AF691307

# The Ecology & Behavior of Amphibians

A FROG-CATCHER

*The lower sort of people in China are not delicate in their eating; they do not scruple to feed heartily upon animals which die a natural death, as is noticed by several voyagers. The poor eat frogs and rats, dried hams of which latter are exposed to sale in the streets; and the middling sort think a young dog no bad food . . . . They have a method of taking frogs in China, during the night, by means of fire, which is carried in a wire net, as reproduced in this plate.* —George Henry Mason, *The Costume of China* (London: W. Miller, 1800).

Traditional frog catchers in China probably had detailed knowledge of the natural history of the animals they sought as food. Here the image of the frog-catcher represents the many biologists who have ventured into the field, often at night, to study the ecology and behavior of amphibians. This print is dated May 4, 1799, from a drawing by a Chinese artist, Pu-Qua, of Canton. From the author's collection.

# The Ecology & Behavior of Amphibians

Kentwood D. Wells

The University of Chicago Press
Chicago and London

KENTWOOD D. WELLS is professor in the Department of Ecology and Evolutionary Biology at the University of Connecticut

The University of Chicago Press, Chicago 60637
The University of Chicago Press, Ltd., London
© 2007 by The University of Chicago
All rights reserved. Published 2007
Printed in the United States of America

16 15 14 13 12 11 10 09 08 07    1 2 3 4 5

ISBN-13: 978-0-226-89334-1 (cloth)
ISBN-13: 978-0-226-89335-8 (paper)
ISBN-10: 0-226-89334-0 (cloth)
ISBN-10: 0-226-89335-9 (paper)

Library of Congress Cataloging-in-Publication Data

Wells, Kentwood David, 1948–
The Ecology and behavior of amphibians / Kentwood D. Wells.
p. cm.
Includes bibliographical references and index.
ISBN-13: 978-0-226-89334-1 (cloth : alk. paper)
ISBN-10: 0-226-89334-0 (cloth : alk. paper)
ISBN-13: 978-0-226-89335-8 (pbk. : alk. paper)
ISBN-10: 0-226-89335-9 (pbk. : alk. paper)
1. Amphibians—Ecology. 2. Amphibians—Behavior. I. Title.
QL641.W35 2007
597.8—dc22

2007001687

♾ The paper used in this publication meets the minimum requirements of the American National Standard for Information Sciences—Permanence of Paper for Printed Library Materials, ANSI Z39.48-1992.

*This book is for Marta, Camila, and Gabriela*

# Contents

# Preface

THE TRAIN from Balboa Station moved slowly at first, making its way past the warehouses, military bases, and red-tiled roofs of Canal Zone houses, past the locks of Miraflores and Pedro Miguel, and alongside the Gaillard Cut. There it picked up speed, traveling parallel to the Panama Canal and eventually reaching stretches of tropical rainforest, dotted with small farms and banana fields. As I gazed out the window at the ponds and lakes that lay alongside the tracks, I thought about all the frogs that must be out there, feeding, calling, defending territories, laying eggs, avoiding snakes, and otherwise going about their lives. I couldn't see them, of course, nor could I hear them at night over the din and clatter of the train, but I knew they were there. On some nights, arriving on the 11:00 o'clock train at Frijoles Station on the way to Barro Colorado Island, I could hear the last few squeaks and clicks of the small treefrogs along the shore, or the last plaintive whines of túngara frogs finishing up a night of calling.

I had come to Panama in 1976 as a postdoctoral fellow of the Smithsonian Tropical Research Institute. Fresh out of graduate school, I had little experience in the tropics, but I knew of the great diversity of frogs to be found there, sometimes with more species breeding at a single pond than can be found in the entire northeastern United States. I am not sure when I first thought of writing a book on the ecology and behavior of amphibians, but certainly ideas began to germinate as I took those long train rides along the Panama Canal and my many boat rides across Gatun Lake to the jungle island of Barro Colorado.

I actually began writing this book in the early 1980s, so to say it has had a long gestation period would be a gross understatement. Indeed, at times the project seemed to take on a life of its own, like one of those never-ending scholarly endeavors usually associated with the humanities, such as editing the correspondence of Charles Darwin or writing the definitive multivolume biography of Herman Melville. Much has changed in the intervening years, both in my life and in the study of amphibian ecology and behavior. I began the book as a young assistant professor and now find myself one of the most senior members of my department. I got married, raised two young daughters, bought and remodeled a house, taught classes, supervised graduate students, conducted field research, and served on more departmental and university committees than I can remember.

At every opportunity, I continued working on the book, sometimes for months at a time, sometimes in fits and starts as time allowed. I spent many fulfilling hours in libraries, not only at the University of Connecticut, but also at the Smithsonian Tropical Research Institute during summer trips for field research. Over the years, much of the technology associated with book writing has changed. The initial version of the first chapter I wrote was composed on a portable electric typewriter, and from there I graduated to an IBM Selectric, a mainframe computer terminal, and a succession of desktop computers. Frequent trips to the library to fill notebooks with handwritten notes have largely been replaced by access to electronic journals and the printing of thousands of pages of recent journal articles. Figures drawn by hand with a lettering set have given way to computerized graphics, while the tedious developing of black and white photographs in the darkroom has been replaced by digitally scanned images.

What also has changed over the years is the amount of information available on the ecology and behavior of amphibians, which has increased at an exponential rate, as has the number of researchers studying amphibians. When I began this book, the most recent scholarly treatment of amphibian biology was G. K. Noble's *Biology of the Amphibia,* published in 1931 and reprinted by Dover Publications in the 1950s. Since then, we have seen the publication of Duellman and Trueb's *Biology of Amphibians* (1986); the multivolume Australian series *Amphibian Biology,* edited by Harold Heatwole (and others), which began in 1994; and volumes dealing with the environmental physiology of amphibians (Feder and Burggren 1992), amphibian hearing and acoustic communication (Fritzsch et al. 1988; Ryan 2001; Gerhardt and Huber 2002; Narins and Feng 2006), reproductive biology (Jamieson 2003; Sever 2003a), tadpole biology (McDiarmid and Altig 1999a), salamander ecology (Hairston 1987; Bruce, Jaeger, and Houck 2000), amphibian distribution (Duellman 1999c), and amphibian conservation (Beebee 1996; Semlitsch 2003a; Lannoo 2005). There also have been many other books and monographs dealing with specialized aspects of amphibian biology, or geographic treatments such as Petranka's *Salamanders of the United States and Canada* (1998) and Savage's *The Amphibians and Reptiles of Costa Rica* (2002). Meanwhile, the explosion of journal literature dealing with amphibian ecology and behavior over the last 25 years often left me feeling like Lewis Carroll's Red Queen, always running faster and faster just to keep up. At some point, the search for that last key paper must come to an end. For this book, that point essentially was reached in the fall of 2005, although a few papers from 2006 did find their way into the references. The impossibility of being completely up-to-date is illustrated by the appearance of a 370-page monograph on the phylogeny and classification of amphibians (Frost et al. 2006) just as I was making final preparations to send this book to the publisher (see chapter 1).

Of course, the amphibians themselves have changed as well. Several species have gone extinct since I began writing this book, and many more are in decline or threatened by assaults on their environment. Consequently, many researchers have shifted their focus from investigating the basic biology of amphibians to research programs aimed at understanding the decline of many amphibian populations and identifying ways to prevent it. Yet a clear understanding of the basic physiology, ecology, and behavior of amphibians remains central to any successful conservation effort, and my hope is that this book will contribute to this body of knowledge.

Although I wrote this book myself, it would not have been possible without the encouragement and assistance of many family members, friends, and colleagues. In acknowledging these people, it seems appropriate to start at the beginning. I owe a tremendous debt to my parents, W. Willeroy Wells and Leora Wood Wells, who unfortunately did not live to see this book completed. My father, having been a Rhodes scholar with a keen interest in history, instilled in me a love of research on all sorts of subjects. My mother, a voracious reader, a writer, and an accomplished editor, conveyed to me her love of books and her sense that writing is the essence of communicating both facts and ideas. My parents never questioned my desire to become a naturalist, or to fill our basement with tanks and cages full of frogs, salamanders, and snakes, to major in zoology in college, or to pursue a PhD in the obscure field of frog behavior. They provided love and financial support, without which this book would not have been possible. I am saddened that they did not get to see the final product.

Next to my parents, perhaps the greatest influence on my development from boy naturalist to college professor has been my teachers, starting with my high school biology teacher, Richard Ailstock, who often accompanied my friends and me on expeditions to various swamps, presented me with my first copies of *Copeia,* and invited me back to his classroom for years after I was a student to show slides of frogs to the next generation of students. At Duke University, I got my first taste of real field research in an ecology class taught by Dan Livingstone, and encouraged by him and the late Joe Bailey, I continued my class project and eventually published two papers from the work. My development as a biologist continued in a PhD program at Cornell University, a wonderful center for research in ecology and behavior. I benefited especially from the guidance, encouragement, and criticism of my major professor, F. Harvey Pough, who taught me the importance of integrating studies of physiology, ecology, and behavior to understand the biology of organisms. Steve Emlen, an inspirational teacher and pioneer in the study of vertebrate social behavior, encouraged me in every aspect of my research with frogs by his sheer enthusiasm for all things behavioral. Kraig Adler gave me access to his incredible library as I was writing my first review of frog mating systems, which was the key to getting my career as a scientist underway. Interactions with other faculty members, graduate students, and postdocs at Cornell enriched my experience and shaped my development as a biologist. There are too many of these people to mention individually, but especially influential were Jack Bradbury, Will Provine, Carl Gerhardt, Peter Narins, John Heiser, Tom Wiewandt, Bill Mautz, Gene Helfman, Gary McCracken, and the late Jasper Loftus-Hills. My formal education was completed with the postdoctoral fellowship that took me to Panama, where I benefited from interactions with many graduate students, postdocs, and staff members at the Smithsonian Tropical Research Institute. Especially important was the influence of my postdoctoral advisor, Stan Rand, field naturalist and tropical herpetologist extraordinaire, who unfortunately also did not live to see this book completed.

During the long gestation of this book, many friends and colleagues have provided me with information, unpublished manuscripts, Master's theses, PhD dissertations, and comments on various drafts of chapters, in some cases so long ago that the final version may not be recognizable. I hope I have not forgotten to acknowledge anyone: Kraig Adler, Stevan J. Arnold, David Bickford, Andrew Blaustein, Ron Brandon, Janalee P. Caldwell, Adão Cardoso, Tun-Yu Dave Chen, Martha Crump, Sharon Emerson, Martin Feder, Robert Gatten, H. Carl Gerhardt, Harry Greene, Tim Halliday, Mark Hayes, Lynne Houck, Richard D. Howard, Victor H. Hutchison, Robert Jaeger, Michael Jennions, Murray J. Littlejohn, Horst Lüddecke, Ralph Mac Nally, Peter Morin, Peter Narins, Carlos Navas, Kiisa Nishikawa, Jim O'Reilly, Neville Passmore, John Phillips, Ken Prestwich, Michael J. Ryan, Andreas Schlüter, Joshua J. Schwartz, Ray Semlitsch, Roger Seymour, Cynthia Kagarise Sherman, Vaughan Shoemaker, Martin P. Simon, David Skelly, Margaret M. Stewart, Stephen G. Tilley, Catherine A. Toft, Daniel S. Townsend, C. Richard Tracy, Joe Travis, Paul Verrell, David Wake, Marvalee Wake, Bruce Waldman, Richard Wassersug, Earl Werner, Henry Wilbur, and Larry Woolbright. Two of my graduate students, Patrick C. Owen and Noah Gordon, did yeoman service one summer in checking all of the references in the text and reference list for the first 15 chapters, finding many errors and omissions.

Walter Hödl and Wayne Van Devender contributed many photographs to the book, and their skill as photographers has enhanced it immensely. Others who also contributed photographs or other illustrative materials include Allen Allison, Milos Andera, Carl D. Anthony, Stevan J. Arnold, David Bickford, David Bradford, Edmund D. Brodie, Jr., Edmund D. Brodie III, Janalee P. Caldwell, Alan Channing, Reginald B. Cocroft, Harold Cogger, Jon Costanzo, Martha Crump, Stephen Deban, Robert Drewes, Sharon Emerson, H. Carl Gerhardt, Mac F. Given, Ulmar Grafe, Margaret Gunzburger, Celio F. B. Haddad, James Hanken, Susan Z. Herrick, Lynne Houck, Richard D. Howard, Kristiina Hurme, Victor H. Hutchison, Carlos Jared, Rafael Joglar, Karl Heinz Jungfer, Y.-C. Kam, Stefan Kaminsky, Joseph M. Kiesecker, Krushnamegh Kunte, Julian C. Lee, William P. Leonard, Marcio Martins, Kerry Matz, Charles W. Myers, Kiisa Nishikawa, David Pfennig, Michael Plummer, Rick Relyea, Stephen J. Ressel, Carl W. Rettenmeyer, Stephen Richter, Mark Oliver Rödel, Rudolfo Ruibal, Ivan Sazima, Nancy Staub, Margaret M. Stewart, Daniel S. Townsend, James Vonesh, Karen Warkentin, and Richard J. Wassersug. Very special thanks go to Mary Jane Spring and Virge Kask, biological illustrators at the University of Connecticut, who showed great skill and unending patience in preparing the graphs and line drawings. Virge Kask scanned many slides and combined these with digital images provided by many different people to create the photographic figures.

My many colleagues in the Department of Ecology and Evolutionary Biology at the University of Connecticut have enhanced my work environment immeasurably. I am particularly grateful to Theodore L. Taigen for our years of collaboration on the calling energetics of frogs. Thanks also to the Vertebrate Lunch Group (Kurt Schwenk, Eric Schultz, Margaret Rubega, Elizabeth Jockusch, Chris Elphick, Bruce Goldman, and many students) for many semesters of stimulating discussion. My department heads, especially William Crepet and Gregory J. Anderson, showed enormous patience and encouraged my work on this book, even though it detracted from fieldwork and from my getting lucrative grants. Working on this book and on amphibians in the field would not have been nearly as much fun without my interactions with the many University of Connecticut students who have shared my enthusiasm for amphibians, especially Joshua J. Schwartz, Ellen M. Dawley, Mac F. Given, Catherine R. Bevier, Stephen J. Ressel, Patrick C. Owen, Roberto Ibañez, L. Kealoha Freidenburg, Jennifer A. O'Brien, Susan Z. Herrick, Kristiina Hurme, Philip Robakiewicz, Carlos Navas, Tobias Landberg, Noah Gordon, Laura Runkle, Katherine Bard, Beverly Greer, Joseph Markow, Dawn McKay, Sal Zimmitti, Sharyn Rusch, and Stacey Lance.

The editors and staff of the University of Chicago Press have shown infinite patience and provided much encouragement as the years of book writing rolled on, often with no end in sight. I am especially grateful to the late Susan Abrams, my original editor, who did so much to get this project off the ground, and to Christie Henry, my current ever-helpful editor. The University of Chicago Press provided financial support for the preparation of figures, as did the University of Connecticut Research Foundation, the Department of Ecology and Evolutionary Biology, and the Biological Illustration Office. The Interlibrary Loan Department of the Homer Babbidge Library at the University of Connecticut was endlessly helpful and surely is one of the most efficient operations in the entire university.

Finally, very special thanks go to my wife, Marta Lucía Martínez Wells, and my two daughters, Camila (who thinks frogs are kind of slimy) and Gabriela (who will grab any amphibian she sees), for enhancing my life in every possible way. This book is dedicated to them.

*Storrs, Connecticut*
*April 2006*

# Chapter 1 Phylogeny, Classification, and Morphological Evolution

*These foul and loathsome animals . . . are abhorrent because of their cold body, pale color, cartilaginous skeleton, filthy skin, fierce aspect, calculating eye, offensive smell, harsh voice, squalid habitation, and terrible venom; and so their Creator has not exerted His powers to make many of them.*
—Carl von Linne (Linnaeus), *Systema Naturae* (1758)

LINNEAUS, THE Swedish naturalist who founded the modern system for classifying animals and plants, took a rather dim view of amphibians (he actually did not distinguish between amphibians and reptiles). He was not unique in holding such negative opinions of these animals. Amphibians often seem to evoke fear and loathing from people who are repelled by their cold, clammy skin and noxious skin secretions. They frequently have been invested with magical powers. European fire salamanders (*Salamandra salamandra*), for example, were long thought to have a miraculous ability to survive in fire, no doubt because they sometimes were seen crawling from logs that had been tossed on the kitchen hearth. Toads have come in for special opprobrium, being considered hideous, frightening, and even dangerous animals that possessed a variety of magical powers and an ability to live for extraordinarily long periods of time, perhaps even for centuries. Toads, newts, and salamanders were frequent ingredients in witches' brews and other magical potions that might be used to poison an enemy or cast a spell on a particularly irritating neighbor. Toads appear frequently in European art and literature as symbols of evil and agents of the Devil (De Graaff 1991). In other cultures, however, frogs and toads were widely considered to be symbols of fertility and life. This was particularly true in places where frogs seemed to appear almost by magic at the start of the rainy season, as in Ancient Egypt and the Maya culture of Central America (see chapter 2).

In more recent times, attitudes toward amphibians have shifted, and frogs in particular rank with pandas, lions, and dinosaurs as among the most popular animals in our culture (J. Gibbons 2003a). By the mid-nineteenth century, frogs and toads were popular subjects for children's stories, as they are today. All manner of toys, knickknacks, and decorative objects for the home have been produced with a frog or toad motif, and many people, including some biologists who study amphibians, are avid collectors of amphibian paraphernalia (Donaldson 1980; Ribuoli and Robbiani 1990). There also is a thriving worldwide pet trade in frogs and other amphibians. Yet the average person seldom sees any kind of amphibian in the wild and knows almost nothing about their natural history. People who can expound at length on the social behavior of lions or the latest discoveries in dinosaur paleontology often do not know anything about the feeding habits or reproductive biology of the toads living in their backyard gardens. Most probably could not identify a salamander if they encountered one in the woods.

Even among biologists, there is considerable ignorance about the vast diversity of amphibian species and their great variety of lifestyles. All too often, biologists are familiar with only the few species of amphibians that are widely used for laboratory studies in physiology, genetics, developmental biology, and molecular biology. This can lead to uncritical use of such animals as model organisms, based on the assumption that animals chosen mainly for their suitability as laboratory animals are representative of amphibians as a whole (Cannatella and de Sa 1993; Shaffer 1993). Yet amphibians are an enormously diverse group. With more than 6,000 species currently known, and more being described every year, they represent a greater proportion of living vertebrates than do the mammals, which often are considered the dominant group of terrestrial vertebrates (Glaw and Köhler 1998; Hanken 1999b; for a current count of species, see Amphibia Web at http://elib.cs.berkeley.edu/aw/index.html). Furthermore, living amphibians have a very long evolutionary history, with many of the modern families having been well established for at least 50 to 100 million years (Savage 1973; Duellman and Trueb 1986; Heatwole and Carroll 2000). This means that different types of amphibians are potentially as different from one another biologically as a horse is from an opossum, a bat, or an anteater.

Amphibians occupy nearly every available habitat on earth except for the open oceans, distant oceanic islands, and the frozen reaches of the Arctic and Antarctic. Some amphibians live entirely in water, others entirely on the land. Some live in freshwater swamps, fens, bogs, ponds, lakes, rivers, and streams, environments where water is nearly always available. Others live in deserts, where water is present only for very brief periods each year. Frogs and caecilians are most diverse in tropical rainforests, which provide an ideal combination of warm temperatures and abundant rainfall. Yet a few frogs live above the Arctic Circle in both North America and Europe. Salamanders are most common in the relatively cool, moist forests of eastern North America and the mountains of Central America, but others have penetrated into lowland tropical rainforests, deserts, and even the tundra of Siberia.

This book is about the diversity of amphibian life and the ecological and behavioral adaptations that have made amphibians a successful, if somewhat inconspicuous, component of terrestrial and aquatic ecosystems. The remainder of this chapter provides an evolutionary context for understanding the ecology and behavior of amphibians. The first section reviews general characteristics of amphibians and then briefly discusses the origins of amphibians and their evolutionary history. The rest of the chapter provides a discussion of phylogenetic relationships of the major clades of amphibians and brief synopses of the classification, geographical distribution, and natural history of all amphibian families, followed by more detailed discussions of major features of morphological evolution that are important for understanding their ecology and behavior.

For the most part, the classification used here follows that of Frost (2004). A monograph that proposes a radically revised phylogeny and classification of all amphibians (Frost et al. 2006) appeared just as I was preparing the final manuscript of this book. It seemed neither possible nor desirable to incorporate these changes into the book, especially because this would result in the use of many genus and family names that differ from those cited in the literature being discussed. Such a major change in phylogeny and classification will require time for other scientists to analyze the implications of the proposed changes before the new scheme is widely adopted, and using the new names here would simply cause unnecessary confusion. One implication of the revised phylogeny of amphibians proposed in Frost et al. (2006) is that many previous phylogentic analyses of ecological or behavioral traits discussed in this book, such as the evolution of feeding modes, reproductive modes, life histories, courtship behavior, or parental care, may need to be reexamined. While I have not adopted the taxonomic changes recommended in Frost et al., I have made note of cases in which their taxonomy would be significantly different from the classification used in this book.

The other chapters of the book are roughly divided into three major themes. Chapters 2 to 6 deal mainly with the interactions of amphibians with the physical environment and therefore focus on physiological ecology. Chapters 7 to 13 are concerned with biological interactions among members of the same species, covering such topics as reproduction, parental care, communication, sexual selection and mate choice, aggressive behavior, and the biology of amphibian larvae. Chapters 14 to 16 discuss interspecific interactions, including interactions of amphibians with their predators, the organization of amphibian communities, and prospects for future survival and conservation.

## General Characteristics of Living Amphibians

The three major groups of amphibians are very different from one another morphologically (fig. 1.1). Frogs and toads (order Anura) are specialized for jumping, with greatly enlarged hind legs, shortened bodies, no tail, and large heads and eyes. Salamanders and newts (order Urodela) are much more elongate, with front and back legs of approximately equal size and a long tail (this clade is called Caudata by some systematists, derived from the Latin rather than the Greek word for "tail"). Caecilians (order Gymnophiona) are specialized for life underground. They have elongated, snakelike bodies that lack legs, and they have greatly reduced eyes. Despite these differences, all amphibians share certain physiological characteristics that together set them apart from other terrestrial vertebrates. One such feature is their scaleless, highly permeable skin, which allows for rapid passage of both water and respiratory gases. This feature prevents them from living in salt water or being active in very dry environments (chapter 2). However, it allows them to make use of the skin for respiration to a much

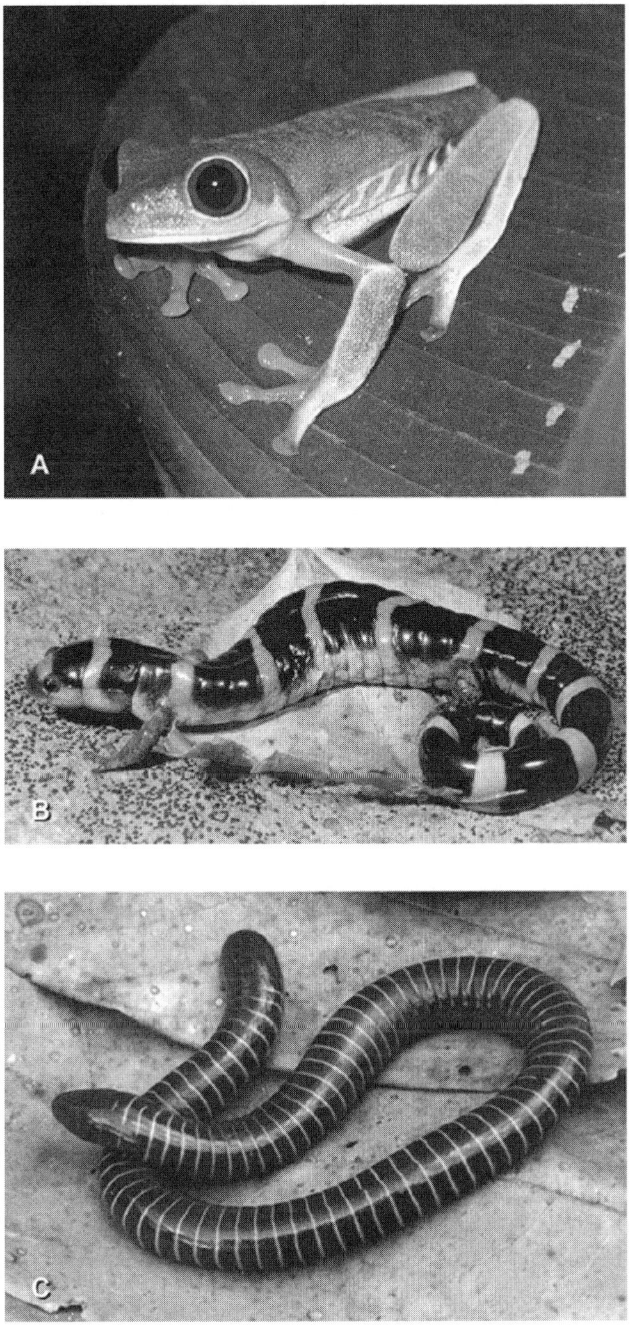

**Fig. 1.1.** Representatives of the three major clades of living amphibians. (A) *Agalychnis callidryas* (Anura, Hylidae). (B) *Ambystoma annulatum* (Urodela, Ambystomatidae). (C) *Siphonops annulatus* (Gymnophiona, Caeciliidae). Photos by Kentwood D. Wells (A), Wayne Van Devender (B), and Janalee P. Caldwell (C).

greater degree than any other vertebrate group (chapter 4). It also has a major effect on their ability to maintain a constant body temperature. Amphibians are ectothermic, a physiological trait inherited from their aquatic ancestors. They are incapable of generating their own body heat (chapter 3) and consequently have much lower metabolic rates than endothermic birds and mammals (chapter 5).

Most amphibians are somewhat dependent on water for reproduction, even when adults are terrestrial, although several of the most successful clades have evolved completely terrestrial reproduction. The egg is never protected by a hard shell, like that of reptiles and birds, and therefore loses and gains water across the egg membrane very rapidly. Even when amphibians breed on land, they must place their eggs in wet microhabitats (chapter 10). Amphibians that lay aquatic eggs, and some that place their eggs out of water, have aquatic larvae that lead lives very different from those of adult amphibians, particularly in frogs (chapter 12). These eventually undergo metamorphosis, a major reshaping of larval morphology into an adult form that is usually adapted for life on land (chapter 13). While the change to a completely terrestrial adult is a new innovation in amphibians, many fishes have larval stages that are very different from the adults.

## The Origin and Evolution of Amphibians

The three living groups of amphibians are descended from a diverse group of tetrapods that first appeared in the Devonian Period, about 400 million years ago (for reviews, see Heatwole and Carroll 2000, and R. Carroll 2001). At this time, the ancient continents were uniting into a single large landmass, Pangaea, much of which was situated in tropical or subtropical latitudes. The climate is assumed to have been relatively warm and equable for terrestrial life. Land communities were characterized by assemblages of relatively primitive plants and arthropods (Benton 1997). We can only speculate about the selective pressures that led a group of fishes to emerge onto land, thus beginning the vast radiation of vertebrates into terrestrial habitats (for general reviews, see Little 1990; Benton 1997; Pough et al. 2004). Some authors have argued that the unexploited food resources in the emerging terrestrial communities and the absence of large predators would have favored the evolution of a more terrestrial lifestyle. Others have suggested that the earliest amphibians evolved from air-breathing fishes in warm, swampy environments where oxygen content of the water was low, favoring the emergence onto land, where oxygen is virtually unlimited. Another theory is that the aquatic habitats of tetrapod ancestors were subject to periodic drought, favoring the ability to leave drying pools to seek out other aquatic habitats.

### Devonian Tetrapods

The precise group of fishes that gave rise to the first tetrapods has been a matter of some debate. All authors agree that the most likely ancestors were sarcopterygian (lobe-finned) fishes, because of the similarities in the structure of the lobed fins and tetrapod limbs. The only living representatives are the lungfishes (order Dipnoi) and the coelacanths (*Latimeria,* order Actinistia), peculiar lobe-finned fishes from the Indian Ocean. There has been considerable disagreement over which of these is the closest living relative of modern amphibians, and both morphological and molecular data have been interpreted differently by various authors (Rosen et al. 1981; Forey 1986, 1988, 1991; Panchen and Smithson 1987; Robineau 1987; A. Meyer and Wilson 1990; Forey, Gardiner, and Patterson 1991; Gorr, Kleinschmidt, and Fricke 1991; Gorr and Kleinschmidt 1993; Zardoya and Meyer 997; Zardoya et al. 1998). Indeed, some workers have maintained that the coelacanth and lungfishes are more closely related to each other than either is to tetrapods (Forey 1988). Regardless of which theory proves to be true, it is clear that neither of these living groups very closely resembles the actual ancestor of tetrapods.

Until relatively recently, the traditional view of most paleontologists has been that the earliest tetrapods evolved from a group of lobe-finned fishes called the osteolepiforms, which had a limb-structure and other skeletal features similar to that of a putative tetrapod ancestor (Schultze 1991; Benton 1997). This was disputed by Rosen et al. (1981), who argued that most of the characters uniting these fishes with tetrapods were ancestral rather than derived; they favored a close relationship of tetrapods to lungfishes. More recent cladistic analyses of fossil morphology have focused on a group of lobe-finned fishes called panderichthyids (fig. 1.2). These resemble osteolepiforms, and sometimes are considered part of that group, but are considered by other paleontologists to be a separate lineage closely related to osteolepiforms (Vorobyeva and Schultze 1991).

The panderichthyids had a number of tetrapod-like features of the skull, axial skeleton, and general body morphology, although the braincase lacked certain key tetrapod features (Ahlberg, Clack, and Lukševičs 1996). These fishes had relatively long snouts, eyes placed on the top rather than on the sides of the head, flattened bodies, and reduced median fins. They probably were predators in shallow water that were capable of limited locomotion on land, similar to that seen in some living fishes such as mud-skippers and walking catfish (Ahlberg and Milner 1994). The pectoral fins of these fishes were more derived and more tetrapod-like than were the pelvic fins, suggesting that tetrapod limb characters did not all evolve at the same rate (Boisvert 2005). Recent phylogenetic analyses place the panderichthyids as the sister group to the tetrapods, with the combined clade formed by these two groups comprising part of a larger clade Tetrapodomorpha, which includes the Osteolepiformes (fig. 1.3). The latter is paraphyletic with respect to tetrapods (Ahlberg and Johanson 1998). This analysis

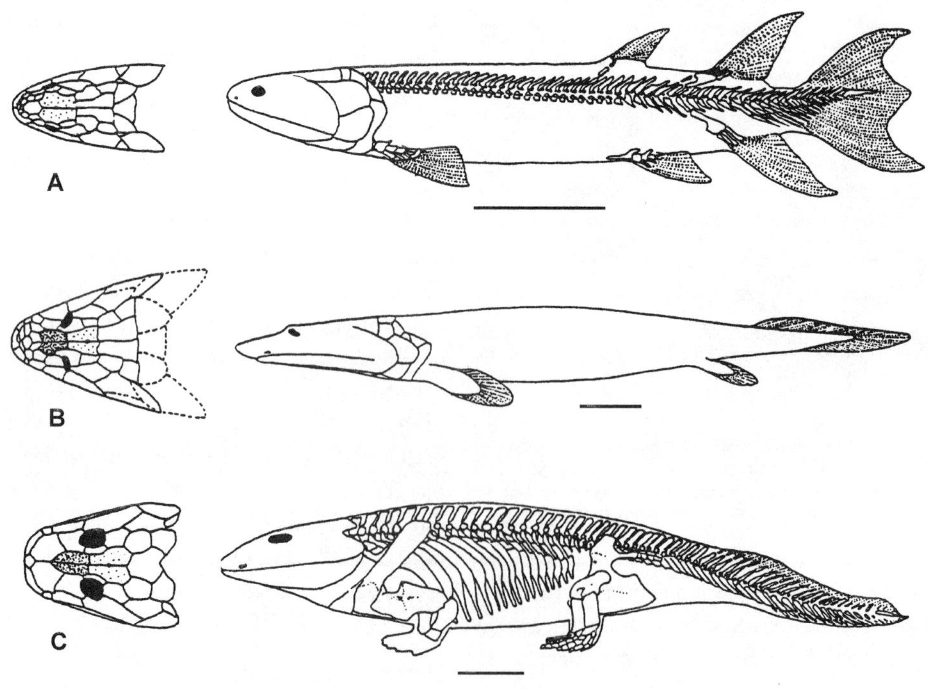


**Fig. 1.2.** A comparison of skull morphology and body form in osteolepiform fishes, panderichthyid fishes, and early tetrapods. (A) Skull and lateral view of the osteolepiform fish *Eusthenopteron.* Scale bar is 100 mm. (B) Skull and lateral view of the panderichthyid fish *Paderichthys rhombolepis.* Scale bar is 10 mm. (C) Skull and lateral view of the Devonian tetrapod *Ichthyostega.* Scale bar is 100 mm. After Ahlberg and Milner (1994).

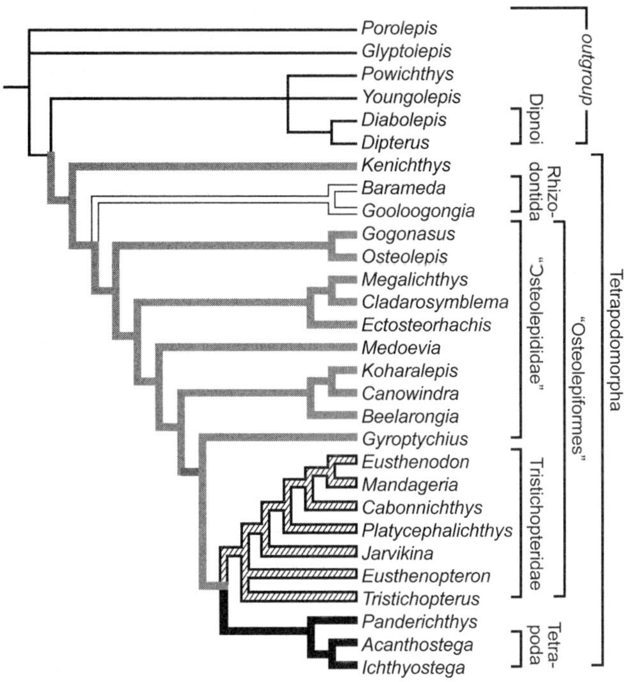


Fig. 1.3. Phylogeny of the tetrapodamorph stem group. Thin black lines represent outgroups. Gray bars indicate members of the paraphyletic family Osteolepididae. Open bars indicate the Rhizodontida. Hatched bars indicate the Tristichopteridae, a family of osteolepiform fishes. Black bars indicate the early tetrapods and their closest relatives. After Ahlberg and Johanson (1998).

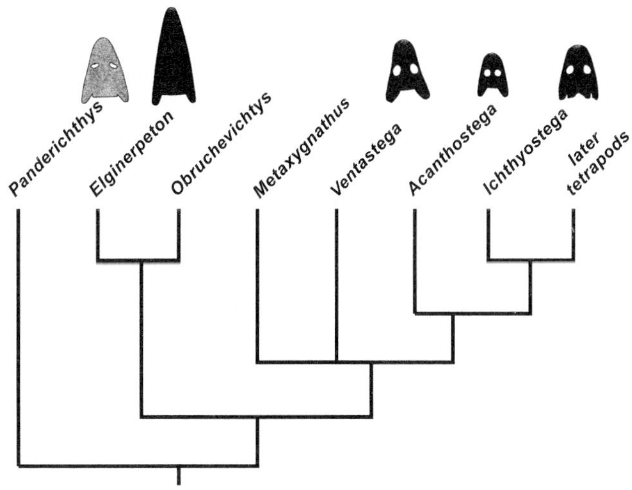


Fig. 1.4. Phylogeny of Devonian tetrapods, with diagrammatic representations of skull structure of major clades. *Ichthyostega* is considered the sister group to the later tetrapods. After Ahlberg (1995).

also suggests that the evolution of certain tetrapod-like morphological features of the skull and reduction or loss of median fins occurred independently in several clades. These apparently represent separate evolutionary experiments leading to large aquatic predators, only one of which survives as the modern tetrapod lineage.

The recent discovery of a new tetrapodomorph fossil from Ellesmere Island in Canada provides an intermediate link between *Panderichthys* and the later Devonian tetrapods. This animal, *Tiktaalik roseae,* had crocodile-like body morphology similar to that of Panderichthys, but had front limbs that were more tetrapod-like than those of panderichthyids (Daeschler et al. 2006). Although this animal had fishlike pectoral fins, the forelimb skeleton had bones apparently homologous to the wrist and finger bones of later tetrapods. These bones gave the animal the ability to adopt a variety of postures, including standing on the substrate while supported by the limbs, as in tetrapods (Shubin, Daeschler, and Jenkins 2006). This intermediate fossil confirms earlier analyses indicating that various tetrapod-like features of the limbs first evolved in aquatic animals with otherwise fishlike morphology (Ahlberg and Clack 2006).

One difficulty in trying to unravel the relationships of various groups of vertebrates is the poor fossil record of Devonian tetrapods (Clack 2000). As of 2001, only eight Devonian tetrapods were known. Two genera, *Elginerpeton* and *Obruchevichthys,* are known mainly from fossil jaws, which were very long and equipped with numerous fishlike teeth. These are thought to be closely related and currently are considered the sister group to all other tetrapods (fig. 1.4). However, these have a relatively derived head and jaw morphology compared to panderichthyids and probably represent an early, specialized lineage. The other Devonian tetrapods, *Ichthyostega, Acanthostega, Ventastega, Metaxygnathus, Hynerpeton,* and *Tulerpeton,* had relatively broad snouts and dorsally placed eyes and were more similar to the panderichthyids (fig. 1.4). Although collectively referred to as "stem tetrapods," this group probably represents several independent lines of evolution, suggesting a considerable diversification of tetrapods by the late Devonian (Ahlberg and Milner 1994; Ahlberg 1995; Jarvik 1996; Clack 2000; R. Carroll 2001; Ruta, Coates, and Quicke 2003). One interesting morphological feature of these fossils is the presence of up to eight toes on the feet, suggesting that a five-toed condition is derived, not ancestral for tetrapods. Other features of the limbs, along with the panderichthyid-like body morphology of these animals, suggest that walking limbs may have evolved in water, rather than for terrestrial locomotion. One possibility is that these limbs facilitated movement through shallow, weed-choked swamps and ponds (Coates and Clack 1990, 1995; Ahlberg and Milner 1994; Clack 2000, 2004; Shubin, Daeschler, and Coates 2004).

### Tetrapods from the Carboniferous to the Triassic

The fossil record for tetrapods is virtually nonexistent over the next 30 million years (R. Carroll 1987; Clack and Car-

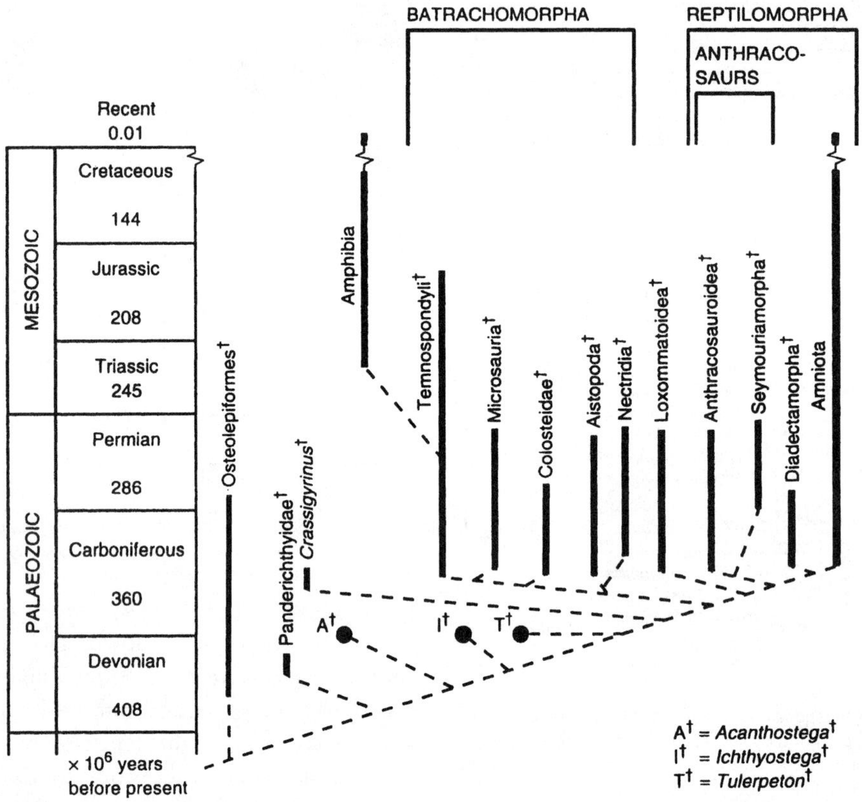


**Fig. 1.5.** Phylogenetic relationships of sarcopterygian fishes and tetrapods, showing the approximate dates of appearance of each clade in the fossil record. Many tetrapod clades were present by the mid-Carboniferous Period, but their relationships to one another are uncertain. The batrachomorph clade, which gave rise to modern amphibians, was clearly differentiated from the reptilomorph clade, which gave rise to the amniotes (reptiles, birds, and mammals). After Pough, Heiser, and McFarland (1996).

roll 2000). Recent discoveries are beginning to fill in this gap, however. In addition to isolated limb and girdle bones from the Lower Carboniferous Period, there is now an articulated skeleton of *Pederpes,* the earliest known tetrapod, that appears to be adapted for walking on land (Clack 2002). This animal, which lived about 350 million years ago, may have had only five toes on the feet, although there are indications of vestigial elements of additional digits. By about 335 million years ago, a host of very distinct tetrapod lineages suddenly appear in the fossil record (fig. 1.5). Indeed, the Carboniferous Period has sometimes been characterized as the Age of Amphibians, but most amphibian fossils come from North American and European deposits, making up only about 1% of the time period encompassed by the Carboniferous (R. Carroll 2000b). Hence, the fossil record for early tetrapods is fragmentary at best.

Many of the Carboniferous tetrapods would not be recognizable as amphibians in the modern sense. While modern amphibians are mostly quite small, and all have highly permeable skin, many Carboniferous tetrapods were large and often protected by scales or even heavy dermal armor. The scaly or armored skin of these animals indicates that they probably lacked the cutaneous respiration characteristic of many living amphibians, and relied instead on well-developed lungs (Holmes 2000). These animals were the first truly successful vertebrate colonists of the land, although most are thought to have retained an aquatic mode of reproduction. At this point, all of the major continents had united into a single landmass, Pangaea, with much of Europe and North America centered near the equator. Land plant communities were much more developed than in the Devonian, with forests that eventually gave rise to the extensive coal deposits that give the period its name. The fossils of early tetrapods have been found in these coal deposits, which probably were originally warm, swampy habitats like those favored by many modern amphibians.

The Carboniferous tetrapods were much more diverse than Devonian tetrapods, and more highly derived relative to their fish ancestors, but the large gap in the fossil record makes their relationship to Devonian tetrapods obscure. They included aquatic, semiaquatic, terrestrial, and fossorial species, some of which had body forms superficially similar to modern crocodilians, lizards, snakes, and salamanders. Others were distinct from any living vertebrates. None of these early tetrapods had a body form resembling the highly derived morphology of frogs. By the late Devonian or early Carboniferous, a "batrachomorph" or "amphibian stem" lineage was well differentiated from a "reptiliomorph" or "reptile stem" lineage, represented by Anthracosaurs and Seymoriamorphs (fig. 1.5). The reptilomorph clade generally is considered to be the sister group of the amniotes (reptiles, birds, and mammals; Smithson 2000). Hence, modern amphibians and reptiles probably have not shared a common ancestor for more than 360 million years (Ahlberg and Milner 1994; Benton 1997; Clack and Carroll 2000; R. Carroll 2001; Ruta, Coates, and Quicke 2003).

The fossil tetrapods of the Carboniferous are much too diverse to discuss in detail here. Relationships among the many groups of early tetrapods are much disputed, with little agreement among various phylogenetic analyses. For recent discussions of phylogenetic relationships and accounts of the various groups, the reader should consult Ahlberg and Milner (1994), R. Carroll (1995, 2002), Laurin and Reisz (1997), J. S. Anderson (2001), Laurin (2002), Ruta, Coates, and Quicke (2003), and the books edited by Panchen (1980), Benton (1988), and Heatwole and Carroll (2000). The group that is most relevant to this book is the order Temnospondyli, considered by many paleontologists to be the likely ancestral group for the modern amphibians (Holmes 2000; Warren 2000; R. Carroll 2001; Ruta, Coates, and Quicke 2003; for contrasting views, see Laurin and Reisz 1997; Laurin 2002). This was an extremely diverse group that first appeared in the late Carboniferous or early Permian (286 million years ago), with some lineages persisting into the early Cretaceous (144 million years ago). Temnospondyls included the very large, fully terrestrial predator *Eryops* (fig. 1.6 A), probably the most frequently illustrated early amphibian. They also included other large terrestrial or semiaquatic forms, with long, crocodilian-like snouts that probably were used to capture fish in freshwater lakes (fig. 1.6 C). Others were smaller and more terrestrial, possibly feeding on insects and other invertebrates. The full extent of the terrestrial tetrapod fauna is unknown, because animals from lowland swamps or lakes are much more likely to be preserved as fossils. Hence, the fossil record is biased toward such habitats, even though tetrapods very likely were present in upland habitats as well (Boy and Sues 2000).

In the late Carboniferous and Permian, many freshwater lakes supported a diverse fauna of temnospondyls. Many of these are known only from larval stages or a few metamorphosed adults. In some cases, impressions of soft tissues, including external gills, are preserved in the fossils. The classification of these forms is somewhat confusing. The term "branchiosaurs" has been used to describe this assemblage, but this appears to represent more than one evolutionary lineage. Two groups of temnospondyls, the families Branchiosauridae and Micromelerpetonidae, are included, as is one unrelated family of Seymouriamorpha (Boy and Sues 2000). The branchiosaurids and micromelerpetonids had a superficial resemblance to modern salamanders (fig. 1.6 D), with short heads, slender trunks, and long tails. They also had three pairs of external gills, like those of modern salamander larvae. In some deposits, hundreds of larval animals have been found, but in most cases, fully metamorphosed adults are very scarce or completely absent. This could mean that adults were fully terrestrial and not preserved in the lake sediments. An alternative explanation is that some of these animals exhibited facultative paedomorphosis like that of modern ambystomatid salamanders (see chapter 13), retaining some larval features as adults when conditions were favorable to fully aquatic life (Boy and Sues 2000).

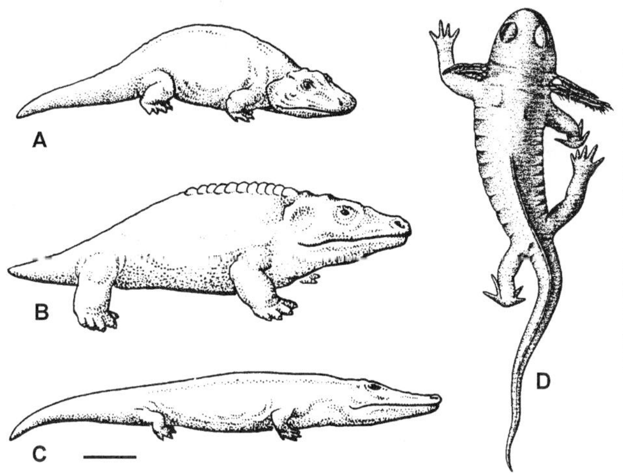


**Fig. 1.6.** Representative temnospondyls. (A) *Eryops*, a terrestrial dissorophid from the Permian. (B) *Cacops*, another terrestrial dissorophid from the Permian. (C) *Cyclotosaurus*, an aquatic crocodile-like capitosaur from the late Triassic. Scale lines indicate 10 cm. (D) *Branchiosaurus*, a paedomorphic or larval temnospondyl from the early Permian, with external gills similar to those of modern larval or paedomorphic salamanders. Scale line for (D) indicates 5 cm. After Pough, Heiser, and McFarland (1996).

Using evidence of food items preserved in some fossils, the functional morphology of the jaws and branchial apparatus, and inferences from modern amphibians, Boy (1993) constructed a hypothetical food web for an Early Permian lake in Germany (fig. 1.7). The smallest branchiosaurids (*Apateon*) probably fed mostly on zooplankton and other small invertebrates, although there is some evidence of cannibalism of smaller larvae as well. Somewhat larger micromelerpetonids (*Micromelerpeton*) probably fed on invertebrates and on smaller amphibians. These amphibians probably competed for food with acanthodian fishes (*Acanthodes*) and actinopterygian fishes (*Parablypterus*). All of these animals in turn were eaten by large crocodilian-like archeogosauroid amphibians such as *Sclerocephalus* and *Archegosaurus*. The smaller amphibians appeared to thrive

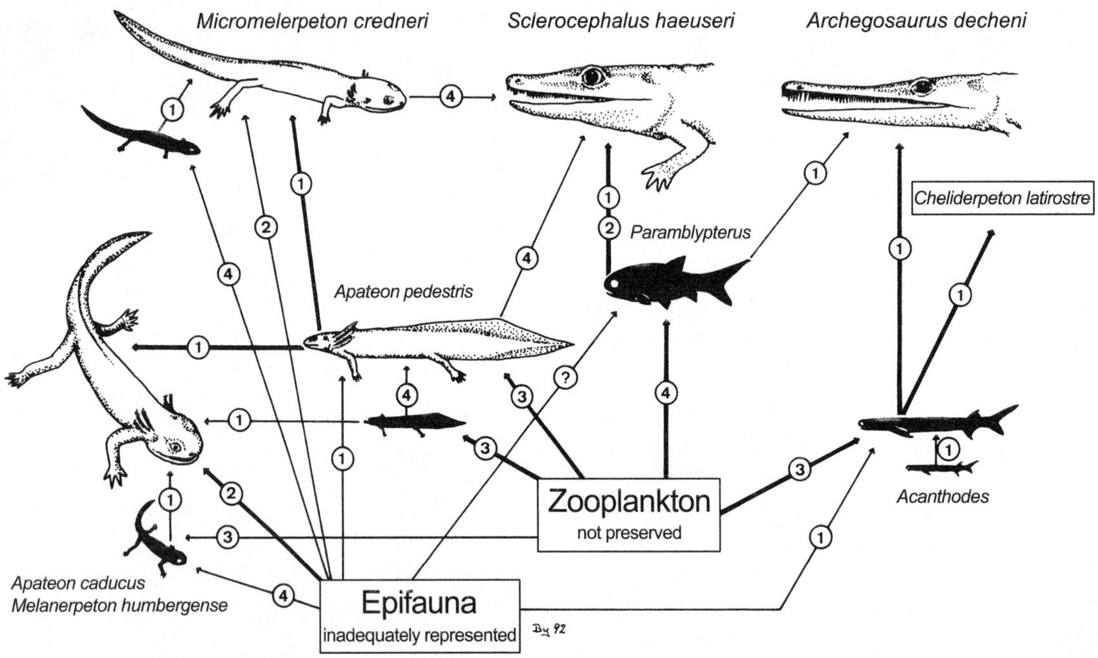


**Fig. 1.7.** Inferred trophic relationships among Early Permian tetrapods and fishes from the Saar-Nahe Basin in Germany. Numbers indicate the type of evidence used to infer trophic relationships: 1 = gut contents or food items stuck in throat region. 2 = coprolites (fossilized dung) that can be unambiguously assigned to their producers. 3 = functional interpretations of features of the jaw and brachial apparatus. 4 = possible predator-prey relationships based on analogies with living species. After Boy and Sues (2000).

mainly in lakes that lacked very large predatory fishes, just as modern ambystomatid salamanders do today (see chapter 14). In some lake deposits, there is evidence that the lake underwent gradual succession, with the smaller amphibians largely disappearing as large predatory fishes became more abundant (Boy and Sues 2000).

By the Triassic Period (245 million years ago), the temnospondyls were represented by a diverse monophyletic clade, the stereospondyls. These were mostly aquatic animals, although evidently derived from terrestrial ancestors. Fully terrestrial temnospondyls by this time had been largely replaced by the enormous diversity of terrestrial reptiles (Benton 1997). Like earlier lineages of temnospondyls, the Triassic forms probably all had aquatic larvae, and some are known to have undergone a rather dramatic morphological metamorphosis. Most probably were covered in some form of scales; well-preserved heavy dermal armor is known for some species. Adult body forms were extremely diverse. Many were superficially crocodile-like in appearance, with both broad-snouted generalist predators and narrow-snouted fish-eaters represented. Others did not resemble any modern vertebrates. The crocodile-like trematosaurids evidently were marine animals, indicating that Triassic nonamniote tetrapods did not all have physiological adaptations like those of modern amphibians. Most stereospondyls were extinct by the end of the Triassic, but some groups persisted in Asia and Australia into the Jurassic (Warren 2000).

A final assemblage of Carboniferous and Permian amphibians that requires brief mention is a group collectively known as lepospondyls (R. Carroll 2000c). These animals were smaller than most temnospondyls and evolved into a variety of body forms that superficially resemble those of modern salamanders, lizards, and snakes (fig. 1.8). Many apparently were terrestrial, at least as adults, and some were burrowing animals, but others were fully aquatic. Hence, in some ways these animals were ecologically more similar to modern amphibians than were the temnospondyls. The relationship of the lepospondyls to the temnospondyls has been disputed, as has the monophyly of this clade (Ruta, Coates, and Quicke 2003). At least six major lineages of lepospondyls are known, of which the most diverse and best known are the microsaurs, aïstopods, and nectrideans. Microsaurs were mostly terrestrial, with a variety of salamander-like and lizardlike body forms (fig. 1.8 A–C). Aïstopods were elongate, snakelike legless amphibians (fig. 1.8 D). They represent one of several clades of lepospondyls that exhibited a tendency toward elongation of the body and reduction of limbs. The nectrideans were a diverse group that included both aquatic and terrestrial forms (fig. 1.8 E). Most were elongate, but this was due to an elongated tail, not an elongated trunk region as in the aïstopods. Some of these amphibians had bizarre flattened heads with projecting hornlike structures, but the functional significance of this morphology is unknown (fig. 1.8 F). Others

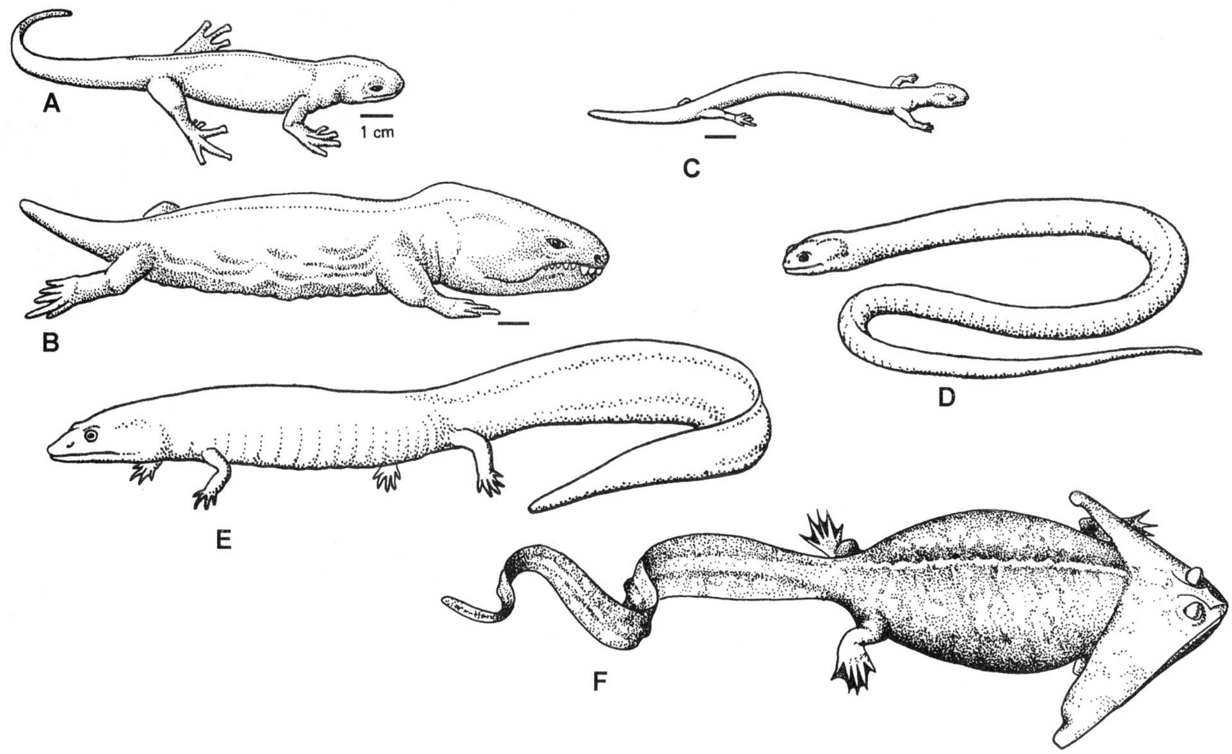


**Fig. 1.8.** Representative Carboniferous and Permian lepospondyls. A–C are microsaurs. (A) *Tuditanus.* (B) *Pantylus.* (C) *Goniorhynchus.* (D) *Ophiderpeton,* a snake-like aïstopod. (E) *Sauropleura,* an aquatic nectridean. (F) *Diplocaulus,* a bizarre horned nectridean from early Permian deposits in Oklahoma and Texas. A–E after After Pough, Heiser, and McFarland (1996); F from Benton (1990).

had elongate, crocodilian-like skulls characteristic of specialized fish-eaters.

## The Origin of Modern Amphibians

The general consensus among most amphibian systematists is that the three groups of living amphibians are descended from a common ancestor and therefore form a monophyletic clade, the Lissamphibia, united by a number of derived morphological characters (Duellman and Trueb 1986; Milner 1988; Bolt 1991; Trueb and Cloutier 1991a; R. Carroll 1995, 2000c, d, 2001; Ruta, Coates, and Quicke 2003). These include the presence of pedicellate teeth and the amphibian papillae, a unique sensory organ in the ear. Molecular evidence also supports the monophyly of the three living groups of amphibians (Hay et al. 1995; Feller and Hedges 1998; Zardoya and Meyer 2001; San Mauro et al. 2005; Zhang et al. 2005). Most systematists argue that this clade is most closely related to a group of terrestrial temnospondyls called dissorophoids, or possibly to the aquatic branchiosaurs, which are included in the superfamily Dissorophoidea (fig. 1.9 A; Trueb and Cloutier 1991a; Milner 1994; Roček and Rage 2000a; R. Carroll 2001; Ruta, Coates, and Quicke 2003). In contrast, Laurin and Reisz (1997) argued that lissamphibians are most closely related to lepospondyls, and these groups together form a sister clade to the amniotes (fig. 1.9 B), but most other workers have not accepted this view. It is difficult to fully resolve this controversy in the absence of key fossils linking lissamphibians with earlier clades of tetrapods. For example, Ruta, Coates, and Quicke (2003) could not identify a specific dissorophoid as the sister group to lissamphibians.

All of the modern amphibian clades are relatively small-bodied animals compared to many Permian and Triassic tetrapods. The largest living amphibians, the Asian giant salamanders (*Andrias*), are about 1.5 meters long, and would be no match for the two-meter long *Eryops* or the crocodile-like capitosaurs from the late Triassic. Indeed, as a group, amphibians are the smallest of all terrestrial vertebrates. It has been suggested that evolutionary reduction in body size was accompanied by truncation of development in many morphological features, a type of paedomorphosis known as progenesis (Bolt 1977, 1979, 1991; Roček and Rage 2000a) that is common in modern amphibian clades (see the following discussions). This process may account for the apparent convergence in morphology of small, salamander-like amphibians in a number of different clades, including the lissamphibians, nectrideans, and microsaurs (fig. 1.8; R. Carroll 1986; Milner 1988). R. Carroll (2000c) argued that the similarities between lissamphibians and lep-

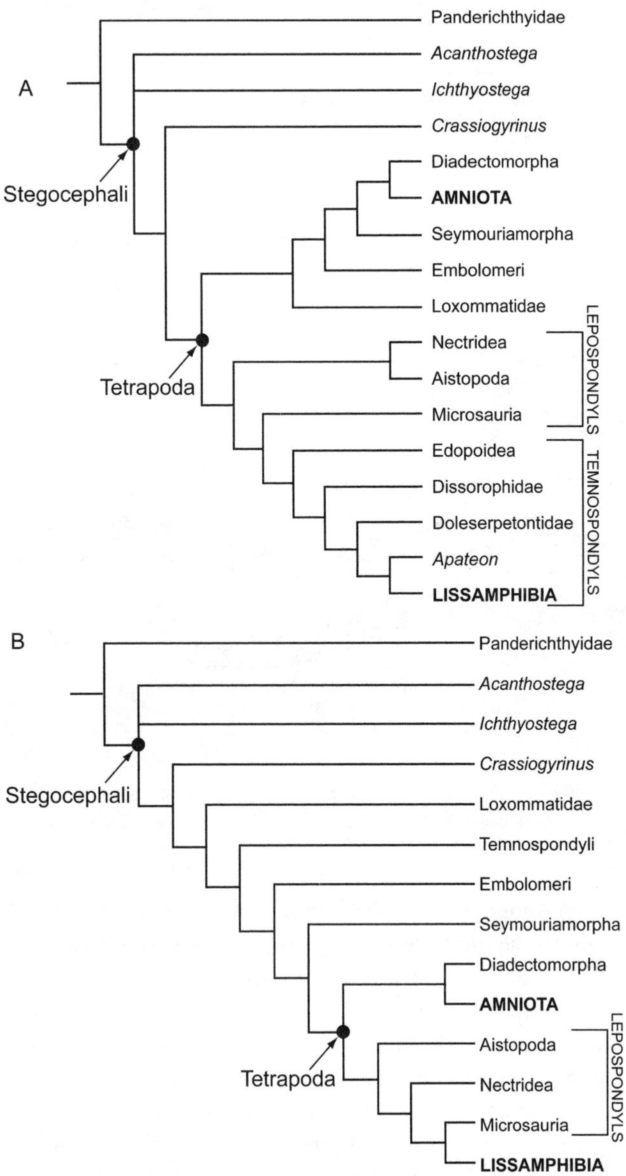


**Fig. 1.9.** Alternative phylogenetic hypotheses for the relationships of lissamphibians, amniotes, and early tetrapods. (A) Traditional hypothesis showing lissamphibians related to temnospondyls. (B) Hypothesis of Laurin and Reisz (1997), showing lissamphibians related to lepospondyls and to microsaurs in particular. After Pough et al. (2004).

ospondyls described by Laurin and Reisz (1997) result from convergent evolution of morphological features related to reduction in body size and therefore would not necessarily be evidence of a close relationship between these groups.

The precise relationship of the modern amphibians to Paleozoic amphibians has been hard to determine, because fossils linking these groups have not been discovered. Most paleontologists agree that anurans show the strongest affinities with Paleozoic temnospondyls, but the enormous gap in the fossil record makes it impossible to determine direct ancestry. There is little doubt that the origins of the three major clades of modern amphibians are very ancient, considerably predating the earliest existing fossils for each group. Recent attempts to apply a molecular clock analysis to genetic data indicate an origin for Lissamphibians between 330 and 367 million years ago, before the breakup of the supercontinent Pangaea (San Mauro et al. 2005; Zhang et al. 2005). This is about 100 million years before the appearance of recognizable fossils of frogs, salamanders, or caecilians, so there obviously is an enormous gap in the known fossil record of these animals.

The first froglike fossil (*Triadobatrachus massinoti*) is a single specimen from early Triassic deposits (245 million years old) in Madagascar (Roček and Rage 2000b). This animal had elongated legs compared to ancestral amphibians, as well as other froglike anatomical features, but lacked the shortened trunk and other anatomical adaptations for jumping found in modern frogs (fig. 1.10 A). A more recently discovered fossil from Poland, *Czatkobatrachus polonicus,* is incomplete, but has froglike characteristics (S. Evans and Borsuk-Białynica 1998). It is about 5 million years younger than *Triadobatrachus,* and its discovery in Europe suggests

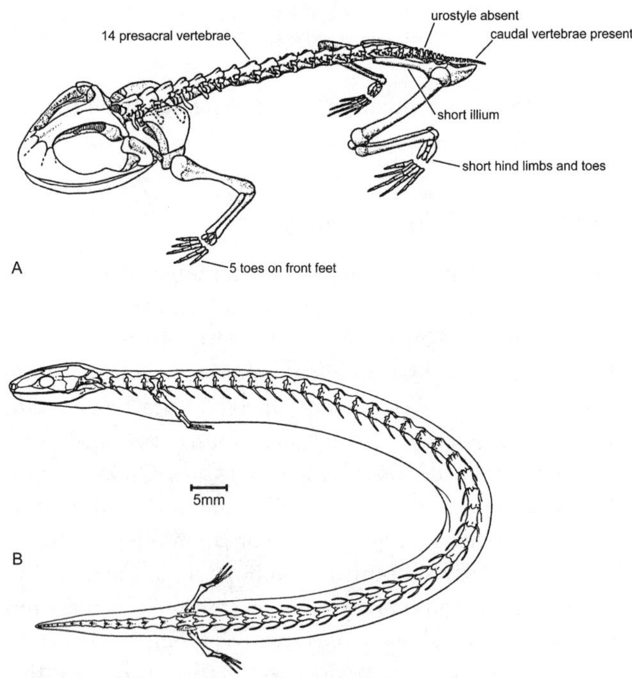


**Fig. 1.10.** Reconstructions of the earliest anuran and caecilian fossils. (A) *Triadobatrachus massinoti,* a froglike amphibian from an early Triassic deposit in Madagascar. This animal had 14 presacral vertebrae, compared to no more than nine for modern frogs. It also had a remnant of a tail and lacked the fused pelvic girdle, elongate illium, and urostyle of modern frogs. It also had five toes on the front feet, whereas modern frogs have four. After Roček and Rage (2000b). (B) *Eocaecilia micropodia,* a caecilian with legs from a Jurassic deposit in Arizona. The skull and lower jaw resemble those of some modern caecilians, but no modern species has limbs. After Carroll (2000c) from Jenkins and Walsh (1993).

that protoanurans were widely distributed in the Triassic (Roček and Rage 2000b).

Shubin and Jenkins (1995) reported the discovery of a fossil frog (*Prosalirus bitis*) from early Jurassic deposits in Arizona that much more closely resembles modern frogs in having a greatly shortened trunk, enlarged hind limbs, and a pelvic girdle adapted for true jumping locomotion. The North American origin of *Prosalirus,* far from the site in Madagascar where *Triadobatrachus* was found, suggests that by the early Jurassic (208 million years ago), the lineage leading to modern frogs was well established in both northern (Laurasian) and southern (Gondwanan) continents. Indeed, another early Jurassic frog, *Vieraella herbsti,* has been found in Argentina. It is sufficiently similar to modern frogs to be referred to the modern family Leiopelmatidae by some systematists, although others do not assign it to any known family. The diversity of frogs increased dramatically during the middle and late Jurassic and the Lower Cretaceous, by which time they were sufficiently modern in appearance for some fossils to be assigned to the most primitive modern clades of frogs, including the Leiopelmatidae, Discoglossidae, Rhinophrynidae, and Pipidae. By the Upper Cretaceous, several other families had made their appearance, including the Leptodactylidae and Pelobatidae, as well as several families that are now extinct (Roček 2000; Gao and Wang 2001).

The early fossil record of the other two groups of modern amphibians is less extensive than the record for frogs, although many new salamander fossils have been found in the last 20 years (Milner 2000). The first clearly recognizable fossil salamanders come from Jurassic deposits in Kazakhstan (*Karaurus*) and Kirghizstan (*Kokartus*), both of which are assigned to the extinct family Karauridae. Although these animals resemble modern salamanders, many features of the skull are more primitive than any living urodeles. Several other salamanders, known only from vertebrae, have been described from Middle Jurassic deposits in Britain. The recent discovery of hundreds of well-preserved adult and larval salamander specimens from Jurassic deposits in China may alter our ideas about salamander relationships (Gao and Shubin 2001, 2003). Two genera described from this assemblage, *Sinerpeton* and *Laccotriton,* appear to be most closely related to cryptobranchids and hynobiids, while *Chunerpeton* appears to be a cryptobranchid. Gao and Shubin argued that these fossils support an Asian origin for all salamanders, although this issue has been debated, as has the exact dating of the deposits in which the Chinese fossils were found (S. Evans et al. 2005). *Iridotriton,* another morphologically primitive salamander from the Late Jurrasic of North America, shares characteristics of both cryptobranchoid and salamandroid salamanders, but may be more closely related to the latter group (S. Evans et al. 2005). Most remaining fossil salamanders can be clearly assigned to modern families or to groups closely related to modern families. Some sirenid and cryptobranchoid salamanders have been found in the Cretaceous, as have primitive salamandroid salamanders. Many modern salamander families probably were well differentiated by the late Cretaceous or early Paleocene (65 million years ago), and fossils of many modern genera from the Paleocene, Eocene, Oligocene, and Miocene (65–23 million years ago; Gardner 2003).

The first caecilian (*Eocaecilia micropodia*) is an elongate amphibian from the same early Jurassic deposit where the anuran *Prosalirus* was found (Jenkins and Walsh 1993; R. Carroll 2000a). Unlike modern caecilians, this animal had well-developed front and back limbs, but these were smaller than in most other amphibians of this period (fig. 1.10 B). The skull, however, has many features in common with modern caecilians, and the lower jaw is nearly identical to that of modern species. There also is evidence that this species had a tentacle, a unique sensory organ found in modern caecilians (see later discussion). The skull of *Eocaecilia* was solidly roofed, as in modern genera such as *Ichthyophis,* whereas some living genera have a gap between the cheek and the rest of the skull. There has been some disagreement among systematists as to which condition is the ancestral condition for the group, but the skull of *Eocaecilia* suggests that the solidly roofed condition is primitive.

R. Carroll and Currie (1975) proposed, long before the discovery of this fossil, that a group of microsaurs in the lepospondyls was the most likely ancestral group for caecilians, and R. Carroll (2000d) argued that the morphology of *Eocaecilia* is consistent with this hypothesis. This would require that caecilians be derived from a different group of Paleozoic amphibians than are modern frogs and salamanders, with a common ancestor for all three groups going back to the Lower Carboniferous. Molecular genetic analyses provide strong support for the monophyly of living amphibians, and they generally support a closer relationship of frogs and salamanders to each other than to caecilians (Zardoya and Meyer 2001; San Mauro et al. 2004, 2005; Zhang et al. 2005). An analysis of fossils also supported this pattern of relationships and failed to find evidence for a close relationship of caecilians to microsaurs (Ruta, Coates, and Quicke 2003).

### Morphological Evolution, Paedomorphosis, and Genome Size

The notion that amphibians represent an intermediate grade of evolution between fishes and reptiles is pervasive even among biologists, but in many respects, this is an inaccurate view of the evolutionary history of vertebrates. The living amphibians and reptiles have not shared a common ancestor for about 360 million years, and the lineage that gave rise to reptiles and other amniotes was well differentiated millions of years before the appearance of the first frogs, salamanders, and caecilians. In many features of their anatomy, amphibians actually have a simplified morphology, even in relation to living bony fishes. This is evident not only in the reduction of various skeletal elements, which can be seen in both fossil and living amphibians, but also in many other organ systems. For example, the nervous systems of amphibians are highly simplified, with salamanders being the most simplified and anurans the least. This is manifested in a reduction in various parts of the lateral line organs, the loss of the tympanum and middle ear cavity and the reduction of middle ear bones in salamanders and caecilians, reduction in the number of retinal cells in salamanders, and simplification of the optic tectum and cerebellum in the brains of all amphibians, especially in salamanders and caecilians (Roth, Dicke, and Nishikawa 1992; Roth et al. 1993; Roth, Blanke, and Wake 1994).

This simplification of organ systems and the loss of elements thought to be present in ancestral tetrapods has been attributed to paedomorphosis, the retention of traits in the adult stage that are present in an earlier stage of development in ancestral species. This is related to two evolutionary trends in amphibians. The first is miniaturization, an evolutionary reduction in body size, which began with the early ancestors of modern amphibians and continues to varying degrees in modern lineages (see subsequent further discussions of miniaturization in the three groups of amphibians). The second trend is the unusually large amount of DNA that amphibians have in their cells compared to other vertebrates. This is particularly true in salamanders, with neotenic forms such as amphiumids, sirens, and proteids having the largest genome sizes. The reasons for this increase in genome size are unclear, but it is correlated with increased cell size, decreased cell metabolism, and decreased rates of cell proliferation and differentiation. In short, amphibians can be expected to have lower metabolic rates, lower growth rates, and slower rates of development than other vertebrates simply because of their larger genomes (Horner and MacGregor 1983; Sessions and Larson 1987; Xia 1995). There also is a positive correlation between genome size and embryonic development time within plethodontid salamanders, after correcting for phylogeny and egg size (Jockusch 1997). Slow rates of cell proliferation and differentiation during development result in simplification of organ systems, with fewer, larger cells packed tightly into a limited space. This trend reaches an extreme in bolitoglossine salamanders, which combine very small body size with relatively large genome and cell size (Roth, Dicke, and Nishikawa 1992; Roth et al. 1993). Large cell size results in simplification of brain structure in both frogs and salamanders. Brain complexity in salamanders is affected by body size as well; larger salamander species have larger and more complex brains (the effect of small body size on brain structure is discussed in more detail in a later section). Patterns of variation in cell size and brain complexity are not correlated in obvious ways with either phylogeny or ecology, making functional interpretations difficult (Roth, Blanke, and Wake 1994).

It might seem that a trait such as large genome size would have only the most tenuous connection to the ecology and behavior of amphibians. However, it may be that many fundamental characteristics of amphibian ecology, life histories, and behavior are direct consequences of having an unusually large genome size (Sessions and Larson 1987; Roth, Dicke, and Nishikawa 1992). For example, the low metabolic rates of amphibians have important consequences for their water relations (chapter 2), temperature relations (chapter 3), modes of respiration (chapter 4), and allocation of energy to activity, growth, and reproduction (chapter 5). Low metabolic rates, low rates of cell proliferation and differentiation, and relatively large cell size could affect the number of eggs produced in a season and a lifetime, and therefore would have a major effect on the evolution of life history strategies (chapter 7). The rather slow rate of development of amphibian eggs, especially in salamanders and species with very large eggs (chapter 7; Bradford 1990), would expose the eggs to predators and other environmental dangers for relatively long periods of time, which might select for extended parental care (see chapter 8). Extended parental care might in turn select for territorial defense of sheltered oviposition sites, where both the eggs and parents are safe from predators (see chapters 8, 9, and 11). Low rates of cell proliferation and differentiation also would affect posthatching development of larvae, and therefore would have a major impact on the biology of larvae (chapter 12) and the timing of metamorphosis (chapter 13; Martin and Gordon 1995). Indeed, large genome size appears to be an evolutionary constraint that affects, directly or indirectly, nearly every aspect of amphibian biology.

## Phylogeny and Classification of Anurans

Most recent taxonomic arrangements recognize about 30 families of anurans, with about 350 genera and nearly 5,300 species. The exact number of families varies because of disagreement among systematists about exactly how higher-order taxa should be arranged. In this book, I generally follow the family level taxonomy of Frost (2004), even though recent phylogenetic analyses suggest that some traditional anuran families are not monophyetic. (A new classification proposed by Frost et al. [2006] recognizes 36 families of anurans, with a substantial number of new or reconfigured genera.) Because different phylogenetic reconstructions often result in different relationships among genera, requiring different family-level classifications, there is little point in radically reconfiguring family names at this point, especially in a book not focused on systematics. One difficulty is that various morphological characters often yield conflicting results, and molecular analyses do not always agree with those dervived from morphology (for more detailed discussions, see Ford and Cannatella 1993; Hillis et al. 1993; Maglia, Púgener, and Trueb 2001; Duellman 2003; García-París et al. 2003; Haas 2003; Púgener, Maglia, and Trueb 2003; Cannatella and Hillis 2004; Frost et al. 2006). Therefore, the phylogeny and classification given here should be treated as a general guide to the major lineages of anurans, subject to future revision. Indeed, the futility of considering any such arrangement the final word on the subject is shown by the recent description of an entirely new family of frogs, the Nasikabatrachidae, currently known from a single species of burrowing frog in India and thought to be most closely related to sooglossid frogs of the Seychelles Islands (Biju and Bossuyt 2003; Frost et al. 2006). The phylogeny shown in fig. 1.11 is similar to that given in Duellman (2003) and Biju and Bossuyt (2003), and reflects a general consensus, as of late 2005, of how anuran families are related to each other.

A group of 27 genera and about 200 species comprising nine anuran families (Ascaphidae, Leiopelmatidae, Discoglossidae, Bombinatoridae, Rhinophrynidae, Pipidae, Pelodytidae, Pelobatidae, and Megophryidae) generally are considered basal frogs. They sometimes have been referred to as "archaeobatrachians," but these clades do not appear to constitute a monophyletic group (Ford and Cannatella 1993; Hertwig, da Sa, and Haas 2004; Hoegg et al. 2004; Roelants and Bossuyt, 2005; Cannatella and Hillis, 2004; Frost et al. 2006). Most systematists have considered *Ascaphus* from the Pacific Northwest region of North America and *Leiopelma* from New Zealand to be the most primitive living frogs. Ford and Cannatella (1993) considered *Ascaphus* to be the sister-group to all other living anurans, whereas Duellman (2003) placed *Leiopelma* as the most basal group, largely on the basis of sperm morphology (Scheltinga and Jamieson 2003). Some systematists have placed *Ascaphus* and *Leiopelma* in the same family (Frost et al. 2006), although these genera appear to be united mostly by ancestral characters rather than shared derived characters.

An analysis based on adult and larval morphology produced a different arrangement, with the Pipoidea (Pipidae + Rhinophrynidae) being the sister group to all other frogs, including Ascaphidae (leiopelmatids do not have aquatic larvae, so were not included in this analysis; Maglia, Púgener,

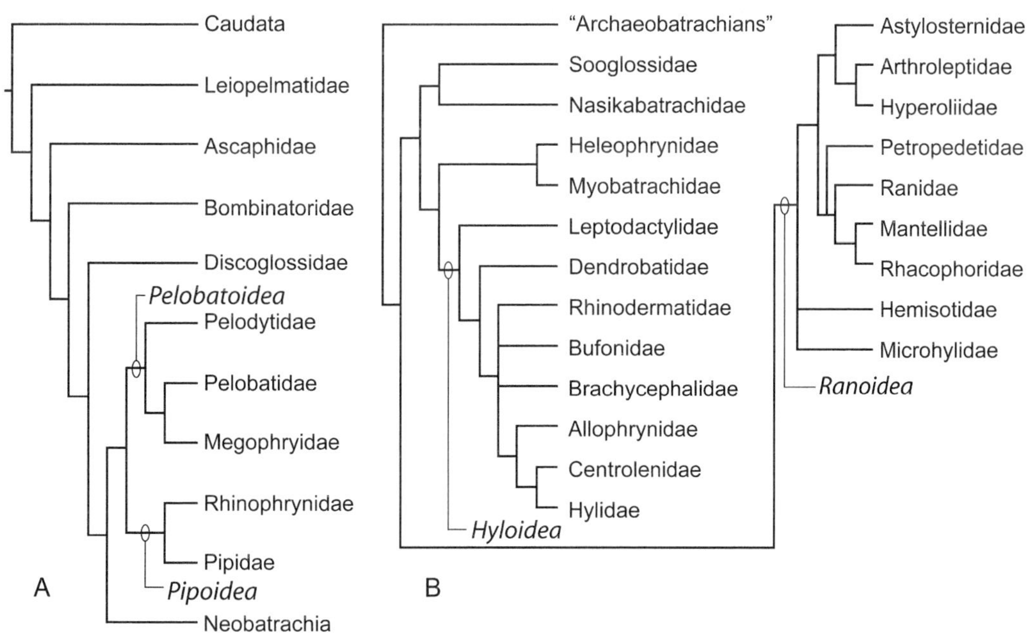


**Fig. 1.11.** Phylogenetic relationships of the major clades of anurans. This figure is meant to show possible relationships of families discussed in this chapter, but other arrangements are possible. Branch lengths are approximate and should not be taken as accurate representations of relative divergence times. (A) Relationships of the "archeobatrachian" anuran families, a nonmonophyletic group. Most phylogenetic analyses agree on the monophyly of the clades Pipoidea and Pelobatoidea, although there is some disagreement over relationships within the latter clade. (B) One of several possible phylogenies for the monophyletic clade Neobatrachia. The monophyly of the clades Hyloidea and Ranoidea is well supported, but relationships of families within these clades are uncertain. Some phylogenies place the hyperoliid subfamily Leptopelinae (not shown here) as more closely related to the Arthroleptidae than to the hyperoliid subfamily Hyperoliinae. After Duellman (2003), with modifications from Biju and Bossuyt (2003).

and Trueb 2001; Púgener , Maglia, and Trueb 2003). Larval morphology alone grouped *Ascaphus* with the pipoids, whereas adult morphology grouped this genus with all the other frog families. In contrast, a different analysis of larval morphology by Haas (2003) placed *Ascaphus* in its traditional position as the sister group to all other anurans, but with pipoids as the next most basal group. Whatever their placement relative to other anurans, the close relationship of pipids and rhinophrynids is well supported by both larval morphology and molecular data, despite their lack of similarity in adult morphology (see the family accounts that follow).

All phylogenetic analysis of adult morphology, larval morphlogy, and molecular characters place the families Discoglossidae and Bombinatoridae as closely related basal groups, in some cases the most basal frogs other than *Leiopelma* and *Ascaphus*. Traditionally, four genera were placed in the family Discoglossidae (*Discoglossus, Alytes, Bombina,* and *Barbourula;* e.g., Duellman and Trueb 1986), but several recent classifications have placed the last two in a separate family Bombinatoridae (Ford and Cannatella 1993; Duellman 2003; Frost et al. [2006] placed *Discoglossus* and *Alytes* in the family Alytidae). I have treated these as two separate families in this book, although studies based on adult morphology, larval morphology, molecular characters, and vocalizations do not give a consistent picture of the relationships among these four genera, in part because *Barbourula* often has been omitted from the analyses (Hay et al. 1995; Bosch and Boyero 2003a; Haas 2003; Púgener, Maglia, and Trueb 2003; Hertwig, da Sa, and Haas 2004; Roelants and Bossuyt 2005; Frost et al. 2006).

The remaining basal families (Pelodytidae, Megophryidae, and Pelobatidae) have been placed together in many phylogenetic analyses as a monophyletic clade, the Pelobatoidea. Both molecular and morphological characters support the close relationship of these groups, although their relationship to other anurans is not fully resolved (García-París et al. 2003; Haas 2003; Púgener, Maglia, and Trueb 2003; Hertwig, da Sa, and Haas 2004; Roelants and Bossuyt 2005). The relationships within this clade have been debated. Traditionally the spadefoot toads from North America (*Scaphiopus, Spea*) have been placed with Eurasian spadefoot toads (*Pelobates*) in the family Pelobatidae, but both morphological and molecular characters indicate that *Pelobates* is more closely related to the Asian megophryids than to the North American spadefoot toads. Roček (1980), Haas (2003), García-París et al. (2003), and Frost et al. (2006) all recommended placing the latter group in a separate family, Scaphiopodidae, but differed on whether the megophryids should be a separate family (García-París et al. 2003; Frost et al. 2006) or included within the Pelobatidae (Haas 2003).

All remaining frog families are placed in an even more derived assemblage, Neobatrachia (fig. 1.11). There is strong support for the monophyly of this clade (Cannatella and Hillis, 2004; Hertwig, da Sa, and Haas 2004; Hoegg et al. 2004; Frost et al. 2006), but relationships within this group, which represents about 95% of all anuran species, are poorly understood (Hillis 1991; Ford and Cannatella 1993; Hass 2003; Cannatella and Hillis 2004; Frost et al. 2006). Ford and Cannatella (1993) recognized a monophyletic clade, Ranoidea, in which they placed the Arthroleptinae, Astylosterninae, Dendrobatidae, Hemisotidae, Hyperoliidae, Mantellinae, Microhylidae, Ranidae, and Rhacophoridae, but relationships among these clades are in dispute. Ford and Cannatella (1993) considered the remaining neobatrachian families not to be a monophyletic group, but others have united these into a group variously called the Hyloidea or Bufonoidea (see later discussion). More recent molecular phylogenies have strongly supported both Ranoidea and Hyloidea as monophyletic groups (Darst and Cannatella 2004; Hoegg et al. 2004), but with the Dendrobatidae removed from Ranoidea.

Emerson et al. (2000) provided a phylogenetic analysis of ranoids using combined morphological and molecular data sets. They found strong support for a monophyletic Ranoidea, excluding the Dendrobatidae. Relationships within the group were not fully resolved, but some apparent relationships differed from traditional classifications. For example, the Arthroleptinae (represented by two species in their analysis) were nested within the Hyperoliidae, but that family did not appear to be monophyletic. Specifically, the genus *Leptopelis,* usually considered the most primitive hyperoliid, appeared to be closely related to *Arthroleptis*. Another novel result is that *Hemisus,* which was previously placed in the Ranidae and now generally treated as a separate family, was grouped with the Microhylidae in their analysis. Finally, the mantelline frogs of Madagascar appeared to be more closely related to rhacophorids than to either ranids or hyperoliids. Unfortunately, some key taxa, including African ranids and astylosternines, were not represented in this analysis, and the great diversity of microhylids also was poorly represented.

A molecular analysis of ranoid frogs from Asia and Madagascar revealed considerable convergence in morphology and ecology, both within and between these regions (Bossuyt and Milinkovitch 2000). This analysis indicates that many traditional subfamilies within the family Ranidae and Rhacophoridae do not represent monophyletic clades. For example, the subfamily Tomopterninae traditionally has been represented by a single genus of burrowing frogs, *Tomopterna,* in Africa, Sri Lanka, and Madagascar. Morphological characters support the monophyly of this genus, despite its disjunct distribution (Glaw, Vences, and Böhme

1998). However, two molecular analyses clearly indicated that the species from Madagascar, *Tomopterna labrosa,* is much more closely related to mantelline frogs from Madagascar than to Asian members of the genus (Richards and Moore 1998; Bossuyt and Milinkovitch 2000). In particular, *T. labrosa* appears to be closely related to the genus *Aglyptodactylus,* and Bossuyt and Milinkovitch (2000) proposed that this species be placed in a new genus, *Laliostoma.* According to their analysis, all of the Madagascar genera traditionally placed in the subfamilies Mantellinae (*Mantella* and *Mantidactylus*), Rhacophorinae (*Aglyptodactylus* and *Boophis*), and Tomopterninae (*Laliostoma*) form a monophyletic clade and should be placed in a separate family, Mantellidae (see also C. Richards, Nussbaum, and Raxworthy 2000). These frogs were separated from ranoid frogs on the Indian subcontinent and the future Seychelles Islands about 87.6 million years ago as the Indian-Seychelles landmass drifted northward (Bossuyt and Milinkovitch 2001). Frogs on Madagascar subsequently evolved into various ecotypes that are very similar to those found among Asian ranid and rhacophorid frogs.

Another persistent problem in interpreting relationships among neobatrachian families has been the position of the family Dendrobatidae. Some of the toxic dendrobatids are very similar to frogs of the genus *Mantella* from Madagascar in morphology, behavior, and ecology (Daly et al. 1996; Heying 2001; Vences et al. 2003; V. Clark et al. 2005), but there is no evidence that these frogs are closely related. Dendrobatids traditionally had been placed within the ranoid frogs (Ford 1993; Ford and Cannatella 1993), although some earlier workers postulated a relationship of dendrobatids to hylodine (cycloramphine) leptodactylid frogs (e.g., Lynch 1971, 1973). Recent evidence does not support a close relationship with ranoid frogs. For example, all recent molecular phylogenies place the dendrobatids outside of the ranoids and within the hyloids (Hedges and Maxson 1993; Hay et al. 1995; Ruvinsky and Maxson 1996; Vences, Kosuch et al. 2000; Darst and Cannatella 2004). Frost et al. (2006) suggested a close relationship to the leptodactylid genus *Thoropa,* which they placed in a new family, Thoropidae.

The hyloids constitute a monophyletic clade that includes the cosmopolitan family Bufonidae, the Sooglossidae of the Seychelles Islands, the Nasikabatrachidae, the Heleophrynidae of southern Africa (not included in Hyloidea by Frost et al. 2006), the Australian Myobatrachidae, and a group of mostly Neotropical families (Allophrynidae, Brachycephalidae, Centrolenidae, Hylidae, Leptodactylidae, and Rhinodermatidae). Dendrobatids appear to fit within this Neotropical group, an arrangement that makes sense biogeographically (Darst and Cannatella 2004). Cranial features of the tadpoles also support a relationship of dendrobatids with the Neotropical hyloid families (Haas 1995), while some features of the finger muscles are consistent with a close relationship between dendrobatids and leptodactylid frogs in the subfamily Cycloramphinae (Hylodinae; T. C. Burton 1998). In contrast, a peculiar morphological feature in some dendrobatids in the genus *Colostethus,* an unusual process on the tongue, supports a relationship with African ranoid frogs, but not leptodactylids (Grant, Humphrey, and Myers 1997). The molecular data indicate that these morphological features almost certainly represent convergent evolution in dendrobatids and ranoids.

There are many other problems yet to be resolved in the phylogeny of anurans. Several of the largest traditional anuran families, including the Hylidae, Leptodactylidae, and Ranidae, clearly are not monophyletic groups united by shared derived characters; these families eventually will be divided into smaller families that are monophyletic (see Frost et al. 2006 for a proposed rearrangement of these groups). Relationships within some of the other large families, including the Bufonidae and Myobatrachidae, are not fully understood, and the monophyly of the latter family has been questioned by some systematists. There also are inconsistencies between some recent molecular analyses and morphological data that must be resolved with further work. A full discussion of anuran systematics is beyond the scope of this book, but some of these issues are discussed briefly in the family accounts that follow.

Throughout the book, I have pursued a relatively conservative approach to names of families, genera, and species, using names that have been widely adopted in the literature. Since this is not a book on systematics, I have chosen not to enter into the current debate over the use of phylogenetically based taxon names, as opposed to traditional taxonomic nomenclature (De Queiroz and Gauthier 1990, 1992; Graybeal and Cannatella 1995; Frost et al. 2006). For that reason, I have retained the use of traditional family names. At lower taxonomic levels, I have largely followed Amphibian Species of the World (Frost 2004) and Amphibia Web (http://elib.cs.berkeley.edu/aw/index.html). My main criterion has been to avoid confusion and to make it possible for readers to find behavioral and ecological literature published on particular species.

## Synopsis of Families of Anura

The following accounts provide a brief summary of the classification, distribution, and natural history of the currently recognized families of anurans. Families are given in alphabetical order for ease of reference and are not grouped according to higher taxonomic levels. The arrangement of families and subfamilies and numbers of genera and species generally follows Frost (2004) and is similar to the classifications in Zug, Vitt, and Caldwell (2001), Duellman (2003), and Pough et al. (2004). Numbers of genera and species should be taken as approximate figures, since the taxonomy of many groups is constantly being revised and new species are discovered every year, especially from the tropics. The taxonomy used here is one of several that are possible and has been chosen as the most widely recognized arrangement. Detailed diagnostic morphological characters of each family are not given; readers should consult Duellman and Trueb (1986), Ford and Cannatella (1993), Zug, Vitt, and Caldwell (2001), and Pough et al. (2004) for this information. For the most part, I have not provided detailed literature citations, except in cases where there has been some dispute over taxonomy. Information on distributions and general natural history come from literature cited elsewhere in the book and from standard reference sources, including Wright and Wright (1949), Liu (1950), Stebbins (1954a, 1985), Cei (1962, 1980), Okada (1966), Stewart (1967), Duellman (1970, 2001), E. N. Arnold and Burton (1978), Duellman and Trueb (1986), Inger and Stuebing (1989), Heyer et al. (1990), Conant and Collins (1991), Glasby, Ross, and Beesley (1993), Glaw and Vences (1994), Passmore and Carruthers (1995), Kuzmin (1999), Schiøtz (1999), Cogger (2000), Lescure and Marty (2000), Channing (2001), Zug, Vitt, and Caldwell (2001), Savage (2002), Pough et al. (2004), and Lannoo (2005).

### Family Allophrynidae

**Distribution:** Guianan region of northeastern South America, including parts of Surinam, Guyana, French Guiana, Venezuela, and Brazil.

**Content:** A single species, *Allophryne ruthveni* (fig. 1.12 A). The relationship of this frog to other anurans has long been a puzzle to systematists. Since it was described in the 1920s, this frog has been variously placed in the families Bufonidae, Centrolenidae, and Hylidae. Ford and Cannatella (1993) failed to find derived characters uniting this species with hylids. They left its placement in a higher taxon unresolved, but Savage (1973) had previously placed it in a monotypic family, and this is the taxonomy currently used by most systematists (Caldwell and Hoogmoed 1998). Analyses using DNA sequences indicated that *Allophryne* is either the sister group to the Centrolenidae (Austin et al. 2002), or is a member of that family (Frost et al. 2006), although it does not resemble centrolenids in general morphology or breeding behavior. An analysis based on hind foot muscle morphology also showed a close relationship to centrolenids, with both families closely related to hylids (T. C. Burton 2004).

**Natural History:** The natural history of this frog is poorly known, but it seems to be a typical treefrog that calls from vegetation around ponds, streams, and rivers, and lays eggs in water.

### Family Arthroleptidae

**Distribution:** Tropical Africa.

**Content:** Three genera, about 50 species. These frogs traditionally were treated as a subfamily of ranids, although Laurent (1986) placed arthroleptines in the family Hyperoliidae. Dubois (1981) united the subfamilies Arthroleptinae and Astylosterninae into the family Arthroleptidae, which was included in Duellman's (1993a) checklist. Ford and Cannatella (1993) could not identify any shared derived characters uniting the two subfamilies and therefore considered the monophyly of the family to be questionable. Following Dubois (1999), Frost (2004) grouped the genera *Arthroleptis* (16 species), *Cardioglossa* (16 species), and *Schoutedenella* (19 species; fig. 1.12 B) in the family Arthroleptidae *sensu stricto,* whereas Duellman (2003) retained the earlier classification uniting arthroleptines and astylosternines into a single family. Frost et al. (2006) placed the arthroleptine and astylosternine genera, along with the genus *Leptopelis* (often placed in the family Hyperoliidae) in an expanded family, Arthroleptidae.

**Natural History:** The natural history of this family is poorly studied. Most arthroleptids are small to medium-size forest-floor or savanna frogs that lay terrestrial eggs that undergo direct development, except for *Cardioglossa,* which has tadpoles.

### Family Ascaphidae

**Distribution:** Pacific Northwest region of North America.

**Content:** One genus, two species. Until recently, only one species, *Ascaphus truei* (fig. 1.12 C), was recognized. Nielson, Lohman, and Sullivan (2001) presented a phylogeographic analysis based on mitochondrial DNA and concluded that inland Rocky Mountain populations should be considered a separate species, *A. montanus.* This genus sometimes has been included in the family Leiopelmatidae (Duellman and Trueb 1986; Duellman 1993a), but *Ascaphus* and *Leiopelma* are united only by shared ancestral characters, not shared derived characters. Ford and Cannatella (1993) considered *Ascaphus* to be a separate clade that is the sister group to all remaining anurans. Mitochondrial DNA data place the ascaphids and leiopelmatids as sister taxa (Hay et al. 1995; Roelants and Bossuyt 2005; Frost et al. 2006). As discussed previously, several other molecular and morphological analyses place these as the most basal frogs.

**Natural History:** These so-called "tailed" frogs live in cold mountain streams and are unique in that males have a copulatory organ that is used to fertilize eggs internally. It

**Fig. 1.12.** Diversity of anurans (Allophrynidae to Brachycephalidae). (A) *Allophryne ruthveni* (Allophrynidae) from South America. (B) *Schoutedenella xenodactyla* (Arthroleptidae) from West Africa. (C) *Ascaphus truei* (Ascaphidae) male, showing copulatory organ, from Washington. (D) *Trichobatrachus robustus* (Astylosternidae) from West Africa. (E) *Bombina orientalis* (Bombinatoridae) from Asia. (F) *Brachycephalus ephippium* (Brachycephalidae) from Brazil; male giving foot-waving display. Photos by Walter Hödl (A, B), William P. Leonard (C), Wayne Van Devender (D), Martha L. Crump (E), and Celio F. B. Haddad (F). For additional photos of these families, see fig. 7.6 E, 14.29 A (*Bombina*), and 10.1 C, D (*Brachycephalus*).

is derived from the cloaca, but has its own skeleton and musculature. Mating sometimes takes place several months before oviposition, with sperm being stored in the female's reproductive tract (Sever et al., 2001, 2003). Eggs are laid in rosary-like strings attached to rocks. The tadpole has an oral disc that allows it to hold onto rocks in swift currents.

### Family Astylosternidae

**Distribution:** Tropical Africa.

**Content:** Five genera, 29 species. These frogs were traditionally treated as a subfamily of ranids, although Laurent (1986) placed astylosternines in the family Hyperoliidae. Dubois (1981) united the subfamilies Arthroleptinae and Astylosterninae into the family Arthroleptidae, which was included in Duellman's (1993a) checklist. Ford and Cannatella (1993) could not identify any shared derived characters uniting the two subfamilies and therefore considered the monophyly of the family to be questionable. Following Dubois (1999), Frost (2004) (http://research.amnh.org/cgi-bin/herpetology/amphibia_tree) grouped the genera *Astylosternus* (11 species), *Leptodactylodon* (15 species), *Nyctibates* (one species), *Scotobleps* (one species), and *Trichobatrachus* (one species; fig. 1.12 D) in the family Astylosternidae, whereas Duellman (2003) retained the earlier classification uniting arthroleptines and astylosternines into a single family. Frost et al. (2006) combined the arthroleptine and astylosternine genera with the genus *Leptopelis* (traditionally placed in the family Hyperoliidae) into an expanded family, Arthroleptidae.

**Natural History:** The natural history of this family is poorly studied. Most astylosternids are riparian, living along rivers and streams. The most famous species is the hairy frog (*Trichobatrachus*), which gets its name from the long hair-like projections that males develop during the breeding season, apparently as auxiliary respiratory organs. All members of this family have aquatic larvae.

### Family Bombinatoridae

**Distribution:** Europe and Asia.

**Content:** Two genera, *Bombina* (eight species) and *Barbourula* (two species). These two genera are sometimes included in the family Discoglossidae. They are united by a number of derived characters that have led taxonomists to place them in a separate family (Lanza, Cei, and Crespo 1976; Cannatella 1985, 1989; Clarke 1987; Ford and Cannatella 1993), although the debate about the relationships among the discoglossoid frogs continues (Haas 2003; Púgener, Maglia, and Trueb 2003; Hertwig, da Sa, and Haas 2004; Roelants and Bossuyt 2005; Frost et al. 2006).

**Natural History:** Species of *Bombina* (fig. 1.12 E) are marked with conspicuous red, yellow, or orange aposematic coloring, mainly on the ventral surface, and have highly distasteful and toxic skin secretions. They lay eggs in shallow ponds and are relatively aquatic as adults, but do venture onto land and readily move between ponds. *Barbourula,* from Borneo and the Philippines, is adapted for living in fast-moving streams, and also has a partially tropical distribution. It feeds in part on freshwater crabs and lays large, unpigmented eggs under rocks in streams.

### Family Brachycephalidae

**Distribution:** Humid coastal forests of southern Brazil.

**Content:** A single genus, *Brachycephalus* (fig. 1.12 F), with eight species, including two species formerly placed in the genus *Psyllophryne.* This small family is endemic to the Atlantic coastal forests of Brazil. Brachycephalids are generally toadlike in appearance and are among the smallest living anurans. The relationships of these very small anurans to other families are unclear, but there is some evidence that they are related to small terrestrial frogs in the genus *Euparkerella,* which are classified as leptodactylids (Giaretta and Sawaya 1998). Darst and Cannatella (2004) found these frogs to be most closely related to eleutherodactylines, and suggested that recognition of the group as a separate family was not warranted. Frost et al. (2006) recommended greatly expanding this family to include the genus *Eleutherodactylus* and its relatives, all previously placed in the family Leptodactylidae. If this change is adopted, the family Brachycephalidae would go from being one of the smallest anuran families to one of the largest, with 16 genera and hundreds of species, all of which apparently produce terrestrial eggs that undergo direct development.

**Natural History:** Very little is known about the natural history of this family. They appear to be typical forest-floor anurans that feed on small insects. One species, *Brachycephalus ephippium,* is bright orange and has highly toxic skin secretions. It is diurnally active and lays eggs in the soil under leaf litter, where they undergo direct development (Pombal, Sazima, and Haddad 1994; Pombal 1999). The other species are cryptically colored, but details of their breeding biology and ecology are unknown.

### Family Bufonidae

**Distribution:** Worldwide, except for Madagascar, Australia, New Guinea, and oceanic islands (*Bufo marinus,* a South American species originally thought to be useful in controlling sugar cane beetles, has been introduced into many Caribbean islands, Australia, and many Pacific islands, including Hawaii and the Philippines).

**Content:** About 33 genera, 480 species. There is no doubt that this family is a monophyletic group, but relationships within the family are poorly understood. The largest genus is *Bufo,* with more than 250 species (fig. 1.13 A, B). Recent analyses based on molecular characters indicate that this genus probably is not monophyletic as presently constituted, because some species share presumed derived characters not found in most members of the genus (Graybeal and Cannatella 1995; Graybeal 1997). On the other hand, some other small bufonid genera actually may belong in the genus *Bufo.* For example, Asian toads from the Himalayas that were placed in the genus *Torrentophryne* because of their unusual morphology and stream-adapted tadpoles appear to be nested within a clade of Asian *Bufo* (Liu et al., 2000). Indeed, Graybeal and Cannatella (1995) found clear evidence of monophyly for only 12 of 33 bufonid genera. The origin of the many species of *Bufo* in North America has been much debated, with various systematists arguing for a close relationship of some or all of these species to African, South American, or Eurasian species. The molecular phylogeny of Pauly, Hillis, and Cannatella (2004) strongly supported a monophyletic clade of North American *Bufo* derived from a Neotropical ancestor.

The Neotropical genus *Atelopus* (fig. 1.13 D), with about 70 species, is the only other genus in the family with more than 25 species. Molecular data suggest that this genus and other Neotropical bufonids not included in *Bufo* (*Osornophryne, Melanophryniscus*) split off relatively early from other bufonids (Graybeal, 1997). Many named genera in this family are very small, with only one to five described species, and have restricted ranges (examples in fig. 1.13 C, F). Several African species originally placed in the genus *Nectophrynoides* (fig. 1.13 E) were placed in separate genera (*Altiphrynoides, Nimbaphrynoides, Spinophrynoides*) by Dubois (1986) because of differences in reproductive biology. These names are used in Frost (2004) but have not been universally adopted by other authors. Nevertheless, egg pigmentation and several morphological characters provide some evidence that the two viviparous species (placed in the genus *Nimbaphrynoides* by Dubois) do in fact form a monophyletic clade, while *Nectophrynoides* in the broad sense does not (Graybeal and Cannatella 1995; Frost et al. 2006). The genera *Altiphrynoides* and *Spinophrynoides* have only one species each, but together appear to form a monophyletic group.

**Natural History:** The so-called true toads (*Bufo*) are found on all major continents except Australia (where one species has been introduced), in both temperate and tropical regions. Despite considerable variation in size, many toads are similar in general morphology and natural history. Most are relatively heavy-bodied, squat anurans with short legs, rough skin, and conspicuous parotoid glands on the head that produce distasteful and sometimes highly toxic secretions containing bufotoxin. These secretions are produced in smaller glands all over the body as well. Many toads feed heavily on ants, which may provide them with at least some of the toxic or distasteful chemicals in their skin. Despite these secretions, most toads are not aposematically colored, probably because of their largely nocturnal habits. Some toads, however, are brightly colored and exhibit marked sexual differences in color, with males being much brighter than females. Most of these species, which include the bright orange *Bufo periglenes* from Costa Rica and *B. canorus* from North America, are active during the day. Most species are terrestrial as adults and are capable of burrowing in soft soils or sand. Toads are moderately active foragers that move by hopping or walking rather than jumping. All species of *Bufo* lay aquatic eggs, usually deposited in long strings in ponds, ditches, and other still bodies of water. Most have aquatic larvae that feed throughout development, but a few species have nonfeeding tadpoles.

The genus *Atelopus* is found mainly along streams in tropical and montane forests of South and Central America (Lötters 1996). These terrestrial anurans have thin bodies, long legs, and slender toes, and generally have bright aposematic markings of red, orange, yellow, and black that advertise their toxic skin secretions. They are largely diurnal and tend to move by walking, rather than hopping like most *Bufo.* They lack a tympanum and a middle ear, yet many species have well-developed vocalizations and can hear airborne sounds. Species of *Atelopus* are stream breeders, usually laying large, unpigmented eggs attached to rocks. The tadpoles have ventral suckers that enable them to hold onto rocks in swift currents. Similar tadpole adaptations are found in several small Old World genera, including the Asian genus *Ansonia* (about 20 species) and the African genus *Werneria* (four species). Arboreal habits are uncommon in bufonids, but some species of *Ansonia, Bufo,* and *Pelophryne* climb onto low vegetation. The brown tree toad of Borneo (*Pedostibes hosii*) is a forest-dwelling arboreal species.

A number of bufonid genera include species with unusual reproductive modes, such as terrestrial oviposition with direct development of eggs in *Oreophrynella,* arboreal oviposition with nonfeeding tadpoles in *Pelophryne,* terrestrial nonfeeding larvae in *Altiphrynoides malcolmi,* ovoviviparity in *Nectophrynoides,* and viviparity in *Nimbaphrynoides liberiensis* and *N. occidentalis.* The last two genera have internal fertilization, as does another African bufonid, *Mertensophryne micranotis.*

### Family Centrolenidae

**Distribution:** Wet tropical forests from southern Mexico through Central America to Argentina.

**Fig. 1.13.** Diversity of anurans (Bufonidae). (A) *Bufo bankorensis* from Taiwan. (B) *Bufo lemur* from Puerto Rico. (C) *Dendrophryniscus minutus* from South America; small male in amplexus with a very large female. (D) *Atelopus varius* from Central America; male giving leg-waving display. (E) *Nectophrynoides tornieri* from East Africa, a viviparous toad. (F) *Schismaderma carens* from South Africa. Photos by Kentwood D. Wells (A), Wayne Van Devender (B), Walter Hödl (C–E), and Alan Channing (F). For additional photos of bufonids, see fig. 1.22 D (*Mertensophryne*); 2.3 B, 8.6 A–C, 14.26 D, 16.5 C, 16.8 A, C, 16.10 C (*Bufo*); and 14.32 D (*Atelopus*).

to which they are stuck. One species, *T. kouniensis,* also feeds on slugs. It has unusually long, recurved teeth that may be useful in grasping these slippery animals and may also be used to penetrate snail shells.

### Correlates of Diet with Head Size and Shape

In frogs with generalized diets, prey size usually is positively correlated with body size and head width, and large frogs usually eat a wider range of prey sizes than small frogs (see chapter 15). Presumably this is because larger frogs can open their mouths wider than small frogs and therefore can accommodate larger prey, but they can still eat small prey as well (Emerson, Greene, and Charnov 1994). This sometimes results in a size-related shift in prey type, with larger individuals or larger species eating categories of prey not available to smaller frogs. For example, the Southeast Asian frog *Limnonectes blythii* eats two general categories of prey: long, thin prey and short, fat prey. Small individuals eat long, thin prey such as millipedes and other elongate invertebrates, but larger individuals also eat long, thin lizards. Beetles constitute the major short, fat prey of small frogs, but larger individuals also eat other frogs, while continuing to eat somewhat larger beetles (Emerson, Greene, and Charnov 1994). In two close relatives of this species, *Limnonectes ingeri* and *L. ibanorum,* males have wider heads and more robust jaw muscles than do females, probably a result of sexual selection (Emerson and Voris 1992; Emerson and Berrigan 1993). Juveniles eat small invertebrates, but never eat crabs, which are too large and too hard-shelled for them to handle. Adult males and females both eat crabs, but the larger heads and more powerful jaw musculature of males allow them to do so at a smaller body size (Emerson and Bramble 1993).

Interspecific comparisons of skull and jaw size and shape have revealed general correlations between diet and morphology in a wide range of anuran species. For example, large-bodied frogs that eat relatively large, slow-moving prey, such as other frogs, typically have relatively wider heads and longer jaws than related species that feed on smaller prey (Scott and Aquino 2005). These are features predicted to increase total gape and the force of jaw closing (Emerson 1985). In contrast, frogs that feed on small, slow-moving prey such as ants tend to have relatively narrow heads, small gapes, and jaws that are shorter than the head, features that facilitate rapid jaw opening and short feeding cycles (the time required to ingest prey).

### Mechanisms of Tongue Protraction

The functional morphology of the tongue and associated musculature and the kinematics of feeding have been studied extensively in frogs, and we now have enough information to place the evolution of feeding behavior in a phylogenetic framework (Nishikawa et al. 1992; Nishikawa 1999, 2000; Deban, O'Reilly, and Nishikawa 2001). Frog tongues all have the same basic morphological structure, but vary in anatomical detail and the degree to which they can be protruded. The ancestral condition, which is found in frogs such as *Ascaphus, Leiopelma, Discoglossus, Bombina,* and most pelobatoid frogs, is a short, disclike tongue broadly attached to the floor of the mouth, with little capacity for projection. This condition also is retained in some neobatrachian groups, including *Telmatobius, Leptodactylus, Lepidobatrachus* (Leptodactylidae), *Limnodynastes* (Myobatrachidae), centrolenids, and all subfamilies of Hylidae except the Phyllomedusinae (fig. 1.28 A). The mechanism of tongue protraction in frogs with this type of tongue is referred to as mechanical pulling. It requires the frog to lunge forward and seize the prey with the tongue and pull it into the mouth (Nishikawa 1999, 2000).

Most of the derived frog lineages have a muscular tongue that is attached in front, with a free posterior flap that can be projected by flipping it forward out of the mouth (fig. 1.28 B; Regal and Gans 1976; Horton 1982; Trueb and Cannatella 1982; Trueb and Gans 1983; Cannatella and Trueb 1988a; Nishikawa et al. 1992; Nishikawa 1999, 2000; Deban, O'Reilly, and Nishikawa 2001). This mode of tongue protraction is called inertial elongation. It allows a frog to capture prey at some distance in front of the mouth without actually lunging toward the prey. Inertial elongation appears to have evolved independently a number of times because it occurs in lineages that are not closely related to one another, including some megophryids (*Megophrys, Leptobrachium*), some leptodacytlids (*Pleurodema, Physalaemus, Eleutherodactylus*), bufonids, phyllomedusine hylids, dendrobatids, and all ranoid frogs except hemisotids and microhylids.

A third mode of tongue protraction is found in the African burrowing frogs in the genus *Hemisus* and many, if not all, microhylid frogs (fig. 1.28 C). These frogs have tongues that are protruded hydrostatically rather than being flipped out of the mouth (Ritter and Nishikawa 1995; Nishikawa, Kier, and Smith 1999; Meyers et al. 2004). As the tongue muscles contract, the thickness of the tongue decreases, while the length increases. This mechanism allows the tongue to be protracted far beyond the tips of the jaws. Although protrusion of the tongue is relatively slow in hydrostatic elongators, this mechanism allows for more precise aiming of the tongue than in species with the other modes of tongue protraction. Precise aiming enhances the ability of these frogs to feed on very small prey, such as mites and termites. The tongue of *Hemisus* is unique in being prehensile, capturing a prey animal by grasping it rather than by adhesion of sticky mucus secretions alone. Contractile properties of the tongue muscles are not dramatically different in species with the three types of tongue protraction

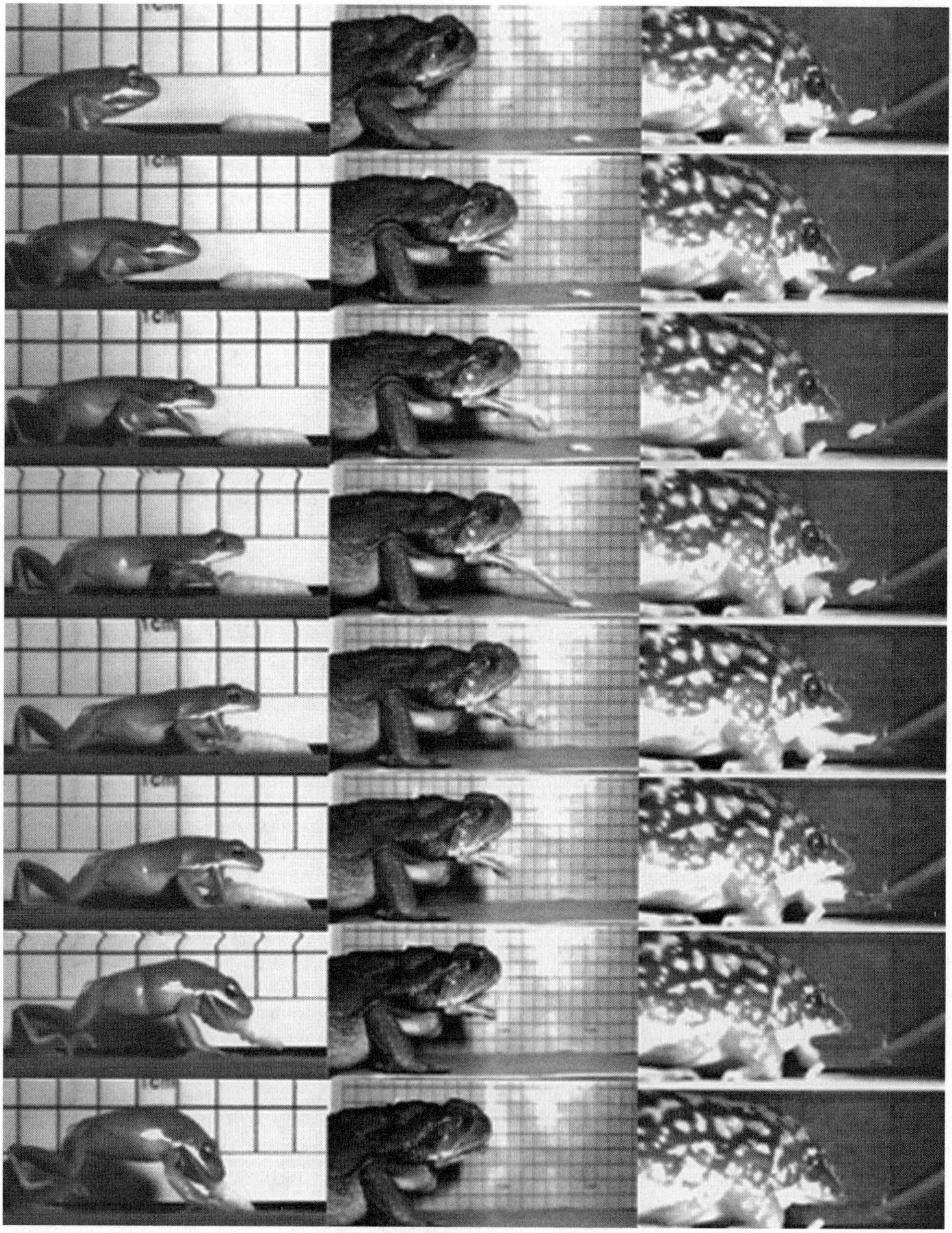

**Fig. 1.28.** Modes of tongue protraction and prey capture in anurans. Each prey capture sequence begins at the top. Left: Mechanical pulling, illustrated by *Hyla cinerea.* The frog lunges toward the prey and pulls it into the mouth with the tongue. Center: Inertial elongation, illustrated by *Bufo marinus.* The tongue is flipped forward out of the mouth and draws the prey into the mouth with little or no movement of the head toward the prey. Right: Hydrostatic elongation, illustrated by *Hemisus marmoratus.* The tongue elongates slowly during protraction, but can be aimed very accurately at small prey animals. Photos from high-speed videos by Kiisa Nishikawa.

mechanisms (Peters and Nishikawa 1999). The tongues do differ in the amount of connective tissue present, with mechanical pullers having the most connective tissue and hydrostatic elongators the least; inertial elongators are intermediate (Nishikawa 2000). The Mexican burrowing frog, *Rhinophrynus dorsalis,* has a tongue that is not attached at the front that can be protruded well beyond the tips of the jaws, but the mechanism of tongue projection has not been investigated in detail. It may represent functional convergence with the tongues of hemisotids and microhylids, but not necessarily convergence in the precise mechanism of tongue elongation (Nishiskawa 2000).

Pipids are unique among anurans in lacking a tongue altogether, but vestiges of the tongue muscles indicate that they evolved from ancestors with functional tongues. The loss of the tongue is related to their aquatic habits. Most pipids feed by using the front feet to shove prey into the mouth and grabbing the prey in the jaws. Members of the African pipid genus *Hymenochirus* are suction feeders both as adults and as tadpoles. Some other frogs have secondarily evolved aquatic feeding as adults, but the lack of suction feeding in the larval stage apparently limits their ability to suction feed as adults. These frogs generally use their jaws to capture prey underwater, because a protrusible tongue would be of little benefit in aquatic feeding (O'Reilly, Deban, and Nishikawa 002).

Differences in tongue protraction mechanisms are correlated with other differences in feeding behavior, but morphological and behavioral traits do not always evolve together in different lineages, and frogs with a derived morphology may retain elements of ancestral feeding behavior. A phylogenetic analysis of the evolution of feeding behavior (fig. 1.29) indicates that frogs with the ancestral tongue morphology typically feed by opening their mouths and bending the jaw downward, lunging forward with the entire body to contact the prey with the short tongue. The head is then bent downward, with the tongue and jaws being used to capture the prey (Nishikawa and Cannatella 1991; Nishikawa and Roth 1991; Nishikawa et al. 1992; Nishikawa 1999, 2000). Because the distance to which the tongue can be protracted is very short (only about 3 mm beyond the tips of the jaws), these frogs must lunge forward with the whole body, and they probably are restricted to feeding on relatively large, slow-moving prey at close range. Indeed, frogs that feed in this manner apparently do not specialize in fast-moving prey, such as flying insects, or very small prey, such as ants, termites, and mites. Instead, they appear to feed mainly on slow-moving prey such as caterpillars, beetles, aquatic insect larvae, spiders, earthworms, ostracods, and snails (Sharell 1966; Nussbaum, Brodie, Jr., and Storm 1983; Nöllert 1984; Herrmann et al. 1987; Nishikawa 2000).

**Fig. 1.29.** Cladogram illustrating the evolution of tongue-protraction mechanisms in anurans. This cladogram also shows relationships among genera within the large family Leptodactylidae and several other families. This analysis suggests that mechanical pulling is the ancestral condition that is found in most primitive frogs and is retained in many neobatrachian clades. Inertial elongation appears to have evolved independently in several different lineages of frogs, whereas hydrostatic elongation apparently evolved only once from an ancestor with inertial elongation. After Nishikawa (1999).

Among more derived frogs, the independent evolution of inertial elongation in several different lineages has enabled the frogs to use the tongue to capture prey without the use of their jaws. Most derived frogs with highly protrusible tongues do not lunge at the prey, but instead project the tongue toward the prey and then pull the prey back into the mouth with the tongue. The tongue can be projected at least 70% the length of the jaw at maximum extension (up to 300% in *Hemisus*). The ability to project the tongue allows for rapid prey capture and probably allows for the evolution of feeding specialization on fast-moving prey and very small prey items. Hydrostatic elongation appears to be derived from inertial elongation. While less suited to capturing rapidly moving prey, it does enable frogs to be more successful in capturing prey on each attempt (Nishikawa 2000).

There is considerable variation in the details of feeding behavior among different lineages. In *Bufo,* for example, there is relatively little forward movement of the head and body (fig. 1.28 B); the toad shoots the tongue forward and then draws the prey into the mouth (Gans and Gorniak 1982a, b). In contrast, phyllomedusine frogs often lunge toward the prey, as in other hylids, but also project the tongue. Sometimes the prey is pinned to the substrate with the tongue and then captured in the jaws when the forward lunge is completed, but small prey may be captured by the tongue and pulled back into the mouth without a lunge (Gray and Nishikawa 1995). The highly protrusible tongue of phyllomedusines may facilitate rapid capture of fast-moving prey such as orthopterans that are common in their rainforest habitats. Spadefoot toads (*Spea*) have relatively long tongues that can be protracted well beyond the tips of the jaws, but they retain the ancestral feeding behavior of lunging forward toward the prey (O'Reilly and Nishikawa 1995).

Some derived frogs exhibit shifts in feeding mode, depending on the type of prey being eaten. For example, when leopard frogs (*Rana pipiens*) are feeding on small, slow-moving prey, they capture the prey by projecting the tongue, with little movement of the body and head. However, when capturing larger prey, they revert to the ancestral pattern of lunging forward and seizing the prey in their jaws (C. W. Anderson 1993; Anderson and Nishikawa 1996). The Australian frog *Cyclorana novaehollandiae* also modulates its mode of prey capture, depending on the size and speed of movement of the prey. Visual assessment of prey characteristics appears to be more important than biomechanical feedback in determining the mode of prey capture employed (Valdez and Nishikawa 1997).

The evolution of tongue morphology and changes in feeding behavior have been accompanied by changes in neural pathways involved in the control of prey capture. In derived anurans, such as *Bufo* and *Rana,* feedback from the hypoglossal nerve running through the tongue is required to trigger proper opening of the mouth, especially when the frogs are feeding on small prey (Nishikawa and Gans 1992; Nishikawa 1999, 2000). Presumably, feeding on small prey captured by the tongue requires tight coordination of mouth opening and tongue projection. However, when *Rana* feeds on large prey by lunging forward, hypoglossal feedback is not required (C. W. Anderson and Nishikawa 1993). The absence of hypoglossal feedback appears to be the ancestral condition for anurans, because it is absent in most species examined so far. It is present in some derived lineages, including most ranoid frogs, but appears to have been secondarily lost in hydrostatic tongue elongators (*Hemisus* and microhylids). In most anurans, withdrawal of the tongue back into the mouth is controlled by another feedback system involving the glossopharyngeal nerve, which allows the frog to determine when the tongue has contacted the prey. Failure to contact the prey often results in a delayed withdrawal of the tongue (Matsushima, Satou, and Ueda 1986, 1987, 1988, 1989; Nishikawa et al. 1992). The presence of glossopharyngeal feedback appears to be the ancestral condition for anurans, but has been secondarily lost in a few anurans and is not present in salamanders.

In addition to use of the tongue and jaws to capture prey, many anurans also use their front feet to capture prey and push it into the mouth, a behavior not seen in urodeles. L. A. Gray, O'Reilly, and Nishikawa (1997) studied the use of forelimbs in feeding in more than 100 species of anurans representing 16 families. They identified several distinct behavior patterns. Two of these, scooping prey into the mouth with the back of the front foot and wiping the prey with the palm of the hand, appear to be ancestral traits that have been retained in many lineages. Other behavior patterns are found only in certain groups. For example, grasping the prey with the fingers is a behavior found only in arboreal species (hylids, hyperoliids, and rhacophorids) that have evolved grasping hands for locomotion in trees. Frogs in several lineages hold prey animals on the substrate with the front feet while pulling the prey into the mouth with the jaws. This behavior appears to have evolved independently in at least a half dozen unrelated clades.

## Phylogeny and Classification of the Urodela

Most recent classifications of the Urodela (also called Caudata; Frost 2004; Frost et al. 2006) recognize 10 families, with about 60 genera and more than 400 species. The family-level classification of salamanders has remained relatively stable, but phylogenetic relationships among the families have been uncertain. One hypothesized phylogeny is shown in fig.1.30. One problem with assessing the relationships of many urodele families is the frequency with which larval characters are retained in reproductive adults (Wiens, Bonett, and Chippindale 2005). In addition, phylogenies based entirely on morphological characters often are not consistent with those based on molecular characters (Mueller et al. 2004; Weisrock, Kozak, and Larson 2005). Some of these problems have been accommodated in recent phylogenetic analyses that combine morphological and molecular data (Larson and Dimmick 1993; Larson, Weisrock, and Kozak 2003).

Many authors place the Sirenidae in a basal position as the sister taxon to all other caudate families. However, the phylogeny presented by Trueb (1993), which was based on an unpublished morphological data set, placed the Sirenidae as the most derived family of salamanders, an arrange-

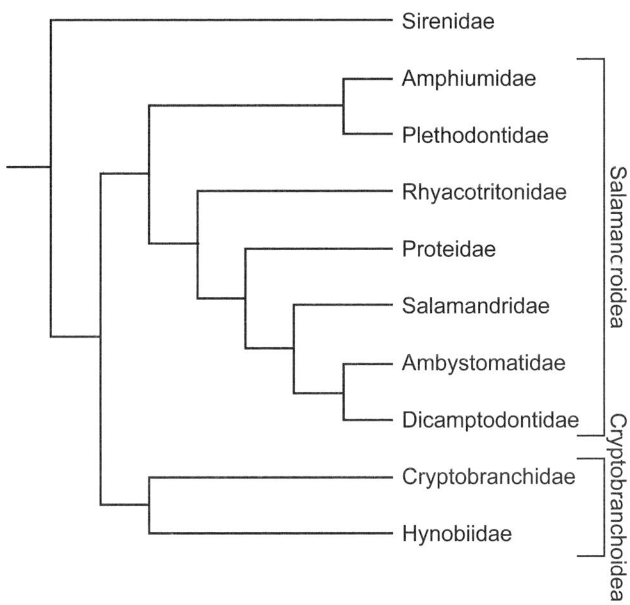


**Fig. 1.30.** Phylogenetic relationships of the urodeles, based on combined molecular and morphological data. Some other recent phylogenies differ in the placement of the family Sirenidae, which may or may not be basal to all other salamander families. There also is some debate over whether Salamandroidea is a monophyletic clade. After Pough et al. (2004), based on A. Larson and Dimmick (1993).

ment that most workers consider unlikely, in part because it would require the secondary loss of internal fertilization in this group (if fertilization actually is external in sirenids). A combined molecular and morphological analysis by Gao and Shubin (2001), which included fossil taxa, did not place sirenids in a basal position, but as the sister group to the Proteidae. Other recent phylogenetic studies based on molecular or combined molecular and morphological data also did not place sirenids in a basal position, but the issue has yet to be fully resolved (Weisrock, Harmon, and Larson 2005; Wiens, Bonett, and Chippindale 2005; Frost et al. 2006).

The remaining families of urodeles usually are placed in two distinct clades, the Cryptobranchoidea (Cryptobranchidae and Hynobiidae) and the Salamandroidea (all remaining families). If Sirenidae is considered a basal group, then this arrangement suggests that internal fertilization probably arose only once in salamander evolution (all salamandroids have internal fertilization, whereas sirenids, cryptobranchids, and hynobiids do not; Weisrock, Harmon, and Larson 2005). Relationships among the salamandroid families are the subject of some debate. In particular, the relationship of the amphiumas to other families is unclear. Most recent phylogenies suggest they are closely related to the plethodontids, as shown in fig. 1.30 (Larson and Dimmick 1993; Larson, Weisrock, and Kozak 2003; Weisrock, Harmon, and Larson 2005; Wiens, Bonett, and Chippindale 2005). The phylogeny presented by Trueb (1993) placed them as the sister group to the Proteidae, an unlikely arrangement that almost certainly results from similar paedomorphic characters (Wiens, Bonett, and Chippindale 2005). There also is conflicting evidence about the position of the plethodontids, the largest and most diverse family of salamanders. Analyses based on morphological characters have identified them as a highly derived group, whereas some molecular phylogenies place them in a position basal to other families (Larson, Weisrock, and Kozak 2003; Weisrock, Harmon, and Larson 2005).

## Synopsis of Families of Urodela

The following accounts summarize the classification, distribution, and natural history of the 10 currently recognized families of urodeles. The classification scheme and numbers of genera and species follow Larson, Weisrock, and Kozak (2003). The family-level classification of the Urodela has been more stable in recent years than that of anurans. The only major change from the classification presented in Duellman and Trueb (1986) is the recognition of the Rhyacotritonidae as a separate family. The genus *Rhyacotriton* was previously included in the family Dicamptodontidae, but it is now clear that *Rhyacotriton* and *Dicamptodon* are not closely related (Good and Wake 1992; Larson and Dimmick, 1993; Larson, Weisrock, and Kozak 2003; Weisrock, Harmon, and Larson 2005; Wiens, Bonett, and Chippindale 2005). *Dicamptodon* appears to be most closely related to the ambystomatids, and with *Rhyacotriton* now removed from the Dicamptodontidae, the dicamptodontids and ambystomatids could be placed in the same family (fig. 1.30). The number of new species of salamanders described each year is much lower than the number of new anuran species, but only slightly lower in proportion to the total number in the two groups (about a 10% increase in caudate species from 1985 to 1992, compared to 14% for anurans; Duellman, 1993a). The difference is due mainly to the low diversity of salamanders in much of South America and their absence from the Old World tropics, the regions where most new frog species have been discovered. It is therefore particularly surprising that a new plethodontid salamander was recently described from Korea, a well-studied region of a continent on which plethodontid salamanders have never been found before (Min et al. 2005).

Information on the general natural history of salamanders is taken from literature cited elsewhere in this book and from standard reference sources, including Bishop (1941b, 1947), Liu (1950), Stebbins (1954a, 1985), Steward (1969), E. N. Arnold and Burton (1978), Duellman and Trueb (1986), Pfingsten and Downs (1989), Conant and Collins (1991),

Griffiths (1996), Petranka (1998), Bruce et al. (2000), and Pough et al. (2004).

## Family Ambystomatidae

**Distribution:** North America from Alaska to the eastern United States and the Mexican Plateau.

**Content:** One genus, *Ambystoma,* 31 species (fig. 1.31 A). Four stream-dwelling species from Mexico formerly were placed in a separate genus *Rhyacosiredon,* but a phylogenetic analysis of electrophoretic data showed *Rhyacosiredon* nested within the genus *Ambystoma* (Shaffer 1984a). Nearly all of the morphological characters thought to distinguish *Rhyacosiredon* from *Ambystoma* are the result of incomplete metamorphosis, a trend that is common in this family. Therefore, *Rhyacosiredon* is now considered a synonym of *Ambystoma* (Reilly and Brandon 1994). Species in the *A. tigrinum* complex show significant genetic divergence with little morphological differentiation (Shaffer 1993; Shaffer and McKnight 1996; Irschick and Shaffer 1997). Several members of this complex traditionally have been treated as subspecies of *A. tigrinum* (*A. t. tigrinum, A. t. diaboli, A. t. mavortium, A. t. melanostictum, A. t. nebulosum, A. t. stebbinsi;* Petranka, 1998). Irschick and Shaffer (1997) recommended that *A. t. tigrinum* and two closely related species, *A. californiense* and *A. velasci,* be recognized

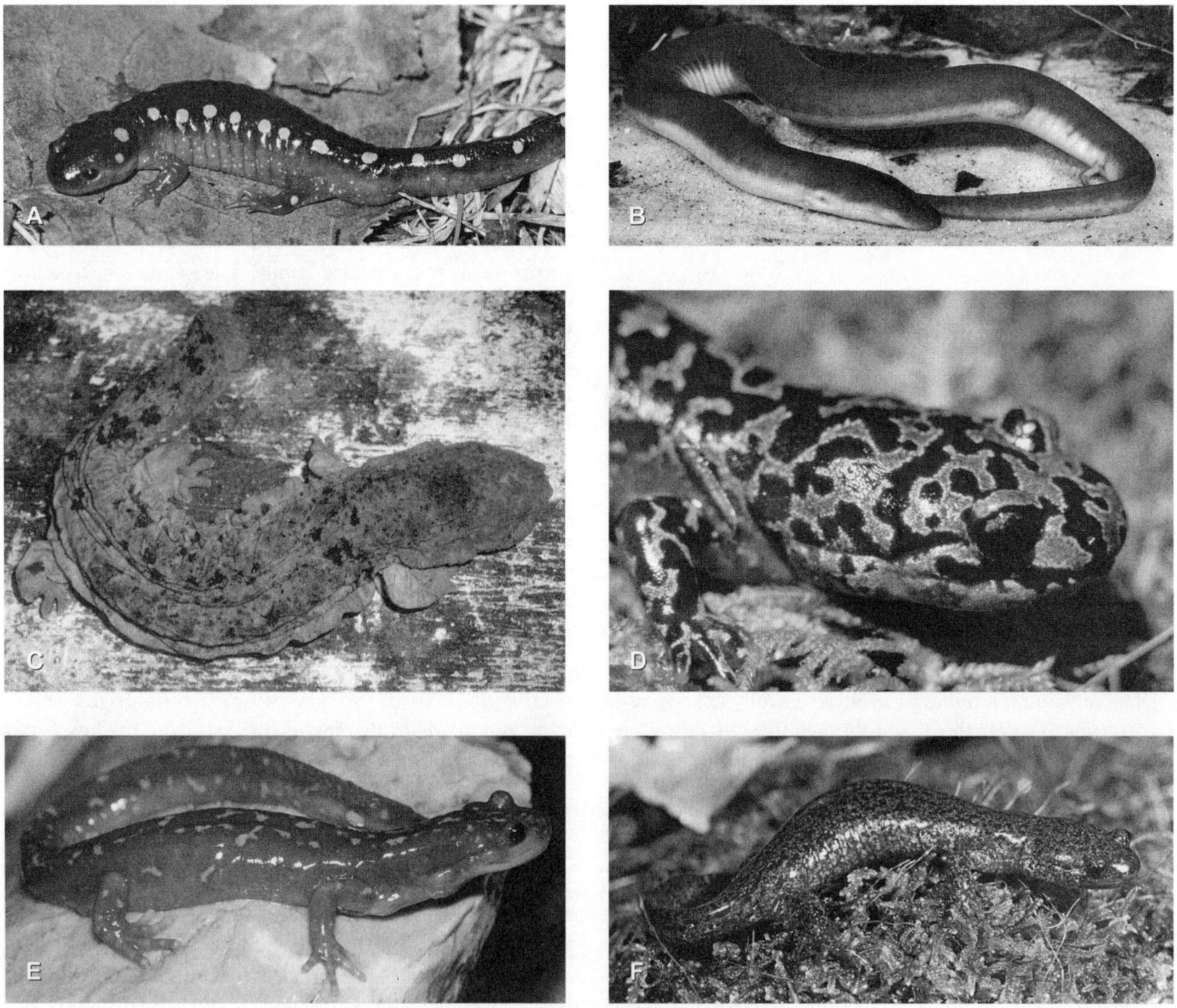

**Fig. 1.31.** Diversity of urodeles (Ambystomatidae to Hynobiidae). (A) *Ambystoma maculatum* (Ambystomatidae) from Connecticut. (B) *Amphiuma tridactylum* (Amphiumidae) from the southern United States. (C) *Cryptobranchus alleganiensis* (Cryptobranchidae) from the Ozark Mountains. (D) *Dicamptodon ensatus* (Dicamptodontidae) from California. (E) Batrachuperus persicus (Hynobiidae) from Iran. (F) *Hynobius* sp. (Hynobiidae) from Asia. Photos by Kentwood D. Wells (A), Wayne Van Devender (B, C, E, F), and William P. Leonard (D). For additional photos of these families, see fig. 1.1 B, 11.2 B, 14.30 F, 16.8 B, 16.9 (*Ambystoma*) and 11.12 A (*Amphiuma*).

as distinct species. The remaining taxa are almost indistinguishable morphologically, but their taxonomy was left unresolved. Frost (2004) and Larson, Weisrock, and Kozak (2003) placed all of these populations in a separate species, *A. mavortium*. Most of the published literature on the ecology of these groups treats them as subspecies of *A. tigrinum*.

**Natural History:** Most ambystomatids are moderately large terrestrial salamanders with lungs. Although commonly called mole salamanders, members of this family have limited burrowing ability and use tunnels constructed by other animals or root channels as retreats outside of the breeding season. Most species are winter or early spring breeders that lay eggs in ponds, either in gelatinous clumps, individually attached to plants, or scattered on the pond bottom (e.g., *A. maculatum, A. tigrinum, A. texanum*). Several Mexican species, including *A. rosaceum, A. ordinarium*, and those formerly placed in *Rhyacosiredon*, are stream breeders that lay eggs in clusters underneath flat stones. The only species in the United States with these breeding habits is *A. barbouri*, a close relative of the pond-breeding species *A. texanum*. Two species, *A. opacum* and *A. cingulatum*, lay eggs on land in depressions that are flooded by winter rains. Females sometimes remain with their eggs in *A. opacum*, but not in *A. cingulatum*. In all species, fertilization is internal by means of spermatophores deposited singly or in groups. Many ambystomatids are paedomorphic and perennibranchiate, retaining external gills and other larval characters after reaching sexual maturity. In some cases, such as the Mexican axolotl (*A. mexicanum*), the capacity for metamorphosis has been lost, but in others, such as *A. talpoideum*, a paedomorphic life history is facultative. This type of life history is especially common at high altitudes or arid regions where the environment surrounding the breeding ponds is inhospitable to terrestrial life (see chapter 13 for a more detailed discussion of this type of life history). Also common is the production of hybrid, polyploid, all female "species" that depend on sperm from a parental species to initiate egg development (gynogenesis). The best known are populations resulting from hybridization between *A. jeffersonianum* and *A. laterale* in the northeastern United States and Canada, but polyploid hybrids that incorporate genomes of *A. texanum* and *A. tigrinum* also occur (Bogart 2003).

### Family Amphiumidae

**Distribution:** Southeastern United States.

**Content:** A single genus, *Amphiuma*, with three species (fig. 1.31 B).

**Natural History:** These are largely aquatic salamanders that have elongate bodies and greatly reduced limbs. They inhabit swamps and ponds that often are poorly supplied with oxygen and therefore depend mainly on their lungs for respiration (see chapter 4). Not much is known about their mating behavior, but fertilization is internal by means of spermatophores. Eggs are laid in burrows constructed at the edges of ponds, under logs, or in the nest mounds of alligators, and are attended by females for several months during dry periods (Fontenot 1999). The larvae hatch at an advanced stage of development and undergo relatively minor morphological change during metamorphosis. The gills are lost in adults, but the gill slits remain open.

### Family Cryptobranchidae

**Distribution:** Eastern Asia and eastern North America.

**Content:** Two genera, the Chinese and Japanese giant salamanders, *Andrias* (two species) and the North American hellbender, *Cryptobranchus alleganiensis* (fig. 1.31 C). These are generally considered to be among the most primitive of living salamanders and are most closely related to hynobiids.

**Natural History:** The three species in this family are the three largest living salamanders. All live in relatively large, well-oxygenated streams and rivers and are completely aquatic. They have flattened bodies with many folds of loose skin that serve as the principal respiratory organ. Male hellbenders are territorial in the breeding season and defend nest sites under rocks, where several females deposit their eggs. In contrast to most salamanders, male cryptobranchids fertilize eggs externally by releasing sperm as they are laid. Males of the Japanese giant salamander (*Andrias japonicus*) build nests in tunnels in riverbanks and defend them as territories, but more than one male sometimes manages to fertilize the eggs of a female in a nest (Kawamichi and Ueda 1998).

### Family Dicamptodontidae

**Distribution:** Pacific Northwest region of the United States and adjacent Canada.

**Content:** A single genus, *Dicamptodon*, with four species (fig. 1.31 D). These salamanders were formerly placed in the family Ambystomatidae and most phylogenetic analyses place them as the sister group to that family. There is considerable genetic differentiation among the four species, but little morphological difference.

**Natural History:** These salamanders are terrestrial as adults, and may even climb a meter or more into vegetation (Stebbins 1954a; Nussbaum, Brodie, and Storm 1983). They usually live near cold, well-oxygenated streams, where eggs are laid in well-hidden locations. Fertilization is internal by means of spermatophores. The larval period is prolonged, sometimes lasting for several years, and some populations exhibit facultative paedomorphosis, retaining larval traits such as external gills as adults (see chapter 13). *D. copei* is permanently paedomorphic.

### Family Hynobiidae

**Distribution:** Eastern Asia, including Siberia, northern China, Korea, and Japan. Disjunct populations in Central Asia.

**Content:** Seven to nine genera, 49 species. This small family comprises some of the most basal of living salamanders and is the sister-group to the cryptobranchids. About two-thirds of the species are in the genus *Hynobius* (fig. 1.31 F). The only other genus with more than two described species is *Batrachuperus* (nine species; fig. 1.31 E). Usually five species are placed in the genus *Ranodon,* but some classifications restrict this genus to the single species *R. sibiricus,* placing the other species in the genera *Liua* (one species) and *Pseudohynobius* (three species).

**Natural History:** Most are terrestrial as adults, but all breed in water, usually in the winter or early spring. *Onychodactylus* is semiaquatic and lives among the rocks of cold mountain streams. It lacks lungs and depends primarily on the skin for respiration. Lungs are reduced in *Ranodon,* another stream-dwelling genus. Hynobiids lay eggs in paired sacs that sometimes resemble the egg clutches of ambystomatids. Eggs are fertilized externally when the male clasps an egg sac after it has been deposited. In one genus, *Ranodon,* the male deposits a large sperm mass, and the female then places her eggs on top of it to achieve fertilization.

### Family Plethodontidae

**Distribution:** North, Central, and South America, with one genus (*Hydromantes*) having a disjunct distribution in northwestern North America and southern Europe. A single species, *Karsenia koreana,* occurs in Korea, the only plethodontid known from Asia (Min et al. 2005).

**Content:** 27 genera, more than 375 species. With more than 70% of all living species, this is the largest and most diverse family of salamanders. Two subfamilies have long been recognized, the Desmognathinae, with two genera (fig. 1.32 A, B), and the Plethodontinae, with all of the rest (fig. 1.32 C–F). The monophyly of the Desmognathinae is strongly supported in most phylogenetic analyses, but several recent studies have shown that this group is derived from within the Plethodontinae, which therefore is not a monophyletic group (Chippindale et al. 2004; Mueller et al. 2004). Frost et al. (2006) recommended eliminating the subfamily name and placed the desmognathine genera in the subfamily Plethodontinae. An analysis based on complete mitochondrial genomes did not support the monophyly of any of the three major clades traditionally recognized within the Plethodontinae (the tribes Plethodontini, Hemidactyliini, and Bolitoglossini; Mueller et al. 2004).

Several of the largest genera of plethodontids, including *Bolitoglossa* (fig. 1.32 C), *Pseudoeurycea, Oedipina* (fig. 1.35 A), and *Nototriton* (fig. 1.35 C), *Chiropterotriton,* are restricted to the Neotropics and are most abundant in montane regions of Mexico and Central America. All of these genera belong to a single clade that is presumed to have invaded the area from North America. Plethodontid salamanders in Central America exhibit a high degree of genetic differentiation among local populations, both within and among species, as well as restricted ranges and sharp elevational limits to their distributions (García-París et al. 2000). This means that there probably are many more species in the region than are currently recognized, and new species are described every year. The only other genera of plethodontids with more than 10 species all are found in North America: *Plethodon* (more than 50 species), *Eurycea* (26 species; fig. 1.32 E), *Desmognathus* (at least 19 species; fig. 1.32 A), and *Batrachoseps* (20 species; fig. 1.35 B).

Plethodontid systematics is somewhat confusing because genetic differentiation and morphological differentiation often have proceeded at different rates. Some genera, including the large genus *Plethodon* and the genus *Aneides,* do not appear to be monophyletic clades as presently constituted (Mahoney 2001). Within genera, there are many cryptic species that are well differentiated genetically, but very similar morphologically. Especially problematic groups are the *Plethodon jordani-P. glutinosus* complex (see fig. 15.11 in chapter 15) of large forest-floor species (Highton 1970, 1990, 1995; Highton and Henry 1970; Highton and Larson 1979; Highton, Maha, and Maxson 1989; Highton and Peabody 2000) and members of the *Desmognathus ochrophaeus* complex of streamside salamanders (see fig. 15.16 in chapter 15; Tilley and Mahoney 1996; Tilley 1997). In these groups, there are many geographic isolates that are morphologically similar, and there are morphologically distinct populations that hybridize to varying degrees where their ranges overlap. There also are cases, as in *Desmognathus imitator,* where extensive variation in color pattern does not correspond to a similar degree of genetic differentiation (Tilley 2000). This has led to disagreement among different authors over how many distinct species should be recognized and what names should be assigned to those species. Similar problems arise in determining species boundaries in other plethodontid genera, including populations of *Eurycea* in Central Texas (Chippindale 2000; Chippindale et al. 2000; Wiens, Chippindale, and Hillis 2003) and populations of *Batrachoseps* (D. Wake and Jockusch 2000) and *Ensatina* on the West Coast of the United States (Jackman and Wake 1994; D. Wake 1997; Highton 1998; D. Wake and Schneider 1998; D. Wake and Jockusch 2000).

**Fig. 1.32.** Diversity of urodeles (Plethodontidae). (A) *Desmognathus marmoratus* from North Carolina. (B) *Phaeognathus hubrichti* from Alabama. (C) *Bolitoglossa pesrubra* from Costa Rica. (D) *Eurycea cirrigera* from North Carolina, with elongated cirri on the male's snout. (E) *Gyrinophilus porphyriticus* from North Carolina. (F) *Hydromantes platycephalus* from California. Photos by Wayne Van Devender (A, B, D–F) and Walter Hödl (C). For additional photos of plethodontid salamanders, see fig. 1.35 A (*Oedipina*); 1.35 B (*Batrachoseps*); 1.35 C (*Nototriton*); 1.42 (*Hydromantes*); 9.13, 14.33 B, 15.16 (*Desmognathus*); 9.19, 14.33 A, 15.11, 16.5 A, B (*Plethodon*); 9.27, 14.25 D (*Aneides*); 11.12 (*Hemidactylium*); 13.24 A (*Gyrinophilus*); 13.24 B, D–F, 14.20 D (*Eurycea*); and 13.24 C (*Haideotriton*).

The complexity of plethodontid systematics reflects a tendency toward morphological conservatism in the group and a history of population interactions at various stages of isolation and speciation (S. J. Arnold 2000). A practical consequence is that different names sometimes have been used in studies of the same or adjacent populations, or names have changed as studies have been conducted. I have attempted to use the currently accepted names for species when discussing studies of ecology and behavior in this book. In some cases, however, this is made more difficult by the lack of agreement on what names should be used. For example, Highton and Peabody (2000), Frost (2004), and Larson, Weisrock, and Kozak (2003) use the name *Plethodon teyahalee* for populations in the *P. glutinosus* complex that were called *P. oconaluftee* by Hairston (1993) and Petranka (1998).

**Natural History:** Plethodontids are lungless salamanders, which depend primarily on the skin for respiration. They generally are small to medium-sized salamanders that are terrestrial, arboreal, or semiaquatic as adults. The desmognathines have wedge-shaped heads (fig. 1.32 A) that may be an adaptation to burrowing under rocks and other cover objects (Dunn 1926c; D. Wake 1966; Schwenk and Wake 1993). One member of this subfamily, *Phaeognathus* *hubrichti* (fig. 1.32 B), is highly fossorial and lays terrestrial eggs that probably undergo direct development (Brandon 1965a). Most other members of this subfamily are associated with streams. The two largest species, *Desmognathus quadramaculatus* (see fig. 15.16 A in chapter 15) and *Desmognathus marmoratus* (fig. 1.32 A), are the most aquatic. Others live along the margins of streams, where eggs are deposited under partially submerged rocks or wet seepage areas and hatch into aquatic larvae (*D. monticola, D. fuscus, D. ochrophaeus, D. ocoee, D. orestes,* and other species; see fig. 15.16 in chapter 15). Two very small species, *D. aeneus* and *D. wrighti,* lay terrestrial eggs that hatch into nonfeeding larvae or undergo direct development. The genus *Desmognathus* thus exhibits a gradient in body size from large aquatic species to very small terrestrial ones. Traditionally this gradient has been viewed as reflecting an ancestral life history of aquatic eggs and larvae, with the smaller and more terrestrial species being more derived (e.g., Dunn 1926c). A molecular phylogenetic study by Titus and Larson (1996) questioned this interpretation. They found that the most terrestrial species represent relatively basal lineages, with some of the more aquatic species being more derived. If this is correct, then it suggests the possible reacquisition of aquatic development in most members of this genus from an ancestor with direct development (Larson, Weisrock, and Kozak 2003; Chippindale et al. 2004). Indeed, within the plethodontids as a whole, the reevolution of an aquatic larval stage from a direct-developing ancestor may have occurred a number of times (Mueller et al. 2004; see fig. 10.14 in chapter 10).

Until recently, nondesmongnathine plethodontids that lay eggs in or near water and have aquatic larvae were considered to be members of a monophyletic clade, the Tribe Hemidactyliini, but a recent analysis indicated that *Hemidactylium* is not closely related to other members of this group (Mueller et al. 2004). Aquatic-breeding plethodontids use a variety of habitats, including springs, streams, and seepage areas (*Pseudotriton, Gyrinophilus,* and most *Eurycea;* fig. 1.32 D, E), as well as ponds, swamps, and ditches (*Stereochilus marginatus, Eurycea quadridigitata,* and *Hemidactylium scutatum*). Many species have become adapted to life in caves. These include some species of *Eurycea* and *Gyrinophilus,* as well as the highly specialized cave salamanders (*Typhlotriton* and *Haideotriton;* cave salamanders previously placed in the genus *Typhlomolge* are now considered synonymous with *Eurycea*). Cave salamanders often have reduced pigment, or lack it altogether, and the eyes often are reduced or absent as well. The most specialized cave-adapted salamanders spend nearly all of their lives in underground streams and pools, and some retain gills throughout life (see further discussion in chapter 13, and fig. 13.24).

Most plethodontids lay nonaquatic eggs that undergo direct development. In North America, most species are terrestrial inhabitants of the forest floor, where they shelter under rocks and logs, or move underground through root channels and tunnels constructed by other animals. Members of the genus *Aneides* have enlarged toe pads for climbing and are arboreal or live in rock crevices. Members of the genus *Hydromantes* (fig. 1.32 F), found in both western North America and southern Europe, live on rock outcrops, cliffs, and in caves (some authors place the European species in the genus *Speleomantes*). Some Neotropical plethodontids also are found under rocks and logs on the ground, but others live in specialized habitats, such as bromeliads or mats of moss and other vegetation on the branches of trees in cloud forests. Within the largest genus *Bolitoglossa,* there are seven well-defined clades, of which only one has made its way into South America from Central America (Parra-Olea, García-París, and Wake 2004).

### Family Proteidae

**Distribution:** Southern Europe and eastern North America.

**Content:** Two genera, *Necturus* from North America (five species; fig. 1.33 A) and *Proteus* from Europe (one species; fig. 1.33 B). Most phylogenetic analyses unite these two genera in a single family, although the analysis of Weisrock, Harmon, and Larson (2005) did not support the monophyly of this group.

**Natural History:** The mudpuppies (*Necturus*) are fully aquatic salamanders with prominent external gills and small eyes. They usually live in relatively large rivers and lakes, but can be found in smaller streams as well, where they prey on invertebrates and small vertebrates. Eggs are laid in clusters under flat stones. *Proteus anguinus,* the European olm, is a blind, cave-dwelling species. Most populations lack skin pigment, but one fully pigmented population has been reported. The natural history in the field is poorly studied, but in captivity, males defend territories centered around rock crevices that are used as oviposition sites by females. All proteids have internal fertilization by means of spermatophores.

### Family Rhyacotritonidae

**Distribution:** Pacific Northwest region of the United States.

**Content:** One genus, *Rhyacotriton,* with four species (fig. 1.33 C). These small salamanders once were all classified as one species, *R. olympicus,* and were placed in the family Dicamptodontidae. The work of Good and Wake (1992) revealed hidden taxonomic diversity in the genus.

**Natural History:** These salamanders live in or near cold, fast running streams. Their lungs are reduced and nonfunc-

**Fig. 1.33.** Diversity of urodeles (Proteidae to Sirenidae). (A) *Necturus punctatus* (Proteidae) from the Coastal Plain of the southeastern United States. (B) *Proteus anguinus* (Proteidae) from Europe. (C) *Rhyacotriton cascadae* (Rhyacotritonidae) from Washington. (D) *Siren lacertina* from the southeastern United States. Photos by Wayne Van Devender (A, C, D) and Walter Hödl (B).

tional for respiration. Their natural history is poorly known, but eggs are laid in well-hidden locations under rocks and in underground springs feeding into the streams where the larvae complete their development. Fertilization is internal by means of spermatophores.

### Family Salamandridae

**Distribution:** Europe, Asia, extreme northern Africa, North America.

**Content:** 15–20 genera, about 70 species (fig. 1.34). Most classifications recognize 15 genera, but new names have been proposed for some species within the genus *Triturus* (*Lissotriton, Mesotriton, Ommatotriton*) and *Salamandra* (*Lyciasalamandra;* see Frost et al. 2006, and Amphibia Web at http://elib.cs.berkeley.edu/aw/lists/Salamandridae.shtml), but I have used the traditional genus names throughout this book. The largest genera are *Triturus* (including the new names listed above [14 species;] fig. 1.34 C), *Salamandra* (up to 13 species, including *Lyciasalamandra*), and *Tylototriton, Cynops* and *Paramesotriton* (seven species each). No other genus has more than four species. Only two genera occur in North America, *Notophthalmus* (fig. 1.34 A) and *Taricha* (fig. 1.34 B), each with three species. Several phylogenetic analyses of the family based on mitochondrial DNA provide strong support for a monophyletic clade composed of *Salamandra, Mertensiella* (fig. 1.34 F), and *Chioglossa* being the sister group to the newts (all remaining genera in the family; Titus and Larson 1995; Weisrock et al. 2001; Larson, Weisrock, and Kozak 2003). Within this group, the two previously recognized species of *Mertensiella* do not appear to form a monophyletic group, and *M. luschani* has been placed in the genus *Salamandra* (or *Lyciasalamandra*) by most recent authors (Weisrock et al. 2001; Larson, Weisrock, and Kozak 2003). Numerous color morphs, described as subspecies of *S. luschani,* may or may not be distinct species. The phylogenetic position of *Salamandrina terdigitata* is uncertain. Its tongue morphology is similar to that of *Salamandra* and *Chioglossa,* both of which feed on land as adults (Özeti and Wake 1969), but it groups with the more aquatic newt clade in mDNA analyses (Titus and Larson 1995). Within the newt clade, the genus *Triturus* appears not to be monophyletic (hence the proliferation of new genus names for members of this group). In addition, the genus *Euproctus* appears to be polyphyletic. Two species, *E. montanus* and *E. platycephalus,* which are found on the islands of Corsica and Sardinia, respectively, are closely related. The Pyrrenean newt (*Euproctus asper*) is more closely related to a clade that includes *Neurergus* and several species of *Triturus* that occur in the Iberian Peninsula and southern Europe. This species has been placed in the genus *Calotriton,* along with a newly de-

**Fig. 1.34.** Diversity of urodeles (Salamandridae). (A) Terrestrial red eft stage of *Notophthalmus viridescens* from Connecticut. (B) *Taricha granulosa* from Washington. (C) *Triturus vittatus* from Asia Minor. (D) *Echinotriton andersonii* from China. (E) *Calotriton* (= *Euproctus*) *asper* from the Pyrenees Mountains of southwestern Europe. (F) *Mertensiella caucasica* from northeastern Turkey and western Georgia in the Caucasus Mountains. Photos by Kentwood D. Wells (A), William P. Leonard (B), and Wayne Van Devender (C–F). For additional photos of salamandrids, see fig. 14.30 A (*Taricha*), 14.30 B (*Paramesotriton*), 14.30 C (*Triturus*); 14.30 E (*Echinotriton*); 14.32 E (*Salamandra*); and 14.32 F (*Tylototriton*).

scribed speceis, *C. arnoldi,* which occurs in an isolated population in Spain (Carranza and Amat 2005). The two North American genera, *Taricha* and *Notophthalmus,* form a monophyletic clade that is closely related to one clade of Old World newts.

**Natural History:** Most salamandrids are newts, semiaquatic salamanders that breed in water, but usually live at least part of their lives on land. *Notophthalmus, Triturus, Cynops, Paramesotriton, Pleurodeles, Tylototriton,* and one species of *Taricha* breed mostly in ponds and attach eggs individually to submerged plants or scatter them individually or in clumps on the pond bottom. After metamorphosis, adults of most species leave the breeding ponds to live on land, although in some populations, adults remain in the water permanently. In most populations of *Notophthalmus*, larvae metamorphose into a terrestrial eft stage (fig. 1.34 A) that lasts up to seven years before adults return to the water to breed. In some populations, some individuals do not lose their gills upon reaching adult body size, but eventually all individuals metamorphose completely. *Chioglossa, Calotriton, Euproctus, Salamandrina,* and two species of *Taricha* are stream-breeders that deposit eggs in hidden locations under rocks. *Euproctus, Calotrion* (fig. 1.34 E), and *Salamandrina* live permanently in or near cold, well-oxygenated streams and have greatly reduced lungs, whereas *Chioglossa* and *Taricha* are more terrestrial as adults. *Salamandra* and *Mertensiella* are fully terrestrial salamanders as adults. Some species lack an aquatic larval stage and give birth to young that develop inside the female's reproductive tract. Most salamandrids are protected by distasteful, and in some cases, highly toxic skin secretions, and they often are aposematically colored. They exhibit a wide range of defensive postures that display their warning colors or direct poison glands toward predators.

### Family Sirenidae

**Distribution:** Southeastern and central United States and northeastern Mexico.

**Content:** Two genera, *Siren* (two species; fig. 1.33 D) and *Pseudobranchus* (two species). Until recently, the genus *Pseudobranchus* was considered to be monotypic, with five recognized subspecies within *P. striatus*. Studies of chromosomes revealed that populations in peninsular Florida belong to a separate species, *P. axanthus* (Moler and Kezer 1993)

**Natural History:** These are fully aquatic salamanders with external gills. They have elongate bodies that lack hind legs. The front limbs are reduced in both genera, and *Pseudobranchus* has only three toes on each front foot. Sirens live mainly in weed-choked ponds, swamps, and ditches, often in habitats with very little oxygen in the water. They utilize gills, skin, and lungs for respiration, but depend mostly on their lungs when oxygen levels are low. They are thought to have external fertilization, because females lack sperm storage organs (spermathecae) characteristic of salamanders with internal fertilization, but mating has not been observed. Eggs are attached to submerged plants.

## Morphological Evolution and Ecology in Salamanders

The salamanders are less diverse morphologically than the anurans. Most species retain the ancestral tetrapod morphology of a relatively small head, elongate body, four limbs of more or less equal size, and a tail. They lack the specialized modes of locomotion seen in anurans, utilizing instead two ancestral modes of locomotion, lateral undulation and walking. Lateral undulation involves bending the body by contracting the trunk and vertebral musculature and is the principal mode of locomotion in water. It also is used when salamanders are slithering across a wet substrate, such as a muddy stream bank. Walking involves alternating movements of the limbs on opposite sides of the body, which are held out from the body in a sprawling gait, along with contractions of the trunk musculature. Despite the general lack of specialization in urodeles compared to anurans, the group does exhibit some morphological variation related to exploitation of different microhabitats. Important themes in salamander evolution include the evolution of very small body size, paedomorphosis, and the evolution of lunglessness.

### Habitat Associations of Adult Salamanders

Four families of salamanders, the Amphiumidae, Cryptobranchidae, Proteidae, and Sirenidae, are largely or strictly aquatic. All are relatively large salamanders that retain some larval features as adults. Some ambystomatids and hynobiids are permanently aquatic as well. Most species in these families are terrestrial as adults, but return to the water to breed. Pond breeding is most common in both families, but some species are stream-breeders. The salamandrids are somewhat more diverse in their habitat associations. The large newts of the genus *Pachytriton* from Asia are permanently aquatic stream-dwellers. Most of the other genera of newts are terrestrial for at least part of their lives, but return to ponds or streams to breed. *Euproctus, Salamandrina,* and *Chioglossa* all breed in streams, but adult *Chioglossa* are more terrestrial than adults of the other two genera. Two genera, *Salamandra* and *Mertensiella,* are fully terrestrial.

The family Plethodontidae is by far the largest lineage of salamanders and also the most diverse. Even within the relatively small subfamily Desmognathinae, there are species with aquatic, riparian, terrestrial, and fossorial habits. In the subfamily Plethodontinae, aquatic, riparian, terrestrial, fossorial, and arboreal habits are represented. All of the species closely associated with water occur in the tribe Hemidactyliini (with *Hemidactylium* removed from this group). These include species mostly associated with streams, a few that are associated with ponds, bogs, and swamps, and a number of cave-dwelling species (see earlier summary under Plethodontidae). All species traditionally placed in the tribe Plethodontini are terrestrial, fossorial, or arboreal, having been freed from dependence on water by the evolution of direct development of terrestrial eggs.

The most diverse plethodontids are the Neotropical bolitoglossines, which include terrestrial, fossorial, and arboreal species. Most species are tropical or subtropical inhabitants of Mexico, Central America, and northern South America. They are particularly diverse in montane cloud forests, where they occupy several unusual microhabitats, including bromeliads and moss mats on trees or on the ground (D. Wake 1987). Some arboreal species have unusual morphological characteristics, including prehensile tails and feet adapted for clinging to vegetation (see the following). This clade also includes the largely terrestrial and fossorial genus *Batrachoseps* from the west coast of North America, and *Hydromantes,* a rock- and cave-dwelling genus from California and Europe (D. Wake 1966) (some authors place European *Hydromantes* in the genus *Speleomantes*).

The bolitoglossines are characterized by a high degree of convergent evolution, with repeated evolution of similar lifestyles in rather distantly related lineages (Parra-Olea and Wake 2001). This has made it extremely difficult to unravel the phylogenetic relationships of the group as a whole. One example of such convergence is the evolution of elongate body form and reduced limbs in fossorial species, including *Batrachoseps* and two Central American genera, *Oedipina* and *Lineatriton* (fig. 1.35 A, B). This morphology also is seen to a lesser degree in some *Plethodon* that live under rocks, logs, and other surface cover. *Oedipina* and *Lineatriton* exhibit the most extreme elongation and are so similar in appearance that they were once placed in the same genus. However, *Oedipina* has evolved an elongate body form by increasing the number of trunk vertebrae, while *Lineatriton* has done so by increasing the length of each individual vertebra (D. Wake 1991; Parra-Olea and Wake 2001).

Another example of convergent evolution in bolitoglossines is found among species adapted for climbing on vegetation or on the walls of caves. Species in a number of different lineages have independently evolved webbed feet, a morphology that greatly increases the surface area that contacts a smooth, wet substrate such as a plant leaf or a cave wall. However, the way in which this morphology has evolved differs, even among members of the same genus. In the genus *Bolitoglossa,* most lowland tropical forest species are arboreal. Larger species have enlarged feet with webbing formed by growth of skin between the toes (fig. 1.36, bottom). These are equipped with muscles that allow the salamander to pull the center of the foot upward, thus creating a suction force that adheres to the leaf (Alberch 1981).

Some small *Bolitoglossa,* many of which live in the leaf axils of *Heliconia* and banana plants, have superficially similar webbed feet, but these are the result of an entirely different evolutionary process. In these species, the feet are not fully developed, but retain the padlike structure that all plethodontid feet have early in development before the toes are fully grown (fig. 1.36, top). These paedomorphic feet cannot generate suction by muscular action, but depend instead on surface adhesion to the leaf (Alberch 1981). This

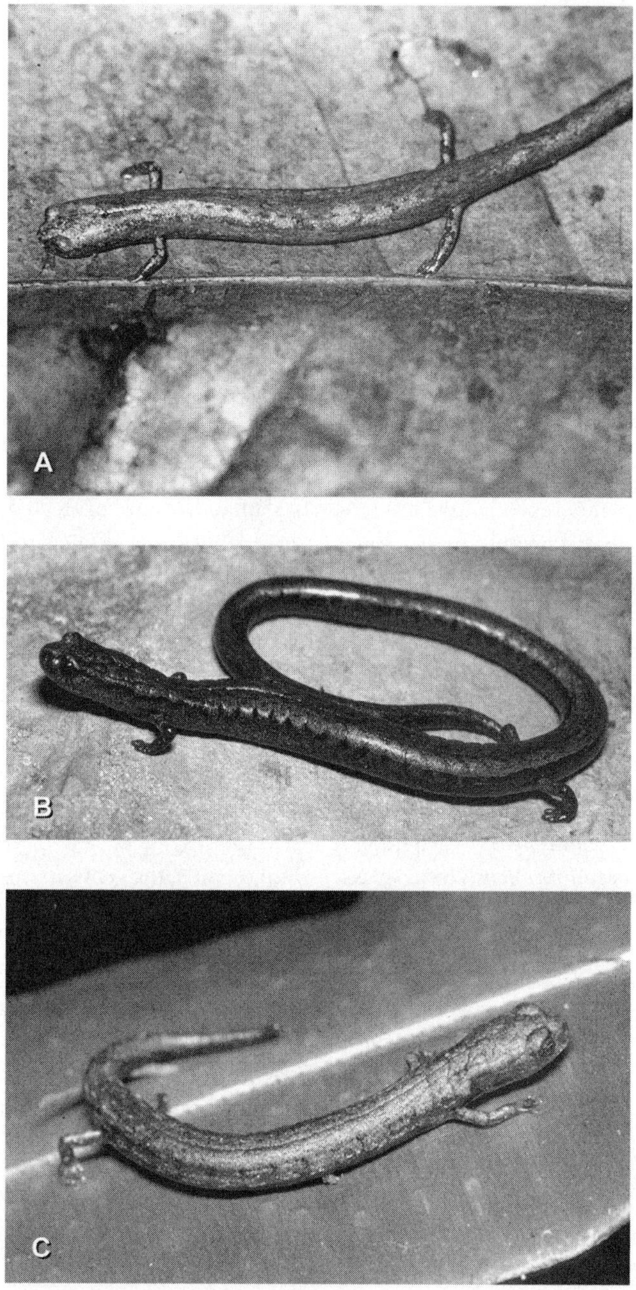

**Fig. 1.35.** Reduction of body size, elongation of the trunk and tail, and reduction of limbs and feet are common themes in the evolution of plethodontid salamanders. (A) *Oedipina parvipes,* a terrestrial salamander from Panama, with elongated trunk and tail and reduced feet. (B) *Batrachoseps attenuatus,* a terrestrial salamander from California, with a highly elongated trunk and tail. (C) *Nototriton picadoi,* a tiny arboreal salamander from Costa Rica, with an elongated trunk and reduced feet. Photos by Kentwood D. Wells (A) and Wayne Van Devender (B, C).

morphology is largely a consequence of reduction in body size through paedomorphosis, not selection for living in a specific habitat; some small species with similar foot morphology are terrestrial rather than arboreal. This sort of convergent evolution in foot morphology apparently has

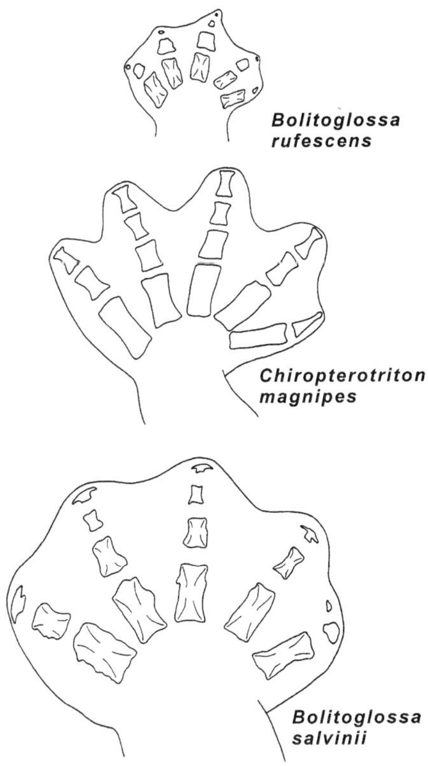


**Fig. 1.36.** Convergent evolution of webbed feet in tropical bolitoglossine salamanders. *Bolitoglossa rufescens* (top) is a small species with padlike feet that are paedomorphic, but without true cutaneous webbing. *Chiropterotriton magnipes* (middle) is a cave-dwelling salamander from northern Mexico with enlarged feet with cutaneous webbing that has grown between the toes. An unrelated species from Guatemala, *Bolitoglossa salvinii* (bottom), also has enlarged feet with cutaneous webbing that has grown between the toes. After D. Wake (1991).

occurred repeatedly in the genus *Bolitoglossa* in different parts of Central America. Some species in other genera, such as *Chiropterotriton,* which also are adapted for climbing, have independently evolved a foot morphology very similar to that of the large *Bolitoglossa,* even though they live in caves, not on vegetation (fig. 1.36, middle; Wake and Brame 1969; D. Wake and Lynch 1976; D. Wake 1987, 1991).

## Body Size and Ecology

As mentioned in the discussion of habitat associations, most of the largest species of salamanders are fully aquatic, while the smallest species tend to be terrestrial or arboreal. This pattern is evident in salamanders as a group, and in particular lineages, such as the desmognathines. There are, however, some exceptions. Members of the genus *Salamandra* are larger than most other salamandrids, yet they are the most terrestrial in the family. Some terrestrial species of *Plethodon* are considerably larger than more aquatic plethodontids such as *Eurycea* or *Stereochilus,* although in this case, the terrestrial and riparian species are in different clades. In general, as terrestrial and arboreal salamanders become smaller, they tend to utilize more cryptic microhabitats that provide protection from desiccation, such as bromeliads, leaf axils, or spaces under bark on fallen logs, whereas larger species are more likely to be seen out and about in the leaf litter, at least on rainy or humid nights.

A trend toward the evolution of very small body size is at least as apparent in salamanders as it is in anurans, and the phenomenon of miniaturization has been studied in considerabe detail in this group (Hanken and Wake 1993). In anurans, miniaturization has occurred in many unrelated lineages, but in salamanders, the evolution of very small body size is limited entirely to plethodontids and occurs most commonly in a single clade, the bolitoglossines. D. Wake (1991, 1992) defined miniaturized salamanders to be those in which at least one sex becomes sexually mature at a body length of less than 30 mm. These include *Desmognathus aeneus* and *D. wrighti, Hemidactylium scutatum, Eurycea quadridigitata, Plethodon websteri, Batrachoseps attenuatus,* a number of species in the Neotropical genera *Bolitoglossa* and *Chiropterotriton,* and all members of the Neotropical genera *Dendrotriton, Nototriton* (fig. 1.35 C), *Parvimolge,* and *Thorius.*

The genus *Thorius* exhibits the most extreme miniaturization, with some species having adult body lengths of less than 20 mm, making them the smallest terrestrial tailed vertebrates (Hanken and Wake 1998). These tiny salamanders typically live in spaces under bark on logs and may even shelter in channels made by insects. They exhibit many features that are paedomorphic relative to more generalized plethodontids. For example, the limbs are greatly reduced, as are the size of the feet and the number of bones in the digits. However, the limb bones and some foot bones exhibit greater ossification than in more typical plethodontids. This had been interpreted by some authors as compensation for a reduction in size by strengthening the bones (D. Wake 1966). However, Hanken (1982, 1993) suggested that this actually is part of a mechanism that restricts bone growth after the salamanders reach sexual maturity (bones cannot increase in length once they are completely ossified). He also found that smaller species in the genus reached this stage of ossification earlier than in larger species, suggesting that differences in adult body size are largely a product of differences in the timing of development.

The most dramatic effects of miniaturization in *Thorius* are seen in the skull, the brain, and the sense organs of the head. The skull is highly paedomorphic, with many features that normally appear late in development being reduced or absent. In fact, even the skull roof is incomplete, with much of the brain being covered with connective tissue rather than bone (fig. 1.37), a most unusual situation in vertebrates. Many other bones of the skull are not fused together, but lie

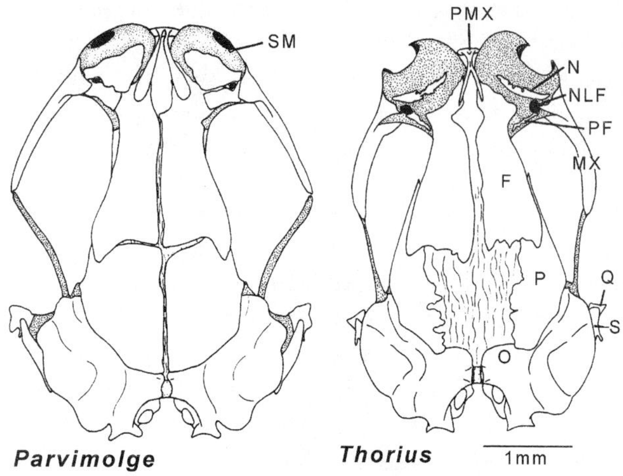


**Fig. 1.37.** Dorsal views of the skulls of two miniaturized species of Neotropical bolitoglossine salamanders. Left: *Parvimolge townsendi* from Veracruz, Mexico. Right: a species of *Thorius* from Mexico. The skulls of both species are somewhat retarded (paedomorphic) in their development. This is especially evident in *Thorius,* in which the braincase is not fully closed and the brain is covered by fibrous connective tissue (wavy lines). Stippled areas represent cartilage. Abbreviations: F = frontal, MX = maxillary, N = nasal, NLF = nasolacrimal foramen, O = otic-occipital, P = parietal, PF = prefrontal, PMX = premaxillary, Q = quadrate, S = squamosal, SM = septomaxillary. After D. Wake (1991).

in a sheet of cartilage (Hanken 1984). The brain and sense organs, especially the eyes, appear enormous in proportion to the size of the skull, although in absolute size, they may be approaching the lower size limit for proper functioning in vertebrates. This results from negative allometry of these organs with body size (that is, relative brain and sense organ size increase among species with decreasing body size). This negative allometry also is evident within the genus *Thorius:* the smallest species have the smallest absolute eye and brain sizes and the largest relative sizes of these organs (Roth, Blanke, and Ohle 1995). Not surprisingly, these relatively large organs impinge on each other. The huge eyes have changed the configuration of the brain, which has been forced downward and backward in the skull, while the large otic capsules that house the inner ear have resulted in major changes in the articulation of the jaw (Hanken 1983).

All of this rearrangement might be expected to have major effects on the functioning of the brain and nervous system. The situation is complicated by the relatively enormous genome of *Thorius,* and of bolitoglossines in general. The main functional consequence of a large genome is large cell size. Thus, as body size becomes smaller, the large cells that make up the brain and sense organs must be packed into smaller and smaller spaces. The inevitable result is a reduction in number of cells. For example, one of the smallest species, *T. narisovalis,* has the smallest number of visual cells in the brain, because its cells are relatively large. However, another small species, *T. pennatulus,* has evolved some evolutionary compensation for small size. It has smaller cells that are more densely packed than in other small species, and consequently has more visual cells than are found in some much larger species. *Thorius* eyes, while large relatively to body size, are still small relative to other amphibian eyes, with only about 6% of the number of photoreceptors found in a ranid frog eye. On the other hand, the number of retinal ganglion cells is approximately equal to the number of photoreceptors, which has the effect of making the entire retina equivalent to the fovea of a frog eye, the region of maximal visual acuity. Thus, while miniaturization has had a major impact on the structure of the eye and brain in these tiny salamanders, they still appear to have the same high visual acuity found in larger plethodontids, a necessity since they use vision to capture prey (Linke, Roth, and Rottluff 1985; D. Wake 1991; Roth, Nishikawa, and Wake 1997).

## Paedomorphosis

The evolution of webbed feet and changes in skull morphology in very small bolitoglossines are part of the broader phenomenon of paedomorphosis that is a dominant theme in the morphological evolution of salamanders (Hanken 1989; D. Wake 1992). Indeed, many morphological features of amphibians in general, and salamanders in particular, are considered highly simplified or paedomorphic relative to ancestral tetrapods (Roth, Dicke, and Nishikawa 1992). In addition to relatively obvious changes in limb and skull morphology discussed previously, paedomorphic simplification can occur in organ systems such as the brain and sense organs. A phylogenetic analysis of the visual, auditory, olfactory, and lateral line systems and several features of the brains of salamanders revealed that most characters exhibit secondary simplification relative to other tetrapod vertebrates, including hagfishes, lampreys, sharks and rays, bony fishes, coelacanths, lungfishes, frogs, and amniotes. Among the salamanders, the greatest degree of simplification is found in the bolitoglossines, considered the most derived lineage in the group (Roth et al. 1993; Roth, Nishikawa, and Wake 1997).

This extensive paedomorphic simplification of the nervous system is correlated with large genome and cell size and slow rates of cell proliferation and differentiation (Roth et al. 1993; Roth, Nishikawa, and Wake 1997). Among salamanders, the large, permanently aquatic families have the largest genomes and cell sizes, while bolitoglossines have the largest genomes and cell sizes among the more terrestrial lineages (Sessions and Larson 1987). In contrast, anurans have smaller genomes than salamanders and exhibit less dramatic examples of paedomorphosis; caecilians are intermediate. If paedomorphosis and simplification of morphological structure, particularly in the nervous system, are a result of slower development and differentiation rates in species with larger genomes and cells, then these morphological patterns may not be adaptive consequences of selection for simplification, but rather constraints placed on morphological evolution by basic characteristics of the salamander lineage. In fact, in some of the very small bolitoglossines with the most simplified nervous systems, there is evidence for the evolution of new structural features that compensate for the overall morphological simplification. Hence, these are among the most specialized of all salamanders in their visual systems and feeding behavior (see the following), in spite of, but not because of, their paedomorphic morphology (Roth and Wake 1985a, b; D. Wake and Roth 1989; Roth et al. 1993).

### The Evolution of Lunglessness

The plethodontids, by far the most successful salamander clade, are unusual among terrestrial vertebrates in having completely lost their lungs, relying entirely on the skin and the lining of the mouth and throat (the buccopharyngeal region) for gas exchange (see chapter 4). Since there are no close relatives of living plethodontids that exhibit an intermediate condition (reduced lungs), it has been difficult to understand exactly what selective pressure led to the initial evolution of lunglessness in this group. Most authors have dismissed the possibility that loss of lungs could be a specific adaptation for terrestrial life, since the switch to cutaneous respiration limits the types of microhabitats in which these animals can live to relatively cool, moist environments (Feder 1983b). The traditional explanation, first proposed by Wilder, Whipple, and Dunn (1920) and elaborated by Dunn (1926c), is that the ancestral plethodontids were highly aquatic, stream-dwelling salamanders from the southern Appalachians, with a lifestyle similar to that of *Desmognathus quadramaculatus* or *D. marmoratus*. The ostensible benefit of lunglessness is to reduce buoyancy, thereby enabling the salamanders to maintain their position in a swift current. Presumably the high oxygen content of cool mountain streams would allow for this because most of the animal's oxygen requirements could be accommodated through cutaneous respiration. Several stream-dwelling salamanders in other families also exhibit moderate to complete loss of lungs, including *Onychodactylus* and *Ranodon* (Hynobiidae), *Rhyacotriton* (Rhyacotritonidae), and *Salamandrina, Euproctus,* and *Chioglossa* (Salamandridae).

Ruben and Boucout (1989) challenged this traditional scenario for the evolution of lunglessness in plethodontids, largely on the grounds that the topography and climate of the southern Appalachian region was not the same as it is today when the plethodontids first appeared. Instead, they believed the geological evidence points to a lowland environment with a warmer, perhaps subtropical climate. In this type of habitat, low oxygen concentration in the water would make lungs essential for fully aquatic salamanders, because they would not be able to take up sufficient oxygen through the skin, and indeed, might lose oxygen to hypoxic water (see chapter 4). They suggested instead that ancestral plethodontids were riparian or terrestrial. Most salamanders in these habitats have narrower heads than more fully aquatic species, and reduction in head width results in reduced capacity to fill the lungs because of a reduction in the throat musculature used to pump air into the lungs. Ruben and Boucout proposed that a reduction in relative head width occurred first, perhaps because of advantages in acquiring terrestrial prey or utilizing cryptic microhabitats under rocks and logs. The reduced ability to use the lungs would then result in increased reliance on cutaneous respiration, a shift that would be easiest in small species with relatively large surface to volume ratios. Reagan and Verrell (1991) speculated that a shift to courtship and mating on land, where oxygen availability is much greater than in the water, might have facilitated the loss of lungs in plethodontids.

Ruben and Boucout's alternative scenario for the evolution of lunglessness was criticized by Beachy and Bruce (1992), who maintained that there were sufficient upland habitats available in the late Mezozoic to allow for the evolution of plethodontids in cool, upland habitats. They also argued that most salamanders, except those that live in hypoxic environments, rely mainly on the skin for respiration, even if they have lungs, and that lungs may have evolved mainly as hydrostatic organs that are used only secondarily as respiratory organs. They stated that desmognathines and hemidactyliines that are closely associated with streamside habitats typically have prolonged larval periods, in some cases lasting for several years (see chapter 13). This suggests that loss of lungs is primarily an advantage for plethodontid larvae in reducing buoyancy in stream habitats. Experiments in which larvae of *Ambystoma maculatum* were raised with and without access to air supported this hypothesis. Those raised without access to air developed smaller and more solidified lungs than did those raised with access to air, and the result was reduced buoyancy and less tendency to be displaced by a current (Bruce et al. 1994).

There clearly is no consensus concerning the origins of lunglessness in the plethodontids. Ruben et al. (1993) challenged the geological data presented by Beachy and Bruce, as well as their assumption that the unusually long larval periods of stream-dwelling plethodontids represent the ancestral condition for plethodontids as a whole. They suggest that the much shorter larval periods characteristic of most ambystomatids and other salamander families are more likely to resemble those of ancestral plethodontids, with the long larval periods of stream-dwelling forms being secondarily derived, or perhaps a proximate result of cold environmental temperatures. They also point out that the larvae of many lunged salamanders live in flowing water and are perfectly capable of walking on the bottom of a stream and reduce buoyancy by releasing air from the lungs. Unfortunately, it may be difficult to resolve this argument definitively. There is considerable disagreement among systematists as to which family is the most likely sister-group to the plethodontids, so it is difficult to infer ancestral larval characteristics from a phylogenetic analysis. Furthermore, the length of the larval period is a notoriously plastic trait in amphibians that tends to be greatly influenced by local environmental conditions (see chapter 13), so the patterns we see in living species may not have been characteristic of their ancestors. Finally, a major morphological change such as the loss of lungs may result from multiple selective pressures and may not be explainable by a simple adaptive advantage such as a reduction in buoyancy or a shift to terrestrial courtship.

## Functional Morphology of Feeding

All salamanders are carnivorous, both as larvae and as adults, although there is evidence that sirens can derive some nutrition from plant material (Pryor, German, and Bjorndal 2006). The mode of prey capture varies among different clades and is affected by habitat, mode of life history, and the presence or absence of lungs (Lombard and Wake 1977, 1986; D. Wake 1982, 1991; D. Wake and Larson 1987; Deban and Wake 2000; D. Wake and Deban 2000; Deban and Marks 2002; O'Reilly, Deban, and Nishikawa 2002). Aquatic salamander larvae have tongues that are little more than thickened pads on the floor of the mouth, a condition that is retained in aquatic salamanders such as *Cryptobranchus* and *Necturus* (fig. 1.39 A). The tongue is not involved in prey capture in salamander larvae or aquatic adults. Instead, they capture prey by suction feeding, an ancestral feeding mode for aquatic vertebrates (Lauder and Shaffer 1993; Deban and Wake 2000; Deban and Marks 2002; O'Reilly, Deban, and Nishikawa 2002). Because lunging toward prey in water tends to displace it away from the salamander's mouth, it is more efficient to capture prey by suddenly opening the mouth and depressing the floor of the buccal cavity, causing the prey to be sucked in (see chapter 12). The hyobranchial apparatus in the floor of the buccal cavity and its associated musculature provide the main force for suction feeding in all aquatic vertebrates. When this apparatus is pulled backward there is a rapid expansion of the buccal cavity. Patterns of muscle activity involved in suction feeding are highly conserved across widely diverse taxa, including fishes and salamanders (Lauder and Shaffer 1985, 1986, 1993; Wainwright et al. 1989). When salamander larvae metamorphose and move onto land, they undergo major morphological changes that affect their feeding behavior. In most species, adults acquire a moveable tongue pad that is supported by a modified hyobranchial apparatus (now the hyolingual apparatus; fig. 1.38). The branchial musculature is modified in such a way that the hyolingual apparatus is protracted, rather than being retracted as it is in suction feeding (Bramble and Wake 1985).

The hyolingual apparatus of adult terrestrial salamanders has a very different structure from that of aquatic larvae. Many skeletal elements, especially those associated with support of the gills, are lost. The ceratohyals, which

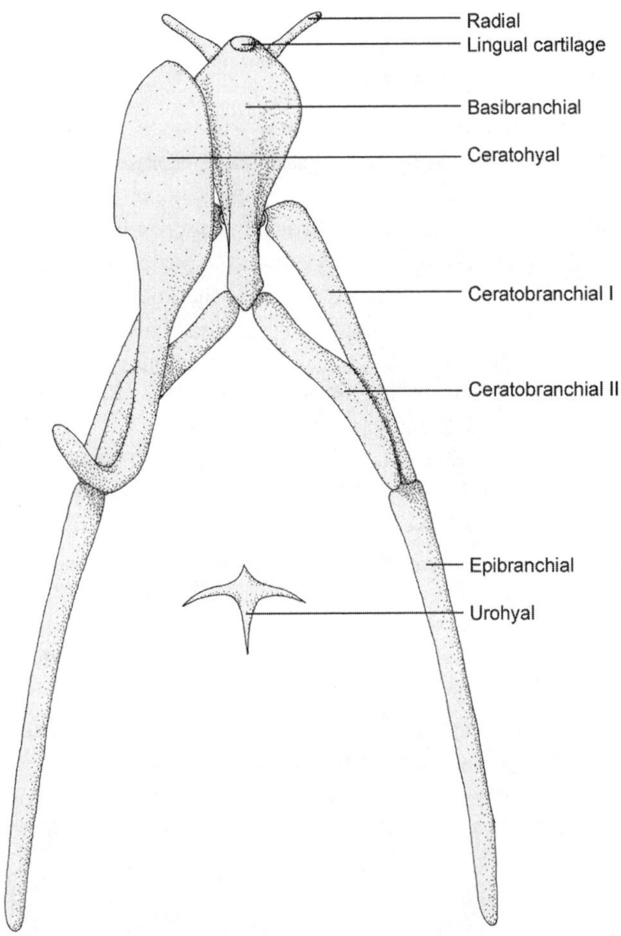


**Fig. 1.38.** Structure of the hyolingual apparatus of an adult plethodontid salamander, *Eurycea bislineata.* The ceratohyals are anchored in the floor of the mouth, while the tongue is supported by the lingual cartilage. During tongue protraction, the whole apparatus slides forward on the ceratohyals. The ceratobranchials are derived from the larval branchial arches that supported the gills. In adult salamanders with lungs, the ceratobranchials are relatively large and are involved in moving the floor of the mouth to fill the lungs with air. In the lungless plethodontids, the ceratobranchials are reduced, but the single pair of epibranchials becomes elongated and the whole hyolingual apparatus becomes specialized for tongue projection. The highest degree of specialization of the hyolingual apparatus is found in plethodontids that undergo direct development, because they are freed from the constraints imposed by larval morphology adapted for suction feeding in water. After Lombard and Wake (1976).

anchor the whole apparatus, become somewhat detached from the other elements that support the tongue pad. This allows those elements to slide forward during tongue protraction. In adult terrestrial salamanders, only two pairs of branchial arches remain from the original three or four present in larval salamanders, represented by the two pairs of ceratobranchials (fig. 1.38). In all families of terrestrial salamanders except hynobiids, these articulate with only a single pair of epibranchials. In salamanders with lungs, the hyolingual apparatus provides the main force to fill the lungs with air by positive-pressure pumping of the throat. In particular, the ceratobranchials are important in depressing the mouth during respiration, as well as during suction feeding in larvae, and they become enlarged early in development. In the lungless plethodontids, the ceratobranchials are reduced, but the epibranchials often become highly elongated (fig. 1.38), and this allows the hyolingual apparatus to become more specialized for tongue projection (D. Wake 1982; Roth and Wake 1989; Deban and Marks 2002).

The hynobiids, ambystomatids, and dicamptodontids, most of which have well-developed lungs, have the most generalized tongue morphology, which is considered to be the ancestral condition for terrestrial salamanders. The tongue is relatively broad and padlike, tightly attached to the floor of the mouth, with only part of the margin of the tongue free (fig. 1.39 B). Consequently, the capacity for tongue projection is limited, with little projection beyond the tip of the snout (Larsen and Guthrie 1975; D. Wake and Deban 2000). Nevertheless, the slight projection of the tongue and the use of the tongue to seize the prey is the principal difference in feeding behavior of terrestrial adults and aquatic larvae. Most other aspects of the kinematics of feeding are remarkably similar (Shaffer and Lauder 1988; Lauder and Reilly 1990; Gillis and Lauder 1994). Although the tongue is used to seize the prey, these salamanders cannot capture prey at long distances, nor are they well adapted to feeding on very small or fast-moving prey (Reilly and Lauder 1989).

Salamandrids generally have a somewhat greater ability to project the tongue than do ambystomatids or dicamptodontids, and there is considerable variation within the family (Özeti and Wake 1969; D. Wake 1982). In *Salamandra salamandra,* eggs are retained in the female's reproductive tract, and larvae are born at a relatively large size and advanced stage of development (see chapter 10). The aquatic larvae use the typical suction feeding seen in other aquatic vertebrates. Motor patterns are very similar to those seen in other aquatic salamander larvae, although larvae of this species capture prey more rapidly than most other species (Reilly 1995). Metamorphosed adults are entirely terrestrial and have a land-tongue morphology similar to that of other terrestrial salamanders. Although their ability to protract the tongue is limited, it can be projected beyond the

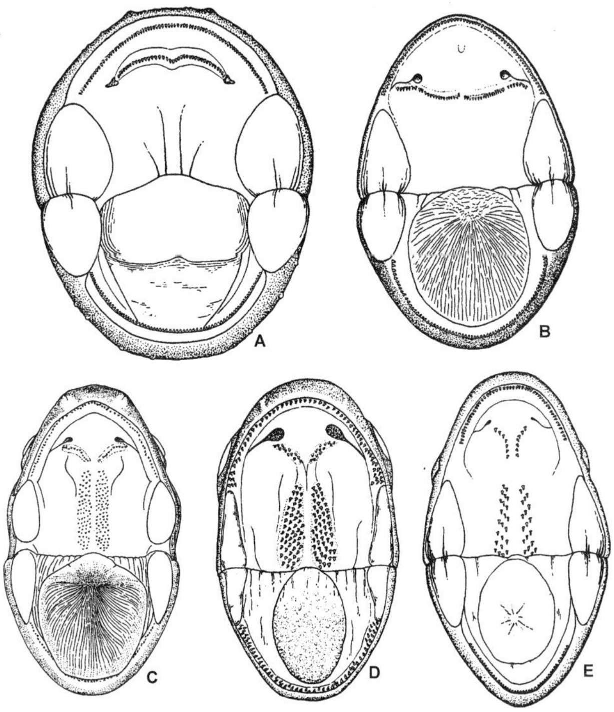


**Fig. 1.39.** Representative salamander tongues. (A) The tongue of the aquatic salamander *Cryptobranchus alleganiensis* is little more than a thickened pad on the floor of the mouth. (B) The tongue of a terrestrial adult of *Ambystoma maculatum* is a padlike structure with only part of the tongue margin detached from the floor of the mouth. (C) *Plethodon glutinosus* has a generalized tongue morphology similar to the ancestral condition for the family Plethodontidae. (D) *Hemidactylium scutatum* has a smaller tongue that is attached in front, but capable of some forward projection. (E) *Eurycea bislineata* has a tongue that is not attached to the tips of the jaws, which allows for greater projection of the tongue to capture prey. After Bishop (1941).

mouth and is used for prey capture (Özeti and Wake 1969; D. Wake 1982). Two other terrestrial genera, *Salamandrina* and *Chioglossa,* have a much greater ability to project the tongue (up to 20% of snout-vent length in *Salamandrina*). In this respect, they are convergent with plethodontids that have highly protrusible tongues (see the following), although the mechanism of tongue projection is different (Özeti and Wake 1969; Findeis and Bemis 1990). It is undoubtedly significant that these genera, like the plethodontids, lack lungs and therefore do not use the hyolingual apparatus for breathing, so it is freed for use in tongue protrusion (D. Wake 1982).

Most of the newts that are largely aquatic as adults have a water tongue morphology with a poorly developed tongue pad and very limited ability to project the tongue (Özeti and Wake 1969; D. Wake 1982). Those that remain in the water for most of their lives, such as the Asian genus *Pachytriton,* retain the labial lobes that aid in suction feeding, but they are lost in species that spend most of their lives on land (they may regrow during the aquatic breeding season). Some newts that are largely terrestrial except during the breeding season have moderately protrusible tongues. One example is *Taricha torosa,* which captures prey on land with a forward lunge coupled with tongue projection; the prey is drawn into the mouth with the sticky tongue. This species also suction feeds in the water, and the need for a hyolingual apparatus that can depress the floor of the mouth for suction feeding and for breathing has apparently limited the evolution of tongue protrusion. In addition, the details of tongue protrusion and the anatomy of the hyolingual apparatus are different from those of plethodontids, clearly indicating an independent evolution of a protrusible tongue (Findeis and Bemis 1990).

All plethodontids have been freed from the morphological constraint of having to use the hyolingual apparatus to fill the lungs with air, although they still pump air into the mouth for buccopharyngeal respiration. This has allowed the hyobranchial apparatus to become specialized for tongue projection. In those species that have aquatic larvae, morphological constraints associated with larval suction feeding remain, and in general, plethodontids with aquatic larvae (the desmognathines and hemidactyliines) have less protrusible tongues than do those that have lost the larval stage altogether. However, even within these groups, there is considerable variation, and it is clear that protrusible tongues have evolved in different ways independently many times in plethodontids (Lombard and Wake 1977, 1986; D. Wake 1982; D. Wake and Larson 1987; D. Wake and Deban 2000; Deban and Marks 2002).

Based on the relationships among plethodontids as they were understood in the 1980s, Wake and Larson (1987) presented a phylogeny of tongue evolution in the family. Recent thinking about phylogenetic relationships within the family have since changed, so some revision of their scenario will be required, particularly regarding the number and direction of evolutionary transitions between different tongue morphologies (Mueller et al. 2004). Because a thorough analysis of tongue evolution based on new ideas of plethodontid phylogeny has yet to be published, I discuss the original ideas of Wake and Larson (1987) here. They argued that the ancestral condition for the family was to have the front of the tongue attached to the jaw by a short, stout genioglossus muscle (fig. 1.41). In species with this type of tongue morphology (the desmognathines, *Plethodon,* and *Aneides*), the tongue is projected by flipping forward a posterior flap of the tongue (fig. 1.40 A, B). This is referred to as an attached protrusible tongue and still would be the ancestral condition under the revised phylogeny of the family (Mueller et al. 2004). Species with this morphology have a limited ability to project the tongue, so they cannot capture prey at long distances (Schwenk and Wake 1993). Several

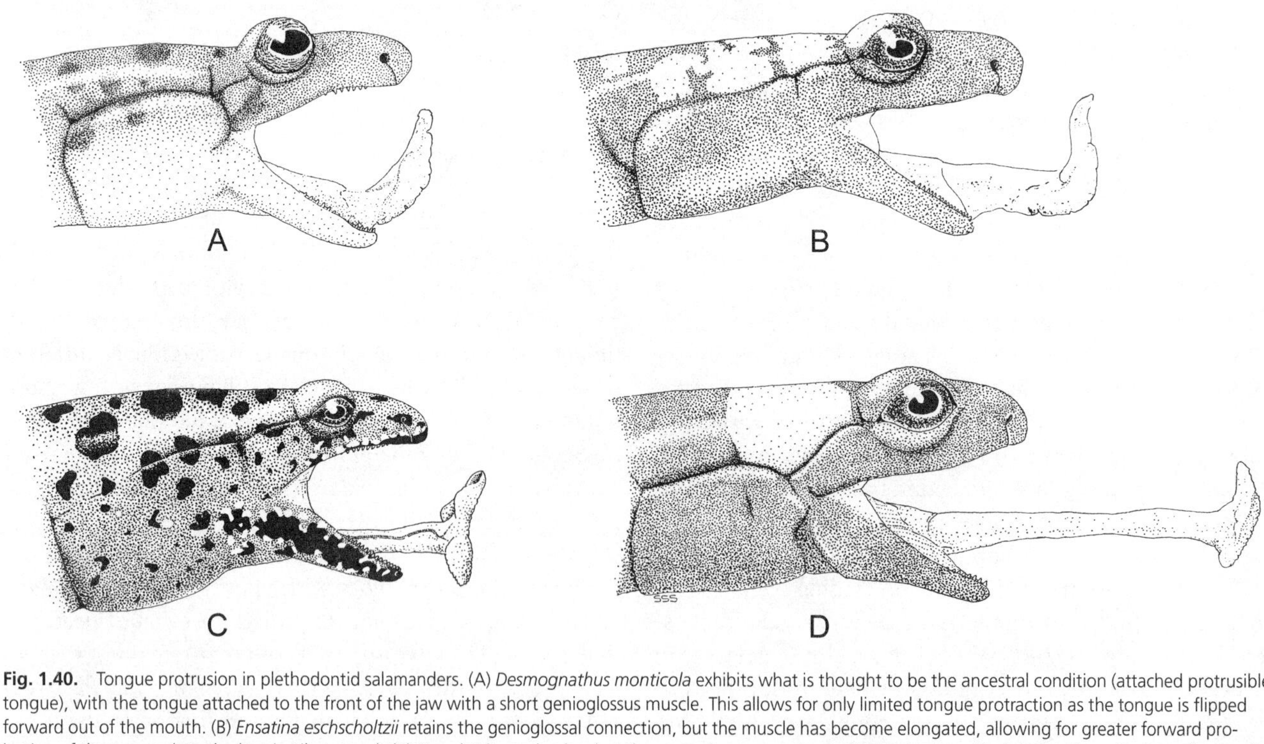


**Fig. 1.40.** Tongue protrusion in plethodontid salamanders. (A) *Desmognathus monticola* exhibits what is thought to be the ancestral condition (attached protrusible tongue), with the tongue attached to the front of the jaw with a short genioglossus muscle. This allows for only limited tongue protraction as the tongue is flipped forward out of the mouth. (B) *Ensatina eschscholtzii* retains the genioglossal connection, but the muscle has become elongated, allowing for greater forward projection of the tongue (attached projectile tongue). (C) *Pseudotriton ruber* has lost the genioglossus muscle, and the tongue pad is detached from the jaw, allowing for much greater forward projection of the tongue (free projectile tongue). (D) A similar tongue morphology has evolved independently in bolitoglossine salamanders, represented here by *Pseudoeurycea bellii,* but these salamanders are capable of even greater projection of the tongue beyond the tips of the jaws. From Lombard and Wake (1977).

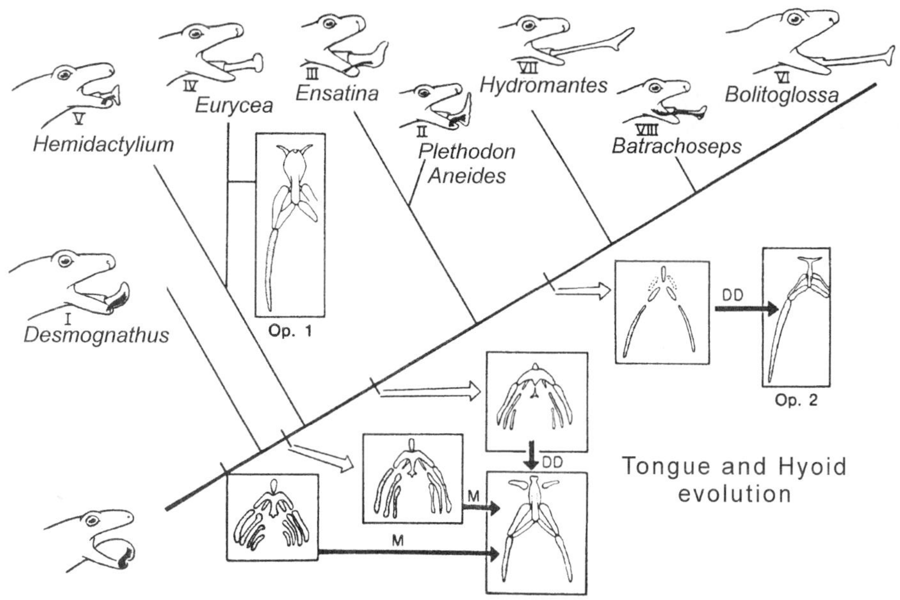


**Fig. 1.41.** Phylogeny of tongue evolution in plethodontid salamanders. Roman numerals refer to eight different modes of tongue projection. In the ancestral condition and in modes I (desmognathines) and II (*Plethodon* and *Aneides*), the tongue is attached to the front of the jaw with a short, stout genioglossus muscle that limits forward tongue projection. In modes III (*Ensatina*), V (*Hemidactylium*), and VIII (*Batrachoseps*), the genioglossal connection is retained, but elongation of the muscle allows for greater tongue projection (attached projectile tongues). This condition appears to have evolved at least three times independently from the ancestral condition. In modes IV (*Eurycea*), VI (bolitoglossines), and VII (*Hydromantes*), the genioglossal connection has been lost, allowing for much greater forward projection of the tongue (free projectile tongues). Tongue projection also is affected by the structure of the hyobranchial apparatus (boxes), which is constrained by larval morphology. Larval desmognathines have four pairs of epibranchials, whereas larval plethodontines have three. In both clades, these are transformed during metamorphosis into a single adult epibranchial on each side. In those species that undergo direct development (tribes Plethodontini and Bolitoglossini), there also is a transformation to a single pair of adult epibranchials. The most derived condition is found in the bolitoglossines, in which the larval epibranchials never form, and only the adult morphology is present. Op. 1 (*Eurycea*) and 2 (bolitoglossines) represent two different options for the evolution of free projectile tongues involving different developmental pathways and different biomechanical properties. After D. Wake and Larson (1987).

lineages that have retained the genioglossal attachment have secondarily achieved a greater degree of tongue protrusion through elongation of the genioglossus muscle. This condition, attached projectile tongues, has evolved independently in *Hemidactylium,* Ensatina, and *Batrachoseps* (fig. 1.41). Of these, only *Hemidactylium* has aquatic larvae, and it has the least protrusible tongue of these genera, while *Batrachoseps,* a genus with direct development, has the most protrusible tongue and probably the greatest ability to feed on small prey at long distances.

In three lineages of plethodontids, the genioglossus muscle has been lost, and the tongue is completely free (fig. 1.41). This condition is found in *Eurycea, Pseudotriton, Hydromantes,* and all of the Neotropical bolitoglossines (fig. 1.40 D). The last two groups, both of which lack aquatic larvae, have the most specialized projectile tongues, capable of extending up to 20–40% of snout-vent length (fig. 1.42). Among the Neotropical bolitoglossines, the most specialized tongues are found in the miniaturized genus *Thorius.* In this group, the different elements of the hyolingual apparatus become disarticulated, allowing them to collapse into a compact linear bundle when the tongue is projected. This results in a highly efficient biomechanical system that probably allows them to capture small, relatively fast-moving prey. The morphological and behavioral specializations of the bolitoglossines are possible because this group is freed from both major morphological constraints on tongue evolution, the use of the hyobranchial apparatus for suction feeding in larvae and its use for breathing in adults. In addition, since these salamanders do not undergo metamorphosis, there is the possibility of novel developmental patterns evolving, and indeed, the structure of the hyobranchial apparatus is different in these species than in other salamanders with projectile tongues (fig. 1.41). However, the fact that lungless salamanders with direct development have not all evolved such specialized tongues indicates that other selective pressures are at work as well (Lombard and Wake 1976, 1977; Roth 1976; D. Wake 1982; D. Wake and Larsen 1987; D. Wake and Deban 2000; Deban and Marks 2002).

**Fig. 1.42.** Extreme tongue projection in the genus *Hydromantes*. Photo by Stephen Deban.

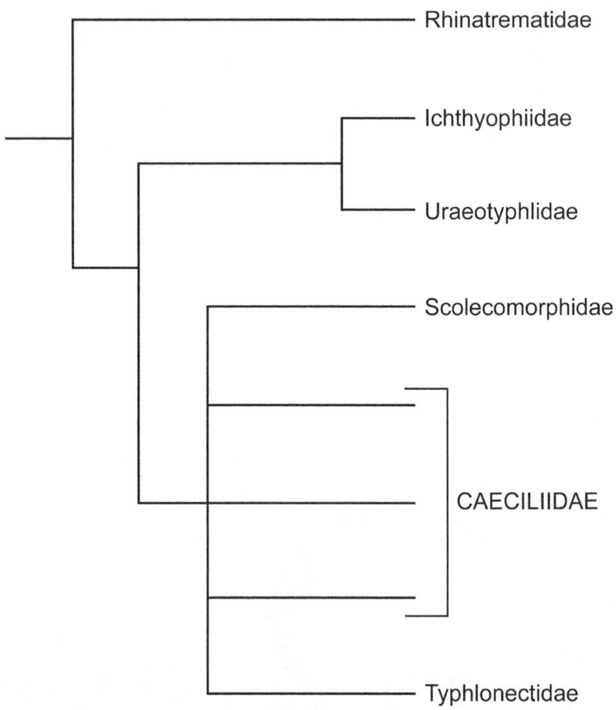


**Fig. 1.43.** Phylogenetic relationships of caecilians, derived from a combination of morphological and molecular data. The family Caeciliidae probably is not a monophyletic group as currently constituted and is shown as several distinct lineages. Some classifications place the typhlonectids in the family Caeciliidae. After Pough et al. (2004), based on Wilkinson and Nussbaum (1999).

## Phylogeny and Classification of the Gymnophiona

The caecilians are elongate, legless amphibians adapted to fossorial or aquatic life. Many species are rarely encountered and have not been studied in great detail. In addition, the highly derived morphology of these animals makes it difficult to determine the phylogenetic relationships of caecilians based entirely on morphological characters (Nussbaum and Wilkinson 1989). Some recent phylogenetic work has been based on molecular data, but the results are not always consistent with the results of morphological studies (Hass, Nussbaum, and Maxson 1993; Hedges, Nussbaum, and Maxson 1993). Even different types of morphological data do not always yield congruent results (M. Wilkinson 1997). There is general agreement that the family Rhinatrematidae is the most primitive group of living caecilians, and the families Ichthyophiidae and Uraeotyphlidae are relatively basal lineages as well and may be closely related to each other (Wilkinson, Drewes, and Tatum 2002; San Mauro et al. 2004; Frost et al. 2006, placed *Uraeotyphlus* within the family Ichthyophiidae). All of these families have free-living aquatic larvae, presumed to be the ancestral condition for caecilians. Some caeciliids also have aquatic larvae, but many caeciliids lay eggs that undergo direct development or are viviparous. The relationships among the more derived families are not yet resolved, but the typhlonectids appear to be closely related to, and possibly derived from, members of the family Caeciliidae, which may not be a monophyletic group (fig. 1.43). Indeed, the major systematists working on caecilians have alternated between treating the typhlonectids as a separate family or as a subfamily of Caeciliidae (Hedges, Nussbaum, and Maxson 1993; M. Wilkinson and Nussbaum 1999).

## Synopsis of Families of Gymnophiona

The family-level classification of caecilians has remained essentially unchanged since the late 1980s, although Frost et al. (2006) proposed collapsing the five generally recognized families into three. Although there have been some minor taxonomic rearrangements at the genus and species level, Duellman (1993a) did not record any net change in the number of described caecilian species between 1985 and 1992. Very little is known about the natural lives of caecilians, so the following summaries of natural history are necessarily brief. The major monograph on the caecilians of the world (E. H. Taylor 1968) contains almost no information on their ecology or natural history. Probably the best-known aspect of their biology is reproduction (M. Wake 1977b, 1992), but only a few species have been studied in detail, almost exclusively in captivity. Sensory systems and locomotion also have been studied in some detail (see the following references cited). Himstedt (1996) provides the best overview of caecilian biology.

### Family Caeciliidae

**Distribution:** Tropical South and Central America, tropical Africa, the Seychelles Islands, the Indian subcontinent.

**Fig. 1.44.** Diversity of caecilians (Gymnophiona). (A) *Oscaecilia ochrocephala* (Caeciliidae) from Panama. (B) *Geotrypetes seraphini* (Caeciliidae) from the Ivory Coast. (C) *Grandisonia alternans* (Caeciliidae) from the Seychelles Islands. (D) *Ichthyophis bannanicus* (Ichthyophiidae) from Vietnam. The tentacle, a sensory organ unique to caecilians, is clearly visible in B and D. Photos by Kentwood D. Wells (A), Mark Oliver Rödel (B), and Wayne Van Devender (C, D). For additional photos of caecilians, see fig. 1.1 C and 11.13 (*Siphonops*).

**Content:** 21 genera, about 100 species (fig. 1.44 A–C). With about 55% of the described species, this is by far the largest family of caecilians and also is the most widely distributed. Most genera have only one to five described species. The largest genus is *Caecilia* (33 species) of northern South America and Panama. Molecular systematic work suggests that this family is paraphyletic, with the genus *Typhlonectes,* often placed in a separate family, being closely related to *Caecilia.* M. Wilkinson and Nussbaum (1999) recommended retaining the Typhlonectidae as a separate family until higher order relationships are further resolved, but Frost (2004) and Frost et al. (2006) treated typhonectids as a subfamily of Caeciliidae. The species found in the Seychelles Islands (*Grandisonia, Hypogeophis, Praslinia;* fig. 1.44 C) appear to form a monophyletic clade that is most closely related to the Indian genus *Gegeneophis* (Hass, Nussbaum, and Maxson 1993; M. Wilkinson et al. 2002).

**Natural History:** The caeciliids are elongate, terrestrial burrowers with blunt heads and no tails. However, they also are capable of living in water, and one species from the Seychelles Islands, *Hypogeophis rostratus,* is commonly found in streams and even engages in aquatic courtship (Nussbaum 1984). As in all caecilians, the eyes are reduced and covered with skin, and specialized tactile and chemosensory organs, the tentacles, are present on the head. Species range in size from less than 100 mm to an impressive 1.5 m. Like all caecilians, they have internal fertilization, which is achieved by use of a special copulatory organ, the phallodeum. Although the reproductive biology of most species is unknown, the family includes species with free-living aquatic larvae (fig. 1.45), species that lay eggs that undergo direct development, and some that give birth to live young (viviparity). Ecological information has been collected on only a handful of species. Densities of some populations can be surprisingly high, even in disturbed habitats such as agricultural land (Oommen et al. 2000; Measey and Di-Bernardo 2003; Measey et al. 2003a, b; Gower et al. 2004; Measey 2004). The few available studies of diets indicate that these animals are generalist predators on soil invertebrates, including earthworms, termites, and other insects (Gaborieau and Measey 2004; Measey et al. 2004).

### Family Ichthyophiidae

**Distribution:** Southeast Asia, the Indian subcontinent, Sri Lanka, Borneo, Sumatra, the Philippines.

Fig. 1.45. Eggs of *Grandisonia alternans* from the Seychelles Islands, a caecilian with aquatic larvae. The eggs normally would be placed in moist areas near water. Photo by Wayne Van Devender.

**Content:** Two genera, *Ichthyophis* (34 species; fig. 1.44 D) and *Caudacaecilia* (five species).

**Natural History:** These are elongate, burrowing animals with longer tails than caeciliids. Eggs are deposited in muddy burrows near water, where they may be attended by the female. Larvae are aquatic and live in both ponds and streams, and adults are perfectly capable of swimming as well (Crapon de Caprona and Himstedt 1985). Recent molecular phylogenetic work suggests that the ichthyophiids of Southeast Asia form a monophyletic clade derived from ancestors that originated on a landmass that included India, Sri Lanka, and the Seychelles, and dispersed into mainland Asia after India became connected to that continent (Gower et al. 2002; M. Wilkinson et al. 2002). This pattern of dispersal out of India also has been reported for some clades of frogs (Bossuyt and Milinkovitch 2001).

### Family Rhinatrematidae

**Distribution:** Tropical South America.

**Content:** Two genera, *Epicrionops* (eight species) and *Rhinatrema* (one species).

**Natural History:** Relatively small, burrowing animals, with short tails. The eyes of these caecilians are less reduced than in more derived families. They lay eggs that hatch into aquatic larvae.

### Family Scolecomorphidae

**Distribution:** Disjunct distribution in tropical West and East Africa.

**Content:** Two genera, *Scolecomorphus* (three species) and *Crotaphatrema* (three species). These two genera together are treated as a subfamily of Caeciliidae by Frost et al. (2006).

**Natural History:** These are moderately large, elongate burrowers with vestigial eyes and no orbits. *Scolecomorphus* probably is viviparous, whereas *Crotaphatrema* may be oviparous (Nussbaum and Wilkinson 1989).

### Family Typhlonectidae

**Distribution:** Disjunct distribution in northern South America and southern Brazil and northern Argentina.

**Content:** Five genera, *Chthonerpeton* (eight species), *Typhlonectes* (two species), *Nectocaecilia* (one species), *Potomotyphylus* (one species), and *Atretochoana* (one species). The genus *Typhlonectes* appears to be closely related to *Caecilia,* based on molecular data, and typhlonectids sometimes are treated as a subfamily of Caeciliidae (Frost 2004; Frost et al. 2006).

**Natural History:** Although often described as strictly aquatic, some typhlonectids exhibit few morphological adaptations for swimming and apparently live in moist soil (e.g., members of the genus *Chthonerpeton;* Nussbaum 1986). The genus *Typhlonectes* is the most aquatic. Although it lacks a true tail, the posterior part of the body is laterally compressed into a fin for swimming. These animals spend the daylight hours in flooded burrows along the margins of streams and rivers. At night, they form large foraging groups in the water (Moodie 1978). All species in this family are assumed to be viviparous. One species of typhlonectid, *Atretochoana eiselti,* was described from a single specimen collected somewhere in South America in the nineteenth century. It was previously placed in the genus *Typhlonectes,* but a reexamination of the specimen revealed that it is completely lungless. It is the only lungless caecilian, and with a length of 725 mm, is the largest known lungless vertebrate (Nussbaum and Wilkinson 1995). After the discovery of a second specimen, the anatomy and systematic relationships of this unusual caecilian were studied in considerable detail (M. Wilkinson and Nussbaum 1997, 1999). It appears to be most closely related to the monotypic genus *Potomotyphylus,* which is fully aquatic. The ecology of *Aretochoana* is completely unknown, but the presence of a fin and laterally compressed body are consistent with an aquatic lifestyle.

### Family Uraeotyphlidae

**Distribution:** Southern India.

**Content:** A single genus, *Uraeotyphlus* (five species).

**Natural History:** These are relatively small caecilians that look similar to ichthyophiids, with a true tail, and appear to be the sister group to that family (M. Wilkinson and

Nussbaum 1996; M. Wilkinson et al. 2002) or part of that family (Frost et al. 2006). They are presumed to be oviparous, and at least one species has aquatic larvae (M. Wilkinson 1992).

## Morphological Evolution and Ecology of Caecilians

All caecilians are elongate animals with large numbers of vertebrae (95–285). They lack any trace of limbs or pelvic and pectoral girdles, and in all but the basal groups, the tail is absent as well (Duellman and Trueb 1986; Himstedt 1996). This morphology presumably is a derived condition related to their fossorial habits. The discovery of well-preserved fossil caecilians from the early Jurassic has shown that the ancestors of modern caecilians probably had well-developed legs and tails (Jenkins and Walsh 1993). The elongation of the body has led to many internal organs, such as the testes and lungs, being elongated as well, a condition also seen in snakes. The left lung is dramatically reduced relative to the right lung, or is lost altogether. This is the same pattern seen in derived families of snakes, which also lack the left lung.

Most caecilians are medium-size to large animals (the largest is *Caecilia thompsoni* at about 1.5 m). However, the miniaturization that has occurred repeatedly in the evolution of anurans and salamanders is evident in caecilians as well. Two of the smallest species, *Idiocranium russeli* of West Africa and *Grandisonia brevis* of the Seychelles Islands, are only slightly more than 100 mm in length (Nussbaum 1984; M. Wake 1986a). *Idiocranium* exhibits a number of paedomorphic features in addition to reduced body size, including retention of some cartilage in the skull and vertebrae and the loss of certain bony elements in the skull that appear in other caecilians relatively late in development (M. Wake 1986a). Unfortunately, nothing is known about the ecology of these tiny caecilians, so functional correlates of reduced body size are unknown.

### Modes of Locomotion

Caecilians are unusual in that the trunk musculature forms a thick band that is closely attached to the skin and relatively unattached to the vertebral muscles. This allows the vertebral column to move independently of the skin and trunk muscles (Naylor and Nussbaum 1980; Nussbaum and Naylor 1982). Caecilians move either by lateral undulation, an ancestral mode for all tetrapods, or by a modified form of concertina locomotion in which the body is thrown into curves to form an anchor to push the front of the body forward. The latter is used when moving inside burrows. Many burrowing caecilians do not bend the entire trunk, but instead bend the vertebral column into curves without bending the trunk musculature, a form of internal concertina locomotion that is unique to caecilians and a few snakes. This type of movement is possible only if there is a loose connection between the trunk and vertebral muscles. Forward movement is enhanced by hydrostatic pressure generated by vertical muscles in the body wall and a helical array of tendons that surrounds the body wall (O'Reilly, Ritter, and Carrier 1997; O'Reilly, Summers, and Ritter 2000). This condition has been lost in the highly derived aquatic typhlonectids, which have a tighter connection between these muscle groups and move mainly by lateral undulation (Gaymer 1971; Gans 1974; Summers and O'Reilly 1997).

### Burrowing

Caecilians burrow by inserting the snout into the ground at an angle, bracing the loosely curved body against the substrate, and elevating the head to push the soil aside. Once the burrow is started, it is enlarged by pushing and compacting the soil with the head, while pushing forward with contractions of the trunk and vertebral muscles (M. Wake 1993b). There are differences in burrowing ability among families of caecilians. Rhinatrematids, which are considered relatively primitive caecilians, spend at least part of their lives on the surface moving through leaf litter and are not as well adapted for burrowing as more derived families. The same is true of ichthyophiids. Ducey et al. (1993) compared the burrowing ability of one ichthyophiid species (*Ichthyophis kohtaoensis*) with several species in the more derived family Caeciliidae. All of the species tested easily burrowed into loosely compacted soil, but had difficulty penetrating more heavily compacted substrates. *Dermophis mexicanus*, a caeciliid, was the only species that could burrow into tightly compacted soil, and only a few individuals succeeded in doing so. When placed on slightly or moderately compacted soil, *Ichthyophis* and one caeciliid (*Schistometopum*) used preexisting artificial tunnels in preference to constructing their own burrows. *Dermophis* did so as well, but was more likely to construct its own burrows. Limited observations in the field indicate that this genus is found in a wider range of soil types than other caecilians.

The skulls of caecilians are highly derived relative to ancestral amphibians, and very different in structure from those of anurans and urodeles. The head is flattened and wedge-shaped, with the mouth underneath a relatively pointed snout. In most species, the whole skull is heavily covered in dermal bone, and the snout region is strengthened (fig. 1.46). The completely roofed skull is similar to that of some elongate fossil tetrapods, especially the microsaurs, leading some authors to conclude that it represents an ancestral condition for caecilians and that this

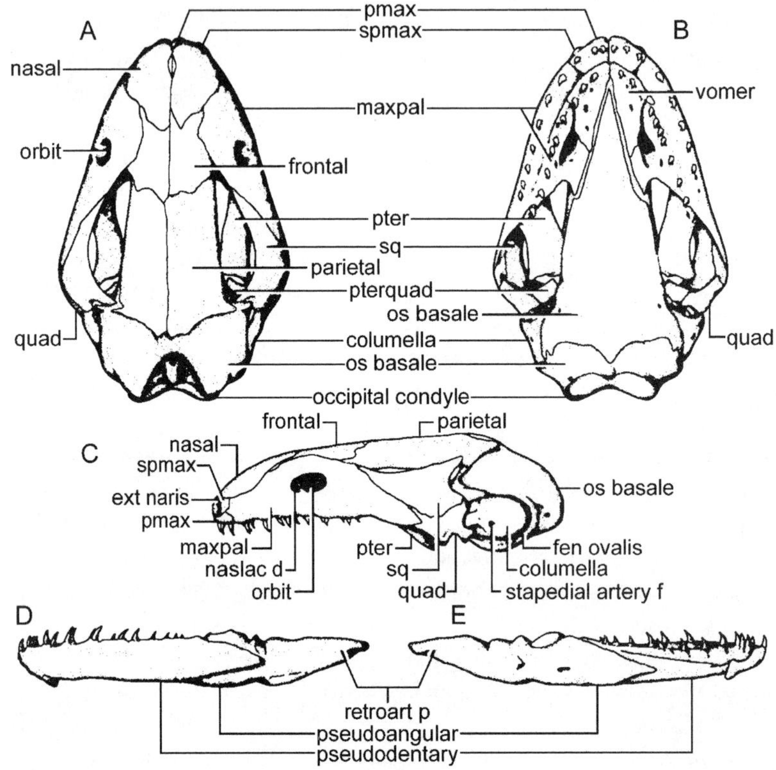


**Fig. 1.46.** Skull of a primitive caecilian, *Epicrionops petersi* (Rhinatrematidae). (A) Dorsal view. (B) Ventral view. (C) Lateral view. (D) Lateral view of mandible. (E) Medial view of mandible. Heavy dermal bone covers the entire skull and snout region. Note the very small size of the orbit. After Trueb (1993).

group is derived from a different clade than anurans and urodeles (Carroll and Currie 1975; see Trueb and Cloutier 1991b for a review of controversies over amphibian evolution). It is more likely a secondary derivation related to fossorial life (M. Wake and Hanken 1982; M. Wake 1986b; Trueb 1993). The skin of the snout and jaw are co-ossified (tightly bound) to the underlying bone, and the skin of the snout is thickly padded (M. Wake and Hanken 1982; Straub 1984; Duellman and Trueb 1986; Trueb 1993; M. Wake 1993b). The skulls of caeciliids appear to be more derived and more specialized for burrowing than those of rhinatrematids and ichthyophiids. The muscles of the head region are modified for burrowing, with powerful muscles that bend the head upward or downward.

### Feeding

Caecilians feed on several types of subterranean prey, including termites, ants, beetles, earthworms, and even an occasional lizard or snake (M. Wake 1986b; O'Reilly 2000; Presswell et al. 2002; Gaborieau and Measey 2004; Measey et al. 2004; the diets of most species are unknown). Until recently, detailed observations of feeding behavior were available only for *Dermophis mexicanus* feeding on earthworms (Bemis, Schwenk, and Wake 1983). These animals seize the prey in their jaws, which are equipped with sharp, recurved teeth. When the mouth is closed, the row of teeth on the lower jaw fit into a space between rows of teeth on the upper jaw (fig. 1.47). This provides an interlocking mechanism that enhances the ability of the animal to hold onto prey. Caecilians lack the protrusible tongue found in most anurans and urodeles, so prey is captured with the jaws. Caecilians have evolved a jaw-closing mechanism that is unique among vertebrates. The skulls of most terrestrial caecilians are heavily roofed with bone, so there are no large openings for expansion of the jaw closing muscles (the temporal fossae) seen in many other tetrapods. Instead, the posterior interhyoideus muscle, which serves a completely different function in other amphibians, runs from the ventral and lateral body walls and inserts on the underside of a vertical process of the jaw (fig. 1.47 A). When this muscle contracts, in conjunction with jaw-closing muscles running from the jaws to the side of the head, the jaw rotates around the quadrate joint (fig. 1.47 C), resulting in rapid closure of the mouth (Bemis, Schwenk, and Wake 1983; Nussbaum 1983).

O'Reilly (2000) studied the kinematics of feeding in sev-

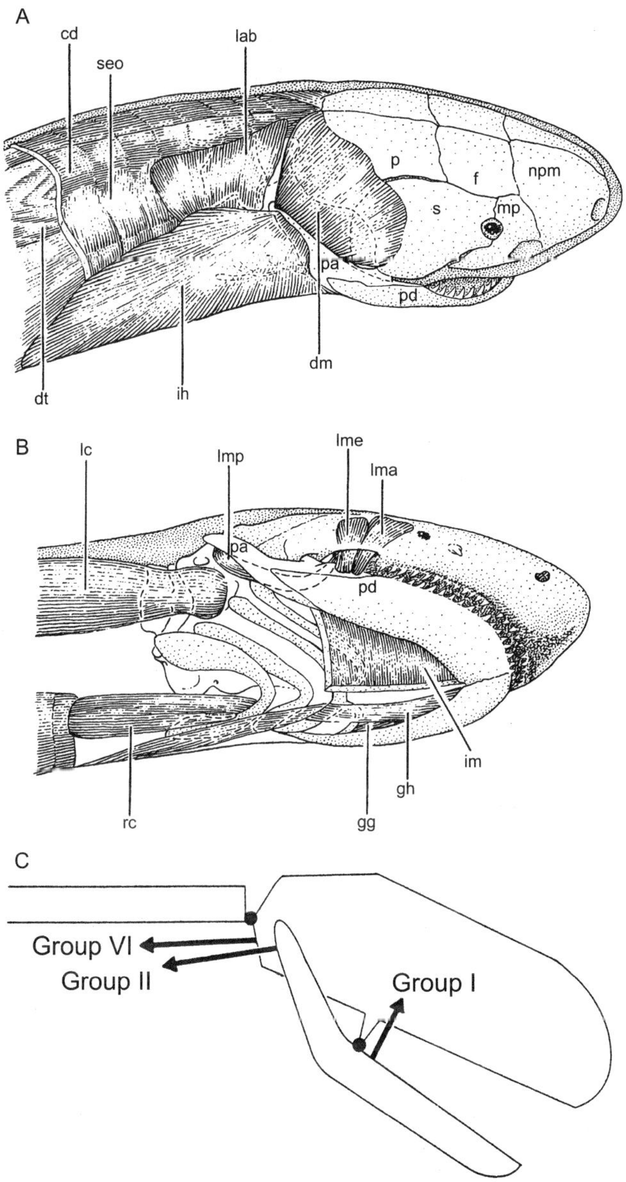


**Fig. 1.47.** Head and jaw muscles of the caecilian *Dermophis mexicanus*. (A) Dorsolateral view, showing superficial muscles. (B) Ventrolateral view with superficial muscles removed. (C) Schematic diagram illustrating mechanism of jaw closing. Group I muscles are the internal adductors (M. levator mandibulae anterior, M. levator mandibulae externus). Group II is the lateral adductor (M. interhyoideus). Group VI is the ventral trunk musculature (M. longus capitis et colli). Arrows indicate direction of movement when these muscles contract. Abbreviations for muscles: dm = depressor mandibulae, dt = dorsalis truncae, gg = genioglossus, gh = geniohyoideus, ih = interhyoideus, im = intermandibularis, lab = levator arcus branchiales, lc = longus capitus, lma = levator mandibulae anterior, lme = levator mandibulae externus, lmp = levator mandibulae posterior, rc = rectus cervicus, rl = rectus lateralis, seo = superficial external oblique. Abbreviations for skull bones: f = frontal, mp = maxillopalatine, npm = nasopremaxilla, p = parietal, s = squamosal. Abbreviations for jaw bones: pa = pseudoangular, pd = pseudodentary. After Bemis, Schwenk, and Wake (1983).

eral families of caecilians. He found that most features of caecilian feeding are relatively conservative in that they do not differ greatly among families. Prey capture in ichthyophiids, which are considered relatively primitive caecilians, is nearly indistinguishable from that of the more derived caeciliids. Typhlonectids differ from these families mainly in being able to close their jaws more rapidly. This may be related to the structure of the skull, which has a temporal opening to accommodate jaw-closing muscles, a feature not seen in most caecilians. This allows for a larger, more powerful muscle that can close the jaw more rapidly. All caecilians open their jaws more slowly than do frogs or salamanders, most of which protrude the tongue to capture prey. O'Reilly (2000) suggested that slow opening of the jaw is necessary in caecilians because they are blind. They must assess the size and activity of their prey with their jaws, whereas frogs and salamanders do so visually before initiating a prey strike. Although some adult caecilians, such as *Hypogeophis* and *Typhlonectes*, are capable of feeding in water, they all capture prey with the jaws rather than employing suction feeding, as seen in many other aquatic vertebrates (O'Reilly, Deban, and Nishikawa 2002).

## Sensory Systems

The sense organs of caecilians have been highly modified for fossorial life. Adult caecilians have reduced eyes, a typical condition for fossorial animals, although many species probably retain some photoreceptive ability, even if they cannot form visual images. Early embryonic development of the eyes proceeds as in other amphibians, but development of various structures is retarded at different stages. Reduction of these parts does not proceed completely in parallel; that is, a species may have a greatly reduced lens or eye musculature, but retain a fully functional retina or optic nerve. However, general evolutionary trends in eye reduction include covering of the eye with skin or both skin and bone, reduction or loss of muscles that control eye movement, reduction in the number of retinal cells, reduction in the optic nerve, reduction or loss of the lens, reduction or loss of the vitreous body of the eyeball, and increased attachment of the lens to the cornea and retina. In general, the least reduced eyes are found in the most primitive families (Rhinatrematidae and Ichthyophiidae), which appear to be adapted to subsurface life under rocks, logs, or leaf litter. The most reduced eyes are found in the more derived families (Caeciliidae and Scolecomorphidae) that are more fully adapted to subterranean life. However, eyes also are somewhat reduced in aquatic typhlonectids, which are thought to have evolved from fossorial caeciliids (M. Wake 1985).

Caecilians have one sense organ that is unique among vertebrates, the tentacle (Fox 1985). This appears to be primarily a chemosensory organ that projects from the side of the head (fig. 1.44 C, D) and has a hollow chamber that connects with the Jacobson's organs, part of the vomeronasal system. The tentacle can be moved and extended or retracted with muscles that originally served to move the eye, but have been taken over for an entirely new function. These muscles are innervated by branches of the oculomotor nerve, which controls eye movements in other amphibians. It also is lubricated with secretions of the Harderian gland, which normally lubricates the eye (M. Wake 1985, 1986b; Billo and Wake 1987). The tentacle is closely associated with the vomeronasal system and may serve to conduct chemicals from the substrate to the vomeronasal organ (Schmidt and Wake 1990). Presumably the tentacle is used to detect prey, perhaps by trailing prey animals in burrows. It also may function in the location and recognition of mates. The only experimental study of tentacle function revealed that blocking the tentacles of *Ichthyophis kohtaoensis* did not impair the animal's ability to locate prey, whereas blocking the nares did interfere with prey location (Himstedt and Simon 1995).

## Summary and Conclusions

The living amphibians are representatives of the first vertebrates to move onto land about 400 million years ago. As such, they must be considered a remarkably successful clade if one measures success in terms of phylogenetic persistence. Amphibians as a group not only survived and prospered through major geological upheavals and rearrangements of the continent landmasses and major changes in climate, but they also survived several major mass extinction events, including the one that wiped out the dinosaurs at the end of the Cretaceous. They also have managed to occupy most of the world's major terrestrial habitats, from the driest deserts to wet tropical rainforests.

Modern amphibians bear only a superficial resemblance to the first tetrapods that emerged onto land, and many morphological features of early tetrapods suggest that they had not yet acquired the distinctive physiological traits of modern amphibians. All living amphibians have a highly permeable integument that is susceptible to desiccation, intolerant of submersion in seawater, and an essential avenue of gas exchange. In contrast, many of the earliest tetrapods were covered with heavy dermal armor, and some evidentially lived in marine environments. By the Carboniferous, a diverse array of tetrapod clades had become the dominant land vertebrates, many of which reached quite large sizes. This contrasts with modern amphibians, most of which are among the smallest of all land vertebrates. Indeed, an evolutionary reduction in body size is a theme that has been repeated many times in the history of amphibians and may account for their success in modern ecosystems, where they often occupy niches not available to larger vertebrates, especially birds and mammals.

The origin of the living groups of amphibians, the frogs, salamanders, and caecilians, is obscure because of a limited fossil record. Nevertheless, by the Jurassic, these three lineages were well differentiated from one another and probably found throughout the world. Animals that we would recognize as modern frogs, salamanders, and caecilians were present by the early Cretaceous, living literally under the feet of the large dinosaurs that dominated terrestrial ecosystems at that time. Indeed, one can speculate that some frogs might have used the water-filled footprints of some of the largest dinosaurs as breeding pools, much as some tropical frogs use the tracks of tapirs and other animals as breeding sites today.

In addition to a general reduction in body size, living amphibians have several general characteristics that have profoundly affected their ecology and behavior. One is a tendency toward paedomorphosis and a simplification of many organ systems. Many living amphibians, for example, have highly simplified sense organs compared to other vertebrates, including fishes. In addition, amphibians have unusually large genomes and large cell size, which in turn affects physiological traits such as metabolism, growth rates, and rates of development. Relatively slow rates of development in turn affect the evolution of parental care, the biology of amphibian larvae, and the timing of metamorphosis.

With about 5,300 of the more than 6,000 species of living amphibians, anurans are by far the most diverse clade of modern amphibians, and this diversity undoubtedly results in part from their highly derived morphology. Frogs in general are morphologically specialized for jumping, which in turn has allowed these animals to radiate into a wide range of aquatic, terrestrial, fossorial, and arboreal habitats. One consequence of this tendency to occupy many different habitats is widespread convergent evolution in morphology. There are, for example, burrowing frogs with very similar morphology in at least a half dozen major anuran families, and there has been repeated evolution of morphologically similar arboreal, semiaquatic, and terrestrial species as well. This tendency toward morphological convergence has made it difficult to resolve the taxonomy and phylogenetic relationships of the major clades of frogs. Some of the largest families of anurans, such as the Ranidae and Leptodactylidae, probably are not monophyletic groups as currently constituted, and relationships among subclades within these families are unclear. The relationships within the large derived ranoid clade (the family Ranidae and its close relatives) are especially problematic, with major conflicts between morphological and molecular data that have resulted in large groups (subfamilies and genera) being placed in different families by different investigators. Recent debates over the relationships of the frogs of Madagascar and the phylogenetic position of the Neotropical dendrobatid frogs are just two examples of the difficulties encountered by systematists in resolving the higher-order relationships of frogs.

In contrast, the urodeles are far less diverse, with about 550 species, and exhibit a tendency toward morphological conservatism. Consequently, the family-level taxonomy of this group has remained relatively stable for many years, although there are continuing debates over the precise relationships among families. The relatively conservative nature of morphological evolution in salamanders has led to different sorts of problems, however. For example, there is a tendency in several groups for genetic differentiation among populations to be much greater than morphological differentiation, and this has resulted in the recognition of many cryptic species and a realization that species-level diversity of some families is greater than previously thought. This is evident to a limited extent in small families such as the Sirenidae, Rhyacotritonidae, and Dicamptodontidae, but is especially characteristic of the family Plethodontidae, which includes more than 70% of all salamander species. This tendency toward a high degree of local genetic differentiation is related to the low vagility of most salamander populations and the very limited geographic ranges of many species, especially those found in mountainous regions that are topographically complex.

With only about 170 described species, the caecilians are the least diverse of the living amphibians, and they also are the most specialized. All caecilians are limited to the wet tropics, and all are either specialized for burrowing or are derived from burrowing ancestors. Their highly derived morphology, which includes a loss of limbs, reduction or loss of the eyes, and major modifications of the skull, has made it difficult to determine the relationships of caecilians to other amphibians. It also has complicated attempts to determine phylogenetic relationships among major clades of caecilians, with morphological data often in conflict with molecular data. The highly derived morphology of caecilians undoubtedly affects nearly every aspect of their ecology and behavior, from modes of locomotion to feeding ecology, physiological ecology, and reproduction. Yet we currently know less about the ecology and behavior of these animals than that of any other amphibians, and indeed, most other vertebrates.

# **Chapter 2** Water Relations

*The amphibians are a defeated group. They were the first vertebrates to emerge from the waters onto the lands, but they were not destined to complete the conquest, and, at first abundant, they have shrunken into insignificance among four-footed vertebrates. . . . The typical amphibian is still chained to the water. In the water it is born; to the water it must periodically return. We have noted various devices among living amphibians which have enabled them to circumvent this difficulty to some extent. But these makeshifts have not been particularly successful. The amphibian is . . . in many respects, little more than a peculiar type of fish which is capable of walking on land.*
—Alfred Sherwood Romer, *The Vertebrate Story* (1959)

For centuries, people have associated amphibians with the margins of the land, viewing them as somewhat loathsome denizens of rivers, swamps, marshes, and ponds. The ancient Egyptians believed that frogs were born of river slime, generated from the coupling of land and water during the annual flooding of the Nile. Frogs were so closely associated with the life-giving properties of water that they became symbols of new life, fertility, and rebirth. The goddess Heqet, who served as a kind of divine midwife, was depicted with the head of a frog and the body of a woman (Keller 1913; Egger 1931; Budge 1969; M. Weber 1972). In other cultures, frogs have been widely associated with rain and fertility of the land. In various parts of India and Nepal, frogs have been hung from trees or presented with offerings of rice and milk to help bring on the rains (Briffault 1927). To the Maya of Central America, frogs were attendants of the thunder gods, and served as musicians who announced the coming of the rains (Wassen 1934; Hamblin 1984). Frogs were considered harbingers of rain by many other tribes in South and Central America, western North America, and the Caribbean. This belief was particularly common in regions subject to periodic drought, where frogs simply disappeared when water was scarce, only to reappear with the onset of rain (Wassen 1934).

Even today, frogs and other amphibians generally are associated in the popular imagination with water and aquatic habitats, although some amphibians are recognized as having made the first tentative steps toward colonization of land. Among biologists, amphibians have assumed the role of transitional animals in progressionist views of evolution that depict the history of vertebrates as a gradual movement from aquatic to terrestrial environments. For example, a review of amphibian endocrinology stated that "Amphibia can be conceived as spanning the phylogenetic step between purely aquatic vertebrates and their terrestrial successors. The contemporary groups contain those that are indeed purely aquatic (the urodeles) and those that are terrestrial as adults (the apodans) and those that essentially dwell in either water or on land (the anurans)" (Balment and Henderson 1987, 490). Actually, the salamanders, caecilians, and anurans all include fully aquatic, fully terrestrial, and semiaquatic species.

The notion that most amphibians remain tied to the water is all too common among biologists unfamiliar with the enormous diversity of amphibian life. Many amphibians are associated with standing water, but some of the most successful lineages are completely independent of standing water, although they do require moist microhabitats. Amphibians can be found in terrestrial habitats from tropical rainforests to some of the driest deserts on earth (Bentley 1966a; Mayhew 1968; Warburg 1972, 1997; Shoemaker 1975, 1988; Heatwole 1984). Although absent from Antarctica and high Arctic latitudes, they are found in nearly all other environments, from below sea level to altitudes above 4,500 m. In North America, one species of toad, *Bufo punctatus,* can be found from the hot floor of Death Valley to above 3,000 m in the Colorado Rockies. Amphibians cannot live in strictly marine environments, but they are found in nearly every other type of aquatic habitat, from brackish estuaries and salty pans to rivers, streams, ponds, lakes, marshes, swamps, and ditches. Some species can survive in acidic or alkaline waters. One of the most interesting aspects of amphibian biology is their ability to tolerate a wide range of environmental conditions, especially variation in the availability of water.

Biologists have known for at least two centuries that amphibians are unusual in several aspects of their water relations. In the late eighteenth century, Robert Townson conducted experiments at Göttingen University on the water relations of *Hyla arborea,* using pet tree frogs named Damon and Musidora (Townson 1794–1795, 1799). He found that his frogs lost water quickly when exposed to dry air, but they could take up water quickly as well, even when only the ventral surface of the skin was in contact with water (C. B. Jørgensen 1994a, 1997a). Subsequent work confirmed the importance of the skin in water exchange (Overton 1904; Maxwell 1913; Adolph 1925, 1932, 1933) and also showed that amphibians can tolerate unusually high levels of water loss compared to other vertebrates (Kunde 1857; Langlois and Pellegrin 1902; F. Hall 1922). C. B. Jørgensen (1997a) provided a detailed and fascinating review of the early history of research on amphibian water relations.

Modern work on water balance, osmoregulation, and nitrogen excretion in amphibians began with the classic work of August Krogh (1939) and continued in the tradition of comparative physiology that he established (Krebs 1975; Bennett 1987; C. B Jørgensen 1997a). This work emphasizes patterns of adaptation to features of the physical environment by comparing selected species with different lifestyles. Generally, certain species have been chosen to represent aquatic, terrestrial, or semiaquatic habits, with relatively little attention to phylogenetic relationships among these species. For example, *Xenopus* and *Siren* often are considered strictly aquatic, *Bufo* and *Salamandra* strictly terrestrial, and various species of *Rana* semiaquatic. Yet all of these animals can face periodic problems of desiccation on land or osmoregulatory stress in water. The "strictly aquatic" *Xenopus* and *Siren* often spend extended periods encased in drying mud, while all "terrestrial" toads (*Bufo*) must return to water to breed, and some spend much of their time in the water. Furthermore, these species differ not only in their habitat requirements, but also in evolutionary history, general morphology, and other aspects of their biology that could influence water relations. Consequently, data from such comparisons must be used cautiously in attempting to explain the evolution of physiological adaptations to the environment (Huey 1987; Bennett and Huey 1990; Harvey and Pagel 1991; Garland and Adolph 1994).

This chapter focuses on the types of problems encountered by amphibians in terrestrial and aquatic environments. I begin with a brief overview of the ecological implications of water availability for amphibians. I then review some aspects of the biology of amphibian skin, because the structure of the skin has important implications not only for problems of water balance, but also for nearly every other aspect of amphibian ecology and behavior. I then discuss adaptations to terrestrial life, since this presents the greatest challenge to the survival of most amphibians. The problems of amphibians living in water are then considered, along with some aspects of kidney function and nitrogen excretion. Although I will touch on cellular-level processes of osmoregulation, transport of water and materials across membranes, and hormonal control of water balance, the main emphasis is on the physiological ecology of whole organisms. For more detailed treatment of physiological mechanisms, readers should consult reviews by Boutilier, Stiffler, and Toews (1992), Shoemaker et al. (1992), and C. B. Jørgensen (1997a).

## Ecological Implications of Water Availability

The availability of water affects nearly every aspect of the lives of amphibians. The availability of aquatic breeding sites undoubtedly is a major determinant of the geographic and local distribution of species that rely on standing water to breed (Rodríguez, Belmontes, and Hawkins 2005). The persistence of aquatic sites can have a major effect on reproductive success (e.g., Pechmann et al. 1989; Rowe and Dunson 1993, 1995; see also chapter 15). Complete reproductive failure is common among amphibians that breed in temporary ponds, especially in arid climates (e.g., Semlitsch et al. 1996; Read 1999). Even in equable climates, periodic drought often decreases the size of local amphibian populations (Dodd 1993, 1994, 1996; Jędrzejewska, Brzeziński, and W. Jędrzejewski 2003; Daszak et al. 2005). Prolonged droughts have been suggested as one of many possible causes for the decline or extinction of populations of aquatic-breeding amphibians in various parts of the world. Such declines can result from either failure of reproduction in several successive years, or drought-induced mortality of juveniles and adults (Corn and Fogleman 1984; W. Osborne

1989; Weygoldt 1989a; M. Crump, Hensley, and Clark 1992; Kagarise Sherman and Morton 1993; Pounds and Crump 1994; Grafe et al. 2004; see also chapter 16).

The availability of suitable moist microhabitats for the terrestrial life-history stages of amphibians is important as well. Some species have successfully colonized deserts and semiarid habitats, but terrestrial amphibians are most diverse and abundant in relatively moist environments, such as cool upland areas and wet tropical rainforests (Wake 1970; Kiester 1971; N. Scott 1976, 1982; D. Wake and Lynch 1976; Pefaur and Duellman 1980; Duellman and Trueb 1986). Comparisons of wetter and drier habitats on a more local scale almost invariably reveal greater diversity and abundance of terrestrial amphibians in the wetter sites (Lloyd, Inger, and King 1968; Barbault 1976a, b; D. Wake and Lynch 1976; Inger 1980; J. Lee 1980; Toft 1980b; Duellman 1988; see chapter 15). On an even smaller scale, soil moisture and the availability of moist retreat sites can be important determinants of the local distribution and abundance of species (Hairston 1949; Stebbins 1954b; Rosenthal 1957; Cunningham 1960; Taub 1961; Heatwole 1962b; Jaeger 1971b, 1980b; Spotila 1972; Pough et al. 1987; Wyman 1988; Petranka, Eldridge, and Haley 1993; Grover 1998, 2000; Vences, Galán et al. 2000).

Problems of evaporative water loss also affect seasonal and daily patterns of activity. Foraging often is limited by dry weather, and this can affect total food intake and allocation of energy to growth and reproduction (Stille 1952; Jaeger 1978, 1980a; Feder 1983b; Feder and Londos 1984; Keen 1984; Stewart 1985; Cree 1989; Duellman 1995). Variation in rainfall can affect the abundance and distribution of prey, which in turn can affect the growth and survival of amphibians (Toft 1980b; Toft, Rand, and Clark 1982; Galatti 1992; Stewart 1995). The availability of moist microhabitats also influences the size and location of individual activity ranges (Dole 1965a, b; Toft 1980a, b; Semlitsch 1981a; Carpenter and Gillingham 1987; Crump and Pounds 1989). When moist retreat sites are scarce, individual amphibians sometimes fight over choice sites (e.g., Wiewandt 1971; Wells 1980a; Keen 1982; Mathis 1989, 1990a).

Movements to and from breeding sites often are affected by rainfall (Blanchard 1930; Baldauf 1952; Packer 1960; Hurlbert 1969; Dole 1971; Fitzgerald and Bider 1974b; Gibbons and Bennett 1974; Semlitsch 1981a, 1983a, 1985a; Gittins 1983a; Griffiths 1984; Semlitsch and Pechmann 1985; Sinsch 1988c; Spieler and Linsenmair 1998; D. Marsh 2000). In both aquatic and terrestrial breeders, reproductive activities such as calling to attract mates, oviposition, and transport of tadpoles, also are affected by rainfall and water availability, especially in strongly seasonal environments (Wells 1980a; Pough et al. 1983; Aichinger 1987a, 1991; Donnelly 1989c; Praderio and Robinson 1990; Moreira and Lima 1991; Donnelly and Guyer 1994; Townsend and Stewart 1994; Moreira and Barreto 1997; Pröhl 1997a; Bertoluci 1998; Arzabe 1999; N. Mitchell 2001, 2002b; Bertoluci and Rodrigues 2002; Kupfer, Nabhitabhata, and Himstedt 2004, 2005; Prado, Uetanabaro, and Haddad 2005; Vaira 2005).

## Structure and Function of Amphibian Skin

The skin of an amphibian is the interface between the animal and its surroundings. It not only provides mechanical protection, but it also serves as a major avenue for the exchange of gases, water, and other materials between the animal and the environment. The maintenance of a relatively stable internal environment (homeostasis) is critical for the proper functioning of various physiological systems and biochemical pathways. The skin is involved in many other facets of amphibian ecology and behavior as well (Heatwole and Barthalmus 1994). Glands in the skin are important for courtship, sex recognition, and other aspects of chemical communication, especially in urodeles (Houck and Sever 1994; chapter 9). Skin glands also produce toxic or distasteful substances that protect amphibians from their predators (Barthalmus 1994; Erspamer 1994; B. Clarke 1997; chapter 14). Changes in the color of the skin are important for communication (chapters 7, 8, 9), temperature regulation (chapter 3), and protection from predators (chapter 14). Structures derived from the epidermis, such as nuptial pads, spines, and claws, are used in courtship and mating (chapters 8, 9, and 10). Hence, an understanding of the basic structure and function of amphibian skin is necessary for understanding the ecology and behavior of these animals.

### Epidermis

The skin of amphibians consists of two major layers, the outer epidermis and the inner dermis (fig. 2.1). The epidermis is derived embryologically from ectodermal cells. The dermis is derived mainly from mesodermal cells, which also give rise to most of the body's other tissues, including bone, muscle, and connective tissues. The pigment cells and glands in the dermis are of ectodermal origin and migrate downward into the dermis during development (Lindemann and Voulte 1976; Whitear 1977; Duellman and Trueb 1986; H. Fox 1986a, b, c, 1994).

The outermost layer of the amphibian epidermis, the *stratum corneum,* generally consists of a single layer of flattened cells. Underlying the *stratum corneum* is a layer of replacement cells known as the *stratum granulosum,* several additional cell layers known as the *stratum spinosum,* and finally the *stratum germinativum,* the site of production of replacement cells for the outer layers of the epidermis

Some small *Bolitoglossa,* many of which live in the leaf axils of *Heliconia* and banana plants, have superficially similar webbed feet, but these are the result of an entirely different evolutionary process. In these species, the feet are not fully developed, but retain the padlike structure that all plethodontid feet have early in development before the toes are fully grown (fig. 1.36, top). These paedomorphic feet cannot generate suction by muscular action, but depend instead on surface adhesion to the leaf (Alberch 1981). This morphology is largely a consequence of reduction in body size through paedomorphosis, not selection for living in a specific habitat; some small species with similar foot morphology are terrestrial rather than arboreal. This sort of convergent evolution in foot morphology apparently has

The hynobiids, ambystomatids, and dicamptodontids, most of which have well-developed lungs, have the most generalized tongue morphology, which is considered to be the ancestral condition for terrestrial salamanders. The tongue is relatively broad and padlike, tightly attached to the floor of the mouth, with only part of the margin of the tongue free (fig. 1.39 B). Consequently, the capacity for tongue projection is limited, with little projection beyond the tip of the snout (Larsen and Guthrie 1975; D. Wake and Deban 2000). Nevertheless, the slight projection of the tongue and the use of the tongue to seize the prey is the principal difference in feeding behavior of terrestrial adults and aquatic larvae. Most other aspects of the kinematics of feeding are remarkably similar (Shaffer and Lauder 1988; Lauder and Reilly 1990; Gillis and Lauder 1994). Although the tongue is used to seize the prey, these salamanders cannot capture prey at long distances, nor are they well adapted to feeding on very small or fast-moving prey (Reilly and Lauder 1989).

Salamandrids generally have a somewhat greater ability to project the tongue than do ambystomatids or dicamptodontids, and there is considerable variation within the family (Özeti and Wake 1969; D. Wake 1982). In *Salamandra salamandra,* eggs are retained in the female's reproductive tract, and larvae are born at a relatively large size and advanced stage of development (see chapter 10). The aquatic larvae use the typical suction feeding seen in other aquatic vertebrates. Motor patterns are very similar to those seen in other aquatic salamander larvae, although larvae of this species capture prey more rapidly than most other species (Reilly 1995). Metamorphosed adults are entirely terrestrial and have a land-tongue morphology similar to that of other terrestrial salamanders. Although their ability to protract the tongue is limited, it can be projected beyond the mouth and is used for prey capture (Özeti and Wake 1969; D. Wake 1982). Two other terrestrial genera, *Salamandrina* and *Chioglossa,* have a much greater ability to project the tongue (up to 20% of snout-vent length in *Salamandrina*). In this respect, they are convergent with plethodontids that have highly protrusible tongues (see the following), although the mechanism of tongue projection is different (Özeti and Wake 1969; Findeis and Bemis 1990). It is undoubtedly significant that these genera, like the plethodontids, lack lungs and therefore do not use the hyolingual apparatus for breathing, so it is freed for use in tongue protrusion (D. Wake 1982).

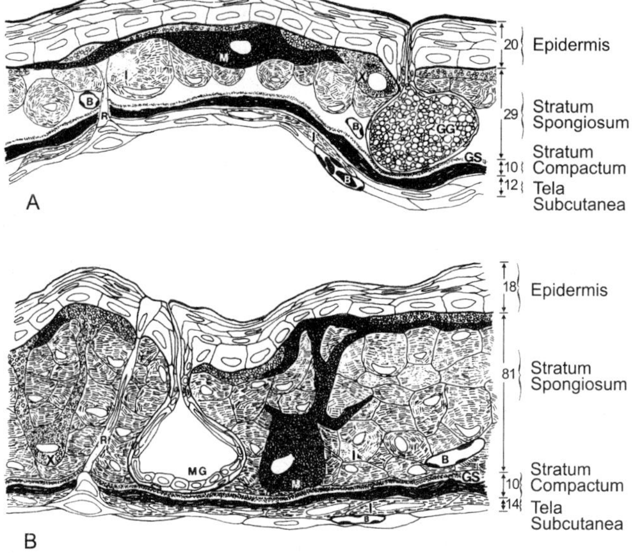


**Fig. 2.1.** Cross-sections of amphibian skin. (A) Dorsal skin of *Hyperolius viridiflavus* in the wet season. (B) Dorsal skin in the dry season, with the number of iridophores having increased by four to six times over the wet season condition. Numbers give thickness of each layer in μm. Abbreviations: B = blood capillary; GS = ground substance layer; GG = granular gland; I = iridophore; M = melanophore; MG = mucus gland; R = radial fiber; X = xanthophore. After Kobelt and Linsenmair (1986).

(H. Fox 1986c, 1994). In most adult amphibians, the epithelial cells forming the *stratum corneum* contain the protein keratin. The degree of keratinization is much less than in other terrestrial vertebrates, and living amphibian species never have keratinized scales on the skin (Lillywhite 2006). Larval amphibians generally do not have keratinized epidermal cells on most of the body. Keratinized cells also are absent from the skin of paedomorphic forms, such as the axolotl (*Ambystoma mexicanum*) and mudpuppies (*Necturus*), which retain some features of larval morphology as reproductive adults (Elias and Shapiro 1957; Spearman 1968; Ling 1972; Duellman and Trueb 1986; Fox 1986c; Warburg, Lewinson, and Rosenberg 1994). In many anuran larvae, cells around the mouth become highly keratinized to form a beak or jaw sheath, as well as rows of denticles or labial teeth. These are used by tadpoles for feeding (chapter 12), but are lost at metamorphosis (Luckenbill 1965; Kaung 1975; Marinelli and Vagnetti 1988; Fiorito de Lopez and Echeverria 1989). Larval amphibians also have a variety of other specialized epidermal cells involved in hatching, adhesion to the substrate, and sensory perception. Adult amphibians retain some of these specialized cells and also have various morphological specializations of the epidermis, such as adhesive toe pads and nuptial pads (see chapters 1 and 9). These specializations are not related to problems of water balance and need not be discussed in detail here (see Fox 1986c, 1994 for reviews).

All amphibians periodically shed the outer layer of the epidermis (*stratum corneum*) and replace it with a new layer of cells. Molting can occur at intervals of less than a week, but the exact timing varies among species. The process begins with increased keratinization of the *stratum corneum* and separation of these cells from the underlying cell layers. Once cornification of the new *stratum corneum* is complete, the old layer of cells is shed intact and usually is swallowed by the animal as molting progresses (Adolph and Collins 1925; Bensden 1956; Taylor and Ewer 1956; Heusser 1958b; C. B. Jørgensen and Larsen 1960, 1961; Ling 1972; H. Fox 1986c). The function of this periodic molting is not entirely clear, but presumably it relates to a need for renewal of the outer layer of skin cells, which are not well protected against injury or mechanical abrasion. Molting cycles do not appear to be directly correlated with periods of increased feeding or growth, but are affected by temperature, metabolic rate, and photoperiod, and are under hormonal control (Ling 1972; H. Fox 1986c). The ability of amphibians to detach the outer layer of cells as one piece has important implications for water balance in burrowing species that form cocoons. These cocoons often are composed of multiple layers of keratinized epidermal cells that have been detached from the underlying layers of skin, but not shed from the body (see the following discussion).

### Dermis

The dermis is considerably thicker than the epidermis and consists of two major layers (fig. 2.1). The *stratum spongiosum* is composed of a network of collagen and elastic fibers, along with a variety of specialized cells, chromatophores, glands, blood vessels, and nerves. Below this is the *stratum compactum,* consisting mainly of compact layers of collagen fibers (H. Fox 1986b, 1994). In some anurans, there is a ground-substance layer between these two, consisting of a thin layer of polysaccharide and calcium granules. Elkan (1976a) noted that this layer is found almost exclusively in terrestrial species and proposed that it acts like a sponge to conserve water. Drewes et al. (1977) disputed this, pointing out that the position of the ground substance layer below the heavily vascularized *stratum spongiosum* puts it in a poor position to retard water loss through the skin. The presence of a ground substance layer in the dorsal skin of *Hyperolius viridiflavus,* and its absence in the highly permeable ventral skin, supports the idea that it plays a role in reducing water loss (Kobelt and Linsenmair 1986), but its precise function is still unclear.

The dermis of caecilians contains unique structures not found in the other groups of amphibians. These are dermal scales, flat discs composed mainly of layers of collagen fibers and minerals that are imbedded in special pockets, with the base of each scale attached to connective tissue (Zylberberg, Castanet, and de Ricqles 1980; H. Fox 1983, 1986c, 1994; Zylberberg and Wake 1990). Neither the evolutionary homology of these scales nor their function is well understood. One possible function is that they provide some structural support that aids in locomotion (Duellman and Trueb 1986). They appear to have no effect on water loss through the skin, which is very high in the few species that have been studied.

Bony osteoderms are present in a number of anuran species from several different families. In some cases, most of the animal's back is covered with this bony armor, which lies beneath the epidermis (Ruibal and Shoemaker 1984). Most of the anurans that have such osteoderms are arboreal, terrestrial, or fossorial, but they occur in a wide variety of habitats, and there is no clear ecological correlation with the presence of osteoderms. Whether osteoderms reduce evaporative water loss is not known, but in most species, the skin between the osteoderm layer and the outer surface of the epidermis is heavily vascularized, suggesting that the osteoderms would provide a poor barrier to water loss. At least one species, *Phyllomedusa bicolor,* applies lipid secretions to its skin to reduce water loss (see the following), so it seems unlikely that the osteoderms serve as a water barrier.

### Dermal Glands

The skin of amphibians is highly glandular compared to that of other vertebrates (Whitear 1977; H. Fox 1986a; B. Clarke 1997). All amphibians have two major types of glands in the dermis: granular glands and mucous glands. The granular glands secrete various proteinaceous substances that serve primarily as defensive chemicals (see chapter 14). They do not play any role in water exchange or moistening of the skin. Mucous glands tend to be much more numerous than granular glands and usually are distributed over the entire surface of the skin, although their precise distribution differs among species (Le Quang Trong 1971, 1975a, b). These glands secrete watery, transparent mucus that keeps the outer surface of the skin moist. This ensures that evaporation of water takes place from the mucus layer rather than the skin itself, protecting the skin from the damaging effects of desiccation (Lillywhite and Licht 1975; Lillywhite and Maderson 1988).

The rate of mucus discharge varies considerably among species of amphibians and is related to their ecology. In anurans that bask in the sun, increased body temperature triggers synchronized release of mucus over the skin surface, thereby cooling the body through evaporation (see also chapter 3). These synchronized discharges are not seen in nocturnal, fossorial, or largely aquatic species, although some of these exhibit a gradual release of mucus over the skin surface in response to desiccation or increased temperature (Lillywhite 1971b, 1975; Lillywhite and Licht 1975). Toads (*Bufo*) generally do not secrete large quantities of mucus except when shedding their skins, giving them their characteristic dry texture, in contrast to the slimy texture of many other anurans. They keep the skin moist via direct movement of water across the skin surface, movement that is facilitated by extensive epidermal sculpturing (Lillywhite and Licht 1974, 1975; see "Water Uptake").

In addition to the two types of glands found in all amphibians, Blaylock, Ruibal, and Platt-Aloia (1976) described a new type of gland in four species of hylid frogs in the genus *Phyllomedusa* from southern South America. These are lipid glands that secrete a waxy coating that the frogs smear over the surface of the skin. This coating provides a remarkably effective barrier against evaporative water loss (see the following discussion). The species in which lipid glands have been described, *P. sauvagii, P. iherengii, P. hypochondrialis,* and *P. boliviana* (= *P. pailona*), live in semiarid regions. Many other species in this genus, including *P. trinitatis, P. lemur,* and *P. tarsius,* live in tropical rainforests, and there is no report of lipid secretion in any of these. Blaylock, Ruibal, and Platt-Aloia (1976) also examined two other species in the same subfamily, *Agalychnis annae* and *Pachymedusa dacnicolor,* and found no evidence of lipid glands. The lipid-producing glands of *Phyllomedusa* appear to be structurally related to the serous (granular) glands that secrete defensive compounds, not to the mucous glands found in all anurans (Delfino et al. 1998; Barbeau and Lillywhite 2005; Lillywhite 2006).

At least one species of rhacophorid tree frog, *Polypedates maculatus,* also produces lipid secretions that are smeared on the skin in a manner nearly identical to that of *Phyllomedusa* (Lillywhite et al. 1997a, b). This behavior undoubtedly evolved independently, because these frogs are not closely related. In contrast to the phyllomedusine frogs, *Polypedates* does not have distinct lipid glands, but secretes a mixture of mucus and lipids from glands that resemble typical mucous glands of other anurans. An Australian tree frog (*Litoria caerulea*) also wipes glandular secretions on its skin, but the structure of the glands that produce these secretions has not been described (Christian and Parry 1997). The glandular secretions of these frogs are composed mostly of protein, but contain lipids as well.

### Regional Differences in Skin Structure and Osmotic Permeability

The skin of amphibians is not uniform throughout the body in either morphology or physiology. Skin in different regions of the body varies in thickness of the various cell layers, vascularization, permeability to water and ions, and in its response to hormones (for a review, see Toledo and Jared

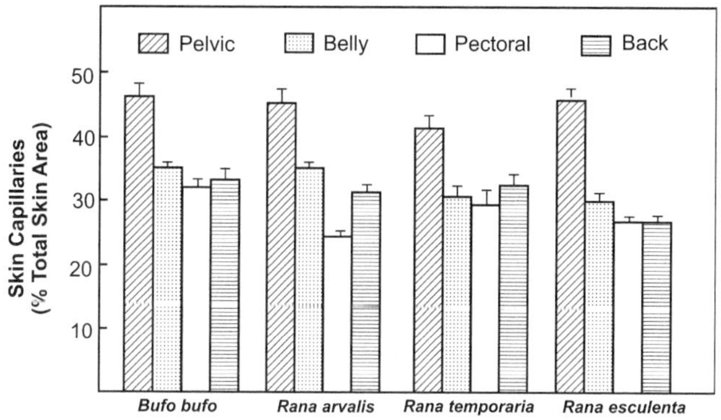


**Fig. 2.2.** Capillary densities of skin from different regions in four species of European anurans, shown as a percentage of total skin area + 1 SE. Pelvic skin has more capillaries than does skin from other regions in all of these species. Plotted from data in Christensen (1974).

1993). It has been known for many years that anurans can take up water from a moist surface through the ventral skin alone, and that the ventral skin is more heavily vascularized than dorsal skin (Overton 1904; Adolph 1933; J. Czopek 1965). In a quantitative study of four species of European anurans, C. Christensen (1974) found that skin from the pelvic region (the pelvic patch) consistently had higher capillary densities than skin from the belly, pectoral region, or back (fig. 2.2). J. Roth (1973) reported that various species of toads (*Bufo*) and spadefoot toads (*Scaphiopus* and *Spea*) have an extensive network of cutaneous veins draining the pelvic region and joining to form large common pelvic veins. This network is largely absent in *Rana catesbeiana* and in *Xenopus laevis*, and the latter also lacks the high density of pelvic skin capillaries found in other anurans (Christensen 1974). The pelvic patches of some toads also have an elaborate complex of cutaneous muscles that press the pelvic patch close to the ground during water absorption behavior (Winokur and Hillyard 1992; Hillyard, von Seckendorff Hoff, and Propper 1998).

Many anurans that take up water from moist substrates rather than bodies of water also have specialized roughened skin either on the entire ventral surface, or concentrated in the pelvic region (fig. 2.3). The rough, granular appearance of the skin derives from flattened polygon-shaped papillae or bumps (termed *verrucae hydrophilica* by Drewes et al. 1977) that are separated by narrow grooves or channels. These *verrucae* are highly vascularized, and along with the network of channels that surrounds them, undoubtedly facilitate rapid uptake of water. This epidermal sculpturing also serves to increase the surface area of the skin available for uptake of water. This structural specialization is common in arboreal frogs such as *Hyla, Phyllomedusa, Litoria, Chiromantis,* and *Hyperolius,* as well as many terrestrial toads (*Bufo*), but it is absent from more aquatic species such as *Rana* and *Xenopus* (Noble 1925; Elias and Shapiro 1957;

**Fig. 2.3.** (A) Ventral surface of the South American treefrog *Phyllomedusa sauvagii,* showing texture of the highly permeable area on the posterior part of the ventral skin. (B) A terrestrial toad, *Bufo fowleri,* showing extensive epidermal sculpturing that conducts water onto the dorsal surface of the skin. Photos by Rudolfo Ruibal (A) and Kentwood D. Wells (B).

Drewes et al. 1977; Duellman and Trueb 1986; Kobelt and Linsenmair 1986). Other epidermal adaptations to facilitate water uptake are described in a later section.

In addition to these structural differences, skin from different parts of the body varies in its osmotic permeability, at least in anurans. Comparisons of isolated samples of skin from ranids, bufonids, hylids, myobatrachids, and leiopelmatids have shown that osmotic flux through the ventral skin generally is somewhat higher than through dorsal skin. Ventral skin from the specialized pelvic region generally exhibits greater osmotic flux than skin from the belly or pectoral area (McClanahan and Baldwin 1969; Bentley and Main 1972b; R. Baldwin 1974; Christensen 1974; Yorio and Bentley 1977; Cree 1988a). Pelvic skin also tends to be much more responsive than skin from other regions to arginine vasotocin (AVT), a hormone that increases skin permeability (fig. 2.4; Bentley and Main 1972a; R. Baldwin 1974; Yorio and Bentley 1977; Cree 1988a, b; see further discussion under "Water Uptake"). In at least one species of treefrog, *Hyla japonica,* males exhibit a greater water absorption response through the ventral skin than do females, suggesting a role for sex hormones in regulation of water

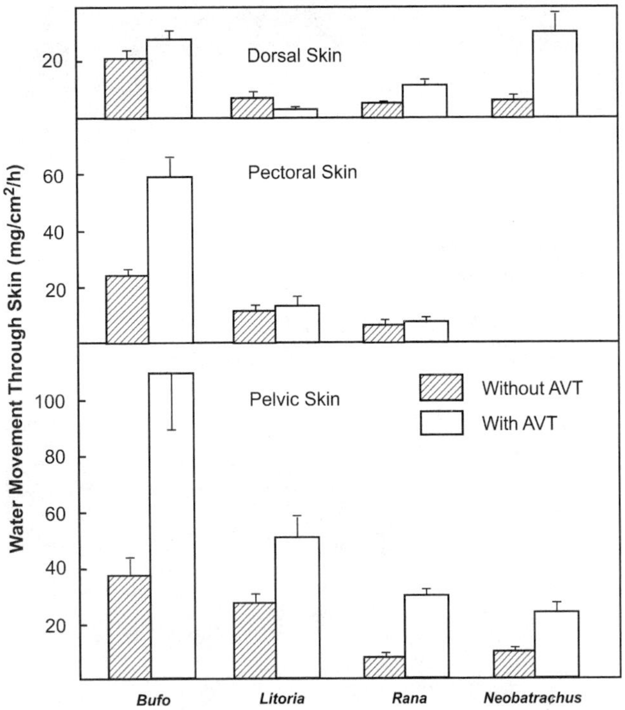


**Fig. 2.4.** Osmotic permeability of skin in different regions and the effects of arginine vasotocin (AVT) on skin permeability in four genera of anurans, including a terrestrial toad (*Bufo*), arboreal frog (*Litoria*), semiaquatic frog (*Rana*), and a fossorial frog (*Neobatrachus*). Pelvic skin exhibits a greater increase in skin permeability to water in response to AVT than does skin from other regions. Plotted from data in Yorio and Bentley (1977).

balance (Kohno et al. 2004). The fossorial Australian frog *Neobatrachus pelobatoides* exhibits a moderate hormonal response in both dorsal and pelvic skin (fig. 2.4). Presumably this is correlated with their habit of taking up water inside a burrow while surrounded by moist soil (Bentley and Main 1972b).

## Determinants of Evaporative Water Loss

The major challenge faced by amphibians in terrestrial environments is evaporative water loss through the skin. Some amphibians have evolved morphological, physiological, or behavioral adaptations that reduce rates of water loss, but many species are limited to moist microhabitats because they have little or no ability to regulate water loss. The variables that affect rates of water exchange between amphibians and the environment have been reviewed by a number of authors (Spotila 1972; Tracy 1975, 1976; Spotila and Berman 1976; Shoemaker et al. 1992; Spotila, O'Connor, and Bakken 1992). An extensive discussion of the biophysics of the process is beyond the scope of this book, and readers should consult these references for further details. In simple terms, the rate of evaporative water loss by an amphibian is a function of the vapor density gradient between the animal and its environment, the resistance of the animal to water loss, and the surface area of the skin exposed to the air. The vapor density gradient is largely a property of the physical environment and cannot be modified by the animal unless it moves to a new microhabitat. It is the difference between the water vapor density at the animal's surface and the water vapor density of the surrounding air. Both temperature and relative humidity of the air affect the vapor density gradient. The gradient will increase as the body temperature of an amphibian increases or as the humidity of the surrounding air decreases. Thus, an amphibian exposed to direct sunlight in relatively dry air will lose water at a much faster rate than the same animal exposed to very humid air at night (Tracy 1976). Nevertheless, even when the air surrounding an amphibian is nearly saturated, as often occurs in a tropical rainforest at night, the animal will continue to lose water at low rates unless its body temperature is lower than that of the air. This is because low levels of metabolic heat production by the animal will be sufficient to raise the temperature of the skin and maintain the vapor density gradient with the environment (Spotila, O'Connor, and Bakken 1992).

The tendency for water to move through a material such as amphibian skin can be expressed in terms of a mass transfer coefficient (measured in centimeters of water movement per second). The reciprocal of this coefficient is termed the resistance to water movement (Shoemaker et al. 1992; Spotila, O'Connor, and Bakken 1992). This represents the combined effects of the inherent resistance of the skin to water movement and the resistance of the boundary layer around the animal. Skin resistance is a function of the skin's morphology and of substances secreted onto its outer surface. Therefore, it can be modified through evolutionary time by changes in the structure of the skin that either increase or decrease resistance to water movement. It also is possible for skin resistance to be modified through physiological responses to water loss (Blaylock, Ruibal, and Platt-Aloia 1976; Kobelt and Linsenmair 1986; Wygoda 1988). Morphological and physiological adaptations that increase skin resistance to water loss are discussed in more detail in a later section.

Boundary layer resistance is more complex. When an amphibian is relatively immobile, water evaporated from the skin's surface forms a thin layer around the animal that reduces the gradient between the animal and the surrounding environment, thereby reducing evaporative water loss. Movement by an animal tends to break up the boundary layer, as does air moving across the animal's surface. Hence, as an animal becomes more active or air speed increases, the resistance afforded by the boundary layer decreases and evaporative water loss increases. For animals with highly permeable skin, boundary layer resistance is a quantitatively more important determinant of water loss than skin resistance (Wygoda 1984; Shoemaker et al. 1992). This means that most amphibians will inevitably have relatively high rates of evaporative water loss when they carry out their normal activities. Boundary layer resistance increases with increasing body mass, so larger amphibians will have greater resistance to water loss than smaller species, even if all other conditions are equal (Shoemaker et al. 1992). In addition, boundary layer resistance can be modified behaviorally by avoidance of windy microhabitats and by changes in the activity of the animal (Heatwole et al. 1969; Pough et al. 1983).

The surface area of skin exposed to the air has a major effect on rates of water loss. Again, large animals are at an advantage over small animals because surface-to-volume ratios decrease as body size increases. This means that for small animals, the area through which water can be lost is increased relative to the volume of water that can be stored inside the body. This makes newly metamorphosed juveniles especially vulnerable to desiccation. Newman and Dunham (1994) found that larger metamorphs of spadefoot toads (*Scaphiopus couchii*) lost water at slower rates than small metamorphs. Because suitable moist microhabitats were patchily distributed, larger metamorphs could survive longer while moving from one moist soil patch to another.

Relative surface area also varies with the shape of an animal. Elongate forms such as plethodontid salamanders have very high surface areas for their body mass and hence are subject to severe water stress (Spotila 1972; Feder 1983b). In contrast, many anurans that inhabit arid regions tend to be globular in shape, with round bodies and relatively short legs. This shape minimizes the surface-to-volume ratio of these animals and therefore reduces rates of water loss. Amphibians also can modify the amount of skin surface exposed to the air by postural changes or by aggregating with other individuals (see "Behavioral Modification of Water Balance").

## Ecological Correlates of Evaporative Water Loss

There have been many studies of evaporative water loss in amphibians, dealing either with single species or with comparisons among species from different habitats. Unfortunately, differences in experimental techniques make quantitative comparisons among studies difficult. Variation in air flow rates in experimental chambers and activity of experimental subjects have a major effect on boundary layer resistance and can lead to differences in rates of water loss that mask any correlation with habitat. Some workers have not corrected data for differences in body mass or accounted for changes in humidity inside experimental chambers (for discussions of methods, see Wygoda 1984; Shoemaker et al. 1992). Estimates of absolute rates of evaporative water loss can be reliably compared only if measurements have been made using exactly the same techniques. Nevertheless, several general patterns now seem clear.

### Ground-Dwelling Amphibians

Amphibians that spend most of their time on the ground generally lack morphological or physiological adaptations for retarding evaporative water loss and do not differ greatly from species that spend most of their time in or near water. For example, plethodontid salamanders have very high rates of evaporative water loss. Differences among species are due almost entirely to differences in body mass and body shape, with larger, more rotund species losing water more slowly than smaller or more elongate species (N. Cohen 1952; Spight 1968; M. Houck and Bellis 1972; Spotila 1972; Feder 1983b). Spotila and Berman (1976) constructed agar models of salamanders for comparison with living salamanders of the same size and shape. They found little difference in water loss from living animals and the models, which essentially were free water surfaces. This indicates that skin resistance to water loss in these animals is negligible. In fact, skin resistance in plethodontids is several hundred times less than that of turtles, lizards, or crocodilians (table 2.1). The only way these salamanders can avoid desiccation is by selecting appropriate microhabitats or by adopting postures that alter surface-to-volume relationships (see "Behavioral Modification").

Having a highly permeable skin allows plethodontids to use the skin as the primary site of respiratory gas exchange (see chapter 4). It also is advantageous in allowing for rapid rehydration following periods of desiccation (see "Water Uptake"). The lack of resistance to water loss is not unique to lungless species. Salamanders such as *Triturus, Salamandra,* and *Ambystoma* all have high rates of water loss, and are no better adapted to desiccating conditions than are plethodontids (Cohen 1952; Spight 1968; Warburg 1971b; Marangio and Anderson 1977; Warburg and Degani 1979; J. Gillis and Breuer 1984).

The same general pattern is found in terrestrial anurans. Wygoda (1984) measured rates of evaporative water loss for a given surface area of skin in two toads (*Bufo terrestris* and *B. marinus*), three ranid frogs (*Rana catesbeiana, R. pipiens,* and *R. utricularia*), a semiaquatic cricket frog (*Acris gryllus*), and a fossorial spadefoot toad (*Scaphiopus holbrookii*). He found that rates of water loss were statistically indistinguishable among these species, and none had rates significantly different from that of the aquatic frog *Xenopus laevis* under the same conditions. Another fosso-

**Table 2.1** Cutaneous resistance to water loss in ectothermic vertebrates

| Species (source no.) | Resistance (s/cm) |
|---|---|
| Lizards, turtles, crocodilians | |
| Xantusiid lizards (9) | 190–3,080 |
| *Mabuya*, 3 species (10) | 1,442 |
| *Sauromalus ater* (1) | 1,400 |
| *Ctenophorus ornatus* (2) | 370 |
| *Anolis carolinensis* (6) | 196 |
| *Terrepene carolina* (6) | 78 |
| *Alligator mississippiensis* (8) | 55 |
| *Apalone spinifera* (11) | 5 |
| Waterproof frogs (Hylidae, Rhacophoridae, Hyperoliidae) | |
| *Phyllomedusa*, 5 species (4, 5, 14) | 200–300 |
| *Chiromantis*, 3 species (3, 7, 14) | 350–400 |
| *Hyperolius*, 3 species (12, 13, 14) | 100–500 |
| Arboreal frogs (Hyperoliidae, Hylidae, Leptodactylidae) | |
| *Litoria*, 4 species (14, 16, 18) | 9–100 |
| *Hyperolius*, 4 species (14) | 25–75 |
| *Pachymedusa dacnicolor* (15) | 14 |
| *Agalychnis*, 2 species (14, 15) | 3–10 |
| *Afrixalus*, 3 species (14) | 3–5 |
| *Osteopilus*, 2 species (15) | 2–3 |
| *Hyla, Pseudacris*, 5 species (14, 15, 17) | 2–4 |
| *Eleutherodactylus*, 2 species (20) | < 1 |
| Other amphibians (Bufonidae, Hylidae, Ranidae, Pelobatidae, Plethodontidae) | |
| *Cyclorana australis* (19) | 8 |
| *Bufo*, 2 species (14, 15) | 1–2 |
| *Rana*, 2 species (14, 15) | < 1 |
| *Scaphiopus holbrookii* (15) | < 1 |
| *Desmognathus* (6) | < 1 |

*Sources:* Adapted from data summarized in references (12)–(16), with additions. (1) Bentley and Schmidt-Nielsen 1966; (2) Dawson, Shoemaker, and Licht 1966; (3) Loveridge 1970; (4) Shoemaker et al. 1972; (5) Shoemaker and McClanahan 1975; (6) Spotila and Bermen 1976; (7) Drewes et al. 1977; (8) J. Davis, Spotila, and Schefler 1980; (9) Mautz 1980; (10) Withers 1981; (11) Robertson and Smith 1982; (12) Withers et al. 1982; (13) Withers, Louw, and Nicholson 1982; (14) Withers, Hillman, and Drewes 1984; (15) Wygoda 1984; (16) Buttemer 1990; (17) Preest, Brust, and Wygoda 1992; (18) Amey and Grigg 1995; (19) Christian and Parry 1997; (20) Rogowitz, Cortés-Rivera, and Nieves-Puigdoller 1999.

rial species, *Gastrophryne carolinensis*, and a tropical rainforest toad, *Atelopus varius*, had slightly higher rates of water loss (table 2.2). Wygoda's work is the best comparative study to date because he immobilized the animals temporarily with the drug curare to control for the effects of activity, but other studies have produced similar results. For example, Thorson (1955) measured evaporative water loss in *Rana catesbeiana*, *R. clamitans*, *Bufo boreas*, and *Spea hammondii* under equivalent conditions and did not find any major differences that could be correlated with habitat.

**Table 2.2** Comparison of area-specific cutaneous water loss of anurans from different habitats, measured under the same experimental conditions

| Species | Habitat | Cutaneous water loss ($mg/cm^2/h$) |
|---|---|---|
| *Xenopus laevis* | Aquatic | 27.5 |
| *Rana catesbeiana* | Semiaquatic | 25.7 |
| *Rana pipiens* | Semiaquatic | 26.8 |
| *Rana sphenocephala* | Semiaquatic | 26.9 |
| *Acris gryllus* | Semiaquatic | 25.2 |
| *Atelopus varius* | Terrestrial | 31.3 |
| *Bufo marinus* | Terrestrial | 27.1 |
| *Bufo terrestris* | Terrestrial | 25.4 |
| *Gastrophryne carolinensis* | Terrestrial, fossorial | 29.9 |
| *Scaphiopus holbrookii* | Fossorial | 26.5 |
| *Hyla cinerea* | Arboreal | 10.4 |
| *Hyla femoralis* | Arboreal | 14.2 |
| *Hyla gratiosa* | Arboreal | 13.7 |
| *Osteopilus septentrionalis* | Arboreal | 13.5 |
| *Pseudacris crucifer* | Scansorial, arboreal | 16.0 |

*Source:* Data from Wygoda (1984).

The same was true for comparisons between two European anurans, *Bufo viridis* and *Rana ridibunda* (Katz and Graham 1980) and two North American species, *Bufo cognatus* and *Rana pipiens* (Withers, Hillman, and Drewes 1984).

### Arboreal Frogs

Although most ground-dwelling amphibians lose water as rapidly as aquatic or semiaquatic species, the same cannot be said for arboreal frogs. Many of these have unusually low rates of evaporative water loss that can be attributed to unusually high skin resistance. Some species are only moderately waterproof, while others are as resistant to evaporative water loss as some squamates, turtles, and crocodilians. Wygoda (1984) compared several North American and Caribbean tree frogs (*Hyla* and *Osteopilus*) with aquatic, semiaquatic, terrestrial, and fossorial anurans from six families. All subjects were immobilized with the drug curare to control for activity. Wygoda found that the arboreal species consistently had rates of evaporative water loss that were 1/3 to 1/2 those of the other species (table 2.2). This was reflected in a much greater skin resistance in the arboreal frogs, while boundary layer resistance was very similar among species (fig. 2.5). Withers, Hillman, and Drewes (1984) reported resistances of the same general magnitude for several New World hylids, including *Hyla gratiosa*, *H. arenicolor*, *Osteopilus septentrionalis*, and *Agalychnis callidryas*, while Bentley and Yorio (1979) reported that the hylid *Pachy-*

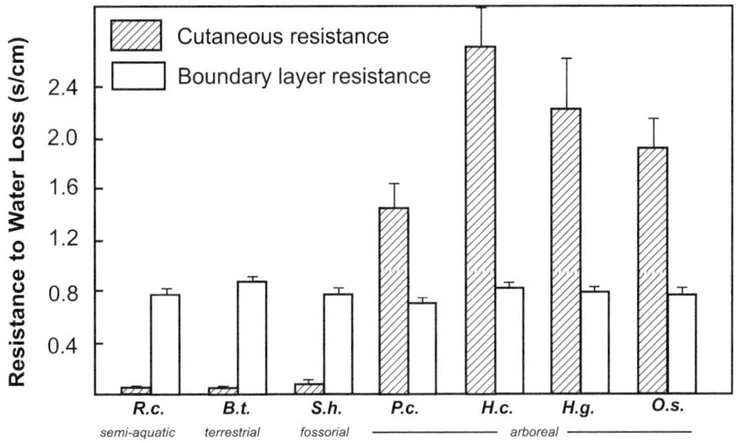


**Fig. 2.5.** Cutaneous and boundary layer resistance to water loss in selected North American anurans. Skin resistance is measured as the reciprocal of the rate of water movement (cm/s) through the skin. Bars show means + 1 SE. Arboreal species have much higher cutaneous resistance than do other anurans, but boundary layer resistance does not differ among species. *R.c.* = *Rana catesbeiana*. *B.t.* = *Bufo terrestris*. *S.h.* = *Scaphiopus holbrookii*; *P.c.* = *Pseudacris crucifer*. *H.c.* = *Hyla cinerea*. *H.g.* = *Hyla gratiosa*. *O.s.* = *Osteopilus septentrionalis*. Plotted from data in Wygoda (1984).

*medusa dacnicolor* had much lower rates of water loss than *Bufo* or *Rana*. Unusually high skin resistance is not characteristic of the whole family Hylidae, because the nonarboreal hylid *Acris gryllus* has a surface-area-specific rate of water loss that is about the same as that of other terrestrial or semiaquatic anurans (Wygoda 1984). In one hylid, *Hyla cinerea*, increased cutaneous resistance to water loss appears at the stage at which metamorphosing juveniles are just emerging from the water, so even the smallest juveniles are somewhat protected from evaporative water loss (Wygoda and Garman 1993).

Arboreal frogs from other regions also have high resistance to evaporative water loss, and some of these are not closely related to hylids. African leaf-folding frogs (family Hyperoliidae), including *Afrixalus fornasinii*, *A. pygmaeus*, and *A. quadrivittatus*, have resistances similar to those of North American hylids (Withers, Hillman, and Drewes 1984). These are small frogs that frequent swamps and grassy shores of pools and ditches. Although they live near water, they often sit in exposed positions on dry vegetation during the day (Stewart 1967). These frogs are not well adapted to extremely dry conditions and generally seek shelter during the dry season in moist locations such as the leaf axils of banana plants (Loveridge 1976).

Reed frogs in the genus *Hyperolius* are found throughout much of sub-Saharan Africa. They often spend the day exposed to the sun on the leaves of plants, and must cope with prolonged dry seasons as well. Some species, such as *H. argus*, *H. pusillus*, and *H. tuberlinguis*, are moderately waterproof, with resistances about 5–10 times those of *Afrixalus* or North American hylids (Loveridge 1976; Withers, Hillman, and Drewes 1984; Geise and Linsenmair 1988). These species spend the dry season sheltered in moist microhabitats such as leaf axils and piles of vegetation (Loveridge 1976). Others, including *Hyperolius nasutus*, *H. marmoratus*, and members of the *H. viridiflavus* complex, are 20 to 250 times more resistant to water loss than North American hylids. These species are some of the most waterproof of all anurans (Loveridge 1976; Withers et al. 1982; Withers, Louw, and Nicholson 1982; Withers, Hillman, and Drewes 1984; Geise and Linsenmair 1986, 1988; Kobelt and Linsenmair 1986; Schmuck and Linsenmair 1988; Schmuck, Kobelt, and Linsenmair 1988). These frogs undergo seasonal changes in skin structure and metabolic physiology (see the following discussion) that enable them to remain in exposed locations during the prolonged dry seasons of the African savannas. Breeding adults appear to be less capable of making these adjustments than juvenile frogs, and few survive for more than one breeding season (Geise and Linsenmair 1986, 1988; Kobelt and Linsenmair 1986).

Rhacophorid tree frogs of the genus *Chiromantis* also are remarkable in their ability to remain in exposed locations during the dry season. The three species in this genus, *C. petersii*, *C. rufescens*, and *C. xerampelina*, are all relatively large tree frogs found in southern and eastern Africa in moderately to extremely dry climates. They have explosive breeding seasons commencing with the onset of heavy rains (Coe 1974). For much of the rest of the year, their environment is hot and dry and seemingly inhospitable to amphibians, yet these frogs can be found in exposed locations during most months of the year (Loveridge 1970; Drewes et al. 1977). All three species have been shown to have extremely low rates of evaporative water loss, with cutaneous resistance more than 200 times that of North American tree frogs and exceeding that of *Anolis* lizards (Loveridge 1970; Drewes et al. 1977; Withers, Hillman, and Drewes 1984). Water conservation is further enhanced by excretion of nitrogenous wastes in the form of uric acid (see "Nitrogen Excretion").

The most unusual of all waterproof arboreal frogs are several species of South American hylids in the genus *Phyllomedusa,* which also are uricotelic. As mentioned previously, several species, including *P. sauvagii, P. iherengii, P. hypochondrialis, P. boliviana,* and *P. bicolor* have glands in the skin that secrete a waxy lipid coating (Blaylock, Ruibal, and Platt-Aloia 1976; Ruibal and Shoemaker 1984). Another species, *P. azurae,* also produces a lipid secretion, but the skin glands have not been investigated (Withers, Hillman, and Drewes 1984). These frogs have an elaborate wiping behavior in which both the front and back legs are used to smear a coating of lipid evenly over the surface of the skin (fig. 2.6). This behavior has been reported not only in adults, but also in newly metamorphosed juveniles of one species, *P. hypochondrialis* (Alamillo and Pramuk 2002). The frogs then sit quietly on an exposed branch or leaf, with the legs folded against the body. In this condition they exhibit rates of evaporative water loss and skin resistance equivalent to those of *Chiromantis* (Shoemaker et al. 1972; Shoemaker and McClanahan 1975; Blaylock, Ruibal, and Platt-Aloia 1976; McClanahan, Stinner, and Shoemaker 1978). Nearly identical wiping behavior has been reported in a rhacophorid tree frog, *Polypedates maculatus* (Lillywhite et al. 1997a, b). Glandular secretions of this frog are composed primarily of mucus, with some lipids as well. This suggests that wiping behavior evolved as a means of spreading mucus on the skin surface, with the evolution of lipid secretions being a later development.

At least one Australian hylid, *Litoria caerulea,* produces lipid secretions and engages in wiping behavior similar to

**Fig. 2.6.** *Phyllomedusa sauvagii* wiping lipid secretions on skin. Photos by Rudolfo Ruibal.

that of *Phyllomedusa* and *Polypedates* (Christian and Parry 1997). This species is widely distributed from relatively mesic to semiarid habitats and sometimes is subjected to hot, dry conditions. Early work by Warburg (1967) had shown that this species has an unusually low rate of evaporative water loss, but the mechanism for retarding water loss was not known (see also Christian and Green 1994). Buttemer (1990) found that skin resistance in this species is two to three times greater than in North American hylids, but not nearly as high as in the lipid-secreting *Phyllomedusa* species. Subsequent work showed that secretions exuded from glands in the skin and wiped over the body are responsible for increased resistance (Christian and Parry 1997). These secretions contain relatively low concentrations of lipids (5%) and a high concentration of protein (85%). Christian and Parry (1997) suggested that the proteins play a significant role in retarding water loss, in addition to the lipids, but this has yet to be demonstrated experimentally. Several hylid treefrogs from Florida that experience prolonged dry seasons also exhibit wiping behavior, but the behavior is less elaborate than that of *Phyllomedusa*. Skin secretions contain a mixture of mucus, lipid, and possibly protein, and resistance to evaporative water loss in much lower than in the true waterproof frogs (Barbeau and Lillywhite 2005).

Several other Australian hylids have moderately high resistance to water loss. *Litoria chloris* has an even higher resistance than *L. caerulea* at temperatures below 40° C, despite the fact that it lives in rain forests (Buttemer 1990). At temperatures above 40° C, skin resistance declines, apparently because of periodic discharge of mucus onto the surface of the skin to facilitate evaporative cooling (fig. 2.7). This reduction in resistance at high temperatures and the frog's small size make it poorly suited to hot, dry conditions. Another tree frog from the same coastal region of Australia, *L. gracilenta,* has skin resistance about twice as high as *L. chloris,* but it is considerably smaller and therefore susceptible to higher water loss under the same conditions (Withers, Hillman, and Drewes 1984). Two other species, *Litoria fallax* and *L. peronii,* also have moderately high resistance to water loss (Amey and Grigg 1995).

A surprising finding by Christian and Parry (1997) is that a terrestrial frog, *Cyclorana australis,* has a resistance to evaporative water loss higher than that of any other terrestrial frog that has been studied, but not as high as arboreal frogs such as *Litoria caerulea* (table 2.1). It had skin secretions with a composition similar to that of *Litoria,* about 10% lipids and 78% protein. This is a burrowing species, and during very dry weather, it retreats underground and forms a cocoon that provides even greater resistance to water loss (see "Cocoon Formation," following). The occurrence of similar strategies for retarding water loss in arboreal frogs and *Cyclorana* without cocoons probably is a function of their close phylogenetic relationship. *Cyclorana* is so unlike other hylids in morphology that it once was placed in the family Myobatrachidae (which was called Leptodactylidae at the time). Later work showed it to be a hylid, a member of the subfamily Pelodryadinae, as are all other Australian hylids. It is possible that in evolving terrestrial and fossorial habits, members of this genus have retained an ancestral mechanism for retarding water loss inherited from arboreal ancestors, or similar mechanisms evolved independently. Several phylogenetic analyses have indicated a close relationship between the subfamily Pelodryadinae in Australia and phyllomedusine hylids in South America, the only New World hylids known to produce concentrated lipid secretions (Ruvinsky and Maxson 1996; Darst and Cannatella 2004; Faivovich et al. 2005).

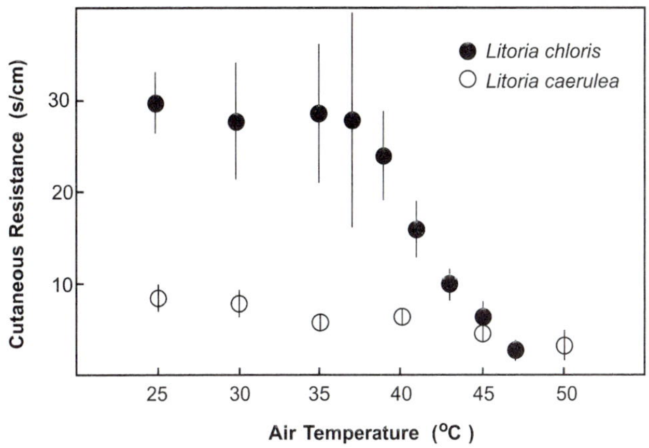


**Fig. 2.7.** Change in cutaneous resistance to water loss as a function of temperature in two Australian treefrogs, *Litoria chloris* and *L. caerulea.* Circles are means; vertical lines show 95% confidence intervals. After Buttemer (1990).

There is no evidence for lipid coatings being applied to the skin in other species of frogs, but it is possible that other frogs will be found with similar behavior (for example, see Barbeau and Lillywhite 2005). Nevertheless, lipids contained within the skin are major barriers to water loss in many other vertebrates (Lillywhite and Maderson 1988), so it seems reasonable to expect the same would be true of frogs. Attempts to correlate lipid content of skin with rates of water loss have produced mixed results. Bentley and Yorio (1979) extracted lipids from the skin of *Pachymedusa dacnicolor* and several nonarboreal species. They did not find any quantitative difference in lipid content that would explain the differences in rates of water loss, but it is possible that differences in the structural configuration of lipids in the skin are responsible for differences in skin resistance. Withers, Hillman, and Drewes (1984) found a variety of lipids in the skin of frogs, but none was unique to arboreal species with high resistance to evaporative water loss. They did not measure lipid content quantitatively, nor did they investigate structural differences in lipid layers. They did find that wiping the skins of two species, *Chiromantis rufescens* and *Hyperolius kivuensis*, with a lipid solvent increased the rate of evaporative water loss. The same was true for *Litoria peronii* (Amey and Grigg 1995). This species has a relatively thick lipid layer located in the *tela subcutanea* of the dorsal skin, but its position below the well-vascularized layers of the skin suggest that it would not provide a major barrier to water loss. *Leptodactylus fallax* has a much thinner lipid layer located just below the epidermis of the dorsal skin, where it would be more effective in retarding water loss. Indeed, this species exhibited a greater reduction in water loss compared to an agar model than did *L. peronii,* but cutaneous resistance to water loss was not calculated.

There is little evidence of unusual lipid layers in the skin of the highly waterproof species of *Chiromantis* and *Hyperolius.* They do have other structural peculiarities that seem to be related to low evaporative water loss, but the precise mechanism is unclear. Drewes et al. (1977) found a nearly continuous layer of specialized pigment cells (iridophores) in the dermis of two species that live in arid environments, *Chiromantis petersii* and *C. xerampelina.* These cells also have been found in *Hyperolius viridiflavus taeniatus* (Kobelt and Linsenmair 1986). In this species, the iridophores undergo a dramatic seasonal change, becoming much more abundant during the dry season (fig. 2.1 B). The iridophores contain many plates of highly reflective purine crystals, which give the frogs a chalky white appearance during the dry season. The main biological role of these iridophores apparently is to increase the reflectance of the skin and decrease radiant heat loads in animals exposed to the sun. This in turn would tend to reduce rates of evaporative water loss. It also is possible that the iridophores provide a physical barrier to water loss, although this has not been conclusively demonstrated. In addition, the synthesis of purines to form the iridophores provides a sink for storage of excess nitrogenous compounds that otherwise would have to be excreted. Temporary storage of nitrogenous wastes in this form saves both water and energy during the harsh dry season (Schmuck, Kobelt, and Linsenmair 1988).

Another possible mechanism for reducing evaporative water loss is formation of a layer of dried mucus on the outer surface of the skin. Geise and Linsenmair (1986) reported that a layer of dried mucus on the skin of *Hyperolius* helped seal the frogs to the substrate and increased cutaneous resistance to water loss. Wygoda (1988) observed increased discharges of mucus onto the skin surface in *Hyla cinerea* in response to desiccation stress. Periodic discharge of mucus is a thermoregulatory response to heat stress in many frogs that remain exposed to the sun, allowing for enhanced evaporative cooling (Lillywhite 1971b, 1975; Lillywhite and Licht 1975; McClanahan, Stinner, and Shoemaker 1978; Withers et al. 1982; Withers, Louw, and Nicholson 1982; Geise and Linsenmair 1986; Kaul and Shoemaker 1989; Shoemaker, Baker, and Loveridge 1989; Buttemer 1990; see also chapter 3). Periodic mucus discharges also could be advantageous in building up a dried layer that could increase cutaneous resistance to water loss. Christian and Parry (1997) suggested that mucus secretions contain proteins that retard water loss.

It is not yet clear how widespread these adaptations for reducing evaporative water loss are among arboreal frogs. Most of the work has concentrated on species known to experience prolonged dry seasons or that sit in exposed positions for long periods of time. There are other tree frogs, such as arboreal species of *Eleutherodactylus,* some tropical microhylids, some centrolenids, and a wide variety of tropical hylids and rhacophorids that might be exposed periodically to desiccation stress and would benefit from increased resistance to evaporative water loss. Wygoda (1984) recalculated data for *Polypedates leucomystax* collected by Shoemaker and McClanahan (1980) and found that this species has moderately high resistance to water loss. In contrast, preliminary data for *Eleutherodactylus coqui* and *E. cooki* from Puerto Rico indicate that these species have low cutaneous resistance to water loss (< 1 s/cm), comparable to those of terrestrial amphibians (Rogowitz, Cortés-Rivera, and Nieves-Puigdoller 1999). Additional studies involving broader comparisons among closely related species from different habitats would be useful in clarifying patterns of evolutionary change in water-conservation strategies.

## Behavioral Modification of Water Balance

For amphibians that lack morphological or physiological adaptations for retarding water loss, the only means of avoiding desiccation is through changes in behavior. These include: (1) selection of moist microhabitats that allow the animal to balance water loss with water uptake; (2) sheltering in moist retreats in leaf litter, piles of debris, or natural cavities; (3) burrowing into soil during dry periods, or making use of burrows constructed by other animals; (4) postural changes; and (5) aggregation with other individuals.

### Microhabitat Selection

One way that amphibians can avoid excessive water loss is to select microhabitats that provide sufficient moisture for water uptake through the skin to balance evaporative water loss (see also "Water Uptake"), or to limit activity to periods with favorable moisture conditions. Every herpetologist knows that amphibians are most easily captured on rainy nights, when they can be found crossing roads or moving about in wet leaf litter. Certainly this is the usual pattern for terrestrial plethodontid salamanders, most of which are nocturnal and seldom seen except in wet weather (Test 1946; Hairston 1949; Ralph 1957; Heatwole 1962b; Jaeger 1980a). Spotila (1972) tested responses of salamanders to humidity in a laboratory gradient consisting of a series of interconnected chambers with screen wire floors, in which relative humidity of the air varied from 0% at one end to 100% at the other. Several species of *Plethodon, Desmognathus,* and *Eurycea* selected relative humidity above 60%, although most species exhibited considerable individual variation in humidity preference. There was some correlation between the level of humidity chosen in the gradient and microhabitats chosen in nature, but the relationship was not exact. Highly terrestrial species such as *Plethodon dorsalis* and *P. serratus* generally selected lower humidity and were found in drier microhabitats than some of the other species. Three species of *Eurycea* that were either semiaquatic (*E. multiplicata*) or associated with microhabitats such as caves and wet ledges (*E. lucifuga, E. longicauda*) consistently selected only the highest humidity.

For most terrestrial salamanders, the moisture content and water-holding characteristics of the soil probably are more important determinants of microhabitat choice than relative humidity of the air, because soil moisture provides the major avenue for water uptake. Field studies and laboratory investigations using soil-moisture gradients have shown that plethodontids, ambystomatids, and salamandrids exhibit specific preferences for soils of certain moisture content, often avoiding both very dry and nearly saturated substrates (Vernberg 1955; Rosenthal 1957; Heatwole 1960, 1962b; Taub 1961; Jaeger 1971b; J. Anderson 1972; Marangio and Anderson 1977; Degani 1984; Wyman 1988; Grover 1998). Soil moisture affects both the horizontal dispersion of salamanders and their vertical distribution in the soil, with individuals moving deeper as the surface layers dry out (Hendrickson 1954; Stebbins 1954b; Rosenthal 1957; P. Anderson 1960; Cunningham 1960; Taub 1961; Heatwole 1962b).

Moisture preferences can be modified by the hydric condition of the salamander, with dehydrated individuals generally showing a stronger preference for wet substrates. Preferences also can be affected by breeding condition. J. Anderson (1972) found that breeding adult *Ambystoma macrodactylum* preferred wetter substrates than did nonbreeding adults, perhaps because they move toward water to breed. In contrast, *A. opacum* adults showed a random preference for soil moisture content during the fall breeding season, perhaps because these animals do not follow a moisture gradient to their terrestrial oviposition sites (Marangio and Anderson 1977).

There have been fewer studies of the role of moisture in microhabitat selection by anurans. Some anurans are strictly nocturnal (Stille 1952; Cree 1989), while others are active both day and night when moisture conditions are suitable (Dole 1965a). Anurans undoubtedly are sensitive to humidity of the air, although this has seldom been investigated experimentally. Stewart (1985) reported that Puerto Rican tree frogs (*Eleutherodactylus coqui*) were seen foraging in the forest canopy most frequently on nights when humidity was high and vegetation was wet. The same pattern was apparent in studies of leiopelmatid frogs in New Zealand, where the number of active frogs was related to relative humidity, rainfall, and wetness of the vegetation (Cree 1989). Juszczyk (cited in Nöllert 1984, p. 44) found more European spadefoot toads (*Pelobates fuscus*) emerging to forage on very humid nights than on drier nights. The activities of ground-dwelling anurans probably are influenced as much by soil moisture as by relative humidity of the air (Tracy et al. 1993). Brekke, Hillyard, and Winokur (1991) showed that *Bufo punctatus,* a desert toad from the southwestern United States, is sensitive to the water potential of the substrate, and preferentially selects those that allow water to be taken up through the skin. Substrate water potential apparently is sensed at least in part through the feet; when the feet were covered, the toads were less selective in choice of substrates. Subsequent work with this species and *Bufo marinus* demonstrated that toads can sense the osmotic concentration of water directly through the ventral skin, and avoid substrates that would tend to draw water out of the body (Hillyard, von Seckendorff Hoff, and Propper 1998; Maleek et al. 1999; Nagai et al. 1999).

Soil moisture content is of critical importance in determining patterns of movement by leopard frogs (*Rana pipiens*). These are more terrestrial than many other temperate-zone ranid frogs and often venture many meters from standing water. Frogs studied during the summer in a Michigan meadow generally limited their activities to portions of the meadow that were visibly wet throughout the summer (Dole 1967). Indeed, Tracy (1975, 1976) found that even the driest soils in the area of the pasture frequented by leopard frogs were wet enough to allow for continuous uptake of water to balance losses through evaporation. For much of the summer, leopard frogs were not very active, often sitting for days in shallow depressions in the moist soil. Movements into drier parts of the pasture were limited to periods of heavy rain (Dole 1965a). Similar rainy-weather excursions outside of normal home ranges have been observed in other frogs (Martof 1953b; Sinsch 1988c; see also chapter 6).

## Sheltering

When conditions on the surface are too dry for amphibians to be active, they often retreat into sheltered sites that protect them from excessive water loss. Most amphibians are smaller than other vertebrates (Pough 1980; Feder 1983b), so they can use microhabitats not available to larger animals. Nearly every conceivable type of retreat is used by some species of amphibian. Many forest-dwelling salamanders make use of surface cover provided by leaf litter, logs, rocks, or piles of debris. Jaeger (1980b) found that *Plethodon cinereus* moved back and forth between litter and retreat sites under logs, with the number in the litter increasing in rainy weather. This species is selective in its choice of retreat sites, generally favoring large cover objects that provide a stable thermal and hydric environment (Mathis 1990a). They also prefer chunky decayed logs to more fibrous or crumbly ones, probably because these provide numerous cracks into which the salamanders can retreat in dry weather (Heatwole 1962b). *Plethodon glutinosus* exhibits a similar preference for chunky logs, with individuals sometimes wedging themselves into cracks and insect tunnels (Wells and Wells 1976). Some members of the genus *Aneides* are highly dependent on logs for shelter, where they are found mainly in cracks under exfoliating bark (T. Davis 2002a). Some small Neotropical salamanders (*Thorius, Chiropterotriton, Bolitoglossa*) live in spaces under the bark of fallen logs (D. Wake and Lynch 1976; Feder 1982b) or under moss mats on rocks, tree trunks, or the ground (D. Wake 1987).

In tropical rain forests, many species of frogs can be found living on or near the ground (N. Scott 1976, 1982; Toft 1980a, b, 1982; Toft, Rand, and Clark 1982). Some of these are terrestrial breeders (see chapter 7), but others must return to water to breed. Ground-dwelling anurans are scarce in most temperate-zone forests, perhaps because the leaf litter is drier for much of the year. In the Neotropics, species diversity of litter anurans tends to increase with increasing rainfall (see fig. 15.2 in chapter 15), suggesting that moisture content of the leaf litter limits the distribution of some species. Many ground-dwelling anurans hide in leaf litter even in relatively dry weather (Toft 1980b; Lynch and Myers 1983), but they use other types of retreats as well. A Puerto Rican frog, *Eleutherodactylus coqui,* is particularly fond of curled, dead leaves of *Cecropia* trees and the rolled petioles of fallen palm fronds, sites that also are used as oviposition sites (Townsend 1989b). The availability of retreat sites apparently limits population size in this species, because addition of bamboo frog houses increased the number of frogs in the forest (Stewart and Pough 1983). Other species of *Eleutherodactylus* hide under rocks, logs, piles of debris, or in crevices along stream banks (Rivero 1978; Stewart 1979; Stewart and Martin 1980; Lynch and Myers 1983).

Terrestrial microhylids from tropical Australia and New Guinea are ecologically similar to *Eleutherodactylus* and often live in the network of roots and tangled vegetation on the forest floor or hide in crevices at the bases of trees (Menzies and Tyler 1977; Zweifel and Tyler 1982; Zweifel 1985). In Queensland, the introduced toad *Bufo marinus* uses rock crevices, hollow logs, and cavities under tree roots or along riverbanks as daytime retreat sites. Toads preferentially select sites with high soil moisture content (Cohen and Alford 1996), and the use of such sites in dry weather can reduce rates of desiccation and increase survivorship (Schwarzkopf and Alford 1996; Seebacher and Alford 2002). In wet weather, the toads often shelter in dense vegetation and often move around in open, grassy areas at night (Seebacher and Alford 1999). The toads also tend to move longer distances in wet weather, when soil moisture levels are high.

A radio-tracking study of a ranid frog (*Hoplobatrachus occipitalis*) in a West African savanna revealed that these frogs often use the same shelters repeatedly for up to two weeks (Spieler and Linsenmair 1998). These frogs are active only at night and spend the daylight hours in crevices under rocks along riverbanks, or under bushes or patches of grass. Shelters that were used by several individuals had more constant temperatures and higher humidity than did similar shelters that were not used by frogs.

Some amphibians find shelter in holes and crevices of rock formations (fig. 2.8 B). Some are rock-dwelling specialists, while others use such sites opportunistically. Specialist crevice-dwelling salamanders include *Aneides aeneus* (R. Gordon 1952), *Eurycea lucifuga* (Hutchison 1958) and *Hydromantes platycephalus* (L. Adams 1942), all of which have flattened heads and bodies that enable them to enter narrow spaces. Other small salamanders, such as *Desmognathus ocoee,* sometimes live in rock crevices in wet areas, but they have few morphological specializations for this habitat (Huheey and Brandon 1973). Two terrestrial-breeding frogs that live in humid crevices and grottos of large boulder fields are *Cophixalus saxatalis,* an Australian microhylid (Zweifel 1985), and *Eleutherodactylus cooki,* a Puerto Rican leptodactylid (Rivero 1978; Joglar 1998). Both *Bufo vertebralis* (Bufonidae) and *Phrynomantis annectans* (Microhylidae) live in rocky, arid regions of southern Africa, and find shelter in spaces under exfoliating layers of granite boulders (Channing 1976a). Both have relatively flattened bodies that facilitate entry into narrow crevices (Passmore and Carruthers 1995). A rock-dweller from a much wetter habitat is *Dendrobates azureus,* which is found only in forest islands in the savannas of Surinam. These frogs are always associated with large boulders, and find shelter in cavities beneath the rocks (Hoogmoed 1969, 1972). Another dendrobatid frog, *Colostethus panamensis,* lives along rocky streams in Panama. Males and females retreat into moist crevices during the dry season (fig. 2.8 A) and defend them vigorously against intruders of both sexes (Wells 1980a).

Arboreal amphibians generally find shelter in spaces un-

**Fig. 2.8.** (A) *Colostethus panamensis* sheltering in a mud bank during dry weather. (B) *Hyperolius orkarkarri* sheltering in a rock crevice during dry season. The whitening of the skin helps to reduce the heat load on the frog when exposed to sunlight. Photos by Kentwood D. Wells (A) and Robert Drewes (B).

der bark or moss on tree trunks, in tree holes, or in the axils of epiphytic plants. *Eleutherodactylus hedricki* of Puerto Rico is a tree-hole specialist, using them for both shelter and reproduction, whereas other species such as *E. coqui* use them occasionally (Rivero 1978). Microhylid frogs from Madagascar in the genera *Platyhyla, Plethodontohyla, Anodonthyla, Platypelis,* and *Paracophyla* live and lay eggs in tree holes, hollow bamboo stems, and the axils of various kinds of epiphytes (Blommers-Schlösser 1975b). An East African microhylid, *Hoplophryne uluguruensis,* also is a bamboo-dweller (Barbour and Loveridge 1928; Lamotte and Lescure 1977), as is *Fritziana ohausi,* a hylid from Brazil (Heyer et al. 1990). Another Brazilian hylid, *Trachycephalus resinifictrix,* is reported to lay eggs in small holes high in trees, including old hymenopteran nests (Goeldi 1907; Lamotte and Lescure 1977). Whether they also use these as permanent shelter sites is not known. Some North American tree frogs use knothole cavities as shelters outside the breeding season (McComb and Noble 1981; Ritke and Babb 1991).

Many tree frogs in the families Hylidae, Leptodactylidae, Rhacophoridae, and Hyperoliidae use the axils of bromeliads, pineapples, bananas, palmettos, and other plants for shelter and sometimes for reproduction (Picado 1913; Dunn 1937; W. Lynn and Grant 1940; W. T. Neill 1951; Slevin 1955; Schiøtz 1967; Stewart 1967; D. Lee 1969; Duellman 1970; Loveridge 1976; Lamotte and Lescure 1977; Rivero 1978; Garrick, Sutton, and Lang 1985; Hedges 1987; Heyer et al. 1990; Ovaska 1991; Duellman and Yoshpa 1996; Meshaka 1996). Retreat sites are located on the ground or in aerial plants high in the forest canopy. An unusual report of a plant used for shelter by frogs is a brief note by W. T. Neill (1946), who found groups of up to two dozen tree frogs (*Litoria* sp.) inside the cavities of ant-plants (*Hydnophytum,* Rubiaceae) in New Guinea. These are epiphytes that grow on the trunks of trees and develop globular swellings on their roots up to 30 centimeters in diameter. They contain a network of interconnected passages inhabited by ants (for an illustration of ant chambers in a related genus, see Hölldobler and Wilson 1990). Neill reported that many ants were present when he discovered the frogs, but how the frogs avoid being bitten is not known.

Some frogs that find shelter in holes have the skin on top of the head tightly co-ossified with the skull (Trueb 1970). This appears to be an adaptation for reducing water loss. Neotropical treefrogs in several genera (*Triprion, Trachycephalus, Corythomantis, Aparasphenodon, Gastrotheca*) use their bony heads to plug holes in logs or tree trunks and spaces in the axils of bromeliads (Lutz 1954; Duellman and Klaas 1964; Bokermann 1966b; Duellman and Trueb 1986; Jared et al. 1999). Seibert, Lillywhite, and Wassersug (1974) reported that evaporative water loss from the head of one species, *Trachycephalus jordani,* was only about a third as high as losses across the dorsal skin and much less than losses from the heads of tree frogs without co-ossified skulls. De Andrade and Abe (1997) reported that in two species, *Aparasphenodon brunoi* and *Corythomantis greeningi,* individuals sheltered in small chambers with the head blocking the entrance had rates of evaporative water loss that were 85–90% lower than those of exposed frogs. Nevertheless, skin co-ossification appears to be less important in reducing water loss than is reduction of evaporative water loss from the skin and behavioral selection of moist retreat sites (Navas, Jared, and Antoniazzi 2002). Cranial co-ossification has evolved in some burrowing anurans as well, including hylid frogs in the genus *Pternohyla* (Duellman 1970; Trueb 1970), some toads in the genus *Bufo* (formerly in a separate genus, *Peltophryne*), and an Old World spadefoot toad, *Pelobates syriacus.* At least one toad, *Bufo empusus,* is known to use its head to block burrow entrances (T. Barbour 1926).

### Burrowing

Many other amphibians spend much of their time below ground. Some species are obligate burrowers that exhibit morphological specializations for digging (see chapter 1), but others are opportunistic burrowers that make use of natural holes, crevices, or tunnels made by other animals. These retreats provide more stable temperature and humidity conditions than do more exposed locations (Denton and Beebee 1993a). Among the most specialized fossorial amphibians are the caecilians. Five of the six living families are specialized for burrowing, with reduced eyes and modified skull morphology (see chapter 1). The other family is largely aquatic, but appears to be derived from fossorial ancestors. Very little is known about the water relations of these animals. The limited data available indicate that rates of evaporative water loss in air are very high (Steele and Louw 1988), and all terrestrial caecilians are restricted to wet soils in the tropics.

Terrestrial salamanders are active on the surface most often on wet nights, and often retreat below ground during dry periods. Most salamanders are not well adapted for digging their own burrows, but generally must rely on previously existing holes and passages in the soil. Terrestrial plethodontids such as *Plethodon jordani, P. glutinosus, P. cinereus,* and *Batrachoseps attenuatus* make use of root tracks, earthworm burrows, and other natural cavities, and are able to enlarge these for their own use (Chadwick 1940; M. Brooks 1946; Cunningham 1960; Heatwole 1960). The desmognathine salamander *Phaeognathus hubrichti* (fig. 1.32 B) spends nearly all of its time in underground burrows. These salamanders have elongate bodies and reduced limbs, suggesting morphological specialization for subterranean life, but they have limited abilities to construct their own burrows in undisturbed soil. Instead, they use natural cavities that they enlarge by wriggling the body from side to side (Brandon 1965a). The same is true of most of the mole salamanders in the family Ambystomatidae, some of which have been found up to a meter below the surface. *Ambystoma tigrinum* is capable of digging its own burrows in loose soil or sand, and *A. talpoideum* has limited burrowing abilities, but *A. annulatum, A. maculatum,* and *A. opacum* can only enlarge preexisting holes (Gruberg and Stirling 1972; Semlitsch 1983b). Loredo, Van Vuren, and Morrison (1996) reported that both adult and juvenile California tiger salamanders (*Ambystoma californiense*) dispersing from a breeding pond spent at least one day in burrows of ground squirrels within 130 m of the pond, while tiger salamanders (*A. tigrinum*) in South Dakota appeared to use prairie dog burrows for more extended periods (Kolbe, Smith, and Browning 2002). Spotted salamanders (*A. maculatum*) also use small mammal burrows as retreats, especially those of shrews (*Blarina brevicauda;* Madison 1997; Regosin, Windmiller, and Reed 2003a).

Burrowing is common in many anuran families (see chapter 1). In contrast to caecilians and urodeles, which are largely restricted to burrowing in moist soils, many frogs spend considerable periods of time in relatively dry soils. Burrowing probably is the single most common adaptation for avoiding excessive water loss among anurans from deserts, savannas, and other dry environments (Main, Littlejohn, and Lee 1959; Bentley 1966a; Main 1968; Mayhew 1968; Shoemaker 1975, 1988; Channing 1976a; Loveridge 1976; Tyler, Watson, and Martin 1981; Lamotte 1983; Heatwole 1984; Warburg 1997), but burrowing is not restricted to anurans from arid regions. There are, for example, numerous burrowing microhylids in the wet forests of New Guinea (Menzies and Tyler 1977; Zweifel and Tyler 1982).

Anurans exhibit a broad spectrum of burrowing behaviors that serve to protect the animals from both desiccation and diurnal predators. Some species, including many toads (*Bufo*), are essentially terrestrial, but often dig temporary shallow burrows to which they retreat in hot, dry weather (Stille 1952; R. Clarke 1974a; Hemmer and Jakobs 1974; Strijbosch 1980; Niekisch 1982; Brinkmann and Podloucky 1987; Hoffman and Katz 1989; Sinsch 1989b; Denton and Beebee 1993a; Rosset and Alcalde 2004). A North American tree frog, *Hyla gratiosa,* escapes from dry conditions by digging shallow burrows in sandy soil; they have been found more than a meter below the surface in winter (W. T. Neill 1952). Other frogs live in more permanent holes or burrows during the day, but emerge at night to forage. Eastern spadefoot toads (*Scaphiopus holbrookii*) behave this way when conditions are sufficiently wet to permit surface activity (S. Ball 1936). Individuals have several relatively permanent burrows within their home ranges (Pearson 1955). European spadefoots (*Pelobates fuscus*) also live in burrows up to two meters deep during the day (Savage 1942), but they often dig a new burrow each morning after a night of foraging (Nöllert 1984). The large Neotropical frog *Leptodactylus pentadactylus* builds relatively permanent burrows, and the same individual can be seen night after night sitting at a burrow entrance waiting for prey to come by (N. Scott 1983; personal observations). Many forest-dwelling microhylids from New Guinea also live in holes and emerge at night to feed (Zweifel and Tyler 1982).

Frogs can save themselves the trouble of digging their own holes by using burrows constructed by other animals. The use of preexisting burrows not only saves energy, but enables anurans to escape from desiccating conditions more rapidly than they would if they dug their own burrows

(M. Parris 1998). Some anurans use the burrows of invertebrates for shelter. For example, some New World microhylid and leptodactylid frogs use the burrows of spiders and crabs as regular daytime retreats (fig. 2.9), apparently coexisting quite peacefully with the original inhabitants (W. Blair 1936; Hunt 1980; Powell, Little, and Smith 1984; Cocroft and Hambler 1989; McIntyre 2003). A South African microhylid, *Phrynomantis bifasciatus,* takes refuge in termite mounds (Loveridge 1976). *Phynomantis microps* has been found in ant nests, as have the hyperoliid frogs *Kassina fusca* and *K. senegalensis* (Dejean and Amiet 1992; Rödel and Braun 1999). Both *Phrynomantis* and *Kassina* have been found in burrows of scorpions as well (Rödel and Braun 1999). A South American leptodactylid, *Lithodytes lineatus,* regularly inhabits and even calls from the nests of leaf-cutter ants (*Atta;* Schlüter and Regös 1981; Regös and Schlüter 1984; Hoogmoed 1986; Lamar and Wild 1995). All of these anurans apparently are protected from attacks by spiders, scorpions, or ants by noxious skin secretions. The association of frogs with noxious animals such as spiders, scorpions, and ants provides antipredator benefits to the frogs (see chapter 14), but they could use the burrows of these animals to maintain a proper water balance as well.

**Fig. 2.9.** Microhylid frogs, *Chiasmocleis ventrimaculata,* from Peru use the burrows of tarantulas, *Xenesthis immanis,* as daytime retreats. The burrows provide the frogs with protection from desiccation, while association with the spider provides protection from predators. (A) A group of frogs emerging from a spider burrow. (B) A single frog foraging on the forest floor underneath a spider. Photos by Reginald B. Cocroft.

Anurans often use the burrows of other vertebrates as retreat sites. European common toads (*Bufo bufo*) occasionally use rabbit holes for shelter (Denton and Beebee 1993a). North American gopher frogs (*Rana areolata, R. capito,* and *R. sevosa*) spend most of their time outside of the breeding season living in the burrows of crayfish, small mammals, and gopher tortoises (D. Lee 1968; Eisenberg 1983; Jackson and Milstrey 1989; Stevenson and Dyer 2002). Gopher tortoise burrows also are frequently used by *Eleutherodactylus planirostris* and occasionally by *Bufo terrestris* and *Gastrophryne carolinensis* (Lips 1991; Witz, Wilson, and Palmer 1991). Narrow-mouthed toads (*Gastrophryne olivacea*) also have been found in the burrows of moles and skinks (Freiburg 1951). The New World treefrogs *Pachymedusa dacnicolor* and *Hyla gratiosa* sometimes use rodent burrows as retreats during dry weather (D. Lee 1968; Wiewandt 1971), and *Rana luteiventris* has been reported in rodent burrows as well (Blomquist and Tull 2002). The burrows of a large South African lizard, *Cordylus giganteus,* are dry season retreats for ranid (*Cacosternum boettgeri*) and hyperoliid frogs (*Kassina wealii*), with up to 20 individuals living in the same burrow (W. Branch and Patterson 1975). Two other South African frogs, *Kassina senegalensis* and *Breviceps verrucosus,* have been found in the tunnels of mole rats (*Cryptomys hottentotus*) and golden moles (*Amblysomus hottentotus;* Loveridge 1976; Poynton and Pritchard 1976).

Many anurans from arid regions remain inactive in deep burrows for many months each year and emerge only after heavy rains for brief periods of breeding and foraging. Frogs with this type of life history include the spadefoot toads (*Scaphiopus, Spea, Pelobates*); Australian myobatrachids (*Notaden, Neobatrachus, Heleioporus,* some *Limnodynastes*) and hylids (*Cyclorana*); African ranids (*Pyxicephalus, Tomopterna, Hildebrandtia*), hemisotids (*Hemisus*), hyperoliids (*Leptopelis*), and microhylids (*Breviceps*); South American leptodactylids (*Lepidobatrachus, Ceratophrys*) and hylids (*Pternohyla*), and the Mexican burrowing frog, *Rhinophrynus dorsalis.* Other species are almost completely subterranean, sometimes even obtaining food underground. These include microhylids from New Guinea in the genera *Barygenys, Choerophryne, Copiula,* and *Xenobatrachus* (Menzies 1976; Menzies and Tyler 1977; Zweifel and Tyler 1982), African shovel-nosed frogs in the genus *Hemisus*

(Wager 1965; Emerson 1976; Passmore and Carruthers 1979), two Australian myobatrachids, *Myobatrachus gouldii* and *Arenophryne rotunda* (J. Roberts 1981, 1984; M. Tyler, Watson, and Martin 1981), and two North American hylid frogs, *Pseudacris streckeri* and *P. ornata* (L. Brown, Jackson, and Brown 1972; Brown and Means 1984). *Pseudacris streckeri* is capable of feeding underground, but feeds on the surface as well (L. Brown 1978). Both species of *Pseudacris* breed in surface bodies of water like all other species in the genus, whereas the other subterranean species listed above lay eggs underground.

### Water Relations of Burrowing Anurans

Studies of water relations in burrowing anurans have focused mainly on desert-dwelling species that experience the most extreme conditions to which amphibians are exposed (Warburg 1997). Burrowing frogs generally store large amounts of water in the bladder while underground (see "Water Storage"). These stocks are resorbed into the tissues as dry-season dormancy progresses. In addition, these frogs have various strategies for avoiding excessive loss of water to the soil. One way is to be highly selective in choice of soils for burrowing, avoiding extremely dry conditions and favoring soils that retain water for much of the dry season (Mayhew 1965, 1968; Heatwole, Cameron, and Webb 1971). Another is to construct burrows near the bases of bushes and other plants, where the soil is wetter than in surrounding areas (W. T. Neill 1952; van Beurden 1984), or next to rocks that channel rainwater into the soil (Bentley, Lee, and Main 1958). Burrowing frogs also could follow soil moisture gradients, digging deeper into the soil as surface layers dry out (Ruibal, Tevis, and Roig 1969).

Many desert anurans, including spadefoot toads, have highly permeable skin that facilitates water uptake when moisture is available (see "Water Uptake"), but this can be a serious problem when the animal is underground. If the soil becomes very dry, water actually can be drawn out of the animal. One way of coping with this problem is for the animals to increase the osmotic concentration of their body fluids, thereby lowering their own water potential (Katz and Hoffman 1990). This not only retards water loss through the skin, but also can reverse the osmotic gradient and allow the animals to take up water from the soil. Spadefoot toads (*Scaphiopus* and *Spea*) from the deserts of the southwestern United States are able to accumulate both urea and electrolytes in their plasma, thereby maintaining a favorable osmotic gradient even during very dry periods (McClanahan 1967, 1972; Shoemaker, McClanahan, and Ruibal 1969). This could be accomplished by a cessation of urine excretion, resorption of urea, acceleration of urea production, or a combination of all three. Accelerated urea production would require a shift from catabolism of stored lipids to catabolism of proteins during periods of osmotic stress (R. M. Jones 1980). McClanahan (1972) demonstrated that the rate of urea accumulation is a direct function of the severity of osmotic stress; spadefoots buried in dry soil accumulated urea more quickly than those in moist soil (fig. 2.10), and he interpreted these results as evidence for increased urea production.

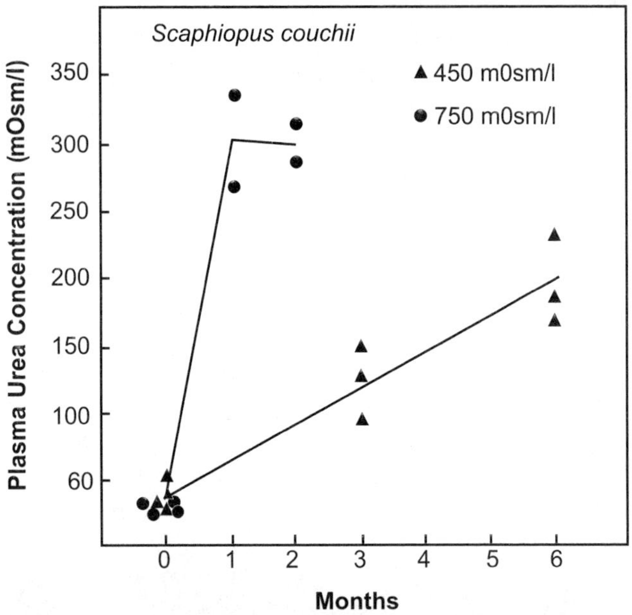


**Fig. 2.10.** Plasma urea concentrations of spadefoot toads (*Scaphiopus couchii*) kept in soils with different water potentials (450 or 750 milliosmoles/liter) as a function of number of months in the ground. The higher number (750 mOsm/l) indicates a lower water availability in the soil. Toads in drier soil accumulated urea more rapidly in the plasma and reached higher total urea concentrations. After McClanahan (1972).

C. B. Jørgensen (1997a, b) questioned this interpretation for spadefoot toads and other amphibians that accumulate urea under osmotic stress. He argued that apparent increased synthesis of urea in water-stressed amphibians is due mainly to a reduction in excretion of urea. He further argued that amphibian species that regularly experience osmotic stress typically are more tolerant of high urea concentrations, and this represents the principal adaptation to dry conditions. Indeed, spadefoot toads can tolerate high osmotic concentrations of urea in part because their muscle tissues are less impaired by high urea concentrations than those of more aquatic species such as *Rana pipiens* (McClanahan 1964). This means they can continue to use their muscles for digging even after many months underground. Old World spadefoot toads (*Pelobates syriacus*) also can raise body fluid concentrations by accumulating urea (Degani, Goldenberg, and Warburg 1983; Degani and Nevo 1986). Their osmoregulatory capacity is more limited than that of some other desert anurans, and they do not live in the driest habitats (Shpun et al. 1993).

This type of response is not limited to pelobatids. Green toads (*Bufo viridis*) from arid regions of Israel can remain buried for several months in relatively dry soil, although they also can remain active under very dry conditions. Green toads develop plasma urea concentrations equaling or exceeding those measured in spadefoot toads (Degani, Silanikove, and Schkolnik 1984; Katz and Gabbay 1986; Hoffman, Eliath, and Katz 1988; Hoffman, Katz, and Eliath 1990; Katz and Hoffman 1990; Hoffman and Katz 1991, 1994, 1997, 1999). This is accomplished by retention of urine and apparently a short-term increase in synthesis of urea (Hoffman and Katz 1994). Three genera of burrowing frogs from Australia, *Neobatrachus, Heleioporus,* and *Cyclorana,* also accumulate urea during estivation, reaching plasma concentrations similar to those of *Scaphiopus* and *Bufo viridis* (Withers and Guppy 1996). The accumulation of urea by *Neobatrachus* and *Cyclorana* is the first to be reported in species that form cocoons. These frogs presumably do not derive an osmotic benefit from urea accumulation because of the impermeable nature of their cocoons, which would prevent water uptake from the soil. Nevertheless, there are other benefits of accumulating urea. One is that increases in intracellular ion concentrations are minimized. Experiments with spadefoot toads have shown that high concentrations of ions (potassium chloride) have a much greater effect in suppressing tissue enzyme activities than do high concentrations of urea, so accumulation of urea actually protects against the more harmful effects of high ion concentrations (Storey 2002).

The highly aquatic frog *Xenopus laevis* sometimes spends dry periods encased in mud at the bottom of drying ponds, and accumulates urea in the plasma (Balinsky et al. 1967). Several other species of anurans exhibit modest increases in plasma urea concentrations when subjected to osmotic stress, such as exposure to salt water (see "Osmoregulation in Saline Environments"). Most of these species have maximum plasma concentrations of urea less than 10% of those found in estivating burrowing anurans (Ackrill, Hornby, and Thomas 1969; Abe and Bicudo 1991; Withers and Guppy 1996). Accumulation of urea in body fluids also has been reported in tiger salamanders (*Ambystoma tigrinum*) left in dry soil in the laboratory for up to nine months (Delson and Whitford 1973a). These salamanders were from a desert-dwelling population that had surface water available only after heavy rains. Etheridge (1990b) reported an increase in plasma urea concentration in *Siren lacertina* estivating in relatively moist soil, but plasma concentrations of urea in estivating sirens were only 5 to 10 percent of those found in estivating desert anurans. The ability to tolerate relatively high levels of urea appears to be widespread in amphibians, at least in anurans, but the ability to actively increase urea concentrations in the body is much less common (Shpun, Hoffman, and Katz 1992).

### Cocoon Formation

The strategy of increasing body fluid concentrations to maintain a positive osmotic gradient with the soil is viable only for species that burrow in soils that have sufficiently high water potentials during the dry season to permit continued uptake of water across the skin. Some types of soils, such as clay, have very low water potentials even when water remains in the soil, making it difficult for amphibians to take up water through the skin. In addition, some species are incapable of digging deep enough into the soil to reach a permanently moist layer.

A number of amphibians deal with this problem by forming relatively impermeable cocoons around their bodies during dormancy (fig. 2.11). Cocoon formation has been reported in anurans from several different families. In Australia, five burrowing hylids in the genus *Cyclorana* form cocoons, including *C. alboguttata,* which sometimes has been classified as *Litoria alboguttata.* Nine myobatrachids in the genus *Neobatrachus* also form cocoons (A. K. Lee and Mercer 1967; van Beurden, 1984; Withers 1995, 1998). Lee and Mercer (1967) reported cocoon formation by another myobatrachid, *Limnodynastes spenceri,* but Withers (1995) was unable to induce cocoon formation in this species or another species in the same genus. Cocoon formation has been reported in an African ranid (*Pyxicephalus adspersus*) and two hyperoliids (*Leptopelis bocagei* and *L. viridis;* Loveridge 1976; Parry and Cavill 1978; Loveridge and Crayé 1979; Loveridge and Withers 1981). Two South American leptodactylids (*Ceratophrys ornata* and *Lepidobatrachus llanensis*) and two New World

**Fig. 2.11.** A leptodactylid frog, *Lepidobatrachus llanensis,* from Argentina, in a cocoon of dried skin layers. Normally the frog would produce such a cocoon while buried in soil. Photo by Rudolfo Ruibal.

hylids (*Pternohyla fodiens* and *Smilisca baudinii*) also form cocoons (McClanahan, Shoemaker, and Ruibal 1976; McClanahan, Ruibal, and Shoemaker 1983; Ruibal and Hillman 1981; McDiarmid and Foster 1987). *Leptopelis, Pternohyla,* and *Smilisca* are largely arboreal during the wet season, whereas the other species are terrestrial. Cocoon formation also has been reported in two urodeles, *Siren intermedia* and *S. lacertina.* These animals are aquatic, but live in swamps that often dry up during the summer. They survive dry periods by burrowing into the mud and forming cocoons around their bodies (Reno, Gehlbach, and Turner 1972; Gehlbach, Gordon, and Weems 1973; Etheridge 1990a, b).

The structure of the cocoon is similar in all of the anurans, being composed of successive layers of stratum corneum that are molted by the animal but not removed from the body. Deposits of dried mucus that are produced during the normal molting cycle are found between the layers of skin (McClanahan, Shoemaker, and Ruibal 1976; McClanahan, Ruibal, and Shoemaker 1983; Loveridge and Crayé 1979; Ruibal and Hillman 1981; McDiarmid and Foster 1987; Withers 1995). McClanahan, Ruibal, and Shoemaker (1983) reported that layers of shed skin were added to the cocoon of *Lepidobatrachus llanensis* at the rate of one per day, with as many as 38 layers being produced. Cocoons with up to 36 layers have been reported in *Pyxicephalus* (Loveridge and Crayé 1979), up to 43 layers in *Pternohyla* (Ruibal and Hillman 1981), up to 40 layers in *Smilisca* (McDiarmid and Foster 1987), and from 50 to more than 100 layers in *Cyclorana* and *Neobatrachus* (Withers 1995).

All cocoon-forming anurans for which data are available exhibit dramatic reductions in evaporative water loss when measured with cocoons intact, ranging from a 77% reduction in *Cyclorana alboguttata* to 97% in *Pyxicephalus adspersus* (table 2.3). McClanahan, Ruibal, and Shoemaker (1983) found that rates of water loss declined rapidly with the addition of new layers of skin to the cocoon in *Lepidobatrachus.* A frog with a 35-layer cocoon exhibited a total resistance to water loss about 10 times that of a frog without a cocoon. Since boundary layer resistance did not change, this was due entirely to an enormous increase in cutaneous resistance (fig. 2.12). Loveridge and Withers (1981) found a continual reduction in evaporative water loss in cocooned *Pyxicephalus* for at least two months after the cocoon was first formed, presumably reflecting a similar addition of skin layers. All of the measurements of evaporative water loss in these frogs were made in dry air, not in soil. Actual water loss would be even lower for species that burrow into soil. The cocoon of *Siren lacertina* appears to provide less resistance to water loss than those of burrowing frogs, and estivating animals continue to lose some water if the surrounding soil has a low water potential (Gehlbach, Gordon, and Weems 1973; Etheridge 1990b).

**Table 2.3** Reduction in evaporative water loss (EWL) with aestivational cocoons

| | EWL (mg/g/h) | | |
|---|---|---|---|
| Species (source no.) | Without Cocoon | With Cocoon | Percent change |
| Hylidae | | | |
| *Cyclorana alboguttata* (1) | | | 77 |
| *Cyclorana australis* (7) | 8.0 | 0.5 | 94 |
| *Cyclorana maini* (8) | 81.7 | 6.2 | 92 |
| *Cyclorana novaehollandiae* (8) | 16.9 | 6.4 | 62 |
| *Cyclorana platycephala* (6) | 12.4 | 0.5 | 96 |
| Hyperoliidae | | | |
| *Leptopelis bocagei* (3) | 25.9 | 5.4 | 80 |
| Leptodactylidae | | | |
| *Lepidobatrachus llanensis* (2) | 8.0 | 0.55 | 93 |
| *Lepidobatrachus llanensis* (5) | 10.0 | 0.90 | 91 |
| Myobatrachidae | | | |
| *Neobatrachus aquilonius* (8) | 33.2 | 10.6 | 68 |
| *Neobatrachus centralis* (8) | 44.2 | 6.8 | 85 |
| *Neobatrachus fulvus* (8) | 13.9 | 3.3 | 76 |
| *Neobatrachus kunapalari* (8) | 34.2 | 2.6 | 92 |
| *Neobatrachus pelobatoides* (8) | 56.7 | 7.0 | 88 |
| *Neobatrachus pictus** (1) | | | 94 |
| *Neobatrachus sutor* (8) | 60.3 | 3.9 | 94 |
| *Neobatrachus wilsmorei* (8) | 21.2 | 5.4 | 75 |
| Ranidae | | | |
| *Pyxicephalus adspersus* (4) | 6.4 | 0.22 | 97 |

*Sources:* (1) Lee and Mercer 1967; (2) McClanahan, Shoemaker, and Ruibal 1976; (3) Loveridge and Craye 1979; (4) Loveridge and Withers 1981; (5) McClanahan, Ruibal, and Shoemaker 1983; (6) Van Beurden 1984; (7) Christian and Parry 1997; (8) Withers 1998.

*Notes:* Evaporative water loss was measured in air, and absolute rates for different species may not be comparable because of differences in techniques. Percent change is either the maximum or the average for each species, depending on the study.

*Evaporative water loss in cocooned frogs estimated from measurements on isolated pieces of cocoon for comparison with water loss by intact frogs without cocoons.

## Postural Changes and Aggregation

Many amphibians reduce rates of evaporative water loss by making postural adjustments that reduce the surface-to-volume ratio of the animal. Salamanders sometimes respond to dry air by coiling the body (Ray 1958; Spotila 1972). However, this probably is relatively ineffective unless the animal can coil up very tightly. Coiling behavior might be used as an emergency response by salamanders unable to escape from desiccating conditions, but most species probably seek shelter before reaching this stage. A more common response is for salamanders to aggregate under cover with other individuals. Often the animals are found tightly intertwined, in some cases forming balls of salamanders that ef-

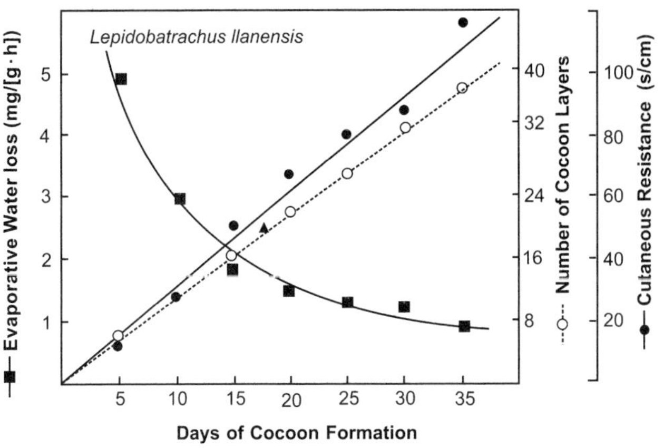


**Fig. 2.12.** Number of cocoon layers, mass-specific rate of evaporative water loss, and cutaneous resistance to water loss in *Lepidobatrachus llanensis* as a function of number of days of cocoon formation. Rate of evaporative water loss and cutaneous resistance to water loss were measured in air. Curves were fitted by eye. Cutaneous resistance to water loss increased as a linear function of number of cocoon layers, resulting in an exponential decrease in mass-specific evaporative water loss. Plotted from data in McClanahan, Ruibal, and Shoemaker (1983).

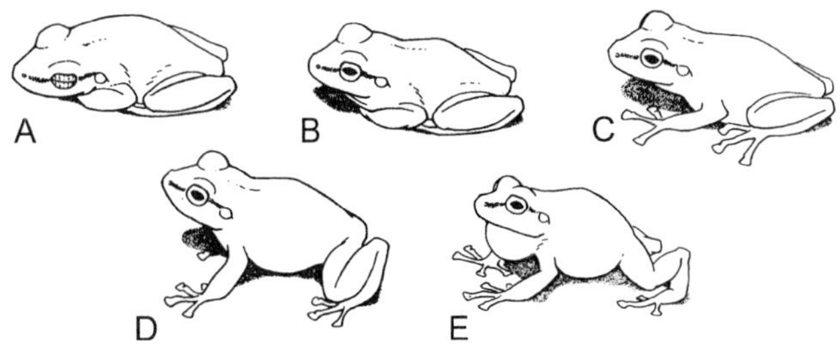


**Fig. 2.13.** Water-conserving and active postures of the Puerto Rican treefrog, *Eleutherodactylus coqui,* on dry and wet nights. (A) Water-conserving. (B) Chin up. (C) Low alert. (D) High alert. (E) Calling. After Pough et al. (1983).

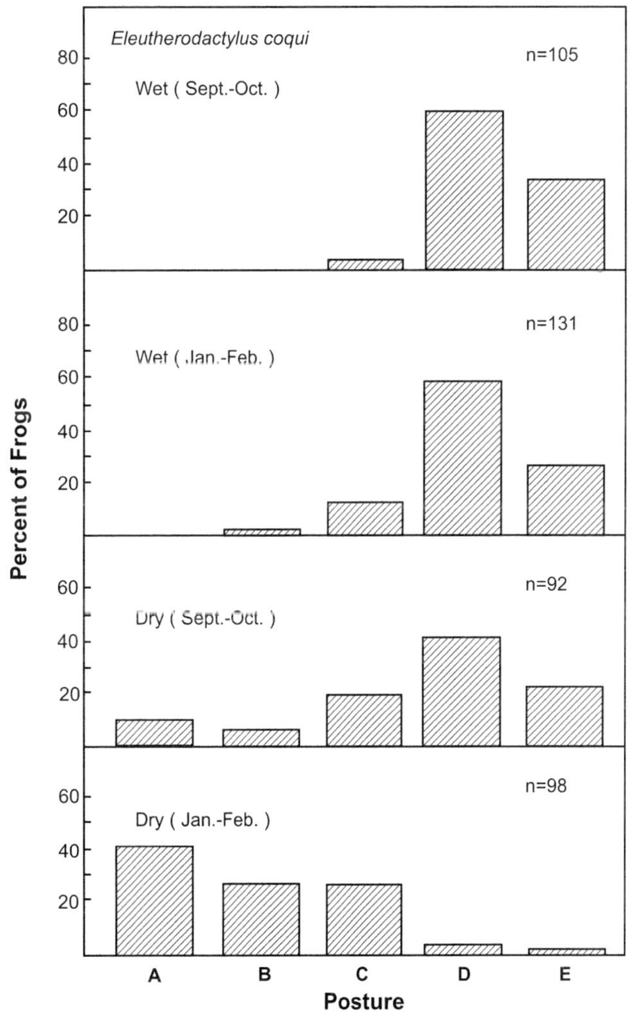


**Fig. 2.14.** Percentage of adult male *Eleutherodactylus coqui* adopting the postures shown in fig. 2.13 on wet and dry nights in the summer and fall (September and October) and in winter (January and February). Frogs are active, alert, and often calling on wet nights in all seasons. On dry nights, frogs are more likely to be active in the summer and fall than in winter, when temperatures are lower and breeding activity is reduced. After Pough et al. (1983).

fectively reduce the surface-to-volume ratio of the whole group. Species observed to aggregate in dry weather include *Plethodon cinereus* (Test 1955; Heatwole 1960), *P. glutinosus* (Humphries 1956; Wells and Wells 1976), *Batrachoseps attenuatus* (J. Cunningham 1960), *Ambystoma macrodactylum* (Alvarado 1967; J. Anderson 1967), *A. tigrinum* (Alvarado 1967; Gehlbach, Kimmel, and Weems 1969), *A. maculatum* (Pough and Wilson 1970), and terrestrial juvenile efts of *Notophthalmus viridescens* (Rohr and Madison 2003).

Arboreal frogs that remain in the open during dry weather adopt water-conserving postures that minimize the area of exposed skin. Generally the legs are folded under the body and the entire ventral surface is pressed tightly to the substrate. Often the eyes are retracted as well. The same basic water-conserving posture is adopted by frogs in several different families, including hylids (Duellman and Trueb 1986; Buttemer 1990), rhacophorids (Loveridge 1976; Drewes et al. 1977), hyperoliids (Withers et al. 1982; Geise and Linsenmair 1986), and leptodactylids (Heatwole et al. 1969; Pough et al. 1983). Some species of *Phyllomedusa* are an exception; they usually remain in a relatively upright, alert posture after coating themselves with lipid secretions, even when conditions are very dry (Blaylock, Ruibal, and Platt-Aloia 1976). The Puerto Rican tree frog *Eleutherodactylus coqui* adopts a variety of postures, depending on the relative humidity of the air and the wetness of the vegetation (fig. 2.13). The full water-conserving posture is adopted during very dry weather, but the frogs become progressively more alert and upright, with more surface area exposed, as hydric conditions improve (fig. 2.14).

Nonarboreal frogs generally do not adopt water-conserving postures when subjected to desiccation stress, although Heatwole, Cameron, and Webb (1971) observed this type of response by *Notaden bennetti* when they were prevented from burrowing in the laboratory. Most terrestrial anurans respond to desiccation by seeking shelter or by burrowing. This often leads to increased activity by frogs subjected to dehydration in laboratory chambers, where they are unable to escape desiccating conditions (Packer 1963a; Heatwole, Cameron, and Webb 1971; Heatwole and Newby 1972; Putnam and Hillman 1977; Schmajuk and Segura 1982). C. Johnson (1969a) and Heatwole, Cameron, and Webb (1971) also observed aggregation responses by *Limnodynastes* and *Notaden* in the laboratory, but it seems likely that these species would simply enter individual burrows in the field. Frogs are sometimes found huddled together under cover objects or in holes, crevices, and other retreats (W. T. Neill 1946, 1951; Slevin 1955; D. Lee 1968; Branch and Patterson 1975; Van Berkum et al. 1982). For example, Cohen and Alford (1996) reported finding large aggregations of *Bufo marinus* in hollow logs and other cavities during dry weather in northern Australia. Gray and Green (2000) reported aggregations of *Dendrobates auratus* in rock crevices and a litter-filled depression in Panama during an unusually dry period in an El Niño year. Newly metamorphosed anurans sometimes aggregate along the shores of ponds. Heinen (1993b) concluded that this type of aggregation in *Bufo americanus* was more important in protecting the toads from desiccation than from predators.

## Water Uptake and Storage

Amphibians are unique among vertebrates in their capacity to take up water through the skin. The highly permeable skin that makes these animals so vulnerable to evaporative water loss also provides a distinct advantage when moisture is available. In fact, nearly all amphibians rely exclusively on the skin for rehydration and never drink water by ingesting it through the mouth (Shoemaker et al. 1992). An exception is the lipid-secreting frog *Phyllomedusa sauvagii,* which responds to raindrops by holding its snout in the air and pumping water into its mouth with the throat muscles. This enables these arboreal frogs to take up water without having to search for pools or wet surfaces (McClanahan and Shoemaker 1987).

### Uptake from Standing Water

When terrestrial amphibians become dehydrated, they often restore body water reserves by moving to streams, ponds, or water-filled tree holes. Carpenter and Gillingham (1987) reported that *Bufo marinus* in Puerto Rico used permanent water holes for rehydration during the dry season, some moving nearly 200 meters from their regular home ranges and visiting water holes about once a week. The toads also rehydrated at ephemeral sites such as water-filled tire tracks. Tevis (1966) observed toads (*Bufo punctatus*) moving several hundred meters up a dry canyon to rehydrate in pools before returning to dry areas to forage. Another canyon-dwelling species, *Pseudacris cadaverina,* made daily trips between daytime resting sites on dry rocks and nightime foraging sites at the edge of a stream, where the frogs could rehydrate before searching for food (R. T. Harris 1975). Studies of a dendrobatid frog in Panama, *Colostethus panamensis,* showed that individuals often left their regular territories on rocky hillsides during the dry season and moved to the few remaining pools in the stream bed to soak up water. After heavy rains, the frogs moved rapidly back up the hill and reoccupied their original territories (Wells 1980a). Some frogs make use of water on the surfaces of leaves, rather than pools or streams. For example, *Rana pipiens* can rehydrate while sitting on dew-covered grass (Dole 1967). *Leiopelma archeyi,* a frog from New Zealand, emerges from daytime retreats only in wet weather and is able to take up water rapidly from wet leaves (Cree 1989).

In some species, periodic rehydration is essential for successful reproduction. For example, in phyllomedusine tree frogs, females carry males on their backs to ponds and absorb water through the skin before ascending into surrounding trees and bushes to lay eggs on the surfaces of leaves. Females often lay eggs in small batches, returning to water as many as six times to rehydrate before completing oviposition (Pyburn 1970, 1980b; Wiewandt 1971; Bagnara et al. 1986). In *Eleutherodactylus coqui,* brooding of eggs by males is critical in providing sufficient water for eggs to complete development (Taigen, Pough, and Stewart 1984; Townsend, Stewart, and Pough 1984). Males periodically leave nest sites and have been seen visiting water sources before returning to the nest (Drewry 1970). Similarly, male *Dendrobates auratus* have been observed sitting in pools or pressing their ventral surfaces against wet leaves immediately before sitting on egg clutches (Wells 1978b; Summers 1989).

All amphibians rehydrate rapidly in fresh water, whether they are completely submerged or have only the ventral skin in contact with the water, but there are interspecific differences in rates of rehydration. In general, terrestrial amphibians take up water more quickly than species that live in or near water (C. B. Jørgensen 1950a; Warburg 1965, 1971a, b, 1972; Spight 1967b; Bentley 1966a, 1971, 1973; Spotila 1972; Mullen and Alvarado 1976; Shoemaker and Nagy 1977; Alvarado 1979). Arboreal anurans usually rehydrate very quickly, even when they live in relatively mesic environments (Shoemaker et al. 1992), and some species of *Hyperolius* have the fastest rehydration rates measured for any amphibian (Geise and Linsenmair 1988). These patterns seem to make adaptive sense because species that are most likely to become dehydrated can replace body water stores quickly. However, such comparisons are complicated by the use of representative aquatic and terrestrial species that are not closely related: *Bufo* and *Xenopus* (Ewer 1952); *Bufo* and *Ascaphus* (Mullen and Alvarado 1976); *Hyla, Rana,* and *Bufo* (W. Schmid 1965); *Neobatrachus, Pelobates, Limnodynastes, Litoria* (Warburg 1972). Not only does phylogenetic history become confounded with habitat differences, but also there often is no replication of species representing different habitat types, making the statistical power of the comparison weak (Garland and Adolph 1994).

Comparisons among closely related species are more convincing, but have produced mixed results. Shoemaker and McClanahan (1980) found a reasonably good correlation between rate of rehydration and degree of terrestrialism in six species of *Rana* from Borneo. Comparisons of Puerto Rican *Eleutherodactylus* from low-altitude (dry) and high-altitude (wet) sites showed that *E. antillensis* took up water more quickly from solutions that varied in water potential than did *E. coqui* (fig. 2.15). The former is found only at low altitudes, whereas the latter is found over a wide range of altitudes and moisture conditions. There was no significant difference in rehydration rates of *E. coqui* from high- and low-altitude sites, suggesting that local populations of this species have not adapted physiologically to different moisture regimes (Van Berkum et al. 1982). A comparison of rehydration rates in four Australian tree frogs did not show a clear correlation with habitat aridity. *Litoria moorei*, a species found only in relatively wet areas, was least tolerant of desiccation, but it also had the highest rehydration rate. *Litoria rubella*, which is found in much drier habitats, had the lowest rehydration rate, whereas *L. caerulea*, another arid-region species, was intermediate (Main and Bentley 1964). In studies of Australian burrowing frogs, Bentley, Lee, and Main (1958) found that rehydration rate was positively correlated with habitat aridity among four species of *Neobatrachus*, but not among species of *Heleioporus*. The latter burrow into deep soil that remains relatively moist during the dry season, whereas the *Neobatrachus* dig relatively shallow burrows and therefore are subjected to greater hydric stress.

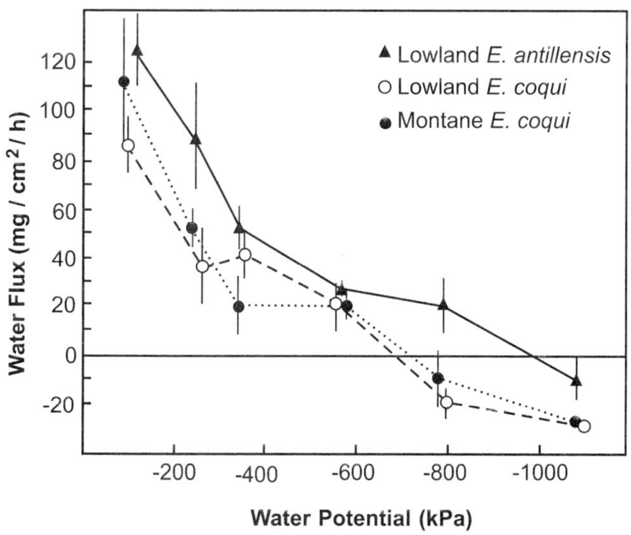


**Fig. 2.15.** Rates of rehydration in Carbowax solutions of different water potentials in three populations of Puerto Rican *Eleutherodactylus*. Data are shown as means ± 2 SE. Larger negative numbers indicate lower water availability. *Eleutherodactylus antillensis*, a grassland species, had higher rehydration rates at all water potentials than did *E. coqui*, which lives in wet forests. There was no difference in rehydration rates of *E. coqui* from different elevations. After Van Berkum et al. (1982).

## Uptake from Soil

Amphibians often do not have ready access to standing water, especially when they are away from aquatic breeding sites, and the ability to take up water from soil or other moist substrates is an important adaptation for terrestrial life. Stille (1952) and Tamsitt (1962) observed toads (*Bufo fowleri* and *Bufo hemiophrys*) moving to lakeshore beaches at night to soak up water from wet sand, but they did not present any quantitative data on rates of water uptake. Other studies have confirmed that both anurans and urodeles can absorb water from moist soil, in some cases at rates nearly equal to those of animals sitting in water (Heatwole and Lim 1961; Packer 1963a; Dole 1967; Spight 1967b; A. K. Lee 1968; McClanahan and Baldwin 1969; Ruibal, Tevis, and Roig 1969; Fair, 1970; R. Walker and Whitford 1970; Spotila 1972; C. Tracy 1976; Bundy and Tracy 1977; Marangio and Anderson 1977; Walters and Greenwald 1977; Degani and Warburg 1980).

Interspecific variation in rates of water uptake from soil seems to be related in part to habitat moisture availability, although there are phylogenetic differences as well. Terrestrial salamanders appear to have lower rates of water uptake than many anurans (Spotila 1972; Marangio and Anderson 1977), although there have not been any studies comparing anurans and urodeles under identical conditions. There is not much information on the relationship between habitat and rate of water uptake from soil among different species of urodeles. Walters and Greenwald (1977) found that the terrestrial eft stage of the red-spotted newt, *Notophthalmus viridescens*, had a somewhat higher rate of rehydration than did aquatic adults. Several studies have shown that fossorial (*Scaphiopus* and *Cyclorana*) and terrestrial anurans (*Bufo*) tend to take up water from soil very rapidly, and can absorb water from drier substrates than can less terrestrial frogs such as *Rana pipiens* (Ruibal, Tevis, and Roig 1969; Walker and Whitford 1970; van Beurden 1984). An exception is the Australian turtle frog (*Myobatrachus gouldii*), which exhibits a less pronounced water balance response than do other burrowing anurans. This species is unusual in that it lives its entire life in habitats where it seldom encounters standing water (Withers and Roberts 1993).

The rate at which an amphibian takes up water from moist soil depends on a number of factors, including the water content and physical characteristics of the soil. Soils vary considerably in the tendency for particles to bind with water. For example, clay soils tend to bind water more tightly than sandy soils, so clay would have lower water potential than sand even when the percentage of water in the soil is the same. In other words, it would be more difficult for an amphibian to extract water from clay than from sand, even if the water content of the soil is identical (C. Tracy 1976; Shoemaker et al. 1992; Spotila et al. 1992). There also are local differences in hydric properties of soil, depending on size of soil particles, degree of compaction, temperature, and water content, so comparisons among species are meaningful only when measurements are made under identical conditions.

The rate of water uptake from soil also is influenced by the condition of the animal. As an amphibian dehydrates, the osmotic concentration of its body fluids increases. In burrowing species, this effect is increased by the retention of urea, as described previously. Consequently, dehydrated amphibians typically take up water more quickly than fully hydrated individuals, and they can absorb water from drier soils (Spight 1967a; Ruibal, Tevis, and Roig 1969; Walker and Whitford 1970; P. Brown, Hastings, and Frye 1977; Marangio and Anderson 1977; Geise and Linsenmair 1988). In *Bufo punctatus,* severely dehydrated individuals show a more pronounced water absorption response than do fully hydrated toads (Brekke, Hillyard, and Winokur 1991; Hillyard, von Seckendorff Hoff, and Propper 1998). Studies of green toads (*Bufo viridis*) showed that the water absorption response also can be induced by injecting toads with saline to increase the osmotic concentration of the body (Hoffman and Katz 1999). The water absorption response of toads involves spreading the hind legs and pressing the pelvic region of the skin (the "pelvic patch" or "seat patch") into the soil. This insures that the highly vascularized pelvic patch is in contact with the substrate (Hillyard von Seckendorff Hoff, and Propper 1998). Similar water-absorbing postures are seen in many anurans, although some species tend to press the entire ventral surface against the substrate (Stille 1958; Dole 1967; C. Johnson 1969b; McClanahan and Baldwin 1969; R. Seymour 1972; Lillywhite and Licht 1974). Salamanders, which lack a specialized pelvic patch, take up water from moist soil by spreading the legs to the sides of the body and flattening the entire ventral surface onto the substrate (Gehlbach, Kimmel, and Weems 1969; Marangio and Anderson 1977).

The rates of water uptake can be affected by morphological features of the skin other than the presence of highly vascularized ventral areas. Hillyard (1976a) found that soil tends to adhere more tightly to skin of burrowing frogs such as *Scaphiopus* than to more aquatic species. This in turn facilitated transfer of water by capillary action. In addition, many terrestrial anurans have elaborate epidermal sculpturing consisting of a network of warts and grooves (fig. 2.3). These channel water by capillary action from the ventral surface onto the dorsal skin (Lillywhite and Licht 1974). Terrestrial salamanders such as *Taricha* and the red eft stage of *Notophthalmus viridescens* also have rough epidermal sculpturing that serves the same function. Most other terrestrial urodeles have costal grooves at intervals along the sides of the body; these function as capillary tubes that draw water onto the dorsal surface of the skin (Lopez and Brodie 1977; S. Brown and P. Brown 1980).

### Hormonal Regulation of Water Uptake

Rehydration from either water or soil is enhanced by physiological adjustments to dehydration, known collectively as the *water balance response.* Early work by Brunn (1921) demonstrated that frogs increased in mass when placed in water after being injected with extracts of the neurohypophysis, a portion of the pituitary gland. Subsequent work confirmed this, and by the 1950s, hormonal control of amphibian water balance had become a major focus of endocrine research (Heller 1974). Some 40 years after Brunn's initial discovery, arginine vasotocin (AVT) was identified as one hormone controlling water uptake (Sawyer, Munsick, and van Dyke 1959; Sawyer 1960a, b; Archer et al. 1960; Heller and Pickering 1961). This hormone is evolutionarily conservative, having the same structure in a variety of vertebrates from agnathans to amphibians, reptiles, and birds (Bentley 1974; Archer, Chauvet, and Rouillé 1997). It is distinct in structure from the mammalian antidiuretic hormone, arginine vasopressin (AVP), but has analogous physiological effects. Many amphibians also are sensitive to AVP, and this hormone has been widely used to investigate the water balance response (e.g., Ewer 1952; Bentley and Heller 1964; Heller and Bentley 1965; Hillman 1974; P. Brown and S. Brown 1977; P. Brown, Hastings, and Frye 1977). Quantitative comparisons among species are difficult because some species respond more strongly to AVT than to AVP, while others exhibit similar responses (Heller and Bentley 1965). There also is considerable variation in dosages of hormones used by different investigators to test the water balance response.

The role of arginine vasotocin in regulating amphibian water balance has been reviewed by many authors (Deyrup 1964; Gorbman 1964; Bentley 1969; 1974; Pang 1977; Sawyer and Pang 1979; Shoemaker et al. 1992; Archer, Chauvet, and Rouillé 1997), and only a brief summary will be given here. Dehydration triggers release of AVT from the neurohypophysis, and this in turn alters water balance by increasing the osmotic permeability of the skin to water (fig. 2.4) and the permeability of the bladder wall to water. AVT also enhances movement of urea across the bladder wall and decreases the rate of urine production by the kidneys (an antidiuretic effect; see "Water Storage" and "Kidney Function"). Some amphibians exhibit all of these responses, while in others AVT affects only the bladder or kidneys. As discussed in the section on skin structure and function, not all regions of the skin are equally sensitive to AVT, at least in anurans. Generally, pelvic skin shows a large increase in osmotic permeability to water in response to AVT, whereas skin from the dorsal surface, legs, or pectoral region does not (Bentley and Main 1972a; R. Baldwin 1974; Yorio and Bentley 1977; Parsons et al. 1978; Cree 1988a, b).

Several investigators have attempted to relate variation in responsiveness to neurohypophyseal hormones to differences in phylogenetic history or ecology. Bentley and Heller (1964) and Bentley (1973) found that the urodeles they tested exhibited little or no cutaneous response to these hormones, but they did show a decrease in urine production. This was taken as evidence that urodeles as a group have a less fully developed water balance response than most anurans, an interpretation that has been repeated in several general reviews (e.g., Heller 1965; Bentley 1969, 1971, 1974; Perks 1987). All of the urodeles used in these studies were either fully or largely aquatic (*Triturus, Siren, Amphiuma, Necturus, Ambystoma mexicanum*). Subsequent work demonstrated that some more terrestrial salamandrids and plethodontids do exhibit a cutaneous water balance response equivalent to that of many anurans (Warburg 1971a; Hillman 1974; P. Brown and S. Brown 1977; P. Brown, Hastings, and Frye 1977; S. Brown and P. Brown 1980). In newts (*Triturus* and *Notophthalmus*), individuals in the terrestrial stage of the life history exhibit stronger water balance responses than do those in the aquatic breeding stage (Warburg 1971a; P. Brown et al. 1983). Bentley and Yorio (1977) reported a modest increase in skin permeability in response to AVT in *Necturus*. These investigators used a more sensitive measurement technique than that employed by Bentley and Heller (1964). It is now clear that lack of a cutaneous response to AVT is not a general characteristic of urodeles.

Heller and Bentley (1965) compared the responses of anurans from several different families to neurohypophyseal hormones. They found little or no cutaneous response in *Xenopus* and *Discoglossus* and a relatively weak response in *Pelobates cultripes*, whereas species of *Hyla, Rana*, and *Bufo* had relatively strong responses. Following Noble's (1931) view of anuran phylogeny, they suggested that a poorly developed water balance response was characteristic of frogs in the more primitive families (Pipidae, Discoglossidae, Pelobatidae), whereas the more derived hylids, ranids, and bufonids had a more highly developed response. Their analysis was confounded by the fact that both *Xenopus* and *Discoglossus* are largely aquatic. The relatively weak response obtained with *Pelobates* might have been due to the condition of the animals. Hillyard (1975, 1976b) found that the skin of *Scaphiopus couchii* was responsive to AVT only in the first few days after emerging from dormancy, but not during the breeding season or after being held in the laboratory for several weeks. Later work by Cree (1985b, 1988a) convincingly refuted the idea that the lack of a cutaneous water balance response is an ancestral trait in frogs. Two species of *Leiopelma*, long considered the most primitive living anurans, show responses to AVT equivalent to those of more derived anuran families.

Many authors have suggested that variation in responsiveness to AVT is related to differences in habitat, with terrestrial species from dry environments showing stronger responses than aquatic species or terrestrial species from more mesic environments. This was first proposed by Steggerda (1937), who compared the responses of *Bufo americanus, Rana pipiens, R. clamitans*, and *Necturus maculosus* to pituitary extracts. The toads showed a rapid increase in water uptake, the frogs a lesser response, and *Necturus* showed no response at all. Subsequent comparisons of *Rana* with *Bufo* (C. B. Jørgensen 1950a; Heller and Bentley 1965; Degani 1985a), *Bufo* with *Xenopus* (Ewer 1952), *Bufo, Rana, Agalychnis*, and *Xenopus* (Yorio and Bentley 1978), and *Crinia* with several other genera (Bentley and Main 1972a) produced similar results. All of these comparisons suffer from the common problem of confounding phylogeny with ecology, since species representing different ecological adaptations also come from different families.

In general, comparisons of water balance responses among closely related species support the idea that terrestrial species are more responsive to neurohypophyseal hormones, but only a few studies have been done. P. Brown, Hastings, and Frye (1977) used AVP to compare responses of *Desmognathus quadramaculatus*, a largely aquatic plethodontid salamander, and four species of terrestrial *Plethodon. Desmognathus quadramaculatus* took up water relatively slowly after dehydration and showed no significant change in response to AVP. All of the other species had higher rehydration rates and responded to AVP by increasing water uptake. Furthermore, two species found in drier microhabitats, *Plethodon hoffmani* and *P. punctatus*, showed stronger responses than did closely related congeners of similar size, *P. cinereus* and *P. glutinosus,* both of which prefer wetter microhabitats. The aquatic and terrestrial species used in this study are in different subfamilies, so it still is possible that phylogenetic history is responsible for the differences in response to the hormone treatments. A comparison of *Desmognathus quadramaculatus* with more terrestrial members of the same genus, such as *D. ochrophaeus* and *D. wrighti,* would be most instructive.

Studies of *Leiopelma* in New Zealand by Cree (1985b, 1988a) are of particular interest because these frogs are thought to represent an ancient lineage. Earlier work by Cameron (1974; cited in Cree, 1988a) had shown that *L. hochstetteri* did not respond to hormone treatments by increasing water uptake, a result supporting Heller and Bentley's (1965) hypothesis that primitive frogs never evolved a water balance response. *Leiopelma hochstetteri* is the most aquatic of the three living species in the genus. Cree showed that the two more terrestrial species, *L. archeyi* and *L. hamiltoni,* not only had much higher rehydration rates than did *L. hochstetteri,* but they also exhibited a cutaneous response to AVT equivalent to that observed in myobatrachids, hylids, bufonids, and ranids. The strength of the water balance response in these species is somewhat surprising, since they are active only in relatively wet microhabitats (Cree 1989) and lack morphological specializations for water uptake, such as epidermal sculpturing and ventral warts (Cree 1985b).

In contrast to these studies, a comparison of several species of Australian frogs in the genus *Litoria* suggests that phylogenetic history sometimes can be more important than ecology in determining the nature of the water balance response. Two species in this genus, *L. aurea* and *L. raniformis,* are ranidlike semiaquatic frogs that frequent the banks of streams, whereas most other species in the genus, including *L. ewingi,* are arboreal. *Litoria aurea, L. raniformis,* and *L. ewingi* all exhibit strong cutaneous responses to AVT, irrespective of the habitat in which they are found (Cree 1985a, 1988b). Furthermore, *L. aurea* retains the well-vascularized, warty ventral skin characteristic of arboreal frogs, despite its semiaquatic habits. Since this species appears to be derived from an arboreal ancestor, Cree suggested that the lack of congruence among ecology, morphology, and hormonal responses is a product of phylogenetic history. Alternatively, she speculated that *L. aurea* spends much of its time basking along the shoreline, resulting in high rates of evaporative water loss that would make rapid rehydration advantageous.

Although most work on hormonal control of water balance in amphibians has focused on the role of arginine vasotocin, this hormone is not the only one that affects water balance. Indeed, C. B. Jørgensen (1992a) argued that dosages of AVT required to elicit a water balance response in experimental studies are many times higher than circulating levels of AVT found in dehydrated anurans. He suggested that AVT is less important in regulating water balance than previously thought, or that it acts in conjunction with other hormones to affect rates of water uptake.

Another class of neurohypophyseal hormones, termed hydrins, has been found in several families of anurans. These are structurally related to vasotocin and have a similar effect on the permeability of the skin and bladder (Rouillé et al. 1989). They do not, however, have the same antidiuretic effect on kidney function as AVT (Archer, Chauvet, and Rouillé 1997). Another hormone, prolactin, has an effect opposite to that of AVT, reducing osmotic permeability of the skin and retarding water uptake. This has been studied mainly in aquatic salamandrids, and is discussed in more detail in the section on adaptations to aquatic life. Epinephrine (adrenaline), which is produced in the medulla of the adrenal gland, can have several β-adrenergic effects on tissues, including dilation of blood vessels. β-adrenergic effectors have been shown to increase the osmotic permeability of the skin in some amphibians (Hillyard 1979; de Sousa and Grosso 1982; Yokota and Hillman 1984). The action of these hormones apparently is independent of AVT and has an even stronger effect on skin permeability. However, very little is known about the physiological significance of these effects or their relationship to amphibian ecology (Shoemaker et al. 1992).

Another hormone, angiotensin II, appears to be of primary importance in regulating the water absorption behavior of anurans (Hillyard, von Seckendorff Hoff, and Propper 1998; Hillyard 1999), as well as water-seeking behavior (Viborg and Rosenkilde 2001). In other vertebrates, including mammals, this hormone regulates drinking behavior. Anurans do not drink by taking water into the mouth, but the behavior of pressing the highly permeable pelvic skin to a wet substrate serves the same physiological function. Hoff and Hillyard (1991) injected fully hydrated toads (*Bufo punctatus*) with different dosages of the human form of angiotensin II. Even very low dosages elicited water absorption behavior. This was inhibited by injection of saralasin, an inhibitor of angiotensin II. In a subsequent study, Hoff and Hillyard (1993) showed that injection of saralasin into dehydrated toads inhibited water absorption behavior, whereas dehydrated toads in the control treatment continued to take up water from a wet substrate. This indicated that angiotensin II is involved in water absorption behavior. Subsequent work has shown that angiotensin II regulates water absorption mainly by inducing anurans to place the pelvic patch onto a substrate (Goldstein, Hoff, and Hillyard 2003). W. Johnson and Propper (2000) reported that dehydration alone is not sufficient to increase circulating levels of angiotensin II in spadefoot toads (*Scaphiopus couchii*), even though dehydrated toads exhibit an increased water absorption response. Indeed, dehydrated toads exhibit a stronger water absorption response than do hydrated toads injected with angiotensin II (Goldstein, Hoff, and Hillyard 2003).

Tran, Hoff, and Hillyard (1992) used a form of angiotensin II derived from bullfrogs (*Rana catesbeiana*) to induce a water absorption behavior in *Bufo woodhousii*. The same form of the hormone induced a water absorption response in *Bufo cognatus* and *Scaphiopus couchii* when injected into the peritoneal cavity (Propper and Johnson 1994) or directly into the brain (Propper, Hillyard, and Johnson 1995). Injection of AVT into the peritoneal cavity failed to induce a water absorption response, however. Angiotensin II also increases blood flow to the pelvic patch, as well as water uptake across the pelvic patch (Slivkoff and Warburton 2001), perhaps in conjunction with regulation of water uptake by bladder volume. Toads with empty bladders initiate water absorption behavior and have a higher rate of water uptake, and when toads with full bladders are injected with angiotensin II, they also take up more water (Parsons et al. 1993). It is possible that stretch receptors monitor bladder volume and trigger production of angiotensin II in toads with decreased bladder volume, but the precise mechanism is not known (Parsons 1994). Actual water uptake by the pelvic skin appears to be triggered by direct contact with water on the substrate (Viborg and Rosenkilde 2004). Anurans have chemical detectors in the pelvic skin that are highly sensitive to the salt concentration of soil, and they avoid sitting on hyperosmotic substrates that would make water uptake impossible (Kostinsky, Miller, and Stewart 2000; Koyama et al. 2001; Hillyard et al. 2004).

### Water Storage

When the explorer and anthropologist Baldwin Spencer traveled through the Australian Outback in the late nineteenth century, the local aborigines demonstrated how they made use of an unusual water-holding frog, *Cyclorana platycephala*, by squeezing the body to bring forth "a wine glass full of sweet water" (B. Spencer 1896, pp. 163–64). Although he later revised his estimate downward to "one or two teaspoons" of water (B. Spencer 1928, vol. 1, p. 45), Spencer remained impressed by the ability of this frog to store water in one of the driest deserts on earth. The water-storing capacity of this species is impressive, with bladder contents sometimes being more than 130% of the remaining body mass in frogs excavated from burrows in the field (van Beurden 1984). This is by no means unique, however; many amphibians store large amounts of water that serve as important reserves during dry periods. Some water is stored in the tissues, and fully hydrated amphibians have somewhat higher body water content than most other vertebrates (Shoemaker and Nagy 1977), but the principal water reserves are stored in the urinary bladder.

Some fishes have urinary bladders, but these are not homologous to the urinary bladder of tetrapods, which is an amphibian innovation. The amphibian urinary bladder is formed as a ventral outpocketing of the cloaca, and urine produced by the kidneys first enters the cloaca before moving into the bladder (Bentley 1966b). Precise measurements of bladder capacity are difficult because it readily expands as it is filled. Most authors agree that aquatic anurans and urodeles have smaller bladder volumes (< 5% of total body mass) than terrestrial and arboreal species (25–50% of total body mass; Bentley 1966b; W. Schmid 1969; Espina and Rojas 1972; Claussen 1974; Shoemaker et al. 1992). The aquatic urodele *Siren lacertina*, which often is forced to estivate in drying mud during the summer, has a maximum bladder volume of about 10% of total body mass (Etheridge 1990b), approaching that of the semiaquatic frog *Rana clamitans* (Schmid 1969; Claussen 1974). The largest bladder capacities are found in burrowing anurans from arid regions, including species of *Bufo, Scaphiopus, Spea, Neobatrachus*, and *Cyclorana*. In some cases, the mass of bladder water in estivating individuals exceeds total body mass with bladder water removed (Shoemaker et al. 1969; van Beurden 1984).

It is assumed that all amphibians with relatively large bladder storage capacities can make use of the stored water by resorbing it across the bladder wall, thereby maintaining a relatively stable body fluid osmotic concentration at least during the early stages of desiccation. Experiments with *Bufo, Scaphiopus, Cyclorana, Phyllomedusa*, and *Ambystoma* have shown that body fluid concentrations rise more rapidly in animals subjected to desiccation after removal of bladder water reserves than in individuals with bladder reserves intact (Ruibal 1962; Shoemaker 1964; McClanahan 1967; Alvarado 1972; Shoemaker and Bickler 1979; van Beurden 1984; Katz and Gabbay 1986). As long as the animals do not lose a substantial amount of water to the surrounding soil, bladder water reserves often are sufficient to sustain burrowed amphibians for many months.

In both anurans and urodeles, the resorption of water from the bladder is enhanced by the action of neurohypophyseal hormones such as AVT. These hormones have several effects, including increasing the permeability of the bladder wall to water, increasing the rate of active transport of sodium across the bladder wall, and increased permeability of the bladder wall to urea (Bentley 1966b; Shoemaker et al. 1992). Active transport of sodium into the tissues from bladder water reserves allows the animal to maintain a positive osmotic gradient for passive diffusion of bladder water into the tissues, counteracting the increase in osmotic concentration inside the bladder as water reserves are drawn down. The accumulation of urea in the plasma also facilitates water resorption by further increasing the osmotic gradient. In some amphibians, the bladder wall appears to be more sensitive to neurohypophyseal hormones than the skin (Bentley 1966b), and a bladder response is present even in some urodeles that exhibit a relatively weak cutaneous water balance response (Warburg 1972). The strength of the bladder response does not seem to be clearly correlated with habitat, at least in urodeles. Some aquatic and terrestrial salamanders, including *Notophthalmus, Triturus, Salamandra,* and *Aneides,* show a substantial increase in bladder water resorption in response to AVT (Bentley and Heller 1965; Warburg 1971b; Hillman 1974; P. Brown and S. Brown 1977). In terrestrial species such as *Plethodon glutinosus,* the effect on the bladder appears to be minimal (P. Brown, Hastings, and Frye 1977).

### The Physiology of Desiccation Tolerance

Despite the various means employed by amphibians to avoid excessive water loss, many species are exposed to dehydrating conditions on a regular basis, and their ability to survive in a particular habitat depends on their ability to tolerate desiccation. Severe dehydration has a detrimental effect on many physiological systems, but changes in osmotic concentration, ion balance, and the circulatory system have received the most attention (Shoemaker et al. 1992). As water is lost from the body, the concentration of solutes in the body necessarily rises. As long as bladder water is available to the animal, the effect on tissues is minimal. Excessive dehydration, on the other hand, can result in sharp rises in concentrations of urea, sodium, potassium, and other solutes in the plasma and body tissues (Shoemaker 1964; Ruibal, Tevis, and Roig 1969; Shoemaker, McClanahan, and Ruibal 1969; Alvarado 1972; McClanahan 1972; Degani and Warburg 1984; Degani, Silanikove, and Schkolnik 1984; Degani 1985b, c; Hillman 1978a, 1982a, 1988; Katz and Gabbay 1986). Physiological regulation of water content in critical sites such as brain tissue minimizes the detrimental effects of increased osmotic concentration (Hillman 1988). Nevertheless, severe dehydration clearly affects organ and tissue function and reduces an animal's capacity for muscle contraction, locomotion, and aerobically supported activity (McClanahan 1964; Hillman 1982a, 1984, 1987; Pough et al. 1983; Beuchat, Pough, and Stewart 1984; Feder and Londos 1984; Gatten 1987a; Gatten and Clark 1989; Moore and Gatten 1989; Stefanski, Gatten, Jr., and Pough 1989). In addition to direct effects on muscle tissue, dehydration reduces heart rate and increases blood viscosity, resulting in a decrease in blood flow to the tissues (Hillman 1978b, 1987; Hillman et al. 1985; Hillman, Zygmunt, and Baustian 1987; Shoemaker et al. 1992). Anurans that have relatively large heart ventricles in relation to body size tend to be more tolerant of dehydration than are those with relatively smaller ventricles (Hillman, Withers, and Drewes 2000). This could be due to the higher stroke volume of species with large ventricles and a greater capacity to pump viscous blood to body organs.

Most comparative studies have focused on the physiological tolerance of amphibians, the point at which a desiccated animal is physiologically incapacitated and likely to die if not removed from the desiccating conditions. Various measures of physiological desiccation tolerance have been used. These include the percentage of body mass or body water lost at death (Thorson and Svihla 1943; Bentley, Lee, and Main 1958; Ray 1958; Main and Bentley 1964; Schmid 1965; Jaeger 1971b) and the percentage lost when the animal can no longer right itself when turned on its back (Ray 1958; Whitford 1968; Farrell and MacMahon 1969; Larson 1971; Ralin and Rogers 1972; Hillman 1980b; Pough, Stewart, and Thomas 1977; Layne, Romano, and Guttman 1989). The second criterion, termed the critical activity point (CAP) by Ray (1958), has proven to be the more easily replicated measure and has become the standard for most such studies.

Broad comparative studies involving anurans from several families, including ascaphids, pipids, ranids, hylids, bufonids, and pelobatids, have shown a positive correlation between desiccation tolerance and degree of terrestrialism (Thorson and Svihla 1943; Thorson 1955; Schmid 1965; Claussen 1973b; Hillman 1980b). Aquatic and semiaquatic species such as *Xenopus, Ascaphus,* and *Rana* generally cannot tolerate a loss of more than 40% of total body mass. Some arboreal, terrestrial, and fossorial species of *Hyla, Bufo,* and *Scaphiopus* can tolerate losses of 45–50% of total body mass, and up to 60% of total body water content. The problem of ecological adaptations being confounded with phylogenetic effects is again evident in these cross-family comparisons.

Studies comparing desiccation tolerance within families, within genera, or even among populations of the same species generally have confirmed that amphibians from drier environments tend to be more tolerant of desiccation, although this is not always true (e.g., Bentley, Lee, and Main 1958). For example, comparisons among plethodontid salamanders have shown that aquatic species of *Desmognathus* and *Eurycea* are less tolerant of desiccation than are more terrestrial species of *Plethodon, Aneides, Ensatina,* and *Batrachoseps* (Littleford, Keller, and Phillips 1947; Ray 1958). M. Houck and Bellis (1972) reported that *Desmognathus ochrophaeus* is more tolerant of desiccation than is the more aquatic *D. fuscus.* Similarly, Jaeger (1971b) found that *P. shenandoah* is more tolerant of dry conditions than is *P. cinereus,* and can occupy rocky habitats unsuitable for the latter species. Even within species, such differences are evident; the terrestrial eft stage of the newt *Notophthalmus viridescens* tolerates greater water loss than does the aquatic adult stage (Walters and Greenwald 1977).

Several studies of leptodactylid and hylid frogs, including many arboreal species, have shown that species from dry habitats tend to tolerate greater loss of water than do those from wet habitats. Pough, Stewart, and Thomas (1977) found that two frogs introduced into Jamaica, *Eleutherodactylus planirostris* and *E. johnstonei,* tolerated much higher levels of water loss than did two native species, *E. cundalli* and *E. gossei,* and were found in correspondingly drier habitats. The Australian tree frog *Litoria caerulea,* which lives in hot tropical areas, is more tolerant than are species living in temperate areas (Main and Bentley 1964). Among North American hylids, semiaquatic species such as *Acris* are the least tolerant of desiccation, whereas semifossorial species such as *Hyla gratiosa* and *Pseudacris streckeri* are the most tolerant; arboreal species are intermediate (Farrell and MacMahon 1969; Ralin and Rogers 1972). A genetic basis for differences in desiccation tolerance was indicated by a comparison of *Hyla gratiosa, H. cinerea* and their hybrids. *Hyla gratiosa* was more tolerant than was *H. cinerea.* Hybrids were intermediate and showed a wide range of variation depending on the relative genetic contribution of the two parental species (Layne, Romano, and Guttman 1989).

Experimental studies of physiological desiccation tolerance provide some information on the ability of amphibians to tolerate extreme environmental conditions, but most amphibians probably retire to shelters or burrows before approaching their physiological limits. It is much more difficult to measure a species' ecological tolerance—the level of dehydration the animals will accept while continuing routine activities such as foraging or advertising for mates. Even submaximal levels of desiccation can depress metabolic rates (Pough et al. 1983; Gatten 1987a; Hillman 1987) and measures of behavioral performance such as stamina, sprint speed, or foraging ability (Beuchat, Pough, and Stewart 1984; Feder and Londos 1984; Gatten and Clark 1989; F. Moore and Gatten 1989; Stefanski, Gatten, and Pough 1989). Most amphibians probably restrict their activities well before they reach a level of dehydration sufficient to produce such effects. For example, the use of water-conserving postures by free-ranging *Eleutherodactylus coqui* limits water loss even on relatively dry nights to about 8% of initial values (Pough et al. 1983), a level that has relatively little effect on either oxygen consumption or jumping ability (Beuchat, Pough, and Stewart 1984). Similarly, cricket frogs (*Acris crepitans*) in the field averaged more than 97% hydration, a level more than sufficient for maximum jump performance (Walvoord 2003). Feder and Londos (1984) found that salamanders (*Desmognathus ochrophaeus*) maintained in the laboratory in relatively dry air would stop foraging and seek out moist retreat sites after losing only 3–4% of their initial body mass. That level of dehydration had little or no effect on foraging ability or several measures of locomotor performance. This suggests that the long-term benefits of maintaining water reserves outweigh the short-term benefits of continuing to forage under less-than-ideal conditions.

## Osmoregulation in Aquatic Environments

Life in the water presents amphibians with fewer challenges than life on land, but it is not free of difficulty. Amphibians in fresh water always have an internal osmotic concentration that is higher than the surrounding environment. Because the skin is highly permeable to water in all amphibians, water tends to be drawn into the body, threatening to upset the water and ion balance of the animal. To compensate for this excess water uptake, amphibians produce large quantities of urine, sometimes voiding the urine as frequently as every 10–15 minutes. Excess water cannot be eliminated without cost, however. Increased production of urine can increase metabolic energy expenditures considerably (Potts 1954). Production of large volumes of urine also creates problems for ion balance, because sodium, potassium, chloride, and other ions are lost with the urine. Further ion losses occur through diffusion across the skin. The principal challenge for amphibians in freshwater habitats is to balance water uptake with water loss while maintaining internal ion concentrations at relatively stable levels. This can be accomplished in several ways: (1) by reducing the rate of water intake, (2) by actively taking up ions from the water to compensate for ion loss, or (3) by reclaiming ions in the urine before they are eliminated from the body. The last two processes further add to the energetic cost of osmoregulation, because they require movement of ions against concentration gradients (Shoemaker and Nagy 1977; Shoemaker 1987; Boutilier, Stiffler, and Toews 1992).

### Skin Permeability and Hormonal Control of Water Balance

As discussed in the section on water uptake in terrestrial amphibians, the skin of aquatic amphibians tends to have a lower osmotic permeability to water than that of terrestrial species. For example, Yorio and Bentley (1978) found that isolated skin from *Xenopus laevis* and *Necturus maculosus,* both highly aquatic species, was 5–20 times less permeable to water than that of *Rana pipiens, Bufo marinus or Pachymedusa dacnicolor.* Both Schmid (1965) and Bentley and Yorio (1976) found that permeability of isolated skin was positively correlated with degree of terrestrialism in a comparison of several ranid and bufonid species. Lower skin permeability in aquatic anurans and salamanders results in lower rates of osmotic water uptake (Mullen and Alvarado 1976; G. Baldwin and Bentley 1982). Comparison of osmotic water flux in a terrestrial caecilian (*Ichthyophis kohtaoensis*) with that of an aquatic species (*Typhlonectes compressicauda*) showed the same pattern; the aquatic caecilian had a much lower rate of water uptake when the animals were submerged in water (Stiffler, DeRuyter, and Talbot 1990).

Pruett, Hoyt, and Stiffler (1991) used data from published literature to compare osmotic flux of water across the skin in aquatic and terrestrial amphibians after correcting for the effects of body size (osmotic flux increases with body size). They could make statistical comparisons only for terrestrial anurans and aquatic salamanders because of small sample sizes, so there is a serious problem of comparing phylogenetically distant groups in this study. Nevertheless, the analysis supported the conclusions of previous workers that aquatic amphibians have lower osmotic flux. Aquatic amphibians also tend to exhibit less change in osmotic permeability in response to neurohypophyseal hormones, presumably because there is little advantage for amphibians living in water to increase rates of water uptake (see further discussion in the earlier section on "Hormonal Control of Water Uptake").

In at least some species that make seasonal movements between water and land, the skin undergoes a corresponding seasonal change in osmotic permeability. This has been studied primarily in newts (Salamandridae). In *Triturus carnifex,* osmotic water flux across the skin is much lower in aquatic animals collected in winter compared to terrestrial animals collected in the summer (fig. 2.16). This reduction in permeability can be induced in terrestrial-phase newts by treating them with the hormone prolactin (Lodi, Biciotti, and Viotto 1982). Similar results have been obtained with North American newts (*Notophthalmus, Taricha;* P. Brown and S. Brown 1973, 1982, 1987; Harlow 1977; S. Brown and P. Brown 1980; P. Brown et al. 1983). A reduction in permeability is seen only after relatively long-term exposure to prolactin, and is assumed to be adaptive in reducing problems of excess hydration in aquatic-phase animals. Several authors have reported that short-term exposure to prolactin actually increases osmotic water uptake in some amphibians (Goldenberg and Warburg 1977, 1983; Warburg and Goldenberg 1978). P. Brown and S. Brown (1987) attributed these results either to pharmacological effects of extremely high dosage levels, or to contamination of hormone preparations used in the experiments. Prolactin also is thought to be responsible for inducing a "water drive" in newts and other amphibians, causing them to search for water at the beginning of the breeding season (Dent 1985), but this has been questioned by some workers (Ishii et al. 1989; Hasumi and Iwasawa 1992).

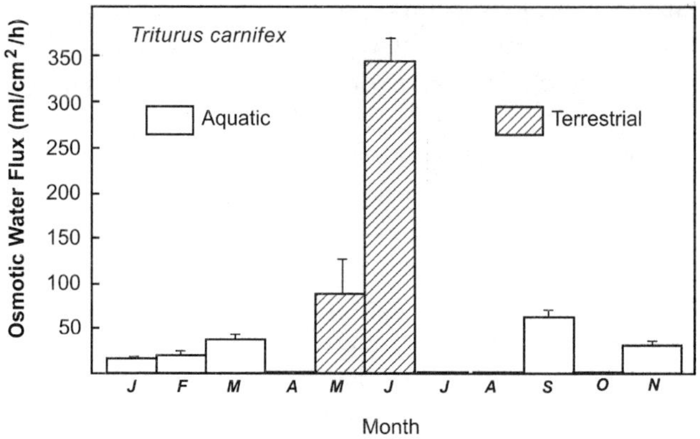


**Fig. 2.16.** Change in osmotic water flux across the isolated skin of Alpine crested newts, *Triturus carnifex,* collected in different months in the terrestrial and aquatic phases of the life history. Bars show means + 1 SE. Terrestrial newts collected in the summer had much higher skin permeability to water than did those collected in water at other seasons. Plotted from data in Lodi, Biciotti, and Viotto (1982).

## Regulation of Ions in Fresh Water

If amphibians were not able to regulate internal ion concentrations by compensating for loss of ions through the skin and urine, then prolonged exposure to fresh water could be fatal. Some species are able to survive long periods of severe hydration stress without apparent detrimental effects. For example, semiaquatic frogs such as *Rana clamitans, R. pipiens,* and *R. esculenta* can survive for several months in distilled or deionized water, even when they are not provided with food (Krogh 1937; Schmid 1965; McAfee 1972). In contrast, Schmid (1965) found that terrestrial or arboreal species of *Bufo, Hyla,* and *Pseudacris* did not do as well under the same conditions, generally lasting less than 10 days. *Rana sylvatica,* a semi-terrestrial species, was intermediate between semiaquatic ranids and terrestrial and arboreal frogs, surviving for about 20 days in distilled water. The deaths of these frogs probably resulted from excessive loss of sodium and other ions and the inability of the animals to compensate their losses by intake of ions in food or through the skin. The greater survival of the more aquatic species could have been due to lower rates of water uptake and sodium loss across the skin, a greater capacity for reclaiming ions in the urine, or a greater capacity to take up ions across the skin.

The ability of amphibian skin to take up ions, particularly sodium and chloride, against a highly unfavorable osmotic gradient was first demonstrated by the pioneering work of Krogh (1937, 1938) and Ussing (1949). Since then, frog skin has been widely used as a model system to understand basic processes of ion exchange across epithelial membranes, and there is a vast literature on the mechanisms of ion exchange. A thorough discussion of ion exchange mechanisms is beyond the scope of a book on amphibian ecology, and the subject has been reviewed in detail several times (Alvarado 1979; Kirschner 1983; Katz 1986; Stiffler 1988; Boutilier, Stiffler, and Toews 1992; Katz and Nagel 1994). Briefly, work on ion exchange has shown that both sodium ($NA^+$) and chloride ($CL^-$) ions are actively transported from the external environment. Transport of these ions is independent, with $NA^+$ being exchanged for ammonium ($NH_4^+$) ions and $CL^-$ being exchanged for bicarbonate ($HCO_3^-$) ions to maintain electroneutrality across the membrane.

All amphibians have some capacity for active ion uptake, even when ion concentrations in the environment are 1/100th those inside the animal. Studies of anurans and urodeles have shown that aquatic species often have a greater capacity for active ion uptake than do terrestrial or arboreal species (Greenwald 1972; Mullen and Alvarado 1976). These data must be interpreted with caution, since different genera and even different families have been used to represent different habitats. A comparison of the aquatic caecilian *Typhlonectes compressicauda* and a terrestrial species, *Ichthyophis kohtaoensis,* did not show the same pattern. The aquatic species had neither a higher apparent affinity for sodium nor a higher maximum capacity for sodium transport than the terrestrial species (Stiffler, DeRuyter, and Talbot 1990). One must be cautious in extrapolating from this two-species comparison, because these caecilians not only live in different habitats, but also are members of different clades (see chapter 1). Furthermore, the aquatic species is thought to have evolved from terrestrial, burrowing ancestors.

There is some evidence for evolutionary adaptation of purely aquatic species to environments with different ionic concentrations. Mullen and Alvarado (1976) found that the skin of *Ascaphus truei,* an inhabitant of mountain streams with very low solute content, could take up sodium from more dilute solutions than could the skin of pond-dwelling anurans. A phylogenetically controlled comparison is virtually impossible in this case, because closely related pond-dwelling species are not available. In general, larval amphibians have lower rates of sodium uptake than do adults. Larvae also tend to have lower rates of water intake and lower rates of sodium loss, so less compensation is required (Alvarado and Kirshner 1963; Bentley and Baldwin 1980). In most amphibian larvae, gills are thought to be more important sites for ion exchange than the skin (Alvarado and Moody 1970; Dietz and Alvarado 1974; Cox and Alvarado 1979; T. Cox 1986; Boutilier et al. 1992).

The ability to take up ions from the water is an important adaptation for aquatic life, but this alone would not be sufficient to prevent disruption of the internal osmotic environment if amphibians lost vast quantities of ions in their urine. Amphibians have a remarkable ability to reclaim ions from the urine before it is eliminated, thereby helping to maintain a stable ion balance in the body. Initial resorption of sodium, chloride, and potassium ions takes place in the kidney tubules and involves active transport of ions against a concentration gradient. Again, there is a large literature on the precise mechanisms of ion resorption that need not be discussed here (see Boutilier et al. 1992 for a review). Quantitatively the resorption of ions by the amphibian kidney is impressive. Studies of various aquatic salamanders have shown that 90–96% of the sodium and chloride and a slightly lower proportion of the potassium ions in the urine are reclaimed, making the final product extremely dilute (Wiederholt et al. 1971; Stiffler and Alvarado 1974, 1980; Boutilier, Stiffler, and Toews 1992). Ion conservation is further enhanced by active transport of ions across the bladder wall, and the anuran bladder has been widely used as a model system for studying ion transport (Bentley 1966b; McKnight, DiBona, and Leaf 1980). Most work on resorption of ions by the bladder has been done on bladders removed from the animals, but Sinsch (1991c) reported results on intact animals demonstrating active transport of ions across the bladder wall in fully hydrated frogs. He also found that resorption of ions was somewhat more efficient in semiaquatic ranids (*Rana ridibunda* and *R. lessonae*) than in a more terrestrial species (*R. temporaria*).

### Osmoregulation in Acidic Environments

A number of environmental variables can affect rates of ionic exchange between amphibians and their surroundings. Of special concern is the effect of acidification of aquatic habitats by acid rain (Dunson, Wyman, and Corbett 1992). Low pH (high acidity) can adversely affect osmoregulation in larval amphibians by substantially increasing rates of sodium ion loss through the skin and by inhibiting the active uptake of sodium from the water (Freda and Dunson 1985a, 1986a; McDonald, Ozog, and Simons 1984; Freda 1986, 1991). For some species, exposure to water with pH below 4.0 can result in the loss of more than 50% of total body sodium, with fatal consequences. Other species are somewhat more tolerant. For example, *Rana clamitans* tadpoles had lower rates of sodium ion loss at low pH than did *R. pipiens* or *R. catesbeiana* tadpoles exposed to the same conditions (fig. 2.17). *Rana clamitans* also had higher initial concentrations of sodium in the body, perhaps allowing it to survive longer in environments that induce rapid sodium loss (Freda and Dunson 1984). A few species of amphibians, including *Hyla andersoni, H. femoralis, Rana virgatipes,* and *Xenopus gilli* have eggs and larvae

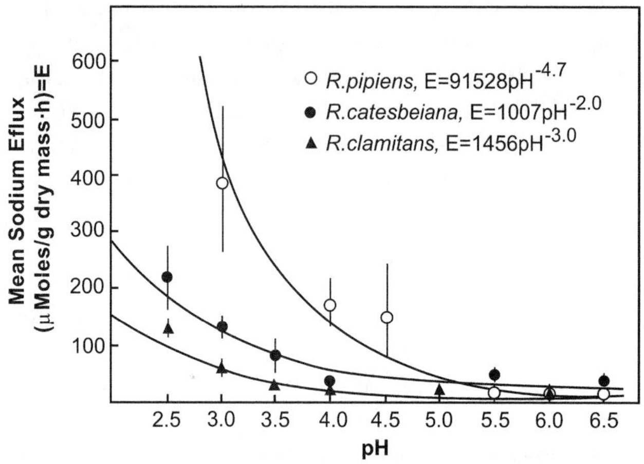


**Fig. 2.17.** Comparison of rates of sodium ion loss at different pH levels in tadpoles of North American leopard frogs (*Rana pipiens*), bullfrogs (*R. catesbeiana*), and green frogs (*R. clamitans*). Increased water acidity (lower pH) results in higher rates of ion loss in all three species. After Freda and Dunson (1984).

with unusually high tolerance to low pH, and consequently can live in acidic swamps and bogs not inhabited by other amphibians (Warner, Dunson, and Travis 1991; Picker, McKenzie, and Fielding 1993).

Ecological consequences of low pH in breeding ponds include direct effects such as delayed hatching of eggs and increased mortality of eggs and larvae. There also can be indirect effects, such as alterations in growth rate, changes in the outcome of competitive interactions among different species of larvae, and changes in predator-prey interactions (Pough 1976; Pough and Wilson 1977; Tome and Pough 1982; Freda and Dunson 1985b, 1986b; Pierce 1985; Cummins 1986, 1989; Warner, Dunson, and Travis 1991; Warner, Travis, and Dunson 1993; Beattie and Tyler-Jones 1992; Karns 1992; Pierce and Wooten 1992; Sadinski and Dunson 1992; K. Grant and Licht 1993; see also chapters 15 and 16). Some amphibian larvae are capable of avoiding highly acidic microhabitats (Freda and Taylor 1992), but it is not clear whether adults actively avoid breeding in such sites (Karns 1992).

Osmoregulatory problems caused by acidic conditions are not limited to aquatic amphibians. Studies of terrestrial salamanders have shown that acidic soil can induce the same osmoregulatory responses seen in aquatic amphibians. *Plethodon cinereus, Eurycea bislineata,* and *Desmognathus ochrophaeus* all suffered rapid loss of sodium ions when exposed to substrates with pH below 4.0 (Frisbie and Wyman 1991, 1992). Terrestrial efts (*Notophthalmus viridescens*), on the other hand, were relatively tolerant of acid conditions and suffered much lower rates of sodium loss. These data are consistent with field and laboratory studies showing a strong tendency for *P. cinereus* and other terrestrial plethodontids to avoid acidic substrates (Vernberg 1955; Mushinsky and Brodie 1975; Wyman and Hawksley-Lescault 1987; Wyman 1988; Wyman and Jancola 1992). Although terrestrial ecosystems are better buffered against the effects of manmade acid rain than are many aquatic habitats, there is a possibility that further acidification of terrestrial environments could have serious consequences for many amphibian species (Wyman 1990; see also chapter 16).

### Osmoregulation in Saline Environments

Amphibians do not normally frequent environments with high salt concentrations, where they would be subjected to excessive water loss and high rates of sodium and chloride uptake through their permeable skin. Most species are killed very quickly by immersion in full-strength seawater. This intolerance of saline environments undoubtedly has been the major barrier to dispersal of amphibians onto distant oceanic islands. However, some amphibians are surprisingly tolerant of high salt concentrations. Some species venture into saline environments only occasionally, perhaps while foraging along a beach or at the margins of a salt marsh (W. T. Neill 1958). Others make use of moderately saline environments on a regular basis and exhibit various physiological adaptations to elevated salt concentrations (Ruibal 1959, 1962; J. Balinsky 1981; U. Katz 1989; Katz and Hoffman 1990; Shoemaker et al. 1992; Markow 1997; Gomez-Mestre and Tejedo 2003).

The best known of these salt-tolerant species is the crab-eating frog (*Fejervarya cancrivora*) of Southeast Asia. It has been known since the nineteenth century that this frog is found in peculiar habitats for an amphibian, including estuarine mudflats and mangrove swamps, although it also is widely distributed in freshwater habitats such as rice fields (Premo and Atmowidjojo 1987). The physiological mechanisms that allow this species to survive in saline environments were first described in classic work by Malcolm Gordon and his colleagues in the early 1960s. Adult crab-eating frogs can live in full-strength seawater and often feed on crabs and other marine invertebrates. These frogs can maintain plasma osmotic concentrations above that of the environment even in nearly full-strength seawater (fig. 2.18). They do so by retaining large quantities of solutes, including urea, sodium and chloride ions, and free amino acids. Urea accounts for about 60% of the increase in osmolality of the plasma, with the rest being mainly sodium and chloride ions. These ions are maintained at relatively low levels inside the cells of muscles and other tissues, but intracellular concentrations of free amino acids can be nearly three times higher in seawater than in fresh water. Urea concentrations can be nearly 10 times higher (M. Gordon, Schmidt-Nielsen, and Kelly et al. 1961; M. Gordon and Tucker 1968).

Because crab-eating frogs move frequently between fresh-

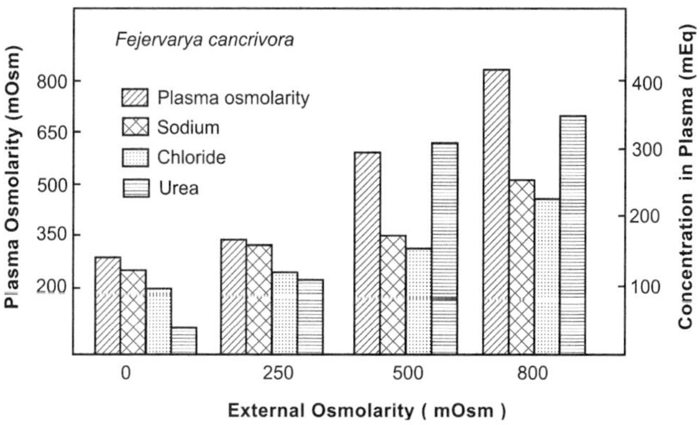


**Fig. 2.18.** Change in plasma urea, chloride, and sodium ion concentration and total plasma osmolality as a function of external osmotic concentration in the Southeast Asian crab-eating frog, *Fejervarya cancrivora*. Urea contributes the most to increased plasma osmolarity when the frogs are in salty water. After Gordon and Tucker (1965).

water and saltwater habitats of different concentrations, they need to respond to changes in salinity relatively quickly. One short-term response is to reduce rates of urine production by the kidneys, thereby retaining more urea as it is produced. Crab-eating frogs show a relatively small reduction in urine production compared to other species subjected to osmotic stress. This frog apparently does not resorb urea in the kidney tubules, an osmoregulatory adaptation found in marine elasmobranchs (Schmidt-Nielsen and Lee 1962). Urea retention is enhanced by increased permeability of the bladder wall to urea, brought about by a rapid increase in circulating levels of the hormone AVT (Chew, Elliot, and Wong 1972; Elliot and Chew 1973). AVT does not increase permeability of the skin to water in this species as it does in many other amphibians, a response that would be maladaptive because it would result in increased water loss across the skin (Dicker and Elliot 1970). Longer-term adaptation to high salinity involves increased synthesis of urea, brought about by increased activity of urea cycle enzymes in the liver (Balinsky, Dicker, and Elliot 1972).

Most amphibian larvae are even less tolerant of high salinity than adults (Shoemaker et al. 1992), but the tadpoles of *Fejervarya cancrivora* are an exception. They can tolerate very high salt concentrations and have been found living in brackish water pools at temperatures approaching 40° C without apparent detrimental effects (M. Gordon, Schmidt-Nielsen, and Kelly 1961; Dunson 1977). This high salinity tolerance is achieved without increasing plasma urea concentrations, because the tadpoles apparently do not acquire the capacity to synthesize urea until just before metamorphosis (Balinsky 1981). The tadpoles do increase plasma concentrations of sodium and chloride ions, but these are maintained below the level of the environment at high salinity. Apparently the tadpoles osmoregulate by taking in salty water with their food and then excreting excess salts by active transport across the gill membranes, a strategy common in saltwater fishes (M. Gordon and Tucker 1965).

Other ranid frogs are found mainly in freshwater habitats, but some species are capable of living in brackish springs, tidal streams, or salt marshes (Annandale 1907; Bertin 1920; Brues 1932; W. T. Neill 1958; Ruibal 1959, 1962; Christman 1974). They survive in such environments by increasing internal solute concentrations. Other frogs in this family are not known to accumulate levels of urea in the plasma and tissues approaching those found in crab-eating frogs. In fact, the concentrations of urea in the muscle tissue of *Fejervarya cancrivora* would be sufficient to incapacitate the muscles of most other frogs (Thesleff and Schmidt-Nielsen 1962; McClanahan 1964). None of the other ranids that have been tested can survive prolonged exposure to very high salinity. These include *Hoplobatrachus tigerinus* (M. Gordon, Schmidt-Nielsen, and Kelly 1961), *R. pipiens* (Homan 1968; Bentley and Schmidt-Nielsen 1971; Jungreis 1971), *R. utricularia* (Christman 1974), *R. temporaria* (Bertin 1920; Duval 1928; Ackrill et al. 1969; Viertel 1999), *R. esculenta* (Ackrill, Hornby, and Thomas 1969), and *R. ridibunda* (Katz 1975; Degani 1985b).

A number of other amphibian species have some capacity to tolerate high salinity, but none habitually lives in marine environments as concentrated as those frequented by *Fejervarya cancrivora.* One such species is *Bufo viridis,* a toad found throughout eastern and southern Europe, western Asia, and the Mediterranean coast of Africa. Although mainly a terrestrial species, it sometimes can be found in brackish marshes and other saline environments. It can tolerate salinity up to about 80% seawater and exhibits many of the same responses seen in *Fejervarya cancrivora.* Concentrations of urea, sodium, and chloride all rise dramatically in the plasma, while intracellular osmotic concentrations are increased mainly by accumulating urea and free amino acids (M. Gordon 1962, 1965; Katz 1973, 1975, 1978; Katz and Gabbay 1984; Hoffman, Eliath, and Katz 1988; Hoffman and Katz 1994; Shpun and Katz 1995). Urine excretion ceases when the frogs are exposed to high salinity (Shpun and Katz 1999). These changes are accompanied by increased activity of urea cycle enzymes, as seen in *Fejervarya cancrivora* (Lee et al. 1982).

The so-called marine toad (*Bufo marinus*) lives mainly in lowland tropical areas, but sometimes can be found along ocean beaches or the margins of mangrove swamps and tidal mud flats (Covacevich and Archer 1975; Zug and Zug 1979; van Beurden and Grigg 1980). It has a relatively poor ability to accumulate urea and other solutes and can tolerate salt concentrations less than half those tolerated by *Bufo viridis* (Liggins and Grigg 1985). This species tends to avoid saline substrates (Maleek et al. 1999). The European natterjack toad (*Bufo calamita*) sometimes inhabits coastal dunes and even breeds in slightly saline water (Andrén and Nilson 1985; Beebee 1985; Gomez-Mestre and Tejedo 2003). It has some capacity to increase plasma urea concentrations, but does not match *Bufo viridis* in its ability to tolerate high salinity (Sinsch and Sherif 1989; Sinsch, Seine, and Sherif 1992). Tadpoles of *B. calamita* can tolerate moderately brackish water, and like *Fejervarya cancrivora,* do so by elevating internal sodium and chloride concentrations, not by accumulating urea (Gomez-Mestre et al. 2004). Tadpoles from the most brackish habitats exhibit higher salinity tolerance than do those from less saline environments, and this variation appears to be genetically based (Gomez-Mestre and Tejedo 2004). The European common toad, *Bufo bufo,* has little ability to tolerate saline environments or to elevate internal solute concentrations (Ferreira and Jesus 1973). Many other species of toads have been reported in brackish water, and a few species, such as *Bufo fowleri,* sometimes breed in moderately saline coastal dune ponds (Markow 1997).

The only other anuran in which salt tolerance has been studied intensively is *Xenopus laevis,* the African clawed frog. This species can tolerate salinity equivalent to about 60% seawater. Responses of *Xenopus* to high salinity are similar to those seen in *Fejervarya cancrivora* and *Bufo viridis*. These include elevation of plasma concentrations of urea, sodium, and chloride and a reduction in rate of urine production (Janssens and Cohen 1968; McBean and Goldstein 1970a, b; Funkhouser and Goldstein 1973; Ireland 1973; Schlisio, Jurss, and Spannhof 1973; Romspert 1976; Seiter, Schultheiss, and Hanke 1978; J. Balinsky 1981). In its natural environment, this species is not normally found near marine habitats, although an introduced population has been reported living in brackish water in California (Munsey 1972). Instead, this species is subjected to periodic osmotic stress when it seeks refuge in drying mud during the dry season. Many of the responses seen in animals exposed to high salt concentrations also are seen in estivating frogs (Balinsky 1981; Katz 1989; Shoemaker et al. 1992). The same is true of *Bufo viridis* (Degani, Silanikove, and Schkolnik 1984; Hoffman, Eliath, and Katz 1988; Hoffman, Katz, and Eliath 1990; Katz and Gabbay 1986; Katz 1989; Hoffman and Katz 1991, 1994), and the ability of these species to tolerate high salinity probably is a by-product of selection for desiccation tolerance.

Relatively few salamanders can tolerate high salinity. However, populations of two terrestrial salamanders, *Batrachoseps attenuatus* and *B. major,* can be found on ocean beaches under piles of salty vegetation (Licht, Feder, and Bledsoe 1975). Plasma concentrations of both sodium and urea are increased in response to high salinity (Jones and Hillman 1978). Some ambystomatid salamanders in western North America breed successfully in ponds and lakes that are both moderately saline and alkaline (Larson 1968; Brandon, Maruska, and Rumph 1981). Some populations exhibit low-level salinity tolerance that is achieved by elevations in plasma concentrations of ions and sometimes urea as well (Kirschner et al. 1971; Gasser and Miller 1986). Moderate salinity tolerance also is seen in several urodeles that do not inhabit saline environments, but are exposed to periodic osmotic stress during estivation (*Ambystoma tigrinum:* Romspert and McClanahan 1981; *Salamandra salamandra:* Degani 1981, 1982, 1985c; *Siren intermedia:* Asquith and Altig 1986).

### Osmoregulation at Low Temperatures

When frogs and toads emerge from hibernation, they often appear bloated, with a large quantity of water present in the tissues and subcutaneous lymph sacs. This increase in water content often is accompanied by a decrease in blood hematocrit and changes in the concentrations of various ions in the tissues and extracellular fluid compartments. These changes are particularly evident in species that hibernate underwater. Excessive hydration also can be induced in terrestrial hibernators such as *Bufo bufo* by placing them in water at low temperatures (C. B. Jørgensen 1950b; Hong 1957; Deyrup 1964; D. Miller, Standish, and Thurman 1968; Zamachowski 1968, 1977; C. B. Jørgensen, Brems, and Geckler 1978; Bradford 1984c; Nielsen and Jørgensen 1990; Sinsch 1991a). Several authors have argued that increased water retention is an adaptive response analogous to that seen in dehydrated anurans, benefiting the animals in some way either during hibernation or during the breeding period immediately following emergence (C. B. Jørgensen et al. 1978; Nielsen and Jørgensen 1990; Sinsch 1991a). The fact that the change in osmotic balance seen in hibernating animals does not impair the functioning of the cardiovascular system (Weathers 1975; Lillo 1980) has been cited as evidence for an adaptive response, but the presumed benefits of excess water accumulation have yet to be identified. An alternative explanation is that the need for precise regulation of body fluid concentration is relaxed at low temperatures, when the animals are physiologically inactive (C. B. Jørgensen 1991; Pinder, Storey, and Ultsch 1992).

### The Lymphatic System and Osmoregulation

Anurans are unusual among vertebrates in having an extraordinarily large lymphatic system, including a series of large lymph sacs located beneath the skin and connected by thin-walled vessels. Fluid accumulates in the lymph sacs and is circulated by lymphatic hearts located throughout the system, although contraction of skeletal muscles probably also plays a role in lymph circulation (Hillman et al. 2004).

Nineteenth-century anatomists described the lymphatic system of European frogs (mainly *Rana esculenta*) in considerable detail, but the function of the system remained unclear. Early workers discovered that enormous volumes of fluid pass freely between the lymphatic system and the blood (Conklin 1930a, b). It was suggested that the lymphatic system might serve as a water reservoir, much like the urinary bladder, that could be used to replenish body water lost through evaporation (Ewer 1952; Deyrup 1964; Middler, Kleeman, and Edwards 1968). Carter (1979) examined the morphology of the lymphatic systems of a large number of anurans with aquatic, terrestrial, fossorial, and arboreal habits. These included Australian myobatrachids and hylids, introduced *Bufo marinus,* a species of semiaquatic *Rana* from New Guinea, *Ascaphus truei,* an aquatic species from North America, and the aquatic African frog *Xenopus laevis.* He found that lymph sacs were larger in aquatic and semiaquatic frogs than in terrestrial, fossorial, or arboreal species. Fossorial species generally had well-developed lymph sacs in the pelvic region, the site of water uptake from moist soil, but dorsal sacs were virtually nonexistent.

The greater development of the lymphatic system in aquatic frogs suggests that it does not serve as a water reservoir to protect against desiccation. Instead, it probably functions mainly to eliminate excess fluid taken up when the frogs are immersed in water, either through the kidneys or by direct diffusion of water across the bladder wall (Carter 1979; A. Baldwin, Rozum, and Gore 1990; Boutilier, Stiffler, and Toews 1992). This hypothesis was supported by Wentzell, McNeil, and Toews (1993), who placed toads (*Bufo marinus*) in a bath containing tritiated water and monitored the activity of the radioactive marker in the lymphatic system, plasma, and urine. Activity of the marker increased more rapidly in the lymphatic system than in the plasma or urine. This indicated that water taken up through the skin probably enters the lymphatic system and is drained by the renal portal system into the kidneys, where excess water is eliminated. Wentzell, McNeil, and Toews also found evidence for some diffusion of water from the lymphatic system directly into the bladder.

## Kidney Function and Nitrogen Excretion

As in all vertebrates, the amphibian kidney performs a number of interrelated functions, including the elimination of nitrogenous and other waste products of metabolism, regulation of internal water, ion, and acid-base balance, and resorption of some nutrients such as glucose. The literature on the physiology of amphibian kidneys is immense because kidneys of frogs have been widely used as a model system for understanding mechanisms of kidney function in vertebrates (Deyrup 1964). I will briefly discuss those aspects of kidney function most directly related to amphibian ecology, but I will not say much about physiological mechanisms. More detailed treatments of several aspects of kidney physiology can be found in Shoemaker and Nagy (1977), Toews and Stiffler (1989), Shoemaker et al. (1992), and especially Boutilier, Stiffler, and Toews (1992).

All amphibians are capable of producing large volumes of dilute urine when they are fully hydrated. This is brought about by a very high glomerular filtration rate (the rate at which fluid is filtered out of the blood) and a high capacity to resorb solutes from the urine in the kidney tubules. Terrestrial amphibians tend to have higher glomerular filtration rates than do aquatic species (Stiffler, DeRuyter, and Talbot 1990; Pruett, Hoyt, and Stiffler 1991), probably because they must rely mainly on the kidneys to eliminate nitrogenous wastes, whereas aquatic amphibians can eliminate most of their nitrogenous wastes across the skin (see the following). One of the first physiological responses of most amphibians to desiccating conditions is to reduce urine production by dramatically decreasing glomerular filtration rate (Schmidt-Nielsen and Forster 1954; Shoemaker and Bickler 1979; Yokota, Benyajati, and Dantzler 1985; Tufts and Toews 1986; Shoemaker et al. 1992). Some species also substantially increase resorption of water by the kidney tubules. Both responses are mediated by the antidiuretic hormone AVT, and renal responses to AVT occur even in some species of amphibians that do not exhibit a pronounced cutaneous water balance response (Bentley 1974; Sawyer and Pang 1975; Pang 1977; Pang, Uchiyama, and Sawyer 1982 (see also "Hormonal Regulation of Water Uptake"). Several other hormones also affect kidney function, and these are discussed by Boutilier, Stiffler, and Toews (1992).

Amphibians are carnivorous or insectivorous as adults, and they produce large amounts of nitrogenous wastes as a byproduct of protein metabolism. The way in which these wastes are eliminated depends on the availability of water. Energetically, the cheapest way to eliminate such wastes is by excreting ammonia, and most fishes and strictly aquatic amphibians are ammoniotelic. In some aquatic amphibians, up to 90% of the ammonia is excreted across the skin, with the kidneys playing a relatively minor role (e.g., *Necturus maculosus;* Fanelli and Goldstein 1964). A study of an aquatic caecilian, *Typhlonectes natans,* is particularly interesting, because typhlonectids are thought to have evolved from fully terrestrial ancestors, which probably excreted most of their nitrogenous wastes as urea. This species excretes ammonia more rapidly than urea, but because urea contains more nitrogen, the contribution of ammonia and urea to total nitrogen excretion is about the same. This is a higher proportion of nitrogen eliminated as urea than in most aquatic amphibians studied so far, perhaps a legacy of the terrestrial ancestry of this species. About 90% of the ammonia and 70% of the urea is eliminated across the skin, so the kidneys are not very important for nitrogen excretion (Stiffler and Manokham 1994).

Ammonia is relatively toxic and cannot be stored in the body in large amounts, so it must be eliminated immediately with large volumes of water. When water availability is restricted, as it is for many terrestrial amphibians, excretion of ammonia becomes impossible. Most semiaquatic and terrestrial amphibians excrete urea, or a combination of urea and ammonia. Excretion of urea requires more energy than excretion of ammonia, and it also requires the presence of appropriate urea-cycle enzymes. Urea is much less toxic than ammonia and therefore can be stored for considerable periods of time when water is scarce, then eliminated when water becomes available (Balinsky 1970, 1981; Jungreis 1976; Shoemaker and Nagy 1977; Shoemaker 1987; Shoemaker et al. 1992). This makes possible the common strategy of retaining urea during periods of osmotic stress, which was discussed in an earlier section.

Amphibians that move between aquatic and terrestrial environments exhibit varying degrees of plasticity in their nitrogen excretion strategies. *Xenopus laevis* is among the world's most aquatic frogs and normally excretes 60–80% of its nitrogenous wastes as ammonia. Most of the ammonia is excreted by the kidneys, not the skin, as in some aquatic salamanders and caecilians (Balinsky and Baldwin 1961). When subjected to desiccation stress during the dry season, it can readily switch almost entirely to production of urea (Balinsky, Cragg, and Baldwin 1961). The Bornean toad *Bufo quadriporcatus* is primarily a forest-floor species. While on land, it excretes approximately 60% of its nitrogenous wastes as urea and the rest as ammonia, but when placed in water, it can excrete up to 91% of its wastes as ammonia (Shoemaker and McClanahan 1980). Other anurans studied by Shoemaker and McClanahan (1980) in Borneo are much less flexible, but the degree of plasticity does not correlate very well with habitat preferences. For example, a toad (*Bufo asper*) and a frog (*Limnonectes blythii*) found along stream banks are ureotelic both on land and in the water, as are several arboreal rhacophorid frogs (fig. 2.19). On the other hand, *Limnonectes kuhlii*, a close relative of *L. blythii*, is mainly ammoniotelic in water and cannot greatly increase production of urea, as is *Rana signata*. Consequently, these species die quickly if completely deprived of water.

Aquatic amphibian embryos and larvae generally excrete mostly ammonia, but those that develop in enclosed environments or under dry conditions face a special problem in eliminating nitrogenous wastes without poisoning themselves. In at least one species with terrestrial eggs, *Geocrinia victoriana*, up to 86% of nitrogenous wastes are in the form

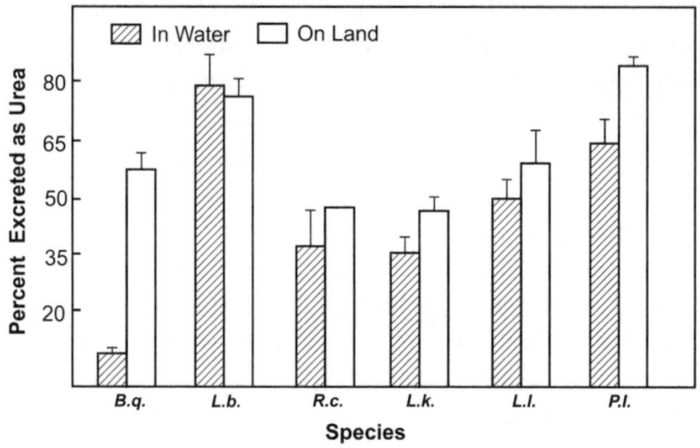


**Fig. 2.19.** Proportion of nitrogenous wastes excreted as urea by various anurans from Borneo in water and on land. The remainder of waste is excreted as ammonia. Bars show means + 1 SE. *B.q.* = *Bufo quadriporcatus. L.b.* = *Limnonectes blythii. L.k.* = *Limnonectes kuhlii. L.l.* = *Limnonectes limnocharis. P.l.* = *Polypedates leucomystax. R.c.* = *Rana chalconota.* Plotted from data in Shoemaker and McClanahan (1980).

of urea (A. Martin and Cooper 1972). Many leptodactylid frogs lay eggs in foam nests placed in terrestrial burrows (see chapter 9), and the larvae remain in the nests for some time. Shoemaker and McClanahan (1973) measured concentrations of urea and ammonia in three foam nests of *Leptodactylus bufonius* from Argentina for six weeks. The tadpoles were primarily ureotelic, especially after the first three weeks of development (fig. 2.20). The tadpoles of an African pig-nosed frog, *Hemisus marmoratus*, which develop inside a burrow, are similarly ureotelic, with about 80% of nitrogren being excreted in the form of urea (Grafe, Kaminsky, and Linsenmair 2005). Urea excretion also occurs in amphibian larvae that develop in pouches on the back of a female or inside the female's reproductive tract (Alcocer et al. 1992; Schindelmeiser and Greven 1981).

Accumulation of ammonium ions to toxic levels also can be a problem in ephemeral breeding pools that dry up as amphibian larvae develop in them. Tadpoles of *Hyperolius viridiflavus*, a species that breeds in ephemeral ponds in West Africa, increased the production of urea in response to increased ammonium ion concentrations. Juveniles that metamorphosed under drying conditions were able to survive dry conditions better than were those that metamorphosed under wet conditions, probably because they had already switched to excretion of urea (Schmuck et al. 1994). Tadpoles of another species, *Hyperolius marmoratus*, did not show this response, but they were able to tolerate higher ammonium ion concentrations. Even bullfrog tadpoles (*Rana catesbeiana*), which are not normally exposed to high ammonia concentrations in their permanent pond habitats, have the capacity to switch to urea excretion when exposed to elevated ammonia levels (Wright and Wright 1996). This plasticity in excretion is retained in adults, which are even

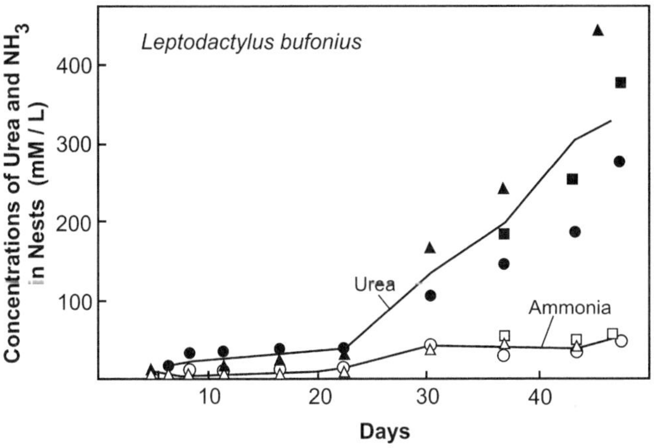


**Fig. 2.20.** Change in the concentration of ammonia and urea in the foam nests of a South American frog, *Leptodactylus bufonius,* during tadpole development. Different symbols indicate data collected from three different nests. The tadpoles exhibit a dramatic increase in urea production as development progresses. After Shoemaker and McClanahan (1973).

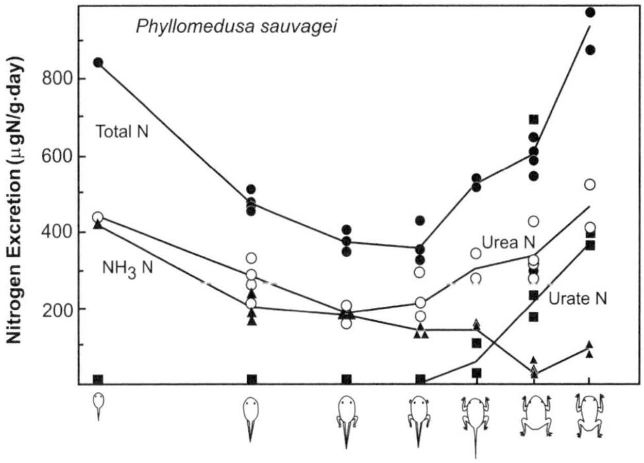


**Fig. 2.21.** Ontogeny of nitrogen excretion in tadpoles and metamorphs of a South American treefrog, *Phyllomedusa sauvagii,* that excretes both uric acid (urate N) and urea as an adult. Total nitrogen excretion increases late in tadpole development, with the proportional contributions of urea and uric acid increasing, while the proportion of ammonia ($NH_3N$) decreases. After Shoemaker and McClanahan (1982).

less likely to be exposed to high ammonia concentrations in nature. This suggests that the ability to switch between ammonia and urea excretion in response to elevated ammonia levels is a general amphibian or anuran trait, and not necessarily a specific adaptation to environments with high ammonia concentrations.

The most unusual form of nitrogen excretion in amphibians is the excretion of uric acid. This is the predominant mode of excretion in squamates and birds, and is advantageous because uric acid is relatively nontoxic and also relatively insoluble in water. This allows much of the water in the excreted material to be resorbed in the kidney, bladder, or cloaca, thereby conserving scarce water in dry environments. For most amphibians, uricotelism would be of limited benefit, because under dry conditions, water loss through the skin greatly exceeds losses in the urine. This is not true for waterproof frogs in two unrelated genera, *Chiromantis* and *Phyllomedusa,* which are uricotelic (Loveridge 1970; Shoemaker et al. 1972; Shoemaker and McClanahan 1975). This shift in mode of excretion is accompanied by a reduction in urea-cycle enzymes in the liver and an increase in uric-acid-cycle enzymes, which usually are not detectable in other amphibians (Balinsky et al. 1976; Shoemaker and McClanahan 1982). These enzymes appear only in metamorphosed frogs, at least in *Phyllomedusa;* tadpoles excrete approximately equal amounts of ammonia and urea early in development, but gradually switch to mostly urea until just before metamorphosis (fig. 2.21). Because excretion of uric acid allows these frogs to resorb most of the water in the urine before it is eliminated, they do not exhibit the dramatic drop in glomerular filtration rates seen in most other anurans that are deprived of water. This means that nitrogen excretion can continue in these frogs even under extremely dry conditions, whereas most amphibians simply stop excreting nitrogen until water becomes available (Shoemaker and Bickler 1979). In many respects, the water balance and excretion strategies of these frogs are more similar to those of squamate reptiles than those of other amphibians.

## Summary and Conclusions

Amphibians usually are associated in the popular imagination with water and aquatic habitats such as ponds, marshes, lakes, and rivers. Even many biologists consider amphibians to be largely tied to water. Yet all three major clades of amphibians (salamanders, frogs and toads, and caecilians), have aquatic, semiaquatic, and fully terrestrial species. Many of the most successful lineages, including frogs in the genus *Eleutherodactylus,* most plethodontid salamanders, and most caecilians, are fully terrestrial and do not even return to water to breed. Much of the research conducted on amphibian water relations has focused on adaptations of amphibians to various sorts of terrestrial environments.

Most aquatic, semiaquatic, and ground-dwelling amphibians lack mechanisms to retard evaporative water loss from the skin, and their activity patterns, movements, and behavior are constrained by the availability of moist microhabitats. One of the most important discoveries in the study of amphibian water balance in the last 30 years is the extent to which arboreal frogs differ from typical aquatic or terrestrial species. Arboreal species in several families have evolved mechanisms to increase the resistance of the skin to evaporative water loss, and the so-called waterproof frogs have skin resistances approaching those of some reptiles. Arboreal habits have evolved independently in many different anuran families (chapter 1), and adaptations that reduce rates of evaporative water loss undoubtedly have evolved independently as well. Rates of evaporative water loss have yet to be investigated in many arboreal leptodactylids, microhylids, centrolenids, bufonids, and ranids, and it seems likely that other moderately waterproof frogs will be discovered in the future.

Nonarboreal amphibians that lack adaptations to retard water loss must depend on behavioral adjustments to maintain the proper water balance in harsh terrestrial conditions. Many species, such as plethodontid salamanders, carefully select microhabitats that remain moist throughout the year, and are active above ground only in relatively wet weather. Amphibians also make use of a wide variety of refugia that provide more stable conditions of temperature and humidity than do more exposed locations. Amphibians find shelter under leaf litter, logs, rocks, or other debris, and they often make use of specialized retreat sites such as the leaf axils of bromeliads and other plants, rock crevices, and burrows constructed by other animals. Burrowing is common in all three major clades of amphibians, although most salamanders are incapable of making their own burrows, and usually rely on root channels or tunnels made by other animals. The frogs and caecilians, on the other hand, include many specialized burrowing species, some of which spend most of their lives underground, emerging occasionally to feed or reproduce.

Burrowing anurans have two distinct strategies that minimize desiccation in underground retreats. Some species, including spadefoot toads (Pelobatidae) and some bufonids, have highly permeable skin that does not provide a barrier to water loss. These species accumulate electrolytes and urea in their plasma, which helps to maintain a favorable osmotic gradient for water uptake, even in relatively dry soil. The aquatic frog *Xenopus laevis* has a similar response when encased in the mud of drying ponds. The alternative strategy is to form a cocoon from layers of shed skin and mucus that provides a physical barrier to water loss. These two strategies are incompatible; species that form cocoons cannot increase water uptake by accumulating urea, because the cocoon would prevent water movement across the skin in both directions.

One of the oddities of amphibian life is that most terrestrial species that live in dry environments have highly permeable skin that does not provide a barrier to water loss. This is advantageous, however, in allowing the animals to rehydrate very rapidly, either by sitting in standing water such as rain pools, or by sitting on moist soil and taking up water through the highly permeable seat patch. Indeed, terrestrial amphibians generally rehydrate more rapidly than aquatic or semiaquatic species, which are less likely to suffer debilitating desiccation. The same is true for arboreal species, which often have low rates of evaporative water loss through the dorsal skin, but high rates of water uptake through the ventral skin, especially in the pelvic region. Excess water is stored in the urinary bladder, and the animals can draw on these water reserves when subjected to desiccation stress.

Work on the hormonal control of water uptake in the past decade has revealed a more complicated picture than is described in most textbooks. For more than 50 years, arginine vasotocin (AVT), a neurohypophyseal hormone, was considered the principal modulator of water balance in amphibians. Many experiments have shown that this hormone, or its mammalian analog, results in increased skin permeability, increased bladder permeability, and increased rates of water uptake. Some workers, such as C. B. Jørgensen (1992a), have questioned the central role of AVT in regulating amphibian water balance, arguing that typical experimental dosages produce pharmacological effects that are much greater than would occur in nature. In addition, a number of additional hormones have been shown to affect various aspects of water balance, especially angiotensin II, a hormone that regulates drinking behavior in other vertebrates, including mammals.

Amphibians that live in or near water are less likely to be subjected to desiccation stress than purely terrestrial species, but they do face osmoregulatory challenges. Because of their highly permeable skin, amphibians in fresh water take up water very rapidly, which is eliminated by production of large amounts of urine. Essential ions tend to be lost in the urine and must be replaced by active uptake of ions across the skin and by resorbing ions before urine is excreted. In general, aquatic amphibians have lower skin permeability and lower rates of water uptake than terrestrial or arboreal species, and their skin is less responsive to the action of hormones such as AVT.

Some aquatic amphibians face special osmoregulatory problems in stressful environments such as acidic or saline water. Many species cannot tolerate highly acidic water, which causes excessive loss of sodium ions to the environment, but a few species have evolved physiological tolerance to acidity that allows them to live in acidic bogs. The sensitivity of most amphibians to high acidity has important conservation implications, because acidification of ponds by acid rain has been cited as a possible cause of declines in amphibian populations in some regions (see chapter 16). Most amphibians are highly intolerant of saline environments as well. A few, such as the crab-eating frog, have managed to penetrate salty mudflats by accumulating urea and other solutes in the plasma, thereby raising their internal osmotic concentration. The crab-eating frog is undoubtedly derived from a typical semiaquatic ancestor, but most of the other anurans that have been shown to tolerate moderate levels of salinity are terrestrial or fossorial species. These species exhibit similar responses to salt stress and desiccation stress, and their ability to tolerate somewhat saline environments probably is a byproduct of selection for tolerance of dry environments.

As with other aspects of water relations, kidney function and strategies of nitrogen excretion by amphibians reflect the environments in which the animals live. Aquatic and semiaquatic species typically excrete most of their nitrogenous wastes in the form of urea, or a combination of urea and ammonia. Because ammonia is toxic in high concentrations, only highly aquatic amphibians excrete most of their nitrogen as ammonia. Some terrestrial species have very plastic nitrogen excretion strategies, switching from excreting mostly ammonia in the water to mostly urea on land, but others are more constrained to one type of excretory product. Excretion of uric acid, which is common in squamate reptiles and birds, is uncommon in amphibians, but has evolved independently at least twice in waterproof frogs in Africa (*Chiromantis*) and South America (*Phyllomedusa*).

The basic features of amphibian water relations have now been rather thoroughly characterized, although it seems likely that there is a considerable amount of natural variation in water balance strategies still to be discovered in the vast array of amphibian species. We also need more explicit connections between physiological traits and the ecology and behavior of amphibians. For example, to what extent is the foraging of terrestrial species constrained by problems of evaporative water loss, as suggested for plethodontid salamanders by Feder (1983b) and others? In arboreal species, what is the relationship between rates of evaporative water loss and the height of foraging sites in the canopy or the choice of calling perches by males? Seasonal calling activity of some frogs that call from elevated perches appears to be limited by problems of evaporative water loss (Pough et al. 1983), but to what extent is nightly calling time constrained by water loss? Are diurnally active amphibians exposed to more severe problems of water loss than more typical nocturnally active species?

A critical need in studies of amphibian water balance, and many other aspects of their physiological ecology, is to ask questions in a more explicit phylogenetic perspective (Huey 1987; Bennett and Huey 1990). The traditional approach to comparative physiological studies has been to choose experimental animals that represent rather broad categories of habitat choice (aquatic, terrestrial, fossorial, or arboreal), with little regard to phylogenetic relationships among the species being studied. Consequently, pairwise comparisons of ranid frogs and bufonid toads or aquatic salamanders and terrestrial anurans are commonplace, even though these animals differ in many features of their biology other than their water relations. While such comparisons have revealed qualitative relationships between physiological traits and habitats, we know almost nothing about the direction of evolution of particular physiological traits that reduce evaporative water loss, facilitate water uptake, or influence osmoregulatory capacities. For example, has the use of concentrated lipid secretions evolved only once in a clade of closely related species within the hylid family Phyllomedusinae, or has it evolved independently several times? How frequently have adaptations to retard evaporative water loss evolved within the diverse family Hyperoliidae? Has there been convergent evolution in physiological traits that reduce evaporative water loss in hylids and other tree frog clades? Is the water balance response that can be elicited by desiccation or by neurohypophyseal hormones in many amphibians an ancestral trait for all amphibians or for all anurans, or has a similar response evolved independently in separately derived clades of terrestrial species? Clearly we are not in a position to answer these questions now, because we lack detailed physiological data on most species of amphibians, and phylogenetic relationships of many major lineages are still unresolved (see chapter 1). Nevertheless, use of phylogenetic information will allow future investigators to make informed choices of experimental animals for comparative studies.

# **Chapter 3** Temperature Relations

*Frogs of various colours are numerous in those parts as far North as the latitude 61°. They always frequent the margins of lakes, ponds, rivers, and swamps; and as the Winter approaches, they burrow under the moss, where they remain in a frozen state till the Spring. I have frequently seen them dug up with the moss (when pitching tents in Winter) frozen as hard as ice; in which state the legs are as easily broken off as a pipe-stem, without giving the least sensation to the animal; but by wrapping them up in warm skins, and exposing them to a slow fire, they soon recover life.*

—Samuel Hearne, *A Journey From Prince of Wales Fort, in Hudson's Bay, to the Northern Ocean* (1795)

IN THE 1760S, the Hudson Bay Company sent a young sailor named Samuel Hearne to the company fort at the mouth of the Churchill River on Hudson's Bay, from which he made a number of exploring trips on foot and by canoe into the interior of the Canadian Arctic. Although not trained as a naturalist, he made detailed notes on the animals of the region, including the description of frozen wood frogs (*Rana sylvatica*) just quoted. His unusual account was mentioned in a number of popular nineteenth-century natural history books (Griffith 1831; Gould 1853; Bingley 1868) and then forgotten. It was not until 200 years later that the physiological mechanism that allows these frogs to survive in a frozen state was understood (Storey and Storey 1984; see discussion at the end of this chapter). Most of the other animals that Hearne observed around the fort were birds and mammals, including large numbers of arctic foxes (*Alopex lagopus*). In contrast to the frogs, the foxes had no difficulty staying active in the coldest winter weather, and in fact, this was the best season for trapping because their coats were in their prime during winter. Arctic foxes can maintain a constant body temperature of 38° C with no increase in metabolic rate, even when exposed to ambient temperatures of −30° C (Scholander et al. 1950).

The ability of these foxes to remain active in even the coldest weather, while the frogs lie frozen under the snow, illustrates one of the key differences between amphibians and mammals. Wood frogs and other amphibians cannot elevate body temperature by metabolic activity, and they lack insulation that would help retain body heat. When cold weather sets in, amphibians are forced to become inactive, either seeking out microhabitats that remain above freezing throughout the winter, or living in a frozen state until warm weather returns. Conversely, as temperatures rise, amphibians gradually become more active until they reach temperatures too high for them to tolerate. At that point, they must again seek out microhabitats that protect them from high temperatures and the associated problem of high evaporative water loss (see chapter 2). The advantage of being ectothermic is that amphibians have very low resting metabolic rates, and hence very low energy requirements compared to birds and mammals (Pough 1980, 1983; see also chapter 5). This low energy lifestyle does come with a cost, however. It means that ambient temperature limits the ability of amphibians to carry on normal activities such as feeding and reproduction.

This chapter reviews the temperature relations of amphibians, with emphasis on the ecological implications of variation in temperature. I first discuss the effects of temperature on various physiological and behavioral processes and how these effects influence the ecology of amphibians. I then discuss aspects of behavioral thermoregulation, temperature acclimation, and responses of amphibians to extreme cold. Details of physiological mechanisms and the biophysics of heat transfer will be mentioned only briefly. Readers should consult the reviews by Hutchison and Dupré (1992); Pinder, Storey, and Ultsch (1992); Rome, Stevens, and John-Alder (1992); and Spotila, O'Connor, and Bakken (1992) for more complete discussions of these topics.

## Effects of Temperature on Physiology and Behavior

Because amphibians are ectotherms, nearly every aspect of their physiology and behavior is affected by temperature. However, the temperature sensitivity of physiological and behavioral processes varies considerably. Some functions, such as metabolic rate or embryonic development, show a positive linear relationship to temperature over most of the natural temperature range encountered by these animals. In contrast, locomotion in some species is relatively temperature independent over a narrow range of temperatures (Rome, Stevens, and John-Alder 1992). Other processes, such as anaerobic metabolism, undergo very little change with changing temperature (Bennett 1978). The following section reviews the effects of temperature on selected physiological and behavioral processes that have a direct effect on amphibian ecology. Some of these topics are covered in more detail in other chapters and will be mentioned only briefly here.

### Metabolic Rates

Aerobic metabolism is temperature dependent in all amphibians and generally increases in a linear fashion as temperature increases (see fig. 5.1 in chapter 5 for a summary). In general, the slopes of regressions of metabolic rate on temperature are similar during rest and exercise, but activity metabolism is several times higher at any given temperature. Phylogeny also influences the relationship between temperature and metabolism. Anurans and urodeles have similar temperature sensitivity (i.e., the slopes of regressions of metabolic rate against temperature are the same), but urodeles have lower metabolic rates at any given temperature than anurans, resulting in lower energy demands (Gatten, Miller, and Full 1992; see chapter 5 for further details).

For temperate-zone amphibians, energetic demands typically increase in spring and summer, but gradually decrease during the fall and are very low during the winter. In addition, some species that are subjected to wide fluctuations in environmental temperature on a daily basis also undergo major fluctuations in metabolic rate (e.g., Carey 1978, 1979a, b). Tropical amphibians, on the other hand, would be expected to have relatively high energetic demands throughout the year, except during periods of dry-season dormancy, and consequently should have higher yearly food intake than most temperate-zone species. Similar differences are expected in populations of the same species from high and low altitudes. Amphibians from high altitudes will tend to have lower metabolic rates, shorter activity seasons, and smaller total energy requirements than will those from lower altitudes, all of which contribute to longer life spans (e.g., Lüddecke 1997).

### Feeding and Digestion

Temperature can affect the ability of amphibians to acquire and process food in several ways. On an ecological level, seasonal and daily variation in temperature sets limits on activity of both amphibians and their prey. For some species, such as plethodontid salamanders, feeding continues at relatively low temperatures even during the winter, but prey availability is low (R. Caldwell and Jones 1973; R. Caldwell 1975; Fraser 1976b). Prey availability for these salamanders also is limited by hot, dry conditions, not necessarily because prey is less abundant, but because salamanders are restricted in their movements (Jaeger 1980b). Anurans in cool climates are limited in both their seasonal and daily activity periods by low temperatures (Mullally and Cunningham 1956; Carey 1978; Bider and Morrison 1981; Smits 1984; Sinsch 1989a). They can compensate by switching from nocturnal to diurnal activity, or by foraging at a wider range of temperatures than species from warmer regions (see "Behavioral Thermoregulation" in a following section). However, species that breed in early spring often depend on stored energy reserves to support reproductive activities because environmental temperatures are too low for their prey to be active (see chapter 5).

Temperature can have direct physiological effects on feeding behavior. Lillywhite, Licht, and Chelgren (1973) provided unlimited numbers of mealworms to juvenile toads (*Bufo boreas*) held at different temperatures and found that toads at 27° C ate more food than those at 14° C and 20° C. Feeding also was suppressed at a temperature above that normally experienced by toads in the field (33° C). Patterns of food ingestion closely paralleled differences in growth rates at these temperatures (growth was maximized at 27° C). The precise reason for the reduced appetite of the toads at low temperatures was not clear, but could have been due to slower rates of food passage through the digestive system. Colder temperatures also could influence rates of movement of their ectothermic prey, which in turn could affect the feeding response of the toads (Ewert 1987). An experiment by Stevens (1988b) avoided this problem by presenting mealworms to toads on a moving circular track at a con-

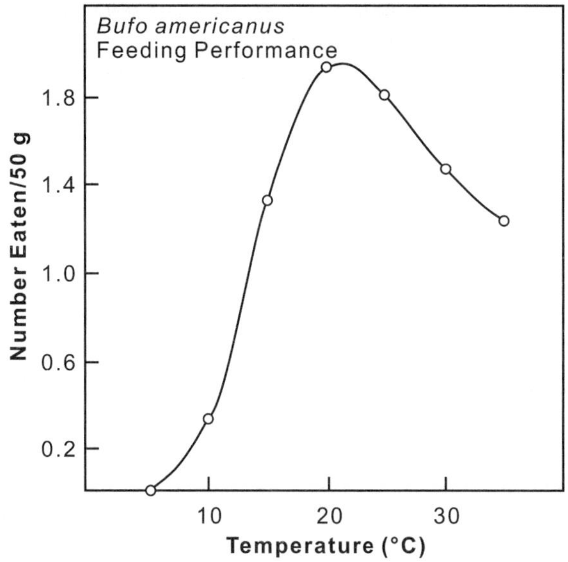


**Fig. 3.1.** Feeding performance of American toads (*Bufo americanus*) at different temperatures, measured as the number of worms eaten per 50 g of body mass. Feeding performance peaks at about 20°C and declines rapidly at colder temperatures. After E. D. Stevens (1988a).

stant velocity, regardless of temperature. The toads still showed a much reduced feeding performance at low temperatures and a slightly reduced response at high temperatures (fig. 3.1). In another experiment, acclimation to a cool temperature (10° C) for several days resulted in lower food intake by the toads, even when they were tested at 20° C (Stevens 1988a). Possibly this was because food remained in the digestive system longer at the lower temperature, resulting in reduced hunger levels. Preest and Pough (2003) reported that low temperatures resulted in some decreased interest in feeding by American toads (*Bufo americanus*) and an increase in the time required to capture prey, but hydration state had a greater effect on feeding performanc than did temperature.

Both the rate of food passage through the digestive system and the efficiency with which energy is assimilated are affected by temperature. For example, red-backed salamanders (*Plethodon cinereus*) fed on fruitflies (*Drosophila*) assimilated about 90% of ingested energy at 10° C, 85% at 15° C, and only about 80% at 20° C. Since metabolic rate increases with temperature while assimilation efficiency decreases, salamanders at high temperatures can be in negative energy balance, especially if dry conditions keep them from foraging (Bobka, Jaeger, and McNaught 1981). In contrast, G. C. Smith (1976) did not find any difference in assimilation efficiency in *Bufo terrestris* over a range of temperatures from 20° to 30° C. Digestive rates increased with increasing temperature, whereas the time required for food to pass through the gut decreased. This probably is a general pattern for amphibians, as for most ectotherms, enabling the animals to process food more rapidly at higher temperatures. This in turn makes more energy available for production of new tissue or for reproduction (Riddle 1909; Seale 1987; Jiang and Claussen 1993; Schabetsberger 1994). Some amphibians appear to behaviorally elevate body temperatures after feeding as a means of increasing the rate of digestion (Lillywhite, Licht, and Chelgren 1973; Freed 1980; Witters and Sievert 2001).

## Development and Growth

### Embryonic Development

Rates of embryonic development are strongly affected by temperature in all amphibians. Embryos develop more rapidly at warm temperatures than at cold temperatures (Moore 1939; Douglas 1948; Volpe 1953, 1957; Herreid and Kinney 1967; H. A. Brown 1967, 1975a, b, 1976b; Zweifel 1968b; Bachmann 1969; Guyétant 1969; Salthe and Mecham 1974; Kuramoto 1975; R. D. Howard 1978b; Bradford 1984a, 1990). Size of the egg also influences development rate, with large eggs taking much longer to develop than small eggs (Salthe and Duellman 1973; Salthe and Mecham 1974; Bradford 1990). Consequently, species such as spadefoot toads (*Scaphiopus*) that lay very small eggs in warm desert pools have the fastest rates of development (only about 12 hours for *S. couchii* at 34° C), an obvious advantage for amphibians using breeding sites that are likely to dry up quickly (Zweifel 1968b). In contrast, amphibians that lay eggs in cold streams or ponds typically have larger eggs and much slower development (up to 6 weeks in *Ascaphus truei* at 4–10° C; H. A. Brown, 1975a). Slow rates of development also are characteristic of amphibians that undergo direct development on land at low temperatures, such as many plethodontid salamanders (Hanken 1979; Bradford 1984a) and microhylid frogs in the cool highlands of New Guinea (M. Simon 1983). Although cold-adapted amphibians typically have faster rates of development at cold temperatures than warm-adapted species, absolute rates of development are faster in warm-adapted species at the temperatures normally encountered during the breeding season (J. A. Moore 1939; Bachmann 1969). Even over short distances, variation in thermal regimes of nearby ponds can result in selection for different rates of embryonic development. In wood frogs (*Rana sylvatica*), embryos from shady, closed-canopy ponds have faster development rates at all temperatures than do those from nearby sunny, open-canopy ponds (Skelly 2004). Phylogeny is important as well; urodele eggs take longer to develop than do anuran eggs, even when egg size and temperature are the same (Bradford 1990; see chapter 10 for further discussion of life history strategies).

### Larval Development

Differentiation and growth of amphibian larvae also are temperature dependent. Differentiation (the passage from one developmental stage to another) is more sensitive to changes in temperature than growth (the production of new tissue). Consequently, low temperatures can inhibit differentiation while growth continues, albeit at a slow rate. The result is that larvae raised at cold temperatures typically are larger at any given stage of development than those raised at warm temperatures (Smith-Gill and Berven 1979; see fig. 13.20 in chapter 13). This has important implications for individual life histories. Some ranid frogs in northern latitudes spend up to three winters in the tadpole stage before metamorphosing, while tadpoles in more southern populations often do not overwinter at all. Overwintering tadpoles typically continue to grow at a slow rate even when differentiation is inhibited by low temperature and eventually transform at a larger size than nonoverwintering tadpoles (Collins 1979a; Collins and Lewis 1979; Bruneau and Magnin 1980b).

Differences in environmental temperature probably account for much of the geographic and altitudinal variation in length of the larval period and size at metamorphosis seen in various species of amphibians (Berven and Gill 1983; Miaud and Guyétant 1998; Morrison and Hero 2003a). For example, green frogs (*Rana clamitans*) at high altitudes have longer larval periods and metamorphose at a larger size than do those at low altitudes. Low-altitude larvae transplanted to high-altitude ponds had even slower rates of development than did larvae from the high-altitude populations, indicating that populations have undergone selection to minimize the effects of the temperature gradient (Berven, Gill, and Smith-Gill 1979). Similarly, environmental temperature accounted for most of the observed variation in life-history traits between lowland Maryland populations of wood frogs (*R. sylvatica*) and montane populations in Virginia, but there were significant genetic differences between the populations as well (Berven 1982a, b). Canadian tundra populations are less temperature sensitive than are those found farther south. They also have shorter larval periods and smaller size at metamorphosis than do their southern relatives. This suggests that selection has favored shorter larval periods in an environment where temperature limits the time available for growth and development (Berven and Gill 1983).

Tadpoles of *Rana temporaria* in northern Sweden actually experience higher water temperatures than do those in southern Sweden because of rapid warming of ponds after the ice melts and longer daylength. Northern tadpoles grew faster at relatively higher temperatures than did southern tadpoles, presumably an adaptation to the much shorter growing season in the north (Ståhlberg, Olsson, and Uller 2001). In both northern and southern populations, growth rate was less variable under optimum conditions for that population (warm in the north, cool in the south) than under less favorable conditions (Olsson and Uller 2002). Northern tadpoles from warmer ponds also experienced higher mortality at low temperatures than did southern tadpoles from cooler ponds (Olsson and Uller 2003). All of these results suggest that selection has favored adaptation in growth and development to local environmental conditions.

Studies of the same species over a latitudinal gradient from southern Sweden to northern Finland showed that northern tadpoles developed faster than did southern tadpoles in the laboratory under identical temperature and food regimes. Hence, there is evidence for genetically based adaptation to short growing seasons in northern populations. In the field, however, variation in pond temperature at the same latitude tended to override genetic variation in development rate, so actual development rates in particular ponds were not closely correlated with latitude (Merilä et al. 2000; Laugen, Laurila, and Merilä 2002; Laugen et al. 2003).

Environmental temperatures can affect the growth and development of different species in different ways, and variation in environmental temperature affects the distribution and abundance of species in nature. For example, some anurans typically breed in open-canopy ponds that receive a lot of direct sunlight and have higher temperatures than do closed-canopy ponds, and these differences could affect rates of larval development. As canopy cover increases over ponds, some amphibians eventually are excluded from such ponds, at least in part because of changes in temperature regime (Skelly, Werner, and Cortwright 1999; Skelly, Friedenburg, and Kiesecker 2002; Halverson et al. 2003). Local populations can show apparent adaptation to differences in environmental temperature even on relatively short time scales. For example, Skelly and Freidenburg (2000) demonstrated that wood frog tadpoles (*Rana sylvatica*) from closed-canopy ponds were less tolerant of high temperatures, as measured by the critical thermal maximum ($CT_{Max}$), than were those from more open ponds formed by beaver dams. Nevertheless, at the cold temperatures normally found in wood frog breeding ponds, tadpoles from closed-canopy ponds actually preferred warmer temperatures than did those from open-canopy ponds, even when raised under identical conditions (Freidenburg and Skelly 2004).

Negovetic et al. (2001) studied the effects of temperature on the performance of tadpoles of the *Rana esculenta* complex in Europe. They found that tadpoles of *R. esculenta*, a hybridogenetic species derived from hybridization between *R. lessonae* and *R. ridibunda*, were more likely to survive and reach metamorphosis at 18° C than were those of the parental species *R. lessonae*. Conversely, *R. lessonae* tadpoles were more successful at 24° C than were *R. esculenta* tadpoles. A survey of a natural pond revealed that the relative abundance of *R. esculenta* tadpoles declined with increasing microhabitat temperature, so there is the potential for at least partial microhabitat segregation of the two types of tadpoles.

In addition to effects on growth, survival, and metamorphosis of larvae, environmental temperature also can affect the expression of physiological traits. For example, tadpoles of *Pseudacris regilla* raised at 15° C had higher burst swimming speed than did those raised at 25° C at test temperatures from 5 to 25° C (Watkins 2000). Tadpoles raised at the cooler temperature also had higher activity of myofibrillar ATPase, an enzyme important for anaerobically supported activity such as bursts of swimming. In the presence of predators, tadpoles that select warm microenvironments are at a disadvantage compared to those that select cooler microenvironments because of reduced swimming speed, which is directly related to the ability to escape predators (Watkins 1996). In predator-free environments, however, selection of warmer microhabitats might be advantageous, especially if the main threat to survival is drying of breeding sites.

### Growth and Development of Juveniles and Adults

Seasonal temperature regimes also have a major impact on patterns of growth and sexual maturation in metamorphosed amphibians. Amphibians at high latitudes or high altitudes grow more slowly and take longer to reach sexual maturity than do populations of the same species in warmer climates (e.g., Bruce 1972b; Tilley 1980; Bury and Whelan 1984; Hemelaar 1988; Ryser 1988b; Caetano 1990; Tilley and Bernardo 1993; Miaud and Guyétant 1998; Marunouchi, Ueda, and Ochi 2000). Presumably this is due to the direct physiological effect of temperature on growth, reduced prey availability at cool temperatures, and the shorter season available for feeding and growth in cool environments (Martof 1956b; C. B. Jørgensen 1992b). For example, bullfrogs (*Rana catesbeiana*) in Quebec grow much more slowly than do those in Louisiana, they take longer to attain sexual maturity, and they reach smaller adult body size (fig. 3.2). Frogs in New York, Illinois, and Missouri have intermediate growth rates (Raney and Ingram 1941; R. Ryan 1953; Durham and Bennett 1963; Schroeder and Baskett 1968; Bruneau and Magnin 1980a).

Some amphibian species reach larger adult body sizes in cool climates than in warm climates (fig. 3.3), a relatively common trend among ectotherms (Partridge and French 1996; Atkinson and Sibly 1997; K. Ashton 2002). For example, amphibians often are considerably larger at high altitudes than are conspecifics at lower altitudes in both temperate and tropical mountains (Narins and Smith 1986; Hemelaar 1988; Lüddecke 1997; Esteban and Sanchiz 2000; Marunouchi, Ueda, and Ochi 2000). Presumably this is due in part to the greater effect of cool temperatures on development than on growth, which in turn would delay

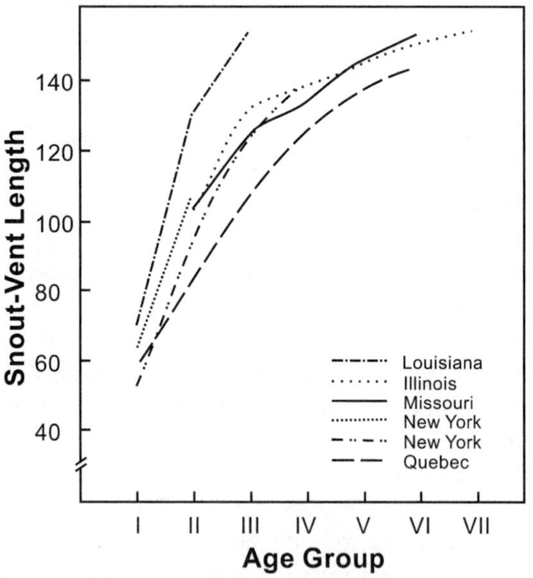


**Fig. 3.2.** Annual growth of bullfrogs (*Rana catesbeiana*) from several North American populations. The two lines for New York are for two different studies conducted in Ithaca, New York, in the 1940s. Growth was fastest in the southernmost population (Louisiana) and slowest in the northernmost population (Quebec). After Bruneau and Magnin (1980a).

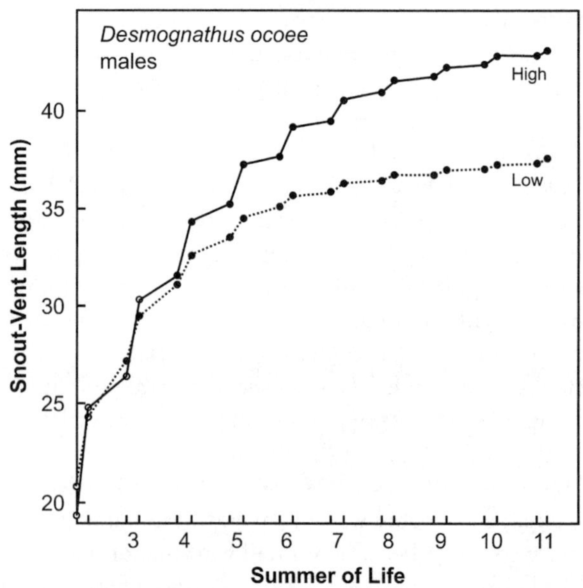


**Fig. 3.3.** Growth of male Ocoee salamanders (*Desmognathus ocoee*) from high-elevation and low-elevation populations in North Carolina. High-altitude males grew faster than did low-altitude males after the fourth summer, probably because of an earlier age of sexual maturity at low altitudes, after which growth rates are reduced. After Tilley (1980).

sexual maturity (Berven 1982a). Once sexual maturity is reached, growth often slows dramatically as the animals shift energy investment into reproduction. Consequently, some early maturing populations have smaller adult body size than do late-maturing populations of the same species (Tilley 1980; Hemelaar 1988; Miaud and Guyétant 1998; Marunouchi, Ueda, and Ochi 2000).

There also is evidence of genetic adaptation for more efficient growth at high altitudes or latitudes. For example, toads (*Bufo bufo*) from Norway or the mountains of Switzerland have shorter growing seasons than do populations from warmer climates, but they actually grow faster during the summer and are larger at sexual maturity (Hemelaar 1988). A similar pattern was found in a comparison of high- and low-altitude populations of *Rana iberica* in Spain. Adult frogs in the high-altitude population grew faster and were larger at any given age than those from a low-altitude population (Esteban and Sanchiz 2000). In this case, age at sexual maturity did not differ between populations. High-altitude populations of amphibians sometimes live longer than do conspecifics at low altitudes (e.g., Lüddecke 1997; Esteban and Sanchiz 2000; Marunouchi, Ueda, and Ochi 2000). This probably is not related directly to temperature, but to shorter activity seasons and reduced exposure to predators.

### Reproductive Cycles and Gametogenesis

Annual cycles of reproduction and gametogenesis in amphibians are strongly temperature dependent. In general, amphibians in wet tropical and subtropical regions have prolonged breeding seasons, in some cases lasting throughout the year (see chapter 10). Males of most of these species probably have continuous production of sperm throughout the year, while females produce single or multiple clutches of eggs during the breeding season (van Oordt 1960; Lofts 1974, 1984; Houck 1977a; Duellman and Trueb 1986; C. B. Jørgensen 1992b; Díaz-Páez and Ortiz 2001). In the Chaco region of Argentina, many anurans show continuous gonadal activity throughout the year, despite some differences in winter and summer temperatures (Cei 1949b, 1980; Pisanó 1957a, b, 1958). However, some species, such as *Lepidobatrachus llanensis, Leptodactylus bufonius,* and *Leptodactylus chaquensis,* have gametogenesis interrupted in winter (Cei 1948, 1949a), and in *L. chaquensis,* gametogenesis also is inhibited by high temperature during the summer dry season. In contrast, *L. ocellatus,* a close relative of *L. chaquensis,* has a continuous gametogenic cycle that is not interrupted in summer or winter (Cei 1948, 1949c, 1980; Rengel 1950). Hot dry seasons also interrupt gametogenesis in *Phyllomedusa sauvagii* (Caruso 1949; Cei 1980).

Most cold-temperate amphibians do not produce gametes during the winter. Females generally complete maturation of their eggs before entering hibernation (C. B. Jørgensen 1973a, 1975, 1992b; C. B. Jørgensen, Hede, and Larsen 1978; C. B. Jørgensen, Larsen, and Lofts 1979; see also chapter 5). In males, spermatogenesis takes place in the late spring or summer. Some species have an internally controlled cycle that results in spermatogenesis stopping during the winter even if the animals are held at warm temperatures. This type of cycle has been found in the northern Eurasian brown frogs (*Rana temporaria, R. arvalis, R. dalmatina*), some plethodontid salamanders, and a cold-adapted South American leptodactylid, *Pleurodema bufonina.* Other amphibians have the potential to continue sperm production throughout the year if temperatures are warm, but sperm production normally is inhibited during winter. Species exhibiting this pattern include European green frogs in the *Rana esculenta* complex, southern European brown frogs such as *R. iberica* and *R. graeca,* the southern European frog *Discoglossus pictus,* and several species of *Triturus.* Among species of leptodactylid frogs in temperate southern South America, warm-adapted populations produce sperm all year, whereas conspecifics in colder climates have spermatogenesis interrupted by winter. Some species that live year-round in cold mountain lakes and streams have continuous spermatogenic cycles, even at temperatures that would inhibit sperm production in other amphibians. Two examples are frogs from far-southern South America, *Telmatobius hauthali* and *Hyla pulchella* (Ifft 1942; Cei 1949d, 1961a, b; van Oordt and van Oordt 1955; van Oordt 1960; J. K. Werner 1969; C. B. Jørgensen 1992b; Lofts 1974, 1984).

### Locomotion

In general, locomotor performance of amphibians decreases at cold temperatures, regardless of whether the animals are jumping, hopping, crawling, or swimming (see Rome, Stevens, and John-Alder 1992, for a more detailed discussion of locomotion and temperature). Locomotor performance is not necessarily a simple linear function of temperature, and the precise shape of the performance curve varies among species and among studies of the same species. For example, Huey and Stevenson (1979) reported that jump performance in *Rana clamitans* was relatively independent of temperature between 10° and 25° C, and a similar pattern was reported for *R. pipiens* by Tracy (1979). However, most other studies of these species and other anurans have failed to find a broad range of temperature independence, although narrower performance plateaus have been observed in some species (fig. 3.4). Locomotor performance generally declines rapidly with decreasing temperature and also falls off abruptly at very high temperatures (Putnam

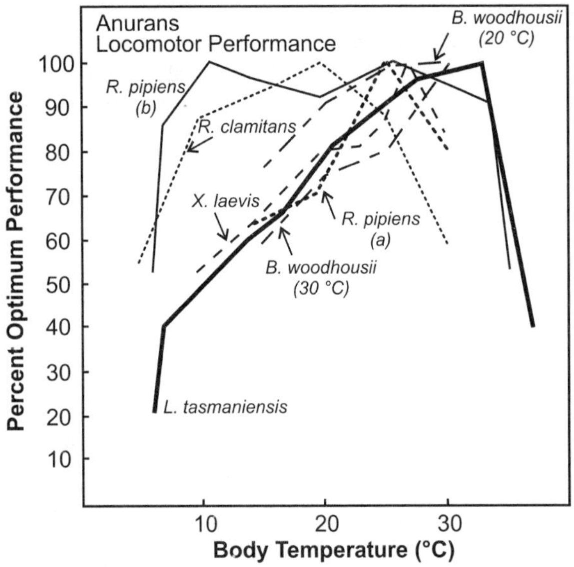


**Fig. 3.4.** Temperature dependence of locomotor performance in selected anurans, measured as percent of optimum performance. Genera: *B* = *Bufo, L* = *Limnodynastes, R* = *Rana, X* = *Xenopus.* Some species, such as *L. tasmaniensis,* exhibit a relatively narrow temperature range for peak performance, whereas others, such as *R. pipiens* and *R. clamitans,* exhibit a relatively broad range of thermal independence. After Whitehead et al. (1989).

and Bennett 1981; K. Miller 1982; Hirano and Rome 1984; Londos and Brooks 1988; Whitehead et al. 1989; Wilson 2001; Gomes, Bevier, and Navas 2002).

The different results obtained in these studies probably relate to differences in experimental techniques. Both Tracy (1979) and Huey and Stevenson (1979) measured the average of three jumps taken by frogs prodded by the investigator. Both of the ranid frogs they studied rely heavily on anaerobic metabolism for bursts of activity, such as leaping away from a predator (see chapter 5). Since anaerobic metabolism is less affected by changes in temperature than aerobic metabolism, this probably accounts for the plateau in jump performance. The other investigators generally measured the average of more jumps or swimming movements during sustained locomotion, which is more likely to be supported by aerobic metabolism and therefore more likely to be affected by changes in temperature. Unfortunately, differences in the way in which locomotor performance has been measured complicate any attempt to generalize about temperature sensitivity in different species. Measures such as maximum jump distance, mean jump distance, and average distance moved per minute often show different patterns of temperature sensitivity, and it is not immediately obvious which type of measurement is the most ecologically relevant. In fact, all can be relevant, depending on whether one is interested in the animals' ability to escape from predators, search for food, or locate mates (Londos and Brooks 1988; Navas et al. 1999; Gomes et al. 2002).

### Communication and Sensory Systems

Any communication signals of amphibians that are produced by muscular action are likely to be strongly temperature dependent. For example, the frequency of tail beating and other courtship displays by alpine newts (*Triturus alpestris*) are strongly influenced by water temperature (Denoël 1998; Denoël, Mathieu, and Poncin 2005). The best-studied communication signals in relation to temperature are the vocalizations of anurans (see chapter 7). Different features of signals are affected to varying degrees by changes in temperature. Features affected by active muscle contraction, such as calling rate and pulse-repetition rate, are highly temperature dependent (fig 3.5), whereas features not involving active muscle contraction, such as dominant frequency, are much less affected by temperature (Zweifel 1968a; Schneider 1977, 1988; B. K. Sullivan 1982c, 1992; Gayou 1984; Wong et al. 2004).

Temperature also affects both the peripheral and central parts of the nervous system involved in call perception. In general, hearing sensitivity in anurans decreases at cold temperatures (Hubl, Mohneke, and Schneider 1977; Hubl and Schneider 1979; Mohneke and Schneider 1979; Walkowiak 1980b; Stiebler and Narins 1990; Narins 1995c). In addition, auditory fibers leading to the amphibian papilla, which detects low-frequency sounds, become selectively tuned to lower frequencies at low temperatures (fig. 3.6; Narins 2001a). Sometimes this can produce a mismatch between the call frequency preferred by females and the frequency of calls produced by males at the same temperature. For example, Gerhardt and Mudry (1980) reported a shift toward lower-frequency preferences in female *Hyla cinerea* at low temperatures when they were presented with calls differing in frequency, even though the dominant frequency of male calls changed very little with an equivalent change in temperature. In contrast, changes in the sensitivity of the brain of gray tree frogs (*Hyla versicolor*) to calls of different pulse rates shows a response to temperature similar to the pulse rates of the calls themselves (Brenowitz, Rose, and Capranica 1985; Rose, Brenowitz, and Capranica 1985). This results in females preferring calls with pulse rates similar to those produced by males at the same body temperature (Gerhardt 1978a).

## Ecological Effects of Temperature

### Adult Thermal Tolerance and Latitudinal Distribution

Amphibians are primarily inhabitants of tropical regions, although some groups, including several salamander families, have distributions centered in the temperate zone. About

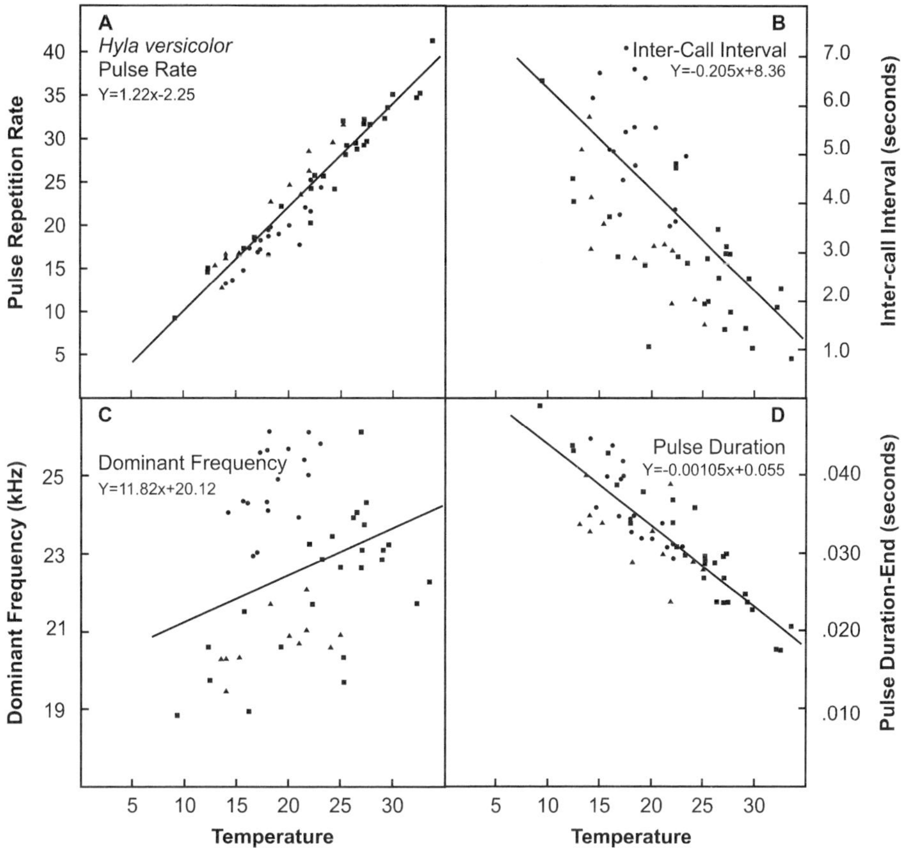


**Fig. 3.5.** Effect of temperature on various features of the calls of the gray treefrog (*Hyla versicolor*). (A) Pulse rate. (B) Intercall interval. (C) Dominant frequency. (D) Pulse duration. Circles are for 19 males recorded in the field. Squares are for 28 males recorded in a temperature-controlled chamber in the laboratory. Triangles are data for a single individual recorded in the laboratory at different temperatures. After Gayou (1984).

80–90% of all anuran species are found in the tropics. A single site in Amazonian Ecuador studied by Duellman (1978) had more species of anurans than do all of the United States and Canada. There are many possible reasons for these patterns of distribution, including past historical events, such as periodic glaciation of high-latitude regions, which undoubtedly reduced amphibian diversity in those areas. Nevertheless, it seems clear that these ectothermic animals do best in warm climates, and the latitudinal distribution of many species probably is limited by their ability to function or survive at low temperatures.

Latitudinal trends in thermal adaptation usually are studied by measuring thermal tolerance limits. The critical thermal maximum ($CT_{Max}$) is most frequently used as an index of high-temperature tolerance. $CT_{Max}$ has been measured in different ways in different studies. Many investigators have slowly heated animals in water until they list over and are no longer able to right themselves. Other investigators have used the onset of spasms as a more accurate measure of the point at which the animal would die if not removed from the high temperature (Hutchison 1961; Hutchison and Maness

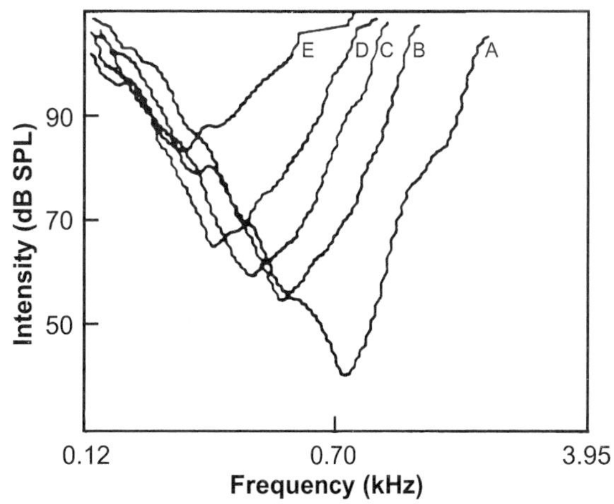


**Fig. 3.6.** Displacement of the tuning curve of an auditory nerve neuron with changing temperature in *Eleutherodactylus coqui*. The Y-axis shows the intensity of sound required to elicit a response from the neuron as a function of frequency. Decreasing temperature decreases the sensitivity of the ear and shifts the tuning curve toward lower frequencies. (A) 25° C, (B) 21° C, (C) 18° C, (D) 15° C, (E) 13° C. After Stiebler and Narins (1990).

1979). The onset of spasms usually occurs at temperatures 0.5–1.5° C above the temperatures at which loss of righting response occurs (V. H. Hutchison, personal communication). Unfortunately, the lack of standardization in measurement techniques makes comparisons among studies difficult. Measurement of critical thermal minima ($CT_{Min}$) is difficult because amphibians simply become more and more sluggish until no longer able to move at very low temperatures. Some authors have used lower lethal temperatures as a measure of cold tolerance. In comparisons of 53 species of anurans from seven families (Bufonidae, Ranidae, Hylidae, Leptodactylidae, Dendrobatidae, Pelobatidae, and Microhylidae), Brattstrom (1968) and Snyder and Weathers (1975) found a significant decline in critical thermal maximum ($CT_{Max}$) with increasing latitude, and an even greater decrease in lower lethal temperatures (fig. 3.7). The result is that tropical species typically tolerate a narrower range of temperatures than temperate-zone species and are particularly intolerant of low temperatures (fig. 3.8). Unfortunately, this comparison was not controlled for phylogenetic effects, so different families were represented in the tropical and temperate-zone samples. When these comparisons are made within families (hylids and bufonids) to control for phylogeny, there is some evidence for greater cold tolerance in temperate-zone species, but the data are limited. For example, a comparison of *Bufo regularis* and *Bufo viridis* from Egypt revealed that *B. viridis* is more tolerant of cold than is *B. regularis*. *Bufo viridis* has a largely temperate distribution, whereas *B. regularis* is a more tropical African species (Katz and Gil 1997).

The possible role of low temperature in limiting latitudinal distributions of some anurans is evident from the distributions of species that barely enter the southern United States. For example, several species of leptodactylid frogs are limited to southern Texas and Florida; all represent northern extensions of genera found mostly in the tropics (Stebbins 1985; Conant and Collins 1991). The introduced marine toad (*Bufo marinus*) is now common in southern Florida, but is limited in its ability to disperse northward because it cannot survive prolonged exposure to temperatures below 10° C (Krakauer 1970). In Guatemala and Texas, this toad is not found in regions with a mean minimum temperature below 15° C (L. C. Stuart 1951; Zug and Zug 1979). Even brief exposure to cold can be lethal for tropical frogs. Heyer et al. (1988) documented a precipitous decline in several species of frogs in the Atlantic coastal forest of southeastern Brazil that was attributed to an unusually severe frost.

The number of amphibian species decreases dramatically at high northern latitudes. In North America, most species of salamanders are not found farther north than southern Canada, although *Ambystoma laterale* reaches Labrador and *Taricha granulosa* is found along the coast in southern Alaska. Many North American anurans also have their northernmost limits in southern Canada, but *Bufo americanus* reaches Labrador, and both *Bufo boreas* and *Rana sylvatica* are found in Alaska. *Rana sylvatica* is the only North American frog found north of the Arctic Circle (Conant and Collins 1991; Pinder, Storey, and Ultsch 1992). In Europe, *Triturus cristatus, T. vulgaris, Bufo bufo,* and *Rana*

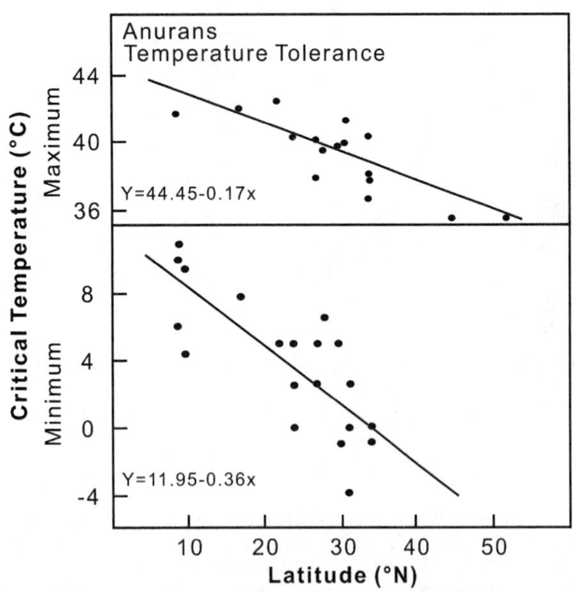


**Fig. 3.7.** Relationship of critical maximum temperature (top) and critical minimum temperature (bottom) to latitude for various species of anurans. The relationships have not been corrected for phylogeny and include tropical and temperate zone species from different families. After Snyder and Weathers (1975).

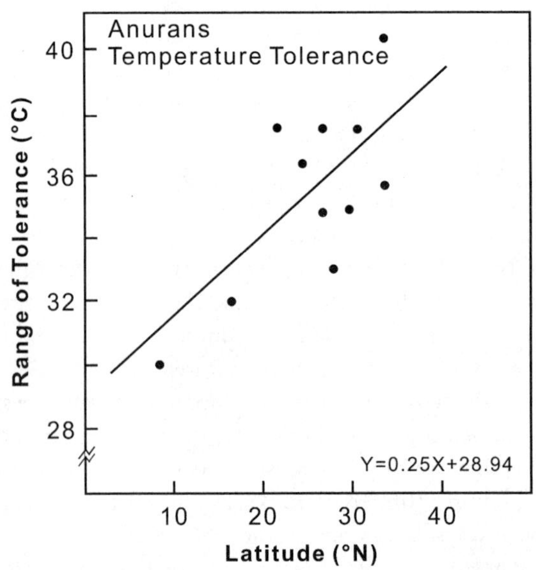


**Fig. 3.8.** Range of temperature tolerance (the difference between critical maximum and minimum temperatures) in anurans as a function of latitude. The data have not been corrected for phylogeny. The increased range of temperature tolerance at high latitudes is due mainly to increased tolerance for cold temperatures. After Snyder and Weathers (1975).

*arvalis* reach the central parts of Scandinavia, while *R. temporaria* lives north of the Arctic Circle. Only one urodele, the Siberian salamander (*Salamandrella keyserlingii*), is found north of the Arctic Circle (Arnold and Burton 1978; Pinder, Storey, and Ultsch 1992).

These far-northern amphibians not only must tolerate or avoid extreme cold in winter, but they also have short activity seasons and often are active at relatively low temperatures even in summer. These conditions can constrain feeding, growth, and reproduction (Forselius 1963; Koskela and Pasanen 1974, 1975; Bider and Morrison 1981; Sjögren, Elmberg, and Berglind 1988; Elmberg 1990). Indeed, several authors have suggested that the northern distribution of amphibians is limited not by low winter temperatures, but by the availability of a sufficient number of warm days during the summer activity period to allow for reproduction, feeding, and larval development (Wynne-Edwards 1952; Darlington 1957).

Among North American hylids, northern species that breed in early spring appear to have undergone adaptation to low temperature that allows them to be active at temperatures that would incapacitate their southern relatives, thereby prolonging their activity seasons (fig. 3.9; John-Alder, Morin, and Lawler 1988). In fact, most of the *Pseudacris* clade consists of species that breed in winter and early spring, a pattern that appears to be phylogenetically derived within North American hylids (Hedges 1986). Within-species comparisons of locomotion in northern and southern populations of *Pseudacris crucifer* did not reveal any difference in temperature sensitivity, however (John-Alder, Barnhart, and Bennett 1989). In contrast, R. S. Wilson (2001) reported that the jumping performance of *Limnodynastes peronii* from lowland tropical rainforests in northern Australia exhibited greater thermal sensitivity at low temperatures than did those from cooler temperate habitats in southern Australia.

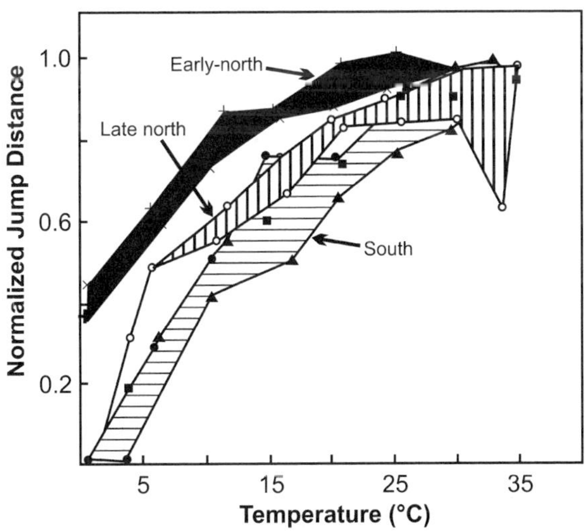


**Fig. 3.9.** Normalized jump distance as a function of temperature for early- and late-breeding northern populations and for southern populations of North American hylid frogs. Black area indicates northern species that breed in early spring (*Pseudacris crucifer* and *P. triseriata*). Vertical striping indicates northern species that breed in late spring and summer (*Hyla andersonii* and *H. versicolor*). Horizontal striping indicates southern species that breed at warm temperatures (*Acris gryllus, Hyla cinerea,* and *H. squirella*). Northern hylids exhibit greater jumping performance at low temperatures than do southern hylids, especially species of *Pseudacris* that breed at low temperatures. Jumping performance is similar for all species at warm temperatures. After John-Alder, Morin, and Lawler (1988).

## Embryonic Thermal Tolerance and Latitudinal Distribution

While the latitudinal distribution of some amphibians is limited by thermal tolerance, or by thermal sensitivity of adults, the effects of temperature on embryos probably is even more important. Early work by J. A. Moore (1939, 1942, 1944, 1949a, b) showed that embryos of species of *Rana* that breed in early spring, such as wood frogs (*R. sylvatica*), are less tolerant of high temperature, and more tolerant of low temperature, than are the embryos of warm-weather breeders such as green frogs (*R. clamitans*) and bullfrogs (*R. catesbeiana*). Differences in lower thermal tolerance are most pronounced, so wood frogs can breed at temperatures that are lethal to green frog or bullfrog embryos. Indeed, wood frog embryos can tolerate brief periods of supercooling to temperatures as low as −2°C, but prolonged exposure to freezing tempertures kills most embryos (Frisbie, Constanzo, and Lee 2000). Among anurans in general, thermal tolerance of embryos tends to be positively correlated with breeding-pond temperature, but thermal tolerance is not necessarily related to latitude in a simple way (R. Douglas 1948; Volpe 1953, 1957; Ballinger and McKinney 1966; L. Licht 1971; H. A. Brown 1967, 1975a, b; Zweifel 1968b, 1977; Bachmann 1969; McLaren and Cooley 1972; Kuramoto 1975, 1978).

In general, intraspecific geographic variation in embryonic thermal tolerance is small relative to interspecific variation (J. A. Moore 1939, 1942, 1949b; Volpe 1953, 1957; Herreid and Kinney 1967; H. Brown 1975b). J. A. Moore (1949a) and others documented geographic variation in embryonic thermal tolerance in *Rana pipiens,* but later taxonomic work showed that these populations comprise a complex of closely related species (Hillis 1988). The same is true for the differences in embryonic thermal tolerance reported by H. A. Brown (1967) for populations of spadefoot toads. In California, *Spea hammondii* breeds after cool winter rains and embryos cannot tolerate high temperatures. In Arizona, populations once thought to be members of the same species breed after warm summer rains, and their embryos can tolerate higher temperatures. This population is now considered to represent a separate species, *Spea multiplicata* (H. A. Brown 1976b). The small differences among populations of the same species, even over wide geographic areas, suggest that evolutionary changes in embryonic thermal tolerance occur slowly compared to the rate at which populations disperse into new habitats. Consequently, it is not unusual for cold-adapted amphibians to breed at temperatures close to their lower thermal limits, while warm-adapted species often breed at temperatures near their upper thermal limits (Bachmann 1969). This means that embryonic thermal tolerance probably limits the latitudinal distributions of some species.

### Altitudinal Distribution

Amphibians at high altitudes face the same problems encountered by those at high latitudes, including long, cold winters and short activity seasons with relatively cool temperatures. In addition, high-altitude environments are subject to wide fluctuations in temperature on a daily basis and often are relatively dry during summer activity seasons (Navas 1997, 2002, 2006). For aquatic breeders, altitudinal distribution also often is limited by the availability of ice-free ponds for larval development, especially for species with prolonged larval periods (Vences, Grossenbacher et al. 2003). Zschokke (1900) reported that anurans in the Alps with relatively short larval periods reach higher altitudes than do those with longer larval periods. All of these factors make high altitudes generally inhospitable to amphibians, and relatively few species live there (Hock 1964). Nevertheless, some amphibians manage to thrive even in these harsh conditions. In the western United States, three pond-breeding salamanders, *Ambystoma tigrinum, A. macrodactylum,* and *A. gracile,* live at elevations of up to 3,000 m. Three species of plethodontid salamanders are found above 2,400 m (*Aneides hardii, Hydromantes platycephalus,* and *Plethodon neomexicanus*). Anurans in this region are somewhat more successful at high altitudes. All five western species of *Pseudacris* and *Hyla* reach elevations of at least 2,300 m, and one, *P. triseriata,* is found above 3,600 m. At least five species of toads (*Bufo*) are found above 2,300 m, and two, *B. canorus* and *B. boreas,* above 3,600 m. Ten species of ranid frogs reach elevations of at least 2,400 m, and four (*Rana sylvatica, R. pretiosa, R. muscosa,* and *R. pipiens*) are found above 3,000 m (Stebbins 1985).

European amphibians generally are found at somewhat lower maximum elevations than are those in North America. In the Pyrenees, the limit appears to be about 2,500 m, whereas in the Alps, some species may live as high as 2,742 m (Vences, Grossenbacher et al. 2003). In Asia, the Himalayas support a rich and diverse amphibian fauna, including hynobiid salamanders and anurans in several families. Although fewer species are found at extreme altitudes than at lower elevations, some Himalayan amphibians are commonly found above 3,700 m, including hynobiid salamanders in the genus *Batrachuperus* and several bufonid, megophryid, rhacophorid, and ranid frogs (Liu 1950; Dubois 1974a, b). One megophryid frog, *Scutiger boulengeri* (= *S. alticola*), is reported to live at nearly 5,200 m, probably the highest elevation for any amphibian in the world (Hock 1964). Tropical mountains also tend to have fewer amphibian species at high elevations than at lower levels, although some groups actually are more diverse at mid-elevations than in the lowlands (Navas 2003 , 2006). Often the faunas of high elevations in the tropics are small subsets of faunas found at lower elevations. For example, one area in Colombia at 3,500 m has one species each of *Atelopus, Colostethus, Eleutherodactylus,* and *Hyla,* genera that are much more diverse at lower elevations (Lüddecke 1997; Navas 1997). Other high-elevation sites in the region have similar small amphibian faunas (J. D. Lynch 2001).

There has been relatively little work on the thermal physiology and ecology of high-altitude amphibians, although there is some work on thermoregulatory behavior (see discussion that follows). A number of authors have reported both interspecific and intraspecific differences in $CT_{Max}$ for amphibians at different altitudes. Low-elevation species or populations generally tolerate slightly higher temperatures than do high-elevation forms (Brattstrom 1968; Delson and Whitford 1973b; K. Miller and Packard 1977; Hoppe 1978; Christian et al. 1988). Rome, Stevens, and John-Alder (1992) argued that amphibians rarely experience temperatures in the field that approach their critical thermal maxima and questioned the significance of these small differences in $CT_{Max}$. For example, J. H. Howard, Wallace, and Stauffer (1983) found that maximum water temperatures in ponds inhabited by *Ambystoma macrodactylum* did not exceed 25° C, but $CT_{Max}$ of several populations was between 33° C and 34° C. Similarly, air temperature in the Puerto Rican lowland forest does not exceed 31° C, considerably lower than the $CT_{Max}$ of *Eleutherodactylus coqui* from low-elevation sites (37.2° C) or *E. portoricensis* from higher elevations (36.3° C; Christian et al. 1988). Nevertheless, individual frogs might be exposed to higher temperatures for short periods of time when sitting in direct sunlight, and it is during these rare but dangerous events that they would benefit from a higher $CT_{Max}$ (V. H. Hutchison, personal communication).

Studies of activities such as calling and locomotion provide a better indicator of how temperature affects the performance of amphibians under natural conditions. Navas (1996c, d) investigated the temperature dependence of swimming in four species of frogs from the high Andes of Colombia and compared their performance with that of congenerics from tropical-lowland populations in Panama. He found

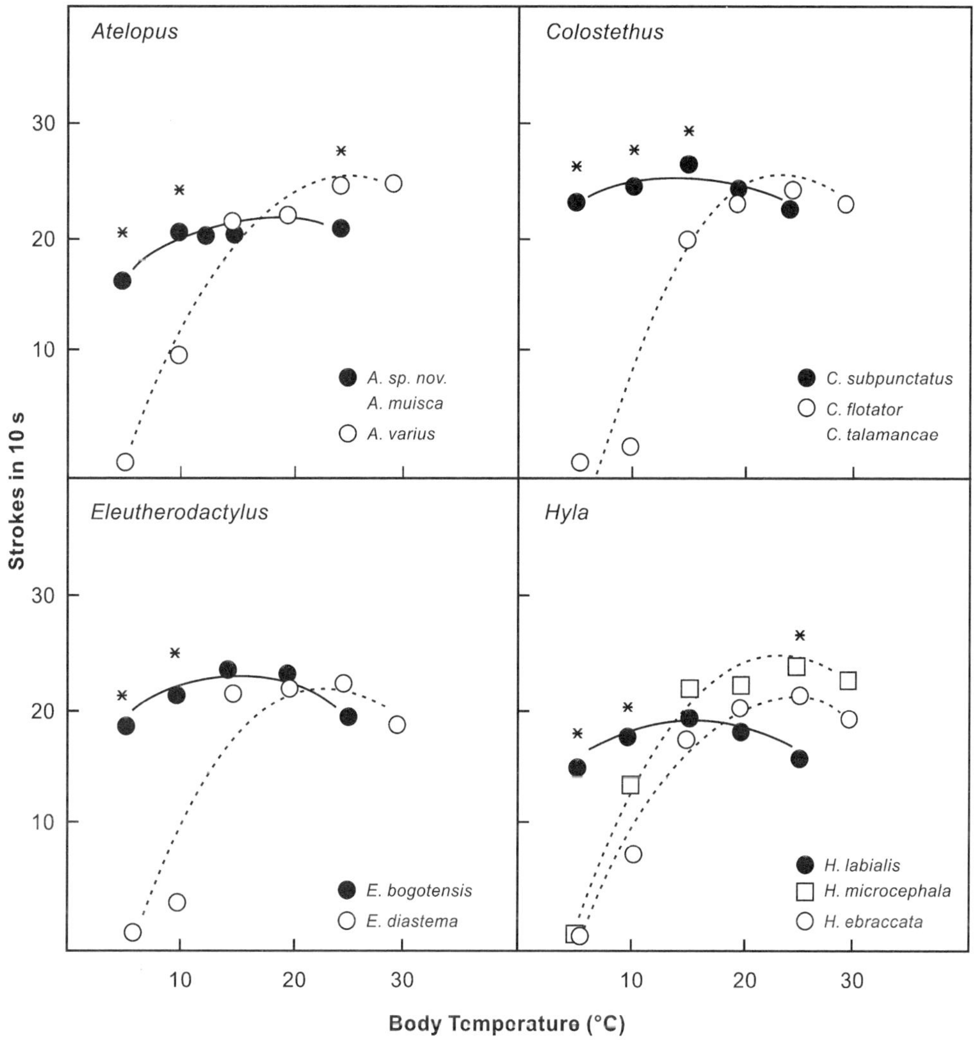


**Fig. 3.10.** Swimming performance as a function of temperature for high-elevation frogs from the Colombian Andes (dark circles) and for low-elevation congeneric species from Panama (open circles). Asterisks indicate temperatures at which significant differences were found between high- and low-elevation species. Data were pooled for *Atelopus muisca* and an undescribed species from the Andes, because they were not significantly different. Data also were pooled for *Colostethus flotator* and *C. talamancae* from Panama. High-elevation species in all four genera exhibit a much greater capacity for locomotion at low temperatures than do low-elevation congenerics. After Navas (1996c).

that populations of all four genera (*Atelopus, Colostethus, Eleutherodactylus,* and *Hyla*) from high-elevation sites could swim at temperatures as low as 5° C, whereas frogs from Panama were largely incapacitated at temperatures below 15° C (fig. 3.10). The enhanced swimming performance of high-elevation frogs was correlated with much higher metabolic scopes at low temperatures. Because the high-elevation species must have had low-elevation ancestors, their enhanced swimming and metabolic performance at low temperatures are derived physiological traits that have evolved independently in the four genera. Calling rates were more difficult to compare because of differences in call structure, but in general, high-elevation frogs called at lower rates than did low-elevation congenerics (Navas 1996a, d). The same was true for different populations of *Hyla labialis,* a tree frog found at a wide range of altitudes in Colombia. There was evidence in this species, however, for reduced thermal sensititivity in pulse-repetition rate at high altitudes (Lüddecke and Sánchez 2002). Daily patterns of calling activity are similar in high- and low-elevation species within genera (largely diurnal in *Colostethus,* nocturnal in *Eleutherodactylus* and *Hyla*). However, males in a population of *Eleutherodactylus bogotensis* from 3,500 m called for only about 1 hour after sunset, because temperatures dropped to 5° C or lower after that. In contrast, males in a low-elevation population of *Eleutherodactylus diastema* in Panama experienced little change in temperature and called from sunset to sunrise.

## Emergence and Timing of Breeding

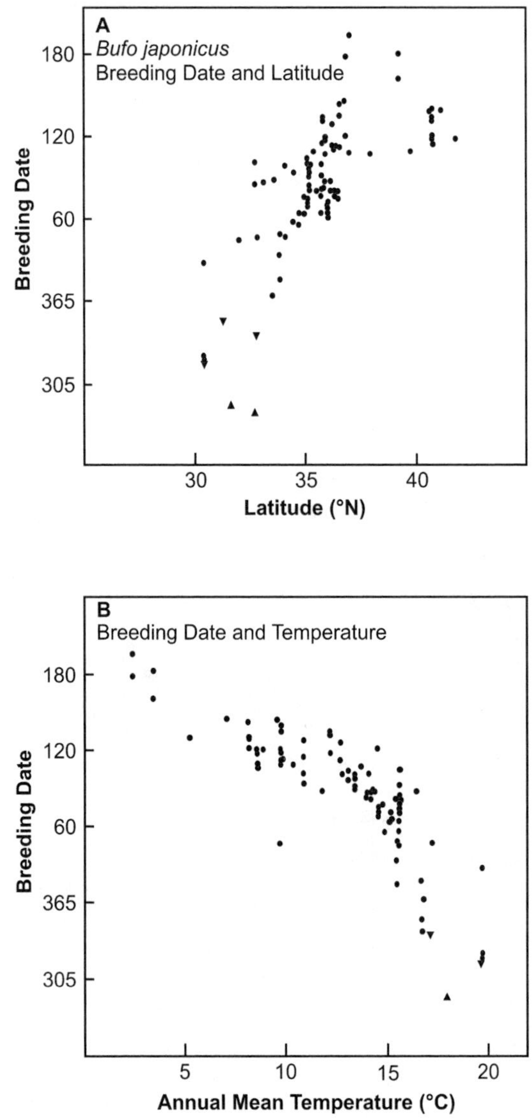


**Fig. 3.11.** Relationship of first breeding date to latitude (A) and temperature (B) for populations of the Japanese toad (*Bufo japonicus*). Toads begin breeding in late winter at low latitudes, but do not breed until early summer at high latitudes (A), presumably because of much lower average temperatures (B). Date 1 = January 1. After Matsui (1989).

For many species of amphibians, emergence from hibernation and the beginning of breeding activity are closely tied to rising temperatures, although rainfall can be an important cue as well (Eibl-Eibesfeldt 1950; Rühmekorf 1958a; W. F. Blair 1960, 1961; Heusser 1960; Packer 1960; Forselius 1963; Hock 1967; L. Licht 1969b; Oldham 1969; Geisselmann, Flindt, and Hemmer 1971; van Gelder and Hoedemaekers 1971; Koskela and Pasanen 1974; Salthe and Mecham 1974; Obert 1975; Uwa et al. 1981; Wisniewski, Paull, and Slater 1981; Sofianidou and Kyriakopoulou-Sklavounou 1983; Bea, Rodriguez-Teijeiro, and Jover 1986; Beneski, Zalisko, and Larsen 1986; Radwan and Schneider 1988; Schneider, Sofianidou, and Kyriakopoulou-Sklavounou 1988; Kusano and Fukuyama 1989; Asimakopoulos, Sofianidou, and Schneider 1990; Elmberg 1990; Kyriakopoulou-Sklavounou and Kattoulas 1990; Le Garff 1998; Reading 2003). There is little evidence for geographic variation in thermal sensitivity within populations of the same species. Consequently, species found over a wide range of latitudes or altitudes usually exhibit marked geographic or altitudinal variation in the timing of breeding activity. For example, spring peepers (*Pseudacris crucifer*) and other members of the genus *Pseudacris* typically start breeding during the winter in southern parts of their ranges, but breed later in the spring in the north (John-Alder, Barnhart, and Bennett 1989). High-altitude populations of *Pseudacris* breed later than do lowland populations (Pettus and Angleton 1967). Wood frogs (*Rana sylvatica*) are winter breeders in the south, but are late spring or summer breeders in the north (Martof and Humphries 1959; Herreid and Kinney 1967). High-altitude populations breed later than do lowland populations at the same latitude (Berven 1982a).

Similar patterns are found in Old World anurans. The Japanese toad (*Bufo japonicus*) is found throughout Japan from about 30° N latitude to 42° N latitude. In southern Japan, this species breeds in winter (November and December), but in northern Japan, some populations do not breed until June (Matsui 1989). Breeding date is significantly correlated with mean annual temperature (fig. 3.11). In Britain, common frogs (*Rana temporaria*) are found over a wide range of altitudes (Beattie 1985). Breeding date is positively correlated with altitude (fig. 3.12 A) and negatively correlated with mean winter temperature (fig. 3.12 B). Common frogs also are found over a wide range of latitudes, from southern Europe—where they live mostly in the mountains—to northern Finland (64° N latitude). Frogs in southern Europe breed much earlier in the year than do those from northern parts of their range (Koskela and Pasanen 1975). In a study based on 140 years of breeding records, Terhivuo (1988) showed that common frogs in southern Finland begin breeding in late April, whereas those in the far north breed at the end of May (fig. 3.13). Similarly, *Rana dalmatina* begins breeding in late January or early February in Greece, late February or early March in Romania, and early April in Sweden (Hartel 2003). Local thermal conditions unrelated to latitude or altitude can affect breeding phenology as well. A population of the rhacophorid frog *Buergeria japonica* from a geothermal hot spring in Taiwan breeds throughout the year, whereas stream-breeding populations do not breed in the winter months (T. Chen, Kam, and Lin 2001). Tadpoles in this population exhibit greater tolerance of higher temperature than do those from cooler habitats (Wu and Kam 2005).

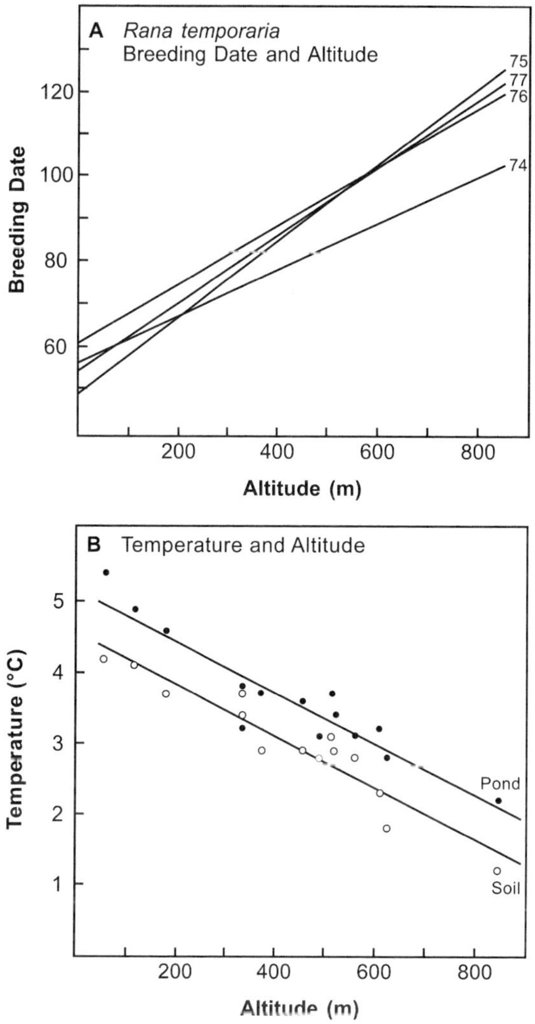


**Fig. 3.12.** (A) Relationship of breeding date to altitude for populations of the common frog (*Rana temporaria*) in Britain. (B) Relationship of average soil and water temperature to altitude. Data in (A) are shown for four different years (1974–1977). Average temperatures decrease with increasing altitude, and this is correlated with much later breeding. After Beattie (1985).

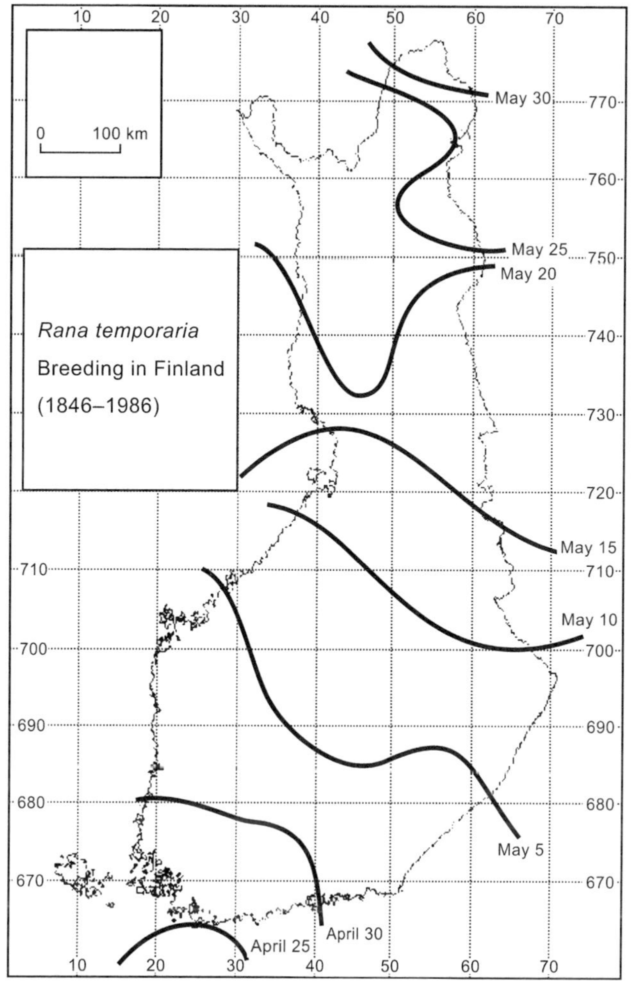


**Fig. 3.13.** Map showing changes in breeding dates for common frogs (*Rana temporaria*) in Finland as a function of latitude. Isobar lines show average first breeding date for a 140-year period (1846–1986). Breeding in northern Finland is delayed by about a month compared to populations in southern Finland. Breeding in most regions of Finland has started earlier in recent decades because of climate warming. After Terhivuo (1988).

With the general warming of the world's climate in recent decades, there is some evidence that the onset of breeding for some temperate-zone amphibians has shifted to earlier dates. For example, Terhivuo (1988) found that breeding of *Rana temporaria* in all parts of Finland has become progressively earlier over the past 140 years due to gradual warming of the climate. Similar trends toward earlier breeding were observed in two species of anurans and three newts over a 17-year period in England, but not for *Rana temporaria* (Beebee 1995a; Forchhammer, Post, and Stenseth 1998). A 19-year study of a population of *Bufo bufo* in southern England showed that year-to-year variation in timing of breeding was correlated with mean winter temperature in the 40 days immediately before breeding began. This study did not find a clear trend toward earlier breeding over the course of the study, although five of the earliest breeding events occurred in the last 10 years (Reading 1998). Other studies have found that some species appear to be breeding earlier, while others are not (Blaustein et al. 2001; Gibbs and Breisch 2001; Tryjanowski, Rybacki, and Sparks 2003).

For many anuran species, the level of calling activity and the number of pairs laying eggs also are affected by daily fluctuations in temperature. Most species appear to have a low-temperature threshold below which they become inactive. This can result in breeding seasons that otherwise would be continuous being divided into a series of short episodes of breeding activity when temperatures are favorable, separated by periods of inactivity in cold weather (Wahl 1969; Obert 1975). This is most likely to occur in early spring breeders, which tend to experience considerable variation in temperature (Sofianidou and Kyriakopoulou-Sklavounou 1983; Schneider, Sofianidou, and Kyriakopoulou-Sklavounou 1988; Kusano and Fukuyama 1989; Asimakopoulos, Sofianidou, and Schneider 1990). It also can occur in the summer, but periods of inactivity tend to be shorter (R. D. Howard 1978a; Radwan and Schneider 1988; Sjögren, Elmberg, and Berglind 1988; Kyriakopoulou-Sklavounou and Kattoulas 1990).

### Choice of Calling and Oviposition Sites

Within breeding ponds, water temperature can have a considerable influence on where males call and where females lay their eggs. For example, *Rana dalmatina* breeds in late January or early February in Greece. Males call mostly underwater early in the breeding season, but move to the surface as air temperature rises (Schneider, Sofianidou, and Kyriakopoulou-Sklavounou 1988). In a population of *Rana lessonae* in Sweden, males consistently chose the warmest locations in the breeding pond to call, and eggs were deposited in the same areas. Males moved from one part of the pond to another depending on the prevailing wind direction and local changes in water temperature (Sjögren, Elmberg, and Berglind 1988). Adams and Frissell (2001) found some clutches of eggs of *Ascaphus montanus,* a cold-adapted stream-breeding frog, in locations near cold groundwater seeps that reduced the temperature of the surrounding water, suggesting active selection of cooler sites for reproduction. In a population of bullfrogs (*Rana catesbeiana*) in Michigan, choruses moved to successively cooler locations as the season progressed; those in early June were in the warmest part of the pond, whereas late-July choruses were in the coolest part (R. D. Howard 1978a). Female bullfrogs avoid depositing eggs in very warm water (above 32° C) that can produce developmental abnormalities in the embryos (R. D. Howard 1978b).

Male bullfrogs are territorial, so choice of calling and oviposition sites are closely linked. In some winter- or spring-breeding ranids, pairs in amplexus move away from calling sites before depositing eggs. They often clump egg clutches together in large communal masses, a behavior generally not seen in warm-weather breeders (Bragg 1953, 1954a; Heusser 1961; Savage 1961; Guyétant 1966; Herreid and Kinney 1967; Zweifel 1968a; L. Licht 1969b, 1971; Hassinger 1970; Beattie 1980). Communal egg masses appear to provide a thermal advantage to the eggs by retaining heat longer than isolated masses as temperatures drop (Seale 1982b; Waldman 1982a; Waldman and Ryan 1983). Communal egg masses often are deposited in very shallow water, exposing them directly to the warmth of the sun during the day (see fig. 10.4 A in chapter 10). Wood frogs (*Rana sylvatica*) consistently place their egg masses in locations where growth and development of tadpoles is enhanced compared to other sites available in the same ponds (Freidenburg 2003). In some species, oviposition behavior changes in response to changes in ambient temperature. For example, J. Caldwell (1986b) found that southern leopard frogs (*R. utricularia*) breeding in warm weather usually deposited egg masses individually, whereas those breeding in cold weather often deposited eggs in communal masses.

## Behavioral Thermoregulation

Ectotherms must rely mainly on behavioral adjustments to regulate their body temperatures. This is particularly true of amphibians, which have very limited abilities to alter rates of heat loss and gain physiologically. Following the publication of the classic paper of Cowles and Bogert (1944) on thermoregulation in desert reptiles, studies of behavioral thermoregulation became a major focus of herpetological research (Avery 1982; Huey 1982). Studies on amphibians have lagged behind those on reptiles, perhaps because overt thermoregulatory behavior is less obvious in amphibians than in reptiles such as desert lizards. In fact, there has been some controversy over whether amphibians engage in active thermoregulation in the field. In part, this results from different definitions of the term thermoregulation. Some workers have defined thermoregulation as an ability to maintain body temperatures above that of the surrounding environment, but probably the best definition is an ability to maintain body temperatures within a range that is narrower than the range of environmental temperatures available to the animal (Hutchison and Dupré 1992).

The precision of thermoregulation varies considerably among species, depending on the ecological circumstances in which they find themselves. Behavioral thermoregulation not only has obvious benefits in enabling an animal to select temperatures that are most favorable for feeding, digestion, growth, locomotion, or reproduction, but it also has costs, such as increased exposure to predation, increased energy expenditure, and decreased time available for other activities. The degree to which a given species engages in active thermoregulation represents a balance between these costs and benefits (Huey and Slatkin 1976; Huey 1982). For example, most desert lizards thermoregulate very precisely. The cost of not doing so can be high, because these animals often are exposed to near-lethal environmental temperatures. In contrast, some tropical rainforest lizards are effectively thermoconformers, allowing their body temperatures to vary with environmental temperature. For these animals, the cost of moving between widely spaced sunspots is greater than the benefit of maintaining temperature within a very narrow range, especially when they can function reasonably well over the entire range of temperatures to which they are normally exposed. The same probably is true for many amphibians. Species that choose a relatively narrow range of temperatures in laboratory thermal gradients often exhibit a much wider range of body temperatures in the field, either because the preferred temperature is not available, or because the cost of shuttling between microhabitats is too great (Hutchison and Dupré 1992).

### Thermoregulation in Water

Because the thermal conductivity of water is much greater than that of air, amphibians that are completely immersed in water seldom have body temperatures that differ significantly from that of the surrounding environment (Spotila, O'Connor, and Bakken 1992). Furthermore, heating and cooling rates tend to be very rapid, especially for small-bodied species. While some reptiles can make physiological adjustments that alter heating and cooling rates (Bartholomew 1982), this does not appear to be true for most amphibians. For example, two relatively large-bodied salamanders, *Amphiuma means* and *Necturus maculosus,* and two frogs, *Rana catesbeiana* and *R. pipiens,* all heat and cool at similar rates in water (Tripp and Lustick 1974; K. A. Anderson and Beitinger 1979; Bohlin and Beitinger 1979; Tracy, Tracy, and Turner 1992). Consequently, the only way that an aquatic amphibian can alter its body temperature is to move between microhabitats in response to local differences in water temperature.

Tadpoles often exhibit clear preferences for certain temperatures when tested in laboratory thermal gradients (Workman and Fisher 1941; Rühmekorf 1958b; Herreid and Kinney 1967; Lucas and Reynolds 1967; de Vlaming and Bury 1970; Casterlin and Reynolds 1978; Hutchison and Hill 1978; Dupré and Petranka 1985; Wollmuth et al. 1987; Wollmuth and Crawshaw 1988; see summary of data in table 9.2 of Hutchison and Dupré 1992). Studies of larval newts and ambystomatid salamanders have produced similar results (P. Licht and Brown 1967; Heath 1975; Keen and Schroeder 1975; Stauffer, Gates, and Goodfellow 1983; Dupré and Petranka 1985). In most laboratory studies tadpoles or salamander larvae were presented with a choice of temperatures from near freezing to between 25° and 35° C. The animals generally chose a range of temperatures somewhere in the middle (between 15° and 25° C), avoiding both very cold and very warm temperatures. However, tadpoles of the tailed frog (*Ascaphus truei*) from cold streams of the northwestern United States selected much lower temperatures (5–9° C) than did the other species tested. Tadpoles confronted with very high temperatures, such as those found in thermal springs, are capable of avoiding near-lethal temperatures (Wu and Kam 2005).

Both acclimation temperature and stage of development can affect the range of temperatures selected by amphibian larvae (Hutchison and Dupré 1992). In bullfrog tadpoles, for example, preferred temperature tends to increase both with increasing acclimation temperature and with stage of development, although the latter trend is not consistent at all acclimation temperatures (Hutchison and Hill 1978; Wollmuth and Crawshaw 1988). Preferred temperature has been reported to increase with stage of development in several other species of amphibian larvae, including *Rana cascadae, R. utricularia, Bufo marinus, B. americanus, Pseudacris triseriata,* and *Ambystoma texanum* (Floyd 1984; Keen and Schroeder 1975; Dupré and Petranka 1985; Wollmuth et al. 1987). This pattern was not seen in *Taricha rivularis,* however (P. Licht and Brown 1967). In some species, preferred temperature increases during most of development, but then drops significantly at metamorphic climax. As a result, the larvae select the warmest temperatures just before metamorphosis, thereby ensuring rapid development when larvae are racing to reach metamorphosis before temporary ponds dry up (see chapter 13) and when they are particularly vulnerable to predators (see chapter 14; Dupré and Petranka 1985; Hutchison and Dupré 1992).

How closely do the temperatures selected by amphibian larvae in laboratory gradients reflect normal activity temperatures and the precision of thermoregulation in the field? Unfortunately, this question is hard to answer, because reliable data on thermoregulation of larvae in the field are scarce. Since these animals generally are too small for temperature to be monitored by implanted thermocouples or temperature-sensitive transmitters, it has been nearly impossible to collect data on changes in body temperature of individual animals over time. Most field data consist of single measurements of body temperatures, or measurements of water temperature in areas where aggregations of larvae were observed (e.g., Rühmekorf 1958b; Brattstrom 1962c, 1963; P. Licht and Brown 1967; J. D. Anderson 1968b; Heath 1975; Wollmuth et al. 1987). Often such measurements reveal a rather wide range of temperatures over which larvae are active, but provide little information on the precision of thermoregulation. The best field evidence for active thermoregulation in aquatic larvae comes from studies showing a tendency for larvae to aggregate in the warmest available parts of the habitat. Nearly all such examples are species that breed at high altitudes in relatively cool environments, or in shallow, temporary ponds (Brattstrom 1962c; Heath 1975; Wollmuth et al. 1987). In either case, rapid growth and development would be advantageous and would be promoted by selection of warm microhabitats.

Information on thermoregulation by adults of aquatic amphibians is meager. There has not been a single detailed study of thermoregulation in a fully aquatic frog, and only a handful of such studies on salamanders and newts. P. Licht and Brown (1967) found little evidence of precise temperature selection by *Taricha rivularis* breeding in a stream, although they did avoid very warm temperatures near hot springs feeding into the stream. Body temperatures of newts in an aquatic gradient in the laboratory were monitored with implanted thermocouples and exhibited rapid fluctuations as the animals moved through the gradient. Although they did not show a strong preference for a narrow range of temperatures, they did avoid raising their body temperatures above 20° C. The newts did not show a clear avoidance of very low temperatures and sometimes climbed onto ice at the cold end of the gradient. In contrast, adult *Ambystoma tigrinum* from high-altitude ponds in Colorado tended to avoid cold water in the field and were consistently found in the warmest sections of the ponds, moving into shallow water during the day and into deeper water at night (Heath 1975). In a laboratory gradient, adults selected a rather wide range of temperatures between about 12° and 26° C and were somewhat less precise thermoregulators than were larvae.

Hutchison and Hill (1976) studied temperature selection by two large aquatic salamander species, *Cryptobranchus alleganiensis* and *Necturus maculosus,* in a laboratory thermal gradient. In both species, acclimation temperature had a significant effect on preferred temperatures. Salamanders acclimated to 5° C tended to select the lowest temperatures available in the gradient (down to about 4° C) and almost always avoided temperatures above 18° C. Those acclimated to 15° C and 25° C selected somewhat higher temperatures, but also were less precise in their thermoregulation. *Necturus* also exhibited a daily shift in preferred temperature, becoming more active and selecting higher temperatures at night than they did during the day. This suggests that when these salamanders are relatively inactive during daylight hours, they retreat to deeper, cooler water and allow their body temperatures to fluctuate with that of the environment (Hutchison and Spriestersbach 1986). Unfortunately, we have little detailed information on temperature selection in the field for either of these species, but both are found mainly in cool environments. Adams and Frissell (2001) found evidence of seasonal movements of *Ascaphus montanus* adults into cool water in streams where summer temperatures exceeded 16° C, but not in streams that remained cool all year.

### Thermoregulation on Land

Thermoregulation on land is considerably more complex than thermoregulation in water because there are more avenues of heat loss and heat gain to be considered (fig. 3.14). Of particular importance is the interaction between temperature and evaporative water loss in animals with a highly permeable skin (Tracy 1975, 1976; Spotila, O'Connor, and Bakken 1992; Tracy, Tracy, and Turner 1992; Bartelt and Peterson 2005). A terrestrial amphibian often is severely constrained in its choice of activity temperatures because preferred temperatures are not available in suitably moist microhabitats. For example, terrestrial salamanders often choose relatively narrow ranges of temperatures in laboratory gradients, but choices can be modified by the availability of cover and hydration state. They choose temperatures higher than those normally preferred if suitable cover is available only at the warm end of the gradient. Dehydrated animals prefer cooler temperatures than those chosen by fully hydrated individuals, thereby reducing rates of evaporative water loss (P. Licht and Brown 1967; Spotila 1972; Feder and Pough 1975).

Terrestrial salamanders have perhaps the simplest form of thermoregulation in the field because they do not engage in overt thermoregulatory behavior such as basking. Moisture probably is more important than temperature preference in selection of microhabitats, but in general, the wet microhabitats favored by salamanders also are relatively cool. Consequently, it is unusual for terrestrial salamanders to be found at temperatures exceeding 25° C, even in the tropics, and typical activity temperatures for most species are below 15° C (Brattstrom 1963; Spotila 1972; Feder 1982b; Feder and Lynch 1982; Feder et al. 1982). Nevertheless, there is limited evidence for active selection of temperatures in the field. For example, Feder (1982b) reported finding individuals of a small Neotropical salamander, *Thorius narisovalis,* under the bark of logs at temperatures up to 6° C warmer than the coolest available temperatures. Even so, the warmest temperature recorded for one of these salamanders was only about 21° C. In contrast, it is not unusual for tropical frogs to have body temperatures above 30° C, and average activity temperatures for many anurans are around 25° C (Hutchison and Dupré 1992). Because of their low activity temperatures and generally slow-moving way of life, terrestrial salamanders, especially plethodontids, have extremely low metabolic rates and the lowest energy requirements of any terrestrial vertebrates (Feder 1983a; see also chapter 5).

Like terrestrial salamanders, many frogs and toads exhibit clear preferences for relatively narrow ranges of temperature when tested in laboratory thermal gradients (Strübing 1954; Cabanac and Jeddi 1971; Lillywhite 1971a; Claussen 1973a; Carey 1978; Gatten and Hill 1984; Smits 1984; Shoemaker et al. 1987; Shoemaker, Baker, and Loveridge 1989; Wollmuth et al. 1987; Malvin and Wood 1991; Tracy et al. 1993; Lüddecke 1995b; Katz, Hoffman, and Gil 1997; see data summary in table 9.2 of Hutchison and Dupré 1992). Preferred temperatures in laboratory thermal gradients

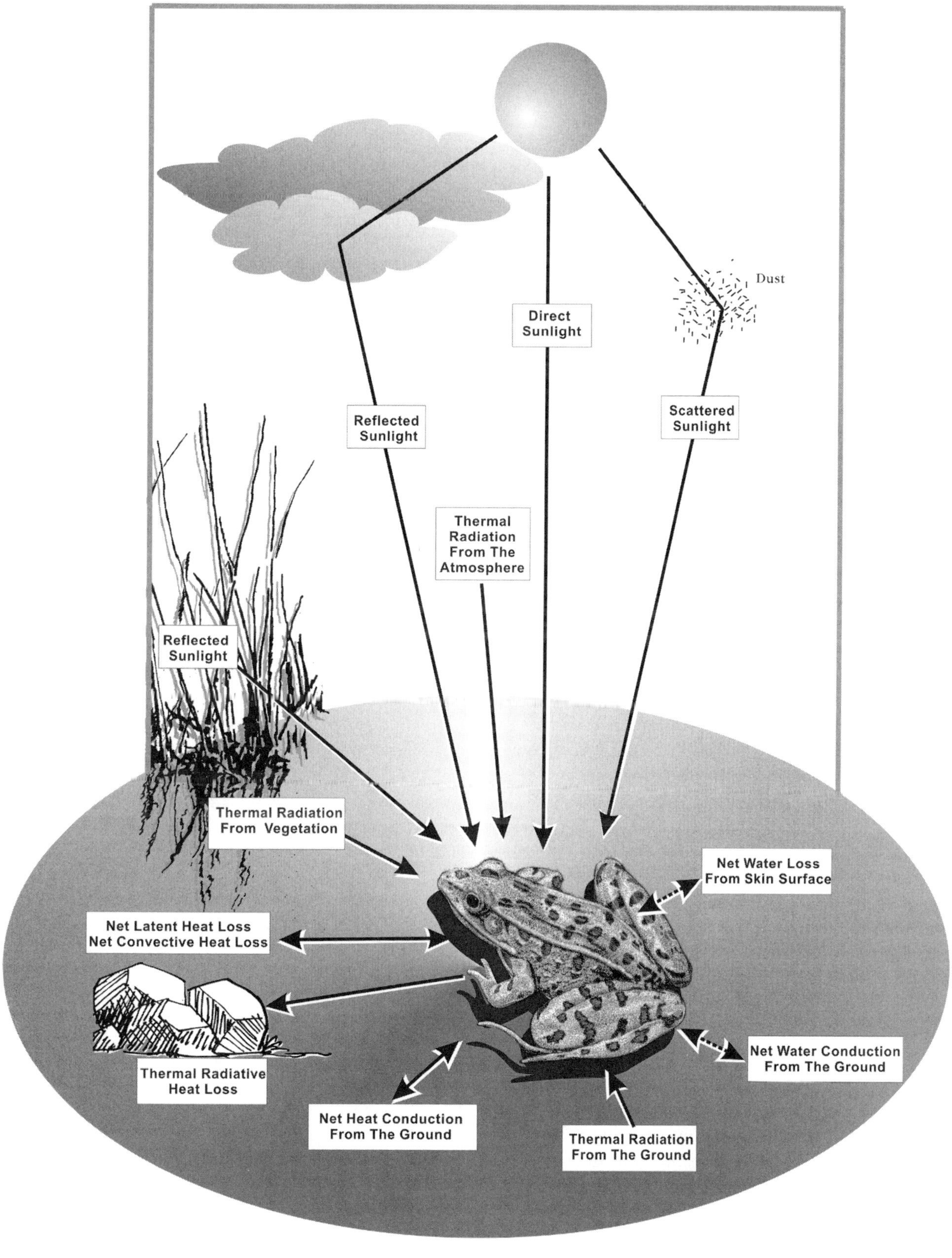


**Fig. 3.14.** Avenues of heat loss and gain (solid arrows) and water loss and gain (dashed arrows) for a leopard frog (*Rana pipiens*) in a meadow. Drawing by Mary Jane Spring after Tracy (1976).

sometimes are correlated with activity temperatures in the field, but often the range of activity temperatures in the field is wider. For example, many anurans select temperatures in laboratory gradients between 26° and 30° C, regardless of the actual temperatures available to these animals in the field (Hutchison and Dupré 1992; Lüddecke 1995b; Katz, Hoffman, and Gil 1997). Most anurans are less secretive than are terrestrial salamanders, which tend to venture from thermally buffered retreat sites only at night or during rainy weather. Consequently, anurans frequently are exposed to major fluctuations in environmental temperature on both a daily and seasonal basis (Carey 1978). Often the body temperatures of these animals simply track ambient temperatures, but many anurans engage in more obvious thermoregulatory behavior than do terrestrial salamanders.

Some anurans regulate their body temperatures by moving between microhabitats that differ in temperature. Sinsch (1984) investigated this type of behavior in three species of European ranid frogs. He put the frogs in a large aquarium with shallow water at one end and soil at the other. The terrestrial section was covered with grass and other plants and provided with boxes for shade. He tested animals during the day and at night. During the day, a light bulb suspended over the soil provided a thermal gradient of about 8° C between the cool aquatic end and the warmer terrestrial end of the tank. The responses of frogs were tested at four different ranges of temperatures (coolest 8–16° C; warmest: 24–31° C).

Sinsch found that *Rana temporaria,* the species with the northernmost distribution, generally preferred the coolest temperatures and could not tolerate the highest range of test temperatures for an extended period. These frogs shifted activity periods from mostly nocturnal at warm temperatures to mostly diurnal at cool temperatures. This is the most terrestrial of the three species outside of the breeding season, and these frogs spent most of their time on land except when retreating to the cooler water at very warm temperatures. The other two species, *R. lessonae* and *R. ridibunda,* have a more southern distribution and both tolerate and prefer higher temperatures than does *R. temporaria. Rana lessonae,* which inhabits the shorelines of ponds, remained on land most of the time, but moved to the water at cool temperatures. While on land, they often sat directly exposed to heat from a lamp. In contrast, *R. ridibunda,* the most aquatic of the three species, spent considerably more time in the water, especially at warmer temperatures, but retreated to terrestrial hiding places at the coolest temperatures. Thus, both the microhabitats and general range of temperatures chosen by these species are correlated with their geographic ranges and habitat associations in nature.

The first detailed fieldwork on thermoregulation in anurans was Lillywhite's (1970) study of bullfrogs (*Rana catesbeiana*) in California. In a pond that lacked vegetation along the shore, the frogs maintained body temperatures within a relatively narrow range by shuttling back and forth between the water and the land. On cool evenings, the frogs were found mainly near the warm water at the center of the pond, but gradually moved closer to shore during the day. As water temperatures rose as high as 35–40° C in the shallow water, close to the critical thermal maximum of 38° C, the frogs emerged onto the shore, where they remained exposed to direct sunlight. On particularly hot days, the frogs adopted an upright posture, with their legs extended, a posture that Lillywhite interpreted as facilitating evaporative cooling. However, in contrast to some lizards, the frogs did not appear to alter their orientation relative to the sun as body temperature changed. At very high temperatures, the bullfrogs periodically discharged mucus onto the surface of the skin, resulting in enhanced evaporative cooling and a temporary reduction in body temperature (Lillywhite 1971b, 1975). These frogs could sit all day in direct sunlight only if a source of moisture (the pond or damp soil) was available to compensate for evaporative water losses. The frogs inhabiting this very exposed pond probably experienced temperatures higher than they would normally prefer, and when artificial shelters were provided, the frogs often moved into the shade.

Although the bullfrogs studied by Lillywhite moved frequently between land and water, such movements are not always necessary to keep body temperatures within a relatively narrow range. Tracy (1976) studied the water and temperature relations of leopard frogs, *Rana pipiens,* using biophysical models to predict the effects of environmental variables such as wind speed, air temperature, substrate temperature, relative humidity, and absorbed solar radiation. During the summer, leopard frogs disperse away from breeding ponds into damp meadows, where individuals are quite sedentary for long periods of time. In an earlier study, Dole (1965a) reported that some frogs spent several days at a time sitting in the same spot in damp soil. Tracy found that frogs sitting in such positions in direct sunlight could maintain body temperatures within a range of less than 5° C, even without extensive movements between microhabitats. He also found that the amount of cover in a frog's immediate surroundings had very little effect on core body temperature, which was more strongly influenced by substrate temperature.

Amount of cover did have a significant effect on rates of evaporative water loss, with frogs in shaded environments losing water at a much slower rate than frogs in direct sunlight. Hence, the shade-seeking behavior observed in bullfrogs and other species probably has more to do with reducing evaporative water loss than with regulating core body temperature. However, even in shaded environments, a frog exposed to relatively high temperatures will desiccate unless it can rehydrate periodically (Tracy 1975, 1976). Therefore, the key to a frog being able to maintain an elevated body temperature is the availability of substrates wet enough to allow the frog to compensate for losses through evaporation. This is consistent with field observations that leopard frogs in meadows are invariably found on wet soil (Dole 1967; Tracy 1976). In contrast to bullfrogs, leopard frogs often are far from standing water during the summer, so shuttling between microhabitats to maintain body temperatures within a narrow range could be costly because movement would increase evaporative water loss. Since these frogs do not heat up excessively as long as they have a source of moisture available, most individuals probably are better off remaining in one spot, even when exposed to bright sunlight, rather than moving through patches of dry, inhospitable habitat.

### Thermoregulation at High Altitudes

Anurans living at high altitudes face particularly difficult problems of thermoregulation. High-altitude environments generally have cooler average temperatures than low-altitude environments, but perhaps more important, they often have much greater fluctuations of temperature during a single day. This is because of the low density and high transparency of air at high altitudes, which produces intense solar radiation during the day, followed by rapid radiative cooling at night. Hence, it is not unusual for anurans living at high altitudes to experience temperature changes of 25° to 30° C in a 24-hour period, and changes of more than 10° C can occur within a few minutes (Carey 1978; Navas 1997, 2003, 2006). Further complicating problems of thermoregulation is the relatively low humidity of many mountain environments during the day, resulting in high rates of evaporative water loss from the exposed skin of amphibians (Bradford 1984b). Anurans at high altitudes have several options for dealing with the combination of low average temperature and high temperature variation. These include (1) being active at a wider range of temperatures than are anurans at lower elevations, (2) shifting activity periods to daylight hours to maximize activity temperatures, and (3) being active in particularly warm microhabitats, provided that suitably moist substrates are available.

The boreal toad, *Bufo boreas boreas,* studied by Carey (1978) in the mountains of Colorado, is one species exposed to such extremes of temperature. Although these toads select a relatively narrow range of temperatures in a laboratory thermal gradient, they are active at a wider range of temperatures in the field, from near freezing to almost 34° C. Toads in her high-elevation population (2,900 m) were active during the day and at night. On very cold nights, they took shelter under logs or in water, but did not use burrows. Carey used implanted radio transmitters to monitor changes in body temperature of several toads released into an enclosure with access to sun, shade, and water. Even in summer, body temperatures of these toads often were only a few degrees above freezing in early morning, but rose rapidly as the sun came out. In direct sunlight, the toads maintained body temperatures between about 27° and 32° C for up to 6 hours in the middle of the day, although extended periods of cloud cover could drop body temperatures by 10° C in an hour. This species has been observed basking on warm rocks around breeding pools, enabling basking individuals to attain body temperatures that were 8° C above those of nonbasking toads in the water (Muths and Corn 1997).

In contrast to high-altitude populations, populations of *B. boreas* studied at low altitudes in California by Carey (1978) and Smits (1984) were primarily nocturnal during the summer, spending hot daylight hours in burrows. Smits found that activity periods changed throughout the year, with toads being much more active during the day in the spring and fall, when air temperatures were cooler. During these cool periods, emergence from burrows was regulated mainly by the time when temperatures at the burrow entrance exceeded that inside the burrow. In contrast, evening emergence was governed mainly by sunset time, although temperature also had an effect (Smits and Crawford 1984). Toads in these populations were active at a narrower range of temperatures (about 10–28°C) than was the high-altitude population, generally avoiding both hot and cold extremes. However, the preferred body temperatures of the high-altitude toads and the two low-altitude populations measured in laboratory gradients were not different, averaging about 24° C in all cases. Thus, it appears that high-altitude populations are active over a wider range of temperatures because that is the only way for them to have enough time to acquire the food they need, not because of inherent differences in preferred body temperature. The increased diurnal activity of high-altitude toads probably is a response to the activity periods of their insect prey rather than a direct physiological response to temperature (Carey 1978).

Physiological differences between high- and low-altitude populations of North American toads are similar to those observed in studies of toads in the Andes of South America. Sinsch (1989a) found that populations of *Bufo spinulosus* at 3,200 m in the Peruvian Andes were active day and night. On sunny days, they warmed up quickly in the morning sun and then remained in the open for several hours. Substrate temperature was more important than air temperature in determining body temperature, so by sitting on warm substrates, the toads could maintain body temperatures above that of the surrounding air, even at night. However, on very sunny days, toads in the open always had body temperatures below air temperature because of evaporative cooling (fig. 3.15).

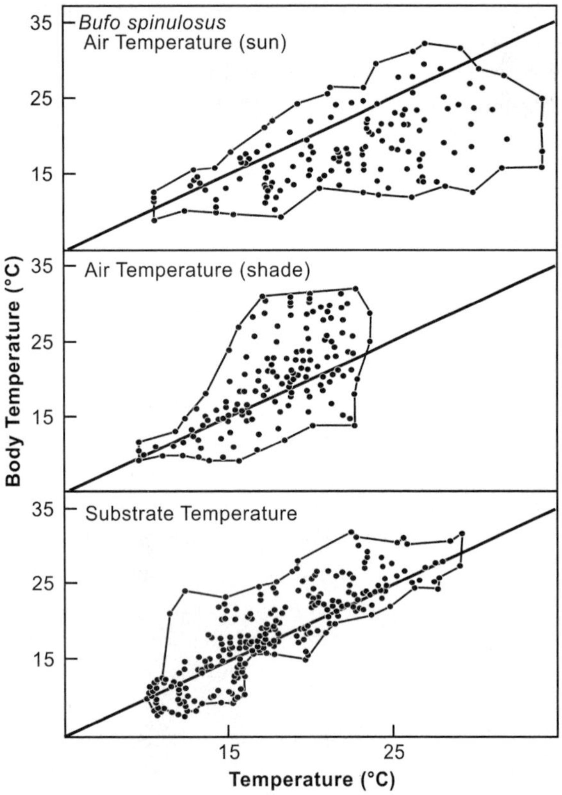


**Fig. 3.15.** Relationship of body temperature to air and substrate temperature in a toad from the high Andes of Peru (*Bufo spinulosus*). Body temperatures of toads typically are lower than air temperature in the sun, but not in the shade, because of increased evaporative water loss in the sun. Body temperatures usually are slightly higher than substrate temperature. After Sinsch (1989a).

Bladder water was gradually depleted on sunny days, so by afternoon, the toads usually retreated to shady sites.

A study of the same species at a higher elevation (4,300 m) by Pearson and Bradford (1976) revealed that adult toads were active only during the day, probably because nighttime temperatures were too low. Lambrinos and Kleier (2003) reported that juveniles at the same altitude also were diurnal, but were active only around midday. Both juveniles and adults were most active on sunny days and had a limited capacity to increase body temperatures by basking. The toads did not reach maximum temperatures as high as those achieved by lizards (*Liolaemus multiformis*) living in the same habitat because evaporation of water from the skin had a cooling effect (fig. 3.16). However, adult toads were active for more hours per day than the lizards and accepted a wider range of activity temperatures.

Navas (1996b) studied the temperature relations of four high Andean frogs in Colombia. He compared microhabitat selection and variation in body temperature in an undescribed species of *Atelopus, Colostethus subpunctatus, Eleutherodactylus bogotensis,* and *Hyla labialis* at 3,500 m. Each of these species used somewhat different microhabitats and consequently had different thermal strategies (fig. 3.17). *Atelopus* was active during the day in relatively exposed locations that received direct solar radiation, but spent the cold nights sheltered in thermally buffered retreat sites under rocks or in dense vegetation. *Colostethus subpunctatus* also was diurnal, but used more sheltered locations and retreated to protected sites at night. *Eleutherodactylus bogotensis* and *Hyla labialis* were nocturnal and therefore exposed to low temperatures for much of their

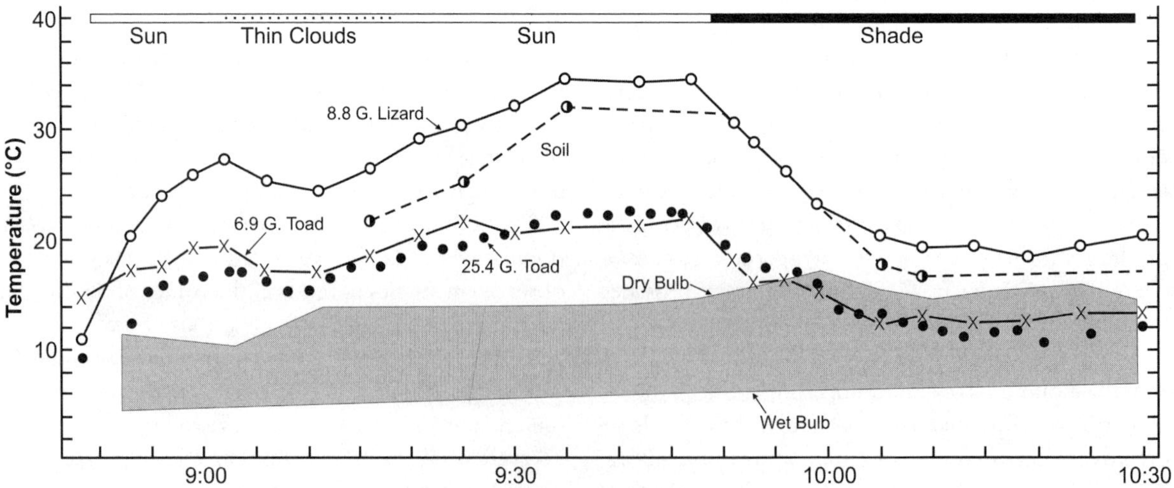


**Fig. 3.16.** Comparison of body temperatures of toads (*Bufo spinulosus*) and a lizard (*Liolaemus multiformis*) for one morning in March at 4,300 m in the Peruvian Andes. Bars at top show sun and cloud conditions. Data are shown for one free-ranging lizard (8.8 g) and a large (25.4 g) and small (6.9 g) toad. Temperatures measured at the soil surface also are shown. The shaded area at the bottom shows the range of air temperatures encompassed by dry bulb (top) and wet bulb (bottom) measurements. The lizard consistently maintained a body temperature significantly above that of the toads and above ambient air temperature in both sun and shade conditions. After Pearson and Bradford (1976).

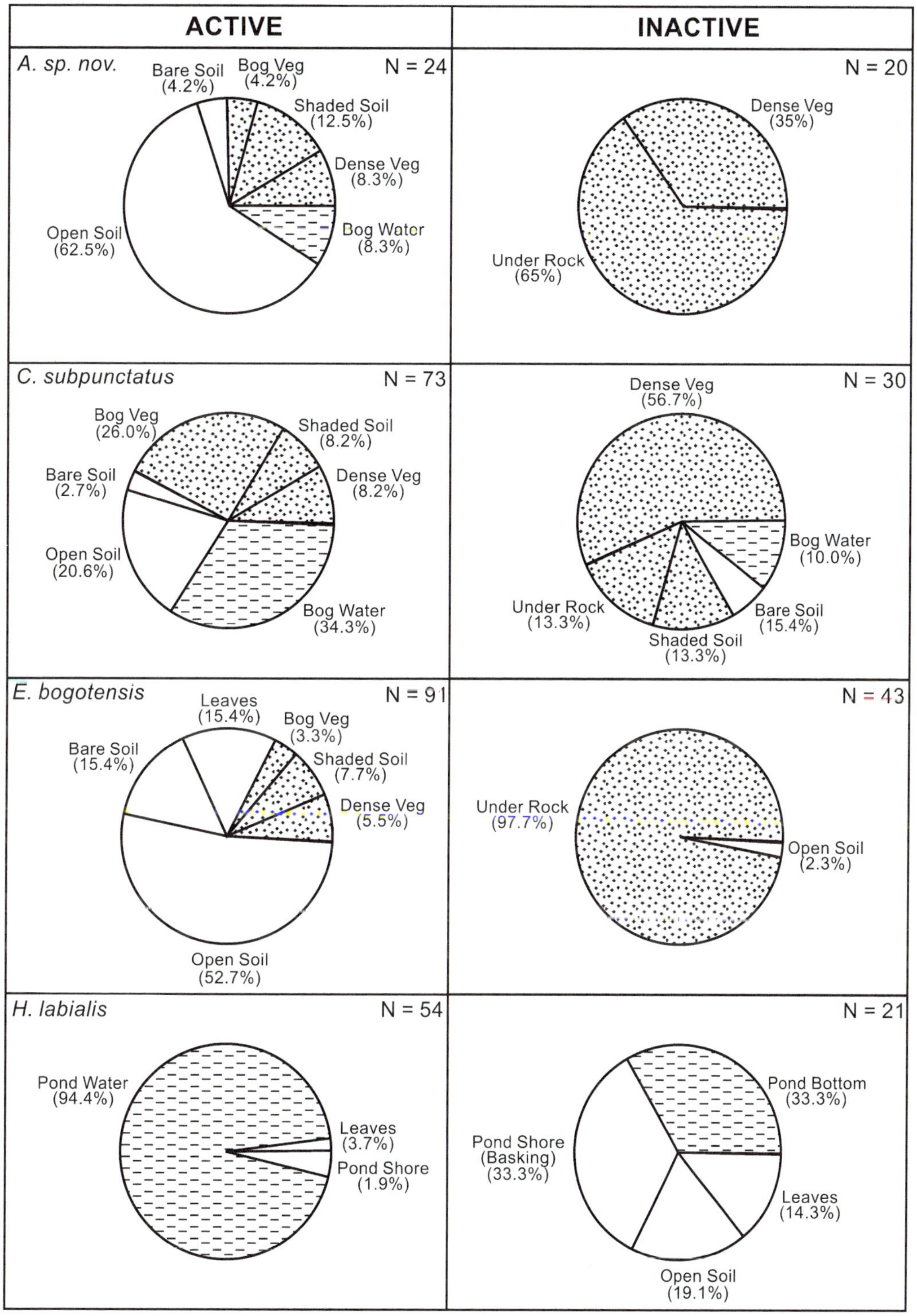


**Fig. 3.17.** Microhabitats selected by active and inactive frogs of four species from the paramo of Colombia. Symbols indicate the thermal stability of the microhabitats. White sections are exposed terrestrial microhabitats with low thermal stability. Dotted sections are sheltered terrestrial habitats with moderate thermal stability. Dashed sections are aquatic habitats with high thermal stability. Genera: *A* = *Atelopus, C* = *Colostethus, E* = *Eleutherodactylus, H* = *Hyla.* After Navas (1996d).

activity periods. However, *E. bogotensis* was terrestrial and often active in exposed locations that experienced the lowest temperatures and the greatest range of temperatures. *Hyla labialis* was found mainly around permanent ponds, where temperatures were much more stable. This species has greater resistance to water loss than the other frogs and can bask in the sun during the day to raise its body temperature above that of the air, especially at lower temperatures. The other three species are thermoconformers, with body temperatures very similar to substrate temperatures. Average activity temperature was lowest for the nocturnal *E. bogotensis* (6.5° C) and highest for the diurnal *C. subpunctatus* (12.5° C). The other two species were intermediate (about 11° C).

Another high-altitude amphibian, *Rana muscosa,* was studied by Bradford (1984b). This species reaches altitudes of more than 3,600 m in the Sierra Nevada of California, as high as any amphibian in North America. These frogs were always in or near water, but spent most of the day exposed to direct sunlight, where body temperatures often exceeded both air and water temperatures by more than 10° C. The frogs moved between microhabitats during the day, consistently selecting the warmest available locations. The frogs sometimes formed dense aggregations at midday, clustering on wet rock surfaces warmed by the sun (fig. 3.18). This allowed the frogs to warm up through contact with a warm substrate, but at the same time minimize evaporative water loss. At night, the frogs were largely inactive, submerging in deep water as temperatures dropped. In a laboratory thermal gradient, frogs tested during the day had a mean selected temperature of about 24° C, almost exactly the same as the toads studied by Carey (1978) and Smits (1984). However, in the field, environmental temperatures were so cool that the frogs seldom reached this temperature except for a few hours on particularly warm afternoons. Thus, these frogs are forced to be active at temperatures well below their preferred body temperature (average daytime activity temperature = 17° C).

**Fig. 3.18.** An aggregation of mountain yellow-legged frogs, *Rana muscosa,* basking on a sunlit rock. Photo by David Bradford.

### Thermoregulation in Arboreal Frogs

In all of the ranids and bufonids discussed so far, the skin offers little or no resistance to water loss (see chapter 2). Hence, heating the skin through increased absorption of solar radiation results in increased evaporation of water across the skin surface, thereby reducing core body temperature. The highly permeable skin of these species therefore limits their ability to use basking as a means of increasing body temperature. This was shown by the inability of *Bufo spinulosus* to achieve body temperatures as high as those of dry-skinned lizards exposed to the same conditions (Pearson and Bradford 1976; see also Tracy, Tracy, and Turner 1992 for a comparison of *Rana pipiens* with a desert lizard, *Sauromalus obesus*). Elevations of body temperature of 2–3° C above air temperature have been reported in mink frogs (*Rana septentrionalis*) sitting partially submerged at the edge of a pond in direct sunlight, but body temperatures were closer to the temperature of the surrounding water (Hedeen 1971). In general, body temperatures of ranids and bufonids sitting in direct sunlight probably track substrate temperature more closely than air temperature (Tracy 1975, 1976; Carey 1978; Sinsch 1989a; Tracy et al. 1993; Vences, Galán et al. 2002). Therefore, these anurans might be more accurately characterized as thigmothermic (thigm [G.] = touch) rather than heliothermic (heli [G.] = sun), since sitting in direct sunlight normally will be correlated with selection of warm substrates.

In contrast to these species, many arboreal frogs have much higher cutaneous resistance to evaporative water loss. This is especially true of the so-called waterproof frogs, including some species of *Phyllomedusa, Chiromantis,* and *Hyperolius* (see chapter 2 for details of water relations). Even some North American tree frogs have resistances to water loss that are several times greater than those of ranids or bufonids (Wygoda 1984). One would expect the temperature relations of arboreal frogs to be different from those of other anurans. Lower rates of evaporative water loss should allow these frogs to make more effective use of basking to raise body temperatures above that of the surrounding environment, but reduced evaporative cooling could lead to heat stress at high temperatures. Furthermore, tree frogs often spend daylight hours perched on dry leaves or branches, making it impossible for them to replace water lost through evaporation without moving to a moist microhabitat.

There are many reports of tree frogs sitting for extended

periods of time in direct sunlight (Brattstrom 1979; Hutchison and Dupré 1992). Examples include hylids from North America (*Hyla cinerea*), South America (*Hyla labialis*), and Australia (*Litoria caerulea, L. chloris, L. xanthomera;* C. R. Johnson 1970; Valdivieso and Tamsitt 1974; Freed 1980; Buttemer 1990; Buttemer and Thomas 2003). Wygoda (1984) reported that several arboreal hylids had body temperatures higher than those of nonarboreal frogs (*Acris, Rana, Bufo, Atelopus, Gastrophryne*) under identical conditions in the laboratory, a difference he attributed to lower rates of evaporative water loss and reduced evaporative cooling. In the field, green tree frogs (*Hyla cinerea*) sitting in direct sunlight have body temperatures as high as 36° C, higher than typical activity temperatures for most frogs (Freed 1980). When water-soaked plaster models of frogs were placed in the same microenvironments as living green tree frogs, the frogs had body temperatures up to 7° C higher than did the models (Wygoda and Williams 1991). In the laboratory, this species heated more rapidly and cooled more slowly than individuals of *Rana sphenocephala* subjected to the same conditions, resulting in higher body temperatures (Wygoda 1988, 1989).

Three Australian tree frogs, *Litoria caerulea, L. chloris,* and *L. xanthomera,* have considerably greater skin resistance to water loss than the North American hylids, but less than the waterproof frogs (see chapter 2). These frogs can tolerate environmental temperatures as high as 50° C and body temperatures up to 38° C. Resistance to water loss is higher in *L. chloris* than in *L. caerulea,* and at any given ambient temperature, *L. chloris* had a somewhat higher body temperature. In both species, there was a linear relationship between air temperature and body temperature at temperatures between 25° C and 40° C, but body temperature was always below air temperature. However, further increases in temperature did not result in a linear increase in body temperature, particularly in *L. chloris* (fig. 3.19). Body temperature leveled off at about 38° C at air temperatures between 43° and 47° C, probably because of increased discharge of mucus onto the skin, resulting in increased evaporative cooling. At these high temperatures, resistance to water loss dropped to a level similar to that of North American hylids (Buttemer 1990). Similarly, cutaneous resistance to evaporative water loss decreased at temperatures above 37° C in *L. xanthomera* (Buttemer and Thomas 2003).

The ability of these frogs to thermoregulate by evaporative cooling is limited by their water storage capacity and the availability of water sources for rehydration (see also chapter 2). A rhacophorid tree frog, *Polypedates maculatus,* exhibits responses to increasing temperature similar to those observed in the Australian species, but this species has a lower temperature threshold for discharge of mucus onto the skin (about 30° C; Lillywhite et al. 1997a, b). *Polypedates*

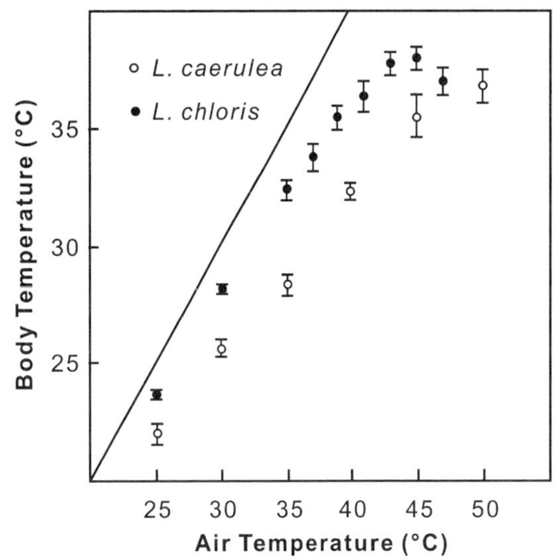


**Fig. 3.19.** Relationship of body temperature to air temperature in two treefrogs from Australia, *Litoria caerulea* and *Litoria chloris.* Symbols and lines indicate means ± 95% confidence intervals. The diagonal line shows body temperature equal to air temperature. After Buttemer (1990).

*maculatus* can tolerate body temperatures as high as 38° C when fully hydrated, but generally avoids high temperatures by seeking shelter in banana plants and other retreat sites.

Waterproof frogs (*Chiromantis* and some species of *Phyllomedusa* and *Hyperolius*) have cutaneous resistance to water loss approaching that of some reptiles (see chapter 2). All of these frogs can sit for days, weeks, or even months at a time in exposed locations during the dry season without rehydrating. Environmental temperatures often exceed 40° C, and all of these frogs can tolerate body temperatures that would be lethal to most semiaquatic and terrestrial anurans: about 40° C in *Chiromantis* and *Phyllomedusa;* up to 43° C in *Hyperolius viridiflavus* (Shoemaker et al. 1987; Geise and Linsenmair 1988). The highest field body temperature ever recorded for an amphibian is 41° C, from *Phyllomedusa sauvagii* in Argentina (McClanahan and Shoemaker 1987; Hutchison and Dupré 1992). The low rates of evaporative water loss in these frogs allows them to maintain body temperatures above that of the surrounding air even after dark. This has been documented in *Hyperolius marmoratus,* which had nocturnal body temperatures averaging about 3° C above air temperature even several hours after sunset (Passmore and Malherbe 1985). This elevation in nocturnal body temperature is considerably greater than the 0.8° C reported for the less waterproof green tree frog (*Hyla cinerea;* Wygoda and Williams 1991).

In all three genera of waterproof frogs, diurnal body temperatures track ambient temperatures rather closely until they approach the critical thermal maximum (fig. 3.20); the frogs then increase evaporative cooling by increasing the

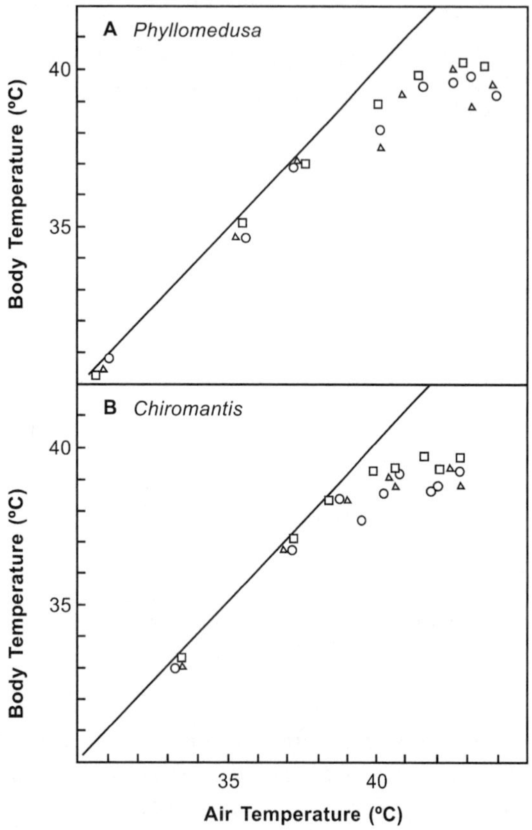


**Fig. 3.20.** Relationship of body temperature to air temperature in two species of "waterproof frogs," *Phyllomedusa sauvagii* and *Chiromantis xerampelina.* The diagonal line shows body temperature equal to air temperature. Symbols show data collected at different air speeds (squares = 160 cm/s, circles = 30 cm/s, triangles = 8 cm/s). After Shoemaker et al. (1987).

discharge of fluid and mucus onto the skin surface (Shoemaker et al. 1987; Geise and Linsenmair 1988). This allows the frogs to maintain a stable body temperature several degrees below ambient temperature. In *Phyllomedusa,* the lipid secretions responsible for very low rates of evaporative water loss begin to lose their effectiveness at temperatures above 35° C, a point at which problems of thermal stress outweigh problems caused by loss of water reserves (McClanahan, Stinner, and Shoemaker 1978). Both *Chiromantis* and juvenile *Hyperolius viridiflavus* further reduce heat loads by increasing the reflectance of their skin, a response that is absent in *Phyllomedusa.* In juvenile *Hyperolius,* changes in skin reflectance occur gradually at the beginning of the dry season through the accumulation of reflective iridophores in the skin (Kobelt and Linsenmair 1986, 1992; Schmuck, Kobelt, and Linsenmair 1988). In *Chiromantis,* individuals undergo a rapid color change from brown to white only at high body temperatures (Shoemaker, Baker, and Loveridge 1989).

Some other tree frogs, such as *Agalychnis callidryas,* assume a chalky white color when exposed to direct sunlight, and this helps reduce radiant heat loads. *Agalychnis* and some other hylid and centrolenid tree frogs also have skin pigments that reflect near-infrared radiation (Schwalm, Starrett, and McDiarmid 1977). This could provide a thermoregulatory advantage to animals sitting in direct sunlight, but the effect of infrared reflectance on body temperature has not been investigated. In nonarboreal frogs that do not have high cutaneous resistance to water loss, changes in skin reflectance are expected to have a minimal effect on body temperature. Higher reflectance could reduce radiant heat gains, but any increased heat gained through darkening of the skin will lead to increased evaporation of water and lowering of body temperature (Tracy 1976, 1979). For example, dark-skinned toads (*Bufo boreas*) sitting in direct sun absorb about 4% more solar radiation than do light-skinned toads, but this is not expected to raise core body temperature by more than about 0.34° C (Carey 1978).

### Adaptive Significance of Basking Behavior

The widespread occurrence of basking in anurans, and the use of warm substrates to raise body temperatures, suggests an adaptive advantage to such behavior, but any benefits must be balanced against potential costs, such as increased risk of desiccation. Only a few studies have examined possible benefits of maintaining elevated body temperatures. At present, the most convincing evidence is that basking is beneficial because it increases rates of digestion and growth, particularly in juvenile anurans. Several studies of toads have shown that newly emerged juveniles often bask on the exposed banks of ponds, while adults in the same populations tend to be active mainly at night (Black and Black 1969; Seymour 1972; Lillywhite, Licht, and Chelgren 1973; Lillywhite and Wassersug 1974). Toad tadpoles generally develop very rapidly and emerge from the water in a relatively underdeveloped state. For example, newly emerged juveniles have a low aerobic capacity for locomotion, but aerobic capacity increases steadily as they grow (Taigen and Pough 1981). Consequently, any increase in growth rate would decrease the time required to reach adult aerobic capacity. Lillywhite, Licht, and Chelgren (1973) demonstrated that juvenile toads (*Bufo boreas*) had a preferred body temperature both in the field and laboratory of about 27° C. Toads that were maintained at 27° C had faster growth rates than did those maintained at either 20° or 33° C, suggesting that the toads preferred a temperature that optimizes growth rate. Furthermore, the toads engaged in basking behavior after feeding, but not after going several days without food. Basking also would be advantageous to juvenile toads in allowing them to attain higher metabolic rates during activity because of the temperature dependence of aerobic metabolism (see chapter 4). Consequently, juvenile toads active at 27° C would have mass-specific metabolic rates approximately equal to those of adults active at 20° C (Taigen and Pough 1981).

The benefits of basking for adult anurans are less obvious. Many adult anurans grow relatively little after reaching sexual maturity, although long-lived species such as *Rana catesbeiana, R. clamitans,* or *R. virgatipes* grow more as adults than do short-lived species, such as many hylids (R. D. Howard 1984; E. Werner 1986; Given 1988a). Freed (1980) found that basking resulted in increased growth rates for juvenile *Hyla cinerea,* but not for adults. However, higher body temperatures produced increased rates of digestion in frogs of all sizes, resulting in more rapid food passage time for both juveniles and adults. Freed suggested that this in turn might make more energy available for activities such as reproduction. Indeed, adult males allowed to bask in the laboratory called in the evening approximately five times more often than nonbasking males. Although these results are intriguing, sample sizes were small, and the experiments need to be repeated. Increased digestion rates and more rapid food passage times also could be important in accelerating accumulation of fat reserves for the winter, particularly in high-altitude populations with short activity seasons (Carey 1978). However, for many anurans, especially arboreal species that sit for long periods exposed to direct sunlight, elevated body temperatures probably are tolerated rather than preferred.

## Thermal Acclimation

Ectotherms typically compensate for seasonal or even shorter-term changes in environmental temperature by acclimation. For example, if a salamander living at 10° C is exposed to a temperature of 20° C, its metabolic rate will rapidly increase, but if the exposure to the warmer temperature lasts for several days, the animal experiences a compensating decrease in metabolic rate. Although such compensation often is not complete (Rome, Stevens, and John-Alder 1992), the result is reduced thermal dependence of metabolism. Other physiological processes undergo acclimation as well. For example, some fish can compensate for reduced muscle performance at low temperatures through physiological and morphological changes in locomotor muscles. This allows cold-acclimated animals to make greater use of their aerobic muscle fibers at low temperatures than warm-acclimated animals (Rome, Loughna, and Goldspink 1985). Many ectotherms also exhibit acclimation of temperature tolerance, with individuals acclimated to high temperatures being able to tolerate higher temperatures than those exposed only to low temperatures. Acclimation of metabolic rate is discussed in more detail in chapter 5, so I will focus on acclimation of locomotor performance and thermal tolerance here.

Data from a limited number of species suggest that most adult amphibians lack a capacity for thermal acclimation of locomotor performance like that seen in many fish, especially compensation for cold temperatures (Rome 1983; Renaud and Stevens 1983b, 1984; Else and Bennett 1987; Whitehead et al. 1989; Knowles and Weigl 1990; Rome, Stevens, and John-Alder 1992; R. Wilson and Franklin 2000b). For example, leopard frogs (*Rana pipiens*) acclimated to 5° C could not jump any farther at temperatures between 5° and 15° C than frogs acclimated to 25° C, and their performance at higher temperatures was worse than that of warm-acclimated frogs (Renaud and Stevens 1983b). Indeed, the only example of acclimation of locomotor performance to cold temperatures is for *Xenopus laevis,* a strictly aquatic frog (R. Wilson, James, and Johnston 2000). In this species, individuals acclimated to 10° C swam 67% faster when tested at 10° C than did those acclimated to 25° C. Individuals acclimated to the higher temperature did not exhibit enhanced performance at that temperature, however. Tadpoles of *X. laevis* also showed acclimation to cold temperatures, but locomotor performance of cold-acclimated tadpoles declined when they were tested at 30° C. Tadpoles of an Australian myobatrachid frog, *Limnodynastes peronii,* exhibited a similar capacity to compensate for cold temperatures (R. Wilson and Franklin 1999), but adults do not. R. Wilson, James, and Johnston (2000) argued that cold acclimation of locomotor performance is limited to aquatic amphibians because aquatic environments are more thermally buffered than are terrestrial environments. The latter can undergo such rapid fluctuations in temperature that acclimation is not possible. Unfortunately, the limited number of species tested for acclimation responses to date makes any such conclusion premature.

Many studies of thermal acclimation in amphibians have focused on acclimation of thermal tolerance (for reviews, see Brattstrom 1970; Hutchison and Maness 1979; Rome, Stevens, and John-Alder 1992). Most temperate-zone amphibians exhibit an increase in $CT_{Max}$ with increasing acclimation temperature, but the magnitude of change varies considerably. For example, in many salamanders, a 20° C change in acclimation temperature (from 5 to 25° C) raises $CT_{Max}$ by only about 1.5–4.5° C. In frogs, a change of similar magnitude (18–20° C) raises $CT_{Max}$ by about 2–8° C. Rates of acclimation vary among species, but generally several days to several weeks are required for acclimation to be complete (Hutchison 1961; Brattstrom 1968; Claussen 1977; Layne and Claussen 1982). Exposure to cold temperatures results in lowered $CT_{Max}$, but the rate of acclimation to cold generally is slower than acclimation to warm temperatures (Rome, Stevens, and John-Alder 1992).

One might expect tropical amphibians that live in relatively equable thermal environments to show less capacity for thermal acclimation, but the limited data available for tropical species are not consistent. For example, some tropical ranids and bufonids tested by Brattstrom (1968) have a capacity for acclimation to warm temperatures similar to that of North American ranids and bufonids, but the tropical species are much less cold tolerant. On the other hand, some tropical hylids showed relatively little acclimation of $CT_{Max}$ (Brattstrom 1968), and the tropical frog *Eleutherodactylus coqui* showed no acclimation, even though this species experiences greater seasonal temperature fluctuation than do many tropical anurans (Christian et al. 1988). Tropical anurans often are killed by low temperatures, so most species have been acclimated over a relatively narrow range of temperatures, making direct comparison with temperate-zone species difficult. Furthermore, comparisons of species in different families often confound latitudinal differences with phylogenetic history. For example, it is possible that *Eleutherodactylus,* a genus largely confined to the Neotropics, lacks a capacity for thermal acclimation, but this is not necessarily true for tropical species in more widespread genera. We need considerably more information on tropical species from a variety of thermal environments, with careful controls for phylogenetic history, before valid comparisons can be made with temperate-zone forms.

Most authors have used $CT_{Max}$ as an index of thermal tolerance, assuming that an increase in $CT_{Max}$ with prolonged exposure to warm temperatures (for example, during summer months) provides added protection from potentially lethal temperatures. Rome, Stevens, and John-Alder (1992) argued that the benefit of the small changes in $CT_{Max}$ seen in many species has yet to be demonstrated. One trivial reason for this is the rarity of observations of amphibians exposed to extreme temperatures in the field (V. H. Hutchison, personal communication). $CT_{Max}$ for most amphibians is 5–15° C higher than their normal activity temperatures. For example, frogs with average activity temperatures around 25° C often have a summertime $CT_{Max}$ approaching or even exceeding 40° C (Brattstrom 1968). Many of the salamander species studied by Hutchison (1961) had a $CT_{Max}$ exceeding 35° C after acclimation to 20° C, and some species, such as *Notophthalmus viridescens,* had a $CT_{Max}$ above 40° C when acclimated to temperatures above 25° C. Both temperate-zone and tropical salamanders typically avoid temperatures above 25° C (Feder 1982b; Feder and Lynch 1982). Nevertheless, amphibians sometimes can be exposed to near-lethal temperatures when trapped in drying pools in hot weather or when exposed to extreme temperatures on land during hot, dry periods. V. H. Hutchison (personal communication) observed a hellbender (*Cryptobranchus alleganiensis*) trapped in a shallow pool that reached 41° C, far above the normal activity temperature of this species. The animal went into spasms, indicating it had reached its $CT_{Max}$. Seasonal acclimation of $CT_{Max}$ could be essential in providing protection against this type of exposure to extreme temperatures, even though such incidents are rare.

Some amphibians appear to adjust thermal tolerance on a daily basis after only brief exposure to unusually high temperatures. For example, Pough and Wilson (1970) studied newly metamorphosed *Ambystoma maculatum* hiding under rocks on the exposed shore of a pond in late summer. In the morning, temperatures under the rocks were between 17° and 21° C, but by afternoon, they were 27–39° C. Salamanders avoided rocks with temperatures above 32° C, and deserted rocks that were heated to this temperature in the laboratory. Nevertheless, many individuals were exposed to unusually high temperatures for several hours each day. The salamanders became dehydrated, but were able to survive. Laboratory studies of animals collected in the field revealed that $CT_{Max}$ increased from about 38.5° to 39.7° C from morning to afternoon. Although the salamanders generally avoided temperatures approaching the $CT_{Max}$, the highest temperature they tolerated in the afternoon (32° C) was high enough to induce some loss of coordination in salamanders collected in the morning. Hence, Pough and Wilson argued that the daily increase in $CT_{Max}$, although small, is correlated with increased resistance to potentially dangerous temperatures. Pough (1974b) reported a similar short-term increase in $CT_{Max}$ in red efts (*Notophthalmus viridescens*) exposed to high temperatures in the field. In both of these studies, changes in thermal tolerance were attributed to rapid thermal acclimation. Other workers have suggested the phenomenon is more accurately described as heat hardening, defined as a rapid increase in thermal tolerance in animals exposed to temperatures close to their $CT_{Max}$ (Hutchison and Maness 1979; Rutledge, Spotila, and Easton 1987a; Rome, Stevens, and John-Alder 1992). One hypothesis to explain increased thermal tolerance in amphibians exposed to short-term increases in temperature is that the animals initially undergo rapid heat hardening that depends on the presence of a heat shock protein (hsp-70) that provides some protection to cells. This is followed by a longer-lasting response that depends on production of new hsp-70 in response to heat shock (Easton, Rutledge, and Spotila 1987; Rutledge, Spotila, and Easton 1987a, b; Near et al. 1990; Yu, Magee, and Spotila 1994; Yu et al. 1998).

## Overwintering and Hibernation

Most studies of amphibian thermal biology have focused on responses to high temperatures, but amphibians living at high latitudes or high altitudes also must contend with periodic exposure to cold. Amphibians generally respond to short-term exposure to low temperatures by retreating to warmer microhabitats and remaining inactive until warm temperatures return. For example, anurans at high altitudes typically seek shelter in burrows or under logs in cold weather (Pearson and Bradford 1976; Carey 1978). However, dealing with many months of cold weather is a different matter. No amphibian can remain active at subfreezing temperatures, and many species spend the winter months in a relatively inactive state, depending on stored energy reserves for survival. Stores of glycogen and lipids accumulate during periods of feeding in the summer and fall and are deposited in fat bodies and the liver to support metabolism, and in some cases, egg development, during the winter (see chapter 5 for additional discussion of overwintering energetics). Because metabolic rates decrease rapidly with falling temperatures, overwintering amphibians often can survive for many months without food (8–9 months for some frogs in northern latitudes or high altitudes; Koskela and Pasanen 1974; Morton 1981; Kuzmin and Maslova 2003). In general, amphibians that are aquatic as adults tend to overwinter in water, whereas terrestrial species usually overwinter on land (Pinder, Storey, and Ultsch 1992). However, there are exceptions, and many species exhibit considerable plasticity in overwintering behavior.

### Urodeles

Fully aquatic salamanders such as *Necturus* and *Cryptobranchus,* which never leave the water even as adults, have no choice but to spend the winter underwater, where they continue to feed, but at a slower rate than in the summer. In the relatively equable climate of the North Carolina mountains, stream-dwelling plethodontids, including species of *Gyrinophilus, Pseudotriton, Eurycea,* and *Desmognathus,* remain active throughout the winter, even when nearby slopes are covered with snow. These species tend to have prolonged larval periods, with larvae spending several winters in streams before metamorphosing (Bruce 1972a, b, 1974, 1978a, b, 1982a, b, 1985a, b, 1988a, b).

Farther north, the semiaquatic plethodontids *Desmognathus fuscus, D. ochrophaeus,* and *Eurycea bislineata* sometimes hibernate on land near streams as adults, but remain active and even feed during warm periods (Bishop 1941b; Ashton 1975; Ashton and Ashton 1978). The more aquatic species *Gyrinophilus porphyriticus* and *Pseudotriton ruber* apparently remain underwater during the winter (Bishop 1941b). Larvae of all of these species spend at least one winter in streams before metamorphosing, but show little growth in winter months (Wilder 1913, 1924; Bishop 1941b; Juterbock 1990). Terrestrial plethodontids always hibernate on land, using root holes and other natural cavities in the soil (Blanchard 1933a; Vernberg 1953; Stebbins 1954a), and they sometimes eat insects underground (R. Caldwell and Jones 1973; R. Caldwell 1975). Adult ambystomatids also hibernate on land and usually are seen above ground only during the breeding season. Spotted salamanders (*Ambystoma maculatum*), for example, typically overwinter in burrows of white-footed mice (*Peromyscus*) and other rodents (Madison 1997). However, in some populations, larval ambystomatids remain in the water throughout the winter (Bleakney 1952; Whitford and Vinegar 1966; Brandon and Bremer 1967), and must do so in *Ambystoma opacum,* which lays eggs in the fall (Noble and Brady 1933; Bishop 1941b).

The natural history of most hynobiid salamanders is so poorly known that little can be said about their winter behavior. In southern Japan (35° N latitude), *Hynobius nebulosus* actually breeds in streams and ditches in midwinter (Tanaka 1989). *Onychodactylus fischeri* apparently hibernates mainly in flowing water that does not freeze in winter (Kuzmin and Maslova 2003). The Siberian hynobiid salamander *Salamandrella keyserlingii,* the only urodele found north of the Arctic Circle, hibernates on land in rotting logs and holes and often forms aggregations of up to 25 individuals (Kuzmin and Maslova 2003). In northern parts of its range, hibernation begins as early as August and lasts until April. This is the only salamander known to be freeze-tolerant (Berman, Leirkh, and Mikhailova 1984; see the following). Hibernation sites are relatively shallow, no more than 7 cm below the surface, where temperatures can be as low as −20° C, but daily fluctuations in temperature are small (Berman 1992; Kuzmin and Maslova 2003).

Adult red-spotted newts (*Notophthalmus viridescens*) in eastern North America spend the winter either on land or in the water. Morgan and Grierson (1932) reported that newts in Massachusetts were active in permanent ponds all winter and continued to feed, mostly on mayfly nymphs, even in the coldest months. In populations of newts from North Carolina to New York, some adults migrate in the fall to breeding ponds, where they spend the winter, while others remain in terrestrial hibernation sites and move to breeding ponds in the spring (Bishop 1941b; Chadwick 1944; Hurlbert 1969). In other populations, such as those studied by Gill (1978a) in the mountains of Virginia, all individuals leave the breeding ponds in the fall and hibernate on land, probably because some ponds freeze all the way to the bottom in very cold winters.

Most species of European newts (*Triturus*) leave their breeding ponds in winter and hibernate on land, and species in the genus *Salamandra* hibernate on land as well. These salamanders use a variety of hibernation sites, including natural cavities in clay deposits and gravel piles; spaces under mats of vegetation, large stones, logs, and piles of timber; and in drain pipes, cellars, and mines (Hecht 1931; M. Smith

1969; Feldmann 1967a, b, 1977; Steward 1969; Kabisch 1971a, b; Kowalewski 1974; Dolmen 1983a; Frazer 1983; Sinsch 1989c; Baumgartner 1999). In regions with mild climates or where permanent water is available, some newts remain in the water all winter. For example, Bell (1977) observed fall migrations of adult *Triturus vulgaris* to permanent breeding ponds, where they spent the winter, a phenomenon not observed in populations that use shallow or temporary ponds for breeding (Griffiths 1984). Van Gelder (1973) reported that a substantial proportion of a population of *Triturus helveticus* in the Netherlands spent the winter in a fen, where breeding also occurred. Some larvae wintered there as well, but metamorphosed juveniles left the water and hibernated on land, returning to the water only when they were sexually mature. Hibernation of newts in ponds and tarns has been reported in other parts of Europe as well (Wolterstorff 1923; Steward 1969; Hagström 1970, 1982).

### Anurans

Most anurans living in cold regions probably hibernate on land, particularly bufonids, pelobatids, bombinatorids, discoglossids, and hylids. Burrowing forms such as toads (*Bufo*) and spadefoot toads (*Scaphiopus, Spea, Pelobates*) often bury themselves in sand or loose soil a meter or more below the surface. They also use natural cavities beneath large stones, cellars, mouse holes, mole tunnels, and other mammal burrows (Breckenridge and Tester 1961; Tester and Breckenridge 1964; M. Smith 1969; J. B. Campbell 1970; Kabisch 1971a; Kowalewski 1974; Frazer 1983; Nöllert 1984; van Gelder, Aarts, and Staal 1986; Sinsch 1989c; Kuyt 1991; Denton and Beebee 1993b; Bosman, van Gelder, and Strijbosch 1996; Kuzmin 1999; Kuzmin and Maslova 2003). European spadefoot toads (*Pelobates fuscus*) and natterjack toads (*Bufo calamita*) hibernate in the nest holes of sand martins (Smith 1969; Frazer 1983; Nöllert 1984), although one account (Thompson 1843) suggested that the martins actually use holes previously constructed by toads. Midwife toads (*Alytes obstetricans*) and common toads (*Bufo bufo*) sometimes hibernate in mine tunnels (Feldmann 1977). There are a few reports of both juvenile and adult *B. bufo* hibernating in water (Waddington 1952; Hagström 1982; Juszczyk et al. 1984; Kwet 1996b), but this appears to be unusual. This species is able to withstand up to a month of flooding of terrestrial hibernation sites, whereas natterjack toads (*Bufo calamita*) suffer high mortality if their burrows are flooded (Bosman, van Gelder, and Strijbosch 1997). In the Russian Far East, *B. gargarizans,* a close relative of *B. bufo,* normally hibernates on land, sometimes in aggregations of more than 60 individuals. Nevertheless, some populations hibernate in shallow rivers and lakes (Kuzmin and Maslova 2003). The Asian toad *B. raddei* hibernates in sandy soil at depths of about 2 m, sometimes in groups of up to 60 individuals (Kuzmin and Maslova 2003).

Fire-bellied toads (*Bombina*) hibernate both on land and in the water, sometimes in aggregations of dozens of individuals. In the Primorye Region of eastern Siberia, *Bombina orientalis* hibernates from September to April or May in rotting logs, leaf litter, rock piles, and occasionally in streams. Frogs in one population were observed hibernating near geothermal springs, where water temperatures in winter can be as high as 24° C, with some individuals remaining active in the warm water (Kuzmin 1999; Kuzmin and Maslova 2003).

North American cricket frogs (*Acris*) have been found hibernating in cracks in soil near breeding ponds, crayfish burrows, cavities under rocks, and gravel banks near streams (W. T. Neill 1948a; A. P. Blair 1951; R. H. Gray 1971; Irwin, Costanzo, and Lee 1999; McCallum and Trauth 2003). Cricket frogs prefer moist sites near water that do not freeze, but are intolerant of prolonged submersion in hypoxic or anoxic pond water. Hibernation sites of the Japanese tree frog *Rhacophorus schlegelii* also appear to be chosen to minimize the chances of freezing and desiccation, with moist sites near the edges of ponds and streams being preferred (Ihara 1999). European tree frogs (*Hyla arborea*) usually hibernate on land in a variety of sites under rocks or clumps of vegetation, in cellars or crevices in brick walls, or buried in piles of leaves or manure (Stumpel 1990). In some regions, these tree frogs also hibernate in mud at the bottom of ponds (Berger and Michalowski 1963; Kowalewski 1974). The closely related *H. japonica* has been found hibernating in leaf litter, cracks in soil, rodent burrows, rock piles, tree holes, and under logs, but not in water (Kuzmin and Maslova 2003). Hibernation sites of most North American tree frogs have not been reported, but *Hyla chrysoscelis,* a freeze-tolerant species (see the following), has been found hibernating in loose soil at the bases of trees (Burkholder 1998) and in tree holes (Ritke and Babb 1991).

The most aquatic North American ranid frogs, such as *Rana catesbeiana, R. clamitans, R. septentrionalis,* and *R. muscosa,* all hibernate in ponds, lakes, swamps, and rivers (Martof 1956a; Willis, Moyle, and Baskett 1956; Bradford 1983; Stinner, Zarlinga, and Orcutt 1994; Matthews and Pope 1999), although occasional terrestrial hibernation has been reported (Bohnsack 1951, 1952; Jenssen 1968). Green frogs (*Rana clamitans*) in upstate New York moved from summer home ranges in ponds to hibernation sites in flowing water, including streams and seeps. These areas remained unfrozen throughout the winter and also were well oxygenated (Lamoureux and Madison 1999). Juveniles of these large ranids appear to be somewhat more tolerant of cold than adults and usually enter hibernation later in the fall and emerge earlier in the spring than adults (Martof 1956a; Willis, Moyle, and Baskett 1956; Jenssen 1968). Juvenile green frogs have been seen active and even feeding in the winter as far north as Illinois (Jenssen and Klimstra 1966). Leopard frogs (*Rana pipiens* complex) are somewhat more terrestrial than the larger North American ranids. They hibernate mainly in ponds and streams (Breckenridge 1944; Manion and Cory 1952; Emery, Berst, and Kodaira 1972; Cunjak 1986; L. Licht 1991). Leopard frogs in Vermont were found hibernating in a well-oxygenated river in the company of map turtles (*Graptemys geographica;* Ultsch, Graham, and Crocker 2000). Leopard frogs also have been found overwintering in caves and rock crevices (Rand 1950; A. P. Blair 1951), as have pickerel frogs (*Rana palustris;* Resetarits 1986). Tadpoles of ranids that breed in permanent ponds often spend one or more winters in the water before metamorphosis (Wright and Wright 1949; Martof 1956b; Collins 1979a; Collins and Lewis 1979; Bradford 1983). Wood frogs (*Rana sylvatica*), among the most terrestrial of North American ranids, generally use shallow ponds for breeding and hibernate in upland forested habitats where the soil is not saturated in winter (Storey and Storey 1986a; L. Licht 1991; Regosin, Windmiller, and Reed 2003b). This is the only North American ranid frog known to tolerate freezing (see following).

In Europe, common frogs (*R. temporaria*)—which resemble wood frogs in morphology, ecology, and behavior—differ from the North American species in that they hibernate mostly underwater (Bannikov 1940; Savage 1961; Ashby 1969; Hazelwood 1969; Kabisch and Engelmann 1971; Koskela and Pasanen 1974; Kowalewski 1974; Feldmann 1977; Frazer 1983; Juszczyk et al. 1984; Verrell and Halliday 1985a; D. Weber 1989b; Pasanen and Sorjonen 1994; Kwet 1996b). Like some North American ranids, such as *Rana clamitans, R. temporaria* often shows a preference for hibernating in sites with running water, including rivers, streams, springs, and ditches (Kuzmin 1999). However, some individuals spend the winter on land, buried in soil or in caves and mine tunnels. Common frogs have even been found hibernating in the dens of poisonous adders (*Vipera berus*) in Finland (Vitanen 1967).

The European moor frog (*Rana arvalis*), which is largely terrestrial outside the breeding season, is mainly a terrestrial hibernator, but some individuals spend the winter in streams or at the bottom of breeding ponds (Juszczyk et al. 1984; Büchs 1987; Hellbernd 1987). Moor frogs have been seen moving about in mild weather during winter (Henle 1999), perhaps shifting between hibernation sites. Another European frog that usually hibernates on land, *Rana lessonae,* sometimes moves between hibernation sites in the winter as well (Holenweg and Reyer 2000). Many other Eurasian ranid frogs, including *Rana dalmatina, R. macrocnemis, R. amurensis, R. asiatica, R. pirica,* and *R. dybowskii,* hibernate either on land or in water, with the latter being the rule in colder regions (Kuzmin 1999). Many of these species hibernate underwater in large groups, with aggregations of 1,000–2,000 being reported for *Rana amurensis,* up to 1,500 for *R. pirica,* and thousands of individuals for *R. dybowskii* (Kuzmin and Maslova 2003). *Rana nigromaculata,* which occurs mostly in warmer climates, usually hibernates in ponds, but sometimes in holes in streams (Kuzmin and Maslova 2003).

Species of water frogs in the *Rana esculenta* complex differ in their hibernation behavior according to their habitat preferences. *Rana ridibunda,* the largest and most aquatic species, hibernates mostly under water (Kowalewski 1974; Berger 1982; Lutschinger 1988; Voituron, Eugene, and Barre 2003). In some populations individuals leave the ponds where they spend most of the year and move to rivers for hibernation, perhaps to take advantage of higher levels of oxygen in the water (Heym 1974). Some *R. ridibunda,* particularly juveniles, spend the winter on land, but this is relatively rare. In contrast, *R. lessonae,* a smaller, more terrestrial species that breeds mainly in temporary ponds, is mainly a terrestrial hibernator. These frogs usually dig individual burrows in the soil or can be found under moss and leaf litter 3–7 cm below the surface (Holenweg and Reyer 2000). Small aggregations under boards and logs also have been observed (Berger 1982). A third "species," *R. esculenta,* is a hybrid between *R. ridibunda* and *R. lessonae* that is perpetuated by hybridogenetic reproduction (Berger 1977; Graf and Polls Pelaz 1989; see chapter 7). Berger (1982) found that populations of *R. esculenta* associated with *R. lessonae* (the most common situation) hibernated on land with the parent species, but populations living in permanent water with *R. ridibunda* hibernated mostly under water. Holenweg and Reyer (2000) also found that *R. esculenta* associated with *R. lessonae* mostly hibernated on land, but they tended to move shorter distances from a pond, and a few individuals were found hibernating underwater. Tunner (1992) reported that thousands of individuals in a mixed population of *R. lessonae* and *R. esculenta* moved up to 15 km from a lake to a terrestrial hibernation site. Some *R. esculenta* spent the winter in a canal running into the lake, but this site was not used by *R. lessonae.* In a study in Switzerland, overwinter survival of mixed populations of *R. lessonae* and *R. esculenta* was negatively correlated with an index of the severity of the winter. This was not due simply to low temperatures, but to variable temperatures. Possibly frogs emerged from retreat sites and moved around during warm periods, thereby depleting some of their stored energy reserves and risking being trapped when freezing winter returns (Anholt et al. 2003).

### Costs and Benefits of Hibernation in Water

For amphibians hibernating in water, there is no danger of desiccation, and little chance of freezing to death unless the water is very shallow. For the most part, spending the winter underwater is simply a matter of slowing metabolism and other physiological processes to very low rates in response to cold temperatures. However, the animals are not in a truly dormant state and sometimes move around under the ice of frozen ponds and lakes. Because of their very low metabolic rates, amphibians wintering underwater generally can take up sufficient oxygen through the skin to supply their metabolic needs. However, in contrast to many freshwater turtles that also hibernate underwater, aquatic amphibians have a poor capacity for anaerobic metabolism in winter. This means that frogs hibernating underwater typically do not bury themselves in mud on the bottom of ponds, which can become anoxic (E. Stewart, Reese, and Ultsch 2004). Some frogs are capable of moving around under the ice (Emery, Berst, and Kodaira 1972) and following oxygen gradients to maintain a capacity for aerobic metabolism. They also can select colder microhabitats, which results in reduced oxygen demands (Boutilier et al. 1997; Tattersall and Boutilier 1997, 1999). The preference of many aquatic hibernators for flowing water enhances survival by ensuring a steady supply of oxygen throughout the winter. Frogs submerged in cold water (3° C) undergo a variety of physiological changes that suppress their metabolic rates by as much as 70% below those measured in air at the same temperature, thereby reducing both energy costs and oxygen demands (Donohoe and Boutilier 1998, 1999; Donohoe, West, and Boutilier 1998, 2000; Ultsch, Reese, and Stewart 2004).

Despite the ability of frogs to lower metabolic rates and seek out microhabitats that provide sufficient oxygen, the chief danger for most aquatic hibernators probably is anoxia caused by freezing over of shallow ponds and lakes and heavy snow cover that blocks sunlight needed for photosynthesis by aquatic plants (Ultsch 1989; Pinder, Storey, and Ultsch 1992; E. Stewart, Reese, and Ultsch 2004; Ultsch, Reese, and Stewart 2004). Mass mortality due to anoxic conditions has been reported in many species of ranid frogs in cold, northern climates, where aggregations of more than 1,000 frogs at a single site have been reported. These include *Rana pipiens* (Manion and Cory 1952), *R. muscosa* (Bradford 1983), *R. temporaria* (Bannikov 1948; Smith 1969; Pasanen and Sorjonen 1994), *R. ridibunda* (Berger 1982; Kuzmin 1999), and the Asian frogs *R. amurensis, R. dybowskii, R.pirica,* and *R. nigromaculata* (Kuzmin and Maslova 2003). More than 3,000 dead *R. dybowskii* were found in one hole in a Russian stream (Maslova 2000). R. D. Howard (1981a) reported that many large adults in a population of bullfrogs (*R. catesbeiana*) overwintering in a shallow marsh died during two especially cold winters, probably having frozen to death because of meager snow cover. Some ranid frogs are better able to tolerate hypoxic conditions than are others. For example, *R. esculenta* is more tolerant of low oxygen levels than is *R. lessonae* (Tunner and Nopp 1979), which might explain its somewhat greater tendency to hibernate underwater.

Predation also can be a significant source of mortality for hibernating frogs, which are particularly susceptible to capture when cold and sluggish, or when they aggregate in large numbers (see chapter 14). European otters (*Lutra lutra*) and water shrews (*Neomys fodiens*) frequently capture hibernating frogs (*Rana temporaria*) in ditches, ponds, and streams (Erlinge 1967; Wolk 1976; J.-M. Weber 1990), and frogs comprise up to 45% of the winter diet of these animals. In the Russian Far East, hibernating *R. dybowskii* are a major component of otter diets as well (Kuzmin and Maslova 2003). European polecats (*Mustela putorius*) also feed heavily on frogs in winter, capturing them while hibernating in ditches or on land (D. Weber 1989a, b). Frogs hibernating in ponds are subject to predation by trout and perhaps other fish as well (Kabisch and Weiss 1968). Breckenridge (1944, p. 85) described a huge aggregation of hibernating leopard frogs (*Rana pipiens*) in Minnesota that "literally paved" the bottom of a sandy stream. Mergansers (*Mergus*) in the area were feeding mainly on hibernating frogs.

### Costs and Benefits of Hibernation on Land

Amphibians hibernating on land do not face a shortage of oxygen, but are susceptible to predation, desiccation, and freezing. Predation on amphibians that hibernate on land is not as well documented as predation on aquatic hibernators. In Finland, however, pine martens (*Martes martes*) were observed feeding on common frogs (*Rana temporaria*) after searching under moss, logs, and stones and in root cavities (Pulliainen and Ollinmäki 1996a). It was not clear, however, whether the frogs were captured during hibernation or killed before hibernation and stored by the martens.

Terrestrial hibernators have only two options available to them to survive freezing temperatures: find microhabitats that remain above freezing throughout the winter, or tolerate freezing of body tissues (Pinder, Storey, and Ultsch 1992). Most species do the former, constructing burrows in deep soil or seeking out natural cavities with sufficient insulation to keep them from freezing. Usually such sites protect the animals throughout the winter, but unusually cold winters can result in significant mortality in terrestrial hibernators. For example, Petranka (1979) found that several populations of terrestrial salamanders (*Plethodon dorsalis* and *P. richmondi*) virtually disappeared from sites in Kentucky after an exceptionally cold winter. Populations of salamanders survived only in rocky areas with extensive networks of underground tunnels. M. Smith (1969) observed mass mortality in a population of *Triturus vulgaris* hibernating under piles of timber, and Berger (1982) found many European water frogs (*Rana esculenta*) dead in their terrestrial hibernation sites. Mass mortality also has been reported at terrestrial hibernation sites of *Hyla arborea* and *Rana arvalis* in very cold winters or years with little snow cover (Kuzmin 1999).

## Freeze Tolerance

Some amphibians have evolved the ability to remain frozen for extended periods of time at temperatures as low as −8° C (Storey 1990). Similar freeze tolerance also is found in many insects, as well as garter snakes (*Thamnophis sirtalis*), and box turtles (*Terrapene carolina;* Storey and Storey 1988; Storey et al. 1988; Costanzo and Claussen 1990; Lee 1990; Storey 1990; Costanzo, Lee, and Wright 1993). There has been some controversy over whether hatchling painted turtles (*Chrysemys picta*) are freeze tolerant. These turtles apparently can survive some freezing at relatively high temperatures (−4° C), but survive lower nest temperatures by supercooling and avoiding freezing (Storey et al. 1988; Churchill and Storey 1992; Costanzo, Iverson, Wright, and Lee 1995; Packard and Packard 1995; Costanzo, Litzgus, Iverson, and Lee 1998; Packard et al., 1997, 1999). The European lizard *Zootoca* ( = *Lacerta*) *vivipara* also employs both freeze tolerance and supercooling to survive the winter (Costanzo, Grenot, and Lee 1995; Grenot et al. 2000; Voituron et al. 2002).

Seven species of amphibians are known to tolerate prolonged freezing (fig. 3.21). Six of these are North American anurans: *Rana sylvatica, Pseudacris crucifer, P. triseriata, P. regilla, H. versicolor,* and *H. chrysoscelis* (Lotshaw 1977; MacArthur and Dandy 1982; Schmid 1982; Storey 1984, 1986, 1987a; Storey and Storey 1984, 1985a, 1986a, b, 1988, 1992; Layne and Lee 1987, 1989; Lee 1988; Costanzo, Wright, and Lee 1992; Lee and Costanzo 1993; Swanson, Graves, and Koster 1996; Packard, Tucker, and Lohmiller 1998; Croes and Thomas 2000; Edwards, Koster, and Swanson 2000). The seventh is a Siberian salamander, *Salamandrella keyserlingii,* a species found well above tree line and farther north than any other salamander (Berman, Leirkh, and Mikhailova 1984; Berman 1992). This species is the most freeze tolerant of all amphibians and can survive temperatures down to −40° C. There are reports of individuals being excavated from depths of 10 m or more in Russian permafrost that appeared from radiocarbon dating to be up to 90 years old (Kuzmin and Maslova 2003). Freeze tolerance is absent in terrestrial hibernators such as *Bufo americanus, B. boreas, B. cognatus, B. woodhousii, Pseudacris streckeri, Rana aurora, Plethodon cinereus,* and *Ambystoma laterale,* as well as the aquatic hibernators *Rana pipiens* and *R. septentrionalis* (Storey and Storey 1986a; Swanson, Graves, and Koster 1996; Packard, Tucker, and Lohmiller 1998). The aquatic hibernator *Rana ridibunda* can tolerate having up to 55% of its body water frozen, but freezing of more than 58% is fatal. This species does not accumulate cryoprotectants like those found in terrestrial freeze-tolerant frogs (Voituron, Eugene, and Barre 2003). Juvenile spadefoot toads (*Spea bombifrons*) can tolerate body temperatures as low as −4° C, but they cannot survive freezing (Swanson and Graves 1995).

**Fig. 3.21.** A frozen wood frog (*Rana sylvatica*), a freeze-tolerant species that hibernates under shallow cover such as moss or leaf litter. Photo by Jon Costanzo.

### Physiology of Freeze Tolerance

Freeze-tolerant amphibians use cryoprotectants to depress the freezing point of their intracellular fluids, much as antifreeze decreases the freezing point of water. In most species, the main cryoprotectant is glucose (Storey 1997; Croes and Thomas 2000), but both glucose and glycerol are produced by *Hyla versicolor* (Schmid 1982; Storey and Storey 1985a; Layne 1999) and the closely related *Hyla chrysoscelis* (Irwin and Lee 2003), and glycerol has been found in *Salamandrella keyserlingii* (Kuzmin and Maslova 2003). *Hyla versicolor* is a tetraploid frog that evolved from ancestral diploid *Hyla chrysoscelis* populations, and one might expect them to have similar physiology. There is a possibility, however, that freeze tolerance has evolved independently several times in this complex of tree frogs, because recent studies of mitochondrial DNA have shown that frogs currently called *H. versicolor* evolved more than once from different lineages of frogs currently called *H. chrysoscelis* (Ptacek, Gerhardt, and Sage 1994). Nevertheless, a study of populations from Missouri, Indiana, and Minnesota failed to reveal any major geographic differences in freeze tolerance or concentrations of cryoprotectants (Irwin and Lee 2003). Frogs from Minnesota did tend to store more glycogen in the liver in preference to storing fat, perhaps as an adaptation to more prolonged periods of cold weather, but this did not translate into higher cryoprotectant concentrations in frozen frogs.

The biochemical and physiological changes that allow amphibians to tolerate subfreezing temperatures are similar in all species studied to date. They have been investigated in most detail for the wood frog (*R. sylvatica*). Frogs can tolerate freezing of extracellular fluids, but not complete freezing of internal organs. As freezing begins, water is drawn out of cells into extracellular spaces, particularly in the abdomen and between the skin and muscles (Storey and Storey 1984; Layne and Lee 1987; Lee et al. 1992). Wood frogs can have more than 60% of their body fluids frozen without ill effect. As a frog cools, internal organs, including the heart, continue to function even after peripheral tissues have started to freeze, allowing cryoprotectant glucose to be distributed to vital organs through the bloodstream (Layne, Lee, and Heil 1989). Thawing occurs throughout the body simultaneously, but organs such as the liver thaw most quickly because they contain high concentrations of cryoprotectant (Rubinsky et al. 1994; Lee and Costanzo 1998).

For frogs to tolerate freezing, it is essential for freezing not to be too rapid and that ice crystal formation occur at a relatively high body temperature. If freezing occurs quickly, it can be fatal (Costanzo, Lee, and Wright 1991) because there is not sufficient time for water to be drawn out of vital organs and for cryoprotectant to accumulate (Costanzo, Lee, and Wright 1992). A frog that has been supercooled (with a body temperature several degrees below 0° C) is at risk of instantaneous freezing (Costanzo, Bayuk, and Lee 1999). Ice formation in the animal generally is initiated by contact with ice crystals in the soil, but also can be promoted by various nucleating agents, such as ice-nucleating bacteria in the gut and skin (Layne, Lee, and Huang 1990; Layne 1995; Costanzo and Lee 1996). Production of cryoprotectant does not begin until ice crystals begin to form in the peripheral tissues (Storey and Storey 1985b).

Conversion of liver glycogen to glucose is supported by anaerobic metabolism facilitated by a rapid increase in the activities of key liver enzymes as soon as ice formation begins (Storey 1987a). Glucose is then rapidly distributed in the blood to vital organs, particularly heart and brain tissue (Storey and Storey 1986b; Storey 1987b). Uptake of glucose by organs is enhanced by a six-fold increase in rates of glucose transport across plasma membranes in frogs collected in the fall, compared to those collected in the spring (P. King, Rosholt, and Storey 1995). A frog's capacity for cryoprotectant production is ultimately limited by the available stores of glycogen in the liver, so frogs that enter hibernation with low glycogen stores should be less capable of surviving subfreezing temperatures (Costanzo and Lee 1993; Jenkins and Swanson 2005).

Freeze-tolerant wood frogs generally do not remain frozen all winter, but undergo repeated cycles of freezing and thawing, with glucose being rapidly reconverted to glycogen and stored in the liver after thawing (Storey and Storey 1986b; Storey 1987b). In fact, these frogs cannot survive prolonged freezing (3–4 weeks), but survival can be improved by injection of supplemental glucose prior to freezing (Layne, Costanzo, and Lee 1998). Wood frogs cannot tolerate freezing at very low temperatures (−7.5° C). Such low temperatures can result in extensive tissue damage, including damage to nerve cells (Costanzo, Allenspach, and Lee 1999) and sperm (Costanzo, Mugnano, Wehrheim, and Lee 1998).

Wood frogs and other freeze-tolerant species retain their capacity to survive subfreezing temperatures even after emerging from hibernation in spring, but low glycogen reserves limit their capacity to produce cryoprotectant. Frogs that have previously thawed lose considerable quantities of glucose through excretion of urine, but some glucose can be recovered by resorption of glucose across the bladder wall (Costanzo, Callahan, Lee, and Wright 1997). Freeze tolerance rapidly declines once the frogs begin to feed (Storey and Storey 1987). Recovery of organ system function is relatively rapid after the frog thaws, with the heart recovering first and peripheral nerves and limb muscles taking somewhat longer (Layne and First 1991; Kling, Costanzo, and Lee 1994). Late freezing does carry some cost, however, because previously frozen males exhibit reduced mate-searching behavior in breeding aggregations for a day or so after freezing (Costanzo, Irwin, and Lee 1997), as well as reduced muscle performance (Irwin, Costanzo, and Lee 2003).

The rapid increase in transport of glucose into cells in freeze-tolerant frogs resembles the physiological responses of many frogs to desiccation. For example, leopard frogs (*Rana pipiens*), which cannot tolerate freezing, can endure losses of up to 50% of total body water content. As desiccation proceeds, a frog maintains the water content of its tissues by greatly increasing intracellular concentrations of glucose, which in turn increase the osmotic concentration of the cells. The glucose is derived from glycogen stored in the liver. Freeze-tolerant wood frogs (*Rana sylvatica*) and spring peepers (*Pseudacris crucifer*) are slightly more tolerant of desiccation than are leopard frogs and have a much greater capacity for transporting glucose into vital organs. Freeze-tolerant chorus frogs (*Pseudacris triseriata*) showed elevated mobilization of liver glycogen in response to desiccation during the winter, whereas freeze-intolerant toads (*Bufo cognatus* and *B. woodhousii*) did not (Edwards, Jenkins, and Swanson 2004). Dehydrated anurans also tend to accumulate urea (see chapter 2), and urea has been shown to have cryoprotectant properties similar to that of glucose. In contrast to glucose, however, urea concentrations do not increase with freezing (Costanzo and Lee 2005). The similarity in responses to desiccation and freezing suggest that the physiological responses of freeze-tolerant frogs are modifications of preexisting physiological responses that are widespread in frogs, although these responses may not be present in all anurans (Churchill and Storey 1993, 1995; Costanzo, Lee, and Lortz 1993).

#### Evolution of Freeze Tolerance

Although a detailed phylogenetic analysis of the evolution of freeze tolerance is not yet possible, the distribution of freeze tolerance among ranids and hylids suggests that this adaptation has evolved independently at least three times and possibly more. *Rana sylvatica* is one of the few terrestrial hibernators among North American ranids, and so far is the only member of its family known to tolerate freezing. The European common frog, *R. temporaria,* can survive at −2° C for 24 hours, but it lacks the prolonged freeze tolerance of *R. sylvatica* (Pasanen and Karhapää 1997). Recent phylogenetic analyses suggest that these frogs are not closely related, despite a general similarity in appearance and breeding ecology (Hillis and Wilcox 2005). The two freeze-tolerant species of *Pseudacris* are closely related (Hedges 1986) and could represent a single origin of freeze tolerance that clearly is independent of wood frogs. *Pseudacris streckeri,* a burrowing species with a more southern distribution, is not freeze tolerant (Packard, Tucker, and Lohmiller 1998). *Pseudacris regilla,* was previously considered not to be freeze tolerant (Storey and Storey 1988), but a recent study demonstrated that individuals from a high-altitude population in northern California have at least a limited capacity for freeze tolerance (Croes and Thomas 2000). Freeze tolerance in *Hyla chrysoscelis* and *Hyla versicolor* could be a single evolutionary event, or several independent events, depending on how many lineages of this complex are freeze tolerant. At least limited freeze tolerance eventually may be found in other amphibians from North America and Eurasia. For example, Berger (1982) found two adult *Rana lessonae* frozen in a block of ice where water had collected in their terrestrial hibernation site. Both individuals were alive when the ice thawed, but six adult *R. esculenta* frozen with them were dead. He also found evidence that *R. lessonae* was more tolerant of cold temperatures in more typical terrestrial hibernation sites than was *R. esculenta.*

### Summary and Conclusions

In addition to their highly permeable skin, no feature of amphibian physiology has a greater impact on their ecology and behavior than ectothermy. Virtually every physiological process in amphibians is temperature dependent, including metabolic rate, the rate at which food can be processed, growth and development, reproductive cycles, locomotion, and sensory biology. This high degree of temperature dependence in turn affects the latitudinal and altitudinal distribution of amphibians, their activity cycles, the timing of breeding and the structure of their mating systems, choice of calling and oviposition sites, and life history strategies. Temperature also can affect the interactions of amphibians with their predators and with other amphibian species in the community.

Because they lack the capacity to generate their own body heat, amphibians are dependent on behavioral thermoregulation to keep their body temperatures within a suitable range. In contrast to many reptiles, however, amphibians are constrained in how precisely they can regulate body temperature by the permeability of their skin. Amphibians can reduce heat loads through evaporative cooling, but only if they have access to sufficient moisture to replace that lost through evaporation. Their ability to increase body temperatures through basking is severely limited, because radiation absorbed by the skin tends to increase evaporative water loss, which in turn lowers body temperature. The only amphibians capable of raising body temperatures significantly through basking are arboreal species that have evolved increased cutaneous resistance to evaporative water loss. Precise thermoregulation is virtually impossible for amphibians in water, because the small size of most amphibians ensures that heat will be conducted away from the body very rapidly. Consequently, amphibians in water are essentially thermoconformers, relying on selection of warm or cool microhabitats to regulate body temperature.

Amphibians at high altitudes face particular severe problems of thermoregulation. Many high-altitude environments undergo major daily temperature fluctuations, with intense solar radiation during the day and rapid cooling at night. High-altitude environments also tend to be relatively dry, which exacerbates problems of evaporative water loss. Some high-altitude populations of amphibians cope with these problems mainly through behavioral adjustments, including being active mainly during the day and at a wider range of temperatures than low-elevation populations. There also is evidence for some tropical frogs that high-elevation populations have evolved physiological traits that enable them to be active at temperatures that would incapacitate their low-elevation relatives.

Temperate-zone amphibians have some capacity to adjust physiologically to seasonal changes in temperature through acclimation of metabolic rate and thermal tolerance, but such acclimation responses appear to be absent in tropical species. Most amphibians exhibit little or no capacity for acclimation of locomotor performance. There have been many experimental studies of acclimation of thermal tolerance, as measured by the critical thermal maximum ($CT_{Max}$), but the adaptive significance of such acclimation is unclear. In many species, a difference of acclimation temperature of as much as 20° C produces change in $CT_{Max}$ of only a few degrees. Because the $CT_{Max}$ typically is well above the upper limit of normal activity temperatures, the benefit of a slight increase in $CT_{Max}$ has yet to be demonstrated. Possibly such acclimation provides survival benefits on rare occasions when amphibians are exposed to unusually high and near-lethal temperatures. Typical acclimation requires a period of days or weeks of exposure to a new temperature regime to produce measurable responses. Such long-term acclimation is unlikely to be beneficial when amphibians are exposed to sudden changes in temperature. Some species exhibit short-term physiological adjustments to elevated temperature in a matter of hours. This type of heat hardening is beneficial in allowing amphibians to survive daily fluctuations in temperature during summer months.

Amphibians are limited in their ability to remain active at very high temperatures, but they are completely incapacitated by very low temperatures. This means that amphibians in winter must seek out microhabitats that remain above freezing, or they must be able to tolerate being frozen on at least a short-term basis. Aquatic hibernators are not freeze tolerant, but often seek out hibernation sites that remain unfrozen and well oxygenated throughout the winter. Most terrestrial hibernators also are not freeze tolerant and must burrow into the ground to a sufficient depth to avoid freezing or seek out caves, rock crevices, and other sheltered sites. A limited number of amphibians have been shown to tolerate freezing for periods of days or weeks. They do so by removing most of the water from their cells, protecting the cells with a cryoprotectant that prevents freezing (glucose or glycerol), and allowing the extracellular spaces to freeze. The phylogenetic distribution of freeze tolerance is poorly understood. At present, all examples of freeze tolerance among anurans are North American hylids (*Pseudacris* and *Hyla*) and one ranid (*Rana sylvatica*). Species that are closely related to these in North America, Europe, and Asia appear not to be freeze tolerant, but most species have not been thoroughly tested.

Another question that remains to be answered is whether freeze tolerance is widespread within species, such as *Pseudacris crucifer, Hyla versicolor,* and *Rana sylvatica,* or is population specific. There was not much geographic variation in freeze tolerance among populations of gray tree frogs from relatively cold regions, but more southern populations have not been studied. It is clear that freeze tolerance has evolved several times independently in North American anurans, possibly multiple times even within widespread species. The question is of particular interest for the gray tree frogs. Both the diploid species, *Hyla chrysoscelis,* and the tetraploid species, *H. versicolor,* have been shown to be freeze tolerant. Recent molecular evidence indicates that tetraploid gray tree frogs have arisen several times independently from different lineages of *H. chrysoscelis.* This raises the question of whether freeze tolerance evolved in all lineages of *H. chrysoscelis* and subsequently was passed on to populations of *H. versicolor,* or arose independently several times within the complex.

# **Chapter 4** Respiration

*The toads,* Bufo, *have a thick-set body, covered with warts or papillae, from which exudes a viscid humor. . . . These hideous and disgusting reptiles ordinarily conceal themselves in shady, humid places, from which they do not go out, except at night, or immediately after warm and abundant rains of summer. . . . In countries where the winter is cold, they pass the season benumbed in holes. Their respiration then becomes extremely limited, and the contact of a very small quantity of air with the skin is sufficient to maintain their existence.*
—W. S. W. Ruschenberger, MD, *Elements of Herpetology and Ichthyology* (1856)

IT HAS LONG been obvious to biologists that there is something unusual about the respiration of amphibians. Not only do they often appear to need very little oxygen to stay alive, but they also can take up most of it through the skin, bypassing the respiratory organs used by other terrestrial vertebrates. These unusual characteristics of amphibians have made them popular subjects for investigations of respiratory physiology since the earliest days of experimental biology. The nineteenth-century anatomist Alexander Ecker stated that the frog "enjoys the doubtful honour of being the physiological domestic animal," being "daily sacrificed upon the altar of science." He added, "the history of the most important physiological discoveries is closely related with the employment of the frog in physiological research" (Ecker 1889, 1). Indeed, studies of frogs were important in the development of modern ideas on animal respiration (for a detailed historical review, see C. B. Jørgensen 2000). In the 17th century, Robert Boyle placed frogs and other animals in chambers evacuated with his newly designed vacuum pump to test the effects of reduced air pressure on living organisms; needless to say, most of these animals died (Boyle 1670). Boyle's interest in respiration was incidental to his research on pneumatics and the chemistry of air, however. At the end of the next century, after the discovery of oxygen by the French chemist Antoine Lavosier, the Italian physiologist Lazzaro Spallanzani performed a number of experiments on frogs in his attempt to understand the nature of respiration. He removed the lungs of frogs and showed that these animals used oxygen and produced carbon dioxide as did intact frogs; he concluded that the skin must be an important avenue of gas exchange (Spallanzani 1803). Spallanzani also was one of the first scientists to suggest that respiration occurs in the tissues and not just in the lungs, as many others had assumed (Perkins 1964; Dolman 1975; Castellani 1984).

In the nineteenth century, the French physiologist William Frederic Edwards (1817) studied the influence of environmental variables on living organisms, including the effects of heat and cold on animals, the role of light in the development of amphibians, and the responses of frogs and other animals to a lack of oxygen. His results were later published in a book (Edwards 1824) that could be considered the first major work on physiological ecology (Allee et al. 1949; Kauta 1971; C. B. Jørgensen 2000). Edwards repeated Spallanzani's experiment of removing the lungs of frogs and confirmed his findings that respiration continued, apparently through the skin. He also showed that frogs without lungs could survive for more than a month in cold water saturated with air. He blocked cutaneous respiration by various means including coating frogs with paint and submerging them in oil. Since many frogs subjected to these treatments died, he concluded that the skin is essential for respiration (Dolk and Postma 1927).

In the 1840s, two French chemists, Victor Henri Regnault and Jules Reiset, developed the first closed respirometer for making quantitative measurements of oxygen consumption and carbon dioxide production in animals. They undertook a detailed comparative study of respiration in "warm-blooded" and "cold-blooded" vertebrates and, in the process, made the first accurate measurements of respiratory quotients in animals (the ratio of $CO_2$ produced to $O_2$ consumed). In experiments with both frogs (probably *Rana esculenta*) and salamanders (most likely *Salamandra salamandra*), they obtained values for $O_2$ consumption that are remarkably close to those reported by later workers (see table 1 in Gatten, Miller, and Full 1992, for comparative data). Their data showed that $O_2$ consumption of frogs increases with temperature, although they did not comment on this point, and they demonstrated that amphibians and reptiles have much lower metabolic rates than do birds and mammals. They followed the example of Spallanzani and Edwards in removing lungs from frogs and found only a slight reduction in the amount of $O_2$ consumed, supporting previous conclusions that most gas exchange takes place through the skin (Regnault and Reiset 1849).

The modern era of amphibian respiration studies was initiated by the Danish physiologist Christian Bohr and his student August Krogh. The former was the first to describe the change in $O_2$ affinity of blood brought about by a change in acidity, now known as the Bohr Effect (Astrup 1972). In his studies of frog respiration, Bohr found that metabolic rates were higher in summer than in winter and that removal of the lungs therefore had a much greater effect on $O_2$ consumption in the summer. He also proposed that the skin of a frog is much more permeable to $CO_2$ than to $O_2$ (Bohr 1900). In 1903, Krogh completed his PhD dissertation on pulmonary and cutaneous respiration in frogs, greatly expanding on Bohr's work. His study was the first to actually quantify the uptake of $O_2$ and release of $CO_2$ through both skin and lungs. He found that most $CO_2$ was eliminated through the skin, while the lungs were more important for uptake of $O_2$. He also confirmed Bohr's finding of seasonal variation in metabolic rates (Krogh 1904). Much of this work was repeated by two Dutch workers, H. E. Dolk and N. Postma (1927), who confirmed Krogh's major results and provided additional evidence for seasonal variation in metabolism. Because all of these early researchers used artificial means to ventilate the lungs, their work provided little information on the relative importance of pulmonary and cutaneous respiration in freely breathing amphibians, and more than 35 years passed before accurate measurements of this type were made on both salamanders and frogs (Whitford and Hutchison 1963; Vinegar and Hutchison 1965).

Much of the subsequent work on amphibian respiration has focused on comparisons of aquatic and terrestrial species. Major research themes have included the relative importance of pulmonary and cutaneous respiration, aspects of acid-base balance as it is affected by gas exchange in the two environments, and chemical influences on the control of respiration (Boutilier, Stiffler, and Toews 1992; Shoemaker et al. 1992; Booth 1994; Malvin 1994). There also has been a revival of interest in the neuromuscular control of lung ventilation (Brainerd 1994), a field that traces back to the work of Robert Townson (1799) in the late eighteenth century (C. B. Jørgensen 2000). Much of this work has been in the tradition of comparative physiology established by Krogh. Hence, most of our knowledge of the details of respiration in amphibians comes from studies of a few species chosen primarily for their usefulness as laboratory animals (e.g., *Rana catesbeiana, Bufo marinus, Xenopus laevis, Ambystoma tigrinum*). Exceptions are a few studies relating respiratory adaptations to environments differing in the availability of oxygen (e.g., Guimond and Hutchison 1976; Ultsch 1976). While the emphasis of most studies has been more on physiological mechanisms than on ecological adaptations, they do provide insights into ways in which amphibians cope with a variety of habitats.

## Environmental Aspects of Gas Exchange

### The Availability of Oxygen

Oxygen is seldom a limiting factor for amphibians in the terrestrial environment. The atmosphere is approximately 21% $O_2$, and its $O_2$ content is not greatly affected by the respiration of animals. While the percentage of $O_2$ in the atmosphere remains constant, the partial pressure of $O_2$ ($PO_2$) declines with increasing altitude, and it is $PO_2$ that determines the availability of $O_2$ to respiring organisms. However, because of their low metabolic rates, terrestrial amphibians do not appear to be limited by low $PO_2$ at high altitudes. In North America, terrestrial salamanders are found as high as 3,300 m (11,000 ft; Hock 1964), while a toad, *Bufo spinulosus,* occurs as high as 4,500 m (14,760 ft.) in the Andes of South America (Pearson and Bradford 1976; Sinsch 1989a). More than a quarter of the anuran fauna of Sichuan Province, China, is found only above 2,200 m (7,200 ft), and some pelobatid frogs in the genus *Scutiger* are regularly found at altitudes over 4,200 m (14,000 ft; Liu 1950; Hock 1964; Zhao, Wu, and Inger 1989). Anurans can easily survive $PO_2$ levels equivalent to those at altitudes of over 20,000 m (Hock 1964), although metabolic rates are depressed under hypoxic conditions (Hou and Huang 1999). Perhaps more pertinent is a study by Withers and Hillman (1983), who demonstrated that maximal $O_2$ consumption during forced exercise in both *Bufo cognatus* and *Rana pipiens* was independent of $PO_2$ down to values as low as 80 Torr (1 Torr = 1 mm Hg). This is equivalent to $PO_2$ at about 4,500 m (14,760 ft). Hence, terrestrial amphibians should be capable of carrying out their normal activities throughout the altitudinal range of all living species.

One situation that might expose terrestrial amphibians to low levels of $O_2$ is burrowing. Indeed, many experimental studies of physiological responses of amphibians to hypoxic conditions (Armentrout and Rose 1971; Boutilier and Toews 1977; D'Eon, Boutilier, and Toews 1978; Boutilier et al. 1979c; Portner, MacLatchy, and Toews 1991) or $O_2$ affinity of the blood (Wood et al. 1975) have been based on the assumption that $O_2$ is depleted inside burrows. The only careful measurements of $O_2$ concentration in soil adjacent to burrowing amphibians are those made by Seymour (1973c) on spadefoot toads (*Scaphiopus couchii*). He found that $O_2$ concentration in soil next to winter burrows (55–80 cm deep) was not significantly different from that at the soil surface. Oxygen concentration in soil adjacent to shallow (8 cm) summer burrows of *S. couchii* was slightly lower than that at the surface, but not enough to have any effect on the animal's metabolism. Hence, one cannot assume that burrowing amphibians are normally exposed to severe hypoxia. Oxygen content of soil decreases with increasing water content, and is lowest in soils with fine particle size, such as those composed mostly of clay. Hypoxic conditions in burrows are most likely to occur in the tropics, where water-logged soils inhibit diffusion of gases. Nevertheless, wet tropical soils can be surprisingly well oxygenated. On Barro Colorado Island in Panama, soil $O_2$ content rarely dropped below 15% during the rainy season, even at depths of 80 cm, and usually was above 17%—much higher than the $O_2$ content of aquatic habitats (Kursar, Wright, and Radulovich 1995).

Aquatic environments can present less favorable conditions for $O_2$ uptake and are more variable in oxygen availability than are terrestrial environments. The $O_2$ content of fully saturated water is only about 1/30th that of air and is strongly influenced by temperature ($O_2$ capacity decreases with increasing temperature; Krogh 1941; Dejours 1976; Randall et al. 1981). In addition, the $O_2$ content of closed bodies of water is affected by the respiration of aquatic organisms, so seasonal and even daily fluctuations in respiration can result in major changes in the availability of $O_2$. Oxygen content also is governed by the degree of mixing in the water column. In the absence of wind-driven or thermally driven mixing in lakes and ponds, deep waters tend to become hypoxic or even anoxic because diffusion of $O_2$ from air into water is exceedingly slow. This problem is less severe in flowing water, where movement of the water usually keeps it well aerated. Hence, amphibians that inhabit rapidly flowing streams or rivers, particularly in cool environments, are less likely to encounter stressful hypoxic conditions than those found in still water.

With the exception of Gordon Ultsch's work on the respiration of sirens (Ultsch 1971, 1973a, b, 1974, 1976), there have been few detailed studies of aquatic $O_2$ availability in relation to amphibian ecology. Scattered data on habitats with low levels of dissolved $O_2$ are summarized in table 4.1. Some of these come from studies of amphibian ecology, while others are taken as representative of conditions in habitats often used by amphibians. In most cases, the values shown are among the lowest recorded and are not necessarily typical of the habitat throughout the year. Open aquatic habitats with exposure to sunlight often have higher $O_2$ levels during the day, when photosynthesizing plants are producing $O_2$, than at night, when photosynthesis stops and the plants and other organisms deplete the $O_2$ through respiration. Some investigators have reported measurements from several depths or different locations within a habitat, but only a few have attempted to document daily or seasonal variation in $O_2$ availability (e.g., Laessle 1961; W. Moore and Burn 1968; Ultsch 1971, 1973a, 1976; Bradford 1983; Wong and Booth 1994; Nie, Crim, and Ultsch 1999).

Low levels of $O_2$ are particularly characteristic of swampy habitats, both in the tropics and the temperate zone (table 4.1). In many areas, such as the "floating meadows" of the Amazon, papyrus swamps of Africa, and lakes invaded by water hyacinths, $O_2$ can be totally depleted in warm weather. This is due to respiration of bacteria breaking down vast amounts of organic material and to a lack of light for photosynthesis resulting from dense mats of vegetation covering the water. Temporary ponds often have very low $O_2$ content as well, but vary considerably depending on depth, density of vegetation, shading from surrounding trees, and so on. Hutchinson, Pickford, and Schnurman (1932) found that shallow pans and vlies in South Africa, the principal breeding sites of many anurans in that region, varied from severely hypoxic to nearly saturated with $O_2$. Consistently low levels were found in shallow seasonal pans with large amounts of decaying organic matter and few photosynthesizing plants.

Hypoxia can be particularly severe in very small bodies of water, where respiration of organisms can quickly deplete the available $O_2$ supply. For example, Kluge (1981) found very low $O_2$ concentrations in the mud basins used by *Hyla rosenbergi* for oviposition (table 4.1), and eggs could survive only if they remained suspended at the surface. Laessle (1961) and Janetsky and Vareschi (1993) made detailed studies of variation in $O_2$ content of water-filled bromeliads in Jamaica. Oxygen levels usually were low, particularly in

**Table 4.1** Examples of aquatic habitats with low dissolved oxygen content inhabited by amphibians, especially larvae

| Habitat/Location (source no.) | Dissolved oxygen (mg/l) | Representative amphibians (source no.) |
|---|---|---|
| Chaco swamp | | |
| Paraguay (1) | 0.0–1.0 | |
| Desert rain pool | | |
| Sudan (8) | 4.3–7.1 | |
| Floating meadow swamp | | |
| Amazon (12, 25) | 0.0–1.5 | *Hyla, Leptodactylus, Sphaenorhynchus** (19) |
| Forest pond | | |
| Louisiana (11) | 0.5–1.5 | |
| Peru (28) | 0.6–6.6 | *Hyla, Phyllomedusa, Leptodactylus* |
| Forest swamp | | |
| Ecuador (22) | 0.3–1.0 | *Hyla, Leptodactylus* (13, 22) |
| Guyana (5) | 0.9 | |
| Grass swamp | | |
| Guyana (5) | 0.3–1.8 | |
| Ice-covered lake | | |
| California (24) | 0.4–9.0 | *Rana muscosa* (24) |
| Mud nest built by frogs | | |
| Panama (23) | 1.3 | *Hyla rosenbergi* (23) |
| Nest in stream | | |
| Australia (29) | 6.5 | *Litoria lesueuri* complex |
| Pans & vlies | | |
| South Africa (8) | 0.9–7.1 | *Phrynomerus, Ptychadena, Pyxicephalus, Hyperolius** (21) |
| River and lake swamp | | |
| Africa (2, 9, 14) | 0.0–7.2 | *Hyperolius, Afrixalus, Ptychadena** (6, 10) |
| Senescent swamp | | |
| India (16) | 1.5–2.7 | *Rana, Rhacophorus** (4) |

*(continued)*

**Table 4.1** (continued)

| Habitat/Location (source no.) | Dissolved oxygen (mg/l) | Representative amphibians (source no.) |
|---|---|---|
| Shallow pond | | |
| England (26) | 1.1–7.9 | |
| Stagnant stream | | |
| Panama (23) | 2.3 | *Leptodactylus, Hyla, Physalaemus, Bufo* (23) |
| Swamp lagoon | | |
| Uganda (31) | 0.3–1.3 | *Hyperolius, Afrixalus, Ptychadena** (6, 10) |
| Temporary pond | | |
| Ecuador (22) | 0.5–3.4 | *Hyla, Leptodactylus* (13, 22) |
| Michigan (27) | < 1.0 | *Hyla, Pseudacris, Ambystoma, Rana** (20) |
| Mississippi (18) | 2.7 | *Ambystoma maculatum* (18) |
| Mississippi (15) | < 2.0 | *Ambystoma opacum* (15) |
| Water-filled bromeliad | | |
| Brazil (30) | 2.6 | *Osteocephalus oophagus* (30) |
| Jamaica (7) | 0.5–3.0 (night) | *Osteopilus brunneus* (7) |
| Water-filled tree hole | | |
| Brazil (30) (30) | 0.2 | *Osteocephalus resinifictrix* |
| Water hyacinth swamp | | |
| Florida (17) | 0.0–4.0 | *Pseudobranchus, Siren** (17) |
| Weedy pond | | |
| India (16) | 1.1–4.2 | *Rana, Rhacophorus** (4) |

*Sources:* (1) Carter and Beadle 1931; (2) Beadle 1932; (3) Hutchinson, Pickford, and Schnurman 1932; (4) McCann 1932; (5) Carter 1935; (6) Schmidt and Inger 1959; (7) Laessle 1961; (8) Rzoska 1961; (9) Bishai 1962; (10) Stewart 1967; (11) Moore and Burn 1968; (12) Junk 1970; (13) Crump 1974; (14) Rzoska 1974; (15) Weigmann and Altig 1975; (16) Dehadrai and Tripathi 1976; (17) Ultsch 1976; (18) Branch and Taylor 1977; (19) Hödl 1977; (20) Collins and Wilbur 1979; (21) Passmore and Carruthers 1979; (22) Crump 1981b; (23) Kluge 1981; (24) Bradford 1983; (25) Junk 1983; (26) Carlvalho 1984; (27) Engle 1985; (28) Magnusson and Hero 1991; (29) S. Richards and Alford 1992; (30) Schiesari, Grillitsch, and Vogl 1996; (31) Chapman et al. 2002.

*Notes:* Most measurements were made at 20–30°C. Water saturated with air at these temperatures would have a dissolved oxygen content of 8–9 mg/l. Species marked with an asterisk (*) are representative of the habitat type but were not necessarily found in the same location in which oxygen measurements were made.

bromeliads filled with organic detritus. When algae were present, water near the surface sometimes was supersaturated in bright sunlight, but the pools became hypoxic, especially deeper in the chamber, when low light prevented photosynthesis by the algae. These bromeliads are inhabited by tadpoles of the hylid frog *Osteopilus brunneus,* which appear to be obligate air-breathers from an early age (Lannoo, Townsend, and Wassersug 1987). Many other species of frogs, including some hylids, microhylids, rhacophorids, and dendrobatids, either breed or deposit their tadpoles in bromeliads or tree holes. The latter are likely to be even more hypoxic because they often are filled with organic detritus and have little exposure to light, but I have been unable to find quantitative measurements of $O_2$ concentrations. Tadpoles in such habitats generally must breathe air to survive and often have specialized morphological adaptations for aerial respiration (Noble 1929a; Wassersug, Frogner, and Inger 1981; Lannoo, Townsend, and Wassersug 1987; Krügel and Richter 1995; see also chapter 12).

Although warm, swampy waters often have low levels of dissolved $O_2$, the same also can be true of cold, clear ponds and lakes when such environments are covered by ice in winter. In shallow lakes with thick ice cover, $O_2$ depletion can be severe enough to cause mass mortality of fish (Greenbank

1945; Barica and Mathias 1979). High winter mortality among predatory fish can actually benefit some amphibians (Eaton et al. 2005), but amphibians that overwinter at the bottom of ice-covered ponds and lakes (Emery, Berst, and Kodaira 1972) sometimes suffer high mortality as well (see chapter 3). One such species, *Rana muscosa*, lives in high-elevation lakes in the western United States (Bradford 1983). Adult frogs were less tolerant of low $O_2$ tensions than were tadpoles and therefore were more susceptible to being eliminated from shallow lakes because of $O_2$ depletion. However, in many lakes, severe $O_2$ depletion probably occurs only in years with exceptional ice cover. Consequently, even if adult frogs are periodically eliminated from the population by winter mortality, the population of most lakes is maintained because overwintering tadpoles are able to survive long enough to reach sexual maturity and reproduce.

## Environmental Carbon Dioxide

In addition to supplying $O_2$ to the tissues, animals also must eliminate $CO_2$ produced as a by-product of metabolism. Carbon dioxide itself is not especially toxic to animals; its main detrimental effects are manifested as changes in the organism's acid-base balance. Terrestrial amphibians have little difficulty eliminating $CO_2$, which readily diffuses into the atmosphere. Probably the only circumstances in which terrestrial species are likely to encounter elevated levels of $CO_2$ are in burrows. A buildup of $CO_2$ inside a burrow could reduce rates of diffusion away from the body, resulting in some respiratory acidosis (Boutilier et al. 1979c). This should be especially true for species that burrow in mud and remain dormant as the mud dries out (e.g., *Siren, Amphiuma*). For species that burrow in loose soil, however, $CO_2$ concentrations in burrows are not likely to be very high because $CO_2$ will diffuse rapidly into the soil. In addition, $CO_2$ released by a burrowing animal will combine with water in the soil, thereby reducing the $PCO_2$ of the burrow.

Aquatic environments often are subjected to major fluctuations in $CO_2$ concentrations. Swampy environments that have very low levels of dissolved $O_2$ usually have high levels of $CO_2$ as well (fig. 4.1). High $CO_2$ concentrations will impede elimination of $CO_2$ through the skin, resulting in severe acidosis. Under some circumstances, $CO_2$ levels in the water can be so high that $CO_2$ actually will move across the skin into the animal instead of being eliminated (Ultsch 1976). Most fish cannot tolerate such conditions for long, but sirens thrive in ponds covered with water hyacinths where $CO_2$ concentrations become very high. Tests of one species, *Siren lacertina,* showed that individuals of all sizes were capable of surviving many days exposure to $CO_2$ levels well above those normally encountered in such ponds (Ultsch 1976).

## Gas Exchange by Adult Amphibians

Amphibians have a number of different avenues of gas exchange in both air and water. Air-breathing amphibians can utilize the lungs, skin, or lining of the mouth and pharynx for gas exchange, while water-breathers use gills or the skin. Some species are biomodal breathers, exchanging gas with

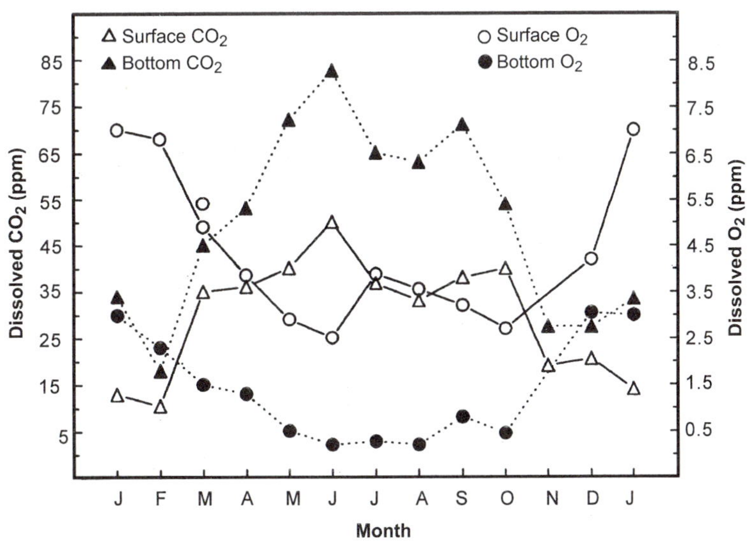


**Fig. 4.1.** Concentrations of dissolved oxygen and carbon dioxide throughout the year in a water hyacinth community in Florida, measured at the water surface and at the bottom of the pond. Oxygen is depleted during the summer, especially on the bottom, while carbon dioxide levels are at their highest during the summer. After Ultsch (1976).

the surrounding water through the skin while surfacing to breathe air with their lungs. The relative importance of the skin for $O_2$ uptake varies among species and within species as a function of temperature and metabolic demand. The skin itself is a metabolically active organ that performs a variety of biological functions (see chapter 2). A significant proportion of the $O_2$ taken up through the skin is used to support the metabolism of the skin itself (Vitalis 1990).

The early work of Krogh (1904) showed that the skin is the major avenue of $CO_2$ excretion in ranid frogs, and subsequent work has confirmed this pattern for most aquatic and terrestrial amphibians (Mendes 1945; Sawaya 1947; Whitford and Hutchison 1963, 1965b; Hutchison, Whitford, and Kohl 1968; Guimond and Hutchison 1968, 1973a, b, 1974, 1976; Bentley and Shield 1973; Shield and Bentley 1973; Gottlieb and Jackson 1976; Ultsch 1976; Mackenzie and Jackson 1978; Emilio and Shelton 1980; Burggren and West 1982; Burggren and Moalli 1984; Feder and Burggren 1985a; Jared et al. 1999). In many species, at least 75% of $CO_2$ exchange takes place through the skin.

## Oxygen Uptake in Air

In the 1960s, investigators studied partitioning of aerial gas exchange in terrestrial amphibians by fitting frogs and salamanders with plastic masks and placing them in a divided chamber. This procedure separates the air surrounding the head from that surrounding the rest of the body (Whitford and Hutchison 1963, 1965a, b, 1966, 1967; Vinegar and Hutchison 1965; Guimond and Hutchison 1968; Hutchison, Whitford, and Kohl 1968). $O_2$ uptake through the skin is then measured in the body chamber, while the combined uptake through the lungs and buccopharyngeal region is measured through the mask. In lungless species, of course, only buccopharyngeal and cutaneous gas exchange are measured. The resulting measurements provide an estimate of the relative importance of various gas exchangers. This procedure is subject to considerable error, however, because the presence of a mask often causes an animal to struggle to free itself, thereby elevating its metabolic rate (Feder 1976a, b, 1977). The result would be an overestimate of the importance of lungs in oxygen uptake, since the lungs account for most of the increase in oxygen uptake in active animals (see the following).

Data for a number of terrestrial salamanders and anurans are summarized in figure 4.2. At 15° C, pulmonary respiration predominates for nearly all of the anurans, whereas only a few species of salamanders take up more than half of their $O_2$ through the lungs. This reflects the more extensive respiratory surface area in the lungs of anurans compared to most salamanders (see further discussion in the following). In plethodontids, the skin is by far the most important av-

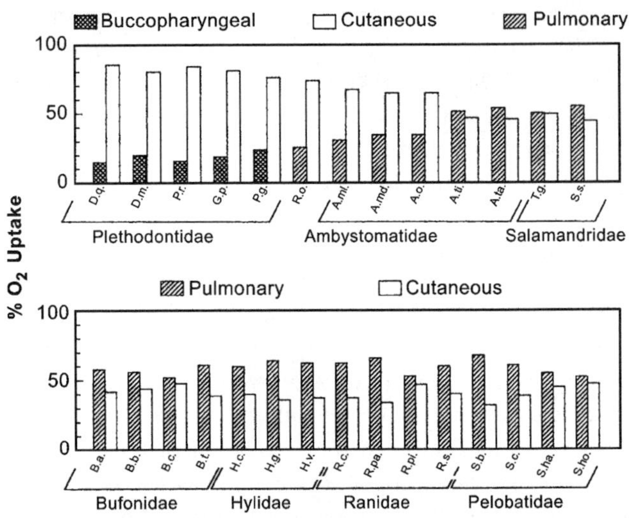


**Fig. 4.2.** Relative importance of pulmonary, cutaneous, and buccopharyngeal oxygen uptake in air for selected terrestrial urodeles (top) and anurans (bottom). Pulmonary and buccopharyngeal respiration could not be measured separately in species with lungs. Bars indicate mean proportions for several individuals of each species. Species abbreviations: *D.q.* = *Desmognathus quadramaculatus; D.m.* = *D. monticola; P.r.* = *Pseudotriton ruber; G.p.* = *Gyrinophilus porphyriticus; P.g.* = *Plethodon glutinosus; R.o.* = *Rhyacotriton olympicus; A.ml.* = *Ambystoma maculatum; A.md.* = *A. macrodactylum; A.o.* = *A. opacum; A.t.* = *A. tigrinum; A.ta.* = *A. talpoideum; T.g.* = *Taricha granulosa; S.s.* = *Salamandra salamandra; B.a.* = *Bufo americanus; B.b.* = *B. boreas; B.c.* = *B. cognatus; B.t.* = *B. terrestris; H.c.* = *Hyla cinerea; H.g.* = *H. gratiosa; H.v.* = *H. versicolor; R.c.* = *Rana clamitans; R.pa.* = *R. palustris; R.pi.* = *R. pipiens; R.s.* = *R. sylvatica; S.c.* = *Scaphiopus couchii; S.ho.* = *Scaphiopus holbrookii; S.b.* = *Spea bombifrons; S.ha.* = *Spea hammondii.* Plotted from data in Whitford and Hutchison (1963, 1965b, 1966), Vinegar and Hutchison (1965), and Hutchison, Whitford, and Kohl (1968).

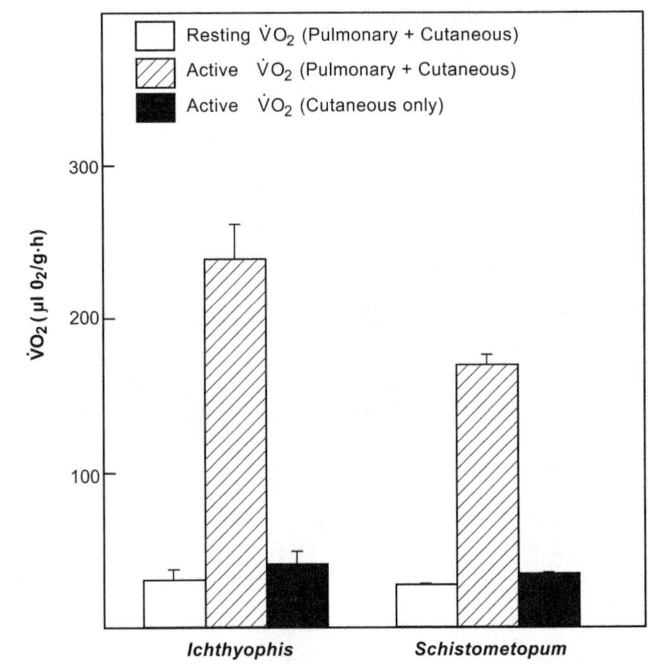


**Fig. 4.3.** Total and cutaneous oxygen consumption measured in two species of terrestrial caecilians, *Ichthyophis kohtaoensis* and *Schistometopum thomense*. Bars show means + 1 SE. After Smits and Flanagin (1994).

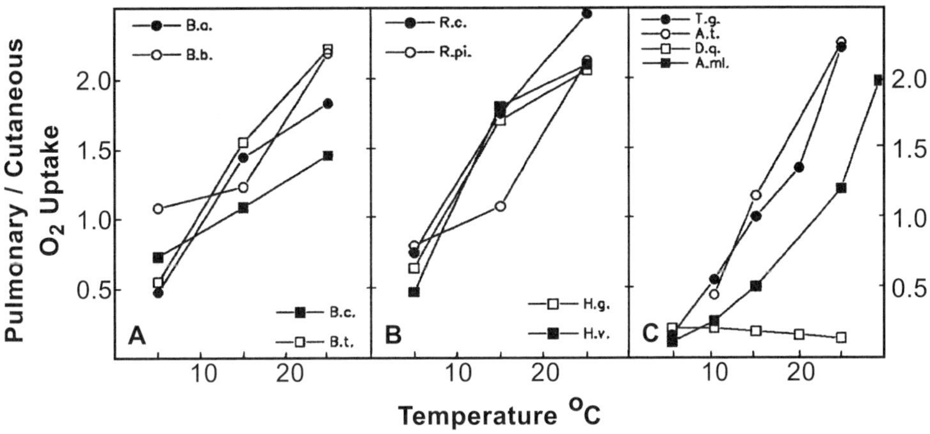


**Fig. 4.4.** Changes in the relative importance of lungs and skin in the uptake of oxygen as a function of temperature in selected amphibians, as shown by the ratio of pulmonary to cutaneous uptake. A positive slope indicates an increase in the importance of pulmonary gas exchange with increasing temperature. (A) Four species of *Bufo.* (B) Two species of *Rana* and two species of *Hyla.* (C) Four species of urodeles. Species abbreviations as in fig. 4.2. Plotted from data in Whitford and Hutchison (1963, 1965b) and Hutchison, Whitford, and Kohl (1968).

enue of $O_2$ uptake, with the buccopharyngeal region seldom accounting for more than 20%. Even in the Ambystomatidae, which have relatively large lungs, most $O_2$ is taken up through the skin. In a few ambystomatids and terrestrial salamandrids, the lungs and skin are about equally important. Data on partitioning of $O_2$ uptake are not available for terrestrial caecilians at rest, but measurements for two species engaged in moderate levels of activity indicate that the skin accounts for only about 20% of $O_2$ uptake (Smits and Flanagan 1994). Combined pulmonary and cutaneous uptake of resting animals was slightly lower than cutaneous uptake in active animals (fig. 4.3), so it is possible that the skin could supply all of the animals' $O_2$ at rest.

The relative importance of different gas exchangers is not fixed, but varies with temperature and level of activity. An increase in temperature leads to a substantial increase in pulmonary $O_2$ uptake in species with lungs, while cutaneous uptake increases more slowly. The net result is a substantial increase in the relative importance of pulmonary respiration (fig. 4.4). This reflects a large effect of increased ventilation on gas exchange through the lungs (Whitford and Hutchison 1965b; Guimond and Hutchison 1968; Hutchison, Whitford, and Kohl 1968; Kruhoffer et al. 1987). Increased activity also can be expected to increase the importance of pulmonary $O_2$ uptake. Although amphibians can increase rates of $O_2$ diffusion through the skin as metabolic demands increase (see Feder and Burggren 1985a and following discussion), the change is less dramatic than the increase in pulmonary gas exchange, and not sufficient to supply all of the needs of an active animal. Lungless salamanders apparently have a limited ability to increase $O_2$ uptake through the buccopharyngeal region, so cutaneous respiration remains the predominant mode even at high temperatures (e.g., *Desmognathus quadramaculatus* in fig. 4.4c). During vigorous exercise, amphibians with lungs greatly increase ventilation rates, and the importance of pulmonary gas exchange presumably increases as well. Large-bodied species that lack lungs are somewhat limited in their capacity to sustain vigorous activity, but this does not seem to be true for very small lungless species (Full 1986; Feder 1988; Feder, Full, and Piiper 1988; Full et al. 1988; see chapter 5).

### Oxygen Uptake in Water

Only a few adult amphibians take up $O_2$ exclusively from the water. Most species are water-breathers only when metabolic demands are relatively low, or $O_2$ content of the water is particularly high. For example, some frogs spend the winter at the bottom of ponds or streams (Emery, Berst, and Kodaira 1972; Bradford 1983). At winter temperatures, metabolic rates are so low that $O_2$ demands can be met through cutaneous respiration (Pinder 1987). However, at other times of year, the same species generally survive only a few days without access to air, even at relatively low temperatures (Holzapfel 1937; Hutchison and Dady 1964; Hutchison and Whitford 1966). A few frogs, such as species of *Telmatobius* from high Andean lakes and *Ascaphus truei* from the streams of western North America, are strictly aquatic throughout their lives and live in cold, $O_2$-saturated water. They probably meet nearly all of their metabolic demands through cutaneous respiration, although they are capable of breathing air if the $O_2$ content of the water declines (Hutchison, Haines, and Engbretson 1976).

Highly aquatic salamanders probably take up most of their $O_2$ through the skin as well. For example, the hellbender (*Cryptobranchus alleganiensis*) is strictly aquatic and is found almost exclusively in relatively cold, well-oxygenated habitats. Its loosely folded skin (fig. 1.31 C in chapter 1) is the primary respiratory organ under most conditions (Guimond and Hutchison 1973a, 1976). The lungs of these salamanders are simple sacs that lack extensive subdivisions. Nevertheless, they are capable of using the lungs to breathe at the surface. They can survive for nearly a week in severely hypoxic water at 20° C by breathing air, but their metabolic rates are greatly reduced under hypoxic conditions and their acid-base balance is disrupted by a buildup of $CO_2$ (Boutilier, McDonald, and Toews 1980; Ultsch and Duke 1990).

Other salamanders that live in relatively cold, well-oxygenated water probably rely mainly on the skin for respiration. This certainly is true of aquatic plethodontids, which lack lungs, such as *Desmognathus marmoratus, Desmognathus quadramaculatus,* and *Gyrinophilus porphyriticus.* Some hynobiids and salamandrids that live in cold mountain streams also have reduced lungs, as do species of *Rhyacotriton* (see chapter 1). All of these species must be mainly skin breathers. Even ambystomatid salamanders, which have well-developed lungs, can remain submerged for long periods of time at cold temperatures when $O_2$ content of the water is high (Whitford and Massey 1970), but come to the surface more often as temperatures rise (Whitford and Sherman 1968).

Some paedomorphic salamanders retain gills throughout adult life and are primarily water-breathers. For example, adult *Necturus* generally surface to breathe only if the $O_2$ content of the water is very low, even though they have fully functional lungs (Lenfant and Johansen 1967; Shield and Bentley 1973; Guimond and Hutchison 1972, 1976; Ultsch and Duke 1990). The gills account for more than half of all $O_2$ uptake and become increasingly important as water temperature increases (Guimond and Hutchison 1972, 1976). At low $O_2$ levels, the lungs can be used as supplementary respiratory organs, enabling these salamanders to survive for up to a week in severely hypoxic water, although with considerably reduced metabolic rates (Ultsch and Duke 1990). When branchial respiration is experimentally blocked, *Necturus* can survive when $O_2$ content of the water is high, but total gas exchange is significantly reduced (Shield and Bentley 1973).

Sirenid salamanders also retain gills as adults, but they are somewhat vestigial and generally account for less than 5% of total $O_2$ uptake. The lungs are much more important than in *Necturus* and are the major sites of $O_2$ uptake at warm temperatures (Guimond and Hutchison 1973b, 1976) or low $O_2$ levels (Duke and Ultsch 1990). These salamanders often live in swampy, oxygen-poor habitats and must rely on aerial respiration to survive under these conditions. Consequently, sirens are essentially air-breathing aquatic salamanders with some capacity for respiration through the skin and gills, in contrast to *Cryptobranchus* and *Necturus,* which are primarily water-breathers that sometimes breathe air.

Amphiumas are even more dependent on their lungs for respiration. These strictly aquatic salamanders lack gills and often live in oxygen-poor habitats. They have the most complex lungs of any of the large aquatic salamanders and cannot survive very low $O_2$ levels without breathing air. The lungs become increasingly important as water temperature increases, accounting for only 8% of total $O_2$ uptake at 5° C, but more than 50% at 25° C (Guimond and Hutchison 1974, 1976). They also have a lower capacity for $O_2$ uptake across the skin than the other "giant" salamanders. Duke and Ultsch (1990) interpreted this as an adaptation for living in hypoxic water or in burrows, where $O_2$ levels are so low that they might actually lose $O_2$ across the skin to the surrounding water.

Most aquatic amphibians do not live in swampy, oxygen-poor habitats like those inhabited by amphiumas and sirens. Most species are bimodal breathers, using the lungs to breathe air and the skin for gas exchange with the water. Partitioning of gas exchange in these species generally is measured by placing the animal in a water-filled chamber with access to air at the top. The water is covered with a layer of oil to minimize diffusion of gases between the aerial and aquatic portions of the chamber (Whitford and Sherman 1968; Guimond and Hutchison 1973b, 1974, 1976; Ultsch 1973b, 1974, 1976; Emilio and Shelton 1974; Gottlieb and Jackson 1976; Wakeman and Ultsch 1976; MacKenzie and Jackson 1978; Burggren and Moalli 1984). Relatively few species have been studied in detail, but variation among species is considerable. In general, salamanders tend to more be dependent on aquatic $O_2$ uptake through the skin than are anurans (see Feder and Burggren 1985a, for summary table). Even a fully aquatic frog such as *Xenopus* takes up about 75% of its $O_2$ via the lungs at temperatures of 20°–25° C (Emilio and Shelton 1974, 1980). These frogs have well-developed lungs that serve as major reservoirs for $O_2$ during prolonged voluntary dives (Boutilier and Shelton 1986b). Measurements of $O_2$ consumption in the aquatic caecilian *Typhlonectes natans* revealed that about 55% of $O_2$ consumed by resting animals is taken up through the lungs, with about 45% being taken up through the skin (Smits and Flanagan 1994). In bimodal breathers, dependence on lungs for $O_2$ uptake generally increases with increasing temperature or activity and decreasing $O_2$ levels, as in amphiumas. In *Typhlonectes* active animals increase $O_2$ consumption by 13 times over resting levels, and about 96% is taken up through the lungs (fig. 4.5).

### Respiratory Surface Area and Capillary Density

A number of workers have made extensive studies of the surface area, capillary density, and thickness of various respira-

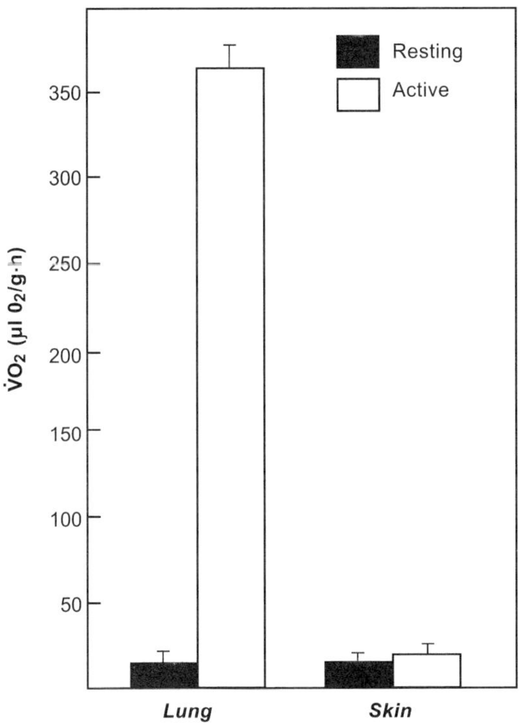


**Fig. 4.5.** Comparison of cutaneous and pulmonary oxygen uptake by the aquatic caecilian *Typhlonectes natans* at rest and after exercise. After Šmits and Flanagin (1994).

tory exchange surfaces in amphibians and have related these morphological variables to the relative importance of each organ in gas exchange (J. Czopek 1955a, b, 1957, 1959a, b, c, 1961, 1962, 1965; G. Czopek and J. Czopek 1955; Szarski 1955, 1964; Strawinski 1956; G. Czopek, J. Czopek, and Szarski 1962; Andrzejewski et al. 1962; Bieniak and Watka 1962; Whitford and Hutchison 1967; Hutchison, Whitford, and Kohl 1968; Ultsch 1973b, 1974). In many of the early studies, data were analyzed as ratios of respiratory surface area or capillary length to body mass, a procedure that can yield misleading interpretations (Packard and Boardman 1986). Therefore, I present the data as allometric regressions. Conclusions drawn from these studies must be treated with some caution because, in many cases, morphological measurements are based on examinations of single specimens of each species. Exceptions are a few studies that examined changes in total respiratory surface area during ontogeny (e.g., Strawinski 1956; Andrzejewski et al. 1962; Ultsch 1973b; Talbot and Feder 1992). There have been few attempts to study intraspecific variation in features such as capillary density or skin thickness, and there has not been a thorough statistical analysis of how morphological features differ among species.

The elongate body form of salamanders often is assumed to favor cutaneous gas exchange more than in anurans because of a more advantageous surface/volume ratio (e.g., Feder 1983b; Duellman and Trueb 1986). However, this assumption is somewhat misleading because anurans generally have much longer legs than salamanders and therefore have considerable skin surface area on the legs and feet that tends to compensate for a reduced surface/volume ratio of the trunk region alone. Data from J. Czopek (1965), Whitford and Hutchison (1967), and Hutchison, Whitford, and Kohl (1968) yield a similar scaling of surface area with body mass in salamanders and anurans (fig. 4.6), with an exponent of about 0.67. There is no clear difference in the intercepts of these regressions, indicating that anurans and salamanders of similar body mass have similar surface areas. A reanalysis of these data by Talbot and Feder (1992) suggests that J. Czopek's (1965) data actually underestimate total skin surface area in anurans because he apparently did not include foot webbing in his measurements.

Some amphibians that are heavily dependent on cutaneous respiration have evolved morphological adaptations for greatly increasing the surface area of the skin relative to their body mass. The most familiar examples are the extremely loose and wrinkled skin of cryptobranchid salamanders and South American aquatic frogs of the genus *Telmatobius* (fig. 4.7). In the African "hairy" frog, *Trichobatrachus robustus* (fig. 1.12 D in chapter 1), males develop extensive

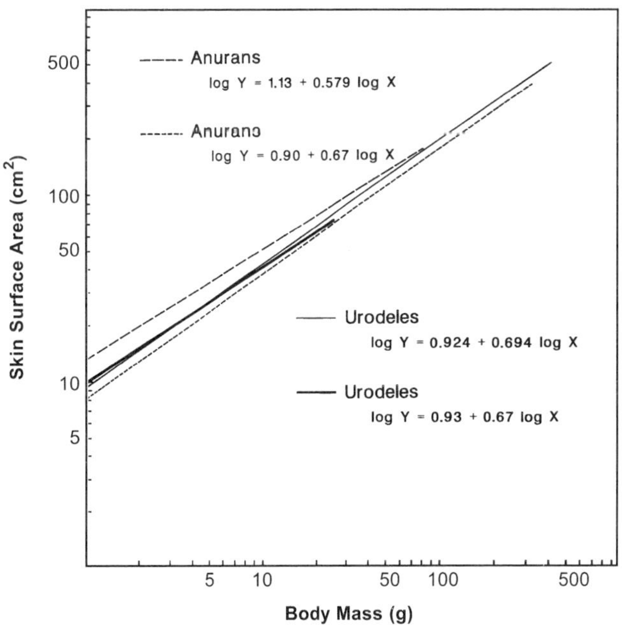


**Fig. 4.6.** Scaling of skin surface area with body mass for anurans and urodeles. Dashed lines show regressions for anurans, plotted from data in Hutchison, Whitford, and Kohl (1968; top equation) and Czopek (1965; bottom equation). Solid lines show regressions for urodeles, plotted from data in Whitford and Hutchison (1967; top equation) and Czopek (1965; bottom equation). The slopes and intercepts of the regressions for anurans and urodeles are not significantly different.

Fig. 4.7. *Telmatobius culeus* from Lake Titicaca, Bolivia, showing extensive skin folds that provide increased surface area for gas exchange. Photo by Victor H. Hutchison.

1989; Maina 1989); exceptions to this trend are *Salamandra, Siren,* and *Amphiuma,* all of which have relatively complex lungs (Wilnow 1964, 1968; Guimond and Hutchison 1976; Goniakowska-Witalinska 1978).

Functional aspects of lung structure have not been studied in detail in caecilians, but most species appear to have well-developed, extensively subdivided lungs that provide a large surface area for gas exchange (Jared et al. 1999; Kc et al. 2000). As in many elongate vertebrates, one lung usually is considerably longer than the other. Usually the right lung is longer and the left lung is vestigial, but the opposite condition is seen in some caecilians (Wilkinson and Nussbaum 1999). Several caecilians exhibit unusual modifications of the respiratory system. The aquatic caecilian *Typhlonectes natans* not only has a long, well-developed right lung, but also a separate tracheal lung between the anterior trachea and the heart that provides additional surface area for gas exchange (Kc et al. 2000). In a related species, *T. compres-*

dermal papillae on the skin, presumably to compensate for increased metabolic demands (Noble 1925).

Skin surface area gives only a rough estimate of an animal's capacity for cutaneous gas exchange because species differ in skin thickness or capillary density (Feder and Burggren 1985a). However, neither variable is closely correlated with dependence on cutaneous respiration. The skin is not consistently thinner nor supplied with more capillaries in plethodontid salamanders, which depend mainly on cutaneous respiration, than in other salamanders or anurans of similar size (Foxon 1964; J. Czopek 1965; Feder 1983b; Feder and Burggren 1985a). Total skin capillary length actually appears to be somewhat higher in anurans than in salamanders of similar mass, particularly in larger species (fig. 4.8). The difference would be even greater if the data were corrected for additional capillaries in the foot webbing (Talbot and Feder 1992). On the other hand, terrestrial anurans often have the most highly vascularized skin on the ventral surface, where it plays little role in gas exchange when the animal is sitting on the ground (J. Czopek 1965; Feder and Burggren 1985a). Such vascularized areas probably are related more to water uptake than to respiration (see chapter 2).

If we consider the total length of capillaries in all respiratory organs (lungs, skin, and oral cavities), anurans consistently have much higher values than salamanders of similar size (fig. 4.9). This is due mainly to the greater development of lungs in anurans, which usually have complex subdivided lungs with many alveoli, compared to the simple saclike lungs of most salamanders (Okada et al. 1962; J. Czopek 1965; Hughes 1970; Smith and Rapson 1977; Goniakowska-Witalinska 1980a, b, 1985, 1986; Burggren

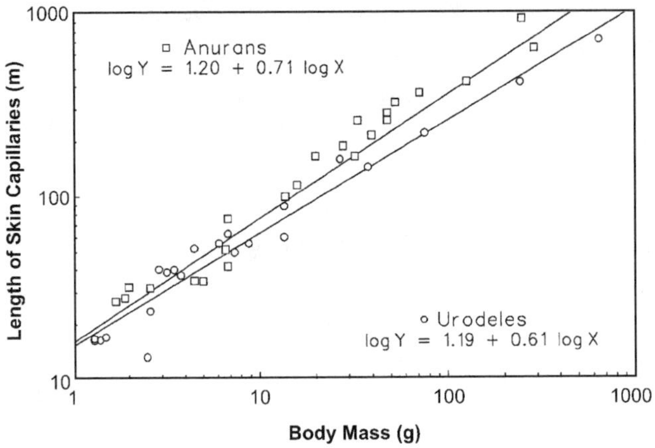


Fig. 4.8. Length of skin capillaries as a function of body mass in anurans and urodeles. Plotted from data in J. Czopek (1965).

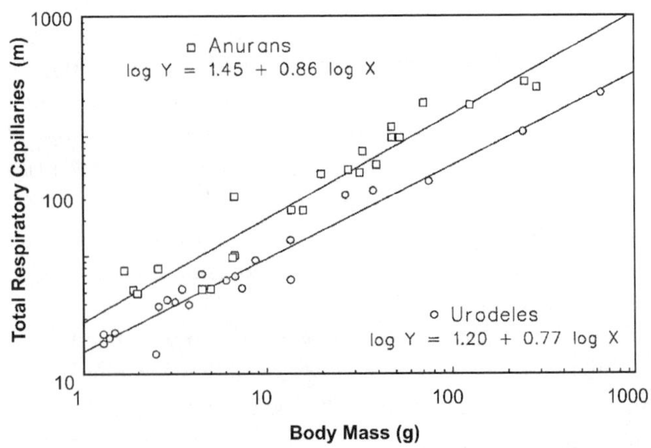


Fig. 4.9. Total respiratory capillary length as a function of body mass in anurans and urodeles. Plotted from data in J. Czopek (1965).

*sicauda,* up to 94% of oxygen uptake is through the lungs, although the skin becomes more important when the animal is completely submerged (Sawaya 1947; Gonçalves and Sawaya 1978).

Another typhlonectid, *Potomotyphlus,* is unique in having the left lung modified into a hydrostatic organ. Pulmonary respiration appears to be reduced in this aquatic caecilian, and the pulmonary circulation is reduced (Wilkinson and Nussbaum 1997). The most unusual modification is found in a strange typhlonectid related to *Potomotyphlus, Atretochoana eiselti.* This caecilian is completely lungless and is the largest known lungless tetrapod (Nussbaum and Wilkinson 1995; Wilkinson and Nussbaum 1997, 1999; Wilkinson et al. 1998). This species is assumed to be aquatic, based on morphological features, such as a tail fin, but the biology of living animals, including their respiratory physiology, is completely unknown. Wilkinson and Nussbaum (1999) hypothesized that both *Potomotyphlus* and *Atretochoana* have abandoned the habit of most typhlonectids of living in mud burrows, which tend to become hypoxic. Their reliance on cutaneous respiration suggests that they probably live in well-oxygenated water.

The lung membranes present a thinner barrier to gas exchange than the skin, so diffusion capacity of the lungs typically is much greater (Withers and Hillman 1988). In general, anurans have somewhat thinner membranes in the lungs than salamanders (Meban 1980), but neither the thinness of the gas-exchange membranes nor capillary density approach those in mammalian lungs (Weibel 1972; Burggren 1989). For most anurans, pulmonary capillaries account for more than half of all respiratory capillaries (fig. 4.10 B). This also is true for some terrestrial salamanders with complex lungs, such as *Salamandra salamandra;* but in other species, cutaneous capillaries predominate (fig. 4.10 A). The skin of plethodontids contains more than 90% of all respiratory capillaries, with the buccopharyngeal region contributing the rest. In most amphibians, capillaries in the lining of the mouth are less than 3% of the total, suggesting that buccopharyngeal respiration is of relatively trivial importance in species other than plethodontids (Foxon 1964; J. Czopek 1965; Duellman and Trueb 1986). Measurements of capillary length are not available for caecilians, but some species appear to have relatively complex, highly subdivided lungs like those of anurans and *Salamandra* (M. Wake 1974; Smits and Flanagin 1994).

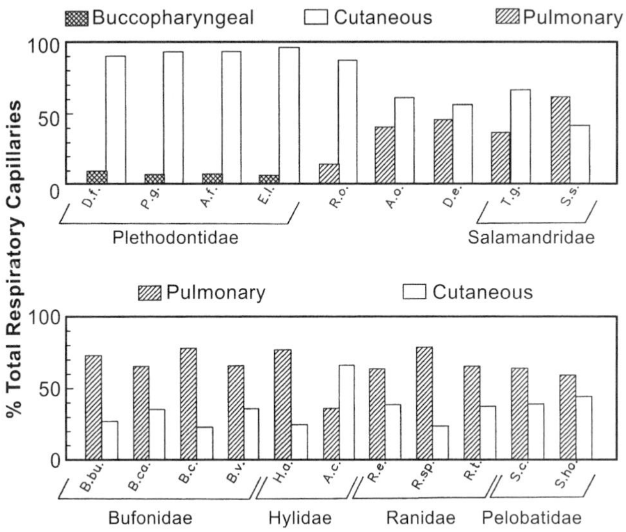


**Fig. 4.10.** Percentage of total respiratory capillaries found in the skin, lungs, and buccopharyngeal regions in (A) selected terrestrial urodeles and (B) anurans. Most anurans have more capillaries in the lungs than in the skin, whereas most urodeles have more capillaries in the skin. Species abbreviations: *D.f. = Desmognathus fuscus; P.g. = Plethodon glutinosus; A.f. = Aneides flavipunctatus; E.l. = Eurycea longicauda; R.o. = Rhyacotriton olympicus; A.o. = Ambystoma opacum; D.e. = Dicamptodon ensatus; T.g. = Taricha granulosa; S.s. = Salamandra salamandra; B.bu. = Bufo bufo; B.ca. = B. calamita; B.c. = B. cognatus; B.v. = B. viridis; H.a. = Hyla arborea; A.c. = Acris crepitans; R.e. = Rana esculenta; R.sp. = R. sphenocephala; R.t. = R. temporaria; S.c. = Scaphiopus couchii, S.ho. = S. holbrookii.* Plotted from data in J. Czopek (1965).

## Ventilation of Respiratory Organs

The surface area, thickness, and capillary density of the various respiratory organs provide a rough estimate of their relative importance in gas exchange, but this also will be affected by ventilation of the respiratory organ with air or water. All air-breathing amphibians employ a positive-pressure buccal pump to force air into the lungs, a mode of ventilation inherited from their sarcopterygian ancestors (Brainerd, Ditelberg, and Bramble 1993; C. B. Jørgensen 2000). Amphibians have a mechanism of buccal pumping that minimizes mixing of inspired and expired air in the buccal cavity (Brainerd, Ditelberg, and Bramble 1993; Brainerd 1994, 1998; Brainerd and Monroy 1998; Kc et al. 2000). Initially, the buccal cavity is partially filled with inspired air, but full expansion of the buccal cavity does not occur until expiration of air from the lungs is complete. Ventilation is accomplished by raising and lowering the floor of the mouth while opening and closing the nares and glottis (De Jongh and Gans 1969). In terrestrial species, a series of shallow ventilatory movements fills the buccal cavity, punctuated by deeper ventilatory movements that fill the lungs (Whitford and Hutchison 1965b; De Jongh and Gans 1969; West and Jones 1975; MacIntyre and Toews 1976). In contrast, fully aquatic species such as *Xenopus, Amphiuma,* and *Siren* ventilate the lungs with each buccal movement, and only a few ventilations are required to fully inflate the lungs. In these species, air is expelled from the lungs as the animal surfaces, thereby reducing the amount of dead space in the lungs and increasing the efficiency of gas exchange (Toews 1971; Toews and McRae 1974; Brett and Shelton 1979; K. Martin and Hutchison 1979; Boutilier 1984, 1988a, 1989). This pattern of rapid emptying and refilling of the lungs enables these animals to spend relatively little time exposed at the surface, where they are vulnerable to predators such as wading birds (Baird 1983). In the aquatic caecilian *Typhlonectes natans*, the pattern of breathing is somewhat different from that seen in other amphibians. Expiration is followed by a series of inspiratory oscillations (M. Gardner, Smits, and Smatresk 2000; Kc et al. 2000). The frequency of breathing episodes and ventilation rates in bimodal breathers are governed mainly by depletion of the body's $O_2$ stores (Boutilier 1984, 1988a, 1989), but are influenced by other variables such as temperature, water depth, and $O_2$ and $CO_2$ content of the water (Shannon and Kramer 1988; Smatresk and Smits 1991; Smatresk 1994; M. Gardner, Smits, and Smatresk 2000).

The three major groups of amphibians differ in the way in which air is expelled from the lungs. In salamanders, the hypaxial body musculature contracts to push air out of the lungs (Brainerd, Ditelberg, and Bramble 1993; Brainerd 1994, 1998; Brainerd and Monroy 1998). Salamanders have three layers of hypaxial muscles, the transverse abdominis, external oblique, and internal oblique, which are closely associated with the rectus abdominis muscle. In *Siren lacertina* and *Necturus maculosus*, only the transverse abdominis muscle is active during expiration. In larval tiger salamanders (*Ambystoma tigrinum*), the transverse abdominis is most active, but the other hypaxial muscles show low levels of activity during expiration. In frogs, the hypaxial muscles have been reduced to two layers. These generally are called the internal and external oblique muscles, but their homologies with salamander muscles are not clear. Frogs do not have active expiration during normal breathing, although the hypaxial muscles are used to force air out of the lungs when males are calling (see chapters 5 and 7). Instead, expiration is passive, relying on elastic recoil of the inflated body wall (Brainerd, Ditelberg, and Bramble 1993). Caecilians also employ passive exhalation (Bennett, Summers, and Brainerd 1999; Kc et al. 2000). These animals are unusual in that they maintain very high pressure in the lungs and peritoneal cavity, apparently to keep the body rigid to aid in locomotion. They also employ an unusually large number of buccal pumping cycles when filling the lungs with air, probably to maintain this high pressure (Carrier and Wake 1995; Kc et al. 2000). Brainerd, Ditelberg, and Bramble (1993) suggested that active exhalation using the trunk musculature is an ancestral trait for all amphibians that has been lost in frogs and caecilians because of their highly derived modes of locomotion.

Aquatic amphibians that live in still water face a special problem in using the skin as a respiratory organ. At low flow rates, the water around the body of the animal tends to form a hypoxic boundary layer that can be a major limitation on gas exchange through the skin (Feder and Burggren 1985a, b; Burggren and Feder 1986; Feder and Pinder 1988). These boundary layers typically are only 1–3 mm thick, but can increase resistance to $O_2$ diffusion up to eight times, even in water with normal $O_2$ content (Feder and Booth 1992; Booth 1994). Even amphibians associated with running water, such as the aquatic plethodontid salamander *Desmognathus quadramaculatus*, often are found in microhabitats in which hypoxic boundary layers are the major impediment to $O_2$ uptake (Booth and Feder 1991). Relatively small movements by the animals can break up hypoxic boundary layers around the skin (Pinder and Feder 1990). Even frogs hibernating under water move periodically (Emery, Berst, and Kodaira 1972), perhaps to break up boundary layers around the skin. The effects of skin ventilation on gas exchange in adult bullfrogs (*Rana catesbeiana*) has been simulated by placing the frogs in stirred and unstirred water, with only the nares exposed to air. Cutaneous $O_2$ uptake declined significantly in water that was not stirred (Burggren and Feder 1986). Other amphibians have been shown to rise to the surface to breathe more frequently in unstirred than in stirred water (Feder and Booth 1992; Booth 1994).

Some highly aquatic amphibians engage in more pronounced movements to ventilate the skin. For example, hellbenders (*Cryptobranchus alleganiensis*) rock the body from side to side, and these rocking movements increase as the $O_2$ content of the water decreases (Boutilier and Toews 1981b; Harlan and Wilkinson 1981). If the animals are prevented from rocking under hypoxic conditions, $O_2$ uptake decreases, but artificial ventilation of the skin restores $PO_2$ to normal levels (Harlan and Wilkinson 1981). The fully aquatic leptodactylid frog *Telmatobius culeus* ventilates its skin by waving the very large flaps of skin in the water (Hutchison, Haines, and Engbretson 1976). In some neotenic salamanders, active movements of the external gills presumably serve to break up any boundary layer formed around the gills in still water. For example, gill movements by *Necturus* are more frequent at high temperatures or when the animal is active (Guimond and Hutchison 1972, 1976), but rates of ventilatory movements and their effect on $O_2$ uptake have not been quantified.

### Perfusion of Respiratory Organs with Blood

Many models of cutaneous gas exchange in amphibians have assumed that exchange is diffusion limited and not affected by changes in blood flow to respiratory exchangers (e.g., Gatz, Crawford, and Piiper 1975; Piiper, Gatz, and Crawford 1976; Piiper 1988). However, for some amphibians, changes in perfusion of respiratory capillaries can affect rates of gas exchange (Feder and Burggren 1985a, b; Burggren 1988) and are under active physiological control (for a review, see Malvin 1994). This has been studied mostly in anurans. When adult bullfrogs (*Rana catesbeiana*) were exposed to air, the number of perfused skin capillaries decreased and the rate of $CO_2$ elimination declined as well; the opposite effect was obtained when the frogs were placed in water (Burggren and Moalli 1984). Poczopko (1957) demonstrated that *Rana esculenta* breathing high levels of $CO_2$ in air increased perfusion of skin capillaries, while Szarski (1959) showed that large changes in $PO_2$ or $PCO_2$ of air in contact with the skin also could affect capillary recruitment. Aquatic hypoxia results in a redistribution of blood from the cutaneous to the pulmonary circulation, while aerial hypoxia has the opposite effect. This has been interpreted as a mechanism to prevent loss of $O_2$ to the hypoxic medium and enhance gas exchange where it is most effective (Boutilier, Glass, and Heisler 1986). Work by Gary Malvin and his colleagues showed that both $PO_2$ and $PCO_2$ at the skin surface can affect cutaneous blood flow and gas exchange. Blood flow increases when skin surface $PO_2$ increases, while it decreases when $PCO_2$ increases. An even greater effect is produced by changes in lung ventilation. When frogs are breathing air through the lungs, blood flow to the skin decreases. Most of this change occurs in the skin on the ventral surface and the legs, but not the dorsal surface (Malvin 1994; Malvin and Hlastala 1986a, b, 1989).

Several unusual features of the circulatory system of frogs, some of which are not present in most salamanders, facilitate the redistribution of blood flow between the pulmonary and cutaneous circulation. Although amphibians have only a single ventricle, oxygenated and deoxygenated blood remain largely separated in the ventricles of anurans and *Amphiuma*. This continues as the blood leaves the ventricle and enters a large vessel, the *conus arteriosus*. Here a spiral valve directs most of the deoxygenated blood from the right atrium into the pulmocutaneous circuit, while oxygenated blood coming from the lungs via the left atrium is selectively distributed to the systemic circulation (fig. 4.11). A similar separation of pulmonary and systemic blood occurs in the caecilian *Typhlonectes compressicauda*, but the anatomy of the heart and associated vessels is somewhat different (Toews and Macintyre 1978). Because amphibians are periodic breathers (Shelton and Boutilier 1982), resistance to blood flow in the lungs changes as a function of breathing activity. When the lungs are inflated, resistance is low and more blood is diverted to the pulmonary circuit; conversely, when pulmonary resistance is high, blood is shunted to the systemic arteries. Hence, perfusion of the lungs is matched to ventilation, thereby increasing the efficiency of gas exchange (Johansen 1963; Johansen and Ditadi 1966; Johansen, Lenfant, and Hanson 1970; Shelton 1970, 1976, 1985; Emilio and Shelton 1974; Johansen 1985; Boutilier, Stiffler, and Toews 1992).

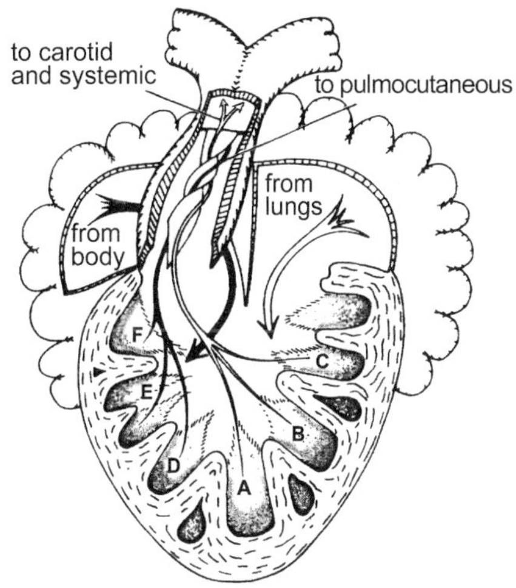


**Fig. 4.11.** Diagram of the heart of a frog (*Xenopus*) and associated blood vessels, showing partial separation of blood flow through the ventricle. Black arrows show the route of deoxygenated blood. White arrows show the route of oxygenated blood. After Shelton (1976).

Perfusion of skin capillaries in anurans is affected by the dual origin of the cutaneous circulation. While some skin capillaries are supplied by the general systemic circulation that carries oxygenated blood, others are supplied by a cutaneous branch of the pulmocutaenous artery, which receives mainly deoxygenated blood from the heart (fig. 4.12). Capillaries supplied by the pulmocutaneous artery will have a more favorable $PO_2$ gradient for cutaneous $O_2$ uptake than those supplied by the systemic arteries because of lower blood $PO_2$. Increased resistance to blood flow through the lungs, coupled with decreased peripheral resistance, will channel more blood through the cutaneous branch of the pulmocutaneous artery. Blood flow through each circuit appears to be regulated in part by the action of muscular sphincters on the two branches of the pulmocutaneous artery (fig. 4.12). These in turn are regulated by mechanoreceptors sensitive to changes in lung inflation and chemoreceptors sensitive to levels of $CO_2$ and $O_2$ (for reviews, see Feder and Burggren 1985a, b; Malvin and Boutilier 1985; Burggren 1988; Malvin 1988, 1994; Boutilier, Stiffler, and Toews 1992; other studies include Shelton 1970; Emilio and Shelton 1972; De Saint-Aubain and Wingstrand 1979; Moalli et al. 1980; De Saint-Aubain 1982; West and Burggren 1984; Boutilier, Glass, and Heisler 1986).

Some amphibians have an additional mechanism to regulate the oxygen-carrying capacity of the blood flowing to the tissues, including the skin. Experiments with European newts (*Triturus carnifex*) showed that, at low temperatures, newts exposed to well-oxygenated water accumulate red blood cells in the spleen. These are released into the circulation when the animals are exposed to lower oxygen levels, thereby increasing the concentration of oxygen-carrying

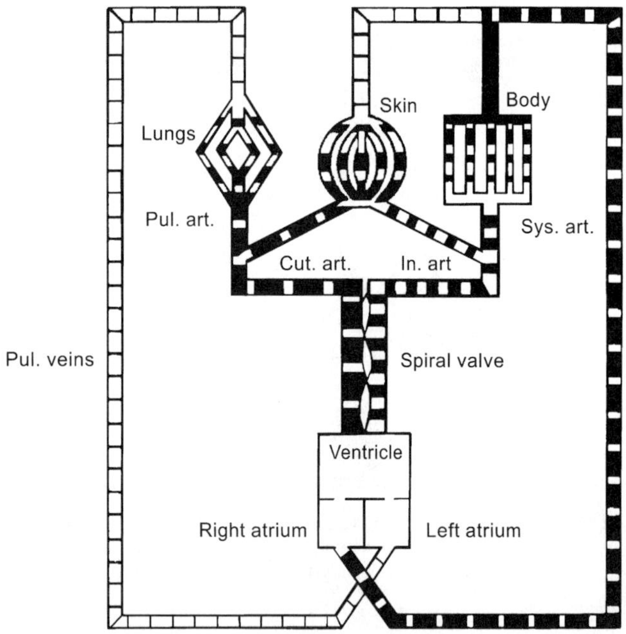


**Fig. 4.12.** Diagram of blood supply to skin and other parts of the body in a frog. Black indicates the route of deoxygenated blood. White indicates the route of oxygenated blood. Mixed colors show partial mixed of oxygenated and deoxygenated blood. Abbreviations: Cut. art = cutaneous artery; In. art. = intersegmental artery; Pul. art = pulmonary arteries; Pul. veins = pulmonary veins; Sys. art = systemic arteries; Sys. veins = systemic veins. After Feder and Burggren (1985a).

cells in the blood (Frangioni and Borgioli 1989, 1991a, b, 1993b). A similar compensatory mechanism has been described in *Rana esculenta,* but in this case, red blood cells accumulate in the liver rather than the spleen (Frangioni and Borgioli 1993a, 1994). Storage of red blood cells does not occur at warm temperatures, presumably because higher oxygen demands at warm temperatures require all available red blood cells to be in circulation.

## Respiratory Properties of the Blood

The availability of $O_2$ to support tissue metabolism is mediated not only by structural features of the respiratory and circulatory systems, but also by oxygen-combining properties of the blood. Many investigators have attempted to correlate respiratory properties of amphibian blood with variation in habitat, $O_2$ availability, mode of respiration, or developmental stage (e.g., Redfield 1933; McCutcheon 1936; McCutcheon and Hall 1937; Lenfant and Johansen 1967; Coates and Metcalfe 1971; Johansen and Lenfant 1972; S. Wood et al. 1975; Hutchison, Haines, and Engbretson 1976; Johansen and Weber 1976; Boutilier and Toews, 1981a; Boutilier, Stiffler, and Toews 1992). The most common approach is to compare $O_2$ dissociation curves and $O_2$ affinity of the blood, as measured by $P_{50}$ (the $O_2$ tension at which the blood is 50% saturated). In general, air-breathing species are thought to have relatively low blood $O_2$ affinity (i.e., the $O_2$ dissociation curve is shifted to the right) compared to higher $O_2$ affinity in the blood of water-breathers, while aquatic species in oxygen-poor environments have higher $O_2$ affinity than those in oxygen-rich environments.

A much-cited example is Johansen and Lenfant's (1972) comparison of *Necturus, Amphiuma,* and *Rana,* which exhibit a shift toward lower $O_2$ affinity as dependence on air breathing with lungs increases (fig. 4.13). This would be advantageous because $O_2$ is readily available to air-breathers, and lower $O_2$ affinity would facilitate unloading of $O_2$ in the tissues. A shift to lower $O_2$ affinity also has been described in the transition from water to air-breathing during metamorphosis and is generally assumed to have the same adaptive function (McCutcheon 1936; McCutcheon and Hall 1937; Riggs 1951; Foxon 1964; Hamada et al. 1966; Gahlenbeck and Bartels 1970; S. Wood 1971; Maclean and Jurd 1972; Hattingh and Bartels 1973; Watt and Riggs 1975; Broyles 1981; Pinder and Burggren 1983; Jurd 1985; Burggren and Just 1992). There is a similar shift in blood $O_2$ affinity in viviparous species at birth (Toews and Macintyre 1977; Garlick et al. 1978). On the other hand, prolonged exposure of adult frogs (*Rana catesbeiana*) to hypoxic water has the opposite effect, resulting in increased blood $O_2$ affinity, blood $O_2$ capacity, and hemoglobin concentration (Pinder and Burggren 1983).

While adaptive explanations relating $O_2$ affinity of the blood to $O_2$ availability in the environment are appealing, caution is required in interpreting these patterns (R. Wells 1990). As pointed out by S. Wood and Lenfant (1987), mea-

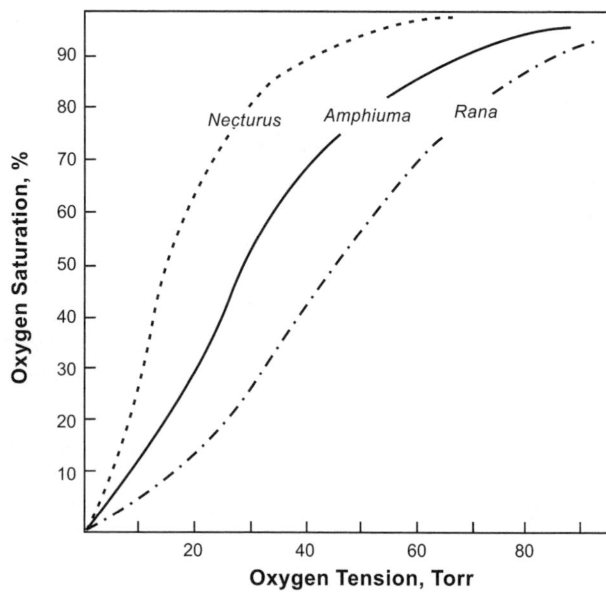


**Fig. 4.13.** Oxygen dissociation curves of blood of *Necturus, Amphiuma,* and *Rana* measured at pH 7.6. The curves show percent saturation of the blood with oxygen at different levels of oxygen availability. After Johansen and Lenfant (1972).

surements of $PCO_2$, arterial and venous $PO_2$, $O_2$ capacity, and blood pH in intact, unrestrained animals are required to properly interpret the significance of the $O_2$ dissociation curve for $O_2$ delivery to the tissues, because other components of the $O_2$ transport system compensate for changes in $O_2$ affinity. Very few such measurements are available, so broad comparisons among species living in different habitats remain largely speculative. Nevertheless, it seems likely that the unusually high $O_2$ affinity of blood in a fossorial caecilian, *Boulengerula taitanus* (S. Wood et al. 1975), and a high-altitude aquatic frog, *Telmatobius culeus* (Hutchison, Haines, and Engbretson 1976), are adaptations to reduced environmental $PO_2$. Actual levels of $O_2$ in the natural habitat have been measured only for the latter species, however. In that case, the lakes in which the frog lives are saturated with $O_2$, but $PO_2$ of the water is relatively low because of the high altitude (nearly 4,000 m).

Another problem with much of the literature on $O_2$ affinity in amphibians is that adaptive explanations sometimes have been based on studies of separated hemoglobin instead of whole blood (e.g., Riggs 1951; Hamada et al. 1966; Taketa and Nickerson 1973; Watt and Riggs 1975; Meirelles et al. 1979; Condo et al. 1981, 1983, 1989; Brunori et al. 1985). Because other components of blood can affect its $O_2$ affinity, functional interpretations based solely on isolated hemoglobin probably are not justified (S. Wood and Lenfant 1987). Much of the work in this field has been done by biochemists concerned more with the structure of the hemoglobin molecule than with physiological implications for the whole animal. This is particularly true of the extensive work on changes in hemoglobin at metamorphosis, where the main focus has been on mechanisms controlling hemoglobin structure and synthesis (for reviews, see B. Sullivan, 1974; Broyles 1981; Jurd 1985).

Interspecific comparisons of blood $O_2$ affinity are complicated by variation in sensitivity to both temperature and pH. Often such comparisons are made at a single temperature and pH that does not always reflect conditions normally encountered by animals in the field. Temperature can affect $O_2$ affinity of blood directly ($O_2$ affinity decreases with increasing temperature) or indirectly through its effect on pH (pH declines with increasing temperature). In some amphibians, there is a relatively small direct effect of temperature on $O_2$ affinity (Kirberger 1953; Straub 1957; Gahlenbeck and Bartels 1968; Jokumsen and Weber 1980; S. Wood 1980; S. Wood and Lenfant 1987). This would be particularly advantageous in species that regularly experience large daily fluctuations in temperature, such as the African tree frog *Chiromantis petersi,* which often is exposed to hot sun during the day and experiences a 15° C temperature change in a few hours (Johansen, Abe, and Weber 1980). Some species exhibit long-term temperature acclimation that reduces the effect of seasonal temperature change on $O_2$ affinity (Kirberger 1953; Straub 1957; Gahlenbeck and Bartels 1968; S. Wood 1980).

Many amphibians, particularly those in aquatic habitats, are subjected to daily fluctuations in environmental pH, in addition to changes in blood pH resulting from metabolism. Consequently, typical pH levels experienced by animals in the field vary among species and will affect $O_2$ affinity of the blood. Interspecific comparisons made at only a single pH, like that of Johansen and Lenfant (fig. 4.15), do not take into account the range of conditions experienced by animals in the field. In addition, many early studies of amphibian blood and even some more recent ones are flawed because they report $O_2$ dissociation curves measured at pH 7.4 (e.g., McCutcheon 1936; McCutcheon and Hall 1937; F. Hall 1966; Lenfant and Johansen 1967; Gahlenbeck and Bartels 1968, 1970), a level that is too low for amphibians at ecologically realistic temperatures; a pH of about 7.8 is more typical for 20° C (S. Wood and Lenfant 1987).

Jokumsen and Weber (1980) reported that $O_2$ affinity of *Xenopus* blood increased during dormancy when the frogs burrow into the mud of drying ponds. They attributed this to the effect of increased concentrations of urea in the blood (see chapter 2) and suggested that the shift might be advantageous because it would inhibit unloading of $O_2$ to the tissues, thereby favoring low metabolic rates and conservation of energy. A similar shift was reported earlier by Seymour (1973c) in dormant *Scaphiopus couchii,* but he did not suggest a mechanism to account for the increase in $O_2$ affinity. His adaptive explanation was somewhat different. He suggested that a high affinity would enhance the storage of $O_2$ in the blood during dormancy, while reduced affinity during the activity season would facilitate unloading of $O_2$ to the tissues when demand is highest. Neither hypothesis has been rigorously tested. Human hemoglobin shows a response to urea similar to that of *Xenopus* blood (Jokumsen and Weber 1980), suggesting that the change in $O_2$ affinity is not a specific adaptation to dormancy in frogs. A more definitive test would be to examine the effect of urea on blood $O_2$ affinity in anurans that do not undergo periodic dormancy and do not accumulate urea in the blood.

When $CO_2$ is released into the blood during respiration, the blood is acidified (see the following), and this in turn decreases the $O_2$ affinity of the blood, a process known as the Bohr Effect. At the level of tissue gas exchange, a moderate Bohr shift can be advantageous because it decreases the tendency for hemoglobin to bind with $O_2$, thereby facilitating release of $O_2$ to the tissues. On the other hand, a pronounced Bohr shift also can reduce the tendency for hemoglobin to take up $O_2$ from the environment and therefore would be disadvantageous under conditions of low $O_2$ availability. Attempts have been made to relate the magnitude of the

Bohr Effect to environmental conditions. For example, S. Wood et al. (1975) suggested that a relatively small Bohr Effect, coupled with high $O_2$ affinity in the blood of the caecilian *Boulengerula taitanus,* is an adaptation that facilitates $O_2$ uptake under hypoxic conditions in burrows. Conversely, Jokumsen and Weber (1980) interpreted the relatively large Bohr Effect in *Xenopus* blood as an adaptation facilitating $O_2$ delivery to tissues during dives. However, there is not a straightforward correlation between the magnitude of the Bohr Effect and habitat (S. Wood and Lenfant 1987), presumably because a number of other variables affect $O_2$ delivery to the tissues.

Several investigators have reported "reversed" Bohr Effects in salamanders (i.e., $O_2$ affinity increases with decreasing pH). This was studied in detail in *Amphiuma means* by Bonaventura et al. (1977), who suggested this might be an adaptation for enhancing $O_2$ uptake during hibernation. Other investigators have described "reversed" Bohr effects in the hemoglobin of other salamanders and have proposed rather vague adaptive explanations related to $O_2$ availability in the environment (Morpurgo, Battaglia, and Leggio 1970; Taketa and Nickerson 1973; Condo et al. 1981, 1983, 1989). However, the reversed Bohr Effect in salamanders appears to be a property of isolated hemoglobin, not whole blood, which exhibits a Bohr Effect similar in direction and magnitude to that of anurans (S. Wood and Lenfant 1987; Condo et al. 1989; Boutilier, Stiffler, and Toews 1992). Consequently, adaptive explanations relating "reversed" Bohr effects to environmental variables probably are unwarranted.

### Elimination of Carbon Dioxide

Carbon dioxide passes through membranes in the gaseous state, but readily goes into solution in water or plasma, where it dissociates into hydrogen ($H^+$) and bicarbonate ($HCO_3^-$) ions. Hence, release of $CO_2$ into the blood can alter the concentration of $H^+$ ions (pH), and this in turn can influence many physiological processes, including the changes in $O_2$ affinity of the blood that were just discussed. Major changes in internal pH can have negative effects on a variety of other physiological processes, and most organisms must maintain pH within a fairly narrow range. Hence, the process of $CO_2$ elimination is tightly linked with acid-base regulation (Howell 1970; Randall et al. 1981; Toews and Boutilier 1986; Stiffler, Ryan, and Mushkot 1987; Boutilier 1988b).

For terrestrial amphibians, excretion of $CO_2$ has little effect on the $CO_2$ content of the surrounding air. Both skin conductance and the magnitude of the $PCO_2$ gradient across the skin determine the rate of $CO_2$ loss. Desiccation of the skin reduces conductance to $CO_2$, again leading to an increase in blood $CO_2$ concentrations and a decrease in blood pH (Boutilier et al. 1979b). A reduction in $CO_2$ permeability is most likely to occur in species with special adaptations to retard cutaneous water loss, such as cocoon formation during dry-season dormancy (see chapter 2). In at least one such species, *Pyxicephalus adspersus,* dormant cocooned frogs had reduced blood pH, apparently resulting from increased resistance to gas diffusion (Loveridge and Withers 1981). In the "waterproof" frog, *Hyperolius viridiflavus,* juveniles spend the dry season sitting on exposed, dry plants. The skin becomes highly impermeable to water, and virtually all cutaneous respiratory exchange stops (Geise and Linsenmair 1986). Whether this results in respiratory acidosis has not been determined, nor is it known whether changes in the acid-base balance of dormant frogs have any detrimental physiological effects.

Nearly all amphibians exposed to high $CO_2$ show an initial increase in plasma $PCO_2$ and a decrease in blood pH. This respiratory acidosis can remain uncompensated for several hours, but eventually an increase in bicarbonate ($HCO_3^-$) ions in the blood gradually raises its pH (Toews and Boutilier 1986; Boutilier 1988b). Some species, including *Amphiuma means, Siren lacertina,* and *Xenopus laevis,* show little or no compensation (Heisler et al. 1982; Toews and Boutilier 1986; Boutilier, Glass, and Heisler 1987), while others, such as the aquatic caecilian *Typhlonectes compressicauda,* exhibit very pronounced compensation (Toews and Macintyre 1978). Still others show intermediate responses (Boutilier et al. 1979a, c; Boutilier and Toews 1981b; Toews and Heisler 1982; Stiffler, Tufts, and Toews 1983). Most amphibians exhibit only partial compensation of extracellular pH under conditions of elevated $CO_2$, independent of the habitat in which they are found (Heisler 1986; Toews and Boutilier 1986; Toews and Stiffler 1990). Intracellular compensation often is much greater, especially in critical tissues such as heart or skeletal muscle, where pH remains largely unchanged even during prolonged exposure to elevated $CO_2$ (Heisler et al. 1982; Toews and Heisler 1982; Toews and Boutilier 1986; Boutilier, Glass, and Heisler 1987; Boutilier, Stiffler, and Toews 1992). This probably facilitates survival in extreme environments such as swamps, but it appears to be a general feature of amphibians and not a specialized adaptation to such habitats.

Terrestrial air-breathing vertebrates such as reptiles, birds, and mammals regulate internal $CO_2$ concentrations and pH mainly by increasing lung ventilation rates in response to high arterial $PCO_2$ (Scheid and Piiper 1986; Shelton, Jones, and Milsom 1986; Tenney and Boggs 1986). In amphibians, most $CO_2$ is rapidly eliminated through the skin and $CO_2$ concentrations in the blood are relatively low. Consequently, it is difficult to maintain a concentration gradient between the blood and lungs that will favor pulmonary excretion of $CO_2$ (Guimond and Hutchison 1976; Shelton, Jones, and Milsom 1986; Toews and Boutilier 1986). Some amphibians increase ventilation rates in response to increased $CO_2$ in inspired air or in the water (Macintyre and Toews 1976; Toews and Macintyre 1978; Boutilier et al. 1979a; Jackson and Braun 1979; Jones 1982; Boutilier and Shelton 1986b; Smatresk 1994; Gardner, Smits, and Smatresk 2000). In terrestrial anurans with relatively well-developed lungs, and probably in caecilians as well, this provides some active regulation of plasma $PCO_2$ and pH; but in *Cryptobranchus,* which has poorly developed lungs, increased ventilation has virtually no effect on these variables. At least one species that lives in high $CO_2$ environments, *Amphiuma tridactylum,* is relatively insensitive to high levels of $CO_2$ and does not increase ventilation rates (Toews 1971). There is an extensive literature on the neural control of breathing in amphibians, but I will not discuss that in detail here (Wakeman and Ultsch 1976; Boutilier 1988a; Smatresk 1994; Kinkead 1997).

## Gas Exchange by Larval Amphibians

Newly hatched anuran and urodele larvae probably are very similar in their respiratory physiology. In both groups, hatchlings have short external gills that are perfused with blood, but not actively ventilated by moving them in the water. The cardiovascular system of newly hatched larvae is relatively poorly developed as well. Because of their very small size, high surface/volume ratios, and low levels of activity, newly hatched larvae probably take up relatively little $O_2$ through the gills, depending instead on simple diffusion through the body wall (Adolph 1979; Burggren 1984; Burggren and Just 1992; Burggren and Infantino 1994; Orlando and Pinder 1995; Ultsch, Bradford, and Freda 1999). Under normal conditions, diffusion provides adequate oxygen for metabolism. Under hypoxic conditions in warm ponds, however, metabolic rates of larvae in early stages of development decline as a function of decreasing oxygen availability. These young larvae are oxygen conformers and have little capacity to increase rates of respiration to maintain constant metabolic rates in low-oxygen environments (Hastings and Burggren 1995; Orlando and Pinder 1995). Early stage tadpoles also have little capacity to use anaerobic metabolism to compensate for reduced aerobic metabolism in hypoxic conditions, and even advanced-stage tadpoles have limited capacities for anaerobic metabolism (Crowder, Nie, and Ultsch 1998; Ultsch, Bradford, and Freda 1999). Aquatic caecilian larvae hatch at a more advanced stage than do those of anurans and urodeles, with larger, more elaborate gills (Sarasin and Sarasin 1887–1890; M. Wake 1977a; Duellman and Trueb 1986). This suggests a greater reliance on gills for respiration early in development, but nothing is known about partitioning of gas exchange or other aspects of respiratory physiology in these animals.

Once the larvae of anurans and salamanders begin feeding, the respiratory systems of the two groups diverge. Urodeles retain external gills throughout larval life, while the gills of anuran tadpoles become encased in a closed gill chamber. The size and complexity of the gills in urodele larvae reflect the environment in which they live. Larvae from well-aerated mountain streams have small gills and probably take up most of their $O_2$ through the skin. Larvae from ponds or other poorly oxygenated waters have large, bushy gills with much greater gas-exchange surfaces (see fig. 12.1 in chapter 12). This suggests a substantial role for the gills in aquatic gas exchange, but quantitative data are scarce. Large ambystomatid larvae can survive when branchial respiration is prevented (Shield and Bentley 1973; Heath 1976), but these experiments tell us nothing about the proportion of $O_2$ normally taken up through the gills in active, unrestrained animals (Burggren 1984).

Nearly all detailed studies of larval respiration and circulation in salamanders have been done on paedomorphic ambystomatids, mainly because their large size makes them suitable for laboratory studies (Gilmore and Figge 1929; Figge 1934, 1936; McIndoe and Smith 1984b; Malvin and Heisler 1988; Malvin 1989). In *Ambystoma tigrinum,* the first three pairs of aortic arches supply blood to the three pairs of gills, while a fourth arch gives rise to the pulmonary artery (fig. 4.14). Because there is some functional separation of systemic and pulmonary blood flow in the heart, the first arch receives the most highly oxygenated blood, and this in turn gives rise to the carotid arteries supplying the brain. The least-oxygenated blood goes to the posterior aortic arches supplying the third pair of gills and the lungs. Studies using radioactive microspheres to trace blood-flow patterns have shown that the major route of flow is not directly from the heart to the lungs, but through the third arch and the *ductus arteriosus* (fig. 4.14). Under hypoxic conditions, the blood can be shunted through the *ductus arteriosus* to the pulmonary artery without passing through the gills. This is advantageous when $O_2$ in the water is so low that the gills either cannot take up $O_2$ effectively or even lose $O_2$ to the water (Malvin and Heisler 1988; Malvin 1989).

There have been very few detailed studies of anuran tadpole gill morphology or function (Malvin 1989). Most studies of the anatomy of the gill region have focused on their role in feeding, not respiration, and comparisons of gill size and complexity have been related to differences in food habits (Wassersug and Hoff 1979; Wassersug 1980; Viertel 1985, 1987; Wassersug and Heyer 1988; Viertel and Richter 1999; see also chapter 12). McIndoe and Smith (1984a, b) described the gill morphology in tadpoles of *Litoria ewingi* and found evidence of shunts between afferent and efferent blood vessels similar to those reported in salamanders. Similar shunts have been described in other species of tadpoles

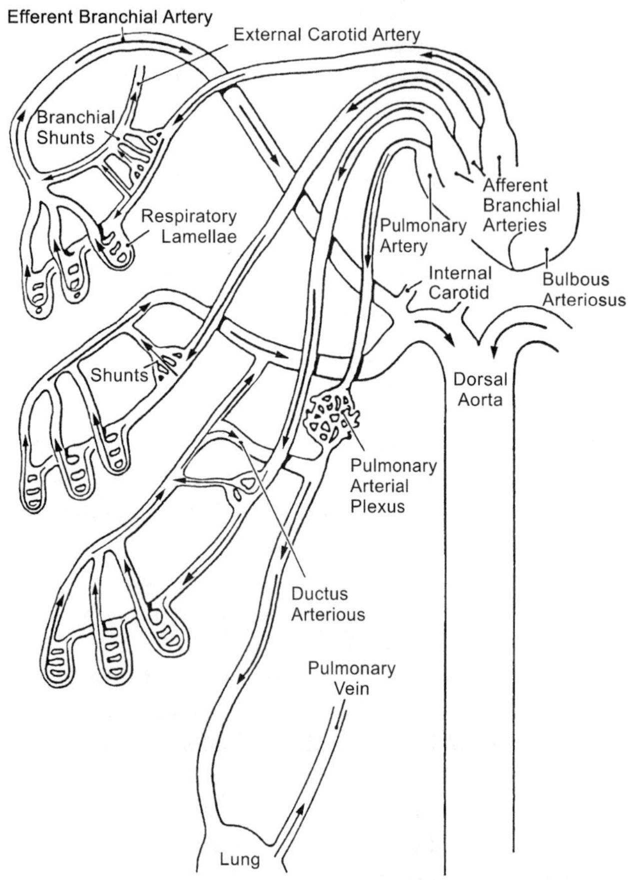


**Fig. 4.14.** Anatomy of the aortic arches, gills, and lungs of the larva of *Ambystoma tigrinum.* Arrows show routes of blood flow. After Malvin (1989).

and West 1982; Burggren and Doyle 1986; Burggren and Just 1992). By cannulating the spiracle of a tadpole placed in an aquatic respirometer, Burggren and West (1982) were able to measure the volume and $PO_2$ of water passing through the gill chambers. Cutaneous respiration was measured by determining the $O_2$ content of water before and after it passed through the chamber in which the tadpole was swimming. Aerial respiration of tadpoles with lungs was measured by allowing them access to air in a closed chamber separated from the aquatic chamber by a layer of mineral oil. In tadpoles without lungs, the skin accounted for about 60% of both $O_2$ uptake and $CO_2$ release (at 20° C), with the gills accounting for about 40%. The contribution of the skin to both $O_2$ and $CO_2$ exchange remained about the same when the lungs first developed, while the gills continued to account for about 40% of $CO_2$ excretion. However, the gills became less important for $O_2$ uptake as the tadpoles surfaced more frequently to breathe. In tadpoles undergoing metamorphosis, the lungs became the most important site of $O_2$ uptake, while $CO_2$ was eliminated mainly through the skin. This pattern remained largely unchanged in adult frogs (fig. 4.15). A later study of bullfrog tadpoles by Crowder, Nie, and Ultsch (1998) showed that even relatively early as well (de Saint-Aubain 1981), and these presumably enable the tadpoles to alter the pattern of blood flow through the gills (de Saint-Aubain 1985).

There has never been a systematic interspecific comparison of gill size and structure in relation to $O_2$ availability in different habitats, but there is evidence of both physiological and evolutionary responses to environments with low levels of $O_2$ (Løvtrop and Pigón 1968). In *Rana catesbeiana* tadpoles, exposure to chronic hypoxia can result in elongation of the gill filaments and thinning of the animal's skin (Burggren and Mwalukoma 1983). However, extreme hypoxic conditions, such as those found in water-filled bromeliads, can favor evolutionary reduction in the size of gill filaments or even a complete loss of functional gills (Noble 1929a). In these environments, aquatic $PO_2$ can be so low that tadpoles actually lose $O_2$ to the water, making a large aquatic respiratory surface disadvantageous (Lannoo, Townsend, and Wassersug 1987). Bromeliad-dwelling tadpoles must therefore depend primarily on aerial respiration.

The only quantitative study of respiratory gas exchange in tadpoles has been done on *Rana catesbeiana* (Burggren

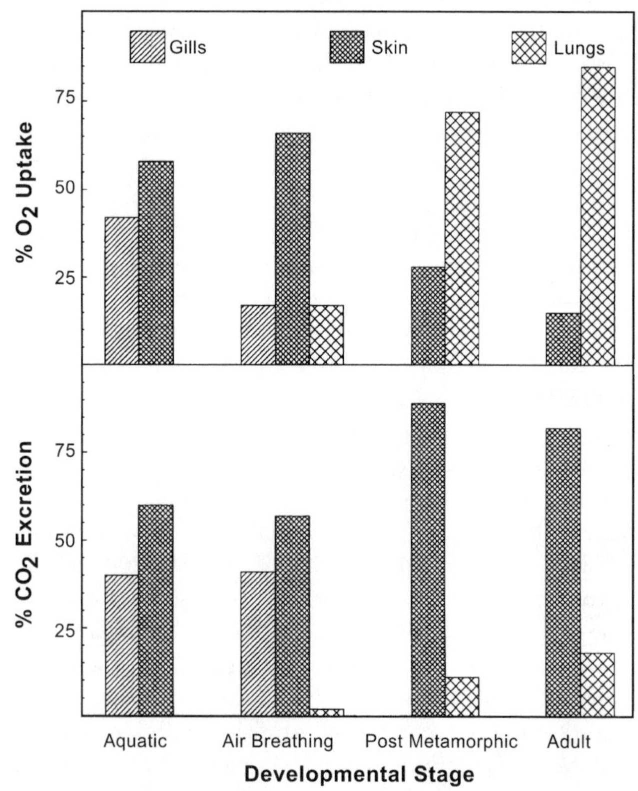


**Fig. 4.15.** Changes in cutaneous, pulmonary, and branchial respiration in tadpoles of the bullfrog (*Rana catesbeiana*) at different developmental stages. Bars show the proportion of $O_2$ and $CO_2$ exchange through the gills, skin, and lungs. Plotted from data in Burggren and West (1982).

stage tadpoles are capable of breathing with their lungs and readily surface to breathe when aquatic $PO_2$ is low. Additional studies of this species and *R. berlandieri* tadpoles showed that partitioning of gas exchange among lungs, skin, and gills was not greatly affected by temperature over a range from 15° to 33° C, even though total $O_2$ consumption increased with increasing temperature (Burggren and Feder 1983). This result contrasts with the increased dependence on pulmonary respiration at higher temperatures in adult frogs (fig. 4.4).

Some tadpoles do not develop lungs until just before metamorphosis (Feder 1983a), while others begin breathing air, presumably using the lungs, almost from the first free-swimming stage (Feder 1984; Ultsch, Bradford, and Freda 1999). Tadpoles that live in fast-moving, well-oxygenated streams do not develop lungs until metamorphosis (Nodzenski, Wassersug, and Inger 1989). In contrast, tadpoles that live in oxygen-poor environments, such as bromeliads, develop lungs very early and are obligate air-breathers (Lannoo, Townsend, and Wassersug 1987). In some species, such as toads of the genus *Bufo,* tadpoles metamorphose with many organ systems, including the lungs, relatively undeveloped, probably because selection favors rapid escape from temporary pond-breeding sites.

Tadpoles with lungs breathe air when the $O_2$ content of the water is low. Air breathing not only provides a more favorable gradient for $O_2$ uptake, but it also compensates for any $O_2$ lost to the water through the gills and skin (Wassersug and Seibert 1975; West and Burggren 1982; Feder 1983d; Wassersug and Feder 1983; Feder and Wassersug 1984). Natural bodies of water often become severely hypoxic during warm weather, and the ability to breathe air probably is critical for survival of tadpoles in such habitats (Wong and Booth 1994; Crowder, Nie, and Ultsch 1998). This is particularly true for species such as bullfrog (*Rana catesbeiana*) tadpoles, which tend to stay near the bottom of ponds, where $PO_2$ is lowest (Nie, Crim, and Ultsch 1999). These tadpoles can survive short periods of submergence in anoxic water by resorting to anaerobic metabolism. They typically die within a few days if deprived of access to air (Crowder, Nie, and Ultsch 1998). Tadpoles also increase rates of air breathing in response to increased activity or disturbance (Feder 1981, 1984). Ranid tadpoles allowed to breathe air have greater stamina during vigorous activity than those restricted to aquatic respiration (Wassersug and Feder 1983). Bufonid tadpoles that do not breathe air appear to be restricted to somewhat more oxygenated water than are ranid tadpoles, although *Bufo* tadpoles actually have a greater capacity to extract $O_2$ from hypoxic water than do *Rana* tadpoles (Feder 1983a; Noland and Ultsch 1981).

The use of lungs also affects the buoyancy of larvae, which in turn is related to characteristics of the habitat or location of food sources. Presumably the absence of lungs in stream-dwelling tadpoles would be advantageous in reducing buoyancy and decreasing the chances of being swept away in a current. *Bufo* tadpoles forage for food on the bottoms of ponds. Since they lack lungs, they remain negatively buoyant throughout development. Tadpoles of *Rana pipiens,* another bottom-dweller, also are negatively buoyant throughout development. Although they breathe air periodically, they usually release a bubble before swimming to the bottom. Other species of tadpoles, including *Rana sylvatica, Hyla versicolor,* and *Pseudacris triseriata,* are neutrally buoyant and can maintain a position in the water column by adjusting the amount of gas in the lungs, thereby allowing them to feed on particles suspended in the water column (Gee and Waldick 1995). Some tadpoles, such as *Xenopus,* are positively buoyant for much of their development and spend most of their time hovering at the water's surface. They must actually use energy to swim downward against their buoyancy, as they might when trying to escape from surface predators (Van Bergeijk 1959). The risk of predation at the surface probably is a major cost of developing lungs early and relying on breathing air for O2 uptake (Feder 1983c; Burggren and Infantino 1994). Larvae of the salamander *Ambystoma tigrinum* reduce air-breathing during daylight hours, when predation at the surface is most likely to occur. They do, however, increase aerial breathing during the day if zooplankton suspended in the water column are particularly abundant (Lannoo and Bachman 1984b).

Some tadpoles increase rates of aerial respiration when they encounter high concentrations of food particles in the water. Because the ventilation of the gills for respiration also draws in food particles that are trapped in part by the gills themselves (see chapter 12), the gill surfaces can become clogged with food and mucus that reduces the area available for gas exchange. Experiments with *Xenopus* tadpoles, which are midwater suspension feeders, demonstrated that high food concentrations caused tadpoles to decrease gill ventilation and increase aerial breathing (Feder et al. 1984). This in turn promotes faster growth rates because tadpoles are able to feed and take up $O_2$ simultaneously. When tadpoles were prevented from breathing air in relatively well-oxygenated water, they reduced food intake rates by expelling particles and mucus from the buccopharyngeal region. This presumably increased the surface area available for gas exchange, but lead to a reduction in energy intake and growth rate (Wassersug and Murphy 1987).

## Gas Exchange by Amphibian Embryos

Amphibian embryos must take up $O_2$ as they develop, but differ from larvae or adults in that they are immobile and therefore cannot breathe air directly or make other behavioral adjustments to enhance respiratory gas exchange. Until the mid-1980s, very little was known about the respiration of amphibian embryos in natural environments. Most measurements of $O_2$ consumption by developing embryos were made on isolated eggs, often removed from their jelly coatings, under ideal laboratory conditions (for references, see Burggren 1984, 1985). These studies tell us very little about gas exchange under field conditions, or the extent to which the availability of $O_2$ is limiting for embryonic development. More recently, however, a series of studies by Roger Seymour and his colleagues and several other investigators has added considerably to our understanding of respiration by embryos (for reviews, see Seymour and Bradford 1995; Seymour 1999).

$O_2$ uptake is limited mainly by diffusion of gas into the egg capsule (Seymour 1994, 1999). In a typical amphibian egg, the embryo and its associated yolk reserves are surrounded by a vitelline membrane, which in turn is surrounded by a jelly capsule (fig. 4.16). The rate of $O_2$ diffusion is limited by the $O_2$ conductance of the egg capsule and the difference in $PO_2$ between the embryo and the outside environment. Oxygen conductance is a function of the physical properties of the egg capsule and its thickness. For individual eggs, the jelly capsule is the principal barrier to gas exchange. Many amphibians, however, lay their eggs in compact jelly masses. This means that the amount of $O_2$ available to an individual embryo also is affected by the size and structure of the egg mass, the distance of individual eggs from the outside of the mass, and the respiration of the surrounding embryos, which tend to deplete $O_2$ inside the mass (fig. 4.16).

The need for $O_2$ to diffuse into eggs is a major constraint on the maximum size of amphibian eggs. Interspecific comparisons (Seymour and Bradford 1995) have shown that egg capsule $O_2$ conductance is well matched to the $O_2$ consumption of the embryos at the hatching stage for both aquatic and terrestrial egg-layers (fig. 4.17). In general, larger eggs have higher total rates of $O_2$ consumption, because they have more metabolically active tissue, even though mass-specific rates of $O_2$ consumption are lower in large eggs. Rates of $O_2$ diffusion are not normally limiting in early developmental stages for very small eggs, either in air or in water, but larger eggs are more prone to oxygen limitation, especially in water or at high temperatures. Comparisons among species of amphibians have shown that rates of $O_2$ consumption at the hatching stage are scaled to egg size with a relatively low exponent (0.52), indicating that $O_2$ becomes increasingly limited at egg volumes of 1 ml or larger (fig. 4.18).

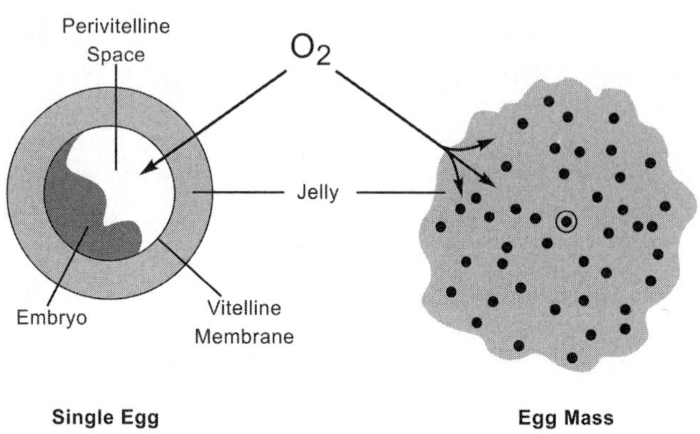


**Fig. 4.16.** Diagrammatic representation of the pathways of oxygen in individual amphibian eggs (left) and jelly egg masses (right). Oxygen uptake is limited by the rate of diffusion through the jelly surrounding the egg mass and the jelly layers and vitelline membrane of individual eggs. After Seymour and Bradford (1995).

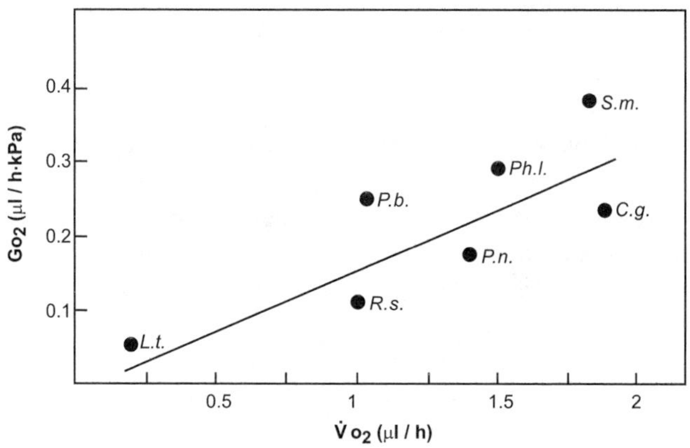


**Fig. 4.17.** Relationship of between oxygen consumption by hatching-stage embryos ($VO_2$) and oxygen conductance of the egg capsule ($GO_2$) for seven species of anurans. Species names: *C.g.* = *Crinia georgiana; Ph.l.* = *Philoria loveridgei; L.t.* = *Limnodynastes tasmaniensis; P.b.* = *Pseudophryne bibronii; P.n.* = *Phryonbatrachus natalensis; R.s.* = *Rana sylvatica; S.m.* = *Spea multiplicata.* After Seymour and Bradford (1995).

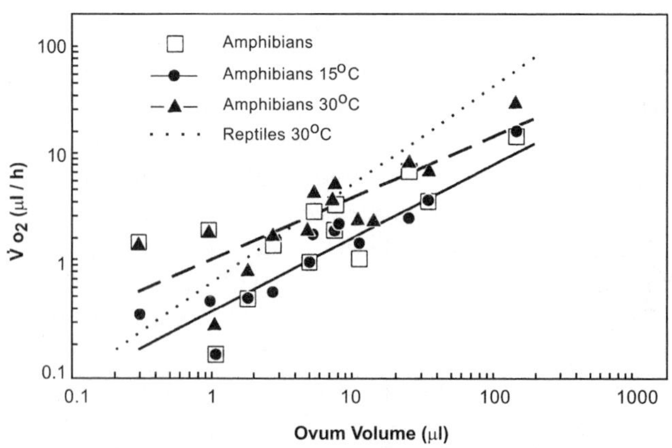


**Fig. 4.18.** Relationship of oxygen consumption of hatching-stage embryos to ovum volume for 14 species of amphibians. Open squares show uncorrected data for measurements made at different temperatures. Closed symbols show data corrected to 15° C and 30° C. Dotted line shows the same relationship for hatching stage reptiles for 30° C from Vleck and Hoyt (1991). Regression equation for amphibians at 15° C: $\dot{V}O_2 = 0.393V^{0.62}$, where V = ovum volume. Regression for amphibians at 30° C: $\dot{V}O_2 = 1.06V^{0.52}$. After Seymour and Bradford (1995).

The rate at which production of new embryonic tissue proceeds is correlated with mass-specific rates of $O_2$ consumption, so large eggs not only have metabolic rates lower than expected, but they also have much longer development times (Bradford 1990; Seymour and Bradford 1995; N. Mitchell and Seymour 2000). The size of amphibian eggs is more constrained in aquatic egg-layers than in terrestrial breeders because of the lower $O_2$ content of water. Egg size also is more constrained in warm, oxygen-poor water than in cold, oxygen-rich water. In both anurans and salamanders, species that breed in warm ponds typically have the smallest eggs and terrestrial breeders have the largest eggs (after correcting for the effect of body size); species breeding in cold streams are intermediate (see chapter 10 for a quantitative analysis of egg size). The metabolic rates of amphibian embryos increase steadily throughout development (H. Brown 1977; Burggren 1984, 1985; Burggren, Infantino, and Townsend 1990; Seymour and Roberts 1991; Seymour, Geiser, and Bradford 1991b; Seymour and Loveridge 1994; Seymour 1995; Seymour and Bradford 1995; Seymour and Roberts 1995; Seymour, Mahony, and Knowles 1995). This means that embryos are most likely to be oxygen-limited during late stages of development.

## Gas Exchange in Air

Many amphibians, including most plethodontid salamanders and tropical frogs in a number of different families, lay eggs out of the water in burrows, in hollow logs, under leaf litter, or in other terrestrial sites (see chapter 7). Terrestrial eggs usually are not surrounded by large masses of jelly, but are laid individually or in small clusters. They sometimes are in partial contact with the wall of a burrow or with other eggs, but they usually are relatively well exposed to open air. Oxygen is unlikely to be limiting, even inside a burrow, but the rate of $O_2$ uptake is limited by the $O_2$ conductance of the egg capsule, especially for large eggs in late developmental stages that are incubating at high temperatures (Seymour, Geiser, and Bradford 1991a, b).

Eggs laid on moist soil or vegetation take up water from the substrate, or from adults providing parental care (A. Martin and Cooper 1972; Bradford 1984a; Taigen, Pough, and Stewart 1984; Bradford and Seymour 1988b; for further discussion of parental care, see chapter 11). This uptake of water can alter surface-to-volume ratios of the eggs and change diffusion distances between the embryo and the environment. This phenomenon has been studied in terrestrial eggs of an Australian frog, *Pseudophryne bibronii*. Eggs of this species are laid in terrestrial nests that eventually are flooded; the larvae complete development in the water. During dry periods, the eggs can remain in the nest for many weeks before hatching (Bradford and Seymour 1985). Uptake of water in the nest causes the volume of the chamber containing the embryo to increase, while the thickness of the jelly capsule surrounding this chamber decreases (Seymour and Bradford 1987). This increases the $O_2$ conductance of the egg at a time when $O_2$ consumption of the embryo is increasing (fig. 4.19). Even under quite dry conditions, $O_2$ diffusion is sufficient to support the embryo's metabolism (Bradford and Seymour 1988b). This change in conductance is a function of developmental stage and is not directly affected by environmental conditions, such as $O_2$ availability and temperature (Seymour, Geiser, and Bradford 1991a).

Once the nests of *P. bibronii* are flooded, the eggs are subjected to decreased ambient $PO_2$. Experimental manipulation of aquatic $PO_2$ in the laboratory showed that below a certain critical $PO_2$, metabolic rates of embryos decreased (Bradford and Seymour 1988b). Very low $PO_2$ induced hatching in eggs that were close to hatching when the experiments began. There was a limit to how much hatching could be accelerated, however, because the embryos must reach a certain stage before they can survive independently. Prolonged exposure to extreme aquatic hypoxia at earlier stages simply retarded development and eventually killed the eggs. Without field data on variation in $O_2$ levels in flooded nests, it is impossible to estimate the importance of this source of mortality in natural populations. However, a field study of three species of *Pseudophryne* over five breeding seasons revealed that embryonic mortality was very low (usually less than 5%) and occurred mostly in the terrestrial rather than the aquatic phase of development (Woodruff 1976b). Hence, mortality due to low $PO_2$ in the nest probably is of trivial importance.

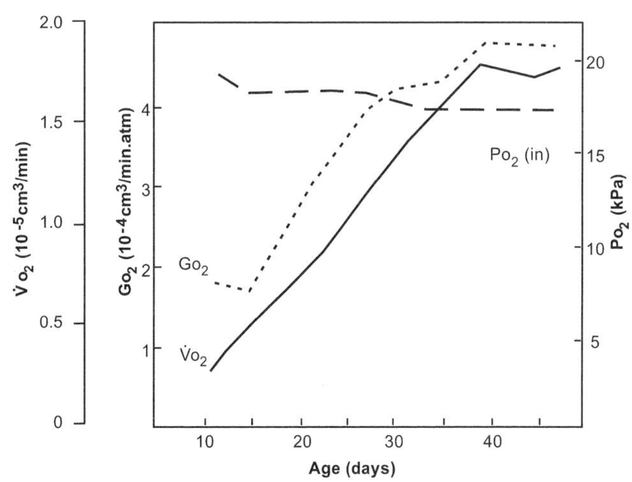


**Fig. 4.19.** Gas exchange in the terrestrial eggs of *Pseudophryne bibronii*. The figure shows the change in capsule wall conductance [$GO_2$], oxygen partial pressure in the perivitelline chamber [$PO_2$ (in)], and oxygen consumption of the embryo [$\dot{V}O_2$] during development. $PO_2$ (in) is calculated assuming a constant $PO_2$ (air) of 20.91 kPa. After Seymour and Bradford (1987).

Another Australian myobatrachid frog, *Crinia* (= *Bryobatrachus*) *nimbus,* deposits small clutches of relatively large eggs in damp moss. Upon hatching, the larvae remain in a gelatinous mass in the nest. This frog has one of the thickest egg capsules known for any anuran, and theoretical calculations indicate that embryos in a clutch of up to 20 eggs should suffer hypoxia at temperatures above 5°C. Field measurements, however, failed to reveal evidence of hypoxic conditions. Apparently the porous nature of the moss nest allows for aeration of the embryos, while photosynthesis by the moss provides supplemental oxygen. The relatively small number of eggs in each clutch and low environmental temperatures, which reduce metabolic demand, also enable the embryos to develop without experiencing hypoxic conditions (Mitchell and Seymour 2003).

Many frogs lay nonaquatic eggs attached to vegetation over water, where the eggs eventually hatch and drop into the water to complete development (see chapter 10). In some species, such as phyllomedusine hylids in the genera *Agalychnis, Phyllomedusa,* and *Pachymedusa,* the eggs are surrounded by abundant jelly, which serves as a water reservoir that reduces the chances of desiccation. Other species, particularly Old World tree frogs in the family Rhacophoridae, lay their eggs in arboreal foam nests. These are made from mucus secretions produced by the female that are beaten into a froth as the eggs are laid (see chapter 10 for details of nest construction).

Seymour and Loveridge (1994) studied the effect of the foam nest on embryonic respiration in the African rhacophorid *Chiromantis xerampelina.* These frogs produce foam nests attached to branches overhanging temporary pools that form after heavy rains. The eggs are relatively small and have very thin jelly capsules. They are scattered through the foam inside the nest. The outer foam layer quickly dries out, forming a relatively hard, protective coating that reduces evaporative water loss, but allows for diffusion of $O_2$ into the nest (fig. 4.20). Before hatching, the eggs could obtain all the $O_2$ they need from the air trapped in the foam bubbles. After the tadpoles hatch, their metabolic rates increase rapidly, and they must depend on $O_2$ that diffuses into the nest from outside. Drying of the foam increases the $O_2$ conductance of the nest. Eventually the tadpoles become very crowded in the small amount of liquid foam remaining in the nest, and $O_2$ becomes limiting, especially at high daytime temperatures. Low $PO_2$ in the nest triggers emergence of tadpoles, which drop into the water below to complete development. The main advantage of laying eggs in a foam nest instead of a jelly mass is that it allows for a much larger egg mass and a larger clutch size because of the air trapped in the foam and the permeability of the drying foam to $O_2$.

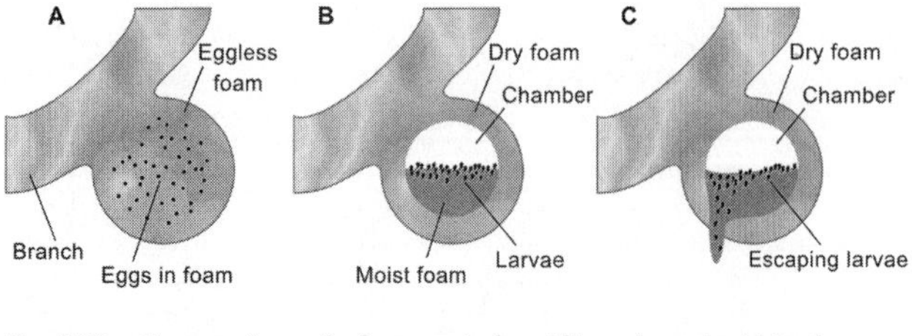


**Fig. 4.20.** Cross-sections of a foam nest of an African rhacophorid treefrog, *Chiromantis xerampelina,* at three stages of its life. (A) Nest immediately after eggs are laid, with eggs dispersed in the center of the foam and surrounded by eggless foam. (B) After tadpoles have hatched and collected in moist foam in the bottom of a chamber formed by a shell of dry foam. (C) Tadpoles escaping from the nest as the pool of moist foam breaks through the outer shell of dry foam. After Seymour and Loveridge (1994).

## Gas Exchange in Water

Eggs laid in water are much more likely to become $O_2$-limited than those laid on land, especially at high temperatures, because aquatic environments often become hypoxic. Amphibians lay eggs in many different types of aquatic sites, from cold, fast-moving streams to warm ponds and small bodies of water in depressions on the ground, burrows, tree holes, and bromeliads (see chapter 10). These are likely to differ considerably in the availability of $O_2$ (see table 4.1). There also is great diversity in the structure of egg masses laid in these sites (see chapter 10), and much of this variation probably is related to problems of $O_2$ uptake.

The first detailed study of respiration of aquatic egg masses was done by Burggren (1985) on *Rana palustris.* This species lays globular egg masses surrounded by a thick jelly coat. Variation in $O_2$ content of water at natural breeding sites has not been studied, but limited field measurements indicated that eggs were in water that was supersaturated with $O_2$ at midday (water temperature about 13 C). Burggren used oxygen electrodes to measure $PO_2$ and $PCO_2$ at different positions in submerged egg masses maintained in the laboratory. He found only a slight decrease in $PO_2$ and a slight increase in $PCO_2$ from the outside to the inside of the egg. At normal ambient $PO_2$, $O_2$ concentrations at the interior of the mass were sufficient to maintain aerobic metabolism. There was no evidence of a buildup of lactic acid in centrally located eggs, and changes in pH were quite small. These results contradicted earlier studies that had found substantially lower $PO_2$ near the center of egg masses in several species, as well as increased levels of lactate that indicated a switch to anaerobic metabolism (R. Savage 1935b; Barth 1946; Gregg 1962).

Given the relatively large size of the egg mass of *Rana palustris,* Burggren (1985) was not able to explain how sufficient oxygen could reach embryos at the center of the mass. Seymour and Roberts (1991) argued that diffusion alone could not account for the relatively high $PO_2$ at the center of the egg masses and hypothesized that water circulating through channels in the jelly must provide oxygen to central embryos. This was subsequently confirmed in studies of wood-frog (*Rana sylvatica*) egg masses. Female wood frogs often lay their eggs in large communal masses that are tightly clustered in shallow water (see fig. 3.14 in chapter 3). This is thought to aid in heat retention during the day, but it also would reduce the surface area available for diffusion of $O_2$ into the egg masses. Indeed, even a single wood frog egg mass exceeds the theoretical limit for egg masses containing a high density of eggs. Nevertheless, the eggs can obtain sufficient $O_2$ throughout development. One reason is that these frogs breed in early spring in relatively cold, oxygen-rich water. In addition, the egg masses are loosely structured, with channels between the eggs that allow water to circulate throughout the mass. Wood frog egg masses start out as compact balls, but rapidly swell as they take up water. This increases the separation of the eggs and the size of the water channels between them. The eggs usually are placed in open sunlight, and as the egg mass warms, convection currents develop that circulate water through the egg mass. Nevertheless, there is a substantial $PO_2$ gradient from the outside to the inside of the mass, with eggs in the center subjected to lower $PO_2$ (Pinder and Friet 1994; Seymour 1995). Individual eggs take up water throughout development, which increases the volume of the space inside the egg and increases the $O_2$ conductance of the egg capsules as the metabolic demands of the eggs increase (fig. 4.21).

Spotted salamanders (*Ambystoma maculatum*) also are early spring breeders that often lay eggs in the same ponds used by wood frogs. Their egg masses are readily distinguished from those of wood frogs because they contain fewer, larger eggs, and the masses are very compact. They lack convection channels that would enhance $O_2$ uptake by the eggs (Pinder and Friet 1994). They do, however, have another source of $O_2$ for the eggs. Many egg masses are colonized by symbiotic green algae (*Oophila ambystomatis*) that apparently live only in association with amphibian eggs (Gatz 1973). A number of early workers had suggested that these algae might produce $O_2$ that could be used by the salamander eggs. In an early study, Hutchison and Hammen (1958) reported that the algae consumed more $O_2$ than they produced, an unlikely result if the algae themselves are growing. Their study, however, was done at 25° C, an unrealistically high temperature for these cold-water breeders. Pinder and Friet (1994) measured $O_2$ consumption and production at 10° C, which revealed that the algae do produce more $O_2$ than they consume, and egg masses exposed to light actually contain more $O_2$ than does the surrounding water. In darkness, when the algae use $O_2$, but do not produce $O_2$ by photosynthesis, steep $PO_2$ gradients developed in salamander egg masses, with central eggs being exposed to nearly anoxic conditions (fig. 4.22). When egg masses were kept in the dark and then exposed to light, $PO_2$ inside the masses increased immediately, indicating an increase in $O_2$ production by photosynthesis (see also Bachmann et al. 1986).

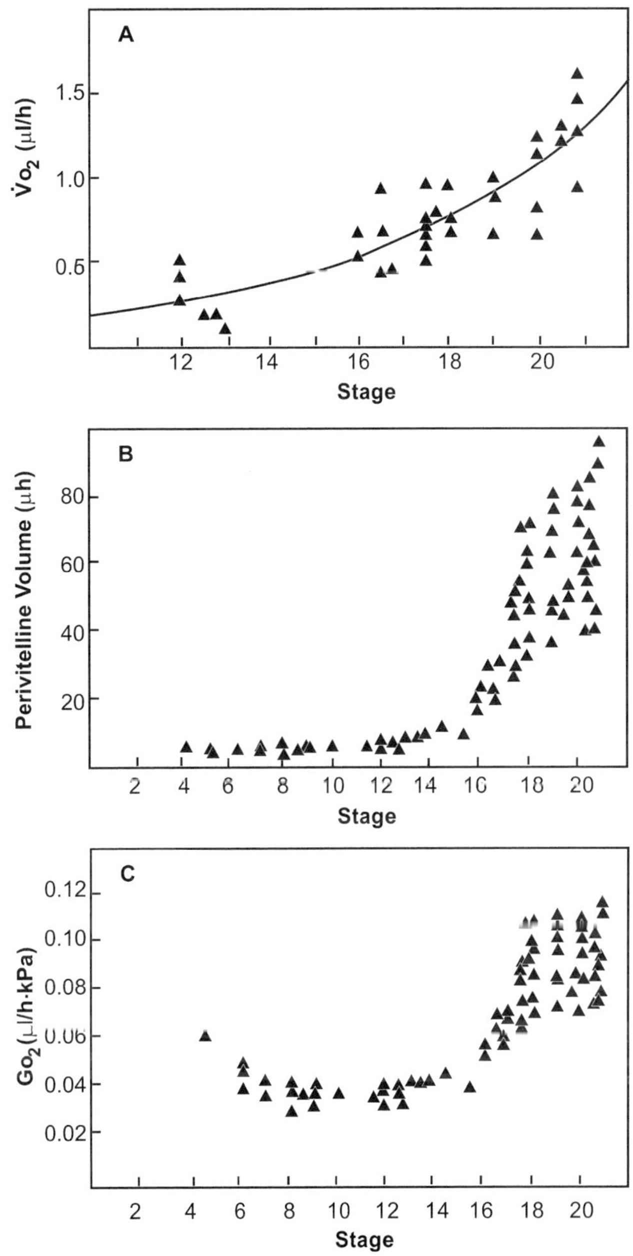


**Fig. 4.21.** Changes in gas exchange and characteristics of wood-frog embryos as a function of Gosner (1960) stage of development. (A) Oxygen consumption ($\dot{V}O_2$) of individual embryos at 15°C. Regression equation: $\dot{V}O_2$ = 0.0793 stage – 1.549. (B) Changes in volume of the perivitelline chamber. (C) Changes in oxygen conductance of the egg capsule ($GO_2$). After Seymour (1995).

In a study of embryonic development in *Ambystoma* and in *Rana palustris*, Petranka, Just, and Crawford (1982) suggested that aquatic amphibian eggs require exposure to low $PO_2$ to induce hatching. Subsequent work has shown this

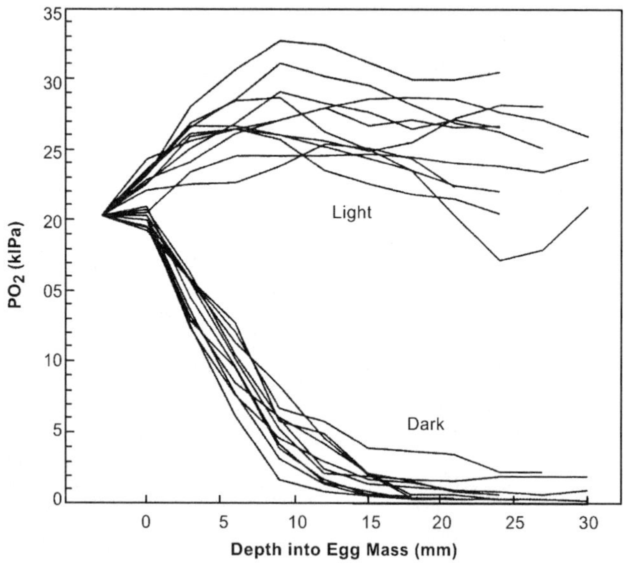


**Fig. 4.22.** Oxygen partial pressure ($PO_2$) gradients in egg masses of *Ambystoma maculatum* that were kept for at least 10 hours in either darkness or light. Egg masses kept in the dark, where photosynthesis by symbiotic algae was not possible, were nearly anoxic at the center of the mass. After Pinder and Friet (1994).

not to be true. For example, Burggren (1985) found that *R. palustris* eggs hatched even in oxygen-saturated water. Mills and Barnhart (1999) found that chronic exposure to hypoxic conditions resulted in slower development and delayed hatching in *Ambystoma maculatum* and *A. annulatum,* and hatchlings were not as well developed as those reared in normoxic conditions. In *A. annulatum,* posthatching survival of larvae reared in hypoxic conditions was poor. In *Rana palustris* and *R. utricularia,* embryos exposed to hypoxic conditions did hatch earlier than did those in normoxic conditions, but the hatchlings were not as well developed. Overall, these results indicate that prolonged exposure to hypoxic water has detrimental effects on the growth, development, and survivorship of amphibian embryos, but embryos probably can survive brief expose to hypoxic conditions without ill effect. There also is evidence for *Ambystoma* that exposure to low levels of dissolved oxygen causes egg capsules to expand, thereby increasing the oxygen conductance of the capsule and partially compensating for low oxygen availability (Mills, Barnhart, and Semlitsch 2001).

Amphibians that breed in warm, oxygen-poor ponds, or in small bodies of water that are likely to be rapidly depleted of $O_2$ by respiring eggs, face the greatest limitations on $O_2$ availability during development. Seymour and his colleagues have studied the respiration of embryos of several Australian frogs that breed in oxygen-poor environments. One is *Limnodynastes tasmaniensis,* which breeds in the spring and summer (October–December) in shallow, temporary ponds. Most populations produce foam egg masses that are produced by the females, which beat the surface of the water with enlarged, paddle-shaped flanges on their front toes (see chapter 10). This creates a bubble nest that floats on the surface, and eggs rise to the surface as they are laid and are suspended in or under the bubbles. As the eggs develop, the foam eventually breaks down, and the eggs sink. Oxygen concentrations were greatest at the foam-water interface and decreased toward the center of the egg masses (fig. 4.23), but most masses had sufficiently high $PO_2$ levels to support the embryos throughout development. Some populations of this species produce nonfoamy egg masses held at the water's surface by vegetation. Most of these also contained sufficient $O_2$ for the embryos, but some parts of the masses became hypoxic if allowed to sink to the bottom (Seymour and Roberts 1991).

Two other Australian myobatrachid frogs lay eggs in small bodies of water. *Crinia georgiana* breeds in relatively cold weather and lays eggs in shallow, water-filled depressions on the ground. The respiratory characteristics of the aquatic eggs of this species were similar to the terrestrial eggs of *Pseudophryne bibronii* (Seymour and Roberts 1995; Seymour et al. 2000). Metabolic rate increased in an exponential fashion as the eggs developed, while $O_2$ conductance increased in a linear fashion as the eggs swelled with water. Consequently, some embryos experienced reduced $O_2$ availability late in development. Despite substantial variation in nest $PO_2$, most embryos survived to hatching, probably because the embryos are unusually tolerant of low $PO_2$. Although, many eggs tend to be placed in sites where trickling water helps to aerate the eggs, there was little evidence that selection of oviposition sites was directly influenced by $PO_2$. Another myobatrachid, *Philoria loveridgei,* lays much larger eggs in a jelly mass in shallow, flask-shaped depressions in moist soil. This is a subtropical frog that breeds at much warmer temperatures than *Crinia georgiana,* and its eggs are subject to more severe $O_2$ limitation. On average, about 2/3 to 3/4 of the eggs hatch, with mortality being greatest in those located at the bottom of the nest, where embryos often are exposed to anoxic conditions (Seymour, Mahony, and Knowles 1995). This species apparently evolved from ancestors that produced foam nests, but the reasons for the loss of foam nesting is unknown.

## Accessory Respiratory Structures in Embryos

While many aquatic and terrestrial embryos probably take up most of their $O_2$ by simple diffusion through the body wall, some species develop elaborate respiratory structures well before hatching. For example, most larvae of salamanders and aquatic caecilians have large branching gills, and some anuran embryos develop large external gills that are lost soon after hatching. Other amphibian embryos have large sheetlike gills or greatly enlarged tail fins that are as-

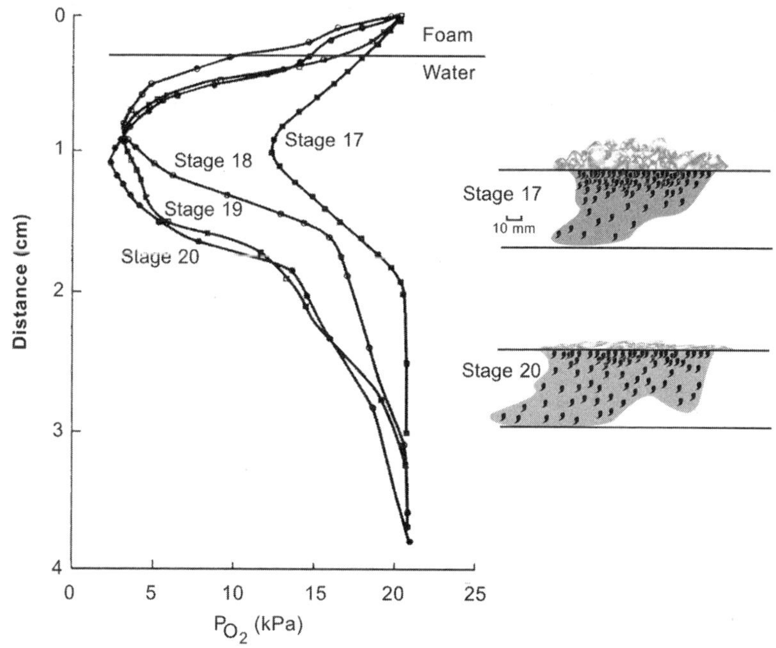


**Fig. 4.23.** Oxygen profiles through the center of a foamy egg mass of *Limnodynastes tasmaniensis* at four Gosner (1960) stages of development. Distance is measured from the top of the egg mass, which is exposed above the water surface, to the bottom of the egg mass about 3.5 cm underwater. Diagram at right shows the spatial distribution of embryos in the egg mass at stages 17 and 20. After Seymour and Roberts (1991).

sumed to have a respiratory function. These are found even in fully terrestrial species that undergo direct development and lack an aquatic larval stage. For example, embryos of terrestrial salamanders such as *Plethodon glutinosus* have large branching gills (Bishop 1941b), while the embryos of *Eleutherodactylus coqui* have wide, thin tail fins (Townsend and Stewart 1985). Such structures are particularly well developed in the embryos of marsupial frogs, which develop in pouches or cavities on the mother's body, and in embryos of viviparous species (see chapter 10). Detailed descriptions and illustrations of these structures are given by Noble (1927, 1931), Lamotte and Lescure (1977), Duellman and Trueb (1986), and Thibaudeau and Altig (1999), as well as numerous papers on embryology and development (Gasche 1939; Lynn 1942; Lynn and Lutz 1946a, b; Jameson 1950; J. Davis 1952; Gorman 1956; Valett and Jameson 1961; Vial 1968; McDiarmid and Worthington 1970; del Pino et al. 1975; M. Wake 1978; del Pino and Escobar 1981; Townsend and Stewart 1985). Functional interpretations of apparent respiratory structures are based almost entirely on their morphology; physiological studies of their role in gas exchange are lacking.

The Neotropical red-eyed treefrog (*Agalychnis callidryas*) lays relatively large eggs on vegetation overhanging water. The embryos develop elaborate branching external gills well that are fully perfused with blood well before hatching (Warkentin 1999b). These gills apparently are important in providing extra surface area for gas exchange for embryos in an egg mass in which diffusion of oxygen is limited. When eggs were removed from masses and exposed to air, the embryos quickly lost the circulation in their external gills, and some lost the gills altogether. Even when gills were retained, they were about 1/10 the length of those of embryos in natural egg masses (Warkentin 2000a). Once the eggs hatch, circulation to the external gills is cut off almost immediately, and gills are quickly lost. Loss of the gills can be retarded, however, by exposing the newly hatched tadpoles to hypoxic water (Warkentin 2000a). These results not only indicate that embryonic gills play an important role in gas exchange, but they also show that development and loss of gills are plastic responses to the availability of oxygen.

## Summary and Conclusions

Amphibians have been used for centuries to answer basic questions about vertebrate respiration, but in many ways, the respiration of amphibians differs from that of other tetrapods. In particular, amphibians rely to a greater extent on cutaneous respiration than do any other vertebrates, and in the case of plethodontid salamanders, reliance on cutaneous respiration has been accompanied by the evolutionary loss of lungs. The use of the skin as a respiratory organ is made possible by the highly permeable nature of the amphibian skin, but is constrained by the need to keep the skin moist at all times. Amphibians cannot achieve dramatic increases in the rate of oxygen uptake through the skin during vigorous activity, so they must either rely on the lungs to supply oxygen during exercise or must depend heavily on anaerobic metabolism.

Amphibians in the terrestrial environment are seldom constrained by the availability of oxygen. Even at the highest altitudes inhabited by amphibians, partial pressures of oxygen ($PO_2$) are more than sufficient to support activity. The only terrestrial amphibians that might be exposed to hypoxic conditions on a regular basis are burrowing species, but the limited data available suggest that even underground burrows usually have sufficient oxygen for normal activity. Aquatic amphibians, on the other hand, often find themselves in environments with limited oxygen supplies. This is particularly true of warm, swampy habitats and shallow, stagnant bodies of water, but also is characteristic of some ice-covered ponds in winter. Some aquatic salamanders, such as sirens, exhibit physiological adaptations to hypoxic environments, including obligate air-breathing with lungs and high tolerance for dissolved $CO_2$. Tadpoles that live in hypoxic environments, such as water-filled bromeliads, exhibit similar adaptations and often develop lungs soon after hatching.

In general, anurans have more complex lungs with greater surface area for gas exchange than do urodeles, and the lungs provide more than half of all gas-exchange surface area in most anurans. There are a few exceptions, however, such as *Siren, Amphiuma,* and *Salamandra,* all of which have complex, well-developed lungs. The structure of most caecilian lungs has not been studied in detail, but the limited data available suggest that caecilians depend primarily on their lungs for gas exchange. In some aquatic species, a separate tracheal lung provides increased surface area for gas exchange, but one species of aquatic caecilian has lost its lungs altogether and is the largest lungless tetrapod. Studies of lung structure of some distantly related giant salamanders have revealed a general correlation between complexity of lung structure and the availability of oxygen in aquatic habitats. There have not been any other comparative studies, however, relating lung structure in other groups, such as anurans, to variation in locomotory activity and degree of dependence on aerobic metabolism.

All amphibians ventilate the lungs with a positive-pressure buccal pump, which involves the throat muscles pumping air into the lungs. Buccal pumping is retained even in lungless species, such as plethodontid salamanders, which also use the lining of the buccal cavity and pharynx for gas exchange. Exhalation is accomplished by contraction of the hypaxial (trunk) muscles in urodeles, but anurans and caecilians rely on elastic recoil of the inflated body for passive exhalation. The exception is in male frogs that call; they have enlarged trunk muscles that contract to force air out of the lungs to produce sound.

Despite having only a three-chambered heart with a single, undivided ventricle, amphibians achieve a considerable degree of functional separation of oxygenated and deoxygenated blood through morphological adaptations of the ventricle and the *conus arteriosus.* At the same time, the undivided circulation allows for shunting of blood away from the lungs when the lungs are compressed and toward the lungs when they are inflated. There is some evidence that oxygen affinity of the blood of amphibians is adapted to the environments in which the animals live, with higher oxygen affinity in aquatic species than in terrestrial air-breathers. Nevertheless, such patterns have not been investigated in a meaningful phylogenetic context, but have been determined for distantly related aquatic and terrestrial species that differ in many other aspects of their physiology and ecology. Ontogenetic studies of anuran tadpoles tend to support this pattern, however, because there is a shift toward lower blood-oxygen affinity at metamorphosis as oxygen becomes more readily available to a terrestrial animal.

Among larval salamanders, there is a general correlation between the surface area of the gills and oxygen availability in the environment, with pond-dwelling larvae typically having larger gills with much more surface area for gas exchange than do stream-dwelling larvae. Again, few quantitative comparisons have been made in a phylogenetic context in clades that occupy diverse habitats, such as the plethodontid salamanders. Virtually nothing is known about the relationship of internal gill morphology to gas exchange capacity in anuran tadpoles, nor have there been any comparative morphological studies relating gill morphology to environmental oxygen availability. Interpretation of gill morphology in anuran tadpoles is complicated by the involvement of the gill region in feeding as well as respiration, but a phylogenetically based comparison of respiratory structures would be worthwhile. We also need additional comparative studies of the ontogeny of respiratory physiology in anuran tadpoles. Limited data for a few taxa suggest that tadpoles that live in oxygen-poor environments often develop lungs sooner than do those that live in well-aerated environments, but again most comparisons have been made between distantly related taxa, such as ranid frogs and bufonid toads.

An area of research that was neglected for many years is the respiration of amphibian embryos. There have been detailed investigations in the last 20 years on a relatively limited number of species found mostly in Australia. These have been chosen to represent a range of oviposition strategies, from terrestrial breeding to laying eggs in ponds, in floating foam nests, or in trees. These studies have shown that the structure and placement of egg masses often is affected by the gas-exchange requirements of the embryos and the availability of oxygen in the environment. Expansion of such comparative studies to a wider range of taxa, such as foam-nesting leptodactylids, stream-breeding frogs, and those that lay eggs in bromeliads and other oxygen-poor environments would be most informative. Many aspects of the evolution of amphibian reproductive modes and life-history strategies, including egg size, clutch size, and the structure and placement of egg masses, traditionally have been interpreted as responses to selective pressures such as predation (see chapter 10). In many instances, however, it appears that responses to the physical environment, and in particular the availability of oxygen, are equally important in shaping the evolution of egg size, clutch size, and clutch structure. We also need more information on the respiratory adaptations of the embryos themselves, particularly those that develop elaborate external gills inside the egg capsule. While such structures appear to provide additional surface area for gas exchange, their actual contribution to embryonic respiration remains almost entirely unknown.

# Chapter 5 Metabolism and Energetics

*Amphibia are cold-blooded; they lack the mechanisms which give the higher types both freedom from environmental change and constancy of chemical activity at the optimum conditions for the expenditure of their energies. . . . Hence their energy sources, the food and oxygen, are made available at a much slower rate in these forms. . . . Amphibia are not able to make use, to the fullest extent, of either their nervous or their motor systems. They remain slaves of their surroundings.*
—G. Kingsley Noble, *The Biology of the Amphibia* (1931)

UNTIL RELATIVELY RECENTLY, biologists tended to view amphibians as somewhat inferior to other terrestrial vertebrates, handicapped by their low body temperatures and metabolic rates, unable to achieve levels of activity equivalent to those of even the most sluggish birds and mammals. Yet amphibians often have been credited with remarkable feats of survival under all sorts of adverse conditions. For centuries, even reputable naturalists thought that toads could survive for many years completely encased in stones or blocks of wood. Typical accounts described toads sealed for years in airtight chambers with no access to food or water, yet miraculously surviving when the chambers were opened (Gosse 1860). Various naturalists tried experiments to determine how long toads could survive under such conditions. For example, the geologist William Buckland performed some experiments in the 1820s in which toads were entombed in stone chambers and buried in a churchyard for a year. Those encased in sandstone all died within a year, whereas some of those in more porous limestone were still alive, but succumbed in the second year (Buckland 1832).

Stories of toads surviving for centuries encased in rock clearly are apocryphal. Nevertheless, amphibians do have unusual metabolic characteristics that allow them to remain underground or underwater during long periods of winter cold or summer heat, often emerging after many months and moving immediately to temporary ponds to breed. Indeed, modern herpetologists have viewed the low metabolic rates of amphibians not as handicaps, but as features that allow them to pursue a low-energy lifestyle not available to endothermic vertebrates (Bennett 1978; Pough 1980, 1983; Feder 1983b; Seale 1987).

This chapter deals with the metabolic characteristics of amphibians. I first discuss sources of variation in resting metabolism and the relative importance of aerobic and anaerobic metabolism during activity. Then I move on to a consideration of the energetics of natural behavior, such as foraging, courtship, aggression, and vocal communication. Finally, I discuss the use of various energy substrates to support metabolic activity and seasonal cycles of energy storage and use.

## Metabolism of Amphibians at Rest

### Sources of Variation in Resting Metabolism

Physiologists have made thousands of measurements of oxygen consumption by amphibians under various conditions and have investigated the relationship between resting metabolism and both temperature and body mass (Tashian and Ray 1957; Whitford and Hutchison 1967; Hutchison, Whit-

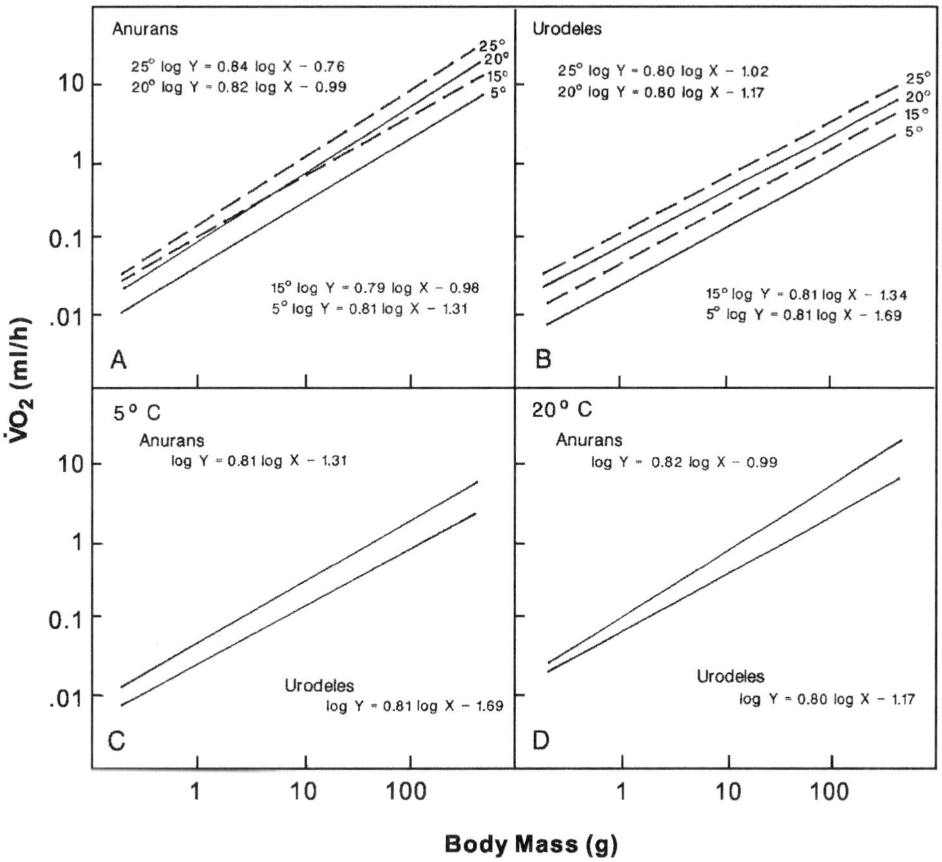


**Fig. 5.1.** Relationship of resting metabolic rates to body mass for anurans and urodeles. (A) Anurans at four different temperatures. (B) Urodeles at four different temperatures. (C) Anurans and urodeles at 5° C. (D) Anurans and urodeles at 20° C. Regressions plotted from equations in Gatten, Miller, and Full (1992).

ford, and Kohl 1968; Ultsch 1973a, 1974; Feder 1976a, b). A comprehensive review by Gatten, Miller, and Full (1992) summarized this literature and tabulated nearly all known measurements for both larval and adult amphibians. The regressions shown in figure 5.1 are plotted from their equations. Many authors have distinguished between "standard" metabolism, measured during the normal period of inactivity, and "resting" metabolism, measured on quiescent animals during the normal activity period (e.g., diurnal measurements on diurnal frogs), but Gatten, Miller, and Full found that the use of these terms was not consistent enough to analyze these measurements separately. Therefore, I have followed their convention in referring to all measurements made on inactive animals as resting metabolic rates.

The scaling of resting metabolism with body mass is identical in adult anurans and urodeles, with slopes of logarithmic regressions averaging about 0.80 (fig. 5.1 A, B). Anurans and urodeles also show a similar response to temperature, with metabolic rate approximately doubling with a 10° C increase in temperature (i.e., $Q_{10} = 2$). At the same temperature, anurans have consistently higher metabolic rates than urodeles (fig. 5.1 C, D). Regressions for caecilians reported by Gatten, Miller, and Full (1992) had intercepts similar to those for anurans and somewhat higher slopes, but these were based on only three or four species and therefore were insufficient for rigorous statistical comparisons. More recent data have shown that caecilians actually have lower metabolic rates than either anurans or urodeles of the same body mass (Smits and Flanagin 1994; Stiffler and Talbot 2000). Data for larval amphibians were more variable than those for adult anurans and urodeles (Gatten, Miller, and Full 1992). The scaling of metabolic rate with body mass was similar for larval and adult urodeles, but larvae appeared to have somewhat lower metabolic rates at any given temperature. The same seems to be true for anuran tadpoles, but the available data are insufficient to draw definitive conclusions.

Gatten, Miller, and Full (1992) compared metabolic rates of amphibians with those of birds and mammals by extrapolating data collected at 20° or 25° C to a more typical endothermic temperature of 38° C, using a $Q_{10}$ of 2.0. They estimated that resting metabolic rates of mammals and nonpasserine birds are about 10 times those of amphibians at 38° C, while passerine birds have metabolic rates about 10 to 30 times higher (bird and mammal data from Lasiewski and Dawson 1967; McNab 1988). Comparisons with data for squamates or for all reptiles revealed that frogs have metabolic rates similar to reptiles at 20° or 25° C, whereas metabolic rates of urodeles are about 30% lower (reptile data from Bennett and Dawson 1976; Andrews and Pough 1985).

These comparisons actually understate the differences in absolute energy requirements between amphibians, reptiles, and endotherms, for several reasons. First, most amphibians are considerably smaller than the smallest birds and mammals, and many also are smaller than most reptiles (Pough 1980; Feder 1983b), so their total energy requirements would be lower even at the same temperature. Second, few amphibians can actually function at body temperatures near 38° C (see chapter 3); meaningful ecological comparisons are best made at normal activity temperatures for each animal. The metabolic rate and total energy requirements of a temperate-zone plethodontid salamander at 15° C probably should be compared to a diurnal frog at 20–25° C (in summer), a lizard at 25–35° C, and a bird or mammal at 38° C. Such comparisons have shown that amphibians in general, and plethodontid salamanders in particular, have the lowest resting metabolic rates and thus the lowest maintenance energy requirements of any terrestrial vertebrates (Pough 1980, 1983; Feder 1983b; Gatten, Miller, and Full 1992).

Broad interspecific comparisons are useful in revealing general relationships between body mass, temperature, and metabolic rate, but can obscure differences among species that are ecologically significant. For example, the regressions in figure 5.1 are based on pooled data for both tropical and temperate-zone species. Several authors have reported lower metabolic rates in tropical than in temperate-zone anurans. Unfortunately, early studies by Tashian and Ray (1957) and Hutchison, Whitford, and Kohl (1968) are flawed because they compared tropical and temperate-zone species in different families and of different body sizes and did not present rigorous statistical comparisons. Weathers and Snyder (1977) compared metabolic rates of three ranid frogs from Borneo with the regression equation for temperate-zone ranids at 25° C given by Hutchison, Whitford, and Kohl (1968) and found that the tropical species had lower metabolic rates. They did not present regression equations for the tropical species that could be compared statistically with the temperate-zone data, however.

Other authors have used these data to support the generalization that tropical anurans have lower metabolic rates than temperate-zone species (Whitford 1973; Duellman and Trueb 1986), but a rigorous statistical analysis that controls for variation in phylogenetic history and body size is needed. Unfortunately, we do not have sufficient data for anurans to make a tropical–temperate-zone comparison among species in the same family (Gatten, Miller, and Full 1992). Navas (1996c) compared the metabolic rates of high-elevation Andean anurans in four different families to those of species in the same genera from low-elevation tropical sites in Panama. He found that low-elevation species of *Colostethus* (Dendrobatidae) and *Atelopus* (Bufonidae) had slightly lower resting metabolic rates than high-elevation species, but this was true only at low temperatures. High- and low-elevation species of *Hyla* and *Eleutherodactylus* did not show this difference, however. Differences in activity metabolism were much more dramatic. All of the low-elevation species had much lower aerobic capacities for locomotion at low temperatures, and in fact, were virtually incapacitated at temperatures regularly experienced by high-elevation species during their normal activity periods.

Feder (1976b) compared resting metabolic rates of 14 Neotropical and 12 North American plethodontid salamanders and found that tropical salamanders had significantly lower metabolic rates than temperate-zone species at both 15° and 25° C. The ecological consequences of these differences in resting metabolism are hard to predict, because we lack detailed information on the amount of time salamanders spend at different temperatures throughout the year. The activity temperatures of tropical and temperate-zone plethodontids overlap considerably, but tropical species do not experience the very low temperatures characteristic of northern winters. Feder and Lynch (1982) found that maximum activity temperatures of tropical and temperate-zone species were not significantly different, but lowland tropical species were poorly represented in their sample. Nevertheless, because of reduced seasonal variation, tropical salamanders probably have higher average temperatures on a yearly basis than temperate-zone species, and might therefore have similar or even higher energy requirements in spite of their lower metabolic rates (Feder, 1976b).

Another potential source of variation in metabolic rates is the relative importance of skin and lungs in gas exchange. Early work by Whitford and Hutchison (1967) had suggested that plethodontid salamanders had lower resting metabolic rates than ambystomatids and salamandrids with well-developed lungs, and that the difference was particularly pronounced at large body sizes. This was attributed to reduced respiratory surface area in lungless species and increasingly unfavorable surface-to-volume ratios in large species. If this pattern were generally true, then the low metabolic rates of urodeles compared to anurans (fig. 5.1) might simply be a consequence of including many plethodontids in the urodele sample. Subsequent work (Feder 1976a) revealed that neither slopes nor intercepts of mass-metabolism regressions differed between lunged and lungless species. Feder attributed the results obtained by Whitford and Hutchison to their use of plastic masks to measure $O_2$ consumption, a treatment that could have excited the animals and elevated metabolism above normal resting levels, particularly in species with lungs.

### Temporal Changes in Resting Metabolism

Early students of amphibian respiratory physiology reported that frogs collected during the summer had higher metabolic rates than those collected in winter or early spring (Bohr 1900; Krogh 1904; Dolk and Postma 1927). Subsequent workers have reported similar patterns in a variety of temperate-zone amphibians, with metabolic rates generally being highest during spring or summer breeding seasons and lowest in fall and winter, even when measured at the same temperature (Vernberg 1952; Fromm and Johnson 1955; Dunlap 1969, 1973, 1980; Harri and Hedenstam 1972; Knapp 1974). Some studies have shown no significant differences in metabolism at different times of year (Davison 1955; Seymour 1973a). The physiological basis for seasonal variation in metabolic rate is poorly understood because several variables that could affect such variation often are not controlled. These include activity of the animal in the metabolic chamber (Fromm and Johnson 1955), thermal history (see the subsequent discussion of acclimation), food intake (Feder et al. 1984), photoperiod (Whitford and Hutchison 1965b; Guimond and Hutchison 1968), and reproductive condition (Fitzpatrick 1971). Seasonal fluctuations in hormone levels, modulated by the central nervous system, can have a direct effect on metabolic rate (Harri and Hedenstam 1972; Lagerspetz 1977), but it is not clear that these effects always are independent of other variables, such as seasonal changes in activity.

Many amphibians exhibit diel variation in resting metabolism, with nocturnal species generally having higher metabolic rates at night and diurnal species having higher rates during the day (Kasbohm 1967; Dunlap 1969; Hutchison and Kohl 1971; Turney and Hutchison 1974; Weathers and Snyder 1977; Carey 1979a; Miller and Hutchison 1979; Bucher, Ryan, and Bartholomew 1982; Taigen and Pough 1983). In some cases, differences between day and night measurements are considerable (e.g., Hutchison and Kohl 1971; Bucher, Ryan, and Bartholomew 1982), but in others, they are quite small and not statistically significant (e.g., Taigen and Pough 1983). Diel variation in metabolic rate appears to be related mainly to changes in activity of the animals and not necessarily to an underlying physiological cycle. Hutchison and Kohl (1971) simultaneously monitored movements and metabolic rates of *Bufo marinus* during a 24-hour cycle. The toads exhibited a marked increase in metabolic rate during the first two to three hours after the onset of darkness and a marked decrease immediately after the onset of light. These changes in metabolism were paralleled by changes in locomotor activity. Similar results were obtained in experiments with *Necturus maculosus* (K. Miller and Hutchison 1979). Both metabolic rates and levels of activity increased dramatically with the onset of darkness as the salamanders increased rates of air-breathing at the water's surface. In both species, changes in activity and metabolic rates were triggered directly by photoperiod and therefore did not represent a true circadian rhythm.

Metabolic rates of amphibians also vary with food intake. Immediately after feeding and for some time thereafter, metabolic rates are elevated because of the energetic cost of digestion (specific dynamic action = SDA). In general, amphibians experience only moderate changes in metabolic rate after feeding. For example, in *Bufo woodhousii,* metabolic rates three hours after feeding were 1.7 times higher than at 120 hours after feeding (Sievert and Bailey 2000), while in *Plethodon metcalfi,* postfeeding metabolism was about 1.8 times resting metabolism (Feder, Gibbs, Griffith, and Tsuji 1984). In *Bufo marinus,* SDA averaged 5.5 times resting metabolism (Secor and Faulkner 2002). Specific dynamic action was similar in *Ceratophrys cranwelli,* averaging two to five times resting metabolism, and it was not affected by the type of diet (earthworms or mice; Powell, Mansfield-Jones, and Gatten 1999; Grayson et al. 2005). Specific dynamic actions in the same general range have been reported for a number of other anurans as well (Busk, Jensen, and Wang 2000; Secor and Faulkner 2002; J. B. Anderson and Wang 2003). For most anurans, the increase in metabolism after feeding is considerably less than that experienced during vigorous exercise, but specific dynamic action varies with both body size and relative meal size.

The limited data available for amphibians suggests that increases in metabolic rate are higher for species that feed infrequently than for those that feed frequently, but it is not clear whether this difference is mainly a result of differences in meal size, differences in seasonal patterns of feeding activity, or both. For example, two anurans that eat large meals at infrequent intervals, *Pyxicephalus adpersus* and *Ceratophrys ornata,* had peak metabolic rates about eight to nine times higher than resting metabolism, with specific dynamic action accounting for about 21–22% of ingested energy. Four unrelated species of anurans that eat smaller meals at frequent intervals increased metabolic rates after feeding by about three to five times resting rates, and specific dynamic action accounted for only 10–17% of ingested energy. These data, along with data on rates of intestinal uptake of amino acids and glucose, indicated that infrequent feeders tend to upregulate digestive function in response to feeding (Secor 2001). Several other studies have shown some degree of seasonal variation in intestinal size and morphology in South American anurans, although these studies provided less detail on digestive physiology (Naya et al. 2003, 2005; Naya and Bozinovic 2004).

In a subsequent study, Secor (2005a, b) attempted to control for phylogenetic effects by making comparions within anuran families, although in some cases, the species used were not very closely related. He reported that *Pyxicephalus, Ceratophrys,* and a toad, *Bufo alvarius,* had much higher specific dynamic action than did three other representatives of the same families, *Rana catesbeiana, Leptodactylus pentadactylus,* and *Bufo marinus.* One major ecological difference between these two groups of species is that the former spend a long dry season estivating underground, whereas the latter do not estivate, and the two tropical species are active throughout the year. The three estivating species shut down digestive function during estivation, thereby saving considerable energy. After feeding, these species exhibited substantial increases in intestinal mass and intestinal nutrient uptake rates, with a 600–1000% increase in total nutrient uptake capacity and a 600–1200% increase in metabolic rate. This contrasted with an increase in nutrient uptake of less than 70% in the nonestivating species and increase in metabolic rate of 300–400%. Nevertheless, postfeeding increases in metabolic rate in these frogs are not as impressive as those observed in snakes that eat large meals at infrequent intervals. In snakes such as *Boa constrictor* and *Python molurus,* metabolic rates after feeding can be as much as 18 times resting rates (1800% increase) and exceed those attained during vigorous exercise (Secor and Diamond 2000; Wang, Busk, and Overgaard 2001).

### Metabolism during Dormancy

Many amphibians undergo a period of dormancy, either to escape desiccating conditions during prolonged dry seasons (see chapter 2) or to survive winter cold (see chapter 3). Dormant amphibians cease all feeding and other activities, and in the case of cocoon-forming species, experience a marked decrease in cutaneous gas exchange capacity (see chapter 4). These changes often are accompanied by reductions in metabolic rates to levels considerably below normal resting levels. In hibernating amphibians, lower metabolic rates are a direct effect of lower body temperature, but this is not the case for estivating animals (Pinder, Storey, and Ultsch 1992). Measurements made during dry-season dormancy are available for nine species in seven families (table 5.1).

All of these species except *Hyperolius viridiflavus* estivate in underground burrows, where they remain for several months, and in extreme cases, for several years. Most of the

**Table 5.1** Metabolic rates of amphibians at rest and during dormancy

| Species (source no.) | Temperature (°C) | $\dot{V}O_2$ rest (ml $O_2$/g h) | $\dot{V}O_2$ dormant (ml $O_2$/g h) | Percent decrease |
|---|---|---|---|---|
| Hylidae | | | | |
| *Cyclorana platycephalus* (3) | 15 | 0.0537 | 0.1056 | 71 |
| Hyperoliidae | | | | |
| *Hyperolius viridiflavus* (6) | 25 | 0.1800 | 0.0750 | 59 |
| Leptodactylidae | | | | |
| *Ceratophrys ornata* (9) | 25 | | | 25[a] |
| *Lepidobatrachus llanensis* (5) | 25 | 0.0970 | 0.0200 | 79 |
| Myobatrachidae | | | | |
| *Neobatrachus wilsmorei* (8) | 25 | 0.0740 | 0.0140 | 80 |
| Pelobatidae | | | | |
| *Scaphiopus couchii* (1) | 15 | 0.0377 | 0.0073 | 81 |
| *Spea hammondii* (adult) (1) | 15 | 0.0600 | 0.0150 | 75 |
| *S. hammondii* (juvenile) (1) | 15 | 0.1720 | 0.0440 | 74 |
| Ranidae | | | | |
| *Pyxicephalus adspersus* (4) | 20 | 0.0370 | 0.0096 | 74 |
| *P. adspersus* (9) | 25 | | | 16[a] |
| Sirenidae | | | | |
| *Siren lacertina* | 23 | 0.0305 | 0.0110 | 64 |

*Sources:* (1) Seymour 1973b; (2) Whitford and Meltzer 1976; (3) Van Beurden 1980; (4) Loveridge and Withers 1981; (5) McClanahan, Ruibal, and Shoemaker 1983; (6) Geise and Linsenmair 1988; (7) Etheridge 1990b; (8) Flanigan et al. 1993; (9) Secor 2005b.

[a]Measurements after only one month of simulated estivation in laboratory.

burrowing anurans had metabolic rates during dormancy about 70–80% below resting rates, although less impressive reductions in metabolic rate, averaging 16 and 25%, were obtained for *Ceratophrys ornata* and *Pyxicephalus adspersus* induced to estivate in the laboratory for only one month (Secor 2005b). *Pyxicephalus, Lepidobatrachus, Ceratophrys, Neobatrachus* and *Cyclorana* all form cocoons to reduce evaporative water loss (Loveridge and Withers 1981; McClanahan, Ruibal, and Shoemaker 1983; Van Beurden 1984; Flanigan et al. 1991; Flanagan, Withers, and Guppy 1993; Withers 1993), but the reduction in the metabolism of those species is not any greater than that of *Scaphiopus,* which does not form cocoons (table 5.1). *Hyperolius viridiflavus* remains exposed on vegetation during the dry season, but becomes completely inactive and greatly reduces evaporative water loss (see chapter 3). This species experiences less reduction in metabolism than do burrowing species. The only salamanders for which data are available are sirens. These form thin cocoons and sometimes spend several months encased in drying mud. The reduction in mass-specific metabolic rate is less pronounced than in the anurans (table 5.1), but sirens also suffer greater loss of body mass from desiccation, so the reduction in total energy demand is comparable (Etheridge 1990a, b).

The greatly reduced metabolic rates of dormant frogs cannot be attributed solely to a reduction in gas exchange capacity, because species that do not form cocoons undergo reductions in metabolism equivalent to those of cocoon-forming species. Furthermore, the dramatic decline in metabolism occurs before cocoons are completely formed (Loveridge and Withers 1981; McClanahan, Ruibal, and Shoemaker 1983). Reduced metabolism also does not appear to be a direct result of starvation, although this certainly can reduce metabolic rates (e.g., Merkle and Hanke 1988). The drop in metabolic rate begins almost immediately and generally is complete within a few days, long before lipid reserves and other energy stores are significantly reduced (Seymour 1973a; Loveridge and Withers 1981; McClanahan, Ruibal, and Shoemaker 1983; Geise and Linsenmair 1988; Etheridge 1990a; Pinder, Storey, and Ultsch 1992).

In *Neobatrachus,* the reduction in metabolism is due in part to a reduction in tissue metabolism that is evident even in isolated muscle tissue (Flanigan, Withers, and Guppy 1991; Flanigan et al., 1993). About 85% of the reduction in whole animal metabolism in dormant *Neobatrachus* is due to a reduction in the metabolic rate of muscle tissue, although other tissues, including the brain, have reduced metabolism as well (Flanigan and Guppy 1997). Metabolic depression is affected by changes in enzyme activities and alterations in rates of glycolysis and other metabolic processes in cells (Flanigan et al. 1990; Pinder, Storey, and Ultsch 1992; Guppy, Fuery, and Flanigan 1994; Fuery et al. 1998), as well as down-regulation of organ systems, especially the digestive system (Secor 2005a, b). The reduction in metabolism can prolong life underground for many months and reduces the possibility that the animals will begin to metabolize their own protein (Seymour 1973a; Van Beurden 1980; Loveridge and Withers 1981; Etheridge 1990a). Indeed, the extremely low metabolism of dormant anurans prevents significant atrophy of skeletal muscle during estivation, an important trait for animals that must dig their way out of the soil once dormancy ends (Hudson and Franklin 2002a, b).

### Metabolic Responses to Temperature Acclimation

Because amphibians tend to be rather imprecise thermoregulators (Carey 1978; Feder 1982b; Sinsch 1989; Hutchison and Dupré 1992; see also chapter 3), they often are exposed to pronounced temperature fluctuations on a seasonal, daily, or even hourly basis. Since metabolic rates are temperature dependent, these changes will result in substantial temporal variation in resting metabolism. Amphibians probably cannot make physiological adjustments to short-term temperature changes on the scale of a few hours or a few days (Feder 1985a, b), but they do exhibit shifts in temperature-metabolism relationships on a seasonal basis through thermal acclimation. Physiological ecologists have spent a considerable amount of time studying patterns of metabolic acclimation in the laboratory in the hope of understanding the functional significance of acclimation responses in the field (for a review, see Rome, Stevens, and John-Alder 1992). Typically, acclimation is studied by holding animals at a fixed temperature for periods of several days to a few weeks and then measuring resting metabolic rates at a range of experimental temperatures. Usually animals are not fed during acclimation experiments, and they often are housed in darkness. Both feeding (Feder et al. 1984) and photoperiod (Vinegar and Hutchison 1965; Kasbohm 1967; Guimond and Hutchison 1968; Hofer 1972; Dunlap 1989) can affect the direction and rate of metabolic acclimation. The effects often are inconsistent among species and even among acclimation temperatures and are therefore difficult to interpret in an ecological context.

Although acclimation responses of amphibians vary among species, a number of patterns appear to be relatively common. For example, plethodontid salamanders typically exhibit partial compensation of metabolic rate when acclimated to temperatures at which the species is normally active (Fitzpatrick, Bristol, and Stokes 1971, 1972; Fitzpatrick 1973a, b; Fitzpatrick and Brown 1975; A. Brown and Fitzpatrick 1981a, b; Feder 1978b, 1985a, b; Feder et al. 1984). If a salamander initially housed at 15° C is transferred to 20° or 25° C, there is an immediate increase in metabolic rate due to the acute effect of temperature, followed by a

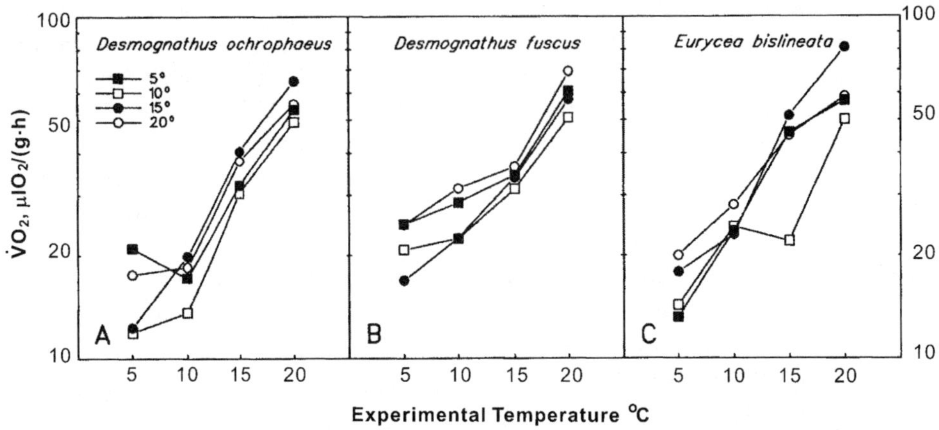


**Fig. 5.2.** Acclimation of metabolic rates in three plethodontid salamanders. Symbols indicate acclimation temperatures. Experimental temperatures are temperatures at which metabolic rates were measured after acclimation. Plotted from data in Fitzpatrick, Bristol, and Stokes (1971, 1972) and Fitzpatrick (1973a).

gradual decrease in metabolism over a period of days. If the salamander is transferred to 10° C, there is an immediate decrease in metabolic rate, followed by a gradual increase. The result is an inverse relationship between acclimation temperature and metabolic rate at a given measurement temperature (fig. 5.2).

Often this pattern is not apparent in animals acclimated to temperatures below the normal activity range. For example, in many species, acclimation to 5° C either produces a decrease in metabolic rate, or very little change (fig. 5.2). Similar patterns are found in anurans that are active at relatively cool temperatures. *Pseudacris triseriata* exhibits an inverse relationship between acclimation temperature and metabolic rate at measurement temperatures of 20–30° C (Dunlap 1980). When these frogs are acclimated to very low temperatures, they exhibit a marked decrease in metabolic rate, particularly when measured at low temperatures (Packard 1972). *Bufo woodhousii* shows the same pattern at 10–20° C, but warm-acclimated animals have reduced metabolic rates when measured at 25° C (Fitzpatrick and Atebara 1974). Tadpoles of the Australian frog *Limnodynastes peronii,* which often overwinter, do not show acclimation of resting metabolic rate to low temperature (E. Marshall and Grigg 1980). Nevertheless, they do exhibit acclimation of both aerobic and anaerobic metabolic enzymes, perhaps allowing tadpoles to maintain slow but steady growth during the winter (Rogers, Seebacher, and Thompson 2004).

Amphibians that are active at higher temperatures often do not exhibit the classic pattern of partial metabolic compensation seen in plethodontid salamanders. For example, in a number of species, including *Triturus alpestris, Salamandra atra, Ambystoma tigrinum, Acris crepitans, Bufo boreas,* and several species of *Rana,* acclimation of summer-collected animals to warm temperatures actually results in higher metabolic rates at moderate to warm measurement temperatures than in animals acclimated to cooler temperatures. In some cases, just the opposite response is seen in winter-collected animals (Dunlap 1969, 1971, 1972, 1980; Harri 1973; Holzman and McManus 1973; Knapp 1974; Weathers 1976; Carey 1979b; Burggren and Wood 1981; Feder 1982a). Tropical amphibians generally exhibit little or no acclimation response, although only a few species have been studied in detail (Feder 1978b, 1982a, 1987; Navas 1996c).

The results of laboratory acclimation studies often are used as a basis for discussing the adaptive significance of acclimation in the field. Putative benefits usually include energy conservation and metabolic efficiency. For example, most authors assume that a reduction in metabolic rate in animals exposed to warm temperatures helps to conserve energy during the summer. Increased metabolism in animals exposed to cool temperatures would enable them to be active and acquire food in the spring or fall at temperatures that would incapacitate them in summer. The reduction in metabolism in animals exposed to cold, such as that seen in *Pseudacris* and many plethodontids, usually is considered a mechanism to conserve energy during winter dormancy (Fitzpatrick, Bristol, and Stokes 1971, 1972; Packard 1972; Fitzpatrick 1973a, b; Fitzpatrick and Brown 1975; A. Brown and Fitzpatrick 1981a, b; Feder 1978b, 1985a, b; Feder et al. 1984). On the other hand, the higher metabolic rates seen in some warm-acclimated animals at moderate temperatures have been interpreted as adaptations to reduce the temperature-sensitivity of metabolism in the face of fluctuating environmental temperatures (Dunlap 1971, 1972).

Although these adaptive explanations often seem plausible, few of the proposed fitness benefits of acclimation have been rigorously quantified. Furthermore, many laboratory studies have not controlled for differences in activity inside metabolic chambers; hence, the higher metabolic rates seen in some warm-acclimated animals could be due to increased movement at higher temperatures (Rome et al. 1992). Functional interpretations of acclimation responses also are complicated by inconsistencies among species that appear to have similar ecological requirements. For example, Fitzpatrick and his colleagues studied metabolic responses to temperature acclimation in *Desmognathus fuscus, D. ochrophaeus,* and *Eurycea bislineata,* all collected in the same ravine. All three species showed similar types of acclimation responses, but the magnitude of the responses varied, being greatest in *D. ochrophaeus,* intermediate in *D. fuscus,* and least in *E. bislineata* (fig. 5.2). These differences were attributed to the last two species being more aquatic than the first and therefore subject to less pronounced fluctuations in temperature (Fitzpatrick, Bristol, and Stokes 1971, 1972; Fitzpatrick 1973b), but this interpretation is largely speculative. Feder (1985a) found similar differences in the rate and magnitude of thermal acclimation in two fully terrestrial plethodontids, *Plethodon cinereus* and *P. metcalfi.*

Another problem is that temperature regimes to which amphibians are acclimated in the laboratory do not always resemble those encountered in the field. Carey (1979b) compared acclimation responses of *Bufo boreas* subjected to a fluctuating temperature regime with those of animals acclimated to constant temperatures of 10°, 20°, and 30° C. She found that the fluctuating regime produced an acclimation response quite different from the constant regimes, but did not necessarily reflect the metabolic responses of toads subjected to natural temperature fluctuations in field enclosures. Furthermore, responses to the different experimental treatments were not consistent even between two populations of the same species. In a rather complicated experiment, Feder (1985a, b) subjected several species of plethodontid salamanders to acclimation regimes that differed in the magnitude and period of temperature fluctuation. Acclimation responses differed considerably among experimental temperatures and among species, making it difficult to place the results in a simple ecological context. In many cases, the apparent energy savings resulting from metabolic acclimation were minimal under a fluctuating temperature regime, calling into question the standard adaptive explanations for such responses.

In general, acclimation of metabolism is viewed as a means of compensating for the effects of temperature variation in a seasonal environment, thereby enabling amphibians to engage in normal activities over a relatively wide range of temperatures without suffering major reductions in performance. If this is true, then one might expect acclimation in the laboratory to have similar effects on resting and activity metabolism. In fact, thermal acclimation often appears to have different effects on resting metabolism, activity metabolism, and locomotor performance, even in the same species (Rome et al. 1992). For example, acclimation of *Desmognathus ochrophaeus* to 21° C reduced both resting and activity metabolism at 13° C, but had no effect on the maximum sustainable speed of locomotion measured on a treadmill. When animals were measured at 21° C, acclimation did not affect metabolic rate, but did influence stamina (Feder 1986).

In contrast, acclimation temperature had no effect on resting metabolism, activity metabolism, or maximum speed of locomotion at 13° C in a tropical plethodontid, *Bolitoglossa subpalmata* (Feder 1987). Studies of other urodeles, including *Necturus* and *Ambystoma,* generally have shown little or no effect of acclimation on locomotor performance or muscle activity (K. Miller 1982; Else and Bennett 1987). In anurans, thermal acclimation generally has little effect on activity metabolism (Carey 1979b), muscle contraction (Rome 1983; Renaud and Stevens 1981, 1984), or locomotor performance (Putnam and Bennett 1981; K. Miller 1982; Renaud and Stevens 1983b; K. Miller and Zoghby 1986; Whitehead et al. 1989). When acclimation effects are evident, they often vary in magnitude and even in direction, depending on the species involved, test temperature, and the measure of locomotor performance used (Renaud and Stevens1983b; K. Miller and Zoghby 1986; Londos and Brooks 1988). Hence, there does not appear to be a consistent relationship between metabolic acclimation in the laboratory and other measures of behavioral or physiological performance.

## Metabolism of Active Amphibians

Early studies of amphibian energetics focused almost exclusively on resting or routine metabolic rates and ignored increases in metabolism needed to support activity. Indeed, some authors constructed detailed energy budgets for amphibians based entirely on resting metabolic rates, with no allowance for increases due to activities such as foraging, searching for mates, territorial defense, or calling (e.g., Merchant 1970; Fitzpatrick 1973a; Burton and Likens 1975a, b; G. Smith 1976). For relatively sedentary species such as plethodontid salamanders, time spent in vigorous activity is such a trivial proportion of an animal's annual time budget that such costs can be safely ignored in calculations of energy budgets. For active foragers, or frogs that call vigorously for many days or weeks, however, the metabolic costs of activity can be considerable. If these are not included in an energy budget, the result can be a gross underestimate of respiratory costs and an overestimate of the amount of energy available for production of new tissue (Seale 1987). Since the early 1970s, physiological ecologists have paid more attention to measuring metabolic rates of active amphibians. We now have a substantial number of measurements of metabolic rates during forced activity (reviewed in Gatten, Miller, and Full 1992), and a small but growing set of metabolic measurements made on animals engaged in natural activities (reviewed in Pough et al. 1992).

### Methods for Measuring Activity Metabolism

Several methods have been used to induce bouts of vigorous activity in amphibians. These were reviewed in some detail by Gatten, Miller, and Full (1992) and will be discussed only briefly here. Some workers have stimulated aquatic amphibians to swim by prodding them manually (Hillman and Withers 1981; Withers and Hillman 1981; Pinder and Burggren 1986) or subjecting them to mild electric shocks (Hutchison, Turney, and Gratz 1977; Preslar and Hutchison 1978; K. Miller and Hutchison 1979; Boutilier, McDonald, and Toews 1980). Cushman, Packard, and Boardman (1976) exercised salamanders (*Ambystoma tigrinum*) by having them swim in a moving current in a water tunnel, but they measured only the accumulation of lactic acid, not rates of oxygen consumption. Jiang and Claussen (1992) developed an aquatic respirometer with a circular tube in which a moving magnetic stirrer was used to prod a newt into continuous swimming around the chamber.

Many early studies of activity metabolism in terrestrial amphibians relied on mild electric shocks to elicit vigorous exercise (Bennett and Licht 1973; Bennett and Wake 1974; Turney and Hutchison 1974; Feder 1977, 1978a, b; K. Miller and Hutchison 1980). This technique was largely abandoned after Hillman et al. (1979) showed that it could lead to serious underestimates of aerobic metabolism and overestimates of the importance of anaerobic metabolism in supporting activity. This was particularly true when electric shock stimulation was combined with manometric methods for measuring oxygen consumption that are based on changes in the volume of gas in a chamber. Heating of the air in the chamber by the electric stimulator and by the activity of the animal causes the gas to expand, thereby underestimating the amount of oxygen consumed. The use of these techniques led to the erroneous conclusion that vigorous activity is supported mainly by anaerobic metabolism, with increases in oxygen consumption occurring only during a postactivity recovery period (Bennett and Licht 1973; Bennett and Wake 1974; Feder 1977, 1978a). Later work showed this not to be true (Hillman et al. 1979; Withers 1980; Gatten 1987a).

Most investigators have exercised animals in rotating chambers that are turned manually (Hillman 1976, 1980b, 1982b, 1987; Hillman et al. 1979; Hillman and Withers 1979, 1981; Withers 1980; Loveridge and Withers 1981; M. Ryan, Bartholomew, and Rand 1983; Withers and Hillman 1983; Sullivan and Walsberg 1985; Walsberg, Lea, and Hillman 1986; Withers, Hillman, Simmons, and Zygmunt 1988; Withers et al., Lea, Solberg, Baustian, and Hedrick, 1988; Prestwich, Brugger, and Topping 1989; Stefanski, Gatten, and Pough 1989) or with a variable-speed motor (Seymour 1973c; Harlow 1978; Carey 1979a, b; McDonald, Boutilier, and Toews 1980; K. Miller and Hutchison 1980; Taigen and Pough 1981, 1983; Taigen, Emerson, and Pough 1982; Pough and Kamel 1984; Taigen and Beuchat 1984; Wells and Taigen 1984; Taigen and Wells 1985; Taigen, Wells, and Marsh 1985; Feder 1986, 1987; Mitton, Carey, and Kocher 1986; Gatten 1987a; Walton 1988; Navas 1996c; Rogowitz and Sánchez-Rivoleda 1999). Rotation of the chamber causes the animal to hop, crawl, or make continual attempts to keep itself upright.

There has been some debate about whether manual or motorized rotation yields a more accurate estimate of maximum aerobic capacity for locomotor activity. Walsberg (1986) reported that manual rotation produced higher estimates of aerobic metabolism than motorized rotation for three species of anurans, but Rogowitz and Sánchez-Rivoleda (1999) reported the opposite pattern for *Eleutherodactylus cooki*. The motorized chambers used by both of these investigators were rotated at a constant speed throughout the exercise period. Most other investigators have adjusted rotation speed to ensure that the animal continues to right itself and does not simply tumble passively inside the chamber (e.g., Seymour 1973c; Taigen and Pough 1981, 1983; Taigen, Emerson, and Pough 1982; Pough and Kamel 1984; Wells and Taigen 1984; Taigen and Wells 1985; Taigen, Wells, and Marsh 1985; Gatten 1987a; Navas 1996c). Therefore, the constant-speed rotation method has not been widely used by other workers, and Walsberg's results do not necessarily indicate that these studies have underestimated aerobic capacity for locomotor activity. Both Walsberg (1986) and M. Ryan (1988b) suggested that problems with methods used to elicit locomotor activity might account for reports that metabolic rates during forced exercise in several species of frogs are lower than metabolic rates during peak calling activity (Taigen and Wells, 1985; Taigen, Wells, and Marsh 1985). Prestwich, Brugger, and Topping (1989) refuted this hypothesis by showing that peak metabolic rates of calling frogs were substantially higher than those obtained when the animals were exercised by manual rotation (see subsequent discussion).

Exercising animals in rotating chambers has proved to be a reliable way of measuring maximum aerobic capacity during nonsustainable locomotion, but it is difficult to quantify precisely the animal's rate of movement with this method or to elicit submaximal levels of locomotion. Some investigators have used motorized treadmills enclosed in metabolic chambers to study the energetics of sustained locomotion (Full 1986; Full et al. 1988; Walton 1988, 1993; Walton and Anderson 1988; B. Anderson, Feder, and Full 1991; Longphre and Gatten 1994; Walton, Peterson, and Bennett 1994). These studies have employed an electrochemical oxygen analyzer to measure continuous changes in oxygen consumption in a flow-through respirometry system, whereas the studies with rotating chambers have relied on a single pair of pre- and postexercise air samples taken from a closed respirometer. The treadmill system allows the investigator to follow the time-course of oxygen consumption as the animal exercises and to calculate the maximum speed at which locomotion can be supported entirely by aerobic metabolism. It also makes possible measurements at relatively low speeds that more closely approximate natural rates of movement in the field. Data collected in this way probably provide a more reliable estimate of the energetic costs of activities such as foraging than data collected from animals exercised to exhaustion (Gatten, Miller, and Full 1992). Animals can be exercised to exhaustion on a treadmill as well. Most measurements of maximum metabolic rates using this technique are similar to those obtained using chambers rotated by hand or with a motor (Walton 1988; Walton and Anderson 1988; Prestwich, Brugger, and Topping 1989), but Walton, Peterson, and Bennett (1994) obtained lower values for toads exercised on a treadmill than for those exercised in rotating chambers.

### Sources of Variation in Activity Metabolism

Gatten, Miller, and Full (1992) tabulated all of the available data on maximum metabolic rates of amphibians during forced activity and calculated regressions of metabolic rate on body mass using only those data collected with rotating chambers or treadmills. At 20° C, the scaling of aerobic activity metabolism with body mass is similar to the scaling of resting metabolism for both urodeles and anurans, but anurans have consistently higher metabolic rates during vigorous activity. In general, most salamanders cannot elevate their metabolic rates to more than three or four times resting metabolism at 20–23° C, although factorial scopes (the ratio of activity metabolism to resting metabolism) greater than 10 have been reported for some species at 13–15° C (Feder 1986, 1987; Mitton, Carey, and Kocher 1986). Many anurans can achieve rates of metabolism more than 10 times resting levels at 20° C, and some have factorial scopes greater than 35 (fig. 5.3), considerably higher than those of most mammals and reptiles (Bennett and Dawson 1976; Lechner 1978; Taigen 1983). Large species tend to have higher factorial scopes than small species when all anurans are combined (Gatten, Miller, and Full 1992), but this is not necessarily true within families. For example, *Pseudacris*

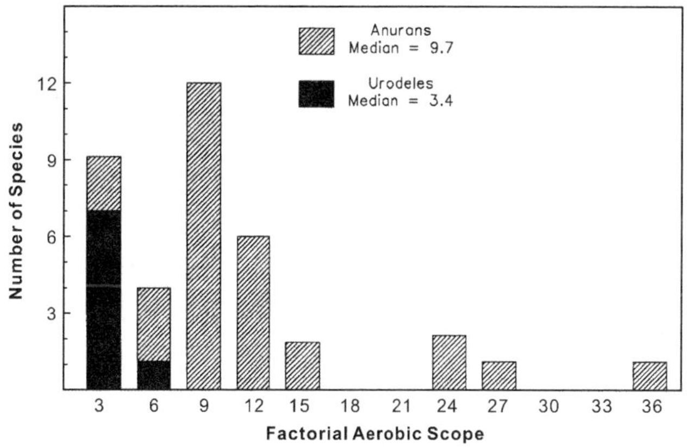


**Fig. 5.3.** Distribution of factorial aerobic scopes for eight species of urodeles and 30 species of anurans measured at 20–22° C. Numbers under each bar are midpoints for each interval. Urodele data are combined for one species each of *Ambystoma*, *Amphiuma*, and *Taricha*, two species of *Desmognathus*, and three species of *Plethodon*. Anuran data include one species each of *Agalychnis*, *Bombina*, *Dendrobates*, *Discoglossus*, *Eleutherodactylus*, *Gastrophryne*, *Hyperolius*, *Kaloula*, *Odontophrynus*, *Osteopilus*, *Pseudacris*, *Pternohyla*, *Pyxicephalus*, and *Spea*. Also included are two or more species of *Bufo*, *Colostethus*, *Hyla*, *Kassina*, and *Rana*. Plotted from data in tables 12.7 and 12.13 in Gatten, Miller, and Full (1992).

*crucifer* (1.3 g), *Hyla arenicolor* (3.4 g), and *Hyla versicolor* (6.1 g) all have factorial scopes between 9 and 10 at 20° C (Taigen, Emerson, and Pough 1982; Kamel, Marsden, and Pough 1985). Aerobic capacity increases with temperature in both anurans and urodeles, but the precise degree of temperature sensitivity has not been determined for urodeles because of insufficient data. For anurans, $Q_{10}$ values for both aerobic capacity and aerobic scope range from about 1.5 to 4.5, with sensitivity being greatest at low temperatures (Gatten, Miller, and Full 1992).

One might expect that the greater aerobic capacity of anurans would allow them to move at higher speeds than urodeles, but an analysis of published and unpublished data by Gatten, Miller, and Full (1992) showed that maximum aerobic speeds did not differ significantly for walking salamanders and hopping frogs and toads. Apparently this is because the cost of locomotion is much higher at any given speed for anurans than for salamanders. For example, a toad (*Bufo fowleri*) moving at its maximum aerobic speed of 0.27 km/h has a metabolic rate about nine times its resting rate (Walton and Anderson 1988). A salamander (*Ambystoma laterale*) moving at a similar speed (0.20 km/h) would have a metabolic rate only about 36% as high, or about four times its resting rate (Full et al. 1988). Indeed, the cost of transport (the energy required to move a given amount of body mass) is relatively high in anurans compared to other vertebrates. For example, the cost of hopping by toads is about twice as high as the cost of running in lizards of the same body mass (Walton and Anderson 1988), whereas the cost of walking by salamanders is considerably lower than that for lizards (Full et al. 1988). The cost of transport also is higher for a hopping or swimming toad (*Bufo marinus*) than for a walking or swimming turtle (*Emydura macquari;* Baudinette, Miller, and Sarre 2000). In part, this is due to a difference in body size, because it is more expensive for a small animal than for a large animal to move a given amount of body mass.

Compared to other terrestrial vertebrates, amphibians do not have much capacity to sustain rapid locomotion. Maximum aerobic speeds of the few species studied to date are one to three orders of magnitude lower than those of mammals or lizards of similar size (Feder, 1986; Full et al. 1988; Walton and Anderson 1988; Gatten, Miller, and Full 1992). Similar differences have been found in comparisons of maximum rates of movement by lizards and frogs engaged in natural foraging behavior (Pough et al. 1992).

The energetic cost of locomotion can be affected by changes in the gaits of animals. Some amphibians rely exclusively on one type of locomotion, such as walking, whereas others, including some toads, switch from walking to hopping at faster movement speeds. B. Anderson, Feder, and Full (1991) measured the cost of locomotion in toads using mixed gaits and those using only walking or hopping gaits. They found that individual hops are more expensive than individual strides during walking, but the distance covered by a hop is greater. This resulted in the cost of transport being about half as great for hopping toads as for walking toads. Toads can therefore save energy by switching to the more economical hopping gait at faster speeds. Some toads, such as *Bufo boreas halophilus,* never hop, but switch from walking to running at high speeds. Walton, Peterson, and Bennett (1994) predicted that this would result in a higher cost of transport at high speeds and perhaps a higher maximum aerobic capacity, but in fact, this species was not very different from other toads that hop. Thus, walking does not appear to be an especially costly form of locomotion, at least in toads, and species that move mainly by walking do not necessarily have higher aerobic capacities than those that combine walking and hopping.

All frogs have lungs, and in most species, lungs are the principal site of oxygen uptake (see chapter 4). Many salamanders, including the entire family Plethodontidae, lack lungs, and one might expect the reduced respiratory surface area in lungless species to limit their ability to increase oxygen consumption during vigorous exercise. The problem would be particularly acute for large individuals, which have higher oxygen demands but relatively lower surface-to-volume ratios. Indeed, comparisons of metabolic rates of ambystomatid salamanders and one large plethodontid, *Desmognathus quadramaculatus,* exercised on a treadmill showed that the ambystomatids had higher maximum metabolic rates, higher factorial aerobic scopes, and higher maximum aerobic speeds. *Desmognathus quadramaculatus* was exhausted in less than 20 minutes by rates of movement that ambystomatid salamanders could sustain for one to two hours. Smaller lungless species, such as *D. ochrophaeus,* appear to be less constrained by their lack of lungs and achieve higher maximum aerobic speeds than *D. quadramaculatus.* The degree to which the activities of lungless salamanders are limited in the field is uncertain, because it is not clear how frequently they are called upon to move as quickly as the speeds used in the treadmill studies (Full et al. 1988).

### Individual Variation in Activity Metabolism and Behavior

Because adaptive evolutionary change depends on natural selection operating on variation among individuals in a population, studies of individual variation in physiological capacities are essential for understanding the relationships between physiological characteristics and individual fitness (Pough 1989). Unfortunately, the study of individual variation has been largely neglected by physiologists (Bennett 1987). Several investigators have begun to document individual variation in metabolic physiology or morphology in reptiles and have attempted to relate this to variation in behavioral performance (Garland 1984, 1985, 1988; Garland and Else 1987; Garland, Hankins, and Huey 1990), but relatively few such data are currently available for amphibians.

Replicated laboratory measurements of aerobic capacity, sprint speed, or stamina have revealed consistent differences among individual salamanders and frogs from the same population (data of Taigen, cited in Pough 1983; Wells and Taigen 1984; Sullivan and Walsberg 1985; Walsberg, Lea, and Hillman 1986; Walton 1988; Bennett, Garland, and Else 1989; Austin and Shaffer 1992; Rogowitz and Sánchez-Rivoleda 1999; Gomes et al. 2004). Other investigators have assumed that such differences among individuals exist, but did not present formal statistical analyses of intra- and interindividual variation (Marker and Gatten 1993; Longphre and Gatten 1994; Watkins 1996). Several studies also have documented individual variation in the behavior of animals in the field, based on such measures of performance as movement at a breeding site (Wells and Taigen 1984), calling rate (B. K. Sullivan 1983b; Wells and Taigen 1984; B. K. Sullivan and Walsberg 1985; Runkle et al. 1994) or foraging velocity (Walton 1988; Pough and Taigen 1990).

Only a few studies have attempted to correlate individual variation in behavioral performance with laboratory measures of physiological capacity. In some cases, aerobic capacity was not correlated with field measurements of behavioral performance. In the study of toads by Walton (1988), locomotor velocities of foraging toads in the field were much lower than were those of animals exercised to exhaustion in laboratory chambers. This suggests that toads engaged in routine foraging probably do not approach their physiological limits, and a close correlation between aerobic capacity and behavioral performance would not be expected. Wells and Taigen (1984) found that movements of male toads in breeding ponds were not correlated with laboratory measures of aerobic capacity. Again, velocities of searching males probably were too low to stress the animals physiologically. Both Wells and Taigen (1984) and Sullivan and Walsberg (1985) reported that calling rates of toads were not correlated with aerobic capacities measured during forced exercise. Subsequent work revealed that the aerobic capacities of muscles used for call production are much higher than those used for locomotion (Pough et al. 1992), so one would not necessarily expect a close relationship between aerobic capacity for locomotion and calling rate (see subsequent discussion of calling energetics).

### The Role of Anaerobic Metabolism

Most routine, low-intensity activities in amphibians probably are supported mainly by aerobic metabolism. Anaerobic metabolism can be important during bursts of high-intensity activity, when the demand for adenosine triphosphate (ATP) production in the muscles exceeds the capacity of the respiratory and circulatory systems to supply oxygen to the tissues (Bennett 1978; Gatten 1985; Gatten, Miller, and Full 1992). Most work has focused on anaerobic glycolysis, yielding lactic acid as an end product, but another anaerobic pathway contributes to the first few seconds of burst locomotion. This involves the direct utilization of stored high-energy phosphates, particularly phosphocreatine, to generate ATP. K. Miller and Sabol (1989) found that *Xenopus laevis* forced to swim at maximum speeds depended mainly on breakdown of phosphocreatine to supply energy in the first few seconds of swimming, with rates of ATP production through this pathway being about nine times the rate of ATP production from anaerobic glycolysis. Phosphocreatine stores in muscles are quickly exhausted, so anaerobic glycolysis is essential for continued activity. The breakdown of phosphocreatine also is an important source of energy during rapid burst swimming in some anuran tadpoles (Gatten, Caldwell, and Stockard 1984).

The role of anaerobic glycolysis in supporting vigorous activity in amphibians generally has been studied by exercising animals at maximal rates for several minutes and then homogenizing the animal to determine whole-body lactate levels. Measurements of lactate concentrations in blood do not provide a satisfactory quantitative estimate of anaerobic metabolism because of variable rates of lactate accumulation in different tissues (Gatten, Miller, and Full 1992). Unfortunately, this means that lactate accumulation cannot be measured in the same individual before and after activity, and investigators must rely on separate pre- and postactivity samples. This approach has shown that capacities for anaerobic ATP production vary considerably among species of amphibians subjected to similar exercise regimes. Species that move mainly by rapid jumping, such as *Rana pipiens*, or by flipping the body to escape from predators, such as *Plethodon jordani*, can derive more than 60% of their energy during vigorous exercise from anaerobic glycolysis. In contrast, species that move at slower, more sustained rates, such as *Bufo americanus* or *Physalaemus pustulosus*, generally have much lower capacities for anaerobic ATP production and rely more heavily on aerobic metabolism. Some of the variation in dependence on aerobic and anaerobic metabolism during vigorous exercise appears to be related to ecological differences among species, and these are discussed in more detail later.

The principal advantage of anaerobic metabolism for amphibians is that it provides high power output more rapidly than is possible through aerobic pathways. In addition, the temperature sensitivity of anaerobic metabolism generally is lower than for aerobic metabolism, making it especially useful to amphibians when they are cold (Bennett 1978; Gatten 1985; Gatten, Miller, and Full 1992). Anaerobic glycolysis is not, however, completely independent of temperature. In *Rana pipiens*, for example, total whole-body lactate accumulation, an index of anaerobic glycolysis, increased with both temperature and duration of exercise. The rate of lactate formation was much greater during short bursts of activity than during longer bouts, especially at high temperatures, and jumping performance improved at warm temperatures (Wine and Gatten 1992).

The disadvantage of anaerobic metabolism is that animals fatigue very quickly and require up to an hour of rest before recovering full locomotor capabilities after a bout of vigorous exercise (Rand 1952; Bennett and Licht 1973; Feder and Olson 1978; Putnam 1979; Zug 1985; Gatten and Clark 1989; Rogowitz and Sánchez-Rivoleda 1999). The physiological basis of fatigue is not fully understood. Fatigue does not appear to be due solely to accumulation of lactate in muscle tissue, because recovery of locomotor abilities precedes complete elimination of lactate (Bennett and Licht 1973; Feder and Olson 1978; Putnam 1979; Gatten and Clark 1989; Gatten, Miller, and Full 1992). It also cannot be attributed to depletion of glycogen stores, which sometimes show little change after brief bouts of exhaustive locomotion (Putnam 1979). Whatever the proximate basis of fatigue, the limited endurance of animals utilizing anaerobic metabolism means that only those activities that are supported by aerobic metabolism can be sustained for long periods of time.

Amphibians use anaerobic glycolysis during short bursts of activity when escaping from predators (Feder and Arnold 1982) or in fighting over territories (Wells 1978a), but it seems unlikely that anaerobic metabolism plays an important role in supporting other routine activities such as migratory movements, foraging, or courtship. Significant increases in whole-body lactate have been reported following bouts of courtship in a salamander, *Desmognathus ochrophaeus,* and nest-building in a frog, *Physalaemus pustulosus,* but in both cases, anaerobic glycolysis accounted for less than 10% of total ATP production (Bennett and Houck 1983; M. Ryan, Bartholomew, and Rand 1983). Anaerobic metabolism also makes an insignificant contribution to calling activity (see the following discussion). Early studies of diving in anurans suggested that frogs might depend on anaerobic metabolism during prolonged submersion (e.g., D. Jones 1967, 1972; Emilio and Shelton 1974). Subsequent work with *Xenopus* showed that significant lactate accumulation occurs during forced dives, whereas voluntary dives appear to be almost entirely aerobic (Boutilier and Shelton 1986a, c; Boutilier 1989). These results parallel those obtained for voluntarily diving lizards, turtles, and snakes (Seymour 1982; Gatten 1985).

In addition to using anaerobic metabolism to support locomotor activity, some animals rely on anaerobic pathways for survival under hypoxic or anoxic conditions. This is particularly common in many aquatic invertebrates, but occurs in some vertebrates as well (De Zwaan and Thillart 1985). For example, some freshwater turtles that hibernate in anoxic mud at the bottom of ponds can survive entirely by means of anaerobic glycolysis and accumulate enormous quantities of lactate (see Gatten 1987b and Ultsch 1989 for reviews). A number of studies have shown that amphibians placed in anoxic conditions in the laboratory can survive for extended periods on anaerobic metabolism alone and exhibit significant accumulations of lactate (F. Rose and Drotman 1967; Armentrout and Rose 1971; F. Rose, Armentrout, and Roper 1971; Christiansen and Penney 1973). It has been suggested that this might be important during winter dormancy, but in fact, we do not have any data on lactate accumulation in overwintering amphibians in the field. The very low metabolic rates of dormant amphibians and their ability to take up oxygen through the skin probably reduce their dependence on anaerobic metabolism. Nevertheless, anaerobic glycolysis is essential in species of anurans that tolerate freezing, including *Rana sylvatica* and *Hyla versicolor,* because it provides the energy for synthesis of glucose and glycerol used as cryoprotectants (K. Storey and J. Storey 1984, 1986b; J. Storey and K. Storey 1985a; K. Storey 1987a; see chapter 3 for further discussion).

### Ontogenetic Changes in Energy Metabolism

When amphibian larvae first hatch from their eggs, their capacities for movement are severely limited because muscles, the cardiovascular system, and metabolic capacities are poorly developed. Both aerobic and anaerobic metabolic capacities probably develop gradually as the larvae grow, but changes in activity metabolism have not been well documented for larval amphibians. The metabolic rates of actively swimming *Bufo fowleri* tadpoles were only about three times those of tadpoles engaged in routine activities (Feder 1983a). This difference probably underestimates aerobic capacities for locomotion, because it is difficult to obtain a true resting metabolic rate for tadpoles. Nevertheless, the capacity of tadpoles to sustain activity aerobically clearly is far below that of adult toads. Factorial aerobic scope did not change significantly with increasing body size in this species, presumably because they do not develop lungs until just before metamorphosis and therefore can do little to increase rates of oxygen consumption. Hillman and Lea (1983) did find an increase in activity metabolism as a function of developmental stage in *Rana catesbeiana* tadpoles, a species that develops lungs well before metamorphosis. Their experiments were ecologically unrealistic, however, because the tadpoles were exercised in air, with only a small amount of water to keep them moist.

Because of their low aerobic capacities, most tadpoles probably depend on anaerobic metabolism to support vigorous activity and therefore are limited in how long such activity can be sustained. Tadpoles also appear to have rather low capacities for ATP production by means of anaerobic glycolysis during short bursts of swimming. Instead, they must depend heavily on breakdown of stored phosphocreatine, and therefore probably are limited to a few seconds of vigorous activity at a time (Gatten, Caldwell, and Stockard 1984). At slower swimming speeds, tadpoles do accumulate lactate (Quinn and Burggren 1983). They also have some capacity to use anaerobic metabolism when exposed to hypoxic conditions, but generally do so only if they are not able to breathe air (Feder 1983a, d).

Watkins (1996) examined the effect of burst swimming speed on the ability of tadpoles (*Pseudacris regilla*) to escape from predators. He first measured sprint speed in plastic troughs and divided tadpoles into fast and slow groups. Only those that differed considerably in speed were used in subsequent experiments. The tadpoles were marked with dye and released into swimming pools with garter snakes (*Thamnophis sirtalis*) as predators. Tadpoles from the fast group were more likely to survive than were those from the slow group. Those that escaped predation swam away from snakes nearly twice as fast as those that were captured. These results suggest a strong potential for predator-mediated selection on burst swimming speed in tadpoles, assuming that there is some genetic variation for physiological or morphological traits that affect swimming speed.

After metamorphosis, anuran species differ in the rate at

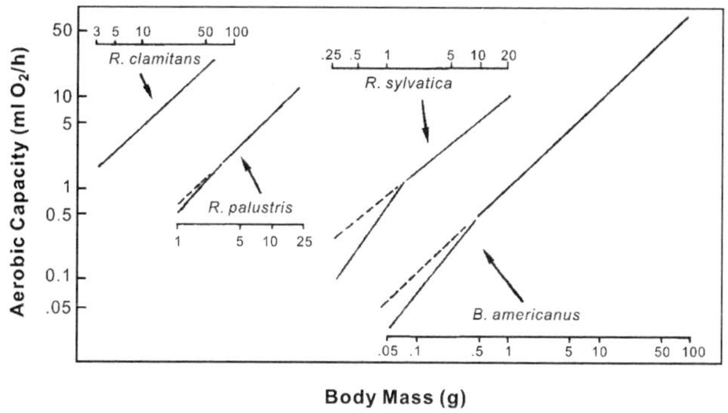


**Fig. 5.4.** Ontogenetic change in aerobic capacity in four species of frogs. Rates of oxygen consumption during exercise are presented on logarithmic scales as a function of body mass. Body mass for each species is shown on a separate horizontal scale. Dashed lines extend slopes for postdispersal individuals to body size range of predispersal juveniles. *Rana sylvatica* and *Bufo americanus* metamorphose with relatively poorly developed aerobic capacities and exhibit a rapid increase before dispersal. After Taigen and Pough (1985).

which their metabolic capacities develop. In species with long larval periods, such as *Rana clamitans,* the lungs and circulatory system are fully functional when the young frogs emerge onto land, and the aerobic capacities of juvenile frogs are indistinguishable from those of adults. In contrast, species that breed in temporary water, such as *Bufo americanus* and *Rana sylvatica,* are under strong selective pressures to leave the water quickly. They emerge in a physiologically underdeveloped state, with poorly developed lungs, small hearts, and low hematocrits. They have relatively low aerobic capacities immediately after metamorphosis, but their performance improves with age (fig. 5.4). *Rana palustris* is somewhat intermediate, with juveniles having slightly lower aerobic capacities than adults, but similar endurance (Taigen and Pough 1981, 1985; Pough 1983; Pough and Kamel 1984). It seems likely that similar patterns of ontogenetic change in physiological performance correlated with differences in breeding ecology and timing of metamorphosis occur in other families of anurans and would be a productive area for future research.

## Morphological Correlates of Activity Metabolism

Vertebrate muscles typically exhibit morphological features that are closely related to the type of work the muscles must perform. Species that engage in sustained periods of exercise have muscles with fiber types, contractile properties, mitochondrial densities, and capillary supplies that enable them to meet demands for high levels of sustained aerobic activity. This correspondence between muscle structure and function is evident in comparisons of anurans with different modes of locomotion. The leg muscles of anurans are composed primarily of fast glycolytic fibers (about 75–80% of total muscle area), but some fast oxidative fibers are present as well (R. S. Smith and Ovalle 1973; Putnam and Bennett 1983; R. Marsh and Taigen 1987; De Costa et al. 1989; Gans and De Gueldre 1992; Wilson et al. 2004). The former depend on anaerobic glycolysis to supply their energy requirements, whereas the latter make use of aerobic pathways. Comparisons among species of *Hyla, Rana, Bufo,* and *Xenopus* have revealed only minor variation in gross fiber type that can be related to differences in mode of locomotion (Putnam and Bennett 1983; R. Marsh and Taigen 1987). However, Mendiola, De Costa, Lozano, and Agulleiro (1991) reported that leg muscles of *Bufo* have higher proportions of fast oxidative and tonic fibers than do leg muscles in *Rana.* The muscles of these two genera also differ in contractile properties. Leg muscles in *Rana pipiens* have relatively fast shortening velocities and high mechanical power under low loads. Muscles of *Bufo americanus* have slower shortening velocities and greater tetanic tension, providing more mechanical power under high loads, such as the animal might experience while burrowing (Renaud and Stevens 1983a).

## Biochemical Correlates of Activity Metabolism

The activities of enzymes involved in catalyzing reactions in anaerobic glycolysis or the Krebs cycle often are used as indicators of an animal's dependence on anaerobic or aerobic metabolism to support muscular work (Alp, Newsholme, and Zammit 1976; Holloszy and Booth 1976; Hochachka and Somero 1984). In studies of amphibians, the glycolytic enzymes phosphofructokinase (PFK), pyruvate kinase, lactate dehydrogenase (LDH), phosphorylase, and hexokinase (HK) have been used to investigate the relationship between enzyme activities and anaerobic capacity. The Krebs cycle enzymes citrate synthase (CS) and oxoglutarate dehydrogenase (GDH) have been used as indicators of aerobic metabolism (Bennett 1974; J. Baldwin, Friedman, and Lillywhite 1977; Cummings 1979; K. Miller and Camilliere 1981; Newsholme and Paul 1983; Putnam and Bennett 1983; Taigen, Wells, and Marsh 1985; Walsberg, Lea, and Hillman 1986; R. Marsh and Taigen 1987; Molina, Mendiola, and De Costa 1987; Given and McKay 1990; Mendiola, Molina, and De Costa 1991; Choi and Park 1996; James et al. 2005). In addition, activities of lipases and β-hydroxylacyl-CoA dehydrogenase (HOAD) have been used as indicators of capacities to use lipids as an energy substrate during muscular activity (Newsholme and Paul 1983; R. Marsh and Taigen 1987).

In comparisons among species, aerobic enzyme activities in muscle tissue generally are correlated with dependence on aerobic metabolism. For example, Putnam and Bennett (1983) found that *Bufo boreas* and *Xenopus laevis* had higher leg-muscle CS activities than did *Rana pipiens.* These differences parallel differences in mode of locomotion in the three species: hopping in *Bufo*, swimming in *Xenopus*, and jumping in *Rana*. In contrast, Newsholme and Paul (1983) reported slightly higher CS activity in sartorius muscles of *Rana temporaria* than in the same muscles of *Xenopus*, but they found that GDH activities were much higher in leg muscles of *Xenopus*. They attributed this difference to the greater aerobic demands of swimming compared to jumping. Mendiola, Molina, and De Costa (1991) reported higher activities of glycolytic enzymes (HK, PFK, LDH) in leg muscles of *Rana perezi* than in the muscles of *Bufo calamita*, again correlating with the greater reliance on anaerobic metabolism in *Rana*. A comparison of LDH and CS activity in leg muscles of *Rana nigromaculata*, *Rana rugosa*, and *Bombina orientalis* did not yield a clear correlation with mode of locomotion. The two ranid species are jumpers, whereas *Bombina* is a walking or hopping species. The three species did not differ in LDH activity, but CS activity was higher in *R. rugosa* and *Bombina* than in *R. nigromaculata* (Choi and Park 1996). All of these comparisons suffer from a lack of phylogenetic control, because putative differences in mode of locomotion are confounded by differences in phylogenetic history. Comparisons among closely related species that differ in mode of locomotion would be preferable, but difficult in practice, because closely related species usually have similar morphology and behavior.

There has been very little work on individual variation in muscle enzyme activities among members of the same species. Walsberg, Lea, and Hillman (1986) did not find any correlation between CS activity and aerobic capacity for locomotion among individual bullfrogs (*Rana catesbeiana*) from a single population. Like most ranid frogs, this species probably relies heavily on anaerobic metabolism for short bouts of jumping, so one would not necessarily expect a correlation between CS activity and locomotor performance. James et al. (2005) also did not find a significant relationship between individual variation in CS activity and jumping performance in *Hyla multilineata*. There was a positive relationship between jumping performance and the glycolytic enzyme pyruvate kinase, as well as total hindlimb muscle mass. Longphre and Gatten (1994) worked with a more aerobic species, *Bufo marinus*. They found a positive correlation among individuals between maximum aerobic speed of locomotion and leg-muscle CS activity, as well as ventricle mass, and iliofibularis muscle mass. Rate of oxygen consumption, however, was not related to any of these variables or to several other morphological traits. In *Rana pipiens*, which relies heavily on anaerobic metabolism, individual variation in sprint speed was not related to variation in glycolytic enzyme activities (Marker and Gatten 1993).

### Ecological and Behavioral Correlates of Activity Metabolism

Physiological ecologists have had some difficulty correlating laboratory measures of physiological performance with behavioral or ecological variables that affect individual fitness in the field (for general discussions, see Taigen and Pough 1985; Arnold 1988; Pough 1989; Pough et al. 1992; and several chapters in Feder et al. 1987). Laboratory measurements of aerobic capacity, anaerobic capacity, or stamina often require animals to perform tasks that do not accurately reflect modes or intensities of locomotor activity in the field. Nevertheless, a correlation between field and laboratory measurements can provide a starting point for understanding the interrelationships among physiology, morphology, ecology, and behavior.

The first attempt to compare the activity metabolism of several species of amphibians in an ecological context was the work of Bennett and Licht (1973, 1974) on a limited number of North American anurans (*Bufo boreas*, *Rana pipiens*, and *Pseudacris regilla*) and urodeles (*Batrachoseps attenuatus*, *Notophthalmus viridescens*, *Aneides flavipunctatus*, and *A. lugubris*). These investigators found a general dichotomy between species with relatively high aerobic capacities and low anaerobic capacities and those with relatively low aerobic capacities and high anaerobic capacities. Although the species studied differed in a number of respects, including body size and shape, mode of locomotion, foraging behavior, and antipredator behavior, Bennett and Licht focused mainly on differences in antipredator behavior. They suggested that species that depend mainly on passive defenses, such as toxic skin secretions, tend to move at relatively slow but steady rates and rely more on aerobic than anaerobic metabolism for locomotion. Conversely, those species that use rapid escape as their principal defense against predators were found to move mainly in short bursts of activity and relied more on anaerobic than aerobic metabolism.

The apparent relationship between antipredator behavior and dependence on aerobic or anaerobic metabolism remained the principal focus of studies of activity metabolism in amphibians for several years (e.g., J. Baldwin, Friedman, and Lillywhite 1977; Hutchison, Turney, and Gratz 1977; Bennett 1978, 1980; Hutchison and Miller 1979a, b; K. Miller and Hutchison 1979). Seymour (1973c), however, had previously suggested that any sustained activity, such as burrowing or active foraging, would be facilitated by a high aerobic capacity. Carey (1979a) argued that reliance on passive defenses and a lack of selection for high anaerobic capacity was an inadequate explanation for the high aerobic capacities of toads. She suggested that positive selection for high aerobic capacities was more likely, but did not specify what the selective pressures might be. Earlier, Emerson (1976) had pointed out that those species of anurans used by previous investigators as exemplars of the passive defense strategy (*Bufo* and *Scaphiopus*) also tended to be burrowing species that forage actively on small prey such as ants. Consequently, differences in mode of locomotion, foraging behavior, or burrowing might be just as important as mode of defense in determining metabolic characteristics of these animals.

These considerations led Taigen, Emerson, and Pough (1982) to examine ecological correlates of activity metabolism in a wider range of species from seven families. Using the same techniques for all species, they measured both aerobic and anaerobic capacities during brief bouts of vigorous exercise. These data were used to calculate an aerobic dependence index, an estimate of the relative importance of aerobic metabolism in supporting activity. Species also were classified as having active or passive antipredator behavior, active or sit-and-wait foraging behavior, jumping or nonjumping locomotion, and terrestrial or arboreal habits. Species placed in each ecological or behavioral category were then tested for statistical associations between the ecological variables and the aerobic dependence indexes.

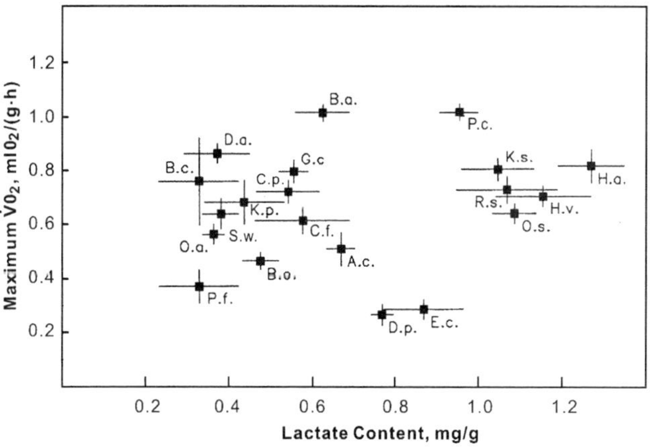


**Fig. 5.5.** Relationship of maximum aerobic capacity during exercise to whole-body lactate content after maximum exercise for 20 species of anurans. Species abbreviations: *A.c.* = *Agalychnis callidryas; B.a.* = *Bufo americanus; B.c.* = *Bufo calamita; B.o.* = *Bombina orientalis; C.f.* = *Colostethus flotator; C.p.* = *Colostethus panamensis; D.a.* = *Dendrobates auratus; D.p.* = *Discoglossus pictus; E.c.* = *Eleutherodactylus coqui; G.c.* = *Gastrophryne carolinensis; H.a.* = *Hyla arenicolor; H.v.* = *Hyperolius viridiflavus; K.p.* = *Kaloula pulchra; K.s.* = *Kassina senegalensis; O.a.* = *Odontophrynus americanus; O.s.* = *Osteopilus septentrionalis; P.c.* = *Pseudacris crucifer; P.f.* = *Pternohyla fodiens; R.s.* = *Rana sylvatica; S.w.* = *Semnodactylus wealii.* After Taigen, Emerson, and Pough (1982), with additional data for dendrobatid frogs from Taigen and Pough (1983).

Two major findings of this study failed to support the earlier conclusions of Bennett and Licht. First, the apparent negative correlation between aerobic and anaerobic capacities that Bennett and Licht found in a limited number of North American anurans disappeared in the larger sample of species. In fact, there was no clear relationship between these two variables (fig. 5.5), indicating that anurans do not cluster into two distinct groups of high aerobic/low anaerobic and low aerobic/high anaerobic species. Second, Taigen, Emerson, and Pough (1982) found that dependence on aerobic metabolism was not significantly associated with antipredator behavior, but was related to both foraging behavior and mode of locomotion. In general, active foragers that depend on walking rather than jumping are more aerobic in their metabolic characteristics than sit-and-wait predators that move mainly by rapid jumps. They also found that burrowing species form a subset of the walking, actively foraging anurans, and accounted for most of the difference in metabolic characteristics related to mode of locomotion. Taigen, Emerson, and Pough suggested that high dependence on aerobic metabolism also might be associated with the occurrence of intensive male combat and competition for territories, but they did not have sufficiently detailed data on aggressive behavior to test this hypothesis statistically. Phylogenetic differences did not explain much of the variation in metabolic characteristics, since variation within families was nearly as great as variation between families.

The study by Taigen, Emerson, and Pough (1982) provided a new ecological framework for understanding the evolution of metabolic characteristics that emphasized the complexity of relationships between morphology, ecology, behavior, and physiology. This was subsequently adopted by other authors (e.g., Pough 1983; Duellman and Trueb, 1986; Seale 1987), but it should be considered only a starting point for future work. Much of the information on behavior and ecology available to these authors was based on descriptive studies and scattered natural history notes, so the quality of the ecological data did not match the detailed quantitative physiological data. They were forced to classify species into general ecological categories, but real ecological variation among species is more like a graded continuum. Anurans exhibit a spectrum of foraging modes rather than falling into two distinct categories of sit-and-wait and active foragers (Pough 1983; Strussmann et al. 1984; Pough et al. 1992; Toft 1995), and correlated morphological, behavioral, and dietary characteristics of species probably are similarly graded. Some species also exhibit considerable seasonal, annual, or geographic variation in diet that could cause them to be placed in different categories depending on when and where the data were collected (e.g., Toft 1980a, b, 1981; Strüssmann et al. 1984). Even antipredator behavior varies among populations or in response to different types of predators (Marchisin and Anderson 1978; Bajger 1980; Arnold 1982; Ducey and Brodie 1983; see also chapter 14) in ways that make very different demands on an animal's metabolic physiology (Pough 1983; Taigen and Pough

1985). Another limitation of the study by Taigen, Emerson, and Pough (1982) is the inclusion of a phylogenetically diverse array of species into a single analysis, leading to possible confounding of adaptive patterns with phylogenetic history (a common problem in nearly all comparative physiological studies of amphibians).

To address some of these problems, Taigen and Pough (1983) conducted a comparative study of three species of dendrobatid frogs (*Dendrobates auratus, Colostethus panamensis,* and *C. nubicola*) and one leptodactylid (*Eleutherodactylus coqui*). The dendrobatids are all relatively active, diurnal foragers, whereas *E. coqui* is a nocturnal sit-and-wait predator. By using previously published data on the diets of these species, Taigen and Pough showed that aerobic capacity during forced locomotion was positively correlated with a tendency to feed preferentially on ants. Although all three dendrobatids were much more dependent on aerobic metabolism than was *E. coqui,* the three species formed a graded series (fig. 5.6 A). *Dendrobates auratus* tended to specialize on ants and was the most aerobically dependent species, whereas the other two dendrobatids had more generalized diets and were intermediate in metabolic characteristics between *D. auratus* and *E. coqui.* Subsequent studies with these species and a fourth dendrobatid, *Colostethus talamancae,* demonstrated positive relationships between aerobic capacity, rates of movement, and number of prey capture attempts per hour (fig. 5.6 B). These data confirmed that the most active foragers are the most dependent on aerobic metabolism (Taigen and Pough 1985; Pough and Taigen 1990).

These studies, along with investigations of functional morphology in anurans (Emerson 1976, 1978, 1979, 1982, 1985, 1988a; Emerson and de Jongh 1980; Zug 1985), have shown that various aspects of morphology, mode of locomotion, burrowing behavior, foraging ecology, and activity metabolism are intercorrelated in a complex manner (see chapter 1). It is much more difficult to establish a clear cause-and-effect relationship among these variables. For example, when we find that an actively foraging species such as *Dendrobates auratus* has a high aerobic capacity, it is not clear whether that is because active foraging has favored the evolution of enhanced aerobic metabolism, or because active foraging is possible because of preexisting metabolic characteristics. When we find that a species such as *Eleutherodactylus coqui* has a very poor capacity to sustain activity aerobically, is this because there has been no selection for enhanced aerobic metabolism, or because the high cost of maintaining metabolic machinery (Taigen 1983) has favored dependence on anaerobic metabolism? At present, it is hard to answer these questions, precisely because we lack detailed information on the direction of evolutionary change in metabolic characteristics within clades of amphibians. Additional studies that place the evolution of locomotor modes, foraging behavior, and metabolic physiology in an explicitly phylogenetic context (Huey 1987) might help to resolve these issues.

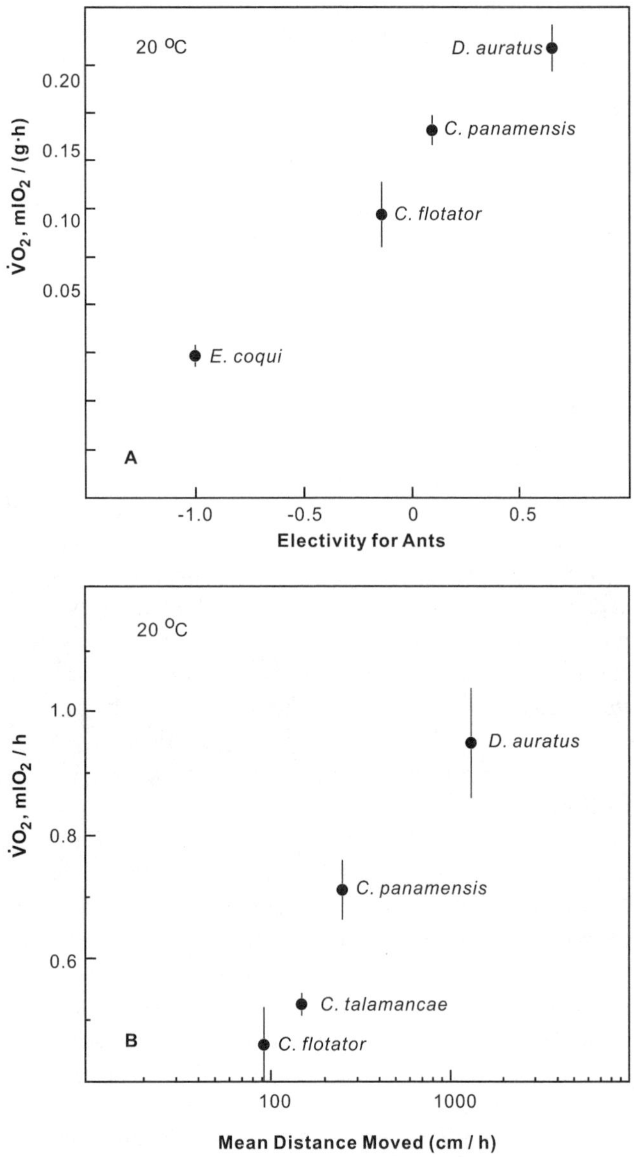


**Fig. 5.6.** Relationship of aerobic capacity to foraging behavior in several species of frogs. (A) Mass-specific aerobic capacity as a function of degree of specialization on ants for *Eleutherodactylus coqui* and three species of dendrobatid frogs. Positive electivity values indicate that frogs take ants in higher proportions than expected from their abundance in the environment. An electivity value of −1.0 indicates avoidance of ants. Metabolic measurements were made at 20° C. Plotted from data in Taigen and Pough (1983). (B) Aerobic capacity as a function of distance moved during natural foraging bouts for four species of dendrobatid frogs. Movement data were collected at natural field temperatures (about 23–27° C). Metabolic measurements were made at 20° C and were adjusted to a grand mean body mass for all four species of 0.95g. Plotted from data in Pough and Taigen (1990).

One also must be cautious in attributing differences in foraging behavior among species directly to physiological constraints imposed by their metabolic physiology. Taigen and Pough (1985) argued that a species such as *E. coqui,* with its poor aerobic capacity, is physiologically incapable of adopting the active foraging mode of *D. auratus,* whereas the latter species is less effective at making sudden lunges to capture large prey. It is not at all clear, however, that the three species of *Colostethus* studied by Pough and Taigen (1990) are incapable of moving as fast as *D. auratus* during normal foraging bouts. Even the fastest moving frogs in the field did not match the rates of locomotion achieved by frogs exercised in rotating chambers. Similarly, Walton (1988) found that natural foraging velocities of toads (*Bufo fowleri*) were only about 20% of the lowest speed at which oxygen consumption was measured in laboratory experiments. It seems unlikely that foraging is an activity that pushes amphibians to their physiological limits, so it is hard to document behavioral differences among species that can be attributed solely to differences in metabolic physiology.

### Evolutionary Consequences of High Activity Metabolism

If frogs or other animals are to achieve high metabolic rates during vigorous activity, then they must have the metabolic machinery to support these high metabolic rates in the form of higher concentrations of mitochondria and higher densities of capillaries to carry oxygen to exercising muscles. This machinery cannot be maintained without cost, however, and several authors have suggested that selection for high levels of activity results in higher resting metabolic rates and consequently higher maintenance costs. It also has been suggested that the evolution of endothermy was a byproduct of increased resting metabolic rates brought about by this type of selection for high activity metabolism (Regal 1978; Bennett and Ruben 1979; Taigen 1983; Walton 1993; Hayes and Garland 1995). Ectotherms such as amphibians lack insulation that would allow them to retain metabolic heat, but increased resting metabolic rates can be viewed as an ecological cost of supporting high levels of aerobic metabolism.

Taigen (1983) analyzed the relationship between maximum aerobic metabolism during exercise and resting metabolism for a number of anurans in several different families. He found a positive correlation, which was consistent with the aerobic capacity model for the evolution of high metabolic rates. Walton (1993) refined Taigen's analysis by making comparisons within the family Hylidae and using modern comparative methods to remove the effect of phylogeny on the relationship between resting and activity metabolism. Again, there was a positive correlation between resting and activity metabolism, regardless of which model of evolutionary change was adopted.

Limited data also have shown a positive correlation between resting and activity metabolism among individuals within a population (Walton 1988), but information on the genetic correlation between these variables is lacking for amphibians. Gomes et al. (2004) conducted a comprehensive survey of metabolic rates in South American anurans. They found that activity metabolism and resting metabolism were not correlated for many species, but were positively correlated in species with the most active lifestyles, especially those with high calling rates. This suggested that maintenance costs are likely to be higher for amphibians that depend heavily on aerobic metabolism to support activity than for species that depend mainly on anaerobic metabolism. Similar results have been obtained for insects (Reinhold 1999). Data for squamate reptiles, on the other hand, have been less supportive of a relationship between resting and activity metabolism. Hayes and Garland (1995) argued that data of the type presented by Taigen (1983) and Walton (1993) are consistent with the aerobic capacity model for the evolution of high metabolic rates, but do not necessarily falsify a number of other hypotheses advanced to explain the evolution of endothermy.

## Energetics of Reproductive Behavior in Male Urodeles

The search for relationships between metabolic capacities of animals exercised in the laboratory and ecological or behavioral traits can, at best, provide only a general correlation between physiology and behavior. This approach cannot be used to estimate the energetic costs of specific types of behavior, because the mode and intensity of locomotion utilized in a forced exercise regime usually does not very closely resemble patterns of activity during foraging, courtship, fighting, burrowing, or other natural activities. Fortunately, amphibians are small enough to enable investigators to make direct measurements of energetic costs, particularly those related to reproduction (for reviews, see Seale 1987; M. Ryan 1988b; Pough 1989; Pough et al. 1992; Prestwich 1994).

The only attempt to directly measure the energetic cost of reproductive behavior in male salamanders is Bennett and Houck's (1983) study of *Desmognathus carolinensis.* They measured oxygen consumption of noncourting pairs of males and females, courting pairs, and pairs of males engaged in aggressive interactions. Courtship bouts lasted more than an hour on average, but metabolic rates of courting pairs were only slightly higher than those of noncourting pairs and less than three times resting rates measured in a closely related species, *D. ochrophaeus* (fig. 5.7). By comparison, the metabolic rates of inactive desmognathine salamanders can increase by as much as 77% simply as a result of feeding (Feder et al. 1984). Metabolic rates of males engaged in aggressive behavior were slightly higher, but courting and fighting pairs of *Desmognathus* did not approach levels of

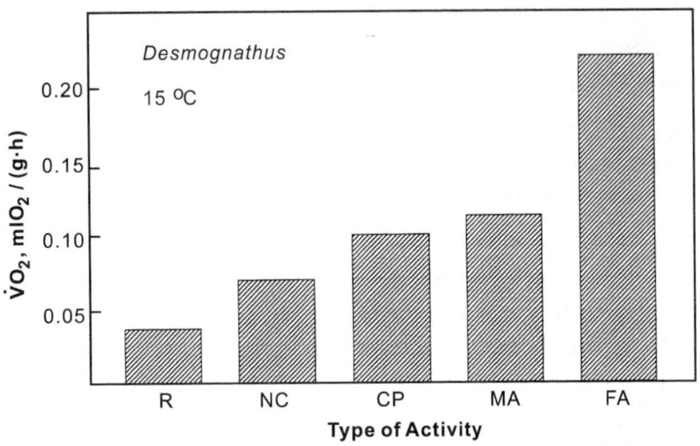


**Fig. 5.7.** Metabolic rates of several closely related species of *Desmognathus* engaged in various activities at 15° C. R = resting metabolism. NC = noncourting pairs. CP = courting pairs. MA = male-male aggression. FA = forced activity. Resting metabolic rate is for *Desmognathus ochrophaeus* from Ohio (Fitzpatrick 1973b). Forced activity is for *D. ocoee* from North Carolina (Feder 1986). All others are for *D. carolinensis* from North Carolina (Bennett and Houck 1983). All of these populations were considered to be *D. ochrophaeus* when the studies were done.

oxygen consumption achieved by individuals during forced exercise. Bennett and Houck (1983) found that at least 90% of ATP production during courtship or aggression was supported by aerobic metabolism. Total energy expenditure during a typical courtship bout was very low, amounting to only about 1–2% of a salamander's daily energy intake. In fact, the energy expended by a male *Desmognathus carolinensis* during courtship was less than the energetic cost of spermatophore production in a closely related species, *D. ocoee,* even though the latter also represents a trivial proportion of the animal's daily energy budget (Marks and Houck 1989).

The urodeles most likely to have energetically costly courtship are aquatic newts (*Triturus, Notophthalmus,* and *Paramesotriton*). Newts have prolonged breeding seasons lasting from several weeks to several months (Verrell and Halliday 1985b; Verrell, Halliday, and Griffiths 1986; Verrell and McCabe 1988). Breeding males of most species develop elaborate secondary sexual characters, such as bright coloration and enlarged tail fins, and the development of these display characters apparently depends on previous foraging success (A. Green 1991b; Baker 1992). Males engage in vigorous courtship that can last up to several hours (Halliday 1974, 1975a, 1976, 1977a; Verrell 1982; Sparreboom 1983, 1984a; Giacoma and Sparreboom 1987). A male's ability to court females for long periods of time is directly related to his reproductive success (see chapter 9 for further discussion). Males of some species also engage in aggressive interactions that can lead to prolonged struggles for possession of females (Verrell 1983, 1984a, 1986a; Sparreboom 1984b; Zuiderwijk and Sparreboom 1986). Although metabolic rates of courting and fighting newts have not been measured, the length and vigor of courtship and aggressive interactions suggest that such behavior is supported mainly by aerobic metabolism.

Several observations suggest that physiological constraints limit the courtship activities of newts (Halliday 1987). In *Triturus vulgaris,* the frequency with which males go to the surface to breathe is directly related to level of activity (Halliday and Worsnop 1977). Breathing frequency increases during sexual activity, and the need to breathe can significantly limit the duration of courtship bouts (Halliday and Sweatman 1976; Halliday and Houston 1991). Breathing rates increase if dissolved oxygen in the water, or oxygen in the atmosphere over the water, is decreased, whereas breathing rates decrease when the water is saturated with oxygen (Spurway and Haldane 1953; Halliday 1977b). Courtship activity also is limited by the need to breathe in *Notophthalmus viridescens* (Verrell 1985a). These results do not demonstrate that newts are energy limited during courtship, but they do suggest that metabolic rates during vigorous activity exceed levels that can be supported solely by extrapulmonary respiration.

A. Green (1991a) provided several lines of indirect evidence that displays of great crested newts (*Triturus cristatus*) are energetically constrained. He found that the number of tail beats in a bout of fanning display decreased as the number of beats per minute increased. Furthermore, tail beat rates decreased as a display bout progressed, suggesting the animals become fatigued. Display bout lengths also decreased over time. Male newts increased both the number of fan beats and the number of display bouts in the presence of competing males, suggesting that they increase their use of energetically costly displays when competition is high, but conserve energy by reducing display activity when competition is low. Green argued that if displays given by male newts are energetically costly, then they can serve as indicators of male condition and possibly male quality to females searching for a mate (see also chapter 9).

## Energetics of Calling in Anurans

Male anurans often call for many hours per night during a breeding season that can last from a few days to several months (table 5.2). Their calls are very loud compared to sounds produced by animals such as birds and insects (Pough et al. 1992; see also table 7.1 in chapter 7). Production of such signals appears to be the most expensive activity performed by male anurans during their lifetimes (Pough et al. 1992; Prestwich 1994; Wells, 2001a). This means that energetic constraints limit the ability of individual males to produce calls, and this in turn can directly affect male mating success (Halliday 1987; M. Ryan 1988b). Limitations

**Table 5.2** Estimates of calling effort for selected anuran species

| Species (source no.) | Temperature (°C) | Measure of calling effort | | | |
|---|---|---|---|---|---|
| | | Notes/h | s/h | h/Night | Notes/Night |
| Bufonidae | | | | | |
| *Bufo woodhousii* (3, 5) | 20 | 240 | 360 | | |
| Centrolenidae | | | | | |
| *Hyalinobatrachium fleischmanni* (9) | 18 | 580 | 120 | 4 | 2,320 |
| Hylidae | | | | | |
| *Agalychnis callidryas* (23) | 26 | 180 | 7 | | |
| *Hyla arborea* (25) | 16 | 11,136 | 890 | | |
| *H. cinerea* (19) | 27 | 3,100 | 430 | 2–4 | 9,300 |
| *H. ebraccata* (7, 23) | 26 | 860 | 142 | 5 | 4,300 |
| *H. gratiosa* (19) | 29 | 3,600 | 650 | | |
| *H. microcephala* (20, 23) | 26 | 3,100 | 200 | 4 | 12,400 |
| *H. rosenbergi* (1) | 26 | 7,200 | 430 | 4 | 28,800 |
| *H. squirella* (19) | 27 | 6,600 | 1,320 | | |
| *H. versicolor* (16) | 19 | 1,000 | 700 | 2–4 | 3,000 |
| *Litoria ewingi* (18) | 10 | 2,200 | 450 | | |
| *Phrynohyas venulosa* (23) | 26 | 1,585 | 580 | | |
| *Pseudacris crucifer* (12) | 19 | 4,500 | 500 | 2–4 | 13,500 |
| *Scinax boulengeri* (23) | 26 | 475 | 173 | 3 | 1,425 |
| *S. rubra* (23) | 26 | 2,065 | 340 | 5 | 10,175 |
| Hyperoliidae | | | | | |
| *Hyperolius marmoratus* (high density) (24) | 25 | 3,100 | 260 | 4 | 12,400 |
| *H. marmoratus* (low density) (24) | 25 | 1,200 | 100 | 4 | 4,800 |
| *H. viridiflavus* (21) | 25 | 5,400[a] | 330 | 4 | 10,000 |
| Leptodactylidae | | | | | |
| *Eleutherodatylus coqui* (13, 14) | 23 | 1,450[b] | 350 | 3 | 4,350 |
| *Leptodactylus fragilis* (23) | 25 | 4,700 | 690 | 3–4 | 16,450 |
| *Physalaemus pustulosus* (2, 4, 22) | 25 | 1,680[b] | 635 | 5 | 8,400 |
| Myobatrachidae | | | | | |
| *Crinia georgiana* (26) | 12 | 346 | | | |
| *C. signifera* (6, 11) | 12 | 7,800 | 940 | | |
| *Geocrinia victoriana* (10) | 13 | 3,840 | 350 | | |
| *Uperoleia laevigata* (15) | 14 | 1,200 | 580 | 4–5 | 5,400 |
| Ranidae | | | | | |
| *Rana clamitans* (22) | 22 | 175 | | | |
| *R. sylvatica* (22) | 8 | 557 | | | |
| *R. virgatipes* (17) | 26 | 180 | 20 | 7 | 1,260 |

*Sources:* (1) Kluge 1981; (2) Bucher, Ryan, and Bartholomew 1982; (3) Sullivan 1982c; (4) Ryan, Bartholomew, and Rand 1983; (5) Sullivan 1983b; (6) Mac Nally 1984b; (7) Wells and Schwartz 1984a; (8) Dyson 1985; (9) Jacobson 1985; (10) Littlejohn and Harrison 1985; (11) Littlejohn, Harrison, and Mac Nally 1985; (12) Taigen, Wells, and Marsh 1985; (13) Woolbright 1985b; (14) Zelick and Narins 1985a; (15) Robertson 1986a; (16) Wells and Taigen 1986; (17) Given 1987; (18) Harrison 1987; (19) Prestwich, Brugger, and Topping 1989; (20) Wells and Taigen 1989; (21) Grafe, Schmuck, and Linsenmair 1992; (22) Ressel 1993; (23) Bevier 1995a; (24) Grafe 1996a; (25) Grafe and Thein 2001; (26) M. J. Smith, Withers, and Roberts 2003.

*Notes:* Data are based on average calling rates and call duration in the field. Calling times (s/h) are periods of actual sound production and do not include inter-call intervals.

[a]Measurements for frogs in laboratory.

[b]Number of calls per hour; calls composed of two distinct types of notes.

on stored energy reserves or foraging time can have a major impact on how long a male frog remains in a chorus (C. Murphy 1994a, b) and how much he calls while in the chorus (Lucas, Howard, and Palmer 1996). Consequently, the study of calling energetics provides a unique opportunity to integrate studies of metabolic physiology and sexual selection (see also chapter 8).

In all anurans studied so far, maximum $O_2$ consumption while calling is more than 10 times resting metabolism, and even average levels for males in the field are 6 to 20 times resting rates (table 5.3). These values probably are not representative of frogs as a whole, because all of these species have relatively high calling rates compared to other frogs (table 5.2). Oxygen consumption increases as calling rate or calling effort increase (fig. 5.8). For species such as *Pseudacris crucifer, Hyla microcephala, Hyla arborea, Hyperolius viridiflavus, Hyperolius marmoratus,* and *Crinia georgiana,* there is little variation in call note duration, and note rate alone explains most of the variation in $O_2$ consumption (Taigen, Wells, and Marsh 1985; Wells and Taigen 1989; Grafe, Schmuck, and Linsenmair 1992; Grafe 1996a; Grafe and Thein 2001; M. J. Smith, Withers, and Roberts 2003). In contrast, *Hyla versicolor* gives relatively long trills that vary in duration, and the best predictor of $O_2$ consumption is the product of note rate and duration (Taigen and Wells 1985; Wells and Taigen 1986), or the total number of pulses produced per hour (fig. 5.8 C; McLister 2001).

Interspecific variation in calling rate is positively correlated with the cost of calling, but absolute costs vary considerably (Prestwich 1994). In most species, maximum $O_2$ consumption during calling equals or exceeds $O_2$ consumption during forced exercise (table 5.3), so the latter cannot be considered the maximum aerobic capacity of these animals. Bucher, Ryan, and Bartholomew (1982) reported that noncalling *Physalaemus pustulosus* males exposed to a chorus had metabolic rates elevated above resting levels, but there is little evidence that this is generally true. In nearly all subsequent studies, predicted $O_2$ consumption at zero calling effort was not significantly different from resting metabolism (Taigen and Wells 1985; Taigen, Wells, and Marsh 1985; Prestwich, Brugger, and Topping 1989; Wells and Taigen 1989; Grafe, Schmuck, and Linsenmair 1992; Grafe 1996a; Grafe and Thein 2001; McLister 2001; M. J. Smith, Withers, and Roberts 2003). The results reported by Bucher, Ryan, and Bartholomew (1982) very likely were the result of noncalling males moving around inside the metabolic chamber.

**Table 5.3** Oxygen consumption of male frogs at rest, during forced exercise, and while calling

| Species (source no.) | Temperature (°C) | $\dot{V}O_2$ (ml $O_2$/g h) Rest | Exercise | Call (avg.) | Call (max.) |
|---|---|---|---|---|---|
| *Crinia georgiana* (13) | 12 | | | 0.49 | 1.80[a] |
| *Hyla arborea* (12) | 16 | 0.08 | | 1.44 | 1.78 |
| *H. cinerea* (6) | 27 | 0.14 | 1.02 | 0.92 | 1.20 |
| *H. gratiosa* (6) | 29 | 0.10 | 1.25 | 1.22[b] | 1.21 |
| *H. microcephala* (7) | 26 | 0.15 | | 1.70 | 2.80 |
| *H. squirella* (6) | 27 | 0.17 | 1.79 | 2.27[b] | 2.10 |
| *H. versicolor* (3, 5) | 19 | 0.08 | 1.09 | 1.70[b] | 1.67 |
| *Pseudacris crucifer* (10) | 23 | 0.15 | | 1.82 | 2.11 |
| *Hyperolius marmoratus* (high density) (9) | 25 | 0.13 | | 0.92 | 1.95 |
| *H. marmoratus* (low density) (9) | 25 | 0.13 | | 0.44 | 1.95 |
| *H. viridiflavus* (8) | 25 | 0.13[c] | 1.10[c] | 1.33 | 1.50 |
| *Physalaemus pustulosus* (1, 2) | 26 | 0.15 | 1.82 | 0.91 | 1.18 |
| *P. pustulosus* (11, 14) | 25 | 0.16 | | 1.30 | 1.83 |

*Sources:* (1) Bucher, Ryan, and Bartholomew 1982; (2) Ryan, Bartholomew, and Rand 1983; (3) Taigen and Wells 1985; (4) Taigen, Wells, and Marsh 1985; (5) Wells and Taigen 1986; (6) Prestwich, Brugger, and Topping 1989; (7) Wells and Taigen 1989; (8) Grafe, Schmuck, and Linsenmair 1992; (9) Grafe 1996a; (10) K. Wells, Taigen, and O'Brien 1996; (11) Bevier 1997b; (12) Grafe and Thein 2001; (13) M. J. Smith, Withers, and Roberts 2003; (14) Wells and Taigen, unpublished.

*Notes:* $\dot{V}O_2$ call (avg.) is the estimated metabolic rate at the average calling rate measured in the field. $\dot{V}O_2$ call (max.) is the highest metabolic rate for a frog calling in a metabolic chamber.

[a]Maximum value estimated from graphical data.

[b]Average calling rate in field exceeded those of males in metabolic chambers.

[c]Data from Taigen, Emerson, and Pough 1982, adjusted to 25° C assuming $Q_{10}$ of 2.

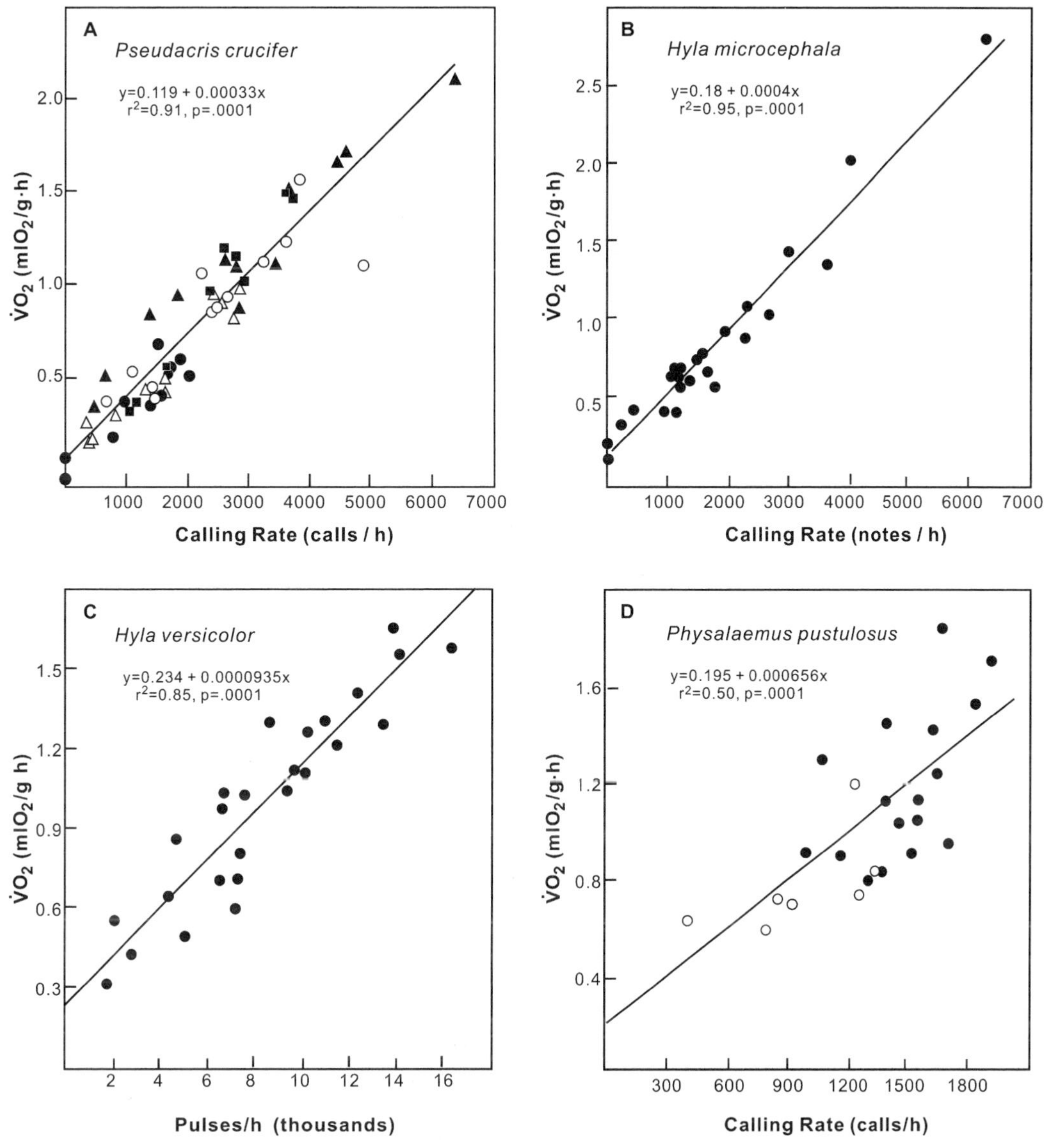


**Fig. 5.8.** Oxygen consumption as a function of calling effort for several species of frogs. (A) *Pseudacris crucifer* from Connecticut measured at five different temperatures (Wells, Taigen, and O'Brien 1996). (B) *Hyla microcephala* from Panama (Wells and Taigen 1989). (C) *Hyla versicolor* from Connecticut (Taigen and Wells 1985; Wells and Taigen 1986). (D) *Physalaemus pustulosus* from Panama (open circles are data from Bucher, Ryan, and Bartholomew 1982, with the two lowest values omitted; closed circles are data collected by T. L. Taigen and K. D. Wells, from Pough et al. 1992).

## Social Interactions and the Cost of Calling

Social interactions among males can have a major effect on calling energetics, but not necessarily in the same way in all species. Male frogs often respond to the calls of other individuals by increasing calling rate, call duration, or the number of notes in calls (see chapter 10). In *Hyla microcephala,* males add notes to their calls as vocal competition increases, leading to a linear increase in metabolic rate (fig. 5.8 B). Males appear to conserve energy by maintaining a low calling effort when only a few males are present, but increasing calling effort and energetic expenditures when competition among males is intense (Wells and Taigen 1989). This species calls in distinct bouts in which several males respond to each others' calls, followed by periods of silence (Schwartz 1991). These periodic pauses appear to be essential for conserving energy because of the extremely high calling rate of this species. Males regularly achieve calling rates of 6,000 notes per hour, and some individuals produce 10,000 notes per hour for short periods of time. Schwartz, Ressel, and Bevier (1995) estimated that males would deplete most of the glycogen reserves in their trunk muscles in less than two hours if they called continuously at such high rates (see the following further discussion of energy substrates).

In *Hyla versicolor,* males increase call duration in response to the calls of other males, but they simultaneously decrease calling rate (fig. 5.9). The result is a relatively constant level of energy expenditure, regardless of chorus density (Wells and Taigen 1986). Playback experiments with females have shown that long calls are preferred over short calls, and high calling rates over low rates (Klump and Gerhardt 1987). More importantly, as predicted by Wells and Taigen (1986), females prefer long calls delivered at slow rates to short calls delivered at fast rates, even when total calling effort is equal (Klump and Gerhardt 1987). Therefore, males in dense choruses alter their calls in ways that enhance their attractiveness to females. Most individuals appear to be calling near maximum sustainable levels, even when calling in isolation, and males probably cannot increase calling rate and call duration simultaneously because of energetic constraints (fig. 5.10).

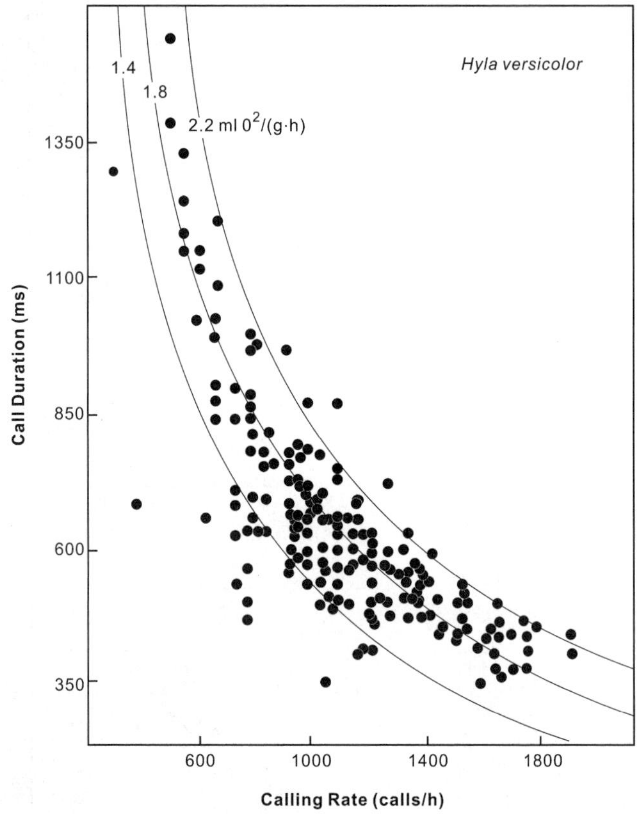


**Fig. 5.10.** Relationship of calling rate to call duration for male gray treefrogs (*Hyla versicolor*) calling in Connecticut. Lines show estimated rates of oxygen consumption for different combinations of rate and duration, calculated from metabolic rates of males measured in chambers placed in a natural chorus. Males cannot produce very long calls at high rates. After Wells and Taigen (1986).

Runkle et al. (1994) studied individual and seasonal variation in calling behavior and energy expenditure in *Hyla versicolor.* Differences in calling behavior among males in a chorus sometimes persisted for several nights. Some males consistently gave relatively long calls at low rates, whereas others consistently gave relatively short calls at higher rates (see Gerhardt, Dyson, and Tanner [1996] for similar results for another population of this species). The behavior of individuals was plastic, however, and this consistency appeared to reflect local differences in chorus density. Neighboring males often responded to each other's calls and therefore tended to have similar levels of energy expenditure. As the season progressed, males dropped out of the chorus, and overall chorus density decreased. This resulted in a significant reduction in average call duration, but a coincident increase in calling rate; average hourly energy expenditures changed very little. Nevertheless, there was a substantial decrease in total energy expenditure for call production over the course of the season, because chorus duration decreased from nearly four hours per night at the beginning of the season to only an hour and a half at the end of the season. This decrease in chorusing time was not related to changes in temperature and did not appear to be a result of decreased energy reserves (Wells et al. 1995), but probably was related to decreased availability of females.

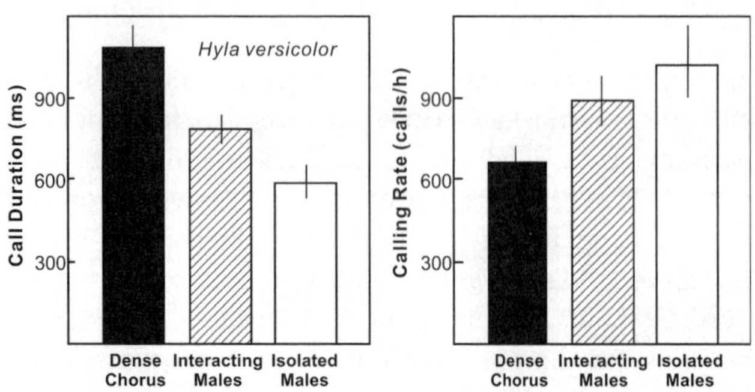


**Fig. 5.9.** Changes in call duration and calling rate as a function of chorus density in gray treefrogs (*Hyla versicolor*) from Connecticut. Males increase call duration but decrease calling rate as chorus density increases. Bars show means ± 1 SE. After Wells and Taigen (1986).

The effect of social interactions on calling energetics also has been examined in *Physalaemus pustulosus.* This species has an unusual call consisting of a frequency-modulated "whine" note that sometimes is combined with one or more secondary "chuck" notes. The latter are given simultaneously with the whine note, and are not added onto the end of the call as in other species with complex multinote calls (M. Ryan 1985a). Males calling in low-density choruses generally give calls with relatively few chucks. In dense choruses, males call at faster rates and add more chuck notes to their calls. Females prefer high calling rates to low rates and calls with chuck notes to whines alone (Rand and Ryan 1981). Although energetic expenditures increase with increasing calling rate, there is no evidence from the data of Bucher, Ryan, and Bartholomew (1982) or the data collected by Taigen and Wells (unpublished data) that number of chucks is related to the cost of calling. Probably this is because the chuck is produced by passive vibration of a fibrous mass in the airstream of the larynx, not by additional contractions of the trunk muscles (Drewry, Heyer, and Rand 1982; see also chapter 10).

## Temperature and Calling Energetics

Most of the frogs studied to date are tropical species or summer-breeding temperate-zone species that call at a relatively narrow range of temperatures. Consequently, variation in temperature explains little of the variation in $O_2$ consumption while calling (Bucher, Ryan, and Bartholomew 1982; Wells and Taigen 1985; Wells and Taigen 1989; Grafe, Schmuck, and Linsenmair 1992). Spring peepers (*Pseudacris crucifer*), however, call at a wide range of temperatures during a breeding season that can last from late March through late May in New England. Early in the season, males call at air temperatures as low as 3° to 4° C, but late in the season, temperatures can be as high as 25° C. Calling rate is tightly correlated with ambient temperature (Lemon and Struger 1980), and the cost of calling is a function of calling rate (Taigen, Wells, and Marsh 1985). One would expect an increase in the cost of calling as the season progresses. Wells, Taigen, and O'Brien (1996) measured metabolic rates of calling spring peepers at temperatures from 5° to 23° C. Intercepts of regressions for different temperatures changed because of changes in resting metabolism, but the relationship between calling rate and metabolic rate was similar at all temperatures (fig. 5.8 A). Temperature had relatively little effect on the energetic cost of calling independent of its effect on calling rate. Both air and water temperatures increase during a typical breeding season, and the estimated cost of calling at the end of a season was double the cost at the beginning (fig. 5.11). Males of *Hyla versicolor* call at a much narrower range of temperatures, but temperature can still affect the energetic cost of calling through its effect on calling rate. As in *P. crucifer,* temperature does not affect the cost of calling independent of calling rate. It does appear to affect the efficiency of call production (the conversion of metabolic energy to sound energy), however. Calling efficiency increased with increasing temperature, although the reasons for this are unclear (McLister 2001).

## Contributions of Anaerobic Metabolism

Because calling is such a vigorous activity that is sustained for many hours, it seems unlikely that anaerobic pathways could make a substantial contribution to calling metabolism. Measurements of lactate accumulation by calling frogs support this prediction. Pough and Gatten (1984) reported

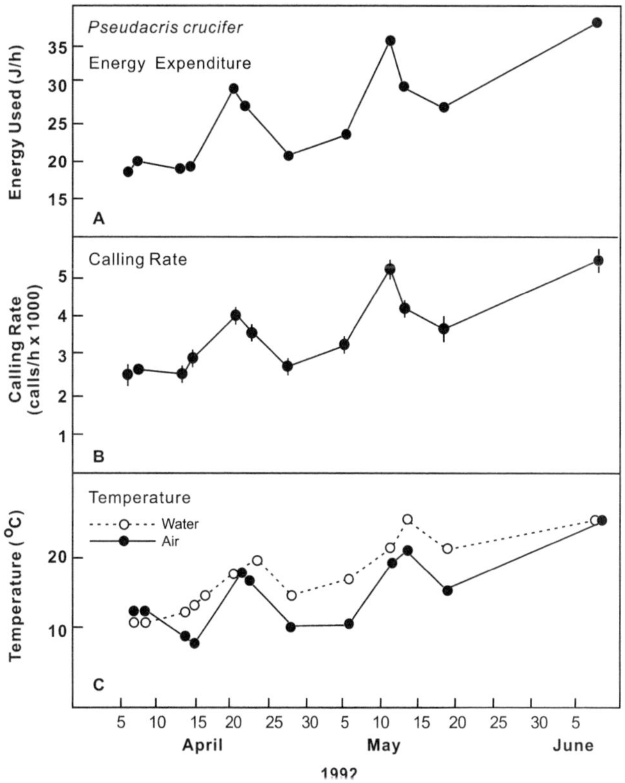


**Fig. 5.11.** Seasonal changes in temperature, calling rate, and estimated energy expenditure by calling spring peepers (*Pseudacris crucifer*) in Connecticut. Energy expenditure was estimated from measurements of metabolic rates of calling frogs at five different temperatures. After Wells, Taigen, and O'Brien (1996).

levels of whole-body lactate in calling male spring peepers that were double those of animals resting in the laboratory and slightly above those of noncalling males in the field. They concluded that calling animals sometimes exceed their maximum aerobic capacity and obtain some of their energy for calling through anaerobic pathways. Nevertheless, whole body lactate levels in calling frogs were less than 10% of those found in animals exercised to exhaustion and would make a trivial contribution to total ATP production. In *Pseudacris regilla* and *Physalaemus pustulosus,* whole-body lactate levels of calling males were not significantly different from those of resting animals (Whitney and Krebs 1975a; M. Ryan, Bartholomew, and Rand 1983), and in *Hyla versicolor,* lactate levels actually decreased after several hours of calling (Taigen and Wells 1985). Prestwich, Brugger, and Topping (1989) and Grafe, Schmuck, and Linsenmair (1992) found that elevated whole-body lactate levels were due almost entirely to increased lactate in the leg muscles, not trunk muscles, and probably were due to frogs moving around while calling. Therefore, it appears that anaerobic metabolism does not make a significant contribution to ATP production to support routine calling behavior and can be ignored in calculations of the cost of calling.

## Total Energetic Costs of Calling

Estimates of total energetic costs of calling for several species of frogs are given in table 5.4. These estimates are based on metabolic rates of average size males calling at average rates in the field and do not indicate the range of variation within species. Such variation can be considerable if males differ substantially in size or calling rate. For example, a *Hyla microcephala* male calling at the highest rate measured in the field would have a net cost of calling nearly double the average figure for this species (Wells and Taigen 1989). Total energy expenditure of calling *Hyperolius marmoratus* males in a high-density chorus was nearly three times that of males in a low-density chorus. A large *Hyla versicolor* male can have a body mass more than double that of the smallest male in the chorus. If these frogs have the same calling effort, the larger male will use more energy, and this will be reflected in the greater sound power output of large males (Taigen and Wells 1985).

Interspecific differences in net cost are due in part to the louder calls and greater sound power output of larger species. For example, *Hyla versicolor* males have much louder calls than those of *Physalaemus pustulosus* and a net cost of calling about 10 times higher. Nevertheless, some very small species, such as *Hyla microcephala* and *Pseudacris crucifer*, have remarkably loud calls for their body size, but net costs of calling are not proportionally higher. This suggests that these frogs are more efficient at converting metabolic energy to sound energy. Prestwich (1994) estimated that *Pseudacris crucifer* has a sound production efficiency of about 5%, nearly five times higher than in *Physalaemus pustulosus*. One must be cautious in interpreting these comparisons, however, because they do not take into account intraspecific variation in calling effort or call intensity. In general, frogs are relatively inefficient at converting metabolic energy into sound energy, with efficiencies ranging from 0.5 to 5%. The efficiency of the muscles that actually produce the sound, however, would be much higher (Prestwich 1994; McLister 2001).

Also shown in table 5.4 is the cost per minute of actual sound production. With the exception of *Hyperolius*, all of the small species (body mass = 1–2 g) have similar costs per minute of calling, despite major differences in the structure of their calls. *Pseudacris crucifer* has a simple tonelike call, whereas *Physalaemus pustulosus* has a much longer and more complex frequency-modulated call. *Hyla microcephala* and *H. squirella* both have rather broad-spectrum calls, but the former has distinct pulses in the calls, while the latter does not. Most of the differences in call structure reflect differences in the structure and action of the larynx and laryngeal muscles (M. Ryan 1985a; Schneider 1988), and these probably contribute relatively little to total energetic costs.

## Energetic Costs of Other Activities

Information on the energetic cost of natural activities other than calling in amphibians is very limited, but the available data indicate that calling is likely to be the most expensive behavior performed by these animals (tables 5.4 and 5.5).

**Table 5.4** Net cost of calling in selected anuran species, listed in order of decreasing body mass

| Species (source no.) | Temperature (°C) | Mass (g) | Net cost of calling J/h | J/g·h | J/min |
|---|---|---|---|---|---|
| *Hyla arborea* (11) | 16 | 6.3 | 172 | 27 | 11.6 |
| *H. gratiosa* (6) | 29 | 12.5 | 280 | 22 | 25.9 |
| *H. versicolor* (3, 5) | 19 | 8.6 | 280 | 33 | 24.0 |
| *H. cinerea* (6) | 27 | 5.1 | 80 | 16 | 11.2 |
| *H. squirella* (6) | 27 | 2.2 | 93 | 42 | 4.2 |
| *Hyperolius marmoratus* (high density) (9) | 25 | 2.0 | 36 | 18 | 8.3 |
| *H. marmoratus* (low density) (9) | 25 | 2.0 | 13 | 7 | 7.8 |
| *H. viridiflavus* (8) | 25 | 2.0 | 48 | 24 | 8.7 |
| *Physalaemus pustulosus* (1, 2) | 26 | 1.7 | 25 | 15 | 3.8 |
| *P. pustulosus* (10, 12) | 25 | 1.8 | 41 | 23 | 3.9 |
| *Pseudacris crucifer* (4) | 19 | 1.2 | 25 | 21 | 3.0 |
| *Hyla microcephala* (7) | 26 | 0.6 | 20 | 33 | 3.4 |

*Sources:* (1) Bucher, Ryan, and Bartholomew 1982; (2) Ryan, Bartholomew, and Rand 1983; (3) Taigen and Wells 1985; (4) Taigen, Wells, and Marsh 1985; (5) Wells and Taigen 1986; (6) Prestwich, Brugger, and Topping 1989; (7) Wells and Taigen 1989; (8) Grafe, Schmuck, and Linsenmair 1992; (9) Grafe 1996a; (10) Bevier 1995b; (11) Grafe and Thein 2001; (12) Taigen and Wells, unpublished data from fig. 5.8.

*Notes:* Cost (J/h) = $\dot{V}O_2$ call – $\dot{V}O_2$ rest × mass × 20.1 J/ml $O_2$; J/min = Cost per minute of actual sound production.

**Table 5.5** Mass-specific metabolic rates and net cost of activity for selected amphibians engaged in natural behavior and at rest

| Species (source no.) | Temperature (°C) | Mass (g) | $\dot{V}O_2$ rest (ml $O_2$/g·h) | $\dot{V}O_2$ activity (ml $O_2$/g·h) | Net cost (J/h) |
|---|---|---|---|---|---|
| *Spea hammondii* (1) | | | | | |
| Burrowing | 30 | 11.8 | 0.11 | 0.51 | 95 |
| *Bufo fowleri* (4, 5) | | | | | |
| Foraging | 21 | 25.8 | 0.13 | 0.30[a] | 85 |
| *Physalaemus pustulosus* (3) | | | | | |
| Nest building | 25 | 1.7 | 0.16 | 0.77 | 20 |
| *Hyla versicolor* amplexus (6) | 25 | | | 0.16 | |
| *Desmognathus carolinensis* (2) | | | | | |
| Aggression | 15 | 2.0 | 0.07 | 0.11 | 1.6 |
| Courtship | 15 | 2.0 | 0.07 | 0.10 | 1.2 |

*Sources:* Modified from Pough et al. (1992). (1) Seymour 1973; (2) Bennett and Houck 1983; (3) Ryan, Bartholomew, and Rand 1983; (4) Walton 1988; (5) Walton and Anderson 1988; (6) McLister 2003.

*Notes:* Net cost = $\dot{V}O_2$ activity – $\dot{V}O_2$ rest × mass × 20.1 J/ml $O_2$.

[a]Estimated from field measurements of movement speed and laboratory measurements of oxygen consumption.

The only activity that entails elevations in oxygen consumption approaching that of calling males is construction of foam nests, which has been studied in *Physalaemus pustulosus*. The male uses his back legs to beat the mucus coating around the eggs into froth as the eggs are laid, a process that requires about one hour (M. Ryan 1985a). Metabolic rates of nest-building pairs of males and females are about five times resting rates (M. Ryan, Bartholomew, and Rand 1983), and the net cost of nest building is about 70 to 80% of the cost of calling (tables 5.4 and 5.5). M. Ryan, Bartholomew, and Rand (1983) estimated that frogs actually kick their legs only about 20% of the time, so the net cost per minute of activity would be about 1.7 J, or about 45% of the cost of a minute of sound production.

Another vigorous activity performed by many species of anurans is burrowing (see chapters 1 and 2). Seymour (1973c) measured the energetic cost of burrowing in a spadefoot toad, *Spea hammondii*. Metabolic rates of burrowing toads were about four and a half times resting rates, and the cost of an hour of burrowing for a 12-g toad was about 95 J (table 5.5). The cost of calling is unknown, but in two hylids of about the same size, *Hyla gratiosa* and *Hyla versicolor*, the cost of calling is about three times the cost of burrowing in *Spea* (tables 5.4 and 5.5). Seymour estimated that the toads were actually burrowing only about 22% of the time, or about 13 minutes per hour, so the cost per minute of burrowing is approximately 7.3 J at 30° C. By comparison, a minute of sound production by *H. gratiosa* at a similar temperature costs about three and a half times as much (table 5.4).

Other routine activities by amphibians probably are much less expensive. For example, Walton (1988) reported that foraging toads (*Bufo fowleri*) typically move at a very slow rate (mean = 0.018 km/h). From a regression of metabolic rate versus movement speed (Walton and Anderson 1988), estimated $O_2$ consumption during foraging is about 2.3 times resting metabolism, yielding a net cost of 85 J/h for a 26-g toad (table 5.5). This is a downward extrapolation from the metabolic data, because natural foraging velocities were only 20% of the lowest speed at which $O_2$ consumption was measured. This figure is well below maximum $O_2$ consumption during forced exercise, indicating that toads probably do not approach their physiological limits during routine foraging (Walton 1988). The costs of calling by frogs are much greater than the costs of reproductive activities in salamanders. For example, the cost of courtship in *Desmognathus* is only about 5% of the cost of calling by a frog of similar size (tables 5.4 and 5.5).

### Comparisons to the Cost of Signaling in Other Animals

The acoustic signals used by frogs are analogous in function to those of some fishes, insects, and birds. There is abundant evidence for both insects and birds that high singing rates make males attractive to females (Andersson 1994; Gerhardt and Huber 2002), so sexual selection is likely to favor the production of energetically expensive signals, just as it does in frogs. On the other hand, mechanisms of sound production differ in different kinds of animals, and some are capable of producing acoustic signals at relatively low cost. For example, the cost of calling appears to be low in the oyster toadfish (*Opsanus tau*). Like many fishes, this species produces sound by exceedingly rapid contractions of sonic muscles associated with the swim bladder. These muscles are tiny compared to the trunk muscles of frogs—1% of male body mass, compared to up to 15% in some male frogs. Consequently, the elevation of metabolic rate above resting levels is minimal (Amorim et al. 2002; these results are based on electrical stimulation of sonic muscles, not spontaneous calling, so they are not strictly comparable to the data on frogs).

Several studies have been done on the energetic cost of sound production in insects, and discussions of the energetics of sound production by frogs and insects can be found in Pough et al. (1992), Prestwich (1994), and Gerhardt and Huber (2002). Insect songs typically have lower sound pressure levels than the calls of frogs, but total sound power output is greater for some insects, especially species that call more or less continuously (Pough et al. 1992). Comparisons both within and among species have shown that metabolic rates of calling insects increase as a linear function of syllable rate, just as the cost of calling in frogs increases with calling rate. Because of the much greater number of syllables produced per minute, mass-specific metabolic rates of calling insects often are much higher (0.4–13 ml $O_2$/g·h, mean = 5.2) than those of calling frogs (0.78–4.20 ml $O_2$/g·h, mean = 1.8; Prestwich 1994).

Insects have higher mass-specific resting metabolic rates than do frogs, so factorial aerobic scopes for calling average about 1.5 times higher in frogs (Prestwich 1994). Two species of mole crickets, for example, had factorial scopes of about five to eight times resting metabolism (Prestwich and O'Sullivan 2005). The Australian bladder cicada (*Cystosoma saundersii*), on the other hand, had a metabolic rate during singing that was 21 times resting metabolism, comparable to the highest factorial metabolic scopes measured in frogs (Mac Nally and Young 1981). This species produces a loud, continuous call, sometimes for a half an hour or more at a time, so the energetic cost of signaling is considerable. Not all insects produce such expensive signals, however. A wax moth (*Achroia grisella*), which produces ultrasonic signals while fanning its wings, has an average factorial scope of only about 2.75 (Reinhold et al. 1998), while the tock-tock beetle (*Psammodes striatus*), which produces sound by tapping on the ground, has a factorial scope of about 2.5 (Lighton 1987). A chirping cricket (*Requena verticalis*), which has a very low calling rate, has a factorial scope of only 2 (Bailey et al. 1993). Despite the relatively low cost of calling in *R. verticalis*, males appear to be under some energy stress. Individuals that were fed a high-quality diet devoted more of their time to calling than those fed a low-quality diet (Simmons et al. 1992). Frogs appear to be more efficient in converting metabolic energy into sound energy (0.5–5% efficiency) than insects (0.03–6%), but the highest estimates are comparable for the two groups (Prestwich 1994; Reinhold et al. 1998; Prestwich and O'Sullivan 2005).

Several lines of evidence suggest that birds are stressed energetically by singing. In some species, singing rates decline in cold weather, perhaps because of the increased energetic costs of thermoregulation (Garson and Hunter 1979; Higgins 1979; Gottlander 1987; Reid 1987; Santee and Bakken 1987). Some birds increase singing rates when food is abundant, or when they are provided with supplemental food (Searcy 1979; Wilhelm, Comtesse, and Phlumm 1980; Davies and Lundberg 1984; Gottlander 1987; Radesäter et al. 1987; Reid 1987; Strain and Mumme 1988; Lucas, Schraeder, and Jackson 1999). Several authors have suggested that females might use song rate as an indicator of male condition or quality. If males sing at times when food is scarce or energy reserves are limited, then only males in good condition could sing at peak rates (McNamara, Mace, and Houston 1987; Reid 1987; Hutchinson, McNamara, and Cuthill 1993). Nevertheless, the main constraint imposed by singing may be a time constraint on foraging rather than the energetic cost of singing per se (Thomas et al. 2003; Ward, Lampe, and Slater 2004).

There have been a number of measurements of the energetic cost of singing in birds. Horn, Leonard, and Weary (1995) measured $O_2$ consumption in crowing roosters. Metabolic rate increased as a function of crowing rate, but even at peak rates, the cost of crowing was relatively low, with metabolic rate elevated only about 15% above resting levels. Roosters produce, at most, only 200–250 crows per day, equivalent to the number of calls produced in one hour by frogs with relatively low calling rates (table 5.2). Mass-specific metabolic rates during crowing were only 1.25 ml $O_2$/(g·day), less than the hourly metabolic rate of a calling spring peeper. In zebra finches (*Taeniopygia guttata*), canaries (*Serinus carnaria*), and starlings (*Sturnus vulgaris*), metabolic rates during singing averaged 1.1–3.4 times resting metabolism, depending on singing rate (Oberweger and Goller 2001; Ward, Speakman, and Slater 2003; Ward and Slater 2005). For pied flycatchers (*Ficedula hypoleuca*), the cost of singing was about 2.7 times resting metabolism and only 1.12 times the cost of standing (Ward, Lampe, and Slater 2004). This species produces only about 90–240 songs per hour, much lower than the calling rates of many frogs. More comparable to calling in frogs is the singing of Carolina wrens (*Thryothorus ludovicianus*), which produce up to 2100 songs per hour. Metabolic rates of singing males were three to nine times resting rates and increased as a linear function of singing rate. At typical singing rates, singing would account for 10–25% of the daily energy budget for male wrens during the breeding season (Eberhardt 1994).

Although factorial increases in metabolic rate by many calling frogs appear to be considerably higher than for birds, this is largely a result of the very low resting metabolic rates of frogs. Mass-specific comparisons of the absolute amount of oxygen consumed for each second of vocal signal showed that costs are similar for birds and frogs (Oberweger and Goller 2001). In birds, both the respiratory muscles and muscles of the syrinx are involved in song production. Birds may save energy during singing because the syrinx modulates a one-way airstream produced during normal respiration, whereas frogs use large trunk muscles to force air out of the lungs and through the larynx only for calling (see chapter 7). The relative contribution of these two sets of muscles to the energetic cost of singing in birds is unknown, but the cost of contracting muscles of the syrinx is likely to be small. These muscles constitute a tiny proportion of a bird's body mass compared to the trunk muscles of frogs. For example, the trunk muscles of a 1.25 g male spring peeper (*Pseudacris crucifer*) weigh about 0.185 g, or nearly 15% of total body mass (Taigen, Wells, and Marsh 1985). In contrast, the syrinx muscles of a male zebra finch weigh about 0.027 g (Wade and Buhlman 2000), about 1% of its 13 g body mass (Nudds and Bryant 2000).

It appears that flying, particularly flights involving vertical takeoffs, is considerably more expensive than singing. In zebra finches, for example, the cost of short flights is 10 times the cost of singing and 28 times resting metabolism (Nudds and Bryant 2000; Ward, Lampe, and Slater 2004). This means that some nonvocal displays of birds are like to be more energetically costly than singing, because they often require moving the total body mass of the bird. Many species, especially those that display on leks, perform displays that involve vigorous and repeated jumping or flying. Norberg (1991) estimated that the flight display of the flappet lark (*Mirafra rufocinnamomea*) requires a metabolic rate 16 times resting metabolism. Male sage grouse (*Centrocercus urophasianus*) perform repeated strut displays to attract females. Instantaneous metabolic rates of displaying males, measured with the doubly labeled water technique, were 14–17 times resting metabolism, comparable to the factorial scopes of some calling frogs. Daily energy expenditure was about twice as high for displaying males as for nondisplaying males on the same lek and about four times as high as for males at rest (Vehrencamp, Bradbury, and Gibson 1989). Total energetic expenditures for all of these bird species are an order of magnitude higher than for frogs because of their much higher metabolic rates and larger size. Nevertheless, singing and other sexual displays probably represent a much smaller proportion of the daily energy budget of a bird than that of a frog simply because so much of a bird's energy budget is consumed by heat production and by flight powered by massive flight muscles (> 20% of total body mass in finches; Nudds and Bryant 2000).

### Morphological Correlates of Calling Energetics

The muscles involved in call production by frogs are quite distinct in morphology and histochemistry from anuran leg muscles. The action of these muscles is described in detail in chapter 7. Briefly, the power for call production is provided by muscles in the trunk region, including the rectus abdominus, external oblique, and internal oblique (= transverse) muscles. The trunk muscles of males generally are much larger (3–15% of body mass) than those of females (1–3%; Pough et al. 1992; Bevier 1995b; Girgenrath and Marsh 2003). Relative trunk muscle mass is greater in species with high calling rates than in species with low calling rates (Bevier 1995b). The laryngeal muscles are involved in opening and closing the larynx during call production, and in some species are actively involved in fine-scale amplitude modulation of the call. These muscles also are larger in males than in females. Both sets of muscles are composed mainly of fast oxidative fibers (> 90%), along with a few tonic fibers, and have rapid shortening velocities (Eichelberg and Schneider 1973, 1974; Schneider 1977, 1988; R. Marsh and Taigen 1987; McLister, Stevens, and Bogart 1995; Boyd et al. 1999; Girgenrath and Marsh 1999, 2003; R. Marsh 1999).

Elastic fibers associated with the trunk muscles increase the efficiency of call production. In male frogs, the rectus abdominus muscle is separated from the other trunk muscles by a band of fibrous tissue, the linea masculina. In at least one species, *Physalaemus pustulosus*, this band of tissue is highly elastic and provides elastic recoil during deflation of the lungs, which increases the efficiency of sound production (Jaramillo, Rand, Ibáñez, and Dudley 1997). Elastic fibers also are found in the vocal sac of *Physalaemus*, and these might increase the efficiency with which air is pumped back into the lungs after a call is produced.

Trunk and laryngeal muscles are dramatically different from leg muscles in ultrastructural characteristics. The muscles involved in call production are well supplied with sarcoplasmic reticulum and have several times as many mitochondria per unit volume as the leg muscles (fig. 5.12). Capillary densities also are much higher in trunk and laryngeal muscles than in leg muscles, reflecting the much higher oxygen demand of the trunk muscles (Eichelberg and Schneider 1973, 1974; Schneider 1977, 1988; R. Marsh and Taigen 1987; Ressel 1993, 2001; McLister, Stevens, and Bogart 1995). In a detailed analysis of muscle ultrastructure in relation to calling behavior, Ressel (1996) found that both mitochondrial volume density and capillary density were positively correlated with calling rate among tropical frogs from two families that call at similar temperatures. Indeed, comparison of mitochondrial volume density for a variety of tropical and temperate-zone species reveals a positive correlation with calling rate that does not appear to be related to phylogeny (Ressel 2001). Ranid frogs with very low calling rates have mitochondrial volume densities similar to those of hylids with low calling rates, such as *Agalychnis callidryas* (fig. 5.13 A). Two leptodactylid frogs, *Physalaemus pustulosus* and *Eleutherodactylus coqui*, have muscle characteristics similar to those of hylids with comparable calling rates.

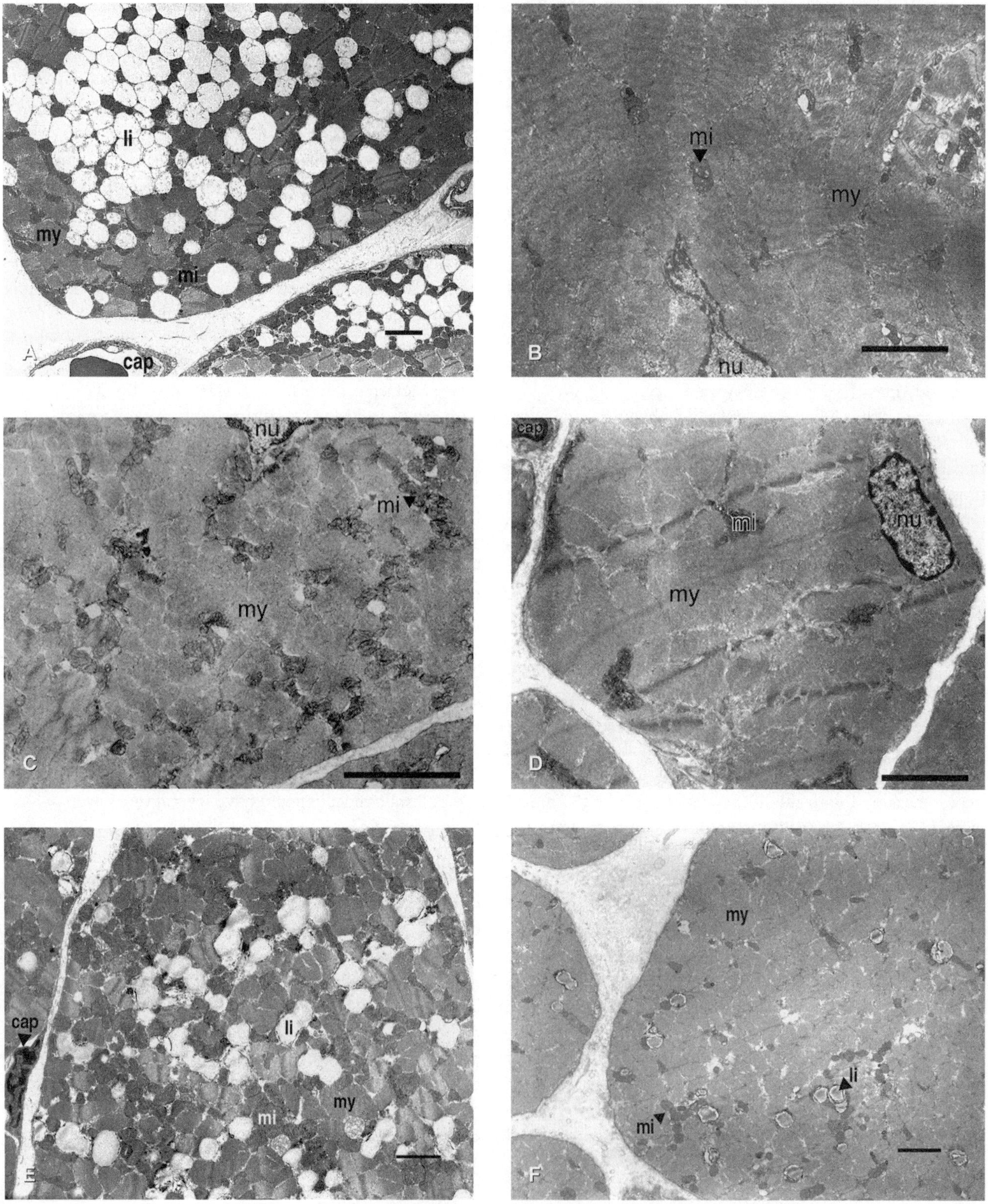


**Fig. 5.12.** Transmission electron micrographs of trunk and leg muscles of frogs. (A) Trunk muscle of *Pseudacris crucifer.* (B) Gastrocnemius (leg) muscle of *P. crucifer.* (C) Trunk muscle of *Rana sylvatica.* (D) Gastrocnemius muscle of *R. sylvatica.* (E) Trunk muscle of *Hyla microcephala.* (F) Trunk muscle of *Agalychnis callidryas.* Abbreviations: cap = capillaries; li = lipid droplets; mi = mitochondria; my = myofibrils; nu = nucleus. Scale lines = 1 μm. Photos by Stephen J. Ressel.

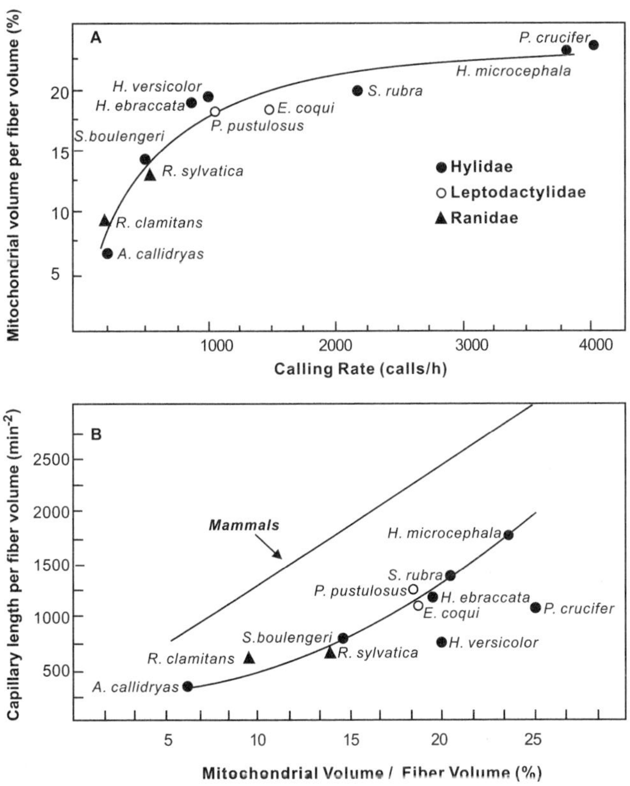


**Fig. 5.13.** (A) Relationship of mitochondrial volume density of trunk muscle tissue to calling rate for several species of anurans in three families. The curve is for Neotropical hylids and leptodactylids only (After Ressel 1996). Data for two temperate zone hylids (*Hyla versicolor* and *Pseudacris crucifer*) and two temperate zone ranids (*Rana clamitans* and *R. sylvatica*) are shown for comparison (data from Ressel 1993). (B) Relationship of trunk-muscle capillary length to mitochondrial volume density for several anurans and for mammals. The curve is for Neotropical hylids and leptodactylids (after Ressel 1996), with other species shown for comparison (data from Ressel 1993).

In mammals, capillary supply to skeletal muscle is closely correlated with mitochondrial volume density, because oxygen supply must increase with metabolic demand. This also appears to be true for frogs, but frogs have fewer capillaries for a given mitochondrial volume than do mammals (fig. 5.13 B). Presumably this reflects the lower activity temperatures of frogs, which in turn result in lower metabolic demands (Ressel 1996). Indeed, points for some temperate-zone frogs that call at lower temperatures than most tropical frogs fall below the curve for tropical frogs alone (fig. 5.13 B). This is particularly true for species with high calling rates, such as *Pseudacris crucifer.*

## Biochemical Correlates of Calling Energetics

Both citrate synthase (CS) and β-hydroxyacyl-CoA-dehydrogenase (HOAD) activity have been used as indicators of aerobic metabolism and lipid metabolism in the trunk muscles of frogs (Taigen, Wells, and Marsh 1985; R. Marsh and Taigen 1987; Given and McKay 1990; Bevier 1995a), as they have been for sexually dimorphic sound-producing muscles in other animals (Rice and Lobel 2002). Activities of both enzymes are consistently higher in trunk muscles than in leg muscles (fig. 5.14), and variation in the activities of these enzymes is closely correlated with variation in calling rate (fig. 5.15). The activities of these enzymes are closely correlated with each other in both trunk and leg muscle. This suggests that species with high calling rates not only have high aerobic capacities, but also depend more heavily on lipids to fuel call production (Bevier 1995a; see subsequent discussion of energy substrates). The enzymes involved in aerobic metabolism are located in the mitochondria, and a comparison of Bevier's (1995a) data on enzyme activities with Ressel's (1993) data on mitochondrial volume density shows a strong positive correlation for both tropical and temperate-zone species (fig. 5.16).

A comparison of two Neotropical hylids in same genus, *Scinax rubra* and *S. boulengeri* (fig. 5.17), nicely illustrates the relationship between calling behavior and trunk muscle

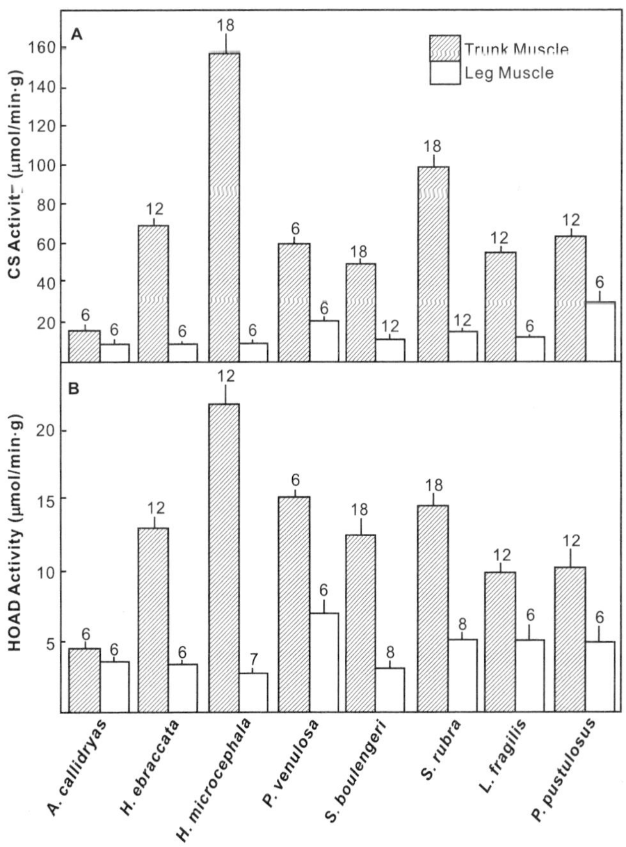


**Fig. 5.14.** Comparison of activities of the enzymes citrate synthase (CS) and β-hydroxyacyl-CoA-dehydrogenase (HOAD) at 25° C in trunk muscles and leg muscles of tropical hylids and leptodactylids. Abbreviations: *A = Agalychnis; H = Hyla; L = Leptodactylus; P. pustulosus = Physalaemus pustulosus; P. venulosa = Phrynohyas venulosa; S = Scinax.* After Bevier (1997b).

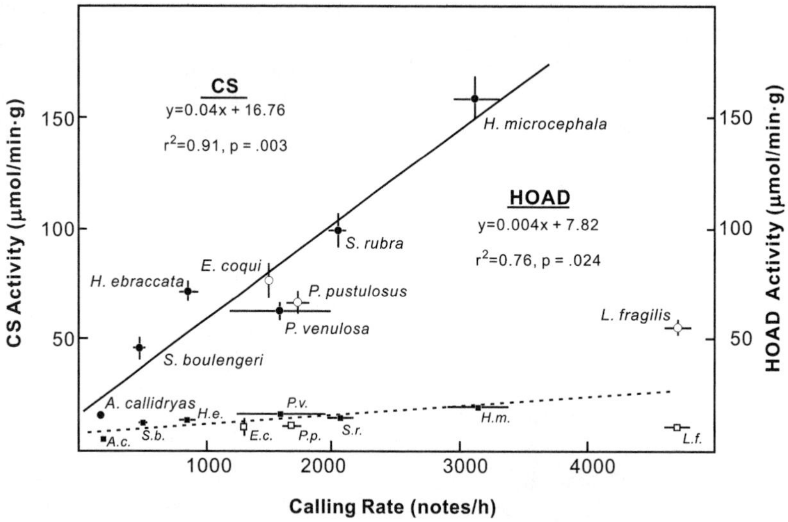


**Fig. 5.15.** Activity of the enzymes citrate synthase (CS) and β-hydroxyacyl-CoA-dehydrogenase (HOAD) in relation to calling rate for tropical frogs. Calling rates are estimated average calling rates at natural field temperatures. The regression line is for the hylid species only (filled symbols), but two leptodactylids (open symbols) are included for comparison. Abbreviations and species names as in fig. 5.14, with the addition of *Eleutherodactylus coqui*. After Bevier (1995b).

characteristics (Ressel 1996; Bevier 1997a). These species are about the same size (40 mm snout-vent length) and mass (3.5 g), and they call at the same temperatures. *Scinax rubra* males have larger trunk muscles (10% of body mass) than do *S. boulengeri* males (6.6%), and their muscles are more aerobic, with much higher mitochondrial volume densities, greater capillary length, and higher enzyme activities, especially CS activity (table 5.6). They differ dramatically in calling behavior as well. *Scinax rubra* is an explosive breeder that calls only after heavy rains, and individual males are seldom present in a chorus for more than one or two nights. Males call at relatively high rates (more than 2,000 calls/hour) for up to seven hours per night, and they move actively about the chorus area in search of females. *Scinax boulengeri*, on the other hand, is a continuous breeder, with some males present in choruses for weeks at a time. Males call at relatively low rates, usually for only about three hours per night, and they reduce calling rates after the first hour. The trunk muscles of *S. rubra* are adapted for very high levels of calling activity during the short periods when females are available, whereas those of *S. boulengeri* are designed for slow but sustained activity over much longer periods of time. This difference in behavior is reflected in their use of energy substrates as well. *Scinax rubra* has much larger stores of lipids and glycogen in the trunk muscles, but also depletes them more rapidly than does *S. boulengeri*.

A study of spring peepers (*Pseudacris crucifer*) by Zimmitti (1999) was the first to examine variation in trunk muscle enzyme activities among males in the same population. He collected pairs of individuals on the same night,

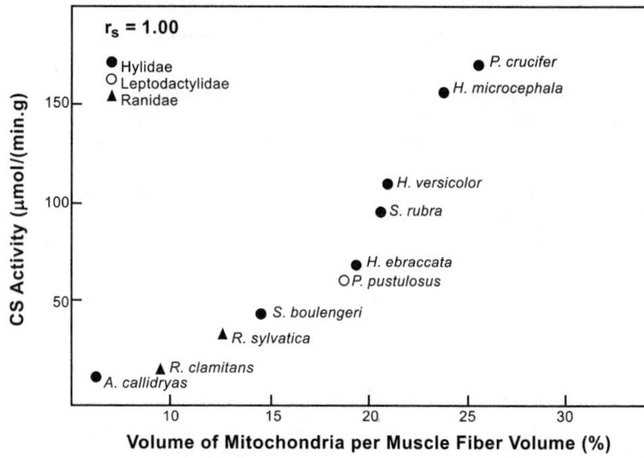


**Fig. 5.16.** Relationship of citrate synthase (CS) activity at 25° C to mitochondrial volume density in anurans. Abbreviations as in fig. 5.14, with the addition of *R* = *Rana* and *P. crucifer* = *Pseudacris crucifer*. $r_s$ = Spearman rank correlation coefficient. Plotted from data in Marsh and Taigen (1987), McKay (1989), Pough et al. (1992), Ressel (1993, 1996), and Bevier (1995b).

choosing males with unusually high or low calling rates. He found that males with high calling rates consistently had higher activities of both CS and HOAD in their trunk muscles than did males with low calling rates, and the activities of the two enzymes were positively correlated within males. In a multiple regression analysis, trunk muscle CS activity was positively correlated with calling rate after accounting for the effects of temperature and body mass. Differences in calling rate among males of this species are consistent from night to night (B. K. Sullivan and Hinshaw 1990), and females are most likely to be attracted to males

**Fig. 5.17.** Two Panamanian treefrogs with different calling behavior and different strategies of energy use. (A) *Scinax rubra* is an explosive breeder. Males call at high rates for only one or two nights after heavy rains and deplete intramuscular glycogen reserves very quickly. (B) *Scinax boulengeri* breeds continuously during the rainy season. Males call at low rates and deplete their energy reserves relatively slowly, even though they start out with lower concentrations of glycogen in their muscles. Photos by Kentwood D. Wells.

with high calling rates (Forester, Lykens, and Harrison 1989). If at least some of the variation in physiological traits that underlies variation in calling performance is heritable, then a female preference for high calling rate would result in selection for morphological and physiological traits that enable males to call at high rates.

## Energy Substrates for Call Production

The enzymatic profiles of muscles involved in call production indicate that both lipids and carbohydrates are important sources of energy for calling frogs. In other vertebrates, sustained aerobic exercise generally is supported by a mixture of lipid and glycogen stored in the muscles. Glycogen provides energy at a faster rate than lipids, and therefore provides greater power output. Glycogen stores are more limited, however, and provide less energy per gram of storage product, so glycogen reserves are depleted much more rapidly. Lipids provide more energy per gram, so stores of lipids are depleted more slowly. Depletion of muscle glycogen reduces muscle performance, even when abundant lipid stores remain, so glycogen represents a short-term constraint on muscular activity (Holloszy and Booth 1976; Hochachka and Somero 1984; Guppy 1988; J.-M. Weber 1988). The relative importance of glycogen and lipids as substrates to support activity varies among species, with lipids being most important in animals that sustain high levels of activity over long periods of time. For example, highly aerobic mammals that are capable of long-distance running, such as dogs, rely much more on intramuscular lipid stores to support vigorous exercise than do less aerobic species that do not move rapidly over long distances, such as opossums (J.-M. Weber 1992). The same is true for acoustic insects. Species with high calling rates typically derive most of their energy for sound production from lipids, whereas those with low calling rates sometimes rely entirely on carbohydrates (Prestwich 1994; Prestwich and O'Sullivan 2005).

The laryngeal and trunk muscles of frogs are well supplied with lipid stores in the form of droplets closely associated with clusters of mitochondria (fig. 5.12; Eichelberg and Schneider 1973; Eichelberg and Obert 1976; Schneider 1977, 1986; R. Marsh and Taigen 1987; Ressel 1993; McLister, Stevens, and Bogart 1995). A comparative study of several Neotropical hylids and leptodactylids (fig. 5.18) showed that lipid content of trunk muscles is positively correlated with calling rate, and this appears to be true for frogs in general (Ressel 1996). One Neotropical hylid with a very low calling rate, *Agalychnis callidryas*, has virtually no lipid in its trunk muscles, and the same is true for two North American ranids that also have low calling rates, *Rana sylvatica* and *R. clamitans*. In contrast, male spring peepers (*Pseudacris crucifer*) have more lipid in their trunk muscles than is found in most other highly aerobic vertebrate muscles (Ressel 1993, 2001).

In species with abundant lipid reserves, changes in lipid content of trunk muscles generally are not detectable on a nightly basis. Of the seven Neotropical hylids and leptodactylids studied by Bevier (1997b), only *Phrynohyas venulosa* exhibited a significant reduction in trunk-muscle lipid stores after two to three hours of calling, and this species started with relatively small lipid stores. *Hyla microcephala*, which had initial lipid stores about three times as large, showed no change after two hours of calling. Spring peepers (*Pseudacris crucifer*) undergo a dramatic seasonal reduction in trunk-muscle lipid reserves. Males collected early

**Table 5.6** Comparison of behavioral, morphological, and biochemical characteristics of two Neotropical hylid frogs

| Characteristics | *Scinax boulengeri* | *Scinax rubra* |
|---|---|---|
| Behavioral characteristics | | |
| Calling rate (calls/h) | 474 | 2,064 |
| Hours of calling/night | 3 | 5 |
| Morphological characteristics | | |
| Body length (mm) | 40 | 41 |
| Body mass (g) | 3.4 | 3.8 |
| Trunk muscle (% body mass) | 7 | 10 |
| Trunk muscle mitochondria (% fiber volume) | 15 | 21 |
| Trunk muscle capillary length per fiber volume (mm) | 774 | 1,389 |
| Trunk muscle lipid (% fiber volume) | 2.5 | 8.7 |
| Biochemical characteristics | | |
| Trunk muscle lipid (% dry tissue) | 4.7 | 16.4 |
| Trunk muscle glycogen (mg/g) | 3.6 | 7.9 |
| Trunk muscle CS activity (U/g) | 47 | 98 |
| Trunk muscle HOAD activity (U/g) | 13 | 15 |

*Source:* Modified from Bevier (1997a).

*Notes: Scinax boulengeri* breeds continuously for much of the rainy season. *Scinax rubra* breeds opportunistically after heavy rains.

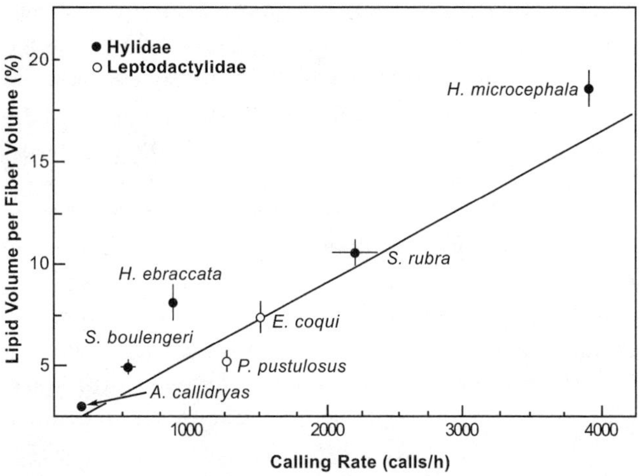


**Fig. 5.18.** Relationship of trunk muscle lipid reserves to calling rate in tropical frogs. Lipid volumes were estimated from transmission electron micrographs like those in fig. 5.12. Calling rates are estimated average calling rates at natural field temperatures. Genera as in fig. 5.14 and 5.15. After Ressel (1996).

in the season have larger reserves than any other species of frog that has been studied, but most of these reserves are gone by the end of the season (McKay 1989; Ressel 1993). Because this species breeds in cold weather, males do not begin feeding until relatively late in the breeding season (Wells and Bevier 1997). This is not true for *Hyla versicolor,* however. This species breeds in warmer weather and feeds throughout the breeding season, so it probably is less dependent on stored lipid reserves than are species that breed in cold weather (S. Walker 1989; Wells 2001a; see subsequent discussion of feeding). Less is known about depletion of energy substrates in laryngeal muscles. Eichelberg and Obert (1976) reported that lipid reserves in the laryngeal muscles of *Bombina bombina* were not significantly reduced by several hours of electrical stimulation, but were reduced after nearly two weeks of calling activity.

Both the trunk and laryngeal muscles of male frogs contain substantial quantities of glycogen (Schneider 1977; Bevier 1997b; Schwartz, Ressel, and Bevier 1995; Wells et al. 1995; Wells and Bevier 1997). Among Neotropical hylids and leptodactylids, initial trunk-muscle glycogen stores were positively correlated with both hourly calling rates and the total number of calls produced in an evening (Bevier 1997b). Rates of glycogen depletion also are correlated with calling rate. Species with relatively high calling rates, such as *Hyla microcephala, H. versicolor, Scinax rubra* and *Physalaemus pustulosus,* can deplete 50–60% of their trunk-muscle glycogen reserves after only two hours of calling (fig. 5.19). Species with lower calling rates, such as *Scinax boulengeri,* use up their trunk-muscle reserves more slowly (Bevier 1997b; Schwartz, Ressel, and Bevier 1995; Wells et al. 1995). In most of these species, liver glycogen reserves show little or no change after two hours of calling, suggesting that these reserves are not used to support call production.

The rapid depletion of glycogen reserves relative to lipid reserves might lead one to conclude that glycogen is a more important source of energy for call production. For many

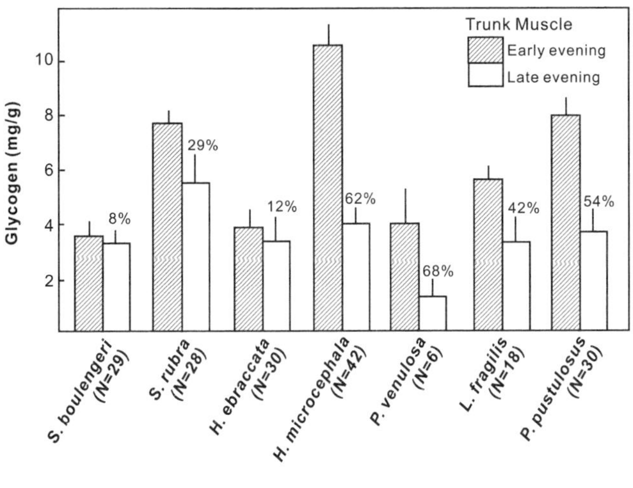


**Fig. 5.19.** Glycogen concentrations in trunk muscles of males for seven species of frogs from Panama. Bars show means ± 1 SE for individuals collected early in the evening and those collected after two to three hours of calling. Percentages show percentage decrease in glycogen concentration after calling. Genera as in fig. 5.14. After Bevier (1997b).

**Table 5.7** Estimated contributions of carbohydrates and lipids to call production in frogs (%)

| Species (source no.) | Carbohydrates | Lipids |
|---|---|---|
| *Hyla arborea*[a] (6) | 69 | 31 |
| *H. microcephala*[b] (2) | 14 | 86 |
| *H. versicolor*[b] (1) | 25 | 75 |
| *H. versicolor*[a] (4) | 40 | 60 |
| *Pseudacris crucifer*[b] (5) | 10 | 90 |
| *Physalaemus pustulosus*[b] (3) | 7 | 93 |
| *Hyperolius marmoratus* | | |
| Before feeding[a] (2) | 13 | 87 |
| After feeding[a] (2) | 63 | 37 |

*Sources:* (1) Wells et al. 1995; (2) Grafe 1996a; (3) Bevier 1997b; (4) Grafe 1997b; (5) Wells and Bevier 1997; (6) Grafe and Thein 2001.

[a]Estimates derived from measurements of respiratory quotients.

[b]Estimates derived from measurements of metabolic rates and depletion of trunk muscle glycogen reserves.

species with high calling rates, however, this is not the case. Because lipids yield much more energy per gram of stored substrate, even a small amount of lipid can account for most of the energy used for call production. The relative importance of glycogen and lipids has been calculated for several species by measuring rates of glycogen depletion and calculating the amount of oxygen used to oxidize glycogen. The remaining oxygen consumed by a calling frog is used to oxidize lipid, so lipid utilization can be estimated by subtraction. Such calculations have shown that most frogs with relatively high calling rates derive 75–95% of their energy from lipids (table 5.7), even though glycogen reserves can be depleted on a single night. These values are mostly consistent with estimates derived from measurements of respiratory quotients (RQ = the ratio of $CO_2$ produced to $O_2$ consumed). Grafe (1997b) measured RQs for calling *Hyla versicolor* males and estimated that they derive at least 60% of their energy for calling from lipids. In *Hyperolius marmoratus,* RQs varied dramatically, depending on whether the frogs had recently fed. Calling males derived 37–89% of their energy from lipids, and they relied much more on stored lipids if they had not fed recently (Grafe 1996a). Grafe and Thein (2001) measured RQs for the European treefrog (*Hyla arborea*) and found that males derived nearly 70% of their energy for call production from carbohydrates, a result that is at odds with data for other frogs with high calling rates. Possibly the males used in the study had been feeding recently, resulting in a shift to greater use of carbohydrates, as in *Hyperolius marmoratus.*

## Feeding and Strategies of Energy Use by Calling Frogs

For species that breed in warm weather, feeding probably is essential if males are to replace depleted energy reserves. Males either feed during hours when choruses are not active, or they leave a chorus periodically to feed. Males of summer breeding frogs in the temperate zone, such as *Rana clamitans* and *Hyla versicolor,* typically have food in their stomachs throughout the breeding season (Jenssen and Klimstra 1966; Wells et al. 1995; Wells 2001a). The foraging time of some calling males probably is limited by the amount of time devoted to calling or defending territories (see the following discussion). Daily food intake probably is converted mostly into glycogen and glucose, which can be utilized for calling within a few nights after a male has fed. This would explain the relatively low reliance on stored lipids for call production by *Hyperolius marmoratus* males that had recently fed.

Several investigators have experimentally supplemented the food of individual male frogs to determine whether food availability limits call production. For example, Sinsch (1988b) studied a high-altitude population of *Gastrotheca marsupiata* in Peru and found that males given supplemental feedings in the laboratory called at higher rates than did those that were not fed. In a study of *Physalaemus pustulosus* in Panama, A. Green (1990) did not find any effect of supplemental feeding on calling activity or the time required for males to return to a chorus. In contrast, Marler and Ryan (1996) found that males of this species given supplemental food were more likely to be calling during subsequent census periods than were males that were not fed. Unfortunately, neither of these studies quantified calling rates or total call production in males that were fed and not fed. Although *Physalaemus pustulosus* breeds more or less continuously throughout the rainy season in Panama, the most active choruses occur on nights after heavy rains (M. Ryan, 1985b). Males probably depend mainly on stored energy reserves to support calling. This species depletes a large proportion of its carbohydrate reserves on a single night of calling, but it apparently derives most of its energy for calling from stored lipids (Bevier 1997b). Feeding might enable males to conserve their lipid reserves by decreasing reliance on stored lipids, as shown for *Hyperolius* (Grafe 1996a), while having only a small effect on call output.

Another frog with a prolonged breeding season is *Hyla gratiosa,* a summer-breeding treefrog from the southeastern United States. Choruses are formed from March through August, and some males remain in choruses for a month or more. Median chorus tenure, however, is much shorter, only two to three days in most years; about 75% of males are present for less than a week. This pattern of chorus attendance suggests that males are energy-limited. Indeed, C. Murphy (1994a, b) found that males suffered reductions in body mass throughout their time in a chorus. Males that remained longer in the chorus were in better condition and lost mass more slowly than those that were present for only a few nights. Males in good condition also were more successful in acquiring mates, both because they were present on more nights, and because they were more likely to attract mates when they were present. Feeding during the breeding season was important for maintaining condition, and males that received supplemental food returned to the chorus more quickly and stayed longer than those that were not fed (C. Murphy 1994b).

In contrast, neither food intake nor body condition had much effect on the number of hours that a male *H. gratiosa* called on any given night (C. Murphy 1999). This suggests that many males probably stop calling well before they have exhausted their energy reserves, perhaps because females are unlikely to continue arriving at the pond. This is consistent with data for *Hyla versicolor* and for many tropical hylids, showing that most males still had significant reserves of glycogen available late in the evening when calling activity was winding down (Schwartz, Ressel, and Bevier 1995; Wells et al. 1995; Bevier 1997b). Food intake in these species probably is important mainly because it allows males to be active in a chorus on many nights, and male chorus attendance is the major determinant of mating success in many species (see chapter 8).

Two studies have examined the impact of supplemental feeding on ranid frogs with low calling rates and long chorus tenure. Judge and Brooks (2001) provided male bullfrogs (*Rana catesbeiana*) with a single meal (two chicken hearts) early in the breeding season. Feeding did not have any effect on chorus tenure or the length of time males occupied territories, but there was a tendency for males in better condition to have longer chorus tenure. Possibly the amount of supplemental food provided to the frogs was not sufficient to produce a significant change in their behavior. N. Gordon (2004) provided male green frogs (*R. clamitans*) with supplemental meals of insects throughout the breeding season. He did not find a significant effect on the amount of time males occupied territories or the vigor with which they responded to playbacks of calls. He did find that males that had been fed tended to move around the pond less than did those that had not been fed. This suggested that males responded to supplemental feeding mainly by reducing the time spent foraging rather than by increasing energy expenditure for territory defense.

Frogs that breed in cold weather face a special energetic challenge, because temperatures often are too low for their prey to be active or for frogs to digest food. Wood frogs (*Rana sylvatica*) and other early spring breeders do not feed at all (see discussion of "Energy Cycles of Males in Explosive Spring-Breeding Anurans"). Wood frogs lack significant lipid reserves, either in the trunk muscles (Ressel 1993) or in abdominal fat bodies (Wells and Bevier 1997). Consequently, they must rely entirely on glycogen reserves to support calling. At the beginning of the breeding season, trunk-muscle glycogen reserves of wood frogs are larger than are those of any other species of frog that has been studied. Despite a relatively low calling rate (table 5.2), these reserves are quickly depleted (fig. 5.20), and in some years, all trunk-muscle reserves probably are gone within a week (Wells and Bevier 1997). Nevertheless, wood frogs can afford to depend entirely on carbohydrates for calling because of their low calling rates and explosive breeding periods.

Spring peepers (*Pseudacris crucifer*) also begin breeding at temperatures too low for feeding, but in contrast to wood frogs, males collected early in the season have very little glycogen stored in either their trunk muscles or livers. Early-season males probably depend almost entirely on their enormous

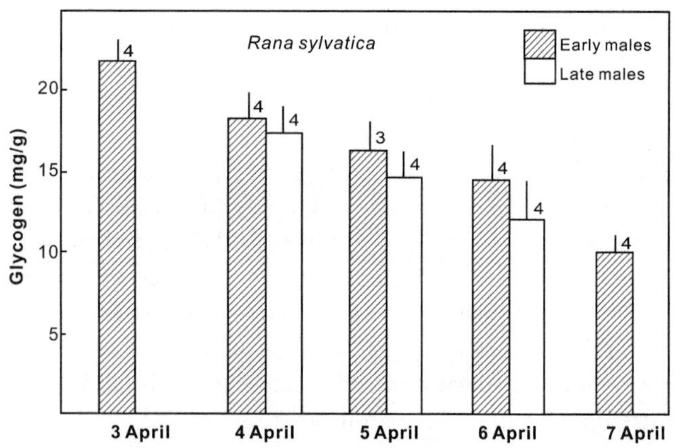


**Fig. 5.20.** Glycogen concentrations in trunk muscles of male wood frogs (*Rana sylvatica*) from Connecticut collected early in the evening and after 2 hours of calling. Bars show means + 1 SE. Numbers show sample sizes. After Wells and Bevier (1997).

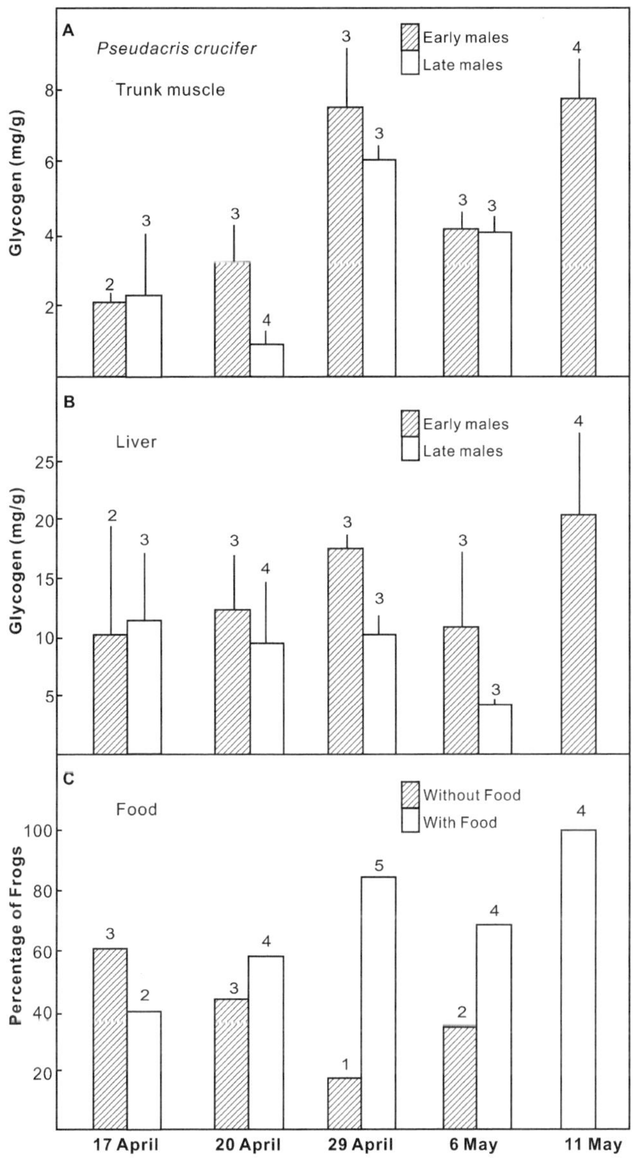


**Fig. 5.21.** Seasonal changes in (A) trunk-muscle glycogen concentration, (B) liver glycogen concentration, and (C) percentage of males with food in the digestive tract for calling male spring peepers (*Pseudacris crucifer*) in Connecticut. Bars show means + 1 SE. Numbers show sample sizes. After Wells and Bevier (1997).

trunk-muscle lipid reserves, and indeed, Wells and Bevier (1997) could not detect any decrease in glycogen reserves after two hours of calling on some nights. As the season progresses, and temperatures increase, calling rates and energy expenditure by male peepers increase (Wells, Taigen, and O'Brien 1996), probably resulting in an accelerating rate of lipid depletion. Insects also are more readily available, and males can partially compensate for lost energy reserves by feeding. Wells and Bevier (1997) found that the amount of food in the stomachs of males increased late in the season, and this was paralleled by an increase in trunk-muscle and liver glycogen reserves (fig. 5.21). It seems likely that late-season males shift to a greater dependence on carbohydrates for calling as their lipid reserves are depleted, much as *Hyperolius* males shift to using carbohydrates after feeding (Grafe 1996a).

Male frogs probably employ a range of energy use strategies during the breeding season similar to that described for breeding females in many kinds of animals. In the literature of life-history theory, animals often are classified either as capital breeders or income breeders (Bonnet, Bradshaw, and Shine 1998). Capital breeders depend on stored energy reserves to support breeding activities, whereas income breeders rely heavily on current food intake to support reproduction. It seems likely that many male frogs pursue a mixed strategy, relying both on stored energy reserves and food intake to support call production (Wells 2001a).

Table 5.8 summarizes the information on strategies of energy use currently available for two temperate-zone ranids and two hylids. Although this summary is somewhat speculative, it provides a useful framework for understanding strategies of energy use in other frogs. Two species that breed in cold weather, *Rana sylvatica* and *Pseudacris crucifer*, are primarily capital breeders, because they either do not feed at all, or begin feeding late in the breeding season. The stored energy substrates are different, however, with *R. sylvatica* relying entirely on glycogen reserves, and *P. crucifer* relying mostly on lipids. Two summer-breeding frogs, *Rana clamitans* and *Hyla versicolor*, are closer to the income breeder end of the spectrum, because both species are known to feed throughout the breeding season. The precise contribution of food intake to their reproductive energy budgets is not known. Despite the lack of effect of supplemental feeding on the territorial behavior of male green frogs studied by N. Gordon (2004), it seems likely that *R. clamitans* males are heavily dependent on food intake, because they have limited glycogen reserves and apparently do not use fat bodies to support call production (Wells 2001a).

**Table 5.8** Hypothesized strategies of energy use in four species of temperate-zone frogs

| | Source of energy | | | |
|---|---|---|---|---|
| Species | Lipids | Glycogen | Food | Energy-use strategy |
| *Rana sylvatica* | | ++++ | | Rely on stored glycogen |
| *Pseudacris crucifer* | ++++ | ++ | + | Mostly use stored lipids |
| *Hyla versicolor* | +++ | ++ | ++ | Mixed strategy: stored lipids, glycogen, and food |
| *Rana clamitans* | | + | ++ | Mostly rely on current food intake |

*Source:* From Wells (2001).

*Notes:* The number of plus signs gives a qualitative estimate of the relative importance of each source of energy for call production. A blank cell indicates that a particular source does not contribute significantly to call production.

### Hormones, Calling, and Energetic Constraints

Both the calling behavior of frogs and the physiological and morphological traits that support call production ultimately are under hormonal control. For example, both testosterone and arginine vasotocin (AVT) are known to stimulate calling behavior in frogs (Boyd 1994; Marler, Chu, and Wilczynski 1995; Solis and Penna 1997; Emerson and Boyd 1999; Burmeister, Somes, and Wilczynski 2001; Wilczynski and Chu 2001; Ten Eyck 2005). Androgens also affect the size and function of muscles involved in call production. For example, experimentally administered testosterone can increase the size of both laryngeal muscles (Kelley 1996) and trunk muscles (Emerson et al. 1999), and castration of males can decrease muscle size. These physiological responses are correlated with a greater expression of androgen receptors in the laryngeal and trunk muscles of males than in females (Kelley et al. 1989; Emerson et al. 1999). Testosterone also regulates seasonal changes in contractile properties of the trunk muscles involved in call production (Girgenrath and Marsh 2003).

Despite the obvious links between hormones and calling behavior, we know relatively little about how hormones affect the energetics of calling in frogs. Measurements of androgen and corticosteroid levels of calling males in natural choruses typically show wide variation both within and among species (Emerson and Hess 1996). Such variation appears to be related to variation in morphological and physiological traits that in turn lead to variation among individuals or among species in calling performance. Emerson (2001a) proposed a model that links levels of androgens and corticosteroid hormones to the energetics of call production. She hypothesized that seasonal development of trunk muscles involved in call production is under androgen control. High levels of testosterone are known to increase muscle mass and also alter contractile properties of muscle tissue in some vertebrates. As testosterone levels rise early in the breeding season, the mass of both trunk and heart ventricle muscle probably increases. As males begin calling, testosterone levels increase further in response to the calling of other males and their own calling behavior. At the same time, levels of corticosteroid hormones, which typically respond to stress, begin to rise as well. Some corticosteroid hormones, specifially glucocorticosteroids, can stimulate gluconeogenesis, thereby providing additional energy for call production. At very high levels, corticosteroid hormones counteract the effects of testosterone and AVT and inhibit reproductive behavior, perhaps causing males to stop calling and resume foraging. This could occur even before stored energy reserves are completely exhausted. Indeed, males of some species appear to stop calling well before energy reserves are completely gone (Wells et al. 1995).

The limited data available on hormone levels of frogs in natural choruses are consistent with this model. For example, there typically is a positive correlation between testosterone and corticosteroid hormone levels among calling individuals of the same species (Emerson and Hess 1996, 2001), even though very high levels of corticosteroids can inhibit calling (Marler and Ryan 1996; Burmeister, Somes, and Wilczynski 2001). In at least one species, *Pseudacris triseriata,* levels of both types of hormones increased during a single evening of calling and were positively correlated with each other (Emerson and Hess 2001).

Levels of these hormones also tend to be positively correlated among species, with the highest levels being found in species with high calling rates (Emerson and Hess 1996). In a detailed analysis of several species of Panamanian frogs previously studied by other workers, Emerson and Hess (2001) found that corticosteroid levels were positively correlated with both average calling rate and relative energetic investment in call production. *Hyla microcephala,* which has among the highest energetic investment in calling of any frog studied to date (Wells and Taigen 1989), had the highest hormone levels. *Scinax boulengeri,* which invests much less energy in call production (Bevier 1997a, b), had the lowest levels. *Physalaemus pustulosus* and *Hyla ebraccata* were intermediate. Thus, relative corticosteroid hormone levels not only were correlated with calling performance, but also with the expression of traits related to call production, including mitochondrial volume density (fig. 5.13 A), capillary density (fig. 5.13 B), citrate synthase activity (fig. 5.15), and intramuscular lipid stores (fig. 5.18).

## Patterns of Energy Storage and Use

The energy substrates stored in trunk muscle are just one component of energy storage and use in amphibians. Other major energy storage sites include abdominal, cutaneous, and subcutaneous fat bodies, the tail (in urodeles), and the liver (a depot for storage of both glycogen and lipids; Hadjioloff 1930; Fitzpatrick 1976; G. Smith 1976; C. B. Jørgensen, Larsen, and Lofts 1979; Seale 1987; Wygoda, Garman, and Howard 1987; C. B. Jørgensen 1992b). Because of their relatively low maintenance energy requirements, amphibians can rapidly accumulate stored energy reserves during one part of the year for use at another time. They also can shift allocation of energy reserves between reproduction and growth according to seasonal demands and the availability of food, although not all species show a tradeoff between reproduction and growth (Lardner and Loman 2003).

### Comparison of Energy Requirements of Males and Females

Maintenance requirements of males and females probably do not differ dramatically for most amphibians, but the energetic demands imposed on the two sexes by reproduction differ both qualitatively and quantitatively. For most species, the major cost of reproduction in females almost certainly is the production of eggs, although other activities such as searching for mates, nest building, brooding of eggs, and feeding tadpoles (in some dendrobatid frogs) add to reproductive costs. For males, the cost of sperm production probably is relatively small. Laboratory studies of males deprived of food have shown that moderate levels of starvation have little or no effect on spermatogenesis in urodeles (Ifft 1942) or anurans (Guha, Jørgensen, and Larsen 1980). Probably the major costs for males are associated with locating, attracting, courting, or fighting over females.

Only limited data are available on reproductive costs for males and females in urodeles. In a study of *Desmognathus ochrophaeus,* Fitzpatrick (1973a) estimated that egg production and brooding consumed about 48% of a female's total annual energy budget, with egg production accounting for 65% of all reproductive costs. The cost of courtship, which Fitzpatrick did not consider, is trivial by comparison (measured for the closely related *D. carolinensis* by Bennett and Houck 1983). Finkler and Cullum (2002) found that metabolic rates of gravid females of *Ambystoma texanum* were double those of males, and free glucose levels in the blood were higher in females as well. They concluded that females have considerably higher costs of reproduction than do males, although they did not measure metabolic rates during activity. In general, it seems likely that reproductive costs for male salamanders (fighting with other males, courtship, and spermatophore production) are small compared to those of females (Bennett and Houck 1983; Marks and Houck 1989).

Most attempts to compare reproductive energy budgets for male and female anurans have not been very precise, because most studies lacked critical data for either males or females. Presumably the high cost of calling in frogs means that reproduction consumes a greater proportion of a male's annual energy budget than it does in salamanders. Nevertheless, in some species, reproductive costs probably are higher for females than for males (Pough et al. 1992). In *Physalaemus pustulosus,* the energy content of eggs is approximately 4 kJ per clutch, while the cost to the female of beating the eggs into a foam nest was estimated to be only 3% as much (0.13 kJ; M. Ryan, Bartholomew, and Rand 1983). The number of clutches produced per year by this species is not known, so it is impossible to calculate the proportion of a female's yearly energy budget that goes into reproduction. Nevertheless, M. Ryan, Bartholomew, and Rand estimated that females that lay between one and 10 clutches per year expend about 1.2–12 times more energy on reproduction than males. More recent measurements of oxygen consumption by calling males in this species (Pough et al. 1992) and average calling rates in the field (Bevier 1995a) suggest that M. Ryan, Bartholomew, and Rand underestimated the cost of calling, perhaps by a factor of two, so their energy budgets probably need to be adjusted.

Grafe, Schmuck, and Linsenmair (1992) used laboratory measurements of oxygen consumption in calling males and egg production by females in *Hyperolius viridiflavus* to calculate total reproductive costs per season. They estimated that female expenditures were about two to three times those of males, but they lacked field data on seasonal variation in male calling effort and female egg production. Mac Nally (1981, 1984b) measured metabolic rates and changes in energy reserves in two species of *Crinia* and estimated that reproductive costs were similar for the two sexes (females apparently produce only a single clutch of eggs each year). He did not have measurements of oxygen consumption in calling frogs, so he used an estimate based on work with *Physalaemus* (Bucher, Ryan, and Bartholomew 1982), a risky procedure given the range of variation among species. His energy budget probably underestimated the cost of calling, because he assumed a five-fold increase in metabolic rate for calling males, much lower than that measured for most species studied to date. Woolbright (1985a) attempted to construct energy budgets for males and females of the Puerto Rican treefrog *Eleutherodactylus coqui.* He estimated that males expend more energy per season on reproduction than do females, but his conclusions must be considered tentative, because he did not have complete data on the costs of calling in males, seasonal and individual variation in male calling effort, or frequency of egg production by females.

## Seasonal Cycles of Energy Allocation

Although information on annual expenditures for reproduction remains sketchy at best, it seems likely that for adult amphibians, reproduction is the single largest component of the annual energy budget, probably exceeding energetic investment in growth and routine activities. This does not mean that all amphibians are energetically stressed by their reproductive activities. If food is readily available, amphibians probably have little difficulty in replacing energy reserves depleted by reproduction. Furthermore, the degree to which reproductive activities themselves are limited by energetic constraints will depend on the timing of reproduction, the length of the breeding season, the nature of the mating system, and a variety of other environmental conditions. Limited data on changes in energy reserves throughout the year suggest that several distinct energy allocation patterns can be found among anurans. Few comparable data are available for urodeles.

### Energy Cycles of Explosive Spring-Breeding Anurans

Some anurans from the Northern Hemisphere breed in early spring, often only a few days after emerging from their winter retreats. Breeding in these species tends to be explosive, with males remaining at breeding sites for only a few days (see chapter 8). These frogs must accumulate enough stored energy to survive the winter and carry them through the breeding season. In two early breeding European species, *Rana temporaria* and *Bufo bufo,* males usually do not feed at all during the breeding period and are unresponsive to prey that is placed in front of them (Juszczyk 1950; C. L. Smith 1950; Mazur 1967; Heusser 1968a, 1969c; Ashby 1969; Itämies and Koskela 1970; Blackith and Speight 1974; Pasanen and Koskela 1974; C. B. Jørgensen 1986, 1992b). In fact, individuals in some populations move to hidden retreats and become inactive for several weeks after the breeding period before beginning to feed in late spring and summer (Heusser 1968c; Ashby 1969). It seems likely that populations from far northern latitudes or high altitudes are particularly stressed energetically because cold temperatures often prevent the frogs from starting breeding activities until late spring. This leaves a relatively short postbreeding activity season in which individuals must replenish their depleted energy reserves before entering winter dormancy, which can last up to eight months in some regions (R. Savage 1961; Koskela and Pasanen 1974, 1975; M. Morton 1981; Beattie 1985; Terhivuo 1988; Elmberg and Lundberg 1991; Lardner and Loman 2003).

In these cold-weather breeders, stored energy reserves, including fat bodies (fig. 5.22), carcass lipids, and glycogen stores in the liver are at their lowest point during or immediately following the breeding period (C. L. Smith 1950; Mazur 1967; Krawczyk 1971; Pasanen and Koskela 1974; C. B. Jørgensen, Larsen, and Lofts 1979; M. Morton 1981; Guzik and Schimscheiner 1985; C. B. Jørgensen 1992b; Lu 2004). One might expect that frogs with very short breeding periods would not be subject to energetic constraints on their activities, but there is evidence of rapid depletion of energy reserves in several explosive-breeding species. Reductions in body mass of males during the breeding period have been reported in both *Rana temporaria* (A. Cooke 1981; Ryser 1989b) and *Bufo bufo* (Arak 1983c). In a population of *Rana temporaria* from northern Finland, males collected immediately after spawning had only half as much glycogen in their livers as those collected during the spawning period, and only a fourth as much as those collected immediately after emergence from winter dormancy (Pasanen and Koskela 1974). In another population from Britain, both males and females had virtually exhausted their liver glycogen stores and fat body lipids by the time breeding was completed (C. L. Smith 1950).

As discussed earlier, male wood frogs (*Rana sylvatica*) depleted trunk-muscle glycogen reserves after only a few days of calling. Liver glycogen reserves were not depleted, except at the end of the breeding season in a year when trunk-

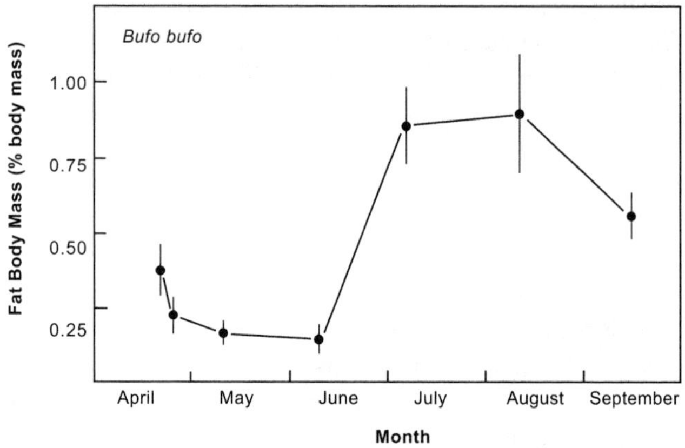


**Fig. 5.22.** Annual cycle of fat bodies in male common toads (*Bufo bufo*) from Denmark. Data show means ± 1 SE for males collected each month. Breeding occurred in April, immediately following emergence from winter dormancy. Plotted from data in Jørgensen, Larsen, and Lofts (1979).

muscle reserves were very low (Wells and Bevier 1997). This species accumulates enormous stores of liver glycogen during the fall. During the winter, these reserves are converted to glucose, which is released in high concentrations into the blood stream and is used as a cryoprotectant to prevent intracellular fluids from freezing (see chapter 3). When the animals thaw, glucose is reconverted to glycogen and stored in the liver. However, repeated cycles of freezing and thawing can deplete liver glycogen reserves (K. Storey and J. Storey 1987; Lee and Costanzo 1993), in part because some glucose is lost through excretion (Layne, Lee, and Cutwa 1996). As a result, frogs entering breeding choruses probably have only a fraction of the liver glycogen reserves that they had the previous fall (Wells and Bevier 1997).

Year-to-year variation in the duration of breeding activity in a given population can be affected by environmental conditions experienced months earlier. Because energy reserves are accumulated during summer and fall feeding periods, poor foraging conditions during these seasons reduce the energy reserves available for breeding. For example, Reading and Clarke (1995) found that toads (*Bufo bufo*) arrived at breeding ponds in unusually poor condition in a year following a hot, dry summer. Conditions immediately before breeding begins can be important as well. If winter and early spring temperatures are unusually warm, males probably arrive at breeding ponds with a substantial proportion of their energy reserves already exhausted (Ryser 1989b). In a 10-year study of a population of *Bufo bufo,* Reading and Clarke (1995) observed an increase in environmental temperatures in the month before breeding, and this was correlated with a decrease in body condition of both males and females at the breeding ponds. This was true even though breeding actually started earlier in the warmer years. Variation in the number of days that individual males spend in a chorus (Davies and Halliday 1979; Reading and Clarke 1983; R. D. Howard and Kluge 1985; Loman and Madsen 1986; Ryser 1989a; Elmberg 1990) also could be related to differences in stored energy reserves, but this has yet to be clearly established.

In early spring breeders, the breeding period is followed by an increase in feeding activity, probably in response to increasing temperatures and a greater availability of food. Feeding also is influenced by endogenous hormonal cycles (C. B. Jørgensen 1986, 1989, 1991; Jørgensen and Wind-Larsen 1987; Wind-Larsen and Jørgensen 1987). In *Rana temporaria,* the digestive tract actually undergoes a partial degeneration during winter dormancy, and this must be reversed before feeding can resume (Juszczyk, Obrzut, and Zamachowski 1966). Early in the feeding period, most of the ingested energy is allocated to growth and tissue repair, but by late summer, energy is increasingly shifted to stored reserves. A pattern found in a number of temperate-zone anurans is for lipid deposition in the carcass and fat bodies to peak in late summer or early autumn, well before the onset of winter dormancy. In contrast, deposition of glycogen reserves in both the liver and fat bodies usually takes place just before hibernation. In some species, glycogen reserves reach their peak even after the frogs become inactive (C. L. Smith 1950; Mizell 1965; Pasanen and Koskela 1974; Schlaghecke and Blüm 1978b; C. B. Jørgensen, Larsen, and Lofts 1979; Farrar and Dupré 1983; C. B. Jørgensen 1986, 1992a).

In environments where food resources are limited, energetic constraints probably are more severe for female anurans than for males, because allocation of energy to developing eggs competes with allocation to stored reserves that enable the animal to survive a period of dormancy. In early spring breeders, a new ovarian cycle of oocyte maturation usually does not begin until two to four months after spawning, when rebuilding of stored energy reserves is well underway (Gallien 1940; Ting and Boring 1940; C. L. Smith 1950; Mizell 1964; Jastrzebski 1968; Krawczyk 1971; Juszczyk and Zamachowski 1973; C. B. Jørgensen 1973a, b, 1975, 1991; Jørgensen, Larsen, and Lofts 1979). In regions where the activity season is very short, females often cannot acquire enough energy to breed every year and have a biennial or irregular reproductive cycle (Turner 1958; Licht 1975; Morton 1981). In many species, the process of vitellogenesis (provisioning of eggs with nutrients) is largely completed before winter dormancy (Su and Yu-lan 1963; C. B. Jørgensen 1973a, b, 1984a; Jørgensen, Hede, and Larsen 1978; Jorgensen, Larsen and Lofts 1979). In some species, the mass of the ovaries continues to increase during the winter (Lu 2004). C. B. Jørgensen, Larsen, and Lofts (1979) argued that this is due mainly to increases in fluid content and does not reflect transfer of energy from stored reserves to the eggs, whereas Lu (2004) argued that in *Rana chinensis,* the increase in ovarian mass is correlated with decreases in the mass of the liver and fat bodies and therefore represents actual transfer of energy reserves to the eggs.

### Energy Cycles of Desert Anurans

Desert-dwelling anurans face many of the same energetic problems that confront early spring breeders, in that breeding often occurs immediately after emergence from dormancy. Furthermore, the frequency, seasonal occurrence, and duration of activity periods generally are less predictable for xeric amphibians than for early spring breeders, because the former respond to sporadic rainfall rather than regular seasonal changes in temperature (Bragg 1945, 1961, 1965; Main, Littlejohn, and Lee 1959; Slater and Main 1963; Mayhew 1965, 1968; Main 1968; Low 1976). Consequently, these animals must store sufficient energy to carry them through dormancy periods of uncertain duration and still have enough reserves to support breeding activities. Most of these species tend to be highly explosive breeders (Bragg 1965; Wells 1977a; B. K. Sullivan 1982b, 1985b; Woodward 1982c), but the energetic costs of reproductive behavior are unknown.

Some desert anurans have extraordinary abilities to survive for long periods without food. Spadefoot toads (*Scaphiopus* and *Spea*) in the southwestern United States sometimes go for a year or more without emerging from their burrows if rainfall is scarce (Bragg 1965; Mayhew 1965). Seymour (1973a) estimated that most individuals in a population of *Scaphiopus couchii* could survive at least nine months on the energy in their fat bodies alone, and some could survive up to two years by metabolizing other energy reserves. Even more remarkable are the survival abilities of water-holding frogs (*Cyclorana platycephala*) from the deserts of central Australia. Using data on stored energy reserves and metabolic rates of dormant frogs, Van Beurden (1980) estimated that all individuals in a population could survive at least 14 months of dormancy, about 20% could survive more than three years, and a few could survive up to nine years without food.

The time available for feeding is unpredictable for desert-dwelling species, so there is a premium on rapid acquisition of food to replenish depleted energy reserves. Couch's spadefoot toads (*Scaphiopus couchii*) can consume insects equivalent to 55% of their own body mass in one feeding. In Arizona, between 57 and 75% of food items in stomachs were winged termites, which emerged after the same heavy rains that triggered breeding by the spadefoot toads (Dimmit and Ruibal 1980). Termites comprised about half of the diet of *Spea multiplicata* as well. In an earlier study, Whitaker, Rubin, and Munsee (1977) reported that termites were the most important prey of *S. couchii* in both Texas and Arizona. Termites are an especially high quality food source because they are rich in lipids and contain relatively small amounts of indigestible material. Dimmitt and Ruibal (1980) estimated that only one or two meals of termites would be required to sustain *Scaphiopus couchii* through very long periods of dormancy in an extremely arid environment.

Most other spadefoot toads live in less arid habitats and do not consistently specialize on termites or other lipid-rich insects. Carabid beetles were the most important prey items for *Spea bombifrons* and *S. multiplicata* in Texas (A. M. Anderson, Haukos, and Anderson 1999a), and *S. bombifrons* from Colorado ate large numbers of these beetles as well (Whitaker, Rubin, and Munsee 1977). *Spea bombifrons* also ate many lepidopteran adults and larvae, and lepidopteran larvae were the most important prey items for *Spea hammondii* in Texas and *Scaphiopus holbrookii* in Indiana (Whitaker, Rubin, and Munsee 1977). In Florida, *S. holbrookii* ate mostly beetles and ants (Pearson 1955). European spadefoot toads (*Pelobates fuscus*) in Romania ate a wide variety of prey, including land snails, beetles, ants, and many other kinds of insects (Cogălniceanu et al. 1998). Much of the variation in diet composition reported in these studies probably is due to local variation in abundance of different kinds of insects, and does not necessarily reflect dietary specialization by particular species.

In contrast to many early spring breeding anurans that overwinter in cold habitats, some desert anurans and other species that are dormant at warmer temperatures continue to provision the eggs during dormancy, drawing on lipid reserves in fat bodies and the carcass (Berk 1938; F. Bush 1963; Seymour 1973a; D. Long 1987a, b). In *Spea multiplicata*, there was no increase in ovary mass or lipid content during dormancy, but individual follicles increased in size. These toads emerged from hibernacula well before breeding, and feeding during the prebreeding period apparently contributed to continued vitellogenic growth of eggs (D. Long, 1989).

### Energy Cycles of Anurans with Prolonged Breeding Periods

In anurans with prolonged breeding seasons, there is considerable evidence that males are energetically stressed by their reproductive activities. Reductions in body mass, lipid reserves, and the mass of energy-storing organs over the course of a breeding season have been reported for males of many species, including *Pseudacris crucifer* (McKay 1989; Lance and Wells 1993; Ressel 1993), *Hyla gratiosa* (C. Murphy 1994a, b), *Rana clamitans* (Jenssen 1972; Wells 1978a), *Rana catesbeiana* (Byrne and White 1975), *Rana virgatipes* (Given 1988a), *Rana palustris* (Given 2005), *Rana epeirotica* (Tsiora and Kyriakopoulou-Sklavounou 2001), *Rana esculenta* (Schlaghecke and Blüm 1978a, b), *Rana ridibunda* (Loumbourdis and Kyriakopoulou-Sklavounou 1991), *Uperoleia laevigata* (J. Robertson 1986b), *Bufo canorus* (M. Morton 1981), *Bufo calamita* (Arak 1983c; Tejedo 1992c), *Bufo fowleri* (Given 2002), *Bufo rangeri* (Cherry 1993), two species of *Crinia* (Mac Nally 1981; Lemckert and Shine 1993), and *Pelobates fuscus* (Eggert and Guyétant 2003).

In *Rana clamitans*, the mass of the fat bodies and liver decrease during the breeding season (fig. 5.23 A, B), with the former being almost depleted by the end of the summer (Jenssen 1972). Reduction in liver mass probably is due in part to a depletion of liver glycogen reserves, and trunk-muscle glycogen reserves are gradually depleted as well (Wells and Bevier, unpublished data). Males take in less food during the breeding season than in spring or fall (fig. 5.23 C). This in turn probably results from the amount of time

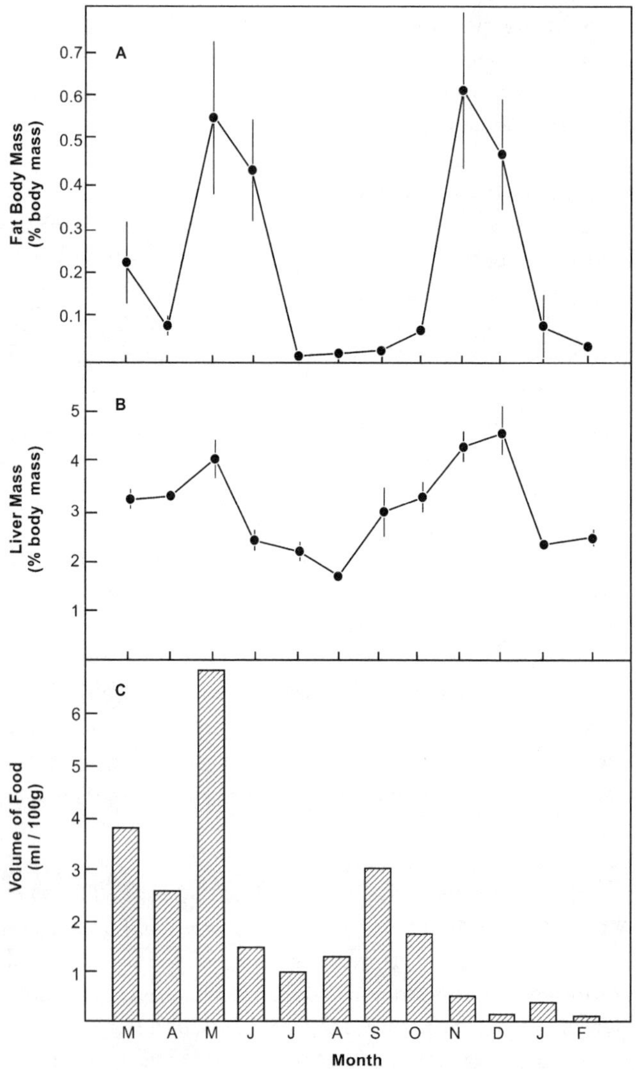


**Fig. 5.23.** Reproductive energetics of male North American green frogs (*Rana clamitans*), a species with a prolonged summer breeding season and territorial behavior. (A) Annual changes in mass of fat bodies. (B) Annual changes in mass of the liver. Data are shown as means ± 1 SE. (C) Mean volume of food in the stomach per 100 g of body mass for male frogs collected each month. Plotted from data in Jenssen (1972) for a population in southern Illinois.

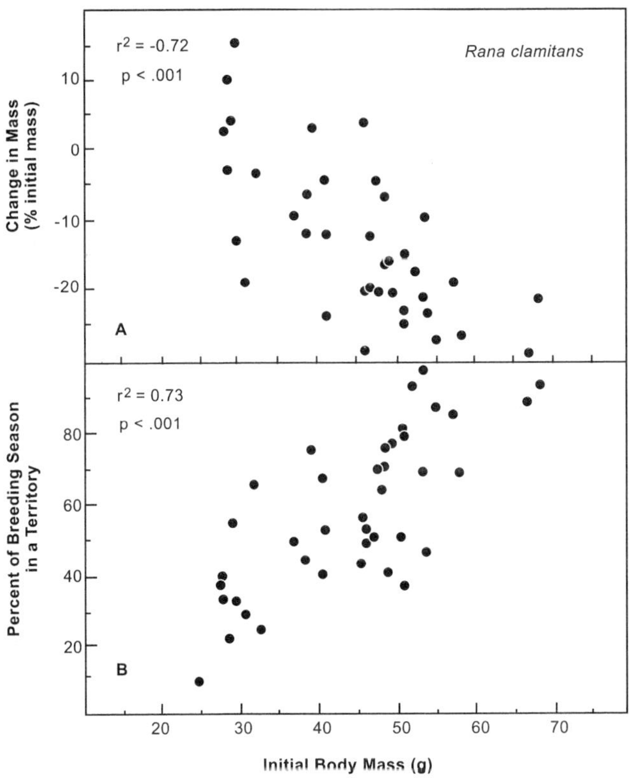


**Fig. 5.24.** Reproductive energetics of male North American green frogs (*Rana clamitans*). (A) Change in body mass as a function of initial body mass. (B) Percentage of the breeding season devoted to territorial defense as a function of initial body mass. Large males spent more time in territories and decreased more in body mass than did small males. Plotted from data in Wells (1976a) for a population in Ithaca, New York.

males invest in territorial defense. In one study (Wells 1976b, 1978a), large males suffered greater reductions in body mass than did small males (fig. 5.24 A), perhaps because they spent more time defending territories (fig. 5.24 B).

In contrast, Tejedo (1992c) reported that small *Bufo calamita* males lost mass at a faster rate than large males. Loss of mass was positively correlated with time spent in a chorus. Nevertheless, because large males had larger initial energy reserves, they remained in the chorus longer than did smaller males and achieved greater reproductive success. Somewhat similar results were obtained in a study of *Pelobates fuscus* in France. Males in good condition tended to arrive earlier and stay longer than those in poor condition, but suffered less total loss of mass (Eggert and Guyétant 2003). Cherry (1993) observed a steady reduction in body mass of breeding males in an African toad, *Bufo rangeri,* but males increased in mass when they left the chorus to feed. Loss of body mass increased as a function of calling rate.

In a summer-breeding frog from North America, *Rana virgatipes,* there was a direct tradeoff between calling rates of small males and growth rate during the breeding season, with some very active callers decreasing in mass, probably because of the combined effects of high energy expenditure and reduced feeding (Given 1988a). Calling also reduced feeding opportunities in a tropical frog from Puerto Rico, *Eleutherodactylus coqui.* Males seldom fed during peak calling times and obtained most of their food in the early morning hours (fig. 5.25). Consequently, food intake by calling males was substantially below that of noncalling males or females (Woolbright and Stewart 1987).

In females of temperate-zone anurans that breed during the summer, cycles of energy storage and use generally are similar to those of spring breeders, with the mass of energy-storing organs being greatest just before winter dormancy and reaching a minimum during or just after breeding. In contrast to early spring breeders, however, these species often emerge well before the breeding season begins and can replenish depleted energy reserves by feeding. In summer-breeding ranids, such as *Rana catesbeiana, R. clamitans, R. esculenta,* and *R. ridibunda,* vitellogenesis begins in the fall, but often is not completed until after spring emergence (Brenner 1969; Jenssen 1972; Byrne and White 1975; Rastogi et al. 1983; Kyriakopoulou-Sklavounou and Loumbourdis 1990). Furthermore, females that breed early in the season sometimes produce a second egg clutch later in the summer, with energy acquired by feeding during the breeding season (Wells 1976a, b; Emlen1977; Rastogi et al. 1983; R. D. Howard 1988a; Kyriakopoulou-Sklavounou and Loumbourdis 1990). Production of multiple egg clutches during summer breeding seasons also has been reported in a number of North American hylid frogs (Godwin and Roble 1983; Perrill and Daniel 1983; Ritke, Babb, and Ritke

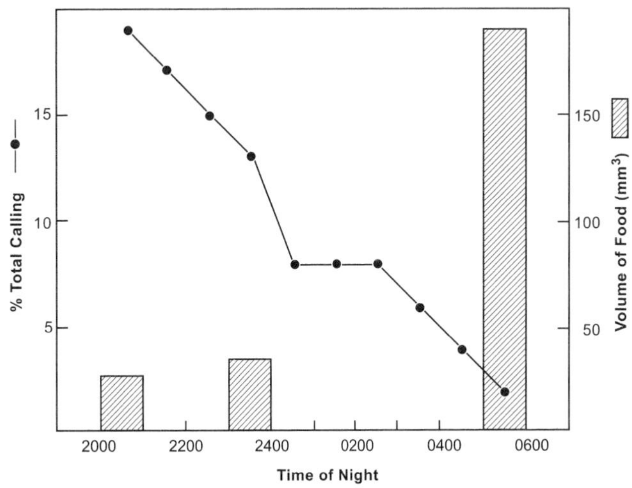


**Fig. 5.25.** Food intake of male *Eleutherodactylus coqui* from Puerto Rico in relation to calling behavior. Line shows proportion of total calling activity at different times of night for 14 males (plotted from data in Woolbright 1985b). Bars show volume of food in stomachs of male frogs collected at three different times of night (plotted from data in Woolbright and Stewart 1987). Males do most of their calling early in the evening and most of their feeding just before dawn.

1990) and a temperate-zone myobatrachid from Australia (Lemckert and Shine 1993). Females of these species probably also rely on feeding during the breeding season to provision their eggs (Ritke and Lessman 1994). Intervals between successive clutches often vary considerably among individual females, perhaps reflecting differences in foraging success (see table 7.15 in chapter 7).

In contrast to the ranid pattern, female cricket frogs (*Acris crepitans*) in Texas did not initiate ovarian development until after they emerged from winter dormancy, apparently utilizing stored lipids in fat bodies and replenishing them by feeding in the spring (D. Long 1987a). Still another pattern was found in *Bufo woodhousii,* which breeds opportunistically from spring through late summer. Ovarian development was initiated in the fall, but vitellogenesis occurred mainly during winter dormancy, drawing on stored energy reserves. Females generally spawned immediately after emergence, but retained some mature follicles throughout the summer, perhaps allowing them to deposit additional egg clutches during unpredictable periods of favorable weather (D. Long 1987a). A somewhat similar pattern was found in populations of *Bufo viridis* from Israel, which usually breed at the onset of winter rains, but also breed opportunistically at other times when water is available. Females feed throughout the year and can initiate vitellogenic growth of eggs at any time (C. B. Jørgensen 1984b).

In the tropics, anurans exhibit a variety of reproductive patterns, ranging from opportunistic breeding after heavy rains to continuous breeding during long rainy seasons (Crump 1974; Duellman and Trueb 1986; Donnelly 1989c; C. B. Jørgensen 1992b; Prado, Uetanabaro, and Haddad 2005; see also chapter 9). In many tropical species, ovarian cycles are asynchronous among females in a population (Church 1960a, b; Inger and Greenberg 1963; Berry 1964; Inger and Bacon 1968; C. B. Jørgensen, Shakuntala, and Vijayakumar 1986). Individual females sometimes lay several egg clutches each year (Senfft 1936; Chibon 1962; Pyburn 1966; Davidson and Hough1969; Crump 1974; Wells 1979, 1980a; Kluge 1981; K. Summers 1989; see table 10.5 in chapter 10 for examples). There is relatively little published information on patterns of energy storage and use in tropical anurans. It seems likely that continuous breeders in the tropics rely heavily on feeding during the breeding season to provision new egg clutches. In contrast, explosive or opportunistic species might be expected to rely at least in part on stored energy reserves. In several tropical species of *Bufo,* the size of fat bodies did not show a clear seasonal pattern. Nevertheless, fat bodies tended to be smallest in females with mature ovarian eggs or those that had just finished spawning, indicating that energy reserves are depleted during egg production (G. Alexander 1933; Church 1960a; Zug and Zug 1979; C. B. Jørgensen et al. 1986). In tropical anurans that breed in areas with pronounced wet and dry seasons, there does tend to be clear seasonal variation in fat body size. Fat bodies typically increase during periods of feeding, but decrease when eggs or sperm are being produced (Pancharatna and Saidapur 1985; Kanamadi et al. 1989; Kanamadi and Jirankali 1991, 1992). This pattern is not always seen in males, however. In some tropical species, variation in fat body size and testis size are not correlated (Saidapur, Kanamadi, and Bhuttewadkar 1989), perhaps because testicular activity is not necessarily directly related to the nutritional condition of males (Kanamadi and Saidapur 1988).

A comparative study of two populations of *Bufo marinus* in a wet and dry site in Venezuela revealed seasonal variation in fat body size, but the pattern differed at the two sites (Lampo and Medialdea 1996). At the wetter site, the toads were dry season breeders, laying eggs in pools along a river. Fat body size of females increased during the wet season. At the drier site, females did not show a clear peak of reproduction in any season, but fat body size did increase slightly in the wetter months. Males showed a similar pattern, but only at the dry site was fat body size inversely correlated with gonad size. In both sexes, fat body size was directly related to rainfall, which in turn was positively correlated with food intake. Toads at the wetter site allocated a larger proportion of their total body mass to fat bodies, indicating that they were more likely than those at the dry site to acquire energy in excess of that needed for reproduction.

In tropical India, frogs typically breed during monsoon rains, which occur either in a single period during the summer or in both summer and winter. In *Euphlyctis hexadactylus* from southern India, females with mature eggs were found throughout the year, but most breeding occurs in three seasonal peaks corresponding to rainy periods (Das 1996b). Fat body size in both males and females increased during the peak feeding months (the dry season), but declined sharply during breeding periods. Depleted fat reserves were rapidly replenished, increasing by up to 300% in about two months. In *Hoplobatrachus tigerinus,* there was only one peak of breeding during the summer monsoon rains (Girish and Saidapur 2000). Again, fat bodies increased during feeding periods and decreased in the breeding season. Supplemental feeding in the laboratory increased both the size of fat bodies and the number and size of ovarian eggs. Experimental removal of fat bodies did not diminish the number of eggs produced, but did result in smaller eggs. Removal of fat bodies from males resulted in regression of the testes (Kasinathan, Gunasing, and Basu 1978).

### Energy Cycles of Urodeles

We know very little about seasonal energy cycles in male salamanders or how energy reserves might limit their reproductive activities, because there has not been much work on seasonal changes in energy stores. Changes in body mass tend to be a poor indicator of nutritional condition, because salamanders often take up substantial amounts of water when they first enter a pond (Aoto 1950; Verrell and Halliday 1985b; Hasumi and Iwasawa 1990). Species that breed during warm weather can feed at breeding sites (Verrell 1985a) and actually can increase in mass if they remain in ponds for long periods of time (Verrell and Halliday 1985b). On the other hand, Verrell, Halliday, and Griffiths (1986) found that male *Triturus vulgaris* showed substantial reductions in the mass of the liver and abdominal fat bodies during the summer breeding season, but replenished their reserves during the fall. How much of their stored energy is used for production of sperm and spermatophores, and how much for aggression and courtship, has yet to be determined, but males do appear to be under some energy stress. In another European salamandrid, *Chioglossa lusitanica,* the mass of the liver and tail decreased in males during the breeding period; both of these organs are important for energy storage (Sequeira, Ferrand, and Crespo 2003). Males of another salamander, *Amphiuma means,* exhibited reductions in fat body mass and carcass lipids during the breeding season (F. Rose 1967), but the reproductive behavior of this species is completely unknown.

Patterns of energy storage and use by female salamanders have not been studied extensively, although there is good information for *Desmognathus ochrophaeus* (Fitzpatrick 1973a, 1976). It is known that lipids stored in both fat bodies and the carcass are mobilized for vitellogenesis (Adams and Rae 1929; Tachi 1963; F. Rose 1967; Rose and Lewis 1968; Lewis and Rose 1969; Fitzpatrick 1973a; D. Scott and Fore 1995). Generally, lipid reserves of temperate-zone salamanders follow the same annual cycle as in anurans, reaching a low point at the end of the reproductive period and increasing until just before winter dormancy. The amount of energy stored as lipids varies with food intake, and this in turn can affect the probability that a female will reproduce in a given year and the number of eggs that she lays (D. Scott and Fore 1995). In some terrestrial salamanders, the tail is a particularly important energy storage organ, sometimes comprising up to 50% of an animal's biomass. In *Chioglossa lusitanica,* the tails of females decrease in mass during the breeding period (Sequeira, Ferrand, and Crespo 2003), and females with autotomized tails are less likely to produce eggs than those with intact tails (Arntzen 1981; see also Maiorana 1977).

The eggs laid by many plethodontid salamanders take weeks or even months to develop and hatch (Houck 1977a). In terrestrial species, the female often remains in close proximity to her eggs, protecting them from predators and providing them with moisture (Crump 1995b; see also chapter 11). This form of parental care can severely limit the foraging of females and lead to depletion of stored energy reserves. Because brooding is relatively prolonged, this can account for a significant fraction of a female's energy budget. Fitzpatrick (1973a) estimated that *Desmognathus ochrophaeus* females used about 16% of their annual energy budget for brooding eggs, mostly derived from stored lipids. The energetic demands of brooding can reduce growth rates as well, and in some species, brooding females decline in overall body condition (Ng and Wilbur 1995). Salamanders that brood their eggs for relatively short periods of time, or at low temperatures, probably incur a lower cost of brooding than those that brood eggs for several weeks during the summer (Kaplan and Crump 1978).

## Summary and Conclusions

Amphibians have the lowest resting metabolic rates and lowest energy requirements of any terrestrial vertebrates. This low-energy lifestyle was once viewed as a handicap that prevented amphibians from exploiting the terrestrial environment as effectively as reptiles, birds, or mammals. Biologists now generally view the low energy requirements of amphibians as a feature that allows them to occupy niches not available to other vertebrates. Metabolic rates of amphibians are temperature-dependent, as they are in all ectotherms. Laboratory studies of thermal acclimation of metabolic rates have revealed a number of different types of response to acclimation at constant temperatures, ranging from complete metabolic compensation to a parallel effect of temperature change on metabolic rate. Many workers have tried to interpret the results of such experiments in terms of adaptations to increase metabolic efficiency or conserve energy, but the connection between laboratory-induced acclimation and acclimitization of metabolism under natural conditions is complex and difficult to evaluate.

Many amphibians rely on anaerobic metabolism to support short bursts of activity, such as rapidly escaping from a predator or lunging at prey. Anaerobic metabolism is less affected by environmental temperature than aerobic metabolism, although it is not completely independent of temperature. Anaerobic metabolism also makes energy available more quickly than aerobic metabolism, but can lead to inefficient use of energy substrates. Studies linking the ecology and behavior of amphibians with their metabolic physiology have revealed a considerable range of variation in metabolic strategies, especially among anurans. Species that engage in long periods of sustained activity, such as walking, swimming, or burrowing, generally have higher aerobic capacities and are less dependent on anaerobic metabolism than species that engage in short bursts of activity. Amphibians as a group have a relatively poor capacity for sustained, aerobically supported locomotion; the aerobic capacities of urodeles generally are lower than those of anurans, even though maximum aerobic speeds are not very different in these groups.

The limb muscles of amphibians that have high aerobic capacities for locomotion have morphological and biochemical characteristics that reflect their capacity for aerobically supported work. Although gross fiber types are not very different in different species of anurans, those with high aerobic capacities, such as toads (*Bufo*), have higher proportions of fast oxidative fibers than do species with lower aerobic capacities, such as frogs in the genus *Rana*. Activities of key enzymes involved in aerobic metabolism also are higher in species with high aerobic capacities. Some studies of individual variation in aerobic capacity, maximum aerobic speed, and sprint speed have revealed a correlation between one or more of these measures of performance capacity and leg muscle enzyme activities, whereas other studies have failed to show such a relationship. Correlation between physiological or behavioral performance variables and muscle biochemistry at the individual level appears to be more likely in species that support locomotion aerobically than in those that rely mainly on anaerobic metabolism.

A major shift in our understanding of amphibian metabolism occurred in the mid-1980s with the discovery that many species of frogs can achieve higher rates of aerobic metabolism while calling than they can during bouts of forced exercise in laboratory chambers. Previous workers had always considered the maximum metabolic rates measured during forced exercise to represent the maximum aerobic capacity of an animal. In some cases, metabolic rates of calling frogs are more than double the maximum aerobic capacity for locomotion. This is not necessarily true for all frogs, however, because nearly all measurements of $O_2$ consumption in calling frogs have been performed on species with relatively high calling rates. The ability of frogs to elevate their metabolism as much as 25 times resting rates became less puzzling with the discovery that the muscles involved in call production are fundamentally different in morphology and biochemistry from those used in locomotion. This is especially true for species with very high calling rates, which have trunk and laryngeal muscles composed of more aerobic fiber types than leg muscles, with much higher concentrations of mitochondria, higher capillary densities, and higher activities of enzymes involved in aerobic metabolism. Recent studies have revealed a wide range of variation in morphological and biochemical characteristics of call-producing muscles, even within the same family of anurans. Species that call at very low rates have muscles that lack many of the features that enhance muscle aerobic capacity, and in some cases, these muscles are not very different from the leg muscles of the same species.

The way in which anurans allocate their energy reserves for call production and other reproductive activities also varies considerably. Anurans with very high calling rates depend heavily on stored lipid reserves to support call production, because available carbohydrate reserves are not sufficient to support calling at high rates for more than a few hours. Some species, such as the spring peeper (*Pseudacris crucifer*) and some tropical hylids, have higher concentrations of lipids in their trunk muscles than are found in most other lipid-rich vertebrate muscle. Stores of lipids are closely associated with the mitochondria, which provide the machinery for aerobic work. Frogs that call at low rates lack extensive intramuscular lipid reserves and apparently depend almost entirely on carbohydrates to support call production. In this respect, the variation in energy-use strategies in calling frogs parallels that of singing insects and exercising mammals.

Annual patterns of energy storage and use by male anurans also vary among species and are related to temporal patterns of reproduction, the availability of females, and the temperature at which males are active. Many frogs and toads that breed in early spring must depend on stored energy reserves to support call production and other reproductive activities, because they emerge from hibernation and move to breeding sites before food is readily available. Some species, such as wood frogs (*Rana sylvatica*), lack extensive lipid reserves and rapidly deplete stored carbohydrate reserves during an explosive breeding period. Frogs with prolonged breeding seasons and relatively low calling rates, such as summer-breeding ranids, appear to deplete energy reserves much more slowly, and they can supplement energy reserves by feeding. Individual males often are present for a substantial proportion of the breeding season. In contrast, frogs with prolonged breeding seasons and high calling rates, such as many hylid treefrogs, deplete energy reserves relatively rapidly, and chorus tenure of individual males often is very short. Desert anurans resemble early spring breeders in that they move to breeding ponds immediately after emerging from dormancy and must rely mainly on stored energy reserves to support reproductive activities. Because favorable conditions for feeding are relatively brief, desert anurans often feed voraciously immediately after breeding and sometimes specialize on lipid-rich prey, such as termites.

Much less is known about patterns of energy use in urodeles. Costs of reproduction, including aggression, courtship, and production of spermatophores, have been measured in only one species of plethodontid salamander, and these appear to be very low compared to the cost of calling in frogs. Newts in the family Salamandridae engage in prolonged, vigorous courtship that is more likely to be energetically costly, but direct measurements of such costs have not been made. Annual patterns of energy use in temperate zone urodeles resemble those of anurans, with energy reserves being depleted during the breeding season and rebuilt before winter dormancy. In some species, however, mating and egg-laying seasons are separated, and courtship takes place both before and after winter hibernation. Females of many plethodontid salamanders remain with their eggs throughout development, which can be prolonged, and the energetic costs of brooding and lost opportunities for feeding can represent a substantial portion of a female's annual energy budget.

The basic features of the metabolism and energetics of amphibians are now well understood, but many details remain to be investigated. For example, there has never been a rigorous, phylogenetically controlled comparison of metabolic rates of anurans from tropical and temperate-zone regions, nor has the temperature sensitivity of metabolism in tropical and temperate-zone species been investigated in detail. One would expect temperate-zone species to exhibit lower temperature sensitivity, especially at low temperatures, because they are more likely to encounter a wide range of temperatures on a daily or seasonal basis. Some low-elevation tropical species have been shown to have lower resting metabolic rates at low temperatures than high-elevation species in the same genera, but the effect was even greater for activity metabolism. Low-elevation species were incapacitated at temperatures commonly experienced by high-elevation forms. Similarly, temperate-zone frogs such as spring peepers (*Pseudacris crucifer*) are active and call at temperatures below the lower lethal limits for some tropical hylids. The capacity for activity at low temperatures is assumed to be a derived trait in hylids (Hedges 1986), and it would be interesting to examine the evolution of physiological traits related to temperature sensitivity in more detail. These might include changes in muscle enzyme activities, the composition of isozymes present in the muscles, or changes in muscle contractile properties.

Ecological correlates of variation in exercise physiology also need to be reexamined, especially in anurans, which are more diverse in their locomotor behavior and foraging ecology than other amphibians. Studies like that of Taigen et al. (1982) revealed a relationship between aerobic capacity and broadly defined ecological and behavioral categories such as foraging mode and burrowing behavior. There is evidence from studies of groups of closely related species, such as dendrobatid frogs, that anurans actually exhibit a continuum of activity strategies correlated with a similar range of variation in metabolic characteristics. There is a need for more detailed, phylogenetically controlled comparative studies that actually quantify behavioral and ecological variables that can be related to metabolic physiology. Investigators also need to pay closer attention to individual variation in behavioral performance and how this is related to variation in physiological traits (Bennett 1987).

Studies of reproductive energetics offer a rich opportunity to examine possible physiological constraints on behavioral performance that directly affects individual fitness. Energetically expensive activities such as calling are more likely to push individual animals to their physiological limits, rather than relatively low-level activities such as foraging, so differences in physiological capacities probably lead to differences in mating success. Although we now have a reasonably good understanding of the cost of calling in some hylids, leptodactylids, and hyperoliids with high calling rates, it is important not to draw sweeping conclusions about the potential cost of calling in all frogs from this limited data set. All of the species for which we have actual measurements of metabolism during calling represent a phylogentically restricted set of species with rather similar breeding ecology. We have some evidence indicating that the cost of calling in other frogs, such as ranids, is much lower than in species with high calling rates, leading to very different strategies of energy allocation. The calling energetics of archeobatrachian frogs has scarcely been investigated, and we know very little about the vast array of tropical forest frogs with unusual reproductive modes. We also need much more detailed information on the natural range of variation in hourly calling rates, nightly calling effort, and seasonal chorus tenure in frogs, so that accurate estimates of total energy expenditures can be made. Until such information is available, it will be impossible to construct reliable annual energy budgets for these species.

# **Chapter 6** Movements and Orientation

*The solitary pool has drawn to itself the entire toad population of the surrounding country. . . . Each toad has his own home or hermitage somewhere in that area, where he spends the greater portion of the summer season. . . . When spring returns he sets out on his annual pilgrimage of a mile or two . . . until he arrives at the sacred pool.*

*The music and revels over, the toads vanish, each one taking his own road, long and hard to travel, to his own solitary home. Their homing instinct . . . is practically infallible.*
—W. H. Hudson, *The Book of a Naturalist* (1919)

One problem with trying to understand the ecology and behavior of amphibians is the difficulty of relating to the spatial scale on which these animals operate. A person walking through a forest inhabited by terrestrial salamanders is much like Gulliver in the land of the Lilliputians. A single stride can encompass the entire home range of a salamander, and several strides will carry the person beyond the area covered by a salamander in a lifetime of movement. Indeed, amphibians almost certainly are the most sedentary of terrestrial vertebrates. A comparison of amphibians to other vertebrates weighing less than 20 g reveals that home ranges of amphibians are 250–10,000 times smaller than those of birds, 125–2,000 times smaller than those of mammals, and 2–500 times smaller than those of small insectivorous lizards (table 6.1). These differences probably are conservative estimates. For reasons that will be discussed, published figures on amphibian home range sizes probably exaggerate the extent of their daily movements. Furthermore, the way in which amphibians utilize their home ranges differs considerably from the behavior of other vertebrates. A bird can traverse its entire home range every few minutes, a small mammal can do so in a few hours, and a lizard in a few days or weeks. An amphibian, on the other hand, probably spends much of its time in a very limited area and makes only occasional movements to other parts of the home range. In addition, the home range of a bird or arboreal lizard often has a substantial vertical component, whereas the home ranges of most amphibians are essentially two-dimensional.

Many aspects of the ecology and behavior of animals are affected by their capacity for movement. Sedentary animals do not have the option of making long-distance movements to escape from periodic changes in weather, and they often are forced to obtain food or mates from a relatively limited area. Daily, seasonal, and lifetime movements affect nearly every other aspect of an animal's behavior and ecology (Swingland and Greenwood 1983), including foraging ecology and energetics (chapter 5), territorial behavior (chapters 8, 9), mating systems (chapters 8, 9), responses to predators (chapter 14), and interactions with other species (chapter 15). Patterns of movement, especially dispersal, affect the dynamics of populations (Gill 1978a, b; Sinsch 1992c; Miaud et al. 1993; Sinsch and Seidel 1995), their genetic structure, (Shields 1982; Breden 1987; Thornhill 1993; Scribner et al. 2001) and the evolution of life history strategies (chapter 10; Horn 1983; Horn and Rubenstein 1984). Conservation decisions also depend on detailed information on pat-

**Table 6.1** Approximate home range sizes of small terrestrial vertebrates (< 20 g)

| Group | *No. of species* | Home range size ($m^2$) Range | Median | Source |
|---|---|---|---|---|
| Birds | 34 | 1,000–44,000 | 10,000 | Schoener, 1968 |
| Mammals | 9 | 500–8,100 | 2,500 | Harestad and Bunnell 1979; Eisenberg 1981 |
| Lizards | | | | |
| Males | 15 | 10–2,000 | 730 | Turner et al., 1969 |
| Females | 10 | 15–1,000 | 450 | Turner et al., 1969 |
| Anurans | 21 | 1–1,900 | 40 | Table 6.5 |
| Urodeles | 13 | .1–90 | 4 | Table 6.4 |

terns of movement within and between habitats (Langton 1989; Reh and Seitz 1990; Dodd 1996; Dodd and Cade 1998; Gibbs 1998a; Hitchings and Beebee 1998; Baker and Halliday 1999; see also chapter 16).

## Types of Movement

Any type of movement carries potential costs for amphibians, including energetic costs, exposure to predation, and exposure to drought or other unfavorable climatic conditions. Consequently, amphibians move only when absolutely necessary. All animal movement is related in some way to acquisition of resources, including food, mates, and shelter from extreme environmental conditions. Some amphibians make repeated short movements away from a central location, such as a diurnal retreat site, while others wander more widely between resource patches. Some also make long distance movements, usually between habitat patches that are used in different seasons. Many amphibians are largely sedentary or even remain underground for much of the year, restricting their movements to certain seasons, when they move to breeding or hibernation sites.

Movements that are directly related to the acquisition of resources such as food, mates, or retreat sites that tend to keep an animal within a home range have been termed station keeping (Dingle 1996). Foraging is one type of station keeping, as is territorial defense of all or part of the home range. Any movement outside of a home to explore new resource patches is termed ranging behavior. Ranging behavior includes occasional forays outside of an established home range by adults, as well as movements of juveniles away from natal areas. Movements of juvenile amphibians away from breeding ponds often have been described as dispersal movements. However, many of these juveniles eventually will return to their natal ponds to breed, so only those that move permanently to new ponds can be considered dispersers (Breden 1987). The rest move some distance from their natal ponds, perhaps searching for new breeding sites, but eventually return without permanently shifting their locations.

The term dispersal often has been used to distinguish one-way movements from bidirectional movement, often called migration. Dingle (1996) defined migration as a specialized kind of movement that is not directly responsive to resources. Typical features of migration include the following: movement that is persistent and of greater duration than during station keeping or ranging; movement that takes place along relatively straight-line paths; temporary suppression of responses to resources that normally terminate movement, such as food; the occurrence of activity patterns related to departure and arrival; and allocation of energy reserves specifically to support migration. Although many migratory animals exhibit bidirectional movement, not all migration is bidirectional. Compared to other terrestrial vertebrates, especially birds and mammals, amphibians are not highly migratory animals. Even the longest migrations of amphibians between winter or summer locations and breeding ponds seldom exceed a few hundred meters.

## Local Movements and Home Range

### Methods for Studying Amphibian Movements

Most amphibians are difficult subjects for studies of movement patterns because of their secretive habits. Continuous observation of marked individuals usually is not feasible, because amphibians often spend much of their time under cover or underground. The most common method for studying movements of amphibians is to mark the animals, release them, and recapture them at variable intervals. A variety of methods has been used to mark and identify amphibians, including toe-clipping, marking with various sorts of waistbands and tags, tattooing, heat branding, freeze branding, marking with silver nitrate, marking with fluorescent powders, skin autografts, and mapping of individual color patterns (Ferner 1979; Donnelly et al. 1994; Kuhn 1994b).

Toe clipping is by far the most common method, because it is the least expensive and easiest method for fieldwork. Some investigators have reported that toe clipping can slightly increase mortality (Clarke 1972) or lead to infected toes (Golay and Durrer 1994), but other studies have not found these detrimental effects (van Gelder and Strijbosch 1996; Lüddecke and Amézquita 1999). Davis and Ovaska (2001) found that toe clipped salamanders (*Plethodon vehiculum*) gained less mass in the field than did unmarked salamanders or those marked with fluorescent paint. There also appeared to be slightly higher mortality among toe-clipped individuals. Some investigators have suggested that toe clipping is responsible for the relatively low recapture rates obtained in many studies (Nishikawa and Service 1988). Nishikawa also argued that toe clipping could cause animals to move farther than they would normally (Nishikawa 1985), but Davis and Ovaska (2001) did not find any effect of marking method on patterns of movement in *P. vehiculum*. The disadvantage of all mark-recapture techniques that have been used in studies of movement is that animals often disappear for weeks or months at a time, then reappear near the original capture site. The apparent movement of the animal is very small, but its actual movements in the intervening period are unknown.

Various techniques have been used for fine-scale studies of individual movement. One is to label animals with small radioactive wires and then locate them on successive days with a detection device (Karlstrom 1957; Breckenridge and Tester 1961; Barbour et al. 1969; Madison and Shoop 1970; Kramer 1973, 1974; Clarke 1974a; Ashton 1975; Ashton and Ashton 1978; Semlitsch 1981b; Kleeberger and Werner 1982). This allows the investigator to obtain many sightings of each individual, thereby providing accurate estimates of day-to-day movements. This method is most useful for small-bodied species with relatively small home ranges (Ashton 1994). It is of limited use for species that spend much of their time deep in subterranean burrows. It also does not uniquely identify individuals, so other identifying marks are required. Unfortunately, the animals sometimes suffer injury from the radioactive wires (Ashton 1975; Semlitsch 1981a), and permission to use radioactive materials can be hard to obtain. The technique also is fairly expensive and has not been widely used in recent years. A technique developed more recently is to mark individuals with tiny implanted passive integrated transponders (PITs) containing computer chips that produce unique identification numbers when activated with a special reading device (Camper and Dixon1988; Sinsch 1992d; Fasola, Barbieri, and Canova 1993). This is an expensive technique, but it allows repeated identification of individuals without a lot of handling. A disadvantage of this technique for studying movements is that the tags can only be detected at close range.

Several investigators have used fluorescent powders of different colors to mark amphibians for long-term or short-term studies of movements. When these are applied to the skin under pressure, they can produce marks that are visible under ultraviolet light and last for more than a year (Nishikawa and Service 1988; Davis and Ovaska 2001). Nishikawa (1990) used this technique to make detailed studies of plethodontid salamanders moving around on the surface at night. Other workers have simply sprinkled fluorescent powders onto the backs of amphibians to monitor nightly movements over relatively short periods of time (Woolbright 1985b; Ovaska 1992). This method is most useful for animals that are readily visible on the forest floor or on vegetation at night, because it allows them to be identified without disturbance. Windmiller (1996) attached strips of yarn dipped in fluorescent powder to the backs of frogs and was able to monitor movements of up to 150 m by following trails of powder left on the ground. He also experimented with capsules filled with luminescent fluid that were attached to the backs of animals and were visible for up to 80 m.

In a study of box turtle movements, Stickel (1950) equipped animals with spools of thread that unwound as the animal moved, leaving a record of its path. This technique has subsequently been used with some success in studies of anuran movements (Dole 1965a, b, 1968, 1972a, c; Tracy and Dole 1969b; Grubb 1970; Neill and Grubb 1971; Sinsch 1987a, c, 1988b, c; Duellman and Lizana 1994; Seebacher and Alford 1999; Tozetti and Toledo 2005). This method provides information on movements over relatively short intervals, but it is impractical for small-bodied species.

Radio tracking has been widely used to study movements of mammals, birds, and reptiles, but until relatively recently, most transmitters were too large to be implanted in or carried by any but the largest amphibians. Transmitters now have been sufficiently miniaturized to be used with many kinds of amphibians. They have been used successfully with toads (*Bufo;* Nuland and Claus 1981; van Gelder et al. 1986; Sinsch 1992b, c; Oldham and Swan 1992; Denton and Beebee 1993a; Kusano, Maruyama, and Kaneko 1995; Seebacher and Alford 1999; Miaud, Sanuy, and Avrillier 2000; Schwarzkopf and Alford 2002; Muths 2003; Bartelt, Peterson, and Klaver 2004), ranid frogs (Fiorito et al. 1994; Spieler and Linsenmair 1998; Lamoureux and Madison 1999; Matthews and Pope 1999; Richter et al. 2001; Bulger, Scott, and Seymour 2003), small rhacophorid frogs (Fukuyama, Kusano, and Nakane 1988; Kusano 1998), hellbenders (*Cryptobranchus;* Stouffer et al. 1983; Gates et al. 1985), ambystomatid salamanders (Madison and Farrand 1998), and newts (Jehle and Arntzen 2000). The transmitters are mounted on a harness or waistband, surgically implanted, or ingested by the animals. Unfortunately, there is considerable potential for either external or implanted transmitters to cause injury even to relatively large species of amphibians (Weick et al. 2005).

### Patterns of Movement

Many amphibians exhibit considerable fidelity to limited areas of suitable habitat, although they sometimes move between habitats at different seasons. Individuals have been found in the same home ranges for up to five years in *Scaphiopus holbrookii* (Pearson 1955, 1957), seven years in *Sala-*

*mandra salamandra* (Joly 1968; Rebelo and Leclair 2003b) and *Plethodon kentucki* (Marvin 2001), and nine years in *Bufo bufo* (Heusser 1968d). Similar fidelity to hibernation sites has been observed in *S. salamandra,* with some individuals hibernating in the same cave for 20 years (Feldmann 1987). Some Manitoba toads (*Bufo hemiophrys*) turned up in the same mounds on a Minnesota prairie every winter for up to six years (Kelleher and Tester 1969). These same toads exhibited very little fidelity to summer home ranges, perhaps because they lived in relatively uniform habitat at that season (Breckenridge and Tester 1961).

Routine movements during activity seasons usually are quite limited, with distances between recapture points averaging less than 10 m for terrestrial salamanders (table 6.2) and less than 15 m for terrestrial anurans (table 6.3). Even the large aquatic salamander *Cryptobranchus alleganiensis* generally moved only 10–30 m between recaptures (Hillis and Bellis 1971; Nickerson and Mays 1973b), although they sometimes moved 1–2 km over a period of months (Gates et al. 1985). Both *Siren lacertina* and *Amphiuma means* in Florida showed similar sedentary patterns of behavior, with no movements of more than 10 m being observed for either species (Sorensen 2004). Nevertheless, recapture rates for these species were extremely low, so the possibility of long-distance movements cannot be ruled out. Daily movements within a home range probably are considerably shorter for most species. Comparisons of movement distances based on mean values can be misleading, because distributions of movement distances usually are negatively skewed (fig. 6.1), with many short movements and only a few long ones (Stebbins 1954b; Bellis 1965; Hillis and Bellis 1971; Wells and Wells 1976). Consequently, median movement distances often are lower than means and provide a better estimate of typical movement (tables 6.2, 6.3).

Amphibians sometimes make brief excursions outside their normal home ranges, possibly in search of mates or better quality home sites. Such excursions usually occur in rainy weather and range from less than 100 m in terrestrial salamanders (Barbour et al. 1969; Wells and Wells 1976; Staub, Brown, and Wake 1995) to several hundred meters in anurans (Raney 1940; Willis, Moyle, and Baskett 1956; Turner 1959, 1960a; Breckenridge and Tester 1961; Bellis 1965; Dole 1965b; Heusser 1968d; Duellman and Lizana 1994; Kam and Chen 2000; Frétey et al. 2004). Long movements often are most apparent with long recapture intervals, so mean movement distance sometimes is positively correlated with mean recapture interval (Ingram and Raney 1943; Turner 1959; Bellis 1965; Hollenbeck 1974; Wells and Wells 1976; Duellman and Lizana 1994). Including these occasional long movements in calculations of home range size (see the following discussion) can lead to serious overestimates of the area in which an animal carries out its daily activities. On the other hand, many studies of movement patterns, especially for terrestrial salamanders, have been conducted on plots that were too small for very long movements to be detected (Staub, Brown, and Wake 1995). Ignoring such movements could result in an underestimate of the amount of habitat that is familiar to an individual animal.

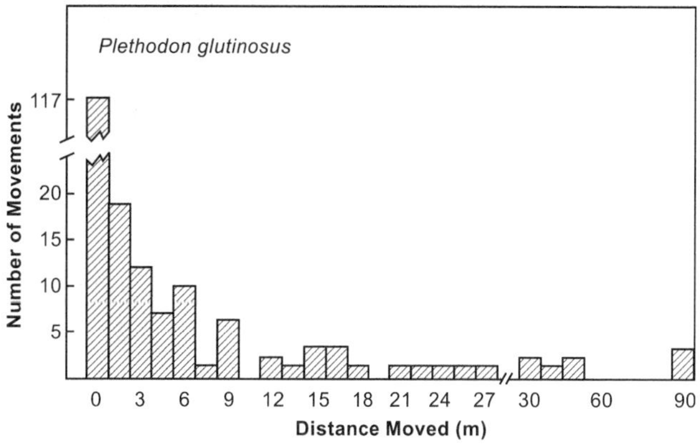


**Fig. 6.1.** Distribution of distances between recaptures for a population of slimy salamanders (*Plethodon glutinosus*) in North Carolina. The pattern of movement is typical of many amphibians, with many short movements and relatively few long ones. After Wells and Wells (1976).

Continuous monitoring of marked individuals throughout their activity periods provides much more useful information on short-term patterns of movement than do mark-recapture studies, particularly in relation to foraging behavior and daily energy budgets (see chapter 5). Unfortunately, such studies are rather tedious and have been attempted for only a few species. One of the most detailed studies of movements in salamanders is Nishikawa's (1990) study of *Plethodon teyahalee* and *P. metcalfi* in the southern Appalachian Mountains. This is a very wet region, but the salamanders were extremely sedentary. Even when visible on the surface, they moved very little. Individuals of *P. teyahalee* that were sighted more than once on the same night did not move more than about one meter between sightings, which was similar to the average distance between recaptures on different nights. *Plethodon metcalfi* was even more sedentary, with individuals moving less than 0.2 m per night, even though individuals of this species actually utilized larger home ranges over a period of several months. In both species, individuals were not active on the surface every night, and many of those that were visible were seen sitting at the entrances to retreat holes. Individuals sometimes used several different retreat holes, but particular holes were never used by more than one salamander. The sedentary behavior of these salamanders, which probably is typical of most terrestrial plethodontids, indicates that they are mainly sit-and-wait foragers.

Short-term movements of anurans have been studied in somewhat more detail. Dole (1965b) observed individual leopard frogs (*Rana pipiens*) for up to 6.5 hours at a time and

**Table 6.2** Movement distances between recaptures for salamanders

| Species (source no.) | $N_i$ | $N_m$ | Duration of study | Distance between captures (m) Range | Mean | Median | Mode |
|---|---|---|---|---|---|---|---|
| Cryptobranchidae | | | | | | | |
| *Cryptobranchus alleganiensis* (8) | 44 | | 4 mths | | 18.8 | 10.7 | 5.0 |
| Males | | | | | | | |
| Females | 29 | | 4 mths | | 18.7 | 8.1 | 5.0 |
| Plethodontidae | | | | | | | |
| *Aneides aeneus*[a] (1) | 12 | 32 | 2 yrs | 0–91 | 5.7 | 0.0 | 0.0 |
| *A. lugubris*[b] (4) | 30 | | 1 yr | 0–5 | 0.9 | 0.0 | 0.0 |
| *Batrachoseps attenuatus*[b] (4) | 80 | | 1 yr | 0–5 | 0.4 | 0.0 | 0.0 |
| *B. attenuatus*[b] (2) | | 101 | 3 yrs | 0–4 | 1.2 | 0.9 | 0.3 |
| *B. pacificus* (5)[b] | 141 | | 6 mths | 0–18 | 0.7 | | |
| *Desmognathus ochrophaeus* (11) | 19 | 74 | 4 mths | | 0.7 | | |
| *D. quadramaculatus* (19) | 73 | | 6 mths | | 1.4 | 0.7 | |
| *Ensatina eschscholtzii* (3) | 54 | 150 | 4 yrs | 0–38 | 7.7 | 6.1 | 3.1 |
| *E. eschscholtzii* (18) | 90 | 115 | 5 yrs | 0–120 | 22.7 | | |
| *Eurycea bislineata* (10) | 20 | 40 | 14 wks | 2–31 | | | |
| *Plethodon cinereus* (15) | | 55 | 6 mths | | 0.4 | | |
| *P. glutinosus* (9) | | 144 | 3 yrs | 0–92 | 6.2 | 0.0 | 0.0 |
| *P. hubrichti* (17) | 8 | 24 | 6 mths | | | 1.0 | |
| *P. jordani* (7) | 29 | 66 | 3 mths | 0–11 | 2.1 | 1.2 | 0.3 |
| *P. kentucki* (21) | 18 | 164 | 7 yrs | 0–4 | 1.5 | | |
| *P. metcalfi* (13, 16) | 32 | 81 | 4 mths | 0–8 | 2.0 | | |
| *P. teyahalee* (13, 16) | 21 | 76 | 4 mths | 0–3 | 1.0 | | |
| *P. vehiculum* (14) | 146 | | 1 yr | 0–8 | < 2.0 | | |
| Rhyacotritonidae | | | | | | | |
| *Rhyacotriton cascadae* (20)[c] | 24 | 129 | 2 mths | 0–6 | 2.4 | 1.7 | |
| Salamandridae | | | | | | | |
| *Salamandra atra aurorae* (22) | 20 | 32 | 4 mths | 0–12 | 5.0[d] | | |
| *Salamandra salamandra* (6) | 89 | 97 | 5 yrs | 0–48 | 11.8 | | |
| *Triturus vulgaris* (12)[e] | | | | | | | |
| Males | | 41 | 1 yr | 1–15 | 2.5 | 1.0 | 1.0 |
| Females | | 123 | 1 yr | 1–16 | 3.2 | 2.0 | 1.0 |

*Sources:* (1) R. Gordon 1952; (2) Hendrickson 1954; (3) Stebbins 1954a; (4) Anderson 1960; (5) Cunningham 1960; (6) Joly 1968; (7) Madison 1969; (8) Hillis and Bellis 1971; (9) Wells and Wells 1976; (10) Ashton and Ashton 1978; (11) Holomuzki 1982; (12) Griffiths 1984; (13) Nishikawa 1985; (14) Ovaska 1988a; (15) Gergits and Jaeger 1990a; (16) Nishikawa 1990; (17) Kramer et al. 1993; (18) Staub, Brown, and Wake 1995; (19) Camp and Lee 1996; (20) Nijhuis and Kaplan 1998; (21) Marvin 2001; (22) Bonato and Fracasso 2003.

*Notes:* All species are largely terrestrial except *Cryptobranchus.* Data are for adult males and females combined unless otherwise indicated. $N_i$ = number of individuals. $N_m$ = number of moves.

[a]Data for males only.

[b]Data include both juveniles and adults.

[c]Data for *Rhyacotriton cascadae* show mean and median lengths for linear ranges along a stream, based on four or more captures per individual.

[d]Data show mean distance moved from first capture point during entire 4-month activity season for adults only.

[e]Movements on land only.

**Table 6.3** Movement distances between recaptures for terrestrial anurans

| Species (source no.) | $N_i$ | $N_m$ | Duration of study | Distance between captures (m) Range | Mean | Median | Mode |
|---|---|---|---|---|---|---|---|
| Bufonidae | | | | | | | |
| *Bufo americanus* (6) | 27 | 86 | 3 mths | | 13.2 | 9.2 | 9.1 |
| *B. boreas* (3) | 13 | 21 | 3 mths | 2–43 | 15.8 | 14.0 | |
| *B. bufo*[a] (15) | 30 | 51 | 12 yrs | 0–70 | | 4.2 | |
| *B fowleri* (13) | | | 5 mths | 0–85 | 32.3 | | 11.4 |
| *B. hemiophrys* (8) | 19 | 276 | 3 mths | 0–229 | | < 9.0 | < 9.0 |
| *B. punctatus* (7) | 40 | 78 | 2 yrs | 0–216 | | < 30.0 | < 30.0 |
| *B. punctatus* (16) | 31 | 83 | 1 yr | | 13.7 | | |
| *B. valliceps* (9) | 32 | 110 | 1 yr | 0–135 | 21.8 | < 24.0 | < 12.0 |
| Dendrobatidae | | | | | | | |
| *Dendrobates granuliferus* (22) | 8 | 71 | 3 wks | 0–6 | 1.4 | | 0.0 |
| *D. histrionicus* (12) | 47 | 128 | 38 days | 0–78 | 7.6 | 5.8 | |
| Hylidae | | | | | | | |
| *Acris crepitans* (5) | | 178 | 10 mths | 0–98 | 38.0 | | |
| *A. crepitans* (6) | 378 | | 2 yrs | | 29.9 | | |
| *Hyla andersonii* (18) | 8 | 93 | 2 mths | 0–102 | | < 20.0 | |
| *H. arenicolor* (14) | 15 | | 3 mths | 0–23 | 7.2 | | |
| Leiopelmatidae | | | | | | | |
| *Leiopelma hamiltoni* (19) | | | | | | | |
| Maud Island A | 9 | 18 | 3 yrs | 0–5 | | 2.2 | |
| Maud Island B | 62 | 142 | 3 yrs | 0–7 | | 0.9 | |
| Stephens Island | 71 | 307 | 3 yrs | 0–6 | | 0.8 | |
| *L. hochstetteri* (20) | 32 | 62 | 9 days | 0–12 | 1.2 | < 0.4 | 0.0 |
| Leptodactylidae | | | | | | | |
| *Eleutherodactylus fitzingeri* (24) | 8 | 12 | 4 mths | 0–35 | 6.9 | | |
| *E. johnstonei* (21) | | | 1 yr | 0–7 | < 2.0 | | |
| Mantellidae | | | | | | | |
| *Mantidactylus microtympanum* (23) | | | | | | | |
| Males | 17 | | 6 mths | 1–24 | 5.3 | 4.0 | |
| Females | 35 | | 6 mths | 0–35 | 4.3 | 3.7 | |
| Myobatrachidae | | | | | | | |
| *Heleioporus australiacus* (26) | 13 | 98 | < 1 yr | 0–463 | 21.7[b] | | |
| Ranidae | | | | | | | |
| *Rana arvalis*[a] (11) | | 51 | 3 mths | | 7.2 | | |
| *R. catesbeiana* (1, 2) | 69 | 190 | 2 yrs | 0–1,600 | 79.2 | 9.1 | 1.5 |
| *R. hecksheri* (4) | 38 | | 6 mths | | 10.3 | | |
| *R. pretiosa*[c] (3) | 17 | 23 | 3 mths | 1–37 | 10.1 | 6.4 | |
| *R. swinhoana* (25) | 243 | | 2 yrs | 0–30 | 15.4 | | |
| *R. sylvatica* (10) | 298 | 339 | 3 mths | 0–71 | 11.3 | 7.0 | 2.7 |
| *R. temporaria* (11) | | 54 | 3 mths | 0–60 | | 4.5 | |

*Sources:* (1) Raney 1940; (2) Ingram and Raney 1943; (3) Carpenter 1954; (4) Hansen 1957; (5) Pyburn 1958; (6) Bellis 1959; (7) Turner 1959; (8) Breckenridge and Tester 1961; (9) Awbrey 1963; (10) Bellis 1965; (11) Haapanen 1970; (12) Silverstone 1973; (13) R. Clarke 1974b; (14) Dole 1974; (15) Haapanen 1974; (16) Weintraub 1974; (17) R. Gray 1983; (18) Freda and Gonzalez 1986; (19) Newman 1990; (20) Tessier, Slaven, and Green 1991; (21) Ovaska 1992; (22) van Wijngaarden and van Gool 1994; (23) Andreone 1998; (24) Höbel 1999a; (25) Kam and Chen 2000; (26) Lemckert and Brassil 2003.

*Notes:* Data are for adult males and females combined unless otherwise indicated. $N_i$ = number of individuals. $N_m$ = number of moves.

[a]Data for both adults and juveniles.

[b]Data show mean weekly movements for adult males and females equipped with radio transmitters.

[c]Data for juveniles only.

found that almost half did not move at all. Some marked individuals remained virtually immobile for more than 24 hours. Clearly this species is quite sedentary on a day-to-day basis, even though some individuals utilize a relatively large home range over a period of several months. Similarly, chorus frogs (*Pseudacris triseriata*) tagged with radioactive wires averaged only 3.5 m of movement per day, but were capable of moving up to 42 m in less than 24 hours (Kramer 1973). Freda and Gonzalez (1986) followed the movements of eight individuals of *Hyla andersonii* and reported that most individuals moved less than 20 m per day and remained within 70 m of a breeding pond; the longest daily movement was about 100 m. In a study of *Pseudacris cadaverina*, a tree frog that lives in rocky canyons in California, R. T. Harris (1975) found that adults spent most of the day sitting on rocks near a stream. In late afternoon, they moved an average of 3 m to the water's edge, where they foraged for insects or engaged in breeding activities until returning to the same diurnal resting sites. These frogs appear to remain within a few meters of one segment of a stream for most of their lives. A high degree of site fidelity and very limited daily movement also is characteristic of several ranid frogs living along tropical and subtropical forest streams (Inger 1969; Kam and Chen 2000).

Movements of terrestrial anurans that are not closely associated with water have not been studied in as much detail, but several species also exhibit a high degree of site fidelity. Woolbright (1985b) studied nightly movements of the Puerto Rican frog *Eleutherodactylus coqui* in the understory of a tropical forest by monitoring marked individuals every 15 minutes. He found that most individuals traveled only 3–5 m per night, usually moving from a diurnal retreat site to a nighttime foraging station at dusk and returning to the retreat at dawn. This low rate of movement is associated with a very low number of prey capture attempts each night (Taigen and Pough 1985). Ovaska (1992) found *Eleutherodactylus johnstonei* in Barbados to be equally sedentary, with individuals moving only 3–4 m per night. Another very sedentary species is *Ceratophrys cornuta*, a large forest-floor frog from Peru. Most individuals moved less than 2 m per night, and many did not move at all (Duellman and Lizana 1994).

All of the anuran species discussed previously are sit-and-wait predators that are expected to be relatively sedentary. In contrast, an actively foraging Panamanian frog, *Dendrobates auratus*, averaged about 160 m of movement (6,000 body lengths) in a 12-hour activity period and made 400–500 attempts to capture prey (Taigen and Pough 1985). This species probably uses a much larger proportion of its home range each day than do species that sit and wait for their prey. Other dendrobatids in the genus *Colostethus* have much lower rates of movement each day, but still are considerably more active than sedentary species such as *Eleutherodactylus coqui* or *Rana pipiens*. Similarly, *Bufo marinus*, an active forager, moves at a faster rate and uses a much larger area each night than does *Leptodactylus ocellatus*, a sit-and-wait predator living in the same habitat (Strüssmann et al. 1984). In Australia, where *Bufo marinus* has been introduced, individuals sometimes moved up to 100 m per night in wet weather and often ended up in diurnal shelters 30–40 m from their previous shelter sites (Seebacher and Alford 1999). At least some individuals of this species appear to be nomadic, wandering over wide areas with little evidence of a fixed home range (Schwarzkopf and Alford 2002).

One aspect of amphibian behavior that has received little attention is vertical movement into trees and other vegetation. Some terrestrial plethodontid salamanders regularly climb into low vegetation on rainy nights, giving them access to insect prey not readily available in the leaf litter and possibly aiding in orientation (Hairston 1949; Madison 1972; Jaeger 1978). Pacific giant salamanders (*Dicamptodon*) have been observed climbing into bushes and trees as high as 1.5 m above the ground (Stebbins 1954a). It is well known that many tree frogs move high into trees after the breeding season and occasionally are heard calling from the tops of trees, but there have been very few attempts to follow the vertical movements of individuals on a daily basis. Stewart (1985) reported that juveniles and females of *Eleutherodactylus coqui* regularly climbed high into the forest canopy to feed at dusk and "parachuted" to the ground at dawn. Males generally remained lower in the vegetation at established calling sites. A reverse pattern was observed in *Hyla rosenbergi* during the breeding season. At dusk, males rapidly descended from diurnal retreat sites to nests on the ground, where they remained throughout the night (Kluge 1981). An unusual example of arboreal behavior was reported in a population of *Bufo valliceps* living in a public park in Austin, Texas. These toads usually are considered to be strictly terrestrial, but Neill and Grubb (1971) observed toads climbing as high as 5 m into live oak trees and repeatedly using tree holes as retreat sites. Several species of European anurans (*Bufo bufo*, *Rana temporaria*, *Alytes obstetricans*) have been observed climbing into oak trees, apparently seeking moisture in wet moss growing on the bark (Gosá 2003).

### Sizes of Home Ranges

Several methods have been used to estimate home range sizes for amphibians. Most investigators have used procedures developed by mammalogists in the 1940s. The most common technique is to draw a convex polygon around all recapture points for each individual (Mohr 1947), or to construct a polygon that encloses the smallest possible area, even if it is not convex (the minimum area polygon method;

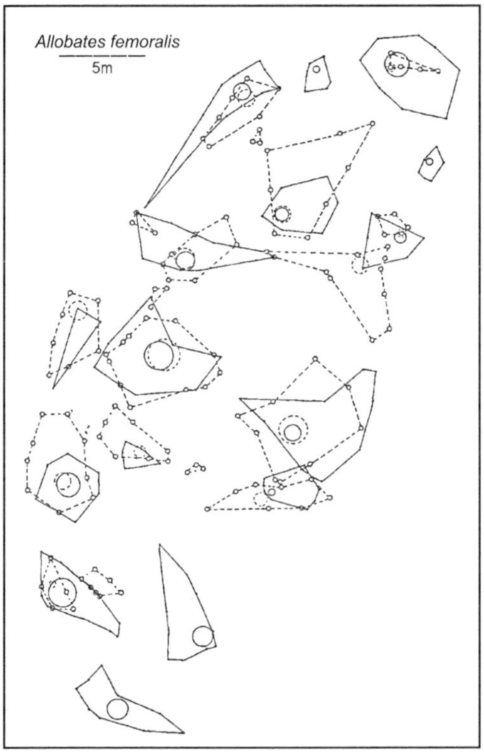


**Fig. 6.2.** Example of polygonal home range boundaries plotted for individual frogs. This figure shows the boundaries of territories defended by males of the dendrobatid frog *Allobates femoralis* in Amazonian Peru. Solid and dashed lines show territory boundaries in two different years. Circles indicate the most frequently used calling sites in those years. After Roithmair (1992).

fig. 6.2). These methods are quite sensitive to number of recaptures, with estimates of home range size generally increasing as the number of recaptures increases (Barbour et al. 1969; Jennrich and Turner 1969; D. Anderson 1982; B. Rose 1982). Polygon area methods tend to exaggerate the importance of occasional long movements. They probably are most accurate for species that travel over much of the home range each day, and least accurate for sit-and-wait predators or species that spend much of their time near retreat sites and make only brief long-distance excursions.

Inger (1969) reported home ranges of several species of *Bufo* and *Rana* as linear transects along tropical streams. This approach is appropriate for frogs that spend all of their lives within a few meters of a stream, but are difficult to compare with estimates of the areas of two-dimensional home ranges of other species. Some authors have determined the size of a circle or ellipse enclosing some arbitrary proportion of all recapture points (Pearson 1955; Madison 1969). This technique has the disadvantage of seriously distorting home range size if the home range deviates significantly from a circular or elliptical shape (Jennrich and Turner 1969; B. Rose 1982). Home range size also can be expressed as a Mean Activity Radius (Hayne 1949; Currie and Bellis 1969; Hillis and Bellis 1971; Holomuzki 1982; Kleeberger and Werner 1982; Kleeberger 1985). The geometric center of all recapture points is determined and the distance from that point to each recapture point is measured. The Mean Activity Radius is the average of these measurements and can be used to calculate the area of a circular home range (e.g., Hillis and Bellis 1971; Nishikawa 1985). This procedure has the advantage of reducing the impact of very long movements. Unfortunately, the statistical center of activity often bears little relationship to the physical activity center of the animal in the field, especially if the animal's movements are asymmetrically distributed around a choice feeding perch or retreat site (e.g., Pearson 1955). There also are various statistical methods that estimate home range use from the distribution of capture points or fixes using radio tracking (e.g., Muths 2003). Often these methods yield larger estimates of home range size than polygon area methods (Clarke 1974a; Healy 1975; Holomuzki 1982; Kleeberger and Werner 1982; Nishikawa 1985), although this is not consistently true (e.g., Watson, McAllister, and Pierce 2003). Unfortunately, there is no way to make statistical comparisons of data obtained by different methods. The data summarized in tables 6.4 and 6.5 include only estimates derived from the minimum area polygon method.

Detailed studies of amphibian home ranges have been done mainly on North American, European, and a few Neotropical species (tables 6.4, 6.5). Home range sizes vary considerably both within and among species. Most authors have presented mean home range sizes, but not medians. In a few cases in which both could be calculated (Hansen 1957; Barbour et al. 1969; Madison and Shoop 1970; Kramer 1974; Ashton 1975), the median was smaller, suggesting that the median would provide a better estimate of typical home range size than the mean.

Given the variability in home range size estimates, few generalizations are possible. Small plethodontid salamanders that center their activities around retreat sites such as logs or burrows tend to have very small home ranges (table 6.4), but the extent of their movements in underground burrows is unknown. Terrestrial salamandrids are more active foragers than plethodontids and tend to have larger home ranges (Joly 1963, 1968; Healey 1975). In the case of *Salamandra*, this probably relates to a much larger body size, whereas red efts (*Notophthalmus viridescens*) are about the same size as small *Plethodon*, but have home ranges up to 20 times larger.

Among anurans, there is no clear relationship between either body size or habitat and home range size, although very small terrestrial dendrobatids, leptodactylids, and leiopelmatids tend to have the smallest home ranges, whereas larger toads and semiaquatic ranids have the largest (table 6.5). The dendrobatid data are not strictly comparable to the

**Table 6.4** Estimate home range sizes for terrestrial salamanders

| Species (source no.) | $N_i$ | Habitat | Duration of study | Home range size (m²) Range | Mean |
|---|---|---|---|---|---|
| Ambystomatidae | | | | | |
| *Ambystoma maculatum* (11) | 6 | Forest floor | 2 mths | 3–29 | 9.8 |
| *A. talpoideum* (8) | 7 | Burrows | 2 mths | 0–23 | 3.6[a] |
| Plethodontidae | | | | | |
| *Desmognathus fuscus* (6) | 14 | Stream banks | 4–10 wks | 0–4 | 1.6 |
| *D. fuscus* (3) | 15 | Stream banks | 2 mths | 0.8–115 | 21.7 |
| *D. ochrophaeus* (9) | 19 | Stream banks | 4 mths | | 0.2 |
| *D. quadramaculatus* (15) | 47 | Rocky streams | 6 mths | 0.01–0.49 | 0.1 |
| *Plethodon cinereus* (10) | 28 | Forest floor | 2–4 mths | 9–20 | 4.4 |
| *P. cinereus* (13) | 20 | Forest floor | 2 mths | | 0.2 |
| *P. hubrichti* (14) | 8 | Forest floor | 6 mths | 0.1–2.4 | 0.6 |
| *P. kentucki* (16) | | | | | |
| Males | 10 | Forest floor | 7 yrs | 0.6–12 | 6.5 |
| Females | 8 | Forest floor | 7 yrs | 1.8–10.4 | 5.5 |
| *P. metcalfi* (1) | | | | | |
| Males | | Forest floor | 3 yrs | | 11.5 |
| Females | | Forest floor | 3 yrs | | 2.8 |
| *P. metcalfi* (12) | | | | | |
| Males | 3 | Forest floor | 4 mths | | 5.0 |
| Females | 8 | Forest floor | 4 mths | | 1.0 |
| Juveniles | 6 | Forest floor | 4 mths | | 0.3 |
| *P. teyahalee* (5) | | | | | |
| Males | | Forest floor | 3 yrs | | 14.4 |
| Females | | Forest floor | 3 yrs | | 6.5 |
| *P. teyahalee* (12) | | | | | |
| Males | 3 | Forest floor | 4 mths | | 0.5 |
| Females | 8 | Forest floor | 4 mths | | 1.0 |
| Juveniles | 6 | Forest floor | 4 mths | | 0.3 |
| Salamandridae | | | | | |
| *Notophthalmus viridescens* (7) (efts) | 10 | Forest floor | 3 mths | 28–153 | 87.0 |
| *Salamandra salamandra* (2) | 8 | Forest floor | 5 yrs | 13–148 | 68.0 |

*Sources:* (1) Merchant 1966; (2) Joly 1968; (3) Barbour et al. 1969; (4) Madison and Shoop 1970; (5) Merchant 1972; (6) Ashton 1975; (7) Healy 1975; (8) Semlitsch 1981b; (9) Holomuzki 1982; (10) Kleeberger and Werner 1982; (11) Kleeberger and Werner 1983; (12) Nishikawa 1990; (13) Mathis 1991b; (14) Kramer et al. 1993; (15) Camp and Lee 1996; (16) Marvin 2001.

*Notes:* Only estimates based on minimum area polygon method are included. Data are combined for males and females unless otherwise indicated. $N_i$ = number of individuals.

[a]Median values given for *A. talpoideum*.

other species, however, because they are based on sizes of territories defended by males. Foraging mode probably has a greater impact on home range size than body size, with active foragers likely to have larger home ranges than sit-and-wait predators of a similar size. Body size and foraging mode tend to be confounded (sit-and-wait predators usually are larger than active foragers), and all of these variables are further confounded by differences in phylogeny. This makes it difficult to do a really sophisticated comparative analysis. It is clear that most anurans spend their lives in quite restricted areas. Even species that make periodic migrations to aquatic breeding sites often return to the same terrestrial

**Table 6.5** Estimated home range sizes for anurans

| Species (source no.) | $N_i$ | Habitat | Duration of study | Home range size ($m^2$) Range | Mean |
|---|---|---|---|---|---|
| Bufonidae | | | | | |
| *Atelopus oxyrhynchus* (8) | 57 | Tropical forest | 13 mths | | 41 |
| *Bufo bufo* (14) | 48 | Pine forest | 2 mths | 490–3,420 | 1,900 |
| *B. fowleri* (7) | 3 | Golf course | < 1 mth | 103–1,245 | 750 |
| *B. japonicus* (20) | 15 | Temple garden | 5 mths | | 220 |
| *B. marinus* (11) | 13 | Clearings | 1 mth | | 168 |
| Dendrobatidae | | | | | |
| *Allobates femoralis* (17) | 34 | Tropical forest | 2 yrs | 0.3–26 | 8 |
| *Colostethus panamensis* (12) (wet season only) | 10 | Rocky stream | 11 mths | 1.5–3.7 | 2 |
| *Dendrobates auratus* (16) | 14 | Tropical forest | 2 yrs | | 3 |
| *D. histrionicus* (18) | 29 | Tropical forest | 3 mths | 2–9 | 4 |
| *D. leucomelas* (18) | 9 | Tropical forest | 4 mths | 1–81 | 20 |
| *D. pumilio* (15) | | | | | |
| Males (home range) | 5 | Tropical forest | 1.5 yrs | 4–10 | 8 |
| Males (defended area) | 7 | Tropical forest | 1.5 yrs | 0.2–5 | 2 |
| Females (home range) | 4 | Tropical forest | 1.5 yrs | 3–6 | 11 |
| *Epipedobates trivittatus* (19) | 21 | Tropical forest | 1.3 yrs | 4–156 | 44 |
| *Mannophryne trinitatis* (13) (territories of females) | 6 | Rocky stream | 2 mths | 0.3–1.0 | 0.6 |
| Hylidae | | | | | |
| *Pseudacris triseriata* (9) | 9 | Farmland | 4 mths | 640–6,024[a] | 2,117 |
| Leptodactylidae | | | | | |
| *Eleutherodactylus marnockii* (2) | 247 | Cliffs | 2 yrs | 210–700 | 435 |
| *Leptodactylus macrosternum* (10) | 10 | Llanos | 1 mth | 10–133 | 43 |
| Pelobatidae | | | | | |
| *Scaphiopus holbrookii* (3) | 88 | Prairie | 1 yr | < 1–33 | 7 |
| Ranidae | | | | | |
| *Rana clamitans* (1) | 29 | Stream bank | 3 mths | 20–200 | 65 |
| *R. hecksheri* (4) | 9 | Stream bank | 6 mths | 4–64 | 16 |
| *R. pipiens* (6) | 18 | Lake shore | 3 mths | 15–270 | 88 |
| *R. pipiens* (6) | 28 | Pasture | 3 mths | 15–615 | 372 |
| *R. pretiosa* (21) | | | | | |
| Breeding season | 9 | Marsh | 3 yrs | | 7,000[b] |
| Dry season | 9 | Marsh | 3 yrs | | 2,000[c] |
| *R. sylvatica* (5) | 17 | Bog | 3 mths | 3–368 | 65 |

*Sources:* (1) Martof 1953b; (2) Jameson 1955b; (3) Pearson 1955; (4) Hansen 1957; (5) Bellis 1965; (6) Dole 1965a; (7) Clarke 1974; (8) Dole and Durant 1974b; (9) Kramer 1974; (10) Dixon and Staton 1976; (11) Zug and Zug 1979; (12) Wells 1980a; (13) Wells 1980c; (14) Sinsch 1987a; (15) Donnelly 1989b; (16) Summers 1989; (17) Roithmair 1992; (18) Summers 1992; (19) Roithmair 1994; (20) Kusano, Fukuyama, and Miyashita 1995; (21) Watson, McAllister, and Pierce 2003.

*Notes:* Only estimates based on the minimum area polygon method are included. Data for aquatic breeders show sizes of terrestrial home ranges during the non-breeding season and combine data for males and females unless otherwise indicated. Home range sizes for dendrobatid frogs are for territories defended by males except where otherwise indicated. Most data are based on at least five captures per individual. $N_i$ = number of individuals.

[a]Activity range for entire year.

[b]Minimum convex polygon for radio-tracked adults in breeding season.

[c]Minimum convex polygon for radio-tracked adults around drying pools.

home ranges each year (Martof 1953a; Turner 1960a; Bellis 1965; Dole 1965b; Heusser 1968d, 1969c; Clarke 1974a; Dole 1974; Yano 1978; Okuno 1985; Fukuyama, Kusano, and Nakane 1988; Kusano 1998).

The factors influencing home range size are unknown for most amphibian species, but presumably availability of suitable shelter sites and abundance of food are important factors. In one of the few experimental studies conducted to date, Kleeberger (1985) examined factors affecting home range size in *Desmognathus monticola* by manipulating both salamander density and the number of cover objects. He found that home range size increased at high population densities, perhaps because of competition for food. Salamanders from high-density plots generally had less food in their stomachs than did those from control plots. When cover objects, or both cover objects and salamanders, were added to plots, there was no effect on home range size. Apparently, as population density increased, individuals wandered more widely in search of food or shelter. Spieler and Linsenmair (1998) observed enormous variation in the size of individual home ranges of a ranid frog (*Hoplobatrachus occipitalis*) in a West African savanna. Frogs inhabiting a sandy riverbank with abundant bushes that were used for shelter sites had much smaller home ranges than did those on more open, rocky riverbanks. These authors speculated that differences in quality and abundance of diurnal retreat sites and prey were responsible for the difference in home range size, but they did not present quantitative data to support their argument.

## Migrations of Adult Amphibians

### Movements to and from Breeding Ponds

Although many terrestrial salamanders and anurans remain in the same areas year round, some amphibians make seasonal migrations between hibernation sites or terrestrial home ranges and aquatic breeding sites. There is a considerable literature on the timing of breeding migrations in relation to time of day, temperature, and rainfall (reviewed in Salthe and Mecham 1974). Most amphibians migrate in wet weather, a pattern to be expected in animals that face severe problems of evaporative water loss (see chapter 2). Migrations of winter and early spring breeders often are strongly influenced by temperature as well, but temperature usually has little effect on patterns of activity in tropical or summer-breeding amphibians (Blanchard 1930; Baldauf 1952; Packer 1960; Oldham 1966; Heusser 1968d; Hurlbert 1969; Fitzgerald and Bider 1974a, b; Gibbons and Bennett 1974; Douglas 1979; Semlitsch 1981b, 1983a, 1985a; Wisniewski et al. 1981; Gittens 1983a; Harrison et al. 1983; Griffiths 1984; Petranka 1984d; Semlitsch and Pechmann 1985; Beneski, Zalisko, and Larsen 1986; Pechmann and Semlitsch 1986; Aichinger 1987a; Sinsch 1988c, 1992c; Donnelly and Guyer 1994; Palis 1998; Spieler and Linsenmair 1998). Most breeding migrations take place at night (Heusser 1968d; Sinsch 1988c; Semlitsch and Pechmann 1985; Pechmann and Semlitsch 1986; Spieler and Linsenmair 1998), and even a bright full moon can reduce activity in some species (Fitzgerald and Bider 1974b). Once a migration has been initiated, however, some amphibians continue to move toward ponds or summer home ranges during daylight hours (Gittins 1983b; Kusano, Fukuyama, and Miyashita 1995).

Information on the distances traveled by adult amphibians to breeding sites is scarce. This is mainly because marked individuals are seldom recaptured away from breeding areas even with frequent and intensive searching. There are scattered reports of adult amphibians being found several hundred meters to several kilometers from the nearest breeding area (Fitch 1958; Landreth and Christensen 1971; Degani and Warburg 1978; Franz, Dodd, and Jones 1988; Dodd, 1996). Displacement studies have shown that some species, such as *Taricha rivularis,* are capable of moving many kilometers, although they probably do not move such long distances on their own (Twitty, Grant, and Anderson 1964; Twitty 1966). One of the most impressive migrations reported so far was observed in a mixed population *Rana lessonae* and *R. esculenta* (a hybridogenetic species) that breeds in a lake on the border of Austria and Hungary. Some individuals moved up to 15 km from breeding sites to hibernation sites in peat fens at the southern end of the lake. Tunner (1992) speculated that this was an ancient hibernation site for these frogs, and that migration distances gradually increased with a northward expansion of the lake in the last few thousand years. In a population of *Hoplobatrachus occipitalis* in a West African savanna, a few individuals moved as much as 6 km from home ranges along a river to flooded areas of the savanna that were used for reproduction. Some frogs moved more than a kilometer on a single night (Spieler and Linsenmair 1998).

The limited data available suggest that migrations of most amphibians are much shorter (Glandt 1986). Some of the best information comes from Heusser's (1968d, 1969c) studies of *Bufo bufo* in Europe. He found that most toads have summer home ranges 500–1,500 m from the breeding pond, with a few individuals moving as much as 3 km. The toads often took the same routes between the pond and their summer home ranges each year. Studies of other populations of the same species yielded similar migration distances, with most individuals occupying home ranges within 1,600 m of their breeding ponds (Blab 1978; Sinsch 1988c). Many individuals of this species begin moving from summer home ranges toward breeding ponds in the fall and hibernate in

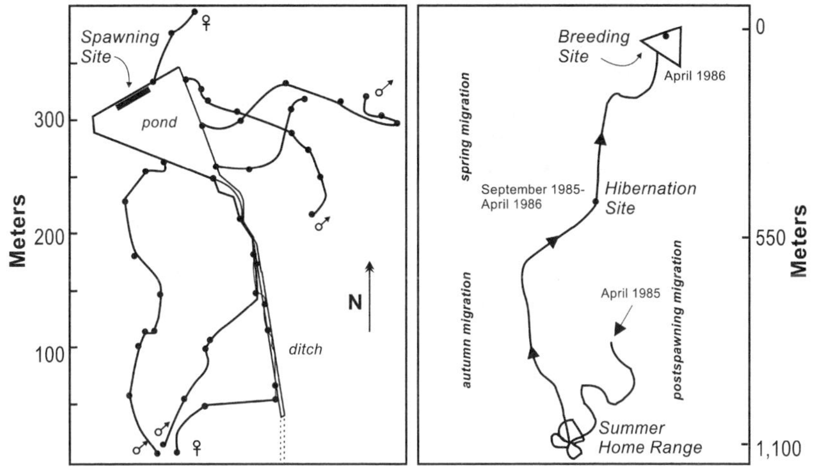


**Fig. 6.3.** Migration behavior of common toads (*Bufo bufo*) in Germany. (A) Paths taken by seven individual toads to a breeding pond in spring. After Sinsch (1988c). (B) Seasonal movements of a single male toad between the summer home range, hibernation site, and breeding pond. After Sinsch (1990).

sites closer to the ponds, so the actual migration distances in the spring are much shorter (Heusser 1968d; Sinsch 1988c, 1989b; fig. 6.3). In a population of the same species in the Austrian Alps, individuals moved 130–1,000 m from breeding ponds, but some also moved nearly 400 m up steep, rocky slopes to reach suitable patches of habitat above the breeding ponds (Sztatecsny and Schabetsberger 2005).

Other toads also migrate a few hundred meters to breeding ponds. Most male natterjack toads (*Bufo calamita*) in a population in Germany remained within 400 m of their breeding sites, but females moved much greater distances, sometimes up to 3 km (Sinsch 1989b, 1992b, c). In a population of the same species in a more arid agricultural region of Spain, individuals moved to sites within 1,000 m of the breeding site, but cumulative movement paths were as long as 4.4 km (Miaud, Sanuy, and Avrillier 2000). American toads (*B. americanus*) usually migrate less than 500 m (Blair 1943; Oldham 1966), and Japanese toads (*B. japonicus*) living in developed areas usually have home ranges within 200 m of breeding ponds (Yano 1978; Okuno 1986; Kusano, Fukuyama, and Miyashita 1995). In a population of *Bufo boreas* in Colorado, most males remained within 500 m of the breeding pond, but some females moved two to three times as far (Muths 2003), a pattern similar to that observed in *Bufo calamita*. The same pattern was seen in a population of *B. boreas* in Idaho, with males moving less than 600 m from breeding sites on average, whereas females averaged over 1100 m of movement (Bartelt, Peterson, and Klaver 2004).

A radio tracking study of *Rana luteiventris* in the mountains of Idaho revealed considerable variation in patterns of movement among different habitats (Pilliod, Peterson, and Ritson 2002). Many individuals remained at breeding sites throughout the summer, long after breeding had ceased, but others moved considerable distances to separate summer habitats. Males were more likely than females to remain in the breeding habitat, and those that did move usually were found within 400 m of the breeding pond, whereas females moved up to 1,033 m from breeding ponds. Similar variation in movement patterns was reported for populations of *Rana draytonii* in California coastal forests and grasslands. Some individuals moved as much as 500 m from breeding sites, either ponds or streams, whereas most remained within 130 m of the breeding area (Bulger, Scott, and Seymour 2003). There was no evidence of a sexual difference in movement in this study.

The reason for shorter movements by males of some species is not entirely clear. Muths (2003) attributed the larger home ranges and longer movements of female boreal toads (*Bufo boreas*) to greater energetic requirements of females compared to males, but there are several possible alternative explanations for this pattern. Often there is strong selection on males to arrive early at breeding ponds (see chapter 8), and remaining close to breeding sites would facilitate rapid return when breeding resumes the next year. This sort of sexual dimorphism in behavior also could reflect differences in locomotor capacities of males and females. In most anurans, including *Bufo calamita, B. boreas,* and *Rana luteiventris,* females are larger than males, and large frogs tend to move farther than small ones (Pilliod, Peterson, and Ritson 2002). Nevertheless, males of all of these species occasionally make considerably longer movements.

Many other anurans have home ranges within a few hundred meters of breeding sites, and often less than 100 m (Martof 1953a; H. Moore 1954; Delzell 1958; Turner 1960a; Kramer 1973; Hollenbeck 1974; Blab 1978; Niekisch 1982;

Freda and Gonzalez 1986; Glandt 1986; Büchs 1987; Hellbernd 1987; Fiorito et al. 1994; Kusano 1998; Richter et al. 2001). In general, one might expect individuals moving toward breeding ponds to move faster and in a more directed manner than those moving away from the pond at the end of the breeding season, especially for explosive early-spring breeders. Few data are available to test this hypothesis, however, mainly because anurans are much more likely to be tracked while leaving ponds than while approaching them. Migrating toads have been observed moving at speeds of about 30 m per hour, or 200–250 m per night (Oldham 1966; Gittins, Parker, and Slater 1980; van Gelder, Aarts, and Staal 1986). Those with home ranges closest to breeding ponds probably could complete the journey in a single night, whereas others might require five or six nights. Movements by *Rana luteiventris* in Idaho were considerably faster, up to 160 m/h, or more than 700 m/day (Pilliod et al. 2003).

Ambystomatid salamanders tagged with radioactive wires or equipped with radio transmitters generally moved less than 250 m from breeding ponds to small mammal burrows and other retreat sites used for much of the year (Williams 1973; Douglas and Monroe 1981; Semlitsch 1981b; Kleeberger and Werner 1983; Madison and Farrand 1998; Faccio 2003; Regosin, Windmiller, and Reed 2003). Most adults in a population of *Hynobius nebulosus* in Japan settled in home ranges less than 100 m from the breeding pond (Kusano and Miyashita 1984). Females of *Salamandra salamandra,* a viviparous species, moved similar distances from their summer home ranges to deposit larvae in the water (Joly 1968). European newts (*Triturus*) sometimes have terrestrial retreat sites only 5–10 m from breeding ponds, although movements of up to 400 m have been recorded (Blab 1978; Dolmen 1981b; Griffiths 1984; Glandt 1986; Jehle and Arntzen 2000). Some striped newts (*Notophthalmus perstriatus*) in a Florida population moved at least 500 m from a breeding pond, although most moved shorter distances (S. Johnson 2003).

### Breeding Site Fidelity

Many amphibians exhibit strong fidelity to particular breeding sites, although occasional movements of 0.5–2 km between breeding sites have been reported (Raney 1940; Blair 1943; Turner 1960a; Miaud, Joly, and Castanet 1993; Sinsch and Seidel 1995). Adult amphibians seldom move between breeding sites within a breeding season, and they often are recaptured in the same pond or stream year after year (Jameson 1957b; Twitty 1959, 1961, 1966; Heusser 1958a, 1960, 1969c; Oldham 1966, 1967; Whitford and Vinegar 1966; Joly and Miaud 1989; Berven and Grudzien 1990; Reading, Loman, and Madsen 1991; Sinsch 1992b, c; Sinsch and Seidel 1995; Lüddecke 1996; Sjögren Gulve 1998b; Kusano, Maruyama, and Kaneko 1999; Pilliod, Peterson, and Ritson 2002; Vasconcelos and Calhoun 2004; Greenberg and Tanner 2005). Some amphibians will return to the same site for several years even after the breeding habitat has been destroyed (Jungfer 1943; P. Anderson 1954; Heusser 1960, 1964, 1969c; McMillan 1963; Shoop 1968). Nevertheless, Schlupp and Podloucky (1994) were able to experimentally shift the fidelity of European common toads (*Bufo bufo*) to a new breeding pond by blocking access to their old pond and transferring them to the new one. After four years, less than 15% of the migration was directed toward the old site, and in later years, less than 1% of the toads moved in that direction.

There is some variation in the degree of site fidelity among species. Griffiths (1984), Joly and Miaud (1989), and Perret et al. (2003) all reported that most adults of European newts (*Triturus*) were captured in only one pond, but some moved between ponds that were relatively close together. Gill (1978a, b) followed the movements of more than 8,500 red-spotted newts (*Notophthalmus viridescens*) and never found an adult that changed breeding ponds, even though some ponds were only a few hundred meters apart and were connected by waterways. A long-term study of *Taricha rivularis* showed that individuals returned to the same 50-yard (46 m) stretch of stream every year and were never recaptured in adjacent steams (Packer 1963b; Twitty 1959, 1961, 1966). Many newts continued to turn up in their home stream segments eleven years after being marked (Twitty, Grant, and Anderson 1967). In a study of *Ambystoma maculatum* in Maine, adults were 100% faithful to their breeding ponds (Vasconcelos and Calhoun 2004).

Berven and Grudzien (1990) marked hundreds of wood frogs (*Rana sylvatica*) in montane ponds in Virginia over a seven year period and found that all adults were completely faithful to the first ponds in which they bred, even though some ponds were only 250–1,000 m apart. Another study of the same species in Maine found that 98% of males and 88% of females were faithful to their breeding ponds (Vasconcelos and Calhoun 2004). Movement of breeding adults between ponds in a population of *Hyla chrysoscelis* was less than 2%, except when one pond dried up (Ritke, Babb, and Ritke 1991). Movement of adult Fowler's toads (*Bufo fowleri*) to new breeding areas was less than 17% in an area with many small ponds separated by sand dunes (Breden 1988). Male natterjack toads (*Bufo calamita*) breeding in areas with networks of small pools showed great fidelity (98%) to specific breeding areas, but were less faithful to individual breeding pools (about 67%). Some males used as many as four different pools in a single breeding season (Sinsch and Seidel 1995). Females showed less site fidelity and sometimes used breeding sites up to 3 km apart in different years (Sinsch 1992c). Some movement between breeding sites within a breeding season also was reported for a population of *B. calamita* in Spain (Miaud, Sanuy, and Avrillier 2000). D. Marsh (2001) studied movements of *Physalaemus pustulosus* in Panama in plots supplied with small artificial pools. Individuals readily moved between pools, which were only five m apart, but seldom moved between plots, which were 200 m apart. These frogs probably do not regularly move between natural breeding ponds that are 200 m apart (D. Marsh, Fegraus, and Harrison 1999).

Okuno (1984) marked about 1,500 Japanese toads (*Bufo japonicus*) at several ponds on a university campus over a period of eight years. Only 25 of these toads changed breeding sites, and nearly all of these moved from the largest pond to smaller ponds with lower density breeding populations. All of the ponds were within 150 m of one another. The degree of site fidelity in various species undoubtedly depends in part on the availability of alternative breeding sites in the area. A comparison of several studies of toads revealed that breeding site fidelity was very high (95–97%) for populations with fewer than four breeding ponds within normal migratory distances, but lower (24–73%) for populations with 6–15 ponds available (Sinsch 1992a). Both *Bufo boreas* and *Rana cascadae* exhibit strong site fidelity to specific breeding sites, even within the same lake. These species often breed in isolated alcoves surrounded by steep cliffs that effectively isolate breeding sites from each other (Olson 1992).

### Movements to Overwintering Sites

Many amphibians that hibernate in terrestrial sites probably spend the winter in sites within their normal home ranges. Some species, however, make regular movements to special hibernation sites. The movements of European common toads (*Bufo bufo*) from summer home ranges to terrestrial hibernation sites closer to their breeding ponds has already been mentioned (Sinsch 1988c). Some amphibians that hibernate in or near water move from summer home ranges to different habitats that provide better conditions for hibernation. For example, a population of green frogs (*Rana clamitans*) in upstate New York moved from permanent ponds, where they spent the summer, to hibernation sites along seeps and streams, where flowing water provided sites that were well supplied with oxygen and unlikely to freeze (Lamoureux and Madison 1999). Movement distances were extremely variable, ranging from 80 to 560 m. In the mountains of California, a population of mountain yellow-legged frogs (*Rana muscosa*) was widely distributed in a system of lakes in the summer. In September, some frogs moved more than 450 m to just a few of these lakes, where they spent the winter in underwater crevices along the lakeshore (Matthews and Pope 1999). Similar migrations to deep lake overwintering sites also have been observed in *Rana luteiventris* in the mountains of Idaho (Pilliod, Peterson, and Ritson 2002; see chapter 3 for a more detailed discussion of hibernation).

## Movements of Juveniles

Data on movements of metamorphosed young from breeding sites are meager. Because mortality rates are very high for juveniles, hundreds or even thousands of individuals must be marked if any are to be recaptured. Furthermore, juveniles tend to disperse in many directions, although they may exhibit preferences for particular terrestrial habitats near breeding sites (e.g., Vasconcelos and Calhoun 2004). It is nearly impossible for investigators to check all possible sites for marked individuals. Consequently, data on movements often are limited to the first few weeks after metamorphosis. If marked juveniles are recaptured at all, they usually are found within 500 m of their natal ponds (Jameson 1956a; Schroeder 1976; Roble 1979; Semlitsch 1981b; Pilliod, Peterson, and Ritson 2002; Vasconcelos and Calhoun 2004). Sometimes a few individuals have been recaptured up to 5 km away (Blair 1953; Dole 1971; Schroeder 1976; Turner 1960a; Gill 1978a; Plytycz and Bigaj 1984; Breden 1987).

There is a general assumption in the literature that most amphibians return to the ponds or streams where they hatched to breed (that is, they exhibit natal philopatry). Indeed, it has been suggested that adult amphibians locate breeding areas by retracing routes followed during dispersal as juveniles (Anderson, in Twitty 1961; D. Ferguson and Landreth 1966). Nevertheless, clear evidence for natal philopatry in amphibians is scarce. In his long-term study of movements in *Taricha rivularis,* Twitty (1966) never marked juveniles dispersing from streams. He did find that hybrid newts reared in tanks and released as larvae into limited areas of a stream tended to return to the same areas as breeding adults. Gill (1978a, b) proposed that the terrestrial eft of the red-spotted newt (*Notophthalmus viridescens*) is a dispersal stage that allows for long-distance migration to new breeding sites. He hypothesized that selection should favor dispersal because of variability and unpredictability in the quality of breeding habitats. Early data supported the hypothesis, because many unknown adults entered his ponds to breed. Subsequent data, however, revealed that the eft stage is much longer (six years or more) than previously suspected. When individuals originally identified as dispersing juveniles finally began moving into breeding ponds, Gill (personal communication) found that nearly 70% of those recaptured returned to their natal ponds, and most of the others moved into the next nearest pond. These data suggest that long-distance dispersal is less important in this species than originally thought. Only about 10% of all juveniles were recaptured as adults, however, so it is possible that some moved longer distances to new breeding areas.

Blair (1953) reported that only 7% of 357 *Bufo valliceps* marked as juveniles were recaptured as adults at their natal pond. Only one other individual was recaptured at all (in another pond), so it is not clear whether the remaining individuals died or dispersed to more distant sites. Spencer (1964) marked over 5,000 juvenile chorus frogs (*Pseudacris triseriata*) in a mountainous region and did not find any movement between breeding sites, some of which were less than 1 km apart. The longest movement by a juvenile frog was only 690 m, and most moved less than 300 m from their natal ponds. In contrast, 18% of 5,000 juvenile wood frogs (*Rana sylvatica*) marked in montane ponds were subsequently recaptured as breeding adults in different ponds (Berven and Grudzien 1990). Breden (1987) found even higher rates of dispersal (27%) by juvenile Fowler's toads (*Bufo fowleri*), but breeding pools were closer together than in the other studies. Juveniles of *Bombina variegata* in Poland were more likely to move to new breeding areas than were adults, although most individuals returned to their natal pond (Płytycz and Bigaj 1984).

Kneitz (1998) presented some of the most detailed information on dispersal of juvenile anurans. He studied populations of anurans in an agricultural landscape in Germany, where a number of ponds were in close proximity to one another, but often separated by relatively inhospitable habitat. He marked several thousand juveniles of three ranid frogs (*Rana temporaria, R. dalmatina, R. esculenta*) and one toad (*Bufo bufo*). Between 80 and 98% of recaptured individuals returned to their natal ponds, and most of the rest dispersed to the next nearest pond. Pool frogs (*Rana lessonae*) in Sweden exhibited less natal philopatry, with about 65% of recaptured juveniles returning to the natal pond. The rest moved to another pond about 270 m away (Sjögren-Gulve 1998b).

It seems likely that juvenile dispersal is more common than is adult dispersal in many aquatic-breeding amphibians. If juveniles move more or less randomly through terrestrial habitat in search of suitable breeding habitat, dispersal distances may be greater for species that take several years to reach sexual maturity than for those that mature early, simply because they have more time to wander from their natal ponds (Palo et al. 2004). Genetic studies of several species have suggested that females often disperse farther than do males, and in these cases, females usually mature at a later age (Austin et al. 2003; Palo et al. 2004). In addition, there may be other reasons for sex-biased dispersal, such as inbreeding avoidance or selection on males to find suitable breeding ponds as early as possible.

## Colonization of New Habitats

If all juvenile and adult amphibians were completely faithful to one breeding pond, then they never would colonize new habitats. Yet such colonization clearly does occur on a regular basis. Some amphibians occur in metapopulations in which many local breeding sites contribute recruits to the overall population, and local extinction is balanced by immigration from other ponds (Gill 1978b; Laan and Verboom 1990; Sjögren 1991a; Olson 1992; Sinsch 1992c; Miaud, Joly, and Castanet 1993; Sjögren Gulve 1994; Sinsch and Seidel 1995). The high degree of philopatry exhibited by adults and juveniles of most amphibian species studied to date probably is due, in part, to a focus on relatively large, isolated breeding sites that are easily encompassed by drift fences.

Amphibians that use more ephemeral habitats are harder to study, but are much more likely to move between breeding sites, because the location of suitable breeding areas changes from year to year. For example, the Neotropical túngara frog (*Physalaemus pustulosus*) breeds in all sorts of temporary water, from rain puddles and water-filled animal footprints to drainage ditches and road ruts. Frogs are readily attracted to shallow pans of water placed on the forest floor (Rand 1983), and even flooded lawns are rapidly colonized by breeding pairs. D. M. Marsh, Fegraus, and Harrison (1999) placed small artificial pools at distances up to 200 m from existing breeding pools on Barro Colorado Island in Panama. They found that ponds close to existing breeding sites were colonized more quickly than were those at greater distances, and more egg masses were deposited in the nearby ponds. They could not determine, however, whether ponds were colonized mostly by dispersing juveniles just starting to breed, or by adults that moved between ponds. In a second set of experiments, D. M. Marsh, Rand, and Ryan (2000) placed artificial ponds in groups of three, with distances of 0–50 m between ponds in a group. Adult male túngara frogs showed strong fidelity to one breeding pond when the ponds were widely spaced, but moved extensively between ponds when they were close together (fig. 6.4). Many other tropical frogs breed in ephemeral habitats ranging from small rock pools to water-filled seed pods (Crump 1974; Gascon 1991, 1992b; Caldwell 1993; Spieler and Linsenmair 1997; see also chapter 10), and these must be colonized by dispersing adults. Even the highly aquatic frog *Pipa arrabali* will colonize small basins placed on the forest floor (Gascon 1992c).

A number of investigators have constructed larger ponds in suitable habitats, and these usually were colonized by a variety of amphibians within a few years (Wilbur and Travis 1984; Laan and Verboom 1990; Resetarits and Wilbur 1989, 1991; Arntzen and Teunis 1993; Banks, Beebee, and Denton 1993; Gressler 1997). Indeed, the construction of new breed-

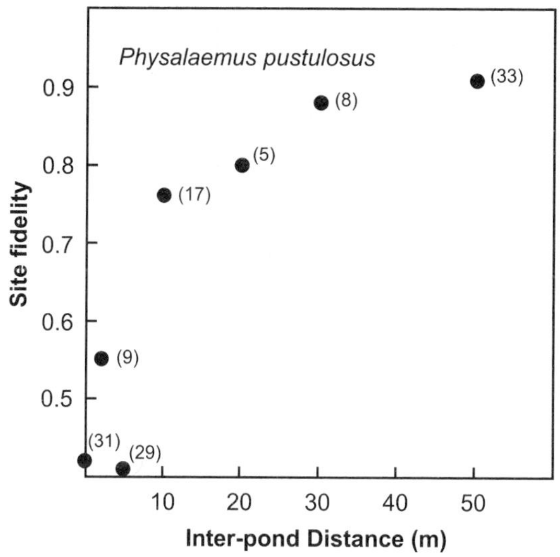


**Fig. 6.4.** Effects of interpond distance on site fidelity of male túngara frogs (*Physalaemus pustulosus*) in Panama. Site fidelity was measured as the probability of a male being recaptured at the previous capture site. Numbers in parentheses show number of recaptures of males at each distance. After D. Marsh, Rand, and Ryan (2000).

ing ponds that are subsequently colonized by existing populations has been a major focus of wetland restoration and amphibian conservation efforts, both in Europe and North America (Thielcke et al. 1985; Pauley and Barron 1995; Beebee 1996; Lehtinen and Galatowitsch 2001; Pechmann et al. 2001). Colonization of new ponds depends not only on the suitability of the ponds as breeding habitats for various species, but also on proximity to established breeding areas, the degree of isolation by unsuitable habitat, and the dispersal abilities of different species (Laan and Verboom 1990; Vos and Chardon 1998; Baker and Halliday 1999; Lehtinen, Galatowitsch, and Tester 1999; Babbitt and Tanner 2000; see subsequent further discussion and chapter 16).

Some amphibians can be considered pioneering species that specialize in invading newly opened habitats. These species can be expected to exhibit greater dispersal tendencies than those that breed in more stable habitats. Pioneering species typically lay large numbers of small eggs, often have short larval periods, and their larvae tend to do poorly in competition with species adapted to more permanent water (see also chapters 10, 13, and 15). Many anurans that breed in temporary ponds exhibit life-history characteristics of pioneering species, but this is more pronounced in some species than in others. For example, many toads (*Bufo*) tend to breed in newly formed ephemeral habitats, but some species show greater tendencies to use such habitats than do others. The European natterjack toad (*Bufo calamita*) is more of a pioneering species than is the common toad (*B. bufo*), with which it shares much of its range. Natterjack toads favor shallow, newly formed, nutrient-poor ponds in sandy regions, and their tadpoles do poorly in more mature, nutrient-rich ponds with abundant populations of predators and competitors (Beebee and Griffin 1977; Beebee 1979b, c, 1983; Banks and Beebee 1986b, 1987a, b, 1988; Griffiths 1991; Griffiths, Edgar, and Wong 1991; Tejedo 1991). The low site fidelity of female natterjack toads presumably facilitates movement into new habitats, although movement of adults between habitat patches appears to be insufficient to account for recolonization of breeding sites after local extinction events (Sinsch 1989b, 1992b). This suggests that dispersal by juveniles also is important in colonization of new breeding sites. Sinsch (1997) found that juvenile natterjack toads dispersed up to two km from their natal ponds within a few weeks. He recaptured only nine tagged individuals, all males, as breeding adults. All of these individuals returned to their natal pond to breed, but this could have been due to the destruction of other nearby breeding ponds.

Another pioneering species is the Neotropical toad *Bufo marinus*. In its native habitats in Central and South America, this species breeds in various kinds of temporary water (Zug and Zug 1979). It was introduced into Queensland, Australia in the 1930s to control sugar cane beetles. It has since dispersed over much of northern Australia and southward to New South Wales, increasing its range by as much as 35 km per year (Van Beurden and Grigg 1980; Easteal 1981; Sabath, Boughton, and Easteal 1981; Easteal et al. 1985; Freeland and Martin 1985; Easteal and Floyd 1986). Transport by humans has been responsible for some of this range expansion, but much of it probably has been due to dispersal of thousands of juveniles from their natal ponds to new breeding sites. Sinsch (1990) cited data on the rate of range expansion given by Freeland and Martin (1985) as evidence that juvenile toads can disperse as much as 35 km per year, but these authors did not present data on juvenile dispersal. Instead, they estimated rates of range expansion from questionnaires distributed to the public. They did report anecdotal evidence that sexually immature toads tended to turn up in new localities before any breeding activity had been reported.

Some amphibians that are not strictly pioneering species nevertheless occupy a wide range of habitats, from relatively ephemeral to more permanent ponds. Many of these species probably are capable of long-distance dispersal, either as juveniles or adults, and this ensures rapid colonization of new breeding sites. For example, the European tree frog (*Hyla arborea*) is widely distributed over much of the continent. In Sweden, Edenhamn (1996) did not find any relationship between degree of pond isolation and the use of ponds for breeding by this species, but Vos and Stumpel (1995) did find an effect of pond isolation in a more fragmented landscape. Juveniles of this species have been recaptured up to 12 km from their natal ponds (Stumpel and

Hanekamp 1986), suggesting an ability to disperse over long distances, but the proportion of long-distance dispersers in a population is unknown.

## Movement and the Genetic Structure of Populations

The genetic structure of populations is governed by current gene flow between populations, selection acting on local populations, and historical events such as expansion and contraction of species' ranges, extinction of local populations, and genetic bottlenecks brought about by small population size (Driscoll 1998; Barber 1999). Gene flow between local populations of amphibians is regulated by dispersal of juveniles and adults between breeding sites. Natal philopatry, coupled with a lack of movement between breeding sites by adults, can lead to increased inbreeding within local populations and a corresponding reduction in genetic variation. It also can lead to genetic differentiation between populations, especially if selection pressures differ between localities. In contrast, movement among populations tends to reduce local genetic differentiation (Merritt et al. 1984; Breden 1987; Berven and Grudzien 1990; Waldman and McKinnon 1993).

Information on adult movements alone is not sufficient to predict patterns of gene flow, because many species that exhibit philopatry as adults often show more extensive movement in the juvenile stage (Shields 1982). Unfortunately, juvenile dispersal has not always been considered in studies of genetic variation. For example, Inger, Voris, and Voris (1974) examined electrophoretic variation in a number of rainforest anurans from Malaysia. They predicted that species in which adults remain in fixed home ranges along streams and do not move to separate breeding sites would show less genetic variation than species that aggregate to breed, because restricted movement would lead to increased inbreeding. This would be true only if juvenile frogs return as adults to breed near their birthplaces. It is equally plausible that juveniles move some distance before finding suitable unoccupied home sites. In fact, they found that patterns of genetic variation in several ranids and bufonids did not support the prediction of low gene flow in streamside species.

### Genetic Variation among Populations

Surveys of anuran populations almost always show some genetic differentiation among populations. Many of these studies involve comparisons of samples from widely separated localities (e.g., Case, Haneline, and Smith 1975; Matthews 1975; Platz 1976; Dunlap and Platz 1981; Easteal 1985; Barber 1999; E. Hoffman, Schueler, and Blouin 2004). Some studies have revealed significant genetic differences among populations only a short distance apart, even when overall genetic similarity among populations is high (Cory 1962; Tordoff, Pettus, and Matthews 1976; Case 1978; Hedgecock 1978; Green 1984; Waldman, Rice, and Honeycutt 1992; Waldman and McKinnon 1993; Scribner et al. 2001; Lampert et al. 2003; Jehle et al. 2005). These data suggest that movement between breeding sites by both juveniles and adults is limited, although some genetic differences could be due to strong local selective pressures that counteract the effects of migration between demes (e.g., Tordoff 1980). Genetic differentiation among populations is likely to be greatest for populations separated by major barriers, such as rivers and mountain ridges (Lougheed et al. 1999; Lampert et al. 2003; Funk et al. 2005). Some amphibians also exhibit a clear correlation between genetic differentiation and distance between populations (e.g., E. Hoffman, Schueler, and Blouin 2004), especially in subdivided habitats (Monsen and Blouin 2003, 2004; Funk et al. 2005).

Genetic structuring sometimes can occur in the absence of obvious habitat barriers to dispersal, especially if individuals have inherently low tendencies to disperse from natal areas. For example, Driscoll (1998) found a high degree of genetic differentiation among populations of two Australian myobatrachid frogs, *Geocrinia lutea* and *G. rosea,* only 1–4 km apart. Genetic differentiation was not related to habitat features such as forest type, the presence of dry ridges, or rivers. Driscoll hypothesized that repeated expansion and contraction of ranges, coupled with very small population size, was responsible for the pattern of genetic structuring in these populations. Once isolated populations were established, there was little or no movement between nearby populations, and hence little or no gene flow. In two closely related species, *G. alba* and *G. vitellina,* there was virtually no dispersal between local populations, although data were available only for adult males. Genetic neighborhoods were very small, less than 40 m in diameter, and were limited to single breeding sites (Driscoll 1997, 1999a). By comparison, Berven and Grudzien (1990) estimated a genetic neighborhood averaging 1.26 km for *Rana sylvatica,* an area that encompassed several different breeding sites. For populations of *Bufo bufo* in England, there is considerable gene flow among ponds less than two km apart, whereas those separated by greater distances show more genetic differentiation (Scribner et al. 2001).

Some amphibian populations exhibit surprisingly low levels of genetic differentiation, despite apparent barriers to dispersal or long distances between populations. Genetic differentiation was low among low-elevation populations of Columbia spotted frogs (*Rana luteiventris*) separated by relatively long distances, whereas genetic differentiation was greater in high-elevation populations (Funk et al. 2005). Populations of *Rana temporaria* and *Bufo bufo* occur on very small, rocky islands in the Baltic Sea, on the coast of Finland. Despite the presence of salt water between the islands, Seppä and Laurila (1999) found little evidence of inbreeding in island populations. Genetic differentiation among islands was small, and most of that was due to variation among groups of islands. Toads have been seen swimming in the sea, and frogs were reported as prey of sea gulls (Kilpi and Byholm 1995). These observations, along with the genetic evidence, suggest that dispersal between islands is relatively common. A study of water frogs (*Rana saharica*) in Morocco also revealed low levels of genetic differentiation among populations, most of which are separated by inhospitable desert (Buckley et al. 1996). In this case, frequent dispersal between local populations seems unlikely. The lack of genetic differentiation is more likely a result of repeated expansion and contraction of local populations during cycles of wetter and drier climates over the last two million years. Recent expansion of the Sahara probably has increased the fragmentation of populations, but these events probably were too recent for substantial genetic differences to appear.

In contrast to these frogs, most populations of salamanders exhibit high levels of genetic differentiation among local populations. Plethodontid salamanders in particular often show greater genetic differentiation among local populations than do most other terrestrial vertebrates (Highton and Webster 1976; Highton 1977; Feder, Wurst, and Wake 1978; Larson and Highton 1978; Duncan and Highton 1979; Tilley et al. 1978; Tilley and Schwerdtfeger 1981; Jacobs 1987; D. Wake 1997). This is especially true of Central American plethodontids, which exhibit high levels of genetic variation within populations, but also remarkably high levels of differentiation between nearby populations (García-París et al. 2000). As discussed at the beginning of this chapter, plethodontid salamanders have some of the lowest rates of movement and smallest home ranges of any terrestrial vertebrates, and this low vagility undoubtedly contributes to low gene flow and high levels of genetic differentiation among local populations. One result has been a proliferation of closely related species in both North America and Central America.

### Genetic Variation within Populations

Several studies have revealed unexpectedly low levels of heterozygosity within breeding populations of amphibians (Tabachnick and Underhill 1972; Guttman and Wilson 1973; Case, Haneline, and Smith 1975; Tabachnick 1977; Hedgecock 1978; Christein, Guttman, and D. H. Taylor 1979; Nevo and Yang 1979; Merritt et al. 1984; Reh and Seitz 1990; Sjögren 1991b; Scribner et al. 2001). There are several possible reasons for this: (1) spatial or social structuring within breeding populations that results in frequent matings of close relatives; (2) selection against heterozygotes; (3) reduced genetic variation during the initial founding of isolated populations; (4) a lack of adult or juvenile dispersal between breeding sites; (5) periodic population declines, resulting in small effective population size; or (6) local extinctions that produce genetic bottlenecks, as sites are recolonized from a single source population.

Substructuring of breeding populations seems unlikely for most species. Two studies of *Bufo americanus* found little evidence for frequent matings among close relatives (Christein, Guttman, and Taylor 1979; Waldman, Rice, and Honeycutt 1992). Waldman, Rice, and Honeycutt (1992) even suggested that female toads might recognize and avoid mating with close male relatives, perhaps from differences in call structure. They reported a significant correlation between genetic relatedness of male toads, as determined by DNA fingerprints, and degree of similarity in temporal features of calls. Unfortunately, they did not present correlation coefficients, but most such relationships appear to be relatively weak. For females to use information from call structure to avoid mating with close relatives, they would need to have a genetic template for call perception that would allow them to identify potential brothers. Waldman (2001) presented results of playback experiments indicating that females can discriminate between the calls of closely related and distantly related males. He hypothesized that females might possess a genetic template for call production that would allow them to recognize the calls of related males, but direct evidence for such a template is currently lacking.

The possibility of selection against heterozygotes often has been inferred from patterns of genetic variation (e.g., Guttman and Wilson 1973; Waldman and McKinnon 1993), but direct evidence for such selection is scarce. Most such evidence consists of interspecific correlations between levels of heterozygosity and environmental variables that are thought to be characteristic of fluctuating or constant environments. For example, Nevo and his colleagues conducted broad-scale electrophoretic surveys of amphibian populations and searched the literature for additional information on genetic variability. They found considerable variation in levels of heterozygosity among families and among species within families (Nevo, Dessauer, and Chauang 1974; Dessauer, Nevo, and Chuang 1975; Nevo 1976a, b; Nevo and Yang 1979, 1982; Nevo, Beiles, and Ben-Shlomo 1984; Nevo and Beiles 1991). They attributed most of this variation to ecological variables—specifically, to differing selective pressures in constant and variable environments.

Nevo and Beiles (1991) reported that terrestrial amphibians (some of which breed in water) exhibited the highest levels of heterozygosity, followed by arboreal, aquatic, fossorial, and subterranean species. They argued that terrestrial and arboreal environments are more variable than aquatic or subterranean environments, and that high levels of heterozygosity represent adaptations to this environmental variation. They also reported that tropical amphibians exhibited higher levels of heterozygosity than did temperate-zone species, despite the fact that many biologists tend to view the tropics as a more equable environment than the temperate zone. They discounted the importance of demographic variables in determining levels of heterozygosity, although they did report that species living in isolated populations or those with a patchy distribution generally exhibited less heterozygosity than those with more widely distributed populations.

The analysis of Nevo and Beiles (1991) provides an unparalleled compilation of data on genetic variation in amphibian populations, but their ecological interpretations of the data are problematic for several reasons. First, they failed to take phylogenetic history into account. Five of the six groups with the highest levels of heterozygosity were anuran families, whereas the three with the lowest levels of heterozygosity were urodele families. Urodeles and anurans were combined in a single analysis, with some families represented by many congeneric species that are ecologically similar (e.g., *Plethodon, Rana, Bufo*), perhaps because of shared ancestry. Other families were represented by only a few species. By using individual species as independent data points in multiple regression analyses, sample sizes were greatly inflated, increasing the probability that significant trends will be apparent (Brooks and McClennan 1991; Harvey and Pagel 1991). In addition, some of the ecological comparisons were biased by heavy representation of certain clades in the analysis. For example, 85% of the tropical species were plethodontid salamanders, all members of a single tribe, the Bolitoglossini. Thus, the comparison of tropical and temperate-zone species is largely a comparison between these salamanders and a variety of other amphibians with diverse life histories and ecology.

Several demographic features of populations related to dispersal probably are closely correlated with the broad ecological categories used in Nevo and Beiles's study. For example, species that were classified as fossorial were mostly desert-dwelling spadefoot toads, which emerge to breed in temporary ponds. Because their ability to move long distances probably is limited by the aridity of their environment, one might expect relatively little dispersal between populations. In addition, local breeding pools often fail to produce any juveniles, so recruits into the population probably come from a small subset of breeding adults. Both of these factors are likely to reduce heterozygosity, and pelobatids as a group had the lowest levels of heterozygosity of any anuran family. Anurans that breed in well-established ponds, such as many ranids, are likely to exhibit greater breeding site fidelity and natal philopatry than are species that breed mostly in temporary water, such as many toads (*Bufo*). Indeed, ranids as a group exhibited lower levels of heterozygosity than did bufonids. Other differences were unexplained. For example, rhacophorid tree frogs had higher levels of heterozygosity than did hylid tree frogs with very similar ecology. Some of these differences probably were related to variation in dispersal patterns of the particular populations used in the analysis.

Other studies have shown that the combined effects of small population size, lack of movement among populations, and periodic local extinction can lead to reduction in heterozygosity within populations and increased genetic differentiation among populations. For example, very low levels of heterozygosity also were reported for local populations of *Rana lessonae* in an area of Sweden not greatly affected by human development (Sjögren 1991a). These frogs reproduced in a network of small pools, many in close proximity to one another. Each pool supported a rather small breeding population, and reproductive failure, usually due to cold weather, was common. In some years, all pools in the area failed to produce metamorphosing frogs, leading to dramatic reductions in recruitment into the breeding population in subsequent years. Migration between pools occurred, but at low levels, and extinction of local populations was frequent. The repeated bottlenecks created by small population sizes, low rates of migration between populations, and periodic extinction probably account for the low level of genetic variation (Sjögren 1991a, b).

## Patterns of Movement and Amphibian Conservation

Information about patterns of movement between terrestrial habitats and aquatic breeding sites, as well as dispersal between breeding sites, is essential for making decisions about protection of critical habitats for threatened or endangered species of amphibians. The importance of dispersal for conservation is discussed in more detail in chapter 16, so only a few points will be mentioned briefly here. One concern is the effect of man-made barriers to dispersal, such as roads, on the genetic structure of populations. High levels of inbreeding and low heterozygosity do not always result in reduced fitness of individuals (Sjögren 1991a), but artificial isolation of populations by roads and other forms of human land use could reduce effective population sizes and increase the effects of inbreeding depression on reproductive success (Waldman and McKinnon 1993; Scribner et al. 2001). For example, Reh and Seitz (1990) studied the genetic structure of populations of *Rana temporaria* in an area of Germany that was highly dissected by roads and agricultural land. All local populations exhibited low levels of heterozygosity, probably because of increased inbreeding.

Heterozygosity decreased with increasing isolation of populations and was especially low for a population that was completely surrounded by roads. Other studies also have shown decreased levels of genetic variation within populations and increased differentiation among populations of anurans isolated by urban development (Hitchings and Beebee 1997, 1998). In *Bufo bufo,* decreased heterozygosity was correlated with an increase in developmental abnormalities in tadpoles and a decrease in tadpole survival, but these effects were not seen in *Rana temporaria*. Negative effects of decreased heterozygosity can be particularly severe under stressful conditions, such as unusually cold winters or very hot, dry summers (Samollow and Soule 1983).

The effects of low levels of genetic variation, combined with small population sizes, also can make such artificially isolated populations especially vulnerable to extinction. Subsequent recolonization becomes impossible because of a lack of suitable dispersal corridors (Laan and Verboom 1990). Not all fragmented populations exhibit genetic effects of isolation, however. For example, a comparison of populations of red-backed salamanders (*Plethodon cinereus*) in a fragmented suburban landscape and undisturbed forest revealed relatively small differences in the degree of genetic differentiation between populations (Gibbs 1998c). Probably this is due to the large population sizes and very low rates of movement in natural populations of this salamander. Even relatively isolated populations probably can persist for a long time in patches of suitable habitat, with few genetic effects. Such isolated populations could be vulnerable to extinction, however, if events such as prolonged drought reduce populations below viable numbers and immigration is prevented because of habitat barriers.

Roads not only prevent movement between populations, but also can have a direct effect on population size, simply because amphibians are killed when attempting to cross them, especially during migrations to and from breeding areas (Fischer 1969; Karthaus 1985; Kuhn 1987; Cooke 1988, 1995; Fahrig et al. 1995; Hels and Buchwald 2001). Two approaches have been widely adopted in Europe, and to a lesser extent in North America, to alleviate this problem (see chapter 16). The first is the construction of various types of tunnels under roads along known migration routes of amphibians (Langton 1989). The second approach is the construction of substitute breeding ponds to intercept amphibians before they cross over roads. Such ponds often are readily colonized by amphibians, although not necessarily by exactly the same assemblage of species found in natural breeding ponds (Podloucky 1989; Schlupp et al. 1989). Migrating amphibians also can be redirected to new breeding ponds by blocking access to old ones (Schlupp and Podloucky 1994). An important consideration is the proximity of suitable terrestrial habitats or dispersal corridors that would allow amphibians access to newly constructed ponds. Use of ponds as breeding sites by amphibians often decreases when ponds are far removed from wooded habitats used for summer home ranges or hibernation (Loman 1988; Wederkinch 1988; Laan and Verboom 1990; Baker and Halliday 1999).

## Homing Behavior

If amphibians make regular movements between breeding sites and terrestrial habitats, or occasional excursions out of their normal home ranges, they must have some means of finding their way home again (Sinsch 1992a). Indeed, many amphibians can find their way back to a breeding pond or home range after being experimentally displaced (tables 6.6, 6.7). Studies of homing behavior are useful in determining the ability of individuals to make long distance migrations to and from breeding sites, and as a means of investigating mechanisms of orientation. The methods used to study homing vary considerably. Some studies have involved rather haphazard displacements of individuals at different distances, with little attempt to quantify homing success as a function of displacement distance. Many investigators have failed to include proper controls (animals marked and released at their home sites), so it is impossible to compare recapture rates of displaced animals with rates expected for animals that are not displaced. In some studies (e.g., Madison 1969), recapture rates actually were higher for animals released at a new site than for those released at home. Presumably this is because displaced animals immediately headed for home on the surface, while nondisplaced animals disappeared into retreat sites in their home ranges. Barthalmus and Savidge (1974) suggested that the time required for individuals to return home as a function of displacement distance would provide a better measure of homing performance than the proportion of animals homing, but few workers have used this method.

### Homing by Urodeles

Homing to nests by displaced females has been reported for several species of plethodontids (table 6.6), but displacement distances in these studies were very short. F. Rose (1966) found that the number of females of *Desmognathus auriculatus* homing successfully from a 6 m displacement was very low (8%), perhaps because these animals were released outside of their familiar foraging ranges. Other studies of terrestrial plethodontids have shown that many individuals can home from locations up to 30 m from their home ranges (table 6.6), a distance considerably greater than that covered in normal daily movements (table 6.3).

**Table 6.6** Studies of homing in salamanders

| Species (source no.) | Goal | Distance (m) | *N* | Percent recaptured | Percent at home: Recaptured | Percent at home: Total |
|---|---|---|---|---|---|---|
| Ambystomatidae | | | | | | |
| *Ambystoma maculatum* (4) | BP | 30 | 15 | | | 68 |
| | | 49 | 12 | | | 25 |
| | | 96 | 11 | | | 82 |
| | | 99 | 11 | | | 36 |
| | | 128 | 15 | | | 67 |
| *A. maculatum* (5) | BP | 20–175 | 18 | | | 89 |
| | | 200 | 13 | | | 46 |
| | | 250 | 9 | | | 31 |
| | | 500 | 4 | | | 25 |
| Plethodontidae | | | | | | |
| *Aneides aeneus* (1) | NE | 0.4–9 | 13[a] | | | 92 |
| *Desmognathus auriculatus* (2) | NE | 3.0 | 60 | | | 90 |
| | | 4.6 | 60 | | | 63 |
| | | 6.0 | 60 | | | 8 |
| *D. ocoee* (10) | NE | 2.0 | 107 | | | 73 |
| *D. ocoee* (9) | HS | 0.5–4 | 86 | 35 | 87 | 33 |
| *D. ochrophaeus* (12) | HR | 30 | 201 | 56 | 45 | 25 |
| *D. fuscus* (8) | HZ | 0 | 87 | 55 | 98 | 54 |
| | | 30 | 120 | 48 | 60 | 28 |
| *D. fuscus* (6) | HZ | 0 | 30 | 47 | 86 | 40 |
| Upstream displacement | | 30 | 79 | 49 | 18 | 9 |
| Downstream displacement | | 30 | 73 | 51 | 41 | 21 |
| Landward displacement | | 30–45 | 71 | 63 | 64 | 41 |
| *Plethodon cinereus* (13) | HR | 8 | 11 | 91 | 100 | 91 |
| | | 15 | 10 | 100 | 100 | 100 |
| | | 30 | 10 | 90 | 89 | 80 |
| | | 90 | 8 | 63 | 40 | 25 |
| *P. jordani* (7) | HR | 0 | 83 | 43 | | |
| | | 6 | 74 | 72 | 96 | 69 |
| | | 9 | 63 | 67 | 86 | 57 |
| | | 15 | 40 | 83 | 91 | 75 |
| | | 30 | 35 | 69 | 83 | 57 |
| | | 60 | 44 | 64 | 61 | 39 |
| | | 150 | 35 | 43 | 20 | 9 |
| | | 300 | 94 | | 9 | |

*(continued)*

Several species exhibited poor homing performance when displaced 60–300 m (table 6.6).

Joly and Miaud (1989) collected groups of alpine newts (*Triturus alpestris*) from three breeding ponds that were 25–150 m apart. The animals were released at four different sites that were either 30 or 100 m from the ponds. Most individuals that were recaptured returned to their native ponds, but a higher percentage returned from short-distance displacements. Individuals displaced 30 m to a wooded site returned to their ponds within two weeks, but those displaced the same distance into a pasture took up to seven weeks to return. Those displaced 100 m into either woods or pasture

**Table 6.6** (continued)

| Species (source no.) | Goal | Distance (m) | *N* | Percent recaptured | Percent at home: Recaptured | Percent at home: Total |
|---|---|---|---|---|---|---|
| Salamandridae | | | | | | |
| *Notophathalmus viridescens* (11) | BP | 0 | 1,206 | 34 | 100 | 34 |
| | | 400 | 1,776 | 33 | 65 | 21 |
| *Taricha rivularis* (3) | HZ | 1,600[b] | 564 | 64 (5 yrs) | 89 | 57 |
| | | 3,200[c] | 692 | 81 (5 yrs) | 95 | 78 |
| | | 4,025[c] | 747 | 38 (2 yrs) | 89 | 34 |
| | | 8,050[c] | 730 | 44 (2 yrs) | 41 | 18 |

*Sources:* (1) R. Gordon 1961; (2) F. Rose 1966; (3) Twitty 1966; (4) Whitford & Vinegar 1966; (5) Shoop 1968; (6) Barthalmus and Bellis 1969; (7) Madison 1969; (8) Barthalmus and Bellis 1972; (9) Huheey and Brandon 1973; (10) Forester 1979c; (11) Gill 1979; (12) Holomuzki 1982; (13) Kleeberger and Werner 1982.

*Notes:* Data are combined for males, females, and juveniles, except for homing to nests, which are for females only. Both the percentage of individuals recaptured and the percentage of animals that exhibited a homing response are reported whenever possible. All recaptures were made in the same season except as indicated. A distance of zero indicates control animals released at the home site. BP = breeding pond; HR = home range; HS = home site; HZ = home zone in stream; NE = nest with eggs.

[a]Number of displacements of four individuals.

[b]Salamanders displaced to a location in home stream.

[c]Salamanders displaced to a different stream.

also took up to seven weeks. An even longer delay was observed by Gill (1979) after he moved adult red spotted newts (*Notophthalmus viridescens*) from one breeding pond to another 400 m away. Many of the newts remained in the new pond for the rest of that breeding season. About a third of the animals were recaptured a year later, and 65% of these (21% of those displaced) had moved back to the original pond. These animals probably delayed their return because the cost in lost breeding opportunities outweighed any advantage of returning in the same season.

The most impressive homing ability exhibited by any amphibian is that of the newt *Taricha rivularis,* studied for many years by Victor Twitty and his colleagues in California. In initial studies, newts were displaced about 1,600 m to a new location in their home stream. Over a five year period, about 64% of the animals were recaptured, and nearly 90% of those returned to their home segments of the stream (Twitty, Grant, and Anderson 1964; Twitty 1966). In subsequent experiments, newts were displaced to different streams, either in an adjacent valley, or in an area separated from the home stream by several ridges. The longest displacement was about five miles (8 km). In only two years, 44% of the animals were recaptured, with about 41% of these returning to the home stream (table 6.6; Twitty 1966). Records of animals recaptured in drift fences placed along ridge tops indicated that the animals moved overland on a relatively straight compass course to reach their home stream. Homeward bound individuals had to pass through a third stream which was ideal newt habitat, but few of them established residence there. Since these newts generally do not wander far from their home stream segments (Twitty, Grant, and Anderson 1967), there is no doubt that they were able to return home even when released in completely unfamiliar areas.

### Homing by Anurans

There have been a number of anecdotal reports of individual anurans being displaced and returning to their home sites (e.g., Durham and Bennett 1963), but quantitative data are scarce. Most studies have involved displacements of less than 300 m, with the highest percentages of successful homing obtained from individuals displaced from breeding ponds (table 6.7; Jameson 1957b; Oldham 1966, 1967; Lüddecke 1996; Holenweg Peter 2001). Presumably return rates were high in these species because animals were strongly motivated to return to the breeding area. One of the few detailed studies of homing from longer distances is Bogert's (1947) work on *Bufo terrestris.* He found that homing success decreased with displacement distance (table 6.7), although the tendency for animals to establish home ranges at the release sites also depended on the suitability of the habitat in the area. In a two week study of homing by *Rana esculenta, R. ridibunda,* and *R. lessonae,* frogs were displaced up to 225 m from their home ponds (Holenweg Peter 2001). Overall, slightly less than half were recaptured, of which 90% were back in the home ponds. Homing performance declined dramatically at distances of more than 75 m, and none of the frogs released more than 150 m from the home pond returned (table 6.7). Adult mountain yellow-legged frogs (*Rana muscosa*) that were moved 144–630 m to new ponds showed

**Table 6.7** Studies of homing in anurans

| Species (source no.) | Goal | Distance (m) | *N* | Percent recaptured | Percent at home: Recaptured | Percent at home: Total |
|---|---|---|---|---|---|---|
| Bufonidae | | | | | | |
| *Atelopus varius* (11) | HR | 10 | 21 | 100 | 86 | 86 |
| *Bufo americanus* (7) | BP | 0 | 577 | 42 | 88 | 37 |
| | | 104[a] | 594 | 29 | 41 | 12 |
| | | 126[b] | 619 | 45 | 67 | 30 |
| *B. americanus* (9) | HR | 60–120 | 11 | 100 | 10 | 10 |
| *B. bufo* (1) | BP | 0 | 30 | | | 60 |
| | | 50–100 | 44 | | | 7 |
| | | 101–200 | 101 | | | 37 |
| | | 201–300 | 202 | | | 8 |
| | | 301–400 | 128 | | | 14 |
| | | 401–500 | 129 | | | 9 |
| | | 501–600 | 50 | | | 2 |
| | | 70–950 | 102 | | | 13 |
| *B. marinus* (4) | HS | 6–70 | 15 | 73 | 100 | 73 |
| *B. terrestris* (2) | HR | 91 | 13 | | | 54 |
| | | 274 | 15 | | | 60 |
| | | 411 | 19 | | | 37 |
| | | 686 | 92 | | | 21 |
| | | 777 | 18 | | | 17 |
| | | 1,609 | 43 | | | 19 |
| *B. valliceps* (5) | BP | 37–223 | 36 | 78 | 100 | 78 |
| Dendrobatidae | | | | | | |
| *Dendrobates pumilio* (10) | HR | 3–20 | 32 | 88 | 61 | 53 |
| Hylidae | | | | | | |
| *Pseudacris regilla* (3) | BP | 0 | 166 | 62 | 100 | 62 |
| | | 115 | 36 | 64 | 87 | 56 |
| | | 275 | 83 | 67 | 100 | 67 |

*(continued)*

a mixed response, with about half moving back to or toward their original capture sites and the rest remaining at the new release site (Matthews 2003).

Gonser and Woolbright (1995) displaced male *Eleutherodactylus coqui* 20–100 m from their calling sites in a Puerto Rican rainforest. These are terrestrial breeders, so they do not move to ponds, but they exhibit considerable fidelity to calling sites. Nearly all males returned to their home sites after displacements of 20 m, more than half returned from 40 m, but only a quarter returned from 100 m (table 6.7). The time required for the frogs to return home increased with displacement distance as well, from an average of four nights at 20 m to nearly 13 nights at 100 m.

The ability of anurans to return home from displacements exceeding 2 km has not been investigated in detail, but a few studies have examined accuracy of initial homeward orientation at different distances. Dole (1968) displaced leopard frogs (*Rana pipiens*) up to 10.5 km from their home ranges. Although he did not report the number of frogs that returned home, he did find that nearly all frogs displaced 200–800 m moved in a homeward direction, while the orientation of frogs displaced 3.2–10.5 km was random. Furthermore, those displaced shorter distances averaged more than twice as much movement per rainy night as those displaced longer distances, suggesting that some frogs in the latter group accepted the release site and established new home ranges. Sinsch (1987a, c) reported that toads (*Bufo bufo*) displaced up to 250 m showed strong initial home-

**Table 6.7** (continued)

| Species (source no.) | Goal | Distance (m) | *N* | Percent recaptured | Percent at home: Recaptured | Percent at home: Total |
|---|---|---|---|---|---|---|
| Leptodactylidae | | | | | | |
| *Eleutherodactylus coqui* (12) | HR | 20 | 20 | | | 95 |
| | | 40 | 20 | | | 65 |
| | | 100 | 20 | | | 25 |
| Ranidae | | | | | | |
| *Rana clamitans* (8) | BP | 0 | 58 | 76 | 100 | 76 |
| | | 155[b] | 271 | 54 | 99 | 54 |
| | HR | 50–114 | 184 | 58 | 80 | 47 |
| *R. esculenta* complex[c] (13) | BP | 0 | 26 | 75 | 100 | 75 |
| | | 38 | | | | 85 |
| | | 75 | | | | 70 |
| | | 100 | | | | 20 |
| | | 150 | | | | 25 |
| | | > 150 | | | | 0 |
| *R. temporaria* (6) | BP | 15–55[a] | 218 | 68 | 64 | 44 |

*Sources:* (1) W. Jungfer 1943; (2) Bogert 1947; (3) Jameson 1957b; (4) Brattstrom 1962c; (5) Awbrey 1963; (6) Oldham 1963; (7) Olham 1966; (8) Oldham 1967; (9) Dole 1972a; (10) McVey et al. 1981; (11) Crump 1986a; (12) Gonser and Wooldbright 1995; (13) Holenweg Peter et al. 2001.

*Notes:* Data are combined for males, female, and juveniles. Both the percentage of individuals recaptured and the percentage of displaced animals that exhibited a homing response are reported whenever possible. All recaptures were made in the same season. A distance of zero indicates control animals that were released at the home site. BP = breeding pond; HR = home range; HS = home site.

[a]Displaced to a different pond.

[b]Displaced on land.

[c]Combined data for *R. ridibunda, R. esculenta,* and *R. lessonae.*

ward orientation, whereas those displaced 500 or 1,000 m were much less accurate. However, seven of ten toads that were displaced 3 km from one breeding pond to a point within 200 m of another pond showed a strong tendency to move toward their home pond.

### The Adaptive Significance of Homing

Several hypotheses have been proposed to explain the evolution of homing abilities in amphibians. For species such as *Taricha rivularis* and *Bufo bufo,* which spend most of their lives on land and return to the water to breed, homing allows the animals to reliably locate a suitable breeding site and then return to a familiar summer home range (Twitty 1966; Heusser 1968d). For species that live along streams, an ability to home allows individuals to return to a familiar area after being accidentally displaced by floods or other chance events. For example, dendrobatid frogs of the genus *Colostethus* live on boulders in tropical streams that are transformed from little more than seepage areas during the dry season to raging torrents during the wet season. It is not uncommon for frogs to be knocked into the water during fights and be washed several meters downstream, but they apparently return to their home sites with little difficulty (Wells 1980a). During dry periods, males abandon their territories and move into streambeds; some individuals return to their former territories in the next rainy season.

Still another possibility is that homing allows individuals to make forays outside of their normal home ranges to search for better feeding areas or retreat sites, but to return if good quality habitat is not found. The tendency for many amphibians to make occasional long-distance movements and then turn up in their old home ranges has already been mentioned. Phillips (1987) suggested that for aquatic breeders such as *Notophthalmus,* the evolution of homing is related to dispersal. Juveniles leaving their home ponds can explore the surrounding area for suitable breeding sites, but an ability to home gives them the option of returning to the natal pond if alternative sites are not available. Probably no single hypothesis will suffice to explain the evolution of homing ability in all amphibians. An ability to return home at least over short distances probably is a general characteristic of most amphibians, and indeed, most vertebrates; the precise advantage derived from the behavior will depend on the ecology and life history of each species.

## Orientation

The information presented so far shows that many amphibians remain in a relatively restricted area for most of their lives, evidently becoming very familiar with their own home ranges. Occasionally they make long excursions into new areas to feed or reproduce and are able to find their way home from distances ranging from several hundred meters to several kilometers. Individuals of some species clearly can home from distances that greatly exceed any natural movements the animals normally make. The remainder of this chapter considers the sensory mechanisms that allow amphibians to find their way home during natural wanderings or after displacement to unfamiliar terrain.

Two types of orientation seem to be most important in guiding the movements of amphibians. The first is piloting, the ability to steer a course using familiar landmarks (Adler 1976; Papi 1990; Sinsch 1990, 1991b, 1992a). Piloting probably is the principal means of orientation used by amphibians for daily movements within their normal home ranges. Many individuals commonly return to the same retreat site, burrow, or feeding station every day, sometimes for many years in succession. Presumably an animal learns the characteristics of the local area and uses visual or chemical cues to relocate its home site after relatively short-distance movements.

Amphibians also are capable of compass orientation without local landmarks. A compass sense is important in maintaining a proper orientation relative to features in the environment, even when these are not visible to the animal. For example, many amphibians are found near the shores of ponds and streams and often escape predators by jumping directly into the water or into surrounding vegetation. When placed in a featureless arena open only to the sky, these animals usually orient as if they were moving perpendicular to their home shore, rather than orienting in a true homeward direction (fig. 6.5; Adler 1970, 1976). This behavior, termed *Y-axis orientation* (D. Ferguson and Landreth 1966), has provided a useful tool for investigating sensory mechanisms of orientation. Animals can be trained to a shoreline in a laboratory tank and then tested for orientation after experimental manipulation of external cues or sense organs.

Amphibians probably also use a compass sense for orientation during movements to and from breeding sites or during occasional excursions out of their normal home ranges. For example, amphibians collected at drift fences around breeding ponds tend to enter and leave a pond at approximately the same point (Shoop 1965, 1968; Shoop and Doty 1972; Williams 1973; L. Hardy and Raymond 1980; Douglas and Monroe 1981; Stenhouse 1985b; Beneski, Zalisko, and Larsen 1986; Phillips and Sexton 1989; Dodd and Cade 1998; Palis 1998). This could be evidence of orientation along a fixed compass axis. However, in many cases, the animals also could be responding to local environmental cues, or they might use such cues in conjunction with a compass sense. Studies of European frogs have shown consistent orientation toward particular types of habitats by both juveniles and adults leaving breeding ponds (Sjögren-Gulve 1998b). Different species showed somewhat different patterns of orientation, with each tending to orient toward habitats typically used by that species (Sjögren-Gulve 1998a). Experiments with North American wood frogs (*Rana sylvatica*) and spotted salamanders (*Ambystoma maculatum*) showed that juveniles and adults exhibited strong orientation toward forest habitats and away from clear-cut areas and open areas near power lines (deMaynadier and Hunter 1999). Similar results were obtained in a study of European newts (*Triturus cristatus* and *T. vulgaris;* Malmgren 2002). The fact that juveniles emerging from ponds showed the same orientation preferences as adults in these studies suggests that individuals are responding to local environmental cues, since the juveniles have not previously moved along any particular compass axis. It is possible that juveniles use local environmental cues to initially learn the direction of suitable habitats, but then rely in part on compass information to return to a pond in subsequent years. A similar multistage process of orientation has been proposed for juvenile sea turtles leaving the beaches where they hatch (Lohmann et al. 1997).

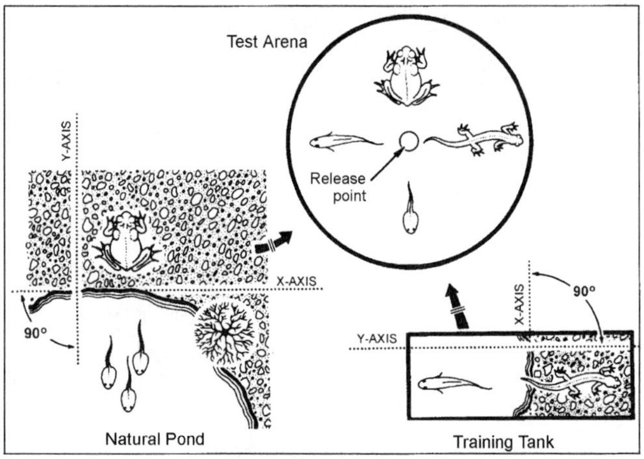


**Fig. 6.5.** Diagram of compass orientation in aquatic and terrestrial vertebrates. When toads and tadpoles collected near a shore are placed in a test arena under a clear sky, they orient along an axis parallel to their natural Y-axis. The toad orients toward land, whereas the tadpoles orient toward water. When a salamander is trained to a shoreline in a training tank and then placed in the test arena, it orients parallel to the trained Y-axis in a shoreward direction. A fish trained to the same shoreline would orient toward water. After Adler (1976).

There is evidence for consistent compass orientation

among adult amphibians moving to and from breeding ponds. Data on movements of salamanders tagged with radioactive wires indicate that most individuals follow a straight-line course when moving from breeding ponds to summer home ranges and simply reverse course to return to the pond the following year (Williams 1973; Semlitsch 1981b, 1983a). Although animals could follow a fixed dispersal path by learning local landmarks, evidence from several anuran species indicates that migrating individuals will maintain their original compass headings even when placed in arenas that obscure local landmarks (Tracy and Dole 1969a; Dole 1972c).

A third type of orientation is true navigation, the ability to orient in the absence of familiar local cues. Navigation requires some type of compass sense. It also requires that the animal be able to determine where it is relative to the goal; that is, it must have a "map sense." Map information could be acquired if an animal integrates information about the distance and direction of its displacement from home and uses this to orient in the opposite direction from its net direction of displacement ("route-based" information). Alternatively, an animal might be able to detect gradients of environmental cues that vary predictablly in different directions and derive a bicoordinate "map" from this information (Phillips 1987; Rodda and Phillips 1992; Fischer et al. 2001).

The sensory basis of orientation can be investigated either by manipulating environmental stimuli and testing the responses of intact animals to these cues, or by eliminating one or more sensory modes and testing an animal's ability to orient without them. Many amphibians probably use several sensory modes and several types of environmental cues for orientation. This redundancy makes it difficult to determine precisely which mechanisms are most important in the natural environment. Experimental manipulation can reveal whether a given environmental cue or sensory system is essential for orientation, but it does not necessarily indicate the relative importance of different cues for orientation in intact animals. By eliminating one sensory mode or altering one cue at a time, investigators often are left with a list of cues that the animal can detect, but none that appears absolutely essential for successful orientation.

### Orientation to Local Environmental Cues

Many amphibians behave as if they can use local landmarks for orientation. For example, frogs often return to exactly the same feeding perch or spot on the shore of a pond night after night, even after relatively long excursions (e.g., Dole 1965b). Male green frogs (*Rana clamitans*) appear to recognize natural features such as clumps of vegetation or rows of bullrushes as boundaries of defended territories (Wells 1977b). Fire salamanders (*Salamandra salamandra*) displaced from their home sites do not return home on a straight-line course, but instead make a series of short movements between natural landmarks such as logs or trees (fig. 6.6; Joly 1963, 1968; Plasa 1979). Despite the obvious importance of visual landmarks for orientation, only Plasa (1979) has done controlled laboratory experiments to test the ability of amphibians to orient using such cues. Salamanders were trained to a home box in an arena with the entrance to the box marked with a variety of striped patterns or a group of small blocks. When the location of the landmarks was switched, the salamanders oriented toward the learned pattern, not in the actual compass direction of the home box.

A number of other types of local orientation cues have been proposed for amphibians. Several early workers suggested that amphibians might locate breeding sites by following humidity gradients (hygrotaxis) or by moving downhill (geotaxis; B. Cummings 1912; Czeloth 1930). Several other studies showed that these cues have little influence on orientation when animals are in a familiar area, but become important when they are in unfamiliar surroundings (Chapman and Chapman 1958; Savage 1961; Oldham 1966, 1967; Twitty 1966; Heusser 1964, 1969c). Omland (1998) found that red-spotted newts (*Notophthalmus viridescens*) tend to orient downhill (geotaxis) when placed in unfamiliar surroundings, presumably because moving downhill generally leads them to water. Adult newts were collected during spring migrations and placed in an apparatus that allowed the slope angle to be varied. They exhibited a strong downhill orientation. In contrast, terrestrial juveniles (efts) did not orient downhill and showed no tendency to move toward ponds. His results are consistent with laboratory studies of magnetic orientation in newts that revealed a high degree of sensitivity to the slope of the test arena, requiring careful leveling of the test apparatus (Phillips and Borland 1994; Deutschlander, Phillips, and Borland 2000).

Although some amphibians use rivulets, streambeds, or other depressions as migration routes (Hurlbert 1969; Hershey and Forester 1980), others appear to move along relatively straight compass courses unrelated to topographic features (Douglas and Monroe 1981; Pilliod, Peterson, and Ritson 2002). A number of investigators have suggested that amphibians have a kinesthetic sense that enables them to remember a path by sensing changes in topography through movements of muscles and joints (Buytendijk 1918; Eibl-Eibesfeldt 1950; Twitty 1959). There is evidence for a kinesthetic sense in newts (Endler 1970) and frogs (Adler 1980), but this probably is more important for making minor corrections in a travel path after avoiding obstacles than for orientation over long distances. Orientation to flowing currents of water (rheotaxis) has been reported in the cave-dwelling salamander (*Proteus anguinus*) and a surface-dwelling relative, *Necturus maculosus* (Durand and Parzefall 1987), as

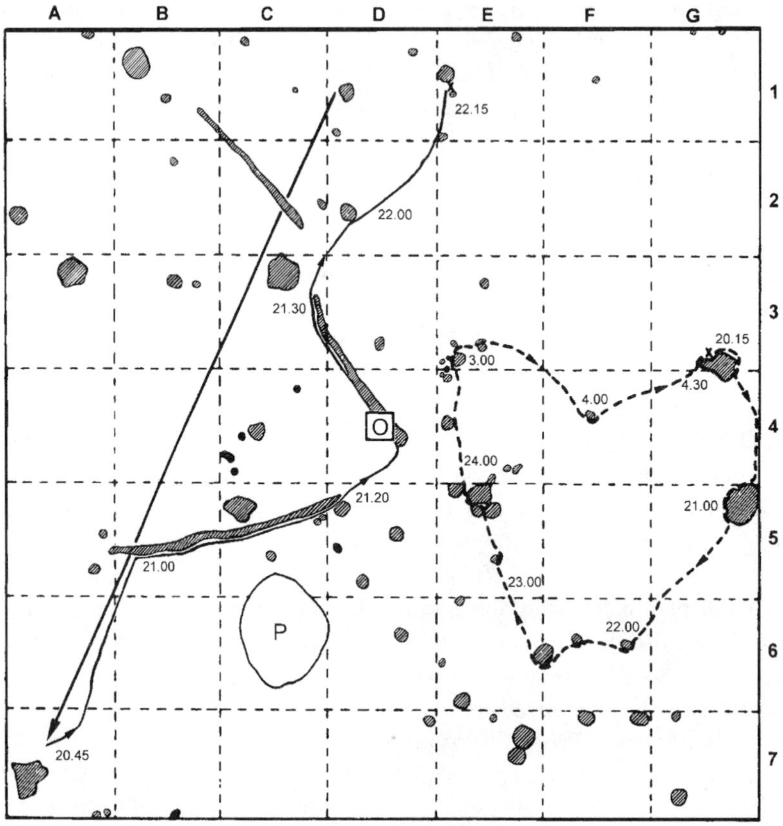


**Fig. 6.6.** Map of movements of adult fire salamanders (*Salamandra salamandra*) in Germany during natural wanderings (dashed line) on a single night and after linear displacement (long arrow) from a home range. These salamanders appear to use natural landmarks such as trees and logs to orient during normal movements and when finding their way home after displacement. Shaded areas are trees and logs. O = observation platform. P = pond. Numbers show the times of night. Each square in the grid is 2 m. After Plasa (1979).

well as *Triturus vulgaris* (Dolmen 1981b). The biological role of this type of orientation is unclear.

## Olfactory Orientation

Local habitat features such as breeding ponds, hibernation sites, or diurnal retreat sites probably have unique chemical signatures that could be used by amphibians for orientation, although the distance at which such cues can be detected is not known for most species. Several lines of evidence suggest that olfactory cues are important for orientation in some anurans and urodeles, but the available data must be interpreted cautiously. Savage (1935b, 1961) provided indirect evidence for olfactory orientation toward breeding ponds by *Rana temporaria*. He found that migration from hibernacula north of a breeding site took place mainly when the wind was blowing from the south across the ponds. He suggested that the odor of algae in the water was important in guiding the movements of migrating frogs. He did not have any experimental evidence that the animals actually were detecting these odors, and studies of other species have failed to detect a relationship between wind direction and migration or the ability to return home after displacement (Eibl-Eibesfeldt 1950; Moore 1954; Heusser 1964, 1969c; Dole 1968). Sinsch (1987c) reported that wind blowing from the direction of home ponds improved the directional orientation of toads (*Bufo bufo*), but orientation was not eliminated when wind was blowing away from the ponds.

Several investigators have used various types of experimental chambers (olfactometers) to present amphibians with a choice of odors from different sources. Early work on *Triturus* by Czeloth (1930) and *Bufo bufo* by W. Jungfer (1943) showed that adults would move toward the odor of a breeding site in a wind tunnel. Martof (1962b) gave *Pseudacris triseriata* adults a choice between the odor of a breeding pond and leaf litter from a forest; the frogs preferred the pond odor. *Pseudacris crucifer* tested in a similar manner did not show a preference for home pond odors. Grubb (1973a, c, 1975) later demonstrated that some anurans preferred the odors of their own breeding areas to those of other bodies of water, including other suitable breeding ponds. Not all test groups showed this discrimination, perhaps because of chance similarity in the odors of different ponds. Grubb (1973b) did not find evidence of olfactory orientation in *Hyla chrysoscelis* or *Scaphiopus hurterii,* but attributed this to differences in the motivation of the test animals rather than an inability to discriminate odors. Metamorphosing juveniles of *Rana lessonae* preferred natal pond odors to odors of tap water or unfamiliar ponds, and exposure to artificial chemical odorants during larval development resulted in a preference for those odors in juveniles (Ogurtsov and Bastakov, 2001). Forester and Wisnieski (1991) reported that *Dendrobates pumilio,* a terrestrial breeder, preferred odors emanating from their own cages to those coming from a blank cage, and they preferred the odor of bromeliads, which normally were placed in their cages, to unfamiliar mint plants. Whether olfactory orientation plays a significant role in the biology of this species in the field is open to question, because these diurnal frogs are very visually oriented in their foraging and social behavior.

Relatively few experimental studies of this type have been done with salamanders, which is surprising, given the importance of chemical senses in the lives of these animals (see chapter 9). A test with *Plethodon metcalfi* failed to demonstrate an ability to discriminate the odor of leaf litter collected in an animal's home area from that collected elsewhere, although other evidence suggested that olfaction might be important in the orientation of this species (Madison 1972). McGregor and Teska (1989) used paper towels impregnated with material from different ponds to test a preference for home pond cues in spotted salamanders (*Ambystoma maculatum*). They scored a choice by determining the relative amount of time the animals spent on marked towels on either side of a test chamber. The salamanders exhibited a strong preference for home pond substrates. The design of this experiment does not clearly distinguish the chemical sense involved in detecting home pond odors (olfactory or vomeronasal system), and it does not necessarily show that the salamanders are capable of detecting airborne odors of ponds at a distance.

Joly and Miaud (1993) presented evidence for olfactory orientation in European newts (*Triturus alpestris*) moving to breeding ponds. Newts were captured while migrating to a breeding pond from different directions. When they were transferred to arenas on the opposite sides of the pond, they showed strong orientation toward the pond. When the pond was covered with a tarpaulin to reduce odor emissions, the newts changed their orientation and moved away from the pond. In laboratory olfactometer experiments, the newts showed a preference for odors emanating from water collected in their own breeding ponds.

Olfactometer tests have the advantage of allowing the investigator to determine preferences for particular odors in animals that have not been traumatized by surgery. Without information about the distance over which odors can be detected in the field, however, they do not necessarily demonstrate an important role for olfaction in orientation. Other workers have surgically impaired visual or olfactory systems in experimental animals to determine whether olfactory cues can be used for homing orientation. The effects of these experimental treatments often appear to be quite traumatic for the animals, and they often exhibit much less movement than control animals (e.g., Sinsch 1987c). Often animals subjected to these treatments eventually find their way home, but in fewer numbers than for control animals, and the time required to make the trip usually is much greater. This could be due to impaired sensory abilities, distress caused by the treatments, reduced survivorship during homing, or a combination of all three.

In several species, including *Bufo boreas, Rana clamitans, Notophthalmus viridescens,* and *Taricha rivularis,* homing ability or homeward orientation was significantly reduced in blinded animals, and in two species, *Salamandra salamandra* and *Bufo spinulosus,* it was eliminated. In many other species, the homing performance of blinded animals was as good as that of normal individuals (table 6.9). In *Bufo bufo,* animals with their eyes covered with tape showed initial homeward orientation as good as that of control animals, but their paths of movement toward home were less straight (Sinsch 1987c). Experiments with blinded animals suggested that olfaction was important for orientation. Unfortunately, many of these studies were done before the possibility of extraocular photoreception or magnetic orientation was recognized, so depriving animals of sight alone does not provide definitive evidence of olfactory orientation.

Many of the same workers also tested the orientation of animals rendered anosmic through surgical severing of the olfactory nerves, perfusing the nostrils with formalin or other chemicals, or blocking the nares with wax or vaseline. Among anurans, only *Bufo boreas* appeared to be completely disoriented when rendered anosmic, although the homing orientation of anosmic *B. bufo* and *R. clamitans* was significantly impaired (table 6.9). Other anurans showed little or no reduction in homing orientation. This suggests that animals simply switch to olfaction or extraocular photoreception if vision is impaired, and to visual cues if olfaction is eliminated. In two studies of toads (*Bufo valliceps* and *B. bufo*), animals rendered both blind and anosmic were completely disoriented (Grubb1970; Sinsch 1987c).

In contrast to the results with anurans, most studies of urodeles have shown that homing ability is reduced or eliminated in anosmic individuals (table 6.9), suggesting a more important role for olfaction in this group. This is to be expected, given the dominant role of chemical senses in the foraging and social behavior of these animals (see chapters

I and II). Most homing studies of salamanders have involved relatively short displacements (< 30 m) from home ranges or breeding sites, so it is conceivable that animals could pick up odor cues from the home area. A notable exception was *Salamandra salamandra,* which was fully capable of homing when nostrils were blocked with Vaseline (Plasa 1979). This species finds prey mainly by visual detection, so it probably relies heavily on visual cues for orientation as well (Himstedt 1994).

The results with *Taricha rivularis* are somewhat puzzling. Anosmic newts were displaced up to 3.2 km (2 miles) to another stream. Very few found their way home, even though both normal and blinded newts homed from the same release point. Most of those that did home were found to have regenerated olfactory tracts (Twitty 1966; Grant, Anderson, and Twitty 1968). These experiments seemed to provide strong evidence for the use of olfaction in homing orientation. Nevertheless, it is hard to imagine how newts could detect odors unique to a particular stream at such great distances, and all of Twitty's (1966) experiments designed to test odor discrimination in this species failed. Alternative interpretations are that survival of anosmic newts was greatly reduced because of an inability to find food or locate moist retreat sites (Twitty 1966), or that the surgery also damaged extraocular photoreceptors important for celestial orientation (Adler 1976) or magnetic receptors (Phillips and Borland 1994). Overall, the available data suggest a role for olfaction in enabling urodeles and perhaps some anurans to find their way home when they are in a familiar area, but its importance for long distance homing is questionable.

### Acoustic Orientation

There has been considerable debate about the importance of vocalizations in guiding the movements of anurans to breeding sites. Many authors have assumed that both females and late-arriving males are attracted to the sound of a chorus (Salthe and Mecham 1974), but males and females of many species find their way to breeding sites before chorusing begins (Heusser 1958a, 1960, 1964, 1969c; Savage 1961; Oldham 1966, 1967; Wells 1979). In fact, European toads (*Bufo bufo*) apparently begin moving from their summer home ranges toward breeding ponds in the fall, and overwinter only a few hundred meters away (Heusser 1968d, 1969c). In experiments with *Bufo americanus* and *Rana clamitans,* Oldham (1966, 1967) found that males from the local area oriented toward familiar breeding sites independent of the sound of a natural or tape-recorded chorus, whereas foreign individuals placed in an arena showed some orientation to broadcast calls. Several other studies have shown that migrating individuals placed in an arena can be diverted from their original compass course by the sound of a chorus. Usually this occurs only when other orientation cues are absent (D. Ferguson and Landreth 1966; Landreth and Ferguson 1966; Tracy and Dole 1969a; Landreth and Christensen 1971). Sinsch (1992b) reported that displaced females of *Bufo calamita* often oriented toward the nearest pond with calling males, whereas males usually oriented toward their home ponds. This suggests that for females, finding a mate quickly is more important than breeding in a particular pond, and is consistent with the much lower level of site fidelity of females compared to males in this species. It is possible that species that breed opportunistically in temporary water are more likely to be attracted to the sound of a chorus than those that breed year after year in the same location.

A surprising example of interorder acoustic orientation was reported by Diego-Rasilla and Luengo (2004). They found that marbled newts (*Triturus marmoratus*) oriented toward playbacks of the calls of natterjack toads (*Bufo calamita*), which often breed in the same ponds, but not to those of green toads (*B. viridis*), which are not sympatric with these newts. Because salamanders lack external ears, they traditionally have been considered deaf to airborne sounds while on land (Wever 1985). Recently, however, Hetherington (2001) found that the lateral body wall of salamanders with lungs, including newts, vibrate in response to airborne sounds, providing a possible extratympanic route for sound transmission similar to that of some frogs that lack external ears (see chapter 7).

### Celestial Orientation

Orientation using celestial cues, particularly the position of the sun, appears to be of primary importance for many amphibians (D. Ferguson 1966, 1971). Celestial cues provide amphibians with a means of compass orientation that is independent of local environmental cues. Most tests have been done on aquatic or semiaquatic species from North America. It is not clear how important this type of orientation is for species that spend all of their lives in forested habitats with little exposure to celestial cues (although these species could use polarized light for orientation—see the following discussion). Most studies have focused on Y-axis orientation. Animals are collected along a natural shoreline, or trained to a shoreline in an arena. The animals are then placed in an aquatic or terrestrial arena that provides a view of the sky, but positioned so that all cues from the home site are excluded. A positive response is scored if an animal moves in a compass direction perpendicular to the shoreline at the home site or training arena. Most species tested to date exhibit Y-axis orientation under a clear sky with the sun visible, but show random orientation under overcast conditions (table 6.8). Several anuran species and one species of newt also can orient along a Y-axis when the moon is

**Table 6.8.** Studies of Y-axis or homeward orientation in amphibians under different light conditions.

| Species (source no.) | Stage tested | Day conditions: Clear | Day conditions: Cloudy | Night conditions: Moon | Night conditions: Stars | Night conditions: Dark |
|---|---|---|---|---|---|---|
| Ascaphidae | | | | | | |
| *Ascaphus truei* (5) | Adults | yes | no | | | no |
| Bufonidae | | | | | | |
| *Bufo americanus* (18) | Juveniles | yes | no | | no | |
| *B. boreas* (10) | Adults | yes | | no | | no |
| *B. boreas* (14) | Juveniles | yes | no | yes | no | |
| *B. fowleri* (2, 9) | Juveniles | yes | no | yes | yes | |
| *B. fowleri* (9) | Tadpoles | yes | no | no | no | |
| *B. quercicus* (11) | Adults | yes | | | | |
| *B. valliceps* (19) | Juveniles | yes | no | no | no | |
| Hylidae | | | | | | |
| *Acris crepitans* (4) | Adults | yes | | yes | yes | |
| *A. gryllus* (1) | Adults | yes | | yes | yes | no |
| *Pseudacris triseriata* (3) | Adults | | | yes | | no |
| Microhylidae | | | | | | |
| *Gastrophryne carolinensis* (9) | Juveniles | yes | | | | |
| *G. carolinensis* (9) | Tadpoles | yes | | | | |
| Pelobatidae | | | | | | |
| *Spea bombifrons* (13) | Adults | yes | | | | |
| Ranidae | | | | | | |
| *Rana catesbeiana* (7) | Adults | yes | no | yes | | |
| *R. catesbeiana* (7, 12) | Juveniles | yes | no | yes | | |
| *R. catesbeiana* (9) | Tadpoles | yes | no | | yes | |
| *R. clamitans* (9) | Tadpoles | yes | | | | |
| *R. pipiens* (17) | Juveniles | yes | | no | no | |
| *R. spenocephala* (8) | Adults | yes | no | | | |
| Ambystomatidae | | | | | | |
| *Ambystoma maculatum* (16) | Juveniles | yes | no | | | |
| *A. opacum* (16) | Juveniles | yes | no | | | |
| *A. talpoideum* (16) | Juveniles | yes | no | | | |
| *A. tigrinum* (15) | Adults | yes | no | | | |
| Salamandridae | | | | | | |
| *Taricha granulosa* (6) | Adults | yes | no | yes | | |
| *Triturus marmoratus* (20) | Adults | | | yes | | no |

*Sources:* (1) D. Ferguson, Landreth, and Turnipseed 1965; (2) D. Ferguson and Landreth 1966; (3) Landreth and Ferguson 1966; (4) D. Ferguson, Landreth, and McKeown 1967; (5) Landreth and Ferguson 1967a; (6) Landreth and Ferguson 1967b; (7) D. Ferguson et al. 1968; (8) Jordan, Byrd, and Ferguson 1968; (9) McKeown 1968; (10) Gorman and Ferguson 1970; (11) Goodyear 1971; (12) Goodyear and Altig 1971; (13) Landreth and Christensen 1971; (14) Tracy 1971; (15) D. Taylor 1972; (16) Tomson and Ferguson 1972; (17) Dole 1972c; (18) Dole 1973; (19) Grubb 1973a; (20) Diego-Rasilla and Luengo 2002.

*Notes:* All animals were tested in aquatic or terrestrial arenas with only the sky visible. Blanks indicate conditions under which animals were not tested. "Dark" indicates a night with complete cloud cover, or a test conducted when the moon and stars were not visible. "Yes" indicates that animals showed orientation. "No" indicates no orientation. Most of these studies were not designed to eliminate the possibility of magnetic orientation.

visible, and a few species show Y-axis orientation when only stars are visible. One study of marbled newts (*Triturus marmoratus*) found similar results in tests for orientation toward a home pond (Diego-Rasilla and Luengo 2002). Species tested under an overcast night sky, or on dark nights when the moon and stars are not visible, invariably display random orientation. In a similar set of experiments, alpine newts (*Triturus alpestris*) showed homeward orientation under a clear night sky, but not when the sky was obscured. In this species, orientation under a clear sky was disrupted by alteration of the magnetic field (see the following discussion; Diego-Rasilla 2003).

When amphibians are collected during migration to a breeding pond and tested in an arena some distance away, they typically maintain their original compass course but do not necessarily orient in the true direction of the breeding pond. This suggests that the mechanism of orientation to breeding sites is similar to Y-axis orientation; that is, the animals move on a fixed compass course until they intersect the shoreline. Tests with migrating animals generally have yielded the same results as tests of animals showing Y-axis orientation to a shoreline. Orientation is maintained under clear day or night conditions when the sun or moon is visible, but orientation sometimes is disrupted when the sky is dark or overcast (D. Ferguson and Landreth 1966; Landreth and Ferguson 1967c; Tracy and Dole 1969a; Landreth and Christensen 1971). On the other hand, Sinsch (1987c) found that European toads (*Bufo bufo*) released under an overcast sky oriented toward breeding ponds nearly as well as those released under a clear sky. This species normally migrates during overcast or rainy weather, when celestial cues are obscured.

These experiments provide strong circumstantial evidence that celestial bodies, particularly the sun, provide cues essential for orientation when only the sky is visible. To use celestial cues for compass orientation, an animal not only must be able to determine the position of the sun or the moon, but also must have a built-in time sense to compensate for the apparent movement of these bodies in the sky. For example, if an animal maintains a fixed compass course, its angle of movement relative to the sun will depend on the time of day, so it must know local time to orient correctly (fig. 6.7). An animal moving south in the northern hemisphere would move at an angle to the right of the sun in the morning, but would move on a course to the left of the sun in the afternoon. If the animal's photoperiod is delayed by six hours, the animal will behave as if the afternoon sun were in the morning position and shift its course 90° in a clockwise direction. Similarly, a six-hour advance in the photoperiod will result in a 90° shift in a counterclockwise direction (Adler 1976).

Several lines of evidence show conclusively that amphibians do have an internal clock synchronized to local solar

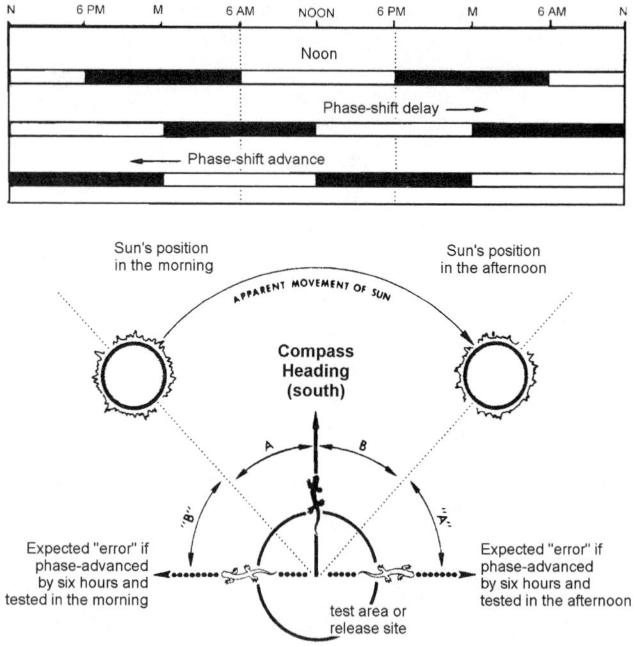


**Fig. 6.7.** The effect of clock-shifting on the orientation of a salamander in a North Temperate location. By changing the photoperiod, animals can be subjected to a six-hour delay phase-shift or a six-hour advance phase-shift (see diagram at top). Animals trained under a normal photoperiod are expected to orient in the correct compass direction (south). Those subjected to phase-shifting are expected to orient in a direction 90° to the right or left of the correct direction, depending on the nature of the phase-shift. Orientation errors are clockwise in phase-delayed animals and counterclockwise in phase-advanced animals. After Adler (1976).

time. First, animals can maintain the correct orientation in an arena independent of time of day (D. Ferguson, Landreth, and McKeown 1967). Second, orientation usually is disrupted if animals are held for several days in the dark (D. Ferguson, Landreth, and Turnipseed 1965; Ferguson, Landreth, and McKeown 1967; Landreth and Ferguson 1967c), indicating that the animals' biological clocks must be reset periodically by exposure to celestial cues. Experiments with *Acris gryllus* showed that exposure to the sun is essential for maintaining synchrony with local time. Animals exposed only to a night sky in a training arena failed to orient to the trained shoreline when tested during the day, but instead continued to orient on the Y-axis of their home shore. In contrast, animals trained under a day sky learned to orient to the new shore (D. Taylor and Ferguson 1969). Finally, all species that have been phase-shifted through altered photoperiods have shown the predicted change in directional orientation (fig. 6.8; D. Ferguson, Landreth, and McKeown 1967; Landreth and Ferguson 1967c; Gorman and Ferguson 1970; D. Taylor and Ferguson 1970; D. Taylor 1972; Justis and Taylor 1976).

The evidence for a lunar or stellar compass sense in amphibians is less convincing than for a solar compass. Be-

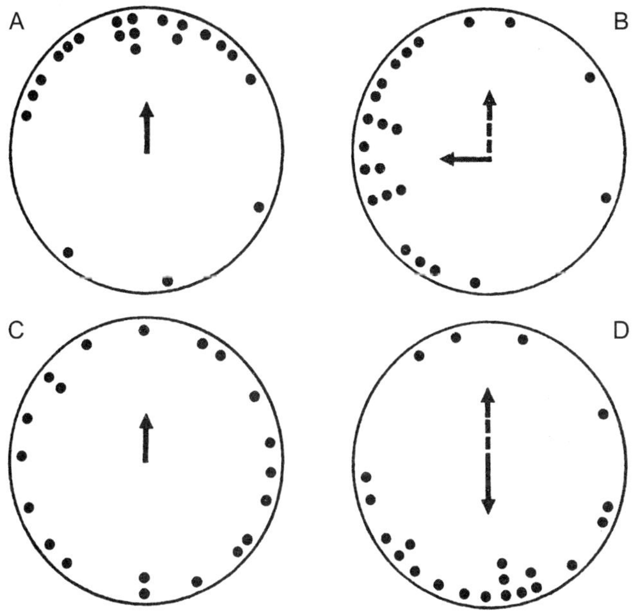


**Fig. 6.8.** Directional responses of tiger salmander larvae (*Ambystoma tigrinum*). (A) Larvae tested immediately after training. Solid arrow shows expected direction of orientation. (B) Larvae subjected to six-hour advance in photoperiod. Solid arrow shows expected direction of orientation; dashed arrow shows original training direction. (C) Larvae tested under complete cloud cover, showing random orientation. (D) Orientation of recently metamorphosed larvae, showing reversal of orientation away from water and toward shore. After D. Taylor (1972).

cause lunar time is not synchronized with solar time, use of a moon compass would require two separate internal clocks (Able 1980). Although several species are reported to show orientation when the moon is visible (table 6.8), the dispersion of individuals in the arena generally is much greater than in daytime tests, suggesting that orientation is less precise at night (D. Ferguson, Landreth, and Turnipseed 1965; Ferguson, Landreth, and McKeown 1967; D. Ferguson and Landreth 1966; Landreth and Ferguson 1966; D. Ferguson et al. 1968; Tracy 1971). There is some indication that animals that tend to move mostly at night, such as juvenile bullfrogs (*Rana catesbeiana*), can orient more accurately under a night sky than primarily diurnal species such as *Bufo fowleri* (D. Ferguson et al. 1968; McKeown 1968), but rigorous statistical comparisons have not been made. There have not been any experiments clearly demonstrating the existence of a lunar clock in amphibians, so the importance of lunar orientation remains unclear. Of seven species tested for orientation under stars alone, only four showed evidence of Y-axis orientation (table 6.8). Again, the responses of individuals tested under stars are not very consistent, and there is a tendency toward bidirectional orientation along the Y-axis in some species (D. Ferguson, Landreth, and Turnipseed 1965; Ferguson, Landreth, and McKeown 1967; D. Ferguson and Landreth 1966; McKeown 1968). The reason for this bidirectionality is not known.

Seasonal and motivational changes can influence orientation responses. For example, D. Ferguson et al. (1968) found that adult bullfrogs (*Rana catesbeiana*) showed Y-axis orientation during a spring breeding season, but the response disappeared in mid-summer after breeding had subsided. However, animals tested in the fall again showed an oriented response. One explanation for this change is that frogs on their summer home ranges learn particular escape routes relative to local features of the environment. When placed in an arena, the frogs often attempt to escape by moving in a preferred direction, rather than on a fixed compass course relative to the shore (D. Ferguson et al. 1968). This interpretation is supported by data showing that individual *Rana* and *Bufo* exhibit consistent differences in their directional orientation in an experimental arena (Adler 1980; Adler and Taylor 1981).

In species that breed in water but return to land as adults, Y-axis orientation is maintained throughout life, but the direction of orientation changes with ontogeny. This was studied in detail by McKeown (1968), who worked with *Bufo fowleri, Gastrophryne carolinensis, Rana catesbeiana,* and *R. clamitans.* He found that Y-axis orientation was well developed in tadpoles once they reached the free-swimming stage. Young tadpoles generally oriented in a direction that would take them toward deep water at the home pond or training tank. Once the tadpoles developed legs and began to emerge from the water, there was a brief period when Y-axis orientation appeared to be random, but eventually the directional preference shifted toward the shallow water end of the Y-axis. Metamorphosed juveniles showed clear orientation toward shallow water when tested in an aquatic arena, suggesting a tendency to disperse away from the breeding pond. When tested on land, some juveniles oriented in a landward direction, while others tended to moved in a direction that would take them back to water. The latter response may have been an attempt to escape from desiccating conditions. Subsequent studies with aquatic larvae and terrestrial juveniles of both anurans and urodeles have produced similar results (D. Ferguson et al. 1968; Goodyear and Altig 1971; Tracy 1971; Dole 1972a, 1973; D.Taylor 1972; Tomson and Ferguson 1972).

### Extraocular Photoreception and Orientation by Polarized Light

A number of studies have shown that blinded amphibians can find their way home after displacement (table 6.9). Originally this was seen as evidence that olfactory cues are important for orientation (see the following discussion). Subsequent work showed that many amphibians have photoreceptors other than their eyes that are used for orientation (Adler 1970, 1976). The importance of extraocular photoreceptors

**Table 6.9.** Sensory impairment studies of homing in amphibians.

| Species (Source) | Goal | Distance (m) | Response | Experimental Groups Normal | Blind | Anosmic |
|---|---|---|---|---|---|---|
| Bufonidae | | | | | | |
| *Bufo americanus* (2) | BP | 36–170 | HO | ++ | | ++ |
| *B. americanus* (12) | HR | 50–235 | HO | ++ | ++ | ++ |
| *B. boreas* (8) | BP | 50–200 | HO | ++ | + | – |
| *B. bufo* (1) | BP | | RH | ++ | + | + |
| *B. bufo* (16) | BP | 145–200 | HO | ++ | + | – |
| *B. calamita* (18) | BP | 130–150 | HO | ++ | – | + |
| *B. spinulosus* (17) | BP | 90–145 | HO | ++ | – | + |
| *B. valliceps* (9) | BP | 80–206 | HO | ++ | ++ | ++ |
| Ranidae | | | | | | |
| *Rana clamitans* (2) | BP | 230 | RH | ++ | | + |
| | | > 90 | HO | ++ | + | + |
| *R. pipiens* (6, 11) | HR | 50–300 | HO | ++ | ++ | ++ |
| Plethodontidae | | | | | | |
| *Desmognathus fuscus* (10) | HR | 3–30 | RH | ++ | ++ | – |
| *Plethodon jordani* (7) | HR | 15–30 | RH | ++ | ++ | – |
| *P. metcalfi* (13) | HR | 15–30 | RH | ++ | ++ | – |
| Salamandridae | | | | | | |
| *Notophthalmus viridescens* (15) | BP | 20 | HO | ++ | + | – |
| *Salamandra salamandra* (14) | HR | < 20 | RH | ++ | – | ++ |
| *Taricha granulosa* (4) | BP | 228 | HO | ++ | ++ | |
| *Taricha rivularis* (3) | SS | 400–3200 | RH | ++ | + | – |

*Sources:* (1) Heusser 1964; (2) Oldham 1966; (3) Twitty 1966; (4) Landreth and Ferguson 1967c; (5) Oldham 1967; (6) Dole 1968; (7) Madison 1969; (8) Tracy and Dole 1969b; (9) Grubb 1970; (10) Barthalmus and Bellis 1972; (11) Dole 1972a; (12) Dole 1972b; (13) Madison 1972; (14) Plasa 1979; (15) Hershey and Forester 1980; (16) Sinsch 1987b; (17) Sinsch 1988b; (18) Sinsch 1992a.

*Notes:* Goal: BP = breeding pond; HR = home range; SS = home stream section. Response: HO = homeward orientation; RH = return home. Symbols: ++ strong orientation response; + some orientation; – little or no orientation.

was first suggested by the work of Landreth and Ferguson (1967c) with *Taricha granulosa.* They found that blinded newts tested in arenas where olfactory cues were not available showed Y-axis orientation similar to that of normal animals. They also found that blinded newts displaced about 230 m from breeding ponds in view of the sun showed accurate homeward orientation, whereas many normal newts displaced under cloud cover or in light-tight containers were disoriented. Finally, blinded newts could orient in a trained direction relative to an artificial light source in the laboratory, but individuals lacking an optic tectum could not.

The principal site of extraocular photoreception in amphibians is the pineal complex, located on the dorsal surface of the diencephalon (fig. 6.9). In anurans, an extension of the diencephalon known as the frontal organ, or pineal end organ, lies in the dermis on the top of the head and probably is the principal extraocular photoreceptor. Urodeles and caecilians lack the frontal organ, but the pineal body itself is sensitive to light transmitted through the skull, as it is in anurans (Eakin 1973; Adler 1976).

The importance of extraocular photoreceptors for orientation has now been demonstrated for several amphibians, including *Acris gryllus* (D. Taylor and Ferguson 1970), *Rana catesbeiana* tadpoles (Justis and Taylor 1976), and *Ambystoma tigrinum* (D. Taylor 1972; Taylor and Adler 1978). In a typical series of experiments, D. Taylor and Ferguson (1970) found that blinded cricket frogs failed to orient properly when the top of the head was covered with opaque plastic, but they could orient when the head was uncovered. Animals with intact eyes could orient properly even with the top of the skull covered. Similar results were obtained with both adult and larval tiger salamanders (D. Taylor 1972). Subsequent work by D. Taylor and Adler (1978) showed that removal of the pineal body in blinded

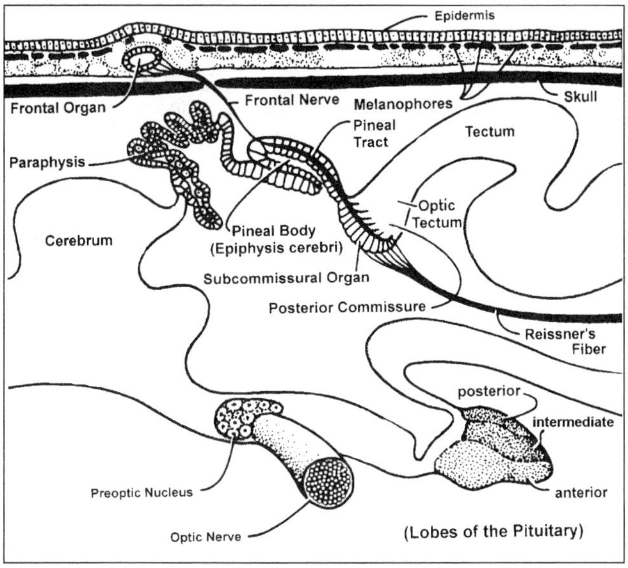


**Fig. 6.9.** Anatomy of pineal complex in a frog (*Rana*), showing the extra-optic photoreceptor (frontal organ) and its connections to the pineal tract and other parts of the brain. Salamanders and caecilians lack the frontal organ, but otherwise are similar to frogs. After Adler (1976).

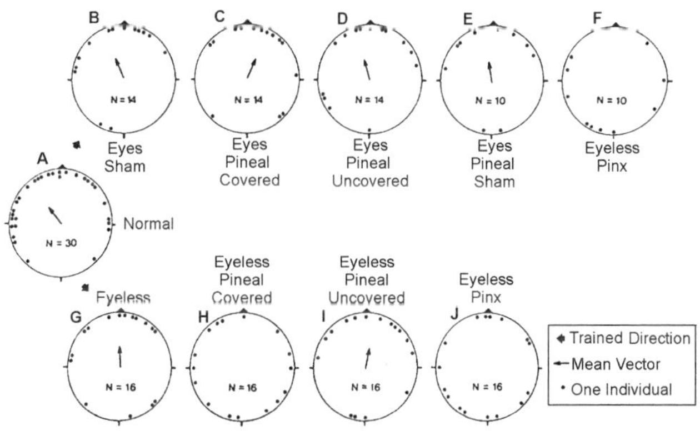


**Fig. 6.10.** Orientation of normal, eyeless, and pinealectomized (pinx) tiger salamanders (*Ambystoma tigrinum*). Thirty normal animals were tested and then divided into two series and tested in sequential order (B–F and G–J). Salamanders with either eyes or pineal intact oriented normally, but blinded animals with the pineal covered or removed did not. After D. Taylor and Adler (1978).

animals produced the same results as covering the top of the head with opaque plastic, thus implicating the pineal as the photoreceptive organ (fig. 6.10). These experiments showed that frogs and salamanders can orient using the pineal alone, but the pineal is not essential for orientation if the eyes are intact. Experiments with bullfrog tadpoles showed that tadpoles can orient correctly as long as one of their photoreceptive organs (eyes, pineal, frontal organ) is intact, but not if all three are impaired (Justis and Taylor 1976).

In most of the daylight orientation experiments discussed so far, the sun either was directly visible to test animals, or its position could be determined from brightness patterns. Some amphibians can determine the position of the sun by detecting polarized light. Polarized light is produced by atmospheric scattering of sunlight, which causes light waves to vibrate primarily in a plane perpendicular to the direction of propagation. At any particular time of day, the direction of vibration (i.e., the bearing of the "e-vector" of polarized light) is directly related to the position of the sun. If an animal can detect the bearing of the e-vector, it can use the information for compass orientation (Auburn and Taylor 1979).

The ability of amphibians to perceive polarized light was first demonstrated by Taylor and Adler (1973) in experiments with *Ambystoma tigrinum*. They placed salamanders in a training tank illuminated by polarized light, with the plane of polarization perpendicular to the long axis of the tank and parallel to the artificial shoreline (fig. 6.11). The animals were then tested in an aquatic arena lacking a shoreline. As expected, the animals oriented perpendicular to the

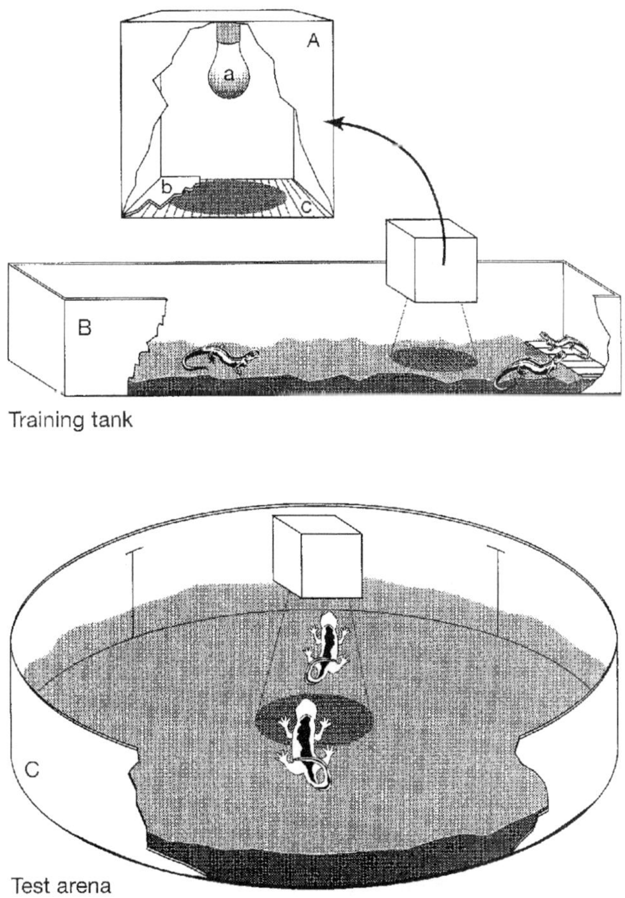


**Fig. 6.11.** Diagram of apparatus used to train and test salamanders under polarized light. (A) Light source, with 25-watt bulb (a), frosted glass (b), and polarizing filter (c) between glass and circular opening in an opaque shield. (B) Training tank, showing position of water, shoreline, and polarizing light source. Parallel lines represent e-vector of polarized light. (C) Test procedure, with e-vector of polarized light shifted 90° relative to its position during training. The salamanders orient in the trained direction relative to the e-vector. After D. Taylor and Adler (1973).

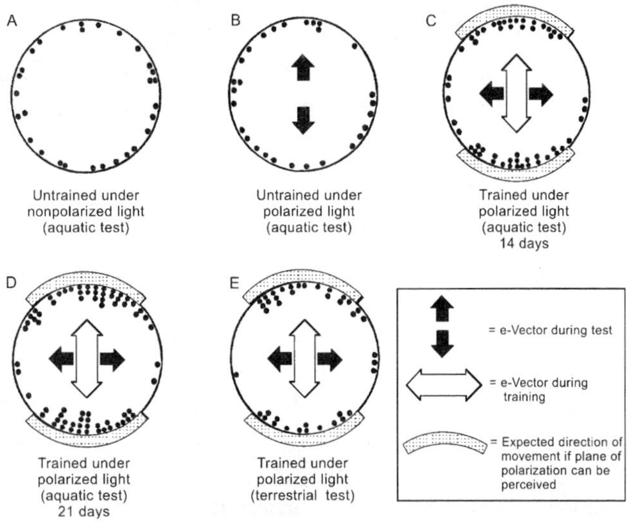


**Fig. 6.12.** Orientation of tiger salamanders (*Ambystoma tigrinum*) under polarized light. (A) Salamanders not trained to a particular direction and tested under nonpolarized light showed random orientation. (B) Salamanders exposed to polarized light but not trained to a shoreline also showed random orientation. Animals trained to move perpendicular to a shoreline under polarized light for 14 days (C) or 21 days (D) were placed in a test arena under polarized light with the e-vector rotated 90° relative to the training direction. These animals oriented in a direction perpendicular to the e-vector, as expected if they could perceive the plane of polarization. Animals in groups C and D were tested multiple times. Those tested under similar conditions only once (E) showed similar orientation. After D. Taylor and Adler (1973).

plane of polarization, a direction that also would have been perpendicular to the learned shoreline (fig. 6.12). Subsequent experiments with bullfrog tadpoles and both efts and adults of red-spotted newts (*Notophthalmus viridescens*) showed a similar ability to orient by polarized light (D. Taylor and Auburn 1978; Auburn and Taylor 1979). Experiments involving impairment of photoreceptors similar to those described above indicated that extraocular photoreceptors are involved in perception of polarized light, but eyes are not (fig. 6.13; Adler and Taylor, 1973; D. Taylor and Auburn 1978).

The ability to use polarized light for orientation is advantageous when animals are deprived of a direct view of the sun. Polarized light is readily perceived under water, and hence is an important cue for orientation by aquatic amphibian larvae. For terrestrial amphibians, the ability to perceive polarized light is particularly useful in forested areas, where the sun often is obscured, but patches of clear sky are visible. Finally, polarization of sunlight is greatest directly overhead at twilight, when many amphibians are moving (Adler 1976). Experiments with bullfrog tadpoles and adult newts showed that the animals could orient under a clear sky at sunrise and sunset, but not under a completely overcast sky, when skylight polarization is absent. Animals trained in an outdoor arena under a clear sky and tested indoors under artificial polarized light oriented in a direction relative to the plane of polarization that was appropriate for the time of day (D. Taylor and Auburn 1978; Auburn and Taylor 1979). These results clearly show that at least some amphibians can use the plane of polarization in natural sunlight for directional orientation.

In addition to perception of directional cues from polarized light, the pineal complex is important for orientation in amphibians because it is involved in synchronizing the animals' internal clocks, a process that is essential for sun-compass orientation. Experiments with several species have shown that blinded animals can be phase-shifted by altering the timing of their photoperiods, or entrained to a new activity pattern by changing the duration of the light and dark portions of the cycle (Adler 1969, 1970, 1971, 1976; Demian and Taylor 1977). If the frontal organ of frogs (Adler 1971) or the pineal body of salamanders (Demian and Taylor 1977) is removed, entrainment to a new photoperiod does not occur.

### Magnetic Orientation and Navigation

For animals that can detect it, the earth's magnetic field is a potential source of information for compass orientation (Wiltschko and Wiltschko 1995; Phillips 1996). Detection of magnetic fields is the least understood sensory system of animals, but recent research has provided evidence of a magnetic sense in birds (reviewed in Able 1980; Phillips 1987, 1996), mammals (Mather and Baker 1981), alligators (Rodda 1984; Rodda and Phillips 1992), and sea turtles (Lohmann, Hester, and Lohmann 1999). The first demonstration of a magnetic sense in amphibians was the work of Phillips (1977) with cave salamanders (*Eurycea lucifuga*). He trained the salamanders to move in a dark corridor of a training box. Some animals were trained to move parallel to the earth's magnetic field. A second group was trained with the corridor parallel to the earth's field, but with the magnetic field shifted by 90° with a magnetic coil. The animals were subsequently tested in a cross-shaped device both with and without an altered magnetic field. Although the test groups did not all exhibit nonrandom orientation, those that did were oriented in the expected direction relative to the magnetic field during training. These responses were independent of other possible directional cues. Hence Phillips concluded that cave salamanders are capable of detecting the earth's magnetic field and using it for directional orientation. This would be particularly useful during those times of year when the salamanders retreat into the completely dark parts of their caves.

Subsequent work showed that red-spotted newts (*Notophthalmus viridescens*) also are capable of detecting slight changes in the earth's magnetic field (Phillips and Adler 1978) and exhibit compass orientation derived from magnetic cues

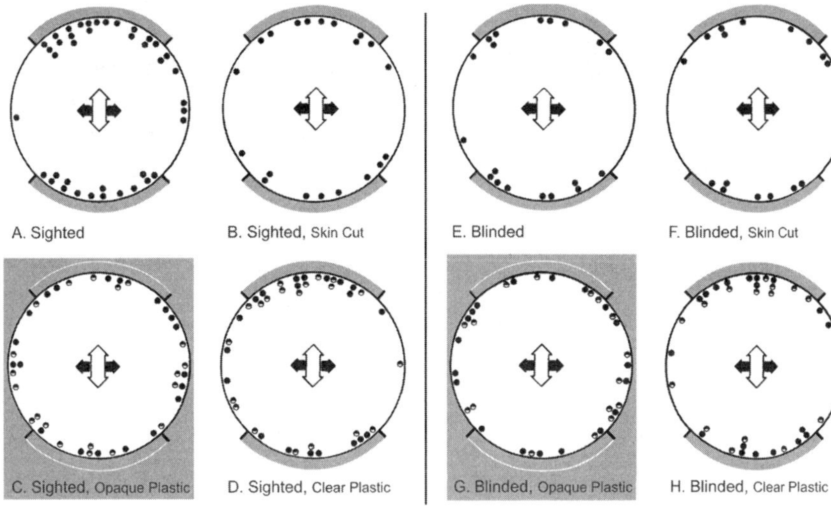


**Fig. 6.13.** Directional orientation of sighted and blinded tiger salamanders (*Ambystoma tigrinum*) under polarized light. Shaded areas at edges of circles indicate quadrants in which animals were expected to score if the plane of polarization is perceived. Black and white arrows as in Fig. 6.12. All blinded or sighted animals could perceive polarized light unless the head was covered by opaque plastic. After Adler (1976).

(Phillips 1986a, b). Newts can orient along a trained magnetic axis (Y-axis orientation), and they are capable of learning the direction of the Y-axis in relation to a magnetic field within 12 to16 hours (Deutschlander, Phillips, and Borland 2000). They also can orient in a true homeward direction after being displaced up to 45 km, even when tested indoors after being deprived of external cues during displacement (Phillips, Adler, and Borland 1995; Fischer et al. 2001). This indicates that newts can perform true navigation using some sort of map sense. Homeward orientation is affected by season. Newts exhibited homeward orientation during the spring migration season, but oriented along the magnetic axis of the training tank at other times (Phillips 1986a, b, 1987). There appear to be two distinct sensory systems involved in magnetic orientation. One is used for simple compass orientation, such as orientation to a trained direction in a tank. When the vertical component of the magnetic field was inverted, this form of compass orientation was reversed. This demonstrated sensitivity to axial components of the magnetic field. The treatment did not affect homeward orientation, suggesting that homing involves a response to the polarity of the magnetic field (Phillips 1986b).

The question that still must be resolved is the precise nature of the sensory system used for magnetic detection and orientation in these newts. Additional experiments have demonstrated a link between the visual system and the magnetic detection system (Deutschlander, Phillips, and Borland 1999). Specifically, changes in the wavelengths of light present during testing can alter the orientation responses of newts. When newts were tested under near-infrared light, with visible light not present, they oriented at random with respect to the magnetic field in a tank (Phillips and Borland 1992b). Orientation to a trained shoreline direction was maintained under full-spectrum visible light and short wavelengths, but was rotated by 90° under long wavelengths. This was shown to be the result of a direct effect of light on the magnetic compass that is sensitive to the inclination of the magnetic field (Phillips and Borland 1992a, c). In contrast, homing orientation is thought to depend on the ability to detect the polarity of the magnetic field. When homing newts were tested under long wavelengths, homeward orientation disappeared, rather than being shifted as in shoreline orientation. This indicated that newts possess two different systems for magnetic reception, one sensitive to light, the other not (Phillips et al. 2002).

Newts exhibiting homeward orientation probably receive inputs from both magnetic receptors (Phillips and Borland 1994). Exactly where these receptors are located, and how information from them is integrated in the brain, have yet to be determined. Ferromagnetic particles that could serve as magnetic detectors have been found in newts, although the precise location and function of these particles has not been determined (Brassart et al. 1999). There also is evidence of a link between polarized light detection by the pineal organ and a light-dependent magnetic compass sense (Phillips et al. 2001).

At present, the degree to which the results for red-spotted newts apply to other amphibians is unknown, but it seems unlikely that this species would have evolved a unique mechanism of orientation. Sinsch (1987c, 1988b) placed small magnets on the heads of toads in an attempt to disrupt the animals' magnetic compass. In *Bufo bufo*, the magnets altered the orientation response, but did not completely eliminate homeward orientation, presumably because other sensory systems were intact (Sinsch 1987c). In *Bufo spinulosus,* magnets shifted the direction of orientation, but they did not have this affect on toads rendered anosmic by blocking the nostrils with Vaseline. This lead Sinsch (1988a, 1992a) to speculate that the olfactory system was somehow linked to magnetic reception, but experimental evidence for this is not definitive. It is possible that with two senses disrupted, the toads simply switched entirely to visual orientation. In a test of celestial orientation in *Triturus marmoratus,* Diego-Rasilla and Luengo (2002) also used magnets to disrupt the ambient magnetic field, but this treatment did not affect orientation under a moonlit sky. In contrast, alteration of the magnetic field did disrupt orientation of *T. alpestris* under similar conditions (Diego-Rasilla 2003), and this species exhibited homeward orientation in response to magnetic cues alone (Diego-Rasilla , Luengo, and Phillips 2005). The reason for the apparent difference in these two congeneric species is unknown. There also is evidence for Y-axis orientation being influenced by the magnetic field in bullfrog (*Rana catesbeiana*) tadpoles (Freake, Borland, and Phillips 2002), and the response to magnetic cues is light dependent, as it is in newts (Freake and Phillips 2005).

## Summary and Conclusions

A variety of techniques have been used to study the natural movements of amphibians. Most workers have used data on recaptures of marked individuals to estimate the magnitude of daily movements and the sizes of home ranges. Unfortunately, recaptures of individuals generally are infrequent and overall recapture rates are low, so the true extent of movements between recapture points cannot be determined. This is particularly true for species that spend most of their time under logs, in rock crevices, or in underground burrows. Use of mark-recapture data to calculate home ranges can exaggerate the area over which daily foraging activities take place. Data derived from continuous observation of individuals generally yield much lower estimates of daily movement distances. It is difficult to make quantitative comparisons of home range size among species, because a variety of methods have been used to calculate home range size, and these methods are not comparable statistically. In general, it appears that most amphibians are quite sedentary, except for seasonal migrations to breeding sites, although active foragers appear to move farther and occupy larger home ranges than sit-and-wait predators. Individuals often exhibit strong fidelity to a home area, and those that breed in water usually return to the same site each year. There is limited evidence that dispersing juveniles return as adults to natal ponds to breed, but data on juvenile dispersal for most species is meager at best.

Choice of a method for analyzing movement patterns and estimating home range sizes should depend in part on the questions being asked. All too often data on amphibian movements have been collected with no clear ecological question in mind, or have been incidental to more general natural history studies. If the goal is to determine the total area familiar to an individual as a prelude to studies of homing and orientation, then it is appropriate to combine recaptures made over a prolonged period at infrequent intervals. On the other hand, if the goal is to understand the normal foraging movements of the animal or estimate the area over which it searches for mates, then fine-grained data collected over shorter periods of time are appropriate. Unless the purposes of such studies are clearly defined in advance and methods of data acquisition and analysis standardized, it will be impossible to make detailed comparisons between species (see also B. Rose 1982).

There are a number of interesting questions relating to the adaptive significance of amphibian movement patterns that have yet to be addressed for most species. For example, the limited evidence available suggests that actively foraging species have larger home ranges than sit-and-wait predators, but we know very little about how such foraging species make decisions about when and how far to move while searching for food. When a dendrobatid frog is foraging for ants on the forest floor (see chapter 15), does it move at a relatively fixed rate, regardless of the number of food items encountered, or does it forage opportunistically, pausing at particularly rich food patches? At what point does it abandon a food patch to search for another? A variety of "optimal foraging" models have been proposed to explain the movement patterns of animals searching for food (reviewed in Pyke 1983). The relevance of these models to amphibians is not clear. Studies of plethodontid salamanders have shown that these animals are able to assess the density and profitability of prey in a simplified laboratory environment and adjust their behavior to take the most profitable prey (Jaeger and Barnard 1981; Jaeger, Joseph, and Barnard 1981, Jaeger, Barnard, and Joseph 1982; Jaeger and Rubin 1982). The movements of individuals between patches of prey in a natural environment and the relationship of these movements to prey profitability have not been investigated.

Another area that has received little attention from amphibian biologists is the question of dispersal strategies (for reviews of theoretical and empirical aspects of dispersal, see Wiens 1976; Greenwood 1983; Horn 1983; Shields 1982, 1983; Stenseth 1983). The lack of reliable information on juvenile dispersal for most species makes it impossible to draw broad conclusions about amphibian dispersal strategies at this point, but it is possible to pose general questions.

For example, how is dispersal related to environmental stability and predictability? At one extreme, that of a suitable breeding area surrounded by inhospitable habitat, the answer is obvious: dispersal is impossible, and in some amphibians this seems to have led to the evolution of neotenic populations that do not even undergo metamorphosis (see chapter 13). In other situations, the answer is not as clear. Certainly one would expect greater dispersal by toads emerging from unpredictable desert rain pools than those hatched in a garden pond in England. On the other hand, a species that uses temporary, but spatially predictable, desert pools for reproduction might show a greater degree of natal philopatry than those using pools surrounded by any other suitable breeding sites (Sinsch 1992d). Interspecific differences in dispersal strategies will have a major impact on the genetic structure and dynamics of populations.

Still another area of potential interest that has scarcely been explored for amphibians is the question of sex differences in dispersal. Among animals in general, one sex often shows a greater tendency to disperse than does the other. These differences are related to several factors, including the probability of breeding with close relatives, the benefits of remaining in groups of close kin, and the probability of finding an unoccupied breeding site or becoming established as an independent breeder in a social group (Greenwood 1980, 1983; Swingland 1983). Sex-biased dispersal has been reported for *Bufo calamita,* but the selective pressures favoring longer movements by females are unclear (Sinsch 1992b). In some species of frogs, such as the bullfrog (*Rana catesbeiana*), males are territorial during the breeding season (see chapter 9), and a young male's chances of obtaining a choice territory depend on the density of older males and the quality of available sites (R. D. Howard 1981a, 1984). It is conceivable that under some circumstances, young males would be better off dispersing to new ponds where competition might be reduced, rather than waiting several years to breed in a natal pond, provided that females visited those ponds as well.

There is limited evidence that either males or females of some species of terrestrial salamanders are territorial (Mathis et al. 1995; see also chapter 9), and competition for choice territory sites might force some individuals to disperse to new areas. This would be particularly true for species utilizing specialized retreat sites, such as rock crevices, that often are in short supply. How do individual males and females of these species trade off the risks of dispersal over unfamiliar ground against the potential benefits of finding uninhabited territories? Only additional fieldwork will provide answers to this and other questions about adaptive patterns of movement in amphibians.

Studies of homing and orientation in amphibians were much in vogue in the late 1960s and early 1970s, inspired in part by the pioneering work of Victor Twitty and his delightful account of the excitement and frustration of fieldwork (see "The Pepperwood Creek Affair" in Twitty 1966). Recently work in this area has diminished, and only a few laboratories remain active in the field. Possibly this is due in part to the technical problems involved in identifying the sensory mechanisms of orientation. Twitty spent years attempting to identify the sensory basis of homing in newts and never obtained a satisfactory result. Partly this was because he focused on senses that are most familiar to humans, mainly vision and olfaction, but was not aware of the potential importance of extraocular photoreceptors, polarized light, or magnetic orientation. Even when all of the potential sensory systems are considered, it often is impossible to determine the relative importance of each system for the orientation of animals in the field. For example, red-spotted newts (*Notophthalmus viridescens*) can orient by means of olfactory cues (Hershey and Forester 1980), celestial cues perceived by the eyes (Hershey and Forester1980), polarized light perceived by the pineal body (D. Taylor and Auburn 1978), magnetic cues (Phillips 1986a, b, 1987), and possibly local landmarks (Hurlbert 1969). How an animal integrates all of this information to find its way around its natural environment remains a mystery.

Perhaps the greatest technical challenges remaining for students of amphibian orientation are to determine how amphibians navigate using a "map sense," and to identify the organs or structures involved in magnetic detection. It seems likely that magnetic cues would be required for such a map sense, but a very high degree of sensitivity would be needed to detect slight changes in the local magnetic field (Phillips 1996). The short distances that most amphibians move would preclude using celestial cues for accurate navigation, although they still could be used to obtain basic compass directions. Recent work has shown that newts are sensitive to both the axis of a magnetic field and its polarity, suggesting that these animals have two distinct types of magnetic receptors that might interact to provide information on an animal's geographical position. In addition, there is evidence for other vertebrates that magnetic detection is involved in calibration of other compass information derived from celestial cues or polarized light, but the way in which these systems are integrated physiologically is not understood.

# **Chapter 7** Anuran Vocal Communication

*The cry of frogs is one of the most wearying, croaking sounds possible, and we have only to place ourselves near some dirty pool in the spring, to convince ourselves of their deep, guttural voices; but bad as this is, it is music compared to the long shrieks, shrill whistlings, snorings, and bellowings of those in other parts of the world.*
—Mrs. R. Lee, *Anecdotes of the Habits and Instincts of Birds, Reptiles and Fishes* (1855)

*The toad's lullaby note comes from the far margin, sweeter than all others. . . . This song has been compared to the slow opening movement of Beethoven's "Moonlight Sonata."*
—Mary C. Dickerson, *The Frog Book* (1906)

*The sound, which the scientific books describe as "croaking," floats far and wide, and produces a beautiful, mysterious effect on a still evening when the last heavy-footed labourer has trudged home to his tea, leaving the world to darkness and to me.*
—W. H. Hudson, *The Book of a Naturalist* (1919)

WHATEVER THEIR musical qualities, vocalizations are conspicuous features of the behavior of most frogs and toads. Although the functions of frog calls were not well understood until relatively recently, most early naturalists realized that calls are given almost exclusively by males, are associated with reproduction, and probably serve to attract mates. Julian Huxley (1927, 223) suggested that a frog's calls are analogous to the songs of a male bird, being in part "an advertisement of his maleness" and in part "an advertisement of the breeding grounds, a boasting of choice real estate." Nevertheless, even a relatively reliable observer such as Mary Dickerson (1906, 35) thought that frogs "feel physical joy and express it in song."

Much of the early literature on anuran vocalizations consisted of brief accounts of calling behavior and onomatopoeic descriptions of calls. For example, Wright (1914) listed seven different characterizations of the bullfrog's call: blood 'n' nouns, br-wrum, be-drowned, more-rum, jug-o-rum, knee-deep, and bottle-o-rum. With few exceptions, conclusions about the functions of calls were based on anecdotal evidence. Only with the advent of portable tape-recording equipment was it possible to analyze calls quantitatively or perform controlled playback experiments to investigate call functions (Rand 2001). Two early pioneers of anuran communication were Arthur A. Allen, an ornithologist at Cornell University, and Peter Paul Kellogg, who produced the first album of frog-call recordings in 1948 (Allen and Kellogg 1948). Even more important were the contributions of Frank Blair and his students at the University of Texas and Charles M. Bogert at the American Museum of Natural History in New York. Blair did extensive research on the role of vocalizations in species isolation, and the work of his students continues to this day. Bogert did less original research on frog calls, but he made major synthetic contributions in the form of the second album of frog-call recordings (Bogert 1958) and a comprehensive review of acoustic communication in amphibians and reptiles (Bogert 1960). This review was the first attempt to place information on anuran calls in an evolutionary and ecological framework. Since then the literature has expanded enormously, and many additional reviews have appeared (Blair 1963, 1968; Schneider 1966;

Lescure 1968; Paillette 1971; Schiøtz 1973; Straughan 1973; Salthe and Mecham 1974; Keister 1977; Littlejohn 1977; Wells 1977b, c, 1988; Gerhardt 1988, 1994a; Narins and Zelick 1988; Rand 1988; M. Ryan 1988a, 2001; Gerhardt and Schwartz 1995; Gerhardt and Huber 2002; Wells and Schwartz 2007).

This chapter focuses on the structure and function of anuran calls. I begin with a discussion of the production of vocal signals. This is followed by a review of the vocal repertoires of anurans, the way in which the physical environment affects the transmission of acoustic signals, and the influence of predators on calling behavior. Next, there is a discussion of the structure and dynamics of anuran choruses, the vocal interactions among frogs in a chorus, the interactions among species in multispecies choruses, and the responses of females to calls of conspecifics and heterospecifics. Finally, I briefly discuss the anatomy and neurobiology of call recognition. The role of vocalizations in mate choice and sexual selection is discussed in more detail in chapter 8.

## Mechanisms of Sound Production

The basic mechanism of sound production in most anurans is a relatively simple system. During the respiratory cycle, air is pumped into the lungs by positive pressure from muscles of the buccal cavity (De Jongh and Gans 1969; Gans 1973). When the frog calls, air is forced from the lungs by contraction of muscles in the trunk region and moves through the larynx into the buccal cavity (fig. 7.1). As the air passes through the larynx, vibration of the vocal cords and associated cartilages produce sound. Actions of the laryngeal muscles shape the sounds into a variety of forms, from frequency-modulated peeps to noisy squawks and musical trills (fig. 7.2). The family Pipidae is an exception to this pattern. Males lack vocal cords and use a very different mechanism (see following) to produce calls that consist of a series of short clicks. In most species, sound is produced only as the air is expired from the lungs (Schneider 1988), but sound is produced during inspiration in *Bombina* (Lörcher 1969) and during both inspiration and expiration in *Discoglossus* (E. Weber 1974). Usually the sound is transmitted to an inflated vocal sac that couples the buccal cavity to the outside environment and serves as a sound radiator (Gans 1973; Duellman and Trueb 1986; Schneider 1986, 1988). The sound-producing system thus involves three major functional units: (1) the trunk muscles that power the system, (2) the laryngeal apparatus that produces the sound, and (3) the buccal cavity and vocal sac that transmit the sound. Although the mechanism of sound production is similar in most anurans, the details of call structure vary and are related to differences in the functioning of these three major components.

**Fig. 7.1.** Calling cycle of the Panamanian túngara frog *Physalaemus pustulosus* with (A) body inflated, and (B) vocal sac inflated. This small tropical frog has become a model organism for the study of anuran communication and behavior. Recent taxonomic rearrangments place this species in the genus *Engystomops* (Ron, Santos, and Cannatella 2006), but I use the name *Physalaemus* throughout this book. Photos by Kentwood D. Wells.

### Trunk Muscles

The power for sound production is provided mainly by contraction of two pairs of broad muscular sheets that enclose much of the thoracic and abdominal cavities (fig. 7.3). The outer layer is composed of the external oblique muscle. The inner layer has been called the internal oblique muscle by many authors (W. Martin 1972; Martin and Gans 1972; Gans 1973; Taigen, Wells, and Marsh 1985; R. Marsh and Taigen 1987), while others have referred to it as the transverse muscle (Duellman and Trueb 1986). The confusion arises because most urodeles have three layers of flank muscles, with the transverse being the deepest. There is considerable variation in the number of muscle layers that are present in

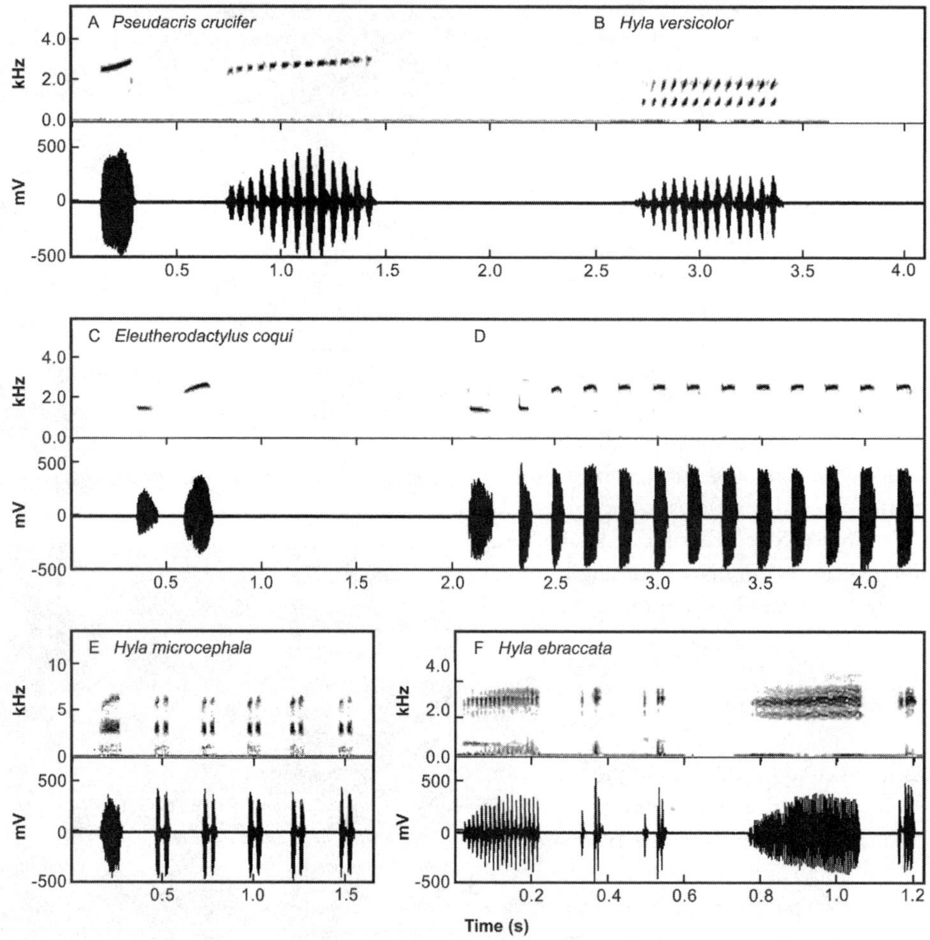


**Fig. 7.2.** Representative anuran calls. The sonagrams at the top of each part show changes in frequency (kHz) over time. The oscillograms at the bottom of each part show changes in amplitude (mV), over time. (A) Tonelike peep of *Pseudacris crucifer* (left) and amplitude-modulated aggressive call (right). (B) Amplitude-modulated advertisement call of *Hyla versicolor.* (C) Tonelike "co" note and frequency-modulated "qui" note of *Eleutherodactylus coqui* advertisement call. (D) Multinote aggressive call of *E. coqui.* (E) Multinote advertisement call of *Hyla microcephala,* composed of an introductory note followed by several biphasic secondary click notes. (F) Three-note advertisement call of *Hyla ebraccata* (left) and three-note aggressive call (right), with much higher pulse rate in the introductory note. From recordings by the author.

urodeles, even within the same family, with some having all four layers and others having only two or three (R. S. Simons and Brainerd 1999; Brainerd and Simons 2000). Anurans have lost the middle layer, leaving only two. One layer is composed of the external oblique muscles, but the homologies of the second layer are uncertain. These muscles have been called both the internal oblique and the transverse muscles by various authors (Ecker 1889; Duellman and Trueb 1986). These layers of muscles are attached to the rectus abdominus muscle that runs longitudinally along the belly and also is involved in call production. These muscles are highly sexually dimorphic, being much larger in males than in females. The muscles of male frogs also exhibit many morphological and biochemical adaptations for call production, including highly aerobic muscle fibers, high concentrations of mitochondria, high activities of aerobic enzymes, heavy vascularization, and ample supplies of energy substrates (see chapter 5 for a more detailed discussion).

The action of the trunk muscles during calling is poorly understood for most anurans. In species that produce a series of simple single-note calls, each call probably is produced by a single contraction of the trunk muscles. For example, each short peep produced by a male *Pseudacris crucifer* (fig. 7.2 A) is preceded by a conspicuous contraction of the body wall. The same is true of the much longer whine call of *Physalaemus pustulosus* (fig. 7.25); males of this species exhibit a contraction of the body wall and a dramatic expansion of the vocal sac as each call is produced (Drewry, Heyer, and Rand 1982; Dudley and Rand 1991). In species that produce a long series of distinct notes, such as *Hyla microcephala* (fig. 7.2 E), each note represents a single contraction of the trunk muscles (Wells and Taigen 1989). In species with

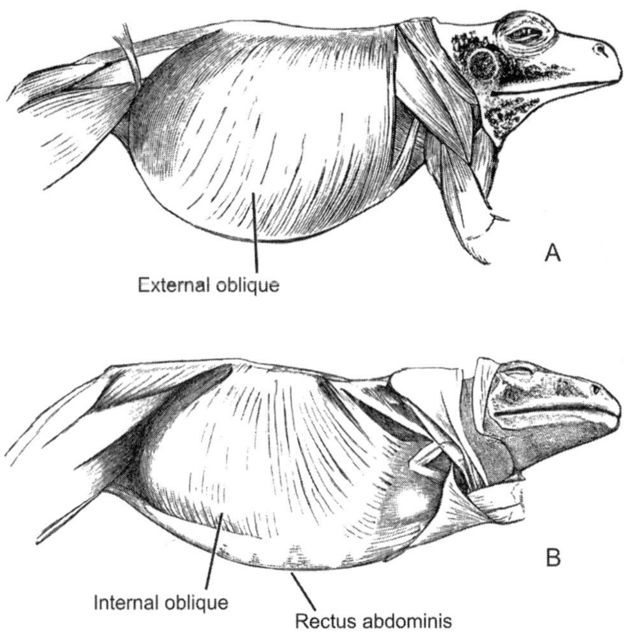


**Fig. 7.3.** Trunk muscles of a male frog, *Rana esculenta,* showing muscles involved in call production. (A) First layer of muscles, viewed from the right side. (B) Second layer of muscles, viewed from the right side and below. The external oblique, internal oblique, and rectus abdominis muscles are the principal call-producing muscles. After Ecker (1889).

long, amplitude-modulated calls consisting of closely spaced pulses, trunk-muscle activity is more complex. W. Martin (1972) speculated that some bufonids with very long calls (sometimes 30 seconds or more) have pulsatile contractions of the trunk muscles. With only a single contraction, the length of a call would be limited by the volume of air in the lungs, but a series of short contractions would allow for alternating periods of expiration and inspiration (Gans 1973). Pulsatile contractions of the trunk muscles have been observed during production of release calls in bufonids (W. Martin and Gans 1972; R. Schmidt 1972b).

In other families, low rates of amplitude modulation are produced by periodic contractions of the trunk muscles alone, or coordinated activity of the trunk and laryngeal muscles (W. Martin 1972). For example, the aggressive calls of several species of European hylids consist of an irregular series of short pulses (Schneider 1967, 1968, 1977; Schneider and Nevo 1972). E. Weber (1976) found that *Hyla savignyi* males could produce these pulses even when the laryngeal muscles were removed, indicating that the trunk muscles are involved. In the two species of North American gray tree frogs (*Hyla versicolor* and *H. chrysoscelis*), each pulse in the call (fig. 7.2 B) corresponds to a separate cycle of contraction and relaxation of the trunk muscles, as demonstrated by electromyographic recordings of muscle activity (Girgenrath and Marsh 1997). *Hyla chrysoscelis,* a diploid species, has a pulse-repetition rate about twice that of *H. versicolor,* a tetraploid species, at the same temperature, and this is reflected in differences in the frequency of trunk-muscle contraction (R. Marsh 1999). The muscle contraction frequencies that maximize the power output of the muscles at 20–25° C (44 Hz in *H. chrysoscelis* and 21 Hz in *H. versicolor*), are very similar to the contraction frequencies observed in calling males (Girgenrath and Marsh 1999).

### Laryngeal Apparatus

The anuran laryngeal apparatus lies between the lungs and the buccal cavity and therefore is involved in both respiration and sound production. Both the laryngeal skeleton and the laryngeal muscles are much larger in male frogs than in females (Wahl 1969; Schneider 1970, 1977, 1986, 1988; Eichelberg and Schneider 1973, 1974; R. Marsh and Taigen 1987; McClelland and Wilczynski 1989; McClelland, Wilczynski, and Ryan 1990; McClelland, Wilczynski, and Rand 1997; Boyd et al. 1999). The development of laryngeal muscles has been studied in detail only in *Xenopus laevis,* which has a mechanism of sound production that is atypical for frogs in general. In this species, sexual dimorphism in laryngeal muscles develops after metamorphosis under the control of male hormones (Kelley 1986, 1996; Sassoon and Kelley 1986; Sassoon, Gray, and Kelley 1987). Not only are the laryngeal muscles of *Xenopus* females smaller than those of males, but they also have different fiber types and are incapable of contracting very rapidly. Hence, females cannot give calls like those of males even when the appropriate parts of the brain are stimulated (Tobias and Kelley 1987), although they do give simple calls that will be discussed in more detail later (Tobias, Viswanathan, and Kelley 1998).

The laryngeal muscles of male frogs resemble the trunk muscles in their fiber composition, high mitochondrial and capillary density, and biochemical features (see chapter 5). Laryngeal anatomy varies among species of anurans (for comprehensive morphological studies, see Ridewood 1897, 1900; Blume 1931, 1932, 1933; Trewavas 1933; Sedra and Michael 1957; Paulsen 1965; Lörcher 1969; Wahl 1969; Heinzmann 1970; Schneider 1970, 1986, 1988; W. Martin 1972; E. Weber 1974; E. Schmid 1977, 1978; McClelland and Wilczynski 1989; M. Ryan and Drewes 1990; Steinwarz 1990; McClelland, Wilczynski, and Ryan 1990, 1998; McClelland, Wilczynski, and Rand 1997).

In most anurans, the sound-producing apparatus consists of two arytenoid cartilages that articulate to form a closed, dome-shaped structure that houses the vocal cords (fig. 7.4). The vocal cords can be activated only when the arytenoid cartilages are moved apart to allow passage of air through the larynx. These moveable cartilages are supported by the cricoid cartilage, which attaches the larynx to the hyoid apparatus and serves as a point of attachment for various

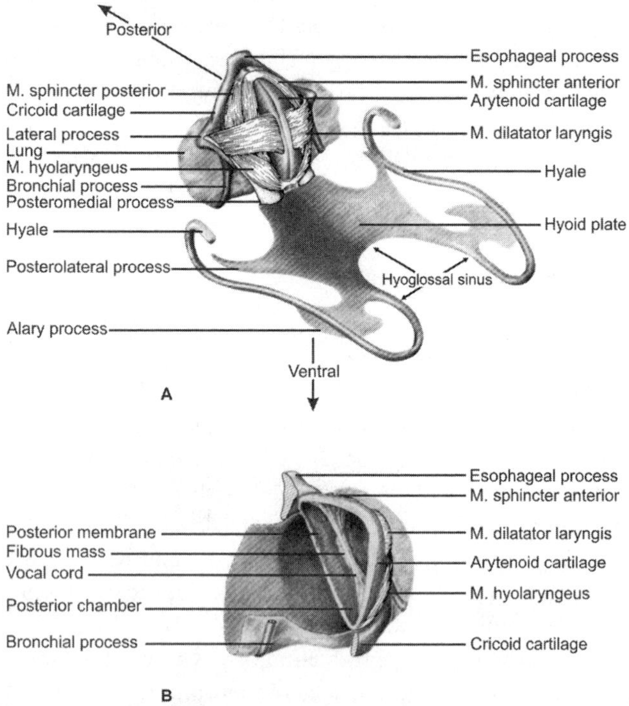


**Fig. 7.4.** The laryngeal apparatus of a typical anuran. (A) Diagram of the laryngeal apparatus of an anuran, showing the hyoid apparatus and the cartilages and muscles of the larynx. The dilator muscles open the cartilages, allowing an airstream to pass over the vocal cords, which are not visible in this figure. The sphincter and hyolaryngeus muscles close the larynx. (B) Cross section of the larynx of an anuran, showing one of the paired vocal cords suspended between the cartilages. After Duellman and Trueb (1986).

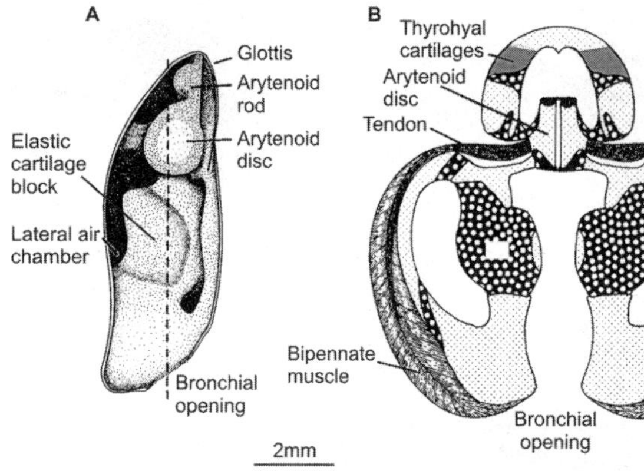


**Fig. 7.5.** Larynx of a male pipid frog, *Xenopus borealis.* (A) Midsagittal view of the interior of the larynx. Dorsal is to the right; anterior to the top. Dashed line indicates the plane of the horizontal section in part B. (B) Horizontal section through the larynx. After Yager (1992b).

laryngeal muscles. In most derived frogs, such as hylids, ranids, leptodactylids, and bufonids, the arytenoid cartilages are large and the cricoid cartilage forms a narrow ring, whereas in *Bombina* and *Discoglossus,* the cricoid cartilage is flattened and considerably larger than the arytenoids (Lörcher 1969; E. Weber 1974; Schneider 1988).

In pipids, the laryngeal skeleton is highly modified (fig. 7.5), with an enormous cricoid cartilage that forms a boxlike structure that partially encloses the greatly reduced arytenoids (Ridewood 1887, 1900). The arytenoid cartilages are held together by fluid surface tension. These frogs lack vocal cords; calls consist of sharp clicking sounds produced when the arytenoid cartilages are suddenly pulled apart, and the sounds are amplified by the boxlike cricoid cartilage (Rabb 1960; Yager 1992a, b, 1996). Hence, sound production in pipids requires contraction of the laryngeal muscles, but does not involve contraction of the trunk muscles or the production of vibrations in an air stream.

In anurans other than pipids, sound is produced only when three conditions are met: (1) the arytenoid cartilages are opened to allow air to pass from the larynx through the glottis to the buccal cavity; (2) the vocal cords are brought into the proper position in the airstream to cause them to vibrate; and (3) pulmonary air pressure is sufficient to set the vocal cords and associated structures in motion. The various laryngeal muscles (fig. 7.4) are involved in the control of all three aspects of sound production, as well as the control of normal respiration. The laryngeal dilator muscles originate on lateral processes of the hyoid and insert on the medial edges of the arytenoid cartilages; they serve to open the arytenoids during normal respiratory movements and vocalization. Two pairs of constrictor muscles (also called the anterior sphincter and hyolaryngeal muscles) originate on the hyoid and wrap around the anterior edge of the arytenoids to insert on the opposite member of each pair. When these muscles contract together, they slide over the arytenoids toward the midline and serve to close the larynx. The posterior sphincter muscles insert on the dorsal ends of the arytenoids and in some species serve to bring the vocal cords together into the air stream (R. Schmidt 1972a; Duellman and Trueb 1986; Schneider 1988).

The action of the various laryngeal muscles during vocalizations varies among species, but has been investigated in detail only for ranids, bufonids, and hylids. In most species, removal of the laryngeal dilators makes vocalization and even normal respiration impossible. These muscles are required to open the larynx during inspiration because the elasticity of the larynx causes it to remain closed unless actively opened by muscle contraction. Frogs lacking these muscles cannot fill the lungs with air and therefore cannot generate the pulmonary pressure required for vocalization (E. Weber 1975a, 1976, 1977, 1978a; Schneider 1988). In some species, these muscles also open the larynx during vocalization. For example, in both hylids and ranids, bursts of sound are preceded by active opening of the larynx by the laryngeal dilators (R. Schmidt 1965, 1973b; Schneider 1977, 1988). In bufonids, on the other hand, the larynx apparently is opened passively by pulmonary air pressure; the call is terminated when the laryngeal dilators contract and move the arytenoid cartilages out of the air stream (W. Martin 1972; Martin and Gans 1972). In some frogs, individual pulses of sound are terminated by rapid closure of the larynx when the constrictor muscles contract (R. Schmidt 1965, 1972a, 1973b). This enables the frog to produce pulses that are highly stereotyped in shape and duration, features that often are important for species recognition (Gerhardt 1988). Removal of these muscles disrupts the spacing and duration of the pulses (E. Weber 1975a, 1976, 1977, 1978a).

In addition to the active amplitude modulation of calls produced by contractions of the laryngeal muscles, some anurans exhibit passive amplitude modulation resulting from oscillations of various elements of the sound-producing apparatus. In many bufonids, the edges of the arytenoids vibrate as pulmonary pressure forces air through the glottis, producing a series of short pulses with very fast rise times. In some species, including *Bufo bufo, B. calamita, B. gariepensis,* and members of the *B. regularis* group in Africa, these pulses are the only type of amplitude modulation in the call. In primitive bufonids, pulses produced by vibration of the arytenoids are imbedded in longer pulses produced by active contraction of the laryngeal muscles. Calls that have only passive amplitude modulation presumably are derived from this pattern through loss of active modulation. Some derived species of toads have gone the other direction and have lost passive amplitude modulation, but retain active modulation (e.g., *Bufo americanus,* W. Martin 1971, 1972). Passive amplitude modulation also can be produced by oscillations of the vocal cords rather than the arytenoids. For example, the release calls of *Rana pipiens* consist of a series of major pulses produced by contraction of the laryngeal muscles; these are subdivided into minor pulses produced by air pressure suddenly pushing apart the paired vocal cords (R. Schmidt 1973b).

The mass and tension of the vocal cords are the major determinants of the dominant frequency of a frog's call. As the mass of the vocal cords increases, the frequency at which they vibrate decreases, thereby decreasing the frequency of the call. Since vocal-cord mass often is positively correlated with overall body size, larger frogs typically produce lower-pitched calls, both within and between species (Zweifel 1968a; W. Martin 1972; Schneider 1977, 1988; M. Ryan 1988a; Penna and Veloso 1990; M. Ryan and Wilczynski 1991; Keddy-Hector, Wilczynski, and Ryan 1992; Wilczynski, Keddy-Hector, and Ryan 1992). The morphology of vocal cords varies among families (Paulsen 1965; E. Schmid 1978), as does the relationship between dominant frequency and body size. In many species, there are fibrous masses imbedded in the membranous vocal cords or attached to them; these increase the mass of the vocal cords and lower the frequency of the call (W. Martin 1972). If the mass of the vocal cords evolves somewhat independently of body size, then the relationship between body size and call frequency will be be relatively weak, as it is in the genus *Physalaemus* (M. Ryan 1985b), or it can differ among populations of the same species (M. Ryan and Wilczynski 1991).

Frequency modulation of a call is most likely achieved by active changes in the tension of the vocal cords, brought about by changes in the shape of the larynx or direct muscular action on the vocal cords. Many frogs of the family Leptodactylidae have frequency-modulated calls (fig. 7.2 C), as do members of several other families. In these groups, the vocal cords insert on flexible tissue at the tips of the arytenoid cartilages, and the posterior sphincter (= constrictor) muscles insert in the same area. Contraction of these muscles increases the tension on the vocal cords, causing an increase in the frequency of the call. In bufonids and most of their relatives, this mechanism for changing the tension of the vocal cords has been lost, and most species show little frequency modulation of the call. W. Martin (1972) suggested that the bufonid lineage shows an evolutionary trend in which the laryngeal muscles have a reduced role in determining the frequency structure of the call, while the role of vocal-cord mass becomes more important. Nevertheless, the tension of the vocal cords still can be changed by changes in the shape of the larynx and play some role in determining the dominant frequency of the call.

### Buccal Cavity and Vocal Sacs

Most anurans have vocal sacs that are inflated during vocalization (fig. 7.6). These are connected to the buccal cavity through openings in the floor of the mouth that range in size from narrow slits to substantial holes. Vibrations produced by the larynx are transmitted through the buccal cavity to the air-filled vocal sac; the vocal sac in turn radiates the sound to the external environment. In general, the efficiency with which frogs convert metabolic energy into radiated sound energy is quite low, due mainly to a mismatch between the size of the vocal sac and the wavelength of the sound being radiated (M. Ryan 1985a; Prestwich, Brugger, and Topping 1989; Prestwich 1994; McLister 2001). Most frogs have vocal sacs that are much smaller than the wavelength of the call and cannot radiate sound very efficiently, although the efficiency is greater than it would be if the larynx served as the only sound radiator. This problem is most acute for species with very low-pitched calls, because low-pitched sounds have long wavelengths. Small species with high-pitched calls and large vocal sacs relative to the size of the head (and larynx) can be expected to be more efficient sound radiators and have unusually loud calls for their body size (e.g., *Hyla microcephala;* see table 7.1).

It had been suggested that vocal sacs might alter the

**Fig. 7.6.** Vocal sac morphology is highly variable among anurans. Many frogs that call from elevated perches have relatively large single medial vocal sacs like that of *Hyperolius concolor* (A). The small paired vocal sacs of *Hylodes asper* (B) produce relatively low-intensity calls, which are supplemented by the visual leg-waving display. Paired vocal sacs that project above the water, like those of *Rana virgatipes* (C) and *Phrynohyas coriacea* (D), radiate most of the acoustic energy of the call to the air. *Bombina bombina* (E) has a relatively primitive vocal sac formed by a loose pocket of the lining of the mouth. The internal vocal sac of *Rana catesbeiana* (F) radiates sound energy to both the air and the water and is typical of many frogs that call from the water's surface. Photos by Walter Hödl (A, B, D), Mac F. Given (C), H. Gasser, courtesy of Walter Hödl (E), and Kentwood D. Wells (F).

**Table 7.1** Peak sound pressure levels of anuran calls measured at 50 cm in front of the animal

| Species (source no.) | SVL (mm) | Peak sound pressure level (dB SPL) Mean | Range | *N* |
|---|---|---|---|---|
| Bufonidae | | | | |
| *Bufo americanus* (1) | 54–85 | 106 | 102–109 | 10 |
| *B. garmani* (2) | 90–97 | 112 | 110–113 | 3 |
| *B. gutturalis* (2) | 80–81 | 115 | 114–115 | 2 |
| *B. terrestris* (1) | 42–82 | 107 | 106–108 | 2 |
| Centrolenidae | | | | |
| *Hyalinobatrachium fleischmanni* (5) | 21–23 | 100[a] | 92–104 | 9 |
| Discoglossidae | | | | |
| *Alytes muletensis* (14) | | 75 | | 24 |
| Hylidae | | | | |
| *Acris gryllus* (1) | 15–29 | 103 | 99–106 | 4 |
| *Agalychnis callidryas* (12) | 42–47 | 93 | | 18 |
| *Hyla avivoca* (1) | 28–39 | 99 | 96–101 | 10 |
| *H. chrysoscelis* (1) | 36–43 | 105 | 101–108 | 10 |
| *H. cinerea* (1) | 37–59 | 103 | 97–107 | 6 |
| *H. ebraccata* (7) | 23–28 | 100 | 96–105 | 12 |
| *H. gratiosa* (1) | 49–68 | 108 | 106–109 | 3 |
| *H. microcephala* (7) | 21–25 | 106 | 101–109 | 12 |
| *H. phlebodes* (7) | 20–24 | 103 | 101–105 | 5 |
| *H. versicolor* (1) | 32–51 | 109 | 105–112 | 9 |
| *Pseudacris crucifer* (1) | 18–29 | 103 | 100–106 | 10 |
| *P. regilla* (11) | | 96[b] | 93–101 | 20 |
| *Scinax boulengeri* (12) | 38–45 | 100 | | 8 |
| *S. rubra* (12) | 38–43 | 94 | | 6 |
| Hyperoliidae | | | | |
| *Afrixalus brachycnemis* (2) | 16–19 | 97 | 91–102 | 8 |
| *A. fornasinii* (2) | 34–36 | 106 | 104–109 | 7 |
| *Hyperolius argus* (2) | 33–38 | 108 | 105–111 | 11 |
| *H. marmoratus* (2) | 28–30 | 104 | 102–105 | 12 |
| *H. pusillus* (2) | 20–23 | 110 | 105–113 | 14 |
| *H. tuberilinguis* (2) | 28–32 | 110 | 105–113 | 15 |
| *Kassina maculata* (2) | | 114 | 112–116 | 6 |
| Leptodactylidae | | | | |
| *Eleutherodactylus coqui* (4) | | | | |
| High elevation | 35–50 | 100[c] | 96–105 | 20 |
| Low elevation | 26–37 | 97[c] | 90–103 | 16 |
| *Eupsophus emiliopugini* (13) | | 86 | 72–88 | 9 |
| *Leptodactylus labialis* (12) | 30–35 | 84 | | 9 |
| *Physalaemus pustulosus* (9) | 27–32 | 90 | | 5 |
| Microhylidae | | | | |
| *Gastrophryne carolinensis* (1) | 20–30 | 95 | 87–101 | 6 |
| Myobatrachidae | | | | |
| *Geocrinia victoriana* (8) | 21–33 | 105 | 101–107 | 7 |
| *Uperoleia laevigata* (6) | 24–32 | 85 | | 15 |
| Petropedetidae | | | | |
| *Cacosternum boettgeri* (2) | 18–19 | 108 | 105–112 | 3 |
| Ranidae | | | | |
| *Rana areolata* (1) | 62–90 | 110 | 109–113 | 2 |
| *R. clamitans* (9) | 54–67 | 96 | 92–102 | 5 |
| *R. esculenta* (3) | 62–76 | 110 | | |
| *R. lessonae* (3) | 54–59 | 105 | | |
| *R. ridibunda* (3) | 77–97 | 116 | | |
| *R. virgatipes* (10) | 43–55 | 109 | 102–112 | 16 |
| *Tomopterna cryptotis* (2) | 38–42 | 106 | 104–108 | 6 |

*Sources:* (1) Gerhardt 1975; (2) Passmore 1981; (3) Brzoska 1982; (4) Narins and Hurley 1982; (5) Wells and Schwartz 1982; (6) J. Robertson 1984; (7) Schwartz and Wells 1984a; (8) Littlejohn and Harrison 1985; (9) Ryan 1985a; (10) Given 1987; (11) Brenowitz 1989; (12) Bevier 1995a; (13) Penna and Solis 1996; (14) Bush 1997.

*Notes:* Data show the overall means for several individual males unless otherwise indicated. Snout-vent lengths (SVL) for most species are ranges for breeding males from standard taxonomic works or original papers. *N* = Number of individuals.

[a]Median value given for *Hyalinobatrachium fleischmanni.*

[b]Fast RMS SPL given for *Pseudacris regilla.*

[c]SPL of "qui" note of *Eleutherodactylus coqui.*

frequency composition of calls by acting as cavity resonators. Experiments with frogs (*Physalaemus pustulosus*) calling in helium showed that dominant frequency was not altered, as would be expected with a cavity resonator (Rand and Dudley 1993). Nevertheless, frequency modulation of calls in this species is affected by the nature of the vocal sac. As the frog calls, the rate of vocal sac inflation gradually decreases as the expansion of the sac reaches its elastic limits. This decrease in inflation rate corresponds with a downward sweep in call frequency and a decrease in call amplitude (Dudley and Rand 1991). Consequently, frequency modulation can be a result of changes in the rate of air flowing over the vocal cords, and not just changes in vocal-cord tension. The size of the head and the capacity of the buccal cavity also can affect the frequency of the call. In *Hyla microcephala,* a frog with a small head and buccal cavity, high frequencies around 6 kHz are emphasized in calls, whereas in *H. ebraccata,* a species with a larger buccal volume, the dominant frequency is around 3 kHz (McClelland, Wilczynski, and Rand 1997).

The morphology of vocal sacs varies considerably among

species of anurans (fig. 7.6). Most accounts of vocal sac anatomy have focused on taxonomic implications rather than the functional significance of structural variation (C. Liu 1935b; Liem 1970; M. Tyler 1971, 1972, 1974, 1985a; M. Hayes and Krempels 1986). We currently know very little about how the structure of a vocal sac influences the way in which sound is transmitted. These sacs are not simply formed by the skin of the throat, but are located between the superficial mandibular muscles and the deeper muscles of the throat (M. Tyler 1971, 1972). This gives the sac a high degree of elasticity and, in some cases, muscle contraction pumps air back into the lungs after a call is produced. The vocal sac of *Physalaemus pustulosus* increases in volume by a factor of 20 to 40 during calling (fig. 7.1; Dudley and Rand, 1991). The vocal sac is highly elastic, due to a layer of elastic fibers located between the skin and the outer surface of the vocal sac muscles. This provides for substantial elastic recoil that facilitates deflation of the sac and refilling of the lungs (Jaramillo, Rand, Ibáñez, and Dudley 1997). This in turn probably reduces the energetic cost of vocalization.

The most common type of vocal sac is a single subgular sac that inflates like a balloon when the frog calls (fig. 7.6 A). Some species have paired subgular sacs (fig. 7.6 B) that are partially separated by connective tissue running between the superficial and deep muscle layers (Duellman 1970). Others have paired lateral sacs that are completely separate chambers on either side of the throat (fig. 7.6 C, D). These are formed by extensions of the interhyoideus muscle (M. Tyler 1971). In at least one species, *Rhinophrynus dorsalis,* the muscles forming the walls of the lateral vocal sacs are quite thick (M. Tyler 1974), suggesting an active role in a vocalization cycle. It seems likely that the biochemical and morphological characteristics of vocal sac muscles are similar to those of the trunk muscles, but this has not been investigated in any species.

The three basic types of vocal sacs (median subgular, paired subgular, and paired lateral) often are further differentiated as internal or external, following the terminology of Noble (1931) and C. Liu (1935b). However, these terms are somewhat misleading, since all sacs actually are internal, with the major difference being the thickness of the vocal sac wall. Small frogs that call in air often have very thin vocal sacs (fig. 7.6 A), whereas those that call in water, particularly very large species, often have thick-walled vocal sacs that appear swollen when inflated (fig. 7.6 E, F). Some ranid frogs have balloonlike extensions of the vocal sacs, called gular sacs, which project beyond the vocal sacs themselves when inflated. These gular sacs are invaginations of the gular skin surrounding the interhyoideus muscle and are everted during sound production. These sacs are found in African ranids of the genus *Ptychadena* and several Old World species of *Rana,* including *R. esculenta, R. ridibunda,* and *R. nigromaculata,* but are not found in any New World ranids (Inger 1956). These gular sacs probably serve to radiate sound to the air when a frog is calling in the water, as do the lateral vocal sacs of species such as *Rana virgatipes* (fig. 7.6 C) and *Phyrnohyas coriacea* (fig. 7.6 D)

Although vocal sacs probably are the main sound-radiating organs in most frogs, they are not the only ones. Purgue (1997) discovered that a significant proportion of the sound energy produced by a calling male bullfrog (*Rana catesbeiana*) is radiated not from the vocal sacs, which sit in the water in this species, but from the very large tympanic membranes. Male bullfrogs, and a number of other species of *Rana,* have tympana up to 50% larger than those of females (fig. 7.7). This difference in size could be related to

**Fig. 7.7.** Comparison of tympanum size and structure in a female (A) and a male (B) bullfrog (*Rana catesbeiana*). The tympanum of the male is much larger in proportion to the size of the head and has a thickened central area. Photos by Susan Z. Herrick.

greater hearing sensitivity in males (Hetherington 1994), but it also appears to relate to the use of the ears as sound radiators. Furthermore, the tympana of male bullfrogs and other ranids with enlarged tympana have an unusual structure, with a thickened central patch that accounts for most of the mass of the eardrum. This patch apparently reduces displacement of the center of the membrane and serves to decouple the auditory and sound-broadcasting functions of the eardrum. In an unusual African frog, *Petropedetes parkeri*, the thickening of the tympanum is carried even further. Males have a conspicuous spongy papilla projecting from the tympanum, but the structure is offset from the center of the membrane. There is some evidence that the ears of these frogs are used for both sound reception and sound transmission, as in North American bullfrogs (Narins et al. 2001).

Various species of frogs lack vocal sacs, but the selective pressures favoring this condition are not always clear. Presumably the pipids never evolved vocal sacs because they do not use a vibrating air stream for sound production. Primitive aquatic species such as *Ascaphus* and some discoglossids also lack vocal sacs (C. Liu 1935b; M. Tyler 1980). *Bombina bombina* lacks a vocal sac, but can expand the floor of the buccal cavity to perform the same function as a vocal sac (fig. 7.6 E). This also is true of the Australian myobatrachids *Heleioporus* and *Neobatrachus* (M. Tyler 1972). Some more-derived frogs have lost vocal sacs secondarily. Loss of vocal sacs and partial or complete loss of long-distance vocal communication sometimes is associated with breeding near cascades or torrential streams. Examples include Australo-Papuan hylids such as *Litoria lesueuri, L. booroolongensis*, and *L. nannotis;* Neotropical hylids such as *Hyla altipodens*, the *H. mixomaculata* group, some members of the *H. bistincta* group, and various species of *Plectrohyla;* the Australian myobatrachid *Taudactylus diurnis;* and some species in several genera of leptodactylid frogs (*Alsodes, Insuetophrynus, Telmatobufo, Telmatobius;* Zweifel 1958; Duellman 1970; M. Tyler 1971, 1972; Menzies 1976; Trueb 1979; Penna, Veloso, and Contreras 1983; Penna and Veloso 1987, 1990; Duellman and Campbell 1992; Barker, Grigg, and Tyler 1995). The usual explanation for this is that the background noise of waterfalls makes vocal communication difficult. However, this argument is weakened by the fact that many anurans that breed in or near torrents have well-developed vocal signals (for example, centrolenid frogs). Some of the species previously listed lack vocal sacs, but still produce vocalizations that presumably are used in close-range communication (e.g., *Litoria nannotis;* S. Richards 1992) or serve as release signals (Penna and Veloso 1987). In some genera, some species have vocal sacs and long-distance vocal signals, but close relatives breeding in similar habitats do not (Duellman 1970; Penna and Veloso 1990).

Various bufonids also lack functional vocal sacs, and some apparently do not have long-distance vocal signals. However, loss of vocalizations generally is not associated with breeding near noisy streams and waterfalls. Rather, it seems to be associated with explosive breeding in permanent water or sites that do not change in location from year to year (Wells 1977b). Males do not need calls to attract females to breeding sites because they are in traditional locations; once at the site, males generally locate females by active searching. Examples include *Bufo alvarius, B. boreas, B. exsul, B. nelsoni, B. holdridgei, B. periglenes, B. rosei*, and most populations of *B. bufo* (W. Blair and Pettus 1954; Heusser 1961, 1969a; Schuierer 1962; J. Savage 1966; Hotz 1970; Black and Brunson 1971; W. Martin 1972; Tandy and Keith 1972; Novak and Robinson 1975; Kagarise Sherman 1980). A number of voiceless toads are active during the day and are sexually dichromatic, with males more brightly colored than females (e.g., in *B. periglenes* males are bright orange, whereas females are dark, with a mottled pattern). The bright color of the males could serve as a visual signal to females that partially replaces vocal signaling (J. Savage 1966; Tandy and Keith 1972; Kagarise Sherman 1980).

In ranids, loss of vocal sacs and reduction of vocal communication has evolved independently in several different lineages, apparently in response to different selective pressures. In western North America, several ranid frogs lack vocal sacs (*R. aurora, R. cascadae, R. muscosa, R. pretiosa*), whereas others retain them (*R. draytonii, R. boylii*). Those lacking vocal sacs do vocalize, but generally do so under water and have relatively weak calls, whereas those with vocal sacs usually call in air (Zweifel 1955; F. Turner 1958; Licht 1969a, b; Morris and Tanner 1969; Altig and Dumas 1971; Hayes and Krempels 1986). Some large African ranids, including *Conraua goliath* and *Aubria subsigillata*, lack vocal sacs and apparently do not call, but their breeding habits are very different. The former breeds in or near river rapids, whereas the latter breeds in temporary ponds (Perret 1966; Lamotte and Perret 1968; Sabater-Pi 1985). Several so-called voiceless ranid frogs also are found in Borneo (*Limnonectes leporinus, L. ibanorum, L. kuhlii, L. laticeps, L. luctuosa, L. microdisca, L. paramacrodon;* Inger 1966). These frogs breed in a variety of habitats from small streams to pools and swampy areas, but their modes of communication are poorly understood (Emerson 1992; Emerson and Inger 1992). The behavior of these frogs varies geographically, probably reflecting hidden taxonomic diversity in the group (Emerson et al. 2000). For example, males of *L. leporinus* (formerly *L. blythii*) in Borneo apparently do not call, but Matsui (1995) reported that males of a *L. blythii*-like species in Malaysia did produce short, single-note calls, as do males of a species from Vietnam, which may be *L. blythii* (Orlov 1997). Males of *L. kuhlii* from Taiwan also produce calls (Tsuji and Lue 1998), but this almost certainly is not the same species as in Borneo; *L. kuhlii* sensu stricto occurs only on Java and differs from those on both Borneo and Taiwan (Iskandar 1998).

## The Vocal Repertoires of Frogs and Toads

Bogert (1960) classified anuran calls into six categories based on the context in which they occur: (1) mating calls, (2) territorial calls, (3) male release calls, (4) female release calls, (5) distress calls, and (6) warning calls. He added a tentative category of rain calls, vocalizations given sporadically by males away from the breeding site during rains or in high humidity. The function of rain calls is unknown and will not be discussed further. Bogert's classification was useful in drawing attention to the diversity of call types and summarizing the available information on their functions. It was later modified by Littlejohn (1977) and Wells (1977a, b, 1988), and the terminology used in this book is derived from those papers.

### Advertisement Calls

Bogert used the term *mating call* to describe the principal signals given by males during the breeding season. These calls are now generally termed advertisement calls (Wells 1977b) because they often serve more than one function or convey more than one type of message (Gerhardt 1992b). Playback experiments have shown that female frogs will approach conspecific calls presented alone or in choice tests with heterospecific calls (see the following and table 7.2). In species in which males call over long periods of time and females are in the same habitat, advertisement calls also could stimulate hormone production in females and maintain reproductive condition in females, although this has rarely been demonstrated experimentally (Lea, Dyson, and Halliday 2001). Advertisement calls also advertise a male's position to other males and help to maintain spacing between calling individuals (Whitney and Krebs 1975b; Wells 1977a, b, 1978b). The distance between males is a function of perceived call intensity (Fellers 1979a; Brzoska, Schneider, and Nevo 1982; J. Robertson 1984; Brenowitz, Wilczynski, and Zakon 1984; Wilczynski and Brenowitz 1988; Brenowitz 1989; Gerhardt, Diekamp, Ptacek 1989; Murphy and Floyd 2005; Owen and Gordon 2005).

Males of some species can use the pitch of other males' advertisement calls to assess the body size of potential opponents in agonistic encounters over females or calling sites (Davies and Halliday 1978; Arak 1983b; Ramer, Jenssen, and Hurst 1983; J. Robertson 1984; Given 1987; Wagner 1989a). Males are more likely to approach or attack speakers playing high-pitched calls of small males, but retreat from the low-pitched calls of large males (fig. 7.8). Males of some species alter the pitch of their calls in response to those of neighboring males (e.g., *Rana catesbeiana;* Bee and Bowling 2002), but it is not always clear that this provides more accurate information about male body size to opponents (Bee, Perrill, and Owen 2000; Bee 2002; see further discussion of aggressive interactions in the following).

There also is evidence for bullfrogs (*Rana catesbeiana*) that features of the advertisement call can be used by males to identify their neighbors individually (M. Davis 1987; Bee and Gerhardt 2001a, b, c, 2002). This allows males to accommodate to the vocalizations of familiar neighbors that do not represent a significant threat to their own territories. Both bullfrogs and green frogs (*R. clamitans*) exhibit short-term habituation to familiar calls (Owen and Perrill 1998; Bee and Gerhardt 2001a; Bee 2003a). This could provide a mechanism for individual recognition of calls (see also Bee and Schachtman 2000; Owen and Perrill 2000). It also appears that there is sufficient individual variation in call properties of both species to enable males to discriminate among neighbors (Bee et al. 2001; Bee 2004). Individual recognition probably is common in territorial frogs that maintain relatively stable spatial relationships within a pond (e.g., *Rana dalmatina;* Lesbarrères and Lodé 2002), but convincing experimental evidence is lacking for most species. Playback experiments with a highly territorial frog from Panama, *Dendrobates pumilio,* failed to demonstrate discrimination between calls of neighbors and those of unfamiliar males (Bee 2003b).

Although the advertisement calls of most anurans consist of a single note, a series of identical repeated notes, or a long trill, some have complex advertisement calls with more than one kind of note (Littlejohn 1977; Wells 1977a, 1988). Most frogs have only a few kinds of notes in their calls, but some rhacophorid treefrogs have extraordinarily complex calls, with a dozen or more distinct kinds of notes. The functions of the different note types in species with exceptionally complex calls are not fully understood, but at least some call components appear to be used in aggressive interactions among males (Narins, Lewis, and McClelland 2000; Christensen-Dalsgaard, Ludwing, and Narins 2002; Feng, Narins, and Xu 2002).

In some species, different types of notes apparently convey separate messages to males and females. For example, in *Eleutherodactylus coqui* (fig. 1.17 F in chapter 1), the beginning co note (fig. 7.2 C) elicits calling and aggressive responses from males, but they show little response to the qui note (Narins and Capranica 1978). In contrast, females are attracted to the qui note, played with or without the co note, but show little response to the co note alone (Narins and

**Table 7.2** Results of two-choice playback experiments testing call discrimination by female anurans

| Species tested (source no.) | Heterospecific alternative | No. of choices | | No. of females | | |
|---|---|---|---|---|---|---|
| | | CS | HS | CS | HS | *P* |
| Hylidae | | | | | | |
| *Hyla andersonii* (13) | *H. cinerea* | 43 | 4 | 12 | 0 | < 0.001 |
| *H. cinerea* (13) | *H. andersonii* | 18 | 0 | 6 | 0 | 0.032 |
| *H. cinerea* (15) | *H. gratiosa* | 323 | 7 | 61 | 3 | < 0.001 |
| *H. gratiosa* (15) | *H. cinerea* | 48 | 0 | 19 | 0 | < 0.001 |
| *H. chrysoscelis* (4) | *H. versicolor* | 32 | 1 | 7 | 1 | 0.124 |
| *H. ebraccata* (20) | *H. microcephala* | 10 | 0 | 10 | 0 | 0.002 |
| *H. ebraccata* (20) | *H. phlebodes* | 14 | 4 | 14 | 4 | 0.180 |
| *H. femoralis* (10) | *H. squirella* | 14 | 0 | 6 | 0 | 0.032 |
| *H. squirella* (10) | *H. cinerea* | 34 | 0 | 14 | 0 | < 0.001 |
| *H. meridionalis* (17) | *H. arborea* (allopatric) | 15 | 1 | 10 | 1 | 0.022 |
| *Pseudacris crucifer* (12) | *P. ornata* | 21 | 0 | 14 | 0 | < 0.001 |
| *P. cadaverina* (16) | *P. regilla* | 53 | 3 | 53 | 3 | < 0.001 |
| *P. regilla* (16) | *P. cadaverina* | 55 | 4 | 55 | 4 | < 0.001 |
| *P. clarkii* (6) | *P. t. triseriata* | 39 | 5 | | | |
| *P. t. triseriata* (6) | *P. clarkii* | 29 | 7 | | | |
| *P. t. triseriata* (3) | *P. feriarum* (allopatric) | 22 | 1 | | | |
| *P. t. triseriata* (3) | *P. nigrita* (allopatric) | 11 | 10 | | | |
| *P. streckeri* (2) | *P. ornata* (allopatric) | 35 | 4 | 7 | 2 | 0.454 |
| *P. streckeri* (1) | *P. clarkii* | 46 | 1 | | | |
| *P. clarkii* (6) | *P. streckeri* | 28 | 1 | | | |
| *Litoria ewingi* | *L. verreauxi* | | | 10 | 0 | 0.002 |
| *L. verreauxi* | *L. ewingi* | | | 9 | 0 | 0.004 |
| Hyperoliidae | | | | | | |
| *Hyperolius argus* (18) | *Hyperolius* (3 species) | 84 | 1 | 15 | 1 | 0.001 |
| *H. marmoratus* (18) | *Hyperolius* (3 species) | 105 | 0 | 20 | 0 | < 0.001 |
| *H. pusillus* (18) | *Hyperolius* (3 species) | 37 | 0 | 8 | 0 | 0.008 |
| *H. tuberilingus* (18) | *Hyperolius* (3 species) | 15 | 0 | 1 | 0 | |
| Leptodactylidae | | | | | | |
| *Physalaemus pustulosus* (21) | *P. coloradorum* | 10 | 1 | 10 | 1 | 0.021 |
| *P. pustulosus* (21) | *P. pustulatus* | 11 | 0 | 11 | 0 | < 0.002 |
| *P. pustulosus* (21) | *P. petersi* | 10 | 0 | 10 | 0 | 0.002 |
| *P. pustulosus* (21) | *P. "pustulatus"* Peru | 12 | 0 | 12 | 0 | < 0.002 |
| *P. pustulosus* (21) | *P. enesefae* | 10 | 0 | 10 | 0 | 0.002 |
| *P. pustulosus* (21) | *P. ephippifer* | 10 | 0 | 10 | 0 | 0.002 |
| *P. pustulosus* (21) | *P. "roraima"* | 10 | 0 | 10 | 0 | 0.002 |

(*continued*)

**Table 7.2** (continued)

| Species tested (source no.) | Heterospecific alternative | No. of choices CS | No. of choices HS | No. of females CS | No. of females HS | P |
|---|---|---|---|---|---|---|
| Myobatrachidae | | | | | | |
| *Crinia signifera* (7) | *C. parinsignifera* | 11 | 0 | 11 | 0 | < 0.002 |
| *C. signifera* (7) | *C. tinnula* | 11 | 0 | 11 | 0 | < 0.002 |
| *C. parinsignifera* (7) | *C. signifera* | 8 | 0 | 8 | 0 | 0.008 |
| *C. parinsignifera* (7) | *C. tinnula* | 8 | 0 | 8 | 0 | 0.008 |
| *C. tinnula* (7) | *C. parinsignifera* | 14 | 0 | 14 | 0 | < 0.002 |
| *C. tinnula* (7) | *C. signifera* | 14 | 0 | 14 | 0 | < 0.002 |
| *Geocrinia laevis* (14) | *G. victoriana* | 5 | 0 | 5 | 0 | 0.062 |
| *G. victoriana* (14) | *G. laevis* | 8 | 1 | 7 | 0 | 0.016 |
| Pelobatidae | | | | | | |
| *Scaphiopus couchii* (8) | *S. hurterii* | 23 | 23 | | | |
| *S. hurterii* (8) | *S. couchii* | 10 | 12 | | | |
| *Spea hammondii* (11) | *S. bombifrons* | 27 | 0 | 9 | 0 | 0.004 |
| *S. bombifrons* (11) | *S. hammondii* | 23 | 0 | 6 | 0 | 0.032 |
| Pipidae | | | | | | |
| *Xenopus laevis* (19) | *X. muelleri* | 268 | 0 | 20 | 0 | < 0.001 |

*Sources:* (1) Littlejohn and Michaud 1959; (2) W. Blair and Littlejohn 1960; (3) Littlejohn 1960; (4) Littlejohn, Fouquette, and Johnson 1960; (5) Littlejohn 1961; (6) Michaud 1962; (7) Straughan and Main 1966; (8) Awbrey 1968; (9) Littlejohn and Loftus-Hills 1968; (10) Gerhardt 1970; (11) Forester 1973; (12) Gerhardt 1973; (13) Gerhardt 1974a; (14) Littlejohn and Watson 1974; (15) Oldham and Gerhardt 1975; (16) Straughan 1975; (17) Gerhardt and Schneider 1980; (18) Telford and Passmore 1981; (19) Picker 1983; (20) Backwell and Jennions 1993; (21) Ryan and Rand 1993a.

*Notes:* Both the number of choices of conspecific (CS) and heterospecific (HS) calls and the number of females choosing each type of call are shown. *P* values are for two-tailed binomial tests comparing the number of females choosing each type of call.

Capranica 1976). In the Australian myobatrachid *Geocrinia victoriana,* the long, pulsatile introductory note conveys an aggressive message to males, while the shorter secondary notes are attractive to females (Littlejohn and Harrison 1985). A somewhat similar system is found in the African hyperoliid *Afrixalus brachycnemis,* which has a rapidly pulsed zip note that serves as an aggressive signal and a longer trill component that is attractive to females (Backwell 1988). In *Afrixalus fornasinii,* males give long trains of very short pulses (trills) that often grade into a series of repeated pulsed notes that are given mainly in response to other males (Schneichel and Schneider 1988). These distinct call components probably convey separate attractive messages to females and aggressive messages to males, but playback experiments needed to determine the functions of the different parts of the call have not been performed. Some other species, such as European frogs in the genus *Pelodytes,* have multinote calls, but the function of the different call notes is unclear (Márquez, Pargana, and Crespo 2001). There also are frogs that add components to their calls during chorusing interactions; these are discussed in a later section.

## Male Courtship Calls

Male frogs often alter their vocal behavior when females are nearby, producing calls that render the male more conspicuous in a chorus (Wells 1977a, 1988). Males of some species respond to females by lengthening their advertisement calls. For example, male spring peepers (*Pseudacris crucifer*) give longer peeps when females are nearby (M. Rosen and Lemon 1974). Male gray treefrogs (*Hyla versicolor*) respond to approaching females by giving trills that can be several times the length of normal advertisement calls (Fellers 1979a; Wells and Taigen 1986; Klump and Gerhardt 1987). Similar behavior is seen in a dendrobatid frog, *Mannophryne trinitatis* (Wells 1980c). Males normally give two-note advertisement calls, but combine these into a continuous trill when females are approaching (fig. 7.9). The result is a 50% increase in the number of notes delivered per minute.

In many species, males simply respond to approaching females by increasing calling rates (P. Anderson 1954; M. J. Coe 1974; Wells 1977a; Fellers 1979a; Wells and Greer 1981; Schwartz and Wells 1984a, b, 1985; Wells and Schwartz

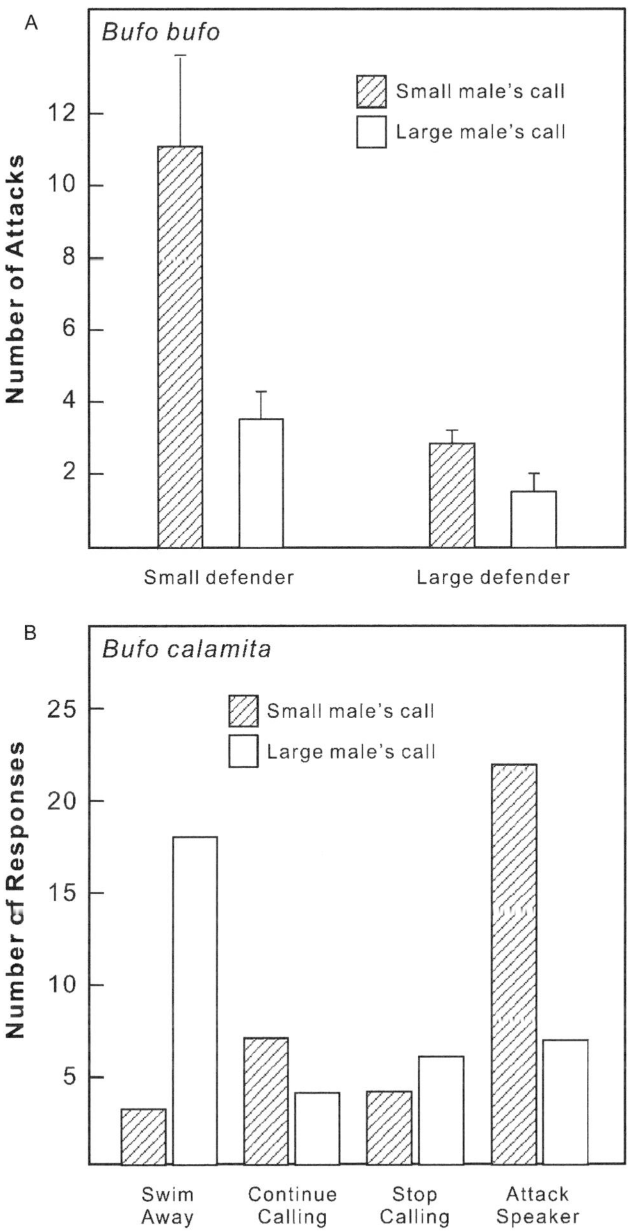


**Fig. 7.8.** Use of call pitch by toads to assess body sizes of opponents in aggressive encounters. (A) In *Bufo bufo,* medium-sized males allowed to attack either large or small males in amplexus with females. Defending males were prevented from calling by rubber bands in their mouths. Tape-recorded calls of large or small males were played during the experiment. Males were more likely to attack when they herd the calls of a small male, but they seldom attacked large defending males. After N. Davies and Halliday (1978). (B) Responses of males of *Bufo calamita* to a speaker playing calls of a large or small male in a chorus. Males were most likely to retreat from calls of large males, but often approached and attacked a speaker playing the calls of a small male. After Arak (1983c).

1984a; Akef and Schneider 1985; Martins and Haddad 1988; Ibáñez 1991) or by giving calls that are modified versions of the advertisement call (Owen and Tucker 2006). In others, males give courtship calls that are quite distinct from the advertisement call. Distinctive courtship calls appear to be

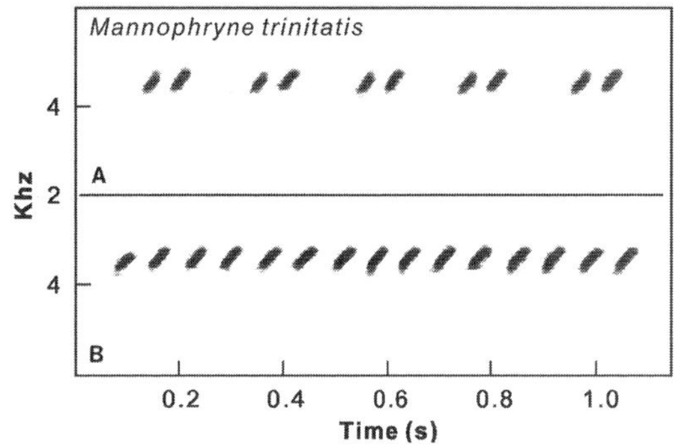


**Fig. 7.9.** Sound spectrograms of (A) the advertisement call and (B) a courtship trill of *Mannophryne trinitatis.* From recordings by the author.

particularly common in species in which the male leads the female to a concealed oviposition site during a prolonged bout of courtship (Jameson 1954, 1955a, b; Dixon 1957; Pengilley 1971a; Lüddecke 1976; Wells 1977a, 1980c; H. and E. Zimmermann 1981, 1985; K. Jungfer 1985; Townsend and Stewart 1986; Hoskin 2004), although they sometimes occur in species in which the male calls from a fixed location to attract the female (Greer and Wells 1980; Kluge 1981; J. Robertson 1986b). In midwife toads (*Alytes obstetricans*), males give courtship calls while moving toward females (Heinzmann 1970), and females sometimes respond with calls of their own (see the following). Courtship calls sometimes are given at lower intensity than advertisement calls, perhaps to avoid alerting other males to the presence of a female. Ovaska and Caldbeck (1997b) showed that males of *Eleutherodactylus antillensis* respond to playbacks of courtship calls by approaching the speaker playing the call, a response that was not seen during playbacks of advertisement or aggressive calls.

Some species have more than one type of courtship call. In *Hyalinobatrachium fleischmanni,* a male gives long, frequency-modulated calls when a moving frog is first detected nearby. This appears to serve as an aggressive call when directed at males, but probably also provides direction cues to females. Once a female begins to approach the male, he switches to a series of short chirps (Greer and Wells 1980). In *Mannophryne trinitatis,* a male gives a long trill when courting a female at a distance (fig. 7.9), but switches to quiet chirps as he leads the female to a hidden oviposition site (Wells 1980c). Males of the Australian microhylid *Cophixalus ornatus* give long courtship calls that are distinct from advertisement calls while leading females to nest sites, but give a shorter call when in the nest with a female (Hoskin 2004). Both types of calls are delivered at a high rate, but low intensity. Short-range courtship calls also have been reported in *Dendrobates speciosus* (K. Jungfer 1985) and several species of *Eleutherodactylus* (Ovaska and Hunte 1992; Michael 1996; Bourne 1997; Ovaska and Caldbeck

1997a, 1999). Courtship calls probably are widespread in anurans, but have not been reported for many species because of the lack of detailed observations of courtship behavior.

### Female Courtship Calls

Some female frogs give calls in response to the calls of males, and these often are called reciprocation calls (Littlejohn 1977; Roy, Borah, and Sarma 1995; Schlaepfer and Figeroa-Sandí 1998; Emerson and Boyd 1999). Such calls are unusual, but perhaps are more common than generally recognized because they tend to be given at very low intensity and therefore are hard to hear. The best-studied species are midwife toads in the genus *Alytes*. Female midwife toads call in response to male calls, and these calls elicit soft courtship calls from males (Heinzmann 1970; Márquez and Verrell 1991; S. Bush 1997; Bosch and Márquez 2001b). The calls given by females probably enhance the ability of males and females to find each other, but do not appear to function in competition among females (Bosch 2001, 2002a). In some instances, female midwife toads even call spontaneously to advertise themselves to males. Male midwife toads often call from hidden locations in rock crevices or burrows, but sometimes move toward females and engage in vocal duets with them (S. Bush, Dyson, and Halliday 1996; Bush 1997).

Females of a number of other anuran species call in response to male vocalizations, and in some cases, this elicits a change in vocal behavior by the male. Dixon (1957) described a call given by females of *Eleutherodactylus angustidigitorum* that elicited a switch by males from advertisement calls to courtship trills. Very similar behavior was observed in *E. podiciferus* from Costa Rica, but in this case, the normal advertisement call is a trill. Males switched to squeak calls after hearing similar squeaks given by approaching females (Schlaepfer and Figeroa-Sandí 1998). Females of *Leptodactylus fragilis* from Panama are reported to give very soft calls during courtship interactions with males, but the calls have not been recorded (Bernal and Ron 2004). Females sometimes call in response to male calls in *Hyla microcephala* (J. Schwartz, personal communication) and two species of *Pelobates* (Andreone and Piazza 1990; Lizana, Márquez, and Martin-Sanchez 1994), but the reactions of males to these calls have not been studied. Duetting between males and females occurs in the strictly aquatic pipid frog *Xenopus laevis* (Tobias, Viswanathan, and Kelley 1998). As in midwife toads, female *Xenopus* sometimes give calls spontaneously, and this elicits calling from nearby males.

Courtship calling by females has been reported in several species of ranid frogs. Given (1987) recorded low-intensity calls given by female *Rana virgatipes* as they approached territorial males; these elicited an increase in calling from the males (Given 1993a). Similar calls are given by female bullfrogs (*R. catesbeiana;* Judge, Swanson, and Brooks 2000), and females of *Rana ridibunda* also are reported to call when approaching territorial males (Frazer 1983). In a so-called voiceless frog from Borneo, *Limnonectes blythii,* and its relatives, males do not give advertisement calls to attract females to their nests, but females sometimes give soft calls when approaching males (Emerson 1992; Orlov 1997). Reciprocal calling by females has been reported in several other Asian ranids that are not closely related to one another, including *Limnonectes limnocharis, Euphlyctis cyanophlyctis,* and *Rana erythraea* (Roy, Borah, and, Sarma 1995), as well as a ranid treefrog from Fiji, *Platymantis vitiensis* (Boistel and Sueur 1997).

A number of functions of female courtship calling have been proposed, including facilitation of mate location by eliciting more calling from nearby males, identification of territorial and satellite males, and identification of females as receptive potential mates rather than territorial competitors (Emerson and Boyd 1999). To date, only a handful of experimental studies of male responses to female calls have been done (e.g., Given 1993a; S. Bush, Dyson, and Halliday 1996; S. Bush 1997; Tobias, Viswanathan, and Kelley 1998; Bosch 2001, 2002a). All of these hypotheses remain as plausible explanations for females calling in some species. A major difficulty in testing hypotheses about the function of female calling is that the distribution of such behavior even among closely related species is not well understood, because the presence of female calling can be documented with much greater certainty than can its absence.

### Aggressive Calls

Many male frogs defend their calling sites against conspecifics (see chapter 8) and often have distinctive aggressive vocalizations. Several terms have been applied to these calls. Bogert (1960) referred to all aggressive calls as territorial calls, whereas Salthe and Mecham (1974) used the term territorial call for calls given in long-range interactions and aggression or aversion call for those given in close-range encounters. McDiarmid and Adler (1974) and Wells (1977b) used the term encounter call for the latter. Littlejohn (1977) retained Bogert's territorial call for calls given in long-range encounters, but adopted encounter call for close-range signals.

I believe the more general term aggressive call is preferable to territorial call to describe vocalizations given in any kind of aggressive encounter. These calls can then be subdivided into long-range and short-range signals, although the two functions often are combined in a single call. Unfortunately the term territorial call has been applied to vocalizations given by many species that clearly are not territorial (see chapter 8). For example, males of some species in the *Rana esculenta* complex are aggressive toward one another, but do not defend long-term mating territories (Blankenhorn 1974, 1977; Tunner 1976; Klöckner 1982). Instead, they maintain a small space between themselves and neighboring males and attempt to clasp any noncalling individuals moving nearby. Hence use of the term territorial call to describe their aggressive vocalizations (e.g., Wahl 1969; Blankenhorn 1974; Brzoska 1980, 1982) seems inappropriate. The term territorial call also has been used for calls given by males of *Rana temporaria* when other males approach them (Brzoska, Walkowiak, and Schneider 1977; Walkowiak and Brzoska 1982). Males of this explosive breeder locate mates by active searching and show no sign of territorial behavior (R. Savage 1961; Arak 1983a; Elmberg 1986).

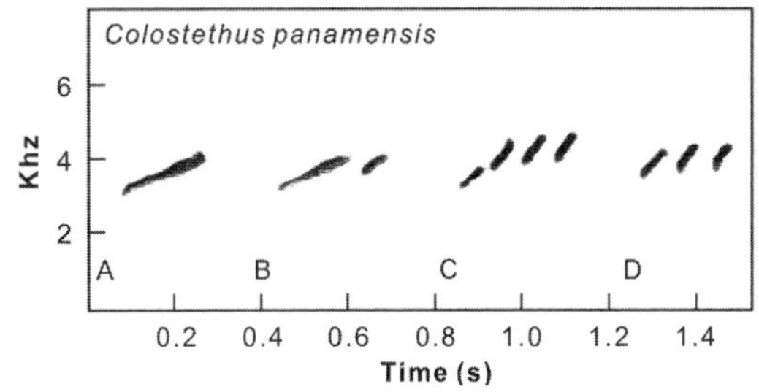


**Fig. 7.10.** Sound spectrograms of the calls of *Colostethus panamensis.* (A) Single-note aggressive call. (B) Two-note aggressive call. (C) Transition from the aggressive call to an advertisement call. (D) Three-note advertisement call. From recordings by the author.

### The Structure of Aggressive Calls

In most anurans, aggressive and advertisement calls have similar dominant frequencies, but differ in temporal structure (fig. 7.2), but there is no unique temporal structure common to all aggressive calls. There are, however, physical constraints on call production that limit how different advertisement and aggressive calls can be. For example, frogs that produce calls with a wide frequency range typically produce relatively short pulses of sound and probably are incapable of producing notes of long duration, whereas frogs that produce narrowly-tuned calls typically have longer notes (Gerhardt and Huber 2002). Consequently, aggressive calls often are modifications of advertisement calls. For example, many North American ranid frogs have wide-spectrum calls, and aggressive calls are subtle variations on the advertisement calls (Wells 1978b; Given 1987; Bee and Perrill 1996; Owen and Perrill 1998; Bee, Perrill, and Owen 1999; Owen 2003; Bevier et al. 2004; K. Larson 2004). In *Pseudacris crucifer,* the advertisement call is a tone-like peep (fig. 7.2 A), whereas the aggressive call is a long trill consisting of a series of relatively long pulses (fig. 7.2 B). In other species of *Pseudacris,* the advertisement call consists of trains of very short pulses, and aggressive calls are longer trains of short pulses, sometimes delivered at faster rates (Owen 2003).

In *Colostethus panamensis,* the call structures are reversed, with the advertisement call being a short trill and the aggressive call a long tone-like peep (Wells 1980a). These calls sometimes grade into one another as a male makes the transition from aggressive to advertisement calling (fig. 7.10). Males of another species in the same genus, *C. subpunctatus,* sometimes respond to calls of other males by grouping call notes into bouts of two or three notes. This does not appear to enhance the attractiveness of males to females, but does function as an aggressive signal (Lüddecke 2002a). The Neotropical hylid frog *Scinax rizibilis* has an advertisement call consisting of a series of repeated notes, but aggressive calls consist of irregular trains of short pulses. This species also has a close-range aggressive call (encounter call) consisting of short bursts of pulses (Bastos and Haddad 2002). In *Eleutherodactylus urichi,* the advertisement call is a short, tone-like peep, the aggressive call an even shorter click (Wells 1981b). On the other hand, the advertisement calls of the African frog *Leptopelis viridis* are short clicks, whereas the aggressive calls are about twice as long and have a slightly lower dominant frequency (Grafe, Steffan, and Stoll 2000).

Aggressive calls with a structure similar to that of advertisement calls, but with a much higher pulse repetition rate, are characteristic of several South and Central American hylids, including *Hyla ebraccata* (fig. 7.2 F), *H. microcephala, H. minuta,* and *H. phlebodes* (Cardoso 1981b; Wells and Greer 1981; Schwartz and Wells 1984a, b, 1985; Wells and Schwartz 1984b). Several hyperoliid frogs from Africa have aggressive calls with a similar structure (Telford 1982). The hylids all have pulsed or trilled advertisement calls, whereas many of the hyperoliids produce insect-like clicks. Some anurans have aggressive calls that are essentially more rapidly repeated versions of the advertisement call (e.g., *Hyla rosenbergi,* Kluge 1981; *Rana virgatipes,* Given 1987; *Leptodactylus albilabris,* Lopez et al. 1988) or have more notes than do advertisement calls (Perrill and Shepherd 1989; Wagner 1989c). On the other hand, some species have several distinct aggressive calls with very different temporal patterns (Capranica 1968; Wells 1978b; Kluge 1981; Given 1987; Martins and Haddad 1988). Because aggressive calls do not function in species recognition, one might expect such calls to be less stereotyped than are advertisement calls. Indeed, in several anuran species, temporal features such as pulse repetition rate and number of pulses are much more variable in aggressive calls than in advertisement calls (Schwartz and Wells 1984a; Littlejohn 2001; Owen 2003).

Some frogs lower the dominant frequency of their advertisement calls when responding to the calls of other males. This type of behavior has been reported in *Leptodactylus albilabris* from Puerto Rico (Lopez et al. 1988) and in several North American frogs, including cricket frogs (*Acris crepitans;* Wagner 1989a, 1992), green frogs (*Rana clamitans;* Bee and Perrill, 1996; Bee, Perrill, and Owen 1999, 2000), carpenter frogs (*R. virgatipes;* Given 1999), and bullfrogs (*R. catesbeiana;* Bee and Bowling, 2002). Lowering the dominant frequency of calls has been interpreted as an aggressive response, perhaps a means of conveying information about the size of the caller, although in bullfrogs, such changes do not appear to be correlated with fighting ability (Bee 2002). This behavior is discussed in more detail in a later section (see "Changes in Call Intensity and Frequency Structure").

## Graded Aggressive Calls

When a species has both long-range and short-range aggressive calls, these generally have been interpreted as discrete call types (hence the terms encounter call and territorial call). Some frogs, however, have graded signaling systems in which long and short-range aggressive calls represent two ends of a continuum. This has been studied in most detail in *Hyla ebraccata* from Panama (Wells and Schwartz 1984b; Wells and Bard 1987; Wells 1989). This species has compound advertisement calls consisting of an introductory note and a series of shorter secondary notes, which have the same dominant frequency and pulse repetition rate. Aggressive calls have introductory notes with much higher pulse rates than do advertisement calls and are more variable in duration (fig. 7.2 F, G). As males approach one another, the introductory notes of their aggressive calls become progressively longer (fig. 7.11 B) and the secondary click notes are dropped (fig. 7.11 A). The secondary notes make the calls more attractive to females, which are not strongly attracted to aggressive calls with only the high-pulse-rate introductory note (Wells and Bard 1987). Hence, males gradually adjust the relative aggressiveness and attractiveness of their calls, depending on the proximity of their opponents. In playbacks of aggressive calls, males increased the duration of aggressive call introductory notes as playback intensity or stimulus duration increased (Wells and Schwartz 1984b; Wells 1989); the number and length of secondary click notes were simultaneously decreased. Similar results were obtained with *H. microcephala* (Schwartz and Wells 1985) and *H. phlebodes* (Schwartz and Wells 1984b), two other Panamanian species that have aggressive calls of variable duration with pulse rates that are higher than those of advertisement calls. Playback experiments with females of *H. microcephala* showed that as in *H. ebraccata,* females prefer the lower pulse rates of advertisement calls to the higher pulse rates of aggressive calls (Schwartz 1987b).

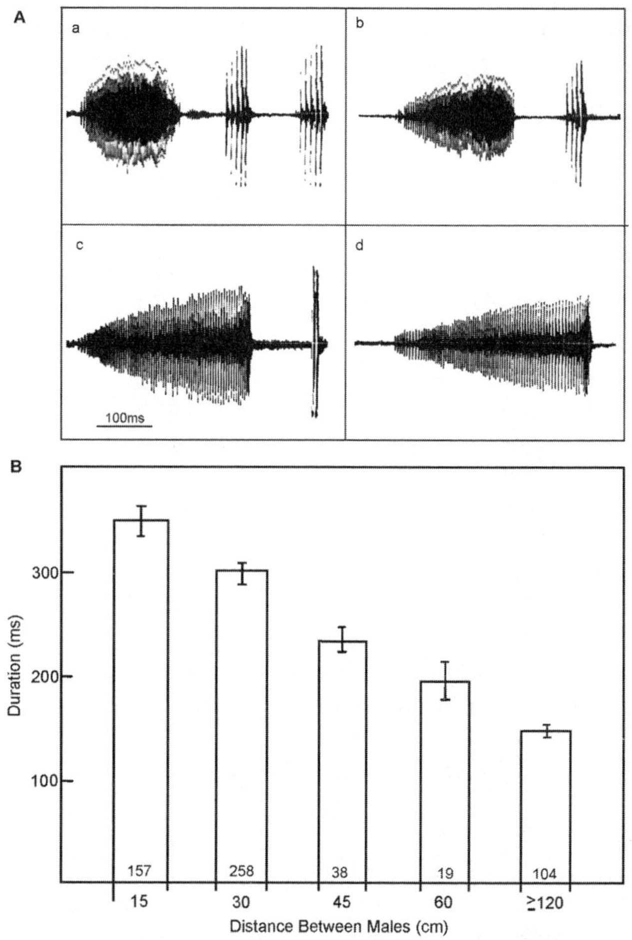


**Fig. 7.11.** Graded aggressive calls of the Panamanian treefrog *Hyla ebraccata.* (A) Oscillograms of four aggressive calls recorded from the same male, showing a gradual increase in the duration of the introductory note and reduction in the duration and number of secondary click notes. Call (a) was given at the longest distance between males; call (d) was given at the shortest distance. (B) Duration of introductory notes of aggressive calls as a function of the distance between interacting males. Numbers at the bottom of each column are sample sizes. After Wells and Schwartz (1984b).

Most North American hylid frogs have less complex vocal repertoires than those of the tropical hylids discussed above, with relatively simple, discrete aggressive calls. A few species have graded aggressive calls, however. Male spring peepers (*Pseudacris crucifer*) have trilled aggressive calls that are easily distinguished from the advertisement peep call (fig. 7.2 A, B). Males increase the duration of the trilled aggressive call in response to both increases in call intensity and the duration of an aggressive call stimulus (Schwartz 1989). Several other species in the genus *Pseudacris* also have trilled aggressive calls, and some of these show evidence of graded variation similar to that seen in spring peepers (Owen 2003). Cricket frogs (*Acris crepitans*) have a somewhat simpler system. Males produce calls with progressively more pulses as they approach one another, but they do not have structurally distinct aggressive calls (Perrill and Shepherd 1989; Wagner 1989b, c, 1992; Burmeister, Wilczynski, and Ryan 1999; Burmeister et al. 2002).

In the Old World, graded aggressive calls have been described in myobatrachid, hyperoliid, ranid, and rhacophorid frogs. In the Australian frog *Geocrinia victoriana,* males increase the duration of the introductory notes of their multinote advertisement calls in aggressive encounters. At close range, the calls have very long introductory notes, but lack secondary notes (Littlejohn and Harrison 1985; Scroggie and Littlejohn 2005). There is limited evidence for graded calls in *Geocrinia laevis,* but the behavior of this species has not been studied in as much detail as that of *G. victoriana* (Harrison and Littlejohn 1985). Australian frogs in the genus *Pseudophryne* give long aggressive calls in close-range encounters, but give shorter calls in more distant interactions. The temporal structure of short and long calls is very similar, with calls that overlap or nearly overlap in duration (Pengilley 1971a). This strongly suggests that the calls are part of a graded system.

In *Hyperolius marmoratus broadleyi*, an African treefrog with a mating system and chorus structure similar to those of the Panamanian hylids, males lengthen their aggressive calls by increasing the number of pulses as opponents approach one another (Grafe 1995). In *Afrixalus brachycnemis,* the duration of aggressive notes does not change, but males gradually increase the proportion of aggressive notes in their calls as chorus size increases (Backwell 1988). Several other species of *Afrixalus* have aggressive calls that are similar in structure to those of the Neotropical hylids, and some of these very likely have graded aggressive calls as well (Schiøtz 1967, 1999).

Some frogs exhibit an extraordinary range of variation in call structure, with graded variation in both advertisement and aggressive calls. Males of *Rana nicobariensis* from Malaysia produce multinote advertisement calls that usually have 1–6 click notes, but when males are calling very close to each other, advertisement calls can have as many as 25 notes (Jehle and Arak 1998). These long calls have shorter notes and higher note repetition rates than do short calls. Males in dense choruses give distinctly different squawk-like aggressive calls, and these sometimes are followed by a series of click-like notes similar to those in the advertisement calls. These compound calls apparently represent transition calls that convey an aggressive message to other males while retaining notes attractive to females, similar to the calls of *Hyla ebraccata.* Males interacting at very close range give yet another type of aggressive call, a short click.

Complex graded vocal repertoires seem to be particularly characteristic of some Old World treefrogs. Several species have vocal repertoires composed of a number of distinct call types, but in most cases, the functions of these calls are not well understood (Kuramoto 1986; Kasuya, Kumaki, and Siato 1992; Andreone 1993; Matsui and Wu 1994; Narins et al. 1998). Males of *Philautus leucorhinus,* a rhacophorid treefrog from Sri Lanka, give longer aggressive calls in close range encounters than in more distant interactions, and the nearly overlapping duration of these call types suggests they are part of a graded aggressive call system (Arak 1983c). Males of another rhacophorid, *Polypedates leucomystax* from Thailand, have a repertoire of at least 12 distinct call types. Many of these calls consist of trains of pulses or clicks that appear to function as aggressive signals, but there is little evidence that the different call types grade into one another (Christensen-Dalsgaard, Ludwing, and Narins 2002). A mantellid treefrog from Madagascar, *Boophis madagascariensis,* has an even more variable vocal repertoire. Narins, Lewis, and McClelland (2000) described 28 different types of calls in this species, although most of these appear to be variants of a single call type that differ in the number of notes and pulses produced (fig. 7.12). The most variable call types, which were termed "iambic" notes, often were given in response to playbacks of similar notes at high intensities and have many of the characteristics of the graded aggressive calls seen in other species.

Males of the Puerto Rican frog *Eleutherodactylus coqui* produce multinote aggressive calls (fig. 7.2 D), but in this case, variation in number of call notes appears to be more related to the types of resources being defended than to the calls given by other males or proximity of competitors. Males of this species seldom defend their calling sites with overt aggression, but when they do give aggressive calls in this context, the calls usually have relatively few notes (O'Brien 2002). Males produce more aggressive calls as the intensity of advertisement call playbacks is increased (Stewart and Bishop 1994) and in response to playbacks of aggressive calls. They do not, however, exhibit a clear tendency to give longer calls to playbacks of longer aggressive calls (O'Brien 2002). On the other hand, males give very long calls when defending retreat sites, and the longest calls of all are given in defense of nest sites containing eggs (Stewart and Rand 1991). Two closely related species, *E. antillensis* and *E. cochranae,* have very similar multinote aggressive calls (Michael 1997b; Ovaska and Caldbeck 1997b). In all three species, the repeated notes resemble the second note of the advertisement call. Michael (1997b) suggested this is at odds with the interpretation of Narins and Capranica (1978) that the second "qui" note of the advertisement call *of E. coqui* is used to attract females, whereas the "co" note functions in male-male interactions. The contradiction is more apparent than real, however, because the co note is a long-range signal that probably mediates spacing between

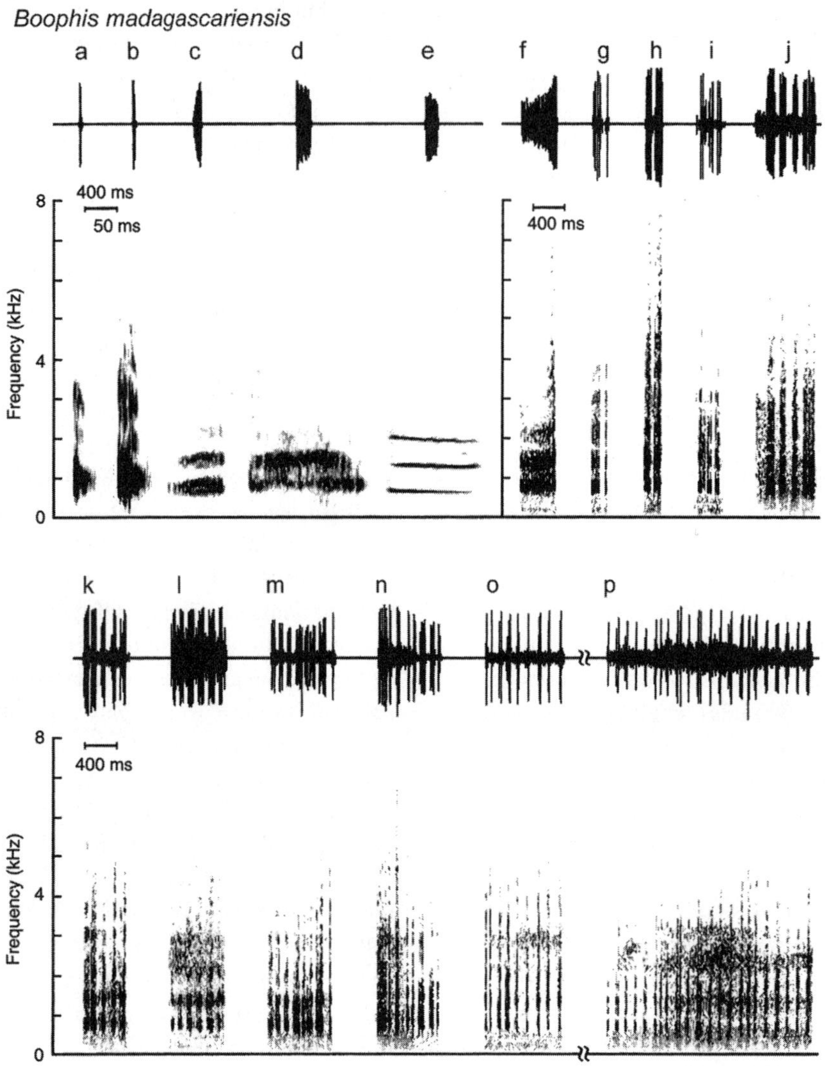


**Fig. 7.12.** Complex vocal repertoire of *Boophis madagascariensis.* Oscillograms are shown on top of each part, sound spectrograms on the bottom. For the first set of calls (a–e), the time scale on the sound spectrograms has been magnified to show details of call structure. Time scales are the same for oscillograms and sound spectrograms for all other calls. (a) toc note. (b) short click note. (c) short rip note. (d) loud click note. (e) tonelike note. (f) long rip note. (g) creak note. (h–p) iambic notes with increasing number of pulses. Males give iambic notes more frequently in response to playbacks of conspecific calls, and these may represent a graded aggressive call system. After Narins et al. (2000).

males in a chorus, whereas the multinote aggressive call is used mainly at close range.

### Behavioral Responses to Aggressive Calls

Playbacks of aggressive calls can inhibit calling, cause males to retreat from speakers or make aggressive advances toward them, or result in increased levels of aggressive calling (W. Snyder and Jameson 1965; Emlen 1968; Wiewandt 1969; Allan 1973; M. Rosen and Lemon 1974; Passmore 1978; Wells 1978b, 1988, 1989; Gambs and Littlejohn 1979; Whitney 1980; Wells and Greer 1981; Brzoska 1982; Brzoska, Schneider, and Nevo 1982; Perrill, Gerhardt, and Daniel 1982; Ramer, Jenssen, and Hurst 1983; Schwartz and Wells 1984b, 1985; Wells and Schwartz 1984a, b; Telford 1985; Given 1987; Backwell 1988; Lopez et al. 1988; Brenowitz 1989; Schwartz 1989; Wagner 1989b; G. Rose and Brenowitz 1991, 1997, 2002; Dyson and Passmore 1992b; Brenowitz and Rose 1994; Grafe 1995). Males usually give more aggressive responses to playbacks of aggressive calls than to playbacks of advertisement calls, and they respond to aggressive calls at lower intensities. These behavioral responses can be relatively plastic, with males habituating to the local density and proximity of other males in the chorus (Brenowitz and Rose 1994; G. Rose and Brenowitz 1997;

V. Marshall, Humfeld, and Bee 2003; see "Spatial Organization of Choruses").

The precise message conveyed by aggressive signals is not entirely clear. Long-range aggressive calls usually are interpreted as warnings that an intruder is trespassing too closely on an occupied calling site (Littlejohn 1977; Wells 1977b). Close-range signals often are considered indicators of imminent attack if the intruder fails to withdraw. Graded aggressive signals can be used by an animal to gradually escalate an encounter and provide increasingly accurate information about the probability that it will attack an opponent (D. Nelson 1984; Grafen 1990a; Capp and Searcy 1991; Hauser and Nelson 1991). In a graded system such as that of *Hyla ebraccata* or *Hyperolius marmoratus broadleyi*, the use of graded signals probably allows the male to signal increased levels of aggressiveness without making the call completely unattractive to females (Wells and Schwartz 1984b; Wells 1989; Grafe 1995).

The same seems to be true of the graded signals of cricket frogs, in which males respond to opponents by modifying the basic advertisement call used to attract females (Wagner 1989b, 1992; Burmeister, Wilczynski, and Ryan 1999). Burmeister et al. (2002) found that lowering of dominant frequency by males in response to playbacks of calls was a good predictor of their likelihood of attacking the speaker. The function of changes in temporal features of calls was less clear. Males that did not respond behaviorally to a simulated intruder gave relatively short calls with fewer pulses and few groups of pulses per call than did those that abandoned their calling sites or attacked the speaker, but there was no difference in call characteristics of those that abandoned a calling site or attacked. None of the changes in dominant frequency or temporal features was clearly related to body size. Hence, while graded changes in frequency and temporal features could signal a male's intention to respond or not respond to an opponent, it is not clear that such variation is related either to actual fighting ability. Adult male cricket frogs are small and exhibit relatively little variation in size, so size-related variation in fighting ability probably is small as well.

Some authors have argued that animals should not give reliable information about their motivational states or the likelihood of attacking an opponent (Dawkins and Krebs 1978; Clutton-Brock and Albon 1980). Instead, aggressive signals might be used mainly to signal the size and strength of an animal to its opponent, thereby allowing animals to avoid costly escalated contests when they have little chance of winning a fight (G. Parker 1974; Maynard Smith and Parker 1976; Dawkins and Krebs 1978; Riechert 1978; Clutton-Brock and Albon 1980; Enquist et al. 1990). This does not seem to be the case in cricket frogs (Burmeister et al. 2002), but might be true for other frogs. It also is possible that physiological condition is a better indicator of fighting ability than is body size alone, especially in frogs with high calling rates and high levels of energy expenditure.

Wells and Schwartz (1984b) speculated that the ability of *H. ebraccata* males to sustain production of long aggressive calls, which are expensive to produce, might be a reliable indicator of size or physical vigor. Aggressive calls of most frogs are exchanged between males in an alternating fashion, a pattern seen in the aggressive signaling of other animals as well (Greenfield 1994a). This would enable males to assess the call structure of their opponents and perhaps evaluate their fighting ability. In a preliminary test of this idea with *Hyla microcephala*, Schwartz (1994) used a computerized playback system to broadcast aggressive calls to males in a chorus in response to spontaneous aggressive calls given by the males. The responses were programmed to be either longer, shorter, or the same length as the initial call of the test male. When Schwartz presented males with calls that were of similar duration to their own calls, they tended to increase call duration in response to the playbacks. They did not increase call duration as much when they were presented with calls that were longer or shorter than their own calls. This suggests that males might escalate their aggressive responses to elicit information about whether the opponent is capable of producing even longer calls (see Greenfield and Minckley 1993, for a similar system in grasshoppers).

### Release Calls

Male anurans usually give release calls when clasped by other males, either when the male being clasped is alone or in amplexus with a female. Unreceptive females, or those that have completed oviposition, often give similar calls. Usually these calls consist of a series of rapidly repeated broad-spectrum notes. They indicate to a clasping male that he has grabbed an inappropriate mate. Male release calls have been described in bombinatorids, discoglossids, pipids, pelobatids, leptodactylids, bufonids, hylids, and ranids (Eibl-Eibesfeldt 1950; Bogert 1960; R. Savage 1961; Rabb and Rabb 1963a; Capranica 1968; R. Schmidt 1972b, 1976; Heusser 1969a; Lörcher 1969; Wahl 1969; Heinzmann 1970; J. Pierce and Ralin 1972; Salthe and Mecham 1974; E. Weber 1974; Brzoska, Walkowiak, and Schneider 1977; Schneider 1977; Wells 1978b; Cei 1980; Picker 1980; Penna and Veloso 1981, 1987; Schneider and Brzoska 1981; Odendaal, Bull, and Telford 1983; Akef and Schneider 1985; R. Schmidt, Kemnitz, and Hudson 1990; Castellano et al. 2002). Females probably give release calls in many of these species as well. In some species, males and females give release vibrations when clasped by other frogs, but do not produce audible sound (Bogert 1960; Penna and Veloso 1982). It has been suggested that the advertisement calls of anurans originally evolved from simple release signals (R. Schmidt 1966, 1968b; Rand 1988; R. Schmidt, Kemnitz, and Hudson 1990). There has not been any phylogenetic analysis that would support this hypothesis, and such a test might be impossible because of the prevalence of advertisement calling in extant anurans.

### Distress Calls, Alarm Calls, and Defensive Calls

Bogert (1960) used the term distress call to describe vocalizations given by frogs being attacked by predators. Usually these are relatively loud screams, often delivered with the mouth open. Many such calls have been described since Bogert's paper appeared (Wahl 1969; Heinzmann 1970; J. Hoff and Moss 1974; E. Weber 1974, 1975b, 1978b; Sazima 1975; Lescure 1977; Tunner and Hödl 1978; Kluge 1981; Hödl and Gollmann 1986; Penna and Veloso 1987; Leary and Razafindratsita 1998), but their function remains obscure. Possibly they serve to warn neighbors of danger or startle the predator, thereby allowing the victim to escape. There is little direct evidence to support either hypothesis, other than reports than humans often are startled by such cries (e.g., Hödl and Gollmann 1986). Leary and Razafindratsita (1998) reported an unusual response to distress calls given by a treefrog, *Trachycephalus* (= *Phrynohyas*) *venulosus,* which was seized by a snake. In less than a minute, 17 other frogs emerged from retreat sites in a tree and oriented toward the sound, but did not do anything else.

Some large frogs actually threaten or attack predators while giving hisses or screams (Barrio 1963; Villa 1969; Vaz-Ferreira and Gehrau 1975; Veloso 1977; Hödl and Gollmann 1986). Although some authors have referred to these as aggressive calls, the term defensive call is more appropriate to distinguish them from calls used in intraspecific aggressive encounters. The effectiveness of the calls in deterring predators is unknown.

## Call Structure and the Physical Environment

The physical environment can distort a signal before it reaches a receiver, making it difficult to predict what an animal will hear when it receives a signal broadcast by another individual some distance away. The most important environmental effects are: (1) attenuation of signals due to spherical spreading, atmospheric absorption, scattering, and boundary interference; (2) degradation of signal structure due to reverberations or irregular amplitude fluctuations brought about by atmospheric turbulence; and (3) masking of signals by background noise (Michelsen 1978; Wiley and Richards 1982; Gerhardt 1983, 1994a; Gerhardt and Klump 1988a; Narins and Zelick 1988; M. Ryan 1988a; Forrest 1994).

Attenuation of signals refers to the reduction in signal intensity over distance. All sounds radiating from a point source suffer attenuation due to spherical spreading of sound waves. Sound pressure level decreases by 6 decibels (dB) for each doubling of distance from the sound source, and this attenuation is independent of the frequency of the sound. Additional reductions in sound intensity due to atmospheric absorption, scattering, or boundary interference constitute excess attenuation. Excess attenuation is greater for high frequencies than for low frequencies in all habitats, so in general, low-frequency sounds are more suitable for long-distance communication, especially when both the signaler and receiver are near the ground (Michelsen 1978; Wiley and Richards 1982; Gerhardt 1983; 1994a).

In addition to suffering excess attenuation in certain habitats, the temporal features of calls are degraded by reverberation of sound through reflecting surfaces such as vegetation or irregular amplitude fluctuations caused by atmospheric turbulence. These processes tend to distort features of calls that are important for species recognition or message transmission. Reverberations are most severe in densely vegetated habitats and obscure high rates of amplitude modulation in calls. Consequently, either tone-like signals or signals with gradual frequency modulation would be best suited for long-distance communication in such habitats (D. Richards and Wiley 1980; Wiley and Richards 1982). Irregular amplitude fluctuations are particularly significant in open environments, especially during the daytime when temperature gradients increase atmospheric turbulence. The environment tends to impose amplitude fluctuations on calls that are not amplitude-modulated, such as pure tones, although this is not likely to affect frogs that call at night, when temperature gradients are minimal.

Several early studies of bird songs (Chappuis 1971; Morton 1975) indicated that birds in tropical forests have lower-pitched songs than those in open savannas, and it was suggested that low frequencies are better suited for propagation through forest habitats. However, D. Richards and Wiley (1980) did not find consistent differences in the frequency structure of bird songs from forested and open habitats in the temperate zone, and there seems to be little evidence that excess attenuation of high frequencies is greater in forested habitats (Wiley and Richards 1978, 1982). Instead, the use of low-frequency sounds by some tropical forest birds seems to be related to their tendency to sing from the ground rather than elevated perches. Differences in the temporal features of bird songs are related to habitat preferences as well. Many forest birds have tone-like songs, while many open country birds have songs with rapid frequency modulation (analogous to rapid amplitude modulation in frog calls; Chappuis 1971; Morton 1975; D. Richards and Wiley 1980). These analyses also are complicated by phylogenetic differences in the birds that live in open and forested habitats, which in turn are correlated with differences in song structure (M. Ryan and Brenowitz 1985).

Several investigators have attempted to apply the ideas derived from studies of birds to studies of call structure in frogs. Some studies have examined the relationship between call structure and habitat in a range of anuran species, while others have used experimental playbacks of recorded calls to determine the effect of habitat structure on call attenuation and degradation.

### Observational Studies of Call Structure in Relation to Habitat

Several authors have proposed that habitat structure is related to the frequency of frog calls (Littlejohn 1977; Telford 1982), but there have been few systematic attempts to examine this question. Any attempt to identify best frequencies for certain habitats is complicated by differences in acoustic properties of habitats that appear superficially similar. Two open habitats can have very different characteristics, depending on the type and density of vegetation, the presence or absence of standing water, temperature gradients, humidity, and atmospheric turbulence (Wiley and Richards 1982). Even more important are differences in calling site preferences of different species. If a male frog calls near the ground, and if females receiving his signals also are on the ground, then high frequency components of calls will suffer much greater attenuation than low frequency components (e.g., Odendaal, Bull, and Telford 1986; Kime, Turner, and Ryan 2000), particularly if the frog is surrounded by vegetation (Zimmerman 1983). However, if either the caller or receiver is elevated, then differential loss of high frequencies will be reduced (Gerhardt 1981b, 1983, 1994a; Forrest 1994; Gerhardt and Huber 2002; Parris 2002).

#### Frequency Structure

Zimmerman (1983) compared frogs living in Amazonian forests with those in floating meadow habitats and found that forest species have significantly lower-pitched calls. Unfortunately, the analysis is complicated by the effect of body size and phylogeny. The forest species generally are larger than those in open areas, and call frequency decreases with increasing body size. Since the relationship of body size to dominant frequency was the same in the two habitats, there is no evidence that habitat per se influences dominant frequency. This is consistent with Wiley and Richards' (1982) conclusion that the best frequency for long-distance transmission does not differ between habitats. Furthermore, the calls of many species probably are not adapted for long-range propagation, but for ease of localization at short distances. Phylogenetic differences probably are important as well. Frogs that are common in forests, such as *Eleutherodactylus,* are uncommon in open areas and tend to have a very different call structure than open-habitat frogs such as hylids. Studies of leptodactylid frogs in marsh and bog habitats in southern Chile also failed to find a clear relationship between habitat and call dominant frequency (Penna and Solis 1998), as did a study of several families of frogs in open and forested habitats in Panama (Kime, Turner, and Ryan 2000). Bosch and De la Riva (2004) found a relatively weak tendency for Bolivian frogs in closed microhabitats to have frequency-modulated calls, but the effect of taxonomy (family and genus) was much stronger. They did not, however, perform a detailed phylogenetic analysis using modern comparative methods.

#### Temporal Structure

Many open-habitat frogs have calls with rapid amplitude modulation (e.g., Schwartz and Wells 1984a, b, 1985; Wells and Schwartz 1984a), but there have been few systematic surveys of call types in different habitats. Zimmerman (1983) found that forest species tend to have more narrowly tuned calls with lower pulse repetition rates than open habitat frogs. In Panama, hylid frogs that call in open, grassy areas tend to have noisy, broad-spectrum calls, but this also is true of hylids that call along forested streams, such as species of *Smilisca.* Leptodactylids that call from forested habitats, such as *Eleutherodactylus,* have narrowly tuned calls, but so do species of *Leptodactylus* and *Pleurodema* that call from open habitats (Kime, Turner, and Ryan 2000). Clearly these comparisons are confounded by differences in the phylogenetic distribution of species that live in forested and open habitats. Furthermore, many open habitat frogs, such as those breeding in marshes, actually call in the midst of dense vegetation, so the acoustic environment does not differ dramatically from that of a densely vegetated forest. Until more precise characterizations of anuran calling sites are available, it will be difficult to discern relationships between habitat types and call structure.

#### Call Intensity

The distance over which a call can be detected will be determined in part by its initial intensity; calls of high sound pressure level will be transmitted over greater distances than low-level calls of the same frequency. Schiøtz (1973) and Dubois (1977a) reported their impressions that the calls of open-habitat species are much louder than are those of forest species, but they did not present actual measurements of sound pressure level. Most of the available data (table 7.1) are for species calling in open habitats such as ponds or flooded fields (Gerhardt 1975; Passmore 1981; Schwartz and Wells 1984a). The only data on forest species are for *Eleutherodactylus coqui* in Puerto Rico (Narins and Hurley 1982) and *Hyalinobatrachium fleischmanni* in Panama

(Wells and Schwartz 1982). The data reveal few general patterns. There is no evidence that interspecific differences in call intensity are consistently related to body size, calling site elevation, or habitat. For example, several species of leptodactylid frogs from southern Chile that call in bogs, where sound transmission is relatively poor, have calls of unusually low intensity (Penna and Solis 1998). Perception of calls in this environment can be enhanced by acoustic properties of the frogs' burrows, however (Penna and Solis 1996, 1999; see following). Call intensity is related to factors other than habitat structure per se. For example, frogs that call in open habitats tend to call in much denser choruses than do those in forested habitats, and competition among males should favor the evolution of very loud calls.

### Experimental Studies of Call Degradation

Only a few experimental studies have examined the effects of vegetation and other environmental factors on the degradation of calls produced by frogs. Wells and Schwartz (1982) showed that excess attenuation of calls of *Hyalinobatrachium fleischmanni* was greater for males calling in the midst of dense vegetation than for those calling from more open, elevated sites. M. Ryan and Wilczynski (1991) found that two subspecies of the cricket frog *Acris crepitans* have calls that differ in temporal and frequency structure. One lives in a forested habitat, the other in more open areas. When the calls of both subspecies were broadcast through each habitat, both suffered more degradation in frequency structure in a forested habitat than in an open habitat, but the subspecies native to forests suffered less degradation in that habitat (M. Ryan, Cocroft, and Wilczynski 1990). Perception of calls by females in the forested habitat was enhanced by more precise tuning of the ear, thereby improving the signal-to-noise ratio of the call (Witte et al. 2005). These studies did not examine degradation of temporal structure in detail. However, in a study of two species of toads, *Bufo woodhousii* and *B. valliceps,* M. Ryan and Sullivan (1989) found that their amplitude-modulated trills were degraded as transmission distance increased in both open and heavily vegetated habitats. The calls of *B. valliceps* suffered greater degradation, possibly because their calls have longer pulses.

Penna and Solis (1998) compared the sound propagation properties of five species of leptodactylid frogs in southern Chile. In two of these species, *Hylorina sylvatica* and *Pleurodema thaul,* males call from open water in marshes. The other three species, *Eupsophus emiliopugini, Batrachyla antartandica,* and *B. leptopus,* males call from hidden locations in bogs and sometimes from burrows. Calls of all of these species exhibited less excess attenuation in the more open marsh habitat than in the bog habitat, regardless of which habitat they actually use. Species with call dominant frequencies below 2 kHz were better adapted for long-distance propagation of calls, but the two species of *Batrachyla* had the highest dominant frequencies, and they called from the habitat most likely to result in attenuation of calls. Castellano, Giacoma, and Ryan (2003) compared attenuation degradation of calls of diploid and triploid species in the *Bufo viridis* complex in different habitats. Although the propagation of calls differed in different habitats, there was no evidence of adaptation to local habitat conditions.

Kime, Turner, and Ryan (2000) obtained similar results in a broad survey of call transmission in open and forested habitats in Panama. They tested the calls of 22 species of bufonids, hylids, centrolenids, leptodactylids, and dendrobatids, using an index of call degradation that combined excess attenuation and degradation of the temporal structure of the call. Degradation was greater in the forested habitat than in the open habitat, and calls with high dominant frequencies suffered the greatest degradation. The height at which calls were broadcast had a major effect on call degradation, being much greater near the ground than when calls were broadcast from elevated sites. Call structure clearly was strongly affected by phylogeny, although this was not analyzed in detail in the study. Most of the hylids had relatively noisy, broad-spectrum calls, whereas many of the leptodactylids, centrolenids, and dendrobatids had more narrowly tuned, tone-like calls or frequency-modulated calls, regardless of the habitat in which they called. Overall, both observational and experimental studies provide little support for the idea that the structure of frog calls is adapted for particular habitats.

### Background Noise

A serious problem faced by many animals is the presence of background noise that masks their calls or makes detection of conspecifics difficult (Gerhardt and Klump 1988a; Narins and Zelick 1988). Such noise can be derived from the physical environment or the signals of other animals. The influence of animal sounds is discussed in later sections, so here I will concentrate on the physical environment. Although high wind and rustling vegetation generate considerable background noise, this probably is not a serious problem for frogs. Many stop calling in heavy wind, especially in dry weather, probably to reduce the risk of desiccation (personal observations). Rain beating down on vegetation also could create considerable background noise, but its effect on frog calling has not been studied, mainly because investigators cannot make recordings during rainstorms and tend to retreat indoors. Although high levels of background noise often inhibits calling by frogs (e.g., Schwartz and Wells 1983b), low background noise may induce frogs to call at higher rates or produce louder calls, thereby improving the signal-to-noise ratio of their calls (Penna, Pottstock, and Velasquez 2005).

One source of environmental noise that could influence the evolution of both call structure and calling behavior is running water. Dubois (1977a, b) found that frogs calling near waterfalls in Nepal generally produce a series of click-like calls delivered in rapid succession, with long pauses between calling bouts. The discontinuous production of very repetitive calls enhances contrast with the continuous background noise of the water. South American frogs of the genus *Hylodes* generally live near torrents and produce calls consisting of a repeated series of frequency-modulated peeps, also separated by long periods of silence (Haddad, Pombal, and Batistic 1996; Vielliard and Cardoso 1996; Haddad and Giaretta 1999). Other stream-breeding frogs that tend to call in discontinuous bouts include some species of centrolenid frogs and *Smilisca sila*, a Central American hylid that often calls near waterfalls and has click-like calls (Ibáñez 1991). Possibly a similar explanation applies to these frogs, although the tendency of *Smilisca sila* to call in synchronized bouts also has been explained as an adaptation to reduce predation by bats (Tuttle and Ryan 1982).

Another way to maximize contrast with the wide-spectrum noise of running water would be to have calls with energy concentrated into a narrow frequency band. Some frogs and birds that call near torrents have narrowly tuned calls (Greer and Wells 1980; Wells 1980a; Passmore 1981; Dubois 1977a, b; Dubois and Martens 1984; Haddad and Giaretta 1999; Feng, Narins, and Xu 2002), but whether this behavior represents a specific adaptation to that environment is not known. In some small stream-dwelling frogs, the dominant frequency of calls is well above that of the sound produced by waterfalls, so acoustic interference is minimal (Haddad and Giaretta 1999). Some frogs that breed in streams near noisy waterfalls deal with background noise in another way; they have evolved visual signals such as foot-flagging displays that either supplement or replace acoustic signals (Haddad and Giaretta 1999; Hödl and Amezquita 2001). Presumably these visual displays make the frog conspicuous to other males or to females at a distance, while vocalizations function mainly in short-distance communication.

### Transmission of Signals from Holes and Burrows

While many anurans call from elevated perches or in the water, others typically call from underground burrows or the surface of the ground. Such behavior is particularly common in leptodactylids (Heyer 1969), myobatrachids (A. K. Lee 1967; Pengilley 1971a; Roberts 1981, 1984), and some burrowing microhylids from New Guinea and Australia (Zweifel 1985). Some species call while completely hidden in a burrow or crevice. Examples include *Leptodactylus albilabris* (E. Lewis and Narins 1985; Lopez et al. 1988; Narins 1990), *L. syphax* (Cardoso and Heyer 1995), *Eupsophus emiliopugini* (Penna and Solis 1996), *Alytes muletensis* (Bush 1997), and various species of *Heleioporus* and *Pseudophryne* (A. K. Lee 1967; Pengilley 1971a; Bailey and Roberts 1981). Males of the Australian turtle frog (*Myobatrachus gouldii*) often call with only the posterior half of the body buried in sand, while *Arenophryne rotunda* males call either underground or on the surface (Roberts 1981, 1984). Males of the leptodactylid frog *Lithodytes lineatus* often call inside nests of leaf-cutter ants (*Atta*), from burrows, or from sheltered sites in leaf litter (Lamar and Wild 1995). Calling from burrows and rock crevices can be disadvantageous for transmission of sound through air because of problems of excess ground attenuation, but the burrow would help to protect the frogs from predators that might home in on their calls.

Frogs that call in burrows have the potential to communicate through two distinct channels, the airborne channel and through ground vibrations. The latter would provide a transmission channel relatively free of background noise (E. Lewis and Lombard 1988; Narins 2001b). There is no evidence that females of any species of frog approach males in their burrows by tunneling through the ground, so the airborne channel probably is essential for attracting mates. Bailey and Roberts (1981) used plaster models of burrows to investigate burrow acoustics in several species of *Heleioporus*. They found that the resonant frequencies of the burrows corresponded to the dominant frequencies of the calls of the different species. They also found that transmission of the calls was maximized when males call near the entrance to the burrow, which serves as a resonator. These frogs lack vocal sacs, and the burrow might compensate for this by increasing the sound pressure level of the call.

A somewhat analagous situation has been reported in a microhylid frog from Borneo, *Metaphrynella sundana*, which calls from water-filled treeholes. Males of this species have vocal sacs, but they appear to adjust the fundamental frequency of their calls to match the resonance frequency of the chamber, thereby enhancing the transmission of the call (Lardner and Lakim 2002). Because both the resonance properties of different treeholes and the dominant frequencies of male calls vary considerably, the result is a wide range of variation in the perceived calls of males in the population (Lardner and Lakim 2004).

A burrow or chamber also can affect the reception of calls produced by other males calling nearby. Penna and Solis (1996, 1999) investigated the burrows of *Eupsophus emiliopugini* affected the ability of the burrow inhabitat to perceive calls of neighboring males. They found that the burrows increased the sound pressure level of the calls of nearby males inside a male's burrow. Males alternate calls with one another and respond to other males calling close to

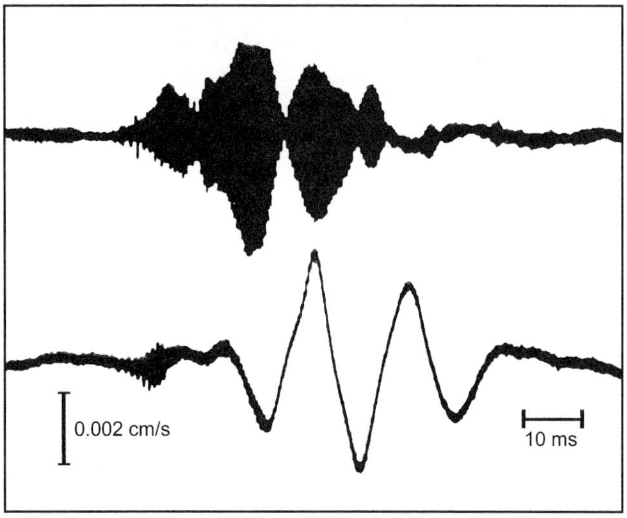


**Fig. 7.13.** Seismic communication in *Leptodactylus albilabris.* The oscillogram on top shows the airborne signal recorded with a microphone. The one on the bottom shows the ground-borne signal recorded with a geophone 1 meter from the calling male. Horizontal time bar = 10 ms. After Narins (1990).

their burrows with aggressive calls. The acoustic properties of the burrow are therefore likely to affect spacing among males in the chorus.

The use of substrate-born signals has been studied in detail only in the Puerto Rican frog *Leptodactylus albilabris.* E. Lewis and Narins (1985) showed that calling males produce a distinct thump that can be detected with a geophone (fig. 7.13). Apparently the vocal sac hitting the walls of the burrow produces this sound. This frog is extremely sensitive to substrate-born vibrations that apparently are detected by the sacculus of the ear (Narins and Lewis 1984; E. Lewis and Narins 1985; E. Lewis et al. 2001). Males alter their vocal behavior in response to both airborne and substrate-born signals (Lopez et al. 1988; Narins 1990). Substrate-born signals appear to be important in maintaining spacing between calling males. Because sound travels more slowly through the ground than through air, there would be a difference in the time a signal arrives at the receiver through the two channels. The frogs could use this information to estimate the distance between themselves and their neighbors.

Cardoso and Heyer (1995) described another form of seismic communication in *Leptodactylus syphax.* Males calling in burrows responded to playbacks of advertisement calls by giving aggressive calls, emerging from the burrow, and pounding their front feet on the ground. This produced an audible clicking sound and very likely produces a seismic signal to which the frogs are sensitive. Females of the Asian rhacophorid frog *Polypedates leucomystax* respond to calling males by tapping their toes on a reed or blade of grass, and males respond by moving toward the females (Narins 1995a).

### Transmission of Signals Underwater

Many frogs call while floating on the surface of the water or while partially submerged. The calls produced by these species probably are audible underwater as well as in air. Dudley and Rand (1992) recorded the underwater sounds of *Physalaemus pustulosus* in Panama and found that the calls were very similar to those recorded in air at close range, but at long distances, high frequencies were more attenuated in water than in air. The calls of bullfrogs (*Rana catesbeiana*) also propagated effectively in both air and water, but it is not known whether either males or females make use of underwater acoustic cues (Boatright-Horowitz, Cheney, and Simmons 1999). In shallow water, frequencies below 1,400–2,000 Hz were greatly attenuated, especially in ponds with soft bottoms. Frogs calling while floating on the surface produce calls that can be detected at least 4 m away underwater. If they were fully submerged, propagation of calls would be reduced to about 1 m.

Some species of anurans typically call while completely submerged. These include not only the aquatic pipids (Osterdahl and Olsson 1963; Rabb and Rabb 1963a; Weygoldt 1976a; Vigny 1979; Picker 1980; Yager 1992a, b, 1996), but also frogs in several other families, including the leptodactylid genus *Telmatobius* (Cei and Roig 1965), the ranid frogs *Rana aurora* (L. Licht 1969a; M. Hayes and Miyamoto 1984), *R. palustris* (personal observations), *R. dalmatina* (Christensen-Dalsgaard and Elepfandt 1995), and *R. subaquavocalis* (Platz 1993), and possibly the highly aquatic bombinatorid genus *Barbourula* (G. Myers 1943). Species that call in shallow water often produce a signal that is audible both underwater and in air, but *Rana subaquavocalis,* a species from Arizona, calls in deep pools, and its calls are not audible above the water.

Probably all anurans are capable of hearing sounds underwater, both those that normally call above the surface (Lombard, Fay, and Werner 1981; Hetherington and Lombard 1982) and those that call exclusively underwater (Christensen-Dalsgaard, Breithaupt, and Elepfandt 1990; Christensen-Dalsgaard and Elepfandt 1995; Elepfandt 1996; Yager 1996). Water is an excellent medium for transmitting sound, which travels about five times faster in water than in air. Also, a sound produced at a given energy level produces a sound pressure level that is much higher in water than in air (Elepfandt 1996). In deep water, sound can be transmitted over long distances with very little distortion. In shallow water, which is more typical of the habitats inhabited by amphibians, the transmission of signals is affected by reflection of sound from both the bottom substrate and the surface of the water, resulting in a complex set of pathways between the signaler and receiver (Rogers and Cox 1988). This complicates the job of females attempting to localize a signaler.

Shallow water environments also tend to filter out very low frequencies, with the cutoff frequency varying depending on the physical characteristics of the substrate (Forrest, Miller, and Zagar 1993). Most frogs are small and tend to produce calls with dominant frequencies above those likely to be filtered out by shallow water environments. Ponds and other bodies of water generally seem relatively quiet to human observers, but some ponds can be surprisingly noisy. For example, Elepfandt (1996) reported that acoustic signals produced by aquatic insects in some African ponds are potential sources of interference with the calls of pipid frogs. Despite the complexities of sound transmission and detection in shallow water, the calling behavior of aquatic species such as *Xenopus laevis* is not markedly different from the behavior of anurans that call out of water (Yager 1992a, 1996). Female pipids readily approach males calling underwater (Picker 1980; M. D. Picker 1983; Yager 1996; Kelley and Tobias 1999; Kelley, Tobias, and Horng 2001) and even exchange vocal signals with them (Tobias, Viswanathan, and Kelley 1998). They appear to have little difficulty locating calling males (Elepfandt 1996).

### Predation and Calling Behavior

Although predation is frequently invoked as a potential cost of acoustic signaling (e.g., Schiøtz 1973), there are only a few studies demonstrating that predators home in on acoustic signals of their prey. Jaeger (1976) provided circumstantial evidence that *Bufo marinus* uses sound to locate calling males of *Physalaemus pustulosus.* Other large anurans, such as *Leptodactylus pentadactylus,* commonly prey on smaller frogs (M. Ryan, Tuttle, and Taft 1981), but whether they use the calls to locate prey is not known. A Neotropical opossum, *Philander opossum,* can use frog calls to locate prey (Tuttle, Taft, and Ryan 1981), and a Neotropical bat, *Trachops cirrhosus,* regularly locates prey by auditory cues and will approach speakers playing anuran calls (Tuttle and Ryan 1981). Other bats are known to eat frogs, and mammals such as raccoons and European polecats often take large numbers of frogs from breeding choruses (see chapter 14). It is not known, however, whether any of these predators home in on the calls of frogs.

The presence of bats and other predators could affect many aspects of a frogs' calling behavior, including selection of calling sites, calling rates, the structure of advertisement calls, and the formation of choruses. Bats are readily attracted to the calls of *Hyalinobatrachium fleischmanni* (Tuttle and Ryan 1981), and this species' habit of calling upside down from the undersides of leaves probably is a defense against such predation (Greer and Wells 1980; Wells and Schwartz 1982). Many other tropical forest frogs call from well-concealed sites under dead leaves or in the axils of bromeliads, probably as a defense against predation (Zimmerman and Bogart 1984, 1988). Some frogs exhibit lunar phobia, reducing their calling activity on bright moonlit nights, perhaps to avoid predation (personal observations).

In contrast, the neotropical hylid *Smilisca sila* is more likely to call in the open on moonlit nights than on dark nights. Apparently they can detect bats more easily on bright nights and take evasive action or reduce calling rates in response to the presence of the bats (Tuttle and Ryan 1982; Nunes 1988b). This species typically calls in the vicinity of waterfalls. Bats are less likely to approach speakers playing calls near waterfall sounds than speakers farther from such sounds, suggesting that the waterfall noise partially masks the frog's calls and makes it hard for bats to locate them. *Smilisca sila* produces complex multinote calls, and males respond very rapidly to the calls of their neighbors, producing a burst of synchronized calling by groups of males, which are followed by long periods of silence. Males overlap their calls with those of their neighbors, but tend to alternate individual notes within calls, a means of avoiding acoustic interference (see the following). Males also tend to increase the number of notes in their calls as the number of notes in a neighbor's calls increase (Ibáñez 1991). The overlapping of calls in synchronized bursts make the calls difficult for bats to locate (Tuttle and Ryan 1982). M. Ryan (1986b) maintained that males of this species respond to the calls of neighbors so quickly that the normal neural pathways used in such vocal interactions would be too slow to account for the behavior. In a more thorough study of response latencies, however, Ibáñez (1991) found that while the responses of this frog are fast, they are within the range of response times observed in a number of other hylids. Very similar synchronized chorusing behavior has been observed in a stream-breeding centrolenid frog, *Cochranella granulosa,* and this probably represents an anti-predator adaptation as well (Ibáñez 1991).

Tuttle and Ryan (1981) showed that bats are more attracted to speakers playing frog calls at high rates than at low rates. Many tropical forest frogs call at relatively low rates, often with long pauses between calling bouts (Duellman 1967b; Dubois 1977a; Lynch and Myers 1983; Zimmerman and Bogart 1984, 1988). This could make it difficult for predators to locate the calls, although low calling rates also could be related to low levels of vocal competition among males. Some acoustic insects exhibit similar adaptations against bat predation (Bellwood and Morris 1987; Bellwood 1990; G. Morris et al. 1994). *Smilisca sila* males reduce their calling rates when bats pass by (Tuttle and Ryan 1982). In another Panamanian frog, *Physalaemus pustulosus,* males often stop calling completely when bats or models of bats are detected overhead (Tuttle, Taft, and Ryan 1982). When two males are calling, they tend to remain silent for longer periods than when only a single male is calling, but the duration of silent periods in response to bat models gradually decreases as chorus size increases from two to five males (Jennions and Backwell 1992). This indicates that males probably assess not only the presence of the predator, but also the relative risk of being eaten, which decreases with chorus size (M. Ryan, Tuttle, and Taft 1981). As in *Smilisca sila,* males of *Physalaemus pustulosus* produce complex multinote calls consisting of an introductory whine note followed by one or more secondary chuck notes (M. Ryan 1985b). Males calling alone often give single-note calls, but they give more complex calls in dense choruses. Complex calls are more attractive to females than are single-note calls (Rand and Ryan, 1981), but they also are more easily located by bats (M. Ryan, Tuttle, and Rand 1982), so there is a cost associated with producing calls that are especially attractive to females.

## The Structure and Dynamics of Anuran Choruses

The mating systems of chorusing frogs and the criteria used by females to select mates are discussed in chapter 8. Here I will focus on the spatial structure and behavioral dynamics of an anuran chorus, especially acoustic interactions among calling males. Reviews of these topics are provided by Schwartz (2001), Gerhardt and Huber (2002), and Wells and Schwartz (2006). There are a number of reasons why males might aggregate at suitable breeding sites. In many cases, such sites are limited, forcing males to call in close proximity to one another. Aggregations also can reduce individual exposure to predation. This would occur if the number of calling males overwhelms the ability of predators to feed on them (M. Ryan, Tuttle, and Taft 1981), but only if the aggregation does not attract substantially more predators to the area. Calling in groups also can decrease rates of predation if acoustic interactions among males, such as the synchronized calling of *Smilisca sila* and *Cochranella granulosa,* confuses predators and makes the frogs more difficult to locate (Tuttle and Ryan 1982; Ibáñez 1991).

Aggregation by calling males can increase the proportion of time that the group as a whole produces signals (the duty cycle of the chorus) or increase the total sound level emanating from the chorus. Changes in the amount of signaling time or chorus intensity do not, however, increase in direct proportion to the number of males in the chorus. It is unlikely that these acoustic properties will lead to an increase in the number of females available for each male (Bradbury 1981; Greenfield 1994a, b). Males also might aggregate because females prefer groups that enable them to compare several potential mates (R. Alexander 1975; Bradbury 1981; Bradbury and Gibson 1983; Greenfield and Shaw 1983; Thornhill and Alcock 1983; Höglund and Alatalo 1995). Several studies of insects and frogs have shown that more females are attracted to multiple sound sources than to single sound sources (G. Morris, Kerr, and Fullard 1978; Cade 1981; Aiken 1982; Walker 1983; Schwartz 1994). Females also are more likely to be found in large choruses than in small choruses (Doolan and Mac Nally 1981; M. Ryan, Tuttle, and Taft 1981; Tejedo 1993a; C. Murphy 2003). Most of these studies have not demonstrated an increase in the per capita availability of females to males in larger choruses, however.

Aggregations of calling male frogs can be formed either because of independent responses of many individuals to the same environmental cues or because the calling of males elicits calling by other males competing to attract mates. There is a large literature showing that numbers of calling males often are correlated with environmental variables such as temperature, rainfall, humdity, wind speed, and ambient light intensity (Blankenhorn 1972; Obert 1975; Wells 1978b; Woolbright 1985a; Banks and Beebee 1986a; Henzi et al. 1995; Navas 1996b; Brooke, Alford, and Schwarzkopf 2000; Friedl and Klump 2002; C. Murphy 2003). Often the abundance of females at a breeding site parallels variation in numbers of males, suggesting either that females are differentially attracted to large aggregations, or that males and females respond to similar environmental cues. For one species, *Hyla gratiosa,* there is evidence that females are attracted to chorus sounds from a distance (Gerhardt and Klump 1988b), but variation in the number of females in a chorus is better predicted by environmental variables (C. Murphy 2003).

There is limited evidence that the presence of calling males attracts other males to calling sites (Oldham 1966, 1967), and calling by males clearly stimulates males in a chorus to increase calling rates or call for longer periods each night (Wells 1988; Schwartz 1991; Brooke, Alford, and Schwarzkopf 2000; Friedl and Klump 2002). Once such choruses are formed, males are faced with the problem of competing for the attention of females in a noisy and acoustically complex environment, and they employ a variety of behavioral tactics to enhance their ability to communicate effectively.

### Spatial Organization of Choruses

To a naive observer, a very large chorus of frogs, like the deafening choruses of spring peepers (*Pseudacris crucifer*), seems like a disorganized cacophony produced by randomly spaced individuals. In fact, most anuran choruses exhibit some degree of spatial structuring. Individual males typically maintain some minimum distance between themselves and their neighbors. They use the perceived sound-pressure level of other males' calls to assess inter-male distance (Fellers 1979a; Brzoska, Schneider, and Nevo 1982; Narins and Hurley 1982; J. Robertson 1984; Brenowitz, Wilczynski, and Zakon

1984; Telford 1985; Wilczynski and Brenowitz 1988; Brenowitz 1989; Gerhardt, Diekamp, and Ptacek 1989; Schwartz 1989; Wagner 1989c; Stewart and Bishop 1994). The distance at which males tolerate other calling males is relatively plastic, and tends to decrease as the number of males in a chorus increases. This is because males accommodate to the close proximity of their neighbors in dense choruses and do not waste time and energy trying to repel them.

In *Pseudacris regilla* and *P. crucifer,* males with close neighbors had a higher threshold for giving aggressive calls than did males with more distant neighbors (G. Rose and Brenowitz 1991; V. Marshall, Humfeld, and Bee 2003), and the threshold was correlated with the sound-pressure level of the neighbors' calls (fig. 7.14 A). When near neighbors were removed, the threshold for aggressive calling decreased (that is, a lower intensity playback of advertisement calls was required to elicit aggressive calling). On the other hand,

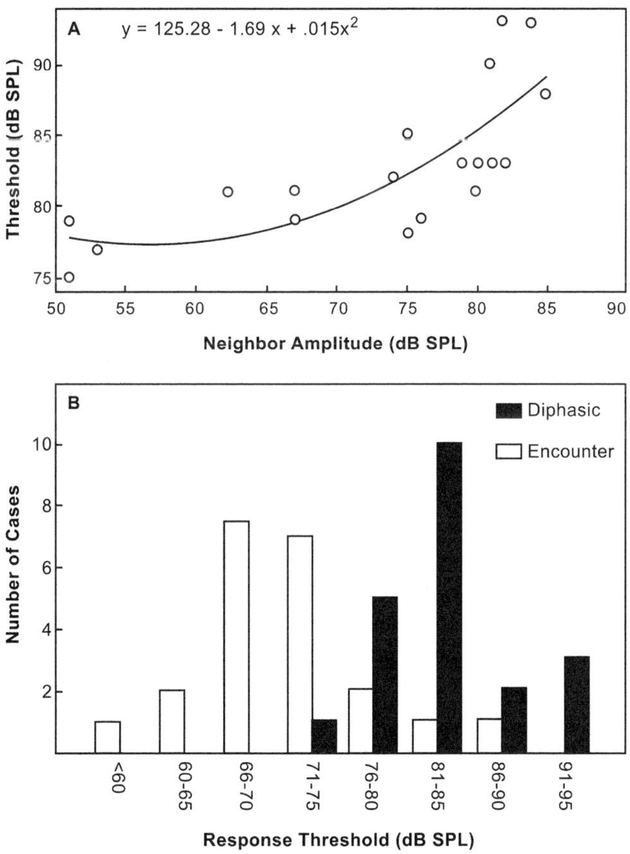


**Fig. 7.14.** Vocal responses of male Pacific treefrogs (*Pseudacris regilla*) in a chorus. (A) Change in aggressive threshold as a function of the amplitude of the nearest neighbor's call. The threshold indicates the playback intensity (SPL) required to elicit aggressive calls in response to playbacks of advertisement calls. Males required a louder stimulus to elicit aggressive responses when other males were nearby. (B) Aggressive thresholds of males in response to playbacks of the advertisement call (diphasic call) and the aggressive call (encounter call). Males gave aggressive responses at lower SPL's in response to aggressive calls that to advertisement calls. After G. Rose and Brenowitz (1991).

when loud advertisement calls were played to males, the threshold for responding with aggressive calls eventually increased. As in most frogs, males give aggressive responses to aggressive calls at lower playback intensities than those required to elicit aggressive calls in response to advertisement calls (fig. 7.14 B). These thresholds were not changed after repeated presentations of advertisement calls at a high level (Brenowitz and Rose 1994). The frogs did accommodate to loud playbacks of aggressive calls by increasing the threshold for aggressive responses, however (G. Rose and Brenowitz 1997; Brenowitz, Rose, and Adler 2001).

The advantage of this sort of plastic response to the call of near neighbors is that males can adjust their aggressive responses to changes in chorus density. As long as a neighboring male remains in one place and does not continue to approach a calling male, it probably represents only a minimal threat to that male's calling territory or his ability to attract mates. If males did not accommodate to the presence of close neighbors, they might spend much of their time giving aggressive calls and spend relatively little time advertising for mates. Experiments with several species have shown that females strongly prefer advertisement calls to conspecific aggressive calls (Oldham and Gerhardt 1975; Schwartz 1987b; Wells and Bard 1987; Grafe 1996b; Brenowitz and Rose 1999; Brenowitz, Rose, and Adler 2001), so there is a clear cost to producing aggressive calls. V. Marshall, Humfeld, and Bee (2003) found that female spring peepers (*Pseudacris crucifer*) preferred advertisement to aggressive calls, but this preference could be reversed by relatively modest increases in the intensity of the aggressive calls. They suggested that energetic constraints also might be important in limiting production of aggressive calls, since these are much longer than are advertisement calls, and production of aggressive calls resulted in a temporary increase in calling effort.

In most anuran choruses, males sort out spatial relationships early in the evening. Once the positions of calling males are relatively stable, males concentrate their efforts on advertising for mates (fig. 7.15). Often there is an intense period of aggressive calling in the first hour or so of chorus formation. This period usually is followed by several hours of advertisement calling in which aggressive calls are heard only occasionally (Jameson 1954; Pengilley 1971a; Allan 1973; Garton and Brandon 1975; Wells 1978b, 1988; J. Robertson 1986b; Wells and Bard 1987; Stewart and Rand 1992; Grafe 1995). One advantage of establishing spatial relationships relatively early in the evening is that the cost of producing aggressive calls that are less attractive to females is minimized because most disputes over calling space are settled before females arrive in the chorus (Wells 1988). Males in very dense choruses tolerate neighbors at relatively close proximity simply because the tradeoff between giving aggressive calls and producing calls that are more attractive

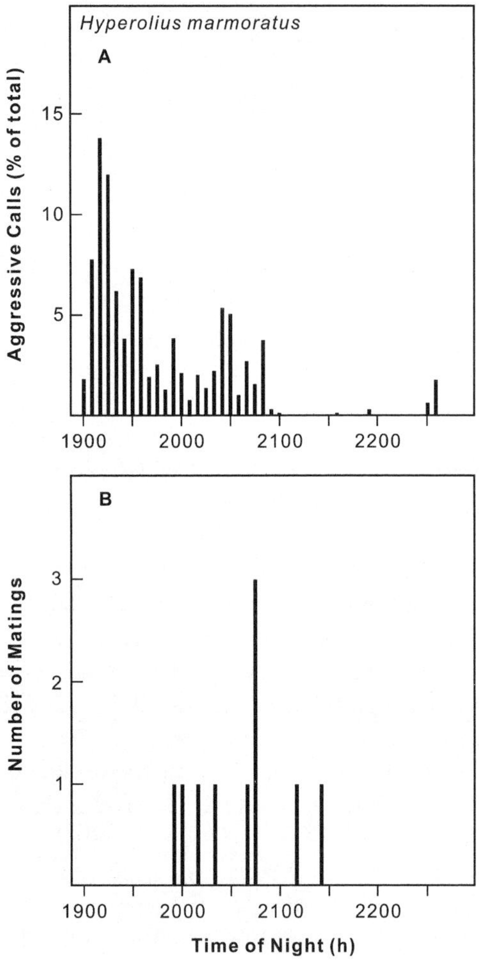


**Fig. 7.15.** Timing of aggressive calling and mating by males of an African frog, *Hyperolius marmoratus broadleyi.* (A) Temporal occurrence of aggressive calls as a percentage of total calls given in a chorus of 11 males on one night. (B) Temporal pattern of mating, based on males seen in amplexus for 18 consecutive nights. Males contest calling territories early in the evening, but exhibit relatively low levels of aggressive calling later in the evening when females are most likely to mate. After Grafe (1995).

to females becomes more costly as the evening progresses and more females are searching for mates. At very high densities, the costs of defending a calling site becomes so high that males simply abandon aggressive calling altogether and even become unresponsive to playbacks of aggressive calls at high intensities (Wells 1988; Brenowitz and Rose 1994; Grafe 1995; G. Rose and Brenowitz 1997).

There seem to be two main advantages to males that maintain exclusive calling spaces in a chorus. First, it reduces the chances that nearby males will interfere with a male's courtship of females. In very dense choruses, this tactic is only moderately successful, because of the tendency for males to adopt satellite tactics and attempt to intercept females that are attracted to calling males (see chapter 8). Spacing in a chorus also provides acoustic advantages by reducing the amount of interference from the calls of other males (Telford 1985). Often the background noise in a frog chorus is very loud and can mask the calls of individual males and make it difficult for females to locate them (Ehret and Gerhardt 1980; Gerhardt and Klump 1988a; Schwartz and Wells 1983b; Wollerman 1999; Gerhardt and Huber, 2002; Wollerman and Wiley 2002a). The most important sources of interference are likely to be a male's immediate neighbors, because their calls will be perceived by females are being the loudest relative to that male's own calls. In experiments with *Hyla cinerea,* Schwartz and Gerhardt (1989) presented females with a choice of an attractive call (advertisement call) and an unattractive call (aggressive call) in the presence of background noise. The frogs were able to detect the calls more readily when the speakers broadcasting the calls were separated from the noise by 45° or 90° than when the speakers were immediately adjacent to those broadcasting noise. The improvement in call detection was relatively small, however, so spatial separation alone probably is less effective than are adjustments in the timing of vocalizations to avoid overlap with those of near neighbors.

## Vocal Interactions among Chorusing Frogs

In addition to the exchange of aggressive calls to facilitate spacing in a chorus, male anurans engage in a number of other kinds of vocal interactions that serve either to reduce acoustic interference from nearby males or to make their signals more attractive to females. These include changes in the timing of calls to reduce call overlap with neighbors and changes in the rate, duration, or complexity of calls. In dense, noisy choruses, these types of vocal interactions usually involve only a male and a few of his closest neighbors, with other males simply contributing to the overall background noise (Narins 1992, 1995a, b; Schwartz 2001). For example, in choruses of *Eleutherodactylus coqui* and *Physalaemus pustulosus,* males usually respond to only one or two near neighbors and ignore the calls of more distant males (Brush and Narins 1989; Greenfield and Rand 2000). In a study of *Hyla microcephala,* Schwartz (1993) used a computerized system to simultaneously monitor groups of four to six males calling in close proximity to one another. He found that a male usually responded only to the calls of his loudest neighbor, even when the perceived intensities of other males' calls were only a few decibels lower. These results indicate that males tend to interact mainly with those males that represent the most important source of acoustic interference and the most immediate competition for any females that are nearby. Nevertheless, the remaining males in a chorus contribute to general background noise that can interfere with a male's ability to attract mates.

Not all frogs exhibit this type of strict selective attention to nearest neighbors. For example, in a chorus of *Hyla ver-*

*sicolor* in an artificial pond, males avoided overlapping calls when only two males were calling, but call overlap increased in larger choruses (up to eight males). Furthermore, males were more likely to overlap at least part of their calls with those of their nearest neighbors than expected by chance (Schwartz 2001; Schwartz, Buchanan, and Gerhardt 2002). Hence, there is no evidence for selective avoidance of call overlap with nearest neighbors in this species. Males of *H. versicolor* respond to other males' calls by lengthening their own calls (see following), and this could partially compensate for parts of their calls overlapping those of other males (Schwartz, Buchanan, and Gerhardt 2001). In addition, spatial separation among males in a chorus improves the ability of females to locate individuals, even when calls overlap (Schwartz and Gerhardt 1995).

Boatwright-Horowitz, Horowitz, and Simmons (2000) reported that male bullfrogs (*Rana catesbeiana*) were more likely to respond to the calls of distant males than to those of close neighbors. They argued that calling by near neighbors tends to inhibit calling by a male, whereas calling by more distant frogs tends in stimulate calling. Nevertheless, male bullfrogs callling within a few meters of one another often alternate calls or call notes with one another (personal observations). Male bullfrogs tend to be widely separated in a chorus, calling rates are low, but individual calls are loud, so males can readily hear the calls of distant males. This probably is not true for very dense choruses, like those of *Pseudacris crucifer* or *Hyla microcephala*, where general background noise masks the calls of distant individuals (Schwartz and Gerhardt 1998; Schwartz 2001).

#### Unison Bout Calling

In very dense frog choruses, with many males competing to be heard by females, calling is not always continuous. Instead, there often are bursts of calling activity in which many males call at the same time, followed by periods of silence. This waxing and waning of chorus activity can continue for much of the evening (Rosen and Lemon 1974; Wickler and Seibt 1974; Garton and Brandon 1975; Whitney and Krebs 1975b; Arak 1983c; Schwartz and Wells 1983a, b, 1984b, 1985; Schwartz 2001). Similar behavior occurs in many acoustic insects (Alexander 1975; Otte 1977; Otte and Loftus-Hills 1979; Greenfield 1983, 1990, 1994 a, b; Greenfield and Shaw 1983) and has been termed unison bout singing. This type of calling pattern probably results from mutual acoustic stimulation by males in the chorus. It could be advantageous for males to call together because the resulting cacophony is confusing to predators. Males also are likely to call in response to the calls of other males simply because they must compete for the attention of females (Brooke, Alford, and Schwarzkopf 2000; Schwartz 2001). Unison bout calling has clear disadvantages as well, especially the high level of background noise and acoustic interference that results from synchronized calling. Individual males can minimize acoustic interference by adjusting the timing of their calls within bouts (see following).

One question that arises from unison bout calling is why frogs stop calling periodically instead of calling continuously throughout the evening. This has been investigated in a series of experiments with *Hyla microcephala* in Panama. In this species, additional males soon join a male that begins calling during a silent period. The males tend to gradually increase the number of repeated notes in their calls as they compete for the attention of females (Schwartz and Wells 1985; Schwartz 1986, 1987b, 1991; Wells and Taigen 1989). Each bout of calling usually lasts about 30 seconds and then subsides. Three hypotheses have been tested to explain this pattern of calling: (1) females prefer males that call in a cyclic pattern to those that call continuously; (2) males stop calling because acoustic interference becomes too great for them to continue; or (3) males stop calling periodically to reduce the high cost of call production (Schwartz 2001).

Schwartz (1991) presented females with a choice of two speakers playing advertisement calls at the same average rate, but one broadcast calls continuously, while the other broadcast calls grouped into bouts separated by silent periods. Females did not exhibit a significant preference for one pattern over the other, so the first hypothesis was refuted. Schwartz (1991) tested the second hypothesis by presenting males in the field with several types of stimuli that were stored on a computer and triggered by the test male's own calls. In one experiment, males were presented with a synthesized chorus at several intensity levels. These generally stimulated more calling by the males rather than inhibiting their calling or causing them to shift their calls to silent periods. When males were presented with simulated choruses of 1, 2, or 12 frogs, they increased the duration of their calling bouts in response to the 12-male chorus, just the reverse of the pattern predicted from the acoustic interference hypothesis. Schwartz (1991) concluded that the energy conservation hypothesis was the most likely explanation for why males stop calling. This was tested in another series of experiments in which rates of glycogen depletion in the trunk muscles of males (see chapter 5) were measured (Schwartz, Ressel, and Bevier 1995). These measurements showed that depletion of glycogen reserves was very rapid. If males attempted to call continuously at rates observed during calling bouts, they would not have sufficient reserves to sustain calling through an entire evening. Because females arrive relatively late in the evening, but over an extended period of time, those males that can sustain several hours of calling probably have the best chance of attracting a mate. Preliminary data using a simulated chorus supported this hypothesis. In this experiment, one speaker broadcast calls in cycles and persisted for five hours, while the others broadcast continuously and stopped earlier. Females were more likely to enter a trap at the cyclical speaker than those at the other speakers (Schwartz 2001). Thus, it appears that periodic pauses in calling represent a way of conserving energy and extending a male's calling period.

### Changes in the Timing of Calls

Male frogs, especially those that call in dense choruses or in unison bouts, often shift the timing of their calls to reduce call overlap with neighboring males (Foster 1967; Paillette 1967; Schneider 1967, 1968; Lemon 1971; Loftus-Hills 1971, 1974; Rosen and Lemon 1974; Wickler and Seibt 1974; Wells 1977b, 1988; Awbrey 1978; Narins and Capranica 1978; Lemon and Struger 1980; Narins 1982a, b, 1992, 1995a, b; Zelick and Narins 1982, 1983, 1985a; Wells and Schwartz 1984a; Akef and Schneider 1985; Sullivan 1985a; Sullivan and Leek 1986; Forester and Harrison 1987; Schwartz 1987a, 1993, 1994; Schneider, Joermann, and Hödl 1988; Walkowiak 1988b, 1992; Brush and Narins 1989; Ibáñez 1991, 1993; Klump and Gerhardt 1992; Given 1993b; Gerhardt 1994a; Greenfield 1994a, b; Grafe 1996b, 2003; Ovaska and Caldbeck 1997b; Penna 1997; Bosch and Márquez 2001a; Márquez and Bosch 2001; Schwartz 2001; Gerhardt and Huber 2002). Similar behavior occurs in many acoustic insects (Alexander 1975; Otte 1977; Greenfield and Shaw 1983; Greenfield 1990, 1994a, b, 2002; Gerhardt and Huber 2002) and singing birds (Todt 1970; Wasserman 1977; Gochfeld 1978; Hultsch and Todt 1982; Dabelsteen et al. 1996).

The precise pattern of vocal interaction varies among species. In some frogs, the calls of two males are precisely alternated (fig. 7.16 A). This not only minimizes call overlap, but also results in neighboring males giving approximately the same number of calls per minute. In other species, the calls of the second male begin immediately after the end of the first male's calls (fig. 7.16 B). Again, call overlap is minimized, and calling effort is very similar. In species that produce long sequences of repeated notes, two males sometimes overlap long calls, but alternate individual notes, resulting in little or no acoustic interference (fig. 7.16 C). Alternatively, some frogs produce a series of repeated notes, with the calls of the second male falling in the intervals between the calls of the first male (fig. 7.16 D). In either case, there is the potential for relatively precise matching of the number of call notes. Finally, in some anurans, males respond to the calls of other males by initiating their own calls so that the calls of two males overlap extensively, with no apparent attempt to avoid acoustic interference (fig. 7.16 E). In the example shown in fig. 7.16 E, the calls are trills consisting of repeated trains of short pulses, but similar call overlap can occur in species in shorter, single-note calls. Examples of each of these types of interactions are discussed in more detail in the following discussion.

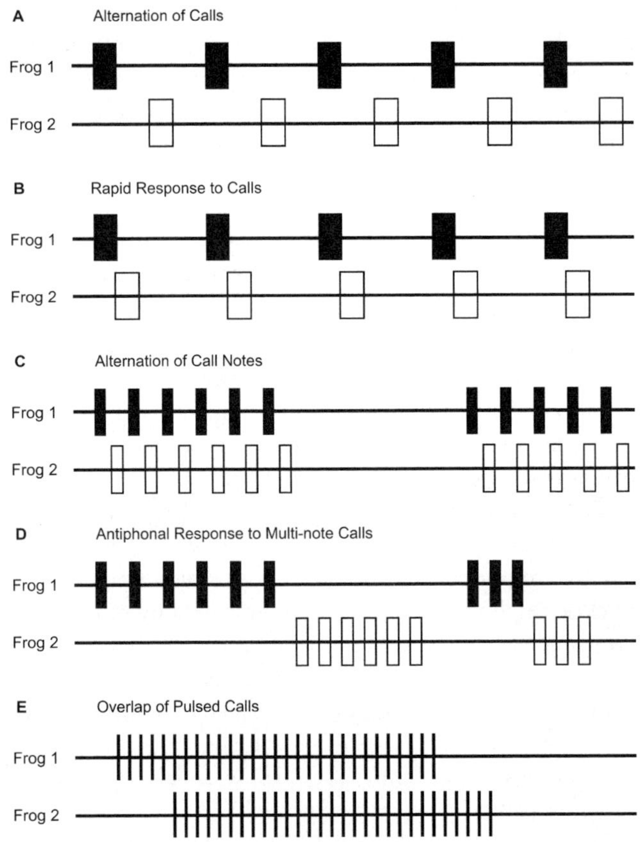


**Fig. 7.16.** Diagram of several possible types of vocal interaction between neighboring males in a frog chorus. (A) Calls consist of relatively long notes given at regular intervals, with the calls of the second male precisely alternated with those of the first. (B) Calls consist of relatively long notes given at regular intervals, with the calls of the second male starting immediately after the end of the first male's calls. (C) Calls consist of a variable number of closely spaced short notes, with individual notes of the second male's calls alternating with those of the first male. The result is minimal acoustic interference and relatively precise matching of the number of call notes. (D) Calls consist of a variable number of short notes. Calls of the second male are given immediately after the entire sequence of notes of the first male has ended, with fairly precise matching of the number of call notes. (E) Calls are trills made up of a rapid series of short pulses. Calls are overlapped with no attempt to avoid acoustic interference.

Most frogs probably have a neural pacemaker or oscillator that sets the spontaneous calling pattern of an individual male (Loftus-Hills 1974; Wickler and Seibt 1974; Lemon and Struger 1980; Zelick and Narins 1985a; Walkowiak 1988b, 1992; S. Moore et al. 1989; Klump and Gerhardt 1992; Greenfield 1994a, b). In most species, especially those with relatively high calling rates, males calling without interruption from other males usually call at a relatively constant rate, which varies with temperature. The call oscillatory period is the time between the onset of one call and the onset of the male's next call (fig. 7.17). This can be divided up into several sub-periods that describe the responsiveness

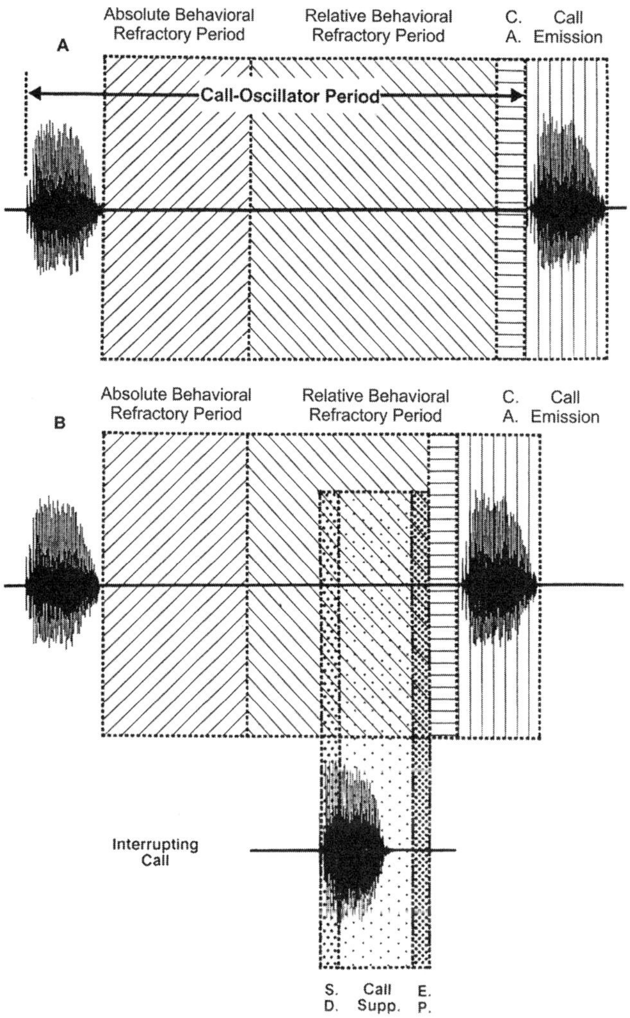


**Fig. 7.17.** Schematic representation of the phases of a calling cycle in frogs. (A) Internally triggered call generation. The period of the calling cycle is set by an internal oscillator, which sets the timing of the next call. (B) Externally triggered call generation. When the frog hears the call of a neighboring male, he resets his calling period so that calls do not overlap. Interrupting calls that occur during the absolute behavioral refractory period do not elicit vocal responses. Interrupting calls falling with the relative behavioral refractory period initially suppress calling and then stimulate calling after a delay. During the call activation phase (C.A.), calling cannot be suppressed, and calls other males are less likely to be perceived while a male is calling (emission phase, E.P.) S.D. indicates suppression delay that occurs when the frog is listening to the other male's call. After Klump and Gerhardt (1992).

of the male to external acoustic stimuli. Most frogs probably have an absolute behavioral refractory period immediately after the production of their own calls, during which they will not call even when they hear the call of another male. The length of this refractory period varies among species (Narins 1982a). After the absolute behavioral refractory period is a period in which males do not normally call spontaneously, but do respond to the calls of another male. In other words, hearing the call of another male causes an individual to call sooner than it otherwise would. Once a male's neural circuits have initiated call production (the call activation phase in fig. 7.17), it cannot be suppressed by hearing another call, and males probably have difficulty hearing calls produced during their own call emission. Consequently, frogs often do not respond to the calls or other males or to playbacks of calls that overlap all or part of their own calls.

Some frogs are capable of responding very rapidly to changes in background noise. For example, males of *Eleutherodactylus coqui* can accurately place their calls in gaps of only 750 milliseconds in continuous background noise, a time only slightly longer than that of their own calls. Furthermore, they can do so even if the timing of the gap is unpredictable, or if the sound level in the gap is only 4–6 dB below that of the overall background noise (Zelick and Narins 1982, 1983). This means that they must initiate production of their own calls within a few milliseconds of perceiving the reduction in background noise. In *Hyperolius marmoratus broadleyi,* males can insert their calls into gaps as short as 600 milliseconds, but not into 300 millisecond gaps (the calls themselves are about 83 milliseconds long). Responses are very fast, with most males initiating calls within 40–80 ms after a stimulus call ended (Grafe 1996b).

In *Hyla microcephala,* males not only alternate individual notes in longer overlapping multinote calls (as in fig. 7.16 C), but they also lengthen the gaps between the notes in their own calls to further reduce the chances of acoustic interference (Schwartz and Wells 1985). Males can respond to gaps in background noise as short as 20 milliseconds, which allows them to make this sort of note-by-note adjustment in the timing of their calls (Schwartz 1993). As a result, less than 10% of a male's calls usually suffer acoustic interference from his loudest neighbor, and even in groups of up to six males, 34–92% of a male's call notes are free of interference. In response to synthetic stimuli of variable duration, males lengthened the gap between their own call notes in response to stimuli up to 200 ms long (Schwartz 1993). They could not lengthen the gap sufficiently to avoid note overlap in response to 400 ms stimuli (fig. 7.18).

Acoustic interference between neighboring males could be detrimental to males either because it reduces a female's ability to locate a male, or because it obscures features that are critical for species identification or are attractive to females. Acoustic interference also could affect communication between males, because males do not hear the calls of other males that overlap their own calls, thereby making it difficult to judge the distance between them (Lemon 1971). Passmore and Telford (1981) found that females of *Hyperolius marmoratus* could locate speakers broadcasting simultaneous or alternating calls equally well. However, since

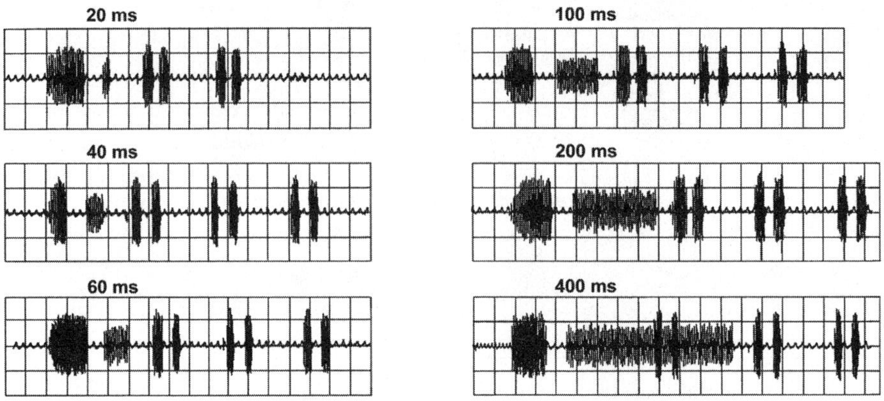


**Fig. 7.18.** Variation in call timing in *Hyla microcephala* in response to interruptions of the call by a synthetic stimulus of variable duration. The call of this species consists of a single introductory note followed by a variable number of diphasic secondary notes. The call notes with lower amplitude in each panel are the interrupting stimulus calls. Males responded to interrupting calls of 20–200 ms by increasing the gap between their call notes, thereby avoiding overlap of individual notes. They were not able to lengthen the gap sufficiently to avoid overlap in response to a 400 ms call. After Schwartz (2001).

calls were broadcast precisely in phase, species-specific temporal patterns were not obscured. Hence, their experiment did not address the second possible consequence of acoustic interference (Wells and Schwartz 1984a).

A definitive test of hypotheses advanced to explain the adaptive significance of call alternation was performed by Schwartz (1987a) using a four-speaker playback design. He found that in species with pulsed calls (*Hyla microcephala* and *H. versicolor*), females preferred alternated calls to those that overlapped so that the temporal structure of the call was obscured. No such discrimination occurred when the overlapped calls were completely in phase. Subsequent experiments with *H. microcephala* showed that the degrading effects of an interfering call were eliminated if the interfering call was only 6 dB lower than a male's own calls. These experiments support the hypothesis that call alternation is advantageous in allowing males to preserve species-specific temporal information in the call. In *Pseudacris crucifer,* a species with tone-like calls lacking temporal structure, females showed no preference for alternated vs. overlapped calls. Schwartz (1987a) also found that males gave more aggressive responses to advertisement calls broadcast in alternation with their own calls than to those that overlapped their calls. This supports the hypothesis of Lemon (1971) and Passmore and Telford (1981) that call alternation is advantageous in allowing interference-free communication with neighboring males. In short, call alternation can improve the effectiveness of communication with individuals of either sex, depending on the structural features of the call.

Experiments with other species generally are consistent with these interpretations. In *Physalaemus pustulosus,* which has a frequency-modulated call, females did not discriminate between overlapped and alternating calls (Schwartz and Rand 1991). In studies with *Hyperolius m. marmoratus,* a species with a tone-like advertisement call, Dyson and Passmore (1988) found that females did not discriminate between overlapping, simultaneous, and alternating calls. They did tend to prefer the leading call when stimulus calls overlapped, or one followed immediately after the other, even when the lead call was broadcast at a lower intensity. Females of *H. m. broadleyi,* which also have tone-like calls, also did not discriminate against overlapping calls, but showed a consistent preference for leading calls when the second call followed shortly afterward, perhaps because the beginning of the second call was masked by the first call. Results of experiments with *Alytes obstetricans* and *A. cisternasii* are not entirely consistent with Schwartz's (1987a) results, however. These frogs have very simple, tone-like calls and show only moderate precision of call alternation (Bosch and Márquez 2001a). Females did discriminate against overlapped calls in favor of nonoverlapped calls, but did not discriminate between leader and follower calls when they were overlapped (Bosch and Márquez 2000).

### Synchronized Calling

In some species, a very rapid response to calls of other males produce a synchronizing effect that results in partial overlap of calls of neighboring males. This represents an extreme form of the unison bout calling described earlier and would seem to be disadvantageous to individual males in attracting mates. This type of chorusing is less common in anurans than in acoustic insects, where synchronized chorusing appears to have evolved as a byproduct of males attempting to out-signal one another (Greenfield 1990, 1994 a, b; Greenfield and Roizen 1993). Some examples of call overlap in anurans appear to be related to anti-predator behavior. For example, males of *Smilisca sila* typically call within 60–80 milliseconds of the onset of other males' calls. Their advertisement call notes are about 136 milliseconds long, so this results in the beginning of the second male's calls overlapping the end of the first male's calls (Tuttle and Ryan 1982; M. Ryan 1986b; Ibáñez 1991). In natural choruses, the calls of individual males overlapped those of their neighbors much more than expected by chance (Ibáñez 1991). Similar behavior was observed in a centrolenid frog, *Cochranella granulosa* (Ibáñez 1991). As in *Smilisca*, a male *C. granulosa* usually responds to the call of another male before that call has ended, resulting in partial overlap among calls of neighbors. This species breeds along streams where frog-eating bats are present, so this behavior probably is an antipredator strategy as well. Protection from predators comes at a cost in reduced attractiveness to females, however. Females of *C. granulosa* chose alternated calls over overlapped, out-of-phase calls in which the temporal structure of overlapped calls was obscured, but did not discriminate between alternated and in-phase overlapped calls (Ibáñez 1993).

Other species exhibit partial call overlap that is not related to protection from predators. For example, males of the Panamanian treefrog *Hyla ebraccata* usually respond to the multinote calls of their neighbors by placing their calls immediately after the end of the first male's introductory note. This results in the long introductory note of the second male overlapping one or more of the short secondary notes of the first male (fig. 7.2 E), effectively masking those notes (Wells and Schwartz 1984a). Experiments with females showed that they prefer the second masking call to the lead masked call when the calls are otherwise identical. It is not clear whether this represents an evolved interference tactic (Greenfield 1994b; Greenfield, Tourtellot, and Snedden 1997) or is simply a byproduct of timed responses to avoid overlap of introductory notes.

Males of the African frog *Kassina fusca* typically respond to calls of their neighbors very rapidly, so that more than 80% of all calls overlap. On average, about 21% of a lead call is overlapped by a follower call. Females exhibit a clear preference for follower calls when the degree of overlap is low, but switch to a preference for leader calls when the degree of overlap is high (Grafe 1999). These results suggest that males actively regulate the degree of overlap to correspond to female preferences. This was confirmed by experiments in which the degree of overlap in calls given in response to playbacks was nearly identical to that in natural interactions.

In some toads, the long trills given by one male stimulate calling by other males, which overlap their calls with those of their neighbors (as in fig. 7.16 E). This appears to be most common in species in which males call to attract females to the breeding site, but locate mates mainly by active searching (see chapter 9) and therefore are less likely to suffer the negative consequences of acoustic interference (Wells 1977b; Sullivan 1985a). In the only experiment performed with of a species that overlaps long calls, females of *Bufo americanus* preferred leader calls to follower calls in an overlapping pair, although this preference was affected by frequency differences as well (Howard and Palmer 1995).

### Changes in Calling Rate

Male frogs usually call at faster rates in choruses than in isolation, and increases in calling rate have been demonstrated experimentally for many species (Wells 1988; Bendix and Narins 1999; Penna, Narins, and Feng 2005). In many cases, such changes result from antiphonal calling between males that can lead to relatively precise matching of calling rates of two interacting males. Such call matching also is seen in species with multinote calls. In *Hyla microcephala*, for example, there is a tendency for males to match the number of notes in their calls to those of their neighbors through note-by-note alternation (Schwartz and Wells 1985; Wells 1988; Wells and Taigen 1989). In the Asian rhacophorid frog *Philautus leucorhinus* and the Australian myobatrachid frog *Crinia georgiana*, males give a series of repeated notes before being answered by a nearby male (as in fig. 7.16 C). In both species, there is relatively precise matching of call note number, but addition of call notes by the responding male tends to level off at higher note rates (Arak 1983c; Gerhardt et al. 2000). Somewhat less precise call note matching was reported for *Kassina kuvangensis* (Grafe 2003). In contrast to the species discussed previously, males of species that do not call in dense choruses often show little or no response to playbacks of conspecific calls (Dubois 1977a; Greer and Wells 1980).

Although energetic costs (chapter 5) or risk of predation increase at high calling rates, females generally show a preference for high calling rates in two-choice experiments (Gerhardt 1988; Wells 1988; Gerhardt and Huber 2002; see chapter 8 for further discussion). Hence, the high calling rates of many species probably are a result of sexual competition. They derive in part from the tendency of many frogs to alternate their calls with those of other males, which leads to matching of calling rates by close neighbors. When males are presented with broadcast stimuli that greatly exceed their own spontaneous calling rates, they often cannot keep up. They tend to respond only to every second call, producing a calling rate about half that of the stimulus (Zelick and Narins 1985a; S. Moore et al. 1989; Klump and Gerhardt 1992; Bendix and Narins 1999; Grafe 2003).

### Changes in Call Duration and Complexity

Frogs that have trilled calls often increase the duration of their calls in response to other males. This behavior has been demonstrated experimentally in *Hyla versicolor.* Males sometimes double call duration in response to the calls of other males, but they simultaneously reduce their calling rates, probably because high energetic costs make it impossible for males to produce long calls at high rates (Wells and Taigen 1986; Schwartz, Buchanan, and Gerhardt 2002; see also chapter 5). Females prefer both high calling rates and long calls, but prefer long calls at low rates to short calls at high rates (Klump and Gerhardt 1987). In natural choruses, background noise reduces the ability of females to discriminate among calls of different duration, and in such conditions, females tend to discriminate against only the shortest calls (Schwartz, Buchanan, and Gerhardt 2001). Consequently, as long as males calling together in one part of a pond roughly match one another in calling rate and call duration, they are likely to be equally attractive to females.

In many other species, males give single-note calls when calling alone, but add notes to their calls when calling in choruses (Wells 1988). In some species, a single type of note is repeated, as in *Rana catesbeiana* (Capranica 1968), *R. clamitans* (Wells 1978b; Ramer, Jenssen, and Hurst 1983; Bee and Perrill 1996), *R. virgatipes* (Given 1987), *R. septentrionalis* (Bevier et al. 2004), *Kassina kuvangensis* (Grafe 2003), and *Eupsophus emiliopugini* (Penna, Pottstock, and Velazquez 2005), although some species vary the structure of individual notes (Bevier et al. 2004; Suggs and Simmons 2005). Other species produce calls with distinct primary and secondary notes, and males tend to increase the number of secondary notes in their calls in response to the calls of other males or to experimental playbacks of calls (Wells 1988). This type of call is found in several Neotropical hylids, including species already mentioned, such as *Hyla ebraccata, H. microcephala, H. phlebodes, H. minuta,* and *Smilisca sila* (Fouquette 1960; Duellman 1970; Cardoso 1981b; Wells and Greer 1981; Tuttle and Ryan 1982; Schwartz and Wells 1984a, b, 1985; Wells and Schwartz 1984a). Similar calls are found in *Physalaemus pustulosus* (Rand and Ryan 1981; M. Ryan 1985b), several Australian myobatrachids (Ayre et al. 1984; Littlejohn and Harrison 1985; Gerhardt et al. 2000), a rhacophorid from Sri Lanka (Arak 1983c), and several ranid and hyperoliid frogs from Africa (Schiøtz 1967, 1999; Passmore 1977, 1978; Telford 1982; Backwell 1988; Pallet and Passmore 1988).

In two-choice playback experiments, females often exhibit a preference for complex, multinote calls over simple, single-note calls. This type of response has been documented in *Physalaemus pustulosus* (Rand and Ryan 1981), *Geocrinia victoriana* (Littlejohn and Harrison 1985), *Hyla ebraccata* (Wells and Schwartz 1984a), *H. microcephala* (Schwartz 1986), and *Hyperolius tuberilinguis* (Pallett and Passmore 1988). Females might prefer multinote calls because they indicate superior genetic quality in males that can sustain high levels of note production. Alternatively, a repetitive series of stereotyped click-like notes provides the best contrast with the continuous background noise of a multispecies frog chorus and therefore makes males more conspicuous to females (Schleidt 1973; S. Green and Marler 1979; D. Richards and Wiley 1980; Wells and Schwartz 1984a; Pallett and Passmore 1988). It also is possible that females are simply attracted to the longest calls, or those with the greatest total sound energy (G. Morris, Kerr, and Fullard 1978; Aiken 1982; Arak 1983a; Wells and Schwartz 1984a; Schwartz 1986, 1987b; Wells and Bard 1987). Males of these species probably would benefit from producing long, complex calls all of the time, but the high energetic cost of producing such calls and increased risk of predation probably makes this impossible (M. Ryan 1988a; Wells and Taigen 1989; Schwartz, Ressel, and Bevier 1995). An interesting question is why so many chorusing frogs tend to produce discontinuous calls consisting of repeated notes instead of long, continuous trills like those of many toads (*Bufo*). In addition to the energetic cost of continuous calling, Greenfield (1990) suggested that this pattern of calling produces acoustic windows of silence that enable a male to monitor the calling activity of his neighbors. This would be especially important in species that call in very dense choruses. Schwartz's (1987a) work on call alternation suggests that this explanation is applicable to frogs as well.

#### Changes in Call Intensity and Frequency Structure

Several studies have shown that males can alter the frequency structure of their advertisement or aggressive calls in response to the calls of other males, but the way in which frequency is altered differs among species. In *Leptodactylus albilabris,* males shifted the dominant frequency of their calls toward that of a call broadcast to them, either by raising or lowering dominant frequency, and they also increased the intensity of their calls in response to both advertisement and aggressive calls (Lopez et al. 1988). The shift in call frequency toward that of an opponent was interpreted as a possible mechanism to match the signal of an intruding male, possibly as a way of directing the signal toward that particular male. The change in call intensity could be advantageous in making a male more attractive to females or in delivering a stronger aggressive message to intruding males. The changes in call intensity and frequency can be constrained by the structure of the sound producing apparatus. Lopez et al. (1988) found that all of the males that showed an increase in call intensity also decreased call dominant frequency, so possibly changes in frequency are not in themselves adaptive.

Changes in call frequency also have been reported in several ranid frogs, including *Rana virgatipes* (Given 1999), *R. clamitans* (Bee and Perrill 1996), and *R. catesbeiana* (Bee and Bowling 2002; Bee 2002). In *R. virgatipes,* males give aggressive calls that are very similar in structure to advertisement calls, as they are in *R. clamitans,* but aggressive calls have secondary frequencies that are lower than those of advertisement calls. Males respond to playbacks of calls by lowering the secondary frequencies of their advertisement calls, which makes them more like aggressive calls (Given 1999).

In *R. clamitans,* males lowered the dominant frequency of their calls in response to playbacks of both single-note and multinote advertisement calls, and they also called at higher rates, produced longer notes, and decreased call amplitude (Bee and Perrill 1996). Bee, Perrill, and Owen (1999) demonstrated that males lower the dominant frequency of their calls more in response to low-frequency calls (350 Hz) than to calls with higher fundamental frequencies (400 and 450 Hz). Fundamental frequency is negatively correlated with body size, so these results suggest that males can use the frequencies of other males' calls to assess their body size. These results are consistent with earlier work by Ramer, Jenssen, and Hurst (1983), which showed that the level of aggressive responses by male green frogs differed in response to calls of large and small males.

The functional significance of this change in frequency is not yet clear. Bee, Perrill, and Owen (2000) tested two hypotheses to explain the significance of lowering the dominant frequency of calls: (1) The lower frequencies provide more accurate information about the size of the caller, or (2) the lower frequency provides deceptive information by making a male appear to be larger than he actually is. These authors found that calls given in response to playbacks were not more strongly correlated with male body size than were those given spontaneously, so the first hypothesis was not supported. Small males tended to reduce the dominant frequency of their calls more than did large males, so the possibility of deception remains a viable hypothesis. It also is possible that the change in frequency is correlated with some other change, such as the production of longer notes or changes in call intensity, and does not have an adaptive explanation. The magnitude of frequency alteration (about 20 Hz on average) is considerably smaller than the 50 or 100 Hz differences used by Bee, Perrill, and Owen (1999) to determine whether males can perceive differences in call frequency. It remains to be determined whether such small differences in dominant frequency elicit different behavioral responses from males. Bullfrogs (*R. catesbeiana*) exhibit similar changes in dominant frequency in response to other males, but lowering of dominant frequency does not appear to be related to fighting ability (Bee 2002).

Cricket frogs (*Acris crepitans*) also change the dominant frequency of their calls in response to calls of other males (Wagner 1989a, 1992). Larger males consistently won fights, and call dominant frequency was negatively correlated with body size. Males tended to retreat from broadcasts of low-frequency calls, but were more likely to attack speakers playing higher-frequency calls. Males also responded to low-frequency calls by lowering the dominant frequency of their own calls. This change in frequency gave males an advantage in aggressive interactions, but the changed frequency actually produced a less accurate correlation between body size and dominant frequency than did the unaltered frequencies. This suggests that the males do not alter call frequency to provide an honest indicator of body size. Wagner (1992) did find a positive correlation between the extent to which males lowered their call frequencies and the probability that they would initiate an attack (see also Burmeister et al. 2002). He therefore suggested that males alter the frequency of their calls to signal their fighting ability to opponents.

### Chorus Leadership

There is a persistent idea in the anuran literature that certain individuals serve as leaders in choruses. An early proponent of this view was the Swedish naturalist Peter Kalm. Traveling in New Jersey in 1749, he reported that bullfrogs seemed to have "a captain among them. . . . When he begins to croak, all the others follow; and when he stops, the others are all silent" (Benson 1966, vol. 1, p. 297). Charles C. Abbott, a later naturalist who lived in the same region, disagreed with Kalm. Having carefully observed bullfrogs calling near his home, he reported that "the croaking of the 'captain' is not always the same individual. At times the initial croak would come from one side of the pond, then the other. . . This shows that not any one individual started and stopped the croaking of its companions" (Abbott 1885, 322–23).

Goin (1949) reported that *Pseudacris crucifer* choruses are organized into duets and trios, with certain individuals serving as chorus leaders. Subsequent authors have reported similar phenomena in other anurans (P. Anderson 1954; Hardy 1959; Bogert 1960; Brattstrom 1962a; Duellman 1967b; Brattstrom and Yarnell 1968; Wickler and Seibt 1974; Whitney and Krebs 1975b). Chorus leaders sometimes have been considered dominant individuals, and ordered sequences of calls have been characterized as hierarchies (Duellman 1967b; Brattstrom and Yarnell 1968; E. O. Wilson 1975). There is no real evidence that call-order is determined by aggressive interactions among individuals, nor is there evidence that choruses are analogous to dominance hierarchies in other animals.

Brattstrom and Yarnell (1968) reported that chorus leaders in *Physalaemus pustulosus* obtain more mates than other males, but subsequent work has refuted this (M. Ryan 1983b, 1985b). In fact, there is little convincing evidence that certain individuals consistently serve as chorus leaders. Most of the reports cited above are descriptive and provide little or no quantitative analysis of calling patterns. All are based on very brief observations, usually on a single night. In many choruses, certain males consistently call at higher rates than do others (Sullivan 1982a, 1983b; Wells and Taigen 1984; Given 1988a; Sullivan and Hinshaw 1990, 1992; Runkle et al. 1994; Jennions, Backwell, and Passmore 1995; Gerhardt, Dyson, and Tanner 1996; Zimmitti 1999; Bee et al. 2001; Tárano 2001; Friedl and Klump 2002; Schwartz, Buchanan, and Gerhardt 2002). These males sometimes tend to start calling bouts during short-term interactions with neighboring males (Whitney and Krebs 1975b), but in many species, leader-follower roles change frequently (e.g., Grafe 2003). Indeed, it has not been shown for any species of anuran that certain individuals serve as leaders of an entire chorus over long periods of time.

### Acoustic Interactions between Males of Different Species

Although some species of anurans call in monospecific choruses, many call in aggregations of as many as 10 to 15 species, especially in the tropics (Duellman 1967a, 1978; Hödl 1977; Passmore 1978; Schlüter 1979, 1980a, b, c, 1981; Passmore and Telford 1981; Telford 1982; Schwartz and Wells 1983a, b, 1984a; Cardoso and Haddad 1992; Arzabe, de Carvalho, and Goes Costa 1998; Chek, Bogart, and Lougheed 2003). The presence of so many species calling together can lead to two problems for intraspecific communication. First, the calls of other species constitute noise that masks the vocalizations of conspecifics or obscure features of the calls needed for species recognition. Second, if calls of different species are sufficiently similar, then females might choose males of the wrong species as mates (Gerhardt 1994a, c). The problem of reproductive isolation in mixed-species choruses is discussed in the next section. Here I focus on problems of acoustic interference (see also Gerhardt and Schwartz 1995; Gerhardt and Huber 2002).

Littlejohn (1977) listed several ways in which acoustic interference between species could be minimized. One is spectral stratification, with each species using a different frequency band for communication. Another is spatial separation, either through aggregation of males into species-specific assemblages, or through use of different microhabitats for calling. A third is partitioning of calling time. Mistakes in species identification could be minimized by use of different coding patterns in advertisement calls, but this would have little effect on acoustic interference.

#### Spectral Separation

There are a number of examples in the literature of sympatric species using distinctly different frequency bands for broadcasting calls. For example, several species of *Eleutherodactylus* in Puerto Rico have calls with different dominant frequencies, so overlap between species is slight (Drewry and Rand 1983). Similar spectral stratification has been reported in anuran communities from South America, Australia, Asia, and Africa (Duellman 1967a, 1978; Dubois 1977a; Hödl 1977; Elzen and Kruelen 1979; Humphries 1979; Schlüter 1979, 1980a, b, c, 1981; Telford 1982; Garcia-Rutledge and Narins 2001). One interpretation of this pattern is that dominant frequencies of sympatric species have diverged, thereby enhancing the ability of each species to communicate in a noisy environment (Littlejohn 1977; W. Smith 1977; Drewry and Rand 1983). According to this view, each species has a unique "acoustic niche," the dimensions of the niche being shaped by acoustic competition among species (Duellman 1978; Duellman and Pyles 1983; Garcia-Rutledge and Narins 2001).

There are several problems with this interpretation. First, the argument implies that assemblages of breeding frogs are stable enough to allow selection to change frequency characteristics of calls. In fact, the number of species occurring together often differs from one site to another, even within a restricted geographic area (Crump 1974; Duellman 1990; Duellman and Thomas 1996; Arzabe, de Carvalho, and Goes Costa 1998), so it is not clear that any given assemblage represents a co-adapted community. Second, interspecific differences in dominant frequency are correlated with differences in body size, and many selective forces affect the evolution of body size. Third, there are many species in these assemblages that exhibit broad frequency overlap (Dubois 1977a; Hödl 1977; Humphries 1979; Schlüter 1979, 1980a, b, c, 1981; Telford 1982; Zelick and Narins 1982; Schwartz and Wells 1983a, b, 1984a; Garcia-Rutledge and Narins 2001). Finally, many species of frogs have broad auditory tuning curves (see discussion that follows) and are sensitive to sounds outside the frequency range of their own calls. Therefore, spectral separation of calls does not necessarily ensure a noise-free communication channel (Narins and Zelick 1988; Garcia-Rutledge and Narins 2001).

Most studies of spectral separation or "acoustic niche partitioning" in multispecies assemblages have not demonstrated statistically that spectral stratification occurs more frequently than expected by chance. Chek, Bogart, and Lougheed (2003) reanalyzed data from a number of these studies, along with previously unpublished data on Amazonian frog assemblages. They found statistical evidence for greater than expected call separation in only three of 11 assemblages studied. In most cases, separation was most apparent in dominant frequency, which probably is related to body size differences among species, but there was some evidence for separation by pulse rate in two assemblages. The evolutionary processes that have resulted in acoustic separation in these assemblages are not known. Overall, the analysis of Chek, Bogart, and Lougheed (2003) provides only weak support for the concept of acoustic niche partitioning in frogs.

### Spatial Separation

Formation of species-specific aggregations within mixed-species choruses is frequently mentioned as a way of improving the efficiency of acoustic communication by polarizing the acoustic environment (Mecham 1961; Dubois 1977a; Hödl 1977; Littlejohn 1977; Passmore 1978; Telford 1982; Passmore and Carruthers 1995). Unfortunately, rigorous statistical analyses of spatial distributions of calling males are hard to find. Some clustering by species results from distinct preferences for calling sites in different microhabitats, on different types of plants, or at different heights in vegetation, but different species usually are within hearing range of each other. Often they are distributed in a mosaic, with heterospecific males calling only a few centimeters from one another (Fouquette 1960; Duellman 1967a; Hödl 1977; Littlejohn 1977; Passmore 1978; Bowker and Bowker 1979; Mac Nally 1979; Telford 1982; Schwartz and Wells 1983a, b, 1984a; Etges 1987; Garcia-Rutledge and Narins 2001).

Drewry and Rand (1983) reported that two Puerto Rican species, *Eleutherodactylus locustus* and *E. brittoni*, tended to use similar calling microhabitats where each species was found alone, but used different microhabitats when found in the same place. Unfortunately, they did not present a quantitative analysis of calling sites. Ptacek (1992) reported that the two species of North American gray treefrogs, *Hyla versicolor* and *H. chrysoscelis*, used a similar range of calling sites when each species was found alone in a pond. In ponds where both occurred, there was a tendency for males of *H. versicolor* to be found at more elevated sites and *H. chrysoscelis* closer to the ground or water. Even with these changes in spatial distribution, the calling sites of the two species overlapped broadly when they occurred in the same pond, and males sometimes respond to the calls of heterospecific males in mixed choruses. Males of *Hyla cinerea* were somewhat more likely to call from elevated perches in ponds where *H. gratiosa* was present than in ponds where it was not; *H. gratiosa* generally calls from the water surface (Höbel and Gerhardt 2003). In a study of two territorial ranids occupying the same pond, Given (1990) reported that choice of territories by males of the larger species, *Rana clamitans*, was not influenced by the presence of the smaller species, *R. virgatipes*, but there was some tendency for males of the latter species to avoid calling close to male green frogs. This behavior probably was related more to defense of oviposition sites than to acoustic interference. Acoustic interaction between the two species was rare.

### Temporal Separation

When there is little spatial separation between species and call frequencies overlap, there can be considerable acoustic interference between species that can affect a male's ability to attract mates. The vocal repertoires of three sympatric Panamanian tree frogs, *Hyla ebraccata, H. microcephala,* and *H. phlebodes*, are very similar in both structure and function, and there is broad frequency overlap in their calls. Choruses of both *H. microcephala* and *H. phlebodes* can inhibit calling by *Hyla ebraccata* (Schwartz and Wells 1983a, b). Both species call in distinct bouts, and *H. ebraccata* calls mainly in the silent periods between bursts of heterospecific calls (fig. 7.19). Experimental playbacks of both a recorded chorus of *H. microcephala* calls and filtered noise of the appropriate frequency showed that *H. ebraccata* males were more likely to give aggressive calls and multinote calls during silent periods than during playbacks (fig. 7.20). Since

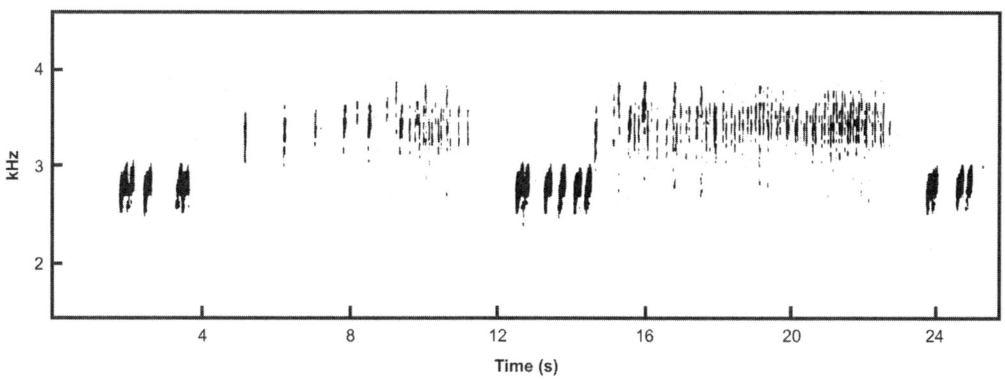


**Fig. 7.19.** Interaction between a chorus of *Hyla phlebodes* and several males of *H. ebraccata*. The calls of higher frequency are those of three or four *H. phlebodes* males calling in a group. The calls of lower frequency are those of several *H. ebraccata* males calling nearby. Note the lack of overlap in calls of the two species. From a recording by Kentwood D. Wells and Joshua J. Schwartz.

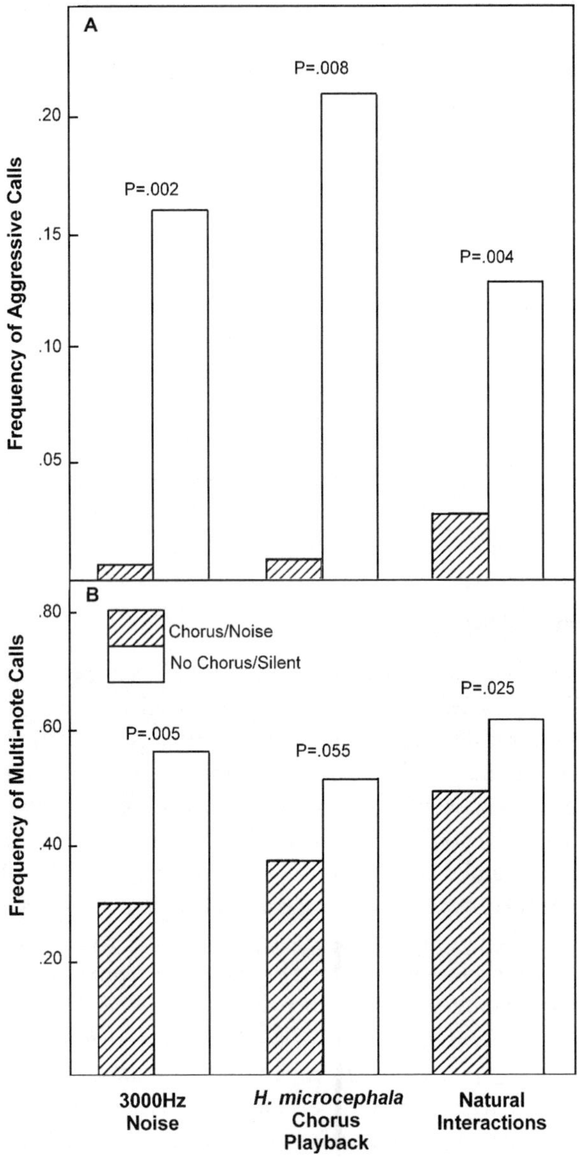


**Fig. 7.20.** Effect of acoustic interference on the calling of *Hyla ebraccata.* (A) Frequency of aggressive calls given by *H. ebraccata* during silent periods and during natural choruses of *H. microcephala,* during playbacks of a *H. microcephala* chorus, and during playbacks of filtered noise centered on the dominant frequency of the calls of *H. ebraccata.* (B) Frequency of multinote calls given by males of *H. ebraccata* in the same situations. The data are combined for all playback intensities between 80 and 104 dB SPL. Both aggressive calls and multinote calls are given in response to other males in the chorus, so these results show that background noise masks the calls of neighboring males. Plotted from data in Schwartz and Wells (1983a).

both types of calls are given in response to other males in the chorus, these experiments showed that background noise masks the calls of neighboring males.

All three species also give synchronized responses to individual heterospecific calls (Schwartz and Wells 1984a, b, 1985). Males of *H. microcephala* sometimes alternate individual call notes with those of *H. phlebodes;* when their calls are interrupted, they increase the interval between notes to avoid note overlap, just as they do in response to conspecific calls. *Hyla ebraccata* males are most responsive to sounds with a dominant frequency and duration similar to that of their own calls, although there was some response calls of different frequency and duration as well. Playback experiments with females of *H. ebraccata* showed that a chorus of *H. microcephala* calls makes it more difficult for females to orient to conspecific calls (Schwartz and Wells 1983b). In addition, females prefer conspecific calls free of interference to those overlapped by individual calls of *H. microcephala* (Schwartz and Wells 1984a). Ehret and Gerhardt (1980) and Gerhardt and Klump (1988a) obtained similar results by using synthetic masking noise in choice experiments with females of *Hyla cinerea.*

There are several other examples of one species shifting the timing of its calls to avoid overlap with those of other species, or males being completely inhibited by chorusing of heterospecifics. Males of *H. ebraccata* sometimes respond to the calls of *Scinax boulengeri,* which completely overlap the frequency range of their calls (personal observations). Males of *Eleutherodactylus coqui* are responsive to a range of frequencies outside those of their own calls and sometimes engage in vocal interactions with males of *E. portoricensis* (Zelick and Narins 1982). The Australian frog *Pseudophryne semimarmorata* is inhibited from calling by the much longer calls of *Geocrinia victoriana* (Littlejohn and Martin 1969). Similarly, the long calls of *Crinina parinsignifera* cause *C. signifera* males to stop calling and move away. The shorter calls of *C. signifera* have a less dramatic effect on *C. parinsignifera;* the latter shift the timing of their calls to avoid overlap with those of *C. signifera,* but they do not reduce overall calling rates (Mac Nally 1982; Littlejohn, Harrison, and Mac Nally 1985). On the other hand, Odendaal, Bull, and Telford (1986) found that *C. signifera* calls were much louder than those of sympatric *C. riparia,* and the former species often calls in dense choruses that produce continuous high background noise. They speculated that the noise generated by *C. signifera* prevents *C. riparia* from calling in the same habitats.

Similar examples of interspecific acoustic interference have been reported in birds (Cody and Brown 1969; Wickler 1972; Ficken, Ficken, and Hailman 1974; Bremond 1978; Popp, Ficken, and Reinartz 1985; Popp and Ficken 1987) and insects (Samways and Broughton 1976; Samways 1977a, b; Latimer 1981; Latimer and Broughton 1984; Bailey and Morris 1986; Greenfield 1988, 1990, 1993, 1997; Schatral 1990; Schatral and Yeoh 1990). There is even a report that calling by a diurnal tropical frog, *Dendrobates pumilio,* is inhibited by the loud sounds of cicadas (Páez, Bock, and

Rand 1993). In both birds and insects, the tendency to shift calls or songs in response to those of other species depends on the similarity of the heterospecific signals, especially the degree of frequency overlap. Furthermore, in orthopterans, males will alternate calls with males of other species only if they respond in this way to conspecific calls (Latimer and Broughton 1984). Hence, interactions between species probably are a by-product of intraspecific interactions. Consequently, discontinuous singers, which produce repeated chirps, are inhibited by continuous singers, which produce trills; only the former are "programmed" to produce their songs after those of a neighbor have ended (Greenfield 1990; Schatral 1990).

### Interspecific Use of Aggressive Calls

Closely related species often have very similar aggressive calls, and in some cases these are used in interspecific communication. Defense of calling sites against heterospecific males has been reported in leptodactylids (Drewry 1970), myobatrachids (Pengilley 1971a; Mac Nally 1982, 1984a; Littlejohn, Harrison, and Mac Nally 1985) dendrobatids (Wells 1980a), and hylids (Paillette 1970b; Humphries 1979; Cardoso 1981b; Schwartz and Wells 1984a, b, 1985). Males of some species give aggressive calls in response to playbacks of heterospecific advertisement or aggressive calls (Gambs and Littlejohn 1979; Humphries 1979; Brzoska 1982; Mac Nally 1982; Littlejohn, Harrison, and Mac Nally 1985; Gerhardt 2001).

The three Panamanian tree frogs discussed earlier, *Hyla ebraccata, H. microcephala,* and *H. phlebodes* (fig. 7.21 top), have very similar vocal repertoires. In particular, the long aggressive calls of these species are much more similar to each other than are their advertisement calls (Schwartz and Wells 1984a). The similarity in the structure of aggressive calls in these three species (fig. 7.21 bottom) could result from descent from a common ancestor. *Hyla microcephala* and *H. phlebodes* are considered to be members of the *H. microcephala* species group, whereas *H. ebraccata* is a member of the *H. leucophyllata* species group. Cladistic analysis suggests that these species groups are relatively closely related (Duellman and Trueb 1983; Kaplan 1994; Duellman 2001). In playback experiments, males of *H. ebraccata* gave more aggressive calls in response to aggressive calls of *H. microcephala* and *H. phlebodes* than to advertisement calls of these species, although they were most responsive to conspecific aggressive calls (fig. 7.22). This indicates that males of *H. ebraccata* recognize the calls of all three species as aggressive calls, and indeed, males have been observed engaging in interspecific fights over calling sites (Schwartz and Wells 1984a). The high pulse repetition rate that characterizes the aggressive calls of all three species is the most important feature in eliciting aggressive responses from males (Wells and Schwartz 1984b).

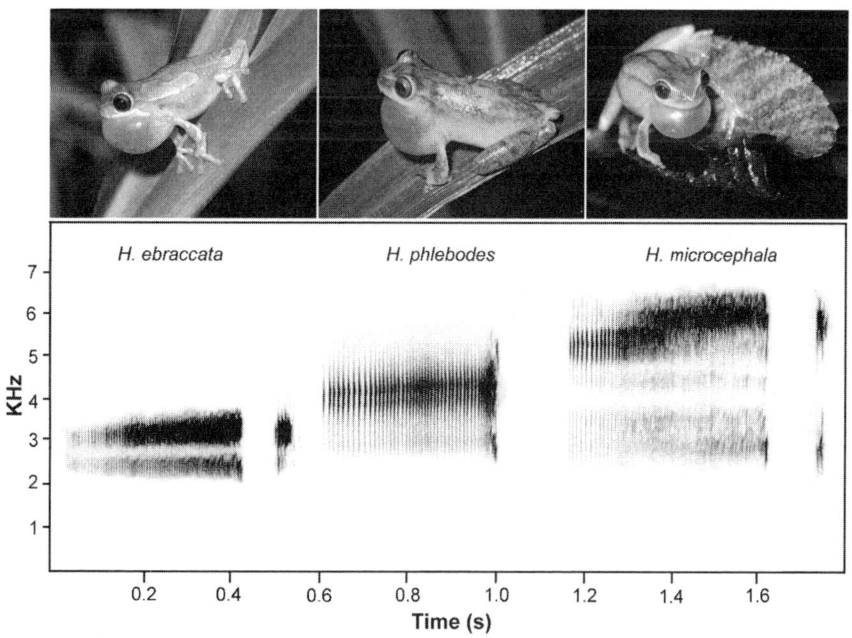


**Fig. 7.21.** Top: Calling males of three sympatric hylid frogs from Panama. *Hyla ebraccata* (left), *H. phlebodes* (center), and *H. microcephala* (right). Photos by Kentwood D. Wells. Bottom: Sound spectrograms of long aggressive calls of *H. ebraccata* (left), *H. phlebodes* (center), and *H. microcephala* (right). Note overlap in both frequency and pulse rate among species. From recordings by the author.

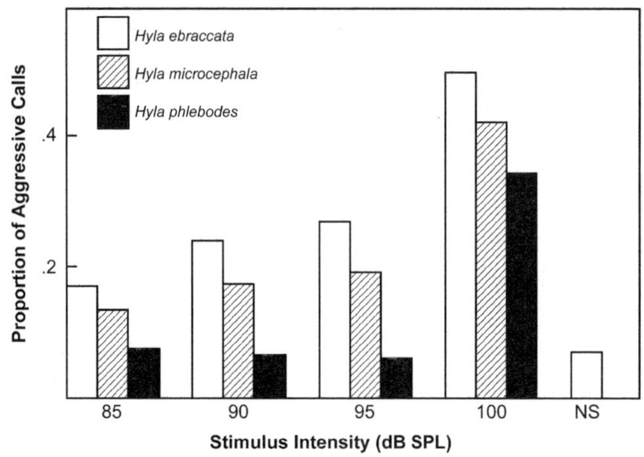


**Fig. 7.22.** Proportion of aggressive calls given by males of *Hyla ebraccata* to long aggressive calls of *H. ebraccata, H. microcephala,* and *H. phlebodes* at four different playback intensities. NS = no stimulus (silent) periods. Males of *H. ebraccata* were most responsive to conspecific calls and least responsive to those of *H. phlebodes*. After Schwartz & Wells (1984a).

## Responses of Females to Conspecific and Heterospecific Calls

A major focus of the early work on anuran vocalizations was the role of calls in maintaining reproductive isolation between species, and this remains an active area of research today (see reviews by Gerhardt 1982, 1988, 1994 a, c; Gerhardt and Schwartz 1995; Gerhardt and Huber 2002). Traditionally, species recognition has been treated as a process distinct from sexual selection. Some authors, however, have treated species recognition and reproductive isolation as incidental effects of sexual selection (Thornhill and Alcock 1983; West Eberhard 1983). They argue that communication signals of two related species diverge in allopatry because of female preferences for certain types of calls. The species-specific nature of the resulting signals would be an incidental by-product of this preference, but the species would be reproductively isolated if they become sympatric. Others, most notably Paterson (1978, 1980, 1981, 1982, 1985), have argued that each species has a set of signals that constitute a Specific Mate Recognition System. Interspecific differences in call structure have been attributed to independent adaptation of signal structure to different environments, with sexual selection and interactions among species playing a relatively minor role (Passmore 1981).

From the point of view of a female frog trying to find a suitable mate, intraspecific mate choice (sexual selection) and species recognition are part of the same process of signal recognition (M. Ryan and Rand 1993b; Gerhardt 1994a, c; Gerhardt and Schwartz 1995). When a female enters a mixed-species chorus, she is confronted by an array of males calling to attract mates. Some of these are conspecifics, but many are not. The female's sensory system is selectively tuned to a certain range of values of each call feature, and those calls that provide the greatest sensory stimulation are most likely to elicit phonotaxis. Often the types of features that distinguish closely related species from one another, particularly differences in frequency and fine-temporal features such as pulse repetition rate, tend to be subject to stabilizing selection. That is, females generally exhibit preferences for average values of these traits over extreme values. These usually are relatively static call features that exhibit little variation within and among conspecifics in a chorus (Gerhardt 1991, 1994a, c; Castellano and Giacoma 1998; Giacoma and Castellano 2001). In contrast, as discussed in chapter 8, females often exhibit directional preferences for gross temporal features such as call intensity, calling rate, call duration, or call complexity. All of these features are quite variable within and among individuals and are subject to sexual selection. The net result of females exhibiting preferences for average values of some call traits and for extreme values of others is that they usually pick males of the same species. Within that subset of males in a mixed chorus, they often prefer individuals that invest a considerable effort in call production.

Under some circumstances, heterospecific calls, or certain features of heterospecific calls, can elicit phonotaxis by females (Gerhardt 1994a). When females have been presented only with heterospecific calls, they sometimes exhibit phonotaxis toward those calls (Martof 1961; Gerhardt 1974a, 1981b, 1994a; Gerhardt and Doherty 1988; Backwell and Jennions 1993; M. Ryan and Rand 1993b; Gerhardt and Schwartz 1995; Gerhardt 2001). Even filtered noise of the appropriate frequency can elicit phonotaxis in no-choice experiments (Ehret and Gerhardt 1980).

In a natural chorus, females might orient toward heterospecific calls if conspecific males are scarce. When many conspecifics are present, females can be expected to exhibit nearly perfect preferences for conspecific calls. Females are most likely to be attracted to heterospecific calls when those calls closely resemble conspecific calls in one or more key features. Calls that are less similar to conspecific calls are less likely to reach a threshold for sensory stimulation required to elicit female phonotaxis. It is possible that some heterospecific calls are so different from conspecific calls that they are simply ignored by females. Females also might actively avoid heterospecific calls although recent experimental studies have shown that females usually do not exhibit this type of avoidance behavior (Gerhardt, Dyson, Tanner, and Murphy 1994). There also have been experiments showing that females sometimes are more attracted to a conspecific call to which an element of a heterospecific call has been added than they are to normal conspecific calls (M. Ryan and Rand 1993a, c); this is discussed in more detail in the following.

**Evidence for Species Recognition**

Many of the pioneers of anuran acoustic research devoted much of their effort to documenting differences in call structure among closely related species and showing that these differences constitute important barriers to gene exchange between species (A. P. Blair 1941a, b, 1942; W. Blair 1955, 1956a, 1958a, b, d, 1964; Littlejohn 1959; Bogert 1960; Martof 1961; Mecham 1961). There have been many descriptive and statistical comparisons of the calls of sympatric anuran species (W. Blair 1958a, b, d; Bogert 1960; Littlejohn 1959, 1965; Fouquette 1960; Littlejohn and Martin 1965a; Duellman 1967a, 1978; Günther 1969; Channing 1976b; Hödl 1977; Passmore 1978; Schluter 1979, 1980a, b, c, 1981; Telford 1982; Drewry and Rand 1983; Duellman and Pyles 1983; Schwartz and Wells 1984a; Wollerman and Wiley 2002b). All of these have shown that sympatric species never have identical advertisement calls. In most cases, the calls of two species can be distinguished by more than one spectral or temporal feature.

There are many examples of cryptic or morphologically similar species of anurans that can be distinguished mainly by differences in their calls, or were first identified as separate species by differences in calls. These include several members of the *Rana pipiens* complex in North America (Hillis 1988), two treefrogs from Puerto Rico, *Eleutherodactylus coqui* and *E. portoricensis* (Thomas 1966), sympatric morphs of *Polypedates leucomystax* in Malaysia (Narins et al. 1998), morphologically similar species of *Breviceps* and *Hyperolius* in Africa (Minter 1997; Channing, Moyer, and Burger 2002), and many others. In some cases, groups of morphologically similar species are complexes of diploid and polyploid species, with the polyploids presumably being derived from the diploids. The latter include *Hyla chrysoscelis* (probably consisting of several separate diploid species) and *H. versicolor* (tetraploid) in North America (Ptacek, Gerhardt, and Sage 1994; Keller and Gerhardt 2001), diploid and tetraploid populations of green toads (*Bufo viridis*) in Europe (Castellano et al. 1998; Castellano, Giacoma, and Dujsebayeva 2000; Stöck 1998), species of *Xenopus* in Africa (Tymowska 1991), species of *Neobatrachus* in Australia (Roberts 1997), and species of *Phyllomedusa* in South America (Haddad, Pombal, and Batistic 1994).

When Charles M. Bogert published his 1960 review, the evidence that anuran calls were important for species recognition was largely circumstantial and based on qualitative comparisons of calls. The only attempt to test the hypothesis had been an experiment in which female *Pseudacris streckeri* were given a choice of conspecific calls and those of *P. clarki* (Littlejohn and Michaud 1959). The experiment showed conclusively that females were attracted only to the calls of their own species. Since then, many similar experiments have been conducted in both the field and laboratory (table 7.2). Some have suffered from a lack of statistical independence because multiple responses of the same females were included in the analysis. Whenever possible, I have reanalyzed the data to compare numbers of females choosing conspecific and heterospecific calls. The few cases in which females did not clearly discriminate between conspecific and heterospecific calls (Martof 1961; Awbrey 1968) probably failed because the test females were not yet ready to mate or were simply unresponsive to the test stimuli. Most investigators have tested females collected while already in amplexus to ensure a high degree of receptivity. Females collected in amplexus do not appear to be any less discriminating than those collected while approaching calling males (C. Murphy and Gerhardt 1996; Witte et al. 2000). In at least one species, gravid females that actually are ovulating are more receptive than those that are not (Lea, Halliday, and Dyson 2000). Gerhardt (1992a) and Gerhardt and Huber (2002) provide thorough reviews of techniques used in playback experiments.

**Identification of Features Used in Species Recognition**

Bogert (1960) cautioned against assuming that those features of calls that seem most distinct to human listeners are necessarily those most important to female frogs. Even when there are significant differences in various temporal features of calls, one cannot assume that all differences are equally important in species discrimination. Similarly, one cannot assume that calls differing in only one or two features are less likely to be distinguished by females than calls differing in many features. Duellman and Pyles (1983) performed a principal component analysis that combined seven spectral and temporal features of the calls of several Neotropical hylid species; they concluded that calls of related species are most similar when the species are not sympatric. Without knowing which features females use for call recognition, however, one cannot make any assumptions about their importance in preventing cross-species matings.

The best way to identify features used for call recognition is to alter calls experimentally, or use synthetic calls to systematically vary spectral and temporal features (Gerhardt 1988, 1992a, 2001; Gerhardt and Huber 2002). In an early experiment, Martof and Thompson (1964) found that major changes in call repetition rate, call intensity, and the temporal structure of call notes had relatively little effect on choice of calls by female *Pseudacris triseriata*. Females did discriminate against calls that deviated significantly from the average dominant frequency of conspecific calls, and they preferred a natural call to white noise. Unfortunately, Martof and Thompson did not use females captured in amplexus, so some of their results are questionable. More recent studies generally have found that females are relatively tolerant of small deviations in frequency or temporal features of calls, but major deviations from typical features of conspecific calls make synthetic calls less attractive to females (e.g., Kime et al. 1998; Witte, Ryan, and Wilczynski 2001; Bosch 2002b).

Loftus-Hills and Littlejohn (1971) were the first to use synthetic calls to investigate the basis of call discrimination in a pair of sympatric species. They presented females of two Australian frogs, *Litoria ewingii* and *L. verreauxii,* with calls differing in pulse repetition rate, but identical in all other temporal properties and dominant frequency. The difference in pulse rate was sufficient for effective discrimination by females of both species, even though this is not the only difference in the natural calls of these species (Littlejohn 1965). In a more detailed study, Straughan (1975) presented female *Pseudacris regilla* and *P. cadaverina* with synthetic calls differing in frequency, duration, pulse duration, and pulse repetition rate, all of which differ in natural calls of the two species. Only pulse repetition rate was important for species discrimination. Additional experiments with synthetic calls have established the importance of pulse repetition rate for species recognition in many species, including *Bufo woodhousii* (Sullivan and Leek 1987) and *Hyla microcephala* (Schwartz 1987b). In the latter species, differences in pulse repetition rate also are sufficient for females to distinguish advertisement calls from conspecific aggressive calls, a result obtained in experiments with *H. ebraccata* as well (Wells and Bard 1987).

### Species Recognition in the *Hyla versicolor* Group

Carl Gerhardt and his colleagues have conducted detailed studies of call recognition in *Hyla versicolor* and *H. chrysoscelis* (Gerhardt 1978c, 1982, 1988; Gerhardt and Doherty 1988; Diekamp and Gerhardt 1995; Gerhardt, Dyson, and Tanner 1996; Gerhardt and Schul 1999; Schwartz, Buchanan, and Gerhardt 2001; Bush, Gerhardt, and Schul 2002; Schul and Bush 2002; see Gerhardt 2001 and Gerhardt and Huber 2002 for reviews). This system is an interesting one in which to study the evolution of call preferences, because the evolutionary history of the two species is well understood. *Hyla versicolor* is a tetraploid species derived from the diploid *H. chrysoscelis,* apparently more than once (Ptacek, Gerhardt, and Sage 1994).

At the same temperature, the calls of sympatric populations of these species have similar dominant frequencies and duration, but differ in pulse shape, pulse duration, and pulse repetition rate. The slower pulse rate of *H. versiolor* is a direct effect of polyploidy, as shown by the lower pulse rates of triploid and tetraploid males artificially produced in the laboratory (Keller and Gerhardt 2001). Differences in pulse rate alone are sufficient for females of each species to discriminate conspecific from heterospecific calls. Females also can use differences in pulse shape and pulse rise time to discriminate between conspecific and heterospecific calls, but these differences are of relativley minor importance compared to the large differences in pulse rate that females normally would encounter in the field. The responses of femaled are temperature-dependent. Pulse rate changes with temperature, and a female prefers calls with a pulse rate corresponding to that of conspecific calls at her own body temperature (Gerhardt 1978c). If the pulse-rate selectivity of females were not adjusted to changing temperature, then they might choose males of the wrong species, because the pulse rate of a warm *H. versicolor* is similar to that of a cool *H. chrysoscelis.*

The two species use somewhat different mechanisms to discern differences in pulse rate, with *H. chrysoscelis* using absolute pulse rate, whereas *H. versicolor* uses a combination of pulse duration and interpulse interval (Schul and Bush 2002). There also are differences in the responses of females of the two species to call frequency structure. Calls of both species have two distinct frequency bands, and females of both species prefer these calls to those lacking either the high or low frequency band (Gerhardt and Doherty 1988 Gerhardt 2001). Females of *H. versicolor* prefer calls with only the higher frequency present to those with only the lower frequency, but this is not true for *H. chrysoscelis* (Gerhardt, 2001). Hence, despite the fact that populations of the tetraploid *H. versicolor* clearly are derived from populations of diploid *H. chrysoscelis,* there has been evolutionary change in the mechanisms of call recognition in the derived species.

### Species Recognition in the *Hyla cinerea* Group

Among the most detailed work on call recognition is the work of Gerhardt and others on *Hyla cinerea* (Gerhardt 1974b, 1976, 1978a, b, 1981b, 1982, 1987, 1988; Gerhardt and Mudry 1980; Megela Simmons 1988; Megela Simmons, Buxbaum, and Mirim 1993; Gerhardt, Allan, and Schwartz 1990; Allan and Megela Simmons 1994) and *H. gratiosa* (Gerhardt 1981a; Bodnar 1996; see Gerhart 2001 and Gerhardt and Huber 2002 for reviews). These species often breed at the same time and place as several other hylids of similar size (fig. 7.23), and *H. cinerea* and *H. gratiosa* are capable of producing viable and fertile hybrids. These species have similar calls, although each one differs from the others in both frequency and temporal structure (fig. 7.24). Gerhardt varied both spectral and temporal properties of calls and tested the responses of females to "standard" and altered synthetic signals. This allowed him to establish which features are important for call recognition and to determine how much each feature can be changed without reducing the effectiveness of the signal. The results can then be related

**Fig. 7.23.** Four mophologically similar treefrogs that sometimes occur together in breeding ponds in the southeastern United States. (A) *Hyla cinerea*. (B) *H. gratiosa*. (C) *H. squirella*. (D) *H. andersonii*. Photos by H. Carl Gerhardt (A, C) and Kentwood D. Wells (B, D).

to differences in the calls of sympatric species, as well as to differences between calls within the species' repertoire.

The advertisement calls of *H. cinerea* have a waveform periodicity of about 300/s. Females prefer standard calls to those with very high (900/s) and very low (below 100/s) periodicity, but do not discriminate between 300 and 450/s (Gerhardt 1978b). The calls of one sympatric species, *H. squirella*, overlap those of *H. cinerea* in frequency, but have a waveform periodicity of around 100–130/s; the difference in this fine temporal property apparently is sufficient for species discrimination. The calls of two other species, *H. andersonii* and *H. gratiosa*, have a periodicity of about 300 and 450/s respectively, so differences in this property alone would not allow *H. cinerea* females to distinguish the calls of these species from those of conspecifics.

The advertisement call of *H. cinerea* has a bimodal spectrum, with one energy peak at around 900 Hz and a second between 2,700 and 3,300 Hz. At low intensities, females do not discriminate between calls with both frequency peaks present and calls with only the 900 Hz component, but at higher intensities, they discriminate against calls with the high frequency component attenuated or missing (Gerhardt 1981b). Even at very low playback intensities, females prefer standard calls to those with the low frequency component attenuated by 12 dB (Gerhardt 1976, 1981b). Hence, the presence of both high and low frequency components is important for species recognition at close range, but low frequency sound is sufficient to attract females to a chorus from some distance away. When the low frequency peak of the call is varied, females prefer calls of average frequency (about 900 Hz) to those with lower (600–700 Hz) or higher (1100–1200 Hz) frequency peaks (Gerhardt 1982), except at very low temperatures (Gerhardt and Mudry 1980). If intermediate frequencies are added to stimulus calls, they be-

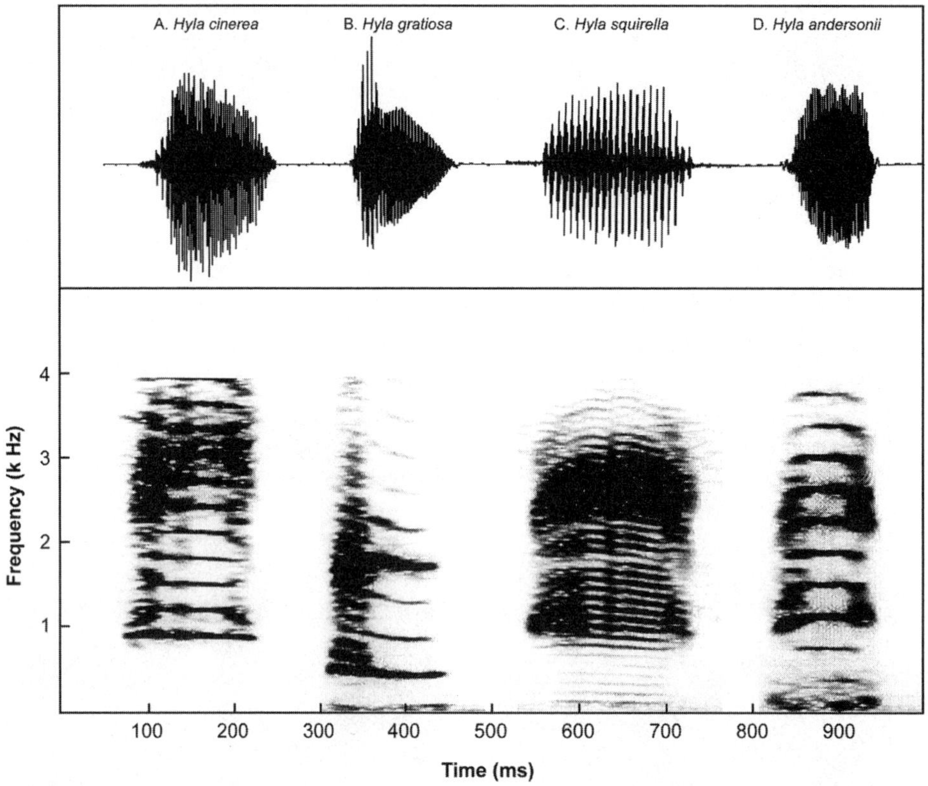


**Fig. 7.24.** Spectral and temporal features of the calls of four sympatric species of treefrogs. Sound spectrograms and sections showing relative amplitude at different frequencies are on the left; oscillograms showing temporal features are on the right. (A) *Hyla cinerea.* (B) *H. gratiosa.* (C) *H. squirella.* (D) *H. andersonii.* Sonagrams and oscillograms courtesy of H. Carl Gerhardt.

come less attractive to females (Gerhardt and Höbel 2005). These preferences are related to the tuning of the peripheral auditory system and have the effect of excluding calls falling within the frequency range of sympatric species (Gerhardt 1987; Gerhardt and Höbel 2005).

Experiments with *H. gratiosa* produced results qualitatively similar to those for *H. cinerea* (Gerhardt 1981a). Again, fine temporal properties alone are not sufficient to allow females to discriminate between conspecific calls and those of *H. cinerea.* The presence of a pulsed component at the beginning of the call is important to elicit phonotaxis in *H. gratiosa,* but not in *H. cinerea.* The presence of two spectral peaks also is important, but the optimum frequencies in these peaks differ from those of *H. cinerea.* Addition of components between the two spectral peaks in synthetic *H. gratiosa* calls has no effect on the attractiveness of the calls, whereas the attractiveness of *H. cinerea* calls is reduced when extra components are added. In fact, the optimum high frequency peak for *H. gratiosa* falls in the range of frequencies that reduces the attractiveness of *H. cinerea* calls when added to synthetic signals (Gerhardt 1974b). Alterations in the frequency and harmonic structure of calls can affect call recognition in this species as well, but their effect depends on how different call characteristics are combined (Bodnar 1996).

A number of general points can be made from the work on these two species. First, some species have very distinctive features in their calls that are irrelevant for species recognition, such as the pulsed component at the beginning of the call in *H. cinerea.* This demonstrates the need for careful experimental studies to confirm the functions of different call components. Second, there are differences in the way species-specific information is encoded in calls, even between closely related species. Hence one cannot generalize from one species to another and assume that certain spectral or temporal properties are always important for call recognition. Third, females probably recognize calls of their own species by responding to several different call characteristics in combination, and the way the characteristics are combined can affect the responses elicited. Finally, different properties can be responsible for the lack of phonotactic responses by females to different heterospecific calls. For example, the properties that lead females of *H. cinerea* to prefer their own species' calls over those of *H. squirella* are not the same as those that elicit greater phonotaxis to conspecific calls than to calls of *H. andersonii* or *H. gratiosa.*

### Species Recognition in the Genus *Physalaemus*

Another anuran that has become a model system for studies of call recognition and the evolution of vocal signals is the leptodactylid frog *Physalaemus pustulosus* (M. Ryan 1985b; Wilczynski, Rand, and Ryan 1995; M. Ryan and Rand 2001). As discussed previously, this species has a call consisting of a whine note that decreases in frequency, sometimes followed by one or more noisy chuck notes (fig. 7.25). The whine alone is both necessary and sufficient to elicit phonotaxis by females, but females prefer calls with one or more chucks (Rand and Ryan 1981). The position of the chuck makes little difference to females; calls in which the chuck preceded the whine were as attractive as normal calls in which the chuck followed the whine (M. Ryan 1983a). In fact, experiments with synthetic calls showed that the chuck could be placed in any number of positions before or after the whine and still be as attractive as the normal whine-chuck call (Wilczynski, Rand, and Ryan 1999). If the frequency sweep of the whine was reversed, however, the call was less attractive to females. Frequency-modulated whines also were more attractive than were constant-frequency tones with frequencies encompassed by the whine. The direction of frequency modulation also affects the responses of males to calls; males gave relatively few responses to reversed whines over a wide range of playback intensities (G. Rose, Zelick, and Rand 1988; G. Rose and Rand 1991).

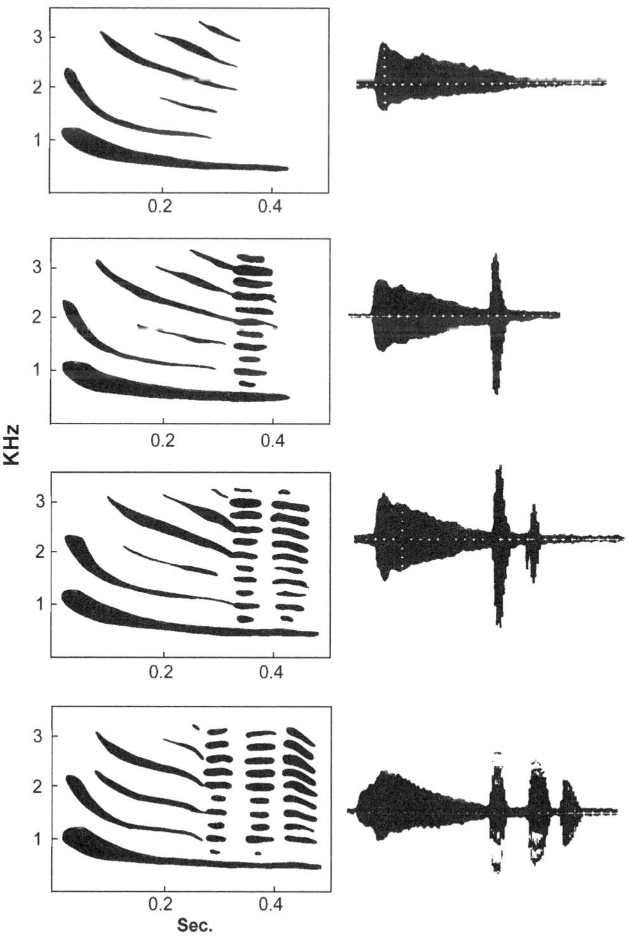


**Fig. 7.25.** Sound spectrograms (left) and oscillograms (right) of the whine advertisement calls of *Physalaemus pustulosus,* with increasing numbers of secondary chuck notes (top to bottom). After Ryan and Drewes (1990).

Other experiments showed that females responded strongly to calls that included the fundamental frequency of the whine (that is, the frequency determined by vibrations of the vocal cords), but not to calls containing only higher harmonics of the fundamental (Rand, Ryan, and Wilczynski 1992). Even a short segment of the whine was sufficient to elicit phonotaxis, although complete whines were more attractive than most partial whines. The chuck, which makes calls more attractive to females, can be altered in various ways without decreasing the responses of females. For example, females responded equally well to chucks with only the top or bottom half of the frequency range present, and they also responded to calls in which the chuck was replaced with a burst of white noise encompassing the same range of frequencies (Rand, Ryan, and Wilczynski 1992).

Tests of *Physalaemus pustulosus* females with calls of a number of other species in the genus demonstrated that they almost always chose conspecific whine notes over those of other species, despite considerable similarity in call structure within the genus (M. Ryan and Rand 1993b, 1995; Tárano and Ryan 2002). Furthermore, most heterospecific calls did not elicit phonotaxis even when paired with filtered noise, but a few heterospecific calls did elicit phonotaxis under these circumstances. Females also exhibited phonotaxis to synthetic calls arrayed along a gradient of intermediates between those of conspecifics and those of several other species, and the strength of their responses was related to how different the synthetic call was from a conspecific call (Ryan et al. 2003). This study and others (Kime et al. 1998) indicate that females will accept a relatively wide array of calls as long as their components are similar to those of conspecific calls. Nevertheless, when given a choice between such intermediates and conspecific calls, females consistently prefer the latter (M. Ryan and Rand 1995, 1999a,b, 2001). The evolution of call structure in this group has been unusually conservative, and many species have very similar whine calls (many of these are not sympatric with one another). This high degree of conservatism in call structure is paralleled by a high degree of similarity in the auditory tuning curves of the various species (Wilczynski, Rand, and Ryan 2001).

Members of the genus *Physalaemus* differ in the types of components added to the basic whine note. One species, *P. pustulatus,* has a pulsed component that precedes the whine note. Addition of this component to a *P. pustulosus* call rendered the call more attractive to females than a whine note

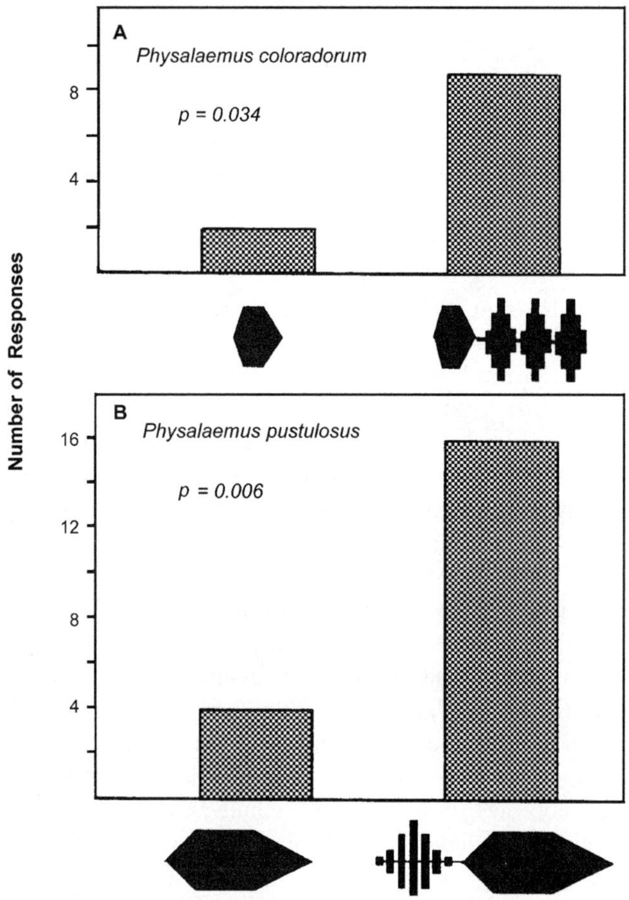


**Fig. 7.26.** Two examples of female responses to call elements from heterospecific calls in members of the *Physalaemus pustulosus* species group. (A) Responses of females of *P. coloradorum* to a conspecific call (left) and a conspecific call to which three chuck notes of *P. pustulosus* were added (right). (B) Responses of females of *P. pustulosus* to a conspecific call (left) and a conspecific call to which the amplitude-modulated prefix of another species, *P. pustulatus,* had been added (right). In both cases, females preferred the calls that had been made more complex by addition of heterospecific call elements. After Ryan and Rand (1993b).

alone, even though such calls are never given by this species (fig. 7.26 B). This suggests that the female sensory system has a pre-existing bias that would favor the evolution of this type of prefix if it were ever to arise by chance (M. Ryan and Rand 1993c). A similar response was obtained with females of another species, *P. coloradorum,* which does not produce chucks. Addition of a chuck from the call of *P. pustulosus* to the call of *P. coloradorum* made the latter more attractive to *P. coloradorum* females (fig. 7.26 A). In contrast, females of another species, *P. enesefae* did not exhibit the same type of pre-existing bias in favor of components of other species' calls when these were added to conspecific advertisement calls. They simply ignored these added components and were equally likely to choose modified and unmodified calls (Tárano and Ryan 2002). *Physalaemus enesefae* is rather distantly related to *P. pustulosus, P. pustulatus,* and *P. coloradorum,* all of which are in one clade within the genus. This suggests that the preference for complex calls has evolved within that clade, but not necessarily in the genus as a whole.

### Hybridization and Hybrid Calls

Many closely related species of anurans exhibit a high degree of genetic compatibility. Because reproductive isolation is not always complete, interspecific hybridization is a regular, although not common, occurrence in anuran populations. Hybridization sometimes occurs in scattered locations throughout the ranges of two broadly sympatric species when the species breed in the same habitats (A. Blair 1941b, 1942; Cory and Manion 1955; Thornton 1955; Wasserman 1957; Mecham 1960, 1965; Zweifel 1968a; L. Brown 1970; Forester 1973; Jones 1973; Gerhardt, Guttman, and Karlin 1980; Green 1983; Sattler 1985; Anderson and Moler 1986; Lamb and Avise 1986; Schlefer et al. 1986; Haddad, Cardoso, and Castanho 1990; Oliveira et al. 1991; Gerhardt 1994a; Gerhardt Ptacek, Barrett, and Torke, 1994; Gerhardt and Schwartz 1995; Gergus, Malmos, and Sullivan 1999; D. Green and Parent 2003; Vines et al. 2003). Usually heterospecific matings are relatively rare in such situations, and hybrid individuals constitute only a small proportion of the adult population.

Hybridization sometimes also occurs in narrow zones where the boundaries of two largely allopatric species come into contact (W. Blair 1955; Henrich 1968; L. Brown and Guttman 1970; Littlejohn, Watson, and Loftus-Hills 1971; A. Martin 1972; Littlejohn and Watson 1973, 1983; Littlejohn and Roberts 1975; McDonnell, Gartside, and Littlejohn 1978; Woodruff 1978; J. Feder 1979; Sage and Selander 1979; Gartside 1980; D. Green 1983; Sullivan 1986a, 1995; Szymura and Barton 1986; Hillis 1988; Sullivan and Lamb 1988; Barton and Hewitt 1989; Sanderson, Szymura, and Barton 1992; Littlejohn 1993, 1999; Szymura 1993; Gerhardt and Schwartz 1995; Nürnberger et al. 1995; MacCallum et al. 1998; Kruuk, Gilchrist, and Barton 1999). Usually these zones occur when two relatively closely related species that have been separated geographically come back into contact, but postmating isolation (genetic incompatibility) is incomplete. Such parapatric hybrid zones (Woodruff 1973) can be stable for long periods of time.

Even if hybrid individuals are competitively inferior to both parental species, a hybrid population can be maintained in a narrow zone of overlap through continual mismatings between species. When two species come into contact in an ecotone between two different habitats, hybrid individuals might actually be better adapted to the ecotone than either parental species, although this has never been demonstrated conclusively for anurans. In that case, a self-maintaining hybrid population can develop in a zone separating allopatric populations of the parental species (Littlejohn and Watson 1973; Woodruff 1973; Green 1983; Littlejohn 1993, 1999). Sometimes hybridization can be favored if hybrid offspring are competitively superior in particular environments. For example, in ephemeral desert pools in Arizona, females of *Spea bombifrons* that mate with males of *S. multiplicata* produce tadpoles that develop more rapidly than do pure *S. bombifrons* larvae, an advantage in a drying pond. In contrast, matings between female *S. multiplicata* and male *S. bombifrons* produce offspring with longer development times and lower survival than pure *S. multiplicata* offspring (K. Pfenning and Simovich 2002).

Some instances of hybridization between broadly sympatric species occur when species that were previously isolated by habitat differences have been brought into contact by man-made habitat disturbance (Cory and Manion 1955; Thornton 1955; Wasserman 1957; Mecham 1960; L. Brown 1970; Schlefer et al. 1986; Lamb 1987). This is not necessarily true for parapatric hybrid zones. In either case, the chances of hybridization are greatest if one or more of the following conditions are met: (1) The advertisement calls of the two species are similar to one another, thereby increasing the probability that a female will move toward a heterospecific male; (2) males of one or both species typically obtain mates by active searching (see chapter 8) and are therefore likely to seize heterospecific females by mistake; (3) males of one species greatly outnumber males of the other species, so females have some difficulty locating conspecific males.

In some instances, hybridization can lead to formation of polyploid species and unsual reproductive systems. The best-known example is in a complex of European water frogs, the *Rana esculenta* complex. Hybridization between *Rana lessonae* and *R. ridibunda* in many separate locations has produced the *R. esculenta* phenotype, which often is competitively superior to the parental species (Berger 1977; Graf and Polls- Pelaz 1989). Usually the *R. esculenta* genotype occurs with *R. lessonae*. The *lessonae* genome is discarded before meiosis when *R. esculenta* females form eggs, and the remaining *ridibunda* genome is then duplicated. Females produce haploid eggs of the *ridibunda* genotype that then combine with *R. lessonae* sperm to reconstitute the *R. esculenta* genome. In populations where *R. esculenta* is associated with *R. ridibunda,* the *lessonae* genome usually is maintained in the eggs. In addition, some *R. esculenta* females produce both haploid and diploid eggs, and subsequent matings with a parental species yield some triploid offspring. Matings between males and females of the *R. esculenta* genotype normally produce inviable offspring of the *R. ridibunda* genotype, but production of viable *R. ridibunda* genotypes occurs occasionally. These are exclusively female and normally mate with males of *R. lessonae,* thereby reconstituting the *R. esculenta* genotype (Vorburger 2001).

The widespread success of the *Rana esculenta* genotype appears to be due mainly to competitive superiority of the hybrid phenotype. Theoretically, hybrids should be at some disadvantage because they are dependent on one of the parent species for successful reproduction. Yet in many populations, *R. esculenta* individuals outnumber the parental species. *Rana esculenta* tadpoles benefit from having an "all purpose" genotype that can they tolerate a wide range of physical conditions in ponds (Tunner and Nopp 1979). They also are superior competitors to the tadpoles of the parental species with which they most often are associated, *Rana lessonae* (Semlitsch and Reyer 1992b; Semlitsch 1993; see also chapter 15). If the *R. esculenta* genotype were to completely outcompete *R. lessonae* in a pond, then *R. esculenta* eventually would disappear, because they cannot reproduce themselves. Both forms continue to persist, however, through a combination of mating preferences by females for *R. lessonae* males and patterns of migration between neighboring populations (Hellriegel and Reyer 2000; Som, Anholt, and Reyer 2000). In addition, the sperm of hybrid *R. esculenta* males are less successful at fertilizing eggs than are those of parental genotypes, so the fitness advantage of the *R. esculenta* genotype is reduced (Reyer, Niederer, and Hettyey 2003).

#### Calls of Hybrid Males

The calls of hybrid males generally are intermediate in both spectral and temporal characteristics to calls of the parental species, although some features are more similar to calls of one parent than to those of the other (W. Blair 1955, 1956b, 1958b; Bogert 1960; Littlejohn and Oldham 1968; Zweifel 1968a; Littlejohn, Watson, and Loftus-Hills 1971; A. Martin 1972; Forester 1973; Littlejohn and Watson 1973; Gerhardt 1974c; Littlejohn and Roberts 1975; Littlejohn 1976; Brzoska 1980, 1982; Doherty and Gerhardt 1983; Schneider and Joermann 1988; Scroggie and Littlejohn 2005). In the *Rana esculenta* complex, however, calls of the hybrid form are not intermediate between those of the parental species, but have frequency components not found in either parental call (Wycherley, Doran, and Beebee 2002). This probably reflects the unusual genetic system in these frogs.

Several playback experiments have demonstrated that calls of hybrid males are less attractive to females of the parental species than are calls of conspecific males (Awbrey 1965; Gerhardt 1974c; Kruse 1981b; Höbel and Gerhardt 2003). However, the results are not as clear-cut as are those for the species discrimination tests described earlier. Some females are readily attracted to calls of hybrid males in the absence of conspecific calls, and they often fail to discriminate between hybrid and conspecific calls that are very similar (Gerhardt 1974c; Littlejohn and Watson 1976). Hybrid

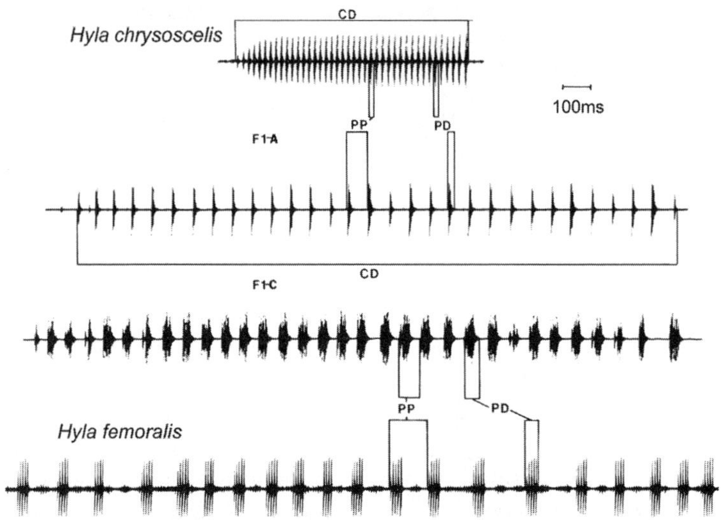


**Fig. 7.27.** Oscillograms of the advertisement calls of *Hyla chrysoscelis* (top), *H. femoralis* (bottom), and hybrids between the two species (middle). Hybrid calls like the one in the top middle figure were more similar in pulse structure to calls of *H. chrysoscelis* than to those of *H. femoralis*. Other calls, like that shown in the bottom middle figure, were more similar to *H. femoralis*. Abbreviations: CD = call duration; PD = pulse duration; PP = pulse period. After Doherty and Gerhardt (1984a).

males presumably are at a disadvantage when trying to attract mates in a large chorus of nonhybrid males, but the disadvantage is not absolute.

There is some evidence that hybrid females prefer the calls of hybrid males to those of the parental species. Mecham (1960) found several pairs of hybrid males and females in amplexus in mixed *Hyla cinerea-H. gratiosa* choruses, even though the number of hybrids in the choruses was small. Doherty and Gerhardt (1983, 1984a) showed that hybrid females produced by crossing *H. femoralis* and *H. chrysoscelis* preferred calls of hybrid males to those of *H. chrysoscelis*. They did not discriminate between hybrid calls and those of *H. femoralis,* which sometimes resembled hybrid calls in pulse repetition rate (fig. 7.27). However, when given a choice of synthetic calls with pulse repetition rates corresponding to the mean values for hybrids and *H. femoralis,* the hybrid females clearly preferred hybrid calls. Again, the situation in the *Rana esculenta* complex differs from other hybrid systems in that both *R. lessonae* females and *R. esculenta* females actively avoid mating with *R. esculenta* males. In this system, *R. lessonae* females have reduced fitness if they mate with *R. esculenta* males, because the *lessonae* genome eventually is discarded. *Rana esculenta* females avoid mating with *R. esculenta* males because matings between hybrids yield offspring with low viability (Abt and Reyer 1993; Roesli and Reyer 2000). Nevertheless, males of *R. esculenta* often acquire mates because they are very aggressive in competing for females (see chapter 8).

## Character Displacement

If two species of frogs with very similar advertisement calls occur in sympatry, females might make frequent mistakes in choosing mates. If hybrid offspring have lower fitness than offspring of conspecific matings, then selection would favor individuals with calls least similar to those of the other species. The result might be divergence in the characteristics of calls in the zone of sympatry, a process known as reproductive character displacement or reinforcement (Littlejohn 1969, 1977, 1981, 1988, 1993, 1999; W. Blair 1974; Gerhardt and Schwartz 1995; Gerhardt and Huber 2002). If the calls of two species were more similar in allopatry than in sympatry, then this would be evidence of possible character displacement. It is possible that differences in the calls of many sympatric species of anurans have evolved through past reinforcement of slight differences in call characteristics, thereby enhancing their effectiveness in restricting interspecific gene exchange.

Despite extensive work on anuran vocalizations, convincing examples of reproductive character displacement have been elusive (W. Blair 1974; Gerhardt and Schwartz 1995; Littlejohn 1999; Gerhardt and Huber 2002). Many of the examples proposed by early workers can be explained as the result of clinal variation in call characteristics, incidental effects of differences in body size, or differences in the temperatures at which calls of different populations were recorded. Furthermore, the lack of evidence of character displacement among sympatric anurans could mean that the calls of different species diverged sufficiently in allopatry to ensure complete reproductive isolation when the species become sympatric. The difficulty in documenting reinforcement of call characteristics has led some workers to suggest that the process occurs very rarely, if at all (Paterson 1978, 1980, 1981, 1982; Passmore 1981; Telford 1982).

Nevertheless, there are several well-documented examples of character displacement in frog calls (Gerhardt and Schwartz 1995; Littlejohn 1999; Gerhardt and Huber 2002). The best known is the *Litoria ewingii* complex in eastern Australia (Littlejohn 1965, 1976). The calls of *L. ewingii* and *L. verreauxii* differ in pulse repetition rate both in allopatry and sympatry, but the differences are much greater in sympatry (fig. 7.28). In fact, the calls of *L. verreauxii* from populations sympatric with *L. ewingii* not only have much faster pulse repetition rates, but they also have less pronounced and more variable depths of amplitude modulation. Although females of both species readily discriminate between conspecific calls and those of sympatric congeners, *L. ewingii* females from a zone of sympatry cannot discriminate calls of sympatric *L. ewingii* from those of allopatric *L. verreauxii* (Littlejohn and Loftus-Hills 1968). On the other hand, the calls of sympatric and allopatric *L. verreauxii* are sufficiently different from each other that *L. verreauxii* females from a zone of sympatry prefer calls from their own population to allopatric calls. In fact, there appears to have been a shift in the way in which females recognize their own species' calls from a quantitative trait (pulse repetition rate) to a qualitative one (pulsed vs. unpulsed signals). The reduction in depth of amplitude modulation in populations sympatric with *L. ewingii* appears to be a result of increasing pulse rate (that is, the frogs have difficulty producing fully modulated calls at high pulse rates Gerhardt and Davis 1988).

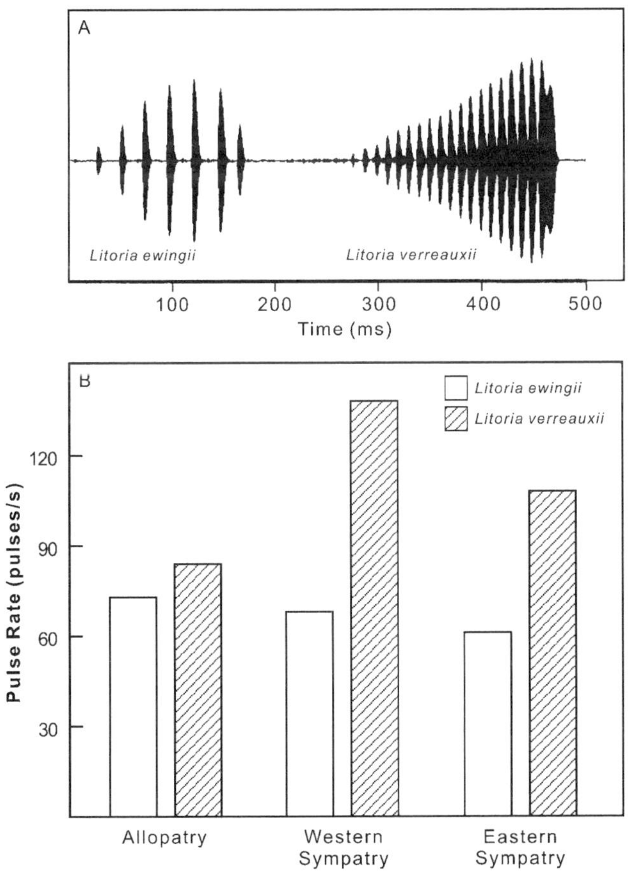


**Fig. 7.28.** (A) Oscillograms of the advertisement calls of two Australian treefrogs, *Litoria ewingii* and *L. verreauxii*. After Littlejohn (2001). (B) Mean pulse rates for the two species in areas of allopatry and western and eastern areas of sympatry at 10° C. The difference in pulse rate between the two species is much greater in sympatry than in allopatry, indicating the character displacement has occurred. Plotted from data in Littlejohn (1999).

Character displacement in pulse repetition rates also has been described in chorus frogs (*Pseudacris*) of the southeastern United States. Calls of *P. nigrita* have a relatively slow pulse repetition rate. Calls of *P. feriarum* sympatric with *P. nigrita* have much faster pulse rates. Allopatric populations have only slightly faster rates (Littlejohn 1960, 1969; Fouquette 1975), suggesting that character displacement has occurred. Furthermore, the calls of *P. feriarum*, an eastern species, are very different from those of *P. triseriata*, a western species. Female *P. triseriata* readily discriminate between calls from their own population and those of *P. feriarum*, but cannot discriminate between conspecific calls and those of allopatric *P. nigrita*, which have a very similar pulse rate (Littlejohn 1960). Unfortunately, the responses of females from sympatric populations of *P. feriarum* and *P. nigrita* have not been tested.

Ralin (1977) argued that character displacement has occurred in calls of *Hyla chrysoscelis* in sympatry with *H. versicolor*. He based this conclusion on differences in pulse repetition rate in sympatric and allopatric populations. Gerhardt (1982) found no differences in the relationship of pulse rate to temperature in *H. chrysoscelis* sympatric with *H. versicolor* in Missouri and allopatric populations, and therefore, no evidence of character displacement. Furthermore, the possibility of an east-west cline in pulse repetition rate in the calls of *H. chrysoscelis*, independent of the presence of *H. versicolor*, cannot be dismissed. The absence of obvious character displacement in these and other pairs of anuran species could be a result of character displacement having occurred in the responses of females rather than the structure of male calls. Gerhardt (1994b) reported that females of *H. chrysoscelis* from populations that were sympatric with *H. versicolor* were more selective in their preferences for conspecific calls than were females from far allopatric populations. In particular, preferences of females from far allopatric populations for calls with pulse repetition rates characteristic of conspecific calls could be weakened by making the calls longer, but this did not occur when females from sympatric populations were tested.

Several other studies also have shown that females are more discriminating in sympatry than in allopatry, and this form of character displacement actually could be more common than evolutionary changes in male calls. Females of two species of midwife toads, *Alytes obstetricans* and *A. cisternasii,* are more discriminating in sympatry than in allopatry, even though the calls of males do not differ between sympatric and allopatric populations (Márquez and Bosch 1997b). In spadefoot toads (*Spea multiplicata*), females from populations sympatric with *S. bombifrons* preferred average calling rates to high rates, whereas females from allopatric populations preferred high rates. Because the calling rate of *S. bombifrons* is higher than that of *S. multiplicata,* females of *S. multiplicata* in sympatry appear to avoid calling rates that are likely to overlap those of *S. bombifrons.* (K. Pfenning 2000). This appears to have resulted in reinforcement of call differences in areas of sympatry and a decrease in the frequency of hybridization over the last quarter century (K. Pfennig 2003). In *Hyla cinerea,* there were subtle differences in male calls in populations sympatric with *H. gratiosa* and allopatric populations, but the main difference was in responses of females. Again, females showed stronger discrimination against heterospecific calls in sympatric populations, although not all populations showed this pattern (Höbel and Gerhardt 2003).

### The Evolution of Call Structure

As discussed in the previous sections, each species of anuran has a call that is distinct in one or more features from those of closely related species, and females consistently prefer the calls of conspecifics to heterospecifics. At the same time, it has long been recognized that the calls of closely related species often share certain basic structural characteristics. Presumably because these characteristics have been inherited from a common ancestor (Blair 1958a, c, 1962; McAlister 1959; Duellman 1970; A. Martin 1972; Nelson 1973; Zweifel 1985; Straughan and Heyer 1976; Tandy and Tandy 1976; Schneider, Hussein, and Akef 1986; Cocroft et al. 1990; M. Ryan and Drewes 1990; Cocroft and Ryan 1995; Gerhardt 2001; Gerhardt and Huber 2002). These similarities in call structure sometimes have been used to infer phylogenetic relationships among species or to place species within a genus into distinct species groups (Blair 1958a; Barrio 1965; Tandy and Keith 1972; C. Myers and Daly 1976; H. Zimmermann and Zimmermann 1988; Cocroft 1994). More recently, some investigators have used phylogenies derived independently from nonacoustic characters to map the evolution of call features as a means of unraveling patterns of call evolution (M. Ryan 1988a; M. Ryan and Drewes 1990; M. Ryan and Rand 1993a, c, 1995; Cocroft and Ryan 1995). I will briefly discuss three examples that have been particularly well studied.

The calls of North American species of toads (*Bufo*) are well known, and W. Martin (1971, 1972) studied morphological correlates of call production in detail. Some of these toads have traditionally been divided into two species groups. The *B. americanus* group includes *B. americanus, B. fowleri, B. hemiophrys, B. houstonensis, B. microscaphus, B. terrestris,* and *B. woodhousii;* the *B. cognatus* group includes *B. cognatus, B. compactilis,* and *B. speciosus* (W. Blair 1972; Maxson, Song, and Lopata 1981). Examination of the calls of these species, placed in a phylogenetic context, reveals major differences in call structure between the two species groups (fig. 7.29). All of these toads produce relatively long trills composed of regularly repeated pulses (fig. 7.29 A), but they differ in pulse repetition rate, call duration, and calling rate. There also are differences in dominant frequency as a result of difference in body size (Cocroft and Ryan 1995). The most distinctive difference is in the internal structure of the calls. The members of the *B. cognatus* group have pulses with internal amplitude modulation (fig. 7.29 B). These are produced by vibrations of arytenoid valves, which are absent in the *B. americanus* group. Because the outgroup species, *B. valliceps,* also has internal amplitude modulation, it appear that this is the ancestral condition for this clade of toads, with the loss of internal modulation being a derived condition.

A similar phylogenetic analysis of the calls of chorus frogs in the genus *Pseudacris* revealed considerable congruence between phylogeny and call structure (fig. 7.30). Cocroft (1994) found that *Pseudacris crucifer,* a species that often has been placed in the genus *Hyla,* appears to be a basal member of the *Pseudacris* clade, an interpretation consistent with electrophoretic work by Hedges (1986). It has a tone-like call, similar in basic structure (fig. 7.30 A) and pulse shape (fig. 7.30 B) to the calls of *P. ornata* and *P. streckeri,* two species that have long been considered close relatives. Most other *Pseudacris* have trilled calls (fig. 7.30 A), with a distinctly different pulse shape (fig. 7.30 B). *Pseudacris ocularis,* a tiny species that is morphologically distinct from other members of the genus and has been placed in various genera, including its own genus, appears to arise between these two major clades and has a two-part call with notes similar to both groups (Cocroft and Ryan 1995). One interesting pattern that has not been analyzed in detail is the evolution of aggressive calls. *Pseudacris crucifer* has a trilled

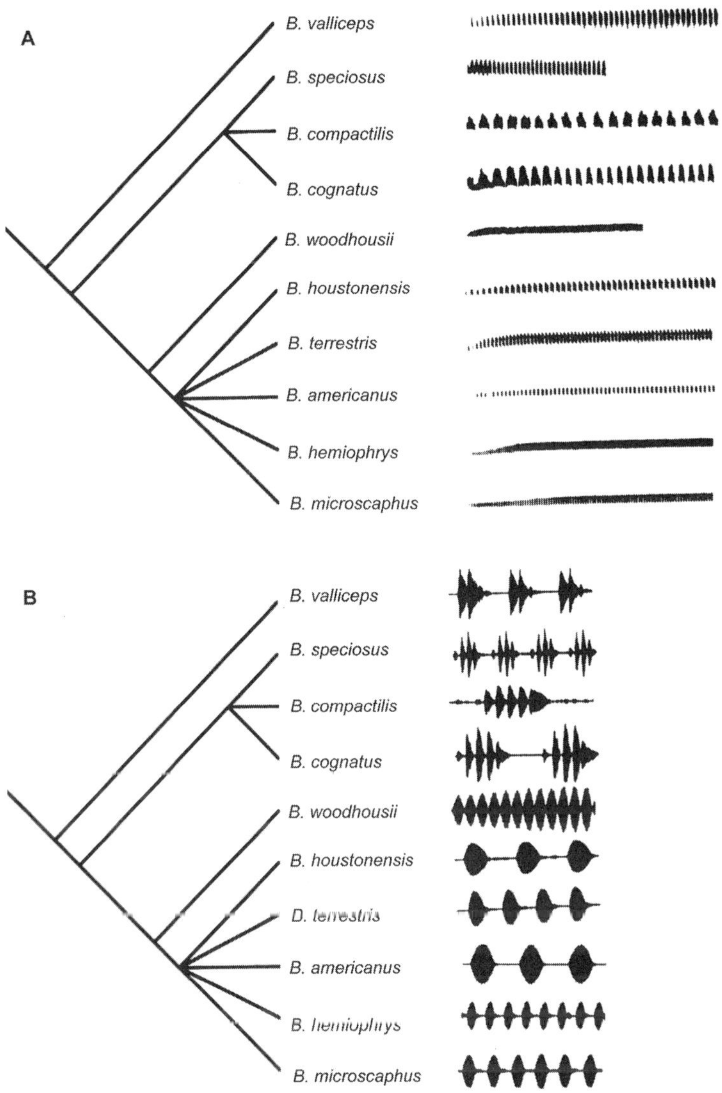


**Fig. 7.29.** Phylogenetic analysis of call evolution in North American toads (*Bufo*). (A) Sound spectrograms of the first second of calls of each species at about 18° C. Frequency scale omitted to facilitate comparison of overall call structure. (B) Oscillograms showing waveform and pulse structure of calls. The *Bufo americanus* group (bottom half) lacks internal amplitude modulation of pulses. From Cocroft and Ryan (1995).

aggressive call similar to the advertisement calls of other *Pseudacris* (Schwartz 1989), as does *Pseudacris regilla* (Allan 1973). Owen (2003) documented the occurrence of trilled aggressive calls in *P. feriarum* from North Carolina, *P. triseriata* from Indiana, *P. brachyphona* from West Virginia, and *P. streckeri illinoiensis* from Illinois, suggesting that this type of aggressive call is widespread in the genus.

A third group that has been investigated in considerable detail is the Neotropical leptodactylid genus *Physalaemus*. Members of this genus have similar frequency-modulated calls with a pronounced harmonic structure. These vary in duration and in the shape of the call envelope, but many species have rather similar dominant frequencies (fig. 7.31). Most species produce whine-like notes alone or in doublets, but a few add other components to the call. Both *P. pustulosus* and *P. petersi* produce secondary chucks. These are short notes with a wide frequency spectrum, but the chucks have a somewhat different structure in the two species (M. Ryan and Drewes 1990). The hypothesized phylogeny of the *P. pustulosus* species group, to which both of these species belong, has changed several times as new data have become available (Cannatella and Duellman 1984; M. Ryan and Rand 1993c, 1995), making it somewhat difficult to interpret patterns of call evolution. If *P. pustulosus* and *P. pe-*

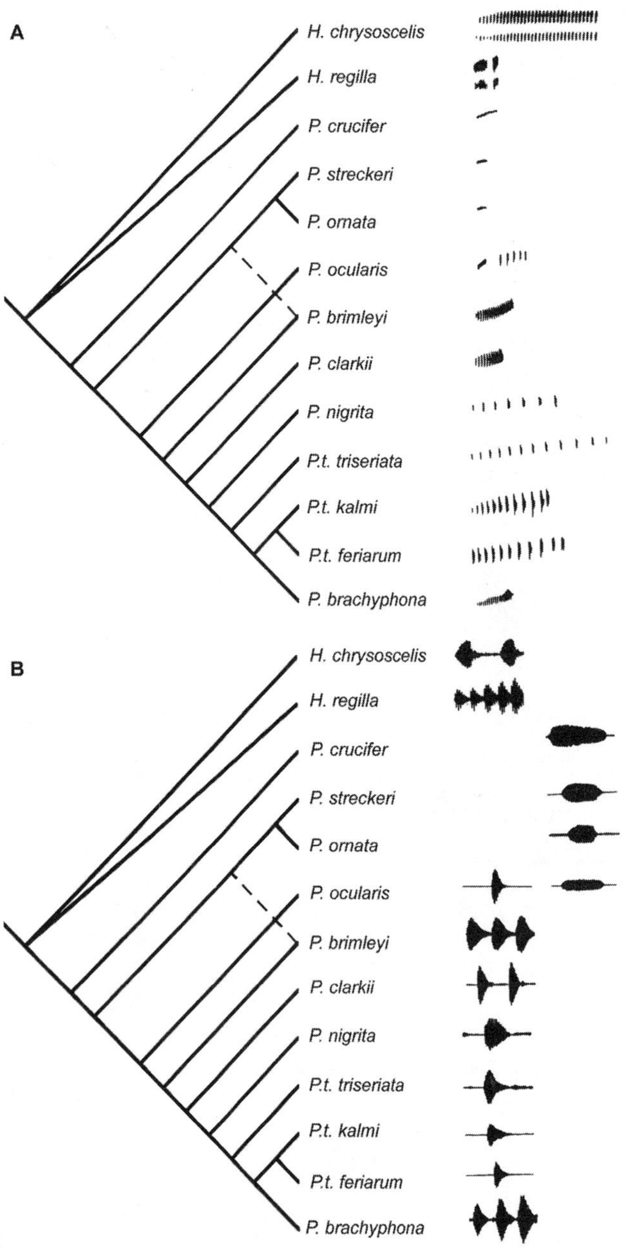


**Fig. 7.30.** Phylogenetic analysis of call evolution in North American chorus frogs (*Pseudacris*) and related species of *Hyla*. (A) Sound spectrograms of the calls of each species, not all recorded at the same temperature. Time scale is the same for each call. Frequency scale omitted to facilitate comparison of overall call structure. (B) Oscillograms showing waveforms and pulse structure of calls. Calls in the first column represent 30 ms in the middle of the call. The second column shows the amplitude envelope for pure-tone calls. Time scale is 200 ms for *Pseudacris crucifer,* 50 ms for *P. streckeri, P. ornata,* and *P. ocularis*. From Cocroft and Ryan (1995).

*tersi* are sister species (Cannatella and Duellman 1984; M. Ryan and Rand 1995), then the chuck notes could have evolved once in a common ancestor; if they are not, then evolution of the chuck would have occurred twice (fig. 7.32). *Physalaemus pustulatus* adds a pulsed prefix to its call (fig. 7.31). This appears to be a uniquely derived trait that is absent in its closest relative, *P. coloradorum,* and all other members of the genus (M. Ryan and Drewes 1990; M. Ryan and Rand 1993a, c).

M. Ryan and Rand (1995, 1999b) went beyond mapping of calls of *Physalaemus* onto cladograms to examine the evolution of call structure in this genus. They also used the phylogeny of the group to estimate characteristics of the calls of the ancestors of extant species (fig. 7.33). They then tested responses of *Physalaemus pustulosus* females to the calls of both the extant and "ancestral" species. In general, they found that females showed strong discrimination between conspecific calls and those of other extant species, but showed less discrimination between conspecific calls and those of their immediate ancestors. Many females did respond to heterospecific calls and even to the hypothetical calls of ancestral species when given a choice between heterospecific calls and white noise, even when the calls presented were those of fairly distantly related species. These results indicate that females recognize a broad range of calls as appropriate signals, but they show strong discrimination against most of these when conspecific calls are available. The tendency of females to respond to heterospecific calls in the absence of conspecific calls was related to the quantitative similarity in call characteristics, but not necessarily to phylogenetic similarity. The estimates of ancestral call characteristics varied depending on the exact model of phylogeny reconstruction that was used (Losos 1999; M. Ryan and Rand 1999a), but most of these differences did not appear to be important to females (M. Ryan and Rand, 1999a).

Major changes in call structure within clades of frogs are correlated with morphological changes in the structures that produce the calls. For example, all members of the *Physalaemus pustulosus* group have fibrous masses in their vocal cords, but only those that have a firm attachment between the fibrous mass and the cricoid cartilage produce chuck notes when the fibrous mass vibrates in the airstream passing through the larynx. In addition, the fibrous mass is much larger in *P. pustulosus* and in other species that produce complex calls than in species that lack chuck notes (Boul and Ryan 2004). Other morphological changes in the larynx can have major effects on call structure as well. One example already mentioned in the loss of arytenoid valves in the *Bufo americanus* species group, resulting in the loss of internal amplitude modulation of pulses (W. Martin 1972). Differences in call note duration and pulse repetition rate could be correlated with differences in the size of the laryngeal muscles (McClelland, Wilczynski, and Rand 1997), as well as physiological changes in aerobic capacity, enzyme activities, muscle fiber type, contractile properties, and blood supply (McLister, Stevens, and Bogart 1995; R. Marsh 1999; see also chapter 5). Presumably the neural programs controlling call production and the timing of muscle contrac-

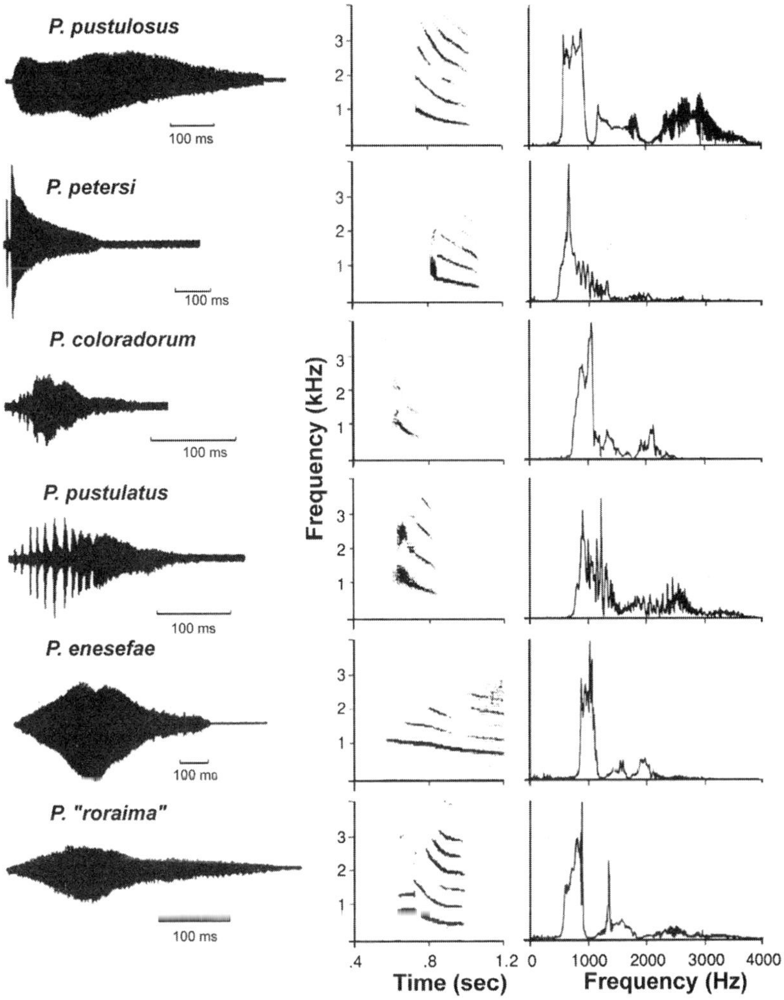


**Fig. 7.31.** Oscillograms (left), sound spectrograms (middle), and power spectra (right) of calls the genus *Physalaemus*. Most species have similar frequency-modulated calls with similar dominant frequencies, but the pattern of frequency modulation (middle) and the amplitude envelope of the call (left) differ among species. From Ryan and Rand (1993c).

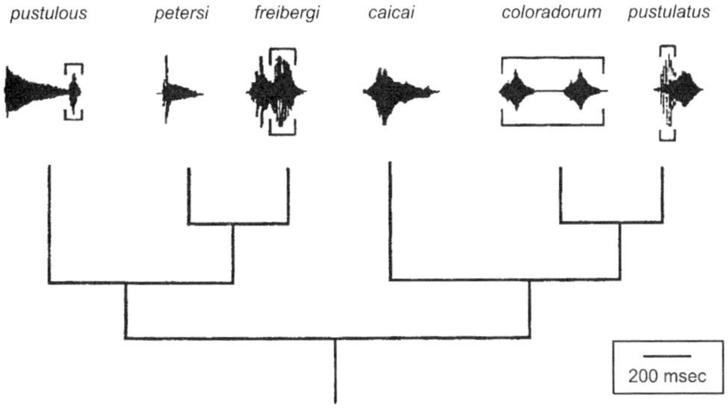


**Fig. 7.32.** Phylogeny of the *Physalaemus pustulosus* group, showing evolution of chucks and other call components. An oscillogram of the call of each species is shown (time scale at lower right). Bracketed figures show derived call characters, including a secondary chuck note in *P. pustulosus*, a "squawk" at the end of the call of *P. freibergi*, double-note call in *P. coloradorum*, and amplitude-modulated prefix in *P. pustulatus*. After Ryan and Rand (1993b).

tion must undergo changes as well, although this has not been thoroughly studied in a phylogenetic context.

## The Anatomy and Neurobiology of Call Recognition

To understand how sounds influence the behavior of anurans, it is important to know something about how sounds are perceived by these animals. The perceptual world of a frog is not the same as that of a human observer and vocal signals are not necessarily interpreted by the anuran auditory system in the same way as they would be by our own. Morphological or physiological constraints imposed by the auditory system can be important in shaping the evolution of anuran vocal communication (M. Ryan 1986a, 1988a, 1990; Gerhardt 1987, 1988; M. Ryan and Rand 1990;

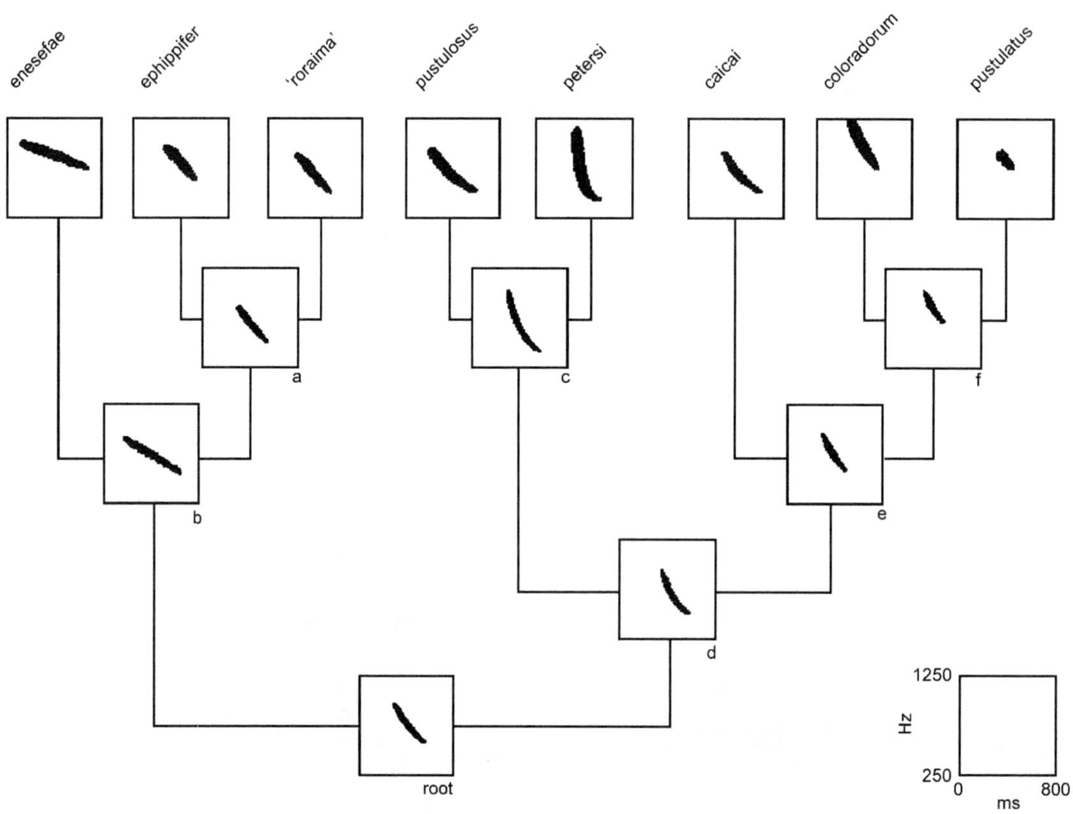


**Fig. 7.33.** Basic structure of whine calls of the genus *Physalaemus,* with inferred ancestral calls (a–f). Sound spectrograms show synthetic models of calls used in playback experiments with females. After Ryan and Rand (1999).

M. Ryan, Fox, Wilczynski, and Rand 1990; M. Ryan and Keddy-Hector 1992). The following section reviews the anatomy and physiology of the anuran auditory system, especially as it relates to behaviorally relevant aspects of sound perception and call recognition. This has been an active field of research for decades, but an exhaustive review of anuran neuroanatomy and neurophysiology is beyond the scope of this book. Several books (Fritzsch et al. 1988; Gerhardt and Huber 2002; M. Ryan 2001; Narins and Feng 2007) and a review by Smotherman and Narins (2004) provide detailed reviews of work in this field, and several earlier reviews are useful as well (Capranica 1976a, b, 1977; Capranica and Moffat 1983; Capranica and Rose 1983; Wilczynski and Capranica 1984).

## Anatomy and Function of the Middle Ear

The anuran ear consists of two major parts, the middle ear and the inner ear (fig. 7.34). The tympanic membrane is flush with the outside of the frog's head and is not recessed in a cavity as in mammals, so there is no outer ear. The middle ear provides the mechanical linkage between the external environment and the fluid-filled cavities of the inner ear, while the inner ear contains the sensory structures that translate mechanical stimulation into nerve impulses that then travel to the brain. While the basic principle of sound reception in anurans is similar to that in other vertebrates, the details of anatomical structure are quite different in some respects. Anurans have more than one system for picking up acoustic stimuli from the environment and more than one type of sensory organ to translate these inputs into nerve impulses (E. Lewis and Narins 1999; Smotherman and Narins 2004).

### The Tympanic Ear

The tympanum, or eardrum, provides the major input of airborne sound into the auditory system of most anurans (fig. 7.34). The tympanum is a thin membrane supported by a bony or cartilage ring. Variation in the size of the tympanic membrane appears to be correlated with differences in auditory sensitivity, both within and between species, with frogs having larger tympana also exhibiting greater sensitivity (Fox 1995). In addition, larger tympana are sensitive to lower frequency sounds, both within and between species (Hetherington 1992a, b). This can lead to sexual differences in frequency sensitivity, as in the bullfrog (*Rana catesbeiana*), a species in which males have much larger tympana (Hetherington 1994).

The inner surface of the tympanum is in contact with the

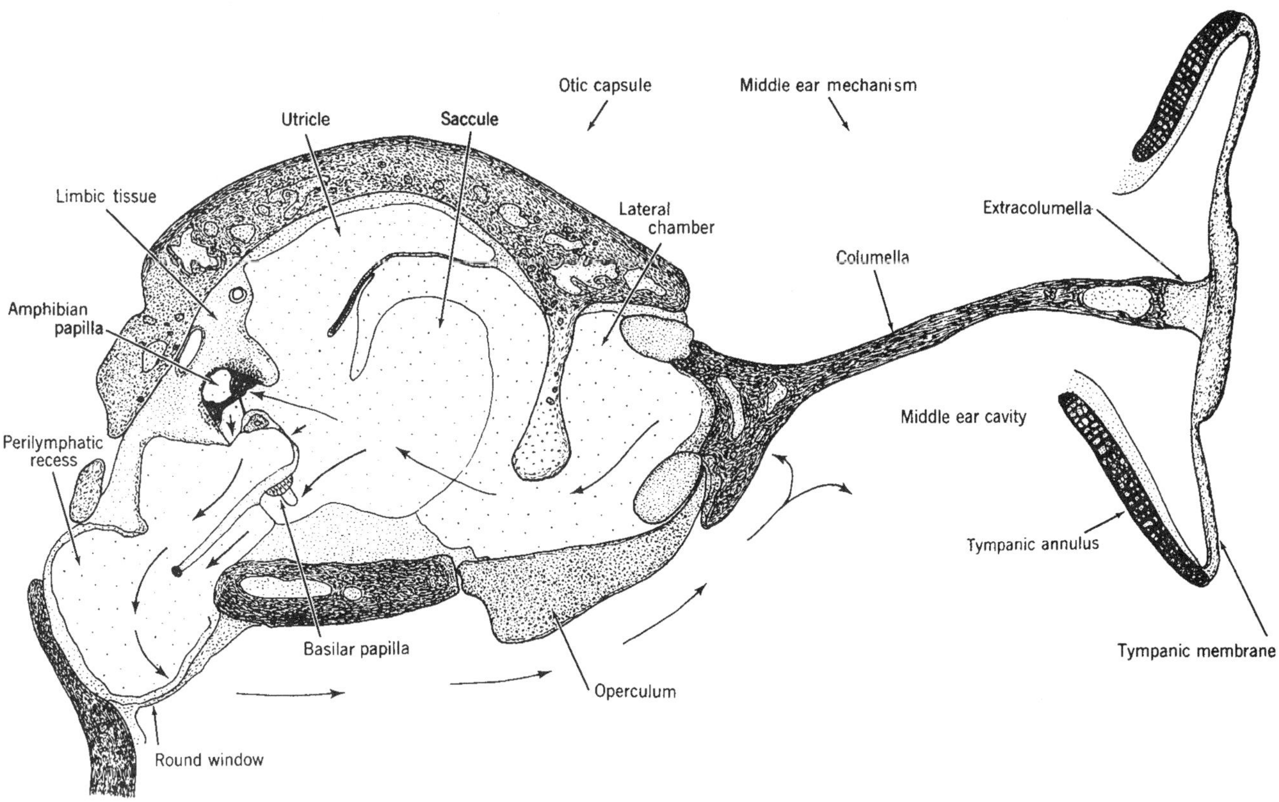


**Fig. 7.34.** Cross-section of the anuran middle ear, showing important structures involved in sound transmission and perception. From Wever (1985).

extracolumella (also called the extrastapes or plectrum), a structure usually composed of cartilage that links the tympanum with the bony columella (stapes). The columella in turn serves as the mechanical link to the inner ear (fig. 7.34). The outer part of the columella passes through an air-filled chamber that has a direct open connection to the buccal cavity through the eustachian tube. Hence, pressure changes in the buccal cavity can affect the pressure on the inner surface of the tympanum. In most anurans, the medial portion of the columella passes through connective tissue and muscle before reaching its connection with the inner ear. This condition contrasts with that of mammals, in which the entire middle ear cavity is filled with air (Wever 1985; Jaslow, Hetherington, and Lombard 1988; Smotherman and Narins 2004).

The medial end of the columella (columellar or stapedial footplate), usually composed of cartilage, rests in the oval window of the otic capsule, the bony structure that protects the inner ear. Vibrations of the tympanum are transmitted by the stapes through this window to the fluid in the inner ear (M. Jørgensen and Kanneworff 1998). This sets up currents that stimulate the sensory structures in the inner ear (Purgue and Narins 2000). Measurements of tympanic vibrations by means of laser interferometry in several anuran species have shown that the tympanum is not finely tuned to particular frequencies, but shows a relatively flat response over a broad range of frequencies. In *Rana pipiens,* sensitivity falls off rapidly below 650 Hz, and the tympanum shows very little response to frequencies below 300 Hz. Sensitivity also decreases at frequencies above 1,200 Hz, but high frequencies are transmitted if the intensity of the sound is sufficiently high (Wilczynski, Resler, and Capranica 1987). Removal of the tympanum causes a substantial decrease of hearing sensitivity at high frequencies (Lombard and Straughan 1974). In frogs with tympanic ears, the tympanum appears to be the major route for acoustic input underwater as well as in air, at least at relatively high frequencies in the range of most anuran advertisement calls (Hetherington and Lombard 1982). Hearing sensitivity underwater is similar to that in air (Lombard, Fay, and Werner 1981), so frogs that call underwater, such as *Rana aurora, R. cascadae,* or *R. palustris,* should have little difficulty detecting conspecific calls.

Although the tympanic ear is the major means of detecting airborne sounds in most anurans, it is not essential for effective sound reception, even at relatively high frequencies. In fact, reduction of various elements of the tympanic ear has occurred repeatedly in anuran evolution (Jaslow, Hetherington, and Lombard 1988). Some species, such as the microhylid *Koloula pulchra,* lack a tympanum and part of the extrastapes, but retain the stapes and middle ear cavity. Others, such as the leptodactylid *Telmatobius exsul*, have lost the tympanum and middle ear cavity, but retain most of the middle ear bones. Still others have lost all of the tympanic ear structures. This condition is found in representatives of at least 11 anuran families, including all of the Sooglossidae and various species of bufonids (*Atelopus* and some *Bufo, Nectophryne*, and *Melanophryniscus*), myobatrachids (*Pseudophryne, Crinia*), and in *Hemisus*. Loss of tympanic ears has not occurred only in voiceless frogs (discussed in a previous section). In fact, many earless frogs have advertisement calls with frequencies in the same range as species with normal tympanic ears (i.e., dominant frequencies above 1 kHz). Behavioral and neurophysiological studies of earless frogs and those lacking a tympanum have shown that acoustic communication and frequency sensitivity are similar to those of normal anurans (Pengilley 1971a; Loftus-Hills 1973a, b; Jaslow 1979; Hill and Robertson 1981; Nussbaum, Jaslow, and Watson 1982; Robertson 1984, 1986b; Crump 1988; Jaslow, Hetherington, and Lombard 1988; Cocroft et al. 1990; Jaslow and Lombard 1996; Lindquist and Hetherington 1996; Lindquist, Hetherington, and Volman 1998). Hence, there must be alternative routes for acoustic input in earless frogs and perhaps other anurans as well (see the following).

### The Opercularis System

One extratympanic pathway for sound reception that is sensitive mainly to low frequency input is the opercularis system, which is unique to amphibians. This system is always present in anurans, even in species that lack a tympanic ear. It is present in most urodeles as well, but is absent from some neotenic groups (proteids, amphiumids, and cryptobranchids; Wever 1985; Jaslow, Hetherington, and Lombard 1988). The operculum is a cartilage disk that sits in the oval window adjacent to the stapedial footplate (fig. 7.34). It is attached to the opercularis muscle, which inserts on the suprascapula (Hetherington, Jaslow, and Lombard 1986). The operculum articulates with the stapedial footplate, often sitting in a notch within the footplate. Lombard and Straughan (1974) suggested that interlocking of the operculum and stapes by action of the opercularis muscle would modify the frequency response of the tympanic ear by increasing its sensitivity to low frequency sound. When they removed the opercularis muscle, sensitivity to frequencies below 1 kHz was greatly reduced. They proposed a dual input system, with frequencies below 1 kHz being received mainly by the opercularis system and those above 1 kHz being received mainly by the tympanic system. Changes in the tension of the opercularis muscle would modify the sensitivity of the combined system to high and low frequencies. More recent work by Hetherington, Jaslow, and Lombard (1986) largely refuted this hypothesis by showing that the two systems articulate in such a way that the operculum and stapes move independently.

A second hypothesis to explain the function of the opercularis system was proposed by Wever (1979, 1985). He suggested that contraction of the opercularis muscle when the frog is vocalizing would cause the operculum to lock the stapedial footplate in place, reducing movement of the stapes and thereby greatly reducing the sensitivity of the tympanic ear. This would serve as a mechanism to protect the auditory system from the loud sound of the frog's own voice. Recent authors have offered several lines of evidence to refute Wever's hypothesis. First, the independent movement of the two systems suggests a lock mechanism is unlikely. Second, the opercularis muscle is made up mainly of tonic fibers, not the fast twitch fibers required for rapid contraction. In fact, it is a very slowly contracting muscle that appears to remain in tonic contraction whenever a frog is sitting on the ground (Becker and Lombard 1977; Hetherington and Lombard 1983; Hetherington, Jaslow, and Lombard 1986; Hetherington 1987).

Hetherington (1988a) argued that the opercularis system functions primarily in detection of vibrations at frequencies below 200 Hz. The tonic contraction of the muscle ensures a rigid connection between the operculum and the pectoral girdle that is well suited for transmission of low-frequency vibrations. This explanation is applicable to amphibians in general; many urodeles are sensitive to low frequency vibrations even though they lack a tympanic ear (Wever 1978, 1985; Ross and Smith 1979). The hypotheses of Lombard and Straughan (1974) and Wever (1979, 1985) do not provide a functional explanation for the opercularis system in nonvocalizing amphibians. In anurans, the opercularis muscle has a vertical orientation when the frog is sitting in a normal posture, and this probably enhances transmission of vertical displacements of the pectoral girdle. If the opercularis muscle is removed, responses of inner ear receptors to vibrations are greatly reduced (Hetherington 1985). Recent work by Mason and Narins (2002) has shown that greater coupling occurs between the operculum and the footplate than previously thought. They revived Weaver's protective hypothesis and argued that the operculum is not the major avenue for detection of ground-borne vibrations afterall. They also suggested that the role of the operculum in amphibians remains unresolved.

### Other Extratympanic Pathways

Although Lombard and Straughan's original concept of separate input channels for high and low frequency sounds seems to be an oversimplification, there is considerable evidence of extratympanic pathways for airborne sound and not just ground-borne vibrations. Wilczynski, Resler, and Capranica (1987) found that both tympanic and extratympanic pathways are important for sound reception in *Rana catesbeiana.* At frequencies below about 900 Hz, auditory nerve fibers were about equally sensitive to tympanic and extratympanic input, but at higher frequencies, sensitivity to tympanic input was much greater. However, over a fairly broad frequency range (about 300 to 1,200 Hz), inner ear responses probably reflect an interaction between tympanic and extratympanic transmission. The precise nature of extratympanic inputs of airborne sounds is unknown. The opercularis system could be involved, as suggested by Lombard and Straughan's (1974) experiments, but this has not been definitively demonstrated (Wilczynski, Resler, and Capranica 1987; Jaslow, Hetherington, and Lombard 1988). Another possibility is that sound is simply transmitted through the tissues of the head, a pathway that is likely to be more effective underwater than in air.

Still another route was suggested by work with *Eleutherodactylus coqui* by Narins, Ehret, and Tautz (1988). They found that the body wall adjacent to the lungs vibrates in response to frequencies within the range of the advertisement call. Since the lungs have an open connection to the middle ear cavity through the buccal cavity and eustachian tubes, these vibrations could be directly transmitted to the auditory system through changes in air pressure in these cavities (fig. 7.35). Subsequent work showed that sounds of about 1 kHz, the frequency of the introductory co note of this species, are transmitted via the lung channel more effectively than are sounds around 2 kHz, the frequency of the qui note (Ehret et al. 1990). Other species of frogs exhibit significant transmission of sound via the lung-ear pathway, suggesting this is an important component of hearing in many anurans (Hetherington 1992a, b; Ehret, Werth, and Kamada 1994), including those that lack a tympanic ear (Lindquist, Hetherington, and Volman 1998; Hetherington and Lindquist 1999) and those that call underwater (Christensen-Dalsgaard and Elepfandt 1995). This is particularly true of small-bodied species, in which the movements of the body wall in response to sound can actually exceed those of the tympanic membrane (Hetherington 1992a). This allowed for the retention of vocal communication in small-bodied anurans that have undergone a reduction or loss of the tympanic ear (Hetherington 1992b). In some species, the body wall over the lungs vibrates in response to lower frequency sound than does the tympanum, but in others, frequency-sensitivity is similar (Ehret, Werth, and Kamada 1994). Input from the lung pathway is thought to be particularly significant for detecting the direction of a sound source (see "Sound Localization" in the following).

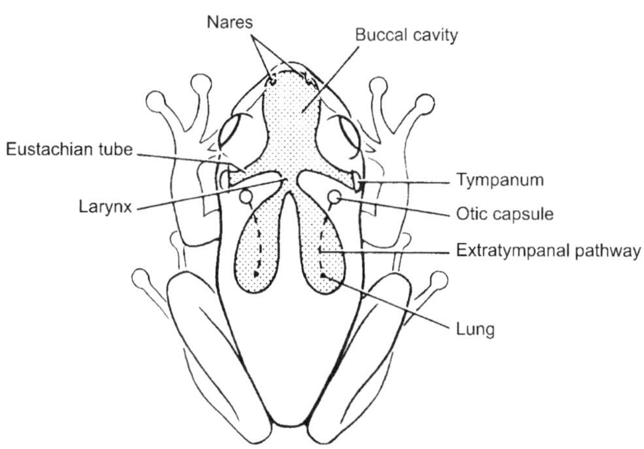


**Fig. 7.35.** Diagram of possible pathways for sound transmission in *Eleutherodactylus coqui.* Vibrations of the body wall over the lungs are transmitted to the ears via the larynx, buccal cavity, and Eustachian tubes. Sound also may be transmitted through the body to the otic capsule. After Narins et al. (1988).

### Anatomy and Function of the Inner Ear

The anuran inner ear is contained within the otic capsule (fig. 7.34) and consists of a number of fluid-filled chambers (the endolymphatic or membranous labyrinth) within which the sensory organs are located. The bony labyrinth of the otic capsule surrounds these sacs, and the space between the membranous and bony labyrinth is filled with fluid as well (the perilymphatic system). The perilymphatic system is important for hearing because it receives vibrations from the middle ear and transmits them to the fluid-filled chambers of the membranous labyrinth (E. Lewis and Narins 1999; Purgue and Narins 2000). The dorsal parts of the otic labyrinth make up the vestibular system for detecting motion and maintaining balance. The ventral part contains four sensory organs, each of which is innervated by a branch of the auditory nerve (VIIIth cranial nerve; fig. 7.36). One of these, the lagena, is mainly a gravity receptor involved in maintaining equilibrium (Caston, Precht, and Blanks 1977). It does not appear to play a role in detecting communication signals (Feng, Narins, and Capranica 1975; Baird and Lewis 1986; M. Jørgensen and Christensen-Dalsgaard 1991). The three remaining organs, the saccule, amphibian papilla, and basilar papilla, are sensitive to vibrations, to airborne sounds, or to both (Wilczynski and Capranica 1984; Wever 1985; E. Lewis and Lombard 1988; Zakon and Wilczynski 1988; M. Jørgensen and Chistensen-Dalsgaard 1991; Yu, Lewis, and Feld 1991; Christensen-Dalsgaard and Narins 1993; Christensen-Dalsgaard and Jørgensen 1996; E. Lewis and Narins 1999).

#### Saccule

The function of the saccule varies among the major vertebrate groups. In fish, it is primarily an acoustic organ, while in birds and mammals it is mainly involved in vestibular function (Fay and Popper 1985). The saccule of amphibians

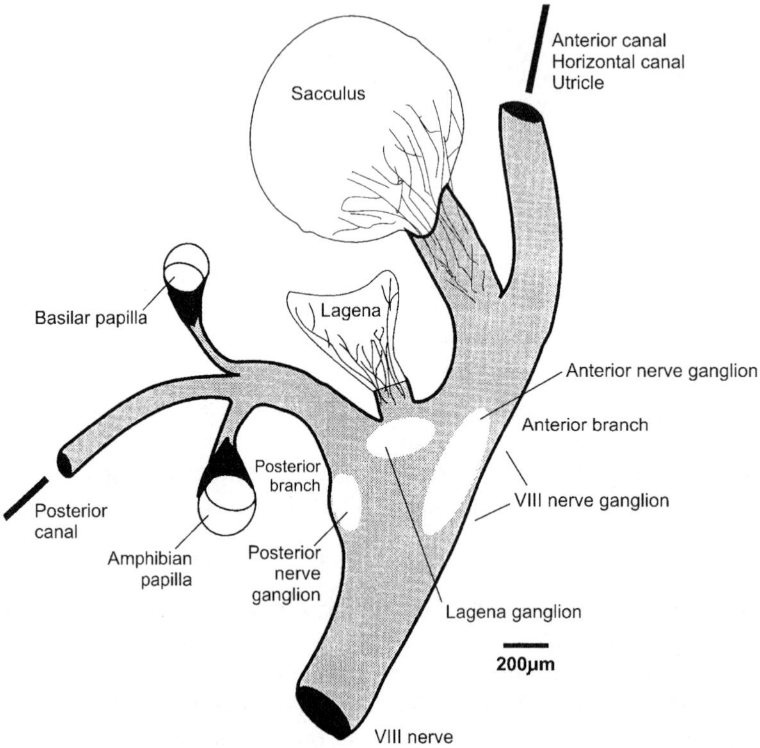


**Fig. 7.36.** Diagram of peripheral innervation of the right VIII nerve of *Rana pipiens* in ventromedial view, showing major sound-receiving organs. After Simmons et al. (1992).

is extraordinarily sensitive to substrate-borne vibrations, particularly frequencies below 100 Hz (Koyama et al. 1982; E. Lewis, Leverenz, and Koyama 1982; E. Lewis 1983; Narins and Lewis 1984; Christensen-Dalsgaard and Jørgensen 1988, 1996; E. Lewis and Lombard 1988; M. Jørgensen and Christensen-Dalsgaard 1991; Yu, Lewis, and Feld 1991; Christensen-Dalsgaard and Narins 1993; Smotherman and Narins 2004). Central processing of vibrational information takes place in the midbrain in the same region where acoustic cues are processed (Walkowiak 1980a; Comer and Grobstein 1981; Fuzessery and Feng 1982; Christensen-Dalsgaard and Jørgensen 1989).

This seismic sense probably is used by most anurans to detect the approach of predators. However, it also functions in the detection of substrate-borne signals produced by frogs that call underground (E. Lewis and Narins 1985) and perhaps vibrations produced by frogs calling in the water (Walkowiak and Munz 1985). The saccule also is capable of detecting low-frequency airborne sounds, and the best excitatory frequencies of neurons innervating the saccule are similar for vibrations and sound (Moffat and Capranica 1976; M. Jørgensen and Christensen-Dalsgaard 1991; Yu, Lewis, and Feld 1991; Christensen-Dalsgaard and Narins 1993; Christensen-Dalsgaard and Jørgensen 1996). Because the tympanum does not transmit sound effectively at such low frequencies, it is possible that the saccule is an important receptor for sound transmitted through extra-tympanic pathways, such as the bones of the skull and the lungs.

### Amphibian Papilla

The amphibian papilla is an organ located in a chamber called the amphibian papillar recess (fig. 7.34). It is found in all three groups of extant amphibians and is unique to these groups, providing evidence for the monophyly of modern amphibians (Wever 1985; E. Lewis and Lombard 1988). Amphibian papillar fibers in the VIIIth nerve are sensitive mainly to frequencies between 100 Hz and 1,200 to 1,400 Hz. Most species have two distinct populations of fibers innervating the amphibian papilla. One group is tuned to frequencies below about 500 Hz and can be inhibited by simultaneous presentation of sounds just above the best frequency of the neuron, a phenomenon known as two-tone inhibition (Capranica and Moffat 1980). The other is a group of fibers tuned to midfrequency sounds (500–1,200 Hz), which cannot be inhibited by higher frequencies. These two types of fibers have been found in various ranids, bufonids, hylids, and leptodactylids (Feng, Narins, and Capranica 1975; Narins and Capranica 1980; Capranica and Moffat 1983; Wilczynski and Capranica 1984; Zakon and

Wilczynski 1988), but only the lower frequency population was present in *Scaphiopus couchii* (Capranica and Moffat 1975). Recent work has shown that the amphibian papilla also is sensitive to vibrations at frequencies overlapping and slightly above the sensitivity of the saccule (M. Jørgensen and Christensen-Dalsgaard 1991; Yu, Lewis, and Feld 1991; Christensen-Dalsgaard and Narins 1993; Christensen-Dalsgaard and Jørgensen 1996).

The amphibian papilla shows a tonotopic organization, with sensory hair cells sensitive to different frequencies distributed on different parts of the papilla. This change in frequency sensitivity is correlated with morphological differences in hair cells from different regions of the organ (Simmons, Bertolotto, and Narins 1994). Hair cells are found in a patch at the rostral end of the papilla and on a long, S-shaped caudal extension (fig. 7.37 A). In general, neurons tuned to the lowest frequencies innervate cells at the rostral end of the papilla; this is the inhibitable population of low-frequency fibers described above. The range of the amphibian papilla extends into higher frequencies in a caudal direction (fig. 7.37 B), with the S-shaped extension giving rise to the population of noninhibitable, midfrequency fibers (E. Lewis, Leverenz, and Koyama 1982; E. Lewis, Baird, Leverenz, and Koyama 1982; Wilczynski and Capranica 1984; E. Lewis and Lombard 1988; Zakon and Wilczynski 1988). In primitive anurans, such as *Ascaphus*, the caudal extension is absent and sensitivity of the amphibian papilla to frequencies above 500 Hz is poor. In pelobatids, this part of the papilla is small, perhaps accounting for the lack of midfrequency, noninhibitable fibers. In more derived anurans, the caudal extension is longer, extending the range of the amphibian papilla to frequencies between 700 and 1,400 Hz (E. Lewis 1981). There is considerable morphological diversity among species of derived anurans that parallels the diversity of call frequencies used by different species (M. Ryan 1986a). Even within closely related species, such as the monophyletic assemblage of frogs in the genus *Eleutherodactylus* on Puerto Rico, there is considerable diversification in the form of this organ, suggesting that changes in the structure of the auditory system have accompanied divergence in the structure of advertisement calls (E. Lewis, Hecht, and Narins 1992).

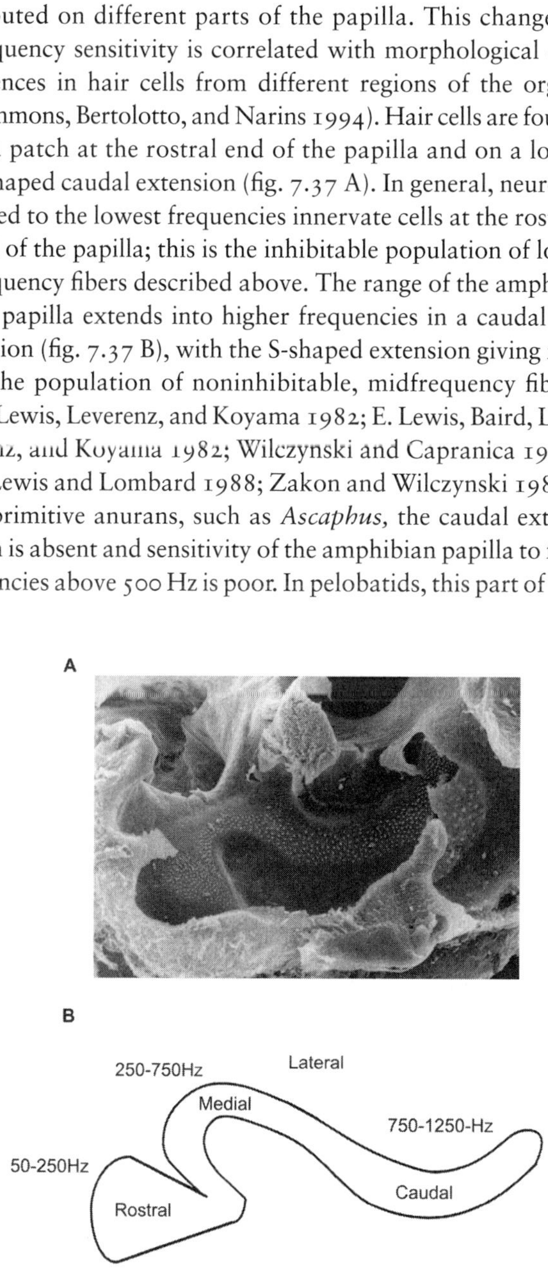


**Fig. 7.37.** (A) Scanning electronic micrograph of the amphibian papilla of *Eleutherodactylus coqui*, showing clusters of hair cells on the surface of the organ. Photo by E. R. Lewis. (B) Diagram of the amphibian papilla showing three regions of different frequency sensitivity. After Smotherman and Narins (1999).

The tuning of the amphibian papilla appears to be based on a mechanism similar to that found in the cochlea of the mammalian ear, although the system in amphibians has evolved completely independently. In mammals, a basilar membrane runs the length of the coiled cochlea and supports the sensory hair cells. Mechanical stimulation of the basilar membrane near its base sets up a traveling wave that moves along the membrane to its apex. Because the thickness and mass of the membrane change along its length, both the velocity and wavelength of the traveling wave change. Maximum excitation of hair cells at different places along the basilar membrane is a function of the frequency of sound being transmitted. This provides a mechanical filter that allows the ear to make fine-scale frequency discriminations (Fay and Popper 1985). An analogous traveling wave filter has been proposed for amphibians, possibly supported by the tectoral membrane that overlies the hair cells in the amphibian papilla (Capranica and Moffat 1977; Hillery and Narins 1984; E. Lewis and Lombard 1988). Like the mammalian basilar membrane, the thickness of the tectoral membrane varies, with the most massive portion being over the low-frequency hair cells at the rostral end of the papilla (Wever 1973; E. Lewis 1976; Shofner and Feng 1983, 1984). Although the details of how such a traveling wave filter actually works is not yet fully understood (E. Lewis and Lombard 1988; Zakon and Wilczynski 1988; E. Lewis, Hecht, and Narins 1992), it appears that the amphibian papilla shows a remarkable degree of functional convergence with the mammalian auditory system.

### Basilar Papilla

The basilar papilla is a small organ that sits in a chamber (the basilar papillar recess) opening to the saccule (fig. 7.34). There has been some debate about whether or not the basilar papilla in amphibians is homologous to auditory organs in reptiles, birds and mammals, or is an independently derived structure (E. Lewis and Lombard 1988; Smotherman and Narins 2004). All anurans have a basilar papilla, as do primitive urodeles, but it is absent in most derived urodeles and most caecilians. Basilar papillar fibers in the VIIIth nerve are most sensitive to high frequency sounds above 1,000 Hz (Feng, Narins, and Capranica 1975; E. Lewis, Leverenz, and Koyama 1982; Wilczynski and Capranica 1984; Zakon and Wilczynski 1988). All fibers innervating the basilar papilla usually have best frequencies within about 100 Hz of each other and effectively are tuned to the same frequency. The basilar papilla functions as a mechanical resonator, with tuning characteristics that depend in part on the size of the organ. The size of the basilar papilla generally increases with body size both within and between species, and larger basilar papillae are tuned to lower frequencies. Consequently, larger frogs generally are most sensitive to lower frequencies (within the range above 1,000 Hz) than are smaller frogs (Wilczynski and Capranica 1984; Zakon and Wilczynski 1988). This is true for comparisons between species (Loftus-Hills 1973a; Walkowiak, Capranica, and Schneider 1981; Wilczynski, McClelland, and Rand 1993), between sexes that differ in body size within species (Narins and Capranica 1976, 1980; Wilczynski, Zakon, and Brenowitz 1984; Wilczynski, Keddy-Hector, and Ryan 1992; McClelland, Wilczynski, and Rand 1997), and between same-sex individuals of different body size (Shofner and Feng 1981, 1984; Wilczynski, Zakon, and Brenowitz 1984; Shofner 1988; Zakon and Wilczynski 1988; Keddy-Hector, Wilczynski, and Ryan 1992).

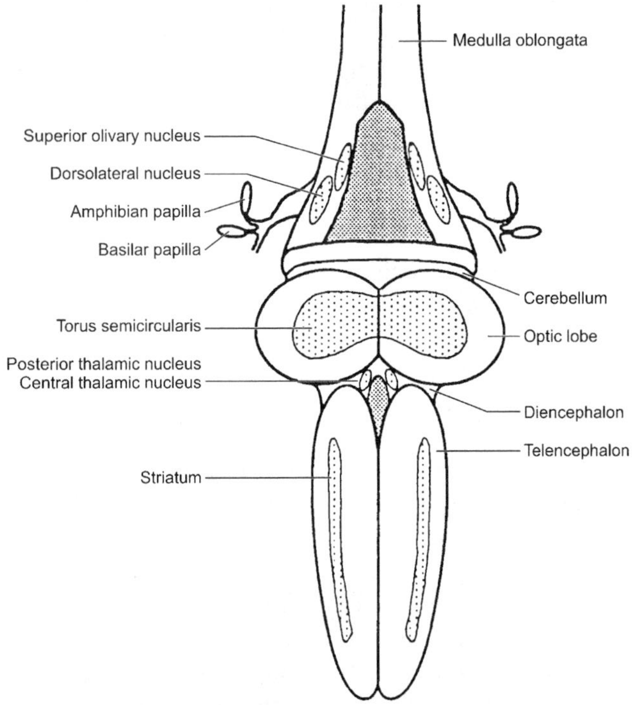


**Fig. 7.38.** Schematic diagram of the brain of a frog, showing approximate locations of principal auditory centers. Anterior is toward the bottom. After J. C. Hall (1994).

### Central Auditory Pathways

Most of our understanding of central auditory pathways in anurans comes from studies of ranid frogs, with limited information on certain parts of the brain from other families. The nonneurobiologist can easily become lost in the complex interconnections of different regions of the brain, and I will not attempt a detailed summary here. Instead, I will simply describe the main features of auditory pathways in ascending order from the medulla (hindbrain) to the diencephalon (midbrain) and telencephalon (forebrain), with a brief mention of features that are important for understanding neural processing of acoustic signals. Detailed descriptions of anatomy and wiring diagrams of connections between major brain centers can be found in E. Lewis, Leverenz, and Koyama (1980), Wilczynski (1981, 1988), Wilczynski and Capranica (1984), Feng (1986a, b), Neary (1988), Will (1988), Will and Fritzsch (1988), and J. C. Hall (1994).

Fibers from the VIIIth nerve enter the medulla in the posterior part of the brain (fig. 7.38) and terminate in the dorsolateral nucleus (also called the dorsal medullary nucleus or dorsal acoustic nucleus). Fibers from the basilar papilla and amphibian papilla project to different regions of this nucleus, and there is separation of the low and midfrequency fibers from the amphibian papilla as well. Cells in the dorsolateral nucleus receive inputs from both ears through connections between the nuclei on each side of the brain (Feng and Capranica 1976; Wilczynski and Capranica 1984; Fuzessery 1988). From the dorsolateral nuclei there are nerve projections to the superior olivary nuclei, located at the base of the medulla, and to both sides of the torus semicircularis in the midbrain. The latter seems to be homologous to the inferior colliculus in the mammalian brain and is a major center for processing of acoustic signals and the integration of other sensory information (Wilczynski 1981, 1988; Wilczynski and Capranica 1984; J. C. Hall 1994; Walkowiak and Luksch 1994). The torus becomes sensitive to conspecific acoustic signals shortly after metamorphosis. As the frog develops and grows, neurons in the torus exhibit a gradual downward shift in the frequencies to which they are most sensitive, as well as a sharpening of frequency tuning (Boatright-Horowitz and Simmons 1995) and changes in their response to amplitude-modulated (Boatright-Horowitz, Garabedian, Odabashian, and Simmons 1999).

The main ascending connections from the torus terminate in the central thalamic nucleus of the diencephalon. Auditory sensitivity also is present in several other nuclei in this region, receiving inputs from the central thalamic nucleus and directly from the torus as well (Neary 1974, 1988; Mudry, Constantine-Paton, and Capranica 1977). Several areas of the forebrain (telencephalon) exhibit acoustic sensitivity, including the striatum, medial pallium, and preoptic area. These receive connections from several nuclei in the diencephalon and directly from the torus (Wilczynski and Northcutt 1977, 1983; Gruberg and Udin 1978; Mudry and Capranica 1980; Neary 1988). Some of these areas are linked to motor centers of the brain and provide a link between the perception of calls and call production (R. Schmidt 1966, 1968b, 1973a, 1974, 1984; Knorr 1976; Wada and Gorbman 1977; Urano and Gorbman 1981; Aitken and Capranica 1984; Wetzel, Haerter, and Kelly 1985; Neary 1988; Schneider 1988; Walkowiak and Luksch 1994), as well as phonotactic responses by females (R. Schmidt 1969, 1971; Walkowiak and Luksch 1994). There are at least three pathways between the torus, auditory centers of the diencephalon, and the hypothalamus. These pathways provide a link between the auditory system and the center for hormonal control of reproductive behavior (Urano and Gorbman 1981; Brzoska and Obert 1980; Ball 1981; Neary 1988; Neary and Wilczynski 1986; Wilczynski and Allison 1989; Allison and Wilczynski 1991; Wilczynski 1992; Wilczynski 1993).

## Processing of Acoustic Signals

So far, I have discussed only the basic structure of the auditory system and the major peripheral and central auditory pathways, but not the way in which this system actually processes acoustic signals. Processing of acoustic information involves at least three major steps: (1) detection of the signal, (2) recognition of key properties of signals that distinguish them from other sounds in the environment, and (3) accurate localization of the signaler. Behavioral experiments have shown that anurans can localize the source of acoustic signals and generally are quite selective in their responses to acoustic stimuli. Selectivity in females usually is tested by presenting them with a choice of two or more sounds and observing their phonotactic responses, as discussed in a previous section (Gerhardt 1992a). In most experiments, male frogs are presented with recorded calls. Either an evoked vocal response or a reduction in calling during the stimulus is then measured (Capranica 1965; Paillette 1970a; Loftus-Hills 1973b; Narins and Capranica 1978; Narins 1982a; Walkowiak and Brzoska 1982; Zelick and Narins 1982; Megela and Capranica 1983; Schwartz and Wells 1983a, b, 1984; Megela Simmons 1984; Wells and Schwartz 1984a, b; Sullivan and Leek 1986; Allan and Megela Simmons 1994). These experiments have demonstrated that anurans are capable of fine-scale discrimination between conspecific and heterospecific calls and between different types of calls within their own species' repertoire. Discrimination between calls can be based on differences in frequency structure, temporal structure, or both (Gerhardt 1988; Walkowiak 1988b; Brenowitz, Rose, and Adler 2001).

### Processing in the Frequency Domain: Peripheral Tuning

Because the anuran tympanum has a relatively flat frequency response over a broad range of frequencies, the middle ear provides little frequency selectivity. It simply acts as a low-pass filter, cutting off frequencies above a certain range, but transmitting those below that range at similar amplitudes. In general, cut-off frequencies at the high end tend to decrease as body size and tympanum size increase. Hence, in a relatively small species, *Hyla cinerea,* there is rapid attenuation of frequencies above 3,500 Hz. For larger species, such as *Bufo americanus* and *Rana pipiens,* the cut-off is between 1,200 and 1,800 Hz (Moffat and Capranica 1978; Capranica and Moffat 1983; Wilczynski, Resler, and Capranica 1987).

The inner ear exhibits a greater degree of frequency selectivity. As previously discussed, nerve fibers innervating the basilar and amphibian papillae are tuned to different frequency ranges. Individual fibers have V-shaped tuning curves; that is, each fiber has a characteristic best frequency to which it responds at the lowest sound intensity threshold. Different populations of nerve fibers are tuned to different frequencies (fig. 7.39 A). This type of tuning is not unique to anurans, but is characteristic of all vertebrate auditory systems. The sharpness of tuning of individual nerve fibers in anurans is not very different from that found in mammals (Wilczynski and Capranica 1984). When responses of fibers from several individuals are pooled, the range of frequency sensitivity in the population can be determined (fig. 7.39 B). Because of structural characteristics of the two auditory organs discussed above, individual amphibian papillar fibers show a wide range of best frequencies within and between individuals, whereas basilar papillar fibers exhibit little variation within individuals, but differ in best frequencies among individuals of different size (Zakon and Wilczynski 1988).

In general, peaks of sensitivity in the inner ear roughly correspond to the distribution of energy in the species' own calls (Gerhardt and Schwartz 2001). This was first demonstrated in Capranica's (1965) studies of bullfrogs (*Rana catesbeiana*). Calls with energy in two bands, one at around 200 Hz and another at 1,500 Hz, were most effective in eliciting vocal responses from males, and this represents the natural distribution of energy in bullfrog calls. Similarly, in *Hyla cinerea,* advertisement calls have two major bands of energy at about 900 Hz and 3,000 Hz, and calls with both frequency bands present are most effective in eliciting phonotactic responses from females (Gerhardt 1981b, 1987). In both cases, the low frequency energy in the call corresponds to peak sensitivities of amphibian papillar fibers, while the high frequency bands correspond to peak sensitivities of

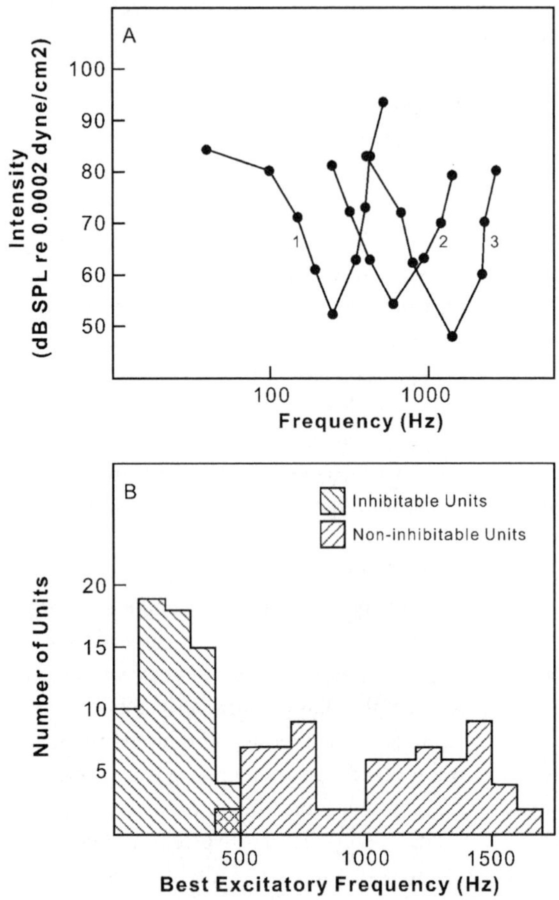


**Fig. 7.39.** (A) V-shaped tuning curves of three populations of nerve fibers of the bullfrog (*Rana catesbeiana*). Frequencies are plotted on a logarithmic scale. The low-frequency population has a best excitatory frequency (BEF) of about 250 HZ, the mid-frequency population has a BEF of about 600 HZ, and the high-frequency population has a BEF of about 1400 HZ. The low and mid-frequency populations are derived from the amphibian papilla, while the high-frequency population is derived from the basilar papilla. (B) Histogram of best excitatory frequencies of individual eighth nerve fibers. After Feng, Narins, and Capranica (1975).

basilar papillar fibers (Frischkopf, Capranica, and Goldstein 1968; Feng, Narins, and Capranica 1975; Capranica and Moffat 1983). Behavioral responses to different frequencies in *H. cinerea* are closely matched to sensitivity peaks in behaviorally derived audiograms of total hearing sensitivity for this species (Megela Simmons, Moss, and Daniel 1985). This is less true for bullfrogs, which exhibit relatively poor sensitivity to 200 Hz sounds, even though these are effective in eliciting vocal responses (Megela Simmons, Moss, and Daniel 1985). Furthermore, these frogs are not deaf to sounds outside the frequency range of their calls, but in fact are sensitive to a relatively broad range of frequencies (Capranica 1992).

In *Rana catesbeiana* and *Hyla cinerea,* co-activation of both the amphibian and basilar papilla seems to be necessary to elicit a maximum behavioral response to a call, but this clearly is not the case for many anurans. Most small frogs have relatively high-frequency calls that fall outside the range of the amphibian papilla and therefore excite only the basilar papilla (e.g., Hödl, Amezquita, and Narins 2004). In contrast, a few species, such as *Physalaemus pustulosus* and *Eleutherodactylus coqui,* have calls or call components falling only within the range of the amphibian papilla (Narins 1983; M. Ryan 1985b; G. Rose, Zelick, and Rand 1988). In species with high frequency calls, there generally is a good match between the tuning of basilar papillar fibers in the VIIIth nerve and the dominant frequency of the call (Capranica 1976a, b; Wilczynski, Zakon, and Brenowitz 1984; Zakon and Wilczynski 1988; Bosch and Boyero 2003a). The tuning of the basilar papilla is related to body size, so there often is parallel geographic or altitudinal variation in body size, dominant frequency of the call, and frequency sensitivity of the ear (Capranica, Frischkopf, and Nevo 1973; Nevo and Schneider 1976; Nevo and Capranica 1985; Narins and Smith 1986; M. Ryan 1988a). There is evidence from studies of *Acris crepitans* that the evolution of call frequency and basilar papilla sensitivity has been independent of body size. The result is a closer match between call frequencies and tuning of the auditory system than expected from interpopulation differences in body size alone (M. Ryan and Wilczynski 1988; Keddy-Hector, Wilczynski, and Ryan 1992; Wilczynski, Keddy-Hector, and Ryan 1992).

In some species, there is evidence of sexual dimorphism in peripheral auditory tuning that is related at least in part to differences in body size (Zakon and Wilczynski 1988). For example, in *Eleutherodactylus coqui,* amphibian papillar fibers of both males and females exhibit sensitivity to frequencies in the range of the "co" note (about 1,000 Hz). Excitation of this organ alone is sufficient to elicit a vocal response from males (Narins 1983). Males seem to pay little attention to the higher frequency "qui" note, which falls within the range of the basilar papilla (Narins and Capranica 1976, 1978, 1980). Females, on the other hand, are attracted to this note, but are little influenced by the co note. Females have a population of basilar papillar fibers that are well tuned to the frequency of the qui note (about 2,000 Hz). Males, on the other hand, are smaller than females and have many basilar papillar fibers tuned to frequencies above 3,000 Hz, well above the frequency range of any conspecific calls (fig. 7.40).

Similar sexual dimorphism has been found in the tuning of basilar papilla fibers in *Pseudacris crucifer* (Wilczynski, Zakon, and Brenowitz 1984) and other hylids (McClelland, Wilczynski, and Rand 1997). In these cases, the result is that ears of females are well matched to the average frequency of the call, whereas males have peak sensitivity at higher frequencies. In many species, however, the match between basilar papilla tuning and call frequency is not perfect. In several species, including *Physalaemus pustulosus* and *Acris*

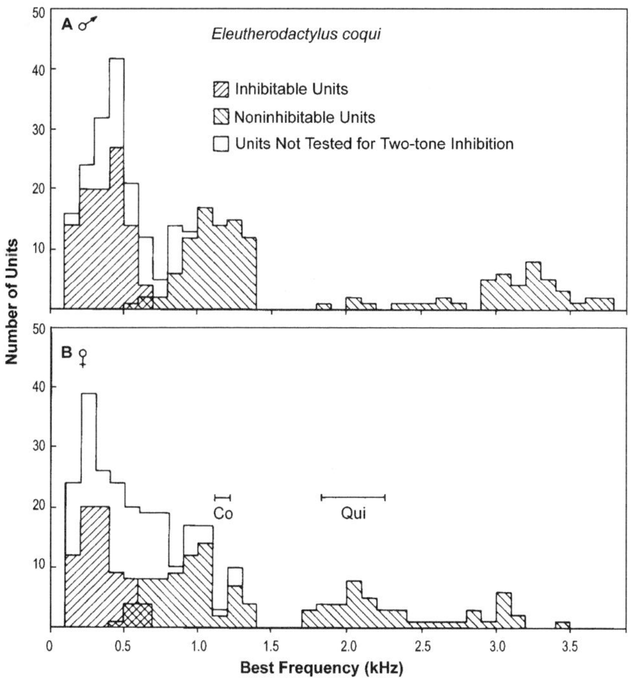


**Fig. 7.40.** Sexual dimorphism in tuning of peripheral auditory system in *Eleutherodactylus coqui.* Nerve fibers with best excitable frequencies below 1.5 kHz are derived from the amphibian papilla, while those above 2 kHz are derived from the basilar papilla. The latter exhibit sexual differences in tuning, with males having fibers tuned to higher frequencies. After Narins and Capranica (1976).

*crepitans,* the basilar papilla of females is tuned to slightly lower frequencies than the average frequency of male calls in the same population (Gerhardt and Schwartz 2001). One possible explanation for this is that tuning of the auditory system represents a trade-off between long-distance and short-distance communication (Sun et al. 2000). This is because high frequencies in calls exhibit greater attenuation with distance than do low frequencies, so an ear tuned to lower than average frequency could be adapted to detect calls at a distance. It also is possible that auditory tuning and call frequency structure are subject to different evolutionary constraints, and any deviation from a theoretical optimum or perfect match does not reflect an adaptive response.

It is important not to read too much into a match, or lack of a match, between the average dominant frequencies of male calls in a population and average best frequencies in the tuning curves of females. The tuning curves of frogs are not very sharp, but exhibit a fairly broad range of frequency sensitivity. The match between the tuning curves of females and male calls could be sufficient to bias choices of females toward conspecific calls and away from heterospecific calls that differ significantly in frequency structure. It is not clear, however, that this can account for choices that females make among conspecific calls differing by a few hundred Hertz in dominant frequency. This is especially true for species with high-frequency calls. Because of the structure of the basilar papilla, female frogs are expected to have relatively poor intensity-independent discrimination of frequencies that stimulate only that organ (Zakon and Wilczynski 1988; Gerhardt and Schwartz 2001).

Schwartz and Gerhardt (1998) found relatively poor frequency discrimination in female spring peepers (*Pseudacris crucifer*), especially in the absence of background noise. Furthermore, the choices of individual females that were tested several times often were not consistent. This is the only study in which the tuning curves of individual females have been determined after testing their behavior in playback experiments. The results showed that a female's preference for call dominant frequency was not correlated with best excitatory frequency, as determined from multi-unit recordings in the brain. This calls into question adaptive explanations for female preferences based on average best excitatory frequencies for whole populations (e.g., Sun et al. 2000; see further discussion of female choice in chapter 8).

### Processing in the Frequency Domain: Central Nervous System

Tuning of the peripheral auditory system is only the first step in frequency processing. Further processing and integration of frequency information takes place in the auditory centers of the brain. Most of our knowledge of frequency processing in the brain is derived from studies of the response characteristics of individual neurons. We know a good deal about tuning characteristics of these neurons and how they respond to changes in frequency, but we know less about how information from neurons with different response characteristics is integrated. We also know relatively little about where this integration takes place, how the nervous system actually recognizes different types of calls, or how behavioral responses to calls are regulated (Fuzessery 1988). Therefore, only a brief outline of frequency processing in the central nervous system will be given here.

Most neurons in the dorsolateral nucleus seem to reflect the tuning characteristics of the VIIIth nerve fibers that terminate there. Cells in this region exhibit the V-shaped tuning curves characteristic of VIIIth nerve fibers, and the same three populations of fibers tuned to low, middle, and high frequencies can be identified. Hence, it appears that relatively little additional processing occurs at this level (Feng and Capranica 1976; Fuzessery and Feng 1981, 1983a; Fuzessery 1988). Many of these characteristics also are found in neurons of the superior olivary nucleus, but there also is evidence of neuronal interactions at this level that result in a sharpening of frequency sensitivity (Fuzessery 1988).

This sharpening of frequency tuning continues in the torus, where the majority of neurons exhibit single best frequencies like those at lower levels. However, other neurons in the torus exhibit a variety of responses not seen at lower levels. Some show little evidence of narrow tuning, but respond to a wide range of frequencies. Others have a bimodal response to two distinct frequencies, but are inhibited by or unresponsive to frequencies in between. Still others exhibit maximal responses to specific combinations of frequencies (Loftus-Hills 1971; Walkowiak 1980a; Hermes et al. 1981, 1982; Fuzessery and Feng 1982; Fuzessery 1988; Diekamp and Gerhardt 1992). Hence, in the torus we find the first evidence of an ability to selectively respond to sounds with complex frequency structure.

Further processing almost certainly occurs in higher brain centers as well, but our knowledge of frequency processing in the diencephalon and telencephalon is very incomplete and is based mostly on studies of *Rana pipiens*. In this species, some cells in the thalamus are highly selective in their responses to tone combinations, giving maximum responses to combinations corresponding to the distribution of energy in the advertisement call (Mudry, Constantine-Paton, and Capranica 1977; Fuzessery and Feng 1983b; J. C. Hall 1994). Mudry (1978) found evidence of similar selective responses to tone combinations in *Rana catesbeiana* and *Hyla cinerea* (cited in Capranica and Moffat 1983; Capranica and Rose 1983). In all three of these species, the excitatory tone combinations result in stimulation of both the amphibian and basilar papilla. How applicable these results are to anurans in general, particularly those with only a single frequency band in the call, is not yet known.

### Processing in the Time Domain

The early work of Capranica and his colleagues focused almost exclusively on the frequency-tuning characteristics of the anuran auditory system. They were especially concerned with identifying potential "mating call detectors" in the brain that might function as a matched frequency filter to recognize biologically relevant sounds (e.g., Capranica and Moffat 1983). Textbook accounts often stop at this point, giving the misleading impression that processing of acoustic signals is simply a matter of recognizing the distribution of sound energy in the calls (e.g., Hopkins 1983; Duellman and Trueb 1986). While much of this early neurobiological work was being done, however, a large body of behavioral evidence was accumulating showing that recognition of conspecific calls and discrimination between different calls within a species' repertoire often is based on temporal features of the calls and not just frequency structure (see Gerhardt 1988, 1994a, c; Gerhardt and Schwartz 1995; Gerhardt and Huber 2002; and previous sections of this chapter). Consequently, in the 1980s investigators began to consider ways in which temporal features of calls are processed (Capranica and Rose 1983; Walkowiak 1988a; J. C. Hall 1994; Brenowitz, Rose, and Alder 2001; Gerhardt and Huber 2002; Narins and Feng 2006).

In contrast to the frequency selectivity of the peripheral auditory system, VIIIth nerve fibers are not tuned to particular rates of amplitude modulation (Feng, Hall, and Sidique 1991). Instead, the nerve simply fires in synchrony with amplitude modulated signals up to a certain rate (Walkowiak 1988a). To date, most species tested have been those with relatively low rates of amplitude modulation in their calls (less than 150 pulses/s). These species exhibit a cut-off rate at about 100–150 Hz (Capranica and Moffat 1975; Capranica and Rose 1983; G. Rose and Capranica 1983, 1985; Walkowiak 1984, 1988a). In the green treefrog (*Hyla cinerea*), however, rates of amplitude modulation within calls are much higher (260–300 Hz), and the auditory neurons of this species are capable of tracking such modulation (Klump et al. 2004). This is not surprising, because we know from behavioral experiments that some species with much higher pulse rates can distinguish between calls on the basis of pulse rate alone. For example, *Hyla microcephala* females exhibit a clear preference for calls of 225 pulses/s over those with pulse rates of 95, 150, or 325 pulses/s (Schwartz 1987b). It would be interesting to know the cut-off rate for VIIIth nerve synchrony in this species. It should be noted that these tropical frogs are active at a higher temperature (about 26° C) than temperate zone species tested previously (usually at around 20° C), and the ability to synchronize with an amplitude-modulated signal is very likely to be temperature dependent.

Most neurons in the acoustic nuclei of the medulla (dorsolateral nucleus and superior olivary nucleus) exhibit the same low-pass characteristics found in VIIIth nerve fibers, but are unselective in their responses to AM rates within the appropriate range. However, some cells appear to be selectively tuned to specific AM rates (Walkowiak 1988a; Condon, Chang, and Feng 1991), a trend that becomes more pronounced in the torus and probably continues in high brain centers such as the thalamus (J. C. Hall and Feng 1986, 1987; Walkowiak 1988a). The torus has been the site of the most intensive studies of temporal processing. Single-cell recordings in ranids, bufonids, hylids, and leptodactylids have all yielded similar results. Several types of neurons have been identified in the torus, each with a distinct response to AM signals (fig. 7.41). Some cells are tuned to specific rates of amplitude modulation, often approximating the rate of the species' advertisement call. However, the occurrence of AM-tuned units in the midbrain does not appear to be a specific adaptation of anurans for detecting conspecific calls, but rather a general property of vertebrate brains (G. Rose 1986). Furthermore, not all toral cells are AM-tuned. Many are nonselective, responding equally well to a wide range of amplitude-modulated signals. Still others act as high or low-pass filters, responding only to AM rates above or below a certain level. Finally, some units are suppressed

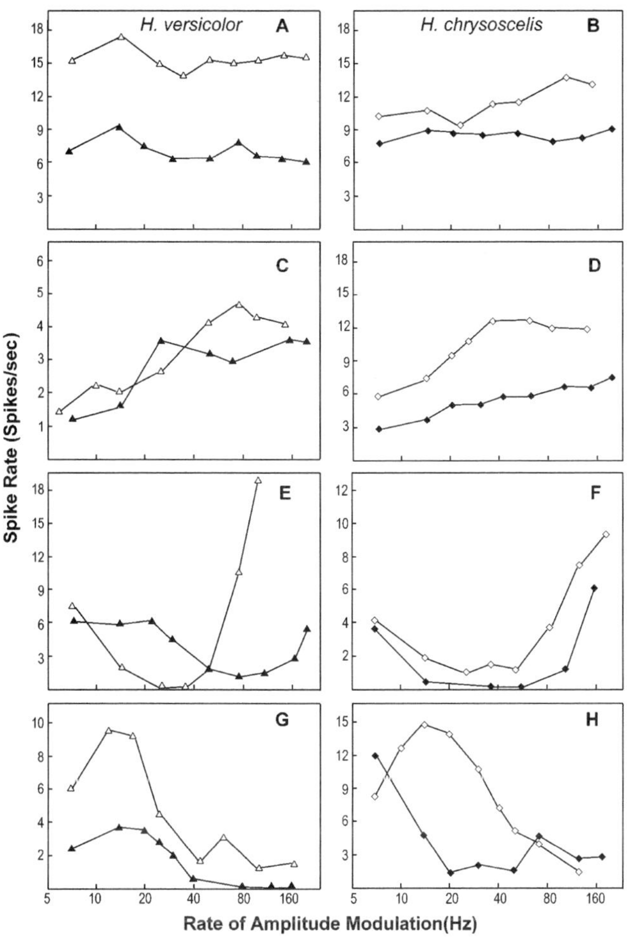


**Fig. 7.41.** Iso-intensity spike rates of single units of the torus semicircularis of two species of treefrogs, *Hyla versicolor* and *H. chrysoscelis,* as a function of the rate of amplitude modulation of a white-noise sitimulus. The two types of symbols indicate recordings from different neurons. Some units show nonselective responses to amplitude-modulated stimuli (A, B). AM high-pass units (C, D) respond strongly to high rates of amplitude modulation. Band-suppression units (E, F) respond to high and low rates of modulation, but less strongly to intermediate rates. Low-pass units (G, H) respond primarily to low rates of amplitude modulation. Not shown are band-pass units that respond to a range of intermediate AM rates. After G. Rose, Brenowitz, and Capranica (1985).

by certain AM rates (Capranica and Rose 1983; G. Rose and Capranica 1983, 1984, 1985; Walkowiak 1984, 1988a; Capranica, Rose, and Brenowitz 1985; G. Rose, Brenowitz, and Capranica 1985; Epping and Eggermont 1986a, b; J. C. Hall 1994; Penna, Lin, and Feng 1997). Responses to rates of amplitude modulation in the torus are temperature-dependent (Brenowitz, Rose, and Capranica 1985; G. Rose, Brenowitz, and Capranica 1985), which probably accounts in part for temperature-dependent changes in behavioral preferences for pulse rate observed in some treefrogs (Gerhardt 1978c; Gerhardt and Doherty 1988). Neurons in the torus also are selectively responsive to intrapulse amplitude modulation, pulse duration, pulse shape, and call duration (Penna, Lin, and Feng 1997; Alder and Rose 2000).

### Long-term Temporal Integration

Recent work in which synthetic calls were played to *Pseudacris regilla* has shown that cells in the torus integrate the temporal characteristics of calls, especially pulse repetition rate, over relatively long periods of time (on the order of hundreds of milliseconds). This type of long-term temporal integration enables the frogs to discriminate between advertisement and aggressive calls that differ mainly in pulse rate (Alder and Rose 1998; Brenowitz, Rose, and Aldler 2001; G. Rose and Brenowitz 2002). Long-term temporal integration also has been demonstrated in *Rana pipiens*. Specifically, neurons in the torus that were most sensitive to relatively high pulse repetition rates (40–50 pulses/s in this species) required a threshold number of pulses to be presented before they could respond to the stimulus. Once the neurons began to respond, responses continued as long as the stimulus continued, so responses were greater to longer stimuli. At pulse rates above those to which the neurons were most sensitive, neural responses were reduced, apparently because they did not have sufficient time to recover between pulses.

The need for female frogs to perform long-term integration of male signals to detect the appropriate pulse rate of conspecific advertisement calls could drive the evolution of call structure and calling behavior in males. In amplitude-modulated calls consisting of trains of repeated pulses, long calls contain more pulses repeated at the appropriate rate than do short calls, and females might prefer males that give longer calls. Such preferences could be considered to result from sensory exploitation (M. Ryan 1990), with males exploiting a pre-existing sensory mechanism (long-term integration of pulse rate) to make their calls more attractive (Alder and Rose 2000; G. Rose et al. 2002).

Female preferences for long calls have been demonstrated in a number of species, including *Hyla versicolor* (Klump and Gerhardt 1987; see also chapter 8). In that species, males respond to the presence of females by greatly lengthening their calls beyond the duration of normal advertisement calls. As discussed in a previous section of this chapter, males of other species make similar adjustments in their calls when females are detected nearby, often giving longer trains of pulses or repeating a pulsed call at a higher rate, as in *Hyla ebraccata* and *H. microcephala*. In *H. ebraccata,* females prefer calls with the largest number of pulses repeated at the pulse rate of the advertisement call. Hence, they prefer single-note advertisement calls to aggressive calls of the same duration; the latter have much higher pulse rates. They also prefer aggressive calls to which secondary notes have been added to those consisting only of the introductory note. Secondary notes have the same pulse rate as do advertisement calls. Multi-note advertisement calls with long primary notes and short secondary notes of the same pulse rate were about as attractive to females as a continuous single-note call of the same duration calls (Wells and Bard 1987), a result that supports the temporal integration model of Brenowitz, Rose, and Adler (2001).

### Processing of Signals in Noise

Most of the neurobiological studies discussed above have involved testing of individuals under ideal experimental conditions, with pure tones or AM stimuli being delivered directly to the auditory system. However, frogs in the real world frequently communicate in a very noisy environment, in which conspecific calls, calls of other species, and nonbiological sources of noise all reduce their ability to detect and recognize calls of other individuals (Narins and Zelick 1988; Wollerman 1999). Frogs exhibit a variety of behavioral strategies to reduce the effects of background noise, including a remarkable ability to detect even very brief periods of quiet and insert their own calls in those periods (Schwartz and Wells 1983a, b; Zelick and Narins 1983). These behavioral responses were discussed in more detail in the section on chorusing interactions.

The main effect of noise on the auditory system is to reduce the sensitivity of individual nerve fibers and to saturate the auditory system so that signals of interest cannot be distinguished from background noise. Such physiological masking is not limited to simultaneous exposure to the signal and to noise; exposure to loud noise can reduce the sensitivity of the ear for time period after the exposure ceases (Megela and Capranica 1982; Zelick and Narins 1985b). Hence a major difficulty for animals trying to communicate in noise is to make accurate intensity discriminations that allow them to pick out relevant signals from the background. One way of accomplishing this is to increase the thresholds of individual neurons to avoid overloading the system. An auditory nerve fiber fires at an increasing rate as the sound pressure level of a stimulus increases, up to a certain point at which the fiber is saturated and firing at its maximum rate. Hence, differences in firing rate can be used to code differences in signal intensity. In *Eleutherodactylus coqui,* Narins (1987) found a shift in the relationship between intensity and firing rate in the presence of broadband noise, effectively decreasing the sensitivity of the lowest threshold fibers and avoiding complete saturation of the auditory system. This would allow the frog to make intensity discriminations needed for signal detection even at high noise levels. Signal detection also is enhanced because individual neurons tuned to the same frequency have a wide range of intensity thresholds. Hence, at high noise levels, the most sensitive fibers become saturated, but higher thresholds units are still able to detect the signal (Capranica and Moffat 1975; Feng 1982; Narins and Zelick 1988).

The tuning of peripheral auditory nerve fibers to frequencies near those of conspecific calls, and the sharpening of that tuning in the central nervous system, enable a frog to extract signals from noise more effectively than would be possible if all fibers were broadly tuned. However, this alone is not sufficient, because background noise in a chorus often has energy concentrated around the frequency of the frog's own call (Gerhardt and Klump 1988a; Narins and Zelick 1988). Nerve fibers differ in their capacity to reject noise near the frequency of the signal being detected, so the behavioral effects of masking depend on both the frequency and intensity of the masking sound (Ehret and Gerhardt 1980; Narins 1982b). In general, frequency selectivity is greatest at low frequencies, where tuning curves of auditory nerve fibers are relatively narrow, and gradually decreases at higher frequencies (Ehret and Capranica 1980), a pattern found in many vertebrates. However, using a behavioral technique, Moss and Megela Simmons (1986) found that the ability of *Hyla cinerea* to extract signals from noise is greatest at frequencies similar to those in the advertisement call. In contrast, *Rana catesbeiana* exhibits its greatest capacity to reject noise at around 900 Hz, a frequency not present in the advertisement call. This suggests that matching of frequency selectivity to the spectrum of the advertisement call is not a general feature of all anuran auditory systems (Megela Simmons 1988).

Chorus noise often obscures key temporal features of calls used in call recognition, so even if the call can be detected, discrimination between calls is impaired (Schwartz and Wells 1983a, b; Schwartz 1987a; Gerhardt and Klump 1988a; Schwartz and Gerhardt 1989). This masking effect is somewhat reduced if the signal and interfering noise come from different directions, but the improvement is relatively small (Schwartz and Gerhardt 1989). Narins and Zelick (1988) suggested that the ability to detect specific temporal features such as pulse rate in the presence of background noise should be enhanced by AM-selective neurons, because they are keyed to a particular temporal pattern not present in the random fluctuations of the chorus sound.

### Sound Localization

Once a female frog has detected and recognized a conspecific call, she still is faced with the task of locating the calling male. The same is true for a male trying to locate an intruder in his territory. Often this is made more difficult by the types of calling sites chosen by males, which often are hidden in dense vegetation or located in tree holes or burrows that distort the directional cues available to the receiver. Yet frogs clearly do manage to locate one another, and often do so very quickly in a natural chorus. Several behavioral studies have demonstrated that frogs can accurately locate a speaker broadcasting conspecific calls and can do so in both the horizontal and vertical planes (Feng, Gerhardt, and Capranica 1976; Rheinlaender et al. 1979;

Gerhardt and Rheinlaender 1980; Passmore et al. 1984; Rheinlaender and Klump 1988; M. Jørgensen and Gerhardt 1991). In some species, the female will move her head from side to side before orienting her body toward the sound source, but other species do not exhibit head-scanning behavior (M. Jørgensen and Gerhardt 1991). The frog then jumps toward the sound source, often with some error that is corrected for on the next jump. The result is a characteristic zigzag pattern of approach to the speaker. Rheinlaender et al. (1979) showed that scanning with the head before jumping improves the accuracy of approach to the speaker. Both ears are required for sound localization, and covering one tympanum with grease causes the frog to turn in a circle (Feng, Gerhardt, and Capranica 1976). These early studies demonstrated that a frog could determine whether a sound was coming from the left or right of its head, but did not demonstrate an ability to discriminate between different angles of sound incidence. Later work by Klump and Gerhardt (1989) showed *that Hyla gratiosa* females are capable of true angle discrimination.

All of the frogs tested so far for sound localization ability are very small species, and one, *Colostethus flotator,* is one of the world's smallest frogs, with a distance between the ears of less than 5 mm (Gerhardt and Rheinlaender 1980). This presents a serious problem for sound localization, because the small distance between the ears makes it nearly impossible for frogs to make use of the two types of cues usually employed by vertebrates to locate sounds: differences in the intensity of sound reaching the two ears and differences in time of arrival. The paradox of sound localization by such small animals is partially resolved if one assumes that localization is not based simply on comparisons of input from the two ears, but also on comparison of input to the two sides of the tympanum. In other words, the anuran ear seems to function as a pressure gradient system that compares sound waves reaching the two sides of the tympanum through different pathways (Rheinlaender, Walkowiak, and Gerhardt 1981).

Research on the directionality of hearing in frogs is highly technical, much of it focused on the effects of sound source direction on responses of neurons in the peripheral auditory system or in higher brain centers. Only a brief summary will be given here (for a more detailed review, see Gerhardt and Huber 2002). The middle ear cavity is open to the buccal cavity through the Eustachian tube, so several routes to the inside of the tympanic membrane are possible. Potential routes include the ear on the other side of the head, the skin over the lungs, and other as yet unknown extratympanic pathways such as the tissues of the head (Rheinlaender, Walkowiak, and Gerhardt 1981; Pinder and Palmer 1983; Vlaming, Aertsen, and Epping 1984; Aertsen et al. 1986; Michelsen, Jørgensen, and Christensen-Dalsgaard 1986; Wilczynski, Resler, and Capranica 1987; Eggermont 1988; Narins, Ehret, and Tautz 1988; M. Jørgensen 1991; M. Jørgensen, Schmitz, and Christensen-Dalsgaard 1991; Ehret, Werth, and Kamada 1994). The tympanum is inherently directional in its responses to sound. Directionality varies with sound frequency, but in most species studied to date, peak directionality does not correspond precisely to the dominant frequency of the advertisement call (M. Jørgensen 1991; M. Jørgensen and Gerhardt 1991; M. Jørgensen, Schmitz, and Christensen-Dalsgaard 1991; Wang, Ludwig, and Narins 1996). The directionality of tympanic responses appears to be enhanced by input from extratympanic pathways, such as the lungs (M. Jørgensen 1991; M. Jørgensen, Schmitz, and Christensen-Dalsgaard 1991; Wang, Ludwig, and Narins 1996), but it is not clear whether this actually enhances the ability of a frog to locate a sound source. The responses of neurons in the auditory nerve and higher brain centers also are sensitive to changes in the direction of a sound source. This sensitivity can affect both their frequency response and responses to temporal patterns such as amplitude modulation (Feng 1981; Feng and Capranica 1976, 1978; Schmitz, White, and Narins 1992; White, Schmitz, and Narins 1992; Gooler et al. 1993; Xu, Gooler, and Feng 1994, 1996; Schwartz and Gerhardt 1995; Wang, Ludwig, and Narins 1996). What is not yet fully understood is how directional information from the two ears is integrated in the brain to allow a frog to locate a calling male.

## Summary and Conclusions

In most anurans, sound is produced when the trunk muscles force air through the larynx into the buccal cavity and vocal sac. Both the trunk muscles and laryngeal muscles of male frogs exhibit morphological, physiological, and biochemical features that enhance their aerobic capacity and make them well suited for long periods of sustained activity. Furthermore, variation in morphological, physiological, and biochemical features of muscles parallels variation in calling activity among species (see chapter 5). Electromyographic studies have shown that in some species with slow pulse repetition rates, the trunk muscles are involved in production of individual call pulses. In two of these species, *Hyla versicolor* and *H. chrysoscelis,* differences in the biochemical and contractile properties of the trunk muscles parallel differences in pulse rate. We also know next to nothing about the physiological characteristics of vocal sac musculature or the role of these muscles in sound production, although there is evidence that these muscles are unusually elastic in some species. The size and shape of the vocal sac in relation to the wavelength of sound being radiated affect the efficiency with which metabolic energy is converted into acoustic energy, but again, little comparative work has been done on calling efficiency. Some frogs lack vocal sacs and use other modes of communication in aggressive interactions or in attracting mates, but the behavior of most voiceless anurans has not been studied in the field.

Anuran vocal repertoires can be divided into several functional types of calls, including advertisement calls, courtship calls, aggressive calls, release calls, and distress or defensive calls. Advertisement calls are either simple, single-note calls, or complex calls consisting of one or more types of repeated notes. Complex calls of some species appear to have attractive and aversive elements combined into one call, so a male can convey separate messages to males and females simultaneously. In other species, males alter the temporal structure of part of the advertisement call, turning it into a call that functions mainly in male-male aggressive interactions. Aggressive calls of most frogs are readily distinguished from advertisement calls of the same species, but there is no temporal structure that is common to aggressive calls of all species. Some frogs have more than one type of aggressive call used in long-range and short-range encounters, or they have graded aggressive calls that change gradually as frogs approach one another. The function of graded aggressive calls is not yet clear, but they could enable a male to convey a more precise message about his aggressive motivation, or they might be used in assessing the qualities of an opponent. Graded calls also could allow a male to gradually shift from attractive to aversive calls by dropping elements that are attractive to females only when immediately threatened by a competitor.

Several features of the physical environment potentially influence the evolution of anuran calls and calling behavior, but this aspect of vocal communication is only beginning to be explored. Although there have been suggestions that the frequency structure of bird songs is adapted for transmission through particular habitats, there is little evidence that this is true for anurans. Nevertheless, the height of the caller above the ground, the density of vegetation surrounding his calling site, and other aspects of the physical environment will affect the attenuation of different frequencies in a frog's call and thereby affect the way a signal is perceived by other individuals. Some authors have suggested that frogs breeding in open country have louder calls than do those breeding in forests, but there has not been a systematic attempt to test this hypothesis.

The evolution of calls also can be affected by background noise derived from the physical environment. For example, the sound of running water, particularly near waterfalls or torrents, can mask frog calls; highly repetitive calls or narrowly tuned calls can be effective in counteracting acoustic masking. Problems of sound transmission through air are most severe near the ground, particularly for species calling from burrows. However, at least one species, *Leptodactylus albilabris,* is highly sensitive to substrate-born vibrations and apparently can use the ground as a relatively noise-free channel for acoustic communication. The extent to which this is true for other burrowing frogs needs to be investigated.

The biological environment also has a major effect on the evolution of vocal communication. Some bats and several other types of predators can use the calls of frogs to locate prey, and the presence of these predators could affect the structure of a frog's call, the temporal pattern of calling behavior, and the nature of the calling site. Many frogs call in dense choruses of both conspecific and heterospecific individuals. Often these are incidental aggregations around choice breeding sites, but individuals also could derive various benefits from being in an aggregation, including greater protection from predators or a greater chance of attracting females. However, the costs and benefits of chorusing behavior are poorly understood for most species.

One major cost of calling in either a group of conspecifics or a mixed-species chorus is acoustic interference from other calling males. Such interference reduces a female's ability to perceive and locate individual calling males, and it makes advertisement of a male's territory to other males more difficult. At the intraspecific level, most chorusing frogs exhibit adaptations to reduce acoustic interference and increase the distinctiveness and detectability of their own calls. These include matching the calling rate of neighboring males, timing calls to avoid overlap with the calls of other individuals, and changes in the complexity or duration of advertisement calls in response to those of other males. Experiments with females of a few species have shown that males usually alter their vocalizations in choruses in ways that enhance their ability to attract mates, but the dynamics of chorusing behavior need to be studied in much more detail. Most playback experiments with females have been conducted in highly simplified acoustic environments, and the ways in which females detect signals in the complex environment of a natural chorus is not well understood. Consequently, the relationship between socially mediated changes in the calling behavior of males and their mating success has scarcely been explored (see chapter 8).

Many of the behavioral adaptations for avoiding acoustic interference at the intraspecific level extend to interspecific interactions as well. For example, frogs often are inhibited from calling by the calls of other species with a similar frequency structure. Individuals can avoid interference by calling in silent intervals between bouts of calling by other species, or they alter the timing of their calls in response to individual calls. The effectiveness of heterospecific calls in eliciting such responses is a function of the similarity of spectral and temporal features of calls. Some frogs also engage in aggressive interactions with other species and use aggressive calls for interspecific communication. Again, the effectiveness of calls in repelling heterospecific males depends on the similarity of the calls of the two species.

Mixed-species choruses also present problems to females searching for mates, because they must correctly identify males of their own species from a cacophony of different advertisement calls. The role of vocalizations in reproductive isolation and the identification of properties important for call recognition are perhaps the best studied aspects of anuran communication. Choice experiments with many species have shown that female anurans generally have no difficulty distinguishing conspecific calls from those of other species. The use of synthetic calls to systematically vary individual properties of calls has been an effective way to identify call properties critical for species recognition. These experiments have revealed that some species-specific features of calls are irrelevant to females, and that females of closely related species often use quite different properties to discriminate against calls of other species. A major challenge for future investigators is to determine how females integrate information about different properties of calls when choosing a mate in a natural chorus.

Despite the species-specific nature of anuran calls, closely related species often share basic structural characteristics of their calls that have been inherited from a common ancestor. These similarities have been used to identify higher-order relationships among groups of species, or to reconstruct the phylogeny of a particular clade. The use of independently generated phylogenies to examine the evolution of call structure promises to be an important approach for future investigators. Limited evidence for a few groups, including North American toads (*Bufo*) and chorus frogs (*Pseudacris*) suggests that changes in call structure often are congruent with phylogenies based on morphological or molecular characters. As with any study based on phylogenetic evolution, however, our understanding of call evolution will only be as good as the phylogenetic hypothesis used to examine it. Changes in the proposed phylogenetic relationships of closely related species, such as those that have occurred several times for the genus *Physalaemus*, can change conclusions about the evolution of call structure. Studies that couple phylogenetic studies of call evolution with investigations of changes in the morphology and physiology of the call-producing apparatus would be particularly useful.

The structure and function of the anuran auditory system has been a major focus of research for many years, but our understanding of how acoustic signals are processed is still very incomplete. We know a good deal about frequency response characteristics of the peripheral auditory system and the role of the amphibian and basilar papillae in sound reception. We also have a considerable amount of information on the frequency selectivity of individual neurons in the medulla and midbrain acoustic centers, but considerably less about frequency processing at higher levels. Although many behavioral studies have shown that species recognition and discrimination among call types within a species' repertoire often is based on temporal features of the calls, there were few studies of temporal processing until the 1980s. These studies have shown relatively little temporal selectivity in the peripheral auditory system, but some selectivity for specific rates of amplitude modulation in the midbrain and perhaps in higher auditory centers as well. However, much of our understanding of sound processing in the central nervous system is based on studies of a limited number of ranids, bufonids, and hylids. We have no information on temporal processing in species that have calls with extremely high rates of amplitude modulation. Unfortunately, many anurans with calls of this type are very small and not ideal subjects for neurophysiological work. Nevertheless, the diversity of call structures in anurans presents a challenge for future studies of the neurophysiological correlates of call recognition. We also know very little about how frequency and time information are integrated in the brain, but the early concept of a single "mating call detector" in the brain seems to be an oversimplification.

The ability of female frogs to locate calling males in a chorus or artificial sound sources has been well established. It is now generally accepted that sound localization involves a pressure-gradient system in which inputs from two sides of each tympanic membrane are compared. These inputs come not only from sound impinging directly on the tympanum, but from extratympanic pathways such as the ear on the opposite side of the head, the lungs, and perhaps the bones and tissues of the head. Both the peripheral and central auditory systems exhibit inherent directional responses, but the way in which directional information from the two ears is integrated to allow for accurate sound localization is not yet understood.

# **Chapter 8** Mating Systems and Sexual Selection in Anurans

*It is surprising that these animals have not acquired more strongly marked sexual characters; for, though cold-blooded, their passions are strong. . . . An unfortunate female toad [is sometimes found] dead and smothered from having been so closely embraced by three or four males. Frogs have been observed . . . fighting all day long during the breeding season.*

*Among almost all animals there is a struggle between the males for possession of the female. . . . Females have the opportunity of selecting one out of several males, on the supposition that their mental capacity suffices for the exercise of a choice.*
—Charles Darwin, *The Descent of Man and Selection in Relation to Sex* (2nd ed., 1874)

IN 1871, Charles Darwin published his second most famous book, *The Descent of Man, and Selection in Relation to Sex*. Although his discussion of human evolution was the most controversial part of the book, more than half the book was devoted to animal behavior and Darwin's theory of sexual selection. This has been one of his most lasting contributions to biology, and was the starting point for all modern investigations of sexual selection and the evolution of mating systems. Darwin first proposed the theory of sexual selection to explain the evolution of morphological traits in animals that did not appear to be products of natural selection. He noted that male animals often are larger than females, more brightly colored, or endowed with special weapons used in competition for mates. He argued that such traits, which he called secondary sex characters, often decrease the survivorship of males, either because they are more vulnerable to predators, or because they kill or injure each other in fights. Nevertheless, if such traits enhance a male's ability to acquire mates, then they will be favored by sexual selection.

Darwin recognized two components of sexual selection that could lead to the evolution of secondary sex characters or differences in body size between males and females. The first was the relatively uncontroversial observation that males often fight among themselves for possession of females, leading to selection for weapons and fighting ability. The second was female choice, which involved females comparing the traits of potential mates and selecting those that were the most attractive. Many of Darwin's contemporaries were skeptical that female choice occurred in nature, in part because Darwin attributed to animals a sense of beauty like that of humans. Although a few investigators studied sexual selection and female choice, especially in birds, the subject was largely neglected until the second half of the 20th century (Cronin 1992). Sexual selection, and specifically female choice, is now among the most active areas of research in behavioral ecology and evolutionary biology (Andersson 1994).

Darwin had relatively little to say about amphibians in his book, although he did note that male frogs sometimes fight among themselves for possession of females, and he suggested that the calls of frogs are analogous to the songs of birds, and probably evolved through sexual selection. One of the few investigators in the early twentieth century who addressed questions of sexual selection and mate choice in anurans was G. Kingsley Noble. He published a pioneering study of the behavior of *Hyla andersonii* in the Pine Barrens of New Jersey (Noble and Noble 1923), and he contributed to studies of the mating behavior of other frogs (Noble and Aronson 1942). He also devoted a chapter of his book on *The Biology of the Amphibia* (1931) to secondary sex characters. Serious study of the mating systems of anurans began with the publication of several studies of territoriality and mate choice in frogs (Emlen 1968, 1976; Heusser 1969d; Wiewandt 1969; Wells 1977c, 1978b; R. D. Howard 1978a; Fellers 1979a, b) and a review of anuran social behavior (Wells, 1977b). Since then, anurans have become important model organisms for testing hypotheses about sexual selection and mate choice (Arak 1983a; M. Ryan 1985b, 1991; Gerhardt 1988, 1994a; Halliday and Tejedo 1995; B. K. Sullivan, Ryan, and Verrell 1995; Grafe, Spieler, and König 1999; Gerhardt and Huber 2002).

## Mating Systems and Sexual Selection

The process of sexual selection takes place in the context of particular mating systems. These encompass the array of behavioral tactics used by individuals to acquire mates. Most studies have focused mainly on the mating systems of male animals, but sexual selection can operate on traits of females as well. Females also can vary in their choosiness, depending on the relative costs and benefits of selecting particular males as mates (Jennions and Petrie 1997). In general, males are subject to more intense sexual selection than are females because of differences in the way each sex invests energy in the production of gametes. Females usually produce eggs that are provisioned with large yolk reserves to support the growth and development of the embryos, and some provide additional nutritional resources to embryos retained inside the body (chapter 10) or to young after they have hatched (chapter 11). Males, on the other hand, produce sperm that are supplied with only enough energy to get them to the eggs. This means that the cost of sperm production generally is lower than the cost of egg production, even when males produce millions of sperm, and they usually have sufficient sperm supplies to fertilize the eggs of many females. This difference in parental investment in gametes means that the reproductive success of males usually is limited by the number of mates they can acquire, whereas the reproductive success of females tends to be limited by energy availability (Trivers 1972). This disparity in parental investment means that the potential reproductive rate of males usually exceeds that of females, and females then become a limiting resource and the object of competition among males (Clutton-Brock and Vincent 1991; Clutton-Brock and Parker 1992; G. Parker and Simmons 1996). Traits that enhance the ability of males to locate, attract, capture, or retain possession of females will be favored by sexual selection, as will traits that reduce the amount of time between successive matings (Arnold and Duvall 1994).

In general, male animals adjust their mate-locating tactics to the spatial and temporal distribution of females (Emlen and Oring 1977; Arnold and Duvall 1994). When females are highly dispersed, males often have little choice but to search over wide areas for mates. This type of mating system is common in reptiles that have large home ranges, such as widely foraging lizards and snakes (Wells 2001b), and it occurs in some pond-breeding salamanders and newts as well (see chapter 9). Mate searching is uncommon in anurans, which are relatively sedentary (chapter 6) and usually adopt the more efficient tactic of calling to attract females. Although males of some tropical forest frogs call from widely separated locations to attract females, most anurans are moderately to highly aggregated at terrestrial calling sites or aquatic breeding sites. The form of the mating system that results is partly a product of the temporal availability of females.

## Temporal Patterns of Reproduction in Anurans

Most anurans can be classified as either explosive or prolonged breeders, although these actually represent two ends of a continuum from single night breeding in some desert species that use ephemeral rain pools to year-round breeding in some tropical anurans (Wells 1977b). Variation in the temporal pattern of breeding appears to be related mainly to abiotic factors, such as the seasonal distribution of rainfall and the permanence of bodies of water used for reproduction, but biotic factors such as predation, competition, and foraging opportunities can be important as well (Lucas, Howard, and Palmer 1996; McCauley et al. 2000).

### Breeding Patterns in Temperate-Zone Frogs

Most temperate-zone anurans live in seasonal environments in which breeding activities are restricted to warm months with sufficient rain to fill breeding ponds. Many temperate-zone ranids and bufonids breed explosively in early spring, often in temporary ponds that disappear by late summer. The precise duration of the breeding period varies both within and among species, with some populations completing breeding activity in only three or four nights. In other cases, males may be present in a pond for several weeks, but with actual breeding activity concentrated into a few peak nights (e.g., Lodé, Holveck, and Lesbarrères 2005).

In cold-water breeders, explosive breeding enables females to clump eggs into huge communal masses that reduce heat loss (J. Moore 1939; R. Savage 1961; Herreid and Kinney 1967; Hassinger 1970; Licht 1971; Beattie 1980; R. D.

Howard 1980; Caldwell 1986a) and increase development rate (Waldman 1982a). Early synchronous breeding also allows tadpoles to develop before populations of aquatic predators become well established. For example, ambystomatid salamanders often breed in the same ponds as *Rana sylvatica* and lay eggs at the same time. The wood frog eggs develop more quickly, so tadpoles are mobile before the salamander larvae are large enough to eat them (Wilbur 1972). In other areas, the major predators are aquatic insect larvae, and populations do not reach high levels until summer (see chapters 13 and 14). Heusser (1970) suggested that synchronous breeding in some European ranids is advantageous because late eggs are likely to be consumed by early-hatching conspecific tadpoles. The same was found to be true for North American wood frogs (*Rana sylvatica;* Petranka and Thomas 1995; see Crump 1992 and chapter 12 for a general discussion of cannibalism in tadpoles).

Many temperate-zone frogs in both the northern and southern hemispheres have spring or summer breeding periods lasting two to three months (Wells 1977b). Breeding activity often is more or less continuous, although there are some nights on which activity is greater than on others (R. D. Howard 1978a; Ritke, Babb, and Ritke 1992; Henzi et al. 1995). Others have long breeding seasons divided into short breeding periods, producing a pattern similar to that of explosive breeders that breed only once each year. For example, a population of southern leopard frogs (*Rana sphenocephala*) breeding in a temporary pond in Mississippi formed several short breeding aggregations over a period of several months, with breeding episodes sometimes separated by a month or two. Eggs were laid only when water in the pond was high, and each period of egg laying lasted less than three days (Doody and Young 1995). Similar episodic breeding triggered by rainfall has been reported for *Bufo calamita* in Europe, with some populations having as many as nine distinct calling periods during a three-month breeding season (Banks and Beebee 1986; Sinsch 1988d; Denton and Beebee 1993a).

Northern populations of the European water frogs (the hybrid *Rana esculenta* and its associated parental species, usually *R. lessonae*) generally have one short breeding period (Forselius 1963). Southern populations have a series of short breeding periods (Rühmekorf 1958a; Heusser 1961; Günther 1969; M. Smith 1969; Wahl 1969; van Gelder and Hoedemaekers 1971; Blankenhorn 1974, 1977). Some discoglossids and bombinatorids have a series of short breeding periods (Knoepffler 1962; Lörcher 1969; Heinzmann 1970; Akef and Schneider 1985; Reading and Clarke 1988), but the proximate control of reproductive activity differs from that of ranids. In the ranids, periodic reproduction appears to be a facultative response to changing temperature (Obert 1975), whereas in the discoglossids and bombinatorids, it is controlled by an endogenous hormonal cycle that is independent of temperature (Obert 1973, 1974, 1977). Nevertheless, populations of midwife toads (*Alytes*) at high elevations or in arid regions often have only a single short breeding season, imposed by environmental conditions (Angelier and Angelier 1964; Márquez 1992).

### Breeding Patterns in Tropical Frogs

In tropical and subtropical regions, conditions often are suitable for anuran reproduction for many months, even for species that breed in temporary ponds. Many species breed more or less continuously during long rainy seasons, but the number of frogs at a breeding site varies with rainfall (Berry 1964; Inger and Bacon 1968; Crump 1974; Wells 1977b; M. Ryan 1985b; Aichinger 1987a; Cardoso and Haddad 1992; Donnelly and Guyer 1994; Moreira and Barreto 1997; Pröhl 1997a; Arzabe et al. 1998; Bastos and Haddad 1999; Bertoluci and Rodrigues 2002; Vaira 2005). This pattern is characteristic of Neotropical species such as *Hyla ebraccata, H. microcephala, Physalaemus pustulosus,* and many others (M. Ryan 1985b; Schwartz and Wells 1985; Donnelly and Guyer 1994; Arzabe, Carvalho, and Goes Costa 1998; D. Marsh 2000). Sometimes closely related species have very different breeding patterns. For example, in Panama, *Scinax boulengeri* breeds in relatively permanent water and is active throughout the rainy season, whereas a congeneric species, *S. rubra,* forms explosive mating aggregations in temporary pools only after heavy rains (Bourne 1992; Bevier 1997a). Terrestrial breeders that do not depend on standing water for reproduction sometimes breed for most of the year in wet tropical climates, or they reproduce mostly in the rainy season (Donnelly 1989c, 1999; Praderio and Robinson 1990; Moreira and Lima 1991; Townsend and Stewart 1994; Pröhl 1997a; Brooke, Alford, and Schwarzkopf 2000).

Many aquatic-breeding tropical frogs form explosive mating aggregations after heavy rains, as described previously for *Scinax rubra.* The large canopy-dwelling hylids *Trachycephalus* (=*Phrynohyas*) *venulosus* and *Phrynohyas coriaceus* often appear in ponds at the onset of the rainy season, engage in a brief burst of reproduction, and then disappear for the rest of the year (Rivero and Esteves 1969; Zimmerman and Hödl 1983), although in the Cerrado of Brazil, *T. venulosus* engaged in repeated bouts of reproduction after heavy rains (Rodrigues, Uetanabaro, and Lopes 2005). Other species form explosive mating aggregations in shallow rain pools, including the Costa Rican hylid *Hyla pseudopuma* (M. Crump and Townsend 1990), the Australian hylids *Litoria chloris* and *L. xanthomera* (Morrison, Hero, and Smith 2001), the Brazilian microhylid *Chiasmocleis shudikarensis* (Zimmerman and Bogart 1988), a variety of anurans from the Brazilian Pantanal (Prado, Uetanabaro, and Haddad 2005), and the golden toad of Costa Rica, *Bufo periglenes* (now apparently extinct; Jacobson and Vandenberg 1991). Two other Neotropical toads, *Bufo alatus* (formerly *B. typhonius*) and *B. dapsilis,* are explosive breeders, but are not restricted to temporary rain pools. The former breeds in pools in permanent streams, while the latter breeds in shallow pools at the edges of streams (Wells 1979; Zimmerman and Bogart 1988). Explosive breeding in *B. alatus* probably swamps potential predators of eggs and tadpoles, including tadpoles of another frog, *Leptodactylus pentadactylus* (Wells 1979).

### Breeding Patterns in Desert and Savanna Frogs

Most anurans from arid or semiarid regions have explosive breeding periods because they use ephemeral rain pools for reproduction. The short life of the breeding pools places a premium on rapid oviposition, so males and females typically arrive at ponds immediately after heavy rains, engage in a frenzy of mating activity for a few nights, and then leave. In some species, males attend several different choruses in one year, but females usually breed only once. This type of breeding behavior is characteristic of desert-dwelling anurans in many families, including bufonids, pelobatids, microhylids, hylids, ranids, leptodactylids, and myobatrachids (Poynton 1964b; Wager 1965; Main 1968; Channing 1976b; Wells 1977b; Cei 1980; B. K. Sullivan 1989a; Krupa 1994), as well as temperate-zone anurans from more mesic areas with irregular rainfall patterns (Greenberg and Tanner 2004). Some anurans from tropical savannas also have a single explosive breeding period, or a series of explosive breeding periods, depending on the persistence of aquatic breeding sites (Balinsky 1969; Barbault 1972; Grobler 1972; Channing, du Preez, and Passmore 1994; Rödel 1996).

### Geographic Variation in Breeding Phenology and Mating Systems

Temporal patterns of breeding appear to be relatively flexible in many anurans, and it can be misleading to consider a particular temporal pattern as being a fixed species-specific trait. Variation in temporal patterns of breeding can, in turn, result in considerable variation in mating systems, even among populations of the same species. In some North American toads, the length of the breeding season is quite variable, depending on local weather conditions (Kagarise Sherman 1980; B. K. Sullivan 1982a, 1985b, 1986b, 1989a; Woodward 1982b, c; Wagner and Sullivan 1992). For example, some populations of American toads (*Bufo americanus*) complete their breeding activities in less than three days (R. D. Howard 1988b; Howard and Young 1998), but others remain at a breeding site for up to five weeks (Forester and Thompson 1998). The behavior of males in some choruses is like that of typical explosive breeders, but in others it more closely resembles that of prolonged breeders. Geographic variation has been reported in the mating systems of both *Bufo calamita* and *B. bufo* in Europe. In most regions, the former has a long breeding season (Rühmekorf 1958a; Heusser and Miesterhans 1969; Beebee 1979c, 1983; Sinsch 1988d) and a mating system that has been compared to a lek (Arak 1983b; 1988a, b). In Spain, the same species has an explosive breeding season and exhibits behavior typical of other explosive-breeding toads (Tejedo 1988). The northern subspecies of the common European toad, *B. bufo bufo,* has an explosive breeding season throughout its range (R. Savage 1934; W. Jungfer 1943; Eibl-Eibesfeldt 1950; Heusser 1960, 1961, 1963, 1968b, d, 1969a; N. Davies and Halliday 1977, 1978, 1979; Gittins, Parker, and Slater 1980; Hemelaar 1983; Loman and Madsen 1986; Höglund and Robertson 1987). The southern subspecies, *B. bufo spinosus,* has a long breeding season and a mating system very different from that of its explosive-breeding relatives (Hotz 1970; Dolce 1976; Sofianidou and Schneider 1985). Some geographic variation in length of the breeding period has been reported in the European spadefoot toads, *Pelobates fuscus* and *P. cultripes,* but most populations have relatively long breeding periods (> three weeks; Nöllert 1984; Lizana, Márquez, and Martin-Sanchez 1994; Eggert and Guyétant 2003). Geographic variation probably is common in widely distributed temperate-zone anurans, with breeding seasons becoming shorter with increases in latitude, altitude, or habitat aridity.

Geographic variation probably occurs in widely distributed tropical species as well, but has not been as thoroughly documented. In Panama, *Leptodactylus pentadactylus* is a prolonged breeder, with males often being found in low-density choruses or calling alone (Breder 1946; personal observations), but Rivero and Esteves (1969) described an explosive aggregation of the same species in Venezuela. *Physalaemus cuvieri,* which is common throughout much of Brazil, calls more or less continuously for six to eight months in the relatively mild climate of southern Brazil. In the hotter, drier regions of central Brazil, the same species calls for only about four months and does so sporadically after heavy rains in semiarid regions of northeastern Brazil (Moreira and Barreto 1997). Again, as in temperate-zone frogs, increasing habitat aridity or decreased pond hydroperiod tends to produce a shorter and less predictable breeding season in tropical frogs. Some species even exhibit different breeding patterns in different ponds in close proximity to one another. For example, in Brazil, *Scinax eurydice* was found calling for six months in a permanent pond, but for only one month in a temporary pond about one km away (Arzabe, Carvalho, and Goes Costa 1998).

### Implications of Prolonged and Explosive Breeding

Variation in the temporal pattern of reproduction has important implications for the operation of sexual selection in anuran mating systems and the way in which males allocate their time and energy for reproduction. One predicted difference between prolonged and explosive breeders is that the opportunity for female choice should be more limited in explosive breeding aggregations because of the high densities of competing males. In general, male-male competition is expected to be the most important determinant of variation in male mating success in explosive breeders. In contrast, when the breeding period is prolonged, females arrive asynchronously, and direct competition among males for individual females is expected to be rare. Females should be able to compare males before selecting their mates (Wells 1977b). Recent research has largely supported these predictions (Arak 1983a; B. K. Sullivan et al. 1995). Nevertheless, females sometimes can choose mates in low-density choruses, even when the breeding season is relatively short (B. K. Sullivan and Sullivan 1985; B. K. Sullivan, 1989b, 1992; Wagner and Sullivan, 1992; K. Pfennig 1998, 2000).

Because of the highly synchronized nature of explosive breeding aggregations, males are expected to have relatively few opportunities to compete for mates, because many females will be mating simultaneously (Wells 1977b). Indeed, males of explosive-breeding species seldom obtain more than three matings in a season, and usually less than 5% obtain more than one (B. K. Sullivan, Ryan, and Verrell 1995). Multiple mating opportunities are more common in prolonged breeders (fig. 8.1), and in some species, a single male can mate more than 15 times in a breeding season of three to six months (Murphy 1994a; Dyson, Bush, and Halliday (1998). The major determinant of the intensity of sexual selection operating on male traits is variation in male mating success (Wade 1979; Wade and Arnold 1980; S. Arnold 1983a; S. Arnold and Wade 1984a, b; S. Arnold and Duvall 1994). Consequently, the opportunity for sexual selection to operate on male traits should be more limited in explosive breeders than in prolonged breeders.

This prediction is complicated by variation in the operational sex ratio, defined as the number of males relative to the number of females in breeding aggregations. In explosive mating aggregations, males often outnumber females by as much as 10 to 1 on any given night, and the total number of males visiting a pond throughout the breeding period usually exceeds the total number of females. This means that many males fail to mate at all. Consequently, the opportunity for sexual selection to operate on male traits can be high even when only a few males mate more than once (B. K. Sullivan, Ryan, and Verrell 1995). Studies of two species of toads that have short breeding periods, *Bufo woodhousii* and *B. valliceps,* showed that the opportunity for sexual selection was correlated with the seasonal operational sex ratio for different populations (fig. 8.2). When the sex ratio was strongly biased in favor of males, the opportunity for sexual selection on male traits increased. In species with prolonged breeding periods, sex ratios also tend to be biased toward an excess of males (B. K. Sullivan, Ryan, and Verrell 1995; Lüddecke 2001; Eggert and Guyétant 2003). This bias, coupled with the ability of some males to obtain multiple matings, will tend to produce even greater variance in male mating success and further increase the intensity of sexual selection on male traits.

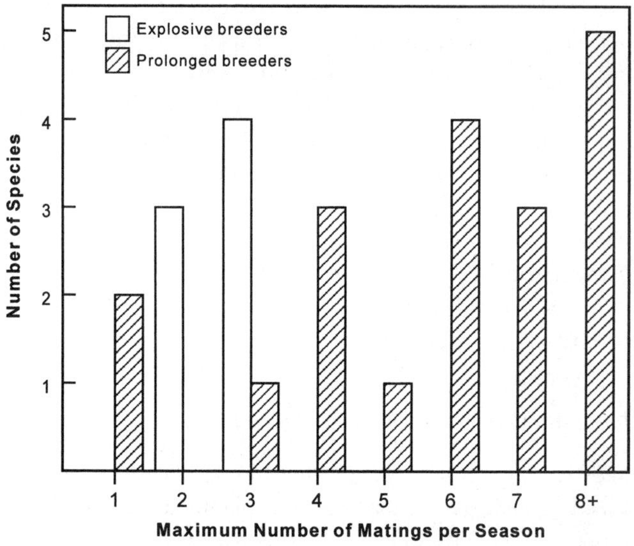


**Fig. 8.1.** Maximum number of matings obtained by individual males in a single breeding season in explosive and prolonged breeding anurans. Each species is represented only once. The two prolonged-breeding species with only a single mating per season are both from studies with very small sample sizes. Plotted from data summarized in Kluge (1981) and B. K. Sullivan, Ryan, and Verrell (1995) and additional data from Roithmair (1992), Dyson, Henzi, Halliday, and Barrett (1998), and Pröhl and Hödl (1999).

Variation in the duration of breeding seasons also has implications for the way in which males allocate their time and energy to reproduction (Wells 1977b; Arak 1983a). Explosive breeding places a premium on rapid location of mates, because females are available for such a short time. Males can afford to use a lot of energy searching for mates and struggling over females, even if this leaves them largely drained of energy reserves at the end of the breeding period. In contrast, prolonged breeders must husband their resources to last much longer, especially if a male's mating success is determined in part by how long he can remain at the breeding site. Male frogs can conserve their energy reserves in several ways. One is to use inexpensive forms of mate-locating behavior, such as assuming a satellite role near a calling male (Arak 1983a). A second is to call mostly when females are likely to be present (Woolbright 1985a; Bevier

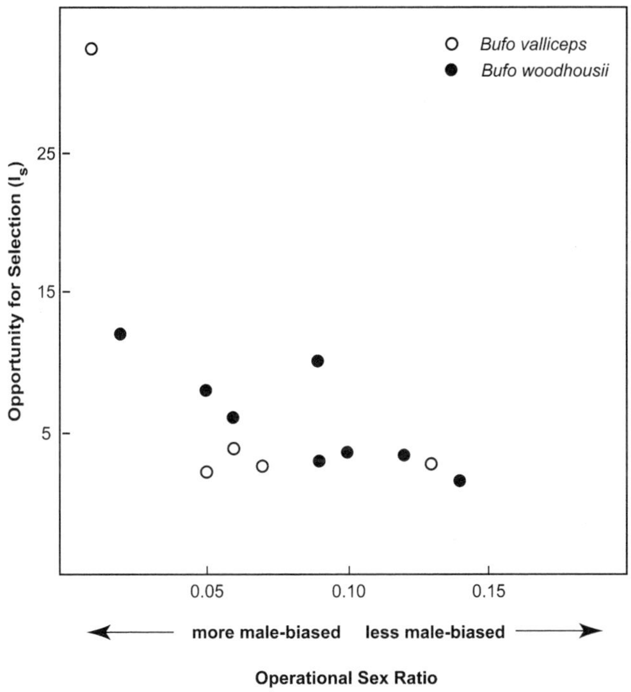


**Fig. 8.2.** Relationship of the operational sex ratio to the opportunity for sexual selection in males for two species of toads, *Bufo valliceps* and *Bufo woodhousii.* The operational sex ratio was measured as the ratio of the number of females to the number of males in a chorus. The opportunity for sexual selection ($I_s$ = variance in male mating success/squared mean mating success) is an indicator of the potential strength of sexual selection on male traits that are related to mating success. After B. K. Sullivan, Ryan, and Verrell (1995).

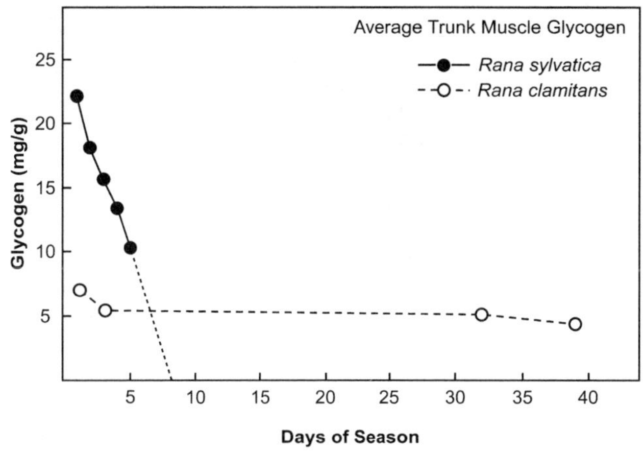


**Fig. 8.3.** Relative rates of depletion of glycogen reserves in the trunk muscles of males of the explosive-breeding wood frog (*Rana sylvatica*) and the prolonged-breeding green frog (*Rana clamitans*). Wood frogs exhaust their glycogen reserves within a few days, whereas green frogs maintain low levels of glycogen in their muscles throughout the breeding season. From Wells (2001).

1997a). A third is to call in bouts separated by silent periods (Schwartz et al. 1995). Alternatively, males can replace depleted energy reserves by leaving the breeding area periodically to feed (Fellers 1979b; Greer and Wells 1980; Kagarise Sherman 1980; Arak 1983a; Woodward 1982b; Godwin and Roble 1983; M. Ryan 1983b). Nevertheless, male frogs often deplete a significant proportion of their energy reserves by the end of the breeding season (Jenssen 1972; Wells 1978b; Mac Nally 1981; Arak 1983a; Given 1988a; Ressel 1993; Eggert and Guyétant 2003; see also chapter 5). In some cases, the ability of individual males to remain in a chorus is limited by energy reserves (C. Murphy 1994a, b).

As expected, explosive breeders tend to deplete their energy reserves more quickly than do prolonged breeders (Halliday and Tejedo 1995). Daily changes in body mass in the explosive-breeding toad *Bufo bufo* were greater than in *B. calamita,* a prolonged breeder (Arak 1983a). Explosive-breeding wood frogs (*Rana sylvatica*) deplete their trunk-muscle glycogen reserves more quickly than do prolonged-breeding green frogs (*Rana clamitans*), even though wood frogs begin the season with much larger reserves (fig. 8.3). Male wood frogs usually spend only a few nights at the breeding pond, whereas male green frogs can be present every night for up to two months. Green frogs conserve their energy reserves in part by calling at much lower rates than do male wood frogs. They also feed throughout the breeding season to supplement stored energy reserves, something that wood frogs cannot do because of the low temperatures at their breeding ponds (Wells and Bevier 1997; Wells 2001a). In *Scinax rubra,* a tropical hylid with an explosive breeding period, males call at higher rates and for more hours per night than do males of *S. boulengeri,* a prolonged breeder, and they deplete their energy reserves much more quickly. These differences are reflected in the much longer chorus attendance of *S. boulengeri* males. This species further reduces energy expenditures by calling at high rates only early in the evening, when females are most likely to be present, and then sharply reducing calling rates for the rest of the night (Bevier 1997a).

## Scramble Competition Mating Systems

The distinction between explosive and prolonged breeders does not, in itself, determine the precise tactics used by males to acquire mates, but the temporal pattern of breeding provides the framework for the evolution of mating systems. The mating systems of most anurans fall into one of three categories: (1) scramble competition, (2) choruses and leks; and (3) resource defense (Wells 1977b; Arak 1983a; B. K. Sullivan, Ryan, and Verrell 1995). Most anurans with very short breeding periods fall into the first category, but one must keep in mind that the distinctions among these mating systems are somewhat artificial. In reality, the mating tactics of male anurans exhibit a considerable degree of plasticity and tend to grade into one another, even within a single species (see the following "Plasticity in Anuran Mating Sys-

**Fig. 8.4.** A dense explosive mating aggregation of American toads (*Bufo americanus*), with many pairs in amplexus and single males searching for females. Photo by Kentwood D. Wells.

tems"). Some anurans, especially those that do not lay their eggs in water, have mating systems that do not fit conveniently into any of these categories. Rather than engaging in debates about whether a particular species fits an arbitrary set of criteria used to define a particular type of mating system, it is more informative to describe the full range of variation seen in anuran mating systems. The following discussion focuses mainly on the behavior of males and its effect on male mating success. The role of female choice is discussed in a later section.

### Scramble Competition in Aquatic Mating Aggregations

A number of well-studied species of anurans in Europe and North America form explosive mating aggregations and exhibit classic scramble competition. These include *Rana temporaria, R. dalmatina, R. sylvatica, Bufo americanus, B. bufo,* and several other species of *Bufo, Scaphiopus,* and *Spea.* Similar mating systems are found in anurans in these and other families around the world. The typical pattern for most explosive breeders is for males to gather in very dense choruses (fig. 8.4). They tend to arrive at breeding ponds somewhat earlier than females and often greatly outnumber them (fig. 8.5). Males typically search for females in the water or on land around the breeding site (R. Savage 1961; Wells 1977b, 1979; N. Davies and Halliday 1979; R. D. Howard 1980, 1988b; Kagarise Sherman 1980; Berven 1981; Kruse 1981a; R. D. Howard and Kluge 1985; B. K. Sullivan 1985b; Elmberg 1986; Loman and Madsen 1986; Okuno 1986; Höglund and Robertson 1987, 1988; Tejedo 1988, 1992c, 1993a; Ryser 1989a; Telford and van Sickle 1989; Woolbright, Greene, and Rapp 1990; Jacobson and Vandenberg 1991; Lizana, Márquez, and Martin-Sanchez 1994; Tarkhnishvili 1994; Tsuji and Kawamichi 1996a, b; Forester and Thompson 1998; R. D. Howard and Young 1998). Apparently males cannot distinguish the sex or even the species of other individuals visually. Mate acquisition is a trial-and-error process, with little evidence that males discriminate among females of different sizes (Hettyey, Török, and Hévizi 2005). Indeed, males often attempt to clasp almost any moving object, including, on occasion, members of other anuran species (Marco et al. 1998; Marco and Lizana

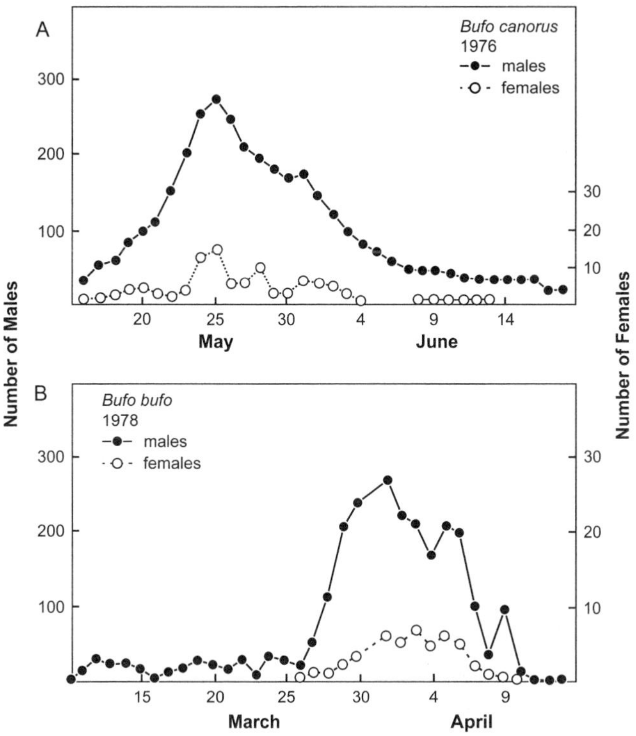


**Fig. 8.5.** Temporal pattern of male and female arrival at breeding ponds in explosive-breeding toads. (A) *Bufo canorus* from Yosemite National Park, California. (B) *Bufo bufo* from Great Britain. Females generally arrive later than males and in fewer numbers. (A) after Kagarise Sherman (1980). (B) after N. Davies and Halliday (1979).

2002; Hettyey and Pearman 2003; Ficetola and De Bernardi 2005). Other males are frequently clasped, but are released after giving release calls or vibrations (see chapter 7).

Males in explosive mating aggregations often struggle for possession of individual females (fig. 8.6). Unpaired males try to displace those already in amplexus, sometimes resulting in large mating balls in which a half dozen or more males clasp a female simultaneously. Contests among males sometimes are fatal to females. One often finds female wood frogs (*Rana sylvatica*) crushed or drowned after they were clasped by groups of struggling males (R. D. Howard 1980), and dead females have been found in mating balls of *Bufo bufo* as well (Verrell and McCabe 1986). Mating balls are common in many ranids and bufonids (Verrell and McCabe 1986; Forester and Thompson 1998), but occur in other families as well. For example, as many as 16 males have been observed clasping a single female *Hyla pseudopuma* during an explosive breeding event (Crump and Townsend 1990). Scramble competition and multiple males clasping females also have been observed in another hylid, *Triprion petasatus* (fig. 8.6 C), a highly explosive breeder (J. Lee and Crump 1981).

In most species, males compete by pushing and kicking one another, but in some, males use specialized weapons that can injure other males. For example, male African bullfrogs (*Pyxicephalus adspersus*) engage in prolonged wrestling bouts and bite one another with tooth-like projections on the lower jaw (B. Balinsky and Balinsky 1954; T. Hayes and Licht 1992; Channing, du Preez, and Passmore 1994). In some leptodactylid frogs, such as *Leptodactylus pentadactylus*, males are equipped with sharp spines on their front feet that are used to jab their opponents (Rivero and Esteves 1969; Shine 1979).

### Competition among Males and Mating Success

In explosive breeding aggregations where male densities are high, unpaired males frequently attack mated pairs and attempt to displace males already in amplexus. Generally paired males have an advantage in such contests, so rates of displacement are low for most species, often less than 5% (Wells 1979; Gatz 1981a; Kagarise Sherman 1980; R. D. Howard 1988b; Tejedo 1988; Forester and Thompson 1998). Displacement rates of 10–39% have been reported in dense populations of *Bufo bufo* (N. Davies and Halliday 1979; Loman and Madsen 1986; Reading 2001). Usually large males displace smaller males (N. Davies and Halliday 1977, 1978, 1979; Wells 1979; Gittins et al. 1980; Kagarise Sherman 1980; Berven 1981; Lamb 1984a; R. D. Howard and Kluge 1985), but there are exceptions (e.g., Loman and Madsen 1986; Höglund and Robertson 1987).

The intense direct competition among males for possession of females generally results in large males being more successful in acquiring mates than small males (fig. 8.7; N. Davies and Halliday 1977, 1979; Wells 1979; Gittins, Parker, and Slater 1980; R. D. Howard 1980; Kagarise Sherman 1980; Berven 1981; Gatz 1981a; Reading and Clarke 1983; R. D. Howard and Kluge 1985; Loman and Madsen 1986; Olson, Blaustein, and O'Hara 1986; R. D. Howard and Young 1998). Some studies have revealed no such size advantage, however (Kruse 1981a; Arak 1983a; B. K. Sullivan 1983a, 1985b; B. K. Sullivan and Sullivan 1985; Olson, Blaustein, and O'Hara 1986; Elmberg 1987; Höglund and Robertson 1987; Ryser 1989a), or have found only a small difference between successful and unsuccessful males (R. D. Howard 1988b). One study of explosive-breeding hylid frogs in Australia (*Litoria chloris* and *L. xanthomera*) unexpectedly showed that small males were more likely to acquire mates than were large males, but this was related more to differences in chorus tenure than to competitive ability (Morrison, Hero, and Smith 2001). Size-related mating patterns are by no means fixed for a particular species. Large males enjoy a mating advantage in some populations but not in others, or the advantage varies among years in the same population (B. K. Sullivan 1985b; Olson, Blaustein,

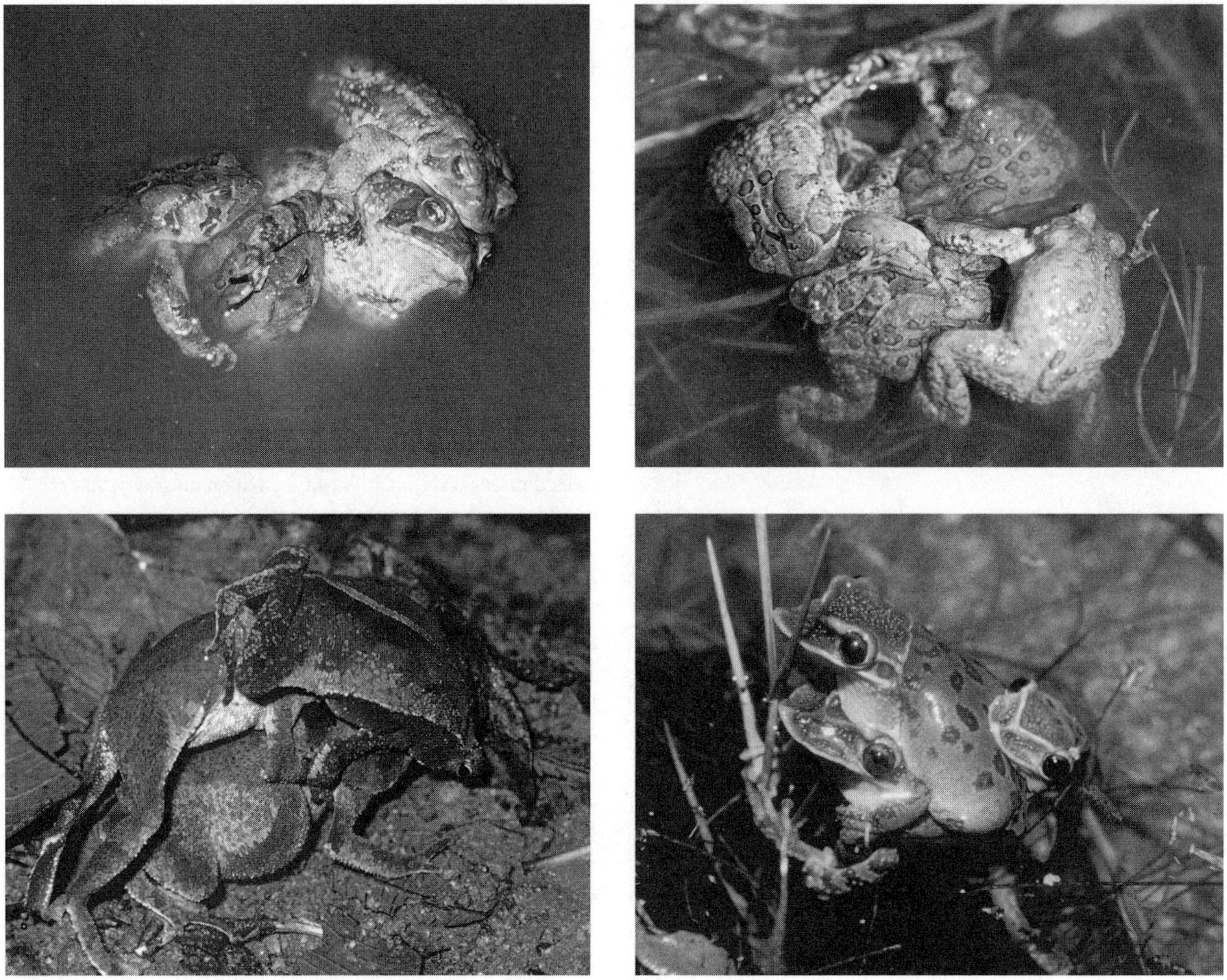

**Fig. 8.6.** Scramble competition among males of explosive breeding anurans. (A, B) Several male American toads (*Bufo americanus*) competing for a large female. (C) Two males of the Panamanian toad *Bufo alatus* struggling for possession of a female (D) Two males of the Central American treefrog *Triprion petasatus* in amplexus with a female. Photos by Kentwood D. Wells (A–C) and Julian Lee (D).

and O'Hara 1986; R. D. Howard 1988b; Tejedo 1992c; Halliday and Tejedo 1995). In some species, an advantage of large size is apparent at high densities but not at low densities (Arak 1983a).

There is relatively little evidence for size-assortative mating for most explosive breeding anurans (Wells 1979; Gittins Parker, and Slater 1980; Kagarise Sherman 1980; Berven 1981; Gatz 1981a; Kruse 1981a; Woodward 1982c; Arak 1983a; Hemelaar 1983; Reading and Clarke 1983; B. K. Sullivan 1983a; R. D. Howard and Kluge 1985; Loman and Madsen 1986; Olson, Blaustein, and O'Hara 1986; Elmberg 1987; Höglund and Robertson 1987; R. D. Howard 1988b; Ryser 1989a; Wogel Abrunhosa, and Pombal 2005). The occurrence of assortative mating sometime varies among years or among populations (Olson, Blaustein, and O'Hara 1986). Licht (1976) and N. Davies and Halliday (1977) proposed an adaptive explanation for assortative mating in toads, arguing that a close match in sizes of males and females in amplexus would increase the probability of eggs being successfully fertilized. Subsequent work with several explosive breeders has shown that fertilization rates in the field almost always are very high and generally are not related to relative sizes of males and females in mated pairs (Kruse 1981a; Halliday and Verrell 1984; R. D. Howard and Kluge 1985; Höglund and Robertson 1987; Krupa 1988).

The variable occurrence of assortative mating in different populations probably results from differences in male-male competition at different densities and does not require an explanation based on adaptive advantages to females. Small males clasping large females are likely to be displaced, whereas large males on large females are not (fig. 8.8). This

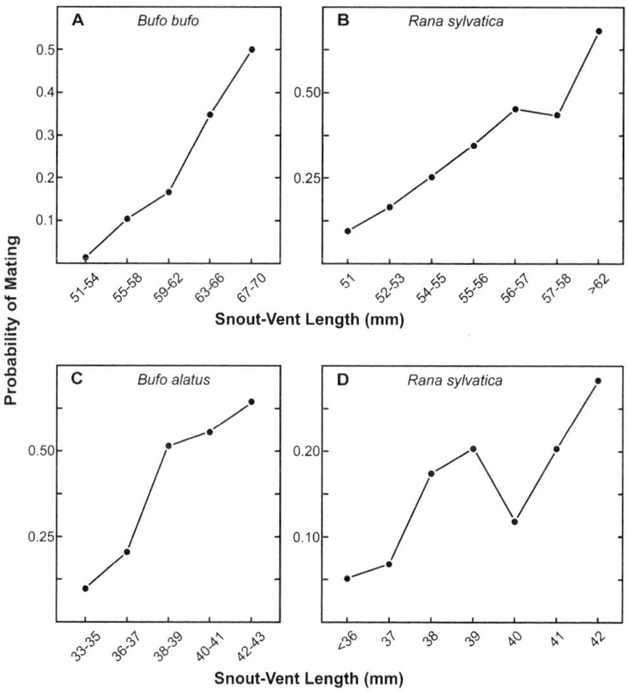


**Fig. 8.7.** Effect of male size on the probability of mating for several explosive-breeding anurans. (A) Common toads (*Bufo bufo*) from Great Britain. (B) Wood frogs (*Rana sylvatica*) from three ponds in Maryland and Virginia. (C) Toads, *Bufo alatus,* from Panama. (D) Wood frogs (*Rana sylvatica*) from Michigan. Plotted from data in (A) N. Davies and Halliday (1977), (B) Berven (1981), (C) Wells (1979), (D) R. D. Howard (1980).

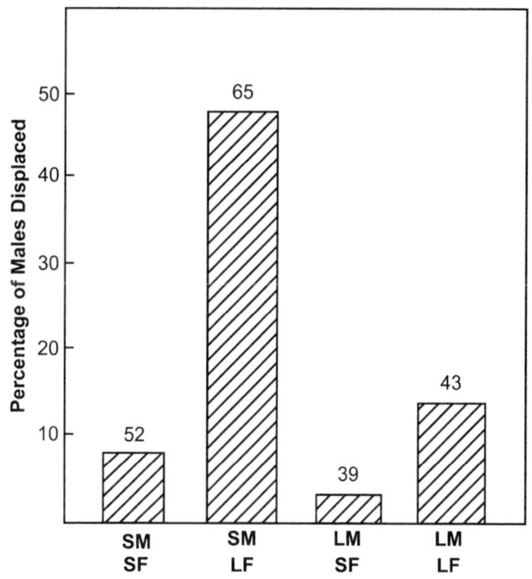


**Fig. 8.8.** Effect of relative size of paired males and females on the probability of a male being displaced from amplexus in wood frogs (*Rana sylvatica*). LF = large female; SF = small female; LM = large male; SM = small male. Small males have difficulty holding on to large females and therefore are more likely to be displaced than are small males on small females or large males on either larger or smaller females. Plotted from data in R. D. Howard and Kluge (1985).

would produce a positive correlation between male and female sizes in mated pairs. At very high densities, male-male competition often is so intense that small males are displaced even from small females. In those cases, large males will enjoy a mating advantage, but assortative mating will not be apparent (Gittins, Parker, and Slater 1980; Arak 1983a).

## Chorus Tenure and Mating Success

In many explosive breeders, the arrival of males at the breeding site is so synchronized that variations in length of attendance are hard to detect (e.g., Wells 1979; Berven 1981). In species that have a series of relatively explosive breeding events in a single season, the total number of nights present at breeding sites probably is a major determinant of male mating success (Wogel Abrunhosa, and Pombal 2005). Even in species with a single short breeding period, slight differences in time of arrival can have a major impact on male mating success, particularly if many females arrive on the first few nights of breeding activity. This in turn can select for early arrival by males. In some early spring breeders, males can be seen making their way to breeding ponds over snow and ice, while females show up a few days later. Gatz (1981a) found that successful *Bufo americanus* males arrived slightly earlier and stayed slightly longer than did unsuccessful males; many of the early arrivals were relatively large. Kagarise Sherman (1980) reported similar results for *B. canorus*. Large males arrived earlier than small males in two of three years, and males that eventually mated arrived at ponds about 2.5 days earlier than unsuccessful males. More important, in each year, 30 to 60% more females were available during the tenure of successful males than during the tenure of unsuccessful males. In other words, mating success in this species was determined in part by males being in the right place at the right time.

In a population of *Rana dalmatina* in France, males arrived in breeding ponds earlier than females and stayed longer, and the largest males were the first to arrive. There was no information on individual mating success in this study, although there was evidence for some multiple paternity of egg clutches (Lodé, Holveck, and Lesbarrères 2005). In one study of *Rana sylvatica,* there was a slight tendency for large males to arrive earlier, and large males were more likely to obtain mates (R. D. Howard and Kluge 1985). Large males also tended to arrive earlier than small males in a population of *Bufo bufo* studied by Loman and Madsen (1986), but mating success appeared to be more related to size than to time of arrival. In another study over a 20-year period, large males with previous breeding experience consistently arrived earlier in the breeding period than did smaller, inexperienced breeders (Reading 2001). In contrast to these results, Morrison, Hero, and Smith (2001) reported that small males had longer chorus tenure than did large males in *Litoria xanthomera,* and this probably was the reason for greater mating success of small males. These authors attributed the longer chorus tenure of small males to lower energetic investment in calling.

The advantage of arriving early at a breeding site seems clear for species in which the breeding season is short and female arrival is relatively synchronous. Because many females are mating at the same time, males that arrive late risk missing the peak of female availability. The reason why large males often arrive earlier than small males is less obvious. There are several possible and nonexclusive explanations for this pattern. On a proximate level, it may be that large males, which often are older as well, are more familiar with the breeding site than are first-year breeders and therefore arrive first (Reading 2001). It also is possible that large males are more likely to move in the fall from summer home ranges to overwintering sites relatively close to breeding ponds, and therefore have less distance to cover in the spring. Autumn movements to overwintering sites close to ponds has been reported in a number of species (see chapter 6), but the influence of body size has not been examined.

Another possibility is that large males save energy by securing the available females early before small males arrive. In anurans with scramble competition mating systems, it almost certainly is more difficult to displace another male from amplexus than to clasp an unmated female, even when the initial clasping male is smaller. Indeed, as mentioned previously, mated males of most species are rarely displaced from amplexus (< 5% of mated pairs), although takeover rates as high as 30% have been reported in some populations of toads. A model developed for mate-guarding crustaceans with internal fertilization, in which males guard females until they are ready to molt, predicted that when takeovers are difficult, large males should begin guarding females early to ensure that they monopolize the available females. In cases where takeovers are easy, large males are expected to simply usurp smaller males guarding females, and prolonged mate guarding is less likely (Härdling, Kokko, and Elwood 2004; see also Candolin and Voigt 2003). Because takeovers appear to be difficult in most anurans, the usurpation tactic is not likely to be a viable one except for extremely large males, especially when there is relatively little delay between males entering amplexus and females beginning to lay eggs (a feature that differs from that of crustaceans, where males may have to wait for some time for females to become receptive).

### Male Mobility and Mating Success

Because males in explosive breeding aggregations often obtain mates by active searching, one might expect male mating success to be related to mobility or activity at the breeding site. Males that move more frequently and attempt to clasp many other individuals should be those most likely to obtain mates (Wells 1977b). Unfortunately, detailed data on movements of individual males are scarce, and unless all males in a population can be monitored simultaneously, it can be difficult to relate movement patterns to mating success because many of the animals under observation do not mate (e.g., Wells and Taigen 1984).

Gatz (1981a) reported that successful male *B. americanus* did not move more often than unsuccessful males, but he considered only large-scale movements between different parts of the pond. Kagarise Sherman (1980) quantified movements of *B. canorus* and *B. exsul* males through focal-animal observations, but she was not able to relate movements to mating success. In a low-density population of *Rana sylvatica,* Woolbright, Greene, and Rapp (1990) found that males that moved were more likely to clasp females than were those that remained stationary, but they did not relate clasping success to quantitative measures of movement. Furthermore, final mating success was not greater for the more mobile individuals because several males were displaced by other males before they could fertilize a clutch of eggs. R. D. Howard (1988b) made the most concerted effort to date to measure individual variation in searching behavior. He found that male *Bufo americanus* differed in rates of movement and number of clasping attempts, but there was no relationship between these variables and mating success. The whole question of individual variation in mate searching behavior and its relationship to mating success in explosive breeders is one that needs to be addressed in more detail.

### Multiple Matings and Sperm Competition

In most aquatic-breeding frogs, a single male usually succeeds in defending a female against other males and eventually fertilizes her eggs without interference, even in species that form large "mating balls." Nevertheless, this type of scramble competition provides an opportunity for multiple males to release sperm simultaneously, leading to direct sperm competition among males (Halliday 1998). In many cases, the male with his vent in closest proximity to that of the female is likely to fertilize most of her eggs, because sperm typically is released as soon as the eggs emerge from the female's vent. In some species, however, fertilization by multiple males has been documented. Genetic data from *Rana temporaria* tadpoles that hatched from single egg clutches have shown that more than one male can fertilize the eggs of a single female (Laurila and Seppa 1998). More recent work has shown that unpaired males of this species often engage in clutch piracy, clasping and fertilizing egg masses after they have been laid (Vieites et al. 2004). This is similar to behavior reported for some hynobiid salamanders (see chapter 9).

There also is evidence of multiple paternity in a mixed

population of *Rana esculenta* and *R. lessonae* (Berger and Rybacki 1992), but these frogs were studied only in the laboratory. The behavioral mechanism that resulted in multiple paternity is not known. These species often lay eggs communally, so it is possible that sperm being released by several males in amplexus simultaneously become mixed and fertilize more than one batch of eggs. Other possibilities are that more than one male remains in amplexus with a single female, or unpaired males release sperm in mating aggregations. A relatively low frequency of multiple paternity has been reported in *Rana dalmatina,* a species in which males sometimes are territorial, but may engage in scramble competition when males greatly outnumber females (Lodé and Lesbarrères 2004; Hettyey, Török, and Hévizi 2005; Lodé, Holveck, and Lesbarrères 2005).

Two forms of multiple paternity have been reported in explosive-breeding species of *Leptodactylus* from Brazil (Prado and Haddad 2003). In *L. chaquensis,* up to seven satellite males placed themselves in foam nests being constructed by pairs in amplexus. The paired males tried to evict the satellite males with aggressive attacks, but often without success. Satellite males beat the foam in synchrony with the mated pairs and presumably released sperm as well. In *L. podicipinus,* a small male was seen to insert himself between a larger male and a female, with both males beating their legs to construct the foam nest and presumably releasing sperm. An analysis of testis size showed that these two species have larger testes relative to body size than do other *Leptodactylus* or other foam-nesting leptodactylids.

Multiple paternity has been investigated in some detail in *Crinia georgiana,* a small myobatrachid frog from Australia. Although this species has a relatively prolonged breeding season, breeding tends to be episodic, with large numbers of individuals aggregated into small temporary pools (P. Byrne 2002). Sex ratios are heavily biased in favor of males. Consequently, males often engage in scramble competition for access to females. As many as eight males have been found in amplexus with a single female, and groups of two or three males are common. Genetic analysis of egg clutches laid by multiply mated females showed that more than one male often fertilized the eggs of a single female (Roberts, Standish, Byrne, and Doughty 1999). Multiple paternity appears to result entirely from competition among males for access to females. Multiply mated females suffer a cost in reduced fertilization success, probably because of jostling by competing males (Byrne and Roberts 1999). There is no evidence of any genetic benefits to females in mating with more than one male that would compensate for the cost of reduced fertilization. This was determined by comparing the growth, development, and survivorship of tadpoles that hatched from eggs sired by single males or multiple males (Byrne and Roberts 2000).

### Scramble Competition in Arboreal Frogs

Although intense scramble competition among males is commonly associated with large aggregations of males at aquatic breeding sites, this type of mating system is found in some nonaquatic egg-layers as well. Some phyllomedusine tree frogs that lay eggs in vegetation over water form explosive mating aggregations after heavy rains. Males actively search for females, and high densities of males can result in vigorous pushing and kicking among males competing for females. In some cases, males in amplexus fail to dislodge competing males, and several males participate in fertilizing the eggs, resulting in possible sperm competition (Halliday and Verrell 1984; Halliday 1998). N. Scott and Starrett (1974) observed an aggregation of several thousand *Agalychnis spurrelli* around a temporary pond in Costa Rica. Males clambered over one another while searching for females in vegetation overhanging the water, and some males were observed scraping eggs of other males off of leaves.

W. Roberts (1994) observed several dense aggregations of another Costa Rican species, *A. saltator.* Hundreds of individuals were climbing around in vines over a swamp, and eggs were deposited on moss that covered the vines and nearby tree trunks. Males kicked and pulled at other males in amplexus, attempting to dislodge them, and a few cases of multimale matings were observed. Direct competition for females and multimale matings also have been observed in *Pachymedusa dacnicolor* and *Phyllomedusa rohdei,* which sometimes forms explosive mating aggregations, and in *Agalychnis callidryas,* which has a prolonged breeding season, but sometimes occurs at high densities (Pyburn 1970; Wiewandt 1971; Wogel, Abrunhosa, and Pombal. 2005). In *A. callidryas,* multiple paternity of two egg clutches was confirmed by DNA analysis (D'Orgeix and Turner 1995), and it probably occurs in other phyllomedusines as well.

Several species of rhacophorid frogs that deposit eggs in foam nests on trees overhanging ponds or in shallow burrows at the edges of ponds form dense mating aggregations in which males attempt to displace others from amplexus. Multimale matings have been observed in several of these species (Kato, 1956; M. J. Coe 1967, 1974; Kasuya, Shigehara, and Hirota 1987; Kasuya, Hirota, and Shigehara 1996; Feng and Narins 1991; Fukuyama 1991; Jennions, Backwell, and Passmore 1992). In some cases, explosive mating aggregations formed after heavy rains (M. J. Coe 1974), but in others, the breeding season lasted for several weeks or months. In those cases, the specialized nature of the oviposition sites produces local aggregations of males around the best sites. In addition, the long period of time required to complete a foam nest (up to several hours in *Chiromantis xerampelina;* Jennions, Backwell, and Passmore 1992) allows males to intrude on mating pairs and attempt to fertilize eggs as peripheral or satellite males (see fig. 10.7 B). Males assist in the construction of the foam nest by beating mucus secretions produced by the female with their hind legs (see chapter 7). Coe (1967) originally interpreted multimale mating in *Chiromantis* as a form of cooperative behavior, suggesting that more than one male might be needed to build a satisfactory nest, an interpretation that was repeated by E. O. Wilson (1975). It now seems clear that this behavior is simply the result of competition among males for access to females.

Two lines of evidence suggest that direct sperm competition is a regular feature of the mating systems of these rhacophorid frogs. First, Jennions and Passmore (1993) enclosed the rear ends of *Chiromantis xerampelina* males with plastic bags, preventing them from fertilizing eggs. In two trials in which peripheral males were not present, none of the eggs was fertilized, whereas in one trial in which peripheral males were present, about 37% of the eggs hatched. Those eggs could only have been fertilized by a peripheral male. Unfortunately, DNA analysis of multiple paternity has yet to be performed with these frogs.

The second line of evidence for sperm competition comes from measurements of the relative size of testes in rhacophorids in which multimale mating has been reported. In a comparative study that was not controlled for phylogeny, Kusano, Toda, and Fukuyama (1991) showed that foam-nesting Asian rhacophorids, and *Rhacophorus arboreus* in particular, have larger testes in proportion to body size than do other Asian anurans, including bufonids, hylids, ranids, and microhylids. In a similar analysis that combined Asian and African foam-nesting frogs with anurans from several different families, Jennions and Passmore (1993) showed that the three foam-nesting genera in which multimale mating has been reported have relatively larger testes than other anurans (fig. 8.9). Based on somewhat meager field data, they also suggested that species in which multimale matings are common (*Chiromantis xerampelina* and *Rhacophorus arboreus*) have relatively larger testes than do foam-nesters in which multimale matings are less common (*Rhacophorus schlegelli* and *Polypedates leucomystax*). More detailed field observations on a larger number of rhacophorid species are needed to rigorously test this hypothesis.

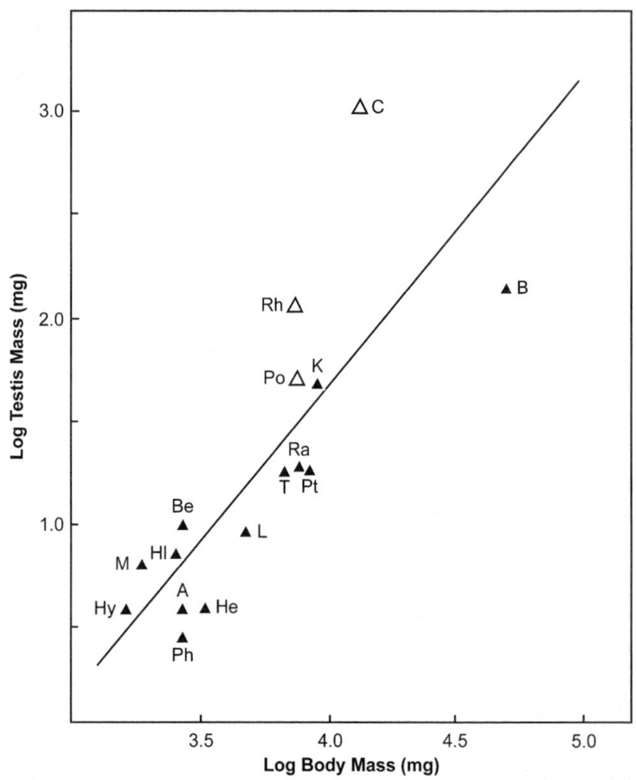


**Fig. 8.9.** Relative testis size in several genera of anurans. Those shown with open symbols are foam-nesting rhacophorids that exhibit multiple mating. Abbreviations: A = *Afrixalus;* B = *Bufo;* Be = *Buergeria;* C = *Chiromantis;* He = *Hemisus;* Hl = *Hyla;* Hy = *Hyperolius;* K = *Kassina;* L = *Leptopelis;* M = *Microhyla;* Ph = *Phrynomantis;* Po = *Polypedates;* Pt = *Ptychadena;* Ra = *Rana;* Rh = *Rhacophorus;* T = *Tomopterna.* This analysis is not corrected for phylogenetic relationships among these genera. After Jennions and Passmore (1993).

## Male Mating Tactics in Prolonged-Breeding Anurans

In most anurans with long breeding seasons, males call from stationary locations to attract females, and do not initiate amplexus until contacted by the female (Wells 1977b). Active searching is rare, but occasionally it occurs in very dense choruses. In most prolonged breeders, however, there is little evidence of direct competition among males for possession of individual females. Instead, competition takes the form of elaborate vocal interactions (chapter 7), defense of calling sites or resource-based territories, or elaborate courtship of females. In this section, I describe the variety of mating systems of prolonged-breeding anurans in some detail. Certain groups of terrestrial-breeding frogs are treated separately, including dendrobatid frogs, members of the widespread tropical genus *Eleutherodactylus,* and the midwife toads (*Alytes*). Aspects of the breeding biology of these frogs, including nonaquatic oviposition and elaborate forms of parental care, make it difficult to fit their mating systems into categories applied to anurans that lay their eggs in or near water.

### Choruses and Leks in Aquatic and Arboreal Frogs

Probably the most common mating system for aquatic-breeding frogs is for males to gather in choruses around suitable breeding sites. Males call from the shores of ponds and streams, from elevated perches on vegetation, rocks, or other prominent sites, from the surface of the water, or, in a few species, from underwater (fig. 8.10 E, F, G). Usually, males maintain some minimum distance between themselves and their neighbors, often defending calling sites with aggressive calls and physical attacks (Wells 1977b; B. K. Sullivan, Ryan, and Verrell 1995). Fidelity of individuals to specific calling sites varies among species. In some species, males defend the same calling sites for days or weeks at a time (Paillette 1970b; M. Rosen and Lemon 1974; Mac Nally 1979, 1984a; Forester and Daniel 1986; Robertson 1986b). In others, males occupy different sites on successive nights and sometimes move between calling sites on a single night (Whitney and Krebs 1975b; J. S. Garton and Brandon 1975; Fellers 1979a, b; K. Wells and Greer 1981; Arak 1983a, b, c, 1988a, b; B. K. Sullivan 1982b; Wells and Schwartz

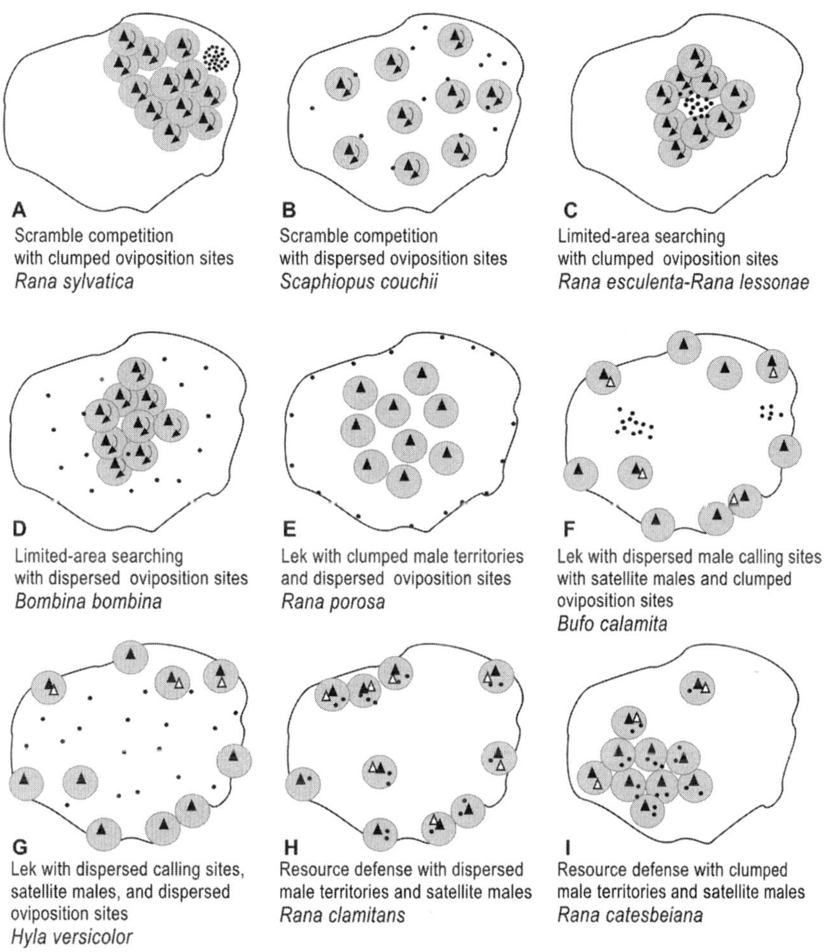


**Fig. 8.10.** Spatial organization of choruses of pond-breeding anurans, showing a variety of possible mating systems. Closed triangles show calling males. Open triangles show satellite males. Curved arrows indicate movements of calling males. Dots indicate distribution of egg clutches. Shaded circles are territories or activity spaces of males. (A) Scramble competition typical of explosive-breeding frogs such as wood frogs (*Rana sylvatica*) that lay eggs in communal masses. (B) Scramble competition typical of explosive-breeding spadefoot toads like *Scaphiopus couchii* that lay eggs in dispersed clumps. (C) Mating system of frogs in the *Rana esculenta* complex. Males maintain a spatial organization similar to a lek, but engage in active searching for mates. Eggs are laid in communal masses. (D) Mating system of *Bombina bombina*. Males defend small territories, but engage in active searching within their territories. Eggs are dispersed. (E) Lek mating system of ranid frogs that breed in rice paddies, such as *Rana porosa brevipoda*. Males defend small territories that are used as mating stations, but eggs are dispersed on the periphery of the chorus. (F) Lek mating system of natterjack toads (*Bufo calamita*). Males call from dispersed territories on the periphery of the pond and sometimes are accompanied by satellite males. Eggs are laid in communal sites in shallow water. (G) Lek mating system of *Hyla versicolor* and many other treefrogs. Males call from elevated perches on the periphery of the pond or in the water and sometimes are accompanied by satellite males. Males approach females at their calling sites, but carry them to the pond to lay eggs. (H) Resource defense mating system of green frogs (*Rana clamitans*). Male territories are located mostly in sheltered sites along the shore of the pond, where females lay their eggs. Satellite males often accompany territorial males. (I) Resource defense mating system of bullfrogs (*Rana catesbeiana*), with male territories clustered in open water, where females lay their eggs. Satellite males often accompany territorial males.

1984b; Telford 1985; Gerhardt et al. 1987; Telford and Dyson 1988; Dyson 1989; M. Morris 1989; Shimoyama 1989; Cherry 1993). This sort of shifting from one calling site to another seems to be common in species that call from the surface of the water in clustered territories, in species that breed in unstable habitats where water levels or temperatures vary from night to night, and in species that utilize abundant and largely interchangeable calling sites such as tree branches or the banks of ponds and rivers.

Females typically approach males at their calling sites, enter into amplexus, and then carry the males away to suitable oviposition sites. This type of mating system resembles the leks of birds and mammals in that male territories are used for mating, but not for oviposition (Höglund and Alatalo 1995). Nevertheless, the evolutionary origins of leklike mating systems probably differ in these groups. In anurans, the locations of aggregations are determined at least in part by the distribution of suitable aquatic breeding sites, whereas in birds and mammals, the location of leks appears to be largely independent of nesting or birthing areas used by females. Some authors have referred to anuran choruses as resource-based leks. This is a somewhat confusing term, because lek mating systems usually are contrasted with resource defense mating systems. There is evidence for some groups of birds and mammals, such as grouse and ungulates, that lek mating systems are most likely to evolve when females occupy very large home ranges that make defense of resource-based territories or groups of females difficult. There is no evidence of any such relationship in anurans. Nevertheless, it does appear that many of the processes of mate choice and sexual selection described for bird and mammal leks also apply to anuran choruses.

Höglund and Alatalo (1995) suggested that lek mating systems are less common in anurans than is defense of resource-based territories. I believe just the opposite is true. Leklike choruses appear to be the predominant mating systems of aquatic-breeding anurans with prolonged breeding seasons in the families Bombinatoridae, Bufonidae, Discoglossidae, Hylidae, Hyperoliidae, Leptodactylidae, Mantellidae, Microhylidae, Myobatrachidae, Pelobatidae, Pelodytidae, Pipidae, Pseudidae, Ranidae, and Rhacophoridae (table 8.1). Many investigators have characterized the choruses of prolonged-breeding anurans as leks (Emlen 1976; Klöckner 1982; Shimoyama 1982; B. K. Sullivan, 1982b 1983b; Robertson 1986a, b; Arak 1988b, c; Márquez-M. de Orense and Tejedo-Madueno 1990; B. K. Sullivan and Hinshaw 1992; Cherry 1993; Giacoma, Kozar, and Pavignano 1993; C. Murphy 1994a, b; Bastos and Haddad 1996, 2002; Grafe 1997a; Friedl and Klump 2002), despite some quibbling over what criteria should be used to define a lek mating system (e.g., Bourne 1992). Many other anurans essentially have the same type of mating system, but were not originally described as lek species (e.g., Godwin and Roble 1983; M. Ryan 1983b, 1985b; Mac Nally, 1984a; Schwartz and Wells 1984b, 1985; Wells and Schwartz 1984a, b; Akef and Schneider 1985, 1990; B. Balinsky 1985; J. Schwartz 1986; Telford 1985; M. Morris 1989, 1991; Perrill and Shepherd 1989; Wagner 1989a, b, c; Runkle et al. 1994; Barreto and Andrade 1995; B. K. Sullivan, Ryan, and Verrell 1995; Grafe, Steffan, and Stoll 2000; Narins, Lewis, and McClelland 2000).

Lek mating systems are characteristic of anurans that use oviposition sites that cannot be easily monopolized by individual males. Many primitive anurans have inguinal amplexus, including bombinatorids, discoglossids, pelobatids, pelodytids, myobatrachids, and pipids. A common mode of egg deposition in these families is for females to attach eggs individually to plants or other substrates, or to scatter them individually or in small packets on the bottoms of breeding pools (see chapter 10). This mode of oviposition would make it difficult for males to monopolize suitable oviposition sites, so defense of resource-based territories is unlikely. Foam-nesting myobatrachids are an exception, because they often lay their eggs in single masses at sites that can be defended by individual males (see the following). Some derived frogs with axillary amplexus, including some species of *Pseudacris* and *Hyla,* also tend to scatter individual eggs or packets of eggs in different locations (see chapter 7), so they are expected to have lek mating systems as well.

Many species of ranid frogs in Asia breed mostly in rice fields, where males form dense aggregations and vigorously defend small territories against other males. This type of mating system has been described in *Rana brevipoda* (= *R. porosa brevipoda*), *R. porosa porosa, R. nigromaculata, Fejervarya limnocharis,* and *F. syhadrensis.* In these species, mated females apparently leave the cluster of male territories in deeper parts of ponds and lay their eggs in vegetation at the water's edge (Dubois 1977b; Inoue 1979; Shimoyama 1982, 1989, 1993, 1996, 1999; Mallick 1986; Kanamadi, Hiremath, and Schneider 1995). Consequently, their mating systems more closely resemble leks than resource defense (fig. 8.10 E). Mixed species choruses have been observed in rice fields inhabited by both *Rana nigromaculata* and *R. brevipoda,* and in such choruses, males of both species defend their territories against both conspecific and heterospecific males (Shimoyama 1999). Fights usually are won by *R. nigromaculata,* the larger species. The similarity of the vocalizations of these two species facilitates their use in interspecific encounters.

The ranid frogs described previously have clumped male territories and dispersed oviposition sites. Some anurans with leklike mating systems exhibit the opposite pattern, with clumped oviposition sites and more dispersed calling sites (fig. 8.10 F). Most toads (*Bufo*) lay eggs in long strings,

**Table 8.1** Examples of aggression and territoriality in anurans with prolonged breeding seasons

| Species (source no.) | Duration of study | Mark | Site | Type | Behavior | | | |
|---|---|---|---|---|---|---|---|---|
| | | | | | AC | CH | FI | SM |
| Bombinatoridae | | | | | | | | |
| *Bombina bombina* (12) | 4 | No | W | CS | + | + | + | |
| *B. orientalis* (87) | 1 | No | W | CS | + | + | + | |
| *B. variegata* (12, 147) | 4 | No | W | CS | + | + | + | |
| Bufonidae | | | | | | | | |
| *Atelopus chiriquiensis* (52) | 1 | Yes | G | CS | + | + | + | |
| *A. varius* (99, 144) | 3 | Yes | G | CL | + | + | + | |
| *A. zetecki* (137) | 1 | No | G | CS | + | + | + | |
| *Bufo calamita* (72, 73) | 4 | Yes | WG | CS | – | + | + | + |
| *B. canorus* (55) | 4 | Yes | WG | CS | + | + | + | + |
| *B rangeri* (124) | 4 | Yes | WG | CS | – | + | + | + |
| *B. regularis* (110) | 4 | No | W | CS | + | + | + | |
| *B. woodhousii* (69) | 4 | Yes | WG | CS | – | + | + | + |
| Brachycephalidae | | | | | | | | |
| *Brachycephalus ephippium* (129) | 1 | No | G | CO | + | + | + | |
| Centrolenidae | | | | | | | | |
| *Centrolene buckleyi* (145) | 1 | No | V | | + | + | + | |
| *C. prosoblepon* (89) | 3 | Yes | V | CO | + | + | + | |
| *Cochranella griffithsi* (33) | 1 | No | V | CO | + | | + | |
| *Hyalinobatrachium collymbiphyllum* (42) | 4 | Yes | V | CO | + | + | + | |
| *H. fleischmanni* (54, 59, 89) | 3 | Yes | V | CO | + | + | + | |
| *H. valerioi* (26) | 4 | Yes | V | CO | + | + | + | |
| Hylidae | | | | | | | | |
| *Acris crepitans* (102, 105, 107, 108, 109, 121) | 4 | Yes | W | CS | + | + | + | + |
| *A. gryllus* (96) | 3 | Yes | G | CL | + | + | + | |
| *Agalychnis callidryas* (157) | 1 | No | V | CS | + | + | + | |
| *Hyla albopunctata* (58) | 3 | No | V | CS | + | | | |
| *H. arborea* (17) | 1 | No | V | CS | + | + | + | |
| *H. arenicolor* (22) | 1 | No | R | CS | + | + | + | |
| *H. avivoca* (21) | 1 | No | V | CS | + | | | |
| *H. chrysoscelis* (74) | 3 | Yes | VG | CS | + | + | + | + |
| *H. cinerea* (30, 44, 46, 66) | 4 | Yes | V | CS | + | + | + | + |
| *H. crepitans* (157) | 1 | No | VW | CS | + | + | | |
| *H. ebraccata* (62, 85, 86) | 4 | Yes | V | CS | + | + | + | + |
| *H. elegans* (134) | 4 | Yes | W | CS | + | + | + | + |
| *H. faber* (2, 101, 125, 126) | 1 | No | NG | CO | + | + | + | |
| *H. meridionalis* (17) | 1 | No | V | CS | + | + | + | |
| *H. microcephala* (94) | 4 | Yes | VG | CS | + | + | + | + |
| *H. minuta* (58, 78, 95, 114) | 3 | Yes | VGW | CS | + | + | + | + |
| *H. pardalis* (24) | 1 | No | NG | CO | | + | + | |
| *H. phlebodes* (83) | 4 | No | V | CS | + | + | + | |
| *H. rosenbergi* (60) | 4 | Yes | N | CO | + | + | + | |
| *H. sanborni* (58) | 3 | Yes | V | CS | + | + | + | |
| *H. savignyi* (64) | 2 | No | GV | CS | + | + | + | |

*(continued)*

**Table 8.1** (continued)

| Species (source no.) | Duration of study | Mark | Site | Type | Behavior AC | CH | FI | SM |
|---|---|---|---|---|---|---|---|---|
| *H. squirella* (46) | 1 | No | V | CS | + | + | + | + |
| *H. veleta* (58) | 3 | Yes | V | CS | + | | | |
| *H. versicolor* (46, 47) | 4 | Yes | VGW | CS | + | + | + | + |
| *Litoria aurea* (50) | 1 | No | V | | + | | | |
| *L. ewingi* (50, 142) | 2 | No | V | | + | + | + | |
| *L. peroni* (50) | 1 | No | V | | + | + | + | |
| *L. raniformis* (50) | 1 | No | W | | + | + | + | |
| *L. verreauxi* (154) | 1 | No | GW | | + | + | + | |
| *Pachymedusa dacnicolor* (20) | 2 | No | B | DS | + | + | + | |
| *Phyllodytes luteolus* (63) | 1 | No | V | CO | + | + | + | |
| *Phyllomedusa hypochondrialis* (151) | 3 | No | V | CS | + | + | + | |
| *P. sauvagii* (152) | 1 | No | V | CS | + | + | | |
| *P. trinitatis* (157) | 1 | No | V | CS | | + | + | |
| *Phrynohyas resinifictrix* (115, 153) | 4 | Yes | TH | CO | + | + | + | |
| *Pseudacris crucifer* (28, 46) | 2 | No | VG | CS | + | + | + | + |
| *P. regilla* (23, 31, 46, 57, 104, 127) | 3 | Yes | VG | CS | + | + | + | + |
| *P. triseriata* (156) | 1 | No | W | CS | + | + | + | |
| *Scinax boulengeri* (157) | 4 | Yes | V | CL | + | + | | |
| *S. rubra* (117, 123) | 4 | Yes | W | CS | + | + | + | + |
| Hyperoliidae | | | | | | | | |
| *Afrixalus brachycnemis* (70, 98) | 4 | Yes | V | CS | + | + | + | |
| *A. delicatus* (113) | 4 | Yes | VW | CS | + | + | + | + |
| *A. fornasinii* (70, 103) | 4 | No | V | CS | + | + | + | |
| *A. vittiger* (138) | 1 | No | V | CS | | + | + | |
| *Hyperolius argus* (70) | 4 | No | VW | CS | + | + | + | + |
| *H. marmoratus* (70, 95, 118, 133) | 4 | Yes | VW | CS | + | + | + | + |
| *H. pusillus* (70) | 4 | No | VW | CS | + | + | + | |
| *H. tuberilinguis* (70) | 4 | Yes | V | CS | + | + | + | |
| *Kassina senegalensis* (70) | 2 | Yes | VW | CS | + | + | | |
| *Leptopelis natalensis* (70) | 1 | No | V | | + | + | | |
| *L. viridis* (149) | 3 | No | V | CS | + | + | + | |
| Leptodactylidae | | | | | | | | |
| *Batrachyla antartandica* (143) | 3 | No | G | | + | | | |
| *B. leptopus* (143) | 3 | No | G | | + | | | |
| *B. taeniata* (143) | 3 | No | G | | + | | | |
| *Crossodactylus gaudichaudii* (122) | 1 | No | G | CS | + | + | + | |
| *Eleutherodactylus achatinus* (75) | 1 | No | GV | | + | + | + | |
| *E. antillensis* (139, 157) | 3 | No | V | | + | + | + | |
| *E. coqui* (16, 84, 116, 120, 130) | 4 | Yes | VN | CL, O | + | + | + | |
| *E. fitizingeri* (75) | 1 | No | G | | + | | | |
| *E. hedricki* (16) | 1 | No | TH | CO | + | + | | |
| *E. johnstonei* (119, 141) | 4 | Yes | V | CS | + | + | + | + |
| *E. portoricensis* (16) | 1 | No | V | CS | + | | | |
| *E. urichi* (61) | 3 | Yes | V | CL | + | + | + | |
| *Eupsophus emiliopugini* (136) | 1 | No | B | | + | | | |

*(continued)*

**Table 8.1** (continued)

| Species (source no.) | Duration of study | Mark | Site | Type | Behavior | | | |
|---|---|---|---|---|---|---|---|---|
| | | | | | AC | CH | FI | SM |
| *Hylodes asper* (146) | 3 | No | R | CL | + | + | + | |
| *Leptodactylus albilabris* (91, 100) | 3 | No | B | CO | + | + | | |
| *L. insularum* (3) | 1 | No | N | O | + | + | | |
| *L. melanonotus* (8) | 1 | No | R | | + | + | + | |
| *L. syphax* (132) | 1 | No | B | CO | + | + | | |
| *Physalaemus cuveri* (131) | 3 | No | W | CS | + | + | + | |
| *P. pustulosus* (76, 93) | 4 | Yes | W | CS | + | + | + | |
| *Thoropa petropolitana* (49) | 1 | No | R | CL | + | + | + | |
| Mantellidae | | | | | | | | |
| *Boophis madagascariensis* (150) | 2 | No | V | CS | + | | | |
| Microhylidae | | | | | | | | |
| *Breviceps adspersus* (35) | 1 | No | B | CO | | + | | |
| *Cophixalus parkeri* (68, 77) | 4 | Yes | G | CO | | + | + | |
| *Megophrys parva* (38) | 1 | No | V | | + | + | + | |
| *Phrynomantis annectans* (32) | 1 | No | GR | CS | + | + | + | |
| *Platypelis grandis* (29) | 1 | No | TH | CO | + | + | | |
| Myobatrachidae | | | | | | | | |
| *Atelotus brevis* (79, 140) | 4 | Yes | BW | CO | | + | + | |
| *Crinia parinsignifera* (53, 81) | 4 | Yes | G | CL | + | + | | + |
| *C. signifera* (53, 81) | 4 | Yes | G | CL | + | + | | + |
| *Geocrinia laevis* (88) | 4 | No | G | CS | + | | | |
| *G. victoriana* (92) | 4 | Yes | V | CS | + | + | + | |
| *Limnodynastes dumerilii* (50) | 1 | No | W | CS | + | + | | |
| *L. peroni* (7, 11) | 1 | No | WB | CO | + | + | + | |
| *L. tasmaniensis* (50) | 1 | No | W | CS | + | + | | |
| *Pseudophryne bibroni* (19) | 3 | No | B | CO | + | + | + | |
| *P. corroboree* (19) | 3 | No | B | CO | + | + | + | |
| *P. dendyi* (19) | 3 | No | B | CO | + | + | + | |
| *Uperoleia laevigata* (82) | 4 | Yes | G | CL | + | + | + | |
| Petropedetidae | | | | | | | | |
| *Phyrnobatrachus mababiensis* (70) | 1 | No | G | | + | | | |
| Pipidae | | | | | | | | |
| *Hymenochirus boettgeri* (4, 5) | 1 | No | W | | + | + | + | |
| *Pipa carvalhoi* (37) | 1 | No | W | | + | + | + | |
| *P. parva* (6) | 1 | No | W | | + | + | + | |
| *P. pipa* (13) | 1 | No | W | | + | + | + | |
| Ranidae | | | | | | | | |
| *Limnonectes blythii* (142) | 1 | No | NW | CO | + | + | | |
| *L. limnocharis* (135) | 1 | No | W | CS | + | + | + | |
| *Platymantis papuensis* (15) | 1 | No | G | | + | + | | |
| *Ptychadena anchietae* (39, 43) | 4 | No | GW | CS | + | + | | |
| *P. mossambica* (39, 43) | 4 | No | GW | CS | + | + | | |
| *P. oxyrhynchus* (39, 43) | 4 | No | GW | CS | + | + | | |
| *P. taenioscelis* (39, 43) | 4 | No | GW | CS | + | + | | |
| *Pyxicephalus adspersus* (128) | 1 | no | W | CS | | + | + | |
| *Rana berlandieri* (48) | 1 | No | W | | + | | + | |

*(continued)*

**Table 8.1** (continued)

| Species (source no.) | Duration of study | Mark | Site | Type | Behavior AC | CH | FI | SM |
|---|---|---|---|---|---|---|---|---|
| *R. catesbeiana* (9, 14, 34, 41, 56) | 4 | Yes | W | CO | + | + | + | + |
| *R. clamitans* (40, 45) | 4 | Yes | W | CO | + | + | + | + |
| *R. dunni* (18) | 1 | No | W | | + | | | |
| *R. esculenta* (25, 36, 65) | 4 | Yes | W | CS | + | + | + | |
| *R. graeca* (111) | 1 | No | W | CO | + | | | |
| *R. grylio* (1, 80) | 3 | No | W | CO | + | + | + | |
| *R. lessonae* (25, 36, 65) | 4 | Yes | WG | CS | + | + | + | |
| *R. montezumae* (18) | 1 | No | W | | + | | | |
| *R. nigromaculata* (67, 148) | 2 | Yes | W | CS | + | + | + | |
| *R. palustris* (157) | 1 | No | W | CO | + | + | | |
| *R. pipiens* (10, 18, 27) | 1 | No | W | | + | | | |
| *R. porosa brevipoda* (51, 106, 148) | 2 | Yes | W | CS | + | + | + | + |
| *R. ridibunda* (36, 90, 112) | 1 | No | W | CO | + | + | + | |
| *R. septentrionalis* (155) | 4 | Yes | W | CO | + | + | + | |
| *R. sphenocephala* (27) | 1 | No | W | | + | | | |
| *R. syhadrensis* (38) | 1 | No | W | CS | + | + | + | |
| *R. virgatipes* (97) | 4 | Yes | W | CO | + | + | + | + |
| Rhacophoridae | | | | | | | | |
| *Philautus leucorhinus* (71) | 3 | Yes | V | CS | + | + | + | + |

*Sources:* (1) Bogert 1960; (2) Lutz 1960a; (3) Sexton 1962; (4) Osterdahl and Olsson 1963; (5) Rabb and Rabb 1963a; (6) Rabb and Rabb 1963b; (7) Clyne 1967; (8) Brattstrom and Yarnell 1968; (9) Emlen 1968; (10) Schmidt 1968a; (11) Clyne 1969; (12) Lörcher 1969; (13) Rabb 1969; (14) Wiewandt 1969; (15) Zweifel 1969; (16) Drewry 1970; (17) Paillette 1970b; (18) Mecham 1971; (19) Pengilley 1971a; (20) Wiewandt 1971; (21) Altig 1972a; (22) Pierce and Ralin 1972; (23) Allan 1973; (24) Lutz 1973; (25) Blankenhorn 1974; (26) McDiarmid and Adler 1974; (27) Pace 1974; (28) Rosen and Lemon 1974; (29) Blommers-Schlösser 1975b; (30) Garton and Brandon 1975; (31) Whitney and Krebs 1975b; (32) Channing 1976b; (33) Duellman and Savitsky 1976; (34) Emlen 1976; (35) Poyton & Pritchard 1976; (36) Tunner 1976; (37) Weygoldt 1976a; (38) Dubois 1977b; (39) Passmore 1977; (40) Wells 1977b; (41) R. D. Howard 1978a; (42) McDiarmid 1978; (43) Passmore 1978; (44) Perrill, Gerhardt, and Daniel 1978; (45) Wells 1978a; (46) Fellers 1979a; (47) Fellers 1979b; (48) Gambs and Littlejohn 1979; (49) Heyer and Crombie 1979; (50) Humphries 1979; (51) Inoue 1979; (52) Jaslow 1979; (53) Mac Nally 1979; (54) Greer and Wells 1980; (55) Kagarise Sherman 1980; (56) Ryan 1980b; (57) Whitney 1980; (58) Cardoso 1981b; (59) Clark 1981; (60) Kluge 1981; (61) Wells 1981b; (62) Wells and Greer 1981; (63) Weygoldt 1981a; (64) Brzoska, Schneider, and Nevo 1982; (65) Klockner 1982; (66) Perrill, Gerhardt, and Daniel 1982; (67) Shimoyama 1982; (68) Simon 1982; (69) Sullivan 1982a; (70) Telford 1982; (71) Arak 1983c; (72) Arak 1983b; (73) Arak 1983a; (74) Godwin and Roble 1983; (75) Lynch and Myers 1983; (76) Ryan 1983a; (77) Simon 1983; (78) Cardoso and Haddad 1984; (79) Giddings 1984; (80) Lamb 1984b; (81) Mac Nally 1984a; (82) J. Robertson 1984; (83) Schwartz and Wells 1984b; (84) Townsend, Stewart, and Pough 1984; (85) Wells and Schwartz 1984a; (86) Wells and Schwartz 1984b; (87) Akef and Schneider 1985; (88) Harrison and Littlejohn 1985; (89) Jacobson 1985; (90) Kuhn and Schneider 1985; (91) Lewis and Narins 1985; (92) Littlejohn and Harrison 1985; (93) Ryan 1985a; (94) Schwartz and Wells 1985; (95) Telford 1985; (96) Forester and Daniel 1986; (97) Given 1987; (98) Backwell 1988; (99) Crump 1988; (100) Lopez et al. 1988; (101) Martins and Haddad 1988; (102) Perrill and Magier 1988; (103) Schneichel and Schneider 1988; (104) Brenowitz 1989; (105) Perrill and Shepherd 1989; (106) Shimoyama 1989; (107) Wagner 1989a; (108) Wagner 1989b; (109) Wagner 1989c; (110) Akef and Schneider 1990; (111) Asimakopoulos, Sofianidou, and Schneider 1990; (112) Kyriakopoulou-Sklavounou and Kattoulas 1990; (113) Backwell and Passmore 1991; (114) Haddad 1991; (115) Hödl 1991b; (116) Stewart and Rand 1991; (117) Bourne 1992; (118) Dyson and Passmore 1992b; (119) Ovaska and Hunte 1992; (120) Stewart and Rand 1992; (121) Wagner 1992; (122) Weygoldt and Carvalho e Silva 1992; (123) Bourne 1993; (124) Cherry 1993; (125) Martins 1993a; (126) Martins 1993b; (127) Brenowitz and Rose 1994; (128) Channing, du Preez, and Passmore 1994; (129) Pombal, Sazima, and Haddad 1994; (130) Stewart and Bishop 1994; (131) Barreto and Andrade 1995; (132) Cardoso and Heyer 1995; (133) Grafe 1995; (134) Bastos and Haddad 1995; (135) Kanamadi, Hiremath, and Schneider 1995; (136) Formas and Poblete 1996; (137) Lindquist and Hetherington 1996; (138) Rödel 1996; (139) Bourne 1997; (140) Katsikaros and Shine 1997; (141) Ovaska and Caldbeck 1997b; (142) Orlov 1997; (143) Penna 1997; (144) Lindquist and Hetherington 1998; (145) Bolívar-G., Grant, and Osorio 1999; (146) Haddad and Giaretta 1999; (147) Seidel 1999; (148) Shimoyama 1999; (149) Grafe, Steffan, and Stoll 2000; (150) Narins, Lewis, and McClelland 2000; (151) Ribeiro Matos, Andrade, and Hass 2000; (152) Rodrigues, de Souza, andFilho 2003; (153) Schiesari, Gordo, and Hödl 2003; (154) Peter Harrison, personal communication; (155) Lew Oring, personal communication; (156) Patrick Owen, personal communication; (157) K. Wells, unpublished observations.

*Notes:* Data on aggressive behavior of dendrobatid frogs are given in table 11.3 along with information on parental care. Duration of study: 1 = scattered observations; 2 = < 1 month; 3 = > 1 month; 4 = > 1 season. Mark = animals marked (yes, no). Site = location of site defended (B = burrow; G = ground; N = nest; R = rocks; TH = tree hole; V = vegetation; W = water). Type = type of site defended (CS = short term calling site; CL = long term calling site; CO = calling and oviposition site; DS = daytime shelter). Classification of type of site defense represents best estimates based on available data. Behavior: AC = aggressive calling; CH = chasing; FI = fighting; SM = satellite males (+ behavior present, – behavior absent, blank cell = behavior not investigated or unknown).

with many clutches clustered together in shallow water. Some ranid frogs, even those with long breeding seasons, tend to cluster their eggs in areas with dense vegetation, or in sunny areas where warm water ensures rapid embryonic development (Blankenhorn 1977; Radwan and Schneider 1988; Sjögren, Elmberg, and Berglind 1988; Doody and Young 1995). Other species, such as the tropical foam-nesting frogs in the genus *Physalaemus*, prefer to attach their eggs to those of other females, creating huge communal masses that reduce the chances of eggs drying out (M. Ryan 1985b; Barreto and Andrade 1995). This sort of spatial clumping of oviposition sites would make it impossible for males to defend resource-based territories, but defense of small mating territories is common.

In many anurans with lek mating systems, suitable calling sites are spatially separated from oviposition sites, making defense of oviposition sites impossible. These include species in which males call on land and are later carried to water by females, such as *Uperoleia laevigata* (J. Robertson 1986a, b). Males of the Asian rhacophorid *Buergeria buergeri* sit on rocks in mountain streams and defend calling sites against other males (Fukuyama and Kusano 1989, 1992). Females approach males at their calling sites and then carry them up to 50 meters before laying eggs under rocks (Fukuyama, Kusano, and Nakane 1988). Many treefrogs call from elevated perches, ranging from emergent aquatic vegetation only a few centimeters above the water to the branches of bushes and trees (Gerhardt and Rheinlaender 1982; Passmore et al. 1984; S. Mitchell and Miller 1991). The principal advantage of such sites is that propagation of calls over long distances is enhanced by reducing excess attenuation due to ground absorption or passage of signals through dense vegetation (Forrest 1994; Gerhardt and Huber 2002). If eggs are laid in water, then the use of elevated calling sites precludes the defense of oviposition sites by territorial males (fig. 8.10 G). Hence most treefrogs are expected to have lek mating systems unless they use specialized oviposition sites, such as nests constructed by males.

### Resource Defense in Aquatic and Arboreal Frogs

Resource defense mating systems are expected to evolve when males can monopolize resources that are needed by females for successful reproduction. In frogs, such resources almost always are oviposition sites. The first observation of frogs defending oviposition sites as territories was by Albert Hazen Wright, a pioneer student of anuran behavior. He reported that male bullfrogs (*Rana catesbeiana*) establish calling stations near fallen logs or stumps "in which they keep a proprietary interest" (Wright 1914). He also observed certain places along the shore of a pond that were consistently occupied by male green frogs (*R. clamitans*). He frequently found eggs in the same locations. Subsequent work on these two species demonstrated that males do indeed defend territories that are occupied for long periods of time (fig. 8.11). Females usually lay their eggs within the territories of their mates, and mate choice is influenced by the quality of male territories (Martof 1953b; Emlen 1968, 1976; Wells 1977a, 1978a; R. D. Howard 1978a,b; M. Ryan 1980a). These remain the two best-documented examples of resource-defense mating systems in anurans (fig. 8.10 H, I). B. K. Sullivan, Ryan, and Verrell (1995) listed only a few examples of resource-defense polygyny in anurans, but similar mating systems can be expected in almost any species that makes use of oviposition sites that are easily monopolized by males (table 8.1).

Territorial defense of oviposition sites is particularly common in pond-breeding ranid frogs with prolonged breeding seasons. Male carpenter frogs (*Rana virgatipes*) defend long-term territories along the shores of ponds, particularly around clumps of submerged vegetation used by females as oviposition sites (Given 1988b). Male pickerel frogs (*Rana palustris*) call underwater from clumps of vegetation and from the bases of submerged bushes, and they respond to advertisement calls of other males with distinct aggressive calls (Given 2005). Egg clumps are laid in similar locations, suggesting that males defend oviposition sites. In other North American ranids, including *R. grylio* and *R. capito*, females apparently lay eggs at male calling sites, and males probably are territorial (Wells 1977b; Lamb 1984b), although the latter species also has been reported to form explosive mating aggregations after heavy rains (Semlitsch, Gibbons, and Turberville 1995). In Europe, *Rana ridibunda* and several related species breed during the summer in large ponds and lakes. Males are widely spaced in choruses and aggressive toward one another. They probably have a territorial mating system similar to that of North American green frogs and bullfrogs, but there is little information on where females lay eggs (M. Smith 1969; Frazer 1983; Kuhn and Schneider 1984; Schneider and Joermann 1988; Kyriakopoulou-Sklavounou and Kattoulas 1990).

Some stream-breeding ranids also have resource-defense mating systems. The southern European frog *Rana graeca* breeds in rivers and streams during the spring. Males call at the surface or under water in cavities under rooks or tree roots, and eggs are laid in these cavities. Males have calls that are used in aggressive interactions, but whether they are territorial is not known (Asimakopoulos, Sofianidou, and Schneider 1990). In Asian stream-breeding ranids such as *Rana holsti* and *Paa* (= *Rana*) *boulengeri*, males exhibit strong fidelity to isolated pools that are used for oviposition and probably are territorial (C. Liu 1950). Another Asian stream-breeding frog, *Vibrissaphora ailaonica* (Megophryidae), also appears to be territorial. Males build nests be-

**Fig. 8.11.** Territorial behavior of male green frogs (*Rana clamitans*). (A) Two males wrestling while floating in the water. (B, C) Two males wrestling while standing on the pond bottom. (D) A large male holding a smaller male under water. Photos by Kentwood D. Wells.

neath rocks and call to attract females. As many as 17 egg masses have been found under one rock, but it is not entirely clear whether this represented the nest of a single male or those of several males. The males of this species remain with the eggs and guard them (Ho et al. 1999).

The fanged ranid frogs of Southeast Asia also breed in streams, and in at least some species, males build nests and defend them against other males. The species that has been studied in most detail is *Limnonectes leporinus* (formerly *L. blythii*) from Borneo, which builds nests along the edges of streams. Females visit the nests to lay their eggs, which are later guarded by the males (Emerson 1992). Males are larger than females and have hypertrophied jaw muscles and fangs on the lower jaw that are used as weapons (Emerson and Voris 1992; see "Sexual Dimorphism"). Males of this species appear not to have advertisement calls, but are territorial (Emerson and Inger 1992). Males of other fanged frogs, including species currently designated as *Limnnonectes blythii* from Vietnam and *L. kuhlii* from Taiwan, have been observed fighting over territories and biting one another with their fangs. Males of the latter species do not always build nests, but defend oviposition sites in shallow pools along streams. Males of both of these species produce advertisement calls, as do related frogs from penisular Malaysia (Matsui 1995), and some have distinct aggressive calls as well (Orlov 1997; Tsuji and Matsui 2002; Tsuji 2004). Both of these species probably will be given different names in the future, because both appear to be members of geographically separated multispecies complexes (Emerson, Inger, and Iskandar 2000.

Nest-building frogs in other families have resource-defense mating systems as well. The nest-building gladiator frogs (*Hyla boans*, *H. faber*, *H. pardalis*, *H. rosenbergi*, and *H. wavrini*) of South and Central America are the best known (fig. 8.12 A). These species have sharp spines on the front feet that are used in territorial battles among males (fig. 8.12 B). Defense of nests has been reported in several of these species, and in others, males often have scratches and wounds that are consistent with territorial fighting (Breder 1946; Lutz 1960b, 1973; Kluge 1981; Martins and Haddad 1988; Hoogmoed 1990b; Martins and Moreira 1991; Martins 1993a, b; Martins, Pombal, and Haddad 1998).

Many other frogs lay aquatic eggs in secluded spots that can be defended by territorial males. For example, males of the Chinese frog *Rana caldwelli* (formerly *R. adenopleura*) call from shallow water or from flooded burrows, where eggs are subsequently laid by females (C. Liu 1950), and it seems likely that males would defend their burrows. Defense of similar burrows used for oviposition has been reported in a species of *Adenomera* (Leptodactylidae) from Brazil (Kokubum and Giaretta 2005). Males of the Australian myobatrachid frog *Limnodynastes peronii* are territorial and defend sites in submerged vegetation where foam nests are subsequently placed (Clyne 1967, 1969). Males of the South American leptodactylid *Thoropa petropolitana* defend territories around oviposition sites on wet rock faces (Heyer and Crombie 1979). In another leptodactylid, *Crossodactylus gaudichaudii,* males aggressively defend calling sites along streams. They appear to select and excavate nearby nest sites under rocks and then close up the site with stones after the female has laid her eggs (Weygoldt and Carvalho e Silva 1992).

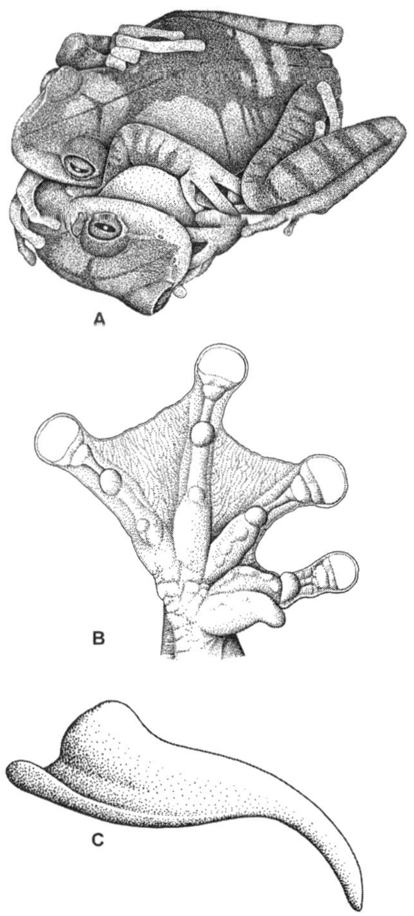


**Fig. 8.12.** (A) Two males of a Neotropical gladiator frog (*Hyla faber*) fighting over a territory. (B) Front foot and (C) prepollical spine of a closely related species, *Hyla boans,* used in fights between males. (A) After Martins, Pombal, and Haddad (1998); (B) after drawings by David M. Dennis in Duellman (1970).

Tree frogs in several families lay eggs in or above small bodies of water in treeholes, water-filled bromeliads, hollow bamboo stems, and other enclosed sites (see chapter 7). When males call from such sites, it is very likely that they defend them against other males, and that females assess the quality of oviposition sites when choosing mates. Males of one bromeliad breeder, *Phyllodytes luteolus,* have been observed fighting for control of suitable oviposition sites (Weygoldt 1981a), although males in one population where bromeliads were abundant were not territorial (Eterovick 1999). In other hylids that breed in treeholes and bromeliads, males often exhibit strong attachment to specific calling sites and sometimes respond to playbacks of calls with distinct, aggressive calls (Hödl 1991b; Jungfer 1996). Aggressive defense of an oviposition site in bamboo has been observed in captivity in a microhylid frog from Madagascar, *Platypelis grandis,* and males in the field have been observed calling from treeholes and water-filled leaf axils (Blommers-Schlösser, 1975b). Apparently not all treehole breeders are territorial, however. In one microhylid frog from India, *Ramanella montana,* up to four males were observed calling from the same water-filled treehole, with no evidence of aggressive interactions (Krishna, Krishna, and Vijayalaxmi 2004).

Males of many centrolenid frogs are territorial and defend calling sites on leaves overhanging streams that are subsequently used for oviposition by females (McDiarmid and Adler 1974; Duellman and Savitsky 1976; McDiarmid 1978; Greer and Wells 1980; M. Clark 1981; Jacobson 1985; M. Hayes 1991; Bolívar-G., Grant, and Osorio 1999). In many cases, the male provides parental care to the eggs and sometimes accumulates several clutches at the same site (see chapter 11). In other centrolenids, such as *Centrolene geckoideum,* eggs are laid on rocks where they are splashed by waterfalls. Aggressive behavior has not been reported, but males have been observed calling while sitting on egg clutches (J. Lynch, Ruiz, and Rueda 1983).

In contrast to the behavior of centrolenids, African hyperoliid frogs in the genus *Afrixalus,* which also place their eggs on vegetation over water, do not appear to defend oviposition sites. Males call on leaves similar to those used as oviposition sites, but mated females often explore several potential sites before laying eggs. Males do not guard eggs and do not exhibit strong site fidelity (Backwell 1991a). Hence, the mating system of these frogs appears to be more like a lek than resource defense. In most tropical hylids that lay eggs on vegetation over water, there is not a close association between male calling sites and oviposition sites, so these frogs apparently have lek mating systems as well. These include the phyllomedusines (*Agalychnis, Phyllomedusa, Pachymedusa*) and species such as *Hyla ebraccata* (personal observations), *H. brevifrons,* and *H. sarayacuensis* (Hödl 1991a). Males have been observed fighting over calling sites in some phyllomedusine frogs, including *Phyllomedusa sauvagii,* but it is not clear whether males defend oviposition sites or only calling sites (Halloy and Espinoza 1999–2000).

## Mating Systems of Terrestrial Egg-Layers

The mating systems of anurans that lay their eggs on land can be difficult to classify, because males are not always aggregated at well-defined breeding areas. Explosive breeding and intense scramble competition have not been reported for terrestrial breeders, although Pombal, Sazima, and Haddad (1994) did observe attempts by males of a small terrestrial brachycephalid toad, *Brachycephalus ephippium,* to displace other males from amplexus. Most terrestrial breeders appear to have lek or resource-defense mating systems, but these two mating systems are not always easily distinguished because of a lack of information on female oviposition behavior for many species. Furthermore, many terrestrial egg-layers have some form of parental care (see chapter 11), which complicates the classification of mating systems. Traditional definitions of lek mating systems exclude cases in which males provide parental care (Höglund and Alatalo 1995). Some frogs with paternal care have mating systems with leklike characteristics, while others clearly appear to be resource defense mating systems. Rather than debating the correct classification of these mating systems, it will be more informative to simply describe them in some detail.

### Defense of Subterranean and Terrestrial Nest Sites

The most obvious cases of resource defense are in species that lay eggs in burrows excavated by the male. Such burrows require a considerable investment in time and energy and therefore are likely to be aggressively defended against intruders. Several species of Australian myobatrachids in the genus *Pseudophryne* lay eggs in burrows that are later flooded. Males defend these sites with aggressive vocalizations and physical attacks (Pengilley 1971a; Woodruff 1976a). Many species of *Leptodactylus* build foam nests in underground burrows. In *L. albilabris* and *L. syphax,* males call from the burrows and respond to the calls of other males with aggressive signals (Narins 1990; Cardoso and Heyer 1995; E. Lewis et al. 2001). In *L. fuscus* males call on the ground near the entrances to burrows and lead females to them. They also respond aggressively to the calls of other males (M. Martins 1988). I have observed similar behavior in a related species, *L. fragilis* (= *L. labialis*), in Panama (unpublished observations). In one instance, a male was observed calling to a female and leading her for several meters to an underground burrow. Bernal and Ron (2004) described similar courtship behavior in this species. A male and female were seen exchanging calls and touching each other while moving toward a hidden burrow. Males of a Brazilian species, *L. troglodytes,* also lead females to burrows that they construct (Arzabe and de Almeida 1997). There is evidence of males defending burrows used for oviposition in the African microhylid *Breviceps adspersus* (Poynton and Pritchard 1976). Males of *Cophixalus parkeri,* a microhylid from New Guinea, defend terrestrial oviposition sites and guard eggs (Simon 1983), and territorial behavior has been observed in *C. ornatus* as well (Zweifel 1985). Aggressive defense of territories that include oviposition sites also has been observed in *Brachycephalus ephippium* (Pombal, Sazima, and Haddad 1994).

Hartmann, Hartmann, and Haddad (2004) described some extraordinarily complex courtship behavior in an unnamed hylid treefrog from Brazil, possibly related to *Hyla ehrhardti.* The behavior of this species resembles that of some species of *Leptodactylus* that breed in burrows. Males called from ground bromeliads and other elavated positions, but led females to previously constructed burrows in mud banks near temporary ponds on the forest floor. The frogs lay unpigmented eggs in these burrows, and tadpoles develop in nearby ponds after the burrows are flooded. Courtship behavior included use of soft courtship calls and a variety of visual and tactile displays, culminating in the male entering the burrow, followed by the female. In one instance, a female inspected a burrow, but did not mate with the owner, which continued to call inside the burrow. This species clearly seems to have a resource-based mating system, but it is not known whether males actively defend burrows against other males. In other species in the *Hyla albosignata* and *H. albofrenata* complexes, which probably are closely related to this frog, males call from subterranean burrows and presumably would defend them against intruding males (Cruz and Peixoto 1984; Gomes and Peixoto 1997).

### Mating Systems of *Eleutherodactylus*

Many terrestrial breeders with direct development have oviposition sites that are some distance from calling sites. This is particularly true of species that lay eggs underground or in leaf litter, but call from elevated perches on logs, low vegetation, bushes, or tree trunks. In some species, females approach males at their calling sites and then carry them in amplexus to a suitable oviposition site. In others, males have elaborate courtship behavior and lead females to oviposition sites before going into amplexus. When males defend calling sites as territories, but females control the selection of oviposition sites, the mating system resembles a lek. If the male leads the female to an oviposition site a short distance from his calling site, it can be located within a defended territory, and the mating system can be considered resource defense. In most cases, we simply do not know enough about the biology of these frogs to clearly describe the mating system. For example, I observed fighting between males of *Eleutherodactylus urichi* from Trinidad (fig. 8.13 A), and males showed strong fidelity to their territories (Wells 1981b). Unfortunately, mating and oviposition were not observed, so it was impossible to determine whether eggs were laid within the areas defended by the males.

The behavior of *Eleutherodactylus johnstonei* from Barbados includes features of both lek and resource-defense mating systems. Males exhibit strong site fidelity and defend small calling territories near the ground against other males (Ovaska and Hunte 1992; Bourne 1997). Females approach males at calling sites, and males then lead them to oviposition sites in small cavities. It is not clear that these nest sites are within the areas defended by males, however. Both Ovaska and Hunte (1992) and Bourne (1997) observed groups of males attempting to court the same female, and Bourne reported instances of unpaired males following courting pairs into nesting cavities. When Ovaska and Hunte (1992) removed a male that was leading a female toward a nesting cavity, the female then followed another male to the same cavity. Both males and females have been reported to attend eggs after they are laid (Townsend 1996), which suggests that males do not necessarily invest heavily in defense of oviposition sites.

In a number of other species of *Eleutherodactylus*, especially those found in Puerto Rico, males call from elevated perches, but lead females to oviposition sites selected by the male (Jameson 1954, 1955a, b; Townsend and Stewart 1986; Michael 1996; Ovaska and Caldbeck 1997a). After oviposition, the males often remain with the eggs until they hatch. The best-studied species is *Eleutherodactylus coqui*. Males prefer open, elevated calling sites, probably because they enhance the transmission of calls (Narins and Hurley 1982; Townsend 1989a). Most oviposition sites are on the ground in sheltered locations, but elevated sites, such as dead *Cecropia* leaves hanging in branches, are actually used more often than expected if males were selecting oviposition sites at random (Townsend 1989a). Males lead females to oviposition sites while giving distinctive courtship calls (Townsend and Stewart 1986; Michael 1996). Males that use elevated oviposition sites tend to call closer to nest sites than those using sites on the ground, thus reducing the time required for courtship. These males also were more likely to acquire

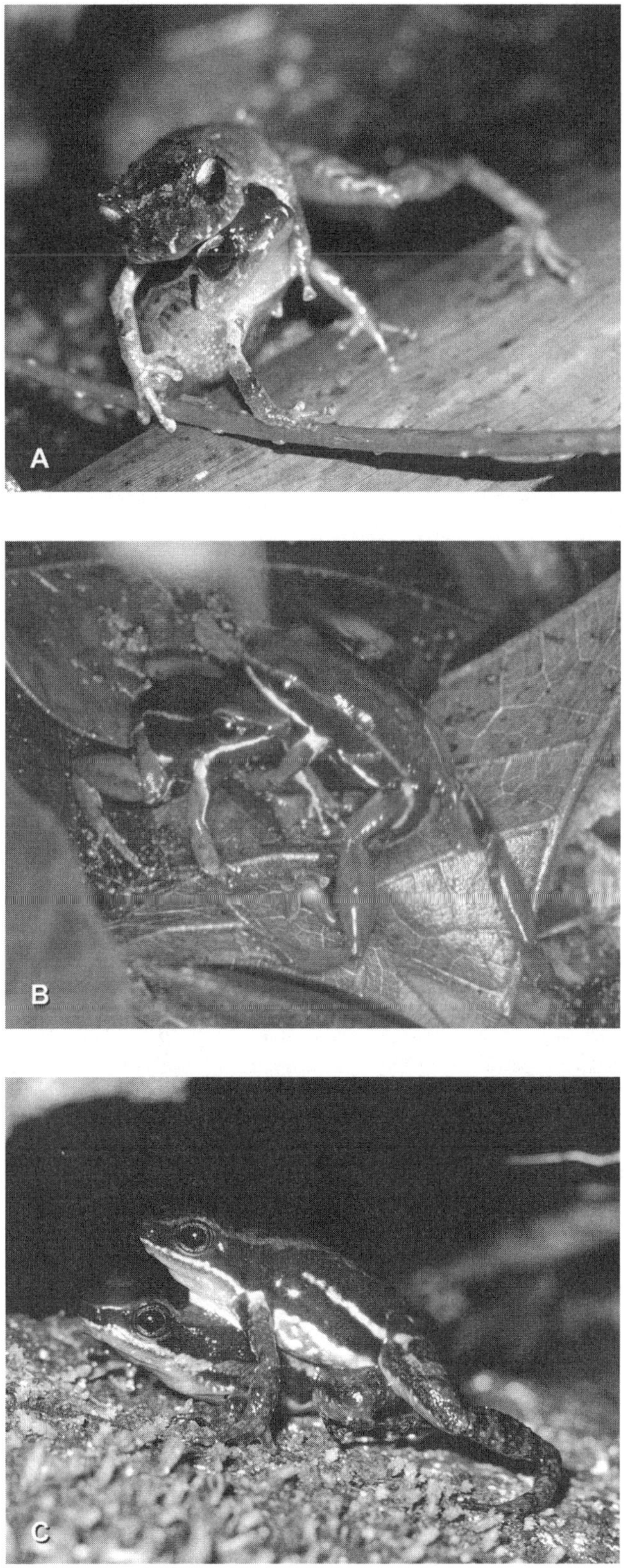


**Fig. 8.13.** Fighting behavior in small terrestrial Neotropical anurans. (A) Two males of the leptodactylid frog *Eleutherodactylus urichi* from Trinidad fighting over a territory. (B) Two males of the dendrobatid frog *Allobates femoralis* from French Guiana fighting in the leaf litter. (C) Two males of the dendrobatid frog *Colostethus panamensis* from Panama fighting over a territory in a rocky stream. The top male is pressing the bottom male onto the rock. Photos by Kentwood D. Wells (A, C) and Walter Hödl (B).

multiple mates at the same site, and had higher hatching success (Townsend 1989a). Males seldom fight over calling sites, but do give aggressive calls in response to calls of nearby males (Stewart and Bishop 1994; O'Brien 2002). They are much more aggressive in defending retreat sites, especially those being used as nest sites (Stewart and Rand 1991, 1992; O'Brien 2002). The mating system of *E. coqui* can be considered a resource-defense mating system, since the quality of nest sites controlled by males probably influences mate choice by females (Townsend and Stewart, 1994).

Other species of *Eleutherodactylus* with male parental care probably have resource-defense mating systems as well. In some species, males call from oviposition sites and aggressively defend them against other males, so resource defense is clear. For example, *E. hedricki* is a tree-hole breeder and exhibits both male parental care (Townsend 1996) and territoriality (Drewry 1970). *Eleutherodactylus cooki* breeds in small caves and rock crevices. Males are aggressive, and often continue to call while attending eggs, sometimes enabling them to obtain additional matings (Joglar, Burrowes, and Rios 1996). Other species, such as *E. antillensis* and *E. cochranae,* resemble *E. coqui* in using courtship calls to lead females from calling sites to oviposition sites (Michael 1996; Ovaska and Caldbeck 1997a). Male parental care is likely in these species, but parental behavior has not been observed in detail (Townsend 1996).

### Mating Systems of Dendrobatid Frogs

Many dendrobatid frogs also lay their eggs on land and exhibit elaborate parental care (see chapter 11). Some species may have resource-defense mating systems, but leklike behavior has been observed as well. The territorial behavior of males differs among species, depending in part on habitat structure and population density (Pröhl 2005b). Members of the genus *Colostethus* often are associated with rocky streams, and some form of territorial behavior probably occurs in most species, especially at high population densities. The best-studied example is the Panamanian species *C. panamensis* (formerly *C. inguinalis;* Wells 1980a). Males occur at high densities on boulders along streams and defend relatively large territories during the rainy season. They call throughout the day and are aggressive toward individuals of both sexes (fig. 8.13 C). Some territories are occupied for up to six months, and appear to contain areas suitable for feeding, shelter, courtship, and oviposition. During the dry season, males abandon their territories and congregate around sources of moisture in streambeds. Here, some males defend small territories, while others simply sit quietly in pools. During the dry season, females also defend territories centered near moist retreat sites or choice feeding sites—but they abandon their territories during the rainy season, when most of the breeding takes place. Females seem to defend small territories only when they need to secure a suitable retreat site during dry weather, whereas the large territories of males during the rainy season are related to competition for mates.

In other species, including *C. pratti, C. talamancae, C. flotator,* and *C. nubicola,* males remain at the same sites for days or weeks and are aggressive toward other males, but their behavior has not been studied in detail (Wells 1977a, 1980a, unpublished observations; K. Summers 2000). These species live on the forest floor and generally occur at lower densities than do the streamside species, so competition for territories probably is not as intense. There are brief reports of territorial behavior for other species of *Colostethus,* and males in captivity often are aggressive toward other males. Species exhibiting some form of aggressive behavior include *C. abditaurantius, C. awa, C. caeruleodactylus, C. degranvillei, C. fraterdanieli, C. marchesianus, C. nexipus, C. sauli,* and *C. subpunctatus* (Lescure 1975a; Heselhaus 1984; Coloma 1995; Fandiño, Lüddecke, and Amézquita 1997; T. Grant and Castro 1998; Lima, Caldwell, and Biavati 2002; Lüddecke 2002a; Pröhl 2005b). Unfortunately, too little is known about these species to characterize their mating systems. One species, *C. beebei,* is unusual in that breeding takes place in water-filled bromeliads, a habit usually associated with other genera of dendrobatid frogs. Males are territorial, aggressively defending their bromeliads against other males (Bourne et al. 2001). Presumably a shortage of suitable oviposition sites has led to the evolution of resource defense in this species. In *C. stepheni,* from Amazonian Brazil, males are territorial and guard oviposition sites on the forest floor (Junca, Altig, and Gascon 1994; Junca 1996, 1998). In contrast to most dendrobatids, tadpoles are not carried to water, but remain in the nest and are guarded by the male until metamorphosis (see chapter 11). Consequently, suitable nest sites might be harder to find than in species that move tadpoles to water.

In several species of *Colostethus* and *Mannophryne,* males are aggressive toward other males, but do not appear to defend long-term territories. Their spatial organization is more similar to a lek than to resource defense, but they do not fit the traditional definition of lekking behavior (Höglund and Alatalo 1995) because males also care for eggs and tadpoles. The best known is *Mannophryne trinitatis,* which has been studied in both Trinidad and Venezuela (Test 1954; Sexton 1960; van Meeuwen 1977; Wells 1980c; Praderio and Robinson 1990; Cummins and Swan 1995). Females are highly territorial and defend suitable feeding and retreat sites along rocky streams. They respond to intruding males and females by pulsating their bright yellow throats and physically attacking other frogs (fig. 8.14 A, B). Males, on the other hand, move frequently between calling sites and do not maintain permanent territories. When males are not calling or defending calling sites, they assume a cryptic brown coloration (fig.

**Fig. 8.14.** Aggressive behavior and communication in a dendrobatid frog from Trinidad, *Mannophryne trinitatis.* (A) Territorial female displaying her conspicuous yellow throat set off by a black band. The female pulsates her throat at intruding frogs of both sexes. (B) Two females fighting over a territory. (C) Male with light brown coloration typical of noncalling males. (D) Calling male with black coloration. Photos by Kentwood D. Wells.

8.14 C). When calling, males turn black and attack other black males calling nearby (fig. 8.14 D). They display from rocks and other prominent sites and often jump up and down while calling, making themselves very conspicuous to females. Females leave their territories and follow males to oviposition sites in leaves some distance from the males' calling sites. They attend the eggs and later carry the tadpoles to water. Two other species, *Mannophryne collaris* and *Colostethus palmatus,* exhibit a similar type of social organization, including female territoriality, darkening of coloration in displaying males, and conspicuous jumping displays by calling males (Dole and Durant 1974a; Durant and Dole 1975; Lüddecke 1974, 1976, 1999). Territorial behavior by females also has been observed in *C. abditaurantius* (Grant and Castro 1998), but nothing is known about the mating system of this species.

Most species of *Dendrobates, Epipedobates,* and *Allobates* are not closely associated with streams, but are either terrestrial or arboreal. Aggressive behavior between males has been observed in many species (fig. 8.13 B), both in the field and in captivity (Duellman 1966; Goodman 1971; Crump 1972; P. Bunnell 1973; Silverstone 1973; Wells 1978a; McVey et al. 1981; H. Zimmermann and E. Zimmermann 1980b, 1981, 1984, 1988, 1990; E. Zimmermann and H. Zimmermann 1982; Jungfer 1985; E. Zimmermann 1990; Forester, Cover, and Wisnieski 1993; Baugh and Forester 1994; Hermans, Pinxten, and Eens 2002; Narins, Hödl, and Grabul 2003). A number of recent field studies with marked individuals have confirmed that males often defend all-purpose territories that usually include both foraging areas and oviposition sites (Summers 1989, 1990, 1992; Roithmair 1992, 1994; van Wijngaarden and van Gool 1994;

Pröhl 1997b; Summers and Amos 1997; Pröhl and Hödl 1999). These species might have resource-defense mating systems, but exactly what resources are being defended is not always clear (Pröhl 2005b).

Most species of *Dendrobates* deposit tadpoles in pools, water-filled treeholes, or bromeliads. Donnelly (1989a, b) reported that males of *Dendrobates pumilio* defend territories centered near trees with bromeliads that serve as tadpole deposition sites. When bromeliads were added to experimental plots, the density of resident males increased, but addition of leaf litter that might serve as oviposition sites did not. This led Donnelly to conclude that bromeliads are the main resources being defended by territorial males. In contrast, other investigators working with the same species have found that tadpole deposition sites often are not located within territories of males, although oviposition sites sometimes are (Pröhl 1997b; Pröhl and Hödl 1999). In fact, territories of males are considerably smaller than home ranges of females and tend to be located where densities of females are highest. Female home ranges are more closely associated with tadpole rearing sites than are territories of males (Pröhl and Berke 2001).

Both males and females of *D. pumilio* were observed to mate with multiple partners, but the degree of polygamy was greater for males than for females, as was variance in mating success. Female densities were higher in a secondary forest, which had many more tadpole rearing sites, than in a primary forest, and average male mating success was consequently higher in secondary forest (Pröhl 2002). Realized reproductive success of males and females depends not only on the number of matings obtained, but also on survival of egg clutches. Loss of egg clutches was a frequent occurrence, and when clutch losses were high, mating frequencies of males were higher, because females were not occupied with caring for tadpoles and therefore could produce new clutches of eggs more quickly. On the other hand, realized reproductive success of males was actually lower when clutch loss was high (Pröhl 2005a).

Overall, the mating system of *D. pumilio* appears to be closer to a lek than to resource defense, with competition among males for calling sites and male calling performance being the main determinants of mating success (Pröhl 2003). Nevertheless, the mating system of this species defies simple classification because of the heavy investment in parental care by both males and females. Although males do not defend tadpole rearing sites, such sites can be limiting for female reproductive success, and females sometimes aggressively defend such sites against other females (Haase and Pröhl 2002). It is not entirely clear whether females are defending rearing sites as resources for future use, or are protecting tadpoles already deposited in these sites against competition or cannibalism.

In species that use treeholes and other large pools as tadpole deposition sites, such as *D. auratus,* such sites often are not located within a male's territory and often are used by more than one male (K. Summers, 1990). Presumably these relatively large pools are scarce, but also difficult for one male to defend against intruders. In contrast, males of *D. ventrimaculatus* and *D. vanzolinii* are territorial and defend water-filled leaf axils or treehole cavities that serve both as oviposition sites and tadpole deposition sites (Caldwell 1997; Summers and Amos 1997; Caldwell and Oliveira 1999). Eggs are laid above pools of water, where they are attended by both parents. Some tadpoles fall into pools without being carried, but others are picked up and carried to other pools. Males carry tadpoles, but they periodically call females to the pools to deposit unfertilized eggs to feed the tadpoles (see chapter 11). In *D. ventrimaculatus,* more than one male sometimes deposited tadpoles in the same axil, and genetic analysis using microsatellite markers revealed that tadpoles inhabiting the same pool often were unrelated. This indicates that the territorial behavior of males is not always successful in preventing other males from using their pools. This can be costly, because if an intruding male deposits large tadpoles in a pool containing small tadpoles from a resident male's clutch, then the larger tadpoles often eat the smaller ones (Summers and Amos 1997).

There is some evidence that territories tend to be somewhat larger in dendrobatids in which males provide parental care than in those with female parental care. For example, territory size is relatively large in *Dendrobates auratus, D. leucomelas, Allobates femoralis,* and *Epipedobates trivittatus,* all of which have male parental care (Summers 1989, 1992; Roithmair 1992, 1994). Territories of males often are quite small in *D. pumilio, D. granuliferus,* and *D. histrionicus,* species in which females assume the primary parental role (Donnelly 1989b; Summers 1992; van Wijngaarden and van Gool 1994; Pröhl 1997b; Pröhl and Hödl 1999; Pröhl and Berke 2001). This difference could simply reflect differences in body size (the species with male parental care are considerably larger) or differences in population density. It also is possible that males of the former group maintain large territories so that egg clutches of different females can be widely separated to prevent egg cannibalism by females (Wells 1978a; Summers 1992). Roithmair (1992, 1994) found that male mating success was positively correlated with territory size in *Allobates femoralis* and *Epipedobates trivittatus,* and suggested that females use territory size as an indicator of male quality. Territory size also was positively correlated with mating success in a population of *Dendrobates pumilio,* but this variable was much less important than the calling activity of the male and the height of his calling perch (Pröhl and Hödl 1999; Pröhl 2003).

Like some other frogs that lay eggs on land at oviposition

sites separated from calling sites, dendrobatid frogs often have elaborate courtship behavior that can last for an hour or more. Usually a female initiates courtship by approaching a calling male (Crump 1972; Polder 1974b, c; Wells 1978a, 1980a; H. Zimmermann 1974, 1978, 1982; Limerick 1980; Weygoldt 1980a, b, c, 1982, 1983; H. Zimmermann and E. Zimmermann 1980a, b, 1981, 1984, 1985; Schülte 1981a, b; E. Zimmermann and H. Zimmermann 1982, 1985; E. Zimmermann 1983; Heselhaus 1984; Jungfer 1985; Lüddecke 1999; Pröhl and Hödl 1999; Bourne et al. 2001). When females are territorial, as in *Mannophryne trinitatis* and *M. collaris,* males initiate courtship at a distance, by using elaborate jumping displays and conspicuous color changes (Dole and Durant 1974a; Wells 1980c). Once a female is nearby, the male leads her to a concealed oviposition site in dead leaves, a rock crevice, or a bromeliad. Sometimes the male and female pause and engage in elaborate tactile interactions. In *Dendrobates histrionicus, D. lehmanni,* and related species, the male follows the female and actively strokes her with his feet (Silverstone 1973; H. Zimmermann and E. Zimmermann 1980b, 1981; E. Zimmermann and H. Zimmermann 1982). In *D. auratus, D. azureus, D. leucomelas, D. tinctorius, Phyllobates terribilis, P. vittatus,* and *Allobates femoralis,* the female takes the more active role (Polder 1974b; Wells 1978a; H. Zimmermann 1974; 1982; H. Zimmermann and E. Zimmermann 1980a, 1981, 1985; Weygoldt 1983; Summers 1989, 1990, 1992). This partial reversal of the usual sex roles in courtship seems to reflect competition among females for preferred males, or attempts by females to prevent tadpoles of other females from being deposited in pools with their own tadpoles, where they suffer cannibalism (Summers 1990, 1992).

When the male and female have found a suitable oviposition site, they often engage in additional tactile interactions, displays, and preparations for mating. These usually include cleaning or wiping motions on the substrate. The details of oviposition and fertilization vary considerably. In some species of *Colostethus, Phyllobates,* and *Dendrobates,* the male clasps the female in cephalic amplexus (see fig. 10.1 D). The male holds the female with the upper surfaces of the front feet pressed against her throat (Myers et al. 1978; Wells 1980a; H. Zimmermann 1982, 1983; E. Zimmermann 1983; Weygoldt 1983; Heselhaus 1984; H. Zimmermann and E. Zimmermann 1985; Bourne et al. 2001). In other species, the male lies loosely across the back of the female during oviposition and fertilizes eggs (Lüddecke 1974, 1999; Silverstone 1976; Schülte 1981b).

In several species in the *Dendrobates auratus* group, males are reported to sit on the eggs and fertilize them after they are laid (Senfft 1936; Polder 1974b, 1976a, b; H. Zimmermann 1974, 1978; H. Zimmermann and E. Zimmermann 1980a; E. Zimmermann 1983). Similar behavior has been reported in *D. histrionicus* and some of its close relatives (H. Zimmermann and E. Zimmermann 1980b, 1981). In *Colostethus beebei,* males were observed to initially clasp females in cephalic amplexus during courtship, but fertilization apparently occurred when males sat on the eggs after they were laid (Bourne et al. 2001). Males of *C. caeruleodactylus* were not observed in amplexus, but did move around on eggs after they were deposited (Lima, Caldwell, and Biavati 2002). In some derived species of *Dendrobates,* such as *D. pumilio, D. speciosus,* and *D. ventrimaculatus,* males apparently deposit sperm on the substrate before eggs are laid (Limerick 1980; Weygoldt 1980a, b; Lescure and Bechter 1982; Jungfer 1985). However, in *D. granuliferus,* the male and female are reported to place their vents in contact, and the male fertilizes the eggs as they are laid (Crump 1972).

There are several possible reasons for the evolution of elaborate courtship in dendrobatids (Wells 1977a, 1980a). First, females could benefit from remaining "coy" until a suitable oviposition site is located. Summers (1989) reported that *Dendrobates auratus* females prefer certain males based on the quality of their territories. They vigorously pursue males and try to prevent other females from mating with them. Males, on the other hand, should persist in courtship only if they receive clear signals that a female is receptive. This could lead to the evolution of a complex exchange of signals between males and females. Second, dendrobatids generally remain in the same habitat throughout the year. Consequently, males must frequently encounter both receptive and nonreceptive females; complex courtship allows males to assess the reproductive condition and receptivity of females that enter their territories. This would be especially important if males and females compete for similar territory sites (Wells 1980a). A male should attempt to determine whether an intruding female is a potential competitor or a potential mate and adjust his behavior accordingly. It also is possible that females require prolonged stimulation to initiate ovulation (Bourne et al. 2001), although it is not clear why this should be more important for dendrobatids than for other frogs.

### The Mating Systems of Midwife Toads

One group of anurans, the midwife toads (genus *Alytes*), has a mating system that does not fit conveniently into the categories described for other anurans. Male midwife toads call on land, either from open ground or from crevices or other retreat sites. Calling sites usually are within 50 meters of pools where tadpoles eventually develop, but are not necessarily immediately adjacent to bodies of water. The breeding period can be prolonged (up to two months) or explosive (less than one week), depending on local weather, but prolonged breeding seasons tend to be broken into a series of short episodes of reproductive activity. Males advertise themselves to females by calling. They are not territorial, although males sometimes maintain a minimum distance between themselves and other individuals. In some cases, however, more than one male calls from the same retreat site (Bush 1997). Compared to other frogs, males exhibit little aggressive competition for calling sites or mates, although occasional struggles over individual females have been reported. Males also will increase their calling rate in response to the calls of other males, a behavior indicative of some level of competition for the attention of females (Bosch and Márquez 1996). Females approach calling males and often answer their calls with their own vocalizations, an unusual occurrence in anurans (see chapter 7). Sometimes a male and female will alternate calls with one another for several minutes before going into amplexus.

The eggs are laid in strings, which the male wraps around his hind legs. He then carries them for several weeks before depositing hatching tadpoles in a pond (Heinzmann 1970; Crespo 1982; Reading and Clarke 1988; Raxworthy 1990; Márquez and Verrell 1991; Márquez 1992, 1993, 1995, 1996; Verrell and Brown 1993; Bush 1996; Bush, Dyson, and Halliday 1996). In *Alytes obstetricans* and *A. cisternasii,* males often mate several times and care for more than one clutch of eggs simultaneously, and large individuals are most likely to obtain multiple matings (Reading and Clarke 1988; Raxworthy 1990; Márquez 1993, 1995; Böll and Linsenmair 1998). Multiple clutches are less common in *A. muletensis* (Bush 1996).

Three features of the mating systems of midwife toads set them apart from most other anurans. First, males and females often gather in mixed-sex groups during the breeding season, and groups of males have been reported to call to the same female. Second, both males and females call. In most cases, the females call in response to males, which stimulates the males to call more quickly (Heinzmann 1970; Márquez and Verrell 1991; Bush 1997; Bosch 2001, 2002). This provides the female with better directional information about the location of the male. In the Majorican midwife toad (*Alytes muletensis*), females were observed advertising for males by calling and even searched for males in rock crevices, but this occurred only when females needed to lay their eggs immediately. Third, females have been observed fighting over males (fig. 8.15; Verrell and Brown 1993; Bush 1997; Bush and Bell 1997). This behavior is predicted to occur in species with a high level of male parental investment, where the availability of males to care for eggs limits the reproductive rate of females (Trivers 1972). Competition among females also is expected when there is a shortage of receptive males for some other reason, or substantial variation in the quality of males available as mates (Jennions and Petrie 1997). In some species, there appears to be variation among males in their readiness to mate, and this could lead to a shortage of mates for females in the population (Lea, Halliday, and Dyson 2003).

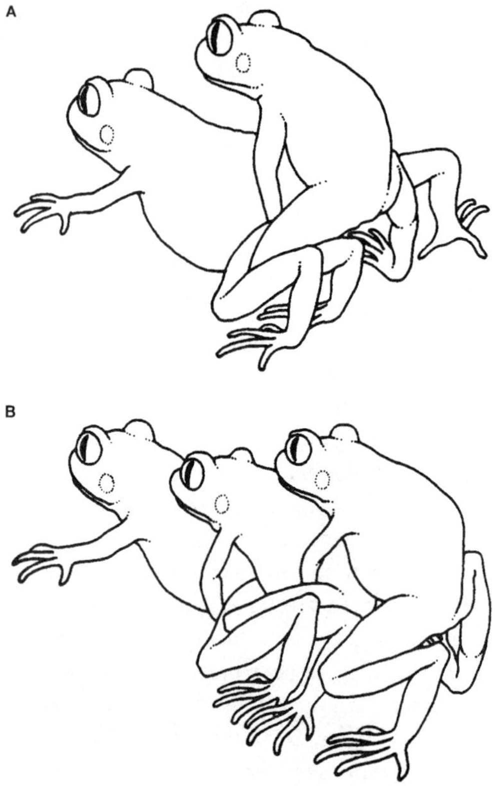


**Fig. 8.15.** Competition between females for access to a male in the Majorcan midwife toad (*Alytes muletensis*). (A) Normal amplexus position, with male on the back of the female. (B) A second female clasping the male from behind, interfering with his mating attempt. After Bush and Bell (1997).

There is some evidence for active choice of mates by both female and male midwife toads. In two species, *A. obstetricans* and *A. cisternasii,* large males obtained more egg clutches than small males. The lack of extensive male-male competition suggests that female choice probably accounts for this variation in male mating success, although the evidence comes mostly from correlational studies of male traits and reproductive success rather than experimental studies (Márquez 1993; Böll and Linsenmai, 1998; Lodé and Le Jacque, 2003). Females did exhibit a preference for low-frequency calls, like those produced by large males, in playback experiments (Márquez 1995), but the preference was not very strong (Bosch and Márquez 2005). Females of *A. cisternasii* showed a stronger preference for high calling rates. In *A. muletensis,* a species in which receptive females appear to outnumber available males, females did not prefer low-frequency calls or long calls, both of which are characteristic of large males (Dyson et al. 1998). Males oriented toward the calls of females and showed a stronger response to loud calls than to less intense calls. They did not, however, prefer low-frequency calls of large females to high-frequency calls of small females (Bush, Dyson, and Halliday 1996). This result was contrary to predictions that males should prefer larger females because they produce larger clutches or larger eggs. The fact that males exhibit any selectivity toward females is unusual among male anurans, which tend to be relatively indiscriminant in their choice of mates (Krupa 1995). Males are expected to be selective only when excess females are available, or individual females differ substantially in the number or quality of eggs they produce.

## Aggressive Displays, Fighting, and Territory Defense

In anurans that defend either calling sites or oviposition sites, displays and aggressive behavior tend to be much more elaborate than in species that engage in scramble competition over females (table 8.1). Presumably the evolution of elaborate displays and aggressive behavior are driven by sexual selection, especially when variance in male mating success is high. In general, aggressive interactions among males appear to be more frequent and more intense in species that defend a valuable resource, such as an oviposition site, than in those that defend only space around their calling sites, although this has not been rigorously quantified.

### Visual Signals

Typically, aggressive encounters are preceded by exchanges of aggressive calls, which usually are structurally distinct from the advertisement call (see chapter 7). Although acoustic signals tend to predominate, many species also use visual displays in close-range encounters. For example, in some species, males display to other males by adopting an inflated or elevated posture (Test 1954; Sexton 1960; Duellman 1966; Emlen 1968; Wiewandt 1969; Goodman 1971; Crump 1972; Durant and Dole 1975; Wells 1977b, 1978b, 1980a, c, 1981b; Howard 1978a; M. Ryan 1980b). In others, males engage in push-up or rocking displays (Brattstrom and Yarnell 1968; McDiarmid and Adler 1974; Greer and Wells 1980) or wave the front feet (Crump 1988; Lindquist and Hetherington 1996, 1998).

In several unrelated species of frogs that call near noisy streams, vocal advertisement is supplemented or has been replaced by foot-flagging displays that involve extension of the hind feet (fig. 8.16). Often, males have light blue or other reflective colors on the webbing of the hind feet that enhances the display (Haddad and Giaretta 1999; Hödl and Amézquita 2001). These displays are directed at other males or potential mates, or both. In *Hyla pardalis,* foot-flagging displays elicited similar displays from other males, suggesting they function mainly in territory advertisement (Amézquita and Hödl 2004). Such displays may be advantageous in an environment characterized not only by high acoustic noise levels, but considerable visual clutter as well. Hindleg waving displays also have been observed during aggressive encounters in *Phyllomedusa sauvagii,* but this species does not call near noisy streams (Halloy and Espinoza 1999–2000). Males of a recently described dendrobatid frog from Amazonian Brazil, *Colostethus caeruleodactylus,* develop bright blue toes during the breeding season, but obvious displays of these markings during aggression and courtship have not been reported (Lima and Caldwell 2001; Lima, Caldwell, and Biavati 2002).

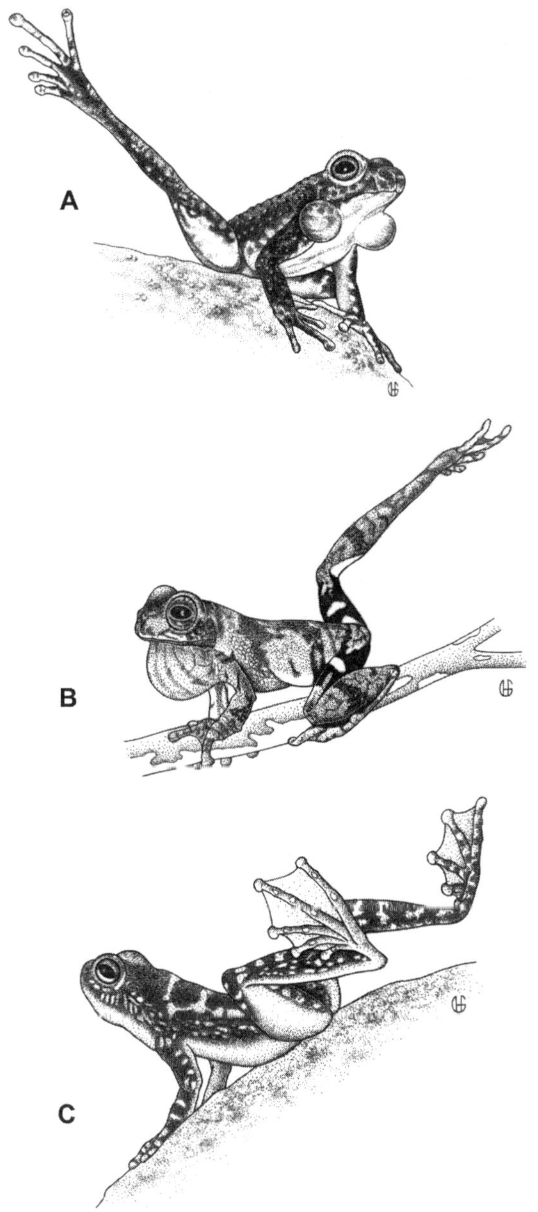


**Fig. 8.16.** Convergent evolution of foot-flagging displays in three species of stream-breeding frogs. (A) *Hylodes asper,* a Brazilian frog that breeds in torrential streams. (B) *Hyla parviceps,* a stream-breeding frog from Venezuela. (C) *Staurois latopalmatus,* a ranid frog that calls from boulders in streams in Borneo. After Hödl and Amézquita (2001); original drawings by H. C. Grillitsch.

Many species of frogs display brightly colored patches on the throat during territorial encounters (Test 1954; Sexton 1960; Emlen 1968; Wiewandt 1969; Durant and Dole 1975; Wells 1978b, 1980a; M. Ryan 1980b). In *Allobates femoralis,* pulsations of the vocal sac are an important component of aggressive displays. Models that were manipulated to pulsate the vocal sac during call playbacks elicited more aggressive responses from territorial males than did acoustic playbacks alone or silent throat movements (Narins, Hödl, and Grabul 2003; Narins et al. 2005). Even in species without distinctly colored vocal sacs, movements of the vocal sac may serve as a visual signal to other males or to females searching for mates (Rosenthal, Rand, and Ryan 2004). In at least two species of dendrobatid frogs, *Colostethus palmatus* and *Mannophryne trinitatis,* calling males turn completely black and are aggressive mainly toward other black males (fig. 8.14 D). Males signal submission by turning brown (Lüddecke 1974, 1976, 1999; Wells 1980c).

Seidel (1999) described an unusual form of territorial display in *Bombina variegata.* In the population that he studied in Austria, choruses have a leklike spatial organization, but males employ a variety of tactics to acquire mates. Large males defended small territories in open water, which they advertised by calling and by producing waves on the surface of the water. They produced the waves by rhythmically striking the surface of the water with their hind legs, sending a series of concentric waves radiating outward. These appeared to demarcate territories that were largely avoided by other males. Smaller males usually did not defend such territories, but engaged in scramble competition for females.

### Fighting Behavior

When vocalizations and displays are ineffective in repelling intruders into territories, male anurans often resort to overt aggression. Physical encounters take a variety of forms, including jumping on opponents, attempting to push opponents off of elevated perches, grappling with the forelimbs, or wrestling with the forelimbs interlocked (fig. 8.12–8.14; Tunner 1976; Wells 1977b, 1978b, b, 1980a, c, 1981b; R. D. Howard 1978a; Fellers 1979a; M. Ryan 1980b; Cardoso 1981b; Weygoldt 1981a; H. Zimmermann and E. Zimmermann 1981, 1984, 1988, 1990; B. K. Sullivan 1982a; Telford 1985; Arak 1983b; Cardoso and Haddad 1984; Akef and Schneider 1985; Robertson 1986b; Crump 1988; Given 1988b; Dyson 1989; Summers 1989, 1990, 1992; Roithmair 1992, 1994; Forester, Cover, and Wisnieski 1993; Baugh and Forester 1994; Halliday and Tejedo 1995; Rödel 1996; Orlov 1997; Summers and Amos 1997; Bolívar-G., Grant, and Osorio 1999; Halloy and Espinoza 1999–2000; Rodrigues, de Souza, and Filho 2003; Wogel, Abrunhosa, and Pombal 2004). In *Atelopus varius,* most fights include a peculiar squashing behavior in which one frog jumps on an opponent's back and presses him to the ground (Crump 1988). This resembles the wrestling behavior of several dendrobatid species, including *Colostethus panamensis* (fig. 8.13 C; Wells 1980a).

**Fig. 8.17.** Skull of a male fanged frog of the genus *Limnonectes* from Borneo, showing sharp fangs used in fighting with other males. Photo by Kerry Matz, courtesy of Sharon Emerson.

More violent aggression occurs as well, including biting (Rabb and Rabb 1963b; Brattstrom and Yarnell 1968; Weygoldt 1981a; Giddings 1984; Townsend, Stewart, and Pough 1984; Katsikaros and Shine 1997). Some frogs have fangs or tusks on their jaws that are used in intraspecific combat (Fabrezi and Emerson 2003). These include the Australian myobatrachid frog *Atelotus brevis* (Katsikaros and Shine 1997) and a group of fanged frogs in the genus *Limnonectes* (fig. 8.17) from Southeast Asia (Emerson 1994, 1996; Emerson and Voris 1992; Emerson and Inger 1992).

Some frogs have spines on the chest or forelimbs that are used to stab or slash opponents in fights. The latter has been studied most thoroughly in the nest-building gladiator frogs, including *Hyla faber, H. pardalis,* and *H. rosenbergi* (Lutz 1960b, 1973; Kluge 1981; Martins and Haddad 1988; Martins, Pombal, and Haddad 1998). In these species, males have a curved, sharp prepollical spine on each front foot, which is used to slash opponents during prolonged wrestling bouts (fig. 8.12 C). Males are frequently injured in fights, sometimes fatally (Kluge 1981). Similar spines are found on the forelimbs of males in several species of *Leptodactylus* and in several other families of anurans (Shine 1979), often associated with other sexually dimorphic characters such as enlarged forelimbs (see "Sexual Dimorphism," following). Probably such structures are used in fighting, but the social behavior of most species is unknown. Males of centrolenid frogs in the genus *Centrolene* have humeral spines on their forelimbs that often are very sharp. In at least one species, *C. buckleyi,* males suffered injuries after grappling with other males, and injuries that could have been inflicted by these spines have been observed in other species as well (Bolívar-G., Grant, and Osorio 1999).

## Male Behavior and Mating Success in Prolonged-Breeding Anurans

### Competition between Males

Most of the literature on aggressive interactions of anurans is still relatively descriptive. Only a few investigators have attempted to measure the factors that determine the winners of contests. In territorial species, encounters usually are won by larger individuals or by residents of territories, a pattern found in most territorial animals (R. D. Howard 1978a; Wells 1978b; 1980a; Kluge 1981; Robertson 1986b; Given 1987, 1988b; Crump 1988; Dyson 1989; Pröhl and Hödl 1999; Seidel 1999; Shimoyama 1999; Pröhl and Berke 2001; Toledo and Haddad 2005). In *Rana clamitans,* an intruder had virtually no chance of evicting a resident unless the intruder was considerably larger (Wells 1978b). Contests were more likely to be escalated when intruders were at least as large as resident males. Otherwise intruders were chased away after a few seconds. In *R. virgatipes,* residents won most interactions. Body size was more important in determining the winner when contests were escalated to physical fights rather than being settled by exchanges of aggressive calls (Given 1988b). Again, escalated fights were more likely to occur when males were relatively evenly matched in size, a trend also reported in *Uperoleia laevigata* by Robertson (1986b). In that species, large males won most contests settled either by vocal interactions or escalated fights. Large males also won more fights in *Bufo woodhousii* (B. K. Sullivan 1982b) and *B. calamita* (Arak 1983b), both of which defend calling sites, but not oviposition sites. In a territorial dendrobatid frog, *Dendrobates pumilio,* there often is little variation in body size among males in a population. Levels of aggression and the outcome of fights in this species are affected by prior residence in a territory, by the length of time a male has been in his territory, and by relative territory quality (Forester, Cover, and Wisnieski 993; Baugh and Forester 1994; Prohl and Hödl 1999). All of these are variables that commonly influence the outcome of fights in other territorial animals.

It seems likely that most aggressive interactions among male frogs are settled by exchanges of vocal signals (chapter 7). This is particularly true of lek-breeding species that do not defend oviposition sites. In many such species, escalated contests are rare compared to those seen in frogs such as *Rana clamitans* and *R. catesbeiana.* In species such as *Hyla versicolor,* for example, males give aggressive calls in response to other males calling nearby, but only occasionally attack another male (Fellers 1979a; Wells, personal observations). Physical aggression seems to be somewhat more frequent in hylids that form very dense choruses, such as *Hyla ebraccata* and *H. microcephala,* but even in these species, most contests are settled with aggressive calling (Wells and Schwartz 1984b; Schwartz and Wells 1985). There is evidence for several species that males can use the pitch of the advertisement call to assess the body size of opponents (N. Davies and Halliday 1978; Arak 1983b; Ramer, Jenssen, and Hurst 1983; Robertson 1986b; Given 1987; Bee and Perrill 1996; Bee, Perrill, and Owen 1999, 2000). Presumably they could do so from aggressive calls as well, and this could have a major effect on the outcome of aggressive encounters (see chapter 7 for a more detailed discussion of aggressive calling).

Despite the fact that large males are likely to win most fights over territories, the largest males in a population do not necessarily achieve the highest mating success. In *Rana clamitans* and *R. catesbeiana,* for example, large males generally were more successful than small males in defending high-quality territories and in obtaining mates, but the correlation between male size and mating success in different years varied from weak to very strong (Wells 1977c; R. D. Howard 1978a, 1981a, 1983). Mating success was more a function of territory quality than male size per se (see the following discussion of female choice). Nevertheless, if females choose good territories, and the largest males occupy the best territories, then this will result in indirect selection for large body size in males.

Large body size probably provides an advantage to males in many other species that defend resource-based territories, even if only indirectly, through female choice based on territory quality. This was true for *Rana virgatipes* (Given 1988b), *Atelotus brevis* (Katsikaros and Shine 1997), and *Limnonectes kuhlii* from Taiwan (Tsuji 2004). On the other hand, large males did not enjoy a mating advantage in the nest-building frog *Hyla rosenbergi* (Kluge 1981). The range of body sizes of males actively defending nests was relatively small for such a large frog, and Kluge (1981) suggested that the lack of effect of body size on mating success might have been due to an inability of females to clearly distinguish between large and small males. Mating success also was not related to body size in two species of centrolenids that defend oviposition sites, *Hyalinobatrachium fleischmanni* (Greer and Wells 1980; Jacobson 1985), and *Centrolene prosoblepon* (Jacobson 1985), or in *Dendrobates pumilio* (Pröhl and Hödl 1999; Pröhl 2003). In all of these species, the range of body sizes of males in the population was relatively small. A large-male mating advantage has been reported in a number of other species, including *Physalaemus pustulosus* (M. Ryan 1980a, 1983b), *Bufo calamita* (Arak 1983b), some populations of *Bufo woodhousii* (Woodward 1982b, c), *Hyla arborea* (Márquez-M. de Orense and Tejedo-Madueno 1990), and *Hyla elegans* (Bastos and Haddad 1996). These species have lek mating systems, but do not defend resource-based territories.

In many lek-breeding anurans, large males do not consistently have a mating advantage over small males. This was the case for *Hyla cinerea* (Gerhardt 1982; Gerhardt et al. 1987), *H. versicolor* (Fellers 1979b), *H. chrysoscelis* (Godwin and Roble 1983), *H. gratiosa* (C. Murphy 1994b), *H. ebraccata* (except for one year; M. Morris 1991), *H. labialis* (Lüddecke 2001; Gutiérrez and Lüddecke 2002), *Pseudacris regilla* (Perrill 1984), some populations of *Bufo woodhousii* (B. K. Sullivan 1982b, 1983b, 1987), *Bufo rangeri* (Cherry 1993), *Bufo verrucossisimus* (Tarkhnishvili 1994), *Bombina variegata* (Barandun 1990), *Hyperolius marmoratus* (Passmore and Telford 1983; Telford and Dyson 1988; Dyson et al. 1992; Dyson, Bush, and Halliday 1998), *Chiromantis xerampelina* (Jennions, Backwell, and Passmore 1992), and *Polypedates leucomystax* (Arak 1988b). In some cases, this was simply due to a very narrow range of body sizes in the male population, but in others, body size had little effect on mating success, despite significant size variation among males.

There is only limited evidence for size assortative mating in anurans with prolonged breeding seasons. R. D. Howard (1981a, 1983, 1988a) found a weak positive correlation between sizes of males and females in mated bullfrog (*Rana catesbeiana*) pairs in two of three years, and Katsikaros and Shine (1997) reported a similar relationship for *Atelotus brevis*. Positive assortative mating also was reported for a population of *Hyla arborea* (Márquez-M. de Orense and Tejedo-Madueno 1990) and on some nights for *H. labialis* (Lüddecke 2001; Gutiérrez and Lüddecke 2002). Robertson (1986a) reported a positive correlation between the mass of male and female *Uperoleia laevigata* in amplexus, even though body lengths were not correlated. In general, the body mass of paired males was about 70% of the body mass of their mates. Robertson originally speculated that females might use body mass as an indicator of male physical condition, but he later showed that females with mates about 70% of their own body mass had the highest rate of fertilization of eggs (Robertson 1990). He suggested that larger males hampered the females in their oviposition behavior, whereas smaller males might have had insufficient sperm reserves. The relative sizes of males and females are reported to affect fertilization success in several other species as well, including *Scinax rubra* (Bourne 1993), *Hyla elegans* (Bastos and Haddad 1996), and *H. labialis* (Gutiérrez and Lüddecke 2002). Two studies of *Hyperolius marmoratus,* on the other hand, showed no evidence of assortative mating or any effect of relative male and female size on fertilization success (Dyson et al. 1992; Grafe 1997a), and there is little evidence for assortative mating in most other frogs. Another possibility is that larger females prefer larger males because the frequency to which their ears are tuned decreases as they grow, so large males with low-frequency calls are likely to be heard better by large females (Márquez-M. de Orense and Tejedo-Madueno 1990).

### Chorus Tenure

In anurans with prolonged breeding seasons, receptive females often arrive asynchronously over a period of several weeks or months. Consequently, males present on many nights are more likely to encounter females than are those that are present on only a few nights (Halliday and Tejedo 1995). A positive correlation between duration of male attendance and mating success has been reported for many species, including *Bufo calamita* (Arak 1983b, 1988b; Denton and Beebee 1993a), *Bufo fowleri* (Given 2002), *B. gutteralis* (Telford and van Sickle 1989), *B. rangeri* (Cherry 1993), *B. woodhousii* (Woodward 1982b), *Hyalinobatrachium fleischmanni* (Greer and Wells 1980; Clark 1981; Jacobson 1985), *Centrolene prosoblepon* (Jacobson 1985), *Physalaemus pustulosus* (M. Ryan 1980a, 1983b, 1985b), *Eleutherodactylus coqui* (Townsend and Stewart 1994), *Hyla rosenbergi* (Kluge 1981), *H. chrysoscelis* (Godwin and Roble 1983; Ritke and Semlitsch 1991), *H. cinerea* (Gerhardt et al. 1987), *Hyla faber* (Martins 1993a), *H. gratiosa* (C. Murphy 1994a, b), *H. versicolor* (B. K. Sullivan and Hinshaw, 1992; Bertram, Berrill, and Nol 1996), *Epipedobates trivittatus* (Roithmair 1994), *Hyperolius marmoratus* (Dyson et al. 1992), *Chiromantis xerampelina* (Jennions, Backwell, and Passmore 1992), *Buergeria buergeri* (Fukuyama and Kusano 1989), *Rana catesbeiana* (R. D. Howard 1978a) *R. clamitans* (Wells 1977c), *R. virgatipes* (Given 1988a, b), and *Limnonectes kuhlii* (Tsuji 2004). In a few studies, male chorus tenure was not correlated with mating success, or was positively correlated with mating success in only some years of the study (Fellers 1979b; B. K. Sullivan 1987; M. Morris 1989).

Male chorus tenure often explains a large fraction of the total variation in male mating success. This reduces the amount of variation to be explained by mechanisms of male-male competition and female choice, and makes identification of other determinants of mating success difficult without large sample sizes. The causes of individual differences in chorus tenure are not fully understood. Two factors that seem most important are depletion of energy reserves (Wells 1978b; Mac Nally 1981; Godwin and Roble 1983; A. Green 1990; C. Murphy 1994a, b; Dyson et al. 1998; Eggert and Guyétant 2003; see also chapter 5) and death due to predation (R. D. Howard 1978a; M. Ryan, Tuttle, and Rand

1982; M. Ryan 1985b, 1988b; Pough et al. 1992). In some studies, large males had longer chorus tenure than small males, perhaps because they have lower mass-specific metabolic rates and larger energy reserves, and this can lead to a positive correlation between body size and mating success, independent of competitive interactions among males (Tejedo 1992c; Tsuji 2004). In other studies, there was not a correlation between body size and chorus tenure (Halliday and Tejedo 1995). In *Pelobates fuscus,* younger males remained in the chorus longer than did older males (Eggert and Guyétant 2003). This could be a result of older males obtaining mates more quickly and then leaving the chorus, or younger males employing mating tactics that require less expenditure of energy.

C. Murphy (1994a, b) found a strong correlation between chorus attendance and male mating success in *Hyla gratiosa* over several years (fig. 8.18). Variation in chorus attendance appeared to be in part to variation in energy reserves of males in the chorus. Males suffered reductions in body mass throughout their time in a chorus, but those that remained longer in the chorus were in better condition and lost mass more slowly than those that were present for only a few nights (see chapter 5 for a more detailed discussion of this work).

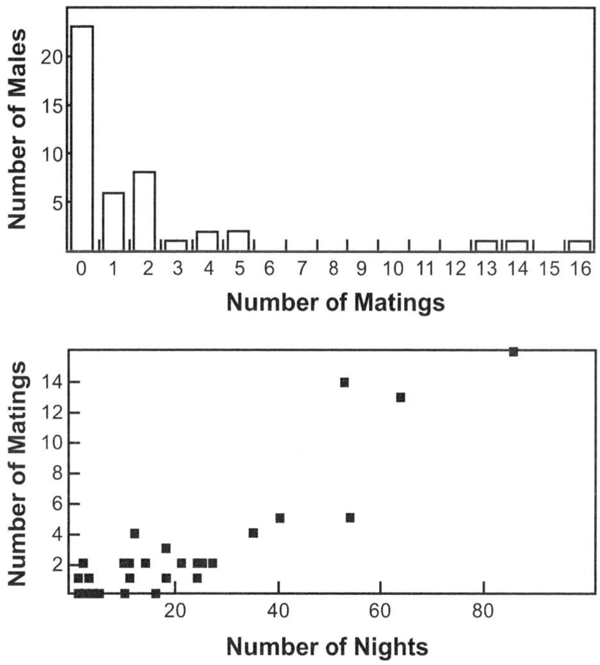


**Fig. 8.19.** Mating success and chorus attendance in an African frog, *Hyperolius marmoratus.* (Top) Variation in number of matings in a chorus of 45 males during a breeding season lasting 167 nights. (Bottom) Relationship of male mating success to male chorus attendance. Successful males spent more time in the chorus than did unsuccessful males, and this was especially true for three males with very high mating success. After Dyson, Henzi, Halliday, and Barrett (1998).

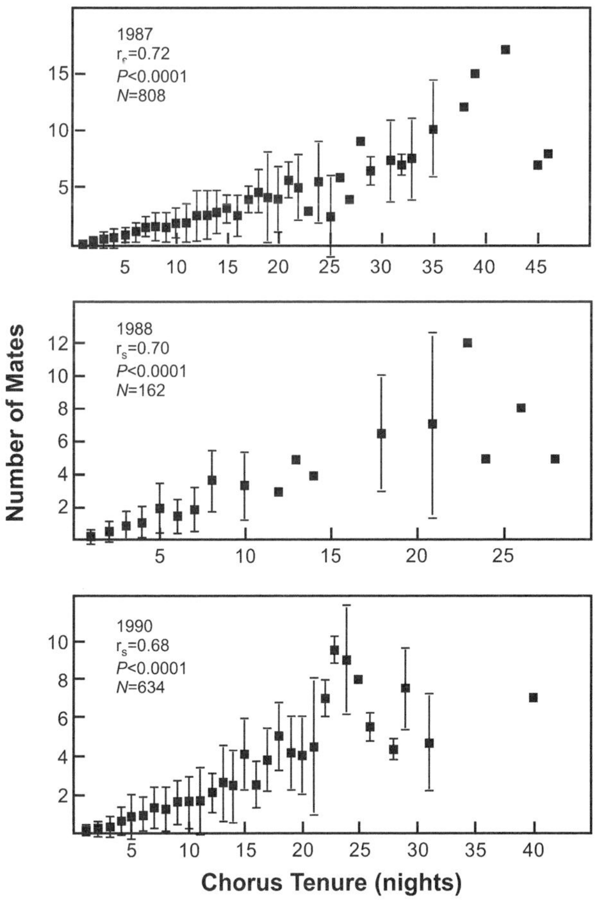


**Fig. 8.18.** Relationship of chorus attendance by male *Hyla gratiosa* to male mating success in three breeding seasons. In all three years, chorus tenure accounted for a substantial proportion of variation in male mating success. After C. Murphy (1994b).

Dyson, Bush, and Halliday (1998) monitored the chorus attendance of a small population (45 males) of *Hyperolius marmoratus* for 167 days at a natural pond in South Africa. Chorus tenure of individual males was extremely variable (1–86 nights), but the median was low (five nights). Eighty-six females mated in the chorus, but mating success of males was unevenly distributed. More than half the males did not mate at all, but some mated as many as 16 times (fig. 8.19 top). Male mating success was strongly correlated with number of nights that males were present in the chorus (fig. 8.19 bottom). Furthermore, a male that mated on one night was twice as likely to mate on the next night as a male that had not mated previously. In this species, most matings occur within the first 1.5 hours of a chorus. Males that acquire mates typically call for much shorter periods and produce about a third the number of calls produced by males that do not mate (Passmore, Bishop, and Caithness 1992). Because the cost of calling in this species is very high (Grafe 1996b), this reduction in total calling effort could result in substantial energy savings, thereby prolonging a male's stay in a chorus and increasing his chances of acquiring additional mates.

It seems unlikely that this explanation can account for the very long chorus attendance (50–86 nights) of the few males that mated many times (13–16 mates; fig. 8.19). The disproportionate success of these few males could have been related to other factors, such as superior foraging ability or behavioral attributes such as calling rate that were not measured in the study.

## Plasticity in Anuran Mating Systems

The discussion of mating systems so far has focused on relatively discrete categories of mating systems, including scramble competition, leks, and resource defense. In fact, it has become clear that mating systems of anurans often are quite variable, both within and between populations. Within breeding aggregations, males sometimes shift between calling and searching tactics as a function of chorus density. They also shift the location of searching activity as a function of body size, or alternate between calling and satellite behavior. In addition, some species of anurans have mating systems that combine elements of active searching and territorial behavior. The result is a tendency for scramble competition in high-density choruses to grade into leklike behavior at lower densities, with leklike mating systems in turn grading into resource defense as densities decrease even further. Similar density-dependent shifts from scramble competition to lek behavior to resource defense have been reported in other vertebrates (Langbein and Thirgood 1989; Balmford et al. 1993; Thirgood, Langbein, and Putnam 1999). In addition, some reptiles, birds, and mammals have mixed mating systems within a single population, with some males visiting leks while others defend isolated territories (Gosling 1986; Gosling and Petrie 1990; Wikelski, Carbone, and Trillmich 1996; Lanctot and Weatherhead 1997). Satellite behavior and other alternative mating tactics are common in other vertebrates as well (Andersson 1994).

### Density-Dependent Shifts between Searching and Calling

The tactics used by males to acquire mates in explosive mating aggregations often vary with male density (Wells 1977b; Arak 1983a; Olson, Blaustein, and O'Hara 1986; R. D. Howard 1988b; B. K. Sullivan 1989a, 1992; Wagner and Sullivan 1992). In very dense choruses, males devote more of their time to active searching and attempting to displace other males from amplexus than they do to calling. They move over relatively large areas in search of females, but searching sometimes becomes concentrated around a communal oviposition site (fig. 8.10 A) as the breeding period progresses (Blankenhorn 1974; N. Davies and Halliday 1979; R. D. Howard 1980; Ryser 1989a). At low densities, males often abandon searching altogether and switch to stationary calling. This reduces the amount of direct competition among males for individual females and provides more opportunities for females to select their mates (Wells 1977b). Density-dependent shifts in male behavior are apparent in comparisons between species, between populations of the same species, or between years in the same population.

Kagarise Sherman (1980) studied the effects of male density on the behavior of *Bufo canorus* by making observations of focal animals. Within years, male density was negatively correlated with individual calling rate and the proportion of time that individuals spent calling. Frequency of movement and frequency of clasping attempts were positively correlated with male density, and males also tended to move farther in a given period of time as density increased. Hence, males clearly shifted from active searching to stationary calling as density decreased. Similar differences were apparent in comparisons between years. A comparison of choruses of *B. canorus* and *B. exsul* revealed that the latter always had higher density aggregations and showed a greater tendency to locate mates by active searching.

A density-dependent shift from searching to calling also has been observed in *Bufo bufo*. This species lacks a vocal sac, and in dense choruses, males seldom call (Heusser 1969a). In a low-density chorus, however, some males remained stationary and called to attract females (Höglund and Robertson 1988). The proportion of males calling increased as density decreased. Other toads that exhibit density-dependent changes in mate-locating tactics include *Bufo americanus* (B. K. Sullivan 1992), *B. woodhousii* (B. K. Sullivan 1986b), *B. cognatus* (L. E. Brown and Pierce 1967; B. K. Sullivan 1982a, 1983a; Krupa 1989, 1994), and *B. valliceps* (Wagner and Sullivan 1992). Similar changes in behavior were reported for wood frogs (*Rana sylvatica*) by Woolbright, Greene, and Rapp (1990). This species normally breeds in dense choruses with very active searching by males (R. D. Howard 1980; Berven 1981; R. D. Howard and Kluge 1985). In the low-density chorus studied by Woolbright et al., males spent less time searching and more time calling than in more typical wood frog choruses. Experimental reduction of male density resulted in even less searching and more calling.

Density-dependent shifts between calling and searching are less common in species with longer breeding periods, probably because densities seldom reach levels seen in explosive mating aggregations. Nevertheless, I have seen males of *Hyla versicolor* swimming about on the surface of a pond searching for mates when densities were very high, even though they normally call from fixed locations. A shift to active searching at high densities has been reported for several toads with relatively long breeding periods, including *Bufo calamita* (Arak 1988c) and *Bufo valliceps*

(Wagner and Sullivan 1992). In *Rana porosa,* an inhabitant of rice fields that has a long summer breeding season, males shift from defense of individual territories at low densities to active searching and attempts to displace males from amplexus at high densities (Shimoyama 1989). The European agile frog (*Rana dalmatina*) also shows some plasticity in behavior. This species has a relatively short, early spring breeding season, with elements of both scramble competition and lek behavior. At low densities, most males call and sometimes defend calling sites as territories, but as a densities increase, many males stop calling and apparently shift to satellite behavior. It is during these periods of highest male density that multiple paternity of egg clutches is most likely to occur, although it is not clear how fertilization by multiple males is achieved (Lodé, Holveck, and Lesbarrères 2005).

### Size-Dependent Variation in Searching Behavior

Males also vary in where they choose to search for mates. This was first studied in *Bufo bufo* by N. Davies and Halliday (1979), who found that males searched either at the spawning site or around the periphery of the breeding area, distributing themselves so that the probability of mating was approximately equal. Males on the periphery were able to intercept females early and suffered less competition from other males. However, when the pairs moved to the spawning site, small males suffered an increased probability of being displaced by males searching there. Large males are more likely than small males to do well in areas where competition for females is intense (Arak 1983a; Loman and Madsen 1986). In *Rana temporaria,* small males tended to search for mates around the edges of a pond or on land, whereas larger individuals were more likely to search at the spawning site (Arak 1983a; Ryser 1989a). In a population of *Bufo americanus* with a somewhat longer breeding period, small males tended to search for females on land around the breeding pond, and they intercepted 70% of the females approaching the pond (Forester and Thompson 1998). Nearly two-thirds of these males remained paired with females after an hour, even though larger males often tried to displace them once they reached the pond, indicating the advantage of being the first male to clasp a female. These mating tactics are by no means fixed; both large and small males were observed to switch mate-locating tactics in all of these studies.

### Gradation between Scramble Competition and Lek Behavior

In some anurans with relatively short breeding seasons, males engage in active searching behavior, but restrict their searching to a relatively small area, resulting in some degree of spatial organization within an aggregation (e.g., *Rana aurora:* Calef 1973b; *R. esculenta:* Wahl 1969; *R. temporaria:* Elmberg 1986; *Bufo boreas:* Black and Brunson 1971; *B. regularis:* Tandy and Keith 1972; *Pelobates fuscus:* Eibl-Eibesfeldt 1956; *Discoglossus pictus:* Knoeppffler 1962; Weber 1974; *Bombina bombina* and *B. variegata:* Heusser 1961; Lörcher 1969; *Bombina orientalis:* Akef and Schneider 1985; *Pyxicephalus adspersus:* Channing, du Preez, and Passmore 1994). Males sometimes defend a small area around themselves, but they usually do not defend permanent territories. Such aggregations have a spatial organization similar to that of a lek (fig. 8.10 D, E), but females often are intercepted by searching males, or have their initial mates displaced by larger males. This leaves females with few opportunities to mate with males of their own choosing (Wahl 1969; Blankenhorn 1974, 1977; Tunner 1976; Kagarise Sherman 1980; Klöckner 1982; B. K. Sullivan 1982b; Radwan and Schneider 1988; Channing, du Preez, and Passmore 1994; Bergen, Semlitsch, and Reyer 1997).

The tendency for scramble competition to grade into lek-like behavior is illustrated by the behavior of the Neotropical hylid frog *Scinax rubra.* Bourne (1992) studied this species in Guyana and characterized its mating system as a lek, but in contrast to the relatively continuous breeding patterns of most lekking anurans, this species breeds in a series of explosive mating aggregations after especially heavy rains (Bourne 1992; Bevier 1997a). Males call for many hours on peak nights of breeding and defend small areas around their calling sites with aggressive vocalizations. Yet males also tend to move frequently while calling, especially small individuals (Bourne 1992; Wells, personal observations). Females move through the chorus and attempt to choose mates, but sometimes are intercepted by noncalling satellite males. More than two-thirds of the matings observed by Bourne (1992) resulted from large males displacing other individuals from amplexus. Hence, any possible advantage to females in choosing particular males as mates often is thwarted by direct competition among males for possession of females (Bourne 1993). While the spatial organization of choruses in this species resembles a lek mating system, the mating tactics used by males are similar to those of many other explosive-breeding frogs.

The African bullfrog (*Pyxicephalus adspersus*), another species that breeds only after heavy rains, has a somewhat similar mating system, but males are much more aggressive toward one another than are males of *Scinax rubra.* Males defend small territories around themselves in a densely packed chorus by attacking and even biting other males (B. Balinsky and Balinsky 1954; Grobler 1972; Picker 1983). The mating system has been described as a lek (Channing, du Preez, and Passmore 1994), but females arriving in the chorus often are chased by several males until one achieves amplexus, so females probably have few opportunities to choose their mates.

Another mating system that combines characteristics of both scramble competition and lek behavior is that of the water frogs of Europe (fig. 8.10 C). The edible frog, *Rana esculenta,* is a hybrid originally produced by matings between *R. lessonae* and *R. ridibunda* (Berger 1977; Graf and Polls Pelaz 1989; Hellriegel and Reyer 2000; see also chapter 7). Matings between male and female *R. esculenta* produce nonviable or sterile offspring with the *R. ridibunda* phenotype. Hybrids therefore must mate with individuals of one of the parent species to reproduce. Most populations of *R. esculenta* are maintained by this hybridogenetic mode of reproduction. Most commonly hybrids are found in mixed populations with *R. lessonae.* The latter predominates in the ancestral breeding areas for this species, such as pools in marshes and forests, whereas *R. esculenta* tends to be more common in disturbed areas such as gravel pits. These frogs can be active at breeding sites for a month or more, but peaks of mating activity often last only a few days. Males gather on display arenas, sometimes at densities up to 25 males per square meter. They establish small territories and constantly jockey for position in the chorus, using both aggressive calls and physical attacks (Wahl 1969; Blankenhorn 1974, 1977; Klöckner 1982; Radwan and Schneider 1988).

Females swim among the calling males, but often are clasped by searching males. Males sometimes attempt to displace other individuals from amplexus. Females of both phenotypes are expected to prefer to mate with *R. lessonae* males. *Rana lessonae* females should do so because they are conspecifics, whereas *R. esculenta* females should do so because only matings with *R. lessonae* males can produce viable offspring. Females apparently are not able to assess the relative densities of the two male phenotypes from long-distance acoustic cues emanating from choruses, but both *R. lessonae* and *R. esculenta* females are attracted to the calls of individual *R. lessonae* males (Roesli and Reyer 2000). When *R. esculenta* females were given a choice of male phenotypes in situations where male-male competition was prevented, they preferred *R. lessonae* males (Abt and Reyer 1993). In a competitive situation with different numbers of the two types of males, however, females mated with them in proportion to their occurrence, suggesting that male-male competition nullifies any preference for *R. lessonae* males (Bergen, Semlitsch, and Reyer 1997). Nevertheless, there is evidence of "cryptic female choice" in this system, because females that were seized by *R. esculenta* males released fewer eggs than did those seized by *R. lessonae* males (Reyer, Frei, and Som 1999). Some females subsequently laid the remainder of their egg clutches later in the same season, but most females apparently resorbed unlaid eggs, thereby improving their body condition and increasing reproductive success the following year. Males are indiscriminant in their choice of mates when presented with parental and hybrid females, at least in laboratory experiments, so *R. lessonae* males often end up mating with hybridogenetic females, even though they do not derive any genetic benefit from such matings (Engeler and Reyer 2001).

Several workers who have studied this system have reported that *R. esculenta* males are more aggressive toward other males than are the *R. lessonae* males. The latter mostly engage in searching behavior and attempts to clasp other frogs (Blankenhorn 1974, 1977; Klöckner 1982; Radwan and Schneider 1988; Abt and Reyer 1993; Bergen, Semlitsch, and Reyer 1997). Blankenhorn (1977) reported that the largest *R. esculenta* males were so aggressive that they did relatively poorly in acquiring mates, even though they outnumbered the *R. lessonae* males, but Bergen, Semlitsch, and Reyer (1997) concluded that the more aggressive males actually enjoyed a mating advantage in mixed populations. A possible explanation for the greater aggressiveness of *R. esculenta* males is the hybrid origin of this phenotype. These frogs are intermediate between the two parental species in morphology (Berger 1977; Schneider and Brzoska 1981; Brzoska 1982; Schneider and Joermann 1988). Patterns of aggressive and sexual behavior are likely to have a genetic basis and to be intermediate between that of the two parental species. *Rana lessonae* males appear to have a typical scramble competition mating system, with little aggression among males and much searching for females. *Rana ridibunda,* on the other hand, is a prolonged breeder, and appears to have a lek or resource-defense mating system (M. Smith 1969; Heym 1974; Frazer 1983; Kuhn and Schneider, 1984; Schneider and Joermann 1988; Kyriakopoulou-Sklavounou and Kattoulas 1990). Males are aggressively territorial and have a variety of aggressive calls. The peculiarly aggressive behavior of *R. esculenta* males could result from their behavior combining elements of two parental species that have different mating tactics.

### Gradation between Lek Behavior and Resource Defense

There is some evidence that pond-breeding ranids that normally defend resource-based territories shift to lek behavior when choruses are unusually dense. This could be the reason for the leklike spatial organization of choruses of Asian frogs that breed in rice fields, where frogs tend to be very abundant. Emlen (1976) originally described the mating system of a dense population of bullfrogs (*Rana catesbeiana*) in Michigan as a lek. Males defended closely packed territories in well-defined choruses that changed position throughout the summer. Emlen observed three females in amplexus that left the chorus before laying eggs, and five of six egg masses that he found were outside the chorus area.

Subsequent work by R. D. Howard (1978a) on the same population in years when population densities were somewhat lower revealed that most egg masses were deposited in the territories of males, and he concluded that these frogs have a resource-defense mating system. However, it is possible that the conflicting results of these studies reflect a density-dependent shift in mating tactics.

In my study of green frogs (*Rana clamitans*), some males abandoned territories along the shore of a pond during dry weather and formed a chorus with males much closer together near the center of the pond, suggesting a shift to leklike behavior (Wells 1977c). The chorus broke up after heavy rain raised water levels in the pond, so I was not able to observe any mating behavior. Some European ranids seem to shift from defense of individual territories to lek behavior or even active searching at high densities, and some also exhibit seasonal shifts in spatial organization (Radwan and Schneider 1988; Kyriakopoulou-Sklavounou and Kattoulas 1990). This sort of intraspecific variation in mating systems needs to be investigated more thoroughly in anurans.

## Satellite Behavior

In addition to population-level shifts in the structure of the mating system, individual males also adopt alternative mating tactics within a chorus. Often this involves some males adopting a satellite role and parasitizing the calling efforts of other males. Such behavior is common in many other vertebrates and invertebrates, and there is a large theoretical and empirical literature on alternative mating tactics (Dawkins 1980; Rubenstein 1980; Waltz 1982; Austad 1984; Dominey 1984; Waltz and Wolf 1984). Alternative mating tactics can be expressions of a conditional strategy in which disadvantaged males adopt less preferred mating tactics. They also can represent mixed evolutionarily stable strategies, in which different males adopt different tactics that yield roughly equal payoffs in mating success (Arak 1983a; Gardner, Morris, and Nelson 1987; Lucas and Howard 1995; Lucas, Howard, and Palmer 1996; McCauley et al. 2000).

### Explosive Breeders

Satellite behavior in explosive-breeding anurans is somewhat similar to the shift between calling and searching discussed earlier, except that satellite males tend to focus their search for mates around particular calling males instead of particular parts of the breeding site (Leary et al. 2004). Again, this appears to be largely a density-dependent phenomenon, occurring mostly at intermediate densities as males shift between calling and searching tactics. Although satellite males sometimes are smaller than callers, this is not always true (Kagarise Sherman 1980; Krupa 1989), and switching between calling and satellite roles probably is common. In *Bufo cognatus*, satellite males tend to be smaller than calling males, although not necessarily younger (Krupa 1989; Leary et al. 2005), and they were most likely to gather around large males that had long calls. In this species, the satellite tactic was clearly inferior to calling, with more than 90% of all matings going to calling males. Satellite males also were smaller, but not younger, in explosive-breeding populations of *Bufo woodhousii* in Oklahoma (Leary et al. 2005). In both of these species, individuals that experience slow growth early in life appear to be more likely to adopt a satellite tactic than do faster-growing males.

### Prolonged Breeders

Satellite behavior is much more common in prolonged breeders with leklike or resource-defense mating systems than in explosive-breeding anurans. Such behavior represents either a mixed evolutionarily stable strategy or a conditional strategy adopted by smaller, younger, or weaker males excluded from choice calling sites (Gardner, Morris, and Nelson 1987; Halliday and Tejedo 1995). Large size differences between territorial and satellite males appear to be most common in anurans with resource defense mating systems. In *Rana clamitans* and *R. catesbeiana*, for example, satellite males often are considerably smaller and younger than territorial residents (fig. 8.20; Wells 1977c, 1978b; R. D. Howard 1978a, 1981a, 1984). In bullfrogs, satellite males sometimes intercept females approaching territorial males, but the mating success of satellite males is relatively low, probably accounting for less than 20% of all matings in the population (R. D. Howard 1978a, 1981a). Satellite males probably also intercept females occasionally in *R. clamitans*, although it has not yet been observed (Wells 1977c, 1978a). In both species males are opportunistic, and readily abandon the satellite role and occupy territories deserted by other males. Young male bullfrogs also attempt to set up territories in areas not occupied by larger males, but quickly abandon them if challenged (R. D. Howard 1981a 1984). Similarly, in *R. virgatipes*, smaller males sometimes assume satellite roles and sometimes are territorial, but quickly retreat when challenged by a larger male or when they are presented with recordings of large male calls. Large males in turn give more aggressive responses when presented with the calls of small males (Given 1987). Small males also adopted satellite tactics in the Asian fanged frog *Limnonectes kuhlii*, and a few succeeded in acquiring mates (Tsuji 2004).

Satellite males have been observed in several species of prolonged-breeding toads (Wells 1977b; Arak 1988c). In most cases, satellites are somewhat smaller than calling males, but quantitative data are scarce. In *Bufo calamita*, three or four individuals often sit near one another in a group, with the largest male calling most of the time and the

**Fig. 8.20.** Satellite behavior in frogs with resource defense mating systems. (A) A green frog (*Rana clamitans*) satellite male in low posture next to a territorial resident in inflated high posture. (B) A large male bullfrog (*Rana catesbeiana*) accompanied by a much smaller satellite male. This male sometimes had as many as three satellites around him at the same time. Photos by Kentwood D. Wells.

others acting as satellites. When large calling males were removed, the smaller satellite males generally dispersed. Removal of satellite males had no effect on the behavior of other males in the group. When calls were played through a speaker, small males often adopted satellite roles around the speaker, but large males were more likely to move away from the speaker. Satellite males in this species sometimes intercept females, but their mating success is lower than that of calling males, even though the latter do not defend resource-based territories (Arak 1983a; 1988a).

Satellite males are common in hylids (table 8.1) and probably other tree frogs (e.g., rhacophorids, Arak 1983c). Satellite males sometimes are slightly smaller than calling males (Garton and Brandon 1975; Arak 1983c; Forester and Lykens 1986; Backwell and Passmore 1991; Haddad 1991), but often there is little or no difference in body size (Perrill, Gerhardt, and Daniel 1978, 1982; Fellers 1979a; Roble 1985; Perrill and Magier 1988; Lance and Wells 1993). Consequently, several authors have suggested that these species exhibit a mixed evolutionarily stable strategy, with calling and satellite strategies yielding equal payoffs (Arak 1983a; Gardner, Morris, and Nelson 1987). Satellite males have been observed intercepting females as they approach calling males (Miyamoto and Cane 1980; Roble 1985; Perrill and Magier 1988; Haddad 1991). Nevertheless, mating success of satellite males often appears to be much lower than that of calling males, and in some studies, satellite males have not been observed to mate at all (Backwell and Passmore, 1991). Perrill et al. (1978) released *Hyla cinerea* females near pairs of calling and satellite males. In 43% of the tests ($N = 30$), the satellite male clasped the female. These data do not necessarily show that the overall payoff of calling and satellite behavior is nearly equal. To determine this, one would require data on the proportion of all matings during the breeding season obtained by calling and satellite males.

In many tree frogs, males readily switch between calling and satellite roles, although some individuals tend to spend more time calling and others more time as satellites. When calling males are removed from caller-satellite associations, satellite males often begin calling, or they reorient toward other calling males (Perrill, Gerhardt, and Daniel 1982; Haddad 1991). In one series of experiments with *Hyla cinerea,* calling males were removed from associations and replaced with a speaker playing recorded calls; all satellite males became satellites on the speaker. Synthetic advertisement calls were effective in eliciting satellite behavior, but synthetic aggressive calls were not (Perrill, Gerhardt, and Daniel 1982). In similar experiments with *Acris crepitans,* some calling males switched to a satellite role when presented with a loud recorded call, but switching was less common than in *H. cinerea* (Perrill and Magier 1988).

These experiments demonstrate the flexibility of mating behavior in these species, but the factors determining which role a particular male will adopt on any given night are unknown. Several theoretical models have predicted that a male's tendency to adopt a satellite role will depend on several factors, including his age, body size, and energy reserves (Lucas and Howard 1995; Lucas, Howard, and Palmer 1996; McCauley et al. 2000). For example, the model by McCauley et al. (2000) predicted that males with low energy reserves should switch to satellite behavior to save energy. This prediction is not yet supported by empirical data from natural populations, but very few studies have compared body condition or energy reserves of callers and satellites in the same population. Lance and Wells (1993) did not find any difference in body condition of callers and satellites in spring peepers (*Pseudacris crucifer*), but satellite males were present only early in the breeding season, when energy reserves of males are expected to be high. On the other hand, calling males collected late in the season did have lower body mass for any given length than did those collected early in the season, probably because of depletion of lipid reserves (McKay 1989; Ressel 1993; see chapter 5).

Emerson (2001b) proposed that switches from calling to satellite behavior are mediated by decreasing energy reserves leading to increased production of the stress hormone corticosterone. This in turn was expected to reduce circulating levels of testosterone, perhaps resulting in males switching to satellite behavior. Leary et al. (2004) tested the model with two species of toads, *Bufo woodhousii* and *B. cognatus*. They found that calling males had higher levels of corticosterone than did satellite males, but there was not a negative relationship between corticosterone and testosterone. In both species, satellite males were significantly smaller than calling males, and in *B. woodhousii*, satellite males also were in poorer condition. Their results suggest that differences in both body size and body condition play a role in determining behavioral tactics, but there is not a simple hormonal control mechanism mediating behavior. Leary et al. (2004) suggested that switches between calling satellite behavior are mediated not by testosterone, but by interactions between corticosterone and arginine vasotocin (AVT), which is known to mediate calling behavior in anurans (Marler, Chu, and Wilczynski 1995; Ten Eyck 2005)

## Mating Tactics of Female Anurans

So far, the discussion of anuran mating systems has focused almost entirely on the behavioral tactics used by males to acquire mates, and we tend to classify mating systems based largely on the behavior of males. Yet as Darwin clearly recognized, females often play an active role in choosing their mates, by assessing variation in quality of the males themselves or the resources they control. Much of the recent work on anuran mating systems has focused on the tactics used by females to assess mate quality and the effect of female choice on the evolution of male traits. B. K. Sullivan, Ryan, and Verrell (1995) and Gerhardt and Huber (2002) provide detailed reviews of recent literature on female choice in anurans.

Females could derive either direct or indirect benefits from choosing particular males as mates. Direct benefits include anything that increases the female's production of offspring, whereas indirect benefits refer to anything that enhances the genetic quality of those offspring (Kirkpatrick and Ryan 1981; Kirkpatrick 1982; Andersson 1994). The most obvious direct benefit that females might derive from mate choice would be access to resources controlled by males, such as high-quality oviposition sites, because this is likely to directly affect the survivorship of her offspring. Similarly, if males provide parental care, then choice of males with superior parental abilities would yield immediate fitness benefits to the female. Even when males do not provide the female with resources or parental care, as in lekking species, females could benefit by selecting the most fertile males. If a male's size relative to that of the female affects his ability to fertilize her eggs, then selection of males of the appropriate size would yield direct benefits as well. Females also might benefit from avoiding males infected with parasites or diseases that could be transmitted to her (K. Pfennig and Tinsley 2002). Finally, if searching for a mate is costly because of energetic constraints or the risk of predation, then females will benefit by choosing males that produce especially conspicuous signals that can be easily located (Grafe 1997a; Jennions and Petrie 1997). In some cases, males might evolve signals that exploit some preexisting sensory bias in females, thereby making their signals more easily detected by females (M. Ryan 1990; M. Ryan and Rand 1990, 1993b; see also chapter 7).

Indirect benefits of mate choice depend on genetic benefits provided by males. One form of indirect sexual selection is the so-called runaway sexual selection process first proposed by R. A. Fisher (1930). He proposed that an initial preference for a certain male trait might exist because of some benefit it provides to the female. Once the preference becomes established in the population, the preference and the trait become genetically correlated, so that selection for more exaggerated traits leads to increased preference for the trait, leading to runaway sexual selection. This process has been difficult to demonstrate in nature, and there is no clear evidence that any trait in male anurans evolved by runaway sexual selection. Consequently, nearly all discussion of indirect benefits of mate choice has focused on good genes models of sexual selection. These models postulate that females select mates on the basis of traits that are indicators of overall genetic quality, or traits that reveal the presence of genes that enhance particular aspects of fitness, such as resistance to parasites, high growth rate, or a high physiological capacity for activity (Andersson 1994). For example, suppose females prefer males with high calling rates. If a high calling rate provides reliable information about some fitness-related trait, then a genetic correlation between the trait and the preference cause both the frequency of the correlated trait and the strength of the female preference to increase in the population (B.K. Sullivan, Ryan, and Verrell 1995).

Parker (1983) argued that many apparent examples of active female choice actually could be the result of passive attraction to male displays. Variation among individual males in display behavior might or might not be correlated with genetic differences that would affect female fitness. This is particularly relevant to anurans, because females often must locate a chorus, and individual males within a chorus, before choosing a mate. If females simply go to a male producing the most conspicuous signal, it does not necessarily imply that females actively compare the qualities of different individuals or derive a fitness benefit from choosing one male over another (Arak 1988a; Dyson and Passmore 1988). In practice it often is difficult to distinguish between active choice and passive attraction. B. K. Sullivan (1989b) argued that the distinction is not very useful because it confuses questions about the mechanisms by which females select mates and the ultimate fitness benefits of mate choice (see also Ryan 1990; Forrest and Raspet 1994; B. K. Sullivan, Ryan, and Verrell 1995; Grafe 1997a; Jennions and Petrie 1997). There is considerable evidence that female anurans approach calling males by following a gradient of call intensity, and are more likely to approach louder or closer calls than softer or more distant calls (Beckers and Schul 2004; Castellano, Rosso, and Giacoma 2004). It is less clear whether females are simply reacting to the intensity of stimulation of the auditory system or are discriminating among males on the basis of cues related to male quality.

Most work on female choice in anurans has focused on the proximate cues used by females to select their mates. There is convincing evidence for females selecting mates on the basis of resources controlled by males, as well as phenotypic traits such as body size or calling activity. It has been much more difficult to demonstrate the fitness benefits to females making such choices, especially indirect genetic benefits (B. K. Sullivan, Ryan, and Verrell 1995; Grafe, 1997a; Doty and Welch, 2001; Welch, 2003). A largely neglected area of research is the cost of mate choice by females and variation in the nature and strength of female preferences (Wagner 1998). There is some evidence for anurans that choices made by females are repeatable (Jennions, Backwell, and Passmore 1995; Gerhardt, Dyson, and Tanner 1996; Gerhardt and Huber, 2002), but there also is evidence that the criteria used by females to choose mates sometimes vary within and among individuals and among populations of the same species, even when the general direction of preference is similar for all females (Jennions and Petrie 1997; R. D. Howard and Young 1998; Schwartz and Gerhardt 1998; C. Murphy and Gerhardt 2000; Gerhardt and Huber 2002). The effect of such individual variation in female preference for certain male traits is to weaken potential selection on those traits. Hence, to fully understand how sexual selection acts on male traits, we need replicated studies of mate choice both within and between populations.

### Female Choice in Explosive-Breeding Anurans

Many authors have speculated about criteria that might be used by females to select mates in explosive-breeding aggregations (Licht 1976; N. Davies and Halliday 1977; Wilbur, Rubenstein, and Fairchild 1978; Gatz 1981a; J. Lee and Crump 1981). Nevertheless, there is relatively little convincing evidence that female choice is a major determinant of male mating success in most species that form very dense breeding choruses and engage in scramble competition (Arak 1983a; Halliday 1983a; Halliday and Tejedo 1995; B. K. Sullivan, Ryan, and Verrell 1995). Upon entering a chorus, females usually are mobbed by searching males. Nevertheless, in low-density choruses, some females can approach individual calling males and initiate amplexus without being intercepted by other individuals. For example, Kagarise Sherman (1980) found that approximately half of all matings observed in a population of *Bufo canorus* were initiated when females contacted calling males. B. K. Sullivan (1982a) reported that *B. cognatus* females sometimes approached and contacted calling males and attempted to shake off satellite males that intercepted them. Studies of several other explosive-breeding toads at low densities also have revealed some opportunities for female choice (R. D. Howard 1988b; B. K. Sullivan 1992; Tejedo 1992c).

The criteria used by females to select mates in explosive-breeding species are largely unknown. These species do not defend resources, so any mate choice would have to be based on phenotypic or behavioral characteristics of males. B. K. Sullivan (1983a) found that successful *Bufo cognatus* males called more frequently than did unsuccessful males, and he suggested that females might choose mates on the basis of differences in calling rates. However, he measured only the percentage of observation periods in which males were calling, not hourly calling rates, so these data do not necessarily demonstrate a preference for high calling rates.

B. K. Sullivan (1992) showed that females of *Bufo americanus* preferred a higher calling effort (calling rate × call duration) in controlled playback experiments, but he did not find any evidence that females discriminated between calls of high and low frequency. He was not able to correlate mating success in the field with any aspect of calling behavior, but his sample of matings was very small. R. D. Howard and Palmer (1995) tested females of the same species with high- and low-frequency calls presented in alternation or overlapped with one another (the more common situation in the field). Females preferred calls of low frequency if the calls were alternated or if the low-frequency call preceded the high-frequency call in an overlapped pair, but they showed no preference when the high-frequency call was presented first. Although large males of this species often have a mating advantage (R. D. Howard 1988b), it is not known whether females can assess differences in male body size from the relatively small differences in call frequency usually present in natural choruses (R. D. Howard and Palmer 1995).

Later work on the same population revealed repeatable differences between males in call characteristics (R. D. Howard and Young 1998). Call-dominant frequency was negatively correlated with male body size, and females initially tended to pair with larger than average males. Howard and Young removed some females from amplexus and allowed them to pair repeatedly with different males, and these individuals showed little consistency in their tendency to pair with males of a particular size. Both calling rate and call duration were highly variable, both within and between males, but these traits were not related to body size. Howard and Young did not investigate the effect of these variables on mate choice by females in natural choruses, however. In this population, only about half of the males called actively in the chorus; the remainder mainly engaged in active searching for females. This pattern is typical for this species (Forester and Thompson 1998). Overall, the results of these studies indicate that female American toads sometimes have an opportunity to select individual males in a chorus, but female preferences often are negated by the activities of searching males that seize females as soon as they are encountered.

### Female Choice in Prolonged-Breeding Anurans

Most of the work on female choice in anurans has focused on species with relatively long breeding seasons, in which males form leklike choruses or defend resource-based territories. Females typically have ample opportunities to move among calling males and compare the quality of potential mates or their territories. Unfortunately, we know relatively little about the sampling tactics used by females to assess males in large choruses. In laboratory experiments, females often choose between alternative acoustic stimuli in less than a minute, but they often spend much more time sitting and listening to calling males in the field before beginning their approach. J. Schwartz, Huth, and Hutchin (2004) designed a field experiment to determine how much time is required for females of *Hyla versicolor* to assess differences in number of pulses in calls, a criterion known to be important in mate choice. They found that females probably need only about two minutes to assess subtle differences between calls, even though they appear to take much longer to approach males in the field. This experiment presented females with a somewhat simplified task, however, because they only were required to choose between two sound sources in a relatively quiet environment, whereas females in the field often are confronted with multiple sound sources and high levels of background noise.

One of the few studies in which the behavior of individual females was monitored in detail was that of M. Morris (1989) on *Hyla chrysoscelis*. She found that females typically visited a relatively small subset of calling males in a chorus before choosing a mate. Presumably the potential costs of wider searching, including exposure to predation, outweigh the benefit of sampling a larger number of males. Females of *H. gratiosa* exhibit even more restricted mate sampling (Murphy and Gerhardt 2002). Most individuals mated with the first male encountered upon entering the chorus, suggesting that females evaluate males from a distance before approaching them. The specific criteria used by females to choose mates vary with the type of mating system. Some evaluation of territory quality almost certainly is important for any species in which males defend oviposition sites, because the quality of the oviposition site can have an immediate effect on a female's reproductive success. When males do not defend resources, however, it becomes more difficult to determine how females assess the quality of potential mates.

#### Female Choice and Resource Quality

The easiest form of female choice to understand is choice based on resource quality. If males defend territories centered on oviposition sites and these sites vary in quality, then females are expected to choose mates on this basis. There is good evidence for female choice based on territory quality in *Rana clamitans* and *R. catesbeiana* (Wells 1977c; R. D. Howard 1978a). In both species, physical features of territories were correlated with mating success of territory owners. Although large males tended to acquire the best territories, small males with good territories were more likely to mate than large males with poor territories. R. D. Howard (1978b) demonstrated that bullfrog oviposition sites varied in temperature, resulting in variable hatching success of eggs (fig. 8.21). In one year, hatching success was strongly correlated with male size (R. D. Howard 1978a, b), but in two other years it was not (R. D. Howard 1983). Apparently this was due to a reduction in the number of males competing for territories, leading to greater availability of high-quality sites.

Males of other species, such as *Hyla rosenbergi* and several centrolenid frogs, also defend oviposition sites, but it is not clear whether females choose mates on the basis of territory quality. Kluge (1981) did not find consistent differences between nests defended by successful and unsuccessful *H. rosenbergi* males; he suggested that female choice was more likely to be based on some aspect of male behavior, especially aggressiveness. Greer and Wells (1980) found no differences in survivorship of *Hyalinobatrachium fleischmanni* eggs at sites used frequently for oviposition and those used infrequently, so it was not clear that female choice of mates at particular sites enhanced female fitness.

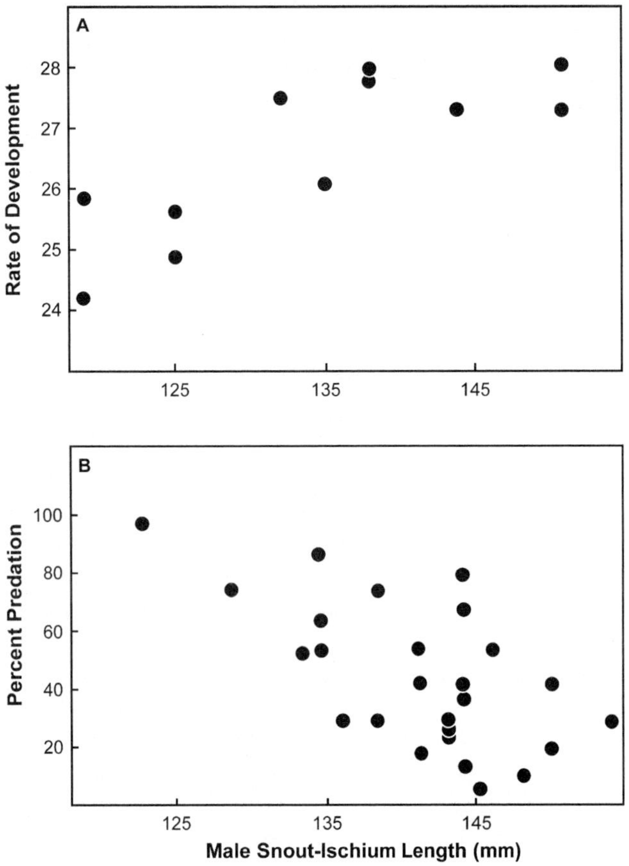


**Fig. 8.21.** Two measures of territory quality as a function of male body size in bullfrogs (*Rana catesbeiana*) from Michigan. (A) Relationship between body size and rate of embryonic development for eggs deposited in male territories. Rate of development is an index of the time required for eggs to reach the hatching stage. (B) Relationship between body size and rate of predation on eggs by leeches. Larger males defended better-quality territories, where embryos developed more quickly than in poor-quality territories, thereby escaping predation by leeches. After R. D. Howard (1978b).

Nevertheless, there could have been differences in post-hatching survival of tadpoles related to placement of oviposition sites. Some egg clutches, for example, were laid on leaves that were not directly over water, increasing the risk that tadpoles would fall onto dry land when they dropped out of the egg mass. It is not known, however, if females can accurately assess whether oviposition sites are properly positioned over water.

### Female Choice and Male Parental Ability

Male parental ability is frequently cited as a possible criterion for female choice in a variety of animals (Clutton-Brock 1991; Andersson 1994; Jennions and Petrie 1997), but there are relatively few convincing demonstrations that females can accurately assess the parental abilities of potential mates. Summers (1989) observed females of *Dendrobates auratus* competing among themselves for access to males. There always were plenty of males advertising for mates, so he interpreted this as competition for particularly high-quality males, or males controlling high-quality territories. He did not, however, have any direct evidence that females assess differences in parental abilities of males. Felton et al. (2006) reported that females of a small microhylid frog in Australia, *Cophixalus ornatus*, preferred nest cavities constructed by older males (some up to 14 years old), but the amount of time males spent attending their eggs after oviposition was not significantly correlated with male age, body size, or condition. These results suggest that females probably have few cues available to accurately predict male parental ability.

Márquez (1993) studied parental care and female choice in two species of midwife toads, *Alytes obstetricans* and *A. cisternasii*. Males often carry more than one clutch at a time, and large males were more likely to obtain multiple clutches than smaller males. Hatching success of the eggs, however, was not correlated with either male body size or the number of clutches a male carried. An earlier study by Reading and Clarke (1988) produced similar results. This suggests that females do not assess the parental ability of males. In these studies, hatching success was measured only once for each male, so it was not possible to measure the repeatability of male parental ability to determine whether there are consistent differences among males that would affect female fitness. It still is possible that females prefer to mate with males already carrying eggs, if this is an indicator of male parental ability. Females of *A. obstetricans* actually appear to suffer some cost if they choose males already carrying eggs. Males with large egg complements retain the eggs longer and release the tadpoles into pools later than do males with small egg complements, thereby exposing the eggs to possible desiccation, or the tadpoles to low water levels late in the summer (Márquez 1996). In an experimental study of another species, *A. muletensis*, Bush (1996) reported that females did not exhibit a significant preference for either brooding or nonbrooding males. Her field studies agreed with those of Márquez in finding no relationship between the number of clutches carried by males and hatching success.

### Female Choice and Call-Dominant Frequency

Many authors have suggested that female anurans might choose mates on the basis of call features that reflect differences in male body size, especially the dominant frequency of male calls (see Gerhardt and Huber 2002, for a review). Unfortunately, there was confusion in some of the early literature about which features of a frog's call are correlated with male body size (Licht 1976; Wilbur, Rubenstein, and Fairchild 1978; Fairchild 1981), despite a clear discussion of the subject by Zweifel (1968b; see also B. K. Sullivan 1982c; Sullivan and Sullivan 1985). In general, interspecific comparisons of closely related species reveal a clear negative correlation between male body size and the dominant frequency of the call. A negative correlation between size and dominant frequency often is found within species as well, but can be more difficult to detect because of the narrower range of body sizes. Even when body size and call frequency are negatively correlated, the correlation sometimes is relatively weak (Oldham and Gerhardt 1975; M. Ryan 1980a, 1983a; Kagarise Sherman 1980; B. K. Sullivan 1982b, 1992; Arak 1983b; B. K. Sullivan and Sullivan 1985; Krupa 1990; B. K. Sullivan and Hinshaw 1992; Cherry 1993; Wagner and Sullivan 1995; Burmeister et al. 2002; Esteban et al. 2002). This means that even if females are capable of fine-scale frequency discrimination, there is a high probability of error in estimating body sizes of males (Gerhardt 1982; J. Robertson 1986a; Gerhardt and Huber 2002). Furthermore, body size sometimes is a relatively poor predictor of age (Lykens and Forester 1987; Halliday and Verrell 1988; R. D. Howard and Young 1998), so the common assumption that females should prefer larger males because they are older is not always valid. In one species, *Cophixlus ornatus*, call-dominant frequency not only was negatively correlated with body size, but also with male age after correcting for body size, even though size and age were not correlated with each other (Felton et al. 2006).

In a number of studies, two-choice playback experiments have been used to test the ability of females to discriminate frequencies within the range found in the population. In some early studies, females were tested mainly with calls at the high and low ends of the population frequency distribution, but high-and low-frequency calls were not tested against those of average frequency. For example, M. Ryan (1980b, 1983b, 1985b) showed that female *Physalaemus pustulosus* could discriminate between calls differing by 60 Hz in fundamental frequency; all females tested approached the speaker playing the lower-pitched call. These frequencies were near the extremes found in natural populations of this species, so the calls of most males would fall between those tested. Females also could discriminate between calls differing by 40 Hz, but not between calls differing by only 30 Hz (M. Ryan 1983b). Ryan's results show that females can make subtle distinctions between calls based on frequency alone, and that some information about male body size might be gained. The degree to which this contributes to the nonrandom mating pattern observed in this species is not clear, because male mating success also is affected by a variety of other factors (M. Ryan 1983b, 1985b). Later work on this species showed a bias toward low-frequency calls, but not a strong preference (Wilczynski, Rand, and Ryan 1995).

Forester and Czarnowsky (1985) presented female spring peepers (*Pseudacris crucifer*) with a choice of calls differing in dominant frequency by about 750 Hz; the stimuli were near the extreme values for their population. As in *Physalaemus*, females preferred the lower frequency call. Unfortunately, this experiment was marred by multiple tests of the same females and a failure to equalize playback intensities at the release point. Furthermore, Forester and Czarnowsky did not test preferences of females for average versus high- or low-frequency calls. In another study of the same species, Doherty and Gerhardt (1984b) found that females preferred calls with average dominant frequencies to those with dominant frequencies above or below the average frequency in the population. J. Schwartz and Gerhardt (1998) found that female spring peepers did not exhibit a preference for any frequency in the absence of background noise, and when noise was present, they actually preferred higher-than-average to lower-than-average frequency calls. They also found that females were not very consistent in their choice of frequencies when they were tested more than once. Furthermore, when auditory tuning curves were determined for the same females that had been tested in playback experiments, there was little or no relationship between frequency sensitivity of individual females and their behavioral responses. Schwartz and Gerhardt concluded that females of this species do not exhibit strong frequency discrimination and are unlikely to choose mates on the basis of differences in call dominant frequency. Bosch, Márquez, and Boyero (2003) reported a similar low level of consistency in choice of dominant frequency for two species of midwife toads (*Alytes*).

Studies with several species have shown either stabilizing selection on call frequency, or weak directional selection for low-frequency calls (Gerhardt and Huber 2002). For example, females preferred calls of average frequency to either high- or low-frequency calls in *Hyla cinerea* (Gerhardt 1982; Gerhardt et al. 1987), *H. gratiosa* (Gerhardt 1981b; C. Murphy and Gerhardt 2000), *H. versicolor* (Gerhardt 1991), *Bufo viridis* (Castellano and Giacoma 1998), *Hyperolius marmoratus broadleyi* (Grafe 1997a), and *Crinia georgiana* (M. J. Smith and Roberts 2003b).

Relatively weak preferences for lower than average frequency calls have been found in several other species. In *Hyla chrysoscelis,* M. Morris and Yoon (1989) found that females preferred low-frequency calls to high-frequency calls, but they did not discriminate between either high- or low-frequency calls and those of average frequency. In *Hyperolius marmoratus marmoratus,* females showed a slight preference for low- over high-frequency calls, but this preference could be altered by changing the timing of the two calls (Dyson and Passmore 1988). Females of *Physalaemus enesefae* preferred low-frequency or average-frequency to high-frequency calls, but did not discriminate between low and average frequencies. The preference for low-frequency calls was strongest when the low-frequency call preceded the high-frequency call and disappeared when calls partially overlapped (Tárano and Herrera 2003). Two species of midwife toads (*Alytes obstetricans* and *A. cisternasii*) showed small biases toward lower-than-average frequency calls as well (Márquez 1995; Márquez and Bosch 1997a), and in one population, mating success was negatively correlated with call frequency (Lodé and Le Jacques 2003). In *Cophixalus ornatus,* males that acquired mates had calls with lower dominant frequencies than did unmated males nearby, but not compared to the population as a whole (Felton et al. 2006).

A few species have shown somewhat stronger preferences for low-frequency calls. Wollerman (1998) reported that female *H. ebraccata* preferred low-frequency calls (2960 Hz) to average-frequency calls (3240 Hz), but did not discriminate between average calls and those of slightly lower frequency (3170 Hz) or higher frequency (3330, 3480 Hz). Nevertheless, four of 20 individuals responded to the average-frequency call when tested with the lowest-frequency call. A significant preference for low-frequency calls also was found in *Acris crepitans* (M. Ryan, Perrill, and Wilczynski 1992), but only in two-thirds of the preference tests. Overall, the evidence suggests that there is not strong directional selection on call-dominant frequency in many anurans, and in some cases, selection is strongly stabilizing in favor of average-frequency calls.

There is considerable evidence that frequency preferences observed in playback experiments with various species roughly correspond to average frequency tuning of the ears of females in the population (see chapter 7). In those experiments in which females preferred calls of average frequency, the preferred call was close to the average best frequency response of the ear of females in that population (Doherty and Gerhardt 1984b). In species in which females preferred lower-frequency calls, the ears of females were tuned to frequencies somewhat lower than the average frequency of calls in the population (e.g., M. Ryan et al. 1992; Sun et al. 2000). Preferences for low-frequency calls over high-frequency alternatives could simply reflect the steeper decline in sensitivity of the ear at frequencies above the best frequency (Gerhardt 1991; Schwartz and Gerhardt 1998; Gerhardt and Huber 2002). In other words, given a choice of calls 200 Hz above and below the best frequency of the female's ear, a female should hear the lower frequency better. In some species, large females show a stronger preference for low-frequency calls than do small females (M. Ryan et al. 1992; Jennions, Backwell, and Passmore 1995). This probably reflects the lower-frequency tuning of the ears of larger females, particularly frequencies detected by the basilar papilla (see chapter 7). One must be cautious, however, in extrapolating from average responses of populations of females to choices made by individual females. The work of Schwartz and Gerhardt (1998) showed considerable variation in behavioral responses both within and among females in a population that was not easily predicted from neural frequency responses.

Further evidence that preferences for low-frequency calls are related to the frequency sensitivity of the ear comes from experiments showing that a preference for low-frequency calls can be reversed by increasing either the intensity or rate of the less-preferred call (Gerhardt 1988, 1994a; Gerhardt and Huber 2002). For example, when M. Morris and Yoon (1989) increased the sound pressure level of a high-frequency call of *Hyla chrysoscelis* by 3 dB, the preference for a low-frequency alternative disappeared. When high-frequency calls were presented at twice the rate of low-frequency calls, the preferences were reversed. On the other hand, *H. versicolor* females continued to prefer calls of average frequency over higher- or lower-frequency calls even when the sound pressure level of the average-frequency calls was reduced by 6 dB (Gerhardt and Doherty 1988). Gerhardt, Dyson, and Tanner (1996) found that in natural choruses of this species, variation in dominant frequency of calls among males in the population was rarely large enough to elicit intensity-independent preferences by females.

All of these experiments. taken together. suggest that in most frogs, any selection on dominant frequency is more likely to be weakly directional or stabilizing than strongly directional (Gerhardt 1991, 1994a; Gerhardt and Huber 2002). In some anurans, however, there appears to be little frequency discrimination by females. Most experiments with toads, for example, have found little evidence of frequency selectivity by females (Arak 1983b, 1988b; B. K. Sullivan 1983b; Höglund and Robertson 1988; B. K. Sullivan 1992; Cherry 1993; Wagner and Sullivan 1995). When such selectivity was evident, it was affected by other variables, such as degree of call overlap (R. D. Howard and Palmer 1995). Lopez and Narins (1991) also failed to find evidence of frequency discrimination in *Eleutherodactylus coqui.*

Other studies have shown that any frequency discrimination evident in two-speaker playback experiments often disappears or is modified in more complex acoustic environments. This was first demonstrated by Gerhardt (1982), who found that the ability of *Hyla versicolor* females to discriminate between calls of different dominant frequency diminished considerably when they were tested in a four-speaker arena. Márquez and Bosch (1997a) used a computer to broadcast calls of different frequencies from an array of seven speakers to females of two species of midwife toads. One species, *Alytes obstetricans,* showed a weak preference for lower-than-average frequency calls. Many females chose the other alternatives, however, and responses of individual females were not consistent. In the multispeaker array, the other species, *A. cisternasii,* did not show a significant preference for low-frequency calls, and again, repeated choices by the same females were not consistent.

In *Hyperolius marmoratus,* females tested in two-speaker playback experiments exhibited a clear preference for low-frequency over high-frequency calls (Dyson and Passmore 1988; Jennions, Backwell, and Passmore 1995), but detailed field studies showed no evidence of a consistent mating advantage for large males (Telford and Dyson 1988; Dyson et al. 1992). Studies of animals in experimental enclosures revealed that large males enjoyed a mating advantage in small choruses, but male mating success was random with respect to size in larger choruses (Telford, Dyson, and Passmore 1989). Other experimental studies showed frequency discrimination by female *Hyperolius* diminished if males were very close together (Dyson and Passmore 1992a; see also Schwartz and Gerhardt 1989). Discrimination also can be affected by the temporal relationship of calls (Dyson and Passmore 1988; Dyson, Henzi, and Passmore 1994). The tendency for females to approach a speaker in playback experiments can be influenced by the presence of suitable perches near the speaker, especially on moonlit nights (Backwell and Passmore 1990). Possibly these frogs prefer not to approach males on the ground or in the water, where they are vulnerable to predators such as snakes (Grafe 1997a). Clearly, caution is required in extrapolating from simplified playback experiments to patterns of mate choice in more complex acoustic environments and situations where mate choice can be costly. Overall, the evidence that female anurans use differences in the dominant frequencies of calls to assess the body size of males when selecting mates is relatively weak. Frequency discrimination could be important in enabling females to avoid mating with males of the wrong species, but it might not contribute very much to variation in male mating success in natural choruses (Gerhardt 1994a; Schwartz and Gerhardt 1998; Gerhardt and Huber 2002).

### Female Choice and Male Calling Performance

Although many investigators have examined the importance of dominant frequency as a determinant of male mating success, other aspects of male calling performance probably are more important. In a natural chorus, females must locate males in a noisy and acoustically complex environment. Therefore, males that produce the loudest, longest, or most complex calls, or those that call most frequently, probably have an advantage in attracting mates (Wells 1977a 1988; Gerhardt 1982, 1988). This advantage need not be related to any genetic benefits accruing to females, but simply reflect the ease with which females perceive and locate males in a noisy chorus (Arak 1983a, 1988b; Halliday 1983a; G. Parker 1983; Wells and Schwartz 1984a; Grafe 1997a). Nevertheless, calling at high rates, or producing loud, long, or complex calls usually increases the energetic costs of calling (see chapter 5), so females might use differences in calling performance to assess differences in the quality of males in a chorus.

Females of many anurans will move toward speakers playing the louder call in playback experiments (Fellers 1979b; Arak 1983a, b, 1988b; Forester and Czarnowsky 1985; Gerhardt 1988, 1994a; Schwartz 1993; Bishop, Jennions, and Passmore 1995; Beckers and Schul 2004; Castellano, Rosso, and Giacoma 2004; and this would be expected of most species. Perceived call intensity depends on a female's proximity to a male, as well as the acoustic properties of the male's calling site (chapter 7). Call intensity per se probably provides the females with little information about the male, although in some species, larger males produce louder calls (Gerhardt 1975; Given 1987). Fellers (1979a) suggested that *Hyla versicolor* males calling from unobstructed sites might have an advantage in attracting females, but he failed to show that successful males occupied superior sites more often than unsuccessful males. Godwin and Roble (1983) found no relationship between mating success and the physical features of calling sites in the closely related species, *H. chrysoscelis*. On the other hand, males of *Hyalinobatrachium fleischmanni* that occupied elevated, unobstructed sites obtained more mates than did those at low, densely vegetated sites (Greer and Wells 1980); the higher sites were those most likely to enhance the broadcasting of calls (Wells and Schwartz 1982). Males of the Puerto Rican frog *Eleutherodactylus coqui* also exhibit a strong preference for elevated sites. Not only do elevated sites affect the propagation of calls (Narins and Hurley 1982), but such sites also are closer to preferred nest sites than are sites closer to the ground (Townsend 1989a).

The vocal interactions among males in a chorus also can influence their ability to acquire mates (Wells and Schwartz 1984a; Wells 1988). Males often increase calling rates, lengthen their calls, or add secondary notes to their calls when they hear the calls of other individuals or detect females nearby (see chapter 7 for a detailed discussion). Female frogs almost always prefer high calling rates to low rates (Whitney and Krebs 1975a; B. K. Sullivan 1983b; J. Schwartz 1986; Gerhardt et al. 1987; Wells and Bard 1987; Arak 1988a; Gerhardt 1988; Wells 1988; Lopez and Narins 1991; Passmore, Bishop, and Caithness 1992; B. K. Sullivan 1992; B. K. Sullivan and Hinshaw 1992; Cherry 1993; Jennions et al. 1995; Polakow et al. 1995; Wagner and Sullivan 1995; Kasuya et al. 1997; Smith and Roberts 2003a, b; Tárano and Herrera 2003). Females sometimes exhibit a preference for calling rates well above the normal range of variation for their species (Gerhardt 1987, 1991), but a preference for high calling rates can be eliminated if another species with a higher calling rate is present (K. Pfennig 2000). Forester, Lykens, and Harrison (1989) tested reactions of female *Pseudacris crucifer* in the field to variation in calling rate by setting up several speakers equipped with traps that captured females. There was an almost linear increase in number of females captured as calling rate increased, indicating that males investing more energy in calling (Taigen et al. 1985; Wells, Taigen, and O'Brien 1996) are more likely to acquire mates.

This is true for other species as well, because the energetic cost of calling generally is a linear function of calling rate (Pough et al. 1992; Prestwich 1994). In a study of *Hyla microcephala* in Panama, Schwartz, Ressel, and Bevier (1995) monitored the calling of several males in a chorus simultaneously with a computer system. Females selected the males with the highest calling rates first. Because females often arrive at the breeding site relatively late at night, males must continue to call for several hours to maximize their chances of mating. Calling at a high rate rapidly depletes the frogs' reserves of glycogen in their trunk muscles, so they must pause periodically to conserve energy. Frogs that were stimulated to call at unusually high rates, which would be most attractive to females, depleted even more of their energy reserves.

In *Hyla versicolor,* males increase call duration but they reduce calling rate in response to other males' calls (Wells and Taigen 1986; Schwartz, Buchanan, and Gerhardt 2002). This does not seem to increase aerobic costs, but could decrease the amount of time a male can call by depleting some types of energy reserves more quickly (Taigen and Wells 1985; Wells and Taigen 1986; Grafe 1997b). Females of this species prefer long calls to short calls, even when total calling effort (seconds of calling per hour) is equalized (Klump and Gerhardt 1987; Gerhardt, Dyson, and Tanner 1996). Females also prefer faster calling rates if call duration is equal, and longer calls if calling rate is equal. The same is true for *H. chrysoscelis* (Gerhardt 2005). Thus, females prefer calling patterns that require increased expenditures of energy. However, the ability of females to discriminate among calls of different durations is considerably reduced in a large chorus with high levels of background noise, but they do tend to avoid males giving very short calls (Schwartz, Buchanan, and Gerhardt 2001).

Females prefer complex calls (i.e., those with distinct secondary notes) in *Physalaemus pustulosus* (Rand and Ryan 1981), *Geocrinia victoriana* (Littlejohn and Harrison 1985), *Hyla ebraccata* (Wells and Schwartz 1984a), and *H. microcephala* (Schwartz 1986). In cricket frogs (*Acris crepitans*), males do not produce distinct secondary notes, but they respond to calls of neighboring males by increasing the number of call groups and the number of pulses per call. These changes have been interpreted as aggressive responses (see chapter 7), but they also make the calls more attractive to females (Kime, Burmeister, and Ryan 2004). It is not entirely clear whether females typically prefer longer or more complex calls because these calls provide females with information about male quality (M. Ryan 1983b), or because complex calls are more easily detected in a noisy chorus. In some species, such as *H. microcephala,* the energetic cost of calling is a linear function of the number of call notes produced, so the number of call notes may provide reliable information about male quality (Wells and Taigen 1989). In other species, such as *Physalaemus pustulosus,* addition of secondary notes does not appear to increase energetic costs (Bucher, Ryan, and Bartholomew 1982), so females would not necessarily gain information about male quality, but might be able to locate calls more effectively.

In most species tested with multinote calls, or calls of different durations, the preferences for more longer or more complex calls is open ended, so that calls that are longer than normal elicit a stronger response from females than do more typical calls. One species from Taiwan, *Rhacophorus moltrechti,* did not show this pattern. This species produces calls with an introductory note with a slow pulse rate, sometimes followed by a number of secondary notes with a faster pulse rate. Females preferred multinote calls of normal duration to calls with just the introductory note alone, but addition of a larger number of secondary notes did not make calls any more attractive (T.-U. D. Chen 2002).

While playback experiments often have demonstrated preferences by females for loud calls, high calling rates, long calls, or more complex calls, it has been surprisingly difficult to correlate individual variation in these aspects of calling behavior with male mating success in the field (B. K. Sullivan, Ryan, and Verrell 1995). In some cases, this probably is because the number of males observed mating for which detailed behavioral information was available was relatively small (e.g., B. K. Sullivan 1992). In other cases, preferences for certain call characteristics, such as high calling rate, can be overridden by stronger preferences for another characteristic, such as call intensity (Schwartz 1986; Arak 1988a; Gerhardt 1988, 1991, 1994a; M. Ryan and Rand 1990). Nevertheless, several studies have demonstrated significant correlations between male mating success and variation in calling rate or total number of calls produced (B. K. Sullivan 1983b; Arak 1988a; M. Morris and Yoon 1989; B. K. Sullivan and Hinshaw 1990; Passmore et al. 1992; Cherry 1993; Grafe 1997a; Kasuya et al. 1997; Pröhl and Hödl 1999; Pröhl 2003; Smith and Roberts 2003a).

In many natural choruses, females probably select their mates from a relatively small subset of males that are calling (M. Morris 1989), so comparisons of successful and unsuccessful males are best made within these subsets of males. Because neighbors often interact with one another and have closely matched calling performance (e.g., Wells 1988; Brush and Narins 1989; Bosch and Márquez 1996; Bendix and Narins 1999; Grafe 2003), variation within local subsets of males often is less than that of the chorus as a whole. Finally, the types of call characteristics that appear to be most subject to directional sexual selection, such as calling rate, call duration, call complexity, and perhaps even call intensity, are features that often vary considerably even within a single male (that is, they are dynamic call properties). Consequently, even when a clear preference can be demonstrated in playback experiments, the potential for sexual selection on these traits is limited by intramale variability. Correlations between male mating success and calling behavior can be difficult to detect, or can vary from one year to the next. Nevertheless, most recent studies of dynamic properties of anuran calls have revealed sufficient among-male variation for females to discriminate among individuals (Gerhardt 1991, 1994a; Runkle et al. 1994; B. K. Sullivan, Ryan, and Verrell 1995; Wagner and Sullivan 1995; Gerhardt, Dyson, and Tanner 1996; Welch, Semlitsch, and Gerhardt 1998; Docherty, Bishop, and Passmore 2000; Bee et al. 2001; Tárano 2001; Friedl and Klump 2002; Given 2002).

### Genetic Benefits of Female Choice

While there is abundant evidence that female anurans prefer certain male traits both in laboratory tests and under natural conditions, there is relatively little evidence establishing a clear link between such preferences and genetic benefits to their offspring. Several investigators have used laboratory experiments to determine whether choice of large males by females yields genetic benefits to their offspring, such as enhanced growth rates or higher survivorship of tadpoles fathered by large males. Although these experiments produced some evidence for paternal effects on fitness-related traits of tadpoles, there was no consistent tendency for offspring of larger males to have faster growth rates, shorter larval periods, larger size at metamorphosis, or higher survivorship. Therefore, these experiments do not provide strong evidence that indirect selection favors choice of large males by females (Woodward, Travis, and Mitchell 1988; S. Mitchell 1990; Howard, Whiteman, and Schueller 1994; Semlitsch 1994). Two studies that investigated the relationship between parasite infection, male calling performance, and male mating success found little effect on male fitness (Hausfater, Gerhardt, and Klump 1990; Tinsley 1990).

*Hyla versicolor* is one of the few frogs in which a preference for some aspect of calling performance has been linked to putative genetic benefits to offspring. Welch, Semlitsch, and Gerhardt (1998) reported that tadpoles fathered by males that gave long calls generally had higher growth and survivorship than did those fathered by males that gave short calls, but the performance of tadpoles varied with food availability and between years. Subsequent work showed that offspring of males with longer than average calls tended to feed more as tadpoles than did offspring of males with shorter calls (Doty and Welch 2001). The genetic mechanism that links male calling performance with tadpole performance is not known, however, and any genetic benefits accruing to offspring are likely to be heavily dependent on environmental conditions that they encounter as larvae. For example, Welch (2003) found that tadpoles fathered by males with long calls were slightly (8%) heavier than those fathered by males with short calls under low-density conditions, but this advantage disappeared at high density. Hence a consistent preference by females for long calls would not necessarily be advantageous. While females theoretically could alter their preferences for male traits if they could assess future environmental conditions for their offspring, it is not clear that female gray treefrogs can make such assessments.

## Sexual Selection on Male Traits

The mating systems of anurans described previously affect the way in which sexual selection operates on male traits, because different traits are likely to be favored depending on whether males compete directly for females, defend resources attractive to females, or call in choruses or leks to attract females. Sexual selection is just one component of selection operating on traits that affect overall male reproductive success or fitness (S. Arnold and Duvall 1994). The lifetime fitness of an individual male is determined by the number of offspring produced by all of his mates (fecundity) and the male's probability of dying in any given year (mortality; fig. 8.22). The number of offspring fathered by a male depends on the fecundity of each female with which he mates (fecundity per mate) and the number of mates he acquires (mating success). Any traits that determine the number of offspring a male produces through their effect on mating success can be considered sexually selected traits.

Males can influence the fecundity of females either by providing a nuptial gift of food (a behavior unknown in amphibians), or by providing parental care for her offspring (fig. 8.22). Male parental care would normally have a positive effect on survivorship of a male's offspring, but it can have either a negative or positive effect on a male's mating success. If the time devoted to parental care reduces a male's ability to advertise or search for additional mates, then the effect is negative. On the other hand, if females actually prefer to mate with males that have demonstrated their parental abilities by caring for the offspring of other females, then the effect can be positive. Even if females do not prefer males that are already caring for offspring, males would not suffer a major cost of parental care if they can continue to attract additional mates while performing parental duties, such as guarding eggs (see chapter 11).

Male mating success is affected by the amount of time required for a male to mate with one female and then locate another (mate cycling time) and by the male's ability to per-

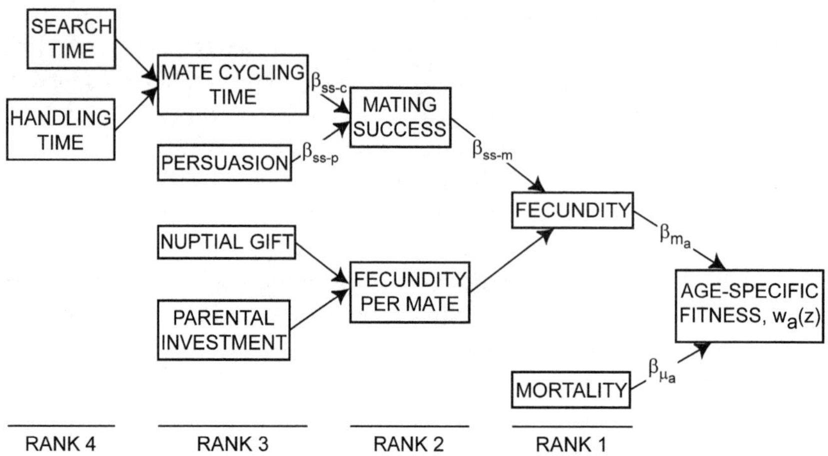


**Fig. 8.22.** Diagram of types of traits affected by sexual selection. Search time and handling time together constitute mate cycling time, a measure of how frequently a male can mate. Traits that influence mate cycling time and traits that influence a male's ability to attract or retain a female (persuasion) together affect mating success and are subject to sexual selection. Other traits combine to affect fecundity, or the number of eggs fertilized by a male. Fecundity and mortality together determine age-specific reproductive success. After Arnold and Duvall (1994).

suade each female to mate with him (persuasion; fig. 8.22). Mate cycling time is influenced both by the time required to locate mates (search time) and the time used for courtship and mating (handling time). Let's consider the behavior of an explosive breeding toad, *Bufo bufo,* as an example (N. Davies and Halliday 1979). Males actively search for females, and a male sometimes spends many hours or even several days searching for a mate. He will then spend several additional hours fertilizing the eggs as they are laid. The long search time and relatively long handling time imposed by the slow rate of oviposition by the female often means that a male will have few opportunities to mate again in a very short breeding season.

In a species with this type of explosive breeding behavior, any traits that reduce either search time or handling time would be favored by sexual selection. Search time might be reduced in a number of ways. One is for males to arrive early at the breeding site to intercept the first females to arrive. Males also might increase their rates of searching at the breeding pond, search a wider area to intercept females before they reach the pond, or concentrate their searching around oviposition sites or other areas where females are most likely to be found. Males also might attempt to displace males already in amplexus. Traits associated with enhanced physiological capacity for movement or calling, or those related to the size and strength of the male, would be favored. Handling time would include not only the time spent fertilizing eggs, but also any mate guarding, to prevent competitors from displacing a male from amplexus. In a species with external fertilization, a male probably has little control over how much time is devoted to fertilizing eggs, since this depends on how rapidly females lay their eggs. In contrast, traits correlated with male size and strength might reduce handling time by enhancing mate-guarding ability. For many explosive breeders, persuasion probably is of little significance in determining male mating success, because most females do not choose their mates.

Now let's consider a frog with a resource-defense mating system, such as *Rana clamitans.* In this species, search time can be considered equivalent to time spent advertising for mates, since males do not actually search for females. Any traits that reduce the amount of time needed to advertise for a mate will thus be favored by sexual selection. For example, any morphological or physiological traits that allow a male to produce louder calls or call at a higher rate would be favored. Once a female is attracted to a male's territory, handling time can be considerable if females spend time visiting several males before choosing a mate (Wells 1977c). Fertilization takes less than 15 minutes in this species (Aronson 1943b), so this adds little to handling time. Defense of an oviposition site can be considered a form of persuasion. A male with a high-quality territory is likely not only to persuade a female to mate with him relatively quickly, but he also is likely to acquire more mates than a male with a low-quality territory. In species with resource-defense mating systems, male traits associated with successful defense of territories, such as size and strength, probably are more important than traits related to calling ability. In a lek species, such as *Hyla versicolor,* persuasion is likely to take the form of enhanced calling ability, but traits associated with defense of mates or territories, such as large body size, enlarged forelimb muscles, or elaborate nuptial pads and spines, are less likely to be important. In terrestrial breeders, such as dendrobatid frogs, a male's ability to persuade a female to mate with him could depend both on his calling or courtship behavior and his ability to provide a suitable oviposition site.

### Measuring Sexual Selection in Anuran Populations

Until recently, most studies of sexual selection were designed to demonstrate that males with certain phenotypic characters are more successful than others at acquiring mates (Andersson 1994), and many studies of anurans have followed this approach. Most investigators have not collected data that allow them to directly estimate the strength of sexual selection acting on particular traits, although this approach has been used more frequently in recent studies (B. K. Sullivan, Ryan, and Verrell 1995). To understand how selection operates, we need measurements of the fitness of individuals in natural populations, as well as methods for estimating the effects of selection on various components of fitness. These statistical methods will not be discussed in detail here, but a number of good summaries are available (Arnold 1983a, b; R. D. Howard 1983; Arnold and Wade 1984a, b; W. Sutherland 1985; D. Brown 1988; Clutton-Brock 1988b; Grafen 1988; Koenig, Albano, and Dickinson 1991; Arnold and Duvall 1994).

In considering how sexual selection operates in nature, it is important to distinguish between selection and the evolutionary response to selection (Wade and Arnold 1980; Arnold and Wade 1984a). The former refers to changes in the statistical distribution of phenotypic characters within a generation. For example, if the size distribution of males that acquire mates differs from that of the entire male population, then selection has occurred. However, the evolutionary effect of selection on a phenotypic character (e.g., body size) depends on the way in which that character is inherited. In some cases, as when mating success is correlated with male body size, much of the variance in body size is related to differences in age or to environmental effects, such as the amount of food acquired by each male. Knowledge of the genetic basic of most phenotypic characters, particularly those directly related to fitness, is virtually nonexistent for amphibians, so my discussion will refer mainly to the opportunity for sexual selection. It also is important to distinguish the effects of sexual selection and natural selection, since these work either in the same direction or in opposite directions. The net effect of successive episodes of sexual or natural selection acting on a phenotypic character is referred to as the *total opportunity* for selection (Arnold and Wade 1984a, b).

The key to measuring the opportunity for sexual selection acting on male traits is to collect data on variation in male mating success (Clutton-Brock 1988). Such data can be obtained either through cross-sectional or longitudinal studies (Arnold and Wade 1984b). In some studies of anuran mating systems, only cross-sectional data have been collected. The usual approach is to collect samples of mated and unmated males on a few nights of the breeding season. The samples are then compared with respect to some phenotypic character, such as body size or calling rate (Licht 1976; Wilbur, Rubenstein, and Fairchild 1978; Gittins, Parker, and Slater 1980; Fairchild 1981; Gatz 1981b; J. Lee and Crump 1981; Woodward 1982c; Hemelaar 1983; Crump and Townsend 1990, Lopez and Narins 1991). Such data are relatively easy to collect, but they provide little information about the mating system. It is not possible to estimate total variation in reproductive success from cross-sectional data alone. This type of data can be used to estimate opportunities for sexual selection only if the relative fitness of different phenotypic classes (e.g., different size classes) does not change over time and the phenotypic character being considered does not change with age (Arnold and Wade 1984b). Both assumptions seem unrealistic for most anuran populations, because large and small males often arrive at different times, and body size generally increases with age.

The most useful data on variation in reproductive success are those obtained from longitudinal studies of marked individuals throughout a breeding season. There are problems associated with this approach as well. First, many species are not very good subjects for studies of mating success because one cannot monitor the behavior of individuals with sufficient accuracy to be sure that all matings are observed. It is very difficult to correlate male mating success with other variables when the number of matings observed in a population is very small (e.g., Fellers 1979b; Lopez and Narins 1991; B. K. Sullivan 1992; Bertram, Berrill, and Nol 1996). Second, there is the problem of choosing an appropriate measure of reproductive success. For most anurans, the appropriate measure is the number of matings obtained by each male, since variation in mating success is the best measure of the intensity of sexual selection. Other measures, such as the number of eggs fertilized, number of hatchlings produced, or the number of offspring surviving to reproductive age (R. D. Howard 1979) often are not practical, and they measure variables affected by both sexual selection and other components of selection, such as fecundity selection (Arnold and Wade 1984b). In species in which direct sperm competition is a possibility, as in rhacophorid or phyllomedusine frogs with multimale matings, measuring mating success accurately is difficult without genetic analyses of parentage.

A third problem is the time over which estimates of reproductive success are made. Most studies of reproductive success in anurans and most attempts to calculate selection intensities or opportunities for sexual selection have considered only selection within breeding seasons (Kluge 1981; B. K. Sullivan 1983b, 1986b; Arnold and Wade 1984b). Opportunities for selection vary considerably from pond to pond or from year to year in the same population (B. K. Sullivan 1986b), but the intensity of sexual selection over the lifetime of an individual cannot be estimated from these data. Consequently, data on variation in lifetime reproductive success of both males and females are needed for a complete understanding of sexual selection (Wade 1979; Wade and Arnold 1980; Clutton-Brock 1983, 1988; Arnold and Wade 1984b). Unfortunately, such estimates are difficult to obtain and are available for only a few animal species (Clutton-Brock 1983, 1988). These data indicate that seasonal and lifetime estimates of reproductive success are not always tightly correlated, especially if reproductive success is strongly affected by age (Clutton-Brock 1983).

Actual data on lifetime reproductive success are not available for any anuran species, but R. D. Howard (1983, 1988a) used data on age-specific survivorship and reproductive success to estimate lifetime reproductive success for *Rana catesbeiana* and *R. sylvatica* from computer simulations. In both species, seasonal reproductive success of males was heavily skewed, with many individuals producing no offspring, and a few males producing many. The number of unsuccessful males was especially high in *R. sylvatica* because of a heavily male-biased sex ratio at the breeding pond (R. D. Howard and Kluge 1985; R. D. Howard 1988a). In *R. catesbeiana,* some males mated several times in a season while others never mated, and male reproductive success was much more variable than that of females (fig. 8.23).

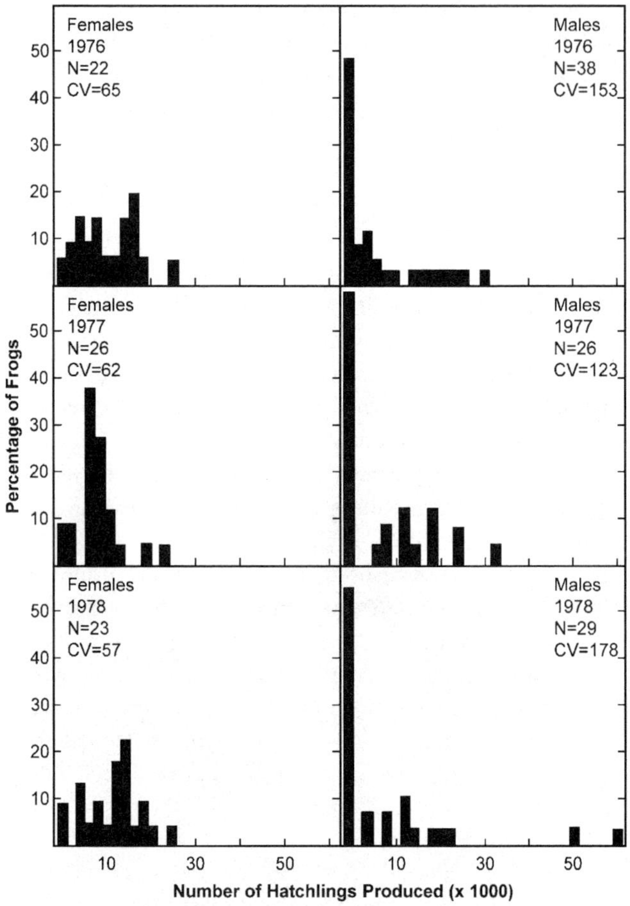


**Fig. 8.23.** Comparison of seasonal estimates of the number of offspring produced by male and female bullfrogs (*Rana catesbeiana*) in Michigan. The higher variance in males was due primarily to some individuals acquiring several mates in one season and only secondarily to the fecundity of their mates. After R. D. Howard (1983).

In both species, estimates of variation in lifetime reproductive success, measured as the number of zygotes produced, were much greater in both sexes than seasonal variation in reproductive success. This was due to the strong influence of survivorship on lifetime estimates. However, the difference between males and females decreased in both species when lifetime estimates were used. This does not necessarily mean that the intensity of sexual selection acting on the two sexes is similar. Since lifetime reproductive success in both species is strongly influenced by adult survivorship (a component of natural selection), variation in lifetime reproductive success alone provides a poor estimate of the opportunity for sexual selection (R. D. Howard 1988a). The importance of sexual selection over the lifetime of the animal can be estimated only if the relative effects of natural and sexual selection are partitioned in the same way as the data from a single season (Arnold and Wade 1984b; Koenig, Albano, and Dickinson 1991).

An additional problem in assessing the potential effect of sexual selection on phenotypic characters is a lack of information on how different phenotypic traits are correlated with one another. Most studies of sexual selection in anurans and other animals have tended to focus on single characters, particularly body size (Arnold and Wade 1984b; Koenig, Albano, and Dickinson 1991). Occasionally, behavioral traits such as calling rate are considered as well (e.g., B. K. Sullivan 1983b). Different phenotypic traits often are either positively or negatively correlated with one another, and selection can affect one trait indirectly by acting on a correlated character (Arnold and Wade 1984b). This makes statistical analysis of the data much more complicated, because covariance between phenotypic traits must be considered. To date, very few studies of sexual selection in anurans have adequately documented the relationship of reproductive success to morphological characters other than body size.

Arnold and Wade (1984b) used R. D. Howard's (1979) data on one season's mating success in bullfrogs (*Rana catesbeiana*) to estimate the relative contribution of sexual and natural selection to total selection on male body size. Variation in number of matings provided an estimate of the opportunity for sexual selection, whereas variation in number of eggs per mate and hatching success provided estimates of fecundity selection. Arnold and Wade concluded that sexual selection was the major force favoring large body size in male bullfrogs, accounting for 75% of the total selection differential (i.e., the change in mean body size after selection). In contrast, fecundity selection, acting through number of eggs per mate and hatching success, accounted for only 25% of the total selection differential. To estimate relative contributions to the total opportunity for selection on male body size, it also was necessary to consider covariation between the different measures of fitness (number of mates, number of eggs per mate, hatching success). Arnold and Wade found that sexual selection accounted for 59% of the total opportunity for selection on male body size. Lifetime reproductive success would be heavily influenced by size- or age-specific survivorship of males, a component of selection not considered in this single-season analysis.

## Sexual Selection and Sexual Dimorphism

Darwin (1871) originally proposed his theory of sexual selection to account for the evolution of sexually dimorphic characters that he believed could not be explained by ordinary natural selection. He recognized that polygynous species (those with high variance in male reproductive success) generally exhibit greater sexual dimorphism than monogamous species. This pattern has been supported by subsequent studies of birds and mammals (Selander 1972; Clutton-Brock, Harvey, and Rudder 1977; Clutton-Brock, Albon, and Harvey 1980; Alexander et al. 1979; Clutton-Brock 1983; Payne 1984), although there are exceptions (Ralls 1976, 1977). Several investigators have discussed the relationship of sexual dimorphism to social behavior in amphibians and reptiles (Shine 1978, 1979; Wells 1978b; Berry and Shine 1980; R. D. Howard 1981b; Woolbright 1983, 1989; R. D. Howard and Kluge 1985; Halliday and Verrell 1986; J. Lee 1986; Halliday and Tejedo 1995; Monnet and Cherry 2002), but much remains to be done. Most anurans exhibit some type of sexual difference in morphology or coloration that allows males and females to be readily distinguished, but in many cases, the functional significance of these differences is not well understood. The most noticeable difference in most species is the presence of vocal sacs in males, a trait obviously related to calling behavior. Many species also are sexually dimorphic in body size or exhibit sexual differences in limb morphology, muscle structure, skin color, skin texture, and other traits (Noble 1931; Duellman and Trueb 1986). I will focus primarily on traits that seem most likely to be related to sexual selection, even when the functional significance of the traits is unclear.

### Sexual Dimorphism in Body Size

Because many anurans continue to grow throughout their lives, interpreting patterns of sexual size dimorphism presents some problems. Apparent differences in body size in a population can reflect different age structures brought about by differential predation on the two sexes or delayed maturity in one sex (R. D. Howard 1981b; Márquez, Esteban, and Castanet 1997; Monnet and Cherry 2002; Gramapurohit, Shanbhag, and Saidapur 2004a). If so, then sexual size dimorphism would not be due to selection for different growth rates. For example, Monnet and Cherry (2002) presented a statistical analysis of size and age, controlled for phylogeny, in a variety of anuran species. They found a strong correlation between the average difference in age between males and females in a population and degree of sexual dimorphism. In most populations included in their analysis, females were older than males and reached larger sizes. This suggests that females often delay sexual maturity and continue to grow, or suffer lower rates of mortality. Monnet and Cherry (2002) concluded that sexual selection is of minor importance in determining patterns of sexual dimorphism in anurans. Nevertheless, early sexual maturity in males can itself be a sexually selected trait that could result in males being smaller than females. Furthermore, their sample included very few species in which males are larger than females, so a role for sexual selection in the evolution of large male body size cannot be ruled out.

In contrast to many birds and mammals, females of most anuran species are considerably larger than males (Shine 1979; Monnet and Cherry 2002). Selection probably favors large body size in females because larger females produce larger clutches of eggs (see chapter 10). In some species, however, males are as large as or larger than females. In some populations of *Bombina bombina* and *B. variegata*, males are as large as or larger than females and exhibit sexual dimorphism in other traits, such as hind leg length and head width. The functional significance of these differences is not clear, however (Radojičić et al. 2002). In temperate-zone ranids, females of explosive breeders such as *Rana sylvatica* and *Rana temporaria* are considerably larger than males (Wells 1978b; R. D. Howard 1980; Berven 1981; R. D. Howard and Kluge 1985; Ryser 1986). In contrast, males and females of territorial species such as *Rana catesbeiana* and *R. clamitans* grow at similar rates and reach similar maximum sizes (Martof 1956a; Durham and Bennett 1963; Schroeder and Baskett 1968; Wells 1978b; R. D. Howard 1981b). In these species, large males win more fights than small males, and spend more time defending territories (fig. 8.24). The same is true for *Hyla rosenbergi* and related species that engage in territorial combat, and again, males tend to be as large as or larger than females in these species (Kluge 1981; Martins and Haddad 1988). Males of the hylid genus *Plectrohyla* also tend to be as large as or slightly larger than females (Duellman and Campbell 1992). These frogs have not been observed fighting, but they have several other morphological features that suggest that male combat is common (see the following). In the African frogs *Pyxicephalus adspersus* (fig. 8.25; Stewart, 1967; T. Hayes and Licht 1992) and *Petropedetes johnstoni* (H. Parker 1936), males are much larger than females. The size differences in *Pyxicephalus* results from higher growth rates in young males than in young females (T. Hayes and Licht 1992). Males are known to fight vigorously among themselves during the breeding season (B. Balinsky and Balinsky 1954; Channing, du Preez, and Passmore 1994). In several species of *Leptodactylus,* some of which are known to exhibit male combat, males are at least as large as females (Shine 1979).

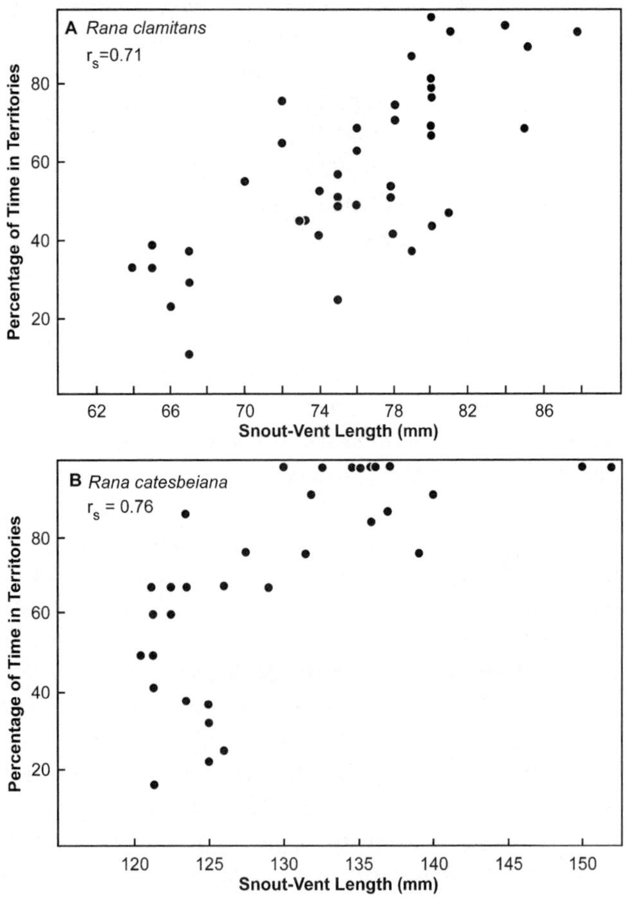


**Fig. 8.24.** Amount of time spent defending territories as a function of male size in two prolonged-breeding North American ranid frogs. (A) *Rana clamitans* from upstate New York. (B) *Rana catesbeiana* from Michigan. Plotted from data in (A) Wells (1976a) and (B) Emlen (1976).

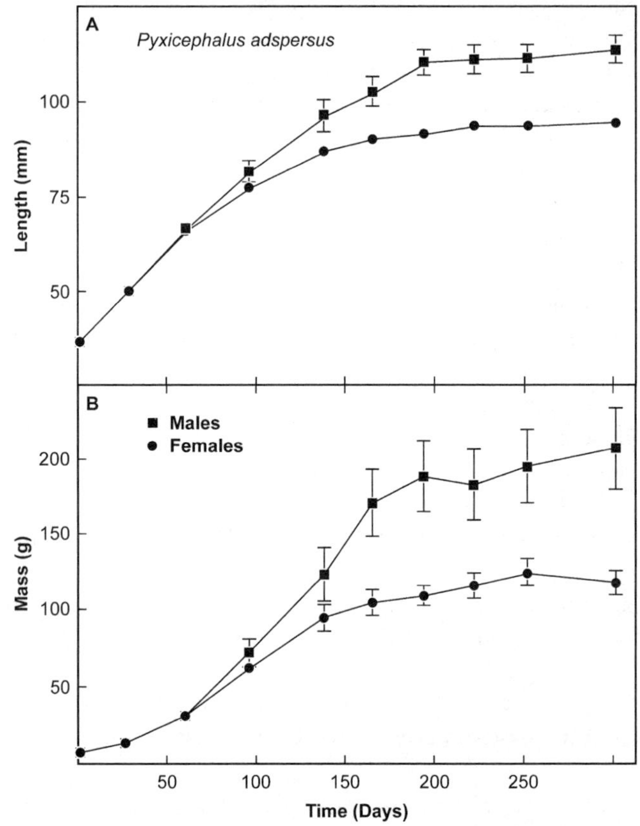


**Fig. 8.25.** Sexual dimorphism in the African ranid frog *Pyxicephalus adspersus.* (A) Rate of growth in length in males and females. (B) Rate of growth in mass in males and females. After about two months, males grow at much faster rates than do females and reach larger asymptotic sizes. After T. Hayes and Licht (1992).

Similar sexual dimorphism in body size is found in the fanged ranid frogs (*Limnonectes*) of Southeast Asia (the current taxonomy of this group is confusing, with numerous unnamed species currently grouped together under the same names; Emerson and Ward 1998; Emerson et al. 2000; Emerson 2001a; B. Evans et al. 2003). In most species, males are larger than females, have wider heads, and are equipped with hypertrophied jaw muscles and fangs on the lower jaw (fig. 8.26 A). Males of several species of fanged frogs have been observed biting one another in fights over territories around nest sites (Orlov 1997; Tsuji and Matsui 2002). There is variation among species, however, in the degree of sexual dimorphism. The ancestral condition for the clade appears to be male body size larger than female body size, with large fangs in males, but both smaller fangs and larger female body size have evolved multiple times in the group. Although males are larger than females in all species with large fangs, males also are larger in some species with smaller fangs (Emerson 2001a). The evolution of larger body size in males and changes in the size and shape of the head probably are related to sexual selection for fighting ability in males and not to ecological differences such as

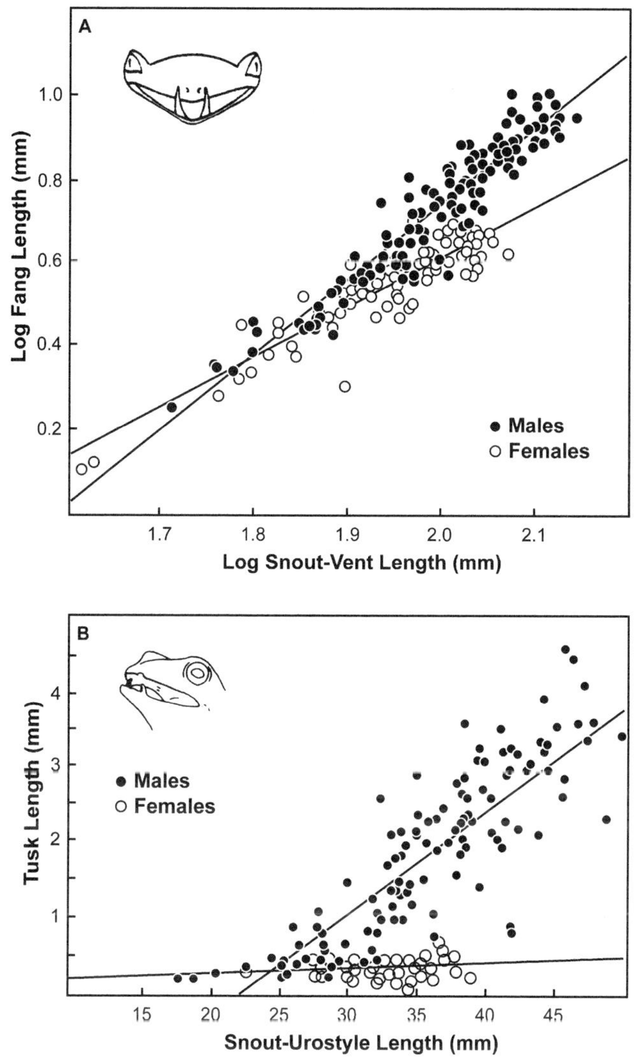


**Fig. 8.26.** Sexual dimorphism in weapons used for fighting by male frogs. (A) Sexual dimorphism in fang length in *Limnonectes leporinus*. After Emerson and Voris (1992). (B) Sexual dimorphism in tusk length in *Atelotus brevis*. After Katsikaros and Shine (1997).

prey selection (Emerson and Inger 1992; Emerson and Voris 1992; Emerson 1994, 1996; Emerson and Ward 1998). Conversely, the evolution of female body size larger than male body size appears to be associated with the evolution of small fangs in males, small body size, and a switch to terrestrial modes of reproduction (Emerson 2001a).

An analysis of sexual dimorphism in the Australian myobatrachid *Atelotus brevis* similarly suggested that sexual selection rather than ecological differentiation is responsible for differences between males and females (Katsikaros and Shine 1997). Males grow larger than females, have larger heads, and are equipped with tusks in the lower jaw, used for fighting (fig. 8.26 B). Males and females eat somewhat different types of prey, but this appears to reflect differences in microhabitat selection rather than differences in body morphology. Males reach larger sizes than do females in several other groups of frogs in which males probably fight with one another, but their behavior has been poorly studied. These include the Asian ranids *Paa* (= *Rana*) *boulengeri* and *P. yunnanensis* (= *Rana phrynoides;* C. Liu 1950) and some Asian stream-breeding megophryid frogs, including some species of *Scutiger* and species in the genus *Vibrissaphora* (Ho et al. 1999). Males of the latter genus are known to build nests under rocks in streams and almost certainly are territorial, but the behavior of *Scutiger* is not well known.

The relationship between male combat and large male size is not very robust when all anurans are considered. Shine (1979) suggested that males should be as large as or larger than females in species with male combat, but this certainly is not true for most small hylids. Many of these were classified as nonfighting species by Shine, but in fact, most probably exhibit aggressive defense of calling sites (table 8.1). Actual physical combat tends to be infrequent, however, with most aggressive encounters being resolved through exchange of vocalizations. Consequently, body size could have relatively little effect on a male's ability to maintain a territory, and therefore little influence on male reproductive success. Patterns of sexual dimorphism cannot be predicted from information on mating systems alone; data on the way in which phenotypic characters influence reproductive success in the two sexes are needed as well, because it is these characters that are subject to sexual selection (Clutton-Brock 1983). Furthermore, even when sexual selection favors large body size in one sex or the other, other selective pressures can favor small body size, so the evolutionary outcome is difficult to predict (Arnold 1983b; Arnold and Wade 1984b; Halliday and Verrell 1986; Halliday and Tejedo 1995).

Woolbright (1983) proposed an explanation for patterns of sexual size dimorphism in anurans based on the combined effects of sexual selection and energetic constraints on males and females. He suggested that the ratio of male to female size should be related to the proportion of the growing season used for reproductive activities, as well as the relative advantage enjoyed by large males in obtaining mates. Woolbright's model is based on several assumptions: (1) that sexual selection always favors large body size in male anurans, (2) that selection for increased clutch size always favors large body size in females, and (3) that energetic constraints on males limit growth rates and hence adult body size.

Woolbright predicted that females should be larger than males in explosive breeders from the temperate zone because the intensity of sexual selection acting on males should be relatively weak, so selection for increased body size would not be very strong. He predicted that in prolonged breeders from the temperate zone, males would be as large as or larger than females, because sexual selection for large body size would be more intense. However, Woolbright argued that in tropical species, a very long breeding season would impose severe energetic constraints on males, so growth rates would be limited and males could not reach a size comparable to females.

Woolbright (1989) demonstrated that *Eleutherodactylus coqui* males raised in the laboratory grew at rates equivalent to those of females and reached a similar size, whereas males in the field had much slower growth rates and were smaller than adult females. The energy limitation hypothesis also is consistent with a study by Given (1988a) showing a direct tradeoff between calling rates and growth rates of small *Rana virgatipes* males. Nevertheless, a number of the assumptions of Woolbright's model do not appear to be valid for all anurans, and therefore the model probably is too simplistic to provide a general explanation of sexual dimorphism for all species (B. K. Sullivan 1984).

One problem is that Woolbright failed to consider increased energetic investment in multiple egg clutches by females of prolonged breeding species, so it is not necessarily true that males of these species expend more energy relative to females than do males of explosive breeding species. Furthermore, the long breeding seasons of many tropical species do not necessarily translate into greater energetic constraints on male growth; some males simply abandon reproductive activities for extended periods to replenish energy reserves. Males of different species also exhibit a variety of calling tactics, some probably very expensive and some less so (see chapter 5). Finally, the assumption that sexual selection always favors large male body size is not necessarily valid, nor is there any clear evidence that sexual selection on male body size is stronger in tropical species with long breeding seasons. In fact, the most convincing evidence for a large-male mating advantage comes from studies of explosive breeding species from the temperate zone, just those species for which Woolbright predicted sexual selection on body size should be the weakest. Even patterns of sexual dimorphism reported in the literature are not always consistent with the model. For example, in some tropical frogs with long breeding seasons, males are smaller than females, while in others they are similar in size (Crump 1974).

## Sexual Dimorphism in Limb Morphology

In addition to differences in body size, anurans exhibit many other forms of sexual dimorphism. One common type is a difference in the proportions of the forelimbs, with those of males generally being larger or more robust than those of females. In many explosive breeding ranids, including *Rana sylvatica, R. pipiens,* and *R. temporaria,* the male's arms completely encircle the female during amplexus, with his front toes entwined under the female's body (Banta 1914; Noble and Aronson 1942; R. Savage 1961). This enables a male to hold on firmly while repelling other males attempting to steal his mate. R. D. Howard and Kluge (1985) hypothesized that sexual selection should favor elongation of the arms in males, and measurements of *Rana sylvatica* revealed that males have longer arms relative to body length than females (fig. 8.27). In contrast, males of the territorial species *R. clamitans* do not completely encircle the female during amplexus, and there is no evidence that males have longer arms than females (fig. 8.28 bottom). Males do wrestle with each other, and arm strength probably is a major determinant of fighting success. Measurements of forelimb proportions revealed that both upper arm diameter (fig. 8.28 top) and forearm diameter are greater in males than in females (Wells, unpublished data).

Tests of Howard and Kluge's hypothesis with other species have yielded mixed results. R. D. Howard (1988b) found no evidence that forelimb length influenced male mating success in *Bufo americanus,* and there was no sexual dimorphism in limb length. J. Lee (1986) found that in two explosive breeding toads, *Bufo terrestris* and *B. marinus,* males in amplexus had relatively longer forelimbs than unmated males, but only in *B. marinus* was this reflected in a slight sexual dimorphism in limb length. No such relationship was found in two other toads, *B. quercicus* and *B. valliceps,* species in which direct competition among males for possession of females was thought to be rare.

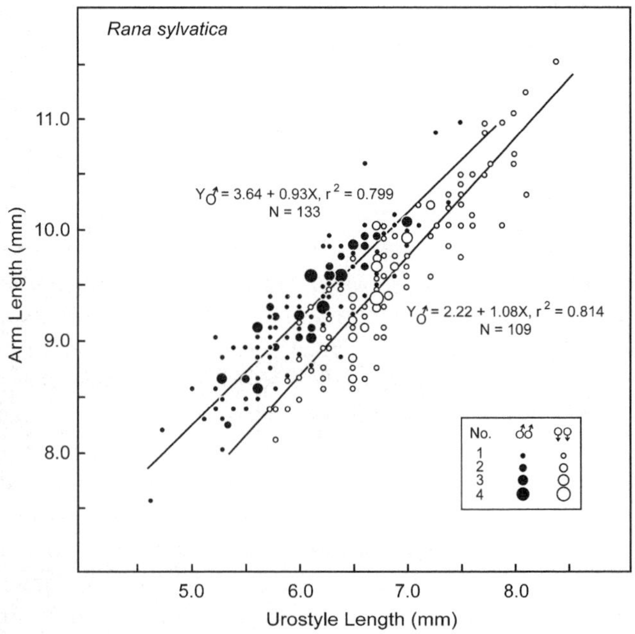


**Fig. 8.27.** Sexual dimorphism in arm length in *Rana sylvatica.* Males have longer arms relative to body length than do females. Here, the length of the urostyle is used as an index of body size. After R. D. Howard and Kluge (1985).

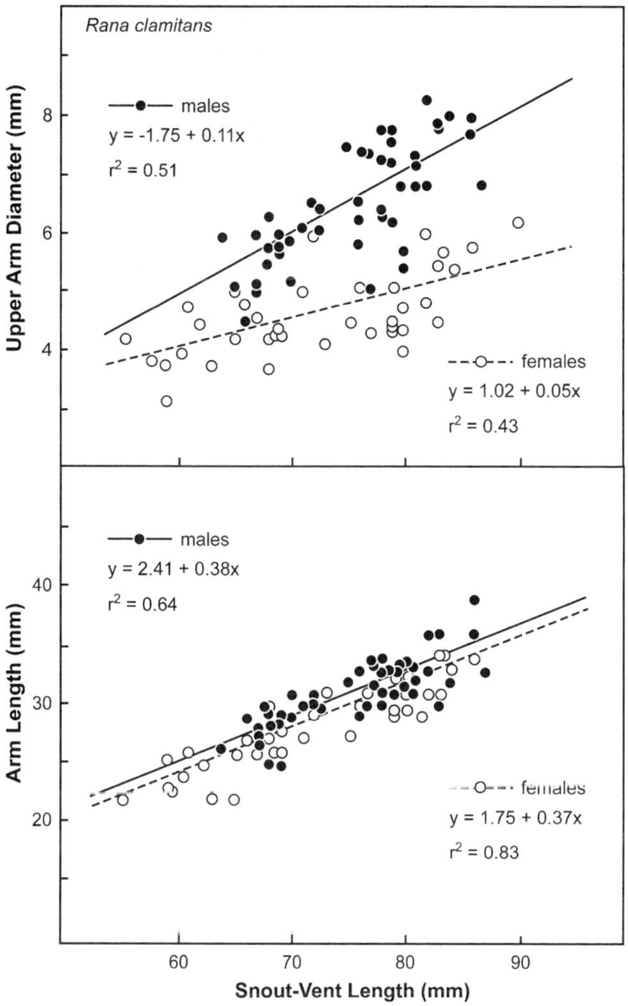


**Fig. 8.28.** Sexual dimorphism in upper arm diameter (top) and arm length (bottom) in *Rana clamitans.* Males have more robust but not longer arms than do females. Data are based on measurements of preserved specimens in the Cornell University collection (K. D. Wells, unpublished data).

There also was no evidence that arm length influenced mating success in an explosive-breeding hylid, *Triprion petasatus,* nor was there any evidence of sexual dimorphism in limb length. The same was true for another hylid, *Hyla marmorata,* but the mating system of this species has not been studied in detail.

One difficulty with these tests is that classifying the species simply by their mating systems obscures major differences in clasping behavior that in turn reflect major differences in the functional morphology of the forelimbs. Most authors simply categorize amplexus in anurans as being axillary (clasping the female under the forelimbs) or inguinal (clasping the female in front of the hind limbs; Duellman and Trueb 1986). Variation in clasping behavior actually is much more complex (Jameson 1955a). In most species of *Bufo,* for example, the male does not completely encircle his mate with the forelimbs, but instead presses his "fists" into the axillae of the female (R. Savage 1934; M. Smith 1969). Consequently, a male's ability to retain possession of a female probably is related more to forelimb strength than to its length, and this in turn is reflected in more massive development of certain muscles involved in holding the female (Emerson 1991). Indeed, subsequent work by Lee (2001) on *Bufo marinus* showed that forelimbs of males are much more robust than are those of females, and males with large forelimbs were more likely to be found in amplexus. Males of this species also have more robust hind limb muscles than do females, but it is not clear whether this difference is a result of sexual selection for more powerful leg muscles or a side effect of hormonally controlled hypertrophy of muscles in general (Lee and Corrales 2002).

In many anurans, the arm of the male has a greater diameter relative to its length than that of the female, and the front feet often are bent inwards (C. Liu 1936). This reflects greater development of the *flexor carpi radialis* muscle in males, as well as several other muscles involved in flexing the hand and forearm (Muller, Galavazi, and Szirmai 1969; Melichna et al. 1972; Thibert and Nicolet 1972; Kirby 1983; Rubenstein, Erulkar, and Schneider 1983; Oka et al. 1984; Yekta and Blackburn 1992; Peters and Aulner 2000; Lee 2001). Not only are these muscles larger in males than in females, but they also have more slow-tonic fibers and a greater capacity for prolonged tonic contraction, which enables the male to hold onto the females for hours or even days (Wacholder and von Ledebur 1930; Rubenstein, Erulkar, and Schneider 1983; Peters and Aulner 2000). The role of various forelimb muscles in clasping has scarcely been investigated, and indeed, comparative studies combining observations of clasping behavior with detailed functional analyses of limb morphology offer a wide open field for future investigators (Emerson 1991).

Even more dramatic examples of sexual dimorphism in limb morphology can be found in some tropical anurans, but often we have only sketchy information on their behavior. In most large species of *Leptodactylus,* including *L. pentadactylus, L. insularum, L. ocellatus,* and *L. chaquensis,* males are as large as or larger than females (Gallardo 1964; Heyer 1979b; Cei 1980) and have greatly enlarged arms compared to those of the female (fig. 8.29). The huge size of the arms in males mainly results from hypertrophy of the forearm flexor muscles. For example, in *L. ocellatus,* the *flexor carpi radialis* muscle can be nearly 20 times more massive in males than in females of the same body size, but there is little or no dimorphism in forearm extensor muscles (Scow 1965). These muscles are anchored on wide flanges that develop on the humerus of the male that are absent in females (Lynch 1971; Duellman and Trueb 1986). Similar enlargement of forelimb muscles and humeral flanges are

**Fig. 8.29.** Sexual dimorphism in the size of the arm in *Leptodactylus insularum* from Panama. (A) Male. (B) Female. Males develop hypertrophied arms in the breeding season and probably wrestle with one another. Photos by Kentwood D. Wells.

found in hylid frogs of the genus *Plectrohyla* as well (Duellman 1970; Duellman and Campbell 1992). Hypertrophied forelimbs are characteristic of males in several other groups of stream-breeding hylids, including *Hyla armata* and the *Hyla pulchella* group (Duellman 1970). These hylids are not closely related to one another, so the enlarged forelimbs of males are thought to have evolved independently several times (Duellman, de la Riva, and Wild 1997). Unfortunately, the behavior of these species is poorly known, so the function of the enlarged forelimbs is unknown, but it seems likely that males fight among themselves in at least some of these species.

In many frogs with hypertrophied forelimbs in males, there also are various kinds of sharp spines on the front feet, usually in the "thumb" region. For example, many species of *Leptodactylus* have sharp spines on the front feet, forelegs, or chest. Early workers (e.g., Noble 1931; C. Liu 1936) interpreted such structures as adaptations for enhancing a male's ability to hold onto a female during amplexus. In many species, however, enlargement of the forearms probably is associated with male combat, and males of several species are known to use spines to jab opponents in fights (Rivero and Esteves 1969; Wells 1977b; Shine 1979). The naturalist W. H. Hudson (1903) described the enormous arms and sharp spines of a male *Leptodactylus* and referred to the species as "the wrestler," but he believed these structures served mainly to protect the frog from its enemies. Large male size, enlarged arms, and spines also are characteristic of the Australian myobatrachid *Limnodynastes peronii*. J. Moore (1961) believed the male used the spines to clasp the female, but this species is known to be territorial and to engage in wrestling bouts during the breeding season (Clyne 1967, 1969).

Enlarged forelimbs, often armed with spines, are found in male frogs in other families around the world. For example, enormous forelimbs and a variety of spines are found in males of several Asian ranids, including *Paa boulengeri*, *P. phrynoides* (fig. 8.30), *P. spinosa*, *Rana holsti*, and *R. subaspersa* (Pope 1931; C. Liu 1935a, 1936, 1950; Inger

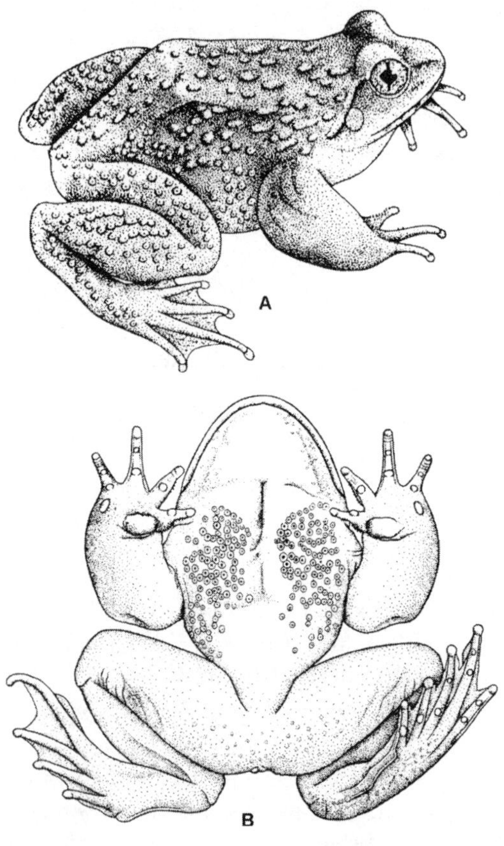

**Fig. 8.30.** Male of a large stream-breeding ranid frog, *Paa phrynoides*, from western China, showing (A) enlarged arms and (B) ventral spines. After C. Liu (1950).

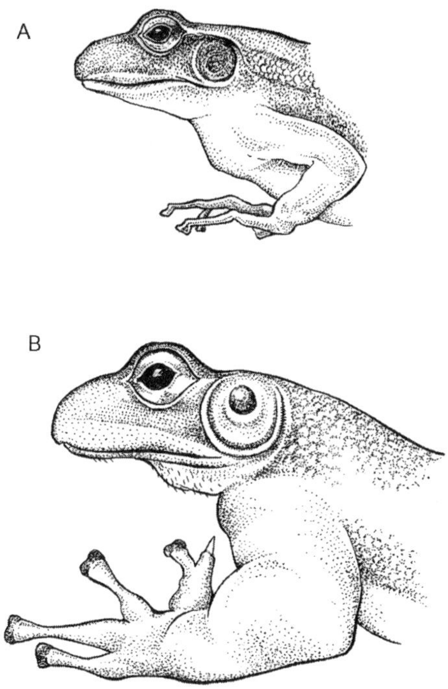


Fig. 8.31. Sexual dimorphism in the African frog *Petropedetes johnstoni.* (A) Female, showing slender arms. (B) Breeding male, showing robust arms and well-developed metacarpal spine. Many males also have scars that probably are derived from wounds received from the spines of other males during fights. Drawing by Mary Jane Spring after H. Parker (1936).

1947; Okada 1966). Unfortunately, their social behavior has not been studied. However, C. Liu (1950) reported that *P. boulengeri* males tend to remain in one place, with only one individual inhabiting a given pool; this suggests that males are territorial. *Rana subaspersa* lays eggs in mud nests similar to those constructed by *Hyla rosenbergi* (Utsunomiya and Utsunomiya 1983) and might be expected to have a similar mating system. Mature males of the African frog *Petropedetes johnstoni* also have enlarged forelimbs and sharp metacarpal spines (fig. 8.31). These characters apparently develop seasonally and are absent in mature, nonbreeding males (H. Parker 1936). Again, the social behavior of this species is unknown, but Sanderson (1936) reported that many nonbreeding males showed "laceration on the gular region, presumably relics of the preceding breeding period."

Sharp spines also are found in a number of groups of treefrogs. These include the nest-building gladiator frogs (*Hyla boans, H. faber, H. pardalis, H. rosenbergi,* and related species). The use of spines in combat by males in *Hyla rosenbergi* and related species was discussed earlier (see fig. 8.12). The forelimbs of the male are not greatly enlarged in these species (Kluge 1981). Prepollical spines are found in several other groups of Neotropical hylids that do not have enlarged forelimbs, including the *Hyla albomarginata* group, the *H. albopunctata* group, the *H. geographica* group, and the *H. granosa* group, but the function of the spines has not been determined in these frogs.

The *Hyla pulchella* group of Andean treefrogs and the *Hyla circumdata* group of southeastern Brazil both have sharp, prepollical spines as well as enlarged forelimbs in males, and on the basis of these characters, were combined into a single clade by Duellman, de la Riva, and Wild (1997). Again, the behavior of males and the use of the spines are unknown in these groups. The only other hylids that combine prepollical spines with enlarged forelimbs are frogs of the genus *Plectrohyla.* The prepollical processes in these frogs vary from blunt to very sharp in different species, suggesting they serve different functions. In two species with sharp spines, males in museum collections have scars on the head and forelimbs that suggest the use of the spines in combat (Duellman and Campbell 1992). Males of one South American hylid, *Hyla armata,* are unique in having clusters of spines on both the inner surface of the hand and on the humerus, as well as greatly hypertrophied forelimbs (fig. 8.32 D). It is not known whether these are used in male combat, to hold the female in amplexus, or both (Duellman, de la Riva, and Wild 1997). Sharp humeral spines also are found in centrolenid frogs of the genus *Centrolene,* and as discussed earlier, these apparently are used in fighting between males (Bolívar-G., Grant, and Osorio 1999).

Many male anurans that lack obvious weapons such as

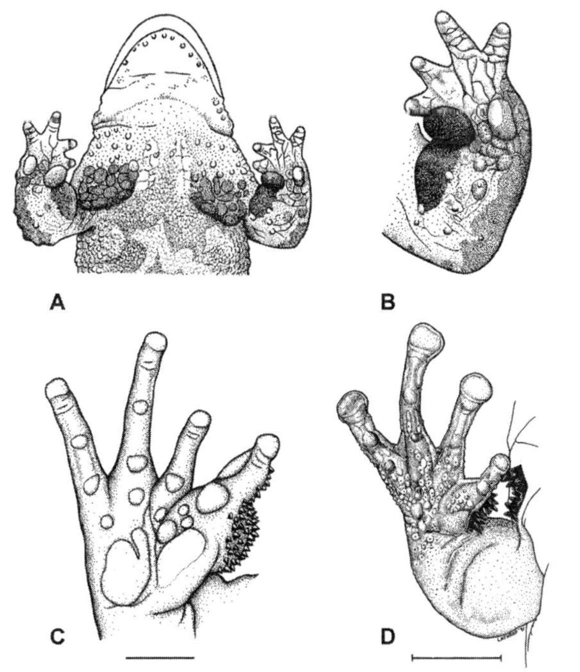


Fig. 8.32. Nuptial pads and spines on the feet of male anurans. (A, B) *Bombina maxima,* a bombinatorid frog that breeds in streams and pools in western China. (C) *Telmatobius truebae,* a stream-breeding leptodactylid frog from Peru. (D) *Hyla armata,* a stream-breeding hylid frog from the Andes of South America. (A, B) After Liu (1950); (C) After J. J. Wiens (1993); (D) After a drawing by Amy Lathrop from Duellman, de la Riva, and Wild (1997).

thumb spines do have nuptial pads or keratinized excrescences on the forelimbs or chest, which usually have modified mucus glands associated with them (fig. 8.32 A; C. Liu 1936; Duellman and Trueb 1986; Epstein and Blackburn 1997). Most authors have assumed that these structures help the male hold onto the female and prevent the female from escaping (e.g., Noble 1931; Duellman and Trueb 1986). Some have argued that well-developed nuptial pads are associated with breeding in water, with less-developed structures being found in terrestrial breeders (Duellman and Trueb 1986). Presumably this is because females would be more difficult to hold in water, particularly running water. Indeed, Paul Kammerer's famous experiments on the development of nuptial pads in the midwife toad (*Alytes obstetricans*) were based on the premise that breeding in water should induce pad development. By forcing the normally terrestrial toads to mate in water, he expected to produce heritable increases in the size of nuptial pads after several generations. Kammerer's career was destroyed when G. K. Noble showed that the nuptial pads on a key specimen had been faked (Koestler 1971; Aronson, 1975). Ironically, the original premise of the experiment was flawed, because the association between nuptial pads and aquatic reproduction seems to be partly spurious. A more likely reason for the lack of nuptial pads in midwife toads is the absence of intense male-male competition for females in these frogs.

R. M. Savage (1961) probably was the first to suggest that nuptial pads are not primarily adaptations for preventing the female from escaping, but rather enable the male to hold the female while fending off rival males. Consequently, the evolution of nuptial pads and other devices for grasping the female should be viewed in the context of mate guarding behavior, rather than mating behavior per se. Viewed in this context, one would predict that nuptial pads will be well developed in explosive breeders with extensive male-male competition for females, but not necessarily in prolonged-breeding, territorial species. This seems to be true for North American ranids. Explosive breeders such as *Rana sylvatica* typically have very large nuptial pads, but in territorial species such as *R. clamitans* and *R. catesbeiana,* the thumb of the male is somewhat enlarged, but there is no keratinized nuptial pad (Wright and Wright 1949; Wells, unpublished observations). Nevertheless, the elaborate nuptial excrescences of some species probably are adaptations for holding onto females in running water. For example, in some Chinese ranids that breed in mountain streams, males have extensive excrescences on the forelimbs, and much of the ventral surface of the body is covered with short spines (fig. 8.30 B), providing a rough texture for holding the female (C. Liu 1936, 1950). Very large, spiny nuptial pads also are characteristic of many other stream-breeding frogs (fig. 8.32).

### Sexually Dimorphic Integumentary Structures

In addition to nuptial pads and the various kinds of spines that develop on the forelimbs and integument of some male frogs, there are a number of examples of sexual differences in integumentary structure. In many cases, the function of these structures is unknown. Males of the so-called hairy frog (*Trichobatrachus robustus*) from West Africa develop strange fringe-like projections from the skin during the breeding season (see fig. 1.12 D in chapter 1). These have been interpreted as supplementary respiratory structures (see chapter 4), but the biology of this frog has not been investigated in detail, and its behavior during the breeding season is unknown. In megophryid frogs of the genus *Vibrissaphora* from Southeast Asia and China, males have rough pustules on the skin that are lacking in females, and they develop bizarre keratinized spines on the upper jaw during the breeding season (fig. 8.33). Males apparently defend oviposition sites under rocks in streams and guard their eggs after they are laid. Ho et al. (1999) suggested that the keratinized spines are used to fend off predators that threaten the eggs, but the actual use of these structures has not been observed in the field. It seems equally likely that they are involved in some aspect of courtship or aggressive interactions among males.

Many other anurans exhibit sexual differences in skin texture, with either the male or female having rough or spiny skin that is not found in the other sex. These differences might function as tactile cues for sex recognition (Duellman and Trueb 1986), but in no case has the function of these differences been studied in detail. The same is true for sexual differences in coloration, which is especially common in toads that are diurnally active (Kagarise Sherman 1980). In some species, such as *Bufo canorus,* females are more brightly colored than are males, whereas in others,

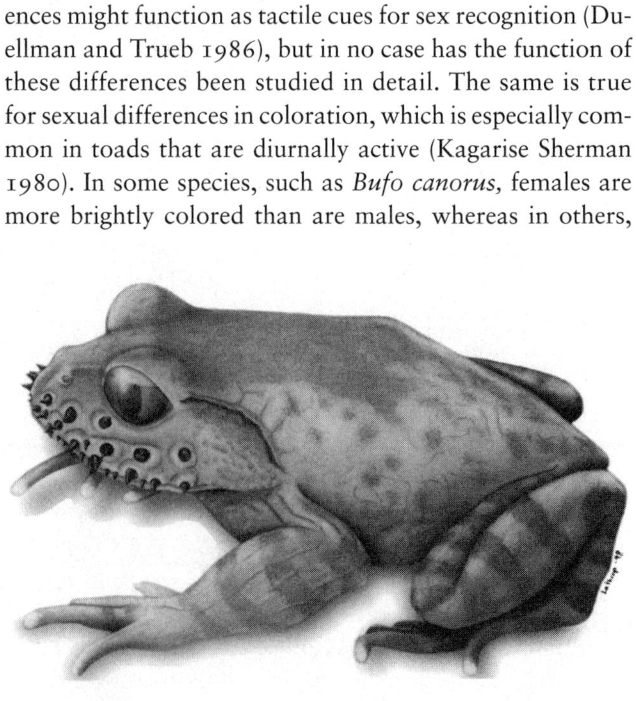

**Fig. 8.33.** Male of the megophryid frog *Vibrissaphora ailaonica* from Vietnam, showing keratinized spines on upper lip. These develop during the breeding season, but their function is unknown. Drawing by Amy Lathrop from Ho et al. (1999).

such as *Bufo periglenes,* the male is more brightly colored. Marked sexual differences in coloration also are found in some hyperoliid frogs (Schiøtz 1967, 1999; T. Hayes and Menendez 1999). Again, these differences could be used for sex recognition, although this has not been demonstrated experimentally. The use of visual cues in mate choice has been demonstrated in a species of dendrobatid frog, *Dendrobates pumilio,* but in this case, females discriminated between different color morphs in a polymorphic species, rather than recognizing a characteristic male color (Summers, Symula, Clough, and Cronin 1999).

Some frogs have glands of unknown function on various parts of the skin, and in many cases, these are sexually dimorphic or found only in one sex (Duellman and Trueb 1986; Thomas, Tsang, and Licht 1993; Brizzi, Delfino, and Jantra 2003). For example, glands are found on the ventral surface of the body or legs in males of some microhylid, hylid, and mantelline frogs (Duellman and Trueb 1986). Ventral glands are known to produce mucus secretions that glue the male to the female during amplexus in some rotund microhylids (see chapter 10), but the function of the glands is not known in other frogs. They could assist the male in adhering to the female, or they could produce pheromones that are transferred to the female during courtship or amplexus. Sexually dimorphic glands are found in the dorsal skin of males of several European ranids (*Rana dalmatina, R. iberica,* and *R. italica*) that do not have vocal sacs. Brizzi, Delfino, and Pellegrini (2002) suggested that these glands produce pheromones that compensate for reduced acoustic signalling in these frogs, but experimental evidence is lacking.

Treefrogs of the genus *Plectrohyla* have unusual skin glands that might be used in sexual behavior. During the breeding season, the upper lip of the male becomes swollen, and in some species, unusually long maxillary and premaxillary teeth actually protrude through the lips. These could be used to scrape the skin of the female, to inoculate her with pheromones, as in some plethodontid salamanders (see chapter 9). Indeed, females of at least one species have been found with scratches on the head and back that could have been caused by the teeth of males (Duellman, de la Riva, and Wild 1997). Clearly, the whole question of possible chemical communication and its role in courtship in anurans needs further investigation.

Pheromonal communication has not been well documented in anurans, but in one species of *Litoria* from Australia, secretions produced by skin glands of the male and released into the water appear to enhance the receptivity of the female (Wabnitz et al. 1999, 2000). Males of the strictly aquatic African pipid frog *Hymenochirus* also produce chemical signals that attract females (Pearl et al. 2000). Pheromonal communication also has been reported in *Leiopelma hamiltoni,* although this is not necessarily connected with reproduction (J. S. F. Lee and Waldman 2002; Waldman and Bishop 2004).

### Sexually Dimorphic Fangs and Tusks

In some anuran species, males lack spines on the forelimbs, but have bony, tusklike projections on the lower jaw. In *Ceratobatrachus, Hemiphractus,* and *Pyxicephalus,* tusks are present in both sexes and might be used for predation (all of these are frog-eaters; Duellman and Trueb 1986). *Pyxicephalus adspersus* males also are known to use their tusks in combat (B. Balinsky and Balinsky 1954). In a small hylid frog, *Phyllodytes luteolus,* males have larger tusks than females, and use them when fighting over bromeliads that serve as oviposition sites (Weygoldt 1981a). Very conspicuous tusks also are found in males of the African petropedetid frog *Dimorphognathus africanus* (Noble 1931; Duellman and Trueb 1986) and in several species that have already been discussed, including the Australian myobatrachid *Atelotus brevis* (J. Moore 1961; Katsikaros and Shine 1997) and fanged frogs of the genus *Limnonectes* (fig. 8.26), which are known to use fangs in combat.

### Sexual Differences Related to Calling

A largely neglected aspect of sexual dimorphism is the relationship of sexual differences in morphology to differences in physiology, especially as related to calling behavior. Calling is energetically expensive and requires suitable metabolic machinery to supply the muscles used in call production with large amounts of energy and oxygen (see chapter 5). Most of the work involved in producing the call is done by the abdominal muscles, primarily the internal and external oblique muscles (chapters 5, 7). Noble (1931) stated that the abdominal muscles of male frogs are more powerful than those of females, but did not cite specific examples. Recent work has shown that trunk muscles of males are much larger than those of females, and have higher activities of enzymes involved in aerobic metabolism. They also differ from muscles of females in fiber type, mitochondrial density, capillary density, lipid concentration, and contractile properties (Taigen, Wells, and Marsh 1985; R. Marsh and Taigen 1987; Given and McKay 1990; Bevier 1995b; Ressel 1996; Girgenrath and Marsh 2003; see chapter 5 for a more detailed discussion).

In addition to morphological and physiological differences in abdominal muscles, there are other physiological differences between the sexes that probably are related to call production. The laryngeal apparatus generally is much larger in males than in females. The laryngeal muscles exhibit morphological and physiological adaptations for high aerobic capacity similar to those of the trunk muscles (Manz

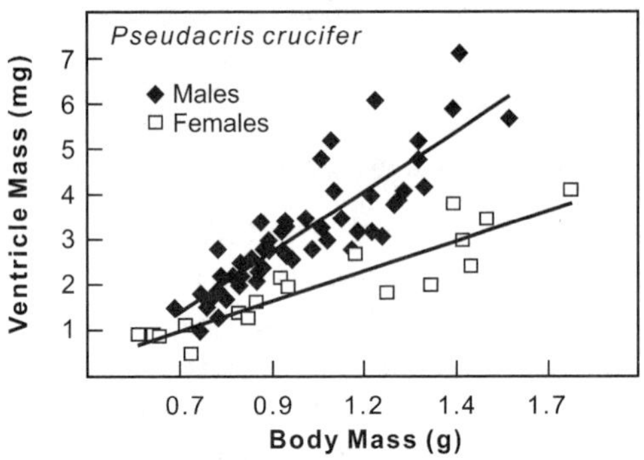


**Fig. 8.34.** Comparison of heart ventricle size in relation to body size in male and female spring peepers (*Pseudacris crucifer*). Males have much more robust ventricles for any given body size than do females, probably because of the greater demands for oxygen during call production. After Zimmitti (1999).

1975; Eichelberg and Schneider 1973, 1974; Eichelberg and Obert 1976; Schneider 1977; R. Marsh and Taigen 1987; Kelley and Tobias 1989; McClelland and Wilczynski 1989; McLister, Stevens, and Bogart 1995; Kelley 1996; McClelland, Wilczynski, and Rand 1997; Boyd et al. 1999). Noble (1931), citing several old references, reported that male frogs also have larger lung capacities and higher concentrations of red blood cells than do females. Zimmitti (1999) found that male spring peepers (*Pseudacris crucifer*) have relatively larger heart ventricles and higher hemoglobin concentrations than do females (fig. 8.34), features that appear to be related to the greater oxygen demands of calling males. There could be other physiological traits in frogs that differ between males and females that have yet to be investigated. For example, nesting male largemouth bass (*Micropterus salmoides*) exhibited a substantial increase in cardiac output and heart rate during the breeding season, and both variables were higher than in females, despite a lack of difference in relative ventricle mass (S. Cooke 2004).

## Hormones and Sexual Dimorphism

Sexually dimorphic traits in anurans are under hormonal control, but the way in which hormones mediate the expression of sexually dimorphic traits varies (Emerson 2000). Some traits are affected by hormones early in development, including alterations in body size and body proportion and development of the vocal tract. Sexual differences generally first appear at or soon after sexual maturity, and are irreversible later in life (T. Hayes and Licht 1992; Kelley 1996). The development of sexually dimorphic traits, especially those related to calling, has been studied in great detail in *Xenopus laevis* by Kelley (1996) and her colleagues. Castration of males early in development results in an underdeveloped larynx, with muscles composed mostly of slow fibers characteristic of females. Subsequent exposure of the muscles to androgens results in proliferation of fast-twitch fibers characteristic of male laryngeal muscles. Castration of sexually mature males does not result in regression of the larynx, however, and castrated males are fully capable of calling, but do not do so unless the appropriate brain center is stimulated. If females are treated with androgens early in development, they will develop malelike laryngeal muscles, but hormone treatments after sexual maturity have no effect. The laryngeal muscles of male bullfrogs (*Rana catesbeiana*), which are larger than those of females and have more aerobic fiber types, also have androgen receptors, and probabably are androgen-sensitive (Boyd et al. 1999). Other permanent sexually dimorphic traits that have been shown to be under androgen control early in development include differences in body size (T. Hayes and Licht 1992), sexual differences in coloration (T. Hayes and Menendez 1999), and development of the vocal sac (T. Hayes and Menendez 1999).

Most other sexually dimorphic traits of male anurans are expressed during the breeding season in response to increases in circulating levels of androgens, but later regress during the nonbreeding period. These include hypertrophy of forelimb muscles (Muller, Galavazi, and Szirmai 1969; Melichna et al. 1972; Oka et al. 1984; Dorlochter, Astrow, and Herrera 1994; Blackburn et al. 1995), development of nuptial pads (Rastogi and Chieffi 1971; Kanamadi and Saidapur 1982; Kao et al. 1994; L. Lynch and Blackburn 1995; Epstein and Blackburn 1997; L. Harvey and Propper 1997), and expression of throat coloration (Greenberg 1942; Kao et al. 1994). In many cases, however, seasonal hypertrophy of muscles or other male sexual traits is superimposed on preexisting differences in underlying structure in males and females that originate early in development (Emerson 2000). For example, while forearm musculature and nuptial pads of males typically regress outside of the breeding season, they generally do not return to a female-like morphology.

Another trait that is likely to be affected by male hormones both early in development and on a seasonal basis is the size of the trunk muscles involved in call production. Male frogs typically have much larger trunk muscles than do females (see chapters 5 and 7), and trunk muscles in some species appear to be partly wasted away by the end of the breeding season (personal observations). In part, this is due to the depletion of energy stores (lipids and glycogen) that contribute substantially to the mass of the muscles, and it seems likely that seasonal accumulation of energy stores in muscles is under hormonal control. It also is likely that the muscle tissue itself undergoes seasonal cycles of hypertrophy and regression. Indeed, males of *Hyla chrysoscelis* exhibit a 50% reduction in trunk muscle mass during the postbreeding period, and contractile properties of muscles are altered as well. Treatment of postbreeding males with testosterone restored both muscle mass and contractile properties to breeding season levels (Girgenrath and Marsh 2003). As in most frogs, females of this species have much smaller trunk muscles with very different contractile properties than those of males. Again, treatment with testosterone produced marked changes in female muscles, with muscle mass and contractile properties similar to those of breeding males.

Species of frogs with high calling rates typically have greater relative trunk muscle mass than do those with low calling rates (Pough et al. 1992; Bevier 1995a, 1997b). Most frogs that have been examined exhibit substantial elevations in circulating androgen levels during the breeding season (Mendonça et al. 1985; Houck and Woodley 1996; L. Harvey et al. 1997), and one might expect higher androgen levels to be found in species with large trunk muscles and high calling rates. A preliminary study of a phylogenetically diverse group of frogs from Southeast Asia provided some support for this hypothesis. Species with high calling rates or loud calls had higher androgen levels than did those with low calling rates or soft calls, but differences in relative trunk muscle size were not examined (Emerson and Hess 1996).

Some frogs have unusually low androgen levels during the breeding season, and in some cases, this is correlated with a reduction in the occurrence of sexually dimorphic traits found in other frogs. For example, *Limnonectes leporinus,* a fanged frog from Southeast Asia, has unusually low androgen levels compared to other ranid frogs, and this species lacks many of the sexual dimorphic traits that commonly appear during the breeding season in other ranids, such as hypertrophied forelimbs and nuptial pads. This condition has been related to the evolution of extensive parental care in this group of frogs (Emerson, Rosewemitt, and Hess 1993; Emerson 1996). Nevertheless, these frogs retain the physiological responses to hormones that were present in their ancestors. When supplemental doses of androgens were administered to males of *L. leporinus,* they showed some development of thumb pads that normally are absent (Emerson, Carroll, and Hess 1997). Low hormone levels are not always associated with parental care or the absence of secondary sexual characters in males, however, because two other Southeast Asian species, *Occidozyga laevis* and *Bufo asper,* also have low hormone levels, but have nuptial pads and other sexually dimorphic traits and lack parental care. Clearly, more comparative data on the hormonal control of sexually dimorphic traits are needed before we will fully understand the processes involved.

## Summary and Conclusions

When I first reviewed anuran mating systems nearly 30 years ago (Wells 1977b), information on male mating tactics, competition for females, and territoriality was almost entirely descriptive. There were no published studies exploring the factors that affect male mating success. Nevertheless, it was possible to make some generalizations from the descriptive natural history literature available at the time. In particular, I suggested a dichotomy between explosive breeders, characterized by synchronous arrival of females, and prolonged breeders, characterized by asynchronous availability of mates. The timing of reproductive activity appears to be influenced mainly by abiotic factors such as variation in temperature and rainfall, but it has a major effect on the tactics used by males to acquire mates and the degree of variation in male mating success in a breeding population. Because a high degree of synchrony tends to result in dense aggregations at breeding sites, I predicted that male-male competition would prove to be more important than female choice in explosive breeders, whereas female choice would play a major role in determining the mating success of males in prolonged breeders. I also predicted that the variance in male mating success should be greater in prolonged breeders because of greater opportunities for males to compete for many different females.

These predictions have been largely supported by subsequent studies. Direct competition for females is common in explosive breeders, and large males tend to be more successful than are small males. In prolonged breeders with lek-like mating systems, females often appear to select mates on the basis of male traits that are not strongly correlated with body size. In species that defend territories centered on oviposition sites, female choice seems to be based more on territory quality than on male phenotype, although these often are correlated, because large males are more likely to acquire good-quality territories than are small males. The prediction that male mating success should be more variable in species with prolonged breeding periods than in explosive breeders also has been largely supported by more recent work. Multiple mating by males is uncommon in explosive breeders, except in unusual cases where females outnumber males (e.g., Denton and Beebee 1993a), but often occurs in prolonged breeders.

In my original review, I pointed out that the dichotomy between explosive and prolonged breeders is somewhat artificial, since these categories actually represent two ends of a continuum of temporal breeding patterns. Indeed, more recent work has revealed a considerable degree of plasticity in the timing and length of the breeding period in many anurans. In general, species living in relatively variable environments, such as deserts or regions with pronounced temperature variation, tend to exhibit more plasticity in the timing of breeding than do species living in more stable environments. In the tropics, anurans in wet regions often breed more or less continuously during the rainy season, although there are some species that time their breeding to coincide with especially heavy rainfall. In drier tropical habitats, such as savannas, breeding seasons are likely to be more abbreviated, or to be broken up into a series of short episodes of breeding. In the temperate zone, summer-breeding anurans that utilize relatively permanent ponds almost always have prolonged breeding periods, whereas spring breeders tend to be more variable and are more likely to have breeding interrupted by cold spells.

This variability in the timing and duration of reproductive activity in response to changes in temperature and rainfall can lead to variation in population densities at breeding sites and in the relative numbers of males and females. Many anurans exhibit density-dependent shifts in mate-locating tactics. For example, males of explosive breeding species often shift from calling to active searching as densities increase, and some males adopt satellite tactics at intermediate densities. These changes in mating tactics are not unique to anurans, but are similar to density-dependent changes in mating tactics observed in insects (Thornhill and Alcock 1983; Finke, Waage, and Koenig 1997; Zuk and Simmons 1997) and other animals. Prolonged-breeding anurans exhibit density-dependent shifts in mating tactics as well, from resource-defense to lekking behavior, or from lekking to active searching. Again, these changes in behavior parallel similar density-dependent shifts in mating tactics seen in lek-breeding insects (Shelly and Whittier 1997) and vertebrates (Gosling 1986; Langbein and Thirgood 1989; Gosling and Petrie 1990; Apollonio et al. 1992; Balmford et al. 1993; Wikelski, Carbone, and Trillmich 1996; Lanctot and Weatherhead 1997; Thirgood, Langbein, and Putnam 1999).

Most work on anuran mating systems has focused on the tactics used by males to acquire mates. Much less attention has been devoted to the mating tactics of females. In part this is because female anurans tend to be cryptic in their behavior and difficult to observe. In nocturnal species that breed in heavily vegetated ponds or swamps, females usually are not seen until already in amplexus with males, so it is difficult to observe their mate-locating behavior in detail. Only a few studies have specifically focused on the movements of females (e.g., Wells 1977c; R. D. Howard 1978a; M. Morris 1989; Grafe 1997a). Despite the difficulty of observing females, their behavior deserves more attention. Just as males adjust their mating tactics to changes in male density, females do so as well. For example, females might spend less time comparing potential mates or their territories if males are highly aggregated, especially if females are subjected to harassment by males or are exposed to predators attracted to the aggregation. In explosive breeders, females in amplexus sometimes leave areas of high male density, perhaps to escape from intense competition among males, which can be fatal to females.

Harassment by males also could explain differences in oviposition behavior by females, which has led investigators to classify some mating systems as resource defense and others as leks. For example, Emlen (1976) interpreted the mating system of a population of bullfrogs (*Rana catesbeiana*) in Michigan as a lek largely because he observed females leaving the territories of males to lay their eggs. R. D. Howard (1978a) studied the same population and concluded the frogs had a resource-defense mating system because most females laid eggs in the territories of their mates. Possibly the higher population density during the years when Emlen studied the population caused many females to leave the chorus because it became difficult for them to lay their eggs without being disturbed by neighboring territory holders. The fundamental strategy of males, the defense of exclusive territories, did not change, however. Several investigators who have studied frogs that breed in rice fields, where densities often are very high, have observed males defending clustered territories, but females leaving the main chorus and moving to the periphery of the breeding area to lay their eggs (e.g., Shimoyama 1982, 1989). The mating system of these frogs resembles a lek, but it would be interesting to determine whether females ever lay eggs in the territories of their mates when competition among males is less intense. In other animals, avoidance of male harassment can affect both the mating and egg-laying behavior of females (e.g., Clutton-Brock et al. 1992; Finke et al. 1997).

Because of the difficulty of observing the behavior of females while simultaneously monitoring the behavior of large numbers of males, it is not a trivial matter to identify all of the factors that contribute to variation in male mating success in a natural population. Even for well-studied species such as the red-winged blackbird, years of work on many different populations were required to determine the relative importance of male behavior, male morphology, or resource quality as determinants of female choice (Searcy and Yasukawa 1995; Beletsky and Orians 1997). Similar problems plague studies of mate choice in anurans. Usually only one population of a given species has been studied intensively, although studies of multiple populations are becoming more common. Even when species have been studied more than once, results and interpretations often are quite different (e.g., *Rana catesbeiana,* Emlen 1976; R. D. Howard 1978a; M. Ryan 1980b; *Bufo americanus,* Gatz 1981a; Kruse 1981a; R. D. Howard 1988b). Often, the determinants of male mating success vary from year to year even in the same population.

The mating success of a male frog probably is determined by many factors. These include the length of his stay in the breeding pond, his ability to obtain a suitable calling site, his ability to displace other males from amplexus or avoid being displaced, his level of activity at the breeding site, and the quality of his territory, as well as phenotypic traits such as the dominant frequency of his calls, calling persistence, call duration, or call intensity. The relative importance of each factor depends on the details of the mating system. Consequently, any search for a single factor to explain most of the variation in male mating success in a population is likely to be futile. Furthermore, individuals with similar mating success sometimes follow different behavioral tracks to achieve the same results. One male might arrive at the pond early, another might be particularly active in searching for females, and a third might be particularly successful in displacing other males from amplexus. Without detailed behavioral observations of each individual, such differences would be obscured, and the investigator would be left with weak but possibly significant relationships between mating success and various behavioral and morphological characteristics of males. Different correlates of mating success often are correlated with each other as well, so the use of multivariate statistics to partition variation in reproductive success is desirable (Arak 1983b). Unfortunately, data on variation in male behavior has not been sufficiently detailed to allow most investigators to use this approach.

These problems are compounded when one considers lifetime reproductive success of long-lived species. R. D. Howard's (1978a, 1981a, 1983, 1984) work with bullfrogs showed that males attempt to obtain mates by different means at different ages, and a young male that has relatively poor reproductive success in one season might obtain several mates the next year. Therefore, size-related mating patterns evident on a seasonal basis do not necessarily reflect the distribution of matings over the lifetimes of all individuals in the population. In addition, many chance events can affect a male's lifetime reproductive success. For example, a harsh winter can greatly reduce the number of males competing for females, so that young males are more successful than in years when population densities are high (R. D. Howard 1983). Changes in the configuration of the breeding site, in the distribution of vegetation, or in the abundance of egg predators also can change the availability of high-quality territories and alter the distribution of matings in the male population. Without long-term observations over several breeding seasons, such events would go undetected (see also Beletsky and Orians 1997).

The importance of female choice in anuran mating systems varies because selection favors different mate choice tactics in different species, depending on the potential costs and benefits of discriminating between potential mates (Janetos 1980; Halliday 1983b; G. Parker 1983; Wittenberger 1983; Jennions and Petrie 1997). If the variance in male quality is low, and the costs of comparing males are high (e.g., Grafe 1997a), then selection should favor females mating with the first male encountered. This hypothesis has yet to be falsified for most anuran species. Alternatively, females might mate with any male that meets some minimum criterion of acceptability, but not make fine-scale discriminations among males. In many populations, very low-quality males are eliminated by male-male competition, so all individuals participating in a chorus are acceptable. Finally, if there is sufficient variation in the quality of potential mates, due either to variation in territory quality or male phenotype, then selection should favor active mate choice by females, providing the costs of such behavior are not too high. Even so, it seems likely that females of most species make comparisons among only a small subset of the males in the population (Janetos 1980; Wittenberger 1983), and one should not expect every female to choose the best possible mate. Furthermore, even if mate quality has a substantial effect on female fitness, the ability of females to distinguish minor differences in quality from male phenotype or behavior is limited.

Perhaps the most controversial problem yet to be resolved in studies of anuran mating systems is the extent to which females can use differences in call characteristics to assess differences in body size or even genetic quality of males. As discussed previously, there has been much speculation on this subject, but relatively little convincing experimental work. Several steps are required to demonstrate active female choice based on characteristics of male calls: (1) It must be shown that some males obtain more matings than expected by chance. This can be done only if complete data on male mating success throughout the breeding period are available. (2) Behavioral data showing that female frogs can actively assess male behavior and deliberately move toward particular males must be presented. Anecdotal reports of a small subset of females approaching calling males (e.g., L. Licht 1976) are not convincing evidence of female choice. (3) Characteristics of calls must be shown to vary in a predictable manner. For example, if dominant frequency is only weakly correlated with male body size, then females will not be able to derive reliable information about the size of males from their calls. (4) It must be demonstrated that females can discriminate between calls falling within the natural range of variation in the population. If females do not distinguish between the calls of most males in the population, then differences in calls probably will have little effect on male mating success.

Two-choice playback experiments must be interpreted cautiously. They demonstrate the ability of females to distinguish between calls in a simplified acoustic environment, but they cannot show conclusively that females do so in a chorus of several dozen or several hundred males. These experiments also cannot demonstrate that such discrimination is the principal determinant of male mating success. One difficulty in extrapolating from playback experiments to a natural situation is that females probably sample only a few males when choosing a mate. In many species, mate choice probably is a multistep process (Felton et al. 2006). Initially, a female probably is attracted to any continuous sound of the correct frequency, even choruses of other species (Gerhardt 1982). After arriving at the chorus, a female then moves to an area where one or more males of her own species are calling. Any comparisons among males probably involve only a few individuals in the local area, with others simply contributing to background noise (Wells and Schwartz 1984a). Consequently, evidence of active mate choice in the population as a whole can be difficult to detect.

Finally, even if active female choice has been documented, it will be impossible to determine whether mate choice is adaptive for females until additional genetic information is available. Some models of sexual selection indicate that females need not use adaptive criteria in choosing mates for active mate choice to be maintained in a population (Lande 1981; Andersson 1982; Arnold 1983b). G. Parker (1983) suggested that adaptive female choice based on criteria yielding only genetic benefits is not likely to be maintained because of the costs of exercising such choice, and Arnold (1983b) has cautioned against viewing female choice as a process of "shopping for good genes." Nevertheless, much of the speculation about why female anurans should prefer large males to small ones is based on adaptive arguments. Common assumptions are that (a) large males are older and therefore have demonstrated an ability to survive, or (b) large males have grown more rapidly and therefore probably are genetically superior in their ability to acquire and use energy (Arak 1983a; M. Ryan 1983b).

Unfortunately, evidence that traits affecting mate choice by females yield genetic benefits to their offspring is weak or nonexistent for most anurans. There is virtually no information on the genetic basis of fitness-related traits in adults (for example, differences in growth rates of juveniles and adults, physiological differences). Several investigators have used laboratory experiments to determine whether choice of large males by females yields genetic benefits to their offspring, such as enhanced growth rates or higher survivorship of tadpoles fathered by large males. Although these experiments produced some evidence for paternal effects on fitness-related traits of tadpoles, there was no consistent tendency for offspring of larger males to have faster growth rates, shorter larval periods, larger size at metamorphosis, or higher survivorship. Therefore, these experiments do not provide strong evidence that indirect selection favors choice of large males by females (Woodward, Travis, and Mitchell 1988; S. Mitchell 1990; Howard, Whiteman, and Schueller 1994; Semlitsch 1994). One study did suggest a link between female choice of males with high calling performance and larval fitness, although the genetic mechanism linking adult and larval performance is not known (Welch, Semlitsch, and Gerhardt 1998; Doty and Welch 2001). Two studies that investigated the relationship between parasite infection, male calling performance, and male mating success found little effect on male fitness (Hausfater, Gerhardt, and Klump 1990; Tinsley 1990).

The relationship of sexual dimorphism to mating systems and sexual selection has scarcely been explored. Many types of sexual dimorphism, such as large body size, enlarged forelimbs, and development of spines and other weapons in males often seem to be related to male combat. Nevertheless, information on aggressive behavior and the contribution of specific phenotypic traits to fighting ability is available for only a few species. Furthermore, the tradeoffs between sexual selection and natural selection acting on characters such as male body size (e.g., Wikelski and Trillmich 1997) are unknown for most species, so it is difficult to predict patterns of sexual dimorphism from comparative studies alone. The functions of many sexually dimorphic traits, such as differences in skin texture or coloration and the presence of cutaneous glands in some male frogs, remain unknown. Finally, physiological correlates of sex differences in morphology can be as important in determining reproductive success as the morphological traits themselves, but they have not been explored in detail. Studies of sexual selection in anurans thus provide an opportunity for interdisciplinary research combining approaches from diverse fields such as behavioral ecology, functional morphology, and physiological ecology.

# Chapter 9 Communication and Social Behavior of Urodeles and Caecilians

*A salamander . . . will remain absolutely motionless for an hour on or under some dead leaf, in the trickling waters that wend their way riverward from a mossy spring. To sit or stand for an hour, and watch this immovable creature, is both painful and monotonous, and when, at last, you disturb it, perhaps accidentally, away it goes to some similar spot near by, and resumes its motionless attitude. . . . Studies of salamander life soon become a bore.*
—Charles C. Abbott, *A Naturalist's Rambles About Home* (1884)

*Among the Newt tribe [sexual] characters take the form of frills and crests and vivid colours. They are intended to stimulate through the sense of sight, and arouse emotion, as a city is beflagged to welcome those it may delight to honour.*
—W. P. Pycraft, *The Courtship of Animals* (1913)

UNTIL RELATIVELY RECENTLY, most biologists, even those working on amphibians, probably gave little thought to the behavior of salamanders, perhaps sharing Charles C. Abbott's opinion that these small, secretive animals have little to offer the inquiring naturalist. Most early published reports of salamander communication and social behavior consisted of descriptive studies of courtship behavior (e.g., Noble and Brady 1933; Bishop 1941b; Stebbins 1954b; Organ 1958, 1960a; Joly 1966; Salthe 1967; Brandon 1970b; Garton 1972; Salthe and Mecham 1974) and brief reports of aggressive interactions (Kerbert 1904; Bishop 1941b; R. Gordon 1952; W. Grant 1955; Hutchison 1959; Brandon and Huheey 1971; Licht 1973; Thurow 1976). Although Darwin (1871) had noted the striking sexual dimorphism of European newts (fig. 9.1), most of the early work on salamander behavior did not relate observations of courtship and aggression to theories of sexual selection or use them as a basis for understanding the evolution of mating systems.

This changed in the mid-1970s, with a renewed interest in evolutionary interpretations of animal communication and social behavior. Of particular importance was the publication of several key papers on the courtship of newts by Halliday (1974, 1975a, b, 1976, 1977a) and North American salamanders by Arnold (1976, 1977). These pioneering studies were the first to place the behavior of salamanders in a conceptual framework of sexual selection theory, and they set the stage for all subsequent work in the field. The literature in this field has expanded at an accelerating rate in the last three decades, and a number of reviews have been published (Arnold and Houck 1982; Halliday and Verrell 1984; Houck 1986; Jaeger 1986; Verrell 1989c; Halliday 1990; Houck and Verrell 1993; Jaeger and Forester 1993; Houck and Sever 1994; Halliday and Tejedo 1995; Mathis et al. 1995; Sullivan, Ryan, and Verrell 1995; Griffiths 1996; Dawley 1998; Denoël 1999; Houck and Arnold 2003).

Caecilians have received much less attention. Although certain aspects of their reproductive biology have been studied in the laboratory (see chapter 10), we know next to nothing about their communication and social behavior in the field. In this chapter, I briefly discuss what little we know about caecilian communication and then review different modes of communication in salamanders and newts and the

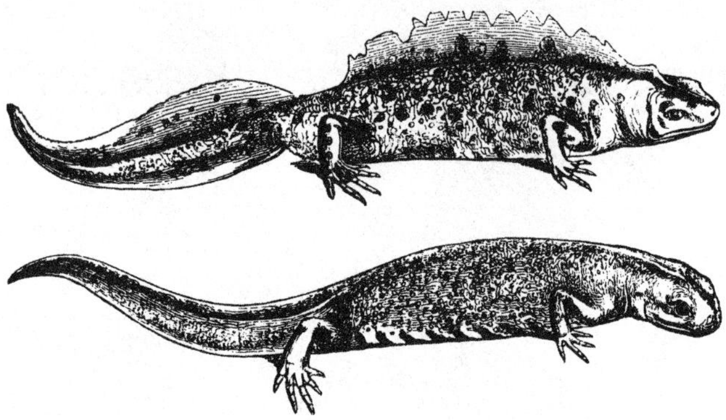

**Fig. 9.1.** Sexual dimorphism in the European crested newt (*Triturus cristatus*), from Charles Darwin (1871), *The Descent of Man, and Selection in Relation to Sex.* Male is on top, female on bottom.

use of communication signals for species, sex, and individual recognition. I then discuss aggression, spacing, and territoriality, the evolution of mating systems, and the action of sexual selection on traits of males and females.

## Communication by Caecilians

Caecilians are mostly burrowing animals that spend their lives underground, although some have secondarily evolved aquatic habits. They are nearly or completely blind, so use of visual signals can be ruled out. There is no evidence that caecilians use sound for communication, so communication in this group almost certainly depends on chemical or tactile cues. The only species of caecilian for which we have any experimental evidence about modes of communication is the aquatic species *Typhlonectes natans.* Animals kept in captivity tend to be gregarious, sheltering under rocks in the water. They appear to mark shelter sites with chemical signals and are attracted to chemicals emanating from aggregations of conspecifics (Warbeck, Breiter, and Parzefall 1996). In this respect, they are remarkably similar to females of the European olm (*Proteus anguinus*), a blind aquatic cave salamander (Guillaume 2000a, 2002) and the gregarious terrestrial salamander *Salamandra luschani* (Gautier et al. 2006). Breeding male *Typhlonectes* in captivity are aggressive toward one another, but not necessarily territorial. Laboratory tests showed that nonreproductive females preferred chemical cues from other females to those from males, whereas reproductive females preferred cues from males. Reproductive males preferred cues from receptive females to those from nonreceptive females, but did not discriminate between related and unrelated females (Warbeck and Parzefall 2001). Because natural population densities and tendencies to aggregate in the field have not been studied, the nature of this animal's mating system remains unknown.

## Modes of Communication in Urodeles

### Acoustic Communication

In contrast to the dominant role of acoustic communication in the lives of anurans, communication by sound is of little importance to salamanders. A number of species of salamandrids, ambystomatids, and plethodontids are known to produce a variety of low intensity hisses, barks, clicks, squeaks, and whistles. In most cases, the function of these sounds is not known, although they seem to be associated with antipredator behavior (Maslin 1950; Bogert 1960), courtship (Wyman and Thrall 1972), or aggressive behavior (Licht 1973; J. Davis and Brattstrom 1975; Ovaska 1987; K. Nishikawa personal communication). The only attempt to play recorded vocalizations to salamanders failed to elicit a response (Wyman and Thrall 1972). Some of the sounds produced by salamanders contain energy at frequencies well above 2 kHz (Bogert 1960; Wyman and Thrall 1972; J. Davis and Brattstrom 1975). These animals lack external tympanic membranes and most have degenerate columellas (Wever 1985; Jaslow, Hetherington, and Lombard 1988). They can detect airborne or waterborne sounds and ground vibrations, but peak sensitivities generally are below 500 Hz, and hearing sensitivity falls off rapidly at frequencies above 1 kHz (Ferhat-Akat 1939; Ross and Smith 1978; Wever 1985). The relative insensitivity of salamander ears to anything but low-frequency sounds and vibrations makes it unlikely that the sounds produced by these animals function in intraspecific communication. Nevertheless, sound production is a regular feature of aggressive interactions in some species, such as *Plethdon vehiculum,* which makes clicks by snapping the jaws together (Ovaska 1987). Further investigation of the hearing capabilities of sound-producing species is needed.

### Visual Communication

Many salamanders are nocturnal and often carry out their activities under cover objects or in other dark environments. They have eyes that are well adapted to functioning in dim light conditions and can capture prey even on relatively dark nights (Roth 1987). Nevertheless, dim light conditions probably select against the use of color in visual displays and select for displays consisting of relatively simple movements or postural changes that can be detected even when an animal is in silhouette. For example, red-backed salamanders (*Plethodon cinereus*) signal aggression by looking toward an opponent and raising the trunk off the ground, sometimes with the back arched (fig. 9.2). Changes in posture appear to form a graded signaling system that indicates gradual changes in aggressiveness (Jaeger 1981b, 1984; Jaeger and Schwarz 1991). Submission is signaled by looking away from opponents and by holding the body flat against the substrate (fig. 9.2). The condition

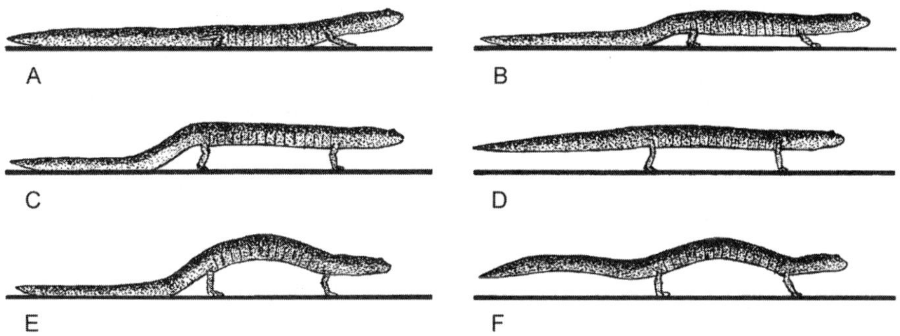


**Fig. 9.2.** Aggressive and submissive postures of the red-backed salamander (*Plethodon cinereus*). (A) Resting posture. (B) All trunk raised, low stance (ATR 1). (C) All trunk raised, high stance (ATR 2). (D) All trunk raised with tail raised (ATR 3). (E) All trunk raised with back arched (ATR 4). (F) All trunk raised with tail raised and back arched (ATR 5). After Jaeger and Schwarz (1991).

of a resident's tail also affects the responses of an intruder to a resident. Red-backed salamanders sometimes lose all or part of their tails in fights or to predators. Residents that have lost their tails are less successful in repelling intruders, even though they are no less aggressive than are individuals with complete tails (Wise and Jaeger 1998). Presumably this difference in behavioral response to the resident is mediated by visual inspection and assessment of the resident's tail condition by the intruder.

Some of the displays described in *P. cinereus* also have been observed in other plethodontids, including *Plethodon vehiculum* (Ovaska 1987), *P. serratus, P. websteri* (Camp 1999), *Aneides flavipunctatus* (Staub 1993), and *Ensatina eschscholtzii* (Wiltenmuth 1996). *Plethdon vehiculum* and *P. websteri* also exhibit tail-waving behavior, in which an animal moves its tail in a high arc over the body, usually immediately before attacking an opponent (Ovaska 1987; Camp 1999). Similar behavior has been observed during aggressive interactions in *Desmognathus fuscus* and *D. monticola* (Keen and Sharp 1984). Both larval and adult ambystomatids are reported to change body postures during aggressive encounters (Ducey and Ritsema 1988; Ducey 1989; Walls 1990, 1991), but it has yet to be clearly established that these constitute aggressive and submissive signals like those described for plethodontids (Mathis et al. 1995).

Visual signals also are used in courtship by some salamanders. In *Rhyacotriton olympicus*, the male wags the tip of his tail during courtship, presumably providing some sort of visual stimulus to the female (Arnold 1977). During the breeding season, males of *Hynobius nebulosus* develop a white throat patch that is pulsated in the presence of females (Thorn 1967). In most species of *Desmognathus*, males swing their forelimbs forward in a butterfly stroke pattern while approaching females. This behavior occurs in both the initial orientation and subsequent persuasion phases of courtship and sometimes is accompanied by undulation of the tail, another visual signal. Butterfly movements sometimes grade into stroking, a tactile signal in which the male strokes the female's head and neck with one forelimb, but not all species of *Desmognathus* engage in stroking behavior (Organ 1961a; Arnold 1977; Maksymovitch and Verrell 1992; Houck and Verrell 1993; Uzendoski and Verrell 1993; Brock and Verrell 1994; Verrell 1994a, b, 1995, 1997; Herring and Verrell 1996; Verrell 1999; Mead and Verrell 2002; Mabry and Verrell 2003).

The precise function of these visual and tactile displays is not known. In *D. ocoee*, there was not a significant correlation between male mating success in laboratory trials and the frequency of individual courtship acts, such as butterfly. Successful males did exhibit a higher composite score of persuasive ability that combined several courtship behaviors, including butterfly (Vinnedge and Verrell 1998), but it is not clear that frequency of this display actually affects mating success (Gershman and Verrell 2002). The butterfly display appears to have been secondarily lost in *D. ochrophaeus*, but is retained in its sister species, *D. orestes*. Mead and Verrell (2002) suggested that the loss of this display and another courtship behavior might account in part for the sexual incompatibility between *D. ochrophaeus* and *D. orestes*.

Visual signals are more prevalent in salamandrids, especially newts in the genera *Triturus, Paramesotriton, Pachytriton, Cynops, Euproctus, Calotriton, Neurergus*, and *Notophthalmus*, which can be active either at night or during daylight hours. The use of complex visual displays appears to be a derived behavior within the family Salamandridae. In most genera, males approach females and seize them in some form of amplexus, either from above or below (fig. 9.3). In species that have amplexus, visual signals seem to be less important than are chemical and tactile signals (Halliday 1990; Houck and Arnold 2003). In some species, such as the North American red-spotted newt (*Notophthalmus viridescens*), males sometimes perform a relatively simple lateral display (the hula) before seizing the female. This can be sufficient to elicit a response from the female, but most courtship bouts proceed to an extended period of chemical stimulation (Verrell 1982, 1983, 1990a).

In newts of the genera *Cynops, Paramesotriton, Pachytriton*, and *Neurergus*, amplexus is absent, but the visual displays of the males are relatively simple (Sparreboom 1983, 1984b, 1993, 1994; Sparreboom and Faria 1997; Thiesmeier and Hornberg 1997; Sparreboom and Thiesmeier 1999; Sparreboom, Sreinfartz, and Schultschik 2000). Males of these genera lack the conspicuous crests and tail fins found in some *Triturus,* but males of some species do develop breeding coloration. In *Paramesotriton,* males change from a dull brown color like that of the female to a much lighter yellow. Many male newts, including *Paramesotriton,* court females most actively after dusk, so the light yellow coloration probably serves to make the male more conspicuous in dim light. In *Neurergus strauchii,* males develop silvery-blue coloration on the sides of the tail in the breeding season, a pattern that also is highly reflective and probably makes the tail conspicuous in dim light (Sparreboom, Sreinfartz, and Schultschik 2000). Similar light blue coloration is seen in breeding males of some populations of alpine newts (*Triturus alpestris;* Bovero, Giacoma, and Andreone 1997).

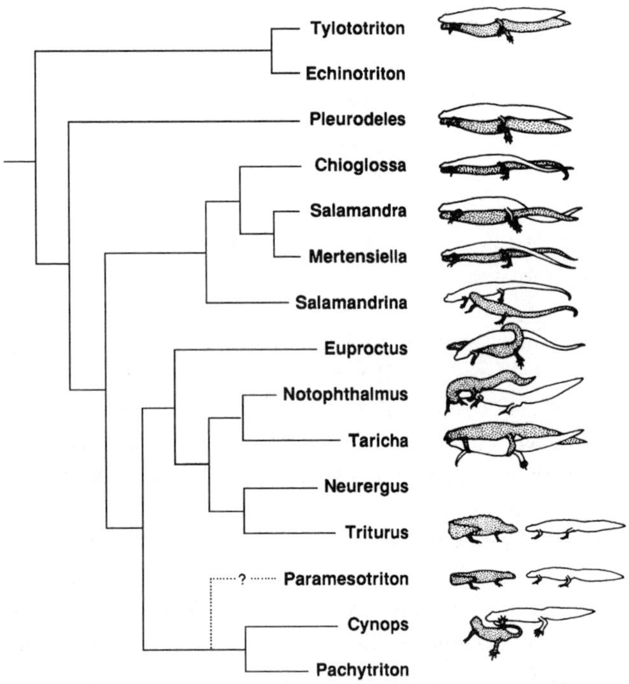


**Fig. 9.3.** Phylogeny of the genera of salamandrids, based on morphological characters, showing different types of amplexus during courtship. In each case, the male is the shaded individual. This figure illustrates the taxonomic distribution of different modes of amplexus among genera of salamandrids, but some more recent phylogenies differ from this one in the details of relationships among the groups. From Halliday (1990).

In *Paramesotriton,* males also have a bright spot or stripe on the tip of the tail, set off with dark lines on either side, a pattern likely to enhance contrast in dim light. Males move the tip of the tail during courtship, a behavior that has been interpreted as mimicking the movements of prey as a way of attracting the attention of females (Sparreboom 1993). Similar caudal luring has been reported in the courtship of several of the smaller species of *Triturus* (*T. alpestris, T. boscai, T. helveticus, T. italicus, T. vulgaris kosswigi*), but it is absent in large species (Rafiński and Czaja 1984; Wambreuse and Bels 1984; Pecio and Rafiński 1985; Giacoma and Sparreboom 1987; Arntzen and Sparreboom 1989; Faria 1993). In *T. boscai, T. italicus,* and *T. pygmaeus,* there is an additional visual display, the flamenco, in which the male raises his tail and waves the tip (fig. 9.4). This attracts the female, which sometimes even snaps at the tail tip (Arntzen and Sparreboom 1989; Faria 1993; Hidalgo-Vila, Pérez-Santigosa, and Díaz-Paniagua 2002). These displays can be considered a form of sensory exploitation (Ryan 1990), in which males have evolved a display that exploits the tendency of females to orient toward small moving prey animals. Female newts often orient toward the male's tail during spermatophore deposition, apparently following the movements of the tail until

**Fig. 9.4.** Flamenco courtship display of *Triturus boscai.* The male wiggles the white tip of the tail from side to side, which attracts the attention of the female. The female orients toward the tail tip and sometimes snaps at it, suggesting that this courtship display exploits the tendency for females to orient toward moving prey. From Faria (1993).

they encounter the spermatophore (Arntzen and Sparreboom 1989). In the genera *Euproctus* and *Calotriton,* the male often holds his tail vertically, apparently as a display to attract the attention of the female, and then uses the tail to seize the female (Guillaume 1999). Males of this genus also restrain females by biting them (Houck and Arnold 2003), but use of chemical signals appears to be minimal (Guillaume 1999).

Most of the courtship displays of *Triturus* combine visual, tactile, and chemical signals in ways that make it difficult to unravel the functional components of the displays. A display that appears to be an ancestral trait for the whole genus *Triturus* is the fan. The male faces head-on toward the female, with the tail folded back along the side of the male. He rapidly beats or vibrates the tail, which directs a current toward the female's snout (fig. 9.5). This conveys pheromones toward the female, but the display probably provides tactile stimulation from the water movement and a visual signal as well (Halliday 1974, 1975b, 1990; Arntzen and Sparreboom 1989; Houck and Arnold 2003). It appears that there has been a gradual shift in the relative importance of chemical and visual display components in the evolution of newt courtship. The fan display is present in several genera that lack broad, visually conspicuous tail fins, including *Neurergus* (Fleck 1982; Sparreboom, Sreinfartz, and Schultschik 2000), *Paramesotriton* (Sparreboom 1983, 1984b), *Pachytriton* (Sparreboom and Thiesmeier 1999), and *Cynops* (Tsutsui 1931; Sparreboom and Faria 1997), and the smaller species of *Triturus* (Arntzen and Sparreboom 1989; fig. 9.6). Presumably the transfer of chemical signals to the female is the most important feature of the display in these species. In

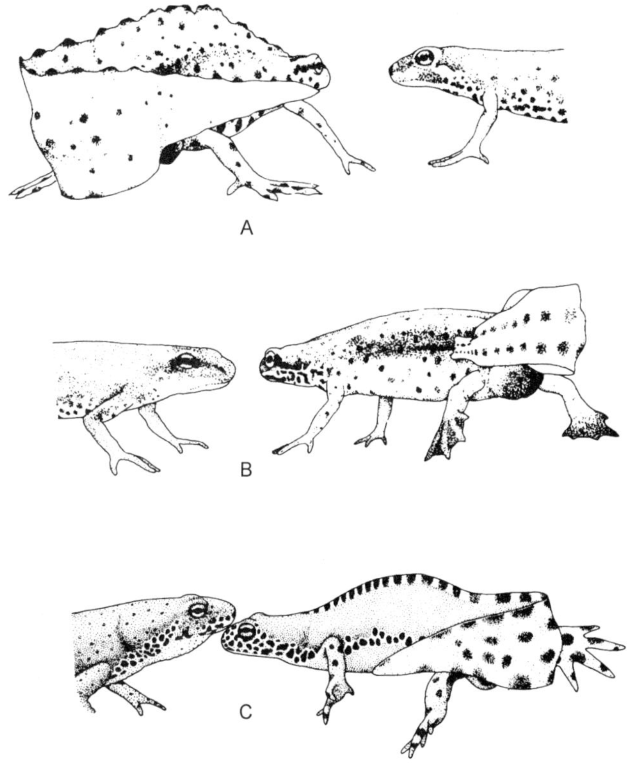


**Fig. 9.5.** Fan courtship displays of three species of European newts. In each case, the male directs a current of water toward the female with movements of his tail. This display appears to be ancestral for the entire genus, but has been secondarily lost in several large-bodied species. (A) *Triturus vulgaris.* (B) *Triturus helveticus.* (C) *Triturus alpestris.* After Halliday (1977a).

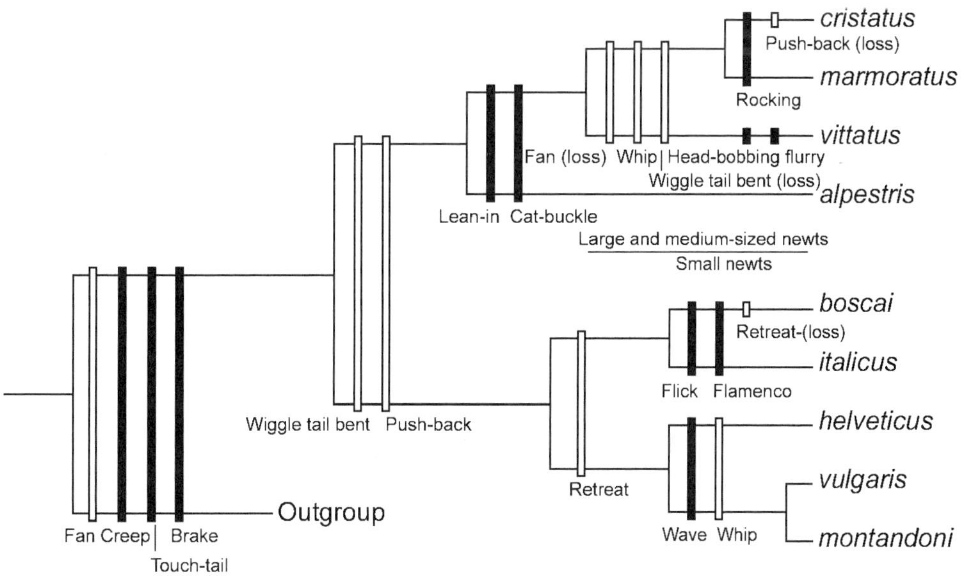


**Fig. 9.6.** Phylogeny of the genus *Triturus,* based on an analysis of biochemical and behavioral data, showing the evolution of various display traits. Solid bars are apomorphies. Open bars are homoplasous characters. This phylogeny and several other analyses of molecular characters support a division of large-bodied and small-bodied newts into distinct monophyletic clades, although some more recent phylogenies differ in details of relationships among species in the genus. After Arntzen and Sparreboom (1989).

species with broader tail fins, such as *Triturus vulgaris,* the visual component probably becomes more important. In *Triturus vulgaris, T. helveticus,* and *T. montandoni,* all of which have moderately broad tail fins and crests, there is another display, the wave, which appears to be primarily a visual signal. The male holds his tail up for a few seconds, presenting a full lateral view of the body and tail to the female (Halliday 1974; fig. 9.7). This display is absent in small species with inconspicuous tail fins, and therefore appears to be a derived signal in the more visually elaborate species.

Visual displays reach their greatest development in the very large species with extremely conspicuous tail fins and colorful crests, including *T. cristatus, T. marmoratus,* and *T. vittatus* (fig. 9.8A). Because these species are in a different clade than the small species of newts, the evolution of complex visual displays appears to have evolved at least twice independently (fig. 9.6). The visual displays of large newts are quite distinct from those of small newts. In the large newts, the male usually does not face the female head-on, and the ancestral fan display has been lost (fig. 9.7). Instead, the male presents himself laterally to the female, showing off his elaborate crest and fin, which would be visible in silhouette even in relatively dim light. Often the male will lean his body toward the female (the lean-in display). A somewhat similar but less intense version of this display is found in *T. vulgaris,* probably an independent derivation (Halliday 1974), but it is best developed in the largest, most brightly colored species. It often is accompanied by the cat-buckle display, in which the male arches his back and kinks the body around the female, showing off the crests and tail fins to maximum effect (fig. 9.8B). The shift toward these more elaborate visual displays is associated with a shift in mating systems, from mate-searching to lekking (Zuiderwijk and Sparreboom 1986; Arntzen and Sparreboom 1989; A. Green 1989; Hedlund and Robertson 1989; Raxworthy 1989; Halliday 1990; Hedlund 1990a, b; Sparreboom and Teunis 1990; see subsequent discussion of Mating Systems). Some of these same displays are used in aggressive interactions among males.

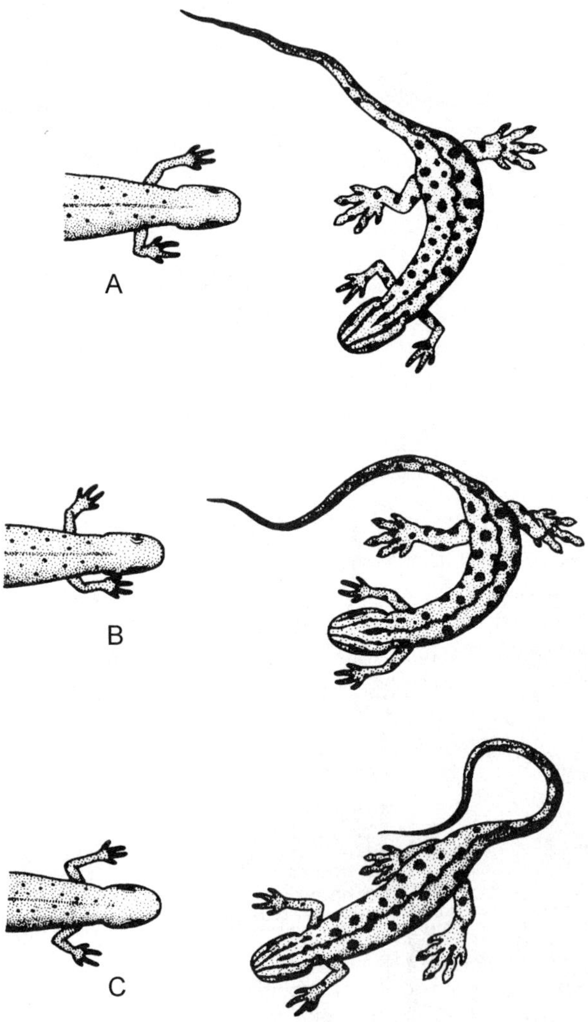


**Fig. 9.7.** Three courtship displays of the common newt (*Triturus vulgaris*). (A) The wave display, in which the male presents the female with a full lateral view of the tail. (B) The whip, a sudden lashing of the tail against the male's body, which directs a blast of water toward the female. (C) The fan, a delicate movement of the tail that directs a current of water toward the female (see also fig. 9.6). After Halliday (1977a).

### Tactile Communication

Tactile communication probably is important in the courtship of many species of salamanders, but it is often hard to separate from visual and chemical communication. As already mentioned, tail fanning displays of European newts probably serve a dual function of directing pheromones toward the female and providing tactile stimulation that is detected by the lateral line system (Halliday 1974; 1975a; 1977a). Some species of newts have an even more intense tactile display, the whip, in which the male beats his tail toward the female with enough force to move the female backward (Arntzen and Sparreboom 1989). This display might provide cues that enable a female to assess the size and strength of the male, especially in dim light or darkness, when visual signals might be difficult to discern. In two small species, *T. boscai* and *T. italicus,* males perform a flick display, which is somewhat similar to the whip. In *T. boscai,* the tip of the male's tail sometimes brushes the female's cheek or snout, probably providing a tactile signal (Pecio and Rafiński 1985; Giacoma and Sparreboom 1987; Arntzen and Sparreboom 1989; Faria 1993, 1995).

Many forms of behavior observed in plethodontid courtship, such as slapping the female's snout with a gland on the male's chin or raking the female with premaxillary teeth, might provide some tactile stimulation, but their primary function is delivery of chemical signals (Arnold 1977; Arnold

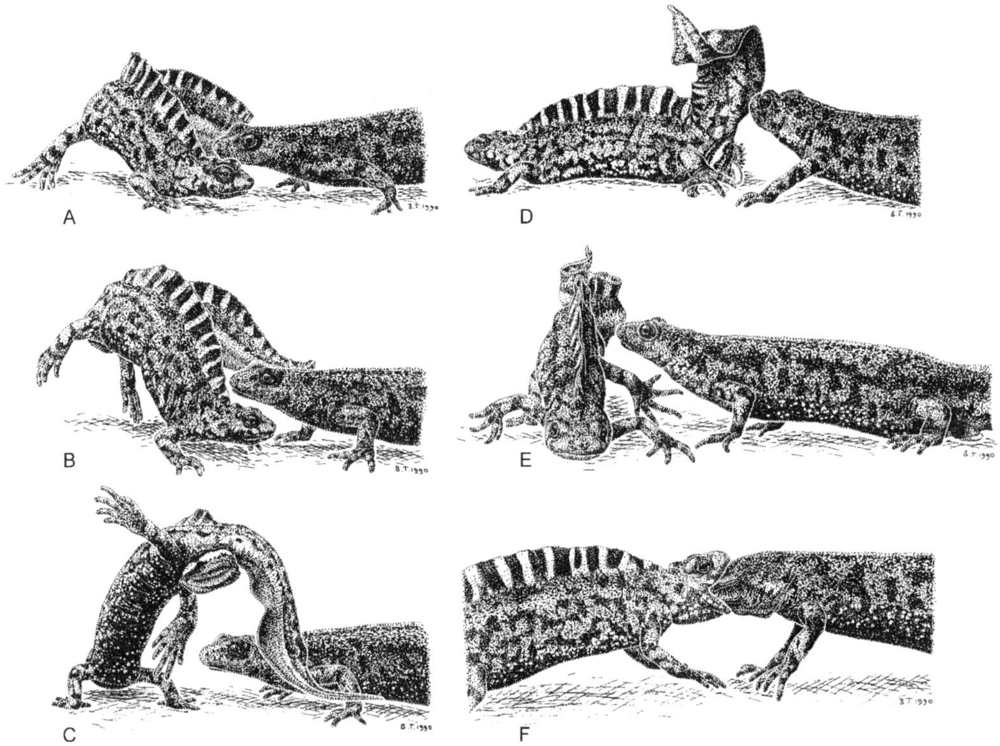

**Fig. 9.8.** Courtship behavior of the marbled newt (*Triturus marmoratus*). (A) The male takes up a position in front of the female, bends his tail in her direction, and arches his back (lean in). (B) The catbuckle display, with the male's back strongly arched and the hind legs off the bottom, sometimes accompanied by irregular waving movements of the tail. (C) Male standing on his front feet, delivering a powerful tail-lash toward the female, sometimes touching her head and body. (D) Male creeping ahead of the female. After the female touches the male's tail, he deposits a spermatophore, which can be seen emerging from the male's cloaca. (E) The male then turns 90° and blocks the female's path, sometime pushing her back with his body or tail (brake and push back). (F) After spermatophore deposition and pick-up, the male sometimes bites the head of the female. This form of post-mating mate-guarding behavior is not seen in most other species of newts. After original drawings by Bas Teunis in Sparreboom and Teunis (1990).

and Houck 1982; Houck and Sever 1994; see subsequent further discussion). Some plethodontids, such as *Gyrinophilus porphyriticus,* lack courtship glands on the chin, but during courtship, the male repeatedly rubs his snout on the snout, head, and body of the female (Beachy 1997a). Similar rubbing behavior occurs during courtship in ambystomatids and amphiumids, which lack courtship glands on the head. Beachy (1997a) suggested that such behavior evolved as a form of tactile stimulation and secondarily assumed the function of delivering chemical signals to the female. In contrast, Houck and Sever (1994) hypothesized that courtship glands and behavior to transfer chemicals to females evolved together, with the courtship glands being secondarily lost in some plethodontids, including *Gyrinophilus.*

In most urodeles, tactile signals transmitted from the female to the male are critical for the successful completion of courtship. In *Triturus,* a male will not deposit a spermatophore until the female touches his tail with her snout (fig. 9.9). Similarly, in many ambystomatids, the female signals her readiness to receive a spermatophore by nudging the base of the male's tail with her snout. In plethodontids, the female signals her receptivity by resting her chin on the dorsal surface of the male's tail and engaging in a tail-straddling walk (fig. 9.10). In all of these groups, males can be induced to deposit spermatophores if they are provided with artificial tactile stimulation (Halliday 1975a; Arnold 1976).

## Chemical Communication

Chemical signals seem to provide the principal cues for individual, sex, and species recognition in most species of salamanders (Arnold 1977; Halliday 1977a, 1990; Madison 1977; Arnold and Houck 1982; Houck and Sever 1994; Dawley 1998; Houck and Arnold 2003; Sever 2003b; Gautier et al. 2006). These signals can be derived from pheromones produced by specialized glands, from odors emanating from unspecialized epidermal secretions, or from fecal material, which sometimes are supplemented with glandular secretions. In many species, chemical signals seem to be transmitted mainly through direct contact between individuals,

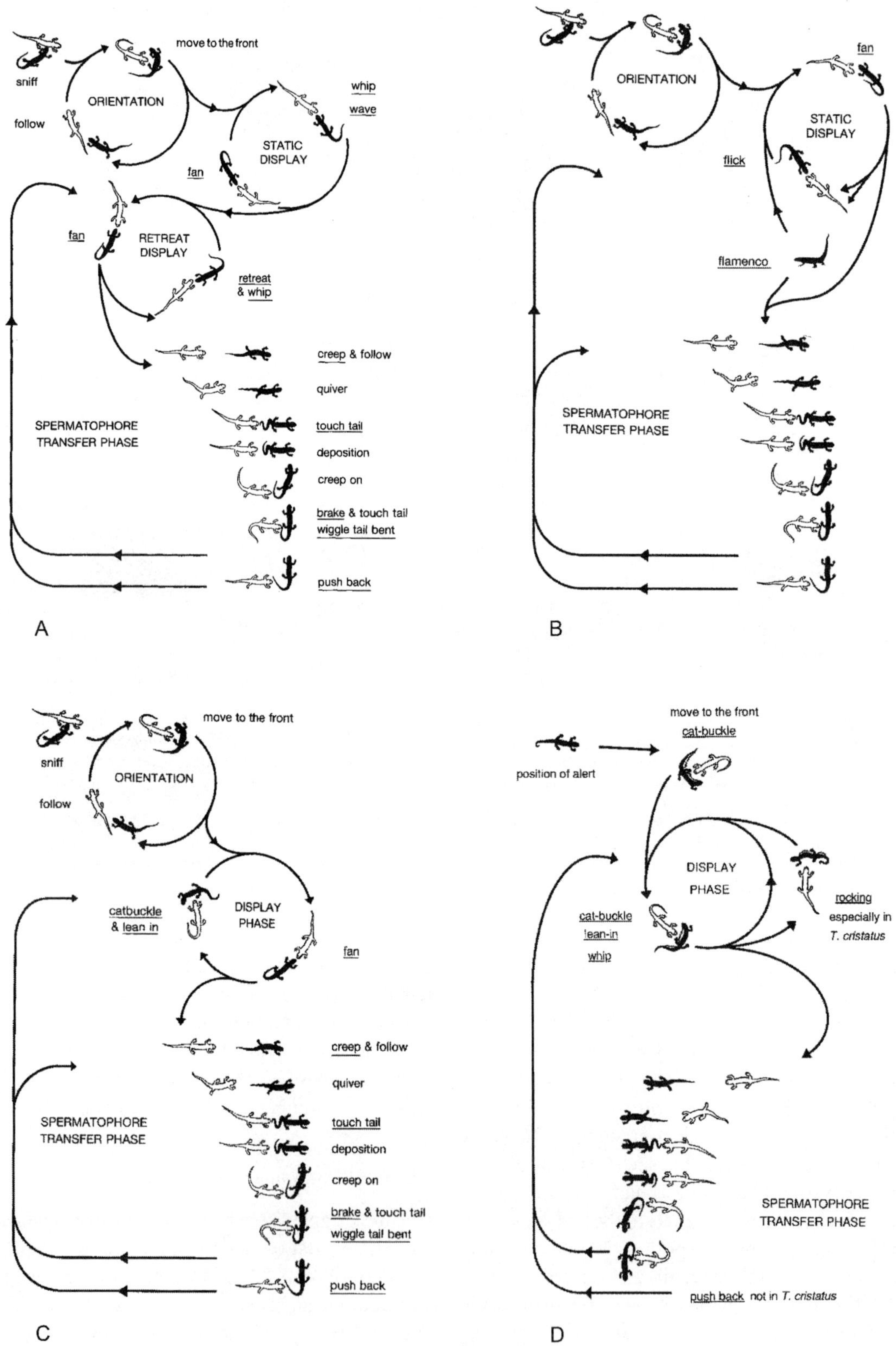


**Fig. 9.9.** Diagrams of courtship behavior of four species of European newts. (A) Courtship of *Triturus vulgaris,* a relatively small species. Courtship in this species and several others includes distinct orientation, static display, and retreat display phases. In the retreat display, the male backs away from the female while fanning or whipping his tail toward her. (B) Courtship of *T. boscai,* another small-bodied newt. This species lacks the retreat display, but has a distinct flamenco display (see fig. 9.5) that is not seen in *T. vulgaris.* (C) Courtship of *T. alpestris,* a medium-sized newt. This species also lacks the distinct retreat display, but has the derived lean in and catbuckle displays that also are seen in larger species. (D) Courtship of the largest species of newts, including *T. cristatus, T. carnifex,* and *T. marmoratus.* These species have lek mating systems and courtship displays that include frequent lateral displays of sexually dimorphic markings and coloration. After Arntzen and Sparreboom (1989).

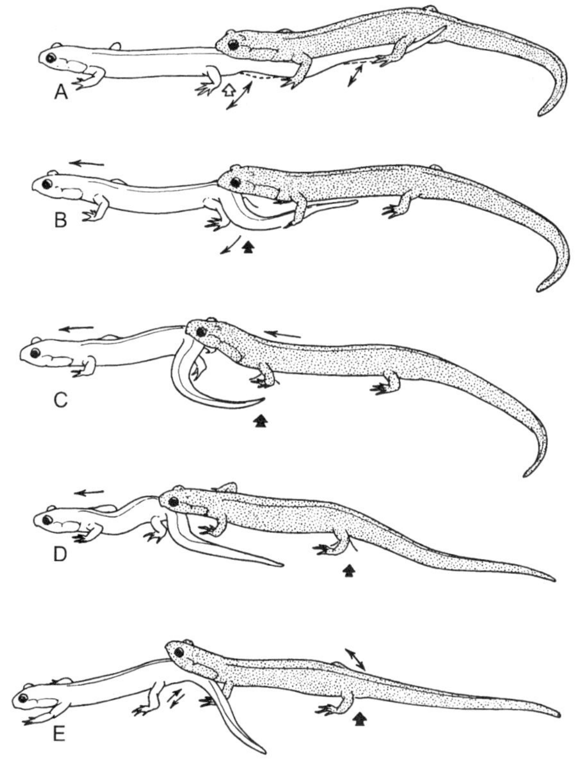


**Fig. 9.10.** Tail-straddling walk during the courtship of *Plethodon shermani.* The shaded individual is the female. (A) The female places her chin on the base of the male's tail and follows him as he moves. (B) The male deposits a spermatophore (black arrow) and moves his tail to the side. (C, D) The male leads the female forward until her cloaca is positioned over the spermatophore. (E) The male lifts the base of his tail as the female is positioned over the spermatophore, and she picks up the spermatophore with the lips of the cloaca. After Arnold (1976).

but some salamanders can obtain information from chemicals deposited on a substrate or from air- and water-borne chemicals. This section focuses on the production, delivery, and detection of chemical signals. The functions of these signals are described in more detail in later sections.

### Production and Delivery of Chemical Signals

Male urodeles generally do not produce signals that attract females, although some species may exhibit a general attraction to conspecific chemical cues (Secondi, Haerty, and Lodé 2005). Typically males search for females in appropriate habitats (see Mating Systems). In some aquatic species, females facilitate mate location by males by producing chemical sex attractants. Behavioral experiments have shown that males of some species move toward chemical cues emanating from females, even without making direct contact (Zippelius 1948; Twitty 1955, 1966; W. C. Davis and Twitty 1964; Dawley 1984a; Malacarne and Giacoma 1986; Verrell 1986b; Andreoletti et al. 1988; Belvedere et al. 1988; Rowland, Robb, and Cortwright 1990; Cogălniceanu 1992, 1994; Park and Propper 2001; Park et al. 2004; Rohr et al. 2005; see further discussion under "Sex Recognition"). The origin of these sex attractants is not known. One likely source is material exuded from unspecialized glands in the skin, but in salamandrids, it also is possible that the ventral cloacal glands of females produce attractant pheromones (Houck and Sever 1994).

Once a male salamander has located and identified a female, he often courts the female using pheromones produced by specialized glands in the cloacal region, at the base of the tail, and on the head. The anatomy and histology of cloacal glands have been studied in all families of salamanders (Sever 1978, 1980, 1981, 1983, 1991a, 1992a, 1994; 2003b; Sever et al. 1990; Brizzi, Calloni, and Delfino 1990, 1992; Brizzi et al. 1995, 1996; Brizzi and Calloni 1992), but their role in chemical communication has been investigated only in salamandrids. The ancestral condition for salamanders with external fertilization is to have a relatively simple gland in the cloacal region of both males and females that produces chemical sex attractants (Sever 1991a). With the evolution of internal fertilization, cloacal glands of males evolved the secondary function of producing the spermatophore (see chapter 10).

In most urodele families, there are three sets of glands involved in spermatophore production: the pelvic glands, ventral glands, and Kingsbury's glands. One pelvic gland, which sometimes has more than one part, appears to be involved in production of courtship pheromones. In salamandrids, this has been called the dorsal or abdominal gland, although the latter term is now considered an inappropriate description of the gland's location (Sever et al. 1990). In all other salamander families, the putative pheromone-producing gland is called the vent gland, which appears to be homologous to the dorsal gland of male salamandrids. The term *dorsal gland* is somewhat confusing, because there are glands of the same name in female salamanders that are not homologous to the dorsal glands of males (Sever 1991a). Guillaume (2000b, c) described an entirely new type of clocal gland in *Proteus anguinus* that secretes onto the ventral skin around the cloaca and may be involved in producing chemical signals used to mark substrates (Guillaume 1999, 2000a). These glands are found in both males and females, but are larger and more active in males, which are territorial when breeding.

In all salamanders that have been studied, the cloacal glands of males involved in both spermatophore and pheromone production become enlarged in response to increases in male hormone levels during the breeding season; these are particularly conspicuous in salamandrids and ambystomatids. In salamandrids, secretory activity of these glands increases as well (Sever et al. 1990). Cedrini and Fasolo (1971) reported that dorsal gland extracts elicited strong responses in the olfactory bulb of female *Triturus,* and these glands appear to be the source of pheromones that help to bring females into a state of sexual receptivity (Malacarne et al. 1984; Malacarne and Giacoma 1986). Behavioral experiments with *Triturus carnifex* demonstrated that females are attracted to water containing courting male-female pairs in preference to water containing sexually inactive pairs or water alone (Malacarne and Vellano 1987). Females were not attracted to water containing males alone, however, suggesting that males do not produce a sex attractant. Instead, they probably release pheromones only when actively courting females. Kikuyama et al. (1995) reported that females of an Asian newt, *Cynops pyrrhogaster,* will orient toward water flowing from a chamber containing actively courting pairs, or to sponges treated with extracts from the cloacal glands. The pheromone that elicits responses from females is a peptide called sodefrin. A closely related species, *C. ensicauda,* produces a pheromone called silefrin, which has a similar structure. It is attractive to conspecific females, but not to females of *C. pyrrhogaster* (Kikuyama et al. 2002).

Pheromones probably are delivered to females during the tail fanning display of male newts (Halliday 1977a). In a study of the red-spotted newt (*Notophthalmus viridescens*), Verrell (1982) placed dye in the water near the male's cloaca; the dye moved toward the female's snout during courtship. In most salamandrids, secretory products of the dorsal gland empty into the cloaca, but in *Salamandra,* which has terrestrial courtship, glandular products discharge into skin furrows outside the cloaca (Brizzi et al. 1996). This suggests that males deposit glandular secretions on the ground during courtship. In another salamandrid with terrestrial courtship, *Echinotriton chinhaeiensis,* the male deposits a trail of mucus from his cloaca, but it is not yet clear whether the male deposits a pheromone that stimulates the female (Sparreboom, Xie, and Liang 2001). In *Euproctus* and *Calotriton,* the male clasps the female and deposits the spermatophore directly into her cloaca (see chapter 10). The delivery of chemical signals to the female seen in the courtship of other salamandrids seems to have been lost, and the dorsal gland is absent as well (Brizzi et al. 1995; Guillaume 1999). A similar and apparently independent loss of the dorsal gland has occurred in *Chioglossa lusitanica,* which also exhibits increased use of tactile signals and clasping during courtship (Brizzi et al. 1999).

The vent glands of other salamanders probably produce courtship pheromones as well, but behavioral experiments have not been performed with other families. The size and presumably the activity of vent glands vary among different groups of salamanders, and this probably reflects variation in the importance of vent gland secretions in courtship. Vent glands are relatively well developed in most salamanders that breed in water, such as proteids and ambystomatids. The female signals her receptivity by nudging the cloacal region of the male, but whether the female is orienting to pheromones produced by the male is not known (S. Arnold 1976, 1977). Vent glands tend to be smaller in most plethodontids, perhaps reflecting the increased importance of courtship glands in other regions, and they have been completely lost in some neotenic hemidactyliines (Sever 1991a, 1994, 2003b). On the other hand, vent glands are large in some plethodontids, including *Aneides lugubris, Pseudotriton montanus,* and a number of bolitoglossine salamanders (Sever 1994). Some of these species, such as *Pseudotriton,* lack pheromone-producing glands on the chin (Houck and Sever 1994), but others with enlarged vent glands do not. Until more behavioral work is done on these salamanders, the functional correlates of relative vent gland size will remain unclear.

Among urodeles as a whole, the occurrence of specialized glands on the head or chin that are used in courtship is a derived condition confined to some salamandrids and to plethodontids (Houck and Sever 1994; Sever 2003b). Noble (1931) referred to a variety of anatomically unrelated glands as hedonic glands, but Arnold (1977) proposed that these be termed courtship glands, and that individual glands be named by their anatomical position rather than their putative functions. In salamandrids, use of courtship glands differs among genera (Halliday 1990). In *Notophthalmus,* the male seizes the female around her neck with his hind legs and rubs her snout with genial glands on his cheeks (fig. 9.11 B). In *Taricha,* the male clasps the female with both pairs of legs and rubs her snout with a submandibular gland (W. C. Davis and Twitty 1964; Propper 1991). This seems to be an obligatory stage that can last for several hours before the male leads the female and deposits spermatophores (Arnold 1977). In the initial stages of courtship in the terrestrial genus *Salamandra,* the male approaches the female from above and rubs his throat on her snout, possibly to deliver pheromones from a submandibular gland (Joly 1966;

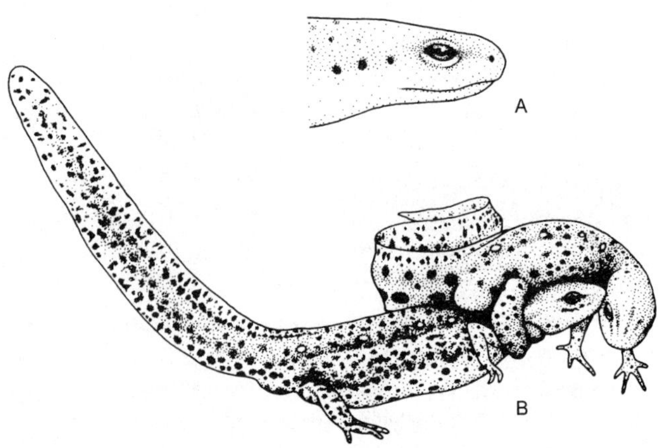


**Fig. 9.11.** Courtship of the North American red-spotted newt (*Notophthalmus viridescens*). (A) Secretory pits of courtship gland on side of the male's head. (B) Amplexus, with the male rubbing the courtship gland against the snout of the female while grasping her with the hind legs. After Halliday (1990).

Häfeli 1971). The male then moves under the female and clasps her in ventral amplexus. In *Triturus,* males lack courtship glands on the head, and, as discussed earlier, males do not clasp females during courtship (Halliday 1977a).

The structure and function of courtship glands have been studied in some detail in *Notophthalmus viridescens* (Pool and Dent 1977; Pool, Dent, and Kemphues 1977). Males and females have structurally similar glands on the sides of the head, but those of females do not normally produce secretory products. The courtship glands of males become functional during the breeding season, secreting a glycoproteinaceous product that accumulates in small epidermal pits on the side of the head (fig. 9.11 A). Secretory activity is induced by a combination of prolactin and testosterone. Administration of these hormones to females can induce glandular secretion as well, but normally females will not have circulating levels of testosterone high enough to produce secretions. The hormonal induction of courtship gland activity coincides with the development of secondary sexual characters such as broad tail fins and nuptial pads on the hind feet of males (Singhas and Dent 1975). Early experiments by Rogoff (1927) showed that a female would not follow a male or pick up spermatophores if the pores of the courtship glands were blocked, or if the female's nares were closed. If the female's snout was held against the male's glands, she would follow the male even if she had not gone through the preliminary stages of courtship. Subsequent work by Arnold (1972) and Verrell (1982) showed that use of the glands during courtship is facultative. When females are particularly receptive, the male skips this stage and proceeds directly to leading the female and depositing spermatophores.

In plethodontids, courtship glands are located on the chin and are termed mental glands (Sever 1976, 2003b). As in newts, seasonal enlargement of the glands occurs during the mating period and is correlated with high levels of circulating androgens (Woodley 1994). The structure of mental glands differs among different groups of plethodontids and roughly corresponds to the mode of pheromone delivery to females (fig. 9.12). The most widespread modes of pheromone delivery, which are considered ancestral for plethodontids, have been termed "pulling" and "snapping" (Arnold 1977; Arnold and Houck, 1982; Houck and Sever 1994). A male either rakes his premaxillary teeth across the female's dorsal skin, or he snaps his head down to puncture the skin (fig. 9.13 A, B). Both actions serve to vaccinate the female with mental gland secretions. Species with this mode of pheromone delivery typically have small mental glands and well-developed premaxillary teeth. In the two smallest species of desmognathines, *Desmognathus aeneus* and *D. wrighti,* males have an unusual mode of pheromone delivery derived from this ancestral mode. The male actually bites the female during courtship and holds on for several

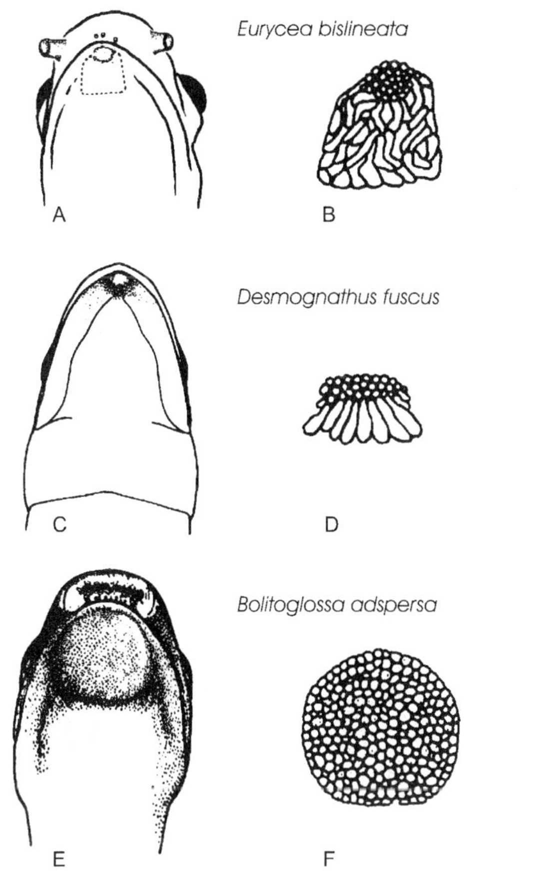


**Fig. 9.12.** Mental glands of plethodontid salamanders. (A) "Fan" type mental gland of *Eurycea bislineata.* The dashed line shows the part of the gland beneath the skin. The small circular area is the exposed secretory portion of the gland. The male also has enlarged cirri on the snout and elongated premaxillary teeth that protrude through the upper lip; these teeth are used to rake the skin of the female to deliver pheromones from the gland. (B) Fine structure of gland, showing horizontally oriented tubules that fan out from the terminal secretory pores. (C) Mental gland typical of the genus *Desmognathus,* a small protrusion at the tip of the chin. (D) Fine structure of the gland, showing vertical tubules and secretory pores, which are located only at the anterior portion of the protrusion. (E) Pad-shaped mental gland of *Bolitoglossa adspersa.* Four premaxillary teeth protrude through the upper lip of the male and are used with the gland to deliver pheromone to the female. Large species of *Plethodon* also have pad-shaped glands, but lack the protruding premaxillary teeth. (F) Cross-section of the gland, showing vertical tubules. The entire surface of the pad contains secretory pores. After Houck and Sever (1994).

minutes while delivering mental gland secretions (fig. 9.13 C, D; Arnold and Houck 1982; Promislow 1987; Verrell 1999). In these species, secretions of the mental glands are released through pores at the base of the male's elongated teeth, in contrast to the usual condition of being released through pores in the skin (Sever 1976). *Desmognathus wrighti* and *D. aeneus* are not sister species, so their unusual mode of pheromone delivery appears to have evolved independently (fig. 9.14). Similar behavior also occurs in *D. imitator,* which is not closely related to either of these species (Verrell 1994b; Mead and Verrell 2002).

In most small-bodied species of *Plethodon,* such as mem-

**Fig. 9.13.** Courtship behavior in the genus *Desmognathus.* (A) Male *D. orestes* (left) in position to initiate a snap, with premaxillary teeth in contact with the female (right). (B) Position of the male (left) after delivering a snap; the snapping action has thrown his body away from the female. (C) Male *D. wrighti* (left) biting the hind foot of the female (right), possibly delivering pheromones into her bloodstream. (D) Male *D. wrighti* biting the female's body, lifting her off of the ground. Photos by Stevan J. Arnold.

bers of the *P. cinereus* group, males transfer pheromones by vaccination before progressing to the tail-straddling walk. In large-bodied species of *Plethodon,* such as members of the *Plethodon jordani* and *P. glutinosus* complexes, males have very large padlike mental glands (fig. 9.12 E). They use a derived mode of pheromone delivery. During the tail-straddling walk, the male turns around and slaps his mental gland on the snout of the female (Arnold 1976; Marvin and Hutchison 1996; fig. 9.15). *Plethodon dorsalis* is a relatively small species from western North America that is phylogenetically intermediate between the small eastern *Plethodon* and the large-bodied species. In this species, the male applies pheromones by slapping or rubbing the mental gland on the female's snout, as in the large-bodied species, but does so before the tail-straddling walk, as in the *P. cinereus* group (Picard 2005).

A number of plethodontids in different clades lack both mental glands and pheromone delivery during courtship (fig. 9.14). Houck and Sever (1994) considered these to be instances of secondary loss of both glands and the behavior associated with pheromone delivery. In contrast, Beachy (1997a) suggested that the ancestral condition for all plethodontids was the absence of mental glands, and this condition has been retained in some groups, such as *Gyrinophilus, Pseudotriton, Ensatina,* and *Phaeognathus.* In this scenario, mental glands would have evolved independently several times within the family, perhaps accounting for the striking differences in mental gland structure in different lineages (fig. 9.12).

Arnold and Houck (1982) reported experiments designed to determine the function of mental gland secretions in *Desmognathus ocoee* (this population was formerly considered

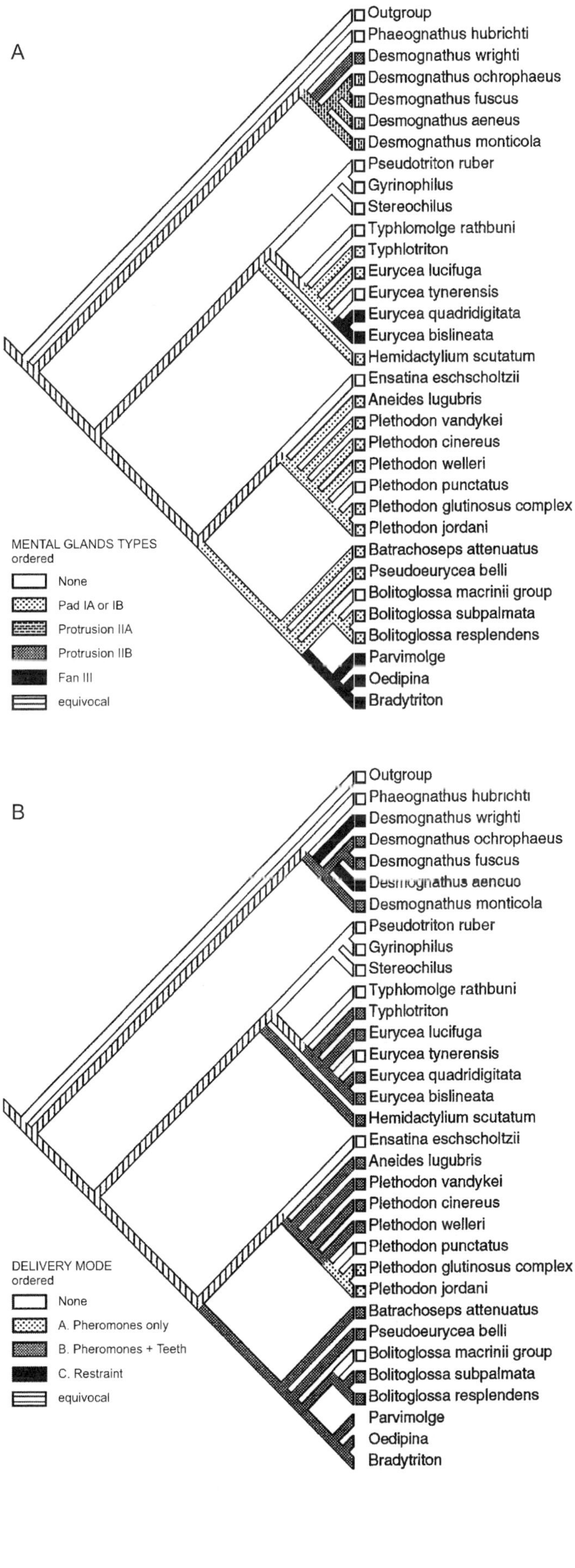


**Fig. 9.14.** Phylogeny of (A) mental gland types and (B) modes of pheromone delivery in plethodontid salamanders. The phylogeny is based on morphological and molecular characters and was derived independently of the mental gland and behavioral traits. The four major clades, from top to bottom, are the Desmognathinae, the Hemidactyliini, the Plethodontini, and the Bolitoglossini. More recent molecular phylogenies differ from this one in that desmognathines are thought to have evolved from within plethodontines, and the monophyly of the Hemidactyliini is doubtful. Consequently, the evolution of pheromone delivery modes may be more complex than depicted here. (A) Protrusion-type mental glands predominate in the Desmognathinae. All other groups have pad- or fan-type glands. Mental glands have been secondarily lost in several genera. (B) Most plethodontids use the premaxillary teeth to deliver pheromones from mental glands. Two species of *Desmognathus* have independently evolved a form of courtship in which the male bites the female and restrains her (see fig. 9.13). In large species of *Plethodon,* the male delivers pheromones by slapping the mental gland on the snout of the female, but does not use premaxillary teeth to rake the female's skin (see fig. 9.15). After Houck and Sever (1994).

Fig. 9.15. Male *Plethodon shermani* slapping his mental gland on the snout of a female during the tail-straddling walk. After Arnold (1976).

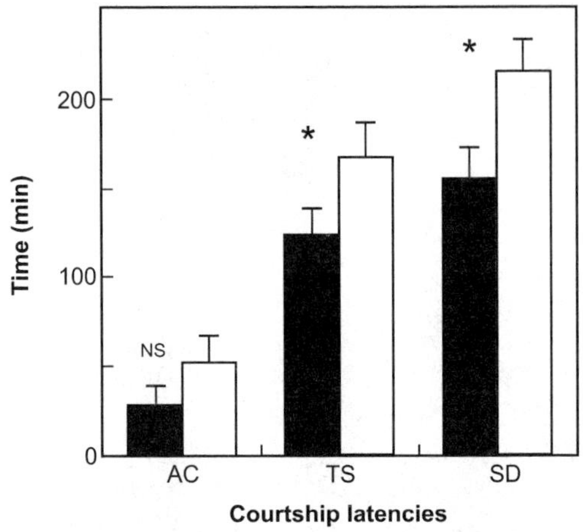


Fig. 9.16. Differences in timing of courtship events for 17 females of *Desmognathus ocoee* treated with pheromones (black bars) or saline (white bars). Data are mean latencies ± 1 SE. (AC) Latency to active courtship. (TS) Latency to tail-straddling walk. (SD) Latency to spermatophore deposition. Females treated with pheromones initiated the tail-straddling walk and deposited spermatophores significantly sooner than did those treated with saline. After Houck and Reagan (1990).

to be *D. ochrophaeus;* see Tilley and Mahoney 1996). They surgically removed mental glands, premaxillary teeth, or both from male salamanders and compared their success in inseminating females with that of control groups. Surprisingly, the absence of teeth and mental glands did not seem to reduce male mating success. Subsequent work by Houck and Regan (1990) on the same species did demonstrate that mental gland secretions can have an effect on courtship behavior. They removed mental glands from males and used extracts from the glands in courtship tests with females. Those females that received glandular extracts applied to the skin on filter paper progressed more rapidly to the tail-straddling walk stage of courtship than did control females, and the time to spermatophore deposition was shorter as well (fig. 9.16).

In a study of *Plethodon shermani,* a member of the *P. jordani* complex, Houck et al. (1998) applied mental gland extracts to the external nares of females, which were then courted by males from which the glands had been removed. Treated females were more receptive than control females, as shown by the shorter time to initiation of courtship and to active courtship by males. The pheromone involved consists mainly of two proteins. One of these, called plethodontid receptivity factor, is sufficient to increase female receptivity (Rollmann, Houck, and Feldhoff 1999), which is mediated by specific neurons in the female vomeronasal system that respond to the pheromone (Wirsig-Wiechmann et al. 2002). Modes of pheromone delivery are evolutionarily conservative in plethodontids, with courtship behavior being nearly identical among closely related species. In contrast, there is considerable variation in the biochemical structure of courtship pheromones among populations of the same species and between closely related species (Rollmann, Houck, and Feldhoff 2000), indicating strong directional selection on pheromones (Watts et al. 2004). Nevertheless, divergence in pheromone structure is not complete, and females of some species are stimulated by the pheromones of closely related species. For example, the pheromones of *P. montanus* and *P. yonahlossee* were as effective as conspecific pheromones in increasing female receptivity in *P. shermani* (Rollmann, Houck, and Feldhoff 2003). These species exhibit relatively low levels of divergence in the gene that codes for the courtship pheromone (Palmer et al. 2005). Indeed, species recognition is not perfect, and there is considerable genetic evidence of both past and present hybridization among closely related species (Highton and Henry 1970; Highton 1995; Highton and Peabody 2000; Weisrock, Kozak, and A. Larson 2005; see the following further discussion of "Species Recognition").

Plethodontids have glands located in other parts of the body that are used for chemical communication, but the functions of these glands are not as well understood as that of mental glands. Plethodontids engage in a tail-straddling walk during courtship, with the female placing her chin at the base of the male's tail (Arnold 1977; Marvin and Hutchison 1996; fig. 9.10). In several genera, including *Desmognathus, Eurycea, Plethodon,* and possibly *Hydromantes,* males have small glands on the dorsal surface of the tail base, where they probably deliver secretions to the nasolabial grooves of the female during the tail-straddling walk (Noble 1929c, 1931; Sever 1985, 1989; Brizzi, Calloni, and Delfino 1991; Houck and Sever 1994; Sever 2003b). These glands often are not visible on the surface of the skin, or form a slightly raised bump. They have been studied histologically in only a few species, and there have not been any experiments demonstrating the signal function of their secretions. Given the ubiquity of the tail-straddling walk in plethodontids, it is likely that such glands occur in many other species that have yet to be examined.

Some plethodontids also have glands used in chemical communication on the ventral side of the tail, just behind the cloaca (Noble 1929c). In *Plethodon cinereus,* the postcloacal gland appears to be used to mark fecal pellets with chemicals, which in turn are used to advertise ownership of a territory (Simons and Felgenhauer 1992; Jaeger and Gabor 1993; Simons, Felgenhauer, and Jaeger 1994). These glands are found in males and females, both of which are territorial (Horne and Jaeger 1988), and active glands have even been found in juvenile red-backed salamanders (Simons, Jaeger, and Felgenhauer 1995; see the subsequent further discussion of territoriality in this species). Postcloacal glands also have been described in both males and females of *Aneides lugubris* (Staub and Paladin 1997). Although present in both sexes, they are larger in males. Their precise function is not known, but it is possible that they are used to mark substrates for territorial advertisement. Members of this genus are known to be very aggressive toward conspecifics (Cupp 1980; Staub 1993). Substrate marking with fecal pellets, presumably carrying glandular secretions, also has been observed in the genus *Salamandra* (Gautier and Miaud 2003). Postcloacal glands appear to be modified granular glands that originally produced defensive chemicals (see chapter 14), and courtship glands on the dorsal surface of the tail are modified granular glands as well (Sever 1989; Brizzi, Calloni, and Delfino 1991; Staub and Paladin 1997). In contrast, mental courtship glands appear to have evolved from mucous glands (Houck and Sever 1994).

## Detection of Chemical Signals

There are several ways in which chemical signals can be detected by salamanders: by unspecialized receptors in the skin, by specialized receptor organs in the nasal cavity, or possibly by brain receptors responding to pheromones that are presumed to be delivered directly into the bloodstream (as in the courtship of plethodontids that involves pulling, snapping, or biting with the teeth). Dawley (1998) provided a detailed review of the structure and function of nasal receptor organs in salamanders, as well as the physiology and neurobiology of chemoreception. Only a brief summary will be given here.

The chemosensory organs in the nasal cavities of salamanders are of two types—the main olfactory chamber and the vomeronasal organ. The chemosensory epithelium in the main olfactory chamber is used primarily to detect volatile airborne odors (in terrestrial salamanders) or waterborne chemicals (in aquatic species). In aquatic salamanders, the sensory epithelium is located in grooves in the nasal epithelium that run parallel to the direction of water flow through the nasal cavity, a structure that presumably facilitates detection of chemicals in the water (Dawley 1998). In contrast, the sensory epithelium is arranged in a flat sheet in terrestrial species, an arrangement that probably maximizes the surface area exposed to air flowing through the nasal cavity. In terrestrial salamanders that have aquatic larvae, the larvae have grooves lined with sensory epithelium, but the epithelium becomes more flattened at metamorphosis (Arzt et al. 1986; Dawley 1998). In some aquatic species, such as *Necturus,* the olfactory epithelium is much thicker than that of most other vertebrates (Farbman and Gesteland 1974; Graziadei and Monti Graziadei 1976). It is not clear whether this reflects a greater sensitivity to chemicals, or is simply a consequence of the large cell size characteristic of salamanders.

The vomeronasal organ is located in a special pocket, or diverticulum, off of the main olfactory chamber. In some terrestrial salamanders, such as *Ambystoma maculatum,* only a small portion of the wall of this chamber is lined with sensory epithelium, suggesting a relatively minor role for the vomeronasal organ. In contrast, a larger area of sensory cells is found in *Triturus* and especially in plethodontids, which have most of the diverticulum lined with sensory cells (Dawley 1998). In *Plethodon cinereus,* the vomeronasal chamber is distinctly separate from the main olfactory chamber (fig. 9.17; Dawley and Bass 1988). In addition, plethodontids have unique structures, the nasolabial grooves, which deliver chemicals directly to the vomeronasal organs, bypassing the main olfactory chamber (Dawley and Bass 1989). The nasolabial grooves are equipped with glands that secrete a lipid compound (C. Brown and Martof 1966; Sever 1975b, 1980). This fluid picks up chemicals when the sala-

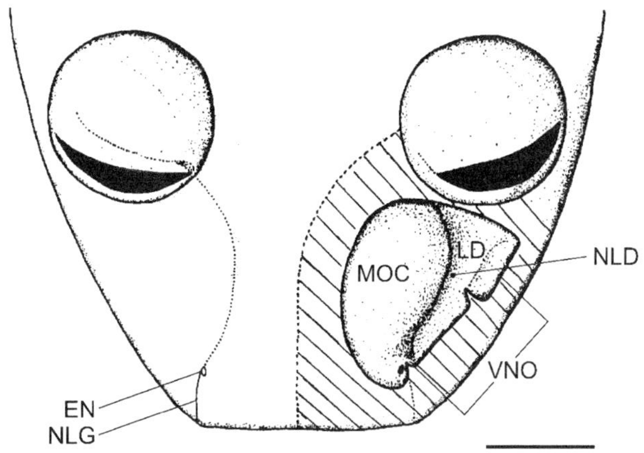


**Fig. 9.17.** Nasal chemosensory organs of the red-backed salamander (*Plethodon cinereus*). Dashed line indicates where skin and parts of the skull were removed to reveal the nasal organ. Abbreviations: EN = external nares; LD = lateral diverticulum; MOC = main olfactory chamber; NLD = nasolacrimal duct; NLG = nasolabial groove; VNO = vomeronasal organ. After Dawley and Bass (1988).

mander touches its snout to the substrate (nose-tapping) or to another salamander, and the fluid and chemical molecules are then carried by capillary action to the vomeronasal organs (C. Brown and Martof 1966; Dawley and Bass 1989).

The relative sizes of the chemosensory organs of salamanders are correlated to some extent with habitat and behavior, but they reflect phylogenetic history as well. Some paedomorphic, fully aquatic salamanders, such as cryptobranchids and amphiumids, have greatly reduced vomeronasal organs (Jurgens 1971). In contrast, secondarily aquatic species with terrestrial ancestors have vomeronasal organs as large or larger than those of terrestrial relatives (Dawley 1992a), a pattern seen in aquatic caecilians as well (A. Schmidt and Wake 1990). In salamandrids and plethodontids, species that have an aquatic larval stage tend to have larger vomeronasal organs, with a more complex projection of nerves from the vomeronasal organ to the accessory olfactory bulb than do strictly terrestrial species with direct development or viviparity (Schmidt, Naujoks-Manteuffel, and Roth 1988; Dawley 1992a, 1998). For example, among plethodontids, semiaquatic species such as *Eurycea wilderae, E. guttolineata,* and *Desmognathus quadramaculatus* have larger vomeronasal organs in relation to body size than do either *Plethodon cinereus* or *Desmognathus wrighti,* both of which have direct development. On the other hand, bolitoglossine salamanders, all of which have direct development, not only tend to have relatively large vomeronasal organs relative to body size, but have large main and accessory olfactory bulbs in the brain as well (Dawley 1998). This suggests that there are phylogenetic patterns in vomeronasal organ evolution for which ecological or behavioral correlates have yet to be identified.

Plethodontid salamanders also exhibit sexual dimorphism in vomeronasal organs that is correlated with sexual differences in behavior. Both the vomeronasal and main olfactory organs are larger in males than in females (Dawley, 1992a, b), probably reflecting the tendency for males to search for and identify potential mates using chemical cues. In addition, *P. cinereus* exhibits seasonal changes in the size of the vomeronasal organ. The organ increases in size in both males and females during the summer, prior to the fall and spring mating season (Dawley and Crowder 1995), but at a time when both males and females tend to be highly territorial (Jaeger 1986). This is due to a seasonal proliferation of new sensory cells (Dawley et al. 2000). The ability to proliferate new cells is an unusual feature of vertebrate olfactory systems and probably reflects a need to regenerate sensory cells that have been damaged by chemicals (Dawley 1998). In the case of *P. cinereus,* and probably other plethodontids, this capacity to regenerate new cells has been adapted to seasonal changes in the need to detect chemical cues related to mate location, courtship, or territorial advertisement. Rather than being triggered by damage to sensory cells, the process leading to the production of new cells is initited by a regular seasonal cycle of events (Dawley et al. 2000).

Seasonal changes in the olfactory system and concomitant changes in sensitivity to chemical signals very likely are hormonally regulated. Thompson and Moore (2000) found that male newts (*Taricha granulosa*) treated with arginine vasotocin (AVT) spent more time with models of females impregnated with female sex pheromones than did untreated males. This is consistent with previous work showing that blocking AVT interferes with courtship behavior in this species (F. L. Moore and Miller 1983). Levels of AVT vary seasonal and increase during the courtship season (F. L. Moore and Zoeller 1979), so it seems reasonable to assume that the sensory system of the male becomes primed by AVT to respond to chemical signals produced by females.

## Species, Sex, and Individual Recognition in Urodeles

The evidence that chemical signals are of primary importance in salamander communication comes mainly from observations of courtship and aggressive behavior. In contrast to the large literature on the ability of anurans to discriminate between different types of calls (chapter 7), there have been fewer experimental studies of chemical discrimination in salamanders. Such experiments are essential for determining the ability of either male or female salamanders to distinguish the odors of different species, sexes, or individuals. To ensure that visual and tactile cues have been eliminated, discrimination experiments must be designed so that the animals being tested do not actually contact the animals providing odor cues. Generally, odor discrimination has been tested by presenting test animals with a choice of previously marked substrates, or by presenting air- or water-borne odors from living salamanders in an olfactometer. Unfortunately, if the species being tested normally detects chemical signals of other individuals only through direct contact, then negative results in these experiments will not necessarily indicate an inability to discriminate odors. Nevertheless, these experiments can provide information on minimum discriminatory abilities.

### Species Recognition

Many investigators have reported that male salamanders usually do not attempt to court females of the wrong species, especially if they have access to females of their own species (e.g., W. C. Davis and Twitty 1964; Arnold 1976, 1977). In *Taricha,* even blinded males easily distinguished between females of different species. In laboratory tests, however, males could be induced to attempt courtship with heterospecific females if water previously inhabited by conspecific females was added to test aquaria (W. C. Davis and Twitty 1964). This clearly indicates that chemical signals are important in species recognition.

Dawley (1984b, 1986b, b, 1987) investigated the ability of several large species of *Plethodon* to distinguish between conspecifics and heterospecifics on the basis of airborne odors alone. Test animals were placed in a glass Y-tube apparatus (fig. 9.18), which received odors from jars containing salamanders of different species. Dawley scored responses to odors by observing whether the test animal moved down the right or left arm of the apparatus. She tested the responses of males in several pairs of closely related species that do not hybridize at the collection sites: *Plethodon teyahalee* and *P. aureolus* from Monroe County, Tennessee; *P. teyahalee* and *P. jordani* from Madison County, North Carolina; and *P. glutinosus* and *P. kentucki* from Wise County, Virginia. These salamanders have undergone a series of confusing name changes as various cryptic species have been identified within what once were considered two widespread species, *P. glutinosus* (Highton 1989, 1995) and *P. jordani* (Highton and Peabody 2000). The Tennessee and North Carolina populations of the *P. glutinosus* complex were referred to as *P. glutinosus* in Dawley (1984b, 1986b), *P. teyahalee* in Dawley (1986a, 1987), and *P. oconaluftee* in Hairston (1993) and Petranka (1998). *Plethodon teyahalee* is name currently in use (Highton and Peabody 2000). *Plethodon aureolus* (Highton 1983) was called "Species A" in Dawley (1984b), while a population of *P. jordani* from Lemon Gap in Madison County, North Carolina, tested by Dawley (1986a), is now called *P. montanus* (Highton and Peabody 2000). Despite this taxonomic confusion, there is no doubt that the pairs tested in all of these experiments represent distinct species.

In all cases, males showed a clear preference for odors of their own species over those of heterospecifics. When males were given the opportunity to court a conspecific and a heterospecific female, they seldom showed any interest in heterospecific females (Dawley 1986b). The responses of females of these species to odors of conspecific and heterospecific males were less consistent. Most females tested during the breeding season showed a preference for odors of conspecific males over heterospecifics, but those tested outside of the breeding season either did not show a preference or actually avoided conspecific odors. The latter response could reflect avoidance of aggressively territorial males at times when females are not interested in finding mates. When salamanders were tested from populations known to hybridize in the field, the strength of behavioral preferences for conspecifics over heterospecifics was weaker in at least one of the species, suggesting that hybridization results from weak behavioral isolation (Dawley 1987). Recent work on the chemical structure of courtship pheromones in the *Plethodon jordani* complex has revealed considerable variation among geographically separated populations (Rollmann, Houck, and Feldhoff 2000; Watts et al. 2004). Nevertheless, the structure of the pheromones remains sufficiently similar that females of *P. shermani* were stimulated by courtship pheromones of closely related species (Rollmann, Houck, and Feldhoff 2003).

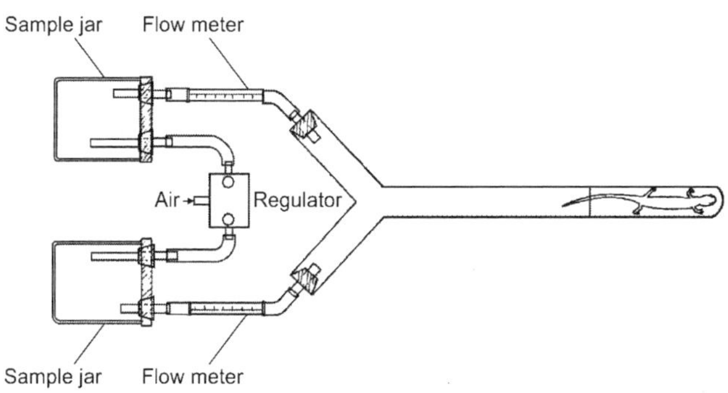


**Fig. 9.18.** Y-tube apparatus used to test for odor discrimination in experiments with members of the *Plethodon glutinosus* complex. The salamander was placed at the end of the tube, facing away from the arms of the apparatus to prevent it from running quickly down one arm of the Y-tube. After Dawley (1984a).

A considerable amount of work has been done on species recognition, sexual compatibility, and speciation in small streamside salamanders in the genus *Desmognathus* (Arnold, Reagan, and Verrell 1993; Verrell and Mabry 2000, 2003; Verrell 2003). This genus contains many very similar species, some of which have overlapping ranges. Genetic studies, based mostly on electrophoretic markers, have shown a wide range of variation in the degree of genetic differentiation among populations (Tilley et al. 1978; Tilley, Verrell, and Arnold 1990). Populations originally assigned to the widespread species *D. ochrophaeus* are especially confusing and have recently been partitioned into four cryptic species with ranges that show little or no overlap. Levels of genetic differentiation among these species (*D. ochrophaeus, D. ocoee, D. carolinensis,* and *D. orestes*) suggest that they have been evolving independently for considerable periods of time (Tilley and Mahoney 1996). Two other members of the *D. ochrophaeus* complex, *D. imitator* and *D. apalachicolae,* were previously separated as distinct species (Tilley 1981, 1985; Means and Karlin 1989). This pattern of differentiation into many very similar species is somewhat like the pattern seen in other complexes of cryptic salamander species (Highton 1989, 1995, 1997; Good and Wake 1992, 1993). The resulting changes in nomenclature mean that many behavioral studies refer to older names; careful attention to the geographic location of animals used in these studies is required to determine which species were being investigated (see Tilley and Mahoney 1996, for a guide to the literature on *Desmognathus*).

Behavioral experiments with many different populations in this complex have shown that sympatric populations of distinct species usually are strongly isolated from one another and do not mate with heterospecifics, even when not given a choice of mates. For example, in tests of sympatric populations of *D. ocoee* and *D. imitator* from the Great Smoky Mountains in Tennessee, Verrell (1989a) found that only 5% of heterospecific matings resulted in insemination of females. In most cases, males ignored heterospecific females and did not even initiate the preliminary stages of courtship. When males were tested in a Y-tube apparatus, both species showed a strong preference for conspecific female odors, indicating that chemical cues are used for species recognition. More detailed studies have been done on sympatric populations of *D. orestes* and *D. fuscus* from Mt. Rogers, Virginia. Houck, Arnold, and Hickman (1988) found nearly complete behavioral isolation between these populations in no-choice tests with heterospecifics. In a subsequent study, Uzendoski and Verrell (1993) found that only one of 60 courtship trials between heterospecifics resulted in deposition of a spermatophore by the male, and none resulted in insemination of the female. Again, males tended to ignore heterospecifics, and Y-maze tests showed a strong preference of males for conspecific female odors. Males of *D. orestes* that were reared in isolation in the laboratory showed the same lack of interest in females of *D. fuscus* as did field-collected males, indicating that prior experience is not necessary for species recognition (Verrell 1994c).

In addition to tests of sympatric populations that clearly belong to distinct species, there has been extensive investigation of behavioral isolation among allopatric populations of the *D. ochrophaeus* complex. Studies involving thousands of courtship trials have shown that behavioral isolation tends to be strongest in populations that are most widely separated geographically (Tilley et al. 1990; Verrell 2003). Although degree of genetic differentiation also was correlated with geographic distance between populations, genetic differentiation did not account for a significant amount of variation in degree of sexual isolation when geographic distance was held constant. It appears that gradual divergence of allopatric populations within the *D. ochrophaeus* complex has led to gradual evolution of reproductive isolation. Those populations now considered to be distinct species (*D. ocoee, D. orestes, D. carolinensis*) generally exhibit the strongest degree of behavioral isolation. Geographically close populations now considered to be the same species (Tilley and Mahoney 1996) generally exhibit low to moderate levels of isolation (Arnold, Verrell, and Tilley 1996; Herring and Verrell 1996), but may exhibit preferences for mates from their own population over those from other populations (Verrell 2003). Significant behavioral isolation also has been found among allopatric populations of *D. santeelah*, suggesting that these are in the process of splitting into distinct species as well (Maksymovitch and Verrell 1993).

In all of these salamander complexes, it appears that the complex topography of the southern Appalachians has resulted in populations becoming geographically isolated from one another for periods of time sufficient for gradual genetic divergence and divergence of mate recognition systems. Although the courtship behavior of most species of *Desmognathus* is very similar (Houck and Verrell 1993; Herring and Verrell 1996; Verrell and Mabry 2000, 2003), there are quantitative differences in the frequencies of different courtship behaviors. In addition, gradual divergence in chemical signals probably is responsible for much of the behavioral isolation among these populations (Verrell 2003). Not all plethodontids show this pattern, however. For example, there appears to be relatively little behavioral isolation among widely separated populations of *Eurycea wilderae* from different watersheds in North and South Carolina (Kozak 2003).

There has been some work on reproductive isolation in newts of the family Salamandridae. Although sympatric species normally breed only with members of their own species, hybridization between different species of newts is widespread but relatively infrequent in most regions (T. R. Halliday 1990; Arntzen et al. 1998; Michalak and Rafiński 1999; Babik, Szymura, and Rafiński 2003). Tests with allopatric populations of different species of Asian newts (*Cynops*) showed that interspecific courtship sometimes proceeded to spermatophore deposition, but usually the female failed to pick up the spermatophore (Kawamura and Sawada 1959; Sawada 1963). This was attributed to differences in the temporal patterning of courtship behavior. Michalak, Grzesik, and J. Rafiński (1997) tested for degrees of isolation between allopatric populations of two very closely related species, *Triturus vulgaris* and *T. montandoni*. Males in heterospecific tests were less likely to display to females than those in homospecific tests, and when display did occur, courtship was less likely to proceed to spermatophore deposition or transfer in hetersospecific pairings. Females were even less likely to mate with heterospecific males if they were given a choice between conspecific and heterospecific males (Michalak and Rafiński 1999). The overall courtship behavior of these two species is very similar, but color patterns of males are quite different and might be important for species recognition. It also is possible that chemical signals delivered during courtship displays differ, but this has not been investigated.

Much less work has been done on species recognition in other families of salamanders. In general, salamanders in other families tend to be well differentiated from sympatric species, but occasional hybridization has been reported. Present or past hybridization among different species of *Ambystoma* has resulted in the production of triploid hybrids, which usually live in association with one of the parental species. Reproduction by hybrid females depends on matings with males of a parental species. However, the male genome is not permanently incorporated into the offspring, so males that mate with hybrid females instead of conspecifics suffer lost reproductive success (Bogart 1989; Licht 1989; Lowcock 1989). In many populations, hybrid females outnumber diploid females of the parental species, so such matings are likely to be common (Bogart and Klemens 1997). However, males of one parental species, *Ambystoma jeffersonianum,* can distinguish the odors of conspecific females from those of triploid females associated with them and show a strong preference for conspecifics (Dawley and Dawley 1986). This would ensure greater fitness of males and help maintain males in the population, because hybrid offspring almost always are females.

### Sex Recognition

It is possible that salamanders could distinguish between their own odors and those of other species, but not obtain any information about the sex of these individuals. Olfactory cues are widely assumed to be important in sex recognition in these animals, but it is surprising how few experiments to test chemical sex recognition have been performed. Twitty (1955, 1966) placed sponges impregnated with skin secretions from females of *Taricha rivularis* in streams and found that the sponges attracted males from long distances downstream. Control sponges without secretions did not attract males. Unfortunately, sponges bearing male secretions were not tested. Similar results were obtained in experiments with *T. granulosa* using pheromone-soaked realistic rubber models of female newts (Thompson et al. 1999). Verrell, Strand, and Hanson (2001) impregnated sponges with chemicals from females of *Ambystoma macrodactylum;* the sponges were then placed in traps in ponds. Males were attracted to sponges with female cues, but not to sponges alone, and males also entered traps containing live females. Females, on the other hand, were not attracted to males. Gauss (1961) reported that males of *Triturus* mark branches and other surfaces in ponds by pressing their cloacal glands to the substrate. Females passing through the area investigated these marks, and Gauss suggested that females are attracted to odors of males. This has not been confirmed by any other investigators, however.

In *Notophthalmus,* males can distinguish between odors of males and females and are strongly attracted to female odors. However, females did not show discrimination in laboratory experiments, nor did they tend to move toward male odors (Dawley 1984a). This is consistent with behavioral observations on newts that suggest males initiate courtship after chemically identifying females (Verrell 1982). When several male newts are attempting to court a single female, the combined pheromones of the males actually repel other males, even though males are attracted to chemical cues from noncourting males (Park and Propper 2001). It is not clear, however, whether males in the field use chemical cues from other males to avoid situations where competion for females would be intense. Rohr et al. (2005) suggested that males assess local operational sex ratios from chemical cues. They found that males were more attracted to cages housing females alone than to those with both males and females, especially early in the breeding season. Later in the season, males were equally attracted to caged females with and without courting males. They suggested that this indicated a tendency for late-season males to be more likely to accept competitive situations, but in fact the lack of difference in response was due mainly to a reduction in overall responsiveness and not to increased responsiveness to cages that included males.

In another aquatic species, the European cave salamander (*Proteus anguinus*), neither males nor females exhibited clear differences in responses to chemical cues emanating from males or females in the water, nor did they discriminate between substrates previously inhabited by males or females. Nevertheless, a male evidently can determine the sex of another individual within a few seconds of making contact with his snout. Males and sexually immature individuals evoked an aggressive response, but females did not (Parzefall 1976). Breeding males are territorial and use chemical signals to mark their courtship territories, but nonbreeding males and females often are gregarious and apparently use chemical cues to locate conspecifics under suitable shelters (Guillaume 2000a, 2002). In contrast, the stream-breeding salmander *Calotriton* (= *Euproctus*) *asper,* which sometimes lives in caves, shows little site attachment and does not mark substrates with chemical cues. They appear to tolerate other individuals under shelter sites, but are not positively attracted to them (Guillaume 2002) and apparently do not use chemical cues for sex recognition (Guillaume 1999).

Dawley (1984b) investigated chemical sex recognition in *Plethodon montanus* and several members of the *P. glutinosus* complex. She found that both nonbreeding and breeding males and females could distinguish sex by airborne odors in an olfactometer. When given a choice between male and female odors, or between female odors and a blank, males oriented toward female odors. Males showed no preference in tests of male odors versus blanks. Females consistently avoided odors of conspecific males. They even chose odors of heterospecifics over conspecific males, perhaps because of a strong tendency to avoid territorial males during the nonbreeding season. Neither males nor females differed in their reactions to substrate odors of nonbreeding males and females. These results contrast with those of Jaeger and

Gergits (1979), who found that adults of *Plethodon cinereus* could distinguish sex from substrate odors. Both males and females tended to avoid male-marked substrates, and subsequent work has shown that both males and females use chemicals to mark territories (see the following discussion). Avoidance of male-marked substrates also has been reported in *P. vehiculum,* a species that does not appear to be strongly territorial in the field (Ovaska 1988a). A. Evans, Forester, and Masters (1997) reported that females of *Desmognathus ochrophaeus* preferred substrates marked by males from a different population to those marked by males from the same population. Their experimental design did not, however, distinguish between attraction to foreign male odors and avoidance of familiar male odors.

### Individual Recognition

Many salamanders probably can distinguish their own chemical signals from those of other individuals, and in some cases, they can distinguish between the signals of different individuals (Mathis et al. 1995). In the blind, aquatic European cave salamander, *Proteus anguinus,* nonbreeding adults could distinguish between cover objects that had never been occupied and those previously occupied by other salamanders, and they preferred the occupied sites even when odor cues were five days old (Parzefall 1976). When given a choice between cover objects marked with their own odors and those of other individuals, they often preferred those of other individuals. This species is gregarious outside of the breeding season and apparently uses substrate odors to locate communal hiding places (Briegleb 1962; Gillaume 2000a). Adults preferred water that had been inhabited by at least three other salmanders over fresh water, but they showed no difference in response to water inhabited by one or two salamanders compared to control water (Parzefall 1976).

Adult *Proteus* apparently can learn the odors of particular individuals if the odor is associated with negative reinforcement. Parzefall (1976) introduced males into the defended territories of other males and allowed the intruders to be attacked. These males were then placed in a new aquarium and given a choice of cover objects marked with their own odors or those of their attackers. Even on the day after being attacked, the salamanders showed a strong avoidance of the odors of the males that had attacked them. Males given a choice of odors from an unknown sexually immature individual and a previous attacker also avoided the attacker's odor. In contrast, test males that had not been previously attacked showed no difference in their reactions to odors of mature males and sexually immature individuals. Odor discrimination has been investigated in less detail in another proteiid, *Necturus maculosus.* Adults could distinguish between control water and water inhabited by other individuals and preferred the latter. They also preferred substrates marked with odors of their own species to unmarked substrates, but their ability to distinguish the odors of different individuals is unknown (Parzefall, Durand, and Richard 1980).

Some larval ambystomatids apparently use chemical cues to identify close relatives. Such kin recognition can mediate aggressive or cannibalistic interactions among larvae (Walls and Roudebusch 1991; Pfennig and Collins 1993; Pfennig, Sherman, and Collins 1994; Walls and Blaustein 1995; Hokit, Walls, and Blaustein 1996); this is discussed in more detail in chapter 12. There is no evidence that aquatic salamander larvae can identify particular individuals, as opposed to classes of individuals, from chemical cues. However, in laboratory tests with recently metamorphosed juvenile marbled salamanders (*Ambystoma opacum*), Walls (1991) found that when animals were siblings, they showed different behavioral responses to familiar and unfamiliar individuals. This was interpreted as evidence for possible neighbor recognition, but the spatial dispersion of juveniles of this species in nature is unknown. There is some evidence for discrimination among familiar and unfamiliar individuals in *Notophthalmus viridescens* as well (Wise et al. 1993), but how this relates to spatial interactions among individuals in the field is unclear.

Terrestrial salamandrids of the genus *Salamandra* often exhibit strong site attachment to particular shelter sites, moving relatively little and sometimes using the same sites for many years (see chapter 6). Laboratory experiments with two species, *S. lanzai* and *S. atra,* showed that these salamanders can use fecal pellets to identify shelter sites. They also seem to be able to identify the individual residents of particular sites and therefore use fecal pellets as territorial markers (Gautier and Miaud 2003). Territorial behavior and scent-marking in this species develop in older juveniles before they reach sexual maturity, which occurs at about eight years. Young juveniles show little reaction to odors of other salamanders, and other salamanders do not react to odors of young juveniles (Gautier, Léna, and Miaud 2004). In contrast to the territorial behavior of these species of *Salamandra,* the closely related species *S. luschani* is gregarious and uses chemical cues to locate suitable rock crevices, which are used as shelters. The more gregarious nature of this species apparently is related to the aridity of its habitat and a shortage of suitable retreat sites (Gautier et al. 2006).

The most detailed work on individual odor recognition has been done with plethodontids. Tristram (1977) reported that *Plethodon cinereus* given a choice of substrates marked with their own odors or those of unfamiliar conspecifics tapped their naso-labial grooves to the substrate more frequently in response to their own odor. Jaeger and Gergits (1979) subsequently showed that nonbreeding males spent more time on substrates marked with their own odors than on those marked with odors of conspecific males. Females exhibited a weaker tendency to avoid odors of conspecific males, but did not differ in their reactions to their own odors and those of unfamiliar females. Nevertheless, *P. cinereus* females are territorial in the field and mark substrates with fecal pellets and pheromones secreted by glands on the underside of the base of the tail, as do males (Horne and Jaeger 1988; Simons and Felgenhauer 1992; see further discussion that follows). Although salamanders in the field frequently investigate fecal pellets by contacting them with their snouts, they also are capable of detecting airborne components of these territorial markers (S. Martin, Jaeger, and Prosen 2005). Avoidance of substrates marked by conspecifics has been reported in *P. dunni* and *P. serratus* as well, two small species that both appear to be territorial (Ovaska and Davis 1992; Mathis, Deckard, and Duer 1998). Dawley (1984b) performed a similar experiment with *P. glutinosus* from three different localities in Virginia and Tennessee. Salamanders were tested during the nonbreeding season, and most males and females preferred substrates marked by other individuals of the same sex to substrates marked with their own odors. Attraction to substrates marked by conspecifics also has been reported in males of two species from the Ozarks, *P. caddoensis* and *P. ouachitae* (Anthony 1993), but the reason for this behavior is unclear. Individuals of *Batrachoseps attenuatus* also exhibit discrimination between substrates marked with their own odors and those of other individuals (Gillette 2002).

In addition to discriminating between their own odors and those of other individuals, some terrestrial plethodontids also can discriminate between odors of different individuals. McGavin (1978) tested odor preferences of nonbreeding *P. cinereus* that had been exposed to the odors of other individuals for two weeks. Both males and females performed significantly more nose-taps in response to the odors of familiar salamanders than to the odors of unfamiliar ones. Jaeger (1981b) exposed adults of the same species to the odors of other individuals for a week and then staged encounters between test animals and animals with both familiar and unfamiliar odors. The test animals were significantly less aggressive toward familiar conspecifics. This type of "dear-enemy" recognition has been reported in many territorial birds and probably provides a mechanism to reduce the cost of aggression toward individuals that do not represent an immediate threat to the owner's territory (E. O. Wilson 1975). Females also exhibit "dear-enemy" recognition in interaction with other females and are more aggressive toward unfamiliar individuals (Jaeger and Peterson 2002).

Madison (1975) reported that *P. metcalfii* from Highlands, North Carolina, also can distinguish odors of familiar and unfamiliar individuals. However, his experiments differed from those with *P. cinereus* in that test animals were not exposed to odors of other individuals in the laboratory. Instead, animals collected within three m of one another were considered "neighbors," and those collected at least 30 m apart were considered "non-neighbors." Salamanders were given a choice of airborne odors of neighbors and non-neighbors in an olfactometer. Outside the breeding season, both males and females showed a significant preference for neighbors, but females shifted to a preference for non-neighbors in the breeding season. It is not clear how these results relate to the mating system of this species.

## Aggression, Spacing, and Territoriality in Urodeles

Salamanders exhibit aggressive behavior in a number of different contexts. Sometimes individuals fight over discrete resources, such as prey items. Such behavior has been observed in plethodontids, salamandrids, and both adult and larval ambystomatids, mostly in captivity (Licht 1973; Thurow 1976; Walls and Jaeger 1987; Ducey 1989; Walls 1990; Ducey and Heuer 1991; Walls and Semlitsch 1991; Mathis et al. 1995; V. Townsend and Jaeger 1998). Evidence for such behavior in the field is largely anecdotal, because researchers are unlikely to encounter two individuals just at the moment they find the same prey item. A special case of resource-item competition is fighting over females by males during the mating season. This is common in many salamanders and is discussed in more detail in the section on "Mating Systems," following.

Some salamanders also exhibit strong attachment to specific sites and advertise and defend them against other individuals. This type of site defense is similar to that seen in lizards that are not otherwise territorial, such as skinks, lacertids, teiids, and some varanids (Stamps 1977; E. Martins 1994). In salamanders, the sites most commonly defended are retreat sites, such as logs, rock crevices, or tunnels, and nest sites where females are brooding eggs (Mathis et al. 1995). Much of the evidence for site defense comes from studies of captive animals, although brief observations have been made of salamanders in the field defending either retreat sites or nest sites against conspecifics (R. Gordon 1952; Hutchison 1959; Brandon and Huheey 1971; Hillis and Bellis 1971; Nickerson and Mays 1973a; Nussbaum, Brodie, and Storm 1983). Some salamanders might go further and defend all or part of the home range as an exclusive territory. Evidence for home range defense comes mainly from studies showing lower than expected overlap among home ranges of different individuals (Nishikawa 1990; Mathis 1991b) and a few observations of aggressive interactions away from obvious retreat sites (Gergits and Jaeger 1990a). In practice, it often is difficult to distinguish between complete or partial home range defense and site defense. In some species, these two forms of aggression grade into one another, with individuals defending home ranges to some extent, but showing much stronger defense of specific retreat sites or nests. In the following discussion, both site defense and home range defense are considered forms of territoriality.

### Territorial Behavior of the Red-backed Salamander

Due to the efforts of Robert Jaeger and members of his laboratory, we now know more about the territorial behavior of the red-backed salamander (*Plethodon cinereus*) (fig. 9.19) than that of any other salamander (Jaeger 1986; Jaeger and Forester 1993; Mathis et al. 1995). Red-backed salamanders are inhabitants of deciduous forests throughout much of eastern North America and can reach densities of several individuals per square meter. Burton and Likens (1975b) estimated that the biomass of these salamanders in the Hubbard Brook Experimental Forest in New Hampshire exceeded that of birds during the breeding season. The high density of some populations means that competition for food and suitable retreat sites is likely to be intense.

Red-backed salamanders have three major microhabitats available to them: the forest leaf-litter, surface cover objects such as logs and rocks, and underground tunnels constructed by other animals (Jaeger 1980b). In wet weather, the salamanders move through the leaf litter in search of small invertebrates, particularly soft-bodied animals such as springtails (Collembola; Jaeger 1972, 1980a; Burton 1976). In dry weather, the salamanders retreat under logs and other cover objects that protect them from desiccation. This restricts their ability to forage and limits the types of food available to them (Jaeger 1980a). They often feed on prey associated with logs, such as ants and termites. The latter are greatly preferred, even by individuals that previously fed on ants (Jaeger, Schwartz, and Wise 1995), because their less chitinous bodies are more easily digested and therefore are more profitable (Jaeger and Barnard 1981; Jaeger and Rubin 1982; Gabor and Jaeger 1995). In very dry weather, salamanders are forced underground, where food is much less abundant (Taub 1961; Fraser 1976a; Jaeger 1980b), or they aggregate under cover objects, thereby avoiding desiccation, but reducing their ability to search for food (Test 1955). Presumably individuals that can maintain territories with adequate, moist retreats and feeding sites will be those most likely to remain in good condition in dry weather (Jaeger 1980a).

**Fig. 9.19.** The red-backed salamander (*Plethodon cinereus*), the subject of many studies of territorial behavior. Photo by Wayne Van Devender.

Several lines of evidence from field studies in Virginia indicate that both male and female red-backed salamanders are territorial for much of their activity season, although this is not necessarily true in all parts of the species' range (Maerz and Madison 2000). In an early study, Jaeger (1979) found that adults often did not share the same retreat sites. He also found that individuals were spaced farther apart in the summer than in the spring, perhaps because warmer, drier weather intensifies competition for choice foraging sites. During the summer, when the salamanders are not courting, it is rare for more than one salamander to be found under the same cover object, and when this does occur, the log usually is large enough to allow salamanders to use different foraging areas with little overlap (Mathis 1989). Adults are relatively sedentary, occupying small home ranges for periods of weeks or months, to which they readily return if displaced (Kleeberger and Werner 1982; Gergits and Jaeger 1990b; Jaeger et al. 1993). Many individuals even returned to precisely the same cover object after being displaced.

The most convincing evidence of territoriality in the field comes from a series of studies by Alicia Mathis. In one study, she removed resident salamanders from cover objects, and these sites were reoccupied by new individuals more often than were cover objects that had not had a salamander in residence (Mathis 1990a). This indicates that the best-quality sites were those already occupied by salamanders. Mathis also found that the original residents of these sites were larger than the individuals that moved into the vacated sites, suggesting that large adults had monopolized the best territories, and smaller individuals either were in lower-quality sites or were nonterritorial floaters. A survey of natural cover objects showed that large animals were more likely to be found under large logs than were small animals (fig. 9.20), even though small individuals preferred large cover objects in both laboratory and field choice experiments. Large cover objects probably provided better microhabitats because soil temperatures were slightly cooler, and soil moisture presumably was higher.

Other studies have shown that large salamanders do not necessarily defend the largest cover objects (Gabor 1995; Faragher and Jaeger 1997). This suggests that the quality of

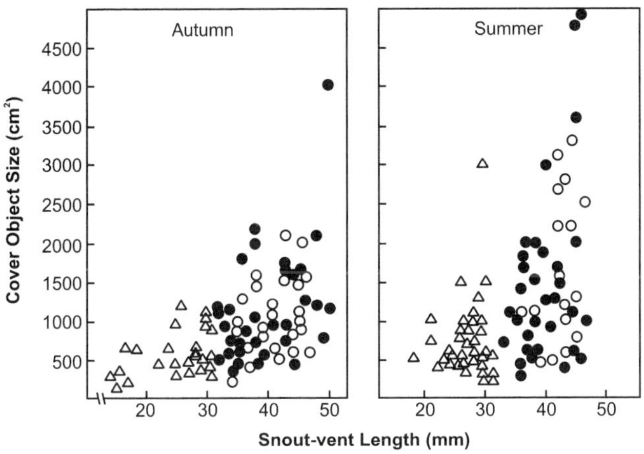


**Fig. 9.20.** Relationship between cover object size and snout-vent length for *Plethodon cinereus* in Virginia during the autumn courting season and the summer noncourting season. Open circles are adult males; closed circles are adult females; triangles are juveniles. In both seasons there was a similar positive relationship between body size and cover object size. After Mathis (1990a).

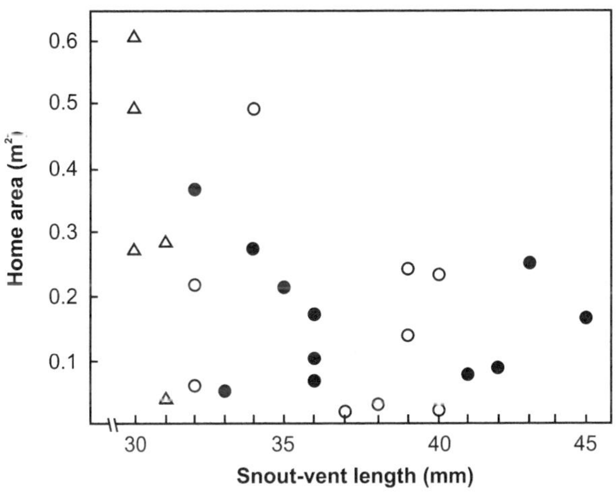


**Fig. 9.21.** Relationship of home range size to snout-vent length in red backed salamanders (*Plethodon cinereus*) in Virginia for males (closed circles), females (open circles), and juveniles (triangles). After Mathis (1991b).

food resources rather than size of a cover object is the most important determinant of territory quality. Indeed, Gabor (1995) found that territories held by large salamanders also provided more prey than did territories held by smaller individuals, especially soft-bodied prey (Diptera) of the type favored by the salamanders. Mathis (1991b) found that large salamanders actually had smaller home ranges than did small salamanders (fig. 9.21). This could be because food is patchily distributed and large salamanders occupy the best sites, or smaller individuals could be floaters that have not yet settled in a permanent home range. Those individuals that appeared to have established permanent territories were in better condition than were presumed floaters, as indicated by the length of the tail, an important energy-storage organ in salamanders (Maiorana 1977; Fraser 1980; see also chapter 5).

Mathis (1991b) also examined the spatial relationships of males and females in her population. Previous studies had shown that both males and females are aggressive toward conspecifics and defend territories in laboratory containers (Horne 1988; Horne and Jaeger 1988). Nevertheless, there also are indications from laboratory studies that males are somewhat more tolerant of intrusions by females than by males (Wrobel, Gergits, and Jaeger 1980; Jaeger 1984; Horne 1988; Thomas, Jaeger, and Horne 1989), and males are less aggressive toward females with which they are familiar (Guffey, MacKinster, and Jaeger 1998). This is to be expected if females must enter territories of males to find mates (Hom, Jaeger, and Willits 1997) or if they remain more or less permanently within male home ranges. In one field site, most adults of both sexes had home ranges that did not overlap those of other individuals, suggesting that both males and females were territorial. When home ranges did overlap, their residents were likely to be of different sexes, and individuals of both sexes were most likely to have a member of the opposite sex as a nearest neighbor (Mathis 1991b). Large males appear to have an advantage over small males in their access to females. In the field, males found within 10 centimeters of a female were larger than those not associated with females. In laboratory trials, females tended to associate more with large males than with small males, and large males dominated small males in contests over females. Subsequent field studies showed that males and females of this species were much more likely to share the same cover object than were same-sex pairs (Jaeger et al. 1995), and these associations persist even outside the reproductive season (Jaeger et al. 2001). There also is evidence from laboratory studies that females prefer to associate with particular males with which they are familiar (Gillette et al. 2000).

Work on captive animals by Lang and Jaeger (2000) suggested that males and females sometimes codefend a common territory during the breeding season, although they do not do so cooperatively. Males were more aggressive toward other males than toward intruding females, whereas females were more aggressive toward other females. Intruders were introduced into territories, in the laboratory, that were defended by a pair, by a single male, or by a single female. Male intruders did not leave pair-defended territories any faster than they left territories defended by single individuals, whereas females did leave pair-defended territories more quickly. The spatial strategies of males and females are not identical, however. Males appear to dominate preferred feeding territories (Gabor 1995). Some females are able to live within the territories of males, but others apparently act as nonterritorial floaters or are relagated to lower-quality territories. When individuals were removed from territories in the field, the vacated cover objects were more likely to be invaded by females than by males, regardless of the sex of the original inhabitant (Toll, Jaeger, and Gillette 2000). Thus, females probably spend more time moving around in search of high quality retreat sites and feeding territories than do males, and they may use fecal pellets of other individuals, either males or females, to assess the quality of potential foraging sites (Karuzas, Maerz, and Madison 2004).

Juvenile red-backed salamanders do not exhibit the territorial behavior seen in adults. Instead, they move between the leaf litter and adult territories under logs and other cover objects, depending on the weather and the amount of moisture in the litter. In laboratory tests, juveniles were actually attracted to pheromones of adults, rather than being repelled by them, even though adults are large enough to injure or kill juveniles. In laboratory encounters, adults displayed aggressively to juveniles by raising the trunk off of the ground, but they did so less frequently than in encounters with other adults, and they almost never bit juveniles. These results suggest that adults are relatively tolerant of intrusions by juveniles into their territories. There was a difference in response depending on whether the adults were familiar with the juveniles; they showed greater tolerance to familiar individuals. Jaeger, Wicknick, Griffis, and Anthony (1995) hypothesized that such responses reflect recognition and tolerance of close kin, but the genetic structure of natural populations and the proximity of close relatives remains to be investigated.

The postural displays by which red-backed salamanders signal aggression and submission during agonistic encounters have already been discussed (fig. 9.2). Overt aggression consists mainly of biting attacks directed at an opponent's tail, trunk, and snout (Jaeger 1981b, 1984; Jaeger, Barnard, and Joseph 1982). Attacks are directed disproportionately to the snout region, often injuring the nasolabial grooves (fig. 9.22). This can interfere with an animal's ability to find food (David and Jaeger 1981) or mates, and it makes it difficult for a salamander to monitor chemical territory markers of other individuals or find its way back to a home site (Graves 1994). Thus, even though fighting is rarely fatal, it can significantly reduce an individual's lifetime fitness (Jaeger 1981b). Territorial defense also can be costly if it reduces foraging time or causes animals to choose less profitable prey (Jaeger, Nishikawa, and Barnard 1983). Unfortunately, because overt aggression is so rarely observed in nature (Gergits and Jaeger 1990a), it probably is impossible to estimate the frequency of escalated fights or injury in natural populations. In laboratory encounters, a number of factors influence the outcome of aggressive interactions. Not surprisingly, body size is important, with large salamanders dominanting smaller individuals, thereby gaining access to more food

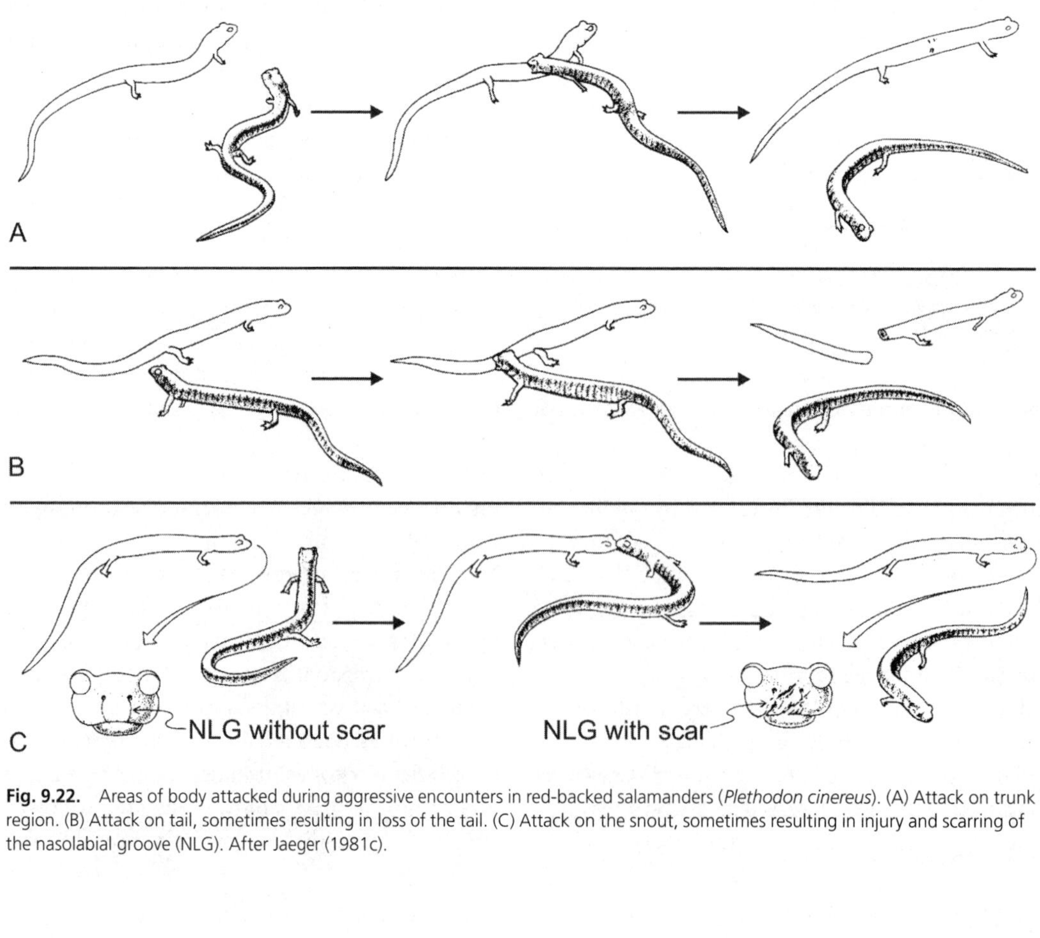


**Fig. 9.22.** Areas of body attacked during aggressive encounters in red-backed salamanders (*Plethodon cinereus*). (A) Attack on trunk region. (B) Attack on tail, sometimes resulting in loss of the tail. (C) Attack on the snout, sometimes resulting in injury and scarring of the nasolabial groove (NLG). After Jaeger (1981c).

(Townsend and Jaeger 1998). In addition, individuals with territories containing their preferred food (termites) initiated more aggressive displays and were more likely to bite intruders than those in territories with lower quality food (ants; Gabor and Jaeger 1999). Threat displays appear to be relatively inexpensive to produce and are not always followed up by biting attacks, suggesting that residents sometimes bluff their opponents in territorial encounters.

An important component of the territorial behavior of red-backed salamanders is the use of chemical signals in advertising ownership of territories and possibly conveying information about the size of territory owners and the quality of their territories. Early experiments demonstrated that red-backed salamanders can distinguish between their own substrates and those previously inhabited by other individuals, as well as the odors of familiar and unfamiliar individuals (Tristram 1977; McGavin 1978; Jaeger and Gergits 1979; Jaeger 1981b). Later, G. Simon and Madison (1984) tested the hypothesis that either fecal pellets or cloacal odors were the source of chemical cues used for identification. They scored odor preferences by the position of an animal in a test arena and found that salamanders preferred their own fecal or cloacal odors to those of other individuals. They hypothesized that one or more of the cloacal glands were the source of pheromones in fecal pellets.

In a more sophisticated experiment, Jaeger et al. (1986) placed salamanders in containers with artificial burrows. The investigators marked burrow entrances by placing a fecal pellet of the test animal, a fecal pellet of another animal, or a paper surrogate fecal pellet on the substrate. Males spent more time in burrows marked with their own fecal pellets than in those marked with either conspecific or surrogate pellets, and they spent more time in surrogate-marked burrows than in those marked with another individual's pellets. This indicated that males not only were attracted to their own fecal pellets, but also avoided those of other salamanders. Test males also were more likely to assume a flattened submissive posture (fig. 9.2) in the presence of a pellet from another individual. Like Simon and Madison, Jaeger et al. (1986) suggested that cloacal glands, probably the vent glands, were responsible for chemical signals deposited with the fecal pellets.

Subsequent work demonstrated that female red-backed salamanders behave in much the same way as do males when confronted with their own fecal pellets or those of other individuals (Horne and Jaeger 1988). They spent more time in burrows marked with their own pellets and showed submissive behavior to pellets of other salamanders. They also showed more interest in fecal pellets than did males, not only investigating them by nose-tapping, a behavior that conveys odorants to the vomeronasal organ, but actually squashing the pellets with their snouts. Some females also displayed aggressive postures in the presence of other salamanders' pellets. Mathis (1990b) demonstrated that females deposited fecal pellets most rapidly when confronted with their own pellets, whereas males deposited pellets in response to the odors of females. She suggested that substrate marking serves different functions in the two sexes, with females using chemical signals when advertising ownership of territories and males using them in both territory advertisement and attraction of females.

The fact that females exhibit many of the same forms of territorial behavior as do males, and apparently mark their territories with fecal pellets, indicated that the vent gland probably is not the source of pheromones deposited with fecal pellets, because female salamanders do not have vent glands. More recent work has shown that the likely source is a gland or set of glands located on the underside of the tail, just behind the cloaca. Presumably pheromones can be deposited on a fecal pellet by dragging or pressing the gland over the pellet after it has been deposited. Salamanders responded to extracts of postcloacal glands by nose-tapping (Simons and Felgenhauer 1992), and residents were more likely to assume aggressive postures when they had marked a substrate with secretions of the postcloacal glands (Simons, Felgenhauer, and Jaeger 1994). Test animals showed more interest in substrates marked with postcloacal gland secretions of other animals than unmarked substrates.

A surprising finding of these experiments was that salamanders did not avoid marked substrates when a resident salamander was absent. This suggested that chemical markers do not repel intruders, but instead provide information about the identity of the resident. When a resident is present, an intruder can attempt to match chemicals on the substrate with the odor of the resident to determine if that animal is, in fact, the owner of the territory. Intruders into territories in laboratory tests often contacted residents and investigated the region of the tail where marking glands are located (Jaeger and Gabor 1993), sometimes even pushing their snouts under the tail of the resident. This sort of detailed investigation probably enables a salamander to assess the resident's status, individual identity, size (Mathis 1990b), and perhaps fighting ability of opponents, allowing them to avoid costly escalated fights when they are unlikely to prevail (Simons, Jaeger, and Felgenhauer 1997). It also appears that residents are more aggressive when they are in territories marked with their own pheromones (Jaeger 1981b; Nunes and Jaeger 1989; Simons, Jaeger, and Felgenhauer 1997), and indeed, they probably use the presence of these pheromones to identify their own territories. Residents also tend to be more aggressive if they are in good-quality territories (Nunes 1988a), if they have been in a territory for a long time (Nunes and Jaeger 1989), and if they are large individuals (Mathis and Simons 1994).

The amount of information conveyed by the seemingly simple behavior of placing pheromones and fecal pellets in a territory is quite remarkable. In addition to information about a male's identity, status, and fighting ability, fecal pellets apparently provide females with a considerable amount of information about the qualities of males as potential mates. When females were given a choice of artificial burrows marked at the entrance with fecal pellets of males that had eaten termites (a high-quality food) or ants (a lower-quality food), they spent more time in the burrows with pellets containing termite remains (Walls et al. 1989). A field survey also revealed that females were more likely to be associated with males that had been feeding on termites, even though ants were much more abundant in the area. Gravid females are more likely than nongravid females to inspect fecal pellets closely by squashing them with their snouts (Jaeger and Wise 1991). It is not clear why females should prefer to associate with males that feed on a higher quality diet. One possibility is that females make use of food resources in the male's territory to produce eggs. Another is that they lay eggs in the male's territory, and their offspring gain access to the male's resources (see the following discussion of Mating Systems). Whatever the reason, males are more efficient foragers on termites, the prey type that makes them more attractive to females (Jaeger, Schwarz, and Wise 1995).

There is evidence that the benefits of territoriality extend to competitive interactions between species. Jaeger and Gergits (1979) reported that when red-backed salamanders were given a choice of substrates marked with their own odors or those of a closely related species, *Plethodon shenandoah,* they spent more time on their own substrates. Similar results were obtained when *P. shenandoah* was tested. These results were interpreted as evidence that salamanders avoid the pheromones of other species. In the field, *P. cinereus* and *P. shenandoah* have parapatric distributions, with the latter limited to drier talus slopes in areas surrounded by wetter habitats favored by *P. cinereus* (Jaeger 1970). *Plethodon cinereus* cannot invade the drier habitats, but they are able to exclude the other species from moist habitats (Jaeger 1971a, b, 1972; see also chapter 15). *Plethodon cinereus* is more aggressive than is *P. shenandoah* in laboratory encounters (Wrobel, Gergits, and Jaeger 1980), and *P. cinereus* can defend territories against the other species in preferred habitat (Griffis and Jaeger 1998). The parapatric distributions of the two species result from the ability of *P. shenandoah* to tolerate drier conditions, combined with aggressive interspecific territoriality of *P. cinereus* in preferred habitats (Jaeger 1974). Experimental removal of resident *P. cinereus* from rocks near talus slopes inhabited by *P. shenandoah* allowed the latter species to invade the surrounding habitat (Griffis and Jaeger 1998).

Populations of red-backed salamanders also have parapatric distributions with several other salamanders of similar size (*P. wehrlei, P. hubrichti, P. nettingi, P. hoffmani*), suggesting that they are aggressive toward these species as well (Pauley 1978; Jaeger and Forester 1993). In fact, *P. cinereus* is highly aggressive toward *P. hoffmani,* and is aggressively superior to that species (Jaeger, Prosen, and Adams 2002). Red-backed salamanders also are aggressive toward similar-size juveniles of a much larger species, *P. glutinosus,* and apparently exclude them from their territories in the field, but they do not interact with larger adults of the same species (Lancaster and Jaeger 1995; Jaeger, Gabor, and Wilbur 1998). In laboratory encounters, *P. cinereus* dominated adults of *Eurycea bislineata,* a relatively nonaggressive species that sometimes uses similar retreat sites (J. Markow, personal communication), as well as the closely related species *E. cirrigera* (Jaeger, Gabor, and Wilbur 1998). In contrast, adults of *Desmognathus ochrophaeus* and *D. fuscus* were able to displace red-backed salamanders from retreat sites in laboratory encounters, even when they were somewhat smaller than the residents (E. Smith and Pough 1994; Jaeger, Gabor, and Wilbur 1998). This reflects the even more aggressive nature of many desmognathine salamanders (see the following). Red-backed salamanders probably avoid contact with species of salamanders large enough to prey on them, such as large *Desmognathus* and *Ambystoma maculatum* (Ducey, Schramm, and Cambry 1994; Jaeger, Gabor, and Wilbur 1998).

A caveat to the story of territoriality in red-backed salamanders is that their behavior is not necessarily uniform across their very broad geographic range. For example, studies of a population in northern Michigan revealed that individuals were much less likely to be distributed in a uniform spatial pattern typical of territorial animals than were those from Virginia, where most of the previous work has been done (Quinn and Graves 1999). Michigan salamanders also were more likely to be found in aggregations under the same cover objects than were those from Virginia. Furthermore, when salamanders from these two regions were housed in the laboratory under identical conditions, the Michigan salamanders showed a much greater tendency to aggregate than did Virginia salamanders. The reasons for these differences in behavior are not known, but they could be related to differences in the abundance or predictability of food resources, which appear to be less seasonally variable in Michigan than in Virginia.

### Aggression and Territoriality in Other Plethodontines

Other species of salamanders have not been studied in as much detail as *P. cinereus,* but there is evidence that many other species in the subfamily Plethodontinae are aggressive and probably territorial (Mathis et al. 1995). Several small eastern species of *Plethodon* that are ecologically similar to *P. cinereus* exhibit very similar territorial behavior, including *P. serratus* (Mathis, Deckard, and Duer 1998; Camp 1999), *P. websteri* (Camp 1999), and *P. angusticlavius* (Mathis and Britzke 1999; Mathis, Schmidt, and Medley 2000). These species share many of the elements of aggression and display behavior seen in *P. cinereus,* but their behavior and spatial relationships in the field have not been studied in detail. In western North America, *P. dunni,* a small species similar in appearance and ecology to *P. cinereus,* exhibits many of the same types of territorial behavior, including scent-marking (Ovaska and Davis 1992) and aggressive interactions (Ovaska 1993). In contrast, another very similar sympatric species, *P. vehiculum,* does not appear to be territorial, because groups of individuals sometimes use the same retreat sites (Ovaska 1988b), and aggressive interactions between residents and intruders are rare (Ovaska 1993). Males do exhibit some aggressive behavior, including biting, during the mating season (Ovaska 1987). The more aggressive nature of *P. dunni* enables that species to exclude *P. vehiculum* from choice microhabitats (Dumas 1956), either through overt aggression (Ovaska 1993) or avoidance of chemically marked substrates by the less aggressive species (Ovaska and Davis 1992).

Both intra- and interspecific aggression and territoriality have been reported for a number of larger species of *Plethodon* as well. The best-studied example is the interaction between sympatric members of the *Plethodon glutinosus* and *P. jordani* complexes in the southern Appalachians (see also chapter 15). Both *P. metcalfi* (a member of the *P. jordani* complex) and *P. teyahalee* appear to be intraspecifically territorial (the latter was called *P. glutinosus* by Nishikawa [1985, 1987], *P. teyahalee* by Nishikawa [1990], and *P. oconaluftee* by Hairston [1993]). Adult males, females, and juveniles all occupy relatively fixed home ranges that overlap those of other individuals, particularly of the same age and sex, less than expected by chance (fig. 9.23). The home ranges of *P. teyahalee* in the Great Smoky Mountains were smaller and more exclusive than were those of *P. metcalfi* in the Balsam Mountains, and always were associated with retreat holes that probably were defended against other individuals. Those of *P. metcalfi* were not always associated with retreat holes or cover objects, suggesting that this species defends territories in the leaf litter. Members of both species complexes were highly aggressive toward conspecifics in laboratory encounters (Nishikawa 1985, 1987), and adults in the field often had conspicuous bite wounds (Nishikawa 1990).

Individuals of *Plethodon teyahalee* from the Great Smoky Mountains, where the two species experience intense interspecific competition (see chapter 15), were more aggressive toward both conspecifics and heterospecifics than were those from the nearby Balsam Mountains, where interspecific competition was less intense (fig. 9.24). Members of the *Plethodon jordani* complex did not show a similar level of variation. Adult *P. jordani* from the Great Smoky Mountains and *P. metcalfi* from the Balsam Mountains were aggressive toward both conspecifics and heterospecifics in both populations (see also Selby, Winkel, and Petranka 1996). Members of the *P. jordani* complex generally occur at much higher densities than does *P. teyahalee,* suggesting that intraspecific competition for choice territories is likely to be intense in both populations (Nishikawa 1985). *Plethodon teyahalee* tended to treat both conspecific and heterospecific intruders as equally important intruders in all populations, whereas members of the *P. jordani* complex were less aggressive toward *P. teyahalee* than toward conspecifics where intraspecific competition was intense, but interspecific competition was not (Nishikawa 1987). Members of these two complexes apparently are not aggressive toward one another in all localities, however. For example, Rissler, Barber, and Wilbur (2000) did not observe any aggressive behavior between conspecific or heterospecific adults of *P. glutinosus*

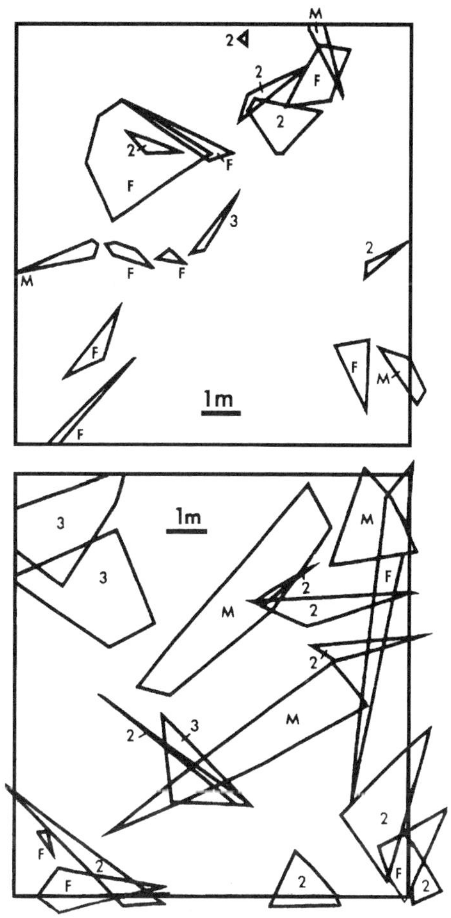


**Fig. 9.23.** Spatial relationships of home ranges of large *Plethodon* on plots in the Smoky Mountains (top) and Balsam Mountains (bottom) of North Carolina. The top figure shows home ranges of *Plethodon teyahalee.* The bottom figure shows home ranges of *P. metcalfi.* Home ranges were larger and overlapped more in the Balsam Mountains. M = male. F = female. 2 = 2-year-old juvenile. 3 = 3-year-old juvenile. After Nishikawa (1990).

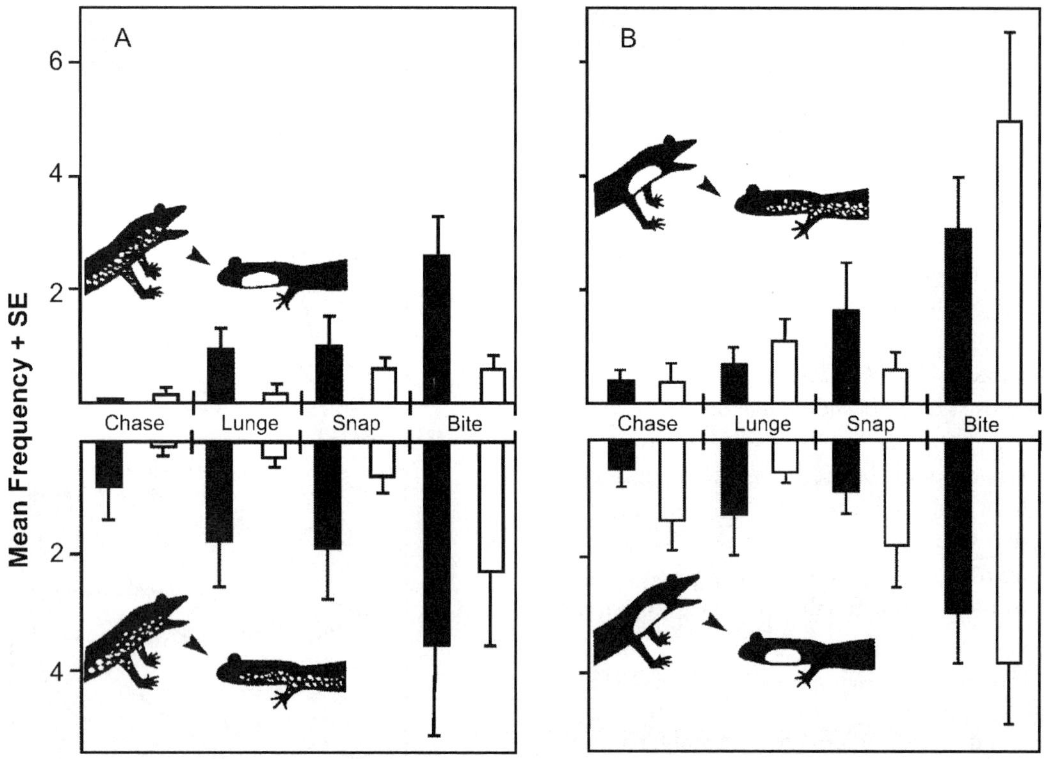


**Fig. 9.24.** Intra- and interspecific aggressive interactions between members of the *Plethodon jordani* and *P. glutinosus* complexes. (A) Average frequency ± 1 SE of aggressive behavior patterns exhibited by resident *Plethodon teyahalee* from the Smoky Mountains (solid bars) and Balsam Mountains (open bars) during encounters with heterospecifics (top) and sympatric conspecifics (bottom). (B) Average frequency ± 1 SE of aggressive behavior patterns exhibited by resident members of the *P. jordani* complex from the Smoky Mountains (solid bars) and Balsam Mountains (open bars) during encounters with sympatric *P. teyahalee* (top) and sympatric conspecifics (bottom). *Plethodon teyahalee* is shown with small white spots. Members of the *P. jordani* complex, shown with white cheeks, are *P. jordani,* a red-cheeked species, in the Smoky Mountains, and *P. metcalfi,* a gray-cheeked species, in the Balsam Mountains. Behavioral categories increase in level of aggressiveness from left to right. Individuals of *P. teyahalee* from the Smoky Mountains were more aggressive to both conspecifics and heterospecifics than were those from the Balsam Mountains, but this was not true for the *P. jordani* complex. After Nishikawa (1985).

and *P. montanus,* an introduced member of the *P. jordani* complex, at Mountain Lake in Virginia.

Anthony (1993) studied intra- and interspecific interactions in two relatively large salamanders from the Ouachita Mountains in Arkansas, *Plethodon ouachitae* and *P. albagula* (a member of the *P. glutinosus* complex). He found that both species investigated fecal pellets of conspecifics, but were not repelled by them. Some individuals squashed fecal pellets with their snouts, a behavior similar to that of *P. cinereus,* and some also appeared to mark over conspecific pellets by pressing postcloacal glands over the pellets. Adults of *P. ouachitae* tend not to co-occur under the same cover objects in the field and are highly aggressive toward one another in laboratory encounters (Thurow 1976; Anthony and Wicknick 1993). The fights observed in this species were much more violent than those reported for other species of *Plethodon,* with prolonged biting and wrestling bouts in which the opponents rolled over one another and became twisted into knots (fig. 9.25). Adults also tried to bite juveniles, but the latter usually ran away. Juveniles did not avoid substrates marked by adults, but cannibalism of juveniles was not observed. The very aggressive nature of *P. ouachitae* enabled this species to compete with and sometimes displace adults of the larger, but less aggressive *P. albagula* (fig. 9.26; Anthony, Wicknick, and Jaeger 1997). The aggressive interspecific territoriality of *P. ouachitae* probably accounts in part for its greater abundance in favorable habitat where the two species are sympatric.

Both intra- and interspecific territoriality has been reported in *P. kentucki* and *P. glutinosus* that live together in the mountains of Kentucky (Marvin 1998a, b). In the field, adult *P. kentucki* maintained relatively fixed home ranges that did not overlap with those of same-sex individuals. Home ranges of males and females, however, often overlapped, especially when females were gravid. Those of juveniles were randomly distributed, but juveniles did exhibit some aggression toward other salamanders. In laboratory encounters, males were more aggressive toward other males in the

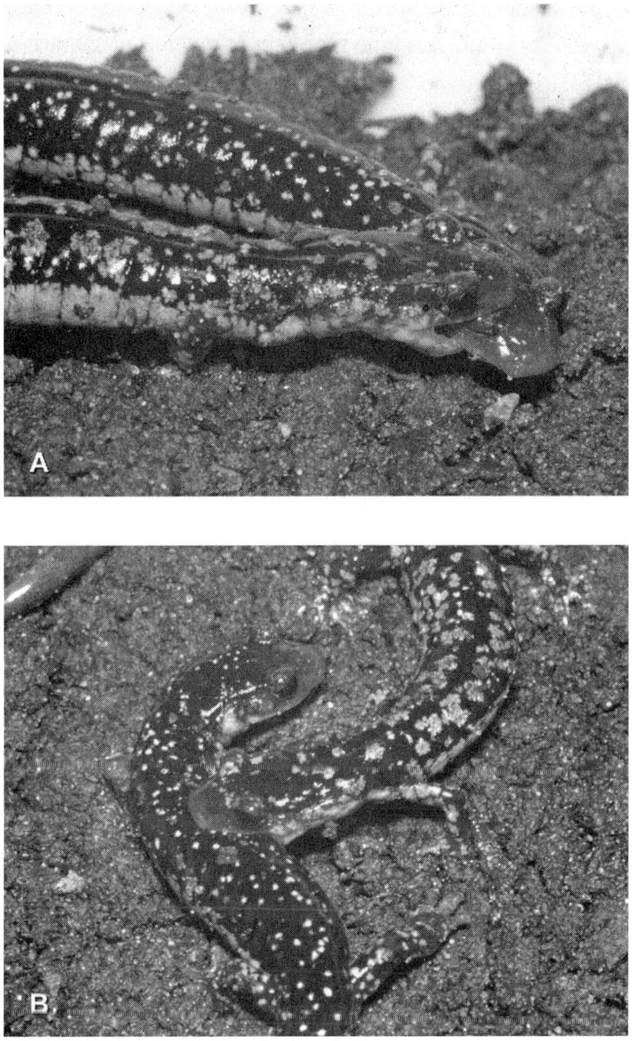


Fig. 9.25. Aggressive behavior in *Plethodon ouachitae*. (A) One male biting another male on the head. (B) One male biting another male in the trunk region. Photos by Carl D. Anthony.

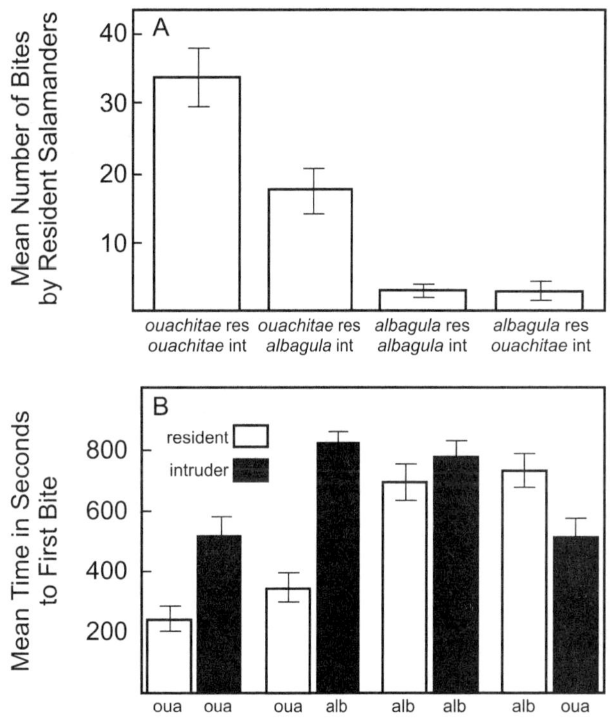


Fig. 9.26. Intraspecific and interspecific aggression in *Plethodon ouachitae* and *P. albagula*. (A) Mean number of bites delivered by the resident salamander (res) to the intruding salamander (int) in intra- and interspecific encounters. *Plethodon ouachitae* residents were much more aggressive toward conspecific and heterospecific intruders than were *P. albagula* residents. (B) Time to first bite by residents and intruders. *Plethodon ouachitae* residents attacked intruders of both species quickly, while *P. albagula* residents did not. Both graphs show that *P. ouachitae* was more aggressive toward conspecifics than toward heterospecifics. After Anthony, Wicknick, and R. G. Jaeger (1997).

breeding season than at other times of year, whereas females and juveniles showed the same level of aggressiveness toward males, females, and juveniles at all times of year. The overall level of aggressiveness by females and juveniles was much lower, however, than that of breeding males. In a high-density population, large males had territories that overlapped more home ranges of gravid females than did the territories of smaller males, but this was not true for a low-density population. Both *P. kentucki* and *P. glutinosus* were equally likely to attack conspecifics and heterospecifics in the laboratory, although the intensity of aggression was greater in conspecific encounters. These salamanders appear to compete for choice retreat and foraging sites both intra- and interspecifically, with *P. glutinosus*, the larger species, being dominant over *P. kentucki*.

Two other genera of plethodontine salamanders exhibit relatively high levels of aggression and evidence of territoriality. Members of the genus *Aneides* from both eastern and western North America are aggressive toward conspecifics and defend retreat sites such as rock crevices and tree holes as territories (Gordon 1952; Cupp 1980; Staub 1993; T. Davis 2002a). These salamanders have exceptionally large jaw-closing muscles, especially in males, long, sharp teeth, and various anatomical features that strengthen the skull (fig. 9.27). Fights include prolonged biting similar to that seen in *Plethodon ouachitae*, but not in most other species of *Plethodon*. Males in particular are capable of inflicting serious wounds on opponents, and bite scars are common in animals collected in the field, especially in three western species in the *A. lugubris* complex (*A. lugubris*, *A. ferreus*, and *A. flavipunctatus*; Staub 1993).

A relatively large salamander from the West Coast of the United States, *Ensatina eschscholtzii*, also appears to be territorial (Wiltenmuth 1996). In laboratory encounters, established residents were more aggressive than were intruders, and males were more aggressive than were females. Overt aggression included chasing, pushing opponents with

**Fig. 9.27.** Male *Aneides lugubris,* showing very large jaw muscles used in fighting. Photo by Nancy Staub.

the head, and biting. *Ensatina* lacks the hypertrophied jaw muscles and long teeth found in *Aneides,* and serious wounds were not observed in this species. *Ensatina* is a classic example of a "ring species," in which lineages have diverged as they moved from a common point of origin, with genetically divergent populations then coming into secondary contact where their ranges come together. Divergence of lineages has been accompanied by some divergence in aggressive behavior, particularly among females (Wiltenmuth and Nishikawa 1998). Females in coastal populations tend to be more aggressive and exhibit greater avoidance of other females than do those in inland populations. Presumably this divergence is related to ecological differences between populations, but these have not been studied in detail.

### Aggression and Territoriality in Desmognathines

As in the genus *Aneides,* desmognathine salamanders have unusually large jaw-closing muscles and are capable of inflicting serious injuries on other salamanders. Males and females of several species of *Desmognathus* defend retreat sites against conspecifics, and biting attacks have been observed both in the field and in the laboratory (Brandon and Huheey 1971; Keen and Sharp 1984; Keen and Reed 1985; Jaeger 1988; Verrell and Donovan 1991; Camp 1996; Camp and Lee 1996). In some species, an attacker sometimes bites off part of the tail of an opponent, and large adults sometimes cannibalize smaller juveniles (Keen and Sharp 1984; Camp 1996). Scarring of the snout region, probably resulting from intruders being bitten while investigating occupied retreat sites, also is common. Several experiments have shown that large individuals can exclude smaller salamanders from retreat sites. The social behavior of one desmognathine, *Phaeognathus hubrichti,* a large burrowing species, is largely unknown. Males are larger and are more muscular than females, and they often exhibit extensive scarring, apparently from bites. This suggests that aggressive behavior among males may be common (Bakkegard and Guyer 2004).

The spatial distribution of desmognathines in a population is influenced by both the density of suitable cover objects and the presence of resident salamanders (Keen and Sharp 1984; Keen and Reed 1985; Kleeberger 1985; Jaeger 1988; Colley, Keen, and Reed 1989). Aggression and territorial behavior extend to interspecific interactions as well. Large species, such as *D. quadramaculatus* and *D. monticola,* can exclude smaller species, such as *D. fuscus* and *D. ocoee,* from preferred retreat sites. This can affect the microhabitat use and spatial distribution of individuals in a multispecies community, although there is variation among populations in the frequency of interspecific aggression (Southerland 1986a; see chapter 15 for a more detailed discussion of interspecific competition).

There also is some evidence from laboratory studies that desmognathine salamanders can recognize substrate-borne or airborne odors of other salamanders. In some experiments, salamanders preferred their own odors to those of other individuals, but there is not a consistent tendency for individuals to avoid the odors of conspecifics (Keen, McManus, and Wohltman 1987; Evans and Forester 1996). When females of a population of *D. ochrophaeus* from Maryland were given a choice of substrates marked by males from their own population and those from a nearby population, they spent more time on the substrates from the other population (A. Evans, Forester, and Masters 1997). The experimental design did not, however, discriminate between attraction to odors of males from other populations and avoidance of males from the same population. There is some evidence that juveniles of *Desmognathus* avoid the odors of adults, which sometime prey on them (Evans and Forester 1996). Roudebush and Taylor (1987b) reported that *D. monticola* avoided substrates marked by the larger *D. quadramaculatus,* although they did not avoid feces or cloacal gland odors of that species (Jacobs and Taylor 1992). At present, evidence for chemical advertisement of territories is not nearly as strong for desmognathines as it is for plethodontines. In part, this is simply because they have been studied less intensively. It also seems likely that the semiaquatic habitats used by many desmognathines are less suitable for deposition of long-lasting chemical markers than are the drier terrestrial habitats used by *Plethodon, Aneides,* and *Ensatina.*

### Aggression in Ambystomatids and Dicamptodontids

Aggressive behavior has been reported in both the larvae and adults of ambystomatid salamanders. Aggression in larval ambystomatids includes apparent visual displays and biting, and it sometimes grades into cannibalism. Both aggressive competition and cannibalism tend to increase as population density increases, and represent forms of interference competition for limited food resources (Walls and Jaeger 1987; Semlitsch and Reichling 1989; Petranka 1989a; C. K. Smith 1990; Van Buskirk and Smith 1991; Walls and Semlitsch 1991; Mathis et al. 1995; Walls 1998; Wildy et al. 2001). Larval competition is discussed in more detail in chapter 13. There is no convincing evidence that larvae of ambystomatids or any other family of salamanders are territorial.

Aggressive behavior of terrestrial juveniles and adults of ambystomatid salamanders has been observed in the laboratory. Outside of the breeding season, these animals spend most of their time in hidden retreat sites, especially small mammal tunnels through the soil and leaf litter (Semlitsch 1981a; Madison 1997; Madison and Farrand 1998). Marked individuals often move to the same general area each year after leaving breeding ponds (see chapter 6), but it seems unlikely that they occupy precisely the same tunnels, because these probably are somewhat ephemeral. It is conceivable that animals advertise ownership of particular tunnels or burrows with chemical markers and defend them aggressively against intruders, but such behavior has not been observed in the field.

Ducey and Ritsema (1988) housed adult spotted salamanders (*Ambystoma maculatum*) in cages with artificial plastic tunnels. When two individuals were housed together, they tended to occupy different tunnels. They also defended tunnels by elevating the body, advancing toward or touching intruders, and by biting intruders on the snout. Subsequent work with several other species (*A. opacum, A. talpoideum, A. macrodactylum,* and triploid hybrids of the *A. jeffersonianum* complex) housed in the same type of cages with artificial tunnels revealed similar types of aggressive behavior, although the frequency of biting varied among species (Ducey 1989). Biting by captive adults was previously reported in *A. gracile* by Licht (1973), and has been observed in newly metamorphosed juveniles of several species (Walls 1990, 1991). There also are anecdotal observations of similar aggression in *Dicamptodon,* the sister group to the ambystomatids (Nussbaum, Brodie, and Storm 1983). Ducey and Heuer (1991) reported that aggression in several ambystomatid species was reduced when the animals were well fed. They interpreted this as evidence that aggression serves mainly to defend feeding territories rather than tunnels used as retreat sites, but the absence of field observations makes this conclusion highly speculative. If ambystomatids do defend retreat sites in the field, such defense appears to be relatively ineffective. Multiple spotted salamanders (*Ambystoma maculatum*) sometimes are found in the same burrows in the field, and they readily occupied the same artificial burrows in the laboratory (Regosin, Windmiller, and Reed 2004). Whether these salamanders actually are attracted to conspecifics, as in *Salamandra luschani* (Gautier et al. 2006) or simply tolerate them, is not yet known.

There is relatively little evidence for pheromonal marking of territories by ambystomatids. Ducey and Ritsema (1988) and Ducey (1989) did not report any behavioral acts that seem clearly related to substrate marking, such as pressing the cloaca to the substrate. Furthermore, none of the species tested showed a tendency to avoid substrates marked by other conspecifics (see also D. Martin, Jaeger, and Labat 1986) or to distinguish between odors of familiar and unfamilar conspecifics (Ducey, 1989). Indeed, in some cases, individuals spent more time on substrates from cages of conspecifics than on their own substrates. Mathis et al. (1995) interpreted these results as evidence that resident individuals were attempting to confront intruders or mark over their chemical signals, but it also is possible that the observed behavior simply reflects a tendency for salamanders to investigate unfamiliar odors in their environment. Ambystomatids do not have naso-labial grooves to conduct odorants from the substrate to the vomeronasal organs, but they do tend to press their snouts against substrates and other salamanders when investigating them, a behavior that probably has a chemosensory function. Nevertheless, the question of scent marking and its effect on the spacing behavior of ambystomatids requires further investigation.

### Aggression in Salamandrids

Territorial behavior appears to be rare among salamandrids. J. Davis and Brattstrom (1975) reported that captive individuals of *Taricha torosa* defended rocks in aquaria with elaborate visual and vocal displays. However, aggressive behavior has never been reported in the field, so the significance of the behavior is not known. Terrestrial adults of European fire salamanders (*Salamandra*) occupy limited home ranges (Degani and Warburg 1978; Plasa 1979) and often return to familiar retreat sites that appear to be marked with chemical signals (Gautier and Miaud 2003). Kastle (1986) observed wrestling bouts in *S. salamandra,* but not biting, between males, which appeared to be defending prominent landmarks, such as piles of rocks and tree roots, as territories. Whether such behavior occurs only during the breeding season, and whether it represents true territorial behavior, is unclear. Andreone (1992) reported similar behavior in another member of the same genus, *S. lanzai.* Males of this species also occupied elevated perches at night, and two males were found clasped together in an apparent fight.

Males of some aquatic salamandrids are aggressive toward other males during the breeding season, and several large species of European newts have leklike mating systems in which males defend small mating territories (see "Mating Systems and Sexual Selection," following). Most other newts are nonterritorial and exhibit relatively little aggressive behavior. There is mixed evidence about the possible use of scent marking by salamandrids. An early paper by Gauss (1961) reported that males of *Triturus cristatus* advertised their territories by placing chemical markers on the substrate. Subsequent work has shown that males in some populations of this species form leks and defend small mating territories, but scent-marking has not been reported in these studies (Zuiderwijk and Sparreboom 1986; Hedlund and Robertson 1989). Apparent cloacal scent marking also was reported in another aquatic salamandrid, *Calotriton asper,* by Lengvenus and Parzefall (1992), who also observed males biting one another. It is not clear whether the aggression observed in this species represents territorial behavior or sexual interference during courtship (Thiesmeier and Hornberg 1990).

### Aggression in Other Families

Aggression in salamanders was first described in aquatic cryptobranchids (Kerbert 1904; S. Bishop 1941b; Hillis and Bellis 1971). Males establish territories under large rocks in streams or rivers and vigorously defend them by biting intruders, especially other males. In the Japanese giant salamander (*Andrias japonicus*), very large males hold the best territories, which serve as both retreat and oviposition sites. Intruding males are vigorously attacked, but multiple sneaker or satellite males often enter a nest when females are laying eggs. Cryptobranchids have external fertilization, so these intruding males probably have opportunities to fertilize some of the female's eggs (Kawamichi and Ueda 1998). Once the eggs have been laid the female leaves, but the male remains with the eggs and guards them after they are deposited (see chapter 11). These salamanders essentially have a resource-defense mating system similar to that of some frogs (see the following discussion and chapter 8). Some hynobiids, which also have external fertilization, are reported to defend territories under rocks in the breeding season (Thorn 1962, 1963, 1967; Tanaka 1986, 1987, 1989), and they probably have resource-defense mating systems as well (Nussbaum 1985).

Some sirenids, the only other group of salamanders with external fertilization, also might be territorial, but the evidence is circumstantial (Mathis et al. 1995). Individuals have been seen fighting over retreat sites (Asquith and Altig 1987), and both males and females often have bite wounds (Godley 1983). Unfortunately, the behavior of sirenids is so poorly known that even mating behavior has never been observed. Nevertheless, the occurrence of bite scars only in sexually mature individuals, and primarily in the breeding season, suggests that the scars result from aggressive interactions among males, although it also is possible the some females acquire scars during courtship or while guarding eggs against conspecific predators (Fauth and Resetarits 1999). Observations of proteids in captivity suggest that males defend territories under rocks, but observations of aggression are anecdotal (Briegleb 1962; Parzefall 1976; Parzefall, Durand, and Richard 1981; R. Ashton 1985). These salamanders have internal fertilization, and guarding of eggs, if it occurs, is more likey to be done by the female (see chapter 11).

## Mating Systems and Sexual Selection in Urodeles

The mating systems of salamanders exhibit many similarities to those of anurans (chapter 8) and can be analyzed according to the same principles (Verrell 1989c; Sullivan, Ryan, and Verrell 1995). Nevertheless, the details of courtship, reproductive competition, and sexual interference differ substantially from those of anurans. This is because most salamanders have internal fertilization by means of spermatophores, which contrasts with the external fertilization found in most anurans. This can affect the evolution of mating systems in several ways. First, it allows for the temporal and spatial separation of courtship and mating from oviposition. In some plethodontids, for example, courtship and mating occur in the fall and spring, but oviposition does not take place until summer. This makes it unlikely that males will adopt a mating strategy of defending oviposition sites that are attractive to females (resource defense polygyny), although this does seem to occur in the externally fertilizing cryptobranchids (S. Bishop 1941b) and hynobiids (Tanaka 1989; Kawamichi and Ueda 1998). The use of spermatophores also requires relatively close coordination between males and females to ensure successful pickup of spermatophores by females. This seems to have led to the evolution of much more elaborate courtship behavior in urodeles than that seen in most anurans (Arnold 1977; Halliday 1990). Finally, internal fertilization changes the nature of sperm competition (Halliday and Verrell 1984; T. Halliday 1998). Simultaneous fertilization of a female's eggs by several males, like that seen in some arboreal frogs (see chapter 8), is not possible with internal fertlization, although sequential fertilization by several males is possible. This is expected to lead to the evolution of various forms of sexual interference in which males try to disrupt the mating attempts of other males before mating takes place, and to mate guarding as a means of denying rival males access to a female. Such mate guarding can occur both before and after a male completes courtship with a female (Arnold 1976; Halliday and Verrell 1984; Verrell 1989c; Halliday 1990, 1998; Halliday and Tejedo 1995). Postmating mate guarding is likely to have limited effect, however, if females can store sperm for long periods of time (e.g., E. Adams, Jones, and Arnold 2005).

The range of mating systems described for urodeles is similar to that seen in anurans. These include explosive mating aggregations with scramble competition, competitive mate searching and mate guarding, leks, and resource defense (Verrell 1989c; Halliday and Tejedo 1995; B. Sullivan, Ryan, and Verrell 1995). It is important to recognize that these mating systems represent points along a continuum of mating strategies, with the distinct categories being defined largely for the convenience of investigators. For example, the type of scramble competition seen in explosive mating aggregations tends to grade into competitive mate searching and mate guarding, depending on the degree of synchrony in the mating season and the density of competing males. Similarly, active mate searching grades into lekking behavior largely as a function of population density, characteristics of the breeding site, and the degree of site fidelity exhibited by displaying males.

### Temporal and Spatial Patterns of Reproduction

Like anurans, salamanders can be divided into species with prolonged and explosive courtship periods. Highly explosive breeding is relatively rare in salamanders, perhaps because few species breed in temporary water. This pattern is found mainly in ambystomatids, most of which breed in late fall, winter, or early spring, including *Ambystoma annulatum, A. cingulatum, A. jeffersonianum, A. macrodactylum, A. maculatum, A. opacum,* and *A. texanum* (Arnold 1976, 1977; McWilliams 1992; Krenz and Scott 1994; Verrell and Pelton 1996). Explosive breeding also occurs in some hynobiid salamanders that breed in winter or early spring (Hasumi 1994). Most other salamanders have courtship periods lasting from several weeks to several months, and many tropical species probably court throughout the year (Houck 1977a). So little is known about sirenids and amphiumids that their courtship patterns cannot be characterized.

Arnold (1976, 1977) emphasized the impact of temporal variation in female availability on patterns of male courtship. Salamanders appear to have a limited supply of spermatophores available each year (Halliday 1976; Verrell 1986a, 1988a; Verrell, Halliday, and Griffiths 1986), so temporal variation in female arrival will determine how a male can best allocate his sexual resources (W. Harris and Lucas 2002). As in frogs, short breeding seasons and intense male-male competition probably select for rapid courtship of females and expenditure of considerable effort in courting the first female encountered, even at the expense of reduced fertilization efficiency. On the other hand, if the male has the opportunity to court many females each year, then selection should favor husbanding of resources and maximum efficiency in courtship and mating. Patterns of sperm allocation are expected to be affected not only by the availability of females, but also by the intensity of competition among males, since reproductive payoffs will be relatively low when many males are present (W. Harris and Lucas 2002). The duration of the courtship season also affects the form of male-male competition. In dense aggregations of explosive breeders such as some *Ambystoma,* one would expect various forms of scramble competition to predominate, whereas in low-density populations of more prolonged breeders, other forms of competition, including guarding of individual females and territoriality, might be employed (Wells 1977a; Verrell 1989c; Halliday and Tejedo 1995; B. Sullivan, Ryan, and Verrell 1995). Finally, in species with short courtship seasons, male-male competition restricts opportunities for female choice to operate, as it does in frogs. In more prolonged breeders, females can compare potential mates and actively choose among them. Consequently, one might expect greater sexual dimorphism in display structures and more complex courtship behavior in prolonged breeders.

### Explosive Mating Aggregations and Scramble Competition

Explosive mating aggregations occur when the arrival of males and females at a breeding site is highly synchronized and the duration of the mating season is short. Often this results in high densities of males competing for access to limited numbers of females. The predominant form of competition is scramble competition, in which males attempt to mate with females as rapidly as possible (Verrell 1989c; Halliday and Tejedo 1995; B. Sullivan, Ryan, and Verrell 1995). Males typically search for females in breeding ponds and probably are attracted to chemical cues emanating from females. There is little evidence that females are attracted to males (P. Verrell, Strand, and Hanson 2001).

#### Ambystomatidae

Most ambystomatid salamanders make synchronized movements to breeding sites after warm rains in winter or early spring. Males usually arrive before females and often outnumber them, sometimes by 10 to 1. The effective bias in sex ratio is increased by the much shorter residence time of females, which probably engage in courtship only once and leave the ponds soon after oviposition (Douglas 1979; Semlitsch 1985a; Beneski, Zalisko, and Larsen 1986; Sexton et al. 1986; McWilliams 1992; Verrell and Pelton 1996; Verrell and Krenz 1998; W. Harris and Lucas2002). In spotted salamanders (*Ambystoma maculatum*), courtship lasts no more than a month, and often lasts only a few days in any given pond. Males greatly outnumber females at the breeding site. Males move about the pond searching for females, and several males often attempt to court the same female simultaneously, each depositing many spermatophores scattered over a large area. Courtship behavior is relatively simple, with little direct interaction between individual males and females. Males do not attempt to monopolize females, but they do compete by placing their spermatophores near those already present, resulting in clusters of spermatophores on the bottom of the pond. Males also sometimes place their spermatophores directly on top of those already deposited by other males (fig. 9.28), a form of sexual interference (see the following "Alternative Mating Tactics"). Each male deposits an average of 40 spermatophores per courtship—nearly half his total complement for the year—but there is a low probability that any particular spermatophore will be located by a female (Arnold 1976). This species exhibits an extreme form of scramble competition, with little or no opportunity for females to choose mates. Genetic studies have shown that multiple paternity is common, but there is a clear advantage to males that mate first with females, and this probably favors early arrival of males in ponds (Tennessen and Zamudio 2003; E. Myers and Zamudio 2004).

Similar behavior has been described in *A. texanum* (Garton 1972; Petranka 1982a; McWilliams 1992; W. Harris and Lucas 2002). A male sometimes deposits up to 128 spermatophores in a single courtship, suggesting that males enhance their chances of fertilizing a female mainly by depositing as many spermatophores as possible, instead of investing a lot of time in courtship before spermatophore deposition. The precise allocation of sperm to each courtship event will be affected by the size of a male's sperm reserves and the expected future availability of females as the number of females declines during the breeding season (W. Harris and Lucas 2002). Males are aware of the presence of females, probably relying on chemical cues for sex identification, and they deposit spermatophores most rapidly when females are nearby. There is little direct interaction among males in a courting group, but as in *A. maculatum,* males sometimes cover previously deposited spermatophores with their own. Females pick up large numbers of spermatophores (average of 26 spermatophores per courtship), so there is the potential for sperm competition among males if several males court a female simultaneously. In laboratory experiments, males deposited more spermatophores if other males were present than if they were alone (McWilliams 1992). The effect of this behavior on actual male reproductive success remains to be determined.

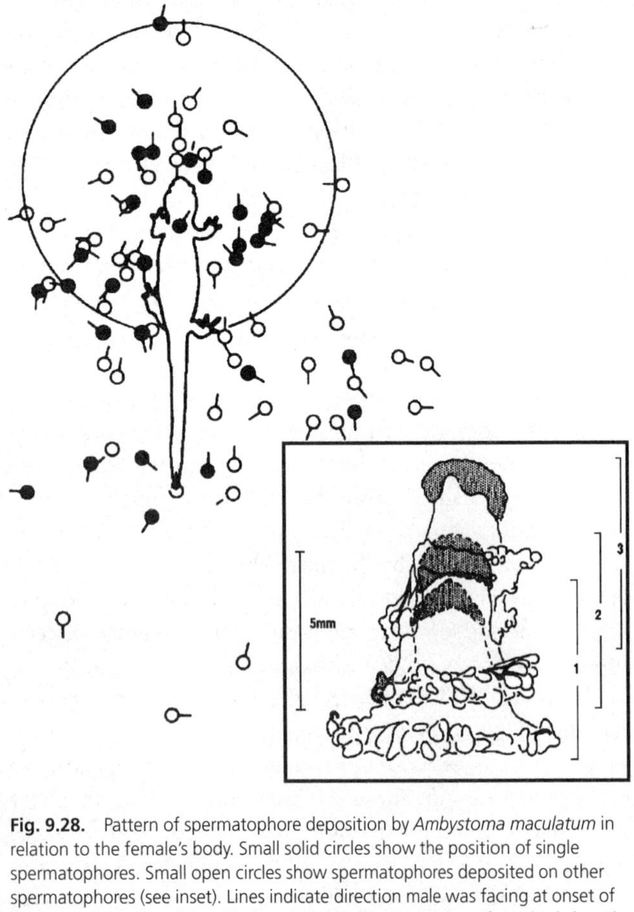


**Fig. 9.28.** Pattern of spermatophore deposition by *Ambystoma maculatum* in relation to the female's body. Small solid circles show the position of single spermatophores. Small open circles show spermatophores deposited on other spermatophores (see inset). Lines indicate direction male was facing at onset of spermatophore deposition. The large circle indicates a radius of one body length around the female's snout. Large numbers of spermatophores are randomly distributed around the female's body. After Arnold (1976).

Marbled salamanders (*A. opacum*) and ringed salamanders (*A. annulatum*) both have similar mating systems, but in these species, courtship takes place on land after adults have migrated to dry or partially filled pond basins in the fall (Noble and Brady 1933; Spotila and Beumer 1970). Males typically outnumber females and engage in scramble competition for mating opportunities. Krenz and Scott (1994) discovered that some females in a population of marbled salamanders in South Carolina arrived at the breeding area having already mated, suggesting that some males intercept and court females as they are moving toward the breeding area. This would enable males to avoid competition from other males at the breeding site, reducing the likelihood of sexual interference, such as covering of spermatophores.

### Hynobiidae

Some hynobiids also have explosive breeding periods, but in contrast to ambystomatids, these salamanders have external fertilization. Females lay eggs in paired jelly sacs, which are then fertilized by males as they clasp the sacs with their front legs. In explosive-breeding species, including *Hynobius retardatus, H. dunni,* and *H. nigrescens,* males form mating balls around newly deposited egg sacs, competing for the chance to fertilize the eggs (Sasaki 1924; Hasumi and Iwasawa 1990; Sato 1992; Hasumi 1994). In *H. nigrescens,* a male clasps a female around her pelvic region as soon as she begins to deposit an egg sac, but shifts his grasp to the egg sac itself as it emerges, releasing sperm immeditately. Other males cluster around the sacs, jostling one another and competing to clasp the egg sac, with as many as eight males participating in a mating ball. Males also bite at competing males and twist their bodies and tails from side to side to dislodge competitors. In general, the first male to clasp the egg sacs was the most successful in maintaining a hold long enough to fertilize eggs (Hasumi 1994). As in explosive-breeding ambystomatids, there is little or no opportunity for females to select mates. In *H. leechii,* multiple insemination has been observed when different males clasp each of the two paired egg sacs produced by a female (S.-R. Park, Park, and Yang 1996; Park and Park 2000). This species also exhibits postmating guarding of egg sacs. Such guarding is relatively brief, averaging only about half a minute, but presumably is sufficient to reduce the chances of a second male attempting to fertilize the eggs immediately after the first male has mated.

#### Salamandridae

Scramble competition and formation of balls of males competing for access to a single female also has been reported in *Taricha granulosa,* but in this species fertlization is internal by means of spermatophores. A male clasps a female in amplexus from above, and other males sometimes attempt to wrestle the female away from the original male (Arnold 1977; Janzen and Brodie 1989). This behavior resembles that of explosive-breeding frogs, but presumably courtship cannot proceed to spermatophore deposition until all of the intruding males have been repelled. Males allowed to court females in the absence of competitors normally spend a considerable amount of time in courtship before depositing a spermatophore (Propper 1991). Janzen and Brodie (1989) reported that males that were found paired with females or participating in mating balls were larger than unpaired males, but they did not present data on the relative sizes of males that were ultimately successful in completing courtship. Females are known to mate with more than one male, but the first male to mate with a female normally fertilizes the largest proportion of her eggs (A. Jones et al. 2002). In the field, sex ratios normally are heavily skewed toward an excess of males, and this is likely to lead to increased mating skew and an increased intensity of sexual selection acting on males (A. Jones, Adams, and Arnold 2004).

### Mate Searching and Mate Guarding

Mate searching is defined here as a mating system in which males move about the breeding area or their normal year-round habitats in search of females, rather than remaining in one place and waiting for females to come to them. Once a male locates a female, he sometimes guards her against rival males. Mate guarding behavior includes seizing females in amplexus during courtship, transporting females away from rival males, or aggressively repelling competitors. Competitive mate searching and mate guarding are widespread among salamanders and probably is the most common mating system in this group.

#### Ambystomatidae

As the length of the breeding season increases, or as population densities at breeding sites decrease, the various forms of scramble competition seen in explosive-breeding ambystomatids tend to grade into a system of competitive mate searching and mate guarding. In a study of tiger salamanders equipped with radio transmitters, Madison (1998) found that males initially gathered in the deepest, clearest part of a breeding pond, presumably to search for females where visibility was highest. In mate-searching species, a male typically invests time and energy in monopolizing females and courting them individually before depositing spermatophores, instead of scattering groups of spermatophores around females. For example, male tiger salamanders (*A. tigrinum*) push females away from competitors (fig. 9.29); each female is then courted individually and led over a spermatophore in a tail-nudging walk (Arnold 1976). Although a male deposits only about 20 spermatophores per courtship, the chances of each one being found are higher than in *A. maculatum.* Rival males interfere with courtship by interposing themselves between a courting male and his female. By mimicking female behavior in the tail-nudging walk, the intruding male causes the leading male to deposit a spermatophore, which he then caps with his own (fig. 9.30). This tactic is used mainly by males that are smaller than those already courting females; larger individuals are more likely to push females away from rivals (R. D. Howard, Moorman, and Whiteman 1997). There is some opportunity for females to choose or reject particular males as mates, but male-male competition probably is a major determinant of male mating success.

Other ambystomatids with mating systems resembling that of *A. tigrinum* include *A. jeffersonianum, A. laterale, A. dumerili, A. talpoideum, A. macrodactylum,* and *A. gracile.* In all of these species, males court females individually, deposit relatively few spermatophores, and lead females over spermatophores (Mohr 1931; J. Anderson 1961; Licht 1969b; Storez 1969; Uzzell 1969; Brandon 1970b; Arnold 1977; Verrell and Pelton 1996; Verrell and Krenz 1998). In several of these species, mate guarding is carried further than in *A. tigrinum.* Males actually seize females in amplexus, clasping them with their forelimbs. A male often engages in a prolonged period of courtship stimulation, rubbing his chin against the cheek of the female. Evidentally this is primarily a tactile signal, because mental courtship glands have not been described in ambystomatids (Verrell and Pelton 1996). The female signals her receptivity by

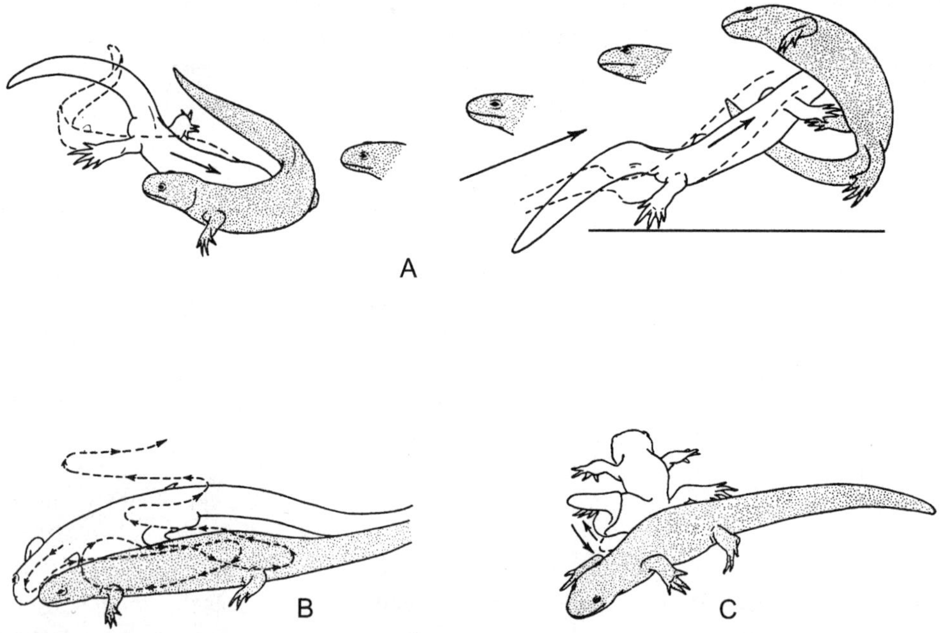


**Fig. 9.29.** Initiation of the tail-nudging walk during courtship in *Ambystoma tigrinum.* The male is shown in white, the female in gray. (A) The male transports the female away from rival males by shoving her through the water. (B) The male rapidly slides his snout over the back of the female and then moves away from her on a path indicated by the dashed line. (C) The male pauses while facing away from the female and repeatedly taps his tail on her dorsum. After Arnold (1976).

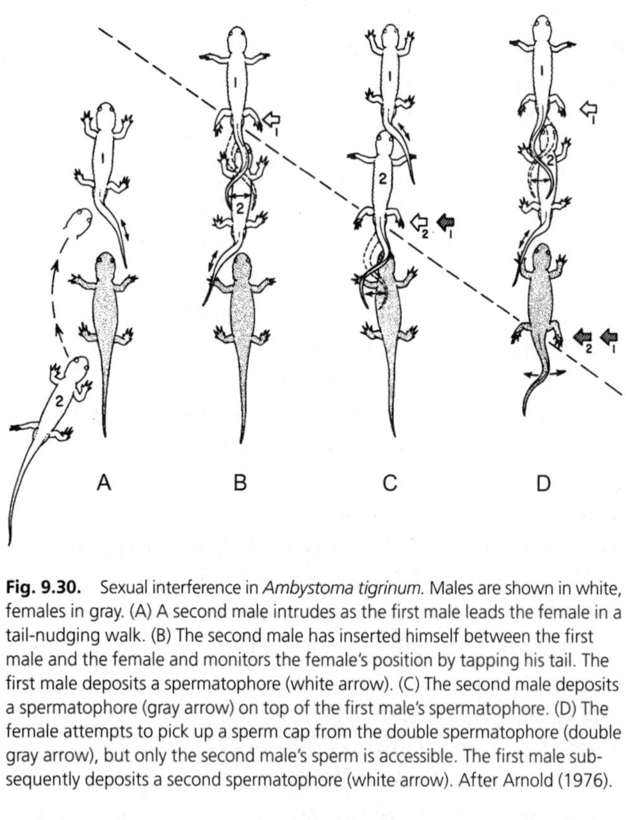


**Fig. 9.30.** Sexual interference in *Ambystoma tigrinum.* Males are shown in white, females in gray. (A) A second male intrudes as the first male leads the female in a tail-nudging walk. (B) The second male has inserted himself between the first male and the female and monitors the female's position by tapping his tail. The first male deposits a spermatophore (white arrow). (C) The second male deposits a spermatophore (gray arrow) on top of the first male's spermatophore. (D) The female attempts to pick up a sperm cap from the double spermatophore (double gray arrow), but only the second male's sperm is accessible. The first male subsequently deposits a second spermatophore (white arrow). After Arnold (1976).

lifting her head, and the male responds by proceeding to spermatophore deposition.

### Salamandridae

Many genera of Old World salamandrids engage in some sort of amplexus during courtship (Houck and Arnold 2003). In *Salamandra, Mertensiella, Chioglossa, Pleurodeles, Echinotriton,* and *Tylototriton,* the male seizes the female from below by looping his front legs over those of the female (fig. 9.3). Presumably amplexus functions as mate guarding behavior, allowing the male to court the female without interference from other males. Nevertheless, the precise nature of the mating systems of these salamanders is not clear, because observations of courtship have been made mostly in the laboratory. There is some evidence that males of *Salamandra* are territorial (Kastle 1986; see previous discussion), but whether females mate in the territories of males is not known. In laboratory observations, males maintain an alert posture and actively pursue any moving salamanders, suggesting that males search for mates (Himstedt 1965; Joly 1966; Arnold 1987). Similar behavior has been seen in *Chioglossa lusitanica,* which probably is closely related to *Salamandra* (Arnold 1987; Houck and Arnold 2003). In both of these genera, the male holds the female tightly throughout amplexus, but moves aside after depositing a spermatophore to lower the female's cloaca over the spermatophore. This close guarding of females suggests that attempted interference by other males is common in these species, something that seems unlikely if males courted females in exclusively held territories. Unfortunately, courtship in the presence of rival males has not been investigated, nor is there any information on spatial relationships and movements of males and females in the field.

Males of *Mertensiella caucasica* exhibit courtship amplexus like that of *Salamandra* and *Chioglossa* (fig. 9.3), clasping females throughout courtship and moving aside to lower them over the spermatophore. In this species, and in another previously placed in the genus *Mertensiella*, *Salamandra luschani*, males have an unusual morphological feature not found in any other salamanders. This is a fleshy protuberance at the base of the tail, just above the cloaca (fig. 9.31). This is placed in the cloaca of the female during amplexus (Schultschik 1994) and probably serves to align the female's cloaca with that of the male, ensuring that she will be lowered directly over the spermatophore (Arnold 1987). There are subtle differences in the histology of the glandular structure of the tubercles in the two species, which suggests the possibility of independent evolution of this trait (Sever et al. 1997). This interpretation is supported by recent phylogenetic work showing *S. luschani* to be more closely related to other species of *Salamandra* than to *M. caucasica* (Larson, Weisrock, and Kozak 2003).

Males of the genus *Pleurodeles*, which court females in water, also seize females from below, but the mode of sperm transfer is different. After depositing a spermatophore, the male releases one of the female's forelimbs and pivots around her so he is facing toward her. The female is then turned around to align her cloaca with the spermatophore (Arnold 1977; Houck and Arnold 2003). In laboratory observations, rival males sometimes attempted to wrestle males off of females or clasped females during spermatophore transfer, thereby disrupting the mating attempt. Again, this suggests that the tight amplexus of these salamanders helps a male monopolize a female.

In the European genera *Euproctus* and *Calotriton*, males

**Fig. 9.31.** Breeding male of *Salamandra luschani*, showing the dorsal protuberance at the base of the tail, which is used in courtship. This type of structure also occurs in *Mertensiella caucasica* and formerly was used as a character to unite these two species in the genus *Mertensiella*. Molecular phylogenetic studies suggest that *S. luschani* is more closely related to other species of *Salamandra* than to *M. caucasica*, in which case the tubercle probably evolved independently in the two species. This conclusion is supported by differences in the histology of the structure in the two species. From Halliday (1990).

**Fig. 9.32.** Amplexus and mating in *Calotriton* (= *Euproctus*) *asper*. The male has his body wrapped under that of the female, with his tail over her tail. Drawing by Bas Teunis. Reprinted from: R.A. Griffiths. *Newts and Salamanders of Europe* (p. 48). Academic Press, 1996, with permission from Elsevier.

have evolved a completely different form of mate guarding and spermatophore transfer. The female is seized from below, as in the previous genera, but she is held tightly by the male's tail, which is looped around her body, rather than his forelimbs (fig. 9.32). Sometimes the male also holds the female with his hind legs or his jaws. Amplexus can be a prolonged affair in these salamanders, lasting up to 30 hours in *Calotriton asper* (Thiesmeier and Hornberg 1986). Rival males often attempt to displace males from amplexus with aggressive attacks, and more than one male has been observed clasping the same female (Thiesmeier and Hornberg 1986, 1990). The mode of spermatophore transfer is unique to this genus; spermatophores are placed directly into the cloaca of the female in what amounts to copulation. The cloaca of the male is formed into a conical protuberance that facilitates placement of sperm into the female's cloaca. Again, the precise nature of the mating systems of these salamanders is unclear. Observations by Thiesmeier and Hornberg (1990) suggest that males of two species, *Euproctus platycephalus* and *E. montanus*, actively search for females and sometimes seize them aggressively with their jaws before clasping them with the tail. In contrast, males of *C. asper* are reported to assume a conspicuous display posture that attracts females, with the tail elevated above the body. Males of this species also seem to seize females more gently. These observations suggest that the mating system of *C. asper* more closely resembles lekking behavior than mate searching, but additional field observations are needed.

North American newts of the genus *Notophthalmus* also search for mates and seize them in amplexus as a means of guarding them from rival males (Verrell 1982, 1983, 1990a; Massey 1988). In this genus, however, males capture females from above with their hind legs (fig. 9.11 B). Other males often try to insert themselves between the male and the female, or displace the courting male by wrestling with him (fig. 9.33). Usually the original male retains possession of the female, but occasionally a larger intruder will displace

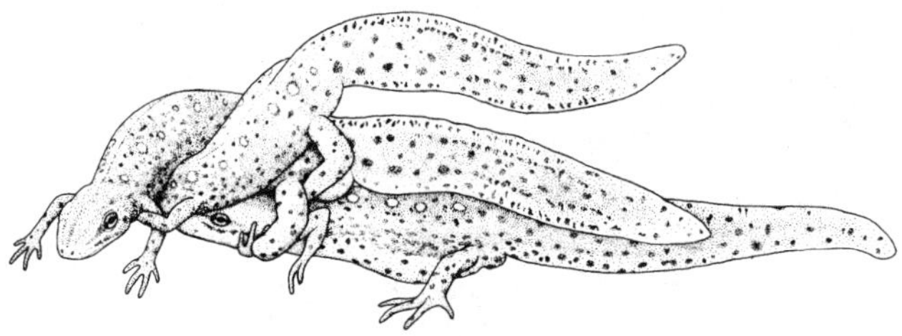

**Fig. 9.33.** Males of *Notophthalmus viridescens* wrestling over a female. The original male has his hind feet wrapped around the neck of the female. The intruder is attempting to insert his body between the male and female. Drawing by Tim Halliday. Reprinted from: P.A. Verrell. 1986. Wrestling in the red-spotted newt (*Notophthalmus viridescens*): Resource value and contestant asymmetry determine contest duration and outcome. *Anim. Behav.* 34:398–402. Academic Press, 1986, with permission from Elsevier.

the courting male (Verrell 1986c). Males sometimes skip the amplexus stage of courtship when rival males are absent and proceed directly to a brief display before initiating spermatophore deposition (Verrell 1983). Male-male competition clearly influences mating success, but females also have opportunities to accept or reject particular males. In a natural population, most courtship attempts were unsuccessful, either because of interference from rival males or because the female simply left after being released by the male (Massey 1988).

Males of most Old World newts (*Triturus, Neurergus, Paramesotriton, Cynops,* and *Pachytriton*) probably search for mates in breeding ponds, approaching anything that resembles a female newt, although some species of *Triturus* form leks (see the following discussion). These genera lack amplexus, but males engage in elaborate courtship of individual females. The details of these displays have already been discussed in the section on "Modes of Communication." The absence of amplexus and the use of elaborate visual, chemical, and tactile signals during courtship appear to be derived conditions within the family Salamandridae (Halliday 1990; fig. 9.3). Because males do not physically monopolize females, rival males frequently interrupt courtship bouts and attempt to insert themselves between the courting male and female, often just before the original male would deposit a spermatophore (fig. 9.34). The intruder then attempts to lure the female away from the orignal male and complete the courtship sequence with her. Males of several species also have been observed engaging in female mimicry (fig. 9.35). This involves a male inserting himself between a courting male and female, following the lead male while touching his tail. This sometimes induces the first male to deposit a spematophore, which often is not picked up by the female and is wasted. Sometimes the female moves away from the original courting male and follows the intruder and completes courtship with him (Verrell 1984a, 1988c; Giacoma and Crusco 1987; Verrell and McCabe 1988; Halliday 1990; Faria 1995; Halliday and Tejedo 1995; Sparreboom 1996; Sparreboom and Faria 1997). Often both males fail to inseminate the female, because females tend to move away from competing males. Insemination of females by multiple males appears to be relatively common in some species of *Triturus* (Rafiński 1981; Rafiński and Osikowski 2002; Garner and Schmidt 2003) and very likely occurs in other newts as well. There is evidence that in one species, *T. alpestris,* females benefit from multiple matings by having increased fertilization of eggs (Osikowski and Rafiński 2001). Given a choice, females of this species also appear to discriminate against closely related males and mate differentially with less related males (Garner and Schmidt 2003).

Many of these newts, especially the smaller species of *Triturus,* are notable for their lack of aggression and overt mate guarding behavior by males. Males compete for females by chasing them (Verrell and McCabe 1988), by displaying more intensively to females in the presence of rival males (Verrell 1984b), or by altering the temporal pattern of some courtship elements (Sparreboom 1996; Denoël, Poncin, and Ruwet 2001a). Two closely related genera of Asian newts, *Cynops* and *Paramesotriton,* exhibit more aggressive forms of mate guarding that are seldom seen in most species of *Triturus.* Male Chinese fire-bellied newts (*Cynops orientalis*) have been observed biting females for up to five minutes after spermatophore deposition, a behavior that usually immobilizes the female (Sparreboom and Faria 1997). The function of this behavior is unclear, but it could serve as a form of postmating guarding behavior that prevents the female from immediately mating with another male. Japanese sword-tailed newts (*Cynops ensicauda popei*) also have been observed biting females, but only before spermatophore deposition. In most cases, the male actually picked up the female and transported her away from a rival male (Sparreboom1996). Biting is more aggressive in *Paramesotriton.* Males of several species often violently bite one another and engage in prolonged fights for possession of females (Sparreboom 1984b). The courtship behavior of these two genera is similar to that of *Triturus,* but with less complex display elements than those seen in species of *Triturus* that lack aggressive mate guarding behavior (Sparreboom 1993; Sparreboom and Faria 1997). If the courtship of *Triturus* represents a more derived condition, then competitive display might have replaced more aggressive forms of competition for mates. In larger species of *Triturus,* however, both display behavior and aggression are well developed (see the following leks discussion).

### Plethodontidae

The courtship behavior of plethodontid salamanders has been studied in considerable detail, but most of these studies have been done in the laboratory, and their mating sys-

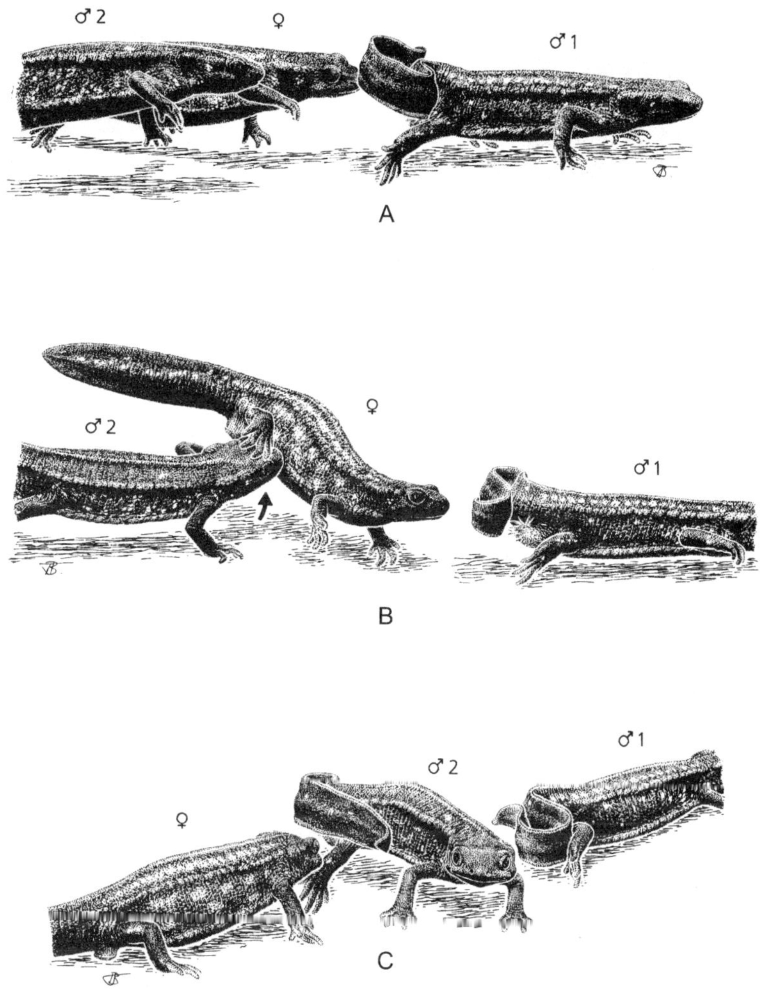


**Fig. 9.34.** Sexual Interference in Japanese newts (*Cynops ensicauda popei*). (A) The courting male (♂1) is creeping and being followed by the female, while the interfering male (♂2) approaches from beside the female. (B) The interfering male (♂2) nudges the belly and cloaca of the female. (C) The original courting male (♂1) remains in creeping position, while the interfering male (♂2) turns in front of the female and gives fan displays. From Sparreboom (1996).

tems are not well understood (Houck and Verrell 1993). For most plethodontids, we have so little information on the daily movements of males and females (chapter 6) that it is impossible to determine whether males search for females or females search for males. Both lekking and resource defense have been proposed as possible mating systems in plethodontids (Mathis et al. 1995; Sullivan, Picard, and Madison 1995), but in many cases, competitive mate searching and mate guarding are equally plausible explanations for their behavior. In several species, there appears to be greater home range overlap between males and females than between individuals of the same sex (Nishikawa 1990; Mathis 1991b; Marvin 1998b). In some species, males and females are more likely to be found under the same cover object than are individuals of the same sex (Wells 1980c; Lang and Jaeger 2000), or members of one sex are more likely to have members of the opposite sex as nearest neighbors (Mathis 1991b). Nevertheless, it is not always clear whether females move into male home ranges or males leave their home ranges to search for females (Toll, Jaeger, and Gillette 2000).

Most plethodontids have relatively long courtship seasons, and courtship behavior is remarkably uniform in the family. Males never gather in dense aggregations or engage in mass courtship. In every species, males court individual females with an elaborate behavioral repertoire involving both chemical and tactile stimulation, as described earlier in this chapter. Invariably this ends in a tail-straddling walk. The female rests her chin on the base of the male's tail, and the male leads her over a single spermatophore (Arnold 1977; Marvin and Hutchison 1996). Males of the *Plethodon jordani* complex invest considerable time in each courtship (average = 56 minutes), but the probability of each

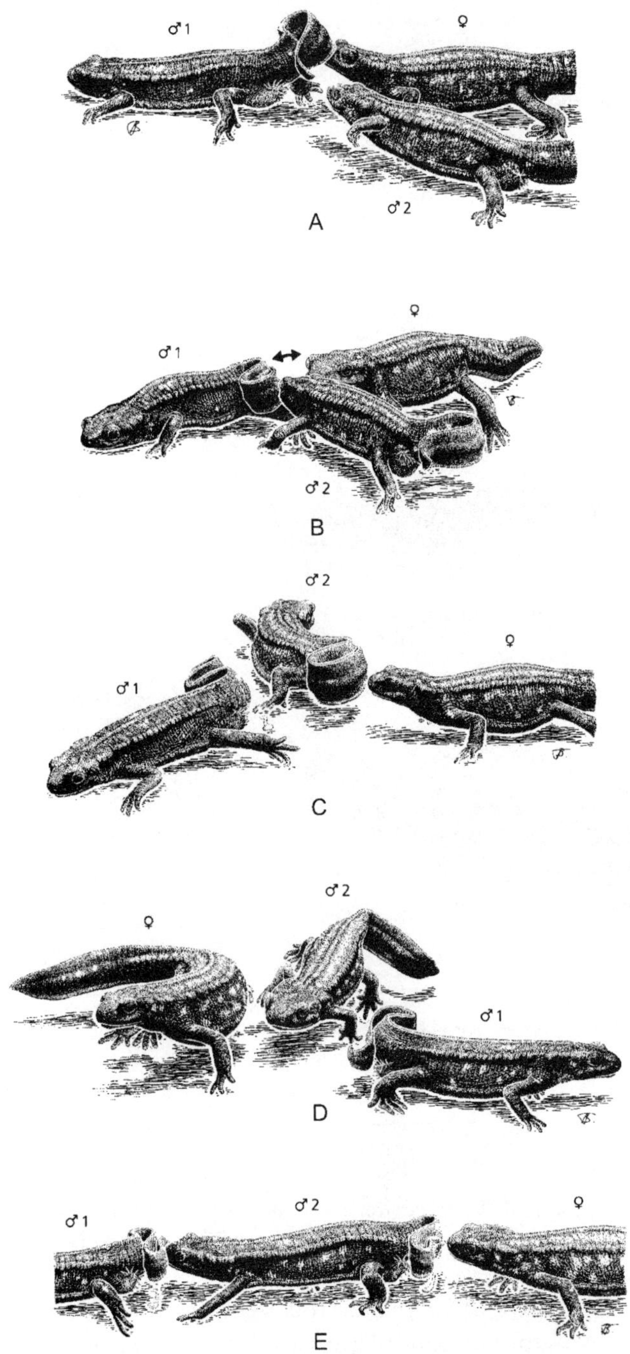


**Fig. 9.35.** Examples of sexual mimicry in Japanese newts (*Cynops ensicauda popei*). (A) A courting male (♂1) is creeping, with the female following; an interfering male (♂2) interrupts the courtship. (B) The interfering male (♂2) then inserts himself between ♂1 and the female, while swinging his head and moving the female away from the original male. (C) The original male (♂1) has deposited a spermatophore, but ♂2 has led the female away. (D) In another interaction, an interfering male (♂2) has interrupted a courtship, and the female moves away. (E) An interfering male (♂2) has induced ♂1 to deposit a spermatophore by touching his tail; ♂2 then deposits a spermatophore while being followed by a female. From Sparreboom (1996).

spermatophore being picked up is very high. A male seldom produces more than one spermatophore per courtship (Arnold 1976), although multiple spermatophore production has been observed in *P. kentucki* (Marvin and Hutchison 1996). Postmating guarding of mates does not seem to be common in plethodontids. This might be because females become unreceptive to further courtship soon after mating and remain unreceptive for several days (Verrell 1998a, 1991a). Some females do eventually court with additional males. Because they can store sperm for some time, this sets the stage for direct sperm competition (Tilley and Hausman 1976; Houck, Tilley, and Arnold 1985; Halliday 1998; E. Adams, Jones, and Arnold 2005).

Various forms of sexual interference and male-male competition are common in plethodontids. Female mimicry has been observed in several genera, including *Plethodon, Pseudotriton, Ensatina,* and *Desmognathus* (Arnold 1977). Usually this involves an intruding male engaging in a tail-straddling walk with another male, causing the leading male to deposit a spermatophore. Although the second male does not cover the spermatophore, it is not available to females and therefore represents wasted sperm (Arnold 1976). Aggressive competition among males, including biting of rivals, has been observed in *Plethodon, Eurycea, Pseudotriton, Gyrinophilus, Aneides,* and *Desmognathus* (Organ 1958, 1960a, b; Arnold 1972, 1976, 1977; Jaeger 1984; Ovaska 1987; Houck 1988; Verrell and Donovan 1991; Staub 1993; Halliday and Tejedo 1995; Mathis et al. 1995; Bernardo and Arnold 2002). Sometimes this is related to defense of territories, but in other cases, males compete for access to individual females (Arnold 1976; Houck 1988; Gergits and Jaeger 1990a; Verrell and Donovan 1991).

### Leks

Lek mating systems are distinguished from other types of mating aggregations and mate searching in that males exhibit site fidelity, they tend to defend courtship sites against other males, and they perform elaborate courtship displays that attract females and help to bring them into receptivity. Although there are hints that some species of *Desmognathus, Plethodon,* and *Calotriton* might have leklike mating systems (Thiesmeier and Hornberg 1990; Sullivan, Ryan, and Verrell 1995), the best documented examples of lekking behavior are several of the largest and most brightly colored species of *Triturus.* These include *T. vittatus, T. marmoratus, T. cristatus, T. carnifex, T. karelini,* and *T. dobrogicus* (the last three were formerly considered subspecies of *T. cristatus*). Lekking has not been described in detail for all of these, and does not necessarily occur in all populations. *Triturus marmoratus* and the *T. cristatus* group appear to form a well-defined clade allied to *T. vittatus,* although the phylogenetic position of the latter is not entirely clear (Arntzen and Sparreboom 1989; Halliday 1990; MacGregor, Sessions, and Arntzen 1990; Halliday and Arano 1991). These phylogenetic relationships make it likely that leklike mating systems have evolved only once or twice in the genus. Exactly why such a mating system might evolve is not clear, however.

Males of these species establish small territories on the bottoms of ponds after dark. Often they spend daylight hours hiding in dense vegetation, but move to more open areas for courtship display. This can lead to aggregation of males in suitable display areas. In contrast to many lekking birds (Höglund and Alatalo 1995), male newts exhibit only temporary attachment to specific territories or aggregations, often moving to new display sites on successive nights. Males of all of these species have elaborately marked crests and tail fins, making them the most sexually dimorphic species in the genus (fig. 9.36). As discussed previously, courtship displays emphasize lateral display of crests and tail fins to females, often accentuated by the lean-in and cat-buckle displays, in which the male leans toward the female and kinks his body around hers (fig. 9.8). Aggressive interactions among males are relatively common, although this varies among populations, and many of the displays used in courtship, especially the lateral displays, are used in aggressive encounters as well (Zuiderwijk and Sparreboom 1986; A. Green 1989; Hedlund and Robertson 1989; Raxworthy 1989; Hedlund 1990b; Sparreboom and Teunis 1990). The principal function of aggressive territoriality in these lekking species presumably is to allow a male to court females while minimizing sexual interference by neighboring males. Hence, it serves much the same function as mate guarding in other salamanders.

### Resource Defense

Defense of resources that are attractive to females is an uncommon mating system in salamanders, for reasons discussed earlier. The mating system of the red-backed salamander (*Plethodon cinereus*) has been described as resource defense because there is evidence that females associate with males that have high-quality food resources in their territories (Mathis et al. 1995; Sullivan, Ryan, and Verrell 1995; Lang and Jaeger 2000; Toll, Jaeger, and Gillette 2000). It is not clear, however, how females might benefit from choosing males on the basis of territory quality. Three possibilities are: (1) females benefit directly by foraging in male territories, (2) the offspring of females grow up in the territories of their mates and have access to high-quality food resources, or (3) females use the quality of territories to assess the over-

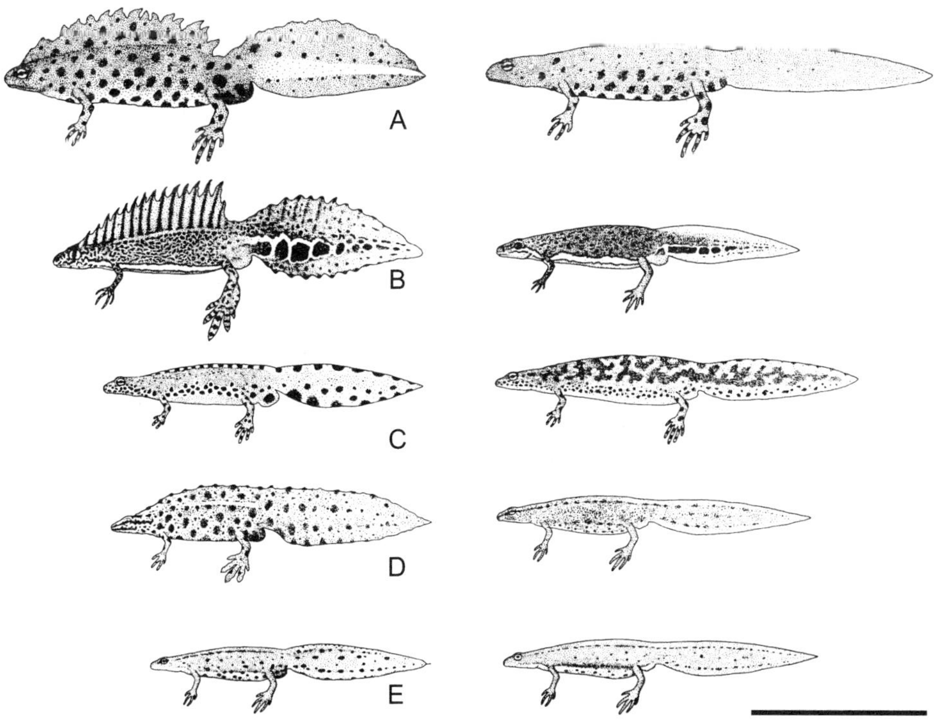


**Fig. 9.36.** Sexual dimorphism in five species of European newts. Males are on the left, females on the right. (A) *Triturus cristatus* and (B) *T. vittatus* have lek mating systems and exhibit the greatest sexual dimorphism in ornamentation. (C) *T. alpestris.* (D) *T. vulgaris.* (E) *T. boscai.* Scale bar represents 5 cm. From Halliday and Arano (1991).

all genetic quality of potential mates (Jaeger et al. 2000). At present, all of these hypotheses remain to be tested.

The best-documented examples of resource defense mating systems are in the cryptobranchids and hynobiids, salamanders in which external fertilization sets the stage for defense of oviposition sites by males. In *Cryptobranchus alleganiensis,* males have been observed defending nest sites under large flat rocks against other males, but accepting females into their territories, where they lay eggs attached to the underside of the rocks. Males guard the eggs after they are laid, but they continue to accept females into their territories, and some males have been observed tending several egg clutches simultaneously (B. G. Smith 1907; Bishop 1941b; Nickerson and Mays 1973a). The Asian giant salamanders (*Andrias*) have a similar mating system, with males defending suitable oviposition sites as territories (Kerbert 1904; Stejneger 1907; Tago 1929; Kuwabara et al. 1989; Kawamichi and Ueda 1998). In these salamanders, small males have been observed entering the nests of large males when females are laying eggs (Kawamichi and Ueda 1998). This type of sexual interference is analogous to the satellite behavior seen in many frogs (chapter 8) and fishes (Taborsky 1994), with intruding males attempting to fertilize some of the eggs deposited in another male's territory.

Males of some hynobiid salamanders also defend territories centered around suitable oviposition sites (Tanaka 1986, 1987, 1989). In Japan, *Hynobius nebulosus* has a prolonged breeding season, from November through February. In a population studied in a ditch by Tanaka (1989), adults migrated overland to the breeding site, with the largest males arriving somewhat earlier than smaller males. The largest males established territories around the roots of horsetails, under fallen leaves, or near holes in mud, and sometimes occupied them for several weeks. Territories with abundant horsetails or other vegetation suitable for attachment of egg sacs were preferred, and received the largest number of egg sacs (up to 11 sacs in some territories). Smaller, nonterritorial males adopted a satellite or sneaker tactic. They entered the territories of large males as females were depositing eggs and attempted to clasp and fertilize egg sacs. Satellite males were most abundant around the highest-quality sites. Satellite males frequently moved between territories, hiding in mud while waiting for females to appear, and some males switched between territorial and satellite behavior. The mating system of this salamander is remarkably similar to those of resource-defending frogs, such as *Rana clamitans* and *R. catesbeiana* (see chapter 8).

### Determinants of Male Mating Success

In contrast to the voluminous literature on determinants of male mating success in anurans, there has been relatively little work on this topic with urodeles. In part this is because these animals are much more difficult to observe in the field and less suitable for experimental work than the highly vocal anurans. Consequently, any conclusions about factors that influence variation in male mating success must be somewhat tentative. As in anurans, the amount of time a male spends engaged in reproductive activities is likely to have a major effect on male mating success, and this probably accounts for the early arrival of males at breeding sites in many species (e.g., Douglas 1979; Tanaka 1989). The amount of time a male spends at a breeding site or the length of his courtship season might be constrained by depletion of energy reserves (Halliday 1987). Unfortunately, information on individual variation in energy reserves, male tenure at breeding sites, and individual reproductive success is not available for any species of urodele. Both tenure at the breeding site and number of eggs obtained in territories were positively correlated with male body size in *Hynobius nebulosus,* but reproductive success was clearly influenced by male territory quality as well (Tanaka 1989).

Physiological constraints on male performance could limit male mating success on a shorter time scale as well (Halliday 1987). In newts, repeated courtship of a female increases the probability of successful spermatophore transfer (Teyssedre and Halliday 1986), and males that court many females are likely to have the highest reproductive success. One constraint on courtship time is the need for aquatic animals to breathe. Time spent surfacing to breathe reduces time available for courting (Halliday and Sweatman 1976; Halliday and Worsnop 1977; Halliday and Houston 1991), and males engage in longer bouts of courtship when aquatic oxygen concentrations are increased (Halliday 1977b). It is possible that males with large crests and tail fins relative to body size enjoy some increase in cutaneous respiration (see chapter 4) that would enable them to remain longer under water, but this has not been demonstrated experimentally. Energetic investment in courtship display also could be constrained by depletion of energy reserves, forcing males to reduce the frequency of costly display elements. This might have a significant effect on male mating success in competitive situations, because male newts typically increase display rates in the presence of other males (Verrell 1984b; Green 1991a). Consequently, males with depleted reserves would be at a competitive disadvantage in trying to court females. In other species, however, the energetic cost of courtship is relatively trivial (Bennett and Houck 1983; see chapter 5).

Physiological traits related to locomotor ability or stamina might influence male mating success when males search for females. For example, two studies of red-spotted newts (*Notophthalmus viridescens*) have shown that male mating success is positively correlated with tail fin size. There is no evidence that females prefer males with large tails in this species, but larger tail fins appear to enhance the ability of males to capture females in amplexus (Able 1999; Gabor, Krenz, and Jaeger 2000). Direct scramble competition for females also can select for traits that enhance a male's ability to hold on to females once they are encountered. In *Taricha granulosa,* males engage in scramble competition for females, and both a large tail and a bulky body appear to enhance male mating success (Janzen and Brodie 1989; Halliday 1998). Males of this species increase in bulk during the breeding season due to hormonally regulated uptake of water (Propper 1991; Moore 1994). The same is true of some hynobiid salamanders, which engage in scramble competition for egg sacs rather than females. In some species, uptake of water results in enlargement of the head, which confers an advantage in competition to fertilize egg sacs (Hasumi and Iwasaw 1990; Hasumi 1994; see also "Patterns of Sexual Dimorphism," following).

Aggressive interactions among males probably influence male mating success in many species of salamanders (Halliday and Tejedo 1995). In several species, large males have been shown to dominate small males in contests over females or territories (Tanaka 1987; Houck 1988; Janzen and Brodie 1989; Mathis 1990b; Verrell and Donovan1991; Wiltenmuth 1996; R. D. Howard, Moorman, and Whiteman 1997; Marvin 1998b; Mathis and Britzke 1999), although this is not true for some species (Camp 1999). This is likely to be true for most species in which males fight over females, but not necessarily for species that engage in relatively nonaggressive scramble competition (Hasumi 1994). The various forms of sexual interference that commonly occur in urodeles also are likely to affect male mating success, because many courtship attempts are disrupted before mating is completed. In some cases, the use of interference tactics by small males reduces the competitive advantage of large males attempting to monopolize females (Tanaka 1989; R. D. Howard, Moorman, and Whiteman 1997).

Female choice probably has some effect on male mating success in most salamanders except those with explosive breeding periods. In both newts and terrestrial salamanders, many courtship attempts end with the female leaving before mating is completed. Again, information about the criteria used by female salamanders to choose mates is meager compared to the large literature for anurans (Sullivan, Ryan, and Verrell 1995). In salamanders that have resource-based mating systems, variation in territory quality might influence female choice of mates. The possibility that female red-backed salamanders select mates based on the quality of food found in their territories has already been discussed. Although both laboratory and field observations tend to support this hypothesis (Mathis et al. 1995), we do not yet have much evidence that females move from one male territory to another to compare them. Variation in resource quality probably influences female choice in cryptobranchids and hynobiids in which males defend oviposition sites. In cryptobranchids, the size or other features of retreat sites under rocks probably influence the choice of mates by females. In *Hynobius nebulosus,* the density and quality of vegetation available for egg attachment appears to have a major influence on female choice (Tanaka 1989).

In most other salamanders, female choice would necessarily be based on morphological features of males, aspects of their behavior, or a combination of the two. This question has been addressed in most detail in studies of European newts. In both *Triturus cristatus* and *T. vulgaris,* females pick up more spermatophores deposited by males with large crests than by those with small crests (Malacarne and Cortassa 1983; Hedlund 1990b; A. Green 1991b; Gabor and Halliday 1997). It is not known whether females make visual assessments of relative crest height or choose mates based on features of courtship displays that are correlated with crest height. In either case, the height of the crest appears to provide females with reliable information about male quality. Only males in good condition develop large crests, and those provided with extra food develop larger crests (A. Green 1991b; J. M. R. Baker 1992). On the other hand, neither crest height nor body size appears to be a reliable indicator of the inherent mating capacity of a male, as measured by the number of spermatophores he produces (J. M. R. Baker 1990).

Female newts also might select mates on the basis of differences in courtship behavior of males. However, Michalak (1996) found little consistency within males for several elements of courtship behavior in *T. montandoni,* and little consistency in preferences of females for particular courtship elements. Breeding coloration, perhaps in combination with other sexually dimorphic traits, might influence mate choice in some newts. In a study of a population of alpine newts (*Triturus alpestris apuanus*) that included both paedomorphic and metamorphosed males, females preferred to mate with metamorphosed males. These males develop bright blue breeding coloration and other sexually dimorphic traits that are absent in paedomorphic males (Bovero, Giacoma, and Andreone 1997).

There is some evidence that females change their selectivity toward males, depending on whether they have mated previously. In an experimental laboratory study, Gabor and Halliday (1997) found that female smooth newts (*Triturus vulgaris*) did not discriminate between males with large and small crests when they were the first males encountered. In contrast, females that had already mated once would remate only with males with large crests. This suggests that females might mate relatively quickly with the first male encountered, as long as he meets some minimum standard for quality, to ensure that her eggs would be fertilized, but would remain receptive to other males of higher apparent quality. Because female newts can store sperm, there is the potential for direct sperm competition to occur (A. Jones, Adams, and Arnold 2002; Garner and Schmidt 2003). In the laboratory, males sometimes are selective in their choice of mates as well. Males that were presented with more than one female simultaneously reduced the amount of time spent in amplexus with any one female (Verrell 1985b) and were more likely to initiate courtship with the larger of two females, presumably because this will result in the male fertilizing more eggs. Verrell (1985b, 1986b) proposed that males discriminate between large and small females using either visual or chemical cues, but Rowland, Robb, and Cortwright (1990) argued that discrimination of female body size based on chemical cues is unlikely. In any case, it probably is rare for males to encounter multiple females simultaneously in natural populations, where sex ratios usually are heavily skewed toward an excess of males.

Information on female choice of mates for other salamanders is so fragmentary that little can be said about it. Neither size nor age was correlated with courtship success of male *Desmognathus* in noncompetitive situations (Houck 1988; Houck and Francillon-Viellot 1988; Verrell 1991a). Donovan and Verrell (1991) reported that pairs of *D. ocoee* that were familiar with one another were less likely to complete spermatophore transfer than were unfamiliar pairs, but it was not clear whether this was due to male rejection of females or the reverse. Several studies of *Desmognathus* have shown that males exhibit siginificant differences in mating success in laboratory trials, but in most cases, information on variation in male behavior was not available (Houck, Tilley, and Arnold 1985; Shillington and Verrell 1996). Vinnedge and Verrell (1998) and Gershman and Verrell (2002) addressed this problem by examining the mating success of *D. ocoee* males paired with females in noncompetitive trials. They then examined variation in several aspects of male courtship behavior. They found that males that had achieved a high level of mating success had significantly higher frequencies of persuasive courtship displays, based on a composite score of several individual courtship elements. In particular, males that performed the highest frequency of "snap" behavior involved in transferring pheromones to females achieved the highest reproductive success.

Females also might use differences in body size to assess mate quality. Females of *Plethodon cinereus* were more likely to associate with large males than with small males in laboratory experiments, but it is not clear whether females were assessing male body size directly or were responding to some other cue that is correlated with body size (Mathis 1991b). Male body size did not appear to influence female choice of mates in *Ambystoma tigrinum,* but those males that achieved successful spermatophore transfer had slightly longer tails (R. D. Howard, Moorman, and Whiteman 1997). Again, the way in which females might assess differences among males is not known. In some salamanders, there is evidence for males discriminating among females on the basis of body size. For example, male *Desmognathus* that were offered access to a large and a small female simultaneously preferred to court the larger individuals (Verrell 1989b, 1994b). How often males might encounter two females simultaneously in the field is not known. A preference for larger females might be expected in species in which males search for females and invest considerable time and energy in courtship, and in those in which males are likely to have access to many females. Such a preference seems less likely in species in which the time and energy costs of courtship are small, or males seldom encounter females.

### Patterns of Sexual Dimorphism

The various problems associated with interpreting patterns of sexual size dimorphism were discussed in chapter 8. Differences in body sizes of males and females result from different selection pressures acting on the two sexes, and it often is difficult to predict exactly how these will affect relative adult body sizes (Halliday and Tejedo 1995). In most species of urodeles, fecundity selection is likely to favor large body size in females because of size-related increases in clutch size, egg size, or both (see chapter 10). Indeed, the typical pattern for most urodeles, as in anurans, is for females to reach larger sizes than do males. In some species, aggressive competition among males probably favors large body size as well, because large males are almost certain to win more fights than are small males. In a preliminary analysis, Shine (1979) reported a significant association between reports of male combat in urodeles and males being either larger than or about the same size as females. His analysis was flawed, however, because he lacked information on aggressive behavior in many species, some of which have subsequently been shown to be aggressive (Halliday and Verrell 1986). He also relied on literature records for data on body sizes, and in many cases, only average body sizes were available. In most ectothermic animals, including salamanders, growth tends to be asymptotic, and the asymptotic or maximum body size typically gives a better estimate of sexual dimorphism than does average body size (Stamps 1993). Finally, Shine's (1979) analysis did not incorporate phylogenetic information, making his statisical inferences suspect.

It seems clear that there is not a simple relationship between sexual size dimorphism, measured as a difference in body length, and the occurrence of male combat in urodeles. In cryptobranchids and hynobiids, males of some species are reported to be as large as or larger than females, and some of these species also exhibit male combat (Nickerson and Mays 1973a; Shine 1979). However, there has not been a comparison of closely related hynobiids that have different mating systems. Large body size apparently influences male mating success in *Hynobius nebulosus,* a territorial species (Tanaka 1989), but not in *H. nigrescens,* a species with scramble competition (Hasumi 1994). In the latter species, the heads of males become swollen during the breeding season, apparently through hormone-mediated water uptake, and this swelling apparently confers some advantage in scramble competition for control of egg sacs.

Shine (1979) listed several species of salamandrids in which males are at least as large as females, including *Calotriton asper* and *Pleurodeles waltl,* both of which compete by clasping females in amplexus, but in most other genera, males reach shorter body lengths than do females. This is true even for species of *Triturus* that defend territories on leks and for genera in which males have been observed fighting, such as *Paramesotriton* and *Salamandra.* In the genus *Plethodon,* females of most species reach greater body lengths than do males, even in species that are territorial and sometimes violently aggressive. The genus *Desmognathus* is unusual in that males of many species reach sexual maturity at a younger age than do females, but ultimately achieve a larger body size (Organ 1961a; Tilley 1977, 1980; Juterbock 1978; Bruce 1993). Aggressive competition among males is common in this genus, and large males typically dominate smaller individuals (Houck 1988). Nevertheless, the pattern of sexual dimorphism probably has more to do with the very large energetic investment in egg production by females and the dramatic slowing of growth after sexual maturity (Hom 1988; Bruce 1993). Males also are larger than females in the only other desmognathine salamander, *Phaeognathus hubrichti* (Bakkegard and Guyer 2004).

Features of male morphology other than body size can confer an advantage in aggressive contests, and degrees of sexual dimophism in these traits need to be investigated in more detail. For example, biting is common in many plethodontid salamanders, and one might expect this mode of fighting to favor relatively wider heads or stronger jaw musculature in males. Juterbock (1978) reported that the heads of male *Desmognathus* are wider and more massive than those of females, but he did not present a detailed analysis of head size allometry in relation to body size in the two sexes. Most species of *Plethodon* do not exhibit obvious differences in head size or shape between males and females, perhaps because aggression and territoriality are common in both sexes. Members of the genus *Aneides* have conspicuously hypertrophied jaw muscles and reinforcement of parts of the skull that are related in part to fighting (Staub 1993). Both males and females are aggressive in several species, and sexual dimorphism is not pronounced (Staub and Paladin 1997), but in *A. hardii,* males have much more massive jaw musculature and skulls than do females, a feature that is very likely related to aggressive behavior.

Salamanders have many other sexually dimorphic traits that are related to mating behavior or reproductive competition among males, even in species that lack overt aggression (Halliday and Tejedo 1995). Males of species that clasp females in amplexus usually have enlarged hind or forelimbs, sometimes equipped with horny nuptial pads to facilitate holding onto females. Such features are particularly conspicuous in salamandrids, such as *Notophthalmus.* These morphological features typically develop during the breeding season in response to elevated levels of sex hormones (Houck and Woodley 1995). The same is true for the many sexually dimorphic display organs, such as wide tail fins in salamandrids, and courtship glands used for female persuasion; their role in courtship and reproductive competition has already been discussed in some detail. The evolution of sexually dimorphic morphological traits in urodeles has yet to be studied thoroughly or placed in an explicitly phylogenetic context. There have been attempts to examine the phylogenetic evolution of behavioral traits that are related to morphology, such as courtship in newts (fig. 9.6; Arntzen and Sparreboom 1989). Further investigations along these lines should be a promising area of research.

## Summary and Conclusions

Urodeles tend to be much less conspicuous than anurans, so information about their communication and social behavior is more limited. Nevertheless, research over the last 30 years or so has revealed a rich repertoire of communication signals and behavioral interactions in both aquatic and terrestrial species. In contrast to anurans, urodeles do not use sound to communicate. The ears of salamanders are sensitive mainly to low frequencies well below those found in the hisses, squeaks, and clicks that are produced by some species, so the importance of these vocalizations as communication signals is unclear. Salamanders do make use of visual, tactile, and chemical signals, which are given primarly during courtship or aggressive interactions.

Many salamanders are most active at night, so visual signals often take a form that can be perceived in dim light, such as changes in body posture and movements of the tail. Color markings that develop during the breeding season in some salamanders often appear to increase the reflectivity of the animal or a particular body part, or to increase contrast between the animal and its background. Highly reflective signals that enhance the conspicuousness of displaying males include white patches on the throat in *Hynobius,* light yellow body coloration and bright tail spots in *Paramesotriton,* and silvery blue body coloration in *Neurergus* and some species of *Triturus.* Many species of *Triturus* also have bright coloration on belly and tail, often with contrasting black spots or other markings that tend to make the tail especially conspicuous. Within the family Salamandridae, the ancestral form of courtship involves males clasping females in amplexus and stimulating them with chemical signals, with visual signals being less important. In more derived species, especially in the genus *Triturus* and its relatives, amplexus has been lost, and visual signals have become more important in the initial stages of courtship, although chemical signals also are used. In the largest, most derived species of *Triturus,* many of which have lek mating systems, visual signals have become elaborated, with males using displays that present females with a lateral view of their sexually dimorphic body coloration, crests, and tail fins.

Many of the displays of newts combine visual, tactile, and chemical signals in ways that make it difficult to distinguish between these modes of communication. For example, the tail-fanning display found in most Old World newts not only conveys chemical signals to the female, but probably also provides tactile stimulation through water movements and visual stimulation as well. In some newts, new display elements have appeared that seem to emphasize either visual signaling (the wave display) or tactile signaling (the whip and flick displays). In plethodontids and ambystomatids, many forms of courtship behavior appear to combine tactile signals with chemical signals, with visual signals being of secondary importance. These include the application of pheromones from courtship glands by slapping the female with a chin gland or raking her with sharp teeth in plethodontids. Tactile signals are widely used by female salamanders to signal receptivity to males during courtship and usually involve a female nudging or touching the male.

Chemical signals probably are of primary importance for most urodeles and convey information about the species, sex, and sexual receptivity of individuals, as well as serving as persuasive signals that increase female receptivity during courtship. Chemical signals are produced by a variety of specialized glands located mostly in the cloacal or head region. In general, male salamanders do not appear to produce chemical signals to attract females. Males often are attracted to chemical cues emanating from females, but it is not known whether females of most species produce specialized male-attracting pheromones. Males of many salamander species produce special courtship pheromones that are either applied directly to the female's body or wafted toward the female during courtship displays. It is assumed that these pheromones serve to increase female sexual receptivity, although this has been demonstrated experimentally in only a few species of salamandrids and plethodontids.

Chemical signals produced by males can be detected by receptors in the skin of females, by specialized receptor organs in the nasal cavity, or, in some plethodontids, by brain receptors responding to pheromones injected directly into the bloodstream of the female. Chemosensory organs in the nasal cavity include the main olfactory chamber, which detects mainly volatile airborne odors or waterborne chemical cues, and the vomeronasal organ, which detects less volatile chemicals. The vomeronasal organ is best developed in plethodontids, which have nasolabial grooves that convey chemicals from the substrate by capillary action directly to the vomeronasal organ. In some plethodontids, the vomeronasal organ is sexually dimorphic and is much larger in males than in females. This probably relates to the tendency for males to search for females during the breeding season.

Numerous experimental studies have demonstrated that salamanders can use chemical cues to identify the species, sex, and even individual identity of other individuals. Chemical signals probably are of primary importance in reproductive isolation of closely related species, many of which have nearly identical courtship behavior. The evolution of reproductive isolation apparently has been a gradual process, at least in closely related species of *Desmognathus.* Populations generally considered to be distinct species generally exhibit strong behavioral isolation, whereas geographically close populations of the same species tend to show less behavioral isolation than those that are geographically separated.

Many terrestrial salamanders use pheromones to mark substrates in their territories, and other individuals can use these chemical signals to determine the identity of the territory holder. The red-backed salamander (*Plethodon cinereus*) has been used as a model system for understanding territorial behavior in terrestrial salamanders, although not all populations of this species are territorial. Territories in this species and in other terrestrial plethodontids are centered on retreat sites such as fallen logs, which provide moist microhabitats for shelter and foraging. Both males and females are territorial, and the largest indivduals tend to acquire the best sites. Territories that contain highly profitable, easily digested prey such as termites are preferred over those containing less profitable prey, such as ants. There is some evidence that females use the fecal pellets of males to assess the quality of their diets, and they use this information to select males with the best territories as mates. In some cases, males and females appear to defend the same territory, although they do not act cooperatively. Females are most aggressive toward other females, whereas males are most aggressive toward other males. Possible benefits to a female in selecting a male with a high-quality territory include access to preferred prey for herself and access to prey for her offspring, which might grow up in the vicinity of the male's territory. These putative benefits have yet to be demonstrated experimentally, however.

Red-backed salamanders and a number of other terrestrial plethodontids are interspecifically territorial and readily attack other species of salamander of similar size. Indeed, this form of interference competition appears to be the most important mechanism of interspecfic competition in salamander communities (see chapter 15). In some cases, aggression by a larger or more dominant species limits the microhabitats available to a competitively subordinate species. A classic example is the restriction of *Plethodon shenandoah* in Virginia to relatively dry talus slopes that cannot be inhabited by the more widespread species *Plethodon cinereus,* which occupies most of the surrounding wetter microhabitats. In at least one instance, a local population of *P. shenandoah* was driven to extinction by its inability to move into moister habitats during a drought. Both intra- and interspecific aggression and territoriality are common in the genus *Desmognathus* as well. These salamanders have very robust jaw musculature, which probably evolved originally for prey capture, but which also equip them with powerful weapons for fighting. Larger species often exclude smaller species from microhabitats nearest to streams through aggression and occasional predation (see chapter 15). Species of *Desmognathus* also have been found to be aggressive toward species of *Plethodon* and *Eurycea* of similar size, and can dominate these salamanders in aggressive encounters. Aggression is relatively uncommon in adults of other families of salamanders, and evidence for territoriality is mostly lacking or anecdotal.

Salamanders have an array of mating systems similar to that seen in anurans, including scramble competition in explosive mating aggregations, mate searching and mate guarding, leks, and resource defense. Scramble competition is characteristic of ambystomatids and some hynobiids that have very short breeding seasons. Males often gather in large numbers and compete vigorously to fertilize the eggs of females, either by deposition of many spermatophores, as in *Ambystoma,* or by monopolizing egg sacs, as in *Hynobius.* Various forms of sexual interference, including covering the spermatophores of rival males, and shoving females away from rival males, are common in explosive-breeding ambystomatids. Some salamandrids, such as *Taricha granulosa,* also engage in scramble competition, in which males compete for females by clasping them in amplexus, a form of mate guarding that prevents rival males from courting females and depositing spermatophores.

This type of scramble competition grades into various forms of mate searching and mate guarding behavior, with males seeking out individual females, sequestering them from rival males, and courting them individually. Mate searching and mate guarding are common in ambystomatids with somewhat longer breeding periods than those of explosive breeders, as well as many newts in the family Salamandridae. In the latter family, the ancestral form of competition among males involves males clasping females in amplexus to physically prevent other males from gaining access to them. Rival males often attempt to displace other males from amplexus, so sexual selection favors enlarged feet and other structures that allow a male to hold tightly to a female. Most Old World newts, apparently a derived clade within the family Salamandridae, have abandoned amplexus and court females using visual, tactile, and chemical cues. Sexual interference is common in these species as well, but males do not physically restrain females and usually do not attack rival males.

Some of the largest, most derived species in the genus *Triturus* have evolved lek mating systems in which males are aggressive toward other males and establish small mating territories on the bottoms of ponds. This form of competition can be seen as an alternative to guarding individual females or physically transporting females away from rival males. Newts with lek mating systems exhibit the greatest degree of sexual dimorphism, especially in traits related to visual courtship displays. This type of mating system provides the greatest opportunities for females to choose among available males, and there is evidence that females of some species prefer males with visually conspicuous traits, such as large tail fins.

Resource defense mating systems are much less common in urodeles than in anurans. This probably relates to the mode of fertilization, which in most salamanders involves transfer of sperm into the female's reproductive tract by means of spermatophores. This allows for the separation of mating and oviposition in both space and time, making it difficult for males to defend resources as a way to attract females. The exceptions are cryptobranchids and some hynobiids, which have external fertilization. Males often defend nest sites as territories, and it seems likely that females select mates in part on the basis of nest site quality. As in anurans with resource defense mating systems, satellite males are common, often clustering around the territories of the largest males and attempting to fertilize eggs of females entering the nest. Some terrestrial plethodontids may have resource defense mating systems as well, but in these cases, the benefits to females in choosing certain territories are less clear. They could gain benefits for themselves in the form of good foraging sites, or, if their offspring remain in the territories of their mates, they could gain valuable resources for their young.

Patterns of sexual dimorphism in salamanders are not clearly related to differences in aggressive behavior. Early analyses suggested that males may be larger relative to females in species that engage in male combat. This probably accounts for the large size reached by males in some territorial species, such as the cryptobranchids. Nevertheless, more recent work has revealed that aggression among males is common in many salamander species in which males reach considerably smaller adult sizes than do females. Many sexually dimorphic traits in salamanders undoubtedly are related to competition among males to attract or monopolize females. These include morphological structures used to clasp females in amplexus, such as enlarged hind legs, as well as traits that enhance a male's ability to chase after females, such as enlarged tail fins. In species of newts with lek mating systems, the elaboration of visually conspicuous crests, tail fins, and colorful markings on the body and tail very likely is due in part to preferences among females for such traits.

# Chapter 10 The Natural History of Amphibian Reproduction

*When with their croaking voyces the male provoketh the female to carnal copulation, which he performeth not by the mouth (as some have thought) but by covering her back: the instrument of generation meeting in the hinder parts, and this they perform in the night season, nature teaching them the modesty and shamefastness of this action. The land frogs are ingendered out of Eggs.... When the Egge breaketh or is hatched, there cometh forth a little black thing ... having no visible part of a living creature upon them, besides their eyes and their tails, and within a short space after their feet are formed, and their tail divided into two parts, which tail becometh their hinder-legs.*

—Edward Topsell, *The History of Serpents* (1658)

Reproduction is the most conspicuous feature of amphibian biology, and many species are seldom seen except during the breeding season. Frogs have been considered symbols of fertility in many human cultures, from ancient Egypt to India to the Mayan region of Central America. Their association with reproduction and fertility derives in part from their connection with life-giving water (see chapter 2) and in part because of the prolific nature of their spawning (Briffault 1927; Egger 1931; Wassen 1934; M. Weber 1972; Hamblin 1984). Yet for centuries, the details of amphibian reproduction were poorly understood. The peculiar description given by Topsell actually is a translation of a Latin work by Conrad Gesner, published in the previous century, but Gesner's account in turn was taken almost verbatim from that of Pliny the Elder, a Roman writer of the first century AD (Bastock and Riley 1900). Not until the eighteenth century were accurate descriptions of amphibian reproduction published. The best were those of August Johann Rösel von Rosenhof (1705–1759), whose spectacularly illustrated book, *Historia Naturalis Ranarum Nostratium* (1753–1758), provided accurate and detailed descriptions of fertilization, egg-laying, and development of most of the common European species (Adler 1989; Tunner 1996).

Most European amphibians have a "standard" amphibian life history, with aquatic eggs and aquatic larvae that metamorphose into terrestrial adults. However, even by the mid-eighteenth century, it was becoming clear that not all amphibians have this type of life history. The only European frogs with an unusual mode of reproduction, the midwife toads (*Alytes*), were first discovered in the 1740s (Boulenger 1912), and the live-bearing habits of European fire salamanders (*Salamandra*) were well known in Rösel von Rosenhof's day as well. Furthermore, as European explorers ventured into the tropics, all sorts of frogs with peculiar reproductive habits were discovered. For example, Linnaeus described the bizarre Surinam toad (*Pipa pipa*) in 1758. Soon afterward, the French physician Philipp Fermin, who spent some time in Surinam, provided the first, albeit somewhat inaccurate, description of its odd reproductive biology, with eggs undergoing development in cavities in the back of the female (Fermin 1765; cited in Rabb and Snedigar 1960). Similar discoveries of species with foam nests, nonaquatic oviposition, direct development, and various forms of parental care were made throughout the nineteenth century (Sampson 1900).

This chapter presents a broad overview of the natural history of amphibian reproduction, with a particular focus on life-history traits such as modes of egg deposition and development, egg size, clutch size, and reproductive effort. The role of energy storage in reproductive energetics is discussed in chapter 5, while behavioral aspects of courtship, mating, parental care, and the evolution of mating systems are discussed in chapters 8, 9, and 11. The first part of this chapter reviews modes of fertilization in the three major amphibian groups. The next section presents a detailed descriptive review of modes of egg deposition and development and discusses possible selective pressures affecting oviposition behavior in different environments. The third section discusses interspecific variation in egg size and clutch size as a function of body size, reproductive mode, and taxonomic group, as well as possible selective pressures affecting the evolution of these life-history traits. The final section examines intraspecific variation in egg size and clutch size and the relative contribution of genotype and environment to this variation, with a particular focus on the fitness consequences of variation in life history traits both within and between populations.

## Mating Behavior and Modes of Fertilization

There is little doubt that external fertilization is the ancestral condition in amphibians, but internal fertilization has evolved independently in all three groups. This is most developed in the caecilians, all of which have internal fertilization, and least developed in anurans, most of which retain the ancestral condition.

### Amplexus Position and External Fertilization in Anurans

Nearly all species of anurans fertilize their eggs externally. Usually the male clasps the female in amplexus, and eggs are fertilized as they emerge from the female's cloaca. At some point in the evolution of modern frog families, a major shift occurred in amplexus position. The ancestral condition appears to be inguinal amplexus, with the male clasping the female around the waist (fig. 10.1 A), while the derived condition is axillary amplexus, with the male clasping the female behind the front legs (fig. 10.1 B). Groups with inguinal amplexus include all of the "primitive" frogs: *Ascaphus, Leiopelma, Rhinophrynus,* as well as the bombinatorids, discoglossids, pelobatids, pelodytids, pipids, myobatrachids, and sooglossids (Lynch 1973; Nussbaum 1980; Duellman and Trueb 1986; Ford and Cannatella 1993). Some South American leptodactylids also have inguinal amplexus. These frogs (*Eupsophus, Pleurodema, Batrachyla, Somuncuria*) are found in relatively cold regions in the southern beech (*Nothophagus*) forests at the southern end of the continent, thought to be where the family originated (Formas 1976; Duellman and Veloso 1977; Cei 1980; Formas and Vera 1980; Ortiz, Ibarra-Videl, and Formas 1989). Heyer (1975) suggested that these leptodactylids are descended from ancient leiopelmatid stocks, and they apparently have retained the form of amplexus that is the ancestral state for all frogs (Lynch 1978). The small toadlike frogs in the family Brachycephalidae also have inguinal amplexus (fig. 10.1 C), but males shift to a more dorsal position (fig. 10.1 D) during oviposition (Pombal, Sazima, and Haddad 1994), a behavior similar to that seen in the midwife toads (*Alytes*). Because the relationships of brachycephalids to other anurans are unclear (Ford and Cannatella 1993; Frost et al. 2006), it is hard to know whether inguinal amplexus in this family represents retention of ancestral behavior, or secondary derivation of this form of amplexus.

Inguinal amplexus has been reported in several bufonids from South and Central America. These include *Osornophryne bufoniformis* and *O. percrassa* from Colombia and Ecuador and two toads from Central America, *Bufo holdridgei* and *B. fastidiosus* (Novak and Robinson 1975; Ruiz-Carranza and Hernandez-Camacho 1976; Graybeal and de Queiroz 1992). The last two species probably are closely related to each other, but the relationship of *Osornophryne* to other bufonids is unclear. Ruiz-Carranza and Hernandez-Camacho (1976) suggested these toads might be related to the primitive leptodactylids that retain inguinal amplexus, whereas Hoogmoed (1987) considered this to be a secondary derivation of inguinal amplexus from the axillary amplexus that is typical of other bufonids. Graybeal and de Queiroz (1992) suggested that inguinal amplexus probably is secondarily derived in the two species of *Bufo* as well, but left open the possibility that these species actually are more closely related to *Osornophryne* than to other species of *Bufo* with axillary amplexus. Since amplexus position is an evolutionarily conservative character in most anurans, the reasons for one or more reversions to inguinal amplexus in bufonids are unknown. However, in most of these species, sexual dimorphism is extreme, with the female much larger than the male, perhaps making it difficult for the male to clasp the female in axillary amplexus (Duellman and Trueb 1986).

Frogs in several other families that normally have axillary amplexus also appear to have evolved inguinal amplexus secondarily (Duellman and Trueb 1986). Inguinal amplexus in these species generally is related to an unusual mode of reproduction or physical constraints of body size and shape. For example, in the burrowing microhylid *Myersiella microps* from Brazil, the male grasps the female around the waist. She then burrows headfirst through leaf litter to lay eggs, dragging the male with her. Presumably, inguinal amplexus in this species makes it easier for the female to get the male into her burrow (Izecksohn, Tenorio de Albuquerque, and Furtado de Mendoca et al. 1971). The same

**Fig. 10.1.** Amplexus positions of anurans. (A) Inguinal amplexus in *Rhinophrynus dorsalis* from Belize. (B) Axillary amplexus in *Hyla ebraccata* from Panama. (C, D) The male of the *Brachycephalus ephippium* shifts from inguinal (left) to cephalic (right) amplexus when the eggs are about to be laid. (E) Cephalic amplexus in *Colostethus panamensis* from Panama. The elongated middle finger of the male is aligned along the jaw of the female. (F) Reverse hind leg clasp in *Eleutherodactylus coqui,* a Puerto Rican frog with internal fertilization. The female places her hind legs over those of the male, locking the male's cloaca in contact with her cloaca. Photos by Kristiina Hurme (A), Kentwood D. Wells (B, E), Marcio Martins (C), Ivan Sazima (D), and Daniel S. Townsend (F).

probably is true for the African frog, *Hemesis guineensis*, another headfirst burrower with inguinal amplexus (Amiet 1991). In a burrowing microhylid, *Breviceps adspersus*, from southern Africa, the male is much smaller than the female and is unable to reach around her rotund body; during amplexus, the male becomes glued to the female with mucus skin secretions (Wager 1965). In several other burrowing microhylids, the male clasps the female in axillary amplexus, but he also becomes glued to the female for a stronger hold. This has been reported in two species of *Kaloula* from the Philippines (Inger 1954), two North American species of *Gastrophryne* (Fitch 1956; Conaway and Metter 1967), and a South American species of *Elachistocleis* (Scrocchi and Lavilla 1990).

### Functional Significance of Amplexus Position

The functional significance, if any, of the evolutionary switch from inguinal to axillary amplexus in the transition from primitive to advanced frogs has never been fully explained. Rabb and Rabb (1963b) suggested that inguinal amplexus is "less efficient" than axillary amplexus because the male's vent is not positioned directly above that of the female, so the male must move into position as eggs are deposited by the female (Eibl-Eibesfeldt 1956). There is no evidence, however, that eggs of species with inguinal amplexus are any less likely to be fully fertilized than those of more advanced frogs with axillary amplexus. Lynch (1973) argued that the somewhat awkward mating posture of species with inguinal amplexus is best suited to oviposition in quiet waters and suggested, without further explanation, that axillary amplexus allows for reproduction in a greater diversity of habitats, including moving water. While it is true that most frogs that lay eggs in flowing streams are members of derived families with axillary amplexus (Leptodactylidae, Bufonidae, Hylidae, Ranidae), there are some frogs with inguinal amplexus that have evolved similar breeding habits. These include megophryids of the genus *Scutiger* from the mountains of China, which attach eggs to rocks and vegetation in small streams (Liu 1950). Some Australian myobatrachids, including *Crinia riparia* and several species of *Taudactylus*, also inhabit moderate to fast-flowing streams and lay their eggs at the edge of a stream or attached to rocks underwater (Littlejohn and Martin 1965a; Barker and Grigg 1977; Hero, Littlejohn, and Marantelli 1991; Cogger 2000).

I believe a major advantage of axillary amplexus is that it enables the male to keep his vent closely aligned with that of the female, and this in turn often allows the entire clutch of eggs to be fertilized at once. In species with inguinal amplexus, the male must arch his back to bring his vent close to that of the female each time she releases eggs, and it probably is difficult to maintain this posture for very long. Consequently, frogs with inguinal amplexus typically fertilize eggs in a series of bouts in which only a small proportion of the eggs is released and fertilized. Each bout lasts only a few seconds and is separated by intervals of several seconds to several minutes. Most commonly, eggs are laid individually or in small packets or strings floating on the surface, attached to submerged vegetation or stones, or lying on the bottom of shallow pools.

This appears to be the typical pattern of oviposition and fertilization in many basal clades of anurans. These include *Discoglossus* (Knoeppffler 1962; Herrmann et al. 1987), *Bombina* (R. Savage 1932; C. Liu 1950; Birkenmeir 1954; Heusser 1961; Akef and Schneider 1985; Herrmann et al. 1987), *Scaphiopus* and *Spea* (Wright and Wright 1949; Stebbins 1954a; Bragg 1965), *Pelobates* (Eibl-Eibesfeldt 1956; Nöllert 1984), *Pelodytes* (Elzen 1976; Raehmel 1983; Hartley 1990; Toxopeus 1985; Toxopeus, Ohm, and Arntzen 1993; Guyétant, Temmermans, and Avrillier 1999), *Xenopus* (Bles 1906; Rabb 1973), *Hymenochirus* (Österdahl and Olsson 1963; Rabb and Rabb 1963b), and the myobatrachid genera *Neobatrachus, Crinia,* and *Uperoleia* (Barker and Grigg 1977; Hero, Littlejohn, and Marantelli 1991; Cogger 2000). Foam-nesting myobatrachids are exceptional in that they deposit eggs in a series of bouts, but all of the eggs are combined into a foam nest that floats on the water's surface (Tyler and Davies 1979; see the following further discussion of foam-nesting behavior). One consequence of this mode of fertilization and oviposition is that oviposition can be a lengthy process, lasting for many hours in some species (table 10.1).

Some frogs with axillary amplexus also deposit eggs individually or in small isolated packets. This is the typical mode of oviposition in many temperate-zone hylids (*Hyla, Acris,* and *Pseudacris;* Wright and Allen 1908; Noble and Noble 1923; Wright and Wright 1949; Livezey 1952; Eibl-Eibesfeldt 1956; Martof and Thompson 1958; Gosner and Rossman 1959; Fouquette and Littlejohn 1960; Redmer 1998), as well as some hyperoliid frogs (Wager 1965; Backwell 1991). However, in these species, oviposition usually lasts only two to three hours and sometimes is completed in less than one hour (table 10.1). Intervals between bouts of oviposition and fertilization often are much shorter than in species with inguinal amplexus. For example, Wright and Allen (1908) reported intervals of 21 seconds between emissions of sperm in *Pseudacris triseriata,* while Livezey (1952) observed intervals of only seven seconds in *Pseudacris nigrita.* Fouquette and Littlejohn (1960) reported that a pair of *Pseudacris streckeri* deposited 72 separate masses of two to ten eggs each in only 25 minutes. Redmer (1998) reported that a female *Hyla avivoca* released small groups of three to nine eggs at intervals of one to four minutes, with up to 180 eggs being laid in 40 minutes. An Asian fanged frog, *Limnonectes kuhlii,* lays eggs in shallow streams and trickles.

**Table 10.1** Time required for fertilization and deposition of eggs in anurans with inguinal and axillary amplexus

| Species (source no.) | Type of egg mass | No. of eggs | Oviposition time |
|---|---|---|---|
| *Species with inguinal amplexus* | | | |
| Bombinatoridae | | | |
| *Bombina bombina* (29) | Individual eggs or small packets | 100 | 4 hrs |
| *B. orientatlis* (26, 31) | Individual eggs or small packets | 200 | up to 24 hrs |
| *B. variegata* (5, 12) | Individual eggs or small packets | 150 | 5–12 hrs |
| Discoglossidae | | | |
| *Discoglossus pictus* (17) | Small packets | 900 | up to 24 hrs |
| Myobatrachidae | | | |
| *Limnodynastes peroni* (20) | Foam nest | > 1,000 | > 2 hrs |
| Pelodytidae | | | |
| *Pelodytes punctatus* (37) | Short strings | 350 | 2 hrs |
| Pipidae | | | |
| *Hymenochirus boettgeri* (18) | Individual eggs | 450 | 2–7 hrs |
| *Xenopus laevis* (1) | Individual eggs | | 12 hrs |
| *Species with axillary amplexus* | | | |
| Bufonidae | | | |
| *Bufo americanus* (3, 9) | Long strings | 7,000 | 3–4 hrs |
| *B. bufo bufo* (21) | Long strings | 5,000 | several hrs |
| *B. bufo spinosus* (28) | Long strings | 7,240 | 6–8 hrs |
| *B. cognatus* (39) | Long strings | 9,376 | 3–6 hrs |
| *B. fowleri* (9) | Long strings | 7,000 | 5 hrs |
| *B. microscaphus* (41) | Long strings | 4,500 | 10 hrs |
| *B. raddei* (4) | Long strings | | 2 hrs |
| *B. torrenticola* (40) | Long strings | 3,500 | 75 mins |
| Hylidae | | | |
| *Hyla arborea* (10) | Small packets | | < 1 hr |
| *H. brevifrons* (34) | Film on leaves | 100 | 5 mins |
| *H. chrysoscelis* (15) | Small packets | 1,500 | 26 mins |
| *H. cinerea* (22) | Surface film | 700 | 30 mins |
| *H. sarayacuensis* (34) | Film on leaves | 150 | 20 mins |
| *H. rosenbergi* (24) | Surface film | 2,350 | 10 mins |
| *Litoria vereauxi* (23) | Small packets | 885 | 2–4 hrs |
| *Phyrnohyas spilomma* (19) | Surface film | 3,000 | < 1 min |
| *Pseudacris nigrita* (11, 13, 14) | Small packets | 100 | 2–3 hrs |
| *P. streckeri* (15) | Small packets | 350 | 25–50 mins |
| *P. triseriata* (2) | Small packets | 100 | 2 hrs |
| Hyperoliidae | | | |
| *Afrixalus delicatus* (33) | Packets on leaves | 60 | 3 hrs |
| Leptodactylidae | | | |
| *Edalorhina perezi* (32) | Foam nest | 35 | 30–60 mins |
| *Lithodytes lineatus* (25) | Foam nest | 60 | 3 hrs |
| *Physalaemus ephippifer* (30) | Foam nest | 410 | 30–50 mins |
| *P. pustulosus* (27) | Foam nest | 234 | 1 hr |
| *Pleurodema diplolistris* (36) | Foam nest | 650 | 12 mins |

*(continued)*

**Table 10.1** (continued)

| Species (source no.) | Type of egg mass | No. of eggs | Oviposition time |
|---|---|---|---|
| Microhylidae | | | |
| *Chiasmocles ventrimaculata* (35) | Surface film | 380 | 30 mins |
| *Microhyla ornata* (42) | Surface film | 225 | 3–5 mins |
| Ranidae | | | |
| *Pyxicephalus adspersus* (38) | Surface film | 3,500 | 3–28 mins |
| *Rana boylii* (43) | Attached to rocks | | 8 mins |
| *R. catesbeiana* (8) | Surface film | 12,000 | 12 mins |
| *R. clamitans* (8) | Surface film | 4,000 | 15 mins |
| *R. nigromaculata* (4) | Globular mass | 4,000 | 4 mins |
| *R. pipiens* (3, 6) | Globular mass | 4,000 | 4 mins |
| *R. septentrionalis* (7) | Globular mass | 1,000 | 10 mins |
| *R. sylvatica* (3) | Floating mass | 1,750 | 10 mins |
| *R. temporaria* (16) | Floating mass | 2,000 | < 1 min |
| *R. palustris* (3) | Globular mass | 2,500 | 3 mins |
| Rhacophoridae | | | |
| *Buergeria japonica* (44) | Small packets | 38–53 | 1 min[a] |

*Sources:* (1) Bles 1906; (2) A. Wright and Allen 1908; (3) A. Wright 1914; (4) C. Liu 1931; (5) R. Savage 1932; (6) Noble and Aronson 1942; (7) Aronson 1943a; (8) Aronson 1943b; (9) Aronson 1943c; (10) Eibl-Eibesfeldt 1952; (11) Livezey 1952; (12) Birkemeir 1954; (13) Martof and Thompson 1958; (14) Gosner and Rossman 1959; (15) Fouquette and Littlejohn 1960; (16) R. Savage 1961; (17) Knoeppfler 1962; (18) Rabb and Rabb 1963a; (19) Pyburn 1967; (20) Clyne 1969; (21) M. Smith 1969; (22) Garton and Brandon 1975; (23) Anstis 1976; (24) Kluge 1981; (25) Regös and Schlüter 1984; (26) Akef and Schneider 1985; (27) Ryan 1985a; (28) Sofianidou and Schneider 1985; (29) Herrmann et al. 1987; (30) Hödl 1990a; (31) Marshall, Doyle, and Kaplan 1990; (32) Schlüter 1990; (33) Backwell 1991; (34) Hödl 1991a; (35) Schlüter and Salas 1991; (36) Hödl 1992; (37) Toxopeus, Ohm, and Arntzen 1993; (38) Channing, du Preez, and Passmore 1994; (39) Krupa 1994; (40) Tsuji and Kawamichi 1998; (41) A. Brown et al. 2000; (42) Schleich and Kästle 2002; (43) Wheeler, Welsh, and Heise 2003; (44) Haramura 2005.

*Note:* Only species that lay eggs in water are included.

[a]Pairs lay several small clusters of 1–6 eggs each, with each oviposition bout lasting about 1 minute; total elapsed time about 1 hour.

Females remained in amplexus for about 50 minutes on average, but actual oviposition lasted only half this time. They laid eggs in small batches at intervals of about three minutes, eventually producing about 50 eggs altogether (Tsuji and Lue 1998). Some of these species of frogs remain in amplexus for several hours before they begin laying eggs. Nevertheless, the limited quantitative data in table 10.1 suggest that one advantage of axillary amplexus is that it reduces the amount of time required to actually lay the eggs, a period when the pair is likely to be vulnerable to predation.

The evolutionary shift to axillary amplexus also is correlated with a greater tendency for the more derived frogs to lay their eggs in a single large mass that is fertilized very rapidly. For example, many microhylids, hylids, ranids, and hyperoliids lay all of their eggs as a single unit, consisting of a globular cluster, floating jelly raft, or surface film, while toads (*Bufo*) typically lay eggs in a continuous jelly tube. Such egg masses often are produced by a single convulsive expulsion of eggs by the female, accompanied by simultaneous release of sperm by the male. In some species, a series of egg-deposition and fertilization bouts are required (Wright 1914; C. Liu 1931; R. Savage 1934, 1961; Noble and Aronson 1942; Aronson 1943a, b, 1944; Eibl-Eibesfeldt 1952, 1956; Fouquette and Littlejohn 1960; Pyburn 1967; Garton and Brandon 1975; Kluge 1981; Hödl 1990a, 1991a; Schlüter and Salas 1991).

In some cases, fertilization is extraordinarily fast (table 10.1). For example, R. Savage (1961) reported that less than five seconds is required for deposition and fertilization of the entire egg clutch in *Rana temporaria. Phrynohyas spilomma* deposits its eggs in a surface film in about the same amount of time (Pyburn 1967). Even species that lay thousands of eggs in a surface film or globular cluster, such as *Rana catesbeiana* (fig. 10.2), *R. clamitans,* or *R. pipiens,* often complete oviposition in 10 minutes or less. Oviposition this fast has not been reported for any frog with inguinal amplexus, most of which lay only a few hundred eggs (table 10.1). Toads, which have axillary amplexus, are an exception to this pattern, often taking several hours to deposit their strings of eggs. Much of this time is spent resting between egg-laying bouts and moving around to drape the eggs onto vegetation and other supports; actual oviposition

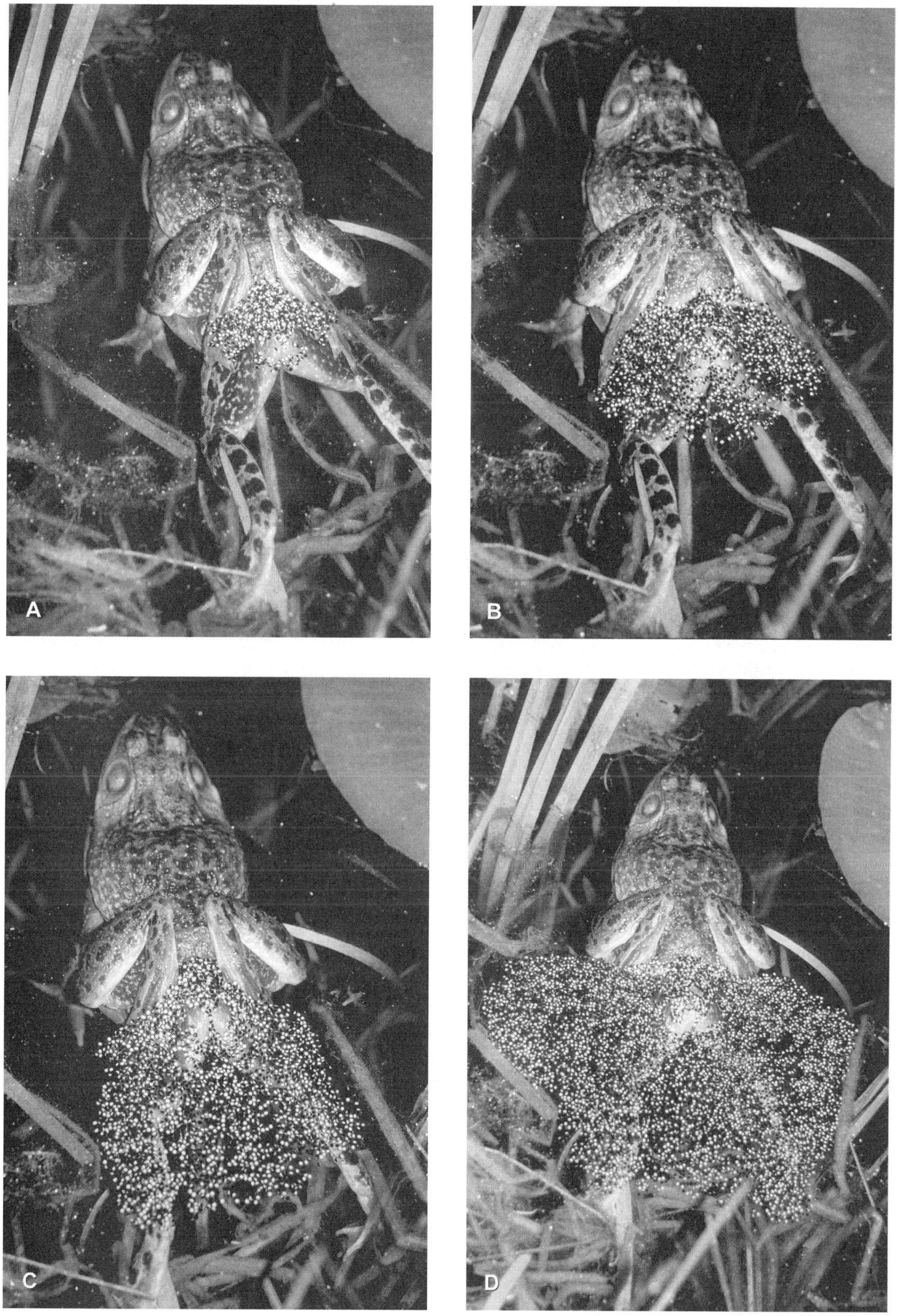


**Fig. 10.2.** Pair of American bullfrogs (*Rana catesbeiana*) depositing a surface film of eggs. The male forms a basket with his hind legs near the female's cloaca to receive and fertilize the eggs. The sequence from A to D took less than 3 minutes. Photos by Kentwood D. Wells.

and fertilization takes only a few minutes (Aronson 1944). For example, in *Bufo microscaphus,* bouts of fertilization were 42–75 seconds, with long resting periods in between. More than 10 hours would be required for a female to deposit all of her 4,500 eggs (A. Brown et al. 2000). Females of a stream-breeding toad (*Bufo torrenticola*) in Japan had spawning periods lasting about 75 minutes on average, and laid eggs in three or four separate batches. Fertilization of each batch required only about 10–15 seconds, however, with the remaining time being used for resting or moving between spawning sites (Tsuji and Kawamichi 1998). Pairs remained immobile for about 10 minutes after each spawning event, perhaps to ensure complete fertilization of the eggs. Some species of toads also remain in amplexus for several hours before egg laying begins, either because the female is seized by a male before she is ready to lay, or because she has not yet located a suitable oviposition site (J. Krupa 1994; Tsuji and Kawamichi 1998).

**Fig. 10.3.** Unusual mode of oviposition and fertilization in a ranid frog from India, *Nyctibatrachus petraeus.* The frogs sometimes gather on vegetation over water in mixed-sex groups, and eggs are fertilized after being deposited by the female, without any form of amplexus. Photo by Krushnamegh Kunte.

#### Modification and Loss of Amplexus

In some frogs, the male hardly clasps the female at all, but simply rests on her back or clasps her loosely during egg deposition. Most of these are species that lay eggs out of water. In *Rhinoderma darwinii,* a peculiar South American frog that broods its tadpoles in the vocal sac, fertilization is accomplished with only brief contact between the male and female (Pflaumer 1936; Cei 1962). In several species of *Mantidactylus* (Mantellidae) from Madagascar, the female deposits eggs on a leaf. The male fertilizes them while sitting on the female's head and shoulders, apparently releasing sperm to run down her back onto the eggs (Blommers-Schlösser 1975b, 1979). In another species in the same genus with aquatic oviposition, *M. curtus,* amplexus is very brief (Arnoult and Razarihelisoa 1967). Very short amplexus also has been observed in species of *Mantella,* terrestrial breeders in the same family (Arnoult 1966; H. Zimmermann 1992). Mantellid frogs are unusual in that males have femoral glands on the undersides of their thighs (Duellman and Trueb 1986; Glaw and Vences 1994). The function of these glands is unknown, but possible functions include facilitating adhesion of the male to the back of the female, tactile stimulation of the female with the rough surface of the gland, and delivery of chemical pheromones to the skin of the female (see chapter 8 for a discussion of other sexually dimorphic glands in anurans).

Dendrobatid frogs, terrestrial breeders that are remarkably similar to *Mantella* in morphology and behavior, have some unusual forms of amplexus as well. In some species, males clasp females by placing the upper surfaces of their front feet under the female's jaw, a position known as cephalic amplexus (fig. 10.1 C). Males sometimes clasp one another the same way during fights (Myers, Daly, and Malkin 1978; Wells 1980a). This appears to be the ancestral form of amplexus in the family. In derived dendrobatids (*Phyllobates* and *Dendrobates*), cephalic amplexus has been lost during mating, but is used by males of some species in aggressive interactions (Myers 1987). The male either fertilizes the eggs by sitting on them after they have been laid, or eggs are laid while the male and female sit facing away from one another with their vents in close proximity (H. Zimmermann and E. Zimmermann 1988; E. Zimmermann 1990).

An unusual case of loss of amplexus has been observed in a ranid frog from India, *Nyctibatrachus petraeus* (Kunte 2004; Das and Kunte 2005). This arboreal frog deposits eggs on leaves overhanging small streams or on bare rocks. Males called from vegetation and defended suitable oviposition sites as territories. Females visited males in their territories and deposited eggs at the male's calling site without amplexus or other physical contact between the male and female. Once the eggs were laid, the male sat on the eggs and fertilized them. Males sometimes attracted several females in succession to the same oviposition site, and in at least one instance, two females approached a male simultaneously and one deposited eggs while the other waited nearby (fig. 10.3).

### Internal Fertilization in Anurans

Internal fertilization has evolved independently a number of times in frogs (Sever et al. 2003). In *Ascaphus,* the "tailed frogs" from western North America, the cloaca of the male has been modified to form a copulatory organ that is inserted into the cloaca of the female for mating. These frogs are the only species with internal fertilization that have true inguinal amplexus. The poor alignment of the male and female cloaca during amplexus makes the copulatory organ particularly advantageous, especially in the swift-water habitats where these frogs breed. During mating, the copulatory organ is bent forward toward the female's cloaca while the male clasps her around the waist. Mating is a prolonged affair, lasting for up to 90 hours (Noble and Putnam 1931; Slater 1931; Metter 1964; B. Stephenson and Verrell 2003). The reason for such prolonged amplexus is unknown. It could represent a form of mate-guarding behavior (see chapter 8), but competition among males for access to females has not been observed in these frogs. Female tailed frogs have the capacity to store sperm in the oviducts for long periods of time, which allows for the separation of the mating season and the egg-laying season (Sever et al. 2001), but it also would tend to make prolonged mate guarding ineffective in reducing sperm competition. The eggs are deposited in water and the tadpoles undergo normal aquatic development.

Internal fertilization is known in two species of *Eleutherodactylus* from Puerto Rico. *Eleutherodactylus jasperi,* which was not discovered until 1976, is ovoviviparous, retaining fertilized eggs in the oviducts and giving birth to small froglets (Drewry and Jones 1976; M. Wake 1978). It must have internal fertilization, but mating behavior has not been observed, and unfortunately this species appears to be extinct. Mating leading to internal fertilization has been observed in *E. coqui,* a species that retains the normal mode of terrestrial oviposition and direct development seen in other members of the genus (Townsend et al. 1981). In this species, the male rests on top of the female in a sheltered nest, but does not actually clasp her in amplexus. The pair can remain in this position for more than seven hours. Eventually, the female loops her hind legs over those of the male in a "reverse hind leg clasp" (fig. 10.1 D). This apparently serves to press the male's cloaca downward against the vent of the female. Mating can last more than two hours before oviposition begins. The female normally lays her eggs with the male still resting on her back, but fertile eggs are laid even if the male is removed (Townsend and Stewart 1986). Mating behavior has not been observed in great detail in most other species of *Eleutherodactylus.* It is possible that other examples of internal fertilization will be found, particularly in other Puerto Rican species, but it apparently does not occur in *E. antillensis* (S. Michael 1997a).

All other known examples of internal fertilization in anurans occur in African bufonids. African toads in the genus *Nectophrynoides* and several related genera are of particular interest because this group includes species with a continuum of reproductive modes, from strictly oviparous to fully viviparous (Grandison 1978; M. Wake 1980c; Xavier 1986; see the following further discussion). In some species, the male's cloacal opening faces downward to facilitate internal fertilization (Angel and Lamotte 1948; Lamotte and Xavier 1972; Grandison 1978). Based on reports of field collectors, Grandison (1978) reported that male *Altiphrynoides malcolmi* sometimes clasp females from below (belly-to-belly amplexus) and bend the cloacal opening upward to contact the female's cloaca. Similar behavior has been observed in *Mertensophryne micranotis,* an oviparous African bufonid with internal fertilization (Grandison 1980; Grandison and Ashe 1983), but it is not clear whether this is the typical mating position in either species. *Mertensophryne* males have unusual spines on the cloaca, which are thought to aid in copulation with the female.

### External Fertilization in Urodeles

In contrast to anurans, more than 90% of urodeles have internal fertilization. Only the most primitive families (Cryptobranchidae, Hynobiidae, Sirenidae) have external fertilization (Duellman and Trueb 1986). In *Cryptobranchus,* males defend potential oviposition sites under rocks in streams and rivers as territories. Females enter these sites and lay their eggs, which are then fertilized by the male (B. Smith 1907; S. Bishop 1941b). Similar behavior has been observed in the giant Asian salamanders (*Andrias*) (Kerbert 1904; Tago 1929) and in *Hynobius* (Sasaki 1924; Thorn 1962, 1963, 1967; Kusano 1980; Tanaka 1989). Male cryptobranchids fertilize eggs as they are laid by releasing a milky cloud of sperm over them. In contrast, male *Hynobius* seize the egg sacs and release sperm onto them after the female has spawned. Although sirens are assumed to have external fertilization, actual mating behavior has never been observed, and repeated attempts to induce mating in captivity have failed (Hanlin and Mount 1978; Godley 1983).

### Spermatophores and Internal Fertilization in Urodeles

The major innovation in the evolution of urodele reproductive modes was the evolution of the spermatophore and internal fertilization (Salthe 1967; Salthe and Mecham 1974; Nussbaum 1985). The selective forces that led to the initial evolution of spermatophore production are unknown. Salthe (1967) believed that spermatophores evolved mainly as an adaptation to flowing water that allowed sperm transfer to take place without the sperm being washed away in a current. Nussbaum (1985) disagreed, arguing instead that the main advantage is that it allows the male to deposit sperm before the female spawns, thus enabling the male to more effectively prevent other males from fertilizing his female's eggs. This contrasts with the situation in explosive breeding species of *Hynobius,* in which several males release sperm simultaneously onto a previously deposited egg mass (Sasaki 1924; Nussbaum 1985; Hasumi and Iwasawa 1990; Hasumi 1994). However, even in species that fertilize eggs internally by means of spermatophores, males have evolved various means of interfering with sperm transfer, sometimes resulting in direct sperm competition or fertilization of a female's eggs by more than one male (Arnold 1976; Halliday and Verrell 1984; Verrell 1989a; see chapter 9).

Whatever initial advantage accrued to salamanders that produced spermatophores, it is clear that the evolution of the spermatophore has had a profound effect on the reproductive biology and social behavior of urodeles. One consequence was the evolution of internal fertilization in all of the derived families of urodeles. Females of some species can retain viable sperm in their reproductive tracts for long periods of time (Boisseau and Joly 1975; Halliday and Verrell 1984; Houck and Schwenk 1984). This sets the stage for the separation of mating from oviposition, something that is virtually nonexistent in the anurans (except for *Ascaphus;* Metter 1964; Sever et al. 2001, 2003). For example, many North American plethodontid salamanders engage in courtship in both the fall and spring, but generally lay eggs only in late spring or early summer (Houck 1977a). This separation of courtship and oviposition allows females to make use of hidden oviposition sites that are not necessarily good courtship sites (Salthe 1967; Nussbaum 1985). The fact that males and females are not together when eggs are laid probably also accounts for the fact that care of the eggs is performed exclusively by the female in salamanders with internal fertilization (Nussbaum 1985; see also chapter 8). Finally, the relatively high cost of spermatophore production (compared to the cost of sperm production alone) limits the rate at which spermatophores can be produced (Verrell, 1986c 1988b; Halliday 1987; Marks and Houck 1989). This has resulted in selection for increased efficiency of spermatophore transfer, which in turn has led to the evolution of elaborate forms of courtship seen in many salamanders, particularly salamandrids and plethodontids (Arnold 1976, 1977; Halliday 1977a, 1990; Verrell 1989a; see also chapter 9).

The spermatophores of salamanders consist of a gelatinous base that supports a whitish or yellowish cap containing the spermatozoa (Jordan 1891; B. G. Smith 1910; Wilder 1923; Noble and Weber 1929; Noble 1931; S. Bishop 1932, 1941b; Blanchard 1933b; Noble and Brady 1933; R. E. Smith 1941; Organ 1958, 1960a, 1968; Briegleb 1961; Organ and Lowenthal 1963; Organ and Organ 1968; Uzzell 1969; J. Anderson 1970; Spotila and Beumer 1970; Arnold 1976; Labanick and Davis 1978; Russell et al. 1981; T. Mann 1984). All spermatophores share a basic similarity in structure, but there are differences in size and shape of the spermatophore, mode of attachment of the sperm mass to the base, and the structure of the sperm cap (see Zalisko, Brandon, and Martan 1984 for a comparative review). The stalk and the cap containing the sperm are produced by male cloacal glands (Sever and Houck 1985). Salamanders with external fertilization have cloacal glands, but not the ones involved in spermatophore production. In these species, the glands apparently function only in production of pheromones to attract females. This ancestral function is retained, and in some cases amplified, in derived urodele families, but the cloacal glands also have taken on the secondary function of spermatophore production (Sever et al. 1990; see also Sever 1991a, 1992a, 1994, 2003a for detailed reviews of the anatomy and evolution of cloacal glands).

In most urodeles with internal fertilization, the male courts the female and deposits one or more spermatophores on the ground, the bottom of a pond, a rock, or other substrate. The female then moves over the spermatophore and picks up the sperm with the lips of her cloaca (see chapter 9). In ambystomatids and salamandrids, the sperm are located near the upper surface of the spermatophore cap. When the female moves over the spermatophore, she removes only the top of the spermatophore cap into her cloaca, leaving part of the cap and the base behind. In contrast, plethodontid spermatophores have sperm concentrated near the center of the cap, which sits on the pointed tip of the gelatinous base. Normally the entire cap is taken into the female's cloaca, leaving only the base behind (Arnold 1976; Zalisko, Brandon, and Martan 1984; Arnold, Reagan, and Verrell 1993).

In the salamandrid genera *Euproctus* and *Calotrion,* the male grabs the female either by biting her, or by encircling her with his tail. The male deposits spermatophores on the female's body, sometimes placing them directly in the cloaca, or pushing them into the cloaca with his feet (Alcher 1981; Thiesmeier and Hornberg 1986, 1990). Often several spermatophores are transferred in a copulation that can last for several hours. Salthe (1967) suggested that this mode of sperm transfer represents a primitive state within the Salamandridae, but it seems more likely that it is secondarily derived from the more widespread pattern of depositing spermatophores on the substrate, perhaps as an adaptation to mating in flowing water. This interpretation is supported by morphological specialization of *E. montanus* and *E. platycephalus.* In both species, the cloaca of the male is elongated into a conical shape that facilitates placement of spermatophores in the female's cloaca (Carranza and Amat 2005). Direct transfer of spermatophores to the cloaca of the female apparently also occurs in *Amphiuma* (C. Baker, Baker, and Caldwell 1947), a species that does not breed in flowing water, but the details of mating behavior are not clearly understood.

Once sperm enter the female's cloaca, they are stored in the spermathecae, structures consisting of a number of tubules that empty directly into the cloaca, or in plethodontids, into a common duct connected to the cloaca (Kingsbury 1895; Wahlert 1953; Sever 1991a, 1992a, 1994, 1995, 1997, 2002; Sever and Kloepfer 1993; Sever et al. 1995; Sever and Bart 1996; T. Halliday 1998; Sever and Brizzi 1998; Sever et al.

2004a, b). The structure of the spermathecae varies, being relatively simple in most families, but more complex in plethodontids. The more complex structure is presumed to be the derived condition (Sever and Brizzi 1998). In some species, including *Notophthalmus viridescens, Triturus vulgaris,* and several species of *Desmognathus,* sperm often is packed into spermathecae in tangled masses (M. Hardy and Dent 1986; Sever and Hamlett 1998; Sever et al. 1999). This could represent an adaptation to restrict movement of sperm from rival males into the spermathecae, thereby reducing the potential for sperm competition.

In some urodeles, the walls of the spermathecal tubules provide nutritional secretions that sustain the sperm for some time (Benson 1968; Dent 1970; Pool and Hoage 1973). The length of time that sperm remain viable has not been precisely determined for most species. There are reports that viable sperm are stored for more than two years in *Salamandra salamandra* (Joly 1960) and several other species (Sever and Brizzi 1998), but more recent work suggests that most species do not retain sperm capable of fertilizing eggs for more than a few months (Sever1995; Sever, et al. 1996; Sever, Rania, and Krenz 1996; Halliday 1998; Sever and Brizzi 1998). In some species, such as ambystomatid salamanders, sperm remains viable for only a few hours to a few days, and eggs normally are laid shortly after mating (Humphrey 1977; Sever 1995; Sever and Brizzi 1998). However, Krenz and Scott (1994) reported that female *Ambystoma opacum* sometimes arrive at oviposition sites with eggs already fertilized, indicating that courtship occurred somewhere else. In plethodontids, which often lay eggs long after mating, sperm must remain viable for several months (Marynick 1971; Pool and Hoage 1973). Since females can mate with several different males during the long mating period, there is the possibility for direct sperm competition among males having a major impact on their reproductive success (Halliday and Verrell 1984; Halliday 1998). Fertilization of egg clutches by more than one male has been detected in *Desmognathus* (Tilley and Hausman 1976; Houck, Tilley, and Arnold 1985; E. Adams, Jones, and Arnold 2005). In some *Desmognathus,* viable sperm are retained even after eggs are laid, setting the stage for sperm competition among males over more than one breeding season (Houck and Schwenk 1984). In contrast, Sever (1991b, 1992b) found that sperm in the spermathecae of *Eurycea cirrigera* degenerate soon after egg-laying, making it unlikely that sperm are viable for more than one season. This seems to be the pattern in most salamanders (Sever and Brizzi 1998).

### Internal Fertilization in Caecilians

All caecilians have internal fertilization, but the mode of fertilization is quite different from that of urodeles and clearly is independently derived. In male caecilians, the end of the cloaca is modified into a copulatory organ, the phallodeum. In contrast to *Ascaphus,* which has its copulatory organ permanently extended, the phallodeum is normally folded back inside the cloaca, but is everted during mating by contractions of muscles in the body wall and in the cloaca (M. Wake 1977b, 1992). Fluid for transporting the sperm and nutrients to support them are provided by accessory glands, called Müllerian glands, which are not found in anurans, urodeles, or other vertebrates (M. Wake 1981). Actual mating behavior has seldom been observed in caecilians, even in captivity, so details of courtship and copulation are unknown. We also know nothing about how long viable sperm can be retained in the female's reproductive tract, although female caecilians lack any sort of specialized sperm storage organ like that found in urodeles (M. Wake 1972; Sever 2002)

## Modes of Egg Deposition and Development in Amphibians

Amphibians are unique among vertebrates in the variety of ways in which they reproduce. Bony fishes approach amphibians in the diversity of their reproductive modes (Balon 1975, 1981), but lack fully terrestrial reproduction (although some fishes do place their eggs out of water). Anurans, urodeles, and caecilians all include species with aquatic, partially aquatic, and terrestrial reproduction, as well as live-bearing species (Duellman and Trueb 1986). There have been many attempts to classify amphibian reproductive modes by site of egg deposition, type of egg, and patterns of egg and larval development. For example, Boulenger (1886) divided anurans into species with small eggs and aquatic larval development and those with large eggs and direct development. These categories were further subdivided into species with aquatic oviposition, terrestrial oviposition, and parental transport of eggs. Later workers provided general reviews of amphibian reproductive modes (Sampson 1900; Wunder 1932; Lutz 1947; Goin 1960a; Salthe 1969; Salthe and Duellman 1973; Salthe and Mecham 1974; Lamotte and Lescure 1977; Duellman 1985, 1989a; Lescure 1986; Haddad and Prado 2005). Others have published family-level surveys (Noble 1927, 1931; Jameson 1955a, 1957a) and summaries of reproductive modes of amphibians from particular regions (A. Martin 1967; Crump 1974; Duellman 1978, 1988, 1989b; W. Brown and Alcala 1983; Tyler 1985b; Amiet 1989; Hödl 1990b; Rand and Myers 1990; Donnelly 1994; Perotti 1997; Haddad and Prado 2005).

Duellman (1985) listed 29 reproductive modes in anurans, and these were discussed in considerable detail by Duellman and Trueb (1986). Their classification was based on features such as site of egg deposition (in water, on land, or retained in the oviducts), type of eggs (small aquatic eggs, large terrestrial eggs, foam nests, etc.), and mode of development (feeding or nonfeeding aquatic larvae, nonaquatic larvae, or direct development of eggs into small froglets). This scheme has been widely used, with some modifications, by subsequent authors (Hödl 1990b; Donnelly 1994; Perotti 1997; Haddad and Prado 2005). Duellman (1988) used a different numbering scheme in a discussion of reproductive modes in Neotropical anurans, and Duellman (2003) added two new reproductive modes to the list, thereby renumbering some modes. More recently, Haddad and Prado (2005) added 10 newly discovered reproductive modes to the list, bringing the total to 39, with some of Duellman and Trueb's modes being renumbered (table 10.2). Salthe (1969) developed a somewhat different classification of reproductive modes for urodeles. However, all of the modes of egg deposition and development found in urodeles have been described in anurans, so I will use the classification given in table 10.2 for both groups. The reproductive biology of most caecilians is so poorly known that it is impossible to determine where many species fit in this scheme, so I will discuss their reproductive modes in more general terms. Attempts to classify amphibian reproductive modes have been complicated by variation in how different authors define certain reproductive modes. For example, Lamotte and Lescure (1977) and Duellman and Trueb (1986) used the term "direct development" for species in which eggs hatch into fully formed juveniles with no feeding or nonfeeding larval stage. They included species in which the eggs develop in terrestrial or arboreal nests, on the body of a parent, or in a vocal sac or special pouch. They distinguished direct development from modes of reproduction in which eggs are retained in the oviducts for all or part of development. M. Wake (1980c, 1982, 1989b) used a broader definition of direct development to include any species that lacks a free-living, feeding larval stage, such as those with eggs that hatch into nonfeeding larvae and those that retain encapsulated eggs throughout development in the reproductive tract. I will use the more restricted definition in my discussion. Salthe and Mecham (1974) and M. Wake (1982) also used the term "ovoviviparity" to refer to any species that carries eggs that develop entirely from yolk reserves and hatch into larvae or juveniles. They included species that carry eggs on the back, in pouches, in the vocal sac, or in the stomach. I prefer the more traditional use of the term "ovoviviparity" to refer only to species that retain eggs in the oviducts, but do not provide them with any nutrition other than the yolk. "Viviparity" is used to refer only to species that retain eggs in the oviducts and provide them with additional sources of nutrition. The term "paraviviparity" has been proposed to describe development of embryos from yolk reserves in sites outside of the mother's reproductive tract, such as brood pouches (Altig and Johnston 1989; Thibaudeau and Altig 1999), but this term has not been widely used in the literature.

Most authors assume that some form of aquatic oviposition is the ancestral condition for all three living orders of amphibians, but the primitive mode of egg placement and the structure of egg clutches in ancestral amphibians are unknown. Salthe and Mecham (1974) speculated that the ancestral condition might be the deposition of eggs in rosary-like strings, with individual eggs connected by jelly strands constricted between the eggs. This type of egg mass is characteristic of some relatively primitive salamanders (*Cryptobranchus, Andrias, Amphiuma*) and several tropical plethodontids, in which it probably is secondarily derived. All of the oviparous caecilians for which eggs have been described also produce this type of egg mass, as do some relatively primitive frogs, including *Leiopelma, Ascaphus,* and *Alytes* (Salthe and Mecham 1974; Duellman and Trueb 1986; Funk et al. 2004), as well as some terrestrial-breeding microhylid frogs of the genus *Cophixalus* (Hoskin 2004). It is not clear whether clutch structure is homologous in all of these groups or simply the result of superficial similarity in appearance, because the morphology of the eggs has not been studied in detail.

The evolution of nonaquatic oviposition has long been recognized as a recurring theme in amphibian evolution (Lutz 1947; Jameson 1955a; Goin 1960a; Crump 1974; Salthe and Mecham 1974; Lamotte and Lescure 1977; McDiarmid 1978; Bogart 1981; Duellman 1985, 1989a; Duellman and Trueb 1986). In some cases, the shift to nonaquatic oviposition involves little more than placing aquatic eggs in a saturated environment where larvae eventually have access to water. In other groups, adaptation to the terrestrial environment has gone much farther, culminating in the complete loss of the aquatic larval stage and direct development of eggs. Although most authors have considered the most likely direction of evolution in reproductive modes to be from aquatic to terrestrial reproduction, there is evidence for plethodontid salamanders that evolution has gone in the opposite direction, with reacquisition of an aquatic life history from direct-developing ancestors (Titus and Larson 1996; Chippindale et al. 2004; Kozak et al. 2005; see the following further discussion).

For more than 60 years, there has been speculation about the selective pressures favoring the evolution of nonaquatic reproduction, especially in anurans. The most common explanation is that nonaquatic oviposition removes the most vulnerable life-history stages, the eggs and early larvae, from the dangers presented by a host of aquatic predators. This advantage would be most pronounced in species with direct development, a strategy that completely eliminates aquatic stages (B. Lutz 1947; Tihen 1960, 1965; Poynton 1964b;

**Table 10.2** Modes of egg deposition and development in anurans

| Egg deposition site | D&T mode | Tadpole development | Selected examples |
|---|---|---|---|
| *Aquatic eggs* | | | |
| 1. Still water | 1 | Feeding in pond | Many *Rana, Hyla, Bufo* |
| 2. Flowing water | 2 | Feeding in stream | *Ascaphus, Atelopus* |
| 3. Subaquatic chamber in stream | | Feeding in stream | *Crossodactylus, Hylodes* |
| 4. Streamside basin | 3 | Feeding in stream | *Hyla boans, Hyla vasta* |
| 5. Subterranean nest | | Feeding in water after flooding | *Aplastodiscus, Hyla albofrenata* group |
| 6. Treehole, bromeliad | 4 | Feeding in water | *Anotheca, Nyctimantis, Phyllodytes* |
| 7. Water-filled depression | 5 | Nonfeeding in water | *Pelophryne, Eupsophus, Leiopelma hochstetteri* |
| 8. Treehole, leaf axil | 6 | Nonfeeding in water | *Anodonthyla, Frostius, Platypelis* |
| 9. On ground or in water | 7 | Develop in stomach | *Rheobatrachus* |
| 10. Bubble nest on pond | | Feeding in pond | *Chiasmocleis leucosticta* |
| 11. Foam nest in pond | 8 | Feeding in pond | *Adelotus, Physalaemus,* some *Leptodactylus* |
| 12. Foam nest in pool | 9 | Feeding in stream | *Megistolotis lignarius* |
| 13. Foam nest in basin | | Feeding in pond | *Leptodactylus labyrinthicus* |
| 14. Foam nest in ground bromeliad | | Feeding in pond | *Physalaemus spiniger* |
| 15. Dorsum of female | 10 | Feeding in pond | *Pipa carvalhoi* |
| 16. Dorsum of female | 11 | Direct development | *Pipa pipa* |
| *Nonaquatic eggs* | | | |
| 17. Terrestrial nest | 12 | Feeding in pond | *Pseudophryne, Geocrinia* |
| 18. Ground, rock, burrow, depressions, nest | 13 | Tadpoles move to water | *Hemisus, Leptopelis, Batrachyla,* some *Centrolene, Phrynomedusa, Paratelmatobius* |
| 19. Humid rocks, rock crevices, tree roots above water | | Feeding semiterrestrial tadpoles on rocks in water film or splash zone | *Cycloramphus, Thoropa, Petropedetes* |
| 20. Terrestrial nest | 14 | Tadpoles carried to water | *Dendrobates, Colostethus* |
| 21. Terrestrial nest | 15 | Nonfeeding in nest | *Thoropa, Leiopelma, Zachaenus* |
| 22. Terrestrial nest | 16 | On dorsum, in vocal sac, in pouches | *Sooglossus, Rhinoderma, Assa, Cycloramphus stejnegeri* |
| 23. Terrestrial nest | 17 | Direct development | *Brachycephalus, Eleutherodactylus, Myersiella, Platymantis* |
| 24. Leaves over water | 18 | Feeding in pond | *Agalychnis, Phyllomedusa,* some *Hyla* |
| 25. Leaves over water | 18 | Feeding in stream | *Hyalinobatrachium, Phasmahyla* |
| 26. Walls of tree holes or hollow bamboo | 19 | Feeding in water | *Acanthixalus, Nyctixalus, Chirixalus* |
| 27. Arboreal nest | 20 | Direct development | *Eleutherodactylus, Platymantis* |
| 28. Foam nest on ground | | Feeding in pond after flooding | *Physalaemus signifer* group |
| 29. Foam nest in basin | 21 | Feeding in pond after flooding | *Heleioporus* |
| 30. Foam nest in burrow | | Feeding in pond after flooding | *Adenomera bokermanni, Leptodactylus fuscus* group |
| 31. Foam nest in burrow | | Feeding in stream after flooding | *Leptodactylus cunicularius* |

*(continued)*

**Table 10.2** (continued)

| Egg deposition site | D&T mode | Tadpole development | Selected examples |
|---|---|---|---|
| 32. Foam nest in burrow | 22 | Nonfeeding in burrow | Some *Adenomera, Stumpffia* |
| 33. Foam nest in tree | 23 | Feeding in water | *Chiromantis, Rhacophorus, Polypedates* |
| 34. Carried on male's legs | 24 | Feeding in pond | *Alytes* |
| 35. Dorsal pouch on female | 25 | Feeding in pond | *Gastrotheca* (some) |
| 36. Dorsum of female | 26 | In bromeliads or bamboo stems | *Flectonotus, Fritziana* |
| 37. Dorsum of female | 27 | Direct development | *Gastrotheca* (some), *Stefania, Hemiphractus* |
| 38. Eggs retained in oviducts | 28 | Nutrition provided by yolk | *Eleutherodactylus jasperi* |
| 39. Eggs retained in oviducts | 29 | Nutrition provided by female | *Nectophrynoides occidentalis* |

*Sources:* Modified from Haddad and Prada (2005). D&T mode indicates reproductive modes of Duellman and Trueb (1986).

A. Martin 1967; Heyer 1969; Salthe 1969; Van Dijk 1971; Crump 1974; Salthe and Mecham 1974; McDiarmid 1978; Duellman and Trueb 1986). Nevertheless, terrestrial and arboreal eggs are preyed upon by snakes, spiders, insects, and other arthropods, often at rates similar to those reported for aquatic breeders (see chapter 14). Very high survivorship often is achieved only if nonaquatic oviposition is coupled with parental care (see chapter 11). A variety of other putative benefits of nonaquatic reproduction have been proposed, including reduced larval competition (B. Lutz 1947; A. Martin 1967; Crump 1974; Duellman and Trueb 1986), escape from problems of desiccation in temporary ponds (A. Lee 1967; Crump 1974; McDiarmid 1978; Duellman and Trueb 1986), escape from low oxygen levels in tropical ponds (Van Dijk 1971), and avoidance of oviposition in streams with swift currents (Alcala 1962; Goin and Goin 1962; W. Brown and Alcala 1983).

Magnusson and Hero (1991) correlated the diversity of reproductive modes of frogs at various Amazonian sites with environmental variables thought to affect the survival of eggs and larvae, including abundance of predators, water temperature, oxygen availability, and probability of desiccation. They found a positive correlation between the proportion of species with nonaquatic reproductive modes and the abundance of invertebrate predators. Sites where fish were abundant actually had a higher proportion of species with aquatic oviposition, perhaps because fish tend to prey on invertebrate egg predators but represent only a minor threat to amphibian eggs. Physical factors such as oxygen, temperature, and chances of the habitat drying up were not clearly related to the frequency of nonaquatic reproduction. This study helps to explain why anurans with certain modes of reproduction do better than others in particular habitats, but it tells us very little about the origins of these modes.

In fact, any attempt to find a general explanation for the evolution of nonaquatic reproduction probably is a fruitless exercise, for several reasons. First, species with different reproductive modes often coexist in the same habitat, particularly in the tropics (e.g., Crump 1974), and represent equally successful alternatives for dealing with certain kinds of environments. Second, many amphibians with nonaquatic reproduction represent ancient lineages that evolved their modes of reproduction in ecological conditions quite different from those in which they are currently found. Other groups, such as centrolenid frogs that lay eggs on leaves overhanging streams, probably have remained closely associated with the habitats in which their reproductive modes evolved. In any case, an assemblage of species in a particular locality, like those included in the study by Magnusson and Hero (1991), almost certainly did not evolve together, but instead represents a diverse array of lineages with independent evolutionary histories.

Amphibians with nonaquatic oviposition are exposed to selective pressures quite different from those affecting aquatic breeders. Oxygen availability is seldom a serious problem, although under some circumstances, such as high temperature, metabolic rates and development of embryos are limited by rates of gas diffusion (Seymour and Bradford 1987; Seymour, Geiser, and Bradford 1991a, b; Seymour and Loveridge 1994; see also chapter 4). In contrast, a high rate of evaporative water loss probably is the most important constraint on successful development of terrestrial eggs (A. Martin and Cooper 1972; Taigen, Pough, and Stewart 1984; Bradford and Seymour 1988a). Consequently, adaptations for terrestrial or arboreal oviposition generally include choice of suitable wet substrates for oviposition, placement of eggs in sheltered locations, and parental attendance of eggs (Salthe and Mecham 1974; McDiarmid 1978; Wells 1981a; Taigen, Pough, and Stewart 1984; Nussbaum 1985; Townsend 1989a; N. Mitchell 2001, 2002a, b).

## Reproductive Modes of Anurans

A summary of anuran reproductive modes, modified from Duellman and Trueb (1986) and Haddad and Prado (2005), is presented in table 10.2. The various features of reproductive modes need not have evolved in parallel. There are, for example, frogs with terrestrial eggs and aquatic larvae, frogs with aquatic or terrestrial eggs and nonfeeding tadpoles or direct development, and frogs that lay their eggs in foam nests placed either in the water or on land. Reproductive modes are far more diverse in the tropics than in other regions, and most of this diversity is found in wet tropical forests, especially in the New World. For example, Crump (1974) identified 10 distinct modes of reproduction at a single rainforest site in Ecuador, while the Atlantic rainforest of Brazil has 27 different modes represented (Haddad and Prado 2005). Overall, 31 of 39 modes (80%) listed by Haddad and Prado (2005) are represented in the Neotropical region. In contrast, 90% of the frogs in the United States lay eggs in permanent or temporary ponds, as do nearly all species in Europe. Africa, southern Asia, and the Australo-Papuan region have fewer reproductive modes represented than do the Neotropics, but more than in temperate regions (Duellman and Trueb 1986). Arid regions typically have relatively few reproductive modes represented, because most anurans in these regions are dependent on standing water. For example, in a semiarid region of Argentina, half of the anuran species laid eggs in ponds, and only four reproductive modes were represented (Perotti 1997).

One difficulty with the classification scheme of Duellman and Trueb (1986), as well as that of Haddad and Prado (2005), which is derived from it, is that some reproductive modes encompass a wide variety of egg-laying strategies that represent fundamental adaptive differences, while others are very specialized reproductive modes that differ only slightly from others. For example, Mode 1 (oviposition in still water with feeding aquatic larvae) includes species that lay eggs in many different types of aquatic habitats that differ in size, depth, temperature, and permanence. Variation in habitat structure results in different selective pressures shaping the evolution of oviposition behavior (see the following additional discussion). On the other hand, species with Mode 4 reproduction (oviposition in a streamside basin constructed by the frogs) do not differ in any fundamental way in their egg and larval development from many anurans with Mode 1 reproduction. Species in the genus *Pipa* all share a unique mode of egg deposition in which eggs develop in cavities in the back of the female, but some retain the eggs longer than others, giving birth to small froglets (Mode 16) rather than aquatic tadpoles (Mode 15). Other reproductive modes group together species that are only superficially similar to one another. Some anuran species exhibit considerable plasticity in oviposition behavior, resulting in a single species exhibiting several of the reproductive modes listed in table 10.2. Some frogs, for example, construct basins in which eggs are laid, but sometimes also place them in natural rock pools or streams. Other species usually are pond-breeders, but sometimes will place their eggs in quiet stream pools.

Consequently, while these classification schemes provide a useful starting point for further discussion, the various modes of reproduction should not be viewed as clearly differentiated grades of evolution or the final word on the diversity of reproductive modes in anurans. Indeed, if one were to clearly differentiate among various types of oviposition sites, larval development, and amount of parental care, the list in table 10.2 could easily be expanded to 50 or 60 distinct modes of reproduction. The list in table 10.2 also should not be interpreted as an evolutionary progression from ancestral to derived modes of reproduction, even though it begins with the most generalized mode and ends with modes generally considered to be highly derived. The numbers in table 10.2 provide a convenient shorthand for referring to reproductive modes, but I do not follow the order of reproductive modes precisely as listed in the table. Instead, I group them roughly by ecological similarity.

### Oviposition in Standing Water (Mode 1)

Oviposition in standing water such as permanent and temporary ponds has long been considered the ancestral condition in anurans (Noble 1927; Goin 1960a, Salthe and Duellman 1973; Salthe and Mecham 1974; McDiarmid 1978; Duellman and Trueb 1986), although some authors have questioned this (Bogart 1981). Certainly this appears to be the most generalized reproductive mode, and is by far the most common, being found in about 80% of anuran families (Duellman and Trueb 1986). Mode 1 is a very broad category, encompassing species that lay eggs in all sorts of standing water, from permanent lakes and ponds to vernal pools, swamps, flooded fields, temporary rain forest pools, desert rain pools, and small rock pools on the shores of rivers and lakes. The physical and biological characteristics of these habitats vary considerably (table 10.3) and almost certainly have shaped the evolution of both clutch structure and oviposition behavior. Clutch sizes, egg sizes, placement of eggs, and development times all vary enormously among Mode 1 anurans. Most studies have tended to focus on a single selective pressure, such as predation or oxygen availability, but the type of oviposition found in any given species or clade probably represents a compromise among conflicting selective pressures. At present we can do little more than speculate about the adaptive significance of various types of egg clutches, because quantitative measurements of selection gradients are virtually nonexistent.

**Table 10.3** Selected physical and biological features of various kinds of aquatic sites commonly used by breeding amphibians

| Habitat type | Water temperature | Dissolved oxygen | Probability of drying | Food for larvae | Predation pressure |
|---|---|---|---|---|---|
| Vernal pool | Low | Moderate | Moderate | Variable | Low |
| Permanent pond (in summer) | High | Low | Low | High | High |
| Swamp (in summer) | High | Very low | Moderate | High | High |
| Mountain lake | Low | High | Low | Low | Variable |
| Desert rain pool | Very high | Variable | High | Variable | Low |
| Lowland stream | Variable | Variable | Low | Variable | Variable |
| Mountain stream | Low | High | Variable | Low | Low |
| Streamside basin (tropics) | High | Low | Moderate | Variable | Low |
| Tree hole (tropics) | Moderate | Low | Low | Low | Low |
| Bromeliad (tropics) | High | Low | Low | Low | Low |

### Eggs Laid in Compact Masses

Many ranid frogs that deposit eggs in cold ponds and lakes lay their eggs in compact jelly clumps attached to submerged vegetation in relatively deep water. Placing the eggs well below the surface presumably reduces their chances of being frozen or drying out if water levels drop (J. Moore 1939, 1940, 1949b; Licht 1971; Salthe and Mecham 1974). At the same time, clumping the eggs in a compact mass is feasible because levels of dissolved oxygen generally will be high enough in cold water to allow adequate diffusion of oxygen into the center of the mass (Burggren 1985; Seymour and Bradford 1995; Seymour 1999; see also chapter 4). Both the thick jelly layer surrounding the egg masses and the relatively tough egg capsules of these species provide protection against some types of predators, such as fishes and chewing insects, but are less effective against newts and sucking insects (Salthe 1963; Grubb 1972; Werschkul and Christensen 1977; see also chapter 14).

### Eggs Laid in Floating Rafts

An alternative mode of egg deposition practiced by some cold-water breeders, such as the North American wood frog (*Rana sylvatica*) and the European common frog (*Rana temporaria*), is to lay eggs in floating jelly rafts at the surface, usually in exposed locations that receive direct sunlight. This increases rates of embryonic development by exposing them to higher temperatures than those experienced by submerged egg masses, a distinct advantage for species breeding in vernal pools or other temporary sites that undergo rapid drying. The thermal advantage of laying eggs exposed at the surface is enhanced by selecting the warmest available sites for egg deposition (Seale 1982b; Sjögren, Elmberg, and Berglind 1988) and by clumping eggs into large communal masses, sometimes composed of hundreds of clutches (fig. 10.4 A). On sunny days, temperatures inside these communal egg masses tend to be higher than in isolated clutches or the surrounding water. The communal masses also retain heat longer than do isolated masses (Heusser 1961; R. Savage 1961; Guyétant 1966; Herreid and Kinney 1967; Zweifel 1968b; Licht 1969a, 1971; Hassinger 1970; Beattie 1980; Seale 1982b; Waldman 1982a; Waldman and Ryan 1983; Caldwell 1986b; Håkansson and Loman 2004; see chapter 3 for further discussion of embryonic thermal requirements). In some cases, clustering of egg masses also may provide some defense against predators such as leeches, because isolated masses may be more vulnerable to attack from the periphery (Håkansson and Loman 2004). A clear disadvantage of placing eggs at the surface is the possibility of freezing or desiccation, but laying eggs in communal masses reduces the rate of evaporative water loss when the eggs are stranded above water and prolongs survival of the eggs (Forester and Lykens 1988).

### Eggs Laid in Strings

Most toads in the genus *Bufo* and some pelobatids lay eggs in long jelly strings (10.4 B). These strings typically are draped over vegetation in the water or are deposited on the bottom of a pond. Presumably such egg masses enhance gas exchange for the embryos by ensuring that each embryo is exposed to open water (Zweifel 1968b; Duellman and Trueb 1986; Seymour 1999; see chapter 4 for a discussion of respiration by amphibian embryos). Nevertheless, toad eggs often are piled up into large aggregations in shallow water, a pattern of egg deposition that would tend to decrease oxygen availability, particularly in very warm water. In some species, such as *Bufo quercicus,* egg strings are broken into short bands of five or six eggs that are scattered on the pond bottom, and a similar pattern of oviposition is seen in some spadefoot toads (*Spea*) as well.

### Egg Scattering

Many pond-breeding anurans scatter eggs individually or in small packets on the bottom of a pond. This mode of

**Fig. 10.4.** Aquatic egg masses of anurans. (A) Communal egg mass of *Rana sylvatica.* (B) String of eggs of *Bufo americanus.* (C) Small packet of eggs of *Pseudacris feriarum* laid at the water's surface. (D) Diffuse surface film of a microhylid frog, *Kalophrynus interlineatus,* with an ovipositing pair in amplexus. (E) Tiny pigmented eggs of *Hyla wavrini* in a streamside mud basin. (F) Large unpigmented eggs of *Hyla leucopygia* in a flooded subterranean chamber. Photos by Kentwood D. Wells (A–C), Wayne van Devender (D), Walter Hödl (E), and Celio F. B. Haddad (F).

oviposition is typical of frogs in the genera *Bombina* and *Discoglossus,* some species of *Paratelmatobius* from Brazil (Pombal and Haddad 1999), some myobatrachid frogs, and many others. Eggs also are attached individually or in small groups to submerged plants, as in some species of *Pseudacris, Acris, Hyla, Spea,* and *Xenopus,* or released individually at the water's surface, as in *Hymenochirus.* The behavior of scattering eggs rather than laying them in a single mass generally is interpreted as an antipredator adaptation that makes eggs inconspicuous and therefore difficult for predators to locate (Rabb 1973; Salthe and Mecham 1974; C. Marshall, Doyle, and Kaplan 1990). Antipredator benefits of egg scattering have been demonstrated more clearly for newts than for anurans (see the following discussion). This type of oviposition also could enhance gas exchange in oxygen-poor water by maximizing the exposed surface area of the eggs (Zweifel 1968b; Seymour 1999).

#### Surface Films

Many anurans that breed in warm water lay eggs in films only a single egg thick on the surface of the water (fig. 10.4 C–E). Following the work of J. Moore (1940), most authors have interpreted this as an adaptation to poorly oxygenated water. The surface film allows the eggs to obtain oxygen directly from the air, even when the water is nearly anoxic (Zweifel 1968b; Van Dijk 1971; Salthe and Mecham 1974; Kluge 1981). M. Ryan (1977) and Kluge (1981) also found that surface films tend to be slightly cooler than the surrounding water, a possible advantage if high temperatures result in developmental abnormalities (R. D. Howard 1978b), but it seems unlikely that this is the principal selective advantage of such clutches. Poynton (1964a) suggested that surface films are an adaptation to reduce predation because they keep the eggs off the bottom, where most predators are found. However, this does not deter leeches, a major predator on clutches of bullfrog eggs (R. D. Howard 1978b). This explanation also does not account for the strong association of surface films with warm, poorly oxygenated breeding sites.

### Oviposition in Flowing Water (Mode 2)

Many Mode 1 frogs that normally breed in ponds also breed in slow-moving streams and stagnant pools at the edges of streams. For example, pond-breeding anurans such as green frogs (*Rana clamitans*), pickeral frogs (*R. palustris*), and American toads (*Bufo americanus*) often lay eggs in the quiet backwaters of streams. There also are stream-breeding species, such as *Bufo torrenticola* in Japan, that have oviposition behavior, egg clutch structure, and larval development that are very similar to those of pond-breeding species (Tsuji and Kawamichi 1996b, 1998). Other anurans have become more specialized for breeding in flowing water from moderately fast streams to mountain torrents (Mode 2), with modifications in oviposition behavior, egg size, clutch structure, and larval morphology that enhance survival in rapidly flowing water. For example, stream-breeding bufonids in the genus *Atelopus* lay large and unpigmented eggs in double strings attached to rocks (Starrett 1967). Parallel adaptations for stream breeding are found in the bufonid genera *Werneria* from Africa and *Ansonia* from Asia (Amiet 1972; W. Brown and Alcala 1983; Lamotte and Lescure 1989b). The tailed frog (*Ascaphus truei*) lays large, unpigmented eggs in rosary-like strings attached to the undersides of rocks in streams (H. Brown 1989). Megophryids in the genus *Scutiger* from the mountains of China lay large, unpigmented eggs in clusters attached to the undersides of stones or plant roots in streams (C. Liu 1950). All of these anurans have highly modified tadpoles adapted to holding onto rocks in fast-moving water (see chapter 12).

Ranid frogs from many parts of the world breed in streams. These include species from Africa (*Conraua*), Asia (*Staurois, Amolops, Paa,* some *Rana*), Europe (*Rana graeca, R. iberica, R. italica, R. pyrenaica*), North America (*Rana subaquavocalis*), and Central America (*Rana warszewitschii, R. maculata;* C. Liu 1950; Lamotte and Perret 1968; Kuramoto 1973; Villa 1979a; Sabater-Pi 1985; Kusano and Fukuyama 1989; T. Matsui and Matsui 1990; Platz 1993; Donnelly 1994; Kam et al. 1998; Schleich and Kästle 2002). Similar stream-breeding habits have evolved independently in many other families, including mantellids from Madagascar, rhacophorids from Asia (*Buergeria*), astylosternids from Africa (*Trichobatrachus*), *Heleophryne* from southern Africa, South American leptodactylids (*Insuetophrynus, Megaelosia, Telmatobius, Hylodes,* and others), several groups of Neotropical hylids (*Plectrohyla, Ptychohyla,* several species groups of *Hyla*), some hylids from Australia and New Guinea (*Litoria booroolongensis, L. eucnemis, L. genimaculata, L. lesueri, L. pearsoniana, L. spenceri, L. subglandulosa,* and the genus *Nyctimystes*), and some Australian myobatrachids (*Taudactylus, Rheobatrachus,* some *Crinia;* Duellman 1970; Visser 1971; Boycott 1972, 1988; Formas, Diaz, and Valencia 1980; Blommers-Schlösser 1981; Sinsch 1985; Fukuyama, Kusano, and Nakane 1988; Amiet 1989; M. Davies 1989; Fukuyama and Kusano 1989, 1992; Lamotte and Lescure 1989a; McDonald and Davies 1990; Duellman and Campbell 1992; Giaretta, Bokermann, and Haddad 1993; Anstis and Littlejohn 1996; Duellman, de la Riva, and Wild 1997; Anstis, Alford, and Gillespie 1998; Cogger 2000; Andreone et al. 2002; Vences et al. 2002).

Most stream-breeding frogs lay relatively large eggs compared to closely related species that breed in standing water, often in relatively compact masses attached to rocks in relatively sheltered sites. Presumably this type of oviposition site reduces the chances of eggs being swept away in a current. A common pattern among stream-breeding frogs in the tropics is for breeding to occur mostly in the dry season, when stream flow is low and eggs are less likely to be swept away in torrential floods (Menzies 1963; C. B. Jørgensen, Shakuntala, and Vijayakumar 1986; Fukuyama and Kusano 1992; Kam et al. 1998). Eggs laid in hidden locations often are unpigmented, presumably because their exposure to ultraviolet radiation in sheltered oviposition sites is minimal (Salthe and Mecham 1974). In all of these genera, the tadpoles have special adaptations for adhering to rocks in stream currents (see chapter 12). Reproductive Mode 2 tends to grade into Mode 18 among stream-breeding frogs that lay eggs on wet rock faces, often in the splash zones of waterfalls. Tadpoles of those species either drop into streams to complete development, or remain attached to the rock faces (see the following "Non-Aquatic Eggs with Feeding Tadpoles"). In some cases, eggs of such species also are unpigmented (e.g., *Amolops cremnobatus;* Inger and Kottelat 1998). Mode 2 also grades into Mode 4 (eggs laid in streamside basins), because some basin-building frogs also place eggs directly in streams (e.g., *Smilisca sordida;* Malone 2004).

In at least some lineages, stream breeding appears to be derived from ancestral pond-breeding habits. For example, two subfamilies of mantellid frogs in Madagascar, Mantellinae and Boophinae, include extensive radiations of stream-breeding species. In both groups, stream breeders appear to form monophyletic groups that probably are derived from pond-breeding ancestors. In the genus *Boophis,* stream breeding also seems to have resulted in higher rates of speciation than pond breeding, perhaps because of lower rates of movement in stream breeders and a greater tendency for populations to become isolated from one another (Vences et al. 2002). One species, *B. occidentalis,* lays relatively large eggs attached in a single layer to rocks. Their larvae exhibit typical morphological adapations for stream life (see also chapter 12). On the other hand, this species often forms relatively large breeding aggregations and lays large clutches, features that are more characteristic of pond-breeding members of the genus (F. Andreone et al. 2002).

### Aquatic Eggs in Constructed Nests, Chambers, or Basins (Modes 3, 4, and 5)

The habit of laying aquatic eggs in basins or nests constructed by a parent has evolved independently several times in the family Hylidae and probably several times in the family Ranidae as well. Presumably the main advantage to laying eggs in a nest is to remove the eggs from habitats where aquatic predators are abundant, or to avoid periodic flooding events that could sweep away eggs.

#### Eggs Laid in Constructed Subaquatic Nests (Mode 3)

Several stream-breeding hylodine leptodactylid frogs from the Atlantic rainforest of Brazil place their eggs in chambers constructed by the males, a reproductive mode with some similarities to that of Mode 2 species that attach their eggs to natural cavities under rocks in streams. In *Crossodactylus gaudichaudii,* males call along the edges of streams and excavate nesting chambers under submerged rocks, apparently located within defended territories (Weygoldt and Carvalho e Silva 1992). In *Hylodes asper,* males use a complex array of acoustic and visual signals, including foot-waving displays, to advertise their territories and attract females. Males periodically dive underwater to excavate nesting chambers by pushing sand between submerged rocks, and mated pairs lay their eggs in these locations (Haddad and Giaretta 1999).

#### Nest-Building Hylids

In several species of Neotropical gladiator frogs (*H. boans, H. circumdata, H. faber, H. pardalis, H. rosenbergi,* and *H. wavrini*), males construct large basins beside streams (Mode 4), and females are then attracted to the nest by the male's calls. In *H. boans,* however, males call mostly from trees rather than from their nests (Magnusson et al. 1999). Eggs are laid in a surface film, probably an adaptation to low oxygen levels in the nests (Breder 1946; Lutz 1960a, b, 1973; Crump 1974; Lamotte and Lescure 1977; Kluge 1981; Martins and Haddad 1988; M. Martins 1993a, b; fig. 10.4 E).

In some regions, nest building appears to be a facultative response to habitats that lack suitable natural pools for egg laying, but nest building is abandoned if oviposition sites are plentiful. In Brazil, Caldwell (1992) found that *Hyla boans* breeding in sandy areas constructed nests, while those breeding along rocky streams usually deposited their eggs in natural rock pools. In *H. wavrini,* an Amazonian species that until recently was considered conspecific with *H. boans* (Hoogmoed 1990b), males make nests by pushing debris out of natural water-filled depressions, but have not been observed making elaborate nests like those of *H. boans* (Martins and Moreira 1991; Martins 1993a). Breder (1946) reported that males of *Hyla rosenbergi* in Panama sometimes did not build complete mud nests, but instead constructed mud walls to dam up natural rock pools. In Costa Rica, this species used cattle footprints and other small puddles as nests, with less than a third of all nests being constructed by males (Höbel 1999b, 2000). *Hyla crepitans,* another member of this group, usually breeds in ponds, flooded areas, and stream pools without building nests. However, Caldwell (1992) found a single nest constructed by this species on a sandy spit in a stream in Venezuela, even though most of the frogs at this locality did not build nests. At other sites, this species is reported to make use of water-filled depressions (Martins 1993a). *Hyla faber* in Brazil normally constructs mud nests, but during periods of high water, they sometimes deposits eggs as surface films directly in ponds (Haddad and Sawaya 2000).

A large treefrog from Hispaniola, *Osteopilus vastus,* lays eggs in shallow basins at the edges of mountain torrents. Eggs are laid in clumps attached to rocks at the bottom of the nest. Whereas the newly hatched tadpoles of *H. rosenbergi* and *H. faber* have large external gills, presumably an adaptation to very low oxygen levels in their nests (Kluge 1981; Martins 1993a), those of *O. vastus* lack these gills and complete their development in well-oxygenated streams (Noble 1927). An Australian hylid in the *Litoria lesueuri* complex lays its eggs in nests at the edges of streams, but nest construction has not been observed. Eggs are laid in a globular cluster that rests on the bottom of the nest (S. Richards and Alford 1992; S. Richards 1993). Levels of dissolved oxygen in nests of this species were considerably higher than those reported for *H. rosenbergi* in Panama (Kluge 1981).

Several Brazilian treefrogs have similar breeding habits, but show some plasticity in the placement of eggs. For example, a clutch of eggs of *Hyla luctuosa* was found in a water-filled depression at the edge of a pond, while other eggs were found in a pond (Pombal and Haddad 1993). *Hyla ibitipoca* males have been observed calling in natural cavities formed by tree roots at the edges of streams (Caramaschi and Feio 1990). Martins (1993a) suggested that the use of these types of natural, water-filled cavities gave rise to the more elaborate nest-building behavior of species such as *H. boans, H. faber,* and *H. rosenbergi.* A thorough phylogenetic analysis of the entire complex is needed before the direction of evolution in reproductive modes will be clear, but unfortunately, even the species-level taxonomy of these frogs is poorly understood at this point (Kluge 1979; Cardoso 1983; Hoogmoed 1990b).

Other Brazilian hylids have somewhat similar reproductive modes, except that aquatic eggs are placed in subterranean nests that are subsequently flooded, with tadpoles completing development in temporary ponds or streams (Mode 5). These frogs (*Hyla cavicola, H. leucopygia,* other members of the *H. albosignata* group, and *Aplastodiscus perviridis*) lay aquatic eggs in burrows constructed at the edges of temporary ponds or streams (Cruz and Peixoto 1984; Haddad and Sazima 1992; Haddad and Sawaya 2000; Haddad, Faivovich, and Garcia 2005; Haddad and Prado 2005). Because the eggs of these species are laid in a covered chamber, they generally lack pigmentation (fig. 10.4 F), in contrast to eggs of hylids that construct exposed nests. After the nests of *H. leucopygia* are flooded, their roofs often collapse, and the resulting basins sometimes are used as oviposition sites by *H. luctuosa* (Haddad and Sawaya 2000).

In Costa Rica, hylids in the genus *Smilisca* exhibit a variety of reproductive modes and also exhibit some plasticity in oviposition behavior (Malone 2004). *Smilisca puma* has the most generalized mode, placing eggs in temporary ponds (Mode 1), as do other members of the genus elsewhere in Central America. These frogs tend to be explosive breeders that use temporary ponds formed by heavy rains. In contrast, *Smilisca sila* and *S. sordida* are associated with forest streams, and place their eggs in constructed nests near the water (Mode 4). Both also use natural stream pools for oviposition. *Smilisca sordida* exhibits unusually plastic oviposition behavior. They sometimes attach eggs to the substrate, bury them beneath the substrate, or deposit them in a floating film on the water's surface. This species was most likely to place constructed basins on sandbars and small islands in streams, rather than on streambanks. Basin construction was most common during the middle of the breeding season, whereas eggs were sometimes placed in natural pools at the beginning of the breeding season.

#### Nest-Building Ranids

Nest building also has been reported in a number of ranid frogs from Asia. In two species, *Rana holsti* and *R. subaspera* (both sometimes placed in the genus *Babina*), eggs are laid as a surface film in natural basins or mud nests constructed at the edge of a stream (Utsunomiya and Utsunomiya 1983; Utsunomiya et al. 1983). Males and females sometimes work together to build a nest of gravel along a stream in fanged frogs from Borneo (*Limnonectes leporinus*) and Vietnam (*L. blythii*). The eggs are then mixed with sand or gravel in the bottom of the nest and tadpoles develop in the stream (Dring 1979; Emerson 1992; Orlov 1997).

### Oviposition in Very Small Bodies of Water (Modes 6, 7, and 8)

Frogs with reproductive Mode 6 lay eggs in very small bodies of water such as treeholes, the axils of aerial plants, and water-filled seedpods and fruit capsules (phytotelmata; Lehtinen, Lannoo, and Wassersug 2004). The eggs hatch into feeding tadpoles that subsist on organic material that accumulates in their tiny pools. Mode 6 represents a somewhat artificial grouping of unrelated frogs, with repeated independent evolution of superficially similar reproductive modes that actually encompass a wide variety of oviposition sites. A bufonid from East Africa, *Mertensophryne micranotus,* breeds in crevices in rotten trees (Grandison 1980). A toad from Brazil, *Bufo castaneoticus,* lays its eggs in water-filled, woody fruit capsules of the Brazil nut tree (*Bertholletia excelsa,* Lethycidaceae) after rodents have removed the seeds from the capsules. The tadpoles subsist on detritus that collects in the water (Caldwell 1991, 1993; Caldwell and Araújo 2004).

The Jamaican hylid frog *Calyptahyla crucialis* has been observed calling from hollow trees, but it appears to breed mainly in bromeliads. All of the other Jamaican hylids (*Hyla wilderi, H. marianae,* and *Osteopilus brunneus*) also are bromeliad-breeders (Dunn 1926a; Lynn and Grant 1940; A. Schwartz and Fowler 1973; Garrick, Sutton, and Lang 1985; Hedges 1987). Other Neotropical hylids that breed in bromeliads include *Anotheca spinosa* and members of the *Hyla bromeliacia* group from Central America (Duellman 1970) and the South American genus *Phyllodytes* (Bokermann 1966a; Weygoldt 1981a; Peixoto, Caramaschi, and Freire 2003; Lehtinen, Lannoo, and Wassersug 2004). At least some members of the South American genus *Osteocephalus* breed both in bromeliads and the water-filled leaf axils of forest palms (Zimmerman and Bogart 1988; Hödl 1990b). The tadpoles of some bromeliad-breeders, such as *Phyllodytes luteolus,* are generalist feeders on organic material, and have relatively unspecialized mouthparts (Weygoldt 1981a). The tadpole of *Phyllodytes gyrinaethes* has a bizarre enlarged oral disc, and older tadpoles eat frog eggs (Peixoto, Caramaschi, and Freire 2003). Other bromeliad breeders, such as *Osteopilus brunneus* and *Osteocephalus oophagus,* have reduced mouthparts, and subsist partly or entirely on the eggs of their own or other species of frogs (Lannoo, Townsend, and Wassersug 1987; Jungfer and Schiesari 1995; see also chapter 12).

A large hylid from lowland Amazonian forests, *Phrynohyas resinifictrix,* breeds in stagnant water in tree holes. Some of these are the abandoned nests of bees and are lined with resin, leading Goeldi (1907) to incorrectly conclude that these frogs actually lined their nests with gum (Lamotte and Lescure 1977; Zimmermann and Hödl 1983; Hödl 1990b, 1991b). Tadpoles of this species often eat conspecific eggs, but females apparently do not deliberately feed eggs to tadpoles, as occurs in some species that breed in phytotelmata (Schiesari, Gordo, and Hödl 2003). Tadpoles also lack the specialized mouth morphology typical of obligate egg-eating species (Schiesari, Grillitsch, and Vogl 1996).

Several frogs in the family Mantellidae from Madagascar also breed in phytotelmata. One species of *Mantella* breeds in bamboo stumps or treeholes (Heying 2001, 2004), whereas at least six species of *Mantidactylus* breed in *Pandanus* plants (Lehtinen, Richards, and Nussbaum 2004). All of these appear to be descended from pond-breeding ancestors, and there has been one apparent reversal to pond breeding within the clade of phytotelm-breeding *Mantidactylus.* A small West African frog, *Phrynobatrachus guineensis,* breeds in treeholes, water-filled fruit capsules, and empty snail shells (Rödel 1998; Rödel, Frohschammer, and Linsenmair 2004). Most members of this genus breed in temporary ponds, so it seems likely that *P. guineensis* had a pond-breeding ancestor. Males of a small microhylid frog from Borneo, *Metaphrynella sundana,* call from small pools in tree holes, but the placement of eggs has not been described (Inger and Stuebing 1997). Most species in another genus of Asian microhylids, *Ramanella,* breed in water-filled treeholes or buttress roots, laying eggs either in the water (Mode 6) or just above the water (Mode 26) (Manamendra-Arachchi and Pethiyagoda 2001; Krishna, Krishna, and Vijayalaxmi 2004), although some species use temporary rain pools (Dutta, Sushree, and Mohanthi-Hejmadi 1990–1991).

Because food tends to be in short supply in small bodies of water such as those found in bromeliads and tree holes, some frogs that breed in such sites provision their eggs with sufficient yolk to allow their aquatic tadpoles to develop without feeding (Mode 8). These endotrophic tadpoles typically lack mouthparts that would allow them to feed. This unusual reproductive mode is found in three genera of microhylids from Madagascar (*Anodonthyla, Platypelis, Plethodonthyla;* Blommers-Schlösser 1975a). It also has been reported in a Neotropical microhylid, *Syncope antenori,* which breeds in terrestrial or arboreal bromeliads (Hödl 1990b; Krügel and Richter 1995), and two Brazilian bufonids, *Dendrophrynicus brevipollicatus* and *Frostius pernambucensis* (Carvalho 1949; Lamotte and Lescure 1977; Lehtinen, Lannoo, and Wassersug 2004).

There also are frogs that breed in small water-filled depressions on the ground and have nonfeeding aquatic tadpoles (Mode 7). Leptodactylid frogs in the genus *Eupsophus* from southern Chile lay their eggs in cold water that collects in cavities under moss, logs, or rocks; the nonfeeding tadpoles complete development in these small pools (Formas and Pugin 1978a, b; Formas and Vera 1980; Formas 1989). A very similar reproductive mode is found in the most aquatic species of *Leiopelma* from New Zealand, *L. hochstetteri* (Archey 1922, 1935; N. Stephenson and Thomas 1945; Turbott 1949; N. Stephenson 1951, 1955; E. Stephenson and Stephenson 1957; B. Bell 1978, 1982, 1985; B. Bell and Wassersug 2003). The larvae complete most of their development in the eggs and do not feed after hatching. Indeed, they have lost all vestiges of feeding structures typically found in most free-living tadpoles and lack functional gills and lungs as well. Nonfeeding tadpoles in other species usually retain more indications of feeding structures than do those of *Leiopelma.* The tadpoles of *L. hochstetteri* do retain other morphological features typical of tadpoles, including a muscular tail that enables them to swim. The life history of this species clearly appears to be derived from one with a free-living tadpole stage, perhaps resembling the stream tadpoles found in other families of frogs (B. Bell and Wassersug 2003).

The golden toad of Costa Rica (*Bufo periglenes*), now thought to be extinct, bred in small pools on the forest floor and had tadpoles that could survive to metamorphosis without feeding, although they could feed if food was available (Crump 1989b). An Australian myobatrachid, *Crinia georgiana,* breeds in very shallow seeps where the tadpoles remain until metamorphosis. As in *B. periglenes,* they are capable of feeding, but their large yolk reserves allow them to reach metamorphosis without feeding (Doughty 2002).

Most bufonids in the genus *Pelophryne,* from high-altitude forests in the Philippines, have nonfeeding tadpoles that develop in water-filled depressions on the ground (Inger 1960; Duellman and Trueb 1986). However, the eggs and nonfeeding tadpoles of *Pelophryne brevipes* have been found three to five meters above the ground in leaf axils of screw pines (*Pandanus;* Alcala and Brown 1982) and in tree holes (Denzer 1994), a mode of reproduction closer to Mode 8 than Mode 7. Indeed, these two categories tend to grade into one another. For example, tadpoles of *Kalophrynus pleurostigma,* a microhylid from Borneo, have been found not only in water-filled holes in logs and in pitcher plants, but also in a forest-floor pool and a burrow in a stream bank (Inger 1985).

### Oviposition above Water-Filled Cavities (Mode 26)

A number of arboreal frogs attach their eggs to the walls of water-filled tree holes, buttress tanks, or cavities in plants, and their tadpoles drop into the water and feed on organic material or frog eggs (Mode 26). This reproductive mode tends to grade into Mode 6 (eggs placed in the water in tree holes or bromeliads) and Modes 24 and 25 (oviposition on vegetation above water). For example, the Asian rhacophorid *Chirixalus eiffingeri* usually places its eggs on the walls of water-filled tree holes, leaf axils, or bamboo stumps. Some eggs are placed in the water, but are soon consumed, because tadpoles of this species feed mainly on conspecific eggs that are provided by the female (Kuramoto 1973; Ueda 1986; Kuramoto and Wang 1987); see also chapter 11. Wassersug, Frogner, and Inger (1981) described tadpoles of a treehole-breeding rhacophorid from Thailand, identified as a species of *Philautus.* They also are mainly egg-eaters that probably are fed by the female. Whether fertilized eggs are normally placed in the water or on the walls of the tree hole is unknown. Other Asian rhacophorids in the genera *Theloderma* and *Nyctixalus* also are thought to have Mode 26 reproduction, although egg-laying sites are not known for all species (C. Liu 1950; Inger 1966; Wassersug, Frogner, and Inger 1981; W. Brown and Alcala 1983). Most phyllomedusine hylid frogs lay arboreal eggs on vegetation above ponds or streams (Modes 24 and 25), but *Agalychnis craspedopus* in Peru placed eggs on the sides of tree buttresses above small pools, a mode of reproduction closer to Mode 26 (Block et al. 2003).

The African hyperoliid *Acanthixalus spinosus* lays its eggs on the walls of large water-filled tree holes, where the tadpoles apparently feed on algae (Perret 1962; Lamotte and Lescure 1977; Schiøtz 1999). A second recently described species, *A. sonjae,* has similar breeding habitats, but the tadpoles will eat other species of tadpoles as well as vegetation (Rödel et al. 2003). Two other hyperoliid frogs, *Hyperolius acutirostris* and *H. mosaicus,* also are treehole breeders that deposit eggs just above the water (Schiøtz 1999). Tree hole breeding also has been reported in an African petropedetid frog, *Phrynobatrachus guineensis,* but other members of this genus breed in temporary ponds (Rödel 1998). African microhylids of the genus *Hoplophryne* lay their eggs on the walls of hollow bamboo stems and rolled banana leaves; their tadpoles subsist at least in part on conspecific eggs (Noble 1929a). A microhylid from India, *Ramanella montana,* lays eggs on the walls of tree holes just above the water line, or on floating dead leaves in the water (Krishna, Krishna, and Vijayalaxmi 2004), so reproduction in this species grades between Modes 6 and 26.

### Oviposition in Bubble or Foam Nests (Modes 10–14, 29–33)

The habit of laying eggs in a frothy or foamy nest on the water's surface (Modes 10, 11, and 12), in water-filled basins (Mode 13), in phytotelmata (Mode 14), on the ground (Mode 28), in burrows (Modes 29–32), or in trees (Mode 33) has evolved independently in several major clades of frogs. Duellman and Trueb (1986) and Haddad and Prado (2005) placed foam-nesting frogs in several different reproductive modes, but in many cases, these modes tend to grade into one another. For example, some frogs lay eggs in foam nests in both still (Mode 11) and slowly moving water (Mode 12), and foam nesting in ponds (Mode 11) tends to grade into foam nesting in basins or burrows (Modes 29–32). Some frogs lay eggs both on the surface or at the edge of ponds (Mode 11) and in trees (Mode 33).

#### Bubble Nest on Water's Surface (Mode 10)

An apparently unique mode of reproduction has been described in the Neotropical microhylid frog *Chiasmocleis leucosticta* (Haddad and Hödl 1997). Amplexed pairs laid eggs on the surface in small groups, with a total of up to 70 bouts of oviposition. After deposting the eggs, the male and female submerged together and released air bubbles from their nostrils from underneath the floating eggs, creating a frothy bubble nest floating on the water's surface. This is the only known example of an amphibian nest being constructed by exhaling air. Presumably the bubbles help to hold the eggs at the water's surface, and perhaps provide supplementary oxygen in warm water. It is somewhat reminiscent of the bubble nesting of a variety of tropical fishes that live in poorly oxygenated water, although in those cases, the males construct their bubble nests and defend them prior to ovipostion (Hostache and Mol 1998; Jaroensutasinee and Jaroensutasinee 2001). The genus *Chiasmocleis* includes 18 species, all of which breed in warm, poorly oxygenated pools, but the frequency of bubble nesting in the genus is not known.

### Foam Nesting in Nonarboreal Frogs (Modes 11–14, 28–32)

Foam-nesting is characteristic of most frogs in the subfamily Leptodactylinae, including *Leptodactylus, Physalaemus, Pleurodema, Lithodytes, Edalorhina,* and *Vanzolinius* (Fernandez and Fernandez 1921; Eisentraut 1932; Breder 1946; Cei 1949a, 1980; Gallardo 1958; Heyer 1969; Rivero and Esteves 1969; Lescure 1973, 1979; M. Crump 1974; Philibosian et al. 1974; Vaz-Ferreira and Gehrau 1975; Heyer and Rand 1977; Lamotte and Lescure 1977; Cardoso 1981a; Lescure and Letellier 1983; Regös and Schlüter 1984; M. Ryan 1985b; Hödl 1986, 1990a, b, 1992; M. Martins 1988; Duellman and Morales 1990; Schlüter 1990). It also has been reported in one Neotropical hylid species (Haddad, Pombal, and Gordo 1990).

Foam nests of leptodactylids are derived from mucus secretions produced by the female as she lays her eggs (fig. 10.5). These are whipped into foam when the male kicks his legs (fig. 10.5 A). Hödl (1990a) provided a detailed description of foam nest construction in *Physalaemus ephippifer.* Construction of a complete nest takes about half an hour to an hour and involves an average of 85 individual bouts of foam beating by the male. Each bout of activity is initiated by a female laying a small number of eggs, which are then beaten into the foam by rotating movements of the male's hind legs. Individual bouts of fertilization and foam beating are very short, averaging only three to four seconds, with intervals of 20 to 30 seconds between bouts.

In *Pleurodema, Physalaemus,* and *Edalorhina,* foam nests are usually placed on the surface of the water in shallow, temporary pools (Mode 11). Often the foam nests are attached to bits of vegetation, dead leaves, or mud at the edge of the pools (Sexton and Ortleb 1966). In some species, pairs attach their eggs to previously laid foam nests, resulting in large communal egg masses of several hundred clutches (Hödl 1992; fig. 10.5 B). The smaller frogs in this group, such as *Physalaemus* and *Edalorhina,* exhibit considerable flexibility in choice of oviposition sites. They use almost anything from moderately large temporary ponds to roadside ditches, tire tracks, water-filled animal footprints on the forest floor, water-filled seed pods, and pools between tree roots (Rand 1983; Hödl 1986; Tárano 1998; Dillon and Fiaño 2000; D. Marsh and Borrell 2001). Schlüter (1990) even reported observations of *Edalorhina perezi* building a foam nest in the axil of a palm tree, and tadpoles were found in a water-filled depression on a tree trunk. More typical oviposition sites for this species are water-filled holes left by the roots of large fallen trees (P. Murphy 2003). Another member of this genus, *Physalaemus signifer,* also exhibits considerable plasticity in reproductive behavior, with foam

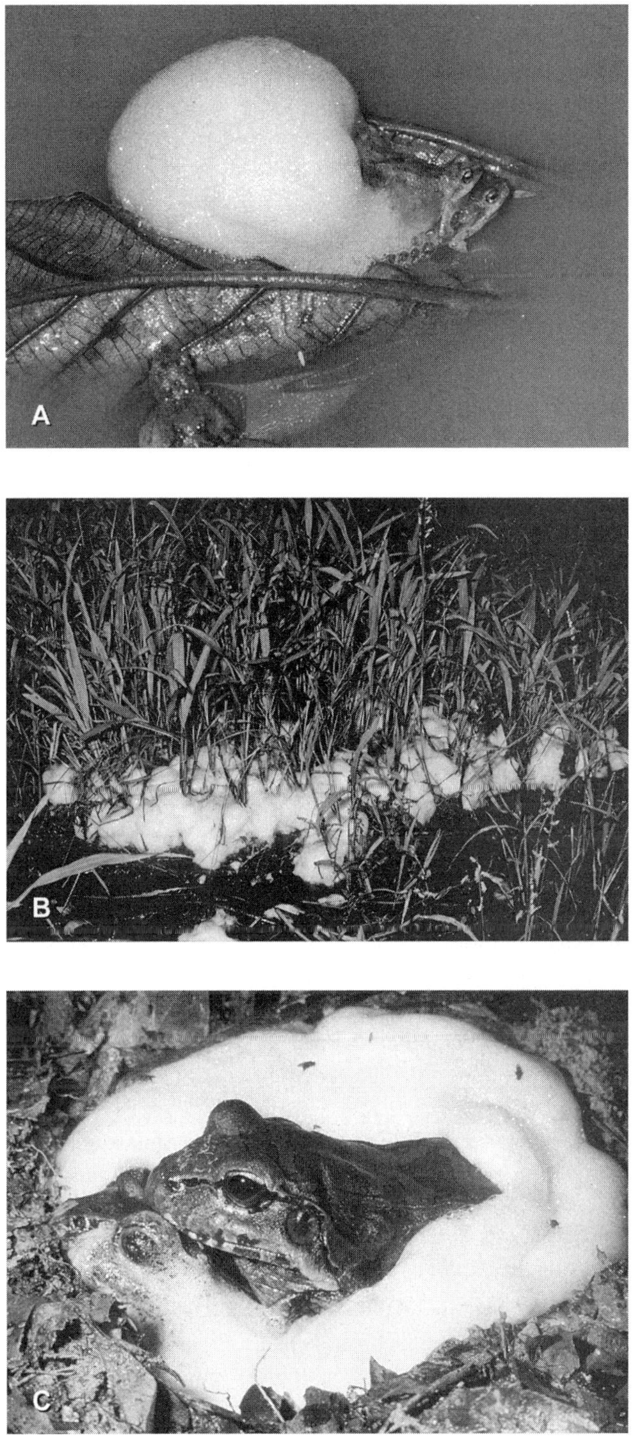


**Fig. 10.5.** Foam nesting in leptodactylid frogs. (A) Male and female of *Physalaemus pustulosus* laying eggs attached to a floating leaf. (B) Communal nest composed of the egg masses of many different pairs of *P. pustulosus.* (C) Pair of *Leptodactylus knudseni* in amplexus. The foam nest is placed in a shallow depression near the edge of a pond. Photos by Kentwood D. Wells (A, B) and Walter Hödl (C).

nests being placed in shallow ponds, on moist ground, or in water accumulated in the axils of terrestrial bromeliads; tadpoles develop in water once the area is flooded (Haddad and Pombal 1998). The plasticity of behavior in this species led Haddad and Prado (2005) to place this species in three distinct reproductive modes (Modes 11, 14, and 28), but all of these clearly are variations on the same mode of reproduction. Several closely related species also place foam nests on the ground or in terrestrial bromeliads and may show similar plasticity in behavior.

Most members of the *Leptodactylus ocellatus* and *L. melanonotus* species groups deposit eggs in foam nests floating on the water's surface or at the edge of ponds (Mode 11), whereas members of the *L. pentadactylus* group deposit their large foam nests in open cavities at the edges of ponds (Mode 13; fig. 10.5 C; Breder 1946; Heyer 1969; Rivero and Esteves 1969; Hödl 1990b; Prado, Uetanabaro, and Lopes 2000). In one Brazilian member of the *L. melanonotus* group, *L. podicipinus,* males construct water-filled basins near small ponds. The frogs call from the basins, which often are covered by a roof of leaves, and attract females that lay foamy egg clutches in the basins (Prado, Uetanabaro, and Haddad 2002).

In frogs of the *Leptodactylus fuscus* group, the male builds an underground burrow in saturated soil and leads the female to the nest site, where a foam nest is produced (Heyer 1969; Lescure 1973; Martins 1988; Perotti 1997). Tadpoles remain in the nest until it is flooded, and complete development by feeding in the water (Mode 30). *Leptodactylus mystaceus,* another member of this group, places its foam nests in open mud basins under logs or other cover (Caldwell and Lopez 1989). In one species, *L. bufonius,* foam nests are constructed in an open burrow at the margin of a pond, and the nest is then covered over with soil after eggs are laid (Pisanó 1962; Philibosian et al. 1974; M. Crump 1995a; Reading and Jofré 2003). This type of nest may help protect the developing embryos from extreme temperatures in the hot, dry habitat of this species.

In *Leptodactylus fallax,* which is a member of the *L. pentadactylus* group, tadpoles remain in a foam nest in an underground burrow throughout development. Most authors have stated that the tadpoles do not feed (Lescure and Letellier 1983; Duellman and Trueb 1986) and this frog has been placed in a different reproductive mode (Mode 32) than other leptodactylids. S. Davis et al. (2000) reported that even relatively young tadpoles of this species appeared to lack substantial yolk reserves, but they were not able to determine whether the tadpoles took in supplemental food. Gibson and Buley (2004) subsequently reported that females of this species actually feed tadpoles by laying unfertilized eggs in the foam nest, a form of parental care previously known mostly for species that lay eggs in bromeliads or tree holes (see chapter 11).

A similar mode of reproduction has been reported for the genus *Adenomera,* although there is no evidence that females feed the tadpoles. These frogs typically place their foam nests in open terrestrial chambers in which the tadpoles develop without feeding (Heyer and Silverstone 1969; Lescure 1979; Heyer et al. 1990; Kokubum and Giaretta 2005). In a population of this genus in Bolivia, however, females laid eggs at the edges of temporary ponds, and tadpoles completed their development in the ponds (De la Riva 1995), and this also may be true for some other species in the genus (Kokubum and Giaretta 2005). Members of this genus therefore exhibit both Mode 11 and Mode 32 reproduction, but whether this represents species-specific differences in mode of development, or perhaps hidden taxonomic diversity within the genus, is not yet clear. Microhylid frogs of the genus *Stumpffia* from Madagascar lay eggs in foam nests in leaf litter and apparently have nonfeeding tadpoles, so this represents a completely independent evolution of this type of reproduction in another family (Glaw and Vences 1994).

Australian myobatrachids in the subfamily Limnodynastinae exhibit a striking convergence with leptodactyline frogs in their reproductive biology (A. Martin 1970). Most species of *Limnodynastes,* as well as *Adelotus* and *Lechriodus,* build foam nests on the surface of shallow pools (fig. 10.6 A, B), where tadpoles complete their development (Mode 11; Main, Littlejohn, and Lee 1959; J. Moore 1961; A. Martin 1967; Tyler 1985b). However, the way in which the nest is constructed is completely different from that seen in leptodactylines, and really should be considered a distinct mode of reproduction. As the eggs are laid, the female beats the water in front of her with flattened, paddle-shaped structures on her front toes (fig. 10.6 C). This causes bubbles to rise underneath the female, where they combine with mucus secretions produced by the female to form a mass of foam that traps the eggs (H. Parker 1940; A. Martin 1967; Tyler and Davies 1979). In some populations of *Limnodynastes tasmaniensis* in southern Australia, females do not build foam nests, but instead produce a surface film of eggs. Females in these populations lack the enlarged flanges on the front feet (J. Roberts and Seymour 1989). Some species of *Limnodynastes* are flexible in their choice of oviposition sites. For example, *L. dorsalis* is reported to place foam nests in standing water (Mode 11), flowing water (Mode 12), or even in burrows at the edges of streams (Mode 31; A. Martin 1967).

The burrowing frog *Megistolotis lignarius,* which lives around boulder fields and rocky ledges near small streams in northern Australia (Cogger 2000), also is a foam-nest builder, but it places nests in pools that become flowing streams after heavy rains (Mode 12). Females lack flanges on the front toes, and the mode of nest construction is un-

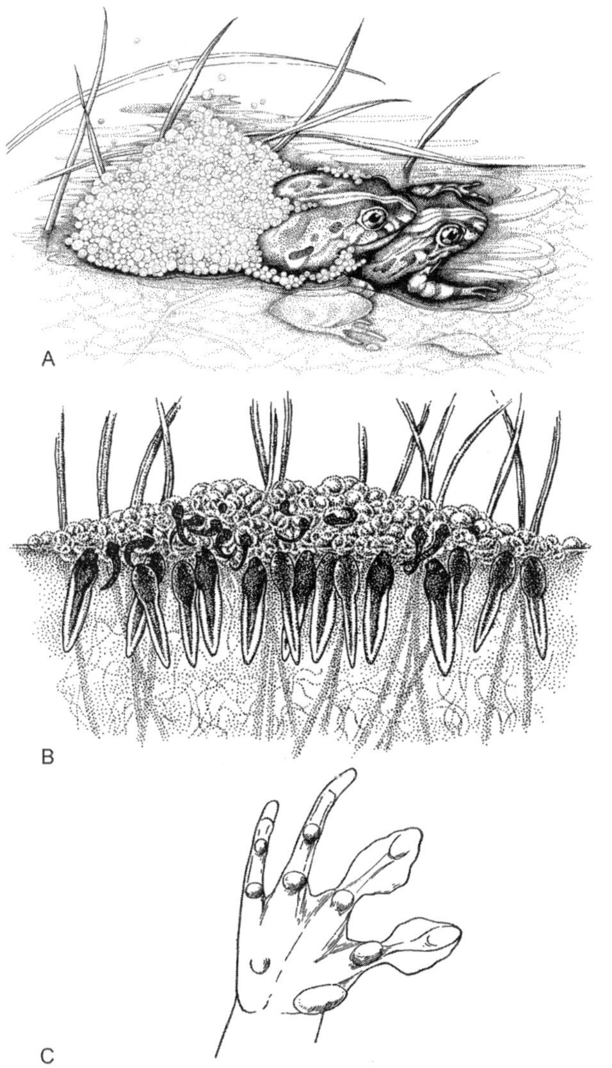


**Fig. 10.6.** Foam nests of limnodynastine frogs. (A) Pair of *Limnodynastes ornatus* laying eggs in a foam nest. (B) Hatched tadpoles suspended from the foam nest of *Limnodynastes tasmaniensis.* (C) Flanges on the front feet of the female of *Limnodynastes.* (A, B) from Littlejohn et al. (1993), (C) from Noble (1931).

known. Another group of foam-nesters that lacks enlarged front toes is the genus *Heleioporus*. These are frogs that usually place their eggs in burrows at the edge of standing water or in places where they are flooded after eggs are laid, allowing tadpoles to complete development in the water (Mode 31; A. K. Lee 1967; A. Martin 1967). However, some members of the genus place their foam nests in standing water (Mode 11; M. Tyler 1985b). Frogs in the limnodynastine genus *Philoria* produce foam nests that are placed in terrestrial situations such as moist depressions or sphagnum moss. The tadpoles develop completely in the nest and apparently do not feed (Mode 32; J. Moore 1961; Littlejohn 1963; A. Martin 1967; Tyler 1985b).

Construction of aquatic foam nests is characteristic mainly of species that breed in warm, poorly-oxygenated water, and many foam-nesting species utilize temporary ponds or very small bodies of water such as animal footprints. Several benefits of foam nesting have been proposed, and these are not necessarily mutually exclusive. Some authors have suggested that placing eggs in a foam nest reduces exposure to aquatic predators (A. Martin 1967; Heyer 1969; M. Ryan 1985b), while terrestrial or flying insects become entangled in the sticky foam (Villa, McDiarmid, and Gallardo 1982). Presumably species that place foam nests in burrows or depressions away from standing water would derive the greatest benefit. Nevertheless, foam nests are subject to predation by tadpoles, aquatic insects, and a variety of terrestrial arthropods (Villa, McDiarmid, and Gallardo 1982; M. Ryan 1985b; Downie 1988). Unfortunately, there are no quantitative estimates of egg survivorship in species that build foam nests and those in the same habitat that do not.

Another common suggestion is that foam nests protect the eggs from desiccation when temporary pools in which they are laid dry up, keeping the eggs viable until the next rain (Breder 1946; A. Martin 1967; Heyer 1969; Salthe and Mecham 1974; Duellman and Trueb 1986; Hödl 1986; Duellman 1989a). Downie (1988) found that some eggs in foam nests of *Physalaemus pustulosus* could survive for two or three days if stranded, although those close to the surface died quickly under dry conditions. This is not much better than the survivorship of *Rana sylvatica* eggs left out of the water in the laboratory, but the latter were not exposed to the high temperatures experienced by *Physalaemus* eggs in the field (Forester and Lykens 1988). However, aquatic foam nests probably provide enough protection to get the eggs through brief dry periods during the rainy season, an effect that probably is enhanced by grouping the eggs together in large communal masses (M. Ryan 1985b). The limited protection from desiccation provided by foam nests presumably preadapts foam-nesting species to move their eggs out of the water altogether, especially if they are placed in a relatively moist burrow (Heyer 1969). Lynch (1971) suggested that foam-nesting in leptodactylines evolved during a period of increasingly dry climate, while Heyer (1975) argued that foam-nesting originated in wet forests, but preadapted leptodactylines for later invasion of drier savanna habitats. In some species that produce foam nests in burrows, hatched tadpoles generate their own foam from mucus mixed with water from the substrate. This enables the tadpoles to survive for a number of days in a state of developmental arrest before the burrow floods and the tadpoles move into a pond (Downie 1984, 1989, 1994; Caldwell and Lopez 1989; Downie and Weir 1997; Downie and Smith 2003).

Other authors have suggested that foam nests provide some thermal advantage to the eggs, but results of several studies are inconsistent. Gorzula (1977) reported that temperatures at the center of *Physalaemus* foam nests were considerably cooler than the surrounding water on a sunny day and suggested that the foam reflects solar radiation and helps to prevent the eggs from overheating. In contrast, Dobkin and Gettinger (1985) found that *Physalaemus* foam nests reflected relatively little short-wave radiation and warmed more rapidly than the surrounding water, sometimes exceeding air temperatures by 4–8° C. They interpreted this as an advantage for species breeding in temporary pools, where rapid embryological development would be favored. Downie (1988) found little difference between nest and water temperatures in the same species. He found that foam nests usually were slightly cooler than the surrounding air, but still rose as high as 35° C without any apparent damage to the eggs. Hence, the evidence that foam nests provide any heating or cooling advantage to the eggs is weak.

A fourth possibility that seems more promising is that foam nests ensure a supply of oxygen by surrounding eggs with air-filled bubbles and elevating them above the warm, hypoxic water typical of their breeding sites (Downie 1988; Seymour 1999). This has not been thoroughly tested for New World species. However, Seymour and Roberts (1991) measured oxygen profiles of foamy and nonfoamy egg masses of an Australian frog, *Limnodynastes tasmaniensis.* The structure of the foam nest in this species, and presumably in other limnodynastine frogs, is different from that of leptodactyline frogs. In the latter, most of the eggs are held above the water's surface by the foam for most or all of development. In contrast, those of *Limnodynastes* were mostly submerged in the water, suspended under a thin layer of foam floating on the surface. Oxygen levels in the foam and near the surface, where the majority of eggs were concentrated, remained relatively high throughout development, but eggs that were a centimeter or more below the surface experienced much lower levels of oxygen, especially late in development. Nonfoamy egg masses suspended near the surface showed a similar oxygen profile, but with an even steeper decline in oxygen levels with increasing depth. When either type of egg mass was allowed to sink to the bottom, at least part of the mass became anoxic and many eggs died. The foamy egg masses of this species apparently serve mainly to keep the eggs suspended near the surface, and therefore function much like the surface films of other warm-water breeders. In fact, at high temperatures, the foam nests of limnodynastine frogs tend to break up into surface films (Tyler 1985b; Seymour and Roberts 1991). This has not been reported for leptodactyline frogs, although their nests disintegrate in heavy rains (M. Ryan 1985b).

### Foam Nesting in Arboreal Frogs (Mode 33)

Some rhacophorid and hyperoliid frogs build foam nests, often in trees, but sometimes in the water or on the ground. African treefrogs in the genus *Chiromantis* live in a variety of habitats, from moist forests to savannas. They breed in the rainy season, sometimes quite explosively after heavy rains. *Chiromantis xerampelina* and *C. rufescens* typically deposit their eggs in foam nests overhanging temporary pools or slow-moving rivers and streams (fig. 10.7 A), where the tadpoles complete their development (Wager 1926, 1965; Lamotte and Perrett 1963a; Coe 1974; Jennions, Backwell, and Passmore 1992). *Chiromantis petersi,* a savanna species, often breeds in areas with very few trees, and attaches its eggs to almost any suitable object, including grass stems, logs, tree trunks, rocks, or branches of shrubs (Coe 1974). The foam nest is constructed by rapid leg movements of both the male and female, and it is not uncommon for several males to participate in building a nest with a single female (Coe 1967, 1974; Jennions, Backwell, and Passmore 1992; see chapter 8). In *C. xerampelina,* females lay eggs in bouts interrupted by periodic trips to the water to rehydrate. When a female resumes her egg laying, she does not necessarily mate with the same male or group of males. Females of this species also have been observed returning to nests on the night after egg-laying and adding an additional layer of foam (Jennions 1992). The nests of these frogs can survive prolonged exposure to hot, dry conditions. They develop a hard crust on the outside, but remain moist on the inside. The tendency of these frogs to deposit eggs in large communal nests probably helps to reduce water loss as well. The hardening of the surface of the egg mass actually increases diffusion of oxygen into the egg mass, but early in development, air trapped in the bubbles of foam provide most of the oxygen needed by the developing embryos (Seymour and Loveridge 1994; see also chapter 4).

Foam nesting also is characteristic of Asian rhacophorids in the genera *Rhacophorus* (fig. 10.7 B), *Polypedates,* and some species of *Chirixalus* (fig. 10.7 C, D). These frogs exhibit considerable plasticity in placement of foam nests, especially in disturbed habitats. Some species place their eggs in trees over water. Others place them on almost any sort of substrate, including vegetation at the surface of ponds, mud banks, or man-made walls. Eggs also are laid under stones or logs, in ditches, floating on the surface of shallow ponds, in depressions on muddy ground, or in terrestrial burrows (Ikeda 1897; Siedlecki 1909; Pope 1931; Bhaduri 1932; Liu 1950; Kato 1956; Okada 1966; Mallick and Mallick 1982; Maeda and Matsui 1989; Girish and Saidapur 1999b; Kadadevaru and Kanamadi 2000). Indeed, the distinction among reproductive modes becomes somewhat arbitrary for these frogs. For example, a large treefrog from Borneo, *Rhacophorus harrissoni,* places its foam nests on the trunks of trees over tanks formed by the fusing of buttress roots, a mode of reproduction similar to that of some treehole breeders that lay nonfoamy egg masses in much smaller bodies of

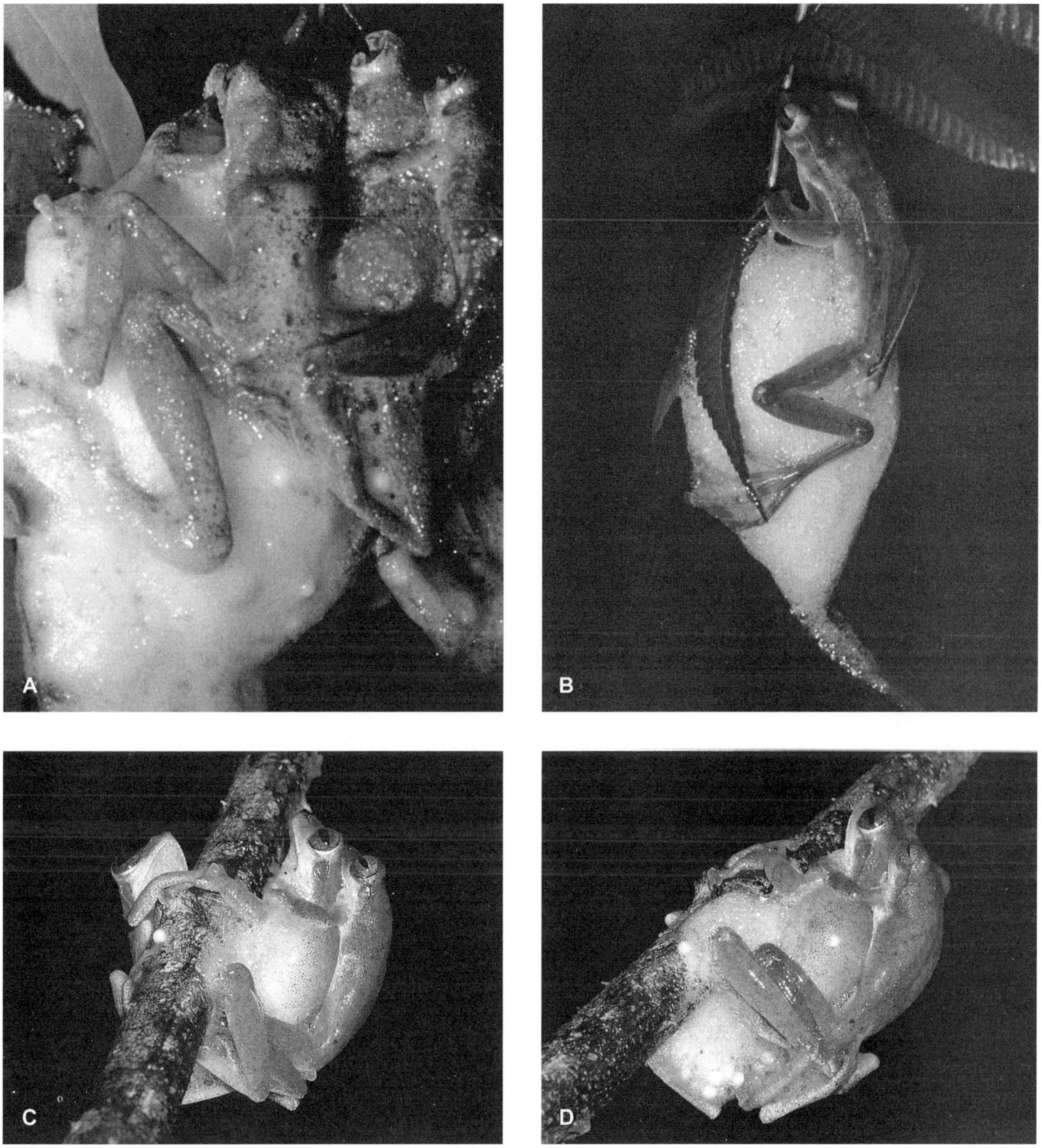

**Fig. 10.7.** Foam nests of arboreal rhacophorid frogs. (A) Multiple males fertilizing the eggs of a single female of *Chiromantis rufescens* from West Africa. (B) Female of *Rhacophorus dulitensis* from Borneo smoothing the outside of a foam nest. (C) Pair of *Chirixalus nonkhorenis* from Vietnam depositing a foam nest on a branch, with a satellite male nearby, possibly attempting to fertilize some eggs. (D) Same pair with unpigmented eggs visible in the foam. Photos by Mark Oliver Rödel (A), Walter Hödl (B), and Wayne Van Devender (C, D).

water (Mode 26; Inger and Stuebing 1997). In one Japanese species, *Rhacophorus viridis,* eggs have been found several meters above the ground in trees (Mode 33), in grassy ponds (Mode 11), between rocks, on mud (Mode 28), and in terrestrial burrows (Mode 29) (Kuramoto 1973; Utsunomiya and Utsunomiya 1983). Several Asian rhacophorids build communal nests similar to those of *Chiromantis.* Simultaneous matings by several males with a single female have been observed in both arboreal and terrestrial breeders (Kasuya et al. 1987; Feng and Narins 1991; Fukuyama 1991; Kusano et al. 1991; see chapter 8 for additional discussion of the significance of group mating in these frogs).

Foam nesting also has been reported in one hyperoliid frog, *Opisthothylax immaculatus* from West Africa. It produces a foam nest folded inside a leaf overhanging a body of water (Amiet 1974b, 1989). The eggs are unusually large (4.6 mm in diameter) compared to those of most other foam-nesting species, and only a few eggs are laid. This suggests that foam nesting in this species is secondarily derived from arboreal egg-laying in nonfoamy masses, like that seen in some species of *Afrixalus* (see the following), rather than from aquatic foam nesting. The function of the foam is unknown, but it could provide a reservoir of water for the eggs or simply serve to cement the surrounding leaves together.

### Arboreal Oviposition without Foam Nests (Modes 24 and 25)

In addition to arboreal foam-nesters discussed earlier, many other arboreal frogs deposit eggs on leaves, branches, or trunks of trees above water. The egg-laying adaptations in these frogs are extremely diverse and parallel those found in terrestrial breeders (see the following). The very different structure of the egg masses produced by different groups of arboreal frogs clearly reflects the repeated independent evolution of arboreal reproductive modes. As with many of the reproductive modes already discussed, Modes 24 and 25 could easily be subdivided into several distinct reproductive modes. Some Neotropical hylids, such as *Hyla ebraccata* and *H. brevifrons,* place their eggs on the upper surfaces of leaves, where they develop rapidly, and tadpoles drop into the water to complete development. Egg masses in these species are little more than aquatic egg masses attached to vegetation (fig. 10.8 A), although eggs tend to be somewhat larger than those of aquatic breeders of similar body size (Duellman 1970; Salthe and Duellman 1973; Crump and Kaplan 1979; Aichinger 1987a; Hödl 1991a). Somewhat similar egg masses are laid on leaves by some species of *Mantidactylus,* a group of mantellid frogs from Madagascar.

All of the phyllomedusine hylid frogs of South and Central America lay eggs in sites over water, and their tadpoles drop into the water to complete development, but the types of oviposition sites used vary considerably. Most members of this subfamily attach eggs to leaves overhanging ponds, but eggs also can be attached to tree trunks, stems, vines, and rocks (Caldwell 1994). Some species of *Phyllomedusa* place their eggs on vegetation over streams or attach them to rocks (A. Lutz and Lutz 1939; Cannatella 1980a; Langone, Prigioni, and Venturino 1985; Hödl 1990b). Members of the genus *Phrynomedusa,* formerly placed in the *Phyllomedusa fimbriata* group, deposit eggs in cavities, rock crevices, or hollows in fallen trees (Cruz 1982, 1990; Weygoldt 1984b, 1985, 1991; Jungfer and Weygoldt 1994). *Agalychnis craspedopus* lay eggs on vegetation over small forest pools or water-filled hollow logs (Hoogmoed and Cadle 1991), or on tree buttresses above small pools (Block et al. 2003). A closely related species, *A. calcarifer,* apparently breeds exclusively in small pools formed in cavities formed by the buttresses of large fallen trees. Eggs are deposited on vegetation over the pools or on the sides of the fallen trees just above the water (Marquis, Donnelly, and Guyer 1986; Donnelly et al. 1987; Caldwell 1994; Donnelly 1994). The reproductive modes of these species therefore grade into Mode 26 (see the following). Another species, *A. saltator,* lays eggs in huge communal masses imbedded in moss and epiphyte rootlets growing on vines over swamps (W. Roberts 1994). Egg capsules of this species are unusually hard and sticky and lack the large jelly mass typical of leaf-breeding members of the genus.

In *Pachymedusa dacnicolor,* most species of *Agalychnis, Phyllomedusa lemur,* and probably other members of the *Phyllomedusa buckleyi* group, the eggs are laid on the exposed surfaces of leaves, but the leaves are not wrapped around the eggs. Sometimes the female provides the eggs and surrounding jelly with water by periodically returning to the water with the male to rehydrate between bouts of egg deposition (Pyburn 1970; Wiewandt 1971; N. Scott and Starrett 1974; Bagnara et al. 1986; Jungfer and Weygoldt 1994; Leary and Parker 1998). This behavior does not occur in *A. saltator* (W. Roberts 1994). Many species of *Phyllomedusa* surround their eggs with one or more leaves that are wrapped around the eggs by the female, or formed into a funnel-shaped nest before eggs are laid. These frogs do not return to the water to rehydrate during oviposition. Instead, eggless jelly capsules are incorporated into the nest as water reservoirs. As the eggs develop, water passes by osmosis from the eggless capsules to the adjacent eggs (compare fig. 10.8 C and 10.8 D; Agar 1909; Kenny 1966; Pyburn 1970, 1980a; Pyburn and Glidewell 1971; Crump 1974; Lescure 1975b; Lamotte and Lescure 1977; Weygoldt 1981b; Cannatella 1982; Langone, Prigioni, and Venturino 1985; Hödl 1990b; Langone 1993; Jungfer and Weygoldt 1994; Vaira 2001).

Most centrolenid frogs deposit relatively large, unpigmented eggs on the undersides of leaves overhanging streams

**Fig. 10.8.** Nonaquatic egg masses of anurans. (A) Male of a Central American treefrog, *Hyla ebraccata* (Hylidae), calling above an egg clutch on a leaf over a pond. (B) Egg mass of *Hyalinobatrachium fleischmanni* (Centrolenidae) placed on the underside of a leaf over a stream in Panama. (C) Egg mass of *Phyllomedusa trinitatis* (Hylidae) from Trinidad at an early stage of development. The embryos are wrapped in leaves and lack pigment. Empty egg capsules serve as water reservoirs. (D) Eggs of *P. trinitatis* in a later stage of development, with water-filled empty capsules largely depleted. (E) Eggs of the South African Kloof frog, *Natalobatrachus bonebergii* (Ranidae) on a branch. (F) Egg mass of *Cycloramphus dubius* (Leptodactylidae) from Brazil, placed on a tree root near the splash zone of a waterfall. Photos by Kentwood D. Wells (A–D), Alan Channing (E), and Marcio Martins (F).

(fig. 10.8 B), and in many species, a parent guards or broods the eggs (Duellman and Tulecke 1960; McDiarmid 1975, 1978; Greer and Wells 1980; Villa 1984; S. Jacobson 1985; M. Hayes 1991; see also chapter 11). Tadpoles drop into streams where they burrow into mud at the bottom (Villa and Valerio 1982). Other frogs that deposit eggs on vegetation over streams include *Hyla calypsa* and *H. thorectes* (Hylidae) from Central America (Duellman 1970; Lips 1996, 2001), *Litoria iris* (Hylidae) from the mountains of New Guinea (Zweifel and Tyler 1982), *Rana leytensis* (Ranidae) from the Philippines (Alcala 1962), and *Nyctibatrachus petreua* (Ranidae) from India (Kunte 2004).

Many hyperoliid frogs lay eggs out of water, with eggs being placed at various heights from just above the water line to four or five meters above the water (Wager 1965; Schiøtz 1967, 1999; Stewart 1967; Rödel 2000). In one species that lays eggs close to the water, *Hyperolius tuberilinguis,* pairs in amplexus return to the water after laying eggs, rehydrate, and then return to deposit additional layers of jelly on the eggs (M. Mason and Passmore 1998). This behavior is reminiscent of the behavior of some phyllomedusine frogs, which either rehydrate during egg laying or deposit empty water-filled jelly capsules in the egg masses. In another species, *H. spinigularis,* females have been observed returning for several nights after eggs have been deposited to moisten them with water (R. Stevens 1971). In most species of leaf-folding frogs in the genus *Afrixalus,* eggs are laid above water and are wrapped in leaves that are glued around the egg mass, a behavior remarkably similar to that of some phyllomedusine hylid frogs (Schiøtz 1999). A phylogenetic analysis of reproductive modes in hyperoliid frogs suggested that nonaquatic oviposition was derived from aquatic oviposition, probably more than once (fig. 10.9; Duellman 1989a).

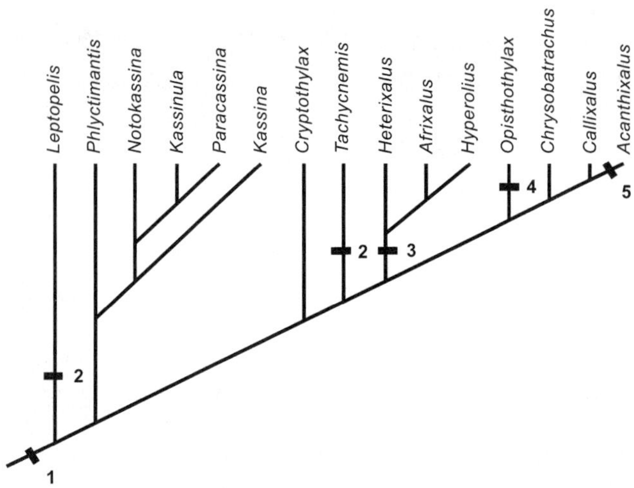


**Fig. 10.9.** Hypothesized phylogenetic relationships of hyperoliid frogs, based on morphological characters. Numbers show evolutionary changes in reproductive modes. 1 = aquatic eggs and larvae. 2 = terrestrial eggs and aquatic larvae. 3 = eggs on vegetation above water with aquatic larvae. 4 = arboreal foam nest with aquatic larvae. 5 = eggs and larvae in water-filled cavities in trees. After Duellman (1989a).

## Terrestrial Oviposition with Aquatic Tadpoles (Modes 17–21)

Many groups of frogs have evolved reproductive modes that involve deposition of eggs on land, but they nevertheless retain the aquatic larval stage. In some cases, the eggs are laid in sites that are later flooded, thereby allowing the tadpoles to complete their development in water. There also are species that lay their eggs at some distance from water, and the tadpoles either make their way to water on their own, or are carried there by a parent. Finally, there are species that lay eggs in terrestrial nests in which the tadpoles complete development without feeding at all.

### Nonaquatic Eggs with Feeding Tadpoles (Modes 17, 18, and 19)

Perhaps the least specialized mode of nonaquatic oviposition is to place the eggs in a terrestrial nest or natural cavity in a saturated area that is likely to be flooded by future rains. The tadpoles remain in the nest until it is flooded and then complete their development in temporary ponds (Mode 17). This mode of reproduction is characteristic of several Asian species of *Rana,* including *R. caldwelli* in China (described as *R. adenopleura* in C. Liu 1950) and *R. psaltes* in the Ryukyu Islands of Japan (Kuramoto 1985). In Taiwan, *R. psaltes* places eggs in small nests (about 30 mm in diameter) in very shallow water (personal observations), so the reproductive mode of this species grades into Mode 4. The eggs of these species are clearly derived from those of aquatic-breeding ancestors, being surrounded by a thick jelly mass characteristic of many pond-breeding frogs. This reproductive mode resembles that of the Brazilian hylid frog *Hyla leucopygia* and other members of the *H. albosignata* group, which was discussed earlier under Mode 4. A rhacophorid frog from Taiwan, *Chirixalis idiootocus,* lays pigmented eggs in shallow depressions or holes at the edges of temporary ponds. Hatching apparently is stimulated by heavy rain, and the tadpoles complete their development in a pond (Kuramoto and Wang 1987).

A similar reproductive mode is found in Australian myobatrachids in the genus *Geocrinia,* which lay eggs at the base of grass tussocks, and *Pseudophryne,* which lay eggs in burrows. The eggs hatch when the nests flood after heavy rains. These frogs have relatively large eggs that are not surrounded by a jelly mass. The eggs are morphologically and physiologically more similar to eggs of other terrestrial breeders than those of aquatic breeders (Jacobson 1963; A. Martin and Cooper 1972; Watson and Martin 1973; Woodruff 1976b, 1977; Bradford and Seymour 1985, 1988a, b; Seymour and Bradford 1987; Gollmann and Gollmann

1991; Seymour, Geiser, and Bradford 1991a, b). Newly hatched tadpoles of these species have considerable yolk reserves that allow them to develop rapidly in areas flooded for only a few weeks. One species, *P. australis,* has more specialized reproductive habits. It breeds throughout the year and lays eggs in terrestrial sites through which water flows after heavy rains (Thumm and Mahony 2002a). Eggs generally hatch at a more advanced stage of development than do those of other members of the genus, but there is considerable plasticity in both the timing and stage of hatching (Thumm and Mahony 2002b).

Some frogs deposit their eggs in terrestrial sites that are close enough to water to allow the tadpoles to drop into the water or wriggle to the water to complete development (Mode 18). This reproductive mode lumps together a number of unrelated anurans and does not fully capture the wide range of variation in both the types of sites used for oviposition and the way in which tadpoles reach the water. There is no doubt that Mode 18 (Mode 13 of Duellman and Trueb 1986) represents a somewhat artificial grouping of many independently evolved reproductive modes.

Some species in the genus *Leptopelis* (Hyperoliidae) from western and southern Africa lay their eggs in excavated depressions or natural cavities close to temporary savanna ponds, forest pools, or streams. When tadpoles hatch, they wriggle into the water (Lamotte and Perret 1961; Schiøtz 1963, 1967, 1999; Wager 1965; Largen 1977; Oldham 1977; Amiet 1989; Rödel 2000). Some species of the African ranid genus *Cardioglossa* apparently have a similar reproductive mode (Amiet 1989; Rödel, Schorr, and Ernst 2001). The large myobatrachid frogs of the genus *Mixophyes* are reported to lay eggs on rocks or roots above streams (Watson and Martin 1973; Duellman and Trueb 1986), although Hero, Littlejohn, and Marantelli (1991) reported that *M. balbus* lays eggs directly in streams and rock pools (Mode 2). *Paratelmatobius poecilogaster,* a small leptodactylid frog from the Atlantic forests of southeastern Brazil, lays eggs on rocks overhanging small pools on the forest floor. When the eggs hatch, the tadpoles drop into the pools, where they complete their development. The tadpoles sometimes are found in streams as well, having been washed out of the forest pools by heavy rains (Pombal and Haddad 1999). *Rana ishikawae,* a frog endemic to the Ryukyu Islands of Japan, lays unpigmented eggs in burrows in stream banks a meter or more above the water level. Tadpoles have been found in streams, but how they get from the burrow to the stream is unknown (Utsunomiya, Utsunomiya, and Katsuren 1979). Nests of an Indonesian frog, *Rana arathooni,* are located on steep banks above mountain streams, and tadpoles eventually drop into the streams to complete development (R. Brown and Iskandar 2000). Terrestrial eggs are laid in an underground chamber by African burrowing frogs in the genus *Hemisus* (Hemisotidae), sometimes with little or no water (Amiet 1991). After the eggs hatch, the female opens the burrow and constructs a channel through the mud, and the tadpoles follow her through the channel to open water (Wager 1965; Rödel et al. 1995; Spieler 1997; Kaminsky, Linsenmair, and Grafe 1999; Rödel 2000; see also chapter 11).

Many frogs with Mode 18 reproduction are inhabitants of wet, cool montane regions. For example, the three species in the leptodactylid genus *Batrachyla* are found only in the cool mountain forests and moors of southern Chile. Eggs are deposited on rocks or under rocks, logs, and leaf litter close to cold streams and moorland pools. When the tadpoles hatch, they find their way into the streams or pools (Capurro 1958; Cei and Capurro 1958; Cei 1962; Barrio 1967; Busse 1971; Formas and Pugin 1971; Formas 1976). The Philippine frog *Rana magna,* an inhabitant of cool mountain streams, lays its eggs on rock faces and moss in shaded locations, and the tadpoles complete development in streams (Alcala 1962). Duellman and Trueb (1986) classified the *Heleophryne* of the mountains of southern Africa as Mode 2 stream-breeders, but there is some evidence for at least one species, *Heleophryne purcelli,* that eggs are laid on rocks or gravel beside mountain streams and not always directly in the water (Visser 1971; Channing 2001). Two centrolenid frogs, *Centrolene geckoideum* and *C. petrophilum,* from the mountains of Colombia and Ecuador, also attach their eggs to rocks overhanging the water (Lynch, Ruiz, and Rueda 1983; Ruiz-Carranza and Lynch 1991b). Reproductive Mode 18 sometimes grades into arboreal modes of reproduction, such as Modes 24 and 25 (eggs attached to leaves over water). For example, the Kloof frog of southern Africa (*Natalobatrachus bonebergi*), a mountain stream breeder, lays eggs on vertical rock faces, leaves, or branches (fig. 10.8 E), usually about 1 m above the water, but sometimes as much as 4 m high (Wager 1965).

A number of frogs have a reproductive mode similar to those described previously, with eggs being laid on the vertical surfaces of rocks, roots, and other substrates near streams and waterfalls (fig. 10.8 F). However, instead of dropping into the water, the tadpoles of these frogs are adapted to adhering to the rock surface, where they feed on algae and bacteria growing in the thin film of water bathing the rocks (Mode 19). This mode is characteristic of South American leptodactylids in the genera *Thoropa* and *Cycloramphus* (oviposition sites of most *Cycloramphus* have not been reported). A similar reproductive mode is characteristic of African frogs in the genus *Petropedetes* (Lamotte 1954; Lamotte, Perret, and Dzieduszycka 1959; Lamotte and Lescure 1977, 1989b; Heyer and Crombie 1979; Heyer 1983; Wassersug and Heyer 1983; Cocroft and Heyer 1988; Amiet 1989, 1991). Both eggs and tadpoles of a ranid frog, *Amolops cremnobatus,* have been found adhering to rocks near a waterfall in Laos, but it is not known whether the tadpoles complete development on the rock faces (Inger and Kottelat 1998).

#### Nonaquatic Eggs with Larval Transport (Modes 20 and 22)

Another way of removing eggs from the water while retaining an aquatic larval stage is for a parent to carry tadpoles to water once they hatch (Mode 20). In most dendrobatid frogs, either the male or female, depending on the species, carries the tadpoles from a terrestrial oviposition site, either transporting the whole clutch of tadpoles at once or only one or two at a time. The tadpoles are placed in streams or in small bodies of water in tree holes and bromeliads, where they feed on organic matter or conspecific eggs (Weygoldt 1987). The parental behavior of these frogs is discussed in more detail in chapter 11. Males of the Bornean ranid frogs *Limnonectes finchi* and *L. palavanensis* also carry tadpoles on their backs to water (Inger 1966; Inger, Voris, and Frogner 1986; Inger and Stuebing 1997). Free-swimming tadpoles have been found in quiet stream pools, a tree buttress tank, and a water-filled hole in a log.

Frogs with Mode 22 reproduction are similar to those discussed previously, except that tadpoles complete development on the body of the parent and are not always transported far from the nest. This reproductive mode actually combines several distinct modes of tadpoles transport in unrelated clades of frogs. Tadpoles hatched from terrestrial eggs are carried on the back of a parent in *Sooglossus seychellensis,* where they develop without feeding. The three most terrestrial species of *Leiopelma* from New Zealand, *L. hamiltoni, L. archeyi,* and *L. pakeka,* lay eggs in moist terrestrial sites under rocks, logs, or vegetation. These hatch at an advanced stage and remain closely associated with the male or complete development on his back (B. Bell 1978, 1982, 1985; B. Bell and Wassersug 2003). In the myobatrachid frog *Assa darlingtoni,* tadpoles hatching from terrestrial eggs make their way into inguinal brood pouches on the sides of the male, where they complete their development (Ingram, Anstis, and Corben 1975; Ehmann and Swan 1985). Another Mode 22 frog is *Rhinoderma darwinii* from Chile, in which nonfeeding tadpoles complete development in the male's vocal sac (Cei 1962). However, in the only other species in the genus, *R. rufum,* the tadpoles are carried in the vocal sac to water (Mode 20; Formas, Pugin, and Jorquera 1975), suggesting that Mode 22 reproduction is relatively easily derived from Mode 20. All of these are discussed in more detail in chapter 11.

#### Nonaquatic Eggs with Nonfeeding Tadpoles (Mode 21)

Some terrestrial-breeding frogs supply their eggs with enough yolk to support larval development all the way to metamorphosis. Eggs are laid in natural cavities, excavated burrows and nests, or the surfaces of leaves and rock walls. After hatching, tadpoles remain at the nest site and develop without feeding. This mode of reproduction, termed nidicolous (= nest inhabiting) by some authors (Thibaudeau and Altig 1999), has evolved independently in a number of families. The best-studied example of this mode of development is the Tasmanian myobatrachid frog *Crinia nimbus.* Eggs are laid in small cavities in wet moss or lichens, where they undergo a prolonged development that can last up to a year. Hatched tadpoles remain in a thick soup formed by the remains of the jelly that initiatially surrounded the egg mass, apparently without feeding (N. Mitchell and Swain 1996; N. Mitchell and Seymour 2000; N. Mitchell 2002a, b). Although their development is potentially limited by lack of sufficient oxygen, the embryos and tadpoles do not normally experience hypoxic conditions because of the very low temperatures of their nests and extra oxygen supplied by photosynthesis of the plants surrounding the eggs (Mitchell and Seymour, 2003; see also chapter 4).

Terrestrial reproduction with nonfeeding tadpoles has evolved repeatedly in the family Microhylidae. One example is *Synapturanus salseri,* a burrowing microhylid from the Amazonian region of Colombia. The frogs live in burrows just below a matted layer of roots in the forest floor. Eggs are deposited in the burrow. Tadpoles apparently hatch at an advanced stage and complete development in the nest without feeding (metamorphosed froglets still had substantial yolk reserves; Pyburn 1975). Another species in the same genus, *S. rabus,* also has large terrestrial eggs, but the complete mode of development has not been observed (Pyburn 1976). The genus *Myersiella,* thought to be closely related to *Synapturanus* (Nelson and Lescure 1975; Zweifel 1986), was listed by Duellman and Trueb (1986) as having direct development (Mode 23), but in fact, it appears to have a reproductive mode similar to that of *Synapturanus* (Izecksohn et al. 1971). Several Old World microhylids have similar modes of reproduction, including the Asian frog *Calophrynus pleurostigma* and *Madecassophryne truebae* from Madagascar, but their reproduction is poorly studied (Glaw and Vences 1994; Thibaudeau and Altig 1999). Microhylid frogs in the genus *Stumpffia* from Madagascar also are believed to have nonfeeding tadpoles, but these frogs lay eggs in foam nests in leaf litter (Glaw and Vences 1994), so are more appropriately placed in Mode 32.

A burrowing leptodactylid, *Cycloramphus stejnegeri,* from the Atlantic coastal forests of Brazil, has nonfeeding terrestrial larvae. Tadpoles have been seen on the back of a female, but whether the female carries them around or simply remains with them in the nest is unknown (Heyer and Crombie 1979). Terrestrial, nonfeeding tadpoles also have been described in *Zachaenus parvulus* (Lutz 1944), a closely related species (Heyer 1983). Another Brazilian leptodactylid, *Thoropa lutzi,* which is associated with rock walls, also has terrestrial, nonfeeding tadpoles (Bokermann 1965). This mode of reproduction has recently been described in three species of *Colostethus* (Dendrobatidae) from South America (*C. degranvillei*, *C. stepheni*, and *C. nidicola*) and one species from the island of Martinique (*C. chalcopis*; Junca, Altig, and Gascon 1994; Kaiser and Altig 1994; Kaiser, Coloma, and Gray 1994; Caldwell and Lima 2003).

*Altiphrynoides malcolmi*, a bufonid from the highlands of East Africa, lays terrestrial eggs in moist locations at the bases of plants. Tadpoles hatch with large yolk reserves and complete development in the nest (Grandison 1978). Although considered by M. Wake (1980c) to be a form of direct development, it is more similar to Mode 21. Several small African petropedetid frogs also lay terrestrial eggs and have nonfeeding, terrestrial tadpoles. Two of these, *Arthroleptella hewitti* and *A. lightfooti*, lay eggs in or under wet moss near cold mountain streams. The tadpoles remain in wet mud or moss throughout development and apparently do not feed (Wager 1965). Another species, *Phrynodon sandersoni*, is unusual in laying its eggs on the surface of leaves of rainforest plants, where the eggs are attended by the female. The ecology of the tadpoles has not been studied in nature, but the location of the oviposition sites and the morphology of the tadpoles suggest that they develop on land without feeding (Amiet 1981, 1991). *Phrynobatrachus alticola* lays eggs in wet leaf litter. Tadpoles remain in the eggs until just before metamorphosis and then hatch into nonfeeding tadpoles that quickly lose their tails (Rödel and Ernst 2002).

## Nonaquatic Oviposition with Direct Development (Modes 23 and 27)

The most complete form of adaptation to terrestrial life in frogs is direct development of terrestrial eggs (Mode 23). Typically these eggs are very large compared to those of frogs with other reproductive modes and are enclosed in thick capsules, but are not surrounded by a mass of jelly, which would be of little use in retarding evaporative water loss. This mode of reproduction enables frogs to sever all ties to standing water and move into habitats that otherwise might not be available to them. Several major lineages of frogs and a few minor ones have evolved direct development independently, and most of these are associated with wet tropical forests. The most successful group is the leptodactylid genus *Eleutherodactylus* (fig. 10.10 A), the largest genus of vertebrates, with over 500 named species. Direct development also is found in other members of the tribe Eleutherodactylini, including the genera *Adelophryne*, *Euparkerella*, *Ischnocnema*, *Holoaden*, *Phyllonastes*, *Phyzelaphryne*, and *Phrynopus*. Hedges (1989) placed several other genera with direct development in the genus *Eleutherodactylus*, including *Halactophryne*, *Sminthillus*, *Syrrhophus*,

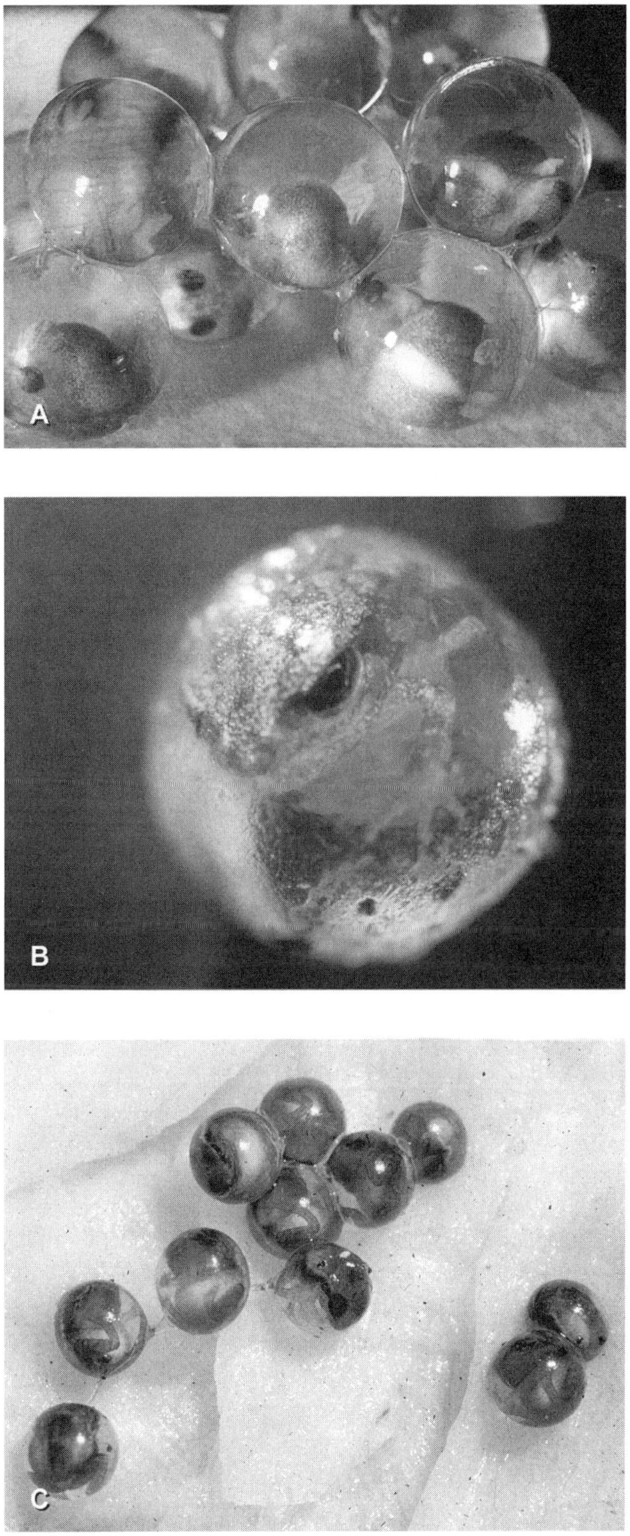


**Fig. 10.10.** Direct development in terrestrial frogs. (A) Developing embryos of *Eleutherodactylus coqui* (Leptodactylidae) from Puerto Rico inside their egg capsules. Light areas on the bellies of the embryos are yolk reserves. (B) Late-stage embryo of *Athroleptis stenodacylus* (Arthroleptidae) from West Africa. (C) Eggs of *Sooglossus gardineri* (Sooglossidae) from the Seychelles Islands, with developing embryos inside. Photos by James Hanken (A), Ulmar Grafe (B), and Wayne Van Devender (C).

and *Tomodactylus,* but some authors have continued to treat them as separate genera.

Species of *Eleutherodactylus* are the dominant terrestrial anurans of most tropical forests in South and Central America and the Caribbean islands, where they inhabit a variety of microhabitats, including leaf litter, piles of debris, the branches and trunks of trees and shrubs, tree holes, bromeliads, and the margins of streams. Most species of *Eleutherodactylus* use terrestrial nest sites (Mode 23), but some use arboreal nests (Mode 27). For example, *E. hedrecki* of Puerto Rico lays eggs in tree holes, sometimes up to 8 meters above the ground (Rivero 1978; Joglar 1998). *Eleutherodactylus coqui* lays eggs mostly in rolled up sierra palm leaf petioles or the curled dead leaves of *Cecropia* trees. Although most nests are on the ground, the frogs actually show a preference for arboreal sites, using suitable elevated nest sites at frequencies much higher than their occurrence in the forest (Townsend 1989a). Other species, such as *E. cochranae* of Puerto Rico, live and lay their eggs in bromeliads (Rivero 1978; Joglar 1998). Some species have elaborate parental care (e.g., *E. coqui:* Townsend, Stewart, and Pough 1984; see also chapter 11), but others simply lay their eggs in moist leaf litter, under logs, or in rock crevices and abandon them. Embryos undergo complete development within the egg capsule and emerge as tiny froglets ready for fully independent terrestrial life (Sampson 1904; Dunn 1926b; Noble 1926b, 1927; Martin del Campo 1940; Lynn 1942; Lynn and Lutz 1946a, b; Goin 1947; Jameson 1950, 1955b; Adamson, Harrison, and Bailey 1960; Chibon 1960; Heatwole 1962a; Salthe and Mecham 1974; Lamotte and Lescure 1977; Lescure 1980; Duellman and Trueb 1986; Hayes-Odum 1990; Bourne 1997, 1998).

The developmental biology of *Eleutherodactylus coqui* has been examined in more detail than that of any other frog with direct development (Elinson 1990; Hanken, Jennings, and Olsson 1997). Thibaudeau and Altig (1999) provided a detailed review of the developmental biology of *Eleutherodactylus* and other direct-developing frogs, and I will give only a brief summary here. This genus has a relatively derived type of direct development. The embryo does not simply undergo larval development inside the egg, but instead exhibits major changes in the timing and patterning of development compared to that of aquatic-breeding frogs. For example, bone appears in the developing embryo, as do paired limbs. The pattern of tissue development is like that of adult frogs, with most larval features having been lost. For example, larval mouthparts do not develop in *Eleutherodactylus,* but there are vestiges of larval mouthparts in some other direct-developing embryos, such as *Arthroleptella* (Thibaudeau and Altig 1999). Most direct-developing embryos lack gills or have only vestigial structures, but some embryos develop a tail that can serve as an accessory respiratory structure (see chapter 4). The embryos of *Eleutherodactylus* also differ from those of aquatic-breeding frogs in the timing of thyroid axis development. The mother provisions the embryos with some thyroid hormone, a pattern seen in some other frogs as well. The thyroid gland becomes active about two-thirds of the way through embryonic development in *E. coqui,* whereas this is delayed until after hatching in aquatic-breeding frogs with tadpoles (Jennings and Hanken 1998). The onset of thyroid activity corresponds with a number of morphological changes similar to those seen in metamorphosis, including regression of the tail, elongation of the limbs, remodeling of the cranium, and development of the hypaxial musculature. Thyroid hormone activity is necessary for these features of development to proceed normally (Callery and Elinson 2000).

The closest ecological equivalents of *Eleutherodactylus* in the Old World are microhylid frogs in two closely related subfamilies, the Genyophryninae (= Sphenophryninae) and the Asterophryninae of northern Australia, New Guinea, and islands in the Australo-Papuan region northward to the southern Philippines. Recent molecular systematic work by David Bickford (2001) suggests that the latter subfamily is not monophyletic, and he advocated placing all New Guinea microhylids in the subfamily Genyophryninae. As far as is presently known, all of the frogs (more than 150 species) in these subfamilies have direct development of eggs into small froglets, and most, if not all species also have parental care (Simon 1983; Bickford 2001; see also chapter 11). Like *Eleutherodactylus* in many parts of the Neotropics, these frogs are the dominant terrestrial anuran group in the Australo-Papuan region, comprising nearly 50% of the frogs of New Guinea (Zweifel and Tyler 1982). No doubt their mode of reproduction is responsible for much of their ecological success, especially in the highlands of New Guinea, where ponds are scarce.

Microhylids from New Guinea have radiated into a wide variety of ecological niches, and they use a correspondingly wide array of oviposition sites (Bickford 2001). Burrowing frogs in the genera *Callulops* place eggs in burrows as much as 140 cm deep, while other burrowing frogs (*Copiula, Xenobatrachus*) use shallower enclosed soil cavities or depressions under logs. Terrestrial frogs such as *Austrochaperina, Sphenophryne, Hylophorbus, Liophryne,* and some species of *Cophixalus,* use shallow cavities in soil, burrows near streams, or sites in leaf litter and detritus. Arboreal frogs, including other species of *Cophixalus, Albericus,* and *Oreophryne,* use cavities in tree ferns, the crotches of trees, or the surfaces of leaves. Unfortunately, little or nothing is known about the reproductive biology of most species in this group (M. Tyler 1963a; Zweifel 1972, 1985; Zweifel and Tyler 1982; W. Brown and Alcala 1983; Hoskin 2004).

Ranid frogs in the genera *Platymantis* and *Ceratobatra-*

*chus* overlap these terrestrial-breeding microhylids geographically, being found in New Guinea and nearby islands, the Bismark Archipelago, the Solomon Islands, the Philippines, and Fiji. They occupy similar habitats and have the same direct development mode of reproduction (W. C. Brown 1952; Zweifel 1969, 1975; W. Brown and Parker 1970; Zweifel and Tyler 1982; Thibaudeau and Altig 1999). There are both terrestrial and arboreal frogs in the genus *Platymantis*, so both Mode 17 and Mode 20 (direct development in an arboreal nest) are represented (W. Brown and Alcala 1983).

Other examples of direct development of terrestrial eggs are scattered in various unrelated groups around the world (Thibaudeau and Altig 1999). In some cases, complete development of the eggs has been observed, but in others, direct development has been inferred from the size and appearance of the eggs and site of egg deposition. Unfortunately, without complete information on egg development, it is difficult to distinguish Mode 23 (direct development into froglets) from Mode 21 (development into nonfeeding tadpoles). The large, unpigmented ovarian eggs of frogs in the small Neotropical family Brachycephalidae suggest that they have direct development (Izecksohn 1971; Duellman and Trueb 1986), and this has been confirmed for one species, *Brachycephalus ephippium*. In this species, eggs are laid in soil under leaf litter and take up to two months to develop (Pombal, Sazima, and Haddad 1994; Pombal 1999). Several genera of Neotropical bufonids (*Oreophrynella, Osornophryne,* and *Rhamphophryne*) have large terrestrial eggs and probably have direct development (Duellman and Trueb 1986), although this has been confirmed only in *Oreophrynella* (McDiarmid and Gorzula 1989). Two obscure monotypic genera of African bufonids, *Laurentophryne* and *Didynamipus,* also have direct development (Grandison 1981).

A number of other Old World frogs have direct development, including *Sooglossus gardinerii* (fig. 10.10 C) from the Seychelles Islands (Duellman and Trueb 1986), members of the rhacophorid genus *Philautus* from the Philippines (Alcala and Brown 1982), *Rhacophorus microtympanum* from Sri Lanka (Kirtisinghe 1946), the African genus *Arthroleptis* (fig. 10.10 B), and African petropedetids in the genera *Anhydrophryne* and *Arthroleptella* (Guibe and Lamotte 1958; Lamotte and Perret 1963b; Wager 1965; Amiet 1989; Thibaudeau and Altig 1999). Several Australian myobatrachids also have direct development, including *Geocrinia rosea* and *G. lutea*, possibly *Metacrinia nichollsi,* and two peculiar burrowing frogs from the western Australian deserts, *Arenophryne rotunda* and *Myobatrachus gouldii.* The two species of *Geocrinia* and *Arenophryne* all deposit their eggs in burrows. The latter is very unusual in that breeding pairs remain together in an underground burrow for many months. Burrows where eggs are deposited are deeper (80 cm) than those of most other terrestrial breeders, no doubt an adaptation to its desert environment (Roberts 1984). *Myobatrachus* is an even more peculiar frog, being highly modified for subterranean life. It has a reproductive mode very similar to that of *Arenophryne,* but it lives in drier habitats and lays its eggs even deeper (more than 1 meter; Roberts 1981).

### Transport of Eggs by a Parent (Modes 9, 15, 16, 34, 35, 36, and 37)

Reproductive Modes 9, 15, 16, 34, 35, 36, and 37 all involve parental transport of eggs that either hatch into tadpoles or undergo complete development in the egg capsule (table 10.2). This type of parental care is discussed in more detail in chapter 11, so only a brief summary will be given here. Undoubtedly the most bizarre form of egg transport is found in gastric-brooding frogs of the genus *Rheobatrachus* from Australia (Mode 9). Eggs are swallowed by the female and develop entirely in her stomach (Corben, Ingram, and Tyler 1974; Tyler and Carter 1981; Tyler 1983; Leong, Tyler, and Shearman 1986). The midwife toads of the genus *Alytes* are the only frogs in which males are known to carry eggs. The eggs are wrapped around the hind legs of the male and carried for up to a month before they hatch into aquatic tadpoles (Mode 34; Boulenger 1912; Heinzmann 1970; Lopez Jurado, Caballero, and Dos-Santos Freitas 1979; Crespo 1982; König and Schlüter 1991; Márquez and Verrell 1991; Márquez 1992, 1993).

In several genera of hemiphractine frogs from South and Central America, eggs are carried on the back of the female, either attached externally (*Stefania, Cryptobatrachus, Hemiphractus*), or in partially or fully enclosed dorsal pouches (*Flectonotus, Fritziana, Gastrotheca;* see fig. 11.4 in chapter 11). Eggs undergo direct development (Mode 37) in *Cryptobatrachus, Stefania, Hemiphractus,* and some *Gastrotheca.* In *Stefania evansi,* a species that carries eggs exposed on the back, the froglets remain on the back of the female even after the egg membranes have ruptured (Gorzula, Morales, and Hernandez 1983; Jungfer and Boehme 1991). In *Flectonotus* and *Fritziana,* tadpoles are placed in water-filled leaf axils or tree holes (Mode 36). Duellman and Gray (1983) reported that these tadpoles develop to metamorphosis without feeding. However, subsequent investigations revealed that tadpoles of at least one species, *Fritziana goeldii,* do feed (Weygoldt 1989b; Weygoldt and Carvalho e Silva 1991). In some species of *Gastrotheca,* eggs hatch into feeding tadpoles that are deposited in ponds (Del Pino 1980, 1989a, b; Duellman and Maness 1980; Del Pino and Escobar 1981; Auber-Thomay and Letellier 1986; Gipouloux 1986; Duellman and Hillis 1987; Duellman 1989a; Juraske and Sinsch 1992).

The only fully aquatic frogs that carry their eggs are the bizarre South American frogs in the genus *Pipa*. In *P. parva,*

*P. carvalhoi,* and *P. myersi,* eggs are carried in cavities on the back of the female and hatch into aquatic tadpoles that complete their development in a pond (Mode 15). In *P. pipa, P. arrabili,* and *P. snethlageae,* eggs hatch into froglets (Mode 16; Rabb and Rabb 1960, 1963a; Weygoldt 1976a, b; Trueb and Cannatella 1986; Duellman 1989a). While the origins of this bizarre mode of reproduction seem mysterious, some aspects of mating and oviposition in *Pipa* appear to have been inherited from ancestors with other modes of oviposition. Using a phylogeny for pipid frogs derived from morphological characters (Cannatella and Trueb 1988a), Duellman (1989a) concluded that the ancestral mode of reproduction in the family Pipidae was the deposition of small aquatic eggs on underwater plants, as seen in all species of *Xenopus* (fig. 10.11). In two genera, *Silurana* and *Hymenochirus*, small eggs are deposited on the water's surface, a process that involves elaborate mating behavior in which the pair turn upside down in the water as each egg is released. This peculiar mating behavior is retained in *Pipa,* but the eggs are pushed into spongy tissue on the female's back by the male after the eggs are fertilized (fig. 10.12).

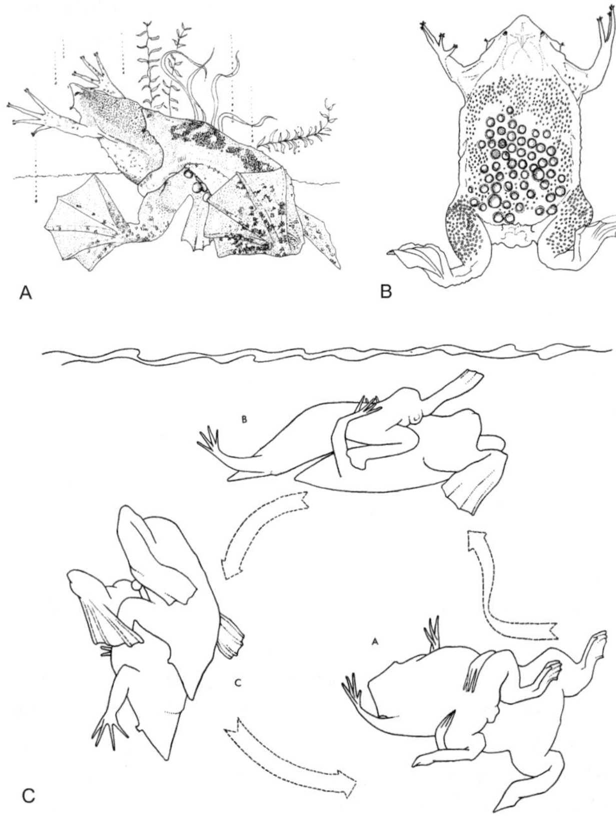


**Fig. 10.12.** Mating and oviposition behavior in *Pipa pipa.* (A) Male and female in amplexus. (B) Female with eggs embedded in the dorsal skin. (C) Midwater turnover that accompanies the release of each egg by the female. The released egg then catches between the male and female and is pressed into the dorsal skin of the female by the male in amplexus. Other pipids exhibit similar midwater turnovers during egglaying, but lack parental care. After Lamotte and Lescure (1977; A, B) and Rabb (1973; C).

### Retention of Eggs in the Oviducts (Modes 38 and 39)

Because of the rarity of internal fertilization in anurans, it is not surprising that only a few species have evolved ovoviviparous (Mode 38) or viviparous (Mode 39) reproduction. Ovoviviparous species retain eggs in the oviducts throughout development, but nutrition to support embryonic development comes entirely from yolk reserves. This reproductive mode has been described in *Eleutherodactylus jasperi,* a bromeliad-dweller from Puerto Rico that unfortunately now appears to be extinct (Joglar 1998). This species has very large eggs typical of members of this genus and a very small clutch size (three to five eggs) that probably is due to its small body size. Development is relatively rapid, taking less than one month. Although eggs are retained in the oviducts, most features of embryonic development are very similar to that of other *Eleutherodactylus* (Drewry and Jones 1976; M. Wake 1978). It is possible that a similar reproductive mode will be discovered in other members of this genus, since the reproductive biology of most species is largely unknown.

All of the other frogs that are known to retain eggs in the oviducts are African bufonids. Traditionally all of these species were placed in a single genus, *Nectophrynoides,* along with *N. osgoodi,* an oviparous species. Dubois (1986) divided the genus into four genera on the basis of differences in reproductive mode. In his arrangement, the name *Nectophrynoides* is used only for ovoviviparous species. Currently five species are recognized—*N. viviparus, N. tornieri, N. cryptus, N. wendyi,* and *N. minutus,* but evidence for monophy of the group is weak (Graybeal and Cannatella

**Fig. 10.11.** Hypothesized phylogenetic relationships of pipid frogs, based on morphological characters. Numbers show evolutionary changes in reproductive modes. 1 = acrobatic turnovers during amplexus; eggs and larvae aquatic. 2 = acrobatic turnovers during amplexus; eggs on back of female, with aquatic larvae. 3 = acrobatic turnovers during amplexus; eggs on back of female, which undergo direct development. After Duellman (1989a).

1995). The name *Nimbaphrynoides* was applied to two viviparous species, *N. occidentalis* and *N. liberiensis*), but other workers have continued to used the name *Nectophrynoides* for these frogs (Lamotte and Sanchez-Lamotte 1999). Dubois (1986) placed *Nectophrynoides osgoodi*, which has aquatic oviposition and free-living tadpoles, in the monotypic genus *Spinophrynoides*. He placed one species, *N. malcolmi*, in the monotypic genus *Altiphrynoides*. These frogs retain eggs in the oviducts for only a short time before being deposited on land in wet areas, where the tadpoles develop without feeding (Grandison 1978; M. Wake 1980c).

As originally conceived by Grandison (1978, 1981), these frogs were a group of closely related frogs that exhibited a graded continuum of reproductive modes, from a presumed ancestral condition of oviparity with feeding tadpoles (*N. osgoodi*) to oviparity with nonfeeding tadpoles (*N. malcomi*) to ovoviviparity and finally viviparity. M. Wake (1980c), on the other hand, suggested that the brief retention of eggs in *N. malcomi* actually is derived from an ovoviviparous ancestor that retained eggs throughout development, but did not provide them with supplemental nutrition. Unfortunately, the phylogeny of bufonids is not fully understood, and molecular data are not available for any of the African frogs with unusual reproductive modes (Graybeal 1997; some of these species are considered in the analysis of Frost et al. 2006). Morphological data suggest that all of the species with unusual reproductive modes belong to one clade, but this group also includes a number of other genera with "normal" oviparous reproduction. Thus, it is possible that the species with unusual reproductive modes are not most closely related to each other (Graybeal and Cannatella 1995; Graybeal 1997). These toads also appear to be relatively basal within the family Bufonidae, even though their reproductive modes appear to be derived.

The most derived reproductive mode appears to be that of *N. occidentalis* and *N. liberiensis*, both of which are viviparous, but it is not clear whether these evolved from oviparous or ovoviviparous ancestors (Graybeal and Cannatella 1995). The tadpoles developing inside the oviducts first use up their yolk reserves, which are less abundant than in the other species, and then feed off of "uterine milk" secretions produced by glands in the oviduct walls (Vilter and Lugand 1959; Lamotte and Xavier 1972; M. Wake 1980c; Xavier 1977, 1986). Eggs of all of the ovoviviparous and viviparous bufonids are smaller than are those of *Eleutherodactylus jasperi*, and clutch sizes are considerably larger, probably a remnant of their aquatic ancestry. Development times are long, due in part to the cold temperatures of their high-altitude habitats (M. Wake 1978, 1980c). For example, *N. occidentalis* females sometimes are pregnant during a prolonged dry season that is spent underground, and development of the young can take up to nine months (Angel and Lamotte 1944, 1948; Lamotte and Xavier 1972; Lamotte and Sanchez-Lamotte 1999).

## Reproductive Modes of Urodeles

The reproductive modes of urodeles are much less diverse than those of anurans, with only 10 of the 39 modes of Haddad and Prado (2005) represented (table 10.4). In part, this reflects the much lower species diversity of urodeles, with

**Table 10.4** Modes of egg deposition and development in urodeles

| Egg deposition site | D&T mode | Salthe mode | Larval development | Selected examples |
|---|---|---|---|---|
| *Aquatic eggs* | | | | |
| 1. Still water | 1 | I | Feeding in ponds | *Siren, Ambystoma, Triturus* |
| 2. Flowing water | 2 | II | Feeding in streams | *Dicamptodon, Eurycea, Cryptobranchus, Necturus* |
| *Nonaquatic eggs* | | | | |
| 17. Terrestrial nest | 12 | III | Feeding in ponds | *Ambystoma opacum, Ambystoma cingulatum* |
| 18. Terrestrial nest | 13 | III | Larvae move to water | *Hemidactylium*, some *Desmognathus* |
| 21. Terrestrial nest | 15 | III | Nonfeeding in nest | *Desmognathus aeneus* |
| 23. Terrestrial nest | 17 | III | Direct development | *Desmognathus wrighti, Plethodon, Batrachoseps* |
| 27. Arboreal nest | 20 | III | Direct development | *Aneides lugubris*, many Bolitoglossini |
| 38. Eggs retained in oviducts | 28 | | Nutrition provided by yolk | *Salamandra salamandra, Mertensiella caucasica* |
| 39. Eggs retained in oviducts | 29 | | Nutrition provided by mother | *Salamandra atra, Mertensiella luschani* |

*Sources:* D&T mode indicates corresponding anuran reproductive modes from Duellman and Trueb (1986). Salthe mode indicates urodele reproductive modes of Salthe (1969).

*Note:* Modes are numbered as in table 10.2, but modes not found in urodeles are omitted.

only about 560 species, compared to nearly 5,300 species of anurans. Some types of reproduction, such as foam nesting, are completely absent. In addition, there are fewer arboreal species of salamanders, and they lack the diversity of reproductive modes found in arboreal frogs. All arboreal salamanders are in a single subfamily, the Plethodontinae, and most are in one tribe, the Bolitoglossini, which has undergone a large radiation in Mexico and Central America (D. Wake and Lynch 1976; D. Wake 1987; D. Wake and Elias 1983; García-Paris et al. 2000). The entire tribe is characterized by direct development of eggs laid out of water, either on land (Mode 23) or in arboreal nest sites, such as bromeliads (Mode 27). Hence, the great variety of reproductive modes seen in arboreal frogs, including hanging eggs on vegetation over water (Modes 24, 25, 26, 33) or placing them in tree holes, leaf axils, or bromeliads to hatch into feeding or nonfeeding larvae (Modes 6, 8) are simply absent in urodeles. Also not represented is any reproductive mode involving transport of eggs (Modes 9, 15, 16, 34–37) or larvae (Mode 20) on the body of a parent.

### Oviposition in Still Water (Mode 1)

Oviposition in still water, such as ponds, swamps, and stream pools (Mode 1 in table 10.4; Mode I of Salthe 1969) is characteristic of many hynobiids, ambystomatids, and salamandrids, all sirenids, and a few species of plethodontids. Eggs of pond-breeding hynobiids typically are laid in paired egg sacs resembling the jelly clumps of some ambystomatid salamanders. These are laid on the pond bottom or attached to submerged logs, branches, and plant stems (Kunitomo 1910; Sasaki 1924; Thorn 1962, 1963, 1967; Kusano 1980; Saito 1987; Takahashi and Iwasawa 1988a, b, 1989b; Sato 1989; Tanaka 1989). Most pond-breeding species of *Ambystoma* have relatively short breeding seasons. Eggs are laid soon after mating, often as a single clump or several clumps attached to submerged branches and plant stems (fig. 10.13 A). Species with this mode of oviposition include *Ambystoma maculatum, A. jeffersonianum, A. texanum, A. gracile, A. annulatum* and eastern populations of *A. tigrinum* (Noble and Marshall 1929; Henry and Twitty 1940; S. Bishop 1941b, 1947; Hamilton 1948; Stebbins 1954a, 1985; Knudsen 1960; P. W. Smith 1961; Uzzell 1964; Spotila and Beumer 1970; Eagleson 1976; Petranka 1982b; Hutcherson, Peterson, and Wilkinson 1989; Kraus and Petranka 1989). As in anurans, deposition of eggs in a compact clump under water is characteristic mainly of species that breed in relatively cold water and possibly protects eggs from freezing or predators (Salthe and Mecham 1974; Ward and Sexton 1981; Morin 1983b). Thick jelly also provides some protection to eggs that are temporarily stranded in drying ponds (Marco and Blaustein 1998; Greven 2003a).

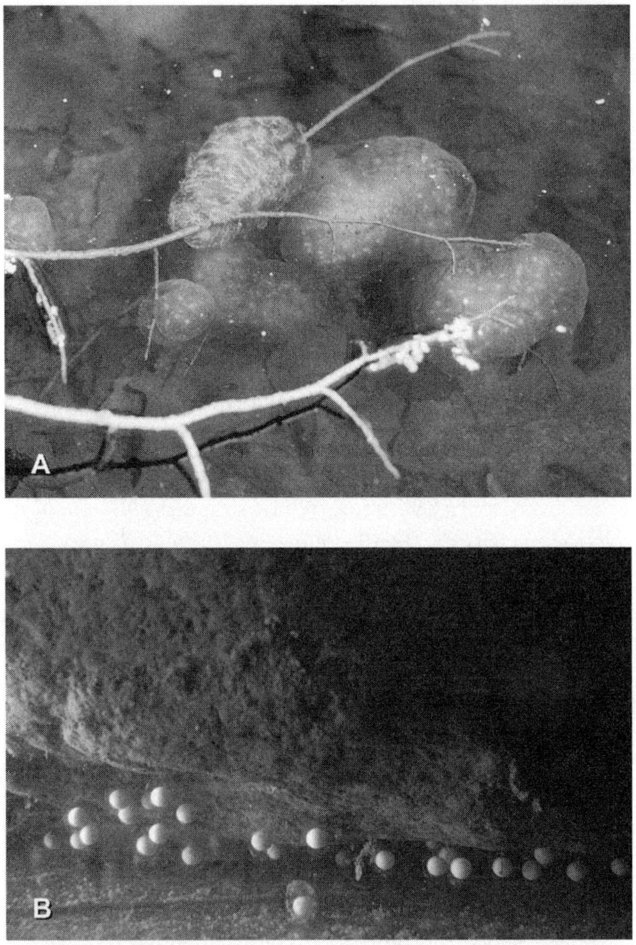


**Fig. 10.13.** Representative aquatic salamander eggs. (A) Eggs of *Ambystoma maculatum,* a pond-breeding salamander. Many ambystomatids produce egg masses with a thick jelly coat, usually attached to branches or other objects in the water. (B) Eggs of *Eurycea bislineata* attached to a rock. Each egg is individually attached by a short stalk. The lack of pigment is characteristic of eggs laid in hidden locations where the eggs are not exposed to direct sunlight. Photos by Kentwood D. Wells.

Other ambystomatids scatter individual eggs on the bottom of a pond or attach them to grass stems and other substrates. Species with this type of oviposition include *A. mabeei, A. laterale,* and some hybrid combinations of *A. laterale, A. jeffersonianum, A. texanum,* and *A. talpoideum.* In western North America, *A. californiense,* some populations of *A. tigrinum,* and low-altitude populations of *A. macrodactylum* also lay eggs individually (Bishop 1941b, 1947; Stebbins 1954a, 1985; H. M. Smith 1956; Uzzell 1964; J. Anderson 1967; J. Hardy 1969; Bogart et al. 1987). High-altitude populations of *A. macrodactylum* lay eggs attached in clusters to the undersides of logs, rocks, and other objects in relatively deep water, a possible adaptation to avoid late-spring freezes (J. Anderson 1967). Populations in Oregon tend to lay eggs mostly in small clumps of 5–25 eggs (Kezer and Farner 1955; Ferguson 1961; J. Anderson 1967; Walls et al. 1993). Variation in egg-laying behavior has been reported in other species as well. For example, *A. talpoideum* has been reported laying eggs either in small clumps of up to 40 eggs (Bishop 1947; Mosimann and Uzzell 1952; Shoop 1960) or scattered individually on the pond bottom (Semlitsch 1985b). Some females of *A. texanum* also lay eggs singly, even though oviposition in clumps is more common (Petranka 1982b; Kraus and Petranka 1989). Some of this variation can be attributed to differences in the availability of suitable objects such as branches to which egg masses can be attached, variation in movements of females during egg deposition, and amount of disturbance of egg-laying females by other salamanders (Bragg 1957b; Spotila and Beumer 1970). Scattering of eggs also appears to be more common in species breeding in warm water, where compact egg masses might become hypoxic.

Most pond-breeding salamandrids have relatively long breeding seasons; individual females sometimes lay eggs over a period of days or weeks, a habit that allows them to disperse their eggs in both time and space. Two species of newts from western North America, *Taricha torosa* and *T. granulosa,* breed in ponds and slow-moving streams. *Taricha torosa* lays eggs in spherical jelly clumps of about 15–20 eggs, whereas *T. granulosa* scatters individual eggs on the bottom of a pond (Twitty 1942, 1966; Stebbins 1954a; Brame 1968; Marshall, Doyle, and Kaplan 1990). The European salamandrid *Pleurodeles waltl* lays eggs either singly or in small clumps attached to plants and stones (Steward 1969). One member of the genus *Neurergus,* found in the Middle East, breeds in ponds and stagnant stream pools. The other members of the genus are all stream breeders, so pond breeding appears to be derived in this group (Steinfartz et al. 2002). Most European newts of the genus *Triturus* breed in ponds, although some, such as *T. boscai,* sometimes breed in streams as well (Orizaola and Braña 2003a).

Many pond-breeding salamandrids, including eastern North American newts (*Notophthalmus*) and Eurasian newts (*Triturus, Paramesotriton, Cynops*) attach eggs to underwater plants. They either wrap leaves around individual eggs, or place rows of eggs between two grass blades (P. H. Pope 1924; Tsutsui 1931; S. Bishop 1941b, 1947; Romer 1951; Steward 1969; Salthe and Mecham 1974; G. Bell and Lawton 1975; Sparreboom 1984c; Diaz-Paniagua 1986, 1989b; Miaud 1995; Orizaola and Braña 2003a; Norris and Hosie 2005). Experiments with *Triturus* showed that wrapped eggs enjoyed higher survivorship than unwrapped eggs. There was a reduction in predation by female newts and by one species of water beetle (*Acilius sulcatus*), but another water beetle, *Dytiscus marginalis,* fed heavily on both wrapped and unwrapped eggs (Miaud 1993, 1994). Other experiments with dragonfly naids (*Aeshna cyanea*) also showed that wrapping protects eggs from predation (Orizaola and Braña 2003a).

All sirenids breed in still or slow-flowing water and typically inhabit heavily vegetated, swampy ponds and ditches. The eggs of *Pseudobranchus* are attached individually to leaves, stems, and roots of underwater plants (Noble and Marshall 1932; Carr 1940; S. Bishop 1941b), whereas those of *Siren* are deposited in clumps in tangled mats of roots and other vegetation (Noble and Marshall 1932; Godley 1983). A few species of plethodontid salamanders also are found mainly in swampy habitats. One example is *Eurycea quadridigitata,* which is unusual for members of this genus, because it migrates seasonally to ponds and swamps to breed (Semlitsch and McMillan 1980). Eggs have been found attached to the undersides of a submerged log in a pond (Carr 1940), among leaves and debris in seepage areas (Brimley 1923), and on wet logs several centimeters above the water (Goin 1951). Hence, both reproductive Mode 1 and Mode 18 have been observed in this species. The same is true for *Stereochilus marginatus,* another inhabitant of swamps and ditches. Noble and Richards (1932) reported that eggs of this species are attached to logs, bark, or other objects in the water (Mode 1). Both A. Schwartz and Etheridge (1954) and Rabb (1956) found eggs under logs at the edges of ponds and above the water line (Mode 18), and J. Wood and Rageot (1963) found eggs both in the water and above the water line. Another plethodontid, *Pseudotriton montanus,* is found mainly in swampy bottomlands, sluggish streams, and ponds at low elevations in the southeastern United States (Bruce 1975). Eggs have been found attached to rootlets, dead leaves, and other objects in muddy seepage areas (Fowler 1946; S. Bishop 1947).

### Oviposition in Flowing Water (Mode 2)

Most other salamanders with aquatic eggs breed in some sort of flowing water, from mountain torrents to large rivers (Mode 2 in table 10.4; Mode II of Salthe 1969). These include all cryptobranchids, proteids, dicamptodontids, and rhyacotritonids, and some hynobiids, ambystomatids, salamandrids, and plethodontids. In contrast to many pond-breeding urodeles that lay their eggs in exposed locations (some newts being an exception), most Mode 2 species hide their eggs, usually under rocks (Salthe 1969). Cryptobranchids lay eggs in rosary-like strings in cavities under rocks in rivers and large streams (Kerbert 1904; B. G. Smith 1907; Tago 1929; Bishop 1941b; Nickerson and Mays 1973b). Eggs of stream-breeding hynobiids are similar to those of pond-breeding species, being laid in large jelly sacs attached to rocks or lying in quiet stream pools (Oyama 1929; Liu 1950; Sparreboom 1979; Steward 1969; Akita 1989; Griffin and Solkin 1995; Serbinova and Solkin 1995).

Other large-bodied stream-breeding salamanders typically attach individual eggs in clusters to the undersides of flat stones, logs, branches, or roots. Species with this type of oviposition behavior include *Necturus, Proteus, Ambystoma barbouri, A. ordinarium, A. rosaceum, A. rivularis,* and several other species of *Ambystoma* from the mountains of Mexico (S. Bishop 1926, 1941b, 1947; K. Fitch 1959; Briegleb 1962; H. Campbell and Simmons 1962; Shoop 1965; J. D. Anderson and Worthington 1971; Brandon and Altig 1973; J. D. Anderson and Webb 1978; Ashton and Braswell 1979; Durand 1986; Kraus and Petranka 1989; Sih and Maurer 1992). Some small stream-breeders also attach their eggs in clusters to the undersides of rocks, roots, or dead vegetation. These include *Eurycea* (fig. 10.13 B), *Gyrinophilus, Pseudotriton, Desmognathus marmoratus, Desmognathus quadramaculatus,* and *Taricha rivularis* (C. H. Pope 1924; Wilder 1924; Bishop 1941b, 1947; Twitty 1942, 1966; Mohr 1943; J. Wood 1950, 1953a; J. Wood and McCutcheon 1954; Organ 1961a; Franz 1964; Spotila and Ireland 1970; Ireland 1974, 1976).

Other stream-breeding salamanders lay their eggs in even more cryptic locations. For example, some Old World salamandrids, including *Paramesotriton caudopunctatus, Euproctus,* and *Calotriton,* attach their eggs individually to the undersides of rocks or place them in crevices, inserting them with a protruding cloaca that the female develops in the breeding season (Sparreboom 1983; Thiesmeier and Hornberg 1986, 1990). Two genera of salamanders found only in streams in the Pacific Northwest region of North America, *Dicamptodon* and *Rhyacotriton,* often lay eggs in holes and crevices well underground in springs and seepage areas, often several meters from the main streams where the larvae complete their development (Nussbaum 1969a, b; Nussbaum and Tait 1977). Some species of *Rhyacotriton* also lay eggs under rocks in the main channels of streams (Karraker 1999; MacCracken 2002; Russell et al. 2002). Small *Dicampton* larvae are very cryptic, sometimes living in the spaces between rocks 30 centimeters or more below the beds of streams (Feral, Camann, and Welsh 2005). In at least some populations of the Japanese hynobiid salamander *Hynobius naevius,* large yolky eggs are laid in underground streams, where larvae develop without feeding (Tanabe 2002, cited in Tominaga, Nishikawa, and Sato 2003).

The antipredator advantages of cryptic oviposition in stream-breeding salamanders seem clear (Nussbaum 1985), but other factors, such as current velocity or silt formation, also might be important. In experimental studies with stream-breeding populations of *Ambystoma barbouri,* Sih and Maurer (1992) exposed some clusters of eggs by overturning rocks and left others in cryptic locations on the undersides of rocks. About 93% of the hidden eggs survived to hatching, whereas almost 98% of the exposed eggs died, mostly because of desiccation or being washed away in the current. Accumulation of silt on exposed egg clutches did not appear to have much effect on survivorship, and the effects of predation were not considered.

### Nonaquatic Oviposition with Aquatic Larvae (Modes 17, 18, and 21)

Salamanders in several families have independently evolved nonaquatic modes of oviposition, although some retain an aquatic larval stage. Presumably, the selective pressures favoring this shift are the same as those already discussed for anurans. Salthe (1969) had difficulty placing species with terrestrial eggs and aquatic larvae in his scheme of reproductive modes, but generally assigned them to either Mode I (pond-breeders) or Mode II (stream-breeders) on the basis of larval adaptations, regardless of where their eggs are laid. Several urodeles lay eggs in nests that are later flooded, allowing larvae to complete their development in the water (Mode 17). These include *Ambystoma opacum, A. cingulatum,* and probably all three species of *Amphiuma.* The two ambystomatids resemble Australian frogs in the genus *Pseudophryne* in that eggs are laid during the fall in low-lying areas that will later be flooded by winter rains. *Ambystoma opacum* females remain with the eggs until they are flooded (Noble and Brady 1933; Petranka and Petranka 1980; M. Jackson, Scott, and Estes 1989), whereas *A. cingulatum* females scatter their eggs under logs, leaf litter, sphagnum mats, or on the bare ground and desert them (J. Anderson and Williamson 1976). In *Amphiuma,* eggs are laid in depressions in dried-up swamps, holes under logs near ponds, and similar locations that are flooded by spring and summer rains. The female apparently stays with the eggs until they are flooded, sometimes for several months (J. A. Weber 1944; Bishop 1947; C. Baker 1945; Fontenot 1999). Larvae hatch at an advanced stage of development once the eggs are flooded, and retain their gills for only about two weeks before undergoing metamorphosis (Ultsch and Arceneaux 1988; Gunzburger 2001, 2003).

A somewhat similar mode of reproduction (Mode 18) is found in several species that lay their eggs just above the water line or at the edges of ponds or streams. The larvae hatch in the terrestrial nest and wriggle or flip themselves into the water to complete development. In addition to *Stereochilus marginatus* and *Eurycea quadridigitata,* which appear to exhibit both Mode 1 and Mode 18 reproduction (see the previous discussion), this mode is found in *Hemidactylium scutatum,* which lays eggs in sphagnum moss over or adjacent to standing water (see fig. 11.12 C in chapter 11). Larvae hatch at a relatively early stage (with only the front legs well developed) and drop into the water to complete development (Blanchard 1933b, 1936; Bishop 1941b; P. Gilbert 1941; J. Wood 1953a; R. Harris and Gill 1980).

Mode 18 reproduction also is found in *Echinotriton andersoni,* a salamandrid from the Ryukyu Islands of Japan. Eggs are laid under leaf litter at the edges of ponds, puddles, and springs. The larvae creep about in the mud until they are able to make their way into the water, where they undergo typical aquatic larval development (Utsunomiya and Utsunomiya 1977; Utsunomiya, Utsunomiya, and Kawachi 1978; Nussbaum and Brodie 1982). In the only other member of this genus, *E. chinhaiensis* from China, eggs are laid in moist detritus up to 180 cm from the edges of ponds. The larvae apparently complete development on land, although it is possible that some larvae make their way into the ponds or are washed into the water by rain (Xie et al. 2000).

This mode of reproduction also is characteristic of species of *Desmognathus* that lay their eggs in seepage areas or cavities under rocks, moss, roots, or leaf litter at the edges of streams and swamps. These include *D. fuscus, D. ochrophaeus, D. ocoee, D. orestes, D. monticola,* and several other species. The larvae hatch at a relatively advanced stage (all four legs well developed) and remain with the mother in the nest for several days before eventually moving to water (Bishop 1941b, 1947; Hairston 1949; Organ 1961c; Petranka 1998). Another species, *D. aeneus,* is even less dependent on standing water, and lays eggs that hatch into advanced larvae that complete development in the nest without feeding or moving to water (Mode 15; Valentine 1963; J. R. Harrison 1967).

### Nonaquatic Oviposition with Direct Development (Modes 23 and 27)

Complete emancipation from standing water has been achieved by salamanders that lay nonaquatic eggs that undergo direct development within the egg capsule (Mode III of Salthe 1969). Direct development occurs only in plethodontids. It has been reported in one desmognathine species (*Desmognathus wrighti;* Organ 1961b) and is presumed to occur in a fossorial desmognathine, *Phaeognathus hubrichti,* although complete egg development has not been observed in this species (Brandon 1965a; Petranka 1998). All of the other salamanders with direct development are members of the plethodontine tribes Plethodontini and Bolitoglossini, and all members of these tribes have this mode of reproduction. This probably accounts for their ecological success in invading a variety of habitats, from relatively dry regions to tropical cloud forests (Wake and Hanken 1996). Indeed, more than 300 of the approximately 377 species of plethodontid salamanders are members of these two tribes, and the proportion is sure to increase as additional tropical species are described.

All salamanders with direct development lay large, unpigmented eggs in hidden locations. The large size of eggs, along with a very large genome size, especially in bolitoglossines (Jockusch 1997), results in prolonged periods of embryological development and probably favors the evolution of parental care (see chapter 11). Females remain with the eggs in most species, but some species of *Batrachoseps* and some Neotropical bolitoglossines apparently lack parental care. In terrestrial breeders (Mode 23), eggs usually are laid in small clusters in underground cavities, rock crevices, or inside hollow logs (Bishop 1941b, 1947; Kessel and Kessel 1942; L. Miller 1944; Stebbins 1949, 1954a; J. Davis 1952; R. Gordon 1952; Dumas 1955; A. Schwartz 1955; Highton 1956, 1962; Johnston and Schad 1959; Livezey 1959; Organ 1960b; McDiarmid and Worthington 1970; Houck 1977a, b).

Arboreal species (Mode 27), which are found mostly in Mexico and Central America, deposit their eggs under mats of mosses and liverworts on tree limbs, in bromeliads, and other cryptic sites, but the oviposition sites of most tropical salamanders remain unknown (Robinson 1976; Houck 1977b; D. Wake 1987). Some species, such as *Nototriton picadoi,* deposit eggs in mats of moss in both terrestrial and arboreal sites (Bruce 1998). Clutches of *N. barbouri* eggs have even been found in abandoned hummingbird nests, which were lined with moss like that normally used as oviposition sites by these salamanders (McCranie and Wilson 1992). At least some species of *Nototriton* and perhaps some other Neotropical salamanders appear to lack parental care (Bruce 1998). *Aneides lugubris,* an arboreal salamander from western North America, lays eggs in terrestrial sites, decaying logs, or hollow trees as much as 10 meters above the ground (Bishop 1947; L. Miller 1944; Stebbins 1954a), while the eggs of *A. vagrans* have been found in hollow spaces in fallen logs (T. Davis 2003).

It is not entirely clear how many times plethodontid salamanders have evolved reproductive modes that lack an aquatic larval stage. It has long been assumed that aquatic reproduction represents the ancestral condition for all plethodontids (Dunn 1926c). Most plethodontids with fully aquatic reproduction breed in streams, and it has been proposed that the relatively large eggs of stream breeders is a prerequisite for the eventual evolution of fully terrestrial reproduction (Collazo and Marks 1994; D. Wake and Hanken 1996). Nevertheless, phylogenetic analyses suggest that nonaquatic reproduction could have evolved more than once in the family, and indeed, there could have been some reversals from direct development back to reproductive modes with an aquatic larval stage.

The subfamily Desmognathinae is especially problematic. Traditionally, most authors have followed Dunn (1926c) and Organ (1961c) in assuming that aquatic oviposition and aquatic larvae, as exhibited by *Desmognathus marmoratus* and *D. quadramaculatus,* represent the ancestral condition for the entire subfamily. Most members of the genus *Desmognathus* lay eggs in wet seeps or along the sides of streams, but retain aquatic larvae. This has been considered the next step toward fully terrestrial reproduction. Nonaquatic oviposition and loss of aquatic larvae would then represent the most derived condition, perhaps having evolved independently in two species of *Desmognathus* (*D. wrighti* and *D. aeneus*) and in *Phaeognathus hubrichti*.

Molecular phylogenetic studies of the subfamily by Titus and Larson (1996) and Chippindale et al. (2004) suggested that this view of the evolution of reproductive modes is incorrect. Titus and Larson (1996) found that *Phaeognathus* is a basal desmognathine, and both *D. wrighti* and *D. aeneus*, the two species in the genus *Desmognathus* that lack aquatic larvae, are basal species within that genus, not the most derived species as often assumed (see also Kozak et al. 2005). One implication of this phylogeny is that nonaquatic reproduction is ancestral for the entire desmognathine clade, with metamorphosis and aquatic larvae having re-evolved secondarily. The analysis of Chippindale et al. (2004) also supported the idea of an aquatic life history having re-evolved in *Desmognathus*, because the entire desmognathine clade appears to have been derived from within a clade of direct-developing plethodontines (fig. 10.14). Traditionally the tribes Plethodontini and Bolitoglossini have been considered closely related, with a single origin of direct development (D. Wake and Hanken 1996), but the analysis of Chippindale et al. (2004) suggests direct development could have evolved independently in these two clades. An alternative is that a reversal to an aquatic life history occurred more than once, with most of the hemidactyline salamanders having reverted to this type of life history from a direct-developing ancestor.

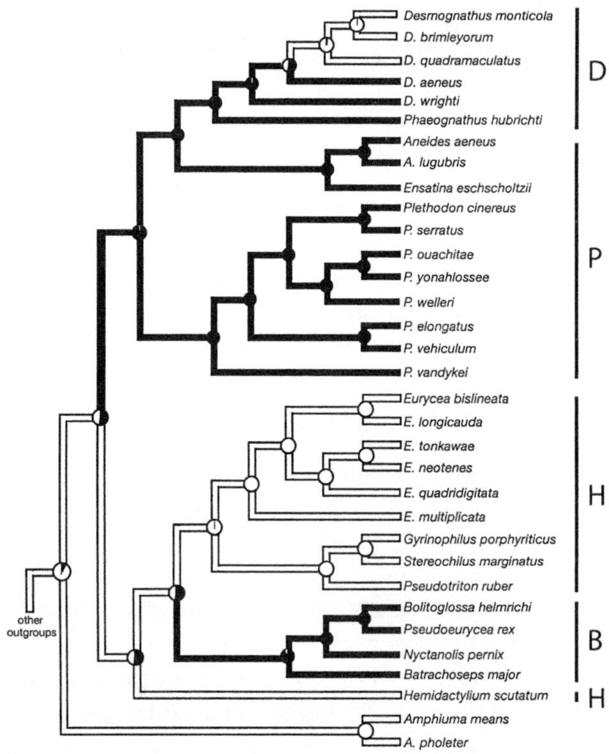


**Fig. 10.14.** Phylogeny of plethodontid salamanders, based on the single most parsimonious tree dervived from both nonmolecular characters and mitochondrial and nuclear DNA sequences. Branch shading shows modes of development (light = free-living aquatic larvae; dark = direct development). This phylogeny suggests the reevolution of an aquatic life history in desmognathines, which appear to be derived from within a clade of direct-developing salamanders. Abbreviations indicate traditional subfamilies and tribes, some of which appear not to be monophyletic. D = Subfamily Desmognathinae; P = Tribe Plethodontini; H = Tribe Hemidactyliini; B = Tribe Bolitoglossini. From Chippindale et al. (2004).

Additional phylogenetic analyses and comparative studies of embryological development are needed to resolve the question of how many times direct development and aquatic larval development have evolved in the plethodontids. There are major differences in patterns of development in the Desmognathinae, Plethodontini, and Bolitoglossini that suggest the possibility of independent origins of direct development in these three clades. In general, desmognathines show much greater retention of some features of larval development in direct-developing embryos than do the other two clades (Marks and Collazo 1998). The bolitoglossines exhibit the most fully derived pattern of development and the greatest loss of larval structures, including the hyobranchial apparatus (D. Wake and Hanken 1996).

The loss of larval structures has, in turn, led to extensive repatterning of development. For example, the loss of larval hyobranchial structures involved in larval feeding in bolitoglossines has resulted in extensive repatterning of the entire hyobranchial skeleton and the precocious development of the tongue well before hatching occurs. It also has allowed for the evolution of a freely projectile tongue and increased reliance on the tongue for prey capture (Lombard and Wake 1977; D. Wake 1982; see also chapter 1). There have been major changes in patterns of development of the nervous and sensory systems as well, and these changes are associated with the evolution of enhanced visual acuity and changes in motor performance associated with prey capture (G. Roth and Wake 1985b; D. Wake and Hanken 1996). Finally, there have been changes in patterns of limb development associated with loss of free-living larvae. Specifically, direct-developing salamanders exhibit accelerated hind limb development, with hind limb buds appearing at about the same time as forelimb buds (D. Wake and Marks 1993; D. Wake and Shubin 1994; D. Wake and Hanken 1996).

### Retention of Embryos in the Oviducts (Modes 38 and 39)

Despite the nearly universal occurrence of internal fertilization in urodeles, only a few species have evolved reproductive modes in which embryos are retained in the lower part of the oviduct (uterus) for most or all of development. Indeed, all of these species are members of a single clade consisting of two genera of terrestrial salamandrids, *Salamandra* and *Mertensiella* (Griffiths 1996; Greven 2003b). Nevertheless, ovoviviparous and viviparous reproduction apparently has evolved independently a number of times within this clade, because there is considerable variation in reproductive modes even within a single species. The distinction between ovoviviparous and viviparous reproduction becomes somewhat blurred in these salamanders, because the two modes tend to grade into one another because of the unusual number of ways in which developing embryos receive nutrients.

*Mertensiella caucasica* exhibits the least specialized mode of reproduction because it retains the oviparous habits of most salamanders. Eggs are deposited in sheltered locations in streams and hatch into larvae that complete development in the water (Muskhelishvili 1964; Schultschik 1994). The other species traditionally placed in this genus, *M. luschani,* is considered by recent authors to be a member of the genus *Salamandra* (Weisrock et al. 2001). This species is viviparous, so placement of this species in *Salamandra* means that viviparity occurs only in that genus. Only two fertilized eggs develop inside the female, one in each oviduct, where the embryos remain for 5–12 months before emerging as fully metamorphosed young (Özeti 1973, 1979). The mode of nutrient provisioning has not been studied in detail, but the developing young presumably rely mainly on yolk reserves early in development. Later, the larvae are known to feed on the remains of unfertilized eggs in the oviducts (Guex 1994).

The genus *Salamandra* currently includes six species, *S. salamandra, S. atra, S. lanzai, S. algira, S. corsica,* and *S. luschani,* but there is considerable debate about the taxonomic status of many described subspecies (Griffiths 1996). *Salamandra salamandra* exhibits considerable geographic variation in size, color pattern, and mode of reproduction. Over much of its range, this species is ovoviviparous. Females retain relatively large numbers of developing embryos in an enlarged portion of the oviduct (the uterus) that serves as little more than a brood chamber. The embryos are surrounded by jelly like that of aquatic-breeding salamanders and derive all of their nutrition from yolk reserves. Once these reserves are exhausted, the larvae are born and complete their development in a pond or stream (Joly 1968, 1971, 1986; Greven and Guex 1994; Joly, Chesnel, and Boujard 1994; Warburg 1994; Degani and Warburg 1995; Greven 1998, 2003b; Dopazo and Korenblum 2000; Rebelo and Leclair 2003a; Weitere et al. 2004).

Some populations of *S. salamandra* have evolved a more derived form of viviparous reproduction. Females retain much smaller numbers of embryos in the uterus and give birth to nearly or fully metamorphosed young (Joly 1968, 1971, 1986; Fachbach 1969, 1976; Greven 1976, 1977, 1998, 2003b; Depazo and Korenblum 2000). As in ovoviviparous populations of this species, the developing embryos initially depend on yolk reserves, but after these are exhausted, the larvae hatch and remain in the uterus. There they feed on the remains of unfertilized eggs, as in *S. luschani,* or on smaller siblings (Joly 1968, 1986; Fachbach 1969, 1976; Dopazo and Alberch 1994; Greven 1998, 2003b). Some authors have argued that viviparity in this species is found mostly in high-altitude populations, and the production of a few large young is thought to be an adaptation to relatively harsh terrestrial environments (Gasser and Joly 1972; Joly, Chesnel, and Boujard 1994; Veith et al. 1998). In Spain, however, both modes of reproduction occur in populations of *Salamandra salamandra* that are genetically similar and geographically close to one another, and viviparity does not appear to be associated with altitude (Alcobendas, Dopazo, and Alberch 1996; Dopazo, Boto, and Alberch 1998; Dopazo and Korenblum 2000; García-París et al. 2003).

*Salamandra atra* and *S. lanzai* are viviparous and give birth to fully metamorphosed terrestrial juveniles. Both of these species are inhabitants of cool montane habitats and have an extraordinarily long gestation period (up to five years in some populations). In *S. atra,* only a single egg is fertilized in each oviduct. The remaining unfertilized eggs break apart and are fed upon by the two developing larvae. Eventually, additional nutrition is provided by secretions of the oviduct walls. The larvae use specialized teeth to scrape material from the epithelium of the uterus (Vilter and Vilter 1960, 1962, 1964; Häfeli 1971; Guex and Chen 1986; Vilter 1986; Guex 1994; Guex and Greven 1994; Greven 1998, 2003b).

## Reproductive Modes of Caecilians

The reproductive biology of caecilians is poorly understood because most species are fossorial and seldom encountered in the field. The basic reproductive mode of many species is unknown. Most information on caecilian reproduction comes from studies of captive individuals, and the eggs, larvae, and juveniles of most species have never been seen in the field. However, there is enough information to suggest general distributions of reproductive modes in the group (M. Wake 1977b, 1992; Wilkinson and Nussbaum 1998). The Rhinatrematidae of South America are generally considered the most basal living caecilians (Duellman and Trueb 1986; Nussbaum and Wilkinson 1989; Frost et al. 2006), and all appear to be oviparous. Free-living larvae of several species in the genus *Epicrionops* have been found in streams, and *Rhinatrema* is assumed to have aquatic larvae as well. However, oviposition sites have not been reported for any members of this family. The family Ichthyophiidae of southeastern Asia is also considered to be a relatively basal group and appears to be exclusively oviparous with aquatic larvae. Eggs are not laid in the water, but most commonly are placed in burrows or under mats of vegetation at the margins of bodies of water, where they are attended by the female. Larvae eventually make their way to water after hatching (Sarasin and Sarasin 1887–1890; M. Wake 1977b; Kupfer, Nabhitabhata, and Himstedt 2004). Hence, reproduction in these species most closely resembles reproductive Mode 18. The largest family of caecilians is the Caeciliidae, a largely terrestrial and fossorial group. This family exhibits the widest range of reproductive modes, with both oviparous and viviparous species. Larval stages have been reported for a few species in this family, but many others apparently lay terrestrial eggs that undergo direct development (Mode 23). Genera known or assumed to have direct development include *Idiocranium, Grandisonia, Hypogeophis, Siphonops* (Goeldi 1899; Gans 1961; M. Wake 1977b, 1989b; Delsol 1986), and at least some species of *Caecilia* (Funk et al. 2004).

Viviparity (Mode 39) apparently has evolved independently several times in caecilians, but the details are hard to unravel because of major gaps in our knowledge of reproductive modes and uncertainty about the phylogenetic relationships of the major caecilian clades (Nussbaum and Wilkinson 1989; Hass, Nussbaum, and Maxson 1993; M. Wake 1993b; Frost et al. 2006). Viviparity is known in some caeciliids, and one species, *Dermophis mexicanus,* has been the subject of detailed studies of reproduction and embryological development (M. Wake 1977a, b, 1980a, b, 1992; M. Wake and Hanken 1982). Viviparity is thought to be characteristic of all typhlonectids as well, a group of aquatic caecilians found only in South America, as well as members of the small African family Scolecomorphidae, a terrestrial group (M. Wake 1977a, b; both typhlonectids and scolecomorphids were placed in the family Caeciliidae by Frost et al. 2006). In all of these viviparous species, early embryological development is supported by yolk reserves, but these are soon exhausted. Lipid-rich secretions produced by the walls of the oviducts (uterus) provide additional nutrition. These secretions are directly ingested by the developing larvae, which scrape cells from the oviduct wall with special fetal teeth. The teeth become fully developed at about the same time that the oviduct begins secreting nutrients (M. Wake 1977a, 1980a, b, 1989b, 1992; Welsch, Muller, and Schubert 1977; M. Wake and Hanken 1982; Exbrayat, Delsol, and Lescure 1983). Gestation lasts for many months, and the newborn young typically are very large relative to the size of the female, indicating a considerable energetic investment by the mother (M. Wake 1977a; Exbrayat, Delsol, and Flatin 1981, 1982). *Typhlonectus* larvae develop enormous, highly-vascularized, sac-like gills that probably are used for gas exchange across the mother's uterine wall and possibly is used for uptake of nutrients as well (Delsol, Flatin, Exbrayat, and Bons 1981, 1983; Delsol, Exbrayat, Flatin, and Gueydan-Baconnier 1986; M. Wake 1992). In contrast to viviparous salamandrids, developing caecilian larvae are not known to consume undeveloped eggs in the oviducts (M. Wake 1989b).

## Interspecific Variation in Egg Size, Clutch Size, and Reproductive Effort

Since the nineteenth century, biologists have recognized that different reproductive modes in amphibians are associated with different numbers and sizes of eggs. Boulenger (1886) noted that aquatic egg-layers typically have smaller eggs and lay more of them than do terrestrial egg-layers. Several later workers attempted to synthesize information on variation in egg and clutch size, most notably Salthe (1969) and Kaplan and Salthe (1979) on urodeles, Crump (1974) and Duellman (1978) on Neotropical anurans, and Salthe and Duellman (1973) and Duellman and Trueb (1986) on all anurans. Hödl (1990b) tabulated data for 130 species of lowland Amazonian anurans, but did not attempt a detailed analysis of the data. Tilley (1968) examined both intra- and interspecific size-fecundity relationships in desmognathine salamanders, while Kuramoto (1978a) did the same for 12 species of Japanese urodeles and anurans. M. Wake (1978) analyzed the relationship of egg size and clutch size to body size in the genus *Eleutherodactylus.* Barbault (1984) presented an interspecific analysis for 17 species of anurans from West Africa, contrasting aquatic breeders with species that lay eggs out of water. Lang (1995) examined the relationship of body size to egg size and clutch size in several genera of stream-breeding hylids from Middle America. Many other studies have examined size-fecundity relationships in individual species (see the following citations). The following discussion reexamines some of the patterns revealed by these studies and presents some new statistical analyses of egg and clutch sizes for a limited number of reproductive modes.

Duellman and Trueb (1986) summarized several conclusions about the relationship of body size, egg size, and clutch size that emerged from the studies cited previously: (1) Within a given reproductive mode, there is a positive correlation between clutch size and female body size. (2) Within a given reproductive mode, there is a positive correlation between ovum size and female body size. (3) Regardless of reproductive mode, there is a negative correlation between clutch size and ovum size. (4) Regardless of reproductive mode, there is a positive correlation between ovum size and size of hatchlings. In addition, total clutch volume does not appear to be related to reproductive mode, but is tightly correlated with body size for all modes.

While these generalizations probably are correct in some instances, most of the studies from which these conclusions have been drawn are flawed in one way or another. None of the papers presents a statistical analysis of differences in egg size and clutch size within or among reproductive modes. Some of the studies, including the major reviews by Salthe (1969) and Salthe and Duellman (1973), do not present either the original data or regression equations relating egg and clutch size to body size. Instead, conclusions are drawn from visual inspection of scatter plots. Some studies have not accounted for variation in body size in comparisons among reproductive modes. For example, Duellman (1978) and Hödl (1990b) presented a comparison of mean egg size and clutch size for anurans with several different reproductive modes. However, since both of these variables covary with body size, such comparisons are not very informative.

Finally, all of these studies grouped together amphibians with similar reproductive modes without regard to phylogeny, making it impossible to discern patterns of evolutionary change within particular clades. In Kuramoto's (1978a) study, for example, urodeles and anurans were combined into one regression analysis, obscuring any differences between the two orders. In other studies, apparent differences among reproductive modes could simply reflect the influence of a single family or genus on the data set. I have attempted to partially control for taxonomic differences by presenting analyses both for individual families or genera and for several families or genera combined. This procedure cannot be considered a true phylogenetic analysis (P. Harvey and Pagel 1991), because it simply uses a hierarchical classification of families, genera, and species as currently accepted by taxonomists. This classification does not necessarily reflect true phylogenetic relationships, which are constantly being revised (see Frost et al. 2006). For example, several of the largest anuran families that are well represented in the data set, including the ranids and leptodactylids, probably are not monophyletic groups, and the relationship of genera within these groups is uncertain (Ford and Cannatella 1993; Frost et al. 2006). The analysis presented here should be treated as a descriptive summary of the data, not a definitive analysis of evolutionary trends. My hope is that these patterns will provide raw material for more detailed phylogenetic analyses of life-history traits, something that is not yet possible for most groups because of incomplete information on reproductive modes, or uncertainties about phylogenetic relationships. Once we have more robust phylogenies of large families such as Hylidae, Ranidae, and Leptodactylidae, these data should be reanalyzed using modern comparative methods (see further discussion in the last section of the chapter).

### Interspecific Variation in Egg and Clutch Size in Anurans

The data used in my analyses are heavily weighted toward Neotropical species, because these have been studied in more detail than other groups (Crump 1974; Duellman 1978; Hödl 1990b) and North American anurans, taken mainly from A. Wright and Wright (1949). Since this book did not give mean snout-vent lengths, I used the midpoint of the body size range to approximate the mean. Most measurements of egg diameters are based on mature ovarian eggs and do not include jelly capsules surrounding the eggs. In some cases, only diameters of deposited eggs were available, but data were used only when the diameter of the egg itself, without the jelly capsules, was given. Some of the variation among species probably is due to differences in measurement procedures in the original studies. Certain common genera such as *Bufo, Hyla, Rana,* and *Eleutherodactylus* tend to be overrepresented in the data set, whereas little information is available for many other taxa. I have focused on several of the most common reproductive modes: pond-breeding species (Mode 1), species with terrestrial or arboreal eggs that undergo direct development (Modes 23 and 27), and species that lay nonaquatic eggs that hatch into tadpoles (Modes 18–21 and 24–25), because these are the only reproductive modes with sample sizes large enough for statistical analysis. In all of the regression analyses in the following discussions, I used natural logarithms of snout-vent length, egg size, and clutch size to improve the fit to a linear model (data tables and statistical analyses are available from the author).

For some reproductive modes, both egg size and clutch size are positively correlated with body size among species. For Mode 1 anurans as a group and for three families individually (Bufonidae, Hylidae, and Ranidae), clutch size was much more strongly correlated with body size than was egg size (fig. 10.15). The same was true for anurans with direct development (Modes 23 and 27). There were significant differences among families in both egg size and clutch size after adjusting for differences in body size. Bufonids have smaller eggs than do hylids, hyperoliids, and ranids (fig. 10.15 top), but the other families did not differ from each other statistically. This pattern reflects a general tendency for egg size to vary much less than clutch size among species of anurans. This suggests that the evolution of egg size is constrained within fairly narrow limits within reproductive modes, whereas clutch size is free to vary with changes in body size.

Mode 4 hylids were not analyzed separately because only

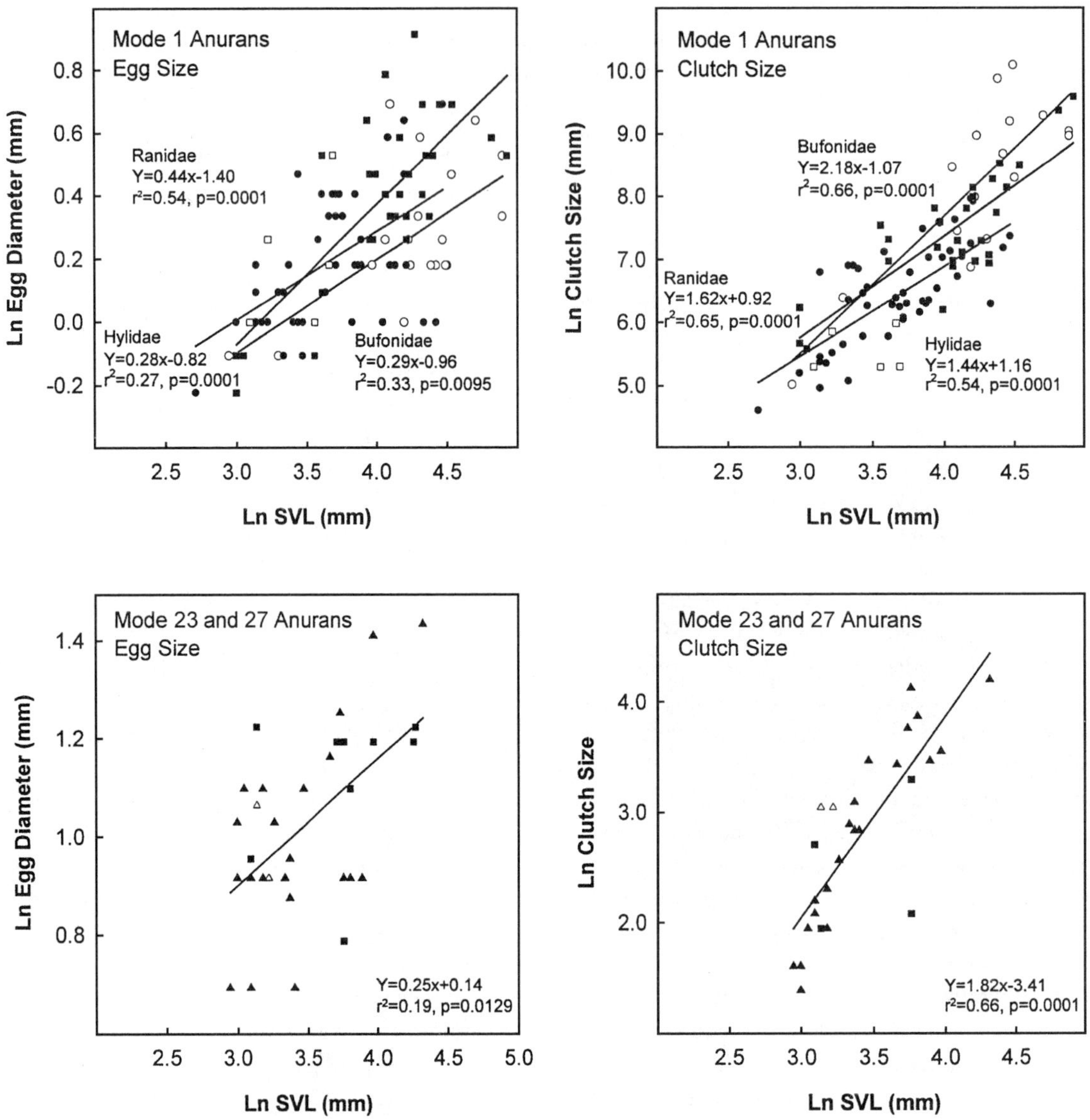


**Fig. 10.15.** Relationship of egg size and clutch size to body size for pond-breeding anurans (Mode 1) and those with direct development (Modes 23 and 27), plotted as natural logarithms. Mode 1 anurans include Bufonidae (open circles), Hylidae (closed circles), Hyperoliidae (open squares), and Ranidae (closed squares). Regression lines are not shown for Hyperoliidae. Mode 23 and 27 anurans include Arthroleptidae (open triangles), Leptodactylidae (closed triangles), and Ranidae (closed squares). Regression lines are for combined data. This analysis is not corrected for phylogenetic relationships.

three species were available, but egg and clutch sizes in these frogs are similar to those of Mode 1 hylids of the same size (fig. 10.16). Clutch size was not related to body size in hylids that lay eggs out of water (Modes 24 and 25; all squares in fig. 10.16). Hylids with these reproductive modes include two distinct groups that have quite different egg-laying strategies. Species of *Hyla,* such as *H. brevifrons, H. leucophyllata,* and *H. ebraccata,* lay relatively large clutches of small eggs on the surfaces of leaves (fig. 10.8 A), and these eggs hatch very quickly. These are little more than aquatic egg masses transferred to terrestrial oviposition sites. In contrast, the phyllomedusines (*Agalychnis, Phyllomedusa*) generally are larger species, and have larger eggs (fig. 10.8 C, D) with longer development times, but proportionately smaller clutches (hence, the lack of correlation between clutch size and body size when the two groups are pooled). An analysis of covariance for the Mode 24–25 hylids revealed a significant effect of taxonomic group on both egg size and clutch size. The phyllomedusines also have much larger eggs and smaller clutches than do Mode 1 hylids of similar body size (see closed squares in fig. 10.16), whereas Mode 24–25 *Hyla* have only slightly larger eggs and slightly smaller clutches (open squares in fig. 10.16). Although often considered to have the same reproductive mode, these groups clearly represent multiple independent origins of nonaquatic oviposition and quite different life-history strategies.

For all reproductive modes combined, egg size was negatively correlated with clutch size, but this is not true for any reproductive mode alone or for individual families. In fact, for Mode 1 hylids and ranids, egg size and clutch size were positively correlated, reflecting the positive correlation of each variable with body size. To eliminate the effect of body size, I used the residuals of regressions of both egg and clutch size on body size. When this was done for all frogs combined, there was a negative relationship between size-adjusted egg and clutch size (fig. 10.17). Within reproductive modes, only Mode 24–25 hylids exhibited a negative relationship, which reflects the dichotomy between the small egg/ large clutch strategy of the genus *Hyla* and the large egg/ small clutch strategy of the phyllomedusines. Hence, it appears that the negative relationship between egg size and clutch size reported by Salthe and Duellman (1973), Duellman and Trueb (1986), and others is largely a result of combining frogs of many different reproductive modes. Within reproductive modes, egg size varies relatively little among species, and there is not necessarily a direct tradeoff between these two life-history traits.

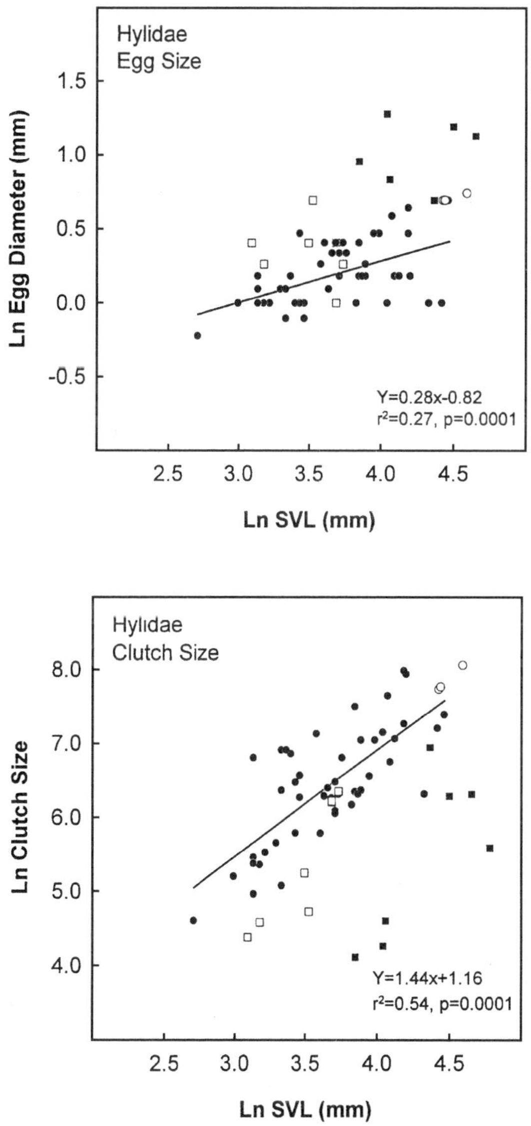


**Fig. 10.16.** Relationship of egg size and clutch size to body size for anurans in the family Hylidae, plotted as natural logarithms. Data include hylids that place eggs in ponds (Mode 1; closed circles), streamside nest-builders (Mode 4; open circles), *Hyla* that place eggs on vegetation over water (Mode 24; open squares), and Mode 24 phyllomedusines (closed squares). Regression lines are for Mode 1 hylids only. This analysis is not corrected for phylogenetic relationships.

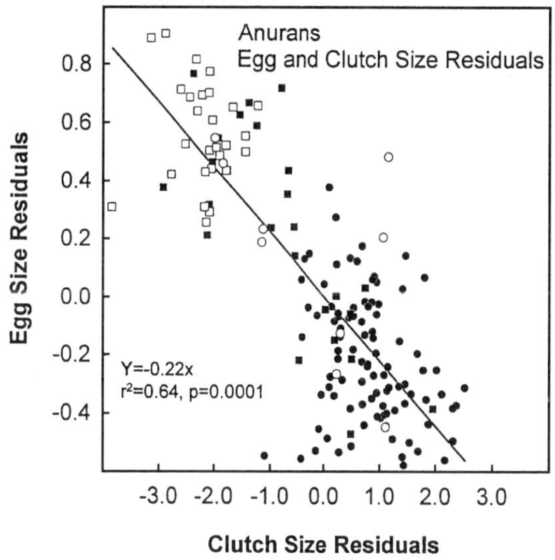


**Fig. 10.17.** Relationship of egg size to clutch size in anurans, after accounting for the effect of body size on both egg and clutch size. Data include pond breeders (Mode 1; closed circles), pond-breeding foam nesters (Mode 11; open circles), frogs that place eggs on vegetation over water (Mode 25; close squares), and those with direct development of terrestrial or arboreal eggs (Modes 23 and 27; open squares). Regression is for all reproductive modes combined and is not corrected for phylogenetic relationships.

## Interspecific Variation in Egg and Clutch Size in Urodeles

Kaplan and Salthe (1979) provided the most detailed analysis of the relationship of egg and clutch size to body size and reproductive mode in urodeles. However, they did not present regression equations for all analyses, nor did they attempt to account statistically for family differences. Therefore, I have done a new analysis, using original data kindly supplied by Robert Kaplan. In this data set, both body size and egg size were expressed as volumes, which were calculated from linear dimensions. I used natural logarithms rather than the base-10 logarithms used by Kaplan and Salthe (1979); all statistical results are identical for the two types of transformed data. I also made some modifications to the data set. Kaplan and Salthe (1979) grouped all species

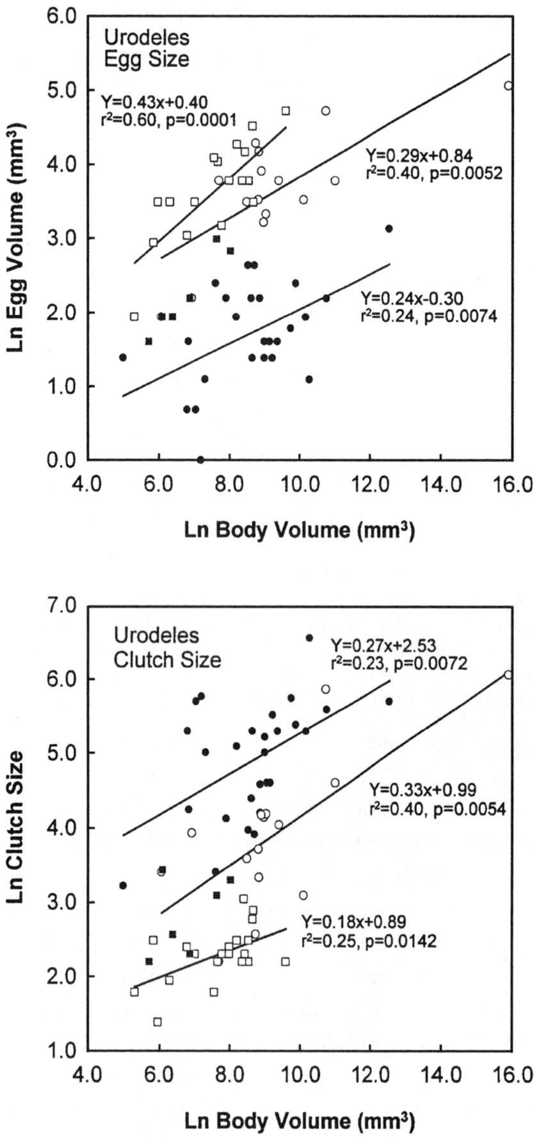


**Fig. 10.18.** Relationship of egg size and clutch size to body size in salamanders, plotted as natural logarithms. Data include pond breeders (Mode 1; close circles), stream breeders (Mode 2; open circles), species with terrestrial eggs and aquatic larvae (Mode 18; closed squares), and those with terrestrial eggs and direct development (Mode 23; open squares). Regression lines are not shown for Mode 18 species.

of *Desmognathus* except *D. wrighti* with the Mode 2 stream breeders because of their larval adaptations. However, I have removed all of the species that have terrestrial eggs and aquatic larvae from Mode 2 and placed them in Mode 18, leaving only *D. quadramaculatus* and *D. marmoratus* in Mode 2. For the analysis, the Mode 18 desmognathines were combined with *Hemidactylium scutatum,* which also has Mode 18 reproduction, and *D. aeneus,* which has Mode 21 reproduction (terrestrial eggs with nonfeeding larvae). *Stereochilus* and *Eurycea quadridigitata,* which were included in Mode 2 by Kaplan and Salthe, are placed in Mode 1 because they are commonly associated with ponds and swamps. Although data on egg size and clutch size are now available for some species not included in Kaplan and Salthe's analysis, I have used only the original data set, because the new information is unlikely to affect the results.

All of Kaplan and Salthe's major conclusions are supported by the reanalysis of the data. Both egg size and clutch size were positively correlated with body size for Mode 1, 2, and 23 urodeles (fig. 10.18). For plethodontids that have terrestrial eggs that hatch into larvae (Modes 18 and 21), egg size was positively related to body size, but clutch size was not. The lack of correlation between clutch size and body size could be a result of small sample size, but the relationship of egg size and body size was very robust despite the sample size. In general, plethodontids with terrestrial eggs (Modes 18, 21, 23) exhibited a stronger correlation between body size and egg size than between body size and clutch size. This generally was not the case in aquatic egg layers (Modes 1 and 2), where egg size appears to be more conservative. An analysis of covariance for all urodeles combined revealed a significant effect of both body size and reproductive mode on both egg size and clutch size. Mode 1 species had the smallest eggs and the largest clutches, while Mode 23 species had the largest eggs and the smallest clutches (fig. 10.18).

Regressions of egg size residuals on clutch size residuals revealed a strong negative relationship between size-adjusted egg size and size-adjusted clutch size for all urodeles combined (fig. 10.19) and for Mode 1, Mode 2, and Mode 23 spe-

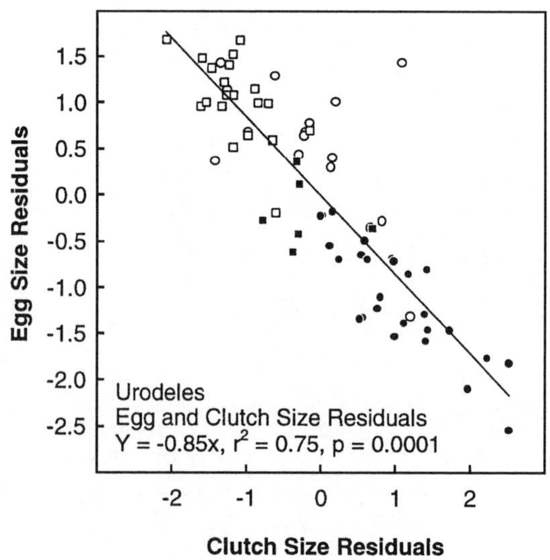


**Fig. 10.19.** Relationship of egg size to clutch size for salamanders, after accounting for the effect of body size on both egg and clutch size. Data include pond breeders (Mode 1; closed circles), stream breeders (Mode 2; open circles), species with terrestrial eggs and aquatic larvae (Mode 18; closed squares), and those with terrestrial eggs and direct development (Mode 23; open squares). Regression is for all reproductive modes combined and is not corrected for phylogenetic relationships.

cies. Within the Mode 1 species, only salamandrids showed a significant tradeoff between egg and clutch size after adjusting for body size, but sample sizes for all within-family comparisons were small (six to eight species per family). Mode 2 plethodontids exhibited a near perfect negative correlation between egg size and clutch size residuals, even though only six species were included in the analysis.

### Egg Size, Hatchling Size, and Development Rate

I did not attempt a new analysis of the relationship between egg size and hatchling size, but in general, species with large eggs have larger hatchlings than species with small eggs in both anurans and urodeles (fig. 10.20). Large eggs also take longer to develop and usually hatch at a more advanced stage than small eggs, but both rate and length of development are affected by temperature and phylogenetic differences as well (Bradford 1990). Rate of development was positively correlated with temperature when the effects of egg size and phylogeny (urodeles or anurans) were removed, and negatively correlated with egg size when the effects of temperature and phylogeny were removed (fig. 10.21). In

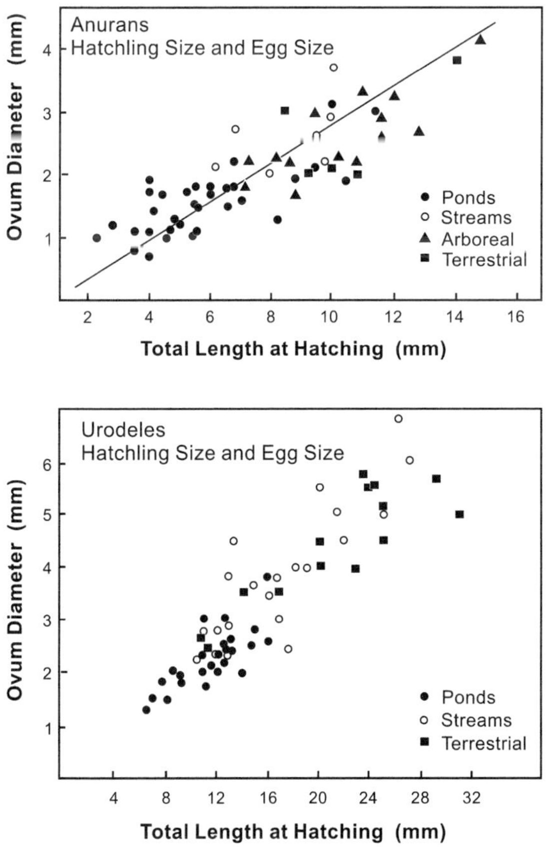


**Fig. 10.20.** Relationship of hatchling size and egg size for anurans (top) and urodeles (bottom) with different reproductive modes. Anuran data after Duellman and Trueb (1986). Urodele data after Salthe (1969).

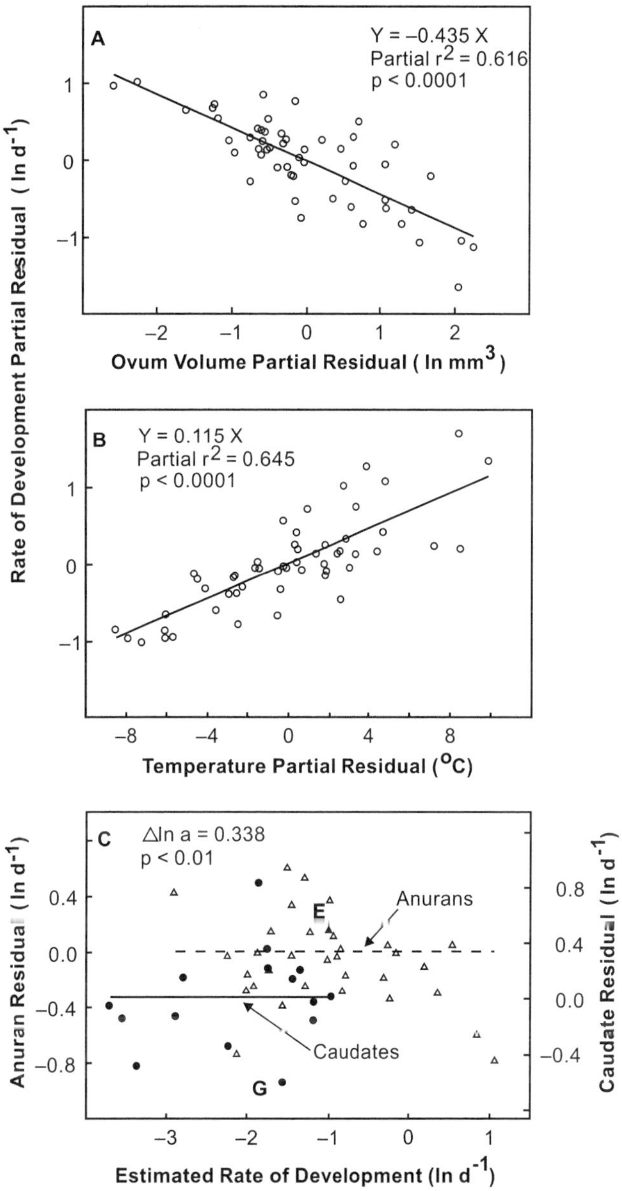


**Fig. 10.21.** The effect of egg size, temperature, and phylogeny on rate of egg development for anurans and urodeles. (A) Partial residual plot for rate of development as a function of egg volume, with effects of temperature and taxonomic group removed. This shows that large eggs develop more slowly than small eggs. (B) Partial residual plot for rate of development as a function of temperature, with the effects of egg volume and taxonomic group removed. This shows that eggs develop more rapidly with increasing temperature. (C) Partial residual plot showing difference in rate of development of anurans and urodeles as a function of predicted rate of development for both groups. This shows that anuran eggs develop more rapidly than urodele eggs of the same size and at the same temperature. Two species, *Eleutherodactylus coqui* (E), a tropical frog with direct development, and *Gastrotheca riobambae* (G), a "marsupial" frog, were outliers that were not included in the analysis. After Bradford (1990).

addition, urodele eggs have slower embryonic development than anuran eggs, take longer to hatch, and hatch at a more advanced stage, even when egg size and temperature are the same. The slower development times of urodele eggs probably are due at least in part to larger genome size, which tends to be negatively correlated with development rate (see chapter 1). Indeed, within plethodontid salamanders, which have the largest genomes of any vertebrates, development rate is negatively correlated with genome size, but is not related to egg size after accounting for genome size and temperature (Jockusch 1997).

In both anurans and urodeles, species that lay terrestrial eggs have longer development times, larger hatchlings, and more advanced hatchlings than aquatic egg-layers. Differences in development time generally have been attributed to differences in the size of the eggs, but differences in genome size could be more important. This has not been examined for anurans, but in plethodontid salamanders, species that lay terrestrial eggs with direct development generally have larger genomes than do aquatic-breeding species (Jockusch 1997). Egg size undoubtedly does influence hatchling size because of the much greater yolk reserves in terrestrial eggs, which allow for more prolonged development inside the egg. The disadvantage of having a long incubation period is the increased exposure of the eggs to predators and other sources of mortality. Hence, it is not surprising that many species that lay terrestrial eggs also have parental care, and survivorship of eggs often is poor if the attending parent is removed (see chapter 11).

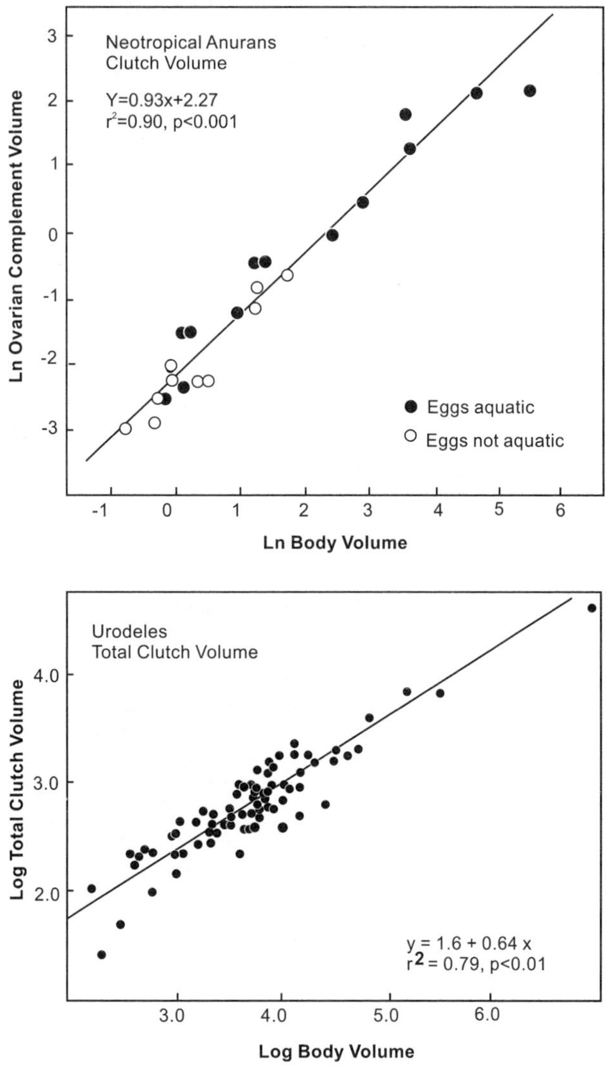


**Fig. 10.22.** Relationship of total clutch volume to body size for anurans (top) and urodeles (bottom). Anuran data after Duellman and Trueb (1986). Urodele data after Kaplan and Salthe (1979), with regression equation calculated from original data supplied by Robert Kaplan (personal communication).

## Reproductive Mode and Energetic Investment

Reproductive mode does not appear to have much effect on the energetic investment that females make in each clutch of eggs, but instead alters the way in which energy is partitioned. In urodeles (Kaplan and Salthe 1979) and anurans (Crump 1974; Barbault 1984; Duellman and Trueb 1986), total clutch volume is tightly correlated with body size, regardless of reproductive mode (fig. 10.22). This relationship does not appear to be strongly affected by phylogeny, but this has not been explicitly tested using phylogenetically-based comparative methods. Because of the allometric relationship between clutch volume and body size, smaller species tend to have proportionately larger clutch volumes than large species, regardless of reproductive mode (Crump 1974; Kaplan and Salthe 1979). Many species of anurans and urodeles with nonaquatic reproductive modes are smaller than species with fully aquatic reproduction, so these relationships can give the impression that terrestrial-breeders invest more heavily in reproduction than do aquatic-breeders (e.g., Tilley 1968). When related species of similar size are compared, however, total energetic investment per clutch is similar, but aquatic egg-layers divide their investment into a large number of small packages, while terrestrial egg-layers divide theirs into a small number of large packages (M. Crump and Kaplan 1979). Unfortunately, without knowing something about rates of energy intake and energy expenditure, it is impossible to determine whether species allocate different proportions of their total energy budget to reproduction, the most accurate measure of reproductive effort (Seale 1987; Stearns 1992).

Total annual reproductive effort is even more difficult to estimate, because in most cases we have inadequate information on the number of clutches a female lays each year. Most urodeles probably produce only a single clutch per year, and females of some temperate-zone species do not breed every year (R. Harris and Ludwig 2004). Some tropical salamanders probably produce several clutches per year, but information on these species is limited (Houck 1977a; Tilley and Bernardo 1993). In anurans, cold-weather breeders reproduce only once a year at best, but some warm-weather breeders in both the tropics and temperate zone produce multiple clutches per year (table 10.5). Unfortunately, most of the species for which quantitative information is available are Mode 1 pond-breeders, so the effect of reproductive mode on frequency of reproduction is unknown. However, two lines of evidence suggest that production of multiple clutches is common in both aquatic and

**Table 10.5** Examples of multiple egg clutch production by anuran females in a single breeding season

| Species (source no.) | $N_f$ | Months of observation | Largest no. of clutches | Days between clutches: Range | Days between clutches: Mean |
|---|---|---|---|---|---|
| Temperate zone species | | | | | |
| *Bufo calamita* (14) | | 3 | 2 | | 42 |
| *B. calamita* (17) | 5 | 4 | 2 | 43–96 | 63 |
| *B. cognatus* (10) | 2 | 2 | 2 | | 44 |
| *B. gutturalis* (9) | 6 | 6 | 2 | 55–123 | 81 |
| *Crinia signifera* (15) | 8 | 12 | 2 | | 74[a] |
| | 18 | 12 | 3 | 6–304 | 162[a] |
| *Hyla chrysoscelis* (11) | 19 | 4 | 3 | 12–73 | 36 |
| *H. cinerea* (7) | 37 | 2 | 3 | 6–40 | 19 |
| *Limnonectes kuhlii* (18) | 11 | 3 | 9 | 3–28 | 13 |
| *Pseudacris regilla* (7) | 16 | 2 | 5 | 13–69 | 37 |
| *Pseudophryne australis* (19) | 11 | 48 | | | < 30[b] |
| *Rana catesbeiana* (3) | 5 | 3 | 2 | 15–22 | 19[b] |
| *R. clamitans* (1, 2) | 15 | 3 | 2 | 15–46 | 25 |
| Tropical species | | | | | |
| *Bufo alatus* (4) | 9 | 3 | 2 | 38–42 | 42 |
| *Colostethus panamensis* (5) | 22 | 11 | 2 | 61–260 | 133 |
| *Eleutherodactylus coqui* (16) | 7 | 3 | 2 | 50–71 | 58 |
| *Hyla ebraccata* (20) | 13 | 3 | 2 | 10–41 | 18 |
| *H. rosenbergi* (6) | 16 | 5 | 5 | 21–33 | 25 |
| *Hyperolius viridiflavus* (13)[c] | 15 | 5 | 12 | | 16 |
| *H. marmoratus* (13)[c] | 29 | 5 | 12 | | 16 |
| *H. marmoratus* (12) | 13 | 4 | 2 | 18–30 | 24[d] |
| | 21 | 4 | 2 | 60–103 | 78[d] |
| *Physalaemus pustulosus* (8) | | 8 | | 28–35 | |

*Sources:* (1) Wells 1976a; (2) Wells 1976b; (3) Emlen 1977; (4) Wells 1979; (5) Wells 1980a; (6) Kluge 1981; (7) Perrill and Daniel 1983; (8) Ryan, Bartholomew, and Rand 1983; (9) Balinsky 1985; (10) Krupa 1986; (11) Ritke, Babb, and Ritke 1990; (12) Telford and Dyson 1990; (13) Grafe, Schmuck, and Linsenmair 1992; (14) Arak, cited in Silverin and Andren 1992; (15) Lemckert and Shine 1993; (16) Townsend and Stewart 1994; (17) Denton and Beebee 1996; (18) Tsuji and Lue 2000; (19) Thumm and Mahony 2002a; (20) K. Wells, unpublished data.

*Note:* $N_f$ = number of females.

[a]Some *Crinia signifera* females produced a small second clutch 1–3 months after the first; others produced larger clutches at longer intervals.

[b]Based on sightings of gravid females.

[c]Based on frogs breeding in laboratory.

[d]Some *Hyperolius marmoratus* females produced a second clutch within 30 days; others produced second clutches after more than 60 days due to differences in rainfall. Mean for all females was 57 days (range 18–103).

terrestrial-breeding frogs with prolonged breeding seasons: the simultaneous occurrence of more than one size of oöcytes in the ovaries (Church 1960b; Crump 1974; Barbault and Trefaut Rodrigues 1978a, b, 1979a, b; Barbault and Pilorge 1980; Rastogi et al. 1983; Shimoyama 1986; Donnelly 1989c; Kyriakopoulou-Sklavounou and Loumbourdis 1990; Silverin and Andren 1992) and observations of individual females breeding at frequent intervals in captivity (Senfft 1936; Chibon 1962; Davidson and Hough 1969; Grow 1977; Richards 1977; Schülte 1981a; Weygoldt 1982, 1984a; E. Zimmermann and Zimmermann 1982; Jungfer 1985; Lüddecke 1995a; Thumm and Mahony 2002a). In the field, intervals between clutches can be affected by a variety of environmental variables, including temperature, patterns of rainfall, and availability of food (Telford and Dyson 1990; Townsend and Stewart 1994; Harris and Ludwig 2004). While different species vary considerably in total annual reproductive output, it is not clear that this variation is related in any systematic way to differences in reproductive mode.

### Adaptive Explanations for Interspecific Variation in Egg and Clutch Size

How can we explain the differences in egg size and clutch size in different reproductive modes? Many discussions of this problem published in the 1970s and 1980s were framed explicitly or implicitly in terms of r-and-K selection (M. Crump 1974, 1981b; Barbault and Trefaut Rodrigues 1978a, b, 1979a, b; Duellman 1978; Barbault and Pilorge 1980; Barbault 1984; Duellman and Trueb 1986). Aquatic habitats, especially temporary ponds, generally have been characterized as unstable, unpredictable, or fluctuating environments where eggs and larvae are subjected to high levels of mortality from predation or other causes. In contrast, terrestrial habitats, particularly in the wet tropics, have been characterized as stable, equable, or predictable environments where mortality of eggs and larvae is lower. Amphibians with large clutches of small eggs have been viewed as classic examples of "r-selected" species that produce large numbers of offspring, each at relatively low energetic cost. Terrestrial breeders, particularly those with direct development and various forms of parental care, have been viewed as "K-selected" species that invest heavily in each individual to enhance its competitive ability, but at a cost in reduced numbers of offspring. In some cases, r-and-K selection has been invoked as an explanation for the evolution of egg size and clutch size, even when other life-history traits, such as age at maturity, frequency of reproduction, age-specific mortality, and length of reproductive life were unknown (Crump 1974; Duellman 1978). In other studies, some of these variables were measured as well (Barbault and Trefaut Rodrigues 1978a, b, 1979a, b; Barbault and Pilorge 1980).

The use of r-and-K selection as an explanatory framework in studies of life-history evolution has now been largely abandoned. The scheme fails to account for the range of variation in life-history strategies seen in nature and does not make precise predictions about the types of life-history traits to be expected in r- and K-selected species (Stearns 1992). It is now clear that the evolution of life-history strategies is far more complex than originally thought. We need several kinds of data to fully understand the evolution of life-history traits in any group, including amphibians. We need detailed information not only on egg size and clutch size, but also on demographic features of populations (age at first reproduction, age-specific reproduction and mortality, reproductive life span), the genetic correlation between life-history traits, physiological costs, and the effects of phylogeny of specific life-history traits (Stearns 1992). These data are simply not available for most amphibian species, so a comprehensive discussion of the evolution of reproductive strategies is not possible at this point. Nevertheless, it is possible to identify some selective pressures that are likely to affect the evolution of egg size and clutch size in predictable ways.

All of the environmental variables listed in table 10.3 that affect the evolution of clutch structure and oviposition behavior probably also affect the evolution of egg and clutch size. Clearly these variables are not completely independent, but covary in different habitats. For example, aquatic environments with low temperatures typically have high levels of dissolved oxygen, while warm habitats have low oxygen levels. Physical factors such as water temperature, dissolved oxygen content, and probability of the habitat drying do not necessarily select for specific egg or clutch sizes, but they could act as constraints that limit the range of variation possible in a given habitat, particularly variation in egg size. In general, small eggs have high surface-to-volume ratios, short development times, and hatch into relatively small, poorly developed larvae. Large eggs have low surface-to-volume ratios, longer development times, and hatch into larger, more fully developed larvae. These differences are magnified by differences in water temperature. Species with the shortest development times and the most underdeveloped hatchlings typically are those that lay small eggs in warm aquatic environments, such as desert pools, whereas those with the longest development times and most fully developed hatchlings lay large eggs in cool aquatic or terrestrial habitats.

Small eggs probably are necessary for successful reproduction in any aquatic habitat that combines warm temperatures with low oxygen availability, such as warm ponds, swamps, desert pools, streamside basins in the tropics, or water-filled bromeliads. Oxygen demand increases with temperature, but diffusion of oxygen into the egg decreases as egg size increases (Bradford 1990; Seymour 1999). Hence, metabolic demands and oxygen availability set upper limits on the size of eggs that can survive in these types of habitats (see also chapter 4). This probably accounts in part for the small range of interspecific variation in egg size among Mode 1 anurans. In addition, the relatively large clutch sizes of most warm-water Mode 1 amphibians probably are indirect consequences of selection for small egg size, and not necessarily the result of selection on clutch size per se.

Small eggs would be particularly advantageous in warm, ephemeral habitats such as desert rain pools and shallow tropical ponds, where very rapid development should be favored because of rapid drying of the habitat (Zweifel 1968b). This is consistent with the smaller eggs of Mode 1 *Bufo*, which tend to breed in ephemeral habitats, compared to Mode 1 hylids or ranids, which usually breed in larger temporary or permanent ponds. Even within the genus *Rana*, species that breed in warm water have smaller eggs than do cold-water species with the same reproductive mode. The advantage of small eggs in warm, ephemeral habitats is enhanced by physiological changes that result in increased development rates at a given temperature (see also chapter 3). For example, spadefoot toads (*Scaphiopus* and *Spea*) from North American deserts not only have very small eggs, but they have faster rates of development at any given temperature than do other anurans from the same region (Zweifel 1968b; Bradford 1990).

In cooler, well-oxygenated aquatic environments, such as cold ponds, lakes, and mountain streams, egg size would be less constrained by oxygen availability, particularly because embryonic metabolic rates would be lower at cold temperatures (H. Brown 1977; Bradford 1990; Seymour 1999). This would allow cold-water breeders to produce larger eggs than do warm-water breeders. In fact, this is the pattern seen in both urodeles and anurans, particularly in stream-breeding species (Salthe 1969; Salthe and Duellman 1973; Kaplan and Salthe 1979; Nussbaum 1985; Bradford 1990). However, greater oxygen availability and lower metabolic demands would not necessarily select for larger eggs, since this is accompanied by a decrease in clutch size.

Nussbaum (1985, 2003) provided a thorough review of the various adaptive explanations for the correlation between cold aquatic environments and large eggs that have been proposed for urodeles, anurans, and fishes. Many authors have simply noted the correlation and have argued that large eggs are an adaptation for breeding in cold environments, without specifying the selective advantage of large eggs (J. Moore 1949b; Salthe 1969; Licht 1971). Others have argued that cold environments favor the production of large, relatively advanced larvae, and this can be achieved only by having large eggs. Large larvae might be advantageous if competition for food were particularly severe in cold environments with low productivity (an argument made mainly by ichthyologists cited in Nussbaum 1985). Large, yolky eggs also might provide sufficient energy reserves for larvae to survive for a considerable period of time without food, until they reach a size that allows them to feed effectively (H. Brown 1977, 1989). In flowing water, large size enhances the ability of larvae to hold their position in a current (Dunn1923; Noble 1927; Salthe 1969; Salthe and Duellman 1973; Salthe and Mecham 1974; S. Richards 2002). This idea was tested experimentally by Petranka, Kats, and Sih 1987), who compared the eggs and larvae of a stream-breeding ambystomatid, *Ambystoma barbouri*, with those of a closely related pond-breeder, *A. texanum* (originally the stream- and pond-breeders were thought to be two different races of *A. texanum*). *Ambystoma barbouri* has larger eggs and larger larvae, which were displaced shorter distances in the stream current than were those of *A. texanum*.

Nussbaum (1985) argued that the correlation between large eggs and cold environments is largely spurious, at least for urodeles, with the real correlation being between large eggs and flowing-water habitats, most of which also happen to be relatively cold. He proposed that large eggs in stream-breeding salamanders are favored mainly because of the types of food available for their larvae. Most pond-dwelling salamander larvae feed on small, relatively abundant prey, such as zooplankton, which tend to be evenly distributed in the environment. In contrast, stream-dwelling salamander larvae eat mostly larger prey, such as aquatic insects, that are more patchily distributed and less abundant, because zooplankton is not available. Consequently, large, relatively advanced hatchlings that can handle large prey would have an advantage over smaller, less advanced hatchlings that can eat only small prey. This argument does not seem applicable to stream-breeding anurans, because these have completely herbivorous larvae that usually scrape algae off of rocks or intercept small particles in the current (see chapter 12). Nevertheless, a simulation study of reproductive strategies in fishes concluded that production of a few large larvae was superior to production of many small larvae in all environments with low food abundance at all levels of small-scale prey patchiness (Winemiller and Rose 1993). Because productivity generally is lower in streams than in ponds, large larvae are favored in stream-breeding anurans as well.

Indeed, large eggs are likely to be favored in any aquatic environment that provides poor food resources for the larvae. For example, frogs that place their eggs in small bodies of water, such as bromeliads, tree holes, water-filled bamboo, and water-filled depressions on the ground often have larger eggs than do close relatives that breed in larger bodies of water. These types of sites generally provide little food for developing larvae. In some species, the tadpoles have become obligate or facultative egg-eaters, subsisting on eggs deliberately or accidentally deposited by females in the same bodies of water (Lannoo, Townsend, and Wassersug 1987). Others supply their eggs with enough yolk to allow the tadpoles to develop without feeding (e.g., *Eupsophus, Pelophryne, Kalophrynus, Syncope, Platypelis, Plethodonthyla, Anodonthyla*).

Several species of toads that breed in small bodies of water with low larval food supplies have much larger eggs than more typical pond-breeding toads. One such species is *Bufo periglenes,* from montane cloud forests of Costa Rica (unfortunately, now thought to be extinct), which bred in small water-filled depressions on the forest floor. These sites can be very crowded and are largely devoid of suitable food for the tadpoles. Eggs of this species were about 3 mm in diameter, larger than those of two other Costa Rican species that breed in ponds (both < 2 mm in diameter). These eggs are about twice the size expected for a toad with a body size of *B. periglenes* (SVL = 160 mm; predicted egg diameter from bufonid regression in fig. 10.15 = 1.7 mm). Tadpoles were able to feed if provided with suitable food, but also could complete metamorphosis without feeding if deprived of food (Crump 1989b). In contrast, tadpoles of the two pond-breeding species with small eggs, *B. marinus* and *B. coniferous,* could not survive to metamorphosis without food.

Another species that is closely related to *B. periglenes,* and has similar habits, is *B. holdridgei.* This toad also has much larger eggs (2.8 mm) than do congenerics of similar size (SVL = 46 mm; predicted egg diameter from regression in fig. 10.15 = 1.2 mm). Tadpoles probably can reach metamorphosis without feeding (Novak and Robinson 1975). Yet another Neotropical species with unusually large eggs is *B. castaneoticus,* which lays eggs in water-filled fruit capsules of the Brazil nut tree, a microhabitat likely to be poorly supplied with food (Caldwell 1991, 1993). The Cape Mountain toad of southern Africa, *B. rosei,* has exceptionally large eggs (2.6 mm) for its body size (37 mm; predicted egg diameter from regression in fig. 10.15 = 1.1 mm). These are laid in rosary-like strings similar to those of *B. holdridgei* in shallow, water-filled depressions. The tadpoles hatch at an advanced stage of development, with rear leg buds already visible. Wager (1965) characterized the mouth as "feebly developed," suggesting the tadpoles do little or no feeding before metamorphosis. Presumably the habit of laying unusually large eggs has evolved independently several times in the genus *Bufo,* but a detailed phylogenetic analysis of the genus is needed to test this hypothesis rigorously. Unfortunately, none of these species was included in the most recent phylogenetic analysis of bufonids, so their relationships to other species of *Bufo* with smaller eggs are unclear, as are their relationships to each other (Graybeal 1997). Similar arguments have been advanced to explain unusually large egg size in *Crinia georgiana* compared to other members of the genus. Tadpoles of this species also can reach metamorphosis without feeding (Doughty 2002).

Nonaquatic egg-layers, particularly those with direct development, have the largest eggs in relation to body size in both anurans and urodeles. Most authors have argued that the principal advantage of large eggs in these species is the production of large, advanced larvae or juveniles that can cope with the rigors of the terrestrial environment. Usually predation and competition for food are invoked as the major selective pressures favoring large, advanced hatchlings (Salthe 1969; Salthe and Duellman 1973; Crump 1974; Salthe and Mecham 1974; Duellman and Trueb 1986), but the selective pressures that favor particular sizes of eggs are varied and complex (Bernardo 1996a, b). In fact, terrestrial microhabitats do not provide any suitable food for aquatic larvae. Consequently, unless the larvae are carried to water by the parent, terrestrial eggs must be provided with large yolk reserves. In some cases, large yolk reserves allow for prolonged development until the nest site is flooded (Mode 17). In others, large eggs develop into larvae large enough to make their way to water independently (Modes 18, 24, 25, and 26), or complete development without feeding, either inside the egg capsule (Modes 23 and 27) or in the nest (Modes 18 and 32). In addition, problems of desiccation set a lower limit on egg size in terrestrial environments. Large eggs lose water more slowly than do small eggs because of their more favorable surface-to-volume ratios. Nonaquatic egg-layers must select relatively saturated oviposition sites, or provide water to the eggs in some other way. Some species surround their eggs with water-filled jelly (a strategy seen in many anurans that lay eggs on leaves overhanging water). In others, parental attendance of the eggs provides the necessary moisture (Taigen, Pough, and Stewart 1984; see also chapter 11).

In general, oxygen is readily available to terrestrial eggs, so problems of gas diffusion probably would not place an upper limit on their size, within the range of sizes normally seen in amphibians (see also chapter 4). Nevertheless, an adequate rate of oxygen diffusion depends on adequate hydration of the eggs (Seymour and Bradford 1987; Bradford and Seymour 1988a; Seymour 1999). Furthermore, under some circumstances, such as high temperatures or flooding, oxygen can become limiting even for terrestrial eggs (Bradford and Seymour 1988b; Seymour, Geiser, and Bradford 1991a). Most nonaquatic eggs are laid in relatively wet, sheltered locations that would reduce exposure to extreme conditions that might limit the availability of oxygen. However, the arboreal foam nests produced by some tropical rhacophorids often are placed in direct sunlight, and temperatures inside the nest can reach levels higher than those tolerated by many amphibian eggs. These high temperatures would increase metabolic demands of the embryos and tend to favor relatively small eggs. In one such species, *Chiromantis xerampelina,* the eggs are about 1.7 mm in diameter, smaller than those of many arboreal or terrestrial egg-layers of similar size. The wet foam surrounding the eggs actually impedes diffusion of oxygen into the nest, but as the outer surface of the nest dries, gas conductance increases. The air-filled bubbles making up the nest supply most of the oxygen needed by the embryos. As metabolic rates of the embryos increase during development, oxygen demand can exceed rates of oxygen uptake, and this serves as a stimulus for the larvae to hatch and drop out of the nest (Seymour and Loveridge 1994).

## Intraspecific Variation in Egg and Clutch Size

Studies of interspecific variation in egg size and clutch size have revealed clear differences among reproductive modes that appear to be consistent across different clades of amphibians. For example, terrestrial egg-layers in many families have larger eggs and smaller clutches than do closely related aquatic egg-layers. Within aquatic egg-layers, stream-breeders consistently have larger eggs and smaller clutches than their pond-breeding relatives. All of these analyses are based on species averages for body size, egg size, and clutch size, and therefore mask a considerable amount of intraspecific variation in life-history traits. Such variation is evident in comparisons among populations, among individual females in the same populations, and even within clutches of eggs produced by individual females. Intraspecific variation in egg size and clutch size provides a potentially rich source of information on adaptation of life histories to local conditions, as well as the relative contribution of genotype and environment to variation in life-history traits.

### Variation within Populations

Many investigators have reported a positive correlation between body size and clutch size within populations in both anurans and urodeles (table 10. 6), and none has reported a significant negative correlation. Not all species exhibit this type of relationship, however. For example, Crump (1974) reported that only 11 of 41 species of tropical frogs for which she had adequate data showed a positive correlation between body size and clutch size, despite robust sample sizes for many of the species. Lips (2001) did not find any correlation between clutch size and body size in the Neotropical treefrog *Hyla calypsa,* which lays eggs on vegetation over water. Kuramoto (1978a) examined intraspecific relationships for 12 species of Japanese amphibians. In four of these (two frogs and two salamanders), clutch size was not related to body size, but in two cases, this could have been due to small sample sizes ($N = 10$). Many authors have cited interspecific relationships between body size and clutch size, such as those reported by Salthe and Duellman (1973), as evidence that within species, larger females should produce larger clutches. While this certainly is true for many species, one cannot simply assume that it is true for all amphibians. There have been fewer attempts to document variation in egg size, but a positive correlation has been reported in anurans and urodeles from several different families (table 10. 6). Other species show no relationship between body size and egg size (e.g., Lips 2001), while a few exhibit negative relationships (Crump 1974; Kuramoto 1978a).

There are several reasons why a positive intraspecific correlation between body size, clutch size, and egg size is not always apparent. Several studies have shown a significant effect of female age on clutch size, independent of body size (table 10. 6). For example, Berven (1988) reported that both clutch size and egg size were positively correlated with body size within three age classes of female wood frogs (*Rana sylvatica*), but one-year-old females produced larger clutches of smaller eggs than did two- or three-year-olds, after adjusting for differences in body size. He also found a slightly higher total investment in egg production (total egg volume) by two-year-old females than by one- or three-year-olds (fig. 10.23). M. Gibbons and McCarthy (1986) reported a similar increase in relative egg size with age in *Rana temporaria,* while Elmberg (1991) reported an increase in relative clutch size (adjusted for body size) with increasing age in the same species. In a West African frog, *Hyperolius nitidulus,* some individuals became sexually mature within a few weeks of metamorphosis and reproduced in the same season. These rapidly maturing individuals were smaller than those breeding earlier in the year, and females produced smaller eggs. They also laid smaller egg clutches than first clutches of later breeding females (Lampert and Linsenmair 2002). In a facultatively paedomorphic salamander, *Ambystoma talpoideum,* egg size, adjusted for body size, increased with age, but clutch size did not (Semlitsch 1985b). Fully metamorphosed terrestrial adults, which included females that were one to five years old, produced larger eggs than did paedomorphic individuals, but the latter were all only a year old. Among one-year-olds, egg size was similar in the two types of morphs, but terrestrial females produced more eggs for their body size.

Another factor that influences the relationship between body size and both clutch and egg size is the number of clutches produced by an individual female in one season. In *Rana catesbeiana,* second clutches were smaller and had smaller eggs than first clutches (R. D. Howard 1978b, 1988a). Second clutches also were smaller in *Hyla rosenbergi* (Kluge 1981), but not in *Hyperolius marmoratus* (Telford and Dyson 1990). In the Australian frog *Crinia signifera,* which has a very long breeding season, some females produced a second, smaller clutch within three months of laying their

**Table 10.6** Examples of intraspecific correlation between clutch size (CS) or egg size (ES) and body size in amphibians

| Species (source no.) | CS | ES | Other variables correlated with egg or clutch size |
|---|---|---|---|
| Bombinatoridae | | | |
| *Bombina bombina* (60) | + | | |
| *B. orientalis* (52) | | + | Population, year |
| *B. variegata* (48) | + | | |
| Bufonidae | | | |
| *Bufo alatus* (8) | + | | |
| *B. bufo* (25, 33, 35) | + | + | Population |
| *B. calamita* (6, 33, 49) | + | + | Age, population, breeding date |
| *B. canorus* (22) | + | + | Number of clutches laid |
| *B. cognatus* (50) | + | | |
| *B. exsul* (22) | + | 0 | |
| *B. japonicus* (43) | + | + | Latitude, temperature |
| *B. marinus* (19) | + | | |
| *B. viridis* (29, 59) | + | + | |
| Dendrobatidae | | | |
| *Dendrobates pumilio* (42) | + | | Month, food availability |
| *Epipedobates pictus* (8) | + | | |
| Hylidae | | | |
| *Hyla bifurca* (8) | + | | |
| *H. bokermanni* (8) | 0 | – | |
| *H. boans* (8) | + | | |
| *H. brevifrons* (8) | + | | |
| *H. calypsa* (54) | 0 | 0 | |
| *H. granosa* (8) | + | | |
| *H. labialis* (56) | + | 0 | Altitude |
| *H. leucopygia* (53) | + | | |
| *H. marmorata* (8) | 0 | – | |
| *H. parviceps* (8) | + | | Month |
| *H. sarayacuensis* (8) | + | | |
| *H. japonica* (16) | + | 0 | |
| *Litoria chloris* (57) | 0 | 0 | |
| *L. fasciolatus* (57) | 0 | 0 | |
| *L. iteratus* (57) | + | 0 | |
| *L. lesueuri* (57) | + | 0 | |
| *L. pearsoniana* (57) | 0 | 0 | |
| *Pseudacris triseriata* (3) | + | + | Population (altitude) |
| Leptodactylidae | | | |
| *Eleutherodactylus altamazonicus* (8) | + | | |
| *E. coqui* (51) | + | | Month, rainfall |
| *E. lacrimosus* (8) | + | | |
| *E. lanthanites* (8) | + | | Month |
| *Physalaemus pustulosus* (31) | + | | |

(*continued*)

**Table 10.6** (continued)

| Species (source no.) | CS | ES | Other variables correlated with egg or clutch size |
|---|---|---|---|
| Petropedetidae | | | |
| *Phrynobatrachus accraensis* (17) | + | | |
| *P. calcaratus* (20) | + | | |
| *P. plicatus* (14) | + | | Month |
| Ranidae | | | |
| *Limnonectes limnocharis* (16) | + | 0 | |
| *Ptychadena maccarthyensis* (13) | + | | |
| *P. oxyrhynchus* (13) | + | | |
| *Rana catesbeiana* (15, 21, 40) | + | + | Clutch (2nd < 1st) |
| *R. curtipes* (61) | + | | |
| *R. dalmatina* (28) | + | + | |
| *R. japonica* (16) | + | + | |
| *R. nigromaculata* (16) | 0 | + | |
| *Rana porosa brevipoda* (16) | + | 0 | |
| *R. ridibunda* (46) | + | + | |
| *R. sylvatica* (39, 40) | + | + | Age, latitude, altitude |
| *R. tagoi* (16) | + | 0 | |
| *R. temporaria* (7, 9, 34, 35, 41, 47, 55, 58) | + | + | Age, year, latitude, altitude, food intake |
| *R. tsushimensis* (16) | + | + | |
| Rhacophoridae | | | |
| *Buergeria buergeri* (16) | 0 | + | |
| *Rhacophorus schlegelii* (16) | + | 0 | |
| Ambystomatidae | | | |
| *Ambystoma barbouri* (30) | + | | |
| *Ambystoma maculatum* (18, 26) | + | + | Type of breeding pond |
| *A. opacum* (18, 37) | + | + | Population |
| *A. talpoideum* (32) | + | + | Age, morph |
| *A. tigrinum* (18) | + | 0 | |
| Hynobiidae | | | |
| *Hynobius nebulosus* (23) | + | | Breeding date |
| *H. nigrescens* (44) | + | + | Population, altitude |
| *Onychodactylus japonicus* (16) | 0 | + | |
| Plethodontidae | | | |
| *Bolitoglossa rostrata* (12) | + | | |
| *Desmognathus aeneus* (2) | + | | |
| *D. fuscus* (4, 38) | + | | Population, altitude |
| *D. monticola* (4) | + | | Population |
| *D. ocoee* (4) | + | | |
| *D. quadramaculatus* (4) | + | | |
| *D. wrighti* (4) | + | | |
| *Eurycea quadridigitata* (24) | + | | |
| *Gyrinophilus porphyriticus* (5) | + | | |

*(continued)*

**Table 10.6** (continued)

| Species (source no.) | CS | ES | Other variables correlated with egg or clutch size |
|---|---|---|---|
| *Hemidactylium scutatum* (1) | + | | |
| *Plethodon websteri* (27) | + | | |
| *Pseudoeurycea goebeli* (11) | + | | |
| *Pseudotriton montanus* (5) | + | | |
| *P. ruber* (5) | + | | |
| Salamandridae | | | |
| *Cynops pyrrhogaster* (16) | + | – | |
| *Triturus vulgaris* (10) | + | | Age |
| Ichthyophiidae | | | |
| *Ichthyophis kohtaoensis* (62) | + | +[a] | |

*Sources:* (1) Blanchard, 1936; (2) J. R. Harrison, 1967; (3) Pettus and Angleton, 1967; (4) Tilley, 1968; (5) Bruce 1969; (6) Hemmer and Kadel 1971; (7) Kozłowska 1971; (8) Crump 1974; (9) Koskela and Pasanen 1975; (10) Bell 1977; (11) Houck 1977a; (12) Houck 1977b; (13) Barbault and Trefaut Rodrigues 1978a; (14) Barbault and Trefaut Rodrigues 1978b; (15) R. D. Howard 1978b; (16) Kuramoto 1978; (17) Barbault and Trefaut Rodrigues 1979b; (18) Kaplan and Salthe 1979; (19) Zug and Zug 1979; (20) Barbault and Pilorge 1980; (21) Bruneau and Magnin 1980b; (22) Kagarise Sherman 1980; (23) Kusano 1980; (24) Semlitsch and McMillan 1980; (25) C. B. Jørgensen 1981; (26) Woodward 1982b; (27) Semlitsch and West 1983; (28) Sofianidou and Kyriakopoulou-Sklavounou 1983; (29) C. B. Jørgensen 1984b; (30) Petranka 1984d; (31) Ryan 1985a; (32) Semlitsch 1985b; (33) Banks and Beebee 1986b; (34) Cummins 1986; (35) Gibbons and McCarthy 1986; (36) Reading 1986; (37) Walls and Altig 1986; (38) Hom 1987; (39) Berven 1988; (40) R. D. Howard 1988a; (41) Ryser 1988a; (42) Donnelly 1989c; (43) Matsui 1989; (44) Takahashi and Iwasawa 1989a; (45) Bruce and Hairston 1990; (46) Kyriakopoulou-Sklavounou and Loumbourdis 1990; (47) Elmberg 1991; (48) Rafinska 1991; (49) Tejedo 1992a; (50) Krupa 1994; (51) Townsend and Stewart 1994; (52) Kaplan and King 1997; (53) Haddad and Sawaya 2000; (54) Lips 2001; (55) Laugen, Laurila, and Merilä 2002; (56) Lüddecke 2002; (57) Morrison and Hero 2002; (58) Lardner and Loman 2003; (59) Castellano et al. 2004; (60) Cogalniceanu and Miaud 2004; (61) Gramapurohit, Shanbhag, and Saidapur 2004b; (62) Kupfer, Nabhitabhata, and Himstedt 2004.

*Notes:* The correlation can be positive (+), negative (–), or nonsignificant (0). Blank cells indicate that the correlation was not reported.

[a]Hatchling size positively correlated with female body size; egg size was not measured.

first clutch, whereas other individuals produced a relatively large second clutch after a much longer interval (five to six months; Lemckert and Shine 1993). Egg clutches of *Hyla calypsa* laid early in the breeding season were larger than those produced later in the season, but data were not presented for different clutches of individual females (Lips 2001). In *Bufo canorus,* a toad that breeds at high altitudes in the mountains of California, many females do not breed every year. Kagarise Sherman (1980) found that females that laid eggs at intervals of one or two years produced significantly smaller clutches than did females that reproduced every three or four years, even though the former were larger individuals.

The tendency for second clutches to be smaller than first clutches suggests that females are under considerable energy stress and do not have sufficient energy reserves, or have not been able to consume sufficient food, to produce a full complement of eggs shortly after producing their first clutches. Seasonal variation in clutch sizes in some tropical frogs supports this interpretation as well. For example, Donnelly (1989c) found that clutch sizes in *Dendrobates pumilio* varied seasonally, apparently in response to changes in food availability. Townsend and Stewart (1994) reported that clutches produced by female *Eleutherodactylus coqui* in the dry season were significantly smaller than were clutches produced in the wet season, even though clutch size generally was positively correlated with body size, and the females in the wet season sample were slightly smaller. Seasonal variation in clutch size also has been documented in some tropical African frogs (Barbault and Trefaut Rodrigues 1978b, 1979a). In one of these species, *Arthroleptis poecilonotus,* clutch size was not related to body size, but clutches were much larger in a wet year than in a dry year.

Some temperate-zone anurans, such as *Rana temporaria, R. sylvatica, Bufo bufo,* and *Bombina orientalis* also exhibit year-to-year variation in egg size, clutch size, or both, that sometimes equals or exceeds variation among populations (C. B. Jørgensen, Larsen, and Lofts 1979; Cummins 1986; Berven 1988; Kaplan and King 1997). All of these species breed in the spring and depend at least in part on stored energy reserves accumulated during the previous year to support egg production and maturation. This makes it difficult

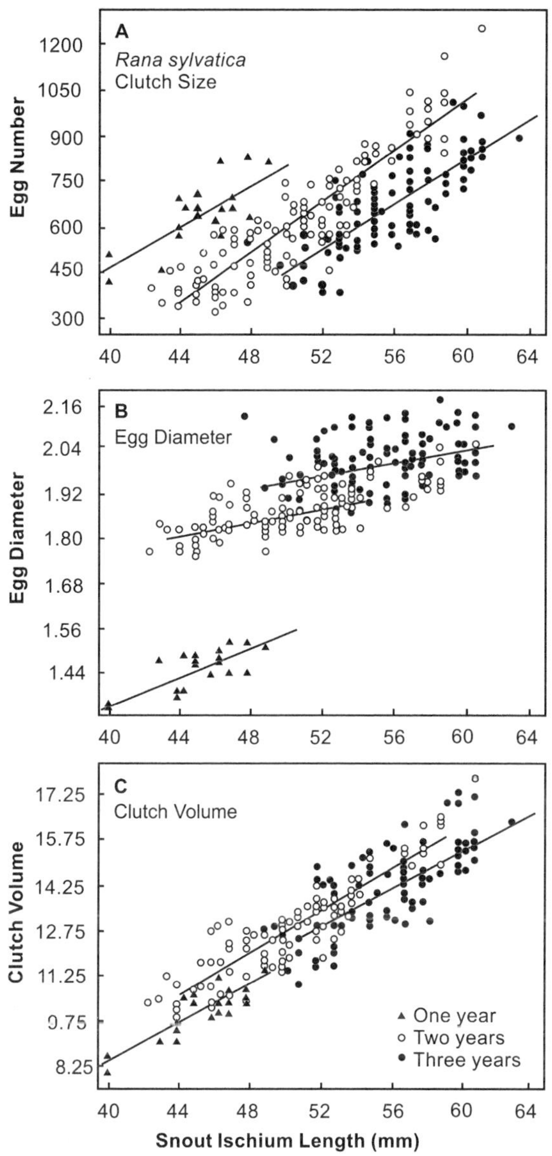


**Fig. 10.23.** Relationship of clutch size, egg size (mm), and total clutch volume (mm ) to body size for *Rana sylvatica* females at one, two, and three years of age. Second- and third-year females have smaller clutches and larger eggs than do first-year females. After Berven (1988).

to identify the causes of year-to-year variation in egg and clutch size, but presumably these would include variation in food abundance, length of the activity season, temperature, and rainfall, which might affect time available for foraging.

Laboratory studies of several species of amphibians have shown that food availability can affect both the number and size of eggs produced by a female. The number of eggs is determined relatively early in the ovarian cycle and is a function of the number of small oöcytes available for recruitment. The nutritional condition of the female at this stage of the cycle can affect the number of oöcytes recruited, but once this is set, clutch size does not undergo any further increase. Yolk is deposited later, in the vitellogenic growth period, which generally corresponds to the main period of feeding and body growth. The nutritional condition of the female during this period has a profound effect on egg size and probably accounts for much of the variation observed in natural populations (C. B. Jørgensen 1973a, b, 1974a, 1975, 1981, 1982, 1984a, b, 1992b; C. B. Jørgensen, Larsen, and Lofts 1979; Fraser 1980; Smalley and Nace 1983; R. Kaplan 1987; Kaplan and King 1997). Any effect on clutch size, such as that observed by Kaplan (1987; clutch size was smaller in females provided with small amounts of food), probably is due to loss of oöcytes in energetically stressed females, rather than recruitment of new oöcytes in well-fed individuals (C. B. Jørgensen 1992b).

Many authors have assumed that egg size and clutch size are inversely related to each other within populations. Indeed, this type of trade-off between offspring size and number is a central assumption of much recent work on life-history evolution (G. Bell and Koufopanou 1986; Stearns 1989, 1992; Roff 1992). Evidence for an intraspecific trade-off of egg size and number is mixed. A number of studies have reported a positive correlation between egg size and clutch size (e.g., Sofianidou and Kyriakopoulou-Sklavounou 1983; Semlitsch 1985b; Matsui 1989; Kyriakopoulou-Sklavounou and Loumbourdis 1990), a pattern to be expected when both egg and clutch size increase with body size. However, some investigators have found a negative correlation between egg and clutch size after eliminating the effect of body size, indicating that females with relatively large clutches for their body size produce relatively small eggs (e.g., *Ambystoma jeffersonianum* complex: Wilbur 1977c; *A. tigrinum:* Kaplan and Salthe 1979; *Bufo bufo:* Reading 1986; *Bufo viridis:* Castellano, Cucco, and Giacoma 2004; *Rana temporaria:* Cummins 1986; Gibbons and McCarthy 1986; *Hyla calypsa:* Lips 2001; *Hyla labialis:* Lüddecke 2002b). Other studies have failed to show such a relationship (e.g., *Ambystoma opacum* and *A. maculatum:* Kaplan and Salthe 1979; *Bombina orientalis:* Kaplan 1987; *Bufo calamita:* Tejedo 1992b; five species of *Litoria:* Morrison and Hero 2002). Rico et al. (2004) reported a negative correlation between egg size and clutch size in field-deposited clutches of *Scinax trapicheiroi*, but they did not know the body sizes of the females that laid the eggs.

Presumably a trade-off between egg size and clutch size is most likely to be observed when most females in a population have been able to recruit near-maximum numbers of oöcytes and abundant food has enabled them to fully supply them with yolk. Because the number of oöcytes is determined first, females that have large clutch sizes relative to body size will face constraints imposed by the total space available for developing eggs. Consequently, they will be able to allocate less yolk to each egg than will females with smaller clutches relative to body size. For example, Kaplan and Salthe (1979) observed a trade-off between egg and clutch size only in a population of *Ambystoma tigrinum* that had a larger total clutch volume than predicted from an interspecific regression for several species in the genus. They suggested that this population was more constrained by available body space than were populations of other species that did not exhibit such a trade-off. Gibbons and McCarthy (1986) attributed the negative relationship between size-adjusted egg and clutch size in *Rana temporaria* in Ireland to an unusually large total investment in eggs compared to populations of the same species in other parts of Europe.

Lüddecke (2002b) reported that females of a tropical treefrog, *Hyla labialis,* that were supplied with unlimited food in the laboratory showed a significant trade-off between clutch size and egg size. Females in the best condition tended to produce larger clutches of smaller eggs than did those in poorer condition. Nevertheless, there was not a tendency for females with the largest total volume of eggs to have the strongest trade-off between egg and clutch size. Studies of repeated spawnings by the same females in captivity also suggested that certain females are simply more physiologically efficient at converting energy into eggs. Some females consistently had higher total volumes of eggs than did other females, yet these females often returned to a better body condition than did females that laid smaller egg clutches (Lüddecke 1995a).

### Variation within Clutches

In addition to variation in egg size among females in the same population, egg size also varies within a single clutch. Such intraclutch variation has been documented in a number of anurans and urodeles, including tropical hylids (Crump 1981b; Lips 2001), *Pseudacris crucifer* (Crump 1984b), *Crinia signifera* (Williamson and Bull 1989), *Pseudophryne australis* (Thumm and Mahony 2005), *Bufo calamita* (Tejedo and Reques 1992), *Bombina orientalis* (Kaplan and King 1997), two species of *Hynobius* (Takahashi and Iwasawa 1988b), and several ambystomatid and plethodontid salamanders (Beachy 1993a). Several authors have suggested that such variation is an adaptation to fluctuating or unstable environments, where the relative advantage of large and small eggs is unpredictable. The hypothesis has been variously referred to as "adaptive coin flipping" (Cooper and Kaplan 1982; Kaplan and Cooper 1984) or "bet hedging" (Crump 1981b, 1984b; Lips 2001; Thumm and Mahony 2005). It suggests that females produce a variety of egg sizes within a single clutch to ensure that at least some are well adapted to the prevailing conditions encountered that year. This model has been questioned on theoretical grounds, mainly because fluctuating environments alone are not sufficient to result in selection for variable egg size (McGinley, Temme, and Geber 1987). Others have disagreed, arguing that variation in egg size is a widespread if not yet fully understood phenomenon that could have an adaptive explanation (Bernardo 1996b).

### Consequences of Variation in Egg Size

The hypothesis that within-clutch variation in egg size represents an adaptive response to fluctuating environments was invoked to account for ranges of variation that are larger than expected if there is a single optimum egg size for a given species or population. To test this idea, it is first necessary to examine fitness consequences of variation in egg size among and within clutches. Attempts to do so have produced mixed results. Some of the differences in results reported for different species probably reflect differences in experimental techniques, sample sizes, or methods of measuring larval performance, while others reflect real differences among species. In most studies, variation in egg size among clutches in the same population has been correlated with variation in larval life-history traits, particularly the time required to reach hatching, feeding, and metamorphosis, and the sizes of larvae at each of those stages. Only two studies have examined relationships within individual egg clutches (Crump 1984b; Williamson and Bull 1989). Unfortunately, most authors have not measured all of these indicators of larval performance, making interspecific comparisons difficult.

In most species, egg size appears to have little or no effect on time until hatching or metamorphosis (table 10. 7). Loman (2002a) reported that tadpoles of *Rana temporaria* hatching from large eggs developed faster and metamorphosed earlier than did those from small eggs raised in the same conditions. Most other studies have not found such effects. The most consistent effect of large egg size is on the size of larvae at hatching and at the first feeding stage. Whenever a significant effect has been detected, it has been positive (table 10. 7), indicating that large eggs tend to give rise to large hatchlings. It is not clear that this effect always carries over to metamorphosis, however. Only a few workers have examined size at metamorphosis as a function of egg size, but positive effects were found in only two species, *Taricha torosa* (R. Kaplan 1985) and *Rana sylvatica* (Berven and Chadra 1988). In *Taricha,* the correlation was measured indirectly; size at metamorphosis was positively correlated with size at first feeding, which in turn was positively correlated with egg size. In *Rana sylvatica,* the positive effect was seen only when larvae were raised at high densities; at low densities, the pattern was reversed, with smaller eggs actually producing larger metamorphs than did large eggs, particularly at high food levels (Berven and Chadra 1988).

**Table 10.7** Effect of egg size on larval life-history traits in amphibians

| | Effect on time | | | Effect on size | | |
|---|---|---|---|---|---|---|
| Species (source no.) | H | F | M | H | F | M |
| Among-clutch variation | | | | | | |
| *Ambystoma barbouri* (5) | | | 0 | | 0 | 0 |
| *A. maculatum* (1) | +[a] | – | | + | + | |
| *A. maculatum* (4) | | | | + | | |
| *A. opacum* (4) | 0 | – | | + | + | |
| *A. talpoideum* (4) | | | | 0 | | |
| *A. texanum* (5) | | | 0 | | + | 0 |
| *A. tigrinum* (1) | | | – | | + | |
| *Taricha torosa* (3) | | | –[b] | | + | + |
| *Bombina orientalis* (7, 10) | 0 | | | + | | |
| *Bufo calamita* (9) | | + | 0 | | + | 0 |
| *Crinia signifera* (8) | – | 0 | | + | + | |
| *Pseudophryne australis* (13) | 0 | | | + | | |
| *Rana sylvatica* (6) | | | | | | |
| Low density, high food | | | – | | | – |
| High density, low food | | | 0 | | | + |
| *Rana temporaria* (11, 12) | | | –[c] | | | +[c] |
| Within clutch variation | | | | | | |
| *Crinia signifera* (8) | 0 | 0 | | + | + | |
| *Pseudacris crucifer* (2) | 0 | | 0 | + | | 0 |
| *Pseudophryne australis* (13) | 0 | | | + | | |

*Sources:* (1) Kaplan 1980; (2) Crump 1984b; (3) Kaplan 1985; (4) Walls and Altig 1986; (5) Petranka, Sih, L. B. Kats, and J. R. Holomuzki 1987; (6) Berven and Chadra 1988; (7) Kaplan 1989; (8) Williamson and Bull 1989; (9) Tejedo and Reques 1992; (10) Parichy and Kaplan 1995; (11) Laugen, Laurila, and Merilä 2002; (12) Loman 2002; (13) Thumm and Mahony 2005.

*Notes:* Effects of increasing egg size can be positive (+), negative (–), or nonsignificant (0). Blank cells indicate effects that were not measured. H = hatching; F = feeding; M = metamorphosis.

[a]Time to hatching was positively correlated with egg size in only one of three experimental temperature regimes for *Ambystoma maculatum*.

[b]Time to metamorphosis was negatively correlated with feeding size in *Taricha torosa* when larvae were provided with unlimited food, but was positively correlated when food was limited.

[c]Egg size affected time to metamorphosis in one study of *Rana temporaria* and size at metamorphosis in another.

The reasons for this are unclear, but it points up the complex nature of relationships between egg size and other measures of offspring fitness (Bernardo 1996b). Most studies have examined this relationship under ideal laboratory conditions with larvae kept at relatively low densities and supplied with abundant food. In the field, these conditions are likely to vary considerably from year to year and among ponds. This makes predictions about exactly what size of egg will yield the highest fitness difficult. The only attempt to test the prediction that egg size should be more variable in species that experience unpredictable environments was M. Crump's (1981b) study of five Neotropical treefrogs. She predicted that frogs breeding in temporary ponds, which were more variable in several ways, should have greater intraclutch variation in egg size than permanent-pond breeders, but this was not the case. However, temporary-pond breeders exhibited a somewhat more even distribution of egg sizes, which Crump interpreted as partial support for the hypothesis.

### Variation among Populations

Variation among populations in both egg size and clutch size has been documented for both anurans and urodeles (Pettus and Angleton 1967; Kozlowska 1971; Koskela and Pasanen 1975; Berven 1982a, 1987; Berven and Gill 1983; Banks and Beebee 1986b; Cummins 1986; Reading 1986; Beattie 1987; Takahashi and Iwasawa 1988a, 1989a; Matsui 1989; Bruce and Hairston 1990; Elmberg 1991; Lai, Kam, and Lin 2003). Usually differences among populations have been attributed to differences in temperature, altitude, or latitude (Morrison and Hero 2003b). Populations from cool regions, high altitudes, or high latitudes often have larger eggs and smaller clutches than do populations from warmer environments. This trend is not universal, however (Berven and Gill 1983; Elmberg 1991), and in some cases simply reflects differences in female body size (Laugen, Laurila, and Merilä 2002). Furthermore, the populations in question often differ in other respects as well, such as the nature of the breeding site. Year-to year variation within populations has seldom been compared with variation among populations, but in some cases, it is nearly as great (e.g., Cummins 1986; Kaplan and King 1997), raising questions about adaptive interpretations of differences among populations. Studies of geographic variation in life-history traits frequently suffer from a lack of statistical replication, with comparisons made between single populations or pooled samples from different altitudes or latitudes (e.g., Pettus and Angleton 1967; Banks and Beebee 1986a; Beattie 1987; Laugen, Laurila, and Merilä 2002; Lai, Kam, and Lin 2003). Studies comparing egg size or clutch size along a gradient of altitude, latitude, or temperature (e.g., Matsui 1989; Takahashi and Iwasawa 1989a; Elmberg 1991) provide more convincing evidence of a relationship between life-history traits and environmental variables. At best, even these studies simply reveal a pattern of correlation, but not the underlying causes of the relationship. Only experimental studies can reveal whether geographic variation is due mainly to direct environmental effects on life-history traits or genetic differences among populations.

The most comprehensive experimental analysis of geographic variation in egg and clutch size for any amphibian is Keith Berven's work on North American wood frogs (*Rana sylvatica*). He initially studied wood frogs from nine ponds along an altitudinal gradient from near sea level to over 1000 m (Berven 1982a). Females from montane populations reached sexual maturity at a later age and at a larger size than lowland females. Both egg and clutch size were positively correlated with body size within and between populations, resulting in larger absolute egg and clutch sizes in montane populations. However, relative to body size, montane females produced larger eggs and smaller clutches than did lowland females, reflecting a trend reported in other frogs. In an experiment that ran for several years, Berven reciprocally transferred thousands of newly metamorphosed juveniles between lowland and montane ponds and then monitored their survival, age, and size at sexual maturity, and the size of eggs and clutches laid by females.

The experiment showed that both genetics and the environment in which the juveniles grew up influenced age and size at first reproduction. For example, lowland females that remained in their home ponds reached sexual maturity in two years, whereas individuals from the same population transferred to the mountains matured in three years. However, this was still faster than montane females reared in mountain ponds, which usually took three to four years to reach maturity (montane females transferred to the lowlands took two to three years). Age at first reproduction was more sensitive to environmental influences in montane females, resulting in considerable variation among individuals. Size at first reproduction for montane females was less plastic and was similar in both environments, although there was considerable year-to-year variation in both environments as well. These results suggest that the cooler temperatures and shorter growing season in the mountains are largely responsible for the later maturity of montane females, but there are genetic differences among populations that reinforce this trend.

In contrast to the strong environmental effect on age and size at maturity, egg size and clutch size were entirely determined by the population of origin, indicating genetic differences between populations that presumably result from differential selection in the two environments. Montane females had larger eggs and smaller clutches, relative to body size, than lowland females, regardless of which environment they were in (fig. 10.24). Berven speculated that selection acts mainly on egg size, with other life-history traits changing secondarily. He suggested that large eggs, and the large energy reserves associated with them, would be advantageous in cool montane environments, where larval periods are relatively long and pond productivity is low. Presumably, the additional energy provided by large eggs should enable the larvae to reach metamorphosis more rapidly or at a larger size under the adverse conditions of high-altitude ponds than larvae hatched from small eggs. In fact, experiments on larval growth and development showed that rate of development was largely determined by environmental temperature, regardless of where the eggs came from, whereas growth rate and size at metamorphosis were affected by both the environment and genetic differences between populations (Berven 1982b). Montane larvae grew faster and were larger at all developmental stages than lowland larvae. Reciprocal

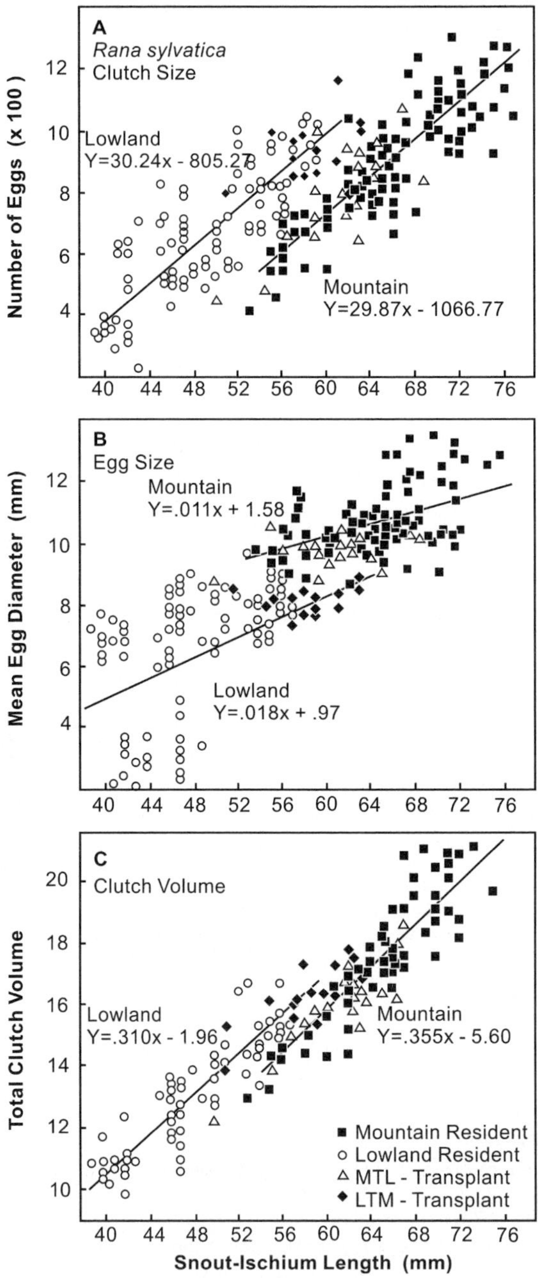


**Fig. 10.24.** Relationship of clutch size, egg size (mm) and total clutch volume (mm$^3$) to body size for *Rana sylvatica* females from lowland and mountain populations. Also shown are data for females transplanted as juveniles from mountain to lowland habitats (MTL) and lowland to mountain habitats (LTM). Mountain females are larger, on average, than lowland females. Mountain females also have larger eggs and smaller clutches than do lowland females of the same body size. After Berven (1982a).

matings between lowland and montane males and females also revealed a significant nongenetic maternal effect on length of the larval period and size at metamorphosis. Larvae derived from montane mothers had shorter larval periods and larger size at metamorphosis than larvae derived from lowland mothers when they were reared under the same conditions in the laboratory. Presumably this maternal effect was due mainly to larger eggs being produced by montane females.

In contrast to the general expectation that latitudinal variation should parallel altitudinal variation, wood frogs from the far northern parts of their range tend to be smaller than are those from lower latitudes, and they lay smaller eggs. High latitudes resemble high altitudes in having short growing seasons, but they also have longer days, and both pond temperatures and productivity often are higher than in high-altitude ponds. In this type of environment, rapid development is favored over large size at metamorphosis, and the advantage of having large eggs is reduced because of relatively high larval food availability (Berven 1982a; Berven and Gill 1983). It is difficult to determine precisely the fitness advantages of large and small eggs in different environments because of the large effect of larval density and food availability (Berven and Chadra 1988). These do not vary in predictable ways among habitats, and indeed, year-to-year variation within one habitat can equal or exceed differences among habitats.

## Summary and Conclusions

Amphibians exhibit some of the most varied modes of reproduction of any group of vertebrates. All three major clades of living amphibians have species with internal and external fertilization, but the distribution of modes of fertilization varies from nearly all external in anurans to mostly or entirely internal in urodeles and caecilians, respectively. In all three clades, external fertilization is considered an ancestral trait, although the morphologically primitive frog *Ascaphus* is one of the few anurans with internal fertilization. In most anurans, external fertilization occurs while the male and female are in amplexus, but amplexus has been secondarily lost a number of times in anuran evolution, mostly in species with nonaquatic reproduction. Early in anuran evolution, there was a shift from inguinal amplexus to axillary amplexus, with the latter being far more common today. The adaptive significance of this shift in amplexus position is not entirely clear, but it has major implications for the evolution of oviposition behavior and anuran mating systems. One important development in the oviposition behavior of anurans with axillary amplexus is a tendency for egg-laying times to be reduced and for eggs to be laid in compact masses. This in turn may have set the stage for the evolution of resource-based mating systems, in which males defend oviposition sites (see chapter 8).

In urodeles, the ancestral external mode of fertilization is retained only in the most primitive families, the cryptobranchids, hynobiids, and sirenids, and sometimes is accompanied by male defense of oviposition sites, a type of mating sytem that is not known in salamanders with internal fertilization. The evolution of the spermatophore was a key event in the evolution of urodeles, because it allows mating and oviposition to be separated in time. This in turn makes parental care by males unlikely; parental care by females is the rule in urodeles (see chapter 11). Internal fertilization by means of spermatophores, coupled with a capacity of females to store sperm, reduces the effectiveness of male mate guarding and allows for multiple paternity of egg clutches. The length of time that sperm remains viable inside the female varies among salamanders from a few days to several years, but the reasons for this variation are largely unknown. All caecilians have internal fertilization, accomplished by means of a special copulatory organ and not by means of spermatophores. In contrast to both anurans and urodeles, in which viviparous reproduction is rare, the evolution of internal fertilization has been followed by the evolution of viviparity in many caecilian lineages.

Reproductive modes, defined in terms of egg deposition sites and mode of embryonic and larval development, are extremely diverse among anurans. Currently nearly 40 different reproductive modes have been described for anurans, but these could easily be subdivided into many more distinct modes of reproduction. The most common modes, deposition of eggs in either still or flowing water with an aquatic larval stage, are found throughout the world in many different families. These reproductive modes acutally encompass a wide range of variation in the location of oviposition sites, the structure of egg masses, and the details of embryonic and larval development. The significance of this variation is just beginning to be explored.

The large number of more specialized and apparently more derived modes of reproduction, often involving partial or complete removal of the embryos and larvae from the water, are found almost exclusively in the wet tropics. It is striking how much still remains to be learned about the basic features of anuran reprodutive biology, with as many as ten new modes of reproduction having been described in recent years just from one area, the Atlantic rainforest of Brazil, a hotspot for endemic species of frogs. It seems likely that many more unusual modes of reproduction eventually will be discovered in tropical forests around the world. One feature of anuran reproductive modes that deserves more attention from researchers is the high degree of plasticity in choice of oviposition sites in many species. Some tropical frogs use a wide range of temporary pools, sites on or near the ground, or elevated sites to lay eggs, resulting in some species exhibiting several different modes of reproduction even in the same population. In other species, males build nests in some locations, but not in others. The fitness consequences of such extreme variation in oviposition site choice have scarcely been addressed.

One widespread evolutionary theme among frogs in the humid tropics is the evolution of various mechanisms to avoid placing eggs directly in oxygen-poor environments, such as warm, swampy pools, or to mediate the effects of such environments on embryonic development. These mechanisms include production of very small eggs with rapid development, placement of eggs in surface films in contact with the air, placement of eggs in bubble or foam nests that provide supplemental oxygen supplies, placement of eggs in well-oxygenated streams, or placement of eggs out of water altogether. Each of these diverse modes of oviposition entails unique costs and benefits that have yet to be studied in detail for most species.

The reproductive modes of urodeles and caecilians are much less diverse than are those of anurans, in part because both of these clades contain only a small fraction of the number of species of anurans. Furthermore, the urodele clade is dominated by a single family, the Plethodontidae, most of which exhibit direct development of nonaquatic eggs. Indeed, this mode of reproduction, and the lack of dependence on standing water, may have allowed this one lineage to colonize a variety of temperate and tropical habitats. There is evidence from phylogenetic studies, however, for one or more reversals of life-history evolution, with a reacquisition of an aquatic larval stage in some desmognathine salamanders. This form of life history may be particularly suitable for salamanders in environments with abundant, well-oxygenated aquatic habitats, such as streams, springs, and caves, where food resources for newly hatched offspring are more abundant or more accessable in water than on land.

Along with the great diversity of reproductive modes found in amphibians, there also is considerable inter- and intraspecific variation in both egg size and clutch size, and some variation in the frequency of reproduction. In general, egg size appears to be dictated by reproductive mode in both anurans and urodeles. Eggs laid in oxygen-poor aquatic environments invaribly are small, regardless of the size of the adult female, whereas clutch sizes tend to be large and highly correlated with female body size. In well-oxygenated aquatic habitats, eggs are invariably larger, with clutch sizes correspondingly smaller. For eggs laid outside of water, oxygen limitations are less important, but problems of evaporative water loss tend to favor large eggs, regardless of the size of the female. Hence, tiny species of *Eleutherodactylus* with direct development produce enormous eggs relative to their body size, but cannot produce very many of them (in the most extreme cases, producing only one at a time). Larger ranids or bufonids with direct development can produce even larger eggs, but they are smaller relative to adult body size, and clutch sizes are larger.

Arboreal egg-layers exhibit a wide range of egg sizes and clutch sizes, and the significance of this variation has yet to be explored in detail. Foam-nesting rhacophorid treefrogs tend to have relatively small eggs and large clutches, presumably to facilitate rapid development in hot environments, where eggs are placed over ephemeral rain pools. Some hylid frogs that lay nonfoamy egg masses on leaves over water have relatively small eggs with little jelly protection, as in *Hyla ebraccata*. Others, such as frogs in the genus *Agalychnis*, have relatively large eggs with abundant jelly surrounding the clutch. Presumably the first mode of reproduction is advantageous in environments where rapid embryonic development is favored, whereas the latter may be advantageous where longer development time and larger hatchlings are favored.

In many amphibians, intraspecific variation in egg size or clutch size, or both, is related to variation in body size, with larger females usually producting larger clutches and often producing larger eggs. Indeed, selection for increased reproductive output probably is a major determinant of female body size and probably accounts for the fact that females of most amphibians are larger than males. Not all species show a positive relationship between body size and egg or clutch size, however. In some cases, this is because age of the female also affects these variables, regardless of body size. In some facultatively paedomorphic salamanders, alternative life history strategies are correlated with differences in egg size or clutch size. In some anurans that produce multiple egg clutches each season, second or third clutches often are smaller than first clutches, although this depends on the time intervals between clutches. At high altitudes, some females do not reproduce every year, and those that have longer intervals between clutches sometimes produce larger clutches. All of these patterns suggest that female amphibians often are energy limited, and allocate less energy to reproduction when food supplies are low or intervals between egg-laying events are short. When females are not energy limited, they may allocate the maximum possible amount of energy to reproduction, which can result in a statistical tradeoff between number of eggs and egg size that is not apparent when females lack the energy reserves to develop a full complement of eggs.

Some amphibians exhibit considerable variation in egg size even within clutches. The physiological basis for such variation, which presumably results from differential allocation of energy reserves to different eggs, has not been thoroughly investigated. Some investigators have suggested that this variation represents a "bet-hedging" strategy that is best suited for variable and unpredicatable environments in which small eggs with rapid development may be favored in some circumstances, and larger eggs with slower development in other circumstances. There have been relatively few rigorous tests of this hypothesis. Most studies have shown that egg size has little or no effect on time until hatching or time until metamorphosis, but a number of studies have shown a positive effect of egg size on size of larvae at hatching or at the first feeding stage. These studies indicate positive selection on egg size, but there is relatively little evidence of selection for variation in egg size. Many amphibians also exhibit geographic variation in egg size and clutch size, although again, the adaptive significance of such variation is poorly understood and deserves further experimental investigation.

# **Chapter 11** Parental Care

*Wandering over the meadow, now sodden with a recent rain, I observed a small frog. . . . To my astonishment, I saw that its back was covered with little black bodies, set close together like paving-stones on a street. . . . I very quickly saw that these were tadpoles, so crowded in the small space that the tails and part of the bodies were hidden.*
—Herbert H. Smith, "On oviposition and nursing in the Batrachian genus *Dendrobates*" (1887)

NINETEENTH-CENTURY naturalists who were familiar with the rather uniform breeding habits of North American and European frogs often were amazed by the variety of reproductive strategies encountered in the tropics. Some tropical frogs lay their eggs in water and abandon them, but others have remarkably complex forms of parental behavior. Although the occurrence of parental care in amphibians is no longer astonishing, it still holds considerable interest for behavioral ecologists. Parental care refers to any type of parental investment in offspring after the eggs are laid or the young are born, but does not include the initial energetic investment in producing gametes (Trivers 1972; Clutton-Brock 1991). I exclude from my discussion of parental care those species that have evolved viviparous reproduction; these modes of reproduction were discussed in chapter 10. For parental behavior to be considered true parental care, it is assumed that such behavior increases the survivorship of the young, although in practice this has been demonstrated experimentally only in a small subset of species that exhibit parental care. Parental care usually entails some cost to the caregiving parent, although again, only a few studies have thoroughly documented such costs.

Some form of parental care has been reported in most of the major amphibian families, having evolved independently many times. The eggs of amphibians lack protective shells or membranes and are exposed to a variety of dangers. Terrestrial eggs are susceptible to desiccation, whereas aquatic eggs suffer from floods, droughts, or lack of oxygen (see chapter 4). The eggs are subjected to attacks by pathogenic bacteria, fungi, and a host of predators, such as leeches, insects, spiders, snails, fishes, amphibians, and snakes (see chapter 14). Parental care presumably has evolved as a response to the dangers posed by these physical and biological threats to survival. Amphibian parental care has been reviewed a number of times (Salthe and Mecham 1974; Lamotte and Lescure 1977; McDiarmid 1978; Wells 1981a; Nussbaum 1985; M. Hayes 1991), with the most comprehensive reviews being those of M. Crump (1995b, 1996), Nussbaum (2003), and Lehtinen and Nussbaum (2003). Crump's papers include very complete tables summarizing reports of parental care in individual species and a thorough discussion of the evolution of parental behavior, but some interesting examples of parental care in amphibians have been described since those reviews were written. I will first describe the basic patterns of parental care in amphibians and then discuss experimental work on the benefits and costs of parental care and major trends in the evolution of parental care. Whenever possible, I will try to interpret the evolution of parental care in an explicitly phylogenetic context.

## Patterns of Parental Care

The frequency of parental care varies among the major amphibian groups. McDiarmid (1978) estimated that parental care is present in about 10–15% of anuran species. Crump (1996) reported that parental care has been observed in about 6% of anuran species, representing 15–17 different families, depending on the family-level classification that is used. This figure undoubtedly underestimates its actual occurrence, because the reproductive biology of many species is unknown, especially in the tropics, where parental care is most common. Parental care has not been reported in the families Ascaphidae, Bombinatoridae, Brachycephalidae, Heleophrynidae, Pelobatidae, Pelodytidae, Pseudidae, or Rhinophrynidae. In several of the largest families of anurans, including the Bufonidae, Hylidae, Leptodactylidae, Myobatrachidae, and Ranidae, parental care either is uncommon, or is restricted to a few lineages within the family (some of these families are not monophyletic groups as presently constituted; see chapter 1). Parental care is more common in salamanders, occurring in at least 18% of all species. Parental care is particularly common in the family Plethodontidae, which includes more than 70% of living urodele species, and probably occurs in many species for which it has not been reported, simply because eggs of these species have never been observed. Parental care has not been reported in the family Rhyacotritonidae, and it is uncommon in hynobiids and ambystomatids (Nussbaum 1985, 2003). Crump (1995b, 1996) did not list any examples in the family Salamandridae, but more recent work on Asian newts of the genus *Pachytriton* in captivity indicates that females aggressively guard their eggs (Thiesmeier and Hornberg 1997, 1998). Egg attendance by females also has been reported in *Euproctus montanus* (Nussbaum 2003). The ancestral mode of reproduction in caecilians is oviparity, whereas most of the derived clades are viviparous (see chapter 10). Most oviparous caecilians are thought to exhibit parental care, but well-documented examples are rare (less than 10 species; M. Wake 1977a, 1992; Crump 1995b, 1996).

## Parental Care in Anurans

Several classifications of anuran parental care have been proposed. McDiarmid (1978) recognized 12 categories, based on oviposition habitat (aquatic or terrestrial), site of parental investment (nest, burrow, or on the parent), nature of the larval stage (free-swimming or direct development), and sex of the caregiving parent. In a subsequent review, I combined his categories into four major types of parental care: attendance of eggs, attendance of larvae or young, transport of eggs, and transport of larvae (Wells 1981a). Because of more recent work, it is now necessary to add several additional categories, including attendance and transport of terrestrial juveniles and feeding of young (Crump 1995b, 1996; Lehtinen and Nussbaum 2003). In some anurans, either the male or female builds some sort of nest to receive the eggs, but they do not necessarily remain with the eggs after they are laid. These nests include mud basins constructed at the edges of streams, underground burrows, foam nests placed on the water or in vegetation, and nests made by wrapping leaves around the eggs. These modes of reproduction were discussed in chapter 10 and will not be included in the discussion of parental care unless parents remain with the eggs.

### Attendance of Eggs

This is the most common form of parental behavior, and occurs in 13–15 anuran families, depending on which classification is used (tables 11.1–11.3). Attendance of aquatic eggs is relatively rare. In the large Neotropical tree frog *Hyla rosenbergi*, males construct mud nests in which eggs are laid. Males sometimes guard the nests after eggs are laid (Kluge 1981), although Höbel (2000) reported that males in a population in Costa Rica exhibited only facultative nest construction and did not guard their eggs. Other closely related species build similar nests (M. Martins 1993a, b). Parental attendance of nests has not been reported in most of these, but in *H. faber*, males guard their nests when densities of calling males are high (fig. 11.1). This suggests that the main benefit of parental care in this species is to protect the eggs from being damaged by intruding males (M. Martins,

**Fig. 11.1.** Male *Hyla faber* from Brazil guarding a clutch of eggs in a streamside nest. Photo by Celio F. B. Haddad.

**Table 11.1** Examples of egg attendance by frogs

| Species (source no.) | Attendant | Oviposition site | Calling site |
|---|---|---|---|
| Astylosternidae | | | |
| *Trichobatrachus robustus* (9) | Male | Rocks in stream | Rocks in stream |
| Bufonidae | | | |
| *Nectophryne afra* (12, 25, 56) | Male | Shallow pool | Shallow pool |
| *N. batesi* (54) | Male | Shallow pool | Shallow pool |
| *Nectophrynoides malcolmi* (28) | Male | Terrestrial nest | |
| *Oreophrynella nigra* (55) | Male or female[a] | Burrow or tunnel | Rock crevice |
| Centrolenidae | | | |
| *Centrolene geckoideum* (40, 66) | Male | Rock near waterfall | Rock near waterfall |
| *C. notostictum* (58) | Male | Leaf over stream | Leaf over stream |
| *C. petrophilum* (58) | Male | Rock in stream | Rock in stream |
| *C. prosoblepon* (47, 58) | Male[b] | Rock or leaf | Rock or leaf |
| *Cochranella savagei* (33) | Male | Leaf over stream | Leaf over stream |
| *Hyalinobatrachium bergeri* (59) | Male | Leaf over stream | Leaf over stream |
| *H. colymbiphyllum* (29, 57, 79) | Male | Leaf over stream | Leaf over stream |
| *H. fleischmanni* (35, 47, 57) | Male | Leaf over stream | Leaf over stream |
| *H. orientalis* (49) | Male | Leaf over stream | Leaf over stream |
| *H. valerioi* (29, 57) | Male | Leaf over stream | Leaf over stream |
| *H. vireovittata* (57) | Male | Leaf over stream | Leaf over stream |
| Hemisotidae | | | |
| *Hemisus guineensis* (54, 56) | Female | Burrow | Ground |
| *H. guttatus* (8, 63) | Female | Burrow | Ground |
| *H. marmoratus* (63, 68) | Female | Burrow | Ground |
| Hylidae | | | |
| *Hyla faber* (67) | Male | Mud nest | Mud nest or vicinity |
| *H. rosenbergi* (36) | Male | Mud nest | Mud nest or vicinity |
| Hyperoliidae | | | |
| *Alexteroon obstetricans* (18, 56) | Female | Leaf over water | Leaf over water |
| *Hyperolius spinigularis* (15) | Female | Leaf over water | Leaf over water |
| Leiopelmatidae | | | |
| *Leiopelma archeyi* (45) | Male | Under rocks or vegetation | Under rocks or vegetation |
| *L. hamiltoni* (45) | Male | Under rocks or vegetation | Under rocks or vegetation |
| *L. hochstetteri* (45) | Male | Small pools | Ground near pools |
| Leptodactylidae | | | |
| *Cycloramphus boraceiensis* (74) | Male | Rock crevice | Wet rocks |
| *C. stejnegeri* (31) | Female | Depression in soil | Unknown |
| *Geobatrachus walkeri* (30) | Female | Terrestrial nest | Unknown |
| *Leptodactylus chaquensis* (69) | Female | Aquatic foam nest | Shallow water |
| *L. fallax* (11, 32, 39, 77) | Female[c] | Terrestrial foam nest | Ground |
| *L. fuscus* (17, 52) | Female | Burrow built by male | Near burrow |
| *L. insularum* (53, 73) | Female | Aquatic foam nest | Shallow water |
| *L. ocellatus* (19, 22) | Female | Aquatic foam nest | Shallow water |
| *L. podicipinus* (69) | Female | Aquatic foam nest | Shallow water |
| *L. validus* (64) | Female | Aquatic foam nest | Shallow water |

(*continued*)

**Table 11.1** (continued)

| Species (source no.) | Attendant | Oviposition site | Calling site |
|---|---|---|---|
| *Lithodytes lineatus* (43) | Female | Terrestrial foam nest | Ground |
| *Thoropa petropolitana* (31) | Male | Wet rock surface | Wet rock |
| *Zachaenus parvulus* (31) | Female | Terrestrial nest | Unknown |
| Mantellidae | | | |
| *Mantidactylus argenteus* (61) | Male | Leaf over stream | Unknown |
| *M. bicalcaratus* (75) | Male or female | Leaf over water-filled leaf axil | Leaves |
| *M. majori* (81) | Male | Leaf over stream | Leaves |
| *M. punctatus* (75) | Male or female | Leaf over water-filled leaf axil | Leaves |
| *M. webbi* (61) | Male | Rock in stream | Rock in stream |
| Megophryidae | | | |
| *Brachytarsophrys intermedia* (80) | Male | Under rock in stream | Under rock in stream |
| Microhylidae | | | |
| *Albericus darlingtoni* (6) | Unknown | Terrestrial nest | Ground |
| *A. sp. "brownie"* (70) | Male, female | Tree fern, tree crotch detritus | Arboreal vegetation |
| *A. sp. "greenie"* (70) | Male | Tree crotch detritus | Arboreal vegetation |
| *Anodonthyla boulengerii* (20) | Male | Tree hole | Tree hole |
| *A. montana* (61) | Male | Pool in rocks or wet moss | Unknown |
| *A. nigrigularis* (61) | Male | Tree hole | Tree trunk |
| *A. rouxae* (61) | Male | Hollow bamboo | Bamboo trunk |
| *Aphantophryne pansa* (37, 38) | Male | | |
| *Austrochaperina blumi* (70) | Male | Fissure in clay bank | Unknown |
| *A. brevipes* (48, 69) | Male | Forest floor | Forest floor |
| *A. derongo* (70) | Male | Cavity in soil | Ground |
| *A. fryi (73)* | Male | Under debris | Ground |
| *A. macrohynca* (70) | Male | Cavity in stream bank | |
| *A. mehelyi* (10) | Female | Forest floor | Unknown |
| *A. novaebritanniae* (69) | Female? | Small depressions | Unknown |
| *A. palmipes* (37, 70) | Male | Streamside burrow | Ground |
| *A. robusta* (73) | Unknown | Under log | Ground |
| *Barygenys flavigularis* (50) | Male | Burrow | Unknown |
| *Breviceps adspersus* (8) | Female | Burrow | Ground |
| *B. gibbosus* (5, 8) | Female | Burrow | Ground |
| *B. sylvestris* (8) | Female | Under rock | Ground |
| *Callulops robustus* (16) | Male | Tree stump | Burrow |
| *C. sp. cf. "robusta"* (75) | Male | Burrow | Burrow |
| *C. wilhelmanus* (16) | Male | Log, moss, litter | Burrow |
| *Cophixalus aenigma* (78) | Male | Hole in bank | Hole in bank, under litter, debris, and rocks |
| *C. bombiens* (78) | Unknown | Under log | Unknown |
| *C. cheesmani* (72) | Male | Topside of leaf | Leaves |
| *C. cocinnus* (78) | Unknown | Under log | Hidden elevated sites in tree holes, rock crevices, or vegetation |
| *C. exiguus* (78) | Male | Under epiphyte | Unknown |
| *C. hosmeri* (78) | Male | Under leaf litter | Under leaf litter |

*(continued)*

**Table 11.1** (continued)

| Species (source no.) | Attendant | Oviposition site | Calling site |
|---|---|---|---|
| *C. infacetus* (48, 78) | Male | Under fern, under leaf litter | Ground, logs, shrubs, under litter |
| *C. mcdonaldi* (78) | Male | In palm axil | In palm axil |
| *C. monticola* (78) | Male | In palm axil | In palm axil |
| *C. neglectus* (48, 78) | Male | Rotting log, under palm frond, under leaves | Ground |
| *C. neglectus* (78) | Female | Under rock | |
| *C. ornatus* (48, 78) | Male[d] | Under moss, under leaf in tree hole, under rock slab | Ground or elevated sites; under leaf litter in tree hole |
| *C. parkeri* (41) | Male | Moss, litter, log | Low plants |
| *C. riparius* (23, 37) | Male, female | Tree fern | Tree fern, palm |
| *C. saxitalis* (78) | Male | Rock crevice | Unknown |
| *C. sphagnicola* (37, 38) | Male | | |
| *C. sp. "b"* (72) | Male | Cavity in soil | Unknown |
| *C. sp. "big & green"* (72) | Male | Tree cavity above ground | Arboreal vegetation |
| *Cophyla phyllodactyla* (62) | Male | Tree hole, bamboo | Leaves |
| *Copiula sp.*(72) | Male | Enclosed cavity in soil | Ground |
| *Hylophorbus rufescens* (72) | Male, female | Shallow cup in soil | Ground |
| *Liophryne rhododactyla* (37) | Male | Forest floor | Unknown |
| *L. schlaginhaufeni* (72) | Male | Soil depression, leaf litter | Ground |
| *Myersiella microps* (13) | Female | Terrestrial nest | Unknown |
| *Oreophryne anthonyi* (2) | Male, female | | |
| *O. flava* (6) | Unknown | Terrestrial nest | Unknown |
| *O. sp.* (61) | Male | Underside of leaf | Leaf clusters |
| *O. sp. "rattler"* (72) | Male | Underside of leaf | Leaves |
| *Oxydactyla brevicrus* (6) | Unknown | Forest floor | Forest floor |
| *O. stenodacylus* (71) | Unknown | Moss tussock | Unknown |
| *Platypelis grandis* (20) | Male | Tree hole | Tree hole |
| *Plethodontohyla mihanika* (76) | Male | Tree hole | Tree hole |
| *P. notosticta* (20) | Male | Tree hole | Tree hole |
| *P. tuberata* (20) | Male | Burrow | Unknown |
| *P. sp.* (61) | Male | Water-filled leaf | Tree trunk |
| *Probreviceps rhodesianus* (6) | Female | Burrow | Open ground |
| *Sphenophryne cornuta* (72) | Male | Leaf litter | Ground |
| *Synapturanus salseri* (21) | Male | Burrow | Burrow |
| *Xenobatrachus sp. "blue spot"* (72) | Male | Cavity in embankment | Cavity |
| *X. sp. cf. "mehelyi"* (72) | Male, female | Soil depression under log | Unknown |
| Myobatrachidae | | | |
| *Adelotus brevis* (4) | Male | Hole or crevice | Hole or crevice |
| *Arenophryne rotunda* (44) | Male[e] | Burrow | On ground or underground |
| *Assa darlingtoni* (46) | Female[f] | Terrestrial nest | Ground |
| *Limnodynastes dorsalis* (4) | Male | Burrow | Burrow |
| *Philoria sphagnicola* (4) | Female | Wet moss | Ground |
| *Pseudophryne bibroni* (14, 27) | Male | Burrow | Burrow |
| *P. coriacea* (60) | Male | Under grass mat | Ground |

(*continued*)

**Table 11.1** (continued)

| Species (source no.) | Attendant | Oviposition site | Calling site |
|---|---|---|---|
| *P. corroboree* (14) | Male | Burrow | Burrow |
| *P. dendyi* (14, 27) | Male | Burrow | Burrow |
| *P. douglasi* (7) | Male | Shallow pool | Shallow pool |
| *P. semimarmorata* (14, 27) | Male | Burrow | Burrow |
| Petropedetidae | | | |
| *Petropedetes cameronensis* (34, 54) | Male | Leaf near stream | Unknown |
| *P. newtoni* (34, 54, 56) | Male | Rock near stream | Rock near stream |
| *P. parkeri* (54) | Male | Rock near stream | Rock near stream |
| *P. parreti* (34, 54) | Male | Rock near stream | Rock near stream |
| *Phrynodon sandersoni* (34, 56) | Female | Leaf near stream | Vegetation |
| Ranidae | | | |
| *Limnonectes finchi* (51) | Male | Under leaves | Under leaves |
| *Limnonectes palavanensis* (51) | Male | Under leaves | Under leaves |
| *Pyxicephalus adspersus* (1, 3, 8) | Male | Shallow water | Shallow water |
| *R. ishikawae* (26) | Male | Water in burrow | Burrow |
| Rhacophoridae | | | |
| *Chirixalus eiffingeri* (65, 70) | Female, male[g] | Bamboo stumps | Bamboo stumps |
| *Chiromantis xerampelina* (8) | Female | Arboreal foam nest | Trees |
| Sooglossidae | | | |
| *Sooglossus gardinearii* (42) | Female | Terrestrial nest | Ground |
| *S. seychellensis* (42) | Female | Terrestrial nest | Ground |

*Sources:* (1) B. Balinsky and Balinsky 1954; (2) Zweifel 1956; (3) Poynton 1957; (4) J. Moore 1961; (5) W. Rose 1962; (6) Tyler 1963a; (7) Main 1964; (8) Wager 1965; (9) Perret 1966; (10) Tyler 1967; (11) G. Brooks 1968; (12) Scheel 1970; (13) Izecksohn et al. 1971; (14) Pengilley 1971a; (15) Stevens 1971; (16) Zweifel 1972; (17) Lescure 1973; (18) Amiet 1974a; (19) Vaz-Ferreira and Gerhau 1974; (20) Blommers-Schlösser 1975b; (21) Pyburn 1975; (22) Vaz-Ferreira and Gerhau 1975; (23) Menzies 1976; (24) Poyton and Pritchard 1976; (25) Lamotte and Lescure 1977; (26) Katsuren, Tanaka, and Ikehara 1977; (27) Woodruff 1977; (28) Grandison 1978; (29) McDiarmid 1978; (30) Ardila-Robayo 1979; (31) Heyer and Crombie 1979; (32) Lescure 1979; (33) Cannatella 1980a; (34) Amiet 1981; (35) M. Clark 1981; (36) Kluge 1981; (37) Simon 1982; (38) Zweifel and Allison 1982; (39) Lescure and Letellier 1983; (40) Lynch, Ruiz, and Rueda 1983; (41) Simon 1983; (42) Nussbaum 1984; (43) Regös and Schlüter 1984; (44) Roberts 1984; (45) B. Bell 1985; (46) Ehmann and Swan 1985; (47) Jacobson 1985; (48) Zweifel 1985; (49) Cannatella and Lamar 1986; (50) Stocks 1987; (51) Inger and Voris 1988; (52) Martins 1988; (53) Wells and Bard 1988; (54) Amiet 1989; (55) McDiarmid and Gorzula 1989; (56) Amiet 1991; (57) M. Hayes 1991; (58) Ruiz-Carranza and Lynch 1991b; (59) Ruiz-Carranza and Lynch 1991c; (60) Cogger 1992; (61) Johnston and Richards 1993; (62) Glaw and Vences 1994; (63) Passmore and Carruthers 1995; (64) Downie 1996; (65) Kam et al. 1996; (66) Grant et al. 1998; (67) Martins et al. 1998; (68) Kaminsky et al. 1999; (69) Prado et al. 2000; (70) Kam, Chuang, and Yen 2000; (71) Zweifel 2000; (72) Bickford 2001; (73) Ponssa 2001; (74) Giaretta and Facure 2003; (75) Lehtinen 2003; (76) Vences, Raxworthy, Nussbaum, and Glaw 2003; (77) Gibson and Buley 2004; (78) Hoskin 2004; (79) Drake and Ranvestel 2005; (80) B. Stuart 2005; (81) Vences and de la Riva 2005.

*Notes:* Species in the genus *Eleutherodactylus* are listed separately in Table 11.2. Dendrobatid frogs, which also carry tadpoles, are listed in Table 11.3. In cases where both males and females have been observed with egg clutches, the most frequent attendant is listed first.

[a]Both males and females of *Oreophrynella nigra* were found near egg clutches laid communally in underground tunnels. One female was found with a single clutch.

[b]Jacobson (1985) reported that *Centrolene prosablepon* females remained with eggs for up to 2 hours, but this may not be parental care.

[c]*Leptodactylus fallax* females reported to feed tadpoles.

[d]Several adults of *Cophixalus ornatus* of undetermined sex found under slabs of exfoliating rock with multiple egg clutches.

[e]One male *Arenophryne rotunda* was found sitting on an egg clutch; another clutch had no attendant.

[f]Females of *Assa darlingtoni* guard eggs, but males carry tadpoles.

[g]Males of *Chirxalus eiffingeri* attend eggs in bamboo stumps; females return to feed eggs.

Pombal, and Haddad 1998), as has been suggested for *H. rosenbergi* as well (Kluge 1981).

Females of several species of *Leptodactylus* remain with eggs in an aquatic foam nest or in a burrow near water (Lescure 1973, 1979; Vaz-Ferreira and Gehrau 1975; Lescure and Letellier 1983; Downie 1996; Prado, Uetanabaro, and Lopes 2000; I. Martins 2001; Ponssa 2001), and they often remain with the tadpoles after they hatch (see the following "Attendance of Tadpoles and Young"). In *L. fallax*, a species that places foam nests in cavities on the ground, both males and females remain near the nest and are aggressive toward human intruders and, presumably, potential predators (Gibson and Buley 2004). Attendance of foam nests by males has been reported in the Australian myobatrachid *Adelotus brevis* (J. Moore 1961) and the African bufonid *Nectophryne afra* (Scheel 1970). Males of the hairy frog of Africa, *Trichobatrachus robustus* (Astylosternidae), are reported to attend their eggs, which are laid in streams (Perret 1966).

Attendance of nonaquatic eggs is more common, and has been reported in bufonids, centrolenids, dendrobatids, hemisotids, hyperoliids, leiopelmatids, leptodactylids, mantellids, microhylids, myobatrachids, petropedetids, rhacophorids, rhinodermatids, and sooglossids (Crump 1995b, 1996; Lehtinen and Nussbaum 2003). The widespread occurrence and repeated evolution of egg attendance in different families suggests that the transition from unattended eggs to attended eggs is a relatively simple process, since it merely requires a parent to remain at the oviposition site after eggs are laid. Parental care is not essential for successful nonaquatic oviposition, however. Many frogs that lay eggs on land or on vegetation do not care for them, but simply abandon them at the oviposition site (Lamotte and Lescure 1977; see also chapter 10). There are, for example, many species of hylids that lay eggs on leaves overhanging bodies of water, but parental attendance of eggs has not been reported in any of these species (J. Hayes 1991). Many species of *Eleutherodactylus* that lay their eggs on the ground or in leaf litter also abandon them (Townsend 1996).

The details of parental care are not known for most of the species listed in tables 11.1–11.3; most accounts consist of brief observations of adult frogs found in association with eggs at oviposition sites. Nevertheless, detailed studies of a few species have revealed a considerable amount of variation in the nature and persistence of parental behavior. In some frogs, a parent sits on the egg mass, with the body pressed against the eggs. Species of frogs that exhibit this form of parental care include two species of *Leiopelma;* most, if not all of the microhylids from New Guinea (fig. 11.2 A, B); the African petropedetid *Phrynodon sandersoni;* many species of *Eleutherodactylus* (fig. 11.2 C); and some species of *Colostethus* and *Epipedobates* (Dendrobatidae; Lüddecke 1974; McDiarmid 1978; Amiet 1981, 1989; Simon 1983; Townsend, Stewart, and Pough 1984; B. Bell 1985; Weygoldt 1987; Johnston and Richards 1993; Crump 1995b; Diesel, Baurle, and Vogel 1995; Joglar, Burrowes, and Rios 1996; Burrowes 2000; Bickford 2001, 2004; Lehtinen and Nussbaum 2003). In some of these species, the parent remains with the eggs most of the time during development, but in others, parents attend eggs only at night and seek more protected retreat sites during the day (McDiarmid 1978; M. Hayes 1991; Bickford 2004). Attendance can be prolonged, especially in species with direct development (about three weeks in *Eleutherodactylus coqui;* up to three months in *Cophixalus parkeri*). Female African shovel-nosed frogs (*Hemisus*) have been observed in close contact with eggs when they were unearthed from underground burrows, but the details of their parental behavior underground are not known (Rödel et al. 1995; Rödel 1996; Kaminsky, Linsenmair, and Grafe 1999).

Other species are less attentive and do not always sit directly on the eggs. For example, males of the centrolenid frog *Hyalinobatrachium fleischmanni* and several closely related species do not attend their eggs continuously, but return periodically to sit on them for brief periods of time, apparently releasing water onto the eggs (M. Hayes 1991). In *H. fleischmanni,* such behavior occurs mostly after midnight and is so infrequent that Greer and Wells (1980) failed to observe it. M. Hayes (1991) reported that attendance of eggs was more frequent in dry weather than in wet weather, suggesting a degree of plasticity in parental behavior. Males of this species spend relatively little time attending eggs when they are not providing them with water, whereas males of some centrolenids attend their eggs for longer periods of time. For example, McDiarmid (1978) reported that males of *H. valerioi* attended eggs during the day and at night, although they usually did not sit on the egg masses (fig. 11.2 D). In contrast, males of *H. colymbiphyllum* attended eggs only at night. They also spent most of their time sitting next to, but not on, egg masses. In *Centrolene geckoideum,* which lays eggs attached to rocks near waterfalls, males have been observed sitting on or near egg masses, but the amount of time that males spend attending eggs has not been determined (Lynch, Ruiz, and Rueda 1983; T. Grant, Bolívar-G., and Castro 1998). Males of some hyperoliid frogs also remain near eggs laid on vegetation, but do not sit on the eggs (fig. 11.2 E).

Males of several African frogs in the genus *Petropedetes* are reported to attend eggs only at night, and they sit next to the eggs, not directly on them (Amiet 1981). Males of the Taiwanese rhacophorid frog *Chirixalus eiffingeri* remain at nest sites in bamboo stumps, sometimes in contact with eggs laid on the walls of the bamboo (fig. 11.2 F), but females return periodically to feed the tadpoles (see the following; Kam, Chuang, and Yen 1996). In *Colostethus stepheni,* a dendrobatid frog that does not transport its tadpoles, males remain within about 20 cm of the oviposition site after eggs are laid, but visit the site only occasionally after the first two to three days (Junca, Altig, and Gascon 1994; Junca 1996). In other species that do transport tadpoles, such as *C. beebei,* males visit egg clutches repeatedly to moisten them, but do not remain with them continuously (Bourne et al. 2001). Observations of two other dendrobatids in captivity, *Epipedobates pulchripectus* and *E. pictus,* revealed that males typically sat on their egg clutches at night, but deserted them for variable intervals during the day, either to rehydrate or to feed. Males of *E. pictus* were reported to become

**Table 11.2** Examples of egg attendance in the genus *Eleutherodactylus*

| Species (source no.) | Attendant | Oviposition sites | Calling sites |
|---|---|---|---|
| *E. abbotti* (26) | Male[a] | Leaf litter | Not reported |
| *E. amadeus* (17) | Male | Under rocks, logs | Ground or vegetation |
| *E. amplinympha* (23) | Male[a] | Rock crevice on cliff | Rocks |
| *E. antillensis* (33) | Male[b] | Leaf litter | Shrubs |
| *E. augusti* (3) | Male[a] | Under rock | Rocks, ground |
| *E. auriculatoides* (26) | Male[a] | Bromeliads | Not reported |
| *E. bakeri* (26) | Male | Bromeliads | Bromeliads? |
| *E. cochranae* (30) | Male | Bromeliads | Bromeliads |
| *E. cooki* (6, 25, 30, 31) | Male | Under rocks | Tops of rocks |
| *E. coqui* (11, 30) | Male | Dead leaves, petioles | Shrubs, trees |
| *E. diastema* (26) | Male[a] | Bromeliads, leaf axils | Bromeliads, leaf axils |
| *E. glandulifer* (26) | Male | Under rocks | Ground |
| *E. guantanamera* (19) | Male | Bromeliad axils | Bromeliads |
| *E. hedricki* (9) | Male | Tree holes | Tree holes, trunks |
| *E. inoptatus* (26) | Male | Under rocks, burrows | Not reported |
| *E. intermedius* (20) | Male[a] | Hole in soil | Hole in soil |
| *E. montanus* (26) | Male[a] | Under rocks | Not reported |
| *E. portoricensis* (13) | Male | Dead leaves, petioles | Shrubs, trees |
| *E. tetajulia* (24) | Male | Hole in tree fern | Hole in tree fern |
| *E. unicolor* (26) | Male[a] | Not reported | Not reported |
| *E. ventrilineatus* (26) | Male | Not reported | Not reported |
| *E. wightmanae* (26, 36) | Male | Leaf litter | Leaf litter |
| *E. alticola* (15 | Either | Under rocks | Not reported |
| *E. johnstonei* (12, 27) | Either | Leaf litter | Shrubs, trees |
| *E. affinis* (26) | Female | Under rock | Not reported |
| *E. albipes* (35) | Female | Under rock | Not reported |
| *E. angelicus* (15) | Female | Terrestrial cavity | Not reported |
| *E. anomalus* (13) | Female[a] | Depression in sand | Ground or rocks |
| *E. atkinsi* (18) | Female[a] | Cavity in ground | Not reported |
| *E. caryophyllaceus* (5) | Female | Leaf surface | Shrubs |
| *E. cundalli* (11, 22) | Female | Leaf litter, caves | Shrubs, trees, caves |
| *E. decoratus* (7) | Female | Under rocks | Not reported |
| *E. dimidiatus* (16) | Female[a] | Leaf litter | Not reported |
| *E. fitzingeri* (13, 32) | Female | Leaf litter | Ground, shrubs, logs |
| *E. glaucoreius* (26) | Female | Soil cavities, banks | Not reported |
| *E. lymani* (28) | Female | Under rocks | Not reported |
| *E. martinicensis* (1, 10) | Female | Under rocks | Shrubs, walls |
| *E. maussi* (5) | Female | Leaf litter | Not reported |
| *E. nicefori* (26) | Female[a] | Under rocks | Not reported |
| *E. noblei* (34) | Female | Under leaves in tree buttress | Not reported |
| *E. nubicola* (1) | Female | Under rocks | Rocks, banks |
| *E. pantoni* (26) | Female | Soil cavities, banks | Not reported |
| *E. pentasyringos* (2) | Female | Under rocks | Not reported |
| *E. percultus* (26) | Female[a] | Bromeliad axil | Not reported |

(*continued*)

**Table 11.2** (continued)

| Species (source no.) | Attendant | Oviposition sites | Calling sites |
|---|---|---|---|
| *E. psephosypharus* (21) | Female[a] | Hole under stump | Not reported |
| *E. richmondi* (9) | Female | Decaying log | Logs, ground |
| *E. simoterus* (26) | Female[a] | Under rock | Not reported |
| *E. terraebolivaris* (6) | Female | Soil and leaf litter | Not reported |
| *E. varleyi* (16) | Female[a] | Leaf litter | Not reported |
| *E. zygodactylus* (26) | Female[a] | Depression in soil | Not reported |

*Sources:* (1) Lynn and Grant 1942; (2) Lynn and Dent1943; (3) Jameson 1950; (4) Adamson, Harrison, and Bailey 1960; (5) Heatwole 1962a; (6) Heatwole 1963; (7) Bogert 1969; (8) Myers 1969; (9) Drewry 1970; (10) Lemon 1971; (11) Stewart 1979; (12) Lescure 1980; (13) J. D. Lynch and Myers 1983; (14) Townsend, Stewart, and Pough 1984; (15) M. Hayes 1985; (16) Estrada 1987; (17) Hedges et al. 1987; (18) Novo, Estrada, and Moreno 1987; (19) Estrada 1990; (20) Estrada 1992; (21) J. A. Campbell, Savage, and Meyers 1994; (22) Diesel, Baurle, and Vogel 1995; (23) Kaiser, Green, and Schmid 1995; (24) Estrada and Hedges 1996; (25) Joglar, Burrowes, and Rios 1996; (26) Townsend 1996; (27) Bourne 1997; (28) J. D. Lynch and Duellman 1997; (29) Bourne 1998; (30) Joglar 1998; (31) Burrowes 2000; (32) Quijano et al. 2002; (33) Ovaska and Estrada 2003; (34) Whitfield and Pierce 2003; (35) Alonso, Rodríguez, and Hernandez 2005; (36) Joglar et al. 2005.

*Notes:* Species are arranged by the sex of the care-giving parent. Information based on data summarized in Wells (1981a), Crump (1995), and Townsend (1996) and later papers, with additional data supplied by Daniel Townsend.

[a]Indicates species for which only single observations of an attending parent are available, or the extent of parental care is uncertain.

[b]One male observed near egg clutch; other clutches found without attending parent.

**Table 11.3** Parental care and aggressive behavior in dendrobatid frogs

| Species (source no.) | Tadpole habitat | Sex caring for: Eggs | Sex caring for: Tadpoles | No. of tadpoles carried | Aggressive behavior ♂ | Aggressive behavior ♀ |
|---|---|---|---|---|---|---|
| *Allobates femoralis* (22, 27, 36, 43, 56, 79)[a] | Palm leaves, fruit husks | Male | Both | 7–22 | T | – |
| *Colostethus abditaurantius* (96) | Streams | | | | T | T |
| *C. awa* (90) | Swamps | | Male | 11 | + | – |
| *C. beebei* (104) | Bromeliads | Male | Male, female[b] | 1–2 | T | – |
| *C. chalcopis* (87, 88) | Stay in nest | Male | Male | | | |
| *C. delatorreae* (90) | Ditches | | Male | | | |
| *C. degranvillei* (94) | On male | | Male | | T | |
| *C. elachyhistus* (90) | Streams | Male | Male | 19 | | |
| *C. exasperatus* (70, 90) | Streams | | Male | 14 | | |
| *C. fascianiger* (97) | Streams | | Male | 4–5 | | |
| *C. flotator* (29, 49, 66, 93) | Streams | | Male | 2–10 | T | – |
| *C. fraterdanieli* (97) | Streams | | Male | 8–12 | + | – |
| *C. infraguttatus* (90) | Streams | | | 15 | | |
| *C. jacobuspetersi* (90) | Streams | | | 4 | | |
| *C. machalilla* (90) | Pools | | Male | 15 | | |
| *C. marchesianus* (77, 90) | Bamboo, fruit husks | | Both[c] | 5–34 | | |
| *C. marchesianus* (105) | Streams | Male | Male | 8–37 | T | |
| *C. mcdiarmidi* (78) | | | Male | 3 | | |
| *C. nubicola* (29, 49, 66, 93, 102) | Streams | | Male | 2–15 | T | – |
| *C. olfersoides* (67) | Streams | | Male | 2–3 | | |
| *C. palmatus* (14, 23) | Streams | Male | Male | 25–30 | + | + |
| *C. panamensis* (38, 40) | Streams | | Female | 20–35 | T | T |

*(continued)*

**Table 11.3** (continued)

| Species (source no.) | Tadpole habitat | Sex caring for: Eggs | Tadpoles | No. of tadpoles carried | Aggressive behavior ♂ | ♀ |
|---|---|---|---|---|---|---|
| *C. patitae* (106) | Streams | | Male | 7 | | |
| *C. pratti* (29, 49, 97) | Streams | | Female | 1–8 | T | – |
| *C. pulchellus* (90, 91) | Ditches, swamps, streams | | Male | 4 | | |
| *C. sauli* (62)[a] | Pools | Male | Male | 12–16 | + | – |
| *C. steveni* (86, 94) | Stay in nest | Male | Male | | | |
| *C. subpunctatus* (5, 95) | Pools | Male | Male | 1–21 | T | |
| *C. talamancae* (21, 27, 49, 96, 102) | Streams | | Male | 8–29 | T | – |
| *C. toachi* (90) | Streams | | Male | 16 | | |
| *C. vertebralis* (90) | Swamps, pools | | Male | 2 | | |
| *Cryptophyllobates azureiventris* (102) | Unknown | Male | Male | 12–18 | | |
| *Dendrobates auratus* (1, 2, 3, 18, 31, 32, 73) | Tree holes, palm fronds | Male | Male | 1–3 | T | + |
| *D. azureus* (10, 15–17)[a] | | Male | Male | 1–2 | – | + |
| *D. bombetes* (37) | | | Male | 1–2 | | |
| *D. castaneoticus* (74) | Fruit husks | | | | | |
| *D. fantasticus* (101) | Bromeliads | Male | Both[b] | 1–2 | | |
| *D. granuliferus* (7, 9, 21, 59, 61, 62, 82, 89, 103) | Bromeliads, leaf axils | Female | Female[d] | 1–2 | T | – |
| *D. histrionicus* (12, 20, 45, 50, 54, 61, 81) | Bromeliads, leaf axils | Female | Female[d] | 1–3 | T | – |
| *D. imitator* (71, 76) | Bromeliads | Male | Male | 1 | | |
| *D. lehmanni* (45, 50, 61)[a] | Bromeliads, leaf axils | Female | Female[d] | 1 | + | – |
| *D. leucomelas* (44, 59, 81) | Tree holes | Male | Male | 1–2 | T | + |
| *D. minutus* (21, 66, 101) | Bromeliads | | Male | 1–3 | | |
| *D. occultator* (24) | | | | | + | – |
| *D. opisthomelas* (37) | Bromeliads | | Male | 1 | | |
| *D. pumilio* (11, 34, 35, 41, 42, 46, 83) | Bromeliads, leaf axils | Male | Female[d] | 1–4 | T | – |
| *D. quinquevittatus* (74, 77) | Stumps, logs, fruit husks | | | | | |
| *D. reticulatus* (52, 57, 63)[a] | Bromeliads | Male | Both[b] | 1 | + | – |
| *D. speciosus* (61, 62, 64, 101)[a] | Bromeliads, leaf axils | Female | Female[d] | 1 | + | – |
| *D. tinctorius* (53, 62)[a] | Bromeliads | Male | Male | 1–2 | + | + |
| *D. vanzolinii* (98) | Tree holes | Both | Both[b] | 1–2 | T | – |
| *D. ventrimaculatus* (Ecuador) (96, 99, 100) | Bromeliads | Male | Male | 1–2 | T | + |
| *D. "ventrimaculatus"* (Peru) (36, 47, 51, 63, 101) | Bromeliads | Both | Both[b] | | | |
| *D. virolensis* (80) | Bromeliads | | Male | 1–2 | | |
| *Epipedobates andinus* (68) | Bromeliads | | | 2 | + | – |
| *E. bassleri* (48, 62) | | Male | Male | | | |
| *E. bilinguis* (72)[a] | | Male | Male | 8 | + | – |
| *E. boulengeri* (27) | | | Male | 4–9 | | |
| *E. braccatus* (85) | | | Male | 7 | | |
| *E. erythromus* (69) | | | Male | | | |
| *E. espinosai* (27) | | | Male | 2–3 | | |
| *E. flavopictus* (85) | Streams | | Male | 18 | | |
| *E. hahneli* (85) | Streams | Male | Male | 2–8 | | |
| *E. macero* (84) | Streams | | Male | 2–4 | | |

(*continued*)

**Table 11.3** (continued)

| Species (source no.) | Tadpole habitat | Sex caring for: Eggs | Tadpoles | No. of tadpoles carried | Aggressive behavior ♂ | ♀ |
|---|---|---|---|---|---|---|
| *E. parvulus* (58)[a] | | Male | Male | 9–16 | + | – |
| *E. petersi* (27) | | | Both | 1–8 | | |
| *E. pictus* (22, 27, 58, 85)[a] | | Male | Both[c] | 2–16 | + | – |
| *E. pulchripectus* (58)[a] | Tree holes | Male | Male | 9–16 | + | + |
| *E. silverstonei* (8, 33, 62)[a] | Pools | Male | Male | 15–17 | + | – |
| *E. smaragdinus* (27) | | | Male | 7 | | |
| *E. tricolor* (27, 60, 62)[a] | | Male | Male | 15–30 | + | – |
| *E. trivittatus* (25, 26, 27, 62) | Palm leaves | Male | Male | 3–46 | + | – |
| *Mannophryne collaris* (13, 19)[e] | Streams | | Male | | + | T |
| *M. herminae* (64)[e] | Streams | | Male | 4–15 | | |
| *M. oblitterata* (64)[e] | Streams | | Male | 7 | | |
| *M. trinitatis* (4, 6, 28, 39, 75, 92)[e] | Streams | Male | Male | 2–12 | + | T |
| *Phyllobates lugubris* (25, 26, 32, 55)[a] | Pools | Male | Male | 10–15 | + | + |
| *P. terribilis* (30, 66)[a] | Pools | Male | Male | 1–23 | + | + |
| *P. vittatus* (27, 55, 62, 102) | Tree holes | Male | Male | 7–21 | T | + |

*Sources:* (1) Senfft 1936; (2) Dunn 1941; (3) Eaton 1941; (4) Test 1954; (5) Stebbins and Hendrickson 1959; (6) Sexton 1960; (7) Goodman 1971; (8) Lüling 1971; (9) Crump 1972; (10) Hoogmoed 1972; (11) Bunnell 1973; (12) Silverstone 1973; (13) Dole and Durant 1974a; (14) Lüddecke 1974; (15) Polder 1974a; (16) Polder 1974b; (17) Polder 1974c; (18) H. Zimmermann 1974; (19) Durant and Dole 1975; (20) Polder 1975; (21) Silverstone 1975; (22) Lescure 1976; (23) Lüddecke 1976; (24) C. Myers and Daly 1976; (25) Polder 1976a; (26) Polder 1976b; (27) Silverstone 1976; (28) Van Meeuwen 1977; (29) Wells 1977a; (30) C. Myers, Daly, and Malkin 1978; (31) Wells 1978b; (32) H. Zimmermann 1978; (33) C. Myers and Daly 1979; (34) Graeff and Schülte 1980; (35) Limerick 1980; (36) Meede 1980; (37) C. Myers and Daly 1980; (38) Wells 1980a; (39) Wells 1980c; (40) Wells 1980b; (41) Weygoldt 1980a; (42) Weygoldt 1980b; (43) Weygoldt 1980c; (44) H. and E. Zimmermann 1980a; (45) H. and E. Zimmermann 1980b; (46) McVey et al. 1981; (47) Schülte 1981b; (48) Schülte 1981a; (49) Wells 1981a; (50) H. and E. Zimmermann 1981; (51) Lescure and Bechter 1982; (52) C. Myers 1982; (53) Weygoldt 1982; (54) E. and H. Zimmermann 1982; (55) H. Zimmermann 1982; (56) Hödl 1983; (57) Myers and Daly 1983; (58) Weygoldt 1983; (59) E. Zimmermann 1983; (60) H. Zimmermann 1983; (61) H. and E. Zimmermann 1983; (62) Heselhaus 1984; (63) H. and E. Zimmermann 1984; (64) Dixon and Rivero-Blanco 1985; (65) Jungfer 1985; (66) H. and E. Zimmermann 1985; (67) Weygoldt 1986; (68) C. Myers and Burrows 1987; (69) Weygoldt 1987; (70) Duellman and Lynch 1988; (71) H. and E. Zimmermann 1988; (72) Jungfer 1989; (73) Summers 1989; (74) Caldwell and Myers 1990; (75) Praderio and Robinson 1990; (76) H. and E. Zimmermann 1990; (77) Aichinger 1991; (78) Reynolds and Foster 1992; (79) Roithmair 1992; (80) Ruiz-Carranza and Ramirez-Pinilla 1992; (81) Summers 1992; (82) Van Wijngaarden and Bolanos 1992; (83) Brust 1993; (84) Rodriguez and Myers 1993; (85) Haddad and Martins 1994; (86) Junca, Altig, and Gascon 1994; (87) Kaiser and Altig 1994; (88) Kaiser, Coloma, and Gray 1994; (89) Van Wijngaarden and Van Gool 1994; (90) Coloma 1995; (91) Crump 1995b; (92) Cummins and Swan 1995; (93) Ibáñez and Smith 1995; (94) Junca 1996; (95) Fandiño, Lüddecke, and Amézquita 1997; (96) Summers and Amos 1997; (97) Grant and Castro 1998; (98) Caldwell and de Oliveira 1999; (99) Summers 1999a; (100) Summers and Earn 1999; (101) Summers et al. 1999; (102) Lötters, Jungfer, and Widmer 2000; (103) Summers 2000; (104) Bourne et al. 2001; (105) Lima and Keller 2003; (106) Lötters, Morales, and Proy 2003.

*Notes:* Many species use more than one type of tadpole deposition site; only the most commonly used sites are listed. For the genera *Allobates, Colostethus, Epipedobates, Mannophryne,* and *Phyllobates,* small numbers of tadpoles carried probably do not represent the entire complement. Aggressive behavior: (+) present, (–) absent, T = frogs are territorial in the field. Blank cells indicate that there is no information.

[a]Species in which aggressive behavior has been observed only in captivity.

[b]Males carry tadpoles; females feed tadpoles with unfertilized eggs.

[c]Males usually carry tadpoles, but occasional transport by females has been reported.

[d]Females feed tadpoles with unfertilized eggs.

[e]These species of *Mannophryne* were previously in the genus *Colostethus.*

progressively less attentive to their eggs and started advertising for additional mates before their first clutches hatched (Weygoldt 1987). Many other dendrobatids do not remain for long periods of time at the oviposition site, but return periodically to moisten the eggs and eventually to transport the tadpoles to water. Sometimes a parent can care for more than one clutch in different locations simultaneously (Wells 1978a; H. Zimmermann and Zimmermann 1981, 1988; Weygoldt 1987; K. Summers 1989). It is possible that some dendrobatids do not care for their eggs at all, but return to the oviposition site only to carry their tadpoles (Roithmair 1992).

### Transport of Eggs

Frogs in four distantly related families have evolved the habit of carrying eggs on a parent's body. In each family a single clade exhibits a unique form of parental care. In the midwife toads, *Alytes* (Discoglossidae), pairs mate on land, and after the eggs are fertilized, the male wraps the strings of eggs around his hind legs. The eggs are carried for about

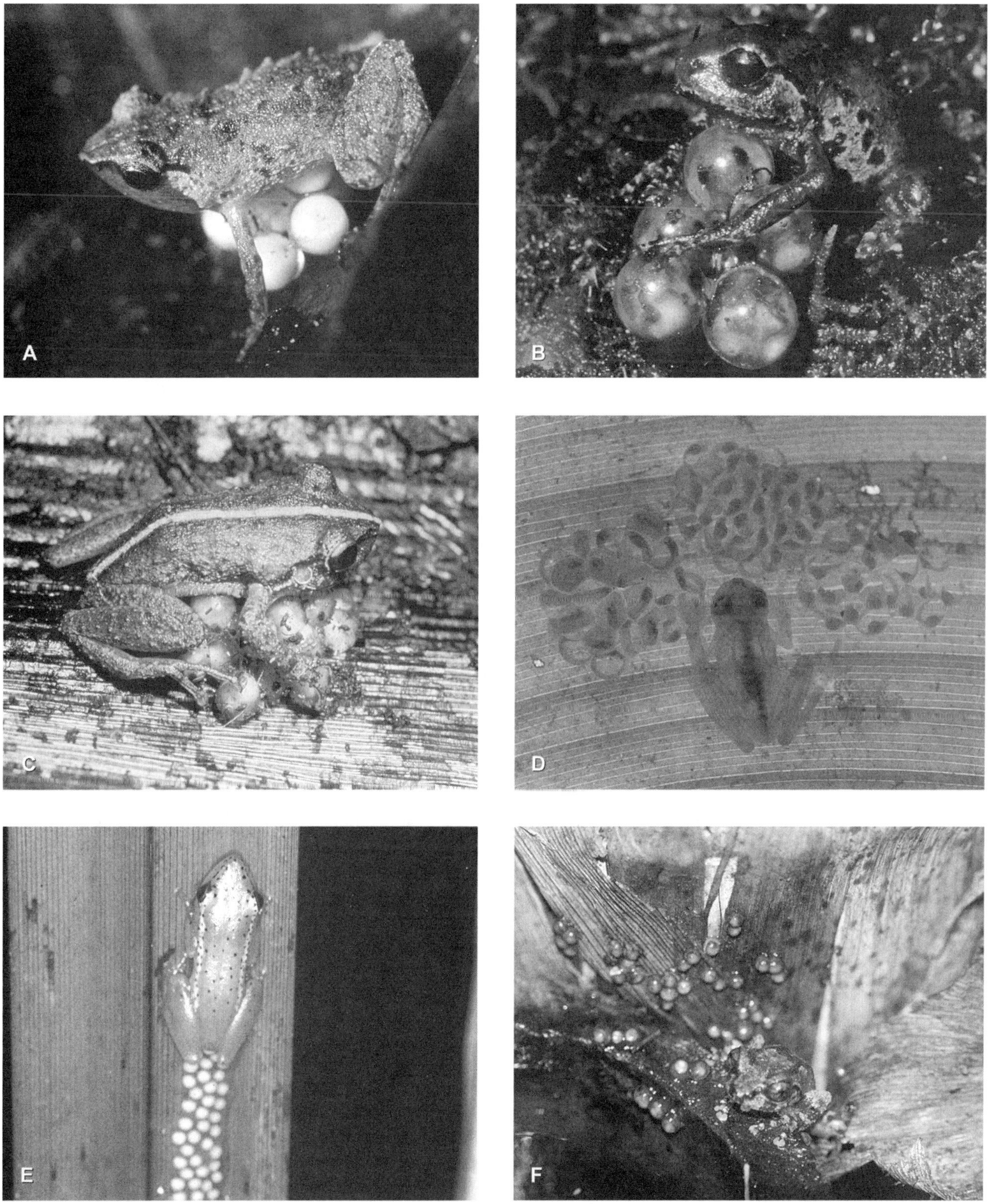


**Fig. 11.2.** Attendance of nonaquatic eggs by male anurans. (A) A male of a species of *Albericus* (Microhylidae) from New Guinea brooding a clutch of unpigmented eggs. (B) A small male of a species of *Oreophryne* (Microhylidae) from New Guinea brooding a clutch of very large eggs. (C) A male of *Eleutherodactylus coqui* (Leptodactylidae) from Puerto Rico brooding eggs. (D) A male of *Hyalinobatrachium valerioi* (Centrolenidae) from Costa Rica guarding three clutches of eggs laid on a leaf, photographed from below the leaf. (E) A male of *Hyperolius parkeri* (Hyperoliidae) from East Africa guarding a clutch of eggs on a leaf. (F) A male of *Chirixalus eiffingeri* (Rhacophoridae) guarding eggs at the edge of a hollow bamboo stump. The eggs usually are laid just above the water line, but some eggs may become submerged. Photos by David Bickford (A, B), Margaret M. Stewart (C), Walter Hödl (D, E), and Kentwood D. Wells (F).

a month until they hatch and tadpoles are released into ponds (Boulenger 1912; Lopez-Jurado, Caballero, and Dos-Santos Freitas 1979; Crespo 1982; König and Schlüter 1991; Márquez and Verrell 1991; Márquez 1992). Males are capable of carrying up to three clutches of eggs simultaneously, and some males carry more than 100 eggs at a time. Large males tend to have greater mating success and therefore carry more eggs than small males (Reading and Clarke 1988; Raxworthy 1990; Márquez 1993; Verrell and Brown 1993). The origin of this unique form of parental care, which is shared by all four species in the genus (*A. obstetricans, A. cisternasii, A. muletensis, A. dickhilleni*), is obscure. All living species that are closely related to *Alytes* lay eggs in water and lack parental care (see chapter 10). Discoglossid frogs are an ancient lineage, and the living species represent a small remnant of a much more diverse group (Sanchiz and Roček 1996) that might have included other species with parental care. The genus *Alytes* is estimated to be at least 15 million years old (Maxson and Szymura 1984), so this form of parental care could be at least that ancient.

In all other species that transport eggs, it is the female that carries the eggs. In the family Pipidae, one genus, *Pipa,* exhibits parental care; all others lay eggs in water (see chapter 10). After each egg is released by the female and fertilized by the male, the pair performs a complex acrobatic maneuver in the water that causes the egg to stick to the back of the female (see fig. 10.12 in chapter 10). Eventually the tissues on the female's back swell into a spongy matrix that engulfs the eggs. The eggs complete development in separate capsules within this tissue. In some species, including *P. parva, P. myersi,* and *P. carvalhoi,* the eggs hatch into tadpoles, which complete their development in the water. In others, including *P. pipa, P. arrabali,* and *P. snethlageae,* the eggs undergo direct development and hatch as miniature froglets (Rabb and Snedigar 1960; Rabb and Rabb 1960, 1963a; Weygoldt 1976a, b; Trueb and Cannatella 1986). Species with direct development are thought to represent the most derived condition (Duellman 1989a; see fig. 10.11 in chapter 10). As in midwife toads, it is hard to imagine the evolutionary steps leading to such an unusual form of parental care, but again, this behavior could have evolved millions of years ago. Fossil pipids in South America go back at least 95–100 million years, but the age of the genus *Pipa* is unknown (Báez 1996).

One of the strangest modes of egg transport was described in two species of Australian myobatrachids, *Rheobatrachus silus* and *R. vitellinus.* The female would ingest relatively late-stage embryos and then carry the eggs and larvae in her stomach; this behavior therefore grades into tadpole transport (Corben, Ingram, and Tyler 1974; K. McDonald and Tyler 1984; Leong, Tyler, and Shearman 1986). The female would give birth through her mouth (Tyler and Carter 1981). This peculiar behavior was possible because secretion of digestive enzymes is inhibited during gastric brooding, apparently in response to hormones secreted by the embryos (Tyler 1983; Tyler et al. 1983). These frogs were of great interest to digestive physiologists because of possible insights they might provide into mechanisms to shut down production of digestive enzymes, with possible applications to treating gastric ulcers (Tyler 1983). Unfortunately, it now appears that both species are extinct, part of a general decline in rainforest frogs in northern Australia (K. McDonald 1990; S. Richards, McDonald, and Alford 1993). Consequently, the opportunity to learn more about this unique form of parental care probably has been lost. The origins of this form of parental care are a mystery, and even the relationship of *Rheobatrachus* to other myobatrachids is uncertain (Ford and Cannatella 1993; Littlejohn et al. 1993).

Egg transport is characteristic of one group of Neotropical frogs, the Hemiphractinae. This group traditionally has been considered part of the family Hylidae (Sheil, Mendelson, and da Silva 2001), but both molecular studies (Ruvinsky and Maxson 1996; Faivovich et al. 2005) and tadpole morphology (Haas 2003) suggest that the group actually is rather distantly related to other hylids. Indeed, the group as a whole may represent as many as three separate clades that are not closely related to each other, in which case various forms of egg transport would have evolved independently (Frost et al. 2006). The embryos of these frogs are unique in developing large bell-shaped gills that provide enhanced surface area for gas exchange and sometimes completely envelop the developing embryos. In all species the female carries the eggs, but the site of egg transport and the structure of brood pouches varies considerably among different genera. In three largely terrestrial genera, *Hemiphractus, Cryptobatrachus,* and *Stefania,* large eggs are carried exposed on the back of the female, without an enclosed brood pouch (Duellman 1970; Trueb 1974; Duellman and Hoogmoed 1984; Jungfer and Boehme 1991; Crump 1995b). Del Pino (1980) thought that mucus glands on the back of the female produce secretions that glue the eggs in place, but mating and egg deposition has not been observed. Possibly the eggs are held onto the back of the female by mucus produced in the oviducts and extruded with the eggs, as has been reported for *Flectonotus goeldii* (Weygoldt and Carvalho e Silva 1991). The eggs of all of these genera undergo direct development and hatch into tiny froglets. In at least one species of *Stefania,* the hatchlings remain on the back of the female after hatching (fig. 11.3) and will even try to climb back on if they fall off (Gorzula, Morales, and Hernandez 1983; Jungfer and Boehme 1991), so egg transport grades into offspring transport in these species.

In the arboreal genus *Flectonotus,* eggs are carried in a dorsal brood pouch with an opening down the middle (Duellman and Gray 1983). In *F. goeldii* and *F. ohausi,* there is scarcely

Fig. 11.3. Female of *Stefania evansi* (Hemiphractinae) carrying froglets on her back. Members of this genus carry eggs exposed on the back of the female before hatching. Photo by Karl Heinz Jungfer.

any brood pouch. Eggs are attached to the dorsum of the female, with flaps of dorsal skin forming slight walls around them, but not meeting in the middle. There is little or no structural alteration of the skin to form a brood pouch (Del Pino 1980; Weygoldt and Carvalho e Silva 1991). The male places the eggs on the back of the female. He grasps each egg with his hind foot as it emerges from the female's cloaca. He then rotates the egg onto the female's back and beats the mucus secretions surrounding the eggs into froth. This eventually hardens to form a brood sac, which readily detaches from the female once the eggs hatch. In *F. fissilis*, the lateral skin flaps form a more complete brood pouch, with the edges glued together in the middle with mucus secretions. The most complete brood pouches are found in *F. pygmaeus* and *F. fitzgeraldi*, two closely related species. The brood pouch is completely enclosed (fig. 11.4 A). The pouch tissue swells during incubation and becomes moderately well vascularized, suggesting that there is some gas exchange between the mother and the embryos (del Pino 1980). Mating behavior has been observed in *F. pygmaeus*. The male opens the pouch with his feet and catches each egg as it is laid, pushing it into the pouch with his feet and body (Duellman and Maness 1980).

In all species of *Flectonotus*, eggs develop relatively rapidly (three to four weeks in captivity) and hatch as tadpoles in an advanced stage of development. These are deposited by the female in small bodies of water in bromeliads, the leaf axils of aroids and *Heliconia* plants, and holes in bamboo. The tadpoles do not have well-developed mouthparts (Wassersug and Duellman 1984). Initially, all species were thought to complete development without feeding by living off their abundant yolk reserves (Duellman and Gray 1983), but Weygoldt (1989b) reported that tadpoles of *F. goeldii* feed on detritus that falls into their bromeliads.

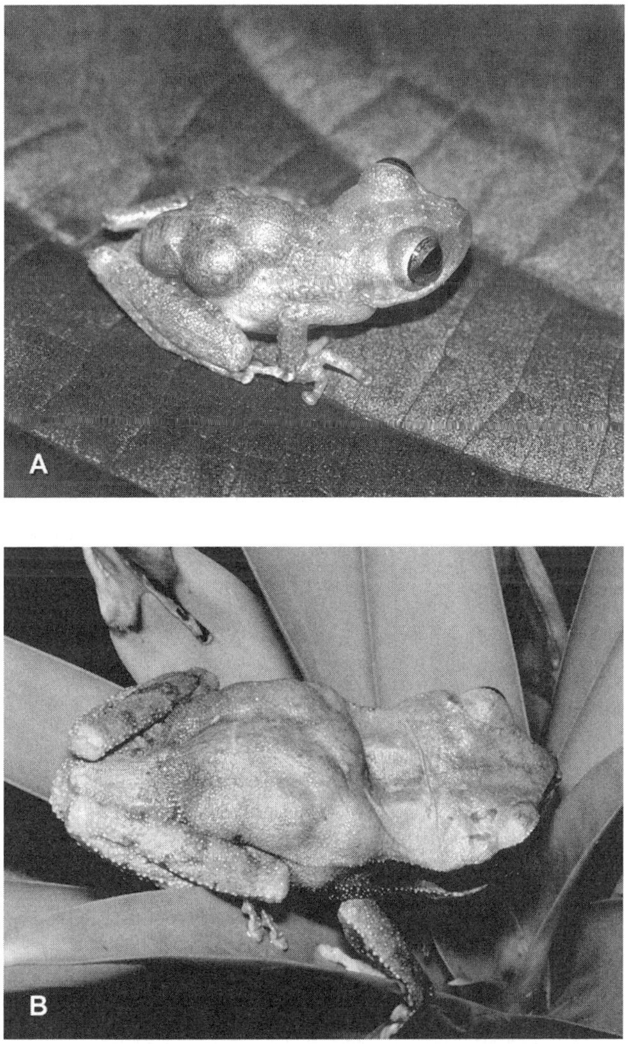


Fig. 11.4. Fully enclosed brood pouches of hemiphractine marsupial frogs. (A) Female *Flectonotus fitzgeraldi* from Trinidad with a clutch of four eggs. The male pushes the eggs into the pouch through the slit down the center of the back, which later closes over the eggs. (B) Female *Gastrotheca fissipes* from Brazil carrying a large clutch of eggs. The pouches of this genus have small posterior openings. Photos by Kentwood D. Wells (A) and Ivan Sazima (B).

The most complex brooding behavior is found in the genus *Gastrotheca,* a diverse group that includes both arboreal and terrestrial species (Duellman and Fritts 1972; Duellman 1974; Duellman and Hillis 1987). These frogs have completely enclosed brood pouches with a relatively small posterior opening (fig. 11.4 B). The male pushes the eggs into the pouch, as in *Flectonotus* (Duellman and Maness 1980). The tissue surrounding the eggs swells and becomes highly vascularized, probably providing for gas exchange between the mother and the embryos, but not exchange of nutrients (del Pino 1980). Incubation can be prolonged, up to 120 days in captivity in some species. More than half the species in the genus have direct development of eggs, and many of these live in lowland rainforest habitats. Other species in the genus have eggs that hatch into feeding tadpoles with well-developed mouthparts (Fitzgerald, Guillette, and Duvall 1979; Wassersug and Duellman 1984). Most species deposit their tadpoles in ponds, but some place them in smaller bodies of water, such as bromeliads (Duellman and Maness 1980). Tadpole-producing species of *Gastrotheca* are limited to high elevations in the Andes (Duellman and Hillis 1987; Duellman, Maxson, and Jesiolowski 1988).

The origins and evolution of egg transport in hemiphractines are difficult to reconstruct. There are no other hylids or leptodactylids that transport their eggs, and relatively few species that provide their eggs or larvae with any kind of parental care. Consequently, it is difficult to determine the intermediate steps leading from more typical aquatic or nonaquatic modes of reproduction to egg transport in this group. Molecular data suggest that this assemblage of frogs is very old, with the various genera diverging from one another as much as 50–100 million years ago (Duellman, Maxson, and Jesiolowski 1988; Frost et al. 2006). Hence, there has been ample time for the evolution of the distinct modes of egg transport seen in the different lineages, with little indication of transitions between them.

The morphology of the brood pouches suggests a transition from transport of eggs on the dorsum of the female as the ancestral condition, as seen in *Hemiphractus, Cryptobatrachus,* and *Stefania,* to the more complex, partially or completely enclosed brood pouches of *Flectonotus* and *Gastrotheca.* On the other hand, all of the species that lack a brood pouch have direct development, usually considered a derived reproductive mode. Duellman (1989a) used a phylogeny of hemiphractine frogs derived from molecular data and developmental characters to trace the evolution of reproductive modes in the group. He argued that direct development is the ancestral mode for the entire subfamily, with deposition of tadpoles, either feeding or nonfeeding, having evolved independently in several lineages. This analysis also suggested that evolution of a partially or completely enclosed brood pouch occurred only once. A basic assumption of this analysis is that the hemiphractines comprise a monophyletic clade, but this conclusion is subject to revision as more modern molecular techniques are used to explore the phylogenetic relationships of the group (Frost et al. 2006).

### Attendance of Tadpoles

Most adult frogs are terrestrial, and only a few species attend aquatic tadpoles. Males of an African bufonid, *Nectophryne afra,* remain with tadpoles for up to two weeks after they hatch, and are reported to aerate the nest by kicking their legs while swimming in place (Scheel 1970). In the African bullfrog, *Pyxicephalus adspersus* (Ranidae), males remain with their tadpoles, which form dense schools after hatching (B. Balinsky and Balinsky 1954; W. Rose 1956; Poynton 1957). Some authors questioned whether this species actually protects the tadpoles (Wager 1956, 1965; Lambiris 1971). More recent observations by Kok, du Preez, and Channing (1989) and C. Cook, Ferguson, and Telford (2001) showed that males sometimes construct channels that allow their tadpoles to move from isolated shallow pools at the edges of ponds to deeper water, where they can complete their development. C. Cook, Ferguson, and Telford (2001) also confirmed the earlier observations of B. Balinsky and Balinsky (1954) that males attack potential tadpole predators and sometimes are killed in the process.

Females of African shovel-nosed frogs in the genus *Hemisus* remain with their tadpoles in an underground burrow (fig. 11.5 A, B). Eventually the female frees the tadpoles by digging her way out of the nest and constructing a slide or channel that allows the tadpoles to reach the water. The tadpoles follow the female closely (fig. 11.5 C) and may even ride on her back as she makes her way to deeper water, where the tadpoles complete development (Wager 1965; Rödel et al. 1995; Rödel 1996; Kaminsky, Linsenmair, and Grafe 1999).

Somewhat similar behavior has been observed in several Neotropical frogs in the genus *Leptodactylus,* but in these species, eggs are laid in foam nests floating in the water. In a Panamanian species, *Leptodactylus insularum,* females lead schools of tadpoles in shallow water and push their way through obstructing mud and vegetation into deeper pools (Wells and Bard 1988; Vaira 1997). They communicate with their tadpoles by means of a "pumping" display in which the rear part of the body was moved up and down in the water, creating a series of concentric waves that moved toward the tadpoles (fig. 11.6). Females also are aggressive toward simulated predators or other frogs (Vaira 1997; Ponssa 2001). Very similar behavior has been observed in a closely related species, *L. ocellatus.* Females attend schools of tadpoles and guard them against predators (Vaz-Ferreira and Gehrau 1975). This species also has been observed leading tadpoles with pumping movements similar to those of *L. insularum* (Jan Caldwell, personal communication). Females of an-

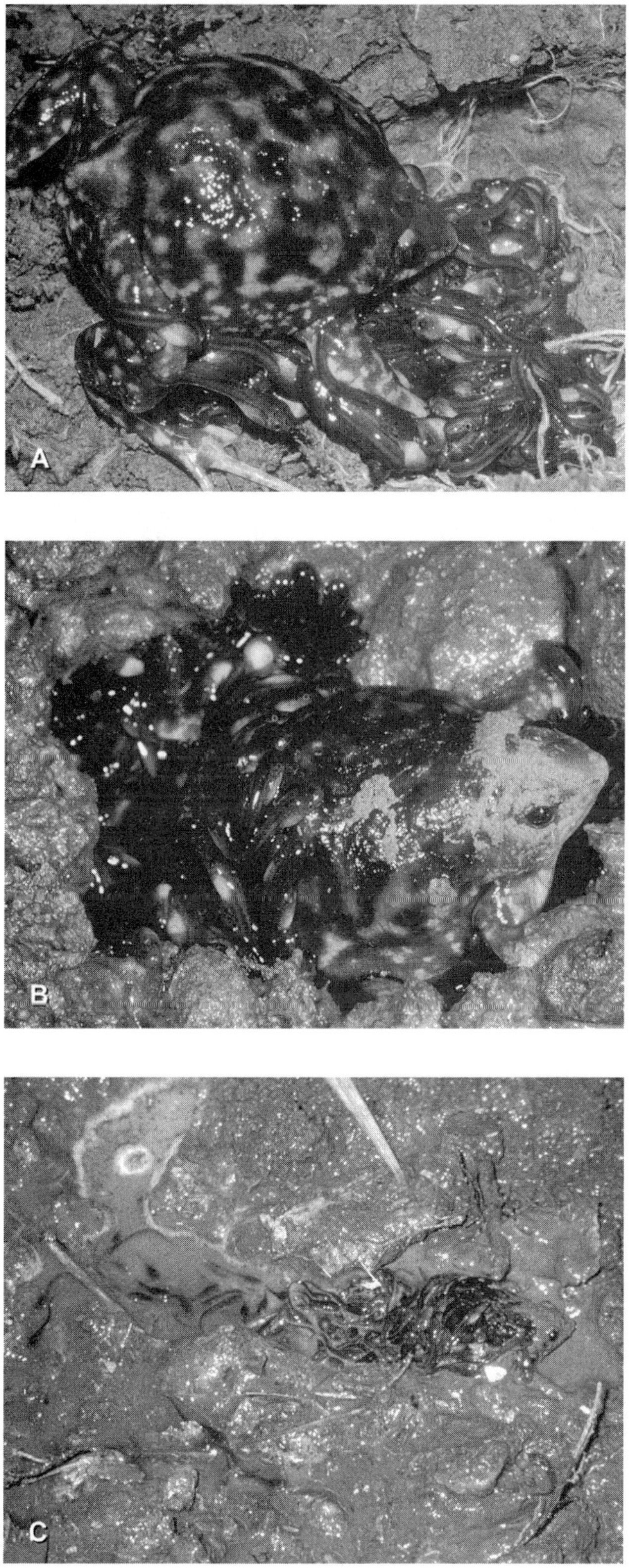


**Fig. 11.5.** Parental care in the West African frog *Hemisus marmoratus.* (A) Female attending a group of tadpoles in a subterranean nest, from which the top has been removed. (B) Tadpoles clustered around the female, with some tadpoles clinging to her back, as she digs out of the burrow. (C) School of tadpoles following the female in a shallow channel in the mud. Photos by Mark Oliver Rödel (A) and Stefan Kaminsky (B, C).

other species in the *L. ocellatus* group, *L. chaquensis,* also attend schools of tadpoles, but their behavior toward the tadpoles has not been observed in detail (Prado, Uetanabaro, and Lopes 2000).

Pumping movements by females attending schools of tadpoles have been reported in three species in the *L. melanonotus* group, which is thought to be closely related to the *L. ocellatus* group (Downie 1996; Prado, Uetanabaro, and Lopes 2000; Prado, Uetanabaro, and Haddad 2002; I. Martins 2001). In two of these species, *L. validus* and *L. leptodactyloides,* aggressive behavior by females has not been reported. Females of the third species, *L. podicipinus,* did attack simulated predators (sticks and hands). These frogs are all smaller than are members of the *L. ocellatus* group, so it is possible that aggressive attacks on predators are less common in this group. In some other species of *Leptodactylus,* a parent remains with nonfeeding tadpoles in an aquatic or terrestrial foam nest, but their behavior has not been studied in detail (G. Brooks 1968; Lescure 1973, 1979).

Parental attendance of tadpoles also has been reported in several species of anurans that have nonfeeding tadpoles that hatch from eggs laid in small bodies of water or in terrestrial nests (see chapter 10). For example, males of several microhylids from Madagascar remain with nonfeeding larvae in tree holes and leaf axils (Blommers-Schlösser 1975a). Males of the South American microhylid *Synapturanus salseri* are reported to remain with eggs and nonfeeding tadpoles in an underground nest (Pyburn 1975). In two species of *Leiopelma* in New Zealand, *L. hamiltoni and L. archeyi,* the larvae remain in the nest with the male through metamorphosis; in a third species, *L. hochstetteri,* larvae are aquatic and do not remain with a parent (B. Bell 1985; B. Bell and Wassersug 2003).

### Transport of Tadpoles

In some anurans, attendance of tadpoles tends to grade into tadpole transport. The example of some tadpoles of *Hemisus marmoratus* riding on the back of the female to water has already been discussed. Heyer and Crombie (1979) found a female of the leptodactylid frog *Cyclorhamphus stejnegeri* under a log with 40 tadpoles on her back. Whether females transport tadpoles to water or simply attend them for some period of time is not known, but the morphology of the tadpoles suggests that they complete development out of water. In *Leiopelma archeyi* and *L. hamiltoni,* the tadpoles climb onto the back of the male, but the male evidently remains at the oviposition site until tadpoles metamorphose, rather than transporting them to another site (B. Bell 1985; B. Bell and Wassersug 2003). Tadpoles of *Sooglossus seychellensis* (Sooglossidae) attach themselves to the back of a parent, where they complete their development by drawing on their large yolk reserves. Originally males were thought to perform pa-

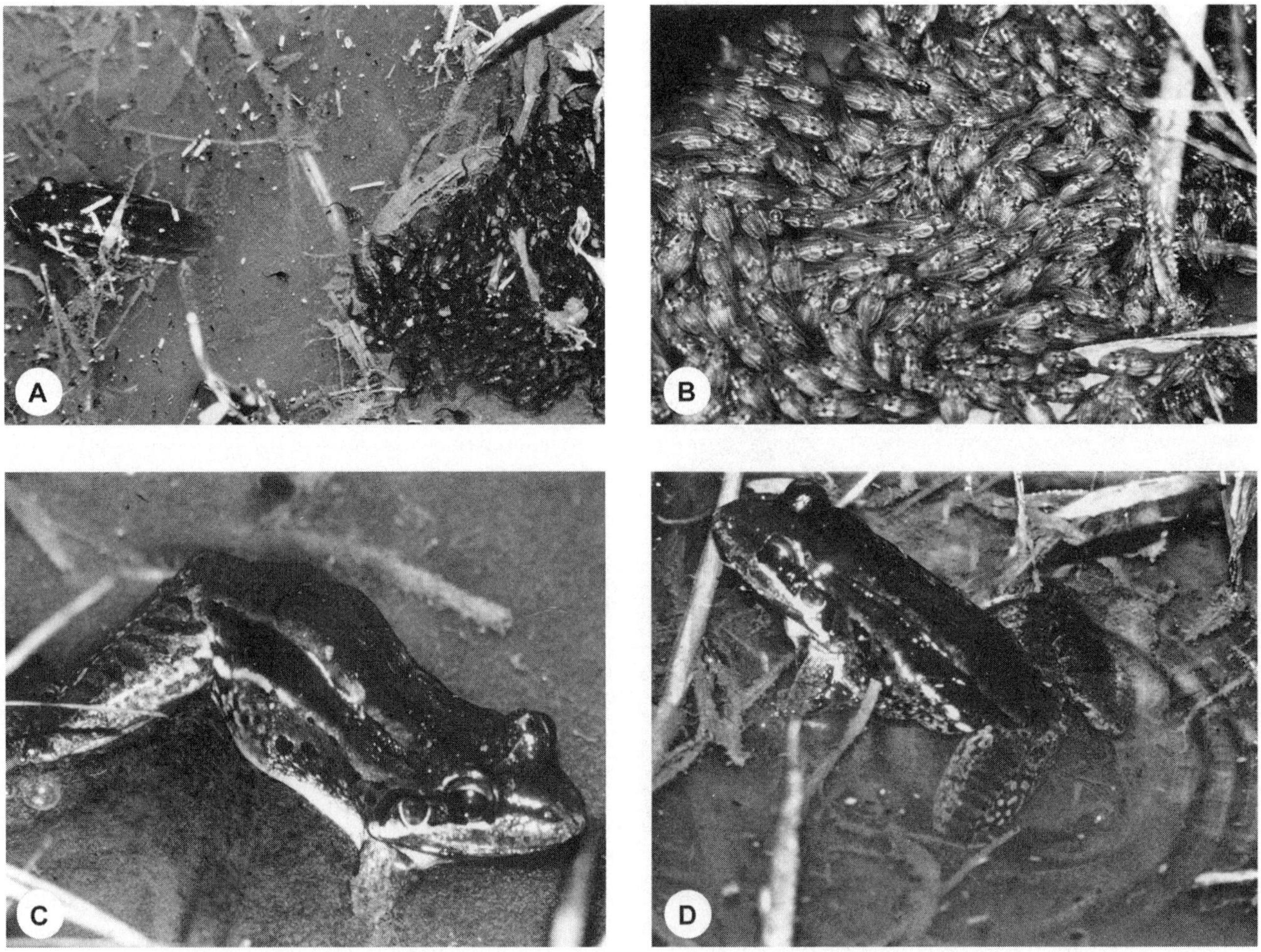

**Fig. 11.6.** Parental care in the Panamanian frog *Leptodactylus insularum*. (A) Female followed by a school of tadpoles. (B) Dense school of tadpoles. (C) Female giving pumping signal to tadpoles, with rear of the body elevated. (D) Female bringing the rear of her body down into the water, producing concentric waves on the water. Photos by Kentwood D. Wells.

rental duties, but more recent observations indicate that the female is the caregiving parent (Nussbaum 1984; Nussbaum, unpublished data cited in Crump 1995b). Tadpoles of the dendrobatid frog *Colostethus degranvillei* also are reported to complete development on the back of the female (Lescure 1984).

Most dendrobatid frogs carry their tadpoles on their backs from a terrestrial oviposition site to water (table 11.3). The only exceptions known at present are *Colostethus degranvillei* and two species with tadpoles that develop at the terrestrial oviposition site without being carried to water, *C. stepheni* and *C. chalcopis* (Junca, Altig, and Gascon 1994; Kaiser and Altig 1994; Kaiser, Coloma, and Gray 1994; Junca 1996). The reproductive biology of many species of *Colostethus*, a large and diverse genus, is unknown, and it is possible that other species that do not carry tadpoles will be discovered. The reproductive biology of the nocturnal species *Aromobates nocturnus*, which is thought to be the sister species to all other dendrobatids, also is unknown, but tadpoles have been found in streams (C. Myers, Paolillo O., and Daly 1991).

In *Colostethus, Mannophryne, Allobates, Epipedobates, Phyllobates,* and *Cryptophyllobates,* a parent carries up to 40 tadpoles at once (fig. 11.7 A). The tadpoles are placed in a variety of aquatic sites (fig. 11.7 B), including streams, streamside pools, swamps, temporary ponds, small pools on the forest floor, or water collected in stumps, logs, fallen bamboo, the petioles of fallen palm fronds, or seed pods (Jowers and Downie 2005). One species, *Mannophryne trinitatis,* even has been observed depositing tadpoles on wet leaf litter when suitable pools were not available (Downie, Livingstone, and Cormack 2001). Tadpoles of one species, *Colostethus bromelicola,* have been found in bromeliads up to five meters above the forest floor, and adults are never seen on the ground (Dixon and Rivero-Blanca 1985). It is not known whether the eggs are laid in bromeliads, or the

**Fig. 11.7.** Parental care in dendrobatid frogs. (A) Female *Colostethus panamensis* carrying a complete clutch of tadpoles. (B) Male *Allobates femoralis* depositing a tadpole in water. This species carries an entire clutch of tadpoles and releases them into small pools. (C) Male *Dendrobates auratus* sitting on eggs. The male returns to the eggs periodically to provide them with moisture, but does not brood them continuously. (D) Male *D. pumilo* carrying a single tadpole on his back. Photos by Kentwood D. Wells (A, C), Walter Hödl (B), and Edmund D. Brodie III (D).

tadpoles are transported to bromeliads from other oviposition sites. Another bromeliad-breeder, *C. beebei*, uses giant terrestrial bromeliads. Females lay eggs on leaves above the water, and males subsequently transport tadpoles to water-filled leaf axils (Bourne et al. 2001). Males also were reported to move tadpoles from pools that lacked sufficient food resources to other pools with accumulations of detritus and algae.

In these genera, tadpoles either are released simultaneously, or they drop off gradually over several days (Luling 1971; Lüddecke 1974, 1976; Lescure 1976; Silverstone 1976; Van Meeuwen 1977; Wells 1977a, 1980a, c, 1981a; C. Myers, Daly, and Malkin 1978; C. Myers and Daly 1979; Weygoldt 1980c, 1983, 1987; H. Zimmermann and E. Zimmermann 1981, 1985, 1988, 1990; H. Zimmermann 1983; Aichinger 1991; Coloma 1995; Fandiño, Lüddecke, and Amézquita 1997; Lötters, Jungfer, and Widmer 2000; Jowers and Downie 2005). In *Colostethus panamensis*, tadpoles grow while on the back of the female, drawing on their large yolk reserves. Individual females that were recaptured several times had fewer tadpoles on their backs each time, and there was a strong negative correlation between number of tadpoles carried and tadpole size (Wells 1980b). A similar negative correlation between number and size of tadpoles was reported for *Mannophryne trinitatis* (Cummins and Swan 1995), but in this case, it is the male that carries the tadpoles. These authors suggested that this correlation might be a result of smaller clutches having larger eggs, although they could not rule out the possibility of tadpoles growing while being transported. Subsequent work on the same species by Downie et al. (2005) showed that the tadpoles do grow while on the back of the male. After about eight days, all yolk reserves were exhausted and the larvae did not continue to grow.

In *Dendrobates,* tadpoles are shuttled from a terrestrial oviposition site (fig. 11.7 C) to water and are carried only briefly. Usually only one or two tadpoles are carried on each trip (fig. 11.7 D). Transporting parents often place individual tadpoles in separate bodies of water in tree holes, stumps, logs, bromeliads, or the leaf axils of plants (Senfft 1936; Dunn 1941; Eaton 1941; H. Zimmermann 1974, 1978; Silverstone 1976; Wells 1978a, 1981a; Graeff and Schülte 1980; Weygoldt 1980a, b, 1982, 1983, 1984a, 1987; H. Zimmermann and E. Zimmermann 1980a, b, 1981, 1984, 1988, 1990; Lescure and Bechter 1982; E. Zimmermann and H. Zimmermann 1982; Jungfer 1985; Summers 1989, 1990, 1992; 1999b; Van Wijngaarden and Bolanos 1992; Brust 1993; Summers and Amos 1997; Caldwell and Oliveira 1999; Summers and McKeon 2004). At least two species, *Dendrobates castaneoticus* and *D. quinquevittatus,* place their tadpoles in the water-filled woody fruit capsules of the Brazil nut tree and in pools formed in the petioles of fallen palm fronds (Caldwell and Myers 1990; Caldwell and Araújo 2004).

The transport of only a few tadpoles at a time is thought to be a derived form of parental care within the family (Weygoldt 1987; H. Zimmermann and E. Zimmermann, 1988, 1990; Summers and McKeon 2004). Placing tadpoles in separate bodies of water probably is advantageous because of the limited food supplies in the small bodies of water used by these frogs. Summers (1990) demonstrated experimentally that growth rates of *Dendrobates auratus* tadpoles decreased as tadpole density increased, while mortality increased. The larval period also was longer in more crowded sites. Larger bodies of water, such as water-filled tree holes, could support more tadpoles than small bodies of water. Many dendrobatid tadpoles compensate for limited food supplies in bromeliads and tree holes by eating other tadpoles, so placement of several tadpoles in the same small body of water makes them susceptible to cannibalism. This would favor the evolution of behavior to disperse tadpoles in different deposition sites (Summers 1990, 1999b; Summers and Amos 1997; Caldwell and Araújo 1998; Caldwell and de Oliveira 1999; Summers and Earn 1999; Summers and Symula 2001; Summers and McKeon 2004).

Two ranid frogs from Borneo, *Limnonectes palavaensis* and *L. finchi* (formerly considered subspecies of *L. microdisca*) have evolved a form of parental care with remarkable parallels to that of dendrobatid frogs. These are small forest-floor frogs about the size of the largest species of dendrobatids. Terrestrial egg clutches have been found in leaf litter, either attended by a male or unattended. Males have been seen carrying up to 52 tadpoles at once, and these are deposited in pools along the margins of streams, in the main body of slow-moving streams, or rain pools on the forest floor (Inger 1966; Inger , Voris, and Frogner 1986; Inger and Voris 1988). Other details of oviposition and parental care are unknown.

Other frogs have unusual ways of transporting their larvae. In South American *Rhinoderma* (Rhinodermatidae), the male picks up the eggs before they hatch and carries the hatched tadpoles in an enlarged vocal sac (see fig. 1.20 E in chapter 1), sometimes carrying more than one cohort of tadpoles at the same time (M. Crump 2002). In *R. darwinii,* the tadpoles remain in the vocal sac for up to 52 days before completing metamorphosis (Cei 1962; Busse 1970; Jorquera, Pugin, and Goicoechea 1972; Goicoechea, Garrido, and Jorquera 1986), whereas tadpoles of *R. rufum* emerge before metamorphosis and complete development in the water (Jorquera, Garrido, and Pugin 1982). Tadpoles of *R. darwinii* apparently receive nutrition through the membranes of the male's vocal sac, so tadpole transport grades into tadpole feeding. The internal structure of the sac indicates that secretory activity is present (Garrido, Pugin, and Jorquera 1975). The larvae are not capable of intestinal absorption until just before metamorphosis (Jorquera, Garrido, and Pugin 1982), suggesting that any nutrient passage would be through the skin. Goicoechea, Garrido, and Jorquera 1986) tested this hypothesis by injecting horseradish peroxidase and labeled amino acids into the circulatory system of the brooding male. These labeled tracers found their way into the tissues of the tadpoles, indicating that transfer of materials from the male's circulation to the larvae does occur. This type of nutrient transfer is analogous to processes found in some viviparous amphibians (see chapter 10), but involves a completely independent evolution of a different organ system.

In the Australian myobatrachid *Assa darlingtoni,* the female apparently remains with the eggs, but whether she provides active parental care is not known. When the eggs are ready to hatch, the male sits on them and the tadpoles wriggle into inguinal brood pouches through slits near the junction of the hind legs with the body (fig. 11.8). The tadpoles remain there up to 80 days before emerging as small froglets (Ingram, Anstis, and Corben 1975; Ehmann and Swan 1985). In both *Assa* and *Rhinoderma,* there is no obvious transitional stage from ancestral modes of reproduction to these highly specialized forms of parental care. The relationships of *Rhinoderma* to other anuran families are unclear (see chapter 1), and no other myobatrachid has a reproductive mode similar to that of *Assa.*

### Attendance of Froglets

In some anurans with direct-developing eggs, a parent remains with the juveniles after they hatch. In *Cophixalus parkeri,* a microhylid frog from New Guinea, the young remain in the nest with a parent for 30–40 days after hatching (Simon

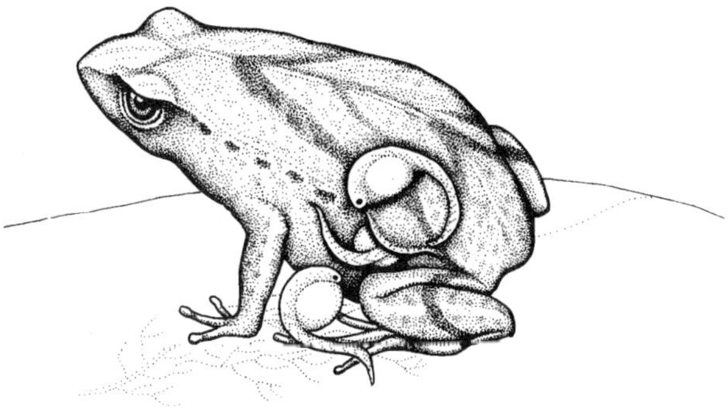

**Fig. 11.8.** Tadpoles entering the inguinal pouch of the male in the Australian myobatrachid frog *Assa darlingtoni*. Drawing by T. Wright from Roberts (1993).

1983). Similar behavior has been observed in many other genera of New Guinea microhylids, including *Xenobatrachus, Copiula, Hylophorbus, Callulops, Austrochaperina, Sphenophryne,* and *Liophryne,* although in most cases the duration of attendance apparently is shorter than in *C. parkeri* (Bickford 2001). In at least some species of *Sphenophryne, Liophryne,* and *Cophixalus,* males actually transport hatched young on their backs (see the following discussion). Juveniles of *Eleutherodactylus coqui* sometimes stay with a male in the nest for several days after they hatch (Townsend, Stewart, and Pough 1984), as do juveniles of a closely related species, *E. cooki* (Joglar, Burrowes, and Rios 1996). Another species in the same genus, *E. cundalli,* has made the transition from attending juveniles to transporting them; the tiny froglets are carried on the back of the female (Diesel, Baurle, and Vogel 1995; see the following.

### Transport of Froglets

Transport of froglets is a form of parental care that was discovered only recently. It has been reported in several unrelated species of frogs with eggs that undergo direct development. In *Stefania evansi,* a hemiphractine frog, the eggs are carried on the back of the female, as described earlier, and the froglets remain on her back after hatching. They still have abundant yolk reserves (fig. 11.3) and presumably can remain on the female for some time and continue to grow (Gorzula, Morales, and Hernandez 1983; Jungfer and Boehme 1991). In *Eleutherodactylus cundalli* (Leptodactylidae), a Jamaican species, terrestrial eggs are laid on small terraces, crevices, and depressions in caves and attended by the female. They undergo direct development and hatch into small froglets that climb onto the back of the female. Between 30 and 72 froglets can be carried at once. As in *Stefania,* froglets still had yolk reserves when they initially climbed on the female, but eventually these were depleted as the froglets continued to grow. Most females carrying young were observed near the entrance to a cave, and one was seen outside, about 100 meters from the nearest cave entrance. Transport of the young in this species apparently serves to carry them out of caves to habitats with more abundant food supplies (Diesel, Baurle, and Vogel 1995). Similar behavior evidently occurs in *E. glaucoreius,* a species closely related to *E. cundalli* (Townsend 1996).

Froglet transport has been described in several genera of microhylid frogs from New Guinea (Bickford 2001, 2002). All of these species have direct development. Some species inhabit terrestrial or arboreal habitats similar to those used by Neotropical *Eleutherodactylus,* but others are largely subterranean. Males have been observed carrying groups of froglets in both terrestrial and arboreal species in several genera, including *Liophryne, Sphenophryne,* and *Cophixalus* (fig. 11.9). Froglets gradually jumped off of the parent's back into the leaf litter as they were carried for several days for distances of up to 55 meters from the oviposition site.

**Fig. 11.9.** Transport of froglets by microhylid frogs from New Guinea. (A) A male of an arboreal species of *Cophixalus* with young froglets on his back. (B) A male of the terrestrial frog *Liophryne schlaginhaufeni* carrying froglets. Photos by David Bickford.

The main function of froglet transport in these species appears to be dispersal of juveniles into suitable microhabitats. Bickford (2001) also conducted a phylogenetic study based on mitochondrial and nuclear gene sequences. He found that the current genus-level taxonomy of these frogs does not reflect true phylogenetic relationships. The species in which froglet transport has been described appear to be closely related, despite being placed in different genera. It seems likely that such behavior will be found in many other New Guinea microhylids, so the complete phylogenetic distribution of the behavior has yet to be determined.

### Feeding of Tadpoles

Perhaps the most unusual form of parental care reported in frogs is the feeding of tadpoles with fertilized or unfertilized eggs. The behavior is always associated with the use of oviposition or tadpole deposition sites that have limited food resources. Several species of dendrobatid frogs feed their tadpoles. All of these species place their tadpoles in water-filled leaf axils of bromeliads and other plants. Tadpole feeding was first observed in *Dendrobates pumilio* in captivity (Graeff and Schülte 1980; Weygoldt 1980a, b) and subsequently confirmed in field observations and experiments (Brust 1993). The eggs are laid in leaf litter and attended by the male, which returns periodically to moisten them. It is the female that carries the tadpoles, apparently summoned to the oviposition site by the male's calls. The female places the tadpoles in the leaf axils of bromeliads or other water-collecting plants and returns every few days to feed them with unfertilized eggs (Graeff and Schülte 1980; Weygoldt 1980a, b; Brust 1993). The tadpoles are obligate egg-eaters and eventually die unless fed by the female. Similar behavior has been observed in *D. histrionicus, D. lehmanni, D. granuliferus,* and *D. speciosus* in captivity (H. Zimmermann and E. Zimmermann 1981, 1983, 1988, 1990; E. Zimmermann and H. Zimmermann 1982, 1985; E. Zimmermann 1983; Heselhaus 1984; Jungfer 1985; Weygoldt 1987), and in the field for *D. histrionicus* (H. Zimmermann 1989). Tadpoles of *D. granuliferus* have been found in the field in the leaf axils of plants, along with undeveloped eggs presumably deposited by females (Van Wijgaarden and Bolanos 1992). *Dendrobates arboreus* probably feeds its tadpoles as well (Summers et al. 1999). These species and *D. pumilio* all are members of the *D. histrionicus* species group, which is centered in northern South America and Central America. The occurrence of tadpole feeding in most or all of the members of this group suggests a single origin for this type of parental care in this clade. The tadpoles of the *D. histrionicus* group have mouthparts with reduced labial denticles, which is typical of obligate egg-eaters (see chapter 12).

Tadpole feeding also has been observed in several other dendrobatids found in Amazonian South America, including "*D. ventrimaculatus,*" *D. variablis, D. fantasticus, D. reticulatus, D. imitator,* and *D. vanzolinii.* These species are part of a confusing complex of similar-looking frogs often grouped together in the "*D. quinquevittatus* species group" (Silverstone 1975; C. Myers 1982; Caldwell and Myers 1990). Unfortunately, the name *D. quinquevittatus* has been used incorrectly in many accounts of parental care in captive animals (Meede 1980; Schülte 1981b; Lescure and Bechter 1982; H. Zimmermann and E. Zimmermann 1984, 1988, 1990; Weygoldt 1987; Crump 1995b). The species described by these authors actually is "*D. ventrimaculatus,*" which itself probably comprises a number of very similar species that have yet to be named (Caldwell and Myers 1990). How many of these have this unusual form of parental care is not known, but field studies of "*D. ventrimaculatus*" in Amazonian Ecuador indicated that frogs in that population do not feed their tadpoles (Summers and Amos 1997; Summers 1999b; Summers and Earn 1999; Summers et al. 1999). The name *D. quinquevittatus* is properly used for an entirely different species that does not deposit tadpoles in bromeliads or plant leaf axils, but in water-filled holes in stumps and logs, in leaf petioles of falling palm fronds, and in Brazil nut fruit capsules on the forest floor. The complex of tadpole-feeding species in Amazonia is more accurately described as the "*Dendrobates ventrimaculatus* complex" (Summers et al. 1999). Recently, two instances of females providing unfertilized eggs to tadpoles were reported for *Colostethus beebei* from Guyana (Bourne et al. 2001). This species is unusual among members of this genus in using terrestrial bromeliads as oviposition and tadpole deposition sites. *Colostethus beebei* exhibits several features of parental behavior that are convergent with the behavior of the Amazonian species of *Dendrobates.*

The details of parental care and tadpole development in most of these species have been observed in captivity, although there are field studies of *D. vanzolinii* from Brazil, "*D. ventrimaculatus*" from Ecuador, and *Colostethus beebei* from Guyana (Caldwell 1997; Summers and Amos 1997; Caldwell and de Oliveira 1999; Bourne et al. 2001). Parental care in Amazonian species of *Dendrobates* and in *C. beebei* differs in several respects from that of the *D. histrionicus* group. Eggs are laid on the leaves of bromeliads, *Heliconia,* or other plants, and in some cases, only a single egg is laid. Eggs are tended by the male, which transports the tadpoles to a water-filled leaf axil. Males call to and court females, either during tadpole transport or afterward, thereby inducing the female to lay additional eggs in or above pools where tadpoles reside. Hence, these species exhibit biparental care, in contrast to members of the *D. histrionicus* group, in which females transport tadpoles and lay trophic eggs without the male being present. In the population of "*D. ventrimaculatus*" studied by Summers and Amos (1997) in Ecuador, the basic reproductive biology is similar to that described previously, but there was no evidence of females laying trophic eggs. Instead, males performed all parental care, including carrying tadpoles from leaf axils where eggs are deposited to different leaf axils where the tadpoles develop. Many tadpoles simply hatched from egg clutches deposited on the sides of leaf axils and dropped into the pool below, or were deposited there by males. Although the frogs avoided depositing eggs above pools containing large tadpoles, tadpoles of different sizes often ended up in the same pool, and cannibalism of eggs and tadpoles was common (Summers 1999b).

Egg eating appears to be facultative in the Amazonian species, and the tadpoles have mouthparts typical of detritus-feeding species, not obligate egg-eaters (Kneller 1982; H. Zimmermann and E. Zimmermann 1984, 1988, 1990; Weygoldt 1987). *Colostethus beebei* also is a facultative egg-eater, with most of their nutrition coming from detritus, algae, or small insect larvae that accumulate in bromeliad pools (Bourne et al. 2001). Both Weygoldt (1987) and H. and E. Zimmermann (1988, 1990) suggested that this form of tadpole feeding is the ancestral condition leading to the more specialized behavior of frogs in the *D. histrionicus* group. An analysis of dendrobatid phylogeny based on mitochondrial DNA makes this scenario unlikely, however, because the Amazonian *Dendrobates* do not appear to be closely related to the *D. histrionicus* complex (fig. 11.10). The most likely hypothesis is that tadpole feeding has evolved independently in these two groups (Summers et al. 1999) and again in *C. beebei.* In the Amazonian clade, males transport tadpoles, a behavior that appears to be the ancestral condition for the entire genus *Dendrobates.* This necessitates cooperation between males and females if females are to locate tadpole deposition sites to feed the tadpoles. In the *D. histrionicus* group, however, transport of tadpoles by females and feeding of tadpoles appear to have evolved together, thereby releasing the male from further parental duties.

It has been known for more than half a century that tadpoles of some other frogs that live in bromeliads and similar small bodies of water eat frog eggs and sometimes have mouthparts adapted for egg eating (Noble 1929a; Dunn 1937; E. Taylor 1954; Lannoo, Townsend, and Wassersug 1987). Several of these species, quite unrelated to one another, have been shown to feed their tadpoles with eggs in a manner reminiscent of the behavior of dendrobatids de-

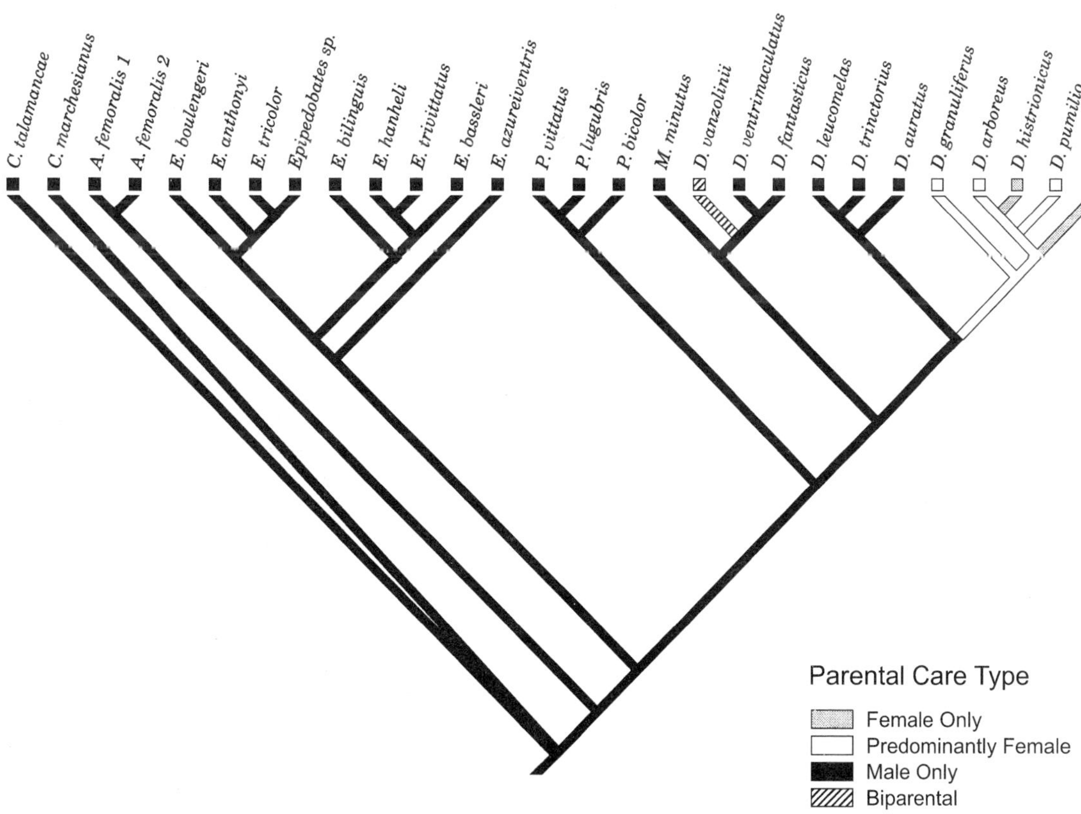


**Fig. 11.10.** Phylogeny of dendrobatid frogs, showing the evolution of male, female, and biparental care. The phylogeny is based on mitochondrial DNA sequences. This analysis indicates that biparental care in *Dendrobates vanzolinii* probably evolved from male parental care, which is ancestral for the whole family. A few species of *Colostethus* that are not included in this analysis have female parental care. Genera: *A* = *Allobates* (= *Epipedobates femoralis*); *C* = *Colostethus; D* = *Dendrobates; E* = *Epipedobates; M* = *Minyobates; P* = *Phyllobates.* From Clough and Summers (2000).

scribed previously. The best known of these tadpole-feeding frogs is *Osteopilus brunneus* from Jamaica. Dunn (1926a) reported finding eggs in the stomachs of tadpoles and described morphological adaptations of their mouthparts for egg eating. Lannoo, Townsend, and Wassersug (1987) conducted a much more detailed study of tadpole morphology and behavior (see chapter 12) and presented indirect evidence of females feeding their young, but they did not actually observe parental behavior. A study by Thompson (1996) confirmed that females do feed their tadpoles. The eggs of this species are very small, probably an adaptation to the extremely low oxygen levels in bromeliad tanks (see chapter 4). Clutch sizes are much larger than are those of dendrobatids, and some bromeliads have as many as 170 tadpoles in them. The eggs hatch within two days, but the tadpoles can take more than two months to reach metamorphosis. Females begin feeding their tadpoles immediately after they hatch and return, on average, every four days to deposit eggs. Early in the development period, females bring males to the bromeliads and deposit only fertilized eggs. Tadpoles eat some of these eggs, but others survive and join the developing brood. After some tadpoles have metamorphosed, the female usually switches to providing only unfertilized eggs for the remaining tadpoles. Presumably the tadpoles are now large enough to eat all of the eggs, so there would be no advantage to depositing fertilized eggs (some large tadpoles can have as many as 180 eggs in their stomachs; Lannoo, Townsend, and Wassersug 1987). As in some dendrobatids, the tadpoles are obligate egg-eaters and die if they are not fed.

Very similar behavior has been described in *Anotheca spinosa,* which has been studied mostly in captivity. E. Taylor (1954), D. Robinson (1961), and Duellman (1970) all reported finding frog eggs in the stomachs of tadpoles that were living in bromeliads and water-filled tree holes, but they did not observe any parental behavior. Jungfer (1996) placed a pair of adults in a terrarium and provided them with artificial containers designed to resemble water-filled tree holes. The frogs deposited large numbers (50–300) of small eggs on the walls of the containers, just above the water's surface. About a week after the first eggs were laid, the female returned to lay additional eggs. If the male was present, she attached fertilized eggs to the walls of the container, but if the male was not present, she laid unfertilized eggs directly in the water. The tadpoles typically swarmed around the female and began eating the eggs as soon as they were laid (fig. 11.11 A). The female returned every four to five days to lay additional eggs, averaging about 135 eggs per clutch. This continued for two to four months until the tadpoles reached metamorphosis. Larger tadpoles appeared to out-compete smaller tadpoles for nutritive eggs and transformed more quickly.

Yet another hylid with egg-eating tadpoles is a species from Brazil, *Osteocephalus oophagus.* This species lays eggs in a variety of small bodies of water, including bromeliads on tree branches or on the ground, the leaf axils of palms, the fallen palm bracts, and tree holes as much as 35 m above the ground. Females return to lay more eggs about every five days, very similar to the interval reported for the other two hylids. Females of this species deposit only fertilized eggs, some of which escape predation and hatch into tadpoles that join the developing brood; most are eaten by tadpoles already present (fig. 11.11 B). The tadpoles cannot complete development without being fed several times by the female (Jungfer and Schiesari 1995), and females do not deposit eggs at new sites while they have tadpoles developing in another site (Jungfer and Weygoldt 1999). Tadpoles of a recently described species from Ecuador, *O. deridens,* have been observed in the field with eggs in their digestive tracts, suggesting that this species has the same type of parental care (Jungfer et al. 2000).

A rhacophorid tree frog, *Chirixalus eiffingeri,* from Taiwan and Japan exhibits parental behavior very similar to that of these hylids, especially *Anotheca spinosa* (fig. 11.11 C, D). Eggs are laid just above the water's surface in a variety of sites, including hollows in the forks of trees, leaf axils, and cups formed when the hollow stems of bamboo are cut. Ueda (1986) reported egg attendance by males, which sometimes sat on the eggs to moisten them (fig. 11.2 F). He also reported that tadpoles of this species are obligate egg-eaters. Eggs that accidentally fell below the water line were eaten. In addition, females were observed laying unfertilized eggs directly in the water, and the tadpoles subsequently consumed these eggs. Kam, Chuang, and Yen (1996) reported that males remained with some egg clutches in bamboo cups in the field, but initially did not observe parental behavior by either males or females.

Through a series of laboratory and field experiments, Kam, Chuang, and Yen (1996), Kam et al. (1997, 1998, 2000, 2001, 2002), Kam and Yang (2002), Liang, Huang, and Kam (2002), and Huang, Huang, and Kam (2003) demonstrated that egg eating is necessary for the tadpoles to grow. They also showed that females rely on the location of bamboo stumps to relocate their offspring, and that there is complex tactile communication between tadpoles and females. When tadpoles were transferred from their original bamboo stumps to new ones that did not contain tadpoles, they were not fed by females and did not grow. When tadpole densities were high, growth was slower than when densities were low, but females evidently can increase their production of trophic eggs as the biomass of tadpoles in a pool increases. Kam, Chuang, and Yen (1996) estimated that each tadpole would consume about 87 eggs during a two-month development period. Since each clutch contains about 80 eggs, a female would have to lay several thousand eggs to successfully rear one clutch of tadpoles. This imposes a significant

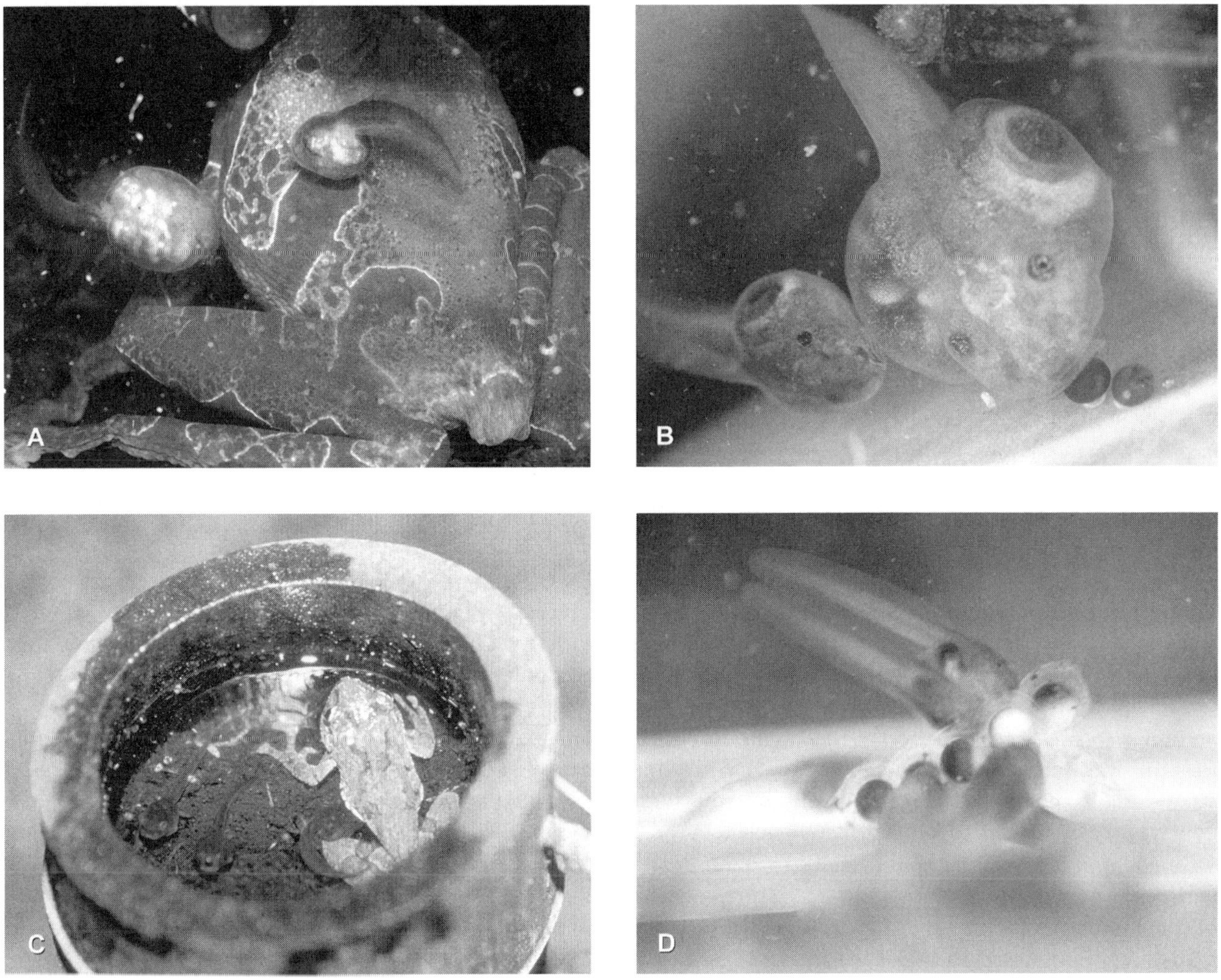


**Fig. 11.11.** Feeding of tadpoles by frogs that breed in small bodies of water. (A) Female of *Anotheca spinosa* (Hylidae), a bromeliad breeder, depositing unfertilized eggs in water. The stomachs of the tadpoles are full of eggs. (B) Tadpoles of *Osteocephalus oophagus* (Hylidae) feeding on newly deposited eggs. (C) Female of *Chirixalus eiffingeri* (Rhacophoridae) visiting a bamboo stump to deposit eggs to feed developing tadpoles. (D) Tadpoles of *C. eiffingeri* feeding on newly deposited eggs. Photos by Karl Heinz Jungfer (A), Walter Hödl (B), and Y.-C. Kam (C, D).

energetic constraint on females, which do not normally produce a new clutch of fertilized eggs until they have finished rearing their current clutch of tadpoles (Kam et al. 2000). The necessity for females to provision their tadpoles also limits the growth of the tadpoles. Normally females feed their young about once every eight days. Experimental treatments in which tadpoles were fed on chicken egg yolk showed that tadpoles fed more frequently than this grew faster, while those fed only every two weeks grew more slowly (Liang, Huang, and Kam 2002). All tadpoles eventually reached metamorphosis, however, and tadpoles of this species are better able to tolerate periods without food than are herbivorous tadpoles in the same genus.

Perhaps the most surprising report of tadpole feeding is for *Leptodactylus fallax*, a frog found only on the islands of Dominica and Montserrat. As in other members of this genus, eggs are laid in foam nests, but these are placed away from water under rocks or logs, in tree stumps, or in underground burrows. Males defend oviposition sites and attract females to them. Pairs take up to 14 hours to construct the foam nest and lay eggs. The clutch size is relatively small (25–45 eggs), but individual eggs also are small and lack sufficient yolk reserves for tadpoles to complete development without feeding. Both males and females guard the nest, and the female periodically enters the nest to lay clutches of unfertilized eggs that are rapidly eaten by the tadpoles. Development is relatively slow, requiring up to two months to metamorphosis in captivity at temperatures similar to those in the field. Females provisioned eggs on average about every three days, with tadpoles in a single nest consuming an estimated 10,000–25,000 eggs (Gibson and Buley 2004). In a species from the Cerrado of Brazil, *L. labyrinthicus*, a substantial proportion of the eggs deposited in a foam nest do not develop and apparently are not fertilized, and these serve as food for developing tadpoles. There is no evidence, however, of females returning to the nest to provision the tadpoles (Shepard and Caldwell 2005).

I have described the behavior of tadpole-feeding frogs in some detail, because it represents a remarkable degree of convergence in parental behavior of frogs in several different families that use similar sites to rear tadpoles. Providing either fertilized or unfertilized eggs for tadpoles to eat seems like an expensive way for female frogs to rear their young, but this behavior should be viewed as only one alternative strategy for making use of sites that provide little or no food for the young. The other alternative is to provision the eggs with sufficient yolk reserves to allow the young to develop without feeding, also an expensive option. This strategy has evolved in a number of frogs that lay eggs in tree holes, leaf axils, and small pools (reproductive Mode 6; see chapter 10). The disadvantage of this strategy is that eggs must be large, so clutch sizes are small. If eggs or larvae are subject to high predation, then having small clutch sizes probably is not a viable option. On the other hand, if eggs are to be placed directly in small, oxygen-poor bodies of water, as they are in *Osteopilus brunneus* and *Osteocephalus oophagus*, then the eggs must be small because of gas exchange constraints (see chapter 4). Feeding the tadpoles probably is the only way the offspring can complete development.

In addition to convergence in overall parental behavior in these frogs, there also has been convergence in the system of communication between the tadpoles and their mothers. Studies of several species of *Dendrobates*, as well as *Anotheca spinosa* and *Chirixalus eiffingeri*, revealed that tadpoles respond to the arrival of the female by swimming vigorously and poking or chewing at the female's skin, especially on her back and cloacal region. Tadpoles of *Leptodactylus fallax* also showed increased activity and moved toward the female when she arrived at the foam nest (Gibson and Buley 2004). The activity of tadpoles appears to elicit egg deposition by the female, and therefore is functionally equivalent to the begging behavior of young birds in a nest (Weygoldt 1980a, Jungfer 1985, 1996; Ueda 1986; E. Zimmermann 1990). Tadpoles of *Osteocephalus oophagus* also swim and bump against the female as she lays eggs, but in this case, such stimulation is not necessary to elicit oviposition by females (Jungfer and Weygoldt 1999).

Frogs are not the only animals that have adapted to bromeliads by providing parental care to their young. In the same area of Jamaica where *Osteopilus brunneus* lives, a tiny crab (*Metopaulias depressus*) lives in the water-filled leaf axils of large ground bromeliads. This species produces large, yolky eggs and have a larval period of about two months, shorter than that of many marine crabs. The female remains with her larval brood in the bromeliads. She provides a number of important services, including cleaning leaf litter out of the leaf axils, depositing bits of broken snail shells that reduce the acidity of the water, and feeding the larvae with prey captured on the bromeliad leaves (Diesel 1989, 1992a, b). These crabs are generalist predators and might be expected to prey on the eggs or larvae of frogs. These two species seldom encounter one another, however, because the crabs prefer bromeliads of the genus *Aechmea*, which are avoided by *O. brunneus*. The frogs prefer bromeliads of the genus *Hohenbergia*, which are used less frequently by the crabs. In addition, the frogs use central leaf axil tanks almost exclusively, whereas the crabs prefer peripheral axils formed by older leaves (Diesel 1989; Thompson 1996).

## Parental Care in Urodeles

Salthe and Mecham (1974), Nussbaum (1985, 2003), and Crump (1995b, 1996) have reviewed the parental behavior of urodeles. In many ways, the parental behavior of urodeles is less complex than that of anurans. Salamanders do not transport eggs or larvae, nor do they feed their young, so parental care consists entirely of attendance of eggs and larvae. Attendance of aquatic eggs by males has been reported in cryptobranchids in both North America (*Cryptobranchus*) and Asia (*Andrias;* Kerbert 1904; Stejneger 1907; B. Smith 1907; Tago 1929; S. Bishop 1941b; Nickerson and Mays 1973a; Kuwabara et al. 1989). Two stream-breeding hynobiid salamanders, *Hynobius kimurai* and *H. naevius*, are reported to have male parental care as well (Tago 1931; cited in Nussbaum 1985). Thorn (1962, 1967) reported that males of a pond-breeding species, *H. nebulosus*, also attend their eggs. This has not been observed in the field (Kunitomo 1910; Kusano 1980), and Nussbaum (1985, 2003) considered Thorn's interpretations questionable. There is considerable confusion about parental care in the European olm, *Proteus anguinus* (Proteidae). Reports based on observations in captivity have variously described egg attendance by males, females, or both (Vandel and Bouillon 1959; Briegleb 1962; Durand and Bouillon 1964; Durand and Vandel 1968; Durand 1986). Some of this confusion could be due to captive animals seeking shelter under a limited number of stones in aquaria and sometimes choosing sites with eggs. This species lives in deep, dark caves, and its behavior in nature has not been observed.

There are brief reports of females attending eggs in streams, usually attached to the undersides of rocks, in *Necturus* (S. Bishop 1926, 1941b; Ashton and Braswell 1979), *Desmognathus* (C. Pope 1924; Organ 1961c; Martof 1962a), *Eurycea* (Richmond 1945; Wood 1951, 1953a; Franz 1964; Baumann and Huels 1982; Bruce 1988a; Marshall 1996), *Gyrinophilus* (S. Bishop 1924; Bruce 1978a), *Pseudotriton*

(Goin 1947; Bruce 1978c), and *Dicamptodon* (Nussbaum 1969a; Nussbaum, Brodie, and Storm 1983; L. Jones, Bury, and Corn 1990; Steele, Brodie, and McCracken 2003a). Most of these reports are based on only a few observations of females with eggs, and the details of parental care in these genera remain largely unknown. Nevertheless, they do indicate that attendance of eggs by females is common in stream-breeding salamanders. Observations of stream-breeding Asian newts of the genus *Pachytriton* in captivity suggest that females aggressively guard their eggs, which are attached to the undersides of stones. Females also have been observed eating undeveloped eggs in their clutches and depositing secretions from the skin on their eggs, perhaps to inhibit fungal or bacterial growth (Thiesmeier and Hornberg 1992, 1997, 1998). The behavior of these newts has not been studied in the field, however.

Pond-breeding salamanders typically do not attend their eggs. Females of *Siren intermedia* have been discovered near clumps of aquatic eggs in mats of vegetation (Noble and Marshall 1932; Godley 1983), but these reports are anecdotal and do not provide any details of parental care. Other sirenids scatter their eggs individually and therefore probably do not have parental care (Nussbaum 1985). In *Amphiuma,* females lay eggs in burrows on the bottoms of shallow ponds (fig. 11.12 A). If a pond dries up, the female remains coiled around her eggs (J. Weber 1944; Baker 1945; Gunzburger 2001, 2003).

In several species of salamanders that lay eggs out of water, but have aquatic larvae, females remain with the eggs. In *Ambystoma opacum* (fig. 11.12 B), females lay eggs in the dry bed of a temporary pond and usually remain with the eggs until the pond fills (Noble and Brady 1933; Kaplan and Crump 1978; Petranka and Petranka 1981; Jackson, Scott, and Estes 1989; Petranka 1990; Figiel and Semlitsch 1995; Croshaw and Scott 2005). Petranka (1990) reported that about 25% of females in a population in North Carolina deserted their eggs within a month of oviposition. Eggs that were deserted suffered higher mortality than did those that were attended. Jackson, Scott, and Estes (1989), who studied a population in South Carolina, also found that eggs that were deserted early suffered higher mortality. Croshaw and Scott (2005) performed a series of laboratory and field experiments in which attending females were removed. Their results indicated that protection from predators and removal of eggs infected by fungi increased survivorship of eggs. In contrast to the behavior of *A. opacum,* there is no evidence of parental care in *A. cingulatum,* a species with similar breeding habits. Females of this species simply scatter their eggs in leaf litter at the bottom of dry ponds, where the eggs eventually are flooded by winter rains (J. Anderson and Williamson 1976).

In *Hemidactylium scutatum* (fig. 11.12 C), eggs are laid in sphagnum bogs above the water level, and larvae make

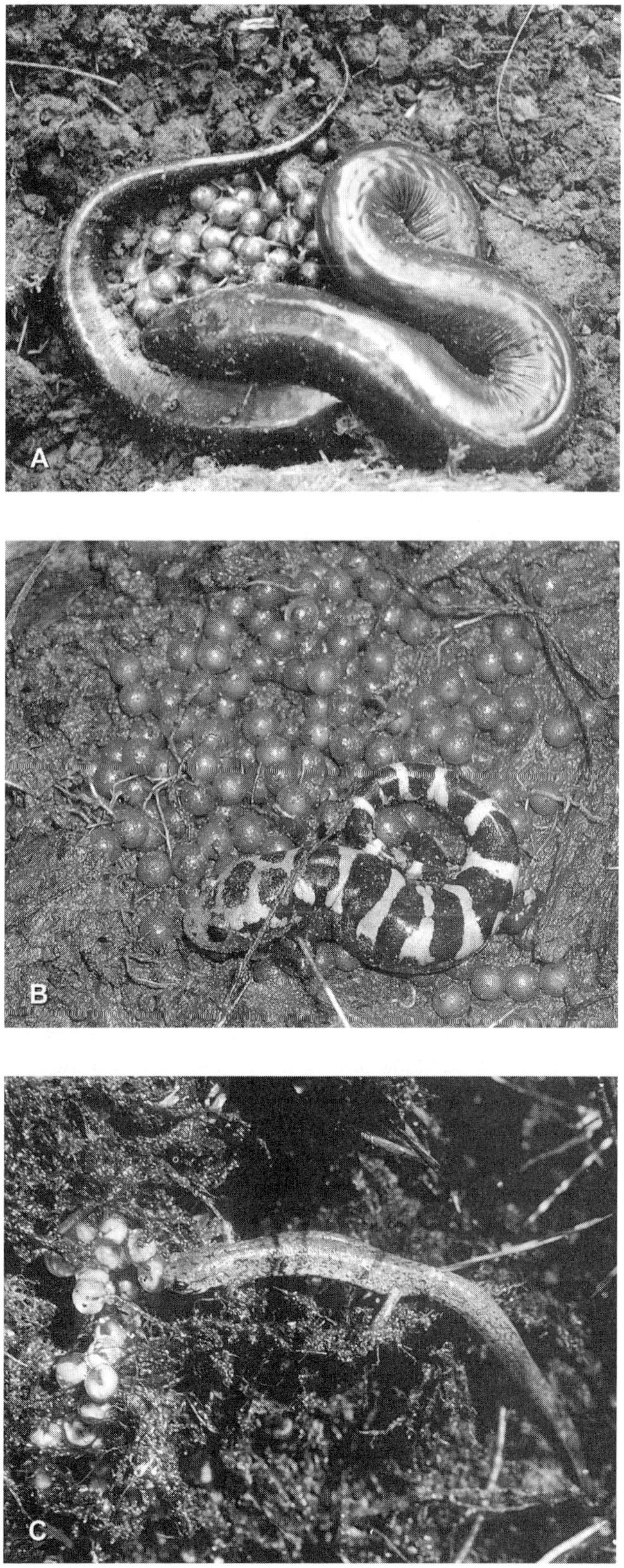

**Fig. 11.12.** Attendance of eggs and larvae by female salamanders. (A) Female *Amphiuma means* (Amphiumidae) with eggs after being excavated from a subterranean nest. (B) Female *Ambystoma opacum* (Ambystomatidae) with eggs. This species lays eggs in the fall on land in places likely to be flooded by snow or rain. (C) Female *Hemidactylium scutatum* (Plethodontidae) with eggs in a sphagnum moss nest. Maternal care occurs in most plethodontids that lay terrestrial eggs. Photos by Margaret Gunzburger (A), Wayne Van Devender (B), and Kentwood D. Wells (C).

their way into the water to complete development. Females usually attend their eggs, but sometimes they are abandoned (S. Bishop 1919; Blanchard 1934, 1936; Blanchard and Blanchard 1931; J. Wood 1953b; R. N. Harris and Gill 1980; Breitenbach 1982; R. N. Harris et al. 1995). In contrast to some other plethodontids, female *Hemidactylium* do not attempt to protect their eggs from attacks of predators, suggesting that parental care is poorly developed in this species (Carreño and Harris 1998). Variable egg attendance also has been reported in *Stereochilus marginatus,* a species with a somewhat similar mode of reproduction. Females attend some egg clutches, but they abandon others (A. Schwartz and Etheridge 1954; Rabb 1956; J. Wood and Rageot 1963).

An unusual phenomenon reported in *Ambystoma opacum, Hemidactylium scutatum,* and *Stereochilus marginatus* is communal nesting, with several females placing eggs in the same nest, which often is attended by only a single female. In the population of *A. opacum* studied by Petranka (1990), joint nesting was rare, with only 6.5% of more than 300 nests containing eggs of two or three females. In some cases, more than one female remained with the eggs, but in others only one female was present. Petranka's data suggested a slight decrease in survivorship for joint nests attended by only one female, but sample sizes were too small for rigorous statistical analysis. Very little is known about communal nesting in *Stereochilus,* because only a brief anecdotal report is available (J. Wood and Rageot 1963).

Communal nesting has been studied in more detail in *Hemidactylium.* On the basis of preliminary observations, R. N. Harris and Gill (1980) hypothesized that the presence of eggs from more than one female might be the result of successive females supplanting one another at the nest site. Each female would lay her own clutch of eggs and then remain with it, along with other eggs that were already present. They also suggested that later-arriving females might eat some of the eggs of other individuals, providing themselves with extra energy while attending their eggs. A more detailed five-year study refuted this hypothesis (R. N. Harris et al. 1995). There was no evidence that females supplanted one another, or that the last arriving female was the one most likely to attend the eggs. In fact, in more than half the cases, the first female remained with the eggs even as other clutches were laid nearby. There also was no evidence of aggression among females or egg eating. Communal nesting did not appear to be a form of brood parasitism, as seen in some birds, because females had approximately equal reproductive success when they attended their own eggs or contributed to a joint nest.

Another possibility is that in *Hemidactylium,* and in the other species that sometimes nest communally, joint nesting is a result of prime nest sites being in short supply. This did not appear to be the case for *Hemidactylium* in the Virginia populations studied by R. N. Harris et al. (1995), because the percentage of joint nesting was not related to population density (about a third of nests were communal, on average, regardless of population size). These investigators also manipulated densities in artificial habitats and found no effect on the proportion of joint nests; again, about a third of nests were communal in both low- and high-density treatments. They did find, however, that as many as seven females in the high density treatment sometimes laid eggs in the same nest, whereas all joint nests in the low-density treatment contained the eggs of only two females. Some females that laid eggs in either single nests or joint nests deserted their eggs before they hatched, and these suffered higher mortality than attended eggs. There was a greater tendency for single nests to be deserted than joint nests, suggesting that nesting with another female provides an advantage for females that desert their eggs. The eggs take about six weeks to develop, and this evidently imposes an energetic cost on brooding females; reduction in body mass was positively correlated with the number of days a female remained with her eggs. It has yet to be determined whether females desert their eggs because their energy reserves are depleted, or for other reasons (see further discussion of the "Costs of Parental Care," following).

Attendance of eggs by females is the rule in most other genera of plethodontid salamanders that lay eggs on land. These include species of *Desmognathus* that have aquatic larvae and lay eggs at the margins of streams and in seepage areas, such as *D. fuscus, D. ochrophaeus,* and *D. ocoee* (Tilley 1972; Forester 1977, 1978, 1979a, c, 1981, 1983, 1984; Forester, Harrison, and McCall 1983; R. L. Jones 1986; Juterbock 1986, 1987; Hom 1987). Females sometimes remain in the nest with hatched larvae for a few days before the larvae make their way to water. Parental care by females is common, perhaps nearly universal, in genera with direct development, including *Aneides, Chiropterotriton, Desmognathus, Ensatina, Hydromantes, Lineatriton, Parvimolge, Plethodon,* and *Pseudoeurycea* (Nussbaum 1985; Crump 1995b). In those species that have been studied in detail, females generally remain with the eggs throughout development; parental care can last two months or more in species that lay eggs in cool microhabitats (Fitzpatrick 1973b; Forester 1983; Bachmann 1984; R. L. Jones 1986; Juterbock 1986, 1987; Hom 1987; Ng and Wilbur 1995). The genus *Batrachoseps* is unusual among plethodontids in that maternal brood care apparently is absent in most species, but communal oviposition is relatively common (Maslin 1939; Jockusch and Mahoney 1997). Unattended clutches of other species are found occasionally, but it is difficult to determine whether this is a result of observers disturbing the nests. There are scattered reports of males of several species being found in or near nests, either alone with eggs or with a female (Storer 1925; Storm 1947; Vial

1968; Jockusch and Mahoney 1997), but there is not one unequivocal observation of a male providing parental care in any plethodontid salamander.

### Parental Care in Caecilians

Nearly all aspects of the biology of caecilians are poorly understood because of their secretive and often subterranean habits, and their parental behavior is no exception. Although the morphology of the reproductive system has been studied in a number of species (reviewed by M. Wake 1977a, b, 1992), information on other aspects of reproduction is scarce. More than half the species for which the reproductive mode is known are viviparous, and their reproductive biology was discussed in chapter 10. Females probably attend eggs in many of the oviparous species (M. Wake 1977a), but most accounts of parental care consist of brief natural history notes. In the Sri Lankan species, *Ichthyophis glutinosus* (Ichthyophiidae), females coil around eggs, which eventually hatch into aquatic larvae (Sarasin and Sarasin 1887–1890; Breckenridge and Jayasinghe 1979). All other reports of females attending eggs are for members of the family Caeciliidae, all species with direct development (fig. 11.13). These include the African species *Idiocranium russeli* and *Afrocaecilia taitana* (Nussbaum, unpublished data cited in M. Crump 1995b), the South American species *Siphonops paulensis* (Goeldi 1899; Gans 1961), and species from three genera found only in the Seychelles islands, *Grandisonia, Hypogeophis,* and *Praslinia* (Brauer 1897; Nussbaum 1984).

**Fig. 11.13.** Parental care by the caecilian *Siphonops annulatus* from Brazil. (A) Female brooding eggs. (B) Female with newly hatched young. Photos by Carlos Jared.

### Benefits of Parental Care

Although the supposed benefits of parental care have been the subject of much speculation, there is a dearth of quantitative information. I have already discussed transport and feeding of tadpoles in some detail, and the benefits of such behavior are obvious. I will limit my discussion here to the benefits of egg attendance, because most experimental studies have focused on this form of parental care. Parental attendance of eggs clearly enhances the survival of offspring (Highton and Savage 1961; M. Bachmann 1964; Vial 1968; Snyder 1971; Tilley 1972; Forester 1979a; R. N. Harris and Gill 1980; Simon 1983; Townsend, Stewart, and Pough 1984; Juterbock 1987; Jackson, Scott, and Estes 1989; Petranka 1990; M. Hayes 1991; R. N. Harris et al. 1995; Thompson 1996; Bickford 2001, 2004; Crespi and Lessig 2004). The precise way in which survival is improved is known for only a few species. In some amphibians, the presence of a parent has little or no effect on offspring survival (Woodruff 1977), and parental attendance is not true parental care. Although observational studies are an important first step in unraveling the adaptive significance of parental care, only controlled experiments are likely to yield definitive conclusions about proximate benefits. The most convincing are those performed in the field, where eggs are exposed to natural environmental conditions. Among the potential benefits that have been proposed for egg attendance by amphibians, the most reasonable are protection against conspecifics, protection against heterospecific predators, protection against desiccation, protection against fungus and other pathogens, and aeration and agitation of eggs.

#### Protection against Conspecifics

Predation of eggs by conspecifics apparently is widespread in aquatic and terrestrial salamanders (Salthe and Mecham

1974; Nussbaum 1985, 2003). Field observations of predatory behavior are limited to scattered reports of adult salamanders invading nests and eating eggs (S. Bishop 1941b; Tilley 1972; Kaplan and Sherman 1980), and there are even fewer observations of parents protecting eggs from predators. Often egg "guarding" is inferred from aggressive responses of females to fingers or other objects introduced into the nest. Brooding females of *Aneides aeneus* have been observed driving intruding males from nests (R. Gordon 1952). D. Snyder (1971) found that very few eggs of this species survived if the attending female was removed, and predation by conspecifics appeared to be a major source of mortality. Several workers have introduced conspecific adults into containers with brooding females, which invariably attack the intruders (Highton and Savage 1961; F. Rose 1966; Forester 1978, 1979a; Bachmann 1984). Forester (1983) found that brooding *Desmognathus ocoee* females became more aggressive as eggs developed and attributed this to the greater difficulty of replacing clutches lost late in the breeding season. However, Bachmann (1984) did not find a similar increase in aggressiveness in *Plethodon cinereus.*

Predation on eggs by conspecifics has been observed in leptodactylid, microhylid, and dendrobatid frogs. In *Eleutherodactylus coqui,* conspecific predation is a major source of embryo mortality, and parental males vigorously defend their clutches with aggressive vocalizations and by pushing and biting intruders (Townsend, Stewart, and Pough 1984; Stewart and Rand 1991). Conspecific predation also appears to account for most egg mortality in the closely related species, *E. cooki* (Burrowes 2000, Rogowitz et al. 2001). In the New Guinea microhylid *Cophixalus parkeri,* many eggs are lost to predators when the parent is removed, but actual defense of eggs has not been observed because of the frogs' secretive habits (Simon 1983). Egg predation by both males and females is common in dendrobatid frogs raised in captivity (Wells 1978a; Weygoldt 1980a, b; H. Zimmermann and E. Zimmermann 1981) and has been observed in the field for *Dendrobates auratus* (Summers 1989). Most dendrobatid frogs do not guard their eggs continuously, but in some species, the territorial behavior of males probably keeps intruders away from a resident's eggs (Wells 1980a; Roithmair 1992; Summers 1989, 1992). In *Dendrobates auratus,* males lead successive females to different oviposition sites (Wells 1978a). This probably makes attendance of eggs less efficient, but it prevents females from destroying or eating clutches laid by previous mates. Females, on the other hand, benefit from eating eggs, not only for their nutritional value, but also because this reduces the time that their mates will spend caring for other females' eggs (Summers 1990; see "Sex Roles in Parental Behavior," following).

Another form of protection against conspecifics has been reported in *Hyla rosenbergi.* Guarding males attack and usually repel other males intruding on their mud nests (Kluge 1981). This is important, because if the surface film of eggs is disturbed, they will sink to the bottom of the nest and die. Although it is not known whether males deliberately try to destroy other males' eggs, it is clear that paternal guarding is essential for egg survival when male densities are high. When male densities and the probability of intrusion were low, eggs were seldom guarded, indicating the facultative nature of parental care. Facultative parental care also was reported for a closely related species, *Hyla faber,* in Brazil. As in *H. rosenbergi,* males abandoned nests when densities of males at the site were low, but guarded their nests when densities were high (M. Martins, Pombal, and Haddad 1998).

### Protection against Heterospecific Predators

It is often assumed that amphibian parents defend their eggs against heterospecific predators, but the evidence for this is mostly circumstantial. Eggs of the salamander *Desmognathus ocoee* quickly disappear when left unattended in the field (Tilley 1972), suggesting that predation is a major source of mortality. Forester (1978, l979a) introduced carabid beetles, ringnecked snakes (*Diadophis punctatus*), and three species of predatory salamanders (*Pseudotriton ruber, Gyrinophilus porphyriticus, Desmognathus monticola*) into laboratory chambers with brooding females of *D. ocoee.* The females successfully defended their clutches against the beetles, but generally could not defend against snakes or large salamanders. Similar results were obtained with *Plethodon cinereus;* most females fled when snakes were introduced into nest cavities, but one female did attack an intruding snake (Bachmann 1984). When eggs of *Ambystoma opacum* were enclosed in mesh predator-exclusion cages, without attending females, survival of eggs was higher than for unenclosed eggs (Croshaw and Scott 2005). Small millipedes (*Oxidus gracilis*) were identified as important predators of A. opacum eggs.

Adults of two large frog species, *Leptodactylus ocellatus* and *Pyxicephalus adspersus,* actually attack birds and other vertebrates feeding near their tadpoles (Balinsky and Balinsky 1954; W. Rose 1956; Poynton 1957; Vaz-Ferreira and Gerhau 1975; C. Cook, Ferguson, and Telford 2001), and other species of *Leptodactylus* have been induced to attack sticks and other model predators (Vaira 1997; Prado, Uetanabaro, and Lopes 2000; Prado, Uetanabaro, and Haddad 2002). The terrestrial eggs of *Eleutherodactylus coqui* are occasionally attacked by invertebrates, but this is a relatively minor source of mortality, and the incidence of invertebrate predation did not increase when parental males were removed (Townsend, Stewart, and Pough 1984). In contrast, predation by invertebrates appears to be a major source of mortality for *Cophixalus parkeri* eggs, and unattended eggs have almost no chance of surviving (Simon 1983). All of the terrestrial and fossorial microhylids (*Xenobatrachus, Copiula, Liophryne, Hylophorbus, Callulops,* and *Austrochape-*

*rina*) studied in New Guinea by Bickford (2001, 2004) exhibited defensive behavior against ants or other arthropod predators, but arboreal species did not. Many of the fossorial species were particularly aggressive in guarding their eggs, and often tried to block entrances to burrows. Some of these microhylids also attacked pencils inserted into nests as model predators. When attending parents (usually males) were removed from egg clutches of two common species, *Hylophorbus rufescens* and *Oreophryne* sp., all of the eggs died or disappeared. Most mortality in *Hylophorbus* was caused by ant predation, whereas unattended eggs of *Oreophryne*, an arboreal species, were more likely to die from desiccation (Bickford 2004).

Eggs of centrolenid frogs are laid in exposed positions on leaves overhanging water and are attacked by a variety of invertebrates, including small crabs, spiders, and insects (McDiarmid 1978; M. Hayes 1983), but apparently not by conspecifics (M. Hayes 1991). Males of several species guard their eggs (McDiarmid 1978; M. Clark 1981; Jacobson 1985; M. Hayes 1991), and removal of the male greatly decreases survivorship of eggs. McDiarmid (1978) attributed most of the mortality in unattended clutches to predation by invertebrates, but in *Hyalinobatrachium fleischmanni*, desiccation seems to be a more important source of mortality than predation (M. Hayes 1991).

### Protection against Desiccation

There is little doubt that protection against desiccation is a major benefit of parental care in many terrestrial species, although it is not important in species that lay eggs in saturated microhabitats (Simon 1983). Most plethodontid females remain in contact with their eggs and usually coil around them during dry weather (Noble and Brady 1933; Organ 1960b; Brode 1961; Highton and Savage 1961; Vial 1968; McDiarmid and Worthington 1970; Tilley 1972; Forester 1979a). Some salamanders moisten their eggs by periodically voiding urine (Highton and Savage 1961; Brode 1961). D. Snyder (1971) found that survivorship of *Aneides aeneus* eggs was poor when the female was removed, even when the eggs were surrounded by predator exclusion cages; much of the mortality was due to desiccation. Similarly, Forester (1984) demonstrated that eggs of *Desmognathus ocoee* lost more water when left unattended than when brooded by the female. In *Plethodon cinereus*, eggs brooded by large females had greater mass and higher survivorship than did those brooded by small females. Furthermore, in dry years, maternal body size had a greater effect on hatchling size than did initial egg size, suggesting that brooding mothers protect eggs from desiccation (Crespi and Lessig 2004).

Many frogs sit directly on their eggs, sometimes covering them with their bodies and limbs (C. Myers 1969; Amiet 1981; Townsend, Stewart, and Pough 1984; B. Bell 1985; M. Hayes 1991; Diesel, Baurle, and Vogel 1995; Joglar, Burrowes, and Rios 1996; Quijano, Santos-Barrera, and Pacheco-Rodríguez 2002; Bickford 2004). This posture clearly is well suited to minimizing evaporative water loss, as well as transferring water from the highly permeable pelvic skin to the eggs. There is indirect evidence of the importance of parental care in preventing desiccation in *Leiopelma*. In two species that lay eggs in relatively dry sites, males sit on their eggs until they hatch; in a third species that lays eggs in semiaquatic sites, males remain near their eggs, but usually do not sit on them (B. Bell 1985). Dendrobatid frogs take up water from wet leaves or small puddles and then transfer it to terrestrial egg clutches (Senfft 1936; H. Zimmermann 1974, 1978; Wells 1978a; Weygoldt 1980a, b; H. Zimmermann and E. Zimmermann, 1980b, 1981, 1984). Whether they do so by releasing urine or by transferring water directly from the skin is not clear. Attending adults sometimes manipulate the eggs with their back legs, perhaps to enhance the uptake of water (fig. 11.7 C).

Similar manipulative behavior has been observed in the centrolenid frog *Hyalinobatrachium fleischmanni*, a species that lays eggs on the surfaces of leaves. M. Hayes (1991) placed water soluble dyes on egg clutches and showed experimentally that manipulation of the eggs caused the dye to diffuse more rapidly and more completely into the egg clutch, suggesting that the behavior would enhance water uptake. Experimental removal of brooding males showed that brooding reduced embryo mortality from desiccation. Hayes also artificially watered some clutches from which males had been removed. Survival of embryos was higher in these clutches than in unattended clutches that were not watered. Parental attendance was most important in the first few days after eggs were laid. Removal of males at that time resulted in higher mortality than removal of males later in development. As mentioned previously, males of this species and several other centrolenids do not attend their eggs continuously, but visit them for short periods of time, usually after midnight. Hayes found that extended bouts of brooding resulted in eggs becoming more susceptible to infection by fungi, and these clutches also were more likely to be invaded by predatory fly larvae (see Villa 1984). In other words, there was a tradeoff between increased likelihood of desiccation with too little brooding and increased mortality from fungi and fly larvae with too much brooding. Intermediate levels of brooding produced the highest survivorship.

Brooding by males is essential for maintaining the proper hydric regime for the eggs of *Eleutherodactylus coqui* (fig. 11.14). Brooding seems to be less important for *E. cooki*, which breeds in very humid caves (Burrowes 2000; Rogowitz et al. 2001). Males of *E. coqui* select oviposition sites, usually in curled dead leaves, rolled palm fronds, or other enclosed microhabitats that provide high-humidity environments for egg development (Townsend 1989a). They

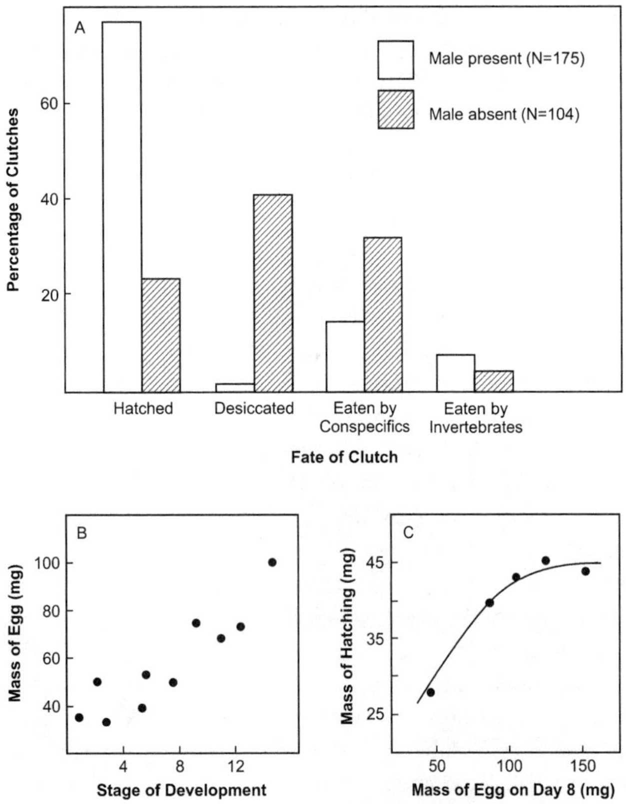


**Fig. 11.14.** Effect of male parental care on egg survival and development in the Puerto Rican treefrog *Eleutherodactylus coqui.* (A) Fate of control clutches with male present and experimental clutches with male absent. (B) Mass of the eggs as a function of stage of development in clutches attended by males. (C) Mass of hatchlings as a function of the mass of eggs at day 8. (A) Plotted from data in Townsend et al. (1984). (B, C) After Taigen, Pough, and Stewart (1984).

lead females to their nest sites (Townsend and Stewart 1986), and after the eggs are laid, the male remains with them until they hatch. When males were removed from eggs, 41% of the clutches died from desiccation, compared to 1% of attended clutches (fig. 11.14 A; Townsend, Stewart, and Pough 1984). Water is transferred by osmosis directly from the male's skin to the eggs. The water not only increases survivorship of eggs, but also the mass of hatchlings (fig. 11.14 B, C), thereby probably improving their chances of surviving outside the nest (Taigen, Pough, and Stewart 1984). Hydration of eggs also appears to be the most important benefit of parental care in an arboreal species of *Oreophryne* from New Guinea, but it appears to be of minor importance for terrestrial and fossorial species (Bickford 2001, 2004).

### Protection against Fungus and Pathogens

Fungi often infect terrestrial amphibian eggs, and these infections are fatal if left unchecked (Villa 1979b; Simon 1983). Fungal infections often begin on eggs that have died from other causes and spread to living eggs (Forester 1979a). Tilley (1972) reported that living *Desmognathus ocoee* eggs could survive heavy fungal infections, but Forester (1979a) found that fungal infections frequently were fatal. Unattended eggs tend to develop fungal infections more rapidly than attended clutches (Storm 1947; Vial 1968; D. Snyder 1971; Salthe and Mecham 1974; Forester 1979a; Simon 1983), but there has been some controversy over the role of the parent in preventing fungal growth. Early literature suggested that adult amphibians might produce antifungal secretions in the skin (reviewed by Salthe and Mecham 1974), but several experimental studies have failed to support this hypothesis (Daniel and Simpson 1954; Vial and Preib 1966; Forester 1979a; Simon 1983). Thiesmeier and Hornberg (1997) speculated that females of the Asian newt *Pachytriton* release antifungal secretions from the skin, but experimental evidence is lacking. R. Austin (2000) found a substantial bacterial community on the skin of *Plethodon ventralis,* and at least some of these bacteria exhibited antifungal and antibacterial properties, suggesting that they might play a role in protecting eggs brooded by females. Periodic movement or abrasion of the eggs by the female also could inhibit fungal growth, but evidence for this is mixed (Tilley 1972; Forester 1979a). Females of *Desmognathus ocoee* selectively eat infected eggs, and this protects the rest of the clutch (Tilley 1972; Forester 1979a). Several species of microhylid frogs from New Guinea were observed to eat egg clutches infected with fungus, and in one species, *Sphenophryne cornuta,* an attending parent selectively ate only infected eggs (Bickford 2001).

### Aeration and Agitation of Eggs

Several types of parental activities in aquatic amphibians have been interpreted as adaptations for aerating eggs. This is a reasonable hypothesis, since fanning eggs is an important component of parental care in many fishes (Blumer 1979; Perrone and Zaret 1979; Baylis 1981). This behavior helps to break up the boundary layer of still water surrounding the eggs, thereby facilitating diffusion of oxygen into the egg mass. Evidence for such behavior in amphibians is inconclusive. *Necturus* females keep eggs constantly agitated in aquaria by waving their gills, and *Cryptobranchus* males achieve the same effect through continuous movements of the body (Salthe and Mecham 1974). Such movements frequently occur when these species are exposed to low oxygen tensions as a means of enhancing the animal's own respiration (Guimond and Hutchison 1976); any advantage to the eggs could be incidental and has yet to be demonstrated. The apparently deliberate agitation of eggs by kicking in *Nectophryne afra* (Scheel 1970) is more reasonably interpreted as parental behavior, although the survival value of the behavior has not been investigated.

Agitation of eggs also occurs in terrestrial amphibians. Dendrobatids, centrolenids, and other frogs move their legs around in the eggs while wetting them (Wells 1978a; H. Zimmermann and E. Zimmermann 1981; M. Hayes 1991). This might ensure an even distribution of moisture, but it also could be important in preventing yolk stratification or adhesion of the embryo to the egg membranes (Salthe and Mecham 1974). Simon (1983) showed that survivorship of terrestrial frog eggs was enhanced by periodic agitation. Bickford (2001) observed many species of microhylids from New Guinea jostling eggs, but he did not present experimental data on the function of this behavior. Survivorship of eggs appears to be enhanced by jostling in *Desmognathus* (Forester 1979a), but not in *Hyalinobatrachium fleischmanni* (M. Hayes 1991) or *Ambystoma opacum* (Croshaw and Scott 2005).

## Costs of Parental Care

In many species, attendance of eggs or larvae undoubtedly entails some cost to the parent, but measurement of such costs is extremely difficult. Theoretically, both benefits and costs should be measured in terms of lifetime fitness (Trivers 1972; Clutton-Brock 1991; Crump 1996), but in practice, one must be satisfied with estimates of proximate costs and benefits. Even this is difficult, because costs and benefits often cannot be measured in the same units. Benefits accrue as increased offspring survival, whereas costs include physical risk to the parent, energetic costs, and decreased opportunities for future reproduction.

### Physical Risk to the Parent

There are a few anecdotal accounts of brooding amphibians being consumed by predators such as ants (Forester 1979c) or snakes (Forester 1978), but these observations mainly involve captive individuals. The best evidence for a direct risk to attending parents comes from observations of male African bullfrogs (*Pyxicephalus adspersus*) being injured or killed when defending schools of tadpoles against water birds (C. Cook, Ferguson, and Telford 2001). In two particularly grisly instances, herons pecked out the eyes of defending males and then ate their offspring. *Hyla rosenbergi* males are sometimes injured while defending their nests (Kluge 1981), but this probably is rare in other amphibians. In *Eleutherodactylus coqui,* intruding males are sometimes injured in fights, but parental males are not (Townsend, Stewart, and Pough 1984). To show that parental behavior is costly, one must demonstrate that risks to caregiving individuals exceed those incurred by individuals engaged in other activities. Indirect evidence that brooding females suffer increased mortality comes from Organ's (1961c) study of the demography of several species of *Desmognathus.* He found that sexually mature females had lower survivorship than males, and attributed this to greater exposure to predation in aquatic or semiaquatic nest sites.

### Energetic Costs

The metabolic costs of egg attendance and other forms of parental care probably are trivial compared to other reproductive activities such as calling, searching for mates, and courtship behavior (see chapter 5). Nevertheless, brooding adults that are confined to nests have few opportunities to feed, and sometimes suffer reductions in body mass or stored energy reserves (Fitzpatrick 1973b; Krzysik 1980; Forester 1981). Brooding *Desmognathus* females frequently have empty or nearly empty stomachs (Tilley 1972) and use up significant amounts of fat. Fitzpatrick (1973b) estimated that about 16% of the annual energy budget of a *D. ochrophaeus* female was allocated to maintenance during brooding, but he did not attempt to estimate the cost of parental care relative to that of alternative uses of energy such as foraging, growth, or production of eggs.

A reduction in growth rates and body mass of brooding females also has been reported for *Plethodon cinereus* (Ng and Wilbur 1995). Brooding females of this species produced fewer eggs the following year than did those that abandoned their eggs or did not brood at all (Yurewicz and Wilbur 2004). It has been suggested that many temperate-zone plethodontids cannot produce eggs every year because they spend so much of the summer growing season brooding eggs without feeding (Sayler 1966; Hom 1988; Crump 1995b). Nevertheless, if the chances of eggs surviving without parental care are sufficiently poor, then females should remain with their eggs even if this entails high costs (Hom, Willits, and Clark 1990). For some salamanders, the energetic costs of parental care are relatively low. For example, Kaplan and Crump (1978) did not find any evidence that brooding entailed substantial energy costs in *Ambystoma opacum.* They attributed this to the brief period of parental care and the fact that most individuals do not feed during the fall anyway.

The New Guinea microhylid frog *Cophixalus parkeri* has an extraordinarily long brooding period, with some males remaining with eggs for nearly five months (Simon 1983). Brooding males consume much less food than nonbrooding adults, and often have empty stomachs. They sometimes resort to eating sphagnum moss from the nest area, a food that probably provides very little nutrition. Brooding adults also have reduced fat bodies compared to nonbrooding frogs (fig. 11.15 A). Hence, the prolonged period of parental care clearly imposes a heavy energetic cost. In *E. coqui,* neither brooding nor calling males do much feeding, but parental males were more likely to have empty stomachs (Townsend

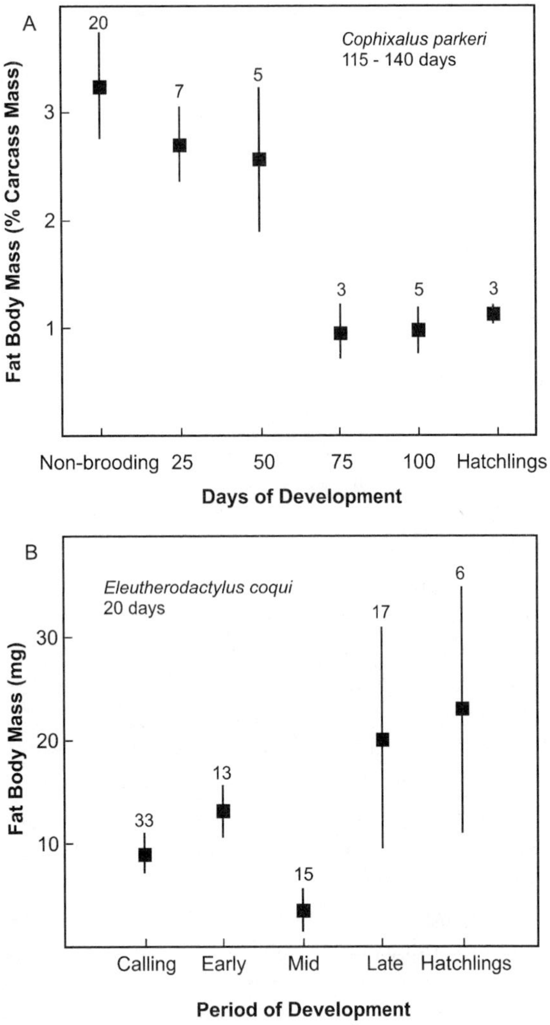


**Fig. 11.15.** The energetic cost of parental care in two terrestrial frogs. (A) Reduction in relative abdominal fat body mass in males of a microhylid frog from New Guinea, *Cophixalus parkeri,* as a function of stage of embryonic development. Fat reserves of males decrease throughout the long period of parental care in this species. (B) Change in absolute abdominal fat body mass in males of the Puerto Rican treefrog *Eleutherodactylus coqui* as a function of stage of embryonic development. Fat reserves of males decrease at the peak period of parental care, but increase late in embryonic development as some males leave the eggs periodically to feed. (A) After Simon (1983). (B) Plotted from data in Townsend (1986).

1986). Nevertheless, fat bodies were significantly reduced only in the middle of the brooding period (fig. 11.15 B), suggesting that the energetic cost of the relatively short (20 day) period of parental care is low compared to the cost for *Cophixalus.*

Almost nothing is known about the details of parental care by caecilians in the field, but Kupfer, Nabhitabhata, and Himstedt (2004) reported that body mass of females of *Ichthyophis kohtaoensis* in Thailand decreased during the breeding season and was lower for females brooding eggs at advanced stages of development. Body condition, adjusted for body length, also was lower for females brooding older eggs than for those brooding young eggs. Parental care is prolonged in this species, with clutches required up to three months to develop, but the constancy of female attendance has not been determined. Nevertheless, the changes in body mass and condition suggest that foraging opportunities for brooding females are limited, making parental care energetically costly. These observations are consistent with an earlier report that females in captivity showed a reduction in body mass during egg brooding, but in that case, captivity may alter natural foraging behavior (Kramer, Kupfer, and Himstedt 2001).

Shorter-term types of parental care, such as transporting larvae from oviposition sites to water, probably have relatively low energetic costs, because the time devoted to the activity is so short. The only attempt to investigate such costs is a study by Downie et al. (2005) on *Mannophryne trinitatis,* a dendrobatid frog in which males carry tadpoles for several days. They did not find any evidence that foraging of tadpole-carrying males was reduced relative to that of calling males, nor did the extra burden of tadpoles affect locomotor performance. They were not able to measure actual metabolic costs of carrying tadpoles, however.

### Reproductive Costs

Prolonged parental duties also could limit opportunities for future reproduction. For females, this is largely an energetic cost, since an inability to feed while brooding eggs would reduce a female's capacity to produce more eggs. Alternatively, parental care might limit reproductive opportunities because environmental conditions are not suitable for reproduction all year. For example, some plethodontids might be limited to annual or biennial reproduction because of the costs of parental care, although another possibility is that ecological factors restrict opportunities for more frequent reproduction, thereby making extended parental care possible. Many salamanders that lack parental care also have annual or biennial reproductive cycles (Salthe and Mecham 1974).

Data on frequency of reproduction in tropical frogs are so scarce that it is impossible to say whether parental care limits reproduction by females. In the dendrobatid frog *Colostethus panamensis,* a species with female parental care, females apparently reproduce no more than twice a year (Wells 1980a), a surprisingly low figure for a species with favorable breeding conditions for nine or ten months a year. Females of some other dendrobatids with male parental care, such as *Dendrobates auratus,* produce clutches every 10 to 14 days (Senfft 1936), but this also is true for species such as *D. pumilio,* in which females assume some parental duties and even feed their tadpoles (Weygoldt 1980a, b).

For males, the chief cost of parental care would be an inability to advertise for additional mates while attending eggs. This seems to be true for *Hyla rosenbergi* (Kluge 1981). In *Eleutherodactylus coqui,* most males remain with their eggs throughout development and do not advertise for additional females; a few leave their eggs late in development and occasionally succeed in attracting additional mates (Drewry 1970; Townsend, Stewart, and Pough 1984). Nevertheless, a reduced opportunity to obtain mates does appear to be a significant cost of parental care in that species (Townsend 1986, 1989b). On the other hand, if males can care for eggs and attract additional mates at the same time, then the cost of parental care is greatly reduced. Attendance of multiple egg clutches by males has been reported in cryptobranchid salamanders (B. Smith 1907; S. Bishop 1947; Nickerson and Mays 1973a; Kuwabara et al. 1989), several species of *Eleutherodactylus* (Drewry 1970; Joglar, Burrowes, and Rios 1996; Rogowitz et al. 2001), some centrolenid frogs (McDiarmid 1978; M. Hayes 1991), some microhylid frogs in the genus *Cophixalus* (Hoskin 2004), and other anurans (Wells 1981a). Male midwife toads (*Alytes*) frequently carry eggs from several different females (Reading and Clarke 1988; Raxworthy 1990; Márquez 1993; Verrell and Brown 1993). Furthermore, the relatively short breeding period of some midwife toad populations reduces the cost of parental care, because females are available for only a short period of time (Márquez 1992).

## Sex Roles in Parental Care

The publication of Trivers' (1972) important paper on parental investment led to increased interest in male and female roles in all aspects of reproductive behavior, including parental care. Once viewed as an example of cooperation between the sexes, parental behavior is now evaluated in terms of potential evolutionary conflicts in the reproductive interests of males and females (Trivers 1972; E. O. Wilson 1975; Maynard Smith 1977, 1978; Grafen and Sibly 1978) or parents and offspring (Trivers 1974). Amphibians and fishes are particularly useful groups for testing theoretical ideas about parental care, because these groups include species with no parental care, parental care by males alone, parental care by females alone, and biparental care. There now is a considerable theoretical and empirical literature on the origins of maternal and paternal care in these groups (G. Williams 1966, 1975; Trivers 1972; Dawkins 1976; Dawkins and Carlisle 1976; Maynard Smith 1977, 1978; Grafen and Sibley 1978; McDiarmid 1978; Ridley 1978; Blumer 1979; Perrone and Zaret 1979; Barlow 1981; Gross and Shine 1981; Baylis 1981; Wells 1981a; Gross and Sargent 1985; Nussbaum 1985, 2003; Clutton-Brock 1991; Crump 1995b, 1996; Beck 1998; Lehtinen and Nussbaum 2003).

### Mate Desertion and Parental Care

Theoretical models often are designed to predict which parent will desert offspring first. Such behavior would force the remaining parent either to desert the offspring or to care for them alone (Trivers 1972; Dawkins 1976; Dawkins and Carlisle 1976; Maynard Smith 1977, 1978; Grafen and Sibly 1978; Ridley 1978; Gross and Sargent 1985; Clutton-Brock 1991). These models assume that strategies adopted by individuals of one sex depend on activities of both same sex and opposite sex individuals in the population. Although such models show how a set of specified alternative strategies could be maintained in a population, they tell us little about how such strategies originated because this depends largely on initial conditions (Maynard Smith 1977). Since both maternal and paternal care in amphibians almost certainly evolved from a condition of no parental care, it is unlikely that desertion by one parent requires the other to care for eggs. Therefore, it is more appropriate to ask why individuals of either sex remain with the eggs instead of deserting them.

### Mode of Fertilization and Parental Care

Parental care by males alone is more common among animals with external fertilization than among those with internal fertilization (Trivers 1972; Wells 1977b, 1981a; Ridley 1978; Gross and Shine 1981; Clutton-Brock 1991). Among vertebrates, care by the male alone is uncommon in mammals, birds, reptiles and salamanders with internal fertilization, but is relatively widespread in teleost fishes, frogs, and the few salamanders with external fertilization. Two general explanations of this relationship have been proposed. First, with external fertilization, males must be present when eggs are laid and therefore are available to assume parental duties (G. Williams 1966, 1975; Trivers 1972; Wells 1977b, 1981a; Maynard Smith 1978; Crump 1995b, 1996). In species with internal fertilization, male parental care often is not an option, because mating occurs long before eggs are laid. In many temperate-zone plethodontid salamanders, for example, mating can occur in both the fall and spring, but oviposition occurs in the summer (Houck 1977a); parental care by females is nearly universal. External fertilization also increases the probability that all eggs in a clutch are a male's own, thus reducing the probability that a male will care for another's eggs (G. Williams 1966, 1975; Trivers 1972; Maynard Smith 1978; Ridley 1978; Blumer 1979). The importance of the certainty of paternity hypothesis in the evolution of paternal care has been questioned on theoretical grounds (Maynard Smith 1978; Werren, Gross, and Shine 1980; Gross and Shine 1981). It cannot explain the widespread occurrence of either male or female parental care in species with external fertilization (Wells 1981a; Crump 1995b, 1996). In addition, there are cases of multiple paternity of egg clutches in both anurans and urodeles with external fertilization (see chapters 8 and 9), so males cannot always be certain of their paternity.

Beck (1998) used phylogenetically based comparative methods to examine the relationship between mode of fertilization and sex of the caregiving parent in anurans. He concluded that mode of fertilization did not explain the occurrence of male or female parental care in this group. There are several problems with his analysis, however. First, he limited his analysis to anurans, nearly all of which have external fertilization. When salamanders are included in the analysis, it is clear that most salamanders have female parental care, with the only exceptions being species with external fertilization. In anurans, on the other hand, paternal care is at least as common as is maternal care, suggesting that external fertilization facilitates the evolution of paternal care. Second, the sample of anuran species with internal fertilization is very small and biased by reproductive mode. Of the genera known to have internal fertilization, two (*Ascaphus* and *Mertensophryne*) do not have any parental care and are therefore irrelevant to the analysis. Five other African bufonids have viviparous reproduction, as does *Eleutherodactylus jasperi.* In these species, parental care consists of embryos being retained in the reproductive tract of the mother, so male parental care is not possible, and they should be omitted from the analysis (Crump 1996). The sole remaining species is *Eleutherodactylus coqui.* All of the closest relatives of this species within the genus *Eleutherodactylus* have male parental care, regardless of mode of fertilization (Townsend 1996).

There is evidence for amphibians that the proximity of the male during oviposition is more important than mode of fertilization in determining the pattern of parental care, especially if the male chooses the oviposition site. In *Eleutherodactylus coqui,* males perform all parental care, but fertilization is internal (Townsend et al. 1981; Townsend, Stewart, and Pough 1984). Nevertheless, mating occurs immediately before oviposition at a site chosen by the male, so the male is available to assume parental duties. A Central American species, *E. angelicus,* also might have internal fertilization, although the evidence is circumstantial, and nothing is known about the timing of mating and oviposition. Females attend eggs in this species (M. Hayes 1985).

Among salamanders, the best-documented cases of male parental care are in cryptobranchids, which have external fertilization (S. Bishop 1941b; Nussbaum 1985, 2003). Oviposition takes place at a nest site previously chosen by the male, and the males then simply stay with the eggs. Sirens are assumed to have external fertilization (see chapter 10), yet in one species, females are reported to attend their eggs. In this species, however, eggs are laid in tangles of vegetation probably chosen by the female. Unfortunately, mating has never been observed in sirens, either in the field or in captivity. Male parental care might occur in the European olm (*Proteus anguinus*), although the details of parental care are unclear (see the previous discussion). Even though this species has internal fertilization, mating apparently takes place immediately before oviposition, so the male is available for parental duties. For most salamanders, the evolution of fertilization by means of spermatophores probably set the stage for the evolution of female parental care, not because fertilization is internal, but because courtship and oviposition generally are separated in both space and time, and oviposition sites are chosen by females alone. This enables females of many species to make use of cryptic, protected oviposition sites that probably are not suitable for courtship (Nussbaum 1985).

### Territoriality and Parental Care

If a male defends an oviposition site that enhances his attractiveness to females, he might remain in his territory and continue to attract females while giving his eggs some protection. In the initial evolution of parental care, the reproductive cost to the male, in the form of decreased opportunities for future matings, would be relatively low (Trivers 1972; G. Williams 1975; Wells 1977b, 1981a; Woodruff 1977; McDiarmid 1978). This argument has been applied to the evolution of male parental care in fishes (G. Williams 1975; Ridley 1978; Blumer 1979; Perrone and Zaret 1979; Barlow 1981; Baylis 1981; Gross and Sargent 1985; Clutton-Brock 1991). There is little doubt that it applies to amphibians as well, but several additional factors must be considered (Wells 1981a). First, one cannot assume that a male remaining at an oviposition site protects the eggs. In *Rana clamitans* and *R. catesbeiana,* males remain in their territories and continue to call after fertilizing eggs (Wells 1977c; R. D. Howard 1978a), but there is no evidence that this is advantageous to the eggs. The same is true for Australian frogs of the genus *Pseudophryne,* which lay eggs in burrows (Woodruff 1977). Second, a correlation between male parental care and territoriality is expected only if males defend oviposition sites. Many species of frogs defend calling sites separated from oviposition sites (chapter 8). Finally, some forms of parental care make territory defense difficult, so the advantages of parental care have to be balanced against the disadvantages of losing the territory.

In an earlier review, I suggested that frogs in the genus *Eleutherodactylus* would be useful for comparative studies of parental strategies, because the genus contains species with maternal care, paternal care, or no parental care at all (Wells 1981a). Since that paper was published, additional examples of parental care in this group have been reported, although most such reports are anecdotal, and many are based on unpublished information (Townsend 1996). Although the data are still very incomplete, several trends are evident. In general, parental care by females appears to be strongly associated with oviposition in nests on the ground, including spaces under rocks, holes in sand or soil, subterranean burrows constructed by the female, or in leaf litter. These sites also are the ones most commonly used by species that apparently lack parental care.

Some species with male parental care also nest on the ground or in leaf litter, but many others use elevated sites in such as tree holes, bromeliads, curled dead leaves hanging in vegetation, the axils of palm trees, and the sides of rocks in caves (table 11.2). The evolution of arboreal habits potentially exposes frogs and their eggs to drier conditions than those experienced by ground-dwelling frogs. This requires the use of specialized oviposition sites in sheltered locations that provide high-humidity environments for egg development (Townsend 1989a), just the types of sites that are likely to be in short supply and defensible as territories by males. Indeed, aggressive defense of oviposition sites has been reported in several species of *Eleutherodactylus*. Townsend (1996) also reported that males have been observed calling from oviposition sites in seven species, and accumulation of multiple egg clutches at a defended site is relatively common. In some species, such as *E. coqui*, males usually do not call directly from the oviposition site, but instead lead the female there from an elevated calling site (Townsend and Stewart 1986). Once a male has acquired an egg clutch, however, he sometimes calls from the nest during late stages of embryonic development.

Townsend (1996) also identified geographical correlates of parental care in *Eleutherodactylus*. Parental care by males appears to be quite uncommon among species from the mainland of South and Central America. Townsend could find only two examples of possible male parental care out of 28 species for which some information was available. One was for an *E. augusti* male found under a rock with a clutch of eggs (Jameson 1950); the other was for an *E. diastema* male found in the leaf axial of a bromeliad with an egg clutch (Dunn 1937). The former is known to call from the oviposition site, and the latter exhibits a high degree of site fidelity and very likely is territorial (Wells, unpublished observations; Wilczynski and Brenowitz 1988). All other mainland species for which egg clutches are known (a very small proportion of the total) either have no parental care or female parental care. In contrast, parental care by males is much more common in the West Indies, especially in Hispaniola, Puerto Rico, and the Lesser Antilles. This pattern reflects a fundamental difference in the ecology of *Eleutherodactylus* in these areas. Many mainland species are terrestrial, forest-litter inhabitants, with arboreal habitats being dominated by hylids and centrolenids. Many West Indian species are terrestrial as well, but others have moved into arboreal microhabitats, especially on islands where hylids are uncommon or absent. Many of these make use of elevated oviposition sites and have male parental care.

Phylogenetic history appears to be important as well, in part because phylogeny and ecology are correlated to some degree. For example, all of the *Eleutherodactylus* of Puerto Rico and the Lesser Antilles are thought to be closely related, and indeed, appear to form a monophyletic group that is largely arboreal, but with some terrestrial, cave, and rock-dwelling species (Hedges 1989, 1996; Joglar 1989, 1998). The common occurrence of male parental care in these frogs could represent a single origin from a common ancestor. A distinct clade, the subgenus *Euhyas*, is endemic to the western Caribbean (Cuba, Jamaica, and Hispaniola) and is primarily terrestrial. Every example of female parental care among West Indian species cited by Townsend (1996) comes from this lineage. This includes all Jamaican species for which parental care is known, except for *E. alticola*, which is reported to have either male or female parental care. There are Cuban and Hispaniolan species in this subgenus with male parental care as well (Townsend 1996), and all of these probably evolved arboreal habits secondarily (Hedges 1989).

Other anuran species show a similar association between male territoriality, defense of specialized oviposition sites, and male parental care (table 11.1). In species in which females attend their eggs, oviposition usually occurs in burrows constructed by the female, foam nests formed by pairs in amplexus, and natural cavities on the ground under logs, rocks, or leaf litter. In species with male parental care, males usually call from tree holes, burrows, rock cavities, nests constructed by the male, leaves overhanging streams, or other types of oviposition sites likely to be in short supply or variable in quality and defensible as territories. Active territorial defense of these sites has been reported in several species, and in some, males have been observed attending more than one clutch, or are known to continue to call while attending eggs. Many dendrobatid frogs exhibit a slightly different strategy. Males often are territorial, but instead of defending a single oviposition site, they defend an area containing several suitable sites and then court females and lead them to oviposition sites (Wells 1978a; Summers 1989, 1992; Roithmair 1992).

As in the genus *Eleutherodactylus*, the phylogenetic history of other anurans appears to affect the role of the two sexes in parental care, although this also reflects the phylogenetic distribution of territorial defense of oviposition sites by males. In centrolenids, for example, only male parental care has been well documented (table 11.1). The only report of females remaining with eggs is Jacobson's (1985) account of *Centrolene proseblepon* females remaining near their eggs for several hours after completing oviposition. It is not clear whether this is an example of parental care, or simply the result of exhausted females resting after laying eggs. Parental care apparently is rare in hyperoliids, some of which lay eggs on leaves over water in a manner similar to that of centrolenids. Only female parental care has been observed in this family. Male parental care appears to be much more common than female parental care in microhylid frogs, the main exceptions being burrowing or ground-nesting species. Male parental care clearly is the dominant pattern in several microhylid genera from Madagascar (*Anodonthyla, Cophyla, Platypelis,* and *Plethodontohyla*), all in the subfamily Cophylinae. The same is true for microhylids from New Guinea and Australia (Bickford 2001, 2004). Female parental care predominates in leptodactylid genera other than *Eleutherodactylus,* perhaps reflecting a correlation with oviposition in foam nests or nest sites selected by the female.

### Sex Roles in Egg and Tadpole Transport

Why is egg transport almost exclusively a female activity in frogs? McDiarmid (1978) suggested that females are more likely to be able to carry a complete clutch because they often are larger than males. This explanation seems unlikely, because it should apply equally well to tadpole transport. Yet there are many species of dendrobatid frogs in which males have no difficulty transporting entire clutches of tadpoles. I believe several other factors are more important. First, there would be severe problems in transferring eggs from the female to the body of the male if fertilization is external. In pipids and egg brooding hylids, the actions involved in placing the eggs on the female's back or in a brood pouch are modifications of normal amplexus, in which the male is on top of the female. Although specific movements vary considerably among species, all allow the eggs to pass the male's cloaca and be fertilized before being positioned on the female (Rabb and Rabb 1960, 1963a; Weygoldt 1976a; Duellman and Maness 1980). Rather complicated gymnastics would be required for eggs to end up on the back of the male. In *Alytes,* the only genus with male transport of eggs, the eggs are laid in strings, thus facilitating their attachment to the male's legs. Another factor that would mediate against the evolution of egg transport by males is territoriality; a frog carrying eggs probably cannot defend a territory because of potential injury to the eggs. Furthermore, if a male could carry only one clutch at a time, then a prolonged period of parental care would prevent a male from obtaining additional mates (Wells 1981a). Again, *Alytes* is an exception; males can carry several clutches simultaneously, and there is no evidence that males are territorial (Márquez 1992; Verrell and Brown 1993).

Carrying tadpoles on the back probably interferes with aggressive defense of territories, and this could influence the sex of the care-giving parent in dendrobatid frogs. When male-male competition for mating territories is intense, then one might expect selection to work against males carrying tadpoles. On the other hand, when competition for territories is less intense, or when males are not territorial, then either sex might carry tadpoles. These predictions are most relevant to species that carry tadpoles for several days before releasing them, as in some species of *Colostethus* and *Mannophryne* (table 11.3). Tadpole transport by males is the predominant form of parental care in these genera and is assumed to be the ancestral condition for all dendrobatids with parental care (Summers et al. 1999). In *Colostethus panamensis, C. pratti,* and *C. talamancae,* however, females carry tadpoles and males defend long-term territories (Wells 1980a, b, 1981a, unpublished observations). In *Mannophryne trinitatis, M. collaris,* and *Colostethus palmatus,* females defend territories and males carry tadpoles. Males are aggressive toward other calling males, but apparently do not defend fixed territories (Test 1954; Sexton 1960; Duellman 1966; Lüddecke 1974, 1976; Durant and Dole 1975; Van Meeuwen 1977; Wells 1980c, 1981a). In some other species, such as *C. flotator,* males are aggressive toward one another and apparently are territorial (Wells 1981a; unpublished observations). Males also carry tadpoles, but these are transported to pools and released very quickly (Wells, unpublished observations). Whether a male subsequently returns to the same territory is not known. Patterns of male and female parental care in *Colostethus* and *Mannophryne* probably also reflect phylogenetic history, but a complete phylogeny of these genera is not yet available.

Males carry tadpoles in most species of *Dendrobates, Allobates, Epipedobates, Phyllobates,* and *Cryptophyllobates,* apparently reflecting the ancestral condition for the family (Summers et al. 1999; Lötters, Jungfer, and Widmer 2000; Lehtinen and Nussbaum 2003). In most species, tadpole transport requires only a few minutes to a few hours. Even when males are territorial, as in *Dendrobates auratus, D. leucomelas,* and *Allobates femoralis,* they must stop defending their territories only for short periods of time to carry tadpoles to water, and then are free to return and resume calling and territorial behavior (Summers 1989, 1992; Roithmair 1992). Parental care by females has evolved in two different lineages within the genus *Dendrobates.* In members of the *Dendrobates histrionicus* complex, including *Dendrobates pumilio, D. granuliferus, D. histrionicus, D. lehmanni, D. arboreus,* and *D. speciosus,* males are very aggressive in defending territories, and tadpoles are carried by females and later fed with trophic eggs (table 11.3). In *D. pumilio, D. granuliferus,* and probably *D. arboreus,* males attend the eggs before females take over all other parental duties, but they do not remain continuously with the eggs and are free to defend territories and advertise for additional mates. In the more derived species, such as *D. histrionicus* and *D. speciosus,* however, females perform all parental care (fig. 11.14), and males can return immediately to territorial defense and calling to attract mates.

In members of the "*Dendrobates ventrimaculatus* complex" in Amazonian South America, biparental care predominates. In contrast to the members of the *D. histrionicus* complex, males not only attend the eggs, but also carry the tadpoles. As described earlier in the chapter, males of most species then call the females and lead them to the tadpole deposition sites, where the females feed the tadpoles. In at least one species, *D. vanzolinii,* males and females form long-term monogamous pair bonds while caring for their young (Caldwell 1997; Caldwell and Oliveira 1999). The ancestral condition for this clade appears to be parental care performed entirely by the male. This is seen in a population of "*D. ventrimaculatus*" from Ecuador, which exhibits egg attendance and tadpole transport by males, but not feeding of tadpoles by females (Summers et al. 1999). Additional behavioral data on other populations of Amazonian frogs and further resolution of the taxonomy and phylogeny of the group are needed before the evolution of male and female parental care can be completely understood.

Summers and Earn (1999) suggested that differences in parental strategies within the genus *Dendrobates* are related in part to the size of pools used for tadpole deposition and the amount of food available in these pools. Species that have only male parental care often deposit tadpoles in relatively large pools in tree holes, stumps, fallen logs, or seed husks on the ground, where organic detritus tends to accumulate. While tadpoles in these pools often are cannibalistic, they often can survive without feeding on other tadpoles and without being fed by the female. Males sometimes deposit more than one tadpole in the same pool, although more commonly they disperse them among different pools. Summers and Earn (1999) suggested that placement of small tadpoles in pools that already contain larger offspring of the same male could be a form of facultative tadpole feeding.

Most of the species that have biparental or female parental care with tadpole feeding use very small pools in bromeliads or *Heliconia* leaf axils, or in the case of *D. vanzolinii,* tiny pools in tree saplings and vines (Caldwell and de Oliveira 1999). Such pools are more abundant in the rainforest than are larger pools, but they also lack sufficient food for tadpole development. Hence, the use of these small pools favored the evolution of parental feeding of tadpoles and led to the evolution of either female or biparental care (Caldwell and de Oliveira 1999; Summers and Earn 1999). Biparental care is most likely to evolve when males and females remain in habitats near their tadpole deposition sites, which would enable males to lead females to these sites. Parental care by the female alone is most likely to evolve when tadpole deposition sites are a considerable distance from the normal home ranges of males and females. In *Dendrobates pumilio* and related species, for example, the adults are primarily terrestrial, but tadpole deposition sites often are in bromeliads high in the canopy.

### Parental Care and Sex Role Reversal

G. Williams (1966), Trivers (1972), Knowlton (1982) and others have hypothesized that high male parental investment leads to reversal of the usual sex roles in reproductive behavior, with females competing among themselves for access to males. This type of mating system is found in some pipefishes (Vincent et al. 1992), and in insects that exhibit high levels of male parental investment, such as giant waterbugs (Heteroptera: Belostomatidae; R. L. Smith 1979; Kruse 1990) and katydids (Orthoptera: Tettigoniidae; Gwynne 1985; Gwynne and Simmons 1990; L. Simmons 1992). Dendrobatid frogs have been cited repeatedly as possible examples of sex role reversal because males often provide parental care, and females of some species are more aggressive or more colorful than males (Trivers 1972; E. O. Wilson 1975; Ridley 1978; Kluge 1981). One must be cautious in interpreting these patterns, however, because an extensive male role in parental care is not necessarily equivalent to high male parental investment. In species where males can simultaneously care for eggs and attract additional mates, the cost of parental investment is relatively low. For sex role reversal to occur, male parental care must be so expensive that it inhibits males from obtaining additional mates, thereby decreasing the availability of males in the population and limiting female reproductive success (Trivers 1972; Knowlton 1982; Gwynne 1991; Vincent et al. 1992)

The following types of data are needed to demonstrate a relationship between male parental investment and sex role reversal in frogs: (1) receptive females should outnumber receptive males; (2) the shortage of males must be related to male parental investment and not simply to an unequal population sex ratio; (3) it should be relatively common for more than one female to approach a male simultaneously; (4) female-female aggressive encounters should be related to competition for mates and not solely to competition for ecological resources such as feeding sites or shelters; (5) females should take the more active role in courtship; and (6) males should discriminate between females when given a choice of mates (Wells 1981a). Such behavior is most likely when females produce many small clutches of eggs at short intervals and the time required to care for each clutch is longer than the interclutch interval. Female egg production might outstrip the ability of males to care for them, and a shortage of males would result. One must keep in mind that both males and females can be subject to sexual selection, and it is possible for individuals of both sexes to compete against same-sex individuals for access to mates (Andersson 1994).

The possibility of sex-role reversal has been investigated in some detail in *Dendrobates auratus,* a species originally cited by Trivers (1972) as an example of this phenomenon. Observations of this species in captivity and in the field revealed that more than one female sometimes approaches a calling male. Females sometimes fight with one another over males, females assume an active role in courtship, and they sometimes destroy the eggs of other females (Senfft 1936; H. Zimmermann 1974, 1978; Wells 1978a). All of these observations are consistent with the sex-role reversal hypothesis. Somewhat similar behavior has been observed in several other species with male parental care, including *Dendrobates azureus, D. leucomelas, D. tinctorius, Allobates femoralis, Phyllobates terribilis,* and *P. vittatus* (Polder 1974b; H. Zimmermann and E. Zimmermann 1980a, 1981, 1985; H. Zimmermann 1982; Weygoldt 1983).

In a detailed field study of *D. auratus* in Panama, Summers (1989) found that the availability of receptive males did not limit the mating success of females. There always were numerous males defending territories and calling to attract females. Females were observed to fight with one another and interfere with courtship attempts, as had been observed in the laboratory, but males also competed for access to females. Females took the more active role in courtship. Summers concluded that aggression among females and courtship of males by females reflected competition for particular males that controlled high-quality tadpole deposition sites. Furthermore, it was advantageous for females to try to prevent males from mating with other females, because they suffered a fitness cost if males mated polygynously. When males cared for the eggs of more than one female simultaneously, they often deposited several tadpoles in the same pools. Because the tadpoles are cannibalistic, this could potentially reduce the fitness of a particular female (Summers 1990). Hence, there appears to be a conflict of interest between males, which maximize their reproductive success by mating with as many females as possible, and females, which maximize their reproductive success by monopolizing the services of particular males. The same interpretation probably applies to the very similar behavior of *D. leucomelas,* but this species has not been studied in as much detail (Summers 1992).

Females also have been observed to compete with one another for access to males in *Alytes obstetricans,* another species with male parental care (Verrell and Brown 1993). In a laboratory colony, females tried to interfere with mating by pushing males that were already in amplexus with other females. Females could not completely control access to particular males, at least in the confines of a laboratory, because one male mated with two females that had competed with one another. Males showed little evidence of aggression or competition for mates other than calling to attract females. Females could not be competing for males with high-quality territories in midwife toads, because males are not territorial, but they might compete for access to males perceived to be superior caregivers for their eggs.

## Ecology and the Evolution of Parental Care

Several authors have proposed complex evolutionary schemes relating ecological variables to the evolution of parental care in amphibians, particularly terrestrial species (Salthe 1969; Salthe and Duellman 1973; Salthe and Mecham 1974; Lamotte and Lescure l977; McDiarmid 1978; Nussbaum 1985, 2003; Crump 1995b; Lehtinen and Nussbaum 2003). Some have suggested that parental care in amphibians evolved as part of a more widespread tendency toward terrestrial modes of reproduction. According to this view, rigors of the aquatic environment, especially predation, favored placement of eggs away from water. This in turn introduced new dangers, such as the problem of desiccation. Consequently, terrestrial reproduction set the stage for increased parental attendance of eggs and eventually the evolution of more elaborate forms of parental care (Salthe and Duellman 1973; Salthe and Mecham 1974; McDiarmid 1978; Duellman and Trueb 1986; Crump 1995b). Some authors have assumed that parental care in anurans and urodeles evolved in response to similar selective pressures, and Salthe and Mecham (1974) called particular attention to the apparent parallelism between terrestrial plethodontids and frogs with direct development, such as *Eleutherodactylus.*

I believe the available evidence indicates that parental care in terrestrial salamanders and terrestrial frogs had quite different origins and evolved in response to different selective pressures, a conclusion also drawn by Nussbaum (1985, 2003) and Crump (1995b, 1996). Salamanders and frogs differ in several respects. First, terrestrial reproduction without parental care is relatively common in frogs (Lamotte and Lescure 1977; Lehtinen and Nussbaum 2003), but is rare in salamanders (Nussbaum 1985, 2003). Second, parental attendance of eggs is rare in aquatic frogs, but is relatively common in aquatic salamanders (Nussbaum 1985, 2003; Crump 1995b, 1996). Finally, both terrestrial reproduction and parental care are virtually nonexistent in temperate zone frogs, but are characteristic of most temperate zone salamanders. These patterns provide some insights into the evolution of parental care.

### The Evolution of Parental Care in Urodeles

Nussbaum (1985, 2003) argued that the relative scarcity of food in many running-water environments has favored large egg size in salamanders because larvae hatching from large eggs will be better able to capture the relatively large but scarce prey items found in streams (see also chapter 10). In contrast, salamanders that breed in ponds produce relatively small eggs, perhaps because small larvae can utilize abundant small zooplankton present in such habitats, and because of constraints on gas exchange in low-oxygen environments (see chapter 4). One cost of large egg size in stream-breeding salamanders is a longer embryonic period, especially in cold water (see chapter 10), and this in turn exposes the embryos to a greater risk of predation. Hence, selection has favored placement of eggs in hidden nest sites and maternal attendance to reduce embryo mortality in stream-breeding salamanders. Pond breeders generally have evolved other strategies for protecting their eggs from predators, including encasing the eggs in thick coats of jelly, scattering eggs individually on the bottoms of ponds, or hiding eggs individually in vegetation (see chapter 10 for more details).

As mentioned earlier, conspecific predation on eggs appears to be widespread among salamanders, particularly those using aquatic or semiaquatic nest sites. Such predation results in part from competition among females for suitable nest sites. Salamanders breeding in streams usually place their eggs in specially selected sites that have the proper orientation to the current and are hidden from potential predators. Such sites probably are in short supply in many habitats, and those available probably vary in quality. Consequently, intraspecific competition for nest sites would be expected. By preying on eggs already present in a nest site, an intruding female might cause the resident female to abandon the site, thereby making it available to the intruder. Such predation would favor parental protection of nest sites, even in the absence of strong predation pressure from other species.

Once parental care had evolved in aquatic salamanders, they were preadapted for reproduction in terrestrial microhabitats. Presumably, the principal benefits of parental care gradually shifted from protection against predators to protection against desiccation in fully terrestrial nest sites, as in many species of *Plethodon.* Parental attendance of eggs probably accounts for the success of plethodontids in colonizing a variety of microhabitats in the temperate zone. Subsequently one lineage invaded the humid tropics and has since undergone extensive radiation (D. Wake 1987; see also chapter 1).

This scheme yields several testable predictions. First, facultative parental care should be relatively common in aquatic salamanders. Specifically, when nest sites are not in short supply, or when population densities are low, females might not guard their eggs. This would not be predicted if eggs were guarded mainly against heterospecific predators. Unfortunately, there are few reliable data on intraspecific variation in parental care. J. Wood (1953a) found large clutches of *Eurycea bislineata* eggs, apparently laid by several females, in masses of dead leaves; most were not attended. In other areas, where females lay eggs under rocks or similar defensible sites, parental attendance is more prevalent (S. Bishop 1941b; Baumann and Huels 1982). The evidence for parental attendance in *Pseudotriton* also is mixed, suggesting facultative variation among populations (Salthe and Mecham 1974; Bruce 1972a, 1975, 1978c). A second prediction is that parental care might become facultative or be lost altogether in terrestrial species that place their eggs in saturated microhabitats where predation is not very severe. For example, the frequent abandonment of eggs by females of *Hemidactylium* and *Stereochilus,* both of which lay eggs in very wet sites, is consistent with this prediction.

One also might expect secondary loss of parental care in some tropical salamanders that lay their eggs in moss mats and similar microhabitats. Unfortunately, the eggs of most tropical plethodontids have not been found, so it is impossible to generalize about the frequency of parental care.

## The Evolution of Parental Care in Anurans

The possible origins of terrestrial reproduction and direct development in frogs were discussed in chapter 10. Although parental care in anurans often is associated with terrestrial reproduction, this is not universally true. What does seem to be generally true is that parental attendance of eggs usually is associated with the use of small, discrete oviposition sites, whether these are on land or in the water. Such sites include small pools, basins constructed by the sides of streams, spaces under rocks, rock crevices, burrows, sheltered sites in leaf litter, and oviposition sites on leaves, in leaf axils of bromeliads and other plants, in hollow bamboo stems, and in tree holes. Not only are these sites generally hidden from predators, but they also are readily defended against intruders, and when on land, often provide the high humidity conditions necessary for embryonic development. Anurans that place their eggs in highly exposed sites, or that group their eggs with those of other individuals, seldom exhibit any form of parental care.

What makes the analysis of the evolution of parental care in anurans very difficult is the frequent independent evolution of parental care in many different lineages of frogs, probably in response to a variety of selective pressures. This contrasts with the situation in salamanders, where the vast majority of species with parental care are in the family Plethodontidae, a group with a much less diverse array of reproductive modes. McDiarmid (1978), Duellman and Trueb (1986), and Crump (1995b, 1996) all provided hypothetical scenarios for the evolution of parental care in anurans, progressing through a series of steps toward increasingly terrestrial reproduction. They envisioned the process beginning with frogs placing their eggs near water, but retaining aquatic larvae and either not providing parental care (as in phyllomedusine hylids) or providing some parental care (as in centrolenids). Other species lay eggs in terrestrial sites that are close enough to water to allow tadpoles to move to water or be transported to water by a parent. The final step is considered to be the evolution of terrestrial oviposition, with eggs giving rise to nonfeeding tadpoles or direct development, or in some cases, transport of eggs by the parent.

This type of scenario can be a useful heuristic tool, but I believe it is somewhat misleading when one considers particular lineages of frogs. In most cases, there is little evidence that either reproductive modes or parental care actually evolved through the proposed intermediate steps. This is particularly true of complex forms of egg and larval transport; in most lineages, the closest relatives of species with these complex forms of parental care do not have similar reproductive modes, and intermediate steps leading to the evolution of parental care cannot be identified. Probably a more productive approach in the future will be to examine in detail certain lineages that exhibit variation in reproductive modes and parental care and attempt to reconstruct the evolution of parental care in a more explicitly phylogenetic context, as has been done for dendrobatid frogs (Summers et al. 1999) and some other clades (Lehtinen and Nussbaum 2003).

## Summary and Conclusions

Amphibians display a remarkably complex array of parental strategies. Although many species lay their eggs in water or on land and abandon them, others exhibit various forms of parental care, including attendance of eggs or larvae at the nest site or transport of eggs, larvae, or juveniles on the body of a parent. The frequency of parental care in caecilians is unknown because their reproductive biology is poorly understood. Parental attendance of eggs, usually by the female, appears to be the rule among urodeles, particularly those in the family Plethodontidae. Parental care probably occurs in only 10–15% of all anuran species, but frogs exhibit the greatest diversity of parental behavior. Several experimental studies have shown that parental care enhances the survivorship of eggs and larvae. Putative benefits include protection from predators, protection from desiccation, protection from fungus and other pathogens, aeration of eggs, and in a few species, provision of food to the larvae. In most species, parental care probably provides more than one type of benefit. Potential costs of parental attendance include physical risk to the parent, energetic costs, and decreased opportunities for future reproduction, particularly in species with male parental care.

The evolution of sex roles in parental care is a topic of considerable interest to behavioral ecologists, and the amphibians can serve as a model system for testing hypotheses about parental investment. Mode of fertilization tends to be associated with the form of parental care, with male parental care being more common in species with external fertilization. This association appears to be derived from the temporal coincidence of mating and oviposition and the role of the male in choosing the nest site, rather than the mode of fertilization per se. Territorial defense of oviposition sites by males probably preceded the evolution of male parental care in some species, because males can continue to attract females while attending previous egg clutches. In species that transport eggs or tadpoles, however, territorial behavior by males could select against parental care if eggs or tadpoles are likely to be injured, or if prolonged parental care reduces a male's chances of retaining his territory. Some dendrobatids exhibit apparent sex role reversal associated with high levels of male parental investment; females are more aggressive in competing with one another than males and take a more active role in courtship. Nevertheless, such competition among females probably has more to do with competition for the best-quality males or oviposition sites than with a shortage of males. brought about by high male parental investment. Possible sex role reversal also occurs in midwife toads, a species in which males provide parental care, but do not defend territories or offer resources to females.

Parental care probably evolved independently many times in anurans, usually in response to the rigors of terrestrial reproduction, whereas parental care in plethodontid salamanders probably originated in an aquatic environment and subsequently contributed to the success of this family in colonizing terrestrial habitats. Among aquatic salamanders, maternal care occurs almost exclusively in stream-breeding species, which usually have relatively large eggs and long embryonic periods. The long embryonic period entails increased risks of predation, and selection has favored parental attendance to increase embryonic survival.

A greater understanding of the evolution of parental care in amphibians will require more detailed phylogenetic information on the relationships of species in particular clades, as well as more comparative information on the occurrence of parental care, particularly in tropical frogs, and the role of the sexes in such behavior. There also is a need for additional quantitative and experimental studies of the costs and benefits of parental care. In particular, the sources of embryonic and larval mortality and the importance of parental care in reducing such mortality need to be carefully investigated. This information, coupled with data on the energetic and reproductive costs of parental care, will provide new insights into relationships between reproductive modes, parental behavior, and mating systems in amphibians.

# Chapter 12 The Ecology and Behavior of Amphibian Larvae

*So still the Tadpole cleaves the watery vale*
*With balanced fins, and undulating tail;*
*New lungs and limbs proclaim his second birth,*
*Breathe the dry air, and bound upon the earth.*
—Erasmus Darwin, *The Temple of Nature* (1803)

THE REMARKABLE transformations that amphibians undergo during metamorphosis have fascinated biologists since the time of Aristotle. Indeed, the name *amphibian* comes from Greek roots meaning "double life," and this unique feature of their biology has long been a distinguishing characteristic of the class. Early naturalists viewed the metamorphosis of tadpoles as evidence that amphibians were intermediate in the scheme of creation between fishes and reptiles, since most species make the transition from water-breathing aquatic larvae to air-breathing terrestrial adults.

Although amphibians played an important role in classical embryology and our understanding of processes of development, the ecology and behavior of amphibian larvae, as independent organisms, was largely neglected until recently. R. Maxwell Savage, a pioneer student of tadpole biology, put it this way: "It often seems that the larvae receive less attention than the adults. This is curious, not only because larvae have a considerable interest in themselves, but because, in the study of the whole life history of an animal, it is often the larval stage that throws much light on features of the ecology that would otherwise remain incomprehensible." He advocated the study of tadpoles, "not as temporary animals in the process of becoming something else, but as animals whose life, although short compared to that of adults, is lived on an entirely different plane, and has called for special larval organs, habits, and responses, to be considered just as if the adult frog had never been discovered" (Savage 1961, p. 24).

This is the approach I have taken in this chapter. I begin by describing morphological adaptations of amphibian larvae for different environments, as well as adaptations for acquiring and processing food, since these are important for understanding other aspects of larval ecology and behavior. Finally, I conclude with a discussion of the social behavior of amphibian larvae and its relationship to population ecology, including the role of kin recognition in larval social behavior. Other aspects of larval biology are discussed elsewhere in the book. The physiological ecology of amphibian larvae was discussed in chapters two through five. Chapter 13 reviews the evolution and maintenance of a complex life cycle and ecological aspects of metamorphosis, particularly environmental and genetic influences on larval growth and development. Responses to predators and the ecology of larval communities are discussed in chapters 14 and 15.

## The Origin and Evolution of Amphibian Larvae

A complex life history with an aquatic larval stage is thought to be the ancestral condition for all three groups of living amphibians, the caecilians, urodeles, and anurans. Indeed, this type of complex life history probably was inherited from the common ancestor of living amphibians (Hanken 1999a). Fossils representing several early tetrapod groups exhibit unmistakable larval features. These include dissorophoid temnospondyls (Boy 1974; Bolt 1977, 1979; Milner 1982; Schoch 1992; Boy and Sues 2000), which generally are considered the closest relatives to modern amphibians (see chapter 1). Fossil salamander and frog larvae resembling those of modern families or genera have been found in deposits as old as the Cretaceous (Estes, Spinar, and Nevo 1978; Naylor 1978; Wassersug and Wake 1995; S. Evans and Milner 1996; S. Evans, Milner, and C. Werner et al. 1996; Roček 2000; Roček and Rage 2000b). These fossils show that an aquatic larval stage is an ancient feature of amphibian biology, extending back at least 250 million years.

## Morphology, Development, and Ecology of Caecilian Larvae

The ancestral mode of reproduction for caecilians is the production of eggs that hatch into aquatic larvae. This mode of reproduction is retained in the two most primitive families, the Rhinatrematidae and Ichthyophiidae, as well as some species in the more derived family Caeciliidae (chapter 10). The development of aquatic caecilian larvae is poorly understood compared to that of urodeles and anurans, and even less is known about their ecology. Only a few species in the family Ichthyophiidae have been studied in any detail (Sarasin and Sarasin 1887–1890; Breckenridge, Nathanael, and Pereira 1987; Himstedt 1991, 1996). Caecilian larvae look more like adults than do the larvae of either urodeles or anurans. They hatch at a relatively advanced stage of development, with lungs already present. The large external gills that are present in the developing embryo degenerate at hatching, but the larvae retain gill slits (either one or two pairs, depending on the family). The body is elongate and similar in form to that of adults, but aquatic larvae often have tail fins that enhance aquatic locomotion. In terrestrial species, the tail is lost at metamorphosis. At metamorphosis, the rather delicate chondrocranium of the caecilian larva undergoes extensive ossification to form the heavily reinforced skull of the burrowing adult (see chapter 1). There also are changes in the morphology of the hyobranchial skeleton and tongue that accompany the shift from aquatic suction feeding to capturing terrestrial prey with the jaws (M. Wake 1989a).

Some of the most significant changes that occur at metamorphosis are alterations of the sensory systems (Himstedt 1996). Caecilian larvae have well-developed lateral line systems with two kinds of receptors. The first are neuromasts, which are mechanical receptors similar to those of other aquatic vertebrates. Caecilians also have ampullary organs, which function as electroreceptors (Hetherington and Wake 1979; Fritzsch et al. 1985; Wahnschaffe, Fritzsch, and Himstedt 1985). These organs are sensitive to very weak electric fields, like those produced by living organisms, and probably aid in location of hidden prey in the water (Himstedt and Fritzsch 1990). Urodele larvae also have ampullary organs that are similar in morphology and function to those of caecilian larvae, but these organs are absent in anuran tadpoles, which lack the capacity for electroreception (Fritzsch 1981; Münz, Claas, and Fritzsch 1982, 1984; Fritzsch and Wahnschaffe 1983; A. Roth and Schlegel 1988). The lateral line system disappears in terrestrial adult caecilians.

Caecilians often are thought to be completely blind, but many species have functional eyes both as larvae and as adults (M. Wake 1985). The eyes are small relative to those of urodeles and anurans, and they do not increase in size as the body grows (Himstedt 1995). The muscles that move the eye in larvae assume a secondary function of retracting the tentacle, a chemosensory organ found only in adult caecilians (Billo 1986; Billo and Wake 1987; Himstedt 1996; see also chapter 1). In both larvae and adults, the eyes are covered with semitransparent skin and probably do not form images. Nevertheless, the eyes are sensitive to differences in the intensity of light and are involved in phototactic behavior. Larvae of the Asian caecilian *Ichthyophis kohtaoensis* exhibit positive phototaxis when they first hatch from the eggs, presumably because they use light cues to find their way out of the terrestrial burrow where the eggs were laid. Shortly after hatching, however, phototactic behavior reverses, suggesting that the larvae use negative phototaxis to locate suitable hiding places during the day (Himstedt 1995).

The ecology and behavior of caecilian larvae are almost completely unknown, but work on *Ichthyophis kohtaoensis* in Thailand has shown that the larvae are not strictly aquatic, as usually assumed, but are amphibious. During the day, they remain hidden in terrestrial retreat sites under rocks, logs, or vegetation at the edges of ponds, sometimes in groups, but emerge at night to forage in the water (Himstedt 1991). This amphibious mode of life is retained to some extent in the adults of this species, which are perfectly capable of swimming in the water, despite their morphological adaptations for burrowing (Crapon de Caprona and Himstedt 1985). Anecdotal evidence for other species of *Ichthyophis* and several other caecilians suggests that this type of amphibious behavior is relatively common in larval caecilians (Himstedt 1996). An analysis of stomach contents of juveniles of the aquatic caecilian *Typhlonectes compressicauda* showed that these animals eat earthworms, frog eggs, tadpoles, and both aquatic and terrestrial insects. These animals were observed foraging in shallow water, and may have captured terrestrial insects that fell into the water (Verdade, Schiesari, and Bertoluci 2000).

## Morphology, Development, and Ecology of Urodele Larvae

Most discussions of urodele larvae emphasize the similarity of general body form of larvae and adults and the lack of dramatic changes at metamorphosis (Salthe and Mecham 1974; Duellman and Trueb 1986). Most salamanders have relatively extended larval periods, and in some stream-dwelling species, the larval period can last for more than a year. At the opposite extreme are amphiumas, which hatch at an advanced stage of development and lose their gills after only two weeks (Gunzburger 2003). Urodele larvae are more similar to adults than are anuran tadpoles, but there are several distinctive larval characteristics. These include a skeleton composed mostly of cartilage, flattened tail fins, external gills, open gill slits, and the absence of moveable eyelids (eyelids serve to clean and moisten the eyes in terrestrial amphibians, but do not serve a useful function in aquatic animals). Some or all of these features are retained in paedomorphic species (see chapter 13), but they change in other species at metamorphosis. Changes occur in many other characters, including modification of the head, mouth, and associated musculature. These changes accompany the switch from aquatic suction feeding to terrestrial prey capture (Lauder and Shaffer 1986, 1988; Lauder and Reilly 1988, 1990; Reilly and Lauder 1988; Shaffer and Lauder 1988; B. Miller and Larsen 1989; Reilly 1994, 1996).

Other changes at metamorphosis are less conspicuous, but no less important as the animals make the transition from aquatic to terrestrial life. The skin, for example, develops the glands that characterize adult skin (see chapter 2). Cell layers are added to the epidermis, which develops a distinct *stratum corneum,* the outer layer of the epidermis that provides some mechanical protection to the skin (H. Fox 1994; Warburg, Lewinson, and Rosenberg 1994). These morphological changes are accompanied by physiological changes in osmoregulation and ion transport (Burggren and Just 1992). Metamorphosing salamanders also develop adult urinary and reproductive systems. Changes in the digestive system and other visceral organs are less dramatic than in anuran tadpoles (see the following), because salamander larvae already are adapted to a carnivorous diet, as are adults, but there is some shortening of the gut (Tilley 1964). Finally, salamanders exhibit a number of changes in their sensory systems at metamorphosis (Just, Kraus-Just, and Check 1981; Burggren and Just 1992). These include loss of the lateral line system in terrestrial species, although this is retained in permanently aquatic, paedomorphic species, such as the European cave salamander (*Proteus anguinus;* Fritzsch and Wahnschaffe 1983; Istenic and Bulog 1984). The shift from aquatic to terrestrial habits also is accompanied by changes in the shape of the eye lens. Aquatic salamander larvae have spherical lenses in their eyes, which are best suited to seeing underwater. At metamorphosis, the lenses become more flattened, a change that enhances vision in air (Wilder 1925; Sivak and Warburg 1980). Highly aquatic species, such as newts of the genus *Triturus,* retain spherical lenses throughout life (Manteuffel, Wess, and Himstedt 1977).

### Functional Morphology of Locomotion

Salamander larvae have the same thin, elongate body morphology found in adults. In contrast to anuran tadpoles, which have a tail with only a flexible notochord for an axial skeleton, salamander larvae have a fully developed vertebral column supporting the tail musculature (Wassersug 1989a). Larvae that live in ponds typically have very large tail fins that extend along the back. Mountain brook larvae have small, fleshy tail fins that do not extend beyond the tail itself; stream larvae are intermediate (fig. 12.1). In all salamander larvae, the front limbs develop before the hind limbs, the opposite of the pattern found in anurans. The timing of limb development varies, but most species have fully functional limbs for much of the larval period, in contrast to anurans, in which limbs emerge shortly before metamorphosis.

Pond-dwelling larvae, including many hynobiids, ambystomatids, and salamandrids, often have balancers when they hatch, but these are absent in sirenids, which also have pond-type larvae. Balancers are extensions of the mandibular arch that produce a sticky mucus secretion that helps to hold the larva in place before the front limbs develop. Once the limbs have developed, the balancers either break off, as in the ambystomatids, or are resorbed, as in the salamandrids. Larvae that live in flowing water typically lack balancers altogether, but hatch with both the front and hind limbs more fully developed. A phylogenetic analysis suggested that balancers evolved in the common ancestor of the derived salamander families, after the divergence of the Sirenidae, but were lost repeatedly in various lineages (Crawford and Wake 1998). It also is possible that balancers evolved twice, once in the Hynobiidae and again in the common ancestor of ambystomatids and salamandrids. Crawford and Wake (1998) argued that this was less likely, because the complexity of balancer morphology suggests a single origin of the structure.

The locomotion of salamander larvae has been described as anguilliform (eel-like) and involves lateral bending of both the body and tail (Hoff et al. 1989; Wassersug 1989a). Salamander larvae do not engage in extended bouts of sustained swimming, but instead move in short bursts, separated by periods of relative inactivity. This mode of locomotion is used both to escape from predators and to capture prey. The presence of large external gills and well-developed limbs increases drag as the animal moves. During rapid lunges toward prey,

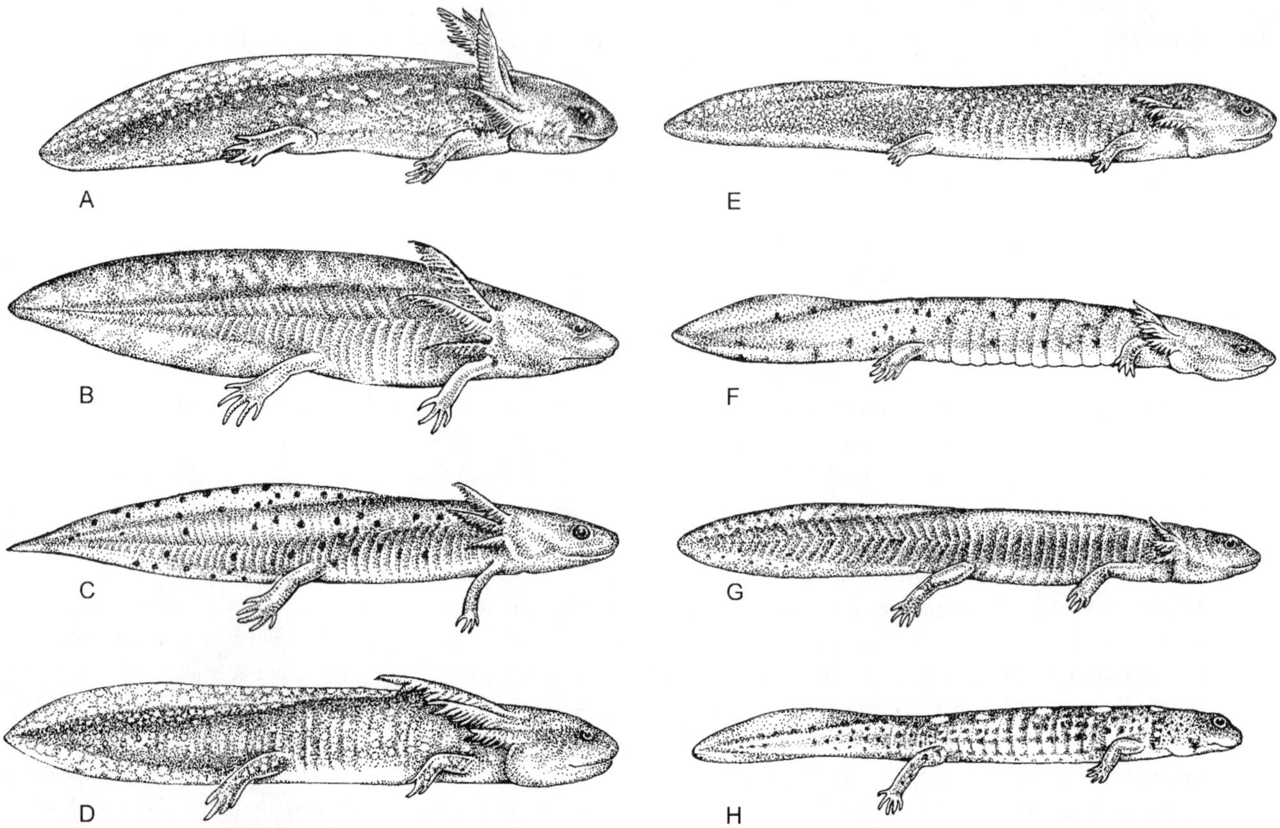


**Fig. 12.1.** Morphological types of salamander larvae (not drawn to same scale). (A–D) Pond type larvae. (A) *Ambystoma tigrinum* (Ambystomatidae). (B) *Pleurodeles waltl* (Salamandridae). (C) *Triturus cristatus* (Salamandridae). (D) *Hemidactylium scutatum* (Plethodontidae). (E–F) Stream type larvae. (E) *Pseudotriton ruber* (Plethodontidae). (F) *Cryptobranchus alleganiensis* (Cryptobranchidae). (G–H) Mountain brook larvae. (G) *Calotriton asper* (Salamandridae). (H) *Rhyacotriton olympicus* (Rhyacotritonidae). (A, D–F) Drawings by Mary Jane Spring after Bishop (1941b). (B, C, G) After D. W. Ovenden in E. Arnold and Burton (1978) and Stebbins (1951).

salamander larvae fold their gills and forelimbs against the body, thereby reducing the amount of drag and increasing the speed of movement. They extend their limbs to slow forward movement (Hoff, Lannoo, and Wassersug 1985).

Pond-dwelling larvae often remain suspended in the water column when they are small, but gradually shift to a more benthic existence as they grow and develop limbs. Large larvae are stronger swimmers than are small larvae, but they also spend more time crawling on the bottom or through vegetation and less time swimming in open water than do small larvae (J. Anderson 1968a; Hassinger, Anderson, and Dalrymple 1970; Leff and Bachmann 1986; McWilliams and Bachmann1989a, b). Stream-dwelling larvae generally do not inhabit open water, but instead spend most of their time resting or crawling among rocks on the bottom. Most of their swimming is confined to lunges at prey or escapes from predators (M. Parker 1994).

## Respiratory Structures

All salamander larvae have three sets of external gills, each consisting of a central stem (ramus) that supports the small respiratory structures, or fimbriae (fig. 12.1). The length of the rami and the number of fimbriae vary with the habitat in which the larvae are found (Valentine and Dennis 1964). Species that live in ponds or lakes with relatively low levels of dissolved oxygen generally have large, bushy gills with many fimbriae that provide a large surface area for gas exchange. Various species of *Ambystoma* have classic pond-type larvae, but similar larval morphology is found in other families as well, including the permanently aquatic proteids, salamandrids such as *Notophthalmus, Taricha,* and *Triturus,* and a few plethodontids such as *Hemidactylium* (fig. 12.1).

At the opposite extreme are larvae found in mountain brooks, where the water often is saturated with oxygen. These larvae have much reduced gills with only a few fimbriae (fig. 12.1). They include the North American genera *Rhyacotriton* and *Dicamptodon* and Asian hynobiids such as *Onychodactylus* and *Batrachuperus* (Valentine and Dennis 1964). Many salamanders, including some *Ambystoma* and plethodontids such as *Gyrinophilus, Pseudotriton,* and *Eurycea,* have intermediate gill morphology and are referred to as stream type larvae, although they may live in a variety of habitats, from fast-moving streams to rivers and still pools. Respiratory surface area exhibits some facultative response to the environment. Low levels of dissolved oxygen produce an increase in fimbrial area, whereas high oxygen levels cause a decrease (Bond 1960; Guimond and Hutchison 1976). Nevertheless, mountain brook and stream larvae never develop the very large gills of pond larvae and generally cannot survive at low oxygen levels. Many salamander larvae probably develop lungs relatively early, but only a few species have been examined in detail (Burggren and Just 1992). Larvae use both the skin and lungs to supplement oxygen uptake through the gills (see chapter 4). They also use the lungs to regulate buoyancy (Lannoo and Bachmann 1984b).

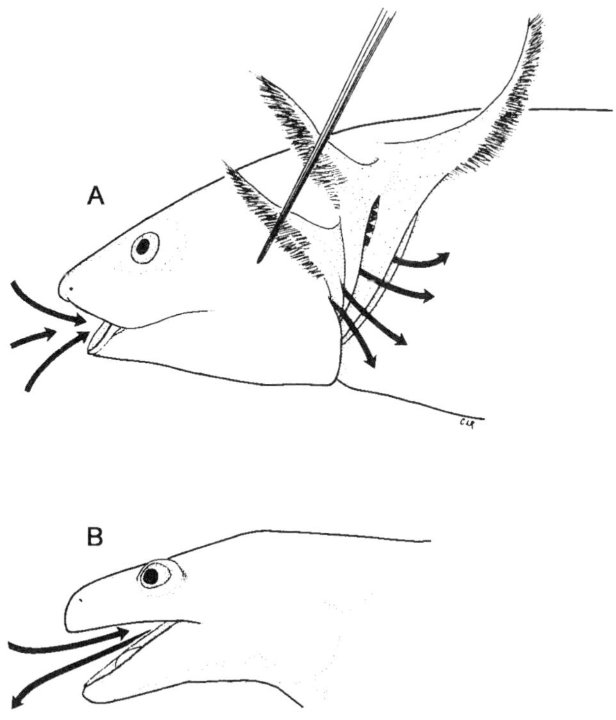


Fig. 12.2. (A) Diagram of one-way flow of water through gill slits of aquatic *Ambystoma tigrinum* larva during feeding. (B) Two-way flow of water in mouth of transformed adult. After Lauder and Shaffer (1985).

## Functional Morphology of Feeding

Some terrestrial salamanders use inertial feeding to capture prey, grabbing the prey with the jaws and then moving the head forward while simultaneously releasing the prey (Bramble and Wake 1985). Other species capture prey by protruding the tongue from the mouth, in some cases for a considerable distance beyond the jaws (Özeti and Wake 1969; Lombard and Wake 1976, 1977; D. Wake 1982; see chapter 1). These modes of feeding are less effective in the aquatic environment, because the movement of the predator's head or tongue would displace the prey away from the mouth. Aquatic salamander larvae feed by suction, moving the snout close to the prey and suddenly opening the mouth while depressing the hyoid apparatus. This produces negative pressure in the mouth, which sucks in the prey (Shaffer and Lauder 1985a, b; Reilly and Lauder 1989; Gillis and Lauder 1994; Reilly 1995, 1996). Both the initial lunge toward the prey and the opening of the mouth can be extremely fast (Hoff, Lannoo, and Wassersug 1985). This mode of feeding is characteristic of all aquatic salamander larvae, and indeed, most aquatic ectothermic vertebrates (Lauder 1985; Wainwright et al. 1989; Sanderson and Kupferberg 1999).

In ambystomatid salamander larvae, the efficiency of suction feeding is greatly enhanced by the one-way flow of water into the mouth and out the open gill slits (fig. 12.2). At metamorphosis, the gill slits close, making a one-way flow of water impossible. The result is that metamorphosed adults are less effective at capturing prey in the water than are aquatic larvae (Lauder and Shaffer 1986). Lauder and Reilly (1988) experimentally tested the role of gill slits in enhancing aquatic feeding in adult axolotls (*Ambystoma mexicanum*), which retain both gills and open gill slits throughout life. They sutured the gill slits to prevent a one-way flow of water during suction feeding. This resulted in a decrease in aquatic feeding efficiency equivalent to that observed in metamorphosing *Ambystoma*. The closing of the gill slits is not solely responsible for the decrease in aquatic feeding efficiency of metamorphosed adults, however. There also are changes in the structure of the hyobranchial apparatus and size and strength of the muscles that open the mouth. The result is a decrease in buccal pressure and in the effectiveness of suction feeding at metamorphosis (Lauder and Reilly 1990).

Decreased aquatic feeding performance also was observed in red-spotted newts (*Notophthalmus viridescens*) at metamorphosis. Terrestrial efts were less successful at capturing prey in the water than were aquatic larvae. In some populations of this species, some adults retain external gills into adulthood, but they lack open gill slits. The presence of the gills alone was not sufficient to increase aquatic feeding performance of branchiate adults over that of fully transformed individuals (Reilly and Lauder 1988). Nevertheless, both types of adults were much more successful at capturing prey in the water than were terrestrial efts. Fully transformed individuals of other salamandrid species that spend all or most of their adult lives in water also are very successful at capturing aquatic prey, even though all of these species lack open gill slits (B. Miller and Larsen 1989). This indicates that while a one-way flow of water may enhance aquatic feeding in some salamanders, it is not necessary for successful feeding in water. Adults of these aquatic species have morphological adaptations of the hyoid apparatus and

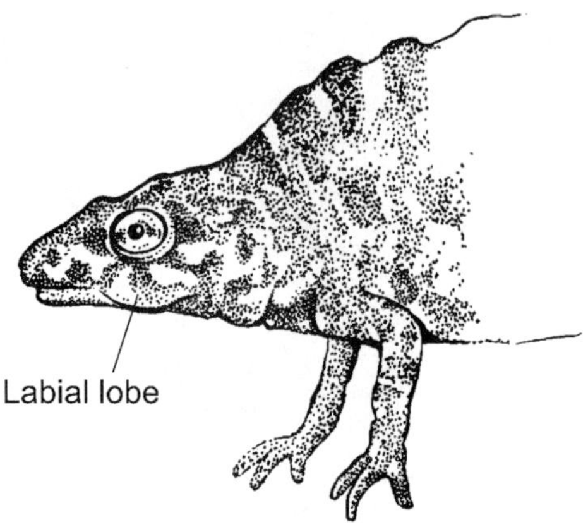


**Fig. 12.3.** Labial lobes in an adult *Triturus marmoratus,* an adaptation for aquatic feeding. Drawing by Mary Jane Spring from a photograph by Günter Diesener in Diesener and Reichholf (1986), p. 27.

mouth that enable them to continue feeding in much the same way as larvae.

The mouths of salamander larvae exhibit several adaptations for suction feeding. The tongue is little more than a glandular pad on the floor of the mouth, a condition that is retained in adults of highly aquatic species (Regal 1966; Özeti and Wake 1969). The mouth is relatively small, and large flaps of skin, or labial lobes, are present on the upper and lower jaws (fig. 12.1). These further reduce the size of the mouth opening and give it a rounded shape, thereby increasing the velocity of water flowing into the mouth and ensuring even acceleration of water moving around the edges of the mouth. The labial lobes also prevent water from flowing out of the mouth when it is closed (Matthes 1934; Özeti and Wake 1969; Lauder 1985). The lobes are lost at metamorphosis in terrestrial species, but may reappear in breeding adults of some species (fig. 12.3), enabling the animals to feed effectively in the water. The labial lobes are retained throughout life in adult salamanders that spend most of their time in water, and in some species, such as *Pachytriton brevipes,* the lobes are unusually large (B. Miller and Larsen 1989).

### Feeding Ecology

All larval salamanders are carnivorous, as are adults. Most species are generalist predators on small invertebrates, and they can have a major impact on the structure and dynamics of freshwater invertebrate communities (Sprules 1972; Burton 1976; Holomuzki and Collins 1987; Taylor et al. 1988; Holomuzki 1989b; Petranka 1989a; Holomuzki, Collins, and Brunkow 1994; Blaustein et al. 1996). Salamander larvae, especially large individuals, also are important predators on anuran tadpoles and other salamander larvae (Metter 1963; Burton 1976; Caldwell, Thorp, and Jervey 1980; Stenhouse, Hairston, and Cobey 1983; Stenhouse 1985b; Cortwright 1988; Gustafson 1993; Griffiths, de Wijer, and May 1994; Morin 1995; Kiesecker 1996; Babik 1998; see also chapters 14 and 15). Larvae of at least one species (*Dicamptodon ensatus*) have been reported to feed on juvenile fish (M. Parker 1993). Most species are sit-and-wait predators that hang in the water column, rest on the bottom of a pond or stream, or hide under rocks and other debris. From these positions, the larvae make rapid lunges at passing prey animals (Hoff, Lannoo, and Wassersug 1985; Leff and Bachmann 1986; McWilliams and Bachmann 1989a, b; M. Parker 1994). In ponds where zooplankton is abundant, salamander larvae sometimes actively pursue prey as well (Hassinger, Anderson, and Dalrymple 1970). The tendency for salamander larvae to sit and wait for prey to come by produces a diet dominated by relatively active, mobile invertebrates. Cryptic, slow-moving invertebrates sometimes are underrepresented in the diet when compared to their abundance in the environment (M. Parker 1994).

Many species exhibit an ontogenetic shift in their diets, taking larger prey as they grow. In part, this is due to changes in the size of the mouth that enable large larvae to eat larger prey (Dodson and Dodson 1971; Collins and Holomuzki 1984; Petranka 1984c; Leff and Bachmann 1986; B. Taylor et al. 1988; M. Parker 1994; Sanderson and Kupferberg 1999). In pond-dwelling larvae, the shift in prey size also is related to the tendency for large individuals to spend most of their time crawling on the bottom rather than swimming in open water. Consequently, they feed mostly on benthic organisms that are larger than the zooplankton consumed by smaller individuals (Dodson and Dodson 1971; Leff and Bachmann 1986; McWilliams and Bachmann 1989a, b). Some pond-dwelling salamander larvae make daily vertical migrations in the water column in response to movements of their prey. Larvae also may change their daily patterns of activity or shift between microhabitats in response to movements of their prey or the activity of their predators (Efford and Mathias 1969; Dodson and Dodson 1971; Anderson and Williamson 1974; Sprules 1974b; L. Branch and Altig 1981; J. Taylor 1983a; Holomuzki 1986b; Leff and Bachmann 1986; McWilliams and Bachmann 1989a; Stangel and Semlitsch 1987; B. Taylor et al. 1988). There can be substantial variation in activity patterns, use of microhabitats, and prey selection among populations of the same species, and even among individuals in the same population (B. Henderson 1973; Efford and Mathias 1969; Collins and Holomuzki 1984; Zerba and Collins 1992). In general, stream-dwelling larvae do not exhibit pronounced ontogenetic shifts in microhabitat use or the type of prey consumed, but they sometimes shift activity patterns in the presence of predators. In most streams, open water areas lack prey that are suitable for salamander larvae, so the larvae tend to feed on benthic prey throughout development (Petranka 1984c; Nussbaum 1985; M. Parker 1994).

### Cannibalism and Cannibalistic Morphs

As salamander larvae grow larger, they often attack smaller conspecifics, especially when population densities are high. Such attacks can produce serious injuries, including the loss of limbs and all or part of the tail. This type of aggressive interference competition can significantly reduce larval survivorship (R. N. Harris 1987b; Walls and Jaeger 1987; Petranka 1989a; Semlitsch and Reichling 1989; D. Scott 1990; C. K. Smith 1990; Van Buskirk and Smith 1991; Walls and Roudebush 1991; Hokit, Walls, and Blaustein 1996; Reques and Tejedo 1996). If size disparities between individuals are large enough, then aggressive attacks sometimes grade into full-scale cannibalism. Opportunistic larval cannibalism has been reported in several species of *Ambystoma* (Stenhouse, Hairston, and Cobey 1983; Lannoo, Lowcock, and Bogart 1989; Semlitsch and Reichling 1989; Walls and Roudebush 1991; M. Crump 1992; Nyman, Wilkinson, and Hutcherson 1993; Walls and Blaustein 1995; Brodman 1996), as well as *Dicamptodon* (C. Johnson and Schreck 1969; Nussbaum and Clothier 1973), *Pseudotriton ruber* (Gustafson 1993), *Salamandra salamandra* (Degani, Goldenberg, and Warburg 1980; Degani,1993), *Notophthalmus viridescens* (R. N. Harris 1987b), *Triturus cristatus* and *T. alpestris* (Dalbik 1990), and several species of *Hynobius* (Kusano, Kusano, and Miyashita 1985; Ohdachi 1994; Wakahara 1995; Nishihara 1996; Fu et al. 2003). Consumption of conspecifics can significantly increase growth rates of larvae and accelerate metamorphosis (Degani, Goldenberg, and Warburg 1980; Lannoo, Lowcock, and Bogart 1989; Warburg 1992; Degani 1993; Nyman, Wilkinson, and Hutcherson 1993; Wakahara 1995; Wildy et al. 1998; Kohmatsu, Nakano, and Yamamura 2001), an advantage that is particularly important for species that breed in temporary ponds.

In some populations, large cannibalistic individuals have proportionately larger heads than do noncannibalistic individuals, but the degree to which larvae develop into distinctive cannibalistic morphs varies among species. In ringed salamanders (*Ambystoma annulatum*), some populations of eastern tiger salamanders (*A. tigrinum tigrinum*), and some Asian salamanders in the genus *Hynobius,* cannibals are unusually large and have wide heads, but otherwise are not very different from noncannibals (Kusano, Kusano, and Miyashita 1985; Lannoo, Lowcock, and Bogart 1989; Nyman, Wilkinson, and Hutcherson 1993; Nishihara 1996). In contrast, populations of barred (*Ambystoma tigrinum mavortium*), blotched (*A. t. melanostictum*), Arizona (*A. t. nebulosum*), and eastern tiger salamanders (*A. t. tigrinum*), as well as long-toed salamanders (*A. macrodactylum columbianum*), sometimes have distinctive cannibalistic and noncannibalistic morphs. These differ not only in the size and shape of the head, but also in the size of the vomerine teeth (Powers 1903, 1907; Glass 1951; Gehlbach 1967, 1969; Rose and Armentrout 1976; Collins 1981; Pierce et al. 1983; Lannoo and Bachmann 1984a; Pedersen 1991; Collins, Zerba, and Sredl 1993; Walls et al. 1993; D. Pfennig, Ho, and Hoffman 1998; K. Larson et al. 1999). These large-headed morphs eat conspecifics much more frequently than do normal morphs (fig. 12.4), but they also eat large invertebrates, tadpoles, fishes, and other prey (Collins and Holomuzki 1984; D. Pfennig, Ho, and Hoffman 1998; K. Larson et al. 1999; Whiteman et al. 2003). In one population of *A. tigrinum* in South Dakota, cannibal morphs were found among both larvae and paedomorphic individuals (the latter are sexually mature individuals that retain larval gills and overwinter in the water before eventually metamorphosing; K. Larson et al. 1999). Paedomorphic cannibals also were described in the earliest study of cannibal morphs by Powers (1907), but in most other populations, the phenomenon is limited to sexually immature larvae. Cannibalistic morphs nearly always coexist in a pond with "typical" omnivorous morphs, with the frequency of cannibalistic morphs varying with environmental conditions (see Wakano, Kohmatsu, and Yamamura 2002, for a theoretical treatment of this phenonmenon).

The proximate determinants and ecological significance of cannibal morph development have been studied in considerable detail in the Arizona tiger salamander (*Ambystoma tigrinum nebulosum*). These salamanders typically are the top predators in the ponds where they live (Holomuzki and Collins 1987). When larval densities are high and competition for food is intense, individuals that develop into cannibals enjoy an advantage, because they have access to large prey, including conspecifics, which are not available to noncannibals (Maret and Collins 1997). This feeding advantage

**Fig. 12.4.** Cannibalistic morph of *Ambystoma tigrinum nebulosum* eating a conspecific. Photo by David Pfennig.

comes from an improved ability to handle large prey, which is related to the increased size of the head and vomerine teeth, and not to improved efficiency of suction feeding (Reilly, Lauder, and Collins 1992).

Larvae can be induced to develop into cannibals in the laboratory when reared at high density (Collins and Cheek 1983). A critical factor in controlling the expression of the cannibal phenotype is the availability of conspecifics as potential prey. When larvae were fed only invertebrates or tadpoles, or a combination of both prey types, they did not develop into cannibals if conspecifics were not present (Loeb, Collins, and Maret 1994). Nevertheless, the availability of tadpoles as prey does play an important role in the development of cannibalism. Salamander larvae that feed on tadpoles grow faster than do those that feed only on small invertebrates, and it is only the largest individuals that reach a size large enough to allow them to prey on conspecifics (Maret and Collins 1996). Similarly, individuals that feed on relatively large invertebrate prey grow faster than those that feed on smaller prey and are more likely to develop the cannibalistic morphology (Whiteman et al. 2003). Initial size differences among young larvae are magnified by interference competition among larvae, with the largest individuals inhibiting the feeding and growth of smaller larvae. This may result in part from avoidance of potentially cannibalistic large larvae by smaller ones, but it also could facilitate the development of cannibalistic behavior by increasing size differences among larvae in a pond (Ziemba and Collins 1999).

Development of cannibalistic morphs also is affected by the prevalence of an infectious bacterium (*Clostridium*), which often causes mass mortality in larval populations (Worthylake and Hovingh 1989). In natural populations, cannibalism is uncommon in lakes that are heavily infested with bacteria, and cannibals are disproportionately represented among dead and dying individuals (D. Pfennig, Loeb, and Collins 1991). Experimental work with another subspecies of tiger salamander, *A. t. tigrinum,* demonstrated that bacterial infections can be transmitted when cannibals eat diseased individuals, and in fact, cannibals were more likely to eat diseased larvae than healthy ones (D. Pfennig, Ho, and Hoffman 1998). Individuals that ate diseased conspecifics suffered reduced growth rates and increased mortality, effects that were not seen when larvae fed on infected individuals of another species. Cannibalism also can increase the probability of being infected by a deadly iridovirus that is known to cause mass mortality in tiger salamanders, but the salamanders do not appear to detect the presence of the virus in a population or avoid eating infected salamanders (Parris et al. 2005).

Cannibalistic salamander larvae actually preferred to feed on larvae of another species if given a choice. This suggests that there is a significant cost to intraspecific cannibalism in the form of increased likelihood of disease transmission, a cost that may account for the relative rarity of cannibalism in nature (D. Pfennig 1997), despite its obvious nutritional advantages (M. Crump 1992). Cannibalistic larvae of *A. t. nebulosum* tend to avoid eating close relatives if alternative prey is available. Such avoidance could be related either to inclusive fitness costs to eating close relatives, or to greater transmission of bacterial infections among close relatives (D. Pfennig, Sherman, and Collins 1994; for further discussion, see later section on "Kin Recognition, Cannibalism, and Aggression in Salamander Larvae").

Experiments with the Asian salamanders in the genus *Hynobius* have shown that the mechanism for induction of cannibalistic morphs is similar to that in *Ambystoma tigrinum.* Cannibalistic morphs are most common in ponds with high densities of larvae and low food availability (Michimae and Wakahara 2001, 2002b; Fu et al. 2003). In ponds with low availability of small food items, larger individuals often feed on tadpoles, and these individuals are most likely to develop into the broad-headed cannibalistic morph (Michimae and Wakahara 2002a). These salamanders often breed very early in the spring in ponds where the only available food consists of relatively large prey such as tadpoles or conspecifics. Under such circumstances, broad-headed cannibals grow faster than do normal morphs. As in *Ambystoma tigrinum,* development of cannibalistic morphs is inhibited by the presence of a lot of closely related larvae (Michimae and Wakahara 2001).

## Morphology and Development of Anuran Tadpoles

The basic internal and external anatomy of tadpoles has been reviewed in exhaustive detail by Altig and McDiarmid (1999), Cannatella (1999), Lannoo (1999), and Viertel and Richter (1999), so I will describe only the essential features of tadpole anatomy that relate directly to their ecology and behavior. In basic body morphology, anuran tadpoles and adults could scarcely be more different. Tadpoles are well adapted to aquatic locomotion, with rounded bodies and long tails with flattened fins. The transformation to a fully terrestrial tetrapod adapted for jumping, walking, or crawling on land and feeding on terrestrial prey is a remarkable one, involving major rearrangements of both internal and external morphology (Hourdry and Beaumont 1985; Wassersug 1997; Viertel and Richter 1999) and major changes in physiology (Burggren and Just 1992; Ultsch, Bradford, and Freda 1999). In tadpoles, the hind limbs emerge first, the opposite of the pattern seen in salamander larvae, with the front limbs appearing shortly before metamorphosis is complete. In many species, lungs develop well before metamorphosis and before the internal gills have been lost. In others, lung development is incomplete when the young frogs emerge onto land (Burggren and Just 1992). Lung development affects not only the mode of respiration in tadpoles (see chapter 4), but the regulation of buoyancy as well (Gee and Waldick 1995). In general, benthic tadpoles, which spend most of their time resting on the bottom of a pond or stream, develop and inflate their lungs later than do tadpoles that feed in the water column. Lung development often is delayed in stream-dwelling tadpoles compared to more generalized tadpoles, probably because the increased buoyancy provided by inflated lungs would be disadvantageous in flowing water (Wassersug and Heyer 1988; Nodzenski, Wassersug, and Inger 1989). Lung development also is delayed in some anurans that undergo rapid development in temporary ponds and emerge in a relatively underdeveloped state (see chapter 4).

Other major morphological alterations that occur during metamorphosis include development of an adult urogenital system and the rearrangement and ossification of skeletal elements (Duellman and Trueb 1986). The skin undergoes the same sorts of changes seen in metamorphosing urodeles, including addition of cell layers to the epidermis, development of a distinct *stratum corneum,* development of adult skin glands, and changes in the physiology of osmoregulation (Burggren and Just 1992; Fox 1994; Warburg, Lewinson, and Rosenberg 1994). There also are major changes in sense organs and other parts of the nervous system (Boatright-Horowitz and Simmons 1997; Lannoo 1999). In some cases, metamorphic changes involve loss of elements of the nervous system that function only in tadpoles, such as the loss of the lateral line system in terrestrial frogs. Other systems, such as an auditory system that can perceive airborne sounds, do not develop until metamorphosis. There also are changes in the sensitivity of some sensory systems as anurans make the transition from water to land. For example, the visual pigments of tadpoles in several different families have been shown to be most sensitive to wavelengths of light found in aquatic habitats. At metamorphosis, there is a shift toward wavelengths characteristic of terrestrial habitats (Hödl 1975).

The most dramatic changes are in the feeding apparatus and digestive system. Aquatic tadpoles are highly specialized for suspension feeding, and the tadpole chondrocranium is arranged to support a branchial skeleton that serves as a pump to move water into the mouth and across the gills (see the following). During metamorphosis, the transition to a terrestrial carnivore requires major reorganization of chondrocranial elements to form the ossified adult skull, loss of the larval mouthparts, and development of the adult jaws and tongue (De Jongh 1968; Wassersug and Hoff 1982; Burggren and Just 1992). Tadpoles have greatly elongated guts, but the morphology of the digestive system is relatively simple. Most tadpoles lack a functional stomach for digesting large food particles, although the larval stomach, the *manicotto glandulare,* is acidic and produces some digestive enzymes (I. Griffiths 1961; Viertel and Richter 1999). The gut lacks papillae or folds to increase absorptive surface area, as well as muscles to move food through the gut. A more distinct stomach develops during metamorphosis. The intestine becomes considerably shorter, but with increased internal surface area for nutrient absorption. At the same time, there are changes in the relative size of other visceral organs, such as the liver and pancreas. These morphological changes are accompanied by major changes in the physiology of digestive enzyme secretion, nutrient absorption, and the functioning of the muscles that move material through the digestive tract (Hourdry, l'Hermite, and Ferrand 1996; Wassersug 1997).

### Functional Morphology of Locomotion

Tadpoles are unique among aquatic vertebrates in having an extremely short body and a relatively inflexible spine composed of a small number of vertebrae. The body of a tadpole is not streamlined like that of a fish, but is globular, with a clear division between the body and the tail (fig. 12.5). This unusual morphology is constrained by the developmental requirements of the terrestrial adult, in which a short, inflexible spine is essential for jumping locomotion (Wassersug 1989a, 1997; Hoff et al. 1999; Hoff and Wassersug 2000). The tail is supported only by a flexible notochord and, except at the base, lacks caudal vertebrae. Most tadpoles have tails that are tapered and pointed at the end, in contrast to the blunt or expanded tails of fishes. This appears to be the optimum design for a tail that is flexible rather than stiff, because the smaller surface area at the end of the tail reduces drag (Wassersug 1989a). The tail fin itself is little more than a double layer of skin, which lacks skeletal support. It has properties that allow it to remain relatively stiff while the tadpole is swimming, but it also is quite fragile (Hoff and Wassersug 2000). Tadpoles often suffer injury to their tails fins from predators, but their fragile tail fins may enable tadpoles to escape by preventing predators from getting a grip on the tail musculature (P. Doherty, Wassersug, and Lee 1998).

The design of a tadpole's body appears to be less suitable for aquatic locomotion than that of a fish, but fluid-dynamic models of tadpole locomotion have shown that they are relatively efficient swimmers (H. Liu, Wassersug, and Kawachi 1996, 1997). When a tadpole swims, the tip of the snout moves from side to side (Wassersug and Hoff 1985; Hoff and Wassersug 1986). These oscillations are under active

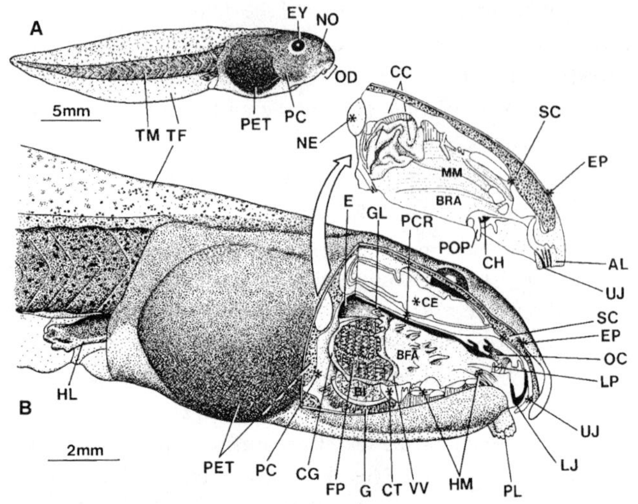


**Fig. 12.5.** Morphology of a tadpole of *Rana temporaria.* (A) External morphology of body. EY = eye; NO = external nares; OD = oral disc; PC = peribranchial chamber; PET = peritoneum; TF = tail fin; TM = tail muscles. (B) Internal anatomy of branchial chamber and filter apparatus. AL = anterior labium; B = branchial arch; BFA = buccal floor area; BRA = buccal roof area; CC = ciliary cushion; CE = brain; CG = ciliary groove; CH = choanae (internal nares); E = esophagus; EP = epidermis; FP = filter plates; G = gills; GL = glottis; HL = hind limb; HM = hyoidean and mandibular muscles; LJ = lower jaw; LP = lingual papilla; MM = mandibular muscles; NE = pronephros; OC = oral cavity; PCR = cartilage of primordial cranium; PL = posterior labium; POP = postnarial papilla; SC = secretory cells; UJ = upper jaw; VV = ventral velum. Reprinted from fig. 1 (p. 347) of Viertel (1985) with kind permission of Springer Science and Business Media.

muscular control and may help to generate thrust (Hoff and Wassersug 2000). This seemingly awkward mode of locomotion also may increase the efficiency of swimming for an animal with a tadpole-shaped body, because it breaks up the boundary layer of fluid around the body. In contrast, a fish-shaped animal forced to swim like a tadpole would experience a substantial increase in drag compared to an animal swimming like a fish (Wassersug 1997).

In contrast to urodele larvae, most anuran tadpoles lose their external gills soon after hatching, before they reach the free-swimming stage. In *Agalychnis callidryas,* which has a plastic hatching period, tadpoles that hatch early lose their external gills within a day of hatching, whereas embryos of the same age that remain in the eggs retain their gills (Warkentin 1999b). Presumably the added surface area of external gills is important for embryonic respiration (see chapter 4), but become a handicap for hatchlings because they tend to increase drag and reduce swimming efficiency (Warkentin 2000b). Hatched tadpoles have plenty of other avenues for gas exchange, including the skin, internal gills, and eventually, the lungs. The internal gills are enclosed in a branchial chamber, an arrangement that provides a hydrodynamic advantage by reducing drag (fig. 12.5).

Tadpoles also differ from urodele larvae in the pattern of forelimb emergence. Most tadpoles develop their front legs inside the atrial chambers that house the gills. This prevents the forelimbs from increasing drag forces as the tadpole swims. Eruption of the legs is delayed until just before metamorphosis, thereby minimizing the amount of time the tadpole spends in the water when it has become a much less efficient swimmer (Wassersug and Sperry 1977; Wassersug 1989a; Dudley, King, and Wassersug 1991). Although the hind limbs emerge before the forelimbs, they nevertheless appear relatively late in development. Just as forelimbs would impede locomotion, the presence of hind limbs would reduce the efficiency of swimming by increasing drag (Gal and Blake 1987; Dudley, King, and Wassersug 1991). If legs appeared early in a tadpole's development, the animal might be much more vulnerable to aquatic predators because of reduced swimming performance. The hydrodynamic cost of having back legs is minimized by their position near the junction of the round body and the flat tail. The globose shape of the tadpole makes this a region of relatively low water flow that is unlikely to be affected by small increases in drag forces. In some tadpoles, the small hind limbs are kept folded against the body early in development, thereby reducing the effects of drag. Later, as the legs grow, they are held out to the sides of the body, and the effects of drag increase (Wassersug 1997).

Changes in the size and shape of a tadpole's body during development and the emergence of limbs may require tadpoles to change their modes of locomotion, especially when attempting to escape from predators. Both burst swimming speed and evasiveness have been shown to increase survival of tadpoles exposed to natural predators (Watkins 1996; see chapter 14). In at least some species of tadpoles, burst swimming speed increases with increasing body size, thereby reducing vulnerability to predators in larger tadpoles (R. Wilson and Franklin 2000a). This advantage can be reduced, however, as hind limbs develop. For example, wood frog tadpoles (*Rana sylvatica*) are relatively poor swimmers when they are small, but swimming performance increases as they grow. Swimming performance declines as the hind limbs develop and even more so once the front legs have appeared (R. Brown and Taylor 1995). This reduces the capacity of older tadpoles to escape from predators by outswimming them. Older tadpoles, especially those with all four legs present, tend to make more turns when subjected to attacks by a simulated predator. Brown and Taylor (1995) interpreted this behavior as a way to compensate for reduced swimming performance by emphasizing maneuverability over speed when escaping from predators. However, maneuverability is a measure of an animal's ability to make its way around obstacles, not simply a measure of turning frequency (R. Wassersug, personal communication). Changes in maneuverability of tadpoles as a function of developmental stage have not been investigated, nor have there been any comparative studies of differences in maneuverability among species.

### Functional Morphology of Feeding

Tadpoles are essentially eating machines, and require a specialized feeding apparatus to maintain a constant intake of food (Wassersug 1975). Despite some adaptive modifications (see the following "Ecological Morphology of Tadpoles"), the basic mechanism is similar in most species and involves pumping water and suspended organic material across structures that trap the food. Water is pumped into the oral cavity with a buccal pump consisting of paired ceratohyals and associated hypobranchial and ceratobranchial structures (fig. 12.6). The ceratohyals articulate medially with the hypobranchial plates, which in turn support the branchial baskets. They also articulate near their distal (lateral) end with the palatoquadrates. When the muscles attached to the distal ends of the ceratohyals contract, those ends are elevated, and the ceratohyals rotate around the articulations with the palatoquadrates (fig. 12.7). This depresses the medial ends of the ceratohyals and expands the buccal cavity (the anterior part of the oral cavity), thereby drawing water into the mouth. When the ceratohyals return to their original position, water is forced from the buccal cavity into the pharynx (the posterior part of the oral cavity) and eventually exits through the spiracle. In most tadpoles, back flow of water is prevented by cyclic opening and closing of the mouth and by a transverse flap in the mouth, the ventral velum, which serves as a one-way valve between the buccal cavity and the pharynx (Wassersug 1972, 1976, 1980; Wassersug and Hoff 1979; Wassersug and Rosenberg 1979).

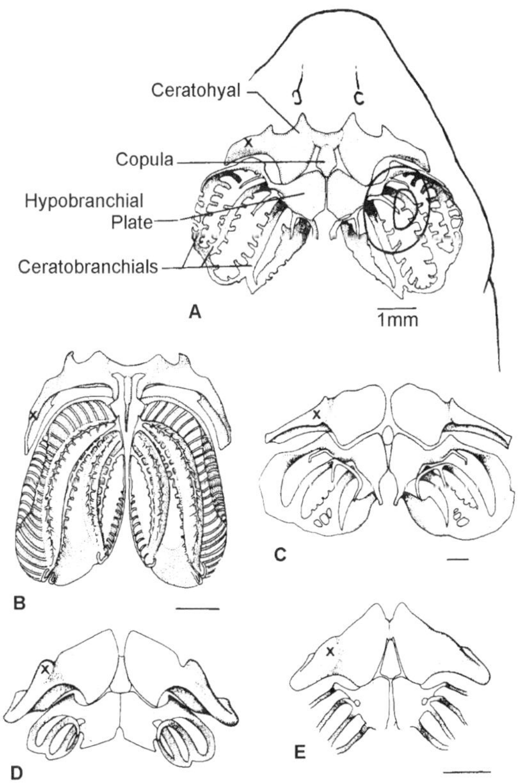


**Fig. 12.6.** Branchial skeletons of anuran tadpoles in dorsal view, showing elements of the buccal pump. The ceratohyals together make up the branchial basket that supports the gill filters. The "x" on each drawing indicates the point of articulation of the ceratohyal with the palatoquadrate bar (not shown). (A) *Rana pipiens* (Ranidae), a pond tadpole with a generalized diet. (B) *Gastrophryne carolinensis* (Microhylidae), an obligate midwater suspension-feeder with very large branchial baskets forming a fine seive. (C) *Heleophryne natalensis* (Heleophrynidae), a benthic tadpole that lives in rapidly flowing water. This tadpole has a suctorial mouth and grazes on periphyton attached to rocks. (D) *Anotheca spinosa* (Hylidae), an egg-eating tadpole that lives in bromeliads. The branchial baskets are reduced to short bars that intercept relatively large food particles. (E) *Hyla microcephala* (Hylidae), a pond tadpole that ingests large filamentous plant fragments and other large food particles. This species also has reduced branchial baskets typical of tadpoles with a macrophagous diet. After Wassersug and Hoff (1979).

As particles suspended in the water pass through the oral cavity, they are sorted by size. Large particles are channeled directly into the esophagus by papillae or ridges in the buccal cavity. Smaller particles pass into the pharynx, where they are entrapped in mucus secreted by the branchial food traps (fig. 12.6) or intercepted by gill filters born on the paired branchial baskets. The movements of cilia then transport cords of mucus and food particles into the esophagus (De Jongh 1968; Gradwell 1968, 1970, 1972a, b; Kenny 1969b, c; Wassersug 1972, 1975, 1980; Seale and Wassersug 1979; Wassersug and Hoff 1979; Wassersug and Rosenberg 1979; Seale, Hoff, and Wassersug 1982; Viertel 1982, 1984a, b, 1985, 1987, 1991; Inger 1985; Wassersug and Heyer 1988).

Although the pumping of water through the buccal cavity appears similar during irrigation of the gills for respiration and during feeding, these actions are not identical. Three muscles (hyoangularis, orbitohyoideus, and antrior interhyoideus) are involved in gill irrigation, but three additional muscles (intermandibularis, suspensorioangularis, and levator mandibulae longus superficialis) are used during feeding. These additional muscles are invoved in the opening and closing of the mouth (P. Larson and Reilly 2002). Tadpoles are unusual in that part of the neurocranium has been modified to function as an upper jaw, and this is powered by a set of muscles that in ancestral amphibians actually inserts on the lower jaw (Haas 2001). In addition, the lower jaw articulates with the condrocranium much farther forward than in other vertebrates, resulting in a feeding apparatus that is unique among tetrapods (Cannatella 1999; Haas 2001; P. Larson and Reilly 2002).

### Ecological Morphology of Tadpoles

In contrast to the relatively minor variation in overall morphology of urodele larvae living in different habitats,

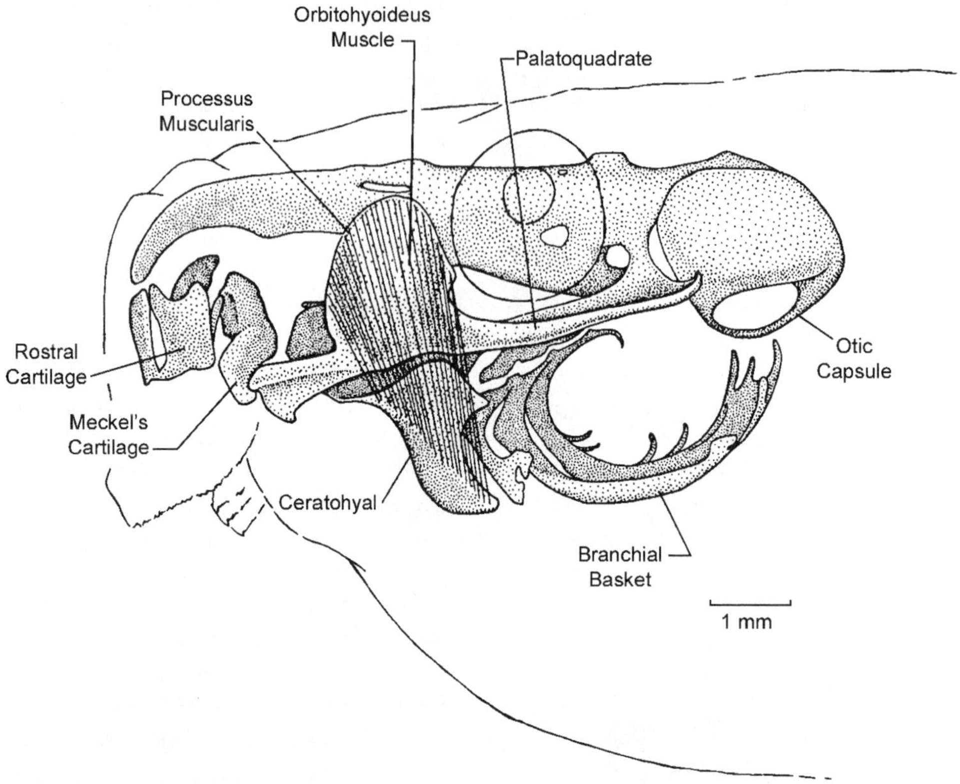


**Fig. 12.7.** Lateral view of the chondrocranium of a generalized pond tadpole (*Alytes obstetricans*). Contraction of the orbitohyoideus muscle causes the ceratohyal to rotate around its articulation with the palatoquadrate, thereby depressing the floor of the buccal cavity and drawing water into the mouth. Elevation of the buccal floor pushes water across the gill filters, located on the branchial baskets. After Wassersug and Hoff (1979).

anuran tadpoles exhibit a wide range of morphological variation that is correlated with differences in microhabitat use and feeding ecology. Most of this variation involves differences in the overall shape of the body, the size and shape of the tail fin, and the morphology of the oral disc, oral cavity, and branchial baskets. Many tadpoles have generalized body morphology and oral discs adapted for feeding on a variety of foods. These tend to be found mostly in ponds and quiet stream pools. Other species are specialized for particular habitats, such as streams, wet rocks, or bromeliads, or for particular types of food, such as microscopic algae, zooplankton, frog eggs, or other tadpoles. There also are tadpoles that do not feed at all, but complete development by drawing on stored energy in yolk reserves (Duellman and Trueb 1986). Some of these exhibit major morphological differences from aquatic tadpoles, whereas others do not (Candioti, Úbeda, and Lavilla 2005). Table 12.1 provides a summary of the major ecological types of tadpoles, based on the classification of Altig and Johnston (1989; see also Altig and McDiarmid 1999).

Despite the efforts of these authors to standardize the terminology used to describe ecological variation in tadpoles, there is some confusion in the literature. Some authors, such as Chou and Lin (1997), have adopted the classification of Altig and Johnston (1989). Others have used slightly different terminology. For example, Passmore and Carruthers (1995) distinguished "pelagic" from "nektonic" tadpoles, whereas Altig and Johnston (1989) used these terms as synonyms. Tadpoles of *Xenopus* were called "pelagic" by Van Dijk (1972), "nectonic filter-feeders" by Duellman and Trueb 1986, "suspension feeders" by Altig and Johnston (1989), and "nektonic" by Passmore and Carruthers (1995). Tadpoles of hylids and hyperoliids with laterally compressed bodies and wide tail fins, which live among dense vegetation, were called "pelagic/hydrophytophilic" by Van Dijk (1972) and "nektonic" by Altig and Johnston (1989). Duellman and Trueb (1986) simply treated them as a subset of pond-type tadpoles without giving them a specific name. Tadpoles that live in fast-flowing streams have been variously described as "rheophilic," "rhyacophilous," or "stream-adapted." Much of the confusion of terminology used to categorize tadpoles results from a lack of detailed behavioral and ecological information to support putative ecomorphological categories. Even the most detailed monographs on tadpoles usually include only brief ecological notes, mostly related to the types of sites where tadpoles

**Table 12.1** Ecomorphological classification of anuran tadpoles

| Tadpole type | Habitat and ecology | Morphology | Examples |
|---|---|---|---|
| *I. Feeding tadpoles in nonflowing water (Lentic habitats)* | | | |
| Benthic | Bottoms of ponds, either in shallow or deep water. Feed by rasping substrates or taking in particulate matter | Globular or somewhat flattened body. Eyes dorsal. Mouth ventral or anteroventral, with well-developed beak and tooth rows on oral disc. Tail fin not extending onto the body. | *Acris, Alytes;* some *Bufo, Colostethus, Dendrobates, Ptychadena, Rana, Smilisca* |
| Carnivorous | Ponds, especially temporary ponds | Similar to benthic tadpoles, but often with terminal mouth. Sometimes with enlarged mouth, beak, and jaw muscles (*Ceratophrys*) or enormous mouth with no oral disc (*Lepidobatrachus*). Some with no hard mouth parts, but with mouth that forms a tube for suction feeding (*Hymenochirus, Pipa carvalhoi*). | *Scaphiopus, Ceratophrys, Lechriodus, Lepidobatrachus, Hymenochirus, Pipa carvalhoi,* some *Hyla* |
| Macrophagous | Bottoms of ponds or in water column in vegetation | Body similar to other pond tadpoles. Oral disc small or absent, but large jaw sheathes present. | *Afrixalus, Cardioglossa, Occidozyga; Hyla leucophyllata, microcephala, parviceps* groups |
| Nektonic (Pelagic) | Live in open water of ponds, often moving through vegetation. Feed by scraping or rasping algae and other food off of rocks and other substrates | Body laterally compressed. Eyes lateral. Mouth anteroventral, with well-developed tooth rows on oral disc. Fins often high and sometimes extend onto body as far as head. Flagellum often present on tail tip. | Many *Hyla;* some *Scinax, Smilisca; Phyrnohyas, Trachycephalus* |
| Neustonic | Feed on organisms and organic matter in surface films of ponds | Similar in body form to other pond tadpoles, but with inverted, umbrella-like oral disc. | *Microhyla achatina* |
| Surface swimmer | Tadpoles swim near the surface of ponds and feed on floating vegetation, such as duckweed (*Lemna*). Can escape from predators by jumping out of water and skittering on surface | Body elongate, with top of head flat. Eyes large and lateral. Mouth terminal and small, bordered by flaplike labia with papillae. Serrated beak. Jaw muscles massive, but buccal muscles small. Long muscular tail with low dorsal fin, less-shallow ventral fin. | *Scarthyla ostinodactyla* |
| Suspension Feeder | Filter-feed on very small particles while hanging in water column | Flattened body with lateral eyes. Low tail fin with terminal flagellum often present. Hard mouth parts absent. Terminal mouth often small or slitlike. Barbels sometimes present at sides of mouth. | *Rhinophrynus, Silurana, Xenopus,* many microhylids (*Chiasmocleis, Ctenophryne, Gastrophryne, Hypopachus,* some *Microhyla, Phrynomerus*) |
| Suspension Rasper | Often feed on suspended particles in mid-water, but also can rasp algae and other food from substrates. Often swim in head-up posture and form schools | Keratinized jaw sheathes and labial teeth present. Eyes lateral. Tail with wide fin and often with terminal flagellum. | Most *Agalychnis; Kassina, Pachymedusa,* most *Phyllomedusa* |
| Phytotelmonous | Tadpoles develop in water in bromeliads, tree holes, fallen logs, bamboo, seed pods, leaf axils, etc. May be carnivorous, obligate egg-eaters, or detritovores | Either elongate or short with stout body. Elongate tadpoles usually have thin tails with little or no tail fin. Mouth and oral cavity often specialized for feeding on large prey or frog eggs. Often obligate air-breathers with well developed lungs. | *Hoplophyrne, Anotheca, Hyla bromeliacia, Hyla zeteki, Osteocephalus oophagus, Osteopilus brunneus, Phrynohyas resinifictrix, Acxanthixalus, Nyctixalus;* some *Dendrobates, Mantidactylus* |
| *II. Feeding tadpoles in flowing water (Lotic habitats)* | | | |
| A. Tadpoles live in flowing portions of streams and rivers | | | |
| Clasping | Live in slow-flowing streams. Clasp substrate with oral disc and feed by rasping with labial teeth | Eyes small and dorsal. Oral disc usually large and ventral, with many marginal papillae. Muscular tail with low fin terminating at junction with body. Tail tip often rounded. | Some *Boophis; Natalobatrachus, Rana palmipes* and *R. pustulosa* groups |

(*continued*)

**Table 12.1** (continued)

| Tadpole type | Habitat and ecology | Morphology | Examples |
|---|---|---|---|
| Adherent | Live in fast-flowing streams and rivers, sometimes in torrents. Maintain position with oral disc and feed by rasping substrates with labial teeth or catching food particles in current. | Often with very large oral discs with many marginal papillae and many labial tooth rows (*Plectrohyla, Ptychohyla*). Others have reduced or absent labial teeth and jaw sheathes that are not robust (*Hyla rufioculis, H. uranochroa, Taudactylus*). Belly sometimes flattened. Muscular tail with low fin and rounded tip. | Many *Hyla; Plectrohyla,* some *Ptychohyla,* some *Mantidactylus, Taudactylus* |
| Suctorial | Live in very fast-flowing streams and torrents. Cling to rocks with highly modified oral discs. | Large oral discs with many marginal papillae. In some species, fleshy flap forms extension of oral disc that enhances suction. Shape of body variable, but often flattened or with anterior end flattened. Muscular tail with low fin and rounded tip. | *Ascaphus, Bufo penangensis, B. torrenticola, Conraua, Heleophryne, Petropedetes, Scutiger, Trichobatrachus,* many hylids |
| Gastromyzophorous | Live in very fast-flowing streams and torrents. Cling to rocks with oral disc and belly sucker | Large belly sucker present, equipped with papillae or keratinized patches. Muscular tail with low fin, ending before junction with body. Tail rounded at tip. | *Amolops, Atelophryniscus, Atelopus,* some *Ansonia, Rana sauteri* |
| Psammonic | Lie buried in sandy bottoms of streams and feed on organisms in spaces between sand grains | Flattened body. Long, fanglike serrations in jaw that keep sand grains out of mouth. Long spiracular tube held with opening above sand layer in current. Oral disc reduced. Eyes dorsal. | *Otophryne robusta* |
| Semiterrestrial | Tadpoles live on wet rock faces, splash zones of waterfalls, and rivulets | Elongate, flattened body with dorsal eyes that are large and protuberant. Belly often flattened. Long thin tail with low find and sharp tip. Ventral mouth. Hind legs often develop early. | *Cycloramphus, Leptodactylus rugosus, Nannophrys ceylonensis, Rana beddomei, Thoropa, Werneria, Petropedetes* |
| B. Tadpoles live in still portions of streams and rivers | | | |
| Benthic | Bottoms of pools, either in shallow or deep water. Feed by rasping substrates or taking in particulate matter | Similar to generalized benthic pond tadpoles. | *Colostethus inguinalis, Hyla arenicolor,* some *Leptopelis, Oreolalax,* some *Rana, Strongylopus grayii* |
| Fossorial | Burrow into mud at bottom of stream pools or live in spaces between rocks | Thin, wormlike, flattened body. Long thin tail with low fin. Eyes dorsal. Small oral disc. Body often red from extensive cutaneous blood flow (in centrolenids). | Centrolenids, *Leptobrachium* |
| Nektonic (Pelagic) | Live in open water of pools | Similar to nektonic pond tadpoles, but with lower tail fins and less lateral eyes. | *Hyla miotympanum, Strongylopus fasciatus* |
| Neustonic | Feed on organisms in surface films of stream pools | Enlarged, umbrella-like oral disc oriented dorsally to funnel surface particles into mouth. Eyes lateral. Tail long with low fins. | *Colostethus flotator, Mantidactylus oviparus, Megophrys montana, Phyllomedusa guttata, P. jandaia* |
| *III. Nonfeeding tadpoles develop in terrestrial nest or small body of water* | | | |
| Nidicolous | Tadpoles hatch from eggs laid in moist terrestrial nests, tree holes, or bromeliads and complete development in nest without feeding, depending on yolk reserves. Aquatic microhabitats similar to those of feeding phytotelmonous tadpoles | Morphology variable, but often with tail fins, mouth parts, and pigment reduced. *Leiopelma hochstetteri, Mertensophryne,* | *Adenomera, Anodontohyla, Colostethus* (3 species), *Craspedoglossa, Eupsophus, Nectophrynoides malcomi, Paracophyla, Pelophryne, Philoria, Phrynodon, Platypelis, Plethodontohyla,* some *Rana, Syncope antenori, Synapturanus, Vanzolinius, Zachaenus* |

*Notes:* Modified from Altig and Johnston (1989). Numbering system and order of tadpole types are different from those of Altig and Johnston.

were collected (e.g., Chou and Lin 1997). The behavior of tadpoles in the field is rarely described, and inferences have been drawn mostly from the morphology of the tadpoles.

## Pond-Dwelling Tadpoles

Generalized pond tadpoles, which are found in most anuran families, typically have oval bodies with broad, muscular tails and wide tail fins. The mouth usually is oriented downward rather than forward. It is bordered by rows of papillae that may have a sensory function, may help to hold the mouth to the substrate, or both. The oral disc usually has several rows of keratinized labial teeth (denticles) used to anchor the oral disc and scrape algae and other organisms off of rocks and other substrates (Wassersug and Yamashita 2001). Usually there are horny jaw sheaths (often called a beak) as well, which are used to rasp food from substrates or bite through large or tough pieces of food (fig. 12.8). The branchial baskets are moderately well developed (fig. 12.6 A), with extensive gill filters and mucus-secreting areas on the food traps, which lie under the velum and above the gill filters. The buccal pump is well adapted for continuous pumping of water across the gills. Generalized pond tadpoles often have opportunistic feeding habits. They can extract phytoplankton and other suspended particles from the water column (Seale 1982a), but they also can feed on algae scraped off the substrate, on decaying animal matter, or on almost anything else that can be broken into small particles by the jaw sheaths and labial teeth. Some species become carnivorous and even cannibalistic under certain circumstances (see the following "Cannibalism and Morphological Polyphenism").

### Benthic and Nektonic Tadpoles

Some pond tadpoles have become specialized for living in particular microhabitats or feeding on particular kinds of food. Many tadpoles are benthic in their habits, spending most of their time on the bottoms of ponds and relatively little time in the water column (table 12.1). These include tadpoles of many species in the families Bufonidae, Discoglossidae, Hylidae, Leptodactylidae, Megophryidae, Myobatrachidae, Pelobatidae, Ranidae, Rhacophoridae, and

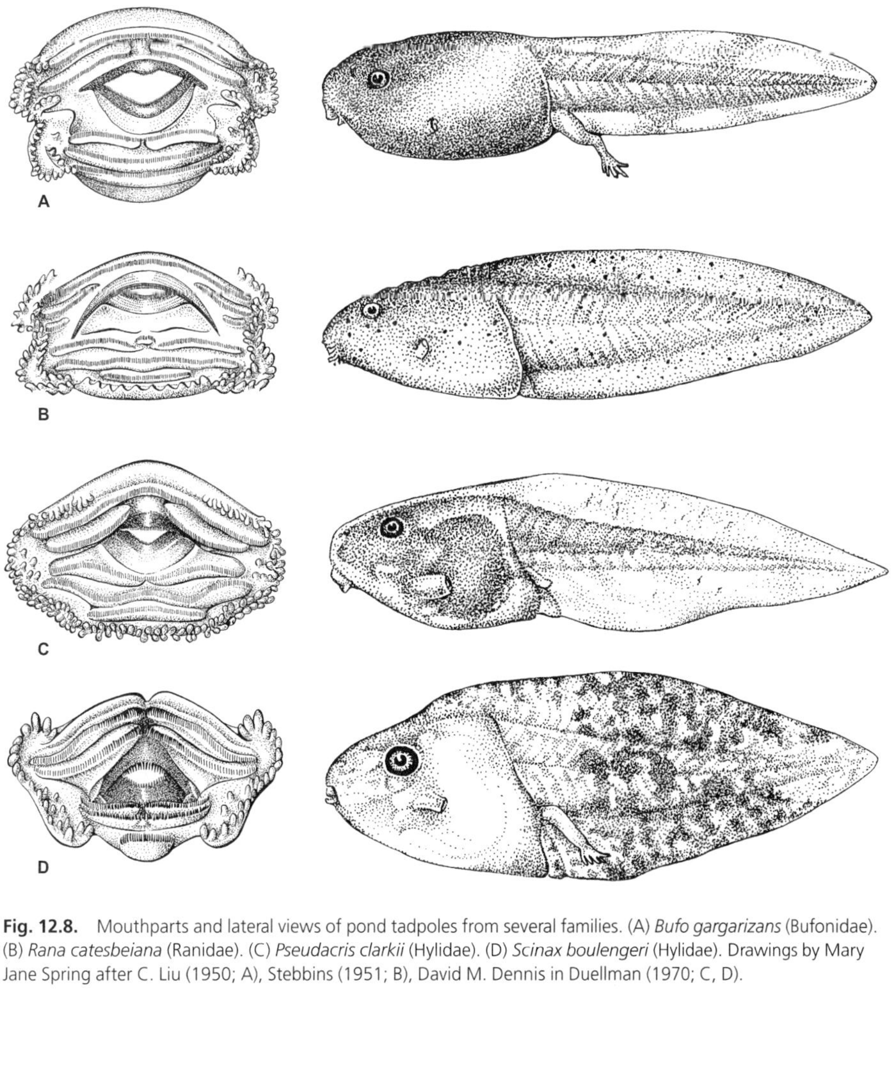

**Fig. 12.8.** Mouthparts and lateral views of pond tadpoles from several families. (A) *Bufo gargarizans* (Bufonidae). (B) *Rana catesbeiana* (Ranidae). (C) *Pseudacris clarkii* (Hylidae). (D) *Scinax boulengeri* (Hylidae). Drawings by Mary Jane Spring after C. Liu (1950; A), Stebbins (1951; B), David M. Dennis in Duellman (1970; C, D).

others (Wassersug 1980; Inger 1985; Wassersug and Heyer 1988). These tadpoles usually have the generalized morphology described previously. Other pond-dwelling tadpoles spend more time in the water column, often moving through patches of vegetation. These nektonic or pelagic tadpoles tend to have bodies that are more laterally compressed than those of benthic tadpoles, with the eyes on the sides of the head instead of the top. Nektonic tadpoles often have much taller tail fins than do benthic tadpoles, sometimes extending along the back all the way to the head (fig. 12.8 D). The tail usually tapers to a narrow point and may have a thin flagellum at the end, a structure that can move independently of the rest of the tail. Many tadpoles in the families Hylidae and Hyperoliidae have this type of morphology.

### Midwater Suspension Feeders

Certain groups of pond-dwelling tadpoles are specialized for feeding in open water on very small particles, such as phytoplankton. These tadpoles typically are found in highly productive ponds in which the density of microscopic phytoplankton is sufficiently high to permit suspension feeding. Most pipid tadpoles (except *Hymenochirus*), tadpoles of the Mexican burrowing frog (*Rhinophrynus dorsalis*), and pond-dwelling microhylid tadpoles (fig. 12.9 B) lack keratinized structures on the oral disc and have become obligate suspension feeders. The morphology of the buccal pump allows large volumes of water to be drawn into the mouth and across the gills with each expansion and contraction of the buccal cavity. The branchial baskets are very large, with dense gill filters well adapted to extracting small particles from the water (fig. 12.6 B). The similarity in morphology and feeding styles in pipids and *Rhinophrynus* probably represents descent from a common ancestor, because these families are very closely related, despite striking differences in adult morphology (Ford and Cannatella 1993; Haas 2003). Midwater suspension feeding evolved independently in microhylids, which are among the most derived frogs and only distantly related to the other two families (Wassersug 1989b; Donnelly, De Sá, and Guyer 1990; Haas 2003). Tadpoles of hylid frogs in the subfamily Phyllomedusinae have hard jaw sheaths and labial teeth that are capable of scraping material from the substrate, but they often behave as midwater suspension feeders and have converged with the pipids and microhylids in the structure of the buccal pump (Wassersug and Hoff 1979). Altig and Johnston (1989) classified these tadpoles as suspension-raspers. Many of these midwater suspension feeders exhibit a strong tendency to form schools, probably as a means of protection against predators in areas that lack protected microhabitats (see the following "Social Behavior").

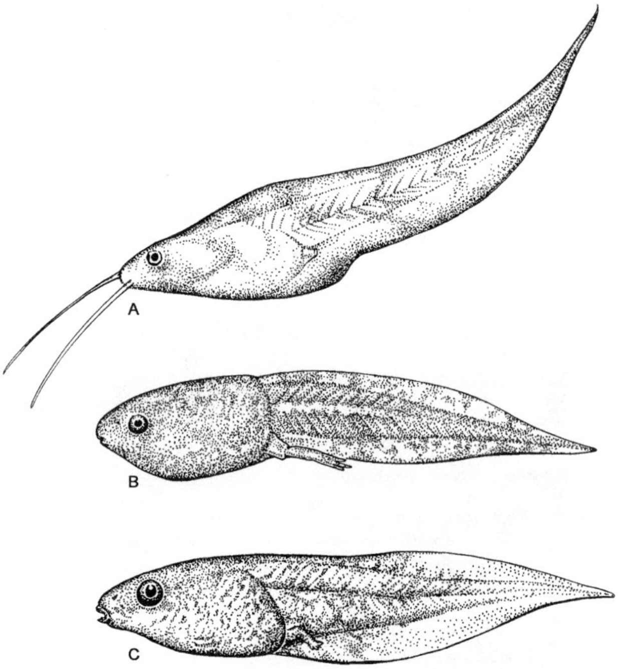


**Fig. 12.9.** Microphagous midwater suspension feeding tadpoles. (A) *Xenopus laevis* (Pipidae), showing typical head-down swimming posture. (B) *Kaloula rugifera* (Microhylidae). (C) *Agalychnis spurrelli* (Hylidae). (A) Drawing by Mary Jane Spring after Duellman and Trueb (1986). (B) After Liu (1950). (C) After David M. Dennis in Duellman (1970).

The morphology and ecology of suspension feeding have been studied in considerable detail in *Xenopus laevis,* which differs from most tadpoles in several respects (Wassersug 1996). *Xenopus* tadpoles spend most of their time feeding while suspended in the water column in a head-down position (fig. 12.9 A). They maintain this position by rapidly oscillating the tip of the tail (Hoff and Wassersug 1986). In contrast to most tadpoles, which have tail muscles composed entirely of relatively anaerobic white muscle fibers, *Xenopus* tadpoles have a band of red-muscle fibers outside the thicker white muscles (Muntz, Hornby, and Dalooi 1989). These more aerobic fibers have higher concentrations of mitochondria and lipids than do white fibers and are responsible for the continuous movements of the tip of the tail.

These tadpoles lack a ventral velum, the membrane that separates the buccal cavity from the pharyngeal cavity in most tadpoles. A one-way flow of water is made possible by valvelike action of the margins of the spiracle, which opens and closes as water is pumped into the mouth (Gradwell 1975). The combined buccopharyngeal cavity has a very large volume. *Xenopus* tadpoles take in much more water with each stroke of the buccal pump than do more generalized tadpoles of similar size (Seale 1982a). In most tadpoles, food trap areas that secrete mucus are found only in the pharynx, but in *Xenopus,* secretory ridges cover the whole floor of the buccopharyngeal cavity.

The modifications of the oral cavity make *Xenopus* tadpoles very efficient suspension feeders. They are capable of trapping extremely small food particles, some as small as viruses. They also can feed in water with particle concentrations that would be too low to allow more generalized tadpoles to extract sufficient energy by suspension feeding (Seale 1982a; Seale, Hoff, and Wassersug 1982; Viertel 1990, 1992). *Xenopus* tadpoles lack true gill filaments, but in well-oxygenated water, oxygen probably is taken up through the skin, the lining of the buccopharyngeal cavity, and the gill filters. The gill filters can become clogged with food when particle concentrations are high, thereby reducing surface area for respiration. The problem of maintaining aquatic respiration when food intake rates are high favors the early development of lungs and aerial breathing, and these tadpoles are obligate air-breathers during much of their development (M. Feder and Wassersug 1984; Feder et al. 1984; Pronych and Wassersug 1994; Orlando and Pinder 1995).

### Surface Feeders

A few species of pond-dwelling tadpoles have unusual modes of feeding, some of which also are found in tadpoles that inhabit quiet pools in streams (see the following "Stream Tadpoles"). The tadpoles of several microhylid frogs from Southeast Asia, including *Microhyla achatina, M. borneensis,* and *M. heymonsi,* have inverted, umbrella-like oral discs that are held at the water's surface and used to capture microorganisms and organic material from the surface film (Altig and Johnston 1989; T. Leong and Chou 1999). These have been termed neustonic tadpoles (the neuston comprises the organisms found in surface films). Similar adaptations are found in tadpoles of some species of *Phyllomedusa* (Duellman and Trueb 1986) and in some stream-dwelling tadpoles (see the following). Another peculiar tadpole is that of an Amazonian hylid, *Scarthyla ostinodactyla* (Duellman and de Sá 1988). This tadpole has an elongate, somewhat flattened body with a very low dorsal fin and a muscular tail (fig. 12.10). It has powerful jaw muscles and an impressive serrated jaw sheath that probably is used to cut pieces of floating plants. It swims just below the surface and apparently feeds on surface vegetation such as duckweed (*Lemna*). It also has an unusual way of escaping from predators—it can use its muscular tail to jump out of the water and skitter across the surface very rapidly.

**Fig. 12.10.** Tadpole of *Scarthyla ostinodactyla* (Hylidae). This tadpole swims just below the water surface and feeds on duckweed (Lemna) and other surface vegetation. The muscular tail is used to jump out of the water and skitter across the surface to escape from predators. From Duellman and de Sá (1988).

### Macrophagous Tadpoles

Some pond-dwelling tadpoles are adapted for feeding on large food particles and are termed *macrophagous* tadpoles (Duellman and Trueb 1986; Altig and Johnston 1989). The buccal cavity often is relatively large, enabling the tadpole to ingest large food items. The buccal pump is capable of generating strong suction for capturing food particles, but usually is not adapted for continuously pumping large volumes of water across the gills (Wassersug and Hoff 1979). Branchial baskets, gill filters, and branchial food traps are reduced and incapable of capturing small particles. Macrophagous feeding appears to be a derived trait in several anuran families, having evolved independently many times (Sokol 1975; Wassersug and Hoff 1979; Wassersug 1980).

Like most tadpoles, many macrophagous species are herbivorous. For example, Neotropical hylids such as *Hyla microcephala, H. phlebodes, H. ebraccata,* and other members of the *H. microcephala* and *H. leucophyllata* species groups apparently feed mainly on filamentous algae (Duellman and Trueb 1986; Duellman 2001), although *H. microcephala* tadpoles sometimes feed on small invertebrates as well (R. Wassersug, personal communication). These tadpoles have large ceratohyals and relatively large buccal cavities for dealing with large food particles, but the ceratobranchials and associated filter apparatus are reduced (fig. 12.6 E; Wassersug and Hoff 1979; Wassersug 1980). These species have rather simple mouths with strong horny jaw sheaths for breaking up filaments of algae (fig. 12.11). Macrophagous tadpoles can be either benthic or nektonic in their habits, and thus overlap in ecology with species that have more generalized diets. An example of a benthic macrophagous tadpole is that of *Occidozyga lima,* which has a flattened body and reduced tail fins, especially the ventral fin. In contrast, nektonic macrophagous tadpoles, like that of *Hyla microcephala,* have rounder bodies and much higher tail fins (fig. 12.11 A).

### Carnivorous Tadpoles

Although the vast majority of tadpole species are entirely or primarily herbivorous, some pond tadpoles also feed on animal matter. Some of these can be considered macrophagous tadpoles that have morphological specializations for feeding on large food items, but others are generalized pond tadpoles that exhibit few adaptations for a macrophagous diet. For example, green frog (*Rana clamitans*) tadpoles, which are quintessential generalized pond tadpoles, are voracious predators on the eggs of wood frogs (*Rana sylvatica;* Petranka and Kennedy 1999). Tadpoles of the Neotropical frog *Leptodactylus pentadactylus* feed on algae scraped from the substrate (Vinton 1951), but they also can be major predators on the eggs and larvae of other anurans (Heyer, McDiarmid, and Weigmann 1975; Wells 1979).

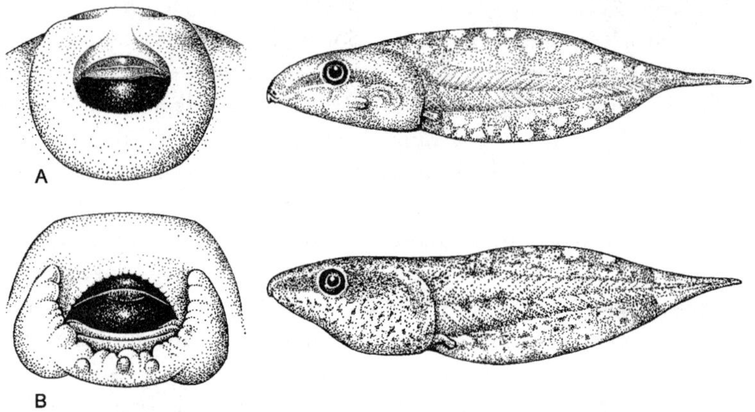


**Fig. 12.11.** Mouthparts and lateral views of pond tadpoles that feed mainly on filamentous algae. (A) *Hyla microcephala.* (B) *Hyla ebraccata.* Drawings by Mary Jane Spring after David M. Dennis in Duellman (1970).

Both the external mouthparts and internal oral structures of this species lack obvious specialization for a carnivorous diet and are characteristic of generalized pond-dwelling tadpoles (Wassersug and Heyer 1988). A variety of other generalized tadpoles in the tropics are opportunistic predators on the eggs or larvae of other species (Hero and Galatti 1990; Magnusson and Hero 1991; Silva, Giaretta, and Facure 2005), as are tadpoles of various species of *Bufo, Rana, Pelobates,* and *Pelodytes* (Heusser 1970; Banks and Beebee 1987b; Tejedo 1991; see chapter 14). Opportunistic carnivores sometimes are cannibalistic, consuming eggs, larvae, or transforming individuals of their own species (see the following "Cannibalism and Morphological Polyphenism"). Tadpoles of *Bufo crucifer* in Brazil ate crustacean eggs, although they did so only when feeding in aggregations, and the eggs did not make up a large proportion of the diet (Eterovick 2000).

Specialized macrophagous carnivores include tadpoles of the Neotropical genus *Ceratophrys,* which are active predators with large heads and massive, serrated jaw sheaths (Fernandez and Fernandez 1921; Noble 1931; La Marca 1986; fig. 12.12). The digestive tract of this species is shorter than those of most herbivorous tadpoles, but it is not known whether the anterior portion of the digestive tract functions like a true stomach or produces pepsin and other enzymes that digest protein (Fry and Kaltenbach 1999). Asian and African ranids in the genus *Hoplobatrachus* also are specialized carnivores that readily prey on eggs and tadpoles of other species or on conspecifics. They have massive serrated jaw sheaths similar to that of *Ceratophrys,* as well as needlelike labial teeth. The branchial baskets and gill filters are reduced in size, and the esophageal opening is very large (Chou and Lin 1997; Grosjean, Vences, and Dubois 2004).

Tadpoles of South American frogs in the genus *Lepidobatrachus,* which is closely realted to *Ceratophrys* (Haas 2003), have enormous heads, wide mouths, and large buccal cavities for ingesting large prey, but lack hard mouthparts for tearing their food. They feed mainly on small invertebrates and tadpoles that are swallowed whole and sometimes are cannibalistic (Cei, 1968; Wassersug and Hoff 1979). Many features of their anatomy are unusual for tadpoles. The stomach is more clearly differentiated from the intestine and the gut is much shorter than in typical tadpoles. The stomach begins to produce pepsin as soon as the tadpoles begin to feed (E. Carroll, Seneviratne, and Ruibal 1991). Many features of the oral apparatus and digestive tract in these tadpoles appear to be more like those of adult frogs than of tadpoles, features that are correlated with their adultlike diet. These frogs breed in ephemeral pools that often dry up quickly. Larval development is accelerated, with some tadpoles leaving the water in only 20 days (Ruibal and Thomas 1988).

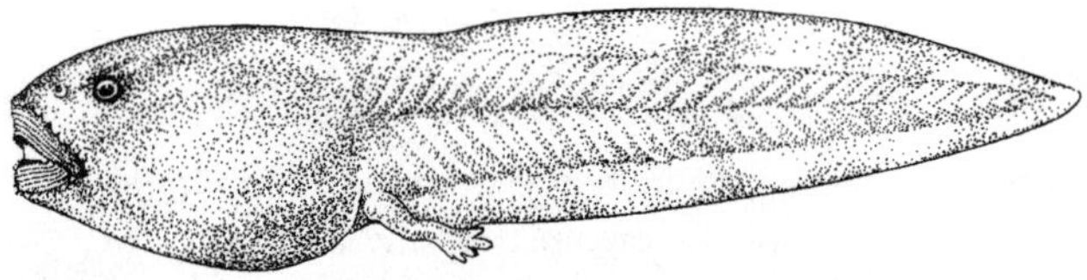

**Fig. 12.12.** A carnivorous, macrophagous tadpole (*Ceratophrys dorsata*) (Leptodactylidae). The well-developed rows of labial denticles are used to rasp flesh from tadpoles and other prey. Drawing by Mary Jane Spring after Noble (1931).

The African pipid frog *Hymenochirus boettgeri* has an unusual tadpole adapted for feeding on small invertebrates. Like all pipids, these tadpoles lack keratinized mouthparts. The front of the mouth is modified to form an oral tube when the jaws are depressed, quite unlike the mouths of *Xenopus* tadpoles (fig. 12.13). The ceratohyals are greatly enlarged, but the ceratobranchials are much reduced and lack any trace of a filter apparatus. There is a common buc-

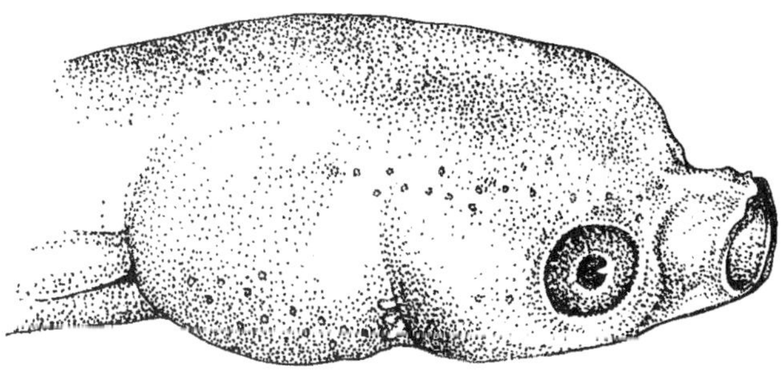

**Fig. 12.13.** The carnivorous tadpole of *Hymenochirus boettgeri* (Pipidae) with a mouth modified into a feeding tube for sucking in zooplankton. Drawing by Mary Jane Spring after O. Sokol (1962).

copharyngeal chamber, and the effective buccal volume is large, allowing the tadpole to ingest large prey whole (Sokol 1962; Ueck 1967; Wassersug and Hoff 1979). The tadpoles actively pursue small invertebrates and suck them into the mouth by rapidly expanding the buccal cavity. The pumping mechanism is similar to that of a microphagous filter feeder, such as *Xenopus*, and clearly is derived from it. The tadpole does not maintain a constant current of water through the buccopharyngeal cavity, but instead pumps water only when a prey animal is encountered. The narrow oral tube apparently serves to accelerate the flow of water into the mouth by constricting it into a small space, while the round opening enhances suction feeding by ensuring even acceleration around the edges of the mouth (Sokol 1962; Deban and Olson 2002).

### Cannibalism and Morphological Polyphenism

Generalized pond tadpoles that normally eat plant material or detritus sometimes kill and eat conspecific eggs or tadpoles (see also chapter 14). There are field observations of cannibalism of either eggs or larvae in *Hymenochirus boettgeri* (Sokal 1962), *Rhinophrynus dorsalis* (Starrett 1960), *Lechriodus fletcheri* (J. Moore 1961), *Ceratophrys* (Cei 1980), *Leptodactylus* (Hero and Galatti1990; Magnusson and Hero 1991; Silva, Giaretta, and Facure 2005), *Rana temporaria* (Heusser 1970, 1971b), *R. sylvatica* (Petranka and Thomas 1995), *Euphlyctis cyanophlyctis* (McCann 1932), *Hoplobatrachus* (McCann 1932, Grosjean, Vences, and Dubois 2004), *Hyla pseudopuma* (Crump 1983), and *Spea* and *Scaphiopus* (S. Ball 1936; Bragg 1957a, 1964, 1965; Pomeroy 1981). There are observations of cannibalism in the laboratory for several other species (M. Simon 1984). Opportunistic cannibalism often is associated with reproduction in ephemeral pools, where food can be scarce, and drying of pools places a premium on rapid development (M. Simon 1984; Polis and Myers 1985; Crump 1992; see also Heusser 1970, 1971b; Kluge 1981; Crump 1983, 1986b; Petranka and Thomas 1995). Usually cannibalistic tadpoles prey on smaller conspecifics, but in some species, tadpoles prey on metamorphosing individuals, especially when the latter have developed all of their legs but have not yet lost their tails. This type of cannibalism has been observed in a Neotropical hylid, *Osteopilus septentrionalis* (Crump 1986b) and in several species of spadefoot toads (*Scaphiopus* and *Spea;* Ball 1936; Bragg 1957a, 1964, 1965; Pomeroy 1981).

Some spadefoot toads in the genus *Spea* also exhibit diet-induced changes in morphology and feeding habits. In three species, *S. bombifrons, S. intermontana,* and *S. multiplicata,* tadpoles exhibit two distinct phenotypes (fig. 12.14). Most individuals are generalized omnivores that feed mainly on organic detritus on the bottoms of temporary desert rain pools. Others develop into carnivores, with enlarged jaw musculature and heavy, serrated jaw sheaths for attacking prey (Bragg 1946, 1956, 1957a, 1962, 1964, 1965; Bragg and Bragg 1959; Pomeroy 1981).

This type of polyphenism has been studied in considerable detail in *Spea multiplicata* (D. Pfennig 1990a, 1992a, b; D. Pfennig, Mabry, and Orange 1991; D. Pfennig and Frankino 1997). In contrast to the situation observed in some species of *Ambystoma,* a high density of conspecifics in the pond is not the main trigger that induces the carnivorous morph in spadefoot toads. Instead, the carnivorous morph develops in response to high densities of tiny freshwater shrimp that are found in ephemeral desert pools. Frequencies of carnivorous tadpoles are correlated with shrimp densities in natural pools, and shrimp density, in turn, is negatively correlated with pond persistence (D. Pfennig 1990a). Feeding shrimp to omnivorous tadpoles induces development of these morphs (D. Pfennig 1992b). The precise physiological mechanism that induces the carnivorous morph is not clear, but it does not appear to be due to increased intake of thyroid hormones in the prey (Storz 2004). In at least some species, the morphology of the tadpoles is reversible; carnivorous tadpoles given only a diet typical of omnivorous tadpoles quickly revert to the omnivore phenotype. The carnivorous tadpoles feed mostly on shrimp, but some-

**Fig. 12.14.** Omnivorous (top) and carnivorous (bottom) morphs of the tadpoles of *Spea bombifrons.* Photo by David Pfennig.

times take other species of tadpoles or conspecifics as well. Thus, they tend to be more opportunistic cannibals than are the cannibalistic morphs in *Ambystoma.* Tadpoles of the carnivorous morph are much more likely to eat conspecifics than are typical morph tadpoles (D. Pfennig and Frankino 1997). Both the behavior of carnivorous tadpoles and the initial development of the carnivore phenotype are mediated by kinship relationships among individuals in a pond (see "Kin Recognition and Cannibalism in Tadpoles" in a later section).

The development and ecology of carnivorous morphs also is influenced by interactions among species. Tadpoles of *Spea bombifrons* are more effective predators on shrimp than are tadpoles of *S. multiplicata,* whereas *S. multiplicata* tadpoles are superior competitors when feeding on detritus. When both species are present in the same pond, tadpoles of *S. bombifrons* are more likely to become carnivorous than when that species is found alone. Conversely, tadpoles of *S. multiplicata* are less likely to become carnivorous in the presence of *S. bombifrons* than in single-species ponds (D. Pfennig and Murphy 2000). Across a range of natural ponds with different densities of the two species, *S. multiplicata* tadpoles were increasingly omnivore-like as the abundance of *S. bombifrons* increased, whereas *S. bombifrons* tadpoles became increasingly carnivore-like as the density of *S. multiplicata* increased. Similar results were obtained when the tadpoles were reared in the laboratory with different numbers of heterospecific tadpoles (D. Pfennig and Murphy 2002).

The adaptive significance of this type of phenotypic plasticity appears to be the same as in cannibalistic *Ambystoma.* Carnivores grow faster and metamorphose earlier, an advantage in pools that dry out quickly. Omnivores, on the other hand, do better in longer lasting ponds, because they reach metamorphosis with larger fat reserves (D. Pfennig 1992a). In *Spea multiplicata,* age and size at metamorphosis are negatively correlated, the reverse of the pattern seen in most anurans (see chapter 13). This is largely because the highest quality food used by these tadpoles, small shrimp, is rapidly depleted. When the first tadpoles begin to metamorphose, the remaining individuals do not enjoy a release from food competition that often occurs in species that feed on algae or bacteria, which continue to increase in abundance over time (e.g., R. Newman 1989). Consequently, late-metamorphosing individuals actually are smaller than are those that leave the pond early (D. Pfennig, Mabry, and Orange 1991).

### Phytotelmonous Tadpoles

Some tadpoles live in water that accumulates in bromeliads, leaf axils, tree holes, hollow logs, seedpods, and other plant parts (Lehtinen 2004). Some of these tadpoles hatch from eggs deposited in these small pools (see chapter 10), but in dendrobatid frogs and some hemiphractine hylids (*Flectonotus*), a parent carries tadpoles to these sites (see chapter 11). These phytotelmonous tadpoles (*phyto* [G.] = plant; *telma* [G.] = pool) have a variety of feeding habits. Some do not feed at all (see the following "Nonfeeding Tadpoles"). Others feed on organic detritus that collects in the water, or they may prey on small invertebrates such as mosquito larvae, and even on conspecific tadpoles (K. Summers 1990; Caldwell 1993; Caldwell and de Araújo 1998). Still others feed opportunistically or entirely on anuran eggs deposited in the same body of water. The South American hylid *Trachycephalus resinifictrix,* which breeds in tree holes, appears to be an opportunistic egg-eater (Schiesari, Grillitsch, and Vogl 1996), as are some species of *Dendrobates* tadpoles (Weygoldt 1987).

Specialized egg-eaters include Neotropical hylids such as *Anotheca spinosa, Osteopilus brunneus, Osteocephalus oophagus,* and *Hyla zeteki* (Dunn 1926a, 1937; E. Taylor 1954; Duellman 1970; Lannoo, Townsend, and Wassersug 1987; Jungfer and Schiesari 1995; Jungfer 1996; Schiesari, Grillitsch, and Vogl 1996; Thompson 1996). Tadpoles of the African microhylid *Hoplophryne uluguruensis,* several Asian rhacophorids (*Chirixalus eiffingeri, Philautus, Theloderma*), and several species of *Dendrobates* also feed largely or exclusively on frog eggs (Noble 1929a; Wassersug, Frogner, and Inger 1981; Ueda 1986; Weygoldt 1987; Kam, Chuang, and Yen 1996). In the case of *Osteopilus brunneus, Osteocephalus oophagus, Anotheca spinosa, Chirixalus eiffingeri,* and some *Dendrobates,* the mother provides the tadpoles with trophic eggs (see chapter 11 for further discussion of this form of parental care). In *Chirixalus eiifingeri,* tadpoles compete with one another in their bamboo pools, and may scavenge dead tadpoles, but they do not cannibalize living tadpoles and appear to be entirely dependent on eggs provided by the mother (Y. Chen et al., 2001; Kam et al. 2001).

Egg-eating tadpoles typically have terminal mouths with large jaw sheaths but reduced labial teeth, relatively large heads, and large buccal cavities (fig. 12.15). Structures associated with a microphagous diet often are reduced (fig. 12.6 D), although this is not true for some egg-eating dendrobatid tadpoles (Weygoldt1987; Caldwell and de Araújo 1998). Tadpoles of *Osteopilus brunneus* do not maintain a continuous flow of water through the buccopharyngeal cavity, but suck food in as it is encountered. Consequently, there is not a continuous stream of moving water to aerate the gills. The lack of gill aeration, together with the low levels of oxygen in water-filled bromeliads, makes aerial respiration obligatory for these tadpoles (Lannoo, Townsend, and Wassersug 1987).

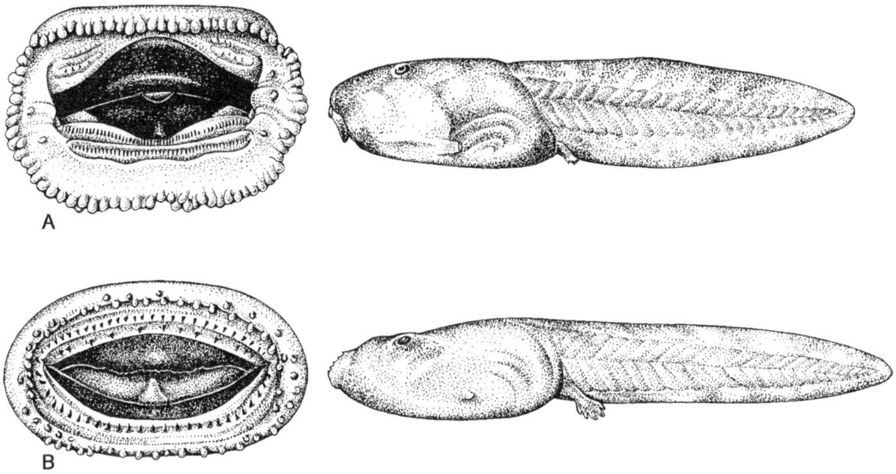

**Fig. 12.15.** Mouthparts and lateral views of two egg-eating Neotropical hylid tadpoles that live in bromeliads. (A) *Anotheca spinosa*. (B) *Hyla zeteki*. These species are not closely related, but have converged in the evolution of diet and morphology. Drawings by Mary Jane Spring after David M. Dennis in Duellman (1970).

## Stream Tadpoles

In areas where ponds and other bodies of standing water are scarce, such as mountainous regions, streams and rivers provide the principal breeding habitats for aquatic-breeding frogs. Streams offer a variety of microhabitats for tadpoles, from deep, quiet pools that are not very different from small ponds, to rapidly flowing riffles and torrents. Different species of stream-dwelling tadpoles have evolved morphological and behavioral adaptations for particular stream microhabitats and often are precluded from living in other microhabitats (Inger, Voris, and Frogner 1986; S. Richards 2002). Indeed, tadpoles living in different stream microhabitats display much greater morphological diversity than do pond-dwelling tadpoles, and tadpoles in different families often exhibit striking convergence in overall morphology and feeding structures (Lamotte and Lescure 1989a, b; Haas and Richards 1998). Some stream tadpoles are fossorial, spending most of their lives buried in gravel, sand, or mud at the bottom of streams. Fossorial life probably is not possible for most pond-dwelling tadpoles, especially those living in warm climates, because of the lack of oxygen on the bottoms of many ponds (see chapter 4).

### Benthic, Nektonic, and Neustonic Stream Tadpoles

Many stream tadpoles are confined to relatively quiet pools and backwater areas, where they are not exposed to the direct action of flowing water. These tadpoles resemble pond-dwelling tadpoles in their morphology and exhibit only minor adaptations to stream habitats. For example, various species of *Rana* in the southwestern United States can be found in stream pools, while others are found mostly in ponds. Tadpoles that live in streams tend to have slightly thicker, more muscular tails and thicker tail fins than do those found in ponds, but otherwise are similar to pond-type tadpoles (Hillis 1982; N. Scott and Jennings 1985; Jennings and Scott 1993). Pond-dwelling ranids sometimes breed in quiet pools of streams, and their tadpoles sometimes exhibit minor variation in body shape and tail morphology, but it is not clear whether such differences have a genetic basis or are due to phenotypic plasticity (Jennings and Scott 1993). Some stream-dwelling ranids in other regions, such as *Rana angolensis* in Africa, *Limnonectes blythii* and its relatives in Borneo and Southeast Asia, and the Neotropical species *Rana palmipes*, have benthic tadpoles that are not much different from pond-dwelling species (Inger 1985; Altig and Johnston 1989; T. Leong and Chou 1999).

Benthic ranid stream tadpoles tend to have generalized feeding habits. They can grow even when feeding on relatively nutrient-poor stream sediments (Flecker, Feifarek, and Taylor 1999), but do better when feeding on more nutritious food, such as diatoms that grow on filamentous algae (Kupferberg, Marks, and Power 1994; Kupferberg 1997a, b). Benthic stream tadpoles can have a major effect on total community structure. Their feeding activities remove significant quantities of diatoms or sediments that tend to inhibit the growth of filamentous algae, and they can have a negative effect on grazing invertebrates that feed on diatoms (Kupferberg 1997a; Flecker, Feifarek, and Taylor 1999).

In addition to ranids, many other anurans have benthic tadpoles that live on the bottoms of quiet stream pools. These include some bufonids (*Bufo asper, B. fastidiosus*), dendrobatids (*Colostethus pamamensis*), hylids (*Hyla arenicolor, Pseudacris regilla*), hyperoliids (*Leptopelis*), leptodactylids (*Alsodes, Telmatobius*), megophryids (*Leptobrachium, Ore-

*olalax*), myobatrachids (*Mixophyes*), rhacophorids (*Polypedates colletti, Rhacophorus gauni*), and others (Inger 1985; Wassersug and Heyer 1988; Altig and Johnston 1989; Zhao et al. 1994; Lips and Krempels 1995; T. Leong and Chou 1999). Most of these tadpoles are very similar to benthic pond-dwelling tadpoles, but may have somewhat more flattened bodies and shallower tail fins.

Tadpoles that live in open water are less common in streams than in ponds, probably because submerged vegetation and other sources of food are scarce in these habitats. Tadpoles of some South African stream frogs in the genus *Rana* (sometimes placed in the genus *Strongylopus*) are nektonic, or intermediate between nektonic and benthic (Altig and Johnston 1989; Passmore and Carruthers 1995), as are tadpoles of the *Hyla miotympanum* group from Central America (Duellman 1970). These tadpoles lack the very high tail fins of nektonic pond tadpoles (Altig and Johnston 1989).

Midwater suspension feeders usually are not found in streams, although Neotropical frogs such as *Agalychnis callidryas* sometimes place their eggs over quiet pools in streams and their tadpoles subsequently develop in those pools. Some stream-dwellers are neustonic feeders. These include several species of *Colostethus* that have funnel-shaped mouths that are oriented upward. Presumably these tadpoles feed either on organisms in the surface film, or perhaps from the undersides of rocks (fig. 12.16 B). Several megophryid frogs in the genus *Megophrys* and *Ophryophryne* from Asia have very peculiar mouthparts adapted to feeding on surface films. The mouth has wide lips forming an upturned funnel (fig. 12.16 C). There are very few labial teeth and no horny jaw sheaths. The tadpoles are found in pools under cascades in small mountain streams, suspended from the surface by the funnel mouth, with the opening directed into the current to catch floating particles (M. A. Smith 1926; Liu 1950; Inger 1985, 1986; Nodzenski, Wassersug, and Inger 1989; T. Leong and Chou 2000; Grosjean 2003).

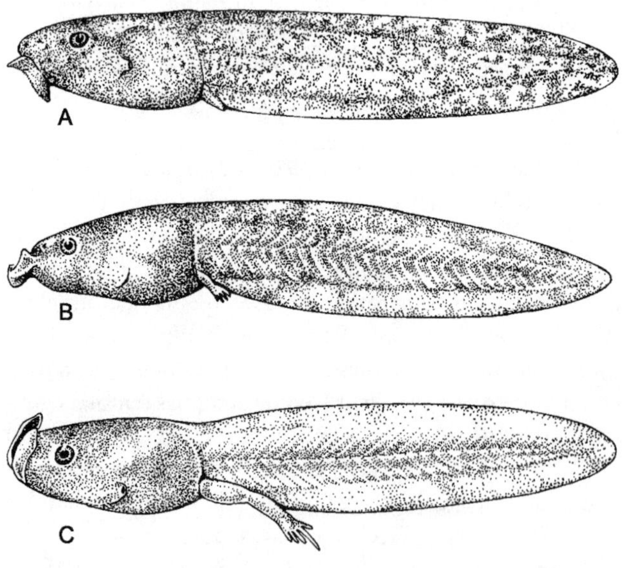


Fig. 12.16. Stream tadpoles with funnel mouths. (A) *Duellmanohyla rufioculis* (Hylidae), which is adapted for feeding on the bottoms of quiet pools. (B) *Colostethus nubicola* (Dendrobatidae), with an upturned funnel for feeding from the undersides of rocks, (C) *Megophrys minor* (Megophryidae), with an upturned funnel for feeding from the water's surface. (A) Drawing by Mary Jane Spring after David M. Dennis in Duellman (1970). (B) After Duellman and Trueb (1986). (C) After C. Liu (1950).

### Fossorial and Psammonic Tadpoles

Centrolenid tadpoles have a very different set of adaptations for stream life. They have elongated bodies with muscular tails and reduced tail fins (fig. 12.17 A). They burrow into gravel or under leaves in streams, and also have been observed burrowed in mud or wet leaves some distance from flowing water. These tadpoles have poorly developed eyes until just before metamorphosis and lack extensive pigmentation in the skin, features that are consistent with their fossorial habits (Villa and Valerio 1982; Mijares-Urrutia 1990; Jaramillo et al. 1997). They often appear to be bright red because of the extensive blood flow through the skin, which may facilitate cutaneous gas exchange.

The tadpoles of some Asian megophryid frogs of the genera *Leptobrachium, Leptobrachella,* and *Leptolalax* also are fossorial (Altig and Johnston 1989). These tadpoles have elongated bodies that are superficially similar to those of centrolenid frogs fig. 12.17 B). They spend most of their time in the spaces between rocks and gravel in quiet stream pools (Liu 1950; Inger 1966, 1985; Inger, Voris, and Frogner 1986). Tadpoles of some species in these genera are more robust

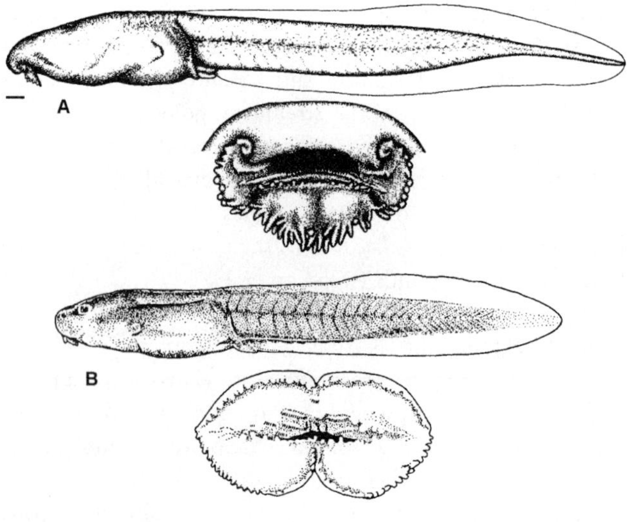


Fig. 12.17. Two fossorial tadpoles, showing convergent evolution in body and tail morphology. (A) *Hyalinobatrachium uranoscopum* (Centrolenidae). (B) *Leptolalax gracilis* (Rhacophoridae). (A) After Heyer (1985), and (B) after Lamotte and Lescure (1989a).

and live in more open habitats, such as the bottoms of quiet pools. One fossorial species, *Leptolalax gracilis,* has a relatively smaller gut and liver than do more generalized tadpoles, probably because of space constraints imposed by their elongate body form. These tadpoles also develop lungs relatively late. This not only alleviates the space constraints for internal organs, but also keeps the tadpoles negatively buoyant (Nodzenski, Wassersug, and Inger 1989).

One of the most bizarre of all tadpoles is that of the South American microhylid *Otophryne pyburni.* The tadpole was first described as that of *O. robusta* by Pyburn (1980b), but the population that he studied is now considered to be *O. pyburni* (J. A. Campbell and Clarke 1998). This tadpole has sharp, needlelike teeth on both the upper and lower jaws (fig. 12.18). Pyburn (1980b) thought the teeth indicated a carnivorous diet, but subsequent work showed this to be incorrect. In fact, these tadpoles appear to be specialized for burrowing into the sandy bottoms of streams, where they are passive filter feeders on organic material in the sand. The teeth appear to be adaptations for preventing sand grains from entering the mouth (Wassersug and Pyburn 1987). The fossorial tadpoles of a centrolenid frog, *Hyalinobatrachium ignioculus,* have somewhat similar serrations on the jaw sheaths, but the function of these structures is not known (Noonan and Bonett 2003). Another unusual feature of the *Otophryne* tadpole is an unusually long spiracle tube (fig. 12.19). Wassersug and Pyburn (1987) speculated that the opening of the tube is held above the sand surface when the tadpole is buried (fig. 12.19 B). Currents passing by the opening would draw water out of the tube, thereby allowing for passive movement of water into the mouth without active pumping. These were called psammonic tadpoles by Altig and Johnston (1989; *psamm* [G.] = sand).

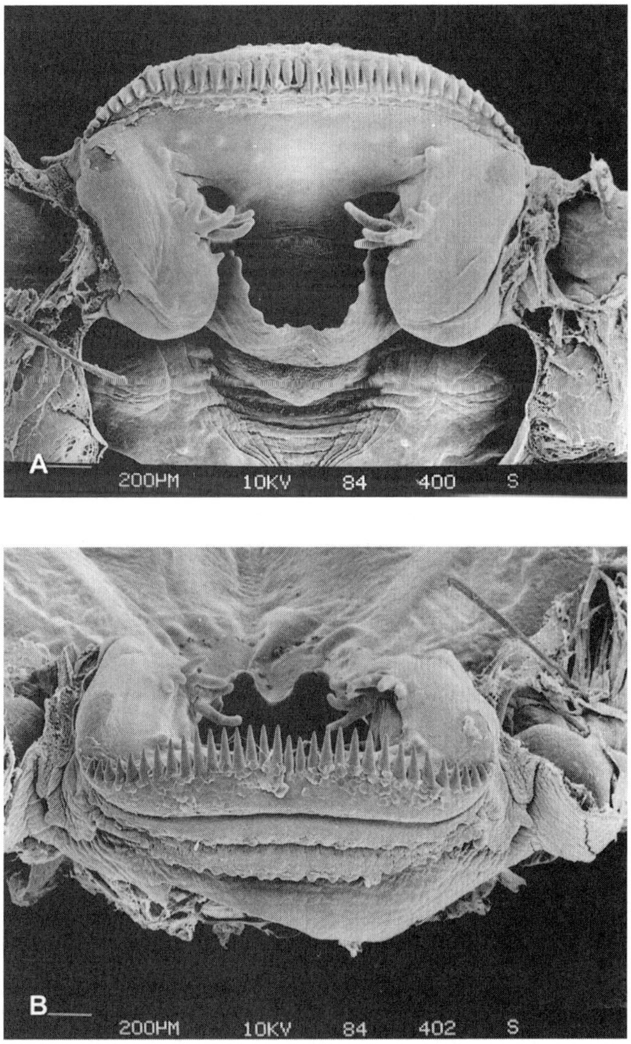


**Fig. 12.18.** Mouthparts of the microhylid frog *Otophryne pyburni* (Microhylidae). (A) Scanning electron micrograph of front of the roof of the mouth, showing sharp teeth and transverse orientation of jaws. (B) Scanning electron micrograph of front of lower jaw, with mouth in closed position, showing sharp teeth. Photos by Richard J. Wassersug.

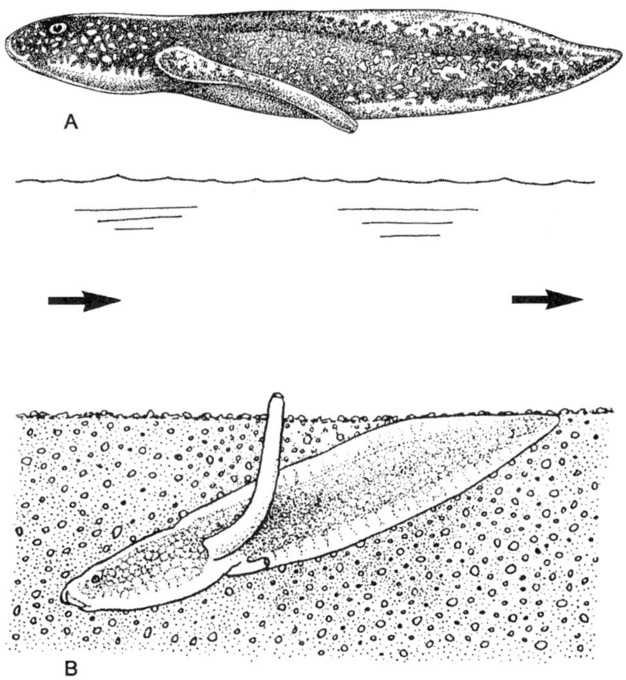


**Fig. 12.19.** (A) Lateral view of *Otophryne pyburni* (Microhylidae), showing elongated spiracle tube on the left side of the body. (B) Hypothesized position of spiracle tube in current when tadpole is buried in sand. Arrows indicate direction of water flow. (A) Drawing by Mary Jane Spring after Pyburn (1980a). (B) After Wassersug and Pyburn (1987).

### Tadpoles Adapted to Fast-Flowing Water

Tadpoles in several families are associated with rapidly flowing streams or torrents. In such habitats, both suspended plankton and larger organic particles are rare, and tadpoles often depend primarily on food attached to the substrate. They face the dual problem of maintaining a position in the current and removing algae that is tightly attached to rocks

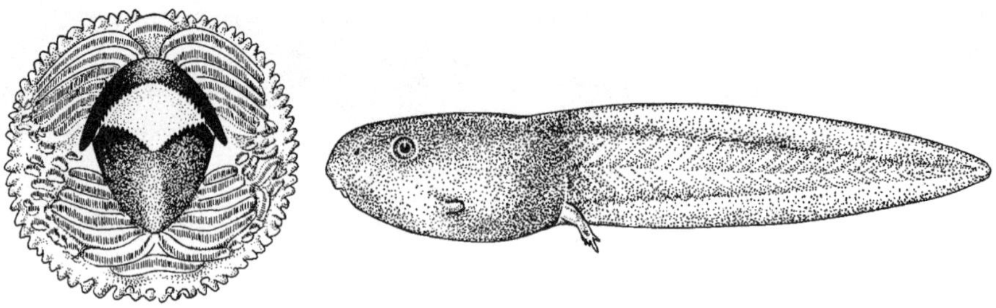

**Fig. 12.20.** Mouthparts and lateral view of *Oreolalax popei* (Microhylidae), a Chinese stream tadpole with a partially carnivorous diet. Drawing by Mary Jane Spring after C. Liu (1950).

(Lamotte and Lescure 1989a, b; S. Richards 2002). Tadpoles that live in rapid currents typically have elongated or flattened bodies, muscular tails, and reduced tail fins. The elongated body places constraints on the development of the feeding apparatus. These species generally have relatively small heads, which precludes having large buccal cavities and macrophagous feeding habits (Wassersug and Hoff 1979). However, some species are predatory. For example, Chinese megophryids of the genus *Oreolalax* live in mountain streams and eat insect larvae and tadpoles in addition to algae; they have powerful, serrated jaws for capturing prey (fig. 12.20).

Many stream-dwelling tadpoles have evolved special morphological adaptations for adhering to rocks, either with enlarged mouthparts and sucking discs, or suction discs on the belly. Altig and Johnston (1989) and Altig and McDiarmid (1999) classified tadpoles that cling to rocks with their mouthparts into several subcategories, including clasping, adherent, and suctorial forms (see table 12.1). These categories reflect apparent differences in the ability of tadpoles to cling to rocks in increasingly fast currents, determined mainly from observed differences in morphology. These categories tend to grade into one another, however, so that even relatively closely related species in the genus *Hyla* may fall into different categories (Duellman 1970).

Tadpoles that live in slow-moving currents may clasp rocks or other substrates and often have somewhat enlarged oral discs. They maintain their position in the current mostly by swimming with their muscular tails. These clasping tadpoles include some petropedetids, such as the South African kloof frog (*Natalobatrachus bonebergi*). Some mantellid treefrogs from Madagascar in the genus *Boophis* also have clasping tadpoles (other tadpoles in this genus live in ponds or quiet stream pools; Altig and Johnston 1989; Lamotte and Lescure 1989b; Glaw and Vences 1994; Passmore and Carruthers 1995).

Stream tadpoles in several families, including hylids, myobatrachids, and mantellids, have enlarged, umbrella-shaped oral discs that enable them to cling to rocks in faster currents. These so-called adherent tadpoles include many species of *Hyla, Duellmanohyla, Plectrohyla* and *Ptychohyla* (fig. 12.21) from South and Central America (Duellman 1970, 2001; Lamotte and Lescure 1989b; Duellman and Campbell 1992). Tadpoles with enlarged oral discs also are characteristic of some Australian hylids in the genus *Litoria,* myobatrachids in the genera *Taudactylus* and *Crinia* (*C. riparia*), and some mantellid frogs from Madagascar in the genus *Mantidactylus* (Altig and Johnston 1989; Lamotte and Lescure 1989b; Davies 1991; Tyler and Davies1993; Glaw and Vences 1994). There also are tadpoles with upturned, funnel-shaped mouths that live in flowing water and apparently use their mouths to adhere temporarily to the undersides of rocks. These include several species of *Phyllomedusa* (Hylidae), *Mantidactylus* (Mantellinae), *Colostethus* (Dendrobatidae), and *Megophrys* (Megophryidae; Lamotte and Lescure 1989b).

Tadpoles that live in even faster currents often have the oral disc developed into a powerful sucker that enables them to cling to rocks more or less continuously. Oral suckers have evolved independently in the North American tailed frog (*Ascaphus truei*) (fig. 12.22 A), South African frogs in the genus *Heleophryne,* many Neotropical species of *Hyla* (fig. 12.22 B–D), hylid frogs from Australia and New Guinea (*Litoria, Nyctimystes*), a number of species of *Bufo,* Asian bufonids in the genus *Ansonia,* Asian megophryids in the genus *Scutiger,* and several genera of ranid, petropedetid, and astylosternid frogs (*Conraua, Petropedetes, Trichobatrachus;* Lutz and Orton 1946; Duellman 1970; Duellman and Trueb 1986; Altig and Johnston 1989; Iwasawa and Saito 1989; Lamotte and Lescure 1989a, b; Inger 1992; Tyler and Davies 1993; Passmore and Carruthers 1995; Haas and Richards 1998; Altig and McDiarmid 1999; S. Richards 2002).

In *Ascaphus,* there is an extensive system of lymphatic canals that provides a hydraulic framework to support the front and sides of the sucker. This allows for tight attachment to the substrate while suction is generated by negative pressure in the buccal cavity. An oral valve that closes when

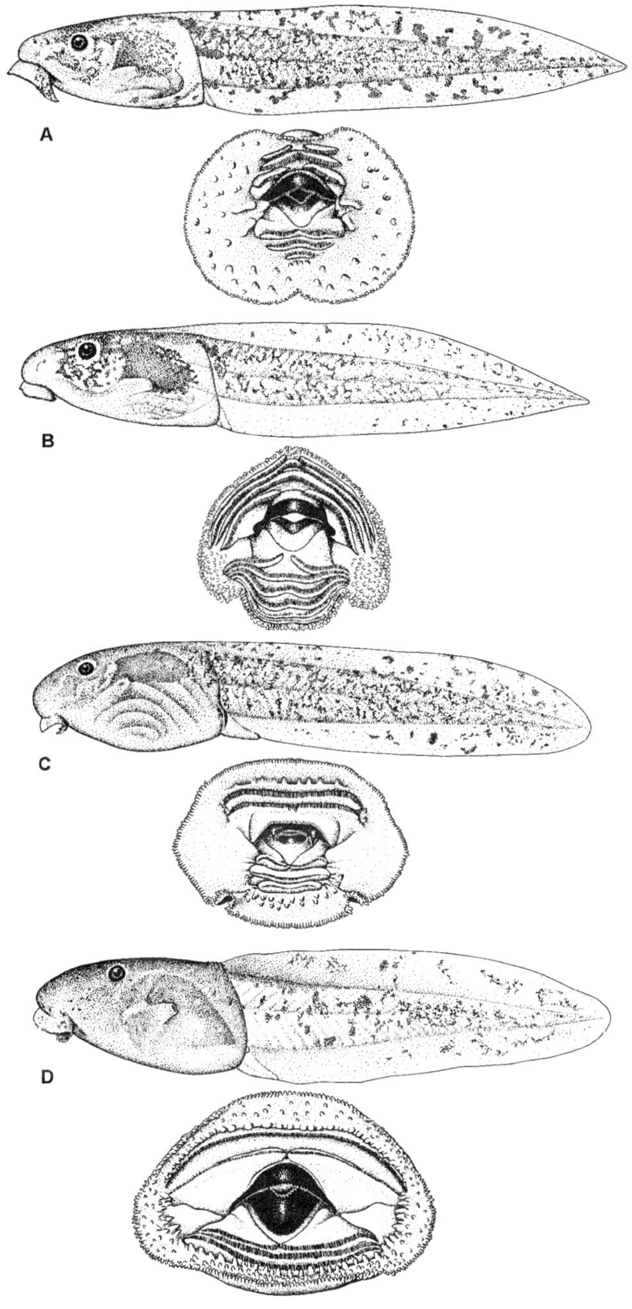


**Fig. 12.21.** Hylid stream tadpoles with adherent mouthparts. (A) *Duellmanohyla schmidtorum.* (B) *Ptychohyla euthysanota.* (C) *Plectrohyla matudai.* (D) *Plectrohyla guatemalensis.* The precise morphology of the mouthparts varies among genera and even within the genus *Plectrohyla,* but all have converged on a morphology that allows the tadpoles to hold onto rocks in flowing water. After drawings by David M. Dennis in Duellman (1970).

the sucker is engaged prevents water flow into the buccal cavity and allows the tadpole to remain attached to rocks for long periods of time without continuous buccal pumping. These tadpoles have an unusual tendon connection between the muscles of the trunk and tail and the lower jaw cartilages, which allows for coordinated movement of the jaw and tail musculature during "suctorial crawling" across the surface of rocks (Gradwell 1973). Suctorial tadpoles in other families appear to move in a similar fashion (Gradwell 1975; C. Taylor and Altig 1995). Feeding in species with suctorial mouths is accomplished by scraping algae off of rocks, and these tadpoles often have well-developed horny jaw sheaths and extra rows of labial teeth on the oral disc (Duellman 1970, 2001; Duellman and Trueb 1986; Lamotte and Lescure 1989a, b; fig. 12.22). The way in which suspended particles are transported through the oral cavity is not entirely clear. One possibility is that the sucker is loosened periodically to allow water to flow into the oral cavity and move food particles into the pharynx (Gradwell 1971; Haas and Richards 1998). Water flow over the gills for respiration is maintained by inflow of water through the nostrils (Altig and Brodie 1972; Gradwell 1971, 1975).

Suctorial tadpoles in several different families exhibit remarkable convergence in morphology, not only of the oral sucker, but also in features on the chondrocranium and muscle system. Haas and Richards (1998) compared the cranial morphology of suctorial tadpoles in the Australian hylid subfamily Pelodryadinae (*Nyctimystes* and the *Litoria nannotis* species group) with that of closely related pond-type tadpoles. Many of the differences in cranial morphology reflect alterations in the position and relative proportions of cranial elements, rather than entirely new structures. The heads of suctorial tadpoles are much wider at the front than are those of pond-type tadpoles, providing a base of attachment for the large oral sucker and the muscles associated with it. Suctorial tadpoles also have relatively smaller branchial baskets, but large ceratohyals. The *rectus abdominis* muscle, which runs along the belly of the tadpole, has a more anterior insertion in suctorial tadpoles. This probably enables the tadpole to draw its body forward after securing a hold with the sucker. Many of the morphological features seen in suctorial hylid tadpoles from Australia are very similar in Neotropical suctorial tadpoles, such as *Hyla armata,* in *Heleophryne* tadpoles from southern Africa, and in mantellid tadpoles of the genus *Boophis* from Madagascar. These groups are not closely related, so the similarity in structure reflects convergent adaptations to similar flowing-water habitats. The basic structure of the cranium also is similar in *Ascaphus,* but this species has some unique morphological features that reflect its much greater phylogenetic distance from other species with suctorial tadpoles, all of which are in neobatrachian families.

Some stream-dwelling tadpoles have an enormous disc posterior to the mouth and extending over much of the ventral surface of the body (fig. 12.23). These so-called gastromyzophorous tadpoles are found in two genera of

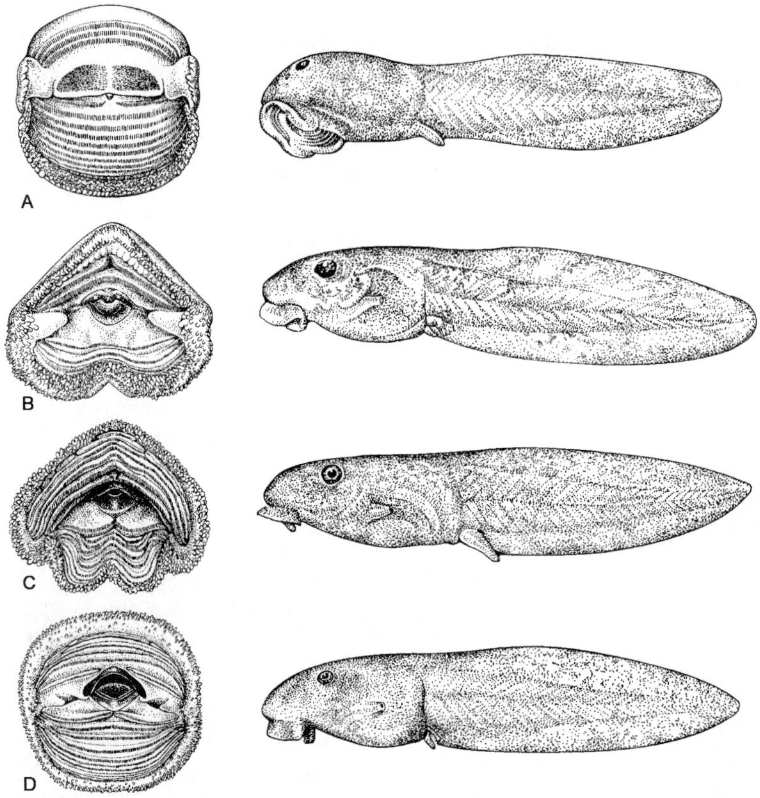


**Fig. 12.22.** Stream tadpoles with oral suction discs. (A) *Ascaphus truei* from North America. (B–D) Three Middle American hylids, showing variation in the number of labial denticles for scraping algae off of rocks. (B) *Hyla rivularis* (2 upper, 3 lower tooth rows). (C) *Hyla colymba* (6 upper, 9 lower rows). (D) *Hyla mixomaculata* (7 upper, 11 lower rows). Drawings by Mary Jane Spring after Stebbins (1951; mouthparts in A), Noble (1931; tadpole in A), David M. Dennis in Duellman (1970; B, C, D).

Neotropical bufonids, *Atelopus* and *Atelophryniscus,* one species of Asian bufonid in the genus *Ansonia,* Asian ranids in the genus *Amolops,* and the Asian ranid *Rana sauteri.* The abdominal sucker enables these tadpoles to adhere to rocks in very strong currents under cascades in mountain streams (Noble 1929a; Hora 1930; Liu 1950; Starrett 1967; Duellman and Lynch 1969; Lescure 1981; P. Gray and Cannatella 1985; Inger 1985, 1992; Duellman and Trueb 1986; Lamotte and Lescure 1989b; Chou and Lin 1997). Hora (1922) reported that an *Amolops* tadpole lifted out of the water by its tail could support a rock attached to its ventral disc that was 60 times the mass of the tadpole. This extraordinary suction is not created by negative pressure in the buccal cavity, but by action of abdominal and branchial muscles. Large pads of connective tissue, to which muscles are attached, support the flaps of skin that form the disc. When the muscles contract, the central part of the disc is elevated while the edges remain attached to the substrate, thereby creating a vacuum inside the disc (Noble 1929a). A similar set of muscles is found in the one species of *Ansonia* that has an abdominal sucker (Inger 1992). The way in which suction is generated by *Atelopus* tadpoles has not been reported. Tadpoles of this genus have less developed keratinized mouthparts than do *Amolops* tadpoles, suggesting that their feeding ecology is different or that they live on different kinds of rocks (Wassersug and Heyer 1983).

### Semiterrestrial Tadpoles

Tadpoles of two genera of leptodactylids, *Cycloramphus* and *Thoropa,* are found on wet rocks along streams in the forests of southeastern Brazil (fig. 12.24 A, B). In most species, tadpoles are adapted for adhering to the surface of rocks splashed by waterfalls or covered by a film of water from seepage areas (Heyer 1983; Heyer and Maxson 1983; Wassersug and Heyer 1983). Their bodies are greatly elongated, with reduced tail fins. The venter is flattened, and in *Cycloramphus,* there is a shelf-like projection extending beyond the body that provides increased area for adhesion to the substrate by means of surface tension, but there is no

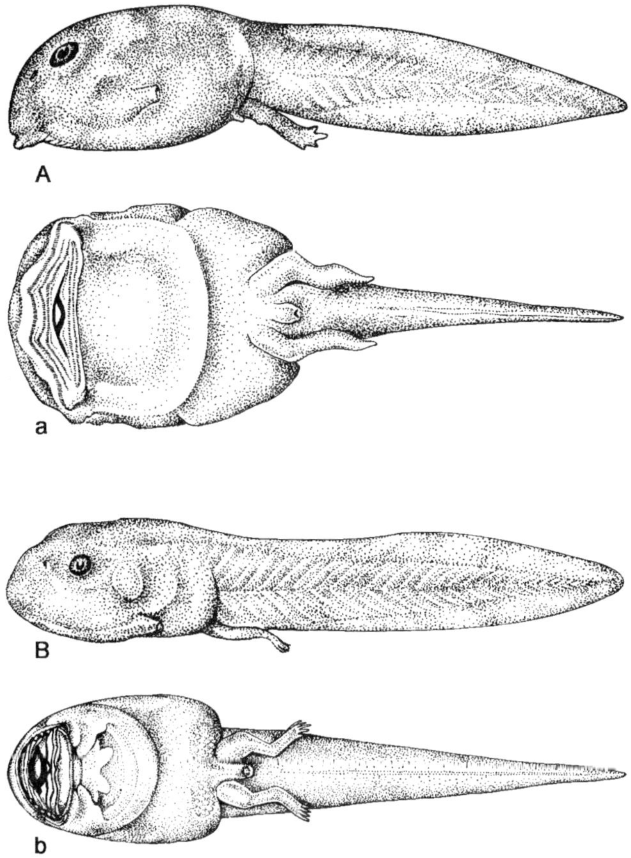


**Fig. 12.23.** Stream tadpoles with ventral sucker discs. (A) *Atelopus certus* (Bufonidae) lateral view; (a) ventral view. (B) *Amolops mantzorum* (Ranidae) lateral view; (b) ventral view. (A) Drawings by Mary Jane Spring after Duellman and Lynch (1969), and (B) after C. Liu (1950).

specialized suction apparatus. The branchial baskets of these tadpoles are relatively flattened and lack the dense filter apparatus needed for feeding on plankton (Wassersug and Heyer 1983). Evidently they feed by scraping algae off of rocks, but their ecology is poorly known.

Some tadpoles in several other families exhibit ecological and morphological adaptations similar to those of *Thoropa* and *Cycloramphus*. For example, some species in the African genus *Petropedetes* have flattened bodies and sometimes adhere to rocks in the splash zones of waterfalls (fig. 12.24 C). These tadpoles apparently spend much of their time out of the water, and even have been observed moving about on the forest floor with only their back legs fully emerged (Lawson 1993; see also fig 9.6 in Hoff et al. 1999). In contrast to the leptodactylids, however, *Petropedetes* tadpoles also have large suctorial oral discs characteristic of other tadpoles found in fast-flowing streams (Lamotte and Lescure 1989a). Some Asian ranid tadpoles in a different subfamily have similar habits and body morphology, and these also have suctorial oral discs. These include *Indirana beddomii* and *Nannophrys ceylonensis* (Lamotte and Lescure 1989a; Wickramasinghe et al. 2005). Finally, there are some West African bufonid tadpoles with similar habits, flattened bodies, and moderately to well developed suctorial oral discs. These include *Bufo perreti* and some species of *Werneria* (Altig and Johnston 1989; Lamotte and Lescure 1989a, b).

### Nonfeeding Tadpoles

Tadpoles in several families of anurans complete development in an enclosed nest, sometimes attended by a parent (see chapters 10 and 11). These were termed nidicolous tadpoles by Altig and Johnston (1989; *nid* [L.] = nest; *-colous* [L.] = inhabiting). Many of these tadpoles do not feed, but live entirely on yolk reserves. For example, tadpoles of *Cycloramphus stejnegeri* develop in a terrestrial nest attended by the female (Heyer and Crombie 1979). They lack the flattened venter characteristic of aquatic members of the genus, as well as most feeding structures (Heyer 1983). Oral structures have been lost or reduced to varying degrees in other tadpoles that complete development in an enclosed nest, but in others, such as *Eupsophus calcaratus*, are not dramatically different from generalized feeding tadpoles (Candioti, Úbeda, and Lavilla 2005). Nidicolous tadpoles are known from species of bufonids, dendrobatids, leiopelmatids, leptodactylids, microhylids, myobatrachids, and ranids (Lutz 1947; Lescure 1973, 1975a, 1979; Salthe and Mecham 1974; Blommers-Schlösser 1975a; Lamotte and Lescure 1977; Altig and Johnston 1989; Junca, Altig, and Gascon 1994; Kaiser and Altig 1994; Caldwell and Lima 2003; see also chapter 10). This type of reproductive mode is somewhat intermediate between aquatic development and fully terrestrial eggs with direct development (see chapter 10). In the case of dendrobatids in the genus *Colostethus*, nidicolous tadpoles probably have evolved independently several times from an ancestral mode of transporting tadpoles to water (Caldwell and Lima 2003). A complete phylogenetic analysis of the genus and its close relatives is needed to test this hypothesis, however,

Some tadpoles that either hatch in or are carried to small pools of water in plant parts, such as bromeliads, leaf axils, and seed husks (phytotelmata), complete development with little or no independent feeding (see also the previous "Phytotelmonous Tadpoles"). For example, the hemiphractine hylid frog *Flectonotus goeldii* carries eggs in a dorsal brood pouch (see chapters 10 and 11). After the eggs have hatched, the mother deposits tadpoles in bromeliads, where they complete their development. The tadpoles appear to be capable of completing development without feeding, although they have been observed to ingest undeveloped eggs and

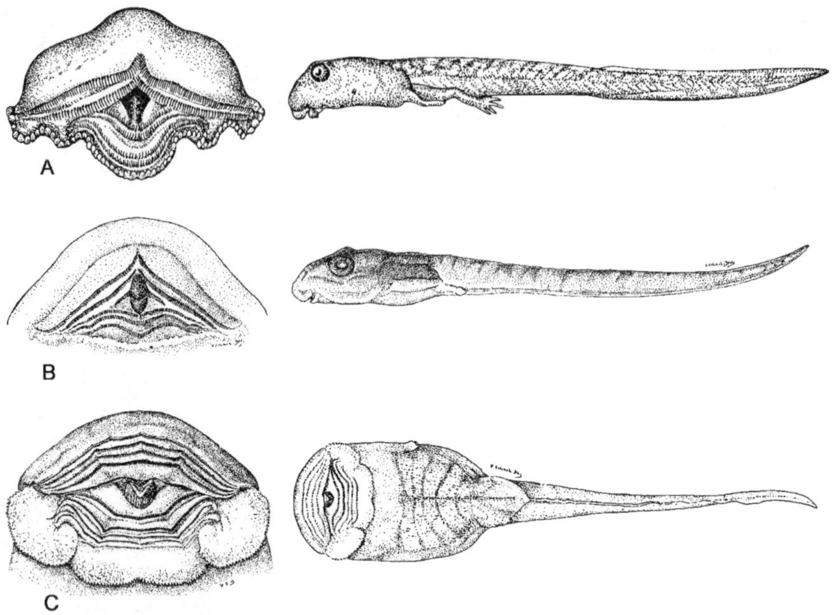

**Fig. 12.24.** Semiterrestrial tadpoles. (A) Lateral view and mouthparts of *Cycloramphus brasiliensis* (Leptodactylidae), a tadpole adapted for living on wet rocks. Lateral projections of skin on the belly provide additional surface for adhering to the surface. (B) Tadpole of *Thoropa miliaris* (Leptodactylidae). (C) Tadpole of *Petropedetes natator* (Petropedetidae). (A) Drawing by Mary Jane Spring after Heyer (1983). (B, C) After Lamotte and Lescure (1989a).

dead or dying tadpoles in captivity (Weygoldt 1989b; Weygoldt and de Carvalho e Silva 1991). Tadpoles of *F. goeldii* lack keratinized labial teeth on the oral disc. The branchial baskets are small and poorly developed, and some branchial muscles are weakly developed or missing. The tadpoles also lack the branchial food traps and gill filter rows that are characteristic of free-living tadpoles in the closely related genus *Gastrotheca* (Haas 1996). The morphological features that are missing or poorly developed in *F. goeldii* are those that normally appear late in the development of *Gastrotheca* tadpoles, suggesting that development has been truncated in *Flectonotus*.

## Social Behavior of Amphibian Larvae

Adult amphibians show little evidence of true sociality, and virtually all behavioral interactions between them, except for courtship and mating, are competitive (see chapters 8 and 9). Most amphibian larvae are relatively solitary as well, and many appear to actively avoid contact with other individuals. This is especially true of salamander larvae, which tend to be aggressive toward other larvae and sometimes are cannibalistic (Mathis et al. 1995; Blaustein and Walls 1995; see also the earlier "Cannibalism and Cannibalistic Morphs"). In some anuran tadpoles, however, there is evidence of social attraction, coordinated behavior among individuals, and even limited cooperation. Often these interactions are mediated by recognition of close kin (Blaustein and Walls 1995; Hoff et al. 1999).

### The Natural History of Tadpole Aggregations

The occurrence of dense aggregations of anuran tadpoles in temporary ponds was reported by a number of early workers (e.g., Abbott 1884; S. Ball 1936; Richmond 1947), but it was mainly the work of Arthur N. Bragg on spadefoot toads in Oklahoma that called attention to the social behavior of tadpoles. In a long series of papers starting in the early 1940s, Bragg described a number of types of aggregations in several species of spadefoot toads (summarized in Bragg 1965). Some of these aggregations were enormous, containing hundreds of thousands of tadpoles. In one pond, for example, he observed a feeding aggregation nearly 7 m long, 1 m wide, and several centimeters deep, with the entire mass of tadpoles moving slowly in the same direction along the bottom of the pond (Bragg 1957a). He also described huge aggregations of thousands of metamorphosing juveniles emerging synchronously from the receding waters of temporary pools (Bragg 1965). Later work revealed that species in at least 10 families of anurans have tadpoles that form aggregations (table 12.2 and reviews by Lescure 1968; Wassersug 1973; Caldwell 1989; Blaustein and Walls 1995).

Much of the early literature on tadpole aggregations

**Table 12.2** Examples of social aggregations in anuran tadpoles.

| Species (source no.) | Pond type | Tadpole color | Type of aggregation |
|---|---|---|---|
| Pipidae | | | |
| *Xenopus laevis* (11, 20) | Permanent and temporary | Light, nearly transparent | Polarized midwater schools; coordinated filter feeding |
| Rhinophrynidae | | | |
| *Rhinophrynus dorsalis* (5) | Temporary | Light, nearly transparent | Polarized schools on pond bottom; coordinated movement in feeding swarms |
| Microhylidae | | | |
| *Chiasmocleis ventrimaculata* (25) | Permanent and temporary | Gray | Flat, wide schools on surface; regular spacing and parallel orientation |
| *Hamptophryne boliviana* (25) | Permanent and temporary | Dark brown above; gray below | Flat, wide schools on surface; regular spacing and parallel orientation |
| *Phrynomantis annectens* (9, 14) | Permanent and temporary | Black and gold | Polarized midwater school; parallel orientation and coordinated movement |
| *P. microps* (32) | Temporary | Black and gold | Aggregations of different sizes; tadpoles not always segregated by size; some parallel orientation in windy conditions |
| Pelobatidae | | | |
| *Scaphiopus* and *Spea* (2, 3, 6) | Temporary | Black | Stationary aggregations of thousands of tadpoles on bottom; parallel orientation and coordinated movement in feeding swarms |
| Bufonidae | | | |
| *Bufo americanus* (12, 16, 17, 23) | Temporary | Black | Large stationary aggregations on bottom with random spacing; some coordinated movement when feeding |
| *B. boreas* (27) | Permanent | Black | Huge stationary aggregations on bottom; some coordinated movement when feeding |
| *B. crucifer* (36) | Temporary | Black | Dense moving aggregations in shallow water |
| *B. maculatus* (35) | Temporary | Black | Dense stationary aggregations in shallow pools and large moving aggregations in deeper water, with some parallel orientation |
| *B. rufus* (33) | Temporary | Dark brown | Stationary or moving aggregations of variable size |
| *B. woodhousii* (21) | Temporary | Black | Large stationary or slow-moving aggregations; some parallel orientation and size segregation |
| *Schismaderma carens* (7, 9) | Temporary | Black | Densely packed tadpole "balls" in midwater; lower density feeding groups on bottom |
| Hylidae | | | |
| *Hyla geographica* (8, 10, 18, 29, 37) | Permanent | Black | Densely packed tadpole "balls" or flat aggregations; rolling or swarming movement without parallel orientation; segregation by size |
| *Phyllomedusa vaillanti* (24) | Permanent | Gold | Dense, highly mobile polarized schools in midwater, on surface, or on bottom; segregation by size |
| Leptodactylidae | | | |
| *Leptodactylus insularum* (28) | Temporary | Brown and gold | Densely packed swarms with coordinated movement led by mother |
| *L. chaquensis* (18) | Temporary | | Dense aggregations |
| *L. melanonotus* (11) | Temporary | Black | Dense aggregations |
| *L. ocellatus* (13) | Temporary | Brown | Densely packed swarms with coordinated movement led by mother |
| *L. wagneri* (25) | Temporary | Dark | Polarized aggregations on bottom with parallel orientation and coordinated movement |

(*continued*)

**Table 12.2** (continued)

| Species (source no.) | Pond type | Tadpole color | Type of aggregation |
|---|---|---|---|
| Ranidae | | | |
| *Pyxicephalus adspersus* (4, 9) | Temporary | Black | Dense aggregations with coordinated movement accompanied by father |
| *Rana cascadae* (27) | Permanent and temporary | Brown | Small stationary aggregations on bottom |
| *R. heckscheri* (1, 18) | Permanent | Black and gold | Dense schools with coordinated rolling movement |
| *R. sylvatica* (26) | Temporary | Dusky, greenish gray | Usually not aggregated; occasional polarized schools at high density |
| *R. temporaria* (34) | Temporary | Brown to black | Moving or stationary aggregations of variable size |
| Rhacophoridae | | | |
| *Philautus cherrapunjiae* (11) | | Black | Compact, irregular schools |

*Sources:* (1) A. Wright 1932; (2) S. Ball 1936; (3) Richmond 1947; (4) B. Balinsky and Balinsky 1954; (5) L. Stuart 1961; (6) Bragg 1965; (7) Wager 1965; (8) Kenny 1969c; (9) Van Dijk 1972; (10) Duellman and Lescure 1973; (11) Wassersug 1973; (12) Beiswenger 1975; (13) Vaz-Ferreira and Gehrau 1975; (14) Channing 1976b; (15) Test and McCann 1976; (16) Beiswenger 1977; (17) Duellman 1978; (18) Cei 1980; (19) Altig and Christensen 1981; (20) Katz et al. 1981; (21) Breden et al. 1982; (22) O'Hara and Blaustein 1982; (23) Waldman 1982) (24) Branch 1983; (25) Schlüter 1984; (26) Waldman 1984; (27) O'Hara and Blaustein 1985; (28) Wells and Bard 1988; (29) Caldwell 1989; (30) Punzo 1991; (31) Punzo 1992; (32) Rödel and Linsenmair 1997; (33) Eterovick and Sazima 1999; (34) Nicieza 1999; (35) Spieler and Linsenmair 1999; (36) Eterovick 2000; (37) Wells, unpublished.

focused on the proper nomenclature for these groups. Bragg (1965) distinguished between asocial and social groups. The former resulted from tadpoles aggregating in response to some feature of the environment, such as temperature gradients or a temporary concentration of food; the latter resulted from the mutual attraction of individual tadpoles. Other workers have used a variety of terms to describe tadpole groups, and Wassersug (1973) attempted to standardize the terminology. He classified groups as simple aggregates and biosocial aggregates, which were equivalent to Bragg's categories. He further subdivided them into polarized and nonpolarized groups, the former being those in which individuals show parallel orientation in the same direction. Wassersug restricted the term "school" to biosocial aggregates showing some degree of coordinated group movement, but the degree of polarization in schools can vary considerably. Some ichthyologists have used the term "shoal" to refer to aggregations of fishes, restricting the term "school" to polarized groups exhibiting coordinated movement (Pitcher and Parrish 1993).

The diversity of tadpole aggregations is evident from table 12.2. This is not an exhaustive list; other species are briefly described in natural history accounts as being gregarious, but the details of group formation and behavior are unknown. The groups that most closely resemble schools of fishes are the highly polarized midwater schools formed by planktivorous tadpoles of *Xenopus* (Wassersug 1973; L. Katz, Potel, and Wassersug 1981; Wassersug, Frogner, and Inger 1981). These form groups characterized by rather even spacing of individuals, parallel orientation, and coordinated movement in one direction. Very similar behavior occurs in the African microhylid *Phrynomantis annectans,* also a midwater planktivore (Van Dijk 1972; Channing 1976b). Other microhylid tadpoles, including *Chiasmocleis ventrimaculata, Hamptophryne boliviana,* and *Phrynomantis microps,* typically hang just below the surface in large aggregations with regular spacing between individuals (Schlüter 1984; Rödel and Linsenmair 1997). Tadpoles of the Mexican burrowing frog, *Rhinophrynus dorsalis,* are morphologically similar to those of *Xenopus* and sometimes form polarized schools. Apparently they feed mostly on small particulate matter, but can feed on small invertebrates as well (Starrett 1960; L. Stuart 1961; Wassersug 1973 and personal communication).

Hylid tadpoles in the subfamily Phyllomedusinae also are similar to *Xenopus* tadpoles in morphology and feeding ecology, being adapted for swimming and hovering in the water column (Wassersug 1973). At least some species also are gregarious. For example, *Phyllomedusa vaillanti* of Amazonian Brazil lays eggs on leaves overhanging ponds. Tadpoles tend to hatch synchronously (Schlüter 1984) and remain in the spot where they land in the water until a group is formed (L. Branch 1983). They travel in polarized schools that are segregated by size and therefore may represent groups of siblings from single egg clutches. Sometimes schools merge during feeding bouts, but quickly separate. This species exhibits a variety of feeding modes, including grazing on the bottom, midwater filter-feeding, and feeding from the surface film, but in all cases tadpoles are gregarious and exhibit coordinated movements (L. Branch 1983).

Tadpoles of spadefoot toads (*Scaphiopus* and *Spea*) and many true toads (*Bufo*) form stationary aggregations on the bottom of shallow ponds; these groups range in size from a few dozen to hundreds of thousands of individuals (fig. 12.25 A). Groups of spadefoot toad tadpoles tend to reach very high densities because they typically breed in ephemeral rain pools that rapidly decrease in size (Bragg 1965). Individuals in stationary aggregations often show little regular-

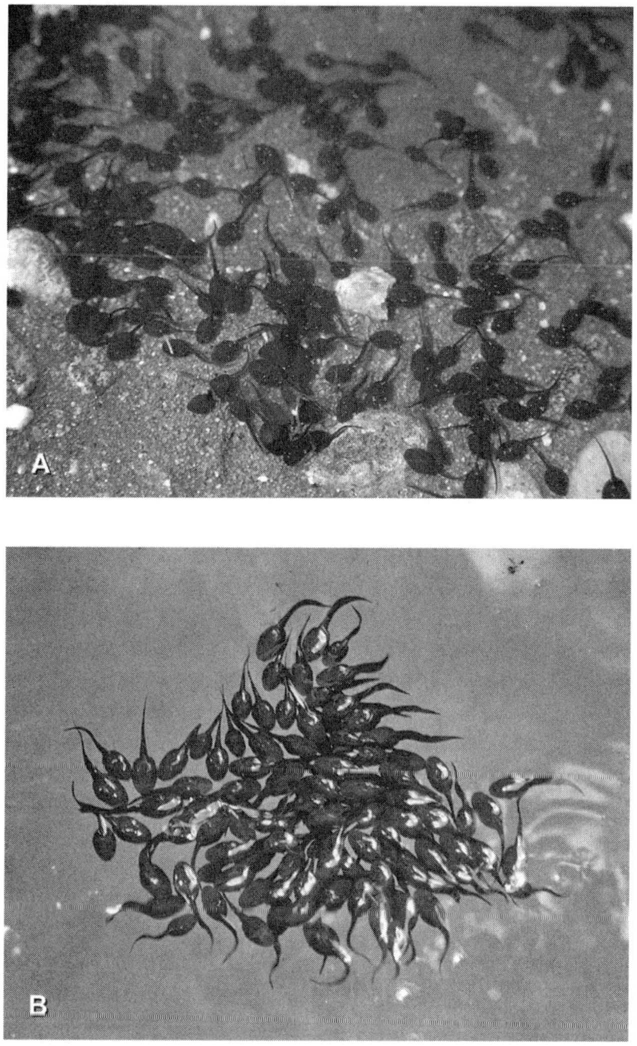

**Fig. 12.25.** (A) Aggregation of American toad tadpoles (*Bufo americanus*) in a shallow pond. These tadpoles often form dense aggregations in shallow water during sunny weather, but then move to deeper water at night. (B) School of *Hyla geographica* tadpoles in a river in Trinidad. These tadpoles form tight balls and swim together in a coordinated manner, probably for protection against predators. Photos by Kentwood D. Wells.

ity in spacing or orientation. In some species, there is a statistical tendency for similar-sized individuals to be associated with one another and to orient in the same direction, but this is not obvious from simple observation (Wassersug, Frogner, and Inger 1981; Breden, Lum, and Wassersug 1982). In both groups, polarized schools are formed when the tadpoles are actively feeding. Usually this involves substrate feeding, with the entire mass of tadpoles moving in the same direction, churning up the pond bottom (Richmond 1947; Bragg 1965; Beiswenger 1975), but occasionally they will feed on suspended particles or plankton in the water column (Bragg 1965; Test and McCann 1976). Many species also are highly gregarious immediately before and immediately after metamorphosis (Bragg 1948, 1950, 1952, 1965; Bragg and Brooks 1958; Black and Black 1969; Lillywhite and Wassersug 1974; S. Arnold and Wassersug 1978).

Some of the most impressive social groups of tadpoles are the densely packed tadpole balls formed by some bufonids and hylids. In these groups, tadpoles do not show parallel orientation, but instead form a seething mass that tends to roll through the water. Often the tadpoles are in direct bodily contact, forming an impressive aggregation that is very conspicuous in open water. This type of aggregation has been described in the African toad *Schismaderma carens* (sometimes placed in the genus *Bufo;* Wager 1965; Van Dijk 1972). Very similar aggregations occur in *Hyla geographica* (fig. 12.25 B; Kenny 1969a; Duellman and Lescure 1973; Duellman 1978; Caldwell 1989). Aggregations of this species observed by Caldwell (1989) in Brazil usually contained tadpoles in a relatively narrow range of sizes, suggesting that each group represents a single clutch of eggs, or offspring of several clutches that hatched at about the same time. When swimming in midwater, the tadpoles form tight balls, but they sometimes spread out into flat schools on the surface. Often the tadpoles in these aggregations were facing inward toward the center of the group, as if trying to swim toward the middle. I have observed similar behavior in the same species in a river in Trinidad. These tadpoles sometimes spread out into a flat school while feeding on the bottom, but formed balls in open water (my unpublished observations).

Several species of leptodactylid frogs have gregarious tadpoles (table 12.2), and in several species, parents accompany the schools. Tadpoles of *Leptodactylus insularum* often form very dense schools with individuals in close bodily contact. These schools are attended by the female, which actively leads the tadpoles through shallow pools, communicating with them by means of tactile or chemical cues (Wells and Bard 1988; see further discussion in chapter 11). Female *L. ocellatus* accompany groups of gregarious tadpoles, and females also have been observed directing the movements of tadpole schools (Vaz-Ferreira and Gehrau 1975; J. Caldwell, personal communication). Tadpoles of the closely related species *L. chaquensis* also are gregarious (Cei 1980), but parental attendance has not been reported. Very similar behavior has been described in the African ranid, *Pyxicephalus adspersus,* with the male rather than the female accompanying the tadpole school (Kok, du Preez, and Channing 1989).

Although most other ranid tadpoles seem to be solitary, there is evidence of limited sociality in some species. Tadpoles of *Rana cascadae* often are found in small stationary aggregations, although there is little indication of coordinated behavior or complex spatial organization (O'Hara and Blaustein 1985; Blaustein and O'Hara 1986). Wood frog (*Rana sylvatica*) tadpoles generally are not gregarious, but Waldman (1984) reported that they sometimes form polarized feeding schools in ponds with very high population densities. The most social North American ranid species appears to be the river frog, *Rana heckscheri,* which forms dense schools and has coordinated feeding behavior (Altig and Christensen 1981; Punzo 1991, 1992). A. Wright (1932), who first described this species, reported that the behavior of the tadpoles was more similar to that of *Bufo* tadpoles than to that of other species of *Rana.* At least two Old World ranids, *Rana chalconota* and *Aubria subsigillata,* are reported to form compact schools (Schiøtz 1963; Wassersug 1973), as do tadpoles of an Asian rhacophorid, *Philautus cherrapunjiae* (Roonwal and Kripalani 1961).

Despite considerable variation in the details of social behavior, gregarious species have several characteristics in common. Most obvious is that many are visually conspicuous, often being either black or bright gold (table 12.1). The exceptions are the midwater schoolers, which tend to have light coloration or are nearly transparent. Some degree of size segregation has been reported in several species (*Bufo woodhousii, Hyla geographica, Phyllomedusa vaillanti*) suggesting a tendency for siblings from the same clutch to group together (see the following discussion). Finally, some species tend to be more gregarious during the day than at night. For example, tadpoles of *Bufo americanus, Phyllomedusa vaillanti, Phrynomantis annectans,* and *Phrynomantis microps* are highly social during the day, but are randomly distributed throughout the pond at night (Channing 1976b; Beiswenger 1977; L. Branch 1983; Rödel and Linsenmair 1997). Aggregation behavior in *Phrynomantis* is affected by water clarity, with larger and denser aggregations occurring in clear water, where tadpoles probably are most vulnerable to predators (Spieler 2003). Other species exhibit tighter parallel orientation or more even spacing during the day than at night (L. Katz, Potel, and Wassersug 1981; Wassersug, Frogner, and Inger 1981). This is not universally true, however. Bragg (1948) observed dense feeding schools of spadefoot toad tadpoles at night, and *Hyla geographica* tadpoles are reported to remain gregarious at night as well (L. Branch 1983). *Leptodactylus* tadpoles attended by a parent also are highly gregarious at all times (Vaz-Ferreira and Gehrau 1975; Wells and Bard 1987).

### Benefits of Aggregation Behavior

Many authors have speculated on the adaptive significance of tadpole aggregations, but evidence for any hypothesis is scarce. Some explanations seem rather implausible. For example, both R. Savage (1961) and Bragg (1965) proposed that grouping is advantageous because large stationary aggregations reduce the oxygen and increase the carbon dioxide content of the water near the tadpoles, and this in turn may help to repel predators such as water beetles. However, the costs of reduced oxygen levels to the tadpoles themselves were not considered. Richmond (1947) suggested that schooling helps reduce the frequency of collision between swimming tadpoles, but the advantage of this is unclear. Most explanations have been more reasonable, focusing on benefits such as increased feeding efficiency or protection against predators, the same benefits proposed for group-living fishes (Pitcher and Parrish 1993).

#### Feeding Benefits

The benefits of cooperative feeding often are invoked to explain the polarized substrate-feeding schools observed in species such as *Bufo, Spea,* and *Scaphiopus.* Such schools frequently form when tadpole densities are high and pond productivity low. In temporary rain pools, macrophytic plants and phytoplankton often are scarce, and the only organic matter available to the tadpoles is in the bottom mud. Cooperative schools might enhance feeding efficiency by churning up the substrate and exposing food not available to individual tadpoles (Richmond 1947; Bragg 1954b, 1965; Bragg and King 1960; L. Stuart 1961; Beiswenger 1975; Waldman 1984, 1991; Blaustein and Walls 1995). Aggregation also might be favored if it enabled tadpoles to displace competing species that do not aggregate. There are anecdotal reports, for example, of small groups of *Pseudacris regilla* tadpoles being displaced by large aggregations of *Bufo boreas* tadpoles (Blaustein and Walls 1995). A similar explanation for schooling behavior has been proposed for some coral reef fishes (Robertson et al. 1976).

Although the hypothesis that aggregation enhances feeding seems reasonable, there is little experimental evidence to either support or refute it. Wilbur (1977a) did find that increased densities of *Bufo americanus* tadpoles had a positive effect on size at metamorphosis, but only at rather low initial densities. In laboratory experiments, tadpoles of this species had the largest mass at metamorphosis when all individuals in a tank had access to a common food resource (Breden and Kelly 1982). This could have been due to social facilitation in breaking up large food particles or to nutritional benefits derived from the feces of other individuals (Steinwascher 1978b). Similar enhancement of growth in interacting tadpoles compared to isolated ones has been observed in *Xenopus laevis* (L. Katz, Potel, and Wassersug 1981) and *Rhynophrynus dorsalis* (Foster and McDiarmid 1982). These experiments do not directly demonstrate that large aggregations enhance feeding efficiency under natural conditions, but they do suggest that limited social interactions benefit individuals even in the absence of predators. Punzo (1991) reported that both learned avoidance of noxious stimuli and coordinated swimming were enhanced in social groups of *Rana heckscheri* tadpoles, but he did not directly address the question of facilitated feeding behavior, which has been suggested, but not demonstrated experimentally for this species (Altig and Christensen 1981). In a study of *Bufo crucifer* tadpoles in Brazil, Eterovick (2000) found that the tadpoles ate more heterotrophic organisms (protozoans, ciliophores, and crustacean eggs) when feeding in aggregations than when feeding alone, but the nutritional benefits of this change in feeding behavior, if any, are not known.

### Protection against Predators

Protection against predators is perhaps the most common explanation for the formation of schools, flocks, and herds by animals (Wittenberger 1981; Pitcher and Parrish 1993). Several types of antipredator benefits have been proposed for tadpole aggregations, but again, there is little experimental support for any of these ideas. Some authors have simply suggested that predators are reluctant to enter tadpole schools, although the reasons for this are not clear. For example, Black (1970) reported that beetle larvae commonly fed on tadpoles of *Spea bombifrons* on the periphery of schools, but they were not seen to enter the aggregations themselves. Bragg (1965) observed beetle larvae turning away when they reached the edge of a tadpole school. Carpenter (1953a) thought that aggregation by *Rana pretiosa* tadpoles might reduce attacks by leeches or somehow facilitate their removal. Both Bragg (1965) and Black (1970) proposed that aggregations provide protection against intraspecific cannibalism, because cannibalistic tadpoles were observed feeding on isolated conspecifics, but not on those in aggregations. This hypothesis seems at odds with other data showing that cannibalism is most likely to occur when population densities are high (see the following "Cannibalism and Kin Recognition").

Some tadpoles form more cohesive aggregations in the presence of predators or chemical cues derived from predators. For example, when tadpoles of European common toads (*Bufo bufo*) were presented with chemical cues from predatory fishes, they formed more cohesive groups than in the absence of such cues (Watt, Nottingham, and Young 1997). Tadpoles of the African microhylid *Phrynomantis microps* formed aggregations in the presence of dangerous predators, such as aquatic insects, but not in the presence of harmless insects or other species of tadpoles (Rödel and Linsenmair 1997). Aggregation behavior also was induced by presenting tadpoles with chemicals derived from injured conspecifics, suggesting a form of alarm response in this species. In another African species, *Bufo maculatus,* tadpoles formed aggregations only in response to both chemical stimuli from injured conspecifics and mechanical stimuli produced by movements of predators (Spieler and Linsenmair 1999). In many other tadpoles, the typical reaction to chemicals released by injured tadpoles is to flee from the source of the chemical or reduce swimming activity, but not to form aggregations (Hrbâçek 1950; Kulzer 1954; Pfeiffer 1966, 1974; Hews and Blaustein 1985; Hews 1988; Nicieza 1999; see chapter 14 for further discussion of alarm responses to predators). *Bufo maculatus* tadpoles showed similar avoidance responses to injured tadpoles in the absence of mechanical stimuli from swimming predators.

Aggregated animals sometimes benefit from a "selfish herd" effect, or from satiation of local populations of predators (W. D. Hamilton, 1971). The best evidence for such an effect in tadpoles is the study by Watt, Nottingham, and Young (1997) of *Bufo bufo* tadpoles. They found that attacks by predatory fish increased as the size of a tadpole aggregation increased, but the per capita attack rate declined (fig. 12.26). Unfortunately, because of ethical concerns, they were not able to observe actual predation rates on groups of different sizes, but the inference from the experiment is that individual tadpoles are safer in larger groups. Kehr and Schnack (1991) reported that tadpoles of *Bufo arenarum* were least vulnerable to predation by giant water bugs (*Belostoma*) at high tadpole densities, but it was not clear whether this was due to predator satiation or an effect of aggregation by tadpoles on prey capture success by the predators.

Aggregated tadpoles might benefit from more rapid detection of predators or confusion effects. Schooling tadpoles approached by a predator typically fan out and swim in several directions, although some degree of coordinated movement is maintained (L. Branch 1983; personal observations on *Hyla geographica*). Tadpoles near the center of an aggregation might benefit from rapid transmission of a fright reaction initiated by those at the periphery swimming away from a predator (Treherne and Foster 1981). In some species, the benefits of grouping may continue beyond metamorphosis. It has been proposed that synchronized metamorphosis

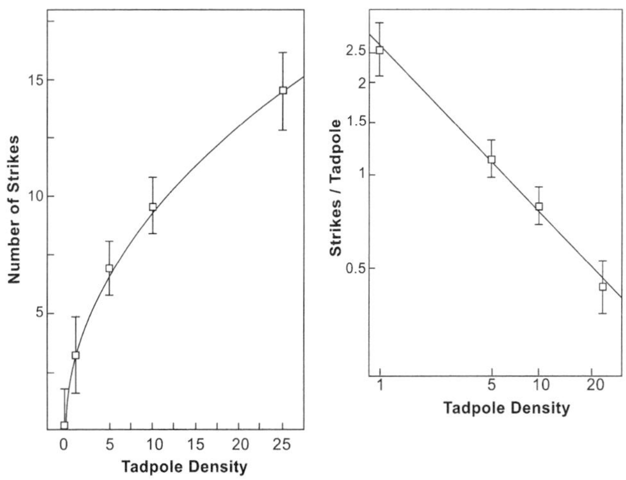


**Fig. 12.26.** The effect of group size on attack frequency by predators of *Bufo bufo* tadpoles. The number of predator attacks increased with increasing tadpole density (left), but per capita risk decreased with increasing density (right). After P. Watt, Nottingham, and Young (1997).

and postmetamorphic aggregations in some species serve to swamp predators and thereby reduce individual risk (Beiswenger 1975; Arnold and Wassersug 1978).

Finally, if tadpoles are distasteful, then aggregation might enhance protection against predators by providing a more conspicuous aposematic signal (Wassersug 1973; Peterson and Blaustein 1991). The conspicuously black coloration of many social tadpoles (table 12.1) has been cited as support for this hypothesis, but direct evidence for an aposematic function of black coloration is scarce. Certainly *Bufo* tadpoles are distasteful to some predators, although the effectiveness of noxious chemicals depends on the developmental stage of the tadpole (Formanowicz and Brodie 1982; Glandt 1984) and the feeding behavior of the predator (chapter 14). The palatability of most schooling tadpoles is not known (Wassersug 1973), but there is little evidence that tadpoles of spadefoot toads are distasteful. In fact, they appear to be preferred prey for some natural predators, and their best protection is for females to lay eggs in temporary ponds where predators are scarce (Woodward 1983).

#### Physiological Benefits

Some species of tadpoles tend to aggregate in the warmest parts of ponds, except when temperatures reach lethal values (Mullally 1953; Brattstrom 1962c; Karlstrom 1962; Bragg 1965; Beiswenger and Test 1966; Beiswenger 1977). Brattstrom (1962c) found that a pan of water containing a large group of tadpoles heated more quickly in the sun than a pan with water alone, leading him to suggest that aggregation serves a thermoregulatory function, which would be enhanced by the black color of many species. However, Brattstrom's experiment is not convincing, because it did not allow for convective cooling resulting from water currents or movements of the tadpoles themselves. Beiswenger (1977) found that screened enclosures in ponds with tadpoles actually heated more slowly than did empty enclosures because of the currents set up by moving tadpoles. Bragg (1962) and Black and Black (1969) also proposed physiological benefits to explain the formation of metamorphic aggregations. They suggested that clustering might reduce desiccation of tadpoles emerging from exposed ponds. Lillywhite and Wassersug (1974) considered this unlikely, but did not present data to refute the hypothesis; they concluded that aggregation in juveniles might be a nonadaptive continuation of larval behavior patterns.

### Costs of Aggregation Behavior

Most authors have given scant attention to the possible costs of group living in tadpoles. In any animal, the frequency of group formation, the sizes of groups, and the spatial and social relationships within groups represent a balance between conflicting selective pressures (Bertram 1978; Pulliam and Caraco 1984; Magurran and Bendelow 1990; Richardson 1994). Furthermore, the costs and benefits of group living often are two sides of the same coin. For example, schooling can be an effective defense against certain types of predators, but it also can increase exposure to other predators or induce a compensating change in their behavior (e.g., Major 1977, 1978). Group foraging may increase feeding efficiency under some circumstances, but result in increased competition if the distribution of resources changes. Consequently, it is impossible to evaluate the costs and benefits of group living solely on the basis of casual natural history observations; careful experimental studies are needed as well. The key issue is whether the per capita benefit to individuals increases with increasing group size, even if larger groups attract predators or suffer increased competition for food.

#### Competition for Resources

One potential cost of group living is increased competition for resources, including oxygen and food. As aquatic oxygen levels decrease, tadpoles switch to aerial respiration (see chapter 4). Synchronized aerial breathing, with many tadpoles swimming simultaneously to the surface, has been reported in a number of social species (L. Stuart 1961; Wager 1965; Wassersug 1971, 1973; L. Branch 1983). However, it is not clear that a social aggregation would deplete the oxygen in a pond more quickly than would an equally large number of dispersed tadpoles. Similarly, the effects of social aggregation on competition for food is not well understood. Increased population density often leads to decreased individual growth rates and smaller size at metamorphosis (see chapter 13). Presumably this is due largely to exploitative competition for food, although there is limited evidence for interference competition as well. For example, large tadpoles may push smaller tadpoles aside and interfere with their feeding (R. Savage 1952; Gromko, Mason, and Smith-Gill 1973; John and Fenster 1975; Wilbur 1977a; Faragher and Jaeger 1998). Probably the total density of tadpoles in a pond affects the intensity of exploitative competition, but not necessarily by the degree of aggregation. In contrast, interference competition probably is most intense in aggregating species (R. Griffiths and Foster 1998).

#### Risk of Predation

The second major cost of group living is likely to be an increased risk of predation. Although aggregation may decrease the risk of predation under some circumstances, conspicuous groups also could attract more predators, and this would be particularly costly if the tadpoles are not distasteful to all types of predators. Information bearing on this point is largely anecdotal. For example, with reference to *Schismaderma carens*, Wager (1965, 110) stated that "Predators love this gregarious habit: dragonfly larvae dive into the mass and emerge immediately with a struggling victim; a bird such as a Hammerhead will eat hundreds in a single meal, and a Water Tortoise will take them by the mouthful." I once observed a turtle feeding on a dense aggregation *of Leptodactylus insularum* tadpoles in Panama (unpublished observations), and several authors have reported predation by birds on conspicuously aggregated tadpoles (Wassersug 1971; Ideker 1976; Beiswenger 1981). In *Pyxicephalus adspersus* and *Leptodactylus ocellatus,* attending parents are reported to attack vertebrate predators (B. Balinsky and Balinsky 1954; Vaz-Ferreira and Gehrau 1975). Such behavior could be an evolutionary response to the increased vulnerability of aggregated tadpoles.

Cannibalism by conspecifics tends to occur mostly in tadpoles that form aggregations (see previous discussion of "Cannibalism"), but it is not clear that this represents a cost of aggregation per se. Black (1970) reported that cannibalistic *Spea bombifrons* tadpoles were present only in ponds where tadpoles were in large feeding schools. He interpreted this as evidence that schools were formed as protection against cannibals. However, it also is possible that cannibalism is an opportunistic response to the availability of densely aggregated prey, or that both cannibalism and schooling are independent responses to food shortage or high population density (see D. Pfennig 1992a, b).

## Kin Recognition in Tadpoles

The occurrence of size-specific schools of tadpoles in species such as *Hyla geographica* and *Phyllomedusa vaillanti* strongly suggests that schools are formed from single egg clutches, and this almost certainly is the case for *Leptodactylus* tadpoles attended by the mother and *Pyxicephalus* tadpoles attended by the father. However, direct evidence that tadpoles can recognize close kin comes not from these tropical species, but from work on North American ranids, bufonids, and pelobatids. Kin recognition was first demonstrated in *Bufo americanus* tadpoles by Waldman and Adler (1979) and subsequently described in several other species by Waldman, Blaustein and O'Hara, and several other investigators. Much of the work by these authors has involved rearing tadpoles under various conditions in the laboratory and testing their ability to distinguish siblings from nonsiblings. Most investigators have used some measure of spatial association to determine whether tadpoles can recognize their kin. In Waldman's early experiments, sibships were colored with different vital stains and then placed in equal numbers in a single tank. Kin recognition was tested by analysis of the spatial distribution of the colored tadpoles. In most other studies, individual tadpoles were placed in a tank with groups of different sibships in chambers at either end. Kin recognition was tested by measuring the amount of time the tadpoles spent in each end of the tank. Later studies included tests for kin association in more realistic field conditions (Waldman 1982b; O'Hara and Blaustein 1985; Hokit and Blaustein 1997), as well as investigations of kin recognition in relation to cannibalism by tadpoles (D. Pfennig, Reeve, and Sherman 1993; D. Pfennig and Frankino 1997; D. Pfennig 1999). There have been several comprehensive reviews of kin recognition in tadpoles (Waldman 1986a, 1991; Blaustein, Beckoff, and Daniels 1987a, b; Blaustein 1988; Blaustein and Waldman 1992; Blaustein and Walls 1995; Hoff et al. 1999). I will give only a brief summary of this work here.

### Development of Kin Recognition

In all of the species that exhibit any sort of kin recognition, tadpoles raised exclusively with siblings associated preferentially with sibs over nonsibs, although in *Rana aurora* this was true only for very young tadpoles. In a few species, including *Pseudacris crucifer, P. regilla, Rana pipiens,* and *R. pretiosa,* tadpoles reared only with siblings did not exhibit any tendency to associate with close relatives (O'Hara and Blaustein 1988; Fishwald et al. 1990). None of these species is known to aggregate with conspecifics in the field.

The effect of exposure to nonsibs during development varies. In *Bufo americanus,* tadpoles reared in mixed groups did not show a preference for familiar sibs over familiar nonsibs in laboratory tests, but did exhibit such preferences in the field (Waldman 1981, 1982b). When tadpoles of *B. americanus* were exposed to sibs before being reared in a mixed group, they retained their ability to recognize siblings. In contrast, when tadpoles were exposed first to mixed groups and later to siblings, they did not exhibit a preference for siblings over familiar nonsibs. These results suggest that experience early in the larval period affects kin recognition. One might predict that when the eggs of several different females are laid in close proximity, as sometimes occurs in this species, tadpoles would exhibit weaker preferences for siblings than those that hatch from isolated egg clutches. Exposure to siblings is not essential for kin recognition, however, because tadpoles reared in isolation also exhibited a preference for associating with sibs (Waldman 1981). In *Bufo boreas,* tadpoles exposed only to siblings exhibited a preference for familiar sibs over unfamiliar nonsibs. In contrast to *B. americanus,* however, subsequent exposure to mixed groups of tadpoles eliminated this preference (O'Hara and Blaustein 1982). This species often occurs in aggregations of hundreds of thousands of individuals, many of which are not related to each other, so it seems unlikely that tadpoles of this species commonly associate with close relatives in the field. Tadpoles of a toad from India, *Bufo melanostictus,* exhibited a preference for associating with sibs in laboratory experiment and could recognize both familiar and unfamiliar kin regardless of rearing conditions (Saidapur and Girish 2000).

The development of kin recognition in two ranids, *Rana cascadae* and *R. sylvatica,* appears to be somewhat different from that seen in the toads. *Rana cascadae* tadpoles could distinguish unfamiliar sibs from unfamiliar nonsibs when reared only in sib groups, in mixed groups, or in social isolation (Blaustein and O'Hara 1981, 1982a, 1983). In laboratory tests, tadpoles reared in mixed groups did not distinguish between sibs and nonsibs from their own rearing groups (O'Hara and Blaustein 1981), but they did exhibit a tendency to associate with familiar sibs over familiar nonsibs in field trials (O'Hara and Blaustein 1985). Taken together, these results suggest that early exposure to siblings is not required for kin recognition, and exposure to nonsibs does not eliminate a preference for associating with sibs. Tadpoles collected from natural aggregations in the field preferred to associate with members of their own aggregations rather than tadpoles from other aggregations (Blaustein and O'Hara 1987). This species forms relatively small aggregations that tend to remain in the same general area for long periods of time. Genetic relationships in these aggregations have not been determined, but it is possible that they are composed largely of siblings from the same egg clutch.

Tadpoles of *R. sylvatica* differ from those of *R. cascadae* in their social behavior. They seldom form aggregations in the field, and when they do so, the aggregations often are composed of large numbers of individuals that could not all be from the same clutch (Waldman 1984). Nevertheless, tadpoles of this species did exhibit some tendency to associate with sibs in laboratory tests, whether they were raised only with sibs or in mixed groups (Waldman 1984; Cornell, Berven, and Gamboa 1989; Fishwald et al. 1990; Rautio et al. 1991). These tadpoles also tended to associate more with familiar sibs with whom they had been reared than with unfamiliar sibs reared in separate tanks (Waldman 1984). These results suggest that familiarity with certain individuals can affect the way in which tadpoles associate with other individuals. Developmental stage can affect kin recognition as well. Rautio et al. (1991) found that 10-day-old *R. sylvatica* tadpoles did not exhibit kin recognition, whereas tadpoles that were at least 17 days old did. These authors also found evidence that tadpoles were positively attracted to siblings, but did not avoid nonsiblings. This contrasts with another set of tests conducted by Waldman (1986a). He observed spacing patterns of *R. sylvatica* tadpoles in laboratory pools and found that they seemed to avoid unrelated conspecifics, rather than being attracted to closely related individuals.

There is limited evidence for some species that kin recognition persists in postmetamorphic anurans. When *Rana cascadae* juveniles were tested in a Y-tube apparatus, they tended to move toward the odors of siblings than toward nonsibs. This preference persisted for up to 47 days after metamorphosis (Blaustein, O'Hara, and Olson 1984). In *R. sylvatica,* juveniles tested within 24 hours of metamorphosis exhibited a tendency to move toward siblings (Cornell, Berven, and Gamboa 1989), but this apparent preference for siblings was not evident several days after metamorphosis (Waldman 1989).

### Sensory Mechanisms of Kin Recognition

It seems clear that chemical cues are the principal means of kin recognition in all of these species. Tadpoles probably use visual cues from other swimming tadpoles to locate aggregations of conspecifics and especially to form polarized schools (Wassersug and Hessler 1971; Wassersug 1973; Foster and McDiarmid 1982). Nevertheless, there is no evidence that visual cues provide sufficient information to enable tadpoles to identify close relatives. In laboratory tests with *Bufo americanus, Rana cascadae,* and *R. aurora,* tadpoles moved preferentially toward water containing siblings, even when all visual cues were excluded (Blaustein and O'Hara 1982a; Waldman 1985a; Blaustein et al. 1993). All of the tests of kin recognition in metamorphosed juveniles also were done with an apparatus that excluded all but chemical cues.

The role of chemical cues in the formation of kin aggregations was revealed indirectly in studies of kin recognition in spadefoot toad tadpoles. In *Spea multiplicata,* tadpoles associated preferentially with siblings over nonsiblings, but they also preferred to associate with nonsiblings that had been raised on the same type of food to siblings raised on different food (D. Pfennig 1990b). These results suggested that tadpoles use chemical cues to select preferred feeding patches in a pond. An apparent preference for aggregating with close kin could be due, in part, to this type of habitat selection. Siblings are likely to remain near where they hatched and feed on similar foods, thereby acquiring chemical cues from their food or surroundings. In an experiment with *S. intermontana,* J. A. Hall et al. (1995) found little evidence that tadpoles tend to aggregate with siblings, but they did find a preference for aggregating with tadpoles raised on the same type of food.

The importance of chemical cues from the environment in the development of kin recognition is supported by some unpublished experiments reported by Waldman (1991). Eggs of *Rana sylvatica* were injected with either lemon or orange extract early in development. Tadpoles hatching from these egg clutches subsequently exhibited a preference for the odors to which they initially had been exposed. Waldman also injected eggs with different cultures of algae, which sometimes grow commensally with the eggs of *R. sylvatica*. Again, tadpoles exhibited a preference for the odors of familiar algal cultures. Hence, even though *R. sylvatica* tadpoles exhibit a preference for siblings when they have not been raised with them, it is clear that environmental cues can affect the choices made by tadpoles in laboratory tests.

### How Often Do Tadpoles Associate with Kin?

Although the evidence for kin-recognition abilities in tadpoles seems compelling, the importance of such behavior in the field is less clear. Waldman (1982b) and O'Hara and Blaustein (1985) reared sibships of *Bufo americanus* and *R. cascadae* separately in the laboratory, colored them with vital stains, and released them together into natural ponds. Analyses of the spatial distribution of colored tadpoles revealed a tendency to associate with sibs. Such experiments have not been performed with other species. The genetic composition of naturally occurring aggregations of tadpoles or juveniles is not known for any species. Since different sibships cannot be distinguished visually in the field, a genetic analysis using DNA fingerprinting or other molecular techniques would be required to investigate this problem. This would be difficult in species that form aggregations containing individuals from many eggs clutches, because any kinship associations are likely to be manifested as slight statistical differences in spatial distribution within a group. Species that form size-specific aggregations, such *as Hyla geographica* and *Phyllomedusa vaillanti,* would be ideal subjects for such studies, as would species in which schools of tadpoles are attended by a parent. These are species that are mostly likely to live in aggregations consisting mostly or entirely of siblings.

Although tadpoles apparently make fine-scale distinctions between conspecifics, the choices made by tadpoles are not perfect. Indeed, they generally do not approach the level of discrimination exhibited by frogs in species recognition tests using acoustic cues (see chapter 7) or salamanders in sex or species recognition tests using chemical cues (chapter 8). In many of those experiments, nearly 100% of individuals being tested make the correct choice. In contrast, tadpoles being tested for sibling recognition often spent only about 55–58% of their time on the sibling side of a test aquarium (O'Hara and Blaustein 1981, 1982; Blaustein and O'Hara 1982a, 1986). These results are statistically significant with large sample sizes, but do not provide evidence of robust or unanimous preferences for associating with siblings. In field tests with *B. americanus,* about two-thirds of the aggregations sampled differed significantly from a random distribution of sibships. In most of these groups, about 60–70% of the individuals were from the same sibship; none was a pure sibling group (Waldman 1982b). In groups of *R. cascadae* tadpoles that differed significantly from a random distribution of sibships, the average proportion of the more common one was about 72%, and pure sibship aggregations were rare (O'Hara and Blaustein 1985). Thus, even in species with well-developed abilities to recognize kin, individual tadpoles are likely to be in contact with both relatives and nonrelatives under natural conditions. In species such as *Bufo boreas,* which form enormous aggregations of hundreds of thousands of individuals, tadpoles probably have few opportunities to interact with siblings in the field, and kin recognition may not be very important in normal social interactions. The same probably is true for species such as *Phrynomantis microps,* which forms swarms composed of many different sizes of tadpoles. Tadpoles of this species readily aggregate with unrelated individuals in the laboratory (Rödel and Linsenmair 1997).

Grafen (1990b) argued that many examples of apparent kin recognition in tadpoles and other animals actually are a by-product of other recognition systems, such as species or group recognition. His argument is that animals use cues emanating from conspecifics to recognize them as conspecifics or members of a social group. They are attracted to conspecifics for protection against predators, enhancement of foraging, or to locate mates. Animals learn the chemical signatures of other individuals early in their development, or they compare cues coming from their own bodies with those of other individuals (phenotype matching). In either case, close relatives are likely to be more similar to one another than are distantly related individuals. Even in the absence of selection for kin recognition per se, responses to such cues could result in apparent kin recognition in laboratory and spatial associations between siblings in the field. Grafen's paper generated some debate (see Blaustein et al. 1991; Byers and Beckoff 1991; R. Stuart 1991; and replies by Grafen 1991a, b, c), but Blaustein et al. (1991) agreed that it often would be difficult to distinguish true kin recognition from recognition that is a by-product of species recognition. Grafen's argument is consistent with the relatively weak kin recognition observed in several species of tadpoles, and the lack of evidence for pure-sibling groups in natural aggregations. His interpretation also is consistent with the results of studies with spadefoot toad tadpoles that suggest that aggregation with kin is a by-product of habitat selection.

### Why Aggregate with Close Kin?

Explanations for the preferential aggregation of tadpoles with siblings generally have involved kin-selection models in which group formation itself is favored by kin selection. However, some species exhibit a preference for siblings in the laboratory, but have few opportunities to preferentially associate with kin in the field (e.g., O'Hara and Blaustein 1982; Blaustein and O'Hara 1986), so such models must be treated cautiously. Most authors have assumed that any benefits realized by individuals living in groups will be enhanced by aggregating with kin, but actual fitness benefits of such behavior have not been well documented (Blaustein 1988; Blaustein et al. 1991; Waldman 1991; Blaustein and Walls 1995). It also has been proposed that an ability to recognize siblings developed in the tadpole stage might carry over into adulthood, possibly to be used in avoiding mating with close relatives (Blaustein, O'Hara, and Olson 1984). Convincing evidence to support this hypothesis is lacking, although other types of kin recognition by adult anurans are possible (Waldman, Rice, and Honeycutt 1992).

### Antipredator Benefits

Two principal kin-selection models of tadpole aggregation both relate to protection from predators. One hypothesis is that distasteful tadpoles benefit from grouping with close kin, because even those individuals sampled by predators will benefit if close relatives are saved (Wassersug 1973; Waldman and Adler 1979; Waldman 1982b; J. Peterson and Blaustein 1991). Alternatively, it has been proposed that tadpoles enjoy inclusive fitness benefits by transmitting warnings of predator attacks to close relatives. This could either occur through physical transmission of a fright reaction, or through release of "warning pheromones" (Blaustein and O'Hara 1981; O'Hara and Blaustein 1981, 1985; Waldman 1982b; Hews and Blaustein 1985). The only rigorous test of this hypothesis with a relatively social species of tadpole was work by Hokit and Blaustein (1995) on *Rana cascadae* tadpoles. They failed to find any evidence for an alarm response to injured tadpoles. The alarm reaction hypothesis assumes that chemical substances eliciting a fright reaction in tadpoles are pheromones that evolved specifically as warning signals. If so, then tadpoles would gain greater fitness benefits if these signals were directed toward relatives. Another possibility is that tadpoles have evolved a behavioral reaction to any chemicals released from injured skin, but the chemicals have other functions and are not produced specifically as warning pheromones. In that case, a kin selection model is not required, because there is no cost associated with production of the signal. At present, the nature of the alarm pheromone in tadpoles is unknown.

Another way of looking at this problem is to assume that aggregation per se has not evolved through kin selection, but when selective pressures such as predation and a shortage of food favor group life, then aggregating with kin may minimize the costs of such behavior. When the benefits of group life greatly exceed the costs, then one might expect tadpoles to form large aggregations of both relatives and nonrelatives, as undoubtedly occurs in many species of *Bufo*. In these species, the ability to recognize kin may not be fully utilized (e.g., *Bufo boreas;* O'Hara and Blaustein 1982). When there are few selective pressures favoring group formation, then any tendency to associate with siblings in laboratory tests are expected to be relatively weak, because the tadpoles derive few benefits from such behavior (e.g., *Rana aurora;* Blaustein and O'Hara 1986).

When selection favors group formation, there can be distinct advantages to aggregating with individuals of a similar size. As pointed out by L. Branch (1983), size-specific schooling is very common in other organisms such as fishes. Individuals that differ in size from most of their companions are more conspicuous to predators. They also are at a disadvantage in fleeing from predators if they cannot move at the same speed (e.g., Mueller 1971, 1975; Milinski 1977; Hobson 1978; N. Wolf 1985; Ranta and Lindström 1990; Ranta, Lindström, and Peuhkuri 1992; Pitcher, Magurran, and Edwards 1985; Pitcher, Magurran, and Allan 1986; Peuhkuri 1997; Peuhkuri, Ranta, and Seppä 1997). There also could be advantages in grouping with similar-sized individuals if large tadpoles out-compete smaller ones through behavioral interference (Faragher and Jaeger 1998).

One way for tadpoles to insure size-specific aggregation is to associate preferentially with conspecifics, which are likely to be more similar in size than are members of other species. Indeed, multispecies aggregations of tadpoles have not been reported in the field, even in situations in which several species live in the same pond. Griffiths and Denton (1992) reported that tadpoles of *Bufo bufo,* a relatively social species, were attracted to tadpoles of both *B. calamita* and *Rana temporaria* in laboratory tests when no other tadpoles were present, but exhibited a somewhat greater attraction to conspecific tadpoles. Nevertheless, when given a choice between an aggregation of conspecifics and one containing *R. temporaria* tadpoles, the *B. bufo* tadpoles did not exhibit a clear preference for conspecifics, even though the *Rana* tadpoles were considerably larger. When given a choice between conspecifics and *B. calamita* tadpoles, the *B. bufo* tadpoles actually showed a slight preference for heterospecific tadpoles. Tadpoles of *B. calamita* are somewhat smaller than are those of *B. bufo,* so the reason for this preference is unclear. Griffiths and Denton (1992) did not present any evidence of mixed-species aggregations in nature.

Given that most tadpoles seem to aggregate mainly with conspecifics, then they could further increase the probability of aggregating with tadpoles of similar size if they associated with siblings from the same egg clutch, because these tadpoles are likely to have hatched at the same time. Preferences for siblings observed in laboratory and field tests could be the result of individuals avoiding nonsiblings because they are more likely to be of a different size, and might have little to do with a preference for kin per se. In a laboratory test of chemical cues used in kin recognition, Waldman (1985b) found that *B. americanus* tadpoles showed no preference for sibling odor over pure water, but preferred pure water to nonsibling odor. This indicates that they are repelled by nonsiblings, but are not necessarily strongly attracted to siblings. Chemical cues need not provide information about size per se. A preference for sibling odors could develop simply if this is most likely to result in tadpoles finding themselves with individuals of similar size.

If preferential aggregation with siblings is mainly a strategy to ensure size-specific grouping, then one might expect kin preferences to be weakly developed in many explosive breeders, because most tadpoles in a pond will be of similar size. Such abilities might be more strongly developed in tadpoles of prolonged breeders, which would be exposed to conspecifics of various sizes. This might explain the relatively imperfect discrimination of many of the species tested to date, most of which are explosive breeders. Unfortunately, in North America, schooling tadpoles tend to be most common among explosive breeders that use ephemeral breeding sites, so these predictions are difficult to test. Again, some of the tropical species that have prolonged breeding seasons and form size-specific aggregations, such as *Hyla geographica,* would be the most appropriate ones for testing this hypothesis.

If predation is the major force favoring group formation, then size-specific aggregation, and possibly kin recognition, also should be most strongly developed in permanent-water species exposed to visually hunting predators that actively search for prey, such as fishes. *Hyla geographica,* which breeds in permanent lakes and rivers with predatory fishes, is one possible example. If the major predators on tadpoles are sit-and-wait predators that do not necessarily select prey by size, then size-specific aggregation may not be favored. For example, tadpoles of the African frog *Phrynomantis microps* are preyed upon mostly by insects, and these tadpoles tend to form aggregations that are not sorted by size (Rödel and Linsenmair 1997). Unfortunately, possible antipredator benefits of size-specific aggregation have not been tested experimentally with tadpoles. For example, there have not been any studies comparable to those with fish to determine whether predators preferentially select the "odd" tadpole in a group. Instead, most studies of predation on tadpoles have focused on the effect of tadpole size or developmental stage on vulnerability to predation, with tadpoles of similar size being presented to predators simultaneously (see chapter 14 for further discussion).

### Effect of Kin Association on Growth and Development

Various authors have suggested that tadpoles aggregate with kin because of beneficial effects on growth and development. Unfortunately, it is difficult to predict exactly how aggregating with kin will affect growth and development toward metamorphosis. Waldman (1991) suggested that in dense aggregations of tadpoles, as might occur in a temporary pond in the process of drying up, smaller tadpoles might gain inclusive fitness benefits by aggregating with kin and allowing their own growth to be stunted in favor of larger, faster growing siblings. Other authors have argued that aggregating with kin might ameliorate the effects of intraspecific competition, thereby allowing all individuals in an aggregation to grow more rapidly or reach a larger size at metamorphosis. In fact, experimental studies of the effect of kin association on growth and development have produced evidence of both positive and negative effects of living with kin.

In a study of chorus frog tadpoles (*Pseudacris triseriata*), a species not known to form kin-based aggregations, D. C. Smith (1990) found that tadpoles reared with siblings grew more rapidly than did those raised with nonsiblings. Similar results were obtained in a study of *Bombina variegata* tadpoles, another species not known to form kin-based aggregations (Jasienski 1988). In contrast, tadpoles of *Rana arvalis,* a species that sometimes forms aggregations of unknown composition, grew larger when reared with nonsiblings than with siblings (Shvarts and Pyastolova 1970a). A preliminary study of *Rana cascadae,* a species that sometimes aggregates with close relatives, showed that siblings raised together at high density had lower body mass than did nonsiblings raised under the same conditions (Hokit and Blaustein 1994). Later, a more thorough study of the same species in which tadpoles were raised in field enclosures revealed complex interactions between the effects of kinship, type of substrate, and larval density on growth and development. Under some conditions, tadpoles reared with siblings had higher survivorship than did those raised with nonsiblings, but they also tended to be smaller, especially at high densities (Hokit and Blaustein 1997). Pakkasmaa and Aikio (2003) found relatively weak effects of kinship on growth and development of *Rana temporaria* tadpoles, and the effect of kinship interacted with density. Tadpoles of an Indian frog, *Rana temporalis,* had faster growth rates and reached larger metamorphic size when reared in single sibships than when reared in mixed groups, but only at high density (Girish and Saidapur 1999a). Sib groups also had shorter larval periods and were more likely to reach metamorphosis at high density than were nonsib groups (Girish and Saidapur 2003). In contrast, developmental and metamorphic traits of another Indian frog, *Tomopterna breviceps,* were unaffected by kinship (Gramapurohit et al. 2004). Waldman (1986b) reported both positive and negative effects of kin association on growth rates of *Bufo americanus* tadpoles, with some families experiencing enhanced growth and others retarded growth. Taken together, these studies demonstrate that kinship can affect the growth and development of tadpoles, but in ways that depend on other ecological factors and on larval density. These studies do not provide convincing evidence that species of tadpoles that form aggregations consistently benefit from aggregating with close relatives.

### Kin Recognition and Cannibalism in Tadpoles

Studies of kin association among tadpoles have produced relatively little clear evidence that tadpoles derive fitness advantages from aggregating with close relatives, as opposed to forming aggregations with randomly related individuals. One type of behavioral interaction for which the benefits of kin recognition seem more predictable is cannibalism by conspecifics. Kin selection theory clearly predicts that when tadpoles or salamander larvae cannibalize other members of their population, they are likely to derive inclusive fitness benefits from recognizing close relatives and refraining from eating them (D. Pfennig 1997). In addition, there are direct benefits from not eating close relatives, such as reduced chances of transmitting diseases through cannibalism.

D. Pfennig, Reeve, and Sherman (1993) tested for kin recognition in tadpoles of *Spea bombifrons,* one of the species that sometimes develops a carnivorous morph. In a test of kin association, omnivores spent about 60% of their time close to siblings, compared to 43% for carnivores. These results suggest that carnivores actually avoid being in close proximity to their relatives. Carnivores that consumed other tadpoles were much more likely to eat nonsiblings (77% of the tadpoles consumed) than siblings (23%). The tendency to avoid eating siblings broke down, however, if the carnivores were very hungry. Subsequent work showed that carnivores of this species are much more likely to eat conspecifics than are omnivores, so the carnivores clearly represent a greater danger to their own siblings than do omnivores (D. Pfennig and Frankino 1997). Tadpoles of *S. bombifrons* also were more likely to develop into carnivores when they were raised alone than when they were raised with siblings, but they were not more likely to do so than when raised in mixed groups of siblings and nonsiblings. Tadpoles of another species, *S. multiplicata,* were more likely to develop into carnivores when reared in mixed groups rather than pure sibling groups.

Tadpoles of *S. bombifrons* exhibit a greater tendency to develop into carnivores than do tadpoles of *S. multiplicata,* and the former also exhibit a stronger avoidance of kin as prey (D. Pfennig 1999). Different sibships *of S. multiplicata* also vary in their propensity to become carnivores. Members of sibships that showed a strong tendency to become carnivores also exhibited the strongest avoidance of kin as prey. When some of these carnivores reverted to the omnivore morph, however, they no longer avoided eating siblings, but as omnivores, they were much less likely to eat any tadpoles, and therefore represented a less serious threat to their relatives. Taken together, these results suggest that the expression of kin recognition in spadefoot toad tadpoles is context-dependent and varies according to the danger that individual tadpoles pose to their relatives.

### Kin Recognition, Cannibalism, and Aggression in Salamander Larvae

Salamander larvae differ from many tadpoles in being relatively solitary and actively avoiding contact with conspecifics, so aggregations of closely related larvae are not expected, except in situations where dispersal from a nest site is minimal (e.g., R. N. Harris et al. 2003). The antisocial behavior of salamander larvae probably is related to widespread aggressive competition among larvae of many species of salamanders, as well as opportunistic cannibalism and the formation of specialized cannibalistic morphs in some populations. As with cannibalistic tadpoles, kin selection is expected to favor kin recognition in aggressive or cannibalistic salamander larvae, because the larvae derive inclusive fitness benefits from directing aggression or cannibalism differentially toward nonrelatives (D. Pfennig 1997). Cannibalistic larvae of *Ambystoma t. nebulosum* recognize close relatives and prefer to feed on unrelated individuals if they are available (D. Pfennig, Sherman, and Collins 1994). The presence of close relatives also can inhibit the development of cannibalistic morphs (D. Pfennig and Collins 1993).

There are two possible advantages to preying preferentially on nonrelatives. The first is a reduction in the inclusive fitness cost that would be incurred if relatives were consumed (the kin selection hypothesis). The second is the possibility that larvae are more susceptible to pathogens carried by close relatives, perhaps because they share similar immune systems (D. Pfennig, Ho, and Hoffman 1998; the disease avoidance hypothesis).

D. Pfennig, Collins, and Ziemba (1999) performed a series of experiments to distinguish between the kin selection and disease avoidance hypotheses for the evolution of kin recognition in tiger salamanders. Contrary to expectation, they found that cannibals were more likely to contract disease after eating nonrelatives than after eating relatives. Larvae did not show any tendency to avoid diseased prey, and in fact, preferred such prey because they were easier to catch. Finally, larvae from populations where disease was prevalent did not exhibit any greater kin discrimination than did those from disease-free populations. All of these results indicate that disease avoidance does not favor the evolution of kin recognition. In contrast, those individuals that did discriminate between kin and nonkin as prey had more surviving siblings than did nondiscriminators, indicating that kin recognition is favored by kin selection. The cost of kin discrimination was low, because growth and survival were similar in discriminators and nondiscriminators.

Kinship influences the behavior of other species of salamander larvae as well, but not always in the way observed in *A. tigrinum nebulosum*. Larvae of *Ambystoma opacum* were less aggressive and more submissive toward siblings of a similar size than toward nonsiblings (Walls and Roudebush 1991). The ability to recognize siblings persisted after metamorphosis, but degree of familiarity with other individuals appeared to have a greater impact on aggressive interactions than did kinship (Walls 1991). Later studies showed that larvae of this species actually preferred to eat siblings rather than nonsiblings when they were much larger than their prey (Blaustein and Walls 1995), but the reason for this unexpected preference is unclear. A complex laboratory experiment did reveal that kinship did not, by itself, influence either aggression between larvae or larval performance, but interactions between kinship, larval density, and food availability did have a significant effect (Hokit, Walls, and Blaustein 1996). These results indicate that the expression of kin recognition and the effect of kinship on larval fitness depend on ecological context, but the fitness consequences of kin recognition in *A. opacum* are less clear than for highly cannibalistic populations of *A. tigrinum*. When larvae were raised in field enclosures, kinship did not affect larval performance, even though growth and size were affected by density (Walls and Blaustein 1994). Another species, *A. gracile*, did not exhibit clear evidence of kin recognition (Blaustein and Walls 1995). Larvae of this species are aggressive toward conspecifics, but cannibalism is relatively rare.

## Summary and Conclusions

Larval amphibians essentially function as independent organisms with adaptations quite different from those of adults. This is particularly true of anuran tadpoles. Although the basic morphological structure of larvae is well known for most groups of amphibians, the study of larval ecology and behavior is still in its infancy. The work of Wassersug and others has shown how much can be learned about the ecology of tadpoles from considerations of functional morphology. Tadpoles adapted for generalized suspension feeding, midwater plankton feeding, or feeding on large food particles exhibit strikingly different configurations of the buccal pump and associated feeding structures. However, many of the inferences about relationships between morphology and feeding ecology are based on anecdotal accounts of feeding behavior. Detailed observational and experimental studies of tadpole feeding behavior in the field and laboratory are needed to supplement the wealth of information available on tadpole morphology. This is particularly true for some of the tropical species with unusual morphological adaptations.

A striking feature of anuran tadpole evolution is the frequent occurrence of convergent evolution of similar morphology in tadpoles from distantly related families that live in similar environments. Striking examples include the repeated evolution of very similar suction devices to hold onto rocks in fast-moving water (Lamotte and Lescure 1989a, b; Haas and Richards 1998), the evolution of similar body and tail fin morphology in nektonic tadpoles in different families (Altig and Johnston 1989), and the repeated reduction of mouthparts and other structures in egg-eating tadpoles that live in bromeliads and other phytotelmata (Lannoo, Townsend, and Wassersug 1987; Schiesari, Grillitsch, and Vogl 1996). Ecomorphological studies of tadpoles also have shown that tadpole morphology can be relatively plastic and can evolve rapidly in response to differences in larval habitats. For example, stream-dwelling tadpoles often exhibit morphological adaptations for life in fast-moving water that are absent in closely related members of the same genus (Iwasawa and Saito 1989; Inger 1992). Similar divergence is seen in egg-eating tadpoles when compared with pond-breeding relatives (Schiesari, Grillitsch, and Vogl 1996). Even within species, there can be considerable morphological plasticity, ranging from relatively minor differences between pond- and stream-dwelling populations (Jennings and Scott 1993) to within-population differentiation of omnivorous and carnivorous morphs and variation among populations in the frequency of these morphs (D. Pfennig and Frankino 1997).

Because salamander larvae are less diverse in their morphological and ecological adaptations, their ecology has received less attention than that of tadpoles. However, the idea that salamander larvae are simply aquatic equivalents of adults because they are at the same trophic level (Duellman and Trueb 1986) is clearly an oversimplification. While it is true that both larval and adult salamanders are carnivorous, the larvae often undergo ontogenetic changes in diet as they grow. In addition, the shift from an aquatic to terrestrial habitat often is accompanied by rather dramatic changes in prey selection. For example, larval *Gyrinophilus porphyriticus* feed mainly on aquatic invertebrates, whereas adults prey on adult salamanders of other species (Bruce 1979, 1980; see chapter 13 for further discussion). In part, this shift in diet reflects the availability of prey in different microhabitats. However, it also is related to morphological differences that make larvae better at capturing prey in water, whereas transformed individuals are more adept at capturing prey on land (Lauder and Shaffer 1986). Additional studies of ontogenetic changes in microhabitat selection and feeding behavior of both urodele and anuran larvae would add considerably to our understanding of their ecology.

The study of social behavior in larval amphibians, particularly anuran tadpoles, also is just beginning. There are now many descriptive accounts of aggregations and schooling behavior in tadpoles, as well as abundant speculation about the functions of such groups, but careful experimental studies of the effects of grouping on feeding behavior and predator avoidance are almost nonexistent. There is intriguing evidence that interaction among close relatives is an important component of tadpole social behavior. The nature of cues used to recognize close relatives, as well as the ontogeny of kin recognition, are now partially understood for a few species, but the prevalence of kin recognition in the thousands of species of tadpoles around the world is completely unknown. Many of the species studied to date have only moderately social tadpoles; tropical species that appear to exhibit close aggregation of kin groups many provide more interesting examples of kin recognition in the future. As in the case of grouping behavior in general, the adaptive significance of kin recognition for tadpoles in the field is poorly understood. Several speculative hypotheses have been proposed, related either to enhanced growth or protection from predators, but as yet there are few definitive tests of these ideas. The most convincing evidence for an adaptive advantage to kin recognition comes from studies of kin discrimination in relation to cannibalism in both tadpoles and salamander larvae.

# Chapter 13 Complex Life Cycles and the Ecology of Amphibian Metamorphosis

Amblystoma's *a giant newt who rears in swampy waters*
*As other newts are wont to do, a lot of fishy daughters:*
*These Axolotls, having gills, pursue a life aquatic,*
*But, when they should transform to newts, are naughty and erratic.*

*They change upon compulsion, if the water grows too foul,*
*For then they have to use their lungs, and go ashore to prowl:*
*But when a lake's attractive, nicely aired, and full of food,*
*They cling to youth perpetual, and rear a tadpole brood.*

*And newts Perennibranchiate have gone from bad to worse;*
*They think aquatic life is bliss, terrestrial a curse.*
*They do not even contemplate a change to suit the weather,*
*But live as tadpoles, breed as tadpoles, tadpoles altogether.*
—Walter Garstang, "The Axolotl and the Ammocoete" (1985)

Walter Garstang's verse nicely illustrates both the complexity and variability of amphibian life cycles. Indeed, amphibians are unique among tetrapods in the complexity of their life cycles and the degree to which their life cycles vary in response to environmental conditions. Amphibians are not so unusual, however, when all groups of animals are considered. Werner (1988) estimated that at least 80% of all animals have complex, multistage life cycles, including the most successful lineages of insects and most marine invertebrates (see also B. Hall and Wake 1999). Most amphibians have a standard life cycle in which eggs are laid in water and hatch into aquatic larvae. These larvae then feed, grow, and develop in the water until they eventually metamorphose into terrestrial or semiaquatic adults. Yet, as discussed in chapter 10, many amphibians have abandoned the aquatic stages of their life cycles and have evolved reproductive modes that make them more or less independent of water. Other species, like the axolotl described in Garstang's poem (the current name for this animal is *Ambystoma*), have lost the terrestrial portion of the life cycle and retain at least part of their larval morphology as sexually mature adults. This variability in amphibian life histories suggests that the advantages of a complex life cycle vary with environmental conditions. In this chapter, I begin by considering the evolution and maintenance of a complex life cycle from an ecological perspective, including interspecific variation in the length of the larval period in relation to environmental conditions. Next, I discuss proximate determinants of larval life history traits, including growth rates, the timing of metamorphosis, and size at transformation. I also discuss the importance of phenotypic plasticity as an adaptive response to variable environments. Finally, I consider the evolutionary loss of metamorphosis in paedomorphic amphibians and the adaptive significance of this type of life history.

## The Adaptive Significance of Complex Life Cycles

In the previous chapter, I reviewed a number of morphological and behavioral adaptations of amphibian larvae, particularly as they relate to feeding ecology. A key question that needs to be considered is why amphibians have a larval stage at all. Certainly part of the answer is related to the evolutionary history of amphibians. A complex life cycle with a free-living aquatic larva is considered by most investigators to be the ancestral condition in all three major lineages of amphibians (Hanken 1999a; R. N. Harris 1999; Reiss 2002). Nevertheless, the course of larval evolution has diverged in these three clades. Both urodeles and caecilians have larvae that resemble adults in basic morphology. The larvae undergo gradual developmental changes as they make the transition from aquatic larvae to terrestrial adults, but they do not undergo the very dramatic metamorphosis seen in anurans. This lack of dramatic metamorphosis is related to the absence of a major shift in diet from the larval stage to the adult. In both of these groups, larvae and adults are carnivorous, very likely the ancestral condition for all amphibians. Individuals often exhibit ontogenetic shifts in size or type of prey as they grow, but the basic diet remains the same. Most anurans, on the other hand, undergo a relatively rapid and dramatic change from a largely herbivorous diet to an entirely carnivorous one. This major shift in diet in turn requires a major rearrangement of organ systems and external morphology.

Caecilians, urodeles, and anurans also differ in their tendency to exhibit evolutionary loss of a complex life cycle. Most caecilians have evolved a specialized lifestyle as terrestrial, burrowing animals. The most successful lineages have lost the aquatic larval stage altogether, shifting either to direct development of terrestrial eggs or to viviparity (see chapter 10). In those caecilians that retain a complex life cycle, the larvae often are not very different from adults ecologically. Most species hatch at a relatively advanced stage of development, and the larvae of some species have become amphibious like the adults, rather than being strictly aquatic (see chapter 12). Urodeles exhibit two opposing evolutionary trends. Among plethodontids, loss of the aquatic larval stage and the evolution of direct development occurred several times, and the lineages that are most successful in terms of numbers of species all have direct development (see chapter 10). Other urodeles have gone in the opposite direction, eliminating the terrestrial adult stage and becoming paedomorphic, retaining at least some larval features throughout life. Direct development has evolved many times among anurans as well. Some lineages have become very successful, possibly as a result of losing the aquatic larval stage (e.g., *Eleutherodactylus*). A few species have become viviparous (see chapter 10). Nevertheless, most anurans have retained a complex life cycle, even in species in which feeding by larvae has been reduced or eliminated. What accounts for the widespread persistence of aquatic larvae in anurans?

The anuran tadpole almost certainly represents a derived type of larval morphology, although a relatively ancient one. The ancestors of modern frogs are assumed to have had a larva much more like that of urodeles, with gradual morphological changes rather than rapid and dramatic metamorphosis (Wassersug and Hoff 1982; R. N. Harris 1999; Reiss 2002). Nevertheless, tadpoles with morphology like that of modern forms have been around at least since the Cretaceous, and probably evolved much earlier (Hanken 1999a). The marked difference in morphology and diet of tadpoles and adult anurans suggests that anurans actually have evolved a more complex life cycle, with selection having favored morphological and ecological divergence of larvae and adults.

In an influential paper, Istock (1967) argued that complex life cycles are inherently unstable. He assumed that populations of larvae and adults were independently regulated, and he developed a model of population growth rates derived from the logistic equation and incorporating information on age-specific mortality and fecundity. He argued that if factors regulating either larval or adult populations are relaxed, then selection should favor the loss of one stage or the other. In the case of amphibians, life histories would tend to move either toward loss of the larval stage and direct development (chapter 10) or retention of a permanent larval morphology (paedomorphosis). Hence, he considered the maintenance of complex life cycles in many amphibians, especially anurans, to be a paradox.

Istock's model was criticized on a number of grounds relating to both the mathematical treatment and its biological assumptions (Bryant 1969; Slade and Wassersug 1975; Wassersug 1975; Strathmann 1978; Wilbur 1980; Werner 1988; Ebenman1992). A number of amphibian biologists have argued that complex life cycles are maintained because of the unique ecological role of the aquatic larval stage, particularly in anurans. Their conclusions are based in part on an analogy with many holometabolous insects, in which nearly all of an individual's growth takes place in the larval stage, with the adult serving mainly for dispersal and reproduction. For example, Wassersug (1975) argued that typical anuran tadpoles are essentially eating machines adapted for rapid utilization of food sources that undergo rapid seasonal fluctuations. Complex life cycles are therefore most likely to be maintained under a regime of short-term environmental fluctuations (e.g., alternating cold and warm or wet and dry seasons). Loss of one stage of the life cycle is more likely when either the aquatic or terrestrial environment becomes more stable. Because morphological constraints prevent the development of a functional reproductive system in an anuran tadpole, only the larval stage can be lost in this group. Similarly, Wilbur (1980) argued that amphibian larvae are adapted for exploiting opportunities for rapid growth by taking advantage of seasonal flushes of primary productivity in aquatic environments. Adults, on the other hand, were viewed as being adapted for survival through unfavorable seasons and dispersal to new breeding sites (see also Wilbur and Collins 1973).

## Amphibian Metamorphosis as an Ontogenetic Niche Shift

The Wassersug-Wilbur model is an appealing one, and for some time was the standard explanation for the maintenance of complex life cycles in amphibians (e.g., Duellman and Trueb 1986). It is, however, heavily influenced by work on frogs that breed in temporary ponds, and probably oversimplifies the evolution of complex life cycles in general. It is not clear, for example, that all larval habitats experience an initial flush of productivity, as originally proposed by Wassersug (R. N. Harris 1999). Furthermore, Werner (1986, 1988) pointed out that most of an individual's growth to adult size usually takes place after metamorphosis, not before. For example, *Rana catesbeiana* has one of the longest larval periods and the largest size at metamorphosis of any North American frog, yet a transforming juvenile is only 4% of the mass of an adult. Werner estimated that in most anurans, more than 90% of growth occurs in the terrestrial phase. So common is this pattern that one species that does complete most of its growth in the tadpole stage has been named the "paradox frog" (*Pseudis paradoxa;* Emerson 1988b). Werner also cited examples of metamorphosed juveniles actually growing more rapidly than aquatic larvae. Hence, the strict dichotomy between larval growth and adult dispersal stages does not seem to apply to most amphibians.

Building on the arguments of Wassersug (1975), Wilbur (1980), and others, Werner (1986, 1988) focused attention on the shift in an amphibian's ecological niche as it makes the transition from the larval to the terrestrial stage. This type of ontogenetic niche shift is widespread even among animals that increase in size during development, even those that do not change habitats or undergo dramatic metamorphosis (Werner and Gilliam 1984). In contrast to previous models of complex life cycles, which focused on the effects of age-specific mortality and growth, Werner proposed a model based on changes in size-specific mortality and growth, because both of these variables tend to be more affected by size than by age in amphibians.

Werner proposed that the costs and benefits of living in the water or on land depend on the size-specific growth and mortality rates experienced in both habitats. If the timing of metamorphosis was simply a function of growth rate, then larvae should transform when size-specific growth rates in the terrestrial environment are expected to exceed those in the aquatic environment (fig. 13.1 A). However, if size-specific mortality rates vary in the two environments as well, then the optimum time for metamorphosis changes, depending on the trade-off between growth and mortality rates (fig. 13.1 B). Werner's model indicates that lifetime fitness is maximized for prereproductive individuals when the ratio of mortality to growth rates is minimized. Therefore, an individual should remain in the larval habitat as long as this ratio is lower in the water, and metamorphose at the size at which this ratio first becomes lower for the terrestrial habitat (fig. 13.1 B). This prediction is independent of the source of mortality. For example, if breeding sites are highly ephemeral, then selection will favor early transformation because the ratio of mortality to growth in the aquatic environment will exceed that in the terrestrial environment, even at small body sizes (i.e., all larvae will perish if they remain in the pond). On the other hand, larvae in a more permanent pond might have prolonged larval periods if aquatic predators are scarce, but short larval periods and smaller sizes at transformation if size-specific predation rates in the aquatic environment exceed those in the terrestrial environment.

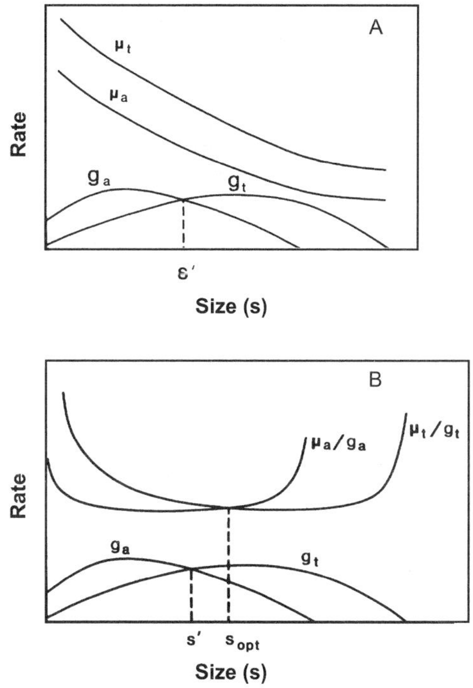


**Fig. 13.1.** Werner's model of amphibian metamorphosis as an ontogenetic niche shift. (A) Hypothetical size-specific rates of growth (g) and mortality (μ) in the aquatic (a) and terrestrial (t) habitats. The size at metamorphosis that maximizes growth across the entire life history is given as s′. (B) Ratios of mortality to growth (μ/g) in the aquatic (a) and terrestrial (t) environments. The optimum size at metamorphosis ($s_{opt}$) is shifted to the right toward a larger size than s′ and represents the size at which the ratio of mortality to growth in the aquatic environment surpasses the ratio in the terrestrial environment. After Werner (1986).

Several authors have modified and expanded upon Werner's original model. Jackson and Semlitsch (1993), in a study of paedomorphosis in *Ambystoma talpoideum,* criticized Werner's (1986) model and earlier ones for not including tradeoffs in reproduction. This is particularly relevant to paedomorphic species, which sometimes can reproduce earlier by staying in the water as paedomorphic adults instead of transforming and moving onto land (see subsequent discussion of paedomorphosis). Ludwig and Rowe (1990) and L. Rowe and Ludwig (1991) expanded Werner's model to include consideration of reproduction in species that always metamorphose, such as anurans. They explored the effects of time constraints on the timing of metamorphosis and the optimum size at metamorphosis. In essence, they argued that metamorphic timing and optimum size should vary, depending on the time between metamorphosis and reproduction. In species with short breeding periods that occur at a fixed time after metamorphosis, individuals that metamorphose at a small size probably will not have time to catch up with those that metamorphose at a large size before reproducing. When reproduction is prolonged, on the other hand, metamorphosing at a small size to escape aquatic predators or enhance growth rates might not be a disadvantage, because small individuals eventually reach a large size before reproducing. As in Werner's model, however, the precise point at which metamorphosis is favored depends on the balance of size-specific growth and mortality rates in aquatic and terrestrial habitats.

Ebenman (1992) incorporated genetic correlation between larval and adult morphological traits into a model of life-history evolution based on ecological niche shifts. His main focus was the initial evolution of a complex life cycle with pronounced metamorphosis, like that of anurans, from a life cycle with less dramatic morphological change between stages. He argued that when larvae and adults make use of very different resources in different habitats, their ability to use resources efficiently often is constrained by genetic correlation between larval and adult morphology. Such constraints are evident in the feeding behavior of salamanders, for example, because morphological features of the hyoid apparatus that are required for suction feeding in water limit the evolution of tongue projection in adults (D. Wake 1982). Ebenman (1992) showed that selection can favor the disruption of such constraining genetic correlation, leading in turn to divergence in larval and adult morphology. Once evolved, this type of complex life cycle, with morphologically distinct larvae undergoing dramatic metamorphosis into adults, is thought to be more stable than the type of life cycle seen in urodeles. Hence, this type of life cycle is less likely to evolve toward reduction or elimination of either the larval or adult stage (see R. N. Harris 1999, for a somewhat similar model). From a mechanistic point of view, the evolution of complex life cycles like those of anurans can result in different developmental programs guiding the formation of larval and adult morphology (Alberch 1987, 1989; Elinson 1990; Hanken, Jennings, and Olsson 1997; Hanken 1999a). From an evolutionary perspective, a complex life cycle with metamorphosis can be seen as an adaptation that permits independent responses of different life history stages to selective pressures in different environments (Werner 1988, Ebenman 1992; R. N. Harris 1999).

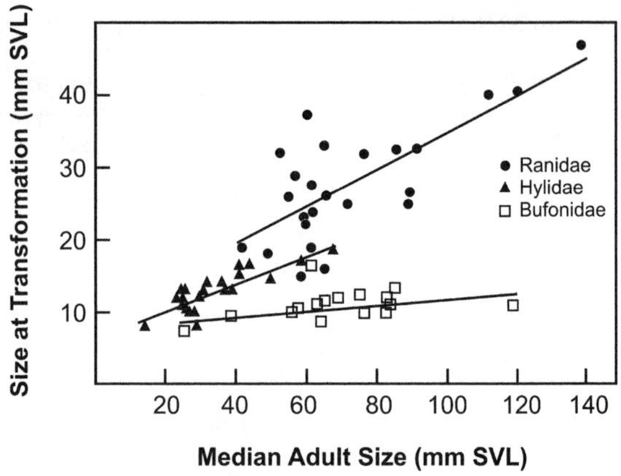


**Fig. 13.2.** Size at metamorphosis as a function of median adult size for three families of North American anurans. Ranids transform at a relatively large size compared to adult body size, and size at metamorphosis is positively correlated with adult body size. Bufonids always transform at a very small body size, regardless of adult body size. Hylids are intermediate. After Werner (1986).

## Interspecific Variation in Length of the Larval Period

Ontogenetic niche shift models are particularly useful in explaining interspecific differences in the length of the larval period and the size of individuals at metamorphosis. They also can be applied to different populations of a species, because selection can favor different life history strategies in different environments. Werner (1986) pointed out that North American ranid frogs show a positive correlation between adult body size and size at metamorphosis (fig. 13.2). The largest ranids tend to breed in permanent water and remain in the larval stage for more than a year (e.g., Collins 1979a). Tadpoles of many large ranids appear to be unpalatable to aquatic predators (Walters 1975; Kruse and Francis 1977; Formanowicz and Brodie 1982; Morin 1983a; Woodward 1983; Kruse and Stone 1984) and eventually grow to a size that allows them to escape most predators. They also exhibit behavioral responses to predators that reduce the chances of being eaten (see chapter 14). On the other hand, they are vulnerable to a variety of predators on land and will be subjected to the physical stresses of a terrestrial environment. Selection will favor remaining in the larval stage as long as the mortality/growth ratio in the aquatic environment is lower than in the terrestrial environment. Smaller species, such as wood frogs (*Rana sylvatica*), typically metamorphose at a smaller size than do large species, and they also are more likely to use temporary ponds that periodically dry up. In these species, selection favors a relatively short larval period, regardless of the mortality risks to terrestrial juveniles and adults (Werner 1986).

In contrast to ranids, North American bufonids show little or no correlation between adult body size and size at metamorphosis. All species transform at a small size (fig. 13.2), and juveniles of a large species, such as *Bufo alvarius*, are similar in size to those of a smaller species, such as *Bufo americanus*. Most toads breed in temporary water, so a short larval period is favored regardless of conditions in the terrestrial environment. Even those that breed in permanent water typically metamorphose at a relatively small size, so pond hydroperiod is not the only determinant of the length of the larval period.

North American spadefoot toads (*Scaphiopus* and *Spea*), which often breed in even more ephemeral pools than do toads, have extremely rapid rates of development (R. Newman 1987, 1988a, b, 1989, 1994; Pfennig, Mabry, and Orange 1991). In contrast, Old World spadefoot toads (*Pelobates*) have more prolonged breeding seasons and breed in more permanent ponds. They have longer larval periods and metamorphose at larger sizes than do tadpoles of *Spea* or *Scaphiopus*. There also are phylogenetic differences between the two North American genera that do not correspond precisely to differences in permanence of breeding habitats. The two species of *Scaphiopus, S. couchii* and *S. holbrookii,* have shorter larval periods and smaller size at metamorphosis than do species of *Spea,* even though some of the latter breed in desert rainpools (Buchholz and Hayes 2000, 2002).

There are costs associated with small transformation size that probably limit how small a metamorphosing anuran can be. Juvenile toads (*Bufo*) have poorly developed lungs and low aerobic capacities compared to adults, but this is not true of ranids that metamorphose at a larger size (Taigen and Pough 1981; Pough and Kamel 1984; see also chapter 5). Short larval periods also can be associated with incomplete morphological development at metamorphosis. For example, in both *Bufo americanus* and a hylid with relatively rapid development, *Pseudacris crucifer,* the middle ear is poorly developed at metamorphosis, whereas in ranids, the ear of metamorphosing juveniles is very similar to that of the adult (Hetherington 1988b). Small metamorphic size also can be disadvantageous in dry environments, such as deserts, because of the more rapid rate of evaporative water loss in small individuals (R. Newman and Dunham 1994).

Werner's model also is consistent with the life histories of various salamanders, although he did not discuss these in detail. For example, the plethodontid salamander *Gyrinophilus porphyriticus* inhabits a variety of permanent aquatic habitats, including springs, seepage areas, and headwater streams in moist forests (Bruce 1979). This species has a remarkably long larval period, lasting up to four years in some locations, and larvae are larger in relation to adult body size than those of most other plethodontid salamanders (Bruce 1972b, 1978b, 1979, 1980). Most larvae metamorphose at about the same size (Bruce 1980). Although both larvae and adults utilize the same aquatic habitats, adult salamanders take some of their food on land and exhibit a marked shift in diet at metamorphosis. Larvae are dietary generalists, feeding mainly on invertebrates and smaller salamander larvae; adults feed heavily on adults of other salamander species. Bruce (1979, 1980) suggested that selection has favored a prolonged larval period because this allows the salamanders to reach a size at which they are effective predators on other salamanders. This fits very well with Werner's view that metamorphosis should occur at an optimum size for shifting from an aquatic to a terrestrial niche, but is hard to explain solely as a function of larval growth. Many salamanders, including one species of *Gyrinophilus,* have gone a step further and have abandoned metamorphosis altogether. These salamanders exhibit a life history known as paedomorphosis, in which larval characters are retained in a reproductive adult. This type of life history has evolved frequently enough in the urodeles (there are no paedomorphic anurans) to suggest a major cost to maintaining a dual life cycle under certain ecological conditions. This topic is discussed in more detail at the end of the chapter.

Werner's ontogenetic niche shift model and those derived from it are essentially evolutionary models that are useful in understanding the selective pressures that have led to major differences in larval life history traits among species or populations. For example, these models allow us to make at least qualitative predictions about the range of sizes at metamorphosis and lengths of larval periods that can be expected in amphibians that breed in permanent ponds, semipermanent ponds, or ephemeral rain pools, and in habitats with different densities of predators. These models are less useful for predicting how individual larvae will respond to different environmental conditions. Aquatic larvae have no way of anticipating rates of growth and mortality on land and comparing these with growth and mortality in the water. They can only respond to conditions as they encounter them in the aquatic environment (R. N. Harris 1999). In some circumstances, deteriorating conditions in the water probably are accompanied by unfavorable conditions on land as well. For example, a rapidly drying pond is likely to be surrounded by relatively hot and increasingly dry terrestrial habitat. Nevertheless, larvae have no chance of surviving if a pond dries up completely, so they often will be forced to move into an unfavorable environment, and not necessarily one in which the growth to mortality ratio improves dramatically.

Although Werner's model postulates relatively fixed ages and sizes at metamorphosis within populations, he readily acknowledged that life-history traits such as timing of metamorphosis and size at metamorphosis often are not fixed for a given species, but show considerable plasticity. Much of the research on amphibian life cycles has been devoted to understanding the proximate determinants of rates of growth and development of larvae, the timing of metamorphosis, and the size of individuals at metamorphosis within populations.

## Divergent Approaches to the Study of Amphibian Metamorphosis

The process of metamorphosis has been studied both by developmental biologists and ecologists, but these two disciplines traditionally have approached the problem from very different perspectives, with minimal interaction between these fields. Developmental biologists have been concerned mainly with processes of morphogenesis, the regulation of gene expression, and the hormonal control of development (Etkin 1968; Dodd and Dodd 1976; B. White and Nicoll 1981; H. Fox 1984; Kikuyama et al. 1993; Denver 1996; Kaltenbach 1996; Tata 1996, 2005; C. Rose 1999; Shi 2000). A few representative species, such as *Rana catesbeiana, R. pipiens, Xenopus laevis,* and *Ambystoma tigrinum,* have been used as model systems for understanding development in all amphibians. This work has led to a very detailed understanding of metamorphosis at the level of genes, cells, and tissues. Generally the process is viewed as an orderly sequence of steps culminating in the transformation of an aquatic larva into a terrestrial adult, with little attention to environmentally induced variation in larval growth and development. Several methods have been devised to describe the stages of larval development (see Just, Kraus-Just, and Check 1981 for a review). The staging table for *Bufo valliceps,* shown in fig. 13.3, was developed by Gosner (1960) and is derived from the embryonic staging table of Shumway (1940) and the widely used staging table for *Rana pipiens* tadpoles published by A. Taylor and Kollros (1946). Staging procedures for salamander larvae are less standardized, and different authors have characterized development in different ways (Just, Kraus-Just, and Check 1981).

While developmental biologists have become increasingly reductionist in their approaches to amphibian metamorphosis (e.g., Shi 2000), ecologists have been more concerned with the way in which the environment impinges upon and modifies the basic program of larval development (Wilbur and Collins 1973). Staging tables such as those in fig. 13.3 are based on groups of tadpoles raised at constant temperatures with unlimited food. These tables often include estimates of larval size and age at each stage, giving the impression of highly predictable rates of development. However, amphibian larvae in nature are exposed to variable conditions that can affect growth rates, length of the larval period, and size at metamorphosis. Laboratory workers have long recognized that rearing conditions, including temperature, larval density, and food abundance, can have major effects on rates of growth and differentiation (see Salthe and Mecham 1974 for an early review). Amphibian larvae often respond to environmental cues such as larval density, food abundance, and pond drying by altering their developmental trajectories, either speeding up or slowing down progression toward metamorphosis in ways that presumably enhance their fitness. Ideally, one should attempt to combine the perspectives of developmental biologists and ecologists to provide new insights into the evolution of amphibian life history strategies and the proximate determinants of plastic developmental responses, but such connections have been made only recently (e.g., Denver 1995, 1997a, b; C. Rose 2005).

## Hormonal Control of Metamorphosis

A detailed discussion of the physiology of metamorphosis is beyond the scope of this book, but a brief overview of hormonal control will be useful for understanding ecological aspects of metamorphosis. This summary is based on several general reviews that can be consulted for additional details (Etkin 1964, 1968, 1970; Frieden and Just 1970; M. Dodd and Dodd 1976; B. White and Nicoll 1981; Hourdry and Beaumont 1985; Kikuyama et al. 1993; Denver 1996; Kaltenbach 1996; Tata 1996; C. Rose 1999; Shi 2000; Denver, Glennemeier, and Boorse 2002; Tata 2005).

### The Role of Thyroid Hormones

The timing of critical events in amphibian development ultimately is under genetic control, but gene expression is mediated by hormone action. Larval development involves both growth (an increase in mass) and differentiation (structural changes in tissues and organs leading to metamorphosis), and different hormones control the two processes. It has been known since early in this century that metamorphosis is initiated by thyroid hormones. Thyroid-mediated metamorphosis probably is an ancestral trait for all amphibians, but we do not have complete data to test this hypothesis. Both anurans and urodeles have thyroid-mediated metamorphosis, but very little is known about metamorphosis in caecilians (C. Rose 1999). Thyroid-mediated metamorphosis also is found in teleost fishes that undergo pronounced metamorphosis (de Jesus, Toledo, and Simpas 1998; Schreiber and Specker 1998; Yamano and Miwa 1998; Solbakken et al. 1999). On the other hand, in lampreys, metamorphosis is not triggered by a rise in thyroid hormones (Youson 1997). The closest living relatives of amphibians, the lungfishes and coelacanths, are not much help. Very little is known about coelacanth reproduction, but they are thought to be live bearers. Hormonal control of lungfish development is not well understood, but these animals do not undergo a dramatic metamorphosis and appear to be neotenic (Joss et al. 1997; Joss 1998).

A simplified diagram of the production pathways and regulatory interactions of thyroid hormones and other

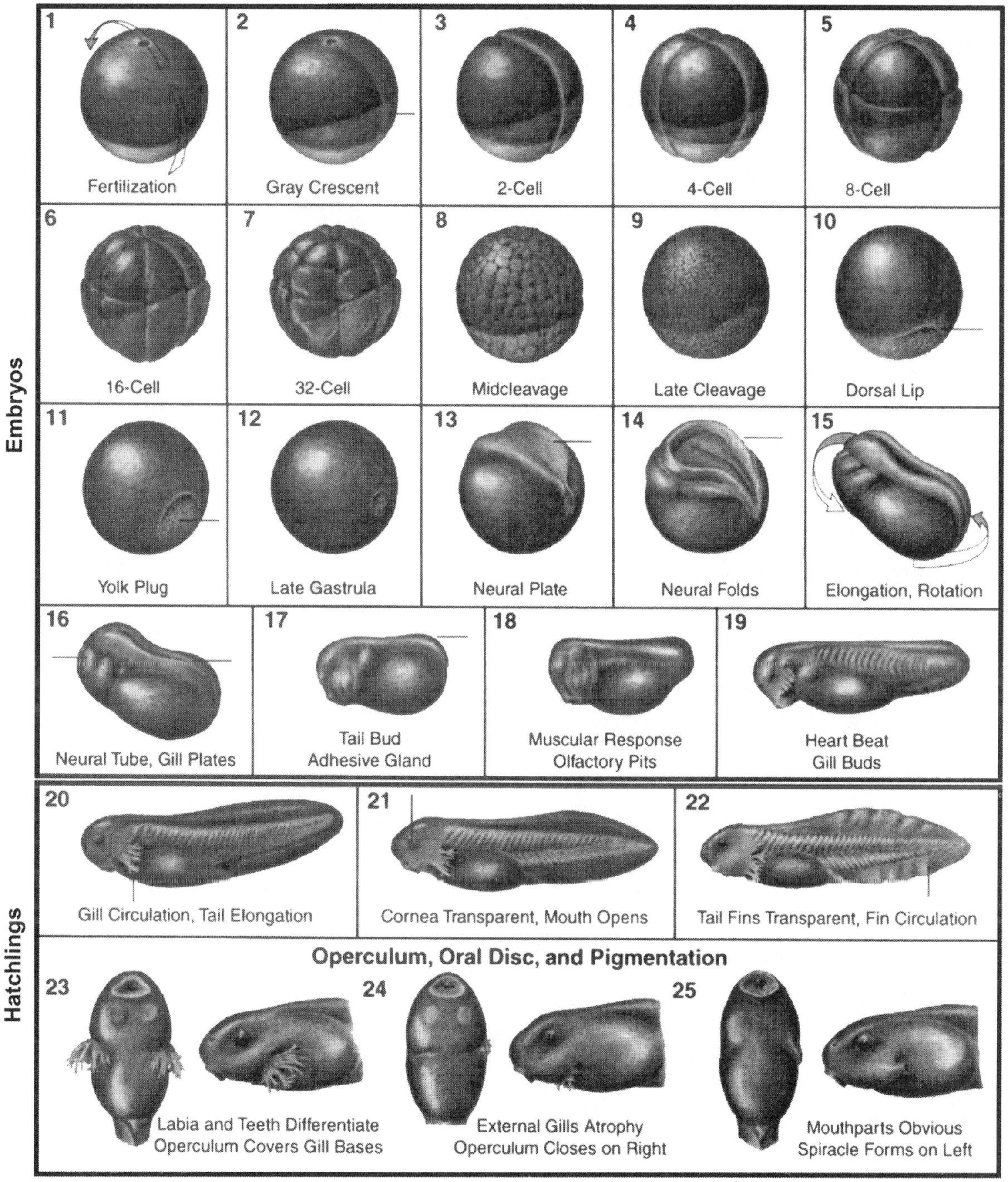


**Fig. 13.3.** Gosner (1960) staging system for anuran tadpoles. These stages are based on Bufo valliceps raised at 25° C. Stages 1–19 are embryonic stages, although hatching can occur as early as stage 16 in some anurans. Hatchling stages (20–25) represent the transition to an active, feeding tadpole. External gills disappear by the end of the hatchling period, and the spiracle (opening from internal gill chamber) appears. Mouthparts used in feeding also appear. Stages 26–41 are tadpoles, which undergo extensive growth and develop limbs. Metamorphs (stages 41–46) have fully developed legs, and eventually reabsorb the tail. After McDiarmid and Altig (1999).

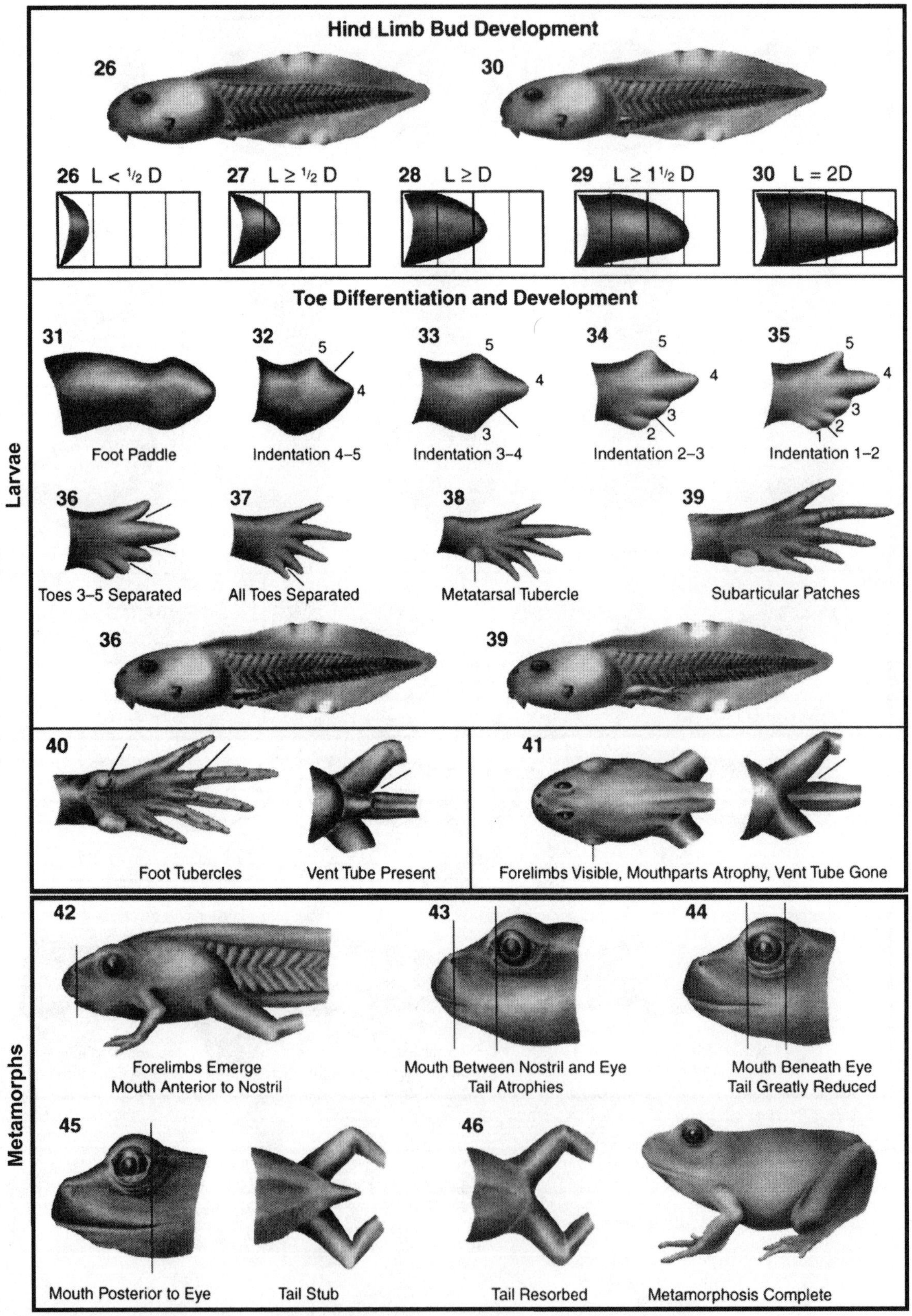


**Fig. 13.3.** (continued)

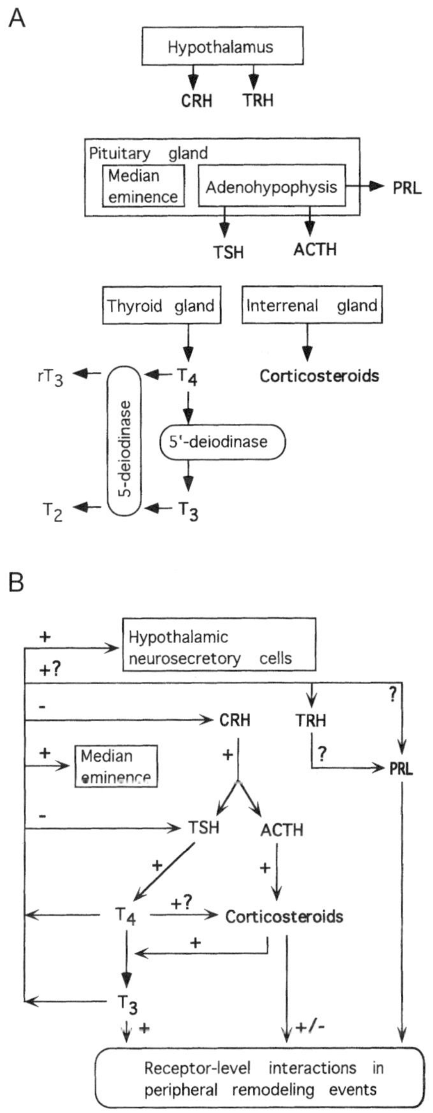


**Fig. 13.4.** Production pathways and regulatory interactions of thyroid hormones, corticosteroid hormones, and prolactin in tadpoles. (A) Endocrine glands involved in hormone production and their products. (B) Stimulatory (+) and inhibitory (–) effects of these hormones on gland maturation, hormone production and processing pathways, and tissue responses. Abbreviations: ACTH = adrenocorticotropin; CRH = corticotropin-releasing hormone; PRL = prolactin; TRH = thyrotropin-releasing hormone; TSH = thyroid-stimulating hormone; $T_2$ = diiodothyronine; $T_3$ = triiodothyronine; $T_4$ = thyroxine. After C. S. Rose (1999).

hormones that affect amphibian metamorphosis is shown in fig. 13.4. Early in development, the thyroid gland is small and relatively inactive. The production of thyroid hormones is regulated by thyroid-stimulating hormone (TSH) produced in the pituitary, which in turn is regulated by production of corticotropin-releasing hormone (CRH) in the hypothalamus (Denver 1996; Denver, Glennemeier, and Boorse 2002). Initially, the part of the hypothalamus that releases CRH is relatively small. It increases in size and activity in response to low levels of thyroid hormones in the blood, and levels of CRH begin to rise. Circulating levels of thyroid hormones are very low in the earliest stages of tadpole development, but they increase during prometamorphosis and peak at metamorphic climax (Kaltenbach 1996). At the same time, the sensitivity of target tissues to thyroid hormones increases and also peaks at metamorphosis, but the timing of sensitivity peaks varies in different tissues. The action of thyroid hormones is mediated by interactions with several other hormones, including prolactin, which is produced by the pituitary gland, and corticosterone, produced by the adrenal glands (T. Hayes 1997).

## The Role of Prolactin

Growth is stimulated by prolactin and possibly other hormones secreted by the pituitary. Apparently this occurs because prolactin increases intestinal absorption of amino acids and glucose from digested food (Frye, Brown, and Snyder 1972). The action of prolactin and thyroid hormones is antagonistic at the level of the target tissues, and treatment of tadpoles with mammalian prolactin can have inhibitory effects on metamorphosis. These results lead Etkin (1968) to propose a model of metamorphosis that predicted that circulating levels of prolactin would be high early in development and decrease as tadpoles progressed toward metamorphosis, resulting in low levels of this hormone at the time when thyroid hormones are high. This model was widely adopted as the standard model of amphibian metamorphosis (e.g., Bentley 1982; Duellman and Trueb 1986; Beachy, Surges, and Reyes 1999). Subsequent work has shown this model to be incorrect, because circulating levels of prolactin increase during development and peak in late metamorphic climax (Kikuyama et al. 1993; Kaltenbach 1996), as does expression of the gene that encodes prolactin (Buckbinder and Brown 1993). Rather than inhibiting production of prolactin, thyroid hormones have been shown to increase circulating levels of prolactin late in development (K. Kawamura, Yamamoto, and Kikuyama 1986; Denver 1996). Prolactin does not appear to affect the timing of metamorphosis, but it does mediate the effects of thyroid hormones on particular target tissues (Denver 1996). For example, prolactin can inhibit some, but not all, of the programs of tail resorption in *Xenopus* tadpoles (Huang and Brown 2000; Tata 2005).

## Interaction between Hormones and Environmental Cues

While this overview is simplified, it indicates how growth and differentiation are regulated by interactions among different endocrine glands and the target tissues. What is less well understood is how environmental factors impinge on this regulatory process. Changes in temperature, food availability, or larval density could affect the production of hormones by the endocrine glands, the regulating function of the hypothalamus, the sensitivity of the target tissues to different hormones, or a combination of all three. For example, low temperature is known to inhibit production of thyroxin and reduce the sensitivity of target tissues to this hormone (Kollros 1961; Etkin 1964, 1968; Smith-Gill and Berven 1979). Starvation can inhibit both growth and differentiation, although not necessarily to the same degree (Pandian and Marian 1985a). Hence, there is ample opportunity for environmental variables to influence both duration of the larval period and body size at metamorphosis.

As discussed in more detail in the next section, many anuran tadpoles exhibit considerable plasticity in the timing of development and size at which individuals metamorphose. A number of species exhibit accelerated metamorphosis under deteriorating environmental conditions, such as pond drying (R. Newman 1992). Ecological models of metamorphosis are based on the assumption that larvae can alter their rates of development in response to environmental conditions such as pond drying, increased competitor density, or decreasing food supplies (Wilbur and Collins 1973). Most ecologists, however, have paid little attention to the link between environmental cues and the hormonal control of metamorphosis, and most development biologists have not considered environmental factors at all.

Corticotropin-releasing hormone (CRH) provides a possible link between certain types of environmental stress, such as pond drying, and the hormonal control of metamorphosis (fig. 13.5). In all vertebrates, CRH functions as a general stress hormone that stimulates production of corticosterone in the adrenal glands. This hormone in turn stimulates escape reactions and other behavioral and physiological responses to stressful conditions. In tadpoles, CRH has been shown to activate both adrenal and thyroid hormone secretion (Denver 1995, 1997a, b; Boorse and Denver 2004; C. Rose 2005). One effect on tadpoles that are close to metamorphosis is an increase in foraging behavior and increased mobilization of energy reserves, which in turn can accelerate growth and development (Crespi and Denver 2004, 2005). In addition, CRH stimulates increased production of thyroid hormones and facilitates conversion of thyroxine to a more potent thyroid hormone, triiodothyronine (T. Hayes and Wu 1995). Circulating levels of CRH and corticosterone have been shown to increase dramatically in tadpoles in response to pond drying (Denver 1998; Boorse and Denver 2004), increased density of competitors (T. Hayes 1997; Glennemeier and Denver 2002), and restricted food availability (Boorse and Denver 2004; Crespi and Denver 2005). When tadpoles of *Bufo, Spea,* and *Rana* were treated with CRH, metamorphosis was significantly accelerated compared to untreated controls (Denver 1993, 1997a, b; Miranda, Affani, and Paz 2000). Similar results were obtained in experiments with *Ambystoma tigrinum*, indicating a common mechanism for anurans and urodeles (Boorse and Denver 2002). Taken together, these studies provide a plausible mechanistic explanation for the ability of tadpoles to assess environmental conditions and alter their development in ways that increase their chances of survival.

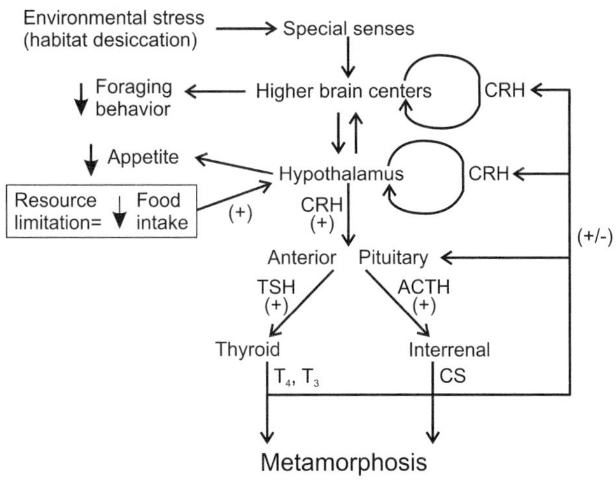


**Fig. 13.5.** A model for the interaction of environmental stress factors, especially pond drying, with the hormonal control of metamorphosis in tadpoles. The stress of pond drying is sensed through special senses, which have yet to be determined. This stress activates extrahypothalamic corticotropin-releasing hormone (CRH) neurons, which then reduce foraging behavior. Activation of CRH neurons in the hypothalamus reduces appetite. Hypothalamic CRH stimulates pituitary thryoid-stimulating hormone (TSH) and adrenocorticotropic hormone (ACTH) secretion, which in turn stimulates the production of thyroid hormones (TH) and corticosteroid (CS) hormones that promote metamorphosis. After Denver (1997a).

Another way in which the environment could affect the hormonal control of metamorphosis is through variation in larval diet (Kupferberg 1997c). When tadpoles are fed on a high protein diet, they grow faster than those fed on low protein diets, and they metamorphose earlier as well. Much of this difference is no doubt due to the direct effect of improved nutrition on growth and development, but it is possible that diet quality has an indirect effect on metamorphosis through its effect on hormone function. In other vertebrates, the activity of an enzyme that converts thyroxine to triiodothyronine is mediated by the availability of dietary protein. Diet quality in natural populations of *Pseudacris regilla* tadpoles in California has been shown to affect larval growth and development. Tadpoles with access to filamentous green algae covered with epiphytic diatoms had shorter larval periods and metamorphosed at a larger size than did tadpoles that did not have access to diatoms (Kupferberg, Marks, and Power 1994). Diet also can affect other aspects of development, such as the development of specialized carnivorous or cannibalistic morphs in spadefoot toad (*Spea*) tadpoles and tiger salamander (*Ambystoma tigrinum*) larvae (see chapter 12). Unfortunately, there has not yet been a study of thyroid hormone levels in relation to variation in tadpole diets that would establish the link between diet and hormone function (C. Rose 2005).

## The Ecology of Amphibian Development and Metamorphosis

Ecologists have long been interested in the way in which environmental factors influence the growth and development of amphibian larvae (e.g., R. Savage 1952, 1961; Salthe and Mecham 1974). It was not until the 1970s, however, that a conceptual framework for understanding larval development and metamorphosis was developed. The starting point for modern discussions of the ecology of amphibian metamorphosis is a model developed by Wilbur and Collins (1973; see R. N. Harris 1999, for a detailed review). Wilbur and Collins were mainly concerned with explaining why individual larvae in the same population and even the same age cohort often differ dramatically in growth rate, the length of the larval period, and size at metamorphosis. They proposed that the timing of metamorphosis is related to larval growth rates and physical condition, these being proximate physiological cues that allow an individual larva to assess the relative costs and benefits of remaining in the larval stage or emerging onto land. Larval growth, in turn, is influenced both by physical factors such as temperature and by larval density and food availability. This model has been very influential in studies of amphibian metamorphosis, but it has been applied to the development of other organisms as well, including insects (Bradshaw and Johnson 1995; Peckarsky et al. 2001), copepods (Twombly 1996), barnacles (Hentschel and Emlet 2000), and fishes (Reznick 1990).

### Elements of the Wilbur-Collins Model

The essential features of the Wilbur-Collins model are as follows. Amphibian larvae raised in isolation exhibit an initial period of nearly exponential growth, followed by a decrease in growth rate and eventually a reduction in mass at metamorphosis. In a natural population, individuals vary in growth rate. Faster-growing individuals will enjoy a head start over their companions, and this translates into competitive superiority throughout the larval period. This produces a skewed distribution of body sizes in the population, with many individuals remaining small while a few reach larger sizes.

Wilbur and Collins postulated that each species has a characteristic minimum size that must be reached before metamorphosis can occur, as well as a maximum size that cannot be exceeded in the larval stage because of increasing inefficiencies in feeding (fig. 13.6). They predicted that when

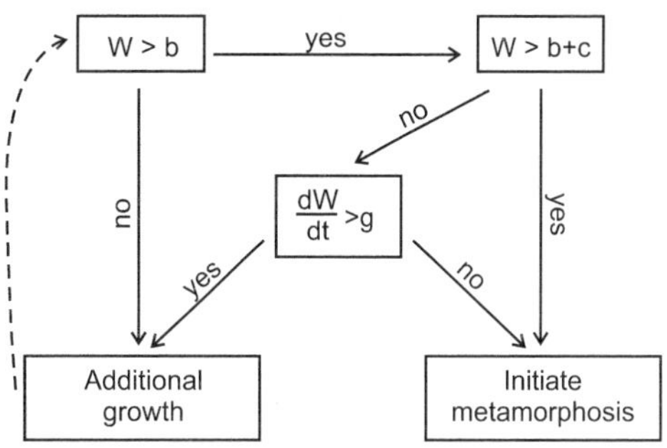


**Fig. 13.6.** Graphical representation of the Wilbur-Collins model of amphibian metamorphosis. The model postulates that larvae must reach a minimum threshold size (b) before they can undergo metamorphosis. Metamorphosis must occur before the larvae reach a maximum size (b + c), at which it is no longer possible to remain in the larval stage. The probability of undergoing metamorphosis depends upon larval body size (W) and size-specific growth rate (dW/dt). If larvae have reached a body size that exceeds the minimum size for metamorphosis (W > b), but have not yet reached the maximum larval size (b + c), then they are expected to continue growing if recent growth rates are high (dW/dt > g). If recent growth rates are low (dW/dt < g), where g is some function of current body mass, then they are expected to metamorphose. After Wilbur and Collins (1973).

resources are abundant, larvae grow quickly, and most reach the maximum size before transforming. Consequently, variation in the length of the larval period and size at metamorphosis will be minimal. When resources are scarce, both the average size at metamorphosis and the range of variation will change with larval density. Competitively superior individuals will be the first to reach the minimum metamorphic size. Because conditions are unfavorable for growth, these larvae are expected to transform at or near the minimum size. Others, having been released from competition with larger conspecifics, eventually reach metamorphic size, but only after a longer larval period. If the breeding pond lasts long enough, some late-transforming larvae reach a larger size at metamorphosis because of reduced competition. However, in temporary ponds, many larvae cannot reach metamorphic size before the water disappears.

Day and Rowe (2002) developed a more formal mathematical model with many of the same elements as the Wilbur-Collins verbal model. Specifically, they examined the effect of developmental thresholds on the response of larvae to changes in environmental conditions. Their results were largely consistent with the Wilbur-Collins model. They pointed out that a developmental threshold, such as the minimum size required for metamorphosis postulated by Wilbur and Collins, has the effect of reversing the growth response to poor resource availability. Below the threshold for metamorphosis, larvae are expected to respond to low food levels by slowing rates of growth and development, thereby lengthening the larval period. Above the size threshold, larvae are expected to respond to slow growth rates brought about by low food availability by accelerating development and metamorphosing as quickly as possible. This model is consistent with the findings of Crespi and Denver (2004) that stress hormones, which increase when food levels are reduced, stimulate foraging and accelerate growth and development in tadpoles nearing metamorphosis, but do not affect foraging by tadpoles in earlier stages of development.

### Empirical Evidence for Variation in Larval Growth Rates

Many studies have established that individual larvae grow at different rates, even when raised under identical conditions (Wilbur 1984). In some species, individuals that initially have slow growth rates never catch up with faster-growing larvae (Travis and Trexler 1986), but in other species, they do (Semlitsch and Caldwell 1982). Life-history traits such as larval growth rate, length of the larval period, and size at metamorphosis all are affected by genetic differences among individuals, leading to consistent differences in performance among sibships (Travis 1980a, 1981, 1983a; Berven and Gill 1983; Pływcz et al. 1984; Travis, Emerson, and Blouin 1987; Blouin 1992; Semlitsch 1993; Girish and Saidapur 1999b; G. R. Smith 1999; Laurila, Karttunen, and Merilä 2002). These phenotypic differences translate into competitive differences among individuals or different capacities to escape predation (Travis 1980b, 1983b; Travis, Keen, and Juilianna 1985a, b; Travis and Trexler 1986). Experiments in which half-sibs were raised under identical conditions have shown that both paternal and maternal genotypes influence phenotypic characters related to larval fitness (Woodward 1986; Travis, Emerson, and Blouin 1987; R. Newman 1988a; Semlitsch 1993). In some studies, maternal effects were of relatively minor importance (Travis 1983b; Travis and Trexler 1986; Travis, Emerson, and Blouin 1987; Laurila, Karttunen, and Merilä 2002) or diminished as the larvae grew (R. Newman, 1988a).

Nongenetic maternal effects can be important as well. Variation in egg size and energy content within and among females has been documented in a number of species (Kuramoto 1978; R. Kaplan 1980, 1985; Crump 1981b, 1984a; Travis 1983a), but the importance of such variation for larval fitness is poorly understood. The effect of variation in egg size is discussed in more detail in chapter 10, and only a brief summary will be given here. Kaplan (1980, 1985) found that *Ambystoma* and *Taricha torosa* larvae that hatched from large eggs grew more rapidly and transformed at a larger size than did those from smaller eggs. In most studies, however, egg size had relatively little effect on the length of the larval period (see examples in table 10.7 in chapter 10). It is more common for egg size to influence the size of larvae at hatching and at first feeding, but differences often disappear by metamorphosis (e.g., Semlitsch and Gibbons 1990). In one study of natterjack toads (*Bufo calamita*) in Spain, egg size was positively correlated with the size of hatchlings when they began to feed, but this variation did not have any effect on survivorship or size at metamorphosis (Tejedo and Reques 1992). Berven and Chadra (1988) examined the effects of egg size, density, and food level on larval development in wood frog tadpoles (*Rana sylvatica*). Tadpoles that hatched from large eggs had shorter larval periods, but only at relatively low densities. There was no direct effect of egg size on size at metamorphosis, but there were complex interactions between egg size and both density and food abundance. At low density, tadpoles that hatched from small eggs metamorphosed at a larger size than did those from large eggs, but the pattern was reversed at high density. Overall, tadpoles derived from large eggs were less sensitive to the effects of density than were those derived from small eggs. In a study of *Salamandra salamandra,* which gives birth to larvae rather than laying eggs, larger size at birth translated into faster development and larger size at metamorphosis, although the effect was mediated by environmental factors such as temperature and food availability (Alcobendas, Buckley, and Tejedo 2004). Taken together, these studies suggest that genetic and nongenetic variation among individuals in a larval cohort is sufficient to produce the asymmetries in body size and competitive ability required by the Wilbur-Collins model, but the sources of such phenotypic variation vary considerably among species.

### Intraspecific Competition and Density-Dependent Growth

#### Studies of Anuran Tadpoles

The effects of intraspecific competition on larval growth and metamorphosis have been investigated in both the laboratory and field for many species of anurans. Many laboratory studies have shown that tadpoles raised under crowded conditions have slower growth rates, longer larval periods, and smaller sizes at metamorphosis than those raised at lower densities (Adolph 1931; Lynn and Edelmann 1936; Richards 1958; S. Rose 1960; Licht 1967; Pourbagher 1969; Heusser 1972a, b; Gromko, Mason, and Smith-Gill 1973; Wilbur 1977a, b; Smith-Gill and Gill 1978; Crump 1981a; Semlitsch and Caldwell, 1982; Berven and Gill 1983; Petranka 1984d; A. Sokol 1984; Travis 1984; Pandian and Marian 1985a, b; Berven and Chadra 1988; Cummins 1989; G. R. Smith 1999; Buchholz and Hayes 2000; Tejedo, Semlitsch, and Hotz 2000; Morey and Reznick 2001; Saidapur and Girish 2001; Girish and Saidapur 2003; Gramapurohit, Veeranacoudar, Shanbhag, and Saidapur 2004). Manipulation of food levels while density is held constant produces similar results, with tadpoles having longer larval periods and smaller size at metamorphosis when provided with small amounts of food (Travis 1984; Berven and Chadra 1988; Semlitsch 1993; Hotz et al. 1999; Laugen, Laurila, and Merilä 2002) or low quality food (Álvarez and Nicieza

2002a). Some of these studies have used densities of tadpoles that seem unrealistically high for natural populations of the species being studied (e.g., G. R. Smith 1999), but others have used more natural densities (e.g., Wilbur 1977a, b). Nevertheless, laboratory studies are somewhat biased toward discovering competitive effects because of the highly simplified environment in which the tadpoles are raised (see Skelly and Keisecker 2001).

Several workers have raised tadpoles at different densities in field enclosures, cattle tanks, or natural pools, with similar results (Brockelman 1969; Wilbur 1971, 1972, 1976; Wilbur and Collins 1973; D. C. Smith 1983; Travis 1983b; Travis, Keen, and Juilianna 1985a; Travis and Trexler 1986; R. Newman 1987, 1989; Warner, Dunson, and Travis 1991; Goater 1994; Tejedo and Reques 1994b; Altwegg 2002). Typically, as initial density increases, larval growth rates decrease (fig. 13.7). Not only does size at metamorphosis decrease with increasing density, but the variance in size at metamorphosis changes as well. At high densities, most larvae remain very small, while a few continue to grow and reach metamorphic size (fig. 13.8; Wilbur and Collins 1973; Wilbur 1976). At lower densities, most of the tadpoles reach a relatively large size, and variation is reduced. In some cases, survivorship of tadpoles is reduced at high densities as well (Warner, Dunson, and Travis 1991). Similar density effects have been found in experiments designed to examine interspecific competition among two or more species of tadpoles; these experiments are discussed in chapter 15.

Semlitsch and Caldwell (1982) found that *Scaphiopus holbrookii* tadpoles raised at high densities in the laboratory actually transformed at a larger mean size than those raised at intermediate densities. They attributed this to lower survivorship at high densities, with the survivors being released from competition and resuming growth. In experiments with *Hyla gratiosa* tadpoles raised individually in separate containers, Travis (1984) showed that low food levels had the same effect on larval growth, length of the larval period, and mean size at metamorphosis as high-density treatments. However, absolute food level did not affect variation in length of the larval period. Travis simulated competitive release by adding increments of food to each tadpole's container in proportion to the number of larvae in the other containers that transformed each day. Under these conditions, tadpoles responded to the food supplement by increasing growth and reaching a larger size at metamorphosis. R. Newman (1987) provided supplemental food to tadpoles of *Scaphiopus couchii* in natural ponds with high and low densities of tadpoles. Many of these ponds dried up before the tadpoles reached metamorphosis. Nevertheless, one high- and one low-density pond with supplemental food produced larger metamorphs than ponds without extra food (fig. 13.7 B).

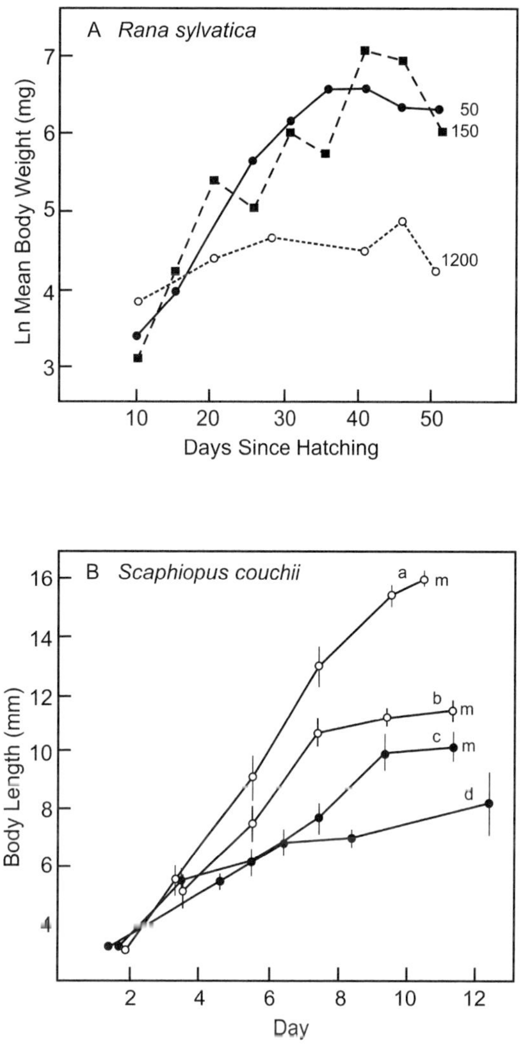


**Fig. 13.7.** Representative growth curves for tadpoles raised at different densities. (A) Growth curves for *Rana sylvatica* tadpoles at initial densities of 50, 150, and 1,200 individuals reared in pens in a natural pond. After Wilbur and Collins (1973). (B) Growth curves for *Scaphiopus couchii* tadpoles in high- and low-density natural ponds. Curve (a) is for a low-density pond to which food was added. Curve (b) is for a low-density pond to which food was not added. Curve (c) is a high-density pond to which food was added that produced metamorphs. Curve (d) is for a high-density pond to which food was not added that failed to produce metamorphs before the pond dried up. After Newman (1987).

In most of the experiments designed to test for density effects on larval growth, environmental conditions have been held relatively constant. However, in natural ponds, density per se has little direct effect on larval survival, but can affect survival through the action of other selective agents, such as predators and a failure of larvae to reach metamorphosis before ponds dry up (e.g., R. Newman 1987; Loman 2002c). The responses of larvae to differences in density and other environmental variables interact in complex ways. For example, predators reduce larval densities, thereby freeing surviving larvae from competition and allowing them to grow at faster rates (D. C. Smith 1983; Wilbur, Morin, and Har-

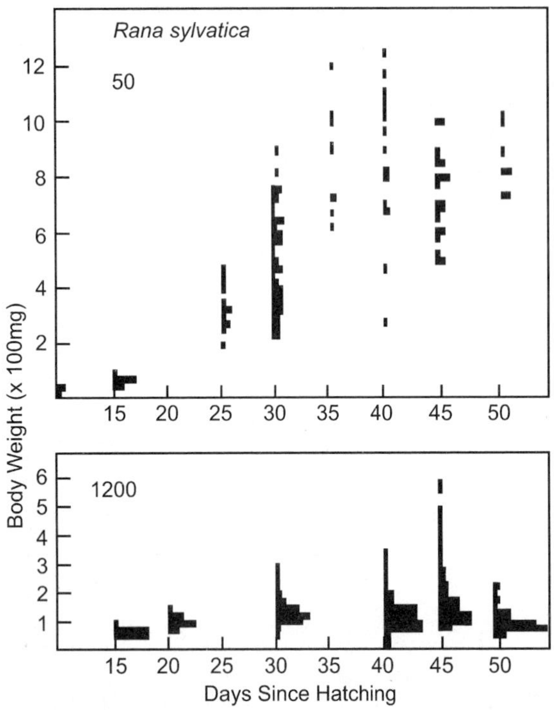


**Fig. 13.8.** Frequency histograms of body mass of *Rana sylvatica* tadpoles at different ages. Tadpoles were raised at low (50) and high density (1,200). Tadpoles in low-density pens grew rapidly and metamorphosed at relatively large sizes. Tadpoles in high-density pens mostly remained very small, with only a few individuals reaching metamorphic size. After Wilbur and Collins (1973).

ris 1983; Travis, Keen, and Juilianna 1985a). On the other hand, the effectiveness of predators in removing larvae from the population is influenced by initial larval density (Travis, Keen, and Juilianna 1985a) and larval size (Caldwell et al. 1980; Woodward 1983; Travis, Keen, and Juilianna 1985b; Cronin and Travis 1986; see chapter 14).

### Studies of Salamander Larvae

Density effects and intraspecific competition for food have not been studied as extensively in salamander larvae as in tadpoles, and most of the work that has been done has been on pond-dwelling *Ambystoma* larvae. In an early study using pens in natural ponds, Wilbur (1972) found that the presence of a predator, *Ambystoma tigrinum,* reduced the survivorship of larvae of other species of *Ambystoma,* but body size of the survivors was larger than when predators were not present. This suggested that predation reduced densities of larvae and enabled the survivors to find more food. In another set of experiments in the same pens, Wilbur (1976) raised larvae of *A. maculatum* and *A. laterale* at different densities. High-density treatments resulted in higher mortality, slower growth, smaller size at metamorphosis, and longer larval periods in both species. Similar results were reported by Wilbur and Collins (1973), and they also found that body sizes of larvae were skewed toward small sizes in high-density pens. Later, Van Buskirk and Smith (1991) manipulated densities of *A. laterale* larvae in natural rock pools along the shore of Lake Michigan, all of which were free of predators. As in earlier pen experiments, high densities decreased both growth rate and survival, and body sizes of larvae in high-density pools were skewed toward small sizes. Studies of larval *A. talpoideum* in artificial ponds (Semlitsch 1987a) and *A. tigrinum nebulosum* in pens placed in a natural lake (Brunkow and Collins 1996) showed similar density-dependent effects on larval growth.

Studies of the marbled salamander (*Ambystoma opacum*) are particularly useful in evaluating the effects of density on larval growth and development because of the variety of techniques that have been used. Experiments have been conducted in laboratory containers (Stenhouse, Hairston, and Cobey 1983), small pens placed in natural ponds (C. K. Smith 1990), artificial ponds (C. K. Smith 1990), large enclosures in natural ponds (D. Scott 1990, 1994; Chazal, Krenz, and Scott 1996), and natural ponds divided in half (Petranka 1989b). The results of these studies have been synthesized into a general demographic model for *Ambystoma opacum* (B. Taylor and Scott 1997). Experiments conducted in small pens or laboratory containers can bias results toward finding density effects (Travis, Keen, and Juilianna 1985a), and often result in considerable variation among replicates of the same treatment (e.g., Wilbur 1976). The same is true for small artificial ponds or cattle tanks, but to a lesser degree (Skelly and Kiesecker 2001). Laboratory experiments and those conducted in small field enclosures sometimes yield unusually high estimates of survivorship, because predators are excluded from the experiments (see D. Scott, 1990, for a comparison of survivorship estimates for *A. opacum*).

Stenhouse, Hairston, and Cobey (1983) raised larvae of both *Ambystoma maculatum* and *A. opacum* at different densities in small laboratory containers and reported that both species exhibited slower growth at high densities than at low densities. Densities in these experiments ranged from those similar to field densities to densities that probably were higher than those commonly observed in the field. Petranka (1989b) attempted to deal with the problems of laboratory and small enclosure experiments by dividing eleven natural ponds in half and stocking larvae of *A. opacum* in each half at natural densities or at one fourth natural densities. Overall, there were few differences in larval growth rates in high- and low-density treatments for the first seven weeks of development, but size differences began to appear by the eleventh week and persisted until metamorphosis, when larvae from low-density treatments emerged at significantly larger sizes. There was considerable variation among ponds, however, with larvae from ponds with low absolute densities generally showing little or no difference between treatments. Nevertheless, density-dependent growth was evident in one low-density pond and was not detected in one high-density pond. Probably these differences were due to variation in characteristics of the ponds, including predator density and food supply. Density dependence was most pronounced in ponds with relatively low populations of cladocerans, the principal food of *Ambystoma* larvae. In most ponds, survival of larvae was higher in the low-density half of the pond.

D. Scott (1990, 1994) conducted a similar set of experiments with the same species, but in large enclosures that allowed for replicated treatments within ponds. These were open-topped enclosures of either 41 m² or 23 m², placed in two different natural ponds. Each enclosure contained natural populations of invertebrate prey and predators. Initial density of larvae was varied to mimic high and low densities in natural ponds. As in Petranka's study, larvae at high densities had lower survivorship, slower growth rates (fig. 13.9), and smaller size at metamorphosis than those at low density. Survival rates were not significantly different from those in natural ponds, however. Mortality was especially high, up to 98%, in high-density enclosures in years when the ponds dried early, because most larvae had not yet reached metamorphic size. The effect of density on length of the larval period in these experiments and others conducted in one of the same ponds (Chazal, Krenz, and Scott 1996) were inconsistent, with high densities resulting in longer larval periods in some years, but not in others. Variation among years accounted for much more of the variance in larval period than did density treatment (D. Scott 1990). Experiments conducted in replicated artificial ponds with the same species also showed density-dependent growth and development (C. K. Smith 1990). Variation in larval performance can have significant effects on adult demographic traits (D. Scott 1994) and population dynamics (B. Taylor and Scott 1997). The specific effects on adult fitness are discussed in more detail in a later section.

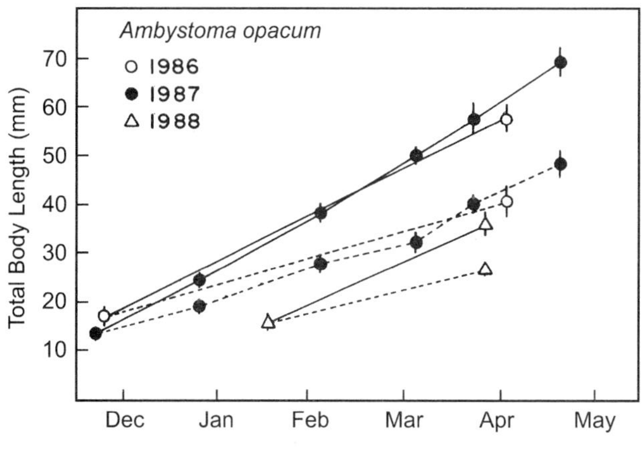


**Fig. 13.9.** Growth curves for *Ambystoma opacum* larvae grown at high densities (dashed lines) and low densities (solid lines) in field enclosures in a temporary pond in three different years. In all three years, growth rates and size at metamorphosis were higher for larvae in low-density enclosures, but all larvae grew slowly in 1988, regardless of density. After D. Scott (1990).

Information on salamanders other than *Ambystoma* is limited, but density-dependent effects on growth and survival have been reported for both *Notophthalmus viridescens* and *Siren intermedia* larvae raised with different densities of conspecifics in cattle tanks (R. N. Harris 1987b; Fauth, Resetarits, and Wilbur 1990). Larval growth and size at metamorphosis also were lower in a natural pond with high larval densities than in the same pond in another year with low densities (R. N. Harris, Alford, and Wilbur 1988). In contrast, larvae of *Hemidactylium scutatum* raised under different food regimes showed little difference in length of the larval period, although slow-growing larvae did metamorphose at a smaller size (O'Laughlin and Harris 2000). These results suggest that this species has a relatively fixed rate of larval development, but growth rate is influenced by food availability.

### Evidence for Food Limitation in Nature

An implicit assumption of all of the studies described in the previous section is that levels of competition detected under experimental conditions reflect those normally encountered in nature. However, this assumption has seldom been explicitly tested (Skelly and Kiesecker 2001; Skelly 2002). In some studies, larval densities have been considerably higher than estimates for field populations, but evidence for competition has been found even at relatively low densities (Travis et al. 1985b; Travis and Trexler 1986). Several field studies of natural ponds have demonstrated a negative correlation between tadpole density and size at metamorphosis, or increased mortality at high densities (R. Savage 1952, 1961; R. Newman 1987; Reques and Tejedo 1995; Reading and Clarke 1999). These studies suggest that intraspecific competition is occurring, although factors such as food abundance and quality, water volume, and temperature variation cannot be controlled in natural ponds, and all of these factors could influence the growth and survivorship of tadpoles, either independently or in conjunction with tadpole density.

Many anuran tadpoles eat periphyton and other plant material that is seasonally superabundant (see chapter 12), and they are subject to predation that can rapidly decrease tadpole numbers (see chapter 14). This has led some authors to conclude that tadpole populations often are below carrying capacity and not food-limited (Herreid and Kinney 1966; Calef 1973a; Heyer 1973, 1974, 1976; Licht 1974; Heyer, McDiarmid, and Weigmann 1975; Cecil and Just 1979). Nevertheless, there is evidence of food limitation in some natural habitats. Food quality for tadpoles can vary annually in the same habitat (Kupferberg, Marks, and Power 1994) or in adjacent habitats (Peterson and Boulton 1999), or as a result of natural succession and changes in leaf litter composition in ponds (Rubbo and Kiesecker 2004). Variation in food quality is reflected by differences in growth rate. For example, both wood frog (*Rana sylvatica*) and spring peeper (*Pseudacris crucifer*) tadpoles grew more slowly in closed-canopy, light-limited ponds with low productivity than in open-canopy ponds with higher productivity (Skelly, Friedenburg, and Kiesecker 2002). Experimental supplementation of food in pond enclosures increased growth rates in both open- and closed-canopy ponds, but the effect was greatest in open-canopy ponds. Tadpoles of two European frogs, *Rana temporaria* and *R. arvalis,* also had slower development and smaller size at metamorphosis in closed-canopy ponds than in open-canopy ponds (Loman 2002b), and tadpoles of the Australian frog *Litoria ewingii* had slower development and decreased growth rates in shaded ponds (Lauck, Swain, and Barmuta 2005). Tadpoles of the stream-breeding frog *Ascaphus truei* had much slower growth rates in a shaded, cool stream than in a nearby less-shaded, warmer stream (Mallory and Richardson 2005).

Dense populations of tadpoles can substantially reduce the standing crop of algae in ponds and streams and alter the composition of the algal community (Dickman 1968; Seale et al. 1975; Seale, Rodgers, and Boraas 1980; Lamberti et al. 1992; Kupferberg 1997b; Graham and Vinebrooke 1998; Holomuzki 1998), whereas lower densities of tadpoles sometimes have little measureable effect on algal biomass (Kiffney and Richardson 2001). In some circumstances, grazing by tadpoles can actually facilitate growth of some algae (Osborne and McLachlan 1985; Kupferberg 1997a). Measurements of organic nitrogen in one pond ecosystem indicated limited availability of food for tadpoles (Seale 1980). Similarly, addition of nutrients to stream channels in British Columbia increased the biomass of both periphyton and grazing *Ascaphus* tadpoles, with tadpoles growing at much faster rates in nutrient-enriched streams (Kiffney and Richardson 2001). In a study of *Pseudacris triseriata* tadpoles in isolated pools, D. C. Smith (1983) found that addition of food to pools partially reversed the negative effects of density on tadpole growth and development. The positive effect of food supplementation on growth of *Scaphiopus couchii* tadpoles (R. Newman 1987) in natural ponds has already been discussed. All of these data indicate that anuran larvae can be food-limited in natural habitats.

There is less direct evidence for food limitation in pond-dwelling salamander larvae, but it seems likely that salamander larvae often are food-limited because of their high trophic position in the community. Kusano (1981) reported that larvae of *Hynobius nebulosus* grew more slowly in a pond than larvae raised in the laboratory at the same temperature, and he found a negative correlation between number of larvae emerging from the pond each year and mean larval size. Stenhouse, Hairston, and Cobey (1983) reported that *A. opacum* larvae appeared to be hungrier in the field than in the laboratory. In some studies, *Ambystoma* larvae have been shown to eliminate preferred arthropod prey from ponds (Dodson and Dodson 1971; Sprules 1972), and these results provide indirect evidence of food limitation. On the other hand, Van Buskirk and Smith (1991) did not find any relationship between abundance of prey and larval growth rate in *A. laterale,* even though both growth and survival were negatively correlated with larval density. They also reported that high densities of larvae did not significantly reduce prey abundance. G. Bell and Lawton (1975) found that larvae of *Triturus vulgaris* were always well fed and grew at near maximum rates in the field, although this could have been related to some peculiarity of their study area, such as unusually high productivity.

Food is likely to be limiting for many stream-dwelling salamander larvae because of the low productivity of most stream environments (Beachy 1995). Stream-dwelling larvae often grow very slowly compared to pond-dwelling larvae, sometimes taking several years to reach metamorphosis. This could be due to low food availability, low temperature, or both, but there have been very few experimental studies designed to separate the effects of temperature and food availability on growth (Bernardo and Reagan-Wallin 2002). Petranka (1984d) reported that *Ambystoma barbouri* larvae in a stream had substantially higher growth rates when provided with supplemental food. Furthermore, size of metamorphosing larvae was inversely related to larval density (Petranka and Sih 1986). After catastrophic flooding of one stream, there was a dramatic decline in larval density, but the few surviving larvae experienced reduced competition and had faster growth rates and larger sizes at metamorphosis than larvae produced in years without floods. Studies of clinal variation in larval life histories of stream-dwelling plethodontids have shown that larvae from cold, high-elevation sites generally grow more slowly and metamorphose at smaller sizes than do those at warmer low-elevation sites (Bernardo and Reagan-Wallin 2002). In one study of *Desmognathus ocoee* from high and low elevations, the abundance of invertebrate prey and the length of the feeding season appeared to be more important that average environmental temperature in determining differences in growth rate (Bernardo and Agosta 2003).

### Mechanisms of Intraspecific Competition

Another assumption of the Wilbur-Collins model is that larger individuals are competitively superior to smaller ones, a prediction supported by some experiments (e.g., Steinwascher 1978b, 1979a; Woodward 1987), but not others. Both exploitative and interference competition could be involved, but the relative importance of the two mechanisms probably differs among species. Exploitative competition operates through individual differences in the ability of larvae to obtain and process food resources. Interference competition involves behavioral or chemical interactions among individuals that prevent some individuals from gaining access to resources, or directly inhibit growth.

### Exploitative Competition in Anuran Tadpoles

Exploitative competition probably is the most important form of competition in anuran tadpoles, but predicting the outcome of such competition is complicated. Larger tadpoles can pump water at a faster rate because buccal volume is a direct function of body size. Therefore, the quantity of suspended particles ingested should be greater for larger tadpoles, and they can monopolize a disproportionate share of the available food. On the other hand, efficiency of food capture is related to the surface area of the food capturing structures (Wassersug 1975). These structures exhibit negative allometric growth, which means that the surface area of the food-capturing structures does not increase in direct proportion to overall body size (Wassersug and Hoff 1979). Consequently, relative feeding efficiency should decrease as tadpoles grow. Wassersug (1975) found that twelve small *Rana pipiens* tadpoles could clear suspended particles from water faster than three large tadpoles of equivalent total mass. He suggested that this is important in setting an upper limit on tadpole size. One of the largest known tadpoles is that of *Pseudis paradoxa* (Emerson 1988b), which has branchial baskets no larger than those of the smaller tadpoles of *Rana catesbeiana*. Perhaps *P. paradoxa* tadpoles get by with a less efficient feeding apparatus by living in habitats with very high productivity.

The outcome of competitive interactions between tadpoles of different sizes also is hard to predict because of plasticity in their feeding behavior. Tadpoles can alter rates of ingestion in response to change in food concentration (Seale and Wassersug 1979). Generalized pond tadpoles also can change their feeding behavior, depending on the availability of different types of food. For example, Steinwascher (1979a) found that *Rana clamitans* tadpoles concentrated on solid food (rabbit chow) when food was abundant, but shifted to feeding on suspended particles when food was scarce. The efficiency of feeding on each type of food was a function of body size. Tadpoles also respond to reduced food levels by increasing the residence time of food in the gut, thereby increasing digestive efficiency (Wassersug 1975), and by increasing the proportion of time spent feeding (Horat and Semlitsch 1994).

### Interference Competition in Anuran Tadpoles

In some species, large tadpoles physically displace smaller ones from food patches (R. Savage 1952; Gromko, Mason, and Smith-Gill 1973; John and Fenster 1975; Wilbur 1977a; Faragher and Jaeger 1998), a form of interference competition. Cannibalism of small tadpoles by larger conspecifics also occurs in some species (see chapter 12). However, the possibility of chemical interference has received more attention in the literature. Early laboratory studies showed that large tadpoles often inhibit the growth of smaller individuals under crowded conditions, even when food is superabundant (Bilski 1921; Goetsch 1924; Adolph 1931; Rugh 1934; Lynn and Edelman 1936). This suggested that large tadpoles release a growth-inhibiting chemical into the water, and considerable effort has gone into identifying such a substance.

Several investigators have shown that the growth of tadpoles can be inhibited by water conditioned by tadpoles, even when other tadpoles are not present (C. Richards 1958; S. Rose 1960; S. Rose and Rose 1961; Akin 1966; Licht 1967; Gromko, Mason, and Smith-Gill 1973). C. Richards (1958, 1962) hypothesized that growth inhibition was associated with "algal-like" cells found in tadpole feces, an explanation supported by other workers (S. Rose and Rose 1961, 1965; Licht 1967). However, L. West (1960) believed growth inhibitors were found in the "slime" growing on the walls of the rearing containers, not cells in the feces. Akin (1966) reported that conditioned water inhibited tadpole growth even when feces and associated material were filtered out. She proposed that an inhibitory chemical produced by the tadpoles became associated with algal cells and was subsequently released into the water, but the nature of the chemical was unknown. Various proteinaceous compounds have been found in the excreta of tadpoles (Runkova, Stepanova, and Koval'chuck 1974; Stepanova 1974). These metabolites inhibit the growth of other tadpoles, but the precise nature of the inhibitory substances remains unknown. A further complication is that kinship alters the effect of growth inhibitors, although how it does so is unclear. Some tests have shown that tadpoles reared with siblings grow more slowly than did those grown with nonsiblings (Shvarts and Pyastolova 1970a; Waldman 1986b), but in other cases, the opposite was true (Waldman 1986b).

Gromko, Mason, and Smith-Gill (1973) found no evidence that any sort of cells was necessary to produce a growth-inhibiting effect, and they questioned whether diffusable substances are responsible for crowding effect in tadpoles. They suggested that the cells observed by Richards and others are parasitic, with the reduction in tadpole growth being a result of increases in the parasite population. A similar suggestion was made by Steinwascher (1978b, 1979b), who found cells in the guts of *Rana clamitans* tadpoles that were identified as a parasitic yeast, *Candida humicola*. At low concentrations, these yeast cells actually enhanced the growth of tadpoles, but at higher concentrations, growth was inhibited. Inhibition was size-specific, with small tadpoles being inhibited more than large ones. Furthermore, the effect of the yeast on tadpole growth depended on the amount of food available and the density of tadpoles (Steinwascher 1979a). The result is a complex interaction in which small tadpoles are consistently at a competitive disadvantage relative to larger individuals.

Several conclusions can be drawn from this brief review of interference competition. First, many authors have assumed that tadpoles produce chemical substances that evolved specifically as growth inhibitors, but at present there is only circumstantial evidence to support this view. Second, crowding effects observed in laboratory experiments can involve several different processes that produce superficially similar results. These include responses to specific growth inhibitors, responses to metabolic waste products (see the following), interactions with parasite populations, or a complex interaction between exploitative and interference competition. Finally, the significance of crowding effects and growth inhibitors for tadpoles in natural populations is unclear. In most studies, tadpoles reared in the laboratory were exposed to densities far greater than those normally encountered in the field. Furthermore, the choice of subjects for such studies was unfortunate, because most have involved ranid tadpoles, particularly *Rana pipiens,* which are not noted for having exceptionally high population densities in nature. Much of the evidence for interspecific growth inhibition comes from laboratory studies in which tadpoles were raised in crowded conditions (e.g., Beebee 1991).

Neither Morin and Johnson (1988) nor Petranka (1989a) found any evidence of such inhibition in artificial or natural ponds. In contrast, workers studying interspecific competition between tadpoles in England reported significant inhibition of growth in relatively small artificial ponds (Griffiths, Edgar, and Wong 1991; Griffiths, Denton, and Wong 1993). They identified an unpigmented unicellular alga (*Prototheca richardsi*) in the tadpole feces that affected tadpole growth. Subsequent genetic work has shown that this organism apparently does not belong in the genus *Prototheca,* but should be placed in a new genus, *Anurofeca* (G. Baker, Beebee, and Ragan 1999). The alga apparently is not toxic to tadpoles (Wong and Beebee 1994). Nevertheless, when algal cells were abundant in the water, tadpoles fed on them, but derived relatively little nutritional benefit. Growth inhibition would then occur because of tadpoles being diverted away from more profitable food (Beebee and Wong 1992). The significance of this mechanism for intraspecific competition is unclear. It has been implicated in interspecific competition (see chapter 15), but there is considerable controversy over the interpretation of experimental results (Beebee 1995b; Griffiths 1995; Petranka 1995; R. N. Harris 1999).

One difficulty with the argument that chemical interference is an important form of intraspecific competition is that it is not clear why individuals that are the source of chemicals would be immune to their negative effects. Some observations and experiments that have suggested a role for chemical interference could be related to general deterioration of the aquatic environment, especially in very shallow ponds in hot climates, where ponds tend to dry up quickly. For example, Schmuck, Geise, and Linsenmair (1994) found that concentrations of nitrogenous waste products, especially ammonia, reached high levels in natural ponds inhabited by dense populations of *Hyperolius* tadpoles in Africa. High tadpole densities in these ponds resulted in longer larval periods and smaller size at metamorphosis, but whether these effects were related to ammonia concentrations or to other density effects is not clear. One species, *H. viridiflavus ommatostictus,* responded to high ammonia concentrations physiologically by increasing the activity of the enzyme cycle that converts ammonia to nontoxic urea. Another species, *H. marmoratus nitidulus,* exhibited an unusually high tolerance for ammonia. Physiological adjustments that enable tadpoles to tolerate high ammonia concentrations also have been reported in leptodactylid frogs in which tadpoles develop inside a foam nest (Shoemaker and McClanahan 1973; see also chapter 2). These physiological adjustments that allow tadpoles to survive in crowded conditions carry costs in reduced growth rates, however.

### Competition in Salamander Larvae

Many studies have shown that growth rates of salamander larvae decrease with increasing density. This could be due either to exploitative or interference competition. Large individuals might enjoy an advantage over smaller individuals in exploitative competition. Because a larva's gape increases with age, larger individuals are expected to exploit a wider range of prey sizes than small individuals (see chapter 12). For the most part, however, experimental studies have failed to support this prediction. For example, Brunkow and Collins (1996) found that larvae of *Ambystoma tigrinum nebulosum* were negatively affected by high densities, but those with similar-sized individuals actually grew more slowly than did those with variable-sized individuals, perhaps because of increased aggression among similar-sized larvae. Large individuals do not necessarily have an advantage over small individuals when all size classes feed on the same small prey items, such as zooplankton.

C. K. Smith (1990) explicitly tested for effects of variation in body size on competition among *A. opacum* larvae in laboratory and field experiments. In laboratory experiments, he found that high densities of larvae resulted in slower growth, but the size of competitors was important only when larvae could physically interact with one another

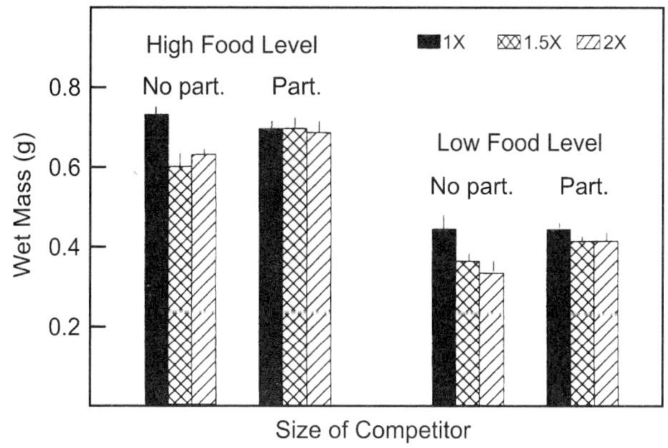


**Fig. 13.10.** Effects of relative size of competitors (1X, 1.5X, 2X) on body mass of *Ambystoma opacum* larvae raised in the laboratory in containers with and without partitions to separate the larvae. The containers without partitions allowed larvae to interact aggressively and resulted in reduced growth of small larvae when competitors of 1.5 or 2 times their body mass were present. This effect was seen at both high and low food levels. Relative body size of competitors did not affect larval growth when the larvae could not interact. After D. C. Smith (1990).

(fig. 13.10). He interpreted this as evidence that behavioral interference is more important than exploitative competition, at least with regard to differences in body size. He did not find the same effect of body size differences in experiments conducted in artificial ponds, where larval densities were much lower. This could have been due to limited opportunities for larvae to interact with one another. Van Buskirk and Smith (1991) did find evidence of competition among *A. laterale* larvae in natural rock pools. High densities of larvae did not significantly reduce prey abundance, but did result in much higher frequencies of tail damage. These results suggested that aggressive attacks on conspecifics constituted the most important type of competition. Evidence of aggressive interactions among larvae and extensive injuries at high densities have been reported for several species of salamanders (R. N. Harris 1987a; Walls and Jaeger 1987; Petranka 1989b; Semlitsch and Reichling 1989; D. Scott 1990; Walls and Roudebush 1991; Hokit, Walls, and Blaustein 1996; Reques and Tejedo 1996). In some extreme cases, large salamander larvae become cannibalistic and eat smaller conspecifics (see "Cannibalism and Cannibalistic Morphs" in chapter 12).

### Phenotypic Plasticity in Larval Growth, Development, and Metamorphosis

The studies of intraspecific competition discussed in the previous sections have clearly shown that high larval densities and a shortage of food can have detrimental effects on growth, survival, and the length of the larval period. The results of many of these studies are consistent with the basic assumptions of the Wilbur-Collins model and other models of amphibian metamorphosis. Nevertheless, a demonstration of density-dependent growth and mortality does not, in itself, test a key prediction of these models. The effects of density observed in studies of intraspecific competition can be attributed to the direct effect of resource depletion or interference competition, and larvae that emerge from the water at small sizes can simply be viewed as losers in competitive interactions.

All of these models predict that amphibian larvae should exhibit adaptive phenotypic plasticity in response to environmental conditions. In other words, larvae are expected to adjust the timing of metamorphosis and size at metamorphosis in ways that enhance their overall fitness. To examine this prediction, we need to know what environmental variables can trigger plastic responses in larvae and the limits of phenotypic plasticity in these animals.

As originally formulated, the Wilbur-Collins model of metamorphosis assumed that tadpoles can respond to environmental conditions at virtually any point in development by altering the timing of metamorphosis and size at transformation. In fact, there are constraints on plasticity related to temporal differences in the sensitivity of growth and development to environmental conditions. For example, the Wilbur-Collins model predicts that declining food supplies late in development will cause tadpoles to speed up development and metamorphose at a size close to the minimum viable size for terrestrial juveniles, or initiate metamorphosis immediately if the minimum size has already been reached. Conversely, if the availability of food actually increases late in development, perhaps because competing larvae have already left the pond or have been eaten by predators, then tadpoles are expected to delay development and continue growing to a larger size. If, on the other hand, rates of differentiation become relatively fixed late in development, then tadpoles might not be able to respond to altered food supplies in this way, because the timing of metamorphosis has already been determined.

Travis (1984) proposed an alternative view of amphibian metamorphosis that postulated that development rate is fixed relatively early in the larval period, but growth remains plastic, continuing throughout development, resulting in variation in size at metamorphosis. Several other authors have proposed modifications of the Wilbur-Collins model that provide for plastic responses in both growth and development rates for part of the larval period, but with development rate becoming fixed at a particular developmental stage (Alford and Harris 1988; Hensley 1993; Leips and Travis 1994; Hentschel 1999). Hentschel's model went further than did the others in attempting to predict the point at which phenotypic plasticity in timing of metamorphosis is lost. Day and Rowe (2002) argued that what appears to be a fixed rate of development after a critical point in the larval period actually may result from developmental thresholds that reverse the effects of changing food resources.

R. N. Harris (1999) proposed a graphical model that is consistent with the other models, but focuses on the relative allocation of energy to growth and development at different stages in the larval period. He argued that when larvae experience low food levels, they will allocate relatively more resources to development than to growth and will metamorphose at the minimum viable size, as proposed by Wilbur and Collins. Those with more food available will allocate relatively more resources to growth and reach a larger size at metamorphosis. The absolute level of resources for both growth and development obviously is higher for well-fed tadpoles, and they not only will tend to be larger at metamorphosis, but will have shorter larval periods than poorly fed individuals.

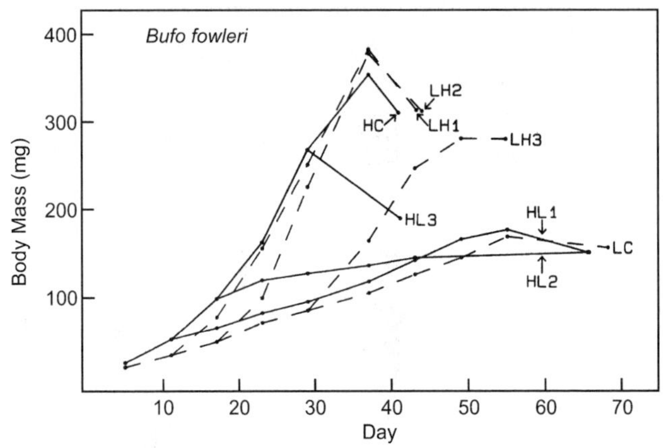


**Fig. 13.11.** Growth curves for *Bufo fowleri* tadpoles raised in small containers at different food levels. Tadpoles raised at constant high food levels (HC) grew rapidly and metamorphosed at a large size, whereas those raised at constant low food levels (LC) grew slowly and metamorphosed at a small size. Tadpoles switched from high to low food rations early (HL1) or in the middle (HL2) of development grew slowly and reached a size similar to those raised at constant low food. Tadpoles switched to low food late in development (HL3) grew rapidly until the food change and then metamorphosed at the same time as those in constant high food, but at a much smaller size. Tadpoles switched to high food early (LH1) or midway through development (LH2) caught up to those raised on constant high food. Tadpoles raised for most of development on low food and switched late to high food (LH3) took longer to reach metamorphosis, but metamorphosed at a size only slightly below that of tadpoles on constant high food. After Alford and Harris (1988).

### Phenotypic Plasticity and Variation in Food Resources

The empirical evidence to support these modified versions of the Wilbur-Collins model comes from experiments in which food rations for tadpoles were changed at different points in development. For example, Alford and Harris (1988) raised tadpoles of *Bufo fowleri* with either high or low food rations and then switched some tadpoles to either lower or higher food rations at different ages. As expected, individuals on high food rations grew faster and reached a larger size at metamorphosis than did those on low food, and they had shorter larval periods as well. When slowly growing tadpoles were switched to high food rations, they had longer larval periods than did those with constant high food, but they all reached about the same size at metamorphosis (fig. 13.11). Conversely, when rapidly growing tadpoles were switched to low food rations relatively early in development, larval periods were similar to those of tadpoles on a constant low food ration, and they metamorphosed at the minimum size. Tadpoles switched to a low food ration late in development metamorphosed at about the same time as those on a constant high food ration, but they were smaller at metamorphosis. These results generally supported the Wilbur-Collins model, but with some limitation on phenotypic plasticity in timing of metamorphosis. Experiments with another species of toad, *Bufo calamita,* showed that tadpoles switched from low to high resources tended to speed up metamorphosis relative to those on constant low food. Those switched from high to low resources did not alter timing of metamorphosis relative to tadpoles on constant high food, but they were smaller at metamorphosis (Tejedo and Reques 1994a). Resources were switched at only one time (17 days) in this study, so the effect of timing of resource change was not investigated.

Hensley (1993) performed a similar experiment with *Pseudacris crucifer* tadpoles. He found that changes in food rations before about Gosner stage 38 affected both the length of the larval period and size at metamorphosis, but changes at later stages did not affect the timing of metamorphosis. Switching tadpoles to low food rations late in development had little effect on size at metamorphosis, but switching them to high food late in development produced larger metamorphs. Leips and Travis (1994) obtained comparable results with tadpoles of *Hyla cinerea,* which breeds in relatively permanent ponds, and *H. gratiosa,* which breeds in temporary ponds. In both species, the length of the larval period was affected by changes in food rations through about 60% of the larval period, but not in the last 40%. *Hyla gratiosa,* the temporary pond species, exhibited greater plasticity in the timing of metamorphosis than did *H. cinerea,* but the period of sensitivity to changes in food level was similar in the two species. The effect of food rations on size at metamorphosis was more complex, but in general, additional food resulted in larger metamorphs, and decreased food resulted in smaller ones. Food ration late in development had the greatest effect on size at metamorphosis. There also were differences in response of tadpoles, depending on which way the food rations were changed. Those that started with high food rations and were switched to low food all were smaller at metamorphosis than were those on constant high food rations. In contrast, those that were switched from low to high rations generally reached a size at metamorphosis similar to tadpoles on constant high food rations, regardless of when the switch occurred (fig. 13.12).

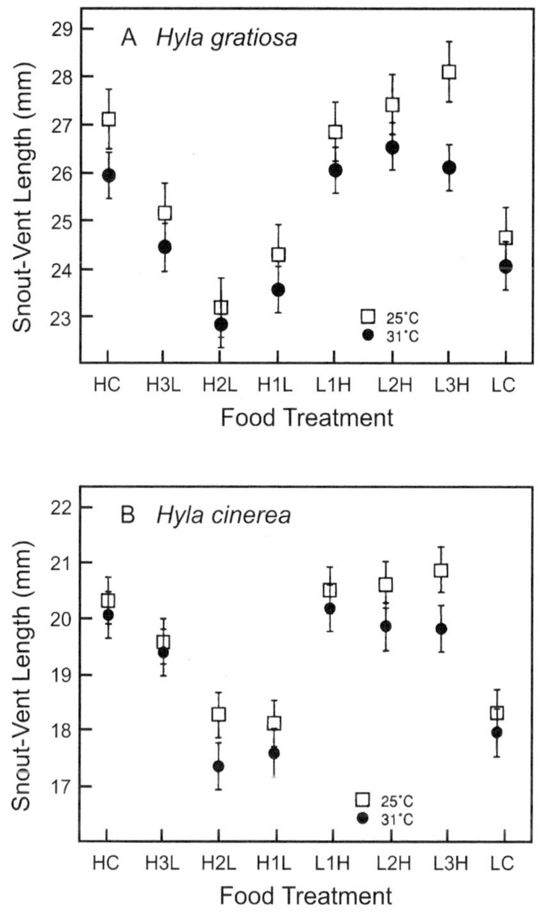


**Fig. 13.12.** Size at metamorphosis for *Hyla gratiosa* and *Hyla cinerea* tadpoles raised at two different temperatures on different food rations. In general, switching from low to high food rations (LH1, LH2, LH3) increased size at metamorphosis relative to tadpoles on constant low food, regardless of when the switch occurred. Switching tadpoles from high to low food rations decreased size at metamorphosis, especially if the switch occurred early in development (LH1, LH2). Tadpoles raised at the lower temperature reached larger body sizes at metamorphosis than those raised at the warmer temperature, but took longer to reach metamorphosis. After Leips and Travis (1994).

A study by Audo et al. (1995) used a different experimental design to investigate the effects of short-term starvation on larval growth and development in *Hyla chrysoscelis*. In their experiments, tadpoles were fed unlimited food, except for a period of ten days in which they were deprived of food. The starvation period occurred in the early, middle, or late stages of development. Tadpoles that were starved early and then given plenty of food were able to catch up with those that were given constant high food. They metamorphosed at a similar size, but had longer larval periods than did those on high food rations. This treatment might mimic a situation in which high tadpole densities early in development result in strong competition for food, but some individuals are later released from such competition by high predation on competing tadpoles (e.g., Semlitsch and Caldwell 1982). Tadpoles that were starved late in development metamorphosed at the same time as those on constant high food, but they were smaller and suffered much higher mortality.

Beck (1997) examined the effect of food resources on metamorphosis in yet another hylid treefrog, *Hyla squirella*. The experimental design was similar to the other studies, except that food rations were adjusted to the changing mass of the tadpoles. This was considered a better test of the Wilbur-Collins model, because it predicts that tadpoles will adjust the timing of metamorphosis in response to changes in mass-specific growth rate. As in the other studies, switching the tadpoles from high to low rations or from low to high late in development (after late limb bud development) did not affect timing of metamorphosis. Tadpoles switched from high to low rations all were smaller at metamorphosis than those maintained on constant high rations. The trend was for the difference to be greatest in tadpoles that were switched early, although results for individual treatments did not differ from one another statistically, probably because of relatively small sample sizes. The opposite pattern was seen in tadpoles switched from low to high rations.

In all of the previous studies, tadpoles that were switched to low resource levels continued to receive some food. Morey and Reznick (2000) used a different approach in a study of three species of spadefoot toads, *Scaphiopus couchii*, *Spea intermontana*, and *Spea hammondii*. *Scaphiopus couchii* breeds in highly ephemeral ponds and has perhaps the shortest larval period of any North American anuran, about seven days (Buchholz and Hayes 2000). The other two species breed in longer-lasting ponds and have larval periods of up to a month. Morey and Reznick raised tadpoles initially at high, medium, or low food levels, and then abruptly cut off all food at different stages of development. At constant food levels, tadpoles of all three species reached larger sizes at metamorphosis on high food rations than on low rations. Length of the larval period was not much affected by food ration in *Scaphiopus couchii*, but in *Spea intermontana*, tadpoles on low food took twice as long to reach metamorphosis as those on high food (fig. 13.13). In all three species, tadpoles deprived of food early in development usually failed to reach metamorphosis, whereas those switched at later stages usually did complete metamorphosis (fig. 13.14). Each species had a characteristic developmental threshold after which metamorphosis was possible when tadpoles were deprived of food. In *Scaphiopus couchii*, tadpoles that had their food removed at Gosner stage 34 could still achieve metamorphosis, whereas in the other two species, the developmental threshold was slightly later. When tadpoles initially were maintained on a high food ration before food was removed, the developmental threshold for successful metamorphosis was slightly earlier than for tadpoles on a low food ration. Presumably this was because they had greater energy reserves available when food was removed.

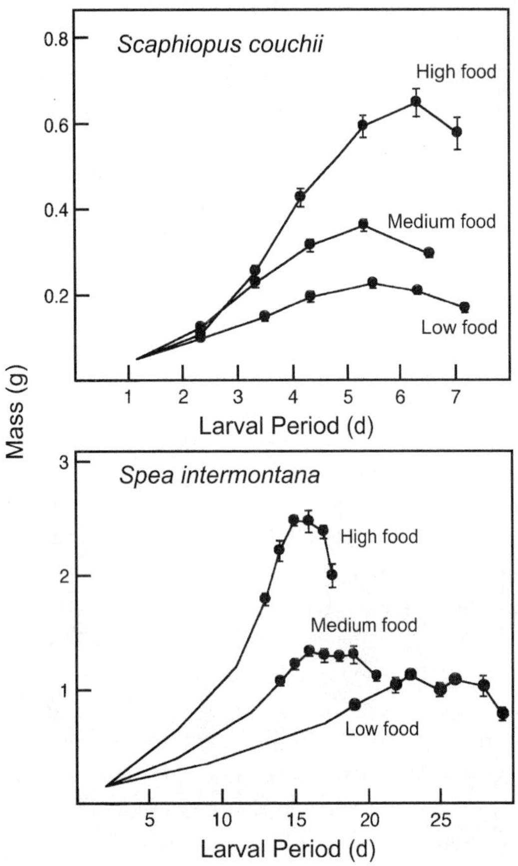


**Fig. 13.13.** Growth of tadpoles of *Scaphiopus couchii* and *Spea intermontana* at three different food rations. Food ration had a large effect on size at metamorphosis in *Scaphiopus couchii,* but a relatively small effect on length of the larval period, which is normally very short. Food ration affected both size at metamorphosis and length of the larval period in *Spea intermontana,* which has a much longer larval period. After Morey and Reznick (2000).

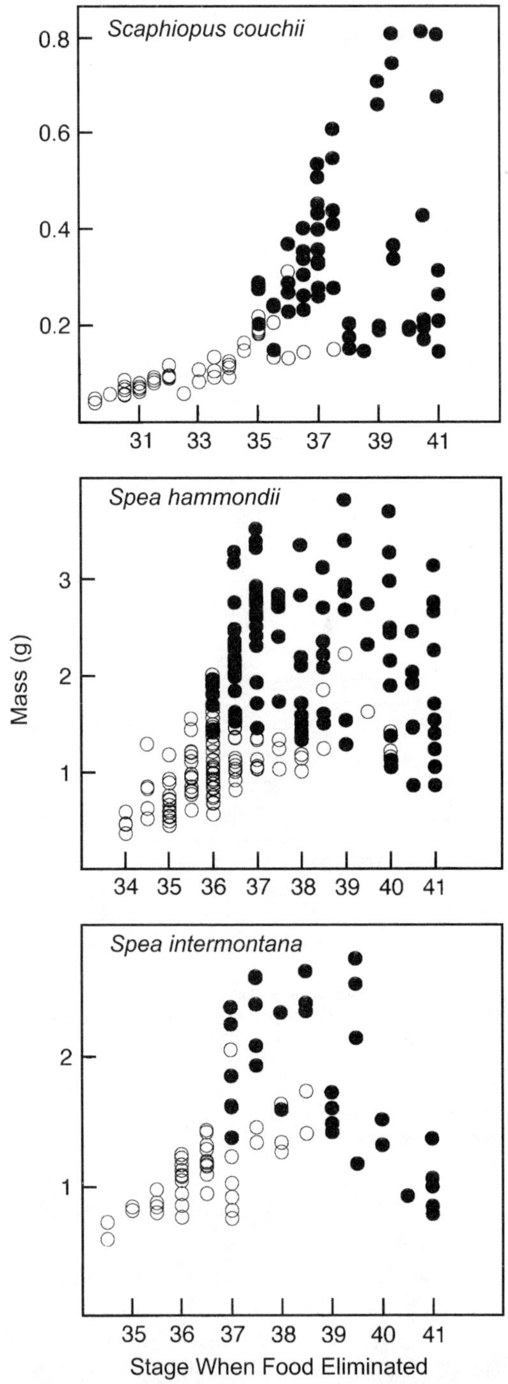


**Fig. 13.14.** Growth of tadpoles of three species of spadefoot toads when food was completely eliminated at different stages of development. Closed circles show tadpoles that reached metamorphosis. Open circles are those that failed to metamorphose. Tadpoles of *Scaphiopus couchii* could reach metamorphosis if food was removed as early as Gosner stage 35. The threshold for successful metamorphosis was later for *Spea hammondii* (stage 36) and *Spea intermontana* (stage 37). After Morey and Reznick (2000).

Nevertheless, all tadpoles that metamorphosed early after being deprived of food did so at much smaller body sizes than those maintained on constant food (see also Boorse and Denver 2004).

Overall, the experiments of Morey and Reznick were consistent with predictions of the Wilbur-Collins model and the developmental threshold model of Day and Rowe (2002). Both timing of metamorphosis and size at metamorphosis were plastic traits that were influenced by food availability and growth rate. In general, *Scaphiopus couchii,* the species that breeds in the most ephemeral habitats, showed the most plastic responses. The methods used Morey and Reznick's experiment do not precisely mimic situations that tadpoles normally encounter in the field. Indeed, in some spadefoot toads, tadpoles have access to increased resources late in development because they either feed on dead or dying conspecifics or actively cannibalize other tadpoles (see chapter 12). Nevertheless, the results of these experiments were consistent with those of R. Newman (1994), who found that decreasing food rations resulted in an acceleration of metamorphosis, whereas a constant low food ration delayed metamorphosis. The sudden deprivation of food in Morey and Reznick's experiments probably induced a stress response in the tadpoles, and such responses have been shown to induce early metamorphosis in these tadpoles (Denver 1997a). In nature, however, the principal environmental stress favoring early metamorphosis in spadefoot toads probably is not a lack of food, but drying up of breeding ponds.

### Phenotypic Plasticity and Pond Drying

For amphibians that breed in temporary ponds, direct mortality of larvae due to starvation or poor-quality food probably is relatively rare. The major threat to survival in these habitats is drying of the ponds before larvae are able to metamorphose. Thus, while competitive effects do not necessarily result in direct mortality, slower growth rates and longer larval periods in competitive environments will decrease the chances that tadpoles will reach metamorphosis before the pond disappears. In such species, selection should favor plastic developmental responses that enable individuals to speed up development in response to pond drying. Indeed, a number of observational and experimental studies have shown that some species do exhibit this type of plasticity, although interpretation of experimental results is not always straightforward.

The best-studied examples are spadefoot toads. Reproductive failure due to pond desiccation is common in these species (R. Newman 1987), so there is strong selection on tadpoles to increase rates of development and minimize the length of the larval period. Tadpole densities often are extremely high in these species, and food availability can be low. In addition, breeding pools often reach high temperatures, especially as they begin to dry out. Selection pressures in this very harsh environment have resulted in spadefoot toads having the fastest rates of development and the shortest larval periods of any North American anurans, as well as the greatest tolerance of high temperatures (Zweifel 1977). Some species of spadefoot toads can accelerate metamorphosis by becoming carnivorous and even cannibalistic, which provides higher-quality food and allows for more rapid development. This type of phenotypic plasticity is discussed in more detail in chapter 12.

In addition, several species of spadefoot toads exhibit considerable plasticity in timing of metamorphosis in response to pond drying (R. Newman 1992). In an experiment with *Scaphiopus couchii*, R. Newman (1988b) reared sibling groups of tadpoles in pens placed in artificial ponds that differed in initial depth and therefore in pond duration. Each sibling group was replicated four times in long- and short-duration ponds. Tadpoles had shorter larval periods in short-duration ponds than in long-duration ponds, and

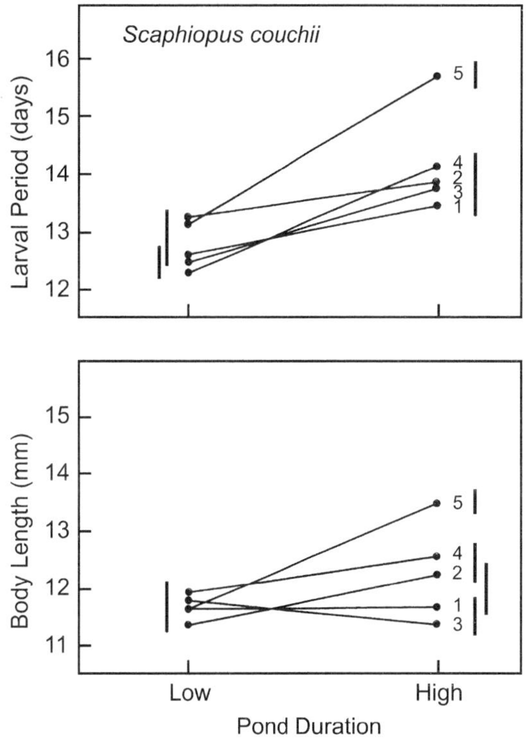


**Fig. 13.15.** Effect of pond duration on length of the larval period and size at metamorphosis for spadefoot toad (*Scaphiopus couchii*) tadpoles. In general, early drying of ponds resulted in shorter larval periods and smaller size at metamorphosis. Different sibships (1–5) differed in timing and size at metamorphosis in ponds of long duration, but mostly were similar in short-duration ponds. Vertical lines indicate values that are not significantly different from one another. After Newman (1988b).

size at metamorphosis was positively correlated with larval period (fig. 13.15). There also were significant differences among sibling groups in development rate, and the sibship with the slowest development suffered high mortality in the short-duration ponds. The different sibling groups did not benefit uniformly from being in long-duration ponds, because tadpoles from some groups reached larger metamorphic sizes than did others in the same ponds.

In another experiment, pond duration, tadpole density, and food rations all were varied in both long- and short-duration ponds. Tadpoles at high density with low food never reached metamorphosis before ponds dried, even in long duration ponds (pond duration varied between 11 and 17 days). In other food and density treatments, tadpoles were more likely to reach metamorphosis in long duration ponds. As in the previous experiments, tadpoles metamorphosed earlier in short-duration ponds than in long-uration ponds. In general, tadpoles with longer larval periods emerged at larger sizes, but food level had the greatest effect on metamorphic size. The patterns seen in the experimental ponds mirrored those seen in nearby natural ponds, where tadpoles metamorphosed earlier from short-duration ponds.

Natural ponds with very high tadpole densities typically did not produce any metamorphs. These results suggest that adult spadefoot toads are not very good at predicting conditions in ponds when they lay eggs, because the very ponds that were most heavily used by breeding adults often resulted in no reproductive success.

These experiments did not directly address the proximate causes of abbreviated larval periods in short-duration ponds. Many features of the ponds changed as the ponds dried, including water depth, pond surface area, resource levels, tadpole density, and temperature. Drying ponds almost certainly change in water chemistry as well, partly as a result of the buildup of nitrogenous waste products of tightly packed tadpoles (e.g., Schmuck, Geise, and Linsenmair 1994). Temperature alone could have accounted for some or all of the accelerated metamorphosis in rapidly drying ponds, because temperature increased substantially in these ponds. A set of laboratory experiments in which both food and tadpole density were varied showed that tadpoles that experience declining growth rates respond by metamorphosing early, and at a relatively small size, as predicted by the Wilbur-Collins model and other models derived from it (R. Newman 1994). Metamorphic size and age were positively correlated when food was abundant, but there was no such correlation when food was scarce. In other words, most tadpoles exposed to low food levels metamorphosed at similar small sizes.

Laboratory experiments with both *Scaphiopus couchii* and *Spea hammondii* directly addressed the effect of decreasing water volume on initiation of metamorphosis (Denver, Mirhadi, and Phillips 1998; Boorse and Denver 2004). The main result of these experiments was that decreasing water volume alone led to shorter larval periods, in the absence of any change in temperature. Furthermore, the response was continuous. Larval periods were shortest and body size smallest in response to the most extreme volume reductions (fig. 13.16). When water volumes were increased before tadpoles reached metamorphosis, they resumed rapid growth and delayed metamorphosis until they had reached a larger size. These responses were not due solely to increases in tadpole density as water volume decreased, because tadpoles raised alone showed a similar response. As discussed in the section on hormonal control of metamorphosis, experimental reductions in water volume resulted in dramatic increases

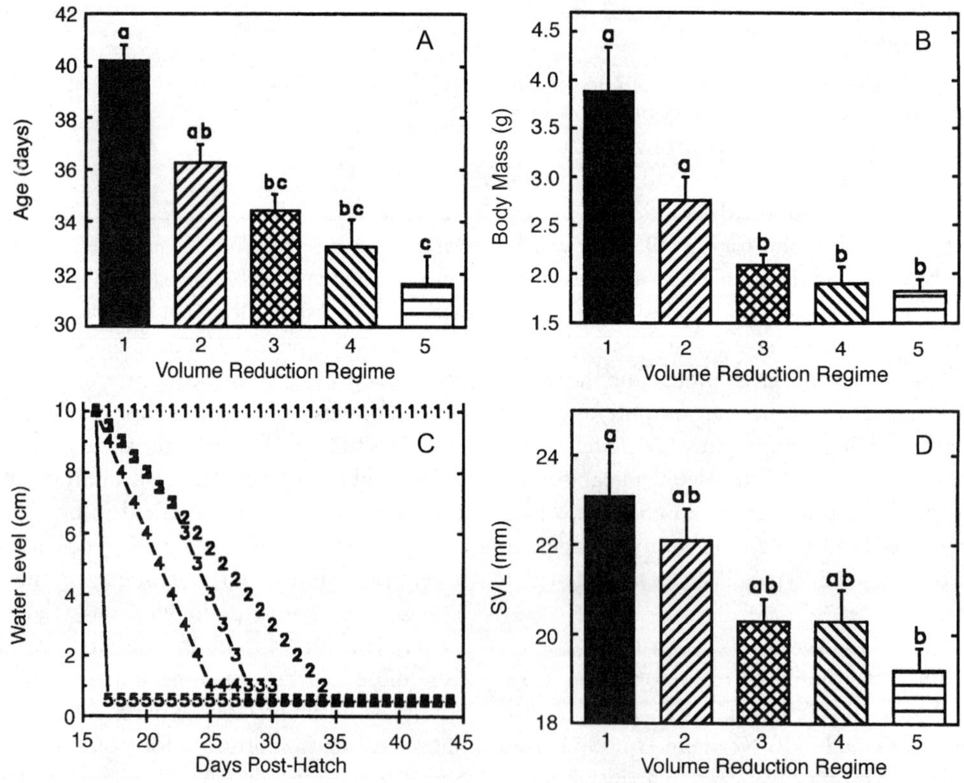


**Fig. 13.16.** Effect of drying regime on age and size at metamorphosis for spadefoot toad (*Spea hammondii*) tadpoles raised in laboratory aquaria. (A) Age at metamorphosis as a function of drying regime. (B) Body mass at metamorphosis as a function of drying regime. (C) Water levels in containers with different drying regimes (1–5). (D) Snout-vent length as a function of drying regime. Bars show means ± 1 SE. Lowercase letters indicate values that are not significantly different from one another. After Denver, Mirhadi, and Phillips (1998).

in levels of corticosterone and thyroid hormones in the body (Denver 1997a, b, 1998). This hormonal response was very rapid, occurring within 24 hours in tadpoles that were exposed to a sudden drop in water volume (Boorse and Denver 2004).

While plasticity in the timing of metamorphosis is particularly dramatic in spadefoot toads, which breed in some of the harshest environments of any amphibian, these are not the only species that show such responses. For example, M. Crump (1989a) reported that tadpoles of a Neotropical treefrog, *Hyla pseudopuma,* which breeds in ephemeral rain pools and other temporary ponds, exhibits accelerated metamorphosis in response to reduction in water volume. A tropical toad (*Bufo maculatus*) from the Ivory Coast also had shorter larval periods and smaller size at metamorphosis, but a ranid frog (*Hoplobatrachus occipitalis*) that also breeds in temporary water did not (Spieler 2000). The natterjack toad (*Bufo calamita*), another temporary pond breeder, also changed the timing of metamorphosis in response to variation in pond duration, density, and food level, but the interaction among these factors was complex (Tejedo and Reques 1994b). Tadpoles in some short-duration ponds metamorphosed early, but only when raised by low density. At high density, size at metamorphosis was greatly reduced, larval periods were long, and the number of tadpoles reaching metamorphosis was reduced, probably because of inadequate nutrition resulting from increased density. Another study of *Bufo calamita, B. bufo,* and *Rana temporaria* in England found no effect of pond drying on timing of metamorphosis, although tadpoles from drying ponds metamorphosed at smaller sizes (Brady and Griffiths 2000). In an experiment designed mainly to look at intra- and interspecific competition, Wilbur (1987) found that tadpoles of *Bufo americanus* at low density metamorphosed several days earlier in ponds that dried after 50 or 100 days than in those that never dried. In *Rana lessonae* and the hybridogenetic species *R. esculenta,* drying regime did not have a significant effect on timing of metamorphosis, but tadpoles emerged at a smaller size in drying ponds (Semlitsch and Reyer 1992b). Experiments with *Rana blairi* and *R. sphenocephala* using a similar experimental design showed a significant reduction in length of the larval period and metamorphic size for both species in drying ponds (M. Parris 2000; T. Ryan and Winne 2001).

In contrast to the results obtained by Brady and Griffiths (2000), several experiments with European common frogs (*Rana temporaria*) exposed to gradual (Laurila and Kujasalo 1999) or rapid (Loman 1999) reduction in water volume showed some acceleration in timing of metamorphosis. A comparison of populations in southern Sweden, where ponds frequently dry up, and northern Sweden, where ponds seldom dry, revealed greater plasticity in timing of metamorphosis in the southern population (Merilä et al. 2000; Laurila, Karttunen, and Merilä 2002). Another study compared tadpoles from ponds in a relatively small geographic area in southern Sweden that differed in hydroperiod. Tadpoles from temporary and permanent ponds showed no difference in degree of plasticity in timing of metamorphosis under controlled conditions (Loman and Claesson 2003), even though tadpoles in natural ponds accelerated metamorphosis if ponds were in danger of drying up (Loman 2002b). There are limits to plasticity that can prevent tadpoles from escaping from drying ponds. Loman (2002c) found that *R. temporaria* tadpoles exposed to a rapid reduction in water volume at an early stage of development mostly failed to accelerate metamorphosis, but instead grew and developed more slowly than those exposed to constant water depth. He attributed this mostly to the effects of crowding, but it also may be that the tadpoles were subjected to the drying regime before reaching the developmental threshold (Day and Rowe 2002) that would enable them to accelerate development.

In all of these species, differences in timing of metamorphosis were considerably less impressive than in spadefoot toads. For example, in *Bufo calamita,* the difference in length of the larval period for tadpoles at low density in long- and short-duration ponds was only one day, or about 3% of a 32-day larval period in constant water. Differences were similar in *Bufo americanus,* with a 4% difference between ponds that dried after 50 days and those that never dried. *Bufo maculatus,* which breeds at the edges of rivers, reduced the larval period by about 10–20%. In *Rana temporaria,* differences ranged from about 1 day to 2.5 days, about 3–6% of the larval period in constant water. Differences in development time for *R. blairi* averaged about 8% of the expected larval period in constant water, while the difference for *R. sphenocephala* was only about 4%. Development times for *Hyla pseudopuma* in decreasing water volume were highly variable, but averaged about 3–5% less than those of tadpoles in constant volumes. This compares with differences in length of the larval period of up to 50% for spadefoot toad tadpoles in natural long- and short-duration ponds (R. Newman 1989), and up to 20% in laboratory drying experiments (Denver, Mirhadi, and Phillips 1998). These differences suggest that phenotypic plasticity in response to pond drying is much greater in species that breed in highly ephemeral desert pools than in those using other types of temporary ponds. A much larger range of species in different habitats needs to be examined to rigorously test this hypothesis. Clades that exhibit considerable interspecific variation in choice of breeding sites would be particularly useful for comparative studies to control for phylogenetic differences that confound interpretation of the available data.

One difficulty in interpreting the results of studies in

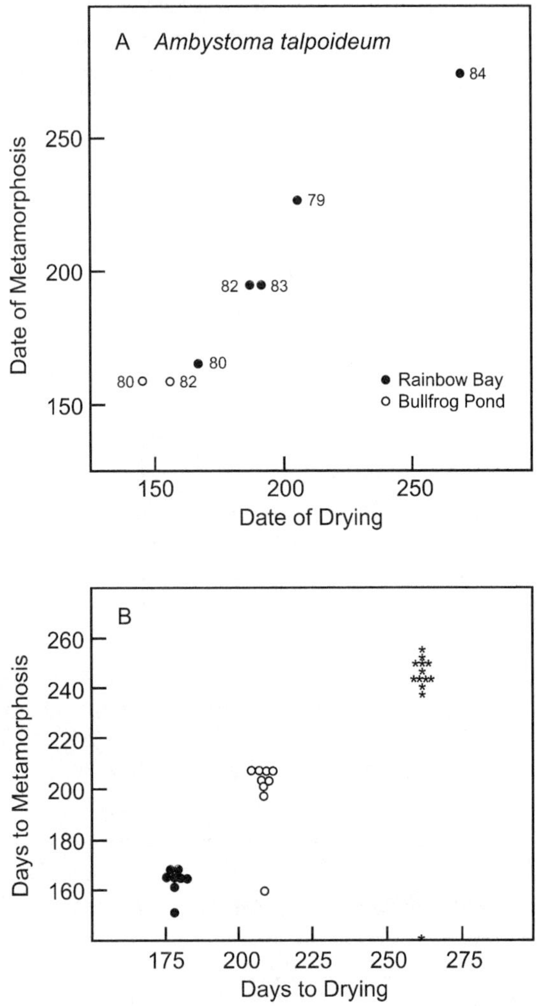


**Fig. 13.17.** Relationship of pond duration to date of metamorphosis in *Ambystoma talpoideum.* (A) Natural variation in date of metamorphosis in two temporary ponds (Rainbow Bay and Bullfrog Pond) between 1979 and 1984. After Semlitsch (1987c). (B) Number of days to metamorphosis as a function of three different drying regimes in artificial ponds. Data points represent individuals at three drying dates: July 20 (solid circles), August 21 (open circles), and October 14 (asterisks). After Semlitsch and Wilbur (1988).

which hydroperiod is varied is that drying treatments produce several effects simultaneously. Experimental drying obviously reduces the volume of water and the persistence of the pond, and hence the time available for tadpoles to develop. In addition, average water temperature also tends to increase, which would tend to increase development rates. On the other hand, shallower ponds experience increased variation in temperature, reaching colder temperatures at night that partially compensate for higher temperatures during the day (Brady and Griffiths 2000). Perhaps most important, tadpole densities are greatly increased in drying ponds. This can result in increased competition for food, as well as possible chemical interactions that inhibit growth and development. For example, in experiments with *Hyla gratiosa* and *H. cinerea,* Leips, McManus, and Travis (2000) found that drying regime did not have a direct effect on length of the larval period, but increased density of tadpoles in drying ponds actually resulted in longer larval periods and smaller sizes at metamorphosis. Hence, neither species exhibited the predicted adaptive shortening of development time in drying ponds. Even though *H. gratiosa* typically breeds in temporary ponds, it actually had a considerably longer larval period than did *H. cinerea,* which tends to breed in more permanent water.

Much less work has been done on responses of salamander larvae to pond drying, in part because most salamanders do not breed in highly ephemeral habitats. S. Freeman and Bruce (2001) studied variation in length of the larval period in *Eurycea guttolineata,* a pond-breeding species. At low elevations, larval periods were several months shorter than at high elevations, where most larvae overwintered before metamorphosing. There was little difference in temperature at the different elevations, but breeding sites at low elevations tended to dry up by late summer. Freeman and Bruce interpreted their results as evidence of adaptive change in length of the larval period in response to increased likelihood of pond drying. The study did not involve experimental manipulation of drying regimes, however, so it is not clear whether this represents a plastic developmental response or genetic differences in life-history traits that have been favored by different selective regimes.

The only detailed experimental studies of the effect of pond drying on larval development in salamanders are on *Ambystoma talpoideum,* which breeds in Carolina bays and other relatively shallow temporary ponds. Reproductive success varies considerably from pond to pond and from year to year within ponds, with most larval mortality resulting from early drying of ponds (Semlitsch 1987d; Pechmann et al. 1989; Semlitsch et al. 1996). In natural ponds, the timing of metamorphosis can vary by more than 100 days, with metamorphosis occurring earlier in ponds that dry early (fig. 13.17 A). Larvae cannot reach metamorphosis in ponds that persist for less than 145 days. Those in ponds of long duration typically reach larger sizes at metamorphosis.

Semlitsch and Wilbur (1988) conducted an experiment in which *A. talpoideum* larvae were housed in artificial ponds that either had constant water levels, or one of three different drying regimes. They found that timing of metamorphosis was strongly correlated with time to pond drying, and the pattern closely mirrored that seen in natural ponds (fig. 13.17 B). The first metamorphs emerged from all ponds at about the same time, however, indicating significant individual variation in timing of metamorphosis within ponds. Individuals that emerged from drying ponds did so at a smaller size than those that emerged from ponds with constant water. Many of the larvae in constant ponds reached very large sizes and became paedomorphic; these responses are discussed in more detail in a later section.

### Fitness Consequences of Variation in Size and Age at Metamorphosis

A general assumption of all models of amphibian metamorphosis is that larvae that metamorphose at a large size will do better on land than those that metamorphose at a small size. Interspecific comparisons suggest that species that metamorphose at small sizes, relative to adult body size, often are morphologically or physiologically underdeveloped. For example, in some species with relatively short larval periods, such as *Bufo americanus,* juveniles emerge from the water with poorly developed lungs and aerobic capacities that are much lower than those of adults. This is not true for some species with long larval periods, such as *Rana clamitans* (Taigen and Pough 1981, 1985; Pough and Kamel 1984; Burggren and Just 1992). Sensory systems sometimes are incompletely developed in species with short larval periods. For example, the development of structures in the ear that enable frogs to detect airborne sounds are less developed in species with short larval periods than those with long larval periods (Hetherington 1988b). Another important physiological consequence of small metamorphic size is greater susceptibility to evaporative water loss, a problem that can be particularly acute for species such as spadefoot toads that emerge into very dry environments (R. Newman and Dunham 1994).

Interspecific comparisons suggest that patterns of larval development can affect juvenile performance, but they do not directly address the question of whether developmental differences among individuals of the same species affect juvenile performance and lifetime fitness. Several laboratory studies have examined the behavioral and physiological performance of conspecific juveniles that experienced different densities as larvae. John-Alder and Morin (1990) reported that juvenile Fowler's toads (*Bufo fowleri*) emerging from high-density larval populations were smaller and achieved shorter jump distances than those emerging from low-density populations. Differences in performance were due entirely to differences in metamorphic body size, not independent effects of larval crowding. Toads from high-density populations also had lower locomotor stamina than did those from low-density populations. Lower jumping performance in small individuals emerging from high-density larval populations also has been reported in another toad, *Bufo bufo* (Goater, Semlitsch, and Baernasconi 1993). Two ranids with longer larval periods, *Rana blairi* and *R. sphenocephala,* showed similar differences between high- and low-density metamorphs, and again, most of the variation in jumping performance was due to differences in body size (Semlitsch et al. 1999). Differences in body size also affected jumping performance in metamorphs of *Discoglossus galganoi,* but in this case, the differences were due to manipulation of food quality and temperature during larval development, rather than larval density (Álvarez and Nicieza 2002b).

In European water frogs (*Rana esculenta* complex), the effect of larval growth history on morphology and jumping performance was not consistent (Tejedo, Semlitsch, and Hotz 2000). In general, frogs derived from low-density ponds were larger and jumped farther than were those from high-density ponds. In one hemiclone of the hybrid *Rana esculenta* genotype, frogs raised at low density had relatively longer hind legs than those raised at high density, and this translated into increased jumping performance. This pattern was not seen in one of the parental species, *R. lessonae,* or two other hemiclones of *R. esculenta.* Surprisingly, larval density did not affect relative hind limb length in the other parental species, *R. ridibunda,* but these frogs did show an effect of larval density on juvenile jumping performance. This suggests that larval growth history affected muscle morphology or physiological traits that are related to jumping performance.

Beck and Congdon (2000) examined the effect of both age and size at metamorphosis on metabolic rate, sprint speed, and endurance in juvenile *Bufo terrestris.* Differences in age and size at metamorphosis were induced by varying food rations and temperature, but not tadpole density (tadpoles were reared alone). Metabolic rate, sprint speed, and locomotor endurance all were negatively correlated with length of the larval period and positively correlated with size at metamorphosis (fig. 13.18). Metabolic rate also was affected by temperature and food rations, after accounting for differences in age and size. The negative effect of age at metamorphosis on performance variables such as sprint speed is related to a negative correlation between age at metamorphosis and relative hind-limb length (Emerson 1986). Variation in metabolic rate is hard to interpret in terms of organismal performance, because the effects of activity on metabolic rate were specifically excluded from the experiment.

In a related study, Beck and Congdon (1999) measured metabolic rates of juveniles that varied in age and size at metamorphosis and then released them into small (0.5 $m^2$) enclosures in a forest. They then counted the number of survivors after one month and determined the mass of each animal. This experiment failed to show any relationship between either body size or metabolic rate and survival of the toads, but unfortunately, sample sizes were too low to detect such an effect. Initial body size was positively correlated with body size at the final census. In a second experiment, newly metamorphosed toads were placed in small field enclosures that were designed to allow some small terrestrial arthropods to move into the enclosures as prey for the toads. After 11 days, the toads were placed in a second set of

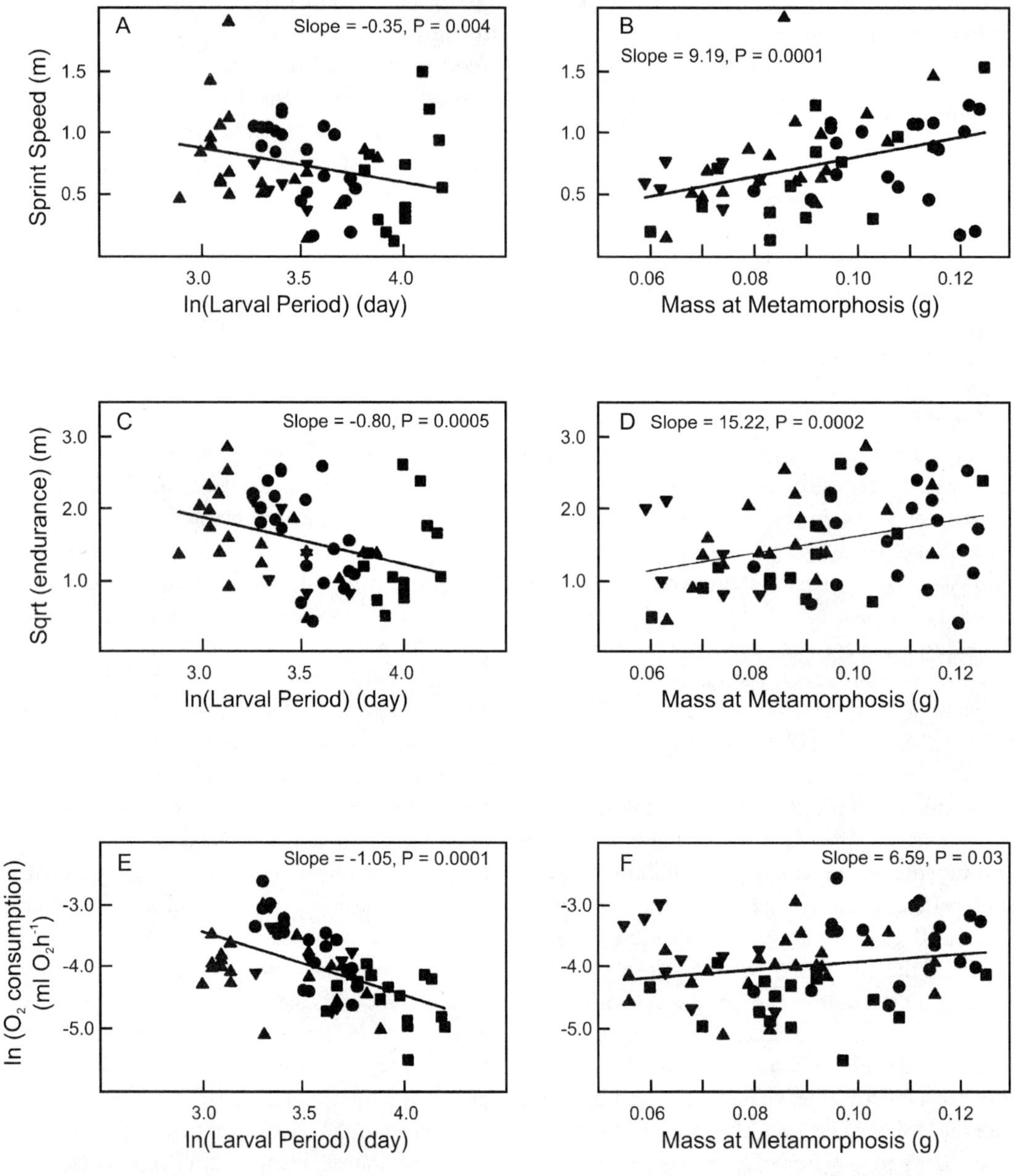


**Fig. 13.18.** Relationship of physiological performance of juvenile toads (*Bufo terrestris*) to length of the larval period and size at metamorphosis. Top: Sprint speed as a function of length of the larval period (A) and mass at metamorphosis. Middle: Endurance as a function of length of the larval period (C) and mass at metamorphosis (D). Bottom: Metabolic rate as a function of length of the larval period (E) and mass at metamorphosis (F). After Beck and Congdon (2000).

enclosures at lower densities to simulate dispersal away from a pond. In this experiment, size at metamorphosis was positively related to survival to day 11, but age at metamorphosis was not. Small individuals grew more rapidly than large ones, but mass at metamorphosis was not related to growth in mass or length. Relatively few toads (24%) were recaptured after about two months in the second, lower-density enclosures. There was a weak positive relationship between age at metamorphosis and survival after two months. Overall, the results of these experiments suggested that age- and size-related differences in behavioral performance of newly emerged toads had relatively little effect on short-term survival. Juvenile toads undergo enormous increases in body size after metamorphosis, so it seems likely that the effects of small differences in initial body size will disappear as the toads grow. Unfortunately, the sources of mortality in these experiments were not known, although it is unlikely that major predators were present in the enclosures. Initial differences in performance variables such as sprint speed and endurance still could have significant effects on early survival if the animals were confronted with predators such as garter snakes soon after emerging from a pond (e.g., Wassersug and Sperry 1977).

Another experiment that used experimental enclosures to examine the effects of larval density on juvenile survivorship was conducted by Morey and Reznick (2001) on spadefoot toads (*Spea hammondii*). Tadpoles raised at low density metamorphosed at a larger size than did those raised at high density, and larger juveniles were more likely to survive their first year on land. Initial differences in body size were maintained or increased during the first year of terrestrial life. These size differences disappeared when juveniles were provided with extra food, either in the laboratory or in field enclosures. Hence, while the larval environment can influence fitness of juveniles and adults, such effects are mediated by conditions in the terrestrial environment. Altwegg and Reyer (2003) also used enclosures to follow survivorship of juveniles of *Rana esculenta* and *R. lessonae* raised in different larval environments. They found that large metamorphs had higher survivorship, faster growth rates, and larger size at maturity than did small metamorphs, regardless of what experimental manipulation produced the differences in body size. In contrast to the work with spadefoot toads, they did not find any tendency for small metamorphs to compensate for small initial size by increasing rates of postmetamorphic growth. Manipulations of juvenile density in enclosures had some effect on juvenile growth rate, but the effect was less than that of larval density. Juvenile density did not affect survivorship (Altwegg 2003b).

Several studies have examined the effect of size at metamorphosis on fitness in adults by following cohorts of individuals emerging from natural ponds and then returning one or more years later to breed. D. C. Smith (1987) followed marked juveniles of *Pseudacris triseriata* emerging from natural rock pools and measured survival and both age and size at sexual maturity. Frogs that metamorphosed early and at a large size had higher survival to maturity than did those that metamorphosed late and at a small size. The larger frogs also tended to reach sexual maturity after one year, whereas smaller frogs often did not reach maturity until the second year. Overall survival to the second year was not affected by size at metamorphosis, however. Given the relatively short life span of these frogs, early sexual maturity probably has a large effect on total lifetime reproductive success.

Berven and Gill (1983) reported a positive correlation between size at metamorphosis and both juvenile survival and size at first reproduction in populations of wood frogs (*Rana sylvatica*). Sample sizes were very small (< 12 individuals) because of high mortality. In another population, cohorts of juveniles emerged from two nearby ponds at different times and at different sizes. On average, the later-emerging cohort was larger at metamorphosis and at sexual maturity. The difference in average size of the two cohorts at metamorphosis was about 2 mm, as was the difference in size at maturity. The latter was about 4.7% of the average size of the larger cohort. Other work on this species suggests that a size difference of this magnitude would translate into a fecundity difference of about 50 eggs in first-year females, or about 10% of average clutch size (Berven 1988). The impact of this difference on overall fitness is unknown, as is the effect of differences in size at maturity for males.

Two studies of salamanders suggest that age and size at metamorphosis can affect adult fitness traits. Semlitsch, Scott, and Pechmann (1988) conducted an eight-year study of *Ambystoma talpoideum* in Rainbow Bay, a temporary Carolina bay that was completed encircled by a drift fence that allowed all immigrating and emigrating salamanders to be captured, marked, and measured. Sample sizes were relatively robust (about 20–111 individuals measured at sexual maturity each year). Survival to maturity was not affected by either age or size at metamorphosis, but both of these variables affected size at maturity. In general, individuals that metamorphosed early were larger at maturity than were those that metamorphosed late. Independent of age at metamorphosis, those that emerged at larger sizes were larger at maturity, and they remained larger in their second year of reproduction. Early metamorphosis and large metamorphic size also resulted in earlier reproduction by females, but this was not true for males.

In contrast to many anurans, *A. talpoideum* metamorphoses at a very large size, up to 95% of size at maturity. This compares with about 37% in wood frogs (Berven and Gill 1983). The limited growth of these salamanders after metamorphosis means that size at metamorphosis is likely to have a greater impact on adult fitness than in many anurans, some of which complete up to 95% of their growth after metamorphosis. Presumably the greater opportunities for postmetamorphic growth in anurans means that at least some individuals that emerge at a small size eventually will catch up to those that emerged at a larger size. This probably is more likely in species with long life spans than in those with relatively short life spans, such as *Pseudacris triseriata* (D. C. Smith 1987).

The second study of salamanders was a nine-year study of *Ambystoma opacum* (D. Scott 1994). Larvae were raised in large enclosures in natural ponds, as described in a previous section. At metamorphosis, the animals were released into the surrounding habitat. Various fitness-related traits, such as body size and clutch size, were then measured in individuals that returned to the ponds to breed over six or

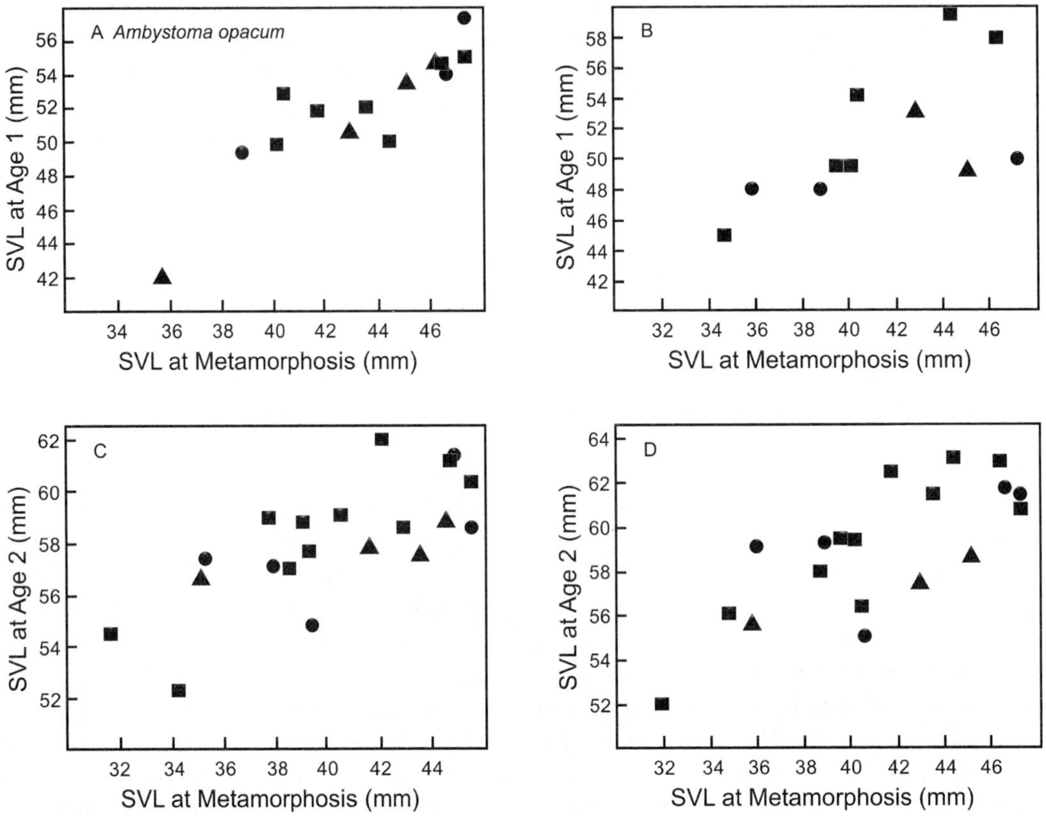


**Fig. 13.19.** Effect of size at metamorphosis on adult size of males (A, C) and females (B, D) at one and two years after metamorphosis in *Ambystoma opacum*. Each point represents a cohort of individuals from different experimental and natural ponds: Ginger's Bay experimental populations (squares), Ginger's Bay natural populations (circles), and Rainbow Bay natural populations (triangles). After D. Scott (1994).

seven breeding seasons. Scott found that metamorphs that emerged from low-density enclosures were larger and had larger lipid stores at metamorphosis than those from high-density enclosures. The former group returned to breed as larger adults than did those raised in high-density enclosures, and this size difference persisted in the second year (fig. 13.19). This species grows more between metamorphosis and sexual maturity than does *A. talpoideum;* metamorphs were about 75–88% of size at maturity. Individuals from low-density enclosures also had higher survivorship and reproduced at a younger age than those from high-density enclosures. Egg clutches of females were up to 70% larger in low-density individuals in one year, but not in another. The effect of body size on foraging success in a natural habitat is unknown, but any size-related difference in the ability of terrestrial adults to obtain food is likely to translate into a significant difference in reproductive performance, especially for females (D. Scott and Fore 1995). Overall, the results of this study provide the most convincing evidence for any amphibian that larval growth and development can have a significant impact on adult fitness and therefore on the dynamics of populations (B. Taylor and Scott 1997).

### Temperature Effects on Larval Life-History Traits

Most field and laboratory studies of larval growth and development have focused on density effects and competition for food, and most have been consistent with the major features of the Wilbur-Collins model and later modifications of that model. In general, larvae reared under conditions that are unfavorable for growth transform at a small size and leave the water as soon as possible; under more favorable conditions, they continue to grow to a larger size. However, some embryologists criticized the model on the grounds that differentiation rate, not growth rate, is the most accurate predictor of the timing of metamorphosis (Smith-Gill and Berven 1979; Just, Kraus-Just, and Check 1981). As discussed in more detail later, this argument is somewhat circular, because differentiation rate is, by definition, the rate at which a tadpole progresses toward metamorphosis. Nevertheless, there is no doubt that temperature can have a major impact on rates of growth, timing of metamorphosis, and size at metamorphosis. For example, *Rana pipiens* tadpoles reared at low temperatures grow more slowly and take longer to reach metamorphosis than those reared at

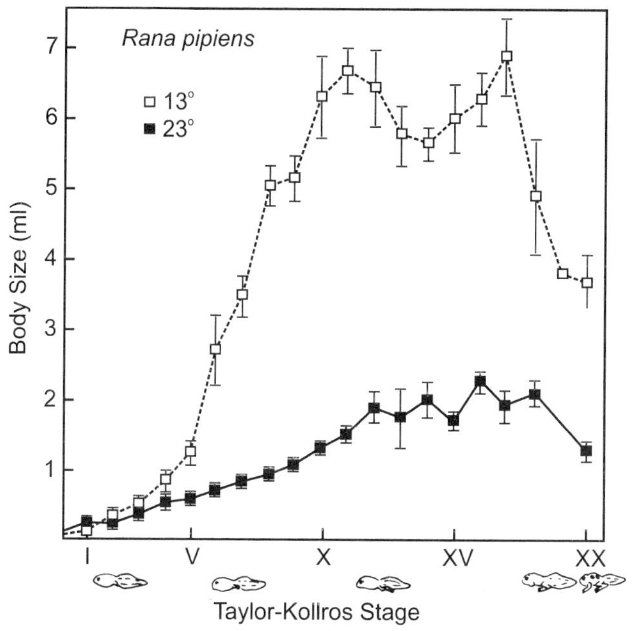


**Fig. 13.20.** Effect of temperature on development and body size in *Rana pipiens* tadpoles. Tadpoles raised at 13° C were larger at any given developmental stage than those raised at 23° C, but took longer to reach those sizes. Data are shown as means ± 2 SE. After Smith-Gill and Berven (1979).

warm temperatures, but they are larger at every stage (fig. 13.20). Hence, slow-growing tadpoles actually transform at a larger size than fast-growing ones, a result that appears to contradict the predictions of the Wilbur-Collins model. The explanation is that low temperatures inhibit differentiation more than growth (Smith-Gill and Berven 1979).

An analogous situation occurs in tadpoles that spend one or more winters in the larval stage. Differentiation virtually ceases during the winter, but growth continues, albeit at a reduced rate. In the spring, growth increases more rapidly than differentiation, so overwintering tadpoles typically transform at larger sizes than individuals that metamorphose in the same year that they hatch (Viparina and Just 1975; Collins 1979a; Berven, Gill, and Smith-Gill 1979; Smith-Gill and Berven 1979). Cold temperatures also reduce growth rates in pond-dwelling salamander larvae, but larvae from cold habitats often transform at a larger size (Bizer 1978; Sexton and Bizer 1978; Petranka 1984d). Similar effects can be produced by experimental manipulation of water temperature (Harkey and Semlitsch 1988 [corrected data for their table 2 supplied by R. Semlitsch]; Beachy 1995).

### Temperature and Geographic Variation in Larval Life Histories

Differences in environmental temperature and the length of the growing season probably account for much of the geographic variation in larval life-history traits reported for many amphibians. Interpreting patterns of geographic variation can be difficult (Berven and Gill 1983). Some differences among populations are genetic responses to selective pressures in different environments, but others represent plastic responses to local environmental variables. Often these variables are confounded with one another in ways that makes analysis of life-history traits even more difficult. For example, habitats of pond-dwelling tadpoles vary in pond permanence, water depth, pond surface area, temperature, dissolved oxygen concentration, pH, habitat productivity, and abundance and types of predators. Closed-canopy ponds have lower light levels, lower temperatures, lower productivity, and often lower numbers of predators than do open-canopy ponds (Loman 2002b; Skelly, Friedenburg, and Kiesecker 2002). All of these variables could affect larval growth and development, either independently or in concert. Latitudinal gradients differ in average temperature, the length of the growing season, and the number of hours of daylight during the summer growing season, all of which can affect pond productivity.

Despite this confounding of variables, there is a clear tendency for many pond-dwelling amphibians to have slower growth and longer larval periods in cold environments than in warm ones. In some cases, however, larvae from cold environments actually metamorphose at a larger size because of the greater effect of cold temperatures on differentiation than on growth. These patterns are particularly evident along altitudinal or latitudinal gradients in similar breeding sites. For example, North American bullfrogs (*Rana catesbeiana*) have much faster growth rates and shorter larval periods in southern populations than in northern ones, but size at metamorphosis often is smaller in southern populations (Viperina and Just 1975; Collins 1979a). Tailed frogs (*Ascaphus*) of western North America have larval periods of up to four years in cold mountain streams (Metter 1967; H. Brown 1990), but only one or two years in coastal and southern populations (Wallace and Diller 1998; Bury and Adams 1999).

Berven, Gill, and Smith-Gill (1979) studied the genetic and environmental components of such variation in high- and low-elevation populations of *Rana clamitans*. High-elevation tadpoles experience colder temperatures and have shorter growing seasons than lowland populations. Consequently, high-elevation tadpoles grow more slowly and have much longer larval periods than lowland tadpoles, but they transform at a larger size because of the greater effect of low temperature on differentiation than on growth. When tadpoles from high and low elevations were raised at low temperatures, those from high elevations actually grew faster than those from low elevations, while the reverse effect was observed at high temperatures. Hence, while differences in growth rate and length of the larval period are largely developmental responses to environmental temperature, natural selection appears to have acted against the selection gradient, so that the negative effect of temperature on growth is reduced. In other words, while high-elevation tadpoles grow more slowly than low-elevation tadpoles in natural environments, they would grow even more slowly if they had the same sensitivity to temperature as low-elevation tadpoles. Although Berven, Gill, and Smith-Gill (1979) did not perform genetic crosses, it is clear that differences in life histories among populations result from interactions of genetic and environmental effects.

A more detailed analysis of altitudinal variation in life history strategies of wood frogs (*Rana sylvatica*) was performed by Berven (1982b). He combined measurements of growth and development under natural conditions with reciprocal transplants of larvae and genetic crosses between high- and low-elevation males and females to unravel genetic and environmental contributions to life-history variation. As in most previous studies, tadpoles from high-elevation natural environments had slower growth rates, longer larval periods, and larger sizes at metamorphosis than those from low-elevation sites (fig. 13.21). Transplant experiments revealed that nearly all the variation in length of the larval period was due to environmental effects, whereas genetic differences between populations accounted for some of the differences in larval growth rates and size at metamorphosis. High-elevation tadpoles had faster growth rates than low-elevation tadpoles in both types of ponds. However, environmental effects still accounted for nearly three-quarters of the variation in size at metamorphosis. Nongenetic maternal effects, such as differences in egg size (Berven 1982a), also had small, but significant, effects on size at metamorphosis. Subsequent crossing experiments showed high heritabilities and considerable additive genetic variance for development time in montane and lowland populations, and for body size at metamorphosis in montane populations (Berven 1987).

Berven and Gill (1983) extended the work on *R. sylvatica* to broader geographic comparisons. Despite the fact that pond temperatures generally decrease with increasing latitude, the length of the larval period of wood frogs actually decreases from south to north, as does size at metamorphosis. In part this is due to genetic differences in temperature sensitivity. Wood frog tadpoles from the Canadian tundra are less sensitive to temperature changes and therefore have shorter larval periods at low temperatures than tadpoles from the mountains of Virginia or low-elevation ponds in Maryland. These populations also differed in their responses to changes in tadpole density. In the Canadian population, the length of the larval period was relatively insensitive to changes in density, whereas size at metamorphosis decreased rapidly with increasing density. In contrast, the size of Maryland tadpoles was less affected by density, whereas the length of the larval period was more sensitive. Montane tadpoles from Virginia were somewhat intermediate (fig. 13.22).

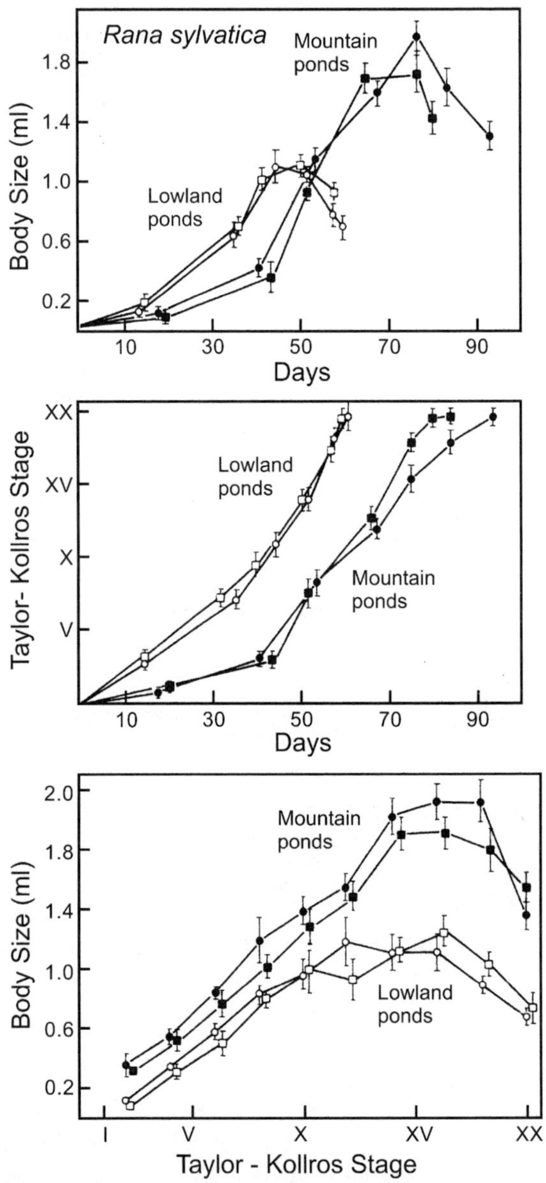


**Fig. 13.21.** Growth and development curves for wood frog (*Rana sylvatica*) tadpoles in natural ponds at high and low elevations. Tadpoles in mountain ponds grew more slowly and had longer larval periods than did those in low-elevation ponds, but reached larger sizes at metamorphosis. Data are shown as means ± 2 SE. After Berven (1982b).

Berven and Gill (1983) interpreted their results as evidence that the strength of natural selection operating on life-history traits differs among the populations. In the lowland population, both juvenile survivorship and size at first reproduction were positively correlated with size at metamorphosis. Selection seems to have favored large size at metamorphosis and has acted to minimize the sensitivity of this trait to environmental variables, particularly density. Be-

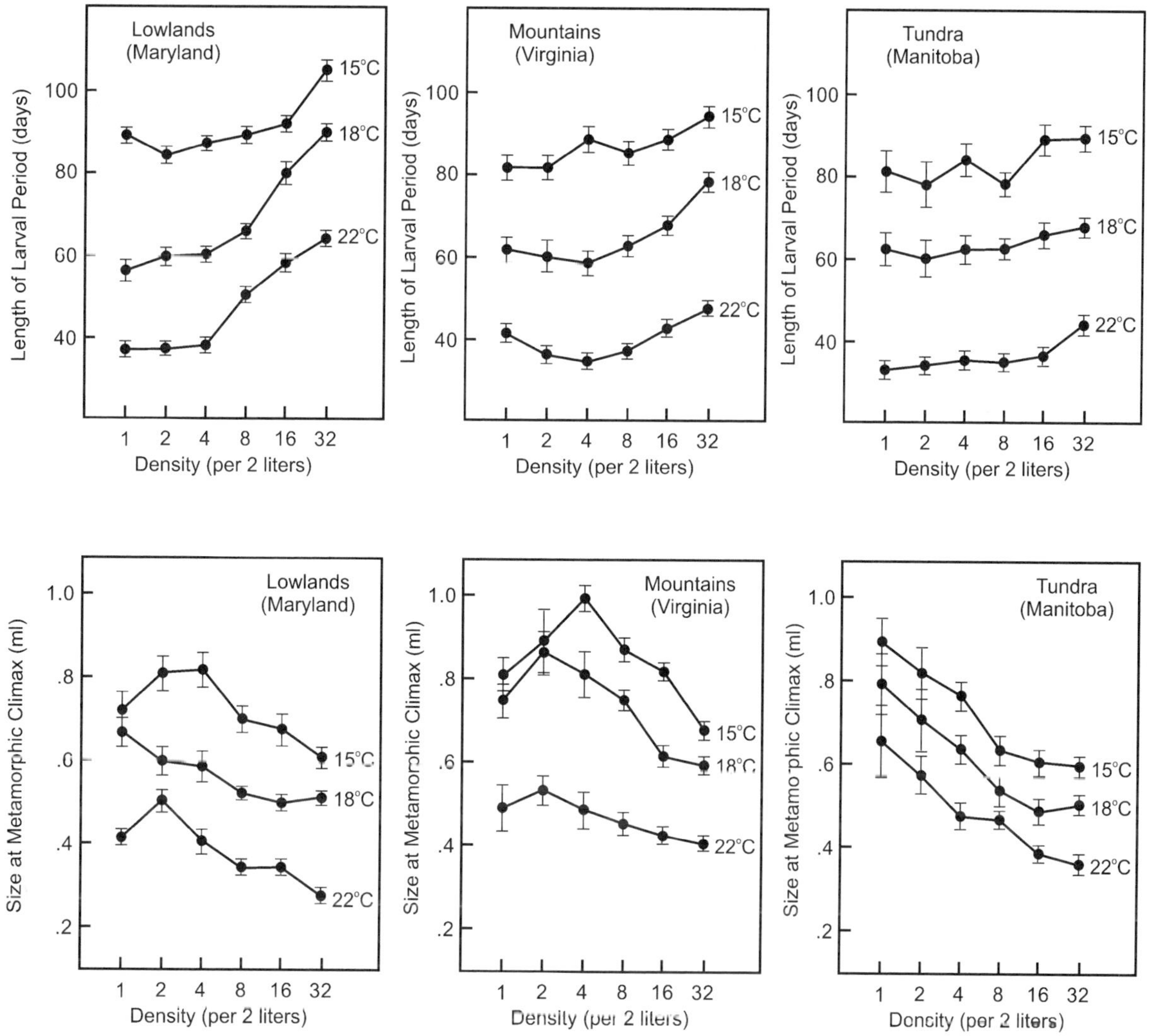


**Fig. 13.22.** Effect of temperature and larval density on length of the larval period and size at metamorphosis in wood frog (*Rana sylvatica*) tadpoles from three geographic locations. Size at metamorphosis was strongly affected by density in tundra tadpoles, but length of the larval period was not. In lowland tadpoles from Maryland, both length of the larval period and size at metamorphosis were affected by density. Tadpoles from the mountains of Virginia were somewhat intermediate, but more similar to those from Maryland than to those from the tundra. Data are shown as means ± 1 SE. After Berven and Gill (1983).

cause the growing season is relatively long, the disadvantage of a long larval period is reduced; selection has favored tadpoles that remain in the pond long enough to reach maximum body sizes. The more intense selection on body size was reflected in low heritability of that trait. In contrast, the very short growing season in the tundra appears to favor rapid larval development, even at the expense of a small size at metamorphosis. Consequently, length of the larval period was relatively insensitive to environmental effects in the Canadian population and exhibited low heritability, suggesting strong selection on that trait. Size at metamorphosis, on the other hand, was strongly influenced by density, and exhibited a much higher level of additive genetic variance. One mechanism that may allow for faster growth rates at high latitudes, or in other environments with short growing seasons, is a higher physiological efficiency in converting food into body tissue (Lindgren and Laurila 2005).

Studies of wood frogs in Michigan revealed a suite of life-history traits that were somewhat different from those in Maryland and Virginia and those in Canada (Riha and Berven 1991). In Michigan, tadpoles in the field grew faster, had shorter larval periods, and were larger at all stages of development than were low-elevation populations in Maryland or high-elevation populations in Virginia. When raised in the laboratory, most of the variation in development rate in populations from all three areas was due to variation in temperature, but development was slower in Michigan and Maryland than in Virginia. In contrast, Maryland and Virginia populations had similar growth rates at the same temperature, but growth rates of Michigan larvae were higher, especially at low temperatures. Some of the differences in larval life histories in these populations were due to differences in breeding time. Wood frogs in Michigan breed later in the spring than do those in Maryland and Virginia, and breeding ponds actually warm up faster than in the more southern populations. Pond productivity appears to be higher in Michigan as well, so more food is available to tadpoles. This could be due to longer days during the growing season, to local differences in nutrient input, or both.

In some amphibians, larvae do not metamorphose at larger sizes in cooler environments, as previously described. For example, some stream-breeding plethodontids exhibit the opposite pattern, with larvae reaching larger sizes at metamorphosis in warm, low-elevation habitats than in cool, high-elevation habitats (Bruce 1988b; Juterbock 1990; Voss 1993; Benardo and Reagan-Wallin 2002; Bernardo and Agosta 2003). Habitats of stream-dwelling salamander larvae vary along altitudinal gradients in a number of respects, including stream order, current velocity, temperature, habitat productivity, abundance of invertebrate prey, and density of predators. For these larvae, the low productivity of cool, high-elevation streams, and particularly the low abundance of suitable invertebrate prey, appear to limit feeding opportunities for salamander larvae and retard rates of growth and development. While prey abundance can be affected by environmental temperature, the effect of prey abundance on larval growth and development appears to be greater than the direct effect of temperature.

Taken together, these studies demonstrate that both environmental effects and genetic differences among populations can be important determinants of variation in life history strategies. In some cases, differences in patterns of larval growth and development that appear to be adaptive responses to natural selection actually are phenotypically plastic responses to the environment. In other instances, there are genetic differences among populations, but these can be identified only through carefully controlled experimental studies. Investigators should resist the temptation to invoke adaptive explanations for differences in life-history traits without adequate data on the genetic basis of such variation. In some cases, natural selection can favor increased phenotypic plasticity as an adaptation to highly variable environments.

The work of Smith-Gill and Berven (1979) and subsequent studies have been useful in demonstrating the importance of temperature as an environmental variable that can profoundly influence growth rate, the timing of metamorphosis, and size at transformation. However, the conclusions of these studies are not inconsistent with the Wilbur-Collins model and other models of amphibian metamorphosis that focus on relative growth rates. These models were designed primarily to explain variation in larval life histories among cohorts of larvae in the same population, where individuals are exposed to similar temperature regimes. They were not designed to account for differences among populations that differ substantially in environmental temperature.

The argument that differentiation rate is more useful than growth rate in predicting amphibian metamorphosis is ecologically uninformative. The stages used by developmental biologists to describe larval differentiation (e.g., the Gosner stages in fig. 13.3) culminate in metamorphosis, so by definition, differentiation rate is the rate at which a larva progresses toward metamorphosis. Therefore, it is a tautology to state that differentiation rate predicts the timing of metamorphosis (Sokol 1984; Wilbur 1984). Furthermore, their model assumes that all larvae metamorphose at exactly the same stage of development. However, Sokol (1984) found that tadpoles of *Litoria ewingi* differed in body proportions, and therefore degree of development, depending on rearing conditions, even though they all metamorphosed at the same Taylor-Kollros stage. Tadpoles reared under crowded conditions were anatomically underdeveloped compared to those reared at low densities, and this could affect their subsequent survival. Interspecific differences in degree of development can be even greater. Indeed, as the discussion in the next section makes clear, the degree of differentiation of larvae at metamorphosis can be just as plastic as are traits such as length of the larval period and size at metamorphosis, especially in urodeles.

## Paedomorphosis as a Life-History Strategy

When the Spanish invaded Mexico in the sixteenth century, they found among the food items of the Aztecs a peculiar creature resembling the salamanders of Europe, but having conspicuous gills on each side of the head (fig. 13.23). The Aztecs called these creatures axolotls, after the exceptionally ugly god Xolotl, who once assumed this amphibian form and retreated into a deep lake to escape the wrath of other deities (H. M. Smith and Smith 1971). Over the next 300 years, various Spanish travelers mentioned axolotls in accounts of Aztec life, and a few preserved specimens found their way to Europe. Naturalists puzzled over the taxonomic status of the animal. Most considered it to be a primitive salamander, allied to the sirens, but others, including the great French naturalist Cuvier, thought it must be the larval form of an undiscovered species. A few, including Henri Milne-Edwards, considered it to be a functionally adult salamander, but one that had remained in an arrested stage of development (Gould 1977).

Scientific interest in the axolotl increased in the mid-nineteenth century, when live specimens were imported to

**Fig. 13.23.** An early engraving of the axolotl (*Ambystoma mexicanum*), a permanently paedomorphic salamander that retains gills throughout life. From Duméril, Bibron, and Duméril (1854).

France from Mexico. The specimens were collected in Lake Xochimilco, near Mexico City, during a short-lived attempt to add Mexico to a resurgent French empire. They were sent by an army officer to the French herpetologist, Auguste-Henri-André Duméril, who soon established a thriving colony in Paris. Several individuals attained sexual maturity and reproduced, but retained their larval morphology. However, to Duméril's astonishment, some of the offspring subsequently lost their gills and transformed into terrestrial adults that were nearly indistinguishable from salamanders of the genus *Ambystoma* (or *Amblystoma*, as the genus was called at the time). This created a considerable stir, since most naturalists had placed the axolotl in the genus *Siredon*. Either Duméril had witnessed the transformation of one genus of salamander into another, or the classification was in error. Eventually the axolotl was recognized as an example of arrested development, an adult form that retained the morphological features of a larva. August Weismann (1878) and others realized that this type of life history was a major blow to the idea that ontogenetic development recapitulates phylogenetic history; the axolotl clearly was derived from a terrestrial ancestor, yet it did not pass through a terrestrial stage during ontogeny (Gould 1977).

The Mexican axolotl, now called *Ambystoma mexicanum*, has become a famous animal, and descendents of Duméril's original stock continue to be used for all manner of experimental laboratory studies (H. M. Smith and Smith [1971] recorded more than 3,300 publications on the axolotl). However, its peculiar life history is far from unique. We now know that similar life histories are widespread among urodeles. All members of four families, the Cryptobranchidae, Sirenidae, Amphiumidae, and Proteidae retain some larval traits as adults and do not undergo further metamorphosis, even when treated with thyroid hormones. They differ, however, in which larval characters are retained in the adult stage (Noble 1924; Dent 1968; Rosenkilde 1979). Cryptobranchids, for example, lack external gills as adults, whereas these are retained in the other families. Retention of larval traits in mature adults occurs in some species in most other urodele families (Ambystomatidae, Dicamptodontidae, Hynobiidae, Plethodontidae, and Salamandridae). In some cases, such life histories have become genetically fixed, but in others they are facultative (Whiteman 1994).

A confusing variety of terms has been used to describe the retention of larval characters in sexually mature salamanders, and different authors often use these terms in contradictory ways. For example, Wilbur and Collins (1973) used the term *paedogenesis* to refer to genetic fixation of larval characters in adults, while the term *neoteny* was used to describe facultative metamorphosis. Salthe and Mecham (1974) used these terms in exactly the opposite way. Other authors have designated as neotenic any individuals exhibiting delayed metamorphosis, whereas those that become sexually mature while retaining larval characters are referred to as paedogenetic (Bizer 1978; Sexton and Bizer 1978). This confusion has led to extensive debate among herpetologists over the proper use of these and other terms (B. Pierce and Smith 1979, 1980; R. Sullivan 1980; D. Wake 1980). One problem is that the term *paedomorphosis* has been used both to describe the specific case of retention of larval characters, such as external gills, in sexually mature salamanders, and a more general evolutionary phenomenon of the retention in adults of any sort of traits in derived species that occur only in earlier developmental stages in ancestral species (see chapter 1). D. Wake (1992) argued that the term should be restricted to the latter situation, but most recent authors have referred to retention of larval characters in adult salamanders as paedomorphosis, and I have adopted that terminology here.

Gould (1977) argued that retention of larval characters in adults can come about either by retardation of somatic development (neoteny) or acceleration of sexual maturation (progenesis). He considered all paedomorphic salamanders to be neotenic, but other authors have disagreed (e.g., Denoël and Joly 2000). More recent workers have identified three processes that can operate alone or together to produce paedomorphic individuals: (1) *neoteny,* defined as a decelerated rate of somatic development, (2) *postdisplacement,* defined as the delayed onset of metamorphic change, and (3) *hypomorphosis,* defined as an early cessation of somatic differentiation that results in early onset of reproduction (Reilly, Wiley, and Meinhardt 1997; T. Ryan and Semlitsch 1998).

Gould's definitions say nothing about genetic fixation of the trait; salamanders can be genetically fixed for paedomorphosis, or facultatively paedomorphic (Whiteman 1994; Denoël, Joly, and Whiteman 2005). Even here there has been confusion. The Mexican axolotl and several other ambystomatids seldom, if ever, metamorphose in nature (see the following), and some authors have categorized them as "obligate neotenes" or "obligate paedomorphs" along with the proteids, sirenids, cryptobranchids, and amphiumids (Duellman and Trueb 1986; Werner 1986). Nevertheless, some individuals in paedomorphic populations of ambystomatids retain the capacity for metamorphosis when treated with thyroid hormones (Dent 1968; Taurog 1974; Rosenkilde 1979). As Duméril's experience showed, they sometimes metamorphose spontaneously in the laboratory (Chauvin 1876; Weismann 1878; H. M. Smith and Smith 1971; Brandon 1976; Brandon, Maruska, and Rumph 1981; Krebs and Brandon 1984). Hence, while some or all individuals are paedomorphic in nature, these species are not considered obligate paedomorphs.

### Examples of Paedomorphosis in Urodeles

Ambystomatid salamanders are quite flexible in their life history characteristics, and paedomorphosis has evolved independently many times in the family. Paedomorphic life histories are particularly common in populations inhabiting the Mexican Plateau, a relatively arid region that includes a number of permanent lakes. Many of the lakes support isolated populations of *Ambystoma*, some of which are entirely paedomorphic. In addition to *Ambystoma mexicanum*, these taxa include *A. andersoni*, *A. dumerilii*, *A. taylori*, and several populations in the *A. tigrinum* complex (Brandon 1976; Brandon 1981; Krebs and Brandon 1984; Shaffer 1984a, b). Other Mexican taxa, including *A. ordinarium*, *A. rosaceum*, *A. amblycephalum*, *A. lermaense*, and some populations of *A. tigrinum*, have variable life histories, as does at least one species formerly placed in the genus *Ryacosiredon* (now considered part of *Ambystoma*). In some cases, paedomorphic and transforming individuals coexist in the same population, while in others, local populations vary in their metamorphic tendencies (J. Anderson and Worthington 1971; Brandon and Altig 1973; J. Anderson and Webb 1978; Collins 1979b; Shaffer 1984a, b). A similar range of variation is found in populations of *A. tigrinum*, *A. talpoideum*, and *A. gracile* north of Mexico (R. Snyder 1956; Gehlbach 1967, 1969; D. Larson 1968; R. Webb and Roueche 1971; Sprules 1974a, b; Eagleson 1976; F. Rose and Armentrout 1976; Patterson 1978; Collins 1981; Semlitsch and Gibbons 1985; Whiteman 1994). The four species of *Dicamptodon*, which are closely related to ambystomatids (see chapter 1), differ in their metamorphic tendencies. Individuals of *Dicamptodon copei* rarely metamorphose in the field (Nussbaum 1970). Most individuals of *Dicamptodon aterrimus*, *D. ensatus*, and *D. tenebrosus* metamorphose, but some individuals are paedomorphic in permanent ponds and streams (Schuierer 1958; Nussbaum 1970; Nussbaum and Clothier 1973; Petranka 1998).

The frequency of paedomorphosis in hynobiid salamanders is unclear. Most discussions (e.g., Wilbur and Collins 1973; Duellman and Trueb 1986) are based on an early report by Sasaki (1924) of paedomorphic *Hynobius retardatus* from a cold montane lake in Japan. He reported that individuals of the same species in temporary ponds always metamorphosed. The population described by Sasaki disappeared from the lake after 1932, apparently eliminated by introduced salmonid fishes, and paedomorphic individuals have never been collected from other localities (Moriya 1983). Sasaki believed that cold temperatures induced paedomorphosis in the lake population. This hypothesis was tested by Moyira (1983); he subjected larvae of transforming populations to low temperatures and found that many never metamorphosed. He did not demonstrate that the larvae ever attained sexual maturity, however, so it is not clear that they were truly paedomorphic.

In plethodontids, paedomorphosis is associated mainly with life in caves, although not all cave-dwelling plethodontids are paedomorphic. The genus *Gyrinophilus* exhibits a range of life-history strategies, from fully transforming to fully paedomorphic (Brandon 1966; Bruce 1979). *Gyrinophilus porphyriticus* is mainly a surface-dwelling species, but it sometimes lives in caves as well (Cooper and Cooper 1968; Brandon 1971). As discussed earlier, it has a very long larval period, but apparently is never paedomorphic. *Gyrinophilus subterraneus*, a species associated exclusively with caves, has larvae that reach nearly adult size that apparently transform just before sexual maturity (Besharse and Holsinger 1977). Although definitive evidence of paedomorphosis in this species has yet to be found, it appears to be evolving toward permanent larval life (fig. 13.24 A). A third species, *G. palleucus*, also is found exclusively in caves, but probably evolved independently from a *porphyriticus*-like ancestor (McCrady 1954; Lazell and Brandon 1962; Brandon 1965b, 1971). It is paedomorphic in nature, although it metamorphoses when treated with thyroid hormones (Dent and Kirby-Smith 1963), and a few spontaneously metamorphosed individuals have been found in the field (D. Simmons 1976; Brandon et al. 1986).

Most of the other paedomorphic plethodontids are in the genus *Eurycea*. Paedomorphic forms are particularly common on the Edwards Plateau of Texas, a region of extensive limestone caves and permanent springs, and the salamanders in this region form a monophyletic clade (Chippindale 2000; Chippindale et al. 2000). Again, this group exhibits a range of life histories from transforming to completely paedomorphic, with both cave- and surface-dwelling populations (Bruce 1976; Sweet 1977b, 1982; Petranka 1998). The latter are associated exclusively with permanent springs; when these fail, populations are forced underground into cave systems. In most populations, all individuals are

**Fig. 13.24.** Adaptations of plethodontid salamanders for life in caves and springs. (A) *Gyrinophilus subterraneus,* a species that sometimes lives in caves. It has a long larval period and reaches a very large size before metamorphosing. (B) *Eurycea tynerensis,* a surface-dwelling salamander from the Ozark Plateau with some paedomorphic populations. (C) *Haideotriton wallacei,* another blind paedomorphic salamander found in deep wells and caves. (D) *Eurycea rathbuni,* a species highly adapted for life in caves and springs, lacks pigment and functional eyes. (E) Pigmented, surface-dwelling larva of *Eurycea spelaea* from the Ozark Plateau. (F) Metamorphosed cave-dwelling adult of *E. spelaea,* with reduced pigment and degenerated eyes. Photos by Wayne Van Devender.

paedomorphic, but can transform when treated with thyroid hormones (Kezer 1952). However, several populations have significant numbers of transformed adults, and in some, paedomorphosis is uncommon or absent (Sweet 1977b, 1982). Metamorphosis is most common in surface-dwelling populations, but occurs occasionally in caves.

Most of the paedomorphic salamanders on the Edwards Plateau have restricted ranges, in many cases being limited to one cave system, stream drainage, or spring (J. K. Baker 1961, 1966; B. Brown 1967a, b, c; Sweet 1977a; Chippindale et al. 2000). These include *E. nana, E. neotenes, E. pterophila, E. sosorum, E. tridentifera, E. troglodytes,* and two species previously placed in a separate genus *Typhlomolge, E. rathbuni* (fig. 13.24 D), and *E. robusta* (Potter and Sweet 1981; Petranka 1998; Chippindale 2000; Chippindale et al. 2000). It is likely that many unnamed species in this complex remain to be discovered. Although these species usually are paedomorphic in nature, some retain the capacity for partial or complete metamorphosis when treated with thyroid hormones (Dundee 1957; Potter and Rabb 1960; Brandon 1971). All live in springs, but exhibit varying degrees of specialization for cave life. Most have some reduction in the size and function of the eyes, and *E. tridentifera, E. rathbuni,* and *E. robusta* are completely blind. These three species also are nearly or completely albinistic, a characteristic of most animals living in dark caves. *Eurycea rathbuni* and *E. robusta* are the most specialized cave dwellers, with spindly legs and flattened snouts that probably are used for aquatic feeding (fig. 13.24 D). These species somewhat resemble another blind paedomorphic salamander, *Haideotriton wallacei* (fig. 13.24 C), known from deep wells and caves in Georgia and Florida, but these species undoubtedly represent independent evolutionary events (Brandon 1971). This species is placed in the genus *Eurycea* by Frost et al. (2006).

Several other hemidactyline salamanders also exhibit varying degrees of paedomorphosis and adaptation to cave life. In the Ozark Plateau and Ouachita Mountains of Missouri, Arkansas, Oklahoma, and Texas, two putative subspecies of *Eurycea multiplicata* are primarily surface-dwelling and usually metamorphosing: *E. m. multiplicata* in the Ouachitas and *E. m. griseogaster* in the Ozarks (Bonett and Chippindale 2004). *Eurycea m. griseogaster* traditionally has been considered very closely related to *E. tynerensis* of the Ozark Plateau (fig. 13.24 B), which is surface-dwelling, but paedomorphic (Dundee 1965; Ireland 1976; Tumlison, Cline, and Zwank 1990). Recent phylogenetic work using mitochondrial DNA showed that *E. tynerensis* is not genetically distinct from *E. m. griseogaster,* and that a paedomorphic life history has evolved independently between three and nine times in this clade. Bonett and Chippindale (2004) recommended that all populations previously designated as *E. tynerensis* and *E. m. griseogaster* be treated as *E. tynerensis.* Another member of the *E. multiplicata* complex is *E. spelaea* (Bonett and Chippendale 2004). Larvae of this species have been found both in and out of caves. These have functional eyes, but transform into adults that have degenerate eyes and live exclusively in caves (Brandon 1970b, 1971; Besharse and Brandon 1974; fig. 13.24 E, F).

The only other urodeles in which paedomorphosis regularly occurs are the newts, particularly species of *Triturus* and *Notophthalmus.* One of the earliest accounts of paedomorphosis in any urodele was de Filippi's (1861) description of sexually mature "larvae" of *Triturus alpestris.* Paedomorphosis frequently occurs in this species (Radovanović 1951; Roček 1974; Ferracin, Lunadei, and Falcone 1980; Dolce and Stoch 1984; Džukić and Kaleszić 1984; Chacornac and Joly 1985; Giacoma 1988; Kaleszić et al. 1996; Denoël and Joly 2000; Denoël et al. 2001; Denoël 2004; Denoël, Schabetsberger, and Joly 2004). Paedomorphosis has been reported in several other European newts, including *T. helveticus, T. italicus, T. vulgaris, T. cristatus,* and *T. carnifex,* but the frequency of paedomorphosis varies among species and among populations of the same species (Kaleszić and Džukić 1985; Džukić et al. 1990; Kaleszić et al. 1994). For example, many paedomorphic individuals have been found in populations of *T. helveticus* in the Netherlands and southern France (van Gelder 1973; Gabrion, Sentein, and Gabrion 1977, 1978), but paedomorphosis is rare in British populations (Frazer 1983). In general, paedomorphosis appears to be most common in isolated mountain lakes, in regions such as the Alps and Balkans, but it sometimes occurs in permanent water at low altitudes.

There are many reports of paedomorphosis in *Triturus vulgaris* (Boettger and Schwartz 1928; Hartwig and Rotmann 1940; Gislén and Kauri 1959; Benl 1965; Dely 1967; Dolmen 1978, 1981a; Dolce and Stoch 1984; Kaleszić and Džukić 1985; Tucić et al. 1985), but usually all individuals eventually metamorphose. Hartwig and Rotmann (1940) reported that about 15% of newts in ponds near Cologne, West Germany were paedomorphic, but most eventually metamorphosed. At least one Swedish population was completely paedomorphic, but it was eliminated by introduced trout (Dolmen 1978, 1981a). There are reports of newts in some populations failing to reach sexual maturity (M. Smith 1969), perhaps as a result of incomplete development. Often such populations are albinistic (Benl 1965; M. Smith 1969; Frazer 1983). This suggests a defect in pituitary function, since both melanin production and metamorphosis are regulated by pituitary hormones.

There are scattered reports of paedomorphosis in *Triturus cristatus,* but Hartwig and Rotmann (1940) never found paedomorphic individuals of this species, even in locations where paedomorphic *T. vulgaris* were found. On the other hand, in Sweden, a permanently paedomorphic population of *T. cristatus* occurred in the same lake as paedomorphic *T. vulgaris* (Dolmen 1978, 1981a). Gabrion, Sentein, and Gabrion (1977) never found paedomorphic *T. marmoratus* in ponds where paedomorphic *T. helveticus* were common, and there appear to be no records of paedomorphosis in this species.

Paedomorphosis is uncommon in North American newts of the genus *Taricha*, many of which are largely terrestrial as adults. However, populations of the most aquatic species, *T. granulosa*, sometimes contain paedomorphic individuals (Reimer 1958), and some populations in high-altitude lakes are nearly all paedomorphic (Marangio 1978). Paedomorphosis appears to be more widespread in *Notophthalmus*, having been reported in both *N. viridescens* and *N. perstriatus* (Noble 1926a, 1929b; S. Bishop 1941a, b; Healy 1970; R. N. Harris 1987a). It is most common in coastal plain populations, but has been reported in permanent ponds in other areas as well (Brandon and Bremer 1966). Life histories of local populations shift in response to climatic variation, with larvae and even some neotenes completing metamorphosis when the aquatic environment deteriorates (Healy 1970; Reilly 1987). In many populations where paedomorphosis occurs, some transformed individuals omit the terrestrial stage and remain permanently in the water (Noble 1929b; Healy 1973, 1974). There has been confusion on this point. Some authors (e.g., Wilbur and Collins 1973) have implied that all individuals that bypass the eft stage are paedomorphic, but aquatic adults can be fully metamorphosed, or they can retain their gills (Noble 1926a, 1929b; S. Bishop 1941a; Healy 1970, 1973, 1974; Reilly 1986, 1987; R. N. Harris 1987a). Often individuals in all stages of metamorphosis occur in the same population.

One difficulty with determining the frequency of paedomorphosis in different groups is that developmental processes leading to retention of larval characters are not necessarily the same in all urodeles. For example, ambystomatids and salamandrids exhibit quite different life-history patterns. Paedomorphic ambystomatids have skulls nearly indistinguishable from those of larvae (Reilly 1986). Hence, these are individuals in which metamorphosis has been delayed beyond the onset of sexual maturity. In some species, such as *Ambystoma talpoideum*, sexually mature paedomorphic individuals can undergo rapid metamorphosis if the pond dries up (Semlitsch 1985b; Reilly 1987). In *Notophthalmus viridescens*, and probably in species of *Triturus* and *Taricha* as well, the skulls of branchiate (gilled) adults are fully metamorphosed, with the only paedomorphic characters being the retention of some or all of the gill structures. Hence, paedomorphic newts are individuals that have initiated but not completed metamorphosis. Reilly (1987) found that permanent paedomorphosis is rare in *Notophthalmus viridescens*. Eventually most branchiate adults apparently complete metamorphosis by losing the gills, even if they remain in the water.

### The Mechanistic Basis of Paedomorphosis

Changes in the tendency of amphibian larvae to undergo metamorphosis must involve alterations at one or more stages of the thyroid hormone axis that affect the production of thyroid hormones, the sensitivity of tissues to these hormones, or a combination of the two. There is some evidence that alterations in the hormonal control of metamorphosis leading to a paedomorphic life history have occurred differently in different lineages of salamanders (Shaffer and Voss 1996). For example, individuals in many paedomorphic populations retain the capacity to metamorphose when they are treated with thyroid hormones, but in other populations, this capacity has been lost. In the Mexican axolotl (*Ambystoma mexicanum*), production of thyroxine appears to be normal, and there is a surge of hormone production early in the larval period, as in other larval salamanders. This surge of thyroxine does not lead to metamorphosis, however, apparently because this species lacks the enzyme that converts thyroxine to the more potent triiodothyronine, which actually initiates metamorphosis. Late in ontogeny, activity of the thyroid axis declines, and these animals never metamorphose in nature (Galton 1992; Rosenkilde and Ussing 1996). These developmental changes appear to be due to changes in a major gene, or quantitative trait locus, and not a simple single genetic locus (Voss and Shaffer 1997). In ambystomatids that are facultatively paedomorphic, the lack of metamorphosis is correlated with changes at higher levels of the thyroid axis, such as the regulation of thyroid hormones by the hypothalamus or pituitary (Shaffer and Voss 1996).

In proteids, which do not undergo metamorphosis even when treated with high doses of thyroid hormones, the loss of metamorphosis is due in part to changes in the sensitivity of target tissues to thyroid hormones. Both *Necturus* and *Proteus* have normal thyroid hormone receptor genes, but one of these genes, TRβ, is expressed at very low levels in these salamanders. This suggests that up-regulation of this gene does not occur in response to thyroid hormone secretion and metamorphosis is not initiated (Safi et al. 1997).

### Ecological Correlates of Paedomorphosis

Most early students of paedomorphosis in salamanders believed that metamorphosis was directly inhibited by some environmental factor, and there were many attempts to identify the proximate causes of this condition. Because paedomorphic ambystomatids and salamandrids are frequently found in high-altitude lakes, many workers suggested that paedomorphosis is caused by low temperatures that either reduce the sensitivity of target tissues to thyroid hormones, or reduce the activity of the thyroid gland (Sasaki 1924; R. Snyder 1956). There is no doubt that environmental temperatures can affect the tendency of individual larvae to metamorphose. For example, Sprules (1974b) raised *Ambystoma gracile* larvae at 12°C and 19°C and found a greater incidence of paedomorphosis at the colder temperature. However, individuals apparently differed in their response to cold. Furthermore, there clearly are genetic differences among populations in metamorphic tendencies. For example, Eagleson (1976) raised *A. gracile* larvae from high- and low-altitude lakes at the same temperature; only 24% of the larvae from the high-altitude lake metamorphosed, compared to 88% of those from the low-altitude lake. In any case, the proximate effects of low temperature are not sufficient to account for paedomorphosis in all urodeles, since many paedomorphic salamanders do not live in cold lakes.

A variety of other direct environmental influences have been invoked to account for paedomorphic tendencies in urodele populations. These include low levels of iodine in high-altitude lakes, resulting in abnormal thyroid activity; insufficient food; an abundance of food; and steep-sided ponds that make it difficult for larvae to emerge from the water (Hartwig and Rotmann 1940; Van Gelder 1973; Salthe and Mecham 1974; Sprules 1974a). Most of these explanations lack generality, particularly for populations in which metamorphosing and paedomorphic individuals coexist. Most recent workers have assumed that paedomorphosis represents a life-history strategy that has evolved in response to natural selection. Presumably selection could operate through: (1) changes in the activity of the thyroid gland, (2) changes in the sensitivity of target tissues to thyroid hormones, (3) changes in the response of the thyroid or target tissues to environmental variables, or (4) a combination of all three. Selection also could favor plasticity in the hormonal control of metamorphosis, allowing facultative paedomorphosis when conditions are favorable, but transformation when they are not. This might occur through changes in hypothalamic activity in response to external environmental cues.

Many authors have proposed that a paedomorphic life history is favored in permanent, relatively equable aquatic habitats surrounded by unstable or inhospitable terrestrial environments (Wilbur and Collins 1973; Healy 1974; Salthe and Mecham 1974; Sprules 1974a). In fact, this was suggested more than a century ago by August Weismann (1878), who attributed the permanent larval life of the axolotl to circumstances that "made life on land more difficult than life in the water," an explanation echoed by Walter Garstang's verse at the beginning of this chapter. Often this hypothesis is stated in terms of a harsh physical environment on land resulting from cold temperatures or arid conditions (e.g., Sprules 1974a; Duellman and Trueb 1986). However, this is an oversimplification, because both physical and biological factors can affect the relative suitability of the aquatic and terrestrial environments.

Werner's niche shift model again provides a useful framework for understanding paedomorphosis. He pointed out that harsh terrestrial conditions would tend to raise the mortality/growth ratio in the terrestrial environment relative to that in the water, while equable conditions in the water would lower the mortality/growth ratio in that habitat. Environmental variables that might shift the balance between growth and mortality in the water or on land include the following: (1) differences in temperature, which could affect growth rates or the length of the growing season; (2) aridity of the terrestrial environment; (3) predation pressure; and (4) availability of food in the two habitats. In addition, the permanence of the aquatic environment will determine whether a paedomorphic life history is possible. There are many examples of paedomorphosis occurring in urodele populations that inhabit permanent ponds, lakes, or streams, whereas populations of the same species in temporary water undergo metamorphosis (Schuierer 1958; Brandon and Bremer 1966; J. Anderson and Worthington 1971; Webb and Roueche 1971; Nussbaum and Clothier 1973; Nussbaum 1976; F. Rose and Armentrout 1976; Patterson 1978; Collins 1981; Semlitsch and Gibbons 1985). Nevertheless, there also are examples of paedomorphic populations occurring in both permanent and temporary habitats, as in *Triturus alpestris* in Europe (Denoël et al. 2001).

The relative importance of these environmental variables depends on the ecology of each species. Many paedomorphic urodele populations are found in permanent lakes or ponds in relatively arid regions. The various species of paedomorphic ambystomatids on the Mexican plateau and elsewhere in western North America (Gehlbach 1967; D. Larson 1968; Webb and Roueche 1971; Collins 1979b, 1981) are obvious examples. Populations of paedomorphic *Triturus helveticus* often are associated with arid limestone tablelands (Gabrion, Sentein, and Gabrion 1977) or inland dunes (van Gelder 1973). The occurrence of paedomorphosis in some coastal plain populations of *Notophthalmus* has been attributed to the relatively arid, sandy terrestrial environment (Healy 1970; Wilbur and Collins 1973). Similarly, high-altitude environments not only have long, cold winters, but also tend to have relatively dry summers, a regime that selects against metamorphosis (Sprules 1974a; Dolmen 1978, 1981a; Marangio 1978). On the other hand, no matter how harsh the terrestrial environment, a paedomorphic life history often is impossible if aquatic predators are abundant. Consequently, paedomorphic urodeles in arid regions or at high altitudes are found mostly in lakes without predatory fish (Carr and Goin 1943; Brandon and Bremer 1967; D. Larson 1968; Sprules 1974a; Dolmen 1978, 1981a; Marangio 1978).

The relative availability of food in aquatic and terrestrial environments has not been investigated as thoroughly. Healy (1973) found that growth of young *Notophthalmus viridescens* was faster in the water than on land in coastal plain areas; this should favor paedomorphosis unless drying of the pond (Healy 1970) or high levels of predation (Werner 1986) result in excessive mortality in the water. The association of paedomorphosis with life in caves also is related to the availability of food, because much of the prey available in a cave is found in the water (Brandon 1971). Although metamorphosed salamanders often feed in the water (P. Smith 1948; Bruce 1979), they probably are less effective at doing so than aquatic larvae (Lauder and Shaffer 1986), and therefore would be at a competitive disadvantage compared to paedomorphic adults. Metamorphosed adults *of Gyrinophilus palleucus* and *G. subterraneus* tend to be emaciated, whereas larvae are robust, with extensive fat reserves (D. Simmons 1976; Besharse and Holsinger 1977; Brandon et al. 1986); this suggests a shortage of food for terrestrial animals. Even aquatic paedomorphs probably are energy-limited in many cave systems, because nearly all energy input comes from outside of the caves. At least one species, *Eurycea spelaea,* even consumes bat guano, which can be as important an energy source as invertebrate prey (Fenolio et al. 2006).

Whiteman, Wissinger, and Brown (1996) examined the foraging ecology of metamorphosed and paedomorphic tiger salamanders (*Ambystoma tigrinum nebulosum*) in Colorado, where the animals are facultatively paedomorphic. Metamorphic adults grew faster in both length and body mass than did paedomorphic adults. Metamorphosed adults also had more caloric value of food in their stomachs than paedomorphic adults. The main reason for this is that metamorphosed adults often moved to temporary ponds, where they had access to fairy shrimp, a preferred type of prey. These prey animals were not available to paedomorphic animals, which are restricted to permanent ponds where fairy shrimp do not occur. Metamorphosed adults also have access to prey in terrestrial habitats that are not accessible to paedomorphic salamanders. The results of this study suggest a strong fitness advantage for metamorphosed adults, raising the question of why paedomorphosis occurs at all in this population. The answer is not entirely clear, although Whiteman, Wissinger, and Brown (1996) suggested that paedomorphosis might be advantageous in permanent ponds with poor growing conditions, because it would allow the salamanders to reach sexual maturity and reproduce at a younger age than those that undergo metamorphosis. In this sense, paedomorphic individuals persist in the population not because they enjoy a clear advantage over metamorphs, but because that is the best option available to them ("making the best of a bad job"' see also Whiteman 1994).

Whiteman's results with *Ambystoma* contrast with studies of mixed populations of paedomorphic and metamorphosed alpine newts (*Triturus alpestris*) in Europe. As with most polymorphic populations of salamanders that have been investigated, paedomorphs and metamorphs are fully sexually compatible (Denoël, Poncin, and Ruwet 2001b) and have nearly identical courtship behavior, despite differences in morphology (Denoël 2002). The two morphs differ ecologically, however. Paedomorphs feed mainly on zooplankton and can be found throughout the water column, including deep water. Metamorphosed individuals feed mostly on terrestrial insects along the shore or in shallow water (Denoël and Joly 2001; Denoël and Schabetsberger 2003; Denoël, Schabetsberger, and Joly 2004). In a number of ponds and lakes, paedomorphs had higher food intake rates, larger fat reserves, and better body condition than did metamorphs, although this was not true for all ponds (Denoël, Hervant, Schabetsberger, and Joly 2002). When paedomorphic newts were provided with extra food in a laboratory experiment, they grew faster and were more likely to remain paedomorphic than were those on low food rations (Denoël and Poncin 2001). All of these results support the hypothesis that favorable growth conditions in the aquatic environment increase the probability that individuals will remain paedomorphic (Denoël, Džukić, and Kalezić 2005).

### Genes, Environment, and Phenotypic Plasticity in Paedomorphic Urodeles

The preceding discussion has treated paedomorphosis as an adaptive life-history strategy that has evolved many times in response to a variety of selective pressures. This assumes that life-history traits such as the tendency to undergo metamorphosis are sufficiently heritable for selection to operate. I have already discussed evidence for genetic variation in larval growth rate among individuals in anuran populations, which in turn leads to differences in the length of the larval period and size at metamorphosis. This section deals with the environmental and genetic determinants of variation in life-history strategies among populations that exhibit facultative paedomorphosis.

The fact that many populations of urodeles are completely paedomorphic in nature suggests that natural selection has shaped the life histories of these animals in response to local environmental conditions. Certainly this is true for those species that have lost the genetic capacity for metamorphosis. However, the quantitative genetics of paedomorphosis have not been worked out in detail. In populations composed of both paedomorphic and metamorphosing individuals, the mixed life-history strategy could be maintained by a genetic polymorphism (Sprules 1974b; R. N. Harris 1987a), but clear evidence for such polymorphism within populations is lacking. Similarly, differences among populations in the frequency of metamorphosis could be due in part to genetic differences. However, the heritability of the tendency to undergo metamorphosis is not known for any species. There have been several studies of interspecific hybridization among species of *Ambystoma,* some of which are paedomorphic (Brandon 1972, 1977; Nelson and Humphrey 1972), but these were not designed to elucidate the genetic basis of paedomorphosis.

Several studies have examined the responses of individuals from different populations raised under the same environmental conditions. Eagleson's (1976) experimental manipulation of temperature and its effect on the frequency of metamorphosis in different populations of *Ambystoma gracile* was discussed already. R. N. Harris (1987a) reported that larval density affected the frequency of paedomorphosis in *Notophthalmus viridescens.* Larvae raised at low densities in artificial ponds were more likely to become paedomorphic than those raised at high densities. Although Harris suggested that a genetic polymorphism might be involved, he did not conduct any genetic experiments. It could be that all individuals have the capacity to retain larval characters, and paedomorphosis is a direct response to environmental cues. Harris also did not determine whether paedomorphic individuals remained permanently in this condition, or eventually metamorphosed after attaining sexual maturity (Reilly 1987).

Studies of the facultatively paedomorphic salamander *Ambystoma talpoideum* provide the most detailed information on the environmental and genetic basis for variation in life histories. In this species, environmental conditions have a strong effect on the tendency of animals to become paedomorphic. Larvae derived from a population that exhibits a high rate of paedomorphosis were raised in cattle tanks under varying conditions of larval density, food abundance, and pond-drying regime (Semlitsch 1987c). Salamanders were much more likely to become paedomorphic when raised at low density and in constant water, but tended to metamorphose in drying ponds or when larval densities were high (fig. 13.25). The presence of bluegill sunfish reduced the tendency of these salamanders to become paedomorphic, not because they increased rates of metamorphosis, but because larvae in tanks with fish tended to remain longer in the larval stage (M. Jackson and Semlitsch 1993).

Despite the strong effects of environmental variation on paedomorphosis, there are clear genetic differences among populations in the tendency to become paedomorphic. In an early experiment, Semlitsch and Gibbons (1985) collected females from two populations that differed in the frequency of metamorphosis. Adults from a relatively permanent pond

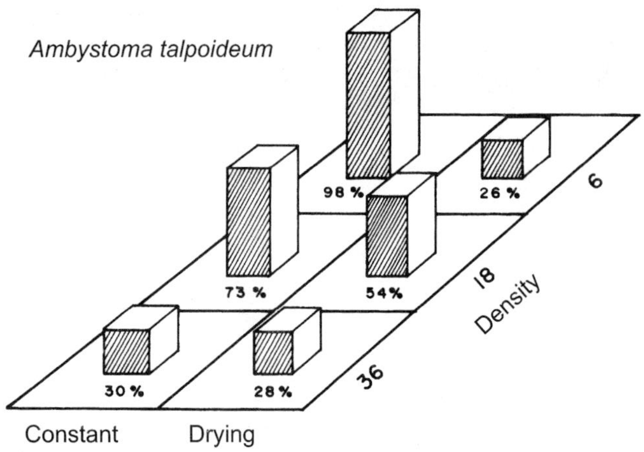


**Fig. 13.25.** Effect of drying regime and larval density on the percentage of *Ambystoma talpoideum* larvae becoming paedomorphic in artificial ponds. Height of each bar shows the average for eight artificial ponds. The larvae were most likely to become paedomorphic in permanent water at low larval densities. After Semlitsch (1987a).

(Ellington Bay) usually were paedomorphic, whereas those from a temporary site (Sun Bay) always transformed. Eggs from these females were raised at different densities and subjected to different rates of pond drying. There were significant differences between the populations in the frequency of paedomorphosis, with larvae from the permanent site being more likely to become paedomorphic under both constant water level and drying treatments (fig. 13.26). This could be the result of genetic differences, but maternal effects could not be ruled out, because only first-generation offspring were considered.

In a subsequent experiment, individuals from these two populations were subjected to different drying regimes over several generations to select for a paedomorphic life history (Semlitsch and Wilbur 1989). Both populations responded to selection for paedomorphosis in permanent water. However, responses of the two breeding lines differed. The line derived from a natural population with a high rate of paedomorphosis responded to selection by decreasing the number of immature larvae and increasing the number of sexually mature paedomorphs. The line derived from a population with a low rate of paedomorphosis responded by decreasing the rate of metamorphosis and prolonging the larval period. The fact that these populations could respond to selection indicated that there is some genetic basis for the tendency to become paedomorphic. This was confirmed in crossing experiments with individuals from three different populations, two from relatively permanent ponds and one from a temporary pond (R. N. Harris et al. 1990).

In another set of experiments, Semlitsch, Harris, and Wilbur (1990) reared larvae from six different populations under the same environmental conditions. Two of these populations were from ponds that seldom dry up, whereas the

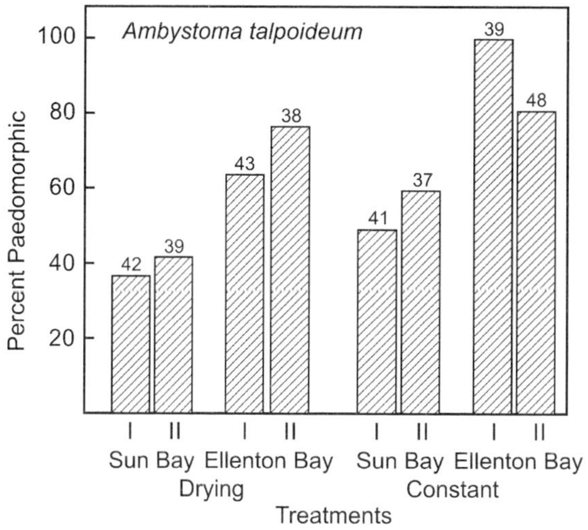


**Fig. 13.26.** Percentage of paedomorphic *Ambystoma talpoideum* from experimental ponds. Source populations were larvae from a relatively permanent pond (Ellenton Bay) and a temporary pond (Sun Bay). Treatments were either tanks that held constant water or those subjected to drying. Roman numerals indicate replicate tanks. Numbers at tops of bars are sample sizes of individuals. After Semlitsch and Gibbons (1985).

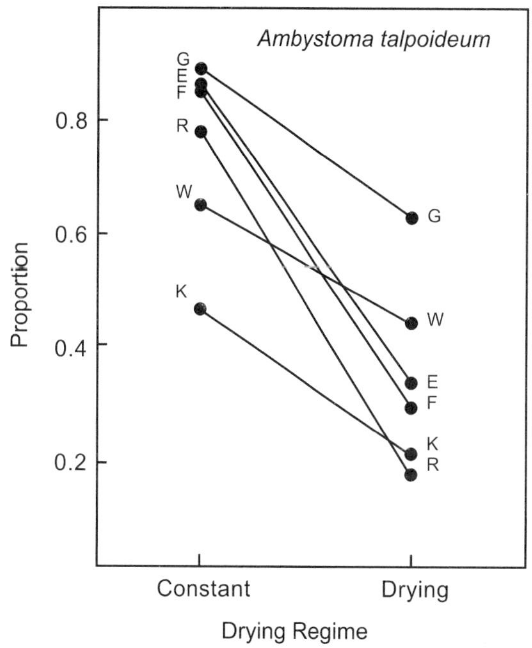


**Fig. 13.27.** Proportion of surviving larvae of *Ambystoma talpoideum* that became paedomorphic in experimental ponds that held constant water or were subjected to drying. Parents of the larvae used in the experiment were obtained from ponds that seldom dry up in the summer (E = Ellenton Bay; F = Flamingo Bay), those that usually dry up (W = Woodland Pond, K = Karen's Bay), and those that always dry up (G = Ginger's Bay; R = Rainbow Bay). After Semlitsch, Harris, and Wilbur (1990).

other four were from ponds that almost always become dry. All of the populations showed a stronger tendency toward paedomorphosis in constant water than in drying tanks, but there were significant differences among populations in the tendency to become paedomorphic (fig. 13.27). These differences did not perfectly reflect differences in natural drying regimes in the source populations, however. For example, one pond that always dries (Ginger's Bay) had the highest frequency of paedomorphs under both drying and constant water regimes. Larvae from another pond that usually dries (Karen's Bay) exhibited a low frequency of paedomorphosis under both drying and constant water regimes. Other populations showed highly plastic responses, with high frequencies of paedomorphosis in constant water and high frequencies of metamorphosis in drying tanks. These results suggest that selection has affected the degree of plasticity in life histories and not just the capacity to remain paedomorphic. These experiments also suggested that recent drying history of the source ponds could not account for all of the variation among populations. Other selective pressures, such as food abundance and larval density, probably are important as well.

Most studies of paedomorphic salamanders have focused on the morphological differences between paedomorphic and transformed adults as they relate to their ability to live in water or on land. However, T. Ryan and Semlitsch (1998) argued that morphological differences are a secondary effect of paedomorphic development, with the main target of natural selection actually being timing of reproduction. By forgoing metamorphosis, individuals have more energy available for growth and sexual maturation, thereby achieving an earlier age at reproduction. Paedomorphic individuals could then reproduce at an earlier age than transformed individuals, or they might reproduce at the same age, but at a larger body size, thus deriving a fecundity advantage from paedomorphosis. Earlier work on this species had shown that paedomorphic females reproduce at a younger age than do metamorphosed females, but metamorphosed females produced somewhat larger clutches for their body size than did paedomorphic individuals of the same age (Semlitsch 1985b).

In the experiments by T. Ryan and Semlitsch (1998), larvae derived from a population that normally metamorphoses were raised at high and low density. In both treatments, sexually mature paedomorphs appeared in ponds before any individuals metamorphosed, although the frequency of paedomorphosis was low at high density (fig. 13.28). None of the newly metamorphosed individuals was sexually mature. Hence a clear advantage of paedomorphosis in this salamander is the decoupling of metamorphosis and sexual maturation, allowing for earlier reproduction by paedomorphic individuals and possibly higher lifetime reproductive success. This can be especially advantageous in an environment with variable rainfall, where movement of metamorphosed individuals into breeding ponds depends on adequate rainfall.

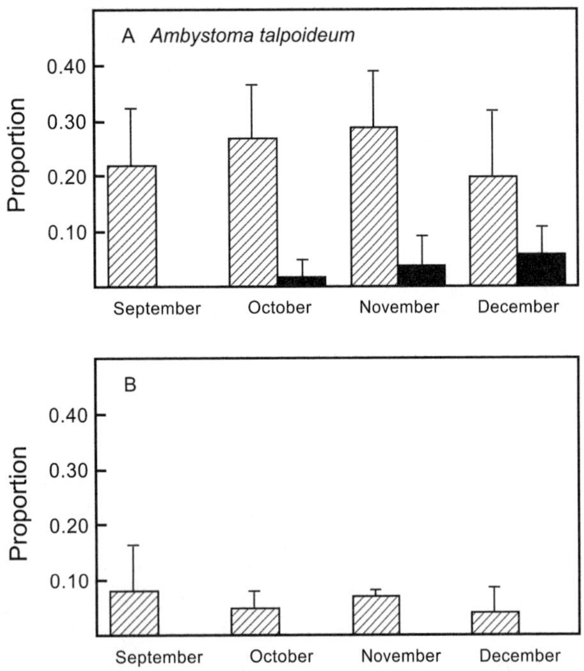


**Fig. 13.28.** Proportion of paedomorphic (hatched bars) and metamorphosed (black bars) adult *Ambystoma talpoideum* in experimental ponds in different months at low (A) and high (B) larval densities. In both high- and low-density ponds, paedomorphic adults appeared before sexually mature metamorphosed adults. All larvae were derived from a population in Rainbow Bay, a pond that always dries up in the summer. After T. Ryan and Semlitsch (1998).

Paedomorphic individuals are already present in permanent ponds and can initiate or even complete reproduction before metamorphosed members of the population even appear in the ponds. Early reproduction, in turn, gives the offspring of paedomorphs a head start in competition with larvae of later arriving metamorphs (T. Ryan and Plague 2004). It also would enable their offspring to complete development early, a major advantage in years when a pond dries up. Because paedomorphic adults retain the ability to metamorphose if conditions deteriorate, they can enjoy the benefits of early sexual maturation without the potential cost of becoming trapped in a drying pond and suffering an early death.

In some populations, any reproductive advantage of paedomorphosis may be greater for one sex than for the other. In a population of *A. tigrinum nebulosum,* for example, metamorphosed females reproduced at shorter intervals than did paedomorphic females, whereas the opposite was true for males. Sex ratios of paedomorphs were biased toward males, but females outnumbered males among metamorphs (Whiteman 1997). A reproductive advantage also can be affected by ecological differences among populations. In European newts, there was no reproductive advantage for paedomorphs in populations of *Triturus vulgaris* and *T. alpestris* from high-altitude lakes (Kalezić et al. 1996; Denoël and Joly 2000), but at lower elevations, paedomorphs of *T. alpestris* grew faster and reached sexual maturity earlier than did metamorphs (Denoël and Joly 2000).

## Summary and Conclusions

A complex life cycle involving an aquatic larval stage and metamorphosis into a terrestrial or semiterrestrial adult is thought to be the ancestral condition for all three major groups of amphibians. Nevertheless, there has been repeated loss of this type of complex life cycle in all three groups. Most caecilians lack an aquatic larval stage, as do most urodele species (but not the majority of families). Many lineages of frogs have partially or completely lost the aquatic larval stage, including some very successful groups, such as the widespread and specious Neotropical genus *Eleutherodactylus.* Some salamanders have evolved in the opposite direction and have lost the terrestrial adult stage of the life cycle. Some salamanders, such as sirenids and proteids, are always paedomorphic, but others, including many species of *Ambystoma,* are facultatively paedomorphic.

The repeated loss of a complex life cycle has led some authors to propose that complex life cycles are inherently unstable and tend to evolve toward loss of either the aquatic or terrestrial stage. Many amphibian biologists, such as Wassersug (1975) and Wilbur (1980), have argued that a complex life cycle is maintained because of the unique ecological role of the larval stage, especially in anurans. In this view, amphibian larvae take advantage of seasonal flushes of productivity in ponds and other bodies of water to acquire abundant resources for growth that often are lacking in the terrestrial environment.

Werner (1986) argued that it is misleading to consider the larval stage of many amphibians, especially anurans, as the principal growth stage, because often more than 90% of total growth takes place in the terrestrial phase of the life cycle. He argued that the transition from an aquatic larval stage to a terrestrial juvenile and adult can be viewed as a shift in ecological niches. He further proposed that the relative costs and benefits of living in the aquatic or terrestrial environment, as measured by changes in the relationship of growth to mortality, will determine the point at which this shift is made by any given species. For many anurans that breed in temporary ponds, mortality becomes unacceptably high if they remain for very long in the aquatic environment, so early metamorphosis at a relatively small body size often is favored. The more ephemeral the breeding habitat, the shorter the larval period must be for successful metamorphosis. In contrast, anurans that breed in permanent ponds often benefit from remaining in the tadpole stage for several years before metamorphosis. Anurans all are constrained, however, by their inability to become reproductive adults while retaining the larval morphology, so eventually all anurans must metamorphose.

Urodeles are less constrained in their development, because they often are fully capable of becoming reproductive adults while retaining many features of larval morphology. The result is the widespread occurrence of either obligate or facultative paedomorphosis in this group. Even salamanders that do metamorphose often have very long larval periods; some achieve up to 90% of their total growth in the larval stage. This type of life history might be favored in species that spend much of their adult life inactive underground, as in many of the mole salamanders (*Ambystoma*), or species that live in caves and other habitats where food availability in the aquatic environment exceeds that of the terrestrial environment.

The process of metamorphosis itself generally has been viewed very differently by developmental biologists and by ecologists. Developmental biologists focus on the genetic and hormonal control of metamorphosis as a relatively orderly and predictable sequence of events, with little consideration of how the environment impinges on the process. Ecologists, on the other hand, have focused on plastic responses of larvae to variation in environmental variables such as temperature, food availability, larval density, and pond permanence, often with little idea of how environmental cues might alter hormonal pathways that ultimately control metamorphosis. Recent work has suggested possible links between the thyroid hormone axis, which regulates the timing of metamorphosis, and environmental variables such as larval density and pond drying, mediated through hormonal stress responses.

Most of the research on the ecology of metamorphosis over the last three decades has focused on proximate determinants of the timing of metamorphosis and the size of individuals at transformation. The conceptual starting point for this work was a model of metamorphosis proposed by Wilbur and Collins (1973), which related the probability of an individual undergoing metamorphosis to the quality of the aquatic environment, as indicated by its own recent growth history. Each species is thought to have a characteristic minimum size at which metamorphosis is possible, as well as a maximum size at which it is no longer efficient or possible to remain in the larval stage. Larvae that fail to reach the minimum size before a pond dries up are doomed, but those that have reached the minimum size for transformation have several options. If growth conditions are good, larvae are expected to remain in the water, acquiring additional resources and increasing body size, which almost always is beneficial once they enter the terrestrial environment. If growth conditions are poor, larvae are expected to metamorphose near the minimum body size, thereby trading an increased probability of escaping deteriorating conditions against the possible costs of being less successful in a harsh terrestrial environment. Different individuals within a population, and even within a single cohort or sibship of larvae, end up metamorphosing at different ages and at different body sizes, because the options available to each larva change as the density of larvae, the abundance of food, and the persistence of the breeding site vary over time.

A basic assumption of the Wilbur-Collins model and related models is that density-dependent competition among larvae will create asymmetries in body size and competitive ability, and these in turn will translate into differences in timing and size at metamorphosis. Many experimental studies have demonstrated such density-dependent competition in both anuran and urodele larvae in laboratory aquaria, cages placed in natural ponds, cattle tanks and other artificial ponds, and in natural bodies of water. Mechanisms of competition differ somewhat between anurans and urodeles. Exploitative competition through resource depletion appears to be most important for anuran tadpoles, with interference competition mediated by behavioral or chemical interactions playing a secondary role. Exploitative competition also occurs in larval urodeles, but interference competition in the form of aggressive interactions or even cannibalism appears to be more important in many species.

Many experiments have shown that the growth and development of amphibian larvae are very plastic and readily respond to short-term changes in resource abundance, intensity of competition, or water level. Diminishing resources or increased competition often induce larvae to speed up development at the expense of growth and metamorphose early at a small size. Drying of ponds has the same effect in some species. Increasing resources or diminished competition resulting from individuals metamorphosing and leaving the water or being eaten by predators often result in increased growth, prolongation of the larval period, and metamorphosis at a large body size. These responses are somewhat constrained by stage of development. Larvae that are deprived of resources relatively early in development often remain at a small size and early developmental stage and never reach metamorphosis. Larvae that experience increased resources late in development allocate those resources to additional growth, but with relatively little effect on the timing of metamorphosis.

The highly plastic nature of amphibian growth and development probably explains much of the success of these animals in colonizing nearly every available habitat, from wet tropical rainforests to cold mountain streams and dry deserts. Indeed, some of the most dramatic examples of developmental plasticity in amphibians can be found in species that inhabit the most unpredictable and inhospitable habitats, such as deserts. Anurans that breed in desert rain pools not only have very short larval periods compared to other anurans, but they exhibit a much greater degree of plasticity in their life histories. Some spadefoot toads, for example, can shorten the larval period by nearly 50% in response to drying of ponds. These species also exhibit extraordinary levels of phenotypic plasticity in morphology and behavior, often switching to carnivorous diets and even becoming cannibalistic, a strategy that greatly increases their rate of growth and development and improves their chances of escaping a drying pond (see chapter 12).

In contrast to anurans, some urodeles have partially decoupled developmental processes such as sexual maturation from growth and transformation. This has enabled many urodeles to become paedomorphic, essentially retaining the ecological advantages of staying in the larval habitat without the cost of delaying reproduction. Indeed, in some facultatively paedomorphic species of *Ambystoma,* individuals actually can initiate reproduction earlier by retaining the larval morphology than by metamorphosing and migrating onto land. For urodeles that live in very predictable environments, such as caves and permanent springs, lakes, and rivers, an obligate paedomorphic life history can be advantageous, especially in regions where such predictable and equable aquatic habitats are surrounded by inhospitable terrestrial habitats. In contrast, paedomorphosis often is a highly plastic trait in urodeles that inhabit unpredictable environments, such as ponds that dry up in some years but not in others. For these animals, a capacity to become paedomorphic when ponds remain full, but to metamorphose, even after reaching sexual maturity, when ponds dry up, probably gives them an advantage over species that have fixed life histories.

The evolution of paedomorphic life histories has implications that go beyond the adaptation of particular populations to local environments. One consequence of paedomorphosis is that gene flow between populations is likely to be reduced, because the terrestrial animals that normally would disperse to new breeding ponds are absent. This is especially true in species that have become obligate paedomorphs, which often become isolated in permanent ponds and lakes that are separated from other suitable aquatic habitats by inhospitable terrestrial habitats. This isolation of local populations probably accounts for the high degree of genetic diversity among permanently paedomorphic populations of *Ambystoma* in the southwestern United States and Mexico (Shaffer 1984a) and the proliferation of species of cave- and spring-dwelling *Eurycea* on the Edwards Plateau of Texas (Sweet 1977b, 1982).

One approach to studies of the ecology of metamorphosis and the evolution of phenotypic plasticity in growth and development that has been largely neglected is the inclusion of a phylogenetic perspective. Experimental work on the plastic responses of spadefoot toads and other anurans to pond drying has shown that spadefoot toads are far more plastic in their responses than other species that have been studied. However, all of these other species are in families only distantly related to pelobatids, so it is not yet possible to conclude that species using ephemeral breeding habitats have always consistently evolved more plastic responses than those using more permanent sites. Even within pelobatids, there is considerable variation in the pattern of metamorphosis that appears more closely related to phylogeny than to ecology (Buchholz and Hayes 2002). Ideally, comparisons should be made among closely related species, as Morey and Reznick (2000) have done for different species of spadefoot toads. Alternatively, multiple comparisons might be made between species in different clades that breed in highly ephemeral habitats and close relatives that use more permanent breeding sites. If species that breed in ephemeral sites are repeatedly shown to have more plastic life histories than close relatives in less ephemeral sites, then this would provide strong evidence for convergent evolution of phenotypic plasticity. A similar approach could be taken to investigate the range of variation in degree of plasticity in paedomorphic salamanders, although in this case, the range of species available for study is much more limited. European and North American newts, which exhibit a wide range of variation in frequency of paedomorphosis, might be a promising group for phylogenetic comparisons.

# **Chapter 14** Amphibians and Their Predators

*Their common enemies are the Weasels, Poul-cats, and Ferrets, for these do gather them together, and lay of them great heaps within their dens, whereupon they feed in Winter. The Heron also and Bittern, is a common destroyer of Frogs, and so likewise some kinds of Kites. . . . The moles also are enemies to Frogs.*

*The Mole is also an enemy to the Toad, for as Albertus writeth, he himself saw a Toad crying above the earth very bitterly, for a mole did hold her fast by the leg within the earth, labouring to pull her in again. . . . They are also at variance with the Lizard, and all kindes of Serpents. . . . The Buzard and the Hawk are destroyers of Toads, but the Stork never destroyeth a Toad to eat it, except in extremity of famine.*
—Edward Topsell, *The History of Serpents* (1658)

THE LITERATURE ON amphibian ecology often conveys an impression of animals beset by enemies on every side, leaving the reader wondering how any manage to survive. As M. Tyler (1976, p. 153) put it, "There are . . . millions of animals seeking out frogs and feeding upon them, so that frogs are the cannon fodder of the animal kingdom. Each stage of development from egg to adult faces its own particular dangers, and at no time of day or night are frogs and their offspring free from attack." Similarly, Porter (1972, p. 330) asserted that "almost anything will eat an amphibian." Certainly, many amphibians appear to be highly vulnerable to predation, being small, slow moving, and relatively defenseless against larger animals. Because amphibians are efficient at converting ingested energy into biomass (see chapter 5), they often reach high population densities, and provide a ready source of protein for animals at higher trophic levels (Burton and Likens 1975a; Pough 1980, 1983). Furthermore, the lack of indigestible materials such as hair, feathers, or chitin makes amphibians a particularly high-quality food.

Predation has been important in the evolution of many amphibian characteristics, such as toxic and distasteful skin secretions, cryptic and aposematic coloration, and a variety of defensive postures and behaviors. In addition, predation is routinely cited as a selective pressure that affects many other aspects of amphibian ecology and behavior (table 14.1). Given the central importance of predation in the lives of these animals, it is important to understand the nature of interactions between amphibians and their predators. This chapter addresses three major questions: (1) What kinds of animals eat amphibians? (2) What is the impact of predation on amphibian population dynamics? (3) How has predation acted as a selective force in shaping the evolution of amphibian ecology and behavior? The impact of predation on larval growth and development and the timing of metamorphosis are discussed in chapter 13, and the importance of predation in the dynamics of amphibian communities is discussed in chapter 15.

## The Predators of Amphibians

The following section reviews what is known about the major predators of amphibians. Whenever possible, I have tried to supplement lists of potential predators with quantitative estimates of the importance of amphibians in the predators'

**Table 14.1** Aspects of the ecology and behavior of amphibians that are affected by predation

| Ecological or behavioral variable | Chapter | Source no. |
|---|---|---|
| Timing and duration of breeding season | 10 | 2, 6, 71 |
| Diel timing of breeding migrations | 10 | 29, 36 |
| Choice of breeding habitats | 14 | 5, 19, 42, 80 |
| Choice of calling sites | 7 | 14, 17, 56 |
| Patterns of calling behavior | 7 | 13, 15, 17 |
| Mate-searching behavior | 9, 14 | 105, 116 |
| Choice of oviposition sites | 10, 14 | 2, 8, 52, 56, 86 |
| Oviposition behavior | 10, 14 | 63 |
| Size and number of eggs | 10 | 3, 11, 41 |
| Timing of egg hatching | 14 | 64, 74 |
| Parental care of eggs and larvae | 11 | 25, 27, 28 |
| Choice of tadpole deposition sites | 14 | 99 |
| Tadpole morphology and coloration | 14 | 78, 83, 91, 102, 103, 104, 108, 109, 110, 112, 114, 115 |
| Timing of aerial breathing by aquatic larvae | 4 | 20, 21, 26 |
| Aggregation of larvae and juveniles | 12, 14 | 4, 7, 18, 85 |
| Diel activity cycles of larvae | 12, 14 | 22, 44 |
| Choice of foraging microhabitats | 14, 15 | 12, 40, 43, 46, 49, 50, 54, 55, 58, 90, 95, 111 |
| Activity and movements of larvae | 14, 15 | 43, 47, 55, 69, 71, 75, 79, 87, 89, 92, 101, 103, 104, 108, 110, 113, 114, 115 |
| Larval swimming speed | 14 | 104 |
| Larval growth rates | 13, 14 | 10, 32, 37, 43, 55, 60, 89, 114 |
| Timing of metamorphosis | 13, 14 | 10, 37, 50, 101 |
| Structure and dynamics of larval communities | 15 | 23, 24, 35, 39, 48, 53, 57, 70, 73, 88, 92, 97, 101 |
| Evolution of foraging behavior | 5, 14, 15 | 16, 30, 31 |
| Spatial orientation and escape behavior | 14 | 1, 9, 34 |
| Alarm responses and predator avoidance | 12, 14 | 45, 51, 54, 62, 65, 67, 68, 76, 77, 82, 93, 94, 98, 106, 107, 116 |
| Distribution and abundance of species | 15, 16 | 33, 38, 61, 66, 81, 84, 96, 112 |
| Sensitivity to pesticides | 16 | 101, 117 |

*Sources:* (1) D. Ferguson et al. 1968; (2) Heusser 1970; (3) Salthe and Duellman 1973; (4) Wassersug 1973; (5) Heyer, McDiarmid, and Weigmann 1975; (6) Wells 1977c; (7) Arnold and Wassersug 1978; (8) R. D. Howard 1978b; (9) Gregory 1979; (10) Wilbur 1980; (11) Crump 1981b; (12) Jaeger 1981a; (13) M. Ryan, Tuttle, and Taft 1981; (14) Tuttle and Ryan 1981; (15) M. Ryan, Tuttle, and Rand 1982; (16) Taigen, Emerson, and Pough 1982; (17) Tuttle and Ryan 1982; (18) Waldman 1982; (19) Woodward 1982; (20) Branch 1983; (21) Feder 1983d; (22) Holomuzki and Collins 1983; (23) Morin 1983a; (24) Morin 1983b; (25) Simon 1983; (26) Lanno and Bachmann 1984b; (27) Townsend, Stewart, and Pough 1984; (28) Nussbaum 1985; (29) Semlitsch and Pechmann 1985; (30) Taigen and Pough 1985; (31) Toft 1985; (32) Travis, Keen, and Juilianna 1985b; (33) Hairston 1986; (34) Licht 1986b; (35) Morin 1986; (36) Pechmann and Semlitsch 1986; (37) Werner 1986; (38) Hom 1987; (39) Morin 1987a; (40) Petranka, Kats, and Sih 1987; (41) Petranka et al. 1987 (42) A. Roth and Jackson 1987; (43) Semlitsch 1987a; (44) Stangel and Semlitsch 1987; (45) Hews 1988; (46) Holomuzki 1989a; (47) Lawler 1989; (48) Fauth 1990; (49) Morey 1990; (50) Skelly and Werner 1990; (51) Brodie, Formanowicz, and Brodie 1991; (52) Crump 1991; (53) Fauth and Resetarits 1991; (54) Kats and Sih 1991; (55) Resetarits 1991; (56) Resitarits and Wilbur 1991; (57) Gascon 1992a; (58) Sih, Kats, and Moore 1992; (59) Kats and Sih 1992; (60) Skelly 1992; (61) Bradford, Tabatabai, and Graber 1993; (62) Elliot, Kats, and Breeding 1993; (63) Miaud 1993; (64) Sih and Moore 1993; (65) Wilson and Lefcourt 1993; (66) Brönmark and Edenhamn 1994; (67) Cupp 1994; (68) Hileman and Brodie 1994; (69) Horat and Semlitsch 1994; (70) Werner and McPeek 1994; (71) Anholt and Werner 1995; (72) Petranka and Thomas 1995; (73) Skelly 1995b; (74) Warkentin 1995; (75) Anholt, Skelly, and Werner 1996; (76) Kiesecker, Chivers, and Blaustein 1996; (77) Lefcourt 1996; (78) McCollum and Van Buskirk 1996; (79) Skelly 1996; (80) Gamradt, Kats, and Anzalone 1997; (81) Hecnar and M'Closkey 1997a; (82) Laurila, Kujasalo, and Ranta 1997; (83) McCollum and Leimberger 1997; (84) Resetarits 1997; (85) Rödel and Linsenmair 1997; (86) Spieler and Linsenmair 1997; (87) Griffiths et al. 1998; (88) Hero, Gascon, and Magnusson 1998; (89) Kiesecker and Blaustein 1998; (90) Kupferberg 1998; (91) Lardner 1998; (92) Laurila, Kujasalo, and Ranta 1998; (93) Petranka and Hayes 1998; (94) Schley and Griffiths 1998; (95) Sih, Englund, and Wooster 1998; (96) Tyler et al. 1998; (97) Azevedo-Ramos, Magnusson, and Bayliss 1999; (98) Kiesecker et al. 1999; (99) Downie, Livingstone, and Cormack 2001; (100) Relyea and Mills 2001; (101) Barnett and Richardson 2002; (102) Relyea 2002a; (103) Relyea 2002b; (104) Relyea 2002c; (105) Richardson 2002a; (106) Rohr, Madison, and Sullivan 2002a; (107) Rohr, Madison, and Sullivan 2002b; (108) Van Buskirk 2002a; (109) Van Buskirk 2002b; (110) Van Buskirk and Arioli 2002; (111) Van Buskirk et al. 2002; (112) Vorndran, Reichwaldt, and Nürnberger 2002; (113) Altwegg 2003; (114) Relyea 2003a; (115) Relyea 2003b; (116) Rohr, Madison, and Sullivan 2003; (117) Relyea 2004.

*Note:* The first column indicates chapters in which topics are discussed in detail.

diets. These data must be evaluated cautiously, because different types of prey are digested at different rates, and analyses of stomach contents, feces, or regurgitated pellets provide only rough estimates of the number of amphibians consumed (Fitch 1960, 1963; Craighead and Craighead 1969; Calef 1973a; Kushlan 1978; Delany and Abercrombie 1986). The data are better for some groups than for others. Predation by invertebrate predators is hard to quantify, and only a few species have been investigated. Information on predation by fishes, amphibians, and most reptiles is largely anecdotal, although there is good information on predation by snakes. Many species of birds and mammals eat amphibians, but good quantitative data are available for only a few species in each group.

## Predation by Invertebrates

Many species of invertebrates prey on amphibians (McCormick and Polis 1982; Toledo 2005). Among the most important predators are leeches, crabs and crayfish, amblypygids, spiders, caddisfly larvae, dragonfly naiads, aquatic bugs, the larvae and adults of aquatic beetles, and several families of flies (table 14.2). Other invertebrates, including free-living flatworms, mosquito larvae, wasps, mantids, and crickets, are less frequently reported as predators of amphibians, but can have a significant impact on amphibian populations in certain habitats. The list in table 14.2 is far from complete, because there are vast numbers of potential invertebrate predators of amphibians in various parts of the world. Regional accounts of amphibians often include invertebrates in lists of natural enemies, without detailed documentation (e.g., Kuzmin 1999), and I have not attempted to include all of these. None of the species of invertebrates that prey on amphibians is an amphibian specialist, with the exception of certain fly larvae that live in egg masses of frogs. Most species are generalist predators that feed opportunistically on amphibians when they are available, but shift to other types of food when they are not. Unfortunately, there is little information on the relative importance of amphibians in the diets of most species; invertebrates usually feed by sucking or chewing their prey, making analyses of stomach contents difficult or impossible. In one of the few studies of the ecological impact of predation on amphibians on populations of invertebrates, Hirai and Hidaka (2002) found that an endangered species of giant water bug in Japan, *Lethocerus deyrollei,* is heavily dependent upon breeding anurans and transforming juveniles during the spring and summer. Recent declines of frogs in rice paddies brought about by changes in agricultural practices therefore may threaten the survival of this insect.

Invertebrate predators undoubtedly take their greatest numerical toll on the eggs and larvae of amphibians. In many temporary ponds that lack fish, invertebrates are the most important predators of amphibian eggs and larvae. Indeed, the assemblage of invertebrate predators present in particular ponds can be one of the most important determinants of the structure and dynamics of amphibian communities (Heyer, McDiarmid, and Weigmann 1975; Van Buskirk 1988; Gascon 1992a; Hero, Gascon, and Magnusson 1998; Azevedo-Ramos, Magnusson, and Bayliss 1999; see also chapter 15).

### Predation on Aquatic Eggs

In many ponds that lack fish, the larvae of caddisflies (Trichoptera) are important invertebrate predators of amphibian eggs (fig. 14.1 A). These insects feed on a broad diet of organic detritus, plant material, and occasional invertebrate prey (Winterbourn 1971a; N. Anderson 1976). Several studies have shown that a diet of animal protein enhances growth of caddisfly larvae and allows them to pupate more rapidly than when they feed entirely on plant material (N. Anderson 1976; N. Anderson and Cummins 1979; Majecki and Majecka 1996). In some species, a diet of animal protein late in development may be necessary for successful pupation (Winterbourn 1971b). Caddisfly larvae readily prey on both frog and salamander eggs, which provide a concentrated source of protein for a brief period during spring breeding seasons. They often swarm over egg masses in large numbers and easily cut through the jelly and tough capsules that surround the eggs (Stout, Stout, and Stihler 1992; C. Rowe, Sadinski, and Dunson 1994; Kwet 1996a; Majecki and Majecka 1998; Richter 2000). In one pond in West Virginia, 82% of all individuals of one caddisfly species, *Banksiola dossuaria,* were found on egg masses of spotted salamanders (*Ambystoma maculatum*), with more than 20 individuals sometimes feeding on a single egg mass (Stout, Stout, and Stihler 1992). When population densities of predatory caddisfly larvae are high, they can consume nearly all of the amphibian eggs laid in some years. Population densities often fluctuate from year to year in the same pond, enabling amphibians to breed successfully in years when caddisfly larvae are scarce (Rowe, Sadinski, and Dunson 1994; personal observations).

Leeches occur in ponds both with and without fish and sometimes prey on amphibian eggs (Cory and Manion 1953; R. D. Howard 1978b; Kwet 1996a), as well as tadpoles and adults. I have seen leeches (*Macrobdella decora*) swarming over egg masses of *Ambystoma maculatum* that also were being attacked by both caddisfly larvae and red-spotted newts (*Notophthalmus viridescens*). R. D. Howard (1978b) estimated that 43% of the eggs in one population of bullfrogs (*Rana catesbeiana*) were destroyed by leeches, with up to 80% mortality in some clutches (fig. 14.1 B). Leeches prey on the eggs of several species of Eurasian amphibians, as do free-living planarians, snails (*Lymnaea*), mites (*Gydrachna,*

**Table 14.2** Some invertebrate predators of amphibians

| Predator | Family | Prey | Stage | Source no. |
|---|---|---|---|---|
| Flatworms (Platyhelminthes) | | | | |
| *Phagocotus* | | *Ambystoma* | L | 88 |
| *Polycelis* | | *Rana* | E | 103 |
| Leeches (Hirudinea) | | | | |
| *Batrachobdella* | Hirudinidae | *Bufo* | L | 19 |
| | | *Rana* | E | 20 |
| *Desserobdella* | Hirudinidae | *Rana* | L | 130 |
| *Haemopis* | Hirudinidae | *Bufo, Rana* | E | 103 |
| *Hirudo* | Hirudinidae | *Bufo, Rana, Triturus* | L, A | 14, 134 |
| *Macrobdella* | Hirudinidae | *Rana* | E | 7, 8, 33, 135 |
| | | *Ambystoma* | E, L, A | 12 |
| Snails (Gastropoda) | | | | |
| *Pomacea* | Ampullariidae | *Hyla* | L | 142 |
| Unidentified | | *Ramanella* | E | 149 |
| Crustaceans | | | | |
| *Astacus* | Astacidae | *Bufo, Hyla, Rana, Triturus* | E, L | 108 |
| *Pacifastacus* | Astacidae | *Bufo, Hyla, Rana, Triturus* | E, L | 108 |
| *Cambarus* | Cambaridae | *Hyla* | L | 69 |
| *Macrobrachium* | Cambaridae | *Agalychnis, Colostethus* | L | 128, 131 |
| *Orconectes* | Cambaridae | *Ambystoma* | E, L | 44, 66 |
| *Procambarus* | Cambaridae | *Hyla* | L | 77 |
| | | *Taricha* | E, L | 100 |
| *Armases* | Grapsidae | *Dendrobates* | L | 127 |
| *Sesama* | Grapsidae | *Cochranella* | E | 48 |
| *Potamocarcinus* | Potamidae | *Physalaemus* | A | 43 |
| *Triops* | Triopidae | *Rana* | L, J | 34 |
| Whip-scorpions (Amblypygi) | | | | |
| *Phrynus* | Amblypigidae | *Eleutherodactylus* | J | 41 |
| Spiders and relatives (Arachnida) | | | | |
| *Cupiennius* | Ctenidae | *Centrolene, Eleutherodactylus, Hyalinobatrachium* | A | 48, 56 |
| *Aganippe* | Ctenizidae | *Crinia* | J | 11 |
| *Lycosa* | Lycosidae | *Leptodactylus* | E | 45 |
| | | *Rana, Microhyla, Pseudacris* | J, A | 21, 24, 30, 113 |
| *Oligoctenus* | Lycosidae | *Eleutherodactylus* | J | 41 |
| *Trochosa* | Lycosidae | *Bufo, Pleurodema* | L | 147 |
| Unidentified | Lycosidae | *Rana* | J | 5 |
| *Ancylometes* | Pisauridae | *Scinax* | A | 144 |
| *Dolomedes* | Pisauridae | *Acris, Hyla, Pseudacris* | A | 3, 72, 148 |
| | | *Bufo, Hyla, Rana* | L | 60, 102, 107 |
| | | Unidentified | L | 2 |
| *Thalassius* | Pisauridae | *Hyperolius* | J | 158 |
| *Trechalea* | Pisauridae | *Hylodes* | J | 95 |
| *Thaumasia* | Pisaurudae | *Scinax* | J | 139 |

*(continued)*

**Table 14.2** (continued)

| Predator | Family | Prey | Stage | Source no. |
|---|---|---|---|---|
| *Olios* | Sparassidae | *Eleutherodactylus* | J | 41 |
| *Selenocosmia* | Theraphosidae | *Helioporus* | | 9 |
| *Sericoplema* | Theraphosidae | *Dendrobates, Physalaemus* | A | 123, 125 |
| *Therophosa* | Theraphosidae | *Oscaecilia* | A | 132 |
| *Atrax* | Dipluridae | *Litoria* | | 6 |
| *Prionostemma* | Phalangidae | *Hyalinobatrachium* | J | 48 |
| Caddisfly larvae (Trichoptera) | | | | |
| *Banksiola* | Phrygaeneidae | *Ambystoma* | E | 86 |
| *Eubasilia* | Phrygaeneidae | *Hynobius* | E | 75 |
| *Oligotricha* | Phrygaeneidae | *Rana, Triturus* | E | 104, 117 |
| *Ptilostomis* | Phrygaeneidae | *Ambystoma, Rana* | E, L | 13, 22, 91 |
| Unidentified | | *Ramanella* | E | 149 |
| Dragonfly naiads (Odonata) | | | | |
| *Aeschina* | Aeshnidae | *Hyla, Pseudacris* | L | 51, 54 |
| *Aeshna* | Aeshnidae | *Bombina, Bufo, Hyla, Limnodynastes, Rana, Triturus* | L | 64, 70, 81, 110, 112, 116, 153, 156 |
| *Anax* | Aeshnidae | *Ambystoma, Bufo, Hyla, Phrynomantis, Pseudacris, Rana* | L | 39, 47, 51, 61, 65, 76, 92, 98, 98, 106, 114, 118, 154 |
| *Gynacantha* | Aeshnidae | *Osteocephalus, Scinax, Phyllomedusa* | L | 80 |
| Unidentified | Aeshnidae | *Epipedobates, Osteocephalus, Phyllomedusa* | L | 82 |
| *Hemicordulia* | Corduliidae | *Limnodynastes, Litoria, Pseudophryne, Ranidella* | L | 73, 74, 85 |
| *Leucorrhinia* | Libellulidae | *Bufo, Rana* | L | 70 |
| *Libellula* | Libellulidae | *Hyla* | L | 63 |
| *Pachidiplax* | Libellulidae | *Hyla* | L | 119 |
| *Pantala* | Libellulidae | *Bufo, Phrynomantis, Phyllomedusa, Physalaemus* | L | 27, 68, 114 |
| *Sympetrum* | Libellulidae | *Hyla* | L | 54 |
| *Tramea* | Libellulidae | *Bufo, Hyla, Phrynomantis, Pseudacris, Rana* | L | 57, 65, 83, 93, 114 |
| *Trapezostigma* | Libellulidae | *Hyperolius* | L | 158 |
| Unidentified | Libellulidae | *Epipedobates, Osteocephalus, Phyllomedusa* | L | 82 |
| True bugs (Heteroptera) | | | | |
| *Abedus* | Belostomatidae | *Hyla* | A | 143 |
| *Belostoma* | Belostomatidae | *Bufo, Hyla, Rana* | L | 16, 47, 61, 78, 99, 111 |
| | | *Physalaemus, Hyla, Scinax Leptodactylus* | A | 137, 145 |
| *Hydrocyrius* | Belostomatidae | *Phrynomantis* | L | 114 |
| *Lethocerus* | Belostomatidae | *Bufo, Hyla, Rana, Rhacophorus, Cynops, Physalaemus* | L, A | 24, 32, 64, 71, 84, 101, 109, 133, 140, 146 |
| *Corixa* | Corixidae | *Bufo, Rana* | E, L | 70 |
| *Cymatia* | Corixidae | *Bufo, Rana* | E, L | 70 |
| *Glaenocorisa* | Corixidae | *Bufo, Rana* | E, L | 70 |

*(continued)*

**Table 14.2** (continued)

| Predator | Family | Prey | Stage | Source no. |
|---|---|---|---|---|
| *Gerris* | Gerridae | *Rana* | L | 122 |
| *Nepa* | Nepidae | *Bufo, Rana* | L | 16, 46 |
| *Nepella* | Nepidae | *Phrynomantis* | L | 114 |
| *Ranatra* | Nepidae | *Bufo, Phrynomantis, Rana* | L | 16, 26, 114 |
| *Notonecta* | Notonectidae | *Bufo, Hyla, Rana* | E, L | 26, 46, 58, 63, 70, 101 |
| Beetles (Coleoptera) | | | | |
| *Catadromus* | Carabidae | *Limnodynastes, Litoria* | J, A | 35 |
| *Chlaenius* | Carabidae | *Litoria, Ranidella, Uperoleia* | J, A | 67 |
| *Maronatus* | Carabidae | *Desmognathus* | E | 39 |
| *Acilius* | Dytiscidae | *Bufo* | L | 18 |
| | | *Triturus* | E | 87 |
| *Colymbetes* | Dytiscidae | *Bufo* | L | 46 |
| *Dytiscus* | Dytiscidae | *Ambystoma, Bufo, Hyla, Pelobates, Rana* | L | 16, 17, 26, 32, 46, 47, 50, 59, 84, 89, 115, 154 |
| | | *Bufo, Rana, Triturus* | E | 70, 87 |
| *Hydaticus* | Dytiscidae | *Hyla* | L | 63 |
| *Ilybius* | Dytiscidae | *Hyla* | E, L | 119 |
| *Rhantus* | Dytiscidae | *Bufo, Hyla, Pseudacris, Rana* | L | 51, 56, 70 |
| *Rhantus* | Dytiscidae | *Physalaemus* | E | 1 |
| *Hydrochara* | Hydrophilidae | *Hyla* | L | 63 |
| *Hydrophilus* | Hydrophilidae | *Bufo, Hyla, Leptodactylus, Lyapsis, Scaphiopus, Scinax* | L | 56, 120 |
| Unidentified | Staphylinidae | *Phyllomedusa* | E | 151, 152 |
| Flies and mosquitoes (Diptera) | | | | |
| *Caiusa* | Calliphoridae | *Polypedates, Rhacophorus* | E | 124 |
| *Lucilia* | Calliphoridae | *Polypedates* | E | 53 |
| *Forcipomyia* | Ceratopogonidae | *Bufo, Leptodactylus, Pseudis, Rana* | A | 136 |
| *Parachironomous* | Chironomidae | *Ambystoma* | E | 42 |
| *Corethrella* | Corethrellidae | *Hyla* | A | 28, 79 |
| *Culex* | Culicidae | *Bufo* | A | 150 |
| *Culiseta* | Culicidae | *Bufo* | L | 90 |
| *Drosophila* | Drosophilidae | *Centrolene, Hyalinobatrachium* | E | 40 |
| *Hirtodrosophia* | Drosophilidae | *Hyalinobatrachium* | E | 31, 40, 55 |
| *Beckeriella* | Ephydridae | *Physalaemus, Leptodactylus* | E | 10, 40, 45, 55, 141, 146, 157 |
| *Tipopsilopa* | Ephydridae | *Hyperolius* | E | 128, 158, 159 |
| *Aphiura* | Phoridae | *Heleioporus* | E | 138, 155 |
| *Megaselia* | Phoridae | *Agalychnis, Eleutherodactylus* | E | 52 |
| | | *Desmognathus* | E | 37 |
| | | *Leptodactylus* | E | 95 |
| Unidentified | Phoridae | *Aneides* | E | 23 |
| | Phoridae | *Dendrobates* | E | 25 |
| | Phoridae | *Hyperolius* | E | 128 |
| | Phoridae | *Phyllomedusa* | E | 151, 152 |
| *Psychoda* | Psychodidae | *Agalychnis, Hyalinobatrachium* | E | 40 |

*(continued)*

**Table 14.2** (continued)

| Predator | Family | Prey | Stage | Source no. |
|---|---|---|---|---|
| Unidentified | Psychodidae | *Centrolene* | E | 40 |
| *Rhysops* | Syrphidae | *Hyalinobatrachium* | E | 4 |
| *Tabanus* (larva) | Tabanidae | *Scaphiopus* | J | 49 |
| Unidentified | Unidentified | *Litoria* | E | 15 |
| Ants and wasps (Hymenoptera) | | | | |
| *Angiopolybia* | Vespidae | *Leptodactylus* | E | 38 |
| *Forselius* | Formicidae | *Scaphiopus* | E | 121 |
| *Polybia* | Vespidae | *Agalychnis* | E | 129 |
| Unidentified | Unidentified | *Hyalinobatrachium* | E | 36 |
| | | *Agalychnis* | E | 45 |
| Mantids (Mantodea) | | | | |
| *Archimantis* | Mantidae | *Litoria* | A | 29 |
| *Choeradodis* | Mantidae | *Bolitoglossa* | A | 94 |
| *Hierodula* | Mantidae | *Litoria* | A | 29 |
| Crickets (Orthoptera) | | | | |
| *Paroecanthus* | Gryllidae | *Hyalinobatrachium* | E | 48 |

*Sources:* (1) F. Williams 1936; (2) McKeown 1942–43; (3) Goin 1943; (4) Costa-Lima 1946; (5) W. T. Neill 1948b; (6) McKeown 1952; (7) Carpenter 1953b; (8) Cory and Manion 1953, (9) Main and Main 1956; (10) Bokerman 1957; (11) Butler and Main 1959; (12) Cargo 1960; (13) T. Murphy 1961; (14) R. Savage 1961; (15) M. Tyler 1962; (16) Wager 1965; (17) Herreid and Kinney 1966; (18) W. E. Neill 1968; (19) Brockelman 1969; (20) Licht 1969c; (21) Bhatnagar 1970; (22) Dalrymple 1970, (23) D. Snyder 1971; (24) Calef 1973a; (25) Silverstone 1973; (26) Licht 1974; (27) Heyer, McDiarmid, and Weigmann 1975; (28) McKeever 1977; (29) Ridpath 1977; (30) Sharma and Sharma 1977; (31) Villa 1977; (32) Brodie, Formanowicz, and Brodie 1978; (33) R. D. Howard 1978b; (34) Knoepffler 1978–79; (35) Littlejohn and Wainer 1978; (36) McDiarmid 1978; (37) Forester 1979a; (38) Lacey 1979; (39) J. Caldwell, Thorp, and Jervey 1980; (40) Villa 1980; (41) Formanowicz et al. 1981; (42) Leclair and Bourassa 1981; (43) Ryan, Tuttle, and Taft 1981; (44) Ward and Sexton 1981; (45) Villa, McDiarmid, and Gallardo 1982; (46) Beebee 1983; (47) Brodie and Formanowicz 1983; (48) M. Hayes 1983; (49) Jackman et al. 1983; (50) Kruse 1983; (51) D. C. Smith 1983; (52) Villa and Townsend 1983; (53) Yorke 1983; (54) Crump 1984a; (55) Villa 1984; (56) Szelistowski 1985; (57) Travis, Keen, and Juilianna 1985a; (58) Cronin and Travis 1986; (59) Leclair, Alarie, and Bourassa 1986; (60) Bleckman and Lotz 1987; (61) Brodie and Formanowicz 1987; (62) Newman 1987; (63) A. H. Roth and Jackson 1987; (64) Hews 1988; (65) Van Buskirk 1988; (66) Holomuzki 1989a; (67) Robertson 1989; (68) Sherratt and Harvey 1989; (69) Fauth 1990; (70) Henrikson 1990; (71) Hinshaw and Sullivan 1990; (72) J. Mitchell 1990; (73) S. Richards and Bull 1990a; (74) S. Richards and Bull 1990b; (75) Sato 1990; (76) Skelly and Werner 1990; (77) Figiel and Semlitsch 1991; (78) Kehr and Schnack 1991; (79) McKeever and French 1991; (80) Azevedo-Ramos et al. 1992; (81) Chovanec 1992b; (82) Gascon 1992a; (83) Gascon and Travis 1992; (84) J. Peterson and Blaustein 1992; (85) A. Peterson, Bull, and Wheeler 1992; (86) Stout, Stout, and Stihler 1992; (87) Miaud 1993; (88) Sih and Moore 1993; (89) Tejedo 1993b; (90) L. Blaustein and Margalit 1994; (91) Rowe, Sadinski, and Dunson 1994; (92) Werner and McPeek 1994; (93) Anholt and Werner 1995; (94) Buttenhoff 1995; (95) Downie et al. 1995; (96) Schiesari, Acuna Junca, and de Mattos Accacio 1995; (97) Skelly 1995b; (98) Anholt, Skelly, and Werner 1996; (99) Babbitt and Jordon 1996; (100) Gamradt and Kats 1996; (101) Kiesecker, Chivers, and Blaustein 1996; (102) S. Johnson, 1996; (103) Kwet 1996a; (104) Lefcourt 1996; (105) Majecki and Majecka 1996; (106) McCollum and Van Buskirk 1996; (107) Rogers 1996; (108) Axelsson et al. 1997; (109) Haddad and Bastos 1997; (110) Laurila, Kujasalo, and Ranta 1997; (111) Mijares-Urrutia, Arends, and Hero 1997; (112) McCollum and Leimberger 1997; (113) Owen and Johnson 1997; (114) Rödel and Linsenmair 1997; (115) Lardner 1998; (116) Laurila, Kujasalo, and Ranta 1998; (117) Majecki and Majecka 1998; (118) Petranka and Hayes, 1998; (119) Resetarits, 1998; (120) Azevedo-Ramos and Magnusson 1999; (121) Dayton and Jung 1999; (122) Eaton and Paszkowski 1999; (123) H. Gray, Green, and Peters 1999; (124) Lin and Lue 1999; (125) Summers 1999b; (126) Warkentin 1999a; (127) H. Gray and Christy 2000); (128) Vonesh 2000 (129); Warkentin 2000a; (130) Berven and Boltz 2001; (131) Downie, Livingstone, and Cormack 2001; (132) Boistel and Pauwels 2002; (133) Hirai and Hidaka 2002; (134) Merilä and Sterner 2002; (135) Schalk, Forbes, and P. J. Weatherhead et al. 2002; (136) Spinelli, Schaefer, and Kehr 2002; (137) Brasileiro, Sawaya, and Giradelli 2003; (138) R. Davis and Disney 2003; (139) Marra et al. 2003; (140) McCoy 2003; (141) Menin and Giaretta 2003; (142) Peltzer and Lajmanovich 2003; (143) Pineda 2003; (144) Prado and Borga 2003; (145) Toledo 2003; (146) Giaretta and Menin 2004; (147) Jara andPerotti 2004; (148) Jeffrey, Waddle, and Maskell 2004; (149) Krishna, Krishna, and Vijayalaxmi 2004; (150) Krishna and Vijayalaxmi 2004; (151) Neckel-Oliveira 2004; (152) Neckel-Oliveira and Wachlevski 2004; (153) Schmidt and Van Buskirk 2004; (154) Storfer and White 2004; (155) Davis and Roberts 2005; (156) Kraft, Wilson, and Franklin 2005; (157) Silva, Giaretta, and Facure 2005; (158) Vonesh 2005a; (159) Vonesh 2005b.

*Note:* Stages: egg (E), larvae (L), juvenile (J), adult (A).

*Hydryphantes*), and the usual assortment of aquatic insects (Kwet 1996a; Kuzmin 1999). Planarians (*Polycelis nigra*) can sometimes consume entire clutches of *Rana temporaria* eggs within a few days (Kwet 1996a).

Several species of freshwater crayfish prey on eggs of *Ambystoma* (Ward and Sexton 1981), *Taricha* (Gamradt and Kats 1996), and *Triturus, Bufo, Hyla,* and *Rana* (Axelsson et al. 1997). Crayfish that have been introduced into nonnative habitats may pose a significant threat to amphibian populations. Gamradt and Kats (1996) discovered that the introduced crayfish *Procambarus clarkii* readily preyed upon egg masses of *Taricha torosa* in California and eliminated the newts from some streams. These newts normally breed in streams that are largely free of egg predators, and they lay their eggs in submerged vegetation or under rocks, where crayfish are most likely to forage. Introduced crayfish are considered a potential threat to amphibian populations in some parts of Europe as well (Axelsson et al. 1997).

### Predation on Nonaquatic Eggs

Many amphibians do not place their eggs directly in water, but lay them in foam nests floating on the water's surface,

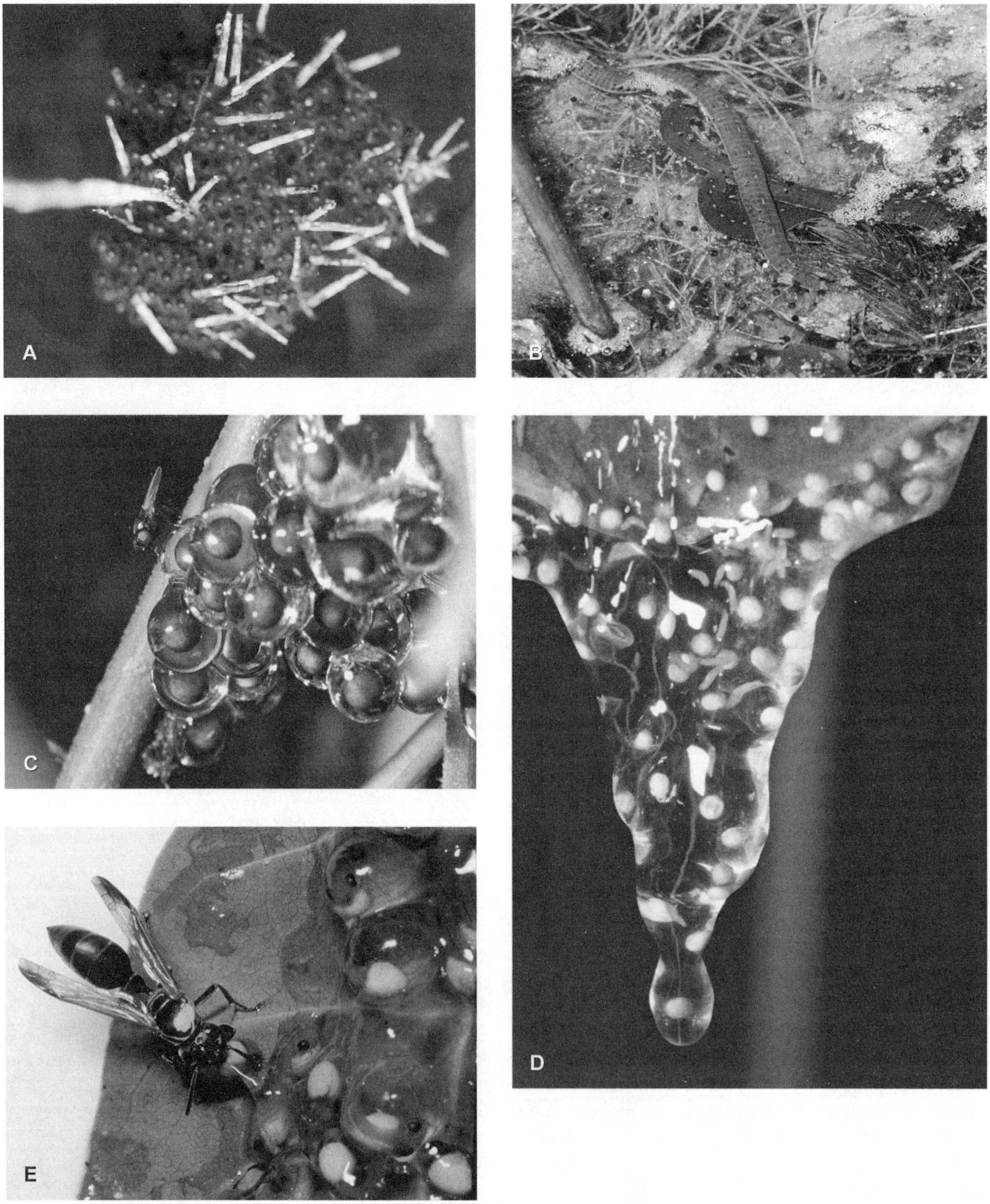

**Fig. 14.1.** Invertebrate predators of amphibian eggs. (A) Egg clutch of *Rana sevosa* infested with the larvae of caddisflies. (B) Leeches (*Macrobdella*) feeding on eggs of *Rana catesbeiana*. (C) Adult female of an ephydrid fly (*Typopsilopa*) depositing eggs on a clutch of frog eggs (*Hyperolius mitchelli*). (D) Egg clutch of *H. mitchelli* infested with fly larvae, which are the small, white animals at the top of the egg clutch. (E) Wasp (*Polybia rejecta*) removing an egg from an arboreal egg mass of *Agalychnis callidryas*. Photos by Stephen Richter (A), Richard D. Howard (B), James Vonesh (C, D), and Karen Warkentin (E).

on leaves overhanging streams and ponds, or in terrestrial nests (see chapter 10). These modes of reproduction probably reduce predation by aquatic invertebrates, but they expose the eggs to attack from a variety of other predators, including crabs, spiders, carabid beetles, wasps, crickets, and several types of fly larvae (Forester 1979a; Villa 1977, 1980, 1984; Villa, McDiarmid, and Gallardo 1982; M. Hayes 1983; Villa and Townsend 1983; Lin and Lue 1999; Warkentin 2000b; R. Davis and Disney 2003; Menin and Giaretta 2003).

The fly larvae are particularly destructive. Up to 80% of the foam nests of *Physalaemus cuvieri* examined by Bokermann (1957) in Brazil were infested with larvae of an ephydrid fly, *Beckeriella niger.* In another study, predation by these fly larvae resulted in an average mortality in *P. cuvieri* egg clutches of 74% (Menin and Giaretta 2003). Even foam nests of *Leptodactylus fuscus* hidden in burrows are attacked by phorid fly larvae (Downie et al. 1995), and flies in this family attack the foam nests of the myobatrachid frogs *Heleioporus* in Australia (R. Davis and Disney 2003). Phorid fly larvae and staphilynid beetles together caused up to 77% mortality in eggs of two species of *Phyllomedusa* in Brazil (Neckel-Oliviera and Wachlevski 2004). In Central America, 70–100% of *Hyalinobatrachium fleischmanni* clutches were infested with larvae of the drosophilid fly *Hirtodrosophila batracida* (Villa 1977, 1984; this fly originally was identified as a species of *Zygothrica* by Villa, but was later described as a new species, *H. batracida,* by Grimaldi 1994). The *Hirtodrosophila* found in *Hyalinobatrachium* egg masses appear to be obligate egg associates, as are various species of *Gastrops,* but some of the other flies listed in table 14.2 are facultative predators (Villa 1980). In Taiwan, up to 80% of the foam nests of several species of rhacophorid treefrogs were infested with fly larvae of the genus *Caiusa* (Lin and Lue 1999). The fly larvae often consumed all of the eggs in masses reared in the laboratory.

Eggs of several species of *Hyperolius* in Uganda were infested with both ephydrid and phorid fly larvae (Vonesh 2000, 2005a, b; fig. 14.1 C). These frogs lay eggs on vegetation over water. Overall, 40% of more than 1,200 egg clutches were infested. The eggs of several frog species (*H. lateralis* and *H. cinnamomeoventris*) seemed to be preferred over those of other species, especially *H. kivuensis,* although egg masses were attacked indiscriminately in the dry season, when frog eggs were scarce. The fly larvae usually destroyed all of the eggs of the preferred species, whereas some eggs of *H. kivuensis* survived infestation by frog fly larvae. This may be due in part to the more rapid development of this species, with eggs often hatching before the fly larvae consumed the entire clutch. In another species, *A. spinigularis,* predation on eggs by flies and adult frogs induced early hatching by the eggs, and the reduced densities of larvae resulting from egg predation increased larval survival and size at metamorphosis (Vonesh 2005a, b). This in turn made the larvae less vulnerable than expected to predation by dragonfly naiads (*Trapezostigma basilaris*), either because the tadpoles grew more rapidly to a less vulnerable size, or because the lower density of tadpoles decreased the hunting efficiency of the insects (Vonesh and Osenberg 2003).

### Predation on Aquatic Larvae

Even if amphibian eggs escape being eaten, there are numerous invertebrate predators ready to attack larvae soon after they hatch. Caddisfly larvae prey on newly hatched tadpoles and salamander larvae that remain near egg masses, but probably are incapable of catching free-swimming older larvae (Rowe et al. 1994). Leeches prey on both large and small tadpoles and sometimes occur in densities sufficiently high to be a major source of mortality (Carpenter 1953b; Brockelman 1969; Berven and Boltz 2001). They also can have nonlethal effects on tadpole fitness by reducing growth rates and prolonging the larval period (Berven and Boltz 2001; Schalk, Forbes, and Weatherhead 2002). A free-living flatworm, *Phagocotus gracilis,* occurs at very high densities (up to 656 worms per square meter) in streams in Kentucky, where it is a significant predator on newly hatched larvae of *Ambystoma barbouri* (Petranka, Kats, and Sih 1987; Sih and Moore 1993). Even aquatic snails have been reported to prey on tadpoles (Peltzer and Lajmanovich 2003).

In most freshwater habitats, the most important invertebrate predators on amphibian larvae are arthropods (fig. 14.2 A–C), including crayfish (Figiel and Semlitsch 1991; Veith 1993; Gamradt and Kats 1996; Axelsson et al. 1997), freshwater shrimp (Warkentin 1995, 1999a; Downie, Livingstone, and Cormack 2001), freshwater crabs (Kuzmin 1999), spiders (Bleckmann and Lotz 1987), dragonfly naiads (Aeshnidae, Corduliidae, Libellulidae), aquatic bugs (Belostomatidae, Nepidae, Notonectidae), and the larvae of aquatic beetles (Dytiscidae, Hydrophilidae; see many references cited in table 14.2). Larval mosquitoes (*Culiseta longiareolata*) fed on newly hatched toad tadpoles (*Bufo viridis*) in desert rain pools in Israel. Conversely, larger *Bufo* tadpoles preyed on early stage mosquito larvae, and the two species competed for food as well (L. Blaustein and Margalit 1994). In aquatic habitats where tadpoles compete with mosquito larvae, mosquitos sometimes avoid depositing eggs in pools with tadpoles present (L. Blaustein and Kotler 1993; Mokany and Shine 2003a). Mosquito larvae (*Culex*) also feed on eggs of the Siberian newt (*Salamandrella keyserlingii;* Kuzmin 1999).

Arthropod predators of amphibian larvae include both active foragers and sit-and-wait predators that use different sensory modes to locate and capture prey. Crustaceans, including crayfish and freshwater shrimp, tend to be active hunters that forage on the bottom of ponds (Figiel and Semlitsch 1991; Gamradt and Kats 1996; Axelsson et al. 1997;

**Fig. 14.2.** Invertebrate predators of larval and terrestrial amphibians. (A) Dragonfly naiad (*Anax*) feeding on a tadpole. (B) Larvae of a water beetle (*Dystiscus*) feeding on a tadpole. (C) A backswimmer (*Notonecta*) feeding on a tadpole. (D) Giant water bug (Belostomatidae) feeding on an adult microhylid frog (*Chiasmocles shudicarensis*). (E) Scorpion feeding on a Puerto Rican frog (*Eleutherodactylus coqui*). (F) Spider feeding on a frog (*Sphaenorhynchus lacteus*). Photos by Rick Relyea (A, B), Karen Warkentin (C), Walter Hödl (D, F), and Rafael Joglar (E).

Warkentin 2000b). At least some species detect prey mainly by chemical cues (Harpaz and Steiner 1990). Some aquatic bugs, including notonectids and some belostomatids (e.g., *Hydrocyrius*), are active hunters in the water column and probably detect prey visually or by means of tactile cues (Rödel and Linsenmair 1997). Other belostomatids (*Belostoma, Lethocerus*) are ambush predators that lurk in vegetation and detect prey mostly by visual and tactile cues, as are nepids such as *Ranatra* (P. Bailey 1986; Rödel and Linsenmair 1997). Larvae of dytiscid beetles are active hunters that use both chemical and tactile cues to locate and capture prey (Formanowicz 1987; Resetarits 1998).

Dragonfly naiads (fig. 14.2 A) include species that use both active foraging and ambush tactics. In general, libellulids and corduliids tend to sit and wait for prey near the bottoms of ponds, whereas aeshnids are more likely to climb into aquatic vegetation, where they use a combination of stalking and ambushing techniques (Pritchard 1965; Resetarits 1998; P. Corbet 1999). Aeschnids have better-developed eyes than do libellulids and corduliids, and often rely on visual cues to detect and orient toward prey (Pritchard 1965, 1966; Sherk 1977; Chovanec 1992a). Nevertheless, they are capable of detecting and striking at prey using tactile cues alone, or in combination with visual cues (Pritchard 1965; Kanou and Shimozawa 1983; S. Richards and Bull 1990a). Either visual or tactile cues allow them to use movements to detect prey, so active amphibian larvae are particularly susceptible to predation by these insects (Caldwell, Thorp, and Jervey 1980; S. Lawler 1989; Azevedo-Ramos et al. 1992; Chovanec 1992b; Skelly 1994).

### Predation on Juveniles and Adults

Predation by invertebrates on juvenile and adult amphibians probably is less common than predation on eggs and larvae, but it still can have an important effect on population dynamics (Toledo 2005, summarizes recent data, including some unpublished examples not included in table 14.2). Juvenile amphibians are subjected to predation by several kinds of invertebrate predators (fig. 14.2 D–F), including centipedes (M. Stewart and Woolbright 1996), scorpions (M. Stewart and Woolbright 1996), amblypygids (Formanowicz et al. 1981), spiders (Neill 1948b; Butler and Main 1959; Sharma and Sharma 1977; Formanowicz et al. 1981; Vonesh 2005b), giant water bugs (Hirai and Hidaka 2002), and carabid beetles (Littlejohn and Wainer 1978; Ovaska and Smith 1988; J. Robertson 1989; Kuzmin 1999). Invertebrate predation on juvenile amphibians probably is underreported, simply because such predation is difficult to observe. A number of kinds of invertebrate predators readily capture adult amphibians, especially when they are aggregated at breeding sites. Among the most important such predators are spiders (fig. 14.2 F; Goin 1943; Bhatnagar 1970; M. Hayes 1983; J. Mitchell 1990; Stewart and Woolbright 1996; R. Owen and Johnson 1997), giant water bugs (fig. 14.2 D; Belostomatidae; Hinshaw and Sullivan 1990; Haddad and Bastos 1997), and leeches (Merilä and Sterner 2002). Giant water bugs can kill between 4% and 8% of adult frogs in small breeding congregations on a single night, and massive mortality of adult anurans and nests from leech infestation has been reported as well (Merilä and Sterner 2002). There are brief reports of attacks on small frogs by mantids (Ridpath 1977; Buttenhoff 1995), crabs (Ryan, Tuttle, and Taft 1981; Stewart and Woolbright 1996; Kuzmin 1999), carabid beetles (Littlejohn and Wainer 1978; Robertson 1989; Kuzmin 1999), and rove beetles (Staphylinidae; Jung et al. 2000), but the impact of these predators on amphibian populations is unknown.

An interesting feature of invertebrate predation on amphibians is the frequent occurrence of trophic loops, in which a species of amphibian is both a predator of and prey for a species of invertebrate. This occurs most frequently when an invertebrate eats eggs, larvae, or juveniles of an amphibian, but adults of the amphibian species prey on the invertebrate (Polis and Myers 1985; Reagan, Camilo, and Waide 1996). For example, many species of fly larvae prey upon amphibian eggs, but some adult frogs probably prey on adult flies as well. Dragonfly naiads are major predators of ranid tadpoles, but large ranid frogs frequently prey on both larval and adult dragonflies, especially those just emerging from the water (Geelen, van Gelder, and Sax 1970; Stewart and Sandison 1972; Corse and Metter 1980; Werner, Wellborn, and McPeek 1995; Kuzmin 1999; personal observations). Adult frogs sometimes eat other tadpole predators as well, including spiders and aquatic bugs.

In the Luquillo Rainforest of Puerto Rico, several kinds of invertebrates prey on eggs or juveniles of *Eleutherodactylus coqui*. These include a centipede (*Scolopendra alternans*) and a crab spider (*Stasina portoricensis*), both of which are eaten occasionally by adult coquis. The frogs can be significant predators on crab spiders, which decreased in experimental plots in which frog densities increased (Stewart and Woolbright 1996). Small individuals of other invertebrate predators, including an amblypygid (*Phrynus longipes*) and a tarantula (*Avicularis laeta*) are potential prey for adult coquis, which live in the same microhabitats as these species (Reagan, Camilo, and Waide 1996). Adult coquis also eat two genera of ants, *Iridomyrmex* and *Paratrechina*, that eat coqui eggs (Stewart and Woolbright 1996), but the impact of frogs on ant populations is likely to be small.

## Invertebrate Parasites

Except for midges and mosquitoes that feed on the blood of frogs, table 14.2 includes only those invertebrate predators that consume all or most of their prey; it does not include parasites of amphibians, some of which can be fatal to their hosts. Internal parasites of amphibians include many types of protozoans (Elkan 1976b), as well as a variety of parasitic worms. The latter include tapeworms (Cestoda), intestinal round worms (Acanthocephala), nematodes that infect the lungs and visceral organs, and many species of monogenean and digenean helminths that infect the lungs, muscles, bladder, and digestive organs (Prudhoe and Bray 1982; Tinsley and Earle 1983; D. Brooks 1984; Duellman and Trueb 1986; Tinsley and Jackson 1986, 1988; Goater, Esch, and Bush 1987; Tinsley 1989; Aho 1990; Yoder and Coggins 1996; McAlpine 1997; McAlpine and Burt 1998; P. Johnson and Sutherland 2003). External parasites include leeches, some of which also live inside the bodies of amphibians as endoparasites (Mann and Tyler 1963), as well as mites, ticks, and chiggers (Rankin 1937; Anthony, Mendelson, and Simons 1994) and parasitic copepods (C. Goin and Ogren 1956). Larvae of sarcophagid, calliphorid, and chloropid flies parasitize adult anurans, and some are thought to be obligate anuran parasites (Askew 1971; Crump and Pounds 1985). A number of freshwater invertebrates attach themselves to amphibians, which then carry them between ponds. These include tiny ostracod crustaceans (Seidel 1989) and bivalve clams (*Pisidium;* Scherff-Norris and Livo 1999). It is not always clear whether this represents a parasitic or commensal relationship. Tadpoles and aquatic salamanders serve as hosts for the parasitic larval stages of freshwater mussels (glochidia; Seshaiya 1941, 1969; A. D. Howard 1951; Watters 1997; Watters and O'Dee 1998). The impact of these animals on their amphibian hosts has not been investigated, however.

### Ecological Impact of Parasites

Most papers on amphibian parasites are brief reports of the occurrence of parasites in particular host species, surveys of parasite communities in amphibians of a particular habitat, or studies of parasite life histories. These studies tend to focus mostly on the biology of the parasites, rather than their effect on the hosts. The importance of parasites as a source of mortality and their impact on amphibian population dynamics has seldom been addressed. A detailed discussion of amphibian parasites is beyond the scope of this book, but I will mention a few cases in which the effect of parasites on the physical condition or survivorship of amphibians has been investigated.

Adult *Atelopus varius* in Costa Rica are parasitized by larval sarcophagid flies (*Notochaeta bufonivora*), particularly when the frogs are aggregated near waterfalls (Crump and Pounds 1985; Pounds and Crump 1987). Fly parasitism could be a major source of mortality in the population, but quantitative data are lacking. Larvae of a tachinid fly, *Lucilia bufonivora*, sometimes infest populations of *Bufo calamita* and other anurans and can cause substantial mortality of adults (Meisterhans and Heusser 1970; Sinsch 1998). Gill (1978a) described heavy infestations of leeches in populations of *Notophthalmus viridescens* and speculated that they might be a major factor controlling the population dynamics of newts, either through direct effects of leech attacks or the effects of blood parasites (*Trypanosoma*) transmitted by the leeches. However, subsequent work showed that the trypanosomes had little or no effect on the survivorship of newts or their rates of reproduction (Mock and Gill 1984; Gill and Mock 1985); the effects of leech attacks alone were not investigated. Two studies of the effects of infection with parasitic nematode lung worms (*Rhabdias*) on growth and survival of newly metamorphosed anurans (*Bufo bufo* and *Rana sylvatica*) failed to reveal any significant effect (Goater 1994; Goater and Vandenbos 1997). However, earlier studies with *Bufo bufo* had shown negative effects of heavier infections by the same parasite on locomotor performance, growth, and survival of metamorphs (Goater and Ward 1992; Goater, Semlitsch, and Baernasconi 1993).

Infection by a species of parasitic trematode (*Ribeiroia ondatrae*) has been implicated in developmental abnormalities seen in several species of anurans, especially the presence of extra limbs (Sessions and Ruth 1990; P. Johnson et al. 1999, 2002, 2003; Sessions et al. 1999; P. Johnson, Lunde, Haight, et al. 2001; P. Johnson, Lunde, Ritchie, et al. 2001; Stopper et al. 2002; Blaustein and Johnson 2003a, b; P. Johnson and Sutherland 2003). Although such deformities have received a lot of news coverage and even were the subject of a popular book (Souder 2000), this is not a new phenomenon. There are reports going back nearly 300 years of amphibians with extra limbs (Ouellet 2000; P. Johnson et al. 2003). Not all limb malformation can be attributed to parasite infections; various chemical pollutants may be important as well (Ouellet 2000; Blaustein and Johnson 2003a, b). Nevertheless, a recent study of preserved specimens of deformed anurans from museum collections turned up evidence of *Ribeiroia* infection in a number of historic populations, some of which are from sites where recent infections have been reported (P. Johnson et al. 2003). As with other trematodes, freshwater snails (*Planorbella*) are the intermediate hosts for *Ribeiroia,* so the prevelence of deformed amphibians may be tightly linked to the population dynamics of the snails. There is evidence that in severely eutrophic ponds impacted by fertilizer runoff or other sorts of pollution, snail populations increase, and this may lead to increased parasite infections in amphibians. There also appears to have been an increase in the number of populations of amphibians infected with these parasites in recent years (Blaustein and Johnson 2003a,b; P. Johnson et al. 2003), and the prevalence of deformed individuals is correlated with trematode abundance (Johnson, Lunde, Ritchie, Reaser, and Launder 2001; P. Johnson et al. 2002).

Presumably the fitness consequences for the affected individuals are severe, since their locomotion is impaired and they are likely to be highly susceptible to predation, especially by water birds that are attracted to the awkward movements of deformed frogs. The ecological impact of parasite infections and the resulting limb malformations is not entirely clear. This phenomenon often has been discussed in the context of declining amphibian populations (e.g., Souder 2000; P. Johnson et al. 2002; Blaustein and Johnson 2003a, b). Certainly in some populations, deformities in up to 90% of individuals could result in local population extinctions, but there is little evidence so far that trematode infections have led to widespread declines of amphibian populations (Ouellet 2000). Indeed, some sites where large numbers of deformed frogs were present more than 50 years ago still support frog populations, with many individuals still infected with parasites (P. Johnson et al. 2003). High parasite infection rates are most common in farm ponds and other bodies of water heavily impacted by human activities or grazing cattle, so these parasites cannot explain the widespread decline of amphibian populations in more pristine habitats (see also chapter 16).

Two studies have examined the effects of parasite infections on calling performance and mating success of adult anurans. Hausfater, Gerhardt, and Klump 1990) quantified total numbers of parasitic nematodes and helminths in various organs of gray tree frogs (*Hyla versicolor*). Although many individuals were heavily parasitized, there was no relationship between parasite load and measures of calling performance, such as call duration and calling rate. Furthermore, when males with high and low levels of calling performance were compared for parasite load, there was not a significant difference for any type of parasite. There also was little evidence that parasite infection affected the time of arrival of males at the chorus, chorus attendance, or mating success. Tinsley (1990) studied spadefoot toads (*Scaphiopus couchii*), which usually are heavily infected by a monogenean parasite, *Pseudodiplorhis americanus,* which lives in the urinary bladder. Heavily infected toads can be debilitated by depletion of energy reserves, especially during hibernation. Nevertheless, males in a breeding chorus that were captured in amplexus had the same level of parasite infection as those not seen in amplexus. Mating success also was not related to body size, body mass, fat deposits, or hematocrit. Tinsley (1990) suggested that those males that have been debilitated by parasite infection simply do not come to breeding assemblages at all, whereas males in choruses are not sufficiently impacted by parasites to affect their reproductive success.

The effect of parasitism by mites (*Hannemania eltoni*) on plethodontid salamanders (*Plethodon angusticlavius*) was examined in laboratory experiments by Maksimowich and Mathis (2000). Even though heavily infested salamanders were not noticeably unhealthy, heavily infested males were less aggressive toward other individuals than were males with low parasite loads. If this effect occurs in the field, it could reduce the ability of heavily infested males to defend territories. Females were tested for the effect of mite infestation on foraging behavior with both fruit flies and termites as prey. Heavily infested females took longer to make a first attack on fruit flies than did females with low parasite loads, but this was not true for females presented with termites. The attack delay averaged only about 75 seconds and was not reflected in lower numbers of prey being consumed or in differences in nose-tapping behavior (an indicator of chemosensory exploration). Hence, the ultimate fitness effects of mite infestation are not known, particularly for animals in the field. Nevertheless, in behavioral experiments, there was a tendency for nonparasitized individuals to avoid fecal pellets of heavily parasitized individuals; these pellets are used to mark the territories of individual salamanders (Maksimowich and Mathis 2001).

### Predation by Fishes

Fishes often are the dominant predators in permanent ponds and lakes, as well as large streams and rivers, but they normally are absent from temporary ponds and very small streams. They can be voracious predators of amphibian eggs and larvae, and some large species can eat adult amphibians as well. An enormous number of temperate-zone fishes readily feed on amphibians. A partial list includes bowfin (*Amia*), mosquitofish (*Gambusia*), sticklebacks (*Gasterosteus*), catfish (*Ictalurus*), pike (*Esox*), goldfish (*Carassius*), carp (*Cyprinus*), bream (*Abramis*), asp (*Aspius*), chub (*Leuciscus, Semotilus*), darters (*Etheostoma*), shiners (*Notropis*), trout, char, grayling, and other salmonids (*Brachymystax, Hucho, Oncorhynchus, Salmo, Salvelinus, Thymallus*), burbot (*Lota*), sculpins and bullheads (*Cotto, Cottus*), sheatfishes (*Parasilurus, Silurus*), snakeheads (*Ophiocephalus*), bass (*Micropterus*), sunfish (*Enneacanthus, Lepomis*), and perch (*Lucioperca, Perca;* Lewis et al. 1961; W. Lewis and Helms 1964; Lowry 1966; Macan 1966; Voris and Bacon 1966; Grubb 1972; Efford and Tsumura 1973; Kruse and Francis 1977; Werschkul and Christensen 1977; Glandt 1983, 1984, 1985; Petranka 1983; Kruse and Stone 1984; Hayes and Jennings 1986; Giacoma 1988; Semlitsch 1987b, 1988; Semlitsch and Gibbons 1988; Bradford 1989; Blouin 1990; Kats and Sih 1992; Feminella and Hawkins 1994; Horat and Semlitsch 1994; Gamradt and Kats 1996; Lefcourt 1996; Kiesecker and Blaustein 1997b; Kurzava and Morin 1998; Tyler, Liss, Ganio et al. 1998; T. Tyler, Liss, Hoffman, and Ganio 1998; Goodsell and Kats 1999; Kuzmin

1999; Lawler et al. 1999; Jordan and Arrington 2001; Monello and Wright 2001). While the impact of fishes on amphibian populations can be severe, the importance of amphibians in the diet of fishes varies widely. There are no species of fishes known to specialize on the eggs, larvae, or adults of amphibians, and amphibians make up a relatively small proportion of the diet of many generalist predators (e.g., Jordan and Arrington 2001).

Predation on amphibians by tropical fishes has not been studied in detail. Various cypriniform, cichlid, and synbranchid fishes have been cited as potential predators of amphibian eggs and larvae, but actual predation is rarely seen in the field (K. Liem 1961; Heyer 1973; Kluge 1981; Branch 1983; Schlüter 1984; Azevedo-Ramos and Magnusson 1999). A fish of the genus *Hoplias* (Erythrinidae) ate a male toad (*Bufo crucifer*) in a breeding chorus in Brazil (Haddad and Bastos 1997). In East Africa, the introduction of food fishes into mountain lakes has virtually eliminated what once were enormous populations of *Xenopus* (Worthington and Worthington 1933; Tinsley 1973; Tinsley, Loumont, and Kobel 1996). Whether the frogs were eliminated by predation or competition has not been established, however. In tropical Queensland, Australia, eels (*Anguilla*) have been seen feeding on adult frogs (Beumer 1979; Harvey, Morrison, and Hero 1999), and barramundi (*Lates calcarifer*) will feed on tadpoles (Lawler and Hero 1997).

Predatory fishes affect the ecology and behavior of amphibians in several ways. They can influence the choice of breeding habitats and the placement of eggs within breeding sites (see the following). Fishes also can influence the relative abundance of amphibian species in aquatic communities by eliminating one or more species of amphibian larvae, or by altering the interactions between amphibian larvae and other predators (Werner and McPeek 1994; Hecnar and M'Closkey 1997b; Resetarits 1997; Kurzava and Morin 1998; see the following and chapter 15). The presence of predatory fish often causes amphibian larvae to change their activity and choice of foraging microhabitats, thereby affecting growth, survivorship, and the timing of metamorphosis (see the following "Defenses of Larvae"). Fish can be especially devastating predators of amphibian larvae when other predators, such as larval or adult salamanders, bullfrogs, or crayfish, also are present (Gamradt and Kats 1996; Kiesecker and Blaustein 1997b; Kurzava and Morin 1998). This may be in part because certain types of antipredator defenses that are effective against one type of predator are not effective against another. This can result in the combined effects of fish and other predators being greater than the effect of each predator alone.

One line of evidence for fish having a significant impact on amphibians comes from studies of distributional patterns and the effect of introduced fishes on amphibian populations (see also chapter 16). Permanently aquatic paedomorphic salamanders are most successful where fish are absent (see chapter 13). Some populations have been wiped out by introduction of nonnative game fishes (Carr and Goin 1943; Brandon and Bremer, 1967; D. Larson 1968; Sprules 1974a; Dolmen 1978; Marangio 1978; Moriya 1983; Sexton and Phillips 1986; see also chapter 16). Even in metamorphosing populations, predation on eggs and larvae may be sufficiently intense to eliminate salamanders from some habitats, or to prevent them from reproducing successfully. In ponds in the western United States, survival of larval *Ambystoma gracile* and *A. macrodactylum* is greatly reduced by the presence of introduced trout (T. Tyler et al. 1998). Nevertheless, the latter species can persist with trout in low numbers, especially in lakes with nonreproducing (recently stocked) fish (T. Tyler, Liss, Ganio, Larson, Hoffman, Deimling, and Lomnicky, 1998). *Ambystoma macrodactylum* is capable of rapidly recolonizing lakes where trout have disappeared. A study in Montana showed that salamanders were present in five of six high-altitude lakes where trout had gone extinct between 1978 and 1998, but salamanders were found in only two of 12 lakes that still had trout (Funk and Dunlap 1999).

In the eastern United States, spotted salamanders (*Ambystoma maculatum*) regularly coexist with bluegill sunfish (*Lepomis macrochirus*), but growth of *Ambystoma* larvae is reduced in the presence of fish. This apparently results from a combination of competition between larvae and fish for the same prey and microhabitat shifts by salamander larvae to avoid predation by fishes (Figiel and Semlitsch 1990). Banded sunfish (*Enneacanthus obesus*) appear to exclude red-spotted newts (*Notophthalmus viridescens*) from ponds in the Pine Barrens of New Jersey, but in this case, competition for food seems to be more important than predation (Bristow 1991).

In the western United States, introduced fish may be responsible for declines of many local populations of amphibians (see also chapter 16). For example, introduced goldfish (*Carassius auratus*) eliminated most amphibians from ponds in Idaho (Monello and Wright 1999). Both California newt larvae (*Taricha torosa*) and Pacific treefrog tadpoles (*Pseudacris regilla*) are subject to predation by mosquitofish (*Gambusia*) that have been widely introduced as mosquito control agents. When salamander larvae were housed with mosquitofish in stream enclosures, the fish ate nearly all of the larvae within 24 hours (Gamradt and Kats 1996). Predation on tadpoles was high even in streams where mosquito larvae were abundant (Goodsell and Kats 1999). Mosquitofish caused newly hatched tadpoles of *Rana draytonii* to reduce their activity, but the fish preferred invertebrate prey to tadpoles and did not have a significant effect on tadpole survivorship in seminatural ponds (Lawler et al. 1999).

Predation by introduced game fishes may be responsible for declining populations of some ranid frogs in the western United States (Hayes and Jennings 1986; Bradford 1989). Tadpoles of red-legged frogs (*Rana aurora*) do poorly in the presence of introduced smallmouth bass (*Micropterus dolomieui*), especially when introduced bullfrogs also are present (Kiesecker and Blaustein 1997). Introduced game fishes also can alter the structure of anuran populations by preventing dispersal between fish-free habitats (Bradford, Tabatabai, and Graber 1993), thereby increasing the chances that local populations will become extinct (see also chapter 16). Similar patterns are seen in Europe, where the introduction of trout into ponds has eliminated some populations of tree frogs (*Hyla arborea;* Brönmark and Edenhamn 1994). The European pool frog (*Rana lessonae*) requires permanent ponds for reproduction, but is largely excluded from very large ponds because of predation by pike (*Esox;* Sjögren 1991a). Many other European ranids, such as the moor frog *(Rana arvalis*), normally breed in fish-free ponds and are threatened by stocking of fish in breeding ponds (Podloucky 1987). In Australia, many frogs are restricted to temporary ponds that lack fish (M. Healey, Thompson, and Robertson 1997), and survival of their tadpoles can be reduced in the presence of introduced fishes (L. Morgan and Buttemer 1996).

In Europe and Asia, the presence of introduced game fishes is considered to be a major threat to stream-dwelling salamanders in the family Salamandridae and Hynobiidae (Picariello 1993; Kuzmin 1999), as well as amphibians that breed in mountain lakes (Mazzotti 1993; Braña, Frechilla, and Orizaola 1996). Effects of fishes may be due to predation, competition, or both. European newts (*Triturus*) generally are found in fish-free ponds, and many populations have been reduced or eliminated by introduced fish (Esher 1972; G. Bell 1979; Dolmen 1981a, 1982b; Giacoma 1988; Reshetnikov and Manteifel 1997; Kuzmin 1999). However, newts sometimes coexist with fish if invertebrates that serve as alternative prey for fish are sufficiently abundant (Dolmen 1982b). Larvae of *T. cristatus* are more vulnerable than are *T. vulgaris* because they feed in open water, whereas *T. vulgaris* larvae generally stay on the bottom (Dolmen 1982b, 1983a).

## Predation by Amphibians

Many species of amphibians prey upon the eggs, larvae, juveniles, or adults of their own or other species of amphibians. Salamander larvae probably are more significant predators of amphibians than are adults, most of which feed mainly on small invertebrates. Conversely, most anuran tadpoles are herbivorous, although even these can be opportunistic predators on eggs and larvae of other amphibians. Large adult anurans, on the other hand, are carnivorous and often take amphibians. Very little is known about the diets of caecilians. Juveniles of an aquatic caecilian, *Typhlonectes compressicauda,* from Brazil eat mostly invertebrates, but also eat the eggs and tadpoles of frogs (*Leptodactylus macrosternum;* Verdade, Schiesari, and Bertoluci 2000).

### Predation by Larval Urodeles

Larval and adult salamanders are the most important vertebrate predators of amphibian eggs and larvae in many ponds and lakes that lack fish (J. D. Anderson 1968a; Wilbur 1972; Calef 1973a; Walters 1975; Burton 1977; Morin 1983a, b; Thiesmeier 1990; Holomuzki, Collins, and Brunkow 1994). Salamander larvae are very effective predators, capable of nearly eliminating some anuran eggs and tadpoles from small ponds (Walters 1975; Morin 1983a, 1995; Cortwright 1988; Skelly 1992; Blaustein et al. 1996; Loeb, Collins, and Maret 1994; Maret and Collins 1996). Predation by salamander larvae can alter the species composition and dynamics of communities of anuran tadpoles by favoring species that are least vulnerable to predation and by reducing competition among surviving individuals (Wilbur, Morin, and Harris 1983; Wilbur 1987; Morin 1995; Blaustein et al. 1996; see also chapter 15). Larval salamanders also eat larvae of other species of salamanders in both ponds (Stenhouse, Hairston, and Cobey 1983; Stenhouse 1985a; Cortwright 1988; Kuzmin 1999; Brodman 2004) and streams (Resetarits 1991; Gustafson 1993). Cannibalism of conspecific larvae is relatively common (Degani, Goldenberg, and Warburg 1980; Collins and Cheek 1983; Lannoo and Bachmann 1984a; Simon 1984; Kusano, Kusano, and Miyashita 1985; Semlitsch and Reichling 1989; Crump 1992; Nyman, Wilkinson, and Hutcherson 1993; Degani 1993; M. Parker 1994; Bressi, Dolce, and Stoch 1996; Kuzmin 1999), and some ambystomatids even develop into specialized cannibalistic morphs (see chapter 12). In most cases, however, conspecific larvae make up a relatively small proportion of the diet.

### Predation by Adult Urodeles

In the eastern United States, adult red-spotted newts (*Notophthalmus viridescens*) are important predators on eggs and larvae of other amphibians. This species sometimes acts as a keystone predator in communities of larval anurans (Morin 1981, 1983a, b, 1986, 1987b; Morin, Wilbur, and Harris 1983; Wilbur 1987; Alford 1989b; Wilbur and Fauth 1990; Fauth and Resetarits 1991; Kurzava and Morin 1994; Kurzava and Morin 1998; Fauth 1999a). By changing the outcome of competition between species, a keystone predator can maintain a higher diversity of species in a community than would persist in the absence of the predator (see chapter 15). Such predators also may eliminate certain prey species altogether, leading to domination of a community by other species (e.g., Kurzava and Morin 1998). The impact of newts on their prey depends in part on the size of the newts. For example, a smaller subspecies, *N. v. dorsalis,* had less impact on populations of *Bufo americanus* tadpoles than did a larger subspecies, *N. v. viridescens* (Kurzava and Morin 1994). The presence of the smaller subspecies can alter competitive interactions among species of tadpoles, favoring those that are least vulnerable to newt predation (Morin 1981, 1983a; Alford 1989b; Fauth and Resetarits 1991). The larger subspecies, on the other hand, tends to reduce survivorship of all species of tadpoles (Morin 1986, 1987a). In one set of experiments, however, *N. v. viridescens* eliminated tadpoles of *Bufo woodhousii,* resulting in a community dominated by *Pseudacris* tadpoles, with several other species surviving in low numbers (Kurzava and Morin 1998). In South Carolina, another subspecies of newt, *N. v. louisianensis,* had less impact on the structure on larval anuran communities than did newts from North Carolina. In South Carolina, a large salamander, *Ambystoma talpoideum,* which was not present in North Carolina, filled the role of keystone predator. Ponds with this salamander present had a higher diversity of anuran species than did those without *A. talpoideum* (Fauth 1999a).

Red-spotted newts often have complex predator-prey interactions with other species of salamanders. In North Carolina, for example, adult newts are major predators on the eggs of tiger salamanders (*Ambystoma tigrinum;* Morin 1983b), and I have seen newts preying on eggs of *Ambystoma maculatum* as well (unpublished observations). High densities of newts are sufficient to eliminate most *Ambystoma* eggs from small ponds before they can hatch. This in turn may indirectly benefit anuran tadpoles, because *Ambystoma* larvae are capable of eliminating most tadpoles from small ponds (Morin 1983a). The larvae of *A. tigrinum* grow much faster than do those of *Notophthalmus* and eventually reach a size that enables them to feed on newt larvae (Morin 1983b). Newts in North Carolina also live in ponds with *Siren intermedia,* a salamander that can be both a competitor and predator (Fauth, Resetarits, Jr., and Wilbur 1990; Fauth and Resetarits 1991). Sirens prey on larval newts, but it also is possible that adult newts prey on larval sirens (Fauth and Resetarits 1991). Red-spotted newts also prey on their own eggs (R. Harris 1987a), and the presence of sirens may actually increase survivorship of newt eggs by reducing cannibalism by adults (Fauth and Resetarits 1991).

Sirens are unselective predators on anuran tadpoles and will even consume toad tadpoles that are distasteful to many other vertebrate predators (Lefcourt 1998), but the diet is composed mostly of invertebrates (Scroggin and Davis 1956). Sirens also will cannibalize the eggs of their own species (Scroggin and Davis 1956; Collette and Gehlbach 1961). In some circumstances, sirens may act as keystone predators in communities of larval anurans, or they may support the keystone role of newts (Resetarits and Fauth 1998; see chapter 15). When *Siren intermedia* lives in the same ponds as *Ambystoma talpoideum,* both species tend to be the top predators of anuran tadpoles, but sirens will prey upon larval *Ambystoma* as well. Hence, sirens are both competitors and predators of *A. talpoideum,* and the presence of sirens has a negative impact on growth and survivorship of *Ambystoma.* In general, the distributions of the two species among ponds are negatively correlated (Fauth 1999b). *Siren intermedia* generally does not live in the same ponds as a larger member of the same genus, *S. lacertina,* perhaps because of competitive or predatory interactions (Snodgrass et al. 1999). The latter species, in turn, tends not to be found with *Amphiuma means,* another predatory salamander of similar size. The impact of these two species on larval amphibian communities has yet to be studied, but *Amphiuma* is known to eat tadpoles, including toad tadpoles that are distasteful to some other predators (Lefcourt 1998).

In the western United States, aquatic habitats lack the diversity of amphibians found in the southeast. Nevertheless, newts of the genus *Taricha* can have a major impact on populations of other amphibians. Calef (1973a) estimated that newts (*Taricha granulosa*) and larval *Ambystoma gracile* together eliminated over 90% of the *Rana aurora* tadpoles from a 10-hectare lake in British Columbia, with newts being the most important predators. Even when tadpoles escape predation by *Taricha,* the presence of newts can result in reduced levels of feeding activity by tadpoles (D. J. Wilson and Lefcourt 1993), which in turn may reduce growth rates and survivorship (Skelly 1992). Adult *Taricha* also can be voracious predators on eggs and larvae of their own species (Kaplan and Sherman 1980; Marshall, Doyle, and Kaplan 1990; Kats, Elliott, and Currens 1992), although they may avoid eating conspecifics if alternative prey is available (Kerby and Kats 1998). Larval *Taricha* tend to react to the presence of adults by hiding (Elliott, Kats, and Breeding et al. 1993; Kerby and Kats 1998).

European newts (*Triturus*) also eat other amphibians. Adults of several species eat frog eggs and tadpoles, but most species tend to avoid toad eggs and tadpoles (Heusser 1971a; Cooke 1974; Griffiths 1986; Banks and Beebee 1988; Reading 1990; Denton and Beebee 1991; Kwet 1996a; Martínez-Solano 2000). Eggs and tadpoles usually are a much less important component of the diet than are various kinds of invertebrates (Griffiths and Mylotte 1987; Kuzmin 1999), but the impact on local populations of frogs can be severe (Kwet 1996a). Adults of *T. alpestris, T. helveticus, T. vittatus,* and *T. vulgaris* eat conspecific eggs, or those of other newts (Miaud 1993; Kuzmin 1999), but the overall importance of eggs in the diet is relatively trivial (Griffiths 1986; Kuzmin 1999). Adult *Triturus* sometimes feed on the larvae of other newts, but again, they make up a tiny fraction of the normal diet

(Dolmen and Koksvik 1983; Griffiths 1986; Kuzmin 1999). Adults of large species, such as *T. cristatus* and *T. marmoratus,* also feed occasionally on adults of smaller species, such as *T. vulgaris, T. helveticus,* and *T. alpestris* (Hagström 1979; Frazer 1983; Kuzmin 1999; Orizaola and del Valle 2000).

Intra- and interspecific predation on eggs by adults is widespread in plethodontid salamanders (Baldauf 1947; J. Wood and Clarke 1955; Snyder 1971; Simon 1984; Nussbaum 1985; Polis and Myers 1985; Crump 1992). Such predation may be a major selective force favoring the evolution of parental care (see chapter 11). Adults of some large plethodontid salamanders, such as *Desmognathus marmoratus, D. quadramaculatus, Gyrinophilus porphyriticus, Pseudotriton ruber,* and *Eurycea lucifuga,* sometimes prey on smaller salamanders (Martof and Scott 1957; Peck and Richardson 1976; Bruce 1979; Hairston 1986; Bock and Fauth 1992; Formanowicz and Brodie 1993; Secki and Queral-Regil 1997; Bernardo 2002), and there is even a report of an adult red-backed salamander (*Plethodon cinereus*) eating a conspecific juvenile (Maerz and Karuzas 2003). Predation by large species, especially *Desmognathus quadramaculatus,* has been considered a major determinant of microhabitat selection and community structure in streamside assemblages of *Desmognathus* (Hairston 1986, 1987; Southerland 1986a, b; Roudebush and Taylor 1987b; Beachy 1994; see chapter 15 for further details). Both *Gyrinophilus* and *Desmognathus quadramaculatus* will attack smaller salamanders in laboratory experiments, but the latter is a less effective predator than the former (Formanowicz and Brodie 1993). Studies of the natural diet of *D. quadramaculatus* indicate that salamanders are seldom eaten, and most of those consumed are conspecific larvae or juveniles. Predation on adults of other species of *Desmognathus* appears to be rare (Davic 1991; Camp 1997).

### Predation by Anuran Tadpoles

Although most anuran tadpoles are herbivores or detritivores, some prey on eggs, tadpoles, or metamorphosing juveniles of their own and other species (Heusser 1970, 1971b; Heyer, McDiarmid, and Weigmann 1975; Crump 1983, 1986b, 1991, 1992; Simon 1984; Polis and Myers 1985; Rödel and Linsenmair 1997; Spieler and Linsenmair 1997; Petranka and Kennedy 1999; Summers 1999b; Babbitt and Meshaka 2000; Vredenburg 2000). A few species have become specialized carnivores or cannibals; these were discussed in chapter 12 (see "Cannibalism and Morphological Polyphenism"). Predation by generalized tadpoles that lack adaptations for a carnivorous diet probably is an underappreciated source of mortality in many amphibian populations (Petranka and Kennedy 1999). For example, most Eurasian anurans have generalized pond tadpoles, but many of these species are known to prey on anuran eggs. Predatory or cannibalistic species include *Rana arvalis, R. dybowskii, R. lessonae, R. ridibunda, R. temporaria, Bufo bufo, B. calamita, Bufo verrucoissimus, Hyla arborea, Pelobates cultripes, Pelobates fuscus, Pelodytes caucasius,* and *Pelodytes punctatus* (Heusser 1971b; Beebee 1977; Sacher 1986; Banks and Beebee 1987b; Tejedo 1991; Kuzmin 1999). Some tadpoles even will eat the relatively distasteful eggs of the natterjack toad (*Bufo calamita*). Predation by tadpoles can result in nearly complete mortality of eggs in some ponds and appears to be a major factor that prevents *Bufo calamita* from successfully colonizing breeding ponds used by other anurans (Banks and Beebee 1987b; Tejedo 1991).

Wood frog (*Rana sylvatica*) tadpoles readily consume conspecific eggs (Petranka and Thomas 1995), and they are voracious predators on the eggs of American toads (*Bufo americanus*) as well, even though the eggs of this species contain toxic chemicals (Petranka, Brannon, Hopey, and Smith 1994). Wood frog tadpoles also prey on eggs and hatchlings of other anurans, including *Rana palustris* and *Hyla chrysoscelis,* as well as eggs of the spotted salamander (*Ambystoma maculatum*). In some ponds, predation on *Ambystoma* eggs by wood frog tadpoles can result in very high levels of mortality (Petranka, Rushlow, and Hopey 1998). This represents an interesting reversal of the usual ecological roles of these two species, because *Ambystoma* larvae often are the most important predators of *Rana sylvatica* tadpoles in temporary ponds (Wilbur 1972; Holomuzki, Collins, and Brunkow 1994). Green frog (*Rana clamitans*) tadpoles, which spend one or more winters in ponds, can sometimes virtually eliminate eggs of wood frogs laid in early spring (Petranka and Kennedy 1999).

Many species of tadpoles in the tropics prey on eggs of other anurans, even when the tadpoles lack specialized mouthparts and other morphological adaptations for a carnivorous diet. For example, the tadpoles of *Leptodactylus pentadactylus* have a generalized morphology, but are voracious predators on the eggs of other frogs, including those of *Bufo alatus* (formerly *B. typhonius*), which may be toxic or distasteful to other predators, including fish (Heyer, McDiarmid, and Weigmann 1975; Wells 1979). Tadpoles of a closely related species, *L. labyrinthicus,* also eat eggs of their own and other species (Silva, Giaretta, and Facure 2005). Tadpoles of several species of frogs from the tropical parts of Australia, including *Limnodynastes ornatus, Litoria alboguttata,* and *Litoria rubella,* prey on the eggs of native species of frogs (Crossland 1998a), as well as those of the introduced toad *Bufo marinus.* The eggs of this species are highly toxic, and consumption of eggs leads to high mortality among native tadpoles (Crossland 1998b; Crossland and Alford 1998). Because these species did not evolve with *Bufo marinus,* they do not exhibit any tendency to avoid these toxic eggs, so the consequences for populations of native frogs could be severe (Crossland 1998b). Tadpoles of *B. marinus* do not, however, appear to be significant predators on eggs, hatchlings, or tadpoles of native Australian frogs (Crossland 1998a).

Predation by tadpoles on the eggs and tadpoles of conspecifics or heterospecifics is particularly common among tropical anurans that breed in ephemeral pools, where rapid drying of pools favors rapid growth. For example, tadpoles of the Neotropical frog *Hyla pseudopuma* often cannibalize eggs and hatchlings of conspecifics (Crump 1983), thereby enhancing their own growth (Crump 1990). Tadpoles of another Neotropical treefrog, *Hyla acreana*, have been observed eating tadpoles of *Hamptophryne boliviana* and a species of *Scinax* in the field, and readily consume tadpoles of these species and those of *Phyllomedusa* and *Allobates* in captivity (McKeon 2005). The Cuban treefrog (*Osteopilus septentrionalis*) breeds in ephemeral drainage ditches in Florida, where tadpoles often occur at very high density. Cannibalism of conspecific tadpoles shortened the larval period and increased size at metamorphosis when tadpoles were fed mostly on a low-quality diet (algae and periphyton). This effect was not seen in tadpoles fed on a high-quality diet (rabbit chow; Babbitt and Meshaka 2000). Similar cannibalism has been described in *Hoplabatrachus occipitalis*, a West African frog that breeds in ephemeral rock pools (Spieler and Linsenmair 1997). Tadpoles of this species are specialized carnivores that also eat insect larvae and tadpoles of other species, in contrast to the generalized morphology and diet of the two hylids. Predation by tadpoles on eggs also is common in restricted habitats such as bromeliads and tree holes, which often harbor obligate egg-eating species (Simon 1984; Lannoo, Townsend, and Wassersug 1987; Crump 1992; Summers 1999b; see chapter 12 for further discussion). Cannibalism also occurs in frogs that lay eggs in shallow aquatic nests, including *Hyla rosenbergi* (Kluge 1981) and *H. faber* (Bernarde and Machado 1999).

### Predation by Adult Anurans

Adults of a number of species of frogs that lay terrestrial eggs will cannibalize the eggs of their own species (Tyler 1963a; Weygoldt 1980a; Wells 1981a; Simon 1983, 1984; H. Zimmermann and E. Zimmermann 1983; Townsend, Stewart, and Pough 1984; Summers 1989; Bonilla and La Marca 1996). Intraspecific cannibalism can be a major selective pressure favoring parental attendance of eggs (see chapter 11). Adults of the African hyperoliid frog *Afrixalus fornasini* eat the eggs of other species of frogs, including *Chiromantis xerampelina* and several species of *Hyperolius* (Drewes and Altig 1996; Vonesh and Osenberg 2003; Vonesh 2005a, b). Both of these species lay eggs on plants or tree branches over water, but do not exhibit parental care. The eggs of *Chiromantis* are laid in foam nests, but *Afrixalus* readily attack these eggs by thrusting their heads into the foam mass and pulling out eggs. *Afrixalus fornasini* also eats eggs of conspecifics. Drewes and Altig (1996) suggested that this species' habit of gluing their eggs inside of a folded leaf might afford some protection against cannibalism by conspecifics.

Adults of some aquatic frogs also eat anuran eggs and tadpoles. Pipid frogs, including *Xenopus* and some species of *Pipa*, prey on conspecific eggs or those of other species (McCoid and Fritts 1980; Gascon 1992b). *Xenopus* also eat tadpoles (Wager 1965) and adults of their own or other species (Inger and Marx 1961), but their role as predators in aquatic communities has not been studied in detail. Adult males of *Bombina orientalis* also consume eggs of conspecifics. These eggs are scattered in loosely connected strings on the bottoms of ponds and have little protection against conspecific predation (Marshall, Doyle, and Kaplan 1990). This species also feeds on the tadpoles of other frogs (Kuzmin 1999). Other Eurasian anurans that sometimes prey on eggs or tadpoles or their own or other species include *Bombina bombina, B. variegata, Bufo viridis, Rana lessonae, R. macrocnemis, R. nigromaculata,* and *R. ridibunda* (Kuzmin 1999).

Many medium and large adult frogs probably are opportunistic predators on smaller anurans (fig. 14.3). For example, ranids such as *Rana blairi, R. cascadae, R. catesbeiana, R. clamitans, R. esculenta, R. lessonae, R. luteiventris, R. muscosa, R. perezi, R. pretiosa, R. ridibunda, R. sylvatica,* and *R. vaillanti* prey on juveniles, adults, and even tadpoles of other species of frogs, as well as juvenile conspecifics. The importance of anuran prey varies geographically and seasonally from a trivial proportion of the diet to more than 30% (Korshgen and Moyle 1955; Frommhold 1956; Korshgen and Baskett 1963; Jenssen and Klimstra 1966; Stewart and Sandison 1972; Dushin 1974; Corse and Metter 1980; Frazer 1983; Lizana, Ciudad Pizarro, and Pérez-Mellado 1986; Werner, Wellborn, and McPeek 1995; Kuzmin 1999; Pilliod 1999; K. Pope 1999; Pearl and Hayes 2002; McCallum, Klotz, and Trauth 2003; Rombough, Jordan, and Pearl 2003; Vaughan 2003; Bolek and Janovy 2004). In North America, the bullfrog (*Rana catesbeiana*) undoubtedly is the most significant anuran predator of other amphibians, simply because it is the largest species. Adult bullfrogs are known to eat tadpoles, juveniles, adults, and even eggs of their own species and other species of frogs (Korshgen and Moyle 1955; Korshgen and Baskett 1963; G. Brooks 1964; M. Stewart and Sandison 1972; McKamie and Heidt 1974; Corse and Metter 1980; Bury and Whelan 1984; McAlpine and Dilworth 1989; J. Stuart and Painter 1993; Werner, Wellborn, and McPeek 1995). They also will eat salamanders (Balfour and Stitt 2003) and even newts (*Taricha torosa*), which are well protected from most predators by toxic skin secretions (M. Jennings and Cook 1998).

**Fig. 14.3.** Large adult *Leptodacylus ocellatus* eating a smaller frog. Photo by Walter Hödl.

The decline of several native ranids in the western United States has been attributed in part to predation by introduced bullfrogs (Moyle 1973; Schwalbe and Rosen 1988; P. Rosen and Schwalbe 1995; P. Rosen et al. 1995), but evidence for this hypothesis is equivocal (M. Jennings and Hayes 1985; Hayes and Jennings 1986). Introduced bullfrogs clearly have a negative impact on some native species, such as the red-legged frog (*Rana aurora*) and yellow-legged frog (*R. boylii*). The impact of bullfrogs appears to result from several factors. These include competition among tadpoles (Kupferberg 1997a), predation by larval and adult bullfrogs on tadpoles of other species (Kiesecker and Blaustein 1997b; Lawler et al. 1999), and the interactive effect of bullfrogs and other introduced predators on tadpoles (Kiesecker and Blaustein, 1998). The impact of predation by adult bullfrogs on the juveniles and adults of native frogs has yet to be determined. Experimental removal of large numbers of adult bullfrogs from a wildlife refuge in Arizona resulted in some increase in the density of *Rana chiriacahuensis* (Schwalbe and Rosen 1988). The natural disappearance of bullfrogs from a park in Canada had a similar effect on the density of green frogs (*R. clamitans;* Hecnar and M'Closkey 1997a). It is not possible to determine from these studies whether the effect of bullfrog removal was due to reduced predation, reduced competition, or a combination of the two (see chapter 16 for further discussion of introduced predators).

A number of large tropical frogs, including the African ranid *Pyxicephalus adspersus*, the South American leptodactylids *Cacophrys, Ceratophrys* and *Lepidobatrachus*, and the South American hylid *Hemiphractus*, have been described as voracious predators of other anurans (Duellman and Trueb 1986). While all are known to eat frogs (A. Loveridge 1947; Wager 1965; Stewart 1967; Duellman 1978; Cei 1980), there is little detailed information on their feeding habits, and the relative importance of anurans in the diet is unknown. *Cacophrys, Ceratophrys* and *Lepidobatrachus* all have very wide mouths and appear to be morphologically specialized for eating large prey, including frogs (Scott and Aquino 2005). Other species that eat anurans at least occasionally include *Bufo marinus* (Jaeger 1976; Tyler 1976), *Leptodactylus chaquensis* (Duré 1999), *L. laticeps* (Scott and Aquino 2005), *L. pentadactylus* (Ryan, Tuttle, and Taft 1981), *L. labyrinthicus* (Vaz-Silva, Silva, and da Silva 2003c), *L. ocellatus* (Gallardo 1964; Kokubum and Rodrigues 2005), and *Trachycephalus venulosus* (Dundee and Liner 1985).

## Predation by Turtles

Many freshwater turtles that are carnivores or omnivores prey on amphibians, especially tadpoles. There are at least 21 species of North American turtles that feed on amphibians (Ernst, Barbour, and Lovich 1994). These represent about 44% of turtle species associated with freshwater habitats (excluding marine turtles and herbivorous tortoises). Several species are too poorly studied for their food habits to be known. Turtles that eat at least some amphibians include snapping turtles (Chelydridae), mud turtles (*Kinosternon*), musk turtles (*Sternotherus*), soft-shelled turtles (*Apalone*), and emydid species such as wood turtles (*Clemmys insculpta*), box turtles (*Terrepene*), and pond turtles (*Clemmys, Chrysemys, Pseudemys, Trachemys, Deirochelys, Emydoidea*). Snapping turtles (*Chelydra serpentina*) even have been seen eating the eggs of spotted salamanders (*Ambystoma maculatum;* Spohr 2005). These turtles are frequently considered major predators of frogs and sometimes can have a significant impact on a local population. R. Howard (1978a) reported that large male bullfrogs (*Rana catesbeiana*) were particularly vulnerable to snapping turtle predation when calling in their territories. Ninety-five percent of snapping turtles captured in Florida had eaten tadpoles (Punzo 1975), but amphibians were a minor component of snapping turtle diets in other areas (M. Alexander 1943; Lagler 1943). Amphibians generally make up less than 10% of the diet of other carnivorous turtles, including *Clemmys marmorata* and several species of *Kinosternon* (Mahmoud 1968; Hulse 1974; Punzo 1974; Bury 1986).

Europe has only two genera of freshwater turtles, *Emys* and *Mauremys*, both of which eat amphibians, especially tadpoles (Ernst and Barbour 1989a; Kuzmin 1999). In Spain, *Mauremys leprosa* is largely herbivorous, and less inclined to eat tadpoles than is the more carnivorous *Emys orbicularis* (Gomez-Mestre and Keller 2003). The genus *Mauremys* also occurs in Southeast Asia and Japan, and several species include amphibians in a varied diet. A number of other Asian pond turtles (*Chinemys, Hardella*) also eat some amphibians, as do Asiatic softshell turtles (*Amyda, Lissemys*), but there is little quantitative information on the importance of amphibians in their diet. Australia has only one family of freshwater turtles, the family Chelidae, found in both temperate and tropical regions. Several genera of these side-necked turtles are omnivorous and may include amphibians in their diet. These include the snake-necked turtles (*Chelodina*), the Australian snapping turtles (*Elseya*), and several species in the genus *Emydura* (Worrell 1963; Ernst and Barbour 1989a; Legler and Georges 1993).

Relatively little is known about predation on amphibians by tropical turtles. Central American turtles are members of families or genera also found in North America, and several of these eat frogs and tadpoles in a varied diet of vertebrate and invertebrate prey (e.g., *Chelydra, Kinosternon, Claudius, Staurotypus* (Vogt and Guzman 1988; Ernst and Barbour 1989a; J. Lee 1996; J. Campbell 1998). In the community of turtles studied by Vogt and Guzman (1988) in southern Mexico, both *Kinosternon leucostomum* and *Staurotypus triporcatus* ate tadpoles, but tadpoles made up only a small portion of the diet. I observed an unidentified turtle feeding on an aggregation of *Leptodactylus insularum* tadpoles in Panama (unpublished observations). *Kinosternon* has a distribution extending into South America, where these turtles probably eat tadpoles. Most of the freshwater turtles of South America are side-necked turtles (families Chelidae and Pelomedusidae). Several genera feed on amphibians as part of a carnivorous or omnivorous diet, including *Pelmomedusa, Pelusios,* and *Podocnemis* (Pelomedusidae) and *Hydromedusa, Phrynops,* and *Platemys* (Chelidae; Ernst and Barbour 1989a). In one species, *Phrynops geoffroanus,* amphibians made up less than 1% of the diet (Fachin Teran, Vogt, and de Fatima Soares Gomez 1995).

Wager (1965) saw turtles feeding on aggregations of toad tadpoles (*Schismaderma carens*) in southern Africa. Hatchlings of side-necked turtles (*Pelomedusa subrufa*) were important predators on several species of tadpoles in temporary ponds on a West African savanna (Rödel 1999; Grafe et al. 2004). Benthic tadpoles that spend most of their time on the bottom of a pond were most vulnerable. These include tadpoles of *Hemisus marmoratus* (Hemisotidae), *Hyperolius nitidulus* (Hyperoliidae), and *Ptychadena maccarthyensis* (Ranidae). Tadpoles that inhabit the water column were less vulnerable, except when drying of ponds crowded tadpoles together. African softshell turtles (*Trionyx, Cycloderma*) also include amphibians in a varied carnivorous diet (Ernst and Barbour 1989a).

### Predation by Crocodilians

Some juvenile crocodilians prey on amphibians, but adult crocodilians seldom do so, probably because amphibians are too small to be profitable prey for such large animals. Anurans were eaten by Nile crocodiles (*Crocodylus niloticus*) in their first three to four years of life, but never by individuals more than 2.5 m long (Cott 1961). A few remains of small frogs were found in stomachs of adult dwarf crocodiles (*Osteolaemus tetraspis osborni*) collected in a market in the Congo, but the precise habitat where these animals had fed was not known (Riley and Huchzermeyer 2000). Juveniles of both the American crocodile (*C. acutus*) and Morelet's crocodile (*C. moreletti*) sometimes eat some frogs (J. Lee 1996; J. A. Campbell 1998). In Australia, juvenile *C. johnstoni* sometimes ate frogs, but they were only about 2% of the diet (Webb, Manolis, and Buckworth 1982). In contrast, 66% of the prey taken by *Caiman crocodilus* during the wet season in Venezuela were frogs (Gorzula 1978). These were captured in shallow water along the shores of temporary savanna lagoons. In a study of Amazonian crocodilian diets, Magnusson, Vieira da Silva, and Lima (1987) found no evidence of predation on anurans by *Caiman crocodilus, Melanosuchus niger,* or *Paleosuchus palpebrosus,* but they were eaten by juvenile *Paleosuchus trigonatus.* American alligators (*Alligator mississippiensis*) seldom eat frogs, even when they are abundant (Valentine et al. 1972), but they sometimes eat large amphibians such as *Siren lacertina* (Delany and Abercrombie 1986).

### Predation by Lizards

Lizards probably are the least significant reptilian predators of amphibians. Most small lizards are insectivorous, while many larger species tend to be herbivorous (Pough 1973). Furthermore, lizards usually are diurnal and often live in relatively dry habitats, whereas amphibians are mainly nocturnal and prefer moist habitats. Nevertheless, a few carnivorous species do eat amphibians occasionally. Wager (1965) reported that frogs are a major food of young monitor lizards (*Varanus*) in South Africa. Several Australian and African species of *Varanus* sometimes eat frogs, but they usually are less than 5% of the diet (King and Green 1979; Shine 1986a; D. Bennett 2002). The same is true for the Bengal monitor (*V. bengalensis*), a largely insectivorous lizard that lives in relatively dry habitats. This species is known to eat nearly 20 species of frog and toads, but anurans comprise less than 1% of its prey (Auffenberg 1994). Another species of monitor lizard (*V. flavescens*) that lives mostly in marshy habitats in India does feed heavily on frogs and toads, with 40% of all identified prey items being anurans (Auffenberg et al. 1989). This species also eats anuran egg masses, a type of food not reported for any other species. *Varanus griseus* feeds occasionally on frogs (*Rana ridibunda*) and toads (*Bufo viridis;* Kuzmin 1999). Other lizards that live near water may take an occasional frog. For example, there are reports of ranid frogs in the diets of *Basiliscus* from Costa Rica (Fleet and Fitch 1974; Cover 1986). Australian water skinks (*Eulamprus quoyii*) have been observed eating both tadpoles and juveniles of the hylid frog *Litoria aurea* (Pyke and Miehs 2001). The large teid lizard *Tupinambis texuixin* lives in a variety of habitats and takes frogs occasionally (J. Murphy 1997). There even are anecdotal reports of small frogs (*Eleutherodactylus*) being eaten by lizards that normally are strictly insectivorous, including *Ameiva exsul, Anolis gundlachi, Anolis cristatellus,* and *Sphaerodactylus klauberi* (Stewart and Woolbright 1996; R. Thomas and Kessler 1996), but these predators probably have little impact on anuran populations. The Eurasian anguid lizard *Ophisaurus apodus* feeds occasionally on toads (Kuzmin 1999).

## Predation by Snakes

In contrast to lizards, snakes are major predators of amphibians; in some areas, they may be more important than any other animals (fig. 14.4). Some snakes are generalist predators, but others have narrow food preferences and have evolved morphological adaptations for eating certain kinds of prey (H. Greene 1983, 1997; Pough and Groves 1983; Savitsky 1983; Mushinsky 1987). Many species in both temperate and tropical regions feed primarily on amphibians, particularly frogs. Nearly 50% of the more than 100 species of snakes in the United States and Canada eat some amphibians (Wright and Wright 1957). Only two or three species of European snakes regularly feed on amphibians (less than 10% of the total snake fauna; Steward 1971; Diesener and Reichholf, 1986). This low number probably is related in part to the low diversity of frogs in Europe. Arnold (1972) found that the number of anuran-eating snakes in the Northern Hemisphere is strongly dependent on the number of anuran species. The scarcity of frog-eating snakes in Europe probably also is related to the low diversity of colubrid snakes in the subfamily Natricinae, the principal clade of frog-eating snakes in the Northern Hemisphere (Greene 1997). Natricines are much more common in Asia and often live near water, where frogs are readily available. Pope (1935) listed 28 species of Chinese snakes that eat frogs, the largest number of species to feed on any type of prey, but not all were anuran specialists. This represents about 25% of the Chinese snake fauna known at that time. In neighboring Japan, 13 of 28 species (46%) feed on amphibians, but none is an amphibian specialist (Mori and Moriguchi 1988). Many of these are natricines, although colubrines and some viperids also eat frogs. Frogs are more abundant in the tropics than in temperate regions, and most tropical areas with high anuran diversity have many species of frog-eating snakes (Barbault 1977, 1987, 1991; Duellman 1978; Vitt 1983, 1987; Vitt and Vangilder 1983; Greene 1988; Cadle and

**Fig. 14.4.** Snake predators of amphibians. (A) Neotropical snake (*Leptodeira*) eating egg mass of *Agalychnis callidryas.* (B) Hognose snake (*Heterodon platyrhinos*) eating a toad. (C) Leptodactylid frog (*Adenomera andreae*) being eaten by a snake. Photos by Karen Warkentin (A), Michael Plummer (B), and Walter Hödl (C).

Greene 1993; Luiselli, Akani, and Capizzi 1998; Luiselli 2003; see the following further discussion of specific regions).

### Primitive Snakes

Phylogenetic history often is an important determinant of diet in snakes. Among the primitive snake families, small species feed mainly on invertebrates, whereas large species eat a variety of vertebrate prey. Snakes in the families Uropeltidae and Xenopeltidae are entirely tropical in distribution and apparently feed mainly on elongate vertebrates, including caecilians. Remains of caecilians have been found in stomachs of a South American species, *Anilius scytale,* and an Asian species, *Cylindrophus rufus* (Greene 1983). There also are field observations of *C. rufus* feeding on caecilians of the genus *Ichthyophis* (Kupfer, Gower, and Himstedt 2003). Some dwarf boas (family Tropidophiidae) eat anurans, mainly frogs of the genus *Eleutherodactylus* (Greene 1983; Henderson and Crother 1989). *Exiliboa placata,* an inhabitant of cloud forests in southern Mexico, apparently feeds exclusively on frogs and salamanders (Campbell and Camarillo 1992). Most members of the family Boidae eat birds and mammals, but the Puerto Rican boa (*Epicrates inornatus*) and other species of *Epicrates* sometimes eat frogs (Reagan 1984; J. Murphy 1997). Most West Indian boids probably take anurans occasionally (Henderson and Powell 1999).

### Colubrid Snakes

The largest family of snakes is the family Colubridae, a group with a worldwide distribution. Colubrids eat almost every available type of prey, including insects, earthworms, slugs, snails, fish, tadpoles, frogs, salamanders, lizards, snakes, birds, and mammals. Variation in diet tends to reflect both phylogenetic history and local abundance of prey. The habit of feeding on small ectothermic vertebrates, such as lizards and frogs, as part of a more generalized diet, is widespread in the family and may represent an ancestral type of diet for at least some clades. A few clades within the family, especially in the Neotropics, include species that tend to specialize on eating frogs (Greene 1988; Cadle and Greene 1993).

Most North American colubrids are members of the subfamily Colubrinae, a large, worldwide assemblage of species that probably is not a monophyletic group. Many colubrines are generalist predators that eat a variety of prey, including amphibians. A good example is the black racer (*Coluber constrictor*), which preys on anurans, reptiles, birds, mammals, and insects. In southern Illinois, anurans accounted for 8% of their annual diet, but up to 22% in midsummer (Klimstra 1959c). Frogs were of minor importance in most parts of Kansas, but in one area where *Rana pipiens* was abundant, frogs made up nearly 13% of the diet (Fitch 1963). Other North American and Eurasian colubrines that sometimes eat amphibians include indigo snakes (*Drymarchon*), North American rat snakes and Eurasian racers (*Elaphe*), coachwhip snakes (*Mastocophis*) Asian banded ratsnakes (*Ptyas*), milk snakes and king snakes (*Lampropeltis*), green snakes (*Opheodrys*), and scarlet snakes (*Cemophora;* Wright and Wright 1957; Klimstra 1959b; Ernst and Barbour 1989b; Trauth and McAllister 1995; Greene 1997; Fitch 1999; Kuzmin 1999; Pauwels 2002; Ryberg and Dayton 2004).

Snakes in the subfamily Natricinae, which is sometimes considered to be a separate family, often are important amphibian predators, although diet tends to vary with local abundance of different prey species. For example, one population of European grass snakes (*Natrix natrix*) in England ate almost nothing but tadpoles, juveniles, and adults of the common toad (*Bufo bufo;* Reading and Davies 1996). Another population in the marshes of southeastern England fed primarily on the introduced frog *Rana ridibunda,* the most abundant anuran in the area (Gregory and Isaac 2004). In Sardinia, the same species fed mostly on *Hyla sarda* (Capula, Rugiero, and Luiselli 1994). In fact, this species feeds on many different species of amphibians throughout Europe and northern Asia (Luiselli and Rugiero 1991; Luiselli and Capula 1997; Kuzmin 1999). Other Eurasian members of this genus, especially *N. maura* and *N. tessellata,* also eat amphibians, as do Asian natricines such as *Amphiesma* and *Rhabdophis* (Santos and Llorente 1998; Kuzmin 1999).

Anurans make up about 90% of the adult diet of the North American water snake *Nerodia erythrogaster* (Diener 1957; Mushinsky and Hebrard 1977; Mushinsky, Hebrard, and Vodopich 1982). Small individuals eat mostly small fish, but shift to anurans as they grow (fig. 14.5). This shift is reflected in changes in preferences for chemical extracts of prey. Young snakes do not prefer one prey type to another, but develop a strong preference for frog extract at eight to nine months of age (Mushinsky and Lotz 1980). *Nerodia fasciata* and *N. sipedon* usually eat a mixture of amphibians and fish (R. Clark 1949; E. Brown 1958; Mushinsky and Hebrard 1977; Camp, Sprewell, and Powders 1980; Fitch 1999), with the former showing an ontogenetic shift from fish to anurans (Mushinsky and Lotz 1980; Mushinsky, Hebrard, and Vodopich 1982). *Nerodia sipedon* in particular is a generalist predator that tends to be quite variable in its dietary habits. Himes (2003) reported that in southern populations, small individuals ate mostly fish, frogs, and salamanders, whereas larger individuals ate mostly fish. E. Brown (1958) found that the proportion of anurans in the diet of *N. sipedon* varied from 17% along streams in New York to over 90% in Michigan bogs. Other North American water snakes, such as *N. taxispilota, N. cyclopion,* and *N. rhombifera,* are primarily fish-eaters and usually avoid amphibians (Mushinsky and Hebrard 1977; Kofron 1978; Camp, Sprewell, and Powders 1980; Manjarrez and Macias Garcia

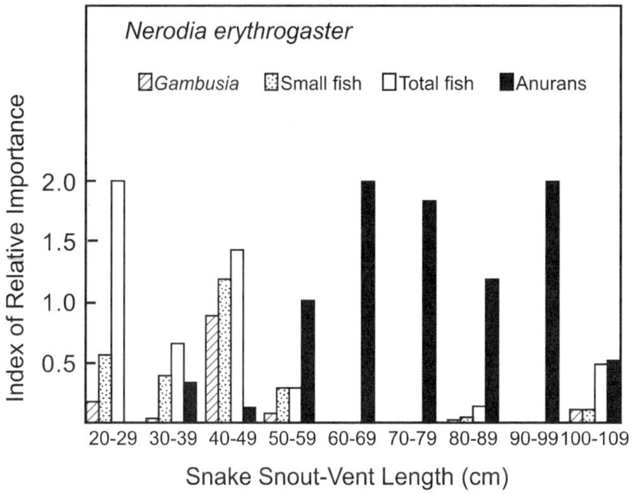


**Fig. 14.5.** Ontogenetic change in the diet of the water snake *Nerodia erythrogaster.* Juveniles eat mostly fish, whereas adults eat mostly frogs. After Mushinsky, Hebrard, and Vodopich (1982).

1991). Nevertheless, anurans were 66% of the prey eaten by *N. cyclopion* in Florida (Van Hyning 1932) and 45% of the prey of *N. rhombifera* in Texas (Bowers 1966).

Many species of garter snakes (*Thamnophis*) prey on amphibians (Rossman, Ford, and Seigel 1996; table 14.3), and some species may depend on amphibians as their principal food. For example, *Thamnophis elegans* in the Sierra Nevada of California feed mostly on anurans, and are not found in areas from which amphibians have disappeared (W. Jennings, Bradford, and Johnson 1992). Tadpoles and metamorphosing anurans tend to be the most common prey, but some garter snakes prey on larval and adult salamanders as well (Dowdey and Brodie 1989; Lind and Welsh 1990; Murray and Jenkins 1999; Painter, Scott, and Altenbach 1999). Geographic variation in diet often reflects local differences in prey abundance. In California, *T. sirtalis* ate mostly anurans and was most common where these were abundant (Kephart 1982). *Thamnophis elegans* ate a mixture of fish, amphibians, and leeches, but ate mostly anurans at sites where they were abundant. *Thamnophis validus,* from upland regions of Baja California, ate only anurans and tadpoles, whereas those from mainland Mexico ate mostly fish as adults, but fed on amphibians as juveniles (De Queiroz, Henke, and Smith 2001). This ontogenetic shift in diet from anurans to fish is the opposite of that seen in many other natricines, and may reflect a shortage of small fish that would be suitable prey for young snakes.

Even very localized differences in prey abundance can affect prey choice. Two species of *Thamnophis* captured in a marsh in California ate mostly frogs, whereas those captured in a nearby stream ate mostly fish (M. White and Kolb 1974; fig. 14.6). A Canadian population of *T. sirtalis* ate mostly larval and adult amphibians at a natural lake, but ate mostly fish in fish hatchery ponds (Gregory and Nelson 1991). Seasonal variation can be important as well. Garter snakes (*T. radix*) in Missouri fed mostly on earthworms in the spring and fall, but switched to feeding almost exclusively on frogs during the summer, when large numbers of metamorphosing juveniles were available (Rossman, Ford, and Seigel 1996). At a lake in California, metamorphosing *Bufo boreas* were the principal summer prey of *T. elegans,* but the snakes shifted to feeding on fish and leeches in years when toad reproduction failed because of dry conditions (Kephart and Arnold 1982; fig. 14.7).

S. Arnold (1981c) suggested that predation on frogs and fish in riparian habitats represents the ancestral condition for all natricine snakes. Some species have subsequently diverged to specialize on other prey, including slugs and earthworms. A genetic basis for variation in slug-feeding tendencies has been demonstrated in populations of *T. elegans,* with marked geographic differentiation between slug-eating coastal populations and slug-avoiding inland populations (Arnold 1981a, b, c). There also is heritable genetic variation in responses to amphibian prey, but little geographic differentiation. Anuran tadpoles and juveniles elicit strong chemoreceptive and feeding responses from most populations of garter snakes, even in species that feed almost exclusively on earthworms (W. Fox 1952; Dix 1968; Burghardt 1969; Arnold 1978, 1981a). *Thamnophis elegans* also showed moderately strong chemosensory responses to plethodontid salamanders (Arnold 1981c), whereas Burghardt (1969, 1970) obtained weak responses to plethodontids from five species (*T. butleri, T. elegans, T. ordinoides, T. radix,* and *T. sirtalis*). *Thamnophis sirtalis* from the mountains of Virginia and Kentucky apparently feed more frequently on salamanders than on frogs (Uhler, Cottam, and Clarke 1939; Barbour 1950). Their chemosensory responses to salamanders would be worth investigating. Other small natricine snakes (*Virginia* and *Storeria*) have evolved more terrestrial habits. They feed mostly on soft-bodied invertebrates, but sometimes take small frogs, salamanders, and amphibian eggs as well (Wright and Wright 1957; Ernst and Barbour 1989b).

The desert night snake (*Hypsiglena torquata*) of the southwestern United States is a member of the subfamily Dipsadinae, which is largely a Central American clade (sometimes referred to as "Central American Xenodontines"). Although it feeds mostly on lizards, this species also feeds on both anurans (*Bufo, Pseudacris, Scaphiopus*) and salamanders (*Batrachoseps;* Rodriguez-Robles, Mulcahy, and Greene 1999). This snake is closely related to a group of Neotropical snakes that mostly eat frogs. Also in the same subfamily are snakes of the genus *Radinaea,* which eat frogs, salamanders, and their eggs, as well as lizards (Myers 1974). Most species are found from Mexico to northern South America, but one

**Table 14.3** Food habits of garter snakes (*Thamnophis*), showing percentage of diet composed of amphibians and other prey

| Species | Locality (source no.) | AM | FI | EW | LE | SL | OT |
|---|---|---|---|---|---|---|---|
| *T. brachystoma* | Pennsylvania (7) | 0 | 0 | 100 | 0 | 0 | 0 |
| *T. butleri* | Michigan (5) | 0 | 0 | 83 | 10 | 0 | 0 |
| *T. cyrtopsis* | New Mexico (9) | 100 | 0 | 0 | 0 | 0 | 0 |
| | Texas (6) | 99 | 0 | 0 | 0 | 0 | 1 |
| *T. couchi couchi* | California (2) | 74 | 26 | 0 | 0 | 0 | 1 |
| *T. couchi hammondi* | California (2) | 89 | 7 | 2 | 0 | 0 | 2 |
| *T. couchi hydrophila* | California (2) | 61 | 37 | 0 | 0 | 0 | 2 |
| *T. elegans elegans* | California (11)[a] | 5 | 63 | 1 | 1 | 0 | 30 |
| *T. elegans terrestris* | California (13) | 3 | 0 | 1 | 0 | 95 | 1 |
| | British Columbia (12)[b] | 1 | 27 | 3 | 0 | 39 | 30 |
| | New Mexico (9) | 78 | 18 | 4 | 0 | 0 | 0 |
| *T. elegans vagrans* | California (13) | 38 | 18 | 0 | 44 | 0 | 0 |
| | California (14) | 33 | 26 | 0 | 41 | 0 | 0 |
| *T. eques* | Mexico (17) | 9 | 11 | 40 | 38 | 0 | 2 |
| *T. melanogaster* | Mexico (16) | 17 | 57 | 3 | 20 | 0 | 2 |
| *T. proximus* | Texas (10) | 94 | 2 | 0 | 0 | 0 | 4 |
| *T. rufipunctatus* | New Mexico (9) | 0 | 100 | 0 | 0 | 0 | 0 |
| *T. sauritus* | Michigan (5) | 90 | 2 | 0 | 0 | 0 | 4 |
| *T. sirtalis* | California (2) | 65 | 3 | 24 | 4 | 1 | 3 |
| *T. sirtalis fitchi* | California (11) | 56 | 36 | 0 | 3 | 0 | 4 |
| | California (14) | 69 | 29 | 0 | 2 | 0 | 0 |
| | California (15) | 90 | 2 | 0 | 8 | 0 | 0 |
| *T. sirtalis parietalis* | Kansas (8) | 93 | 0 | 3 | 0 | 0 | 4 |
| *T. sirtalis pickeringi* | British Columbia (12)[b] | 53 | 0 | 31 | 1 | 0 | 15 |
| *T. sirtalis sirtalis* | Virginia (1) | 56 | 0 | 37 | 0 | 0 | 17 |
| | Michigan (3) | 33 | 5 | 45 | 1 | 0 | 16 |
| | Michigan (5) | 15 | 1 | 80 | 1 | 0 | 3 |
| | New York (4) | 28 | 2 | 57 | 0 | 0 | 13 |
| *T. validus* | Baja California (uplands) (18) | 100 | 0 | 0 | 0 | 0 | 0 |
| | Mainland Mexico (18) | 38 | 60 | 0 | 2 | 0 | 0 |

*Sources:* (1) Uhler, Cottam, and Clarke 1939; (2) Fitch 1941; (3) Lagler and Salyer 1945b; (4) Hamilton 1951; (5) Carpenter 1952; (6) Fouquette 1954; (7) Asplund 1963; (8) Fitch 1965; (9) Fleharty 1967; (10) Clark 1974; (11) White and Kolb 1974; (12) Gregory 1978; (13) Arnold 1981a; (14) Kephart 1982; (15) Kephart and Arnold 1982; (16) Drummond 1983; (17) García and Drummond 1988; (18) De Queiroz, Henke, and Smith 2001.

*Notes:* Numbers show percentage of total prey items in stomachs unless otherwise specified. Data are from Appendix I of Drummond (1983) with additions. Prey categories: AM = amphibians; FI = fish; EW = earthworms; LE = leeches; SL = slugs; OT = other.

[a]Percentage of prey volume.

[b]Percentage of stomachs containing prey item.

species occurs in the southeastern United States (*R. flavilata*) and is known to eat *Acris, Hyla,* and other frogs (Myers 1967). Several other North American snakes that sometimes eat amphibians appear to be related to Neotropical lineages in which frog eating is common, but their precise phylogenetic relationships are unclear (Greene 1997). These include ring-necked snakes (*Diadophis punctatus*), which often feed on small plethodontid salamanders (Uhler, Cottam, and Clarke 1939; Barbour 1950; Fitch 1975; Blanchard, Gilreath, and Blanchard 1979; Cupp 1994; Lancaster and Wise 1996) and small frogs (Myers 1965). There even is one report of this snake feeding on the eggs of a plethodontid sala-

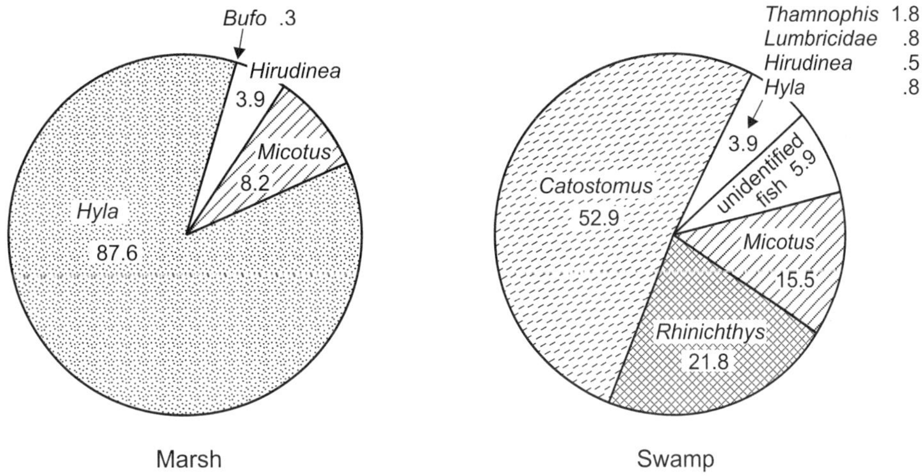


**Fig. 14.6.** Comparison of proportions of prey in the stomachs of garter snakes from adjacent marsh and stream habitats in California. Garter snakes (mostly *Thamnophis sirtalis*) from a marsh habitat ate mostly frogs (*Pseudacris regilla*), whereas *T. sirtalis* and *T. elegans* from a stream habitat ate mostly fish (*Catostomus* and *Rhinichthys*), but took some frogs as well. After White and Kolb (1974).

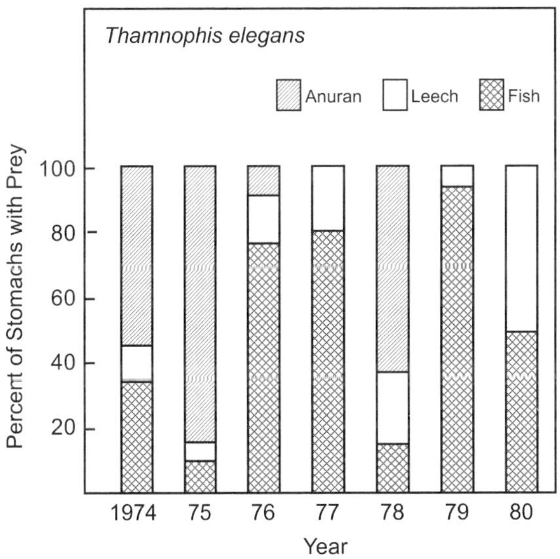


**Fig. 14.7.** Annual variation in proportion of major prey items in stomachs of garter snakes from a pond in California. In years when anurans (*Bufo boreas*) were scarce (1977, 1979, 1980), the snakes switched to eating mostly fish. After Kephart and Arnold (1982).

mander, *Plethodon albagula* (Milanovich, Trauth, and Mary 2005). Another is the mud snake (*Farancia abacura*), which specializes on amphiumas, sirens, and other aquatic salamanders (Wright and Wright 1957; Ernst and Barbour 1989b).

North American hognose snakes (*Heterodon*) generally are considered to be amphibian specialists (fig. 14.4 B). Many textbooks state that hognose snakes subsist almost entirely on toads. In some areas this is true, with over 95% of the diet being composed of toads (Edgren 1955; Platt 1969). However, hognose snakes do vary their diets in response to changes in prey abundance. For example, at one site in Kansas where toads were scarce but *Rana pipiens* was common, 66% of prey items in stomachs of *H. platyrhinos* were frogs, 28% were toads, and 3% were salamanders (*Ambystoma tigrinum;* Platt 1969). The western hognose snake (*H. nasicus*) is less of a toad specialist than the eastern species, and eats substantial numbers of lizards and some small mammals (Platt 1969). The tropical analogs to hognose snakes are the genera *Xenodon* and *Waglerophis,* which specialize on toads and other large-bodied anurans (Duellman 1978; Vitt 1983; Martins, Sazima, and Egler 1993; Greene 1997). They belong to a largely South American clade (the subfamily Xenodontinae). They resemble hognose snakes morphologically (H. M. Smith and White 1955; Kardong 1979), either because of common ancestry or convergent evolution on a common prey type (Cadle 1984, 1987). These snakes also resemble poisonous pit vipers in morphology and color pattern and may be Batesian mimics of these dangerous snakes (Greene 1997).

Other Neotropical colubrids that frequently prey on anurans include nocturnal tree snakes such as *Imantodes, Leptodeira,* and *Thamnodynastes,* which often capture frogs in bushes and trees near breeding ponds (Duellman 1958, 1963, 1978; Test, Sexton, and Heatwole 1966; Kluge 1981; Vitt and Vangilder 1983; Vitt 1996). *Leptodeira* also eats the arboreal eggs of *Agalychnis, Phyllomedusa* and other frogs (Warkentin 1995; Greene, 1997). The first two genera are members of the "Central American Xenodontine" clade (subfamily Dipsadinae), whereas *Thamnodynastes* belongs to the "South American Xenodontine" clade (subfamily Xenodontinae; Cadle and Greene 1993). Despite their names, both of these clades are widely distributed throughout Central and South America, and the Central American clade extends into Mexico and the United States.

Some diurnal colubrine tree and vine snakes, including *Oxybelis, Xenobelis,* and *Leptophis,* also prey on frogs, usually seeking them out in diurnal retreat sites (Duellman 1963, 1978; Henderson, Nickerson, and Hoevers 1977; Lopez, Giraudo, and Arzmendia 2003; Muniz and Silva 2005). Terrestrial Neotropical snakes that feed mainly on anurans or their eggs include *Chironius, Dendrophidion, Drymobius, Simophis* (Colubrinae), *Coniophanes, Pliocercus, Rhadinaea, Urotheca* (Dipsadinae), *Liophis,* and *Lystrophis* (Xenodontinae; Duellman 1963, 1978; Myers 1974; Henderson and Hoevers 1977; Vitt 1983; Seib 1984, 1985; Greene 1988; Michaud and Dixon 1989; Sazima et al. 1992; Cadle and Greene 1993; Martins et al. 1993; Bizerra et al. 1994; Mendes Castanho 1996; Cassimiro and Bertoluci 2003; Da Silva, França e Souza, Vaz Silva, and Rodrigues Silva 2003; M. Lopez, Giraudo, and Arzmendia et al. 2003; Solé and Kwet 2003; Manzanilla et al. 2005). A number of other genera include frogs as at least a minor part of their diets, including *Drymoluber, Mastigodryas, Scaphiodontophis* (Colubrinae), *Amastridium, Nothopsis* (Dipsadinae), *Oxyrhophus, Helicops, Hydronastes, Philodryas, Tropidodryas* (Xenodontinae), and *Sordellina* (lineage uncertain; Cadle and Greene 1993; Da Silva et al. 2003a; M. Lopez 2003; Hartmann and Marques 2005; Menin 2005). Nearly all of the genera that have amphibians as a minor component of the diet feed mostly on other ectotherms, especially lizards and sometimes fish.

In the West Indies, lizards (*Anolis*) and frogs (*Eleutherodactylus*) are the most abundant vertebrate prey for snakes, and as in some South and Central American snake communities, there is an association between predation on lizards and predation on frogs (Henderson and Sajdak 1996; Henderson and Powell 1999). Lizards make up about 75% of the prey of West Indian colubrids, frogs about 21% (Henderson 1984; Henderson and Crother 1989; Henderson and Sajdak 1996). *Alsophis, Chironius, Darlingtonia, Ialtris, Liophis, Mastigodryas,* and *Uromacer* eat frogs, but there are surprisingly few frog specialists. However, on Hispaniola, *Darlingtonia haetina* feeds almost exclusively on *Eleutherodactylus* (Henderson and Schwartz 1986), and *Chironius* is a frog specialist as well (Henderson and Powell 1999). *Uromacer catesbyi* feeds primarily on the hylid frog *Osteopilus dominicensis* (70% of diet), but eats *Anolis* as well (Henderson, Schwartz, and Noeske-Hallin 1987; Henderson et al. 1988). The most widespread genus of colubrid snakes in the West Indies is the genus *Alsophis.* Species throughout the range of this genus feed mainly on *Anolis* lizards, with frogs generally being the second most abundant type of prey (Henderson and Sajdak 1996). The low number of snake species that feed mostly on frogs may be due in part to the distribution of frogs and snakes. *Eleutherodactylus* are most abundant at mid-elevations, whereas snake diversity peaks at lower elevations (Henderson and Crother 1989).

Diets of colubrid snakes in the Old World Tropics are not as well known as those in the New World, but there are many species that eat substantial numbers of anurans. Colubrine and natricine colubrids are found in both the temperate and tropical parts of Asia, and as previously mentioned, a number of species, especially natricines, eat frogs found in rice paddies and natural ponds, or along streams. Some colubrines, such as the mangrove tree snake (*Boiga dendrophila*) include frogs in a varied diet of vertebrate prey (Greene 1997). Several other subfamilies of colubrids are found mostly in southern and Southeast Asia, the Indonesian Archipelago, and northern Australia. These include homolopsine water snakes. These are mostly fish eaters, but some species also eat frogs, tadpoles, and even frog eggs. A study of one species, *Enhydris plumbea,* in rice paddies in Borneo revealed that frogs were the most common prey (Voris and Karns 1996). Also found in Southeast Asia is the small subfamily Xenoderminae, which includes the Javan mudsnake (*Xenodermis javanicus*), a fossorial species that feeds on frogs (Greene 1997).

Many African colubrids also feed on anurans, but the diets of most African snakes have not been studied in detail. A study of the amphibian and reptile community of the Lamto Savanna in West Africa showed that more than a third of the snake species fed mostly on frogs, especially during wet season, when frogs were breeding (Barbault 1974b, 1977). These snakes were sufficiently abundant to have a major impact on anuran populations (Barbault 1987, 1991). Most of the frog-eating snakes were colubrids, but some pythons, vipers, and elapids fed on frogs as well. Frog-eating colubrids included representatives of several distinct clades, including colubrines (*Chlorophis, Philothamnus, Gonionotophis, Dromophis, Crotaphopeltis*), natricines (*Natriceteres*), and psammophiines (*Rhamphiophis*), found in a variety of habitats (terrestrial, arboreal, fossorial). In an assemblage of snakes in a swamp rainforest in Nigeria, there are three semiaquatic genera of colubrids (*Afronatrix, Grayia, Natriciteres*) that eat mostly frogs, a terrestrial snake (*Crotaphytopeltis*) that eats some toads, and two species of filesnakes (*Mehelya*) that eat juvenile toads (Luiselli, Akani, and Capizzi 1998; Akani and Luiselli 2001). Toads also have been found in the stomachs of several species of *Mehelya* from southern Africa, and in two species of a closely related genus, *Gonionotophis* (Shine et al. 1996).

### Viperid Snakes

Most viperid snakes are adapted for feeding on birds and mammals as adults (Pough and Groves 1983), but their broad heads and large gapes make them well suited for preying on large-bodied amphibians as well. Juvenile viperids often feed on ectothermic prey, including invertebrates, lizards, and frogs. As adults, they usually switch to mostly endothermic prey (Sexton 1956–57; Dixon and Soini 1986; Mushinsky 1987; Sazima 1992; H. Greene 1997; Daltry, Wuster, and Thorpe 1998; Nogueira, Sawaya, and Martins 2003). Juveniles often have conspicuous yellow or pink tips on their tails that are used as lures to attract frogs or lizards (W. Neill 1960; Greene and Campbell 1972; Heatwole and Davison 1976; Schuett 1984; Carpenter and Gillingham 1990; Sazima 1991, 1992; Greene 1992; J. Murphy 1997).

The North American cottonmouth (*Agkistrodon piscivorus*) preys mainly on ectothermic vertebrates, and amphibians often are a major component of the diet. In Kentucky, sirens and ranid frogs made up 75% of the diet (Barbour 1956). In southern Illinois, amphibians were about 26% of the diet, but up to 40% in midsummer (Klimstra 1959a). Amphibians made up 20–30% of the food in stomachs of cottonmouths from several southern states (Burkett 1966; Himes 2003, 2004). Copperheads (*A. contortrix*) are found in drier habitats, but they sometimes eat frogs, especially when young (Ernst and Barbour 1989b; Fitch 1999). Bullfrogs exhibit limited resistance to copperhead venom, and greater resistance to copperhead than to cottonmouth venom, probably due to the greater toxicity of the latter (Heatwole, Poran, and King 1999). It is not known, however, whether this resistance is sufficient to allow frogs to escape predation after being struck by one of these snakes. Some Asian crotaline pitvipers, especially *Gloydius* and *Trimeresurus*, feed mostly on frogs (Pope 1935; Mao 1970; Mori and Moriguchi 1988; Kuzmin 1999). Another crotaline, the Malayan pitviper (*Calloselasma rhodostoma*), has a varied diet of mammals, birds, lizards, frogs, and invertebrates, with frogs being eaten most frequently by small individuals in most populations (Daltry, Wuster, and Thorpe 1998). In Java, frogs made up about 63% of the prey of small individuals, whereas mammals made up 75% of the prey of large snakes. Many Neotropical viperids, such as *Bothrops*, *Botriechis*, and *Porthidium*, feed on anurans at least occasionally, with frogs being most important for juvenile snakes (Sexton 1956–57; Duellman 1978; J. Campbell and Solorzano 1992; McCoy and Censky 1992; Sazima 1992; Martins and Gordo 1993; Greene 1997; J. Murphy 1997; Nogueira, Sawaya, and Martins 2003). Several rattlesnakes eat frogs occasionally, but they probably are important prey only for *Sistrurus*, which are frequently found in bogs and swamps (Wright and Wright 1957; Klauber 1972; Ernst 1992). Frogs were the most important prey of a population of *Sistrurus miliarius* in Florida (Roth, May, and Farrell 1999), but *Sistrurus catenatus* in a drier habitat in Wisconsin did not eat frogs (Keenlyne and Beer 1973). The Florida snakes were attracted to chemical cues from frogs in field experiments.

Predation on amphibians by most Old World viperine snakes is not well documented. Most species appear to feed primarily on endothermic prey, but there are exceptions. The African night adders (*Causus*) feed mainly on anurans (Barbault 1974b; Greene 1992, 1997; Luiselli et al. 2004), and anurans are a minor component of the diet of the puff adder (*Bitis arietans;* Barbault 1974b). Both juvenile and adult Eurasian vipers (*Vipera aspis*, *V. berus*, *V. lebetina*, and *V. ursinii*) feed on anurans (Saint Girons 1980; Monney 1993; Kuzmin 1999). Asian saw-scaled vipers (*Echis carinatus*) also prey on frogs (*Rana ridibunda*) and toads (*Bufo viridis;* Kuzmin 1999).

### Elapid and Atractaspid Snakes

In Australia, the snake fauna is dominated by poisonous elapids, and these have radiated into most of the niches occupied by colubrids and viperids on other continents. Since much of Australia is arid, lizards are the predominant prey of many species, but frogs often are the next most important prey (table 14.4). *Denisonia devisi*, *D. maculata*, *Hemiaspis dameli*, *Notechis scutatus*, and *Pseudechis porphyriacus* are

**Table 14.4** The importance of frogs in the diets of some Australian elapid snakes

| Species (source no.) | Total prey (%) | Frogs (%) | Other major prey |
|---|---|---|---|
| *Acanthophis antarcticus* (4) | 34 | 6 | Lizards, mammals |
| *Austrelaps labialis* (10) | 53 | 17 | Lizards |
| *A. ramsayi* (10) | 59 | 17 | Lizards |
| *A. superbus* (10) | 98 | 34 | Lizards |
| *A. superbus* (1) | 43 | 61 | Lizards |
| *Caciogus squamulosus* (3) | 69 | 6 | Lizards |
| *Demansia atra* (2) | 22 | 27 | Lizards |
| *D. psammophis* (2) | 75 | 7 | Lizards |
| *D. devisi* (8) | 42 | 88 | Lizards |
| *D. maculata* (8) | 18 | 95 | |
| *Drysdalia coronata* (5) | 32 | 53 | Lizards |
| *Ecliopsis curta* (6) | 54 | 31 | Lizards |
| *Elapognathus minor* (9) | 6 | 66 | Lizards |
| *Hemiapsis damelii* (11) | 21 | 95 | |
| *H. signata* (11) | 105 | 23 | Lizards |
| *Notechis scutatus* (1) | 90 | 92 | Nestling birds |
| *Pseudechis porphyriacus* (1) | 296 | 83 | Lizards |
| *Pseudonaja textilis* (1) | 32 | 6 | Lizards, mice |
| *Tropidechis carinatus* (7) | 27 | 41 | Mammals |

*Sources:* (1) Shine 1977; (2) Shine 1980b; (3) Shine 1980a; (4) Shine 1980c; (5) Shine 1981; (6) Shine 1982, (7) Shine and Charles 1982; (8) Shine 1983; (9) Shine 1986b; (10) Shine 1987c; (11) Shine 1987b.

*Notes:* Numbers show percentage of total prey items in stomachs. Most data are from preserved museum specimens except data from Shine (1977), which are from freshly collected specimens.

primarily frog predators. However, diets vary with changes in prey availability. For example, *Notechis scutatus* preys on both frogs and lizards at several mainland sites, but feeds mainly on small mammals, nestling birds, and lizards on offshore islands where frogs are scarce (Shine 1987a).

Some elapid snakes in other parts of the world eat amphibians as well. In a savanna community in West Africa, African gartersnakes (*Elapsoidea sundevalli*) fed exclusively on frogs, and frogs were a minor item in the diet of cobras (*Naja nigricollis;* Barbault 1974b). Both *Naja nigricollis* and *N. melanoleuca* in Nigeria included frogs in a varied diet of vertebrates in both swamp forest habitat and in human-inhabited areas (Luiselli, Akani, and Capizzi 1998; Luiselli, Angelici, and Akani 2002). In some parts of central Asia, anurans are important prey of another cobra, *N. oxiana* (Kuzmin 1999). Other Old World elapids that eat some frogs and toads include desert blacksnakes (*Walterinessia*) of the Middle East and tropical Asian kraits (*Bungarus;* Greene 1997). Most New World coral snakes (*Micruroides, Micrurus*) are predators on elongate vertebrates, especially snakes, lizards, and amphisbaenas. They apparently do not eat amphibians very frequently (Greene 1984, 1997), but the diets of most species are not well studied.

Venomous snakes in the genus *Atractaspis* are closely related to elapids, but are generally placed in a separate family. These are burrowing snakes that feed mostly on other elongate, fossorial vertebrates, including snakes and amphisbaenians, as well as small mammals such as shrews and rodents. There is one report of *Atractaspis aterrima* from Tanzania having consumed a caecilian (*Scolecomorphus kirkii*), but it is not known whether caecilians are an important component of the diet (Gower et al. 2004).

## Predation by Birds

Hundreds of species of birds are potential predators of amphibians, but it would be impossible to enumerate all of these without a major survey of the ornithological literature (for lists of avian predators and bibliographies, see Kabisch and Belter 1968; Cook 1987; Shea 1987; Ross 1989; J. Martín and López 1990; Kuzmin 1999). In Europe, more than 90 species of birds have been reported to eat frogs, eggs, and tadpoles (Kabisch and Belter 1968; J. Martín and López 1990). At least 30 species of Australian birds are known to eat frogs or tadpoles, but this probably is an underestimate for this region (Shea 1987). On Barro Colorado Island in Panama, a relatively small area of tropical forest, 16 species of birds were found to have eaten frogs (all members of the genus *Eleutherodactylus*), but generally these were incidental items in diets composed mostly of insects (Poulin et al. 2001).

Birds in at least 15 nonpasserine orders prey on amphibians. These include cassowaries (Casuariiformes), loons (Gaviiformes), grebes (Podicipediformes), cormorants and darters (Pelicaniformes), herons, egrets, bitterns, ibis, storks, and spoonbills (Ciconiiformes), ducks (Anseriformes), sandpipers, gulls, and terns (Charadriiformes), cranes, rails, coots, and bustards (Gruiformes), quails, pheasants, and turkeys (Galliformes), kingfishers, kookaburras, rollers, bee-eaters, motmots, and todies (Coraciiformes), woodpeckers, barbets and puffbirds (Piciformes), cuckoos (Cuculiformes), frogmouths (Caprimulgiformes), hawks, kites, and eagles (Falconiformes), and owls (Strigiformes; Kabisch and Belter 1968; Perrins and Middleton 1985; W. Cook 1987; Shea 1987; J. Martín and López 1990).

Many families of passerine birds (order Passeriformes) include species that sometimes eat amphibians: Atrichornithidae (scrub-birds), Campephagidae (cuckoo-shrikes), Cinclidae (dippers), Corvidae (crows, ravens, jays, magpies), Cracticidae (Australian magpies, butcherbirds, and currawongs), Dendrocolaptidae (woodcreepers), Emberizidae (tanagers), Formicariidae (ant birds), Furnariidae (ovenbirds), Icteridae (American blackbirds), Laniidae (shrikes), Mimidae (catbirds and thrashers), Motacillidae (wagtails), Muscicapidae (thrushes), Orthonychidae (whipbirds), Paridae (tits), Paruliidae (wood warblers), Ploceidae (weavers), Sturnidae (starlings), Troglodytidae (wrens), Tyranidae (kingbirds, kiskadees, and tyrant flycatchers), and Vireonidae (vireos). For most of these groups, the importance of amphibians in the diet is unknown, as is their impact on amphibian populations. Groups of birds that are likely to be important predators of amphibians are summarized in the following.

### Wading and Fishing Birds

Among the nonpasserine birds, herons, bitterns, and storks probably are major predators of amphibians. Hancock and Kushlan (1984) listed 60 species of herons and bitterns (family Ardeidae) around the world; at least 41 include amphibians in the diet. Amphibians are of minor importance for some species, but are a staple food for others. Jenni (1969) found that anurans were the principal prey of Little Blue Herons (*Florida caerulea*) in Florida, comprising 54% of the food delivered to nestlings. Frogs also were a significant part of the diet of cattle egrets (*Bubulcus ibis;* 32%), but were of minor importance for snowy egrets (*Leucophoyx thula;* 5%) and Louisiana herons (*Hydranassa tricolor;* 0.2%) living in the same area. The importance of amphibians in heron diets varies geographically within the same species, no doubt reflecting the habitats used by the birds and the availability of prey. For example, anurans comprised an average of 34% of the food delivered to cattle egret chicks at various sites in Florida (Jenni 1973), with some variation in the species taken most frequently. Cattle egrets in South Africa were much more insectivorous, with only 13% of the diet composed of amphibians (Siegfried 1971), but in another study in South Africa, frogs (juvenile *Pyxicephalus adspersus*) made up a large proportion of the diet (Kopij 2003). In the same region, frogs (*Rana* and *Xenopus*) were a major part of the diets of sacred ibis (*Threskiornis aethiopicus*) and African spoonbill (*Platalea alba*) chicks (Kopij, Kok, and Roos 1996; Kopij 1997).

Mukherjee (1969, 1971a, b, 1974) analyzed stomach contents of water birds from India and found that 12 of 16 species ate some amphibians. These included four species of herons, two egrets, a bittern, a stork, a grebe, a cormorant, an anhinga, and a coot. Amphibians made up 2–28% of the diets of these birds, with large ranid tadpoles being the principal amphibian prey. Amphibians were the most important prey of two species, the chestnut bittern (*Ixobrychus cinnamomeus;* 27% of the diet) and the night heron (*Nycticorax nycticorax;* 28%). In northern Italy, spadefoot toads (*Pelobates fuscus*) made up 24% of the diet of the same species of night heron during the spring breeding season and were important food for nestlings (Vasvari 1935/38; cited in Nöllert 1984). In the Ukraine, fire-bellied toads (*Bombina bombina*) accounted for up to 25% of the diet of night herons (Kuzmin 1999). Various other water birds feed on almost all species of European amphibians, with frogs comprising up to 40% of the diet in some herons (Vasvari 1927/28, 1929/30, 1935/38; Kabisch and Belter 1968; J. Martín and López 1990; Kuzmin 1999). In New Zealand, the Australian bittern (*Botaurus poiciloptilus*) and the White-faced heron (*Egretta novaehollandiae*) feed primarily on the introduced Australian treefrog, *Litoria aurea* (Hancock and Kushlan 1984). The hammerhead (*Scopus umbretta*), a wading bird of African savannas, frequently preys on frogs, including *Xenopus* that the birds captured while skimming along the surface of a lake (Wager 1965). This species also eats the schooling tadpoles of a toad, *Schismaderma carens*.

Amphibians probably are of minor importance in the diets of most nonciconiiform water birds. Mukherjee (1969, 1974) found that amphibians always made up less than 10% of the diet of aquatic birds such as coots, anhingas, cormorants, and grebes. The moorhen (*Gallinula chloropus*), a species of rail, preys on the eggs of *Rana temporaria*, sometimes eliminating most of the eggs from a pond (Savage 1961). Purple gallinules (*Porphyrula martinica*) fed on eggs of *Agalychnis spurelli* laid on vegetation over a pond in Costa Rica (N. Scott and Starrett 1974). *Agalychnis* tadpoles, which form large surface-feeding aggregations (see chapter 12), are preyed upon by green kingfishers (*Chloroceryle americana;* Wassersug 1971; N. Scott and Starrett 1974). Belted Kingfishers (*Megaceryle alcyon*) eat *Rana catesbeiana* tadpoles (Terres 1968), and the Australian azure kingfisher (*Alcedo azurea*) eats tadpoles of the myobatrachid frog *Mixophyes fasciolatus* (Bolitho and Retallick 1996). Gulls, terns, and sandpipers are occasional predators on frogs, tadpoles, and newts (Creutz 1963; Kabisch and Belter 1968; Schardier and Jackson 1982; Beebee 1983; Stangel 1983; J. Martín and López 1990; Sinsch 1998; M. Jones, Goetti, and Livo 1999).

### Ducks and Swans

Several species of swans and ducks prey on tadpoles and small frogs, but in most cases, the importance of amphibians in the diet is unknown. Potential predators in the Northern Hemisphere include mute swans (*Cygnus olor*), mallard (*Anas platyrhynchos*), black duck (*A. rubripes*), gadwall (*A. strepera*), pintail (*A. acuta*), shoveler (*A. clypeata*), red-crested pochard (*Netta rufina*), pochard (*Aythya ferina*), widgeon (*Mareca penelope*), and several species of mergansers (*Mergus;* Kabisch and Belter 1968; Bellrose 1976; J. Martín and López 1990). Widgeon (*Mareca penelope*) prey on *Bufo calamita* in Britain (Beebee 1983), and mallard (*Anas platyrhynchos*) eat frogs, toads, tadpoles, and even spadefoot toad eggs (Kabisch and Belter 1968; Mueller 1980; Steiof and Tietje 1985; Mjelstad and Sactersdal 1989; Kwet 1996a; Sinsch 1998; Jones, Goetti, and Livo 1999; M. Hayes and Rombough 2004). R. Savage (1961) attributed a decline in populations of *Rana temporaria* in several ponds in Britain to mallard predation, but provided no direct evidence. Similarly, Martof (1956a) asserted that mallard and black ducks (*Anas rubripes*) "feed avidly on tadpoles and frogs and can easily deplete the frog populations of small ponds." The hooded merganser (*Mergus cucullatus*) is sometimes called the "frog duck" because of its propensity for preying on anurans (A. Wright 1920). Although mergansers eat mostly fish, they also feed on aggregations of hibernating leopard frogs (*Rana pipiens;* Breckenridge 1944).

### Hawks, Kites, and Eagles

Various species of raptors feed on amphibians, but few rely on them as a major source of food. One exception is an Asian hawk, *Accipiter soloensis,* which feeds on frogs in rice fields and is sometimes called the grey frog hawk (L. Brown and Amadon 1968). Other raptors that sometimes eat frogs and toads include many species of *Accipter* and *Falco,* kites (*Chondrohierax, Elanoides, Elanus, Ictinia, Milvus*), harriers (*Circus*), eagles (*Aquila*), bald eagles (*Haliaeetus leucocephalus*), many hawks of the genus *Buteo,* and the black hawk (*Buteogallus anthracinus*), savanna hawk (*Heterospizias meridionalis*), and caracara (*Polyborus plancus*) of the Neotropics (L. Brown and Amadon 1968; Kabisch and Belter 1968; Ross 1989; Franca e Souza et al. 2003; Machado and Galdino 2005). Another Neotropical species, the crane hawk (*Geranospiza caerulescens*) has very long legs that it uses to extract tree frogs from leaf axils of bromeliads (Bokermann 1978). In Guatemala, frogs were a relatively minor part of the diet of this species (16% of prey items) and were eaten mainly in wet weather when the frogs are particularly active (Sutter et al. 2001).

The importance of amphibians in the diets of raptors varies considerably. In the Lamto Savanna of the Ivory Coast, a region with dense populations of anurans during the wet season (Barbault 1972), frogs formed only a small proportion of the diets of most raptors, with only the black kite (*Milvus migrans*) being a significant predator (Thiollay 1976). Tree frogs (*Hyla cinerea* and *H. squirella*) were the most frequent prey delivered to nestling swallow-tail kites (*Elanoides forficatus*) in Florida (N. Snyder 1974), and frogs also are eaten by hook-billed kites (*Chondrohierax uncinatus*), black-shouldered kites (*Elanus caeruleus*), and Mississippi kites (*Ictinia mississippiensis*) (D. Ross 1989).

North American red-shouldered hawks (*Buteo lineatus*) and broad-winged hawks (*B. platypterus*) take considerable numbers of amphibians, which may comprise more than 20% of the diet in spring and summer (S. Ernst 1945; R. Stewart 1949; Craighead and Craighead 1969; Mosher and Matray 1974; Rosenfield, Gratson, and Carson 1984; D. Ross 1989). Broad-winged hawks eat substantial numbers of toads (*Bufo americanus*), which generally are well protected by toxic skin secretions. In these species, and in other raptors, the importance of amphibians in the diet may be underestimated by analyses of regurgitated pellets. For example, in Massachusetts, anurans were only about 0.5% of the items identified in pellets of red-shouldered hawks, but nearly 20% of the prey carried to the nest (Portnoy and Dodge 1979). Estimates of numbers of anurans consumed by broad-winged hawks derived from direct observations also are higher than those derived from pellet analysis (D. Ross 1989).

Red-tailed hawks (*Buteo jamaicensis*) and marsh hawks (*Circus cyaneus*) probably are important predators of amphibians in some areas (Bent 1961). In Puerto Rico, 42% of the prey items brought to nestlings of red-tailed hawks were frogs (*Eleutherodactylus coqui;* Santana and Temple 1988), but frogs were a minor item in the diet in various parts of North America (Ross 1989). Other North American hawks that eat significant numbers of frogs include black hawks (*Buteogallus anthracinus*), white-tailed hawks (*Buteo albicaudatus*), zone-tailed hawks (*B. albonotatus*), ferruginous hawks (*B. regalis*), and American kestrels (*Falco sparverius;* Ross 1989). In a study of American kestrels in California, the birds captured mostly mammals in a relatively dry, cold winter, but shifted to eating mostly frogs and insects in a warm, wet winter when mammals were scarce (Collopy 1977; Collopy and Koplin 1983). Bald eagles (*Haliaeetus leucocephalus*) normally eat mostly fish, but there are observations of eagles eating pickerel frogs (*Rana palustris*) in Maine (Applegate 1990) and sirens (*Siren lacertina*) in Florida (McEwan and Hirth 1980). Another fish-eating species, the osprey (*Pandion haliaetus*), occasionally takes frogs as well (Ross 1989).

### Owls

One might expect owls to be important predators of amphibians because of their nocturnal habits. However, most species feed primarily on small mammals and birds, and amphibians usually are a minor component of the diet. Amphibians are eaten occasionally by North American barred owls (*Strix varia*), screech owls (*Otus asio*), and great horned owls (*Bubo virginianus;* Hamerstrom and Hamerstrom 1951; Bent 1961; Rising and Schueler 1980; Ross 1989; Beane 2005). Burrowing owls (*Athene cunicularia*) sometimes eat spadefoot toads (*Scaphiopus*) and other amphibians (Sperry 1941; Gleason and Craig 1979; J. Tyler 1983). Frogs are taken occasionally by barn owls (*Tyto alba*), especially those that hunt in marshy areas (Calderón and Collado 1976; Rey, Esteban, and Sanchiz 1994; Pailley and Pailley 1995; Pezzo and Morimando 1995), but they usually are a minor source of food (Bunn, Warburton, and Wilson 1982). However, in one collection of barn owl pellets from Surinam, Haverschmidt (1962) found the remains of 55 *Leptodactylus ocellatus,* the second most numerous type of prey. Puerto Rican screech owls (*Otus nudipes*) are considered to be important predators of frogs, especially the very abundant *Eleutherodactylus coqui* (Stewart and Woolbright 1996; Pérez-Rivera 1997). In one study, *E. coqui* comprised 40% of the prey items brought to nestlings of this species (N. Snyder, Wiley, and Kepler 1987). In some areas, frogs are important prey of European tawny owls (*Strix aluco;* Uttendörfer 1939; Schnurre 1956, 1957; Wehner 1962), and individuals may specialize on frogs in the breeding season (Betts 1927; Wendland 1984). Tawny owls and little owls (*Athene noctua*) also eat newts (Smallcombe 1934; Trimnell 1945). Other European owls that feed on amphibians at least occasionally include the eagle owl (*Bubo bubo*), long-eared owl (*Asio otus*), short-eared owl (*A. flammeus*), and scops owl (*Otus scops;* J. Martín and López 1990).

### Passerine Birds

Although passerine birds in many families prey on amphibians, none relies on them as a principal source of food. Shrikes (*Lanius*) eat frogs, toads, salamanders, and newts, impaling them on thorns along with lizards and other small vertebrates (Ross 1989; Hernández 1995; Jensen 2003). Shrikes will even consume amphibians with toxic or distasteful skin secretions, such as *Salamandra salamandra* and various species of *Triturus* and *Bufo* (Hernández 1995). Corvids and icterids probably are significant predators as well. Many species of crows, ravens, magpies, and jays include amphibians in their diets (Goodwin 1976). Hooded crows (*Corvus corone*) sometimes catch frogs (*Rana tem-*

*poraria*) during the breeding season and tear them to pieces, leaving only the oviducts and eggs of the females (M. Smith 1969). Common ravens (*Corvus corax*) attack toads (*Bufo boreas*) in breeding aggregations, killing more than 20% of the toads in one chorus (Olson 1989). Beasley and Carothers (1974) observed mixed flocks of red-winged and Brewer's blackbirds (*Agelaius phoeniceus* and *Euphagus cyanocephalus*) feeding on young *Rana pipiens*. Crows, jays, and grackles have been seen attacking aggregations of *Bufo* and *Scaphiopus* tadpoles (Ideker 1976; Beiswenger 1981; Harestad 1985) and metamorphosing *Pseudacris* (Tordoff 1980). In Puerto Rico, icterids such as grackles and orioles eat frogs (*Eleutherodactylus;* Wetmore 1916; Pérez-Rivera 1997).

Several other kinds of passerine birds prey on anurans in the tropics. In Puerto Rico, predators of *Eleutherodactylus* include a kingbird (*Tyrannus caudifasciatus*), a flycatcher (*Myiarchus antillarum*), the red-legged thrush (*Turdus plumbeus*), the pearly-eyed thrasher (*Margarops fuscatus*), a vireo (*Vireo altiloquus*), and a warbler (*Dendroica adelaidae*), as well as the nonpasserine tody (*Todus mexicanus;* Wetmore 1916; Stewart and Woolbright 1996; Pérez-Rivera 1997). Red-legged thrushes also eat other frogs, such as *Leptodactylus albilabris,* and in one study, frogs and lizards comprised more than 8% of all prey items (Rolle 1963). On Barro Colorado Island, Panama, the most frequent avian predators of *Eleutherodactylus* were ant birds (family Formicariidae), woodcreepers (family Dendrocolaptidae), and a few species of flycatchers (family Tyrranidae). In contrast to the tendency of some birds to attack amphibians at aquatic breeding sites, predation on the terrestrial-breeding *Eleutherodactylus* peaked in the dry season, when frogs were most abundant in the leaf litter and arthropods were least abundant (Poulin et al. 2001). In Argentina, kiskadees (*Pitangus sulphuratus*) and woodcreepers (*Xiphocolaptes major*) have been seen eating tadpoles and frogs (Carrizo 1990; Crump and Vaira 1991), while a cuckoo (*Guira guira*) was seen eating an adult *Physalaemus fuscomaculatus* (Kokubum and Zacca 2003).

Passerine birds can be important predators of terrestrial salamanders. Plethodontid salamanders (*Plethodon cinereus* and *Desmognathus ochrophaeus*) made up 25% of the food brought to nestlings by hermit thrushes (*Hylocichla guttata;* Coker 1931). There are anecdotal reports of plethodontid salamanders being captured by American robins (*Turdus migratorius;* Brandon and Huheey 1975; A. Wilson and Simon 1985) and a tufted titmouse (*Parus bicolor;* Prescott 1985), and both American robins and catbirds (*Dumetella carolinensis*) consumed salamanders placed in outdoor trays during an experimental study (Brodie and Brodie 1980). California scrub jays *(Aphelocoma coerulescens*) have been seen eating plethodontid salamanders (*Aneides, Batrachoseps;* Rubinoff 1996; Reaser 1997), but their impact on salamander populations is unknown. A number of passerine birds, including blue jays (*Cyanocitta cristata*), common grackles (*Quiscalus quiscula*), and brown thrashers (*Toxostoma rufum*) readily eat salamanders in captivity, and have been used in experimental studies of Batesian mimicry (Huheey 1960; Brodie and Howard 1973; R. R. Howard and Brodie 1973; Hensel and Brodie 1976; Tilley, Lundrigan, and Brower 1982). Unfortunately, information on the frequency of avian predation on salamanders in the wild is not available.

## Predation by Mammals

The list of potential mammalian predators of amphibians probably is not as long as the list of avian predators, but some mammals can have a significant impact on local populations of amphibians. Many mammals in the order Carnivora have catholic diets that include amphibians, but there are few carnivores that rely on amphibians as a principal source of food. Small carnivores, such as weasels, badgers, otters, skunks, and mongooses often are generalist or opportunistic predators that readily switch between different types of prey in response to geographic, yearly, or seasonal variation in prey abundance (Erlinge 1986; Palomares 1993; Roper 1994). Studies of the diets of such predators in a single location or in a single year may suggest that the predator specializes on a particular type of prey, but in fact, the diet often reflects local availability of prey.

### Mustelids

A generalist predator that often eats large numbers of amphibians is the European polecat (*Mustela putorius*). Studies of this species in forested regions of Switzerland and Poland revealed a preponderance of anurans in the diet (Jędrzejewski, Jędrzejewska, and Szymura 1989; D. Weber 1989a), and it has been characterized as a specialist predator of anurans (D. Weber 1989b). In Poland, heavy reliance on amphibians during the winter set the polecat apart from other carnivores living in the same habitat, most of which depended heavily on mammals (Jędrzejewski, Jędrzejewska, and Szymura 1989). Yet this species exhibits considerable variation in diet composition (Lodé 1997). In Switzerland, animals that foraged near farmhouses or trash dumps ate very few anurans (D. Weber 1989b). In Poland, polecats in a mature forest fed almost exclusively on hibernating anurans in winter, which they found by traveling along streams. In very cold weather, however, they switched to feeding on small mammals, because streams were frozen over (Jędrzejewski, Jędrzejewska, and Brzeziński 1993). In Belarus, amphibians were eaten most frequently in the spring, but did not appear in the winter diet at all, when polecats relied almost entirely on small mammals (Sidorovich et al. 1998). In an area of western France that included marshes, meadows, and forest, polecats fed mostly on voles and rats that were captured in meadows and forest (Lodé 1991, 1993). In the spring and early summer, however, the animals spent more time in marshes, where they fed heavily on anurans captured in breeding aggregations (Lodé 1996, 2000). Polecats also used marshes more frequently when rodents were scarce (Lodé 1994), and they shifted their daily patterns of activity to correspond with times when anurans were most active (Lodé 1995). Calling males are most vulnerable, and polecat predation can alter sex ratios of frogs in choruses (Lodé et al. 2004).

Another generalist predator that sometimes eats substantial numbers of amphibians, and sometimes eats very few, is the European badger (*Meles meles*). Badgers eat earthworms, arthropods, amphibians, reptiles, small mammals, and a variety of plant foods, but the proportion of each type of food is highly variable (Long and Killingley 1983; Kruuk 1989; Roper 1994; Roper and Mickevicius 1995; Neal and Cheeseman 1996; Revilla and Palomares 2002). Toads (*Bufo bufo*) were the second most important prey of badgers in central France, making up nearly 30% of the diet overall, and up to 80% at some times of year (C. Henry 1983, 1984a, b). In the spring, badgers fed mostly on toads aggregated at breeding sites, and they also were an important source of food in the summer when earthworms were scarce (C. Henry 1984b). The badgers ate the toads from the ventral side, avoiding the poisonous parotoid glands (C. Henry 1984a). The numbers consumed can be significant; C. Henry (1983) estimated that one badger could eat 56 kg of toads per year, or about 1,360 individuals. Amphibians also were an important component of the diet of badgers in Spain (Ibañez and Ibañez 1980), Finland (Kauhala, Laukanen, and von Rege 1998), and Belarus (Sidorovich and Pikulik 1997). Amphibians were a minor component of the diet in Denmark (Andersen 1954), England (Shepherdson, Roper, and Lüps 1990), and most parts of the former Soviet Union (Roper and Mickevicius 1995). In several other parts of Europe, badgers seldom ate amphibians at all (Ciampalini and Lovari 1985; Roper 1994). In Taiwan, amphibians were the most important prey of the ferret badger (*Melogale moschata;* Chuang and Lee 1997).

Otters in various parts of the world are opportunistic predators of amphibians. Most otters feed primarily on fish, crabs, or crayfish, but their diet varies considerably in different habitats (Jędrzejewska, Brzeziński, and Jędrzejewski 2001). Amphibians often are a lower ranked food source that is consumed when other prey are scarce (K. Greer 1955; Ewer 1973; Rowe-Rowe 1977; Chapman and Feldhamer 1982; Chanin 1985; Mason and Macdonald 1986; Adrian and Delibes 1987; J.-M. Weber 1990; Beja 1991; Reid et al. 1994; Helder-Jose and Andrade 1997; Pardini 1998; Anoop and Hussain 2005). In studies of European otters (*Lutra lutra*) in Poland, however, frogs were as important in the diet as were fish (Brzeziński, Jędrzejewski, and Jędrzejewski 1993; Jędrzejewska, Brzeziński, and Jędrzejewski 2001). Those otters lived along small forest rivers, in contrast to the large rivers or coastal areas studied by many other workers. Otters in Belarus also ate substantial numbers of anurans, especially along fast-flowing forest streams (Sidorovich and Pikulik 1997; Sidorovich et al. 1998). Frogs are particularly vulnerable to otter predation in the winter and spring. Frogs hibernating on the bottoms of ponds, streams, and ditches constitute up to 65% of winter prey of otters in some parts of Europe (Erlinge 1967, 1969, 1972; Callejo Rey et al. 1979; Chanin 1985; Mason and Macdonald 1986; Weber 1990; Brzeziński, Jędrzejewski, and Jędrzejewski 1993; Sidorovich et al. 1998; Jędrzejewska, Brzeziński, and Jędrzejewski 2001). European otters also feed on frogs during the breeding season, and may travel some distance to find breeding aggregations (Jenkins and Harper 1980; Fairley 1984; J.-M. Weber 1990; Jędrzejewska, Brzeziński, and Jędrzejewski 2001). In Scotland, there was a strong positive correlation between the density of anurans in the habitat and the proportion of anurans in the diet of otters (J.-M. Weber 1990).

In areas where both frogs (*Rana*) and toads (*Bufo*) are abundant, European otters tend to eat mostly frogs and avoid toads (Erlinge, 1968; López-Nieves and Hernando 1984; Fairley and McCarthy 1985; Adrian and Delibes 1987; de Jongh 1988; J.-M. Weber 1990; Sidorovich and Pikulik 1997). Nevertheless, otters can be a significant source of mortality for toads in some circumstances. During the spring, a population of otters in the mountains of Spain fed mostly on toads (*Bufo bufo*) captured in breeding aggregations (Lizana and Pérez-Mellado 1990), perhaps because their preferred prey (fish) are scarce at high altitudes (Ruiz-Olmo 1998). Like the French badgers discussed previously, otters in this population have learned to manipulate toads to avoid eating parts where poison glands are concentrated. Otters in Poland also captured toads (*B. bufo*) in breeding aggregations (Jędrzejewska, Brzeziński, and Jędrzejewski 2001), and those in Belarus ate toads (*B. bufo* and *B. viridis*) when preferred prey species were scarce (Sidorovich and Pikulik 1997).

Outside of Europe, otters seem to eat fewer amphibians. North American river otters (*Lutra canadensis*), for example, feed mostly on fish and crustaceans and sometimes do not eat amphibians at all, even in habitats where amphibians are likely to be common (Melquist and Hornocker 1983; Tumlison and Karnes 1987). In some populations, amphibians are a minor component of the diet, appearing mostly during the spring breeding period of frogs (Greer 1955; Reid et al. 1994), although otters were observed feeding on Oregon spotted frogs (*Rana pretiosa*) in both late winter and late summer (M. Hayes et al. 2005). The Neotropical river otter (*Lontra longicaudis*) feeds mostly on fish and crustaceans, with amphibians comprising less than 3% of the diet (Helder-Jose and Andrade 1997; Pardini 1998). In southern Africa, the Cape clawless otter (*Aonyx capensis*) eats mostly crabs, but in some habitats, frogs are as important as fish as a secondary food item (Rowe-Rowe and Somers 1998). In the same region, spotted otters (*Lutra maculicollis*) eat mostly fish and crabs, with frogs being important secondary prey in wet areas (Rowe-Rowe 1977; Rowe-Rowe and Somers 1998), but less so in drier habitats (Somers and Purves 1996). Several species of otters in Asia feed mostly on fish, but take amphibians occasionally (Kruuk et al. 1994; Hussain and Choudury 1998; Anoop and Hussain 2005).

Mink often live in the same habitats as otters and can be significant predators on amphibians as well, but frogs are eaten less frequently than mammals, fish, and crayfish (Gerell 1968; Ewer 1973; Erlinge 1969, 1972; Jenkins and Harper 1980; Wise, Linn, and Kennedy 1981; Chapman and Feldhamer 1982). In Sweden, introduced American mink (*Mustela vison*) preyed on frogs mostly during the spring breeding season, and at times when preferred foods were scarce. There also was evidence of predation on hibernating frogs in some areas (Gerell 1968). In much of Europe, the American mink has become common, while the native European mink (*M. lutreola*) has declined. Where these two species live together in Belarus, European mink ate significantly more anurans than did American mink, with anurans being the most important prey for European mink in fall, winter, and spring (Sidorovich and Pikulik 1997; Sidorovich et al., 1998). In the Białowieża Forest of Poland, American mink ate substantial numbers of anurans, especially in winter (Jędrzejewska, Brzeziński, and Jędrzejewski 2001). The abundance of frogs in the diet of these mink appears to reflect their abundance along the small forested streams of the forest. Indeed, across much of Europe, the proportion of amphibians in the diet of American mink tends to be much higher for populations along rivers and streams than for those living in lakes and ponds.

Other mustelids, such as weasels, martens, and skunks, take amphibians opportunistically, but amphibians usually are not a major component of the diet (Ewer 1973; Corbet and Southern 1977; Chapman and Feldhamer 1982; Reig and Jędrzejewski 1988; Jędrzejewski, Jędrzejewska, and Brzeziński 1989; Redford and Eisenberg 1992; Pulliainen and Ollinmäki 1996b; Sidorovich and Pikulik 1997; Kuzmin 1999). As with otters and mink, however, these predators sometimes can take a considerable toll on hibernating frogs (Pulliainen and Ollinmäki 1996a) or breeding aggregations of anurans (Groves 1980).

#### Mongooses

Mongooses (Herpestidae) are omnivores that can be opportunistic predators of amphibians. In the West Indies, introduced mongooses (*Herpestes*) eat frogs and toads (C. Williams 1918; J. Myers 1931; Pimentel 1955). In Hawaii, they even feed on introduced cane toads (*Bufo marinus;* Baldwin, Schwartz, and Schwartz 1952), a highly toxic species that can be fatal if eaten by dogs, pigs, carnivorous marsupials, varanid lizards, snakes, and other predators (Dryden 1965; Covacevich and Archer 1975; M. Tyler 1976; McCoid, Hensley, and Witteman 1994; McCoid 1995; Burnett 1997). Egyptian mongooses (*Herpestes ichneumon*) living in a wetland habitat in southern Spain ate spadefoot toads (*Pelobates cultripes*) in the wet season, but they never were the most important component of the diet (Palomares 1993). The large African water or marsh mongoose (*Atilax paludinosus*) eats mostly crabs, but also takes significant numbers of frogs (Rowe-Rowe and Somers 1998; Ray and Sunquist 2001). The crab-eating mongoose (*Herpestes urva*) of Taiwan also supplements its diet of crabs with frogs (Chuang and Lee 1997).

#### Other Carnivores

Relatively few other carnivores are known to eat amphibians in large numbers. Cats seem an unlikely type of predator on amphibians, but some small cats do eat them. Up to a third of the scats of Tsushima Island leopard cats (*Felis bengalensis*) in Japan contained remains of amphibians, with the frequency being highest in the winter, and amphibians were part of the diet of Iriomote cats (*Felis iriomotensis*) from Japan as well (Kitchener 1991). Frogs were found in 77% of the scats of servals (*Felis serval*) in the Ngorongoro Crater in Tanzania, and were the second most frequent type of prey after small mammals (Geertsema 1985). Frogs made up about 11% of the diet of ocelots (*Felis pardalis*) in Venezuela (Bisbal 1986), and they may be eaten occasionally by margays (*F. wiedii*) as well (Azevedo 1996).

Even domesticated cats eat amphibians occasionally. In a survey of cat-owning British households during a single spring and summer season, 986 cats were reported to have brought nearly 600 amphibians back to their homes, of which most were ranid frogs (probably mostly *Rana temporaria*). Overall, it was estimated that the approximately 9 million household cats in Britain bring about 5 million amphibians and reptiles back to homes each year, of which 75–80% are frogs (Woods, McDonald, and Harris 2003). This figure undoubtedly underestimates the actual mortality caused by cats, since many prey items probably are consumed away from the home. The estimate also does not include feral cats, which are likely to obtain all of their food by hunting.

In Venezuela, remains of anurans (*Bufo, Eleutherodactylus, Hyla, Leptodactylus, Pseudis*) were found in the diet of crab-eating foxes (*Cerdocyon thous;* Bisbal and Ojasti 1980; Sunquist, Sunquist, and Paneke 1989). The omnivorous South American maned wolf (*Chrysocyon brachyurus*) eats mostly mammals, birds, and fruit, but it also eats toads (Guix 1993). In some areas, the Eurasian raccoon dog (*Nyctereutes procyonides*), a species that frequents the shores of ponds and streams, eats substantial numbers of anurans, especially during the spring breeding season (Stroganov 1969; Ewer 1973; Viro and Mikkola 1981; Kauhala 1996; Kauhala, Laukanen, and von Rege 1998; Kuzmin 1999; Kauhala and Auniola 2001). In other areas, frogs are a minor part of the diet (Jędrzejewski, Jędrzejewska, and Szymura 1989).

Among the procyonids, the North American raccoon (*Procyon lotor*) is frequently cited as a major predator of frogs, but detailed studies of food habits have shown that frogs constitute a surprisingly small proportion of the diet, with more than 50% being composed of plants (Ewer 1973; Chapman and Feldhamer 1982). Nevertheless, raccoons are attracted to breeding aggregations of anurans, or sites where frogs are trapped by drying ponds (Fitch 1958; Schaaf and Garton 1970; M. Jones, Goetti, and Livo 1999). Crab-eating raccoons (*P. cancrivorus*) sometimes eat anurans as well (Guix 1993). Coatis (*Nasua narica*) take an occasional frog or caecilian, but do not seem to be major amphibian predators (Kaufmann 1962). Genets (*Genetta genetta*) are the only viverids found in Europe, and are most common in the Mediterranean region, especially the Iberian Peninsula. They eat mostly small mammals, but other types of prey, including amphibians, are taken occasionally. The frequency of amphibians in the diet is highest on two Mediterranean islands, Ibiza and Cabrera, where alternative prey may be scarce (Virgos, Llorente, and Cortes 1999). Other carnivores that take amphibians occasionally include foxes (*Vulpes, Alopex*), wolves (*Canis lupis*), and bears (*Ursus*) (Jones, Goetti, and Livo 1999; Kuzmin 1999), but it seems unlikely that amphibians are major prey for any of these animals.

### Shrews and Moles

Many insectivores probably feed on amphibians at least occasionally. In Poland, up to 40% of the winter diet of the water shrew (*Neomys fodiens;* fig. 14.8) is composed of hibernating frogs captured in ditches (Wolk 1976). Dead frogs are frequently stored for future use (Buchalczyk and Pucek 1963) and are more completely consumed as winter progresses and food availability declines (Wolk 1976). Several other European shrews include amphibians as a minor item in the diet (Aitchison 1987; Kuzmin 1999). The North American water shrew (*Sorex palustris*) eats salamanders (*Dicamptodon;* Nussbaum and Maser 1969), as does *Blarina brevicauda,* which has been used in tests of amphibian antipredator secretions (Brodie, Nowak, and Harvey 1979; Brodie and Formanowicz 1981). *Blarina* appears to be a minor predator of amphibians in the field, however (Hamilton 1930). Musk shrews (*Suncus*) in India eat frogs and toads (Dharmakumarsinhji 1946; Kunte 1996). There are anecdotal reports of European moles (*Talpa europaea*) capturing anurans (Mellanby 1971; Kuzmin 1999), and the Russian desman (*Desmana moschata*) preys on a number of kinds of amphibians (Kuzmin 1999). The North American shrew mole (*Neurotrichus gibbsi*) is reported to eat salamanders (Nowak and Paradiso 1983).

**Fig. 14.8.** European water shrew (*Neomys fodiens*), a voracious predator on frogs hibernating underwater. This species often caches excess frogs for future use during the winter. Photo by Milos Andera.

### Rodents

Many species of rodents are partially carnivorous (Landry 1970). Muskrats (*Ondatra*) occasionally feed on frogs and salamanders along with other animals such as crayfish and mussels (Landry 1970; Kuzmin 1999). The Australian water rat (*Hydromys*) is known to eat anurans, including introduced *Bufo marinus* (Tyler 1976; Woolard, Vestjens, and MacLean 1978). An African water rat (*Colomys goslingi*) lives in the vicinity of running water and appears to be largely carnivorous; captive individuals eagerly captured and ate tadpoles (Dieterlen and Statzner 1981). There also several aquatic rats in South America (*Ichthyomys, Anotomys, Neusticomys, Daptomys*) that are thought to be carnivorous (Nowak and Paradiso 1983), although captive *Ichthyomys* showed relatively little interest in tadpoles (Voss Sliva L., and Valdeo L. 1982). The common Norway rat (*Rattus norvegicus*) and black rat (*Rattus rattus*), introduced all over the world by humans, are known to prey at least occasionally on frogs and toads (M. Smith 1969; M. Tyler 1976; Frazer 1983; Stewart and Woolbright 1996). There even are reports of chipmunks (*Tamias*), red squirrels (*Tamiasciurus hudsonicus*), and hamsters (*Cricetus*) taking frogs, toads, and salamanders (Landry 1970; B. D. Sullivan 1991; Kuzmin 1999), but their impact is likely to be trivial.

### Bats

Some large bats eat small vertebrates, including frogs, but most also eat insects, and none is known to feed exclusively on frogs (Norberg and Fenton 1988). In the Old World, frogs are eaten by false vampire bats in the family Megadermatidae (*Megaderma, Macroderma, Cardioderma;* Brosset 1962; Douglas 1967; Vestjens and Hall 1977; Advani 1981; Marimuthu, Habersetzer, and Leippert 1995). Radio-tagged false vampire bats (*Megaderma lyra*) in India captured both frogs and insects (Audet et al. 1991). Remains of many ranid frogs (*Ptychadena*) and a few clawed frogs (*Xenopus laevis*) have been found beneath roosts of an African nycterid bat, *Nycteris grandis* (Fenton, Thomas, and Sasseen 1981; Fenton et al. 1990). In the Neotropics, the phyllostomid bat *Trachops cirrhosus* preys on frogs that it locates by homing in on their calls (Tuttle and Ryan 1981; see chapter 7 for further details). At one pond, *Trachops* captured 95 *Physalaemus pustulosus* in slightly more than 14 hours, and were the most important predators on calling males (Ryan, Tuttle, and Taft 1981). This bat shows a strong preference for edible frogs over poisonous species and can distinguish between them by their calls (Page and Ryan 2005). Four species of bats known to eat frogs (*Trachops,* two species of *Megaderma,* and *Cardioderma cor*) all have unusual accessory submandibular salivary glands unlike glands found in other bats. These may produce salivary secretions that neutralize the toxic skin secretions of the frogs eaten by the bats, although this has not been demonstrated experimentally (Tandler, Phillips, and Nagato 1996; Tandler, Nagato, and C. J. Phillips 1997).

### Primates

Amphibians appear to be of little significance in the diets of most primates, but they provide supplemental protein for a few species. Several species of African monkeys (*Cercopithecus mitis, Cercocebus torquatus, Cercocebus diana*) have been observed feeding on the foam nests of rhacophorid tree frogs, *Chiromantis xerampelina* and *C. rufescens,* as well as the gelatinous egg masses of some hyperoliid frogs (Huntly 1985; Rödel et al. 2002). In the Ivory Coast, sooty mangabeys (*Cercocebus torquatus atys*) were observed systematically searching vegetation over water for frog eggs, suggesting that frog eggs could be a regular item in the diet, and monkey predation could have a major impact on egg survivorship (Rödel et al. 2002). Capuchin monkeys (*Cebus apella*) were suspected as predators on arboreal egg masses of *Phyllomedusa bicolor* in Brazil (Neckel-Oliviera and Wachlevski 2004).

Some monkeys also feed on adult anurans. African patas monkeys (*Erythrocebus pats*) were seen feeding on toads in Senegal (Galat-Luong 1991). Japanese macaques (*Macaca fuscata*) sometimes eat both frogs and lizards (Suzuki et al. 1990), as do stumptail macaques (*M. arctoides;* Estrada and Estrada 1977). Frog remains made up 8% of the stomach contents of *Galago alleni,* a nocturnal prosimian studied by Charles-Dominique (1977) in Gabon. All five diurnal monkeys studied by Terborgh (1983) in Peru preyed on frogs, but these made up only 1–3% of all prey items. In Colombia, one of these species, *Cebus apella,* extracted frogs from hiding places in hollow bamboo stalks after enlarging the entrances to the retreats with their teeth (Izawa 1978). The monkeys rubbed the frogs on tree branches after killing them, possibly to remove noxious skin secretions. A closely related species, *C. albifrons,* has been seen capturing frogs from the water-filled axils of plants (Defler 1979). Tamarins (*Saguinus*) and marmosets (*Callithrix*) also eat frogs occasionally, but they make up a very small proportion of the diet (Digby and Barreto 1998; Heymann, Knogge, and Tirado Herrera 2000; Passamani and Rylands 2000; A. Smith 2000; Canale and Lingnau 2003).

### Other Mammals

Mammals in several other orders are potential predators of amphibians. A somewhat surprising one is the nine-banded armadillo (*Dasypus novemcinctus*), which eats a variety of frogs, salamanders, and even amphiumas (Fitch, Goodrum, and Newman 1952; Breece and Dusi 1985; Wirtz, Austin, and Deklee 1985). Another unusual predator is the European wild boar (*Sus scrofa*), which occasionally digs up and consumes spadefoot toads (*Pelobates fuscus;* Briedermann 1976; cited in Nöllert 1984) and other anurans (Kuzmin 1999), as well as highly toxic fire salamanders (*Salamandra salamandra;* Carretero and Rosell 1999). The North American opossum (*Didelphis virginiana*) and a number of South American opossums include amphibians in a varied diet of small vertebrates, invertebrates, and plants (Chapman and Feldhamer 1982; Streilein 1982; Nowak and Paradiso 1983; Redford and Eisenberg 1992). In Panama, the four-eyed opossum, *Philander opossum,* used acoustic cues to locate frogs at a breeding chorus (Tuttle, Taft, and Ryan 1981). Some carnivorous Australian marsupials, such as quolls (*Dasyurus*), eat anurans and consequently are threatened by the spread of the highly toxic introduced cane toad (*Bufo marinus;* Burnett 1997).

## Man the Frog Hunter

In some areas of the world, humans are significant predators of amphibians. Frogs probably have been at least a minor component of the human diet for most of our evolutionary history. Frog bones have been found associated with hominid remains and concentrations of other vertebrate bones at Olduvai Gorge in Tanzania, although it is nearly impossible to determine whether frogs actually were eaten (Isaac and Crader 1981). Amphibian bones often turn up in archaeological digs, but it is not always clear that they represent remains of human food. For example, bones of five genera of frogs have been found at Paleolithic sites in Spain, but they may have been eaten by black kites (*Milvus migrans*) nesting near humans, not by humans themselves (Freeman 1981). Remains of *Bufo marinus* often are found at archaeological sites in Central America (Coe and Diel 1980; R. Cooke 1984; Hamblin 1984). There has been some debate over whether these toads were used as food, or collected for their poisonous skin secretions to be used as psychoactive drugs (Dobkin de Rios 1974; Furst 1976; Kennedy 1982; W. Davis and Weil 1992).

Many hunting and gathering people, as well as those that combine hunting and agriculture, probably take amphibians for food whenever they are encountered. It is not uncommon for such people, especially those living in rain forests, to consider nearly all animals edible (Hayden 1981). For example, various groups of South American Indians eat frogs, toads, tadpoles, and even anuran eggs, but amphibians seldom represent more than a minor source of protein (Wassen 1934; Clastres 1972; Dufour 1983; Hill and Hawkes 1983; Abrams 1987). The Wayapi of French Guiana eat 16 species of frogs and the eggs of eight species, but amphibians constitute a trivial proportion of their total diet (Lescure, Grenand, and Grenand 1980). In Peru, the Cashinahuas eat large tree frogs (*Phrynohyas coriacea*) after roasting them in banana leaves (Duellman and Thomas 1996). Some groups of Indians in Amazonian Peru and Brazil also collect another large tree frog, *Phyllomedusa bicolor,* and use its skin secretions in hunting magic. The secretions are introduced into fresh burns on the skin, resulting in violent illness, followed by a state of euphoria (Daly et al. 1992). The impact of this practice on populations of the frog is unknown. Some North American Indians used bullfrogs (*Rana catesbeiana*), leopard frogs (*Rana pipiens*), wood frogs (*Rana sylvatica*), and even hellbenders (*Cryptobranchus alleganiensis*) as food (Waugh 1916; Wassen 1934; Speck and Dodge 1945; McCoy 1982), but little detailed information is available.

Reports on the use of amphibians as food by humans in other parts of the world are sketchy. David Livingstone reported that people in southern Africa captured *Pyxicephalus adspersus* at breeding ponds after heavy rains and ate them with "great eagerness" (Livingstone 1858, p. 48); this practice has continued in recent times (Mitchell 1946, cited in Stewart 1967). In West Africa, local people prize the giant frog, *Conraua goliath,* in part because it is very difficult to catch (Zahl 1967; Sabater-Pi 1985). In arid regions of the Sudan, where sources of protein are scarce, local people prepare a variety of fermented foods based mostly on sorghum, with added ingredients that include frogs, fish, caterpillars, locusts, and cow urine (Dirar 1994).

In a few areas, amphibians are a more significant component of the diet. Before the introduction of black bass as a food fish into some East African lakes in Uganda, Rwanda, and Zaire, *Xenopus* were used as a substitute for fish by people living near the lakes (Worthington and Worthington 1933; de Witte 1941). Similar use is made of aquatic frogs of the genus *Telmatobius* by people living near high Andean lakes (Crump 2003), and the Mexican axolotl (*Ambystoma mexicanum*) has been used for food since Aztec times (see chapter 13). Frogs are an important resource for some people in the highlands of New Guinea, where mammalian prey is scarce. Both children and adults spend considerable time systematically hunting for frogs along streams, and they have developed an elaborate taxonomy to name them (Bulmer and Tyler 1968). Adult frogs are smoked, baked, or roasted over hot coals, while tadpoles may be roasted or stewed (M. Tyler 1976). In some parts of New Guinea and New Britain, large ranid frogs are hunted with dogs, which are used to drive frogs from hiding places along streams. Human predation on several frogs in New Guinea, including *Rana arfaki* and *R. jimiensis,* is so intense that the species have become rare and may be endangered (M. Tyler 1976). Frogs also are eaten as occasional delicacies by some native Australians (M. Tyler 1976). The Nuaulu people of eastern Indonesia combine swidden agriculture with collection of wild plants and animals in the forest. An exceptionally large treefrog, *Litoria infrafrenata,* is sometimes eaten, but is not a major food (Ellen, Stimson, and Menzies 1976).

Frogs are eaten in many other parts of Asia, although they probably are not a major source of food. In ancient times, people in southern China ate frogs, a habit that was detested by their northern cousins (M. Freeman 1977; Schafer 1977). In some parts of China, a special soup prepared from frogs collected while breeding was associated with sexual power (Eberhard 1968). In modern Asian countries, including China, Japan, the Philippines, and Malaysia, frog consumption is limited mainly to large and abundant ranid frogs (Okada 1927; Angel 1947; Heang 1984), many of which are readily captured in rice fields. China is a major market for frog legs imported from other Asian countries, with up to six million *Hoplobatrachus rugulosus* (Ranidae) shipped from Thailand to China each year (Jensen and Camp 2003). In Okinawa, frogs were once considered sufficiently important as a source of food for frog catching to be a hereditary profession (Okada 1927), and there were professional frog catchers in ancient China as well (see frontispiece; Eberhard 1968). Some tribal people in India use several species of anurans in traditional medicines (Azmi 1990).

Consumption of frogs has a long history in Western and Central Europe. Medieval monks were allowed to eat meat on only a limited number of days each year, but got around the restriction by classifying aquatic animals such as frogs and beavers as fish (Tannahill 1973). The use of frogs as food undoubtedly originated among farmers and peasants who collected wild frogs to supplement the meager protein in their diets, but they soon became established as a regular item on the tables of city dwellers and the aristocracy. At a banquet given in Paris for Elizabeth of Austria in 1571, 1,000 frogs were served along with a wide assortment of fish and shellfish (Tannahill 1973). By the eighteenth century, the use of frogs for food was so commonplace that Linnaeus named one of the most frequently eaten species *Rana esculenta,* the edible frog. In modern times, consumption of frogs has been most common in France, Belgium, Holland, Germany, Switzerland, and Italy (Angel 1947; Dubois 1983a), although frogs are eaten to a limited extent in most European countries. Frogs never have been a popular food in Britain, and during their many long wars with the French, the British often disparaged their enemies as "frog-eaters" or simply "frogs" (Hackwood 1911).

The impact of human consumption of frogs in Europe is hard to estimate because these populations also are threatened by habitat destruction, pollution, and collecting for scientific research. Nevertheless, frog populations have declined in some European countries to the point where conservationists and government agencies are now discouraging their use as food (Dubois 1983a; Honnegger 1981, 1984; Thielcke et al. 1985). The effects of western European consumption of frog legs have been felt in other countries as well. As populations in western and central Europe have declined, restauranteurs and grocers have turned to Greece, Turkey, India, Bangladesh, Indonesia, and other countries as sources of imported frog legs (Jensen and Camp 2003). Thielcke et al. (1985) reported that 96,000 kg of frogs were imported into Switzerland in 1976, representing about a million individuals. Imports into France have been even higher, ranging from 2,700 to 4,500 metric tons per year from 1976 through 1982 (Dubois 1983a). By the 1980s, export of frog legs from India reached as high as 4,400 tons per year (about 4 million kg), representing some 13,000 tons (12 million kg) of frogs. A survey conducted among local farmers suggested that populations of large ranid frogs had declined or disappeared in many parts of India (Abdulali 1985), and there is increasing concern about declining frog populations in India (Oza 1990; Gore and Prayag 1993) and other Asian countries such as Indonesia (Patel 1993; Veith et al. 2000).

In the late nineteenth century, frogs were a regular item on menus in many parts of the United States, and many states had a thriving frog collecting industry. Chamberlain (1898) estimated that at least a million frogs were taken annually for food in the late 1890s, with frogs being a commercially important fisheries product in at least 15 states. By 1900, frogs were the single most valuable product of the fisheries industry in Minnesota (A. Wright 1920). The majority of frogs used for food came from states in the Mississippi River drainage, although states such as New York and California had major frog industries as well. Before 1900, the most heavily exploited species in California was *Rana draytonii,* with harvests totaling nearly 120,000 frogs (22,000 kg) in 1895 (Jennings and Hayes 1985). By 1900, this species had undergone a dramatic decline from which it has never recovered. The scarcity of native frogs may have been the reason for the introduction of bullfrogs (*Rana catesbeiana*) from the eastern United States into California (Jennings and Hayes 1985). The current level of exploitation of frogs for food is difficult to estimate because detailed harvest records are not available for most states, and local populations often are severely threatened by illegal poaching. In addition, collection of frogs for research and educational use is now a substantial enterprise that may have a greater effect on populations in North America than exploitation for food. In some states, ranid frogs, desmognathine salamanders, and ambystomatid larvae are sold in bait shops, but the impact of this largely unregulated trade on amphibian populations is unknown (Jensen and Camp 2003).

## The Impact of Predation on Amphibian Populations

From the preceding survey, it is clear that amphibians are vulnerable to a wide variety of predators at all stages of their lives. However, quantitative data on the demographic impact of predation are scarce, and measuring the impact of specific predators is even more difficult. Studies that provide estimates of mortality in amphibian populations often include lists of potential predators, but they seldom provide quantitative estimates of the impact of those predators on survivorship (e.g., Peterson et al. 1991; Ramirez, Vogt, and Villarreal-Benitez 1998). The study of amphibian demography is complicated by the complex life cycles of most species, with different stages being exposed to very different physical and biological environments. To understand what factors regulate the sizes of amphibian populations, one must study processes occurring at all life history stages (Wilbur 1980; Hellriegel 2000). Many ecologists have divided the typical amphibian life history into two major stages, one occurring in the water and the other on land (Wilbur 1980; Werner 1986). However, when considering the impact of predation on amphibian populations, it is more useful to think of the life history as being composed of four stages: (1) embryos, (2) larvae, (3) metamorphosing juveniles, and

(4) adults. Predation can be a significant source of mortality at any of these stages, but not necessarily at all stages for any given species. At present, we do not have complete information on sources of mortality at all stages of the life cycle for any species of amphibian.

### Embryonic Mortality

Embryonic mortality in aquatic-breeding amphibians often is extremely variable, both within and among species. For example, embryonic mortality ranging from 2% to nearly 100% has been reported for spotted salamanders (*Ambystoma maculatum*) (Shoop 1974; R. Harris 1980; Stangel 1988; Rowe, Sadinski, and Dunson 1994; Brodman 1995; Stenhouse 1987; Petranka et al. 1998). Stenhouse (1987) observed variation from 0 to 100% mortality among different ponds in the same year. In many ponds, freezing was a more important source of mortality than predation, but in some cases, sources of mortality could not be determined. Nevertheless, it is clear that certain predators can have devastating effects on spotted salamander embryos. Survivorship of embryos in ponds containing wood frog (*Rana sylvatica*) tadpoles was 91% for egg masses enclosed in predator exclusion cages, but only 14% for those not in cages (Petranka et al. 1998). Mortality of spotted salamander embryos exceeded 90% in ponds with dense populations of caddisfly larvae, but little or no mortality in ponds that lacked these predators (Rowe, Sadinski, and Dunson 1994).

Other salamanders exhibit similar variation in embryonic mortality. Brodman (1995) found consistently low mortality (< 15%) for *Ambystoma jeffersonianum* in a semipermanent pond in Ohio over several years. Freezing contributed to high mortality (62%) of eggs in a population of *Ambystoma tigrinum* studied by J. Anderson, Hassinger, and Dalrymple (1971), but there was evidence of predation by caddisfly larvae as well. Some *Ambystoma* eggs are vulnerable to predation by adult newts (*Notophthalmus*) and fish, which can virtually eliminate eggs from permanent ponds (Morin 1983b; Petranka 1983; Semlitsch 1988). In stream-breeding populations of *A. barbouri,* embryonic mortality was low (16–20%), and only about 3% of the eggs were eaten by predators (Petranka 1984b). Similar low mortality has been observed in the stream-breeding hynobiid salamander *Ranodon sibiricus* (31%; Bannikov 1958) and the pond-breeding species, *Hynobius nebulosus* (14–33%; Kusano 1980). In contrast, extremely high embryonic mortality (97%) was found in a pond-breeding population of newts (*Triturus vulgaris*), and there was evidence that predation was a major source of mortality (G. Bell and Lawton 1975).

Mortality of aquatic frog eggs is also highly variable. In Marion Lake, British Columbia, only about 5% of all *Rana aurora* eggs died prior to hatching, despite a long period of development (more than one month) in a lake inhabited by predatory newts and ambystomatid salamanders (Calef 1973a). All of the mortality appeared to be due to fungal infection, freezing, and desiccation, not predation. Licht (1974) found similar low mortality (9%) in *Rana aurora* eggs laid in permanent water where invertebrate predators were abundant. Mortality of *R. pretiosa* eggs in the same habitat was higher (29%), but the difference was due to more frequent desiccation of *R. pretiosa* eggs, not to greater predation. Mortality also was low, averaging 3%, in clutches of *Heleioporus albopunctatus* eggs in Western Australia, despite attacks by fly larvae (R. Davis and Roberts 2005). R. D. Howard (1978b) observed very high mortality in egg masses of bullfrogs (*Rana catesbeiana*) that were attacked by leeches, but little or no mortality in other clutches.

Although temporary ponds lack predatory fish, they often are filled with other types of predators, particularly aquatic insects that rapidly colonize new ponds, and many of these prey on amphibian embryos (see table 14.2). In general, invertebrate predator densities are lower in temporary than in permanent water (Heyer, McDiarmid, and Weigmann 1975; Crump 1981b; Woodward 1983), and small pools often have fewer predators than large ones (D. G. Smith 1983; A. H. Roth and Jackson 1987). Usually desiccation is the most important source of embryonic mortality, but predation can be significant in ponds that do not dry up. Amphibian larvae, including conspecifics, can be major predators of anuran eggs in some ponds (see previous discussion of predation by amphibian larvae). For example, Banks and Beebee (1988) found that less than 3% of *Bufo calamita* eggs laid in shallow sand-dune ponds were killed by predators, whereas 26% of the eggs laid in a more permanent heathland pond were eaten. Among the most important predators were tadpoles of *Rana temporaria* and *Bufo bufo,* which are capable of eliminating *Bufo calamita* eggs from some ponds (Banks and Beebee 1987b). An increase in numbers of *B. bufo* in Britain has been accompanied by a dramatic decline in *B. calamita,* possibly because of predation on eggs (Beebee 1979b). The latter species breeds most successfully in very shallow temporary ponds that are not used by *B. bufo,* but these ponds are more vulnerable to desiccation (Banks and Beebee 1987a, 1988). In Spain, predation by tadpoles of *Pelobates cultripes* and *Pelodytes punctatus* resulted in mortality of more than 50% for *Bufo calamita* embryos, and these predators eliminated all toad eggs in one pond (Tejedo 1991). In North Carolina, American toads (*Bufo americanus*) are prevented from breeding in ponds containing wood frog (*Rana sylvatica*) tadpoles, which are capable of consuming several thousand toad eggs in less than an hour at natural tadpole densities (Petranka, Hopey, Jennings, Baird, and Boone 1994).

Nonaquatic oviposition or oviposition in very small pools has been interpreted as a means of reducing predation on eggs (Lutz 1947; Salthe and Duellman 1973; Salthe and Mecham 1974), but rates of predation sometimes exceed those in species that breed in large bodies of water. For example, embryonic mortality was high (82%) in *Ramanella montana,* a microhylid frog from India that lays eggs in tree holes just above the water surface or on floating dead leaves. Several kinds of invertebrates were observed feeding on the eggs, including snails, caddisfly larvae, and millipedes (Krishna, Krishna, and Vijayalaxmi 2004). Mortality of centrolenid eggs, much of it attributable to predation by drosophilid fly larvae, ranged from 14% to nearly 70% (Villa 1977, 1984; McDiarmid 1978; Greer and Wells 1980; Jacobson 1985). Similarly, embryonic mortality was high (72%) in the East African frog *Hyperolius spinigularis,* which lays eggs on emergent vegetation in water or on leaves over water. The main sources of mortality were predation by frogs (*Afrixalus;* 41%), predation by fly larvae (12%), desiccation (10%), and predation by insects (3%; Vonesh 2005a). Predation by *Afrixalus* was particularly high early in the breeding season, with 73% of clutches being attacked, compared to 13–37% later in the season. Embryonic mortality in some species is reduced by parental care, as it is in frogs such as *Eleutherodactylus coqui* (Townsend, Stewart, and Pough 1984) and *Cophixalus parkeri* (Simon 1983; see chapter 11). However, some terrestrial breeders have low embryonic mortality even without parental care. Embryonic mortality was only 3–5% in three species of *Pseudophryne* in Australia, and there was no evidence of predation (Woodruff 1976b). These species lay eggs in a moist, cold environment where predators are scarce.

Marbled salamanders (*Ambystoma opacum*) lay eggs on land in areas that are later flooded. Embryonic mortality is variable, but can be up to 60–90% in some populations (Stenhouse 1987; M. Jackson, Scott, and Estes 1989; Petranka 1990). Freezing and failure of rains to flood nests appear to be the most important sources of mortality, but protection of eggs from predators by attending females also is important (M. Jackson, Scott, and Estes 1989; Croshaw and Scott, 2005). Specific predators have not been identified, however. Embryonic mortality can be relatively high (80–90%) in four-toed salamanders (*Hemidactylium scutatum*), another species that lays eggs on land (Harris and Gill 1980; Harris et al. 1995). Brooding by females enhances embryonic survival, again suggesting that protection against predators might be important. However, females often desert their eggs if they are flooded by heavy rains, and much of the mortality occurs in water, rather than in terrestrial nests. Natural embryonic mortality has not been measured for most salamanders that lay eggs on land, but experimental removal of attending parents in several species of plethodontid salamanders resulted in high mortality, much of it apparently due to predation (see chapter 11).

### Larval Mortality

The larval period of aquatic-breeding amphibians generally is much longer than the egg stage, so the larvae are potentially exposed to many kinds of predators between hatching and metamorphosis. Many species probably are confronted with a choice of oviposition sites along a gradient from highly ephemeral rain pools to permanent ponds, lakes, and streams. There may be an optimum trade-off between the risk of predation in permanent sites and the risk of desiccation in temporary ones (Heyer, McDiarmid, and Weigmann 1975; Wilbur 1980; D. C. Smith 1983; Woodward 1983; Petranka 1984b; Banks and Beebee 1987a, 1988; A. H. Roth and Jackson 1987; Semlitsch 1987d). However, the persistence of ponds and the abundance of predators vary among breeding sites and from year to year at the same site (Shoop 1974; Petranka 1984b; D. C. Smith 1983; Stenhouse 1985a; Semlitsch 1987d; Banks and Beebee 1988). This makes it impossible for amphibians to anticipate the conditions under which larvae will develop (see chapters 13 and 15 for further discussion).

Mortality rates of larvae from hatching to metamorphosis are summarized in table 14.5. With the exception of a population of *Polypedates* from Malaysia and *Hemisus marmoratus* from West Africa, all of the species are from the temperate zone, but there is no reason to expect survivorship of tropical species to be very different. Mortality generally is very high, with less than 10% of hatchlings surviving to metamorphosis in most populations, but there is considerable year-to-year variation in mortality rates. In species with high embryonic mortality as well, survivorship from egg to metamorphosis can be less than 1%. While desiccation obviously is an important source of mortality in temporary ponds, predation also can be high in temporary water (e.g., *Bufo calamita, Hemisus marmoratus, Polypedates leucomystax, Ambystoma maculatum,* and *A. tigrinum*). Furthermore, there is no obvious difference in overall mortality rates among habitats; survivorship is low in permanent ponds, temporary ponds, and in streams.

In many larval populations, there is a dramatic decrease in population size soon after hatching (Herreid and Kinney 1966; Calef 1973a; Licht 1974; Cecil and Just 1979; Viertel 1980; Yorke 1983). This has led many authors to assume that amphibian larvae exhibit Type III survivorship curves—that is, a pattern in which age-specific mortality decreases rapidly with age (Deevey 1947). However, this interpretation results from confusion over arithmetic and logarithmic scales used to plot survivorship curves (Petranka 1985). In absolute terms, much of the mortality in a larval population

**Table 14.5** Estimates of larval mortality in natural populations of amphibians, with the most likely major causes of mortality

| Species (source no.) | Breeding habitat | Mortality (%) | Causes of mortality |
|---|---|---|---|
| Anurans | | | |
| *Bufo bufo* (10) | Permanent pond | 80–98 | Predation |
| *B. calamita* (8) | Temporary pond | 99 | Predation |
| *B. calamita* (18) | Temporary pond | 95–99 | Desiccation, predation |
| *Crinia signifera* (22) | Permanent pond | 90 | Predation |
| *Hemisus marmoratus* (24) | Temporary pond | 1–36 | Predation by turtles |
| *Pelobates fuscus* (23)[a] | Temporary pond | 96–100 | Unknown |
| *Philoria frosti* (15) | Temporary pond | 70 | Desiccation |
| *Polypedates leucomystax* (13) | Temporary pond | 97 | Starvation, predation |
| *Pseudacris triseriata* (12) | Temporary pools | 0–100 | Desiccation (small pools); predation (large pools) |
| *Rana aurora* (4) | Permanent lake | 95 | Predation |
| *R. aurora* (5) | Permanent pond | 95–99 | Predation |
| *R. catesbeiana* (9) | Permanent pond | 82–87 | Predation |
| *R. pretiosa* (5) | Permanent pond | 93–99 | Predation |
| *R. sylvatica* (2) | Temporary pond | 95 | Unknown |
| *R. sylvatica* (20)[a] | Temporary pond | 92–99 | Competition, predation, desiccation |
| Urodeles | | | |
| *Ambystoma annulatum* (21)[a] | Semipermanent pond | > 99 | Predation |
| *A. barbouri* (14) | Permanent stream | 94–99 | Predation, floods, desiccation |
| *A. maculatum* (17) | Temporary pond | 99 | Desiccation, predation |
| *A. maculatum* (6) | Temporary pond | 49–99 | Desiccation |
| *A. maculatum* (19) | Temporary pond | 96 | Unknown |
| *A. opacum* (17) | Temporary pond | 76 | Desiccation |
| *A. talpoideum* (16)[a] | Temporary pond | 96–100 | Desiccation, predation |
| *A. tigrinum* (3) | Temporary pond | 92 | Predation |
| *Hynobius nebulosus* (11) | Permanent pond | 80–99 | Predation |
| *Ranodon sibiricus* (1) | Permanent stream | 75 | Unknown |
| *Triturus vulgaris* (7) | Permanent pond | 91 | Predation |

*Sources:* (1) Bannikov 1958; (2) Herreid and Kinney 1966; (3) Anderson, Hassinger, and Dalrymple 1971; (4) Calef 1973a; (5) Licht 1974; (6) Shoop 1974; (7) G. Bell and Lawton 1975; (8) Kadel 1975; (9) Cecil and Just 1979; (10) Viertel 1980; (11) Kusano 1981; (12) D. C. Smith 1983; (13) Yorke 1983; (14) Petranka 1984a; (15) Malone 1985; (16) Semlitsch 1987a; (17) Stenhouse 1987; (18) Banks and Beebee 1988; (19) Stangel 1988; (20) Berven 1990; (21) C. L. Peterson et al. 1991; (22) Williamson and Bull 1999; (23) Hels 2002; (24) Grafe et al. 2004.

[a]Estimates of total premetamorphic mortality from oviposition to metamorphosis, but most of the mortality usually is in the larval stage.

does occur early in development, but age-specific mortality rates remain relatively constant in most species (fig. 14.9). Hence, most species exhibit a Type II survivorship curve, as defined by Deevey (1947), not a Type III curve. This is true even though many studies have shown that individual larvae become less vulnerable to predation as they grow (see the following further discussion). However, population survivorship curves integrate all sources of mortality, not just predation, and therefore might not reflect a pattern predicted solely from the susceptibility of individuals to predation (Petranka 1985). The survivorship curves in fig. 14.9 are only for the larval portion of the life cycle. Since mortality rates generally are much lower for adults than for larvae, survivorship curves for the entire life span would more closely resemble a Type III curve (see the following).

### Mortality of Metamorphosing Juveniles

Quantitative estimates of mortality rates in metamorphosing juveniles are almost impossible to obtain because the an-

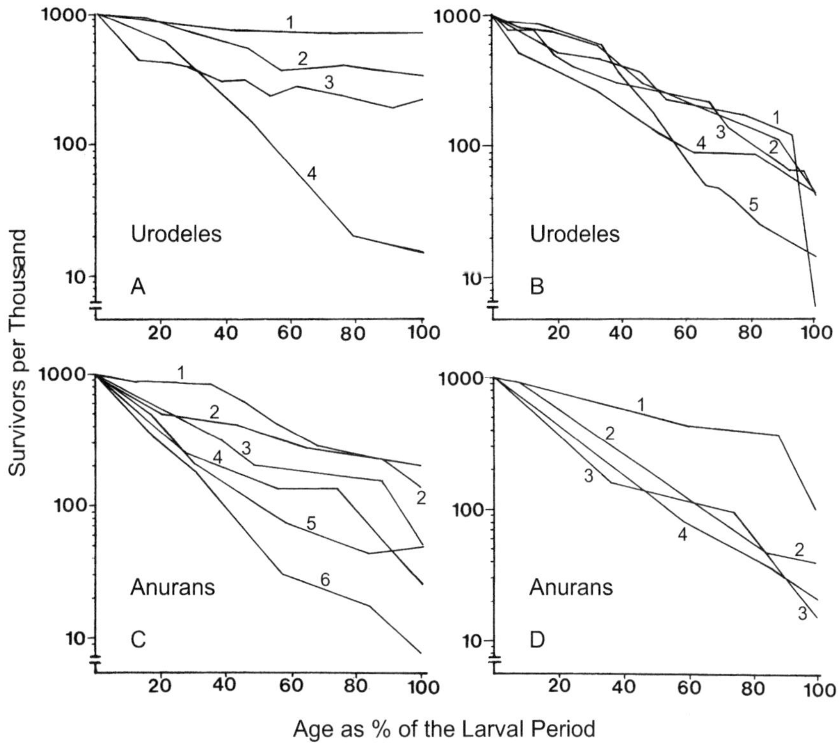


**Fig. 14.9.** Survivorship curves for several larval amphibian populations. (A) *Hynobius nebulosus* in four different years (1–4). (B) *Ambystoma barbouri* in three different years (1, 3, and 5) and two different cohorts of *Triturus vulgaris* from the same pond (2, 4). (C) Four species of anurans: *Polypedates leucomystax* (1), *Rana catesbeiana* (2), *Rana aurora* (3, 4), and *Rana pretiosa* (5). (D) Four populations of *Rana sylvatica* from Alaska. After Petranka (1985).

imals disperse soon after emerging from the water. Nevertheless, the period of metamorphic transformation appears to be a particularly risky stage for many species. The animals are small and slow moving, and aerobic capacities for sustained locomotion may be poorly developed (Taigen and Pough 1981; Pough and Kamel 1984). Metamorphosing anurans that have all four legs, but still retain much of the tail, are inferior to tadpoles in holding a position in the water and inferior to transformed individuals in jumping ability on land (Wassersug and Sperry 1977). When garter snakes (*Thamnophis sirtalis*) were presented with premetamorphic tadpoles and metamorphosing individuals of *Pseudacris triseriata,* they were much more effective at capturing metamorphosing frogs, even though they struck more often at the tadpoles (Wassersug and Sperry 1977).

The greater success of snakes in capturing metamorphosing frogs was reflected in their diets in the field, which consisted mainly of metamorphic stages (fig. 14.10). Similar results were obtained at several sites where *Bufo boreas* tadpoles were transforming; snakes were much more likely to consume metamorphosing frogs than premetamorphic stages (fig. 14.11; Arnold and Wassersug 1978). Metamorphosing frogs and toads also are vulnerable to a variety of other predators, including birds and mammals attracted to aggregations (Fitch 1958; Beasley and Carothers 1974; Tordoff 1980; Harestad 1985), invertebrates (Jackman et al. 1983; Crump 1984a), and even conspecific tadpoles (Crump 1986b). High mortality during metamorphosis may have been a major selective pressure favoring the evolution of short metamorphic transformation in many species (Szarski 1957; Wassersug and Sperry 1977; Downie, Bryce, and Smith 2004). In 12 species of anurans with a mean larval period of 62 days, the time spent in transformation averaged only about 9 days, or 15% of the larval period (Wassersug and Sperry 1977).

### Adult Mortality

In contrast to aquatic larvae, which do not disperse from the breeding area until metamorphosis, adult amphibians are free to move to other areas. Consequently, it often is difficult to distinguish adult mortality from emigration. Even when adult survivorship can be estimated with some precision, it may be impossible to determine the causes of mortality. Predation usually is considered to be important, but the evidence generally consists of scattered observations of animals being eaten by predators, followed by lists of potential predators seen in the area (e.g., Martof 1956a; F. Turner 1960a; Heusser 1968b; Licht 1974). Other sources of adult

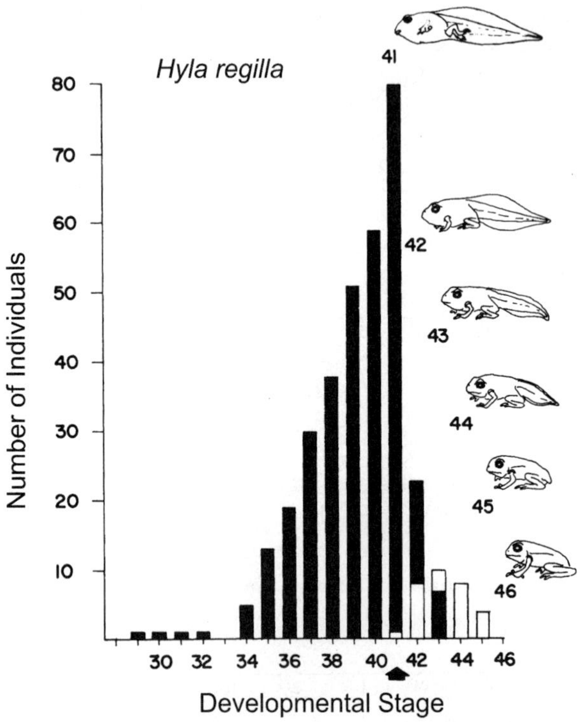


**Fig. 14.10.** Frequency distribution of *Pseudacris regilla* of different metamorphic stages taken from a pond (black bars) and from the stomachs of garter snakes (*Thamnophis elegans* and *T. sirtalis;* open bars) in California. Arrow indicates the last premetamorphic stage. Developmental stages from Volpe (1961). Snakes prey disproportionately on early metamorphic stages. After Arnold and Wassersug (1978).

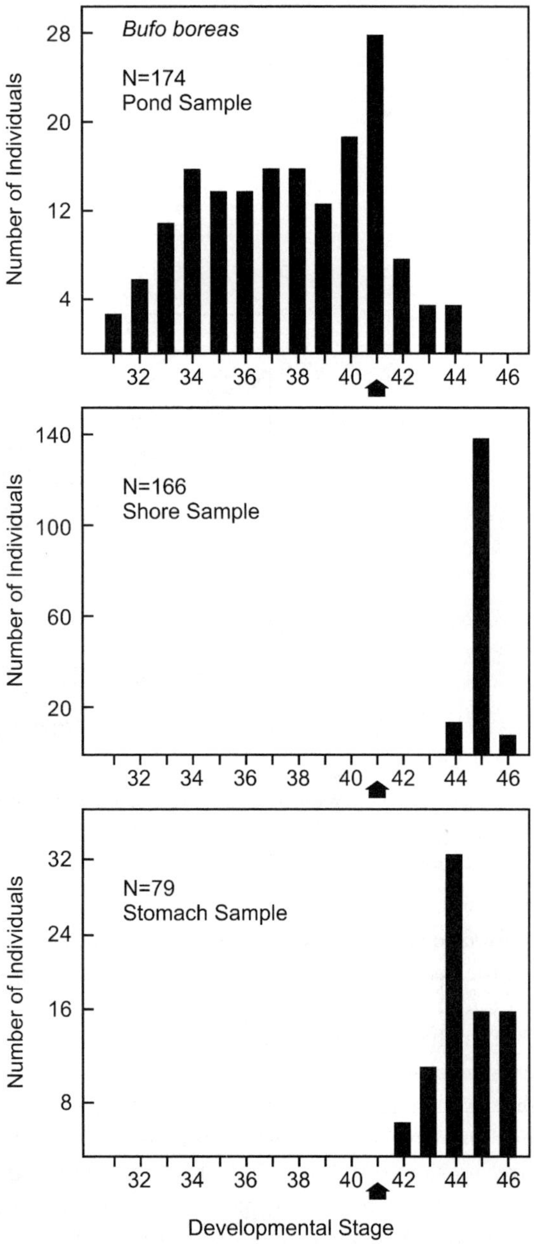


**Fig. 14.11.** Frequency distribution of *Bufo boreas* of different metamorphic stages in a pond, along the shore, and in the stomachs of garter snakes (*Thamnophis elegans* and *T. sirtalis*) in California. Arrows indicate the last premetamorphic stage. After Arnold and Wassersug (1978).

mortality, such as disease and winter kills (Kelleher and Tester 1969; R. D. Howard 1988a), often are ignored.

Survivorship of postmetamorphic individuals can be determined from recaptures of marked individuals, or by examining the age structure of the population. Mark-recapture studies have been most successful with species that return to the same pond each year to breed. Sometimes, most of the population can be collected by hand (e.g., Pearson 1955; F. Turner 1960a; Heusser 1968b; W. Brown and Alcala 1970; Licht 1974; Gittins 1983c) or intercepted with a drift fence surrounding the breeding pond (e.g., Husting 1965; G. Bell 1977; Gill 1985; Caldwell 1987; Kuhn 1994a). There are several potential sources of error in using mark-recapture data to calculate survivorship. The census effort may not be complete, so only part of the marked cohort is recaptured each year. Individuals may not return to the same breeding site and therefore will be lost from the recapture sample, or they may not breed every year, so surviving animals will not be counted in a short-term study. There may be sex differences in frequency of breeding, with one sex (often males) returning every year to breed, while the other returns less frequently, making comparison of male and female survivorship difficult.

Probably the first source of error, incomplete counting, is the one most commonly encountered in such studies. Some workers have simply equated "minimum survivorship" with the proportion of marked individuals recaptured in a subsequent year. However, if the probability of recapturing any surviving individual is substantially less than 100%, then these data almost certainly will underestimate actual survivorship, because marked individuals that are still alive will be missed by chance (Turner 1962). There are various statistical techniques available to adjust mark-recapture data for the probability of recapture and to use estimates of total population size to calculate annual survivorship (Seber 1973; Southwood 1978; Begon 1979), or to use capture-recapture data to model survivorship probabilities (K. Wood et al. 1998; B. Schmidt and Anholt 1999; Henle 2001, 2005; B. Schmidt, Schaub, and Anholt 2002; Schmidt, Feldmann, and Schaub 2005). Unfortunately, many studies of amphibian populations have not employed these techniques. Estimates of adult survivorship based on unadjusted recapture percentages (table 14.6) generally are quite low compared to corrected estimates (table 14.7). Interspecific comparisons of adult survivorship based on figures published in the literature can be quite misleading if the methods used to calculate survivorship are not considered (e.g., Duellman and Trueb 1986). Even the data in table 14.7 are not strictly comparable, because some studies used only relatively simple corrections for the probability of recapture, whereas others used more sophisticated statistical models to estimate survivorship (e.g., B. Schmidt, Schaub, and Anholt 2002).

Keeping in mind the limitations of the data, it appears that annual adult survivorship in medium to large amphibians sometimes is better than 50% and in some cases approaches 90%, although in harsh environments, year-to-year variation in adult survival can be considerable (e.g., Grafe et al. 2004). Estimates of annual survivorship in salamanders derived from the age structure of the population, rather than mark-recapture data, are similar. For example, estimated annual survivorship was 41% for *Eurycea wilderae* (Bruce 1988a), 81% for *Plethodon jordani* (Hairston 1983b), and 79% for *Salamandra luschani* (Olgun, Miaud, and Gautier 2001). Some small amphibians apparently have much lower annual survivorship than the medium to large species shown in table 14.7, but reliable data are scarce. Caldwell (1987) concluded that small *Pseudacris* are essentially annual species once they reach sexual maturity, with only a small fraction of the population surviving to a second breeding season. Age estimates for spring peepers (*P. crucifer*) in breeding choruses suggest that some adults survive to

**Table 14.6** Estimates of minimum annual survivorship of adult amphibians based on uncorrected recapture data

| Species (source no.) | Duration of study (years) | Number marked | Annual survivorship (%) |
|---|---|---|---|
| *Bufo bufo* (12)[a] | 3 | 4,084 | 5–25 |
| *B. fowleri* (9) | 2 | 563 | 30 |
| *B. hemiophrys* (6) | 5 | 385 | 32 |
| *B. valliceps* (1) | 2 | 178 | 11 |
| *B. valliceps* (5) | 2 | 109 | 4 |
| *B. woodhousii* (5) | 2 | 208 | 14 |
| *B. woodhousii* (8) | 2 | 114 | 2 |
| *Eleutherodactylus marnockii* (2) | 4 | 255 | 40 |
| *Gastrophryne olivacea* (4) | 3 | 6 | 17 |
| *Hyla arborea* (13) | 2 | 63 | 30 |
| *Pelobates fuscus* (14) | 4 | 929 | 31 |
| *Pseudacris clarkii* (4) | 3 | 14 | 57 |
| *P. nigrita* (10) | 4 | > 300 | 11 |
| *P. ornata* (10) | 4 | > 800 | 33 |
| *P. regilla* (3) | 2 | 373 | 10 |
| *P. streckeri* (4) | 3 | 61 | 50 |
| *Rana erythraea* (7) | 4 | 1,152 | 5 |
| *R. temporaria* (11)[b] | 10 | 546 | 31 |
| *Scaphiopus couchii* (4) | 3 | 44 | 55 |

*Sources:* (1) W. Blair 1955; (2) Jameson 1955b; (3) Jameson 1956a; (4) Jameson 1956b; (5) Volpe 1960; (6) Kelleher and Tester 1969; (7) W. Brown and Alcala 1970; (8) L. Brown and Brownell 1971; (9) Clarke 1977; (10) Caldwell 1987; (11) Elmberg 1990, (12) Kuhn 1994a; (13) Friedl and Klump 1997; (14) Hels 2002.

*Note:* In these studies, animals that were not recaptured were assumed to be dead.

[a]Data for females only.

[b]Data for males only.

**Table 14.7** Estimates of annual survivorship of adult amphibians based on recapture data corrected for probability of recapture, or on age distributions and life table data

| Species (source no.) | Duration of study (years) | Number marked | Annual survivorship (%) |
|---|---|---|---|
| Anurans | | | |
| *Bufo bufo* (7)[a] | 5 | > 11,000 | 77 |
| *B. bufo* (13) | 4 | > 10,000 | 62 |
| *B. bufo* (33)[b] | 3 | 2385 | 38 |
| *B. calamita* (25) | 7 | > 400 | 40 |
| *B. fowleri* (19) | 5 | < 2,000 | 58 |
| *B. punctatus* (3) | 2 | 34 | 40 |
| *Crinia signifera* (26) | 4 | | 70 |
| *Hemisus marmoratus* (35) | 6 | 821 | 6–53 |
| *Leptodactylus pentadactylus* (21) | 1 | 269 | 15 |
| *Pelobates fuscus* (24) | 7 | 1,632 | 19–50 |
| *Rana aurora* (9) | 4 | > 50 | 69 |
| *R. cascadae* (8) | 2 | 483 | 53 |
| *R. esculenta* (31) | 4 | 1,376 | 53–70 |
| *R. lessonae* (31) | 4 | 861 | 72–84 |
| *R. luteiventris* (9) | 4 | > 100 | 64 |
| *R. luteiventris* (2)a | 4 | > 1,300 | 61 |
| *R. temporaria* (18) | 2 | 2,000 | 45 |
| *R. vaillanti* (27) | 1 | 1,851 | 0 |
| *Scaphiopus holbrookii* (1)[c] | 3 | 153 | 60 |
| Urodeles | | | |
| *Ambystoma californiense* (30) | 8 | | 28–71 |
| *A. maculatum* (5) | 5 | 1,244 | 63–94 |
| *A. talpoideum* (20) | 7 | > 600 | 74 |
| *Cryptobranchus alleganiensis* (11, 16) | | | 80–90 |

*(continued)*

breed for two or three years, but most disappear after their first year (Lykens and Forester 1987; Zimmitti 1999). Small plethodontid salamanders, however, sometimes live longer than do larger species. Estimated annual survivorship of small species of *Desmognathus,* such as *D. wrighti, D. ocoee,* and *D. orestes,* was higher than for larger and more aquatic species such as *D. monticola* and *D. fuscus* (table 14.7).

Most of the data in table 14.7 are for temperate-zone species. Data for tropical species are scarce, but there is some evidence that even relatively large tropical anurans suffer much higher rates of predation than do temperate-zone amphibians. For example, Kluge (1981) never recaptured an adult male *Hyla rosenbergi* in two successive years, and he concluded that many individuals lived for only a few months after entering a breeding chorus. The population appeared to be under heavy pressure from a variety of predators, including snakes, caimans, mammals, and larger frogs. Similarly, most adults survived for less than a year in a population of *Leptodactylus pentadactylus* from Brazil (Galatti 1992) and a population of *Rana vaillanti* from Mexico (Ramirez, Vogt, and Villarreal-Benitez 1998). In contrast, many North American ranids of similar size live up to 10 years (see the following). The reasons for the low survivorship of these relatively large frogs are unclear, but the abundance of tropical snakes that feed mostly on frogs probably contributes to high mortality (Vitt and Vangilder 1983; Greene 1988; Cadle and Greene 1993). Some small African frogs (*Phrynobatrachus* and *Arthroleptis*), which are about the same size as *Pseudacris,* apparently live only two to seven months after reaching sexual maturity (Barbault and Trefaut Rodrigues 1978b, 1979a, b; Barbault and Pilorge 1980). Adult mortality has been attributed mainly to predation (Barbault 1984, 1987), but the evidence is circumstantial. In tropical savannas, adults of some small species have difficulty surviving the severe dry season (Geise and Linsenmair 1986; Rödel 1996), so physi-

**Table 14.7** (continued)

| Species (source no.) | Duration of study (years) | Number marked | Annual survivorship (%) |
|---|---|---|---|
| *Desmognathus fuscus* (4)[a] | 3 | | 58 |
| *D. fuscus* (10) | 5 | | 23–43 |
| *D. monticola* (4)[a] | 3 | | 50 |
| *D. monticola* (23) | 5 | | 50–60 |
| *D. ocoee* (13) | 7 | 758 | 63–74 |
| *D. orestes* (4)[a] | 3 | | 67 |
| *D. wrighti* (4)[a] | 3 | | 79 |
| *Mertensiella caucasica* (22) | 5 | < 500 | 77 |
| *Notophthalmus viridescens* (17) | 10 | > 7,000 | 67–79 |
| *Plethodon jordani* complex (15) | 5 | | 81 |
| *P. kentucki* (32) | 7 | 42 | 84 |
| *Taricha rivularis* (6) | > 10 | > 20,000 | 91 |
| *Triturus alpestris* (29) | | | 81 |
| *T. alpestris* (34) | 5 | 1,385 | 29–82 |
| *T. cristatus* (28) | 8 | | 60 |
| *T. dobrogicus* (24) | 7 | 689 | 34–54 |
| *T. vulgaris* (12) | 3 | 1,000 | 50 |

*Sources:* (1) Pearson 1955; (2) F. Turner 1960a; (3) F. Turner 1960b; (4) Organ 1961a; (5) Husting 1965; (6) Twitty 1966; (7) Heusser 1968d; (8) Briggs and Storm 1970; (9) Licht 1974; (10) Danstedt 1975; (11)Taber, Wilkinson, and Topping 1975; (12) G. Bell 1977; (13) Tilley 1980; (14) Gittins 1983c; (15) Hairston 1983; (16) C. L. Peterson et al. 1983; (17) Gill 1985; (18) Ryser 1986; (19) Breden 1988; (20) Raymond and Hardy 1990; (21) Galatti 1992; (22) Tarkhnishvili and Serbinova 1993; (23) Bruce 1995; (24) Jehle, Hödl, and Thonke 1995; (25) Sinsch and Seidel 1995; (26) Williamson and Bull 1996; (27) Ramirez, Vogt, and Villarreal-Benitez 1998; (28) J. M. R. Baker 1999; (29) Miaud, Miaud, Guyétant, and Faber 2000; (30) Trenham et al. 2000; (31) Holenweg Peter 2001; (32) Marvin 2001; (33) B. Schmidt, Schaub, and Anholt 2002 (based on partial data set of Kuhn 1994a); (34) Perret et al. 2003; (35) Grafe et al. 2004.

*Note:* Number marked shown only for mark-recapture studies.

[a]Survivorship of males only.

[b]Survivorship of females only.

[c]Recaptures not adjusted for probability of recapture, but census is relatively complete.

cal factors alone could result in nearly annual turnover in the population.

The data discussed so far provide an estimate of overall adult survivorship, but not age-specific mortality. The latter can be estimated from mark-recapture data if the ages of marked individuals are known (e.g., F. Turner 1960a; Tilley 1980; Gill 1985). More commonly, age-specific survivorship is estimated by constructing a life table based on the age distribution of the population (e.g., Organ 1961a; Briggs and Storm 1970; Danstedt 1975; G. Bell 1977; Hairston 1983b; Gibbons and McCarthy 1984; Khonsue, Matsui, and Misawa 2002; Marunouchi, Kusano, and Ueda 2002). The disadvantage of this method is that it assumes a stable population size and a stable age distribution. Hence, the method is inappropriate for many aquatic-breeding species that experience boom or bust population cycles, with high recruitment of juveniles in some years and virtually no recruitment in others (e.g., Shoop 1974; Semlitsch 1983b, 1987d; Caldwell 1987; Stenhouse 1987). This method is more appropriate for some terrestrial salamanders that have relatively stable populations (Hairston 1983b, 1987).

For many years, one difficulty in estimating age-specific survivorship was the lack of reliable methods to age adult amphibians (F. Turner 1962; Halliday and Verrell 1988). Traditionally, most workers estimated ages from size distributions, coupled with growth rates of marked individuals (e.g., Briggs and Storm 1970; G. Bell 1977; Hairston 1983b; Shirose et al. 1993; Shirose and Brooks 1995a, b). In some salamanders, males can be crudely aged from the number of lobes on the testes (Organ 1961a; Danstedt 1975; Bruce 1976), but the relationship between age and number of testis lobes is not very precise (Tilley 1977). The best technique is the use of growth rings on bones to age adults. These rings form due to slower growth at certain seasons of the year, either during the winter or in extended dry periods. Bone growth rings can be counted in most temperate-zone species and in tropical frogs that live in regions with distinct wet and dry seasons, but they may not be visible in tropical species that live in continuously wet habitats.

This technique has been available for some time and is now widely used for studies of amphibian populations (Senning 1940; Schroeder and Baskett 1968; Kleinenberg and Smirina 1969; Barbault et al. 1979; Francillon 1979; Castanet and Pilorge 1980; Hemelaar 1981, 1983, 1985; Hemelaar and van Gelder 1980; Dolmen 1982a; Gittins, Steeds, and Williams 1982; Gibbons and McCarthy 1984; and references cited in table 14.8). It is particularly useful when phalanges can be used for aging, because these can be removed without killing the animal, allowing the same animals to be used for aging and for mark-recapture studies. In many amphibian populations, mortality appears to be relatively constant across all age classes of adults (Hairston 1987; Galatti 1992; Jehle, Hödl, and Thonke 1995), indicating that most species have Type II adult survivorship curves (Deevey 1947). In some cases, however, smaller, and presumably younger, individuals suffer higher mortality than do larger, older individuals (Shirose and Brooks 1995a).

If predation is an important source of mortality, then one might expect mortality to be higher in the sex that is most conspicuous to predators or in the poorest condition because of energy expenditures during the breeding season. For example, predators such as bats, possums, and frogs are known to locate male frogs by their calls (see previous discussion and chapter 7), so one might expect to find higher rates of mortality in male than in female frogs. The opposite pattern might be expected in many salamanders, because males of most species do not have conspicuous signals equivalent to frog calls, and females probably expend more energy in reproduction (see chapter 10).

In fact, it is hard to generalize from the existing data. A fairly common pattern in skeletochronological studies of age structure in large ranid frogs is for females to be slightly older than males (e.g., Khonsue et al. 2001a, b; Marunouchi, Kusano, and Ueda 2002), but this is not always the case (e.g., Khonsue, Matsui, and Misawa 2002). The same pattern was found in several subtropical hylid frogs from Australia and a myobatrachid frog (Morrison, Hero, and Browning 2004). It is not always clear, however, that sex differences in maximum age can be attributed to differential predation on males. Similar studies of salamanders often reveal relatively little difference in age, if any, between males and females (e.g., Lima, Arntzen, and Ferrand 2000; Miaud, Guyétant, and Faber 2000), but there also are populations in which females apparently live longer than males (e.g., *Salamandra luschani:* Olgun et al. 2001) and others in which males apparently live longer than females (e.g., *Desmognathus monticola* and *D. quadramaculatus:* Bruce, Castanet, and Francillon-Vieillot 2002; *Euproctus platycephalus:* Bovero et al. 2003). In none of these cases are the causes of sex differences in logevity understood.

Mark-recapture studies have sometimes been used to estimate differences in age-specific mortality between the sexes, although many of these studies have not corrected for possible sex differences in recapture probabilities. Males appeared to have higher age-specific mortality rates in populations of *Rana pretiosa* (Licht 1974), *Rana catesbeiana* (Shirose et al. 1993), *R. grylio* (K. Wood et al. 1998), and *Triturus vulgaris* (G. Bell 1977). Apparent female mortality was higher in *Rana cascadae* (Briggs and Storm 1970), *Ambystoma maculatum* (Husting 1965), some populations of *Notophthalmus viridescens* (Gill 1985), and several species of *Desmognathus* (Organ 1961a). Female mortality also appeared to be slightly higher for populations of *Triturus dobrogicus* and *Pelobates fuscus* in Austria, although not consistently so in all years (Jehle, Hödl, and Thonke 1995). There was no apparent difference in male and female mortality in some populations of *Desmognathus* (Tilley 1980), a population of *Rana luteiventris* (F. Turner 1960a), and a population of *Rana temporaria* (Gibbons and McCarthy 1984). Several studies of frogs provide evidence that reproductive activities expose males to higher levels of predation than females. Caldwell (1987) found that survivorship of male *Pseudacris* was much lower than that of females, but only in years when the population was actually breeding. Calderón and Collado (1976) compared the sizes of adult *Rana ridibunda* captured by barn owls (*Tyto alba*) with sizes in natural populations and found that the owls preyed selectively on male frogs, possibly because they were attracted to their calls. In a study of agile frogs (*Rana dalmatina*) in France, Lodé et al. (2004) reported that polecats (*Mustela putorius*) took a much higher toll on males in breeding choruses than on females, resulting in shifts in the sex ratio that altered male mating tactics.

Many amphibians are surprisingly long-lived for such small animals. Large anurans in the temperate zone, many of which do not become sexually mature until they are several years old, often have maximum life spans of five to ten years, while smaller frogs have maximum life spans of about four years (table 14.8). Even relatively small frogs that are well protected against predators by toxic skin secretions may have very long life spans. For example, Płytycz and Bigaj (1993) estimated that some yellow-bellied toads (*Bombina bombina*) live as long as 20 years in the wild. Highly toxic newts and other salamandrids also tend to have long life spans, with many species reaching maximum ages of 12–15 years, and some living as long as 20 years (table 14.8). More surprising is the relatively long life span of plethodontid salamanders, many of which live at least 10–12 years (table 14.8). Some species, such as members of the *Plethodon jordani* and *P. glutinosus* complexes, are well protected by

**Table 14.8** Estimates of maximum longevity of amphibians in the field, determined from mark-recapture studies or skeletochronology

| Species (source no.) | Location | Method | Maximum longevity (years) |
|---|---|---|---|
| Anurans | | | |
| *Alytes cisternasii* (46) | Spain | Skeletochronology | 6 |
| *A. obstetricans* (46) | Spain | Skeletochronology | 6–7 |
| *Ascaphus montanus* (9) | Montana | Mark-recapture | 14+ |
| *Bombina variegata* (28) | Poland | Mark-recapture | 20 |
| *Bufo americanus* (23) | Virginia | Skeletochronology | 5 |
| *B. bufo* (3) | Switzerland | Mark-recapture | 10 |
| *B. bufo* (19) | Switzerland | Skeletochronology | 12 |
| *B. bufo* (19) | Germany | Skeletochronology | 8 |
| *B. bufo* (19) | France | Skeletochronology | 9 |
| *B. bufo* (19) | Netherlands | Skeletochronology | 9 |
| *B. bufo* (19) | Norway | Skeletochronology | 11 |
| *B. calamita* (36) | Germany | Mark-recapture | 7 |
| *B. calamita* (10) | Ireland | Skeletochronology | 7 |
| *B. hemiophrys* (4) | Minnesota | Mark-recapture | 6 |
| *Bombina bombina* (73) | Romania | Skeletochronology | 5 |
| *Geocrinia alba* (51) | Australia | Skeletochronology | 6 |
| *G. vitellina* (51) | Australia | Skeletochronology | 6 |
| *Hyla arborea* (45) | Germany | Skeletochronology | 6 |
| *Litoria chloris* (76) | Australia | Skeletochronology | 6 |
| *L. lesueuri* (76) | Australia | Skeletochronology | 5 |
| *L. pearsoniana* (76) | Australia | Skeletochronology | 3 |
| *Microhyla ornata* (64) | India | Skeletochronology | 5 |
| *Mixophyes fleayi* (76) | Australia | Skeletochronology | 8 |
| *Nectophrynoides occidentalis* (57) | Ivory Coast | Skeletochronology | 5 |
| *Pelobates fuscus* (33) | Austria | Mark-recapture | 9–10 |
| *Pelobates fuscus* (52) | France | Skeletochronology | 8 |
| *Pseudacris crucifer* (55) | Connecticut | Skeletochronology | 3 |
| *P. crucifer* (18) | Maryland | Skeletochronology | 5 |
| *P. nigrita* (15) | South Carolina | Mark-recapture | 3 |
| *P. ornata* (15) | South Carolina | Mark-recapture | 3 |
| *Rana catesbeiana* (29, 35) | Ontario | Mark-recapture | 9–10 |
| *R. clamitans* (35) | Ontario | Mark-recapture | 6–7 |
| *R. dalmatina* (48) | Germany | Mark-recapture | 5–6 |
| *R. esculenta* complex (73) | Romania | Skeletochronology | 10 |
| *R. iberica* (58) | Spain | Skeletochronology | 4–5 |
| *R. japonica* (69) | Japan | Skeletochronology | 4 |
| *R. latastei* (74) | Italy | Skeletochronology | 4 |
| *R. luteiventris* (1) | Wyoming | Mark-recapture | 9–11 |
| *R. nigromaculata* (62) | Japan | Skeletochronology | 5–6 |
| *R. perezi* (41) | Spain | Skeletochronology | 3–6 |
| *R. pipiens* (17) | Quebec | Skeletochronology | 4–5 |

*(continued)*

**Table 14.8** (continued)

| Species (source no.) | Location | Method | Maximum longevity (years) |
|---|---|---|---|
| *R. porosa brevipoda* (69) | Japan | Skeletochronology | 4 |
| *R. rugosa* (63) | Japan | Skeletochronology | 4–5 |
| *R. saharica* (53) | Morocco | Skeletochronology | 6 |
| *R. sakuraii* (34) | Japan | Skeletochronology | 5 |
| *R. septentrionalis* (43) | Quebec | Skeletochronology | 4 |
| *R. septentrionalis* (35) | Ontario | Mark-recapture | 6 |
| *R. subaquavocalis* (47) | Arizona | Skeletochronology | 11 |
| *R. sylvatica* (49) | Quebec | Skeletochronology | 4 |
| *R. temporaria* (48) | Germany | Mark-recapture, skeletochronology | 6–7 |
| *R. temporaria* (10) | Ireland | Skeletochronology | 7 |
| *R. temporaria* (14, 20) | Switzerland | Mark-recapture, skeletochronology | 8–11 |
| *R. temporaria* (54) | France | Skeletochronology | 12–15 |
| *Scaphiopus couchii* (37) | Arizona | Skeletochronology | 11–13 |
| Urodeles | | | |
| *Ambystoma californiense* (60) | California | Skeletochronology | 10+ |
| *A. maculatum* (26) | Quebec | Skeletochronology | 32 |
| *Batrachoseps attenuatus* (38) | California | Skeletochronology | 8 |
| *Calotriton asper* (24) | Spain | Skeletochronology | 26 |
| *C. asper* (78) | Spain | Skeletochronology | 16–28 |
| *Chioglossa lusitanica* (61) | Portugal | Skeletochronology | 8 |
| *Desmognathus monticola* (39) | North Carolina | Skeletochronology | 11 |
| *D. ocoee* (39) | North Carolina | Skeletochronology | 10 |
| *D. quadramaculatus* (39, 66) | North Carolina | Skeletochronology | 15 |
| *Euproctus platycephalus* (72) | Italy | Skeletochronology | 17 |
| *Mertensiella caucasica* (30, 31) | Georgia Republic | Skeletochronology | 25+ |
| *M. luschani* (66) | Turkey | Skeletochronology | 10 |
| *Notophthalmus viridescens* (25) | Maryland | Skeletochronology | 10 |
| *N. viridescens* (13) | Virginia | Mark-recapture | 12–15 |
| *N. viridescens* (40) | Quebec | Skeletochronology | 9–13 |
| *Plethodon kentucki* (65) | Kentucky | Mark-recapture | 13–16 |
| *P. metcalfi* (11) | North Carolina | Mark-recapture | 25[a] |
| *P. metcalfi* (71) | North Carolina | Skeletochronology | 8–10 |
| *P. teyahalee* (11) | North Carolina | Mark-recapture | 25[a] |
| *Salamandra salamandra* (16) | Germany | Mark-recapture | 20+ |
| *S. salamandra* (32) | Israel | Mark-recapture | 20 |
| *S. salamandra* (56) | Iberian Peninsula | Skeletochronology | 12+ |
| *Taricha rivularis* (2) | California | Mark-recapture | 10+ |
| *Triturus alpestris* (5) | Czech Republic | Skeletochronology | 10 |
| *T. alpestris* (42) | Austria | Skeletochronology | 20+ |
| *T. alpestris* (59) | France | Skeletochronology | 20 |
| *T. boscai* (21, 50) | Portugal | Skeletochronology | 6–8 |
| *T. cristatus* (22, 27) | France | Skeletochronology | 14–16 |
| *T. cristatus* (7, 8) | Sweden | Skeletochronology | 16 |

*(continued)*

**Table 14.8** (continued)

| Species (source no.) | Location | Method | Maximum longevity (years) |
|---|---|---|---|
| *T. dobrogicus* (33, 44) | Austria | Mark-recapture | 9 |
| *T. dobrogicus* (73) | Romania | Skeletochronology | 5 |
| *T. karelinii* (79) | Turkey | Skeletochronology | 8–11 |
| *T. marmoratus* (22) | France | Skeletochronology | 14 |
| *T. marmoratus* (68) | France | Skeletochronology | 9 |
| *T. marmoratus* (12, 21) | Portugal | Skeletochronology | 11–16 |
| *T. marmoratus* (75) | France | Skeletochronology | 9 |
| *T. vittatus ophryticus* (77) | Turkey | Skeletochronology | 10–16 |
| *T. vulgaris* (6) | England | Mark-recapture | 11–12 |
| *T. vulgaris* (7, 8) | Sweden | Skeletochronology | 20 |
| *T. vulgaris* (73) | Romania | Skeletochronology | 6 |

*Sources:* (1) F. Turner 1960a; (2) Twitty 1966; (3) Heusser 1968d; (4) Kelleher and Tester 1969; (5) Smirina and Roček 1976; (6) G. Bell 1977; (7) Hagström 1977; (8) Hagström 1979; (9) Daugherty and Sheldon 1982; (10) Gibbons and McCarthy 1983; (11) Hairston 1983b; (12) Caetano et al. 1985; (13) Gill 1985; (14) Ryser 1986; (15) Caldwell 1987; (16) Feldmann 1987; (17) Leclair and Castanet 1987; (18) Lykens and Forester 1987; (19) Hemelaar 1988; (20) Ryser 1988b; (21) Caetano 1990; (22) Francillon-Vieillot et al. 1990; (23) Kalb and Zug 1990; (24) Montori 1990; (25) Forester and Lykens 1991; (26) Flageole and Leclair 1992; (27) Miaud et al. 1993; (28) Plytycz and Bigaj 1993; (29) Shirose et al. 1993; (30) Tarkhnishvili and Serbinova 1993; (31) Tarkhnishvili and Gokhelashvili 1994; (32) Warburg 1994; (33) Jehle et al. 1995; (34) Kusano et al. 1995; (35) Shirose and Brooks 1995a; (36) Sinsch and Seidel 1995; (37) Tinsley and Tocque 1995; (38) D. Wake and Castanet 1995; (39) Castanet et al. 1996; (40) Caetano and Leclair 1996; (41) Esteban et al. 1996; (42) Griffiths 1996; (43) Leclair and Laurin 1996; (44) Ellinger and Jehle 1997; (45) Friedl and Klump 1997; (46) Marquez et al. 1997; (47) Platz et al. 1997; (48) Kneitz 1998; (49) Sagor et al. 1998; (50) Caetano and Leclair 1999; (51) Driscoll 1999b; (52) Eggert and Guyetant 1999; (53) Esteban et al. 1999; (54) Miaud et al. 1999; (55) Zimmitti 1999; (56) Alcobendas and Castanet 2000; (57) Castanet et al. 2000; (58) Esteban and Sanchiz 2000; (59) Miaud et al. 2000, (60) Trenham et al. 2000; (61) Lima et al. 2000; (62) Khonsue et al. 2001b; (63) Khonsue et al. 2001a; (64) Kumbar and Pancharatna 2001; (65) Marvin 2001; (66) Olgun et al. 2001; (67) Bruce et al. 2002; (68) Jakob et al. 2002; (69) Khonsue et al. 2002; (70) Marunouchi et al. 2002; (71) Ash et al. 2003; (72) Bovero et al. 2003; (73) Cogălniceanu and Miaud 2003; (74) Guarino et al. 2003; (75) Jakob et al. 2003; (76) Morrison et al. 2004; (77) Kutrup et al. 2005; (78) Miaud and Guillaume 2005; (79) Olgun et al. 2005.

[a]Maximum longevity calculated from life table data.

noxious skin secretions, but others, such as *Desmognathus*, do not have such skin secretions and are readily eaten by many predators. Their surprisingly long life spans probably are related to their cryptic behavior, which reduces exposure to predators.

Average life expectancies for all of these amphibians are much shorter than the maximum life spans in table 14.8, because only a small percentage of adults will reach these advanced ages. Nevertheless, when compared to other terrestrial vertebrates, life expectancies of many amphibians are impressive. Annual adult survivorship of temperate-zone passerine birds is about 25–80% and averages about 50%; maximum life spans of many species are between 5 and 15 years, similar to those recorded for amphibians (Dorst 1974). Annual survivorship of adult snakes also is relatively high (35–80%), with an average of about 60% and maximum life spans of 4–20 years (W. Parker and Plummer 1987). Annual survivorship of small lizards (< 50 g) generally is much lower (0–83%, average = 28%, $N = 44$), and it is rare for individuals to live more than five years (F. Turner 1977). This figure may be somewhat biased by an overrepresentation of desert-dwelling lizards in the sample. Compared to amphibians, small mammals such as rodents and insectivores are extremely short-lived. Life spans generally are measured in months rather than years. It is rare for more than 10–15% of an adult population to survive one year (French, Stoddart, and Bobek 1975; Fleming 1979), although some tropical rodents have annual survivorship rates as high as 35–45% (Fleming 1975). Given the relatively high rates of survivorship and long life expectancies of some amphibians, it is worth reconsidering whether amphibians really are the "cannon fodder of the animal kingdom," as suggested at the beginning of this chapter (M. Tyler 1976). R. Savage (1961) argued many years ago that frogs lead relatively safe lives as adults compared to other terrestrial vertebrates, despite many statements in the literature to the contrary. Indeed, the real question is how amphibians manage to live so long despite their apparent vulnerability to the vast numbers of predators that feed on them. The final section of this chapter focuses on this question.

## Evolutionary Responses to Predation

As mentioned at the beginning of this chapter, predation has been an important selective force in the evolution of many behavioral and life-history traits of amphibians (table 14.1). Several of these are discussed in more detail in other chapters. These include the influence of predation on anuran vocal behavior (chapter 7), egg size and clutch size (chapter 10), parental care (chapter 11), aggregation of aquatic larvae (chapter 12), the length of the larval period and the timing of metamorphosis (chapter 13), and the organization of amphibian communities (chapter 15). The final section of this chapter focuses on adaptations that reduce the vulnerability of amphibians to predation at different stages of the life cycle.

## Protection of the Eggs

One might expect amphibian eggs to be particularly vulnerable to predation, because they are small and relatively unprotected, and they cannot move to escape from predators. Yet embryonic survivorship can be surprisingly high in many species. The vulnerability of amphibian eggs to predators depends in part on the types of predators present in the breeding habitat. Some predators may not be very good at finding amphibian eggs because of the sensory modes and foraging tactics used to acquire food. For example, dragonfly naiads that are visually oriented sit-and-wait predators probably are not major predators on amphibian eggs. Actively foraging dytiscid beetle larvae, on the other hand, often use chemical cues to detect prey, and have little difficulty locating and consuming amphibian eggs (Resetarits 1998). Other active predators, such as leeches, caddisfly larvae, crayfish, freshwater shrimp, fish, tadpoles, salamander larvae, and newts, probably can readily find amphibian eggs, especially if they rely on chemical cues to locate prey. Some of these predators may be deterred by protective adaptations of the eggs, including chemical defenses and mechanical defenses of the egg capsule and the surrounding jelly.

### Chemical and Mechanical Defenses

Some amphibian eggs contain toxic or distasteful compounds that deter predators, but the effectiveness of these chemical defenses depends on the predator. For example, eggs of *Rana clamitans* and *R. catesbeiana* are distasteful to newts (*Notophthalmus viridescens*) and larval ambystomatids (Walters 1975), but are readily consumed by leeches (R. D. Howard 1978b). The eggs of some species of *Bufo* contain toxic and distasteful chemicals that repel fishes (Licht 1968), but mosquitofish (*Gambusia*) can eat *Bufo valliceps* eggs with no ill effects (Grubb 1972). Some toad eggs also appear to be distasteful to salamander larvae and newts (Walters 1975; Denton and Beebee 1991), but others are readily consumed by anuran tadpoles (Heusser 1970, 1971b; Heyer, McDiarmid, and Weigmann 1975; Wells 1979; Banks and Beebee 1987b). The eggs of *Bufo marinus,* which has been introduced into Australia, are sufficiently toxic to kill the tadpoles of many native frogs. These species appear to have little behavioral avoidance of these toxic eggs, so *Bufo marinus* represents a serious threat to these frogs (Crossland 1998b; Crossland and Alford 1998). Many invertebrate predators, such as crayfish and insects with sucking mouthparts, seem not to be strongly deterred by the chemical defenses of toad eggs (Henrikson 1990; Axelsson et al. 1997). The chemical defenses of newt eggs (*Taricha*) also seem to be ineffective against crayfish (Gamradt and Kats 1996).

Even when eggs lack chemical defenses, they may derive some protection from egg capsules (Salthe 1963) and the jelly surrounding the egg mass. Grubb (1972) found that frogs breeding in permanent water (*Rana pipiens* and *Acris crepitans*) had larger, tougher egg capsules than species breeding in temporary water (*Scaphiopus couchii, Pseudacris clarki, Gastrophryne olivacea,* and *Hyla chrysoscelis*) and were eaten less frequently by mosquitofish. Unfortunately, this comparison is confounded by phylogenetic differences among the species examined, as well as differences in water temperature and oxygen availability that might affect egg structure (see chapters 4 and 10). The thick jelly coats surrounding the egg masses of some ranid frogs and ambystomatid salamanders provide good protection against predation by fishes and various invertebrates, whereas isolated eggs are quickly consumed (Werschkul and Christensen 1977; Ward and Sexton 1981). Thick jelly coats provide little protection against adult newts, which can rip individual eggs out of the masses (Morin 1983b). Caddisfly larvae also make short work of the jelly surrounding egg mass of *Ambystoma maculatum* (personal observations).

### Choice of Oviposition Sites

Many amphibian eggs lack both chemical and mechanical defenses. In these species, predation may be an important selective force influencing choice of oviposition sites by females. For example, many species of frogs appear to be limited to reproduction in temporary ponds because their eggs and tadpoles are highly palatable to predators in permanent ponds, especially fish (Licht 1969c; Grubb 1972; Walters 1975; Brönmark and Edenhamn 1994; Alford 1999). Even for species that breed only in temporary ponds, eggs may be in danger from a variety of predators, including invertebrates, tadpoles, salamander larvae, and newts. Consequently, females that can detect the presence of potential predators and avoid laying eggs in sites with predators should have an advantage over females that choose oviposition sites at random. Several investigators have tried to assess the ability of females to detect predators in potential oviposition sites.

In one of the first such studies, Resetarits and Wilbur (1989, 1991) studied the choice of oviposition sites by female *Hyla chrysoscelis* and the choice of calling sites by males. They used an array of artificial ponds (cattle tanks) placed in close proximity to one another. Some of the tanks contained predators and others did not. Resetarits and Wilbur found males were less likely to call in tanks inhabited by black-banded sunfish (*Enneacanthus chaetodon*) and by conspecific tadpoles than in control tanks without predators, but did not avoid tanks with *Ambystoma maculatum* larvae, newts (*Notophthalmus viridescens*), ranid tadpoles, or dragonfly naiads (*Tramea*). They also found fewer eggs in tanks with fish and those with *Ambystoma* larvae, and concluded that females avoided laying eggs in sites inhabited by these predators. Resetarits and Wilbur did not observe the movements of females or pairs in amplexus, nor did they observe actual oviposition. Their experimental design was criticized by Ritke and Mumme (1993), who argued that the individual cattle tanks probably were not treated as separate ponds by the frogs, but as subsections of one large breeding area. The experiment also is somewhat flawed because it used an indirect measure of oviposition site choice, the number of eggs seen in the ponds, rather than direct behavioral observations of females. There is a possibility that some eggs laid in tanks containing predators were consumed before the tanks were checked in the morning.

Crump (1991) conducted a similar experiment with a Neotropical frog, *Hyla pseudopuma*. She found that females laid more eggs in artificial pools without predators (conspecific tadpoles) than in those with predators. She also tested for predation of eggs by tadpoles and found that empty jelly capsules always remained as evidence of a predation event. Her results indicate that females can detect conspecific tadpoles in small pools and avoid depositing eggs in those pools. Females of an African ranid frog, *Hoplobatrachus occipitalis,* also avoid laying eggs in pools inhabited by conspecific tadpoles, which are highly cannibalistic (Spieler and Linsenmair 1997). Natural rock pools that already contained tadpoles were not used as oviposition sites, nor were experimental pools supplied with tadpoles. Females of this species not only could detect potentially cannibalistic tadpoles, but they differentiated between these tadpoles and herbivorous tadpoles (*Ptychadena maccarthyensis*) that did not pose a threat to their offspring. The presence of high densities of herbivorous tadpoles did not deter oviposition by females, whereas low densities of conspecific tadpoles did. These investigators had some evidence that predatory tadpoles are detected by chemical cues in the water, but more experimental work is needed to confirm this. Similar avoidance of pools with conspecific tadpoles has been reported for *Pleurodema borellii* from Argentina (Halloy and Fiaño 2000).

Another anuran that tends to avoid laying eggs in pools containing predatory tadpoles is the American toad (*Bufo americanus*). This species normally breeds in pools that lack predators such as fish (Holomuzki 1995), but the same sites sometimes are used by wood frogs (*Rana sylvatica*), which breed earlier in the spring. Wood frog tadpoles are voracious predators of the eggs and young tadpoles of *Bufo americanus.* Petranka, Hopey, Jennings, Baird, and Boone (1994) demonstrated that the toads always avoided laying eggs in experimental ponds that contained wood frog tadpoles, but readily laid eggs in identical nearby ponds that lacked tadpoles. When toad eggs were placed in experimental or natural ponds containing wood frog tadpoles, the toad eggs and tadpoles were virtually eliminated within a few days.

Several studies have tested the ability of female frogs to avoid breeding sites inhabited by predatory fish. Wood frogs (*Rana sylvatica*) never laid eggs in experimental ponds containing predatory sunfish (*Lepomis auritus*), but did use nearby fish-free ponds (Hopey and Petranka 1994). Adult frogs were seen in both types of ponds, but males were much more likely to call in ponds without fish. This suggests that males tend to avoid sites with predators, and female choice of oviposition sites probably is influenced in part by where males are calling. Binckley and Resetarits (2002) found that squirrel treefrogs (*Hyla squirella*) avoided using cattle tanks containing banded sunfish (*Enneacanthus obesus*) as oviposition sites, even though the fish were enclosed and therefore unable to prey on eggs or tadpoles. In a similar set of experiments with *Hyla chrysoscelis,* females chose among cattle tanks without fish or with one of five different species of fish in enclosures. Females avoided tanks with four of the fish species, but not one species that apparently is not a serious threat to their eggs (Binckley and Resetarits 2003). Not all frogs can avoid potential egg and tadpole predators, however. The European common frog (*Rana temporaria*) is ecologically similar to the North American wood frog. On an island in Finland, this species breeds in shallow rock pools. Some of these pools are inhabited by threespine sticklebacks (*Gasterosteus aculeatus*), which prey on tadpoles (Laurila 1998). Nevertheless, the frogs did not show any tendency to avoid pools into which sticklebacks had been introduced (Laurila and Aho 1997).

Among dendrobatid frogs, the presence of predators can affect choice of either oviposition sites or tadpole deposition sites. For example, females of *Dendrobates ventrimaculatus,* which breed in leaf axils, avoided sites that contained cannibalistic conspecific tadpoles in pools below potential oviposition sites (Summers 1999b). Downie, Livingstone, and Cormack (2001) reported that males of *Mannophryne trinitatis* avoided depositing tadpoles in pools containing fish (*Rivulus*) or freshwater shrimp (*Macrobrachium*). In some populations, males that were given only pools with predators actually deposited tadpoles in wet leaf litter in preference to pools. Prior evolutionary history with predators appeared to influence responses of the frogs. In areas where shrimp were uncommon, males deposited tadpoles in pools with shrimp, but not in pools with fish.

Some salamanders are able to detect and avoid predators in potential oviposition sites. Species such as *Ambystoma talpoideum* and *A. texanum* scatter eggs on the bottom of a pond. The eggs are highly vulnerable to fish predation, as are the larvae, and these species appear to be excluded from reproducing in sites inhabited by fishes (Petranka 1983; Semlitsch 1988). *Ambystoma barbouri* is one of the few North American ambystomatids to breed in streams, which often are inhabited by predatory fishes. Surveys of natural stream pools in Kentucky revealed that salamander eggs were much more likely to be laid in stream pools without fish than in those with fish. When the fish in one stream were killed by drought, pools that had contained fish in the previous year subsequently lacked fish and were then readily used by salamanders for oviposition (Kats and Sih 1992). Another stream-breeding salamander is the California newt (*Taricha rivularis*). The presence of introduced crayfish (*Procambarus clarkii*) in streams inhibits reproduction by the newts. In this case, the mechanism of predator detection is clear, because the crayfish aggressively attack adult newts, even though they are well protected by poisonous skin secretions (Gamradt, Kats, and Anzalone 1997).

Few investigators have considered the effects of variation among oviposition sites within breeding areas on predation of embryos. An exception is R. D. Howard's (1978b) study of *Rana catesbeiana* in a pond in Michigan. He found that eggs deposited in warm water developed more rapidly than those deposited in cooler water, and they were subjected to lower levels of leech predation. Hence, the quality of oviposition sites defended by males became a major determinant of mate choice by females (see also chapter 8). In other species, females may reduce predation on their eggs by scattering them in mud on the bottom of a pond or concealing them in vegetation. For example, the eggs of *Pseudacris crucifer* and *P. cadaverina* are attached individually to underwater plants (Gosner and Rossman 1960; Gaudin 1965), a mode of oviposition that presumably makes it difficult for predators to locate the eggs. Both North American and Eurasian newts (*Notophthalmus, Triturus*) wrap their eggs individually in folded leaves (S. Bishop 1941b; Steward 1969; Walters 1975). This has been shown to reduce rates of predation on eggs of *Triturus* (Miaud 1993).

### Timing of Reproduction

The timing of reproduction also may affect the survival of embryos. Heusser (1970, 1971b) suggested that predation by conspecific tadpoles favors synchronized reproduction in some frogs such as *Rana temporaria,* because earlier hatching tadpoles often consume eggs that are deposited late in the season. Petranka and Thomas (1995) tested this hypothesis by exposing wood frog (*Rana sylvatica*) eggs to conspecific tadpoles that had hatched earlier in the season. Many of the eggs were destroyed by cannibalism. Berninghausen (1998) questioned the generality of this explanation for the evolution of explosive breeding. He found relatively little evidence of predation on eggs by tadpoles of several European anurans. In most cases, the eggs seemed to be well protected from tadpole predation by jelly capsules. Nevertheless, as discussed earlier in this chapter, it is clear that tadpoles of many European anurans are fully capable of consuming eggs of their own or other species.

Cannibalism by conspecific tadpoles need not be the only selective pressure favoring explosive breeding in anurans. The presence of heterospecific predators could favor synchronous breeding as well. Walters (1975) proposed that explosive breeding is advantageous in swamping predators, thereby increasing the probability of survival for individual eggs. This would occur if the number of eggs deposited simultaneously were too large to be consumed by the resident predator population. A similar explanation was offered for the highly explosive breeding behavior of *Bufo alatus* (formerly *B. typhonius*), a tropical toad that breeds in permanent water and does not seem to be stimulated by rainfall or other environmental cues (Wells 1979). These tadpoles are subject to predation by the voracious tadpoles of *Leptodactylus pentadactylus.* Explosive breeding in early spring may allow some species to get through the egg stage before populations of predatory invertebrates or salamander larvae reach peak densities (Wilbur 1972; Wells1977a; Majecki and Majecka 1996; Alford 1999; see chapter 8 for further discussion of explosive breeding in anurans).

### Adaptive Plasticity in Hatching of Eggs

Warkentin (1995) described an unusual form of plasticity in hatching time in the eggs of *Agalychnis callidryas,* a tropical tree frog that lays eggs on leaves over water, where they are frequently preyed upon by cat-eyed snakes (*Leptodeira;* fig. 14.4 A). Once the eggs have reached a stage where they can survive independently in the water, they will hatch almost immediately if a clutch is attacked by a snake (fig. 14.12). Many of the tadpoles survive the snake attack by dropping into the water. This comes at a cost, however, because these early-hatching tadpoles are relatively small and not very mobile, and are therefore vulnerable to predation by aquatic predators such as freshwater shrimp (fig. 14.13). Nevertheless, such plasticity in hatching time is favored, because the tadpoles trade almost certain death from a predatory snake against a somewhat higher chance of surviving in the water. *Agalychnis* eggs exhibit similar plasticity in hatching in response to predation by wasps (*Polybia rejecta*), but in this case, the predator attacks individual eggs and usually only

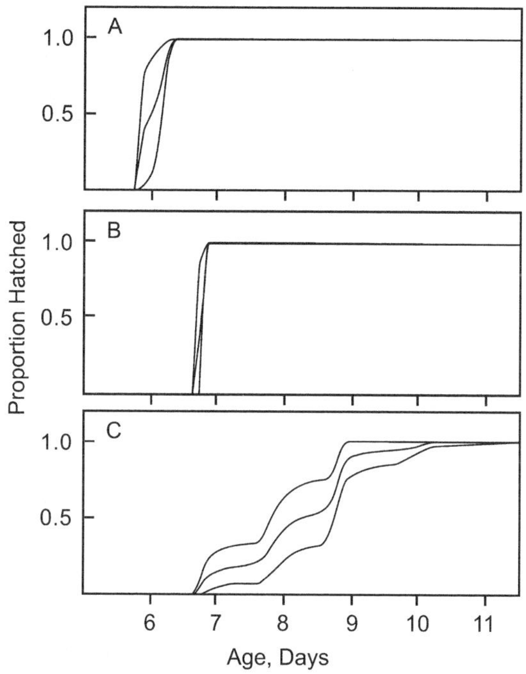


**Fig. 14.12.** Timing of hatching of egg clutches of *Agalychnis callidryas* with and without snake predation. (A) Eggs attacked by a snake (*Leptodeira septentrionalis*) at day 5. (B) Eggs attacked by a snake at day 6. (C) Control egg clutches that were not attacked. Snake attacks induced earlier and more synchronous hatching than that seen in control clutches. After Warkentin (1995).

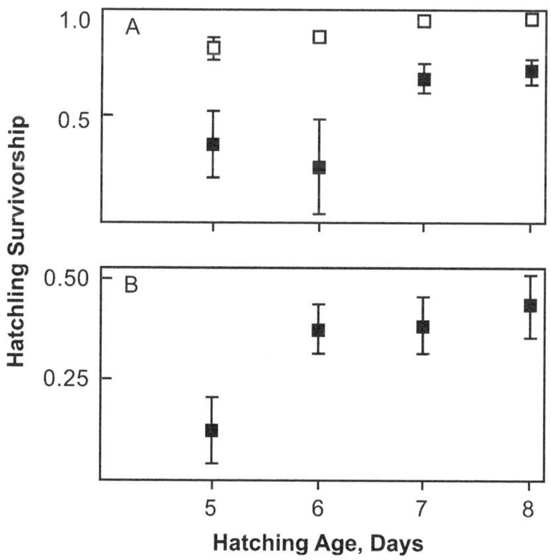


**Fig. 14.13.** Survivorship of hatchling *Agalychnis callidryas* tadpoles as a function of tadpole age in the presence of aquatic predators. (A) Survivorship of tadpoles with small shrimp (*Macrobrachium americana;* open squares) and large shrimp (solid squares). (B) Survivorship with fish (*Brachyraphis rhabdophora*). Large hatchlings are more likely to survive with both types of predators than are small hatchlings. After Warkentin (1995).

single embryos hatch in response to the predation attempt (Warkentin 2000b). This type of response is not limited to animal predators that create a mechanical stimulus that can trigger early hatching. *Agalychnis* eggs also will hatch early in response to infection with a pathogenic fungus. Eggs that are closest to the fungal infection hatch first, presumably responding to some sort of chemical cue (Warkentin, Currie, and Rehner 2001). Eggs of another tropical frog, *Hyperolius spinigularis,* from East Africa, are subject to predation by other frogs (*Afrixalus fornasini*) and ephydrid flies (*Typopsilopa*). They also exhibit plasticity in hatching, with embryos subjected to predator attacks hatching up to three to four days early and at sizes 20–30% smaller than embryos that complete normal development (Vonesh 2005a).

Embryos of the stream-breeding salamander *Ambystoma barbouri* also exhibit adaptive plasticity in time of hatching, but in this case, hatching is delayed in the presence of aquatic predators that threaten newly hatched larvae. One such predator is a free-living flatworm (*Phagocotus gracilis*), which feeds on newly hatched larvae, but cannot catch older larvae. When salamander eggs were exposed to flatworms, or chemical extracts from flatworms, they hatched later than those not exposed to these stimuli. Larvae emerging from delayed-hatching eggs not only were larger than control larvae, but also hatched at a more advanced stage (Sih and Moore 1993). The eggs of *A. barbouri* also hatch later and at a more advanced stage when exposed to chemical cues from green sunfish (*Lepomis cyanellus*), which are even more voracious predators on the larvae (R. Moore, Newton, and Sih 1996). In contrast, exposure to cues from dragonfly naiads (*Anax junius*), a potential larval predator, had little effect on hatching time in spotted salamanders (*Ambystoma maculatum*), wood frogs (*Rana sylvatica*), or southern leopard frogs (*Rana sphenocephala;* A. R. Anderson and Petranka 2003; J. B. Johnson et al. 2003). Embryos of *R. sphenocephala* did exhibit accelerated hatching in response to crayfish (*Procambarus nigrocinctus*), which are efficient egg predators (J. B. Johnson et al. 2003).

### Defenses of Larvae

Once an amphibian hatches from its egg, it often is exposed to a host of aquatic predators. As discussed previously, rates of mortality usually are much higher for larvae than for eggs. In part this is because the larval period is much longer than the period of egg development, but it also is due to the movements of larvae, which are necessary for finding food. The vulnerability and prolonged exposure of amphibian larvae to a wide variety of predators favors the evolution of defensive strategies that reduce the effectiveness of predators. One way to do this is to reduce the likelihood that predators will detect prey in the first place. This can be achieved by means of cryptic coloration that reduces the visibility of a prey species, by active detection and avoidance of dangerous predators, or by behavioral changes that reduce vulnerability to predators. Some tadpoles may simply flee from predators, and natural selection will tend to enhance morphological or physiological traits that are correlated with speed or maneuverability. Rapid growth can be a defense against predators as well, because tadpoles may reach a size that makes them less vulnerable to the most common predators in their habitats. Some larvae also are protected by chemical defenses, and in some cases, advertise their distastefulness with aposematic coloration.

### Color Patterns

For most amphibian larvae, cryptic coloration probably is the first line of defense against visually hunting predators. Amphibian larvae in general are not very colorful. Most species have background coloration in various shades of brown, gray, or green, often with mottling, spots, and other markings that match the background of their environments (see illustrations in chapter 12). Some larvae are countershaded, with lighter coloration on the ventral than on the dorsal surface. Many larvae, especially anuran tadpoles, exhibit general features of disruptive coloration that tend to break up the outline of the body or obscure key features such as the eyes. Disruptive color patterns include vertical bars or saddles on the body and tail, and stripes on the body, tail, or around the eye (Altig and Channing 1993). Midwater suspension feeders, including tadpoles of *Xenopus, Rhinophrynus,* and many microhylids and phyllomedusine hylids, often have light, highly reflective coloration or are nearly transparent. Some amphibian larvae have a limited ability to change color to match the reflectance of the background (Altig 1972b; Fernandez and Collins 1988), but color change is less common than in adult amphibians.

Storfer et al. (1999) examined the adaptive value of cryptic coloration in larvae of a salamander, *Ambystoma barbouri.* This species is found both in relatively permanent streams inhabited by predatory fish and in more ephemeral streams that lack fish. Larvae from streams with fish are lighter in color than those from fish-free streams, and behavioral experiments showed that fish could detect dark larvae more easily against a background of silt-covered rocks (fig. 14.14). There is variation in coloration among larvae in streams with fish, however, with some individuals being darker than are others. Some of this variation is due to periodic gene flow from nearby populations in fish-free streams, where darker coloration seems to be favored. The benefits of dark coloration in the absence of predators are not known, but may include thermoregulatory benefits, reduced exposure to ultraviolet radiation in shallow water, or better protection from terrestrial predators.

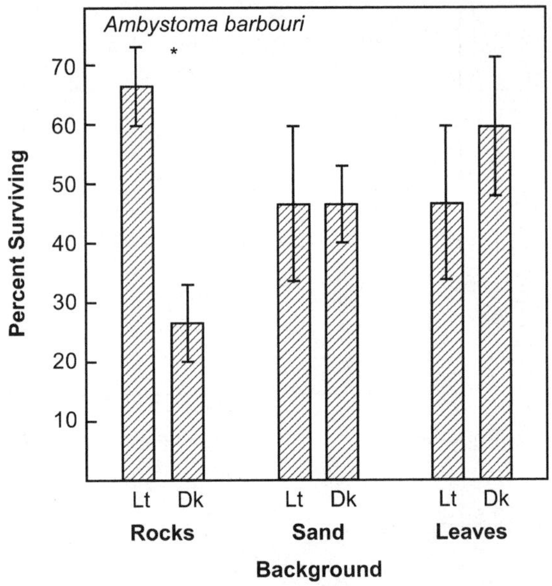


**Fig. 14.14.** Survivorship of salamander larvae (*Ambystoma barbouri*) of light- and dark-color morphs with fish predators against different backgrounds in laboratory aquaria. Light-colored larvae have an advantage over dark-colored larvae against a rock background resembling that of natural streams where the larvae are found. * indicates a significant difference. After Storfer et al. (1999).

Some tadpoles have markings that appear to serve as deflection marks that redirect attacks of predators toward the tail and away from the body and head (Van Buskirk et al. 2004). These include black tail tips, ocelli (fake eyes), and bright coloration (Caldwell 1982, 1986a; Altig and Channing 1993; McCollum and Van Buskirk 1996; McCollum and Leimberger 1997; Van Buskirk and McCollum 1999). Tadpoles often suffer considerable damage to tails without adverse effects on growth and survivorship (Semlitsch 1990; Wilbur and Semlitsch 1990; Figiel and Semlitsch 1991), although this is not always true (Parichy and Kaplan 1992). In many species, the tip of the tail is the area most likely to be injured, and this is the area that has the least effect on swimming performance (J. Blair and Wassersug 2000). Even if tail injury is somewhat costly, it is likely to be less so than attacks on other parts of the body, most of which are likely to be fatal (Van Buskirk et al. 2003).

### Phenotypic Plasticity

Anuran tadpoles exhibit considerable variation in color pattern, even within species. There are a number of examples of ontogenetic changes in color pattern that apparently are related to changes in microhabitat use or vulnerability to predators as tadpoles grow (Caldwell, Thorp, and Jervey 1980; Altig and Channing 1993). In some species, tadpoles differ in color depending on the habitat in which they live. Tadpoles that live in streams may differ in color from conspecifics living in ponds, or those living in clear water may differ from those in murky water. In cricket frogs (*Acris*), black tail tips are most common in populations inhabiting temporary ponds, where dragonfly naiads are the principal predators. Tadpoles living in permanent ponds with fish are more likely to have clear tails (Caldwell 1982). Tadpoles of a Neotropical frog, *Rana palmipes,* have darker coloration on the tail in the presence of predatory water bugs (McIntyre, Baldwin, and Flecker 2004).

Some of this variation in color pattern represents phenotypic plasticity that is induced by direct contact with predators. Tadpoles of gray treefrogs (*Hyla chrysoscelis* and *H. versicolor*) and pinewoods treefrogs (*H. femoralis*) often have deep tail fins with bright orange coloration and dark spots when they live in ponds with dragonfly naiads, but tend to have shallower, unmarked tail fins when such predators are absent (Van Buskirk and McCollum 1999; LaFiandra and Babbitt 2004; fig. 14.15). When tadpoles of *Hyla chrysoscelis* were exposed from an early age to free-ranging dragonfly naiads or to naiads in cages, they gradually developed the phenotypic traits typical of tadpoles in predator-rich ponds (fig. 14.16). In contrast, control tadpoles not exposed to predators retained the phenotype typical of those in predator-poor ponds. Tadpoles that were reared with predators also were less active and spent less time feeding than did those in predator-free treatments, but growth rate was affected only if tadpoles were exposed to free-ranging predators (McCollum and Van Buskirk 1996). Morphological and behavioral responses of *Hyla versicolor* tadpoles change with developmental stage, with younger tadpoles being more likely to alter behavior and older tadpoles being more likely to alter morphology. Nevertheless, there do not appear to be strict developmental windows that constrain the expression of morphological plasticity, because changes in morphology can be induced at almost any developmental stage and are reversible if predators are removed (Relyea 2003c), a pattern seen in other anurans as well (Kishida and Nishimura 2004).

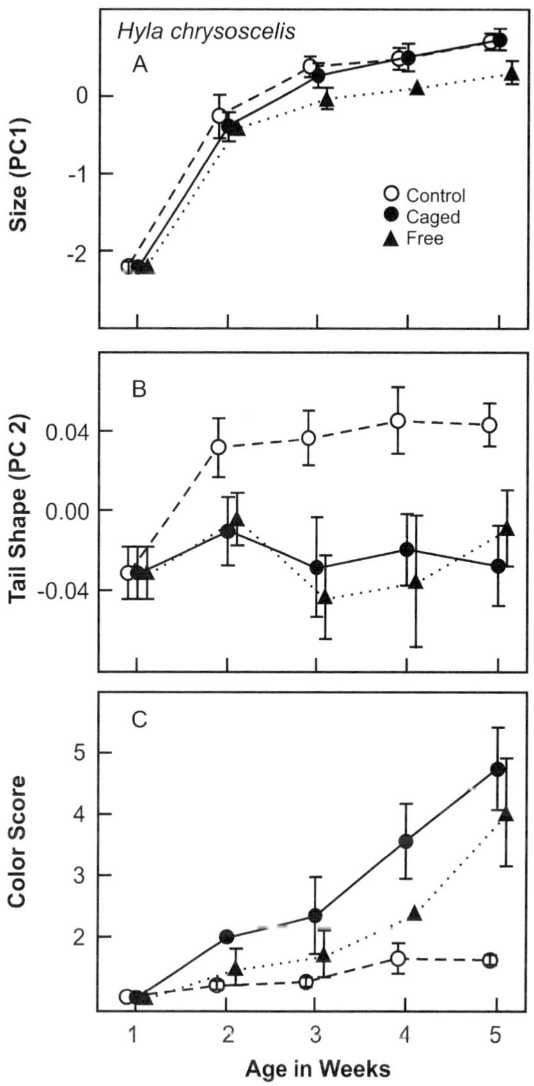


**Fig. 14.16.** Ontogenetic trajectories of the phenotypes of *Hyla chrysoscelis* tadpoles raised with caged or free-ranging predatory dragonfly naiads or in control treatments without dragonflies. (A) Tadpoles exposed to free-ranging predators were significantly smaller (lower PC1 score) than those exposed to caged predators or no predators. (B) Tadpoles exposed to either caged or free-ranging predators had deeper tail fins (lower PC2 scores) than did control tadpoles. (C) Tadpoles exposed to predators had more colorful tails (higher color score) than did control tadpoles. After McCollum and Van Buskirk (1996).

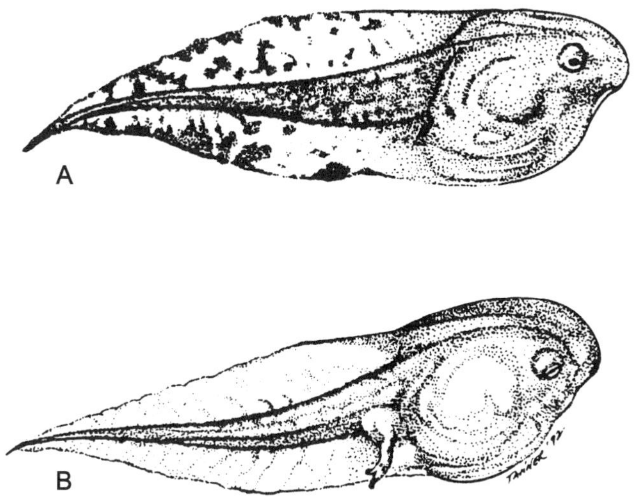


**Fig. 14.15.** Representative tadpoles of *Hyla chrysoscelis* from ponds (A) with predatory dragonfly naiads and (B) without predatory dragonflies. Tadpoles from ponds with dragonflies develop broader and more colorful tail fins. After McCollum and Leimberger (1997).

Direct contact with free-ranging dragonflies was not necessary to induce changes in phenotype, nor were most phenotypic changes affected by experimentally damaging the tails of young tadpoles (McCollum and Leimberger 1997). Exposure to dragonfly naiads that had previously fed on tadpoles induced these phenotypic changes, whereas exposure to starved naiads did not. This suggests that chemical metabolites derived from feeding on tadpoles might be responsible for inducing changes in tadpoles. *Hyla versicolor* tadpoles in natural ponds varied phenotypically in ways predicted from experimental studies similar to those conducted with *H. chrysoscelis.* Tadpoles from a predator-rich pond were more likely to survive in the presence of predators than were those from a predator-free pond (Van Buskirk and McCollum 1999). In laboratory tests, tadpoles were exposed to dragonfly naiads to induce the development of the phenotype with broad tail fins and colorful markings. These were then exposed to both dragonfly naiads and tiger salamander larvae (*Ambystoma tigrinum*) as predators, as were tadpoles that lacked these features. Tadpoles with the predator-induced phenotype had higher survival with both types of predators than did tadpoles that lacked the predator-induced phenotype (fig. 14.17). The greater survival of the predator-induced phenotype was not due to faster swimming, because these tadpole actually were slower swimmers than were those with shallower bodies and tails (Van Buskirk and McCollum 2000a, b). The larger and more conspicuous tail fins of the predator-induced phenotype apparently serve mainly to deflect attacks of predators away from the rest of the body (Van Buskirk et al. 2003). Tadpoles of *Hyla femoralis* exhibit similar responses, which can be triggered by exposure to both dragonfly naiads that have fed on tadpoles and to chemical cues from injured conspecifics (LaFiandra and Babbitt 2004).

Other frogs exhibit plasticity in tail depth in response to predators, but most lack the variation in color pattern seen in *Acris* and several species of *Hyla.* For example, tadpoles of the European treefrog (*Hyla arborea*) exhibited significant increases in tail depth, mainly in the upper tail fin, when exposed to predatory aquatic beetle larvae (*Dytiscus marginalis;* Lardner 2000). D. C. Smith and Van Buskirk (1995) studied populations of *Pseudacris crucifer* and *P. triseriata* in natural rock pools in Michigan. *Pseudacris crucifer* tadpoles are almost always found in pools with predatory dragonflies, whereas *P. triseriata* usually are in predator-free pools. *Pseudacris crucifer* tadpoles have deeper tail fins and more muscular tails than do *P. triseriata* tadpoles. These traits presumably are correlated with swimming speed and predator avoidance. *Pseudacris triseriata* exhibited much more plasticity in tail depth and muscularity than did *P. crucifer.* Tadpoles of *P. triseriata* that were raised in pools with predators developed deep tail fins and muscular tails like those of *P. crucifer,* whereas *P. crucifer* tadpoles showed little response to variation in predator density. A similar pattern was seen in a study of *Bombina* tadpoles in Europe. The species that typically breeds in temporary pools, *B. variegata,* showed more plasticity in morphology in response to predators than did *B. bombina,* which normally breeds in ponds with predators (Vorndran, Reichwaldt, and Nürnberger 2002). Presumably the species that typically live with predators already have phenotypes that have evolved under selection by predators, and their degree of plasticity may be constrained by the costs of plasticity, such as reduced growth rates (Relyea 2002a).

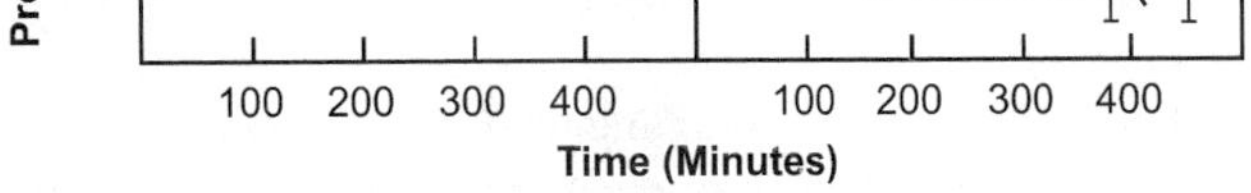


**Fig. 14.17.** Survivorship curves for the two morphological phenotypes of *Hyla versicolor* tadpoles when exposed to predatory dragonfly naiads (*Anax*) and salamander larvae (*Ambystoma*). Tadpoles with the predator-induced morphology had significantly higher survivorship than did those raised without predators. After Van Buskirk and McCollum (2000a).

Several ranid frogs also exhibit phenotypic plasticity in tail morphology in response to free-living predators or chemical cues from caged predators. Tadpoles of several European species, including *Rana arvalis, R. dalmatina, R. ridibunda, R. lessonae,* and *R. temporaria,* all developed deeper tail fins in the presence of cues from several kinds of predators, including diving beetles (*Dytiscus*), dragonfly naiads (*Aeshna* and *Anax*), backswimmers (*Notonecta*), and newts (*Triturus;* Lardner 2000; Van Buskirk 2001; Van Buskirk and Saxer 2001; Van Buskirk 2002b; Van Buskirk and Arioli 2002). Some tadpoles also respond to predators by changing overall body shape. For example, tadpoles of the Australian frog *Limnodynastes peronii* grow excess tissue around the head and body in response to dragonfly naiads (Kraft, Wilson, and Franklin 2005). A somehat similar response is seen in tadpoles of a Japanese frog, *Rana pirica,* in response to hynobiid salamander larvae (Kishida and Nishimura 2005). The function of this response in *Limnodynastes* is not clear, although Kraft, Wilson, and Franklin (2005) suggested that extra tissue on the dorsal side of the tadpole may help protect the spinal cord from predator attacks. In *R. pirica,* the bulgy morphology of tadpoles exposed to *Hynobius* larvae makes it difficult for the salamanders to swallow the tadpoles.

These kinds of plastic responses can be predator specific. For example, many tadpoles develop higher tail fins in the presence of a variety of predators, but respond to the presence of fish by developing more muscular tails that enhance their ability to escape from these active predators (Teplitsky et al. 2005; Wilson, Kraft, and Van Damme 2005). Investment in increased tail muscle is a graded response to different levels of predator risk (Teplitsky, Plénet, and Joly 2005). Indeed, such gradation in morhological plasticty in response to variation predator threat appear to be common (Van Buskirk and Arioli 2002; Laurila et al. 2004), probably because all induced morphological or behavioral responses to predators incur some cost, even after metamorphosis (Van Buskirk 2000; Van Buskirk and Saxer 2001). Given this fine-tuning of plastic responses to predators, one might expect tadpole morphology or degree of plasticity to be closely related to the abundance of predators in natural ponds, but several stuides have shown that this is not necessarily true (Lardner 1998; Van Buskirk and Arioli 2005). One possible explanation for this pattern is that features of local habitats, such as predator abundance, change so rapidly that an evolutionary response to local conditions is precluded.

Several North American ranids, including wood frogs (*R. sylvatica*), leopard frogs (*R. pipiens*), green frogs (*R. clamitans*), and bullfrogs (*R. catesbeiana*), exhibited some plasticity in tail fin morphology in response to dragonfly naiads (*Anax*), but the response of the different species were not identical. Wood frogs exhibited substantial increases in tail fin depth in response to predators, as well as increases in tail muscle depth and width (Van Buskirk and Relyea 1998; Relyea and Werner 2000). Leopard frogs showed a smaller increase in tail fin depth and some increase in tail muscle depth, but not muscle width. Green frogs and bullfrogs showed some changes in overall body shape and size in response to predators, but changes in individual structures such as tail fins generally were not significant (Relyea and Werner 2000). In wood frogs, the predator-induced phenotype had higher survival in the presence of predators than did the noninduced phenotype (Van Buskirk and Relyea 1998), but the adaptive significance of the morphological changes in the other species is not known. The lack of strong responses in tail fin morphology in bullfrogs and green frogs may be related to the fact that these species nearly always live in the presence of predators and have evolved other types of defenses, including behavioral responses and chemical defenses.

Most of the previous studies tested responses of tadpoles to only one type of predator, but morphological plasticity in response to predators is complex and varies as a function of predator type, the number of predator species present, and tadpole species. Some species, such as *Rana temporaria*, exhibit similar morphological responses to several types of vertebrate and invertebrate predators (Van Buskirk 2001). In contrast, wood frogs (*Rana sylvatica*) developed deeper tails in response to dragonfly naiads (*Anax*) and both deeper tails and deeper and shorter bodies in response to mudminnows (*Umbra limi*), but showed no morphological responses to either beetle larvae (*Dytiscus*) or newts (*Notophthalmus viridescens;* Relyea 2001a). When different combinations of predators were presented together, the tadpoles generally responded morphologically as they would to the most dangerous predator alone (Relyea 2003a). American toads (*Bufo americanus*) responded to the presence of mudminnows by developing shallower, longer tails, whereas leopard frogs (*Rana pipiens*) developed deeper tails and shorter bodies in response to the same predator. Even closely related species like bullfrogs (*Rana catesbeiana*) and green frogs (*R. clamitans*) did not respond in exactly the same way to the same predators. This complex variation in morphological responses was accompanied by similar variation in behavioral responses to predators (see the following further discussion).

A surprising result of work by Relyea (2001b) was that overall morphological responses of tadpole species were not closely related to the risk posed to each species by different predators. There were, however, some significant correlations between the risk posed by a particular type of predator and particular morphological traits. For example, wood frogs were more likely to be eaten by mudminnows than were any other species, and they showed the greatest changes in relative tail muscle depth and body length in response to that predator. Green frogs and bullfrogs were the least susceptible to this predator and showed the weakest responses in these morphological traits. Similarly, wood frogs were the most susceptible to predation by dragonfly naiads and exhibited the greatest increase in tail depth in response to those predators, while again green frogs and bullfrogs exhibited little or no response. Other morphological responses were not consistently related to predation risk. One possible reason for such variable results is that prey species do not respond to predators by altering morphology alone; such responses are integrated with behavioral responses such as reduction in activity (see the following). In addition, some species, such as toads, are less palatable to some of the predators than are other species, so they exhibit less pronounced behavioral and morphological changes in response to those predators. Phylogenetic effects clearly have an influence on morphological plasticity, with anurans such as *Bufo* showing relatively little response to predators, and others such as *Hyla* and some *Rana* showing strong reponses. A comparative study of 16 different anuran species that controlled for phylogeny also showed that morphological plasticity tended to be greater in habitats with large temporal variation in predator composition (Van Buskirk 2002a).

There also is evidence for variation among populations of the same species in morphology and degree of morphological plasticity in response to differences in the assemblages of predators present in the habitat. For example, wood frog (*Rana sylvatica*) tadpoles from open-canopy ponds, where predators are relatively abundant, tend to have deeper tails and shorter bodies than to those in closed-canopy ponds, where predators are less abundant (Relyea et al. 2002b).

In addition, there was more variation in morphology among tadpoles from open-canopy ponds, and this reflected greater variation in the abundance and composition of the predator community in these ponds. At least some of this variation was genetic, because phenotypic differences were maintained when tadpoles were reared under the same conditions (Relyea 2002b, 2005a).

Morphological plasticity in response to predators in salamander larvae has not been studied in as much detail. Van Buskirk and Schmidt (2000) exposed larvae of *Triturus alpestris* and *T. helveticus* to dragonfly naiads (*Aeshna*) in experimental ponds. Larvae of both species in ponds with predators developed darker pigmentation on the tail, larger heads, and larger tails than did those in ponds that lacked predators. The predator-induced morph did much better when exposed to free-ranging dragonflies than did those from predator-free ponds, but grew more slowly and took longer to metamorphose. A survey of natural ponds revealed considerable variation in predator density that was correlated with variation in larval head size, as predicted from the experimental results. Not all *Triturus* respond to predators in the same way, however. In a study of six species that live in the same habitats, B. Schmidt and Van Buskirk (2005) found that members of a large-bodied clade (*T. cristatus, T. marmoratus,* and *T. carnifex*) showed somewhat different morphological responses to dragonfly naiads than did members of a more vulnerable, small-bodied clade (*T. vulgaris, T. helveticus,* and *T. alpestris*). Some work also has been done on responses of larval *Ambystoma tigrinum* to diving beetle larvae (*Dytiscus*) and dragonfly naiads (*Anax*). The larvae had shorter bodies and deeper tails in the presence of predator than in predator-free environments, and the response to dragonflies was stronger than the response to diving beetles (Storfer and White 2004).

### Behavioral Responses to Predators

In addition to changes in morphology, the presence of predators can induce amphibian larvae to alter their behavior in ways that reduce their chances of being detected or captured by predators. One type of behavioral response is to form schools, which tend to decrease the chances of an individual larva being captured by predators (Rödel and Linsenmair 1997); this behavior is discussed in more detail in chapter 12. Some amphibians reduce their exposure to predators by shifting their activities to times of day when predators are not very active. For example, *Ambystoma gracile* larvae are active during the day in lakes devoid of predatory fish, but are strictly nocturnal and very wary when fish are present (Efford and Mathias 1969; Sprules 1974b; Taylor 1983a). Larvae of the streamside salamander (*A. barbouri*) also tend to become more nocturnal in the presence of fish (Sih, Kats, and Moore 1992).

Amphibian larvae also commonly shift their activities to protected habitats in the presence of predators. Often these refuge habitats provide less food for the larvae than more exposed locations, so the use of refuges can be costly in terms of reduced feeding and growth. For example, *Ambystoma tigrinum nebulosum* larvae changed their use of foraging microhabitats in the presence of dytiscid beetles (Holomuzki 1986a). In a laboratory experiment conducted at night, salamander larvae were active mostly in shallow water when beetles were absent, but shifted into deeper water when beetles were present. Presumably this is because the beetles prefer to forage in shallow water. Stangel and Semlitsch (1987) found that the presence of sunfish suppressed the nocturnal vertical migration of *Ambystoma talpoideum* larvae into the water column, causing the salamanders to restrict their activity to the bottom of the pond; when fish were removed, activity in the water column resumed (fig. 14.18). Larvae raised in the presence of fish obtained less food and were smaller than those raised without fish, even though densities were lower and the fish were not direct competitors (Semlitsch 1987b). Larvae of *Ambystoma maculatum* living in ponds with predatory marbled salamander larvae (*A. opacum*) did not retreat into refuges and consequently were more vulnerable to predation than were *A. talpoideum* larvae living in the same ponds (Walls 1995). Larvae of *Triturus alpestris* and *T. helveticus* spent more time hiding in leaf litter when exposed to cues from predatory dragonflies (*Aeshna;* Van Buskirk and Schmidt 2000) or fish (Orizaola and Braña 2003b), while young larvae of

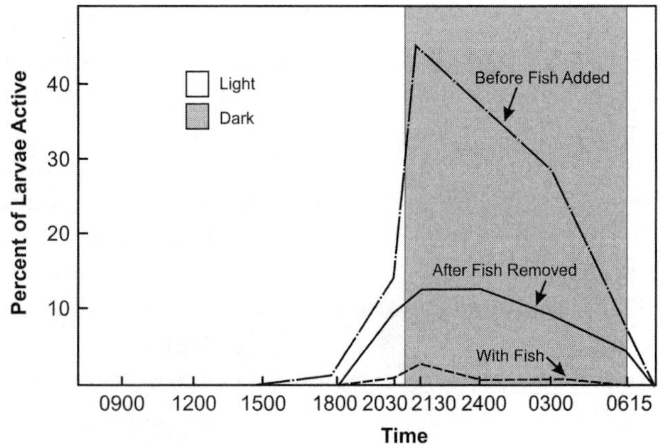


**Fig. 14.18.** Mean percentage of larval *Ambystoma talpoideum* active in the water column in artificial ponds before, during, and after the introduction of predatory fish (*Lepomis macrochirus*). Each line represents averaged data for three 24-hour censuses before fish were added (23 July–4 August), when fish were present (9–14 August), and after fish were removed (16–22 August). The presence of fish dramatically decreased the vertical migration of salamander larvae in the water column, but migration was partially restored when the fish were removed. After Stangel and Semlitsch (1987).

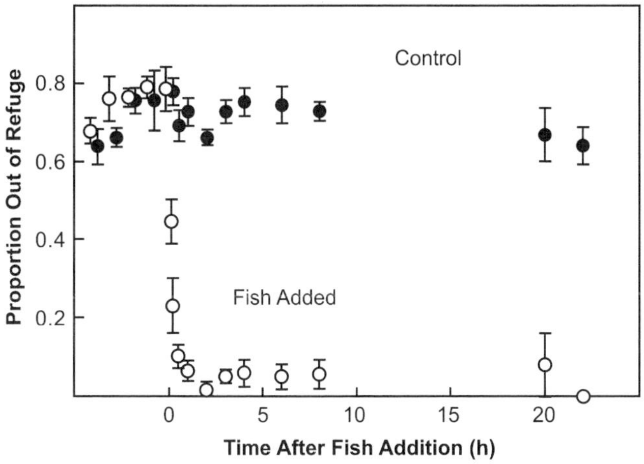


**Fig. 14.19.** The proportion of *Ambystoma barbouri* larvae out of refugia in pools to which bluegill sunfish (*Lepomis macrochirus*) had been added and in control pools that lacked sunfish. After Sih, Englund, and Wooster (1988).

*Dicamptodon tenbrosus* in the laboratory spent more time under artificial shelters in the presence of chemical cues from trout than did those in tanks that lacked such cues (Rundio and Olson 2003).

The use of refuge habitats in response to predatory fish has been studied in considerable detail in the streamside salamander (*Ambystoma barbouri*). This species lives in streams that often dry up in the summer, so larvae must be very active to acquire enough food to reach metamorphic size before the stream dries up (Petranka and Sih 1986, 1987; Maurer and Sih 1996). When confronted with predatory fish, the larvae tend to greatly reduce foraging activity and retreat under cover objects (rocks in natural streams; plexiglass refuges in experimental pools; Petranka 1983). The larvae must emerge periodically from these refuges to feed, and when they do so, they are extremely vulnerable to predation by fish. Experimental studies showed that fewer than 5% of larvae were out of refuges at any given time when fish were present (fig. 14.19), yet predation rates still were relatively high (Sih, Petranka, and Kats 1988). Most larvae begin life in pools with relatively few fish, because females avoid laying eggs in pools with fish (Kats and Sih 1992). Nevertheless, many are carried downstream by currents or floods, which may deposit them in pools with fish. These predators can rapidly eliminate actively foraging larvae from a small pool, so a rapid shift to refuge habitats is required for larvae to survive (Sih, Kats, and Moore 1992). Antipredator behavior is only marginally successful in this species—most larvae exposed to fish are eaten. One reason for this is that the behavior of larvae is shaped by the conflicting need for rapid development, which favors active feeding, and predator avoidance, which favors reduced activity and refuge use (Sih, Kats, and Maurer 2000). Populations of larvae in streams with fish exhibit stronger antipredator responses than do those in streams without fish. Nevertheless, gene flow from populations without fish may reduce the effectiveness of antipredator behavior, which is not as well developed in populations that are close to fish-free populations (Storfer and Sih 1998).

Some anuran tadpoles seem to be less inclined to use sheltered refuges than are salamander larvae, probably because they must have access to algae and other food growing in exposed habitats. Nevertheless, mortality of some species is lower in habitats with extensive cover than in more open habitats (Babbitt and Jordan 1996; Babbitt and Tanner 1997). Some tadpoles alter their use of microhabitats in response to predators. For example, Morin (1986) found that *Pseudacris crucifer* tadpoles used more protected microhabitats in the presence of newts, but he did not find any change in larval growth rates. Tadpoles of *Pseudacris regilla* in a river in California tended to shift from algal mats that contained their principal food, epiphytic diatoms, to lower-quality algal mats and sediments at the bottom of the stream when confronted with predatory garter snakes (*Thamnophis hydrophilus;* Kupferberg 1998). This seemed to be less a result of actively seeking more protected microhabitats than of reduced levels of activity, which caused the tadpoles to sink to the bottom. Tadpoles of various species of *Rana, Hyla,* and *Bufo* increase their use of refuge habitats in response to chemical cues from crayfish (fig. 14.20; Lefcourt 1996; Bridges and Gutzke 1997; Nyström and Åbjörnsson 2000). Tadpoles of *Rana temporaria* also increased their use of refuges when exposed to chemical cues from predatory fish (*Oncorhynchus mykiss*), but tadpoles of *Bufo bufo,* which are distasteful to fish, did not (Nyström and Åbjörnsson 2000). This type of response may be more common than currently recognized, because many laboratory studies have been designed to test the effects of predators on larval activity or spatial avoidance of predators, but have

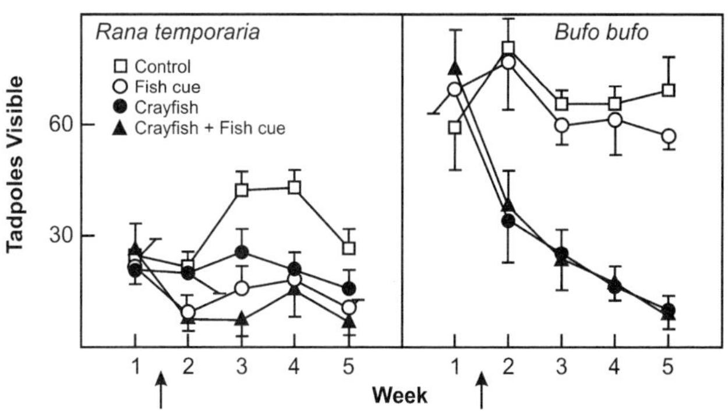


**Fig. 14.20.** Mean number of tadpoles of *Rana temporaria* and *Bufo bufo* visible (out of refugia) in experimental pools without predator cues (controls), with chemical cues from fish (*Oncorhynchus mykiss*), with free-ranging crayfish (*Pacifastacus leniusculus*), and with both fish cues and crayfish. *Rana temporaria* tadpoles actively avoided both fish and crayfish, whereas *Bufo bufo* tadpoles avoided only crayfish. After Nyström and Åbjörnsson (2000).

not included refuge habitats for the larvae (e.g., Mathis and Vincent 2000; Chivers and Mirza 2001; Chivers et al. 2001; Wildy and Blaustein 2001).

Probably the most common behavioral responses of amphibian larvae to the presence of predators are to move away from the predator, reduce foraging activity, and in some cases, become almost immobile. These types of predator avoidance behavior are phylogenetically widespread (Van Buskirk 2002a) and have been observed in tadpoles of *Ascaphus* (Feminella and Hawkins 1994), *Alytes* (Griffiths et al. 1998), *Bombina* (Van Buskirk 2002a), *Pseudacris* (Lawler 1989; Skelly 1995a; Kupferberg 1998; Puttlitz et al. 1999; Chivers et al. 2001), *Hyla* (Lawler 1989; Skelly 1992; Relyea 2001a, b), *Bufo* (Skelly and Werner 1990; Semlitsch and Gavasso 1992; Anholt, Skelly, and Werner 1996; Kiesecker, Chivers, and Blaustein 1996; Laurila, Kujasalo, and Ranta 1998; Lefcourt 1998; Petranka and Hayes 1998; Relyea 2001a, b), and *Rana* (Werner 1991; Stauffer and Semlitsch 1993; Horat and Semlitsch 1994; Werner and McPeek 1994; Anholt and Werner 1995; Manteifel 1995; Lefcourt 1996; Kiesecker and Blaustein 1997b, 1998; Laurila, Kujasalo, and Ranta 1997, 1998; Petranka and Hayes 1998; Van Buskirk and Yurewicz 1998; Relyea and Werner 1999; Eklöv 2000; Laurila 2000a; Nicieza 2000; Thiemann and Wassersug 2000; Babbitt 2001; Chivers and Mirza 2001; Relyea 2001a, b; Van Buskirk 2001). Some salamander larvae, including *Ambystoma* (Semlitsch 1987b; Figiel and Semlitsch 1990; Sih and Kats 1991, 1994; Storfer and Sih 1998; Storfer 1999; Wildy and Blaustein 2001), *Eurycea* (Petranka, Kats, and Sih 1987; Resetarits 1991), and *Notophthalmus* (Mathis and Vincent 2000), also reduce foraging activity or become immobile in response to predators. Larvae of *Hemidactylium scutatum* reduced activity in the presence of newts (*Notophthalmus viridescens*), but only after the first 12 days of development; activity of younger larvae already is very low (C. Wells and Harris 2001). In contrast, small larvae of *Ambystoma annulatum* were more likely to reduce activity in response to cues from newts than were larger individuals that were less susceptible to predation (Mathis, Murray, and Hickman 2003).

There is abundant evidence that movements associated with foraging activity make amphibian larvae especially vulnerable to predators that detect their prey visually or through tactile cues (Caldwell, Thorp, and Jervey 1980; Woodward 1983; Lawler 1989; S. Richards and Bull 1990a; Azevedo-Ramos et al. 1992; Chovanec 1992b; Skelly 1994; M. Moore and Townsend 1998). At the same time, actively foraging tadpoles can process food at a faster rate than less active foragers, and they achieve faster growth rates (Morin 1983a; Lawler 1989; Werner 1991, 1992; Werner and McPeek 1994; Skelly 1995a). A similar relationship between activity and early growth rate has been reported in salamander larvae, although in this case, the comparison was between distantly related species (*Ambystoma jeffersonianum* and *Hemidactylium scutatum;* C. Wells and Harris 2001). This means that reducing activity to escape predation carries a cost in reduced food intake and growth (fig. 14.21; Skelly and Werner 1990; Skelly 1992; Van Buskirk and Yurewicz 1998; Van Buskirk 2000), and these costs may be increased by simultaneous changes in tadpole morphology (Relyea 2002a). There can be other costs as well. Thiemann and Wassersug (2000) reported that tadpoles of *Rana clamitans* that were exposed to caged killifish (*Fundulus*) reduced levels of activity, but were more vulnerable to infection by parasitic trematodes. They attributed this to increased proximity to trematode cercariae in inactive tadpoles that spent more time on the substrate, but this was not tested experimentally.

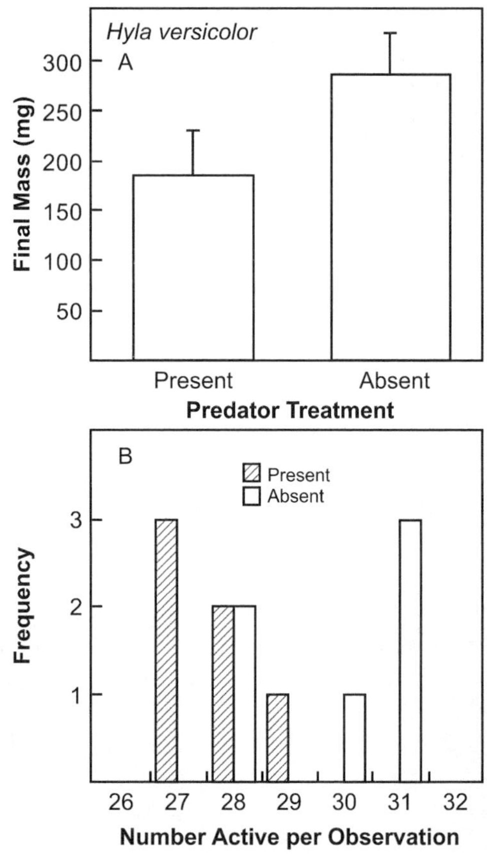


**Fig. 14.21.** Effect of exposure to a predator on growth and activity of *Hyla versicolor* tadpoles. (A) Final body mass of tadpoles growing in enclosures in a natural pond in the presence or absence of a tiger salamander larva (*Ambystoma tigrinum*) housed in a screen cage. (B) Number of active small and large tadpoles in the presence or absence of a tiger salamander larvae in a laboratory experiment. After Skelly (1992).

Because of the tradeoff between vulnerability to predators and foraging efficiency, tadpoles in permanent ponds, where predation is a major threat, often are relatively inactive foragers and tend to have slow growth rates. In contrast, tadpoles that live in temporary ponds, where rapid pond drying favors rapid growth, tend to be much more active, even if this exposes them to predators (Skelly 1995a). This relationship among level of activity, relative growth rate, and habitat appears to hold for several different families of anurans (J. Richardson 2001). One implication of this difference is that temporary pond species often are precluded from invading permanent ponds, especially those with fish, because of ineffective antipredator behavior, whereas permanent pond species are excluded from temporary ponds because of their slow development rates (Woodward 1983; Kats, Petranka, and Sih 1988; Azevedo-Ramos and Magnusson 1999).

A second implication of the tradeoff between feeding and predator avoidance is that amphibian larvae are expected to adjust their responses to predators according to their need for food or levels of food availability. Predators represent an immediate threat to foraging tadpoles, but tadpoles that fail to acquire sufficient food to reach metamorphosis are doomed as well. Tadpoles of *Rana lessonae* and *R. esculenta* reduced their activity and time devoted to foraging in the presence of predators, but hungry tadpoles were more active than well-fed tadpoles, even when predators were present (fig. 14.22;

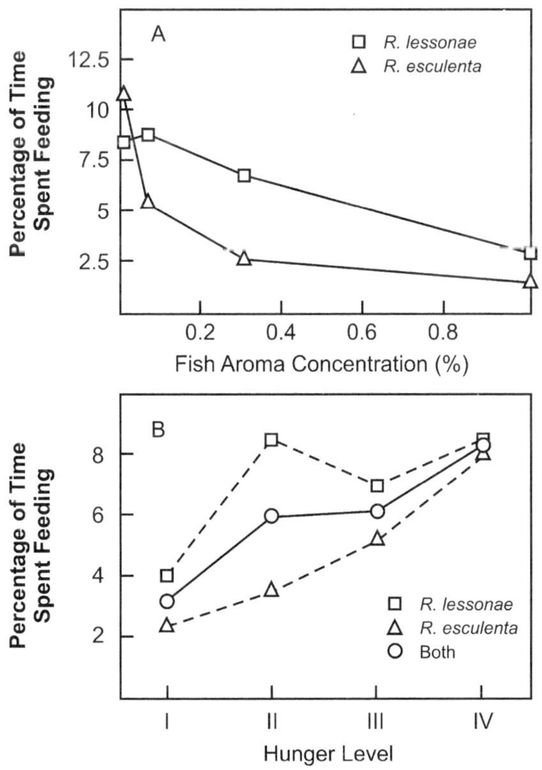


**Fig. 14.22.** Effect of hunger or resource level on feeding activity and mortality of tadpoles in the presence of predators. (A) Effect of fish aroma concentration on time spent feeding by tadpoles of *Rana esculenta* (triangles) and *R. lessonae* (squares). (B) Percentage of time spent feeding by tadpoles of these two species (same symbols) and the two species combined (circles) in the presence of predatory fish as a function of increasing hunger level. After Horat and Semlitsch (1994).

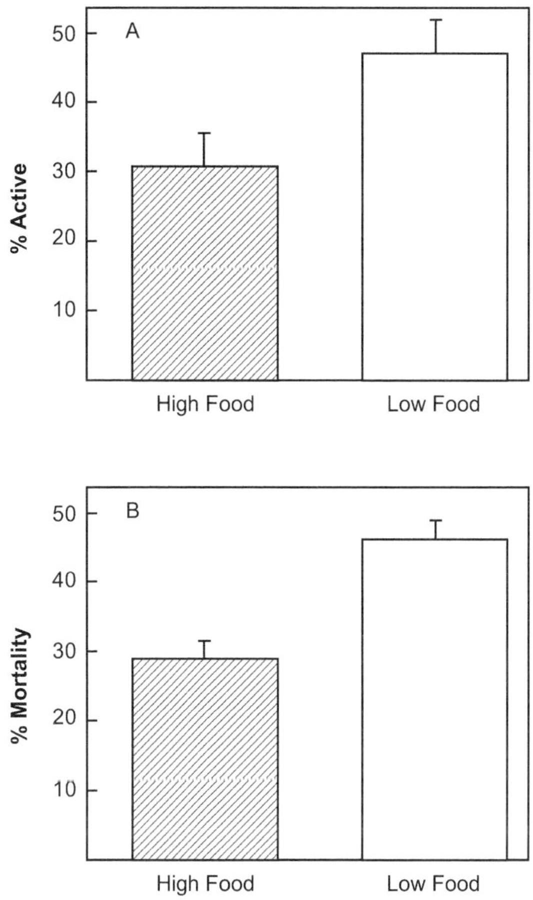


**Fig. 14.23.** (A) Percentage of bullfrog (*Rana catesbeiana*) tadpoles active in the presence of caged dragonfly naiads (*Tramea lacerata*) at high and low food levels. (B) Mortality of bullfrog tadpoles exposed to free-ranging dragonfly naiads at high and low food levels. After Anholt and Werner (1995).

Horat and Semlitsch 1994). Tadpoles of *Rana sphenocephala* were more active when food was abundant than when it was scarce, but also showed a greater reduction in activity in response to predators when food was abundant (Babbitt 2001). Similar experiments with bullfrog (*Rana catesbeiana;* Anholt and Werner 1995) and wood frog (*R. sylvatica*) tadpoles (Anholt and Werner 1998) showed that tadpoles will reduce activity in the presence of predators when resource levels are high, but actually increase activity and suffer higher mortality when resources are low (fig. 14.23). The effectiveness of reduced activity as an antipredator defense also can be affected by the hunger level of the predator. Hungry dragonfly naiads, for example, are more likely to eat inactive tadpoles than are satiated ones (Altwegg 2003a).

In many aquatic environments, amphibian larvae are confronted with an array of predators that use different sensory modes to locate prey. This means that behavioral responses that provide protection against one predator may make them more vulnerable to other predators (Sih, Englund, and Wooster 1998). For example, reducing levels of activity often is an effective defense against visually hunting predators, but not those that detect prey mainly by chemical cues rather than visual or tactile cues. Warkentin (1999a) found that tadpoles of *Agalychnis callidryas* could not escape detection by freshwater shrimp by remaining immobile, because these shrimp detect prey chemically. The only effective defense was to flee, a behavior that became more effective as the tadpoles grew, but this response tended to make the tadpoles more vulnerable to predators such as dragonfly naiads. In some cases, having several kinds of predators present in a pond can actually benefit a species in competition with other amphibians. For example, bullfrog tadpoles (*Rana catesbeiana*) are vulnerable to dragonfly naiads because of their foraging activities, but are relatively well protected from fish by noxious skin secretions. Because fish tend to eat dragonfly naiads, bullfrog tadpoles do better in ponds with fish than do green frog tadpoles (*R. clamitans*), which are vulnerable to both dragonflies and fish (Werner and McPeek 1994; see chapter 15 for further discussion). Similarly, tadpoles of *Hyla cinerea* were less vulnerable to some combinations of predators than to single species of predators alone, apparently because the various predators tend to eat each other (Gunzburger and Travis 2005).

### Sensory Cues for Predator Detection and Recognition

The experiments described previously indicate that amphibian larvae are capable of detecting the presence of dangerous predators and responding with appropriate changes in behavior. A number of sensory cues could be used to detect predators, including tactile, visual, and chemical cues. Several experiments have shown that tactile and visual cues either are not necessary for predator detection, or are not effective. In most cases, chemical cues are used for predator detection and recognition. McCollum and Leimberger (1997) showed that chemical cues are both necessary and sufficient to induce morphological changes. Other experiments have used behavioral responses such as changes in activity and shifts in microhabitat use to determine the cues used by larvae to detect predators. Antipredator responses can be induced by exposing tadpoles (Semlitsch and Gavasso 1992; Semlitsch and Reyer 1992a; Lefcourt and Eiger 1993; Stauffer and Semlitsch 1993; Feminella and Hawkins 1994; Lefcourt and Blaustein 1995; Manteifel 1995; Kiesecker, Chivers, and Blaustein 1996; Lefcourt 1996, 1998; Eklöv 2000; Laurila 2000a) and salamander larvae (Petranka, Kats, and Sih 1987; Sih and Kats 1991) to water conditioned by predators, even when the predators themselves are not present.

In some cases, larvae discriminate between cues derived from dangerous and less dangerous predators (Manteifel 1995; Kiesecker, Chivers, and Blaustein 1996; Lefcourt 1996, 1998). For example, bullfrog (*Rana catesbeiana*) tadpoles exhibit a greater reduction in activity in response to cues from a dangerous predator (dragonfly naiads) than to those from a less dangerous predator (bluegill sunfish; Eklöv 2000). When cues from the two predators were presented together, tadpoles reduced activity to about the same level as they did with bluegills alone. One possible reason for this is that the presence of bluegills reduces activity of dragonfly naiads, which in turn may reduce the cues emanating from the dragonflies to which the tadpoles respond (Eklöv and Werner 2000). Tadpoles of another ranid frog, *Rana temporaria,* show reduced activity to both dragonfly naiads and fish, but show greater spatial avoidance of dragonflies (Laurila 2000a). Tadpoles of the western toad (*Bufo boreas*) avoided chemical cues from backswimmers (*Notonecta*), giant waterbugs (*Lethocerus americanus*), and garter snakes (*Thamnophis sirtalis*), all of which prey on toad tadpoles (Kiesecker, Chivers, and Blaustein 1996). They did not avoid chemical cues from newts (*Taricha granulosa*) or trout (*Oncorhynchus mykiss*), both of which find toad tadpoles distasteful (fig. 14.24). Six different species of anurans tested by Relyea (2001a) differed in their tendency to reduce activity or move away from five different types of predators. All species responded to dragonfly naiads (*Anax*), but they varied in their responses to newts (*Notophthalmus viridescens*), tiger salamander larvae (*Ambystoma tigrinum*), mudminnows (*Umbra limi*), and giant waterbugs (*Belostoma*). Across all anuran species, the decrease in activity in response to mudminnows was proportional to the risk posed by that predator, but this was not true for dragonfly naiads (Relyea 2001b). This may be because dragonflies are relatively dangerous predators for most anuran tadpoles.

Some amphibian larvae respond differently to chemical cues derived from predators that have fed on conspecific larvae compared to those that have fed on other types of food (Lefcourt and Blaustein 1995; Laurila, Kujasalo, and Ranta 1997, 1998; Chivers and Mirza 2001). In some cases, however, they respond equally strongly to predators that have not fed on conspecifics (Petranka and Hayes 1998). Tadpoles also may reduce their antipredator responses as they grow, because large tadpoles often are less vulnerable to certain predators than are small ones (Puttlitz et al. 1999; Eklöv 2000).

Often the ability to recognize predators appears to be genetically based, because previous experience with a predator is not required to elicit a response (Gallie, Mumme, and Wissinger 2001; Mathis, Murray, and Hickman 2003). Nevertheless, some populations of tadpoles can recognize predators from the local area, but not members of the same species from other regions (Griffiths et al. 1998). This suggests that there has been selection at the local level for recognition of dangerous predators. Populations of some species apparently lack the ability to recognize or avoid predators that

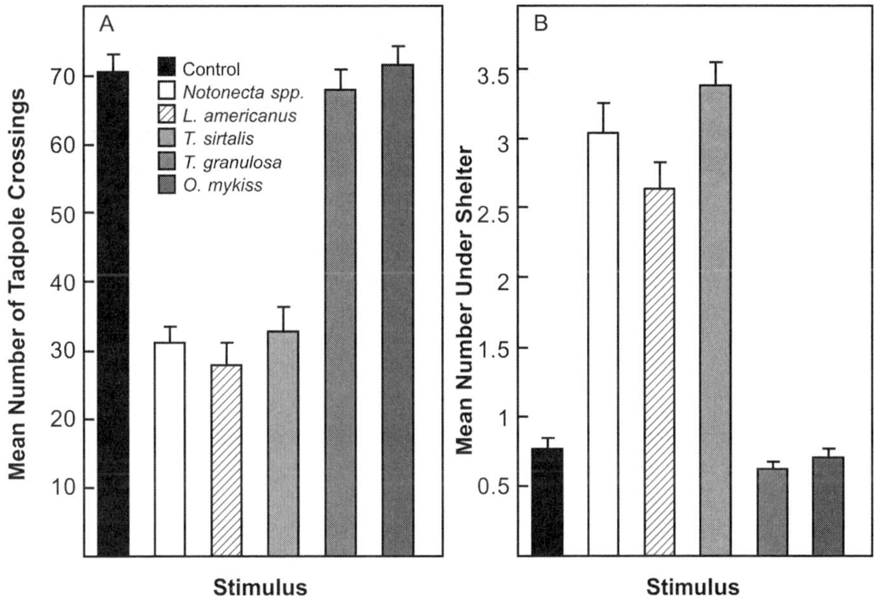


**Fig. 14.24.** Effect of chemical cues from potential predators on activity and use of refugia by tadpoles of *Bufo boreas.* (A) Activity of tadpoles, measured as the number seen crossing the midline of the experimental tank during a five-min trial. (B) Number of tadpoles under a shelter. Tadpoles were exposed to chemical cues from dangerous predators (*Notonecta, Lethocerus,* and *Thamnophis*) and predators that do not feed on toad tadpoles (*Taricha, Oncorhynchus*). The tadpoles exhibited antipredator behavior in response to the dangerous predators, but not to the others. After Kiesecker, Chivers, and Blaustein (1996).

have been introduced into the area very recently (Gamradt and Kats 1996). Similarly, amphibian larvae from temporary pools that lack fish typically do not show antipredator responses to chemical extracts from fish, whereas those from permanent ponds usually do (Kats, Petranka, and Sih 1988).

Antipredator behavior may evolve relatively rapidly, however, in response to strong selection imposed by introduced predators. For example, tadpoles of red-legged frogs (*Rana aurora*) from streams in Oregon inhabited by introduced bullfrogs (*Rana catesbeiana*) reduced their activity and increased their use of refuges when presented with chemical cues from either tadpoles or adult bullfrogs. Tadpoles from streams that lacked introduced bullfrogs did not show these responses, but they did respond to native predators (Kiesecker and Blaustein 1997). A similar difference has been reported in responses of terrestrial juveniles of *Pseudacris regilla* to introduced bullfrogs. Frogs from populations that co-occur with bullfrogs exhibited a much greater avoidance response than did those from populations where bullfrogs are absent (Chivers et al. 2001). In some cases, larvae may learn to avoid predators by responding to cues from injured conspecifics. For example, larvae of *Ambystoma macrodactylum* reduced their activity in response to cannibalistic conspecifics that had been on other larvae, but not those that had fed on *Tubifex* worms, and they responded even more strongly to injured conspecifics (Wildy and Blaustein 2001).

### Alarm Responses

In addition to detecting predators directly, some amphibian larvae respond to cues emanating from injured conspecifics, and use these cues to avoid areas where predators are likely to be found (Hrbâçek 1950; Kulzer 1954; Pfeiffer 1966, 1974; Hews and Blaustein 1985; Hews 1988). Tadpoles of *Bufo americanus* can be induced to abandon a rich food patch if they are exposed to chemical extracts from injured conspecifics (Petranka 1989c). Alarm responses tend to induce behavior that is the opposite of the antipredator behavior discussed previously. Rather than reducing activity, tadpoles tend to increase activity in their attempts to escape from the vicinity of a potential predator. Although increased activity can make a tadpole more conspicuous to predators, this type of reaction can be beneficial when the tadpole already has been detected. Indeed, Hews (1988) showed that tadpoles of *Bufo boreas* that had been exposed to chemicals from an injured conspecific were less likely to be captured by dragonfly naiads. Whether the chemical substances released by injured tadpoles are pheromones that have evolved specifically as alarm signals, or are chemicals in the skin that have other functions, is the subject of some debate, which has yet to be resolved.

Some authors have suggested that alarm responses should be best developed in species of tadpoles that tend to aggregate with kin, because evolution of alarm signaling could be maintained by kin selection (see chapter 12). However, tadpoles of *Bufo boreas,* which forms large aggregations of mixed kin and nonkin, shows relatively strong alarm responses (Hews 1988), whereas tadpoles of *Rana cascadae,* which form small kin aggregations, do not (Hokit and Blaustein 1995). Although responses to alarm substances have been studied mostly in tadpoles that are social, sociality may not be a prerequisite for the evolution of alarm responses. Tadpoles of *Pseudacris regilla,* which are not highly social, will respond to extracts from injured conspecifics or heterospecific tadpoles (*Rana aurora;* M. Adams and Claeson 1998).

### Changes in Growth and Development

For many amphibian larvae, rapid growth probably provides the best protection against predators. Travis (1983b) demonstrated genetic variation in larval growth rate, with high rates being correlated with higher survivorship. Most predators take prey over a limited size range, so some amphibian larvae eventually become too large to be eaten (Heyer, McDiarmid, and Weigmann 1975; Caldwell, Thorp, and Jervey 1980; Brodie and Formanowicz 1983; Crump 1984a; Travis, Keen, and Juilianna 1985b; Alford 1986; Cronin and Travis 1986; Formanowicz 1986; Semlitsch and Gibbons 1988; S. Richards and Bull 1990b; Kehr and Schnack 1991). Although the effect of tadpole size varies with the type of predator and predator size, tadpoles generally become less vulnerable as they grow (Alford 1999). This appears to result from a greater ability of large tadpoles to escape from predators. In general, tadpole sprint swimming speed increases with increasing size (Huey 1980; S. Richards and Bull 1990b), although locomotor performance may decline as a tadpole reaches metamorphosis (R. Brown and Taylor 1995). This means that predators may have to make more strikes at large prey to achieve a successful capture (Formanowicz 1986). Some predators are more adept than others at handling large tadpoles, and most species probably never reach a size that makes them completely invulnerable to attack. Some aquatic predators, and terrestrial predators such as birds, can actually be more successful at capturing large tadpoles than small ones (Crump and Vaira 1991; Tejedo 1993b). Nevertheless, predation probably is an important selective force acting on growth rates of larval amphibians (Travis 1983b; Travis, Keen, and Juilianna 1985a; Cronin and Travis 1986; Semlitsch and Gibbons 1988).

Exposure to predators also can alter rates of development and progress toward metamorphosis. For example, tadpoles of *Bufo boreas* that were exposed to predatory backswimmers (*Notonecta*) or cues from injured conspecifics reached metamorphosis more quickly than those raised with nonpredatory insects (Chivers et al. 1999). This was interpreted as an adaptive change in life history that enabled tadpoles to escape from dangerous conditions in the aquatic environment. Tadpoles of *Bufo boreas, Bufo americanus,* and *Pseudacris regilla* exposed to cues from predatory garter snakes (*Thamnophis sirtalis*) not only showed a tendency to form aggregations, but they also completed metamorphosis more synchronously than did those not exposed to a predator (DeVito et al. 1998, 1999; DeVito 2003). In contrast, larvae of *Ambystoma macrodactylum* that were exposed to cues from cannibalistic conspecifics took longer to metamorphose, presumably because they reduced foraging activity in response to the predator cues (Wildy, Chivers, and Blaustein 1999). A similar cost of antipredator behavior has been reported in several species of tadpoles, especially those from permanent ponds that tend to have long larval periods to begin with (e.g., Relyea and Werner 1999).

### Chemical Defenses

Some amphibian larvae, particularly anuran tadpoles, are distasteful or even toxic to certain predators, but the phylogenetic distribution of distastefulness is poorly understood. Toad tadpoles (*Bufo*) often are assumed to be distasteful (e.g., Waldman 1982b), and the schooling behavior of many toad tadpoles has been interpreted as a behavior that could enhance the aposematic effect of conspicuous coloration in distasteful species (Wassersug 1973; Waldman 1982b; J. Peterson and Blaustein 1991; see chapter 12). However, only a few species have been tested for palatability, and these experiments have produced mixed results. Various *Bufo* tadpoles are distasteful to fishes (Voris and Bacon 1966; Glandt 1984; Kruse and Stone 1984; Kats, Petranka, and Sih 1988; Lawler 1989; Henrikson 1990; Lawler and Hero 1997), salamanders and newts (Cooke 1974; J. Peterson and Blaustein 1991), other tadpoles (Bragg 1960), and some invertebrates (Brodie, Formanowicz, and Brodie 1978). However, other studies have shown toad tadpoles to be palatable to a variety of vertebrate and invertebrate predators (Brockelman 1969; Heusser 1971a; Heyer, McDiarmid, and Weigmann 1975; Brodie, Formanowicz, and Brodie 1978; Morin 1983a; Leclair, Alarie, and Bourassa 1986; Banks and Beebee 1988; Hews 1988; Peterson and Blaustein 1992). Palatability of *Bufo* and other tadpoles to certain predators may depend on the hunger level of the predator and the availability of alternative prey, and it may vary with developmental stage (Brodie and Formanowicz 1987). Newly hatched tadpoles of *Bufo americanus* are distasteful to invertebrate predators, as are metamorphosing individuals, but they appear to be palatable at intermediate stages (Brodie, Formanowicz, and Brodie 1978). Palatable tadpoles may benefit from the distastefulness of newly hatched tadpoles because predators learn to avoid black tadpoles (Brodie and Formanowicz 1987). In contrast, the tadpoles of *Bufo marinus* were unpalatable at all developmental stages to predatory fish, but the later stages were the least palatable (Lawler and Hero 1997).

Closely related predators sometimes differ in their tendency to eat toad tadpoles. For example, adult *Triturus cristatus* readily eat *Bufo* tadpoles, but *T. vulgaris* and *T. helveticus* avoid them (Banks and Beebee 1988). Similarly, *Bufo americanus* tadpoles presented to aquatic beetle larva (*Dytiscus*) were eaten by one species (*D. verticalis*), but only occasionally by another species (*D. harrisii*) from the same pond (Leclair, Alarie, and Bourassa 1986). Failure to eat tadpoles does not necessarily indicate they are distasteful. Leclair, Alarie, and Bourassa (1986) demonstrated experimentally that *D. verticalis* from a pond containing toad tadpoles readily ate them, whereas larvae of the same species from a pond lacking toad tadpoles did not, suggesting that familiarity with the prey or availability of alternative prey may influence feeding preferences.

The relative palatability of tadpoles in other families has scarcely been explored in a systematic manner, although it is clear from the discussion earlier in this chapter that tadpoles are consumed by many kinds of predators. Wassersug (1971) tested leptodactylid, hylid, dendrobatid, and bufonid tadpoles from Costa Rica on human volunteers; only *Bufo marinus* was found to be highly distasteful. Tadpoles of a Neotropical treefrog, *Hyla semilineata,* which form dense schools in ponds and rivers, are black and distasteful to both natural predators (fish and free-ranging birds) and unnatural ones (domestic chickens; D'Heursel and Haddad 1999). Formanowicz and Brodie (1982) found that ranid, hylid, and bufonid tadpoles, as well as larval *Ambystoma maculatum* and *Notophthalmus viridescens,* were palatable to diving beetles (*Dytiscus*), but metamorphic stages of several species were distasteful. This suggests that distasteful skin secretions in some species do not develop until the larvae are about to leave the water. Tadpoles of *Pseudacris regilla* and *Rana cascadae* are highly palatable to both fish and aquatic insects (Peterson and Blaustein 1991, 1992). Liem (1961) found that tadpoles of the Asian ranid *Rana chalconota* had large granular glands and were highly distasteful to fishes, but tadpoles were examined only in late stages of development. Tadpoles of frogs that are highly toxic as adults, such as *Dendrobates, Phyllobates,* and *Atelopus,* have not been tested for palatability to natural predators, but some species of dendrobatid tadpoles readily consume conspecifics (see chapter 12).

Sometimes tadpoles of closely related species differ in their palatability to certain predators. For example, bullfrog (*Rana catesbeiana*) tadpoles are less palatable to fish than are green frog (*R. clamitans*) tadpoles (Werner 1991; Werner and McPeek 1994), although *R. clamitans* tadpoles are less palatable than are wood frog (*R. sylvatica*) tadpoles (Kats, Petranka, and Sih 1988). Tadpoles of the green tree frog (*Hyla cinerea*) are more frequently found in ponds with fish than are those of the barking tree frog (*H. gratiosa*). *Hyla cinerea* tadpoles were less palatable to fish when offered alone, but when the two species were offered in equal numbers, the fish did not discriminate between them (Blouin 1990).

Why has distastefulness or toxicity not evolved in nearly all amphibian larvae? Presumably the reason is that distasteful or toxic skin secretions can be expensive to produce. The cost of producing such compounds may be too great for many temporary pond tadpoles, which must channel most of their available energy into rapid growth to ensure that they leave the pond before it dries up. The cost of producing defensive chemicals probably is not as great for tadpoles that live in permanent ponds, many of which grow slowly anyway and may remain in ponds for months or even for several years. Because of their prolonged exposure to fish and other predators, the enhanced protection afforded by becoming distasteful probably more than repays the energetic cost of producing these defenses. The biochemical pathways leading to synthesis of noxious defensive compounds have not been investigated in tadpoles, but it is possible that this process requires chemical antecedents that cannot be readily obtained from the food available to tadpoles in ponds and other bodies of water. Some highly toxic adult frogs, for example, apparently sequester chemicals obtained by eating insects and use them to manufacture defensive compounds (see the following).

### Defensive Adaptations of Adults

As in larval amphibians, adult amphibians can reduce the chances of being eaten by a predator either by reducing the likelihood of being detected and identified by the predator, or by reducing the chances of being consumed once an encounter takes place (Edmunds 1974; Endler 1986). This can be accomplished through passive defenses such as cryptic coloration, through behavioral avoidance of predators, or through active defenses such as flight, attack, or chemical defenses.

#### Cryptic Coloration

Cryptic coloration provides the first line of defense for many adult amphibians. Often this involves a general matching of dorsal coloration and pattern with that of the environment (Cott 1940; Edmunds 1974). Most amphibians tend to have ground colors that are various shades of green, yellow, red, brown, or gray. Terrestrial salamanders often are dark brown, gray, or black, with very low skin reflectance; this makes them very hard to see at night when they are most active (Norris and Lowe 1964). Many tropical frogs that rest on green vegetation during the day have bright green coloration (e.g., centrolenid frogs, *Agalychnis,* and many species of *Hyla, Litoria,* and *Hyperolius*). Some of these frogs reflect light not only in the visible spectrum, but in the near-infrared as well, perhaps rendering them thermally cryptic to predators such as boid snakes that use infrared radiation to detect prey (Schwalm, Starrett, and McDiarmid 1977; Emerson, Cooper, and Ehleringer 1990). Frogs that are found on dead vegetation near the ground often are colored various shades of yellow or brown (e.g., *Hyla microcephala, H. ebraccata;* Duellman 1970). Dark coloration and mottled patterns are characteristic of many frogs that escape from predators by diving into a pond or stream, where they blend in with algae-covered or muddy substrates (e.g., *Rana muscosa:* Norris and Lowe 1964; *R. septentrionalis:* Hedeen 1972; *R. pretiosa:* Licht 1986a). Frogs that escape by jumping onto land, such as *R. sylvatica, R. aurora,* and *R. pipiens,* often have lighter coloration, which provides a better match with terrestrial vegetation or dead leaves (Heatwole 1961; Hedeen 1972; Licht 1986a).

Some anurans exhibit preferences for certain types of substrates that render them cryptic to predators. For example, juvenile American toads (*Bufo americanus*) preferred visually complex substrates such as sand with holes, vegetation, or gravel to plain sand. The toads were better protected from hunting garter snakes (*Thamnophis sirtalis*) on those substrates than on plain sand (Heinen 1993b). Green and brown color morphs of the Pacific tree frog (*Pseudacris regilla*) exhibited preferences for substrates that best matched their colors, and again, this provided some protection against garter snakes (Morey 1990). Garter snakes probably locate prey largely by chemical cues, so these anurans might be even better protected by background matching from visually hunting predators such as spiders or birds. On the other hand, green and brown color morphs of *Rana pipiens* did not respond differently to green and brown backgrounds, but generally preferred brown to green substrates (Croshaw 2005).

Background matching with the environment can be enhanced by rapid color change. Although amphibians cannot equal the almost instantaneous color changes of flounders, squid, and octopus, some species can change color in a matter of minutes. The eyes probably are involved in the perception of background characteristics, but there is evidence that the skin of some species is sensitive to changes in illumination as well (Iga and Bagnara 1975). In general, frogs exhibit changes in reflectance that result in a lightening or darkening of the skin to approximate the brightness of the substrate, but they do not exhibit precise matching of hue or color purity with that of the background. For example, *Hyla cinerea* exhibits a range of coloration, from lemon yellow to nearly black, but frogs placed on a yellow background may be almost any hue from lemon yellow to green (Nielsen and Dyck 1978). Nevertheless, frogs placed on light substrates are consistently lighter than those on dark substrates. Color change in this species is affected by both background brightness and temperature (R. King, Hauff, and Phillips 1994). Similar results have been obtained with other species placed on colored or black and white backgrounds, including *Pachymedusa dacnicolor* (Iga and Bagnara 1975), *Hyla arborea* (Nielsen 1980), *Pseudacris crucifer* (Kats and Van Dragt 1986), *Rana esculenta* (Nielsen 1979) and a number of other species (references in R. King, Hauff, and Phillips 1994). Background matching may be particularly effective in reducing predation by animals that lack color vision, although it may reduce predation by those with color vision as well.

Some amphibians have evolved cryptic patterns that match particular environments rather precisely. For example, many tree frogs have dorsal patterns resembling lichen-covered tree bark, and others have complex mottled color patterns and dermal appendages that give the frogs the appearance of moss or lichen that grows on trees (fig. 14.25 A–C). Climbing salamanders may resemble lichen-covered rocks (e.g., *Aneides aeneus;* fig. 14.25 D), or bare granite (e.g., *Hydromantes platycephalus*). Mimicry of dead leaves is a particularly common theme among tropical forest frogs (fig. 14.26). In some of these species, the "dead leaf" effect is enhanced by dermal flaps on the eyes and nose that obscure the froglike features of the body, or by flaps along the sides that reduce shadows beneath the body (Cott 1940). *Bufo alatus* reacts to potential predators by freezing and flattening the body against the ground, further enhancing the appearance of a dead leaf (fig. 14.26 D). A more common form of cryptic coloration is a pattern that combines various elements of disruptive coloration (fig. 14.27). These include the light dorsal stripes found in many anurans, dark eye lines or masks that obscure the eyes, and patterns of irregular blotches or spots that break up the outline of the body and limbs. In some species, blotches that appear randomly placed when the animal is active form a contiguous pattern when the animal is at rest, thereby giving a sleeping animal an appearance unlike that of a frog. This type of coincident disruptive coloration is especially common on the limbs of anurans (Cott 1940).

### Color Polymorphism

Many cryptically colored anurans are polymorphic for pattern and coloration (E. Hoffman and Blouin 2000). For example, some ranid frogs are polymorphic for background coloration (Corn 1981), the presence or absence of a light mid-dorsal stripe (Moriya 1952; Moriwaki 1953; Stugren 1966; Fishbeck and Underhill 1971; Ishchenko and Shchupek 1974; Schueler and Cook 1980), or the presence or absence of spots (Merrell 1965, 1973; L. E. Brown and Funk 1977). Cricket frogs (*Acris*) are polymorphic for vertebral stripe color, which may be gray, green, or reddish-brown

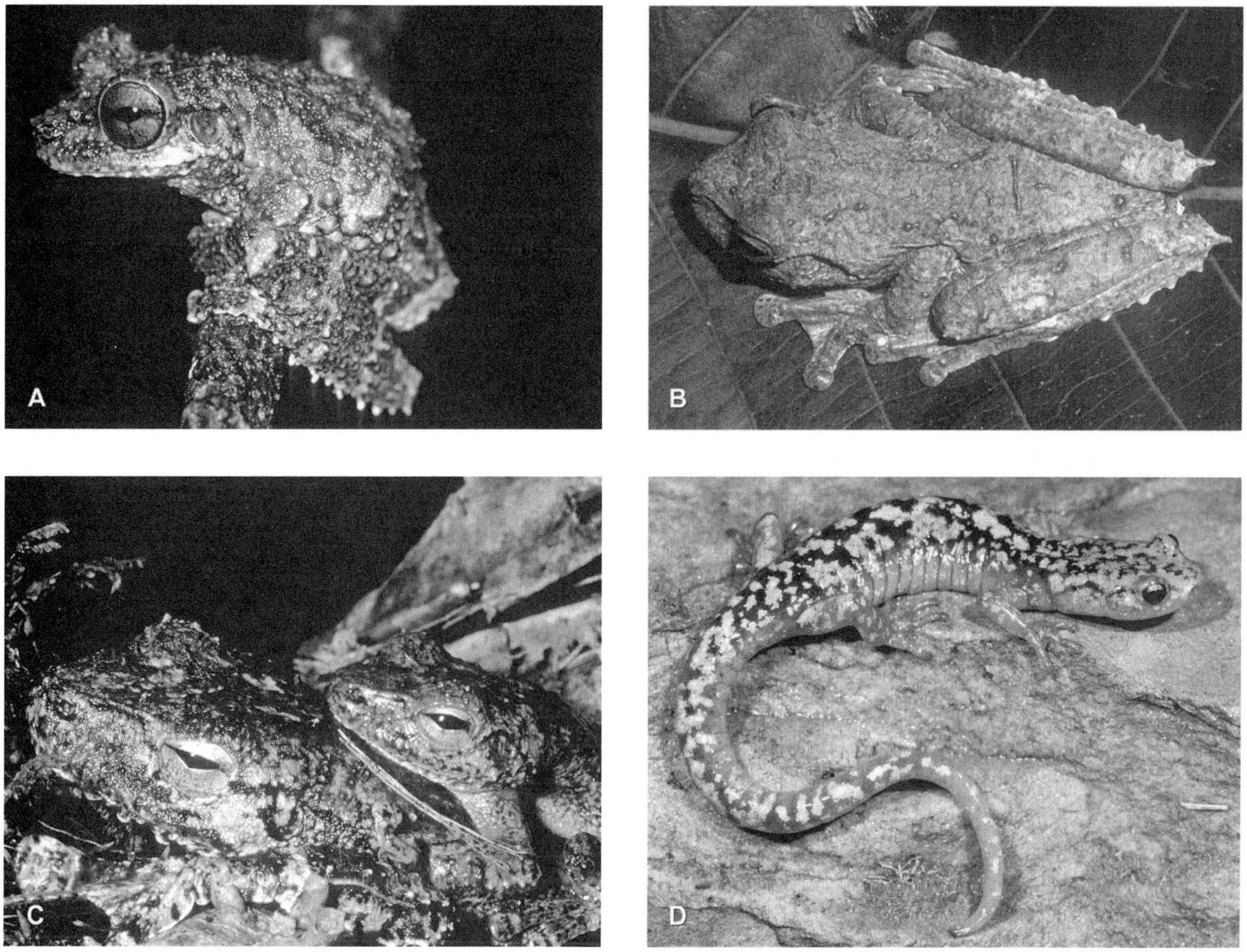


**Fig. 14.25.** A cryptic resemblance to moss or lichen is a common defensive adaptation among amphibians. (A) Mosslike pattern of a Neotropical treefrog, *Osteocephlus buckleyi* (Hylidae). (B) This treefrog from New Guinea, *Litoria genimaculata* (Hylidae) has a similar pattern, which is relatively conspicuous against a plain background. (C) Against a more complex background, this pair of *L. genimaculata* appears more cryptic. (D) The mottled pattern of *Aneides aeneus* (Plethodontidae) blends with the lichen-covered rocks typical of its habitat. Photos by Walter Hödl (A), David Bickford (B, C), and Wayne Van Devender (D).

(Pyburn 1961a, b; Nevo 1973; Milstead, Rand, and Stewart 1974; R. Gray 1983; Gorman 1986). Various species of *Pseudacris* are polymorphic for background coloration and dorsal pattern (Resnick and Jameson 1963; T. Matthews and Pettus 1966; Jameson and Pequegnat 1971; T. Matthews 1971; Tordoff, Pettus, and Matthews 1976; Tordoff and Pettus 1977; Tordoff 1980; Hoppe and Pettus 1984; Travis and Trexler 1984; S. R. Morey 1990). In *P. regilla,* there are both fixed green and brown morphs and a distinctive color-changing morph that can change between green and brown on a time scale of days or weeks (Wente and Phillips 2003). The slowness of color change in this morph would not enable a frog to assume cryptic coloration in response to an immediate predation threat, but would be adaptive in tracking seasonal changes in environmental background coloration. In fact, this species exhibits a seasonal shift in the proportion of brown and green morphs that tracks the availability of brown and green substrates. This shift probably is mediated by differential predation on the two color morphs, as well as the ability of some individuals to change color (Wente and Phillips 2005). Color change in this species is complex, however, and is affected not only by background color, but also by light intensity and temperature. Hence, color change in this species may serve other functions in addition to crypsis, such as thermoregulation (Stegen, Gienger, and Sun 2004).

Color polymorphism seems to be most pronounced in tropical anurans. M. Stewart (1974) estimated that at least 60% of the anuran species found in Malawi exhibit some degree of polymorphism. The most striking example is the genus *Hyperolius,* which exhibits so much variation in color within species that the group is a taxonomist's nightmare (Laurent 1965; Schiøtz 1971, 1999; Wieczorek, Channing, and Drewes 1998; Lötters et al. 2004). Small African frogs

**Fig. 14.26.** A general resemblance to dead leaves is common among terrestrial frogs that live in leaf litter, or those that breed in litter-filled ponds. (A) The Solomon Islands frog *Ceratobatrachus guentheri* (Ranidae) has a variable brown color pattern with skin flaps around the eyes, mouth, and legs that enhance its resemblance to dead leaves. (B) The Neotropical frog *Otophryne pyburni* (Microhylidae) has a leaflike color pattern and adopts an unusual upright, immobile posture to enhance the effect. (C) Another microhylid frog, *Chiasmocles shudicarensis,* has a middorsal line that enhances its resemblance to leaves while floating in a shallow pond. (D) The Panamanian toad *Bufo alatus* (Bufonidae) flattens itself on the ground when threatened, and flaps of skin on the side of the body eliminate shadows and enhance the dead-leaf effect. Photos by Wayne Van Devender (A), Walter Hödl (B, C), and Kentwood D. Wells (D).

of the genus *Phrynobatrachus,* which are rather similar to *Acris* ecologically, are highly polymorphic, with variation in background color, skin texture, and the size, shape, and color of vertebral stripes (Stewart 1974). Similar polymorphism occurs in Neotropical frogs of the genus *Eleutherodactylus,* which vary in background color, striping and spotting patterns, skin texture, and other pattern elements (fig. 14.28). Color patterns often vary in parallel among species in the same region, making species identification difficult (Goin 1947, 1950, 1954, 1960b; Lynch 1966; J. Savage and Emerson 1970; Crombie 1977, 1986; Lynch and Myers 1983; Heyer 1984; J. Savage 1987). Other polymorphic tropical frogs include *Polypedates leucomystax* (Church 1963), *Pleurodema bibroni* (Hellmich 1932; Cei and Capurro 1957; Cei 1958, 1962), *Rhinoderma darwinii* (Cei 1962), various hylid marsupial frogs (Duellman and Fritts 1971; Duellman 1973, 1983; Duellman and Ruiz-C. 1986), other hylid frogs such as *Hyla leucophyllata* and *H. ebraccata* (Duellman 1970), and toads in the *Bufo margaritifer* group (Hoogmoed 1990a; Hass et al. 1995).

Genetic control of color polymorphism has been well established for a number of species. It usually involves a rather simple Mendelian system (Goin 1950, 1954, 1960b; Moriya 1952; Moriwaki 1953; S. Anderson and Volpe 1958; Pyburn 1961a, b; Merrell 1965, 1973; Browder, Underhill, and Merrell 1966; Matthews and Pettus 1966; Fogleman, Corn, and Pettus 1980), although expression of color patterns also can be influenced by environmental conditions during tadpole development (Travis and Trexler 1984; Harkey and Semlitsch 1988) or ontogenetic changes (C. Richards 1976, 1982; Duellman and Ruiz-C. 1986). The selective forces that maintain color polymorphism in a population are not as well understood (see Dubois 1980b and E. Hoffman and Blouin 2000 for reviews). One possibility is that polymorphism is maintained by higher fitness in the heterozygotes (overdominance for fitness), but this is hard to test because heterozygotes generally cannot be identified in the field or laboratory (Merrell, 1969 1973).

Most explanations invoke some form of spatial or temporal variation in selective pressures, with predation often being the putative selective agent. Wood frogs (*Rana sylvatica*) with light vertebral stripes tend to increase from east to west in Minnesota. Fishbeck and Underhill (1971) suggested that striped morphs are more cryptic in areas where pine needles are common on the bottom of ponds, whereas unstriped morphs would be more cryptic in areas of hardwood forest. Schueler and Cook (1980) found a rough correlation between the amount of grass in breeding habitats and the frequency of striped morphs, again suggesting that linear patterns in the environment select for linear patterns on the frogs. However, even in grassy areas, the vast majority of wood frogs lack stripes, suggesting that other forces are important. J. Savage and Emerson (1970) attributed the high degree of polymorphism in *Eleutherodactylus* to the heterogeneous nature of the forest floor, arguing that the variety of colors, textures, and lighting conditions would result in continuously shifting selective advantages for different morphs. Again, predation was assumed to be the principal selective agent.

Other authors have argued that seasonal variation in selective pressures is the main force maintaining polymorphism in anuran populations. Several mechanisms have been proposed. One is that the survivorship of different color morphs changes with seasonal changes in background coloration (Merrell 1965; Matthews 1971; Jameson and Pequegnat 1971; Tordoff 1980). Another is that physiological differences give different morphs an advantage at different seasons (Merrell and Rodell 1968; Hoppe 1978). A third is that differences in rates of larval development favor faster-developing morphs in dry seasons (Merrell 1972; Bull 1977;

**Fig. 14.27.** Many frogs, such as these ranids, exhibit generalized cryptic color patterns that include several elements of disruptive coloration that serve to break up the outline of the body. (A) *Rana palustris* from North America, a species found in all sorts of terrestrial habitats near ponds. Elements of cryptic coloration include spots, transverse bars on the legs, and light dorsal lines. (B) *Rana swinhoana* from Taiwan, which lives among boulders in small streams. Elements of cryptic coloration include brown dorsal color with countershading, blotches on the sides of the body, transverse bars on the hind legs, and a dark line through the eye. (C) *Rana sylvatica* from North America, which breeds in leaf-filled ponds, but spends most of the year in leaf litter on the forest floor. This species exhibits some degree of background matching, adopting a dark color while in breeding ponds, but a lighter color while on land. Photos by Kentwood D. Wells.

**Fig. 14.28.** Color pattern polymorphism in a Puerto Rican frog, *Eleutherodactylus coqui*. All of these color patterns can be found in individuals living in the same habitat. Photos by Margaret M. Stewart (A, D) and Daniel S. Townsend (B, C).

Corn 1981; Hoppe and Pettus 1984). Pyburn (1961a) observed a seasonal shift in color morph frequencies in cricket frogs (*Acris*) that he attributed to changes in substrate coloration. Nevo (1973) reported a correlation between substrate color and morph frequency over a wide geographic area, but he did not detect seasonal shifts in morph frequencies. He thought that both visual selection by predators and physiological differences among color morphs were responsible for the maintenance of color polymorphism. However, subsequent studies have found little evidence of seasonal variation in selection, differential predation, or physiological differences that would favor one morph over another (Isaacs 1971; R. Gray 1977, 1978, 1983; Gorman 1986).

Milstead, Rand, and Stewart (1974) suggested that color polymorphism in frogs is maintained by apostatic selection, a form of frequency-dependent selection (B. C. Clarke 1962; Ayala and Campbell 1974). Predators often form search images of their prey and are likely to focus on the most common prey types, taking them in frequencies greater than expected from their representation in the population, while rare forms are likely to be taken less frequently than expected (Murdoch and Oaten 1975). Switching between prey types as a function of relative abundance has been demonstrated in a number of species (B. C. Clark 1962; Ayala and Campbell 1974; Murdoch and Oaten 1975; Curio 1976; Sargent 1981). Pietrewicz and Kamil (1981) found that blue jays improved their ability to find cryptic prey when they had repeated experience with the same morph, but failed to improve their detection when presented with several morphs in random order. This supports the idea that polymorphism could reduce predation on highly variable frogs such as *Eleutherodactylus*.

Much of the thinking on color polymorphism in anurans has been somewhat simplistic, and many of the proposed modes of selection are at odds with conclusions drawn from theoretical population genetics. The idea that spatial or temporal variation in selective pressures can maintain polymorphism in populations is intuitively appealing, but population genetics models suggest that this will happen only under certain conditions. Fine-grained spatial variation, such as that proposed by J. Savage and Emerson (1970), is unlikely to lead to polymorphism in the absence of overdominance (specifically, unless the geometric mean fitness of the heterozygotes exceeds that of the homozygotes). This is because selective advantages enjoyed by different morphs are never perfectly balanced, and one eventually would become fixed in the population (Felsenstein 1976; Hedrick, Ginevan, and Ewing 1976; Hedrick 1986). Course-grained spatial variation is unlikely to maintain polymorphism unless there is differential choice of appropriate microhabitats by different morphs. Conditions under which polymorphism can be maintained by temporal variation in selective pressures are even more restrictive than for spatial variation, but in general, polymorphism will not be maintained in the absence of overdominance. This is particularly true when the generation time of the organism is a year or more, as it is in frogs (Felsenstein 1976; Hedrick, Ginevan, and Ewing 1976; Hedrick 1986).

At present, some form of frequency-dependent selection, such as the apostatic selection proposed by Milstead, Rand, and Stewart (1974), seems the most likely explanation for the maintenance of color polymorphism, but this has not been tested experimentally for any anuran species. Probably it is unrealistic to expect that a single evolutionary force will explain polymorphism in all anurans. In well-studied polymorphic species such as European banded snails (*Cepaea*), a variety of interacting evolutionary forces have been found to be important in determining morph frequencies in different populations. Heterozygote advantage, climatic selection, visual selection by predators, frequency-dependent selection, density-dependent selection, disruptive selection, stabilizing selection, and various random process all may be operating (J. S. Jones, Leith, and Rawlings 1977). The whole question of color polymorphism in anurans needs to be reexamined, with an eye toward rigorous testing of hypotheses derived from current ideas in population genetics.

### Behavioral Avoidance of Predators

Juvenile or adult amphibians often have characteristic behavioral responses to predators. Perhaps the simplest defense is to avoid locations where predators are likely to be found. Until recently, most studies of antipredator behavior in juvenile or adult amphibians focused on responses to direct contact by predators. It has now become clear that some amphibians are capable of detecting and avoiding chemical cues from predators or injured conspecifics, even when the predator is not present. This is to be expected in aquatic amphibians, which tend to rely heavily on chemical senses for much of their behavior, but perhaps more surprising in terrestrial amphibians, especially anurans, which are not known for their chemosensory abilities. Newts (*Cynops* and *Notophthalmus*) avoided chemical extracts from the skin of conspecifics when these are presented on land (Marvin and Hutchison 1995) or in the water (Woody and Mathis 1997). Adult newts avoided chemical stimuli from predatory fish if these were associated with stimuli from injured conspecifics (newt skin extract), but not to fish stimuli alone (Woody and Mathis 1998). Once conditioned to avoid newt skin extract, the newts also avoided stimuli from an unfamiliar predator (*Siren*) if those stimuli were associated with newt skin extract. Experiments with another aquatic salamander, *Eurycea multiplicata griseogaster*, showed that salamanders could distinguish between predatory fish and nonpredatory tadpoles through chemical, but not visual, cues, and they also could distinguish predatory from nonpredatory fish species through chemical cues (Hickman, Stone, and Mathis 2004).

Several studies have shown that terrestrial and semiaquatic salamanders and anurans also respond to chemical stimuli from injured or stressed conspecifics (Lutterschmidt, Marvin, and Hutchison 1994; Chivers et al. 1996, 1997, 1999; Mathis and Lancaster 1998; B. Graves and Quinn 2000; Hucko and Cupp 2001; Watson et al. 2004), although not all species consistently show this type of response (Marvin et al. 2004). Some salamanders also respond to chemical cues from predatory snakes (Cupp 1994; Flowers and Graves 1997; DeVito et al. 1998; Schley and Griffiths 1998; R. Watson, Mathis, and Thompson 2004). Red-backed salamanders (*Plethodon cinereus*) gave a stronger response to odors from snakes that had fed on salamanders than to those that had fed on earthworms (Madison et al. 1999a, b; McDarby, Madison, and Maerz 1999; Murray and Jenkins 1999). In general, this species also responded more strongly to snakes that had fed on conspecifics than on heterospecific salamanders, although red-backed salamanders that were sympatric with *Eurycea bislineata* also avoided snakes that had fed on that species (A. Sullivan, Picard, and Madison 2005). Red-backed salamanders did not avoid chemical cues from snakes that had fed on *Desmognathus ocrophaeus*, a species that does not use the same microhabitats. *Eurycea bislineata* avoided cues only from snakes that had fed on conspecifics, whereas *D. ochrophaeus* did not avoid any predator cues (A. Sullivan, Madison, and Rohr 2004).

Several amphibians in the western United States showed variable responses to predator cues. Three species (*Rana lutiventris, Pseudacris regilla,* and *Ambystoma macrodactylum*), tested with garter snakes (*Thamnophis elegans*), did not distinguish between chemical cues from predators that had fed on conspecifics or on earthworms, although the two frog species did avoid chemical cues from injured conspecifics (Wirsing, Roth, and Murray 2005). *Rana luteiventris* adults did avoid chemical cues from garter snakes, as did *A. macrodactylum,* whereas *P. regilla* and *Taricha granulosa* did not (Murray, Roth, and Wirsing 2004). Only *R. luteiventris* showed avoidance of chemical cues from bullfrogs (*Rana catesbeiana*), predators that are not native to the region.

### Behavioral Defenses against Predators

Once an amphibian actually encounters a predator, its behavioral responses depend on the morphology and ecology of the animal and the identity and proximity of the predator. In many cases, fleeing from a predator is the most effective defense. Many slender-bodied frogs are agile jumpers (see chapter 1) and rely on rapid flight to escape predators (Marchisin and Anderson 1978; Licht 1986a; Heinen and Hammond 1997). Plethodontid salamanders often exhibit flipping behavior, literally snapping themselves away from the predator with rapid jerks of the body (Brodie 1983; Brodie, Dowdey, and Anthony 1989; Hileman and Brodie 1994; Dowdey and Brodie 1989; Whiteman and Wissinger 1991; Azizi and Landberg 2002). One salamander that lives on rocky slopes, *Hydromantes platycephalus,* coils its body and escapes by rolling away from a predator (García-París and Deban 1995). Similar behavior has been seen in a toad (*Oryophrynella*) that also lives on hillsides (McDiarmid and Gorzula 1989). Alternatively, many amphibians become immobile and flatten themselves against the ground, especially if contacted by a predator such as a snake (Brodie, Johnson, and Dodd 1974; Dodd and Brodie 1976; Brodie 1977; Dodd 1977, 1989; Dodd and Cupp 1978; Dowdey and Brodie 1989; Hayes 1989; Heinen and Hammond 1997). This behavior is found both in species that are well protected by noxious chemicals and those that lack such defenses. Others flip themselves on their backs and remain immobile (fig. 14.29 B), a behavior sometimes called "death-feigning," although the functional significance of the behavior is poorly understood (Toledo 2004b; Vaz-Silva et al. 2004).

Amphibians in close proximity to predators such as snakes often must select an appropriate antipredator behavior very quickly, and choosing the wrong behavior can be fatal. When confronted by birds or mammals, two-lined salamanders (*Eurycea bislineata*) exhibit display postures that present noxious skin glands to the predator (Brodie 1977; Brodie, Nowak, and Harvey 1979; see "Chemical Defenses," following). This type of response is ineffective against garter snakes, which are not repelled by the salamanders' skin secretions. The usual response to snakes is either to remain immobile or to flee, but not to display (Ducey and Brodie 1983). The best response depends on how the snake contacts the salamander. If contacted by the snake's head or body, the most effective defense is to remain immobile, but if contacted by the snake's tongue, then fleeing is most effective. This species exhibits both individual and geographic variation in antipredator behavior. Individuals that fled from garter snakes were faster runners than were those that remained immobile, and salamanders from populations where snakes were common were more likely to flee than those from areas were snakes were scarce (Dowdey and Brodie 1989). Another salamander, *Desmognathus ocoee,* often is confronted by larger species such as *D. quadramaculatus* and *Gyrinophilus porphyriticus,* which are capable of eating it. *Gyrinophilus* is the more dangerous predator (Formanowicz and Brodie 1993), and *D. ocoee* is more likely to flee from that species than from *D. quadramaculatus* (Hileman and Brodie 1994).

Some frogs and salamanders give displays or vocalizations that may serve to startle the predator (Licht 1973; Sazima 1975; Veloso 1977; Brodie 1978; Weber 1978b; see chapter 7). A number of large frogs are capable of inflicting painful bites on attacking predators (Duellman and Trueb 1986), and even small salamanders such as *Desmognathus* can deter predators with persistent bites (Brodie 1978; Brodie, Dowdey, and Anthony 1989; Whiteman and Wissinger 1991). When actually grabbed by a predator, amphibians use a number of methods to reduce their chances of being swallowed. Many anurans inflate the body and stretch out their legs when confronted by predators (fig. 14.29 C), particularly snakes that swallow their prey whole (Hinsche 1928; Hanson and Vial 1956; Marchisin and Anderson 1978).

Tail displays are an important component of antipredator behavior in many salamanders (fig. 14.30), especially in the families Plethodontidae and Salamandridae (Brodie 1977; Brodie, Nussbaum, and DiGiovanni 1984; Brodie, Nussbaum, and Datong 1990). Plethodontids often display the tail to redirect attacks away from the head and body. In some species, the tail is readily autotomized (broken off) when the salamander is attacked (Wake and Dresner 1967). This often allows the salamander to escape while the predator consumes the tail (Arnold 1982; Ducey and Brodie 1983; Labanick 1984; Ducey et al. 1993). The disadvantage of this strategy is that valuable energy reserves may be lost, thereby reducing future growth and reproduction. Maiorana (1977) and Arntzen (1981) suggested that loss of tails in salamanders could result in females skipping a year of reproduction. They worked on two genera (*Batrachoseps* and *Chioglossa*) that have unusually long tails relative to body size. More recent work on two species with shorter tails, *Desmognathus ocoee* and *D. carolinensis,* indicated that even severely injured animals still could reproduce, but the magnitude of tail injury was negatively correlated with reproductive investment (Bernardo and Agosta 2005). Salamanders also use tail displays to direct attacks of predators

**Fig. 14.29.** Defensive postures of anurans. (A) *Bombina orientalis* (Bombinatoridae) has a green and black pattern on the back and bright red and black markings on the feet and belly, probably a means of displaying aposematic coloration to fish attacking from below. When disturbed on land, members of this genus adopt an "unken reflex" posture that reveals the bright coloration on the feet to potential predators. (B) Many anurans become immobile when threatened by predators, adopting a so-called "death feigning" posture, as in *Kassina lamottei* (Hyperoliidae) from the Ivory Coast. (C) Large anurans often repond to predators by elevating the body and infating themselves with air, thereby exaggerating the size of the body, as in *Leptodactylus stenodema* (Leptodactylidae) from South America. (D) Another common defensive posture is to lower the head and elevate the rear part of the body, as in *Leptodactylus pentadactylus.* This posture probably makes the head less vulnerable and often exposes dorsal poison glands to the predator. (E, F) *Physalaemus nattereri* (Leptodactylidae) from Brazil combines this head-down posture with fake eyespots, which are particularly conspicuous from the rear and may startle predators and advertise posterior poison glands. Photos by Martha L. Crump (A), Mark Oliver Rödel (B), Walter Hödl (C, D), and Celio F. B. Haddad (E, F).

**Fig. 14.30.** Defensive postures of urodeles. (A) "Unken reflex" display of aposematic ventral coloration in a North American newt, *Taricha rivularis* (Salamandridae). (B) Display of aposematic ventral coloration by an Asian newt, *Paramesotriton deloustali* (Salamandridae). (C) Rigid posture of the alpine newt, *Triturus alpestris,* with bright orange ventral color on the tail directed toward the predator. (D) Elevated tail display of a North American cave salamander, *Eurycea lucifuga* (Plethodontidae). (E) The Asian newt *Echinotriton andersoni* (Salamandridae) has sharp ribs that protrude through the body wall. This not only injures a predator, but also delivers secretions from poison glands into the predator's mouth. (F) Head-butting posture of a North American salamander, *Ambystoma gracile* (Ambystomatidae), which exposes large parotoid glands to the predator. Photos by Edmund D. Brodie, Jr.

toward aggregations of poison glands on the tail; these displays are discussed in more detail under "Chemical Defenses," following. Some species wrap their tails around a predator and exude copious amounts of mucus, thereby partially immobilizing the predator (Arnold 1982; Brodie 1983).

### Chemical Defenses

Amphibians have a highly glandular skin (see chapter 2) and have a greater array of chemical defenses than any other group of vertebrates. Defensive secretions are produced by granular glands in the skin, so called because of their granular appearance under the microscope. These glands are found in all amphibians and are rather similar in structure, regardless of the types of compounds produced. Lindemann and Voute (1976), Whitear (1977), and Toledo and Jared (1995) provide reviews of amphibian skin and gland structure, and I will not discuss the histology of glands in detail here. Granular glands appear to be an ancestral amphibian trait and originally may have produced compounds other than defensive secretions, with production of highly toxic compounds being a secondary adaptation that has evolved independently a number of times (Neuwirth et al. 1979). For example, many of the skin secretions of amphibians have antibacterial or antifungal properties (Preusser et al. 1975; Zasloff 1987; Barthalmus 1994; Toledo and Jared 1995; Bowie et al. 1999), and protection against these disease agents could have preceded the evolution of chemical defenses against predators.

Granular glands often are distributed throughout the skin, as in many dendrobatids, hylids, ranids, microhylids, pelobatids, leptodactylids, and salamandrids (Lutz 1971; Brodie 1977; Garton and Mushinsky 1979; Neuwirth et al. 1979; Duellman and Trueb 1986; Toledo and Jared 1995). They also may be concentrated in certain regions or aggregated into larger glands. These include the parotoid glands found behind the head in many bufonids and some other anurans. Some bufonids, ranids, pelobatids, leptodactylids, myobatrachids, and discoglossids have large glands, clusters of glandular warts, or glandular ridges on the legs, along the sides of the body, or on the dorsal surface (Toledo and Jared 1995). Parotoid glands are found in many salamandrids and some ambystomatids, hynobiids, and plethodontids (Brodie 1983). Salamanders in these families also have concentrations of granular glands on the dorsal surface of the tail, and salamandrids may have lateral glands as well. In the spiny newts (*Echinotriton*), sharp ribs penetrate the lateral warts (fig. 14.30 E) and serve to injure predators and to deliver poisonous secretions into the wounds (Brodie, Nussbaum, and DiGiovanni 1984). Sharp ribs also are found in *Pleurodeles* and serve a similar function. However, in this genus, poison glands are found all over the body, and are not concentrated around the rib tips (Nowak and Brodie 1978). The European fire salamander (*Salamandra salamandra*) has a double row of very large middorsal glands running from the back of the head to the tip of the tail. These are imbedded in specialized trunk muscles that can be contracted to spray the noxious secretions of the glands toward predators (Brodie and Smatresk 1990).

Defensive postures of amphibians are closely correlated with the distribution of granular glands. Anurans with parotoid glands, or concentrations of dorsal poison glands, respond to predators by bending the head downward and arching the back (fig. 14.29 D), exposing the glands to the attacker (fig. 14.29; Hinsche 1928; Hanson and Vial 1956; Brodie and Gibson 1969; Villa 1969; Altig 1972a; Toledo 2004a). In a number of unrelated species of frogs, conspicuous patches on the flanks or inguinal region are closely associated with concentrations of granular glands. These frogs respond to predators by elevating the rear of the body to expose these bright patches and to orient the glands toward the predator (fig. 14.29 E, F; see the following "Aposematic Coloration"). Anurans that have concentrations of granular glands and bright coloration on the ventral surface, such as *Bombina* and *Melanophryniscus*, respond to predators with the "unken reflex," with the back arched and the ventral surfaces of the feet and throat displayed (fig. 14.29 A).

Some newts have a similar defensive posture (fig. 14.30 A). Salamanders with concentrations of glands in the tail often display the tail, wave it at the predator, or lash it from side to side when attacked (fig. 14.30 B, C). This increases the likelihood that the predator will attack the tail and receive a dose of noxious secretions. It also exposes the least vital part of the animal's body to attack. Salamandrids with granular glands throughout the body often exhibit immobility responses to predators, or adopt postures that expose aposematic patterns to the attacker (Brodie, Johnson, and Dodd 1974; Brodie 1977, 1983). Salamanders that have granular glands concentrated in parotoid glands on the head adopt postures that direct these glands toward a predator (fig. 14.30 F).

Amphibian granular glands produce a wide array of compounds that serve a defensive function (Toledo and Jared 1995). There is an enormous literature on the biochemical structure and pharmacological properties of these secretions, mainly because of their potential application in the development of human drugs. Erspamer (1994) and Daly, Garraffo, and Spande (1999) provided exhaustive reviews of the subject, and I will not go into great detail here (for earlier reviews, see Daly and Witkop 1971; Erspamer 1971; Daly et al. 1978; Daly, Highet, and Myers 1984; Daly, Garraffo, and Spande 1993). Amphibian defensive secretions include two general categories of compounds (Daly and Witkop 1971; Daly, Highet, and Myers 1984). The first are widespread, naturally occurring compounds that may serve other biological functions, such as protection against microbes and fungi, but are effective as defensive secretions when produced in large quantities. These include various biogenic amines and peptides that are found in most of the major amphibian families. Some of these have irritant properties, while others are quite toxic. Other compounds seem to have evolved mainly for defensive purposes and often are very toxic. These include tetrodotoxins and related compounds, found in many species of salamandrids and some anurans (*Atelopus* and *Brachycephalus*); cardiotoxic steroids known as bufodienolides, found mainly in bufonids; and extremely toxic steroidal alkaloids, found in *Salamandra, Dendrobates, Epipedobates, Phyllobates,* and some bufonids (*Melanophryniscus*), myobatrachids (*Pseudophryne*), and *Mantella.* The diversity of compounds that serve much the same function is enormous. Several hundred alkaloids have been described from the Dendrobatidae alone, and altogether, some 400 alkaloids are known from amphibian skin (Daly et al. 1978; Daly, Highet, and Myers 1984; Daly, Garraffo, and Spande 1993). Nearly identical defensive compounds have evolved independently in different families of amphibians (Daly, Highet, and Myers 1984; Daly et al. 1996), yet populations of a single species may produce very different compounds (Daly and Myers 1967; Myers and Daly 1983; V. Clark et al. 2005). Even individuals within the same population often differ in alkaloid profiles (V. Clark et al. 2005).

The origin of many of the alkaloids found in the skin of dendrobatid frogs and other anurans was long a mystery (Daly 1995, 2003; Daly, Garraffo, and Myers 1997). Some of these compounds lack known counterparts in other plants or animals and are presumed to be synthesized by the frogs (B. P. Smith et al. 2002). These include bufadienolides, which are present in laboratory-reared animals raised on a diet of fruit flies (Daly, Garraffo and Myers 1997; Daly, Padgett et al. 1997). Others, however, are derived from chemicals obtained from the arthropods that these animals eat (Daly Garraffo et al. 1994). Many of the toxic dendrobatids feed mostly on ants (Toft 1995; Caldwell 1996b), as do poison frogs (*Mantella*) from Madagascar (Vences, Glaw, and Böhme 1997/98; V. Clark et al. 2005) and many bufonids. Among dendrobatids, the presence of distasteful and toxic alkaloids and aposematic coloration is correlated with the degree of dietary specialization on ants, and has evolved independently in several distantly related clades within the family (Santos, Coloma, and Cannatella 2003; Summers 2003; Darst et al. 2005). Several types of alkaloids found in the skin of these frogs are similar or identical to compounds found in ants (Daly 1998; Spande et al. 1999; T. Jones et al. 1999; Daly et al. 2000, 2002; Saporito et al. 2004). Other probable sources of toxic compounds include beetles and millipedes (Daly 1995; Saporito et al. 2003; V. Clark et al. 2005). When toxic frogs such as *Dendrobates, Atelopus,* and *Mantella* were fed on fruit flies in the laboratory, they lacked most alkaloid skin secretions, indicating that these compounds are derived from the diet. There has been remarkable convergent evolution in aposematic color pattern, the presence of toxic alkaloids, and the ability to acquire defensive chemicals from the diet in dendrobatid and mantellid frogs, two families that are quite distantly related (V. Clark et al. 2005).

When laboratory-reared dendrobatid frogs (*Dendrobates, Phyllobates,* and *Epipedobates*) were fed fruit flies dusted with alkaloids in the laboratory, the chemicals accumulated in the skin and remained active for months (Daly, Secunda et al. 1994). Some of these compounds apparently are sequestered unchanged, but for some species of *Dendrobates,* there is evidence of secondary metabolic enhancement of the toxic effects of ingested alkaloids (Daly et al. 2003). Nontoxic dendrobatids in the genus *Colostethus* do not accumulate alkaloids and apparently lack the biochemical machinery to do so.

The skin secretions of amphibians are noxious to a wide variety of predators, including birds (see references in discussion of aposematic coloration and mimicry), mammals (J. Anderson 1963; Brodie, Nowak, and Harvey 1979; Formanowicz and Brodie 1979; Brodie and Formanowicz 1981; Mason, Rabin, and Stevens 1982), turtles (Hurlbert 1970), snakes (Hurlbert 1970; Lüling 1971; Sazima 1974; C. Myers, Daly, and Malkin 1978; Zielinski and Barthalmus 1989; Barthalmus 1994), amphibians (Brodie 1968b; Hurlbert 1970), fishes (Hurlbert 1970), ants (Fritz, Rand, and de Phaniphilis 1981), and spiders (Szelistowski 1985). The secretions need not be toxic to be effective, but may simply repel predators by irritating mucus membranes of the mouth, causing regurgitation, impairing coordination, or affecting their chemical senses. Nevertheless, amphibians vary in their degree of toxicity to predators, and some have skin secretions that are capable of killing predators (Brodie, Ducey, and Baness 1991). Presumably, many predators that live in the same habitats with highly toxic amphibians learn to recognize them or exhibit innate avoidance of these animals and are seldom actually killed. When toxic prey, such as the toad *Bufo marinus,* are introduced into areas with naïve predators, however, the results can be fatal to the predators (Covacevich and Archer 1975; Burnett 1997; Catling et al. 1999; B. Phillips, Brown, and Shine 2003).

Why do some amphibians evolve extremely toxic skin secretions, while others have only noxious ones? One possible reason is the energetic cost of maintaining the biochemical machinery needed to either manufacture or sequester highly toxic compounds in the skin, but this cost has yet to be investigated. Another is that predators vary in their responses to skin secretions and may evolve resistance to the chemical defenses of amphibians. This in turn will select for increased toxicity in the skin secretions of the amphibian prey. Some tropical snakes, such as *Leimadophis epinephelus,* seem to be remarkably resistant to toxins produced by *Dendrobates, Phyllobates,* and *Atelopus,* and will eat dendrobatid frogs that are rejected by other tropical snakes (Myers, Daly, and Malkin 1978). Snakes such as *Heterodon* and *Xenodon* feed mostly on toads that are distasteful or even toxic to other predators (H. M. Smith and White 1955). Some garter snakes also feed on toads and exhibit enhanced resistance to toad poisons (Macartney and Gregory 1981).

The best-studied example of this type of evolutionary arms race between predators and prey involves garter snakes (*Thamnophis sirtalis*) preying on newts (*Taricha granulosa*), the most toxic of all salamanders (fig. 14.31). Garter snakes in Oregon regularly feed on newts. Snakes from populations that are sympatric with newts and are known to feed on them are more resistant to newt toxins (tetrodotoxin) than are snakes from populations where newts are not found. They also are more resistant than is another species of garter snake, *T. ordinoides,* which never feeds on newts (Brodie and Brodie 1990). On Vancouver Island, British Columbia, newts are much less toxic than are those on the mainland. Garter snakes on Vancouver Island readily feed on these newts, but the snakes are not resistant to the toxins of mainland newts (Brodie and Brodie 1991; Hanifin et al. 1999). Individual snakes appear to assess their own level of resistance and reject newts that are too toxic for them to consume (B. Williams, Brodie, and Brodie 2003).

Tetrodotoxin resistance is derived from evolutionary changes in sodium channels in muscle (Geffeney et al. 2002, 2005). Such resistance appears to carry a cost, because movement speed was negatively correlated with resistance, and this may affect a snake's ability to escape from predators (Brodie and Brodie 1999). Tetrodotoxin resistance has a genetic basis (Brodie and Brodie 1990), and the snakes do not acquire resistance from repeated exposure to the toxin (Ridenhour, Brodie, and Brodie 1999). There is considerable geographic variation in resistance among populations of garter snakes, even on small spatial scales (Brodie, Ridenhour, and Brodie 2002). Phylogenetic analysis of tetrodotoxin resistance across a wide range of colubrid snakes indicated that only *Thamnophis sirtalis* that are sympatric with *Taricha* are highly resistant. However, all species of garter snakes that were tested, and possibly all natricine snakes, exhibit somewhat higher resistance than do other clades of colubrid snakes (Motychak et al. 1999). Species of *Thamnophis* with low resistance exhibit some genetic variation for resistance, indicating that this genus may be predisposed to evolve resistance to toxic prey.

### Aposematic Coloration

Many amphibians are brightly colored, and most either have noxious skin secretions or are Batesian mimics of noxious

**Fig. 14.31.** A garter snake (*Thamnophis sirtalis)* from the Pacific Northwest eating a highly poisonous newt (*Taricha granulosa*). Populations of garter snakes that eat newts have evolved heightened resistance to tetrodotoxin secreted by the newts. Photo by Edmund D. Brodie III.

species (fig. 14.32). Individuals benefit from this aposematic coloration because predators learn to associate certain colors with distastefulness of the prey. This requires that the color patterns of prey species contrast with the prevailing background coloration, making the animals conspicuous to potential predators (Siddiqi et al. 2004). Although most aposematically colored amphibians exhibit contrasting patterns of black, red, orange, or yellow, a variety of other colors, including bright blue and green, can be aposematic if these colors contrast with the background colors of the habitat. Indeed, some species, such as *Dendrobates pumilio,* exhibit extreme polymorphism in aposematic coloration (Summers, Cronin, and Kennedy 2003). Some of this polymorphism may be maintained by sexual selection. Females prefer to mate with males of the same color morph (Summers et al. 1999), even though individuals of different color morphs can successfully interbreed (Summers, Cronin, and Kennedy 2004).

Aposematic coloration has evolved independently many times in amphibians. The most striking examples are the Neotropical dart-poison frogs (*Dendrobates, Allobates, Epipedobates,* and *Phyllobates*), which display various combinations of bright red, orange, yellow, and blue coloration (Myers and Daly 1983). A phylogenetically based comparative analysis of the family Dendrobatidae indicated that level of toxicity, determined by injecting skin secretions into laboratory mice, is positively correlated with brightness of coloration, determined on a ranked scale by human observers (Summers and Clough 2001). Members of the genus *Phyllobates* are the most colorful and the most toxic, whereas members of the genus *Colostethus* are the least colorful and the least toxic. The genera *Allobates, Epipedobates,* and *Dendrobates* are intermediate in both respects. There also is a tendency within lineages for bright coloration to be correlated with large body size, suggesting that selection has favored increased body size in the most aposematically colored species (Hagman and Forsman 2003).

There are many other brightly colored frogs in various parts of the world, especially in the tropics. Frogs of the genus *Mantella* from Madagascar superficially resemble dendrobatid frogs in their morphology, chemical defenses, and coloration, and in their diurnal habits (Daly, Highet, and Myers 1984; V. Clark et al. 2005). Neotropical bufonids of the genus *Atelopus* (Fuhrman, Fuhrman, and Mosher 1969; Kim et al. 1975) often have red, yellow, orange, and black markings. The "golden frog" (*Atelopus zeteki*) of Central America, the Costa Rican "golden toad" (*Bufo periglenes;* J. Savage 1966), and the "gold frog" of southeastern Brazil (*Brachycephalus ephippium*) all are bright orange. The Australian myobatrachids *Notaten bennetti* and *Pseudophryne corroboree* have conspicuous yellow, red, and black markings (Habermehl 1965; Daly, Highet, and Myers 1984; Barker, Grigg, and Tyler 1995). The South American frog *Leptodactylus laticeps* has large red spots on a yellowish-white background (Cei 1980). The African frogs *Kassina senegalensis* (Hyperoliidae) and *Phrynomantis bifasciatus* (Microhylidae) both have patterns combining black with yellow, red, or pink markings (Cott 1940; Jaeger 1971c; Passmore and Carruthers 1995).

The most conspicuously aposematic urodeles are salamandrids such as yellow and black European fire salamanders (*Salamandra*) and the terrestrial red eft stage of the newt *Notophthalmus viridescens* (Cott 1940; Brodie 1968b; Steward 1969). The bright red plethodontid salamander *Pseudotriton ruber* also is distasteful to some predators (Brandon, Labanick, and Huheey 1979b; Brodie, Nowak, and Harvey 1979). The yellow and orange spots of *Ambystoma maculatum* appear to serve an aposematic function (Barach 1951; R. R. Howard 1971; Brodie, Nowak, and Harvey 1979). The same probably is true of the conspicuous black-and-yellow and black-and-white patterns of other ambystomatids such as *A. tigrinum, A. macrodactylum, A. annulatum,* and *A. opacum,* all of which are distasteful to predators (J. Anderson 1963; Brodie 1977; DiGiovanni and Brodie 1981; Mason, Rabin, and Stevens 1982). A number of Neotropical bolitoglossine salamanders have bright red, orange, or yellow markings on a dark background and are quite distasteful to predators (Dodd and Brodie 1976).

Some distasteful species are not brightly colored all over, but have patches of bright coloration that probably serve as aposematic signals. For example, salamanders such as *Plethodon jordani, Bolitoglossa subpalmata,* and *B. flavimembris* have dark dorsal coloration with bright red, orange, or yellow markings on the legs or cheeks; all produce noxious skin secretions (Dodd and Brodie 1976; Hensel and Brodie 1976). A somewhat similar pattern is found in the Asian salamandrid *Tylototriton taliangensis,* which has a dull black dorsum with orange patches on the parotoid glands and toes (C. Liu 1950). Examples among anurans include the Australian myobatrachid *Pseudophryne australis* and the South American leptodactylid *Telmatobufo bullocki,* both of which have red or yellow patches on the head (Cei 1962; Barker, Grigg, and Tyler 1995). Australian myobatrachids such as *Limnodynastes* and *Uperoleia,* several South American leptodactylids (*Pleurodema, Physalaemus*), and African microhylids (*Phrynomantis*) have bright red, yellow, and orange patches on the flanks, groin, and inguinal region (Cei 1980; M. Davies and Littlejohn 1986; M. Davies, McDonald, and Corben 1986; Herrmann 1988a, 1989; Barker, Grigg, and Tyler 1995; Brodie, Williams, and Tyler 1998). All of these species are well endowed with skin glands, which often are closely associated with the color patches (Erspamer, de Caro, and Endean 1966; Erspamer et al. 1975; Cei, Erspamer, and Roseghini 1967; Roseghini, Erspamer, and Endean 1976; Cei 1980).

**Fig. 14.32.** Aposematic coloration in amphibians advertises the presence of toxic or distasteful skin secretions. Most aposematic color patterns consist of either solid red, yellow, or orange, or contrasting patterns of red, yellow, orange, and black. (A) *Phyllobates terribilis* (Dendrobatidae) from Colombia, a lemon-yellow frog with some of the most toxic skin secretions of any anuran. (B) *Mantella aurantiaca* (Mantellidae) from Madagascar, a bright orange frog that exhibits remarkable convergence with dendrobatids in morphology, behavior, and toxic skin secretions. (C) *Dendrobates histrionicus* (Dendrobatidae) from Colombia, a bright red frog with black reticulations, one of many different aposematic patterns found in this species. (D) *Atelopus varius* (Bufonidae) from Ecuador, a highly toxic frog with a contrasting pattern of bright yellow and black markings, one of several aposematic color patterns in this species. (E) *Salamandra salamandra* (Salamandridae) from Europe, a highly toxic salamander with bright yellow or orange markings on a black background. (F) *Tylototriton shanjing* (Salamandridae) from China, a terrestrial salamander with brown-black background coloration with bright orange markings, some of which highlight parotoid glands on the head and rows of poison glands along the body. Photos by Charles W. Myers (A, C), Wayne Van Devender (B, F), Martha L. Crump (D), and Walter Hödl (E).

Some amphibians have aposematic patterns primarily on the ventral surface. When confronted by predators, they often adopt postures that expose these colors to the attacker. A well-known example is the "unken reflex" of *Bombina* (Hinsche 1926; Bajger 1980). These frogs arch the back and raise the legs into the air, exposing the brightly colored ventral surfaces to the predator (fig. 14.28). Nearly identical behavior occurs in the South American bufonid *Melanophryniscus stelzneri* (Fernández 1926; Cei 1980; Herrmann 1988b). Many salamandrids display ventral colors by elevating the anterior part of the body into a vertical position, coiling the body into a circle so the ventral surface is exposed, or displaying the tail in a coil above the body (fig. 14.29; Brodie 1977; Brodie, Nussbaum, and DiGiovanni 1984). In the Asian species *Paramesotriton chinensis,* the display is carried a step further, with the animal turning on its back to reveal a ventral pattern of orange blotches on a black background (Brodie, Nussbaum, and DiGiovanni 1984).

Many anurans have relatively cryptic dorsal coloration, but have bright patches of red, orange, yellow, or blue on the flanks, legs, and feet. These generally are described as "flash colors" that are hidden when the frog is at rest, but serve to startle predators when the frog jumps (Noble 1931; Cott 1940; Duellman 1970; Lutz 1973; Edmunds 1974; Duellman and Trueb 1986; Mattison 1987). However, many of these frogs produce noxious skin secretions, and the "flash colors" may serve primarily as aposematic signals. For example, phyllomedusine tree frogs are noted for their noxious secretions and strong odors when handled, and many species seem to be distasteful to predators (Budgett 1899; Cei and Erspamer 1966; Lutz 1966; Sazima 1974; Cei 1980). Their flank colors often are striking combinations of yellow, orange, and black markings characteristic of many aposematic animals. The Brazilian treefrog *Hyla anceps* has a pattern of dark bands on a coral red background on the flanks and legs. Lutz (1973) suggested that these frogs are Batesian mimics of coral snakes. However, she also stated that they release a sticky secretion and have a strong odor when handled, so the pattern could be aposematic. The same probably is true of the orange "flash colors" of *Hyla versicolor* and *Rana palustris,* both of which are distasteful to some predators (Formanowicz and Brodie 1979; Brodie and Formanowicz 1981). The striking inguinal "eye spots" found in some species of *Physalaemus, Phrynomantis,* and *Pleurodema* (fig. 14.28) probably are aposematic as well as deimatic (startling) signals. The South American hylid *Phrynohyas coriacea* has black spots on the flanks that are revealed during antipredator displays, and this species also has noxious skin secretions (de la Riva 1994). Cott (1940) cited the bright vermilion patches on the thighs and axillae of the African frog *Kassina maculata* as an example of "flash coloration," but this species is a slow-moving frog that does not make sudden leaps to escape from predators (Passmore and Carruthers 1995). It seems more likely that the pattern is aposematic.

Aposematic coloration is advantageous only when predators have color vision; birds generally are assumed to be the principal selective agent favoring aposematic patterns. Birds readily learn to avoid both the bright red efts of *Notophthalmus v. viridescens* (R. R. Howard and Brodie 1971, 1973) and the brown efts of *N. v. louisianensis,* but learning may be more rapid with the brighter color (Brandon, Labanick, and Huheey 1979a). J. Johnson and Brodie (1975) demonstrated that birds learned to avoid *Taricha granulosa* more rapidly when the newts displayed their bright ventral colors, compared to presentations where only the dorsum was visible to the predator. Other experiments have shown that wild-caught passerine birds quickly learn to avoid the red-cheeked and red-legged morphs of *Plethodon jordani,* but are less likely to avoid all-black morphs, even though these also are distasteful (Hensel and Brodie 1976). They also learned to avoid red-cheeked morphs with the red cheeks painted black, suggesting that the red color enhances avoidance learning, but is not essential for the birds to learn the salamanders are distasteful.

Not every species of amphibian that is equipped with chemical defenses is aposematically colored. For example, all species of *Bufo* probably are noxious to predators, but most are cryptically colored. The same is true of large tree frogs such as *Hyla vasta* from Hispaniola and various species of *Phrynohyas* from South and Central America (Duellman 1970; Lutz 1971). Some anurans that are both noxious and cryptically colored probably are too large to be taken by birds, and therefore may not have undergone strong selection for aposematic coloration. Many also are nocturnal and therefore would derive little benefit from conspicuous coloration, but do benefit from chemical defenses against nocturnal predators such as mammals and snakes.

### Mimicry

If predators learn to avoid the aposematic patterns of distasteful species, then edible species can benefit by evolving a pseudoaposematic pattern resembling that of the distasteful model, a form of mimicry known as Batesian mimicry (Cott 1940; Wickler 1968; Edmunds 1974; Huheey 1988; Pough 1988). The occurrence of Batesian mimicry in amphibians was first suggested by Dunn (1927), who pointed out the similarity between the red-cheeked *Plethodon jordani* and red-cheeked *Desmognathus* found together in the mountains of North Carolina, which he called *D. fuscus imitator.* Electrophoretic studies have shown that there actually are two very similar *Desmognathus* with red cheeks in this region (Tilley et al. 1978). The first, *D. imitator,* is endemic to the Great Smoky Mountains, and red-cheeked individuals

**Fig. 14.33.** Examples of mimicry in amphibians. (A) *Plethodon jordani* from Great Smoky Mountains National Park, a distasteful black salamander with reddish-orange cheek markings. (B) *Desmognathus ocoee,* a nondistasteful salamander with a dark body and similar orange cheek markings, a Batesian mimic of *P. jordani.* (C) *Lithodytes lineatus,* a nontoxic leptodactylid frog from the Amazon Basin that may be a Batesian mimic of several toxic dendrobatid frogs, including *Allobates femoralis* (D), *Epipedobates pictus* (E), and *E. trivittatus* (F). These toxic, aposematically colored dendrobatid frogs may form a Müllerian mimicry complex. Photos by Wayne Van Devender (A, B) and Janalee P. Caldwell (C–F).

are found only in sympatry with red-cheeked *P. jordani* (fig. 14.33 A, B). However, red-cheeked morphs of the second species, *D. ocoee,* occur outside of the Smoky Mountains in areas that lack an appropriate Batesian model (these populations are referred to as *D. ochrophaeus* in most of the published literature on mimicry). Some *D. ocoee* have red legs and are found only in sympatry with red-legged *P. shermani* in the Nantahala Mountains, but there are populations of *D. ocoee* with a combination of red cheeks and red legs in areas where appropriate distasteful models are absent (Labanick 1988). Experiments with both domestic chickens and wild-caught passerines have shown that predators conditioned to avoid distasteful *P. jordani* or *P. shermani* also learn to avoid the red-cheeked (*D. imitator*) or red-legged (*D. ocoee*) mimics found in the same area (Brodie and Howard 1973; Labanick and Brandon 1981). However, it is not clear why conspicuous red-cheeked and red-leg/red-cheek morphs of *D. ocoee* are maintained in populations outside the range of appropriate Batesian models, since all *Desmognathus* seem to be palatable to predators (Huheey 1966; Labanick 1988).

Two other examples of mimicry in salamanders both involve red efts of *Notophthalmus viridescens* and plethodontid salamanders. *Plethodon cinereus* lacks chemical defenses and is readily eaten by a variety of predators, including birds. This species has three color morphs: a uniformly slate-gray "lead-backed" morph, a "red-backed" morph with a dull red middorsal stripe, and an "erythristic" morph that is a uniform bright reddish orange. The erythristic morph is found only in the northeastern United States, where red efts are common (Lotter and Scott 1977; Tilley, Lundrigan, and Brower 1982). Experiments with both free-ranging and captive passerines have shown that birds tend to avoid erythristic *P. cinereus* more than the normal red-backed phase, although survival was lower than that of the highly distasteful red efts (Brodie and Brodie 1980; Tilley, Lundrigan, and Brower 1982). This suggests the red *P. cinereus* are mimics of the efts.

The plethodontid salamander *Pseudotriton ruber* also resembles red efts in color and has an overlapping distribution (Huheey and Brandon 1974; Pough 1974a). This species originally was considered a Batesian mimic of red efts because domestic chickens readily ate *Pseudotriton* before being exposed to red efts, but avoided them after experience with the distasteful efts (R. R. Howard and Brodie 1971; Brodie and Howard 1972). Similar results were obtained with wild-caught passerines and naive, hand-reared chicks (R. R. Howard and Brodie 1973). However, Brandon, Labanick, and Huheey (1979b) found that both *P. ruber* and *P. montanus*, another reddish species, were less palatable to chicks than were *Desmognathus*, but more palatable than red efts. Furthermore, these salamanders produce toxic skin secretions (Brandon and Huheey 1981), suggesting that *Pseudotriton* and red efts are part of a Müllerian mimicry system, in which several distasteful species mutually benefit from a similar color pattern. In this case, one species is considerably more noxious than the other. Huheey (1976, 1980, 1988) has argued that a spectrum can exist between pure Batesian and pure Müllerian mimicry, corresponding to a spectrum of relative prey palatability, but this has been disputed by other workers (Benson 1977; Sheppard and Turner 1977; J. Turner 1984a, b).

Another example of apparent mimicry in salamanders is found in California, where the plethodontid salamander *Ensatina eschscholtzii xanthoptica* closely resembles the color pattern of the highly toxic newts of the genus *Taricha*. Experiments with clay models of *Ensatina* painted with aposematic and nonaposematic colors showed that attacks by predators, presumably birds, were more common on the nonaposematically colored models. These results support the hypothesis that *Ensatina* derives some benefit from its resemblance to *Taricha*, although it has not yet been established whether this species is itself distasteful and therefore a Müllerian rather than a Batesian mimic of *Taricha* (Kuchta 2005).

Several examples of mimicry in anurans have been proposed. *Lithodytes lineatus*, a leptodactylid frog of the Amazon Basin, has a pattern of yellow dorsolateral stripes and orange "flash colors" resembling those of *Allobates femoralis* and several species of *Epipedobates* (fig. 14.33 C–F; Nelson and Miller 1971; Duellman 1978; Regös and Schlüter 1984). This resemblance suggests the possibility of Batesian mimicry, while the similarity of patterns in *Allobates femoralis* and *Epipedobates pictus* led Silverstone (1976) to speculate that these species are Müllerian mimics of each other. Although adult *Lithodytes* are largely nocturnal, juveniles are active both day and night, and it is the juvenile pattern that most closely resembles that of the diurnal *Epipedobates* (Duellman 1978; Regös and Schlüter 1984). However, it is not known whether this species also is distasteful and therefore might represent a case of Müllerian mimicry. Adults are associated with leaf-cutter ant nests (Schlüter and Regös 1981; Hoogmoed 1986) and conceivably could produce ant-repelling secretions, like those of *Gastrophryne* (Garton and Mushinsky 1979), which also might be distasteful to vertebrate predators.

Recent work has shown that a species from Amazonian Ecuador that is closely related to *Allobates femoralis, A. zaparo,* is a Batesian mimic of two other species of dendrobatids with very similar aposematic color patterns, *Epipedobates bilinguis* and E. *parvulus*. These two species differ in toxicity, with *E. parvulus* being the more toxic. Where only one model species occurs, *A. zaparo* has a color pattern that mimics that model, but in areas where the two model species overlap, *A. zaparo* actually has a pattern closer to the less toxic model species. The proposed explanation for this counterintuitive result is that birds that have experienced the less toxic model species will avoid only mimics that look very much like that species, whereas an experience with the more toxic species is so distasteful that only a general resemblance to the model is required. Consequently, where the two models overlap, mimics are better off looking like the less toxic species, because they will be protected regardless of which model a predator has encountered (Darst and Cummings 2006; Darst, Cummings, and Cannatella 2006).

Another possible example involves mimicry of one or more species of *Phyllobates* by *Eleutherodactylus gaigeae* in

Panama and Colombia. This species also has light dorsolateral stripes and overlaps the ranges of two dendrobatids with similar patterns, *P. aurotaenia* and *P. lugubris* (Myers and Daly 1983; Lynch 1985). Finally, Symula, Schulte, and Summers (2001) showed that the highly polymorphic species *Dendrobates imitator* has evolved distinct color patterns that closely resemble three other species of dendrobatid frogs in different parts of its range. Because all of the species involved are assumed to be toxic, this is most likely a case of Müllerian mimicry, although as in the case of the interaction between *Allobates zaparo* and the two species of *Epipedobates,* these species could differ considerably in distastefulness and toxicity. This type of mimicry may be relatively common in areas where several dendrobatid frogs are sympatric, but the details of these possible mimicry systems have yet to be explored in detail (Summers 2003).

A highly speculative proposed case of mimicry in another tropical amphibian is the possible mimicry of either banded millipedes or coral snakes by caecilians with banded patterns, such as *Siphonops paulensis* (Vitt 1992). Unfortunately, we have no information on either the efficacy of aposematic coloration in any caecilian or on possible distasteful skin secretions produced by the caecilians themselves, so the nature of this putative mimicry remains unknown.

## Summary and Conclusions

Amphibians are subject to predation by a vast array of invertebrates and vertebrates. Invertebrate predators include leeches, both larval and adult insects, and a variety of other arthropods such as crabs and spiders. Predation by invertebrates falls most heavily on egg and larval stages, and can be the most important source of mortality at these stages for both aquatic and terrestrial breeders. Most invertebrates are generalist predators that take amphibian eggs and larvae along with other types of prey. The only invertebrate predators that seem to specialize on amphibians are a number of genera of dipteran larvae that are obligate associates of amphibian egg masses and various helminth and arthropod parasites of adult amphibians. With few exceptions, the demographic impact of these predators is poorly understood.

Vertebrate predators include many freshwater fishes, both larval and adult amphibians, snakes, turtles, a few lizards and crocodilians, many passerine and nonpasserine birds, and mammals ranging from carnivores and insectivores to bats, rodents, armadillos, monkeys, and humans. The best quantitative data on the importance of amphibians in the diets of predatory vertebrates are for snakes and some kinds of water birds. Relatively few vertebrates are specialists on amphibians. In the temperate zone, the seasonal activity patterns of amphibians probably preclude specialization on amphibian prey by most endothermic predators, although some birds and mammals may concentrate for brief periods on amphibian prey when the latter are aggregated at breeding sites (e.g., hawks, owls, herons, ducks, otters, badgers, raccoons). In the tropics, there are a few hawks, and at least one bat, *Trachops cirrhosus,* which seem to specialize on amphibians, although *Trachops* eats lizards and other vertebrate prey as well as frogs. Only the snakes include large numbers of species that feed exclusively on amphibians. Frog-eating snakes are found in both the temperate zone and the tropics and are most diverse in regions where the diversity of frog species is high. A few snakes, such as the North American water snake *Farancia,* also specialize on salamanders. The fact that both snakes and amphibians are ectothermic, with similar seasonal activity patterns, probably has facilitated the evolution of amphibian eating as a dietary specialization.

While data are available on the importance of amphibians in the diets of certain predators, there are few data on the impact of specific predators or sets of predators on the demography of amphibian populations. In fact, with very few exceptions, it is not possible to identify any particular predator as the major source of mortality in an amphibian population. In part this is due to the rarity of predation events. Field workers seldom observe acts of predation, particularly on adult amphibians. One solution to this problem is for investigators to study the predators rather than the prey. An observer is much more likely to see a water snake eating a frog if he follows the snake while it searches for food rather than sitting and waiting for individual frogs to be eaten. With the help of radio transmitters, predators such as snakes can be located at frequent intervals or followed as they forage. Estimates of food intake by the predators in a relatively restricted area such as a pond, coupled with measurements of amphibian densities, would allow some estimate of the number of prey animals consumed by the predators and hence the demographic impact of predation on the prey population.

One difficulty in estimating the impact of predation on amphibian populations is that methods used to estimate survivorship and mortality in amphibian populations often have not been very sophisticated. Many workers have calculated survivorship from recaptures of marked individuals without adjusting the data for the probability of recapturing marked animals. This provides a rough estimate of "minimum survivorship" in the population, but almost certainly leads to an overestimate of actual rates of mortality. This problem is compounded by the short-term nature of most demographic studies of amphibians. F. Turner (1962) issued a call for more detailed, long-term demographic studies of amphibians, but only a handful of such studies have been conducted in the last 25–30 years, most on North American species.

Even when rates of mortality have been estimated accurately, evidence that predation is a major source of mortality usually consists of occasional observations of predators eating amphibians and sightings of potential predators in the area. Truly integrated studies of predator-prey interactions involving simultaneous long-term observations of amphibian populations and their principal predators are needed if further progress is to be made in this field. Although technically difficult because of the problems involved in accurately censusing amphibian populations, such studies should be possible if species and study sites are chosen carefully.

The means by which amphibians protect themselves from predators are many and varied. Embryos are protected by noxious chemicals, by mechanical properties of the egg capsule or jelly surrounding the egg mass, and by choice of protected oviposition sites by the parents. Some larvae also have distasteful defensive secretions, while others rely on crypsis, choice of protected microhabitats, behavioral avoidance of predators, or rapid growth to escape predation. Primary defensive adaptations of adults include cryptic and aposematic coloration and mimicry. Secondary defensive adaptations decrease the chances of an individual being consumed once it is encountered by a predator. Such adaptations include chemical defenses, defensive postures that expose poison glands or expendable parts of the body to the predator, rapid escape behavior, threatening or startling displays, physical attacks on the predator, immobility responses, inflation of the body, and tail autotomy. Many of these defensive adaptations, such as noxious chemical secretions, aposematic coloration, and defensive postures, have coevolved to provide an integrated system of defense against a variety of predators.

Much of the literature on amphibian defensive adaptations is descriptive. Nevertheless, there have been experimental studies of the effects of aposematic coloration on prey choice by predators (e.g., R. R. Howard and Brodie 1973; Hensel and Brodie 1976), the role of defensive postures in reducing predator attacks (e.g., Johnson and Brodie 1975), the effectiveness of Batesian mimicry in reducing the chances of being eaten (e.g., Brodie and Howard 1973; Labanick and Brandon 1981; Tilley, Lundrigan, and Brower 1982), and the effect of tail autotomy and other types of defensive behavior on salamander survivorship (e.g., Arnold 1982; Ducey and Brodie 1983; Labanick 1984). A number of issues remain unresolved. For example, the presumed Batesian mimicry of dendrobatid frogs by leptodactylids has never been investigated experimentally. In fact, there is only anecdotal evidence that the aposematic coloration of dendrobatid frogs is an effective deterrent to predators, although it is a reasonable assumption. The supposed deimatic function of "flash colors" in many anurans also has not been investigated in any detail. The possibility that such colors actually serve as aposematic signals has been mentioned briefly by a number of authors (e.g., Cott 1940; Formanowicz and Brodie 1979; Brodie and Formanowicz 1981), but this hypothesis has been ignored by most authors and has not been rigorously tested.

Another issue that has seldom been explicitly addressed is the relationship between toxicity and the distastefulness of amphibian defensive secretions, as well as the relationship between these variables and conspicuous coloration. Levels of toxicity generally have been determined by injecting skin extracts into mice or other animals (e.g., Daly and Myers 1967; Brodie 1968a, b; Brodie, Johnson, and Dodd 1974; Myers, Daly, and Malkin 1978; Garton and Mushinsky 1979; Brandon and Huheey 1981). Some animals, such as garter snakes, may be killed by injection of skin extracts, but can eat the same species without permanent damage (Brodie 1968a). Hence, the relationship between level of toxicity and the effectiveness of skin secretions as predator deterrents is not clear. Populations of *Dendrobates pumilio* vary widely in toxicity and coloration, but there is not a clear correlation between conspicuousness and toxicity (Daly and Myers 1967). However, they did not test responses of natural predators to different populations, and it may be that frogs with relatively low levels of toxicity have skin secretions distasteful enough to deter predators. Indeed, it is not clear what selective pressures have favored the evolution of extraordinarily toxic skin secretions in some amphibians, while distasteful chemicals are sufficient to protect other species. In some cases, the evolution of highly toxic skin secretions may be driven by an "arms race" between predators and prey, with predators evolving resistance to amphibian skin secretions and amphibians in turn evolving more potent defenses against these predators.

# Chapter 15 The Ecology of Amphibian Communities

*Brazil is the land of frogs, and almost half the species of the order live here. The naturalist meets with something surprising at every step; every creature is unlike the creatures he knows at home, and he is filled with reverence before such an inexhaustible wealth of forms. But he cannot be content with mere amazement; he longs to understand something of the wealth of species offered by the world about him. No Brazilian animal lives entirely to itself. . . . We shall understand the individual animal only if we consider it in relation to others.*
—Konrad Guenther, A Naturalist in Brazil (1931)

Since the days of the earliest explorer naturalists, European and North American biologists have been impressed by the diversity of tropical life. My first experience with tropical amphibians came on a trip to the summit of El Yunque, Puerto Rico, a mountain topped with a bizarre elfin forest of stunted trees draped with mosses, lichens, and a host of other epiphytes. After a drive up the narrow road to the summit, I emerged from the car on a rainy night and seemed to be surrounded by frogs calling from moss-covered trees, logs, or damp leaves on the forest floor. Nearly a dozen species of frogs are found on El Yunque, most in the genus *Eleutherodactylus* (Joglar 1998; Rivero 1998). Some live in holes and crevices in the gnarled trunks of forest trees; others prefer tree limbs, leaves, forest litter, boulders in mountain streams, or roadside ditches. At night, the sound of calling frogs is almost deafening, and the overall impression is of a forest teeming with amphibians, all sharing the same habitat, but each using the habitat in a slightly different way.

Such diversity of amphibian species in a single locality is not unique to Puerto Rico; indeed, many parts of the tropics have even larger assemblages. More than 30 species of frogs have been observed in a single rain forest pond in Peru (Schlüter 1984), and the area around the village of Santa Cecilia in Amazonian Ecuador once supported more than 80 species (Duellman 1978), approximately the number found in all of the United States. Even in the temperate zone, it is possible to find a number of amphibian species sharing the same habitat. Ponds in the southeastern United States can support up to 17 species of amphibians (Wilbur 1997), all of which contribute larvae to a complex aquatic community, and an equivalent diversity of terrestrial salamanders can be found in some moist forests of the southern Appalachian Mountains.

The occurrence of so many species sharing the same habitat raises a number of questions of interest to community ecologists. What determines the distribution and abundance of species in nature? Why are some species common and others rare, some widespread and others restricted in range, some generalized and others specialized? Do assemblages of species exhibit regular patterns of organization, or are they simply random groupings of species? If some degree of organization is apparent, is this the result of biological interactions among species, or independent adaptations to the environment? Except for chapter 14, which dealt with interactions between amphibians and their predators, most of this book has focused on the ecology and behavior of individual species of amphibians. In this chapter, the emphasis shifts to interactions among species, with an eye toward understanding the structure and organization of amphibian communities.

## A Conceptual Framework for Studies of Amphibian Communities

Studies of the structure and organization of amphibian communities are part of a larger field of community ecology that has undergone gradual evolution over the last 40 years or so. It is not possible to review the complete development of community ecology in this book, but a brief overview of key ideas and approaches will help to provide a framework for understanding the work that has been done on amphibians.

### Defining a Spatial Scale for Community Analysis

Community ecologists traditionally have concerned themselves with local assemblages of species, such as the tadpoles inhabiting a single pond or the salamanders occurring along a mountain stream. Larger-scale patterns of distribution and abundance have fallen mostly within the purview of systematists and biogeographers rather than ecologists (e.g., Duellman 1999a). Nevertheless, community ecologists have recently become interested in species interactions at larger spatial scales, and there have been attempts to unify biogeography and community ecology into a new field of macroecology (Ricklefs and Schluter 1993a; J. Brown 1995). Local species interactions can be affected by processes that occur at larger spatial scales and over long (evolutionary) time spans. These include speciation events, barriers to dispersal, and species-specific responses to features of the environment, such as rainfall patterns, landscape topography, and habitat productivity (Ricklefs 1987; Holt 1993; Ricklefs and Schluter 1993b; Rosenzweig and Abramsky 1993). Macroecological approaches enable ecologists to address questions about regional patterns of species diversity and abundance in relation to environmental variables, the relationship of local abundance to the size of geographic ranges, and the relationship between geographic range and body size. This approach has not been widely used in studies of amphibians, although some studies have addressed ecological determinants of species distribution and abundance at large and intermediate spatial scales (e.g., J. Lee 1980, 1993; Inger and Voris 1993; Murray, Fonseca, and Westoby 1998; Parris and McCarthy 1999; Eterovick and Fernandes 2001; Barr and Babbitt 2002; Eterovick 2003; Poynton 2003). This approach is particularly important for making informed decisions about the conservation of key habitats for amphibians (e.g., García, 2006). Some studies have incorporated phylogenetic information into analyses of speciation and geographic distribution. For example, Graham, Stacy, and Sih (2004) found that closely related species of dendrobatid frogs in the Andes tend to be well separated ecologically along basic habitat axes such as temperature, rainfall, and altitude. If applied to a number of different clades of amphibians with overlapping ranges, this approach has the potential to provide information on how current assemblages of species have come to coexist in particular localities.

### Defining the Species Composition of Communities

Most studies of community structure have focused on only a fraction of the organisms present, with the community being delimited either along taxonomic lines (e.g., the plethodontid salamander community) or by subsets of ecologically similar species that appear to form a unified guild (the forest litter frog community). It is important to keep in mind, however, that such assemblages of species are artificial constructs, because any limited assemblage of species is imbedded within a larger community of organisms that interact with one another as competitors, predators, parasites, or mutualists. For example, the assemblage of frogs (*Eleutherodactylus*) in the El Verde rainforest of Puerto Rico is part of a much larger community composed of a wide variety of invertebrates and vertebrates and the plants on which they feed (fig. 15.1). Interactions among different species of frogs are likely to be influenced, at least in part, by how the frogs interact with all of the other organisms in the forest (Beard et al. 2003).

Many ecologists have assumed that competition is most intense among closely related species. Nevertheless, significant competition can occur among animals as distantly related as ants and rodents (J. Brown and Davidson 1977; J. Brown, Davidson, and Reichman 1979; Davidson, Samson, and Inouye 1985), nectar-feeding birds and insects (F. Carpenter 1979; Kodric-Brown and Brown 1979), insectivorous lizards and spiders (Spiller and Schoener 1988, 1990), and lizards and birds (S. Wright 1979, 1981; Adolph and Roughgarden 1983; Moermond 1983). Interactions among distantly related taxa can be especially complex, because they may interact as both predators and competitors (e.g., Spiller and Schoener 1990).

Insectivorous vertebrates such as birds and lizards are potential competitors of terrestrial amphibians (Barbault 1974a; Duellman and Pianka 1990; Vitt and Caldwell 1994; Caldwell and Vitt 1999), and spiders are potential predators, prey, and competitors (Reagan, Camilo, and Waide 1996; Stewart and Woolbright 1996). A number of investigators have analyzed the distribution, abundance, and resource use of terrestrial amphibians and reptiles living in the same habitats (Lloyd, Inger, and King 1968; Crump 1971; Heyer and Berven 1973; Barbault 1974b, c, 1976a, b, 1977; N. Scott 1976, 1982; Inger and Colwell 1977; Duellman 1978, 1989b, 1990; Inger 1980; Inger et al. 1987; Duellman and Pianka 1990; Vitt and Caldwell 1994; Caldwell and Vitt 1999; Lima, Magnusson, and Williams 2000). While some overlap in resource use is common, especially between insectivorous lizards and frogs, we still know relatively little about how these

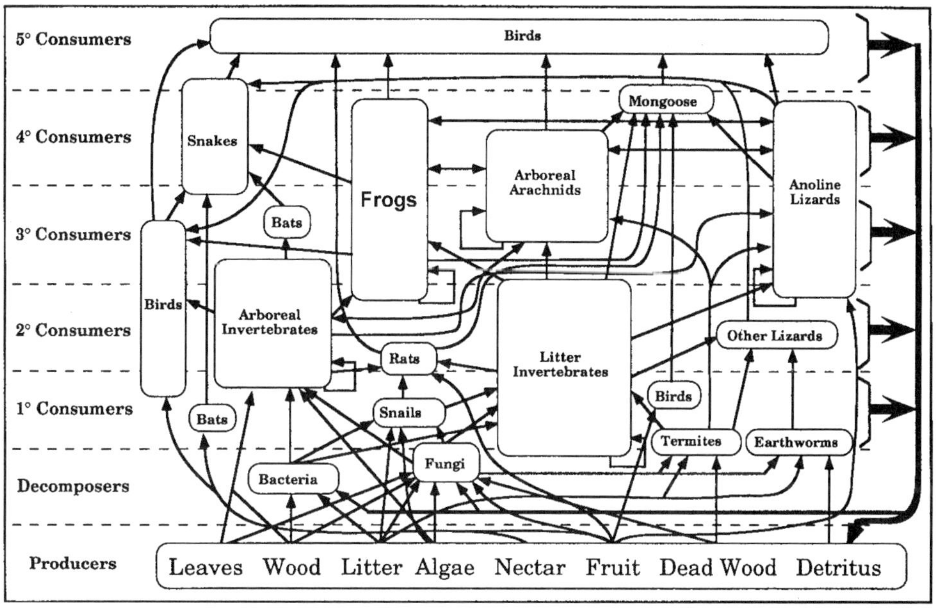


**Fig. 15.1.** The food web of the El Verde Rainforest community in Puerto Rico. Frogs of the genus *Eleutherodactylus* are part of a complex web of trophic interactions among vertebrates, invertebrates, plants, and decomposers. After Waide and Reagan (1996).

animals actually affect one another. The interactions between amphibians and spiders have not been studied, except for the possibility that spiders prey on amphibians (Rubbo et al. 2003). Fishes are both predators (see chapter 14) and potential competitors of amphibian larvae or aquatic adults (Antonelli, Nussbaum, and Smith 1972; Olenick and Gee 1981; Resetarits 1991, 1997; P. Harris 1995). Tadpoles potentially are in competition with other aquatic organisms that feed on algal mats or phytoplankton, such as zooplankton (Alford 1989c), aquatic insects (L. Blaustein and Margalit 1994; L. Blaustein, Friedman, and Fahima 1996; Kuperferberg 1997a; Mokany and Shine 2002, 2003a, b), and snails (Brönmark, Rundle, and Erlandsson 1991). Experimental studies of aquatic communities have provided evidence for competition between amphibians and both aquatic invertebrates (Morin, Lawler, and Johnson 1988; Alford 1989c; Brönmark, Rundle, and Erlandsson 1991; Mokany and Shine 2003a, b) and fishes (Bristow 1991; Resetarits 1991). Some of these studies are discussed in more detail in later sections.

## Defining the Processes That Structure Local Communities

Since the 1950s, most ecologists have assumed that animal communities are not simply random assemblages of species, but rather are organized products of biological interactions among species. Many biological processes may be involved, but it is interspecific competition that has been the principal focus of much of the community ecology literature. This emphasis on competition originated in pre-Darwinian natural history and was central to Darwin's own interpretation of ecological processes (Worster 1977). It was brought to maturity under the influence of G. F. Gause, David Lack, G. Evelyn Hutchinson, Nelson Hairston, Sr., Robert MacArthur, and a host of later workers (Kingsland 1985). The work of the "competition school" of community ecology is based on a number of premises about how natural communities are organized (Schoener 1982). The most important is that communities are resource-limited, with competition for limiting resources being the major determinant of the distribution and abundance of species in the community. Under most circumstances, species using identical resources cannot permanently coexist, so species must exhibit differences in their use of resources sufficient to allow them to live together. These differences, in turn, may be products of interspecific competition, with natural selection favoring divergence among coexisting species when this reduces competition. When species are very similar ecologically, then they may displace one another geographically, so the ranges of some species may be limited, at least in part, by competitive interactions.

In the 1970s and early 1980s, the primacy of interspecific competition as an organizing force in natural communities was challenged on a number of fronts. One criticism was that competition had been overemphasized at the expense of other biological interactions, such as predation, parasitism, or mutualism. Several authors, particularly Connell (1975, 1983), argued that many species do not reach population densities high enough to permit interspecific competition for resources, being limited instead by predation. It is well established that both predation and competition influence the distribution and abundance of species in some systems. In the rocky intertidal, for example, predation on a superior competitor can prevent that species from excluding another from the same habitat (Connell 1971, 1975; Paine 1974). In other cases, predation can actually reverse the outcome of interspecific competition (Morin 1981), so it is important to examine the interactions among a variety of biological processes when studying community structure (see also Quinn and Dunham 1983).

A second criticism was derived from empirical studies that failed to demonstrate interspecific competition among ecologically similar species. One well-known example is work on grassland bird communities, which revealed a broad overlap in resource use by coexisting species, but produced little evidence of competitive interactions (Wiens 1973; Wiens and Rotenberry 1979, 1980; Rotenberry 1980). These results led Wiens (1977, 1984) to question the importance of competition in structuring animal communities. He argued that in a variable environment, populations often are limited by abiotic, density-independent factors, such as weather conditions, resulting in communities that are not at equilibrium. In such communities, resources may become limited during "ecological crunches" brought about by harsh conditions such as droughts, leading to brief episodes of intense interspecific competition. During more benign periods, however, interspecific competition will be relaxed, and the responses of different species to environmental variations should be largely independent of one another. Hence, one would not expect to see the type of tightly structured community predicted by traditional competition theory. These interpretations were strengthened by studies on a variety of other animals, suggesting that interspecific competition is not always an important determinant of community structure (e.g., coral reef fishes: Sale 1977, 1984; insects: Strong 1984; Lawton 1984; Seifert 1984; spiders: Wise 1984).

The most persistent challenge to the "competition school" came from a group of workers centered at Florida State University. These workers questioned not only the importance of interspecific competition in structuring communities, but also the existence of the structural patterns themselves (Connor and Simberloff 1979, 1986; Simberloff 1983; Strong 1983; see also various contributions in Strong et al. 1984). They argued that perceived regularities in patterns of species diversity and abundance, or morphological differences among coexisting species, are more apparent than real. They suggested that many community ecology studies involve circular reasoning: the importance of competition in structuring a community is assumed, differences among species in local distribution, morphology, and resource use are documented, and the results are taken as confirmation that competition has been an important force in community evolution. They advocated a rigorous testing of observed patterns against suitable null hypotheses that assume that biological interactions among species are not important. Such interactions should then be invoked only if the observed patterns differ significantly from those expected under the null hypothesis. Unfortunately, it often is not clear how appropriate null hypotheses should be constructed, or even whether such hypotheses really are "null" models (Harvey et al. 1983; Quinn and Dunham 1983; Roughgarden 1983; Colwell and Winkler 1984; May 1984; other chapters in Strong et al. 1984). Consequently, these criticisms generated much heated debate (Simberloff 1983; Strong 1983; Strong et al. 1984; Connor and Simberloff 1986).

Although the relative merits of various positions can be debated, one useful result of these controversies was to shift the emphasis in community ecology from purely descriptive analyses of community patterns to experimental investigations of the processes that generate these patterns. This shift is readily apparent when one compares the contributions in Strong et al. (1984) with those in an earlier symposium on community ecology (Cody and Diamond 1975). Experimental approaches have come to dominate community ecology since the 1980s (e.g., Hairston 1989a; Resetarits and Bernardo 1998), but it is critical that experimental studies be anchored in solid natural history (Werner 1998). In part, these experimental studies have confirmed earlier views about the nature of community structure. There is now abundant evidence that interspecific competition is an important force in many animal communities (Schoener 1982, 1983; Connell 1983; Sih et al. 1985). This is not always true, however, and its importance cannot simply be assumed. The intensity of competition detected in experimental studies can be influenced by the venue in which the studies are conducted (e.g., Skelly and Kiesecker 2001; Skelly 2002). Furthermore, the way in which competition operates in a community often is mediated by physical factors in the environment or by biological factors such as predation.

Both descriptive and experimental methods have been used in studies of amphibian communities. Many investigators have examined patterns of resource use among species living in the same habitat (e.g., Toft 1985; Kuzmin 1995; Parmelee 1999; Eterovick and Barros 2003). Such studies often include detailed measurements of microhabitat use, activity times, and the sizes and types of prey eaten by different species. These studies often reveal a potential for competitive interactions among species, but do not, in themselves, demonstrate that such interactions are important in determining the relative abundance or patterns of resource use by individual species.

One way to investigate the role or competition or predation in structuring a community is to conduct field experiments in which the abundance of potential competitors or predators is manipulated and the responses of one or more species in the community are observed (Sih et al. 1985; Hairston 1987, 1989a). Field experiments have the advantage of maintaining a degree of realism in the experimental system because the animals are studied in their natural environments (Hairston, 1989b). The disadvantage of field experiments is that many variables than can affect the interactions between species cannot be controlled with great precision, and adequate replication of experimental treatments often is difficult (Morin 1998).

An alternative approach is to perform laboratory experiments that allow the investigator to precisely control all environmental variables except those that are specifically being manipulated. Laboratory experiments afford a high degree of precision in control of variables, but at some cost in realism (Morin 1998). A compromise between laboratory and field experiments is the use of mesocosms, such as cattle tanks used to mimic conditions in temporary ponds (Resetarits and Fauth 1998), or large-scale field enclosures (D. Scott 1990). This maintains some of the precision of control over variables that is possible with laboratory experiments, but also provides a higher degree of realism in the system. The relative merits of these approaches have been hotly debated, particularly with reference to studies of amphibians (e.g., Hairston 1989a, b; Jaeger and Walls 1989; Morin 1989, 1998; Petranka 1989a; Resetarits and Fauth 1998; Wilbur 1989; Skelly and Kiesecker 2001; Skelly 2002). All have advantages and disadvantages, and a research program that integrates information from a variety of approaches is most likely to yield insights into community structure (Werner 1998).

### Global and Regional Patterns of Amphibian Distribution and Abundance

On a global and regional scale, the distribution and abundance of amphibians probably has little to do with biological interactions among species and a great deal to do with historical events and major features of landscape and climate (Duellman 1999a). The global distribution and diversity of amphibian species is strongly affected by latitude. In both the Old and New Worlds, amphibians are far more diverse in the tropics than in temperate latitudes (Duellman 1999b). This is not surprising, since the relatively constant temperatures and wet environments of tropical habitats are ideal for amphibians, especially those that lay eggs out of water. The diversity of amphibians is much higher in the Neotropics than in the Old World tropics. South America, for example, has nearly three times as many species of amphibians as does Africa, which is a much larger continent. In part this reflects the larger expanses of tropical forests in South America compared to tropical Africa, which has extensive grassland habitats. It also reflects the extensive radiation of several large families of anurans in tropical South America, including the Hylidae, Leptodactylidae, Centrolenidae, and Dendrobatidae. Islands tend to have a much lower diversity of amphibians than do mainland habitats, even in the tropics, but species diversity increases with island size. Large tropical islands such as New Guinea, Borneo, Sri Lanka, and Madagascar have high amphibian species diversity, whereas small tropical islands in the Caribbean have much smaller numbers of species.

In both the tropics and temperate zone, amphibian diversity is strongly affected by rainfall. In North America, for example, amphibians are most diverse in the southern Appalachians and the southeastern Coastal Plain, both areas of high rainfall (Duellman and Sweet 1999). A strong positive correlation between amphibian species diversity and rainfall is seen in other regions as well, including Middle America (J. Lee 1980; J. Campbell 1999), South America (Duellman 1999c), and tropical Australia (M. Tyler 1999). Amphibian species diversity also is strongly affected by altitude. In North America, salamanders are more diverse in montane habitats than in lowland habitats, but anurans are not. In South America, the wet slopes of the Andes harbor the highest diversity of anuran species in the world (Duellman 1999c). Montane regions of Africa generally have fewer species of amphibians than do New World mountains, but nevertheless these areas often are rich in endemic species with limited altitudinal distributions (Poynton 1999, 2003).

Certain lineages of amphibians have been particularly successful in montane regions, which often lack permanent or temporary ponds that can be used by amphibians for breeding. In the Neotropics, amphibian clades that deposit eggs out of water are particularly successful in the mountains. These include plethodontid salamanders in Middle America (all tropical plethodontids have direct development) and frogs of the genus *Eleutherodactylus* in South America, as well as dendrobatids (Campbell 1999; Duellman 1999c). Stream-breeding frogs, mostly hylids, centrolenids, and dendrobatids (*Colostethus*), also are very diverse in the mountains. In New Guinea, a very mountainous tropical island, the amphibian fauna is dominated by microhylids with direct development.

Global and regional patterns of amphibian diversity are due largely to independent responses of different lineages to physical factors such as landscape topography and rainfall, as well as historical factors such as speciation and dispersal (Graham et al. 2004). Similar habitats in relatively close proximity to one another often have slightly different assemblages of species present. Some species are distributed across a wide geographic area and live in a variety of habitats. Others have much more restricted ranges or more specialized habitat requirements. The resulting patterns of species diversity set the stage for ecological interactions among species in particular localities (Inger and Voris 1993; Zimmerman and Simberloff 1996; Murray, Fonseca, and Westoby 1998; Bunnell and Zampella 1999; Eterovick 2003; Eterovick and Barros 2003). Such interactions are likely to be more complex in areas with large numbers of species than in communities composed of only a few species. In any given area of tropical rainforest in South America, for example, the number of species in a given locality is likely to be much larger than in areas of similar size in North America. In part this is due to the much greater diversity of reproductive modes in the tropics, where anurans make use of all sorts of microhabitats for laying eggs that are not found in temperate habitats (see chapter 10). Trophic niches may be more diverse or more specialized in the tropics as well, and these are discussed in more detail in the next section. Within North America, assemblages of streamside salamanders, particularly those in the genus *Desmognathus,* are very diverse in the southern Appalachians, with five or six species present in some areas. In more northern habitats, there may be no more than two or three species living in the same type of habitat (Hairston 1987). Ecological interactions may be similar in these two types of assemblages, but clearly the number of possible interspecific interactions is larger in the more diverse assemblage.

## Community Ecology of Adult Anurans

Anurans are diverse both morphologically and ecologically, and many species often can be found sharing the same habitat. This diversity would seem to make these animals ideal subjects for studies of community structure. Unfortunately, the cryptic habits of many species, coupled with relatively high mobility, makes assessment of local densities of anurans difficult, and manipulation of relative abundance even harder. Probably for this reason, studies of terrestrial anuran community ecology have been almost entirely descriptive. Some studies are broad-spectrum analyses of community structure, providing information on species abundance, diet, morphological differences, microhabitat use, activity times, and breeding habits for all of the species in a local assemblage (Crump 1971, 1974; Barbault 1974b, c, 1976b, 1977; Inger and Colwell 1977; Duellman 1978, 1989b, 1990; Toft and Duellman 1979; Pefaur and Duellman 1980; Schlüter 1984; Zimmerman and Simberloff 1996; Eterovick and Sazima 2000; Neckel-Oliveira et al. 2000). The diversity of reproductive modes is discussed in chapter 10, so I will focus on differences in activity time, microhabitat use, and diet.

There have been many descriptive studies of microhabitat use and feeding habits of frogs in local assemblages (Inger and Marx 1961; Inger 1969; Cintron 1970; Pengilley 1971b; Barbault 1974a; Toft 1980a, b, 1981, 1982; Lieberman 1986; Lizana Avia, Ciudad Pizarro, and Pérez-Mellado 1986; Lizana, Perez-Mellado, and Ciudad 1990; Das 1996a, c; A. Lima 1998; A. M. Anderson et al. 1999a; Parmelee 1999; Bellocq, Kloosterman, and Smith 2000; Eniang et al. 2003; Gillespie et al. 2004). There also are more limited studies that describe ecological differences between two or more species, often closely related, that comprise only part of a local community (Heatwole 1963; Inger and Greenberg 1966; Lescure 1971; M. Stewart and Sandison 1972; Heyer and Bellin 1973; V. Brown 1974; M. Stewart 1979; Loman 1978; Stewart and Martin 1980; K. Jones 1982; Mac Nally 1983; Miyamoto 1982; Diaz and Valencia 1985; Licht 1986b; McAlpine and Dilworth 1989; A. K. Howard et al. 1997; Guidali, Scali, and Carettoni 2000; Duré and Kehr 2001; Isacch and Barg 2002; França, Facure, and Giaretta 2004; Lötters et al. 2004). Most studies have shown that nearly all the anurans in multispecies assemblages are generalist predators on insects and other small animals, with broad overlap in diet among species. There may be minor ecological differences among co-occurring species along one or more resource axes, and such differences sometimes have been interpreted as the result of interspecific competition. In most cases, however, the processes underlying patterns of organization in these communities remain largely unknown. I will not review each study in detail, but instead will focus on a few assemblages that exhibit interesting patterns of community organization or are promising subjects for additional investigation.

### Anurans of Tropical Forests

The anurans of most tropical forests can be divided into two distinct assemblages of species. The first are arboreal frogs, which find their food in understory vegetation or even high in the forest canopy (Duellman 1978, 1989b, 1990; Vitt and Caldwell 1994; Parmelee 1999). These frogs tend to be strictly nocturnal, and spend the day in sheltered retreats or adopt water-conserving postures on exposed leaves (see chapter 2). Most species are sit-and-wait predators that eat relatively large prey, such as orthopterans, beetles, and roaches. A few species exhibit unusual feeding modes, however. For example, the Neotropical hylid genus *Sphaenorhynchus* is almost unique among members of that family, and among arboreal frogs in general, in being an ant specialist (Parmelee 1999). These frogs also are unusual in having very narrow mouths, a trait that is more typical of terrestrial ant-eating microhylids than of hylids, which usually have generalized diets.

The second assemblage is one that is virtually unknown in temperate habitats—frogs that spend most of their lives in the leaf litter of the forest floor. Many of these species are active during the day, or both during the day and at night.

Use of this terrestrial microhabitat during the day is made possible by high moisture levels in the litter, and therefore is characteristic only of relatively humid forests. In contrast, dry tropical forests and highly seasonal tropical environments often have terrestrial species, but all are nocturnal (Inger and Colwell 1977; Das 1996c). Use of forest litter habitat is especially prevalent among species with some form of terrestrial reproduction, but it is not restricted to them. In the Neotropics, litter communities are composed mainly of bufonids, dendrobatids, and leptodactylids. The genus *Eleutherodactylus* often is especially common, but other leptodactylids, such as *Physalaemus, Edalorhina, Adenomera, Leptodactylus,* and *Ceratophrys,* may be present as well, and terrestrial or burrowing microhylids are found in some communities. Two of these groups are partially (dendrobatids) or completely (*Eleutherodactylus*) independent of water for reproduction. Diurnal forest floor communities in the Old World are composed mainly of bufonids and ranids, sometimes with microhylids and rhacophorids as well (e.g., Watanabe, Nakanishi, and Izawa 2005). Forest floor communities usually have fewer species in the Old World than in the New World, but the reasons for this are not fully understood (Heyer and Berven 1973; N. Scott 1976, 1982; Voris 1977; Toft 1982). This could be related to local environmental differences, such as the severity of the dry season, or to historical factors, such as the phylogenetic origin of the species in each community. Nevertheless, some Old World tropical forests have high densities of individuals, even when species diversity is low (Watanabe, Nakanishi, and Izawa 2005).

The number of anuran species in Neotropical litter communities is positively correlated with both the number of wet months in a year (fig. 15.2 A) and mean annual rainfall (fig. 15.2 B), with the former being a slightly better predictor of species richness. Similar trends have been found in

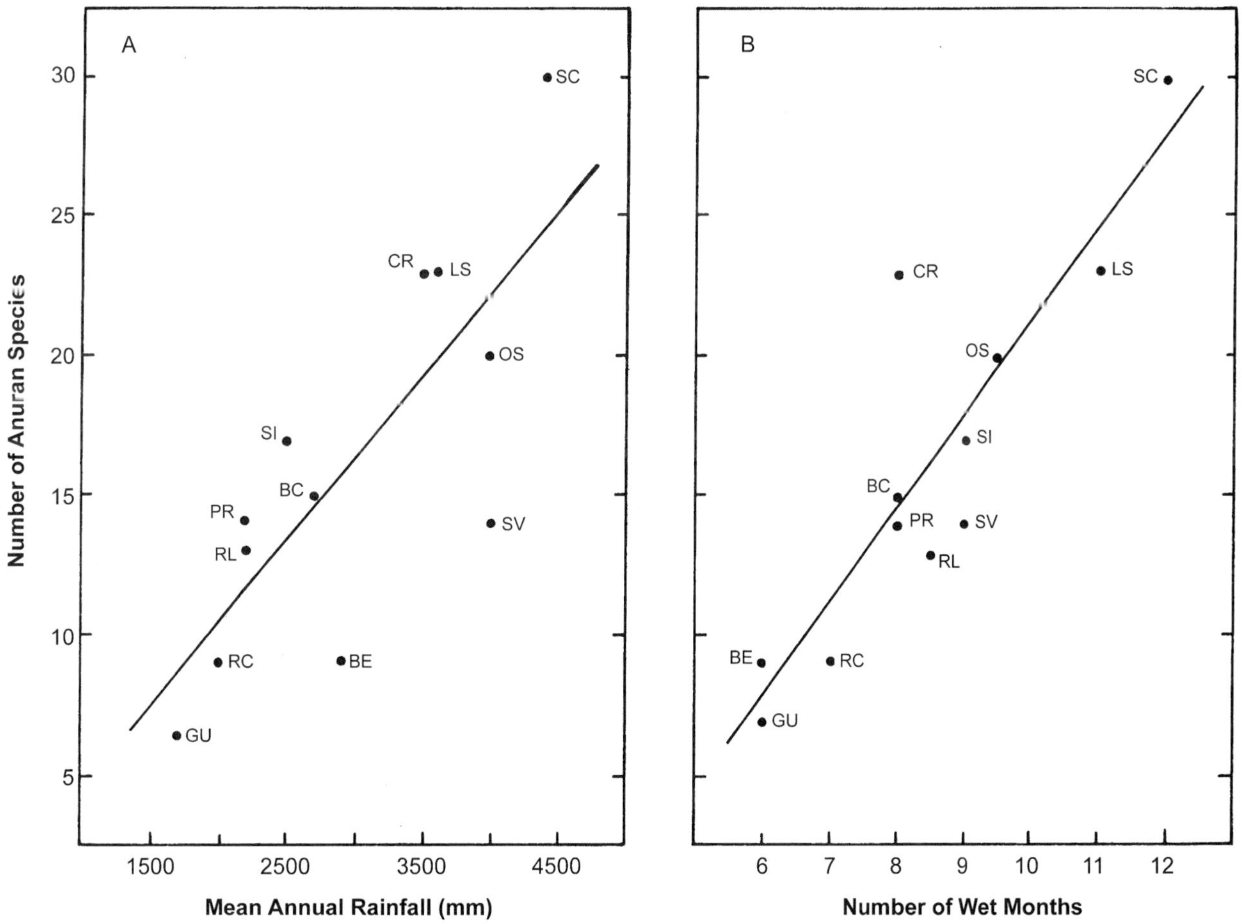


**Fig. 15.2.** Relationship of number of species of litter anurans to mean annual rainfall (A) and number of wet months per year (B) at several sites in South and Central America. Only diurnally active inhabitants of leaf litter are included. Regression equation for (A): $y = -1.42 + 0.006x$, $R^2 = 0.62$, $p < 0.002$; (B): $y = -12.2 + 3.34x$, $R^2 = 0.78$, $p < 0.0001$. Sites: BC = Barro Colorado Island, Panama; BE = Belem, Brazil; CR = Carti Road, Panama; GU = Guanacaste, Costa Rica; LS = La Selva, Costa Rica; OS = Osa Peninsula, Costa Rica; PR = Pipeline Road, Panama; RC = Rio Canclon, Panama; RL = Rio Llullapichis, Peru; SC = Santa Cecilia, Ecuador; SI = Silugandi, Panama; SV = San Vito, Costa Rica. Plotted from data in Heatwole and Sexton (1966), Crump (1971), Scott (1976), Duellman (1978), Toft and Duellman (1979), and Toft (1980a, b, 1981).

comparisons of total amphibian species diversity at several Neotropical localities (Crump 1971; Duellman 1978; J. Lee 1980; Pefaur and Duellman 1980; Duellman 1988; Duellman and Thomas 1996). Although population densities also are higher at wetter sites (N. Scott 1976; Toft 1980a, b), there is little consistency in seasonal patterns of abundance. In some areas, frogs are more abundant in the leaf litter during the wet season (Allmon 1991), but in others they are more abundant in the dry season (Lieberman 1986). These differences probably reflect both seasonal differences in recruitment of new individuals into the population (Toft, Rand, and Clark 1982; Watling and Donnelly 2002) and seasonal shifts in microhabitat preference (Toft 1980b). The numerical distribution of species in the forest litter is very uneven, with a few species usually predominating and others being comparatively rare (N. Scott 1976, 1982; Toft 1980a, b; Heatwole 1982; Toft, Rand, and Clark 1982; Lieberman 1986; Allmon 1991; Watling and Donnelly 2002).

### Feeding Ecology

Neotropical leaf litter anurans differ in the timing of their foraging activity. Some dendrobatids, for example, are relatively inactive at midday, doing most of their foraging in the early morning or late afternoon, whereas other anurans are active throughout the day (Jaeger, Hailman, and Jaeger 1976; Jaeger and Hailman 1981). There also are species of *Eleutherodactylus* that hide in the leaf litter during the day, but usually forage on vegetation above the forest floor at night (Miyamoto 1982; M. Stewart 1985; Woolbright 1985b; Duellman 1990; Parmelee 1999). Each anuran species has a visual system adapted to function best under a specific light regime, and this in turn is reflected in the timing of foraging behavior (Jaeger and Hailman 1973; Hailman and Jaeger 1976). Temporal differences in activity could influence the types of prey available to each species and therefore could be an important component of resource partitioning in the community (Jaeger 1977, 1981c; Jaeger and Hailman 1981). However, at present there is little evidence that differences in diet among sympatric species are due to differences in foraging time, especially among those active mainly during the day (Parmelee 1999).

Although many forest litter anurans have generalized diets, some exhibit a greater degree of taxonomic specialization. Toft (1980a, b, 1981) compared the relative abundance of prey in frog stomachs and leaf litter for Neotropical communities and classified the frogs into three general categories: (1) ant specialists, (2) ant avoiders (= "non-ant specialists"), and (3) generalists. Parmelee (1999) adopted a somewhat similar classification, dividing anurans in Amazonian Peru into an ant/termite guild and a large prey guild. Because ant specialists eat relatively few types of prey other than ants (some eat small mites as well), they tend to have narrow calculated niche breadths. Both the ant avoiders and generalists, on the other hand, tend to eat a wide variety of prey taxa and therefore have much wider niche breadths. Indeed, Toft (1980a) found a strong negative correlation between niche breadth and electivity for ants (a measure of dietary preference). A few species of litter frogs are termite specialists, including *Physalaemus petersi* in Peru. Some ant-eating microhylids take termites as secondary prey (Parmelee 1999), and termites made up the largest volume of prey for the small Brazilian leptodactylid frog *Zachaenus parvulus* (Van Sluys, Rocha, and Souza 2001). Some very tiny forest floor frogs have become even more specialized for feeding on extremely small prey such as mites and collembolans (Simon and Toft 1991). Indeed, the abundance of such tiny prey may have contributed to the independent evolution of miniaturized body size in forest litter frogs in several different families (Clarke 1996; see also chapter 1).

In Panama, Costa Rica, and Peru, bufonids, microhylids, and dendrobatids tend to be ant specialists, whereas most *Eleutherodactylus* avoid ants and feed mainly on large prey, including orthopterans and beetles (Toft 1980a, b, 1981; Lieberman 1986; Parmelee 1999). The dendrobatids are particularly interesting because they exhibit a range of strategies, from active foraging (characteristic of ant specialists) to sit-and-wait predation (characteristic of ant-avoiders; Toft 1995; Caldwell 1996b). This variation is correlated with differences in activity metabolism (Taigen and Pough 1983, 1985), a subject that is discussed in more detail in chapter 5.

Prey size in litter anurans is correlated with body size and mouth width, but ant avoiders tend to take larger prey for a given mouth width than ant specialists (fig. 15.3), while generalists are intermediate. At larger body sizes, ant specialists also have smaller head widths than do ant avoiders of the same size (Toft 1980a; Parmelee 1999). Both generalists and ant specialists exhibit some ontogenetic changes in diet, taking progressively larger prey as they grow. This can result in some shift in types of prey eaten, since different prey taxa typically fall into different size categories (Lima and Moreira 1993; Lima 1998; Van Sluys, Rocha, and Souza 2001). There may be sexual differences as well, with the larger sex (usually females) taking somewhat larger prey than the smaller sex (Bonilla and La Marca 1996).

The feeding ecology of tropical anurans from other parts of the world is poorly known, but the limited data available suggest that general patterns are similar to those of the South and Central American species. In some cases, this similarity may be due to phylogenetic history. For example, bufonids eat large numbers of ants in both tropical and temperate communities. R. Clarke (1974b) reviewed 28 studies of the diets of 14 species of *Bufo* from around the world. In 13 cases, ants were the most important prey items, and in 14 others they ranked second. Many other studies

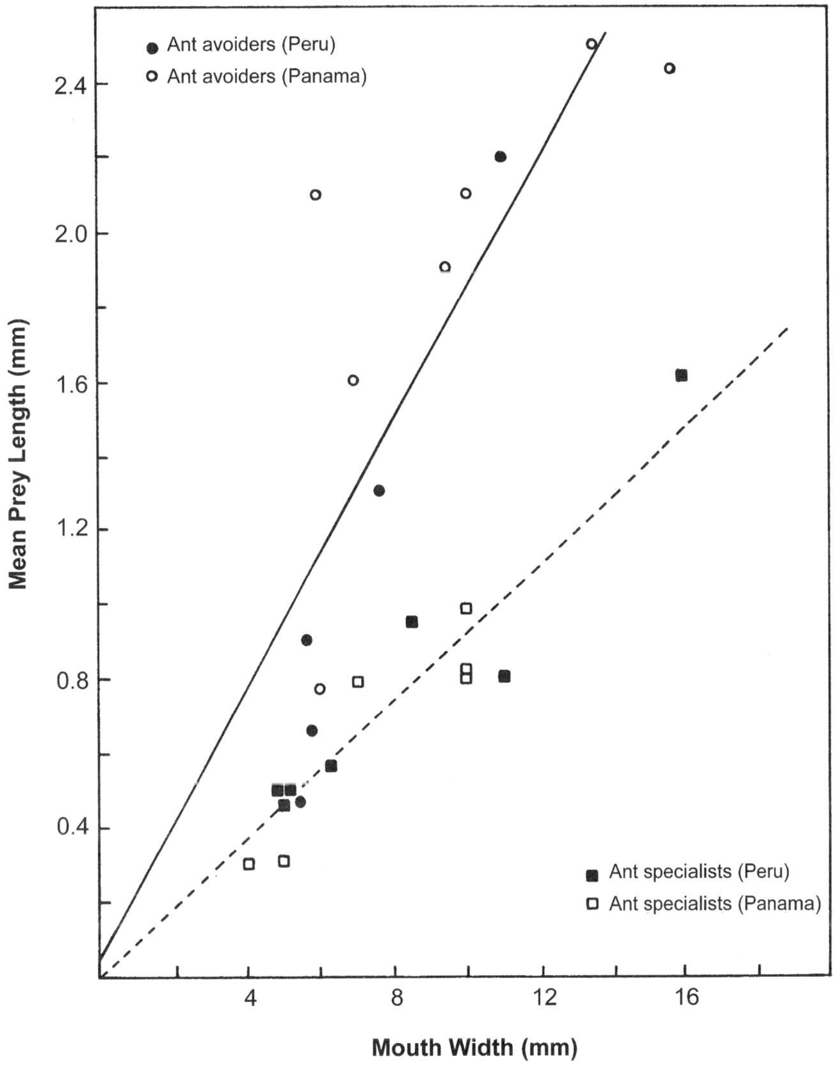


**Fig. 15.3.** Relationship of mean prey size to predator mouth width for litter anurans from Peru (solid symbols) and Panama (open symbols). Circles and solid line are for ant avoiders; squares and dashed line are for ant specialists. Regression equation for ant avoiders: $y = 0.05 + 0.18x$, $R^2 = 0.63$, $p < 0.001$; ant specialists: $y = -0.0008 + 0.09x$, $R^2 = 0.82$, $p < 0.0001$. Plotted from data in Toft 1980a, 1981.

have shown the same pattern (Paulian and Vilardebo 1946; Inger and Marx 1961; P. Berry and Bullock 1962; Lescure 1964, 1971; P. Berry 1970; Forge and Barbault 1978; Toft 1982; Strüssman et al. 1984; Wheater 1986; Vitt and Caldwell 1994; Lajmanovich 1995; F. Meyer, Kutzscher, and Grosse 1999; Parmelee 1999; Isacch and Barg 2002; Vaz-Silva, Silva, and da Silva 2003a, b; Moseley et al. 2005). Ants are important for both diurnal and nocturnal toads, and are heavily used not only by small species, but also by large species such as *Bufo marinus* (N. Weber 1938; Zug and Zug 1979; Strüssman et al. 1984). Terrestrial and burrowing microhylids also tend to be ant or termite specialists in both the Old and New World (Das 1996c; Parmelee 1999). In contrast, other anurans vary considerably in their feeding ecology. For example, while most species of *Eleutherodactylus* are sit-and-wait predators that eat large prey, a few species, such as *E. vocator* in Panama (Toft 1981) and *E. lentus* in the Virgin Islands (K. Jones 1982) are more active foragers and may eat smaller prey, especially ants. Other species of *Eleutherodactylus* include ants in a varied diet of insects, and take them roughly in proportion to their occurrence in the leaf litter (Lieberman 1986).

In several families of anurans, there is an association between specialized feeding on ants and the use of noxious skin secretions as a defense against predators (Toft 1995; Caldwell 1996b; Vences et al. 1997/98; Caldwell and Vitt 1999). For example, most dendrobatids, bufonids, and microhylids are well protected by skin secretions. Among the dendrobatids, the derived, highly toxic species of *Dendrobates* and *Phyllobates* are the most specialized ant predators

(Darst et al. 2005). Species of *Colostethus* (including *Nephelobates;* Bonilla and La Marca 1996) and some species of *Epipedobates* have more generalized diets, although they still eat large numbers of ants (Lima and Moreira 1993; Biavati, Wiederhecker, and Colli 2004). The association between ant-eating and toxic skin secretions derives from the ability of some frogs to sequester chemical compounds acquired from ants in the diet for use in their own defense (see chapter 14 for further details). A specialized diet of ants seems to have resulted in convergent evolution in a number of behavioral and morphological traits in poisonous frogs, especially members of the genus *Mantella* from Madagascar and the Neotropical dendrobatids. In addition to conspicuous aposematic coloration, these frogs share behavioral traits such as diurnal activity, an active foraging mode, and a preference for prey loaded with noxious chemicals (V. Clark et al. 2005). They also share certain morphological features that seem to be associated with feeding on very small prey, including reduction in gape, reduction in teeth, changes in tongue shape, and changes in features of the skull related to the jaw-opening mechanism (Vences, Glaw, and Böhme 1997/98).

#### Niche Overlap and Interspecific Competition

With the exception of the specialized diets discussed previously, litter anurans in the same guild show considerable overlap in prey taxa, as do arboreal frogs in the same communities (Parmelee 1999). In Peru, for example, the ant-avoiding species of *Eleutherodactylus* did not differ significantly in the proportion of different prey taxa in their diets. This is not surprising, since they all take prey taxa other than ants roughly in proportion to their abundance in the litter (Toft 1980a). Similar patterns were found in Panama (Toft 1980b) and in studies of African species (Inger and Marx 1961; Toft 1982). The Neotropical frogs differed considerably in body size, and prey size is correlated with body size (fig. 15.3). Therefore, calculated indices of niche overlap decrease when information on prey size is included, but they are still relatively high for many species pairs (Toft 1980a, b). In a four-species community in Gabon, on the other hand, diets are so similar that inclusion of prey size has little effect on calculations of niche overlap (Toft 1982). In this community, all but one species feed very heavily on termites, and therefore have diets similar in both prey size and prey type. Unfortunately, data for this assemblage are available for only part of one year, so the amount of seasonal variation in diet overlap is unknown.

The importance of competition for food in the organization of leaf litter anuran communities is not at all clear. Experiments with one very abundant species, *Eleutherodactylus coqui,* in Puerto Rico, showed that removal of frogs from small enclosures or from large plots significantly increased the density of herbviorous insects on foliage, but had little or no effect on the abundance of leaf litter insects (Beard et al. 2003). This suggests that leaf litter insects probably are not limiting, even at very high frog population densities.

Specialization on particular types of prey, such as ants, probably is part of a suite of characters resulting from independent feeding adaptations, but interspecific competition may have been more important in the evolution of food size and microhabitat partitioning (Toft 1985). The evidence for this is indirect. In Peru, interspecific overlap in food size was reduced during the dry season, when food was least abundant (Toft 1980a). This is consistent with competition theory, which predicts that niche overlap should decrease as resources become scarce (e.g., Schoener 1982), and with empirical data on other groups (e.g., Boag and Grant 1984). However, the changes in food size also could reflect seasonal shifts in microhabitats. In Panama, food was least abundant late in the wet season. At one site, overlap in prey size decreased in the wet season, but at another site, there was no change (Toft 1980b). Species of *Eleutherodactylus* often differ in their use of microhabitats and in altitudinal distribution, but virtually nothing is known about how biological interactions among species affect habitat use and distribution (e.g., Lynch and Duellman 1997).

### Anurans of Tropical Savannas

Although there is a tendency to think of the tropics as an unbroken expanse of rainforest, there are extensive grasslands in the tropics as well, and many of these support diverse communities of anurans (Rödel 1996, 2000; Neckel-Oliveira et al. 2000). Extensive work on anuran communities in West African savannas provides some interesting parallels with tropical forest litter communities. Most of the studies were carried out in the Ivory Coast, in the Lamto Savanna (Barbault 1972, 1974b, c, 1977) and Comoé National Park (Rödel 1996, 2000). The region is characterized by a three-to-six month dry season and a six-to-nine month wet season, depending on the precise location of the savanna. These savannas exhibit a substantial amount of topographic and habitat complexity, with a relatively dry grassland plateaus, wetter lowlands, woodlands, and stretches of gallery forest.

About 50 species of anurans inhabit the West African savannas (Rödel 2000), with about 40 of these being found in the Lamto Savanna, where most of the detailed ecological work has been done (Lamotte 1967). Some of these are arboreal, but detailed community studies have focused on the terrestrial species, mainly ranids, bufonids, and hyperoliids. In contrast to the Neotropical litter anurans, these anurans are mainly nocturnal, and most depend on standing water for reproduction (a species of *Arthroleptis* is an exception; Barbault and Trefaut Rodrigues 1979a). The density of anurans in the savanna is closely tied to the number of months in which tem-

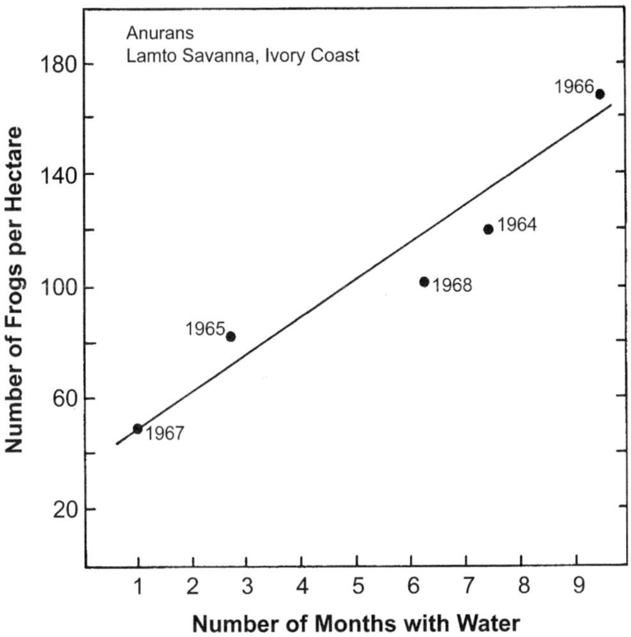


**Fig. 15.4.** Relationship of anuran density to number of months in which temporary ponds were available for breeding in the Lamto Savannah, Ivory Coast. Regression equation: $y = 36.14 + 12.47x$, $R^2 = 0.90$, $p < 0.008$. Plotted from data in Barbault (1972).

porary ponds are available for reproduction (fig. 15.4), and both seasonal and annual fluctuations in population density exceed those observed in humid forested habitats (Barbault 1972, 1974b, c, 1977). As expected, lowland areas with abundant temporary ponds have much higher population densities than drier upland areas (Barbault 1972). In comparisons between two different savannas, the wetter site had both higher species diversity (Barbault 1976b) and higher population densities (Barbault 1976a) than the drier site.

The feeding ecology of these savanna anurans is similar in many respects to that of the forest litter species. Most of the species feed on a variety of insects. However, one arboreal species apparently is an ant specialist, and a burrowing species (*Hemisus marmoratus*) feeds entirely on ants and termites. The bufonids eat mostly ants and beetles, and several ranids also show moderate specialization on ants (Barbault 1974a). As in the forest litter species, dietary diversity in the savanna anurans decreases as the importance of ants in the diet increases (fig. 15.5 A). Although dietary overlap between species is considerable, there is some separation by prey size, which again is strongly correlated with body size of the predator (fig. 15.5 B). As in the other communities discussed earlier, the role of interspecific competition in producing these patterns of resource partitioning is unclear. Barbault (1974a, c) suggested that food probably is not a limiting resource in most years, because population size is controlled mainly by the availability of water for reproduction, not by the abundance of insects. Predation may play an important role as well, by maintaining populations of anurans below levels at which competition for food is likely to occur (Barbault 1987, 1991). Barbault (1991) also speculated that interactions of amphibians with their predators indirectly affect other components of the community, especially insectivorous

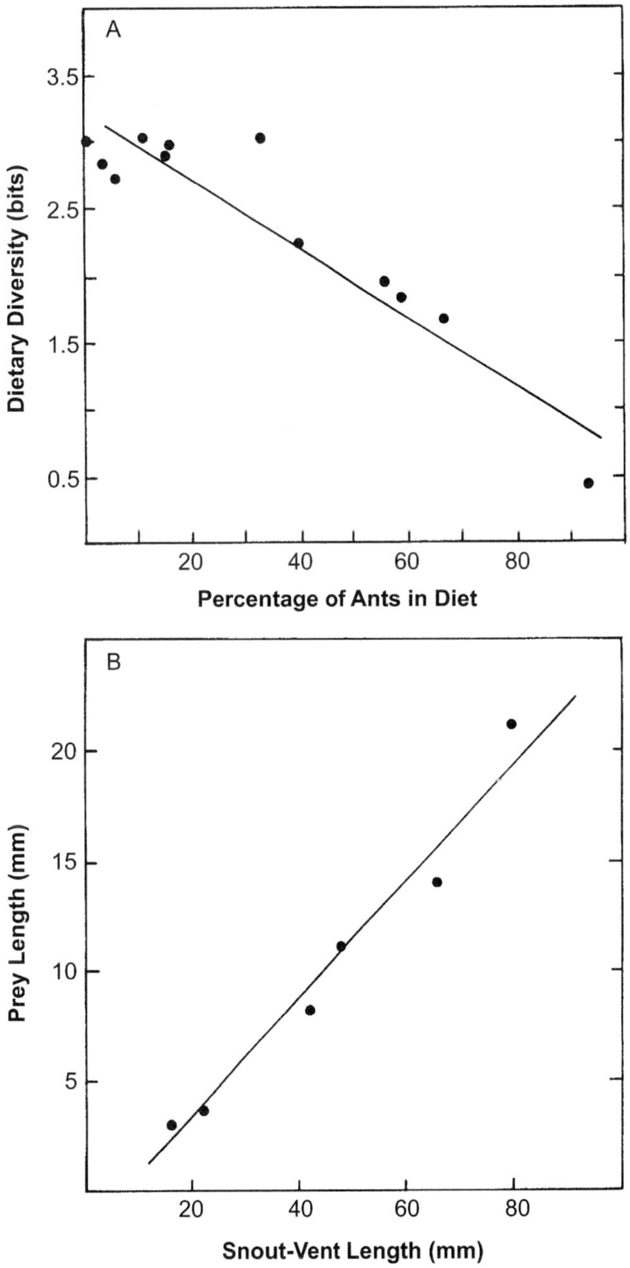


**Fig. 15.5.** (A) Relationship of dietary diversity to the proportion of ants in the diet for 11 species of anurans of the Lamto Savannah, Ivory Coast. Dietary diversity is expressed in bits calculated from the Shannon-Weaver diversity index. (B) Relationship of prey length to predator body size for six species. Prey do not include ants and termites. Regression equation for (A): $y = 3.17 + 0.025x$, $R^2 = 0.85$, $p < 0.0001$; (B): $y = -7.18 + 0.53x$, $R^2 = 0.96$, $p < 0.000$. Plotted from data in Barbault (1974c).

lizards. During dry periods, when amphibians are scarce, their predators may switch to feeding on lizards, which remain abundant during the dry season.

### Tropical Riparian Communities

Although aquatic-breeding frogs are most commonly associated with temporary or permanent ponds, some tropical and subtropical species are found mainly along forested streams (K. Parris and McCarthy 1999). In a relatively dry evergreen forest in Thailand, several species of ranids inhabit riparian habitats. They exhibit minor habitat differences along a horizontal gradient, differing in their affinity for permanent streams and in their tendency to remain in the water. Otherwise, they appear to be rather similar ecologically (Inger and Colwell 1977). In riparian communities of up to 25 anuran species along forest streams in Borneo, there is some separation of microhabitats along both vertical and horizontal axes, with some species being more arboreal or more terrestrial than others (Inger 1969; Inger and Voris 1993; Das 1996a). A similar pattern was seen in a less diverse assemblage in Sulawesi, where stream-breeding frogs differed mostly in the types of perches they occupied along the streams (Gillespie et al. 2004). Most species in these assemblages are generalist predators on small invertebrates. However, bufonids specialize on ants, and two of the largest ranid species feed on crabs not available to the smaller frogs. Closely related species tend to be most similar ecologically, suggesting that phylogenetic history is a more important determinant of resource use than ecological interactions between species (Inger 1969).

An experimental study of three common ranids in these communities provides some evidence of current interspecific competition, although the resource being contested is not known (Inger and Greenberg 1966). There are minor differences among species in microhabitat and prey choice, but all are rather similar ecologically. Inger and Greenberg proposed that the relative abundance of species in the community is regulated by interspecific competition. *Limnonectes leporinus* (formerly *L. blythii*) was very common on all streams, whereas *L. ingeri* (formerly *L. macrodon*) was relatively rare and *L. ibanorum* was intermediate in abundance. When *L. ibanorum* was removed from one stream, the number of *L. leporinus* on the stream doubled over a period of several months. When *L. leporinus* was removed from another stream, *L. ibanorum* showed a small initial increase and then declined slightly, while *L. ingeri* initially showed no change and then increased slightly. These experiments suggest that interspecific competition does limit the numbers of some species on these streams. However, these conclusions are only tentative, because the experiment lacked proper replication and controls, with different streams being used as experimental and control sites. Changes in population size along these streams could have been due to differences between streams that were independent of the experimental treatments. Nevertheless, this study remains, after 40 years, the only attempt to examine competitive interactions among adult anurans experimentally.

### Anuran Communities in the Temperate Zone

Temperate zone anuran communities lack both the structural complexity and high species diversity of tropical assemblages. Although a few species of treefrogs probably do much of their feeding in trees, most temperate zone anurans can be considered either terrestrial or semiaquatic. Indeed, several studies of local assemblages have shown that differences in microhabitat use can account for most of the ecological differences among species. In North America, for example, some ranid frogs, such as *Rana septentrionalis, R. catesbeiana,* and *R. pretiosa,* are more fully aquatic than are species such as *R. clamitans, R. pipiens,* and *R. aurora.* In general, all of these species are generalist predators on arthropods and other invertebrates, but the more aquatic species tend to take larger numbers of aquatic prey than do the species that live along pond shores (Stewart and Sandison 1972; Licht 1986b; McAlpine and Dilworth 1989; Werner, Wellborn, and McPeek 1995). Wood frogs (*Rana sylvatica*) and toads (*Bufo*) tend to be strictly terrestrial outside the breeding season, and can be expected to take mostly terrestrial prey species. Similar patterns are seen in Europe, where anuran communities are largely divided into semiaquatic and terrestrial components (Lizana, Perez-Mellado, and Ciudad 1990). Again, the more aquatic species, such as *Rana ridibunda, R. perezi,* and *Rana iberica,* tend to take more aquatic prey than do the more terrestrial species, such as *R. temporaria, R. dalmatina, R. arvalis,* and various bufonids and discoglossids (Lizana Avia, Ciudad Pizarro, and Pérez-Mellado 1986; Kuzmin 1999; Guidali, Scali, and Carettoni 2000). There have been many studies of the diets of individual species of temperate zone anurans, and these have revealed little evidence of diet specialization, except for the tendency of toads (*Bufo*) to eat large numbers of ants. Nearly all other species are generalist or opportunistic predators that take prey roughly in proportion to their abundance in the habitat, although many species exhibit at least limited aversion to ants (Jenssen and Klimstra 1966; Houston 1973; Blackith and Speight 1974; Labanick 1976; Cogălniceanu et al. 1998; A. M. Anderson, Haukos, and Anderson 1999a; Hirai and Matsui 1999, 2000, 2001; Kuzmin 1999; F. Meyer, Kutzscher, and Grosse 1999; Bellocq, Kloosterman, and Smith 2000).

## Community Ecology of Terrestrial Salamanders

Assemblages of terrestrial plethodontid salamanders are among the most intensively studied of amphibian communities, with the greatest attention having been focused on forest litter species of *Plethodon* or streamside species of *Desmognathus* in the southern Appalachians (Hairston 1987, 1996). Nevertheless, the role of salamanders in terrestrial ecosystems tends to be underappreciated by ecologists in general. The classic study of salamanders at Hubbard Brook in New Hampshire by Burton and Likens (1975a, b) demonstrated that terrestrial salamanders not only can be numerically dominant members of forest ecosystems, but also can account for a large proportion of vertebrate biomass, despite their relatively small size. Subsequent studies, reviewed by Davic and Welsh (2004), have revealed even higher densities and greater biomass of salamanders than those found at Hubbard Brook. Furthermore, there is considerable evidence that salamanders can be dominant predators in a variety of habitats, including the forest floor litter community, and can affect ecosystem dynamics and the population dynamics of leaf-litter invertebrates.

Salamanders tend to be morphologically conservative, with closely related species often being very similar in both appearance and ecology. In some communities, similarity in resource requirements seems to have resulted in competitive exclusion of one species by another from apparently suitable habitat; in others, there is evidence of limited resource partitioning among coexisting species (Toft 1985). Several case studies will be discussed in detail in the following. First, however, I will review the foraging strategies of terrestrial salamanders and the evidence for potential interspecific competition for food, since food is the most commonly measured resource in studies of competition and resource partitioning among vertebrates.

### Foraging Strategies

Most terrestrial plethodontids feed on a variety of small invertebrates, including earthworms, centipedes, millipedes, snails, isopods, pseudoscorpions, spiders, mites, and many types of adult and larval insects (Hairston 1949; Altig and Brodie 1971; Whitaker and Rubin 1971; Donavan and Folkerts 1972; Jaeger 1972; Bury and Martin 1973; Powders and Tietjen 1974; Burton 1976; Fraser 1976a, b; Maiorana 1978a; Sites 1978; Keen 1979; Kryzsik 1979; Lynch 1985; Hairston, Nishikawa, and Stenhouse 1987; Davic and Welsh 2004). Large semiaquatic plethodontids such as *Gyrinophilus* are important predators on smaller salamanders, but eat invertebrates as well (Hairston 1949, 1980c; Martof and Scott 1957; Bruce 1972a; Burton 1976; Kryzsik 1979). Some terrestrial plethodontids also eat smaller conspecific or heterospecific salamanders on occasion, but this appears to be uncommon (Heatwole and Test 1961; Altig and Brodie 1971; Donavan and Folkerts 1972; Powders 1973; Powders and Tietjen 1974).

Although all salamanders are carnivorous or insectivorous, there is no evidence that any species of salamander specializes on particular prey taxa within these broad categories. Only a few workers have compared prey abundance in salamander stomachs and the environment (Martof and Scott 1957; Altig and Brodie 1971; Maiorana 1978a; Sites 1978). Most have simply assumed that salamanders take prey in proportion to their relative abundance, although prey availability also may be affected by the size, mobility, and microhabitat choice of both the predator and the prey (Maiorana 1978a; Jaeger 1980b). Salamanders appear to be more selective in choosing prey sizes than prey taxa. For example, *Desmognathus fuscus* take large prey in higher proportions than their frequency in the environment (Sites 1978). In *Batrachoseps,* diet choice largely reflects prey availability, but some individuals preferentially select larger prey items if they are present (Maiorana 1978a).

Within species, there is a tendency for larger individuals to take larger prey. For example, Whitaker and Rubin (1971) found the same prey taxa represented in all size classes of *Plethodon metcalfi.* However, the relative importance of large prey such as earthworms, millipedes, and centipedes increased with increasing body size, while the importance of small prey such as ants, collembolans, and mites decreased (fig. 15.6). Lynch (1985) also found an ontogenetic decrease in the importance of collembolans and mites with increasing body size in *Aneides flavipunctatus,* and average prey size was positively correlated with salamander body size. Similar relationships (fig. 15.7) have been reported for other species (Fraser 1976a, b; Maiorana 1978a; Sites 1978; Kryzsik

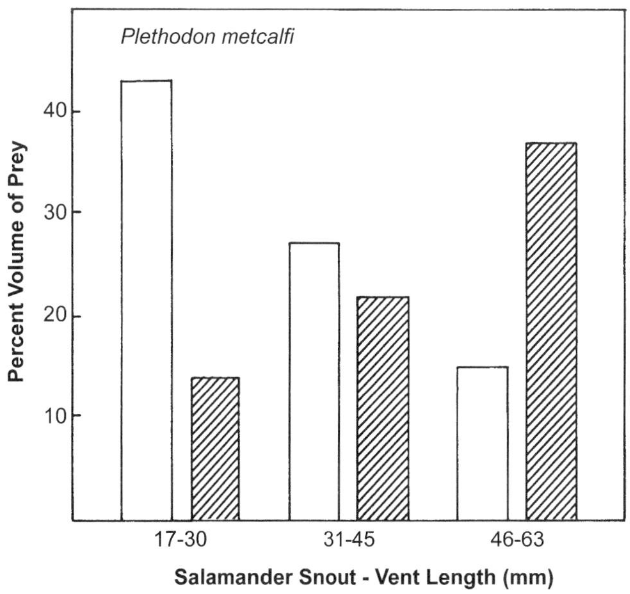


**Fig. 15.6.** Frequency of small prey (open bars) and large prey (striped bars) taken by *Plethodon metcalfi* as a function of salamander body size. Large prey include pooled percentages of spiders, earthworms, millipedes, centipedes, and craneflies; small prey include ants, collembolans, and mites. Plotted from data in Whitaker and Rubin (1971).

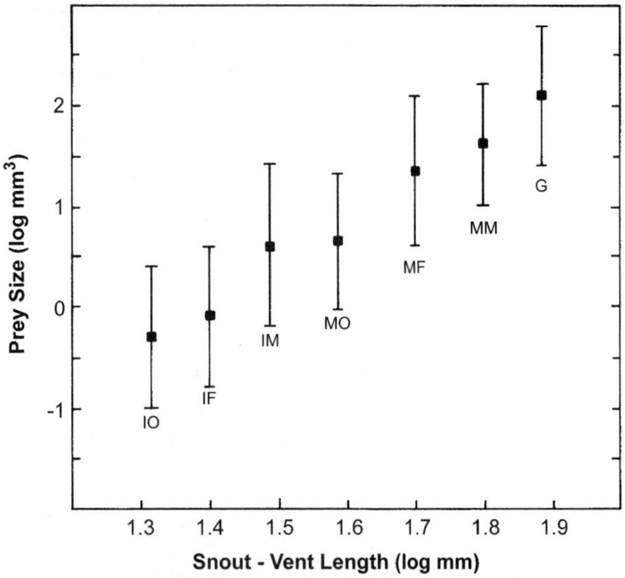


**Fig. 15.7.** Relationship of prey size to salamander body size for adults and juveniles of four species of streamside salamanders in Pennsylvania. Graph shows means ± standard deviations on a log scale. Species abbreviations: IO = immature *D. ochrophaeus,* IF = immature *D. fuscus,* IM = immature *D. monticola,* MO = mature *D. ochrophaeus,* MF = mature *D. fuscus,* MM = mature *D. monticola,* G = *Gyrinophilus porphyriticus.* Plotted from data in Krzysik (1979).

1979). Feeding experiments with *Aneides* showed that maximum prey size probably is determined by physical constraints imposed by the size of the salamander's mouth, while minimum prey size is determined by behavioral avoidance of small prey (Lynch 1985). However, in most species, large individuals eat some small prey (Kryzsik 1979; Lynch 1985), and prey utilization curves for large salamanders may completely encompass those of smaller individuals (fig. 15.8).

There is evidence that salamanders can assess prey profitability and adjust their foraging behavior to concentrate on the most profitable prey when conditions are favorable. A common prediction of optimal foraging models is that animals should become increasingly specialized in their choice of prey as food abundance increases, because a predator can afford to pass up less profitable prey in the expectation that better prey will be available (Krebs 1978). In laboratory experiments, Jaeger and Barnard (1981) presented *Plethodon cinereus* with a choice of large and small flies in equal numbers, but at different total prey densities. Large flies are more profitable because they have proportionately less indigestible exoskeleton. The salamanders specialized on large flies when total prey density was high, but ate large and small flies indiscriminately at low prey densities (fig. 15.9). At high prey densities, salamanders also were more persistent in attacking large flies that were initially missed. They continued to take some small prey if they could do so with relatively little energy expenditure (i.e., by using sit-and-wait foraging tactics), but were less likely to actively pursue small prey than large prey at all but the lowest prey density.

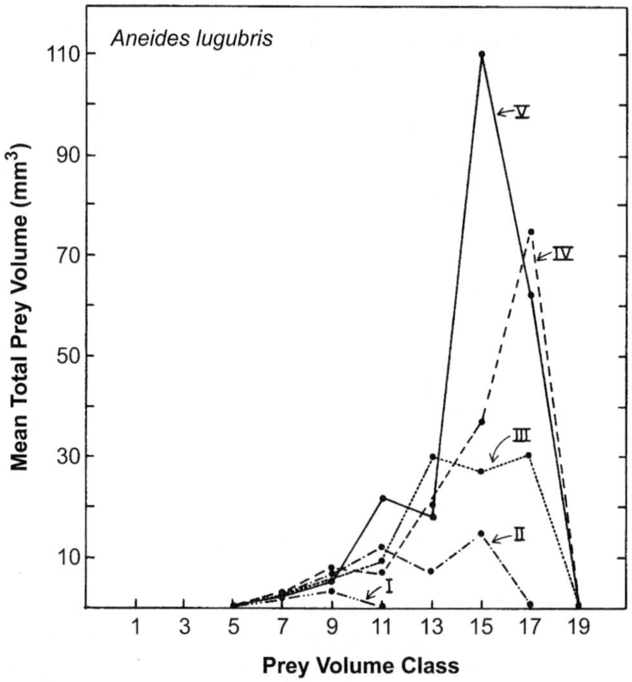


**Fig. 15.8.** Prey utilization curves for different size classes of *Aneides lugubris.* The curves for larger-size classes tend to enclose those for smaller classes. Roman numerals give salamander size classes in increasing order. Prey size classes give relative sizes on a log scale. After Lynch (1985).

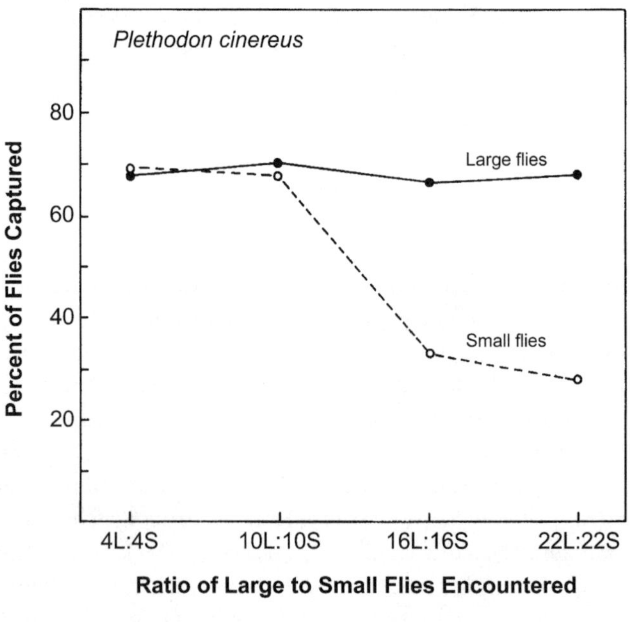


**Fig. 15.9.** Percentage of large and small flies ingested by *Plethodon cinereus* as a function of total prey density. As density increases, the salamanders tend to ignore smaller prey. Plotted from data in Jaeger and Barnard (1981).

Although the foraging behavior of salamanders is difficult to observe in the field, there is some evidence that they use similar "optimal foraging" tactics in nature. Prey availability for terrestrial salamanders is determined mainly by rainfall (Jaeger 1972, 1980b; Fraser 1976a; Keen 1979).

Even when prey abundance remains largely unchanged, the ability of salamanders to leave moist retreat sites to search for food may be severely limited by fluctuations in leaf litter moisture (see also chapter 2). Several authors have reported a tendency for salamanders to specialize on large prey items during relatively wet weather or when feeding in wet microhabitats, but to feed indiscriminately on large and small prey during dry periods or in drier areas (Burton 1976; Sites 1978; Keen 1979; Lynch 1985). Although these results suggest that salamanders are more selective feeders when prey is abundant, it also is possible that salamanders are able to exploit a wider range of microhabitats during wet weather and therefore are exposed to a wider diversity of prey (Jaeger 1978).

### Interspecific Competition for Food

The extent to which food is a limiting resource for salamanders is not clear. Although salamanders may experience periodic food shortages (Jaeger 1972, 1980b; Fraser 1976a), food availability per se may not be the major determinant of population size. Unfortunately, there have been few long-term studies of salamander population dynamics, so the extent and causes of fluctuations in population size are not well understood. Such studies are particularly difficult with salamanders because their cryptic nature makes it difficult for investigators to obtain reliable estimates of fluctuations in population size (Dodd and Dorazio 2004), a necessary prerequisite for determining the causes of such fluctuations. Nevertheless, there is experimental evidence that dense populations of salamanders can have a significant impact on invertebrate populations. For example, Wyman (1998) removed red-backed salamanders (*Plethodon cinereus*) from forest-floor enclosures and found a significant increase in leaf-litter invertebrates. Rooney, Antolik, and Moran (2000) performed a similar set of experiments and found that the presence of *P. cinereus* actually enhanced the abundance of collembolans, which they do not eat, by reducing the abundance of their predators. Although these experiments were designed to investigate the effects of salamanders on invertebrates, rather than the effect of prey abundance on salamanders, they suggest at least the potential for prey to become limiting for some terrestrial salamanders.

In favorable habitats such as the southern Appalachians, salamander populations can be very dense (Hairston 1987; Petranka, Eldridge, and Haley 1993; Petranka and Murray 2001; Davic and Welsh 2004) and may be near the carrying capacity of the environment most of the time (Hairston 1980b). In harsher environments, density-independent factors may maintain populations well below carrying capacity (Dumas 1956). Even when populations are near carrying capacity, the resources that limit population size are not always obvious. Several authors have suggested that space, rather than food per se is the proximate limiting resource for some species. Individuals may compete for access to nesting sites or moist retreat sites that also provide access to food (Jaeger 1974; Fraser 1976a, b; Maiorana 1978b; Hairston 1981). A number of plethodontid salamanders defend territories centered on moist retreat sites against both conspecifics and heterospecifics (see chapter 9). Such retreats may enhance the foraging opportunities of territory holders, even though food is not the direct object of competition.

#### Descriptive Studies

Published studies of the diets of sympatric salamanders (Hairston 1949; Dumas 1956; Altig and Brodie 1971; Jaeger 1972; Bury and Martin 1973; Powders and Tietjen 1974; Burton 1976; Fraser 1976a, b; Krzysik 1979; Holomuzki 1980; Lynch 1985; Hairston, Nishikawa, and Stenhouse 1987) suggest several general conclusions about resource partitioning. Most sympatric species exhibit broad overlap in prey taxa, in some cases reaching levels of similarity of 80–90%. Differences in prey taxa generally reflect differences in microhabitat rather than differences in selectivity within shared microhabitats. For example, in communities of *Desmognathus*, differences in prey taxa generally reflect the proximity of each species to water, with more aquatic salamanders taking more aquatic prey (Hairston 1949; Krzysik 1979).

Differences in prey taxa also reflect morphological differences among salamander species. For example, salamander communities in northern California are composed of several morphologically distinct genera, including one or more species of *Aneides*, which have large bodies and relatively wide heads, *Ensatina*, a generalized terrestrial salamander, and *Batrachoseps*, a specialized burrowing salamander with a very narrow head. Differences in diets among these species are somewhat greater than in communities of salamanders composed of morphologically similar species (Bury and Martin 1973; Lynch 1985). However, differences in prey taxa may be a byproduct of differences in prey size, and may not reflect discrimination among types of prey. Prey size tends to be positively correlated with salamander body size; overlap in prey size is greatest between species with similar body sizes. Even when adults of different species exhibit significant differences in prey size, juveniles of large species often overlap completely with adults of smaller species (Krzysik 1979). Furthermore, as is the case within species, large species often eat both large and small prey, while small species eat only small prey. Although differences in prey size have been interpreted as evidence of resource partitioning among different size classes of salamanders (Fraser 1976a; Krzysik 1979; Toft 1985), there is little direct evidence that prey selection is influenced by the presence of other species. Maiorana (1978b) found that overlap in prey size between sympatric *Aneides* and *Batrachoseps* was positively correlated with overlap in space. This violates the principle of niche complementarity derived from competition theory, which predicts that species that overlap greatly on one resource axis should exhibit reduced overlap on another axis (Schoener 1974, 1982).

Fraser (1976a, b) suggested that both intraspecific and interspecific competition for food might be reduced by individuals having different activity periods, but there is little evidence of major differences in the timing of foraging among sympatric species. Most plethodontids tend to be active on the surface mainly at night, although they may forage under logs, rocks, or litter during the day; all are most active during wet weather. In the genus *Desmognathus*, larger, more aquatic species are reported to be slightly less nocturnal than smaller, more terrestrial species (Hairston 1949; Shealy 1975). Nevertheless, peaks of surface activity usually are similar for sympatric species (Shealy 1975; Holomuzki 1980), and temporal partitioning of resources probably is of little significance in most salamander communities (Toft 1985).

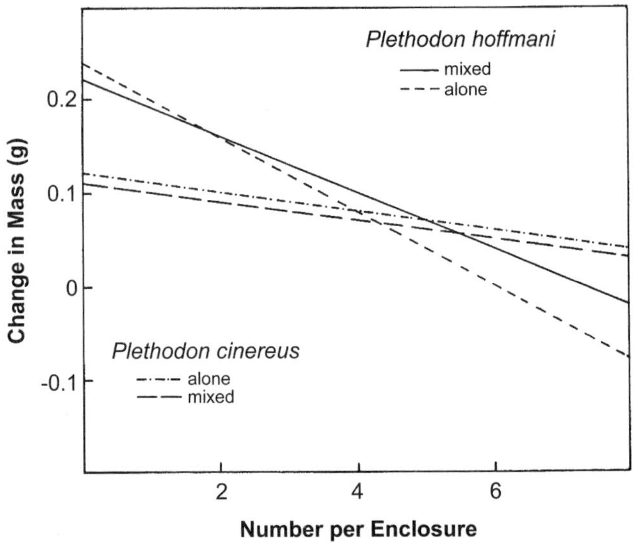


**Fig. 15.10.** Change in mass as a function of density for the salamanders *Plethodon hoffmani* and *P. cinereus* housed in single-species and mixed-species groups in the laboratory. After Fraser (1976b).

### Experimental Studies

Exploitative competition for food can be demonstrated experimentally if the presence of one species significantly reduces the feeding rate, growth rate, energy reserves, or reproduction of a sympatric species, or the presence of one species leads to a shift in the food niche of another species. Only a few such studies have been performed, and most do not strongly support the hypothesis that competition for food is an important determinant of salamander community structure.

Perhaps the best places to look for interspecific competition among plethodontid salamanders are areas in which one species replaces another geographically, often with only a relatively narrow zone of overlap between them. Several small species of *Plethodon* exhibit this type of parapatric or near-parapatric distribution (Highton 1971). For example, the widespread species *Plethodon cinereus* is largely absent from the Valley and Ridge physiographic province of central Pennsylvania, where it is replaced by a very similar species, *Plethodon hoffmani*. Fraser (1976b) placed *P. cinereus* and *P. hoffmani* in laboratory enclosures at different densities in both single-species and mixed-species groups. He found that salamanders of both species lost mass as density increased, but there was no difference in the effects of conspecifics and congenerics (fig. 15.10). Since the highest densities in this experiment were considerably higher than were those in the field, it does not directly test for competition under natural conditions. However, it does show that under intense competition, neither species greatly exceeds the other in its ability to acquire food.

A study of morphological variation and resource use by these two species in sympatric and allopatric populations along a transect in Pennsylvania provides some of the strongest evidence for interspecific competition for food (Adams and Rohlf 2000). Allopatric populations of the two species exhibit few morphological differences in characters of the skull associated with food capture and consumption, but sympatric populations do differ morphologically. Furthermore, sympatric populations exhibited partitioning of prey that was absent in allopatric populations, with *P. cinereus* tending to eat smaller prey items and *P. hoffmani* eating larger prey. Individuals of both species had less food in their stomachs in sympatry than in allopatry, indicating significant competition for food. The morphological differences in sympatric populations involved biomechanical differences in jaw closure that were related to the size of prey that could be most efficiently consumed by each species. Exploitative competition is not the only type of interaction between these species, however. *Plethodon cinereus* tends to be more aggressive toward both conspecifics and heterospecifics than is *P. hoffmani*, suggesting that aggressive interference competition limits the distribution of *P. hoffmani* (Jaeger, Prosen, and Adams 2002).

In studies of another pair of small *Plethodon* species, Jaeger (1971a, 1972) found that *P. cinereus* was competitively superior to *P. shenandoah* in a natural environment, and hypothesized that the former was more effective in finding and capturing prey. However, in mixed-species tests in the laboratory, *P. shenandoah* actually captured more prey than *P. cinereus* (D. Kaplan 1977), the opposite of the result predicted from the field data. Hence, exploitative competition for food does not appear to explain the relative success of the two species in nature. Subsequent experiments by Wrobel, Gergits, and Jaeger (1980) showed that the two species interact aggressively when housed together in the laboratory, but neither species consistently won these encounters, and there was no difference between them in the amount of mass lost during the experiment.

The only field experiment that has specifically tested the hypothesis of food competition is Kleeberger's (1984) work with *Desmognathus*. He added *D. monticola* to unfenced field plots containing both that species and *D. quadramaculatus*. The lipid reserves of *D. monticola* on the plots decreased, but *D. quadramaculatus* was not affected. However, when both *D. monticola* and rocks were added to experimental plots, the lipid reserves of *D. monticola* increased, suggesting that abundance of cover influences the abundance of prey. The addition of *D. quadramaculatus* had no effect on the lipid reserves of either species. These results reveal that food is potentially limiting for *D. monticola* in this area, and intraspecific competition may affect the distribution and abundance of that species, but there was no evidence of interspecific competition for food.

### Interspecific Competition for Space

Although the evidence for exploitative competition among salamanders is relatively weak, it is becoming clear that interference competition is important in many salamander communities. Such competition is mediated by interspecific aggressive encounters (see chapter 9), with the object of competition most likely being suitable sites for feeding, sheltering, or oviposition. In some communities, interspecific competition may result in partitioning of available macro- or microhabitats among species. However, one cannot assume that such partitioning is always the result of interspecific competition, since salamanders might exhibit species-specific habitat preferences regardless of the presence or absence of other species.

#### Altitudinal Separation in Large *Plethodon*

The distribution of large species of *Plethodon* in the southern Appalachians has become a classic example of the role of interspecific competition in determining the limits of species' ranges. In several mountain ranges in the southern Appalachians, relatively large, white-spotted black salamanders tend to co-occur with smaller, unspotted black salamanders with varying degrees of altitudinal overlap and hybridization (Highton and Peabody 2000). The white-spotted species often are placed in the *P. glutinosus* group, a widespread complex with a number of species that occur outside of the southern Appalachians (fig. 15.11). Species in this region include *P. aureolus, P. chattahoochee, P. chlorobryonis, P. cylindraceus, P. glutinosus, P. kentucki*, and *P. teyahalee* (this species has been called *P. oconaluftee* by some authors; e.g., Hairston 1993, 1996; Petranka 1998). The unspotted species, some of which have bright orange markings on their

**Fig. 15.11.** Large plethodontid salamanders of the *Plethodon jordani* and *P. glutinosus* complexes. (A) *P. jordani* from Great Smoky Mountains National Park, North Carolina. (B) *P. metcalfi* from the Nantahala Mountains of North Carolina. (C) *P. teyahalee* from the Nantahala Mountains of North Carolina. Photos by Wayne Van Devender.

cheeks or legs, often are placed in the *P. jordani* group (*P. amplus, P. cheoah, P. jordani, P. meridianus, P. metcalfi, P. montanus,* and *P. shermani*), which is found only in the southern Appalachians. Genetic studies suggest, however, that these two groups may not be monophyletic. For example, *P. teyahalee,* a member of the *P. glutinosus* group, may be more closely related to *P. shermani,* a member of the *P. jordani* group, than to other members of the *P. glutinosus* group, possibly having been derived from hybridization between *P. shermani* and *P. cylindraceus* (Highton and Peabody 2000).

In his pioneering study of salamander ecology, Hairston (1949) reported that *Plethodon metcalfi* in the Black Mountains of North Carolina exhibit a remarkable degree of altitudinal separation from populations of *P. cylindraceus* (Hairston 1996). The former is confined to high elevations, whereas the latter is widely distributed from low to mid-elevations. Later work revealed similar altitudinal separation between spotted and unspotted salamanders in several other mountain ranges (Hairston 1951, 1973, 1980b; Highton and Henry 1970; Highton and Peabody 2000). The unspotted species always are confined to higher elevations, but the precise boundary varies among mountain ranges and with slope exposure. Since the boundaries between these species are not related to obvious changes in climate or vegetation, Hairston (1951, 1973) proposed that competition with unspotted (*P. jordani* complex) species limits the vertical distribution of the spotted (*P. glutinosus* complex) species. On mountains where unspotted salamanders are absent, the local spotted species can be found up to the highest peaks.

This hypothesis was tested by a long-term field experiment in which individuals of the two species complexes were repeatedly removed from experimental plots in zones of sympatry (Hairston 1980b, 1987, 1996; Hairston, Nishikawa, and Stenhouse 1987). Experiments were conducted in both the Balsam Mountains, where species overlap by about 1,200 m, and the Great Smoky Mountains, where the overlap is only about 120 m (fig. 15.12). When these experiments were performed, the lower elevation populations were considered to be *P. glutinosus,* whereas the higher elevation populations were classified as two different color morphs of *P. jordani.* The proper taxonomy of these two salamander complexes has been much debated, but current taxonomy treats all of the lower elevation spotted salamanders used in the experiments as *P. teyahalee.* The two higher elevation experimental populations are *P. jordani* in the Smoky Mountains and *P. metcalfi* in the Balsam Mountains. Despite the confusing changes in taxonomy and some complex genetic interactions among these species (Highton and Peabody 2000), there is no doubt that all of these salamanders are very similar to each other morphologically and ecologically, with the lower elevation species generally being slightly larger and more robust than those at higher elevations.

Hairston predicted that competition should be stronger in the narrower zone of overlap, and this prediction was

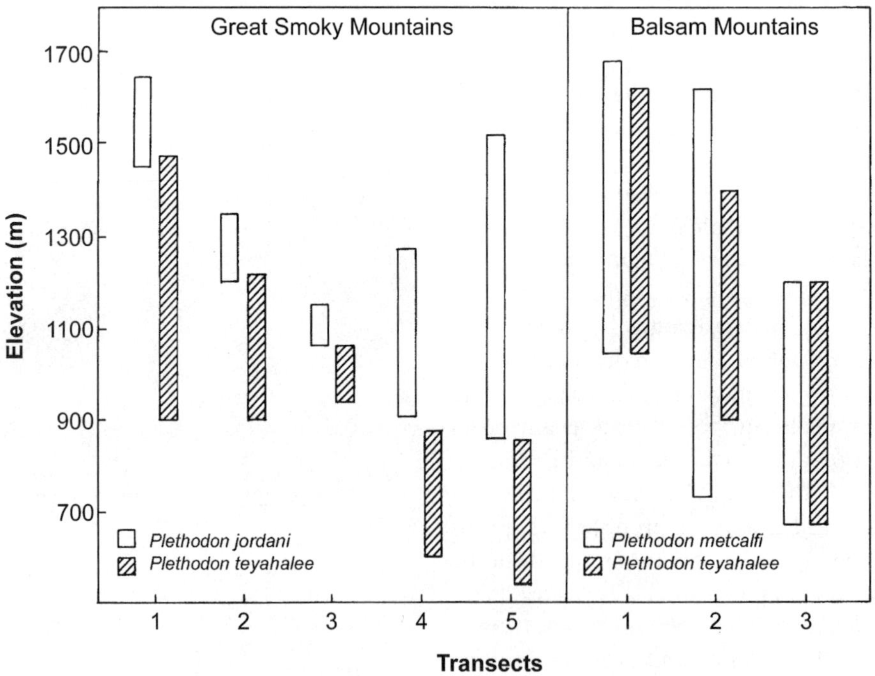


**Fig. 15.12.** Vertical distribution of members of the *Plethodon jordani* and *P. glutinosus* complexes on five transects in the Great Smoky Mountains and three transects in the Balsam Mountains. The distribution of *P. teyahalee* extends to lower elevations than those included in the transects in both mountain ranges. After Hairston (1980b).

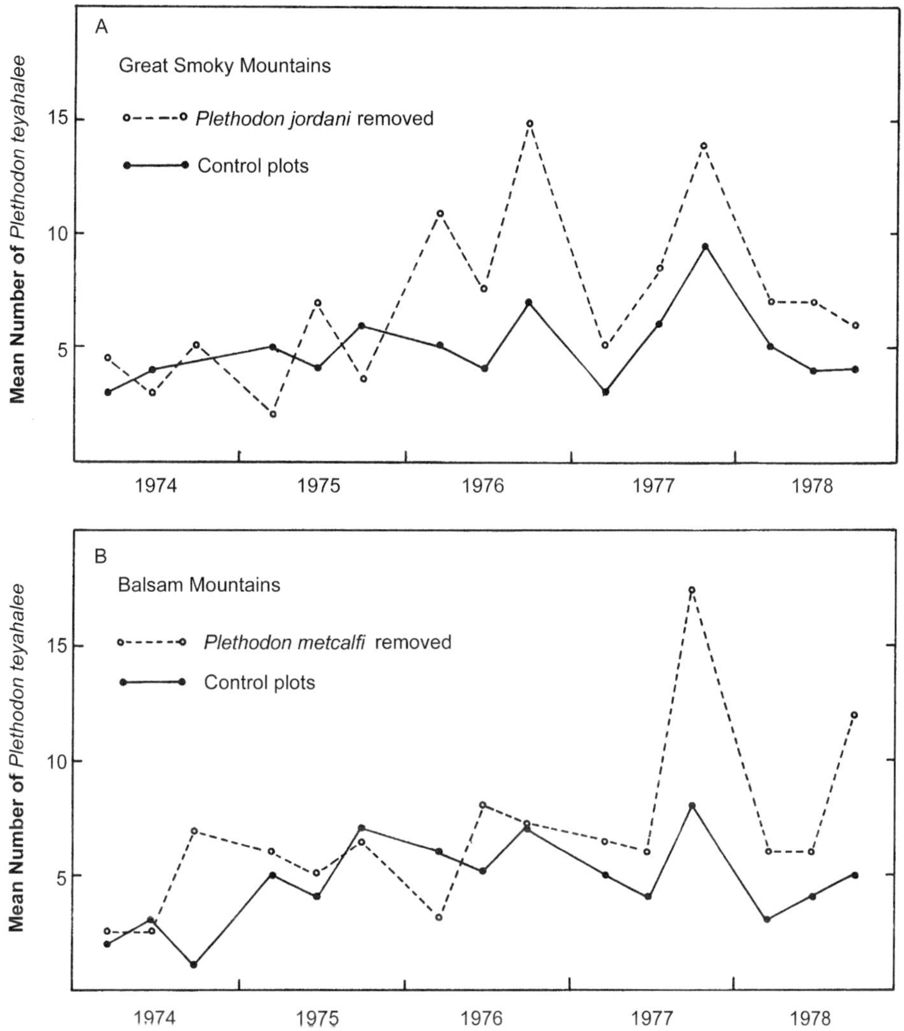


**Fig. 15.13.** Effect of removal of members of the *Plethodon jordani* complex (high-elevation populations) on the number of *P. teyahalee* in plots established in areas of altitudinal overlap. (A) Removal of *P. jordani* from plots in the Great Smoky Mountains produced a strong effect on the population of *P. teyahalee.* (B) Removal of *P. metcalfi* from plots in the Balsam Mountains produced a weaker effect on populations of *P. teyahalee.* After Hairston (1980b).

confirmed. In the Smoky Mountains, removal of *P. jordani* resulted in a significant increase in the abundance of *P. teyahalee* (fig. 15.13 A). In the Balsam Mountains, removal of *P. metcalfi* produced a similar effect, but it was less pronounced and was not evident until several years into the study (fig. 15.13 B). Reciprocal removals of *P. teyahalee* did not increase the abundance of *P. jordani* or *P. metcalfi,* but there was an increase in the proportion of juveniles. Since the unspotted salamanders were numerically dominant when the experiment began, removal of those species would be expected to have a greater effect than removal of *P. teyahalee.* However, the per capita effect of the larger, more robust species, *P. teyahalee,* on *P. jordani* and *P. metcalfi* appeared to be greater than the reverse effect. Several other species of salamanders were present in the experimental plots, but removal of the large *Plethodon* did not affect them, suggesting that not all of the salamanders in the community are competing for the same resources (Hairston 1981, 1984).

Competition between *P. jordani* and *P. teyahalee* is more intense in the Smoky Mountains than is competition between *P. metcalfi* and *P. teyahalee* in the Balsam Mountains. Hairston (1973) had hypothesized that selection may have favored either an increase in the competitive ability of both species in the Smokies, or niche partitioning with a subsequent reduction in competition in the Balsams. These hypotheses were tested by transplanting members of the *P. jordani* complex between the two mountain ranges. This was easily done, because the two species have distinct color patterns. When *P. metcalfi* from the area of lower competition were substituted for *P. jordani* in the Smoky Mountains, the abundance of *P. teyahalee* increased over a period of several years (Hairston 1980a). In contrast, when *P. jordani* from the high-competition population were introduced into the Balsam Mountains, the abundance of *P. teyahalee* decreased (Hairston 1983a). Hence the competitive ability of Smoky Mountain *P. jordani* appears to be greater than that of Balsam Mountains *P. metcalfi* in competition with *P. teyahalee* (Hairston, Nishikawa, and Stenhouse 1987).

These experiments provide strong evidence that these pairs of large *Plethodon* compete for similar resources, but they do not directly assess the mechanism of competitive exclusion. This problem was addressed in additional work on interspecific aggressive behavior by Nishikawa (1985, 1987; (see also Hairston, Nishikawa, and Stenhouse 1987). Nishikawa compared the aggressive behavior of both species in intra- and interspecific encounters in the laboratory, using individuals from both the Balsam and Smoky Mountains.

Differences in the intensity of interspecific aggression paralleled the differences in intensity of competition reported by Hairston. *Plethodon teyahalee* from the Smokies were more aggressive toward heterospecifics than were those from the Balsams (fig. 15.14). There was almost no difference between areas in the per capita effect of *P. jordani* or *P. metcalfi* on the *P. teyahalee* population, and there was no significant difference between *P. jordani* and *P. metcalfi* in the intensity of interspecific aggressive behavior. However, *P. metcalfi* from the Balsams, where interspecific competition is less severe, were more aggressive toward conspecifics than heterospecifics, whereas *P. jordani* from the Smokies showed equal levels of aggression toward conspecifics and heterospecifics (Nishikawa 1987). These experiments provide strong evidence that natural selection has resulted in high levels of interspecific aggressive behavior in both species in the Smoky Mountains. Nishikawa further suggested that because the higher elevation species always are the most common species in zones of overlap, the survival and reproduction of individuals are more likely to be affected by intra- than by interspecific competition. For *P. teyahalee,* interspecific competition probably is more important, and selection for aggressiveness against *P. jordani* in the Smokies may have resulted in greater aggressiveness toward conspecifics as well.

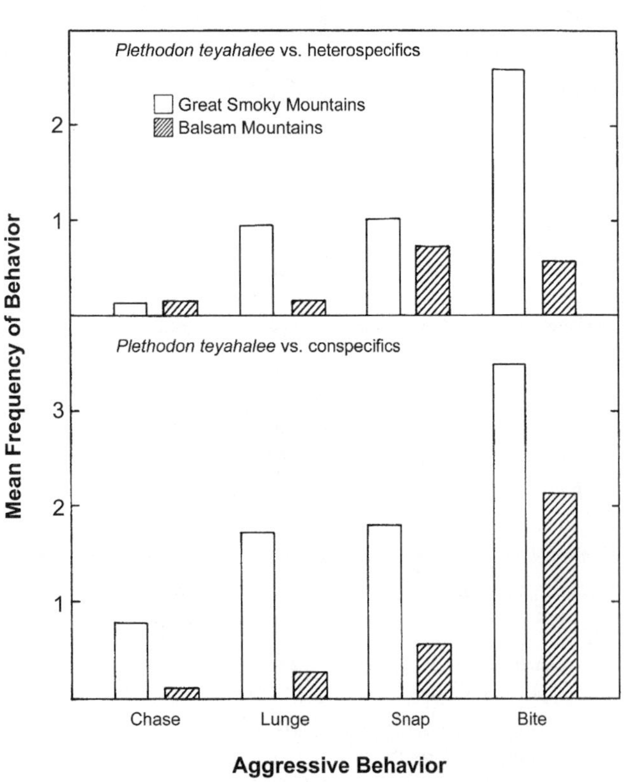


**Fig. 15.14.** Average frequency of aggressive behavior patterns exhibited by *Plethodon teyahalee* in encounters with sympatric conspecifics and heterospecifics. Heterospecific salamanders were *P. jordani* in the Great Smoky Mountains and *P. metcalfi* in the Balsam Mountains. After Nishikawa (1985).

An interesting twist to this story was provided by a study at Mountain Lake Biological Station in Virginia (Rissler, Barber, and Wilbur 2000). *Plethodon glutinosus* is native to this region, but *P. montanus* was introduced to the station in the 1930s after a class field trip to a nearby area (Mt. Rogers) where they occur (this is called *P. jordani* in the paper). In a field survey, Rissler, Barber, and Wilbur (2000) found little evidence of microhabitat partitioning or other indications of competitive interactions between these species. In field enclosures, there were slight differences in foraging behavior of *P. glutinosus* in the presence of *P. montanus,* but neither species showed any effect of interspecific interactions on body condition. In the laboratory, neither species showed much aggression toward conspecifics or heterospecifics. Unfortunately, interspecific interactions have not been studied in the locality from which the introduced *P. montanus* originated (where the *P. glutinosus* is replaced by *P. cylindraceus;* Highton and Peabody 2000), so the level of competition there is unknown. The lack of overt aggression between these species or between conspecifics is perhaps unusual, but Nishikawa's work has shown that levels of aggressiveness can vary among populations. Interpretation of the results is complicated by evidence of hybridization or genetic introgression between the two species at Mountain Lake, although Rissler, Barber, and Wilbur (2000) excluded individuals with apparent hybrid morphology from their experiments. These species apparently do not hybridize where their ranges naturally overlap (Highton and Peabody 2000), but it is possible that the small numbers of *P. montanus* originally released at Mountain Lake forced these individuals to seek matings with the more abundant native species.

Intra- and interspecific aggression and territoriality have been reported in other large species of *Plethodon* (see chapter 8) as well, but in most cases, the species are not as clearly separated spatially as the salamanders in the Balsam and Great Smoky Mountains. For example, in the Ouachita Mountains of Arkansas, *P. ouachitae* can be found in the same habitats as *P. albagula,* but the former species tends to be more abundant. *Plethodon ouachitae* is extremely aggressive toward both conspecifics and *P. albagula,* and it is able to displace the latter species from prime retreat sites (Anthony and Wicknick 1993; Anthony, Wicknick, and Jaeger 1997). Similar interactions occur between *P. glutinosus* and a smaller species, *P. kentucki,* in Kentucky (Marvin 1998a, b). Again, these two species share the same habitat. Both tend to be aggressive and defend retreat sites against conspecifics, but the larger *P. glutinosus* can exclude *P. kentucki* from choice retreat sites. In the Pigeon Mountains of Georgia, *P. glutinosus* is widespread in forested habitats, whereas a similar species, *P. petraeus,* is largely restricted to drier rock outcrops. Both species are aggressively territorial, but *P. glutinosus* is competitively superior to *P. petraeus* and probably limits the latter species to drier patches of habitat (Marshall, Camp, and Jaeger 2004).

### Habitat Separation in Small *Plethodon*

The distributional interactions of small species of *Plethodon* in the southern Appalachians are extremely complex, and taxonomic relationships are confusing as well (Highton 1971). Several species appear to be relict populations restricted to very limited habitats, while others are much more widespread. All of the species are morphologically and ecologically similar, and it is rare for more than two species of similar size to coexist in the same habitat. Species may occasionally hybridize in sympatry, but usually they maintain their genetic identity. The nonoverlapping ranges of some of these small species suggest a role for competition in determining their geographical distributions. For example, Highton (1971) suggested that small, isolated populations of species such as *P. hubrichti, P. richmondi, P. shenandoah,* and *P. welleri* have been prevented from extending their ranges because of competition from the more widespread red-backed salamander (*P. cinereus*).

The interactions between *P. cinereus* and *P. shenandoah* on a microgeographic scale were studied by Jaeger (1970, 1971a, b, 1972, 1974, 1980a) in Shenandoah National Park, Virginia. *Plethodon cinereus* is widespread in forested habitats throughout the park, whereas *P. shenandoah* is restricted to talus slopes consisting of a mixture of loose rocks and patches of soil (fig. 15.15). Jaeger (1971a) placed individuals of the two species in field enclosures in three different types of talus and deep soil in both single-species and mixed-species groups. Neither species was able to survive in bare rocks. *Plethodon shenandoah* had higher survivorship than *P. cinereus* in relatively dry talus, either in single-species or mixed-species groups. Both species did well alone in talus consisting of rocks imbedded in shallow soil and in deep soil, but the presence of *P. cinereus* reduced the survivorship of *P. shenandoah* in both habitats.

Subsequent laboratory experiments revealed that both species preferred soil to rocky substrates, and moist over dry habitats, but *P. shenandoah* had a higher physiological tolerance for dry conditions (Jaeger 1971b). Thus, *P. shenandoah* appears to be excluded from choice habitats by competition from *P. cinereus,* but *P. cinereus* is unable to invade the talus habitat because of physiological intolerance. Because *P. shen-*

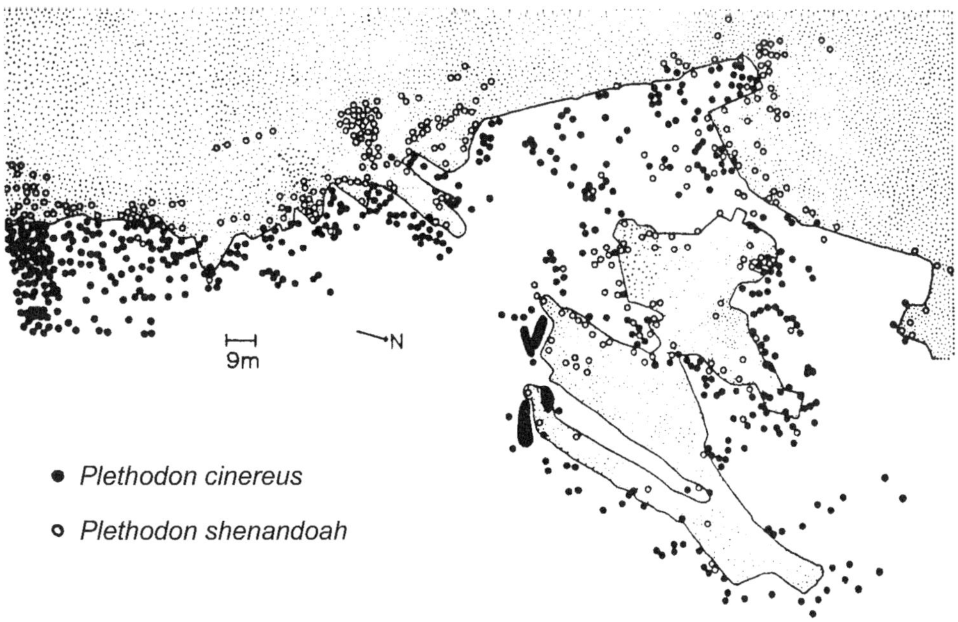


**Fig. 15.15.** Microdistribution of *Plethodon cinereus* and *P. shenandoah* around a talus slope (shaded area) in Shenandoah National Park, Virginia. Each dot represents an individual salamander. After Jaeger (1970).

*andoah* is restricted to relatively poor-quality habitats, its status is precarious, and at least one local deme apparently went extinct during an extended drought (Jaeger 1980a). The microgeographic distributions of these two species are the result of complex interactions between the physiological ecology of each species and interspecific competition.

Exploitative competition for food does not appear to explain the relative success of these species. Instead it appears that interference competition in the form of interspecific territoriality is the principal mechanism of competitive exclusion (Jaeger 1974; Jaeger and Gergits 1979; Wrobel, Gergits, and Jaeger 1980). This hypothesis was tested experimentally by Griffis and Jaeger (1998). They removed individuals of *P. cinereus* from underneath rocks at the edge of a *P. shenandoah* population and left other individuals in place in their retreat sites. Individuals of *P. shenandoah* were much more likely to invade rocks that were not inhabited by *P. cinereus* than those that were inhabited. A somewhat surprising result of the study was that *P. cinereus* was no more aggressive toward territorial intruders than was *P. shenandoah*. This partially contradicted earlier work by Wrobel, Gergits, and Jaeger (1980), which had found evidence of greater aggressiveness by *P. cinereus*. Hence, the behavioral mechanisms that result in competitive exclusion of *P. shenandoah* by *P. cinereus* still are not fully understood. *Plethodon cinereus* was similarly found to be more aggressive than *P. hoffmani,* which has a much more limited distribution (Jaeger, Prosen, and Adams 2002).

Red-backed salamanders are aggressive toward several other species of salamander as well, but in these cases, interspecific competition sometimes results in the exclusion of one species from individual retreat sites, but does not result in separation of species into discrete habitat patches (see chapter 8). For example, adults of *P. cinereus* can defend territories against juveniles of the much larger species, *P. glutinosus,* but not against adults of that species (Lancaster and Jaeger 1995; Jaeger et al. 1998; Price et al. 2002). *Plethodon cinereus* also is aggressive toward *Eurycea cirrigera* (Jaeger, Gabor, and Wilbur 1998) and *E. bislineata* (Joseph Markow, personal communication), both of which tend to be relatively nonaggressive. Possibly red-backed salamanders could influence the choice of retreat sites by *Eurycea,* but this has not been demonstrated conclusively. On the other hand, red-backed salamanders can be displaced from streamside habitats and small seeps by various species of *Desmognathus,* which tend to be even more aggressive (E. Smith and Pough 1994; Fauth 1998; Jaeger, Gabor, and Wilbur 1998; Grover 2000; Grover and Wilbur 2002).

### Microhabitat Separation in *Desmognathus*

Streambank salamanders of the genus *Desmognathus* exhibit a complex array of ecological interactions. Communities of these salamanders consist of as many as five sympatric species in the mountains of North Carolina and southwestern Virginia (Hairston 1949, 1980c; Organ 1961c), but only two or three species farther north (Krzysik 1979) or south (Means 1975). In areas where several species occur together, they often show horizontal separation of microhabitats. Large species are the most aquatic, medium-sized species are semiaquatic, and the smallest species are terrestrial (fig. 15.16). In the mountains of Virginia, for example, the largest species, *D. quadramaculatus,* is restricted to streams and streambanks. *Desmognathus monticola,* a slightly smaller species, is found on streambanks and in seepage areas, while *D. fuscus* is found in the headwaters of streams and in seepage areas. *Desmognathus orestes* is much more terrestrial, being found mainly on the forest floor, while the smallest species, *D. wrighti,* is seldom found near streams and often climbs on forest plants to feed (Organ 1961c). Similar habitat segregation has been reported in other regions (fig. 15.17; Hairston 1949, 1980c; Krzysik 1979), with *D. ocoee* or *D. ochrophaeus* taking the place of *D. orestes* (these species were all considered to be *D. ochrophaeus* when the original ecological studies were done; see Tilley and Mahoney 1996; Petranka 1998).

This gradient from aquatic to terrestrial habits is paralleled by differences in life-history strategies. Since clutch size is positively correlated with body size (see chapter 10), there is a general trend from large to small clutches that parallels the gradient from aquatic to terrestrial life (Organ 1961c; Tilley 1968). While all species lay their eggs in relatively wet sites, the terrestrial species are more likely to lay eggs in seepage areas than in streams. All of the large and medium-sized species have free-living aquatic larvae, but the two smallest species, *D. wrighti* and *D. aeneus,* lack an aquatic larval stage (Organ 1961b, c; J. R. Harrison 1967). The placement of eggs at greater distances from streams and a shorter larval period in the smaller species is paralleled by higher survivorship early in life.

This gradient from aquatic to terrestrial life traditionally has been thought to reflect an evolutionary progression from an ancestral aquatic life history to a more derived terrestrial life history (Hairston 1949, 1987; Organ 1961c). Indeed, this apparent evolutionary progression toward terrestrial life was used by Dunn (1926c) to illustrate a general trend in the evolution of terrestrial habits in the family Plethodontidae. Tilley and Bernardo (1993) argued that in the absence of a rigorous phylogenetic analysis of the subfamily Desmognathinae, it is impossible to determine whether the large aquatic species are most like the ancestor of all *Desmognathus*. Subsequent phylogenetic studies by Titus and Larson (1996) and Chippindale et al. (2004) suggested that the small, fully terrestrial species *D. wrighti* belongs to a lineage that branched off early in the evolution of the genus, while the larger semiaquatic and aquatic species are not basal species, and in fact, are derived from within a clade of

**Fig. 15.16.** Some members of the assemblage of streamside salamanders of the genus *Desmognathus* from the mountains of North Carolina. (A) *D. quadramaculatus.* (B) *D. monticola.* (C) *D. ocoee.* (D) *D. wrighti.* Photos by Wayne Van Devender.

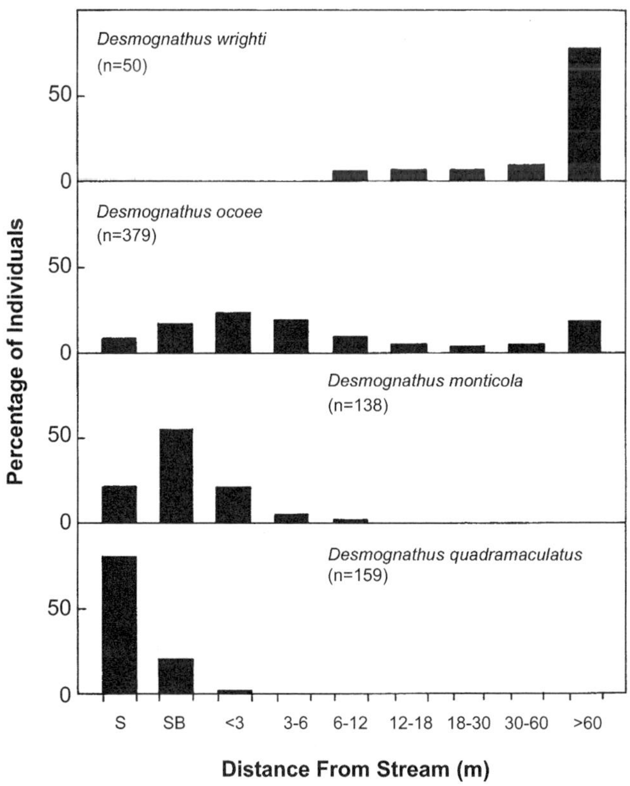


**Fig. 15.17.** Occurrence of four species of *Desmognathus* at different distances from streams in the Black Mountains of North Carolina. S = stream; SB = streambank. Plotted from data in Hairston (1949).

direct-developing terrestrial salamanders. These studies also indicated that the other very small, highly terrestrial species, *D. aeneus*, also represents an early line of desmognathine evolution that may or may not be closely related to *D. wrighti*. One controversial implication of these studies is that most species of *Desmognathus* appear to have secondarily reevolved an aquatic reproductive mode with aquatic larvae from a direct-developing ancestor. This view of desmognathine phylogeny would alter some interpretations of the evolutionary responses of these species to competition or predation (e.g., Hairston 1987), but does not affect interpretation of experiments designed to unravel current ecological interactions among the species.

The ecological separation of these species has been explained by a number of competing hypotheses. Originally, Hairston (1949) attributed their horizontal distribution to interspecific competition, but later suggested that the smaller species were excluded from the vicinity of streams by predators, including fish and larger species of salamanders (Hairston 1980c). Tilley (1968) also supported the view that predation by the larger, more aquatic species, such as *D. quadramaculatus*, tends to exclude smaller species from the most aquatic habitats. Alternatively, the ecological segregation could be the result of habitat preferences that are independent of the presence of other species, although it is difficult to rule out the possibility that competition in the past has favored divergence in the use of microhabitats.

Hairston (1986) tested the predation hypothesis in the Nantahala Mountains of North Carolina. The streamside salamander community included *Desmognathus quadramaculatus, D. monticola,* and *D. ocoee,* with *D. aeneus* living in adjacent terrestrial habitats. Hairston conducted a series of experiments involving removal of either *D. monticola* or *D. ocoee* from plots and measuring the effects of the removals on each other and on the other species. Although it was not possible to completely remove populations from the plots, a sufficient number was removed to have a measurable effect on the other species. Removal of *D. monticola* resulted in a significant increase in the population of *D. ocoee.* This result would be expected if *D. monticola* were either an important predator or competitor of *D. ocoee.* However, the reciprocal experiment of removing *D. ocoee* was interpreted as support for the predation hypothesis. The number of *D. monticola* decreased on the experimental plots, suggesting that *D. ocoee* is an important source of food for this species. There also was evidence, but less conclusive, of a reduction in the number of *D. quadramaculatus* when *D. ocoee* were removed. The more terrestrial *D. aeneus* was not strongly affected by either removal experiment.

The removal experiments also resulted in some shifting in the horizontal distribution of species perpendicular to the stream. When *D. monticola* were removed, *D. ocoee* showed an initial shift toward the stream, perhaps because of reduced predation on this species. However, in subsequent years, *D. ocoee* moved back toward the land. Hairston attributed this to replacement of the missing *D. monticola* by *D. quadramaculatus* and subsequent predation by this species on *D. ocoee.* There was a tendency for *D. quadramaculatus* to shift landward when *D. monticola* was removed, but the pattern was not consistent in all months. Hairston suggested that both predation by the larger *D. quadramaculatus* and competition are important interactions between this species and *D. monticola,* but he provided little direct evidence of predation.

Several other workers have investigated the effect of interspecific interactions on the use of microhabitats by various species of *Desmognathus.* Both laboratory and field experiments have shown that species of *Desmognathus* exhibit preferences for substrate texture, substrate moisture, and size of cover objects that are partly independent of the presence of the other species (Krzysik and Miller 1979; Keen 1982; Carr and Taylor 1985; Southerland 1986a; Grover 2000). For example, *D. monticola* and *D. quadramaculatus* usually are segregated by microhabitat because of the affinity of *D. quadramaculatus* for a stream habitat (Keen 1985). Nevertheless, the larger species may inhibit diurnal activity in *D. monticola* (Carr and Taylor 1985) or prey on juveniles of that species (Southerland 1986b), thereby causing *D. monticola* to shift toward the stream bank. Grover (2000) reported that *D. monticola, D. fuscus,* and *Eurycea cirrigera* were found farther from water at two streams where *D. quadramaculatus* was present than at two where this species was absent. This was taken as evidence that predation by *D. quadramaculatus* resulted in smaller species avoiding microhabitats close to streams, but Grover did not report any observations of actual predation. In experimental enclosures, the presence of *D. quadramaculatus* always resulted in *D. monticola* shifting to more terrestrial microhabitats, where food availability appears to be inferior to that in the aquatic habitat (Rissler, Wilbur, and Taylor 2004). There also is evidence for an evolutionary response to the presence of *D. quadramaculatus* by *D. monticola.* Individuals of *D. monticola* from areas where *D. quadramaculatus* occurs were more terrestrial than those from areas without *D. quadramaculatus,* even in the absence of the latter species.

In some communities, the presence of *D. monticola* can restrict access of *D. fuscus* to streambank microhabitats (Krzysik 1979). This has been attributed mainly to interference competition rather than predation because of the similarity in size of the two species (Keen 1982; Keen and Sharp 1984). Species of *Desmognathus* are known to aggressively defend retreat sites against both conspecifics and heterospecifics, so larger species would be expected to displace smaller ones even if predation is rare (Keen and Sharp 1984; Keen and Reed 1985; Jaeger 1988; Colley, Keen, and Reed 1989; Camp and Lee 1996; see also chapter 9).

Camp (1997) argued that the role of *D. quadramaculatus* as a major predator on smaller species of salamanders has been overstated. He used data on stomach contents from his own work and previously published studies to show that metamorphosed salamanders rarely, if ever, appear in the diet of this species, which eats mostly invertebrates. He also did not find any evidence that *D. monticola,* the next largest species in many streamside communities, feeds regularly on smaller salamanders. In laboratory experiments, Formanowicz and Brodie (1993) found that *D. quadramaculatus* is less effective in capturing salamanders than is another plethodontid that often lives in the same habitats, *Gyrinophilus porphyriticus.* The latter could be an important predator on small species of *Desmognathus,* but its impact on habitat selection by desmognathines has not been studied. Even if predation on adults of smaller species is rare, the threat of predation still could be sufficient to cause smaller species to avoid habitats where the larger species are present. With the data currently available, it is impossible to distinguish between the roles of predation and interference competition in the spatial structuring of *Desmognathus* communities. Another threat to small species of *Desmognathus* may come from the larvae of both *D. quadramaculatus* and *G. porphyriticus,* both of which reach a very large size and prey on the larvae of other plethodontids (Resetarits 1991; Beachy

1993b, 1994, 1997b; Gustafson 1993, 1994). Predation on larvae could affect the relative abundance of smaller species in the community, but is unlikely to have any impact on the spatial distribution of adults.

### Other Terrestrial Salamander Communities

The ecological relationships of other assemblages of terrestrial salamanders, particularly those found in the tropics, are poorly understood (Hairston 1987). The Neotropical plethodontids were once considered to be a relatively minor and uninteresting group, but they are now known to include half of the world's salamander species (D. Wake 1970, 1987; Garcia-Paris et al. 2000). They exhibit a wide range of morphological and ecological adaptations that have scarcely been investigated (D. Wake and Lynch 1976). Of particular interest are microgeographic patterns of distribution along elevation gradients. In the mountains of Guatemala, most species exhibit very narrow vertical distributions (fig. 15.18), so sympatric species often are separated from one another (K. Schmidt 1936; D. Wake and Lynch 1976; Elias 1984). In some cases, this may reflect major changes in vegetation and climate, such as that occurring at the transition from cloud forest to high-elevation woodlands. In other cases, the boundaries between species do not correspond to obvious changes in vegetation or climate, suggesting the possibility of competitive interactions similar to those described for species of *Plethodon* in the Appalachians. Those species that do overlap spatially tend to differ in their use of microhabitats, but the mechanisms responsible for this ecological separation are completely unknown (D. Wake and Lynch 1976).

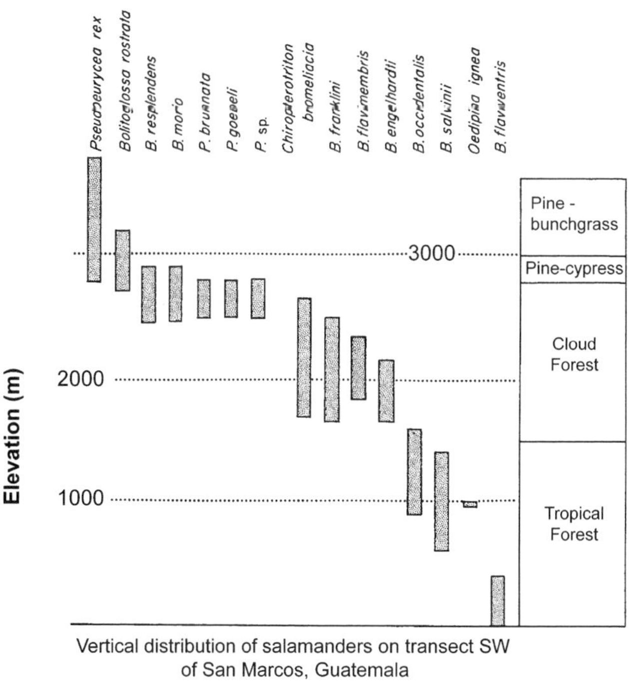


**Fig. 15.18.** Altitudinal distribution of bolitoglossine salamanders along a vertical transect in Guatemala. Many species are restricted to certain vegetation zones and tend to have narrow altitudinal distributions. After D. Wake and Lynch (1976).

Nonplethodontid salamanders have received relatively little attention from community ecologists, except for studies of larval communities (see the following discussion). In North America, adults of several species of *Ambystoma* sometimes occupy the same pond during the breeding season, but they probably do little feeding at that time and therefore are unlikely to compete with one another for food. There is evidence for minor differences in microhabitat selection among adults of several species of *Ambystoma* in Illinois (Parmelee 1993), but there has not been any experimental work to determine whether these differences are due to competitive interactions among the species. Aggressive interactions have been observed between juveniles of different species of *Ambystoma* in the laboratory (Walls 1990), but whether such aggression mediates use of microhabitats in the field is unknown. Most of these species are underground outside of the breeding season, so direct interaction among species is likely to be rare.

### Community Ecology of Pond-Dwelling Adult Salamanders

The importance of ecological interactions among adults of pond-dwelling salamanders is poorly understood. Adults of many salamanders, such as the various species of *Ambystoma*, are present in ponds only during the breeding season, when they probably do not feed, and are terrestrial for the rest of the year, so interaction in ponds probably are minimal. The exceptions are those species that remain in ponds as paedomorphic adults (see chapter 13). Because of their large size, paedomorphic *Ambystoma* are potential predators of other salamander larvae and tadpoles (Morin 1995; Fauth 1999b). The adults of other aquatic salamanders, such as sirens, are potential competitors of paedomorphic *Ambystoma*, and they may prey on them as well (Resetarits and Fauth 1998; Fauth 1999a, b). Adult newts may interact with *Ambystoma* larvae either as competitors or predators (Brophy 1980; Morin 1983b), while newts and sirens can be either predators or competitors with each other (Fauth, Resetarits, and Wilbur 1990; Fauth and Resetarits 1991). Snodgrass et al. (1999) reported that *Siren intermedia* tends to be absent from ponds inhabited by the larger *S. lactertina*, and the latter seldom co-occurs with *Amphiuma means*. These species could interact in either the larval or adult stage as predators, competitors, or both, but the precise nature of the interactions among these species is not known.

Several experimental studies have examined ecological interactions among some of these pond-dwelling species. On the Coastal Plain of North Carolina, red-spotted newts (*Notophthalmus viridescens*) are considered keystone predators in temporary ponds, because they affect the species composition and structure of anuran tadpole communities. In some ponds, lesser sirens (*Siren intermedia*) may play a similar role. Fauth and Resetarits (1991) used experimental ponds (cattle tanks) to investigate the effect of these two species on each other. They housed newts at different densities (two or four adults per pond) with or without a single adult siren. Sirens can eat newt larvae, but they do not eat adults, which are highly toxic. Growth rates of newts were higher at low density than at high density, both with and without sirens, indicating that strong intraspecific competition was occurring in the ponds. Growth rates of newts were even lower in the presence of one siren, especially for those at high density. Newt reproduction also was affected by the presence of sirens, mainly through predation on newt larvae. The newts had no detectable effect on the sirens, but the experiment provided a relatively weak test for such effects because only one siren was present in each pond.

Fauth (1999b) later examined ecological interactions between lesser sirens and paedomorphic adults of *Ambystoma talpoideum,* which often live in the same ponds. Neither of these species is toxic or distasteful, so either potentially could be a competitor or predator of the other, depending on body size relationships. In fact, sirens had a much greater effect on *Ambystoma* than the latter had on sirens. Cattle tanks were set up with zero, four, or eight adult *Ambystoma* per tank and either one siren or no sirens. The presence of one siren significantly reduced the growth rates of adult *Ambystoma,* presumably because of competition. The sirens also virtually eliminated larval *Ambystoma* from the tanks, so predation is important as well. As in the experiment with newts, there was no detectable effect of the treatments on growth or survival of sirens, but again, with only a single siren in each tank, this test is relatively weak. It also is possible that in natural ponds, large *Ambystoma* prey on larval sirens, but this has not been studied. In the field, there was evidence that *Ambystoma* avoided sirens, suggesting that they may be important predators on *Ambystoma.* Nevertheless, the effect is not sufficiently strong to exclude *Ambystoma* from ponds with sirens; in fact, the distribution of the two species in natural ponds was independent.

In Europe, several species of newts in the genus *Triturus* sometimes co-occur in ponds. Competitive interactions have not been studied experimentally, but there have been several descriptive and quantitative analyses of food habits and niche overlap in species sharing the same habitats. Often these species are very similar in size and overall morphology (J. D. Harrison, Gittins, and Slater 1984), suggesting the potential for interspecific competition. In general, results for these species are similar to those for terrestrial salamanders. The newts feed on a broad assortment of invertebrates and show little evidence of taxonomic selectivity (Avery 1968; Griffiths 1986). There often is broad overlap in prey taxa among coexisting species (Avery 1968; Dolmen and Koksvik 1983; Griffiths 1986). The newts feed selectively on larger prey when it is available (Ranta and Nuutinen 1985), as do North American *Notophthalmus* (Attar and Maly 1980; Morin, Wilbur, and Harris 1983). Larger species generally eat larger prey (Avery 1968; Dolmen and Koksvik 1983), leading to some resource partitioning along a prey-size axis, but in some cases, the diet of large species encompasses most of the diet of smaller species (Joly and Giacoma 1992). In Norway and Wales, the relatively large species *Triturus cristatus* tends to feed mainly on the bottom of ponds in deep water, while *T. vulgaris,* a smaller species, spends more time feeding in the water column. This microhabitat partitioning has been interpreted as a mechanism to avoid competition (Dolmen 1983b; Griffiths and Mylotte 1987), but in some ponds, food may be so abundant that competition does not occur (Griffiths 1986). Newt species are not always consistent in their use of microhabitats across their ranges, however. For example, both *T. vulgaris* and *T. helviticus* were primarily bottom dwellers in a pond in Wales and ate very similar prey (Griffiths 1987). In Italy, *T. carnifex,* a large species that is closely related to *T. cristatus,* fed mostly on prey that fell onto the water surface, while even the smallest species, *T. italicus,* consumed prey from the bottom of a pond (Joly and Giacoma 1992).

## Amphibian Larval Communities: General Characteristics

Because of their double life in the water and on land, most aquatic-breeding amphibians are members of two very distinct ecological communities at different times in their lives. Although some amphibians remain in the water after metamorphosis, most adult amphibians have few interactions with larval stages except as predators (see chapter 14). However, since the size of adult populations of many species may be largely controlled by recruitment of metamorphosing individuals into the population, knowledge of interspecific interactions at the larval stage is essential for a complete understanding of the structure and organization of amphibian communities (Wilbur 1980, 1984; Hellriegel 2000; Skelly 2001). There is now a large literature on competitive interactions and predator-prey relationships among species of amphibian larvae. In contrast to the largely descriptive literature on the community ecology of adult amphibians, especially anurans, the literature on larval communities is heavily experimental, in part because larvae are suitable for both laboratory and field manipulation.

The ecology of larval amphibians differs from that of

adult amphibians in several ways. Perhaps most important is the fact that the initial composition of a larval assemblage is not governed solely by interactions among the larvae themselves, but also is a result of choice of oviposition sites by adults (Alford 1999; Skelly 2001). In general, different species of amphibians probably select breeding sites independently, based mostly on physical characteristics of the habitat (see chapter 10), although some amphibians will avoid breeding sites where predators are present (see chapter 14). This means that the composition of the larval community depends on which species breed at a particular site, and this may change from year to year. The timing of breeding by different species may vary as well, as each species responds slightly differently to environmental cues such as temperature and rainfall. Experimental studies have shown that temporal differences in the introduction of various species can have a large effect on interactions among species (see the following). Another major difference between the community ecology of larval and adult amphibians is that larvae are not reproductive animals. Nevertheless, the outcome of competitive interactions among larvae in a pond or stream can affect the timing and success of reproduction by adults, because larval interactions affect the size and competitive ability of metamorphosed adults. The initial number of larvae of each species present in the pond is fixed by the number of eggs laid by females each year, or by the experimenter. Since all populations are expected to suffer mortality, the experimenter can monitor rates at which a population declines, but not population growth rates. Studies of larval competition usually rely on measures of performance thought to be reliable indicators of individual fitness, including (1) survival to metamorphosis, (2) length of the larval period, (3) individual growth rates, and (4) size at metamorphosis.

### Reproductive Input and Output in Pond Communities

In natural ponds, the number of breeding adults and the number of eggs laid by each species often are poor predictors of the final composition of the larval community, especially in temporary ponds (Heyer 1976, 1979a; Seale 1980; Pechmann et al. 1989; Semlitsch et al. 1996). Some species lay thousands of eggs in a pond, but contribute very few metamorphosing young to the next generation; others lay fewer eggs, but have much greater reproductive success. This indicates that processes that occur after eggs have been laid are important in determining the relative reproductive success of different species in the community. These processes can include responses of individual species to physical factors such as temperature and pond drying, as well as biological interactions such as competition and predation. The number of metamorphosing juveniles emerging from a breeding site in any given year may or may not be a good predictor of future adult population size. Adult amphibians typically return to the same pond year after year, although limited migration among sites also occurs (see chapter 6). Juveniles, on the other hand, either return to their natal ponds or disperse to new ponds. In ponds that seldom produce juveniles, populations may be maintained by occasional migration of juveniles from more productive ponds nearby (Sjögren 1991a; Sinsch 1992b; Sinsch and Seidel 1995; Skelly 2001).

The best example of variability in amphibian population dynamics in temporary ponds is a long-term study conducted at Rainbow Bay, a Carolina bay, on the Coastal Plain of South Carolina (Semlitsch et al. 1996). Investigators enclosed the entire pond with a drift fence that trapped adult amphibians as they migrated into and out of the pond during the breeding season, as well as juveniles that left the pond after metamorphosis. The population dynamics of all species using the pond were monitored for 16 years, providing the longest record for any pond in the world used by a large number of amphibian species. A striking result of this study is that most species experienced only a few good reproductive years in the 16 years of the study, and several species failed to reproduce successfully in most years of the study (fig. 15.19).

Much of this variation was due to variation in hydroperiod, the number of days that the pond held water. Four years were drought years in which the pond held water for less than 100 days. This resulted in almost complete reproductive failure for all amphibian species. In other years, the hydroperiod was longer, but not long enough to allow species with relatively long larval periods to reach metamorphosis (e.g., *Rana sphenocephala* and *R. clamitans*). Species with short larval periods, however, did well in those years (e.g., *Pseudacris crucifer* and *P. ornata*). There was evidence for interspecific competition as well. For example, *P. ornata* did relatively poorly when *Rana clamitans* tadpoles were abundant, but did well when *R. clamitans* tadpoles were scarce. Some species, such as *P. crucifer*, showed evidence of density-dependent intraspecific competition as well, while others, such as *R. clamitans*, seemed to do better at high larval densities.

Salamander larvae in this system are predators on some species of tadpoles, and two species, *R. sphenocephala* and *R. clamitans*, did poorly when initial densities of salamander larvae were high. Two species, *Scaphiopus holbrookii* and *Gastrophryne carolinensis*, had only one good reproductive year out of 16. These species breed opportunistically after heavy rains, especially in the summer. In each case, they were successful when the pond dried and then subsequently was refilled by summer rains. The initial drying eliminated other species that were potential predators or competitors of these species. The poor reproductive success of these anurans is especially striking because breeding adults appeared at the pond in most years of the study (fig. 15.19). Overall, the results of this study indicated that the relative

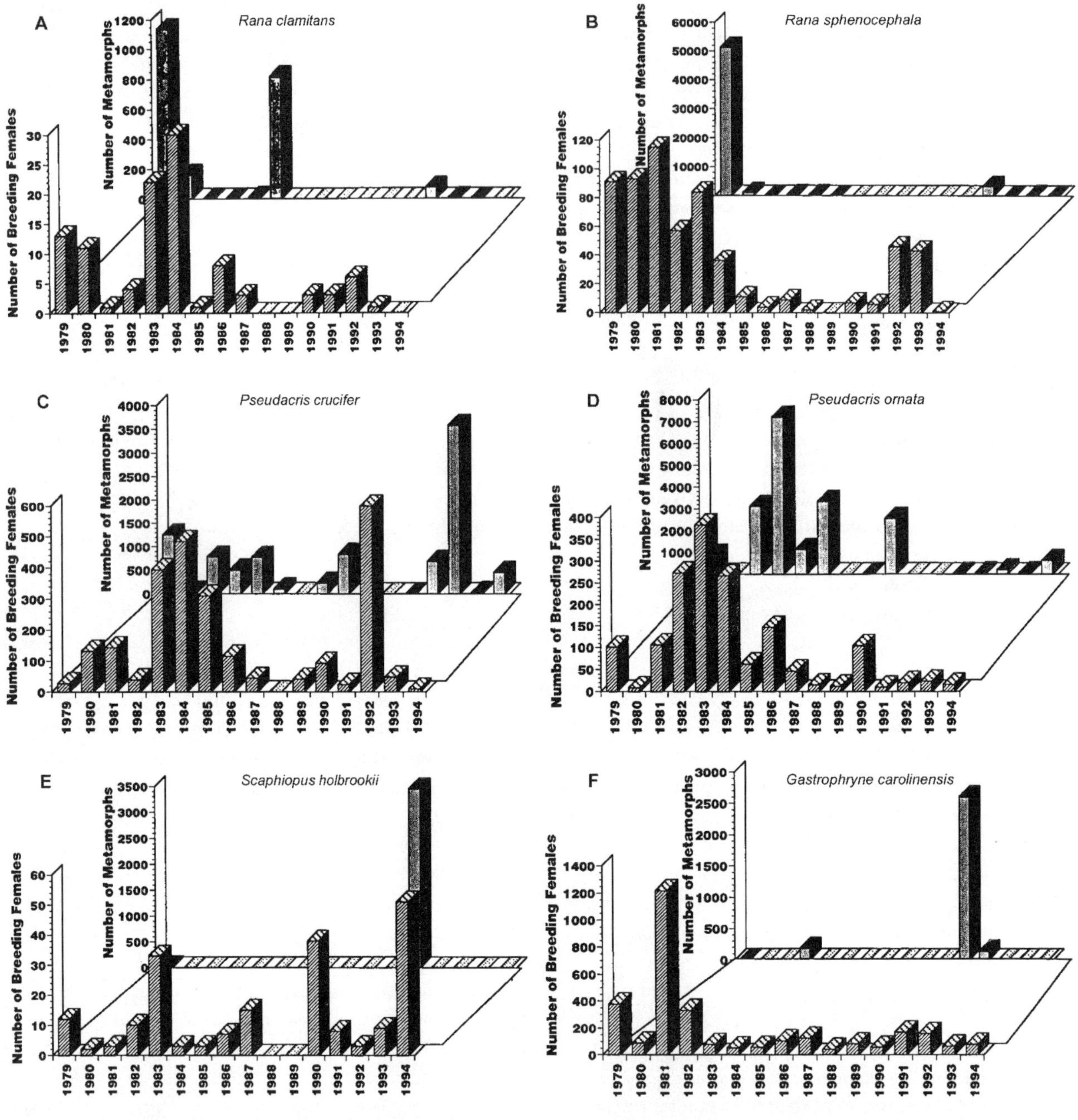


**Fig. 15.19.** Variation in number of breeding females and metamorphosing juveniles for six species of anurans for 16 years at Rainbow Bay, South Carolina, a Carolina bay that dries in most summers. These graphs represent the range of variation in annual reproductive success for a study that included a total of five species of salamanders and eight species of frogs. After Semlitsch et al. (1996).

success of different species in the community is a function of complex interactions between pond hydroperiod, intraspecific and interspecific competition, and predation.

## The Pond Permanence Gradient and the Structure of Larval Communities

The studies at Rainbow Bay illustrate the importance of considering physical factors, such as pond hydroperiod, and the way that they interact with competition and predation to structure larval communities. Amphibians use a spectrum of ponds for breeding, ranging from ephemeral rain pools that often last only a few days to permanent ponds that never dry up (fig. 15.20). In between are vernal pools that typically fill during the winter and spring and often are dry by midsummer, and semipermanent ponds that often hold water for most of the summer, but may dry up every few years (Wellborn, Skelly, and Werner 1996; Skelly 1997, 2001;

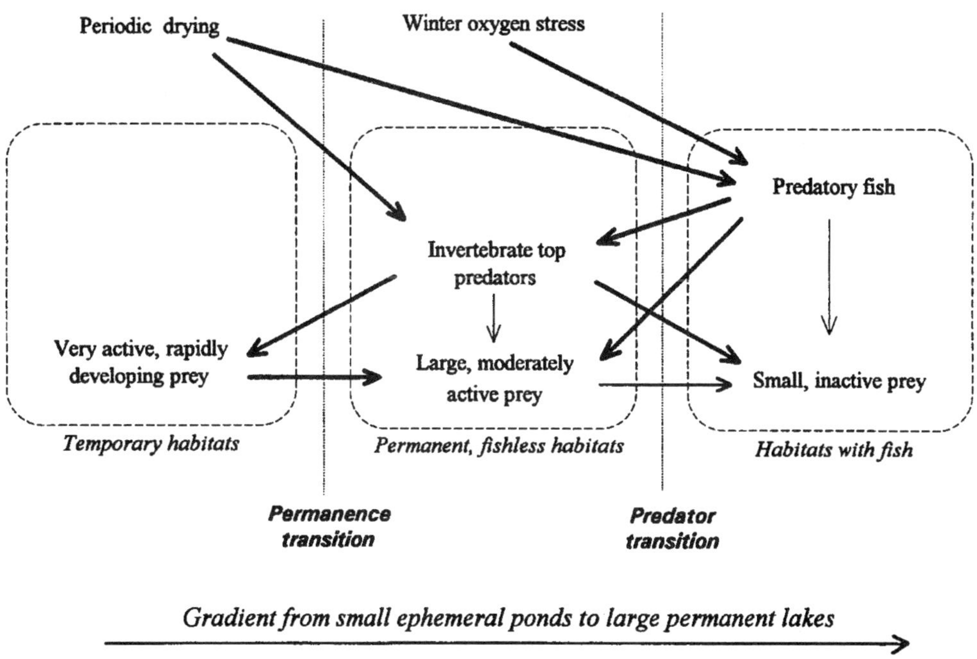


**Fig. 15.20.** Schematic diagram of the gradient in freshwater habitats from ephemeral ponds to permanent lakes. Arrows indicated the direction of negative effects. Thick arrows indicate strong effects that constrain the distribution of species. Thinner arrows indicate weaker effects that do not prevent the coexistence of interacting species. Two major transitions between habitats are indicated. The first is between temporary habitats with few predators and permanent habitats with many invertebrate predators. The second is between permanent habitats that lack fish and those that contain fish. Within temporary ponds, variation in the frequency and timing of drying will effect the composition of both predator and prey communities. From Wellborn, Skelly, and Werner (1996). Reprinted, with permission, from the *Annual Review of Ecology and Systematics,* vol. 27, 1996, by Annual Reviews (www.annualreviews.org).

Baber et al. 2004). The diversity of amphibian species that use ponds along this gradient tends to be low at either end of the gradient, but high in the middle. In other words, relatively few species use either extremely ephemeral pools or permanent ponds for reproduction. In contrast, ponds of intermediate permanence often harbor complex communities of many amphibian species. For example, ponds known as Carolina bays in the southeastern United States, most of which dry periodically, may have the most diverse assemblages of amphibian species of any aquatic habitats, especially when fish are absent from these ponds (Resetarits and Fauth 1998; Snodgrass, Bryan, and Burger 2000).

Amphibians that use different types of ponds along this permanence gradient tend to have very different life-history characteristics (see chapter 13). Highly ephemeral ponds tend to have relatively poor nutrients, and their rapid rate of drying favors rapid larval development and early metamorphosis. These ponds also tend to have fewer predators than do more permanent bodies of water, so the larvae of species that use such sites can afford to be very active in their foraging behavior without exposing themselves to predation (see chapter 14). In contrast, larvae that develop in longer-lasting ponds with many predators tend to be less active, and they often reduce activity in response to the presence of predators, but they also have slower rates of growth and development (Woodward 1983; Skelly 1996; Dayton and Fitzgerald 2001). Although hydroperiod does not vary among permanent ponds, other physical factors such as acidity, dissolved oxygen, maximum depth, surface area, and habitat stucture can influence the distribution of amphibian larvae (e.g., R. Hoffman, Larson, and Brokes 2003).

The presence of predatory fish in many permanent ponds tends to exclude many species of amphibians from using such ponds for breeding (see chapter 14). Ponds with fish tend to have communities composed of fewer species of amphibian larvae than do ponds without fish (e.g., Snodgrass, Bryan, and Burger 2000), although different species of fish have different effects on larval communities (Chalcraft and Resetarits 2003). Different kinds of tadpoles also have evolved different phenotypic responses to predatory fish. Hylid tadpoles in ponds with fish tend to grow more rapidly than do ranid tadpoles, for example (Richardson 2002b). Predation thus assumes a key role in the structuring of permanent pond communities and favors the evolution of antipredator adaptations in species that live in these ponds. Predator avoidance often comes at a cost in reduced rates of growth and development, and some permanent pond species, such as bullfrogs (*Rana catesbeiana*), have tadpoles that spend one or more winters in the pond before reaching metamorphosis. The slower lifestyle of permanent pond species also tends to put them at a competitive disadvantage against more active temporary pond species where they co-occur in semipermanent ponds that lack predatory fish (Skelly 1997). On the other hand, in permanent ponds, the presence of large overwintering tadpoles can have negative effects on the larvae of other species that enter ponds in the spring (Boone, Little, and Semlitsch 2004).

Interactions among different types of predators can directly affect the survivorship of amphibian larvae and can indirectly affect competitive interactions among different species (e.g., Werner and McPeek 1994). In temporary ponds, where many species of amphibian larvae often co-occur, the role of predation may be secondary to interspecific competition in determining the structure of the larval community, although differential vulnerability to predation can influence the outcome of competitive interactions (e.g., Morin 1981, 1983a, 1986, 1987a; Relyea 2000). The Rainbow Bay studies have shown that the rate of pond drying can interact in complex ways with competition and predation to determine the relative success of different species in the community (Pechmann et al. 1989; Semlitsch et al. 1996). The difficulty with such studies is that the relative importance of different factors is hard to determine from statistical correlation alone. Consequently, much of the experimental work on the community ecology of larval amphibians has focused on trying to understand the relative importance of each of these factors and the interactions among them in determining the structure of communities.

## Competition and Predation in Larval Salamander Communities

Salamander larvae generally are the top predators in their communities, especially in temporary ponds, sometimes functioning as keystone predators that affect the structure of the aquatic vertebrate and invertebrate community at lower trophic levels (Holomuzki and Collins 1987; Taylor et al. 1988; P. Harris 1995; Morin 1995; Blaustein et al. 1996; Wissinger et al. 1999; Davic and Welsh 2004). Salamander larvae obviously interact with anuran tadpoles in the same communities, but here I focus on their interactions with one another. The role of larval and adult salamanders as predators of tadpoles is discussed in a later section. Depending on the relative sizes of coexisting species, salamander larvae can affect one another as predators, competitors, or both. Descriptive studies of larval salamander communities have focused mainly on differences in diet and microhabitat use among co-occurring species, whereas experimental studies have mostly dealt with the role of competition and predation in structuring the community.

### Niche Partitioning and Habitat Segregation

Most studies of niche partitioning in salamander larvae have focused on pond-breeding species of *Ambystoma* in the United States. It is common for two or three species of *Ambystoma* to coexist in the same pond, and sometimes as many as five species are present. Larvae and adults of the newt *Notophthalmus viridescens* often occur in the same ponds as well. All of these species tend to be generalist predators on small invertebrates and tadpoles. Overlap in diet among co-occurring species often is high (C. K. Smith and Petranka 1987; B. Taylor et al. 1988; Nyman 1991), as it is for salamander larvae in other parts of the world (Braña, de la Hoz, and Lastra 1986). Some separation in diet may be an indirect effect of the timing of breeding by different species. Typically there is a sequential entry of larvae of different species into the community. For example, in the eastern and southern United States, *Ambystoma annulatum* and *A. opacum* are fall breeders, with the latter species laying eggs in shallow depressions that eventually fill with water to become ponds (Noble and Brady 1933; Petranka and Petranka 1980; Petranka 1998). Larvae hatch and overwinter at a small size, growing steadily during the spring. *Ambystoma tigrinum, A. talpoideum,* and *A. texanum* typically are winter breeders, with the exact date depending on local climatic conditions. *Ambystoma maculatum* breeds in early spring, as do members of the *A. jeffersonianum* complex. Where *A. maculatum* and the *A. jeffersonianum* complex co-occur, the latter often arrives in ponds slightly earlier (Nyman 1991). *Notophthalmus* adults usually enter ponds later in the spring to breed, although in some regions they may remain in the ponds for much of the year.

This staggered breeding pattern has been interpreted as a mechanism for reducing competition, because it leads to separation of species by body size (Worthington 1968, 1969; Hassinger, Anderson, and Dalrymple 1970; Keen 1975; Nyman 1991). There is, however, little evidence to indicate that timing of breeding has evolved as a response to the presence of other species in the community, as opposed to independent responses to factors such as temperature and rainfall. Nevertheless, differences in timing can result in divergence in prey size among species living in the same pond. Small *Ambystoma* larvae typically feed on zooplankton suspended in the water column. As they grow, they shift to larger prey, although they continue to take some small zooplankton (Dodson and Dodson 1971; Collins and Holomuzki 1984; Leff and Bachmann 1986; Holomuzki and Collins 1987; Sredl and Collins 1991). Differences in time of entry into a pond may result in different species using somewhat different prey at any given time, because the earliest arriving species will tend to eat the largest prey (Hassinger, Anderson, and Dalrymple 1970; Nyman 1991).

Species that have similar diets early in life may diverge if they grow at different rates. For example, *A. californiense* larvae in California have faster growth rates than do *A. macrodactylum* larvae, and soon reach a size that enables them to take large prey such as tadpoles, which are not available to the smaller species (J. Anderson 1968a). Nevertheless, there still is substantial overlap in food niches of the two species, and there is no direct evidence that the selection of prey by one species is influenced by the presence of the other. In British Columbia, the slow-growing *A. macrodactylum* often occurs in ponds with a faster-growing species, *A. gracile*. The latter is more affected by increased density of heterospecifics than in the former, whereas *A. macrodactylum* appears to be more sensitive to intraspecific competition (Pearman 2002). Because all salamander larvae are predators, ontogenetic shifts in diet can result in the larger or faster-growing species in a community changing from being primarily competitors to primarily predators of smaller species (Stenhouse 1985b; Cortwright 1988).

Some salamander larvae exhibit differences in use of microhabitats in the same ponds. For example, larvae of the European newt *Triturus vulgaris* spend most of their time on the bottom of ponds, hunting for benthic invertebrates, whereas the larvae of *T. cristatus* spend more time feeding on plankton in the water column (Dolmen 1983a). This difference in microhabitat use is the reverse of that seen in adults, so direct competition between larvae and adults of the same species may be relatively uncommon. In ponds in New Jersey, larvae of *A. maculatum* spent most of their time on the bottom in shallow water, whereas larvae of the *A. jeffersonianum* complex foraged mostly in the water column (Nyman 1991). In some ponds, there is a sequential shift in use of microhabitats that parallels the staggered entry of larvae into the community (Hassinger, Anderson, and Dalrymple 1970). *Ambystoma* larvae typically exhibit a daily cycle of vertical movements in the water column that apparently follow the movements of their prey (J. Anderson and Graham 1967; Hassinger and Anderson 1970; Hassinger, Anderson, and Dalrymple 1970; J. Anderson and Williamson 1974; Petranka and Petranka 1980; Branch and Altig 1981; Holomuzki and Collins 1983). Different species may differ slightly in their patterns of vertical migration, but there is little evidence that species are clearly segregated into distinct strata in the pond. Again, it is not clear whether interactions between the species are responsible for these shifts in microhabitats, or whether they simply reflect the availability of prey of a suitable size for each species.

Habitat segregation at larger spatial scales, such as the use of different types of ponds, results from choices of oviposition sites by adults. Nevertheless, habitat segregation could be an indirect result of interactions among larvae if one species tends to exclude another through competition or predation and adults tend to avoid laying eggs in habitats likely to contain competitors or predators (see chapter 14). Alternatively, such differences in habitat use could result from independent responses of different species to factors such as pond permanence or the presence of predators such as fish. A well-studied example is the microgeographic distribution of *Ambystoma maculatum* and *A. talpoideum* in ponds in the coastal plain of South Carolina. *Ambystoma maculatum* is relatively uncommon in this region, whereas *A. talpoideum* is abundant and widespread. The former is almost entirely restricted to permanent ponds or stream pools that contain predatory fish, whereas the latter is found almost exclusively in fish-free temporary ponds. Although larvae of both species are vulnerable to predation by fish, the larvae of *A. maculatum* have higher survivorship in the presence of fish than do larvae of *A. talpoideum*, provided that spatial refuges are available (Semlitsch 1988). When larvae of the two species were reared together in artificial ponds, there was little evidence that competition affected survivorship. There also was little evidence that *A. maculatum* was incapable of reaching metamorphosis in temporary ponds. Semlitsch (1988) suggested that *A. maculatum* are excluded from temporary ponds because the relatively dry habitats surrounding such ponds are not very suitable for adults. The situation is different in Mississippi and Louisiana, where the two species occur together, usually in ponds without fish. The larvae of *A. talpoideum* are more aggressive than are those of *A. maculatum*, and tend to be competitively superior when the two species are raised together, although the difference is not sufficient to enable *A. talpoideum* to exclude *A. maculatum* from ponds (Semlitsch and Walls 1993).

## Experimental Studies of Competition and Predation

### Pond-Dwelling Salamander Larvae

In addition to the work on *Ambystoma maculatum* and *A. talpoideum* described earlier, there have been a number of other experimental studies of competition and predation in pond-dwelling salamanders (Hairston 1987, 1996). The first detailed work was done by Wilbur (1971, 1972), who studied a community in Michigan that included *Ambystoma tigrinum*, *A. maculatum*, *A. laterale*, and triploid all-female populations of the *A. jeffersonianum* complex. Adult and larval *Notophthalmus viridescens* also were found in natural ponds, as were *Rana sylvatica* tadpoles and a variety of invertebrates

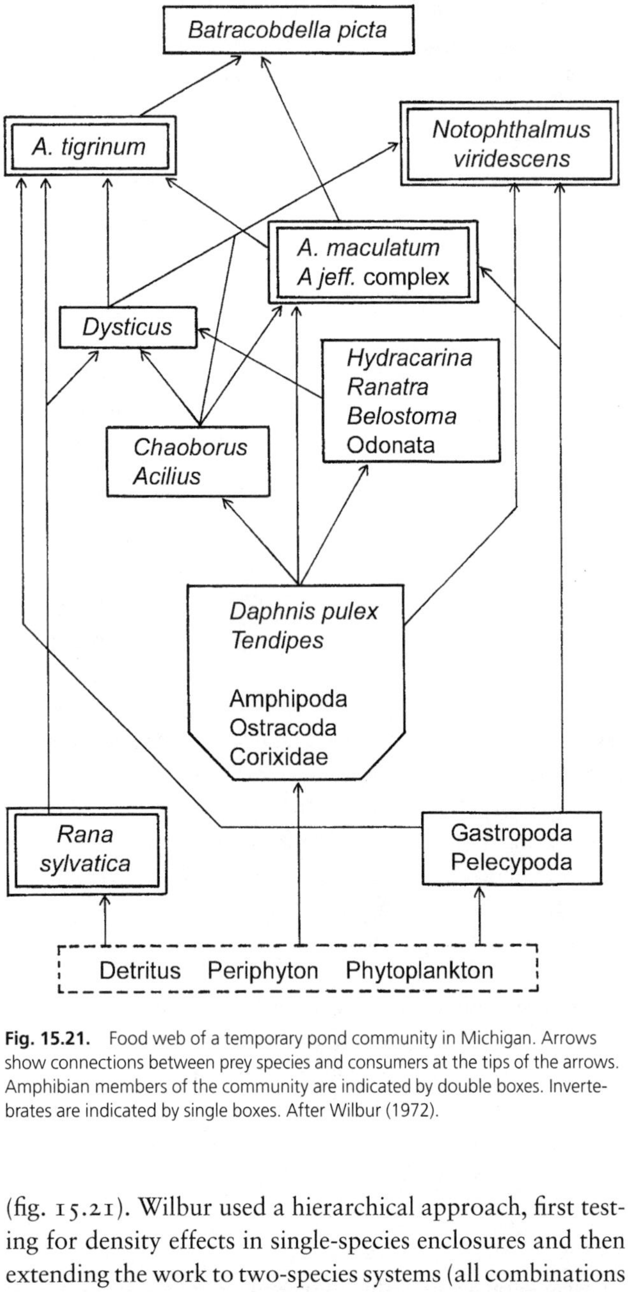


**Fig. 15.21.** Food web of a temporary pond community in Michigan. Arrows show connections between prey species and consumers at the tips of the arrows. Amphibian members of the community are indicated by double boxes. Invertebrates are indicated by single boxes. After Wilbur (1972).

(fig. 15.21). Wilbur used a hierarchical approach, first testing for density effects in single-species enclosures and then extending the work to two-species systems (all combinations of *A. laterale, A. maculatum,* and the triploids) and three-species systems (all three together in different density combinations). The results were variable, with many significant higher-order interactions between treatment effects, some of which are hard to interpret biologically.

Larvae in single-species tests exhibited density-dependent changes in growth and survivorship, but did not respond in exactly the same way. Survivorship of *A. laterale* actually was higher at high than at low densities, but the survivors metamorphosed at a smaller size. *Ambystoma maculatum,* on the other hand, showed greatly reduced survivorship at high densities, but the survivors transformed at a larger size than those raised at low densities, presumably because of reduced competition for food. Density had little effect on survivorship or size at metamorphosis in the triploids, but those at high density had a longer larval period. There also were consistent differences among species in length of the larval period and size at metamorphosis, even when they were raised at the same density. Hence, there appear to be inherent differences between the species that could affect the outcome of competitive interactions among them.

When raised in two- and three-species enclosures, all species showed density-dependent changes in growth and survivorship, and all species were affected by competition from all other species. In some treatments, the differences between species seen in the single-species experiments were maintained or even intensified in the competitive situation. For example, in both two- and three-species systems, there were significant differences among species in the length of the larval period and size at metamorphosis that were consistent with those observed in the single-species treatments. However, interactions in the multispecies systems were not simple functions of adding more larvae to the community; the identity of the species added, their relative proportions in the community, and their absolute abundance all influenced the outcome of competitive interactions. In a natural community, these variables would be governed by a variety of stochastic processes, such as the timing of breeding by each species and the number of adults entering the pond to breed. Consequently, the relative success of each species might vary considerably from year to year and from pond to pond.

The presence of predators in the community can further complicate the picture. Wilbur (1972) compared the performance of the three types of salamander larvae in enclosures with predatory *A. tigrinum* larvae and those without predators. *Rana sylvatica* tadpoles also were added to some enclosures. These served as alternative prey for the *A. tigrinum* larvae, but were seldom eaten by the other species. Again, the results were rather complex. All species had lower survivorship when predators were present. The presence of *R. sylvatica* tadpoles in addition to predators further decreased larval survivorship. This seems paradoxical, since the tadpoles were alternative prey for *A. tigrinum*. However, small *A. tigrinum* larvae cannot prey on the larvae of the other species because they are similar in size. When tadpoles were absent, the *A. tigrinum* larvae evidently preyed mainly on the same small invertebrates eaten by the other salamanders (fig. 15.21). Under these conditions, they had low survivorship, long larval periods, and a small size at metamorphosis, probably because competition limited their food supply. However, small *A. tigrinum* fed preferentially on tadpoles when they were available; this allowed some individuals to gain a rapid size advantage and reach a size that enabled them to prey on the other salamander species. On the other hand, the tadpoles probably reduced the food supply for the invertebrate prey of the other salamander species. Predation by *A. tigrinum* greatly reduced survivorship of the other salamander species. Those that did survive were better off because of reduced competition for their invertebrate prey, and because the *A. tigrinum* larvae also reduced the tadpoles that compete with the invertebrates for food.

Subsequent studies of larval salamander communities have revealed similar complex interactions between competition and predation. In some ponds, fall-breeding *A. opacum* take the place of *A. tigrinum* larvae as top predators. Stenhouse, Hairston, and Cobey (1983) found that the impact of *A. opacum* on *A. maculatum* depended on both the size and density of the predators. *Ambystoma maculatum* larvae were quickly eliminated from laboratory tanks when large *A. opacum* or high densities of predators were present; they were eaten at much slower rates when predators were smaller or present at lower densities. Although there was evidence from field experiments of interspecific competition if *A. opacum* was prevented from reaching a large size, their role as predators appeared to be much more important in determining the success of *A. maculatum*. Because the size and density of *A. opacum* varies from one pond to another and from year to year, the growth and survival of *A. maculatum* varies as well (Stenhouse 1985b). Other experiments in artificial ponds showed that *A. opacum* affects the survivorship of *A. talpoideum* larvae primarily through predation, with early-hatching *A. opacum* larvae having a larger effect than late hatching larvae (Boone, Scott, and Niewiarowski 2002).

In another set of experiments with artificial ponds (cattle tanks), Walls and Williams (2001) found that *A. maculatum* larvae did poorly in the presence of either *A. opacum* or *A. jeffersonianum* larvae, and they did especially poorly when both species were present. The negative impact of these species was due both to direct predation and reduced foraging activity of *A. maculatum* larvae in the presence of the other species. The worst performance was in ponds with both species of predatory salamander larvae and wood frog (*Rana sylvatica*) tadpoles. Although the tadpoles are potential alternative prey for the larger salamander larvae, they also removed most of the filamentous algae from the tanks. This might have removed a source of food for herbivorous invertebrates that serve as prey for the salamander larvae, but it also is possible that tadpoles removed a source of cover for *A. maculatum* larvae. As in the experiments by Stenhouse (1985b), these experiments suggested that *A. maculatum* probably persists in ponds in which the other two salamander species breed because larvae do well in years in which the other two species do poorly. Neither of these studies directly addressed the question of what governs year-to-year differences in larval performance by the predatory species.

The previous experiments ignored the effects of salamanders other than *Ambystoma*, but both larval and adult newts also can be important components of aquatic salamander communities (fig. 15.21). Larval *Notophthalmus* and *Ambystoma* have similar diets and therefore are potential competitors (Brophy 1980). In addition, adult newts prey on *Ambystoma* eggs, while *Ambystoma* larvae eat newt eggs and larvae. Morin (1983b) enclosed *Ambystoma* egg masses in predator-exclusion cages while leaving others unprotected. Survivorship of unprotected eggs was only about 4%, whereas 98% of the caged eggs survived. Hence, newts are capable of nearly eliminating *Ambystoma* from some ponds. In fact, natural ponds with high densities of newts generally had very few *Ambystoma* eggs, whereas ponds with small newt populations had many eggs. The presence of adult newts also reduced the growth rates and size at metamorphosis of *Ambystoma* larvae, indicating an effect of interspecific competition. On the other hand, *Ambystoma* larvae were predators on newt larvae. In the presence of *Ambystoma*, survivorship of newt larvae was reduced, but size at metamorphosis was greater, presumably because of reduced intraspecific competition.

In some ponds in the southeastern United States, newts interact with sirens in both the adult (Fauth and Resetarits 1991) and larval stages. Fauth, Resetarits, and Wilbur (1990) reared larval newts (*Notophthalmus viridescens*) and larval lesser sirens (*Siren intermedia*) in cattle tanks. Each species was raised alone at low (three larvae per tank) or high density (six larvae per tank), or together with three of each species. Both densities were low compared to those in many natural ponds, however (Harris, Alford, and Wilbur 1988). Although these two species are quite different morphologically, they eat similar prey. They were relatively equal competitors in the conditions of the experiment, but each species affected the other in slightly different ways. The presence of siren larvae decreased the survival of newt larvae, but not their growth or the length of the larval period. Newt larvae, on the other hand, decreased growth rates of siren larvae, but not their survival. The presence of three newt larvae in a tank had the same effect on growth rates of larval sirens as did the presence of three additional siren larvae, indicating that the effect is mainly a response to density and not to a specific competitor.

### Stream-Dwelling Larvae

Interactions of stream-dwelling salamander larvae have received less attention than those of pond-dwelling species. One difference between pond and stream larvae is that the latter seldom have either anuran tadpoles or small zooplankton available as food. In the southern Appalachians, stream communities are composed entirely of plethodontid salamanders, including species of *Desmognathus, Eurycea, Gyrinophilus,* and *Pseudotriton.* The identity and relative abundance of different species varies geographically. The structure of the stream community also is related in part to the presence or absence of fish such as brook trout (*Salvelinus fontinalis*), which are potential predators on small salamander larvae and potential competitors with larger larvae and aquatic adults (Resetarits 1991, 1995, 1997). All of the stream-dwelling salamanders tend to be generalist predators on small invertebrates, but the larvae of some species, including *Gyrinophilus porphyriticus, Pseudotriton ruber, Desmognathus quadramaculatus,* and *Desmognathus marmoratus,* reach very large sizes and are capable of consuming smaller salamander larvae. As in communities of pond-dwelling larvae, this means that species may interact as competitors, predators, or both.

Several experimental studies have shown that large species can have a major impact on smaller species as predators, but evidence for strong interspecific competition is mixed. For example, Beachy (1993b 1994) found little evidence for either intra- or interspecific competition among salamander larvae raised at relatively high densities in cages placed in a stream in North Carolina. However, the larvae of both *Desmognathus quadramaculatus* and *Gyrinophilus porphyriticus* consumed larvae of *Eurycea wilderae.*

Resetarits (1991) constructed artificial streams in Virginia and stocked them with two large predators, larvae of the spring salamander (*Gyrinophilus porphyriticus*) and brook trout (*Salvelinus fontinalis*), and two smaller predators, larvae of the two-lined salamander (*Eurycea cirrigera*) and a crayfish (*Cambarus bartonii*). In this experimental system, brook trout had a large effect on both growth and survival of *Gyrinophilus,* but the salamander larvae had little effect on the trout. The precise mechanism leading to reduced growth and survival of the salamander larvae was not clear, but aggressive interference competition and attempted predation are the most likely explanations (the *Gyrinophilus* larvae were mostly too large to be consumed by the trout). Neither trout nor *Gyrinophilus* larvae had a significant effect on the survival of the smaller *Eurycea* larvae, but both predators did reduce their growth rates, especially when both were present together. This effect was largely due to reduced feeding activity in the presence of the predators.

Using the same array of artificial streams, Gustafson (1993, 1994) investigated ecological interactions among larvae of *Gyrinophilus porphyriticus, Pseudotriton ruber,* and *Eurycea cirrigera* in the absence of fish. The first two species are much larger than *Eurycea* and therefore are potential predators on that species and on each other. In fact, Gustafson (1993) found evidence for strongly negative effects of large *Gyrinophilus* and *Pseudotriton* larvae on the survival of smaller *Pseudotriton* larvae and a weaker effect on growth rates of *Eurycea* larvae. As in the experiments by Resetarits, the effect on *Eurycea* growth was largely due to reduced activity in the presence of large predators. A subsequent experiment showed that *Gyrinophilus* also can affect the survival of *Eurycea* larvae, and the effect is greatest when the *Gyrinophilus* larvae are large (Gustafson 1994). This indicates that *Gyrinophilus* larvae become more serious predators on *Eurycea* larvae as they grow. Gustafson's (1993) experiments also showed some reduction in growth rates of large *Pseudotriton* larvae in the presence of *Gyrinophilus,* possibly because of competition. Nevertheless, all of these experiments together provide relatively weak evidence for competition among stream-dwelling salamander larvae, and strong evidence for size-related interspecific predation as a major determinant of community structure.

## Ecology of Anuran Tadpole Communities

Assemblages of anuran tadpoles often are much more diverse than assemblages of salamander larvae. This is especially true in the tropics, where many species of anurans use the same breeding sites, and a dozen or more species of tadpoles can be found at the same site (Heyer 1973; Duellman 1978; Blommers-Schlösser and Blommers 1984; Schlüter 1984; Inger, Voris, and Frogner 1986; Aichinger 1987a; Gascon 1991; Magnusson and Hero 1991; Azevedo-Ramos, Magnusson, and Bayliss 1999). Even in the temperate zone, tadpole communities can be very diverse, with more than a dozen species of tadpoles found in some ponds in the Coastal Plain of the southeastern United States (Wilbur 1997; Resetarits and Fauth 1998; Alford 1999; Snodgrass, Bryan, and Burger 2000). Adult anurans using the same breeding sites often have different temporal patterns of reproduction. Consequently, most sites exhibit a staggered introduction of larvae into the community, and the number of species present varies over time (Blair 1961; Berry 1964; Dixon and Heyer 1968; Heyer 1973, 1976, 1979a; Wiest 1982; Gascon 1991; Semlitsch et al. 1996; Azevedo-Ramos, Magnusson, and Bayliss 1999). This temporal separation has been interpreted as one dimension of resource partitioning among tadpole species. As in the case of salamanders, however, there is little evidence that differences in timing of breeding have evolved in response to interactions among species. It is more likely that each species responds independently to environmental cues, and the resulting pattern of breeding activity then sets the stage for interactions among different tadpole species.

Variation in temporal patterns of reproduction results in different species of tadpoles experiencing different environmental conditions, especially in temporary ponds. Physical factors such as water depth, temperature, and oxygen content often vary over time, as do biological factors such as nutrient availability, fouling of water by waste products from tadpoles and other organisms, and densities of predators and competitors. For species with prolonged breeding seasons, tadpoles may be continuously or episodically added to the community, and different age cohorts of the same species may experience very different conditions (Alford 1999). Furthermore, the nature of interactions with competitors or predators often depends on tadpole body size, so different age cohorts may interact with each other and with other species in different ways. In some cases, this can result in animals switching from being competitors to predators of other species, or even members of their own species, as they grow (Crump 1983; Pfennig 1990a, 1992a, b). All of this variability makes accurate characterization of community structure especially difficult.

## Niche Partitioning and Microhabitat Segregation

### Interspecific Differences in Diet

There have been relatively few studies of dietary differences among tadpoles living in the same habitat (Diaz-Paniagua 1985, 1989a; Inger 1986; Pavignano 1990; Sekar 1992; Alford 1999). This is due in part to the difficulty of determining exactly what these animals are eating. In contrast to salamander larvae, tadpoles are not gape-limited predators, but suspension feeders on algae, phytoplankton, and other particulate matter (see chapter 12). This makes gut contents difficult to identify, especially for those who are not experts on the identification of algae. Furthermore, tadpole body size is not necessarily related to the type or size of food particles ingested (Seale 1980), but it may be related to the rate at which food can be processed (chapter 12). Tadpoles may differ in their ability to capture food particles at various concentrations (Seale and Beckvar 1980; Viertel 1990) and to process different types or sizes of food particles (Altig and McDearman 1975; Altig et al. 1975; Viertel 1992). They also may vary in the efficiency with which food is digested (C. G. Peterson and Boulton 1999) and in the types of nutrients extracted from the food (Steinwascher and Travis 1983).

Tadpoles in a natural environment may be exposed to foods of different quality, and this in turn can impact their growth and development. For example, tadpoles of *Pseudacris regilla* grew faster when fed algae that harbored many epiphytic diatoms than on algae that lacked these organisms (Kupferberg, Marks, and Power 1994). Tadpoles often have been characterized as unselective grazers, but some species are capable of evaluating differences in the quality of food in different habitat patches and choose to feed in higher quality patches (Kupferberg 1997a, 1998). Tadpoles sometimes may be forced by predators to feed in low-quality patches and may grow more slowly than those that have access to high-quality food (Kupferberg 1998). There is no information, however, on how competitive interactions with other species of tadpoles directly affect diet selection in tadpoles. There is evidence from laboratory studies that some species of tadpoles can reduce growth in other species through aggressive interference competition (Faragher and Jaeger 1998), but this effect probably is due to reduced feeding activity rather than changes in diet composition.

### Microhabitat Segregation

In complex assemblages of anuran tadpoles, different species often use slightly different microhabitats, and this can lead to differences in the types of food consumed. Species often differ in their choice of water depth, substrates, or types of vegetation used for shelter and feeding (Inger 1969; Heyer 1973, 1974, 1976, 1979a; Heyer, McDiarmid, and Weigmann 1975; Degani 1986; Inger, Voris, and Walker 1986; Löschenkohl 1986; Diaz-Paniagua 1987; Waringer-Löschenkohl 1988; Zhao et al. 1994; Barreto and Moreira 1996; Wild 1996; Alford 1999). These differences typically are correlated with morphological differences in body form and mouthparts (see chapter 12 for a discussion of the ecomorphology of tadpoles). Although the basic mechanism of suspension feeding is similar for all species, they often differ in the way that food is initially acquired. Some species utilize phytoplankton suspended in the water column, while others feed on particles floating on the water surface. Many species use specialized mouthparts to ingest detritus from the bottom of a pond or scrape algae off of rocks or submerged vegetation. Hence one can find tadpoles feeding in a variety of ways simultaneously in the same pond. Nevertheless, even morphologically distinct tadpoles in different microhabitats in streams often exhibit considerable overlap in diet (Inger 1986; Löschenkohl 1986; Alford 1999). This observation has led some authors to argue that interspecific competition for food is of minor importance in tadpole communities, perhaps because food is always abundant (Heyer 1973, 1974, 1976, 1979a; Heyer, McDiarmid, and Weigmann 1975; Inger 1986; Zhao et al. 1994). There is, however, evidence that tadpoles can substantially alter the abundance and species composition of algae and other vegetation in ponds and streams, suggesting that dense populations of tadpoles are food limited (Dickman 1968; Seale, Rodgers, and Boraas 1975; Seale 1980; Morin and Johnson 1988; Brönmark, Rundle, and Erlandsson 1991; Leibold and Wilbur 1992; Werner 1994; Morin 1995; Kupferberg 1997a; see also chapter 13).

Many of the microhabitat differences seen in assemblages of pond- or stream-dwelling tadpoles probably result from phylogenetic differences in morphology and behavior, not from competitive interactions among species. In fact, there is surprisingly little evidence that the presence of one species of tadpole can cause another species to alter its use of microhabitats. Two examples that do suggest a role for interspecific competition come from Australia. Two species of pond-breeding frogs, *Pseudophryne bibroni* and *P. semimarmorata,* are largely allopatric, but have a narrow zone of overlap in southeastern Australia. *Pseudophryne bibroni* tadpoles consistently out-competed those of *P. semimarmorata* when the two species were reared together, suggesting that competition at the larval stage is one factor preventing greater overlap in these species' ranges (Wiltshire and Bull 1977). A stronger example involves habitat segregation between *Crinia signifera* and *C. riparia.* The former is widely distributed throughout southeastern Australia, while the latter is restricted to streams in the Flinders Range, where it overlaps slightly with *C. signifera* (Odendaal and Bull 1982). *Crinia signifera* tadpoles are typical pond-type tadpoles and have difficulty holding their position in flowing water (Odendaal and Bull 1980). When placed in a rocky stream with tadpoles of *C. riparia,* they tend to stay between the rocks and therefore are excluded from the rock surfaces, where most of the food is located (Odendaal, Bull, and Nias 1982). Although they can survive in fast-moving water, their survival is significantly lower than that of *C. riparia* when the two are housed together in flowing water (Odendaal and Bull 1983). This does not seem to be simply the result of exploitative competition for food, but may result from the more active *C. riparia* tadpoles displacing *C. signifera* from sheltered sites, exposing them to the current and reducing their ability to remain in one place to feed (Odendaal, Bull, and Richards 1984).

## Experimental Methods for Investigating Community Structure and Dynamics

Most studies of tadpole community ecology have been experimental studies in which different combinations of competitors and predators have been raised in artificial ponds and the performance of the various species has been monitored. Most commonly, large cattle watering tanks have been used to simulate conditions in temporary ponds (Wilbur 1989, 1997; Morin 1998; Resetarits and Fauth 1998). These have the advantage of allowing investigators to control tadpole and predator densities and food availability, and treatments can be easily replicated. There are certain features of cattle tanks that do not precisely mimic natural temporary ponds, however. One difference is that cattle tanks typically have straight sides, which removes shallow water habitats that may be important for some species. Some investigators have placed tanks on a slope to simulate a sloping shoreline (Griffiths, Denton, and Wong 1993). When water levels are reduced to simulate pond drying in a straight-sided tank, the surface area of the water remains the same, something that would not be true in natural ponds. The regular size and shape of the tanks also reduces the spatial complexity and diversity of microhabitats typically found in natural ponds.

Investigators have used several other types of containers to perform replicated experiments on competition and predation. These include fiberglass tanks (Semlitsch, Hotz, and Guex 1997), small plastic tubs (Gascon 1992a; G. R. Smith 1999), children's wading pools (Fauth 1999b), plastic barrels cut in half (Lardner 1995), cages placed in natural ponds (Brockelman 1969; Werner 1994; Skelly 1995a, b; Bardsley and Beebee 2000) or streams (Kupferberg 1997a), shallow ponds with plastic liners (Griffiths 1991), large man-made ponds with natural vegetation (Werner and McPeek 1994), natural ponds of similar size (Bardsley and Beebee 1998b), and shallow natural rock pools (D. C. Smith 1983; Laurila 2000b). The various types of experimental systems provide different degrees of realism in replicating conditions in natural ponds and the spatial scale at which interspecific interactions can be examined. Differences in spatial scale can be important in interpreting the results of competition experiments. For example, some studies done in small laboratory containers have yielded very different results from studies conducted on the same species in larger, more natural environments (Werner 1992, 1998; Werner and Glennemeir 1999; Laurila 2000b). The intensity of competition also can differ between artificial ponds such as cattle tanks and natural ponds, with the latter providing weaker evidence of competition (Skelly and Kiesecker 2001; Skelly 2002).

## Tadpole Competition in the Absence of Predators

A number of studies have examined competitive interactions among species of tadpoles in the absence of predators, both in the laboratory (Wilbur 1977b, 1982; Smith-Gill and Gill 1978; Travis 1980b; Alford, 1989b; Werner 1992; Gollmann and Gollmann 1993) and in various kinds of artificial ponds. Most of these studies have revealed competitive effects among species, but the intensity of competition varies considerably.

### Weak Interspecific Competition

Some experiments have produced relatively weak evidence for interspecific competition. For example, in an early study by Wilbur (1977a), density of conspecific tadpoles had a strong negative effect on size at metamorphosis in *Bufo americanus.* When these tadpoles were reared with different densities of conspecifics or tadpoles of pickerel frogs (*Rana palustris*) in pond enclosures, there was not a major difference in the effect of conspecifics and heterospecifics (data from Wilbur 1977, reanalyzed by Alford 1999). A subsequent study by Alford (1989a) in cattle tanks was designed to examine competition between these two species in more detail. Again, competitive effects were relatively weak, with each species having little effect on the performance of the other, regardless of when the two species were introduced into the pond. These results contrasted with earlier experiments with *Bufo americanus* and *Rana sphenocephala* in identical ponds. Those experiments showed strong competition between the two species (Alford and Wilbur 1985). The reasons for the different results with the two species of *Rana* are not clear. Alford (1989a) suggested that *R. sphenocephala* is less likely to share ponds with *B. americanus* than is *R. palustris,* so there might have been weaker selection to evolve differences in microhabitat use or type of food consumed. In the absence of detailed information on resource use, however, there is no direct evidence to support this hypothesis.

Skelly (1995b) examined competitive interactions between tadpoles of spring peepers (*Pseudacris crucifer*) and green frogs (*Rana clamitans*) in pond enclosures stocked at different densities. Green frogs breed mostly in relatively permanent ponds, whereas spring peepers tend to favor more temporary sites. Where the species overlap, green frog tadpoles remain in ponds for one or more winters, so large tadpoles of this species often are present when spring peeper tadpoles hatch. Green frog tadpoles had a relatively small effect on size of spring peepers at metamorphosis, and they did not affect survivorship, growth rate, or length of the larval period. In contrast, green frog tadpoles alone exhibited strong density dependence, with substantial reductions in growth rate and size at metamorphosis at high densities. Skelly concluded that interspecific competition among tadpoles was not sufficient to exclude spring peepers from ponds with large green frog populations. It may be that these species use very different food resources, so competition is weak, despite the much larger size of green frog tadpoles. Skelly (1995a) also found little evidence for interspecific competition between *Pseudacris crucifer* and a much more closely related species, *P. triseriata,* reared in pond enclosures. These results are consistent with those of D. C. Smith and Van Buskirk (1995), who found little evidence of interspecific competition between the same two species in natural rock pools, even though *P. triseriata* exhibits strong intraspecific competition at high densities (D. C. Smith 1983).

Bardsley and Beebee (1998b) did not find evidence of strong interspecific competition in experiments with *Rana temporaria* and *Bufo bufo* tadpoles in natural sand dune ponds. Unfortunately, their experimental design confounded the effects of competition and predation, because ponds used to examine mixed-species interactions contained predatory fish, whereas those used to examine intraspecific competition in single-species treatments contained only invertebrate predators. While survival of *R. temporaria* tadpoles was higher in single-species ponds than in mixed-species ponds, Bardsley and Beebee (1998b) attributed this to differences in predation by fish, not to competition with *Bufo* tadpoles.

### Strong Asymmetric Competition

The results of these experiments contrast with those of many investigators who have reared tadpoles in cattle tanks and other types of artificial ponds, and with some studies in natural ponds. For example, Morin and Johnson (1988) raised spring peeper (*Pseudacris crucifer*) tadpoles with wood frog (*Rana sylvatica*) tadpoles at different densities in artificial ponds. These two species often begin breeding at the same time in early spring. Wood frogs are explosive breeders, with most eggs being laid on just a few nights, whereas spring peepers breed for up to two months. Wood frog tadpoles were clearly competitively superior to spring peeper tadpoles in these experiments. The effect of wood frog tadpoles on mass at metamorphosis for spring peepers was about twice as great as the effect of spring peeper tadpoles on wood frogs. The effect of wood frog tadpoles on members of their own species also was much greater than the effect of spring peeper tadpoles on conspecifics. The competitive superiority of wood frog tadpoles was related to their ability to deplete standing crops of periphyton, the principal food of both species. Strong asymmetric competition also was observed in experiments with wood frog tadpoles and leopard frog (*Rana pipiens*) tadpoles reared in pond enclosures, with wood frog tadpoles again being the superior competitor (Werner and Glennemeier 1999). This result contrasted with an earlier study of these species in the laboratory, which suggested that leopard frog tadpoles were the superior competitor (Werner 1992).

The strong asymmetric competition between wood frog and other tadpoles is typical of the results of many experimental studies of tadpole competition (Alford 1999). Often one species has a very strong effect on the performance of a second species, while the second species has a relatively weak effect on the first. For example, Woodward (1982d) found that tadpoles of *Scaphiopus couchii,* a species that breeds in temporary desert rain pools, was competitively superior to tadpoles of *Bufo woodhousii,* which usually breeds in permanent ponds, and *Rana pipiens,* which always does so. His experiments were conducted in very small (1 $m^2$), shallow (0.4 m) excavated ponds lined with plastic, which would tend to favor temporary pond species. He concluded that both species might be excluded from temporary ponds by competition from *Scaphiopus,* although *R. pipiens* probably would be excluded from such habitats because its larval period is longer than the time such ponds usually hold water. The competitive superiority of *Scaphiopus* tadpoles probably is related to their high levels of foraging activity, which may enable them to process food more efficiently than do other tadpoles. Other experiments have consistently shown that tadpoles with high levels of activity, including other species of *Scaphiopus,* tend to outcompete less active species (Morin 1983a; Wilbur 1987; Dayton and Fitzgerald 2001).

Another example of strongly asymmetric competition comes from studies of natterjack toads (*Bufo calamita*) in Britain. This is a pioneering species that favors temporary ponds in sand dune habitats, especially those that are not used by other species of amphibians. Experiments have shown that tadpoles of this species do poorly in competition with the two most common anurans in the region, *Bufo bufo* and *Rana temporaria,* both of which also breed in temporary ponds, but have more generalized habitat requirements. In addition, *B. calamita* is susceptible to egg predation by these two species. The result is that habitats where the other two species are abundant often are not suitable for natterjack toads (Banks and Beebee 1987a, b; Griffiths 1991; Griffiths, Edgar, and Wong 1991). Populations of natterjack toads are disappearing from dune habitats that have undergone succession, which makes them more suitable for the other species, especially *Bufo bufo* (Bardsley and Beebee 1998a).

Asymmetric competition has been reported in several combinations of ranid tadpoles in both temporary and permanent ponds. Tadpoles of the crawfish frog (*Rana areolata circulosa*) had reduced growth, longer larval periods, and smaller size at metamorphosis when they were raised in artificial ponds with tadpoles of either *R. sphenocephala* or *R. blairi,* even at low densities (Parris and Semlitsch 1998). The effect of *R. blairi* on *R. aerolata* was stronger than that of *R. sphenocephala.* The crawfish frog has a patchy distribution in tallgrass prairie habitats and could be susceptible to local extinction, where the other two species are common. In Europe, there is some evidence that tadpoles of *R. temporaria,* a temporary pond breeder, are competitively superior to those of another species that uses the same types of habitats, *R. arvalis,* but in this case, the effect was relatively weak (Lardner 1995).

Asymmetric competition also occurs between tadpoles of the *Rana esculenta* complex in Europe. *Rana esculenta* is a hybridogenetic species originally produced by hybridization between *R. lessonae,* which tends to inhabit temporary and semipermanent ponds, and *R. ridibunda,* which is found almost exclusively in permanent ponds and lakes. The hybridogenetic species is perpetuated by coexistence and interbreeding with one of the parental species, usually *R. lessonae* (Graf and Polls Pelaz 1989). In populations where *R. esculenta* and *R. lessonae* coexist, the former acts as a sexual parasite on the parental species. *Rana esculenta* contains one haploid genome from each parental species, but the *R. lessonae* genome is discarded in each generation, while the *R. ridibunda* genome is clonally inherited. A population may contain several different hemiclones derived from repeated immigration from populations where hybridization between the two parental species has occurred many times.

*Rana esculenta* often is much more common in natural ponds than is the parental species with which it is associated, especially in disturbed habitats such as gravel pits (Blankenhorn, Heusser, and Notter 1973). One possible reason for this is that *R. esculenta* is competitively superior to the parental species in the tadpole stage, adult stage, or both (Tunner and Nopp 1979). In competition experiments in artificial ponds, *R. esculenta* tadpoles had higher growth rates and tended to emerge earlier from ponds than did *R. lessonae* tadpoles, and the proportion of tadpoles reaching metamorphosis after 60 days was higher for *R. esculenta* as well (Semlitsch 1993). The earlier emergence of *R. esculenta* tadpoles is particularly advantageous in ponds that dry up during the summer (Semlitsch and Reyer 1992b). *Rana esculenta* tadpoles also performed slightly better than did *R. lessonae* tadpoles when competing species of tadpoles (*Hyla arborea* and *Bufo calamita*) were present. The competitive superiority of *R. esculenta* tadpoles derives from their higher level of feeding activity and higher food intake than that of either parental species (Rist et al. 1997).

Competition experiments with three hemiclones from a pond in Switzerland and the two parental species showed that the hemiclones exhibited significant phenotypic variation and performed differently at different densities and combinations of competitors (Semlitsch, Hotz, and Guex 1997). One hemiclone outperformed the other two under most of the conditions of the experiment; this was the most common one in the natural breeding pond, comprising nearly 70% of all tadpoles. The overall competitive superiority of *R. esculenta* over its parental species may derive in part from the diversity of hemiclones inhabiting a pond, each of which may respond slightly differently to environmental variations that occur in all ponds. The three genotypes of water frogs differ physiologically as well, especially in their ability to tolerate low oxygen levels, but experiments to determine whether the hybrid genotype is superior to both parental genotypes in this regard have yielded inconsistent results (Tunner and Nopp 1979; Plénet et al. 2000). Competitive interactions between *R. esculenta* and *R. lessonae* tadpoles also are mediated by different physiological responses to temperature, with *R. lessonae* tadpoles being more tolerant of and performing better at warm temperatures than *R. esculenta* tadpoles (Negovetic et al. 2001).

Bullfrog (*R. catesbeiana*) tadpoles have been shown to be superior competitors to green frog tadpoles (*R. clamitans*) in permanent ponds, where multiple size cohorts of the two species often overlap. One might expect the competitive superiority of bullfrog tadpoles to result from their larger size, but in fact, both large and small bullfrog tadpoles had a negative effect on growth rates and size at metamorphosis of green frog tadpoles (Werner 1994). Small green frog tadpoles had a negative impact on large bullfrog tadpoles, but not on smaller ones. Bullfrog tadpoles tend to be more active foragers than green frog tadpoles, and consequently acquire food at a faster rate (Werner 1991). Werner (1994) argued that the net gain in energy for tadpoles will increase at small body sizes and then begin to decline at larger sizes as the cost of acquiring food approaches and eventually exceeds the gain (fig. 15.22). He also argued that because of their more active foraging, the net gain curve for bullfrogs will be higher than for green frogs at any given body size (fig. 15.23). Even small bullfrog tadpoles may have a net gain that exceeds that of both large and small green frogs, leading to the asymmetric competitive effects observed in the experiments. The interactions of bullfrog and green frog tadpoles in nature are more complicated than indicated by these experiments because competition is mediated by interactions with several kinds of predators (Werner 1991; Werner and Anholt 1993, 1996; Werner and McPeek 1994; Anholt and Werner 1995; see the following discussion).

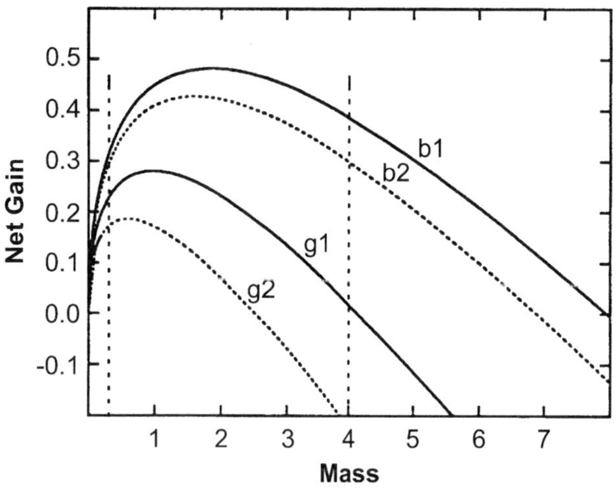


**Fig. 15.23.** Theoretical net gain curves for feeding bullfrog (*Rana catesbeiana*) tadpoles (b1, b2) and green frog (*Rana clamitans*) tadpoles (g1, g2), estimated as in fig. 15.22. At any given size, the more active bullfrog tadpoles are more efficient than green frog tadpoles at collecting food, so the net gain curve for bullfrogs always exceeds that of green frogs. Curves b1 and g1 are for each species in the absence of heterospecific competitors. Curves b2 and g2 are for each species in the presence of competitors. Bullfrog tadpoles have a greater negative effect on energy intake of green frog tadpoles than green frog tadpoles have on bullfrog tadpoles. Bullfrog tadpoles can depress energy intake by green frog tadpoles at both small (left vertical line) and large (right vertical line) body sizes, whereas green frog tadpoles have a significant effect only on larger bullfrog tadpoles. After Werner (1994)

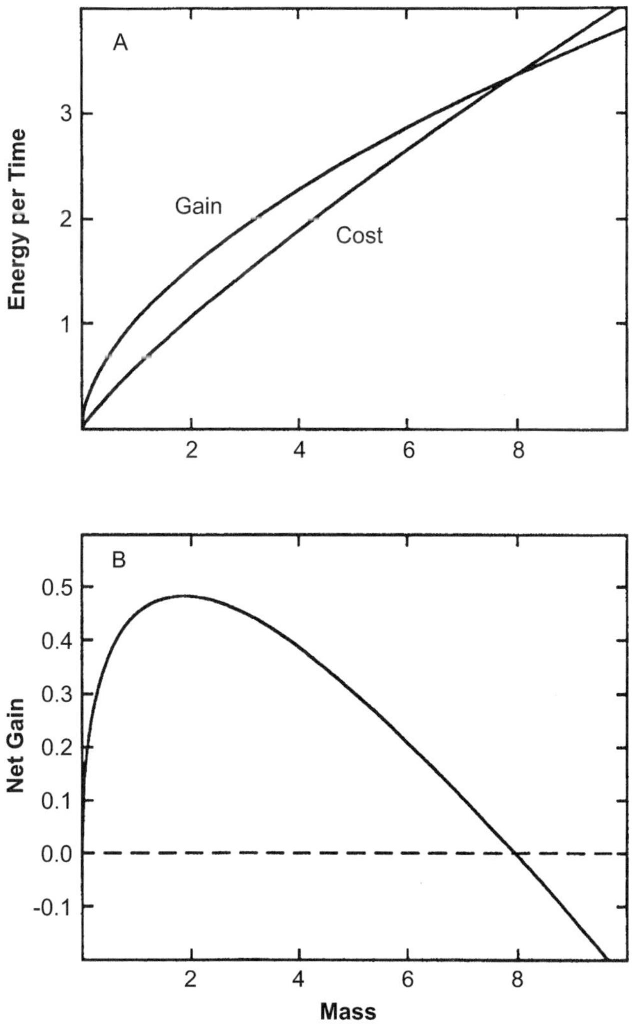


**Fig. 15.22.** (A) Schematic diagram of energetic costs and gains for feeding tadpoles of different body sizes. Gains exceed costs for small-bodied tadpoles, but at larger sizes, costs exceed gains, at which point the tadpole must metamorphose. (B) Net gain calculated by subtracting the cost curve in A from the gain curve. After Werner (1994).

Bullfrog tadpoles can be superior competitors to other species of ranids even when they are introduced into habitats not normally inhabited by bullfrogs. In California, for example, bullfrogs have been introduced into rivers inhabited by tadpoles of a native ranid, *Rana boylii,* and a hylid, *Pseudacris regilla.* In experiments with enclosures in natural streams, both large and small bullfrog tadpoles had a strong negative effect on growth and survivorship of *R. boylii* tadpoles and a significant, but weaker, effect on *P. regilla* (fig. 15.24; Kupferberg 1997a). Bullfrog tadpoles significantly reduced standing crops of algae, suggesting that exploitative competition for food was mainly responsible for the decreased performance of the other species. Bullfrog tadpoles can have other effects as well, such as predation on smaller species of tadpoles. Kiesecker and Blaustein (1997b) reported evidence for predation by introduced bullfrog tadpoles on native tadpoles of red-legged frog (*R. aurora*) in Oregon. It was not clear, however, whether this resulted from deliberate capture of tadpoles or incidental predation when the two species were feeding together. Bullfrog tadpoles also

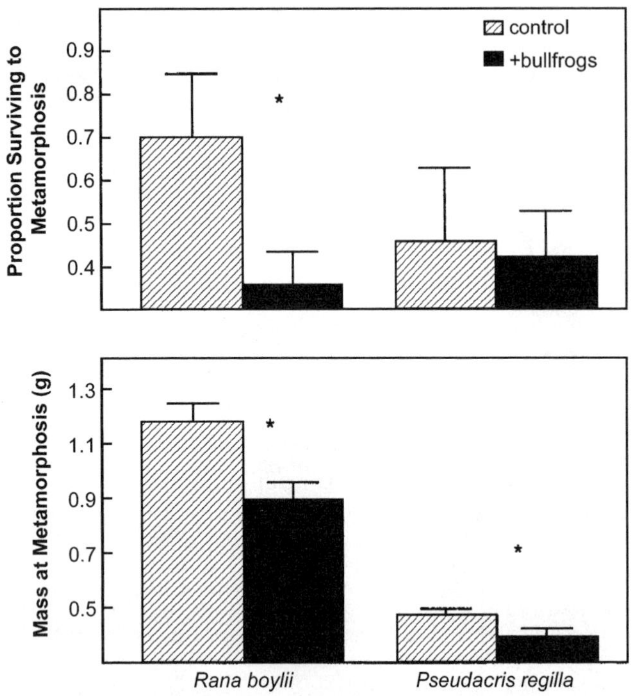


**Fig. 15.24.** Effect of introduced bullfrog (*Rana catesbeiana*) tadpoles on performance of *Rana boylii* and *Pseudacris regilla* tadpoles. Top: Proportion of tadpoles surviving to metamorphosis in stream enclosures with and without bullfrog tadpoles. Bottom: Mass at metamorphosis for tadpoles in enclosures with and without bullfrog tadpoles. * Indicates a significant difference. After Kupferberg (1997b).

reduce the performance of *R. aurora* tadpoles through passive interference competition, especially when food patches are clumped (Kiesecker, Blaustein, and Miller 2001).

### The Effect of Phenology on Interspecific Competition

Although certain species of tadpoles consistently outcompete other species in artificial ponds, the outcome of interspecific competition can be affected by the order in which different species are introduced into the community (Alford 1999). These priority effects are assumed to mimic conditions in natural ponds, in which the order of appearance can vary from year to year because of different responses of breeding adults to environmental cues. Alford and Wilbur (1985) studied priority effects in interactions between *Bufo americanus* and *Rana sphenocephala* tadpoles by varying the presence or absence and time of introduction of each species into cattle tanks. The presence of *Rana* tadpoles had a slightly negative effect on *Bufo,* but the differences were not statistically significant. *Bufo* tadpoles had a strong negative effect on growth and survivorship of *Rana* tadpoles. However, *Rana* tadpoles did better when introduced after *Bufo,* rather than before. A possible explanation for this unexpected result is that *Bufo* tadpoles grow more quickly than *Rana;* when introduced early, they metamorphose early in the season, leaving the *Rana* tadpoles to develop without competition.

Other experiments (Wilbur and Alford 1985) showed that both *Bufo* and *Rana* tadpoles had a negative effect on those of *Hyla chrysoscelis.* The time of arrival of *Rana* tadpoles accounted for more than half the variance in the performance of *Hyla* tadpoles, but the time of arrival of *Bufo* also had a significant effect, even though most of the *Bufo* tadpoles metamorphosed before the *Hyla* were introduced. Evidently, changes in the food supply caused by grazing *Bufo* tadpoles persisted late into the season. In another experiment with *Bufo* and hylid tadpoles, Lawler and Morin (1993) varied the timing of introduction of tadpoles of *Bufo fowleri* and *Pseudacris crucifer.* In natural populations, *Pseudacris* normally breeds earlier than the toads. *Pseudacris* tadpoles were negatively affected by *Bufo* tadpoles, especially when the latter were in the pond at least a week earlier. The negative effect was reduced if *Pseudacris* was in the pond earlier or arrived at the same time as *Bufo.*

Negative effects of one species of tadpole on another sometimes occur even when the species overlap very little in the pond. For example, Morin (1987a) raised spring peeper (*Pseudacris crucifer*) tadpoles in experimental ponds at different densities and then introduced 200 newly hatched gray treefrog (*Hyla versicolor*) tadpoles into each pond shortly before the spring peepers metamorphosed. He found that high densities of spring peeper tadpoles in ponds before the introduction of gray treefrog tadpoles had a negative effect on the growth and survival of the treefrog tadpoles. The two species had little opportunity to interact directly, so this effect must have been due to depletion of food by the spring peeper tadpoles. Morin argued that in some circumstances, an earlier-breeding species might actually exclude a later-breeding species from some ponds. It is not clear how commonly this occurs in nature, however. Many ponds may be so productive that even high densities of tadpoles early in the season do not deplete food resources for later species. For example, large populations of spring peepers and gray treefrogs frequently breed in the same ponds, and may even overlap in breeding seasons (Collins and Wilbur 1979; personal observations).

Very little experimental work has been done with tadpoles in tropical habitats. However, Gascon (1992b) studied interactions among the three most common species of tadpoles in temporary ponds in Amazonian Brazil, including the importance of priority effects. One of the first species to breed is a large hylid, *Osteocephalus taurinus,* which usually lays eggs soon after a pond fills. A second hylid, *Phyllomedusa tomopterna,* lays eggs on vegetation overhanging ponds, and its tadpoles usually appear after *Osteocephalus* is present. Tadpoles of a dendrobatid, *Allobates femoralis,* typically appear later as well, having been carried to water by adults from a terrestrial nest site (Gascon 1992a). In this system, the earliest-arriving species, *Osteocephalus,* did well regardless of when it arrived or what other species were present. *Phyllomedusa,* on the other hand, performed relatively poorly when *Osteocephalus* was already present, while *Allobates* tadpoles always did poorly in the presence of *Osteocephalus.* This study suggests that *Phyllomedusa* may gain some competitive advantage by breeding early, but *Osteocephalus* gains relatively little advantage from its early breeding.

### Mechanisms of Interspecific Competition

The precise mechanisms of competition between species of tadpoles are not fully understood. Most studies have focused on exploitative competition and depletion of food resources by superior competitors; several of the studies already mentioned have found evidence that this occurs. Most of the evidence for exploitative competition is based on the effects of high densities of one species on the performance of another. As discussed in chapter 13, there is considerable evidence that food is often limiting in natural tadpole habitats, with substantial reductions in algal biomass as a result of tadpole grazing (Dickman 1968; Seale, Rodgers, and Boraas 1975; Seale 1980; Lamberti et al. 1992; Kupferberg 1997a; Graham and Vinebrooke 1998; Holomuzki 1998; Mallory and Richardson 2005). In most cases, however, there is little information on the degree to which two or more species in a community actually exploit the same resources.

It also is possible that interference competition occurs in some situations. Faragher and Jaeger (1998) reported that tadpoles of *Rana sphenocephala* inhibited feeding and reduced growth rates of *Hyla cinerea* tadpoles by physically harassing the smaller species in laboratory aquaria. It is not clear from their observations, however, whether such interactions represent aggression by the larger tadpoles or an incidental effect of actively foraging for food in the confines of an aquarium. The behavior of the tadpoles was similar to that interpreted as predatory behavior in interactions between *Rana catesbeiana* and *R. aurora* tadpoles in field enclosures (Kiesecker and Blaustein 1997b). Kupferberg (1997a) reported that she found no evidence of behavioral interactions between bullfrog tadpoles and those of two native species in California. In contrast, Kiesecker, Blaustein, and Miller (2001) reported that interference by *R. catesbeiana* tadpoles decreased survivorship and size at metamorphosis in *R. aurora* tadpoles in ponds in Oregon when food was clumped. The displacement of *Crinia signifera* tadpoles from choice feeding spots in flowing water by *C. riparia* tadpoles is another possible example of interference competition (Odendaal, Bull, and Richards 1984). Overall, however, there is relatively little evidence that behavioral interference is a widespread mechanism of interspecific competition among tadpoles in the field.

Interference competition also could occur through growth inhibition mediated by chemicals released in tadpole waste products, or by organisms such as yeast and algae. The possible role of growth inhibitors in mediating growth of tadpoles crowded together with conspecifics is discussed in chapter 13, and Alford (1999) reviewed the evidence for growth inhibitors being important in interspecific competition. Much of the evidence for interspecific growth inhibition comes from laboratory studies in which tadpoles were raised in crowded conditions (e.g., Beebee 1991). Neither Morin and Johnson (1988) nor Petranka (1989b) found any evidence of such inhibition in artificial or natural ponds. In contrast, workers studying competition between tadpoles of *Bufo calamita* and *Rana temporaria* in England reported significant inhibition of growth in *Bufo* tadpoles that were exposed to *Rana* feces in relatively small artificial ponds (2 meters × 1 meter; Griffiths, Edgar, and Wong 1991; Griffiths, Denton, and Wong 1993) and natural sand dune ponds (Bardsley and Beebee 2001). They identified an unpigmented unicellular alga (*Anurofeca* [= *Prototheca*] *richardsi*) in the tadpole feces that apparently is not toxic to tadpoles (Wong and Beebee 1994). However, when algal cells were abundant in the water, tadpoles fed on them but derived relatively little nutritional benefit. Growth inhibition would then occur because of tadpoles being diverted away from more profitable food (Beebee and Wong 1992). It is not yet clear, however, whether this mechanism of interspecific competition is widespread or important in natural populations of other anurans (Beebee 1995b; Griffiths 1995; Petranka 1995).

### Competition Mediated by Other Organisms

In addition to the organisms implicated as growth inhibitors of tadpoles, other types of organisms can have an effect on competition between species of tadpoles. Among the organisms that can have such effects are predators, and these are discussed in a separate, subsequent section. Less is known about the impact of pathogens and parasites on interspecific competition. Kiesecker and Blaustein (1999) reported an example of competition mediated by a pathogenic fungus. In the western United States, the eggs of amphibians are subject to infection by a water mold, *Saprolegnia ferax* (fig. 15.25), and some species are more susceptible to infection than others. The eggs of Cascades frogs (*Rana cascadae*), for example, are highly susceptible to this fungus, whereas those of the Pacific treefrog (*Pseudacris regilla*) are not (Blaustein et al. 1994). The effect of the fungus on *R. cascadae* is exacerbated by high levels of ultraviolet radiation in shallow water (Kiesecker and Blaustein 1995; Kiesecker,

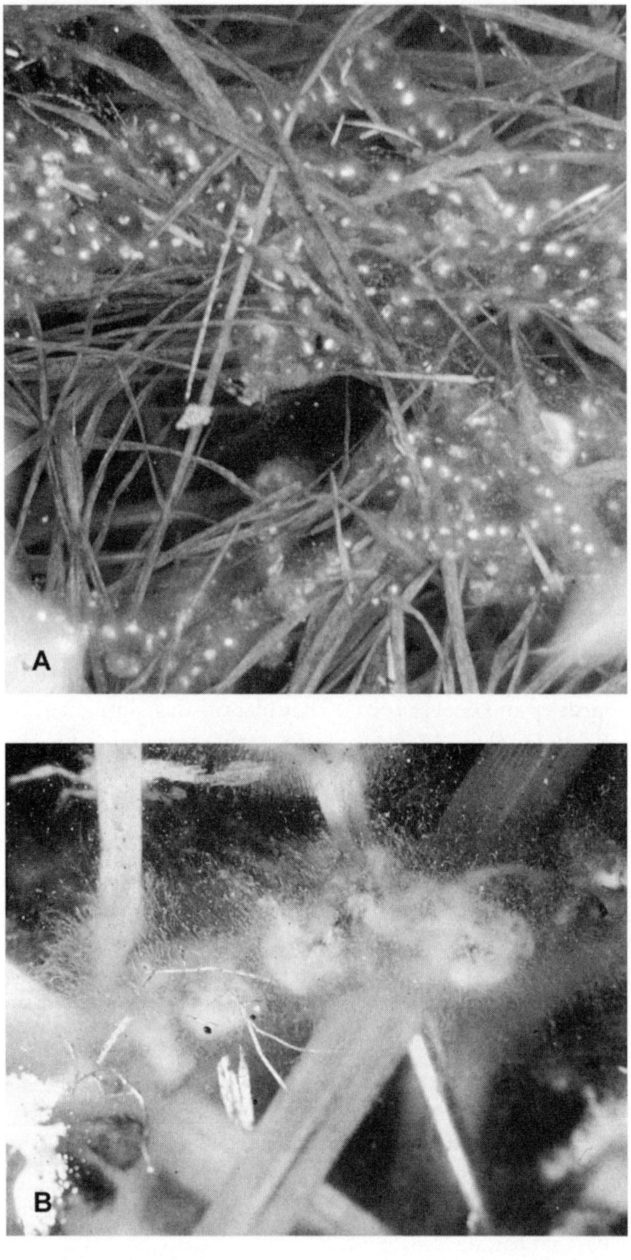


**Fig. 15.25.** Eggs of *Bufo boreas* infected with a water mold, *Saprolegnia ferax*. (A) Communal mass of toad eggs infected with the mold. (B) Close-up of individual infected embryos. Photos by Joseph M. Kiesecker.

Blaustein, and Belden 2001). Kiesecker and Blaustein (1999) raised eggs and tadpoles of these two species through metamorphosis in experimental ponds with and without the fungus, either alone or in competition with each other. In the absence of the fungus, survivorship and mass at metamorphosis were reduced for *Hyla* tadpoles that were raised with *Rana*, while the time to metamorphosis was increased. When the fungus was present, it killed many *Rana* embryos, and this resulted in much lower densities of *Rana* tadpoles and dramatically reduced survivorship to metamorphosis. The *Hyla* tadpoles in these ponds enjoyed reduced competition from *Rana* and had higher survivorship, larger mass at metamorphosis, and shorter larval periods.

### Competition and Pond Hydroperiod

Interspecific competition also can be mediated by physical factors in the environment. Probably the most important of these is pond hydroperiod. Competition among tadpoles generally has the effect of slowing growth rates and increasing the length of the larval period. This can affect survival to metamorphosis if the pond is subject to frequent drying, because tadpoles with long larval periods may not make it out of the pond before the water disappears. Wilbur (1987) investigated the interaction among competition, pond drying, and predation in experimental ponds stocked with four species of tadpoles *(Rana sphenocephala, Scaphiopus holbrookii, Bufo americanus,* and *Hyla chrysoscelis)* at different densities. All species exhibited reduced growth rates, prolonged larval periods, and smaller size at metamorphosis in response to high densities of competitors. Nevertheless, the species differed in their sensitivity to density effects. *Scaphiopus holbrookii,* a species that normally breeds in ephemeral ponds, was the least sensitive, and grew rapidly even at high tadpole densities. Under the pond-drying regime of the experiments, most tadpoles of this species reached metamorphosis before the ponds dried. *Rana* tadpoles, on the other hand, had slow growth rates at all densities and were very sensitive to density effects. These tadpoles never survived in ponds that dried early. *Bufo* tadpoles grew rapidly and could survive in drying ponds, but they tended to be outcompeted by *Scaphiopus* tadpoles. *Hyla* tadpoles did poorly in ponds that did not dry, because of intense interspecific competition. Responses to competition and pond drying were altered when predatory newts were present; that aspect of the study is discussed in more detail in a later section.

Semlitsch and Reyer (1992b) examined the interaction between competition and pond drying in a study of the *Rana esculenta* complex. As mentioned previously, hybrid tadpoles generally are competitively superior to the parental genotypes, and this was especially true when ponds were subjected to a drying treatment. Survivorship of *R. lessonae* tadpoles was considerably lower than that of *R. esculenta* tadpoles in ponds that dried, but the reverse was true for ponds that did not dry. Body size at metamorphosis was smaller for all genotypes for ponds that dried than for those that did not.

### Competition and Pond Acidity

Another feature of the physical environment that could affect competition among tadpole species is water acidity (pH). Some anurans live in naturally acidic water, such as that found in bogs or pine barrens, and certain species may be more tolerant of acidic conditions than others. There also is evidence for variation among populations of the same species in the ability of embryos or larvae to tolerate acid conditions (Andrén, Marden, and Nilson 1989; Glos et al. 2003). Human activities have lowered the pH of some bodies of water because of acid rain, and this has the potential to change the species competition of anuran communities. Warner, Dunson, and Travis (1991), and Warner, Travis, and Dunson (1993) studied the effect of pH on competition between tadpoles of *Hyla gratiosa* and *H. femoralis*. The latter tends to be found in more acidic environments. Tadpoles of *H. gratiosa*, the larger species, usually were competitively superior to those of *H. femoralis* and reduced the size of the latter at metamorphosis. When the tadpoles were raised together in moderately acidic water (pH 4.5), the negative effects of *H. gratiosa* on *H. femoralis* were reduced, suggesting that acidity gave *H. femoralis* a slight competitive advantage. These results contrast with those of Pehek (1995), who studied the Pine Barrens treefrog (*Hyla andersonii*), a species restricted to acidic habitats. Competition from tadpoles of *Rana sphenocephala* reduced both survivorship and mass of *H. andersonii*, while competition from *Hyla versicolor* only reduced body mass. Variation in pH did not have any direct effect on the three species, nor did it affect the outcome of interspecific competition. This means that enhanced competitive ability of tadpoles in acidic environments cannot explain the distribution of *H. andersonii*.

#### Competition and Canopy Cover

Yet another physical factor that could modify competition between species of tadpoles is canopy cover. Certain species of frogs tend to breed in relatively open sites, whereas others often are found in woodland ponds with relatively closed canopies. Natural succession of habitats, such as the growth of woodlands in areas that once were more open, may cause shifts in the relative abundance of different amphibian species (Skelly, Werner, and Cortwright 1999). In particular, species that prefer open habitats may be gradually excluded from ponds as canopy cover and the frequency of drying increase. There are a number of possible reasons for these shifts in species distributions, including changes in the suitability of the ponds for either the eggs or the tadpoles of particular species. For example, differences in canopy cover are likely to be correlated with differences in light availability, water temperature, oxygen content, water pH, algal productivity or species composition, and the type and abundance of predators. Closed-canopy ponds also tend to dry up earlier in the year because of the water demands of the surrounding trees. In general, relatively few North American amphibian species are regularly found in closed-canopy ponds, although some species, such as wood frogs (*Rana sylvatica*) and some species of *Ambystoma* do well in such habitats. Canopy cover may have a strong influence on amphibian species composition in tropical communities as well, but this has not been studied explicitly. There is little overlap in species composition of anuran assemblages in savanna and interior forest ponds in Amazonian Brazil, for example (Gascon 1991; Azevedo-Ramos, Magnusson, and Bayliss 1999), although both types of assemblages include a variety of hylids and leptodactylids. A reasonable hypothesis for future testing is that species distributions are related to relative performance of tadpoles in these very different habitats.

Using the survey of natural ponds by Skelly, Werner, and Cortwright (1999) as background, Werner and Glennemeier (1999) examined the performance of *Rana sylvatica*, *R. pipiens*, and *Bufo americanus* tadpoles in enclosures placed in open-canopy and closed-canopy ponds. The open-canopy ponds were warmer, had higher levels of dissolved oxygen, and were more productive than the closed-canopy ponds, but did not differ in pH. Closed-canopy ponds provided relatively poor-quality food for tadpoles in the form of leaf detritus and the organisms that break down dead leaves, such as fungi and bacteria, but did not support macrophytes. All three species grew rapidly and had high survivorship in the open-canopy ponds, although leopard frog tadpoles had the highest growth rates. Both leopard frog and toad tadpoles grew slowly and had very low survivorship in closed-canopy ponds. Wood frog tadpoles had high survivorship and slightly higher growth rates than the other two species in closed-canopy ponds. In open-canopy ponds, there was strong asymmetric competition between wood frog and leopard frog tadpoles, with the former being the superior competitors. In closed-canopy ponds, leopard frog tadpoles did so poorly that the presence of wood frog tadpoles had little effect. The growth and survival of leopard frog tadpoles in closed-canopy ponds was increased dramatically by the addition of supplemental food, but food addition had no effect on wood frog performance. These experiments demonstrated that both American toads and leopard frogs are effectively excluded from closed-canopy ponds by the poor performance of their tadpoles, probably because of inadequate food resources. Wood frog tadpoles did well in both types of ponds, and in fact, regularly breed in both types of ponds (Skelly, Werner, and Cortwright 1999). Similar observational and experimental studies have shown that spring peeper (*Pseudacris crucifer*) tadpoles do poorly in closed-canopy ponds and are generally excluded from such habitats, whereas wood frog tadpoles are found in both open- and closed-canopy ponds (Skelly et al. 2002; Halverson et al. 2003).

### Interaction between Competition and Predation in Tadpole Communities

The experiments discussed so far were mostly conducted in the absence of major predators, a situation that is unlikely to occur in nature except in the most ephemeral rain pools. Predators can be an important component of any community because they can directly impact the survivorship of prey species and potentially alter their competitive interactions. In addition, predators can have indirect effects on prey species that may affect their competitive ability. For example, some prey species may alter their levels of activity or use of microhabitats in response to predators (see chapter 14), and this in turn can affect the rate at which they acquire food. The direct and indirect effects of predators on prey species can affect the overall diversity of a community by enabling competitively inferior species to prosper at the expense of superior competitors. Predators also can have beneficial effects on individuals lucky enough not to be eaten, because they remove competitors and increase resource availability for the survivors (Wilbur 1997).

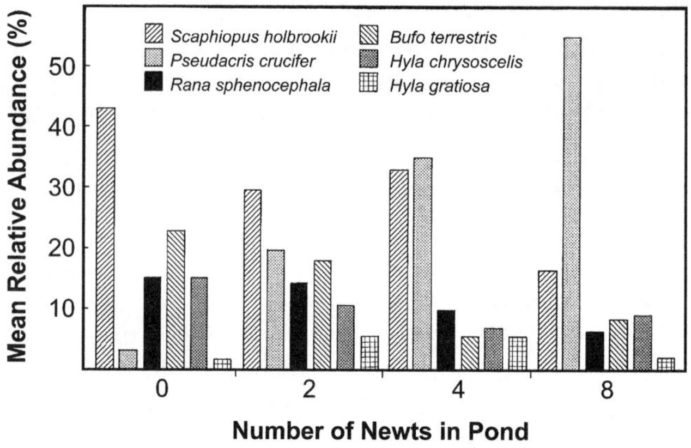


**Fig. 15.26.** Survival of six species of tadpoles in experimental ponds (cattle tanks) as a function of the density of predatory adult *Notophthalmus viridescens*. Tadpoles of each species were initially stocked at densities designed to reflect those found in natural temporary ponds. After Morin (1983a).

### Salamander Predation in Artificial Pond Communities

Starting in the 1980s, Henry Wilbur and his students investigated the role of predators in regulating the structure and dynamics of temporary pond communities, using replicated model communities in cattle tanks to simulate conditions in natural, temporary ponds. These investigators focused on the assemblage of anuran species that occurs on the United States' southeastern coastal plain, with the broken-striped newt (*Notophthalmus viridescens dorsalis*) used as the principal predator (Wilbur 1984, 1987, 1997). A classic series of experiments by Peter Morin set the stage for this line of investigation. In the first set of experiments, six species of tadpoles were introduced into experimental ponds at fixed densities that were thought to reflect typical relative abundance in natural ponds (Morin 1981, 1983a). The tadpoles were introduced in a sequence that preserved the natural phenology of local ponds. Morin varied the intensity of predation by controlling the density of larval salamanders (*Ambystoma tigrinum*) and adult broken-striped newts (*Notophthalmus viridescens dorsalis*). Since the *Ambystoma* larvae eliminated nearly all of the tadpoles from the ponds, only the results with newts as predators were informative.

As expected, the presence of predators greatly reduced overall tadpole survivorship. However, the most dramatic result of the experiments was a reversal in the competitive success of certain species correlated with the density of predators (fig. 15.26). When predators were absent, the species most successful in surviving to metamorphosis were *Scaphiopus holbrookii* and *Bufo terrestris;* these showed dramatic declines with increasing predator density. In contrast, *Pseudacris crucifer* produced almost no metamorphosing young when predators were absent, but showed a dramatic increase in relative success as the numbers of newts increased. Performance of other species, such as *Hyla gratiosa,* was greatest at intermediate predator densities, but was low for all treatments. For several species, size at metamorphosis increased with increasing predator density, while growth rates were inversely related to the density of other tadpole species. This suggested that interspecific competition was occurring. *Pseudacris crucifer* was at a competitive disadvantage compared to other species when predators were absent, but increased predation on the other species enhanced both survivorship and growth in this species. Later experiments showed that *Notophthalmus* did not enhance survivorship of *P. crucifer* tadpoles in the absence of other tadpole species, indicating that the effect of the predator depends on the composition of the tadpole community (Morin 1986). These experiments showed that newts function as keystone predators in the community, because they fundamentally alter the outcome of competition among species of tadpoles. Some species, such as *Pseudacris crucifer,* might not be able to persist with competitively superior species if predators were absent. In other words, the presence of predators is one mechanism that maintains high species diversity in temporary pond communities (Wilbur 1984, 1987, 1997).

In his experiments, Morin (1983a) inferred that competition was occurring by correlating growth performance with final densities of emerging metamorphs. Even stronger evidence for interspecific competition was obtained in another series of experiments, in which both predator density and the density of three tadpole species (*Rana sphenocephala, Bufo terrestris,* and *Scaphiopus holbrookii*) were varied independently (Wilbur, Morin, and Harris 1983). In the absence of predators, *Bufo* tadpoles had strong negative effects on the growth and survivorship of *Scaphiopus* and the growth of *Rana* tadpoles. Predatory newts reduced survivorship in all species, but *Bufo* and *Scaphiopus* tadpoles were virtually eliminated at high newt densities, whereas the larger *Rana* tadpoles were less affected. This freed the *Rana* tadpoles from interspecific competition and resulted in larger sizes at metamorphosis. In fact, the number of newts accounted for nearly two-thirds of the variation in growth rate of *Rana* tadpoles.

The differences in competitive ability and performance in the presence of predators observed in these experiments probably are related mainly to differences in feeding rate and vulnerability to predation. For example, Morin (1983a) found that newts and *Ambystoma* larvae preferred to feed on *Bufo* and *Scaphiopus* tadpoles when they were presented in combination with various hylid and ranid tadpoles, probably because these species are very active and conspicuous when they feed. Most other tadpoles were eaten indiscriminately as they were encountered, but *Rana sphenocephala* tadpoles were actively avoided. This agrees with other observations and experiments indicating that some ranid tadpoles are distasteful to vertebrate predators (Walters 1975; Formanowicz and Brodie 1982; Woodward 1983).

The effects of newts on tadpole communities can be affected by a number of variables, including pond hydroperiod, the phenology of the predator, and interactions with other types of predators. In a rather complex set of experiments, Wilbur (1987) examined competition between *Rana sphenocephala, Bufo americanus, Scaphiopus holbrookii,* and *Hyla chrysoscelis,* at high and low density, with tadpoles subjected to both newt predation and a drying regime. At low density, all species did poorly because of predation. At high density, the competitive dominant, *Scaphiopus,* faired poorly with newts present, while *Bufo* tadpoles did better relative to the other species. *Rana* did relatively poorly as well, because they were the first to be introduced and were subject to predation by newts before alternative prey (zooplankton) were available. When ponds dried at either 50 or 100 days, the two species with short larval periods (*Scaphiopus* and *Bufo*) were the most successful. *Scaphiopus* was the most successful species when newts were absent, but *Bufo* was the most successful when newts were present. Wilbur's experiments showed that pond hydroperiod, competition, and predation all interact to determine the metamorphic success of different species in a community. Rapid pond drying tended to favor species with short larval periods, which also are the most tolerant of high densities (*Scaphiopus* and *Bufo*), but predation determined which species was most successful. *Rana* tadpoles did not survive well in drying ponds because of their slow growth rates, but they also were strongly affected by competition and newt predation. *Hyla* tadpoles also were relatively poor competitors and were highly susceptible to both pond drying and newt predation. Predation can improve the performance of some species, such as *Bufo,* by reducing the density of competitors and increasing growth rates, which would be especially advantageous if ponds dry.

The phenology of predators also can affect tadpole performance. Alford (1989c) examined competitive interactions between the same four species of tadpoles used by Wilbur (1987), with newts present for different periods of time. Some ponds lacked newts, but others had newts present from the beginning of the experiment. Among the latter group, there were ponds from which all newts were removed after nine days or 51 days, and others from which newts were never removed. As in previous experiments, *Scaphiopus* tadpoles tended to dominate ponds that lacked newts. Whenever newts were present early in the larval period of the various anurans the survivorship of the tadpoles was reduced, and the longer newts were present, the greater the effect on tadpole survivorship. Some species, particularly *Rana* and *Scaphiopus,* were virtually eliminated by newt predation when initial tadpole densities were low. At high densities, some tadpoles of each species usually survived when predators were present. The number of *Bufo* and *Hyla* tadpoles was reduced by prolonged exposure to newts, but their relative success increased when predators were present. Later- breeding species (*Bufo, Hyla*) did better when predators were initially present and then removed, because the predators decreased the numbers of tadpoles of earlier-breeding species. Early breeders (*Rana, Scaphiopus*), on the other hand, did better when predators were present throughout the experiment. These species eventually grew to a size at which they were less vulnerable to newt predation, and the continued presence of newts reduced the numbers of their competitors.

Newts are seldom the only predators in temporary ponds, and interactions with other predators could affect the impact of newts on tadpole communities. Fauth and Resetarits (1991) exposed a five-species assemblage of tadpoles (*Rana sphenocephala, Pseudacris crucifer, Bufo americanus, Hyla chrysoscelis,* and *Gastrophryne carolinensis*) to predation from newts at two densities (two or four per tank), either in the presence or absence of one *Siren intermedia.* In this system, *Gastrophryne* and *Rana* tadpoles dominated tanks without any predators, with *Pseudacris* and *Hyla* doing relatively poorly. Even a few newts virtually eliminated *Gastrophryne* tadpoles and greatly increased the relative performance of *Pseudacris.* Sirens were unselective predators on tadpoles, and the presence of one *Siren* in a tank decreased the abundance of all species but did not have a very large impact on the relative proportions of each species (fig. 15.27). Clearly, sirens and newts are not functionally equivalent in their effects on tadpole communities.

A subsequent study by Chalcraft and Resetarits (2003) investigated the effects of three species of fish and three species of salamanders on three species of tadpoles (*Bufo terrestris, Pseudacris crucifer,* and *Rana sphenocephala*). The various types of predators had quite different effects on the

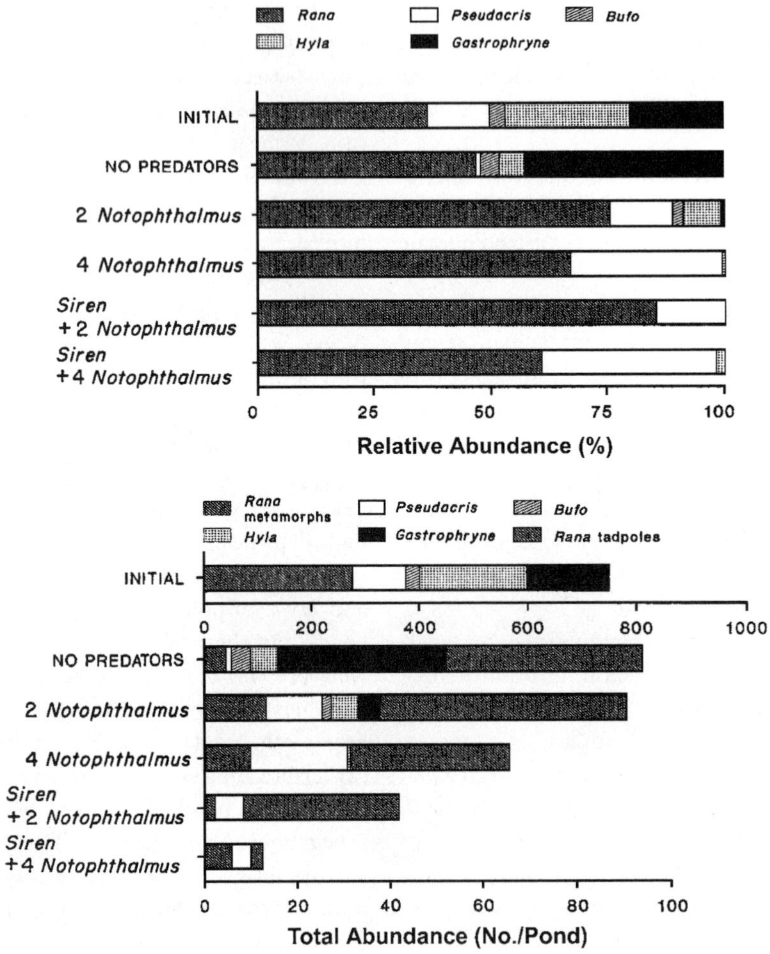


**Fig. 15.27.** Effect of two predators, broken-striped newts (*Notophthalmus viridescens dorsalis*) and sirens (*Siren intermedia*) on the composition of tadpole communities in artificial ponds (cattle tanks). Tadpole species were introduced on a schedule that reflected natural breeding times and in relative abundances that mirror those of natural ponds. Each bar represents the mean of five replicate populations. Top: Relative initial abundance of each species of tadpole and relative abundance of metamorphs or overwintering tadpoles (*Rana sphenocephala* only) in ponds without predators and those with different combinations of the two predators. Bottom: Total absolute abundance of tadpoles. The presence of predators favored tadpoles of *Rana sphenocephala* and *Pseudacris crucifer,* while total abundance of all tadpoles was progressively reduced with increasing numbers of predators. After Fauth and Resetarits (1991).

survivorship and growth rates of the three species of tadpoles. One fish species, the banded sunfish (*Eneacanthus obesus*), reduced survivorship of all tadpole species dramatically, and one other, the eastern mud minnow (*Umbra pygmaea*) strongly affected both *Rana* and *Pseudacris,* but not *Bufo.* There was little consistency in the effects of the other predator species, however, suggesting that these predators were not functionally equivalent in their effects on tadpole community structure.

## Salamander Predation and Tadpole Communities in Natural Ponds

The experiments done with tadpoles and predators in cattle tanks have been designed as model communities to ask basic questions about processes affecting community structure and dynamics. They have not been designed to explain exactly what happens in particular natural ponds (Morin 1998). On the other hand, these experiments are assumed to reflect processes that occur in nature. Resetarits and Fauth (1998) attempted to link studies in experimental ponds with the structure of natural communities by examining correlates of larval species diversity in Carolina bays. Based on previous work in experimental ponds, they made several predictions about the diversity of larval communities in natural ponds. First, when predators are scarce, high larval densities should result in low species diversity, because superior competitors will tend to dominate the ponds. Second, ponds with newts present should have higher species diversity than those without newts, because of the keystone effect of this predator. Third, ponds with sirens should have higher species diversity than those without sirens, because unselective predation will reduce overall competition. Finally, ponds with both predators present should have the highest diversity, because unselective predation by sirens would tend to broaden the range of densities over which the keystone effect of newts could operate. This is because high larval densities will tend to swamp out keystone predators, whereas very low densities could result in the virtual elimination of most species by predators. All of these predictions also should be affected by pond hydroperiod, because species with long larval periods will be excluded from the most temporary ponds.

Resetarits and Fauth (1998) found that pond hydroperiod, the presence of sirens, and the density of tadpoles each explained about a third of the variation in amphibian species diversity. The most surprising result is that the presence of newts had little effect on anuran diversity (fig. 15.28). This is inconsistent with the conclusion from experimental studies that this is a keystone predator in tadpole communities. The reasons for this are not clear. One possibility is that the presence of sirens is necessary for the keystone effect of newts in large ponds like Carolina bays. It also is possible that newts play a keystone role in smaller temporary ponds, which generally lack sirens, and which may be more similar to the conditions in the experimental ponds.

A different subspecies of newt, *N. v. louisianensis,* occurs

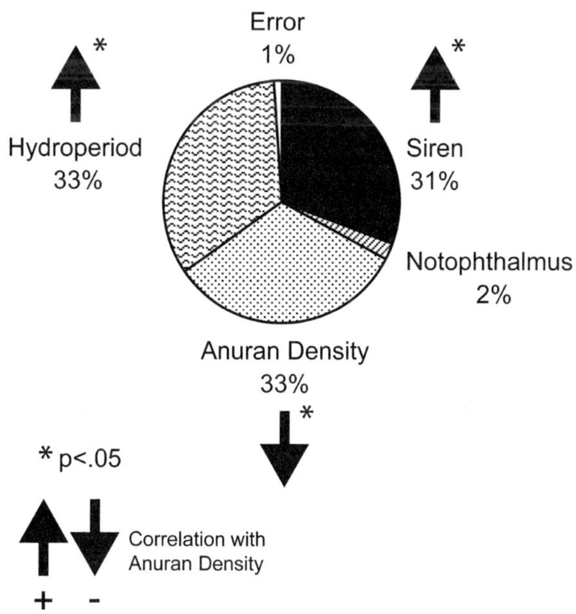


**Fig. 15.28.** Results of a four-factor regression analysis of tadpole species richness in seven Carolina Bays. Each segment of the pie diagram shows the proportion of variance in number of tadpole species explained by each factor (pond hydroperiod, tadpole density, number of sirens present, and numbers of newts). All factors together explained 99.7% of the variance in number of anuran species inhabiting the ponds. After Resetarits and Fauth (1998).

in Carolina bays and other temporary ponds on the coastal plain of South Carolina. Fauth (1999a) conducted a survey of natural ponds similar to that done by Resetarits and Fauth (1998) in North Carolina, and attempted to identify keystone predators by correlating amphibian species diversity with predator presence or absence. Potential predators in these ponds included newts, which were not very abundant, sirens, and either larvae or paedomorphic adults of another salamander, *Ambystoma talpoideum.* There also were several predatory fishes and insects in some ponds. Only *Ambystoma* had a strong effect on amphibian species diversity, suggesting that this species is a keystone predator, whereas newts and sirens are not. Unfortunately, in the absence of experimental work on this system, any explanations for patterns of species diversity in relation to predators remain somewhat speculative. Fauth's interpretation is consistent with experimental work by Morin (1995) in New Jersey showing that larvae of another ambystomatid, *Ambystoma opacum,* can play a role similar to newts in temporary ponds. The experiments by Chalcraft and Resetarits (2003) showed that both *A. opacum* and *Notophthalmus* had a relatively strong effect on community structure by virtually eliminating *Bufo* tadpoles while increasing the relatively success of the less competitive *Pseudacris crucifer* tadpoles. Other experiments by Kurzava and Morin (1994) suggest that geographic variation in the size of newts of different subspecies can result in somewhat different effects on tadpole communities.

### Insect Predation and Tadpole Communities in Artificial Ponds

The survey of natural ponds by Fauth (1999a) suggested that dragonfly naiads of one species, *Tramea carolina,* play a weak keystone role similar to that of salamander larvae. Indeed, predatory insects are common in many kinds of temporary ponds that lack newts or other salamanders. There have been many studies of the responses of individual amphibian species to insect predators (see chapter 14), but fewer dealing with the role of insect predators in mediating interspecific competition in temporary ponds. Van Buskirk (1988) performed experiments with temporary ponds like those used for the experiments with newts. The ponds were stocked with four species of tadpoles (*Pseudacris triseriata, P. crucifer, Rana sphenocephala,* and *Bufo americanus*), either without predators or with two species of dragonfly naiads (*Tramea carolina* and *Anax junius*), alone or together. The dragonflies reduced the abundance of all species, but relative abundance changed when predators were present. Metamorphosing juveniles were predominantly *Bufo* when predators were present, especially when both predators were present. *Rana* tadpoles seldom reached metamorphosis in the absence of predators, largely because competition from other species slowed their growth rates and they did not reach metamorphosis before the ponds dried. Some *Rana* did reach metamorphosis when one predator species was present, presumably because of reduced competition, but very few survived when both species were present. Survivorship of both species of *Pseudacris* was low when either predator was present, and they were eliminated from ponds with both species present. *Pseudacris triseriata* survived in low numbers when one dragonfly species was present, but *Tramea* eliminated *Pseudacris crucifer.* This result contrasts with the results of experiments with newts, which showed that *P. crucifer* is less susceptible to newt predation than are other species.

### Competition, Predation, and the Hydroperiod Gradient

While investigators coming from the Wilbur laboratory have focused most of their attention on competition and predation in artificial pond communities, those from Earl Werner's laboratory in Michigan have used a somewhat different approach. These investigators have focused on differences in the array of predators found in ponds along the gradient from ephemeral vernal pools to permanent ponds and their effects on the structure and dynamics of tadpole communities (Wellborn, Skelly, and Werner 1996; Skelly 1997). Furthermore, the experiments with cattle tanks have been designed to test general predictions about ecological processes and have only secondarily been applied to patterns of species distribution and abundance in natural ponds (Resetarits and Fauth 1998; Fauth 1999a). In contrast, members of the Werner laboratory have started with patterns of species distribution across a gradient of natural ponds and then have designed experiments to determine which factors account for these patterns (Skelly 1997; Werner 1998). Two types of interactions have been investigated in detail: the distribution of two species of *Pseudacris* along the pond permanence gradient, and interactions between green frogs (*Rana clamitans*) and bullfrogs (*R. catesbeiana*) in permanent ponds.

#### Interactions between Spring Peepers and Chorus Frogs

In southern Michigan, both spring peepers (*Pseudacris crucifer*) and striped chorus frogs (*P. triseriata*) breed in early spring. A survey of 22 natural ponds showed that *P. triseriata* tadpoles were found only in temporary ponds (Skelly 1996). *Pseudacris crucifer* tadpoles were found in all types of ponds, but they were least abundant relative to *P. triseriata* in temporary ponds (fig. 15.29 A). These ponds differed not only in drying regime, but also in the array of predators present. Permanent ponds sometimes contained fish, but dragonfly naiads were the most abundant predators. Fish were never found in temporary or intermediate ponds, but both larval salamanders (*Ambystoma tigrinum, A. laterale*) and adult salamanders (*Notophthalmus viridescens*) were found there, along with a variety of insect predators (fig. 15.29 B). In addition, predators tended to be larger in the more permanent ponds, even within predator categories (beetles, dragonflies, and salamanders).

Skelly (1995a) performed a transplant experiment in which tadpoles of each species were placed in enclosures at natural densities in temporary, intermediate, and permanent ponds, with some enclosures stocked with predators normally found in those ponds. The experiment showed that spring peeper tadpoles were more likely to survive to metamorphosis and reached a larger size at metamorphosis in permanent and intermediate ponds, whereas chorus frogs showed the reverse pattern (fig. 15.30). Chorus frogs had faster growth rates and were more likely to metamorphose from temporary ponds before these dried, whereas the slower-growing spring peepers sometimes were killed by desiccation. On the other hand, chorus frog tadpoles fared poorly in permanent ponds with abundant predators, apparently because their higher level of activity makes them more vulnerable to predation (Skelly 1994). In contrast to most of the experiments done in artificial ponds or in the laboratory, there was little evidence of strong interspecific competition between the two species of tadpoles. Hence, it appears that differential susceptibility to drying ponds and predation combine to determine the distribution of these two species.

Skelly's results are consistent with earlier experiments performed with *P. triseriata* in natural rock pools along the shore of Isle Royale, in Lake Michigan. D. C. Smith (1983) found that pools near the lakeshore were relatively small and most likely to dry quickly, but contained relatively few predators. Those close to the forest edge tended to be larger and more permanent, but also contained predators, including dragonfly naiads (*Anax junius*) and salamander larvae (*Ambystoma laterale*). *Pseudacris triseriata* tadpoles were highly susceptible to predation by dragonflies and were effectively eliminated from permanent ponds by predation. They also did poorly in the most temporary ponds because the ponds often dried up. They were most likely to survive in intermediate ponds that usually did not dry, but were too small to support large populations of predators. A later study (D. C. Smith and Van Buskirk 1995) examined interactions between chorus frog tadpoles in the same population and spring peepers, which occurred at low density in the more permanent ponds with predators. Tadpoles of each species were placed in pools along the permanence gradient. Both species had low survivorship in the permanent ponds with predators, but only chorus frog tadpoles did well in more temporary ponds with few predators. Again, this was mainly due to the faster growth rate and shorter larval period of chorus frog tadpoles. As in Skelly's experiment, there was no evidence of strong competition between the two spe-

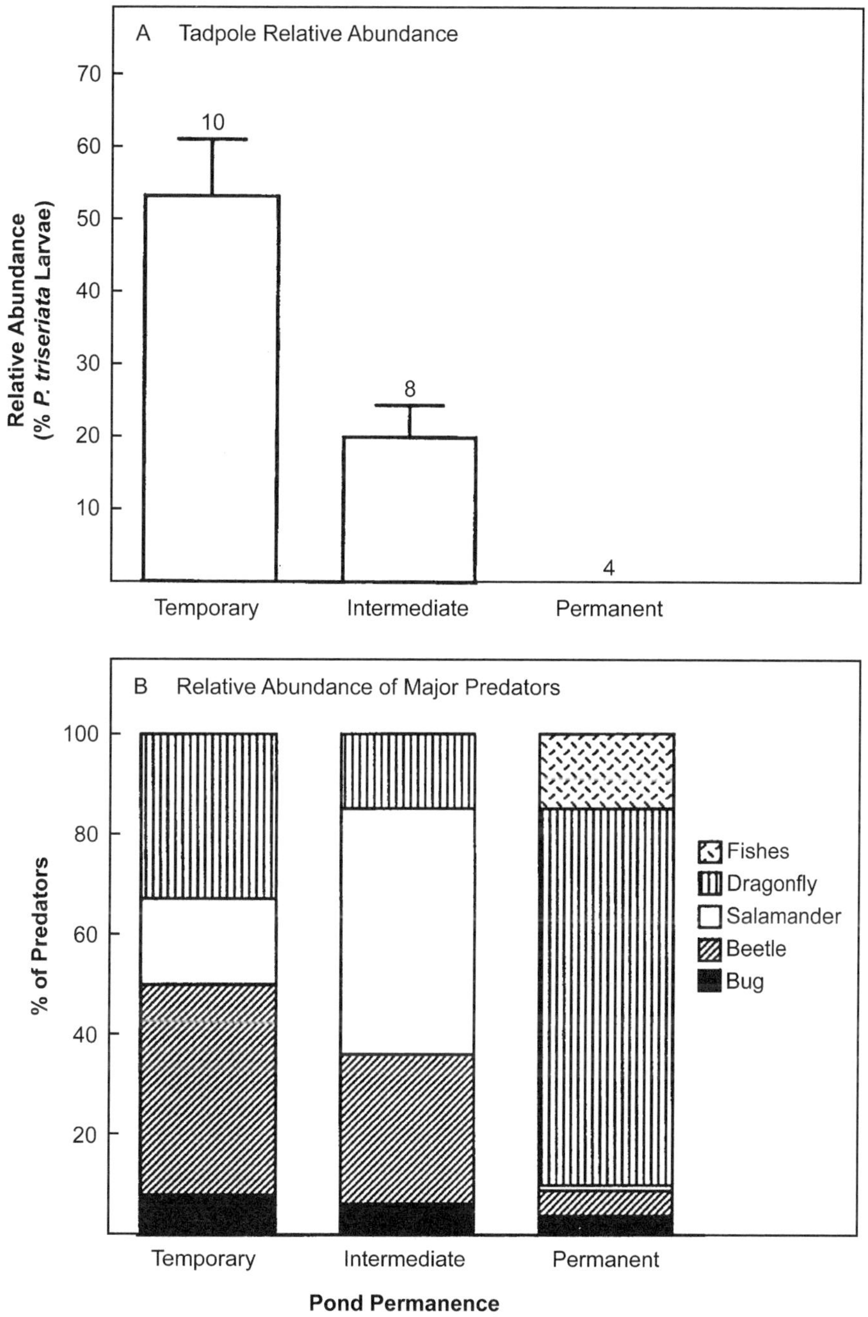


**Fig. 15.29.** Distribution of *Pseudacris* tadpoles and predators in three types of ponds in Michigan. (A) Percentage of tadpoles that were *Pseudacris triseriata* (the remainder were *P. crucifer*). (B) Relative abundance of major predators in three types of ponds. After Skelly (1996).

cies. Overall, these experiments indicate that *P. triseriata* has life-history traits that are well suited to temporary ponds with few predators, including a high level of activity and fast growth. *Pseudacris crucifer*, on the other hand, is less active and grows more slowly, making it poorly adapted to ponds that dry early, but their lower level of activity makes them less susceptible to predators in more permanent ponds.

### Interactions between Bullfrogs and Green Frogs

Interactions between bullfrog (*Rana catesbeiana*) and green frog (*R. clamitans*) tadpoles are different from the interactions between the two species of *Pseudacris*. Both of these species live in permanent and semipermanent ponds, because their long larval periods preclude the use of temporary ponds. Green frogs are more likely than bullfrogs to be

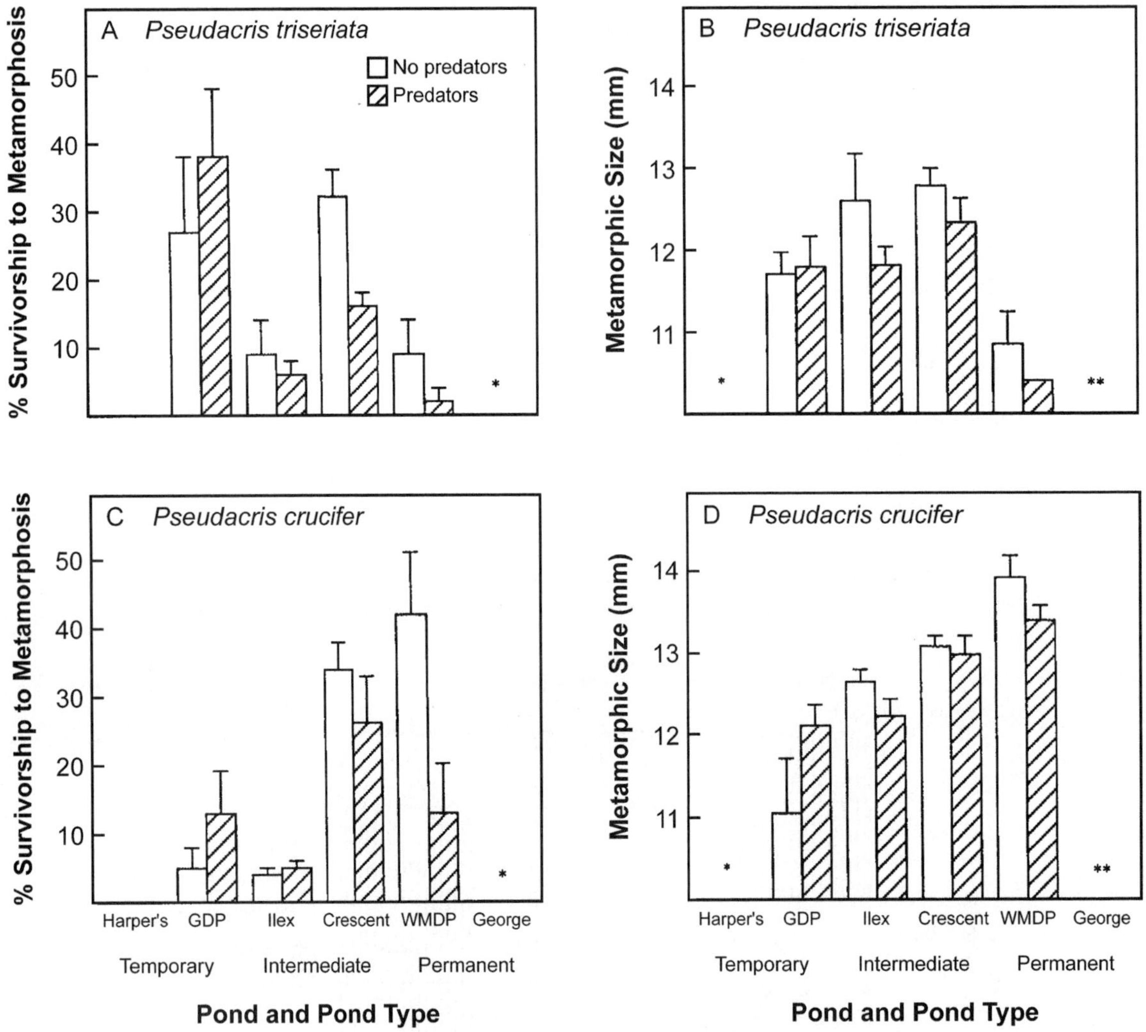


**Fig. 15.30.** Performance of *Pseudacris triseriata* and *P. crucifer* tadpoles raised in enclosures in temporary, intermediate, or permanent ponds with predators (hatched bars) and without predators (open bars). (A) Percentage of *P. triseriata* tadpoles surviving to metamorphosis. (B) Size of *P. triseriata* at metamorphosis. (C) Percentage of *P. crucifer* tadpoles surviving to metamorphosis. (D) Size of *P. crucifer* at metamorphosis. After Skelly (1995b). Ponds marked * and ** were not included because tadpoles invaded enclosures.

found in ponds that dry periodically, however, because they have a shorter larval period. A major effect of periodic pond drying is to exclude fish from the pond. Green frogs are relatively uncommon in ponds with fish, whereas bullfrogs often are abundant in such ponds (Werner and McPeek 1994). As discussed previously in this chapter, there is evidence of strong competition between the two species, with bullfrog tadpoles tending to be the superior competitor because of their higher level of feeding activity (Werner 1991, 1994).

Interactions between these two species are strongly affected by the assemblage of predators present in each type of pond. Werner and McPeek (1994) conducted experiments in subdivided ponds in which one side of each pond was stocked with bluegill sunfish (*Lepomis macrochirus*) and the other was not. The ponds contained a natural assemblage of invertebrate predators, with dragonfly naiads (*Anax*) being the dominant predator. The two sides of each pond were then stocked with more than 3,000 tadpoles of each of the two species, *R. catesbeiana* and *R. clamitans.* The effect of fish on the tadpole community was dramatic. Bullfrog tadpoles never survived in ponds without fish, but dominated ponds with fish. Green frog tadpole survival was low in all treatments, but their survival relative to bullfrog survival was higher in ponds without fish. Similar results were obtained in an unreplicated experiment in which a large pond was partitioned into sides with and without fish. The fish had a major impact on the abundance of other predators in both the experimental ponds and in natural ponds. In both types of ponds, the dominant insect predator (*Anax*) was greatly reduced by fish predation, as were salamander larvae. In laboratory experiments, fish preferred to feed on green frog tadpoles, but both salamander larvae and

*Anax* greatly preferred bullfrog tadpoles. Thus, bullfrogs do better than green frogs in ponds with fish for two reasons: because they are less likely to be eaten by the fish, and because the fish reduce the numbers of more significant predators. Because bullfrog tadpoles are superior competitors to green frog tadpoles, bullfrogs might dominate all ponds, but are excluded from those without fish because of heavy predation by dragonfly naiads. They also are excluded from ponds that dry frequently, by having a much longer larval period than do green frogs.

As discussed in chapter 14, bullfrog tadpoles are protected from fish predation by distasteful skin secretions. Such secretions are present in green frog tadpoles as well, but apparently are less effective at repelling fish. Bullfrog tadpoles are more vulnerable to predation by salamander larvae and dragonfly naiads because of their higher levels of activity. The vulnerability of bullfrog tadpoles to dragonfly naiads is likely to increase at high tadpole densities because depletion of food resources can lead to higher levels of feeding activity (Anholt and Werner 1995). In addition to eating tadpoles, predators can have indirect effects on the growth and competitive ability of tadpoles because of the antipredator behavior of the tadpoles (see chapter 14). For example, in cattle tank experiments, both green frog and small bull frog tadpoles reduced their activity in the presence of caged *Anax* naiads, indicating a response to chemical cues from the predators. This reduction in activity resulted in reduced feeding and slower growth. Large bullfrog tadpoles, on the other hand, were less likely to reduce activity, presumably because they are less vulnerable to dragonfly predation, and actually grew faster in the presence of dragonfly naiads (Werner and Anholt 1996). Small bullfrog tadpoles have a stronger competitive effect on other bullfrog tadpoles and on green frog tadpoles than do large individuals, so the presence of predators produced a greater change in competitive effects of small bullfrog tadpoles. Thus competition and predation interact in complex ways in this system. Increased density of either conspecific or heterospecific tadpoles will tend to reduce growth rates because of competition. This may result in tadpoles becoming more active in their attempts to get enough food, making them more vulnerable to predators. Reduced growth also will increase the amount of time that tadpoles are small and therefore most vulnerable to predators.

These interactions are even more complex when multiple predators are present simultaneously. For example, in laboratory experiments, both green frog tadpoles and small bullfrog tadpoles exhibited large reductions in activity and spatial avoidance of dragonfly naiads. Both species showed little reaction to bluegill sunfish (*Lepomis macrochirus*) and intermediate responses to mud minnows (*Umbra limi*). The strength of behavioral response was directly related to the threat posed by each predator (Relyea and Werner 1999). This means that the most dangerous predator, dragonfly naiads, has two negative effects on tadpole performance. Not only do they reduce survivorship directly by eating the tadpoles, but they also reduce growth rates by reducing feeding activity of tadpoles. Less dangerous predators, such as bluegill sunfish, have only direct negative effects by eating the tadpoles, but they also have indirect benefits by eating the more dangerous predators.

These interactions are further complicated by reactions of predators to each other. For example, when uncaged dragonfly naiads and fish both were simultaneously present with tadpoles, the dragonflies reduced their own activity, apparently in response to the presence of predatory fish. Bullfrog tadpoles, in turn, exhibited less reduction in activity than to dragonflies alone (Eklöv and Werner 2000). Again, bullfrog tadpoles tend to do better when both dragonflies and fish are present than when only dragonflies are present. These changes in behavior can affect competition between bullfrogs and green frogs as well, because green frog tadpoles exhibit greater reductions in activity in response to dragonflies than do bullfrog tadpoles (Werner 1991; Peacor and Werner 1997; Relyea and Werner 1999). Behavioral responses to one type of predator can affect the vulnerability of tadpoles to other types of predators, as well as competitive interactions between tadpoles. For example, when green frog and bullfrog tadpoles in cattle tanks were exposed to free-ranging dragonfly naiads (*Tramea lacerata*) and another type of dragonfly (*Anax*) in cages, the tadpoles showed reduced activity and suffered lower predation than when *Anax* was absent (Peacor and Werner 1997). When large bullfrog tadpoles were present as competitors, these tadpoles depleted resources and caused both green frog and small bullfrog tadpoles to increase their activity. This in turn made these tadpoles more vulnerable to predation by *Tramea*. Taken together, all of these experiments suggest that the distinction between the effects of competition and predation in tadpole communities is somewhat artificial, because each process interacts with the other in complex ways to affect the performance of individual tadpole species (Werner and Anholt 1996).

### Competition and Predation in Tropical Pond Communities

Despite the great diversity and abundance of anurans in the tropics, we know almost nothing about how competition and predation affect the structure of tropical tadpole communities. Tropical tadpole communities can be more complex than are those in the temperate zone. They often have more species, with a wider range of trophic adaptations than those of temperate zone tadpoles. In addition, breeding activity may last for months, with different species having very different temporal patterns of reproduction (Aichinger

1987a). Many temporary ponds may fill, dry, and refill continuously throughout the rainy season, resulting in a constantly changing assemblage of tadpole species at any given site (Gascon 1991).

Most studies of tropical communities have been based on surveys of species occurrence in different bodies of water rather than experimental studies of species interactions. Tropical tadpoles can be found in a variety of habitats, from permanent rivers, lakes, and ponds to large temporary ponds, flooded meadows, streamside pools, and animal wallows. The occurrence of tadpole species in different habitats is largely a result of choice of oviposition sites by breeding adults (Magnusson and Hero 1991), but it also could be affected by interactions among tadpoles or their predators. In a study of tadpole distribution in an Amazonian rainforest, many species of tadpoles were widely distributed across a range of habitats from streamside pools to peccary wallows, and physical factors did not explain the composition of tadpole assemblages (Gascon 1991). The presence of fish was correlated with the composition of tadpole assemblages (Hero, Gascon, and Magnusson 1998). In contrast to earlier reports that most Neotropical tadpoles are found in fish-free habitats (Heyer, McDiarmid, and Weigmann 1975), however, many species used habitats containing a variety of fishes. Some species that were found almost exclusively in streams or streamside pools were consistently found with fish. Temporary ponds in the tropics also vary in the amount of vegetative cover available to tadpoles, and this in turn can affect their vulnerability to different kinds of predators (Kopp, Wachlevski, and Eterovick 2006).

Experimental studies of three common species of tadpoles (*Osteocephalus taurinus, Phyllomedusa tomopterna,* and *Allobates femoralis*) showed that fishes were less effective predators of tadpoles than were dragonfly naiads, although both types of predators reduced tadpole survival in small containers (Gascon 1992a). Both types of predators preferred *Phyllomedusa* tadpoles to *Osteocephalus* tadpoles, but it is not clear whether this is due to differences in palatability, habitat selection, activity, or a combination of all three. A survey of savanna habitats in central Amazonia yielded results similar to the rainforest studies, although the diversity of tadpoles was lower and the species composition was different (Azevedo-Ramos, Magnusson, and Bayliss 1999). Tadpoles were more often found in habitats with insect predators than with fishes, but a number of species consistently co-occurred with fishes. There has not been any experimental work with a tropical assemblage to determine how predation affects competition among species, and, indeed, there is relatively little experimental evidence either for or against competition in tropical tadpole communities.

## Summary and Conclusions

Most studies of amphibian community ecology have focused on interactions among species in local assemblages. Nevertheless, larger-scale patterns of species diversity and distribution can have an impact on local communities, because these patterns set the stage for interactions among co-occurring species. Amphibians are always imbedded in a larger community of primary producers and various types of consumers, including both predators and prey of amphibians. Investigators usually have defined amphibian communities in terms of guilds of ecologically similar species, such as the frogs of the tropical forest floor or the tadpoles of a temporary pond. There has been relatively little attention given to interactions with other organisms that are potential competitors of amphibians, such as insectivorous lizards. What little work has been done in this area is entirely descriptive. Indeed, there are no experimental studies of interactions between amphibians and other terrestrial vertebrates, other than their predators (see chapter 14). There have been a few experimental studies demonstrating potential competitive interactions between aquatic or larval amphibians and other organisms, including fishes and freshwater invertebrates. Given the potential for even very distantly related organisms to interact as competitors, more attention probably should be given to the broader communities in which amphibians live.

Early studies of amphibian community ecology were largely descriptive, with a focus on patterns of microhabitat use and partitioning of food resources in local assemblages of species. For the last 25 years or so, amphibian community ecology has been heavily experimental. Much of this shift to a more experimental approach reflects a general shift toward more experimental work in ecology in response to criticisms that documenting patterns of resource use tells us little about the processes that have created these patterns.

Despite the enormous diversity of anurans, the community ecology of terrestrial anurans is relatively poorly understood as well. Nearly all studies of terrestrial anuran communities have been descriptive studies of microhabitat use and prey selection in assemblages of species sharing the same habitat. Detailed studies have been conducted with frogs of tropical rainforests, especially in the New World, and those of tropical savannas in Africa. Communities of anurans in the temperate zone are much less diverse than in the tropics, often with less than a half-dozen species living in the same habitat, and their community ecology is poorly studied.

The general pattern that emerges from studies of tropical frogs is that most species are generalist predators on insects and other arthropods. Differences in prey consumed by different species in a community often are based on differences in size of prey rather than prey taxon, and these in turn are related to differences in the body size and foraging modes of the anurans. Some tropical frogs are specialist predators, however, and most of these eat large numbers of ants and termites. Certain clades are most likely to include ant specialists, and all of these have toxic or distasteful skin secretions, derived in part from chemicals acquired by eating ants. These include bufonids, toxic dendrobatids, some mantelline frogs from Madagascar, many terrestrial microhylids, and probably some myobatrachids as well. There also are a few large tropical frogs that have become specialist or opportunistic predators on other frogs, an ecological role largely absent from temperate zone communities. Most temperate zone frogs are generalist predators of invertebrates, although bufonids and microhylids in these regions often are ant specialists. Again, partitioning of food resources usually is related to differences in prey size rather than types of prey. Although the tropic niches of terrestrial anurans have been well documented, experimental studies demonstrating competitive interactions among co-occurring species of frogs are lacking. Indeed, there has not been a single well-designed experimental study of terrestrial anuran communities.

This contrasts with extensive experimental work on terrestrial salamander communities. Classic studies of *Desmognathus* salamanders along Appalachian streams and species of *Plethodon* along elevational gradients have shown that competition for space can be intense and can lead to microhabitat partitioning or exclusion of species from some microhabitats. Competition often takes the form of direct interference and aggressive interactions among species of salamanders, many of which are interspecifically territorial. In some cases, interference competition interacts with differences in physiological tolerances to determine the microhabitat distribution of different species, as in the interaction between the widespread species *Plethodon cinereus* and the very similar *P. shenandoah*. The community ecology of terrestrial salamanders outside of the southern Appalachians is poorly understood. Tropical salamander communities offer interesting opportunities for future research, because species in local assemblages often exhibit sharply defined distributional boundaries along elevational gradients. Unfortunately, the relatively inaccessible locations of many of these populations make long-term field studies and, especially, experimental work difficult.

Relatively few salamanders live in aquatic habitats as adults, and the community ecology of aquatic salamanders has not been investigated in as much detail as that of terrestrial species. Nevertheless, temporary ponds in the southeastern United States often harbor either larvae or adults of several species of aquatic salamanders, including newts (*Notophthalmus*), sirens (*Siren, Pseudobranchus*), amphiumas (*Amphiuma*), and paedomorphic mole salamanders (*Ambystoma*). Surveys of natural ponds and experimental studies in artificial ponds have revealed a complex set of interactions among these species, which often function as both predators and competitors of co-occurring salamanders. In some parts of Europe, up to three species of newts (*Triturus*) live together in the same ponds, and they often are very similar in size and ecology. As with terrestrial salamanders, aquatic newts are generalist predators, with no evidence of differences in taxa of prey consumed. Minor microhabitat partitioning has been reported in some ponds, but not in others, but there has not been any experimental work to determine whether choice of microhabitats is influenced by the presence of other species of newts.

The best-studied amphibian communities are assemblages of larval amphibians, especially those that live in temporary ponds. The small size, abundance, and relatively short larval periods of many species have made them ideal subjects for both laboratory and field experiments. Both competitive interactions and the effect predators on interspecific competition have been studied in a variety of laboratory containers, in artificial mesocosms such as cattle tanks, and in small pens and large enclosures in natural ponds. Laboratory studies provide the greatest control over treatment variables and allow for easily replicated treatments, but they sacrifice much of the realism and complexity of natural environments. Mesocosm experiments, such as those using cattle tanks as artificial ponds, provide a greater degree of realism and complexity, while allowing for replicated treatments. Mesocosm experiments have been criticized by some ecologists as being insufficiently realistic to mimic conditions in natural environments such as real temporary ponds. Some investigators have used enclosures in natural ponds to provide a more realistic experimental environment, but in this case, the size of enclosures often must be traded off against the need for multiple replicates of treatments.

All of these types of experiments have revealed significant competition among species of amphibian larvae. However, a survey of experimental studies by Skelly and Kiesecker (2001) showed that laboratory and mesocosm studies are much more likely to produce evidence of strong interspecific competition than are experiments conducted in natural ponds. In some cases, this may be due to higher larval densities in laboratory experiments and in artificial ponds than in natural ponds. In other cases, it may be that the simplified environments of these experimental settings reduce the diversity of microhabitats and food resources available to larvae in natural ponds.

The structure of larval communities is strongly affected by the permanence of breeding ponds. Some species are typically found only in relatively ephemeral ponds that normally completely dry up every year, whereas others are found only in permanent ponds that never dry up. Some species are intermediate, being found in a wide range of ponds. Temporary ponds often have poorer food resources than do permanent ponds, but usually have fewer predators. They are subject to a variety of physical stresses besides rapid desiccation, including high water temperatures and low oxygen levels. Permanent ponds often have abundant food resources but also abundant predators. The largest and most persistent ponds frequently are inhabited by fish, which exclude many species of amphibians from the ponds.

Amphibian larvae that are adapted to temporary ponds typically have rapid growth and development that enables them to escape from drying ponds, but often at a cost of emerging from ponds at a relatively small size. Permanent pond larvae typically grow and develop more slowly, but tend to emerge from ponds at a relatively large size. Temporary pond larvae often have poorly developed defenses against predators, especially fishes, whereas permanent pond larvae often have a variety of behavioral and chemical defenses against both invertebrate and vertebrate predators. When reared in the same ponds, fast-growing species often outcompete slower-growing species in the absence of predators, presumably because they feed at a much faster rate. The presence of predators can alter or even reverse the outcome of interspecific competition by giving an advantage to slower-growing species that are well protected against predators. Predators often have indirect effects on competitive interactions by causing amphibian larvae to reduce their feeding activity or alter their use of microhabitats in ways that reduce rates of growth and development. Interactions between multiple predators, such as fishes and dragonfly naiads, can affect competitive interactions among amphibian larvae as well, because one predator may prey on both the amphibian larvae and the other predators. In addition, defenses that are effective against one type of predator, such as reducing feeding activity, may not be very effective against another predator with a different foraging mode. Since most natural ponds, especially relatively permanent ones, have a rich assortment of potential predators, the interactions among co-occurring species of amphibian larvae can be complex and vary from one pond to another.

Mechanisms of competition appear to differ between communities of larval salamanders and those of anuran tadpoles. Because all salamander larvae are carnivorous, they can function both as competitors and predators of other species, with their effect being determined largely by differences in body size. Direct interference competition in the form of aggressive attacks, biting of tails, and predation are relatively common among salamander larvae. and may be more important than exploitative competition for food resources in some communities. In contrast, most anuran tadpoles are herbivorous, although some species can be opportunistic predators on the eggs or larvae of other species. In most tadpole communities, exploitative competition for food appears to be more important in structuring the community than is aggressive interference competition. There is evidence from laboratory experiments of chemical interference and growth inhibition in tadpoles, but there is limited evidence for this being an important type of interaction in the field.

Our understanding of the structure and dynamics of larval communities is limited by the difficulty of extrapolating from experiments conducted in cattle tanks and other artificial settings to more natural ponds. Much of the work done with cattle tanks and other mesocosms has been designed to answer general theoretical questions about the nature of competitive interactions among species, but with only secondary reference to the structure of particular pond communities in nature. In some cases, results obtained in experimental studies have not been entirely consistent with observations made in natural ponds. For example, there is abundant evidence from cattle tank studies that newts serve as keystone predators in tadpole communities by altering the outcome of competitive interactions among species. Yet some surveys of natural ponds have failed to reveal a strong effect of newts on the structure of local assemblages. Unfortunately, it is hard to compare these results directly, because one set of studies involves experimental manipulation of predator densities, while the other relies on the correlation between predator abundance and species composition of natural communities. Some workers have addressed this problem by starting with natural patterns of species distribution in different types of ponds and then using experiments to tease out the effects of pond permanence, competition, and predation on community structure.

A general problem with many studies of community ecology in both aquatic and terrestrial communities is that seasonal and year-to-year variations in the intensity of competition are largely ignored. Patterns of community organization often are inferred from point samples collected over a relatively short period, but there is considerable doubt as to whether such data are reliable indicators of community structure (Wiens 1981). Studies of other groups, such as lizards (e.g., Dunham 1980, 1983) have shown that the intensity of competition and the degree of niche overlap between species can vary in response to rainfall and food abundance. Given the substantial variation in population size of some amphibians, even in the tropics (Barbault 1972; Toft, Rand, and Clark 1982), as well the variation in numbers (Levings and Windsor 1982) and availability (Jaeger 1980b) of their prey, it seems likely that the intensity of competition between species of amphibians varies as well.

However, only a few investigators have even attempted to document seasonal changes in abundance and resource use of sympatric species (e.g., Toft 1980b).

Studies of community ecology also would benefit from greater consideration of other aspects of amphibian biology. For example, there are many possibilities for integrating information on the physiological ecology of individual species with information on interactions between species. Jaeger's (1970, 1971a, b) work, showing that the distribution of two species of salamanders is governed by interactions between physiological tolerance to desiccation and competition, provides an excellent example. A number of studies have shown that tolerance of water and temperature stress in anurans is correlated with habitat (Ralin and Rogers 1972; Pough, Stewart, and Thomas 1977; Van Berkum et al. 1982; Beuchat, Pough, and Stewart 1984; see also chapter 2). There also is evidence of differences in temperature responses of tadpoles of closely related species (Negovetic et al. 2001). Nevertheless, there have been few experimental attempts to investigate possible interactions between physiological requirements and interspecific competition. There also are possibilities for integrating community ecology studies with studies of behavioral interactions, as has been done in some of the studies of interference competition in salamanders. Interspecific aggressive behavior also has been described in anurans (see chapter 8), but its importance for habitat segregation between species has not been investigated.

Finally, the relationship between community-level processes in aquatic larvae and those of adults remains to be investigated. While it seems clear that the success of different species in producing new recruits into the population will have a major impact on the distribution and abundance of adults in the community, the connections between the two types of communities have not been explored in detail. Most authors have assumed that individuals metamorphosing at a larger body size will have an advantage in both intra- and interspecific competition as adults, and there is some evidence of this from studies of marked juveniles (see chapter 13). It also is clear that both intra- and interspecific competition among larval amphibians can translate into smaller size at metamorphosis and possibly lower survivorship in the terrestrial environment. We know virtually nothing, however, about how juveniles or adults of pond-breeding amphibians interact with one another, or even if they interact at all.

# **Chapter 16** Conservation of Amphibians

Hyla crucifer *is what the biologists call him, but most to of us he is simply the Spring Peeper. . . . Everyone who has ever visited the country in the spring has heard him trilling from the marsh at twilight, and though few have ever caught sight of him most must know that he is a little, inch-long frog, who has just awakened from his winter sleep. . . . The peeper seems to realize, rather better than we, the significance of his resurrection, and I wonder if there is any other phenomenon in the heavens above or in the earth below which so simply and so definitely announces that life is resurgent again.*
—Joseph Wood Krutch, *The Twelve Seasons* (1949)

*These world-filling, mind-altering choruses of spring peepers have no equal in the northeastern landscape. There is talk now of the silence of the frogs, of their striking diebacks, and declines and the disappearance of species globally. . . . It is a silencing that has taken other voices than those of frogs, as well as voiceless presences, all inevitably vanishing with the disappearance of the places in which they must live.*
—David M. Carroll, *Swampwalker's Journal* (1999)

When I began this book, more than 20 years ago, I envisioned a final chapter that might briefly review the highlights of amphibian ecology and behavior discussed in previous chapters, followed by suggestions of critical areas for future research. In the intervening years, however, a greater concern has arisen, a concern that the very animals that amphibian biologists study might be on the verge of extinction. Beginning in the 1980s, amphibian biologists became increasingly convinced that major declines of amphibian populations had occurred in various parts of the world. In some cases, the causes of such declines were clear: destruction of tropical rainforests, drainage of wetlands, conversion of natural habitats to intensive agriculture, urbanization, and other forms of overt habitat destruction. In other cases, the causes were less obvious, with frogs disappearing from seemingly pristine habitats in protected parks and nature reserves and individuals even being found dead or dying in tropical rainforests. Given these alarming developments, it seemed important to use the final chapter to address the issue of declining amphibian populations and what might be done about it.

The literature on declining amphibian populations and amphibian conservation has increased exponentially in recent years. There have been many general reviews (Dodd 1997; Alford and Richards 1999; Corn 2000; Blaustein and Kiesecker 2002; Collins and Storfer 2003; D. Green 2003, 2005a; Storfer 2003; Beebee and Griffiths 2005; Collins and Halliday 2005), regional accounts (D. Green 1997a; Lannoo 1998a; A. Campbell 1999; Minter et al. 2004; Young et al. 2004), and both popular and technical books on the subject (K. Phillips 1994; Beebee 1996; Linder, Krest, and Sparling 2003; Semlitsch 2003a; Lannoo 2005). Given this volume of material, it is fair to ask whether there is anything new to say on the subject. Perhaps not, but I have tried to take a slightly different approach from that of many other authors. Instead of starting with a litany of potential threats to amphibians, I will begin by outlining some features of the basic biology, ecology, and behavior of amphibians that make them either particularly susceptible to environmental perturbations, or not particularly susceptible. Many authors state, almost as a truism, that amphibians are more vulnerable to environmental insults than are other vertebrates and therefore might serve as a kind of "canary in the coal mine," an indicator of environmental degradation (e.g., Carroll 1999). It is therefore worth examining whether this assumption actually is true, or whether there are some features of amphibian biology that make them somewhat less vulnerable to environmental change than are other species and in some instances easier to protect. Whatever the specific causes of decline in a particular population, it is clear that our investigation of these causes and possible solutions to the problem needs to be rooted in a thorough understanding of basic amphibian biology.

## What Makes Amphibians Susceptible to Population Declines?

The idea that amphibians as a group are especially susceptible to environmental degradation derives from several basic amphibian characteristics: their relatively small size; their ectothermic physiology, with limited capacity for activity; their highly permeable skin, which seems vulnerable to various biological or chemical agents; and their dependence on aquatic or moist habitats (Vitt et al. 1990; Murphy, Phillips, and Beasley 2000). Many of these characteristics are shared with other groups of animals that have received less attention or concern, especially in the popular press, including freshwater fishes, terrestrial and freshwater mollusks, aquatic insects, and all sorts of other small aquatic and terrestrial invertebrates. Indeed, there is evidence that many of these groups are threatened as well, at least on a regional or local basis, and the decline of amphibian populations should be viewed in the context of a much larger biodiversity crisis (Lydeard and Mayden 1995; Leidy and Moyle 1998; Chapin, Sala, and Huber-Sannwald 2001; Novacek 2001; Saunders, Meeuwig, and Vincent 2002; Halliday 2005). Nonmarine mollusks, for example, account for more than 40% of all recorded extinctions of animal species since 1500, whereas amphibians comprise the smallest proportion of known vertebrate extinctions (Lydeard et al. 2004).

Currently about 35 species of amphibians are thought to have gone extinct since the 1970s, with another 26 species not having been seen by herpetologists for some years (Stuart et al. 2004; see also The AmphibiaWeb Watch List at http://elib.cs.berkeley.edu/aw/declines/extinct.html). About 425 other species are considered critically endangered, but this list is heavily weighted toward regions where herpetologists have noticed major declines, especially Latin America and Australia, and for many species the data are simply insufficient to determine their current status (fig. 16.1). These numbers need to be considered very rough estimates, because new species continue to be described at a faster rate than species disappear, and for a large proportion of amphibian species, we have insufficient data to determine their status (see Pimenta et al. 2005 and Stuart et al. 2005 for an exchange of views on interpreting the data). Furthermore, quantitative data on population status are most deficient in regions with some of the highest species diversity of amphibians (Africa, Asia, and the New World tropics). Nevertheless, recent estimates suggest that for amphibians, at least in the New World, the proportion of threatened species is considerably higher than for either birds or mammals (fig. 16.2; B. Young et al. 2004). The taxonomic distribution of threatened and declining amphibian species is uneven, with some families having a higher proportion of species at risk

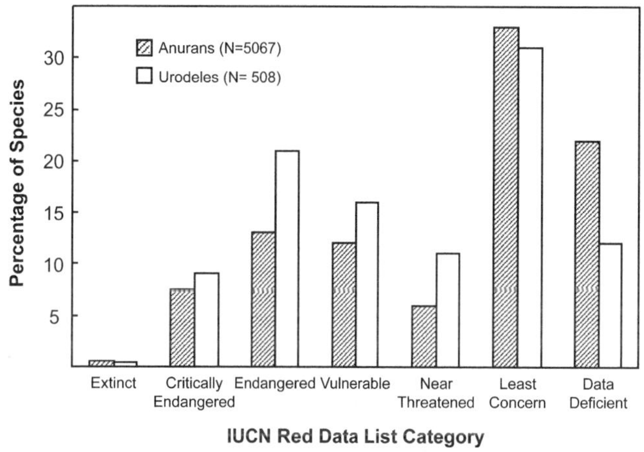


**Fig. 16.1.** Estimated percentages of anurans and salamanders in different Red List Categories designated by the International Union for the Conservation of Nature (IUCN). Caecilians are not included, because sufficient data to determine their conservation status are lacking for nearly all species. Plotted from data in table S3 of Stuart et al. (2004).

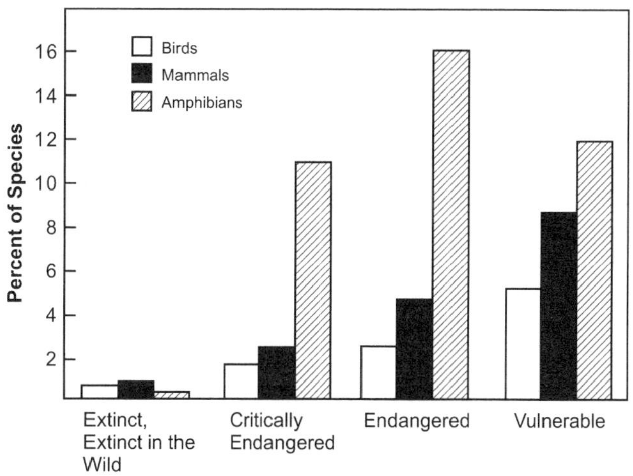


**Fig. 16.2.** Estimated percentages of New World birds, mammals, and amphibians in different IUCN Red List categories. A higher proportion of amphibians appears to be threatened than either birds or mammals. After Young et al. (2004).

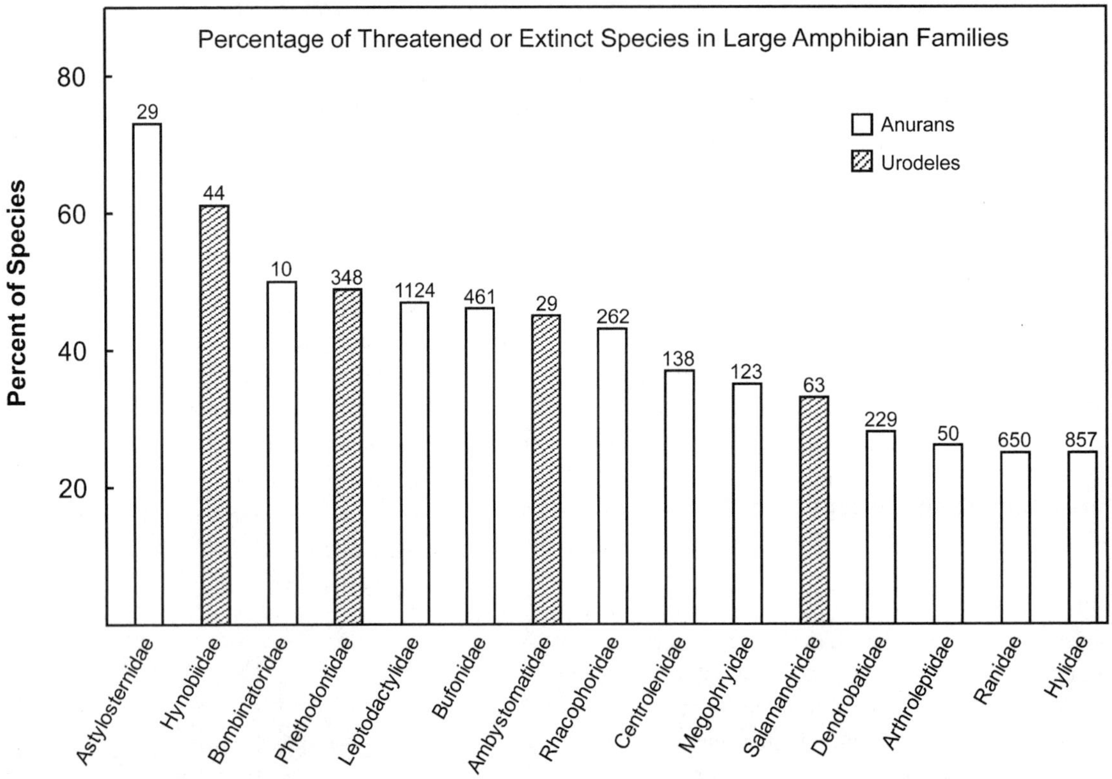


**Fig. 16.3.** Estimated percentages of threatened or extinct species of amphibians in large families of amphibians. Only those families containing at least 10 species, with at least 25% of species threatened or extinct, are included. Quantitative information on the status of these families is available for only a small fraction of all species, so these estimates are tentative. Numbers above each bar show estimated total number of described species in the family. Plotted from data in table S4 of Stuart et al. (2004).

than others, due to differences in ecological adaptations, accidents of geography, or other factors (fig. 16.3).

Declines of reptile populations have received far less attention in the popular press than have declines of amphibian populations, but these animals appear to be susceptible to many of the same environmental changes that threaten amphibians, and many more reptile species are known to have gone extinct (J. Gibbons et al. 2000; Beebee and Griffiths 2005). Some threats, such as human exploitation of eggs or adults for food, are greater for reptiles than for amphibians because of their larger size. Indeed, many of the largest species of reptiles appear to be at greatest risk, including crocodilians, sea turtles, tortoises, freshwater turtles, and large island lizards (G. Webb, Manolis, and Whitehead 1987; J. Ross 1992; Klemens 2000). In addition, there is abundant evidence for widespread declines of many species of birds and mammals, including widely distributed generalist predators such as the eastern spotted skunk (*Spilogale putorius*) (Gompper and Hackett 2005).

Despite the apparent limitations of amphibians, the group as a whole has been remarkably successful. There currently are more described species of living amphibians than species of mammals (Glaw and Köhler 1998). New species are being described at a rapid rate, especially from poorly explored tropical regions such as New Guinea, Sri Lanka, Southeast Asia, and Madagascar, and previously described species often turn out to be complexes of cryptic species (Hanken, Wake, and Savage 2005). Consequently, current estimates of total amphibian diversity probably are too low (Collins and Halliday 2005).

Amphibians as a group have been around for hundreds of millions of years and have experienced dramatic global changes in climate, habitat structure, and even the arrangement of the continents, sometimes with only modest changes in the assemblage of species present. Indeed, many living species of amphibians are very old, in some cases dating back to the Miocene or earlier (Steele et al. 2005), so they not only have experienced major changes in climate, but have witnessed the rise of major mountain ranges as well. In North America and Europe, amphibian species alive today have moved southward with the expansion of the glaciers and then moved back north as the ice retreated, no doubt experiencing changes in breeding phenology and other aspects of their ecology along the way. In many parts of North America and Europe, the same species of amphibians present today were present during the Pleistocene, even though there have been dramatic changes in the mammal faunas of the same regions (Holman 1995, 1998a, b; Delfino, Rage, and Rook 2003; Delfino 2004, 2005). Over even shorter time scales, such as the last 1,000 years, there have been substantial and often quite rapid changes in temperature (Jones, Osborn, and Briffa 2001; Carey and Alexander 2003). Hence the issue is not whether amphibians can adapt to environmental change, but whether the speed at which such changes are now occurring exceeds the rate at which amphibians can adapt, either through evolutionary change or plastic responses to the environment.

It also is important to keep in mind that the decline of amphibian populations probably is not a new phenomenon. It seems likely that as soon as humans began to occupy the natural landscape, populations of many other animals, including amphibians, began to decline and disappear. This process undoubtedly accelerated with the conversion of large tracts of land to agriculture, and accelerated again with the industrialization and urbanization of much of the human landscape and the conversion to more intensive agricultural methods, including the application of vast quantities of synthetic chemicals. A key question is whether the decline of amphibian populations has now accelerated to a new phase or represents a continuation of an ongoing process that simply failed to capture the attention of biologists and the public. A related issue is whether new types of threats to amphibian populations have emerged in recent years (Collins and Storfer 2003).

Data from Dickinson County, Iowa, in the northwestern part of the state, illustrate decades-long declines of amphibian populations. This area contains many wetlands known as prairie potholes that range in size from small, temporary ponds to large lakes, many of which are used by amphibians for breeding. This area is not considered a hotspot of amphibian population declines, but by the 1990s, two of the seven species (28%) present in the 1920s had disappeared (Lannoo et al. 1994). One of these was Blanchard's cricket frog (*Acris crepitans blanchardi*), a species that once was abundant. Furthermore, the abundance of the most common species, the leopard frog (*Rana pipiens*), had decreased by two to three orders of magnitude since the early 1900s. At that time, this species was extraordinarily abundant, with up to 20 million individuals being commercially harvested each year. The decline of these species probably has been underway for decades. The most likely causes appear to be drainage of more than 90% of the wetlands that existed before human settlement and the more recent introduction of nonnative game fishes and bullfrogs into large wetlands. This sort of long-term decline of amphibian populations undoubtedly has occurred in many parts of the world as human populations have expanded and use of the land has become more intensive. It is possible that the apparent acceleration of this process in the past two or three decades has as much to do with changing rates of environmental degradation as with inherent characteristics of amphibians that make them particularly susceptible to such changes.

In this section, I will briefly review some aspects of amphibian morphology, physiology, behavior, and ecology that may affect their vulnerability to environmental change. All of these aspects of their biology were discussed in much more detail in preceding chapters, so I have not included detailed literature citations here.

## Morphological and Physiological Traits

### Small Body Size

As a class, the living amphibians are the smallest of terrestrial vertebrates, and miniaturization has been a recurrent theme in amphibian evolution (see chapter 1). Amphibians range in size from roughly the size of a human thumbnail to the size of a cat, but most have a body mass of less than 20 g. Small body size, especially when coupled with a highly permeable skin, can make amphibians vulnerable to any sort of environmental change that results in drying of microhabitats, from local deforestation to global climate change. On the other hand, small body size provides amphibians with some advantages as well. Small-bodied amphibians are able to occupy microhabitats unavailable to most other vertebrates, including root traces and animal tunnels in the soil, spaces under or in rotting logs, moss mats on trees, cracks in rocks, and other sheltered spots. Even degraded habitats may retain some microhabitats, such as fallen logs, required by terrestrial amphibians for survival, so these animals may be able to persist in areas where larger vertebrates have been extirpated. The small body size of amphibians also means that these animals usually exhibit very limited movement. This can be a disadvantage when habitats are fragmented and separated by areas through which amphibians cannot move, although some small amphibians have surprising abilities to move through inhospitable habitat (D. Marsh et al. 2004). It also can be an advantage, because individuals can persist in relatively small patches of suitable habitat (see the following).

### Ectothermic Metabolism

The low energy demands of amphibians also provide an advantage by allowing large numbers of individuals to live in relatively small patches of habitat. Indeed, both aquatic and terrestrial amphibians sometimes reach extraordinary population densities not realized by any other group of vertebrates. This means that difficulties in locating mates and other effects of population fragmentation may be minimized by the relatively high densities of some amphibian populations. On the other hand, the ectothermic physiology of amphibians and their heavy reliance on anaerobic metabolism may limit their ability to disperse over long distances to new habitat patches. Rowe, Hopkins, and Bridges (2003) argued that amphibians may be susceptible to any environmental changes that decrease food resources available for energy storage used for reproduction, but it is not clear that they would be any more vulnerable to such changes than any other vertebrates. Indeed, they may do better in resource-limited environments than endothermic birds and mammals.

#### Permeable Skin

Of all the traits that can make amphibians vulnerable to environmental perturbations, perhaps none is cited more frequently than their permeable skin (e.g., Boone et al. 2003; Rowe, Hopkins, and Bridges 2003). The free flow of water and other materials across the skin is the key to the success of amphibians, enabling them, among other things, to make use of the skin for respiration. It also enables amphibians to rehydrate quickly when water becomes available. On the other hand, a permeable skin makes amphibians highly susceptible to desiccation. This is likely to be especially true for species that normally live in habitats with abundant moisture, such as tropical rainforests. Those that live in habitats with fluctuating water supplies, such as savannas and deserts, are less likely to be eliminated by prolonged drought because they are physiologically and ecologically adapted to variable water supplies.

One might also expect that a permeable skin would make amphibians susceptible to infectious biological agents such as bacteria and fungi. Indeed, as discussed later in this chapter, there is evidence linking some amphibian population declines to fungal infections. On the other hand, under most circumstances, amphibians are seldom encountered in the field with obvious signs of infection, even when they have suffered serious wounds. The principal reason for this is that amphibians produce a host of biologically active peptides in the skin (Erspamer 1994; B. T. Clarke 1997; Rinaldi 2002). Most work has focused on the toxic or distasteful properties of these secretions as defenses against predators (see chapter 14). Many of these compounds also have antibacterial or antifungal properties and have been shown experimentally to eliminate infections by naturally occurring soil bacteria (Simmaco et al. 1998). Indeed, production of antibacterial peptides appears to be induced by contact with microbes in the environment (Mangoni et al. 2001). Some of these compounds are at least partially effective against infectious agents associated with declining amphibian populations (Rollins-Smith et al. 2002, 2003, 2005; Carey et al. 2003). Antimicrobial compounds have been found in the skin of many amphibians that are not especially toxic or distasteful to vertebrate predators, such as *Plethodon cinereus* (Fredericks and Dankert 2000), and it may be that some distasteful or toxic defensive secretions originally evolved as shields against infection (Daly, Garraffo, and Spande 1993). The presence of such compounds may enable amphibians to live in habitats that normally are filled with a great diversity of bacteria and fungi, including eutrophic ponds and moist soil. Selection imposed by the diversity of microbial agents to which amphibians are exposed presumably has led to the diversification of peptides with activity against particular microbes (Duda, Vanhoye, and Nicholas 2002).

Unfortunately, the amphibian skin provides little or no protection against synthetic chemicals, such as pesticides. Most biologists have assumed that amphibians are unusually vulnerable to such compounds. In fact, in standard laboratory tests, amphibians sometimes are less sensitive to some common chemical contaminants than are other groups of organisms (Fontenot, Noblet, and Platt 1996; M. Harris, Bishop, Struger, Ripley and Bogart 1998; M. Harris, Bishop, Struger, van den Heuvel et al. 1998; Glennemeier and Begnoche 2002), although embryos and larvae may be more sensitive to some chemicals than are adults (Boone and Bridges 2003c). There also is considerable variation both within and among species in sensitivity to pesticides and other contaminants (e.g., Bridges and Semlitsch 2000, 2005a, b; Westerman et al. 2003). Nevertheless, chemical contaminants do accumulate in the tissues of amphibians, having been taken up either through the skin or ingested with their food (e.g., Fontenot et al. 2000; Fagotti et al. 2005), and these compounds can have a variety of detrimental effects on the health and survival of individual animals (see reviews in Sparling, Linder, and Bishop 2000 and Linder, Krest, and Sparling 2003). Pesticides and other chemical contaminants also can affect antipredator behavior and other behavioral traits that may influence survivorship (e.g., Boone and Semlitsch 2001). Conversely, exposure to stressors in the environment such as predators or competitors can increase the lethality of relatively low concentrations of some pesticides (Relyea and Mills 2001). Some chemical contaminants can actually enhance growth of tadpoles by eliminating competitors such as zooplankton that also feed on phytoplankton (Boone and Bridges 2003c).

### Behavioral Traits

#### Limited Dispersal Ability

As mentioned previously, the small body size and limited capacity for activity of most amphibians ensures that they lead relatively sedentary lives (see chapter 6). Indeed, in some groups, such as plethodontid salamanders, individuals can spend their entire lives in an area the size of a small room, and they may limit their daily activities to an area the size of a dining room table. Consequently, these animals can persist for years in relatively isolated patches of suitable habitat, with relatively little impact on either population size or the genetic structure of populations. One study of red-backed salamanders (*Plethodon cinereus*), for example, found slightly more genetic differentiation among fragmented populations in a suburban area than among subpopulations in an undisturbed landscape, but the genetic effects were not of the same magnitude as those found in other fragmented amphibian populations (Gibbs 1998c). Furthermore, despite their relatively sedentary habits, red-backed salamanders are able to cross open ground between patches of forest habitat, so their potential for reoccupying habitats after local extinction events is surprisingly high (D. Marsh et al. 2004).

The small home ranges of many terrestrial amphibians (see chapter 6) may make them less sensitive to the effects of habitat fragmentation than are animals with larger home ranges, especially birds and mammals (e.g., Vallan 2000; Bowne and Bowers 2004; Cushman 2006). Several studies have shown that predictions of extinction risk based on the extent of forest habitat reduction or fragmentation tend to yield overestimates of the number of amphibian species at risk compared to larger, more mobile animals such as birds and mammals (Brook, Sodhi, and Ng 2003; Grelle et al. 2005). Aquatic-breeding amphibians distributed in metapopulations consisting of locally isolated patches of aquatic habitat appear to be more susceptible to the effects of habitat fragmentation, such as increased inbreeding and chance extinction of local populations. Often the ability of aquatic amphibians to reoccupy habitat patches after a local extinction event is constrained by their limited abilities to disperse across the intervening terrestrial habitat and their fidelity to breeding sites (Storfer 2003; L. Andersen, Fog, and Damgaard 2004; Cushman 2006). Nevertheless, some amphibians with relatively poor dispersal abilities do manage to colonize suitable habitats relatively quickly (e.g., *Ambystoma macrodactylum;* Funk and Dunlap 1999). Furthermore, even when dispersal by adults is uncommon, long-distance dispersal by juvenile amphibians probably is much common than generally recognized (Rothermel 2004). Indeed, many amphibian populations inhabiting networks of ponds probably have too much juvenile dispersal between ponds to be considered true metapopulations (M. A. Smith and Green 2005).

### Antipredator Behavior

Various types of antipredator behavior can affect the vulnerability of amphibian populations to decline or extinction. Many amphibians, especially aquatic larvae, have evolved a remarkable ability to detect and identify potential predators and alter their behavior and even their morphology accordingly (see chapter 14). In many cases, these responses are predator-specific, which means that amphibians often do not respond appropriately to exotic predators such as fish to which they have not been previously exposed. Consequently, introduced predators can have a rapid impact on amphibian populations, although there is some evidence that some amphibian species eventually can acquire some capacity to recognize and escape from such predators. Some types of habitat management may interfere with antipredator behavior. For example, removal of aquatic vegetation from ponds, either mechanically or chemically, can remove vital cover into which amphibian larvae normally retreat when exposed to predators (J. E. Murphy, Phillips, and Beasley 2000).

## Life Histories and Reproduction

### Dependence on Water

All amphibians, even those with nonaquatic eggs, depend to some extent on the availability of water for reproduction (see chapter 10), and species that breed in standing or flowing water may be particularly vulnerable to environmental degradation and change. In many parts of the world, freshwater habitats are among the most threatened of all ecosystems, being highly susceptible to biological or chemical pollution, acidification, changes in temperature, silt deposition, and other physical alterations brought about by deforestation and changes to the surrounding terrestrial landscape. Freshwater ecosystems also are among the most heavily impacted by introduced predators and other exotic organisms. The result is an unusually high risk of extinction among freshwater organisms, from plants and invertebrates to fishes and amphibians (Lydeard and Mayden 1995; Benz and Collins 1997; Ricciardi, Neves, and Rasmussen 1999; Dudgeon 2000, 2002; Loh 2000; Warren et al. 2000; Brönmark and Hansson 2002; Saunders, Meeuwig, and Vincent 2002).

Changes in aquatic habitats have the potential to eliminate whole assemblages of amphibians that depend on them for breeding, even when the surrounding landscape appears more or less intact. Probably at high risk are species that breed in clear, cold streams with low nutrient input, because these species are likely to have the narrowest tolerance limits for changes in temperature, nutrient input, or silt load associated with forest clearing, pollution, and other habitat alteration (Corn and Bury 1989; Welsh and Ollivier 1998; Gillespie 2002). Stream-breeding amphibians have declined in several parts of the world (Lips 1998; S. Williams and Hero 1998; Corn, Bury, and Hyde 2003; Young et al. 2004), although it is not clear that these declines all can be attributed to a common cause. Amphibians that inhabit unpolluted, nutrient-poor ponds and lakes, like those found in many montane regions, also are vulnerable because of low tolerance for habitat alteration. For example, stream and lake-dwelling frogs of the genus *Telmatobius* from high altitudes in the Andes are declining, although not necessarily from a single cause (Young et al. 2004). Also at risk are amphibians that depend on vernal pools and other temporary bodies of water. These pools not only provide breeding sites for many amphibians that are intolerant of predators that live in more permanent water, but they also provide important habitats for amphibians moving between larger wetlands. Temporary wetlands often are not afforded the same level of legal protection as more persistent bodies of water and have been disappearing in many parts of the world. They also are highly susceptible to any changes in climate that decrease the amount of time that such pools hold water (Semlitsch and Bodie 1998; Snodgrass, Komorowski et al. 2000; Semlitsch 2002, 2003c; Jakob et al. 2003; R. Brooks 2004). This is especially true for very small vernal pools (< 0.5 hectare), which can be important breeding sites for many amphibians (Egan and Paton 2004).

If freshwater habitats are especially endangered, then amphibians that spend most of their lives on land, particularly those with terrestrial reproduction, might be expected to derive some advantage from their lack of exposure to degraded aquatic habitats. In the rainforests of northern Australia, for example, terrestrial-breeding microhylids and other largely terrestrial frogs have not experienced the declines seen in stream-breeding frogs in the same area (S. Richards, McDonald, and Alford 1993; S. Williams and Hero 1998). A similar trend was seen at Monteverde in Costa Rica, where half the anuran species disappeared in 1987. Species of *Eleutherodactylus* appeared to be less affected than were aquatic-breeding frogs, although three species closely associated with streams did disappear. The only anuran species to reappear in large numbers by 1994 was a species of *Eleutherodactylus,* and several species that had not declined appeared to be increasing in abundance (Pounds et al. 1997). On the other hand, the spectacular golden toad (*Bufo periglenes*) also disappeared, and this was not a stream breeder, although it did lay aquatic eggs. In Puerto Rico and elsewhere in Latin America, there have been declines in some species of *Eleutherodactylus* and other frogs with nonaquatic reproduction, while other species, such as *E. coqui,* have flourished (Woolbright 1997; Lips 1999; Stallard 2001; Young et al. 2001). In Latin America as a whole, anurans with nonaquatic reproduction appear to be more at risk than are aquatic breeders (Young et al. 2004), but this pattern may be confounded by the fact that most terrestrial breeders are forest-interior specialists and therefore are vulnerable to deforestation and habitat fragmentation.

In some areas, pond-breeding and terrestrial-breeding species appear to be somewhat less affected than are stream-breeding species, but the data are not sufficiently robust to allow rigorous statistical analysis (Crump 2003). The picture for terrestrial plethodontid salamanders with direct development is mixed. Hairston and Wiley (1993) failed to find evidence of population declines in six species monitored for 15–20 years in North Carolina. In contrast, Highton (2005) reported that many species were less abundant in the 1990s at sites where they were common in the 1970s. Because most terrestrial plethodontids live in forests, they may be particularly susceptible to destructive forestry practices that eliminate or fragment local populations (Wyman 2003).

Terrestrial amphibians, especially those with nonaquatic reproduction, also are susceptible to any climate changes or habitat alterations that produce changes in temperature, length of the dry season, amount of rainfall in the wet season, soil moisture content, and other environmental variables that directly impact amphibians (Boone et al. 2003). Indeed, amphibians with nonaquatic reproduction generally are confined to moist forest habitats, and may be even more sensitive to habitat disturbance or degradation than are amphibians adapted to more open habitats (Pearman 1997; Wyman 2003). Not only can terrestrial amphibians be affected directly by dry conditions, but they also may experience changes in their food resources that affect population size (Donnelly and Crump 1998). The terrestrial embryos of amphibians with direct development may be especially vulnerable if they are laid in relatively shallow, unprotected sites in leaf litter, or have low tolerance for desiccation (e.g., N. Mitchell 2002a).

### Use of Multiple Habitats

Many amphibians make use of distinctly different habitats, either at different stages of the life cycle, or at different times of year (fig. 16.4). The most common situation is for amphibians to breed in aquatic habitats, but spend the rest

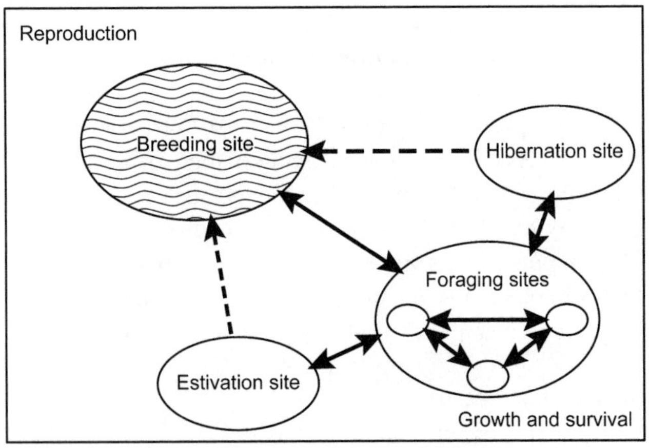


**Fig. 16.4.** Diagram of habitat use by an amphibian that breeds in water, but spends at least part of the year on land. Some species move from aquatic breeding sites to terrestrial foraging areas, where they may spend most of the year and may even hibernate or estivate, making a return trip to water the following year. Other species use distinct terrestrial or aquatic habitats for hibernation and may move directly from these sites to breeding ponds in the spring. Desert-dwelling amphibians often estivate underground and emerge after heavy rains to move directly to temporary water for breeding, followed by a short period of foraging activity. Two-way movements are shown as two-headed, solid arrows. One-way movements are shown as single-headed, dashed arrows. After Semlitsch (2003c).

of the year in surrounding terrestrial habitats (Dodd 1996; S. Pope, Fahrig, and Merriam 2000; Richter et al. 2001; Trenham 2001). Other species move between different types of habitats for reproduction, feeding, and hibernation (Lamoureux and Madison 1999; Matthews and Pope 1999; Pilliod, Peterson, and Ritson 2002; Semlitsch 2003b). In either case, amphibians using multiple habitats are vulnerable to habitat degradation in one type of habitat, even if another remains intact. If wetland management plans fail to include protection of sufficient surrounding terrestrial habitat, then amphibian populations may decline (Semlitsch 1998, 2003b). Deforestation and other forms of habitat degradation may eliminate aquatic-breeding amphibians even if the aquatic habitat appears unaltered. This is much the same problem faced by migratory birds that move from tropical habitats in the winter to temperate habitats during the breeding season, but on a much smaller spatial scale (T. Martin and Finch 1995). Consequently, the problem probably is easier to address from a conservation perspective than is habitat preservation for migratory birds.

### Complex Life Cycles

In most amphibians, movement between habitats is accompanied by metamorphosis from an aquatic larva to a terrestrial adult (see chapter 13). The transition period during metamorphosis is a particular vulnerable one for many species, because they are exposed to predators in two different habitats and exhibit reduced locomotory ability in both habitats (see chapter 14). Most amphibian ecologists have focused on the aquatic stages of the life history and the importance of egg and larval mortality in driving local population dynamics. This is especially true for amphibians that breed in temporary ponds that fluctuate in hydroperiod from year to year. Nevertheless, for some species, mortality of terrestrial juveniles may be at least as important as larval mortality in determining population dynamics and the vulnerability of a population to extinction (Biek et al. 2002; Conroy and Brook 2003), and in long-lived species, adult mortality may be important as well (B. Schmidt, Feldmann, and Schaub 2005). This means that we need to pay close attention to preservation of terrestrial habitats that are essential to juveniles and adults once they leave a pond.

There is evidence that immune responses of amphibians are suppressed during metamorphosis as the larval immune system transitions to the adult immune system. High levels of stress hormones, which are associated with metamorphosis (see chapter 13), also suppress immune responses, so any additional environmental stresses that increase production of these hormones could adversely affect immunity to disease (Rollins-Smith 1998, 2001). Some immune responses, such as the production of antimicrobial peptides in the skin, do not even begin until after metamorphosis is complete.

Any environmental changes that prolong or disrupt the process of metamorphosis are likely to have detrimental effects on amphibian populations. Metamorphosis is regulated by complex hormonal interactions, and there is evidence that some types of chemicals, including certain pesticides, can disrupt or alter hormone action, resulting in retardation of growth and development, incomplete or premature metamorphosis, developmental abnormalities, and altered sexual maturation, all of which are likely to negatively impact survival and reproduction (Carey and Bryant 1995; T. Hayes 2000, 2004; Ouellet 2000; T. Hayes, Collins, Lee, Mendoza, Noriega, Stuart, and Vonk 2002). Because of their complex life cycles, amphibians could be more vulnerable to such endocrine disruptors than are other vertebrates such as reptiles, birds, and mammals (Kloas 2002). Nevertheless, all vertebrates, including humans, share common endocrine controls of reproduction and are potentially at risk of serious endocrine disruption by chemical contaminants, so amphibians are not unique in this regard (D. Kime 1998; Guillette 2000; Fossi and Marsili 2003; Gies 2003; Kawai, Kobayashi, and Kaneko 2003; Hayes 2005).

### Boom and Bust Population Cycles

Many amphibians, especially those that breed in temporary ponds, experience boom or bust population cycles, with a few good years of reproduction often followed by several bad years (Pechmann et al. 1989; Semlitsch et al. 1996; Alford and Richards 1999; Richter and Seigel 2002; Pechmann 2003; Richter et al. 2003). In amphibians with this type of population ecology, the number of breeding adults and the number of eggs produced often are poor predictors of the number of metamorphs emerging from the pond (see also chapter 15). Large fluctuations in population size, which sometimes can result in the complete disappearance of a local population, complicate the task of distinguishing natural population fluctuations from sustained population declines (Pechmann et al. 1991; Pechmann and Wilbur 1994; Pounds et al. 1997; Alford, Dixon and Pechmann 2001; Pechmann 2003; Halliday 2005). In general, one can expect that populations that undergo frequent and dramatic fluctuations in size will be susceptible to chance extinction events during particularly bad years (D. Green 2003). When habitats are fragmented and populations are isolated from one another, the potential for extinction resulting from stochastic events increases. For example, the single known breeding population of the dusky gopher frog (*Rana sevosa*) in Mississippi is vulnerable to extinction because of large year-to-year fluctuations in population size, number of eggs produced, and embryonic and larval survivorship (Richter and Seigel 2002; Richter et al. 2003).

On the other hand, amphibians that experience boom or bust population cycles typically reach sexual maturity relatively quickly and usually produce large numbers of eggs, life-history traits that allow them to take advantage of favorable breeding conditions that occur unpredictably. It also is likely that juveniles of such species tend to disperse more readily from natal ponds than do species with more stable population dynamics, although juvenile dispersal is poorly studied for most amphibians (see chapter 6). Hence, these species may have a greater capacity to recover from events such as droughts than do species with slow growth and development and low reproductive output, especially if dispersal results in recolonization of breeding ponds. Many pond-breeding amphibians also exhibit considerable plasticity in growth and development in response to rapid pond drying and other environmental variables (see chapter 13). Amphibians that do not normally experience major fluctuations in temperature, rainfall, or the persistence of standing water may be less likely to survive prolonged drought or even subtle changes in yearly rainfall patterns or soil moisture that result from deforestation or global climate change.

## Distribution and Ecology

### Limited Geographic Range

Perhaps the single most important ecological factor that can make a species vulnerable to extinction is a limited geographic range. Because of the limited dispersal abilities of amphibians, many species have very small geographic ranges (Beebee 1996; Duellman 1999c). Although they may be locally abundant, species with limited ranges usually exist in relatively few discrete populations. It is the rate at which populations disappear and new ones are founded that ultimately determines the likelihood that a species will go extinct (D. Green 1997b, 2003, 2005a). If a local population of a species with a limited range disappears, it is less likely to be replaced by colonization from other populations than are populations of more widespread species. Consequently, species with limited ranges often receive disproportionate attention from conservation biologists. For example, many of the North American amphibians currently listed or being considered for listing as threatened or endangered species by the federal government are those with historically small ranges, while others have undergone dramatic reductions in range size. Because listing criteria are largely based on the number of viable populations of a given species, species with limited ranges are more likely to receive government protection.

Members of certain clades of amphibians may be particularly vulnerable to extinction because of small geographic ranges. For example, terrestrial salamanders tend to have much smaller geographic ranges than do anurans in the same regions (Beebee 1996). Many plethodontid salamanders in both North America (Highton 1971; Petranka 1998) and Central America (D. Wake and Lynch 1976; García-París et al. 2000) are confined to single mountains or small mountain ranges. Only a few species in the United States, such as *Plethodon nettingi* and *P. shenandoah* (fig. 16.5 A), currently are listed as threatened or endangered. *Plethodon nettingi* occurs in several dozen disjunct populations, and in some places is abundant, but probably always has had a very small geographic range (Pauley et al. 2005). *Plethodon shenandoah* is found in isolated patches of habitat largely surrounded by more competitive species, especially *P. cinereus,* which prevent *P. shenandoah* from extending its range (Jaeger 1971a, b, 1980a). Many other plethodontids have similar small ranges and also may be vulnerable, but have not received federal protection. For example, the Peaks of Otter salamander, *Plethodon hubrichti* (fig. 16.5 B), has a range only about 20 km long in the Blue Ridge Mountains of Virginia, much of it in forest subjected to periodic timber harvesting (Kramer et al. 1993). Although locally abundant in some populations, this species undergoes dramatic population declines after logging (Sattler and Reichenbach 1998).

Small geographic ranges sometimes result from use of specialized habitats that are naturally limited or patchy in their distribution. For example, salamanders in the genus *Eurycea* have undergone extensive evolutionary divergence in caves and springs on the Edwards Plateau of Texas (see fig. 13.24 in chapter 13). Some species are relatively widespread, but others, such as the Barton Springs salamander (*E. sosorum*) and the San Marcos salamander (*E. nana*), are limited to single springs and therefore are highly vulnerable to local extinction (Chippindale 2000; Chippindale et al. 2000). In the case of the Barton Springs salamander, the only known population lives in and around a popular local swimming area, so its habitat is heavily impacted by human use. In Europe, all of the cave- and crevice-dwelling salamanders of the genus *Hydromantes* (sometimes called *Speleomantes*) have small geographic ranges and are considered vulnerable, while the blind cave salamander *Proteus anguinus* (fig. 1.33 B in chapter 1), with a slightly larger range, is endangered (Corbett 1989). Also at risk are populations of European amphibians that live on offshore islands, some of which are endemic and all of which have inherently limited ranges (Corbett 1989; Gibson and Freeman 1997; Andreone and Luiselli 2000).

Small geographic ranges often reflect patterns of climate change during the last few million years that have resulted in advances and retreats of glaciers, alteration of rainfall patterns, and contractions and expansion of forest and grassland habitats that have influenced the current distribution and genetic divergence of amphibian populations. The size and location of a species' current geographic range often reflects a past history of range expansion and contraction. For example, phylogeographic studies of Western Palearctic brown frogs of the genus *Rana* have shown that the re-

**Fig. 16.5.** Many threatened or endangered North American amphibians have small geographic ranges. (A) The Shenandoah salamander, *Plethodon shenandoah,* is found in three isolated populations in Shenandoah National Park. This species is ecologically similar to the more widespread red-backed salamander, *Plethodon cinereus,* and is confined to dry talus slopes by competition from that generalist species. (B) The Peaks of Otter salamander, *Plethodon hubrichti,* probably always had a very small geographic range and is confined to a 16-kilometer section of the Blue Ridge Mountains of Virginia. Populations currently are stable, but are vulnerable to habitat fragmentation and degradation from logging and defoliation of the forest canopy by gypsy moth caterpillars. (C) The Wyoming toad, *Bufo baxteri,* once was abundant along river floodplains of the Laramie Basin of Wyoming. Populations experienced precipitous declines in the 1970s from unknown causes, and the species has not been seen in the wild since 1987. A captive breeding and reindroduction project so far has not been very successful. (D) The relict leopard frog, *Rana onca,* once was considered to be extinct, but survives in small numbers in southern Nevada. Much of its natural habitat has been destroyed by conversion to agriculture, draining of marshes, large-scale dam projects, and invasion of aquatic habitats by introduced bullfrogs (*Rana catesbeiana*) and other alien predators. Photos by Wayne Van Devender (A–C) and Richard D. Bartlett (D).

stricted ranges of many species in southern Europe result from populations being isloated in southern refugia during glacial periods and undergoing subsequent speciation (Veith, Kosuch, and Vences 2003). Similarly, populations of frogs that inhabit the wet tropical forests of northeastern Australia have experienced repeated range expansions and contractions, with populations being restricted to isolated refugia during glacial periods (C. Schneider, Cunningham, and Moritz 1998). In southeastern Brazil, some forest-interior frogs are found in isolated remnants of the Atlantic forest, surrounded by more arid Caatinga habitats. These populations exhibit considerable genetic divergence resulting from contraction of forest habitats that occurred long before the more recent human destruction and fragmentation of the remaining areas of forest (Carnaval 2002). While the processes that resulted in some amphibian populations having small or fragmented ranges are natural parts of the evolution of these species, the resulting patterns of distribution may make such species especially vulnerable to local extinction when isolated habitat patches are destroyed or severely altered.

Many endangered amphibian species with very small geographic ranges represent relict populations that diverged or became isolated from ancestral populations that were more widespread. The Houston toad (*Bufo houstonensis*) is closely related to the American toad (*Bufo americanus*) and is thought to be a Pleistocene relict that became isolated from a more widespread ancestral population because of increasing aridity. Its historic range has always been small, and it currently is found in only a few isolated populations, most of which are in decline (L. E. Brown and Mesrobian 2005). Similarly, the Wyoming toad (*Bufo baxteri;* fig. 16.5 C) is a close relative of the Canadian toad (*Bufo hemiophrys*) and represents another Pleistocene relict population that became isolated in the Laramie Basin and more recently was reduced to a population in a single lake. It now appears to be extinct in the wild, although captive breeding and reintroduction efforts continue (Odum and Corn 2005). Yet another Pleistocene relict is the black toad (*Bufo exsul;* fig. 16.8 A), found only in Deep Springs Valley, California (J. F. Murphy, Simandle, and Becker 2003). The Vegas Valley leopard frog (*Rana fisheri*) was found only in the wetlands in and around what is now the city of Las Vegas, Nevada, and probably has been extinct since the 1940s (R. Jennings 2005). It is a close relative of the aptly named relict leopard frog (*Rana onca;* fig. 16.5 D), which once was considered to be extinct (J. Jaeger et al. 2001). It had a very limited distribution in Arizona, Utah, and Nevada, and the range has recently contracted to a few sites in Nevada (Bradford, Jaeger, Jennings 2004). In Europe, the Mallorcan midwife toad (*Alytes muletensis*) was isolated from ancestral *Alytes* populations on the Balearic Islands. It now is confined to only a few mountain torrents on part of the island of Mallorca, and is therefore highly vulnerable to extinction (Corbett 1989). The European salamanders in the genera *Euproctus* and *Calotriton* all have very small ranges restricted to high elevations, and are thought to be Pleistocene relicts (Beebee 1996; Carranza and Amat 2005).

### Hot Spots for Range-Limited Amphibians

It seems likely that hundreds of species of amphibians around the world are similarly imperiled by very small geographic ranges, but in most cases, the status of these populations is poorly known. Of particular concern to conservation biologists are regions of the world in which severely endangered habitats contain a large number of endemic amphibian species with limited ranges. Indeed, areas of high species diversity and high levels of endemism often contain a high percentage of range-limited species (Duellman 1999b; C. Seymour et al. 2001), many of which also tend to be habitat specialists (Wijesinghe and Brooke 2004). In such situations, habitat destruction has the potential to eliminate whole assemblages of unique amphibians simply because they are found nowhere else. Many areas of the world where concentrations of endemic or range-limited amphibian species occur are relatively small geographically, so protection of a very small proportion of the earth's terrestrial habitats has the potential to protect a large percentage of amphibian species. Some, but not all, of these areas have been designated as biodiversity "hot spots," based largely on analysis of species diversity and endemism of other organisms, especially plant species (N. Myers 1988, 1990, 1996; T. Brooks et al. 2002).

Unfortunately, areas that currently are protected do not always correspond to areas of high species diversity or endemism for amphibians or for other organisms (Prendergast et al. 1993; Young et al. 2004). In subequatorial Africa, for example, where parks and reserves have tended to focus on large, charismatic mammals, most regions with concentrations of endemic or range-limited amphibian species are not currently protected (C. Seymour et al. 2001). In Namibia, which has few endemic amphibian species, but surprisingly high species diversity for a desert region, most of the human population is concentrated near wetlands, which also harbor most of the amphibian species. Most such wetlands are not protected (B. Curtis et al. 1998). In South Africa, on the other hand, the picture is somewhat brighter, with an estimated 96% of amphibian species occurring on protected lands (Drinkrow and Cherry 1995). In general, for sub-Saharan Africa, proposals for land conservation that focus on centers of high species diversity rather than areas already protected or areas chosen to represent particular biomes appear to be most likely to protect the largest proportion of amphibian species (Burgess et al. 2002).

Duellman (1999b) identified over 40 relatively small regions of the world with high amphibian species diversity, a high percentage of endemic species, or both (fig. 16.6). Not all of these regions are equally threatened, but many are in tropical countries where deforestation and other forms of habitat destruction continue at a rapid pace. Of particular concern are montane tropical forests, which are threatened throughout the world. These forests tend to have high amphibian species diversity and large numbers of endemic species. Furthermore, the geographic ranges of many montane amphibian species are quite limited because of physical and altitudinal barriers to dispersal.

High on the list of areas of special concern to amphibian biologists is Madagascar, a very poor country with a high population growth rate (Andreone and Luiselli 2003). Nearly all amphibian species in Madagascar are endemic, and species diversity is high. The total ranges of most species are poorly known, but many are known from only a few localities, in some cases only the type locality where the species was first collected. With increasing fragmentation of the remaining forested regions of Madagascar, there is the potential for a large assemblage of unique amphibians to disappear (Andreone and Luiselli 2003; Lehtinen, Ramanamanjato, and Raveloarison 2003). The amphibian fauna of continental

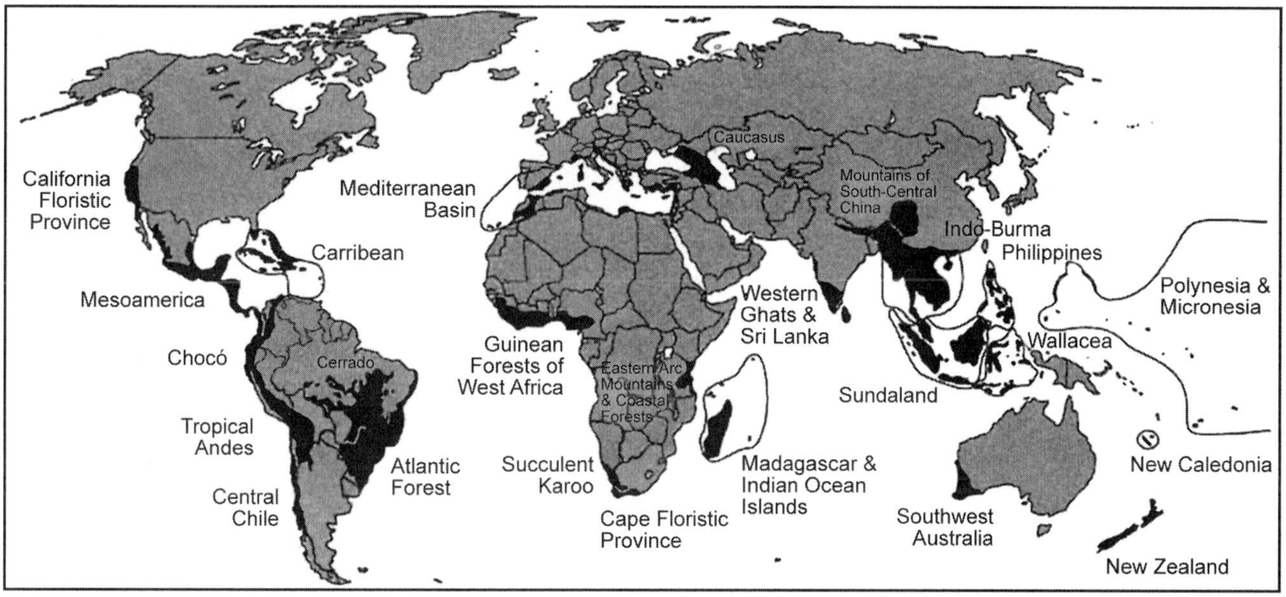


**Fig. 16.6.** Major hotspots of amphibian species diversity or endemism. Names and areas in black show regions with unusually large numbers of amphibian species, a large proportion of endemic species, or both. Species diversity tends to be especially rich in tropical montane regions and in moist forested coastal areas in both the temperate zone and the tropics. Many of these areas also have high proportions of endemic species. Some of these amphibian hotspots unfortunately are located in regions with poor and growing human populations and rapid rates of deforestation and habitat fragmentation. After Hero and Shoo (2003).

Africa has received less attention than that of most parts of the world, but there are a number of relatively small montane regions in subequatorial Africa that are rich in amphibian species, most of them endemic and most with limited ranges. These include highland areas of Tanzania, Kenya, Ethiopia, Zaire and surrounding countries, Cameroon, Malawi, Zimbabwe, Mozambique, Zambia, and South Africa (Poynton 1998, 1999, 2003; Channing 2001; C. Seymour et al. 2001; T. Brooks et al. 2002).

Other tropical or subtropical montane regions with unique assemblages of amphibian species, many with limited ranges, include the Andes of northern South America (Ecuador, Venezuela, and Colombia), the tropical southern Andes (Bolivia and Peru), the Guiana Highlands of Venezuela and the Guianas, the highlands of southern Mexico and Central America from Guatemala to western Panama, the highlands of Haiti, the highlands of Borneo and New Guinea, the highlands of Burma, Thailand, and Vietnam, and the Western Ghats and southern Himalayan Mountains of India (Duellman 1999b; Bossuyt et al. 2004; Dutta et al. 2004; Young et al. 2004).

A recent addition to the list is Sri Lanka, a small island with more than 200 species of amphibians, including many endemics, most of which are concentrated in montane rainforests. This is comparable to the diversity of much larger areas such as Madagascar, Borneo, and the Philippines (Pethiyagoda and Manamendra-Arachchi 1998; Meegaskumbura et al. 2002; Bossuyt et al. 2004). In fact, Sri Lanka may have the highest diversity of amphibian species per unit area of any site in the world. The status of amphibian populations in many of these regions is poorly known, although most are vulnerable to deforestation and other forms of habitat destruction. Sri Lanka already has lost more than 90% of its original rainforest.

Certain coastal regions of the world have unusual combinations of temperature, rainfall, and vegetation that create unique and geographically limited habitats for amphibians. As in many montane regions, these areas tend to have high amphibian species diversity and a large proportion of endemic and range-restricted species. One of the most severely threatened assemblages of amphibians is found in the remaining patches of Atlantic Coastal Forest in Brazil (Morellato and Haddad 2000; Saatchi et al. 2001; Grelle et al. 2005), where 93% of the amphibian species are endemic (Duellman 1999b) and many have quite limited ranges. A number of frog species in this region have declined or disappeared altogether in the last 20 years, and increasing destruction and fragmentation of the forest threatens a unique assemblage of species (Heyer et al. 1988, 1990; Weygoldt 1989a; Young et al. 2004). Also threatened are the many endemic amphibians of the unique Southwestern Cape Region of South Africa, which is subjected to increasing urbanization (Drinkrow and Cherry 1995; C. Seymour et al. 2001; Minter et al. 2004). In Australia, a number of frog species have small ranges limited to acidic habitats such as heathland, *Melaleuca* swamps, sedgeland, and *Banksia* woodlands along the coast of southeastern Queensland and northeastern New South Wales, an area with the fastest growing human population in the country. Other frogs in this region are limited to upland mesic forests that also are under pressure from development. While some species are doing well, others have declined dramatically (Hines, Mahony, and McDonald 1999). Coastal habitats in Australia with somewhat smaller numbers of amphibian species, but with a higher percentage of endemic species, some with limited ranges, are found on the southwestern and southeastern coasts of the continent. Other coastal habitats with relatively large proportions of endemic or range-limited amphibians include the temperate forests of southern Chile and the northwest coast of North America (Duellman 1999b).

### Habitat Specialization

Some amphibians have generalized habitat requirements, while others are quite specialized. All else being equal, one can expect species with specialized habitat requirements to be more vulnerable to extinction, especially if suitable habitats are scarce, patchy in their distribution, or limited geographically. An example of an extreme habitat specialist is the North American green salamander (*Aneides aeneus;* fig. 14.25 D in chapter 14). This species inhabits crevices in rock outcrops in mature forests. Populations tend to be small and widely separated from one another, so are highly vulnerable to local extinction. Consequently, this species is listed as threatened or endangered in most of the states encompassed by its range (Madej 1998). Populations in the southern parts of the species' range, especially in North Carolina, experienced catastrophic declines in the late 1970s, with further declines and some local extinctions recorded in the 1990s (J. Mitchell et al. 1999; Corser 2001; Pauley and Watson 2005). Some local populations were eliminated by housing developments and other forms of habitat destruction, but the causes of declines in other populations are unknown. Because of the patchy distribution of suitable habitat, recolonization of local populations is likely to be extremely slow, so the status of this species, at least in parts of its range, is of major concern. Conservation of these salamanders is further complicated by the likelihood that the single named species actually is a complex of up to four cryptic species. If so, then the geographic ranges of these highly specialized species are even smaller than currently recognized (Pauley and Watson 2005).

Not all habitat specialists have limited ranges. For example, the eastern spadefoot toad (*Scaphiopus holbrookii;* fig. 16.7) can be considered a habitat specialist in that it requires sandy or loamy soils and temporary ponds for reproduction, but it has a wide distribution in the eastern and southern United States (Palis 2005). In regions where such habitats are common, such as the southeastern states, this species can be locally abundant, but at the periphery of its range, where such habitats are rare, it is threatened or endangered. In Connecticut, for example, spadefoot toads historically inhabited areas that mostly have been converted to sand and gravel operations or developed into shopping malls or housing developments. The species is listed as endangered in the state, but it remains relatively common on nearby eastern Long Island and Cape Cod, where sandy habitats are plentiful (Klemens 1993). In Europe, the natterjack toad (*Bufo calamita*) is similarly specialized for sandy habitats. Although it has a wide distribution on the continent, its favored habitats tend to be patchily distributed and increasingly threatened by human activities. Consequently, the number of viable populations has declined over much of the northern part of its range, even though the overall size of the range has not changed very much (Beebee 1977, 1983; Banks, Beebee, and Cooke 1994; Beebee 1996; Sinsch 1998; Buckley and Beebee 2004).

**Fig. 16.7.** The eastern spadefoot toad, *Scaphiopus holbrookii,* requires sandy or loamy soils and temporary ponds for reproduction. It has a large geographic range, extending from southern New England to Florida and Louisiana. It remains abundant in the southeastern United States, but has become rare at the periphery of its range, as in Connecticut, where habitat destruction has eliminated most historic populations. Photo by Wayne Van Devender.

Extreme habitat specialization can make some species vulnerable to natural environmental changes that can lead to the decline or extinction of local populations. For example, a tiny marsupial hylid frog from the Atlantic Coastal Forest of Brazil, *Fritziana ohausi,* was common in the area around Boracéia through the 1970s. Females of this species carry eggs on their backs and deposit tadpoles in small bodies of water, especially in hollow bamboo stems. During the 1970s, the species of bamboo used by this frog flowered synchronously and died, and the frog disappeared from the area, although it may still persist in lower numbers in tree holes and similar breeding sites (Heyer et al. 1990). In a nearby area dominated by ground bromeliads (*Vriesea neoglutinosa*), a fire of unknown origin in 1995 caused a crash in a population of a hylid frog, *Phyllodytes luteolus,* which lives entirely in the bromeliads (Papp and Papp 2000). Such population crashes probably have minimal impact on the survival of a species when that species is relatively widespread, but as populations become increasingly fragmented and the ranges of species are reduced, such events can be catastrophic.

Use of specialized habitats also could make amphibians vulnerable to the secondary effects of selective harvesting of key plant species in the habitat. For example, many species of bolitoglossine salamanders in Mexico and Central America live in bromeliads that grow on the branches of oaks and other trees (D. Wake 1987). Some species of bromeliads in the genus *Tillandsia* are heavily harvested for the home decoration market. A regulated, sustainable commercial harvest of bromeliads has been proposed as a strategy to reduce deforestation by providing a source of income for local people derived from the forest (Wolf and Konings 2001). This seems like an innovative proposal from the standpoint of sustaining populations of bromeliads and protecting the forest, but we have no information on the number, density, or spatial distribution of bromeliads required to sustain populations of animals that depend on them, including salamanders. While any regulated harvest is likely to provide more protection for salamanders and other inhabitants of bromeliads than unregulated harvesting and forest destruction, it may be possible to devise a management plan that is most favorable for populations of both bromeliads and their associated fauna. Cloud forest amphibians also are vulnerable to other kinds of habitat alteration, such as conversion of forested habitats to ornamental fern production (Mendelson et al. 2004).

### Natural Habitat Disturbance and Succession

In considering the effect of habitat alteration on amphibian populations, we must keep in mind that relatively few habitats are completely stable over long periods of time, even in the absence of obvious changes caused by humans. Most habitats are subjected to periodic natural disturbance and alteration. Some changes occur over relatively short time scales, on the order of a few years or decades, whereas others occur over longer time scales, on the order of thousands or tens of thousands of years. Short-term habitat changes can alter the distribution and abundance of amphibian species over relatively small spatial scales. Some short-term changes can be catastrophic, such as hurricanes or fires that cause major damage to habitats, or droughts that temporarily alter soil moisture regimes and eliminate breeding habitat for amphibians. Dry conditions coupled with unusually cold winters may kill some amphibians in terrestrial hibernation sites, even if breeding sites remain intact (Irwin 2005). Such disturbances can alter the species composition and abundance of amphibians in an area, and are likely to impact habitat specialists more than habitat generalists. Eventually, given enough time, an assemblage of amphibian species may return to predisturbance conditions, but there is the potential for rare species to be eliminated altogether by such events. Unfortunately, there has been relatively little research on the long-term effects of natural disturbances such as fire, drought, unusually cold winters, or hurricanes on amphibian populations (Boone et al. 2003; Pilliod et al. 2003).

In Puerto Rico, Hurricane Hugo caused major damage in the Luquillo Experimental Forest in 1989, toppling many trees and opening up the forest canopy. The largest and most common frog, *Eleutherodactylus coqui,* actually increased in abundance following the hurricane, apparently in response to an enormous increase in the abundance of insects feeding on decaying vegetation and new undergrowth (Woolbright 1991, 1996). In contrast, a number of other species either disappeared or became much less common in the immediate aftermath of the hurricane. These included *E. portoricensis, E. wightmanae, E. richmondi,* and *E. eneidae,* which may have suffered from increased dryness of the forest floor following the opening of the canopy and a period of decreased rainfall after the hurricane (Joglar and Burrowes 1996; Woolbright 1997; Joglar 1998). All of these species are smaller and have somewhat more specialized habitat requirements than the habitat generalist, *E. coqui,* and some already were in decline before the hurricane, possibly as a result of increased numbers of days without significant rain in the mid-1980s (M. Stewart 1995). At the same time, populations of a number of species, including *E. coqui,* were declining at higher elevations, not all of which were equally impacted by the hurricane, perhaps in response to other as yet unidentified environmental changes (Stallard 2001).

Other natural environmental changes occur more gradually and can result in changes in the species composition of amphibian assemblages. For example, Skelly, Werner, and Cortwright (1999) examined long-term changes in the distribution of amphibian species in ponds on the Edwin S. George Reserve in Michigan. They found that increased forest growth had resulted in closure of the canopy over some ponds, which in turn shortened the hydroperiod of the ponds. Many of these closed-canopy ponds were abandoned by breeding amphibians, and no new colonization of these ponds occurred. In more open-canopy ponds, on the other hand, the diversity of amphibian species either remained stable or increased. Over the entire mosaic of ponds in the area, new colonization events roughly balanced population extinctions, so the number of breeding populations of most species showed little net change. Beavers may be important in this type of system, because they can reverse the trend toward canopy closure, providing more open habitats for amphibians (Skelly and Friedenburg 2000). Other studies have shown that natural habitat succession can alter the composition of local amphibian assemblages (e.g., Brodman, Cortwright, and Resetar 2002), but where habitat change has been minimal, there has been little change in amphibian populations (e.g., Busby and Parmelee 1996; Crochet et al. 2004; Hossack, Corn, and Pilliod 2005). Changes in regional climatic conditions or other environmental changes also can result in some species in a local assemblage becoming more common while others decline, even as species composition remains stable (Martínez-Solano, Bosch, and García-París 2003).

For some amphibians, populations may be threatened by a combination of human-induced and natural habitat changes. In Britain, for example, the natterjack toad (*Bufo calamita*) originally was a specialist inhabitant of heathland habitats. Most such habitats have been destroyed or fragmented by conversion to intensive agriculture, planting of conifer forests, or urban development. The natterjack toad was largely extirpated from heathland habitats (Beebee 1977, 1983). The remaining patches of heathland are threatened by natural succession to overgrown scrub. Some low-intensity livestock grazing may be required to preserve a more open habitat for natterjack toads and other heathland species (Beebee 1996), which do not appear to be well adapted to moving through complex, enclosed habitats (Stevens et al. 2004). In a similar vein, cattle grazing appears to be important in maintaining temporary wetlands in the Central Valley in a relatively unvegetated state, required by California tiger salamanders (*Ambystoma californiense*) and their principal prey (fairy shrimp; Pyke and Marty 2005). This was due in part to shorter hydroperiods in ungrazed pools.

#### Habitat Isolation and Population Structure

Amphibians that require specific habitats for breeding often use interconnected patches of suitable habitat, and maintenance of stable populations may require continued access to a variety of habitat patches. Many pond-breeding amphibians, for example, exist in metapopulations in which dispersal between breeding sites is relatively rare, but occurs frequently enough to enable recolonization of local ponds after extinction events (see chapter 6). In this type of population, certain highly productive breeding sites may send emigrants to other less productive sites, thereby maintaining populations over the entire area. Amphibians with this type of population structure may be especially vulnerable to habitat fragmentation and artificial barriers to dispersal, such as roads or agricultural land that is unsuitable for amphibians. Such barriers may prevent recolonization of habitat patches following local extinction events. Dispersal among habitat patches also may be essential for maintaining the genetic diversity of populations, and increased barriers to dispersal may result in increased inbreeding and decreased genetic variation. These effects of habitat fragmentation are discussed in more detail in a later section.

For pond-breeding amphibians that live in networks of ponds in relatively close proxmity, extinction of local populations and colonization of habitats by new populations are relatively common events, even in the absence of major habitat changes (Sjögren 1991a; Sjögren-Gulve 1994; Hecnar and M'Closkey 1996, 1997c; Carlson and Edenhamn 2000; D. Green 2003; Trenham et al. 2003). Small populations, in particular, are susceptible to chance extinction due to short-term events such as drought (Jaeger 1980a; Dodd 1993, 1995), killing frosts (Heyer et al. 1988), floods, fires, or disease. For widespread species with generalized habitat requirements, such changes in populations are unlikely to impact the overall status of a species as long as a suitable mosaic of habitats is available and colonization of new habitats roughly balances local extinction events. For species with more specialized requirements, alteration, fragmentation, or isolation of habitats may result in permanent population declines or extinction. The frequency of natural extinction of local populations means that information on the status of a given species at a single locality is of little relevance in establishing whether the species is declining. Only information at larger spatial scales will be useful in determining the long-term prospects for species survival (Hecnar and M'Closkey 1996, 1997c; Bosch, Boyero, and Martínez-Solano 2004).

### How Can We Tell If Amphibian Populations Have Declined or Disappeared?

One difficulty in identifying the causes of declining amphibian populations is actually demonstrating that such declines have taken place. When scientists first became concerned about declining amphibian populations, there was some debate about the quality of the data available to show that widespread declines had occurred. Initially much of the information was anecdotal, with amphibian biologists noting that many species seemed not to be as abundant in areas where they once were common, or had disappeared altogether. In some instances, such as the disappearance of the spectacular *Bufo periglenes* from Monte Verde in Costa Rica, the evidence of decline seemed irrefutable, although even in this case, there was reluctance to concede that the species was permanently gone (M. Crump, Hensley, and Clark 1992).

Several investigators who worked on temporary pond amphibians that exhibit large annual fluctuations in population size argued that the data available for most populations could not be distinguished statistically from random population fluctuations (Pechmann et al. 1991; Pechmann and Wilbur 1994). Others argued that the accumulated weight of such evidence argued for immediate attention to the problem, even if much of the evidence was anecdotal

(Blaustein 1994), or questioned the statistical basis for concluding that populations were not declining (Blaustein et al. 1994). One immediate response to this debate was a call for more long-term monitoring of amphibian populations, to improve the quality of the available data. Since then, through the efforts of the Declining Amphibian Populations Task Force and other organizations, monitoring efforts have been initiated throughout the world (Storfer 2003; Stuart et al. 2004; Young et al. 2004; Heyer and Murphy 2005), and considerable attention has been given to the proper design of monitoring studies (Heyer et al. 1994; Lips et al. 2001; Weir and Mossman 2005).

There really are two ways to go about monitoring amphibian populations (D. Green 1997b). The first is to follow numbers of individuals in particular populations over long periods of time in the hope of discerning long-term trends in population size. The second is to monitor the number of viable populations of a given species over time. The first approach is useful in providing basic information about the demography of amphibian populations and the factors that normally regulate population size, and therefore is important for a complete understanding of amphibian ecology. It is not, however, a very useful approach for determining whether widespread declines of amphibian populations are occurring.

The difficulty is that many years of accumulated data will be required for each population before definitive trends can be identified. Even the few relatively long time-series data sets on amphibian populations may be inadequate to reveal real population declines. In a study of three data sets for the European common frog (*Rana temporaria*), spanning periods of 23–28 years, only one of three populations showed a clear decline, and that was attributable to the introduction of fish into the breeding habitat (Meyer, Schmidt, and Grossenbacher 1998). Some monitoring programs have shown clear population declines when the monitoring effort happened to correspond to a major population crash (e.g., Corn and Fogelman 1984; Semb-Johansson 1991; Bradford 1991; Kagarise Sherman and Morton 1993), but others showed large but nondirectional changes in population size (e.g., Pechmann et al. 1991; D. Green 1997c). One difficulty with most such studies is that population estimates are based on raw counts of numbers of individuals seen, but are not adjusted for the probability that individuals will be detected. Accurate estimates of year-to-year fluctuation in population size require the use of statistical methods to estimate total population size from the number of marked individuals recaptured over time (B. Schmidt 2003). For most amphibian populations, it is simply not possible to obtain accurate estimates of total population size, either because of logistical difficulties, or because characteristics of the species make accurate censuses impossible. As D. Green (1997b, 297) stated, "There is little chance within a human lifetime to identify what is the normal range of fluctuation in population size for many species of amphibians, and therefore no way to know if a population has deviated from this norm."

The second approach is to monitor the occurrence of populations through repeated sampling over long periods of time. All populations, given enough time, are expected to go extinct, but often are replaced relatively quickly by colonists from nearby populations. Consequently, it is the rate at which populations disappear and become reestablished that determines the likelihood of species extinction (D. Green 1997b; Carlson and Edenhamn 2000). An important consideration in monitoring the occurrence of populations over time is the frequency with which sampling is conducted. Long-term studies of amphibian populations in a large number of ponds on the Edwin S. George Reserve in Michigan revealed that infrequent sampling or long intervals between samples are likely to greatly overestimate the number of local population extinctions (Skelly et al. 2003). Comparative surveys of collecting success after intervals of 20 years or more may be useful in documenting the complete disappearance of populations (e.g., Drost and Fellers 1996; Lips et al. 2004), but are less likely to provide reliable estimates of population declines (e.g., Hairston and Wiley 1993; Highton 2005). This is especially true when only two sampling periods are being compared (e.g., Münch and Schröer 1994; Martínez-Solano, Bosch, and García-París 2003), because it is nearly impossible to distinguish between long-term population declines or disappearances and the short-term effects of variation in rainfall or other conditions that affect breeding success.

## Where Are Amphibian Populations Declining?

Declines of amphibian populations have been reported from throughout the world (Stuart et al. 2004), but the data for some areas are better than for others. Information on population size and numbers of viable populations is most reliable for amphibians in North America, Europe, and Australia, where large numbers of biologists are available to monitor population trends. Monitoring efforts are just beginning in some parts of Asia and Latin America and are virtually nonexistent in most of Africa. There are many reports from around the world of apparent declines of individual species or populations of amphibians in *Froglog*, the newsletter of the Declining Amphibian Populations Task Force (DAPTF), but much of this information remains anecdotal and often is based on visits to sites that were studied several decades earlier. The data also are better for some taxonomic groups than for others. Most attention has been given to declines of anuran populations, in part because there are many more of them than other types of amphibians, and because they often are conspicuous and therefore easily monitored. We have much less quantitative information on salamander populations. What we do know often is based on collecting trips 15 or 20 years apart at the same locality, rather than continuous monitoring of populations (e.g., Highton 2005). We have little information on typical population densities of caecilians, although they can be surprisingly abundant, even in human-altered habitats (Oommen et al. 2000; Measey and Di-Bernardo 2003; Measey et al. 2003a, b; Gower et al. 2004; Measy 2004). Although there are some reports of declines in caecilian populations, we lack reliable quantitative data on population trends for these animals (Stuart et al. 2004; Gower and Wilkinson 2005).

In the past 20 years or so, a number of "hot spots" for amphibian population declines have been identified, including montane regions of western North America, Central and South America, and Australia. Amphibian population declines appear to be less severe in other parts of these continents, such as lowland rainforest in South America or eastern deciduous forest in North America. Nevertheless, it is clear that amphibians are vulnerable to declines in any habitat that is under severe pressure, and virtually any amphibian population could be affected by long-term global climate changes. The following sections provide a brief overview of what is known about amphibian declines on all of the major continents and islands.

## North America

### The West

The pattern of amphibian population declines in North America is somewhat paradoxical in that the most dramatic declines have occurred in the West, the least populated part of the continent, while relatively few species have declined in the Northeast, the most densely populated region. Declines of amphibians in the western United States have been of particular concern because many populations have disappeared from apparently pristine and largely uninhabited areas, including national parks. Ranids and bufonids have been most affected, with declines or disappearances of populations reported in Yosemite, Sequoia, Kings Canyon, and Lassen National Parks in California (Bradford 1991; Fellers and Drost 1993; Kagarise Sherman and Morton 1993; Bradford, Graber, and Tabatabai 1994; Drost and Fellers 1996). Populations of native ranids and bufonids have declined elsewhere in the West as well, including the Pacific Northwest, the mountains of Colorado (including Rocky Mountain National Park), Yellowstone National Park in Wyoming, the Central Valley of California, and the desert Southwest (Corn and Fogleman 1984; Clarkson and Rorabaugh 1989; A. Blaustein and Olson 1991; Olson 1992; Carey 1993; Corn 1994, 2003; M. Jennings and Hayes 1994; Fisher and Shaffer 1996; Corn, Jennings and Muths 1997; Patla and Peterson 1999). In Canada, a number of ranids and at least one species of *Bufo* have declined in western Canada (British Columbia and Alberta), but not in eastern provinces. As in the United States, *Rana pipiens* has nearly disappeared from western Canada (D. Green 1997b). M. Hayes and Jennings (1986) estimated that every native ranid species in western North America had experienced either local or regional population declines. A variety of causes have been implicated in these declines, including destruction of breeding habitat, the introduction of exotic predators such as fishes and bullfrogs (*Rana catesbeiana*), use of pesticides and other agricultural chemicals, increased UV radiation, and fungal infections and other diseases. These are discussed in more detail in later sections.

Several of the most severely threatened species of frogs and toads in the West, including those now thought to be extinct, are those with historically limited ranges in specialized habitats, such as canyon streams and springs in otherwise arid environments. These include the Wyoming toad (*Bufo baxteri*), Amargosa toad (*Bufo nelsoni*), Vegas Valley leopard frog (*Rana fisheri*), and relict leopard frog (*Rana onca*). Other species, including *Bufo boreas* (fig. 16.8 C), *B. canorus, Rana pipiens, R. yavapaiensis, R. chiricahuensis, R. pretiosa, R. luteiventris, R. tarahumarae, R. muscosa, R. cascadae,* and *R. draytonii,* were once more widespread, and in some cases were very abundant. All have suffered local or regional population declines, although some species are secure in parts of their ranges (Corn 1994, 2003; M. Jennings and Hayes 1994; Pearl and Hayes 2005). Other western anurans, including hylid treefrogs, the largely fossorial spadefoot toads, and the tailed frogs (*Ascaphus*) apparently have not suffered widespread population declines, although populations of all of these groups are vulnerable to habitat destruction and some have been eliminated from parts of their historic ranges (e.g., *Spea hammondii* in the Central Valley of California; Fisher and Shaffer 1996).

There is little evidence for widespread declines of western species of salamanders, although some species with very small ranges, such as *Batrachoseps aridis,* are endangered (Corn 1994). Some populations of *Ambystoma tigrinum, A. californiense* (fig. 16.8 B), and *A. macrodactylum* have declined or disappeared, in many cases because of habitat destruction or the introduction of predatory fishes (Collins, Jones, and Berna 1988; Ruth 1988; Harte and Hoffman 1989; Corn 1994; Barry and Shaffer 1994). Some of these populations subsequently recovered (Wissinger and Whiteman 1992). *Ambystoma tigrinum* has declined in British Columbia, but not elsewhere in Canada (Weller and Green 1997). Logging and other habitat alteration may threaten local populations of terrestrial and stream-dwelling salamanders, but populations may eventually recover in areas

**Fig. 16.8.** Populations of amphibians in western North America are vulnerable for a number of reasons. (A) The black toad, *Bufo exsul,* has one of the smallest geographic ranges of any North American anuran, being limited to an area of about 15 hectares around four spring systems in Deep Springs Valley, Inyo County, California. The population currently appears to be stable, but is highly vulnerable to habitat alteration. (B) The California tiger salamander, *Ambystoma californiense,* is a California endemic with a range centered on the Central Valley, an area heavily impacted by agriculture, habitat degradation, and chemical pollution. Many populations are considered endangered and others have disappeared. (C) The western toad, *Bufo boreas,* has a wide distribution in western North America and once was very abundant. Some populations have suffered catastrophic declines, especially at high altitudes, but others remain stable. (D) The northern leopard frog, *Rana pipiens,* has one of the largest geographic ranges of any North American anuran. Populations have virtually disappeared from western North America and have suffered major declines in the Midwest, but are relatively secure in eastern North America. Possible factors contributing to the decline of this species include wetland destruction, habitat fragmentation, conversion of natural habitats to agriculture, pollution from pesticides and other chemicals, and introduction of nonnative predators. Photos by Wayne Van Devender (A, B), Carl W. Rettenmeyer (C), and Kentwood D. Wells (D).

that are not completely clearcut (Bury and Corn 1988a, b; A. Blaustein et al. 1995; Diller and Wallace 1996; Grialou, West, and Wilkins 2000; Aubry 2000; Naughton et al. 2000; Wilkins and Peterson 2000).

### The Midwest

The status of midwestern amphibians was reviewed in detail in Lannoo (1998a). Much of this region lacks the vast tracts of national parks and forests and other protected lands that dominate the West, although the Upper Midwest (northern Minnesota, Michigan, and Wisconsin) is heavily forested, as are parts of Indiana, Illinois, and Ohio. Many midwestern amphibians exist in a mosaic of forests, wetlands, and other natural habitats, intensive agricultural lands, and urban and suburban development (Porej, Micacchion, and Hetherington 2004). Major threats to midwestern amphibians appear to be habitat destruction and fragmentation, the intensive use of pesticides, fertilizers, and other agricultural chemicals, and widespread introduction of predatory game fishes and bullfrogs into ponds and lakes. None of the species of amphibians in the Midwest appears to be immediately threatened with extinction, but the abundance of many species, such as the northern leopard frog (*Rana pipiens;* fig. 16.8 D) undoubtedly is a small fraction of what was present before the advent of modern agriculture.

Several of the midwestern states are among those with the greatest loss of original wetland habitat, with estimated losses of at least 90% in Ohio, 89% in Iowa, 87% in Indiana, and 85% in Illinois, and this is likely to have impacted populations of amphibians more than those of some other vertebrates (Leja 1998). Beginning in the Dust Bowl years of the 1930s, many natural wetlands in the Midwest were replaced by hundreds of thousands of artificial farm ponds, many of them stocked with fish. Most of these provide poor habitat for amphibians other than bullfrogs, which can coexist with fish (Lannoo 1996, 1998b). Similar patterns of wetland loss or conversion have occurred in the Great Plains states, including Nebraska (McLeod 2005) and South Dakota (Naugle et al. 2005), with concomitant reductions in the abundance of amphibian species that depend on fish-free temporary ponds for breeding. In some states, loss of amphibian habitats probably has accelerated in recent decades with the decline of cattle grazing and the rise of row crop agriculture. Not only does this tend to eliminate many seasonal wetlands that make planting of crops inconvenient, but it also leads to increased application of fertilizers and pesticides that leach into the remaining wetland habitats (R. Gray and Brown 2005).

In contrast to the western United States, the Midwest has not experienced recent synchronous declines of many species of amphibians in particular regions, nor has there been a differential impact on certain taxa, such as ranids and bufonids. Currently about 30% of the salamanders species and 26% of the anuran species in this region are listed as endangered, threatened, or of special concern in at least one midwestern state, but only a few species are listed in many different states (Lannoo 1998c). Most of these tend to be species with relatively limited ranges or specialized habitat requirements. For example, the four-toed salamander (*Hemidactylim scutatum*) is rare in much of the Midwest and is limited to sphagnum bogs; it is listed in all of the states where it occurs. Similarly, the green salamander (*Aneides aeneus*), a habitat specialist, is at the periphery of its range in this region and occurs in only two states, where it is listed as endangered. The hellbender (*Cryptobranchus alleganiensis*), which requires unpolluted river habitats, is listed as endangered in three of the four midwestern states where it occurs.

Among the anurans, the Illinois chorus frog (*Pseudacris illinoensis*) has perhaps the most specialized habitat requirements, and is limited to isolated patches of once plentiful sand prairie wetlands that mostly have been converted to agriculture (Brandon and Ballard 1998; L. Brown and Cima 1998; Tucker 1998). It is considered rare or threatened in the states where it occurs. An exception to this pattern is the northern cricket frog (*Acris crepitans*), which occurs throughout the Midwest and is abundant in some areas, but has experienced pronounced declines since the 1950s in the upper Midwest (Hay 1998; Lannoo 1998c; R. Gray and Brown 2005). The causes of these declines are unclear, but probably include changes in wetland management that have reduced the availability of fish-free permanent wetlands suitable for cricket frogs, changes in cattle grazing practices that have altered the vegetation on the banks of ponds, and other human-induced habitat alterations. R. Gray and Brown (2005) listed more than 50 possible human-induced environmental changes that may negatively impact cricket frog populations. While many of these factors are likely to affect other amphibians as well, they speculated that cricket frogs are especially vulnerable because of their short adult lifespans and relatively small effective population sizes, which make populations especially vulnerable to events such as prolonged droughts (D. Green 2003).

### The East and South

Less detailed information has been published on the overall status of amphibians in the eastern and southern states than in the West and Midwest, but population monitoring efforts are now underway in most states. These regions have not experienced recent widespread population declines like those in the West and in some species in the Midwest, although local declines have occurred due to habitat alteration or changes in rainfall regimes and pond hydroperiods (Daszak et al. 2005). Many anurans in the eastern United States have large geographic ranges, and many tend to be habitat generalists that can adapt to a variety of breeding sites and terrestrial environments. Some species do well in human-altered environments or are relatively tolerant of the proximity of humans. For example, Fowler's toads (*Bufo fowleri*) favor more open habitats than do American toads (*Bufo americanus*), which share much of their range, and may actually benefit from land clearing (D. Green 2005b). Bullfrogs (*Rana catesbeiana*) do well in fish hatcheries, farm ponds, and other artificial bodies of water that contain fish. In Florida, the introduced Cuban treefrog (*Osteopilus septentrionalis*) has expanded its range, taking advantage of human-altered habitats not used by many other anurans (Meshaka 2001).

Some widespread anuran species that have declined elsewhere remain common in the eastern part of the country. Most notable is the northern leopard frog (*Rana pipiens*), which has disappeared from much of the western part of its range in both the United States and Canada, and has experienced pronounced declines in abundance in the Midwest, but remains common in much of the East (Rorabaugh 2005). For most widespread anuran species in the East and South, local populations remain threatened by habitat destruction and the relentless development of land for housing, shopping centers, roads, golf courses, and other human activities that are detrimental to amphibians, but none appears to be threatened with extinction.

Some eastern anurans have more specialized habitat requirements and may be more sensitive to habitat alteration by human activities. The disappearance of many historical populations of the eastern spadefoot toad (*Scaphiopus holbrookii*) at the periphery of its range in Connecticut has already been mentioned (Klemens 1993). Another habitat specialist is the Pine Barrens treefrog (*Hyla andersonii;* fig. 7.23 D in chapter 7), which is limited to pine barrens habitats and is distributed in a number of disjunct populations in New Jersey, the Carolinas, the Florida Panhandle, and Alabama (Means 2005). While this frog can be locally abundant, populations undoubtedly have been reduced by widespread destruction of suitable habitat. Current populations remain at risk because their habitat often is considered unattractive wasteland and therefore prime land for industrial, commercial, or residential development.

The gopher frogs (*Rana capito* and *R. sevosa*) were once widespread in longleaf pine forests of the Atlantic and Gulf coastal plains, but 95% of this habitat has disappeared, having been replaced by pine plantations or development. These frogs are considered threatened or endangered in all of the states where they occur, and *R. sevosa* currently is known from only a single pond in Mississippi (Richter et al. 2003; Jensen and Richter 2005; Richter and Jensen 2005). Gopher frogs share their specialized habitat with the flatwoods salamander (*Ambystoma cingulatum;* fig. 16.9), which has undergone a similar drastic decline in breeding populations (Dodd 1995; Means, Palis, and Baggett 1996; Palis and Means 2005).

**Fig. 16.9.** The Flatwoods Salamander, *Ambystoma cingulatum,* is endemic to longleaf pine flatwoods habitats in the southeastern United States. This species is threatened by the conversion of most such habitats to managed pine plantations, which often lack the seasonal flooding required by this species for reproduction, as well as herbaceous cover in the terrestrial habitats of adults. Photo by Wayne Van Devender.

Most other eastern and southern species of *Ambystoma* have more generalized habitat requirements and remain relatively common in most areas (Lannoo 2005). Most of these species prefer or require fish-free temporary wetlands for reproduction, so are potentially vulnerable to the introduction of fishes and the usual array of habitat degradation and destruction that affects temporary wetlands throughout the country. Most of the aquatic salamanders of the East and South also appear to be secure (*Necturus, Cryptobranchus, Siren, Pseudobranchus,* and *Amphiuma*), although species with relatively restricted ranges are likely to be at some risk (e.g., *Necturus alabamensis;* Bailey 2005). Local populations of many of these species remain vulnerable to all sorts of habitat degradation that affect aquatic environments, but widespread population declines have not been reported.

### Mexico, Central and South America, and the Caribbean

Much of the recent concern over declining amphibian populations has focused on Latin America, especially montane regions of Mexico, Central America, and South America (Young et al. 2001; Crump 2003; Lips et al. 2004; Young et al. 2004; Lips and Donnelly 2005). Until recently, most information on amphibian populations in this region came from collecting efforts of systematists, not from detailed ecological studies, so quantitative information on the status of populations is scarce. Many parts of Latin America have been poorly studied, and efforts to monitor amphibian populations have only recently begun (Lips et al. 2001). Monitoring efforts will have to continue for many years if long-term population trends are to be detected, but in the meantime, efforts to preserve critical habitats are needed (Young et al. 2001; Lips et al. 2005). Throughout much of this region, amphibian populations are threatened by widespread deforestation and other forms of habitat destruction and degradation associated with rapidly growing human populations. A typical example is Honduras, which has a rich amphibian fauna, but also a poor and expanding human population. Widespread environmental degradation threatens more than half of the amphibian species, some of which already are extinct (McCranie and Wilson 2002; L. Wilson and McCranie 2003).

There also have been disturbing accounts of amphibian species declining or disappearing in apparently pristine protected habitats, such as the cloud forests around Monte Verde in Costa Rica (Pounds and Crump 1994; Pounds et al. 1997), as well as dramatic observations of dead and dying frogs in forests elsewhere in Costa Rica, Panama, and Colombia (Lips 1998, 1999; J. D. Lynch and Grant 1998). Similar declines have occurred elsewhere in Mexico and Central and South America, especially in montane regions above 500–800 meters in Mexico and Central America and 1,000 meters in the Andes of South America (Young et al. 2001; Lips et al. 2004; Lips and Donnelly 2005; Young et al. 2004). Ecuador has been particularly hard hit, with both species diversity and abundance having been reduced at many localities in the Andes (Bustamante, Ron, and Coloma 2005). As mentioned earlier, the Atlantic Coastal Forest of Brazil (fig. 16.10 A, B), which includes some montane habitats, has been especially hard hit (Heyer et al. 1988; Weygoldt 1989a; Guix et al. 1998; Eterovick et al. 2005). Population declines have been less evident in lowland rainforest, including the Amazon Basin (e.g., Duellman 1995; Salas et al. 1999; Azevedo-Ramos and Galatti 2002), but have occurred in some areas where the forest has been destroyed or fragmented (Tocher, Gascon, and Zimmerman 1997; Funk and Mills 2003; Eterovick et al. 2005). Population declines in tropical montane regions are of particular concern, because these areas tend to have the highest species diversity of amphibians, the largest proportion of endemic species, and a high proportion of species with limited ranges (Duellman 1999a; Lips et al. 2004; Lips and Donnelly 2005; Young et al. 2004).

Attention has focused mostly on anurans, which are far more conspicuous and more easily monitored than are urodeles. A wide variety of anuran genera have been affected, but certain groups appear to be particular vulnerable, especially those closely associated with streams. These include bufonids

**Fig. 16.10.** Parts of the Neotropics are considered hotspots for amphibian population declines. The Atlantic Rainforest of Brazil harbors a large number of unusual endemic species of anurans, especially in the family Leptodactylidae. Some of these, including *Paratelmatobius lutzi* (A) and *Holoaden bradei* (B), have not been seen in the wild for many years. Major population declines also have occurred in the cloud forest of Monte Verde in Costa Rica, where the spectacular bright orange toad *Bufo periglenes* (C) apparently has gone extinct. In Puerto Rico, three endemic species of *Eleutherodactylus* appear to be extinct. Other species, including the tiny forest floor frog *E. wightmanae* (D), have declined in some parts of the island, especially at high altitudes. Possible causes for declines of these and other Neotropical species include disease, habitat destruction and fragmentation, and climate change. Photos by Ivan Sazima (A, B), Wayne Van Devender (C), and Kentwood D. Wells (D).

of the genus *Atelopus* (with up to a third of all known species having disappeared and most populations endangered), dendrobatids of the genus *Colostethus*, centrolenids, and several genera of stream-breeding hylids (Coloma 1995; Lötters 1996; Pounds et al. 1997; Young et al. 2001; Lips et al. 2004; La Marca et al. 2005). Anurans with direct development, including some species of *Eleutherodactylus*, have declined in some areas (e.g., southern Mexico; Lips et al. 2004), as have some species that breed in small bodies of water, such as *Bufo periglenes* in Costa Rica, which is thought to be extinct (fig. 16.10 C). In general, however, species that lack aquatic larvae appear to have been less affected than have stream-breeding frogs (Bustamante, Ron, and Coloma 2005; Lips et al. 2005). Most anurans that depend on permanent or temporary ponds for reproduction do not appear to have suffered widespread declines. Nevertheless, there have been localized declines of some species (e.g., *Hyla pseudopuma* at Monte Verde, Costa Rica; Pounds et al. 1997), and whole assemblages of amphibians have been eliminated in areas affected by habitat destruction (Duellman 1978). Although salamander populations in Latin America have not been studied as intensively as have anuran populations, there is evidence for localized declines of some species (e.g., *Bolitoglossa pesrubra* at Cerro de la Muerte, Costa Rica; Lips and Donnelly 2005; *Lineatriton lineolus* in the mountains of southern Mexico; Lips et al. 2004). Because many Neotropical salamanders have narrow altitudinal distributions, future changes in climate may reduce the ranges of many species (Parra-Olea, Martínez-Meyer, and Perez-Ponce de Leon 2005).

The anuran fauna of most Caribbean islands (except for Trinidad and Tobago) differs from that of mainland South and Central America in being dominated by the genus *Eleutherodactylus*, which has direct development of terrestrial eggs. There also are a few species of hylids, dendrobatids, and bufonids, which include riparian, fossorial, and semiarboreal species. Some species of *Eleutherodactylus* also are riparian, although they do not breed in water. Many of these amphibians are associated with forested habitats, although some occur in more open areas. Islands of the Caribbean have suffered enormous reductions in their original forests, with remaining forest cover estimated from a high of about 13% for Cuba to a low of < 1% for Haiti. It therefore seems likely that many Caribbean amphibians are vulnerable to population declines or extinction (Young et al. 2004). Nevertheless, definitive evidence of such declines is limited, largely because of a scarcity of herpetologists searching for rare species in suitable habitats (Hedges 1993).

The best-studied region is Puerto Rico, where there is clear evidence of the decline and possible extinction of some species of *Eleutherodactylus*, while other species have prospered. The largest species on the island, *E. karlschmidti*, has not been found since 1974. It was a conspicuous frog closely associated with stream habitats and is unlikely to be missed by experienced herpetologists, so is presumed to be extinct (Burrowes and Joglar 1991; Hedges 1993; Joglar and Burrowes 1996; Joglar 1998; Burrowes, Joglar, and Green 2004). The cause of the disappearance of this species is unknown, but other species of *Eleutherodactylus* associated with streams have declined elsewhere in the Caribbean (Hedges 1993) and in Mexico and Central America (Lips et al. 2004). Also generally considered to be extinct is the unique ovoviviparous frog *E. jasperi*, last seen in 1981. This species is a bromeliad specialist and therefore more difficult to detect, especially at low population levels. Some of the limited known habitat of this species has been destroyed (Burrowes and Joglar 1991), but definitive evidence that it is extinct is lacking (Hedges 1993). A third species, with more generalized habitat requirements and a wider distribution on the island, *E. eneidae*, also has disappeared. Several other species, including *E. coqui*, *E. portoricensis*, *E. wightmanae* (fig. 16.10 D), *E. richmondi*, *E. locustus*, and *E. gryllus*, have suffered localized population extinctions or declines, especially at higher altitudes, even in protected forests (Joglar and Burrowes 1996; Woolbright 1997; Burrowes, Joglar, and Green 2004). On the other hand, most species found in lowland habitats have not declined, perhaps because they are better able to adapt to human-altered landscapes. *Eleutherodactylus coqui*, in particular, is an extreme habitat generalist and is found at both low and high elevations in Puerto Rico and does well even in plantations of introduced *Eucalyptus* trees (Joglar 1998; Beard et al. 2003; Fogarty and Vilella 2003).

The diversity of habitats used by the declining species in Puerto Rico, and the persistence of some species in areas where others have declined, suggests that there is not a single cause of population declines in this region. Stallard (2001) examined long-term data on environmental variables such as forest cover, temperature, precipitation, rain and stream chemistry, and atmospheric dust transport for eastern Puerto Rico, where high-elevation populations of some frogs have declined. He found significant levels of air pollution in the area, but could not definitively tie pollution to anuran declines, nor were changes in rainfall and temperature thought to be significant enough to contribute to the observed declines. An analysis of remote-sensing data for a number of regions that have experienced amphibian declines, including Puerto Rico, also failed to show definitive correlations between environmental variables and population declines (Carey et al. 2001). Temperatures in Puerto Rico were somewhat higher than normal and precipitation somewhat lower during the periods when some declines of frogs occurred (M. Alexander and Eischeid 2001; Burrowes, Joglar, and Green 2004). Prolonged periods with relatively little rainfall, even when annual rainfall is near normal, can decrease population size in these frogs and could lead to local extinction of rare species (M. Stewart 1995). Stress associated with drought also could weaken defenses of frogs against disease, and there is some evidence for chytrid fungus infections in two species of *Eleutherodactylus* in Puerto Rico in the 1970s, although one of these, *E. coqui,* remains abundant today (Burrowes, Joglar, and Green 2004).

### Australia and New Zealand

For a continent dominated by arid habits, Australia has a remarkably rich anuran fauna (over 200 species), but lacks both caecilians and salamanders. Family-level diversity is low, with the fauna dominated by the families Hylidae and Myobatrachidae. Members of these families have adapted to many different habitats, from extreme deserts to tropical rainforests. A few microhylids and ranids are found in tropical regions of far northern Australia, and one introduced bufonid, *Bufo marinus,* has spread over much of the northeastern part of the country (Lever 2001). Australia has a low human population density over much of the continent, with most of the population concentrated in coastal areas where development may threaten some anurans. The country has a relatively large community of herpetologists, so the status of Australian frogs is reasonably well known compared to some other continents (A. Campbell 1999).

Declines of anuran populations have been reported in several regions of Australia. Some of these clearly seem to be related to human encroachment on natural habitats, as has occurred along the east coast of Queensland and New South Wales (Hines, Mahony, and McDonald 1999) and for some species in areas around Sydney (Thumm and Mahony 1999). Logging and cattle grazing in state and private forests threatens some species in northern New South Wales and southern Queensland (Lemckert and Morse 1999; K. Parris 2001), while human alteration of natural streams has reduced frog habitats on the Southern Tablelands of New South Wales (Hazell, Osborne, and Lindenmayer 2003). Some species also have declined in upland and montane habitats in more protected areas of New South Wales and Victoria (Gillespie and Hines 1999; W. Osborne, Hunter, and Hollis 1999). Many of these species have restricted ranges or utilize specialized habitats (Murray and Hose 2005). The status of anurans in Western Australia is poorly known because relatively few people have worked in this vast, arid region. Three species, *Geocrinia alba, G. vitellina,* and *Spicospina flammocaerulea,* all of which have very small or fragmented ranges, are considered to be endangered (D. Roberts, Conroy, and Williams 1999; Conroy and Brook 2003).

The declines that have received the most attention have occurred in rainforest populations of hylids and myobatrachids in Queensland (K. McDonald 1990; S. S, Richards, McDonald, and Alford 1993; Trennery, Laurance, and McDonald 1993; K. McDonald and Alford 1999). Many of these populations were in national parks and other protected areas, and appear to have undergone relatively synchronous declines starting in the late 1970s. Nearly all of these species were stream breeders, as has been the case for declines in South and Central America (S. Williams and Hero 1998). The relatively rapid declines of many species living in the same habitat suggested the possibility of unusual weather connected with El Niño events or epidemic disease as a major cause of decline (Laurance 1996; Laurance, McDonald, and Speare 1996). Although the latter idea initially was controversial (Alford and Richards 1997; Hero and Gillespie 1997), there is increasing evidence that chrytrid fungal infections were involved in at least some anuran declines in this area (Berger et al. 1998; Berger, Speare, and Hyatt 1999). This issue is discussed in more detail in a later section. A number of frogs, including two species of *Taudactylus* (fig. 16.11 A), the two unique gastric-brooding frogs of the genus *Rheobatrachus* (fig. 16.11 B), and several treefrogs in the genus *Litoria* appear to be extinct (A. Campbell 1999). Some others, such as the cascade treefrog (*Litoria pearsoniana*), seem to have recovered from earlier population declines, although some of the evidence for dramatic declines of this species in the 1970s and 1980s is questionable (K. Parris 2001).

In contrast to Australia, New Zealand has very few frogs. The only native frogs are four species of *Leiopelma* (fig. 1.14 E in chapter 1). Some Australian hylids (*Litoria*) have been introduced, and some populations of these species declined in the 1990s (P. Bishop 1999; B. Bell et al. 2004). This genus *Leiopelma* is now restricted to the northern part of the country and some offshore islands, but probably was once found throughout New Zealand (B. Bell 1994; Holyoake, Waldman, and Gemmell 2001). Several species appear to have gone extinct after the arrival of humans and their associated animals, especially rats (Towns and Daugherty 1994). A number of features of the biology of these frogs make them vulnerable to population declines or extinctions. Most have restricted ranges and limited dispersal abilities. They appear to be unusually long-lived (perhaps more than 30 years; B. Bell, Pledger, and Dewhurst 2004), but have very low reproductive rates. They also are vulnerable to introduced predators, including rats and Australian treefrogs (*Litoria;* Thurley and Bell 1994).

The most widely distributed species is *L. hochstetteri,* but it is found in isolated small populations. *Leiopelma archeyi* has a smaller geographic range, and some populations have experienced significant declines, possibly due to infection by chytrid fungus (B. Bell 1999; B. Bell et al. 2004). This species is now considered critically endangered. Several populations of *Leiopelma* are restricted to two small off-

**Fig. 16.11.** Montane rainforests of Queensland, Australia are hotspots of amphibian population declines. Among the frogs that have disappeared are several stream-dwelling myobatrachid frogs, such as *Taudactylus diurnus* (A) and the unique gastric-brooding frogs of the genus *Rheobatrachus,* including *R. silus* (B). The fungal disease chytridiomycosis has been implicated in the declines of these species. Photos by Harold Cogger.

shore islands, Stephens and Maud Islands. The Stephens Island frogs (*L. hamiltoni*) are very rare, with perhaps no more than 200 individuals (D. Newman 1990). They have been the subject of transplantation efforts to establish a population in a new habitat (D. Brown 1994). The Maud Island population consists of several thousand individuals in a remnant of forest (E. Bell and Bell 1994). In the 1980s, one hundred of these frogs were introduced into a new location on Maud Island and appear to have established a stable population (B. Bell, Pledger, and Dewhurst 2004). The Maud Island frogs were considered conspecific with the Stephens Island *L. hamiltoni* until B. Bell, Daugherty, and Hay (1998) described this population as a new species, *L. pakeka.* More recent DNA sequence data indicated that these two populations are genetically similar and probably should not be considered different species (Holyoake, Waldman, and Gemmell 2001), but the issue remains controversial (B. Bell, Pledger, and Dewhurst 2004). The taxonomic status of the frogs is important for their conservation status, because without the Maud Island population, *L. hamiltoni* has been considered one of the world's rarest and most endangered anuran species. Whatever the taxonomic status of these populations, it is clear that the Stephens Island population cannot survive without active protection and management.

### Europe

Europe has been densely populated by humans for thousands of years and has been heavily altered by agriculture, industrialization, and urbanization. To make matters worse, the species diversity of amphibians in Europe is positively correlated with human population density (Araújo 2003). Partly this reflects a latitudinal effect, with amphibian diversity and human population density both decreasing in the colder northern parts of Europe. Habitats that most North Americans would consider pristine natural areas are scarce; most European amphibians live in a mosaic landscape consisting of small patches of relatively natural habitat interspersed with agricultural land, small towns, suburbs, cities, industrial areas, and roads. Even many national parks and reserves in Europe include human-altered habitats and do not resemble the large tracts of wilderness found in the western parts of North America, or even those in parts of eastern North America such as the Adirondack Mountains in New York, Algonquin Park in Ontario, and the largely uninhabited forests of northern Maine. There are a few exceptions, such as the Białowieża Primeval Forest in Poland, which supports dense populations of amphibians (Pikulik et al. 2001) that provide an essential source of food for the abundant carnivorous mammals in the area (see chapter 14). Unfortunately, only a small portion of this forest is currently protected as a national park, with the remainder subject to logging that may leave the protected forest as an isolated island habitat (Tomiałojć and Wesołowski 2004).

The amphibian fauna of Europe is relatively depauperate, due in large part to Pleistocene glaciation and the east-west orientation of several major mountain ranges that probably blocked movement of some amphibians into southern refugia during glacial periods. Borkin (1999) listed 77 species of amphibians in Europe, from Spain to Russia and Turkey (32 urodeles, 45 anurans), but the number found north of the Mediterranean region is less than half of the total (about 35 species). This means that all of Central and Northern Europe has fewer species of amphibians than the state of North Carolina. Even regions whose local species diversity is relatively high have only 15–20 species of amphibians, fewer

**Fig. 16.12.** Europe has a relatively impoverished amphibian fauna, with the greatest diversity found in southern Europe. The Mediterranean treefrog, *Hyla meridionalis,* shown here, is abundant in southern Europe and is not a major conservation concern. The closely related and morphologically similar northern species, *Hyla arborea,* has suffered declines in many parts of Europe, due mainly to habitat destruction and fragmentation. Habitat protection and restoration have increased population sizes in some countries. Photo by Walter Hödl.

than are found in a small northeastern state like Connecticut. The Mediterranean region, which includes the area from the Iberian Peninsula to Turkey and the northern fringe of Africa, has the most diverse amphibian fauna (fig. 16.12), as well as the largest proportion of endemic species (69% of urodeles and 56% of anurans). Many of these endemics have small geographic ranges resulting from fragmentation of ancestral populations in Pleistocene refugia, and therefore may be at greater risk than are the more widespread species found in northern and central Europe. Northwestern Italy has among the highest local diversity because this is where Italian endemics meet more widespread European species and those found further east in the Iberian Peninsula (Andreone and Sindaco 1998; Andreone and Luiselli 2000). There also are isolated populations of more widespread species, such as *Rana lessonae* in Britain and Sweden, that probably are the result of Pleistocene colonization events at the margins of the species' range (Beebee et al. 2005; Snell, Tetter, and Evans 2005). These peripheral populations are vulnerable to extinction, even if the species is not threatened elsewhere (indeed, the last individual of *R. lessonae* from Britain died in 1999).

The highly altered landscape of Europe, with its mosaic of natural and human-altered landscapes, as well as the large number of species with restricted ranges, has resulted in Europeans having a somewhat different perspective toward amphibian conservation than one finds in North America or in the tropics. North American conservationists tend to focus first on protection of natural habitats and buffer zones around breeding sites and minimizing the impact of human activities on amphibians in mosaic landscapes (Calhoun and Hunter 2003; Semlitsch and Rothermel 2003). European conservationists have been somewhat more focused on active management of landscapes for amphibians, restoration of degraded habitats, creation of new habitats such as breeding ponds, mitigation of wetland loss, and transfer of amphibian populations to newly created wetlands, and captive rearing and release programs to increase the number and size of amphibian populations (Thielcke et al. 1985; Zvirgzds, Stašuls, and Vilnītis 1995; Beebee 1996; Buley and Garcia 1997; Hödl, Jehle, and Gollmann 1997; Directorate of Culture and Cultural and Natural Heritage, Council of Europe 2003; Buckley and Beebee 2004; Edgar, Griffiths, and Foster 2005; Snell, Tetter, and Evans 2005). They have devoted considerable effort to identifying potential amphibian breeding sites in disturbed habitats, such as urban and suburban areas (Beebee 1979c; Dalbeck et al. 1997). European conservationists also have focused attention on individual species that appear to be at elevated risk because of specialized habitat requirements (e.g., the moor frog, *Rana arvalis,* and the natterjack toad, *Bufo calamita;* Beebee 1983, Glandt and Podloucky 1987; Banks, Beebee, and Denton 1993; Sinsch 1998; Buckley and Beebee 2004), fragmented habitats (e.g., *Hyla arborea*) (Stumpel and Tester 1993; L. Andersen, Fog, and Damgaard 2004), or naturally restricted ranges (e.g., island populations; Buley and Garcia 1997; Gibson and Freeman1997; Andreone and Luiselli 2000; Lecis and Norris 2003, 2004).

Most European amphibian species, even those that are not endangered, receive some legal protection (Corbett 1989; Beebee 1996; Beebee and Griffiths 2000; Spellerberg 2002; Directorate of Culture and Cultural and Natural Heritage, Council of Europe 2003). Such laws are aimed primarily at regulating or prohibiting the collection and commercial trading of amphibians. In some cases, there also is some protection against destruction or degradation of critical amphibian habitats, but legal protections are not always effective in protecting amphibians against habitat destruction and fragmentation (McLean, Wight, and Williams 1999). Unfortunately, some European governments have been slow to draft or implement action plans for threatened and endangered species of amphibians that have been identified as priorities by experts in amphibian conservation. Furthermore, despite the large number of herpetologists in Europe and increasing interest in conservation issues, information on the status of many European amphibians is fragmentary (Directorate of Culture and Cultural and Natural Heritage, Council of Europe 2003). There also is considerable variation among European countries in the extent to which natural habitats remain available to amphibians. In Eastern Europe, for example, some countries, such as Romania, have suffered major environmental degradation and undisturbed habitats are scarce. Others, such as Croatia, have many natural areas intact and a large network of protected sites.

The European amphibians at greatest risk for population declines or extinction are mostly those with limited ranges, as discussed previously. These include island endemics such as *Alytes muletensis,* restricted to small habitats on Mallorca, and *Discoglossus sardus, Hyla sarda,* and *Euproctus platycephalus,* which are endemic to Sardinia. Also at risk are habitat specialists such as cave salamanders (*Proteus, Speleomantes* [= *Hydromantes*]), brook salamanders (*Euproctus*), the natterjack toad (*Bufo calamita*), and the moor frog (*Rana arvalis*). Mainland species with relatively small natural ranges are at risk, although not all are currently declining. Many of these species are found in southern Europe, especially the Mediterranean region, including a number of species largely or completely restricted to Italy (*Pelobates fuscus insubricus, Rana italica, R. latastei, Triturus italicus, T. carnifex*) or the Iberian peninsula (*Alytes cisternasii, Rana iberica, Triturus boscai*).

Some relatively widespread European species with generalized habitat requirements have undergone declines in some regions, while remaining stable in others. For example, the common frog (*Rana temporaria*) is found throughout Europe, and has adapted well to human-inhabited landscapes, using garden ponds and other man-made sites for breeding. In Britain, this species declined in the 1960s and 1970s due to intensification of agriculture, but it has rebounded with the increasing popularity of garden ponds in urban and suburban areas and efforts to provide new breeding sites (Beebee 1979c, 1996, 1997; J. M. R. Baker and Halliday 1999; Beebee and Griffiths 2000; Robinson and Sutherland 2002). In Sweden, moderate agricultural land use in the northern part of the country actually provides more favorable habitat for this species than does undisturbed forest. Conversely, in the more intensively used agricultural land of southern Sweden, there is greater fragmentation of forest habitat and poorer quality habitat for common frogs, resulting in lower population densities and lower genetic diversity within populations (Johansson et al. 2005). In Switzerland, long-term population records from the area near Bern indicated that two populations have remained relatively stable, while a third has declined, probabably due to the introduction of predatory fish into the breeding site (Meyer, Schmidt, and Grossenbacher 1998). The common toad (*Bufo bufo*) also is found over much of Europe and is abundant over most of its range, where it has adapted well to human landscapes and often breeds in garden ponds. Nevertheless, recent population declines have occurred for unknown reasons in lowland areas of southeastern England, even in sites where *Rana temporaria* is doing well (J-A. Carrier and Beebee 2003).

Surveys of larval amphibians in a national park in central Spain in the mid-1980s and in 1999 showed that the same overall assemblage of species was present in both surveys, but the relative abundances of species had changed. The population of *Alytes obstetricans* crashed, apparently due to disease, while that of *Rana perezi* increased and most other species remained relatively stable or showed modest declines (Bosch, Martínez-Solano, and García-París 2001; Martínez-Solano, Bosch, and García-París 2003). The reason for the increase in *Rana perezi* is unclear, although the authors suggested that warmer temperatures at higher altitudes enabled this species to expand its range out of the lowlands. Unfortunately, with only two population surveys, it is impossible to determine whether this is a permanent range expansion or a temporary range shift. A resurvey of 56 ponds in southern France originally studied in the 1970s showed little evidence of decline in amphibian populations. There were numerous instances of species disappearing from ponds, but these local extinctions were roughly balanced by colonization of new ponds (Crochet et al. 2004).

In the mosaic landscape of modern Europe, continued habitat destruction and fragmentation are key issues for the conservation of amphibian populations (P. Joly, Morand, and Cohas 2003). Even in widespread species that are not in danger of extinction, there is evidence for negative effects on populations in fragmented landscapes. Most European amphibians breed in ponds, and breeding sites often become surrounded by inhospitable habitat such as agricultural fields, roads, and housing developments that restrict movements between ponds. This type of habitat fragmentation can result in increased mortality from road traffic (A. Cooke 1995; Fahrig et al. 1995; Hels and Buchwald 2001) and increased likelihood of extinction of small populations that are not replenished by migrants from nearby ponds (Balletto and Giacoma 1993; Halley, Oldham, and Arntzen 1996; Vos and Chardon 1998; Joly et al. 2001; Hels and Nachman 2002). There also can be negative genetic effects of fragmented landscapes leading to isolation of populations, including increased inbreeding and decreased heterozygosity (Hitchings and Beebee 1997, 1998; G. Rowe et al. 1999; Garner et al. 2003; Lesbarrères et al. 2003; G. Rowe and Beebee 2003; L. Andersen, Fog, and Damgaard 2004; Johansson et al. 2005), which may or may not result in decreased fitness (Edenhamn, Hoggren, and Carlson 2000; G. Rowe and Beebee 2005).

Many declines of European amphibian populations have been attributed to destruction of essential breeding habitat, fragmentation of habitats, or both. Many widespread European species exist in local metapopulations that depend on a network of breeding ponds in which periodic extinctions of small populations are balanced by recolonization from larger populations nearby (Sjögren 1991a; Sinsch 1992b;

Miaud, Joly, and Castanet 1993; Sjögren Gulve 1994; Sinsch and Seidel 1995; Vos and Stumpel 1996; Carlson and Edenhamn 2000; Ebisuno and Gentilli 2002). Some of these species appear to be particularly susceptible to habitat fragmentation, incuding *Hyla arborea, Bombina bombina,* and several species of *Triturus.* Local or regional declines of these species have been reported in some parts of Europe, while other populations have remained stable.

Several European countries have had considerable success in increasing amphibian populations through a combination of habitat restoration, construction of new breeding ponds, development of habitat corridors to connect breeding sites, and reintroduction of captive-reared animals. In Sweden, for example, habitat restoration and construction of new breeding ponds for *Hyla arborea* has been underway since the 1980s, resulting in an increase in the estimated number of breeding males from about 3,500 to more than 20,000 at more than 500 different sites (J. Baker 1997; Directorate of Culture and Cultural and Natural Heritage, Council of Europe 2003). Similar restoration efforts have increased populations in Denmark as well (L. Andersen, Fog, and Damgaard 2004). Tree frogs collected in Sweden have been used in a program to reintroduce *Hyla arborea* to nearby Latvia, and there is now a stable population in that country (Zvirgzds, Stašuls, and Vilnī 1995). Sweden also has reintroduced *Bombina bombina,* which went extinct in the 1960s, and the current population numbers several thousand individuals in more than 70 breeding ponds. Unfortunately, a reintroduction effort for this species in Latvia in 1988 was not successful (Directorate of Culture and Cultural and Natural Heritage, Council of Europe 2003). The number of populations of natterjack toads (*Bufo calamita*) in Britain has increased by about 10% since 1970, in part because the number of successful translocations to new breeding sites exceeded the number of population extinctions (Buckley and Beebee 2004). Transfers of great crested newts (*Triturus cristatus*) from ponds destroyed by development to newly created wetlands has been used in Britain as a method to mitigate the negative effects of development. Unfortunately, transfers of animals have not been done in a systematic manner, and monitoring of new populations has been spotty and insufficient to determine whether such populations are sustainable (Edgar, Griffiths, and Foster 2005).

Europe has been plagued by the introduction of all sorts of nonnative animals that have the potential to alter the composition of local assemblages of amphibians, as well as intentional or accidental movement of European amphibians to new parts of the continent. Britain, for example, has about as many introduced species of amphibians as native species, but the ecological impact of most of these introductions is not known (Bell and Bell 1995; Beebee and Griffiths 2000). Multiple introductions of some species, such as *Rana lessonae,* have made it difficult to determine whether certain species are entirely introduced or also include native populations (Beebee et al. 2005). Recent research suggests that once a species is introduced into Europe, it has a relatively high probability of becoming established and even expanding its range (Jeschke and Strayer 2005). Some types of introductions clearly have had negative consequences for amphibians. For example, predatory game fish have been widely introduced into previously fish-free habitats. Most European amphibians cannot breed in habitats with fish, and populations often are eliminated when fish are introduced (Brönmark and Edenhamn 1994; Aronsson and Stenson 1995; Braña, Frechilla, and Orizaola 1996; Reshetnikov and Manteifel 1997; Nyström et al. 2002; Böll 2003; Martínez-Solano, Barbadillo, and Lapena 2003; Martínez-Solano, Bosch, and García-París 2003; Reshetnikov 2003; Crochet et al. 2004).

Some large ranids, such as *Rana ridibunda,* do well in permanent lakes, ponds, and rivers inhabited by fish (Teplitsky, Plénet, and Joly 2003). This species was introduced to southeastern England in 1935 and has established large breeding populations in marshes and pasture ditches that are not used by most native amphibian species (Beebee and Griffiths 2000). Throughout Europe, this species tends to be a very successful colonizer of permanent water habitats, whether through natural dispersal or human introduction. Because of its ability to survive in habitats with fish, it can use large rivers as dispersal corridors. Introduced or invasive populations of *R. ridibunda* have been reported in France (Neveu 1997; Pagano, Crochet et al. 2001; Pagano, Joly et al. 2001; Pagano, Lode, and Crochet 2001), Spain (Arano et al. 1995), Switzerland (Vorburger and Reyer 2003), Russia, Ukraine, Kazakhstan, and Kirgizia (Kuzmin 1999). Introduced populations have become established in various parts of Siberia and on the Kamchatka Peninsula in the Russian Far East, aided in part by warming of lake waters by discharges from power plants (Kuzmin and Maslova 2003). In addition to predation and other negative ecological effects on local native species, there is concern that interbreeding between *R. ridibunda* and other European water frogs may result in genetic displacement of other phenotypes by *R. ridibunda* (Arano et al. 1995; Pagano et al. 2003; Vorburger and Reyer 2003).

The North American bullfrog (*Rana catesbeiana*), another fish-tolerant species, also has been introduced into Britain, but populations have not spread as widely as those of *R. ridibunda.* This species has been introduced into parts of Belgium, France, Italy, Germany, the Netherlands, and Greece as well, and some breeding populations have become established (Stumpel 1992; Neveu 1997; Andreone and Marocco 1998). This species is likely to do well in permanent ponds and lakes inhabited by fish, but also could move into other habitats and displace native species. The ecological impact of introduced bullfrogs on native European amphibians has scarcely been studied (Haffner 1997; see http://www.zavod-symbiosis.si/3_en_bullfrog.htm).

### Asia

Because of its enormous size and diversity of climates, Asia has the greatest variety of habitats of any continent, ranging from extreme deserts to tropical rainforests and the highest mountains on Earth. Asia also has a very diverse amphibian fauna, because it represents a contact zone between Palearctic species, Oriental species, and Gondwanan amphibians that entered Asia when India drifted northward and contacted the Asian mainland. With more than 650 species of amphibians, southern Asia has by far the greatest diversity (this area includes Sri Lanka, India, Bangladesh, the countries on the southern flank of the Himalayas, Southeast Asia, Indonesia, and the Philippines; Inger 1999). In contrast, the Palearctic regions of Asia have far fewer species of amphibians, with about 42 species in Western Asia, nine species in Siberia, 20 species in Central Asia, and 55 species in Palearctic Eastern Asia (Borkin 1999). If we consider only the temperate parts of Eastern Asia (China, Taiwan, Korea, the main islands of Japan, Hainan, and the Ryukyu Islands), there are more than 300 species, with far greater diversity in the Oriental realm (southern China, Taiwan, and the Ryukyu Islands) than in the much larger Palearctic realm (Asian Russia, Mongolia, northern China, Korea, and the main islands of Japan; Zhao 1999; Kuzmin and Maslova 2003).

In addition to lowland tropical rainforest areas, which are rich in amphibians, both tropical and temperate montane regions tend to have high species diversity and large numbers of endemic species. Amphibian hot spots of particular interest to conservation biologists, because of high species diversity or large numbers of endemic species with limited ranges, include Sri Lanka, the Western Ghats of India, the Eastern Himalayan region of India, the mountains of southwestern China, montane regions of Southeast Asia, and tropical rainforests of Malaysia and Indonesia. Many of these regions remain poorly explored by herpetologists, and many new species of amphibians continue to be described from Sri Lanka, India, Southeast Asia, and China. In addition, the main islands of Japan and the Ryukyu Islands, while having modest numbers of amphibian species, have a high proportion of endemic species (Zhao 1999; Bossuyt et al. 2004). Several other parts of Asia have relatively low species diversity, but often harbor unique endemic species of amphibians. These include the Caucasus region of Georgia and Russia and several montane areas of Iran, Iraq, Turkey, Afghanistan, Kyrgyzstan, Kazakhstan, and China (Borkin 1999; Tarkhnishvili and Gokhelashvii 1999; Zhao 1999).

As in Europe, human population density in Asia tends to be greatest in the areas of high amphibian species diversity, particularly in India, Sri Lanka, China, and Indonesia. Areas of low amphibian diversity, including Asian Russia, northwestern China, Mongolia, and Central Asia, generally have lower human population densities. This means that deforestation and other human alteration of natural habitats are the chief threats to many species of Asian amphibians. For much of Asia, there is surprisingly little information on the status of amphibian species and almost no quantitative data on declines of amphibian populations, although there have not been reports of large population declines in most regions. There also is wide variation among Asian countries in the level of protection afforded to amphibians or critical amphibian habitats, but in many cases, amphibians are poorly protected in this region and conservation efforts are just beginning.

An appreciation of nature has long been a part of Japanese culture, and Japanese conservationists have paid considerable attention to protecting endangered species of amphibians, despite a shortage of scientists working on the systematics and ecology of amphibians (Matsui 2000). Approximately 83% of the amphibian species or subspecies on the main islands of Japan are endemic, as are 77% of the amphibians of the Ryukyu Islands (Zhao 1999; Matsui 2000; Ota 2000). These include a number of species of hynobiid salamanders, the giant salamander, *Andrias japonicus,* and eleven anuran species. The number of endemic species probably is higher than currently recognized, because genetic studies have revealed hidden taxonomic diversity, especially in the genus *Hynobius* (Matsui et al. 2000, 2001; Nishikawa et al. 2001; Tominaga et al. 2003). Many Japanese species are widespread, but some species with limited habitats are threatened or endangered, especially in the Ryukyu Islands, where deforestation is a major threat (Ota 2000).

On the main islands of Japan, some ranid frogs have declined, including *Rana porosa brevipoda.* The main cause of decline appears to be reduction in natural wetlands due to conversion to agriculture or housing developments, as well as changing agricultural practices in rice fields that provide the main habitat for many species. The traditional method of growing rice involved irrigation of rice paddies by shallow ditches, which provided essential habitat for a variety of aquatic organisms, including frogs. In the last 30 years, most rice field have converted to irrigation by underground pipes that feed deep concrete ditches, which lack the shallow vegetated habitats used by frogs (Fujioka and Lane 1997). The resulting decline in frog populations may impact predators that prey on frogs, such as snakes and wading birds. Also potentially at risk are invertebrate predators that depend heavily on frogs, such as the endangered giant water bug (*Lethocerus deyrollei;* Hirai and Hidaka 2002). The American bullfrog (*Rana catesbeiana*), which was introduced as a source of food in the early twentieth century, is now widespread in Japan and is a potential threat to some native amphibians (Inger 2001). Nevertheless, the only detailed study of the diet of bullfrogs in Japan showed that adults ate mostly crayfish and juveniles ate mostly smaller invertebrates, with little evidence of predation on native anurans (Hirai 2004). Bullfrogs may be competitors of native anurans in this location and may impact populations of other aquatic organisms, but the impact on amphibian populations is unclear.

Taiwan straddles the Tropic of Cancer and has a subtropical climate. Its dense human population is concentrated in the lowland plains of western Taiwan, with the center of the island dominated by high mountains. Amphibian species diversity is modest, with about 30 species, representing a combination of tropical species and temperate zone species that reached the island from mainland China (Zhao 1999). About one third of the species are endemic to Taiwan. Others not currently recognized as endemic, such as Taiwanese populations of *Limnonectes kuhlii,* probably are distinct from populations assigned to the same species on mainland Asia (Emerson, Inger, and Iskander 2000). As many as 20 species have been found in one location in the mountains (Hsu et al. 2003). There is little published information on the status of amphibians on Taiwan, but amphibian numbers have declined and amphibians are threatened by industrialization and deforestation (http://www.tesri.gov.tw/english/content/animal/ani-amphibians). About 12% of Taiwan has been set aside as national parks, nature reserves, and protected areas, mostly in the mountains, and about a third of the amphibian species are protected by law (Hsu and Agoramoorthy 1997). One salamander, *Echinotriton andersoni,* is considered extinct on Taiwan (Zhao 1998).

There is little information published in English on the status of amphibian populations in mainland China, although the status of a few species is discussed on Amphibia Web (http://elib.cs.berkeley.edu/aw/index.html) and in some Chinese publications (Ye, Fei, and Hu 1993; Zhao 1998). This enormous country has a rich and diverse amphibian fauna, particularly in southern China and the southwestern mountains (Zhao and Adler 1993; Fei 1999; Zhao 1999). With the world's largest human population, China also has suffered from enormous environmental degradation (Li et al. 2004), including industrial pollution, extensive deforestation, and loss of wetlands, all of which are likely to have a negative impact on amphibian populations. About 5% of the country has been set aside for parks and nature reserves, although none was specifically designed to protect amphibians (Zhao 1999).

Few detailed surveys of amphibian populations have been conducted in China, but Yiming and Wilcove (2005) estimated that at least 31 species of amphibians in China are threatened or endangered, with two thought to have gone extinct. The major threats to Chinese amphibians are habitat destruction, especially deforestation, pollution, and overexpoitation for food and medicine. Fellers, Yuezhao, and Shaoyin (2003) reported on surveys of three species of amphibians in the Zoige Wetlands at the headwaters of the Yellow River, conducted in 1997 and 2000–2002. All three species showed a reduction in population size, but the number of sites surveyed was too small and the time period too short to demonstrate a significant decline. Anecdotal evidence from local people suggested that these wetlands had decreased in size over the last 20 years and that amphibians had become less common. Currently a number of species of hynobiid and salamandrid salamanders are considered vulnerable or endangered in China, as is the giant cryptobranchid salamander *Andrias davidianus* (R. Murphy et al. 2000; Stuart et al. 2004). At least one salamander, *Cynops wolterstorffi,* is considered extinct in China. Less than a dozen species of anurans in the families Ranidae and Megophryidae are considered endangered or vulnerable, including several very unusual frogs in the genus *Vibrissaphora* (Zhao 1998).

Widespread collection of common ranid species for food and medicine probably threatens some local Chinese populations of amphibians, although not all species are threatened throughout their ranges. According to a recent global assessment of amphibian declines (Stuart et al. 2004), overexploited amphibian species are heavily concentrated in China and Southeast Asia, with ranid frogs being the most frequently overexploited. Yiming and Wilcove (2005) estimated that 40% of the 25 species of Chinese amphibians for which data are available are threatened by overexploitation for food, and 20% are threatened by medicinal use. Illegal harvesting of frogs (*Rana dybowskii*) by Chinese immigrants also is reported to threaten some populations in Far Eastern Russia. Nevertheless, this species and the other ten species of amphibians in this sparsely populated region appear to be secure, and many remain very abundant. The most vulnerable species are forest specialists such as the hynobiid salamander *Onychodactylus fischeri,* but even for this species, threats are largely local. A major threat to some local amphibian populations in both the Russian Far East and in China is the constuction of hydroelectric dams that flood forested gorges. On the other hand, many lowland species that favor open habitats may even increase in human-altered habitats. Potential threats to local populations of amphibians in Asian Russia include the deliberate introduction of nonnative mustelid predators such as mink (*Mustela vison*) and otters (*Lutra lutra*), as well as the European frog *Rana ridibunda,* but the ecological impact of these invasive species has not been well studied (Kuzmin and Maslova 2003). Only a few species of Chinese amphibians are thought to be threatened by alien species, but the problem has been little studied in China (Yiming and Wilcove 2005).

Hong Kong has a surprisingly rich amphibian fauna for such a densely populated area, with 23 species (8% of all Chinese amphibians; Lau and Dudgeon 1999). Despite its large human population, 38% of the land area of Hong Kong is protected. Montane forests are the best-protected habitats, and many amphibians are associated with montane streams. Lowland wetlands, on the other hand, are poorly protected (Yip, Corlett, and Dudgeon 2004). Several anurans closely associated with rice paddies have undergone drastic declines with the disappearance of rice cultivation in Hong Kong, and several other species have quite restricted ranges.

Southeast Asia, including Mayanmar (Burma), Thailand, Laos, Cambodia, Vietnam, Malaysia, Singapore, Indonesia, Brunei, and the Philippines, has an extraordinarily rich amphibian fauna. The greatest species diversity, with a high level of endemism, occurs in Sundaland, which encompasses Sumatra, Java, Borneo, and surrounding smaller islands, as well as the Palawan Islands of the Philippines. This region harbors at least 200 species of amphibians, of which about 60% are endemic. Only about a quarter of the species occur on more than one of the larger islands, although many species are broadly distributed within islands (Inger 1999). The amphibians of Malaysian Borneo have been studied most intensively (Inger and Stuebing 1989, 1997). Those of the other islands are not as well known, although there is a recent guide to amphibians of Java and Bali (Iskandar 1998). Some bufonids, ranids, microhylids, and rhacophorids that favor open habitats have adapted well to rice paddies and other human-altered habitats and often are more common in such habitats than in forests (Inger 1999).

Forest specialists, many of which breed in streams, presumably are at greatest risk because of widespread deforestation, leading to habitat degradation and fragmentation. In some countries, such as Singapore, most of the original forest is gone (Brook, Sodhi, and Ng 2003). In others, large tracts of forest remain, but most Asian forests are at risk from both legal and illegal logging activities. Voris and Inger (1995) found that frog populations were stable from the 1960s through the 1990s along streams in undisturbed forest in two regions of Malaysian Borneo, Sarawak and Sabah, with no evidence of major declines in any species. Nevertheless, stream-dwelling frogs are susceptible to logging that destroys terrestrial habitat and increases silt loads in streams, because most tadpoles of these species cannot survive high silt loads (Inger 1999). Species with very restricted ranges obviously are at risk from logging and other forms of habitat destruction as well. In contrast to the relatively well-protected forests of Malaysian Borneo, deforestation in Indonesian Borneo and Sumatra is proceeding at an alarming rate, even in legally protected forests (Jepson et al. 2001; Curran et al. 2004; Fuller, Jessup, and Salim 2004). This destruction poses an immediate risk to all forest-dwelling amphibians in the region, as well as most of the other wildlife in one of the world's true biodiversity hotspots.

A potential threat to some of the most common ranid frogs in Indonesia is the export of frog legs, mostly to Europe and China. Indonesia has replaced India and Bangladesh as the major source of frog legs after those countries banned frog exports. Nearly all of the frogs being shipped into Europe appear to be crab-eating frogs (*Fejervarya cancrivora*), even though some shipments were declared to be other species, including the introduced American bullfrog (*Rana catesbeiana;* Veith et al. 2000). Estimates of the number of frogs exported from Indonesia range up to a million frogs per year. It seems likely that such numbers will adversely impact frog populations, and marked declines of large ranids in both India and Bangladesh were attributed to overexploitation. Nevertheless, I am not aware of any detailed ecological studies on the impact of frog leg exploitation on anuran populations, the level of harvest that can be sustained, or the amount of time required for a population to recover from overexploitation. Decreases in anuran numbers have been blamed for increased populations of pest insects in India, Bangladesh, and Indonesia, with a concomittant increase in use of chemical pesticides that itself could be harmful to amphibians (Abdulali 1985; Barfield 1986; Oza 1990; Bazilescu 1996; Jacques 1999). Again, I do not know of any rigorous studies designed to demonstrate a causal relationship between anuran declines, insect populations, and pesticide use.

On mainland Southeast Asia, the amphibian fauna of peninsular Malaysia and Thailand is relatively well known (Chan-ard et al. 1999), although a recent guide to amphibians of Thailand (Nutphund 2001) uses outdated taxonomy and has many misidentified photographs (Ohler 2003a). The amphibian fauna of Myanmar (Burma), Laos, Cambodia, and Vietnam is poorly studied. Recent exploration in Vietnam and Laos has resulted in the description of a number of new species of amphibians (Inger, Orlov, and Darevsky 1999; Ohler 2003b; B. Stuart 2005), and there is hidden diversity among those species already described (Bain et al. 2003). Of over 112 species known from Vietnam, only about 10 are officially listed as endangered, threatened, rare, or vulnerable, but there is anecdotal evidence that some species have declined. Some amphibians are at risk from overexploitation for food, medicine, and the pet trade (Nguyen 2000). Others are at risk from forest exploitation that degrades amphibian habitats. In one survey, twice as many species of amphibians were found in undisturbed montane forest as in exploited forest (Nguyen 1998). The amphibian fauna of other Southeast Asian countries is so poorly known that a survey of amphibians from a single mountain range in Cambodia doubled the number of anuran species known from that country (Ohler, Swan, and Daltry 2002), and more than 35 undescribed species of anurans have been discovered in recent surveys in Mayanmar (see http://www.calacademy.org/research/herpetology/mayanmar). As in Vietnam, there appears to be hidden diversity among species already described for Mayanmar, such as the *Fejervarya limnocharis* complex (this group probably contains many cryptic species throughout southern Asia; Toda et al. 1998). Quantitative information on the status or possible decline of amphibian populations is not available for any of these countries.

The countries of South Asia vary in the diversity of their amphibian faunas. The island nation of Sri Lanka is a remarkable amphibian hotspot, with more than 250 species, including more than 100 species of rhacophorid frogs, many still undescribed (Pethiyagoda and Manamendra-Arachchi 1998; Meegaskumbura et al. 2002; Bossuyt et al. 2004). Most of these species are concentrated into about 15,000 km$^2$ of wet forest, giving Sri Lanka the highest density of amphibian species per unit area in the world. Over 220 species of amphibians are found in India and the southern Himalayas, including Nepal and Bhutan, with a number of endemic lineages and especially high diversity in the wet forests of the Western Ghats region and the southern Himalayas (Inger 1999; Bossuyt et al. 2004). New species are regularly described from these areas, and many taxonomic problems remain, so the actual amphibian fauna probably is considerably larger (Dubois 1998). Indeed, the total amphibian fauna of the Western Ghats may be nearly twice as large as currently recognized (Daniels 2003). The region not only harbors a large number of endemic species of frogs, but also at least 14 species of endemic caecilians (Bjatta 1997). Nepal has about 55 species of amphibians, nearly all of which are anurans, with only one caecilian (*Ichthyophis sikkimensis*) and one salamander (*Tylototriton verrucosus;* Schleich, Anders, and Kästle 2002). The amphibian fauna of Bangladesh, much of which is low and wet, is less diverse, probably with fewer than 50 species. The amphibian fauna of relatively arid Pakistan is even smaller, with only about 25 species, mostly bufonids and ranids (Khan 2005).

The status of amphibian populations in these countries is only beginning to be studied (Gupta 1998; Molur and Walker 1998; Daniels 1999). Because the highest species diversity and nearly all of the endemic species in this region occur in wet montane forests, deforestation and forest fragmentation almost certainly are the major threats to amphibian populations. In Sri Lanka, many of the endemic rhacophorid frogs appear to have extremely small geographic ranges, so they are likely to be vulnerable to even relatively localized deforestation. At least 80% of the original wet-zone forests of Sri Lanka are now gone, and the remaining forests are highly fragmented. Museum records from the nineteenth century indicate that at least 15 species of amphibians formerly known from Sri Lanka are now extinct. Given the recent discovery of many new species of frogs, it seems likely that many other species probably went extinct before being described. Along with the direct effects of deforestation, other potential threats to forest-dwelling amphibians in Sri Lanka include increased temperatures and decreased rainfall caused by deforestation, acid rain, and heavy use of pesticides and herbicides in tea plantations that occupy most of the montane regions of the country. In addition, most of the areas of high amphibian diversity are not included in the country's system of parks and nature reserves, which have focused on protecting large mammals in the lower, drier parts of the country (Wijeyamohan, da Silva, and Santiapillai 1994; Pethiyagoda and Manamendra-Arachchi 1998).

Forest-dwelling amphibians face similar threats in other countries with dense human populations, especially India and Nepal. The Western Ghats region of India is a major center of amphibian diversity, where nearly 80% of the amphibians are endemic (Daniels 1992, 1997a, b, c, d; Bhatta 1997; Krishnamurthy 1999, 2003; Krishnamurthy and Hussain 2000; Vasudevan, Kumar, and Chellam 2001; Bossuyt et al. 2004). It also accounts for about 20% of all the tea cultivation in the country. Tea cultivation not only results in removal and fragmentation of natural forest habitats, but it also typically involves heavy application of chemical pesticides that may end up in streams outside of the plantations. Tea plantations can support a surprisingly high diversity of amphibian species (Daniels 2003), but there has not been a carefully designed study to compare the abundance and diversity of amphibians in tea plantations and in nearby natural forests. Amphibians may be threatened even in protected lands, because such parks often include a mosaic of disturbed and undisturbed habitats. A potentially serious threat to amphibians and other animals of the Western Ghats is the increasing frequency of forest fires, even in protected reserves (Kodandapani, Cochrane, and Sukumar 2004). The reason for the increase in fires is not entirely clear, but may result from a combination of land use changes, habitat fragmentation, and climate change, as well as increased human activity in and around forested areas.

In Kudremukh National Park in the Western Ghats, iron mining continues despite the lack of human settlement in most of the park, and this has disturbed and fragmented parts of the natural forest. Of the 42 species of amphibians reported from the park, 22 occurred only in undisturbed habitats, while another 20 occurred in both disturbed and undisturbed habitats (Krishnamurthy 2003). Consequently, species richness was higher in undisturbed sites, and population densities were higher as well. Nepal also has suffered extensive deforestation, despite having a relatively large network of protected areas. Serious habitat degradation in the uplands of Nepal undoubtedly poses a threat to forest-dwelling amphibians, but there is little specific information on the status of amphibian populations in the country (Maskey, Schleich, and Kästle 2002).

A number of South Asian anurans tend to favor open habitats and often occur at higher densities in disturbed habitats around human settlements than in undisturbed forest (Maskey, Schleich, and Kästle 2002; Krishnamurthy 2003). These include the widely distributed toad *Bufo melanostictus* and several pond-breeding microhylids, ranids, and rhacophorids. This appears to be a general pattern in much of Asia, with species that adapt well to agricultural landscapes and other human-altered habitats often becoming very common, while total amphibian species diversity declines as natural forested habitats disappear (Gillespie et al. 2005). Ironically, the abundance of amphibians such as large ranid frogs in human-altered habitats often makes them vulnerable to overexploitation for food or commercial export of frog legs. This apparently occurred in both India and Bangladesh in the 1980s, where *Hoplobatrachus tigerinus* has been the most commercially important frog species. Both countries eventually banned commercial export of frog legs. The large ranid frog *Paa liebigii* is widely collected for food in Nepal (Schleich and Kästle 2002). The ecological impact of harvesting on populations of this species has not been studied in detail, but several members of the genus *Paa* from the Himalayan region are considered to be overexploited (Stuart et al. 2004).

### Africa

Less is known about the amphibian fauna of Africa and the current status of amphibian populations than that of any other continent. Much of the continent remains poorly explored herpetologically, but undescribed species are frequently discovered as new areas are explored. Political instability and lack of basic infrastructure such as roads has limited access of biologists to many parts of Africa. Indeed, only two areas have received much attention in terms of amphibian conservation, southern Africa and Madagascar, although the amphibians of East Africa are beginning to receive more attention (Channing and Howell 2006). Although individual species are known to be at risk elsewhere in Africa, often because of limited geographic ranges (e.g., *Nectophrynoides asperginius* in Tanzania; Weldon and du Preez 2004), there has been little systematic work on the status of African amphibian populations or their even basic habitat requirements (Behangana 2004; Behangana and Arusi 2004). The montane forests of Africa are particularly rich in endemic species of amphibians, many of which have very limited ranges and specialized habitat requirements (Gartshore 1986; Hofer, Bersier, and Borcard 1999, 2000; Poynton 1998, 1999; Channing 2001; J. M. Lea, Luiselli, and Politano 2005; Channing and Howell 2006). We can expect forest-dwelling amphibians, especially stream-breeders and those with nonaquatic reproduction, to be at risk anywhere in Africa where deforestation is proceeding at a rapid rate. Anurans that breed in ponds are more likely to adapt to human-altered habitats than are forest-floor or stream-breeding species (Andreone 1994; J. M. Lea, Luiselli, and Politano 2005). As in Asia, African anurans that adapt well to agricultural landscapes and other human-altered habitats, especially ranids, bufonids, and hyperoliids, are likely to thrive even as forest-dwelling species disappear (e.g., Lawson 1993; S. Lynn and Lindle 2002). In some cases, overall species diversity actually increases in disturbed habitats, even as species' composition changes dramatically (J. M. Lea et al. 2003, 2005).

Civil wars and the movement of large numbers of refugees in various parts of Africa can render even relatively recent work on amphibian populations obsolete and makes the status of protected areas in many parts of the continent uncertain. For example, the Comoé National Park in the Ivory Coast, where extensive studies of amphibian populations have been conducted (Rödel 2000), was invaded by thousands of refugees during the recent civil war and breakdown of government authority in that country. This park is now considered one of the most endangered protected natural areas on the list of UNESCO World Heritage Sites (http://whc.unesco.org/danglist.htm), but the specific effect of recent habitat degradation on amphibian populations is not known. Similar problems afflict several national parks in the Democratic Republic of the Congo and Uganda, and these parks also appear on the list of endangered World Heritage Sites. Other countries simply lack the resources to adequately manage protected areas. The Central African Republic, for example, has about 11% of its land area designated as protected areas. Unfortunately, these areas do not adequately represent all of the ecoregions of the country. Savanna habitats are overrepresented among the protected areas, whereas lowland forests that are likely to harbor many species of amphibians are poorly protected. Furthermore, many areas have little or no active management or protection because of inadequate resources in this very poor country. Poaching of mammals is widespread, but amphibians are more likely to be affected by habitat destruction, and their status is unknown (Blom, Yamindou, and Prins 2004).

By far the best-studied African amphibian fauna is that of southern Africa. The composition of the fauna is well known (Passmore and Carruthers 1995; Channing 2001). This region has a high proportion of endemic species, with about 55% of the species in South Africa, Lesotho, and Swaziland being endemic, and about 66% of all southern African species (J. A. Harrison et al. 2001). A project to compile a detailed atlas of the distribution and status of amphibian species in the region has been underway for a number of years and has now been completed (Minter et al.

2004). In contrast to the heavily forested habitats of the equatorial and montane regions of Africa, much of southern Africa is relatively arid, and amphibians are heavily dependent on wetland habitats for breeding, with temporary wetlands being especially important (Channing and van Dijk 1995). South Africa has lost roughly a third of its original wetlands, and many of the remaining wetland habitats are found in areas of high human population density (J. A. Harrison et al. 2001). Consequently, protection of wetland habitats, especially seasonal wetlands, is a high priority for conservation of southern African amphibians.

Fortunately, a large proportion of amphibian species in South Africa (96%) is represented in protected areas (Drinkow and Cherry 1995), but a much smaller proportion is in protected areas in southern Africa as a whole (C. Seymour et al. 2001). About 72% of the anurans of South Africa, Lesotho, and Swaziland are not threatened or endangered (Minter et al. 2004). Nevertheless, a number of South African species are endangered because of habitat destruction and development. These include several species of *Heleophryne* (see fig. 1.14 C in chapter 1), inhabitants of mountain streams that are found only in this region. Many of the other endangered anurans in South Africa are found in the Southwestern Cape region, where agricultural, urban, and suburban development threaten critical habitats. One highly endangered species is *Xenopus gilli,* which is endemic to fynbos habitats. Most of the known breeding sites of this species have been disturbed or destroyed by agricultural or urban development, and the species also is threatened by hybridization by the much more widespread species *X. laevis* (Picker and de Villiers 1989; Harrison et al. 2001). Other species in the same habitats that have declined due to habitat loss include *Bufo pantherinus, Microbatrachella capensis, Cacosternum capense,* and *Capensibufo rosei* (Harrison et al. 2001; C. Seymour et al. 2001; Minter et al. 2004). There have been local declines of some widespread species, such as *Pyxicephalus adspersus,* due to habitat destruction, but it remains common in much of southern Africa. Because many southern African amphibians breed in seasonal wetlands in savanna or fynbos habitats, a particular threat is the widespread conversion of such habitats to plantations of introduced pine and *Eucalyptus* trees, as well as invasion of wetland habitats by alien plants such as *Acacia* trees, all of which alter the water table and often eliminate breeding sites (Minter et al. 2004).

The importance of Madagascar as a center of amphibian diversity and endemism already has been mentioned. Some species of frogs are widespread in open areas and human-altered habitats, including hyperoliids in the genus *Heterixalus,* the ranid frog *Ptychadena mascareniensis,* and some species of *Mantidactylus.* Different types of agricultural landscapes harbor different assemblages of anuran species. Coffee plantations, for example, often have a relatively high diversity of amphibians and reptiles, whereas rice fields and sugar cane plantations generally have low diversity (Andreone et al. 2003). Many mantellid frogs and nearly all microhylids are forest specialists and hence are susceptible to widespread deforestation and further fragmentation of existing forest habitats in a country that already has lost most of its native forests (Raxworthy and Nussbaum 2000; Vallan 2000, 2002; Andreone and Luiselli 2003; Lehtinen, Ramanamanjato, and Raveloarison 2003). Many of the frogs of Madagascar also have the misfortune of being very colorful and therefore highly desirable in the pet trade. Thousands of individuals of several species of *Mantella,* the unique microhylid tomato frog (*Dyscophus antongili*), and the colorful burrowing frog *Scaphiophryne gottlebei* have been shipped out of the country in recent years. Unfortunately, there is no information on the level of trade in these animals that is ecologically sustainable.

## Why Have Amphibian Populations Declined?

When scientists first became concerned about widespread declines of amphibian populations in the late 1980s, the issue attracted much attention from the press and popular science writers (J. Phillips 1994). At the time, the causes of amphibian population declines were poorly understood, and much of the discussion in the popular press created the impression that some mysterious force was eliminating amphibians around the world. Some supermarket tabloids even suggested that aliens from distant planets were depopulating the world of amphibians! Since then, we have learned a great deal about the general causes of amphibian declines, even though the precise causes of declines of particular species often are not fully understood. Some potential causes of population declines have been discussed briefly in the preceeding review of amphibian populations around the world. This section describes the evidence for some of these causes in more detail. It now seems clear that most amphibian declines share the same basic causes as declines of other groups of organisms in the current biodiversity crisis: habitat destruction and degradation, introduction of alien predators and competitors, environmental pollution, the emergence of infectious diseases (perhaps in conjunction with other environmental stressors), and global changes that can have widespread environmental effects (Semlitsch 2003c). Human exploitation of amphibians for food, medicine, scientific research and education, fishing bait, or the pet trade probably has contributed to declines of some amphibian populations, but there is little evidence that this is responsible for widespread population declines (J. Jensen and Camp 2003). Indeed, the impact of direct human exploitation probably has had a much smaller effect on amphibian populations than has exploitation of many other vertebrates, including fishes, marine mammals, large game mammals, primates, birds, large snakes, crocodilians, and turtles.

### Habitat Destruction

Throughout the world, complete destruction or removal of critical habitats undoubtedly poses the single greatest threat to amphibian populations (Dodd and Smith 2003; Young et al. 2004). The negative effects of replacing natural habitats with agricultural land, roads, industrial sites, cities, suburbs, and other human uses are obvious and have been discussed in the preceding survey of amphibian populations around the world. Nevertheless, some amphibians exhibit a remarkable capacity to adapt to human-altered habitats, including urban and suburban areas (e.g., Beebee 1979c; Chovanec 1994; Delis, Mushinsky, and McCoy 1996; Hermy and Cornelis 2000; Dudgeon 2003; Rubbo and Kiesecker 2005), although amphibians in such environments are threatened by general habitat degradation and loss of connectivity between patches of suitable habitat (Löfvenhaft, Runborg, and Sjögren-Gulve 2004). Amphibians that do well in human-altered habitats tend to be habitat generalists that favor open areas, whereas those that depend on forested habitats tend to fare poorly (Rubbo and Kiesecker 2005). Some species actually are more abundant in agricultural landscapes than in forested habitats, as long as fish-free ponds are available for reproduction and forested habitat is only moderately fragmented (Kneitz 1998; Knutson et al. 1999; Kolozsvary and Swihart 1999; Quinn, Gallie, and Volsen 2001; Guerry and Hunter 2002; J. M. Lea, Politano, and Luiselli 2003; J. M. Lea, Luiselli, and Politano 2005; Hazell et al. 2004; Johansson et al. 2005). There are limits to amphibian tolerance for agricultural landscapes, however. Intensive, industrial-scale farming that involves destruction of small wetlands and heavy application of pesticides has a much greater impact on amphibian populations than do less intensive forms of agriculture such as livestock grazing (Beebee 1977, 1997; P. Joly et al. 2001; R. Gray and Brown 2005). The negative effects of agriculture can be ameliorated to some extent by adjusting agricultural practices such as plowing schedules and application of pesticides to the phenology of local amphibian species, although this may be economically impractical for some crops (Meyer-Aurich et al. 1998).

Two types of habitats are especially critical for a large proportion of the world's amphibian species: forests and wetlands. Although amphibians also can be found in other types of habitats, including tropical savannas, temperate grasslands, and deserts, these areas generally have lower species diversity than do forest and wetland habitats (nearly all grassland and desert amphibians depend on temporary wetlands for reproduction). More arid habitats sometimes face fewer environmental threats than do forests and wetlands because they are less desirable for human use. Nevertheless, conversion of natural grasslands to agriculture and urban and suburban development can threaten amphibians in some regions. I will focus mostly on forests and wetlands here, because the impact of habitat destruction on amphibian populations is best understood for these habitats. For purposes of discussion, it is useful to treat forest and wetland habitats separately, although in reality, many species depend on both types of habitats for survival (Semlitsch 2003b).

#### Deforestation and Forest Fragmentation

Many clades of amphibians are largely or entirely dependent on forested habitats that provide moist terrestrial microhabitats for foraging, and in some cases, for reproduction. These include most plethodontid salamanders, by far the most diverse group of salamanders; many ambystomatids and salamandrids; most eleutherodactyline frogs, one of the largest clades of anurans; many hylid and rhacophorid treefrogs; most arthroleptids, centrolenids, dendrobatids, mantellids, heleophrynids, and microhylids; many bufonids, ranids, megophryids, leptodactylids, and myobatrachids; and many caecilians. In tropical rainforests, there often is an abundant and diverse fauna of forest-floor frogs, some of which are entirely terrestrial and dependent on moist leaf litter for reproduction (see chapter 15). Not only do these anurans depend on moist leaf litter for shelter, but they also depend on the diverse arthropods that inhabit the litter (Vitt and Caldwell 2001). This forest-floor anuran fauna is largely absent from temperate-zone forests, where leaf-litter anurans usually are limited to a few species of *Bufo* and *Rana* that breed in water (*Bufo americanus* and *Rana sylvatica* in Eastern North America; *B. bufo* and *R. temporaria* in Europe; various related species in Asia). In North America, plethodontid salamanders largely occupy the leaf-litter niche, where they can reach extraordinary densities of several thousand individuals per hectare (Davic and Welsh 2004).

All of these groups are at risk in places where deforestation is occurring at a rapid rate. This is particularly true in the tropics, where logging of forests often is followed by replacement of forested habitats by urban development, cattle pastures and farms, tree plantations, roads, mining operations, oil drilling, hydroelectric dams, and other uses that result in the loss of critical amphibian habitats (Vitt and Caldwell 2001; T. Lewis 2002; Brook, Sodhi, and Ng 2003; Crump 2003; Hero and Shoo 2003; Yiming and Wilcove 2005; Channing and Howell 2006). Key areas of special concern for amphibian conservation because of widespread deforestation have been mentioned in the previous section: the lowland and montane rainforests of Central America, South America, the Caribbean, Africa, Madagascar, China,

Indonesia, Southeast Asia, Sri Lanka, Nepal, and India. In general, tropical forest regions in the Old World have less of their original forest remaining than do tropical forests in the New World, and they often are not as well protected (Hero and Shoo 2003). This comparison is somewhat skewed by the very large size of the Amazon rainforest. Although nearly 90% of the Amazon forest remained intact in 2001, some areas, such as Rondônia, are threatened by very rapid deforestation and could lose most of their primary forest within the next decade (Ferraz et al. 2005). Other regions such as Honduras, El Salvador, Jamaica, and Haiti rank among the highest in the world in amount of forest destroyed or rate of deforestation (Crump 2003; Young et al. 2004). The only practical solution to this problem is to find economic incentives to increase protection of old growth tropical forests and promote the restoration of secondary forests in areas of high amphibian diversity, which in many cases also will be areas of high diversity of other taxa.

Even when tropical forests are not completely destroyed, the remaining patches of forest often form a fragmented landscape with a matrix of agricultural land and other disturbed habitats that can be inhospitable to amphibians (Ferraz et al. 2005). Habitat fragmentation can have a number of detrimental effects, several of which have been mentioned previously. Habitat patches can become isolated from one another, leading to reduced gene flow between local populations and decreased genetic diversity within populations. Small, isolated populations often are susceptible to chance extinction, and lack of connectivity between habitat patches reduces the chances that such populations will be replenished by new immigrants. Small habitat patches may be subject to edge effects, with increased exposure to sunlight resulting in warmer, drier microhabitats with the patches that can be inhospitable to amphibians. Finally, forest streams are key breeding sites for many tropical forest amphibians, and logging can affect the hydrodynamics and silt loads of forest streams in surviving forest patches downstream from logged areas.

In some respects, amphibians may be less susceptible to forest fragmentation than are larger vertebrates such as birds and mammals, because of their small home ranges and sedentary habits (see chapter 6). There has been extensive research on the effects of tropical forest fragmentation on a variety of animals, from butterflies to birds and mammals, but comparatively little work on amphibians. Most studies have focused on the effects of forest fragmentation on the abundance and reproductive success of one or two species, but some studies also have addressed the effect of fragmentation on amphibian species diversity. Not surprisingly, most studies of anuran diversity in tropical forest fragments have shown that forest-interior specialists, including tree-hole breeders, bromeliad breeders, species with direct development of terrestrial eggs, and those that depend on clear forest streams are the most vulnerable to forest disturbance and fragmentation (Gascon 1993; Pearman 1997; Tocher, Gascon, and Zimmerman 1997; Tocher, Gascon, and Meyer 2002; Gascon et al. 1999; Funk and Mills 2003). Similar results were obtained in a study of frogs in a temperate rainforest in Australia (Lemckert 1999). Pond-breeding species tend to be much more plastic in their habitat requirements and often can adapt to disturbed habitats. Indeed, many tropical pond-breeding anurans are more abundant and diverse in moderately disturbed habitats than in primary forest (Pearman 1997), possibly because ponds in open-canopy habitats are more productive than are those in closed-canopy forests and therefore may provide better food resources for tadpoles (Skelly, Friedenburg, and Kiesecker 2002; Skelly et al. 2005).

In a fragmented lowland Amazonian forest landscape in Brazil, less than half of the total complement of anuran species found in primary forest was found in pastures, but nearly 80% was found in disturbed and secondary forest (Tocher Gascon, and Meyer 2002). Some species that are characteristic of open, disturbed habitats, such as the hylid frog *Scinax rubra,* apparently used the matrix of disturbed habitats to disperse into forested habitats where they normally would not be found. Other species exhibited considerable plasticity in choice of oviposition sites and were able to breed in all habitat types. For example, *Phyllomedusa tarsius,* which normally lays its eggs on bushes overhanging water, was capable of breeding in primary forest, disturbed forest, and pasture ponds, and actually was most successful in secondary forest ponds (Neckel-Oliveira 2004).

In Amazonian Ecuador, frog species richness again was greatest in undisturbed forest, especially for *Eleutherodactylus,* but was not significantly different from that of selectively logged forest (Pearman 1997). In a study of tropical montane cloudforest in Veracruz, Mexico, there was a significant reduction in anuran species diversity from undisturbed forest (21 species) to shaded coffee plantations (13 species) to pastures (four species), although some coffee plantations had more species than did some forested sites (Pineda and Halffter 2004; Pineda et al. 2005). As in Amazonia, forest-interior specialists were most likely to be absent from disturbed habitats. In Madagascar, only 54% of rainforest anurans were found in secondary forest, only 46% in *Eucalyptus* plantations, and a meager 12% in rice fields (Vallan 2002). Again, forest-interior specialists (mostly microhylids and mantellids) were the first to disappear, whereas pond-breeding ranids and hyperoliids survived in more open habitats. A similar pattern was seen on several small islands in Indonesia, where amphibian diversity was lower in highly disturbed habitats than in undisturbed or moderately disturbed forest (Gillespie et al. 2005). On a river floodplain in Argentina, 18 anuran species, all aquatic breeders, were found in a protected patch of riparian woodland, whereas species diversity in seven disturbed habitats ranged from one to 14 species (Peltzer, Lajmanovich, and Beltzer 2003). In this case, the degree of disturbance was greater than in the other studies, because most plots outside of the protected forest were in suburban areas with only small remnants of woodlands.

The issue of edge effects is closely related to forest fragmentation, because more fragmented forests have more edges adjacent to pasture or other suboptimal amphibian habitats, and the negative effects of such habitat disturbance may penetrate into the interior of small forest fragments. Studies of edge effects in tropical rainforest frogs have been done almost exclusively in the New World and have focused primarily on species of *Eleutherodactylus*, which tend to be forest specialists. In Amazonian Ecuador, species richness of *Eleutherodactylus* increased with distance to the edge of pastures, even though overall frog species diversity did not, because many hylids did well in disturbed habitats (Pearman 1997). At a higher altitude (3,200–3,600 m) site in Ecuador, two species of *Eleutherodactylus* were virtually absent from disturbed habitats surrounding patches of forest, but in only one species was abundance affected by the size of a forest patch (Marsh and Pearman 1997). Similarly, in a cloud forest (2,400–2,800 m) in Ecuador, there was an abrupt transition between forest and agricultural land in the abundance of *Eleutherodactylus*, with this genus being largely absent from nonforested habitats (Toral, Feinsinger, and Crump 2002). There was little evidence of edge effects penetrating into the forest interior, however. In a fragmented forest in Costa Rica, two species of *Eleutherodactylus* were more abundant in interior forest plots than in plots adjacent to pasture, but three other species were not affected by edges (Schlaepfer and Gavin 2001). There was no effect of either fragment size or distance to the nearest edge on abudance of any species of *Eleutherodactylus*. There was a somewhat surprising seasonal effect, with species that did not appear to avoid forest edges becoming more abundant in the forest interior during the rainy season. Overall, these studies suggest that there is not a consistent edge effect on either the diversity or abundance of forest-floor *Eleutherodactylus*, but some species are better than others at tolerating more open-edge habitats.

In North America, the closest equivalents to *Eleutherodactylus* as forest-interior specialists are plethodontid salamanders. Many of these have direct-developing terrestrial eggs, while other species breed in forest streams. Consequently, salamanders, especially plethodontids, have been widely used as indicator species in studying the effects of clearcutting and other forestry practices in North America (deMaynadier and Hunter 1995; Welsh and Droege 2001; Wyman 2003). In contrast to the tropics, where complete deforestation and conversion of forest to agricultural land are the main threats to amphibian populations, the dynamics of forests in North America are more complex. While clear-cutting of forests is widespread and some forests are destroyed for development, many forests are managed for sustainable yield, with continuous regrowth of harvested stands. In some regions, such as New England, overall forest cover is considerably higher than it was 150 years ago, even though timber harvesting or development continues to reduce forest cover in some areas (A. Miller, Bryant, and Birnie 1998). Forestry practices that can affect amphibian diversity and abundance include the type and extent of tree removal, the maintenance of woody debris on the forest floor after tree clearing, rates of forest regeneration, management of riparian zones along forest streams as habitat for salamanders or stream-breeding frogs, conversion of natural forest to tree plantations, and prescribed burns. In addition, fragmentation of forest stands by clearcuts, roads, agricultural land, or development can reduce amphibian population sizes and restrict movements between populations, with some species being more sensitive to fragmentation and edge effects than others (deMaynadier and Hunter 1998, 2000; Gibbs 1998a, b; DeGraaf and Yamasaki 2002; D. Marsh et al. 2004, 2005; Cushman 2006).

There is clear evidence that clearcutting of mature forests has a negative impact on the abundance of many amphibians, presumably because of the warmer, drier conditions that prevail in cutover stands (deMaynadier and Hunter 1995). There was considerable variation among study sites, however, because of differences in forest moisture regimes, amount of woody debris remaining in cleared sites, variation in the age of mature stands used as control plots, and differences among species of amphibians in their ability to tolerate drier conditions. Overall, data from a variety of studies showed that amphibians were about 3.5 times more abundant in mature forest stands than in clear-cut stands, but plethodontid salamanders were, on avearge, about five times more abundant in mature stands.

Nevertheless, in some studies, salamanders were nearly as abundant in clear-cut areas as in uncut forests, especially if abundant woody debris remained after cutting (Cole et al. 1997; Biek et al. 2002). Although terrestrial plethodontids tend to be the most sensitive to forest clearing, other salamanders that use forested habitats outside of the breeding season can be affected as well (Naughton et al. 2000), while more subterranean species may be less sensitive to clear-cutting (Chazal and Niewiarowski 1998). Even less intensive forms of timber harvest can reduce salamander numbers to some extent, although the negative effects may be less persistent (Mitchell, Wicknick, and Anthony 1996; DeMaynaier and Hunter 1998; Messere and Ducey 1998; R. Brooks 1999; Harpole and Haas 1999; Grialou, West, and Wilkins 2000; B. Ross et al. 2000; Bartman et al. 2001; DeGraaf and

Yamasaki 2002; Duguay and Wood 2002; Hicks and Pearson 2003; S. Knapp et al. 2003; Morneault et al. 2004).

Anurans tend to be less sensitive to forest clearing than are salamanders, probably because of the greater tolerance of anurans to desiccation and their greater mobility, which enables individual animals to move out of inhospitable microhabitats. Hylid treefrogs, which are more resistant to evaporative water loss than any other North American amphibians (Wygoda 1984), actually may benefit from forest clearing. Green treefrogs (*Hyla cinerea*), for example, were considerably more abundant in artificial canopy gaps in southern bottomland hardwood forest than in closed-canopy forest (Horn et al. 2005). About 98% of the frogs captured in the gaps were juveniles, which exhibit considerable cutaneous resistance to water loss once they have left the water (Wygoda and Garman 1993).

Clear-cutting also can alter the demographic characteristics of amphibian populations. For example, Ash et al. (2003) reported that individuals of *Plethodon metcalfi* were more massive in 10-year-old clear-cut forest than in mature forest, and immature individuals were largely absent from the clear-cut area. It was not clear, however, whether this was due to poor reproduction by adults in the younger forest or an inability of smaller animals to survive the drier conditions in that area. Estimates of the age of salamanders in the clear-cut forest based on skeletochronology (see chapter 14) indicated that most adults were younger than the age of the forest, suggesting that they were recent immigrants into the area. Quality of prey also may be different in recently cut and mature forests, with soft-bodied prey such as collembolans, a preferred food item of many plethodontids, being less abundant in the drier microhabitats of recently cut forests (Mitchell, Wicknick, and Anthony 1996), even though overall invertebrate abundance in recently cut forests may be as high or higher than in mature forests (Ash 1997; Harper and Guynn 1999).

There has been some controversy over the fate of salamanders in clear-cut forest stands, with some authors arguing that most individuals in such stands probably perish (Petranka, Eldridge, and Haley 1993; Petranka 1994) and others arguing that some individuals move away from disturbed habitats to more hospitable areas (Ash and Bruce 1994). There is little direct evidence for mortality of amphibians in clear-cut forests, simply because dead and dying individuals are seldom found in the field. There is some evidence for individuals of some plethodontid salamanders moving out of clear-cut areas (Sattler and Reichenbach 1998), but also counter-evidence indicating that the abundance of salamanders in areas adjacent to harvested areas did not increase after cutting (Ash 1997; Bartman et al. 2001). Presumably the ability of salamanders and other amphibians to leave clear-cut areas depends on the size of harvested plots, the nature of adjacent forested habitats, and the vagility of individual species.

The amount of time required for salamander populations to return to predisturbance levels also has been a matter of some debate (deMaynadier and Hunter 1995). In deciduous forests of the eastern United States, estimates of the time required for salamander populations to recover to predisturbance levels in previously cut forests range from less than 30 years in relatively moist habitats to as much as 120 years in drier habitats (Pough et al. 1987; DeGraaf and Yamasaki 1992, 2002; Petranka, Eldridge, and Haley 1993; Petranka et al. 1994; Ash 1997; Ash and Pollock 1999; Harper and Guynn 1999; Herbeck and Larsen 1999; W. Ford et al. 2002). Variation in the estimated time required for salamander populations to recover in harvested forests probably depends on overall forest moisture regime, amount of woody debris, rate of decay of logs, leaf litter depth, and other physical factors that affect rates of water loss in amphibians, as well as sampling methods used to estimate salamander numbers (deMaynadier and Hunter 1995). It is not clear that any species of terrestrial salamander is immediately threatened by current forestry practices, although species with naturally restricted ranges, such as *Plethodon hubrichti*, could be at risk if harvest-rotation schedules are too short or if cleared plots are too large (Kramer et al. 1993; Mitchell, Wicknick, and Anthony 1996).

Some studies of Douglas fir forests in the Pacific Northwest have shown relatively little difference between relatively young forest stands (30–75 years old) and very old (up to 450 years old) forest stands in abundance of amphibians (Aubry, Jones, and Hall 1988; Bury and Corn 1988a; Aubry and Hall 1991; Gilbert and Allwine 1991; Steele, Brodie, and McCracken 2002), but most of these studies examined forests that had not been cut, but were regenerating after fires. Other studies in the same region have shown that at least some salamanders are sensitive to stand age, being more abudandant in mature forests than in younger stands. Some terrestrial plethodontids, such as *Plethodon elongatus* and *P. larselli*, are old-growth forest specialists, making them particularly sensitive to logging (Herrington and Larsen 1985; Welsh 1990; Welsh and Lind 1995), whereas habitat generalists such as *P. vehiculum* and *P. dunni* appear to be more tolerant of managed forest habitats (Cole et al. 1997; Dupuis and Bunnell 1999; Aubry 2000; Grialou, West, and Wilkins 2000; Wilkins and Peterson 2000).

The effect of logging on stream-breeding amphibians is somewhat different from the effect on terrestrial-breeding plethodontids. The best-studied species are stream-breeding amphibians of western North America, including *Dicamptodon, Rhyacotriton, Taricha, Ascaphus,* and *Rana aurora,* with only a few studies having been done on species in eastern forests (Lowe and Bolger 2002). Not only is the terrestrial habitat affected by removal of the forest canopy, but the stream environment can be altered by increased silt loads, increased nutrient input, accumulation of woody debris, and changes in stream temperature and solar radiation (Welsh and Ollivier 1998; Corn, Bury, and Hyde 2003). Because stream-breeding species have aquatic larvae, which sometimes remain in streams for several years, changes in the aquatic habitat may have an even greater impact on populations than do changes in surrounding terrestrial areas inhabited by larger adults (Bury 1983; Corn and Bury 1989; Biek, Mills, and Bury 2002). In addition, adults of many stream-breeding amphibians have a limited ability to move overland out of disturbed habitats into new streams (B. Johnston and Frid 2002; Chan-McLeod 2003), and even movement of larvae within streams may be limited (H. Ferguson 2000). The combined effects of reductions in population size and lack of dispersal to undisturbed streams increases the chances of population fragmentation, loss of genetic diversity, and local population extinctions in stream-breeding species (J. Curtis and Taylor 2003).

The negative effects of logging on stream-breeding amphibians can be partially aleviated by retention of forested buffer strips along streams, provided these areas are sufficiently wide (Dupuis and Steventon 1999; B. Johnston and Frid 2002; Vesely and McComb 2002). Unfortunately, widths of buffer strips currently required by some forestry regulations may not be sufficient to protect riparian amphibian populations, and buffer zones are not always required around headwater streams that lack fish (J. Curtis and Taylor 2003). Susceptibility to disturbance of riparian habitats varies among species. Torrent salamanders (*Rhyacotriton*) and tailed frogs (*Ascaphus*), for example, favor cold, rocky headwater streams with steep gradients and low amounts of silt, sand, or woody debris (Diller and Wallace 1996; Dupuis and Steventon 1999; Wilkins and Peterson 2000; Russell, Mabee, and Cole 2004). Their relatively specialized habitat requirements make them very sensitive to forest clearing, especially in the riparian zone (fig. 16.13). Other riparian amphibians, such as *Dicamptodon, Taricha,* and *Rana aurora,* sometimes decrease in abundance in logged areas, but appear to be somewhat less sensitive than are *Rhyacotriton* and *Ascaphus* (Welsh 1990; Cole et al. 1997; Welsh and Olivier 1998; Dupuis and Steventon 1999; Aubry 2000; MacCracken 2002; Vesely and McComb 2002; Steele, Brodie, and McCracken 2003b; Wahbe and Bunnell 2003; Ashton, Marks, and Welsh 2006). In the eastern United States, stream-dwelling salamanders such as *Gyrinophilus porphyriticus* also are susceptible to logging-related increases in sediment loads (Lowe, Nislow, and Bolger 2004). In areas where both aquatic-breeding and terrestrial-breeding salamanders occur, forest buffer zones around streams need to be large enough to accommodate the terrestrial species that tend to

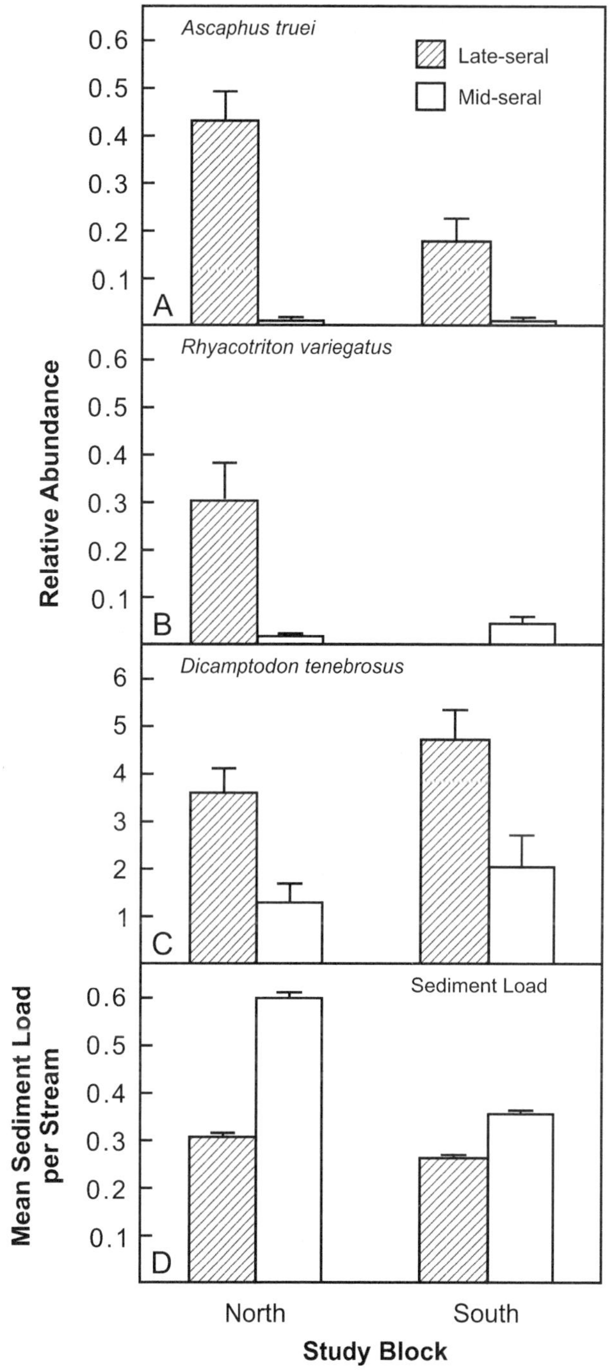


**Fig. 16.13.** Comparison of relative abundance of three stream-associated amphibians in the forests of the Pacific Northwest in late-seral (unharvested) and midseral (37–60 years post harvest) stands in two different study blocks. Populations of *Ascaphus* (A) and *Rhyacotriton* (B) recovered from timber harvesting more slowly than did populations of *Dicamptodon* (C). Sediment loads (D) were significantly higher in midseral stands than in unharvested stands. After Ashton, Marks, and Welsh (2006).

favor more upland habitats while also minimizing the impact of logging on stream habitats (Petranka and Smith 2005).

In many parts of the world, conversion of natural forests to single-species tree plantations may be as great a threat to amphibian populations as is clear-cutting of forests that eventually are allowed to return to their original state. Plantations of spruce, pine, *Eucalyptus,* and teak generally lack the complex understory and thick leaf litter of natural forests, resulting in warmer, drier conditions on the forest floor. This almost always can be expected to have a negative effect on amphibian populations (Waldick, Freeman, and Wassersug 1999; Vallan 2002; Pawar, Rawat, and Choudhury 2004; Yiming and Wilcove 2005). Management strategies in pine plantations that result in removal of the litter layer are likely to decrease soil moisture and expose amphibians to greater risk of desiccation (Moseley, Castleberry, and Ford 2004). In much of the southeastern United States, the dominant forest type was once longleaf pine (*Pinus palustris*), which often occurred in flatwoods habitats with an understory dominated by wiregrass that was flooded in the winter and spring to form shallow inundated savannas (Dodd 1995). Most of this native vegetation has been replaced by plantations of slash pine or loblolly pine, which form a much denser canopy and do not flood because trees are planted in raised beds. The flatwoods salamander (*Ambystoma cingulatum;* fig. 16.9) is a longleaf pine flatwoods specialist that does poorly in pine plantations (Means, Palis, and Baggett 1996). Consequently, this species has declined in much of its range, although it remains common where suitable habitat is present (Palis 1997; Palis and Means 2005). Other amphibians that share this habitat, particular gopher frogs (*Rana capito* and *R. sevosa*), also have declined in some areas (Richter et al. 2003; Jensen and Richter 2005; Richter and Jensen 2005).

Other forestry practices, such as prescribed burns during the wet season, can be detrimental to amphibians because these practices reduce the litter layer, dry out the understory, and tend to eliminate longleaf pine in favor of loblolly pine, which provides less suitable habitat for amphibians (Schurbon and Fauth 2003). Nevertheless, fires are necessary to maintain longleaf pine savannas, essential habitats for a number of southeastern amphibians. Such habitats are quickly replaced by hardwood forest if fire is completely suppressed, but there has been some debate over the optimum interval between prescribed burns (Means et al. 2004; Robertson and Ostertag 2004; Schurbon and Fauth 2004).

The effect of fire on amphibian populations is complex and depends on the type of forest burned, the time of year when burning occurs, and the particular species of amphibians found in those habitats (Pilliod et al. 2003; Bury 2004). Some studies of prescribed burns in North American hardwood forests found few negative effects on amphibian populations, and in some cases found positive effects (Means and Campbell 1981; Kirkland, Snoddy, and Amster 1996; McLeod and Gates 1998; Ford et al. 1999; Russell, Van Lear, and Guynn 1999; Moseley, Castleberry, and Schweitzer 2003). In the Pacific Northwest, natural wildfires appear to have little effect on terrestrial amphibian populations, but result in reduced numbers of stream-dwelling amphibians. Old-growth forests, which provide habitats for a number of amphibians not found in younger forests, are less susceptible to catastrophic fires than are younger stands (Bury 2004). Because much of this region has been heavily logged, younger fire-prone stands now predominate, resulting in reduced habitat for amphibians, especially old-growth specialists. However, there has been relatively little research on the direct effects of prescribed burning on amphibians in this region. Studies of natural and prescribed burns in various habitats in Australia yielded mixed results, with clear negative effects on some anurans, but not on others (Bamford 1992; Driscoll and Roberts 1997). Fire alone is unlikely to threaten many amphibian species with population declines or extinctions, but small isolated populations could be at risk from catastrophic fires (Driscoll and Roberts 1997). Fires also could threaten amphibians in tropical rainforests in regions such as Indonesia, where a combination of increased logging and global climate change may result in increased frequency of catastrophic fires that destroy critical habitats for a wide variety of species.

### Wetland Destruction, Fragmentation, and Management

As mentioned previously, freshwater habitats are under assault throughout the world, and are perhaps the most endangered of all ecosystems. In North America and Europe, which have the strongest wetlands protection programs, many wetlands of importance to amphibians remain highly threatened. In less developed regions, such as South and Central America, wetlands protection is virtually nonexistent or just beginning, and even knowledge of the variety and extent of wetland habitats is rudimentary (Ellison 2004; Junk and Piedade 2004; Loiselle et al. 2004). Threats to stream-breeding amphibians, most of which live in forested habitats, were discussed in the previous section, so I will focus here on amphibians that breed in standing water.

Several characteristics of wetlands that can affect the abundance and diversity of amphibians include wetland size, hydroperiod, the diversity of microhabitats found within a wetland, proximity to other wetland habitats, and proximity to suitable upland habitats. Specific requirements vary among species of amphibians, with some species requiring much longer hydroperiods than others for larvae to reach metamorphosis (Semlitsch 1998, 2000, 2002, 2003b; Paton and Crouch 2002; Semlitsch and Bodie 2003). The array of predators, and particularly the presence or absence of fish, also can affect amphibian diversity and abundance; this is discussed in a separate section that follows.

In most countries where wetland protection has been considered an important conservation issue, including the United States, regulations have focused on protecting relatively large wetlands, such as salt marshes and estuaries, where amphibians do not live, and permanent lakes, ponds, and rivers, which generally have low amphibian diversity. In the United States, Federal protection of small, isolated wetlands and seasonal wetlands is much weaker or nonexistent, although some state and local governments have taken a more active role in protecting such habitats (Semlitsch and Bodie 1998; Russell, Guynn, and Hanlin 2002; Calhoun et al. 2003; Calhoun, Miller, and Klemens 2005; Burne and Griffin 2005b). At the same time, amphibian ecologists have increasingly focused on smaller wetlands with short to intermediate hydroperiods as essential habitats for many pond-breeding amphibians (Dodd 1992; Wellborn, Skelly, and Werner 1996; Griffiths 1997; Babbitt and Tanner 2000; Snodgrass, Komorowski, Bryan, and Burger 2000; Russell, Guynn, and Hanlin 2002; Paton and Crouch 2002; Babbitt, Baber, and Tarr 2003; Babbitt 2005; Burne and Griffin 2005a; Paton 2005; Skidds and Golet 2005). Not only do small isolated wetlands provide fish-free breeding habitats for many amphibians, but they also provide critical stopover points for amphibians moving between larger wetlands (Gibbs 1993; Dodd and Cade 1998; Semlitsch and Bodie 1998) and important foraging sites for amphibians that breed in larger bodies of water (Paton 2005).

Such wetlands go by many names, depending on geographic region, local geology and hydrology, and the nature of the surrounding habitat. Many are simply called seasonal or temporary ponds (Griffiths 1997), but they also include vernal pools (Calhoun et al. 2003; Zedler 2003; Colburn 2004), prairie potholes in the Great Plains (Lannoo et al. 1994; Lannoo 1996), Carolina bays in the southeastern United States (Semlitsch et al. 1996; Sharitz 2003), desert pools and playa wetlands in the arid southwestern United States (A. Anderson, Haukos, and Anderson 1999a, b; M. Gray, Smith, and Brenes 2004), shallow pans in Australia and southern Africa (M. Tyler 1998; Minter et al. 2004), blackwater vlies in South African fynbos habitats (Minter et al. 2004), and shallow seasonal rock pools (D. C. Smith 1983; Spieler 2000; Spieler and Linsenmair 1997). Amphibians also make use of many other types of small depressions that fill temporarily with water, including pools in tropical rainforests (Gascon 1991; Marsh et al. 1999, 2000; Marsh 2000) and animal-generated pools such as bison, hog, peccary, and rhino wallows (Bragg and Smith 1942, 1943; Gascon 1991; Inger and Steubing 1997; Gerlanc and Kaufman 2003) and animal tracks (Rand 1983).

All of these wetlands share certain characteristics, including being filled by precipitation, relatively small size, a lack of permanent connections with other wetlands, a tendency to dry out on a regular basis, and the absence of large predators such as fish that require more permanent water. In addition, their ephemeral nature makes legal protection of such habitats especially difficult. Seasonal pools and even more permanent ponds that form on river floodplains also can be important habitats for amphibians, although such pools are more likely to be invaded by fish than are more isolated wetlands in upland wooded habitats (Healey, Thompson, and Robertson 1997; Burbrink, Phillips, and Heske 1998; Tockner, Schiemer, and Ward 1998; Godreau et al. 1999). Because of river channelization, dams, flood control projects, shoreline development, pollution, and other alterations, floodplain habitats are threatened throughout the world. In Australia, intensive grazing by cattle around floodplain wetlands resulted in degraded habitats and reduced anuran species richness and abundance (Jansen and Healey 2003).

In addition to the size, hydroperiod, and biological communities of temporary wetlands themselves, two critical aspects of such habitats for amphibian conservation are the nature of the habitat surrounding temporary ponds and the degree to which individual ponds are connected by suitable habitat that provides corridors for movement of amphibians between ponds. Because vernal pools and other temporary ponds are, by definition, dry for much of the year, all amphibians that use such ponds must spend considerable time in terrestrial habitats surrounding these ponds, and even amphibians that breed in more permanent aquatic habitats often rely on the surrounding terrestrial habitat outside of the breeding season (e.g., Pope, Fahrig, and Merriam 2000; Pilliod, Peterson, and Ritson 2002). Consequently, amphibian populations are susceptible to degradation of the surrounding terrestrial habitat unless steps are taken to ensure protection of adequate buffer zones around ponds (Dodd and Cade 1998; Semlitsch 1998, 2000, 2002, 2003b; J. Gibbons 2003b; Semlitsch and Bodie 2003). Furthermore, temporary wetlands are by their very nature unpredictable in occurrence. Some vernal pools, for example, may not fill in years due to low winter snowfall or spring rainfall. This makes nearby wetlands essential for amphibians that are displaced from their usual breeding sites, so connectivity between wetlands is critical for amphibian populations (Trenham et al. 2003).

A number of studies have examined the effects of the surrounding habitat on the abundance and diversity of amphibians in breeding ponds, particularly in North America and Europe (Laan and Verboom 1990; Vos and Stumpel 1996; Findlay and Houlahan 1997; J. Mitchell et al. 1997; Hecnar and M'Closkey 1998; Knutson et al. 1999; Kolozsvary and Swihart 1999; Lehtinen, Galatowitsch, and Tester 1999; Pope, Fahrig, and Merriam 2000; Findlay, Lenton, and Zheng 2001; Houlahan and Findlay 2003). Most of these studies have shown that the type of terrestrial habitat surrounding a pond, especially the amount of forest cover, affects species abundance and the composition of wetland amphibian communities. Nevertheless, species differ in their sensitivity to habitat disturbance and alteration. For example, a study of largely agricultural landscapes in Ohio showed that the presence of wood frogs (*Rana sylvatica*), spotted salamanders (*Ambystoma maculatum*), Jefferson's salamander (*A. jeffersonianum*), and smallmouth salamanders (*A. texanum*) was strongly influenced by the amount of forested habitat near breeding sites, whereas tiger salamanders (*A. tigrinum*) were more tolerant of agricultural habitats (Porej, Micacchion, and Hetherington 2004). Several species of amphibians that used temporary playa wetlands in Texas as breeding sites actually were more abundant at ponds surrounded by cultivated agricultural land than at those surrounded by natural grasslands (M. Gray, Smith, and Brenes 2004). Suitable terrestrial habitat is essential for adult amphibians during the non-breeding season (e.g., Pope, Fahrig, and Merriam 2004). In addition, juvenile amphibians emerging from breeding ponds often are reluctant to move across open areas, such as powerline right-of-ways and agricultural fields, that lack the moist microhabitats found in forests (deMaynadier and Hunter 1999; Rothermel and Semlitsch 2002; Vasconcelos and Calhoun 2004). Consequently, dispersal of juveniles between potential breeding sites may be limited by the surrounding habitat. For example, playa wetlands in Texas surrounded by agricultural land could function as population sinks, despite high population abundance, if the ability of metamorphs to move across cultivated land is limited (M. Gray, Smith, and Brenes 2004).

In addition to the presence of suitable terrestrial habitat around breeding ponds, the proximity of ponds to other wetlands and the availability of dispersal corridors between potential breeding sites can be critical for the maintence of healthy amphibian populations. Habitat fragmentation can result either from destruction of wetlands themselves, leading to a reduction in wetland density, or from destruction or alteration of intervening terrestrial habitat that serves to connect wetlands (D. Marsh and Trenham 2001; Trenham et al. 2003). Consequently, it is important to characterize wetland habitats for amphibians on a landscape level, rather than focusing on individual wetlands in isolation. There are many examples of regions in which large numbers of small wetlands persist, but connectivity among wetlands has markedly decreased in recent decades due to increased urban and suburban development, road construction, and other types of habitat alteration (Boothby 1997; Gibbs 2000; Löfvenhaft, Runborg, and Sjögren-Gulve 2004).

Roads can be particularly important in isolating suitable amphibian habitats, because some amphibians are reluctant to move across paved roads, and because of the high mortality of individuals attempting to cross them (Gibbs and Shriver 2005). Consequently, several studies have found negative associations between amphibian abundance or diversity and road density (Findlay and Houlahan 1997; Lehtinen, Galatowitsch, and Tester 1999; Findlay, Lenton, and Zheng 2001; Houlahan and Findlay 2003; Porej, Micacchion, and Hetherington 2004). The effect of roads differs for different species of amphibians, depending on their tendency to move long distances and the speed with which individuals can cross roads (Gibbs 1998c; deMaynadier and Hunter 2000; L. Carr and Fahrig 2001). Certain wetland amphibians, such as the North American red-spotted newt (*Notophthalmus viridescens*), appear to be particularly sensitive to habitat fragmentation because they have difficulty crossing roads and are highly sensitive to loss of forest cover (Gibbs 1998b; deMaynadier and Hunter 2000; Guerry and Hunter 2002; Porej, Micacchion, and Hetherington 2004). In contrast, some inhabitants of open habitats, such as the European natterjack toad (*Bufo calamita*) actually prefer to move through open sandy areas or along roads rather than through forested habitats, although using roads as dispersal corridors still could result in high mortality from cars (Stevens et al. 2004).

Although the focus of this discussion has been on small, temporary wetlands, some amphibians breed in more permanent wetlands and remain at those sites thoughout much of the year. They are susceptible to various sorts of habitat alteration in or near breeding sites. For example, North American green frogs (*Rana clamitans*) are found mostly along the margins of permanent ponds and lakes and are susceptible to the negative effects of shoreline development of lakeside cottages and homes (Woodford and Meyer 2003). These frogs may not be directly affected by the presence of houses per se, but by habitat alteration associated with development, such as removal of submerged woody debris and emergent shoreline vegetation that are considered undesirable because they tend to interfere with boating and swimming. In northern parts of their range, green frogs often are found around peat bog ponds, which apparently are used mainly as nonbreeding habitat, and populations can be negatively affected by peat mining operations that tend to dry out the bogs (Mazerolle and Cormier 2003). Both the behavior and overall abundance of other amphibians can be affected as well (Mazerolle 2001, 2003). Another potential threat to amphibians is mechanical removal of accumulated sediments in recreational ponds and lakes, which often is done to improve conditions for swimming, boating, and fishing. Such drastic habitat alteration not only can cause direct mortality of amphibians that are dredged up with the sediments, but it also can destroy submerged and emergent plants that provide crucial habitat for amphibians (Aresco and Gunzburger 2004). Large salamanders such as *Siren* and *Amphiuma* that use accumlated sediments for estivation in dry weather may be particularly at risk. Streamside habitats also can be degraded by even relatively benign human use, resulting in reductions of habitats for stream-breeding amphibians (Rodríguez-Prieto and Fernández-Juricic 2005).

There is no doubt that wetlands created by human activities, either intentionally or unintentionally, can be important habitats for some amphibians. In agricultural areas, for examples, farm ponds can be essential habitats for amphibians in an otherwise inhospitable landscape (Kneitz 1998; Quinn, Gallie, and Volsen 2001; Knutson et al. 2004). In some cases such ponds may not support all of the species that normally would occur in the region (Hazell et al. 2004), but this is not always true (M. Gray, Smith, and Brenes 2004). There have been extensive efforts in some areas, particularly Europe, to create new habitats for pond-breeding amphibians by building artificial ponds. These often have been colonized by both anurans and urodeles, although the success of these efforts is heavily influenced by the proximity of source populations and the nature of the surrounding habitat (Thielcke et al. 1985; Laan and Verboom 1990; Stumpel and Tester 1993; Chovanec 1994; Hödl, Jehle, and Gollmann 1997; Stumpel and van der Voet 1998; Baker and Halliday 1999; Directorate of Culture and Cultural and Natural Heritage, Council of Europe 2003).

In the United States, there have been numerous attempts to restore degraded wetlands or to build new wetlands to mitigate the loss of natural habitats due to road construction or other projects. Most studies of amphibians in these artificial wetlands have followed populations for less than 10 years after construction of the ponds, so the long-term suitability of constructed wetlands for amphibians is not yet clear. Most studies have shown that amphibians colonize newly constructed wetlands within a year or two of construction and often successfully reproduce (Lehtinen and Galatowitsch 2001; Pechmann et al. 2001; Petranka, Murray, and Kennedy 2003; Vasconcelos and Calhoun 2004). Nevertheless, it is virtually impossible to replicate a natural wetland precisely, because underlying soil permeability and the nature of the surrounding habitat may differ. It can be particularly difficult to replicate the hydroperiod of temporary ponds, because the formation of temporary ponds often is a consequence of local variation in underlying geology and soil conditions. The result is that some species of amphibians that occur in natural wetlands may not use constructed wetlands (Pearl et al. 2005). Nevertheless, some local amphibian populations probably are better off with replacement wetlands than with no wetlands at all. Some anthropogenic wetlands, however, may appear suitable as breeding sites, but actually function as population sinks if amphibians use them as oviposition sites but hydroperiods are too short for larvae to reach metamorphosis in most years (DiMauro and Hunter 2002).

### Introduced and Invasive Species

Aside from habitat destruction, the most clear-cut threat to many amphibian populations is the introduction of exotic predators or competitors into ecosystems where such animals were previously absent (Kats and Ferrer 2003; Kiesecker 2003). A variety of introduced animals have been implicated in declines of amphibian populations in various parts of the world, with the most important being fishes, amphibians, snakes, and crayfish. Introduced mammals, such as the American mink (*Mustela vison*) in Europe and the Russian Far East, also prey on local amphibians (Kuzmin and Mazlova 2003; see also chapter 14). In most cases, these predators target the most common species of amphibians. There is little direct evidence that introduced mammals have had a widespread effect on amphibian populations, but some island populations may be at risk (Ahola et al. 2006). Introduced populations of human commensals such as rats may have contributed to the decline of amphibians such as populations of *Leiopelma* in New Zealand (Towns and Daugherty 1994). Invasive species of birds, such as crows and cattle egrets, often prey on amphibians, but again, there is little evidence that these predators have significantly impacted amphibian populations. Most predatory mammals and birds that feed on amphibians are generalist predators that tend to switch to more abundant prey if amphibians become scarce, or feed on amphibians only as secondary prey items.

#### Introduced Fish and Amphibian Populations

For many amphibians, the aquatic stages of the life history, especially eggs and larvae, are most vulnerable to predators, so it is not surprising that aquatic predators such as fish have had a large impact on amphibian populations. The importance of predatory fishes in limiting the distribution of aquatic-breeding amphibians was discussed in some detail in chapter 14. The larvae of many amphibians, especially those that breed in temporary ponds or in high-altitude lakes that normally are devoid of fish, lack behavioral or chemical defenses against fishes. Consequently, the introduction of predatory fishes into formerly fish-free habitats often has a major impact on local amphibian populations.

The largest literature on the effects of introduced fishes on amphibian populations deals with introductions of trout into streams and lakes. Because of their popularity with recreational anglers, various species of trout have been introduced into streams, rivers, and lakes throughout the world, even in protected national parks and wilderness areas. North American rainbow trout (*Oncorhynchus mykiss*) and brook trout (*Salvelinus fontinalis*) and Eurasian brown trout (*Salmo trutta*) have been the most widely introduced species. Introduced populations of these fish are now well established in many montane habitats in North America, Europe, South America, Australia, and New Zealand (this country, however, lacks native aquatic-breeding amphibians).

Most of these introductions have been initiated by local or national government agencies at the behest of sport fishermen, and generally have been undertaken without prior ecological study of the impact of fish introductions on native species (R. Knapp, Corn, and Schindler 2001). Such introductions have a long history, with nonnative trout having been introduced into Yosemite National Park in California as early as the 1870s (R. Knapp 2005). The impact on amphibian populations probably began soon thereafter (Grinnell and Storer 1924), but was limited by the difficulty of transporting fish into remote wilderness areas. With the advent of fish stocking from airplanes, even the most remote fish-free mountain lakes could be supplied with fish for the entertainment of sportsmen, and the impact on amphibian populations appears to have accelerated in the late twentieth century. Although fish stocking was halted in national parks in the western United States in the 1970s and 1980s, stocking continues in many other lakes in national forests, state forests, and wilderness areas. The result is that an estimated 60% of all montane lakes and 95% of all large lakes in the western United States, nearly all of which originally lacked fish, are now inhabited by introduced salmonids (R. Knapp, Corn, and Schindler 2001). In most western states, stocking of fish continues in 20–75% of all high-altitude lakes (Dunham, Pilliod, and Young 2004).

The impact of these introductions on amphibian populations has been clear and dramatic, as indicated by four lines of evidence: (1) surveys of amphibians in fishless and fish-inhabited lakes, (2) observations of amphibian population declines following the introduction of trout into previously fish-free habitats, (3) experimental studies of predation on amphibians by introduced trout, and (4) recolonization of lakes by amphibians after removal or extinction of introduced trout. Surveys of formerly fish-free lakes in many western states, including California, Oregon, Washington, Idaho, Utah, and Colorado have demonstrated a strong negative correlation between the presence of introduced trout and the presence or abundance of several species of amphibians, including *Ambystoma macrodactylum, A. gracile, Pseudacris regilla, P. maculata, Rana luteiventris, R. cascadae,* and *R. muscosa* (Bradford 1989; Bradford et al. 1998; T. Tyler, Liss, Ganio et al. 1998; T. Tyler, Liss, Hoffman et al. 1998; R. Knapp and Matthews 2000; R. Knapp, Matthews, and Sarnelle 2001; Matthews et al. 2001; Pilliod and Peterson 2001; Bull and Marx 2002; R. Knapp et al. 2003; Dunham, Pilliod, and Young 2004; R. Knapp 2005; Welch and MacMahon 2005). Some species that are distasteful to fish as larvae or adults, such as *Bufo boreas, B. canorus, Taricha torosa,* and *T. granulosa,* do not appear to be as strongly affected by the presence of fish (Corn, Jennings and Muths 1997; Bull and Marx 2002; R. Knapp 2005; Pearl et al. 2005). *Ambystoma gracile,* which often breeds in relatively deep lakes, appears to be more tolerant of fish than are other species of *Ambystoma,* although salamander abundance and patterns of activity can be altered by the presence of trout (Corn, Jennings and Muths 1997; T. Tyler, Liss, Ganio et al. 1998; R. Hoffman, Larson, and Brokes 2004; Pearl et al. 2005).

Studies in other parts of North America and on other continents have shown a similar negative correlation between the presence and abundance of amphibians and the presence of introduced trout or other introduced fish (Brönmark and Edenhamn 1994; Aronsson and Stenson 1995; Braña, Frechilla, and Orizaola 1996; Hecnar and M'Closkey 1997b; Reshetnikov and Manteifel 1997; Funk and Dunlap 1999; Gillespie 2001; Rodríguez 2001; Nyström et al. 2002; Pascual et al. 2002; Böll 2003; Martínez-Solano, Barbadillo, and Lapena 2003; Martínez-Solano, Bosch, and García-París 2003; Reshetnikov 2003; Crochet et al. 2004; Bosch et al. 2006). Introduced trout are present in some streams in South and Central America where toads of the genus *Atelopus* have declined or disappeared, but in some cases, these species coexisted for decades before toad populations declined, suggesting the fish were not a direct cause of the declines (La Marca et al. 2005). In Europe, the impact on both metmorphosing and paedomorphic populations of newts (*Triturus*), many of which occur in isolated populations in formerly fish-free lakes, has been particularly severe (Aronsson and Stenson 1995; Reshetnikov and Manteifel 1997; Denoël, Džukić, and Kalezić 2005). In addition to the effects of fish introductions on previously fish-free lakes, conversion of temporary wetlands to more permanent farm ponds also can encourage stocking of ponds with fish, with negative consequences for amphibians that normally breed in temporary water (Monello and Wright 1999; M. Adams 2000; M. Adams, Pearl, and Bury 2003; Pearl et al. 2005)

For many of these surveys, there are few quantitative data on the occurrence or abundance of amphibians before the introduction of fish, although in some cases, the disappearance of a local amphibian population could be tied to a known fish introduction. The ability of trout and other fishes to prey on amphibians, particularly larvae, is well documented (T. Tyler, Liss, Ganio et al. 1998; T. Tyler, Liss, Hoffman et al. 1998; M. Adams 2000; Kats and Ferrer 2003; see also chapter 14). The virtual exclusion of many amphibian species from lakes inhabited by fish is strong evidence that recent declines in some of these amphibian populations are due to the introduction of fish. The presence of fish also can inhibit recolonization of fish-free lakes by amphibians by making aquatic habitats located between amphibian populations uninhabitable for amphibians (Bradford, Tabatabai, and Graber 1993). Furthermore, the effect of fishes can extend beyond amphibians to impact entire ecosystems of high-altitude lakes. Introduced fishes often alter the composition and abundance of aquatic invertebrates, from small zooplankton to larger invertebrates such as aquatic insects, and this in turn can affect the composition and abundance of algal communities (Bradford et al. 1998; Kats and Ferrer 2003; Dunham, Pilliod, and Young 2004). The effects of introduced fishes can extend up the food chain as well, reducing or eliminating populations of predators such as garter snakes (*Thamnophis elegans* and *T. couchi*) that are heavily dependent on amphibians for food (Matthews, Knapp, and Pope 2002; R. Knapp 2005).

The strongest evidence for introduced trout causing declines or disappearances of amphibian populations comes from studies of recolonization of lakes from which trout have disappeared, either through natural extinction or deliberate removal. In some lakes, introduced trout are not able to establish self-maintaining populations without artificial stocking, and trout populations have subsequently gone extinct. In one study in Montana, Funk and Dunlap (1999) found that long-toed salamanders (*Ambystoma macrodactylum*) recolonized five of six lakes in which trout had gone extinct. The precise timing of recolonization is not known, but it occurred within 20 years, and in some cases, probably within 15 years after fish stocking ceased. In Mount Rainier National Park, introduced brook trout (*Salvelinus fontinalis*) were experimentally removed from one lake, and within a few years, there was a substantial increase in the number of eggs, larvae, and paedomorphic adults of *Ambystoma gracile* seen in the lake (R. Hoffman, Larson, and Samora 2004). Although this experiment lacked replication, the pattern of increasing abudance constrasted with the pattern of random fluctuation seen in one control lake that never contained fish.

A more comprehensive study, with better replication and controls, was conducted by Vredenburg (2004) in Kings Canyon National Park in the Sierra Nevada of California. Introduced trout were removed from five lakes between 1996 and 2003. The abundance of mountain yellow-legged frogs (*Rana muscosa*), a declining species, was monitored in these lakes and in eight fish-control lakes and eight fishless-control lakes. Fish-control lakes were largely devoid of frogs, but there was a rapid increase in numbers of frogs in the fish-removal lakes (fig. 16.14). The key to the rapid recolonization of lakes after the removal of fish was the close proximity of robust frog populations in the remaining fish-free lakes. These results suggest that halting fish stocking, along with removal of the remaining fish, is a viable management strategy to increase populations of amphibians that have declined due to fish introductions. Full recovery of the aquatic ecosystem to conditions that existed before the introduction

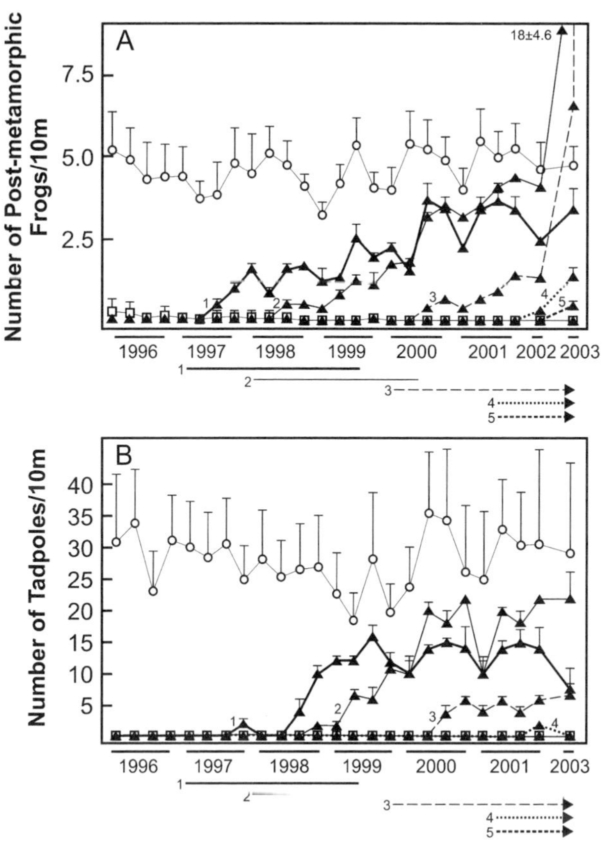


**Fig. 16.14.** Effect of removal of introduced fishes from high-altitude lakes on the abundance of mountain yellow-legged frogs (*Rana muscosa*). (A) Abundance of postmetamorphic frogs. (B) Abundance of tadpoles. Circles show data for eight fishless-control lakes. Triangles show data for five fish-removal lakes. Squares show data for eight control lakes containing fish. Horizonal lines at the bottom show the trout removal period for each of the five fish-removal lakes. After Vredenburg (2004).

of fishes, particularly the invertebrate community, may require 20 years or more (Drake and Naiman 2000; Donald et al. 2001; R. Knapp, Matthews, and Sarnelle 2001).

Although the effects of introduced salmonids have received the most attention from amphibian ecologists, a variety of other fish species have been introduced into nonnative habitats around the world, and some of these also negatively impact amphibian populations. Probably the best documented is the effect of introduced mosquitofish (*Gambusia*) on both pond and stream-breeding amphibians. These highly adaptable fish readily invade temporary wetlands within their natural range, especially in low-lying areas that are frequently flooded by overflow from more permanent water. In central Florida, the native mosquitofish, *Gambusia holbrooki,* was a more voracious predator on anuran tadpoles than was a much larger introduced walking catfish, *Clarias batrachus,* but had about the same impact as did other native fishes (*Fundulus chrysotus* and *Jordanella floridae;* Baber and Babbitt 2003). Hylid and microhylid tadpoles were particularly susceptible to predation by these native fishes, whereas bufonid tadpoles were less vulnerable.

The tendency for mosquitofish to prey on amphibian larvae can have a devastating effect on amphibian populations in habitats where mosquitofish are not native and amphibians presumably have few defenses against them. For example, mosquitofish have been widely introduced into streams in the western United States for mosquito control and have been implicated in declines of native amphibians, including *Pseudacris regilla, Rana aurora,* and *Taricha torosa* (Gamradt and Kats 1996; Goodsell and Kats 1999; Lawler et al. 1999). Introduced mosquitofish also have been implicated in the decline of the Australian golden bell frog (*Litoria aurea*), although here the evidence is somewhat mixed. These fish readily feed on *Litoria* eggs and tadpoles (L. Morgan and Buttemer 1996; Pyke and White 2000). One survey found that *L. aurea* was largely absent from permanent ponds where mosquitofish were abundant, and used mostly ephemeral habitats for breeding (Pyke and White 1996). Another study found that the presence of mosquitofish did not influence the use of aquatic habitats by treefrogs, and tadpoles were found coexisting with the fish (Hamer, Lane, and Mahony 2002a). Nevertheless, *Litoria* tadpoles appear to lack effective antipredator responses to mosquitofish, so reproductive success may be lower in permanent ponds, which once were the preferred habitats of this frog (Hamer, Lane, and Mahony 2002b). Another Australian frog, *Limnodynastes tasmaniensis,* exhibits a greater tendency to avoid bodies of water inhabitated by mosquitofish. This species, and another myobatrachid frog, *Crinia signifera,* both exhibited behavioral avoidance of mosquitofish in experimental tanks, even though neither species evolved in contact with these fish. It is not clear if this represents a general tendency to avoid fish predators, although *Crinia* breeds in ephemeral pools not normally inhabited by fish, whereas *Limnodynastes* breeds in more permanent pools (Lane and Mahony 2002).

In addition to the direct effect of introduced fishes as predators on amphibians, the presence of these species can indirectly facilitate the establishment of other nonnative organisms. One concern is that introduced fishes, especially those from fish hatcheries, can be a source for emerging diseases that attack amphibians (Kiesecker, Miller, and Blaustein 2001; Jancovich et al. 2005; see the following further discussion). Introduced fishes also can improve the chances that other invasive vertebrates become established. For example, in the heavily agricultural Willamette Valley of Oregon, introduced bullfrogs (*Rana catesbeiana*) are widespread. Invasion of ponds by bullfrogs is facilitated by the presence of introduced bluegill sunfish (*Lepomis machrochirus;* M. Adams, Pearl, and Bury 2003). The bluegills are major predators on dragonfly naiads, which in turn are more important predators on bullfrog tadpoles than are the fish themselves (see also chapter 15). Consequently, survival of dragonfly naiads was lower and survival of bullfrog tadpoles was higher in ponds with sunfish than in ponds without fish. M. Adams, Pearl, and Bury (2003) suggested that removing introduced bluegills from ponds may be a more practical way to control densities of introduced bullfrogs than removing the frogs themselves.

## Amphibian Invaders

With the growing concern about declining amphibian populations, it is easy to overlook the potential for some amphibians to become established and even invade new habitats once they are introduced into a new region. One of the most common invasive species is the American bullfrog (*Rana catesbeiana*), which has been widely introduced in western North America, South America, Europe, and Asia. There is evidence that introduced bullfrogs in western North America have had a negative impact on native amphibians (fig. 16.15), especially ranids, through both competition and predation (Kiesecker and Blaustein 1997b, 1998; Kupferberg 1997a; Kiesecker, Blaustein, and Miller 2001; Pearl et al. 2004), although it can be difficult to separate the effects of bullfrogs from other changes in habitat structure and the synergistic effects of other introduced predators, such as fishes (Jennings and Hayes 1985; M. Adams 1999; M. Adams, Pearl, and Bury 2003; Doubledee, Muller, and Nisbet 2003). Experimental studies of predation and competition between introduced bullfrogs and native frogs were discussed in chapter 15 and need not be described in detail here.

As mentioned in the discussion of amphibian declines in

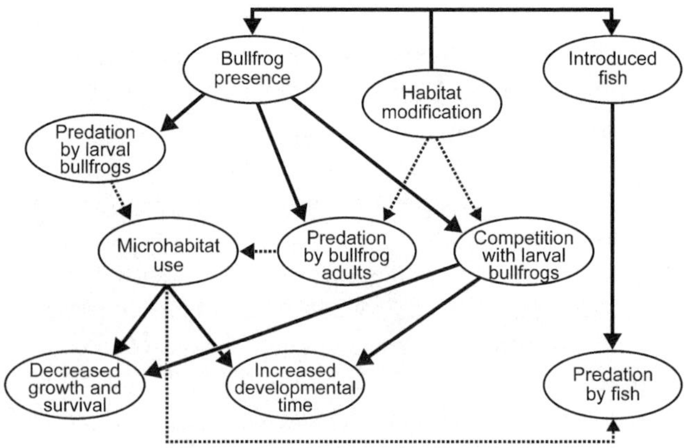


**Fig. 16.15.** Hypothesized interactions between the effects of habitat modification, the presence of introduced bullfrogs (*Rana catesbeiana*), and the presence of introduced predatory fishes on red-legged frogs (*Rana aurora*) in the western United States. Tadpoles and adults of bullfrogs can negatively impact red-legged frogs through both predation and competition, and changes in habitat use induced by bullfrogs can make red-legged frog tadpoles more susceptible to predation by fishes. Solid arrows indicate direct effects; dashed arrows indicate indirect effects. After Kiesecker (2003).

Europe, surprisingly little ecological work has been done on introduced populations of bullfrogs in Europe or other regions of the world. An internet search reveals scores of websites that routinely describe bullfrogs as voracious predators that will eat almost anything. Nevertheless, most studies of the diets of introduced bullfrogs have shown that they eat mostly invertebrates (e.g., Hirai 2004), so the impact on native amphibians is largely unknown. One threat posed by the introduction and international trade in bullfrogs is their potential as reservoirs for disease organisms such as the chytrid fungus (*Batrachochytrium dendrobatidis*), which is thought to be responsible for some amphibian declines. This fungus has been identified in introduced and farmed populations of bullfrogs in South America, although the bullfrogs themselves seem not to be affected by the disease (Mazzoni et al. 2003; Hanselmann et al. 2004). It is not yet known, however, whether bullfrogs actually are responsible for transmission of the disease to native amphibians (Daszak et al. 2004; see the following further discussion of this disease).

The only other species of amphibian that has been as widely introduced as the American bullfrog is the marine or cane toad (*Bufo marinus*). This species has many traits characteristic of an invasive species (Lever 2001). It can reproduce in a wide variety of aquatic sites, from polluted ditches and ponds to pristine rivers. Females produce thousands of eggs, and both egg and tadpole development is very rapid, with metamorphs emerging at a very small size relative to adult size. The toads are highly mobile, with both juveniles and adults being able to move over long distances (Schwarzkopf and Alford 2002). They tend to favor disturbed habitats and often are associated with human habitation. Finally, the eggs, tadpoles, and adults all are highly toxic and distasteful to many predators.

Originally native to South and Central America, this large toad has been introduced accidentally or intentionally, mostly to control insect pests, to Hawaii and many other Pacific islands, Australia, New Guinea, several of the smaller islands of Japan, the Philippines, Diego Garcia in the Indian Ocean, most islands in the Caribbean, Bermuda, and Florida (Lever 2001). Some of these regions, particularly oceanic islands, lack native species of amphibians, although the introduced toads may negatively impact other groups, such as native insects (Matsumoto, Matsumoto, and Miyashita 1984). Other regions, such as the Philippines, where the toad has been widely established for decades (Alcala 1957), have a rich native anuran fauna, but the impact of the toads on native species is largely unknown. Even within South America, *Bufo marinus* may be expanding its range. In Ecuador, for example, the toads now occur at higher altitudes than previously reported (Bustamante, Ron, and Coloma 2005). Although this altitudinal shift has been attributed to increased temperatures at high altitudes, it also could be related to the tendency of this species to spread with human settlement and alteration of natural habitats.

Most ecological studies on the effects of cane toad introductions have been done in Australia, where the species has been considered a major invasive threat to native wildlife. Much of the concern over the cane toad in Australia has focused on its impact on native wildlife species, including mammals, birds, and reptiles, that are poisoned after attacking the highly toxic toads (Covacevich and Archer, 1975; Burnett 1997; Lever 2001; B. Phillips, Brown, and Shine 2003; Aldhous 2004; Webb, Shine, and Christian 2005). The long-term impact of the toads on native predators is unclear (Catling et al. 1999). It is possible that some of these predators eventually will develop learned or evolved behavioral avoidance of adult toads, just as some native fish rapidly learn to avoid their toxic tadpoles (Crossland 2001). There also is evidence of toad-induced selection on morphology in some snakes, with body size and gape size decreasing in a way that makes it more difficult for them to eat large toads (B. Phillips and Shine 2004). Toads also are potential predators on smaller vertebrates, although most studies have shown that insects and other invertebrates comprise most of the diet (Lever 2001). A particularly bizarre impact of cane toads is their ability to prey on the eggs and nestlings of ground-nesting birds, such as the rainbow bee-eater (*Merops ornatus;* Boland 2004).

Despite widespread concern in Australia about the impact of rapidly expanding cane toad populations, there have been surprisingly few detailed ecological studies of the interactions between cane toads and native anurans. Anecdotal observations indicate that cane toads sometimes consume native frogs (M. Tyler 1976), but whether they do so in sufficient numbers to impact frog populations is unknown. Crossland (1998a) found that *Bufo marinus* tadpoles were not significant predators on the eggs or larvae of native frogs, but their own eggs are sufficiently toxic to be a threat to native tadpoles that consume them (Crossland 1998b; Crossland and Alford 1998; Crossland and Azevedo-Ramos 1999; Williamson 1999). The presence of *B. marinus* tadpoles can reduce growth rates and survival of native tapoles, particularly of the genus *Limnodynastes,* in competition experiments (Crossland 1998, 2000; Williamson 1999), but most of the native species do not breed in permanent water, which often is used by cane toads. A study using tadpoles of introduced cane toads in Florida found little impact of this species on growth and development of two native species of tadpoles, *Bufo terrestris* and *Hyla cinerea* (K. G. Smith 2005).

Several other species of amphibians have been introduced into or have invaded new habitats in various parts of the world, but in most cases, the impact on local amphibian populations is poorly understood. The major concerns are competition with or predation on native amphibians and possible genetic consquences of hybridization between introduced species and closely related native species. Populations of African clawed frogs (*Xenopus laevis*) have become established in several parts of the world as a result of releases from laboratory stocks or pets. In California, these frogs are very successful, especially in artificial bodies of water. Although they are voracious predators of invertebrates, predation on amphibians appears to mostly involve cannibalism of conspecifics (McCoid and Fritts 1980, 1993). This species also has become established in Chile, again mostly in artificial ponds (Lobos and Jaksic 2005). As in California, the only amphibians found in the stomachs of *Xenopus* were conspecific tadpoles, so the impact on native amphibians appears to be minimal. Whether these frogs compete with native species has yet to be determined. Introduced populations also are known from Wales, where the only amphibian prey items were conspecific eggs and tadpoles (Measey 1998).

In South Africa, the replacement of natural bodies of water with artificial ponds has facilitated the invasion of *Xenopus laevis* into habitats formerly inhabited by *X. gilli.* The latter species is now endangered because of habitat destruction, predation by *X. laevis,* and hybridization with *X. laevis* in the few remaining habitats that now support *X. gilli* (Picker, Harrison, and Wallace 1996; De Villiers 2004). Hybridization between native species and closely related invasive species also is a concern in Europe, where *Rana ridibunda* has spread into new habitats (Arano et al. 1995; Pagano et al. 2003; Vorburger and Reyer 2003), and in California, where the endangered California tiger salamander (*Ambystoma californiense*) hybridizes with introduced *A. tigrinum* (Riley et al. 2003).

Florida is particularly susceptible to invasive species because of its subtropical climate, and a number of nonnative amphibians have become established there (Meshaka, Butterfield, and Hauge 2004; Meshaka 2005). The best-studied is the Cuban treefrog (*Osteopilus septentrionalis*), which is widespread in the state and favors human-altered habitats (Meshaka 2001). This species is larger than any of the native hylids and generally eats a wider range of prey, including large insects not generally consumed by other species. Although there is considerable overlap in diet between the Cuban treefrog and other hylids, the extent to which competition negatively affects the native species is unknown. Cuban treefrogs also prey on smaller frogs (Meshaka 2001; Wyatt and Forys 2004), but again, the ecological impact of such predation on native frog populations is unknown.

Florida also has been invaded by two species of *Eleutherodactylus* from Caribbean islands. *Eleutherodactylus planirostris* has been widespread in the state for decades, and localized populations of *Eleutherodactylus coqui* from Puerto Rico also have become established near urban areas (Meshaka, Butterfield, and Hauge 2004; Meshaka 2005). These direct-developing frogs have a very different life history from all other Florida anurans and appear to occupy a different niche, so their ecological impact may be minimal. The same two species have become established in Hawaii, presumbly introduced in imported ornamental plants. Since Hawaii lacks native species of amphibians, the main impact appears to be on humans annoyed by their loud calls, although there is a potential for these frogs to negatively impact endemic species of insects, because their populations are expanding rapidly (Kraus and Campbell 2002). Several species of *Eleutherodactylus* also have moved around among various Caribbean islands and from the Caribbean to Central and South America. Invasive species, such as *Eleutherodactylus johnstonei* and *E. planirostris,* appear to be more tolerant of open habitats and dry conditions than are native *Eleutherodactylus* on the same islands. Even if these invasive species do not directly compete with native *Eleutherodactylus,* they eventually may dominate island ecosystems increasingly characterized by heavy deforestation and expanding human occupation of the landscape (Pough, Stewart, and Thomas 1977; M. Stewart 1979; M. Stewart and Martin 1980; Kaiser and Henderson 1994; Kaiser 1997).

### Other Introduced Organisms

A variety of other animals and plants have been introduced all over the world, and many of these species have become highly invasive. For the most part, the impact of these organisms on amphibian populations is poorly studied, although some clearly are potential predators on native amphibians. Introduced crayfish prey on native amphibians, both in North America and Europe, and may exclude amphibians from some habitats (Gamradt and Kats 1996; Axelsson et al. 1997; Gherardi, Renai, and Corti 2001; Beja and Alcazar 2003; Kats and Ferrer 2003). Invasive crayfish also can alter habitat quality by removing aquatic vegetation used for cover by amphibians (Carey et al. 2003). On the Spanish island of Mallorca, the endemic midwife toad (*Alytes muletensis*) is largely excluded from areas inhabited by introduced frogs (*Rana perezi*) and snakes (*Natrix maura*). The distribution of these two species on the island is similar, being limited to lower elevations, and the introduced frogs probabably are the principal prey of the snakes. The midwife toads are now confined to pools in steep, higher-elevation gorges (R. Moore, Griffiths, and Román 2004). The snakes have been present for more than 2,000 years. Both adults and tadpoles of the midwife toads show some behavioral avoidance of chemical cues from snakes, and tadpoles also exhibit morphological changes in areas where snakes are present (Griffiths et al. 1998; Schley and Griffiths 1998; R. Moore et al. 2004). This suggests that the toads have some capacity to adapt to and coexist with the introduced predators. Other introduced species of snakes have the potential to affect amphibian populations, although their impact on amphibian populations has not been studied. For example, large Burmese pythons have become established in the Florida Everglades, but their impact on the native fauna is largely uknown (Lovgren 2004).

Even organisms that are not predators of amphibians have the potential to affect amphibian populations when introduced into nonnative habitats. For example, some invasive aquatic plants, such as water hyacinth (*Eichhornia crassipes*), form dense mats on lakes and ponds that can block light for photosynthesis and reduce the oxygen content of the water. Although some air-breathing amphibians, such as sirens, can tolerate and even thrive in such conditions (Ultsch 1973b), most other vertebrates are excluded from such habitats. Other invasive aquatic plants can alter the hydroperiods, nutrient cycling, and food web dynamics of ponds used by amphibians for reproduction (D. Gordon 1998; Zedler and Kercher 2004). Invasive aquatic snails have the potential to affect larval amphibians, not only as competitors for periphyton (Brönmark, Rundle, and Erlandsson 1991; Cattaneo and Mousseau 1995; P. Harris 1995), but also as reservoirs for trematode parasites that can infect amphibians (Rader, Belk, and Keleher 2003; Johnson and Lunde 2005). Invasive European earthworms, which are gradually spreading in forests of the northeastern United States, could affect amphibians such as plethodontid salamanders by accelerating the removal of leaf litter, thereby depriving salamanders of critical cover and moisture and changing the abundance and distribution of soil invertebrates on which they feed (Bohlen et al. 2004). Even the introduced gypsy moth, which defoliates deciduous trees, and the wooly adelgid, which kills hemlocks, could reduce salamander populations by altering canopy cover and moisture conditions in forests (Carey et al. 2003). None of these interactions has been studied in much detail, and the effect of exotic species on amphibian populations deserves further investigation.

## Emerging Diseases of Amphibians

Although most amphibians have strong natural immunity to all sorts of infectious diseases (Rollins-Smith 2001; Rollins-Smith et al. 2005), there has been great concern in recent years about the emergence of diseases that can decimate amphibian populations, even in well-protected habitats. Indeed, many of the amphibian population declines observed in seemingly pristine areas of montane rainforest in various parts of the world have been attributed to the effects of disease. These effects may act in conjuction with other environmental stressors that weaken the immune defenses of amphibians, such as climate change, human habitat alteration, and environmental pollution (Carey 2000; Carey, Bradford, et al. 2003; Carey, Pessier, and Peace 2003; J. P. Collins et al. 2003; Kiesecker et al. 2004; Rollins et al. 2005). In many cases, emerging infectious diseases are an extension of the problem of introduced organisms, because (1) the pathogens themselves may be newly introduced to an area, (2) other introduced species may serve as reservoirs or vectors for disease organisms, or (3) synergistic effects of disease organisms and other introduced species may precipitate declines of amphibian populations. This problem certainly is not limited to amphibians—emerging diseases are considered major threats to crop plants, wild plants, coral reefs, aquatic and terrestrial wildlife, and even humans (Carey 2000; Daszak, Cunningham, and Hyatt 2000; Cleaveland et al. 2002; Harvell et al. 2002; P. K. Anderson et al. 2004; Sutherland, Porter, and Torres 2004). In some cases, such as the grotesque facial tumors afflicting Tasmanian devils (Bostanci 2005), emerging diseases have wiped out entire populations very rapidly, just as has been observed in some amphibian populations. The effects of disease on key organisms such as corals and forest trees have the potential to alter entire ecosystems by contributing to the decline of other species that depend on them for food or habitat structure (Canterbury and Blockstein 1997; D. Bellwood et al. 2004; G. Jones et al. 2004).

### Chytridiomycosis and Amphibian Declines

Most of the recent literature on the role of disease in amphibian population declines has focused on chytridiomycosis, a fungal disease that attacks keratinized tissue, especially the skin of adult amphibians, but also the keratinized mouthparts of tadpoles. The first indication that amphibians might be dying from an unknown disease was the rapid decline of many species of stream-breeding frogs in the rainforests of Queensland, Australia. On the basis of the timing of declines in different species, Laurance, McDonald, and Speare (1996) argued that the most likely agent of decline was an exotic pathogen, and they suggested a virus was the most likely candidate. This suggestion met with initial skepticism from other Australian ecologists, who argued that there was little or no evidence for the presence of an infectious agent in declining amphibian populations (Alford and Richards 1997; Hero and Gillespie 1997). At about the same time, Lips (1999) was observing massive die-offs of frogs in the mountains of western Panama, even finding dead and dying frogs on the forest floor, certainly a most unusual occurrence. Necropsies of the dead frogs showed that all of them were infected with an unknown fungal disease, and there were signs of fungal infection on the mouthparts of living tadpoles as well. Comparative research revealed that the same fungus was associated with the mortality of frogs in Australia and Central America (Berger et al. 1998). The fungus was subsequently described as a new species, *Batrachochytrium dendrobatidis* (fig. 16.16), a member of a

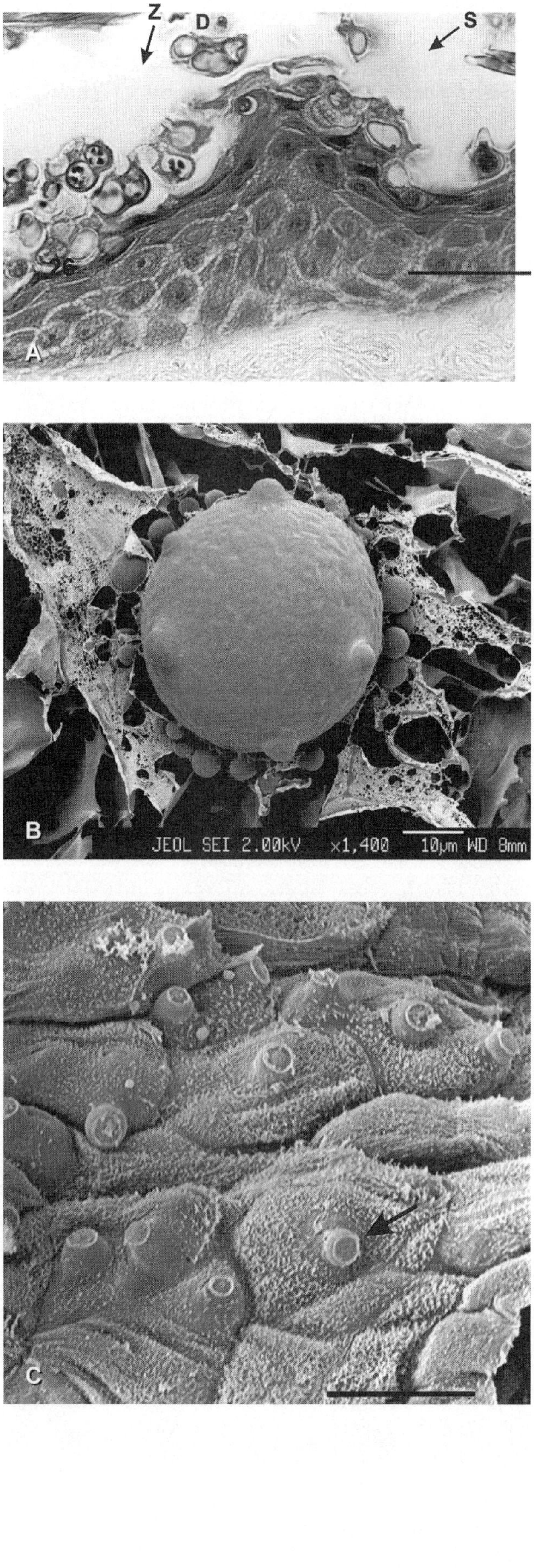


Fig. 16.16. The chytrid fungus *Batrachochytrium dendrobatidis* has emerged as a major threat to amphibian populations in several parts of the world, especially in tropical montane regions. The fungus attacks keratinized cells in the skin, as well as the mouthparts of tadpoles. (A) Cross-section of the skin of a frog infected with *B. dendrobatidis.* Z = zoosporagium containing zoospores. D = discharge tube of the zoosporangium. S = septum in a sporangium. (B) Scanning electron microscope photograph of a large sporangium in culture, showing discharge tubes that release zoospores. (C) Scanning electron microscope photograph of the surface of frog skin infected with *B. dendrobatidis.* Arrow points to discharge tube protruding through the skin. Photos by Lee Berger.

common group of fungi found in soil and water that represents the earliest diverging clade of fungi (Longcore, Pessier, and Nichols 1999; Carey et al. 2003).

There has now been a large body of research on the chytrid fungus and its role in amphibian declines. Amphibians (mostly anurans) infected with *Batrachochytrium* have been reported from every continent except Asia, in both captive and wild populations (Carey et al. 2003; Ouellet et al. 2005; see also a continuously updated website at: http://www.jcu.edu.au/school/phtm/PHTM/frogs/chyglob.htm). In addition to the mass mortality or declines attributed to chytrid fungus infections in Australia and Panama, other population declines associated with chytrid infections have been reported from Costa Rica (Lips, Green, and Papendick 2003), Ecuador (Ron and Merino 2000), Venezuela (Bonaccorso et al. 2003), western North America (D. E. Green and Kagarise Sherman 2001; Bradley et al. 2002; D. E. Green, Converse, and Schrader 2002; Muths, Corn, Pessier, and Green 2003), Europe (Bosch, Martínez-Solano, and García-París 2001), and New Zealand (Waldman et al. 2001). Almost all of these reports come from mid- to high-elevation sites, mostly in stream-breeding frogs. In Costa Rica and Panama, population declines have progressed in a southeasterly direction, having started in the highlands of Costa Rica in the late 1980s, reaching western Panama in the 1990s, and continuing to the present (fig. 16.17). The most recent decline attributable to chytridiomycosis occurred at El Copé, Panama, and was extraordinarily fast, with a decline of more than 50% in density of individuals and number of species within a few months after the first appearance of the fungus in the area (Lips et al. 2006). Most affected were frogs associated with streams, with terrestrial species showing little evidence of decline. Mass die-offs attributable to chytrid infections have not been reported from lowland populations of amphibians in either temperate or tropical regions. This is consistent with laboratory studies showing that the fungus grows best in cool, wet conditions (Longcore, Pessier, and Nichols 1999; Woodhams, Alford, and Marantelli 2003; Berger et al. 2004), and field studies showing that mortality of chytrid-infected Australian frogs was higher in the winter than in other seasons (Berger et al. 2004; Woodhams and Alford 2005). Nevertheless, there are some reports of chytridiomy-

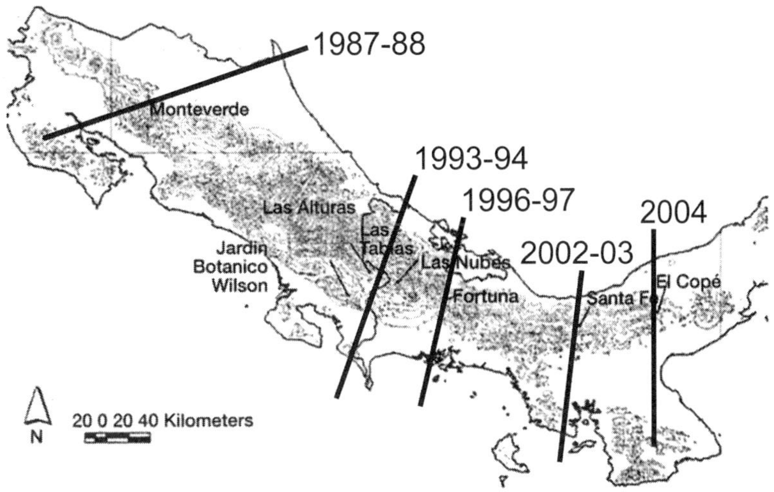


**Fig. 16.17.** Declines of amphibian populations associated with chytridiomycosis have rapidly spread from western Costa Rica, where declines occurred in the late 1980s, to the highlands of western Panama, where declines began in the 1990s and are continuing today. Many species of stream-breeding frogs and some species not associated with streams have been affected. Lines and dates show when declines occurred. After Lips et al. (2006).

cosis from warmer, lower elevation sites as well (Herrera, Steciow, and Natale 2005; Ouellet et al. 2005).

Although chytrid infection clearly has been shown to be the cause of mortality of adult anurans both in the laboratory and in the field (Carey et al. 2003; Berger et al. 2004), the actual mechanism responsible for mortality is not clear. The fungus could interfere with essential functions of the skin, or it could release a toxin that is fatal to many frogs. Infections of tadpoles generally are not fatal, although there is some variation among species in susceptiblity to the fungus (Blaustein et al. 2005). Fungal infections in tadpoles can result in deformed mouthparts that interfere with feeding and can affect rates of growth and development and alter interactions with competitors and predators living in the same pond (Fellers, Green, and Longcore 2001; Parris 2004; Parris and Beaudoin 2004; Parris and Cornelius 2004; Parris, Reese, and Storfer 2006). Nevertheless, the fungus can be transmitted from larvae to adults (Rachowicz and Vrendenburg 2004), so infected tadpoles represent a potential reservoir for the disease, particularly in species with long larval periods.

While many aspects of the basic biology of *Batrachchytrium* are now well understood, many others remain a mystery (McCallum 2005). The three most important unanswered questions are: (1) Where did the fungus originate? (2) Why has it emerged recently as a deadly disease in amphibians? (3) Why are some amphibians highly susceptible to the disease, whereas others can be infected but show few pathological symptoms? Recent genetic analyses of strains from amphibians on different continents indicate that all strains are genetically similar, suggesting a relatively recent common origin of the fungus in widely separated parts of the world (Morehouse et al. 2003). This has led many authors to suggest that the fungus is not native to areas where mass mortality of amphibians has occurred, including Australia and the Western Hemisphere, but was recently spread from another source (Carey et al. 2003). On the other hand, the low level of genetic diversity in tested strains of chytrid fungus is not, in itself, evidence of recent origin, because we lack comparative data on closely related fungi that would provide an expected level of genetic variation (Rachowicz et al. 2005).

If chytridiomycosis is a newly evolved disease, then we must account for its appearance in widely separated parts of the world, and its sudden appearance in areas where it had not previously been present (Lips et al. 2006). One hypothesis is that the fungus has spread around the world through international trade or accidental importation of infected amphibians. Once infected animals are released into the wild, they may encounter species with little or no immunity to the disease. This idea has been supported by reports of nonlethal infections of the fungus in nonnative bullfrogs (*Rana catesbeiana*) in South America, which could serve as reservoirs for the disease (Mazzoni et al. 2003; Hanselmann et al. 2004; Rachowicz et al. 2005).

Weldon et al. (2004) went further in proposing that the fungus most likely originated in Africa and spread to the rest of the world through widespread importation of African clawed frogs (*Xenopus laevis*), which can carry the disease, but are largely resistant to its effects. The evidence for an African origin of the disease is largely circumstantial (McCallum 2005; Rachowicz et al. 2005). Weldon et al. (2004) found infected *Xenopus* specimens in African museum collections dating back to 1938, much earlier than any other reported infection. This frog has been widely transported around the world since the 1930s, initially for pregnancy testing and later as a laboratory animal. Feral populations have become established in a number of locations, including California, Chile, and Wales, and in some cases, have subsequently expanded their ranges. Nevertheless, several lines of evidence argue against this hypothesis. First, the areas in which feral populations of *Xenopus* have become established do not correspond geographically to the hotspots of mortality attributed to chytridiomycosis. Second, feral populations of *Xenopus* almost invariably thrive in warm, manmade bodies of water at low elevations, such as farm ponds (McCoid and Fritts 1980, 1993; Lobos and Jaksic 2005). They are not found in cool, mid- to high-elevation lakes and streams or in montane rainforests, where most instances of the disease have been reported. Although the fungus can survive in water, at least for limited periods of time (Johnson and Speare 2003), we do not yet have a plausible mechanism to explain the dispersal of the fungus from infected to uninfected populations at higher elevations, especially in tropical forest reserves seldom visited by humans.

Until we have more definitive evidence about the origin of this fungus, it will be difficult to determine why the disease has emerged relatively recently as a threat to amphibians. One possibility already mentioned is that the fungus has been transported to areas in which the local amphibians lack defenses against the disease, thereby causing a fast-moving epidemic. Another favorite hypothesis is that other kinds of environmental stressors, such as global climate change, pollution, or habitat alteration have resulted in increased stress for amphibians, leading to impared immune function (Carey, Cohen, and Rollins-Smith 1999; Carey 2000, 2005; Carey and Alexander 2003; Carey et al. 2003; Rachowicz et al. 2005).

Another possibility is that global climate change has altered growing conditions for the fungus itself, making it more of a threat to amphibians. Pounds et al. (2006) reported a strong correlation between episodes of climate warming in Neotropical montane regions and the years in which various species of *Atelopus* disappeared. This stream-breeding genus has been severely impacted by population declines, and chytridiomycosis has been implicated as a cause of these declines. Although the fungus does best at relatively cool temperatures, the effect of a warming climate has been to increase the amount of cloud cover in montane forests. This not only could reduce the frogs' exposure to direct sunlight, but it also decreases maximum daytime temperatures while increasing minimum nighttime temperatures. The result is a narrowing of the daily range of temperatures experienced by the frogs, and by the fungus, with more temperatures falling within the optimum range for fungal growth. Although the data presented by Pounds et al. (2006) show a strong correlation between variation in temperature and declines of *Atelopus* (fig. 16.18), we do not yet have experimental evidence to establish a cause-and-effect relationship between climate change and amphibian declines.

The importance of direct exposure to sunlight as a defense against chytridiomycosis has yet to be explored. Like many organisms, the fungus probably is susceptible to excessive doses of UV-B radiation present in unflitered sunlight, and basking by diurnal amphibians could be one mechanism to combat fungal growth in the skin. Many diurnal anurans that occur at high altitudes, including several species of *Atelopus,* have unusually dark dorsal coloration, and many are nearly all black (Lötters 1996). One likely function of dark coloration is to protect the skin from harmful exposure to UV-B radiation. Many frogs, including some high-altitude species, are known to bask in full sunlight (see chapter 3). It has been suggested that basking could reduce chytrid infections by elevating body temperature (Woodhams, Alford, and Marantelli 2003; Rachowicz et al. 2005). However, neither basking nor dark coloration has a dramatic effect on core body temperature in amphibians, especially small species. Any increase in skin temperature due to increased absorption of sunlight is likely to increase evaporative cooling, thereby lowering core body temperature (see Tracy 1979, and discussion in chapter 3). Nevertheless, it is possible that basking does increase skin temperature, even if core body temperature is relatively unchanged, and dark coloration could enhance this effect. Even a small increase in skin temperature, coupled with direct exposure of the fungus to UV-B radiation, could be important in defending against fungal infections, because the skin is the main target of infection.

A final aspect of chytridiomycosis that remains poorly understood is inherent variation in susceptibility to the disease among different species of amphibians. Reports of mass die-offs of amphibians due to chytrid infection have received wide attention, but there are at least as many reports of amphibians infected with chytrid fungus that do not die from the infection (Carey et al. 2003; Daszak et al. 2005; McCallum 2005; Ouellet et al. 2005; Rachowicz et al. 2005; Woodhams and Alford 2005; Lips et al. 2006). Even in areas where many species of amphibians experienced precipitous declines or disappeared altogether, as in Australia, in Central and South America, and in western North America, other species in the same habitats were unaffected. In Panama, a number of arboreal and terrestrial frogs, includ-

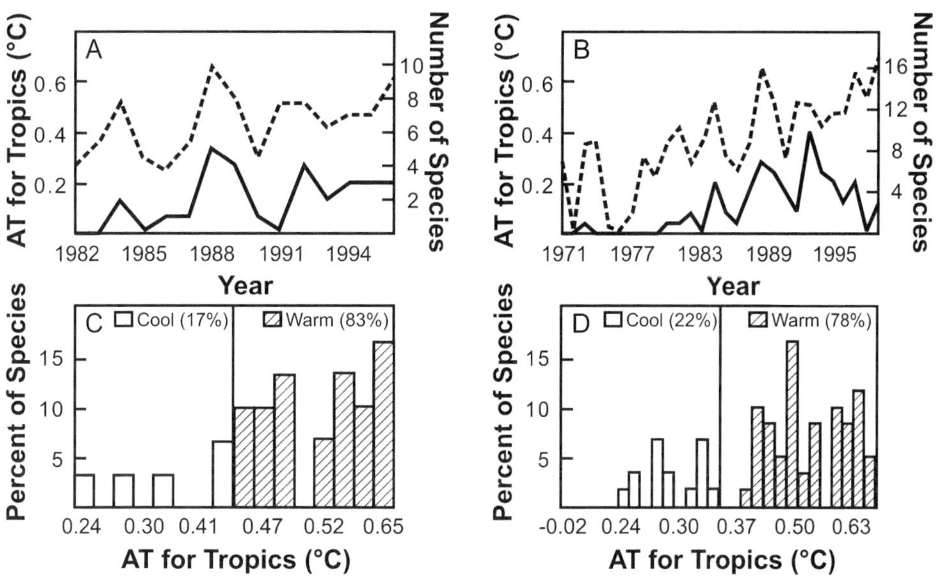


**Fig. 16.18.** Changes in climate appear to be related to the declines of populations of the genus *Atelopus*. (A, B) Relationship of the number of species of *Atelopus* seen for the last time each year (solid lines) to the deviation in air temperature for the tropics (dashed lines. Data in A are for species known to be in decline ($N = 51$); data in B are for those species, plus 49 others suspected of being in decline. Peaks in disappearances of species correspond to unusually warm weather. (C, D) Percentage of species observed for the last time in relative cool and relatively warm years. Data as in A and B. Species of *Atelopus* were much more likely to be seen for the last time following an unusually warm year. After Pounds et al. (2006).

ing *Agalychnis callidryas, Bufo marinus, Hyla microcephala, Leptodactylus pentadacylus, Smilisca phaeota,* and a species of *Rana* appear to be thriving in areas already decimated by the disease, despite the continuing presence of the fungus in the habitat and in the frogs (Lips et al. 2006). In Queensland, one species that suffered drastic declines in the 1980s, *Taudactylus eungellenis,* persists in a number of remant populations. These are infected at low levels with *Batrachochytrium,* but there was little evidence of reduced survival among infected individuals relative to infected animals. A treefrog in the genus *Litoria* in the same area that did not suffer population declines also had low levels of chytrid infection, but with few signs of negative effects on the population (Retallick, McCallum, and Speare 2004). In species such as *Xenopus laevis* and some populations of *Rana catesbeiana,* chytrid infections appear to be endemic, but the frogs show few symptoms of disease, and nonlethal chytrid infections have been reported in many other species of frogs as well (Carey et al. 2003; Davidson et al. 2003; Hopkins and Channing 2003; Daszak et al. 2004; Ouellet et al. 2005). Finally, among salamanders, chytrid infections have been reported in the genus *Ambystoma,* but the disease rarely appears to be fatal (Carey et al. 2003; Davidson et al. 2003; Padgett-Flohr and Longcore 2005).

All of this evidence suggests wide variation among amphibian species in their ability to fight off chytrid infections. What accounts for this variation? The first line of defense against fungal or microbial infections in amphibians is not a cellular immune response, but antimicrobial peptides produced in the skin. Amphibians produce a wide range of peptides that have been intensively studied by biochemists and pharmacologists interested primarily in their potential use as antibiotics in humans (Simmaco et al. 1998; Rinaldi 2002; Nascimento et al. 2003; Sima 2003; Conlon, Kolodziejek, and Nowotny 2004). The importance of these compounds as defenses against disease in wild populations of amphibians is just beginning to be explored (Rollins-Smith et al. 2005).

Recent research has shown that the polypeptide precursors of antimicrobial peptides are remarkably conservative evolutionarily, not only among amphibians, but also among vertebrates in general. These precursors have given rise to thousands of different peptides, all of which may have slightly different antimicrobial properties. Most species of frogs probably have unique combinations of several dozen antimicrobial peptides. The evolutionary strategy in this type of defense is to throw a wide variety of compounds at potential disease organisms, reducing the chances that microbes will evolve immunity to any one compound (Nicolas, Vanhoye, and Amiche 2003; Vanhoye et al. 2003). Not only does the great variety of peptides produced by a given species of frog enable them to defend against a variety of microbes, but the peptides act synergistically, so a combination of peptides has up to 100 times greater antimicrobial activity than does any single peptide (Nicolas, Vanhoye, and

Amiche 2003). This is an ancient evoutionary strategy, with a variety of antimicrobial peptides being present even in the most primitive living anuran, *Ascaphus* (Conlon, Sonnevend et al. 2004).

Most of the antimicrobial peptides that have been tested for activity against chytrid infections have been from ranid frogs, and most have shown strong activity against the fungus (Rollins-Smith, Carey et al. 2002; Rollins-Smith, Reinert et al. 2002; Rollins-Smith et al. 2003, 2005). Recently, additional work with peptides from hylids, leptodactylids, myobatrachids, and pipids turned up other peptides that inhibit growth of the fungus (Rollins-Smith and Conlon 2005; Woodhams et al. 2006). In some cases, such as *Rana tarahumarae,* the frogs produce peptides active against chytrid infections, but infections have been found in both dead and healthy individuals, and the species as a whole has suffered major population declines (Rollins-Smith, Reinert et al. 2002). This has led to the suggestion that other environmental stressors have somehow weakened the defenses of the frogs against the fungus, but evidence that population declines in this species actually were caused by chytridiomycosis is circumstantial (Rorabaugh and Hale 2005).

While much is known about the biochemistry of antimicrobial peptides in amphibians, little is known about the ecology of immune defense or the effect of habitat on the production of such compounds. One alternative to the environmental stressor hypothesis to explain variation among amphibians in susceptibility to chytrid infections is that different species have different innate immunity to the fungus for ecological reasons. In tropical Australia, frog species that have not suffered major declines appear to have skin peptides with greater effectiveness against chytridiomycosis than do species that have suffered declines (fig. 16.19; Woodhams et al. 2006). It seems likely that antimicrobial peptides have undergone rapid evolutionary diversification in response to selection for disease resistance (Duda, Vanhoye, and Nicholas 2002; W. Lee et al. 2005). The presence of a diverse array of bacteria and fungi in the natural environment of amphibians is expected to select for an ever-increasing array of compounds to combat these infectious agents. Furthermore, there is evidence that production of antimicrobial peptides is induced by exposure to soil microbes (Mangoni et al. 2001) or to environments conducive to the growth of fungi and bacteria (Matutte et al. 2000). This suggests that amphibians probably exhibit both plastic phenotypic responses to local assemblages of microbes and long-term evolutionary adaptation to microbes present in the environment.

If this is a general pattern in amphibians, then one would expect to see the evolution of a more powerful and diverse array of antimicrobial skin secretions in amphibians that are regularly exposed to microbe-filled environments. These would include burrowing amphibians in constant contact

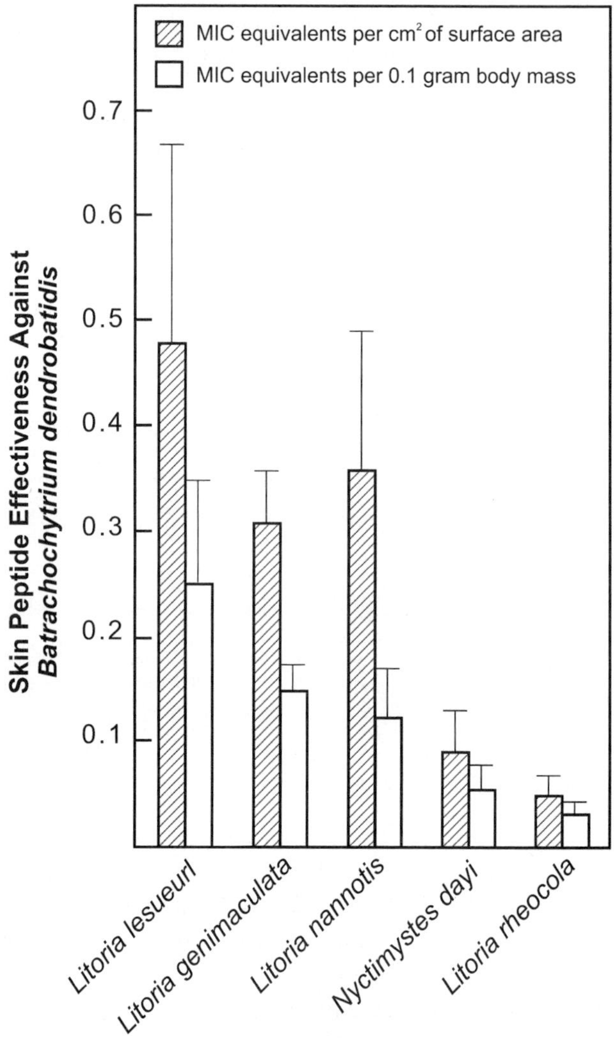


**Fig. 16.19.** Relative effectiveness of skin peptides against the chytrid fungus *Batrachochytrium dendrobatidis* in five species of Australian stream-breeding hylid frogs. Data show Minimal Inhibitory Concentrations (MIC) of mixed skin peptides, expressed in relation to both skin surface area and body mass. The two species on the left (*Litoria lesueri* and *L. genimaculata*) have not experienced population declines, whereas the other three species have declined. The two species on the right have particularly ineffective defenses against the fungus. After Woodhams et al. (2006).

with moist soil, most terrestrial salamanders, leaf-litter frogs in tropical forests, and both salamanders and frogs that breed in warm, eutrophic ponds that are veritable microbial stews. In contrast, amphibians that normally breed in cool, low-productivity environments such as mountain streams or high-altitude lakes would not necessarily be expected to evolve such an array of defenses. If these relatively "clean" habitats are invaded by a newly arrived organism such as a chytrid fungus, the local amphibians may lack the ability to fight off infection, and the result could be an epidemic of disease and rapid declines of local populations, such as those seen at mid- and high-elevation sites in the Western Hemisphere, Australia, and Europe. At the same time, amphibians in the same area that do not breed in streams, such as ground-dwelling *Eleutherodactylus* in South and Central America, might possess greater resistance to infection simply by virture of having a wider array of antimicrobial compounds in the skin. Because the peptides have not necessarily evolved as defenses against specific infectious agents, some terrestrial or pond-breeding amphibians might, by chance, also be susceptible to the disease. Similarly, some stream-breeding frogs might, by chance, be more resistant to the fungus than are others. What is needed to fully understand amphibian defenses against chytrid infection is a comprehensive survey of antimicrobial defenses that combines a rigorous phylogenetic perspective with a realistic ecological context.

The discovery and identification of the organism responsible for chitridiomycosis in amphibians generated a lot of excitement among biologists looking for causes of amphibian declines, and in the popular press. While there is strong evidence associating the presence of chytrid infections in both time and space with certain dramatic amphibian population declines, this disease does not necessarily represent the smoking gun that explains all or even most population declines (McCallum 2005). Nevertheless, it is critical to understand more about this emerging disease and the factors that mediate amphibian susceptibility to it, because there is considerable potential for the disease to spread to areas where it has not yet been encountered (Ron 2005; Lips et al. 2006).

It is important to keep in mind that (1) many amphibian populations have declined in areas where chytrid infection has not been found; (2) not all amphibians infected with the chytrid fungus die or even become sick, (3) there appears to be considerable natural variation in susceptibility to the fungus among amphibian species, and (4) even in species in which chytrid infections have been reported, the disease may not be the principal cause of population declines. A good example is *Rana muscosa* in the western United States. Chytrid infections have been reported from both ranid frogs and bufonid toads in the western United States (D. E. Green and Kagarise Sherman 2001; Bradley et al. 2002; D. E. Green, Converse, and Schrader 2002; Muths, Corn, Pessier, and Green 2003), and *R. muscosa* can be infected with the disease (Rachowicz and Vredenburg 2004), which could be a major threat to now-diminished populations. Nevertheless, it seems clear that the primary cause of population declines in this species up to now has been the introduction of nonnative fish into breeding sites (Vredenburg 2004), and in some areas, pollution from airborne pesticide residues (Davidson, Shaffer, and Jennings 2001, 2002; Davidson 2004; see the following discussion).

### Ranaviruses

Although chytridiomycosis has received the most attention from biologists searching for causes of amphibian population declines, it is not the only disease that can cause mass mortality in amphibians. A survey of reports of dead or dying amphibians in the United States over a six-year period showed that by far the most common cause of mass mortality was infection by ranaviruses (D. E. Green, Converse, and Schrader 2002) in members of the iridovirus family (Chinchar 2002). Molecular genetic techniques have identified a number of different strains of ranaviruses, some of which infect anurans and others that infect mostly urodeles. Different species of salamanders can be susceptible to different strains of virus (Docherty et al. 2003). In contrast to chytridiomycosis, ranaviruses have caused mortality mostly in widely distributed species such as tiger salamanders (*Ambystoma tigrinum;* Collins et al. 2004) and various common ranid frogs, but not in declining species. Nevertheless, viral epidemics are a potential threat to species with restricted ranges, such as the endangered Sonora tiger salamander (*Ambystoma tigrinum stebbensi;* Jancovich et al. 1997) or the California tiger salamander (*A. californiense*).

Viral infections are especially common at high densities, which favor transmission of the disease among individuals. Usually mortality is highest in larvae approaching metamorphosis or in metamorphosed young. The result is that a ranavirus epidemic, sometimes accompanied by a secondary bacterial epidemic, can quickly eliminate a local population, although these often eventually recover. Spread of the virus between populations is less likely because mortality of infected young animals is so high, although there is the potential for sublethally infected metamorphs to harbor the disease and subsequently carry it back to their home ponds or to other nearby ponds (Brunner et al. 2004). There also is the potential for human-assisted transport of infectious agents, or movement of infectious animals though commerce in salamander larvae used for bait (Collins et al. 2003).

Another potential mode of transmission of ranaviruses throughout the world is the widespread commerce in large ranid frogs, including *Rana catesbeiana* and *R. grylio*, for food and for aquaculture. Viral epidemics have been reported in frog farms in China and Thailand, where both North American ranids and Asian ranids (*Hoplobatrachus chinensis*) are raised for food (Chinchar 2002; Weng et al. 2002; Collins et al. 2003). Viral epidemics also may be responsible for die-offs of the most common anuran in Britain, *Rana temporaria* (Cunningham et al. 1996), with imported goldfish in garden ponds being a possible reservoir for the disease (Hyatt et al. 2000; Collins et al. 2003). Because ranaviruses can infect a variety of fish species (Chinchar 2002), stocking of lakes, ponds, and streams with infected game fish provides another potential mode of transfer to amphibian populations (Jancovich et al. 2005). Ranaviruses probably have been endemic in amphibian populations for a long time and cannot be definitively linked to any cases of amphibian population declines, but increased human-assisted transmission of the viruses has the potential to threaten amphibian populations in the future, particularly those that already have restricted ranges or have suffered population declines for other reasons. Isolated populations that have reduced genetic heterozygosity because of natural genetic bottlenecks or habitat fragmentation also may be at risk from newly arriving pathogens such as ranaviruses (Pearman and Garner 2005).

### Other Amphibian Diseases

Amphibians are subject to a variety of other viral, bacterial, and fungal diseases, most of which are endemic in populations and may cause occasional die-offs, but are not thought to be responsible for widespread population declines. One of the most frequently cited diseases is "red-leg," caused by a bacterium, *Aeromonas hydrophila,* but recent research indicates that this is not a major cause of amphibian mass mortality. In many cases, mortality attributed to *Aeromonas* infections may have been misdiagnosed ranavirus infections, or bacterial infections may have been secondary to a viral infection (D. E. Green, Converse, and Schrader 2002) or chytrid infection (Muths and Nanjappa 2005). Some amphibians show little or no sign of disease even when experimentally injected with *Aeromonas* under conditions designed to suppress immune function (Carey, Maniero, and Stinn 1996; Maniero and Carey 1997).

A variety of fungi are found in the digestive tracts of amphibians, although not all of these are pathogenic (R. Nelson et al. 2002). Both eggs and larvae sometimes are attacked by fungal diseases (Czeczuga, Myszyńska, and Krzemińska 1998; Berger et al. 2001), but only *Saprolegnia* fungi have been cited as a major cause of population-wide mortality (Blaustein et al. 1994). Fungal infections of the eggs of *Bufo boreas* (see fig. 15.25 in chapter 15) have been cited as one possible cause of the decline of this species in western North America, but chytridiomycosis also has been found in some populations of this species (Muths and Nanjappa 2005). Susceptibility to fungal infection appears to be related to a number of factors, including water depth, temperature, incident UV-B radiation, a tendency to clump eggs in communal masses, and innate differences in susceptibility of different species to the fungus (Kiesecker and Blaustein 1995, 1997a; Kiesecker, Blaustein, and Miller 2001; J. Robinson, Griffiths, and Jeffries 2003). There is no indication that this fungus is a cause of widespread amphibian population declines.

### Parasites and Amphibian Deformities

At about the same time that public attention first turned to the problem of declining amphibian populations, there also were increasing reports of amphibians turning up with serious deformities, including both missing and extra limbs. There has been a tendency, especially in the popular literature (Souder 2000), to conflate these two issues, but there is little evidence that physical deformities are a widespread cause of amphibian population declines (Carey et al. 2003; P. Johnson and Lunde 2005). Instead, these deformities, usually malformed limbs in metamorphosing frogs, tend to be a local phenomenon that can cause high mortality in particularly ponds (Hoppe 2005), but does not necessarily threaten populations at larger spatial scales. Mortality can be a direct effect of parasite infection, or limb deformites can impact frog locomotion, which in turn affect their foraging behavior or their vulnerability to predators, including wading birds that are the definitive hosts for the trematodes. Although not a direct cause of amphibian population declines, the phenomenon of malformed frogs is of concern to amphibian conservation biologists, because an increased incidence of limb malformations could impact species that already are threatened for other reasons (P. Johnson and Lunde 2005).

When malformed frogs started to appear in large numbers, especially in the midwestern United States, a host of possible causes were proposed, including chemical pollution, increased ultraviolet radiation, and predation (Ouellet 2000; Sessions 2003). Most recent work has focused on the role of trematode parasites in causing limb malformations, although this is not necessarily the only cause of deformities (Bridges et al. 2004). In particular, the trematode *Ribeiroia ondatrae* has been shown to cause limb malformations in a number of anuran species (P. Johnson et al. 1999, 2002, 2003, 2004; P. Johnson, Lunde, Haight et al. 2001; P. Johnson, Lunde, Ritchie et al. 2001; Schotthoefer et al. 2003; Sessions 2003; P. Johnson and Lunde 2005; Sutherland 2005). Although these parasites are a natural part of the ecosystems in which frogs live, there are connections between human-induced habitat changes and the frequency of parasite infection. Limb malformations caused by *Ribeiroia* tend to be most common in eutrophic farm ponds, where increased nutrient input tends to favor snails in the genus *Planorbella,* the first intermediate host of the trematode (Lannoo et al. 2003; P. Johnson and Chase 2004; P. Johnson and Lunde 2005). In addition, widespread, but mostly accidental, transport and introduction of nonnative snails into new habitats has the potential to increase the incidence of trematode infection if the invasive snail species are suitable intermediate hosts. The parasites also can be spread to new habitats by birds that serve as definitive hosts, such as the great blue heron (*Ardea herodias*), which does well in human-altered habitats and is expanding its range in North America (P. Johnson and Lunde 2005).

## Chemical Pollution

Chemical pollution, especially of freshwater habitats, has long been recognized as a potential threat to amphibian populations (Sparling, Linder, and Bishop 2000; Burkhart et al. 2003; Bridges and Semlitsch 2005b). The difficulty in addressing this issue is the enormous number and variety of chemical substances released into the environment by human activities. Heavy metals (Lefcourt et al. 1998), industrial waste products such as PCBs (Fontenot et al. 2000; de Solla et al. 2002), coal ash (Raimondo, Rowe, and Congdon 1998; Rowe et al. 1998; Hopkins, Congdon, and Ray 2000), road salt (Turtle 2000; Kaushal et al. 2005), oil (Mahaney 1994; Lefcourt et al. 1997; Pollet and Bendall-Young 2000), wastewater effluent from factories and sewage treatment plants (Yesilada, Ozmen, and Yesilada 1999; Laposata and Dunson 2000), chemicals leaching from landfills (Bruner et al. 2002), pesticides (see the followng), and fertilizers (see the following) are just some of the many chemical pollutants that have potential negative effects on amphibian populations. It is not possible to individually test the effects of each chemical pollutant on amphibians, and multiple chemical stressors, or chemicals acting in concert with other stressors, often act synergistically to have a greater impact than a single chemical alone (Burkhart et al. 2003). Standard laboratory toxicity assays often provide incomplete or misleading information on the ecological effects of chemical pollutants, because interactions with other environmental factors can either increase or decrease the toxic effects of chemicals on amphibians (Boone and Bridges 2003b, c; Bridges and Semlitsch 2005a, b).

In many cases, amphibians will simply disappear from highly polluted habitats as they avoid using such habitats for breeding or die from exposure to toxic chemicals. In other cases, adults may continue to migrate to polluted habitats to breed, but recruitment of metamorphosing young is so low that these sites constitute population sinks that fail to contribute to the growth of the regional population (Rowe and Hopkins 2003). Low recruitment may result from direct mortality of large numbers of eggs, larvae, or metamorphs, or from sublethal effects of pollutants that increase physiological costs and decrease performance of larvae and metamorphs. There also may be direct effects on adults visiting these sites during the breeding season that further reduce reproductive performance or survival. On a landscape scale, widely dispersed sources of pollution can have a major impact on amphibian populations, whereas the impact of more localized sources probabably is minimal.

Chemical pollution is a symptom of a general degradation of ecosystems by human actitivies. Although high concentrations of chemical pollutants often have localized sources, the problem of chemical pollution is a global one, with many kinds of chemical residues having been found in remote, seemingly pristine areas (Collins and Storfer 2003). Many ecosystems are remarkably resilient and eventually can recover from acute insults, such as dioxin spills, with a variety of animals, including amphibians, eventually recolonizing the area (Garagna et al. 2001). Chronic pollution presents a more difficult problem, both for human health and for ecosystem health. Possible solutions for reducing the effects of pollution on amphibian populations are almost as varied as the number and sources of pollutants themselves, and all come at some economic cost. In some cases, reduced use or elimination of certain chemicals from the environment may be the only solution. In other cases, the effects of pollution may be ameliorated by reducing the number of pollution sources, restricting the dispersal of pollutants, adjusting the timing of chemical use to reduce impacts on amphibians, or preventing amphibians from using polluted sites for breeding. Rather than catalog all possible effects of chemical pollutants on amphibians, I will focus on a few well-studied cases of chemicals that have direct toxic effects on amphibians, those that have sublethal effects on physiology and performance, and those that have indirect effects through interaction with other environmental stressors, both physical and biological (see Sparling, Linder, and Bishop 2000, for a comprehensive review of amphibian ecotoxicology, and Linder, Krest, and Sparling 2003, for a review of multiple stressor effects).

### Pesticides and Amphibian Declines

The toxic effects of pesticides have long been considered a potential cause of amphibian population declines (Boone and Bridges 2003b). Many studies have shown a negative correlation between the amount of agricultural land use and the diversity and abundance of amphibians (see discussion earlier in this chapter), but showing that agricultural pesticides are responsible for these declines has been surprisingly difficult, for a number of reasons. First, agricultural land use itself may produce a habitat structure that is unsuitable for amphibians, independent of chemical applications. Second, there is considerable variation among species of amphibians in sensitivity to pesticides (Bridges and Semlitsch 2000, 2005a), and genetic variation for pesticide resistance among individuals indicates the potential for amphibians to evolve increased resistance to these chemicals (Bridges and Semlitsch 2001; Boone and Bridges 2003b). Third, different pesticides have different effects on amphibians in both aquatic larvae and in terrestrial individuals, and multiple pesticides acting together may have a greater effect than a single pesticide (Relyea 2004a; Bridges and Semlitsch 2005b). Finally, concentrations of pesticides that are lethal to amphibians often are much higher than typical concentrations found in natural habitats or in the tissues of animals from the wild (Fagotti et al. 2005). Standard toxicological tests sometimes show relatively low sensitivity to some pesticides (e.g., Pauli, Coulson, and Berrill 1999; Battaglin and Fairchild 2002), whereas tests of other pesticides have shown strong lethal effects on amphibians (Berrill et al. 1994, 1998). Nevertheless, many pesticides can have sublethal negative effects on amphibians that ultimately affect individual fitness (Boone and Bridges 2003b).

Some of the best evidence for the effect of pesticides on amphibian populations comes not from populations located in agricultural areas, which often are unsuitable for amphibians, but from parks and other seemingly pristine areas that are near agricultural land heavily impacted by pesticides. There is strong circumstantial evidence, for example, that wind-blown pesticides originating in the heavily agricultural Central Valley of California have contributed to the decline of several species of amphibians in the southern Sierra Nevada, especially on the western slopes of the mountains. Severe population declines have been found in several protected areas, including Yosemite National Park (Fellers and Drost 1993; Drost and Fellers 1996). Pesticides are readily transported to higher elevations from the Central Valley by prevailing westerly winds (fig. 16.20 A, B), and pesticide residues have been found in amphibians at considerable distances from major sources of pesticides (Sparling, Fellers, and McConnell 2001; Sparling and Cowman 2003). Furthermore, the geographic pattern of population declines in several species of ranid frogs is strongly correlated with the amount of upwind agricultural land (fig. 16.20 C, D; Davidson, Shaffer, and Jennings 2002) and the amount of pesticides applied to that land (Davidson 2004).

Perhaps the most severely impacted has been the California red-legged frog (*Rana draytonii;* see Shaffer et al. 2004, for recent changes in the taxonomic status of this frog). Originally one of the most common amphibians in California, *R. draytonii* was found in the Central Valley and the Sierra Nevada, as well as many locations along the coast east of the Central Valley, from northern California to Baja California in Mexico. The species is now extinct in the Central Valley and nearly extinct in the Sierra Nevada and southern California, but still occurs in the coastal mountains (Shaffer et al. 2004; Fellers 2005). As with several other ranid frogs, the pattern of decline and extinction in this species is most consistent with the hypothesis that wind-blown pesticides have wiped out populations downwind from the Central Valley (Davidson, Shaffer, and Jennings 2001, 2002; Davidson 2004). Consequently, the prospects for recovery of this endangered species seem slim without major changes in agricultural practices in California.

### Atrazine and Endocrine Disruption

Although one can expect that pesticides designed to kill a variety of plants and animals will have lethal effects on at least some amphibians, sublethal effects of these chemicals also can have dramatic effects on the physiology, behavior, morphology, and ecology of amphibians. One of the best-studied examples is the effect of atrazine on sexual development in anurans. Atrazine is the most widely used herbicide in the world, and is a major contaminant of both surface and groundwater supplies. Atrazine has a variety of negative physiological effects on amphibians, including inhibiting growth and development of larvae, producing morphological deformities, and influencing the behavior and physiology of metamorphosed individuals (Diana et al. 2000; Allran and Karasov 2001; Rohr and Palmer 2005). More disturbing is the fact that at very low, but ecologically realistic doses, atrazine affects gonadal development by producing intersex individuals with both male and female gonadal tissue. In addition, males of some species exhibit reduced larynx size and show other signs of feminization (Reeder et al. 1998; T. Hayes, Collins et al. 2002; T. Hayes, Haston et al. 2002, 2003; Carr et al. 2003).

Although these studies have generated a lot of controversy, primarily from the manufacturers of atrazine, there is no doubt that this widely used chemical can have serious negative effects on amphibian development, even at very low concentrations (T. Hayes 2004). The precise exposure of amphibians to atrazine in the environment depends on a number of factors, including the timing of chemical use and the concentrations used, geology of soils in agricultural fields, the rate of runoff from fields, and the nature of the ponds used for reproduction. These in turn will interact with a variety of biological factors to determine the effects on individuals and populations (fig. 16.21). A direct link between the developmental effects of atrazine and declines of particular populations of amphibians has yet to be established. There is circumstantial evidence that atrazine, or other endocrine disruptors, such as PCBs, could have contributed to the decline of cricket frogs (*Acris crepitans*) in the upper Midwest (Reeder et al. 1998, 2005). Atrazine is only one of several thousand agricultural chemicals that are regularly used in the United States and elsewhere. The effects of most of these chemicals on amphibians are unknown, but the documented effects of atrazine are symptomatic of a broader range of problems caused by many different endocrine-disrupting chemicals in the environment that affect not only amphibians, but many other vertebrates as well (T. Hayes 2000, 2005; Kloas 2002; Mosconi et al. 2002).

### Other Sublethal Effects of Pesticides

Although some pesticides are not directly lethal to amphibians at ecologically realistic doses, these chemicals can have many other effects on amphibians in addition to altering sexual development. Some studies have found that exposure to carbaryl, a widely used insecticide, actually increases rates of growth and size at metamorphosis in some tadpoles, while reducing survival to metamorphosis in others. This occurs when carbaryl kills most zooplankton in ponds, thereby

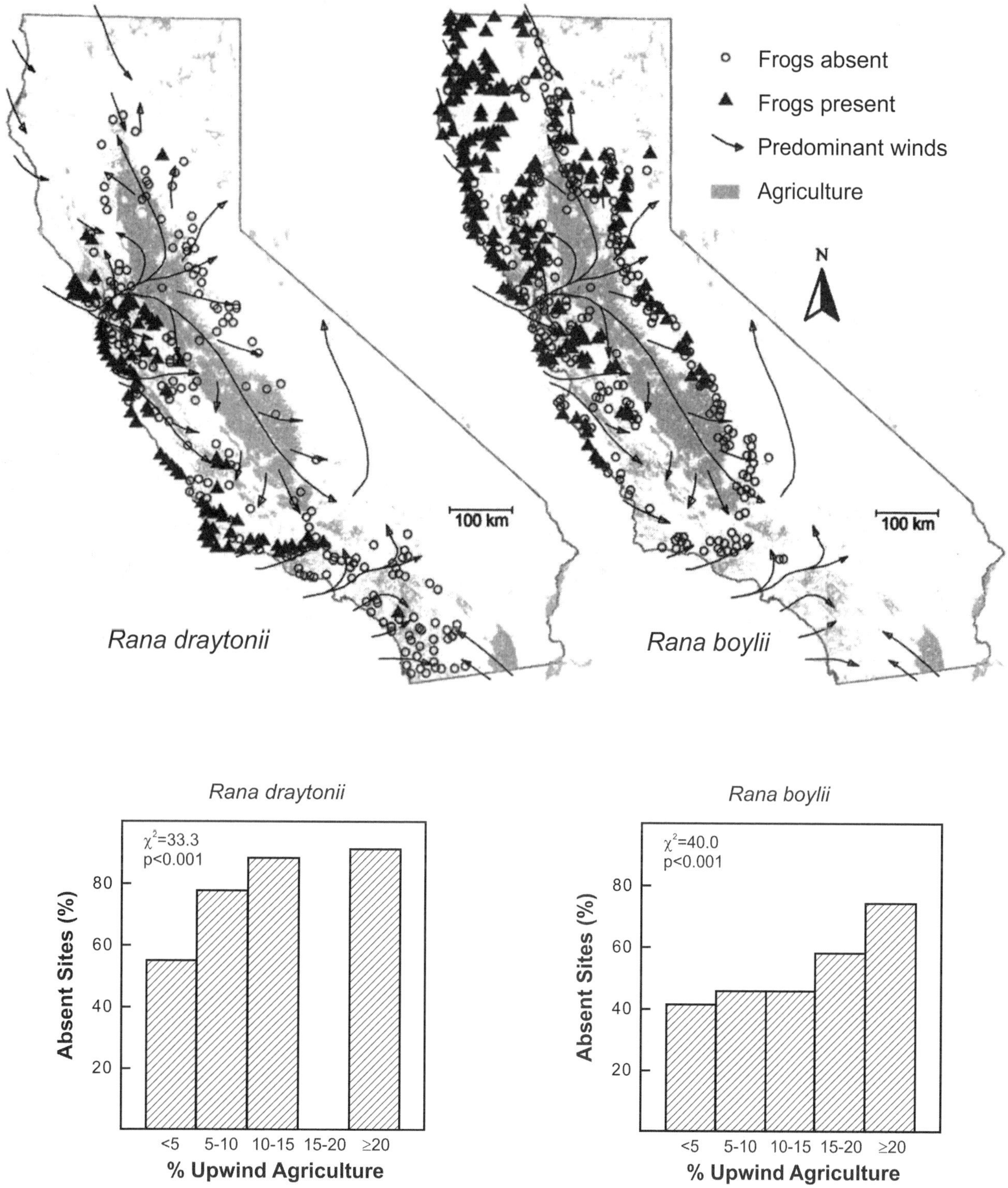


**Fig. 16.20.** Relationship of wind direction and proximity to the heavily agricultural Central Valley of California to the declines of two species of frogs in California. Map on left shows the distribution of historic sites where the California red-legged frog (*Rana draytonii*) is now absent (open circles) and still present (closed triangles). Map on the right shows the distribution of sites for the foothill yellow-legged frog (*Rana boylii*). Graphs show the percentage of sites lacking *R. draytonii* (left) or *R. boylii* (right) in relation to percentage of upwind agriculture. After Davidson et al. (2002).

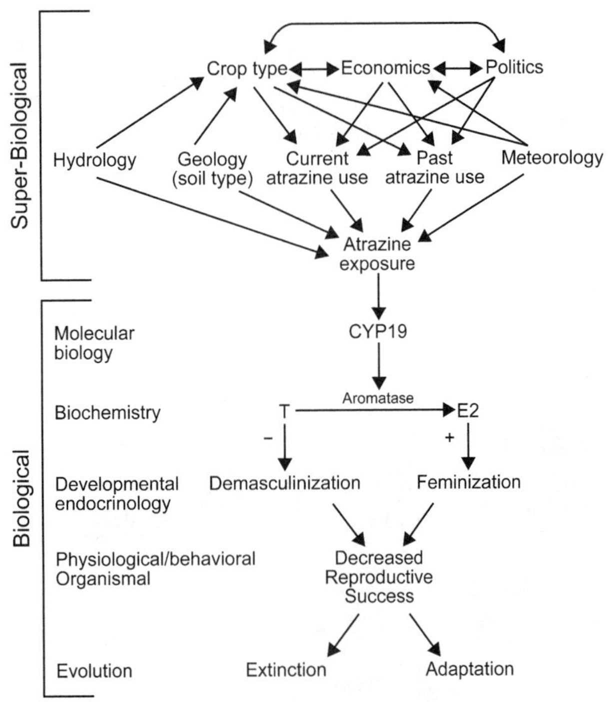


**Fig. 16.21.** Interactions between ecological factors, application of atrazine to agricultural fields, and biological factors that can affect the developmental endocrinology and reproductive success of male amphibians, with potential effects on amphibian populations. After T. Hayes (2005).

increasing the standing crop of algae and providing more food for anuran tadpoles. The effects were most pronounced at higher densities, because tadpoles are most likely to be food-limited at high density (Boone and Semlitsch 2001; Boone et al. 2001; Boone et al. 2004). Although laboratory studies have shown that exposure to UV-B raditation can increase the toxicity of carbaryl to amphibians (Zaga et al. 1998), tadpoles in artificial ponds exposed to both UV-B radiation and carbaryl actually showed increased growth. Again, carbaryl increased the standing crop of algae in ponds, not only providing more food for the tadpoles, but also probably shielding tadpoles from UV-B radiation because of decreased water clarity (Bridges and Boone 2003). Multiple exposure to carbaryl can actually stimulate metamorphosis in green frog (*Rana clamitans*) tadpoles (Boone and Bridges 2003a).

Unfortunately for amphibians, pesticides such as carbaryl are seldom used alone, but often are applied in combination with other agricultural chemicals that can act synergistically to increase mortality of amphibians. For example, Boone and James (2003) examined the effects of exposure to carbaryl (an insecticide), atrazine (a herbicide), and changing water levels on two species of anuran tadpoles and two species of salamander larvae. Carbaryl killed most salamanders in all treatments, and increased mortality of toad (*Bufo americanus*) tadpoles as well. Tadpoles of *Rana sphenocephala* exposed to carbaryl alone increased in size because of an increase in food, but exposure to atrazine decreased body size and development rate because this chemical reduced algal biomass. On the other hand, larvae of spotted salamanders (*Ambystoma maculatum*) did better when exposed to both atrazine and carbaryl than to carbaryl alone. These effects interacted in complex ways with pond hydroperiod and larval density, and also are affected by genetic differences within and among species in sensitivity to the chemicals (Bridges and Semlitsch 2000). Other studies have shown that exposure to multiple pesticides has a greater effect on tadpole mortality than does exposure to single pestcides (Relyea 2004a, 2005b). Synergistic effects of chemicals sometimes can have unexpected results. For example, both carbaryl and ammonium nitrate fertilizer tend to increase algal biomass in ponds, but tadpoles exposed to both of these chemicals together did not show the positive effects on growth and development seen in tadpoles exposed to only one chemical. This suggests that metabolic costs associated with multiple chemical exposure reduce the ability of tadpoles to exploit enhanced food resources (Boone et al. 2005).

Carbaryl also can alter interactions between predators and tadpoles. In one set of experiments, even relatively low doses of carbaryl killed crayfish, which were predators on bullfrog (*Rana catesbeiana*) tadpoles, and higher doses killed predatory sunfish. Consequently, bullfrog tadpoles were released from predation at low carbaryl concentrations, thereby increasing survivorship. As in the other experiments, carbaryl also increased algal biomass, so tadpoles actually were larger with low doses of carbaryl. At higher carbaryl concentrations, however, mortality of tadpoles increased (Boone and Semlitsch 2003). Low doses of carbaryl can inhibit antipredator behavior, such as seeking cover, but also reduce tadpole activity, which makes them less vulnerable to predators (Bridges 1999). Conversely, exposure to predators can affect the toxicity of pesticides to amphibians, presumably because predators increase levels of stress. For example, tadpoles of gray treefrogs (*Hyla versicolor*) suffered higher mortality from exposure to carbaryl in the presence of caged predatory *Ambystoma maculatum* larvae than in the absence of predator cues (Relyea and Mills 2001). Similar increases in mortality from pesticide exposure in the presence of predator cues have been found in several other species (Relyea 2003b, 2004b, 2005c).

Another major concern among amphibian conservation biologists is that pesticides and other chemical pollutants can act as general environmental stressors that may impare immune function in amphibians, thereby making them more susceptible to infection by viruses, bacteria, fungi, or parasites. This indirect effect of chemical exposure has been suggested as one possible reason for the emergence of new diseases that contribute to amphibian population declines (Carey and Bryant 1995). Indeed, a number of studies have shown that exposure to single chemicals, such as malathion or atrazine, or to combinations of chemicals, does diminish immune function in a variety of amphibians (S. K. Taylor, Williams, and Mills 1999; Christin et al. 2003, 2004). Whether this has actually contributed to amphibian population declines has yet to be determined.

### Nitrogen Pollution

The discharge of excess nitrogen into aquatic ecosystems represents one of the major environmental issues of our time, but often receives less attention than the effects of more obviously toxic chemical pollutants. The main sources of excess nitrogen are the heavy application of nitrogenous fertilizers on agricultural lands, managed forests, and lawns (these fertilizers also contain large amounts of phosphorus); runoff from livestock feedlots and slaughterhouses; and the discharge of treated sewage effluent into waterways. Sewage effluent discharged into rivers probably has a minimal impact on amphibians, most of which do not breed in rivers. Nevertheless, wastewater effluent sometimes is used to irrigate agricultural fields, a practice that could contaminate temporary ponds used by amphibians (Laposata and Dunson 2000). Excess nitrogen also can be deposited in both terrestrial and aquatic habitats in the form of airborne residues from the burning of fossil fuels and other human activities (Rouse, Bishop, and Struger 1999). Deposition of atmospheric nitrogen can have negative consequences for ecosystems far from the original source, including relatively remote areas that are considered hotspots of biodiversity (Phoenix et al. 2006).

At low concentrations, nitrogenous fertilizers can enhance tadpole growth and development, particularly at high tadpole densities, by increasing food availability (De Wijer et al. 2003; Boone et al. 2005). Excess nitrogen has a number of negative effects as well, including depeletion of oxygen in eutrophic ponds, inhibition of feeding in tadpoles, and induction of various physical and behavioral abnormalities (Hecnar 1995; Xu and Oldham 1997; Rouse, Bishop, and Struger 1999). At high concentrations, excess nitrogen in the form of nitrates or ammonia can be directly toxic to amphibians (Hecnar 1995; Oldham et al. 1997; P. Watt and Jarvis 1997; Marco, Quilchano, and Blaustein 1999; Marco et al. 2001; Hamer et al. 2004). As with pesticides, sensitivity to nitrogen pollution varies among species of amphibians, and the effects of nitrogen are mediated by interactions with other environmental factors, such as UV-B radiation (Hatch and Blaustein 2003). Diversity and abundance of amphibians often is reduced in agricultural areas where heavy application of fertilizer occurs (Berger 1989; C. Bishop et al. 1999; Oldham 1999; de Solla, Pettit, Bishop, Cheng, and Elliot 2002), and the decline of at least one species of Australian frog, *Litoria aurea,* has been attributed at least in part to heavy use of agricultural fertilizers (Hamer et al. 2004). Nevertheless, the global impact of nitrogen pollution on amphibian population declines has yet to be thoroughly investigated.

### Acid Rain

In addition to direct toxic effects of chemical pollutants, both aquatic and terrestrial habitats can be affected by the increasing acidity of precipitation, which itself is largely a byproduct of air pollution, mostly from coal-burning power plants. Habitat acidity is measured in two ways. First, pH is a measure of instantaneous acidity, with pH values <5.0 being considered acidic. This can vary seasonally with snowmelt or rainfall. Second, habitats differ in acid-neutralizing capacity (ANC), a measure of how well habitats are buffered against changes in acidity. Low pH and low ANC can have a variety of negative effects on amphibians, including inhibiting development and hatching of embryos, retarding growth of larvae, and disrupting the osmoregulation and ion balance of both larvae and terrestrial adults (see chapter 2). These negative effects have been extensively documented in laboratory studies (Beattie and Tyler-Jones 1992; Bradford, Swanson, and Gordon 1992; Dunson, Wyman, and Corbett 1992; Frisbie and Wyman 1992; Rowe and Freda 2000). Acidic habitats also can contain high levels of dissolved metals, such as aluminum, which also negatively impact amphibians (Rowe and Freda 2000).

There is significant variation both within and among species of amphibians in acid tolerance, with those from naturally acidic habitats generally being the most acid tolerant (Pierce 1985; Karns 1992; Pierce and Wooton 1992; Sadinski and Dunson 1992; Räsänen, Laurila, and Merilä 2003a, b). These differences are largely due to maternal effects, but whether they are entirely genetic is not clear. Natural variation in acid tolerance affects the choice of habitats by different species. Acid-tolerant species such as carpenter frogs (*Rana virgatipes*), wood frogs (*Rana sylvatica*), pine-barrens treefrogs (*Hyla andersonii*), and spotted salamanders (*Ambystoma maculatum*), can breed in more acidic water and show less behavioral avoidance of acidic conditions than do less tolerant species such as leopard frogs (*Rana pipiens*) and Jefferson salamanders (*Ambystoma jeffersonianum*). In areas where pond acidity varies on a relatively small spatial scale, amphibians generally avoid the most acidic habitats. In Minnesota peatlands, for example, most amphibian species breed successfully in fens with a pH > 5.0, but not in acidic bogs (pH < 4.5; Karns 1992).

When concern over declining amphibian populations began to grow in the 1980s, acid rain was one of many possible culprits considered by biologists seeking causes for the declines. Of particular concern were montane habitats, where the impact of acid rain can be the most severe, and where negative effects of acidity on fish and invertebrate populations had been well documented. A number of studies in the western United States have documented periodic fluctuations in the acidity of aquatic habitats, largely due to snowmelt. There is little evidence, however, that most aquatic habitats are sufficiently acidic to have caused the decline of amphibian populations (Bradford, Swanson, and Gordon 1992; Corn and Vertucci 1992; Wissinger and Whiteman, 1992; Vertucci and Corn 1996; Muths, Campbell, and Corn 2003). Shenandoah National Park, Virginia, has been heavily impacted by air pollution and increasing stream acidity resulting from forest defoliation by gypsy moth caterpillars. Nevertheless, variation in stream pH and ANC appears to have little effect on the distribution and abundance of plethodontid salamanders (Campbell Grant, Jung, and Rice 2005). All of these studies indicate that acid precipitation is not a significant cause of amphibian population declines, although it could be a threat to some amphibian populations on a local scale (Rowe and Freda 2000).

## Global Changes and Amphibian Population Declines

Most of the potential causes of amphibian declines discussed so far have both local origins and local effects, although in some cases, such as the spread of pesticide residues, effects can be more widespread. Other types of environmental changes take place on a global scale and therefore are much more difficult to combat. Two types of global change that often are conflated in public discourse on the environment are thinning of the ozone layer that protects the earth from ultraviolet radiation, and global warming. The former is caused by the release of certain ozone-destroying industrial chemicals into the atmosphere. The problem is most acute at high latitudes, because so-called "ozone holes" that develop mostly in the summer are centered over the poles. Nevertheless, measurable changes in ultraviolet radiation, and in particular, harmful UV-B radiation, have been detected in equatorial regions as well (Middleton et al. 2001). Global warming is associated with the buildup of greenhouse gases, such as methane and carbon dioxide, in the atmosphere, primarily from the burning of fossil fuels. These gases increase the amount of solar radiation that is retained at the earth's surface, thereby increasing global temperatures. Either of these environmental changes has the potential to affect amphibian populations, although the current impact of both factors remains poorly understood.

### Ultraviolet Radiation

Ultraviolet radiation, and UV-B radiation (wavelengths of 280–320 nm) in particular, is a natural part of the environment of most terrestrial plants and animals and aquatic organisms that live in shallow water exposed to intense sunlight. UV-B radiation in high doses damages DNA and can have a variety of effects on amphibians at all life-history stages, including killing embryos, disrupting embryonic growth and development, producing morphological abnormalities in larvae, making larvae more susceptible to chemical pollutants or disease, and causing cell and retinal damage and altering behavior in terrestrial juveniles and adults (Blaustein et al. 1994, 1997, 1998, 2000; 2003; Kiesecker and Blaustein 1995; L. Long, Saylor, and Soulé 1995; Licht and Grant 1997; Fite et al. 1998; Cummins, Gleenslade, and McLeod 1999; Ankley et al. 2000; Lesser et al. 2001; Formicki, Zamachowski, and Stawarz 2003; Hatch and Blaustein 2003; Pahkala, Laurila, Björn, and Merilä 2003; Pahkala, Merilä, Ots, and Laurila (2003); Blaustein, Romansic, and Scheessele 2005).

Because of these negative effects of UV-B radiation, amphibians have experienced strong selection for various mechanisms to protect themselves from its harmful effects. Larval and adult amphibians generally are well protected against UV-B radiation by melanin in the skin, and amphibians that are most likely to be exposed to high levels of UV-B radiation are the best protected. For example, the black coloration of many tadpoles that habitually live in very shallow, sunny water, often has been interpreted either as a thermoregulatory adaptation (chapter 3) or aposematic coloration (chapter 14), but is just as likely to be an adaptation to avoid harmful UV-B radiation. Similarly, the prevalence of very dark or black coloration in adults of various amphibians at high altitudes generally has been considered an adaptation to enhance thermoregulatory basking (chapter 3), but is more likely an adaptation to avoid the high levels of UV-B radiation present at high altitudes.

Amphibian eggs that are laid in exposed, sunny locations typically are heavily pigmented with melanin, which provides a shield against UV radiation. Consequently, nearly all summer-breeding amphibians that lay eggs in shallow water have black embryos. In contrast, embryos laid in sheltered locations under rocks in streams, in hidden locations on land, or in heavy shade, often have little or no pigment (Licht and Grant 1997; Licht 2003b). Grant and Licht (1995, 1997) and Licht (2003b) reported that the jelly surrounding many aquatic amphibian egg masses also absorbs UV-B radiation and helps to protect embryos, but other studies have reported that egg jelly has no effect on the dosage of UV radiation received by embryos (Hansen, Fabacher, and Calfee 2002; Räsänen et al. 2003). Certainly the protective effect is likely to be much less than that of melanin. Amphibians also can protect their embryos from exposure to UV-B radiation by selecting oviposition sites in deep water, where UV-B radiation is attenuated (Adams, Schindler, and Bury 2001; Kiesecker, Blaustein, and Belden 2001; Licht 2003b), or by wrapping eggs in vegetation to block UV-B radiation (Marco et al. 2001).

Amphibian embryos also have biochemical defenses against UV-B radiation damage. The embryos contain a light-induced enzyme, photolyase, which repairs DNA damaged by UV radiation. Studies of a variety of amphibian species have shown that photolyase activity tends to be highest in species habitually exposed to high levels of sunlight, and consequently, high doses of UV-B radiation. Species that lay eggs in deep water or other sheltered locations tend to have lower photolyase activity, and hence lower capacities for DNA repair (Blaustein et al. 1994, 1996, 1998, 1999; Licht and Grant 1997; van de Mortel et al. 1998; Smith et al. 2002; Licht 2003b; Palen et al. 2005). Not only is there considerable variation among species in photolyase activity, but there also can be adaptive variation in UV-B sensitivity within species as a function of latitude, altitude, or natural exposure to UV-B radiation (Belden and Blaustein 2002a).

Amphibian embryos also may have the capacity to increase photolyase activity in response to UV-B exposure. Smith, Kapron, and Berrill (2000) reported that wood frog (*Rana sylvatica*) eggs exposed to full sunlight had higher photolyase activity than did those kept in the dark. Their experiment did not separate the effects of exposure to UV-B radiation from exposure to longer wavelengths of light, because the embryos were exposed to unfiltered sunlight. Exposure to longer wavelengths, including visible light, is necessary to induce photolyase activity and DNA repair (Palen et al. 2005). In natural environments, intensity of visible light and UV-B radiation are likely to be positively correlated, but it is not yet clear whether amphibian embryos have the capacity to increase DNA repair capabilities if ambient UV-B radiation levels continue to increase in response to ozone depletion, even as visible light levels remain the same. The effects of embryos being exposed to UV-B radiation may carry over to the larval stage. In *Ambystoma maculatum,* larvae that had been exposed to enhanced levels of UV-B radiation as embryos had higher survival when exposed to high UV-B levels than did larvae not previously exposed to high doses of radiation (Calfee, Bridges, and Little 2006).

A study by Blaustein et al. (1994) first raised alarm about the possible effects of increasing exposure to UV-B radiation on amphibian populations. In experiments conducted in Oregon, they found that embryos of *Bufo boreas* and *Rana cascadae* placed in shallow (5–10 cm) water in full sunlight or under filters that transmitted UV-B radiation suffered high mortality, whereas those placed under filters that blocked UV-B radiation had low mortality. Embryos of *Pseudacris regilla,* on the other hand, were not affected by the experimental treatment and had high survivorship with both high and low doses of UV-B radiation. The authors suggested that increasing levels of UV-B radiation could be contributing to the decline of some amphibian species. Subsequent studies reported that exposure to ambient levels of UV-B radiation increased the susceptibility of *Bufo boreas* embryos to a pathogenic water mold, *Saprolegnia* (Kiesecker and Blaustein 1995), but the effect was strongly mediated by water depth. Eggs placed 10 cm under water suffered high mortality, whereas those placed at 50 or 100 cm, where UV-B radiation would be almost completely attenuated, did not (fig. 16.22 A, B). In unmanipulated ponds, mortality of eggs from *Saprolegnia* was much higher in shallow water than in deeper water (fig. 16.22 C). The authors suggested that unsually dry conditions in the Pacific Northwest, and a lack of spring snowmelt, may have resulted in toads laying eggs in unusually shallow water, perhaps accounting for the very high mortality observed in natural populations in the area (Kiesecker, Blaustein, and Belden 2001).

The work of Blaustein and his colleagues on embryonic mortality has generated unusually heated debate among amphibian biologists, with numerous critical articles and reviews and exchanges of comments in the scientific literature (see K. Grant and Licht 1995; Licht 1996, 2003a, b; Licht and Grant 1997 *versus* Blaustein et al. 1996, 1998; Blaustein and Kats 2003). Subsequent studies, all involving experiments with filtered and unfiltered UV-B radiation, have produced mixed results. Some studies have shown higher mortality in embryos of some species exposed to ambient UV-B radiation than in those under filtered sunlight (Anzalone, Kats, and Gordon 1998; Lizana and Pedraza 1998; Broomhall, Osborne, and Cunningham 2000; Häkkinen, Pasanen, and Kukonen 2001; Tietge et al. 2001). Several of these studies tested more than one species under the same conditions and often found differences among species in susceptibility to UV-B radiation. A number of other studies have shown no effect of ambient UV-B radiation on embryonic mortality (K. Grant and Licht 1995; Blaustein et al. 1996, 1999; Corn 1998; Cummins, Gleenslade, and McLeod 1999; Langhelle, Lindell, and Nyström 1999; Merilä, Laurila, and Pahkala 2000; Smith et al. 2000; Starnes, Kennedy, and Petranka 2000; Pahkala, Laurila, and Merilä 2001; Pahkala et al. 2001; Bruner et al. 2002; Pahkala et al. 2003).

In some cases, studies of the same species in different localities have produced different results. Blaustein et al. (1994) reported a strong effect of UV-B radiation on embryonic mortality of *Bufo boreas* in Oregon, but Corn (1998) found no effect in Colorado. Similarly, Häkkinen, Pasanen, and Kukonen (2001) reported an effect of UV-B radiation on embryonic mortality in *Rana arvalis* in Finland, whereas Pahkala, Laurila, Björn, and Merilä (2001) failed to find any such effect in Swedish populations. Lizana and Pedraza (1998) reported a negative effect of UV-B radiation on embryonic survival in *Bufo bufo* at high altitude in Spain, but

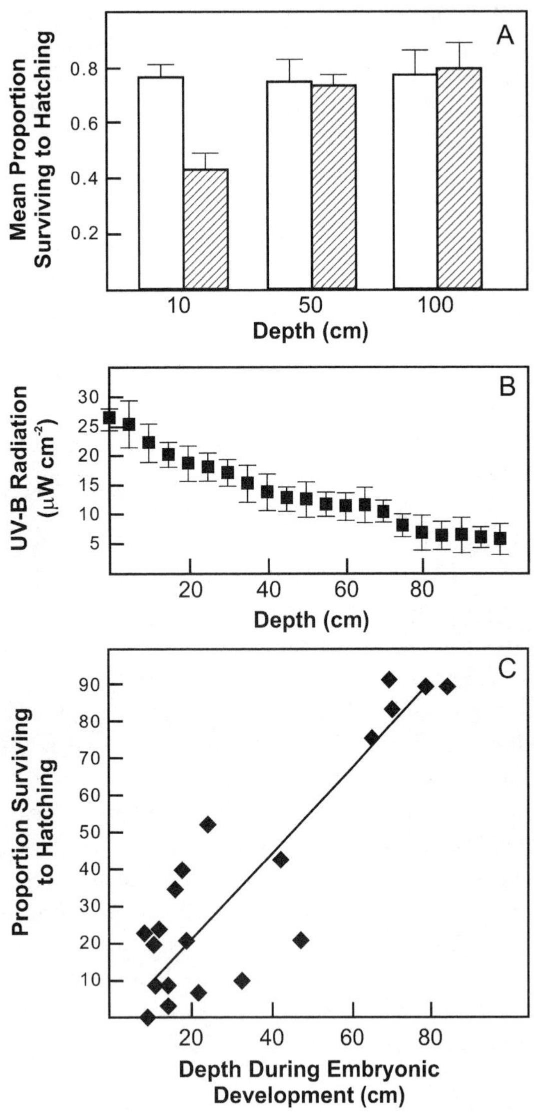


**Fig. 16.22.** Interaction between water depth, intensity of UV-B radiation, and survival of *Bufo boreas* embryos in a clear lake in Oregon. (A) Proportion of embryos surviving without UV-blocking filters (open bars) and with UV-blocking filters (hatched bars). The filters improved survivorship only in shallow water. (B) Attenuation of UV-B radiation with depth in the lake. Only shallow areas are exposed to high doses of radiation. (C) Relationship of water depth to proportion of embryos surviving in egg mass infected with *Saprolegnia ferax,* a water mold. Survivorship of embryos was poor in shallow water, possibly because exposure to increased UV-B radiation increases susceptibility to the water mold. After Kiesecker et al. (2001).

no effect on *B. calamita*. In contrast, Langhelle , Lindell, and Nyström (1999) found no effect of UV-B radiation on embryonic mortality in either *B. bufo* or *B. calamita* in Sweden, and Häkkinen, Pasanen, and Kukonen (2001) found no effect on *B. bufo* in Finland. A number of factors may account for these inconsistent results, including slight differences in experimental design, interspecific and intraspecific differences in photolyase activity and sensitivity to UV-B radiation, variation in water chemistry at the experimental sites, and differences in synergistic interactions with other factors such as pH or the presense of disease organisms.

In natural bodies of water, two physical factors that affect tranmission of UV-B radiation are water depth and the amount of dissolved organic material present in the water. Ultraviolet wavelengths attenuate rapidly with depth. In most bodies of water used for breeding by amphibians, eggs placed near the surface are likely to receive relatively high doses of UV-B radiation, whereas those placed more than 50 cm under the water surface are likely to experience low doses (Licht 2003b). This is especially true when the amount of dissolved organic material in the water is high, which often is the case in the late spring and summer, when levels of ambient UV-B radiation are highest (M. Adams, Schindler, and Bury 2001; Corn and Muths 2002; Diamond et al. 2002; Palen et al. 2002; G. Peterson et al. 2002; Little et al. 2003).

In swamps, marshes, and eutrophic ponds, where many summer-breeding amphibians lay their eggs, levels of dissolved organic material can be high enough to filter out most UV-B radiation within the top few centimeters of the water column (Berrill and Lean 1998; D. Crump, Lean, Berrill, Coulson, and Toy 1999). Amphibians most at risk from UV radiation are likely to be those that breed at high altitudes in very clear lakes, where UV-B penetration is highest. For example, Palen et al. (2002) surveyed 136 ponds and lakes at high altitudes in the Cascades and Olympic Mountains and found that 85% were well protected from UV-B radiation by dissolved organic material, despite being in a region expected to have relatively high UV-B exposure. Lost Lake, Oregon, where the original experiments with *Bufo boreas* were conducted (Blaustein et al. 1994), is one of the clearest lakes, with the highest transmission of UV-B radiation, in the Pacific Northwest (Palen et al. 2004). Although the interpretation of these results has been disputed (Blaustein et al. 2004), there seems to be little doubt that amphibian eggs laid at moderate depths generally are well protected from UV-B radiation. Indeed, there is some evidence to suggest that amphibians with poor inherent protection against UV-B radiation actively choose oviposition sites that minimize exposure to UV-B radiation (M. Adams, Schindler, and Bury 2001; Palen et al. 2005). Another survey of 172 ponds in Glacier National Park failed to support the hypothesis that the distribution of *Bufo boreas* is limited by ambient UV-B radiation (Hossack, Diamond, and Corn 2006).

Breeding phenology clearly has a major impact on the amount of UV-B exposure experienced by amphibian embryos. Temperate-zone amphibians tend to breed later in the year at high altitudes, or high latitudes because of low environmental temperatures (see chapter 3). For example, widely distributed species, such as *Rana temporaria* in Europe, breed in early spring in southern parts of their ranges at low altitudes, but in early summer at higher altitudes or latitudes. Consequently, late-breeding populations will receive higher doses of UV-B radiation and can be expected to have greater inherent protection against UV-B than do early-breeding populations (Merilä, Laurila, and Pahkala 2000). In addition, gradual warming of the climate has resulted in earlier breeding by *Rana temporaria* and some other amphibians in northern latitudes (see chapter 3), and this actually could reduce the amount of UV-B radiation to which embryos are exposed (Merilä, Laurila, and Johanson 2000; Cummins 2003). Breeding phenology also can be affected by other factors, such as rainfall and snowmelt in montane habitats. For example, Corn and Muths (2002) showed that in Colorado, chorus frogs (*Pseudacris maculata*) breeding in dry years with little snowmelt could be forced to breed in relatively shallow water, but they also tend to breed earlier in the year, when levels of UV-B radiation are lower than in late spring. Again, these conclusions have generated some controversy (Blaustein et al. 2004; Corn and Muths 2004), but it is clear that a number of factors interact in complex ways to determine actual exposure of amphibian embryos to UV-B radiation.

Although most research has focused on the effects of UV-B radiation on amphibian embryos, it also is possible for UV radiation to affect both aquatic larvae and terrestrial adults. Unlike embryos, larvae and adults are able to move throughout their habitats and therefore have the potential to select microhabitats that are shielded from UV radiation. Nevertheless, in some situations, such as open shallow pools, it may not be possible to escape such radiation. For amphibian larvae that develop in desert rain pools and other exposed ephemeral sites, exposure to intense sunlight is unavoidable, and indeed, necessary to speed development so that larvae can metamorphose before their ponds disappear (see chapter 13). One can expect larvae in such habitats to be well protected against the harmful effects of UV-B radiation.

The results of experiments with amphibian larvae and UV-B radiation have been somewhat mixed. In some species, exposure to ambient levels of UV-B radiation was lethal to larvae that were prevented from seeking shelter or moving to deeper water (K. Grant and Licht 1995; Häkkinen, Pasanen, and Kukonen 2001; Tietge et al. 2001). Belden et al. (2003) reported a small decrease in survival of *Rana cascadae* tadpoles exposed to full sunlight in shallow water (20–26 cm) for six weeks, even when aquatic vegetation was provided for cover. In other cases, there was no effect of ambient UV-B exposure on larval survivorship, even in clear, shallow water (Crump, Berrill, Coulson, Lean, McGillvray, and Smith 1999; Langhelle, Lindell, and Nyström 1999; G. Smith, Waters, and Rettig 2000; Little et al. 2003). Some of these different results probably reflect inherent differences among species in sensitivity to UV radiation. Belden, Wildy, and Blaustein (2000) reported that larvae of *Ambystoma macrodactylum* from high elevations, where UV-B exposure would be highest, were less sensitive to UV radiation than were larvae from low elevation.

In one experiment with *Rana temporaria* tadpoles, mortality was actually slightly higher in tadpoles shielded from UV-B radiation than in those exposed to normal sunlight, but mortality was even higher in tadpoles exposed to enhanced levels of UV-B (Pahkala, Merilä, Ots, and Laurila 2003). Other studies have demonstrated lethal or sublethal effects of enhanced levels of UV-B on amphibian larvae, indicating a potential for future effects on amphibian populations if levels of UV-B were to increase (K. Grant and Licht 1995; D. Crump et al. 1999; Ankley et al. 2000; Calfee, Bridges, and Little 2006). Nevertheless, larvae of some amphibians appear to be well protected, even against abnormally high levels of UV-B radiation (Little et al. 2003). In some species, larvae can enhance protection by avoiding sunny locations or increasing melanin production in the skin (Nagl and Hofer 1997; van de Mortel et al. 1998; Langhelle, Lindell, and Nyström 1999; Belden, Wildy, and Blaustein 2000; Hofer and Mokri 2000; Belden and Blaustein 2002b; Garcia, Stacy, and Sih 2004). Nevertheless, exposure to ambient UV-B radiation can have a variety of sublethal effects on larvae, with effects on metabolism, growth, development, and behavior that may have fitness consquences for metamorphosed individuals (Belden, Wildy, and Blaustein 2000; G. Smith, Waters, and Rettig 2000; Belden and Blaustein 2002b; Formicki, Zamachowski, and Stawarz 2003; Pahkala, Laurila, and Merilä 2001; Pakhala, Merilä, Ots, and Laurila 2003). These effects are likely to increase if ambient levels of UV-B radiation increase signficantly in the future (Calfee, Bridges, and Little 2006).

In summary, what can we say about the role of ultraviolet radiation in the decline of amphibian populations? (1) It is clear that small increases in UV-B radiation are occurring in various parts of the world, with the greatest impact likely to be at high latitudes in summer. (2) UV-B has a variety of harmful effects on embryos and larvae of amphibians, and some species are sensitive to UV-B in shallow water at current ambient intensities. (3) There is considerable intra- and interspecific variation in how well amphibian embryos and larvae are protected from UV-B, with populations occurring in high-dosage areas having the highest level of protection. (4) Exposure to UV-B in nature is strongly affected by habitat characteristics, with rapid attenuation of UV-B with increasing water depth and levels of dissolved organic material. Many amphibian breeding sites are likely to be well shielded from harmful radiation, although clear ponds and lakes at high altitude may be less well protected. (5) Ambient levels of UV-B, perhaps acting in concert with other factors such as disease, have been implicated in high levels of mortality observed in unmanipulated populations of some amphibian species. (6) There is no evidence that current levels of UV-B radiation have been responsible for widespread declines of amphibian populations in either tropical or temperate regions.

### Global Warming

Despite the claims of some politicians that evidence of human-induced global warming is insufficient, there is little doubt among scientists that human activities are having a significant and accelerating impact on global climate. Indeed, global warming and its effects on the world's ecosystems may be the most important environmental challenge of the twenty-first century, and at the same time, it is one of the most difficult problems to solve, because solutions require widescale international cooperation on a host of issues. The chief cause of global warming is a buildup of greenhouse gases, including carbon dioxide and methane, in the atmosphere, which increases the amount of solar radiation retained at the earth's surface. Although burning of fossil fuels contributes heavily to this accumulation of greenhouse gases, methane emissions from livestock and some types of agriculture, such as rice cultivation (Yan et al. 2005), also make significant contributions, as does the increased use of nitrogen-based fertilizers. Although industrial emissions of greenhouse gases have occurred mostly in the last 150 years, human activities such as burning of forests and grasslands and raising livestock have contributed to greenhouse gas emissions for the last 2,000 years (Ferretti et al. 2005).

The effects of global warming are difficult to predict precisely because of the complexity of interacting factors that affect climate on regional and global scales, but the impact is expected to be widespread and ecologically significant. Average global temperatures are expected to increase by 1.5–5.8°C by the end of the century, but actual temperature changes in different parts of the world will vary (Poff, Brinson, and Day 2002). Some regions are expected to become much warmer, whereas others are expected to become colder. Accompanying these changes will be major alterations in patterns of wind circulation, rainfall, snowfall and snowmelt, and ocean circulation, all of which can affect marine, freshwater, and terrestrial ecosystems.

Global warming is expected to have a major impact on agriculture, forestry, and fisheries, all of which are critical for supporting human life. Some regions, particular northern parts of the temperate zone, actually may become more productive, whereas areas closer to the equator, especially arid regions, may become drier and less productive (MacIver and Wheaton 2005; Maracchi, Sirotenko, and Bindi 2005; Motha and Baier 2005; Salinger, Sivakumar, and Motha 2005; Sivakumar, Das, and Brunini 2005; Zhao et al. 2005). In addition, climate in many parts of the world is expected to become more variable, with increasing frequency of large-scale perturbations such as El Niño events in the Pacific, which impact weather over wide areas. This increased climate variability, coupled with increasing ocean surface temperatures, may lead to a variety of catastrophic events, such as increased drought and fire in some regions, and increased frequency and intensity of hurricanes and typhoons (Salinger 2005; Zhao et al. 2005). These climate changes are expected to alter carbon and nutrient cyles and water cycles, with significant impacts on both agriculture and unmanaged natural ecosystems (Izaurralde et al. 2005; Kim 2005; Thomson et al. 2005). To make matters worse, the negative effects of climate change are likely to be amplified by patterns of human land use, including changing agricultural practices, deforestation and other types of habitat degradation, and urbanization (Siegert et al. 2001; Travis 2003; Zhou et al. 2004; Foley et al. 2005; Tao et al. 2005).

The effects of global climate change on natural ecosystems are expected to be significant and widespread, albeit difficult to predict precisely. Indeed, many such changes already are detectable in response to changes in temperature that have occurred in the last 50–100 years. For example, there is mounting evidence of a poleward extension of the geographic ranges of many warm-adapted marine and terrestrial organisms in the northern hemisphere, with a corresponding reduction in the ranges of many cold-adapted species (Walther et al. 2002; Parmesan and Yohe 2003; Austin and Rehfisch 2005; Hickling et al. 2005; Perry et al. 2005; Hickling et al. 2006). In addition, there is evidence of shifts to earlier breeding in many kinds of organisms, including plants, invertebrates, amphibians, birds, and mammals (Walther et al. 2002; Root et al. 2003). Other potential effects of climate change include either increased or decreased susceptibility of organisms to disease (Harvell et al. 2002), alteration of community structure and food web dynamics (Quinlan, Douglas, and Smol 2005), changes in productivity of high-latitude lakes and ponds (Karlsson, Jonsson, and Jonsson 2005), both positive and negative changes in food supply and growth rates for many kinds of organisms (Yom-Tov and Yom-Tov 2005), and effects on individual reproductive success (Gaston, Gilchrist, and Hipfner 2005). Although some types of organisms will benefit from a warmer climate, the overall impact of global warming is expected to be a rapid increase in the rate of extinction across a wide range of organisms (C. Thomas et al. 2004). Habitat specialists, which often have limited dispersal abilities, are likely to be faced with shrinking habitats and are likely to be the first to disappear.

What is the likely impact of global climate change on amphibians? As with other types of organisms, the effects are hard to predict because different parts of the world are likely to be affected in different ways. Consequently, most of the literature to date is relatively speculative and presents possible scenarios rather than making precise predictions (Donnelly and Crump 1998; Carey et al. 2003). Considerable attention has been given to reports of earlier breeding in various spring-breeding amphibians in the northern hemisphere, because these data provide clear evidence of a biological effect of increasing temperature on amphibian biology. Nevertheless, while some species show a clear trend toward earlier breeding in recent decades, others do not (Terhivuo 1988; Beebee 1995a, b, 2002; Forchhammer, Post, and N. C. Stenseth 1998; Blaustein et al. 2001; Gibbs and Breisch 2001; Blaustein et al. 2002; Tryjanowski, Rybacki, and Sparks 2003). The magnitude of change in breeding date is relatively small, often within the year-to-year range of variation for a given region. Amphibians are notoriously plastic in their breeding phenology, often shifting the date of first activity by several weeks in response to yearly changes in temperature or the timing of snowmelt (see chapter 3). It also is not clear whether earlier breeding is likely to have a negative or a positive effect on these northern amphibians (Merilä, Pahkala, and Johanson 2000; Cummins 2003).

There is relatively little evidence that climate change is currently a major cause of amphibian declines or extinctions in most parts of the world. For example, Alexander and Eischeid (2001) analyzed data on variation in temperature and other weather variables in relation to amphibian declines reported from Colorado, Puerto Rico, Central America, and Queensland, Australia. In some cases, such as Puerto Rico, Central America, and Queensland, amphibian declines corresponded to periods of slightly higher than normal temperatures, but these temperatures were within the normal range of variation for these regions and therefore probably were not responsible for the population declines. In Colorado, temperatures were near normal when amphibian population declines were reported. Similarly, an analysis of climatic data for Puerto Rico alone did not reveal any obvious link to amphibian declines (Stallard 2001). Climate models using these data also suggested that unusual climatic events were not responsible for amphibian declines (Carey et al. 2001). Nevertheless, there is a potential for climate change to impact amphibian populations, especially in conjunction with other factors (Pounds 2001). For example, Kiesecker, Blaustein, and Belden (2001) reported that several years of warm, dry weather in the Pacific Northwest of the United States resulted in western toads (*Bufo boreas*) laying eggs in unusually shallow water. This resulted in increased exposure to UV-B radiation, which in turn increased the susceptibility of the eggs to attack by a water mold (*Saprolegnia ferax*).

Perhaps the habitats in which amphibians are most vulnerable to future climate changes are cool montane forests, both in temperate and tropical regions. Amphibians adapted to cool climates, such as plethodontid salamanders, exhibit marked depression of metabolic rates at temperatures above about 15° C, so populations at lower elevations in mountains may be living near their physiological limits. Any increase in average environmental temperatures may cause a vertical contraction of ranges by driving populations up the mountainsides (Bernardo and Spotila 2006). These effects may be even more pronounced in the tropics, where gradients between high- and low-elevation habitats are steeper than in the temperate zone, and where amphibians generally lack the capacity for long-term temperature acclimation (Navas 2002). As discussed earlier in this chapter, tropical montane forests often are hotspots of amphibian diversity, with large numbers of endemic species, most of which are forest-interior specialists. Tropical montane rainforests also are among the most endangered of all terrestrial amphibian habitats because of accelerating rates of deforestation. Even when high-elevation forests remain intact, amphibians adapted to cool conditions are vulnerable to climate changes that result in shrinking habitat on mountaintops. As warming temperatures push the amphibians higher up into the mountains, they eventually will run out of habitat, and they lack the ability to disperse between mountains. The result could be wholesale extinction of many tropical montane amphibians (S. Williams, Bolitho, and Fox 2003; Parra-Olea, Martínez-Meyer, and Pérez-Ponce de Leon 2005). In addition, as already discussed, there may be synergistic interactions between climate change and the effects of diseases such as chytridiomycosis, which is prevalent in tropical mountains (Pounds et al. 2006).

The threat to montane amphibians comes not only from changes in temperature, but also from changes in the altitude at which clouds form, because mist from clouds often provides the most important source of moisture for these species (P. Foster 2001). The negative effects of rising temperatures may be increased by habitat destruction at lower elevations. For example, there is good evidence that conversion of lowland forest on the Pacific slopes of Costa Rica has resulted in cloud layers moving higher up the mountains, and this may have been responsible, at least in part, for the disappearance of the golden toad (*Bufo periglenes*) and several other high-altitude amphibians in the Monteverde Cloud Forest Preserve in the late 1980s (Pounds and Crump 1994; Pounds, Fogden, and Campbell 1999; Pounds and Puschendorf 2004). Similar effects may be occurring in deforested parts of the Caribbean slope of Costa Rica as well (Lawton et al. 2001) and very likely are occurring in cloud forests and montane rainforests in other parts of the world.

The effect of climate change on amphibians in other habitats is hard to predict. Some climate models suggest that with rising temperatures, rainfall will increase in some wet tropical regions, including Southeast Asia, possibly favoring the expansion of lowland tropical rainforest (Zhao et al. 2005). Nevertheless, the rate at which lowland rainforest is being destroyed by human activity is likely to exceed the rate of forest expansion. This rapid deforestation not only removes critical habitat for many amphibians, but also is likely to increase the incidence of climate-related events, such as catastrophic fires, which until recently were virtually unknown in wet tropical forests (Siegert et al. 2001). In addition, forest fragmentation is likely to result in warmer, drier conditions in the remaining forest fragments, conditions that are likely to be made worse by global warming (Donnelly and Crump 1998). Other wet tropical regions are predicted to experience reduced rainfall (Zhao et al. 2005), a trend that is likely to accelerate with increasing deforestation.

If a warming climate is likely to have the greatest impact on cool-adapted amphibians in the tropics, the reverse is likely to be true in the temperate zone. In these regions, increasing temperature may cause cool-adapted species to expand their ranges toward the poles. In the northern hemisphere in particular, there is a large area for potential expansion of amphibians now found at lower latitudes, particularly those well adapted to forest habitats. In contrast, grassland, semiarid, and desert regions of the northern hemisphere may experience drier conditions than occur today (Kim 2005; Maracchi, Sirotenko, and Bindi 2005; Motha and Baier 2005; A. Thomson et al. 2005). Although many amphibians in such habitats are well adapted to dry climate regimes, most species depend on temporary ponds for reproduction and therefore are highly sensitive to climate changes that result in shorter hydroperiods in breeding ponds and to general water stress caused by reduced soil moisture. In temperate montane regions, such as western North America, increasing temperatures may reduce precipitation or shift precipitation from mostly snow to mostly rain (Kim 2005). The result is likely to be faster runoff of precipitation and alteration of water cycles that could impact amphibians that depend on melting snow to form breeding ponds (Corn and Muths 2002).

### What Difference Does It Make if Amphibians Decline or Disappear?

Since the widespread decline of amphibian populations was first noticed in the 1980s, most research efforts have focused on documenting the extent of these declines and searching for their causes. Very little attention has been given to the impact of amphibian population declines on ecosystems and on other organisms that live with amphibians. For most amphibian biologists and environmentalists, it seems self-evident that the decline of amphibian population is a bad thing. Each amphibian species is a unique product of evolution, and for aesthetic reasons alone, we want to preserve as many of them as possible. Even for the general public, there is a general sense that biodiversity is inherently valuable, if not essential, for human life, and that preservation of biodiversity is a good idea.

Nevertheless, all conservation efforts come with economic, political, and social costs. In wealthy countries, we often are faced with decisions about whether to preserve a swamp or a set of vernal pools or to build yet another supermarket or shopping mall. For most environmentalists, decisions about such tradeoffs are easy: protect the environment and give up the shopping mall. Often public opinion can be rallied to support this point of view. In other parts of the world, decisions usually are more difficult. How do we compare the value of a patch of montane rainforest and all of its inhabitats, including amphibians, with the value of land to grow crops for an impoverished human population? If the Kihansi toad (*Nectophrynoides asperginus*) in Tanzania is endangered by a giant hydroelectric dam (Krajik 2006), how do we compare the value of the toad and other organisms that live in Kihansi Gorge with the value of the dam, which provides one third of Tanzania's electricity to the citizens of one of the world's poorest countries? If the toad is to be saved, how much money should be spent on captive breeding programs or environmental restoration efforts?

These questions do not have easy or obvious answers, and policymakers and the public at large often demand more specific answers to questions about the value of amphibians or any other kind of organism threatened with extinction. Often people want to know not only what amphibians do for the ecosystems in which they live, but also what they do for us. If the Boreal toad (*Bufo boreas*) disappears from Rocky Mountain National Park, will the functioning of the surrounding ecosystem be affected? Will the average visitor to the park even notice the absence of this toad? If most of the frogs along mountain streams in the rainforests of Costa Rica and Panama are wiped out by disease, what impact will this have on the stream ecosystems and the surrounding forest? If frogs disappear from rice paddies in Southeast Asia, will the number of mosquitos that carry human diseases increase?

Unfortunately, despite the vast amount of research that has been done on the ecology of amphibians, biologists often are embarrassingly short of answers to such questions. Efforts to protect amphibians or conserve key ecosystems tend to be justified with fairly generic arguments about the value of particular organisms or the ecosystem services they provide: lost opportunities to cure diseases with chemicals extracted from the skin of frogs, lost services of amphibians as biological control agents for noxious insects, or the importance of amphibians as indicators of overall ecosystem health (the familiar "canary in the coal mine" argument). My feeling is that the canary in the coal mine analogy is overused. After all, humans have demonstrated a remarkable ability to adapt to even the most degraded habitats, from the garbage dumps of India and Colombia to the barren wastelands of the southern Sudan and the eroded hillsides of Haiti. We don't really need a survey of amphibians to tell us these environments are in bad shape; by the time they are uninhabitable by humans, the amphibians probably are long gone.

All too often, evidence for the value of amphibians is anecdotal rather than quantitative. For example, amphibians are commonly cited as important agents in the control of insect pests, and increases in populations of mosquitos and crop-eating insects in various parts of the world have been attributed to declines in frog populations (Abdulali 1985; Barfield 1986; Oza 1990; Bazilescu 1996; Jacques 1999). Yet an official web site of the State of Michigan dealing with emerging infectious diseases and mosquito control flatly states that "There are no amphibians or reptiles that are effective at mosquito control" (http://www.michigan.gov/emergingdiseases). In neither case is there much quantitative information to support these arguments.

Because amphibians tend to be inconspicuous parts of most ecosystems, even when they are very abundant, their role in ecosystem functioning has not been intensively studied. Amphibians generally do not perform ecosystem services that are likely to be clearly important to humans: they do not pollinate crop plants, dispose of decaying carcasses of animals, aerate the soil, or provide oxygen to the atmosphere. If all of the amphibians were to disappear from the Greater Yellowstone Ecosystem, it is unlikely that the effects would be as noticeable as the effects of the disappearance and reintroduction of wolves to the area. Nevertheless, what little research has been done on the role of amphibians in ecosystems suggests that removal of amphibians could have a significant impact on ecosystem structure and function.

### Amphibians as Food

Perhaps the most frequently cited role of amphibians in ecosystems is their role as food for other kinds of animals. The idea that amphibians can be a significant source of energy for animals at higher trophic levels originated with the classic study of Burton and Likens (1975a, b) showing that the biomass of salamanders in the Hubbard Brook Experimental Forest in New Hampshire exceeded that of small birds and mammals. Subsequent work has shown that the biomass of amphibians in other forest ecosystems can be even higher than that reported by Burton and Likens (reviewed in Davic and Welsh 2004). Pough (1980, 1983) pointed out that because of their low metabolic rates, amphibians are far more efficient than are endotherms at converting food into biomass, so they consume a relatively small proportion of the primary productivity of an ecosystem, while providing a large source of energy to higher tropic levels. Studies of the diets of both invertebrate and vertebrate predators have shown that many kinds of animals eat a lot of amphibians (see chapter 14). The key question, however, is what would these animals eat if amphibians were not available? To what extent are animals at higher trophic levels dependent on amphibians for survival, such that the disappearance of amphibian populations would significantly impact populations of these predators?

As discussed in chapter 14, many kinds of invertebrates eat amphibians at all life-history stages, but relatively few are specialist predators of amphibians. Various species of flies with larvae that prey on frog eggs may be an exception, although the extent to which any of these species is dependent upon amphibian eggs for survival is unknown. All sorts of aquatic insects prey on amphibian eggs and larvae, and some species may derive much of their energy from amphibians. For example, caddisfly larvae (Trichoptera) are major predators of ambystomatid salamander eggs, and some species of caddisflies require a diet of protein to complete development (see chapter 14). *Ambystoma* eggs laid in the spring represent a major energy input into vernal pools and other temporary ponds (Regester, Lips, and Whiles 2006), and it is conceivable that populations of caddisflies and other predators would diminish if the number of salamander eggs were reduced. In addition, different species of *Ambystoma* larvae readily prey on each other (see chapter 14), so a decline in one species is likely to impact other species. Nevertheless, we cannot say with certainty that the complete disappearance of a species of *Ambystoma* would have a dramatic effect on any of these predators, because we lack information on the ability of these predators to switch to alternative prey. Most adult aquatic insects are generalist predators that probably can switch from one prey species to another as the relative abundance of different species changes. On the other hand, at least one giant waterbug in Japan is considered sufficiently dependent on tadpoles as prey to be endangered by declining numbers of frogs in rice fields (Hirai and Hidaka 2002).

Most ectothermic vertebrate predators of amphibians also are generalist predators that adjust their prey selection in response to changes in relative abundance of different prey items. For example, there is no evidence that any species of fish or amphibian is solely dependent upon amphibians as a food source (see chapter 14). Even frogs that prey on other frogs generally have relatively broad diets that include a wide range of invertebrate prey (N. Scott and Aquino 2005), although it is conceivable that such predators could suffer declines if smaller species of frogs decline or disappear.

Probably the ectothermic vertebrates most likely to be affected by amphibian declines are snakes, many of which have diets composed largely or exclusively of amphibians (see chapter 14). Indeed, there is good evidence for major declines in several populations of snakes in response to declines in amphibian abundance. For example, garter snakes (*Thamnophis elegans*) virtually disappeared from High Sierra lakes in California where the introduction of nonnative trout had eliminated most frog populations (Matthews, Knapp, and Pope 2002). Neotropical rainforests are particularly rich in snakes that are specialist predators on frogs, tadpoles, or frog eggs. At montane sites in Costa Rica and Panama where dramatic declines of anuran populations have occurred, the abundance of frog-eating snakes has similarly declined (Lips et al. 2006; Whiles et al. 2006). The impact of amphibian population declines is likely to be greater in areas such as these, where whole assemblages of amphibians are declining, than in areas where only the rarest species are threatened.

The impact of amphibian population declines on other vertebrate predators is hard to predict, but most birds and mammals that eat amphibians do not specialize on them. Even though amphibians can comprise a substantial proportion of the diet of some waterbirds, especially food delivered to nestlings (see chapter 14), these predators readily switch to alternative prey if amphibians are scarce. Even those predators that have been characterized as specialist amphibian predators in some parts of their ranges, such as European polecats and badgers, often do not specialize on amphibians in other regions and exhibit considerable plasticity in diet choice both within and between seasons. It seems unlikely that even the complete extinction of amphibians in a particular region would result in the elimination of predatory birds and mammals, although in some cases, they could suffer from reduced food availability and experience decreases in population density.

### Amphibians as Grazers and Predators

Most research on the effects of amphibians on their surrounding ecosystems has focused on their roles as consumers. Most anuran tadpoles are grazers on algae, phytoplankton, and periphyton growing on plants, rocks, or other substrates. Salamander larvae and adult amphibians are significant predators on invertebrates. In some ponds, numbers of larval amphibians can run into the hundreds of thousands or even millions, with a total biomass exceeding that of other animals, even when fish are present (Gehlbach and Kennedy 1978; Deutschmann and Peterka 1988). Even in streams, which generally are less productive than ponds, densities can be high enough to make these animals the dominant consumers in the ecosystem (Huang and Sih 1991; Lips 1999; Davic and Welsch 2004; Whiles et al. 2006). Although these larvae are present in aquatic ecosystems only temporarily, we can expect them to have a major impact on energy flow and nutrient dynamics in bodies of water ranging from high alpine lakes to mountain streams, eutrophic ponds, and vernal pools.

As discussed in chapter 13, dense populations of tadpoles can substantially reduce the standing crop of algae in ponds and streams and alter the composition of the algal community (Dickman 1968; Seale, Rodgers, and Boraas 1975; Seale 1980; Lamberti et al. 1992; Kupferberg 1997a; Graham and Vinebrooke 1998; Holomuzki 1998; Flecker, Feifarek, and Taylor 1999; Ranvestal et al. 2004). Grazing by tadpoles can actually facilitate growth of some algae, thereby increasing primary productivity (Osborne and McLachlan 1985; Kupferberg 1997b; Whiles et al. 2006), so the impact of these animals on ecosystem dynamics can be complex. Comparisons of stream ecosystems in western Panama in areas where anurans already have declined and areas ahead of the spreading front of disease-related declines provide one of the few opportunities to directly assess ecosystem-level impacts of amphibian declines. Although declines of amphibian populations in these areas are relatively recent, there already have been measureable changes in productivity, nitrogen cycling, sediment dynamics, and several other aspects of ecosystem function in streams where tadpoles have disappeared (Ranvestal et al. 2004; Lips et al. 2006; Whiles et al. 2006). These changes, in turn, are likely to impact other grazers in the system, including snails, aquatic insects, and herbivorous fishes, and not just animals that depend on tadpoles as food. In short, the disappearance of a major cohort of consumers from tropical streams has the potential to permanently alter the functioning of these ecosystems.

The role of salamander larvae and adult amphibians such as newts as keystone predators in aquatic ecosystems was discussed in chapter 15 and need not be reviewed in detail here. Not only do these animals affect overall survival of other amphibians and various invertebrates, but they also alter competitive dynamics among species of tadpoles, which in turn can affect the relative abundance of adults of different species. In many of these pond systems, the whole ecosystem may be threatened by habitat destruction or degradation, changes in hydroperiod, or other factors that affect all members of the community together. Nevertheless, one can predict that some future threat to one of these keystone predators alone, such as a newly emergent disease, could have major effects on the whole assemblage of amphibian species, as well as other components of the ecosystem. Similarly, salamander larvae are the dominant vertebrate predators in many forest stream ecosystems in North America. Changes in the terrestrial habitat surrounding these streams, such as logging activities, may differentially impact certain salamander species in ways that eventually alter the aquatic ecosystem as well (Corn, Bury, and Hyde 2003; Davic and Welsch 2004).

Terrestrial frogs and salamanders also can reach extraordinarily high population densities (Davic and Welsch 2004; Woolbright et al. 2006) and can be expected to have a major impact on populations of invertebrate prey animals. Only a handful of studies have documented these effects. Wyman (1998) used small enclosures (3 $m^2$) placed on the forest floor in upstate New York to investigate the effect of red-backed salamanders (*Plethodon cinereus*) on invertebrate densities and leaf litter decomposition rates. The enclosures were stocked with either two or six salamanders, or had no salamanders present. Fewer invertebrates, especially small ones, were collected in enclosures with salamanders than in those without salamanders, although the differences were small. There also was a small decrease in rate of litter decomposition when salamanders were present, presumably because they consumed insects that serve as leaf fragmenters. Although this experiment was conducted at a very small spatial scale, it suggests the potential for a decline in salamander populations to affect both invertebrate populations and the carbon cycle of the leaf-litter community.

A more elaborate set of experiments examined the effects of the Puerto Rican frog *Eleutherodactylus coqui* on leaf-litter invertebrates and forest nutrient dynamics at different spatial scales (Beard, Voget, and Kulmatiski 2002; Beard et al. 2003). Both small enclosures (1 $m^3$) and large (20 × 20 m) removal plots were used to compare the presence and absence of frogs. Densities of aerial insects were lower at both spatial scales when frogs were present, but leaf-litter insects were not affected. The presence of frogs also decreased rates of herbivory on live plants, suggesting that they were consuming significant numbers of leaf-eating insects. The exclusion of frogs also reduced the amounts of several kinds of nutrients in the leaf litter, including dissolved carbon and nitrogen, but the effect was not from changes in invertebrate densities, but apparently was an effect of frog waste products and decomposition. In the smaller enclosures, the presence of frogs increased plant growth rates and decomposition rates, but this effect was not detectable at the more realistic larger spatial scale. Hence, while these experiments demonstrate an effect of frog population on some invertebrates and some nutrient dynamics, one must be cautious in extrapolating from small mesocosm experiments to larger ecosystem-level spatial scales.

In both the experiments with red-backed salamanders and those with coquis, the forest floor community is dominated by a single species of amphibian, neither of which is currently threatened. While the results suggest that the removal of amphibians can have a detectable impact on ecosystem-level processes, they do not address the question of what might happen if most of a large and diverse amphibian community were to disappear, as has been occurring in the montane forests of Costa Rica and western Panama. Even in those areas, some species of frogs have persisted even as most of the stream breeders have succumbed to chytridiomycosis. Clearly the disappearance of the tadpoles of these species has the potential to affect the stream ecosystems in the area, but we do not yet know how the disappearance of the terrestrial juveniles and adults will impact the leaf-litter community, or the extent to which the remaining frogs will take the place of the missing species in terms of ecosystem function. Preliminary analysis of stable isotopes of carbon and nitrogen indicate that the adults of different anuran species in these forests have surprisingly diverse diets, with some feeding in terrestrial upland habitats and others in riparian habitats, so the decline of different species may have different impacts on the surrounding ecosystem (Whiles et al. 2006). Clearly more studies of this type are needed, and in a variety of habitats, from Asian rice fields to tropical forests and North American vernal pools, before we will fully appreciate the role of amphibians in ecosystems.

## The Future of Amphibians

Predicting the future of almost any long-term process, from human population growth and economic trends to the survival of amphibians, is notoriously difficult, because we are limited to observing conditions as they are today. Most such predictions are made with qualifying phrases such as "If current trends continue," but in many cases, we do not know whether current trends will continue as they are, or will improve, or will get worse. This certainly is true for amphibian populations. We have had clear evidence for widespread population declines only in the last two decades, and we are just beginning to understand the principal causes of these declines. In reading through the litany of threats to amphibians described in this chapter, it would be easy to become pessimistic, to assume that amphibians throughout the world face a bleak future. Indeed, one author described his book on amphibians of arid lands as "a requiem for the amphibian class, which is on the verge of extinction" (Warburg 1997). Titles such as *Tracking the Vanishing Frogs* (K. Phillips 1994), *Nature's Fading Chorus* (G. Miller 2000), and *Disappearing Jewels* (Young et al. 2004) have reinforced the view that amphibians throughout the world are under seige. Clearly one function of such titles is to sound a warning, to describe what could happen if current trends continue, not necessarily to precisely predict future events.

I will use three examples to illustrate the difficulty of projecting current trends into the future. In 1813, John James Audubon observed immense flocks of passenger pigeons flying over the woods of Kentucky in numbers so great that "the light of noonday was obscured as by an eclipse." Although the local farmers blasted away at the pigeons with every sort of weapon, Audubon felt confident that no amount of hunting could ever diminish these multitudes of birds. "I have satisfied myself," he wrote, "by long observation that nothing but the gradual diminution of our forests can accomplish their decrease" (quoted in A. B. Adams [1966, 152]). Yet 100 years later, the last passenger pigeon died alone in the Cincinnati Zoo, the rest having succumed to unrelenting commercial slaughter that filled Victorian dinner tables with pigeon pie, but wiped out the species in the process.

About 80 years after Audubon made his prediction about the pigeons, a report from the Canadian Ministry of Agriculture stated that "As to those fishes which, like cod, makerel, herring, etc. are the most important of our sea fishes . . . I say it is impossible, not merely to exhaust them, but even to noticeably lessen their number by the means now used for their capture" (quoted in Kurlansky [1998, 123]). Yet the fish did decline, and Newfoundland's fishermen are now unemployed; both the fish and the fishermen were victims of ever more effective technology that eventually depleted the seemingly limitless stocks. This is an old story of human interactions with animals whose very abundance makes them vulnerable to exploitation—sperm whales, fur seals, elephant seals, sea otters, the North American bison, and even the North American beaver have suffered similar fates, although some of these species have since recovered. Both Audubon and Canadian fisheries officials made the wrong prediction about future trends because they could not conceive of such enormous abundance ever being diminished, nor could they envision future improvements in hunting technology that would make the slaughter of the animals more efficient. Amphibians have largely escaped this sort of indiscriminant slaughter, although some large ranid frogs are threatened by overexploitation, particularly in Asia (Stuart et al. 2004).

In 1962, Rachel Carson made a very different sort of prediction about the future. In her classic book *Silent Spring,* she painted a bleak picture of the possible consequences of environmental degradation, especially from unrestrained use of toxic pesticides:

"There was once a town in the heart of America where all life seemed to live in harmony with its surroundings. . . . The countryside was in fact famous for the abundance and variety of its bird life, and when the flood of migrants was pouring through in spring and fall people traveled from great distances to observe them. . . . Then a strange blight crept over the area and everything began to change. . . . There was a strange stillness. The birds, for example—where had they gone? Many people spoke of them, puzzled and disturbed. The feeding stations in the backyards were deserted. The few birds seen anywhere were moribund; they trembled violently and could not fly. It was a spring without voices. On the mornings that once throbbed with the dawn chorus of robins, catbirds, doves, jays, wrens, and scores of other bird voices there was now no sound; only silence lay over the fields and woods and marsh" (Carson 1962, p. 2).

In fact, there was no silent spring—the birds did not disappear, but not because Rachel Carson was wrong in her prediction, but because her book changed peoples' behavior. Her description, which she called "A Fable for Tomorrow," was not meant to be an accurate prediction of the future, but rather a warning cry, a call to arms to protect the environment before it was too late. Largely because of her book, DDT and other persistent pesticides were banned, and birds such as the Peregrine Falcon, which had been in decline, rebounded. Yet even today, some people insist on drawing the wrong conclusions about her book. One only need type "Rachel Carson Silent Spring" into Google to immediately find blogs for publications like *Capitalism Magazine* arguing that her conclusions, and indeed the entire environmental movement, were based on myths, going on to blame Rachel Carson for the deaths of millions from mosquito-borne diseases such as malaria (the evolution of DDT resistance in mosquitos apparently not entering into the picture; Brockerman 2002).

We are in the warning cry stage in the decline of amphibian populations. Alarm bells have been sounded, both in the technical literature and in the popular press. What we need now is a realistic assessement of what the future holds for amphibians, what conservation measures can be taken, and which species can be saved. Are all amphibians on the brink of extinction, as some writers have suggested? Certainly not. Many amphibians have adapted well to human-altered habitats, and often are more abundant in such habitats than in pristine forests. Others, such as the cane toad (*Bufo marinus*) in Australia have actually spread, with human assistance, to the point of being a threat to other wildlife. Even in areas devasted by chytridiomycosis, some amphibians have survived, apparently enjoying innate immunity to the fungal infection. In Europe, North America, and elsewhere, protection of amphibian habitats, changes in forestry and agricultural practices, restoration of wetlands, creation of new breeding habitats, building of road tunnels for migrating amphibians, and removal of introduced and invasive predators (Vredenburg 2004; Ahola et al. 2006) show great promise in reversing the declines of some species.

Clearly much more needs to be done. We cannot save all species of amphibians, and we are going to lose some pretty spectacular products of evolution, particularly in the tropics. Many frogs with strange and unusual breeding habits, especially those adapted to the interiors of tropical forests, are likely to be lost, as are some species equipped with arsenals of chemical weapons in their skins, some of which might be used to fight human diseases in the future. In many cases, we may lose species of amphibians before we even know they exist. Yet there is some hope.

Despite the frightening nature of emerging diseases such as chytridiomycosis, the most immediate threat to amphibians throughout the world remains the wholesale degradation, fragmentation, and outright destruction of habitats, especially tropical forests (Stuart et al. 2004). Fortunately, the major hotspots of amphibian diversity, and especially areas with a large number of endemic species, represent a relatively small proportion of the earth's land, and these same regions often are rich in endemic species of other kinds of vertebrates (Lamoreux et al. 2006). Some thoughtful allocation of limited resources to protecting such areas has the potential to save many kinds of vertebrates, including amphibians. Saving these hotspots, many of which are in poor tropical countries with expanding human populations, is not just a matter of biology, but also of anthropology, sociology, economics, and politics. Conserving amphibians, or other components of biodiversity, not only means setting aside land, but seeking creative alternatives to forest destruction for people seeking to scrape out a living from the land. It means not only educating people about the value of biodiversity for its own sake, but also for theirs—for the preservation of sources of drinking water, conservation of valuable forest resources, protection against catastrophic floods and landslides in degraded environments, and a host of other reasons. It means encouraging the World Bank and other international agencies to focus on sustainable development, not just rapid development. It also means finding ways to reduce our own use of resources while alleviating the enormous disparity between rich and poor in human societies around the world. It is essential to develop an environmental consciousness and a cohort of local conservation biologists in countries where such concerns now seem mere luxuries for the rich.

Finally, efforts to conserve amphibians and biodiverity in general, from better education to land protection and restoration to the alleviation of poverty—all will be little more than stopgap measures if we fail to recognize the real elephant in the environmental living room, global climate change. The issue here is not relatively small and incremental changes in temperature and rainfall regimes, which already can be detected around the world. Amphibians exhibit a high degree of phenotypic plasticity that enables them to deal with substantial amounts of environmental variation, and there is evidence for a capacity for evolutionary adaptation to changing conditions on very short time scales and small spatial scales (e.g., Skelly 2004). The real danger is that climate change will accelerate, that the melting of glaciers and ice sheets will push climate systems toward a tipping point that will cause major and permanent alteration not only of local climate, but of ecosystems around the world. If individuals and governments continue to ignore problems of global warming, then all of us, amphibians and humans alike, will be in serious trouble.

As Rachel Carson wrote more than 40 years ago, "The road we have been traveling on is deceptively easy, a smooth superhighway on which we progress with great speed, but at its end lies disaster. The other fork of the road—the one 'less traveled by'—offers our last, our only chance to reach a destination that assures the preservation of our earth. The choice, after all, is ours to make" (Carson 1962, p. 277). To assume that individuals and governments will consistently make the wrong choices, as so frequently seems to be the case now, is to give into despair, to conclude that most current efforts to save amphibians and biodiversity in general are doomed to failure. I prefer to assume that humans, the most intelligent species on the planet, eventually will make the right choices, at least part of the time, and that amphibians will continue to be part of our natural world.

It is spring in New England now, as I write the last words of this book. The spring peepers have returned, as they always do, to swamps, marshes, and ponds, with their "world-filling, mind-altering choruses" that so simply and so definitely announce that life is resurgent again. Wood frogs and spotted salamanders have moved to vernal pools, depositing their eggs and starting their offspring on an uncertain journey, an annual race to the land before the ponds dry up. To our south, the rains will soon come again to the lowlands of Panama. A multitude of frogs, large and small, will move to swamps and ponds, feeding, calling, defending territories, laying eggs, avoiding snakes, and otherwise going about their lives, just as they did so many years ago as I rode the train from Balboa Station. If we make the right choices, if we take the correct fork in the road, they will continue to do so for generations of amphibians and humans to come.

# References


Abbott, C. C. 1884. Recent studies of the spadefoot toad. *Am. Nat.* 18:1075–80.

———. 1885. *A naturalist's rambles about home.* New York: Appleton.

Abdulali, H. 1985. On the export of frog legs from India. *J. Bombay Nat. Hist. Soc.* 82:347–75.

Abe, A. S., and J. E. P. W. Bicudo. 1991. Adaptations to salinity and osmoregulation in the frog *Thoropa miliaris* (Amphibia, Leptodactylidae). *Zool. Anz.* 227:313–18.

Able, D. J. 1999. Scramble competition selects for greater tailfin size in male red-spotted newts (Amphibia: Salamandridae). *Behav. Ecol. Sociobiol.* 46:423–28.

Able, K. P. 1980. Mechanisms of orientation, navigation, and homing. In *Animal migration, orientation, and navigation,* ed. S. A. Gauthreaux, Jr., 283–373. New York: Academic Press.

Abourachid, A., and D. M. Green. 1999. Origins of the frog-kick? Alternate-leg swimming in primitive frogs, families Leiopelmatidae and Ascaphidae. *J. Herpetol.* 33:657–63.

Abrams, H. L., Jr. 1987. The preference for animal protein and fat: A cross-cultural survey. In *Food and evolution,* ed. M. Harris and E. R. Ross, 207–23. Philadelphia: Temple University Press.

Abramsky, Z., M. I. Dyer, and P. D. Harrison. 1979. Competition among small mammals in experimentally perturbed areas of shortgrass prairie. *Ecology* 60:530–36.

Abt, G., and H.-U. Reyer. 1993. Mate choice and fitness in a hybrid frog: *Rana esculenta* females prefer *Rana lessonae* males over their own. *Behav. Ecol. Sociobiol.* 32:221–28.

Ackrill, P., R. Hornby, and S. Thomas. 1969. Responses of *Rana temporaria* and *Rana esculenta* to prolonged exposure to a saline environment. *Comp. Biochem. Physiol.* 28:1317–29.

Adams, A. B. 1966. *John James Audubon.* New York: Putnam's.

Adams, A. E., and E. E. Rae. 1929. An experimental study of the fat bodies in *Triturus* (*Diemyctylus*) *viridescens. Anat. Rec.* 41:181–203.

Adams, D. C., and F. J. Rohlf. 2000. Ecological character displacement in *Plethodon:* Biomechanical differences found from a geometric morphometric study. *Proc. Natl. Acad. Sci. USA* 97:4106–11.

Adams, E. M., A. G. Jones, and S. J. Arnold. 2005. Multiple paternity in a natural population of a salamander with long-term sperm storage. *Mol. Ecol.* 14:1803–10.

Adams, L. 1942. The natural history and classification of the Mount Lyell salamander, *Hydromantes platycephalus. Univ. Calif. Publ. Zool.* 46:179–204.

Adams, M. J. 1999. Correlated factors in amphibian decline: Exotic species and habitat change in western Washington. *J. Wildl. Manage.* 63:1162–71.

———. 2000. Pond permanence and the effects of exotic vertebrates on anurans. *Ecol. Appl.* 10:559–68.

Adams, M. J., and S. Claeson. 1998. Field responses of tadpoles to conspecific and heterospecific alarm. *Ethology* 104:955–61.

Adams, M. J., C. A. Pearl, and R. B. Bury. 2003. Indirect facilitation of an anuran invasion by non-native fishes. *Ecol. Lett.* 6:343–51.

Adams, M. J., D. F. Schindler, and R. B. Bury. 2001. Association of amphibians with attenuation of ultraviolet-B radiation in montane ponds. *Oecologia* 128:519–25.

Adams, S. B., and C. A. Frissell. 2001. Thermal habitat use and evidence of seasonal migration by Rocky Mountain tailed frogs, *Ascaphus montanus,* in Montana. *Can. Field-Natur.* 115:251–56.

Adamson, B., R. G. Harrison, and I. Bailey. 1960. The development of the whistling frog *Eleutherodactylus martinicensis* of Barbados. *Proc. Zool. Soc. London* 133:453–69.

Adler, K. 1969. Extraoptic phase shifting of circadian locomotor rhythm in salamanders. *Science* 164:1290–92.

———. 1970. The role of extraoptic photoreception in amphibian rhythms and orientation: A review. *J. Herpetol.* 4:99–112.

———. 1971. Pineal end organ: Role in entrainment of circadian locomotor rhythm in frogs. In *Biochronometry,* 270–87. Washington, DC: U.S. National Academy of Science.

———. 1976. Extraocular photoreception in amphibians. *Photochem. Photobiol.* 23:275–98.

———. 1980. Individuality in the use of orientation cues by green frogs. *Anim. Behav.* 28:413–25.

———. 1989. *Contributions to the history of herpetology.* Oxford, OH: Society for the Study of Amphibians and Reptiles.

Adler, K., and D. H. Taylor. 1973. Extraocular perception of polarized light by orienting salamanders. *J. Comp. Physiol.* 87:203–12.

———. 1981. Toad orientation: variability of response and its relationship to individuality and environmental parameters. *J. Comp. Physiol. A* 144:45–51.

Adolph, E. F. 1925. The passage of water through the skin of the frog, and the relation between diffusion and permeability. *Am. J. Physiol.* 73:85–195.

———. 1931. The size of the body and the size of the environment in the growth of tadpoles. *Biol. Bull.* 61:350–75.

———. 1932. The vapor tension relations of frogs. *Biol. Bull.* 62:112–25.

———. 1933. Exchanges of water in the frog. *Biol. Rev.* 8:224–40.

———. 1979. Development of dependence on oxygen in embryo salamanders. *Am. J. Physiol.* 236:R282–R291.

Adolph, E. F., and H. Collins. 1925. Moulting in an amphibian. *J. Morphol.* 40:575–91.

Adolph, S. C., and J. Roughgarden. 1983. Foraging by passerine birds and *Anolis* lizards on S. Eustatius (Neth. Antilles): Implications for interclass competition and predation. *Oecologia* 56:313–17.

Adrian, M. I., and M. Delibes. 1987. Food habits of the otter (*Lutra lutra*) in two habitats of the Doñana National Park, SW Spain. *J. Zool., London* 212:399–406.

Advani, R. 1981. Seasonal fluctuations in the feeding ecology of the Indian false vampire *Megaderma lyra lyra* (Chiroptera: Megadermatidae) in Rajasthan. *Z. Säugetierkunde* 46:90–93.

Aertsen, A. M. H. J., M. S. M. G. Vlaming, J. J. Eggermont, and P. I. M. Johannesma. 1986. Directional hearing in the grass-frog (*Rana temporaria* L.): II. Acoustics and modelling of the auditory periphery. *Hearing Res.* 21:17–40.

Agar, W. E. 1909. The nesting habits of the tree-frog *Phyllomedusa sauvagii. Proc. Zool. Soc. London* 133:453–69.

Aggarwal, S. J., and A. Riggs. 1969. The hemoglobin of the bull-frog, *Rana catesbeiana:* I. Purification, amino acid composition and oxygen equilibria. *J. Biol. Chem.* 244:2374–83.

Aguiar-Junior, O., A. A. Garda, A. P. Lima, G. R. Colli, S. N. Báo, and S. M. Recco-Pimentel. 2003. Biflagellate spermatozoa of the poison-dart frogs *Epipedobates femoralis* and *Colostethus* sp. (Anura: Dendrobatidae). *J. Morphol.* 255:114–21.

Aguiar-Junior, O., A. P. Lima, A. O. Giaretta, and S. M. Recco-Pimentel. 2002. Cytogenetic analysis of four poison frogs of the *Epipedobates* genus (Anura: Dendrobatidae). *Herpetologica* 58:293–303.

Ahlberg, P. E. 1995. *Elginerpeton pancheni* and the earliest tetrapod clade. *Nature* 373:420–25.

Ahlberg, P. E., and J. A. Clack. 2006. A firm step from water to land. *Nature* 440:747–49.

Ahlberg, P. E., J. A. Clack, and E. Lukševičs. 1996. Rapid braincase evolution between *Panderichthys* and the earliest tetrapods. *Nature* 381:61–64.

Ahlberg, P. E., and Z. Johanson. 1998. Osetolepiforms and the ancestry of tetrapods. *Nature* 395:792–94.

Ahlberg, P. E., and A. R. Milner. 1994. The origin and early diversification of tetrapods. *Nature* 368:507–14.

Aho, J. M. 1990. Helminth communities of amphibians and reptiles: Comparative approaches to understanding patterns and processes. In *Parasite communities: Patterns and processes,* ed. G. W. Esch, A. O. Bush, and J. M. Aho, 157–95. New York: Chapman and Hall.

Ahola, M., M. Nordström, P. B. Banks, N. Laanetu, and E. Korpimäki. 2006. Alien mink predation induces prolonged declines in archipelago amphibians. *Proc. Roy. Soc. London* B 272:1261–65.

Aichinger, M. 1987a. Annual activity patterns of anurans in a seasonal Neotropical environment. *Oecologia* 71:583–92.

———. 1987b. Freilandbeobachtungen zum Fortpflanzungsverhalten von *Hyla brevifrons* Duellman and Crump, 1974 (Anura: Hylidae). *Salamandra* 23:269–76.

———. 1991. Tadpole transport in relation to rainfall, fecundity and body size in five species of poison-dart frogs from Amazonian Peru. *Amphibia-Reptilia* 12:49–55.

Aiken, R. B. 1982. Effects of group density on call rate, phonokinesis, and mating success in *Palmacorixa nana* (Heteroptera: Corixidae). *Can. J. Zool.* 60:1665–72.

Aitchison, C. W. 1987. Review of winter trophic relations of soricine shrews. *Mammal Rev.* 17:1–24.

Aitken, P. G., and R. R. Capranica. 1984. Auditory input of vocal nucleus in the frog *Rana pipiens:* hormonal and seasonal effects. *Exp. Brain. Res.* 57:33–39.

Akani, G. C., and L. Luiselli. 2001. Ecological studies on a population of the water snake *Grayia smythii* in a rainforest swamp of the Niger Delta, Nigeria. *Contr. Zool.* 70:139–46.

Akef, M. S., and H. Schneider. 1985. Vocalization, courtship and territoriality in the Chinese fire-bellied toad *Bombina orientalis* (Anura, Discoglossidae). *Zool. Jb. Physiol.* 89:119–36.

———. 1990. Reproduction, calls and calling of the leopard toad in Egypt (Anura: Bufonidae). *Zool. Anz.* 225:123–32.

Akin, G. C. 1966. Self-inhibition of growth in *Rana pipiens* tadpoles. *Physiol. Zool.* 39:341–56.

Akita, Y. 1989. Breeding cycle of *Onychodactylus japonicus* on Mt. Hodatsu, with special reference to biannual spawning. In *Current herpetology in East Asia,* ed. M. Matsui, T. Hikida, and R. C. Goris, 305. Kyoto, Japan: Herpetological Society of Japan.

Alamillo, H., and J. B. Pramuk. 2002. *Phyllomedusa hypochondrialis azurea* (Orange-legged Leaf Frog). Metamorph behavior. *Herp. Rev.* 33:126–27.

Alberch, P. 1981. Convergence and parallelism in foot morphology in the Neotropical salamander genus *Bolitoglossa.* I. Function. *Evolution* 35:84–100.

———. 1983. Morphological variation in the Neotropical salamander genus *Bolitoglossa. Evolution* 37:906–19.

———. 1987. Evolution of a developmental process—Irreversibility and redundancy in amphibian metamorphosis. In *Development as an evolutionary process,* ed. R. A. Raff and E. C. Raff, 23–46. New York: Alan Liss.

———. 1989. Development and the evolution of amphibian metamorphosis. In *Trends in vertebrate morphology,* ed. H. Splechtna and H. Hilgers, 163–73. Stuttgart: Fischer.

Alberch, P., and J. Alberch. 1981. Heterochronic mechanisms of morphological diversification and evolutionary change in the Neotropical salamander, *Bolitoglossa occidentalis* (Amphibia: Plethodontidae). *J. Morphol.* 167:249–64.

Alberch, P., and E. A. Gale. 1985. A developmental analysis of an evolutionary trend: Digital reduction in amphibians. *Evolution* 39:8–23.

Alcala, A. C. 1957. Philippine notes on the ecology of the marine toad. *Silliman J.* 4:90–96.

———. 1962. Breeding behavior and early development of frogs of Negros, Philippine Islands. *Copeia* 1962:679–726.

Alcala, A. C., and W. C. Brown. 1982. Reproductive biology of some species of *Philautus* (Rhacophoridae) and other Philippine anurans. *Philip. J. Biol.* 11:203–26.

Alcher, M. 1981. Reproduction en élevage de *Euproctus platycephalus* (Urodela, Salamandridae). *Amphibia-Reptilia* 2:97–105.

Alcobendas, M., D. Buckley, and M. Tejedo. 2004. Variability in survival, growth and metamorphosis in the larval fire salamander (*Salamandra salamandra*): Effects of larval birth size, sibship and environment. *Herpetologica* 60:232–45.

Alcobendas, M., and J. Castanet. 2000. Bone growth plasticity among populations of *Salamandra salamandra:* Interactions between internal and external factors. *Herpetologica* 56:14–26.

Alcobendas, M., H. Dopazo, and P. Alberch. 1996. Geographic variation in allozymes of populations of *Salamandra salamandra* (Amphibia: Urodela) exhibiting distinct reproductive modes. *J. Evol. Biol.* 9:83–102.

Alcocer, I., X. Santacruz, H. Steinbeisser, K.-H. Theirauch, and E. M. del Pino. 1992. Ureotelism as the prevailing mode of nitrogen excretion in larvae of the marsupial frog *Gastrotheca riobambae* (Fowler) (Anura, Hylidae). *Comp. Biochem. Physiol.* 101A:229–31.

Alder, T. B., and G. J. Rose. 1998. Long-term temporal integration in the anuran auditory system. *Nature Neurosci.* 1:519–23.

———. 2000. Integration and recovery processes contribute to the temporal selectivity of neurons in the midbrain of the northern leopard frog, *Rana pipiens. J. Comp. Physiol. A* 186:923–37.

Aldhous, P. 2004. The toads are coming! *Nature* 432:796–98.

Alexander, G. 1933. Seasonal variations in a tropical toad. *Univ. Colorado Stud.* 20:195–205.

Alexander, M. A., and J. K. Eischeid. 2001. Climate variability in regions of amphibian declines. *Cons. Biol.* 15:930–42.

Alexander, M. M. 1943. Food habits of the snapping turtle in Connecticut. *J. Wildl. Manage.* 7:278–82.

Alexander, R. D. 1975. Natural selection and specialized chorusing behavior in acoustical insects. In *Insects, science, and society,* ed. D. Pimentel, 356–77. New York: Academic Press.

Alexander, R. D., J. L. Hoogland, R. D. Howard, K. Noonan, and P. W. Sherman. 1979. Sexual dimorphisms and breeding systems in pinnipeds, ungulates, primates, and humans. In *Evolutionary biology and human social behavior: An anthropological perspective,* ed. N. A. Chagnon and W. G. Irons, 402–35. North Scituate, MA: Duxbury.

Alford, R. A. 1986. Effects of parentage on competitive ability and vulnerability to predation in *Hyla chrysoscelis* tadpoles. *Oecologia* 68:199–204.

Alford, R. A. 1989a. Competition between larval *Rana palustris* and *Bufo americanus* is not affected by variation in reproductive phenology. *Copeia* 1989:993–1000.

———. 1989b. Effects of parentage and competitor phenology on the growth of larval *Hyla chrysoscelis. Oikos* 54:325–30.

———. 1989c. Variation in predatory phenology affects predation performance and prey community composition. *Ecology* 70:206–19.

Alford, R. A. 1999. Ecology. In *Tadpoles. The biology of anuran larvae,* ed. R. W. McDiarmid and R. Altig, 240–78. Chicago: University of Chicago Press.

Alford, R. A., P. M. Dixon, and J. H. K. Pechmann. 2001. Global amphibian population declines. *Nature* 412:499–500.

Alford, R. A., and R. N. Harris. 1988. Effects of larval growth history on anuran metamorphosis. *Am. Nat.* 131:91–106.

Alford, R. A., and S. J. Richards. 1997. Lack of evidence for epidemic disease as an agent in the catastrophic decline of Australian rain forest frogs. *Cons. Biol.* 11:1026–29.

———. 1999. Global amphibian declines: A problem in applied ecology. *Ann. Rev. Ecol. Syst.* 30:133–65.

Alford, R. A., and H. M. Wilbur. 1985. Priority effects in experimental pond communities: Competition between *Bufo* and *Rana. Ecology* 66:1097–1105.

Allan, D. M. 1973. Some relationships of vocalization to behavior in the Pacific treefrog, *Hyla regilla. Herpetologica* 29:366–71.

Allan, S. E., and A. Megela Simmons. 1994. Temporal features mediating call recognition in the green treefrog, *Hyla cinerea:* Amplitude modulation. *Anim. Behav.* 47:1073–86.

Allee, W. C., A. E. Emerson, O. Park, T. Park, and K. P. Schmidt. 1949. *Principles of ecology.* Philadelphia: W. B. Saunders.

Allen, A. A., and P. P. Kellogg. 1948. *Voices of the night* (phonograph record). Ithaca, NY: Cornell Laboratory of Ornithology.

Allison, A., and F. Kraus. 2000. A new species of frog of the genus *Xenorhina* (Anura: Microhylidae) from the north coast ranges of Papua New Guinea. *Herpetologica* 56:285–94.

Allison, J. D., and W. Wilczynski. 1991. Thalamic and midbrain auditory projections to the preoptic area and ventral hypothalamus in the green treefrog (*Hyla cinerea*). *Brain Behav. Evol.* 38:322–31.

Allmon, W. D. 1991. A plot study of forest floor litter frogs, Central Amazon, Brazil. *J. Trop. Ecol.* 7:503–22.

Allran, J. W., and W. H. Karasov. 2001. Effects of atrazine on embryos, larvae, and adults of anuran amphibians. *Env. Tox. Chem.* 20:769–75.

Alonso, R., A. Rodríguez, and A. Hernandez. 2005. *Eleutherodactylus albipes* (NCN). Reproduction. *Herp. Rev.* 36:433.

Alp, P. R., E. A. Newsholme, and V. A. Zammit. 1976. Activities of citrate synthase and NAD+-linked and NADP+-linked isocitrate dehydrogenase in muscle from vertebrates and invertebrates. *Biochem. J.* 154:689–700.

Altig, R. 1972a. Defensive behavior in *Rana areolata* and *Hyla avivoca. Q. J. Florida Acad. Sci.* 35:212–16.

———. 1972b. Notes on the larvae and premetamorphic tadpoles of four *Hyla* and three *Rana* with notes on tadpole color patterns. *J. Elisha Mitchell Sci. Soc.* 88:113–19.

Altig, R., and E. D. Brodie, Jr. 1971. Foods of *Plethodon larselli, Plethodon dunni* and *Ensatina eschscholtzi* in the Columbia River Gorge, Multnomah County, Oregon. *Am. Midl. Nat.* 85:226–28.

Altig, R., and E. D. Brodie. 1972. Laboratory behavior of *Ascaphus truei* tadpoles. *J. Herpetol.* 6:21–24.

Altig, R., and A. Channing. 1993. Hypothesis: Functional significance of colour and pattern of anuran tadpoles. *Herpetol. J.* 3:73–75.

Altig, R., and M. T. Christensen. 1981. Behavioral characteristics of the tadpoles of *Rana heckscheri. J. Herpetol.* 15:151–54.

Altig, R., and P. C. Dumas. 1971. *Rana cascadae. Cat. Amer. Amphib. Rept.* (105):1–2.

Altig, R., and G. F. Johnston. 1989. Guilds of anuran larvae: Relationships among developmental modes, morphologies, and habitats. *Herpetol. Monogr.* 3:81–109.

Altig, R., J. P. Kelly, M. Wells, and J. Phillips. 1975. Digestive enzymes of seven species of anuran tadpoles. *Herpetologica* 31:104–8.

Altig, R., and W. McDearman. 1975. Percent assimilation and clearance times of five anuran tadpoles. *Herpetologica* 31:67–69.

Altig, R., and R. W. McDiarmid. 1999. Body plan: Development and morphology. In *Tadpoles. The biology of anuran larvae,* ed. R. W. McDiarmid and R. Altig, 24–51. Chicago: University of Chicago Press.

Altwegg, R. 2002. Trait-mediated indirect effects and complex life-cycles in two European frogs. *Evol. Ecol. Res.* 4:519–36.

———. 2003a. Hungry predators render predator-avoidance behavior in tadpoles ineffective. *Oikos* 100:311–16.

———. 2003b. Multistage density dependence in an amphibian. *Oecologia* 136:46–50.

Altwegg, R., and H.-U. Reyer. 2003. Patterns of natural selection on size at metamorphosis in water frogs. *Evolution* 57:872–82.

Alvarado, R. H. 1967. The significance of grouping on water conservation in *Ambystoma. Copeia* 1967:667–68.

———. 1972. The effects of dehydration on water and electrolytes in *Ambystoma tigrinum. Physiol. Zool.* 45:43–53.

———. 1979. Amphibians. In *Comparative physiology of osmoregulation in animals,* ed. G. M. O. Maloiy, 261–303. New York: Academic Press.

Alvarado, R. H., and L. B. Kirschner. 1963. Osmotic and ionic regulation in *Ambystoma tigrinum. Comp. Biochem. Physiol.* 10:55–67.

Alvarado, R. H., and A. Moody. 1970. Sodium and chloride transport in tadpoles of the bullfrog *Rana catesbeiana. Am. J. Physiol.* 218:1510–16.

Álvarez, D., and G. Nicieza. 2002a. Effects of temperature and food quality on anuran larval growth and metamorphosis. *Funct. Ecol.* 16:640–48.

Álvarez, D., and G. Nicieza. 2002b. Effects of induced variation

in anuran larval development on postmetamorphic energy reserves and locomotion. *Oecologia* 131:186–95.
Amey, A. P., and G. C. Grigg. 1995. Lipid-reduced evaporative water loss in two arboreal hylid frogs. *Comp. Biochem. Physiol.* 111A:283–91.
Amézquita, A., and W. Hödl. 2004. How, when, and where to perform visual displays: The case of the Amazonian from *Hyla parviceps. Herpetologica* 60:420–29.
Amiet, J.-L. 1972. Description de trois Bufonidés orophiles du Cameroun appartenant au groupe de *Bufo preussi* Matschie (Amphibiens, Anoures). *Ann. Fac. Sci. Cameroun* 11:121–40.
———. 1974a. La ponte et la larve d'*Opisthothylax immaculatus* (Boulenger) (Amphibien Anoure). *Ann. Fac. Sci. Cameroun* 17:121–30.
———. 1974b. Le tetard d'*Hyperolius obstetricans* Ahl (Amphibien Anoure). *Bull. IFAN* 36:973–84.
———. 1981. Écologie, éthologie et développement de *Phrynodon sandersoni* Parker, 1939 (Amphibia, Anura, Ranidae). *Amphibia-Reptilia* 2:1–13.
———. 1989. Quelques aspects de la biologie des amphibiens anoures du Cameroun. *Ann. Biol.* 28:73–136.
———. 1991. Images d'Amphibiens camerounais. III. Le comportement de garde des oeufs. *Alytes* 9:15–22.
Amorim, M. C. P., M. L. McCracken, and M. L. Fine. 2002. Metabolic costs of sound production in the oyster toadfish, *Opsanus tau. Can. J. Zool.* 80:830–38.
Andersen, J. 1954. The food of the Danish badger (*Meles meles danicus* Degerbøl) with special reference to the summer months. *Danish Rev. Game Biol.* 3:1–75.
Andersen, L. W., K. Fog, and C. Damgaard. 2004. Habitat fragmentation causes bottlenecks and inbreeding in the European tree frog (*Hyla arborea*). *Proc. Roy. Soc. London B* 271:1293–1302.
Anderson, A. M., D. A. Haukos, and J. T. Anderson. 1999a. Diet composition of three anurans from the playa wetlands of northwest Texas. *Copeia* 1999:515–20.
———. 1999b. Habitat use by anurans emerging and breeding in playa wetlands. *Wildl. Soc. Bull.* 27:759–69.
Anderson, A. R., and J. W. Petranka. 2003. Odonate predator does not affect hatching time or morphology of embryos of two amphibians. *J. Herpetol.* 37:65–71.
Anderson, B. D., M. E. Feder, and R. J. Full. 1991. Consequences of a gait change during locomotion in toads (*Bufo woodhousii fowleri*). *J. Exp. Biol.* 158:133–48.
Anderson, C. W. 1993. The modulation of feeding behavior in response to prey type in the frog *Rana pipiens. J. Exp. Biol.* 179:1–12.
Anderson, C. W., and K. C. Nishikawa. 1993. A prey-type dependent hypoglossal feedback system in the frog *Rana pipiens. Brain Behav. Evol.* 42:189–96.
———. 1996. The roles of visual and proprioceptive information during motor program choice in frogs. *J. Comp. Physiol. A* 179:753–62.
Anderson, D. J. 1982. The home range: A new nonparametric estimation technique. *Ecology* 63:103–12.
Anderson, J. B., and T. Wang. 2003. Cardiorespiratory effects of forced activity and digestion in toads. *Physiol. Biochem. Zool.* 76:459–70.
Anderson, J. D. 1961. The courtship behavior of *Ambystoma macrodactylum croceum. Copeia* 1961:132–39.
———. 1963. Reactions of the western mole to skin secretions of *Ambystoma macrodactylum croceum. Herpetologica* 19:282–84.
———. 1967. A comparison of the life histories of coastal and montane populations of *Ambystoma macrodactylum* in California. *Am. Midl. Nat.* 77:323–55.
———. 1968a. A comparison of the food habits of *Ambystoma macrodactylum sigillatum, A. macrodactylum croceum* and *A. tigrinum californiense. Herpetologica* 24:273–84.
———. 1968b. Thermal histories of two populations of *Ambystoma macrodactylum. Herpetologica* 24:29–35.
———. 1970. Description of the spermatophore of *Ambystoma tigrinum. Herpetologica* 3:304–8.
———. 1972. Behavior of three subspecies of *Ambystoma macrodactylum* in a soil moisture gradient. *J. Herpetol.* 6:191–94.
Anderson, J. D., and R. E. Graham. 1967. Vertical migration and stratification of larval *Ambystoma. Copeia* 1967:371–74.
Anderson, J. D., D. D. Hassinger, and G. H. Dalrymple. 1971. Natural mortality of eggs and larvae of *Ambystoma t. tigrinum. Ecology* 52:1107–12.
Anderson, J. D., and R. G. Webb. 1978. Life history aspects of the Mexican salamander *Ambystoma rosaceum* (Amphibia, Urodela, Ambystomatidae). *J. Herpetol.* 12:89–93.
Anderson, J. D., and G. K. Williamson. 1974. Nocturnal stratification in the larvae of the mole salamander, *Ambystoma talpoideum. Herpetologica* 30:28–29.
———. 1976. Terrestrial mode of reproduction in *Ambystoma cingulatum. Herpetologica* 32:214–21.
Anderson, J. D., and R. D. Worthington. 1971. The life history of the Mexican salamander *Ambystoma ordinarium* Taylor. *Herpetologica* 27:165–76.
Anderson, J. S. 2001. The phylogenetic trunk: Maximal inclusion of taxa with missing data in an analysis of the Lepospondyli (Vertebrata, Tetrapoda). *Syst. Biol.* 50:170–93.
Anderson, K. A., and T. L. Beitinger. 1979. Body heating and cooling in the mudpuppy, *Necturus maculosus. Herpetologica* 35:234–39.
Anderson, K., and P. E. Moler. 1986. Natural hybrids of the Pine Barrens treefrog, *Hyla andersonii* with *H. cinerea* and *H. femoralis* (Anura: Hylidae): Morphological and chromosomal evidence. *Copeia* 1986:70–76.
Anderson, N. H. 1976. Carnivory by an aquatic detritivore, *Clistoronia magnifica* (Trichoptera: Limnephiliidae). *Ecology* 57:1081–85.
Anderson, N. H., and K. W. Cummins. 1979. Influence of diet on the life histories of aquatic insects. *J. Fish. Res. Board Canada* 36:335–42.
Anderson, P. K. 1954. Studies on the ecology of the narrow-mouthed frog, *Microhyla carolinensis. Tulane Stud. Zool.* 2:15–46.
———. 1960. Ecology and evolution in island populations of salamanders in the San Francisco Bay area. *Ecol. Monogr.* 30:359–85.
Anderson, P. K., A. A. Cunningham, N. G. Patel, F. J. Morales, P. R. Epstein, and P. Daszak. 2004. Emerging infectious disease of plants: Pathogen pollution, climate change and agrotechnology drivers. *Trends Ecol. Evol.* 19:535–44.
Anderson, S. C., and E. P. Volpe. 1958. *Burnsi* and *kandiyohi* genes in the leopard frog, *Rana pipiens. Science* 127:1048–50.
Andersson, M. 1982. Sexual selection, natural selection and quality advertisement. *Biol. J. Linn. Soc.* 17:375–93.
———. 1994. *Sexual selection.* Princeton, NJ: Princeton University Press.
Andrén, C., M. Marden, and C. Nilson. 1989. Tolerance of low pH in a population of moor frogs, *Rana arvalis,* from an acid

and neutral environment: A possible case of rapid evolutionary response to acidification. *Oikos* 56:215–23.

Andrén, C., and G. Nilson. 1985. Breeding pool characteristics and reproduction in an island population of natterjack toads, *Bufo calamita* Laur., at the Swedish west coast. *Amphibia-Reptilia* 6:137–42.

Andreoletti, G. E., C. Giacoma, G. Malacarne, and E. Vacca. 1988. Chemical communication in newt courtship: Biochemical and comparative data. *Monit. Zool. Ital.,* n.s. 22:524.

Andreone, F. 1992. Observations on the territorial and reproductive behaviour of *Salamandra lanzai* and consideration about its protection (Amphibia: Salamandridae). *British Herp. Soc. Bull.* 39:31–33.

———. 1993. Two new treefrogs of the genus *Boophis* (Anura: Rhacophoridae) from central-eastern Madagascar. *Boll. Mus. Reg. Sci. Nat. Torino* 11:289–313

———. 1994. The amphibians of Ranomafana rain forest, Madagascar—Preliminary community analysis and conservation considerations. *Oryx* 28:207–14.

———. 1998. Capture-recapture study of *Mantidactylus microtympanum* (Anura, Ranidae, Mantellinae) from the Andohahela low altitude forest, south-eastern Madagascar. *Amphibia-Reptilia* 19:451–55.

Andreone, F., F. Glaw, R. A. Nussbaum, C. J. Raxworthy, M. Vences, and J. E. Randrianirina. 2003. The amphibians and reptiles of Nosy Be (NW Madagascar) and nearby islands: A case study of diversity and conservation of an insular fauna. *J. Nat. Hist.* 37:2119–49.

Andreone, F., and L. Luiselli. 2000. The Italian batrachofauna and its conservation status: A statistical assessment. *Biol. Cons.* 96:197–208.

———. 2003. Conservation priorities and potential threats influencing the hyper-diverse amphibians of Madagascar. *Ital. J. Zool.* 70:53–63.

Andreone, F., and R. Marocco. 1998. *Rana catesbeiana* Shaw, 1802. Rana toro. In *Erpetologia del Piemonte e della Valle d-Aosta. Atlante degli Anfibi e dei Rettili. Monogr. XXVI*, ed. F. Andreone and R. Sindaco, 192–93. Turin, Italy: Museo Regionale di Scienze Naturali.

Andreone, F., and R. Sindaco. 1998. *Erpetologia del Piemonte e della Valle d'Aosta. Atlante degli Anfibi e dei Rettili. Monogr. XXVI*. Turin, Italy: Museo Regionale di Scienze Naturali.

Andreone, F., M. Vences, D. R. Vieites, F. Glaw, and A. Meyer. 2005. Recurrent ecological adaptations revealed through a molecular analysis on the secretive cophyline frogs of Madagascar. *Mol. Phylogen. Evol.* 34:315–22.

Andreone, F., M. Vences, F. M. Guarino, F. Glaw, and J. E. Randrianirina. 2002. Natural history and larval morphology of *Boophis occidentalis* (Anura: Mantellidae: Boophinae) provide new insights into the phylogeny and adaptive radiation of endemic Malagasy frogs. *J. Zool., London* 257:425–38.

Andreone, F., and R. Piazza. 1990. A bio-acoustic study on *Pelobates fuscus insubricus* (Amphibia, Pelobatidae). *Boll. Zool.* 57:341–49.

Andrews, R. M., and F. H. Pough. 1985. Metabolism of squamate reptiles: Allometric and ecological relationships. *Physiol. Zool.* 58:214–31.

Andrzejewski, H., J. Czopek, L. Siankowa, and H. Szarski. 1962. The vascularization of respiratory surfaces in amphibians reaching large body size (*Rana esculenta* L., *R. pipiens sphenocephala* Cope and *Rana grylio* Stejneger). *Studia Soc. Sci. Torunensis Sect. E (Zool.)* 6:163–77.

Angel, F. 1947. *Vie et moeurs des amphibiens.* Paris: Payot.

Angel, F., and M. Lamotte. 1944. Un crapaud vivipare d'Afrique occidentale, *Nectophrynoides occidentalis* Angel. *Ann. Sci. Nat. Zool.* 6:63–89.

———. 1948. Nouvelles observations sur *Nectophrynoides occidentalis* Angel. Remarques sur le genre *Nectophrynoides. Ann. Sci. Nat. Zool.* (Paris) 10:115–47.

Angelier, E., and M.-L. Angelier. 1964. Étude d'une population de crapauds accoucheurs dans un lac de haute montagne. *C. R. Acad. Sci. Paris* 258:701–3.

Anholt, B. R., H. Hotz, G.-D. Guex, and R. D. Semlitsch. 2003. Overwinter survival of *Rana lessonae* and its hemiclonal associate *Rana esculenta. Ecology* 84:39–397.

Anholt, B. R., D. K. Skelly, and E. E. Werner. 1996. Factors modifying antipredator behavior in larval toads. *Herpetologica* 52:301–13.

Anholt, B. R., and E. E. Werner. 1995. Interaction between food availability and predation mortality mediated by adaptive behavior. *Ecology* 76:2230–34.

———. 1998. Predictable changes in predation mortality as a consequence of changes in food availability and predation risk. *Evol. Ecol.* 12:729–38.

Ankley, G. T., J. E. Tietge, G. W. Holcombe, D. L. DeFoe, S. A. Diamond, K. M. Jensen, and S. J. Degitz. 2000. Effects of laboratory ultraviolet radiation and natural sunlight on survival and development of *Rana pipiens. Can. J. Zool.* 78:1092–1100.

Annandale, N. 1907. The fauna of brackish ponds at Port Canning, Lower Bengal. Part I. Introduction and preliminary account of the fauna. *Rec. Indian Mus.* 1:35–43.

Annandale, N., and S. L. Hora. 1922. Parallel evolution in the fish and tadpoles of mountain torrents. *Rec. Indian Mus.* 24:505–9.

Anoop, K. R., and S. A. Hussain. 2005. Food and feeding habits of smooth-coated otters (*Lutra perspicillata*) and their significance to the fish population of Kerala, India. *J. Zool., London* 266:15–23.

Anstis, M. 1976. Breeding biology and larval development of *Litoria verreauxi* (Anura: Hylidae). *Trans. Roy. Soc. South Australia* 100:193–202.

Anstis, M., R. A. Alford, and G. R. Gillespie. 1998. Breeding biology of *Litoria booroolongensis* (Moore, 1961), and *Litoria lesueri* (Duméril & Bibron, 1841) (Anura: Hylidae) and comments on population declines of *L. booroolongensis. Trans. Roy. Soc. South Australia* 122:33–43.

Anstis, M., and M. J. Littlejohn. 1996. The breeding biology of *Litoria subglandulosa* and *L. citropa* (Anura: Hylidae) and a re-evaluation of their geographic distribution. *Trans. Roy. Soc. South Australia* 120:83–99.

Anthony, C. D. 1993. Recognition of conspecific odors by *Plethodon caddoensis* and *P. ouachitae. Copeia* 1993:1028–33.

Anthony, C. D., and J. A. Wicknick. 1993. Aggressive interactions and chemical communication between adult and juvenile salamanders. *J. Herpetol.* 27:261–64.

Anthony, C. D., J. A. Wicknick, and R. G. Jaeger. 1997. Social interactions in two sympatric salamanders: Effectiveness of a highly aggressive strategy. *Behaviour* 134:71–88.

Anthony, C. D., J. R. Mendelson III, and R. R. Simons. 1994. Differential parasitism by sex on plethodontid salamanders and histological evidence for structural damage to the nasolabial groove. *Am. Midl. Nat.* 132:302–7.

Antonelli, A. L., R. A. Nussbaum, and S. D. Smith. 1972. Comparative food habits of four species of stream-dwelling vertebrates (*Dicamptodon ensatus, D. copei, Cottus tenuis, Salmo gairdneri*). *Northwest Sci.* 46:277–89.

Anzalone, C. R., L. B. Kats, and M. S. Gordon. 1998. Effects of solar UV-B radiation on embryonic development in *Hyla cadaverina, Hyla regilla,* and *Taricha torosa. Cons. Biol.* 12:646–53.

Aoto, T. 1950. A remarkable swelling of male skin of a salamander (*Hynobius retardatus* Dunn) in the breeding season. *J. Fac. Sci. Hokkaido Univ., Ser. VI, Zoology* 10:45–54.

Apollonio, M., M. Festa-Bianchet, F. Mari, S. Mattioli, and B. Sarno. 1992. To lek or not to lek: Mating strategies of male fallow deer. *Behav. Ecol. Sociobiol.* 3:25–31.

Applegate, R. D. 1990. *Rana catesbeiana, Rana palustris* (Bullfrog, pickeral frog). Predation. *Herp. Rev.* 21:90–91.

Arak, A. 1983a. Male-male competition and mate choice in anuran amphibians. In *Mate choice,* ed. P. P. G. Bateson, 181–210. New York: Cambridge University Press.

———. 1983b. Sexual selection by male-male competition in natterjack toad choruses. *Nature* 306:261–62.

———. 1983c. Vocal interactions, call matching and territoriality in a Sri Lankan treefrog, *Philautus leucorhinus* (Rhacophoridae). *Anim. Behav.* 31:292–302.

———. 1988a. Callers and satellites in the natterjack toad: Evolutionarily stable decision rules. *Anim. Behav.* 36:416–32.

———. 1988b. Female mate selection in the natterjack toad: Active choice or passive attraction? *Behav. Ecol. Sociobiol.* 22:317–27.

———. 1988c. Sexual dimorphism in body size: A model and a test. *Evolution* 42:820–25.

Arano, B., G. Llorente, M. García-París, and P. Herrero. 1995. Species translocation menaces Iberian waterfrogs. *Cons. Biol.* 9:196–98.

Araújo, M. B. 2003. The coincidence of people and biodiversity in Europe. *Global Ecol. Biogeogr.* 12:5–12.

Archer, R., J. Chauvet, M. T. Lenci, and F. Morel. 1960. Presence d'une vasotocine dans la neurohypophyse de la grenouille (*Rana esculenta* L.). *Biochem. Biophys. Acta* 42:379–80.

Archer, R., J. Chauvet, and Y. Rouillé. 1997. Adaptive evolution of water homeostasis regulation in amphibians: Vasotocin and hydrins. *Biol. Cell* 89:283–91.

Archey, G. 1922. The habitat and life history of *Leiopelma hochstetteri. Rec. Canterbury Mus.* 2:59–71.

———. 1935. Frogs in New Zealand. *Bull. Aukland Zool. Soc.* 2:5–8.

Ardila-Robayo, M. C. 1979. Status sistemático del genero *Geobatrachus* Ruthven 1915 (Amphibia: Anura). *Caldasia* 12:383–495.

Aresco, M. J., and M. S. Gunzburger. 2004. Effects of large-scale sediment removal on herpetofauna in Florida wetlands. *J. Herpetol* 38:275–79.

Armentrout, D., and R. L. Rose. 1971. Some physiological responses to anoxia in the Great Plains toad, *Bufo cognatus. Comp. Biochem. Physiol.* 39:447–55.

Arnold, E. N., and J. A. Burton. 1978. *A field guide to the reptiles and amphibians of Britain and Europe.* London: Collins.

Arnold, S. J. 1972. Species densities of predators and their prey. *Am. Nat.* 106:220–36.

———. 1976. Sexual behavior, sexual interference and sexual defense in the salamanders *Ambystoma maculatum, Ambystoma tigrinum* and *Plethodon jordani. Z. Tierpsychol.* 42:247–300.

———. 1977. The evolution of courtship behavior in New World salamanders with some comments on Old World salamanders. In *The reproductive biology of amphibians,* ed. D. H. Taylor and S. I. Guttman, 141–83. New York: Plenum.

———. 1978. Some effects of early experience on feeding responses in the common garter snake, *Thamnophis sirtalis. Anim. Behav.* 26:455–62.

———. 1981a. Behavioral variation in natural populations. I. Phenotypic, genetic and environmental correlations between chemoreceptive responses to prey in the garter snake, *Thamnophis elegans. Evolution* 35:489–509.

———. 1981b. Behavioral variation in natural populations. II. The inheritance of a feeding response in crosses between geographic races of the garter snake, *Thamnophis elegans. Evolution* 35:510–15.

———. 1981c. The microevolution of feeding behavior. In *Foraging behavior: Ecological, ethological and psychological approaches,* ed. A. Kamil and T. Sargent, 409–53. New York: Garland Press

———. 1982. A quantitative approach to antipredator performance: Salamander defense against shrew attack. *Copeia* 1982:247–53.

———. 1983a. Morphology, performance and fitness. *Amer. Zool.* 23:347–61.

———. 1983b. Sexual selection: The interface of theory and empiricism. In *Mate choice,* ed. P. Bateson, 67–107. New York: Cambridge University Press.

———. 1987. The comparative ethology of courtship in salamandrid salamanders. 1. *Salamandra* and *Chioglossa.* Ethology 74:133–45.

———. 1988. Behavior, energy, and fitness. *Amer. Zool.* 28:815–27.

———. 2000. Systematics at the turn of a century. In *The biology of plethodontid salamanders,* ed. R. C. Bruce, R. G. Jaeger, and L. D. Houck, 167–78. New York: Kluwer Academic/Plenum.

Arnold, S. J., and D. Duvall. 1994. Animal mating systems: A synthesis based on selection theory. *Am. Nat.* 143:317–48.

Arnold, S. J., and L. D. Houck. 1982. Courtship pheromones: evolution by natural and sexual selection. In *Biochemical aspects of evolutionary biology,* ed. M. H. Nitecki, 173–211. Chicago: University of Chicago Press.

Arnold, S. J., N. L. Reagan, and P. A. Verrell. 1993. Reproductive isolation and speciation in plethodontid salamanders. *Herpetologica* 49:216–28.

Arnold, S. J., P. A. Verrell, and S. G. Tilley. 1996. The evolution of asymmetry in sexual isolation: A model and a test case. *Evolution* 50:1024–33.

Arnold, S. J., and M. J. Wade. 1984a. On the measurement of natural and sexual selection: Theory. *Evolution* 38:709–19.

———. 1984b. On the measurement of natural and sexual selection: Applications. *Evolution* 38:720–34.

Arnold, S. J., and R. J. Wassersug. 1978. Differential predation on metamorphic anurans by garter snakes (*Thamnophis*): Social behavior as a possible defense. *Ecology* 59:1014–22.

Arnoult, J. 1966. Écologie et développement de *Mantella aurantiaca. Bull. Mus. Natl. Hist. Nat. Paris* (2) 37:931–40.

Arnoult, J., and M. Razarihelisoa. 1967. Contribution a l'étude des batraciens de Madagascar. Le genre *Mantidactylus:* Adultes et formes larvaires de *M. betsileanus* (Blgr.), *M. curtus* (Blgr.) et *M. alutus* (Peracca). *Bull. Mus. Natl. Hist. Nat. Paris* (2) 39:471–87.

Arntzen, J. W. 1981. Ecological observations on *Chioglossa lusitanica. Amphibia-Reptilia* 1:187–203.

Arntzen, J. W., P. De Wijer, R. Jehle, E. Smit, and J. Smit. 1998. Rare hybridization and introgression in smooth and palmate newts (Salamandridae: *Triturus vulgaris* and *T. helveticus*). *J. Zool. Syst. Evol. Res.* 36:111–22.

Arntzen, J. W., and M. Sparreboom. 1989. A phylogeny for the Old World newts, genus *Triturus:* Biochemical and behavioural data. *J. Zool., London* 219:645–64.

Arntzen, J. W., and S. F. M. Teunis. 1993. A six-year study on the population dynamics of the crested newt (*Triturus cristatus*) following the colonization of a newly created pond. *Herpetol. J.* 3:99–110.

Aronson, L. R. 1943a. The sexual behavior of Anura. IV. Oviposition in the mink frog, *Rana septentrionalis* Baird. *Am. Midl. Nat.* 29:242–44.

———. 1943b. The sexual behavior of Anura. 5. Oviposition in the green frog, *Rana clamitans,* and the bull frog, *Rana catesbeiana. Amer. Mus. Novitates* (1224):1–6.

———. 1944. The sexual behavior of Anura. 6. The mating pattern of *Bufo americanus, Bufo fowleri,* and *Bufo terrestris. Amer. Mus. Novitates* (1250):1–15.

———. 1975. The case of "The Case of the Midwife Toad". *Behav. Gen.* 5:115–25.

Aronsson, S., and J. A. E. Stenson. 1995. Newt-fish interactions in a small forest lake. *Amphibia-Reptilia* 16:177–84.

Arzabe, C. 1999. Reproductive activity patterns of anurans in two different altitudinal sites within the Brazilian Caatinga. *Rev. Bras. Zool.* 16:851–64.

Arzabe, C., and A. C. C. de Almeida. 1997. Life history notes on *Leptodactylus troglodytes* (Anura, Leptodactylidae) in northeastern Brazil. *Amphibia-Reptilia* 18:211–15.

Arzabe, C., C. X. de Carvalho, and M. A. Goes Costa. 1998. Anuran assemblages in Crasto forest ponds (Sergipe State, Brazil): Comparative structure and calling activity patterns. *Herpetol. J.* 8:111–13.

Arzt, A. H., W. L. Silver, J. R. Mason, and L. Clark. 1986. Olfactory responses of aquatic and terrestrial tiger salamanders to airborne and waterborne stimuli. *J. Comp. Physiol. A* 158:479–87.

Ash, A. N. 1997. Disappearance and return of plethodontid salamanders to clearcut plots in the southern Blue Ridge Mountains. *Cons. Biol.* 11:983–89.

Ash, A. N., and R. C. Bruce. 1994. Impacts of timber harvesting on salamanders. *Cons. Biol.* 8:300–1.

Ash, A. N., R. C. Bruce, J. Castanet, and H. Francillon-Vieillot. 2003. Population parameters of *Plethodon metcalfi* on a 10-year clearcut and in nearby forest in the southern Blue Ridge Mountains. *J. Herpetol.* 37:445–52.

Ash, A. N., and K. H. Pollock. 1999. Clearcuts, salamanders, and field studies. *Cons. Biol.* 13:206–8.

Ashby, K. R. 1969. The population ecology of a self-maintaining population of the common frog (*Rana temporaria*). *J. Zool., London* 158:453–74.

Ashton, D. T., S. B. Marks, and H. H. Welsh, Jr. 2006. Evidence of continued effects from timber harvesting on lotic amphibians in redwood forests of northwestern California. *Forest Ecol. Manage.* 221:183–93.

Ashton, K. G. 2002. Do amphibians follow Bergmann's rule? *Can. J. Zool.* 80:708–16.

Ashton, R. E., Jr. 1975. A study of movement, home range, and winter behavior of *Desmognathus fuscus* (Rafinesque). *J. Herpetol.* 9:85–91.

———. 1985. Field and laboratory observations on microhabitat selection, movements, and home range of *Necturus lewisi* (Brimley). *Brimleyana* 10:83–106.

———. 1994. Tracking with radioactive tags. In *Measuring and monitoring biological diversity: Standard methods for amphibians,* ed. W. R. Heyer, M. A. Donnelly, R. W. McDiarmid, L. C. Hayek, and M. S. Foster, 158–63. Waswhington, DC: Smithsonian Institution Press.

Ashton, R. E., Jr., and A. L. Braswell. 1979. Nest and larvae of the Neuse River waterdog, *Necturus lewisi* (Brimley) (Amphibia: Proteidae). *Brimleyana* (1):15–22.

Ashton, R. E., Jr., and P. S. Ashton. 1978. Movements and winter behavior of *Eurycea bislineata bislineata* (Amphibia, Urodela, Plethodontidae). *J. Herpetol.* 12:295–98.

Asimakopoulos, B., T. S. Sofianidou, and H. Schneider. 1990. Reproductive and calling behavior of the Greek Frog *Rana graeca* (Amphibia: Anura) in Greece. *Zool. Anz.* 225:133–43.

Askew, R. R. 1971. *Parasitic insects.* New York: American Elsevier.

Asplund, K. K. 1963. Ecological factors in the distribution of *Thamnophis brachystoma* (Cope). *Herpetologica* 19:128–32.

Asquith, A., and R. Altig. 1986. Osmoregulation of the lesser siren, *Siren intermedia* (Caudata: Amphibia). *Comp. Biochem. Physiol.* 84A:683–85.

———. 1987. Phototaxis and activity patterns of *Siren intermedia. Southwest. Nat.* 32:146–48.

Astrup, P. 1972. Early Danish contributions to oxygen and acid-base research. In *Oxygen affinity of hemoglobin and red cell acid-base status,* ed. M. Rorth and P. Astrup, 809–22. New York: Academic Press.

Atkinson, D., and R. M. Sibly. 1997. Why are organisms usually bigger in colder environments? Making sense of a life history puzzle. *Trends Ecol. Evol.* 12:235–39.

Attar, E. N., and E. J. Maly. 1980. A laboratory study of preferential predation by the newt *Notophthalmus v. viridescens. Can. J. Zool.* 58:1712–17.

Auber-Thomay, M., and F. Letellier. 1986. Observations sur le développement de la rainette marsupiale, *Gastrotheca riobambae* (Hylides). *Rev. Fr. Aquariol.* 13:79–86.

Aubry, K. B. 2000. Amphibians in managed, second-growth Douglas-fir forests. *J. Wildl. Manage.* 64:1041–52.

Aubry, K. B., and P. A. Hall. 1991. Terrestrial amphibian communities in the southern Washington Cascade Range. In *Wildlife and vegetation of unmanaged Douglas-fir forests,* coordinated by L. F. Ruggiero, K. B. Aubry, A. B. Carey, and M. H. Huff, 326–38. USDA Forest Service, Gen. Tech. Rep. PNW-285. Washington DC: Government Printing Office.

Aubry, K. B., L. L. C. Jones, and P. A. Hall. 1988. Use of woody debris by plethodontid salamanders in Douglas-fir forests in Washington. In *Management of amphibians, reptiles, and mammals in North America,* ed. R. C. Szaro, K. E. Severson, and D. R. Patton, 32–37. USDA For. Serv. Gen. Tech. Rep. RM-166. Washington DC: Government Printing Office.

Auburn, J., and D. H. Taylor. 1979. Polarized light perception and orientation in larval bullfrogs, *Rana catesbeiana. Anim. Behav.* 27:658–68.

Audet, D., D. Krull, G. Marimuthu, S. Sumithran, and J. B. Singh. 1991. Foraging behavior of the Indian false vampire bat, *Megaderma lyra* (Chiroptera: Megadermatidae). *Biotropica* 23:63–67.

Audo, M. C., T. M. Mann, T. L. Polk, C. M. Loudenslager, W. J. Diehl, and R. Altig. 1995. Food deprivation during different periods of tadpole (*Hyla chrysoscelis*) ontogeny affects metamorphic performance differently. *Oecologia* 103:518–22.

Auffenberg, W. 1994. *The bengal monitor.* Gainesville: University Press of Florida.

Auffenberg, W., H. Rahman, F. Iffat, and Z. Perveen. 1989. A study of *Varanus flavescens* (Hardwicke & Gray) (Sauria: Varanidae). *J. Bombay Nat. Hist. Soc.* 86:286–307.

Austad, S. N. 1984. A classification of alternative reproductive behaviors and methods for field-testing ESS models. *Amer. Zool.* 24:309–19.

Austin, C. C., and H. B. Shaffer. 1992. Short-, medium-, and long-term repeatability of locomotor performance in the tiger salamander *Ambystoma californiense. Funct. Ecol.* 6:145–53.

Austin, G. E., and M. M. Rehfisch. 2005. Shifting nonbreeding distributions of migratory fauna in relation to climate change. *Global Change Biol.* 11:31–38.

Austin, J. D., J. S. Dávila, S. C. Lougheed, and P. T. Boag. 2003. Genetic evidence for female-biased dispersal in the bullfrog, *Rana catesbeiana* (Ranidae). *Mol. Ecol.* 12:3165–72.

Austin, J. D., S. C. Lougheed, K. Tanner, A. A. Chek, J. P. Bogart, and P. T. Boag. 2002. A molecular perspective on the evolutionary affinities of an enigmatic Neotropical frog, *Allophryne ruthveni. Zool. J. Linn. Soc.* 134:335–46.

Austin, R. M., Jr. 2000. Cutaneous microbial flora and antibiotics in *Plethodon ventralis:* inferences for parental care in the Plethodontidae. In *The biology of plethodontid salamanders,* ed. R. C. Bruce, R. G. Jaeger, and L. D. Houck, 451–62. New York: Kluwer Academic.

Avery, R. A. 1968. Food and feeding relations of three species of *Triturus* (Amphibia: Urodela) during the aquatic phase. *Oikos* 19:408–12.

Avery, R. A. 1982. Field studies in body temperatures and thermoregulation. In *Biology of the reptilia,* vol. 12, ed. C. Gans and F. H. Pough, 93–166. New York: Academic Press.

Awbrey, F. T. 1963. Homing and home range in *Bufo valliceps. Texas J. Sci.* 15:127–41.

———. 1965. *An experimental investigation of the effectiveness of anuran mating calls as isolating mechanisms.* Unpublished PhD dissertation. Austin: University of Texas.

———. 1968. Call discrimination in female *Scaphiopus couchii* and *Scaphiopus hurterii. Copeia* 1968:420–23.

———. 1978. Social interaction among chorusing Pacific tree frogs, *Hyla regilla. Copeia* 1978:208–14.

Axelsson, E., P. Nÿstrom, J. Sidenmark, and C. Brönmark. 1997. Crayfish predation on amphibian eggs and larvae. *Amphibia-Reptilia* 18:217–28.

Ayala, F. J., and C. Campbell. 1974. Frequency-dependent selection. *Ann. Rev. Ecol. Syst.* 5:115–38.

Ayre, D. J., P. Coster, W. N. Bailey, and J. D. Roberts. 1984. Calling tactics in *Crinia georgiana* (Anura: Myobatrachidae): Alternation and variation in call duration. *Aust. J. Zool.* 32:463–70.

Ayyangar, M. 1915. A south Indian flying frog: *Rhacophorus malabaricus* (Jerdon). *Rec. Indian Mus.* 11:140–42.

Azevedo, F. C. C. de. 1996. Notes on the behavior of the margay *Felis wiedii* (Schinz, 1821) (Carnivora, Felidae), in the Brazilian Atlantic Forest. *Mammalia* 60:325–28.

Azevedo-Ramos, C., and U. Galatti. 2002. Patterns of amphibian diversity in Brazilian Amazonia: Conservation implications. *Biol. Cons.* 103:103–11.

Azevedo-Ramos, C., and W. E. Magnusson. 1999. Tropical tadpole vulnerability to predation: Association between laboratory results and prey distribution in an Amazonian savanna. *Copeia* 1999:58–67.

Azevedo-Ramos, C., W. E. Magnusson, and P. Bayliss. 1999. Predation as the key factor structuring tadpole assemblages in a savanna area in Central Amazonia. *Copeia* 1999:22–33.

Azevedo-Ramos, C., M. Van Sluys, J.-M. Hero, and W. E. Magnusson. 1992. Influence of tadpole movement on predation by odonate naiads. *J. Herpetol.* 26:335–38.

Azizi, E., and T. Landberg. 2002. Effects of metamorphosis on the aquatic escape response of the two-lined salamander (*Eurycea bislineata*). *J. Exp. Biol.* 205:841–49.

Azmi, H. K. 1990. Use of poikilothermic vertebrates as traditional drugs in certain tribes of eastern (Purvanchal) Uttar Pradesh. *Uttar Pradesh J. Zool.* 10:80–87.

Babbitt, K. J. 2001. Behaviour and growth of southern leopard frog (*Rana sphenocephala*) tadpoles: Effects of food and predation risk. *Can. J. Zool.* 79:809–14.

———. 2005. The relative importance of wetland size and hydroperiod for amphibians in southern New Hampshire, USA. *Wetlands Ecol. Manage.* 13:269–79.

Babbitt, K. J., M. J. Baber, and T. L. Tarr. 2003. Patterns of larval amphibian distribution along a wetland hydroperiod gradient. *Can. J. Zool.* 81:1539–52.

Babbitt, K. J., and F. Jordan. 1996. Predation on *Bufo terrestris* tadpoles: Effects of cover and predator identity. *Copeia* 1996:485–88.

Babbitt, K. J., and W. E. Meshaka, Jr. 2000. Benefits of eating conspecifics: Effects of background diet on survival and metamorphosis in the Cuban treefrog (*Osteopilus septentrionalis*). *Copeia* 2000:469–74.

Babbitt, K. J., and G. W. Tanner. 1997. Effects of cover and predator identity on predation of *Hyla squirella* tadpoles. *J. Herpetol.* 31:128–30.

———. 2000. Use of temporary wetlands by anurans in a hydrologically modified landscape. *Wetlands* 20:313–22.

Baber, M. J., and K. J. Babbitt. 2003. The relative impacts of native and introduced predatory fish on a temporary wetland tadpole assemblage. *Oecologia* 136:289–95.

Baber, M. J., E. Fleishman, K. J. Babbitt, and T. L. Tarr. 2004. The relationship between wetland hydroperiod and nestedness patterns in assemblages of larval amphibians and predatory macroinvertebrates. *Oikos* 107:16–27.

Babik, W. 1998. Intrageneric predation in larval newts (*Triturus,* Salamandridae, Urodela). *Amphibia-Reptilia* 19:446–51.

Babik, W., J. M. Szymura, and J. Rafiński. 2003. Nuclear markers, mitochondrial DNA and male secondary sexual traits variation in a newt hybrid zone (*Triturus vulgaris* X *T. montandoni*). *Mol. Ecol.* 12:1913–30.

Bachmann, K. 1969. Temperature adaptations of amphibian embryos. *Am. Nat.* 102:115–30.

Bachmann, M. D. 1964. *Maternal behavior of the red-backed salamander Plethodon cinereus.* Unpublished PhD dissertation. Ann Arbor: University of Michigan.

———. 1984. Defensive behavior of brooding female red-backed salamanders (*Plethodon cinereus*). *Herpetologica* 40:436–43.

Bachmann, M. D., R. G. Carlton, J. M. Burkholder, and R. G. Wetzel. 1986. Symbiosis between salamander eggs and green algae: Microelectrode measurements inside eggs demonstrate effect of photosynthesis on oxygen concentration. *Can. J. Zool.* 64:1586–88.

Backwell, P. R. Y. 1988. Functional partitioning in the two-part call of the leaf-folding frog, *Afrixalus brachycnemis. Herpetologica* 44:1–7.

———. 1991. *Aspects of reproduction in the leaf-folding frog,*

*Afrixalus delicatus*. Unpublished PhD dissertation. Johannesburg, South Africa: University of the Witwatersrand.

Backwell, P. R. Y., and M. D. Jennions. 1993. Mate choice in the Neotropical frog, *Hyla ebraccata:* Sexual selection, mate recognition and signal selection. *Anim. Behav.* 45:1248–50.

Backwell, P. R. Y., and N. I. Passmore. 1990. Suitable approach perches affect female phonotaxis in an arboreal frog. *Herpetologica* 46:11–14.

———. 1991a. Satellite behavior in the leaf-folding frog, *Afrixalus delicatus. J. Herpetol.* 25:497–98.

———. 1991b. Sonic complexity and mate localization in the leaf-folding frog, *Afrixalus delicatus. Herpetologica* 47: 226–29.

Baer, J.-G. 1937. L'appareil respiratoire des gymnophiones. *Rev. Suisse Zool.* 44:353–77.

Báez, A. M. 1996. The fossil record of the Pipidae. In *The biology of Xenopus,* ed. R. C. Tinsley and H. R. Kobel, 329–47. New York: Oxford University Press.

Báez, A. M., and L. A. Pugener. 2003. Ontogeny of a new Palaeogene pipid frog from southern South America and xenopodinomorph evolution. *Zool. J. Linn. Soc.* 139:439–76.

Bagnara, J. T., L. Iela, F. Morrisett, and R. K. Rastogi. 1986. Reproduction in the Mexican leaf frog (*Pachymedusa dacnicolor*). I. Behavioral and morphological aspects. *Occ. Pap. Mus. Nat. Hist. Univ. Kansas* (121):1–31.

Bailey, M. A. 2005. *Necturus alabamensis* Viosca 1937. Black Warrior water dog. In *Amphibian declines: The conservation status of United States species,* ed. M. J. Lannoo, 866–68. Berkeley: University of California Press.

Bailey, P. C. E. 1986. The feeding behaviour of a sit-and-wait predator, *Ranatra dispar* (Heteroptera: Nepidae): The combined effect of food deprivation and prey size on the behavioural components of prey capture. *Ethology* 71:315–32.

Bailey, W. J., and G. K. Morris. 1986. Confusion of phonotaxis by masking sounds in the bushcricket *Conocephalus brevipennis* (Tettigoniidae: Conocephalinae). *Ethology* 73:19–28.

Bailey, W. J., and J. D. Roberts. 1981. The bioacoustics of the burrowing frog *Heleioporus* (Leptodactylidae). *J. Nat. Hist.* 15:693–702.

Bailey, W. J., P. C. Withers, M. Endersby, and K. Gaull. 1993. The energetic cost of calling in the bushcricket *Requena verticalis* (Orthoptera: Tettigoniidae: Listroscelidinae). *J. Exp. Biol.* 178:21–37.

Bain, R. H., A. Lathrop, R. W. Murphy, N. L. Orlov, and T. C. Ho. 2003. Cryptic species of a cascade frog from Southeast Asia: Taxonomic revisions and descriptions of six new species. *Amer. Mus. Novitates* (3417):1–60.

Baird, R. A., and E. R. Lewis. 1986. Correspondences between afferent innervation patterns and response dynamics in the bullfrog utricle and lagena. *Brain Res.* 369:48–64.

Baird, T. A. 1983. Influence of social and predatory stimuli on the air-breathing behavior of the African clawed frog, *Xenopus laevis. Copeia* 1983:411–20.

Bajger, J. 1980. Diversity of defensive responses in populations of fire toads (*Bombina bombina* and *Bombina variegata*). *Herpetologica* 36:133–37.

Baker, C. L. 1945. The natural history and morphology of Amphiumae. *Rep. Reelfoot Lake Biol. Sta.* 22:55–91.

Baker, C. L., L. C. Baker, and M. F. Caldwell. 1947. Observation of copulation in *Amphiuma tridactylum. J. Tennessee Acad. Sci.* 22:87–88.

Baker, G. C., T. J. C. Beebee, and M. A. Ragan. 1999. *Prototheca richardsi,* a pathogen of anuran larvae, is related to a clade of protistan parasites near the animal-fungal divergence. *Microbiology* 145:1777–84.

Baker, J. 1997. Stability for Swedish treefrogs? *Froglog* (20):2–3.

Baker, J. K. 1961. Distribution of and key to the neotenic *Eurycea* of Texas. *Southwest. Nat.* 6:27–32.

———. 1966. *Eurycea troglodytes. Cat. Amer. Amphib. Rept.* 23.1–23.2.

Baker, J. M. R. 1990. Body size and spermatophore production in the smooth newt (*Triturus vulgaris*). *Amphibia-Reptilia* 11:173–84.

———. 1992. Body condition and tail height in great crested newts, *Triturus cristatus. Anim. Behav.* 43:157–59.

———. 1999. Abundance and survival rates of great crested newts (*Triturus cristatus*) at a pond in central England: Monitoring individuals. *Herpetol. J.* 9:1–8.

Baker, J. M. R., and T. R. Halliday. 1999. Amphibian colonization of new ponds in an agricultural landscape. *Herpetol. J.* 9:55–63.

Bakkegard, K. A., and C. Guyer. 2004. Sexual size dimorphism in the Red Hills salamander, *Phaeognathus hubrichti* (Caudata: Plethodontidae: Desmognathinae). *J. Herpetol.* 38: 8–15.

Baldauf, R. J. 1947. *Desmognathus fuscus fuscus* eating eggs of its own species. *Copeia* 1947:66.

Baldauf, R. J. 1952. Climatic factors influencing the breeding migration of the spotted salamander, *Ambystoma maculatum* (Shaw). *Copeia* 1952:178–81.

Baldwin, A. L., J. S. Rozum, and R. W. Gore. 1990. Routes of water and small solute transport between blood and lymph in frog. *Physiol. Zool.* 63:1141–56.

Baldwin, J., G. Friedman, and H. Lillywhite. 1977. Adaptation to temporary muscle anoxia in anurans: Activities of glycolytic enzymes in muscles from species differing in their ability to produce lactate during exercise. *Aust. J. Zool.* 25:15–18.

Baldwin, G. F., and P. J. Bentley. 1982. Roles of the skin and gills in sodium and water exchange in neotenic urodele amphibians. *Am. J. Physiol.* 242:R94–R96.

Baldwin, P. H., C. W. Schwartz, and E. R. Schwartz. 1952. Life history and economic status of the mongoose in Hawaii. *J. Mammal.* 33:335–56.

Baldwin, R. A. 1974. The water balance response on the pelvic "patch" of *Bufo punctatus* and *Bufo boreas. Comp. Biochem. Physiol.* 47A:1285–95.

Balfour, P. S., and E. W. Stitt. 2003. *Ambystoma californiense* (California tiger salamander). Predation. *Herp. Rev.* 34:44.

Balinsky, B. I. 1969. The reproductive ecology of amphibians of the Transvaal Highveld. *Zool. Afr.* 4:37–93.

———. 1985. Observations on the breeding of toads in a restricted habitat. *S. Afr. J. Sci.* 20:61–64.

Balinsky, B. I., and J. B. Balinsky. 1954. On the breeding habits of the South African bullfrog, *Pyxicephalus adspersus. S. Afr. J. Sci.* 51:55–58.

Balinsky, J. B. 1970. Nitrogen metabolism in amphibians. In *Comparative biochemistry of nitrogen metabolism. Vol. 2: The vertebrates,* ed. J. W. Campbell, 519–637. New York: Academic Press.

———. 1981. Adaptation of nitrogen metabolism to hyperosmotic environment in Amphibia. *J. Exp. Zool.* 215:335–50.

Balinsky, J. B., and E. Baldwin. 1961. The mode of excretion of ammonia and urea in *Xenopus laevis. J. Exp. Biol.* 38:695–705.

Balinsky, J. B., S. M. Chemaly, A. E. Currin, A. R. Lee, R. L. Thompson, and D. R. Van der Westhuizen. 1976. A comparative study of enzymes of urea and uric acid metabolism in different species of amphibians, and the adaptation to the environment of the tree frog *Chiromantis xerampelina* Peters. *Comp. Biochem. Physiol.* 54B:549–55.

Balinsky, J. B., E. L. Choritz, C. G. Coe, and G. S. van de Schans. 1967. Amino acid metabolism and urea synthesis in naturally aestivating *Xenopus laevis. Comp. Biochem. Physiol.* 22:59–68.

Balinsky, J. B., M. M. Cragg, and E. Baldwin. 1961. The adaptation of amphibian waste nitrogen excretion to dehydration. *Comp. Biochem. Physiol.* 3:236–44.

Balinsky, J. B., S. E. Dicker, and A. B. Elliot. 1972. The effect of long-term adaptation to different levels of salinity on urea synthesis and tissue amino acid concentrations in *Rana cancrivora. Comp. Biochem. Physiol.* 43B:71–82.

Ball, J. N. 1981. Hypothalamic control of pars distalis in fishes, amphibians, and reptiles. *Gen. Comp. Endocrinol.* 44:135–70.

Ball, S. C. 1936. The distribution and behavior of the spadefoot toad in Connecticut. *Trans. Connecticut Acad. Arts Sci.* 32:351–79.

Balletto, E., and C. Giacoma. 1993. Stochastic extinction probability for European populations of *Hyla arborea:* An approach by VORTEX. In *Ecology and conservation of the European tree frog,* ed. A. H. P. Stumpel and U. Tester, 81–90. Wageningen, The Netherlands: DLO Institute for Forestry and Nature Research.

Ballinger, R. E., and C. O. McKinney. 1966. Developmental temperature tolerance of certain anuran species. *J. Exp. Zool.* 161:21–28.

Balment, R. J., and I. W. Henderson. 1987. Secretion of endocrine glands and their relationship to osmoregulation. In *Fundamentals of comparative vertebrate endocrinology,* ed. I. Chester-Jones, P. M. Ingleton, and J. G. Phillips, 413–508. New York: Plenum.

Balmford, A., L. Bartos, P. Brotherton, H. Herrmann, J. Lancingerova, J. Mika, and U. Zeeb. 1993. When to stop lekking: Density-related variation in the rutting behaviour of sika deer. *J. Zool., London* 231:652–56.

Balon, E. K. 1975. Reproductive guilds of fishes: A proposal and definition. *J. Fish. Res. Board Canada* 32:821–64.

Balon, E. K. 1981. Additions and amendments to the classification of reproductive styles of fishes. *Env. Biol. Fish.* 6:377–89.

Bamford, M. J. 1992. The impact of fire and increasing time since fire upon *Heleioporus eyrei, Limnodynastes dorsalis,* and *Myobatrachus gouldii* (Anura: Leptodactylidae) in Banksia woodland near Perth, Western Australia. *Wildl. Res.* 19:169–78.

Banks, B., and T. J. C. Beebee. 1986a. Climate effects on calling and spawning of the natterjack toad *Bufo calamita.* Discriminant analysis and applications for conservation monitoring. *Biol. Cons.* 36:339–50.

———. 1986b. A comparison of the fecundities of two species of toad (*Bufo bufo* and *B. calamita*) from different habitat types in Britain. *J. Zool., London* (A) 208:325–37.

———. 1987a. Factors influencing breeding site choice by the pioneering amphibian *Bufo calamita. Holarc. Ecol.* 10:14–21.

———. 1987b. Spawn predation and larval growth inhibition as mechanisms for niche separation in anurans. *Oecologia* 72:569–73.

———. 1988. Reproductive success of natterjack toads *Bufo calamita* in two contrasting habitats. *J. Anim. Ecol.* 57:475–92.

Banks, B., T. J. C. Beebee, and A. S. Cooke. 1994. Conservation of the natterjack toad *Bufo calamita* in Britain over the period 1970–1990 in relation to site protection and other factors. *Biol. Cons.* 67:111–18.

Banks, B., T. J. C. Beebee, and J. S. Denton. 1993. Long-term management of a natterjack toad (*Bufo calamita*) population in southern Britain. *Amphibia-Reptilia* 14:155–68.

Bannikov, A. G. 1940. The ecological conditions of the hibernation of the grass-frog (*Rana temporaria* L.) in Moscow Distrc. *Sbornik Nauchnykh Studencheskikh Rabot Moskovskogo Gosudarstvennogo Universiteta, Zoologiya* 16:41–64 (in Russian).

———. 1948. On the fluctuation of anuran populations. *Dokl. Acad. Nauk. USSR* 61:131–34 (in Russian).

———. 1958. Die Biologie des Froschzahnmolches, *Ranodon sibiricus* Kessler. *Zool. Jb. Syst.* 86:245–49.

Banta, A. M. 1914. Sex recognition and the mating behavior of the wood frog, *Rana sylvatica. Biol. Bull.* 26:171–83.

Ba-Omar, T. A., J. R. Downie, and W. J. P. Barnes. 2000. Development of adhesive toe-pads in the tree-frog (Phyllomedusa trinitatis). *J. Zool., London* 250:267–82.

Barach, J. P. 1951. The value of the skin secretions of the spotted salamander. *Herpetologica* 7:58.

Barandun, J. 1990. Reproduction of yellow-bellied toads *Bombina variegata* in a man-made habitat. *Amphibia-Reptilia* 11:277–84.

Barbault, R. 1972. Les peuplements d'amphibiens des Savanes de Lamto (Côte-d'Ivoire). *Ann. Univ. Abidjan, Ser. E (Écologie)* 5:61–142.

———. 1974a. Le régime alimentaire des amphibiens de la Savane de Lamto (Côte d'Ivoire). *Bull. IFAN* 36:952–72.

———. 1974b. Les peuplements d'amphibiens et de reptiles de la Savane de Lamto. *Bull. Liaison des Chercheus de Lamto,* special number 1974:1–37.

———. 1974c. Observations écologiques dans la Savane de Lamto (Côte d'Ivoire): structure de l'herpétocén ose. *Bull. Écol.* 5:7–25.

———. 1976a. Etude quantitative des peuplements d'amphibiens et le lezards d'une savane arbustive de la region de Bouake (Cote d'Ivoire): Densites et cycles saisonniers d'abondance. *Ann. Univ. Abidjan, Ser. E (Écologie)* 9:385–92.

———. 1976b. Notes sur la composition et la diversité specifiques d'une herpétocénose tropicale (Bouake, Côte d'Ivoire). *Bull. IFAN* 38:445–56.

———. 1977. Structure et dynamique d'une herpétocénose de savane (Lamto, Côte d'Ivoire). *Geo.- Eco.-Trop.* 4:309–34.

———. 1984. Strategies de reproduction et démographie de quelques amphibiens anoures tropicaux. *Oikos* 43:77–87.

———. 1987. Pression de predation et evolution des strategies démographiques en zone tropicale: Les cas de lezards et des amphibiens. *Rev. Zool Afr.* 101:301–27.

———. 1991. Ecological constraints and community dynamics: Linking community patterns to organismal ecology. The case of tropical herpetofaunas. *Acta Oecol.* 12:139–63.

Barbault, R., J. Castanet, H. Francillon, and A. de Ricqles. 1979. Determination de l'age chez un anoure deserticole *Bufo pentoni,* Anderson 1893. *Terre Vie* 33:129–41.

Barbault, R., and H. B. Lillywhite. 2005. Body wiping behaviors associated with cutaneous lipids in hylid tree frogs of Florida. *J. Exp. Biol.* 208:2147–56.

Barbault, R., and T. Pilorge. 1980. Observations sur la reproduction et la dynamique des populations de quelques anoures tropicaux. V. *Phrynobatrachus calcaratus. Acta Oecol., Oecol. Gener.* 1:373–82.

Barbault, R., and M. Trefaut Rodrigues. 1978a. Observations sur la reproduction et la dynamique des populations de quelques anoures tropicaux. I. *Ptychadena maccarthyensis* et *Ptychadena oxyrhynchus*. *Terre Vie* 32:441–52.

———. 1978b. Observations sur la reproduction et la dynamique des populations de quelques anoures tropicaux. II. *Phrynobatrachus plicatus* (Gunther). *Geo-Eco-Trop.* 2:455–66.

———. 1979a. Observations sur la reproduction et la dynamique des populations de quelques anoures tropicaux. III. *Arthroleptis poecilontus. Trop. Ecol.* 20:64–77.

———. 1979b. Observations sur la reproduction et la dynamique des populations de quelques anoures tropicaux. IV. *Phrynobatrachus accraensis. Bull. IFAN* 41:417–28.

Barber, P. H. 1999. Patterns of gene flow and population genetic structure in the canyon treefrog, *Hyla arenicolor* (Cope). *Mol. Ecol.* 8:563–76.

Barbour, R. W. 1950. The reptiles of Big Black Mountain, Harlan County, Kentucky. *Copeia* 1950:100–107.

———. 1956. A study of the cottonmouth, *Agkistrodon piscivorus leucostoma* Troost, in Kentucky. *Trans. Kentucky Acad. Sci.* 17:33–41.

Barbour, R. W., J. W. Hardin, J. P. Schafer, and M. J. Harvey. 1969. Home range, movements, and activity of the dusky salamander, *Desmognathus fuscus. Copeia* 1969:293–97.

Barbour, T. 1926. *Reptiles and amphibians. Their habits and adaptations.* Boston: Houghton Mifflin.

Barbour, T., and A. Loveridge. 1928. A comparative study of the herpetological faunae of the Uluguru and Usumbara Mountains, Tanganyike Territory, with descriptions of new species. *Mem. Mus. Comp. Zool.* 50:85–265.

Bardsley, L., and T. J. C. Beebee. 1998a. Interspecific competition between *Bufo* larvae under conditions of community transition. *Ecology* 79:1751–59.

———. 1998b. Interspecific competition between larvae is not an important structuring force in mixed communities of *Rana* and *Bufo* on an English sand-dune system. *Ecography* 21:449–56.

———. 2000. Competition between *Bufo* larvae in a eutrophic pond. *Oecologia* 124:33–39.

———. 2001. Non-behavioural interference competition between anuran larvae under semi-natural conditions. *Oecologia* 128:360–67.

Barfield, S. 1986. Indonesia's frog legs. *J. Env. Health* 48:324.

Barica, J., and J. A. Mathias. 1979. Oxygen depletion and winterkill risk in small prairie lakes under extended ice cover. *J. Fish. Res. Board Canada* 36:980–86.

Barker, J., and G. Grigg. 1977. *A field guide to Australian frogs.* Adelaide, Australia: Rigby.

Barker, J., G. C. Grigg, and M. J. Tyler. 1995. *A field guide to Australian frogs.* Chipping Norton, New South Wales, Australia: Surrey Beatty and Sons.

Barlow, G. W. 1981. Patterns of parental investment, dispersal and size among coral-reef fishes. *Env. Biol. Fish.* 6:65–85.

Barnett, H. K., and J. S. Richardson. 2002. Predation risk and competition effects on the life-history characteristics of larval Oregon spotted frog and larval red-legged frog. *Oecologia* 132:436–44.

Barr, G. E., and K. J. Babbitt. 2002. Effects of biotic and abiotic factors on the distribution and abundance of larval two-lined salamanders (*Eurycea bislineata*) across spatial scales. *Oecologia* 133:176–85.

Barreto, L., and G. V. Andrade. 1995. Aspects of the reproductive biology of *Physalaemus cuvieri* (Anura: Leptodactylidae) in northeastern Brazil. *Amphibia-Reptilia* 16:67–76.

Barreto, L., and G. Moreira. 1996. Seasonal variation in age structure and spatial distribution of a savanna larval anuran assemblage in central Brazil. *J. Herpetol.* 30:87–92.

Barrio, A. 1963. Consideraciones sobre comportamiento y "grito aggresivo" propio de algunas especies de Ceratophrynidae (Anura). *Physis (Buenos Aires)* 24:143–48.

———. 1965. El genero *Physalaemus* (Anura: Leptodactylidae) en la Argentina. *Physis (Buenos Aires)* 25:421–48.

———. 1967. *Batrachyla antartandica* n. sp. (Anura, Leptodactylidae)/descripción y estudio comparativo con la especie genotipica, *B. leptopus* Bell. *Physis (Buenos Aires)* 27:101–9.

Barry, S. J., and H. B Shaffer. 1994. The status of the California tiger salamander (*Ambystoma californiense*) at Lagunita: A 50-year update. *J. Herpetol.* 28:159–64.

Bartelt, P. E., and C. R. Peterson. 2005. Physically modeling operative temperatures and evaporation rates in amphibians. *J. Therm. Biol.* 30:93–102.

Bartelt, P. E., C. R. Peterson, and R. W. Klaver. 2004. Sexual differences in the post-breeding movements and habitats selected by western toads (*Bufo boreas*) in southeastern Idaho. *Herpetologica* 60:455–67.

Barth, L. G. 1946. Studies on the metabolism of development. *J. Exp. Zool.* 103–104:463–86.

Barthalmus, G. T. 1994. Biological roles of amphibian skin secretions. In *Amphibian biology. Vol. 1. The Integument,* ed. H. Heatwole and G. T. Barthalmus, 382–410. Chipping, New South Wales, Australia: Surrey Beatty and Sons.

Barthalmus, G. T., and E. D. Bellis. 1969. Homing in the northern dusky salamander, *Desmognathus fuscus fuscus* (Rafinesque). *Copeia* 1969:148–53.

———. 1972. Home range, homing and the homing mechanism of the salamander, *Desmognathus fuscus. Copeia* 1972:632–42.

Barthalmus, G. T., and I. R. Savidge. 1974. Time: An index of distance as a barrier to salamander homing. *J. Herpetol.* 8:251–54.

Bartholomew, G. A. 1982. Physiological control of body temperature. In *Biology of the Reptilia,* vol. 12, ed. C. Gans and F. H. Pough, 167–212. New York: Academic Press.

Bartman, C. E., K. C. Parker, J. Laerm, and T. S. McCay. 2001. Short-term response of Jordan's salamander to shelterwood timber harvest in western North Carolina. *Phys. Geog.* 22:154–66.

Barton, N. H., and G. M. Hewitt. 1989. Adaptation, speciation and hybrid zones. *Nature* 341:497–503.

Bastock, J., and H. T. Riley, eds. 1900. *The natural history of Pliny.* 6 vols. London: George Bell and Sons.

Bastos, R. P., and C. F. B. Haddad. 1996. Breeding activity of the Neotropical treefrog *Hyla elegans* (Anura, Hylidae). *J. Herpetol.* 30:355–60.

Bastos, R. P., and C. F. B. Haddad. 1999. Atividade reprodutiva de *Scinax rizibilis* (Bokermann) (Anura: Hylidae) na Floresta Atlântica, sudeste do Brasil. *Rev. Bras. Zool.* 16:409–21.

———. 2002. Acoustic and aggressive interactions in *Scinax rizibilis (*Anura: Hylidae) during the reproductive activity in southeastern Brazil. *Amphibia-Reptilia* 23:97–104.

Battaglin, W., and J. Fairchild. 2002. Potential toxicity of pesticides measured in midwestern streams to aquatic organisms. *Wat. Sci. Tech.* 45:95–102.

Baugh, J. R., and D. C. Forester. 1994. Prior residence effect in the dart-poison frog, *Dendrobates pumilio. Behaviour* 131:207–24.

Baudinette, R. V, A. M. Miller, and M. P. Sarre. 2000. Aquatic and terrestrial locomotory energetics in a toad and a turtle: A search for generalizations among ectotherms. *Physiol. Biochem. Zool.* 73:672–82.

Baumann, W. L., and M. Huels. 1982. Nests of the two-lined salamander, *Eurycea bislineata. J. Herpetol.* 16:81–83.

Baumgartner, A. 1999. Überwinterung von Teichmolchen (*Triturus vulgaris*) in einer Krauterspirale. *Z. Feldherpetol.* 6:219–38.

Baylis, J. R. 1981. The evolution of parental care in fishes, with reference to Darwin's rule of male sexual selection. *Env. Biol. Fish.* 6:223–51.

Bazilescu, I. 1996. Trading in frog legs. TED Case Studies no. 238 (an online journal available at http://www.american.edu/ted/frogs.htm).

Bea, A., J. D. Rodriguez-Teijeiro, and L. L. Jover. 1986. Relations between meterological variables and the initiation of the spawning period in populations of *Rana temporaria* in the Atlantic region of the Basque country (northern Spain). *Amphibia-Reptilia* 7:23–31.

Beachy, C. K. 1993a. Differences in variation in egg size for several species of salamanders (Amphibia: Caudata) that use different larval environments. *Brimleyana* 18:71–82.

———. 1993b. Guild structure in streamside salamander communities: A test for interactions among larval plethodontid salamanders. *J. Herpetol.* 27:465–68.

———. 1994. Community ecology in streams: Effects of two species of predatory salamanders on a prey species of salamander. *Herpetologica* 50:129–36.

———. 1995. Effects of larval growth history on metamorphosis in a stream-dwelling salamander (*Desmognathus ochrophaeus*). *J. Herpetol.* 29:375–82.

———. 1997a. Courtship behavior in the plethodontid salamander *Gyrinophilus porphyriticus. Herpetologica* 53:289–96.

———. 1997b. Effect of predatory larval *Desmognathus quadramaculatus* on growth, survival, and metamorphosis of larval *Eurycea wilderae. Copeia* 1997:131–37.

Beachy, C. K., and R. C. Bruce. 1992. Lunglessness in plethodontid salamanders is consistent with the hypothesis of a mountain stream origin: a response to Ruben and Boucot. *Am. Nat.* 139:839–47.

Beachy, C. K., T. H. Surges, and M. Reyes. 1999. Effects of developmental and growth history on metamorphosis in the gray treefrog, *Hyla versicolor* (Amphibia, Anura). *J. Exp. Zool.* 283:522–30.

Beadle, L. C. 1932. Observations on the bionomics of some East African swamps. *J. Linn. Soc. London* (*Zoology*) 38:135–55.

Beane, J. C. 2005. *Amphiuma means* (Two-toed Amphiuma). Predation. *Herp. Rev.* 36:295.

Beard, K. H., A. K. Eschtruth, K. A. Vogt, D. J. Vogt, and F. N. Scatena. 2003. The effects of the frog *Eleutherodactylus coqui* on invertebrates and ecosystem processes at two scales in the Luquillo Experimental Forest, Puerto Rico. *J. Trop. Ecol.* 19:607–17.

Beard, K. H., K. A. Voget, and A. Kulmatiski. 2002. Top-down effects of a terrestrial frog on forest nutrient dynamics. *Oecologia* 133:583–93.

Beasley, L. E., and S. W. Carothers. 1974. Unusual feeding habits in two species of blackbirds. *Wilson Bull.* 86:478–79.

Beattie, R. C. 1980. A physico-chemical investigation of the jelly capsules surrounding eggs of the common frog (*Rana temporaria temporaria*). *J. Zool., London* 190:1–25.

———. 1985. The date of spawning in populations of the common frog (*Rana temporaria*) from different altitudes in northern England. *J. Zool., London* (A) 205:137–54.

———. 1987. The reproductive biology of common frog (*Rana temporaria*) populations from different altitudes in northern England. *J. Zool., London* 211:387–98.

Beattie, R. C., and R. Tyler-Jones. 1992. The effects of low pH and aluminum on breeding success in the frog *Rana temporaria. J. Herpetol.* 26:353–60.

Beck, C. W. 1997. Effect of changes in resource level on age and size at metamorphosis in *Hyla squirella. Oecologia* 112:187–92.

———. 1998. Mode of fertilization and parental care in anurans. *Anim. Behav.* 55:439–449.

Beck, C. W., and J. D. Congdon. 1999. Effects of individual variation in age and size at metamorphosis on growth and survivorship of southern toad (*Bufo terrestris*) metamorphs. *Can. J. Zool.* 77:944–51.

———. 2000. Effects of age and size at metamorphosis on performance and metabolic rates of southern toad, *Bufo terrestris,* metamorphs. *Funct. Ecol.* 14:32–38.

Becker, R. P., and R. E. Lombard. 1977. Structural correlates of function in the 'opercularis' muscle of amphibians. *Cell Tissue Res.* 175:499–522.

Beckers, O. M., and J. Schul. 2004. Phonotaxis in *Hyla versicolor* (Anura, Hylidae): The effect of absolute call amplitude. *J. Comp. Physiol. A* 190:869–76.

Bee, M. A. 2002. Territorial male bullfrogs (*Rana catesbeiana*) do not assess fighting ability based on size-related variation in acoustic signals. *Behav. Ecol.* 13:109–24.

———. 2003a. Experience-based plasticity of acoustically evoked aggression in a territorial frog. *J. Comp. Physiol. A* 189:485–496.

———. 2003b. A test of the "dear enemy effect" in the strawberry dart-poison frog (*Dendrobates pumilio*). *Behav. Ecol. Sociobiol.* 54:601–10.

———. 2004. Within-individual variation in bullfrog vocalization: Implications for a vocally mediated social recognition system. *J. Acoust. Soc. Am.* 116:3770–81.

Bee, M. A., and A. C. Bowling. 2002. Socially mediated pitch alteration by territorial male bullfrogs, *Rana catesbeiana. J. Herpetol.* 36:140–43.

Bee, M. A., and H. C. Gerhardt. 2001a. Habituation as a mechanism of reduced aggression between neighboring territorial male bullfrogs, *Rana catesbeiana. J. Comp. Psychol.* 115:68–82.

———. 2001b. Neighbor-stranger discrimination by territorial bullfrogs (*Rana catesbeiana*): I. Acoustic basis. *Anim. Behav.* 62:1129–40.

———. 2001c. Neighbor-stranger discrimination by territorial bullfrogs (*Rana catesbeiana*): II. Perceptual basis. *Anim. Behav.* 62:1141–50.

———. 2002. Individual voice recognition in a territorial frog (*Rana catesbeiana*). *Proc. Roy. Soc. London B* 269:1443–48.

Bee, M. A., C. E. W. Kozich, K. J. Blackwell, and H. C. Gerhardt. 2001. Individual variation in advertisement calls of territorial male green frogs, *Rana clamitans:* Implications for individual discrimination. *Ethology* 107:65–84.

Bee, M. A., and S. A. Perrill. 1996. Responses to conspecific advertisement calls in the green frog (*Rana clamitans*) and their role in male-male communication. *Behaviour* 133:283–301.

Bee, M. A., S. A. Perrill, and P. C. Owen. 1999. Size assessment in simulated territorial encounters between male green frogs (*Rana clamitans*). *Behav. Ecol. Sociobiol.* 45:177–84.

Bee, M. A., S. A. Perrill, and P. C. Owen. 2000. Male green frogs lower the pitch of acoustic signals in defense of territories: A possible dishonest signal of size? *Behav. Ecol.* 11:169–77.

Bee, M. A., and T. R. Schachtman. 2000. Is habituation a mechanism for neighbor recognition in green frogs? *Behav. Ecol. Sociobiol.* 48:165–68.

Beebee, T. J. C. 1977. Environmental change as a cause of natterjack toad (*Bufo calamita*) declines in Britain. *Biol. Cons.* 11:87–102.

———. 1979a. Habitats of the British amphibians (2): Suburban parks and gardens. *Biol. Cons.* 15:241–57.

———. 1979b. Population interactions between the toads *Bufo bufo* (Linnaeus) and *Bufo calamita* (Laurenti): Some theoretical considerations and consequences. *Br. J. Herpetol.* 6:1–5.

———. 1979c. A review of scientific information pertaining to the natterjack toad *Bufo calamita* throughout its geographical range. *Biol. Cons.* 16:107–34.

———. 1983. *The natterjack toad.* Oxford: Oxford University Press.

———. 1985. Discriminant analysis of amphibian habitat determinants in south-east England. *Amphibia-Reptilia* 6:35–43.

———. 1991. Purification of an agent causing growth inhibition in anuran larvae and its identification as a unicellular unpigmented alga. *Can. J. Zool.* 69:2146–53.

———. 1995a. Amphibian breeding and climate. *Nature* 374:219–20.

———.1995b. Tadpole growth: Is there an interference effect in nature? *Herpetol. J.* 5:204–5.

———. 1996. *Ecology and conservation of amphibians.* New York: Chapman and Hall.

———. 1997. Changes in dewpond numbers and amphibian diversity over 20 years on chalk downland in Sussex, England. *Biol. Cons.* 81:215–19.

———. 2002. Amphibian phenology and climatic change. *Cons. Biol.* 16:1454.

Beebee, T. J. C., J. Buckley, I. Evans, J. P. Foster, A. H. Gent, C. P. Gleed-Owen, G. Kelly, G. Rowe, C. Snell, J. T. Wycherley, and I. Zeisset. 2005. Neglected native or undesirable alien? Resolution of a conservation dilemma concerning the pool frog *Rana lessonae. Biodivers. Cons.* 14:1607–26.

Beebee, T. J. C., J. S. Denton, and J. Buckley. 1996. Factors affecting population densities of adult natterjack toads *Bufo calamita* in Britain. *J. Appl. Ecol.* 33:263–68.

Beebee, T. J. C., and J. R. Griffin. 1977. A preliminary investigation into natterjack toad *Bufo calamita* breeding site characteristics in Britain. *J. Zool., London* 181:341–50.

Beebee, T. J. C., and R. Griffiths. 2000. *Amphibians and reptiles. A natural history of the British herpetofauna.* London: HarperCollins.

Beebee, T. J. C., and R. A. Griffiths. 2005. The amphibian decline crisis: A watershed for conservation biology? *Biol. Cons.* 125:271–85.

Beebee, T. J. C., and A. L.-C. Wong. 1992. *Prototheca*-mediated interference competition between anuran larvae operates by resource diversion. *Physiol. Zool.* 65:815–31.

Begon, M. 1979. *Investigating animal abundance: Capture-recapture for biologists.* London: Edward Arnold.

Behangana, M. 2004. The diversity and status of amphibians and reptiles in the Kyoga Lake Basin. *Afr. J. Ecol.* 42 (Suppl. 1):51–56.

Behangana, M., and J. Arusi. 2004. The distribution and diversity of amphibian fauna of Lake Nabugabo and surrounding areas. *Afr. J. Ecol.* (Suppl. 1):6–13.

Beiswenger, R. E. 1975. Structure and function in aggregations of tadpoles of the American toad, *Bufo americanus. Herpetologica* 31:222–33.

———. 1977. Diel patterns of aggregative behavior in tadpoles of *Bufo americanus,* in relation to light and temperature. *Ecology* 58:98–108.

———. 1981. Predation by gray jays on aggregating tadpoles of the boreal toad (*Bufo boreas*). *Copeia* 1981:459–60.

Beiswenger, R. E., and F. H. Test. 1966. Effects of environmental temperature on movements of tadpoles of the American toad, *Bufo terrestris americanus. Proc. Michigan Acad. Sci. Arts Lett.* 51:127–41.

Beja, P., and R. Alcazar. 2003. Conservation of Mediterranean temporary ponds under agricultural intensification: An evaluation using amphibians. *Biol. Cons.* 114:317–26.

Beja, P. R. 1991. Diet of otters (*Lutra lutra*) in closely associated freshwater, brackish and marine habitats in southwest Portugal. *J. Zool., London* 225:141–52.

Belden, L. K., and A. R. Blaustein. 2002a. Population differences in sensitivity to UV-B radiation for larval long-toed salamanders. *Ecology* 83:1586–90.

———. 2002b. UV-B induced skin darkening in larval salamanders does not prevent sublethal effects of exposure on growth. *Copeia* 2002:748–54.

Belden, L. K., I. T. Moore, R. T. Mason, J. C. Wingfield, and A. R. Blaustein. 2003. Survival, hormonal stress response and UV-B avoidance in Cascades frog tadpoles (*Rana cascadae*) exposed to UV-B radiation. *Funct. Ecol.* 17:409–16.

Belden, L. K., E. L. Wildy, and A. R. Blaustein. 2000. Growth, survival and behaviour of larval long-toed salamanders (*Ambystoma macrodacylum*) exposed to ambient levels of UV-B radiation. *J. Zool., London* 251:473–79.

Beletsky, L., and G. H. Orians. 1997. *Red-winged blackbirds: Decision-making and reproductive success.* Chicago: University of Chicago Press.

Bell, B. D. 1978. Observations on the ecology and reproduction of the New Zealand leiopelmid frogs. *Herpetologica* 34:340–54.

———. 1982. The amphibian fauna of New Zealand. In *New Zealand herpetology,* ed. D. G. Newman, 27–89. Wellington, New Zealand: New Zealand Wildlife Service.

———. 1985. Development and parental care in the endemic New Zealand frogs. In *Biology of Australasian frogs and reptiles,* ed. G. Grigg, R. Shine, and H. Ehmann, 269–78. Chipping Norton, New South Wales, Australia:Surrey Beatty and Sons.

———. 1994. A review of the status of New Zealand *Leiopelma* species (Anura, Leiopelmatidae), including a summary of demographic studies in Coromandel and on Maud Island. *New Zealand J. Zool.* 21:341–49.

———. Recent population declines of Archey's frog (*Leiopelma archeyi*) in the central Coromandel range. *New Zealand J. Zool.* 26:255.

Bell, B. D., and A. P. Bell. 1995. Distribution of the introduced alpine newt *Triturus alpestris* and of native *Triturus* species in North Shropshire, England. *Aust. J. Ecol.* 20:367–75.

Bell, B. D., S. Carver, N. J. Mitchell, and S. Pledger. 2004. The recent decline of a New Zealand endemic: How and why did populations of Archey's frog *Leiopelma archeyi* crash over 1996–2001? *Biol. Cons.* 120:193–203.

Bell, B. D., C. H. Daugherty, and J. M. Hay. 1998. *Leiopelma pakeka,* n. sp. (Anura: Leiopelmatidae), a cryptic species of frog from Maud Island, New Zealand, and a reassessment of the conservation status of *L. hamiltoni* from Stephens Island. *J. Roy. Soc. New Zealand* 28:39–54.

Bell, B. D., S. Pledger, and P. L. Dewhurst. 2004. The fate of a population of the endemic frog *Leiopelma pakeka* (Anura: Leiopelmatidae) translocated to restored habitat on Maud Island, New Zealand. *New Zealand J. Zool.* 31:121–31.

Bell, B. D., and R. J. Wassersug. 2003. Anatomical features of *Leiopelma* embryos and larvae: Implications for anuran evolution. *J. Morphol.* 256:160–70.

Bell, E. A., and B. D. Bell. 1994. The local distribution, habitat, and numbers of the endemic terrestrial frog *Leiopelma hamiltoni* on Maud Island, New Zealand. *New Zealand J. Zool.* 21:437–42.

Bell, G. 1977. The life of the smooth newt (*Triturus vulgaris*) after metamorphosis. *Ecol. Monogr.* 47:279–99.

———. 1979. Populations of crested newts, *Triturus cristatus,* in Oxfordshire, England. *Copeia* 1979:350–53.

Bell, G., and J. H. Lawton. 1975. The ecology of the eggs and larvae of the smooth newt, *Triturus vulgaris* (Linn.). *J. Anim. Ecol.* 44:393–494.

Bell, G., and V. Koufopanou. 1986. The cost of reproduction. *Oxford Surv. Evol. Biol.* 3:83–131.

Bellis, E. D. 1959. A study of movement of American toads in a Minnesota bog. *Copeia* 1959:173–74.

———. 1965. Home range and movements of the wood frog in a northern bog. *Ecology* 46:90–98.

Bellocq, M. I., K. Kloosterman, and S. M. Smith. 2000. The diet of coexisting species of amphibians in Canadian jack pine forests. *Herpetol. J.* 10:63–68.

Bellrose, F. C. 1976. *Ducks, geese and swans of North America.* Harrisburg, PA: Stackpole Books.

Bellwood, D. R., T. P. Hughes, C. Folke, and M. Nystrom. 2004. Confronting the coral reef crisis. *Nature* 429:827–33.

Bellwood, J. J. 1990. Anti-predator defenses and ecology of Neotropical forest katydids, especially the Pseudophyllinae. In *The Tettigoniidae: Biology, systematics, and evolution,* ed. W. J. Bailey and D. C. F. Rentz, 8–26. New York: Springer-Verlag.

Bellwood, J. J., and G. K. Morris. 1987. Bat predation and its influence on calling behavior in Neotropical katydids. *Science* 238:64–67.

Belvedere, P., L. Colombo, C. Giacoma, G. Malacarne, and G. E. Andreoletti. 1988. Comparative ethological and biochemical aspects of courtship pheromones in European newts. *Mon. Zool. Ital.* (n.s.) 22:397–403.

Bemis, W. E., K. Schwenk, and M. H. Wake. 1983. Morphology and function of the feeding apparatus in *Dermophis mexicanus* (Amphibia: Gymnophiona). *Zool. J. Linn. Soc.* 77:75–96.

Bendix, J. H., Jr., and P. M. Narins. 1999. Competitive calling behavior by male treefrogs, *Eleutherodactylus coqui* (Anura: Leptodactylidae). *Copeia* 1999:1118–22.

Beneski, J. T., Jr., E. J. Zalisko, and J. H. Larsen, Jr. 1986. Demography and migratory patterns of the eastern long-toed salamander, *Ambystoma macrodactylum columbianum. Copeia* 1986:398–408.

Beniak, A., and R. Watka. 1962. Vascularization of respiratory surfaces in *Bufo cognatus* Say and *Bufo compactilis* Wiegman. *Bull. Acad. Polon. Sci., Ser. Sci. Biol.* 10:9–12.

Benl, G. 1965. Neotenie und Albinismus bei *Triturus vulgaris vulgaris. Salamandra* 1:6–14.

Bennett, A. F. 1974. Enzymatic correlates of activity metabolism in anuran amphibians. *Am. J. Physiol.* 226:1149–51.

———. 1978. Activity metabolism in the lower vertebrates. *Ann. Rev. Physiol.* 40:447–69.

———. 1980. The metabolic foundations of vertebrate behavior. *BioScience* 30:452–56.

———. 1987. Interindividual variability: An underutilized resource. In New *directions in ecological physiology,* ed. M. E. Feder, W. W. Burggren, A. F. Bennett, and R. B. Huey, 147–169. New York: Cambridge University Press.

Bennett, A. F., and W. R. Dawson. 1976. Metabolism. In *Biology of the Reptilia* vol. 5, ed. C. Gans and W. R. Dawson, 127–223. New York: Academic Press.

Bennett, A. F., T. Garland, Jr., and P. L. Else. 1989. Individual correlation of morphology, muscle mechanics and locomotion in a salamander. *Am. J. Physiol.* 256:R1200–R1208.

Bennett, A. F., and L. D. Houck. 1983. The energetic cost of courtship and aggression in a plethodontid salamander. *Ecology* 64:979–83.

Bennett, A. F., and R. B. Huey. 1990. Studying the evolution of physiological performance. *Oxford Surv. Evol. Biol.* 7:251–84.

Bennett, A. F., and P. Licht. 1973. Relative contributions of anaerobic and aerobic energy production during activity in Amphibia. *J. Comp. Physiol.* 87:351–60.

———. 1974. Anaerobic metabolism during activity in amphibians. *Comp. Biochem. Physiol.* 48A:319–27.

Bennett, A. F., and J. A. Ruben. 1979. Endothermy and activity in vertebrates. *Science* 206:649–54.

Bennett, A. F., and M. H. Wake. 1974. Metabolic correlates of activity in the caecilian *Geotrypetes seraphini. Copeia* 1974:764–69.

Bennett, D. 2002. Diet of juvenile *Varanus niloticus* (Sauria: Varanidae) on the Black Volta River in Ghana. *J. Herpetol.* 36:116–17.

Bennett, W. O., A. P. Summers, and E. L. Brainerd. 1999. Confirmation of the passive exhalation hypothesis for a terrestrial caecilian, *Dermophis mexicanus. Copeia* 1999:206–9.

Bensden, J. 1956. Shedding of the skin in the common toad, *Bufo bufo* (L.). *Vidensk. Medd. Naturhist. Foren. Kjobenhavn* 118:211–23.

Benson, A. B., ed. 1966. *Peter Kalm's travels in North America.* 2 vols. New York: Dover.

Benson, D. G., Jr. 1968. Reproduction in urodeles. II. Observations on the spermatheca. *Experientia* 24:853–54.

Benson, W. W. 1977. On the supposed spectrum between Batesian and Müllerian mimicry. *Evolution* 31:454–55.

Bent, A. C. 1961. *Life histories of North American birds of prey.* New York: Dover.

Bentley, P. J. 1966a. Adaptations of Amphibia to arid environments. *Science* 152:619–23.

———. 1966b. The physiology of the urinary bladder of Amphibia. *Biol. Rev.* 41:275–316.

———. 1969. Neurohypophyseal hormones in Amphibia: A comparison of their actions and storage. *Gen. Comp. Endocrinol.* 13:39–44.

———. 1971. *Endocrines and osmoregulation. A comparative account of the regulation of salt and water in vertebrates.* New York: Springer-Verlag.

———. 1973. Osmoregulation in the aquatic urodeles *Amphiuma means* (the Congo eel) and *Siren lacertina* (the mud eel). Effects of vasotocin. *Gen. Comp. Endocrinol.* 20:386–91.

———. 1974. Actions of neurohypophyseal hormones in amphibians, reptiles, and birds. In *Handbook of physiology, Sec.*

*7: Endocrinology, vol. 4,* ed. E. Knobil and W. H. Sawyer, 545–63. Washington, DC: American Physiological Society.

———. 1982. *Comparative vertebrate endocrinology.* 2nd ed. New York: Cambridge University Press.

Bentley, P. J., and G. F. Baldwin. 1980. Comparison of transcutaneous permeability in skins of larval and adult salamanders (*Ambystoma tigrinum*). *Am. J. Physiol.* 239:R505–R508.

Bentley, P. J., and H. Heller. 1964. The action of neurohypophysial hormones on the water and sodium metabolism of urodele amphibians. *J. Physiol.* 171:434–53.

———. 1965. The water-retaining action of vasotocin on the fire salamander (*Salamandra salamandra*): Role of the urinary bladder. *J. Physiol.* 181:124–29.

Bentley, P. J., A. K. Lee, and A. R. Main. 1958. Comparison of dehydration and hydration of two genera of frogs (*Heleioporus* and *Neobatrachus*) that live in areas of varying aridity. *J. Exp. Biol.* 35:677–84.

Bentley, P. J., and A. R. Main. 1972a. Effects of vasotocin on cutaneous water uptake by an Australian frog, *Crinia georgiana. Copeia* 1972:885–86.

———. 1972b. Zonal differences in permeability of the skin of some anuran Amphibia. *Am. J. Physiol.* 223:361–63.

Bentley, P. J., and K. Schmidt-Nielsen. 1966. Cutaneous water loss in reptiles. *Science* 151:1547–49.

———. 1971. Acute effects of sea water on frogs (*Rana pipiens*). *Comp. Biochem. Physiol.* 40A:547–48

Bentley, P. J., and J. W. Shield. 1973. Respiration of some urodele and anuran Amphibia. II. In air, role of the skin and lungs. *Comp. Biochem. Physiol.* 46A:29–38.

Bentley, P. J., and T. Yorio. 1976. The passive permeability of the skin of anuran Amphibia: A comparison of frogs (*Rana pipiens*) and toads (*Bufo marinus*). *J. Physiol.* 261:603–15.

———. 1977. The permeability of the skin of a neotenous urodele amphibian, the mudpuppy, *Necturus maculosus. J. Physiol.* 265:537–47.

———. 1979. Evaporative water loss in anuran Amphibia: A comparative study. *Comp. Biochem. Physiol.* 62A:1005–9.

Benton, M. J., ed. 1988. *The phylogeny and classification of the tetrapods. vol. 1: Amphibians, reptiles, birds.* New York: Oxford University Press.

Benton, M. J. 1990. *Vertebrate palaeontology.* London: Unwin Hyman.

———. 1997. *Vertebrate paleontology.* 2nd ed. London: Chapman and Hall.

Benz, G. W., and D. E. Collins, eds. 1997. *Aquatic fauna in peril: The southeastern perspective. Special Pub. 1, Southeast Aquatic Research Institute.* Decatur, GA: Lenz Design and Communications.

Bergen, K., R. D. Semlitsch, and H.-U. Reyer. 1997. Hybrid female matings are directly related to the availability of *Rana lessonae* and *Rana esculenta* males in experimental populations. *Copeia* 1997:275–83.

Berger, L. 1977. Systematics and hybridization in the *Rana esculenta* complex. In *The reproductive biology of amphibians,* ed. D. H. Taylor and S. I. Guttman, 367–88. New York: Plenum.

———. 1982. Hibernation of the European water frogs (*Rana esculenta* complex). *Zool. Pol.* 29:57–72.

———. 1989. Disappearance of amphibian larvae in the agricultural landscape. *Ecol. Int. Bull.* 17:65–73.

Berger, L., R. Speare, P. Dasak, D. E. Green, A. A. Cunningham, C. L. Goggin, R. Slocombe, M. A. Ragan, A. D. Hyatt, K. R. McDonald, H. B. Hines, K. R. Lips, G. Marantelli, and H. Parkes. 1998. Chytridiomycosis causes amphibian mortality associated with population declines in the rainforests of Australia and Central America. *Proc. Natl. Acad. Sci. USA* 95:9031–36.

Berger, L., R. Speare, H. B. Hines, G. Marantelli, A. D. Hyatt, K. R. McDonald, L. F. Skerratt, V. Olsen, J. M. Clarke, G. Gillespie, M. Mahony, N. Sheppard, C. Williams, and M. J. Tyler. 2004. Effect of season and temperature on mortality in amphibians due to chytridiomycosis. *Aust. Vet. J.* 82:31–36.

Berger, L., R. Speare, and A. Hyatt. 1999. Chytrid fungi and amphibian declines: overview, implications and future directions. In *Declines and disappearances of Australian frog,* ed. A. Campbell, 23–33. Canberra, Australia: Environment Australia.

Berger, L., R. Speare, A. Thomas, and A. Hyatt. 2001. Mucocutaneous fungal diseases in tadpoles of *Bufo marinus* in Australia. *J. Herpetol.* 35:330–35.

Berger, L., and J. Michalowski. 1963. *Keys for the identification of vertebrates of Poland, part 2. Amphibia.* Polish Scientific Publishers (English translation, Central Institute for Scientific, Technical and Economic Information, Warsaw, 1971).

Berger, L., and M. Rybacki. 1992. Sperm competition in European water frogs. *Alytes* 10:113–16.

Berk, L. 1938. Studies in the reproduction of *X. laevis.* 1. The relation of external environmental factors to the sexual cycle. *S. Afr. J. Med. Sci.* 3:72–77.

Berman, D. I. 1992. On places and conditions of hibernation of the Asiatic salamander (*Hynobius keyserlingii*) in the mountain-forest Subarctic zone in the North-East Asia. *Zool. Zh.* 71:75–95 (in Russian).

Berman, D. I., A. N. Leirkh, and E. I. Mikhailova. 1984. Winter hibernation of the Siberian salamander, *Hynobius keyserlingi. J. Evol. Biochem. Physiol.* 1984:323–27 (in Russian with English summary).

Bernarde, P. S., and R. A. Machado. 1999. *Hyla faber* (Smith frog). Larval cannibalism. *Herp. Rev.* 30:162.

Bernardo, J. 1994. Experimental analysis of allocation in two divergent, natural salamander populations. *Am. Nat.* 143:14–38.

———. 1996a. Maternal effects in ecology. *Amer. Zool.* 36:83–105.

———. 1996b. The particular maternal effect of propagule size, especially egg size: Patterns, models, quality of evidence and interpretations. *Amer. Zool.* 36:216–36.

———. 2002. *Desmognathus carolinensis* (Carolina Mountain salmander) and *Plethodon welleri* (Weller's salamander); *Desmognathus monticola* (seal salamander) and *Desmognathus wrighti* (pygmy salamander); *Gyrinophilus porphyriticus* (spring salamander) and *Desmognanthus ocoee* (Ocoee salamander). Intraguild predation. *Herp. Rev.* 33:121.

Bernardo, J., and S. J. Agosta. 2003. Determinants of clinal variation in life history of dusky salamanders (*Desmognathus ocoee*): Prey abundance and ecological limits on foraging time restrict opportunities for larval growth. *J. Zool., London* 259:411–21.

———. 2005. Evolutionary implications of hierarchical impacts of nonlethal injury on reprodution, including maternal effects. *Biol. J. Linn. Soc.* 86:309–31.

Bernardo, J., and S. J. Arnold. 2002. *Gyrinophilus porphyriticus* (spring salamander). Male combat. *Herp. Rev.* 33:121–22.

Bernardo, J., and N. L. Reagan-Wallin. 2002. Plethodontid salamanders do not conform to "general rules" for ectotherm life histories: Insights from allocation models about why simple models do not make accurate predictions. *Oikos* 97:398–414.

Bernardo, J., and J. R. Spotila. 2006. Physiological constraints on organismal response to global warming: mechanistic insights from clinally varying populations and implications of assessing endangerment. *Biol. Lett.* 2:135–39.

Bernal, X., and S. S. Ron. 2004. *Leptodactylus fragilis* (White-lipped Foamfrog). Courtship. *Herp. Rev.* 35:373–73.

Berninghausen, F. 1998. Fressen Kaulquappen europäischer Froschlurche Anurenlaich? *Salamandra* 34:255–68.

Berrill, M., S. Bertram, L. McGillivray, M. Kolohon, and B. Pauli. 1994. Effects of low concentrations of forest-use pesticides on frog embryos and tadpoles. *Env. Tox. Chem.* 13:657–64.

Berrill, M., D. Coulson, L. McGillivray, and B. Pauli. 1998. Toxicity of endosulfin to aquatic stages of anuran amphibians. *Env. Tox. Chem.* 17:1738–44.

Berrill, M., and D. R. S. Lean. 1998. Problems in conducting in situ experiments to evaluate the influence of UV-B radiation on amphibian eggs and larvae. In *Environmental toxicology and risk assessment. Vol. 7. ASTM STP 1333*, ed. E. E. Little, A. J. Delonay, and B. M. Greenberg, 21–30. West Conshoschocken, PA: American Society for Testing and Materials.

Berry, J. F., and R. Shine. 1980. Sexual size dimorphism and sexual selection in turtles (Order Testudines). *Oecologia* 44:185–91.

Berry, P. T. 1964. The breeding patterns of seven species of Singapore anura. *J. Anim. Ecol.* 33:227–43.

Berry, P. Y. 1970. The food of the giant toad *Bufo asper. J. Linn. Soc. London (Zoology)* 49:61–68.

Berry, P. Y., and J. A. Bullock. 1962. The food of the common Malayan toad, *Bufo melanostictus* Schneider. *Copeia* 1962:736–41.

Bertin, L. 1920. Les grenouilles peauveat-elles s'adapter a l'eau saumatre? *C. R. Soc. Biol.* 83:1308.

Bertoluci, J. 1998. Annual patterns of breeding activity in Atlantic rainforest anurans. *J. Herpetol.* 32:607–11.

Bertoluci, J., and M. T. Rodrigues. 2002. Seasonal patterns of breeding activity of Atlantic rainforest anurans at Boracéia, southeastern Brazil. *Amphibia-Reptilia* 23:161–67.

Bertram, B. C. R. 1978. Living in groups. In *Behavioural ecology: An evolutionary approach*, ed. J. R. Krebs and N. B. Davies, 64–96. Sunderland, MA: Sinauer Associates.

Bertram, S., M. Berrill, and E. Nol. 1996. Male mating success and variation in chorus attendance within and among breeding seasons in the gray treefrog (*Hyla versicolor*). *Copeia* 1996:729–34.

Berven, K. A. 1981. Mate choice in the wood frog, *Rana sylvatica. Evolution* 35:707–22.

———. 1982a. The genetic basis of altitudinal variation in the wood frog, *Rana sylvatica*. I. An experimental analysis of life history traits. *Evolution* 36:962–83.

———. 1982b. The genetic basis of altitudinal variation in the wood frog *Rana sylvatica*. II. An experimental analysis of larval development. *Oecologia* 52:360–69.

———. 1987. The heritable basis of variation in larval developmental patterns within populations of the wood frog (*Rana sylvatica*). *Evolution* 41:1088–97.

———. 1988. Factors affecting variation in reproductive traits within a population of wood frogs (*Rana sylvatica*). *Copeia* 1988:605–15.

———. 1990. Factors affecting population fluctuations in larval and adult stages of the wood frog (*Rana sylvatica*). *Ecology* 71:1599–1608.

———. 1995. Population regulation in the wood frog, *Rana sylvatica,* from three diverse geographic localities. *Aust. J. Ecol.* 20:385–92.

Berven, K. A., and R. S. Boltz. 2001. Interactive effects of leech (*Desserobdella picta*) infection on wood frog (*Rana sylvatica*) tadpole fitness traits. *Copeia* 2001:907–15.

Berven, K. A., and B. G. Chadra. 1988. The relationship among egg size, density and food level on larval development in the wood frog (*Rana sylvatica*). *Oecologia* 75:67–72.

Berven, K. A., and D. E. Gill. 1983. Interpreting geographic variation in life-history traits. *Amer. Zool.* 23:85–97.

Berven, K. A., D. E. Gill, and S. J. Smith-Gill. 1979. Countergradient selection in the green frog, *Rana clamitans. Evolution* 33:609–23.

Berven, K. A., and T. A. Grudzien. 1990. Dispersal in the wood frog (*Rana sylvatica*): Implications for genetic population structure. *Evolution* 44:2047–56.

Besharse, J. C., and R. A. Brandon. 1974. Size and growth of the eyes of the troglobitic salamander *Typhlotrion spelaeus. Int. J. Speleol.* 6:255–64.

Besharse, J. C., and J. R. Holsinger. 1977. *Gyrinophilus subterraneus,* a new troglobitic salamander from southern West Virginia. *Copeia* 1977:624–34.

Beshkov, V. A., and D. L. Jameson. 1980. Movement and abundance of the yellow-bellied toad *Bombina variegata. Herpetologica* 36:365–70.

Betts, E. S. 1927. Tawny owls and batrachian food. *British Birds* 21:41–43.

Beuchat, C. A., F. H. Pough, and M. M. Stewart. 1984. Response to simultaneous dehydration and thermal stress in three species of Puerto Rican frogs. *J. Comp. Physiol.* B 154:579–85.

Beumer, J. P. 1979. Feeding and movement of *Anguilla australis* and *A. reinhardtii* in Macleods Morass, Victoria, Australia. *J. Fish Biol.,* 14:573–92.

Bevier, C. R. 1995a. Biochemical correlates of calling activity in Neotropical frogs. *Physiol. Zool.* 68:1118–42.

———. 1995b. *Physiological constraints on calling activity in Neotropical frogs.* Unpublished PhD dissertation. Storrs: University of Connecticut.

———. 1997a. Breeding activity and chorus tenure of two Neotropical hylid frogs. *Herpetologica* 53:297–311.

———. 1997b. Utilization of energy substrates during calling activity in tropical frogs. *Behav. Ecol. Sociobiol.* 41:343–52.

Bevier, C. R., K. Larson, K. Reilly, and S. Tat. 2004. Vocal repertoire and calling activity of the mink frog, *Rana septentrionalis. Amphibia-Reptilia* 25:255–64.

Bhaduri, J. L. 1932. Observations on the urogenital system of the tree-frogs of the genus *Rhacophorus* Kuhl, with remarks on their breeding habits. *Anat. Anz.* 74:336–43.

———. 1935. The anatomy of the adhesive apparatus in the tadpoles of *Rana afghana* Günther, with special reference to the adaptive modifications. *Trans. Roy. Soc. Edinburgh* 58, part II:339–50.

Bhatnagar, R. K. 1970. The spider *Lycosa carmichaeli* Graveley as a predator of small frogs. *J. Bombay Nat. Hist. Soc.* 67:589.

Bhatta, G. 1997. Caecilian diversity of the Western Ghats: In search of the rare animals. *Curr. Sci.* 73:183–87.

Biavati, G. M., H. C. Wiederhecker, and G. R. Colli. 2004. Diet of *Epipedobates flavopictus* (Anura: Dendrobatidae) in a Neotropical savanna. *J. Herpetol.* 38:510–18.

Bickford, D. P. 2001. *The ecology and evolution of parental care* in the microhylid frogs of New Guinea. Unpublished PhD dissertation. Coral Gables, FL: University of Miami.

———. 2002. Male parenting of New Guinea froglets. *Nature* 418:601–2.

———. 2004. Differential parental care behaviors of arboreal and terrestrial microhylid frogs from Papua New Guinea. *Behav. Ecol. Sociobiol.* 55:402–9.

Bider, J. R., and K. A. Morrison. 1981. Changes in toad (*Bufo americanus*) responses to abiotic factors at the northern limit of their distribution. *Am. Midl. Nat.* 106:293–304.

Biek, R., W. C. Funk, B. A. Maxell, and L. S. Mills. 2002. What is missing in amphibian decline research: Insights from ecological sensitivity analysis. *Cons. Biol.* 16:728–34.

Biek, R., L. S. Mills, and R. B. Bury. 2002. Terrestrial and stream amphibians across clearcut-forest interfaces in the Siskiyou Mountains, Oregon. *Northwest Sci.* 76:129–40.

Bieniak, A., and R. Watra. 1962. Vascularization of respiratory surfaces in *Bufo cognatus* Say and *Bufo compactilus* Wiegman. *Acad. Pol. Sci. Ser. Sci. Biol. Bull.* 10:9–12.

Biju, S. D., and F. Bossuyt. 2003. New frog family from India reveals and ancient biogeographical link with the Seychelles. *Nature* 425:711–14.

Billo, R. 1986. Tentacle apparatus of caecilians. *Mem. Soc. Zool. France* 43:71–75.

Billo, R., and M. H. Wake. 1987. Tentacle development in *Dermophis mexicanus* (Amphibia, Gymnophiona) with an hypothesis of tentacle origin. *J. Morphol.* 192:101–11.

Bilski, F. 1921. Über den Einfluss des Lebensraums auf das Wachstum der Kaulquappen. *Archiv. Ges. Physiol.* 188:254–72.

Binckley, C. A., and W. J. Resetarits, Jr. 2002. Reproductive decisions under threat of predation: Squirrel treefrog (*Hyla squirella*) responses to banded sunfish (*Enneacanthus obesus*). *Oecologia* 130:157–61.

Binckley, C. A., and W. J. Resetarits, Jr. 2003. Functional equivalence of non-lethal effects: Generalized fish avoidance determines the distribution of gray treefrog, *Hyla chrysoscelis*, larvae. *Oikos* 102:623–29.

Bingley, W. 1868. *Bingley's history of animated nature.* Cincinnati: C.F. Vent.

Birkenmeyer, E. 1954. Beobachtungen zur Nährungsaufname und Paarungsbiologie der Gattung *Bombina. Verh. Zool.-Bot. Ges. Wien* 94:70–81.

Bisbal, F., and J. Ojasti. 1980. Nicho trofico del zorro *Cerdocyon thous* (*Mammalia*, Carnivora). *Acta Biol. Venezuela* 10:469–96.

Bisbal, F. J. 1986. Food habits of some Neotropical carnivores in Venezuela (*Mammalia*, Carnivora). *Mammalia* 50:329–39.

Bishai, H. M. 1962. The water characteristics of the Nile in the Sudan with a note on the effect of *Eichhornia crassipes* on the hydrobiology of the Nile. *Hydrobiologia* 19:357–82.

Bishop, C. A., N. A. Mahony, J. Struger, P. Ng, and K. E. Pettit. 1999. Anuran development, density and diversity in relation to agricultural activity in the Holland River watershed, Ont., Canada (1990–1992). *Environ. Monit. Assess.* 57:21–43.

Bishop, P. J. 1999. Declining frog populations in New Zealand—The New Zealand Frog Survey and possible future directions. *New Zealand J. Zool.* 26:255–56.

Bishop, P. J., M. D. Jennions, and N. I. Passmore. 1995. Chorus size and call intensity: Female choice in the painted reed frog, *Hyperolius marmoratus. Behaviour* 132:721–31.

Bishop, S. C. 1919. Notes on the habits and development of the four-toed salamander, *Hemidactylium scutatum* (Schlegel). *New York State Mus. Bull.* (219):251–82.

———. 1924. Notes on salamanders. *New York State Mus. Bull.* (253):87–96.

———. 1926. Notes on the habits and development of the mudpuppy, *Necturus maculosus* (Rafinesque). *New York State Mus. Bull.* (268):5–60.

———. 1932. The spermatophores of *Necturus maculosus* Rafinesque. *Copeia* 1932:1–3.

———. 1941a. Notes on salamanders, with descriptions of several new forms. *Occ. Papers Mus. Zool. Univ. Mich.* (451):1–27.

———. 1941b. The salamanders of New York. *New York State Mus. Bull.* (324):1–365.

———. 1947. *Handbook of salamanders.* Ithaca, NY: Comstock Publishers.

Bizer, J. R. 1978. Growth rates and size at metamorphosis of high elevation populations of *Ambystoma tigrinum. Oecologia* 34:175–84.

Bizerra, A. F., R. S. Jordao, and I. Sazima. 1994. *Simophis rhinostoma* (NCN). Diet. *Herp. Rev.* 25:72–73.

Bjatta, G. 1997. Caecilian diversity of the Western Ghats: In search of the rare animals. *Curr. Sci.* 73:183–87.

Blab, J. 1978. Untersuchungen zur Ökologie, Raum-Zeit-Einbindung und Funktion von Amphibienpopulationen. *Schr.-Reihe Landschaftspfl. Naturschutz* 18:1–141.

Black, J. H. 1970. A possible stimulus for the formation of some aggregations in tadpoles of *Scaphiopus bombifrons. Proc. Oklahoma Acad. Sci.* 49:13–14.

Black, J. H., and J. N. Black. 1969. Postmetamorphic basking aggregations of the boreal toad, *Bufo boreas boreas. Can. Field.-Nat.* 83:155–56.

Black, J. H., and R. B. Brunson. 1971. Breeding biology of the boreal toad (*Bufo boreas boreas*) (Baird and Girard) in western Montana. *Great Basin Nat.* 31:109–13.

Blackburn, D. G., R. S. Darrell, K. T. Lonergan, R. P. Mancini, and C. A. Sidor. 1995. Differential testosterone sensitivity of forelimb muscles of male leopard frogs, *Rana pipiens:* Test of a model system. *Amphibia-Reptilia* 16:351–56.

Blackith, R. M., and M. C. D. Speight. 1974. Food and feeding habits of the frog, *Rana temporaria*, in bogland habitats of the west of Ireland. *J. Zool., London* 172:67–79.

Blair, A. P. 1941a. Isolating mechanisms in tree frogs. *Proc. Natl. Acad. Sci. USA* 27:14–17.

———. 1941b. Variation, isolation mechanisms and hybridization in certain toads. *Genetics* 26:398–417.

———. 1942. Isolating mechanisms in a complex of four species of toads. *Biol. Symp.* 6:235–49.

———. 1943. Population structure in toads. *Am. Nat.* 77:363–568.

———. 1951. Winter activities in Oklahoma frogs. *Copeia* 1951:178.

Blair, J., and R. J. Wassersug. 2000. Variation in the pattern of predator-induced damage to tadpole tails. *Copeia* 2000:390–401.

Blair, W. F. 1936. A note on the ecology of *Microhyla olivacea. Copeia* 1936:115.

———. 1953. Growth, dispersal and age and sexual maturity of the Mexican toad (*Bufo valliceps* Wiegmann). *Copeia* 1953:208–12.

———. 1955. Mating call and stage of speciation in the *Microhyla olivaceous-M. carolinensis* complex. *Evolution* 9:469–80.

———. 1956a. Call difference as an isolating mechanism in southwestern toads (genus *Bufo*). *Texas J. Sci.* 8:87–106.

———. 1956b. The mating call of hybrid toads. *Texas J. Sci.* 8:350–55.

———. 1958a. Call difference as an isolation mechanism in Florida species of hylid frogs. *Q. J. Fla. Acad. Sci.* 21:32–48.

———. 1958b. Call structure and species groups in U.S. treefrogs (*Hyla*). *Southwest. Nat.* 5:129–35.

———. 1958c. Mating call and stage of speciation of two allopatric populations of spadefoots (*Scaphiopus*). *Texas J. Sci.* 10:484–88.

———. 1958d. Mating call in the speciation of anuran amphibians. *Am. Nat.* 92:27–51.

———. 1960. A breeding population of the Mexican toad (*Bufo valliceps*) in relation to its environment. *Ecology* 41:165–74.

———. 1961. Calling and spawning seasons in a mixed population of anurans. *Ecology* 42:99–110.

———. 1962. Non-morphological data in anuran classification. *Syst. Zool.* 11:72–84.

———. 1963. Acoustic behaviour of Amphibia. In *Acoustic behaviour of animals,* ed. R. G. Busnel, 694–708. New York: Elsevier.

———. 1964. Isolating mechanisms and interspecies interactions in anuran amphibians. *Q. Rev. Biol.* 39:334–44.

———. 1968. Amphibians and reptiles. In *Animal communication,* ed. T. A. Sebeok, 289–310. Bloomington: Indiana University Press.

———. 1972. *Bufo* of North and Central America. In *Evolution in the genus* Bufo, ed. W. F. Blair, 93–101. Austin: University of Texas Press.

———. 1974. Character displacement in frogs. *Amer. Zool.* 14:1119–25.

Blair, W. F., and M. J. Littlejohn. 1960. Stage of speciation of two allopatric populations of chorus frogs (*Pseudacris*). *Evolution* 14:82–87.

Blair, W. F., and D. Pettus. 1954. The mating call and its significance in the Colorado River toad (*Bufo alvarius*). *Texas J. Sci.* 6:72–77.

Blanchard, F. N. 1930. The stimulus to the breeding migration of the spotted salamander, *Ambystoma maculatum* (Shaw). *Am. Nat.* 64:154–67.

———. 1933a. Late autumn collections and hibernating situations of the salamander *Hemidactylium scutatum* (Schlegel) in southern Michigan. *Copeia* 1933:216.

———. 1933b. Spermatophores and the mating season of the salamander *Hemidactylium scutatum* (Schlegel). *Copeia* 1933:40.

———. 1934. The relation of the female four toed salamander to her nest. *Copeia* 1934:137–38.

———. 1936. The number of eggs produced and laid by the four-toed salamander, *Hemidactylium scutatum* (Schlegel), in southern Michigan. *Pap. Michigan Acad. Sci. Arts Lett.* 21:567–73.

Blanchard, F. N., and F. C. Blanchard. 1931. Size groups and their characteristics in the salamander *Hemidactylium scutatum* (Schlegel). *Am. Nat.* 65:149–64.

Blanchard, F. N., M. R. Gilreath, and F. C. Blanchard. 1979. The eastern ringneck snake (*Diadophis punctatus edwardsii*) in northern Michigan (Reptilia, Serpentes, Colubridae). *J. Herpetol.* 13:377–402.

Blankenhorn, H.-J. 1972. Meteorological variables affecting onset and duration of calling in *Hyla arborea* L. and *Bufo calamita* Laur. *Oecologia* 9:223–34.

———. 1974. *Soziale Organisation einer Mischpopulation von Rana lessonae Camerano und Rana esculenta Linnaeus.* PhD dissertation. Zürich, Switzerland: University of Zürich.

———. 1977. Reproduction and mating behavior in *Rana lessonae - Rana esculenta* mixed populations. In *The reproductive biology of amphibians,* ed. D. H. Taylor and S. I. Guttman, 389–410. New York: Plenum.

Blankenhorn, H.-J., H. Heusser, and P. Notter. 1973. Zur Verbreitung von *Rana esculenta* Linnaeus und *Rana lessonae* Camerano im Zürcher Oberland. *Rev. Suisse Zool.* 80:662–66.

Blaustein, A. R. 1988. Ecological correlates and potential functions of kin recognition and kin association in anuran larvae. *Behav. Genet.* 18:449–64.

Blaustein, A. R. 1994. Chicken Little or Nero's fiddle? A perspective on declining amphibian populations. *Herpetologica* 50:85–97.

Blaustein, A. R., J. J. Beatty, D. H. Olson, and R. M. Storm. 1995. *The biology of amphibians and reptiles in old-growth forests in the Pacific Northwest.* U.S. Forest Service General Technical Report PNW-GTR-337. Washington D C: Government Printing Office.

Blaustein, A. R., M. Beckoff, and T. J. Daniels. 1987a. Kin recognition in vertebrates (excluding primates): Empirical evidence. In *Kin recognition in animals,* ed. D. J. C. Fletcher and C. D. Michener, 287–331. New York: Wiley.

———. 1987b. Kin recognition in vertebrates (excluding primates): Mechanisms, functions, and future research. In *Kin recognition in animals,* ed. D. J. C. Fletcher and C. D. Michener, 333–57. New York: Wiley.

Blaustein, A. R., M. Beckoff, J. A. Byers, and T. J. Daniels. 1991. Kin recognition in vertebrates: What do we really know about adaptive value? *Anim. Behav.* 41:1079–83.

Blaustein, A. R., L. K. Belden, D. H. Olson, D. M. Green, T. L. Root, and J. M. Kiesecker. 2001. Amphibian breeding and climate change. *Cons. Biol.* 15:1804–9.

Blaustein, A. R., D. P. Chivers, L. B. Kats, and J. M. Kiesecker. 2000. Effects of ultraviolet radiation on locomotion and orientation in roughskin newts (*Taricha granulosa*). *Ethology* 106:227–34.

Blaustein, A. R., B. Han, B. Fasy, J. Romansic, E. A. Scheessele, R. G. Anthony, A. Marco, D. P. Chivers, L. K. Belden, J. M. Kiesecker, T. Garcia, M. Lizana, and L. B. Kats. 2004. Variable breeding phenology affects the exposure of amphibian embryos to ultraviolet radiation and optical characteristics of natural waters protect amphibians from UV-B in the U.S. Pacific Northwest: Comment. *Ecology* 85:1747–54.

Blaustein, A. R., J. B. Hays, P. D. Hoffman, D. P. Chivers, J. M. Kiesecker, W. P. Leonard, A. Marco, D. H. Olson, J. K. Reaser, and R. G. Anthony. 1999. DNA repair and resistance to UV-B radiation in western spotted frogs. *Ecol. Appl.* 9:1100–1105.

Blaustein, A. R., P. D. Hoffman, D. G. Hokit, J. M. Kiesecker, S. C. Walls, and J. B. Hays. 1994. UV repair and resistance to solar UV-B in amphibian eggs: a link to population declines? *Proc. Natl. Acad. Sci. USA* 91:1791–95.

Blaustein, A. R., P. D. Hoffman, J. M. Kiesecker, and J. B. Hays. 1996. DNA repair activity and resistance to solar UV-B radiation in eggs of the red-legged frog. *Cons. Biol.* 10:1398–1402.

Blaustein, A. R., D. G. Hokit, R. K. O'Hara, and R. A. Holt. 1994. Pathogenic fungus contributes to amphibian losses in the Pacific Northwest. *Biol. Cons.* 67:251–54.

Blaustein, A. R., and P. T. J. Johnson. 2003a. The complexity of deformed amphibians. *Front. Ecol. Environ.* 1:87–94.

———. 2003b. Explaining frog deformities. *Sci. Amer.* February 2003:60–65.

Blaustein, A. R., and L. B. Kats. 2003. Amphibians in a very bad light. *BioScience* 53:1028–29.

Blaustein, A. R., and J. M. Kiesecker. 2002. Complexity in conservation: lessons from the global decline of amphibian populations. *Ecol. Lett.* 5:597–608.

Blaustein, A. R., J. M. Kiesecker, D. P. Chivers, and R. G. Anthony. 1997. Ambient UV-B radiation causes deformities in amphibian embryos. *Proc. Natl. Acad. Sci. USA* 94:13735–37.

Blaustein, A. R., J. M. Kiesecker, D. P. Chivers, D. G. Hokit, A. Marco, L. K. Belden, and A. Hatch. 1998. Effects of ultraviolet radiation on amphibians: Field experiments. *Amer. Zool.* 38:799–812.

Blaustein, A. R., J. M. Kiesecker, S. C. Walls, and D. G. Hokit. 1996. Field experiments, amphibian mortality, and UV radiation. *BioScience* 46:386–88.

Blaustein, A. R., and R. K. O'Hara. 1981. Genetic control for sibling recognition? *Nature* 290:246–48.

———. 1982a. Kin recognition cues in *Rana cascadae* tadpoles. *Behav. Neur. Biol.* 36:77–87.

———. 1982b. Kin recognition in *Rana cascadae* tadpoles: Maternal and paternal effects. *Anim. Behav.* 30:1151–57.

———. 1983. Kin recognition in *Rana cascadae* tadpoles: Effects of rearing with nonsiblings and varying the strength of the stimulus cues. *Behav. Neur. Biol.* 39:259–67.

———. 1986. Kin recognition in tadpoles. *Sci. Amer.* 254:108–16.

———. 1987. Aggregation behaviour in *Rana cascadae* tadpoles: Association preferences among wild aggregations and responses to non-kin. *Anim. Behav.* 35:1549–55.

Blaustein, A. R., R. K. O'Hara, and D. H. Olson. 1984. Kin preference behavior is present after metamorphosis in *Rana cascadae* tadpoles. *Anim. Behav.* 32:445–50.

Blaustein, A. R., and D. H. Olson. 1991. Declining amphibians. *Science* 253:1467.

Blaustein, A. R., J. M. Romansic, J. M. Kiesecker, and A. C. Hatch. 2003. Ultraviolet radiation, toxic chemicals and amphibian population declines. *Divers. Distrib.* 9:123–40.

Blaustein, A. R., J. M. Romansic, and E. A. Scheessele. 2005. Ambient levels of ultraviolet-B radiation cause mortality in juvenile western toads, *Bufo boreas. Am. Midl. Nat.* 154:375–82.

Blaustein, A. R., J. M. Romansic, E. A. Scheessele, B. A. Han, A. P. Pessier, and J. E. Longcore. 2005. Interspecific variation in susceptibility of frog tadpoles to the pathogenic fungus *Batrachochytrium dendrobatidis. Cons. Biol.* 19:1460–68.

Blaustein, A. R., T. L. Root, J. M. Kiesecker, L. K. Belden, D. H. Olson, and D. M. Green. 2002. Amphibian phenology and climate change. *Cons. Biol.* 16:1454–55.

Blaustein, A. R., and B. Waldman. 1992. Kin recognition in anuran amphibians. *Anim. Behav.* 44:207–21.

Blaustein, A. R., D. B. Wake, and W. P. Sousa. 1994. Amphibian declines: Judging stability, persistence, and susceptibility of populations to local and global extinctions. *Cons. Biol.* 8:60–71.

Blaustein, A. R., and S. C. Walls. 1995. Aggregation and kin recognition. In *Amphibian biology. Vol. 2. Social behaviour*, ed. H. Heatwole and B. K. Sullivan, 568–602. Chipping Norton, New South Wales, Australia: Surrey Beatty and Sons.

Blaustein, A. R., T. Yoshikawa, K. Asoh, and S. C. Walls. 1993. Ontogenetic shifts in tadpole kin recognition: Loss of signal and perception. *Anim. Behav.* 46:525–38.

Blaustein, L., J. Friedman, and T. Fahima. 1996. Larval *Salamandra* drive temporary pond community dynamics: Evidence from an artificial pool experiment. *Oikos* 76:392–402.

Blaustein, L., and B. P. Kotler. 1993. Oviposition habitat selection by the mosquito, *Culiseta longiareolata:* Effects of conspecifics, food and green toad tadpoles. *Ecol. Entomol.* 18:104–8.

Blaustein, L., and J. Margalit. 1994. Mosquito larvae (*Culiseta longiareolata*) prey upon and compete with toad tadpoles (*Bufo viridis*). *J. Anim. Ecol.* 63:841–50.

Blaylock, L. A., R. Ruibal, and K. Platt-Aloia. 1976. Skin structure and wiping behavior of phyllomedusine frogs. *Copeia* 1976:283–95.

Bleakney, S. 1952. The amphibians and reptiles of Nova Scotia. *Can. Field-Nat.* 66:125–29.

Bleckmann, H., and T. Lotz. 1987. The vertebrate-catching behaviour of the fishing spider *Dolomedes triton* (Araneae, Pisauridae). *Anim. Behav.* 35:641–51.

Bles, E. J. 1906. The life history of *Xenopus laevis* Daudin. *Trans. Roy. Soc. Edinburgh* 41:789–821.

Block, J. E., S. L. Unser, J. K. Mooney, and E. R. Wild. 2003. *Agalychnis craspedopus* (Amazon Leaf Frog). Reproduction. *Herp. Rev.* 34:134–35.

Blom, A., J. Yamindou, and H. H. T. Prins. 2004. Status of the protected areas of the Central African Republic. *Biol. Cons.* 118:479–87.

Blommers-Schlösser, R. M. A. 1975a. Observations on the larval development of some Malagasy frogs, with notes on their ecology and biology (Anura: Dyscophinae, Scaphiophryninae, and Cophylinae). *Beaufortia* 24:7–26.

———. 1975b. A unique case of mating behaviour in a Malagasy tree frog, *Gephyromantis liber* (Peracca, 1893), with observations on the larval development (Amphibia, Ranidae). *Beaufortia* 23:15–25.

———. 1979. Biosystematics of the Malagasy frogs. I. Mantellinae (Ranidae). *Beaufortia* 29:1–77.

———. 1981. On endemic Malagasy frogs (Ranidae, Rhacophoridae, and Hyperoliidae). *Monit. Zool. Ital.,* n.s. suppl. 15:217–24.

———. 1993. Amphibiens (deuxieme partie). *Faune de Madagascar* 75 (2):385–530.

Blommers-Schlösser, R. M. A., and L. H. M. Blommers. 1984. The amphibians. In *Key environments. Madagascar,* ed. A. Jolly, P. Oberle, and R. Albinac, 89–104. New York: Pergammon.

Blomquist, S. M., and J. C. Tull. 2002. *Rana luteiventris* (Columbia spotted frog). Burrow use. *Herp. Rev.* 33:131.

Blouin, M. S. 1990. Evolution of palatability differences between closely-related treefrogs. *J. Herpetol.* 24:309–11.

———. 1992. Genetic correlations among morphometric traits and rates of growth and differentiation in the green tree frog, *Hyla cinerea. Evolution* 46:735–44.

Blume, W. 1931. Studien an Anurenlarynx. *Gegenbauers Morph. Jb.* 65:307–464.

———. 1932. Studien an Anurenlarynx II. *Gegenbauers Morph. Jb.* 69:455–622.

———. 1933. Studien an Anurenlarynx III. *Gegenbauers Morph. Jb.* 72:103–52.

Blumer, L. S. 1979. Male parental care in the bony fishes. *Q. Rev. Biol.* 54:149–61.

Boag, P. T., and P. R. Grant. 1984. Darwin's finches (*Geospiza*) on Isla Daphne Major, Galapagos: Breeding and feeding ecology in a climatically variable environment. *Ecol. Monogr.* 54:463–89.

Boatright-Horowitz, S. L., S. S. Horowitz, and A. M. Simmons. 2000. Patterns of vocal interactions in a bullfrog (*Rana catesbeiana*) chorus: Preferential responding to far neighbors. *Ethology* 106:701–12.

Boatright-Horowitz, S. S., C. A. Cheney, and A. M. Simmons. 1999. Atmospheric and undewater propagation of bullfrog vocalizations. *Bioacoustics* 9:257–80.

Boatright-Horowitz, S. S., C. E. Garabedian, K. H. Odabashian, and A. M. Simmons. 1999. Coding of amplitude modulation in the auditory midbrain of the bullfrog (*Rana catesbeiana*) across metamorphosis. *J. Comp. Physiol. A* 184:219–31.

Boatright-Horowitz, S. S., and A. M. Simmons. 1995. Postmetamorphic changes in auditory sensitivity of the bullfrog midbrain. *J. Comp. Physiol. A* 177:577–90.

———. 1997. Transient "deafness" accompanies auditory development during metamorphosis from tadpole to frog. *Proc. Natl. Acad. Sci. USA* 94:14877–82.

Bobka, M. S., R. G. Jaeger, and D. C. McNaught. 1981. Temperature dependent assimilation efficiencies of two species of terrestrial salamanders. *Copeia* 1981:417–21.

Bock, S. F., and J. E. Fauth. 1992. *Pseudotriton ruber ruber* (northern red salamander). Diet. *Herp. Rev.* 23:58.

Bodnar, D. A. 1996. The separate and combined effects of harmonic structure, phase, and FM on female preferences in the barking treefrog (*Hyla gratiosa*). *J. Comp. Physiol. A* 178:173–82.

Boettger, C. R., and E. Schwartz. 1928. Über neotenische Larven des Teichmolches (*Triturus vulgaris* L.). *Zool. Anz.* 78:174–76.

Bogart, J. P. 1981. How many times has terrestrial breeding evolved in anuran amphibians? *Monit. Zool. Ital.*, n.s. suppl. 15:29–40.

Bogart, J. P. 1989. A mechanism for interspecific gene exchange via all-female salamander hybrids. In *Evolution and ecology of unisexual vertebrates*, ed. R. M. Dawley and J. P. Bogart, 170–79. Bulletin 466. Albany: New York State Museum.

Bogart, J. P. 2003. Genetics and systematics of hybrid species. In *Reproductive biology and phylogeny of Urodela*, ed. D. M. Sever, 109–34. Enfield, NH: Science Publishers.

Bogart, J. P., and M. W. Klemens. 1997. Hybrids and genetic interactions of mole salamanders (*Ambystoma jeffersonianum* and *A. laterale*) (Amphibia: Caudata) in New York and New England. *Amer. Mus. Novitates* (3218):1–78.

Bogart, J. P., L. A. Lowcock, C. W. Zeyl, and B. K. Mable. 1987. Genome constitution and reproductive biology of hybrid salamanders, genus *Ambystoma*, on Kelleys Island in Lake Erie. *Can. J. Zool.* 65:2188–2201.

Bogert, C. M. 1947. A field study of homing in the Carolina toad. *Amer. Mus. Novitates* (1355):1–24.

———. 1958. *Sounds of North American frogs. The biological significance of voice in frogs.* Folkways Records, New York (phonograph record). Reissued as compact disc by Smithsonian Folkways Recordings, Washington, DC.

———. 1960. The influence of sound on the behavior of amphibians and reptiles. In *Animal sounds and communication*, ed. W. E. Lanyon and W. N. Tavolga, 137–320. Washington, DC: AIBS.

———. 1969. The eggs and hatchlings of the Mexican leptodactylid frog, *Eleutherodactylus decoratus* Taylor. *Amer. Mus. Novitates* (2367):1–9.

Bohlen, P. J., S. Scheu, C. M. Hale, M. A. McLean, S. Migge, P. M. Goffman, and D. Parkinson. 2004. Non-native invasive earthworms as agents of change in northern temperate forests. *Front. Ecol. Environ.* 2:427–35.

Bohlin, R. G., and T. L. Beitinger. 1979. Heat exchange in the aquatic salamander, *Amphiuma means*. *J. Therm. Biol.* 4:63–67.

Bohnsack, K. 1951. Temperature data on the terrestrial hibernation of the green frog, *Rana clamitans*. *Copeia* 1951:236–39.

Bohnsack, K. 1952. Terrestrial hibernation of the bullfrog, *Rana catesbeiana* Shaw. *Copeia* 1952:114.

Bohr, C. 1900. Über die Haut- und Lungenatmung der Frösche. *Skand. Archiv. Physiol.* 10:74–90.

Boisseau, C., and J. Joly. 1975. Transport and survival of spermatozoa in female Amphibia. In *The biology of spermatozoa*, ed. E. S. E. Hafez and C. G. Thibault, 94–104. Basel, Switzerland: Karger.

Boistel, R., and O. S. G. Pauwels. 2002. *Oscaecilia zweifeli* (Zweifel's caecilian). Predation. *Herp. Rev.* 33:120–21.

Boistel, R., and J. Sueur. 1997. Comportement sonore de la femelle de *Platymantis vitiensis* (Amphibia, Anura) en l'absence du mâle. *C. R. Acad. Sci. Paris, Sci. Vie* 320:933–41.

Boisvert, C. A. 2005. The pelvic fin and girdle of *Panderichthys* and the origin of tertrapod locomotion. *Nature* 438:1145–47.

Bokermann, W. C. A. 1957. Frog eggs parasitized by dipterous larvae. *Herpetologica* 13:231–32.

———. 1965. Notas sobre as especies de *Thoropa* Fitzinger (Amphibia, Leptodactylidae). *Ann. Acad. Brasil. Cien.* 37:525–37.

———. 1966a. O genero *Phyllodytes* Wagler, 1830 (Anura, Hylidae). *Ann. Acad. Brasil. Cien.* 38:335–44.

———. 1966b. Une neuva especie de *Trachycephalus* de Bahia, Brasil. *Neotropica* 12:120–24.

———. 1978. Observações sobre hábitos alimentares do gavião *Geranospiza caerulescens* (Viellot, 1817) (Aves, Accipitridae). *Rev. Brasil. Biol.* 38:715–20.

Boland, C. R. J. 2004. Introduced cane toads *Bufo marinus* are active nest predators and competitors of rainbow bee-eaters *Merops ornatus:* Observational and experimental evidence. *Biol. Cons.* 120:53–62.

Bolek, M. G., and J. Janovy, Jr. 2004. *Rana blairi* (Plains Leopard Frog). Prey. *Herp. Rev.* 35:262.

Bolitho, E. E., and R. W. R. Retallick. 1996. *Mixophyes fasciolatus* (great barred frog). Tadpole predation. *Herp. Rev.* 27:140–41.

Bolívar-G., W., T. Grant, and L. A. Osorio. 1999. Combat behavior in *Centrolene buckleyi* and other centrolenid frogs. *Alytes* 16:7–83.

Böll, S. 2003. Zur Populationsdynamik und Verhaltensökologie einer Rhöner Freilandpopulation von *Alytes o. obstetricans*. *Z. Feldherpetol.* 10:97–103.

Böll, S., and K. E. Linsenmair. 1998. Size-dependent male reproductive success and size-assortative mating in the midwife toad *Alytes obstetricans*. *Amphibia-Reptilia* 19:75–89.

Bolt, J. R. 1977. Dissorophoid relationships and ontogeny, and the origin of the Lissamphibia. *J. Paleontol.* 51:235–49.

———. 1979. *Amphibamus gradiceps* as a juvenile dissorophoid: Evidence and implications. In *Mazon Creek fossils*, ed. M. H. Nitecki, 529–63. New York: Academic Press.

———. 1991. Lissamphibian origins. In *Origins of the higher groups of tetrapods*, ed. H.-P. Schultze and L. Trueb, 194–222. Ithaca, NY: Cornell University Press.

Bonaccorso, E., J. M. Guayasamin, D. Méndez, and R. Speare. 2003. Chytridiomycosis as a possible cause of population declines in *Atelopus cruciger* (Anura: Bufonidae). *Herp. Rev.* 34:331–34.

Bonato, L., and G. Fracasso. 2003. Movements, distribution pattern and density in a population of *Salamandra atra aurorae* (Caudata: Salamandridae). *Amphibia-Reptilia* 24:251–60.

Bonaventura, C., B. Sullivan, J. Bonaventura, and S. Bourne. 1977. Anion modulation of the negative Bohr effect of hemoglobin from a primitive amphibian. *Nature* 265:474–76.

Bond, A. N. 1960. An analysis of the response of salamander gills to changes in the oxygen concentration of the medium. *Develop. Biol.* 2:1–20.

Bonett, R. M., and P. T. Chippindale. 2004. Speciation, phylogeography and evolution of life history and morphology in plethodontid salamanders of the *Eurycea multiplicata* complex. *Mol. Ecol.* 13:1189–1203.

Bonilla, J. P., and E. La Marca. 1996. Hábitos alimentarios de *Nephelobates alboguttatus* (Anura: Dendrobatidae) en una selva nublada andina de Venezuela. *Rev. Biol. Trop.* 44:827–33.

Boone, M. D., and C. M. Bridges. 2003a. Effects of carbaryl on green frog (*Rana clamitans*) tadpoles: Timing of exposure versus multiple exposures. *Env. Tox. Chem.* 22:2695–2702.

———. 2003b. Effects of pesticides on amphibian populations. In *Amphibian conservation,* ed. R. D. Semlitsch, 152–67. Washington, DC: Smithsonian Books.

———. 2003c. A hierarchical approach in studying the effects of an insecticide on amphibians. In *Amphibian decline: An integrated analysis of multiple stressor effect,* ed. G. Linder, S. K. Krest, and D. W. Sparling, 265–70. Pensacola, FL: SETAC Press.

Boone, M. D., C. M. Bridges, J. F. Fairchild, and E. E. Little. 2005. Multiple sublethal chemicals negatively affect tadpoles of the green frog, *Rana clamitans. Env. Tox. Chem.* 24:1267–72.

Boone, M. D., C. M. Bridges, and B. R. Rothermel. 2001. Growth and development of larval green frogs (*Rana clamitans*) exposed to multiple doses of an insecticide. *Oecologia* 129:518–24.

Boone, M. D., P. S. Corn, M. A. Donnelly, E. E. Little, and P. H Niewiarowski. 2003. Physical stressors. In *Amphibian decline: An integrated analysis of multiple stressor effects,* ed. G. Linder, S. K. Krest, and D. W. Sparling, 129–51. Pensacola, FL: SETAC Press.

Boone, M. D., and S. M. James. 2003. Interactions of an insecticide, herbicide, and natural stressors in amphibian community mesocosms. *Ecol. Appl.* 13:829–41.

Boone, M. D., E. E. Little, and R. D. Semlitsch. 2004. Overwintered bullfrog tadpoles negatively affect salamanders and anurans in native amphibian communities. *Copeia* 2004:683–90.

Boone, M. D., D. E. Scott, and P. H. Niewiarowski. 2002. Effects of hatching time for larval ambystomatid salamanders. *Copeia* 2002:511–17.

Boone, M. D., and R. D. Semlitsch. 2001. Interactions of an insecticide with larval density and predation in experimental amphibian communities. *Cons. Biol.* 15:228–38.

———. 2003. Interactions of bullfrog tadpole predators and an insecticide: Predation release and facilitation. *Oecologia* 442:610–16.

Boone, M. D., R. D. Semlitsch, J. F. Fairchild, and B. B. Rothermel. 2004. Effects of an insecticide on amphibians in large-scale experimental ponds. *Ecol. Appl.* 14:685–91.

Bonnet, X., D. Bradshaw, and R. Shine. 1998. Capital versus income breeding: An ectothermic perspective. *Oikos* 83:333–42.

Boorse, G. C., and R. J. Denver. 2002. Acceleration of *Ambystoma tigrinum* metamorphosis by corticotropin-releasing hormone. *J. Exp. Zool.* 293:94–98.

———. 2004. Endocrine mechanisms underlying plasticity in metamorphic timing in spadefoot toads. *Integr. Comp. Biol.* 43:646–57.

Booth, D. T. 1994. Effects of unstirred layers on cutaneous gas exchange. In *Amphibian biology, vol. 1. The integument,* ed. H. Heatwole and G. T. Barthalmus, 168–77. Chipping Norton, New South Wales, Australia: Surrey Beatty and Sons.

Booth, D. T., and M. E. Feder. 1991. Formation of hypoxic boundary layers and their biological significance in a skin-breathing aquatic salamander, *Desmognathus quadramaculatus. Physiol. Zool.* 64:1307–21.

Boothby, J. 1997. Pond conservation: Towards a delineation of pondscape. *Aquat. Cons. Mar. Freshw. Ecosyst.* 7:127–32.

Borkin, L. J. 1999. Distribution of amphibians in North Africa, Europe, Western Asia, and the former Soviet Union. In *Patterns of distribution of amphibians. A global perspective,* ed. W. E. Duellman, 329–420. Baltimore: Johns Hopkins University Press.

Bosch, J. 2001. Female reciprocal calling in the Iberian midwife toad (*Alytes cisternasii*) varies with male call rate and dominant frequency: implications for sexual selection. *Naturwissenschaften* 88:434–37.

———. 2002a. The function of female reciprocal calls in the Iberian midwife toad (*Alytes cisternasii*): Female-female acoustic competition? *Naturwissenschaften* 89:575–78.

———. 2002b. Signal discrimination by female Iberian midwife toads (*Alytes cisternasii*). *Ethol. Ecol. Evol.* 14:217–26.

Bosch, J., and L. Boyero. 2003a. Double stimulation of the inner ear organs of an anuran species (*Alytes cisternasii*) with simple tonal calls. *J. Zool., London* 260:347–51.

———. 2003b. Precopulatory behaviour and the evolutionary relationships of Discoglossidae. *J. Zool. Syst. Evol. Res.* 41:145–51.

Bosch, J., L. Boyero, and I. Martínez-Solano. 2004. Spatial scales for the management of amphibian populations. *Biodivers. Cons.* 13:409–20.

Bosch, J., and I. De la Riva. 2004. Are frog calls modulated by the environment? An analysis with anuran species from Bolivia. *Can. J. Zool.* 82:880–88.

Bosch, J., and R. Márquez. 1996. Acoustic competition in male midwife toads *Alytes obstetricans* and *Alytes cisternasii:* Response to neighbor size and calling rate. Implications for female choice. *Ethology* 102:841–55.

———. 2000. Acoustical interference in the advertisement calls of the midwife toads (*Alytes obstetricans* and *Alytes cisternasii*). *Behaviour* 137:249–63.

——— 2001a. Call timing in male-male acoustical interactions and female choice in the midwife toad *Alytes obstetricans. Copeia* 2001:169–77.

———. 2001b. Female courtship call of the Iberian midwife toad (*Alytes cisternasii*). *J. Herpetol.* 35:647–52.

———. 2005. Female preference intensities on different call characteristics and symmetry of preference above and below the mean in the Iberian midwife toad *Alytes cisternasii. Ethology* 111:323–33.

Bosch, J. , R. Márquez, and L. Boyero. 2003. Behavioural patterns, preference, and motivation of female midwife toads during phonotaxis tests. *J. Ethol.* 21:61–66.

Bosch, J., I. Martínez-Solano, and M. García-París. 2001. Evidence of a chytrid fungus infection involved in the decline of the midwife toad (*Alytes obstetricans*) in protected areas of Central Spain (Anura: Discoglossidae). *Biol. Cons.* 97:331–37.

Bosch, J., P. A. Rincón, L. Boyero, and I. Martínez-Solano. 2006. Effects of introduced salmonids on a montane population of Iberian frogs. *Cons. Biol.* 20:180–89.

Bosman, W., J. J. van Gelder, and H. Strijbosch. 1996. Hibernation sites of the toads *Bufo bufo* and *Bufo calamita* in a river floodplain. *Herpetol. J.* 6:83–86.

———. 1997. The effect of inundation on hibernating *Bufo bufo* and *Bufo calamita. Amphibia-Reptilia* 18:339–46.

Bossuyt, F., M. Meegaskumbura, N. Beenaerts, D. J. Gower, R. Pethiyagoda, K. Roelants, A. Mannaert, M. Wilkinson, M. M. Bahir, K. Manamendra-Arachchi et al. 2004. Local endemism within the Western Ghats-Sri Lanka biodiversity hotspot. *Science* 306:479–81.

Bossuyt, F., M. Meegaskumbura, N. Beenaerts, D. J. Gower, R. Pethiyagoda, K. Roelants, A. Mannaert, M. Wilkinson, C. J. Schneider, M. M. Bahir et al. 2005. Biodiversity in Sri Lanka and the Western Ghats—response. *Science* 308:199.

Bossuyt, F., and M. C. Milinkovitch. 2000. Convergent adaptive radiations in Madagascan and Asian ranid frogs reveal covariation between larval and adult traits. *Proc. Natl. Acad. Sci. USA* 97:6585–90.

———. 2001. Amphibians as indicators of early Tertiary "out-of-India" dispersal of vertebrates. *Science* 292:93–95.

Bostanci, A. 2005. A devil of a disease. *Science* 307:1035.

Boul, K. E., and M. J. Ryan. 2004. Population variation of complex advertisement calls in *Physalaemus petersi* and comparative laryngeal morphology. *Copeia* 2004:624–31.

Boulenger, G. A . 1886. Remarks in connection with the preceding note. *Ann. Mag. Nat. Hist.*, ser. 5, 17:463–64.

———. 1912. Observations sur l'accouplement et la ponte de l'Alyte accoucheur, *Alytes obstetricans. Bull. Class. Sci. Acad. Roy. Belgique* 1912:570–79.

Bourne, G. R. 1992. Lekking behavior in the Neotropical frog *Ololygon rubra. Behav. Ecol. Sociobiol.* 31:173–80.

———. 1993. Proximate costs and benefits of mate acquisition at leks of the frog *Ololygon rubra. Anim. Behav.* 45:1051–59.

———. 1997. Reproductive behavior of terrestrial breeding frogs *Eleutherodactylus johnstonei* in Guyana. *J. Herpetol.* 31:221–29.

———. 1998. Amphisexual parental behavior of a terrestrial breeding frog *Eleutherodactylus johnstonei* in Guyana. *Behav. Ecol.* 9:1–7.

Bourne, G. R., A. C. Collins, A. M. Holder, and C. L. McCarthy. 2001. Vocal communication and reproductive behavior of the frog *Colostethus beebei* in Guyana. *J. Herpetol.* 35:272–81.

Boutilier, R. G. 1984. Characterization of the intermittent breathing pattern in *Xenopus laevis. J. Exp. Biol.* 110:291–310.

———. 1988a. Control of arrhythmic breathing in bimodal breathers: Amphibia. *Can. J. Zool.* 66:6–19.

———. 1988b. Interactions between transcutaneous ion transfer processes and carbon dioxide excretion in amphibians. *Amer. Zool.* 28:1009–18.

———. 1989. Diving physiology: Amphibians. In *Comparative pulmonary physiology: Current concepts,* ed. S. C. Wood, 677–95. New York: Marcel Dekker.

Boutilier, R. G., P. H. Donohoe, G. J. Tattersall, and T. G. West. 1997. Hypometabolic homeostasis in overwintering aquatic amphibians. *J. Exp. Biol.* 200:387–400.

Boutilier, R. G., M. L. Glass, and N. Heisler. 1986. The relative distribution of pulmocutaneous blood flow in *Rana catesbeiana:* Effects of pulmonary or cutaneous hypoxia. *J. Exp. Biol.* 126:33–39.

———. 1987. Blood gases and extracellular/intracellular acid-base status as a function of temperature in the anuran amphibians *Xenopus laevis* and *Bufo marinus. J. Exp. Biol.* 130:13–25.

Boutilier, R. G., D. G. McDonald, and D. P. Toews. 1980. The effects of enforced activity on ventilation, circulation and blood acid-base balance in the aquatic gill-less urodele, *Cryptobranchus alleganiensis:* A comparison with the semi-terrestrial anuran, *Bufo marinus. J. Exp. Biol.* 84:289–302.

Boutilier, R. G., D. J. Randall, G. Shelton, and D. P. Toews. 1979a. Acid-base relationships in the blood of the toad, *Bufo marinus.* I. The effects of environmental $CO_2$. *J. Exp. Biol.* 82:331–44.

———. 1979b. Acid-base relationships in the blood of the toad, *Bufo marinus.* II. The effects of dehydration. *J. Exp. Biol.* 82:345–55.

———. 1979c. Acid-base relationships in the blood of the toad, *Bufo marinus.* III. The effects of burrowing. *J. Exp. Biol.* 82:357–65.

Boutilier, R. G., and G. Shelton. 1986a. The effects of forced and voluntary diving on ventilation, blood gases and pH in the aquatic amphibian, *Xenopus laevis. J. Exp. Biol.* 122:209–22.

———. 1986b. Gas exchange, storage and transport in voluntarily diving *Xenopus laevis. J. Exp. Biol.* 68:99–107.

———. 1986c. Respiratory properties of blood from voluntarily and forcibly submerged *Xenopus laevis. J. Exp. Biol.* 121:285–300

Boutilier, R. G., D. F. Stiffler, and D. P. Toews. 1992. Exchange of respiratory gases, ions, and water in amphibious and aquatic amphibians. In *Environmental physiology of the amphibians,* ed. M. E. Feder and W. W. Burggren, 81–124. Chicago: University of Chicago Press.

Boutilier, R. G., and D. P. Toews. 1977. The effect of progressive hypoxia on respiration in the toad *Bufo marinus. J. Exp. Biol.* 68:99–107.

———. 1981a. Respiratory properties of blood in a strictly aquatic and predominantly skin-breathing urodele, *Cryptobranchus alleganiensis. Respir. Physiol.* 46:161–76.

———. 1981b. Respiratory, circulatory, and acid-base adjustments to hypercapnia in a strictly aquatic and predominantly skin-breathing urodele, *Cryptobranchus alleganiensis. Respir. Physiol.* 46:177–92.

Bovero, S., C. Giacoma, and F. Andreone. 1997. Sexual selection and facultative paedogenesis in the Alpine newt, *Triturus alpestris apuanus* (Bonaparte, 1839). *Geobios* 21:287–92.

Bovero, S., G. Sotgiu, S. Castellano, and C. Giacoma. 2003. Age and sexual dimorphism in a population of *Euproctus platycephalus* (Caudata: Salamandridae) from Sardinia. *Copeia* 2003:149–54.

Bowers, J. H. 1966. Food habits of the diamond-backed water snake, *Natrix rhombifera rhombifera* in Bowie and Red River Counties, Texas. *Herpetologica* 22:225–29.

Bowie, J. H., K. L. Wegener, B. C. S. Chia, P. A. Wabnitz, J. A. Carver, M. J. Tyler, and J. C. Wallace. 1999. Host defence antibacterial peptides from skin secretions of Australian amphibians. The relationship between structure and activity. *Prot. Pep. Lett.* 6:259–69.

Bowker, R. G., and M. H. Bowker. 1979. Abundance and distribution of anurans in a Kenyan pond. *Copeia* 1979:278–85.

Bowman, R. I. 1979. Adaptive morphology of song dialects in Darwin's finches. *J. für Ornithol.* 120:353–80.

Bowne, D. R., and M. A. Bowers. 2004. Interpatch movements in spatially structured populations: A literature review. *Landsc. Ecol.* 19:1–20.

Boy, J. A. 1974. Die larven der rhachitomen Amphibien (Amphibia: Temnospondyli; Karbon-Trias. *Palaontol. Z.* 48:236–68.

———. 1993. Synopsis of the tetrapods from the Rotliegend (Lower Permian) in the Saar-Nahe Basin (SW-Germany). In

*New research on permo-carboniferous faunas,* ed. U. Heidtke, 155–69. Bad Dürkheim, Germany: Pollichia.

Boy, J. A., and H.-D. Sues. 2000. Branchiosaurs: larvae, metamorphosis and heterochrony in temnospondyls and seymouriamorphs. In *Amphibian biology. Vol. 4. Palaeontology,* ed. H. Heatwole and R. L. Carroll, 1150–97. Chipping Norton, New South Wales, Australia: Surrey Beatty and Sons.

Boycott, R. C. 1972. Ghost frog eggs. *Afr. Wildlife* 26 (1):43.

———. 1988. Description of a new species of *Heleophryne* Sclater, 1899 from the Cape Province, South Africa (Anura: Heleophrynidae). *Ann. Cape Province Mus. (Nat. Hist.)* 16:309–19.

Boyd, S. K. 1994. Arginine vasotocin facilitation of advertisement calling and call phonotaxis in bullfrogs. *Horm. Behav.* 28:232–40.

Boyd, S. K., K. D. Wissing, J. E. Heinsz, and G. S. Prins. 1999. Androgen receptors and sexual dimorphisms in the larynx of the bullfrog. *Gen. Comp. Endocrinol.* 113:59–68.

Boyle, R. 1670. New pneumatical experiments about respiration. *Phil. Trans. Roy. Soc., London* 5:2011–32.

Bradbury, J. W. 1981. The evolution of leks. In *Natural selection and social behavior: Recent research and new theory,* ed. R. D. Alexander and D. W. Tinkle, 138–69. New York: Chiron.

Bradbury, J. W., and R. M. Gibson. 1983. Leks and mate choice. In *Mate choice,* ed. P. Bateson, 109–38. New York: Cambridge University Press.

Bradford, D. F. 1983. Winterkill, oxygen relations, and energy metabolism of a submerged dormant amphibian, *Rana muscosa. Ecology* 64:1171–83.

———. 1984a. Physiological features of embryonic development in terrestrially-breeding plethodontid salamanders. In *Respiration and metabolism of embryonic vertebrates,* ed. R. S. Seymour, 87–98. The Hague: Dr. W. Junk.

———. 1984b. Temperature modulation in a high-elevation amphibian, *Rana muscosa. Copeia* 1984:966–76.

———. 1984c. Water and osmotic balance in overwintering tadpoles and frogs, *Rana muscosa. Physiol. Zool.* 57:474–80.

———. 1989. Allotopic distribution of native frogs and introduced fishes in high Sierra Nevada lakes of California: Implications of the negative effect of fish introduction. *Copeia* 1989:775–78.

———. 1990. Incubation time and rate of embryonic development in amphibians: The influence of ovum size, temperature, and reproductive mode. *Physiol. Zool.* 63:1157–80.

———. 1991. Mass mortality and extinction in a high-elevation population of *Rana muscosa. J. Herpetol.* 25:174–77.

Bradford, D. F., S. D. Cooper, T. M. Jenkins, K. Kratz, O. Sarnelle, and A. D. Brown. 1998. Influences of natural acidity and introduced fish on faunal assemblages in California alpine lakes. *Can. J. Fish. Aquat. Sci.* 55:2478–91.

Bradford, D. F., D. M. Graber, and F. Tabatabai. 1994. Population declines of the native frog, *Rana muscosa,* in Sequoia and Kings Canyon National Parks, California. *Southwest. Nat.* 39:323–27.

Bradford, D. F., J. R. Jaeger, R. D. Jennings. 2004. Population status and distribution of a decimated amphibian, the relict leopard frog (*Rana onca*). *Southwest. Nat.* 49:218–28.

Bradford, D. F., and R. S. Seymour. 1985. Energy conservation during the delayed-hatching period in the frog *Pseudophryne bibroni. Physiol. Zool.* 58:491–96.

———. 1988a. Influence of environmental $PO_2$ on embryonic oxygen consumption, rate of development, and hatching in the frog *Pseudophryne bibroni. Physiol. Zool.* 61:475–82.

———. 1988b. Influence of water potential on growth and survival of the embryo, and gas conductance of the egg, in a terrestrial breeding frog, *Pseudophryne bibroni. Physiol. Zool.* 61:470–74.

Bradford, D. F., C. Swanson, and M. S. Gordon. 1992. Effects of low pH and aluminum on two declining species of amphibians in the Sierra Nevada, California. *J. Herpetol.* 26:369–77.

Bradford, D. F., F. Tabatabai, and D. M. Graber. 1993. Isolation of remaining populations of the native frog, *Rana muscosa,* by introduced fishes in Sequoia and Kings Canyon National Parks, California. *Cons. Biol.* 7:882–88.

Bradley, G. A., P. C. Rosen, M. J. Sredi, T. R. Jones, and J. E. Longcore. 2002. Chytridiomycosis in native Arizona frogs. *J. Wildl. Dis.* 38:206–12.

Bradshaw, W. E., and K. Johnson. 1995. Initiation of metamorphosis in the pitcher-plant mosquito: Effects of larval growth history. *Ecology* 76:2055–65.

Brady, L. D., and R. A. Griffiths. 2000. Developmental responses to pond desiccation in tadpoles of the British anuran amphibians (*Bufo bufo, B. calamita* and *Rana temporaria*). *J. Zool., London* 252:61–69.

Bragg, A. N. 1945. The spadefoot toads of Oklahoma with a summary of our knowledge of the group. II. *Am. Nat.* 79:52–72.

———. 1946. Aggregation with cannibalism in tadpoles of *Scaphiopus bombifrons* with some general remarks on the probable evolutionary significance of such phenomena. *Herpetologica* 3:89–97.

———. 1948. Additional instances of social aggregation in tadpoles. *Wasmann Collector* 7:66–79.

———. 1950. Mass movement at metamorphosis in the savannah spadefoot, *Scaphiopus hurterii* Strecker. *Proc. Oklahoma Acad. Sci.* 31:26–27.

———. 1952. A metamorphic aggregation in tadpoles of *Bufo. Proc. Oklahoma Acad. Sci.* 33:71.

———. 1953. A study of *Rana areolata* in Oklahoma. *Wasmann J. Biol.* 11:273–318.

———. 1954a. Aggregational phenomena and feeding reactions in tadpoles of the savannah spadefoot. *Herpetologica* 10:97–102.

———. 1954b. A second season with *Rana areolata* in Oklahoma. *Wasmann J. Biol.* 12:337–43.

———. 1956. Dimorphism and cannibalism in tadpoles of *Scaphiopus bombifrons* (Amphibia, Salientia). *Southwest. Nat.* 1:105–8.

———. 1957a. Aggregated feeding and metamorphic aggregations in tadpoles of *Scaphiopus hurteri,* observed in 1954. *Wasmann J. Biol.* 15:61–68.

———. 1957b. Amphibian eggs produced singly or in masses. *Herpetologica* 13:212.

———. 1960. Aggregational and associated behavior in tadpoles of the plains spadefoot. *Wasmann J. Biol.* 18:273–89.

———. 1961. A theory of the origin of spade-footed toads deduced principally by a study of their habits. *Anim. Behav.* 9:178–86.

———. 1962. Predator-prey relationships in two species of spadefoot toads, with notes on some other features of their behavior. *Wasmann J. Biol.* 10:81–97.

———. 1964. Further study of predation and cannibalism in spadefoot tadpoles. *Herpetologica* 20:17–24.

———. 1965. *Gnomes of the night.* Philadelphia: University of Pennsylvania Press.

Bragg, A. N., and W. Bragg. 1959. Variation in the mouthparts in tadpoles of *Scaphiopus* (*Spea*) *bombifrons* Cope (Amphibia: Salientia). *Southwest. Nat.* 3:55–59.

Bragg, A. N., and M. Brooks. 1958. Social behavior in juveniles of *Bufo cognatus* Say. *Herpetologica* 14:141–47.

Bragg, A. N., and O. M. King. 1960. Aggregational and associated behavior in tadpoles of the plains spadefoot. *Wasmann J. Biol.* 18:273–89.

Bragg, A. N., and C. C. Smith. 1942. Observations on the ecology and natural history of Anura IX. Notes on breeding behavior in Oklahoma. *Great Basin Nat.* 3:33–50.

———. 1943. Observations on the ecology and natural history of Anura IV. The ecological distribution of toads in Oklahoma. *Ecology* 24:285–309.

Brainerd, E. L. 1994. The evolution of lung-gill bimodal breathing and the homology of vertebrate respiratory pumps. *Amer. Zool.* 34:289–99.

———. 1998. Mechanics of lung ventilation in a larval salamander *Ambystoma tigrinum. J. Exp. Biol.* 201:2891–2901.

Brainerd, E. L., J. S. Ditelberg, and D. M. Bramble. 1993. Lung ventilation in salamanders and the evolution of vertebrate air-breathing mechanisms. *Biol. J. Linn. Soc.* 49:163–83.

Brainerd, E. L., and J. A. Monroy. 1998. Mechanics of lung ventilation in a large aquatic salamander, *Siren lacertina. J. Exp. Biol.* 201:673–82.

Brainerd, E. L., and R. S. Simons. 2000. Morphology and function of lateral hypaxial musculature in salamanders. *Amer. Zool.* 40:77–86.

Bramble, D. M., and D. B. Wake. 1985. Feeding mechanisms of lower tetrapods. In *Functional vertebrate morphology,* ed. M. Hildebrand, D. M. Bramble, K. F. Liem, and D. B. Wake, 230–61. Cambridge, MA: Harvard University Press.

Brame, A. H., Jr. 1968. The number of egg masses and eggs laid by the California newt, *Taricha torosa. J. Herpetol.* 2:169–70.

Braña, F., L. Frechilla, and G. Orizaola. 1996. Effect of introduced fish in amphibian assemblages in mountain lakes of northern Spain. *Herpetol. J.* 6:145–48.

Braña, F., M. de la Hoz, and C. Lastra. 1986. Alimentación y relaciones tróficas entre las larvas de *Triturus marmoratus, T. alpestris* y *T. helveticus* (Amphibia: Caudata). *Doñana, Acta Vert.* 13:21–33.

Branch, L. C. 1983. Social behavior of the tadpoles of *Phyllomedusa vaillanti. Copeia* 1983:420–28.

Branch, L. C., and R. Altig. 1981. Nocturnal stratification of three species of *Ambystoma* larvae. *Copeia* 1981:870–73.

Branch, L. C., and D. H. Taylor. 1977. Physiological and behavioral responses of larval spotted salamanders (*Ambystoma maculatum*) to various concentrations of oxygen. *Comp. Biochem. Physiol.* 58A:269–74.

Branch, W. R., and R. W. Patterson. 1975. Notes on the ecology of the giant girdled lizard, *Cordylus giganteus. J. Herpetol.* 9:364–66.

Brandon, R. A. 1965a. Morphological variation and ecology of the salamander *Phaeognathus hubrichti. Copeia* 1965:67–71.

———. 1965b. A new race of the neotenic salamander *Gyrinophilus palleucus. Copeia* 1965:346–52.

Brandon, R. A. 1966. Systematics of the salamander genus *Gyrinophilus. Illinois Biol. Monogr.* 35:1–86.

Brandon, R. A. 1970a. Courtship, spermatophores and eggs of the Mexican Achoque, *Ambystoma (Bathysiredon) dumerili* (Duges). *Zool. J. Linn. Soc.* 49:247–54.

———. 1970b. *Typhlotriton. Cat. Amer. Amphib. Rept.* 84.1–84.2.

———. 1971. North American troglobitic salamanders: Some aspects of modification in cave habitats, with special reference to *Gyrinophilus palleucus. Nat. Speleol. Soc. Bull.* 33:1–21.

———. 1972. Hybridization between the Mexican salamanders *Ambystoma dumerilii* (Dugès) and *Ambystoma mexicanum* under laboratory conditions. *Herpetologica* 28:199–207.

———. 1976. Spontaneous and induced metamorphosis of *Ambystoma dumerilii* (Dugès), a paedogenetic Mexican salamander, under laboratory conditions. *Herpetologica* 32:429–38.

———. 1977. Interspecific hybridization among Mexican and United States salamanders of the genus *Ambystoma* under laboratory conditions. *Herpetologica* 33:133–52.

Brandon, R. A., and R. G. Altig. 1973. Eggs and small larvae of two species of *Rhyacosiredon. Herpetologica* 29:349–51.

Brandon, R. A., and S. R. Ballard. 1998. Status of northeastern Illinois amphibians. In *Status and conservation of Midwestern amphibians,* ed. M. J. Lannoo, 115–24. Iowa City: University of Iowa Press.

Brandon, R. A., and D. J. Bremer. 1966. Neotenic newts, *Notophthalmus viridescens louisianensis,* in southern Illinois. *Herpetologica* 22:213–17.

———. 1967. Overwintering of larval tiger salamanders in southern Illinois. *Herpetologica* 23:67–68.

Brandon, R. A., and J. E. Huheey. 1971. Movements and interactions of two species of *Desmognathus* (Amphibia: Plethodontidae). *Am. Midl. Nat.* 86:86–92.

———. 1975. Diurnal activity, avian predation, and the question of warning coloration and cryptic coloration in salamanders. *Herpetologica* 31:252–55.

———. 1981. Toxicity in the plethodontid salamanders *Pseudotriton ruber* and *Pseudotriton montanus* (Amphibia, Caudata). *Toxicon* 19:25–31.

Brandon, R. A., J. Jacobs, A. Wynn, and D. M. Sever. 1986. A naturally metamorphosed Tennessee cave salamander (*Gyrinophilus palleucus*). *J. Tennessee Acad. Sci.* 61:1–2.

Brandon, R. A., G. M. Labanick, and J. E. Huheey. 1979a. Learned avoidance of brown efts, *Notophthalmus viridescens louisianensis* (Amphibia, Urodela, Salamandridae) by chickens. *J. Herpetol.* 13:171–76.

———. 1979b. Relative palatability, defensive behavior, and mimetic relationships of red salamanders (*Pseudotriton ruber*), mud salamanders (*Pseudotriton montanus*), and red efts (*Notophthalmus viridescens*). *Herpetologica* 35:289–303.

Brandon, R. A., E. J. Maruska, and W. T. Rumph. 1981. A new species of neotenic *Ambystoma* (Amphibia, Caudata) endemic to Laguna Alchichica, Puebla, Mexico. *Bull. Southern California Acad. Sci.* 80:112–25.

Brasileiro, C. A., R. J. Sawaya, and G. Giradelli. 2003. *Physalaemus cuvieri* (Barker frog). Predation. *Herp. Rev.* 34:137.

Brassart, J., J. L. Kirschvink, J. P. Phillips, and S. C. Borland. 1999. Ferromagnetic material in the eastern red-spotted newt *Notophthalmus viridescens. J. Exp. Biol.* 202:3155–60.

Brattstrom, B. H. 1962a. Call order and social behavior in the foam-building frog, *Engystomops pustulosus. Amer. Zool.* 2:394.

———. 1962b. Homing in the giant toad, *Bufo marinus. Herpetologica* 18:176–80.

———. 1962c. Thermal control of aggregation behavior in tadpoles. *Herpetologica* 18:38–46.

———. 1963. Preliminary review of the thermal requirements of amphibians. *Ecology* 44:238–55.

———. 1968. Thermal acclimation in anuran amphibians as a function of latitude and altitude. *Comp. Biochem. Physiol.* 24:93–111.

———. 1970. Amphibia. In *Comparative physiology of thermoregulation,* ed. G. C. Whittow, 135–66. New York: Academic Press.

———. 1979. Amphibian temperature regulation studies in the field and laboratory. *Amer. Zool.* 19:345–56.

Brattstrom, B. H., and R. M. Yarnell. 1968. Aggressive behavior in two species of leptodactylid frogs. *Herpetologica* 24:222–28.

Brauer, A. 1897. Beiträge zur Kenntniss der Entwicklungsgeschichte und der Anatomie der Gymnophionen. *Zool. Jb. Anat.* 10:359–472.

Breckenridge, W. J. 1944. *Reptiles and amphibians of Minnesota.* Minneapolis: University of Minnesota Press.

Breckenridge, W. J., and J. R. Tester. 1961. Growth, local movements and hibernation of the Manitoba toad, *Bufo hemiophrys. Ecology* 42:637–46.

Breckenridge, W. R., and S. Jayasinghe. 1979. Observations on the eggs and larvae of *Ichthyophis glutinosus. Ceylon J. Sci. (Bio. Sci.)* 13:187–202.

Breckenridge, W. R., S. Nathanael, and L. Pereira. 1987. Some aspects of the biology and development of *Ichthyophis glutinosus* (Amphibia: Gymnophiona). *J. Zool., London* 211:437–49.

Breden, F. 1987. The effect of post-metamorphic dispersal on the population genetic structure of Fowler's toad, *Bufo woodhousei fowleri. Copeia* 1987:386–95.

———. 1988. Natural history and ecology of Fowler's toad, *Bufo woodhousei fowleri* (Amphibia: Bufonidae), in the Indiana Dunes National Lakeshore. *Fieldiana: Zool.*, n.s. 49:1–16.

Breden, F., and C. H. Kelly. 1982. The effect of conspecific interactions on metamorphosis in *Bufo americanus. Ecology* 63:1682–89.

Breden, F., A. Lum, and R. Wassersug. 1982. Body size and orientation in aggregates of toad tadpoles, *Bufo woodhousei. Copeia* 1982:672–80.

Breder, C. M., Jr. 1946. Amphibians and reptiles of the Rio Chucunaque drainage, Darien, Panama, with notes on their life histories and habits. *Bull. Amer. Mus. Nat. Hist.* 86:381–435.

Breece, G. A., and J. L. Dusi. 1985. Food habits and home ranges of the common long-nosed armadillo *Dasypus novemcinctus* in Alabama. In *The evolution and ecology of armadillos, sloths, and vermilinguas,* ed. G. G. Montgomery, 419–27. Washington, DC: Smithsonian Institution Press.

Breitenbach, G. L. 1982. The frequency of communal nesting and solitary brooding in the salamander, *Hemidactylium scutatum. J. Herpetol.* 16:341–46.

Brekke, D. R., S. D. Hillyard, and R. M. Winokur. 1991. Behavior associated with the water absorption response by the toad, *Bufo punctatus. Copeia* 1991:393–401.

Bremond, J.-C. 1978. Acoustic competition between the song of the wren (*Troglodytes troglodytes*) and the songs of other species. *Behaviour* 65:89–98.

Brenner, F. J. 1969. The role of temperature and fat deposition in hibernation and reproduction in two species of frogs. *Herpetologica* 25:105–13.

Brenowitz, E. A. 1982. Long-range communication of species identity by song in the red-winged blackbird. *Behav. Ecol. Sociobiol.* 10:29–38.

———. 1989. Neighbor call amplitude influences aggressive behavior and intermale spacing in choruses of the Pacific treefrog (*Hyla regilla*). *Ethology* 83:69–79.

Brenowitz, E. A., and G. J. Rose. 1994. Behavioural plasticity mediates aggression in choruses of the Pacific treefrog. *Anim. Behav.* 47:633–41.

———. 1999. Female choice and plasticity of male calling behavior in the Pacific treefrog. *Anim. Behav.* 57:1337–42.

Brenowitz, E. A., G. J. Rose, and T. Alder. 2001. The neuroethology of acoustic communication in Pacific treefrogs. In *Anuran communication,* ed. M. J. Ryan, 145–55. Washington, DC: Smithsonian Institution Press.

Brenowitz, E. A., G. Rose, and R. R. Capranica. 1985. Neural correlates of temperature coupling in the vocal communication system of the gray treefrog (*Hyla versicolor*). *Brain Res.* 359:364–67.

Brenowitz, E. A., W. Wilczynski, and H. H. Zakon. 1984. Acoustic communication in spring peepers: Environmental and behavioral aspects. *J. Comp. Physiol. A* 155:585–92.

Bressi, N., S. Dolce, and F. Stoch. 1996. Observations on the feeding of amphibians. 4. Larval diet of the fire salamander, *Salamandra salamandra* (Linnaeus, 1758) near Trieste (northeastern Italy) (Amphibia, Caudata, Salamandridae). *Atti Mus. Civico Stor. Natur. Trieste* 47:275–83.

Brett, S. S., and G. Shelton. 1979. Ventilatory mechanisms of the amphibian *Xenopus laevis:* The role of the buccal force pump. *J. Exp. Biol.* 80:251–69.

Bridges, C., E. Little, D. Gardiner, J. Petty, and J. Huckins. 2004. Assessing the toxicity and teratogenicity of pond water in north-central Minnesota to amphibians. *Env. Sci. Poll. Res.* 11:233–39.

Bridges, C. M. 1999. Effects of a pesticide on tadpole activity and predator avoidance behavior. *J. Herpetol.* 33:303–6.

Bridges, C. M., and M. D. Boone. 2003. The interactive effects of UV-B and insecticide exposure on tadpole survival, growth and development. *Biol. Cons.* 113:49–54.

Bridges, C. M., and W. H. Gutzke. 1997. Effects of environmental history, sibship, and age on predator-avoidance responses of tadpoles. *Can. J. Zool.* 75:87–93.

Bridges, C. M., and R. D. Semlitsch. 2000. Variation in pesticide tolerance of tadpoles among and within species of Ranidae and patterns of amphibian decline. *Cons. Biol.* 14:1490–99.

———. 2001. Genetic variation in insecticide tolerance in a population of southern leopard frogs (*Rana sphenocephala*): Implications for amphibian conservation. *Copeia* 2001:7–13.

———. 2005a. Variation in pesticide tolerance. In *Amphibian declines: The conservation status of United States species,* ed. M. J. Lannoo, 93–95. Berkeley, CA: University of California Press.

———. 2005b. Xenobiotics. In *Amphibian declines: The conservation status of United States species,* ed. M. J. Lannoo, 89–92. Berkeley: University of California Press.

Briedermann, L. 1976. Ergebnisse einer Inhaltsanalyse von 665 Wildschweinmagen. *Zool. Gart.* 46:157–85.

Briegleb, W. 1961. Die Spermatophore des Grottenolms. *Zool. Anz.* 166:87–91.

———. 1962. Zur Biologie und Ökologie des Grottenolms (*Proteus anguinus* Laur. 1768). *Z. Morph. Ökol. Tiere* 51:271–334.

Briffault, R. 1927. *The mothers: A study of the origins of sentiments and institutions.* 3 vols. New York: Macmillan.

Briggs, J. L., and R. M. Storm. 1970. Growth and population structure of the cascade frog, *Rana cascadae* Slater. *Herpetologica* 26:283–300.

Brimley, C. S. 1923. The dwarf salamander at Raleigh, North Carolina. *Copeia* 1923:81–82.

Brinkmann, R., and R. Podloucky. 1987. Vorkommen, Gefahrdung und Schutz der Kreuzkröte (*Bufo calamita* Laur.) in Niedersachsen unter besonderer Berucksichtigung von Abgrabungen—Grundlagen fur ein Artenhilfsprogramm. *Ber. Naturhist. Ges. Hanover* 129:181–207.

Bristow, C. E. 1991. Interactions between phylogenetically distant predators: *Notophthalmus viridescens* and *Enneacanthus obesus*. *Copeia* 1991:1–8.

Brizzi, R., and C. Calloni. 1992. Male cloacal region of the spotted salamander, *Salamandra salamandra gigliotii* (Amphibia, Salamandridae). *Boll. Zool.* 59:377–85.

Brizzi, R., C. Calloni, and G. Delfino. 1990. Accessory structures in the genital apparatus of *Salamandrina terdigitata*. IV. Male cloacal glands. A study under light and scanning microscopes. *Z. mikrosk.-anat. Forsch.*104:871–97.

———. 1991. Tail base glands in European plethodontid salamanders with some comments on their biological and phylogenetic significance. *Amphibia-Reptilia* 12:357–72.

———. 1992. Male cloacal region of *Triturus italicus* (Peracca) with reference to the cloacal anatomy and reproductive patterns of the Salamandridae. *Arch. Ital. Anat. Embriol.* 97:121–38.

Brizzi, R., C. Calloni, G. Delfino, and S. Lotti. 1996. The male cloaca of *Salamandra lanzai* with notes on the dorsal glands in the genera *Salamandra* and *Mertensiella* in relation to their courtship habits. *Herpetologica* 52:505–15.

Brizzi, R., C. Calloni, G. Delfino, and G. Tanteri. 1995. Notes on the male cloacal anatomy and reproductive biology of *Euproctus montanus* (Amphibia: Salamandridae). *Herpetologica* 51:8–18.

Brizzi, R., G. Delfino, and S. Jantra. 2003. An overview of breeding glands. In *Reproductive biology and phylogeny of anura*, ed. B. G. M. Jamieson, 253–317. Enfield, NH: Science Publishers.

Brizzi, R., G. Delfino, and R. Pellegrini. 2002. Specialized mucous glands and their possible adaptive role in the males of some species of *Rana* (Amphibia, Anura). *J. Morphol.* 254:328–41.

Brizzi, R., G. Delfino, R. Rebelo, and D. M. Sever. 1999. Absence of dorsal glands in the cloaca of male *Chioglossa lusitanica* and the possible correlation with courtship mode. *J. Herpetol.* 33:220–28.

Brock, J., and P. Verrell. 1994. Courtship behavior of the seal salamander, *Desmognathus monticola* (Amphibia: Caudata: Plethodontidae). *J. Herpetol.* 28:411–15.

Brockelman, W. Y. 1969. An analysis of density effects and predation in *Bufo americanus* tadpoles. *Ecology* 50:632–44.

Brockerman, S. 2002. Rachel Carson's *Silent Spring:* Environmentalist mythology killing us softly. *Capitalism Magazine* (http://www.capmag.com/article.asp?id=1796).

Brode, W. E. 1961. Observations on the development of *Desmognathus* eggs under relatively dry conditions. *Herpetologica* 17:202–3.

Brodie, E. D., Jr. 1968a. Investigations on the skin toxin of the adult rough-skinned newt, *Taricha granulosa*. *Copeia* 1968:307–13.

———. 1968b. Investigation of the skin toxin of the red-spotted newt, *Notophthalmus viridescens viridescens*. *Am. Midl. Nat.* 80:276–80.

———. 1977. Salamander antipredator postures. *Copeia* 1977:523–35.

———. 1978. Biting and vocalization as antipredator mechanisms in terrestrial salamanders. *Copeia* 1978:127–29.

———. 1983. Antipredator adaptations of salamanders: Evolution and convergence among terrestrial species. In *Plant, animal, and microbial adaptations to terrestrial environment*, ed. N. S. Margaris, M. Arianoutsou-Faraggitaki, and R. J. Reiter, 109–33. New York: Plenum.

Brodie, E. D., Jr., and E. D. Brodie, III. 1980. Differential avoidance of mimetic salamanders by free-ranging birds. *Science* 208:181–82.

Brodie, E. D., Jr., P. K. Ducey, and E. A. Baness. 1991. Antipredator skin secretions of some tropical salamanders (*Bolitoglossa*) are toxic to snake predators. *Biotropica* 23:58–62.

Brodie, E. D., Jr., P. K. Ducey, and J. Lemos-Espinal. 1991. Antipredator behavior of the salamander *Bolitoglossa rufescens:* Effects of temperature and location of stimulus. *J. Herpetol.* 25:99–101.

Brodie, E. D., Jr., and D. R. Formanowicz, Jr. 1981. Palatability and antipredator behavior of the treefrog *Hyla versicolor* to the shrew *Blarina brevicauda*. *J. Herpetol.* 15:235–36.

———. 1983. Prey preference of predators: Differential vulnerability of larval anurans. *Herpetologica* 39:67–75.

———. 1987. Antipredator mechanisms of larval anurans: Protection of palatable individuals. *Herpetologica* 43:369–73.

Brodie, E. D., Jr., D. R. Formanowicz, Jr., and E. D. Brodie III. 1978. The development of noxiousness of *Bufo americanus* tadpoles to aquatic insect predators. *Herpetologica* 34:302–6.

———. 1991. Predator avoidance and antipredator mechanisms: Distinct pathways to survival. *Ethol. Ecol. Evol.* 3:73–77.

Brodie, E. D., Jr., and L. S. Gibson. 1969. Defensive behavior and skin glands of the northwestern salamander, *Ambystoma gracile*. *Herpetologica* 25:187–94.

Brodie, E. D., Jr., and R. R. Howard. 1972. Behavioral mimicry in the defensive displays of the urodele amphibians *Notophthalmus viridescens* and *Pseudotriton ruber*. *BioScience* 22:666–67.

———. 1973. Experimental study of Batesian mimicry in the salamanders *Plethodon jordani* and *Desmognathus ochrophaeus*. *Am. Midl. Nat.* 90:38–46.

Brodie, E. D., Jr., J. A. Johnson, and C. K. Dodd, Jr. 1974. Immobility as a defensive behavior in salamanders. *Herpetologica* 30:79–85.

Brodie, E. D., Jr., R. T. Nowak, and W. R. Harvey. 1979. The effectiveness of antipredator secretions and behavior of selected salamanders against shrews. *Copeia* 1979:270–74.

Brodie, E. D., Jr., R. A. Nussbaum, and Y. Datong. 1990. Antipredator behavior of Chinese salamanders. *Zool. Res.* 11:7–18.

Brodie, E. D., Jr., R. A. Nussbaum, and M. DiGiovanni. 1984. Antipredator adaptations of Asian salamanders (Salamandridae). *Herpetologica* 40:56–68.

Brodie, E. D., Jr., B. J. Ridenhour, and E. D. Brodie III. 2002. The evolutionary response of predators to dangerous prey: Hotspots and coldspots in the geographic mosaic of coevolution between garter snakes and newts. *Evolution* 56:2067–82.

Brodie, E. D., Jr., and N. J. Smatresk. 1990. The antipredator arsenal of fire salamanders: Spraying of secretions from highly pressurized dorsal skin glands. *Herpetologica* 46:1–7.

Brodie, E. D., Jr., C. R. Williams, and M. J. Tyler. 1998. Evolution of aposematic behavior and coloration in the Australian frog genus *Uperoleia*. *J. Herpetol.* 32:136–39.

Brodie, E. D., Jr., T. G. Dowdey, and C. D. Anthony. 1989. Salamander antipredator strategies against snake attack: Biting by *Desmognathus*. *Herpetologica* 45:167–71.

Brodie, E. D., III, and E. D. Brodie, Jr. 1990. Tetrodotoxin resistance in garter snakes: An evolutionary response of predators to dangerous prey. *Evolution* 44:651–59.

———. 1991. Evolutionary response of predators to dangerous prey: Reduction of toxicity of newts and resistance of garter snakes in island populations. *Evolution* 45:221–24.

———. 1999. Costs of exploiting poisonous prey: Evolutionary trade-offs in a predator-prey arms race. *Evolution* 53:626–31.

Brodman, R. 1995. Annual variation in breeding success of two syntopic species of *Ambystoma* salamanders. *J. Herpetol.* 29:111–13.

———. 1996. Effects of intraguild interactions on fitness and microhabitat use of larval *Ambystoma* salamanders. *Copeia* 1996:372–78.

———. 2004. Intraguild predation on congeners affects size, aggression, and survival among *Ambystoma salamander* larvae. *J. Herpetol.* 38:21–26.

Brodman, R., S. Cortwright, and A. Resetar. 2002. Historical changes of reptiles and amphibians of northwest Indiana fish and wildlife properties. *Am. Midl. Nat.* 147:135–44.

Brönmark, C., and P. Edenhamn. 1994. Does the presence of fish affect the distribution of tree frogs (*Hyla arborea*)? *Cons. Biol.* 8:841–45.

Brönmark, C., and L.-A. Hansson. 2002. Environmental issues in lakes and ponds: Current state and perspectives. *Env. Cons.* 29:290–306.

Brönmark, C., S. D. Rundle, and A. Erlandsson. 1991. Interactions between freshwater snails and tadpoles: Competition and facilitation. *Oecologia* 87:8–18.

Brook, B. W., N. S. Sodhi, and P. K. L. Ng. 2003. Catastrophic extinctions follow deforestation in Singapore. *Nature* 424:420–23.

Brooke, P. N., R. A. Alford, and L. Schwarzkopf. 2000. Environmental and social factors influence chorusing behaviour in a tropical frog: Examining various temporal and spatial scales. *Behav. Ecol. Sociobiol.* 49:79–87.

Brooks, D. R. 1984. Platyhelminths. In *Diseases of amphibians and reptiles,* ed. G. L. Hoff, F. L. Frye, and E. R. Jacobson, 247–58. New York: Plenum.

Brooks, D. R., and D. A. McLennan. 1991. *Phylogeny, ecology, and behavior. A research program in comparative biology.* Chicago: University of Chicago Press.

Brooks, G. R. 1964. An analysis of the food habits of the bullfrog, *Rana catesbeiana,* by body size, sex, month, and habitat. *Virginia J. Sci.* 15:173–86.

———. 1968. Natural history of a West Indian frog, *Leptodactylus fallax. Virginia J. Sci.* 19:176.

Brooks, M. 1946. Burrowing of *Plethodon jordani. Copeia* 1946:102.

Brooks, R. T. 1999. Residual effects of thinning and high white-tailed deer densities on northern redback salamanders in southern New England oak forests. *J. Wildl. Manage.* 63:1172–80.

———. 2004. Weather-related effects on woodland vernal pool hydrology and hydroperiod. *Wetlands* 24:104–14.

Brooks, T. M., R. A. Mittermeier, C. G. Mittermeier, G. A. B. da Fonseca, A. B. Rylands, W. R. Konstant, P. Flick, J. Pilgrim, S. Oldfield, G. Magin, and C. Hilton-Taylor. 2002. Habitat loss and extinction in the hotspots of biodiversity. *Cons. Biol.* 16:909–23.

Broomhall, S. D., W. S. Osborne, and R. B. Cunningham. 2000. Comparative effects of ambient ultraviolet-B radiation on two sympatric species of Australian frogs. *Cons. Biol.* 14:420–27.

Brophy, T. E. 1980. Food habits of sympatric larval *Ambystoma tigrinum* and *Notophthalmus viridescens. J. Herpetol.* 14:1–6.

Brosset, A. 1962. The bats of central and western India. Part. 2. *J. Bombay Nat. Hist. Soc.* 59:583–624.

Browder, L. W., J. L. Underhill, and D. J. Merrell. 1966. Mid-dorsal stripe in the wood frog. *J. Heredity* 57:65–67.

Brown, A., B. Christensen, T. D. Schwaner, and M. Nickerson. 2000. *Bufo microscaphus* (Arizona toad). Reproduction. *Herp. Rev.* 31:168.

Brown, A. V., and L. C. Fitzpatrick. 1981a. Metabolic acclimation to temperature in the Ozark salamander *Plethodon dorsalis angusticlavius. Comp. Biochem. Physiol.* 69A:499–503.

———. 1981b. Thermal acclimation and metabolism in the grey-bellied salamander, *Eurycea multiplicata griseogaster* (Plethodontidae). *Comp. Biochem. Physiol.* 69A:505–9.

Brown, B. C. 1967a. *Eurycea latitans. Cat. Amer. Amphib. Rept.* 34.1–34.2.

———. 1967b. *Eurycea nana. Cat. Amer. Amphib. Rept.* 35.1–35.2.

———. 1967c. *Eurycea neotenes. Cat. Amer. Amphib. Rept.* 36.1–36.2

Brown, C. E., and B. S. Martof. 1966. The function of the nasolabial grooves of plethodontid salamanders. *Physiol. Zool.* 39:357–67.

Brown, D. 1988. Components of lifetime reproductive success. In *Reproductive success*, ed. T. H. Clutton-Brock, 439–53. Chicago: University of Chicago Press.

———. 1994. Transfer of Hamilton's frog, *Leiopelma hamiltoni,* to a newly created habitat on Stephens Island, New Zealand. *New Zealand J. Zool.* 21:425–30.

Brown, E. E. 1958. Feeding habits of the northern water snake, *Natrix sipedon sipedon* Linnaeus. *Zoologica* 43:55–71.

Brown, H. A. 1967. Embryonic temperature adaptations and genetic compatibility in two allopatric populations of the spadefoot toad, *Scaphiopus hammondi. Evolution* 21:742–61.

Brown, H. A. 1975a. Embryonic temperature adaptations of the Pacific treefrog, *Hyla regilla. Comp. Biochem. Physiol.* 51A:863–73.

———. 1975b. Temperature and development of the tailed frog, *Ascaphus truei. Comp. Biochem. Physiol.* 50A:397–405.

———. 1976a. The status of California and Arizona populations of the western spadefoot toads (genus *Scaphiopus*). *Nat. Hist. Mus. Los Angeles County Contrib.* (286):1–15.

———. 1976b. The time-temperature relations of embryonic development in the northwestern salamander, *Ambystoma gracile. Can. J. Zool.* 54:552–58.

———. 1977. Oxygen consumption of a large, cold-adapted frog egg (*Ascaphus truei* (Amphibia: Ascaphidae)). *Can. J. Zool.* 55:343–48.

———. 1989. Developmental anatomy of the tailed frog (*Ascaphus truei*): A primitive frog with large eggs and slow development. *J. Zool., London* 217:525–37.

———. 1990. Morphological variation and age-class discrimination in overwintering tadpoles of the tailed frog, *Ascaphus truei. J. Zool., London* 220:171–84.

Brown, J. H. 1995. *Macroecology.* Chicago: University of Chicago Press.

Brown, J. H., and D. W. Davidson. 1977. Competition between seed-eating rodents and ants in desert ecosystems. *Science* 196:880–82.

Brown, J. H., D. W. Davidson, and O. H. Reichman. 1979. An experimental study of competition between seed-eating desert rodents and ants. *Amer. Zool.* 19:1129–43.

Brown, L., and D. Amadon. 1968. *Eagles, hawks, and falcons of the world.* New York: McGraw-Hill.

Brown, L. E. 1970. Natural hybrids between two toad species in Alabama. *Q. J. Florida Acad. Sci.* 32:285–90.

———. 1978. Subterranean feeding by the chorus frog *Pseudacris streckeri* (Anura: Hylidae). *Herpetologica* 34:212–16.

Brown, L. E., and J. A. Brownell. 1971. Relative survival of two toad species and their natural hybrids in a disturbed environment. *Am. Midl. Nat.* 86:235–38.

Brown, L. E., and J. E. Cima. 1998. Illinois chorus frogs and the Sand Lake dilemma. In *Status and conservation of Midwestern amphibians,* ed. M. J. Lannoo, 301–311. Iowa City: University of Iowa Press.

Brown, L. E., and R. S. Funk. 1977. Absence of dorsal spotting in two species of leopard frogs (Anura: Ranidae). *Herpetologica* 33:290–93.

Brown, L. E., and S. I. Guttman. 1970. Natural hybridization between the toads *Bufo arenarum* and *Bufo spinulosus* in Argentina. *Am. Midl. Nat.* 83:160–66.

Brown, L. E., H. O. Jackson, and J. R. Brown. 1972. Burrowing behavior of the chorus frog *Pseudacris streckeri. Herpetologica* 28:325–28.

Brown, L. E., and D. B. Means. 1984. Fossorial behavior and ecology of the chorus frog *Pseudacris ornata. Amphibia-Reptilia* 5:261–73.

Brown, L. E., and A. Mesrobian. 2005. Houston toads and Texas politics. In *Amphibian declines: The conservation status of United States species,* ed. M. J. Lannoo, 150–67. Berkeley: University of California Press.

Brown, L. E., and J. R. Pierce. 1967. Male-male interactions and chorusing intensities of the Great Plains toad, *Bufo cognatus. Copeia* 1967:149–54.

Brown, P. S., and S. C. Brown. 1973. Prolactin and thyroid hormone interactions in salt and water balance in the newt *Notophthalmus viridescens. Gen. Comp. Endocrinol.* 20:456–66.

———. 1977. Water balance responses to dehydration and neurohypophyseal peptides in the salamander, *Notophthalmus viridescens. Gen. Comp. Endocrinol.* 31:189–201.

———. 1982. Effects of hypophysectomy and prolactin on the water-balance response of the newt, *Taricha torosa. Gen. Comp. Endocrinol.* 46:7–12.

———. 1987. Osmoregulatory actions of prolaction and other adenohypophysial hormones. In *Vertebrate endocrinology: Fundamentals and biomedical implications, vol. 2. Regulation of water and electrolytes,* ed. P. K. T. Pang, and M. P. Schreibman, 45–84. New York: Academic Press.

Brown, P. S., S. C. Brown, T. Bisceglio, and S. N. Lemke. 1983. Breeding condition, temperature, and the regulation of salt and water by pituitary hormones in the red-spotted newt, *Notophthalmus viridescens. Gen. Comp. Endocrinol.* 51:292–302.

Brown, P. S., S. A. Hastings, and E. B. Frye. 1977. A comparison of the water-balance response in five species of plethodontid salamanders. *Physiol. Zool.* 50:203–14.

Brown, R. M., and D. T. Iskandar. 2000. Nest site selection, larval hatching, and advertisement calls, of *Rana arathooni* from southwestern Sulawesi (Celebes) Island, Indonnesia. *J. Herpetol.* 34:404–13.

Brown, R. M., and D. H. Taylor. 1995. Compensatory escape mode trade-offs between swimming performance and maneuvering behavior through larval ontogeny of the wood frog, *Rana sylvatica. Copeia* 1995:1–7.

Brown, S. C., and P. S. Brown. 1980. Water balance in the California newt, *Taricha torosa. Am. J. Physiol.* 238:R113–R118.

Brown, V. K. 1974. Diets and habitat preferences of selected anurans in southwestern Arkansas. *Am. Midl. Nat.* 91:468–78.

Brown, W. C. 1952. The amphibians of the Solomon Islands. *Bull. Mus. Comp. Zool.* 107:1–64.

Brown, W. C., and A. C. Alcala. 1970. Population ecology of the frog *Rana erythraea* in southern Negros, Philippines. *Copeia* 1970:611–22.

———. 1983. Modes of reproduction of Philippine anurans. In *Advances in herpetology and evolutionary biology,* ed. A. G. J. Rhodin and K. Miyata, 416–28. Cambridge, MA: Museum of Comparative Zoology.

Brown, W. C., and F. Parker. 1970. New frogs of the genus *Batrachylodes* (Ranidae) from the Solomon Islands. *Breviora* (346):1–31.

Broyles, R. H. 1981. Changes in the blood during amphibian metamorphosis. In *Metamorphosis: A problem in developmental biology* (2nd ed.), ed. L. I. Gilbert and E. Frieden, 461–90. New York: Plenum.

Bruce, R. C. 1969. Fecundity in primitive plethodontid salamanders. *Evolution* 23:50–54.

———. 1972a. The larval life of the red salamander, *Pseudotriton ruber. J. Herpetol.* 6:43–51.

———. 1972b. Variation in the life cycle of the salamander *Gyrinophilus porphyriticus. Herpetologica* 28:230–45.

———. 1974. Larval development of the salamanders *Pseudotriton montanus* and *P. ruber. Am. Midl. Nat.* 92:173–90.

———. 1975. Reproductive biology of the mud salamander, *Pseudotriton montanus,* in western North Carolina. *Copeia* 1975:129–37.

———. 1976. Population structure, life history, and evolution of paedogenesis in the salamander *Eurycea neotenes. Copeia* 1976:242–49.

———. 1978a. A comparison of the larval periods of Blue Ridge and Piedmont mud salamanders (*Pseudotriton montanus*). *Herpetologica* 34:325–32.

———. 1978b. Life-history patterns of the salamander *Gyrinophilus porphyriticus* in the Cowee Mountains, North Carolina. *Herpetologica* 34:53–64.

———. 1978c. Reproductive biology of the salamander *Pseudotriton ruber* in the southern Blue Ridge Mountains. *Copeia* 1978:417–23.

———. 1979. Evolution of paedomorphosis in salamanders of the genus *Gyrinophilus. Evolution* 33:998–1000.

———. 1980. A model of the larval period of the spring salamander, *Gyrinophilus porphyriticus,* based on size-frequency distributions. *Herpetologica* 36:78–86.

———. 1982a. Egg-laying, larval periods and metamorphosis of *Eurycea bislineata* and *E. junaluska* at Santeelah Creek, North Carolina. *Copeia* 1982:755–62.

———. 1982b. Larval periods and metamorphosis in two species of salamanders of the genus *Eurycea. Copeia* 1982:117–27.

———. 1985a. Larval period and metamorphosis in the salamander *Eurycea bislineata. Herpetologica* 41:19–28.

———. 1985b. Larval periods, population structure and the effects of stream drift in larvae of the salamanders *Desmognathus quadramaculatus* and *Leurognathus marmoratus* in a southern Appalachian stream. *Copeia* 1985:847–54.

———. 1988a. An ecological life table for the salamander *Eurycea wilderae*. *Copeia* 1988:15–26.

———. 1988b. Life history variation in the salamander *Desmognathus quadramaculatus*. *Herpetologica* 44:218–27.

———. 1993. Sexual size dimorphism in desmognathine salamanders. *Copeia* 1993:313–18.

———. 1995. The use of temporary removal sampling in a study of the population dynamics of the salamander *Desmognathus monticola*. *Aust. J. Ecol.* 20:403–12.

———. 1998. Nesting habits, eggs, and hatchlings of the salamander *Nototriton picadoi* (Plethodontidae: Bolitoglossidae). *Herpetologica* 54:13–18.

Bruce, R. C., C. K. Beachy, P. G. Lenzo, S. P. Pronych, and R. J. Wassersug. 1994. Effects of lung reduction on rheotactic performance in amphibian larvae. *J. Exp. Zool.* 268:377–80.

Bruce, R. C., J. Castanet, and H. Francillon-Vieillot. 2002. Skeletochronological analysis of variation in age structure, body size, and life history in three species of desmognathine salamanders. *Herpetologica* 58:181–93.

Bruce, R. C., and N. G. Hairston, Sr. 1990. Life-history correlates of body size differences between two populations of the salamander, *Desmognathus monticola*. *J. Herpetol.* 24:124–34.

Bruce, R. C., R. G. Jaeger, and L. D. Houck, eds. 2000. *The biology of plethodontid salamanders*. New York: Kluwer Academic/Plenum.

Brues, C. T. 1932. Further studies on the fauna of North American hot springs. *Proc. Amer. Acad. Arts Sci.* 67:185–303.

Bruneau, M., and E. Magnin. 1980a. Croissance, nutrition, et reproduction des ouaouarons *Rana catesbeiana* Shaw (Amphibia Anura) des Laurentides du nord de Montréal. *Can. J. Zool.* 58:175–83.

———. 1980b. Vie larvaire des ouaouarons *Rana catesbeiana* Shaw (Amphibia Anura) des Laurentides, Quebec. *Can. J. Zool.* 58:169–74.

Bruner, M. A., P. A. Shipman, M. Rao, and J. A. Bantle. 2002. Developmental effects of ambient UV-B light and landfill leachate in *Rana blairi* and *Hyla chrysoscelis*. *Ecotox. Env. Safety* 53:73–80.

Brunkow, P. E., and J. P. Collins. 1996. Effects of individual variation in size on growth and development of larval salamanders. *Ecology* 77:1483–92.

Brunn, F. 1921. Beitrag zur Kenntnis der Wirkung von Hypophysen-extrakten auf den Wasserhaushalt des Frösches. *Z. Gesam. Exp. Med.* 25:170–75.

Brunner, J. L., D. M. Schock, E. W. Davidson, and J. P. Collins. 2004. Intraspecific reservoirs: Complex life history and the persistence of a lethal ranavirus. *Ecology* 85:560–66.

Brunori, M., S. Condo, A. Bellelli, B. Giardina, and G. Micheli. 1985. Tadpole *Xenopus laevis* hemoglobin. Correlation between structure and functional properties. *J. Molec. Biol.* 181:237–39.

Brush, J. S., and P. M. Narins. 1989. Chorus dynamics of a Neotropical amphibian assemblage: comparison of computer simulation and natural behaviour. *Anim. Behav.* 37:33–44.

Brust, D. G. 1993. Maternal brood care by *Dendrobates pumilio:* A frog that feeds its young. *J. Herpetol.* 27:96–98.

Bryant, E. H. 1969. A system favoring the evolution of holometabolous development. *Ann. Entomol. Soc. Amer.* 62:1087–91.

Brzeziński, M., W. Jędrzejewski, and B. Jędrzejewski. 1993. Diet of otters (*Lutra lutra*) inhabiting small rivers in the Białowieża National Park, eastern Poland. *J. Zool., London* 230:495–501.

Brzoska, J. 1980. Quantitative studies on the elicitation of the electrodermal response by calls and synthetic acoustical stimuli in *Rana lessonae* Camerano, *R. r. ridibunda* Pallas and the hybrid *Rana "esculenta"* L. (Anura, Amphibia). *Behav. Proc.* 5:113–41.

———. 1982. Vocal response of male European water frogs (*Rana esculenta* complex) to mating and territorial calls. *Behav. Proc.* 7:37–47.

Brzoska, J., and H.-J. Obert. 1980. Acoustic signals influencing the hormone production of the testes in the grass frog. *J. Comp. Physiol. A* 140:25–29.

Brzoska, J., H. Schneider, and E. Nevo. 1982. Territorial behavior and vocal responses in male *Hyla arborea savignyi* (Amphibia: Anura). *Israel J. Zool.* 31:27–37.

Brzoska, J., W. Walkowiak, and H. Schneider. 1977. Acoustic communication in the grass frog (*Rana t. temporaria* L.): Calls, auditory thresholds and behavioral responses. *J. Comp. Physiol.* 118:173–86.

Buchalczyk, T., and Z. Pucek. 1963. Food storage of the European water shrew, *Neomys fodiens* (Pennant 1771). *Acta Theriol.* 7:376–77.

Bucher, T. L., M. J. Ryan, and G. A. Bartholomew. 1982. Oxygen consumption during resting, calling, and nest building in the frog *Physalaemus pustulosus*. *Physiol. Zool.* 55:10–22.

Buchholz, D. R., and T. B. Hayes. 2000. Larval period comparison for the spadefoot toads *Scaphiopus couchii* and *Spea multiplicata* (Pelobatidae: Anura). *Herpetologica* 56:455–68.

———. 2002. Evolutionary patterns of diversity in spadefoot toad metamorphosis (Anura: Pelobatidae). *Copeia* 2002:180–89.

Büchs, W. 1987. Aspekte der Populationsökologie des Moorfrosches (*Rana arvalis* Nilsson): Ergebnisse der quantitativen Erfassung eines Moorfroschbestandes im westlichen Münsterland. *Beih. Schriftenr. Naturschutz Landschaftspfl. Niedersachen* 19:97–110.

Buckbinder, L., and D. D. Brown. 1993. Expression of the *Xenopus laevis* prolactin and thyrotropin genes during metamorphosis. *Proc. Natl. Acad. Sci. USA* 90:3820–24.

Buckland, W. 1832. On the vitality of toads enclosed in stone and wood. *Edinburgh New Phil. J.*, April–October 1832:26–32.

Buckley, D., B. Arano, P. Herrero, and G. Llorente. 1996. Population structure of Moroccan water frogs: Genetic cohesion despite a fragmented distribution. *J. Zool. Syst. Evol. Res.* 34:173–79.

Buckley, J., and T. J. C. Beebee. 2004. Monitoring the conservation status of an endangered amphibian: The natterjack toad *Bufo calamita* in Britain. *Anim. Cons.* 7:221–28.

Budge, E. A. W. 1969. *The gods of the Egyptians*. 2 vols. New York: Dover.

Budgett, J. S. 1899. Notes on the batrachians of the Paraguayan Chaco, with observations upon their breeding habits and development, especially with reference to *Phyllomedusa hypochondrialis* Cope. Quart. *J. Micr. Sci.* 42:305–33.

Buley, K. R., and G. Garcia. 1997. The recovery programme for the Mallorcan midwife toad *Alytes muletensis:* An update. *Dodo* 33:80–90.

Bulger, J. B., N. J. Scott, Jr., and R. B. Seymour. 2003. Terrestrial activity and conservation of adult California red-legged frogs *Rana aurora draytonii* in coastal forests and grasslands. *Biol. Cons.* 110:85–95.

Bull, C. M. 1977. Back pattern polymorphism and tadpole growth rate in two Western Australian frogs. *Aust. J. Zool.* 25:243–48.

Bull, E. L., and D. B. Marx. 2002. Influences of fish and habitat on amphibian communities in high elevation lakes in northeastern Oregon. *Northwest Sci.* 76:240–48.

Bulmer, R. N. H., and M. J. Tyler. 1968. Karam classification of frogs. *J. Polynesian Soc.* 77:333–85.

Bundy, D., and C. R. Tracy. 1977. Behavioral response of American toads (*Bufo americanus*) to stressful thermal and hydric environments. *Herpetologica* 33:455–58.

Bunn, D. S., A. B. Warburton, and R. D. S. Wilson. 1982. *The barn owl.* Vermillion, South Dakota: Buteo Books.

Bunnell, J. F., and R. A. Zampella. 1999. Acid water pond communities along a regional forest to agro-urban ecotone. *Copeia* 1999:614–27.

Bunnell, P. 1973. Vocalizations in the territorial behavior of the frog *Dendrobates pumilio. Copeia* 1973:277–84.

Burbrink, F. T., C. A. Phillips, and E. J. Heske. 1998. A riparian zone in southern Illinois as a potential dispersal corridor for reptiles and amphibians. *Biol. Cons.* 86:107–15.

Burgess, N. D., C. Rahbeck, F. W. Larsen, P. Williams, and A. Balmford. 2002. How much of the vertebrate diversity of sub-Saharan Africa is catered for by recent conservation proposals? *Biol. Cons.* 107:327–39.

Burggren, W. W. 1984. Transition of respiratory processes during amphibian metamorphosis: from egg to adult. In *Respiration and metabolism of embryonic vertebrates,* ed. R. S. Seymour, 31–53. Boston: Dr. W. Junk Publishers.

———. 1985. Gas exchange, metabolism, and "ventilation" in gelatinous frog egg masses. *Physiol. Zool.* 58:503–14.

———. 1988. Role of the central circulation in regulation of cutaneous gas exchange. *Amer. Zool.* 28:985–98.

———. 1989. Lung structure and function: amphibians. In *Comparative pulmonary physiology: Current concepts, vol. 39. Lung biology in health and disease,* ed. S. C. Wood, 153–92. New York: Marcel Dekker.

Burggren, W. W., and M. Doyle. 1986. Ontogeny of regulation of gill and lung ventilation in the bullfrog, *Rana catesbeiana. Respir. Physiol.* 66:279–91.

Burggren, W. W., and M. E. Feder. 1983. Temperature and the balance between aerial and aquatic respiration in larvae of *Rana berlandieri* and *Rana catesbeiana. Physiol. Zool.* 56:263–73.

———. 1986. Effects of experimental ventilation of the skin on cutaneous gas exchange in the bullfrog. *J. Exp. Biol.* 121:445–50.

Burggren, W. W., and R. L. Infantino, Jr. 1994. The respiratory transition from water to air breathing during amphibian metamorphosis. *Amer. Zool.* 34:238–46.

Burggren, W. W., R. L. Infantino, Jr., and D. S. Townsend. 1990. Developmental changes in cardiac and metabolic physiology of the direct-developing tropical frog *Eleutherodactylus coqui. J. Exp. Biol.* 152:129–47.

Burggren, W. W., and J. J. Just. 1992. Developmental changes in physiological systems. In *Environmental physiology of the amphibians,* ed. M. E. Feder and W. W. Burggren, 467–530. Chicago: University of Chicago Press.

Burggren, W. W., and R. Moalli. 1984. 'Active' regulation of cutaneous gas exchange by capillary recruitment in amphibians: Experimental evidence and a revised model for skin respiration. *Respir. Physiol.* 55:379–92.

Burggren, W. W., and A. Mwalukoma. 1983. Respiration during chronic hypoxia and hyperoxia in larval and adult bullfrogs (*Rana catesbeiana*). I. Morphological responses of lungs, skin and gills. *J. Exp. Biol.* 105:191–203.

Burggren, W. W., and N. H. West. 1982. Changing respiratory importance of gills, lungs and skin during metamorphosis in the bullfrog, *Rana catesbeiana. Respir. Physiol.* 47:151–64.

Burggren, W. W., and S. C. Wood. 1981. Respiration and acid-base balance in the salamander, *Ambystoma tigrinum:* Influence of temperature acclimation and metamorphosis. *J. Comp. Physiol.* B 44:241–46.

Burghardt, G. M. 1969. Comparative prey-attack studies in newborn snakes of the genus *Thamnophis. Behaviour* 33:77–114.

———. 1970. Intraspecific geographical variation in chemical food cue preference of newborn garter snakes (*Thamnophis sirtalis*). *Behaviour* 36:246–57

Burkett, R. D. 1966. Natural history of cottonmouth moccasin, *Agkistrodon piscivorus* (Reptilia). *Univ. Kansas Publ. Mus. Nat. Hist.* 17:435–91.

Burkhart, J. G., J. R. Bidwell, D. J. Fort, and S. R. Sheffield. 2003. Chemical stressors. In *Amphibian decline: An integrated analysis of multiple stressor effects,* ed. G. Linder, S. K. Krest, and D. W. Sparling, 111–28. Pensacola, FL: SETAC Press.

Burkholder, G. 1998. *Hyla chrysoscelis* (gray treefrog). Hibernacula. *Herp. Rev.* 29:231.

Burmeister, S. S., A. G. Ophir, M. J. Ryan, and W. Wilczynski. 2002. Information transfer during cricket frog contests. *Anim. Behav.* 64:715–25.

Burmeister, S., C. Somes, and W. Wilczynski. 2001. Behavioral and hormonal effects of exongenous vasotocin and corticosterone in the green treefrog. *Gen. Comp. Endocrinol.* 122:189–97.

Burmeister, S., W. Wilczynski, and M. J. Ryan. 1999. Temporal call changes and prior experience affect graded signalling in the cricket frog. *Anim. Behav.* 57:611–18.

Burne, M. R., and C. R. Griffin. 2005a. Habitat associations of pool-breeding amphibians in eastern Massachusetts, USA. *Wetlands Ecol. Manage.* 13:247–59.

———. 2005b. Protecting vernal pools: A model from Massachusetts, USA. *Wetlands Ecol. Manage.* 13:367–75.

Burnett, S. 1997. Colonizing cane toads cause population declines in native predators: Reliable anecdotal information and management implications. *Pacific Cons. Biol.* 3:65–72.

Burrowes, P. A. 2000. Parental care and sexual selection in the Puerto Rican cave-dwelling frog, *Eleutherodactylus cooki. Herpetologica* 56:375–86.

Burrowes, P. A., and R. L. Joglar. 1991. A survey of the population status and an ecological evaluation of three Puerto Rican frogs. In *Status y distribución de los reptiles y anfibios de la región de Puerto Rico. Publ. Cien. Misc,* ed. J. A. Moreno, 42–46. San Juan, Puerto Rico: Departamento de Recursos Naturales.

Burrowes, P. A., R. L. Joglar, and D. E. Green. 2004. Potential causes for amphibian declines in Puerto Rico. *Herpetologica* 60:141–54.

Burton, T. C. 1998. Pointing the way: The distribution and evolution of some characters of the finger muscles of frogs. *Amer. Mus. Novitates* (3229):1–13.

———. 2004. Muscles of the pes of hylid frogs. *J. Morphol.* 260:209–33.

Burton, T. M. 1976. An analysis of the feeding ecology of the salamanders (Amphibia, Urodela) of the Hubbard Brook Experimental Forest, New Hampshire. *J. Herpetol.* 10:187–204.

———. 1977. Population estimates, feeding habits and nutrient and energy relationships of *Notophthalmus v. viridescens* in Mirror Lake, New Hampshire. *Copeia* 1977:139–43.

Burton, T. M., and G. E. Likens. 1975a. Energy flow and nutrient cycling in salamander populations in the Hubbard Brook Experimental Forest, New Hampshire. *Ecology* 56:1068–80.

———. 1975b. Salamander populations and biomass in the Hubbard Brook Experimental Forest, New Hampshire. *Copeia* 1975:541–46.

Bury, R. B. 1983. Differences in amphibian populations in logged and old-growth redwood forest. *Northwest Sci.* 57:167–78.

———. 1986. Feeding ecology of the turtle, *Clemmys marmorata. J. Herpetol.* 20:515–21.

———. 2004. Wildfire, fuel reduction, and herpetofaunas across diverse landscape mosaics in northwestern forests. *Cons. Biol.* 18:968–75.

Bury, R. B., and M. J. Adams. 1999. Variation in age at metamorphosis across a latitudinal gradient for the tailed frog, *Ascaphus truei. Herpetologica* 55:283–91.

Bury, R.B., and P.S. Corn. 1988a. Douglas-fir forests in the Oregon and Washington Cascades: Relation of the herpetofauna to stand age and moisture. In *Management of amphibians, reptiles and small mammals in North America,* ed. R. C. Szaro, K.E. Severson and D.R. Patton, 11–22. Flagstaff, AZ: U.S.F.S. Rocky Mountain Forest and Range Experiment Station. GTR-RM-166.

———. 1988b. Responses of aquatic and streamside amphibians to timber harvest: a review. In *Streamside management: Riparian wildlife and forestry interactions,* ed. K. J. Raedeke, 165–81. Seattle: Institute of Forest Resources, University of Washington, Contribution 59.

Bury, R. B., and M. Martin. 1973. Comparative studies on the distribution and foods of plethodontid salamanders in the redwood region of northern California. *J. Herpetol.* 7:331–35.

Bury, R. B., and J. A. Whelan. 1984. *Ecology and management of the bullfrog.* Fish and Wildlife Service Resource Publication 155:1–23. Washington, DC: Government Printing Office.

Busby, W. H., and J. R. Parmelee. 1996. Historical changes in a herpetofaunal assemblage in the Flint Hills of Kansas. *Am. Midl. Nat.* 135:81–91.

Bush, F. M. 1963. Effects of light and temperature on the gross composition of the toad, *Bufo fowleri. J. Exp. Zool.* 153:1–13.

Bush, S. L. 1996. Why is double clutching rare in the Majorcan midwife toad? *Anim. Behav.* 52:913–22.

———. 1997. Vocal behavior of males and females in the Majorcan midwife toad. *J. Herpetol.* 31:251–57.

Bush, S. L., and D. J. Bell. 1997. Courtship and female competition in the Majorcan midwife toad, *Alytes muletensis. Ethology* 103:292–302.

Bush, S. L., M. L. Dyson, and T. R. Halliday. 1996. Selective phonotaxis by males in the Majorcan midwife toad. *Proc. Roy. Soc. London B* 263:913–17.

Bush, S. L., H. C. Gerhardt, and J. Schul. 2002. Pattern recognition and call preferences in treefrogs (Anura: Hylidae): A quantitative analysis using a no-choice paradigm. *Anim. Behav.* 63:7–14.

Busk, M., F. B. Jensen, and T. Wang. 2000. Effects of feeding on metabolism, gas transport, and acid-base balance in the bullfrog, *Rana catesbeiana. Am. J. Physiol.* 278:R185–R195.

Busse, K. 1970. Care of the young by male *Rhinoderma darwini. Copeia* 1970:395.

———. 1971. Desarrollo de *Batrachyla leptopus* Bell con observaciones sobre su ecología y comportamiento (Amphibia, Leptodactylidae). *Inv. Zool. Chilenas* 15:5–63.

Bustamante, M. R., S. R. Ron, and L. A. Coloma. 2005. Cambios en la diversidad en siete communidades de anuros en los Andes de Ecuador. *Biotropica* 37:180–89.

Butler, W. H., and B. Y. Main. 1959. Predation on vertebrates by mygalomorph spiders. *West. Aust. Nat.* 11:146.

Buttemer, W. A. 1990. Effect of temperature on evaporative water loss of the Australian tree frogs *Litoria caerulea* and *Litoria chloris. Physiol. Zool.* 63:1043–57.

Buttemer, W. A., and C. Thomas. 2003. Influence of temperature on evaporative water loss and cutaenous resistance to water vapour diffusion in the orange-thighed frog (*Litoria xanthomera*). *Aust. J. Zool.* 51:111–18.

Buttenhoff, P. 1995. *Bolitoglossa rufescens* (Northern Banana Salamander). Predation. *Herp. Rev.* 26:197.

Buytendijk, F. J. J. 1918. Instinct de la recherche du nid et experience chez les crapauds (*Bufo vulgaris* et *Bufo calamita*). *Arch. Neerland. Physiol. l'Homme Anim.* 2:1–50.

Byers, J. A., and M. Beckoff. 1991. Development, the conveniently forgotten variable in 'true kin recognition.' *Anim. Behav.* 41:1088–90.

Byrne, J. J., and R. J. White. 1975. Cyclic changes in liver and muscle glycogen, tissue lipid, and blood glucose in a naturally occurring population of *Rana catesbeiana. Comp. Biochem. Physiol.* 50A:709–15.

Byrne, P. G. 2002. Climatic correlates of breeding, simultaneous polyandry and potential for sperm competition in the frog *Geocrinia georgiana. J. Herpetol.* 36:125–29.

Byrne, P. G., and J. D. Roberts. 1999. Simultaneous mating with multiple males reduces fertilization success in the myobatrachid frog *Crinia georgiana. Proc. Roy. Soc. London B* 266:717–21.

———. 2000. Does multiple paternity improve fitness of the frog *Crinia georgiana*? *Evolution* 54:968–73.

Cabanac, M., and E. Jeddi. 1971. Thermopréférendum et thermorégulation comportementale chez trois poikilotherms. *Physiol. Behav.* 7:375–80.

Cade, W. H. 1981. Field cricket spacing, and the phonotaxis of crickets and parasitoid flies to clumped and isolated cricket songs. *Z. Tierpsychol.* 55:365–75.

Cadle, J. E. 1984. Molecular systematics of Neotropical xenodontine snakes. III. Overview of xenodontine phylogeny and the history of New World snakes. *Copeia* 1984:641–52.

———. 1987. Geographic distribution: problems in phylogeny and zoogeography. In *Snakes: Ecology and evolutionary biology,* ed. R. A. Seigel, J. T. Collins, and S. S. Novak, 77–105. New York: Macmillan.

Cadle, J. E., and H. W. Greene. 1993. Phylogenetic patterns, biogeography, and the ecological structure of Neotropical snake assemblages. In *Species diversity in ecological communities. Historical and geographical perspectives,* ed. R. E. Ricklefs and D. Schluter, 281–93. Chicago: University of Chicago Press.

Caetano, M. H. 1990. Use and results of skeletochronology in some urodeles (*Triturus marmoratus,* Latreille 1800 and *Triturus boscai,* Lataste 1879). *Ann. Sci. Natur. Zool.,* 13th ser., 11:197–99.

Caetano, M. H., M. Castanet, and H. Francillon. 1985. Determination de l'age de *Triturus marmoratus marmoratus* (Latreille 1800) du Parc National de Peneda Geres (Portugal) par squelettochronologie. *Amphibia-Reptilia* 6:117–32.

Caetano, M. H., and R. Leclair. 1996. Growth and population structure of red-spotted newts (*Notophthalmus viridescens*) in permanent lakes of the Laurentian Shield, Quebec. *Copeia* 1996:866–74.

———. 1999. Comparative phenology and demography of *Triturus boscai* from Portugal. *J. Herpetol.* 33:192–202.

Cagle, F. R. 1948. Observations on a population of the salamander *Amphiuma tridactylum* Cuvier. *Ecology* 29:479–91.

Calaby, J. H. 1956. The food habits of the frog *Myobatrachus gouldii* (Gray). *West. Aust. Nat.* 5:93–96.

Calderón, J., and E. Collado. 1976. Selectividad en la predación de la lechuza común (*Tyto alba*) sobre *Rana ridibunda. Doñana Acta Vert.* 3:129–36.

Caldwell, J. P. 1973. *Tropical tree frog communities: Patterns of reproduction, size, and utilization of structural habitat.* Unpublished PhD dissertation. Lawrence: University of Kansas.

———. 1982. Disruptive selection: a tail color polymorphism in *Acris* tadpoles in response to differential predation. *Can. J. Zool.* 60:2818–27.

———. 1986a. A description of the tadpole of *Hyla smithii* with comments on tail coloration. *Copeia* 1986:1004–6.

———. 1986b. Selection of egg deposition sites: a seasonal shift in the southern leopard frog, *Rana sphenocephala. Copeia* 1986:249–53.

———. 1987. Demography and life history of two species of chorus frogs (Anura: Hylidae) in South Carolina. *Copeia* 1987:114–27.

———. 1989. Structure and behavior of *Hyla geographica* tadpole schools, with comments on classification of group behavior in tadpoles. *Copeia* 1989:938–50.

———. 1991. A new species of toad in the *Bufo typhonius* complex from Para, Brazil, with an unusual breeding site. *Pap. Avul. Zool.* 37:389–400.

———. 1992. Diversity of reproductive modes in anurans: Facultative nest construction in gladiator frogs. In *Reproductive biology of South American vertebrates,* ed. W. C. Hamlett, 85–97. New York: Springer-Verlag.

———. 1993. Brazil nut fruit capsules as phytolemata: Interactions among anuran and insect larvae. *Can. J. Zool.* 71:1193–1201.

———. 1994. Natural history and survival of eggs and early larval stages of *Agalychnis calcarifer* (Anura: Hylidae). *Herp. Nat. Hist.* 2:57–66.

———. 1996a. Diversity of Amazonian anurans: The role of sytematics and phylogeny in identifying macroecological and evolutionary patterns. In *Neotropoical biodiversity and conservation,* ed. A. C. Gibson, 73–88. Occ. Pap. Mildred E. Mathias Botanical Garden 1. Los Angeles: University of California.

———. 1996b. The evolution of myrmecophagy and its correlates in poison frogs (family Dendrobatidae). *J. Zool., London* 240:75–101.

———. 1997. Pair bonding in spotted poison frogs. *Nature* 385:211.

Caldwell, J. P. and M. C. de Araújo. 1998. Cannibalistic interactions resulting from indiscriminate predatory behavior in tadpoles of poison frogs (Anura: Dendrobatidae). *Biotropica* 30:92–103.

———. 2004. Historical and ecological factors influence survivorship in two clades of phytotelm-breeding frogs (Anura: Bufonidae, Dendrobatidae). In *Ecology and evolution of phytotelm-breeding anurans,* ed. R. M. Lehtinen, 11–21. *Misc. Publ. Mus. Zool. Univ. Michigan,* no. 193.

Caldwell, J. P., and M. S. Hoogmoed. 1998. Allophrynidae, *Allophryne, A. ruthveni. Cat. Amer. Amphib. Rept.* 666.1–666.3

Caldwell, J. P., and A. P. Lima. 2003. A new Amazonian species of *Colostethus* (Anura: Dendrobatidae) with a nidicolous tadpole. *Herpetologica* 59:219–34.

Caldwell, J. P., and P. T. Lopez. 1989. Foam-generating behavior in tadpoles of *Leptodactylus mystaceus. Copeia* 1989:498–502.

Caldwell, J. P., and C. W. Myers. 1990. A new poison frog from Amazonian Brazil, with further revision of the *quinquevittatus* group of *Dendrobates. Amer. Mus. Novitates* (2988):1–12.

Caldwell, J. P., and V. R. L. de Oliveira. 1999. Determinants of biparental care in the spotted poison frog, *Dendrobates vanzolinii* (Anura: Dendrobatidae). *Copeia* 1999:565–75.

Caldwell, J. P., J. H. Thorp, and T. O. Jervey. 1980. Predator-prey relationships among larval dragonflies, salamanders, and frogs. *Oecologia* 46:285–89.

Caldwell, J. P., and L. J. Vitt. 1999. Dietary asymmetry in leaf litter frogs and lizards in a transitional northern Amazonian rain forest. *Oikos* 84:383–97.

Caldwell, R. S. 1975. Observations on the winter activity of the red-backed salamander, *Plethodon cinereus,* in Indiana. *Herpetologica* 31:21–22.

Caldwell, R. S., and G. S. Jones. 1973. Winter congregations of *Plethodon cinereus* in ant mounds, with notes on their food habits. *Am. Midl. Nat.* 90:482–85.

Calef, G. W. 1973a. Natural mortality of tadpoles in a population of *Rana aurora. Ecology* 54:741–58.

———. 1973b. Spatial distribution and "effective" breeding population of red-legged frogs (*Rana aurora*) in Marion Lake, British Columbia. *Can. Field-Nat.* 87:279–84.

Calfee, R. D., C. M. Bridges, and E. E. Little. 2006. Sensitivity of two salamander (*Ambystoma*) species to ultraviolet radiation. *J. Herpetol.* 40:35–42.

Calhoun, A. J. K., and M. L. Hunter, Jr. 2003. Managing ecosystems for amphibian conservation. In *Amphibian conservation,* ed. R. D. Semlitsch, 228–41. Washington, DC: Smithsonian Books.

Calhoun, A. J. K., N. A. Miller, and M. W. Klemens. 2005. Conserving pool-breeding amphibians in human-dominated landscapes through local implementation of Best Development Practices. *Wetlands Ecol. Manage.* 13:291–304.

Calhoun, A. J. K., T. E. Walls, S. S. Stockwell, and M. McCollough. 2003. Evaluating vernal pools as a basis for conservation strategies: A Maine case study. *Wetlands* 23:70–81.

Calhoun, J. B., and J. U. Casby. 1958. *Calculation of home range and density of small mammals. U.S. Public Health Service Public Health Monogr. 55.* Washington, DC: Government Printing Office.

Callejo Rey, A., J. Guitan Rivera, S. Bas Lopez, J. L. Sanchez Canals, and A. D. E. Castro Lorenzo. 1979. Primeros datos sobre la dieta de la nutria, *Lutra lutra* (L.), en aguas continentales de Galicia. *Doñana, Acta Vert.* 6:191–202.

Callery, E. M., and R. P. Elinson. 2000. Thyroid hormone-dependent metamorphosis in a direct developing frog. *Proc. Natl. Acad. Sci. USA* 97:2615–20.

Calow, L. J., and R. M. Alexander. 1973. A mechanical analysis of the hind leg of a frog (*Rana temporaria*). *J. Zool., London* 171:293–321.

Cameron, M. C. 1974. *Aspects of the osmotic and water balance of the New Zealand native frog Leiopelma hochstetteri Fitzinger, and the Australian whistling frog Litoria ewingi Dumeril and Bibron.* Unpublished M. S. thesis. Palmerston North, New Zealand: Massey University.

Camp, C. D. 1996. Bite scar patterns in the black-bellied salamander, *Desmognathus quadramaculatus. J. Herpetol.* 30:543–46.

———. 1997. The status of the black-bellied salamander

(*Desmognathus quadramaculatus*) as a predator of heterospecific salamanders in Appalachian streams. *J. Herpetol.* 31:613–16.
———. 1999. Intraspecific aggressive behavior in southeastern small species of *Plethodon:* Inferences for the evolution of aggression in terrestrial salamanders. *Herpetologica* 55:248–54.
Camp, C. D., and T. P. Lee. 1996. Intraspecific spacing and interaction within a population of *Desmognathus quadramaculatus. Copeia* 1996:78–84.
Camp, C. D., W. D. Sprewell, and V. N. Powders. 1980. Feeding habits of *Nerodia taxispilota* with comparative notes on the foods of sympatric congeners in Georgia. *J. Herpetol.* 14:301–4.
Campbell, A., ed. 1999. *Declines and disappearances of Australian frogs.* Canberra, Australia: Environment Australia.
Campbell, H. W., and R. S. Simmons. 1962. Notes on the eggs and larvae of *Rhyacosiredon altamirani* (Duges). *Herpetologica* 18:131–32.
Campbell, J. A. 1998. *Amphibians and reptiles of northern Guatemala, the Yucatan, and Belize.* Norman: University of Oklahoma Press.
———. 1999. Distribution patterns of amphibians in Middle America. In *Patterns of distribution of amphibians. A global perspective,* ed. W. E. Duellman, 111–210. Baltimore: Johns Hopkins University Press.
Campbell, J. A., and J. L. Camarillo. 1992. The Oaxacan dwarf boa, *Exiliboa placata* (Serpentes: Trophidophiidae): descriptive notes and life history. *Carib. J. Sci.* 28:17–20.
Campbell, J. A., and B. T. Clarke. 1998. A review of frogs of the genus *Otophryne* (Microhylidae) with the description of a new species. *Herpetologica* 54:301–17.
Campbell, J. A., J. M. Savage, and J. R. Meyers. 1994. A new species of *Eleutherodactylus* (Anura: Leptodactylidae) of the *rugulosus* group from Guatemala and Belize. *Herpetologica* 50:412–19.
Campbell, J. A., and A. Solorzano. 1992. Biology of the montane pitviper, *Porthidium godmani.* In *Biology of the pitvipers,* ed. J. A. Campbell and E. D. Brodie, Jr., 223–50. Tyler, TX: Selva.
Campbell, J. B. 1970. Hibernacula of a population of *Bufo boreas boreas* in the Colorado Front Range. *Herpetologica* 26:278–82.
Campbell Grant, E. H., R. E. Jung, and K. C. Rice. 2005. Stream salamander species richness and abundance in relation to environmental factors in Shenandoah National Park, Virginia. *Am. Midl. Nat.* 153:348–56.
Camper, J. D., and J. R. Dixon. 1988. Evaluation of a microchip marking system for amphibians and reptiles. *Texas Parks and Wildlife Department Research Publication 7100-159:*1–22.
Canale, G., and R. Lingnau. 2003. *Hyla biobeba* (NCN). Predation. *Herp. Rev.* 34:136.
Candioti, M. F. V., C. Úbeda, and E. O. Lavilla. 2005. Morphology and metamorphosis of *Eupsophus calcaratus* tadpoles (Anura: Leptodactylidae). *J. Morphol.* 262:161–77.
Candolin, U., and H. R. Voigt. 2003. Size-dependent selection on arrival times in sticklebacks: why small males arrive first. *Evolution* 57:862–71.
Cannatella, D. 1999. Architecture: Cranial and axial musculoskeleton. In *Tadpoles. The biology of anuran larvae,* ed. R. W. McDiarmid and R. Altig, 52–91. Chicago: University of Chicago Press.
Cannatella, D., and D. M. Hillis. 2004. Amphibians: leading a life of slime. In *Assembling the tree of life,* ed. J. Cracraft and M. J. Donoghue, 430–50. New York: Oxford University Press.
Cannatella, D. C. 1980a. A review of the *Phyllomedusa buckleyi* group (Anura: Hylidae). *Occ. Pap. Univ. Kansas Mus. Nat. Hist.* (87):1–40.
———. 1980b. Two new species of *Centrolenella* from Bolivia (Anura: Centrolenidae). *Proc. Biol. Soc. Washington* 93:714–24.
———. 1982. Leaf-frogs of the *Phyllomedusa perinesos* group (Anura: Hylidae). *Copeia* 1982:501–13.
———. 1985. *A phylogeny of primitive frogs (Archaeobatrachians).* Unpublished PhD dissertation. Lawrence: University of Kansas.
———. 1989. On the monophyly of discoglossid frogs. In *Trends in vertebrate morphology,* ed. H. Splechtna and H. Hilgers, 230–31. Stuttgart, Germany: Gustav Fischer Verlag.
Cannatella, D. C., and R. O. de Sa. 1993. *Xenopus laevis* as a model organism. *Syst. Biol.* 42:476–507.
Cannatella, D. C., and W. E. Duellman. 1984. Leptodactylid frogs of the *Physalaemus pustulosus* group. *Copeia* 1984:902–21.
Cannatella, D. C., and W. W. Lamar. 1986. Synonymy and distribution of *Centrolenella orientalis* with notes on its life history (Anura: Centrolenidae). *J. Herpetol.* 20:307–17.
Cannatella, D. C., and L. Trueb. 1988a. Evolution of pipoid frogs: Intergeneric relationships of the aquatic frog family Pipidae (Anura). *Zool. J. Linn. Soc.* 94:1–38.
———. 1988b. Evolution of pipoid frogs: Morphology and phylogenetic relationships of *Pseudhymenochirus. J. Herpetol.* 22:439–56.
Canterbury, G. E., and D. E. Blockstein. 1997. Local changes in a breeding bird community following forest disturbance. *J. Field Ornithol.* 68:537–46.
Capp, M. S., and W. A. Searcy. 1991. Acoustical communication of aggressive intentions by territorial male bobolinks. *Behav. Ecol.* 2:319–26.
Capranica, R. R. 1965. *The evoked vocal response of the bullfrog: A study of communication by sound.* Cambridge, MA: MIT Press.
———. 1968. The vocal repertoire of the bullfrog (*Rana catesbeiana*). *Behaviour* 31:302–25.
———. 1976a. Morphology and physiology of the auditory system. In *Frog neurobiology,* ed. R. Llinas and W. Precht, 551–75. New York: Springer-Verlag.
———. 1976b. The auditory system. In *Physiology of the amphibia, vol. 3,* ed. B. Lofts, 443–66. New York: Academic Press.
———. 1977. Auditory processing of vocal signals in anurans. In *The reproductive biology of amphibians,* ed. D. H. Taylor and S. I. Guttman, 337–55. New York: Plenum.
———. The untuning of the tuning curve: Is it time? *Semin. Neurosci.* 4:401–8.
Capranica, R. R., L. S. Frischkopf, and E. Nevo. 1973. Encoding of geographic dialects in the auditory system of the cricket frog. *Science* 182:1272–75.
Capranica, R. R., and A. J. M. Moffat. 1975. Selectivity of the peripheral auditory system of spadefoot toads (*Scaphiopus couchi*) for sounds of biological significance. *J. Comp. Physiol.* 100:231–49.
———. 1977. Place mechanism underlying frequency analysis in the toad's inner ear. *J. Acoust. Soc. Am.* 59:S46.
———. 1980. Nonlinear properties of the peripheral auditory system of anurans. In *Comparative studies of hearing in vertebrates,* ed. A. N. Popper and R. R. Fay, 139–65. Berlin: Springer-Verlag.
———. 1983. Neurobehavioral correlates of sound communica-

tion in anurans. In *Advances in vertebrate neuroethology,* ed. J.- P. Ewert, R. R. Capranica, and D. J. Ingle, 701–30. New York: Plenum.

Capranica, R. R., and G. J. Rose. 1983. Frequency and temporal processing in the auditory system of anurans. In *Neuroethology and behavioral physiology,* ed. F. Huber and H. Markl, 136–52. New York: Springer-Verlag.

Capranica, R. R., G. J. Rose, and E. A. Brenowitz. 1985. Time resolution in the auditory systems of anurans. In *Time resolution in auditory systems,* ed. A. Michelsen, 58–73. New York: Springer-Verlag.

Capula, M., L. Rugiero, and L. Luiselli. 1994. Ecological observations on the Sardinian grass snake, *Natrix natrix cetti. Amphibia-Reptilia* 15:221–27.

Capurro, L. F. 1958. Nota sobre la ovipostura y desarrollo de *Eupsophus taeniatus* (Girard). *Inv. Zool. Chilenas* 4:208.

Caramaschi, U., and R. N. Feio. 1990. A new species of *Hyla* (Anura, Hylidae) from southern Minas Gerais Brazil. *Copeia* 1990:542–46.

Cardoso, A. J. 1981a. Biologia e sobrevivência de *Physalaemus cuvieri* Fitz., 1826 (Amphibia, Anura), na natureza. *Ciencia e Cultura* 33:1224–28.

———. 1981b. *Organização espacial e temporal na reprodução e vida larvária em uma comunidade de hilídeos no sudeste do Brasil (Amphibia, Anura).* Unpublished M. S. thesis. Brazil: Universidade Estadual de Campinas.

———. 1983. Descrição e biologia de uma nova especie de *Hyla* Laurenti, 1768 (Amphibia, Anura, Hylidae). *Iheringia, ser. zool., Porto Alegre* 62:37–45.

Cardoso, A. J., and C. F. B. Haddad. 1984. Variabilidade acustica em differentes populações e interações de *Hyla minuta* (Amphibia, Anura). *Ciencia e Cultura* (Sao Paulo) 36:1393–99.

———. 1992. Diversidade e turno de vocalizações de anuros em communidade Neotropical. *Acta Zool. Lilloana* 41:93–105.

Cardoso, A. J., and W. R. Heyer. 1995. Advertisement, aggressive, and possible seismic signals of the frog *Leptodactylus syphax* (Amphibia, Leptodactylidae). *Alytes* 13:67–76.

Carey, C. 1978. Factors affecting body temperatures of toads. *Oecologia* 35:197–219.

———. 1979a. Aerobic and anaerobic energy expenditure during rest and activity in montane *Bufo boreas* and *Rana pipiens. Oecologia* 39:213–28.

———. 1979b. Effect of constant and fluctuating temperatures on resting and active oxygen consumption of toads, *Bufo boreas. Oecologia* 39:201–12.

———. 1993. Hypothesis concerning the causes of the disappearance of Boreal toads from the mountains of Colorado. *Cons. Biol.* 7:355–62.

———. 2000. Infectious disease and worldwide declines of amphibian populations, with comments on emerging diseases in coral reef organisms and in humans. *Environ. Health Perspect.* 108 (Suppl. 1):143–50.

———. 2005. How physiological methods and concepts can be useful in conservation biology. *Integr. Comp. Biol.* 45:4–11.

Carey, C., and M. A. Alexander. 2003. Climate change and amphibian declines: Is there a link? *Divers. Distrib.* 9:111–21.

Carey, C., D. F. Bradford, J. L. Brunner, J. P. Collins, E. W. Davidson, J. E. Longcore, M. Ouellet, A. P. Pessier, and D. M. Schock. 2003. Biotic factors in amphibian population declines. In *Amphibian decline: An integrated analysis of multiple stressor effects,* ed. G. Linder, S. K. Krest, and D. W. Sparling, 153–208. Pensacola, FL: SETAC Press.

Carey, C., and C. J. Bryant. 1995. Possible interrelations among environmental toxicants, amphibian development, and decline of amphibian populations. *Env. Health Perspect.* 103 (Suppl. 4):13–17.

Carey, C., N. Cohen, and L. A. Rollins-Smith. 1999. Amphibian declines: An immunological perspective. *Dev. Comp. Immunol.* 23:459–72.

Carey, C., W. R. Heyer, J. Wilkinson, R. A. Alford, J. W. Arntzen, T. Halliday, L. Hungerford, K. R. Lips, E. M. Middleton, S. A. Orchard, and A. S. Rand. 2001. Amphibian declines and environmental change: Use of remote-sensing data to identify environmental correlates. *Cons. Biol.* 15:903–13.

Carey, C., G. D. Maniero, and J. F. Stinn. 1996. Effect of cold on immune function and susceptibility to bacterial infection in toads (*Bufo marinus*). In *Adaptation to cold: Tenth hibernation symposium,* ed. F. Geiser, A. J. Hulbert, and S. C. Nicol, 123–29. Armidale, New South Wales, Australia: University of New England Press.

Carey, C., A. P. Pessier, and A. D. Peace. 2003. Pathogens, infectious disease, and immune defenses. In *Amphibian conservation,* ed. R. D. Semlitsch, 127–36. Washington, DC: Smithsonian Books.

Cargo, D. G. 1960. Predation of eggs of the spotted salamander, *Ambystoma maculatum,* by the leech, *Macrobdella decora. Chesapeake Sci.* 1:119–20.

Carlson, A., and P. Edenhamn. 2000. Extinction dynamics and the regional persistence of a tree frog metapopulation. *Proc. Roy. Soc. London B* 267:1311–13.

Carlvalho, G. R. 1984. Haemoglobin synthesis in *Daphnia magna* Straus (Crustacea: Cladocera): Ecological differentiation between neighboring populations. *Freshw. Biol.* 14:501–6.

Carnaval, A. C. O. Q. 2002. Phylogeography of four frog species in forest fragments of northeastern Brazil—A preliminary study. *Integr. Comp. Biol.* 42:913–21.

Carpenter, C. C. 1952. Comparative ecology of the common garter snake (*Thamnophis s. sirtalis*), the ribbon snake (*Thamnophis s. sauritus*), and the Butler's garter snake (*Thamnophis butleri*) in mixed populations. *Ecol. Monogr.* 22:235–58.

———. 1953a. Aggregation behavior of tadpoles of *Rana p. pretiosa. Herpetologica* 9:77–78.

———. 1953b. An ecological survey of the herpetofauna of the Grand Teton-Jackson Hole Area of Wyoming. *Copeia* 1953:170–74.

———. 1954. A study of amphibian movement in the Jackson Hole Wildlife Park. *Copeia* 1954:197–200.

Carpenter, C. C., and J. C. Gillingham. 1987. Water hole fidelity in the marine toad, *Bufo marinus. J. Herpetol.* 21:158–61.

———. 1990. Ritualized behavior in *Agkistrodon* and related genera. In *Snakes of the agkistrodon complex: A monographic review,* ed. H. K. Gloyd and R. Conant, 523–32 Oxford, OH: Society for the Study of Amphibians and Reptiles.

Carpenter, F. L. 1979. Competition between hummingbirds and insects for nectar. *Amer. Zool.* 19:1105–14.

Carr, A. F., Jr. 1940. A contribution to the herpetology of Florida. *Univ. Florida Publ. Biol. Sci.* (3):1–118.

Carr, A. F., and C. J. Goin. 1943. Neoteny in Florida salamanders. *Proc. Florida Acad. Sci.* 6:37–40.

Carr, D. E., and D. H. Taylor. 1985. Experimental evaluation of population interactions among three sympatric species of *Desmognathus. J. Herpetol.* 19:507–14.

Carr, J. A., A. Gentles, E. E. Smith, W. L. Goleman, L. J. Urguidi, K. Thuett, R. J. Kendall, J. P. Giesy, T. S. Gross, K. R. Solomon, and G. Van der Kraak. 2003. Response of larval *Xenopus laevis* to atrazine: Assessment of growth, metamorphosis, and gonadal and laryngeal morphology. *Env. Tox. Chem.* 22:396–405.

Carr, L. W., and L. Fahrig. 2001. Effect of road traffic on two amphibian species of differing vagility. *Cons. Biol.* 15:1071–78.

Carranza, S., and F. Amat. 2005. Taxonomy, biogeography and evolution of *Euproctus* (Amphibia: Salamandridae), with the resurrection of the genus *Calotriton* and the description of a new endemic species from the Iberian Peninsula. *Zool. J. Linn. Soc.* 145:555–82.

Carreño, C. A., and R. N. Harris. 1998. Lack of nest defense behavior and attendance patterns in a joint nesting salamander, *Hemidactylium scutatum* (Caudata: Plethodontidae). *Copeia* 1998:183–89.

Carretero, M. A., and C. Rosell. 1999. *Salamandra salamandra* (fire salamander). Predation. *Herp. Rev.* 30:161.

Carrier, D. R., and M. H. Wake. 1995. Mechanisms of lung ventilation in the caecilian *Dermophis mexicanus*. *J. Morphol.* 226:289–95.

Carrier, J-A., and T. J. C. Beebee. 2003. Recent, substantial, and unexplained declines of the common toad *Bufo bufo* in lowland England. *Biol. Cons.* 111:395–99.

Carrizo, G. R. 1990. Un caso de predación de *Xiphocolaptes major* (Passeriformes, Dendrocolaptidae) sobre un anfibio. *Bol. Asoc. Herpetol. Argentina* 6:7.

Carroll, D. M. 1999. *Swampwalker's journal. A wetlands year.* Boston: Houghton Mifflin.

Carroll, E. J., Jr., A. M. Seneviratne, and R. Ruibal. 1991. Gastric pepsin in an anuran larva. *Dev. Growth Differ.* 33:499–507.

Carroll, R. L. 1986. Developmental processes and the origin of lepospondyls. In *Studies in herpetology. Proceedings of the European herpetological meeting,* ed. Z. Roček, 45–48. Prague: Charles University.

———. 1987. *Vertebrate paleontology and evolution.* San Francisco: W. H. Freeman.

———. 1995. Problems of the phylogenetic analysis of Paleozoic choanates. *Bull. Mus. Natl. Hist. Nat. (Paris),* 4th ser., 17 (sec. C):389–445.

———. 2000a. *Eocaecilia* and the origin of caecilians. In *Amphibian biology. Vol. 4. Palaeontology,* ed. H. Heatwole and R. L. Carroll, 1402–11. Chipping Norton, New South Wales, Australia: Surrey Beatty and Sons.

———. 2000b. The fossil record and large-scale patterns of amphibian evolution. In *Amphibian biology. Vol. 4. Palaeontology,* ed. H. Heatwole and R. L. Carroll, 973–78. Chipping Norton, New South Wales, Australia: Surrey Beatty and Sons.

———. 2000c. Lepospondyls. In *Amphibian biology. Vol. 4. Palaeontology,* ed. H. Heatwole and R. L. Carroll, 1198–1269. Chipping Norton, New South Wales, Australia: Surrey Beatty and Sons.

———. 2000d. The lissamphibian enigma. In *Amphibian biology. Vol. 4. Palaeontology,* ed. H. Heatwole and R. L. Carroll, 1270–73. Chipping Norton, New South Wales, Australia: Surrey Beatty and Sons.

———. 2001. The origin and early radiation of terrestrial vertebrates. *J. Paleontol.* 75:1202–13.

Carroll, R. L., and P. J. Currie. 1975. Microsaurs as possible apodan ancestors. *Zool. J. Linn. Soc.* 57:229–47.

Carson, R. 1962. *Silent spring.* Boston: Houghton Mifflin.

Carter, D. 1979. Structure and function of the subcutaneous lymph sacs in the Anura (Amphibia). *J. Herpetol.* 13:321–27.

Carter, G. S. 1935. Results of the Cambridge expedition to British Guiana, 1933. The freshwaters of the rain forest areas of British Guiana. *J. Linn. Soc. London (Zoology)* 39:147–93.

Carter, G. S., and L. C. Beadle. 1931. The fauna of the Paraguayan Chaco in relation to the environment. I. Physico-chemical nature of the environment. *J. Linn. Soc. London (Zoology)* 37:205–58.

Caruso, M. A. 1949. Sobre el ciclo sexual annual de algunos Hylidae del norte Argentino (*Phyllomedusa sauvagei* e *Hyla raddiana*). *Acta Zool. Lilloana* 8:83–103.

Carvalho, A. L. de. 1949. Notas sobre os habitos de *Dendrophryniscus brevipollicatus* Espada. *Rev. Brasil. Biol.* 9:223–27.

Case, S. M. 1978. Electrophoretic variation in two species of ranid frogs, *Rana boylei* and *R. muscosa*. *Copeia* 1978:311–20.

Case, S. M., P. G. Haneline, and M. F. Smith. 1975. Protein variation in several species of *Hyla*. *Syst. Zool.* 24:281–95.

Cassimiro, J., and J. Bertoluci. 2003. *Liophis maryellenae* (Cobra-d'agua). Diet. *Herp. Rev.* 34:69.

Castanet, J., H. Francillon-Vieillot, and R. C. Bruce. 1996. Age estimation in desmognathine salamanders assessed by skeletochronology. *Herpetologica* 52:160–71.

Castanet, J., and T. Pilorge. 1980. Application des techniques squelettochronologiques a l'étude démographique des populations d'amphibiens et de lézards. *Bull. Soc. Zool. France* 105:347–54.

Castanet, J., S. Pinto, M.-M. Loth, and M. Lamotte. 2000. Âge individuel, longévité et dynamique de croissance osseue chez un amphibien vivipare, *Nectophrynoides occidentalis* (Anoure, Bufonidé). *Ann. Sci. Natur.* 21:11–17.

Castellani, C. 1984. Quelques considérations sur les études de Spallanzani a propos de la respiration par rapport a ses manuscrits inédits. *Physis* 26:533–46.

Castellano, S., M. Cucco, and C. Giacoma. 2004. Reproductive investment of female green toads, *Bufo viridis*. *Copeia* 2004:659–64.

Castellano, S., and C. Giacoma. 1998. Stabilizing and directional female choice for male calls in the European green toad. *Anim. Behav.* 56:275–87.

Castellano, S., C. Giacoma, and T. Dujsebayeva. 2000. Morphometric and advertisement call geographic variation in polyploid green toads. *Biol. J. Linn. Soc.* 70:341–60.

Castellano, S., C. Giacoma, T. Dujsebayeva, G. Odierna, and E. Balletto. 1998. Morphometrical and acoustical comparison between diploid and tetraploid green toads. *Biol. J. Linn. Soc.* 63:257–81.

Castellano, S., C. Giacoma, and M. J. Ryan. 2003. Call degradation in diploid and tetraploid green toads. *Biol. J. Linn. Soc.* 78:11–26.

Castellano, S., A. Rosso, and C. Giacoma. 2004. Active choice, passive attraction and the cognitive machinery of acoustic preferences. *Anim. Behav.* 68:323–29.

Castellano, S., L. Tontini, C. Giacoma, A. Lattes, and E. Balletto. 2002. The evolution of release and advertisement calls in green toads (*Bufo viridis* complex). *Biol. J. Linn. Soc.* 77:379–91.

Casterlin, M. E., and W. W. Reynolds. 1978. Behavioral thermoregulation in *Rana pipiens* tadpoles. *J. Therm. Biol.* 3:143–45.

Caston, J., W. Precht, and R. H. I. Blanks. 1977. Response characteristics of frog's lagena afferents to natural stimulation. *J. Comp. Physiol.* 118:273–89.

Catling, P. C., A. Hertog, R. J. Burt, J. C. Wombey, and R. I. Forrester. 1999. The short-term effect of cane toads (*Bufo marinus*) on native fauna in the Gulf Country of the Northern Territory. *Wildl. Res.* 26:161–85.

Cattaneo, A., and B. Mousseau. 1995. Empirical-analysis of the removal rate of periphyton by grazers. *Oecologia* 103:249–54.

Cecil, S. G., and J. J. Just. 1979. Survival rate, population density and development of a naturally occurring anuran larvae (*Rana catesbeiana*). *Copeia* 1979:447–53.

Cedrini, L., and A. Fasolo. 1971. Olfactory attractants in sex recognition of the crested newt. An electrophysiological research. *Monit. Zool. Ital.* 5:223–29.

Cei, J. M. 1948. El ritmo estacional en los fenómenos cíclicos endócrino sexuales de la rana criolla *Leptodactylus ocellatus* (L.) de norte Argentino. *Acta Zool. Lilloana* 6:283–331.

———. 1949a. Costumbres nupciales y reproducción de un batracio characterístico Chaqueño (*Leptodactylus bufonius* Boulenger). *Act. Zool. Lilloana* 8:105–10.

———. 1949b. El ciclo sexual y el prodominio de la espermatogenesis anual continua en batracios chaqueños. *Acta Zool. Lilloana* 7:527–44.

———. 1949c. Factores genético-raciales que differencian la regulación hormonal del ciclo sexual en *Leptodactylus ocellatus* (L.) de la Argentina. "Razas de temperatura" y sus relaciones con algunas características climáticas regionales. *Acta Zool. Lilloana* 7:113–34.

———. 1949d. Sobre la biología sexual de un battracio de grande altura de la región andina (*Telmatobius schreiteri* Vellard). *Acta Zool. Lilloana* 7:467–88.

———. 1958. Polimorfismo y distribución geográfica en poblaciones chilenas de *Pleurodema bibroni* Tschudi. *Inv. Zool. Chil.* 4:300–27.

———. 1961a. Bipolaridad del ciclo espermatogenético autorregulado en los Anfibios. *Revta. Soc. Argent. Biol.* 37:48–51.

———. 1961b. *Pleurodema bufonina* Bell, anfibio australe con ciclo spermatogenetico discontinuo autoregolata. *Arch. Zool. Ital.* 46:167–80.

———. 1962. *Batracios de Chile.* Santiago: Ediciones de Universidad de Chile.

———. 1968. Notes on the tadpoles and breeding ecology of *Lepidobatrachus* (Amphibia: Ceratophryidae). *Herpetologica* 24:141–46.

———. 1980. Amphibians of Argentina. *Monit. Zool. Ital.* (N. S.) Monogr. 2:1–609.

Cei, J. M., and L. F. Capurro. 1957. La distribución de les patrónes de coloración en *Pleurodema bibroni* en relación con la distribución geográfica y el habitat. *Inv. Zool. Chil.* 3:156–61.

———. 1958. Biología y desarrollo de *Batrachyla antartandica* Barrio (Anura, Leptodactylidae). *Inv. Zool. Chil.* 4:150–82.

Cei, J. M,. and V. Erspamer. 1966. Biochemical taxonomy of South American amphibians by means of skin amines and polypeptides. *Copeia* 1966:74–78.

Cei, J. M., V. Erspamer, and M. Roseghini. 1967. Taxonomic and evolutionary significance of biogenic amines and polypeptides occurring in amphibian skin. I. Neotropical leptodactylid frogs. *Syst. Zool.* 16:328–42.

Cei, J. M., and V. G. Roig. 1965. The systematic status and biology of *Telmatobius montanus* Letaste (Amphibia, Leptodactylidae). *Copeia* 1965:421–25.

Chacornac, J. M., and P. Joly. 1985. Activité prédatrice du triton alpestre (*Triturus alpestris*) dans un lac alpin (2125 m, Alpes françaises). *Acta Oecol./Oecol. Gen.* 6:93–103.

Chadwick, C. S. 1940. Some notes on the burrows of *Plethodon metcalfi*. *Copeia* 1940:50.

———. 1944. Observations on the life cycle of the common newt in North Carolina. *Am. Midl. Nat.* 32:491–94.

Chalcraft, D. R., and W. J. Resetarits, Jr. 2003. Predator identity and ecological impacts: Functional redundancy or functional diversity? *Ecology* 84:2407–18.

Chamberlain, F. M. 1898. Notes on the edible frogs of the United States and their artificial propagation. *Appendix, Report of Commissioner of Fish and Fisheries for 1897,* 249–61. Washington, DC: U.S. Government Printing Office.

Chan-ard, T., W. Grossmann, A. Gumprecht, and K.-D. Schulz. 1999. *Amphibians and reptiles of peninsular Malaysia and Thailand.* Würselen, Germany: Bushmaster Pulbications.

Chang, K. C. 1977. Ancient China. In *Food in Chinese culture: Anthropological and historical perspectives,* ed. K. C. Chang, 25–52. New Haven, CT: Yale University Press.

Chanin, P. 1985. *The natural history of otters.* New York: Facts on File.

Chan-McLeod, A. C. A. 2003. Factors affecting the permeability of clearcuts to red-legged frogs. *J. Wildl. Manage.* 67:663–671.

Channing, A. 1976a. Life histories of frogs in the Namib Desert. *Zool. Afr.* 11:299–312.

———. 1976b. Pre-mating isolation in the genus *Kassina* (Amphibia, Anura, Rhacophoridae) in southern Africa. *J. Herpetol.* 10:19–23.

———. 1989. A re-evaluation of the phylogeny of Old World treefrogs. *S. Afr. J. Sci.* 24:116–31.

———. 1993. Observations on the natural history of *Stephopaedes anotis* (Bufonidae). *J. Herpetol.* 27:213–14.

———. 2001. *Amphibians of Central and Southern Africa.* Ithaca, NY: Cornell University Press.

Channing, A., and K. M. Howell. 2006. *Amphibians of East Africa.* Ithaca, NY: Cornell University Press.

Channing, A., D. Moyer, and M. Burger. 2002. Cryptic species of sharp-nosed reed frogs in the *Hyperolius nasutus* complex: advertisement call differences. *Afr. Zool.* 37:91–99.

Channing, A., L. du Preez, and N. Passmore. 1994. Status, vocalization and breeding biology of two species of African bullfrogs (Ranidae: *Pyxicephalus*). *J. Zool., London* 234:141–48.

Channing, A. L., and D. E. van Dijk. 1995. Amphibia. In *Wetlands of South Africa,* ed. G. I. Cowan, 193–206. Pretoria, South Africa: Department of Environmental Affairs and Tourism.

Chapin, F. S., O. E. Sala, and E. Huber-Sannwald, eds. 2001. *Future scenarios of global biodiversity.* New York: Springer-Verlag.

Chapman, B. M., and R. F. Chapman. 1958. A field study of a population of leopard toads (*Bufo regularis regularis*). *J. Anim. Ecol.* 27:265–86.

Chapman, J. A., and G. A. Feldhamer, eds. 1982. *Wild mammals of North America.* Baltimore: Johns Hopkins University Press.

Chapman, L. J., C. A. Chapman, F. G. Nordlie, and A. E. Rosenberger. 2002. Physiological refugia: Swamps, hypoxia tolerance and maintenance of fish diversity in the Lake Victoria region. *Comp. Biochem. Physiol.* A 133:421–37.

Chappuis, C. 1971. Un example de l'influence du milieu sur les émissions vocales des oiseaux: l'évolution des chants en forêt équatoriale. *Terre Vie* 118:183–202.

Charles-Dominique, P. 1977. *Ecology and behaviour of nocturnal primates.* New York: Columbia University Press.

Chauvin, M. von. 1876. Über die Verwandlung der mexikanischen Axolotl in *Amblystoma. Z. Wiss. Zool., Leipzig* 27:522–535.

Chazal, A. C., J. D. Krenz, and D. E. Scott. 1996. Relationship of larval density and heterozygosity to growth and survival of juvenile marbled salamanders (*Ambystoma opacum*). *Can. J. Zool.* 74:1122–29.

Chazal, A. C., and P. H. Niewiarowski. 1998. Responses of mole salamanders to clearcutting: using field experiments in forest management. *Ecol. Appl.* 8:1133–43.

Chek, A. A., J. P. Bogart, and S. C. Lougheed. 2003. Mating signal partitioning in multi-species assemblages: a null model test using frogs. *Ecol. Lett.* 6:235–47.

Chen, T.-C., Y.-C. Kam, and Y.-S. Lin. 2001. Thermal physiology and reproductive phenology of *Buergeria japonica* (Rhacophoridae) breeding in a stream and a geothermal hotspring in Taiwan. *Zool. Sci.* 18:591–96.

Chen, T.-U. D. 2002. Male advertisement call and female choice in Moltrecht's treefrog, *Rhacophorus moltrechti* from Taiwan. Unpublished M.S. thesis. Taiwan: National Cheng Kung University.

Chen, Y-H., Y.-J. Su, Y.-S. Lin, and Y.-C. Kam. 2001. Inter- and intraclutch competition among oophagous tadpoles of the Tawainese tree frog, *Chrixalus eiffingeri* (Anura: Rhacophoridae). *Herpetologica* 57:438–48.

Cherry, M. I. 1993. Sexual selection in the raucous toad, *Bufo rangeri*. *Anim. Behav.* 45:359–73.

Chew, M. M., A. B. Elliot, and H. Y. Wong. 1972. Permeability of urinary bladder of *Rana cancrivora* to urea in the presence of oxytocin. *J. Physiol.* 223:757–72.

Chibon, P. 1960. Développement au laboratoire d'*Eleutherodactylus martinicensis* Tschudi, batracien anoure a développement direct. *Bull. Soc. Zool. France* 85:412–18.

Chibon, P. 1962. Différenciation sexuelle de *Eleutherodactylus martinicensis* Tschudi, batracien anoure a développement direct. *Bull. Soc. Zool. France* 87:509–15.

Chinchar, V. G. 2002. Ranaviruses (family Iridoviridae): Emerging cold-blooded killers. *Arch. Virol.* 147:447–70.

Chippindale, P. T. 2000. Species boundaries and species diversity in the Central Texas hemidactyliine plethodontid salamanders, genus *Eurycea*. In *The biology of plethodontid salamanders,* ed. R. C. Bruce, R. G. Jaeger, and L. D. Houck, 149–65. New York: Kluwer Academic/Plenum.

Chippindale, P. T., R. M. Bonett, A. S. Baldwin, and J. J. Wiens. 2004. Phylogenetic evidence for a major reversal of life-history evolution in plethodontid salamanders. *Evolution* 58:2809–22.

Chippindale, P. T., A. H. Price, J. J. Wiens, and D. M. Hillis. 2000. Phylogenetic relationships and systematic revision of central Texas hemidactyliine plethodontid salamanders. *Herpetol. Monogr.* 14:1–80.

Chivers, D. P., J. M. Kiesecker, M. T. Anderson, E. L. Wildy, and A. R. Blaustein. 1996. Avoidance response of a terrestrial salamander (*Ambystoma macrodactylum*) to chemical alarm cues. *J. Chem. Ecol.* 22:1709–16.

Chivers, D. P., J. M. Kiesecker, A. Marco, E. L. Wildy, and A. R. Blaustein. 1999. Shifts in life history as a response to predation in western toads (*Bufo boreas*). *J. Chem. Ecol.* 25:2455–63.

Chivers, D. P., J. M. Kiesecker, E. L. Wildy, M. T. Anderson, and A. R. Blaustein. 1997. Chemical alarm signaling in terrestrial salamanders: Intra- and interspecific responses. *Ethology* 103:599–613.

Chivers, D. P., J. M. Kiesecker, E. L. Wildy, L. K. Belden, L. B. Kats, and A. R. Blaustein. 1999. Avoidance response of post-metamorphic anurans to cues of injured conspecifics and predators. *J. Herpetol.* 33:472–76.

Chivers, D. P., and R. S. Mirza. 2001. Importance of predator diet cues in responses of larval wood frogs to fish and invertebrate predators. *J. Chem. Ecol.* 27:45–51.

Chivers, D. P., E. L. Wildy, J. M. Kiesecker, and A. R. Blaustein. 2001. Avoidance responses of juvenile Pacific treefrogs to chemical cues of introduced predatory bullfrogs. *J. Chem. Ecol.* 27:1667–76.

Choi, I., and K. Park. 1996. Variations in take-off velocity of anuran amphibians: Relation to morphology, muscle contractile function and enzyme activity. *Comp. Biochem. Physiol.* 113A:393–400.

Choi, I., J. H. Shim, and R. E. Ricklefs. 2003. Morphometric relationships of take-off speed in anuran amphibians. *J. Exp. Zool.* 299A:99–102.

Chou, W.-H., and J.-Y. Lin. 1997. *Tadpoles of Taiwan.* Spec. Publ. 7. Taichung, Taiwan: National Museum of Natural Science.

Chovanec, A. 1992a. Beutewarnemung (Reaktive Distanzen) und Beuteverfolgung (Kritische Distanzen) bei Larven von *Aeshna cyanea* (Müller) (Anisoptera: Aeshnidae). *Odonatologica* 21:327–33.

———. 1992b. The influence of tadpole swimming behaviour on predation by dragonfly nymphs. *Amphibia-Reptilia* 13:341–49.

———. 1994. Man-made wetlands in urban recreational areas—A habitat for endangered species? *Landsc. Urban Plan.* 29:43–54.

Christein, D. R., S. I. Guttman, and D. H. Taylor. 1979. Heterozygote deficiencies in a breeding population of *Bufo americanus* (Bufonidae: Anura): The test of a hypothesis. *Copeia* 1979:498–502.

Christensen, C. U. 1974. Adaptations in the water economy of some anuran Amphibia. *Comp. Biochem. Physiol.* 47A:1036–49.

Christensen, J., and D. Penney. 1973. Anaerobic glycolysis and lactic acid accumulation in cold submerged *Rana pipiens*. *J. Comp. Physiol.* 87:235–45.

Christensen-Dalsgaard, J., T. Breithaupt, and A. Elepfandt. 1990. Underwater hearing in the clawed frog, *Xenopus laevis*. *Naturwissenschaften* 77:135–37.

Christensen-Dalsgaard, J., and A. Elepfandt. 1995. Biophysics of underwater hearing in the clawed frog, *Xenopus laevis*. *J. Comp. Physiol. A* 176:317–24.

Christensen-Dalsgaard, J., and M. B. Jørgensen. 1988. The response characteristics of vibration-sensitive saccular fibers in the grassfrog, *Rana temporaria*. *J. Comp. Physiol. A* 162:633–38.

———. 1989. Response characteristics of vibration-sensitive neurons in the midbrain of the grassfrog, *Rana temporaria*. *J. Comp. Physiol. A* 164:495–99.

———. 1996. Sound and vibration sensitivity of VIIIth nerve fibers in the grassfrog, *Rana temporaria*. *J. Comp. Physiol. A* 179:437–45.

Christensen-Dalsgaard, J., T. A. Ludwing, and P. M. Narins. 2002. Call diversity in an Old World treefrog: Level dependence and latency of acoustic responses. *Bioacoustics* 13:21–35.

Christensen-Dalsgaard, J., and P. M. Narins. 1993. Sound and vibration sensitivity of VIIIth nerve fibers in the frogs *Leptodactylus albilabris* and *Rana pipiens pipiens*. *J. Comp. Physiol. A* 172:653–62.

Christian, K. A., F. Nunez, L. Clos, and L. Diaz. 1988. Thermal relations of some tropical frogs along an altitudinal gradient. *Biotropica* 20:236–39.

Christian, K., and B. Green. 1994. Water flux in the Australian treefrog *Litoria caerulea* under natural and semi-natural conditions. *Amphibia-Reptilia* 15:401–5.

Christian, K., and D. Parry. 1997. Reduced rates of water loss and chemical properties of skin secretions of the frogs *Litoria caerulea* and *Cyclorana australis*. *Aust. J. Zool.* 45:13–20.

Christin, M. S., A. D. Gendron, P. Brousseau, L. Menard, D. J. Marcogliese, D. Cyr, S. Ruby, and M. Fournier. 2003. Effects of agricultural pesticides on the immune system of *Rana pipiens* and on its resitance to parasitic infection. *Env. Tox. Chem.* 22:1127–33.

Christin, M. S., L. Menard, A. D. Gendron, S. Ruby, D. Cyr, D. J. Marcogliese, L. Rollins-Smith, and M. Fournier. 2004. Effects of agricultural pesticides on the immune system of *Xenopus laevis* and *Rana pipiens. Aquat. Tox.* 67:33–43.

Christman, S. P. 1974. Geographic variation for salt water tolerance in the frog *Rana sphenocephala. Copeia* 1974:773–78.

Chuang, S. A., and L. L. Lee. 1997. Food habits of three carnivore species (*Viverricula indica, Herpestes urva,* and *Melogale moschata*) in Fushan Forest, northern Taiwan. *J. Zool., London* 243:71–79.

Church, G. 1960a. Annual and lunar periodicity in the sexual cycle of the Javanese toad, *Bufo melanostictus* Schneider. *Zoologica* 45:181–88.

———. 1960b. The effects of seasonal and lunar changes on the breeding pattern of the edible Javanese frog, *Rana cancrivora* Gravenhorst. *Treubia* 25:215–33.

Church, G. 1963. The variation of dorsal pattern in *Rhacophorus leucomystax. Copeia* 1963:400–5.

Churchill, T. A., and K. B. Storey. 1992. Natural freezing survival by painted turtles *Chrysemys picta marginate* and *C. picta bellii. Am. J. Physiol.* 262:R530–R537.

———. 1993. Dehydration tolerance in wood frogs: a new perspective on development of amphibian freeze tolerance. *Am. J. Physiol.* 265:R1324–R1332.

———. 1995. Metabolic effects of dehydration on an aquatic frog, *Rana pipiens. J. Exp. Biol.* 198:147–54.

Ciampalini, B., and S. Lovari. 1985. Food habits and trophic niche overlap of the badger (*Meles meles* L.) and the red fox (*Vulpes vulpes* L.) in a Mediterranean coastal area. *Z. Säugetierkunde* 50:226–34.

Cintron, G. 1970. Niche separation of tree frogs in the Luquillo forest. In *A tropical rain forest,* ed. H. T. Odum, E51–E54. Washington, DC: U.S. Atomic Energy Commission.

Clack, J. A. 2000. The origin of tetrapods. In *Amphibian biology. Vol. 4. Palaeontology,* ed. H. Heatwole and R. L. Carroll, 979–1029. Chipping Norton, New South Wales, Australia: Surrey Beatty and Sons.

———. 2002. An early tetrapod from 'Romer's Gap.' *Nature* 418:72–76.

———. 2004. From fins to fingers. *Science* 304:57–58.

Clack, J. A., and R. L. Carroll. 2000. Early Carboniferous tetrapods. In *Amphibian biology. Vol. 4. Palaeontology,* ed. H. Heatwole and R. L. Carroll, 1030–43. Chipping Norton, New South Wales, Australia: Surrey Beatty and Sons.

Clark, D. R., Jr. 1974. The western ribbon snake (*Thamnophis proximus*): Ecology of a Texas population. *Herpetologica* 30:372–79.

Clark, M. 1981. *Instinto Parental y Comportamiento Sexual de Centrolenella fleischmanni (Anura, Centrolenidae).* Unpublished thesis. Panama; University of Panama.

Clark, R. F. 1949. Snakes of the hill parishes of Louisiana. *J. Tennessee Acad. Sci.* 24:244–61.

Clark, V. C., C. J. Raxworthy, V. Rakotomalala, P. Sierwald, and B. L. Fisher. 2005. Convergent evolution of chemical defense in poison frogs and arthropod prey between Madagascar and the Neotropics. *Proc. Natl. Acad. Sci. USA* 102:11617–22.

Clarke, B. C. 1962. Balanced polymorphism and the diversity of sympatric species. *Syst. Assoc. Publ.* 4:47–70.

Clarke, B. T. 1987. A description of the skeletal morphology of *Barbourula* (Anura: Discoglossidae), with comments on its relationships. *J. Nat. Hist.* 21:879–91.

———. 1996. Small size in amphibians—Its ecological and evolutionary implications. In *The implications of small body size. Symposia of the Zoological Society of London, no. 69, Miniature vertebrates,* ed. P. J. Miller, 201–24. Oxford: Clarendon Press.

———. 1997. The natural history of amphibian skin secretions, their normal functions and potential medical applications. *Biol. Rev.* 72:365–79.

Clarke, R. D. 1972. The effect of toe clipping on survival in Fowler's toad (*Bufo woodhousei fowleri*). *Copeia* 1972:182–85.

———. 1974a. Activity and movement patterns in a population of Fowler's toad, *Bufo woodhousei fowleri. Am. Midl. Nat.* 92:257–74.

———. 1974b. Food habits of toads, genus *Bufo* (Amphibia: Bufonidae). *Am. Midl. Nat.* 91:140–47.

———. 1977. Postmetamorphic survivorship of Fowler's toad, *Bufo woodhousei fowleri. Copeia* 1977:594–97.

Clarkson, R. W., and J. C. Rorabaugh. 1989. Status of leopard frogs (*Rana pipiens* complex: Ranidae) in Arizona and southwestern California. *Southwest. Nat.* 34:531–38.

Clastres, P. 1972. The Guayaki. In *Hunters and gatherers today,* ed. M. C. Bicchieri, 138–74. New York: Holt, Rinehart and Winston.

Claussen, D. L. 1969. Studies on water loss and rehydration in anurans. *Physiol. Zool.* 42:1–14.

———. 1973a. The thermal relations of the tailed-frog, *Ascaphus truei* and the Pacific treefrog, *Hyla regilla. Comp. Biochem. Physiol.* 44A:137–53.

———. 1973b. Water relations of the tailed frog, *Ascaphus truei,* and the Pacific tree frog, *Hyla regilla. Comp. Biochem. Physiol.* 44A:155–71.

———. 1974. Urinary bladder water reserves in the terrestrial toad, *Bufo fowleri,* and the aquatic frog, *Rana clamitans. Herpetologica* 30:360–67.

———. 1977. Thermal acclimation in ambystomatid salamanders. *Comp. Biochem. Physiol.* 58A:333–40.

Cleaveland, S., G. R. Hess, A. P. Dobson, M. K. Laurenson, H. I. McCallum, M. G. Roberts, and R. Woodroffe. 2002. The role of pathogens in biological conservation. In *The ecology of wildlife diseases,* ed. P. J. Hudson, A. Rizzoli, B. T. Grenfell, H. Heesterbeek, and A. P. Dobson, 139–50. New York: Oxford University Press.

Clough, M., and K. Summers. 2000. Phylogenetic systematics and biogeography of the poison frogs: evidence from mitochondrial DNA sequences. *Biol. J. Linn. Soc.* 70:515–40.

Clutton-Brock, T. H. 1983. Selection in relation to sex. In *Evolution from molecules to man,* ed. S. Bendall, 457–81. New York: Cambridge University Press.

———. 1988. Reproductive success. In *Reproductive success,* ed. T. H. Clutton-Brock, 472–85. Chicago: University of Chicago Press.

———. 1991. *The evolution of parental care.* Princeton, NJ: Princeton University Press.

Clutton-Brock, T. H., and S. D. Albon. 1980. The roaring of red deer and the evolution of honest advertisement. *Behaviour* 69:145–69.

Clutton-Brock, T. H., S. D. Albon, and P. H. Harvey. 1980. Antlers, body size and breeding group size in the Cervidae. *Nature* 285:565–67.

Clutton-Brock, T. H., P. H. Harvey, and B. Rudder. 1977. Sexual dimorphism, socionomic sex ratio and body weight in primates. *Nature* 269:797–800.

Clutton-Brock, T. H., and G. A. Parker. 1992. Potential reproductive rates and the operation of sexual selection. *Q. Rev. Biol.* 67:437–56.

Clutton-Brock, T. H., O. F. Price, and A. C. D. MacColl. 1992. Mate retention, harassment, and the evolution of ungulate leks. *Behav. Ecol.* 3:234–42.

Clutton-Brock, T. H., and A. C. J. Vincent. 1991. Sexual selection and the potential reproductive rates of males and females. *Nature* 351:58–60.

Clyne, D. 1967. Prothalamion in a garden pool. *Victorian Nat.* 84:232–35.

———. 1969. *Australian frogs.* Melbourne, Australia: Periwinkle Books.

Coates, M. I., and J. A. Clack. 1990. Polydactyly in the earliest known tetrapod limbs. *Nature* 347:66–69.

———. 1995. Romer's gap: Tetrapod origins and terrestriality. *Bull. Mus. Natl. Hist. Nat. (Paris)*, 4th ser., 17 (sec. C):373–87.

Coates, M. L. 1975. Hemoglobin function in the vertebrates: An evolutionary model. *J. Mol. Evol.* 6:285–307.

Coates, M. L., and J. Metcalfe. 1971. Evolution of blood oxygen transport in the newt, *Taricha. Respir. Physiol.* 11:94–103.

Cocroft, R. B. 1994. A cladistic analysis of chorus frog phylogeny (Hylidae: *Pseudacris*). *Herpetologica* 50:420–37.

Cocroft, R. B., and K. Hambler. 1989. Observations on a commensal relationship of the microhylid frog *Chiasmocleis ventrimaculata* and the burrowing theraphosid spider *Xenesthis immanis* in southeastern Peru. *Biotropica* 21:2–8.

Cocroft, R. B., and W. R. Heyer. 1988. Notes on the frog genus *Thoropa* (Amphibia: Leptodactylidae) with a description of a new species (*Thoropa saxatilis*). *Proc. Biol. Soc. Washington* 101:209–20.

Cocroft, R. B., R. W. McDiarmid, A. P. Jaslow, and P. M. Ruiz-Carranza. 1990. Vocalizations of eight species of *Atelopus* (Anura: Bufonidae) with comments on communication in the genus. *Copeia* 1990:631–43.

Cocroft, R. B., and M. J. Ryan. 1995. Patterns of advertisement call evolution in toads and chorus frogs. *Anim. Behav.* 49:283–303.

Cody, M. L., and J. H. Brown. 1969. Song asynchrony in neighbouring bird species. *Nature* 222:778–80.

Cody, M. L., and J. M. Diamond, eds. 1975. *Ecology and evolution of communities.* Cambridge: Harvard University Press.

Coe, M. D., and R. A. Diehl. 1980. *In the land of the Olmec. Vol. I. The archaeology of San Lorenzo Tenochtitlan.* Austin: University of Texas Press.

Coe, M. J. 1967. Co-operation of three males in nest construction by *Chiromantis rufescens* Gunther (Amphibia: Rhacophoridae). *Nature* 214:112–13.

———. 1974. Observations on the ecology and breeding biology of the genus *Chiromantis* (Amphibia: Rhacophoridae). *J. Zool., London* 172:13–34.

Cogălniceanu, D. 1992. A comparative ethological study of female chemical attractants in newts (genus *Triturus*). *Amphibia-Reptilia* 13:69–74.

———. 1994. The relative importance of vision and olfaction in mate recognition in male newts (genus *Triturus*). *Herpetologica* 50:344–49.

Cogălniceanu, D., F. Aioanei, C. Ciubuc, and A. Vădineanu. 1998. Food and feeding habits in a population of common spadefoot toads (*Pelobates fuscus*) from an island in the lower Danube floodplain. *Alytes* 15:145–57.

Cognălniceanu, D., and C. Miaud. 2003. Population age structure and growth in four syntopic amphibian species inhabiting a large river floodplain. *Can. J. Zool.* 81:1096–1106.

———. 2004. Variation in life history traits in *Bombina bombina* from the lower Danube floodplain. *Amphibia-Reptilia* 25:115–19.

Cogger, H. G. 2000. *Reptiles and amphibians of Australia.* 6th ed. Sanibel Island, FL: Ralph Curtis Books.

Cohen, M. P., and R. A. Alford. 1996. Factors affecting diurnal shelter use by the cane toad, *Bufo marinus. Herpetologica* 52:172–81.

Cohen, N. W. 1952. Comparative rates of dehydration and hydration in some California salamanders. *Ecology* 33:462–79.

Coker, C. M. 1931. Hermit thrush feeding on salamanders. *Auk* 48:277.

Colburn, E. A. 2004. *Vernal pools: Natural history and conservation.* Blacksburg, VA: McDonald and Woodward Publishing Company.

Cole, E. C., W. C. McComb, M. Newton, C. L. Chambers, and J. P. Leeming. 1997. Responses of amphibians to clearcutting, burning, and gyphosate application in the Oregon coast range. *J. Wildl. Manage.* 61:656–64.

Collazo, A. 1996. Evolutionary correlations between early development and life history in plethodontid salamanders and teleost fishes. *Amer. Zool.* 36:116–31.

Collazo, A., and S. B. Marks. 1994. Development of *Gyrinophilus porphyriticus:* Identification of the ancestral developmental pattern in the salamander family Plethodontidae. *J. Exp. Zool.* 268:239–58.

Collette, B. B., and F. R. Gehlbach. 1961. The salamander *Siren intermedia intermedia* LeConte in North Carolina. *Herpetologica* 17:203–4.

Colley, S. A., W. H. Keen, and R. W. Reed. 1989. Effects of adult presence on behavior and microhabitat use of juveniles of a desmognathine salamander. *Copeia* 1989:1–7.

Collins, J. P. 1979a. Intrapopulation variation in the body size at metamorphosis and timing of metamorphosis in the bullfrog, *Rana catesbeiana. Ecology* 60:738–49.

———. 1979b. Sexually mature larvae of the salamanders *Ambystoma rosaceum* and *A. tigrinum velasci* from Chihuahua, Mexico:Taxonomic and ecological notes. *J. Herpetol.* 13:351–54.

———. 1981. Distribution, habitats and life history variation in the tiger salamander, *Ambystoma tigrinum,* in east-central and southeast Arizona. *Copeia* 1981:666–75.

Collins, J. P., J. L. Brunner, J. K. Jancovich, and D. M. Schock. 2004. A model host-pathogen system for studying infectious disease dynamics in amphibians: Tiger salamanders (*Ambystoma tigrinum*) and *Ambystoma tigrinum* virus. *Herpetol. J.* 14:195–200.

Collins, J. P., J. L. Brunner, V. Miera, M. J. Parris, D. M. Schock, and A. Storfer. 2003. Ecology and evolution of infectious disease. In *Amphibian conservation,* ed. R. D. Semlitsch, 137–51. Washington, DC: Smithsonian Books.

Collins, J. P., and J. E. Cheek. 1983. Effect of food and density on development of typical and cannibalistic salamander larvae in *Ambystoma tigrinum nebulosum. Amer. Zool.* 23:77–84.

Collins, J. P., and T. Halliday. 2005. Forecasting changes in amphibian biodiversity: Aiming at a moving target. *Phil. Trans. Roy. Soc. London B* 360:309–14.

Collins, J. P., and J. R. Holomuzki. 1984. Intraspecific variation in diet within and between trophic morphs in larval tiger salamanders (*Ambystoma tigrinum nebulosum*). *Can. J. Zool.* 62:168–74.

Collins, J. P., T. R. Jones, and H. J. Berna. 1988. Conserving genetically distinctive populations: The case of the Huachuca tiger salamander (*Ambystoma tigrinum stebbinsi* Lowe). In *Management of amphibians, reptiles, and small mammals in North America,* R. C. Szaro, K. E. Severson, and D. R. Patton, technical coordinators, 45–53. Ft. Collins, CO: U.S. Forest Service, Rocky Mountain Forest and Range Experiment Station.

Collins, J. P., and M. A. Lewis. 1979. Overwintering tadpoles and breeding season variation in the *Rana pipiens* complex in Arizona. *Southwest. Nat.* 24:371–73.

Collins, J. P., and A. Storfer. 2003. Global amphibian declines: Sorting the hypotheses. *Divers. Distr.* 9:89–98.

Collins, J. P., and H. M. Wilbur. 1979. Breeding habits and habitats of the amphibians of the E. S. George Reserve, Michigan, with notes on the distribution of fishes. *Occ. Pap. Mus. Zool. Univ. Michigan* (686):1–34.

Collins, J. P., K. E. Zerba, and M. J. Sredl. 1993. Shaping intraspecific variation: Development, ecology and the evolution of morphology and life history variation in tiger salamanders. *Genetica* 89:167–83.

Collopy, M. W. 1977. Food caching by female American kestrels in winter. *Condor* 79:63–68.

Collopy, M. W., and J. R. Koplin. 1983. Diet, capture success, and mode of hunting by female American kestrels in winter. *Condor* 85:369–71.

Coloma, L. A. 1995. Ecuadorian frogs of the genus *Colostethus* (Anura: Dendrobatidae). *Misc.Publ.Univ.Kansas Nat. Hist. Mus.* (87):1–72.

Colwell, R. K., and D. W. Winkler. 1984. A null model for null models in biogeography. In *Ecological communities,* ed. D. R. Strong, Jr., D. Simberloff, L. G. Abele, and A. B. Thistle, 344–59. Princeton, NJ: Princeton University Press.

Comer, C., and P. Grobstein. 1981. Organization of sensory inputs to the midbrain of the frog, *Rana pipiens. J. Comp. Physiol.* 142:161–68.

Conant, R., and J. T. Collins. 1991. *A field guide to reptiles and amphibians, 3rd ed.* Boston: Houghton Mifflin.

Conaway, C.H., and D. E. Metter. 1967. Skin glands associated with breeding in *Microhyla carolinensis. Copeia* 1967:672–73.

Condo, S., A. Bellelli, M. Brunori, M. Corda, M. G. Pellegrini, E. M. Clementi, and B. Giardina. 1989. The functional properties of amphibian hemoglobin: The case of *Salamander salamander* [sic] and *Hydromantes genei. Comp. Biochem. Physiol.* 93A:319–25.

Condo, S., B. Giardina, A. Bellelli, M. Lunadei, A. Ferracin, and M. Brunori. 1983. Comparative studies of hemoglobins from newts: A kinetic approach. *Comp. Biochem. Physiol.* 74A:545–48.

Condo, S., B. Giardina, M. Lunadei, A. Ferracin, and M. Burnori. 1981. Functional properties of hemoglobins from *Triturus cristatus. Eur. J. Biochem.* 120:323–27.

Condon, C. J., S.-H. Chang, and A. S. Feng. 1991. Processing of behaviorally relevant temporal parameters of acoustic stimuli by single neurons in the superior olivary nucleus of the leopard frog. *J. Comp. Physiol. A* 168:709–25.

Conklin, R. E. 1930a. The formation and circulation of lymph in the frog. I. The rate of lymph production. *Am. J. Physiol.* 95:75–90.

———. 1930b. The formation and circulation of lymph in the frog. II. Volume and blood pressure. *Am. J. Physiol.* 95:91–97.

Conlon, J. M., J. Kolodziejek, and N. Nowotny. 2004. Antimicrobial peptides from ranid frogs: Taxonomic and phylogenetic markers and a potential source of new therapeutic agents. *Biochim. Biophys. Acta—Proteins and Proteomics* 1696:1–14.

Conlon, J. M., A. Sonnevend, C. Davidson, A. Demandt, and J. Jouenne. 2005. Host-defense peptides isolated from the skin secretions of the northern red-legged frog *Rana aurora aurora. Dev. Comp. Immunol.* 29:83–90.

Conlon, J. M., A. Sonnevend, C. Davidson, D. D. Smith, and P. F. Nielsen. 2004. The ascaphins: A family of antimicrobial peptides from the skin of the most primitive extant frog, *Ascaphus truei. Biochem. Biophys. Res. Comm.* 320:170–75.

Connell, J. H. 1971. On the role of natural enemies in preventing competitive exclusion in some marine animals and in rainforest trees. In *Dynamics of populations,* ed. P. J. den Boer and G. R. Gradwell, 298–312. Wageningen, The Netherlands: Center for Agricultural Publishing and Documentation.

———. 1975. Some mechanisms producing structure in natural communities: A model and evidence from field experiments. In *Ecology and evolution of communities,* ed. M. L. Cody and J. M. Diamond, 460–90. Cambridge, MA: Harvard University Press.

———. 1983. On the prevalence and relative importance of interspecific competition: Evidence from field experiments. *Am. Nat.* 122:661–96.

Connor, E. F., and D. Simberloff. 1979. The assembly of species communities: Chance or competition? *Ecology* 60:1132–40.

———. 1986. Competition, scientific method, and null models in ecology. *Am. Sci.* 74:155–62.

Conroy, S. D. S., and B. W. Brook. 2003. Demographic sensitivity and persistence of the threatened white-and orange-bellied frogs of Western Australia. *Pop. Ecol.* 45:105–14.

Cook, C. L., J. W. H. Ferguson, and S. R. Telford. 2001. Adaptive male parental care in the giant bullfrog, *Pyxicephalus adspersus. J. Herpetol.* 35:310–15.

Cook, W. E. 1987. Amphibians and birds. *Smithsonian Herpetological Information Service, no. 73.*

Cooke, A. S. 1974. Differential predation by newts on anuran tadpoles. *British J. Herpetol.* 5:386–89.

———. 1981. Amphibian growth rates. *British J. Herpetol.* 6:179–80.

———. 1988. Mortality of toads *Bufo bufo* on roads near a Cambridgeshire breeding site. *British Herpetol. Soc. Bull.* 26:29–30.

———. 1995. Road mortality of common toads (*Bufo bufo*) near a breeding site, 1974–1994. *Amphibia-Reptilia* 16:87–90.

Cooke, R. 1984. Archaeological research in central and eastern Panama: A review of some problems. In *The archaeology of lower Central America,* ed. F. W. Lange and D. Z. Stone, 263–302. Albuquerque: University of New Mexico Press.

Cooke, S. J. 2004. Sex-specific differences in cardiovascular performance of a centrarchid fish are only evident during the reproductive period. *Funct. Ecol.* 18:398–403.

Cooper, J. E., and M. R. Cooper. 1968. Cave-associated herpetozoa. II. Salamanders of the genus *Gyrinophilus* in Alabama caves. *Bull. Natl. Speleol. Soc.* 30:19–24.

Cooper, W. S., and R. H. Kaplan. 1982. Adaptive "coin-flipping": A decision-theoretic examination of natural selection for random individual variation. *J. Theor. Biol.* 94:135–51.

Corben, C. J., G. J. Ingram, and M. J. Tyler. 1974. Gastric brooding: unique form of parental care in an Australian frog. *Science* 186:946–47.

Corbet, G. B., and H. N. Southern, eds. 1977. *The handbook of British mammals.* Oxford, UK: Blackwell Scientific Publications.

Corbet, P. S. 1999. *Dragonflies. Behavior and ecology of Odonata.* Ithaca, NY: Cornell University Press.

Corbett, K., ed. 1989. *The conservation of European reptiles and amphibians.* London: Christopher Helm.

Corn, P. S. 1981. Field evidence for a relationship between color and development rate in the northern leopard frog (*Rana pipiens*). *Herpetologica* 37:155–60.

———. 1994. What we know and don't know about amphibian declines in the West. In *Sustainable ecological systems: Implementing an ecological approach to land management,* W. W. Covington and L. F. DeBano, technical coordinators, 59–67. Ft. Collins, CO: USDA Forest Service Rocky Mountain Forest and Range Experiment Station

———. 1998. Effects of ultraviolet radiation on boreal toads in Colorado. *Ecol. Appl.* 8:18–26.

———. 2000. Amphibian declines: Review of some current hypotheses. In *Ecotoxicology of amphibians and reptiles,* ed. D. W. Sparling, G. Linder, and C. A. Bishop, 663–96. Pensacola, FL: SETAC Press.

———. 2003. Deteriorating status of western amphibians: Can we generalize about causes? In *Amphibian decline: An integrated analysis of multiple stressor effects,* ed. G. Linder, S. K. Krest, and D. W. Sparling, 249–55. Pensacola, FL: SETAC Press.

Corn, P. S., and R. B. Bury. 1989. Logging in western Oregon: Responses of headwater habitats and stream amphibians. *Forest Ecol. Manage.* 29:39–57.

Corn, P. S., R. B. Bury, and E. J. Hyde. 2003. Conservation of North American stream amphibians. In *Amphibian conservation,* ed. R. D. Semlitsch, 24–36. Washington, DC: Smithsonian Books.

Corn, P. S., and J. C. Fogelman. 1984. Extinction of montane populations of the northern leopard frog (*Rana pipiens*) in Colorado. *J. Herpetol.* 18:147–52.

Corn, P. S., M. L. Jennings and E. Muths. 1997. Survey and assessment of amphibian populations in Rocky Mountain National Park. *Northwest. Nat.* 78:34–55.

Corn, P. S., and E. Muths. 2002. Variable breeding phenology affects the exposure of amphibians to ultraviolet radiation. *Ecology* 83:2958–63.

———. 2004. Variable breeding phenology affects the exposure of amphibians to ultraviolet radiation: Reply. *Ecology* 85:1759–63.

Corn, P. S., and F. A. Vertucci. 1992. Descriptive risk assessment of the effects of acidic deposition on Rocky Mountain amphibians. *J. Herpetol.* 26:361–69.

Cornell, T. J., K. A. Berven, and G. J. Gamboa. 1989. Kin recognition by tadpoles and froglets of the wood frog *Rana sylvatica. Oecologia* 78:312–16.

Corse, W. A., and D. E. Metter. 1980. Economics, adult feeding and larval growth of *Rana catesbeiana* on a fish hatchery. *J. Herpetol.* 14:231–38.

Corser, J. D. 2001. Decline of disjunct green salamander (*Aneides aeneus*) populations in the southern Appalachians. *Biol. Cons.* 97:119–26.

Cortwright, S. A. 1988. Intraguild predation and competition: An analysis of net growth shifts in larval amphibian prey. *Can. J. Zool.* 66:1813–21.

Cory, L. 1962. Patterns of geographic variation in Sierra Nevada ranids. *Amer. Zool.* 2:401.

Cory, L., and J. J. Manion. 1953. Predation on eggs of the wood-frog, *Rana sylvatica,* by leeches. *Copeia* 1953:66.

———. 1955. Ecology and hybridization in the genus *Bufo* in the Michigan-Indiana region. *Evolution* 9:42–51.

Cosens, S. E., and J. B. Falls. 1984. A comparison of sound propagation and song frequency in temperate marsh and grassland habitats. *Behav. Ecol. Sociobiol.* 15:161–70.

Costa-Lima, A. 1946. Nova especie do género *Rhysops* Williston (Diptera: Syrphidae). *Bol. Soc. Brasil. Agronomía* 9:155–56.

Costanzo, J. P., and D. L. Claussen. 1990. Natural freeze tolerance in the terrestrial turtle, *Terrapene carolina. J. Exp. Zool.* 254:228–32.

Costanzo, J. P., A. L. Allenspach, and R. E. Lee, Jr. 1999. Electrophysiological and ultrastructural correlates of cryoinjury in sciatic nerve of the freeze-tolerant wood frog, *Rana sylvatica. J. Comp. Physiol.* B 169:351–59.

Costanzo, J. P., J. M. Bayuk, and R. E. Lee, Jr. 1999. Inoculative freezing by environmental ice nuclei in the freeze-tolerant wood frog, *Rana sylvatica. J. Exp. Zool.* 284:7–14.

Costanzo, J. P., P. A. Callahan, R. E. Lee, Jr., and M. F. Wright. 1997. Frogs reabsorb glucose from urinary bladder. *Nature* 389:343–44.

Costanzo, J. P., C. Grenot, and R. E. Lee. 1995. Supercooling, ice inoculation and freeze tolerance in the European common lizard, *Lacerta vivipara. J. Comp. Physiol.* B 165:238–44.

Costanzo, J. P., J. T. Irwin, and R. E. Lee, Jr. 1997. Freezing impairment of male reproductive behaviors of the freeze-tolerant wood frog, *Rana sylvatica. Physiol. Zool.* 70:158–66.

Costanzo, J. P., J. B. Iverson, M. F. Wright, and R. E. Lee. 1995. Cold-hardiness and overwintering strategies of hatchlings in an assemblage of northern turtles. *Ecology* 76:1772–85.

Costanzo, J. P., and R. E. Lee, Jr. 1993. Cryoprotectant production capacity of the freeze-tolerant wood frog, *Rana sylvatica. Can. J. Zool.* 71:71–75.

———. 1996. Mini-review: Ice nucleation in freeze-tolerant vertebrates. *Cryo-Letters* 17:111–18.

———. 2005. Cryoprotection by urea in a terrestrially hibernating frog. *J. Exp. Biol.* 208:4079–89.

Costanzo, J. P., R. E. Lee, Jr., and P. H. Lortz. 1993. Physiological responses of freeze-tolerant and -intolerant frogs: Clues to evolution of anuran freeze tolerance. *Am. J. Physiol.* 265:R721–R725.

Costanzo, J. P., R. E. Lee, Jr., and M. F. Wright. 1991. Effect of cooling rate on the survival of frozen wood frogs, *Rana sylvatica. J. Comp. Physiol.* B 161:225–29.

———. 1992. Cooling rate influences cryoprotectant distribution and organ dehydration in freezing wood frogs. *J. Exp. Zool.* 261:373–78.

———. 1993. Physiological responses to freezing in the turtle *Terrapene carolina. J. Herpetol.* 27:117–20.

Costanzo, J. P., J. D. Litzgus, J. B. Iverson, and R. E. Lee. 1998. Soil hydric characteristics and environmental ice nuclei influence supercooling capacity of hatchling painted turtles, *Chrysemys picta. J. Exp. Biol.* 201:3105–12.

Costanzo, J. P., J. A. Mugnano, H. M. Wehrheim, and R. E. Lee, Jr. 1998. Osmotic and freeze tolerance in spermatozoa of freeze-tolerant and -intolerant frogs. *Am. J. Physiol.* 275:R713–R719.

Costanzo, J. P., M. F. Wright, and R. E. Lee. 1992. Freeze tolerance as an overwintering adapation in Cope's gray treefrog (*Hyla chrysoscelis*). *Copeia* 1992:565–69.

Cott, H. B. 1926. Observations on the life-habits of some batrachians and reptiles from the lower Amazon, and a note on some mammals from Marajo Island. *Proc. Zool. Soc. London* 2:1159–78.

———. 1940. *Adaptive coloration in animals.* Oxford, UK: Oxford University Press.

———. 1961. Scientific results of an enquiry into the ecology and economic status of the Nile crocodile (*Crocodylus niloticus*) in Uganda. *Trans. Zool. Soc. London* 29:211–357.

Covacevich, J., and M. Archer. 1975. The distribution of the cane toad, *Bufo marinus,* in Australia and its effects on indigenous vertebrates. *Mem. Queensland Mus.* 17:305–10.

Cover, J. F., Jr. 1986. *Basiliscus plumifrons* (crested green basilisk lizard). Food. *Herp. Rev.* 17:19.

Cowles, R. B., and C. M. Bogert. 1944. A preliminary study of the thermal requirements of desert reptiles. *Bull. Amer. Mus. Nat. Hist.* 83:261–96.

Cox, C. B. 1966. The Amphibia—An evolutionary backwater. In *Looking at animals again,* ed. D. R. Arthur, 97–118. London: Freeman.

Cox, T. C. 1986. Ion transport across the skin of a larval salamander. *Biochem. Biophys. Acta* 858:142–52.

Cox, T. C., and R. H. Alvarado. 1979. Electrical and transport characteristics of skin of larval *Rana catesbeiana. Am. J. Physiol.* 237:R74–R79.

Craighead, J. J., and F. C. Craighead, Jr. 1969. *Hawks, owls and wildlife.* New York: Dover.

Crapon de Caprona, D., and W. Himstedt. 1985. Das aquatische Verhalten der Blindwühle *Ichthyophis kohtaoensis* Taylor, 1960. *Salamandra* 21:192–96.

Crawford, A. J., and E. N. Smith. 2005. Cenzoic biogeography and evolution in direct-developing frogs of Central America (Leptodactylidae: *Eleutherodactylus*) as inferred from a phylogenetic analysis of nuclear and mitochondrial genes. *Mol. Phylogen. Evol.* 35:536–55.

Crawford, A. J., and D. B. Wake. 1998. Phylogenetic and evolutionary perspectives on an enigmatic organ: The balancer of larval caudate amphibians. *Zoology* 101:107–23.

Cree, A. 1985a. Water balance and nitrogen excretion of two introduced frogs (*Litoria raniformis* and *L. ewingi*). *New Zealand J. Zool.* 12:341–48.

———. 1985b. Water balance of New Zealand's native frogs (Anura: Leiopelmatidae). In *Biology of Australasian frogs and reptiles,* ed. G. Grigg, R. Shine, and H. Ehmann, 361–71. Chipping Norton, New South Wales: Surrey Beatty and Sons.

———. 1988a. Effects of argenine vasotocin on water balance of three leiopelmatid frogs. *Gen. Comp. Endocrinol.* 72:340–50.

———. 1988b. Water balance responses of the hylid frog *Litoria aurea. J. Exp. Zool.* 247:119–25.

———. 1989. Relationship between environmental conditions and nocturnal activity of the terrestrial frog, *Leiopelma archeyi. J. Herpetol.* 23:61–68.

Crespi, E. J., and R. J. Denver. 2004. Ontogeny of corticotropin-releasing factor effects on locomotion and foraging in the western spadefoot toad (*Spea hammondii*). *Horm. Behav.* 46:399–410.

———. 2005. Roles of stress hormones in food intake regulation in anuran amphibians throughout the life cycle. *Comp. Biochem. Physiol.* A 141:381–90.

Crespi, E. J., and H. Lessig. 2004. Mothers influence offspring body size through post-oviposition maternal effects in the redbacked salamander, *Plethodon cinereus. Oecologia* 138:306–11.

Crespo, E. G. 1982. Contribuicão para o conhecimento da biologia dos *Alytes* Ibericos, *Alytes obstetricans boscai* Lataste, 1879 e *Alytes cisternasii* Bosca, 1879 (Amphibia, Discoglossidae): Ovos, posturas (epocas de reprodução). *Arqu. Mus. Bocage,* ser. A, 1:453–66.

Creutz, G. 1963. Ernährungsweise und Aktionsradius der Lachmöwe. *Beitr. Vogelk.* 9:3–58.

Crochet, P.-A., O. Chaline, M. Cheylan, and C. P. Guillaume. 2004. No evidence of general decline in an amphibian community of southern France. *Biol. Cons.* 119:297–304.

Croes, S. A., and R. E. Thomas. 2000. Freeze tolerance and cryoprotectant synthesis of the Pacific tree frog *Hyla regilla. Copeia* 2000:863–68.

Crombie, R. I. 1977. A new species of the genus *Eleutherodactylus* (Amphibia: Leptodactylidae) from the Cockpit Country of Jamaica. *Proc. Biol. Soc. Washington* 90:194–204.

———. 1986. Another new forest-dwelling frog (Leptodactylidae: *Eleutherodactylus*) from the Cockpit Country of Jamaica. *Trans. San Diego Soc. Nat. Hist.* 21:145–53.

Cronin, H. 1992. *The ant and the peacock.* New York: Cambridge University Press.

Cronin, J. T., and J. Travis. 1986. Size-limited predation on larval *Rana aereolata* (Anura: Ranidae) by two species of backswimmer (Insecta: Hemiptera: Notonectidae). *Herpetologica* 42:171–74.

Croshaw, D. A. 2005. Cryptic behavior is independent of dorsal color polymorphism in juvenile northern leopard frogs (*Rana pipiens*). *J. Herpetol.* 39:125–29.

Croshaw, D. A., and D. E. Scott. 2005. Experimental evidence that nest attendance benefits female marbled salamanders (*Ambystoma opacum*) by reducting egg mortality. *Am. Midl. Nat.* 154:398–411.

Crossland, M. R. 1998a. A comparison of cane toad and native tadpoles as predators of native anuran eggs, hatchlings and larvae. *Wildl. Res.* 25:373–81.

———. 1998b. Predation by tadpoles on toxic toad eggs: The effect of tadpole size on predation success and tadpole survival. *J. Herpetol.* 32:443–46.

———. 2000. Direct and indirect effects of the introduced toad *Bufo marinus* (Anura: Bufonidae) on populations of native anuran larvae in Australia. *Ecography* 23:283–90.

———. 2001. Ability of predatory native Australian fishes to learn to avoid toxic larvae of the introduced toad *Bufo marinus. J. Fish Biol.* 59:319–29.

Crossland, M. R., and R. A. Alford. 1998. Evaluation of the toxicity of eggs, hatchlings and tadpoles of the introduced cane toad *Bufo marinus* (Anura: Bufonidae) to native Australian aquatic predators. *Aust. J. Ecol.* 23:129–37.

Crossland, M. R., and C. Azevedo-Ramos. 1999. Effects of *Bufo* (Anura: Bufonidae) toxins on tadpoles from native and exotic *Bufo* habitats. *Herpetologica* 55:192–99.

Crowder, W. C., M. Nie, and G. R. Ultsch. 1998. Oxygen uptake in bullfrog tadpoles (*Rana catesbeiana*). *J. Exp. Zool.* 280:121–34.

Crump, D., D. Lean, M. Berrill, D. Coulson, and L. Toy. 1999. Spectral irradiance in pond water: Influences of water chemistry. *Photochem. Photobiol.* 70:893–901.

Crump, D., M. Berrill, D. Coulson, D. Lean, L. McGillvray, and A. Smith. 1999. Sensitivity of amphibian embryos, tadpoles, and larvae to enhanced UV-B radiation in natural pond conditions. *Can. J. Zool.* 77:1956–66.

Crump, M. L. 1971. Quantitative analysis of the ecological distribution of a tropical herpetofauna. *Occ. Pap. Mus. Nat. Hist. Univ. Kansas* (3):1–62.

———. 1972. Territoriality and mating behavior in *Dendrobates granuliferus* (Anura: Dendrobatidae). *Herpetologica* 28:195–98.

———. 1974. Reproductive strategies in a tropical anuran community. *Misc. Publ. Mus. Nat. Hist. Univ. Kansas* (61):1–68.

———. 1981a. Energy accumulation and amphibian metamorphosis. *Oecologia* 49:167–69.

———. 1981b. Variation in propagule size as a function of environmental uncertainty for tree frogs. *Am. Nat.* 117:724–37.

———. 1983. Opportunistic cannibalism by amphibian larvae in temporary aquatic environments. *Am. Nat.* 121:281–87.

———. 1984a. Intraclutch egg size variability in *Hyla crucifer* (Anura: Hylidae). *Copeia* 1984:302–8.

———. 1984b. Ontogenetic change in vulnerability to predation in tadpoles of *Hyla pseudopuma. Herpetologica* 40:265–71

———. 1986a. Cannibalism by younger tadpoles: Another hazard of metamorphosis. *Copeia* 1986:1007–9.

———. 1986b. Homing and site fidelity in a Neotropical frog, *Atelopus varius* (Bufonidae). *Copeia* 1986:438–44.

———. 1988. Aggression in harlequin frogs: male-male competition and a possible conflict of interest between the sexes. *Anim. Behav.* 36:1064–77.

———. 1989a. Effect of habitat drying on developmental time and size at metamorphosis in *Hyla pseudopuma. Copeia* 1989:794–97.

———. 1989b. Life history consequences of feeding versus non-feeding in a facultatively non-feeding toad larva. *Oecologia* 78:486–89.

———. 1990. Possible enhancement of growth in tadpoles through cannibalism. *Copeia* 1990:560–64.

———. 1991. Choice of oviposition site and egg load assessment by a treefrog. *Herpetologica* 47:308–15.

———. 1992. Cannibalism in amphibians. In *Cannibalism: Ecology and evolution among diverse taxa,* ed. M. A. Elgar and B. J. Crespi, 256–76. New York: Oxford University Press.

———. 1995a. *Leptodactylus bufonius* (NCN). Reproduction. *Herp. Rev.* 26:97–98.

———. 1995b. Parental care. In *Amphibian biology. Vol. 2: Social behaviour,* ed. H. Heatwole and B. K. Sullivan, 518–67. Chipping Norton, New South Wales, Australia: Surrey Beatty and Sons.

———. 1996. Parental care among the Amphibia. *Adv. Stud. Behav.* 25:109–44.

———. 2002. Natural history of Darwin's frog, *Rhinoderma darwinii. Herp. Nat. Hist.* 9:21–30.

———. 2003. Conservation of amphibians in the New World tropics. In *Amphibian conservation,* ed. R. D. Semlitsch, 53–69. Washington, DC: Smithsonian Books.

Crump, M. L., F. R. Hensley, and K. L. Clark. 1992. Apparent decline of the golden toad: Underground or extinct? *Copeia* 1992:413–20.

Crump, M. L., and R. H. Kaplan. 1979. Clutch energy partitioning of tropical tree frogs (Hylidae). *Copeia* 1979:626–35.

Crump, M. L., and J. A. Pounds. 1985. Lethal parasitism of an aposematic anuran (*Atelopus varius*) by *Notochaeta bufonivora* (Diptera: Sarcophagidae). *J. Parasit.* 71:588–91.

———. 1989. Temporal variation in the dispersion of a tropical anuran. *Copeia* 1989:209–11.

Crump, M. L., and D. S. Townsend. 1990. Random mating by size in a Neotropical treefrog, *Hyla pseudopuma. Herpetologica* 46:383–86.

Crump, M. L., and M. Vaira. 1991. Vulnerability of *Pleurodema borelli* tadpoles to an avian predator: Effect of body size and density. *Herpetologica* 47:316–21.

Cruz, C. A. G. da. 1982. Conceituação do grupos de especies de Phyllomedusinae brasiliarias com base em caracteres larvários (Amphibia, Anura, Hylidae). *Arq. Univ. Fed. Rur. Rio de Janeiro* 5:147–71.

———. 1990. Sobre as relações intergenéricas de Phyllomedusinae da Floresta Atlântica (Amphibia, Anura, Hylidae). *Rev. Brasil. Biol.* 50:709–26.

Cruz, C. A. G. da, and O. L. Peixoto. 1984. Especies verdes de *Hyla:* o complexo "*albosignata*" (Amphibia, Anura, Hylidae). *Arq. Univ. Fed. Rural Rio de Janeiro* 7:31–47.

Cummings, B. F. 1912. Distant orientation in Amphibia. *Proc. Zool. Soc. London* 1912:8–19.

Cummings, J. W. 1979. Physiological and biochemical adaptations to training in *Rana pipiens. J. Comp. Physiol.* 134:345–50.

Cummins, C. P. 1986. Temporal and spatial variation in egg size and fecundity in *Rana temporaria. J. Anim. Ecol.* 55:303–16.

———. 1989. Interaction between the effects of pH and density on growth and development in *Rana temporaria* L. tadpoles. *Funct. Ecol.* 3:45–52.

———. 2003. UV-B radiation, climate change and frogs—The importance of phenology. *Ann. Zool. Fennici* 40:61–67.

Cummins, C. P., P. D. Gleenslade, and A. R. McLeod. 1999. A test of the effect of supplemental UV-B radiation on the common frog, *Rana temporaria* L., during embryonic development. *Global Change Biol.* 5:471–79.

Cummins, C. P., and M. J. S. Swan. 1995. Variation in reproductive characteristics of the stream frog *Colostethus trinitatis* on the island of Trinidad. *J. Trop. Ecol.* 11:603–18.

Cunjak, R. A. 1986. Winter habitat of northern leopard frogs, *Rana pipiens,* in a southern Ontario stream. *Can. J. Zool.* 64:255–57.

Cunningham, A. A., T. E. S. Langton, P. M. Bennett, J. F. Lewin, S. E. N. Drury, R. E. Gough, and S. K. MacGregor. 1996. Pathological and microbiological findings from incidents of unusual mortality of the common frog (*Rana temporaria*). *Phil. Trans. Roy. Soc. London B* 351:1539–57.

Cunningham, J. D. 1960. Aspects of the ecology of the Pacific slender salamander, *Batrachoseps pacificus,* in southern California. *Ecology* 41:88–99.

Cupp, P. V. 1980. Territoriality in the green salamander. *Copeia* 1980:263–468.

Cupp, P. V., Jr. 1994. Salamanders avoid chemical cues from predators. *Anim. Behav.* 48:232–35.

Curio, E. 1976. *The ethology of predation.* New York: Springer-Verlag.

Curran, L. M., S. N. Trigg, A. K. McDonald, D. Astiani, Y. M. Hardiono, P. Siregar, I. Caniago, and E. Kasischke. 2004. Lowland forest loss in protected areas of Indonesian Borneo. *Science* 303:1000–1003.

Currie, W., and E. D. Bellis. 1969. Home range and movements of the bullfrog, *Rana catesbeiana* Shaw, in an Ontario pond. *Copeia* 1969:688–92.

Curtis, B., K. S. Roberts, M. Griffin, S. Bethune, C. J. Hay, and H. Kolberg. 1998. Species richness and conservation of Namibian freshwater macro-invertebrates, fish and amphibians. *Biodiversity and Conservation* 7:447–66.

Curtis, J. M. R., and E. B. Taylor. 2003. The genetic structure of coastal giant salamanders (*Dicamptodon tenebrosus*) in a managed forest. *Biol. Cons.* 115:45–54.

Cushman, J. R., G. C. Packard, and T. J. Boardman. 1976. Concentrations of lactic acid in neotenic and transformed tiger salamanders (*Ambystoma tigrinum*) before and after activity. *J. Comp. Physiol.* B 112:273–81.

Cushman, S. A. 2006. Effects of habitat loss and fragmentation on amphibians: A review and prospectus. *Biol. Cons.* 128:231–40.

Czeczuga, B., E. Myszyńska, and A. Krzemińska. 1998. Aquatic fungi growing on the spawn of certain amphibians. *Amphibia-Reptilia* 19:239–51.

Czeloth, H. 1930. Untersuchungen über die Raumorientierung von Triton. *Z. Vergl. Physiol.* 13:74–63.

Czopek, G., and J. Czopek. 1955. The vascularization of respiratory surfaces in *Bombina variegata* (L.). *Bull. Acad. Polon. Sci., Ser. Sci. Biol.* 3:313–18.

———. 1959. Vascularization of respiratory surfaces in *Bufo viridis* Laur. and *Bufo calamita* Laur. *Bull. Acad. Polon. Sci., Ser. Sci. Biol.* 7:39–45

Czopek, G., J. Czopek, and H. Szarski. 1962. The distribution of capillaries in the respiratory surfaces of *Acris crepitans* Baird (Amphibia, Salientia). *Acta Biol. Cracov (Zool.)* 5:165–73.

Czopek, J. 1955a. Vascularization of respiratory surfaces in *Leiopelma hochstetteri* Fitzinger and *Xenopus laevis* Daudin. *Acta Anat.* 25:346–60.

———. 1955b. The vascularization of respiratory surfaces of some Salientia. *Zool. Pol.* 6:101–34.

———. 1957. The vascularization of respiratory surfaces in *Ambystoma mexicanum* Cope in ontogeny. *Zool. Pol.* 8:131–49.

———. 1959a. Skin and lung capillaries in European common newts. *Copeia* 1959:91–96.

———. 1959b. Surface of skin and lung capillaries of amphibians. *Bull. Acad. Polon. Sci.,bSer. Sci. Biol.* 7:479–84.

———. 1959c. Vascularization of respiratory surfaces in *Salamandra salamandra* L. in ontogeny. *Bull. Acad. Polon. Sci., Ser. Sci. Biol.* 7:473–78.

———. 1961. Vascularization of respiratory surfaces of some Plethodontidae. *Zool. Pol.* 11:131–48.

———. 1962. Vascularization of respiratory surfaces in some Caudata. *Copeia* 1962:576–87.

———. 1965. Quantitative studies on the morphology of respiratory surfaces in amphibians. *Acta Anat.* 62:296–323.

da Silva , H. R., M. C. de Britto-Pereira, and U. Caramaschi. 1989. Frugivory and seed dispersal by *Hyla truncata,* a Neotropical treefrog. *Copeia* 1989:781–82.

Dabelsteen, T., P. K. McGregor, M. Shepherd, X. Whittaker, and S. B. Pedersen. 1996. Is the signal value of overlapping different from that of alternating during matched singing in great tits? *J. Avian Biol.* 27:189–94.

Daechler, E. B., N. H. Shubin, and F. A. Jenkins, Jr. 2006. A Devonian tetrapod-like fish and the evolution of the tetrapod body plan. *Nature* 440:757–63.

Dalbeck, L., M. Hachtel, A. Hayd, K. Schäfer, M. Schäfer, and K. Weddeling. 1997. Amphibien im Rhein-Sieg-Kreis und in der Stadt Bonn: Verbreitung, Gewässerpraferenzen, Vergesellschaftung und Gefährdung. *Decheniana* 150:235–92.

Dalrymple, G. H. 1970. Caddis fly larvae feeding upon eggs of *Ambystoma t. tigrinum. Herpetologica* 26:128–29.

Daltry, J. C., W. Wuster, and R. S. Thorpe. 1998. Intraspecific variation in the feeding ecology of the crotaline snake *Calloselasma rhodostoma* in Southeast Asia. *J. Herpetol.* 32:198–205.

Daly, J. W. 1995. The chemistry of poisons in amphibian skin. *Proc. Natl. Acad. Sci. USA* 92:9–13.

———. 1998. Thirty years of discovering arthropod alkaloids in amphibian skin. *J. Nat. Prod.* 61:162–72.

———. 2003. Ernest Guenther Award in Chemistry of Natural Products. Amphibian skin: A remarkable source of biologically active arthropod alkaloids. *J. Med. Chem.* 46:445–52.

Daly, J. W., N. R. Andriamaharavo, M. Andriantsiferana, and C. W. Meyers. 1996. Madagascan poison frogs (*Mantella*) and their skin alkaloids. *Amer. Mus. Novitates* (3177):1–34.

Daly, J. W., G. B. Brown, M. Mensah-Dwumah, and C. W. Myers. 1978. Classification of skin alkaloids from Neotropical poison-dart frogs (Dendrobatidae). *Toxicon* 16:163–88.

Daly, J. W., J. Caceres, R. W. Moni, F. Gusovsky, M. Moos, Jr., K. B. Seamon, K. Milton, and C. W. Myers. 1992. Frog secretions and hunting magic in the upper Amazon: Identification of a peptide that interacts with adenosine receptor. *Proc. Natl. Acad. Sci. USA* 89:10960–63.

Daly, J. W., H. M. Garraffo, G. S. E. Hall, and J. F. Cover. 1997. Absence of skin alkaloids in captive-raised Madagascan mantelline frogs (*Mantella*) and sequestration of dietary alkaloids. *Toxicon* 35:1131–35.

Daly, J. W., H. M. Garraffo, P. Jain, T. F. Spande, R. R. Snelling, C. Jaramillo, and A. S. Rand. 2000. Arthropod-frog connection: Decahydroquinoline and pyrrolizidine alkaloids common to microsympatric myrmicine ants and dendrobatid frogs. *J. Chem. Ecol.* 26:73–85.

Daly, J. W., H. M. Garraffo, and C. W. Myers. 1997. The origin of frog skin alkaloids: an enigma. *Pharm. News* 4:9–14.

Daly, J. W., H. M. Garraffo, and T. F. Spande. 1993. Amphibian alkaloids. In *The alkaloids,* ed. G. F. Cordell, 185–288. San Diego: Academic Press.

———. 1999. Alkaloids from amphibian skin. In *Alkaloids: Chemical and biological perspectives,* ed. S. W. Pelletier, 1–161. New York: Pergamon.

Daly, J. W., H. M. Garraffo, T. F. Spande, V. C. Clarke, J. Ma, H. Ziffer, and J. F. Cover, Jr. 2003. Evidence for an enantioselective pumiliotoxin 7-hydroxylase in dendrobatid poison frogs of the genus *Dendrobates. Proc. Natl. Acad. Sci. USA* 100:11092–97.

Daly, J. W., H. M. Garraffo, T. F. Spande, C. Jaramillo, and A. S. Rand. 1994. Dietary source for skin alkaloids of poison frogs (Dendrobatidae). *J. Chem. Ecol.* 20:943–55.

Daly, J. W., R. J. Highet, and C. W. Myers. 1984. Occurrence of skin alkaloids in non-dendrobatid frogs from Brazil (Bufonidae), Australia (Myobatrachidae) and Madagascar (Mantellinae). *Toxicon* 22:905–19.

Daly, J. W., T. Kaneko, J. Wilham, H. M. Garraffo, T. F. Spande, A. Espinosa, and M. A. Donnelly. 2002. Bioactive alkaloids of frog skin: Combinatorial bioprospecting reveals that pumiliotoxins have an arthropod source. *Proc. Natl. Acad. Sci. USA* 99:13996–14001.

Daly, J. W., and C. W. Myers. 1967. Toxicity of Panamanian poison frogs (*Dendrobates*): Some biological and chemical aspects. *Science* 156:970–73.

Daly, J. W., W. L. Padgett, R. L. Saunders, and J. F. Cover. 1997. Absence of tetrodotoxins in a captive-raised riparian frog, *Atelopus varius. Toxicon* 35:705–9.

Daly, J. W., S. I. Secunda, H. M. Garraffo, T. F. Spande, A. Wisnieski, and J. F. Cover, Jr. 1994. An uptake system for dietary alkaloids in poison frogs (Dendrobatidae). *Toxicon* 32:657–63.

Daly, J. W., S. I. Secunda, H. M. Garraffo, T. F. Spande, A. Wisnieski, C. Nishihira, and J. F. Cover, Jr. 1992. Variability in alkaloid profiles in Neotropical poison frogs (Dendrobatidae): Genetic versus environmental determinants. *Toxicon* 30:887–98.

Daly, J. W., and B. Witkop. 1971. Chemistry and pharmacology of frog venoms. In *Venomous animals and their venoms,* vol. II, ed. W. Bucherl and E. E. Buckley, 497–519. New York: Academic Press.

Daniel, J. D., Jr., and R. W. Simpson. 1954. A negative note on antibiotics. *Herpetologica* 10:16.

Daniels, R. J. R. 1992. Geographical distribution pattern of amphibians in Western Ghats. *J. Biogeogr.* 19:521–29.

———. 1997a. A field guide to the frogs and toads of the Western Ghats, India. Part I. *Cobra* 27:1–25.

———. 1997b. A field guide to the frogs and toads of the Western Ghats, India. Part II. *Cobra* 28:1–24.

———. 1997c. A field guide to the frogs and toads of the Western Ghats, India. Part III. *Cobra* 29:1–13.

———. 1997d. Taxonomic uncertainties and conservation assessment of the Western Ghats. *Curr. Sci.* 73:169–70.

———. 1999. Status of amphibian diversity in India. *Cobra* 35–36:25–28.

———. 2003. Impact of tea cultivation on anurans in the Western Ghats. *Curr. Sci.* 85:1415–22.

Danstedt, R. T., Jr. 1975. Local geographic variation in demographic parameters and body size of *Desmognathus fuscus* (Amphibia: Plethodontidae). *Ecology* 56:1054–67.

Darlington, P. J., Jr. 1957. *Zoogeography.* New York: Wiley.

Darst, C. R., and D. C. Cannatella. 2004. Novel relationships among hyloid frogs inferred from 12S and 16S mitochondrial DNA sequences. *Mol. Phylogen. Evol.* 31:462–75.

Darst, C. R., and M. E. Cummings. 2006. Predator learning favours mimiry of a less-toxic model in poison frogs. *Nature* 440:208–11.

Darst, C. R., M. E. Cummings, and D. C. Cannatella. 2006. A mechanism for diversity in warning signals: Conspicuousness versus toxity in poison frogs. *Proc. Natl. Acad. Sci. USA* 103:5852–57.

Darst, C. R., P. A. Menéndez-Guerrero, L. A. Coloma, and D. C. Cannatella. 2005. Evolution of dietary specialization and chemical defense in poison frogs (Dendrobatidae): A comparative analysis. *Am. Nat.* 165:56–69.

Darwin, C. R. 1871. *The descent of man and selection in relation to sex.* London: John Murray.

———. 1874. *The descent of man and selection in relation to sex* (2nd ed.). London: John Murray.

Darwin, E. 1803. *The temple of nature, or the origin of society: A poem.* London: J. Johnson.

Das, I. 1995. Comparative morphology of the gastrointestinal tract in relation to diet in frogs from a locality in south India. *Amphibia-Reptilia* 16:289–93.

Das, I. 1996a. Folivory and seasonal changes in diet in *Rana hexadactyla* (Anura: Ranidae). *J. Zool., London* 238:785–94.

———. 1996b. Reproductive strategy and lipid mobilization in *Rana hexadactyla* at a seasonal locality. *Ethol. Ecol. Evol.* 8:125–33.

———. 1996c. Resource use and foraging tactics in a south Indian amphibian community. *J. South Asian Nat. Hist.* 2:1–30.

———. 1996d. Spatio-temporal resource utilization by a Bornean rainforest herpetofauna: preliminary results. In *Tropical rainforest research.* Current Issues, ed. D. S. Edwards, W. E. Booth, and S. C. Choy, 315–23. Boston: Kluwer Academic.

Das, I., and M. Coe. 1994. Dental morphology and diet in anuran amphibians from south India. *J. Zool., London* 233:417–27.

Das, I., and K. Kunte. 2005. New species of *Nyctibatrachus* (Anura: Ranidae) from Castle Rock, Karnataka State, Southwest India. *J. Herpetol.* 39:465–70.

Da Silva, N. J., Jr., I. França e Souza, W. vaz Silva, R. S. Ribeiro, M. C. da Costa, N. C. do Valle, and H. L. Rodrigues Silva. 2003a. *Helicops angulatus* (water-snake). Diet. *Herp. Rev.* 34:68.

Da Silva, N. J., Jr., I. França e Souza, W. vaz Silva, and H. L. Rodrigues Silva. 2003b. *Liophis poecilogyrus* (trash-snake). Diet. *Herp. Rev.* 34:69–70.

Daszak, P., A. A. Cunningham, and A. D. Hyatt. 2000. Emerging infectious diseases of wildlife—Threats to biodiversity and human health. *Science* 287:443–49.

Daszak, P., D. E. Scott, A. M. Kilpatrick, C. Faggioni, J. W. Gibbons, and D. Porter. 2005. Amphibian population declines at Savannah River Site are linked to climate, not chytridiomycosis. *Ecology* 86:3232–37.

Daszak, P., A. Strieby, A. A. Cunningham, J. E. Longore, C. C. Brown, and D. Porter. 2004. Experimental evidence that the bullfrog (*Rana catesbeiana*) is a potential carrier of chytridiomycosis, an emerging fungal disease of amphibians. *Herpetol. J.* 14:201–7.

Daugherty, C. H., and A. L. Sheldon. 1982. Age-determination, growth, and life history of a Montana population of the tailed frog (*Ascaphus truei*). *Herpetologica* 38:461–68.

Davic, R. D. 1991. Ontogenetic shift in diet of *Desmognathus quadramaculatus. J. Herpetol.* 25:108–11.

Davic, R. D., and H. H. Welsh, Jr. 2004. On the ecological role of salamanders. *Ann. Rev. Ecol. Evol. Syst.* 35:405–34.

David, R. S., and R. G. Jaeger. 1981. Prey location through chemical cues by a terrestrial salamander. *Copeia* 1981:435–40.

Davidson, C. 2004. Declining downwind: Amphibian population declines in California and historical pesticide use. *Ecol. Appl.* 14:1892–1902.

Davidson, C., H. B. Shaffer, and M. B. Jennings. 2001. Declines of the California red-legged frog: Climate, UV-B, habitat, and pesticides hypotehses. *Ecol. Appl.* 11:464–79.

———. 2002. Spatial tests of the pesticide drift, habitat destruction, UV-B, and climate-change hypotheses for California amphibian declines. *Cons. Biol.* 16:1588–1601.

Davidson, D. W., D. A. Samson, and R. S. Inouye. 1985. Granivory in the Chihuahuan desert: Interactions within and between trophic levels. *Ecology* 66:486–502.

Davidson, E. H., and B. R. Hough. 1969. Synchronous oogenesis in *Engystomops pustulosus,* a Neotropic anuran suitable for laboratory studies: Localization in the embryo of RNA synthesized at the lampbrush stage. *J. Exp. Zool.* 172:25–48.

Davidson, E. W., M. Parris, J. P. Collins, J. E. Longcore, A. P. Pessier, and J. Brunner. 2003. Pathogenicity and transmission of chytridiomycosis in tiger salamanders (*Ambystoma tigrinum*). *Copeia* 2003:601–7.

Davies, M. 1989. Developmental biology of the Australopapuan hylid frog *Litoria eucnemis* (Anura: Hylidae). *Trans. Roy. Soc. South Australia* 113:215–20.

———. 1991. Descriptions of the tadpoles of some Australian

limnodynastine hylid frogs. *Trans. Roy. Soc. South Australia* 115:67–76.

Davies, M., and M. J. Littlejohn. 1986. Frogs of the genus *Uperoleia* Gray (Anura: Leptodactylidae) in south-eastern Australia. *Trans. Roy. Soc. South Australia* 110:111–43.

Davies, M., K. R. McDonald, and C. Corben. 1986. The genus *Uperoleia* Gray (Anura: Leptodactylidae) in Queensland, Australia. *Proc. Roy. Soc. Victoria* 98:147–88.

Davies, M., and S. J. Richards. 1990. Developmental biology of the Australian hylid frog *Nyctimystes dayi* (Günther). *Trans. Roy. Soc. South Australia* 114:207–11.

Davies, N. B., and A. Lundberg. 1984. Food distribution and a variable mating system in the dunnock (*Prunella modularis*). *J. Anim. Ecol.* 53:895–912.

Davies, N. B., and T. R. Halliday. 1977. Optimal mate selection in the toad *Bufo bufo*. *Nature* 269:56–58.

———. 1978. Deep croaks and fighting assessment in toads, *Bufo bufo*. *Nature* 274:683–85.

———. 1979. Competitive mate searching in male common toads, *Bufo bufo*. *Anim. Behav.* 27:1253–67.

Davis, J. 1952. Observations on the eggs and larvae of the salamander *Batrachoseps pacificus* major. *Copeia* 1952:272–74.

Davis, J. E., J. R. Spotila, and W. C. Schefler. 1980. Evaporative water loss from the American alligator, *Alligator mississippiensis:* The relative importance of respiratory and cutaneous components and the regulatory role of the skin. *Comp. Biochem. Physiol.* 67A:439–46.

Davis, J. R., and B. H. Brattstrom. 1975. Sounds produced by the California newt, *Taricha torosa. Herpetologica* 31:409–12.

Davis, M. S. 1987. Acoustically mediated neighbor recognition in the North American bullfrog, *Rana catesbeiana. Behav. Ecol. Sociobiol.* 21:185–90.

Davis, R. A., and R. H. L. Disney. 2003. Natural history and description of *Aphiura breviceps* Schmitz, a scuttle fly (Diptera: Phoridae) whose larvae prey on the eggs of frogs (Anura: Myobatrachidae) in Western Australia. *Aust. J. Entomol.* 42:18–21.

Davis, R. A., and J. D. Roberts. 2005. Embryonic survival and egg numbers in small and large populations of the frog *Heleioporus albopunctatus* in Western Australia. *J. Herpetol.* 39:133–38.

Davis, S. L., R. B. Davis, A. James, and B. C. Pregger Talyn. 2000. Reproductive behavior and larval development of *Leptodactylus fallax* in Dominica, West Indies. *Herp. Rev.* 31:217–20.

Davis, T. M. 2002a. An ethogram of intraspecific agonistic and display behavior for the wandering salamander, *Aneides vagrans. Herpetologica* 58:371–82.

———. 2002b. Microhabitat use and movements of the wandering salamander, *Aneides vagrans,* on Vancouver Island, British Columbia, Canada. *J. Herpetol.* 36:699–703.

———. 2003. *Aneides vagrans* (Wandering Salamander). Reproduction. *Herp. Rev.* 34:133.

Davis, T. M., and K. Ovaska. 2001. Individual recognition of amphibians: Effect of toe clipping and fluorescent tagging on the salamander *Plethodon vehiculum. J. Herpetol.* 35:217–25.

Davis, W. 1988. *Passage of darkness. The ethnobiology of the Haitian zombie.* Chapel Hill: University of North Carolina Press.

Davis, W., and A. T. Weil. 1992. The identity of a New World psychoactive toad. *Ancient Mesoamerica* 3:51–59.

Davis, W. C., and V. C. Twitty. 1964. Courtship behavior and reproductive isolation in the species of *Taricha* (Amphibia, Caudata). *Copeia* 1964:601–10.

Davison, J. 1955. Body weight, cell surface, and metabolic rate in anuran Amphibia. *Biol. Bull.* 109:407–19.

Dawkins, R. 1976. *The selfish gene.* New York: Oxford University Press.

Dawkins, R. 1980. Good strategy or evolutionarily stable strategy? In *Sociobiology: Beyond nature/nurture?,* ed. G. Barlow and J. Silverberg, 331–67. Boulder, CO: Westview Press.

Dawkins, R., and T. R. Carlisle. 1976. Parental investment, mate desertion and a fallacy. *Nature* 262:131–33.

Dawkins, R., and J. R. Krebs. 1978. Animal signals: Information or manipulation? In *Behavioural ecology: An evolutionary approach,* ed. J. R. Krebs and N. B. Davies, 282–309. Sunderland, MA: Sinaue.

Dawley, E. M. 1984a. Identification of sex through odors by male red-spotted newts, *Notophthalmus viridescens. Herpetologica* 40:101–5.

———. 1984b. Recognition of individual, sex, and species odors by salamanders of the *Plethodon glutinosus - P. jordani* complex. *Anim. Behav.* 32:353–61.

———. 1986a. Behavioral isolating mechanisms in sympatric terrestrial salamanders. *Herpetologica* 42:156–64.

———. 1986b. Evolution of chemical signals as a premating isolating mechanism in a complex of terrestrial salamanders. In *Chemical signals in vertebrates 4. Ecology, evolution, and comparative biology,* ed. D. Duvall, D. Müller-Schwarze, and R. M. Silverstein, 221–24. New York: Plenum.

———. 1987. Species discrimination between hybridizing and non-hybridizing terrestrial salamanders. *Copeia* 1987:924–31.

———. 1992a. Correlation of salamander vomeronasal and main olfactory system anatomy with habitat and sex: Behavioral interpretations. In *Chemical signals in vertebrates 6,* ed. R. L. Doty and D. Müller-Schwarze, 403–9. New York: Plenum.

———. 1992b. Sexual dimorphism in a chemosensory system: The role of the vomeronasal organ in salamander reproductive behavior. *Copeia* 1992:113–20.

———. Olfaction. In *Amphibian biology, vol. 3: Sensory biology,* ed. H. Heatwole and E. M. Dawley, 711–42. Chipping Norton, New South Wales, Australia: Surrey Beatty and Sons.

Dawley, E. M., and A. H. Bass. 1988. Organization of the vomeronasal organ in a plethodontid salamander. *J. Morphol.* 198:243–55.

———. 1989. Chemical access to the vomeronasal organs of a plethodontid salamander. *J. Morphol.* 200:163–74.

Dawley, E. M., and J. Crowder. 1995. Sexual and seasonal differences in the vomeronasal epithelium of the red-backed salamander (*Plethodon cinereus*). *J. Comp. Neurol.* 359:382–90.

Dawley, E. M., and R. M. Dawley. 1986. Species discrimination by chemical cues in a unisexual-bisexual complex of salamanders. *J. Herpetol.* 20:114–16.

Dawley, E. M., A. Fingerlin, D. Hwang, S. S. John, and C. A. Stankiewicz. 2000. Seasonal cell proliferation in the chemosensory epithelium and brain of red-backed salamanders, *Plethodon cinereus. Brain Behav. Evol.* 56:1–13.

Dawson, W. R., V. H. Shoemaker, and P. Licht. 1966. Evaporative water loss of some small Australian lizards. *Ecology* 47:589–94.

Day, T., and L. Rowe. 2002. Developmental thresholds and the evolution of reaction norms for age and size at life-history transitions. *Am. Nat.* 159:338–50.

Dayton, G. H., and L. A. Fitzgerald. 2001. Competition, predation, and the distribution of four desert anurans. *Oecologia* 129:430–35.

Dayton, G. H., and R. E. Jung. 1999. *Scaphiopus couchii* (Couch's spadefoot). Predation. *Herp. Rev.* 30:164.

de Andrade, D. V., and A. S. Abe. 1997. Evaporative water loss and oxygen uptake in two casque-headed tree frogs, *Aparasphenodon brunoi* and *Corythomantis greeningi* (Anura, Hylidae). *Comp. Biochem. Physiol.* 118A:685–89.

de Costa, J., M. T. Lozano, B. Agullero, and P. Mendiola. 1989. Histochemical determination of muscle fiber types in locomotor muscles of anuran amphibians. Pflügers Arch. Eur. *J. Physiol.* 414:S27.

De Graaff, R. M. 1991. *The book of the toad.* Rochester, VT: Park Street Press.

de Jesus, E. G. T., J. D. Toledo, and M. S. Simpas. 1998. Thyroid hormones promote early metamorphosis in grouper (*Epinephelus coioides*) larvae. *Gen. Comp. Endocrinol.* 112:10–16.

De Jongh, A. W. J. J. 1988. The lack of toad remains in otter *Lutra lutra* spraints. *Lutra* 31:79–81.

De Jongh, H. J. 1968. Functional morphology of the jaw apparatus of larval and metamorphosing *Rana temporaria* L. *Netherlands J. Zool.* 18:1–103.

De Jongh, H. J., and C. Gans. 1969. On the mechanism of respiration in the bullfrog, *Rana catesbeiana:* A reassessment. *J. Morphol.* 127:259–90.

de la Riva, I. 1994. An undescribed defensive mechanism in the neotropical hylid frog *Phrynohyas coriacea. Amphibia-Reptilia* 15:226–27.

———. 1995. A new reproductive mode for the genus *Adenomera* (Amphibia: Anura: Leptodactylidae): Taxonomic implications for certain Bolivian and Paraguayan populations. *Stud. Neotrop. Fauna Env.* 30:15–29.

de la Riva, I., and J. Köhler. 1998. A new minute leptodactylid frog, genus *Phyllonastes,* from humid montane forests of Bolivia. *J. Herpetol.* 32:325–29.

De Queiroz, A., C. Henke, and H. M. Smith. 2001. Geographic variation and ontogenetic change in the diet of the Mexican Pacific lowlands garter snake, *Thamnophis validus. Copeia* 2001:1034–42.

De Queiroz, K., and J. Gauthier. 1990. Phylogeny as a central principle in taxonomy: Phylogenetic definitions of taxon names. *Syst. Zool.* 39:307–22.

———. 1992. Phylogenetic taxonomy. *Ann. Rev. Ecol. Syst.* 23:449–80.

de Sa, R. O., and D. M. Hillis. 1990. Phylogenetic relationships of the pipid frogs *Xenopus* and *Silurana:* an integration of ribosomal DNA and morphology. *Mol. Biol. Evol.* 7:365–76.

De Saint-Aubain, M. L. 1981. Shunts in the gill filament in tadpoles of *Rana temporaria* and *Bufo bufo* (Amphibia, Anura). *J. Exp. Zool.* 217:143–45.

———. 1982. Vagal control of pulmonary blood flow in *Ambystoma tigrinum. J. Exp. Zool.* 221:155–58.

———. 1985. Blood flow patterns of the respiratory systems in larval and adult amphibians: Functional morphology and phylogenetic significance. *Z. Zool. Syst. Evol.-forsch.* 23:229–40.

De Saint-Aubain, M. L., and K. G. Wingstrand. 1979. A sphincter in the pulmonary artery of the frog *Rana temporaria* and its influence on blood flow in skin and lungs. *Acta Zool. (Stockholm)* 60:163–72.

de Solla, S. R., K. E. Pettit, C. A. Bishop, K. M. Cheng, and J. E. Elliot. 2002. Effects of agricultural runoff on native amphibians in the Lower Fraser River Valley, British Columbia, Canada. *Env. Tox. Chem.* 21:353–60.

de Solla, S. R, C. A. Bishop, K. E. Pettit, and J. E. Elliott. 2002. Organochlorine pesticides and polychlorinated biphenyls (PCBs) in eggs of red-legged frogs (*Rana aurora*) and northwestern salamanders (*Ambystoma gracile*) in an agricultural landscape. *Chemosphere* 46:1027–32.

de Sousa, R. C., and A. Grosso. 1982. Osmotic water flow across the abdominal skin of the toad *Bufo marinus:* Effect of vasopressin and isoprenaline. *J. Physiol.* 329:281–96.

De Villiers, A. L. 2004. *Xenopus gilli* Rose and Hewitt, 1927. In *Atlas and red data book of the frogs of South Africa, Lesotho and Swaziland.* SI/MAB Series #9, ed. L. R. Minter, M. Burger, J. A. Harrison, H. H. Braack, P. J. Bishop, and D. Kloepfer, 260–63. Washington, DC: Smithsonian Institution.

De Vlaming, V. L., and R. B. Bury. 1970. Thermal selection in tadpoles of the tailed-frog, *Ascaphus truei. J. Herpetol.* 4:179–89.

De Wijer, P., P. J. Watt, and R. S. Oldham. 2003. Amphibian decline and aquatic pollution: Effects of nitrogenous fertilizer on survival and development of larvae of the frog *Rana temporaria. Appl. Herpetol.* 1:3–12.

de Witte, G. F. 1941. Batraciens et reptiles. *Explor. Parc. Nat. Albert Miss. G. F. de Witte* 33:1–261.

De Zwaan, and G., Thillart. 1985. Low and high power output modes of anaerobic metabolism: Invertebrate and vertebrate strategies. In *Circulation, respiration, and metabolism,* ed. R. Gilles, 166–92. New York: Springer-Verlag.

Deban, S. M., and S. B. Marks. 2002. Metamorphosis and evolution of feeding behaviour in salamanders of the family Plethodontidae. *Zool. J. Linn. Soc.* 134:375–400.

Deban, S. M., and K. C. Nishikawa. 1992. The kinematics of prey capture and the mechanism of tongue protraction in the green tree frog *Hyla cinerea. J. Exp. Biol.* 170:235–56.

Deban, S. M., and W. M. Olson. 2002. Suction feeding by a tiny predatory tadpole. *Nature* 420:41–42.

Deban, S. M., J. C. O'Reilly, and K. C. Nishikawa. 2001. The evolution of motor control of feeding in amphibians. *Amer. Zool.* 41:1280–98.

Deban, S. M., and D. B. Wake. 2000. Aquatic feeding in salamanders. In *Feeding: Form, function, and evolution in tetrapod vertebrates,* ed. K. Schwenk, 65–94. San Diego: Academic Press.

DeBenedictus, P. A. 1974. Interspecific competition between tadpoles of *Rana pipiens* and *Rana sylvatica:* An experimental field study. *Ecol. Monogr.* 44:129–51.

Deevey, E. S., Jr. 1947. Life tables for natural populations of animals. *Q. Rev. Biol.* 22:283–314.

Defler, T. R. 1979. On the ecology and behavior of *Cebus albifrons* in eastern Colombia. I. Ecology. *Primates* 20:475–90.

Degani, G. 1981. Salinity tolerance and osmoregulation in *Salamandra salamandra* (L.) from different populations. *J. Comp. Physiol.* 145:133–37.

———. 1982. Water balance of salamanders (*Salamandra salamandra*) L.) from different habitats. *Amphibia-Reptilia* 2:309–14.

———. 1984. The selection of hiding places by *Triturus vittatus vittatus* (Jenyns). *Biol. Behav.* 9:235–42.

———. 1985a. Effect of arginine vasotocin on plasma concentration and water tissues in terrestrial toad *Bufo viridis* and semiaquatic frog *Rana ridibunda. Comp. Biochem. Physiol.* 80A:75–77.

———. 1985b. Urea tolerance and osmoregulation in *Bufo viridis* and *Rana ridibunda. Comp. Biochem. Physiol.* 82A:833–36.

———. 1985c. Water balance and body fluids of *Salamandra salamandra* (L.) in their natural habitats in summer and winter. *Comp. Biochem. Physiol.* 82A:479–82.

———. 1986. Growth and behaviour of six species of amphibian larvae in a winter pond in Israel. *Hydrobiologia* 140:5–10.

———. 1993. Cannibalism among *Salamandra salamandra* (L.) larvae. *Israel J. Zool.* 39:125–29.

Degani, G., S. Goldenberg, and M. R. Warburg. 1980. Cannibalistic phenomena in *Salamandra salamandra* larvae in certain water bodies and under experimental conditions. *Hydrobiologia* 75:123–28.

———. 1983. Changes in ion, urea concentration and blood osmolarity of *Pelobates syriacus* juveniles under varying conditions. *Comp. Biochem. Physiol.* 75A:619–23.

Degani, G., and E. Nevo. 1986. Osmotic stress and osmoregulation of tadpoles and juveniles in *Pelobates syriacus. Comp. Biochem. Physiol.* A 83:365–70.

Degani, G., N. Silanikove, and A. Schkolnik. 1984. Adaptation of green toad (*Bufo viridis*) to terrestrial life by urea accumulation. *Comp. Biochem. Physiol.* 77A:585–87.

Degani, G., and M. R. Warburg. 1978. Population structure and seasonal activity of the adult *Salamandra salamandra* (L.) (Amphibia, Urodela, Salamandridae) in Israel. *J. Herpetol.* 12:437–44.

———. 1980. The response to substrate moisture of juvenile and adult *Salamandra salamandra* (L.) (Amphibia: Urodela). *Biol. Behav.* 5:281–90.

———. 1984. Changes in concentration of ions and urea in both plasma and muscle tissue in a dehydrated hylid anuran. *Comp. Biochem. Physiol.* 77A:357–60.

———. 1995. Variations in brood size and birth rates of *Salamandra salamandra* (Amphibia, Urodela) from different habitats in northern Israel. *Amphibia-Reptilia* 16:341–49.

DeGraaf, R. M., and M. Yamasaki. 1992. A nondestructive technique to monitor the relative abundance of terrestrial salamanders. *Wildl. Soc. Bull.* 20:260–64.

———. 2002. Effects of edge contrast on redback salamander distribution in even-aged northern hardwoods. *Forest Sci.* 48:351–63.

Dehadrai, P. V., and S. D. Tripathi. 1976. Environment and ecology of freshwater air-breathing teleosts. In *Respiration of amphibious vertebrates,* ed. G. M. Hughes, 39–72. New York: Academic Press.

Dejean, A., and J.-L. Amiet. 1992. Un cas de myrmécophilie inattendu: La cohabitation de l'anoure *Kassina senegalensis* avec la fourmi *Megaponera foetens. Alytes* 10:31–36.

Dejours, P. 1976. Water versus air as the respiratory media. In *Respiration of amphibious vertebrates,* ed. G. M. Hughes, 1–15. New York: Academic Press.

Del Pino, E. M. 1980. Morphology of the pouch and incubatory integument in marsupial frogs (Hylidae). *Copeia* 1980:10–17.

———. 1989a. Marsupial frogs. *Sci. Amer.* 160:76–84.

———. 1989b. Modifications of oogenesis and development in marsupial frogs. *Development* 107:169–87.

Del Pino, E. M., and B. Escobar. 1981. Embryonic stages of *Gastrotheca riobambae* (Fowler) during maternal incubation and comparison of development with that of other egg-brooding hylid frogs. *J. Morphol.* 167:277–95.

Del Pino, E. M., M. L. Galarza, C. M. de Abuja, and A. A. Humphries, Jr. 1975. The maternal pouch and development in the marsupial frog *Gastrotheca riobambae* (Fowler). *Biol. Bull.* 149:480–91.

Delany, M. F., and L. L. Abercrombie. 1986. American alligator food habits in northcentral Florida. *J. Wildl. Manage.* 50:348–53.

Delfino, G., B. B. Alvarez, R. Brizzi, and J. A. Cespedez. 1998. Serous cutaneous glands of Argentine *Phyllomedusa* Wagler 1830 (Anura: Hylidae): Secretory polymorphism and adaptive plasticity. *Trop. Zool.* 11:333–51.

Delfino, M. 2004. The Middle Pleistocene herpetofauna of Valdemino Cave (Liguria, North-western Italy). *Herpetol. J.* 14:113–28.

———. 2005. The past and future of extant amphibians. *Science* 308:49–50.

Delfino, M., J. C. Rage, and L. Rook. 2003. Tertiary mammal turnover phenomena: What happened to the herpetofauna? *Deinsea* 10:153–61.

Delis, P. R., H. R. Mushinsky, and E. D. McCoy. 1996. Decline of some west-central Florida anuran populations in response to habitat degradation. *Biodivers. Cons.* 5:1579–95.

Delsol, M. 1986. Les types fondamentaux de la reproduction. In *Traite de zoologie. Vol. 14, fasc. I-B (Amphibiens),* ed. P.-P. Grasse and M. Delsol, 321–88. Paris: Masson.

Delsol, M., J. Flatin, J.-M. Exbrayat, and J. Bons. 1981. Développement de *Typhlonectes compressicaudus,* amphibien apode vivipare. Hypotheses sur la nutrition embryonnaire et larvaire par un ectotrophoblaste. *C. R. Acad. Sci., Paris* 293:281–85.

———. 1983. Développement embryonnaire de *Typhlonectes compressicaudus* (Dumeril et Bibron, 1841). Constitution d'un ectotrophoblaste. *Bull. Soc. Zool. France* 108:680–81.

Delsol, M., J.-M. Exbrayat, J. Flatin, and M. Gueydan-Baconnier. 1986. Nutition embryonnaire chez *Typhlonectes compressicaudus* (Dumeril et Bibron, 1841), amphibien apode vivipare. *Mem. Soc. Zool. France* 43:39–54.

Delson, J., and W. G. Whitford. 1973a. Adaptations of the tiger salamander, *Ambystoma tigrinum* to arid habitats. *Comp. Biochem. Physiol.* 46A:631–38.

———. 1973b. Critical thermal maxima in several life history stages in desert and montane populations of *Ambystoma tigrinum. Herpetologica* 29:352–55.

Dely, O. G. 1967. Neure Angaben zur Kenntnis das neotenischer Teichmolchen (*Triturus vulgaris*). *Acta. Zool. Acad. Sci. Hungary* 13:253–70.

Delzell, D. E. 1958. Spatial movement and growth of *Hyla crucifer. Diss. Abstr.* 19:1478.

deMaynadier, P. G., and M. L. Hunter. 1995. The relationship between forest management and amphibian ecology: A review of the North American literature. *Environ. Rev.* 3:230–61.

———. 1998. Effects of silvicultural edges on the distribution and abundance of amphibians in Maine. *Cons. Biol.* 12:340–52.

———. 1999. Forest canopy closure and juvenile emigration by pool-breeding amphibians in Maine. *J. Wildl. Manage.* 63:441–50.

———. 2000. Road effects on amphibian movements in a forested landscape. *Nat. Areas J.* 20:56–65.

Demian, J. J., and D. H. Taylor. 1977. Photoreception and locomotor rhythm entrainment by the pineal body of the newt, *Notophthalmus viridescens* (Amphibia, Urodela, Salamandridae). *J. Herpetol.* 11:131–39.

Denoël, M. 1998. The modulation of movement as a behavioral adaptation to extreme environments in the newt *Triturus alpestris cyreni. J. Herpetol.* 32:623–25.

———. 1999. Le comportement social des urodèles. *Cah. d'Ethol.* 19:221–58.

———. 2002. Paedomorphosis in the alpine newt (*Triturus alpestris*): Decoupling behavioural and morphological change. *Behav. Ecol. Sociobiol.* 52:394–99.

———. 2004. Feeding performance in heterochronic alpine newts is consistent with trophic niche and maintenance of polymorphism. *Ethology* 110:127–36.

Denoël, M., R. Duguet, G. Džukić, M. Kalezić, and S. Mazzotti. 2001. Biogeography and ecology of paedomorphosis in *Triturus alpestris* (Amphibia, Caudata). *J. Biogeogr.* 28:1271–80.

Denoël, M., G. Džukić, and M. Kalezić. 2005. Effects of widespread fish introductions on paedomorphic newts in Europe. *Cons. Biol.* 19:162–70.

Denoël, M., F. Hervant, R. Schabetsberger, and P. Joly. 2002. Short- and long-term advantages of an alternative ontogenetic pathway. *Biol. J. Linn. Soc.* 77:105–12.

Denoël, M., and P. Joly. 2000. Neoteny and progenesis as two heterochronic processes involved in paedomorphosis in *Triturus alpestris* (Amphibia: Caudata). *Proc. Roy. Soc. London B* 267:1481–85.

———. 2001. Adaptive significance of facultative paedomorphosis in *Triturus alpestris* (Amphibia, Caudata): Resource partitioning in an alpine lake. *Freshw. Biol.* 46:1387–96.

Denoël, M., P. Joly, and H. H. Whiteman. 2005. Evolutionary ecology of facultative paedomorphosis in newts and salamanders. *Biol. Rev.* 80:663–71.

Denoël, M., M. Mathieu, and P. Poncin. 2005. Effect of water temperature on the courtship behavior of the alpine newt *Triturus alpestris*. *Behav. Ecol. Sociobiol.* 58:121–27.

Denoël, M., and P. Poncin. 2001. The effect of food on growth and metamorphosis of paedomorphs in *Triturus alpestris aquanus*. *Arch. Hydrobiol.* 152:661–70.

Denoël, M., P. Poncin, and J. C. Ruwet. 2001a. Alternative mating tactics in the alpine newt *Triturus alpestris alpestris*. *J. Herpetol.* 35:62–67.

———. 2001b. Sexual compatibility between two heterochronic morphs in the alpine newt, *Triturus alpestris*. *Anim. Behav.* 62:559–66.

Denoël, M., and R. Schabetsberger. 2003. Resource partitioning in two heterochronic populations of Greek alpine newts, *Triturus alpestris veluchiensis*. *Acta Oecol.* 24:55–64.

Denoël, M., R. Schabetsberger, and P. Joly. 2004. Trophic specialisations in alternative heterochronic morphs. *Naturwissenschaften* 91:81–84.

Dent, J. N. 1968. Survey of amphibian metamorphosis. In *Metamorphosis: A problem in developmental biology*, ed. W. Etkin and L. J. Gilbert, 271–311. New York: Appleton-Century-Crofts.

———. 1970. The ultrastructure of the spermatotheca in the red spotted newt. *J. Morphol.* 132:397–423.

———. 1985. Hormonal interactions in the regulation of migratory movements in urodele amphibians. In *The endocrine system and the environment*, ed. B. K. Follett, S. Ishii, and A. Chandola, 79–84. New York: Springer-Verlag.

Dent, J. N., and J. S. Kirby-Smith. 1963. Metamorphic physiology and morphology of the cave salamander *Gyrinophilus palleucus*. *Copeia* 1963:119–30.

Denton, J., and T. J. C. Beebee. 1991. Palatability of anuran eggs and embryos. *Amphibia-Reptilia* 12:111–14.

———. 1993a. Reproductive strategies in a female-biased population of natterjack toads, *Bufo calamita*. *Anim. Behav.* 46:1169–75.

———. 1993b. Summer and winter refugia of natterjacks (*Bufo calamita*) and common toads (*Bufo bufo*) in Britain. *Herpetol. J.* 3:90–94.

———. 1996. Double-clutching by natterjack toads *Bufo calamita* at a site in southern England. *Amphibia-Reptilia* 17:159–67.

Denver, R. J. 1993. Acceleration of anuran amphibian metamorphosis by corticotropin-releasing hormone-like peptide. *Gen. Comp. Endocrinol.* 91:38–51.

———. 1995. Environment-neuroendocrine interactions in the control of amphibian metamorphosis. *Netherlands J. Zool.* 45:195–200.

———. 1996. Neuroendocrine control of amphibian metamorphosis. In *Metamorphosis. Postembryonic programming of gene expression in amphibian and insect cells*, ed. L. I. Gilbert, J. R. Tata, and B. G. Atkinson, 433–64. San Diego: Academic Press.

———. 1997a. Environmental stress as a developmental cue: Corticotropin-releasing hormone is a proximate mediator of adaptive phenotypic plasticity in amphibian metamorphosis. *Horm. Behav.* 31:169–79.

———. 1997b. Proximate mechanisms of phenotypic plasticity in amphibian metamorphosis. *Amer. Zool.* 37:172–84.

———. 1998. Hormonal correlates of environmentally induced metamorphosis in the western spadefoot toad, *Scaphiopus hammondii*. *Gen. Comp. Endocrinol.* 110:326–36.

Denver, R. J., K. A. Glennemeier, and G. C. Boorse. 2002. Endocrinology of complex life cycles: Amphibians. In *Hormones, brain, and behavior*, vol. 2, ed. D. W. Pfaff, A. P. Arnold, A. M. Etgen, S. E. Fahrback, and R. T. Rubin, 469–513. San Diego: Academic Press.

Denver, R. J., N. Mirhadi, and M. Phillips. 1998. Adaptive plasticity in amphibian metamorphosis: Responses of *Scaphiopus hammondii* tadpoles to habitat desiccation. *Ecology* 79:1859–72.

Denzer, W. 1994. Tree hole breeding in the toad *Pelophryne brevipes* (Peters, 1867). *Amphibia-Reptilia* 15:224–26.

D'Eon, M. E., R. G. Boutilier, and D. P. Toews. 1978. Anaerobic contributions during progressive hypoxia in the toad *Bufo marinus*. *Comp. Biochem. Physiol.* 60A:7–10.

Dessauer, H. C., E. Nevo, and C. K. Chuang. 1975. High genetic variability in an ecologically variable vertebrate, *Bufo viridis*. *Biochem. Genet.* 13:651–61.

Deutschlander, M. E., J. B. Phillips, and S. C. Borland. 1999. The case for light-dependent magnetic orientation in animals. *J. Exp. Biol.* 202:891–908.

———. 2000. Magnetic compass orientation in the eastern red-spotted newt, *Notophthalmus viridescens*: Rapid acquisition of the shoreward axis. *Copeia* 2000:413–19.

Deutschmann, M. R., and J. J. Peterka. 1988. Secondary production of tiger salamanders in three North Dakota prairie lakes. *Can. J. Fish. Aquat. Sci.* 45:691–97.

DeVito, J. 2003. Metamorphic synchrony and aggregation as antipredator responses in American toads. *Oikos* 103:75–80.

DeVito, J., D. P. Chivers, J. M. Kiesecker, L. K. Belden, and A. R. Blaustein. 1999. Effects of snake predation on aggregation and metamorphosis of Pacific treefrog (*Hyla regilla*) larvae. *J. Herpetol.* 33:504–7.

DeVito, J., D. P. Chivers, J. M. Kiesecker, A. Marco, E. L. Wildy, and A. R. Blaustein. 1998. The effects of snake predation on metamorphosis of western toads, *Bufo boreas* (Amphibia, Bufonidae). *Ethology* 104:185–93.

Deyrup, I. J. 1964. Water balance and kidney. In *Physiology of the Amphibia*, vol., ed. J. A. Moore, 251–328. New York: Academic Press.

Dharmakumarsinhji, K. S. 1946. Musk-shrew (*Suncus caeruleus*) attacking bull-frog (*Rana tigrina*). *J. Bombay Nat. Hist. Soc.* 46:180–81.

D'Heursel, A., and C. F. B. Haddad. 1999. Unpalatability of *Hyla semilineata* tadpoles (Anura) to captive and free-ranging vertebrate predators. *Ethol. Ecol. Evol.* 11:339–48.

Diamond, S. A., G. S. Peterson, J. E. Tietge, and G. T. Ankley. 2002. Assessment of the risk of solar ultraviolet radiation to amphibians. III. Predictions of impacts in selected northern midwestern wetlands. *Environ. Sci. Technol.* 36:2866–74.

Diana, S. G., W. J. Resetarits, D. J. Schaeffer, K. B. Beckmen, and V. R. Beasley. 2000. Effects of atrazine on amphibian growth and survival in artificial aquatic communities. *Env. Tox. Chem.* 19:2961–67.

Diaz, N. F., and J. Valencia. 1985. Microhabitat utilization by two leptodactylid frogs in the Andes of central Chile. *Oecologia* 66:353–57.

Díaz-Páez, H., and J. C. Ortiz. 2001. The reproductive cycle of *Pleurodema thaul* (Anura, Leptodactylidae) in central Chile. *Amphibia-Reptilia* 22:431–45.

Diaz-Paniagua, C. 1985. Larval diets related to morphological characters of five anuran species in the biological reserve of Doñana (Huelva, Spain). *Amphibia-Reptilia* 6:307–22.

———. 1986. Selección de plantas para la ovoposición en *Triturus marmoratus. Rev. Española Herp.* 1:315–27.

———. 1987. Tadpole distribution in relation to vegetal heterogeneity in temporary ponds. *Herpetol. J.* 1:167–69.

———. 1989a. Larval diets of two anuran species, *Pelodytes punctatus* and *Bufo bufo,* in SW Spain. *Amphibia-Reptilia* 10:71–75.

———. 1989b. Oviposition behavior of *Triturus marmoratus pygmaeus. J. Herpetol.* 23:159–63.

Dicker, S. E., and A. B. Elliot. 1970. Water uptake by the crab-eating frog, *Rana cancrivora,* as affected by osmotic gradients and by neurohypophyseal hormones. *J. Physiol.* 207:119–32.

———. 1973. Neurohypohysial hormones and homeostasis in the crab-eating frog, *Rana cancrivora. Horm. Res.* 4:224–60.

Dickerson, M. C. 1906. *The frog book.* New York: Doubleday, Page, & Co.

Dickman, M. 1968. The effect of grazing by tadpoles on the structure of a periphyton community. *Ecology* 49:1188–90.

Diego-Rasilla, F. J. 2003. Homing ability and sensitivity to the geomagnetic field in the alpine newt, *Triturus alpestris. Ethol. Ecol. Evol.* 15:251–59.

Diego-Rasilla, F. J., and R. M. Luengo. 2002. Celestial orientation in the marbled newt (*Triturus marmoratus*). *J. Ethol.* 20:137–41.

———. 2004. Heterospecific call recognition and phonotaxis in the orientation behavior of the marbled newt, *Triturus marmoratus. Behav. Ecol. Sociobiol.* 55:556–60.

Diego-Rasilla, F. J., R. M. Luengo, and J. P. Phillips. 2005. Magnetic compass mediates nocturnal homing by the alpine newt, *Triturus alpestris. Behav. Ecol. Sociobiol.* 58:361–65.

Diekamp, B. M., and H. C. Gerhardt. 1992. Midbrain auditory sensitivity in the spring peeper (*Pseudacris crucifer*): Correlations with behavioral studies. *J. Comp. Physiol. A* 171:245–50.

———. 1995. Selective phonotaxis to advertisement calls in the gray treefrog *Hyla versicolor:* Behavioral experiments and neurophysiological correlates. *J. Comp. Physiol. A* 177:173–90.

Diener, R. A. 1957. An ecological study of the plain-bellied water snake. *Herpetologica* 13:203–11.

Diesel, R. 1989. Parental care in an unusual environment: *Metopaulias depressus* (Decapoda, Grapsidae), a crab that lives in epiphytic bromeliads. *Anim. Behav.* 38:561–75.

———. 1992a. Managing the offspring environment: Brood care in the bromeliad crab, *Metopaulias depressus. Behav. Ecol. Sociobiol.* 30:125–34.

———. 1992b. Maternal care in the bromeliad crab, *Metopaulias depressus:* Protection of larvae from predation by damselfly nymphs. *Anim. Behav.* 43:803–12.

Diesel, R., G. Baurle, and P. Vogel. 1995. Cave breeding and froglet transport: A novel pattern of anuran brood care in the Jamaican frog, *Eleutherodactylus cundalli. Copeia* 1995:354–60.

Diesener, G., and J. Reichholf. 1986. *Lurche und Kriechtiere.* Munich, Germany: Mosaik Verlag.

Dieterlen, F., and B. Statzner. 1981. The African rodent *Colomys goslingi* Thomas and Wroughton, 1907 (Rodentia: Muridae)—A predator in limnetic ecosystems. *Z. Säugetierkunde* 46:369–83.

Dietz, T. H., and R. H. Alvarado. 1974. $Na^+$ and $Cl^-$ transport across gill chamber epithelium of *Rana catesbeiana* tadpoles. *Am. J. Physiol.* 226:764–70.

Digby, L., and C. E. Barreto. 1998. Vertebrate predation in common marmosets. *Neotrop. Primates* 6:124–25.

DiGiovanni, M., and E. D. Brodie, Jr. 1981. Efficacy of skin glands in protecting the salamander *Ambystoma opacum* from repeated attacks by the shrew *Blarina brevicauda. Herpetologica* 1981:234–37.

Diller, L. V., and R. L. Wallace. 1996. Distribution and habitat of *Rhyactriton variegata* in managed, young growth forests in north coastal California. *J. Herpetol.* 30:184–91.

Dillon, M. E., and J. Fiaño. 2000. Oviposition site selection by the túngara frog (*Physalaemus pustulosus*). *Copeia* 2000:883–85.

DiMauro, D., and M. L. Hunter. 2002. Reproduction of amphibians in natural and anthropogenic temporary pools in managed forests. *Forest Sci.* 48:397–406.

Dimmitt, M. A., and R. Ruibal. 1980. Exploitation of food resources by spadefoot toads (*Scaphiopus*). *Copeia* 1980:854–62.

Dingle, H. 1996. *Migration. The biology of life on the move.* New York: Oxford University Press.

Dirar, H. A. 1994. Commentary—The fermented foods of the Sudan. *Ecol. Food Nutr.* 32:207–18.

Directorate of Culture and Cultural and Natural Heritage, Council of Europe. 2003. *Convention on the conservation of European wildlife and natural habitats. Report of a group of experts on the conservation of amphibians and reptiles. Strasbourg, France.* 77 pp. (available at http://europa.eu.int/comm/environment/nature/art_17/ memberstates_summary_en.pdf).

Dix, M. W. 1968. Snake food preferences: Innate intraspecific geographic variation. *Science* 159:1478–79.

Dixon, J. R. 1957. Geographic variation and distribution of the genus *Tomodactylus* in Mexico. *Texas J. Sci.* 9:379–409.

Dixon, J. R., and W. R. Heyer. 1968. Anuran succession in a temporary pond in Colima, Mexico. *Bull. Southern California Acad. Sci.* 67:129–37.

Dixon, J. R., and C. Rivero-Blanco. 1985. A new dendrobatid frog (*Colostethus*) from Venezuela with notes on its natural history and that of related species. *J. Herpetol.* 19:177–84.

Dixon, J. R., and P. Soini. 1986. The reptiles of the Upper Amazon basin, Iquitos region, Peru. II. Crocodilians, turtles, and snakes. *Milwaukee Pub. Mus. Contr. Biol. Geol.* 12:1–71.

Dixon, J. R., and M. A. Staton. 1976. Some aspects of the biology of *Leptodactylus macrosternum. Herpetologica* 32:227–31.

Dobkin, D. S., and R. D. Gettinger. 1985. Thermal aspects of anuran foam nests. *J. Herpetol.* 19:271–75.

Dobkin de Rios, M. 1974. The influence of psychotropic flora and fauna on the Maya religion. *Current Anthro.* 15:147–52.

Docherty, D. E., C. U. Meteyer, J. Wang, J. H. Mao, S. T. Case, and V. G. Chinchar. 2003. Diagnostic and molecular evaluation of three iridovirus-associated salamander mortality events. *J. Wildl. Dis.* 39:556–66.

Docherty, S., P. J. Bishop, and N. I. Passmore. 2000. Consistency of calling performance in male *Hyperolius marmoratus marmoratus:* Implications for male mating success. *Afr. J. Herpetol.* 49:43–52.

Dodd, C. K, Jr. 1977. Immobility in juvenile *Bufo woodhousei fowleri. J. Mississippi Acad. Sci.* 22:90–94.

———. 1989. Duration of immobility in salamanders: Genus *Plethodon* (Caudata: Plethodontidae). *Herpetologica* 45:467–73.

———. 1992. Biological diversity of a temporary pond herpetofauna in north Florida sandhills. *Biodiv. Cons.* 1:125–42.

———. 1993. Cost of living in an unpredictable environment: The ecology of striped newts *Notophthalmus perstriatus* during a prolonged drought. *Copeia* 1993:605–14.

———. 1994. The effects of drought on population structure, activity, and orientation of toads (*Bufo quercicus* and *B. terrestris*) at a temporary pond. *Ethol. Ecol. Evol.* 6:331–49.

———. 1995. The ecology of a sandhills population of the eastern narrow-mouthed toad *Gastrophryne carolinensis,* during a drought. *Bull. Florida Mus. Nat. Hist.* 38:11–41.

. 1996. Use of terrestrial habitats by amphibians in the sandhill uplands of north-central Florida. *Alytes* 14:45–52.

———. 1997. Imperiled amphibians: A historical perspective. In *Aquatic fauna in peril: The southeastern perspective.* Special Pub. 1, Southeast Aquatic Research Institute, ed. G. W. Benz and D. E. Collins, 165–200. Decatur, GA: Lenz Design and Communications.

Dodd, C. K., Jr., and E. D. Brodie, Jr. 1976. Defensive mechanisms of Neotropical salamanders with an experimental analysis of immobility and the effect of temperature on immobility. *Herpetologica* 32:269–90.

Dodd, C. K., Jr., and B. S. Cade. 1998. Movement patterns and the conservation of amphibians breeding in temporary wetlands. *Cons. Biol.* 12:331–39.

Dodd, C. K., Jr., and P. V. Cupp. 1978. The effect of temperature and stage of development on the duration of immobility in selected anurans. *British J. Herpetol.* 5:783–88.

Dodd, C. K., Jr., and R. M. Dorazio. 2004. Using counts to simultaneously estimate abundance and detection probabilities in a salamander community. *Herpetologica* 60:468–78.

Dodd, C. K., and L. L. Smith. 2003. Habitat destruction and alteration: Historical trends and future prospects for amphibians. In *Amphibian conservation,* ed. R. D. Semlitsch, 94–112. Washington DC: Smithsonian Books.

Dodd, M. H. I., and J. M. Dodd. 1976. The biology of metamorphosis. In *Physiology of the Amphibia,* vol. 3, ed. B. Lofts, 467–599. New York: Academic Press.

Dodson, S. I., and V. E. Dodson. 1971. The diet of *Ambystoma tigrinum* larvae from western Colorado. *Copeia* 1971:614–24.

Doherty, J. A., and H. C. Gerhardt. 1983. Hybrid tree frogs: Vocalizations of males and selective phonotaxis of females. *Science* 220:1078–80.

Doherty, J. A., and H. C. Gerhardt. 1984a. Acoustic communication in hybrid treefrogs: Sound production by males and selective phonotaxis by females. *J. Comp. Physiol. A* 154:319–30.

———. 1984b. Evolutionary and neurobiological implications of selective phonotaxis in the spring peeper (*Hyla crucifer*). *Anim. Behav.* 32:875–81.

Doherty, P. A., R. J. Wassersug, and J. M. Lee. 1998. Mechanical properties of the tadpole tail fin. *J. Exp. Biol.* 201:2691–99.

Dolce, S. 1976. Distribuzione degli anfibi e biologia del rospo comune (*Bufo bufo spinosus* Daudin, 1803) sul sarso triestino. *Atti Mus.Civ. Stor. Nat. Trieste* 29:175–212.

Dolce, S., and E. Stoch. 1984. Su alcuni di neotenia in *Triturus alpestris alpestris* (Laur.), *Triturus cristatus carnifex* (Laur.), e *Triturus vulgaris meridionalis* (Boul.) nel Friuli-Venezia Giulia (Italia nordorientale). *Atti Mus. Civ. Stor. Nat. Trieste* 36:107–16.

Dole, J. W. 1965a. Spatial relations in natural populations of the leopard frog, *Rana pipiens* Schreber in northern Michigan. *Am. Midl. Nat.* 74:464–78.

———. 1965b. Summer movements of adult leopard frogs, *Rana pipiens* Schreber, in northern Michigan. *Ecology* 46:236–55

———. 1967. The role of substrate moisture and dew in the water economy of leopard frogs, *Rana pipiens. Copeia* 1967:141–49.

———. 1968. Homing in leopard frogs, *Rana pipiens. Ecology* 49:386–99.

———. 1971. Dispersal of recently metamorphosed leopard frogs, *Rana pipiens. Copeia* 1971:221–28.

———. 1972a. Evidence of celestial orientation in newly-metamorphosed *Rana pipiens. Herpetologica* 28:273–76.

———. 1972b. Homing and orientation of displaced toads, *Bufo americanus,* to their home sites. *Copeia* 1972:151–58.

———. 1972c. The role of olfaction and audition in the orientation of leopard frogs, *Rana pipiens. Herpetologica* 28:258–60.

———. 1973. Celestial orientation in recently metamorphosed *Bufo americanus. Herpetologica* 29:59–62.

———. 1974. Home range in the canyon treefrog. *Southwest. Nat.* 19:105–7.

Dole, J. W. and P. Durant. 1974a. Courtship behavior in *Colostethus collaris* (Dendrobatidae). *Copeia* 1974:988–90.

———. 1974b. Movements and seasonal activity of *Atelopus oxyrhynchus* (Anura: Atelopodidae) in a Venezuelan cloud forest. *Copeia* 1974:230–35.

Dolk, H. E., and N. Postma. 1927. Über die Haut- und die Lungenatmung von *Rana temporaria. Z. Vergl. Physiol.* 5:417–44.

Dolman, C. E. 1975. Lazzaro Spallanzani. *Dictionary of Scientific Biography* 12:553–67.

Dolmen, D. 1978. De neotene salamanderne ("skrattabborrene") ved Stensele. *Fauna och Flora* 73:171–77.

———. 1981a. Distribution and habitat of the smooth newt, *Triturus vulgaris* (L.) and the warty newt, *T. cristatus* (Laurenti) in Norway. *Proc. Euro. Herp. Symp. C. W. L. P. Oxford 1980:*127–39.

———. 1981b. Local migration, rheotaxis and philopatry by *Triturus vulgaris* within a locality in central Norway. Brit. *J. Herpetol.* 6:151–58.

———. 1982a. Skeletal growth marks and testis lobulation as criteria for age in *Triturus* spp. (Amphibia) in central Norway. *Acta Zool. (Stockholm)* 63:73–80.

———. 1982b. Zoogeography of *Triturus vulgaris* (L.) and *T. cristatus* (Laurenti) in Norway, with notes on their vulnerability. *Fauna Norv., Ser. A* 3:12–25.

———. 1983a. Diel rhythms and microhabitat preference of the newts *Triturus vulgaris* and *T. cristatus* at the northern border of their distribution area. *J. Herpetol.* 17:23–31.

———. 1983b. A survey of the Norwegian newts (*Triturus,* Amphibia): Their distribution and habitats. *Medd. Norsk Viltforsk.* ser. 3, no. 12:1–72.

Dolmen, D., and J. I. Koksvik. 1983. Food and feeding habits of *Triturus vulgaris* (L.) and *T. cristatus* (Laurenti) (Amphibia) in two bog tarns in central Norway. *Amphibia-Reptilia* 4:17–24.

Dominey, W. J. 1984. Alternative mating tactics and evolutionarily stable strategies. *Amer. Zool.* 24:385–96.

Donald, D. B., R. D. Vinebrooke, R. S. Anderson, J. Sygiannis, and M. D. Graham. 2001. Recovery of zooplankton assemblages in mountain lakes from the effects of introduced sport fish. *Can. J. Fish. Aquat. Sci.* 58:1822–30.

Donaldson, G. 1980. *Frogs.* London: Windward.

Donavan, L. A., and G. W. Folkerts. 1972. Foods of the seepage salamander, *Desmognathus aeneus* Brown and Bishop. *Herpetologica* 28:35–37.

Donnelly, M. A. 1989a. Demographic effects of reproductive resource supplementation in a territorial frog, *Dendrobates pumilio. Ecol. Monogr.* 59:207–21.

———. 1989b. Effects of reproductive resource supplementation on space-use patterns in *Dendrobates pumilio. Oecologia* 81:212–18.

———. 1989c. Reproductive phenology and age structure of *Dendrobates pumilio* in northeastern Cost Rica. *J. Herpetol.* 23:362–67.

———. 1994. Amphibian diversity and natural history. In. *La Selva. Ecology and natural history of a neotropical rain forest,* ed. L. A. McDade, K. S. Bawa, H. A. Hespenheide, and G. S. Hartshorn, 199–209. Chicago:University of Chicago Press.

———. 1999. Reproductive phenology of *Eleutherodactylus bransfordii* in northeastern Costa Rica. *J. Herpetol.* 33:624–31.

Donnelly, M. A., and M. L. Crump. 1998. Potential effects of climate change on two Neotropical amphibian assemblages. *Climatic Change* 39:541–61.

Donnelly, M. A., R. O. De Sá, and C. Guyer. 1990. Description of the tadpoles of *Gastrophryne pictiventris* and *Nelsonophryne aterrima* (Anura, Microhylidae), with a review of morphological variation in free-swimming microhylid larvae. *Amer. Mus. Novitates* (2976):1–19.

Donnelly, M. A., and C. Guyer. 1994. Patterns of reproduction and habitat use in an assemblage of Neotropical hylid frogs. *Oecologia* 98:291–302.

Donnelly, M. A., C. Guyer, J. E. Juterbock, and R. A. Alford. 1994. Techniques for marking amphibians. In *Measuring and montitoring biological diversity. Standard methods for amphibians,* ed. W. R. Heyer, M. A. Donnelly, R. W. McDiarmid, L. C. Hayek, and M. S. Foster, 277–84. Washington, DC: Smithsonian Institution Press.

Donnelly, M. A., C. Guyer, D. M. Krempels, and H. E. Braker. 1987. The tadpole of *Agalychnis calcarifer* (Anura: Hylidae). *Copeia* 1987:247–50.

Donohoe, P. H., and R. G. Boutilier. 1998. The protective effects of metabolic rate depression in hypoxic cold submerged frogs. *Respir. Physiol.* 111:325–36.

———. 1999. The use of extracellular lactate as an oxidative substrate in the oxygen-limited frog. *Respir. Physiol.* 116:171–79.

Donohoe, P. H., T. G. West, and R. G. Boutilier. 1998. Respiratory, metabolic and acid-base correlates of aerobic metabolic rate reduction in overwintering frogs. *Am. J. Physiol.* 274:R704–R710.

———. 2000. Factors affecting membrane permeability and ionic homeostasis in the cold-submerged frog. *J. Exp. Biol.* 203:405–14.

Donovan, A., and P. A. Verrell. 1991. The effect of partner familiarity on courtship success in the salamander *Desmognathus ochrophaeus. J. Herpetol.* 25:93–95.

Doody, J. S., and J. E. Young. 1995. Temporal variation in reproduction and clutch mortality of leopard frogs (*Rana utricularia*) in South Mississippi. *J. Herpetol.* 29:614–16.

Doolan, J. M., and R.C. Mac Nally. 1981. Spatial dynamics and breeding ecology in the cicada *Cystomsoma saundersii:* The interaction between distributions of resources and intraspecific behavior. *J. Anim. Ecol.* 50:925–40.

Dopazo, H., and P. Alberch. 1994. Preliminary results on optional viviparity and intrauterine siblicide in *Salamandra salamandra* populations from northern Spain. *Mertensiella* 4:125–37.

Dopazo, H., L. Boto, and P. Alberch. 1998. Mitochondrial DNA variability in viviparous and ovoviviparous populations of the urodele *Salamandra salamandra. J. Evol. Biol.* 11:365–78.

Dopazo, H. J., and M. Korenblum. 2000. Viviparity in *Salamandra salamandra* (Amphibia: Salamandridae): Adaptation or exaptation? *Herpetologica* 56:144–52.

D'Orgeix, C. A., and B. J. Turner. 1995. Multiple paternity in the red-eyed treefrog *Agalychnis callidryas* (Cope). *Mol. Ecol.* 4:505–8.

Dorlochter, M., S. Astrow, and A. Herrera. 1994. Effects of testosterone on a sexually dimorphic frog muscle: Repeated in vivo observations and androgen receptor distribution. *J. Neurobiol.* 25:897–916.

Dorst, J. 1974. *The life of birds.* 2 vols. New York: Columbia University Press.

Doty, G. V., and A. M. Welch. 2001. Advertisement call duration indicates good genes for offspring feeding rate in gray tree frogs (*Hyla versicolor*). *Behav. Ecol. Sociobiol.* 49:150–56.

Doubledee, R. A., E. B. Muller, and R. M. Nisbet. 2003. Bullfrogs, disturbance regimes, and the persistence of California red-legged frogs. *J. Wildl. Manage.* 67:424–38.

Doughty, P. 2002. Coevolution of developmental plasticity and large egg size in *Crinia georgiana* tadpoles. *Copeia* 2002:928–37.

Douglas, A. M. 1967. The natural history of the ghost bat (*Macroderma gigas*) (Microchiroptera: Megadermatidae) in Western Australia. *West. Aust. Nat.* 10:125–38.

Douglas, M. E. 1979. Migration and sexual selection in *Ambystoma jeffersonianum. Can. J. Zool.* 57:2303–10.

Douglas, M. E., and B. L. Monroe, Jr. 1981. A comparative study of topographical orientation in *Ambystoma (*Amphibia: Caudata). *Copeia* 1981:460–63.

Douglas, R. 1948. Temperature and rate of development of the eggs of British Anura. *J. Anim. Ecol.* 17:189–92.

Dowdey, T. G., and E. D. Brodie, Jr. 1989. Antipredator strategies of salamanders: individual and geographic variation in responses of *Eurycea bislineata* to snakes. *Anim. Behav.* 38:707–11.

Downie, J. R. 1984. How *Leptodactylus fuscus* tadpoles make foam, and why. *Copeia* 1984:778–80.

———. 1988. Functions of the foam in the foam-nesting leptodactylid *Physalaemus pustulosus. Herpetol. J.* 1:302–7.

———. 1989. Observations on foam-making by *Leptodactylus fuscus* tadpoles. *Herpetol. J.* 1:351–55.

———. 1994. Developmental arrest in *Leptodactylus fuscus* tadpoles (Anura: Leptodactylidae) I: Descriptive analysis. *Herpetol. J.* 4:29–38.

———. 1996. A new example of female parental behaviour in *Leptodactylus validus,* a frog of the leptodactylid 'melanonotus' species group. *Herpetol. J.* 6:32–34.

Downie, J. R., R. Bryce, and J. Smith. 2004. Metamorphic duration: An under-studied variable in frog life histories. *Biol. J. Linn. Soc.* 83:261–72.

Downie, J. R., R. H. L. Disney, L. Collins, and E. G. Hancock. 1995. A new species of *Megaselia* (Diptera, Phoridae) whose larvae prey upon the eggs of *Leptodactylus fuscus* (Anura, Leptodactylidae). *J. Nat. Hist.* 29:993–1003.

Downie, J. R., S. R. Livingstone, and J. R. Cormack. 2001. Selection of tadpole deposition sites by male Trinidadian stream frogs, *Mannophryne trinitatis* (Dendrobatidae): An example of anti-predator behaviour. *Herpetol. J.* 11:91–100.

Downie, J. R., E. Robinson, R. J. Linklater-McClennan, E. Somerville, and N. Kamenos. 2005. Are there costs to extended larval transport in the Trinidadian stream frog, *Mannophryne trinitatis* (Dendrobatidae)? *J. Nat. Hist.* 39:2023–34.

Downie, J. R., and J. Smith. 2003. Survival of larval *Leptodactylus fuscus* (Anura: Leptodactylidae) out of water: Developmental differences and interspecific comparisons. *J. Herpetol.* 37:107–15.

Downie, J. R., and A. Weir. 1997. Developmental arrest in *Leptodactylus fuscus* tadpoles (Anura: Leptodactylidae). III: Effect of arrest period on growth potential. *Herpetol. J.* 7:85–92.

Drake, D. C., and R. J. Naiman. 2000. An evaluation of restoration efforts in fishless lakes stocked with exotic trout. *Cons. Biol.* 14:1807–20.

Drake, D. L., and A. W. Ranvestel. 2005. *Hyalinobatrachium colymbiphyllum* (Glass Frog). Egg mass defense. *Herp. Rev.* 36:434.

Drewes, R. C. 1984. A phylogenetic analysis of the Hyperoliidae (Anura): Treefrogs of Africa, Madagascar, and the Seychelles Islands. *Occ. Pap. California Acad. Sci.* (139):1–70.

Drewes, R. C., and R. Altig. 1996. Anuran egg predation and heterocannibalism in a breeding community of East African frogs. *Trop. Zool.* 9:333–47.

Drewes, R. C., S. S. Hillman, R. W. Putnam, and O. M. Sokol. 1977. Water, nitrogen and ion balance in the African treefrog *Chiromantis petersi* Boulenger (Anura: Rhacophoridae), with comments on the structure of the integument. *J. Comp. Physiol.* 116:257–67.

Drewes, R. C., and B. Roth. 1981. Snail-eating frogs from the Ethiopian highlands: A new anuran specialization. *Zool. J. Linn. Soc.* 73:267–87.

Drewry, G. 1970. The role of amphibians in the ecology of Puerto Rican rain forest. In *Puerto Rico Nuclear Center Rain Forest Project Annual Report,* 16–85. San Juan: Puerto Rico Nuclear Center.

Drewry, G. E., W. R. Heyer, and A. S. Rand. 1982. A functional analysis of the complex call of the frog *Physalaemus pustulosus. Copeia* 1982:636–45.

Drewry, G. E., and K. L. Jones. 1976. A new ovoviviparous frog, *Eleutherodactylus jasperi* (Amphibia, Anura, Leptodactylidae), from Puerto Rico. *J. Herpetol.* 10:161–65.

Drewry, G. E., and A. S. Rand. 1983. Characteristics of an acoustic community: Puerto Rican frogs of the genus *Eleutherodactylus. Copeia* 1983:941–53.

Dring, J. 1979. Amphibians and reptiles from northern Trengganu Malaysia with descriptions of two new geckos: *Cnemaspis* and *Cyrtodactylus. Bull. British Mus. Nat. Hist. (Zool.)* 34:181–41.

Drinkrow, D. R., and M. I. Cherry. 1995. Anuran distribution, diversity and conservation in South Africa, Lesotho and Swaziland. *S. Afr. J. Zool.* 30:82–90.

Driscoll, D. A. 1997. Mobility and metapopulation structure of *Geocrinia alba* and *G. vitellina,* two endangered frog species from southwestern Australia. *Aust. J. Ecol.* 22:185–95.

———. 1998. Genetic structure of the frogs *Geocrinia lutea* and *Geocrinia rosea* reflects extreme population divergence and range changes, not dispersal barriers. *Evolution* 52:1147–57.

———. 1999a. Genetic neighborhood and effective population size for two endangered frogs. *Biol. Cons.* 88:221–29.

———. 1999b. Skeletochronological assessment of age structure and population stability for two threatened frog species. *Aust. J. Ecol.* 24:182–89.

Driscoll, D. A., and J. D. Roberts. 1997. Impact of fuel-reduction burning on the frog *Geocrinia lutea* in southwest Western Australia. *Aust. J. Ecol.* 33:334–39.

Drost, C. A., and G. M. Fellers. 1996. Collapse of a regional frog fauna in the Yosemite area of the California Sierra Nevada, USA. *Cons. Biol.* 10:414–25.

Drummond, H. 1983. Aquatic foraging in garter snakes: A comparison of specialists and generalists. *Behaviour* 86:1–30.

Dryden, G. L. 1965. The food and feeding habits of *Varanus indicus* on Guam. *Micronesia* 2:73–76.

Dubois, A. 1974a. Equisse de la zoogéographie des Amphibiens du Népal. *Bull. Soc. Zool. France* 99:788–91.

———. 1974b. Liste commentée d'amphibiens récoltés au Népal. *Bull. Mus. Natl. Hist. Nat. Paris,* ser. 3, 213:341–411.

———. 1977a. Chants et écologie chez les Amphibiens du Nepal. *Colloques. Int. du CNRS* (268):109–18.

———. 1977b. Observations sur le chant et les relations inter-individuelles chez quelques grenouilles du sous-genre *Paa* du Nepal (Amphibiens, Anoures, Genre *Rana*). *Bull. Soc. Zool. France* 102, suppl. 2:163–81.

———. 1980a. Notes sur la systematique et la repartition des amphibiens anoures de Chine et des regions Avoissinantes. IV. Classification generique et gubgenerique des Pelobatidae, Megophryinae. *Bull. Mens. Soc. Linn. Lyon* 49:469–82.

———. 1980b. Populations, polymorphisme et adaptation: Quelques exemples chez les amphibiens anoures. In *Recherches d'ecologie théorique,* ed. R. Barbault, P. Blandin, and J. A. Meyer, 141–58. Paris: Maloine.

———. 1981. Liste des genres et sous-genres nominaux de Ranoidea (Amphibiens Anoures) du monde, avec identification de leurs espèces-types: conséquences nomenclaturales. *Monit. Zool. Ital.* 15:225–84.

———. 1983a. A propos de cuisses de grenouilles. *Alytes* 2:69–111.

———. 1983b. Classification et nomenclature supragenerique des amphibiens anoures. *Bull. Soc. Linn. Lyon* 52:270–76.

———. 1984. La nomenclature supragenerique des amphibiens anoures. *Mem. Mus. Nat. Hist. Natur.,* ser. A 131:1–64.

———. 1986. Miscellanea taxonomica batrachologica (I). *Alytes* 5:7–95.

———. 1992. Notes sur la classification des Ranidae (Amphibiens Anoures). *Bull. Men. Soc. Linn. Lyon* 61:305–52.

———. 1998. South Asian Amphibia: A new frontier for taxonomists. *J. South Asian Nat. Hist.* 4:1–12.

———. 1999. Miscellanea nomenclatorica batrachologica. 19. Notes on the nomenclature of Ranidae and related groups. *Alytes.* 17:81–100.

Dubois, A., and J. Martens. 1984. A case of possible vocal convergence between frogs and a bird in Himalayan torrents. *J. für Ornithol.* 125:455–63.

Dubois, A., and A. Ohler. 1998. A new species of *Leptobrachium* (*Vibrissaphora*) from northern Vietnam, with a review of the taxonomy of the genus *Leptobrachium* (Pelobatidae, Megophryinae). *Dumerilia* 4:1–32.

Ducey, P. K. 1989. Agonistic behavior and biting during intraspecific encounters in *Ambystoma* salamanders. *Herpetologica* 45:155–60.

Ducey, P. K., C. D. Anthony, and E. D. Brodie, Jr. 1991. Thresholds and escalation of antipredator responses in the Chinese salamander *Cynops cyanurus:* Inter- and intra-individual variation. *Behav. Proc.* 23:181–91.

Ducey, P. K., and E. D. Brodie, Jr. 1983. Salamanders respond selectively to contacts with snakes: Survival advantage of alternative antipredator strategies. *Copeia* 1983:1036–41.

———. 1991. Evolution of antipredator behavior: Individual and populational variation in a Neotropical salamander. *Herpetologica* 47:89–95.

Ducey, P. K., E. D. Brodie, Jr., and E. A. Baness. 1993. Salamander tail autotomy and snake predation: role of antipredator behavior and toxicity for three Neotropical *Bolitoglossa* (Caudata: Plethodontidae). *Biotropica* 25:344–49.

Ducey, P. K., D. R. Formanowicz, Jr., L. Boyet, J. Mailloux, and R. A. Nussbaum. 1993. Experimental examination of burrowing behavior in caecilians (Amphibia: Gymnophiona): Effects of soil compaction on burrowing ability of four species. *Herpetologica* 49:450–57.

Ducey, P. K., and J. Heuer. 1991. Effects of food availability on intraspecific aggression in salamanders of the genus *Ambystoma. Can. J. Zool.* 69:288–90.

Ducey, P. K., and P. Ritsema. 1988. Intraspecific aggression and responses to marked substrates in *Ambystoma maculatum* (Caudata: Ambystomatidae). *Copeia* 1988:1008–13.

Ducey, P. K., K. Schramm, and N. Cambry. 1994. Interspecific aggression between the sympatric salamanders, *Ambystoma maculatum* and *Plethodon cinereus. Am. Midl. Nat.* 131:320–29.

Ducibella, T. 1974. The occurrence of biochemical metamorphic events with anatomical metamorphosis in the axolotl. *Develop. Biol.* 38:175–86.

Duda, T. F., Jr., D. Vanhoye, and P. Nicholas. 2002. Roles of diversifying selection and coordinated evolution in the evolution of amphibian antimicrobial peptides. *Mol. Biol. Evol.* 19:858–64.

Dudgeon, D. 2000. The ecology of tropical Asian streams in relation to biodiversity conservation. *Ann. Rev. Ecol. Syst.* 31:239–63.

———. 2002. The most endangered ecosystems in the world? Conservation of riverine biodiversity in Asia. *Verh. Int. Ver. Theor. Angew. Limnol.* 19:1–15.

———. 2003. Clinging to the wreckage: Unexpected persistence of freshwater biodiversity in a degraded tropical landscape. *Aquat. Cons: Mar. Freshw. Ecosyst.* 13:93–97.

Dudley, R., V. A. King, and R. J. Wassersug. 1991. Implications for shape and metamorphosis for drag forces on a generalized pond tadpole (*Rana catesbeiana*). *Copeia* 1991:252–57.

Dudley, R., and A. S. Rand. 1991. Sound production and vocal sac inflation in the túngara frog, *Physalaemus pustulosus* (Leptodactylidae). *Copeia* 1991:460–70.

———. 1992. Underwater sound production in a Neotropical anuran, *Physalaemus pustulosus* (Leptodactylidae). *Bioacoustics* 4:211–16.

Duellman, W. E. 1958. A monographic study of the colubrid snake genus *Leptodeira. Bull. Amer. Mus. Nat. Hist.* 114:1–52.

———. 1963. Amphibians and reptiles of the rainforests of southern El Peten, Guatemala. *Univ. Kansas Publ. Mus. Nat. Hist.* 15:205–49.

———. 1966. Aggressive behavior in dendrobatid frogs. *Herpetologica* 22:217–22.

———. 1967a. Courtship isolating mechanisms in Costa Rican hylid frogs. *Herpetologica* 23:169–83.

———. 1967b. Social organization in the mating calls of some Neotropical anurans. *Am. Midl. Nat.* 77:156–63.

———. 1970. *The hylid frogs of middle America. Univ. Kansas Mus. Nat. Hist. Monogr.* 1. 2 vols. Lawrence, Kansas.

———. 1973. A systematic review of the marsupial frogs (Hylidae: *Gastrotheca*) of the Andes of Ecuador. *Occ. Pap. Mus. Nat. Hist. Univ. Kansas* 22:1–27

———. 1974. A systematic review of the marsupial frogs (Hylidae: *Gastrotheca*) of the Andes of Ecuador. *Occ. Pap. Mus. Nat. Hist. Univ. Kansas* 22:1–27.

———. 1978. The biology of an equatorial herpetofauna in Amazonian Ecuador. *Misc. Publ. Mus. Nat. Hist. Univ. Kansas* (65):1–352.

———. 1983. *Gastrotheca aureomaculata:* Systematic resolution of a case of intrapopulational polychromatism in a hylid frog. *Herpetologica* 39:105–10.

———. 1985. Reproductive modes in anuran amphibians: Phylogenetic significance of adaptive strategies. *S. Afr. J. Sci.* 81:174–78.

———. 1988. Patterns of species diversity in anuran amphibians in the American tropics. *Ann. Missouri Bot. Garden* 75:79–104.

———. 1989a. Alternative life-history styles in anuran amphibians: Evolutionary and ecological implications. In *Alternative life-history styles of animals,* ed. M. N. Bruton, 101–26. Dordrecht: Kluwer Academic.

———. 1989b. Tropical herpetofaunal communities: Patterns of community structure in Neotropical rainforests. In *Vertebrates in complex tropical systems,* ed. M. L. Harmelin-Vivien and F. Bourliere, 61–88. New York: Springer-Verlag.

———. 1990. Herpetofauna in Neotropical rainforests: Comparative composition, history, and resource use. In *Four Neotropical rainforests,* ed. A. H. Gentry, 455–505. New Haven, CT: Yale University Press.

———. 1993a. *Amphibian species of the world: Additions and corrections. Special Publ. no. 21, Univ. Kansas Mus. Nat. Hist.,* Lawrence, Kansas.

———. 1993b. Amphibians in Africa and South America: Evolutionary history and ecological comparisons. In *Biological relationships between Africa and South America,* ed. P. Goldblatt, 200–43. New Haven, CT: Yale University Press.

———. 1995. Temporal fluctuations in abundance of anuran amphibians in a seasonal Amazonian rainforest. *J. Herpetol.* 29:13–21.

———. 1999a. Distribution patterns of amphibians in South America. In *Patterns of distribution of amphibians. A global perspective,* ed. W. E. Duellman, 255–328. Baltimore: Johns Hopkins University Press.

———. 1999b. Global distribution of amphibians: Patterns, conservation, and future challenges. In *Patterns of distribution of amphibians. A global perspective,* ed. W. E. Duellman, 1–30. Baltimore: Johns Hopkins University Press.

———. 1999c. *Patterns of distribution of amphibians. A global perspective,* ed. W. E. Duellman. Baltimore: Johns Hopkins University Press.

———. 2001. *The hylid frogs of middle America.* New and expanded edition. 2 vols. Ithaca, NY: Society for the Study of Amphibians and Reptiles.

———. 2003. An overview of anuran phylogeny, classification

and reproductive modes. In *Reproductive biology and phylogeny of anura,* ed. B. G. M. Jamieson, 1–18. Enfield, NH: Science Publishers.

Duellman, W. E., and J. A. Campbell. 1992. Hylid frogs of the genus *Plectrohyla:* Systematics and phylogenetic relationships. *Misc. Publ. Mus. Zool. Univ. Michigan* (181):1–32.

Duellman, W. E., I. de la Riva, and E. R. Wild. 1997. Frogs of the *Hyla armata* and *Hyla pulchella* groups in the Andes of South America, with definitions and analyses of phylogenetic relationships of Andean groups of *Hyla. Sci. Pap. Nat. Hist. Mus. Univ. Kansas* (3):1–41.

Duellman, W. E., and R. O. de Sá. 1988. A new genus and species of South American hylid frog with a highly modified tadpole. *Trop. Zool.* 1:117–36.

Duellman, W. E., and T. H. Fritts. 1972. A taxonomic review of the southern Andean marsupial frogs (Hylidae: *Gastrotheca*). *Occ. Pap. Mus. Nat. Hist. Univ. Kansas* 9:1–37.

Duellman, W. E., and P. Gray. 1983. Developmental biology and systematics of the egg-brooding hylid frogs, genera *Flectonotus* and *Fritziana. Herpetologica* 39:333–59.

Duellman, W. E., and D. N. Hillis. 1987. Marsupial frogs (Anura: Hylidae: *Gastrotheca*) of the Ecuadorian Andes: Resolution of taxonomic problems and phylogenetic relationships. *Herpetologica* 43:141–73.

Duellman, W. E., and M. S. Hoogmoed. 1984. The taxonomy and phylogenetic relationships of the hylid frog genus *Stefania. Misc. Publ. Univ. Kansas Mus. Nat. Hist.* (75):1–39.

Duellman, W. E., and L. T. Klaas. 1964. The biology of the hylid frog *Triprion petasatus. Copeia* 1964:308–21.

Duellman, W. E., and J. Lescure. 1973. Life history and ecology of the hylid frog *Osteocephalus taurinus,* with observations on larval behavior. *Occ. Pap. Mus. Nat. Hist. Univ. Kansas* (13):1–12.

Duellman, W. E., and M. Lizana. 1994. Biology of a sit-and-wait predator, the leptodactylid frog *Ceratophrys cornuta. Herpetologica* 50:51–64.

Duellman, W. E., and J. D. Lynch. 1969. Descriptions of *Atelopus* tadpoles and their relevance to atelopodid classification. *Herpetologica* 25:231–40.

———. 1988. Anuran amphibians from the Cordillera de Cutucu, Ecuador. *Proc. Acad. Natur. Sci. Philadelphia* 140:125–42.

Duellman, W. E., and S. J. Maness. 1980. The reproductive behavior of some hylid marsupial frogs. *J. Herpetol.* 14:213–22.

Duellman, W. E., L. R. Maxson, and C. A. Jesiolowski. 1988. Evolution of marsupial frogs (Hylidae: Hemiphractinae): Immunological evidence. *Copeia* 1988:527–43.

Duellman, W. E., and V. R. Morales. 1990. Variation, distribution, and life history of *Edalorhina perezi* (Amphibia, Anura, Leptodactylidae). *Stud. Neotrop. Fauna Environ.* 25:19–30.

Duellman, W. E., and E. R. Pianka. 1990. Biogeography of nocturnal insectivores: Historical events and ecological filters. *Ann. Rev. Ecol. Syst.* 21:57–68.

Duellman, W. E., and R. A. Pyles. 1983. Acoustic resource partitioning in anuran communities. *Copeia* 1983:639–49.

Duellman, W. E., and P. M. Ruiz-C. 1986. Ontogenetic polymorphism in marsupial frogs (Anura: Hylidae). *Caldasia* 15:617–27.

Duellman, W. E., and A. H. Savitsky. 1976. Aggressive behavior in a centrolenid frog with comments on territoriality in anurans. *Herpetologica* 32:401–4.

Duellman, W. E., and S. S. Sweet. 1999. Distribution patterns of amphibians in the Nearctic Region of North America. In *Patterns of distribution of amphibians. A global perspective,* ed. W. E. Duellman, 31–109. Baltimore: Johns Hopkins University Press.

Duellman, W. E., and R. Thomas. 1996. Anuran amphibians from a seasonally dry forest in southeastern Peru and comparisons of the anurans among sites in the Upper Amazon Basin. *Occ. Pap. Nat. Hist. Mus. Univ. Kansas* (180):1–34.

Duellman, W. E., and L. Trueb. 1983. Frogs of the *Hyla columbiana* group: Taxonomy and phylogenetic relationships. In *Advances in herpetology and evolutionary biology,* ed. A. G. J. Rhodin and K. Miyata, 33–51. Cambridge, MA: Museum of Comparative Zoology, Harvard University.

Duellman, W. E. and L. Trueb. 1986. *Biology of amphibians.* New York: McGraw-Hill.

Duellman, W. E., and J. B. Tulecke. 1960. The distribution, variation, and life history of the frog *Cochranella viridissima* in Mexico. *Am. Midl. Nat.* 63:392–97.

Duellman, W. E., and A. Veloso M. 1977. Phylogeny of *Pleurodema* (Anura: Leptodactylidae): A biogeographic model. *Occ. Pap. Mus. Nat. Hist. Univ. Kansas* 64:1–46.

Duellman, W. E., and M. Yoshpa. 1996. A new species of *Tepuihyla* (Anura: Hylidae) from Guyana. *Herpetologica* 52:275–81.

Dufour, D. L. 1983. Nutrition in the northwest Amazon: Household dietary intake and time-energy expenditure. In *Adaptive responses of native Amazonians,* ed. R. B. Hames and W. T. Vickers, 329–55. New York: Academic Press.

Duguay, J. P., and P. B. Wood. 2002. Salamander abundance in regenerating forest stands on the Monongahela National Forest, West Virginia. *Forest Sci.* 48:331–35.

Duke, J. T., and G. R. Ultsch. 1990. Metabolic oxygen regulation and conformity during submergence in the salamanders *Siren lacertina, Amphiuma means,* and *Amphiuma tridactylum,* and a comparison with other giant salamanders. *Oecologia* 84:16–23.

Dumas, P. C. 1955. Eggs of the salamander *Plethodon dunni* in nature. *Copeia* 1955:65

———. 1956. The ecological relations of sympatry in *Plethodon dunni* and *Plethodon vehiculum. Ecology* 37:484–95.

Duméril, A. M. C., G. Bibron, and A. Duméril. 1854. *Erpétologie Générale ou Histoire Naturelle Compléte des Reptiles. Atlas.* Paris: Librairie Encyclopédique de Roret.

Duncan, R., and R. Highton. 1979. Genetic relationships of the eastern large *Plethodon* of the Ouachita Mountains. *Copeia* 1979:95–110.

Dundee, H. A. 1957. Partial metamorphosis induced in *Typhlomolge rathbuni. Copeia* 1957:52–53.

———. 1965. *Eurycea tynerensis. Cat. Amer. Amphib. Rept.* 22.1–22.2.

Dundee, H. A., and E. A. Liner. 1985. *Phrynohyas venulosa* (veined treefrog). Food. *Herp. Rev.* 16:109.

Dunham, A. E. 1980. An experimental study of interspecific competition between the iguanid lizards *Sceloporus merriami* and *Urosaurus ornatus. Ecol. Monogr.* 50:309–30.

———. 1983. Realized niche overlap, resource abundance, and intensity of interspecific competition. In *Lizard ecology,* ed. R. B. Huey, E. R. Pianka, and T. W. Schoener, 261–80. Cambridge, MA: Harvard University Press.

Dunham, J. B., D. S. Pilliod, and M. K. Young. 2004. Assessing the consequences of nonnative trout in headwater ecosystems in western North America. *Fisheries* 29:18–26.

Dunlap, D. G. 1969. Influences of temperature and duration of acclimation, time of day, sex and body weight on metabolic rates in the hylid frog, *Acris crepitans. Comp. Biochem. Physiol.* 31:555–70.

———. 1971. Acutely measured metabolic rate-temperature curves in the cricket frog, *Acris crepitans. Comp. Biochem. Physiol.* 38A:1–16.

———. 1972. Latitudinal effects on metabolic rates in the cricket frog, *Acris crepitans:* Acutely measured rates in summer frogs. *Biol. Bull.* 143:332–43.

———. 1973. Latitudinal effects on metabolic rates in the cricket frog, *Acris crepitans:* Seasonal comparisons. *Biol. Bull.* 145:103–18.

———. 1980. Comparative effects of thermal acclimation and season on metabolic compensation to temperature in the hylid frogs, *Pseudacris triseriata* and *Acris crepitans. Comp. Biochem. Physiol.* 66A:243–49.

———. 1989. The effects of differing photoperiods during thermal acclimation on oxygen consumption in the hylid frogs, *Acris crepitans* and *Pseudacris triseriata. Comp. Biochem. Physiol.* 94A:519–23.

Dunlap, D. G., and J. E. Platz. 1981. Geographic variation of proteins and call in *Rana pipiens* from the northcentral United States. *Copeia* 1981:876–79.

Dunn, E. R. 1923. The breeding habits of salamanders and their bearing on phylogeny. *Copeia* 1923:25–28.

———. 1926a. The frogs of Jamaica. *Proc. Boston Soc. Nat. Hist.* 38:111–30.

———. 1926b. The habits of *Sminthillus limbatus. Copeia* 1926:141–44.

———. 1926c. *The salamanders of the family Plethodontidae.* Northampton, MA: Smith College.

———. 1927. A new mountain race of *Desmognathus. Copeia* 1927:84–86.

———. 1937. The amphibian and reptile fauna of bromeliads in Costa Rica and Panama. *Copeia* 1937:163–67.

———. 1941. Notes on *Dendrobates auratus. Copeia* 1941:88–93.

Dunn, J. E., and P. S. Gipson. 1977. Analysis of radio telemetry data in studies of home range. *Biometrics* 33:85–101.

Dunson, W. A. 1977. Tolerance to high temperature and salinity by tadpoles of the Philippine frog, *Rana cancrivora. Copeia* 1977:375–78.

Dunson, W. A., R. L. Wyman, and E. S. Corbett. 1992. A symposium on amphibian declines and habitat acidification. *J. Herpetol.* 26:349–52.

Dupré, R. K., and J. W. Petranka. 1985. Ontogeny of temperature selection in larval amphibians. *Copeia* 1985:462–67.

Dupuis, L., and D. Steventon. 1999. Riparian management and the tailed frog in northern coastal forests. *Forest Ecol. Manage.* 124:35–43.

Dupuis, L. A., and F. L. Bunnell. 1999. Effects of stand age, size, and juxtaposition on abundance of western redback salamanders (*Plethodon vehiculum*) in coastal British Columbia. *Northwest Sci.* 73:27–33.

Durand, J. 1986. Le cycle biologique du Protee. In *Traite de zoologie. Vol. 14, fasc. IB Amphibiens,* ed. P.-P. Grasse and M. Delsol, 515–24. Paris: Masson.

Durand, J., and M. Bouillon. 1964. Observations sur le developpment du Protee, *Proteus anguinus* Laurenti (Batraciens, Urodeles). *C. R. Acad. Sci. Paris* 259:4801–4.

Durand, J., and A. Vandel. 1968. *Proteus:* An evolutionary relic. *Sci. J.* 1968:44–49.

Durand, J. P., and J. Parzefall. 1987. Comparative study of the rheotaxis in the cave salamander *Proteus anguinus* and his epigean relative *Necturus maculosus* (Proteidae, Urodela). *Behav. Proc.* 15:285–91.

Durant, P., and J. W. Dole. 1975. Aggressive behavior in *Colostethus* (= *Prostherapis*) *collaris* (Anura: Dendrobatidae). *Herpetologica* 31:23–26.

Duré, M. I. 1999. *Leptodactylus chaquensis* (NCN). Diet. *Herp. Rev.* 30:92.

Duré, M. I., and A. I. Kehr. 2001. Differential exploitation of trophic resources by two pseudid frogs from Corrientes, Argentina. *J. Herpetol.* 35:340–43.

Durham, L., and G. W. Bennett. 1963. Age, growth, and homing in the bullfrog. *J. Wildl. Manage.* 27:107–23.

Dushin, A. I. 1974. Diet of two frog species in the fisheries of the Mordovian ASSR. *Soviet J. Ecol.* 5:583–85.

Dutta, S. K., J. Sushree, and P. Mohanthi-Hejmadi. 1990–91. Breeding and development of *Ramanella variegata* (Anura: Microhylidae). *J. Zool. Soc. India* 42–43:55–76.

Dutta, S. K., K. Vasudevan, M. S. Chaitra, K. Shanker, and R. K. Aggarwal. 2004. Jurassic frogs and the evolution of amphibian endemism in the Western Ghats. *Curr. Sci.* 86:211–16.

Duval, M. 1928. L'adaptation des grenouilles a l'eau saumatre. *Ann. Physiol. Physiochem. Biol.* 4:181–89.

Dyson, M. L. 1985. *Aspects of social behaviour and mate choice in a caged population of painted reed frogs, Hyperolius marmoratus.* Unpublished M.S. thesis. Johannesburg, South Africa: University of the Witwatersrand.

———. 1989. *Aspects of social behaviour and mate choice in a caged population of painted reed frogs, Hyperolius marmoratus.* Unpublished PhD dissertation. Johannesburg, South Africa: University of the Witwatersrand.

Dyson, M. L., S. L Bush, and T. R. Halliday. 1998. Phonotaxis by female Majorcan midwife toads, *Alytes muletensis. Behaviour* 135:213–30.

Dyson, M. L., S. P. Henzi, T. R. Halliday, and L. Barrett. 1998. Success breeds success in mating male reed frogs (*Hyperolius marmoratus*). *Proc. Roy. Soc. London B* 265:1417–21.

Dyson, M. L., S. P. Henzi, and N. I. Passmore. 1994. The effect of changes in the relative timing of signals during female phonotaxis in the reed frog, *Hyperolius marmoratus. Anim. Behav.* 48:679–85.

Dyson, M. L., and N. I. Passmore. 1988. Two-choice phonotaxis in *Hyperolius marmoratus* (Anura: Hyperoliidae): The effect of temporal variation in presented stimuli. *Anim. Behav.* 36:648–52.

Dyson, M. L. and N. I. Passmore. 1992a. Effect of intermale spacing on female frequency preferences in the painted reed frog. *Copeia* 1992:1111–14.

Dyson, M. L. and N. I. Passmore. 1992b. Inter-male spacing and aggression in African painted reed frogs, *Hyperolius marmoratus. Ethology* 91:237–47.

Dyson, M. L., N. I. Passmore, P. J. Bishop, and S. P. Henzi. 1992. Male behavior and correlates of mating success in a natural population of African painted reed frogs (*Hyperolius marmoratus*). *Herpetologica* 48:2136–46.

Džukić, G., and M. L. Kalezić. 1984. Neoteny in the alpine newt population from the submediterranean area of Yugoslavia. *Alytes* 3:11–19.

Džukić, G., M. L. Kalezić, N. Tvrtković, and A. Djorović. 1990. An overview of the occurrence of paedomorphosis in Yugoslav newts (*Triturus,* Salamandridae) populations. *British Herpetol. Soc. Bull.* 34:16–22.

Eagleson, G. W. 1976. A comparison of the life histories and growth patterns of populations of the salamander *Ambystoma gracile* (Baird) from permanent low-altitude and montane lakes. *Can. J. Zool.* 54:2098–2111.

Eakin, R. M. 1973. *The third eye.* Berkeley: University of California Press.

Easteal, S. 1981. The history of introduction of *Bufo marinus* (Amphibia: Anura): A natural experiment in evolution. *Biol. J. Linn. Soc.* 16:93–113.

———. 1985. The ecological genetics of introduced populations of the giant toad, *Bufo marinus.* III. Geographical patterns of variation. *Evolution* 39:1065–75.

Easteal, S., and R. B. Floyd. 1986. The ecological genetics of introduced populations of the giant toad, *Bufo marinus* (Amphibia: Anura): Dispersal and neighborhood size. *Biol. J. Linn. Soc.* 27:17–45.

Easteal, S., E. K. van Beurden, R. B. Floyd, and M. D. Sabath. 1985. Continuing geographical spread of *Bufo marinus* in Australia: Range expansion between 1974 and 1980. *J. Herpetol.* 19:185–88.

Easton, D. P., P. S. Rutledge, and J. R. Spotila. 1987. Heat shock protein induction and induced thermal tolerance are independent in adult salamanders. *J. Exp. Zool.* 241:263–67.

Eaton, B. R., and C. A. Paszkowski. 1999. *Rana sylvatica* (wood frog). Predation. *Herp. Rev.* 30:164.

Eaton, B. R., W. M. Tonn, C. A. Paszkowski, A. J. Danylchuk, and S. M. Boss. 2005. Indirect effects of fish winterkills on amphibian populations in boreal lakes. *Can. J. Zool.* 83:1532–39.

Eaton, T. H. 1941. Notes on the life history of *Dendrobates auratus. Copeia* 1941:93–95.

Ebenman, B. 1992. Evolution in organisms that change their niches during the life cycle. *Am. Nat.* 139:990–1021.

Eberhard, W. 1968. *The local cultures of south and east China.* The Netherlands: E. J. Brill, Leiden.

Eberhardt, L. S. 1994. Oxygen consumption during singing by male Carolina wrens (*Thryothorus ludovicianus*). *Auk* 111:124–30.

Ebisuno, M., and A. Gentilli. 2002. Reproductive site selection and characteristics of sources and sinks in an Italian tree frog metapopulation (*Hyla intermedia,* Boulenger 1882). *Rev. Ecol. (Terre et Vie)* 57:263–78.

Ecker, A. 1889. *The anatomy of the frog.* Translated by G. Haslam. Oxford, UK: Clarendon Press.

Edenhamn, P. 1996. *Spatial dynamics of the European tree frog (Hyla arborea L.) in a heterogeneous landscape.* Unpublished PhD dissertation. Uppsala, Sweden: Uppsala University.

Edenhamn, P., M. Hoggren, and A. Carlson. 2000. Genetic diversity and fitness in peripheral and central populations of the European tree frog *Hyla arborea. Hereditas* 133:115–22.

Edgar, P. W., R. A. Griffiths, and J. P. Foster. 2005. Evaluation of translocation as a tool for mitigating development threats to great crested newts (*Triturus cristatus*) in England, 1990–2001. *Biol. Cons.* 122:45–52.

Edgren, R. A. 1955. The natural history of the hog-nosed snakes, genus *Heterodon:* A review. *Herpetologica* 11:105–17.

Edmunds, M. 1974. *Defence in animals.* Essex, UK: Longman, Harrow.

Edwards, J. R., J. L. Jenkins, and D. L. Swanson. 2004. Seasonal effects of dehydration on glucose mobilization in freeze-tolerant chorus frogs (*Pseudacris triseriata*) and freeze-intolerant toads (*Bufo woodhousii* and *B. cognatus*). *J. Exp. Zool.* 301A:521–31.

Edwards, J. R., K. L. Koster, and D. L. Swanson. 2000. Time course for cryoprotectant synthesis in the freeze-tolerant chorus frog, *Pseudacris triseriata. Comp. Biochem. Physiol.* A 125:367–75.

Edwards, W. F. 1817. Sur l'asphyxie, considérée dans les Batraciens. *Ann. Chim. Phys.* 5:356–80.

———. 1824. *De L'influence des agents physique sur la vie.* Paris: Crochard.

Efford, I. E., and J. A. Mathias. 1969. A comparison of two salamander populations in Marion Lake, British Columbia. *Copeia* 1969:723–36.

Efford, I. E., and K. Tsumura. 1973. A comparison of the food of salamanders and fish in Marion Lake, British Columbia. *Trans. Am. Fish. Soc.* 102:33–47.

Egan, R. S., and P. W. C. Paton. 2004. Within-pond parameters affecting oviposition by wood frogs and spotted salamanders. *Wetlands* 24:1–13.

Egger, F. 1931. Frosch und Kröte bei den alten Ägyptern. *Mitt. Geogr. Ethnol. Ges. Basel* 4:1–24.

Eggermont, J. J. 1988. Mechanisms of sound localization in anurans. In *The evolution of the amphibian auditory system,* ed. B. Fritzsch, M. J. Ryan, W. Wilczynski, T. E. Hetherington, and W. Walkowiak, 307–36. New York: Wiley.

Eggert, C., and R. Guyétant. 1999. Age structure of a spadefoot toad *Pelobates fuscus* (Pelobatidae) population. *Copeia* 1999:1127–30.

———. 2003. Reproductive behaviour of spadefoot toads (*Pelobates fuscus*): Daily sex ratios and males' tactics, ages, and physical condition. *Can. J. Zool.* 81:46–51.

Ehmann, H., and G. Swan. 1985. Reproduction and development in the marsupial frog, *Assa darlingtoni* (Leptodactylidae, Anura). In *Biology of Australasian frogs and reptiles,* ed. G. Grigg, R. Shine, and H. Ehmann, 279–85. Chipping Norton, New South Wales, Australia: Surrey Beatty and Sons.

Ehret, G., and R. R. Capranica. 1980. Masking patterns and filter characteristics of auditory nerve fibers in the green treefrog (*Hyla cinerea*). *J. Comp. Physiol.* 141:1–12.

Ehret, G., and H. C. Gerhardt. 1980. Auditory masking and effects of noise on responses of the green treefrog (*Hyla cinerea*) to synthetic mating calls. *J. Comp. Physiol.* 141:13–18.

Ehret, G., J. Tautz, B. Schmitz, and P. M. Narins. 1990. Hearing through the lungs: Lung-eardrum transmission of sound in the frog *Eleutherodactylus coqui. Naturwissenschaften* 77:192–94.

Ehret, G., E. K. Werth, and T. Kamada. 1994. The lung-eardrum pathway in three treefrog and four dendrobatid frog species: Some properties of sound transmission. *J. Exp. Biol.* 195:329–43.

Eibl-Eibesfeldt, I. 1950. Ein Beitrag zur Paarungsbiologie der Erdkröte (*Bufo bufo* L.). *Behaviour* 2:217–36.

———. 1952. Vergleichende Verhaltensstudien an Anuren. I. Zur Paarungsbiologie des Laubfrosches, *Hyla arborea* L.). *Z. Tierpsychol.* 9:383–95.

———. 1956. Vergleichende Verhaltensstudien an Anuren. II. Zur Paarungsbiologie der Gattungen *Bufo, Hyla, Rana,* und *Pelobates. Zool. Anz.* Suppl. 19:315–23.

Eichelberg, H., and H. J. Obert. 1976. Fat and glycogen utilization in the larynx muscles of fire-bellied toads (*Bombina bombina* L.) during calling activity. *Cell Tissue Res.* 167:1–10.

Eichelberg, H., and H. Schneider. 1973. Die Feinstruktur der Kehlkopfmuskeln des Laubfrosches, *Hyla arborea arborea* (L.) im Vergleich zu eninem Skelettmuskel. *Z. Zellforsch.* 141:223–33.

———. 1974. The fine structure of the larynx muscles in female tree frogs, *Hyla a. arborea* L. (Anura, Amphibia). *Cell Tissue Res.* 152:185–91

Eisenberg, J. F. 1981. *The mammalian radiations.* Chicago: University of Chicago Press.

———. 1983. The gopher tortoise as a keystone species. *Proc. 4th Ann. Meeting Gopher Tortoise Council.* Gainesville: Florida State Museum.

Eisentraut, M. 1932. Biologische Beobachtungen in Bolivianischen Chaco. V. Ein neuer Fall von Brutfursorge in der Froschgattung *Leptodactylus. Z. Morphol. Ökol. Tiere* 26:317–26.

Eklöv, P. 2000. Chemical cues from multiple predator-prey interactions induce changes in behavior and growth of anuran larvae. *Oecologia* 123:192–99.

Eklöv, P., and E. E. Werner. 2000. Multiple predator effects on size-dependent behavior and mortality of two species of anuran larvae. *Oikos* 88:250–58.

Elepfandt, A. 1996. Underwater acoustics and hearing in the clawed frog, *Xenopus.* In *The biology of Xenopus,* ed. R. C. Tinsley and H. R. Kobel, 177–93. New York: Oxford University Press.

Elias, H., and J. Shapiro. 1957. Histology of the skin of some toads and frogs. *Amer. Mus. Novitates* (1819):1–27.

Elias, P. 1984. Salamanders of the northwestern highlands of Guatemala. *Nat. Hist. Mus. Los Angeles County Contrib. Sci.* (348):1–20.

Elinson, R. P. 1990. Direct development in frogs: Wiping the recapitulationist slate clean. *Sem. Develop. Biol.* 1:263–70.

Elkan, E. 1976a. Ground substance: An anuran defense agent against desiccation. In *Physiology of the Amphibia,* vol. 3, ed. B. Lofts, 101–10. New York: Academic Press.

Elkan, E. 1976b. Pathology in the Amphibia. In *Physiology of the Amphibia,* vol. 3, ed. B. Lofts, 273–312. New York: Academic Press.

Ellen, R. F., A. F. Stimson, and J. Menzies. 1976. Structure and inconsistency in Nuaulu categories for amphibians. *J. d'Agric. Tropic. Bot. Appl.* 23:125–38.

Ellinger, N., and R. Jehle. 1997. Struktur und dynamik einer Donaukammolch-Population (*Triturus dobrogicus,* Kiritzescu 1903) am Endelteich bei Wein: Ein Überblick über neun Untersuchungsjahre. In *Populationsbiologie von amphibien. Stapfia* 51, ed. W. Hödl, R. Jehle, and G. Gollmann, 133–50. Linz, Austria: Oberösterreich Landesmuseum.

Elliot, A. B. and M. M. Chew. 1973. Vasotocin-like activity in the plasma of the euryhaline frog (*Rana cancrivora*) after transfer from fresh water to saline. *J. Endocrinol.* 66:427–34.

Elliott, S. A., L. B. Kats, and J. A. Breeding. 1993. The use of conspecific chemical cues for cannibal avoidance in California newts (*Taricha torosa*). *Ethology* 95:186–92.

Ellison, A. M. 2004. Wetlands of Central America. *Wetlands Ecol. Manage.* 12:3–55.

Elmberg, J. 1986. Apparent lack of territoriality during the breeding season in a boreal population of common frogs *Rana temporaria* L. *Herpetol. J.* 1:81–85.

———. 1987. Random mating in a boreal population of European common frogs *Rana temporaria. Holarc. Ecol.* 10:193–95.

———. 1990. Long-term survival, length of breeding season, and operational sex ratio in a boreal population of common frogs, *Rana temporaria* L. *Can. J. Zool.* 68:121–27.

———. 1991. Ovarian cyclicity and fecundity in boreal common frogs *Rana temporaria* L. along a climatic gradient. *Funct. Ecol.* 5:340–50.

Elmberg, J., and P. Lundberg. 1988. Navigation in breeding-migrating common frogs *Rana temporaria:* A simple translocation experiment. *Amphibia-Reptilia* 9:169–73.

———. 1991. Intraspecific variation in calling, time allocation and energy reserves in breeding male common frogs *Rana temporaria. Ann. Zool. Fennici* 28:23–29.

Else, P. L., and A. F. Bennett. 1987. The thermal dependence of locomotor performance and muscle contractile function in the salamander *Ambystoma tigrinum nebulosum. J. Exp. Biol.* 128:219–33.

Elzen, P. van der. 1976. Remarques sur la biologie de *Pelodytes punctatus* Daudin 1802 (Amphibia, Pelobatidae) en Camargue. *Rev. Fr. Aquariol.* 3:99–104.

Elzen, P. van der, and D. A. Kreulen. 1979. Notes on the vocalizations of some amphibians from the Serengeti National Park, Tanzania. *Bonn. Zool. Beit.* 30:385–403.

Emerson, S. B. 1976. Burrowing in frogs. *J. Morphol.* 149:437–58.

———. 1978. Allometry and jumping in frogs: helping the twain to meet. *Evolution* 32:551–64.

———. 1979. The ilio-sacral articulation in frogs: form and function. *Biol. J. Linn. Soc.* 11:153–68.

———. 1982. Frog postcranial morphology: identification of a functional complex. *Copeia* 1982:603–13.

———. 1983. Functional analysis of frog pectoral girdles. The epicoracoid cartilages. *J. Zool., London* 201:293–308.

———. 1985. Skull shape in frogs—Correlations with diet. *Herpetologica* 41:177–88.

———. 1986. Heterochrony and frogs: The relationship of a life history trait to morphological form. *Am. Nat.* 127:167–83.

———. 1988a. Convergence and morphological constraint in frogs: Variation in postcranial morphology. *Fieldiana (Zool.)* 43:1–9.

———. 1988b. The giant tadpole of *Pseudis paradoxa. Biol. J. Linn. Soc.* 34:93–104.

———. 1991. A biomechanical perspective on the use of forelimb length as a measure of sexual selection in frogs. *J. Evol. Biol.* 4:671–78.

———. 1992. Courtship and nest-building behavior of a Bornean frog, *Rana blythi. Copeia* 1992:1123–27.

———. 1994. Testing pattern predictions of sexual selection: A frog example. *Am. Nat.* 143:848–69.

———. 1996. Phylogenies and physiological processes—The evolution of sexual dimorphism in Southeast Asian frogs. *Syst. Biol.* 45:278–89.

———. 2000. Vertebrate secondary sexual characters—Physiological mechanisms and evolutionary patterns. *Am. Nat.* 156:84–91.

———. 2001a. A macroevolutionary study of historical contingency in the fanged frogs of Southeast Asia. *Biol. J. Linn. Soc.* 73:139–51

———. 2001b. Male advertisement calls: behavioral variation and physiological processes, pp. 36–44. In *Anuran communication,* ed. M. J. Ryan, 36–44. Washington, DC: Smithsonian Institution Press.

Emerson, S. B., and D. Berrigan. 1993. Systematics of southeast Asian ranids: Multiple origins of voicelessness in the subgenus *Limnonectes* (Fitzinger). *Herpetologica* 49:22–31.

Emerson, S. B., and S. K. Boyd. 1999. Mating vocalizations of female frogs: control and evolutionary mechanisms. *Brain Behav. Evol.* 53:187–97.

Emerson, S. B., and D. M. Bramble. 1993. Scaling, allometry, and skull design. In *The skull. Vol. 3. Functional and evolutionary mechanisms,* ed. J. Hanken and B. K. Hall, 384–421. Chicago: University of Chicago Press.

Emerson, S. B., L. Carroll, and D. L. Hess. 1997. Hormonal induction of thumb pads and the evolution of secondary sexual characteristics of the Southeast Asian fanged frog, *Rana blythii. J. Exp. Zool.* 279:587–96.

Emerson, S. B., T. A. Cooper, and J. R. Ehleringer. 1990. Convergence in reflectance spectra among treefrogs. *Funct. Ecol.* 4:47–51.

Emerson, S. B., and H. J. de Jongh. 1980. Muscle activity at the ilio-sacral articulation of frogs. *J. Morphol.* 166:129–44.

Emerson, S. B., and D. Diehl. 1980. Toe pad morphology and mechanisms of sticking in frogs. *Biol. J. Linn. Soc.* 13:199–216.

Emerson, S. B., H. W. Greene, and E. L. Charnov. 1994. Allometric aspects of predator-prey interactions. In *Ecological morphology,* ed. P. C. Wainwright and S. M. Reilly, 123–39. Chicago: University of Chicago Press.

Emerson, S. B., A. Greig, L. Carroll, and G. S. Prins. 1999. Androgen receptors in two androgen-mediated, sexually dimorphic characters of frogs. *Gen. Comp. Endocrinol.* 114:173–80.

Emerson, S. B., and D. L. Hess. 1996. The role of androgens in opportunistic breeding tropical frogs. *Gen. Comp. Endocrinol.* 103:220–30.

———. 2001. Glucocorticoids, androgens, testis mass, and the energetics of vocalization in breeding male frogs. *Horm. Behav.* 39:59–69.

Emerson, S. B., and R. F. Inger. 1992. The comparative ecology of voiced and voiceless Bornean frogs. *J. Herpetol.* 26:482–90.

Emerson, S. B., R. F. Inger, and D. Iskandar. 2000. Molecular systematics and biogeography of the fanged frogs of Southeast Asia. *Mol. Phylogen. Evol.* 16:131–42.

Emerson, S. B., and M. A. R. Koehl. 1990. The interaction of behavioral and morphological change in the evolution of a novel locomotor type: "Flying" frogs. *Evolution* 44:1931–46.

Emerson, S. B., C. Richards, R. C. Drewes, and K. M. Kjer. 2000. On the relationships among ranoid frogs: A review of the evidence. *Herpetologica* 56:209–30.

Emerson, S. B., C. Rosewemitt, and D. Hess. 1993. Androgen levels in a Bornean voiceless frog, *Rana blythi. Can. J. Zool.* 71:196–203.

Emerson, S. B., J. Travis, and M. A. R. Koehl. 1990. Functional complexes and additivity in performance: A test case with "flying" frogs. *Evolution* 44:2153–57.

Emerson, S. B., and H. Voris. 1992. Competing explanations for sexual dimorphism in a voiceless Bornean frog. *Funct. Ecol.* 6:654–60.

Emerson, S. B., and R. Ward. 1998. Male secondary sexual characteristics, sexual selection, and molecular divergence in fanged ranid frogs of Southeast Asia. *Zool. J. Linn. Soc.* 122:537–53.

Emery, A. R., A. H. Berst, and K. Kodaira. 1972. Under-ice observations of wintering sites of leopard frogs. *Copeia* 1972:123–26.

Emilio, M. G., and G. Shelton. 1972. Factors affecting blood flow to the lungs in the amphibian, *Xenopus laevis. J. Exp. Biol.* 56:67–77.

———. 1974. Gas exchange and its effect on blood gas concentrations in the amphibian *Xenopus laevis. J. Exp. Biol.* 60:567–79.

———.1980. Carbon dioxide exchange and its effects on pH and bicarbonate equilibria in the blood of the amphibian, *Xenopus laevis. J. Exp. Biol.* 85:253–62.

Emlen, S. T. 1968. Territoriality in the bullfrog, *Rana catesbeiana. Copeia* 1968:240–43.

———. 1976. Lek organization and mating strategies in the bullfrog. *Behav. Ecol. Sociobiol.* 1:283–313.

———. 1977. "Double clutching" and its possible significance in the bullfrog. *Copeia* 1977:749–51.

Emlen, S. T., and L. W. Oring. 1977. Ecology, sexual selection, and evolution of mating systems. *Science* 197:215–23.

Endler, J. 1970. Kinesthetic orientation in the California newt (*Taricha torosa*). *Behaviour* 37:15–23.

Ender, J. A. 1986. Defense against predators. In *Predator-prey relationships,* ed. M. E. Feder and G. V. Lauder, 109–34. Chicago: University of Chicago Press.

Engeler, B., and H.-U. Reyer. 2001. Choosy femlaes and indiscriminate males: Mate choice in mixed populations of sexual and hybridogenetic water frogs (*Rana lessonae, Rana esculenta*). *Behav. Ecol.* 12:600–606.

Engle, D. L. 1985. The production of haemoglobin by small pond *Daphnia pulex:* Intraspecific variation and its relation to habitat. *Freshw. Biol.* 15:631–38.

Eniang, E. A., R. King, J. Lea, D. Capizzi, and L. Luiselli. 2003. Trophic niches of four sympatric rainforest anurans from southern Nigeria: Does resource partitioning play a role in structuring the community? *Rev. Ecol. (Terre Vie)* 58:321–35.

Enquist, M., O. Leimar, T. Ljungberg, Y. Mallner, and N. Segerdahl. 1990. A test of the sequential assessment game: Fighting in the cichlid fish *Nanancara anomala. Anim. Behav.* 40:1–14.

Epping, W. J. M., and J. J. Eggermont. 1986a. Sensitivity of neurons in the auditory midbrain of the grassfrog to temporal characteristics of sound. I. Stimulation with acoustic clicks. *Hearing Res.* 24:37–54.

———. 1986b. Sensitivity of neurons in the auditory midbrain of the grassfrog to temporal characteristics of sound. II. Stimulation with amplitude modulated sound. *Hearing Res.* 24:55–72.

Epstein, M. S., and D. G. Blackburn. 1997. Histology and histochemistry of androgen-stimulated nuptial pads in the leopard frog, *Rana pipiens*, with notes on nuptial gland evolution. *Can. J. Zool.* 74:472–77.

Erlinge, S. 1967. Food habits of the fish otter (*Lutra lutra* L.) in south Swedish habitats. *Viltrevy* 4:371–43.

———. 1968. Food studies on captive otters, *Lutra lutra* L. *Oikos* 19:259–70.

———. 1969. Food habits of the otter (*Lutra lutra*) and the mink (*Mustela vison*) in a trout water in southern Sweden. *Oikos* 20:1–7.

———. 1972. Interspecific relations between otter (*Lutra lutra*) and mink (*Mustela vison*) in Sweden. *Oikos* 23:327–35.

———. 1986. Specialists and generalists among the mustelids. *Lutra* 29:5–11.

Ernst, C. H. 1992. *Venomous reptiles of North America.* Washington, DC: Smithsonian Institution Press.

Ernst, C. H., and R. W. Barbour. 1989a. *Turtles of the world.* Washington, DC: Smithsonian Institution Press.

———. 1989b. *Snakes of eastern North America.* Fairfax, VA: George Mason University Press.

Ernst, C. H., R. W. Barbour, and J. E. Lovich. 1994. *Turtles of the United States and Canada.* Washington, DC: Smithsonian Institution Press.

Ernst, S. G. 1945. The food of the red-shouldered hawk in New York State. *Auk* 62:452–53.

Ernst, V. 1973. The digital pads of the tree frog *Hyla cinerea.* I. The epidermis. *Tissue and Cell* 5:83–96.

Erspamer, V. 1971. Biogenic amines and active polypeptides of the amphibian skin. *Ann. Rev. Pharmacol.* 11:327–50.

———. 1994. Bioactive secretions of the amphibian integument. In *Amphibian biology, vol. 1. The integument,* ed. H. Heatwole and G. T. Barthalmus, 178–350. Chipping Norton, New South Wales, Australia: Surrey Beatty and Sons.

Erspamer, V., A. Anastasi, G. Bertaccini, and J. M. Cei. 1964. Structure and pharmacological actions of physalaemin, the main active polypeptide of the skin of *Physalaemus fuscumaculatus. Experientia* 20:489–90.

Erspamer, V., G. Bertaccini, and J. M. Cei. 1962. Occurrence of an eledoisin-like polypeptide (physalaemin) in skin extracts of *Physalaemus fuscumaculatus. Experientia* 18:562.

Erspamer, V., G. de Caro, and R. E. Endean. 1966. Occurrence of physaleamin-like polypeptides (uperolein) and other active polypeptides in the skin of *Uperoleia rugosa. Experientia* 22:738.

Erspamer, V., G. Erspamer, G. Mazzanti, and R. Endean. 1984. Active peptides in the skins of one hundred amphibian species from Australia and Papua New Guinea. *Comp. Biochem. Physiol.* 77C:99–108.

Erspamer, V., L. Negri, F. Erspamer, and R. Endean. 1975. Uperolein and other active peptides in the skin of the Australian leptodactylid frogs *Uperoleia* and *Taudactylus. Arch. Pharmacol.* 289:41–54.

Esher, K. 1972. Die Amphibien des Kantons Zürich. *Vierteljarshr. Naturforsh. Ges. Zürich* 117:335–80.

Espina, S., and M. Rojas. 1972. A comparison of the size of the urinary bladder of two South American anurans of different habitat. *Comp. Biochem. Physiol.* 41A:115–20.

Esteban, M., M. García-París, D. Buckley, and J. Castanet. 1999. Bone growth and age in *Rana saharica,* a water frog living in a desert environment. *Ann. Zool. Fennici* 36:53–62.

Esteban, M., M. García-París, and J. Castanet. 1996. Use of bone histology in estimating the age of frogs (*Rana perezi*) from a warm temperate climate area. *Can. J. Zool.* 74:1914–21.

Esteban, M., M. J. Sánchez-Herráiz, L. J. Barbadillo, J. Castanet, and R. Márquez. 2002. Effects of age, size and temperarture on the advertisement calls of two Spanish populations of *Pelodytes punctatus. Amphibia-Reptilia* 23:249–58.

Esteban, M., and B. Sanchiz. 2000. Differential growth and longevity in low and high altitude *Rana iberica* (Anura, Ranidae). *Herpetol. J.* 10:19–26.

Estes, R., Z. V. Spinar, and E. Nevo. 1978. Early Cretaceous pipid tadpoles from Israel (Amphibia: Anura). *Herpetologica* 34:374–75.

Estrada, A., and R. Estrada. 1977. Patterns of predation in a free-ranging troop of stumptail macaques (*Macaca arctoides*): Relations to ecology II. *Primates* 18:633–46.

Estrada, A. R. 1987. Los nidos terrestres de dos especies de anfibios cubanos del genero *Eleutherodactylus* (Anura: Leptodactylidae). *Poeyana* (352):1–9.

———. 1990. Las puestas de *Eleutherodactylus varians* (Gundlach y Peters). *Rev. Biol.* 4:163–67.

———. 1992. Comentarios sobre *Eleutherodactylus intermedius* Barbour et Shreve, 1937 de la Region Oriental de Cuba. *Comm. Brev. Zool.* 1992:12–14.

Estrada, A. R., and H. B. Hedges. 1996. At the lower limit in tetrapods: A new diminutive frog from Cuba (Leptodactylidae: *Eleutherodactylus*). *Copeia* 1996:852–59.

Eterovick, P. C. 1999. Use and sharing of calling and retreat sites by *Phyllodytes luteolus* in a modified environment. *J. Herpetol.* 33:17–22.

———. 2000. Effects of aggregation on feeding of *Bufo crucifer* tadpoles (Anura, Bufonidae). *Copeia* 2000:210–15.

———. 2003. Distribution of anuran species among montane streams in south-eastern Brazil. *J. Trop. Ecol.* 19:219–28.

Eterovick, P. C., and I. S. Barros. 2003. Niche occupancy in south-eastern Brazilian tadpole communities in montane-meadow streams. *J. Trop. Ecol.* 19:439–48.

Eterovick, P. C., and G. W. Fernandes. 2001. Tadpole distribution within montane meadow streams at the Serro do Cipó, south-eastern Brazil: Ecological or phylogenetic constraints? *J. Trop. Ecol.* 17:683–93.

Eterovick, P. C., A. C. O. de Queirox Carnaval, D. M. Borges-Nojosa, D. L. Silvano, M. V. Segalla, and I. Sazima. 2005. Amphibian declines in Brazil: An overview. *Biotropica* 37:166–79.

Eterovick, P. C., and I. Sazima. 1999. Description of the tadpole of *Bufo rufus* with notes on aggregative behavior. *J. Herpetol.* 33:711–13.

———. 2000. Structure of an anuran community in a montane meadow in southeastern Brazil: Effects of seasonality, habitat, and predation. *Amphibia-Reptilia* 21:439–61.

Etges, W. J. 1987. Call site choice in male anurans. *Copeia* 1987:910–23.

Etheridge, K. 1990a. The energetics of estivating sirenid salamanders (*Siren lacertina* and *Pseudobranchus striatus*). *Herpetologica* 46:407–14.

———. 1990b. Water balance in estivating sirenid salamanders (*Siren lacertina*) *Herpetologica* 46:400–406.

Etkin, W. E. 1964. Metamorphosis. In *Physiology of the Amphibia,* vol. 1, ed. J. A. Moore, 427–68. New York: Academic Press.

———. 1968. Hormonal control of amphibian metamorphosis. In *Metamorphosis: A problem in developmental biology,* ed. W. Etkin and L. J. Gilbert, 313–48. New York: Appleton-Century-Crofts.

Etkin, W. E. 1970. The endocrine mechanism of amphibian metamorphosis, an evolutionary achievement. In *Hormones and the environment,* ed.G. K. Benson and J. G. Phillips, 137–55. Cambridge: Cambridge University Press.

Evans, A. L., and D. C. Forester. 1996. Conspecific recognition by *Desmognathus ochrophaeus* using substrate-borne odor cues. *J. Herpetol.* 30:447–51.

Evans, A. L., D. C. Forester, and B. S. Masters. 1997. Recognition by population and genetic similarity in the mountain dusky salamander, *Desmognathus ochrophaeus* (Amphibia: Plethodontidae). *Ethology* 103:865–75.

Evans, B. J., R. M. Brown, J. A. McGuire, J. Supriatna, N. Andayani, A. Diesmos, D. Iskandar, D. J. Melnick, and D. C. Cannatella. 2003. Phylogenetics of fanged frogs: Testing biogeographical hypotheses at the interface of the Asian and Australian Faunal Zones. *Syst. Biol.* 52:794–819.

Evans, B. J., D. B. Kelley, R. C. Tinsley, D. J. Melnick, and D. C. Cannatella. 2004. A mitochondrial DNA phylogeny of African clawed frogs: Phylogeography and implications for polyploid evolution. *Mol. Phylogen. Evol.* 33:197–213.

Evans, S. E., and M. Borsuk-Białynicka. 1998. A stem-group frog from the Early Triassic of Poland. *Acta Palaeont. Polon.* 43:573–80.

Evans, S. E., C. Lally, D. C. Chure, A. Elder, and J. A. Maisano. 2005. A Late Jurassic salamander (Amphibia: Caudata) from the Morrison Formation of North America. *Zool. J. Linn. Soc.* 143:599–616.

Evans, S. E., and A. R. Milner. 1996. A metamorphosed salamander from the early Cretaceous of Las Hoyas, Spain. *Phil. Trans. Roy. Soc. London, Ser. B* 351:627–46.

Evans, S. E., A. R. Milner, and C. Werner. 1996. Sirenid salamanders and a gymnophionan amphibian from the Cretaceous of the Sudan. *Palaeontology* 39:77–95.

Ewer, R. F. 1952. The effects of posterior pituitary extracts on water balance in *Bufo carens* and *Xenopus laevis,* together with some general considerations of anuran water economy. *J. Exp. Biol.* 29:429–39.

———. 1973. *The carnivores.* London: Weidenfeld and Nicholson.

Ewert, J.-P. 1987. Neuroethology of releasing mechanisms: Prey-catching in toads. *Behav. Brain Sci.* 10:337–405.

Exbrayat, J.-M., M. Delsol, and J. Flatin. 1981. Premieres remarques sur la gestation chez *Typhlonectes compressicaudus* (Dumeril et Bibron, 1841), amphibien apode vivipare. *C. R. Acad. Sci., Paris* 292:417–20.

Exbrayat, J.-M., M. Delsol, and J. Flatin. 1982. Observations concernant la gestation de *Typhlonectes compressicaudus* (Dumeril et Bibron, 1841), amphibien apode vivipare. *Bull. Soc. Zool. France* 107:486.

Exbrayat, J.-M., M. Delsol, and J. Lescure. 1983. La viviparité chez *Typhlonectes compressicauda,* amphibien apode. *Bull. Soc. Herpetol. France* 26:23–24.

Fabrezi, M., and S. B. Emerson. 2003. Parallelism and convergence in anuran fangs. *J. Zool., London* 260:41–51.

Faccio, S. D. 2003. Postbreeding emigration and habitat use by Jefferson and spotted salamanders in Vermont. *J. Herpetol.* 37:479–89.

Fachbach, G. 1969. Zur Evolution der Embryonal-bzw. Larvelentwicklung bei *Salamandra. Z. Zool. Syst. Evolut.-forsch.* 7:128–45.

———. 1976. Biologie, Taxonomie und phylogentische Beziehungen der verschiedenen Unterarten von *Salamandra salamandra* im Bereich der Iberischen Halbinsel. Part II. *Z. Zool. Syst. Evolut.-forsch.* 14:81–103.

Fachin Teran, A., R. C. Vogt, and M. de Fatima Soares Gomez. 1995. Food habits of an assemblage of five species of turtles in the Rio Guapore, Rondonia, Brazil. *J. Herpetol.* 29:536–47.

Fagotti, A., L. Morosi, I. Di Rosa, R. Clarioni, F. Simoncelli, R. Pascolini, R. Pellegrino, G.-D. Guex, and H. Hotz. 2005. Bioaccumulation of organochlorine pesticides in frogs of the *Rana esculenta* complex in central Italy. *Amphibia-Reptilia* 26:93–104.

Fahrig, L., J. H. Pedlar, S. E. Pope, P. D. Taylor, and J. F. Wegner. 1995. Effect of road traffic on amphibian density. *Biol. Cons.* 73:177–82.

Fair, J. W. 1970. Comparative rates of rehydration from soil in two species of toads, *Bufo boreas* and *Bufo punctatus. Comp. Biochem. Physiol.* 34:281–87.

Fairchild, L. 1981. Mate selection and behavioral thermoregulation in Fowler's toads. *Science* 212:950–51.

———. 1984. Male reproductive tactics in an explosive breeding toad population. *Amer. Zool.* 24:407–18.

Fairley, J. S. 1984. Otters feeding on breeding frogs. *Irish Nat. J.* 21:372.

Fairley, J. S., and T. K. McCarthy. 1985. Do otters prey on breeding natterjack toads? *Irish Nat. J.* 21:544–45.

Faivovich, J., C. F. B. Haddad, P. C. A. Garcia, D. R. Frost, J. A. Campbell, and W. C. Wheeler. 2005. Systematic review of the frog family Hylidae, with special reference to Hylinae: phylogenetic analysis and taxonomic revision. *Bull. Amer. Mus. Nat. Hist.* 294:1–240.

Fandiño, M. C., H. Lüddecke, and A. Amézquita. 1997. Vocalisation and larval transportation of male *Colostethus subpunctatus* (Anura: Dendrobatidae). *Amphibia-Reptilia* 18:39–48.

Fanelli, G. M., and L. Goldstein. 1964. Ammonia excretion in the neotenous newt, *Necturus maculosus* (Rafinesque). *Comp. Biochem. Physiol.* 13:193–204.

Faragher, S. G., and R. G. Jaeger. 1997. Distribution of adult and juvenile red-backed salamanders: Testing new hypotheses regarding territoriality. *Copeia* 1997:410–14.

———. 1998. Tadpole bullies: examining mechanisms of competition in a community of larval anurans. *Can. J. Zool.* 76:144–53.

Farbman, A. I., and R. C. Gesteland. 1974. Fine structure of the olfactory epithelium in the mudpuppy, *Necturus maculosus. Am. J. Anat.* 139:227–43.

Faria, M. M. 1993. Sexual behaviour of Bosca's newt, *Triturus boscai. Amphibia-Reptilia* 14:169–85.

———. 1995. A field study of reproductive interactions in Bosca's newt, *Triturus boscai. Amphibia-Reptilia* 16:357–74.

Farrar, E. S., and R. K. Dupré. 1983. The role of diet in glycogen storage by juvenile bullfrogs prior to overwintering. *Comp. Biochem. Physiol.* 75A:255–60.

Farrell, M. P., and J. A. MacMahon. 1969. An eco-physiological study of water economy in eight species of tree frogs (Hylidae). *Herpetologica* 25:279–94.

Fasola, M., F. Barbieri, and L. Canova. 1993. Test of an electronic individual tag for newts. *Herpetol. J.* 3:149–50.

Fauth, J. E. 1990. Interactive effects of predators and early larval dynamics of the treefrog *Hyla chrysoscelis. Ecology* 71:1609–16.

———. 1998. Investigating geographical variation in interspecific interactions using common garden experiments. In *Experimental ecology,* ed. W. J. Resetarits, Jr. and J. Bernardo, 394–415. New York: Oxford University Press.

———. 1999a. Identifying potential keystone species from field data—An example from temporary ponds. *Ecol. Lett.* 2:36–43.

———. 1999b. Interactions between branchiate mole salamanders (*Ambystoma talpoideum*) and lesser sirens (*Siren intermedia*): Asymmetrical competition and intraguild predation. *Amphibia-Reptilia* 20:119–32.

Fauth, J. E., and W. J. Resetarits, Jr. 1991. Interactions between the salamander *Siren intermedia* and the keystone predator *Notophthalmus viridescens. Ecology* 72:827–38.

———. 1999. Biting in the salamander *Siren intermedia intermedia:* Courtship component or agonistic behavior? *J. Herpetol.* 33:493–96.

Fauth, J. E., W. J. Resetarits, Jr., and H. M. Wilbur. 1990. Interactions between larval salamanders: A case of competitive equality. *Oikos* 58:91–99.

Fay, R. R., and A. N. Popper. 1985. The octavolateralis system. In *Functional vertebrate morphology,* ed. M. Hildebrand, D. M. Bramble, K. Liem, and D. B. Wake, 291–316. Cambridge, MA: Harvard University Press.

Feder, J. H. 1979. Natural hybridization and genetic divergence between the toads *Bufo boreas* and *Bufo punctatus. Evolution* 33:1089–97.

Feder, J. H., G. Z. Wurst, and D. B. Wake. 1978. Genetic variation in western salamanders of the genus *Plethodon,* and the status of *Plethodon gordani. Herpetologica* 34:64–69.

Feder, M. E. 1976a. Lunglessness, body size, and metabolic rate in salamanders. *Physiol. Zool.* 49:398–406.

———. 1976b. Oxygen consumption and body temperature in Neotropical and temperate zone lungless salamanders (Amphibia: Plethodontidae). *J. Comp. Physiol.* 110:197–208.

———. 1977. Oxygen consumption and activity in salamanders: effect of body size and lunglessness. *J. Exp. Zool.* 202:403–14.

———. 1978a. Effect of temperature on post-activity oxygen consumption in lunged and lungless salamanders. *J. Exp. Zool.* 206:179–190.

———. 1978b. Environmental variability and thermal acclimation in Neotropical and temperate-zone salamanders. *Physiol. Zool.* 51:7–16.

———. 1981. Effect of body size, trophic state, time of day, and experimental stress on oxygen consumption of anuran larvae: An experimental assessment and evaluation of the literature. *Comp. Biochem. Physiol.* 70A:497–508.

———. 1982a. Environmental variability and thermal acclimation of metabolism in tropical anurans. *J. Therm. Biol.* 7:23–28.

———. 1982b. Thermal ecology of Neotropical lungless salamanders (Amphibia: Plethodontidae): Environmental temperatures and behavioral responses. *Ecology* 63:1665–74.

———. 1983a. Effect of hypoxia and body size on the energy metabolism of lungless tadpoles, *Bufo woodhousei,* and air-breathing anuran larvae. *J. Exp. Zool.* 228:11–19.

———. 1983b. Integrating the ecology and physiology of plethodontid salamanders. *Herpetologica* 39:291–310.

———. 1983c. The relation of air breathing and locomotion to predation on tadpoles, *Rana berlandieri,* by turtles. *Physiol. Zool.* 56:522–31.

———. 1983d. Responses to acute aquatic hypoxia in larvae of the frog *Rana berlandieri. J. Exp. Biol.* 104:79–95.

———. 1984. Consequences of aerial respiration for amphibian larvae. In *Respiration and metabolism of embryonic vertebrates,* ed. R. S. Seymour, 71–86. Boston: Dr. W. Junk.

———. 1985a. Acclimation to constant and variable temperatures in plethodontid salamanders. I. Rates of oxygen consumption. *Comp. Biochem. Physiol.* 81A:673–82.

———. 1985b. Acclimation to constant and variable temperatures in plethodontid salamanders. II. Time course of acclimation to cool and warm temperatures. *Herpetologica* 41:241–45.

———. 1986. Effect of thermal acclimation on locomotor energetics and locomotor performance in a lungless salamander, *Desmognathus ochrophaeus. J. Exp. Biol.* 121:271–83.

———. 1987. Effect of thermal acclimation on locomotor energetics and locomotor performance in a tropical salamander, *Bolitoglossa subpalmata. Physiol. Zool.* 60:18–26.

———. 1988. Exercising with and without lungs. II. Experimental elimination of pulmonary and buccopharyngeal gas exchange in individual salamanders (*Ambystoma tigrinum*). *J. Exp. Biol.* 138:487–97.

Feder, M. E., and S. J. Arnold. 1982. Anaerobic metabolism and behavior during predatory encounters between snakes (*Thamnophis elegans*) and salamanders (*Plethodon jordani*). *Oecologia* 53:93–97.

Feder, M. E., A. F. Bennett, W. W. Burggren, and R. B. Huey, eds. 1987. *New directions in ecological physiology.* New York: Cambridge University Press.

Feder, M. E., and D. T. Booth. 1992. Hypoxic boundary layers surrounding skin-breathing aquatic amphibians: Occurrence, consequences and organismal responses. *J. Exp. Biol.* 166:237–51.

Feder, M. E., and W. W. Burggren. 1985a. Cutaneous gas exchange In vertebrates: Design, patterns, control and implications. *Biol. Rev.* 60:1–45.

———. 1985b. The regulation of cutaneous gas exchange in vertebrates. In *Circulation, respiration, and metabolism,* ed. R. Gilles, 101–13. New York: Springer-Verlag.

———. 1985c. Skin breathing in vertebrates. *Sci. Amer.* 253:126–42.

———, eds. 1992. *Environmental physiology of the amphibians.* Chicago: University of Chicago Press.

Feder, M. E., R. J. Full, and J. Piiper. 1988. Elimination kinetics of acetylene and freon 22 in resting and active lungless salamanders. *Respir. Physiol.* 72:229–40.

Feder, M. E., A. G. Gibbs, G. A. Griffith, and J. Tsuji. 1984. Thermal acclimation of metabolism in salamanders: Fact or artifact? *J. Therm. Biol.* 9:255–60.

Feder, M. E., and P. L. Londos. 1984. Hydric constraints upon foraging in a terrestrial salamander, *Desmognathus ochrophaeus* (Amphibia: Plethodontidae). *Oecologia* 64:413–18.

Feder, M. E., and J. F. Lynch. 1982. Effects of latitude, season, elevation, and microhabitat on field body temperatures of neotropical and temperate zone salamanders. *Ecology* 63:1657–64.

Feder, M. E., J. F. Lynch, H. B. Shaffer, and D. B. Wake. 1982. Field body temperatures of tropical and temperate zone salamanders. *Smithsonian Herp. Infor. Service Publ.* 52:1–23.

Feder, M. E., and L. E. Olsen. 1978. Behavioral and physiological correlates of recovery from exhaustion in the lungless salamander *Batrachoseps attenuatus* (Amphibia: Plethodontidae). *J. Comp. Physiol.* 128:101–7.

Feder, M. E., and A. W. Pinder. 1988. Ventilation and its effect on "infinite pool" exchangers. *Amer. Zool.* 28:973–83.

Feder, M. E., and F. H. Pough. 1975. Temperature selection by the red-backed salamander, *Plethodon c. cinereus* (Green) (Caudata: Plethodontidae). *Comp. Biochem. Physiol.* 50A:91–98.

Feder, M. E., D. B. Seale, M. E. Boraas, R. J. Wassersug, and A. G. Gibbs. 1984. Functional conflicts between feeding and gas exchange in suspension-feeding tadpoles, *Xenopus laevis. J. Exp. Biol.* 110:91–98.

Feder, M. E., and R. J. Wassersug. 1984. Aerial versus aquatic oxygen consumption in larvae of the clawed frog, *Xenopus laevis. J. Exp. Biol.* 108:231–45.

Fei, L. 1999. *Atlas of amphibians of China.* Zhengzhou, China: Helan Publishing House of Science and Technology. (in Chinese).

Feldmann, R. 1967a. Nachweis der Ortstreue des Feuersalamanders, *Salamandra salamandra terrestris* Lacépède, 1788, gegenüber seinem Winterquartier. *Zool. Anz.* 178:42–48.

———. 1967b. Winterquartiere des Feuersalamanders, *Salamandra salamandra terrestris,* in Bergwerkstollen des südlichen Westfalen. *Salamandra* 3:1–3.

———. 1977. Bergwerkstollen als Winterquartiere von Amphibien. *Natur und Heimat* 37:23–28.

———. 1987. Überwinterung, Orstreue und Lebensalter des Feuersalamanders, *Salamandra salamandra terrestris.* Schlussbericht einer Langezeituntersuchung. *Jb. Feldherpetol.* 1:33–44.

Feller, A. E., and S. B. Hedges. 1998. Molecular evidence for the early history of living amphibians. *Mol. Phylogen. Evol.* 9:509–16.

Fellers, G. M. 1979a. Aggression, territoriality, and mating behaviour in North American treefrogs. *Anim. Behav.* 27:107–19.

———. 1979b. Mate selection in the gray treefrog *Hyla versicolor. Copeia* 1979:286–90.

———. 2005. *Rana draytonii* Baird and Girard, 1852. California red-legged frog. In *Amphibian declines: Conservation status of United States species,* ed. M. J. Lannoo, 552–54. Berkeley: University of California Press.

Fellers, G. M., and C. A. Drost. 1993. Disappearance of the Cascades frog *Rana cascadae* at the southern end of its range, California, USA. *Biol. Cons.* 65:177–81.

Fellers, G. M., D. E. Green, and J. E. Longcore. 2001. Oral chytridiomycosis in the mountain yellow-legged frog (*Rana muscosa*). *Copeia* 2001:945–53.

Fellers, G. M., W. Yuezhao, and L. Shaoyin. 2003. Status of amphibians at the Zoige Wetlands, Sichuan Province, China. *Froglog* (58):1.

Felsenstein, J. 1976. The theoretical population genetics of variable selection and migration. *Ann. Rev. Genetics* 10:253–80.

Felton, A., R. A. Alford, A. M. Felton, and L. Schwarzkopf. 2006. Multiple mate choice criteria and the importance of age for male mating success in the microhylid frog, *Cophixalus ornatus. Behav. Ecol. Sociobiol.* 59:786–95.

Feminella, J. W. and C. P. Hawkins. 1994. Tailed frog tadpoles differentially alter their feeding-behavior in response to nonvisual cues from four predators. *J. N. Amer. Benthol. Soc.* 13:310–20.

Feng, A. S. 1981. Directional response characteristics of single neurons in the torus semicircularis of the leopard frog (*Rana pipiens*). *J. Comp. Physiol.* 144:419–28.

———. 1982. Quantitative analysis of intensity-rate and intensity-latency functions in peripheral auditory nerve fibers of northern leopard frogs (*Rana p. pipiens*). *Hearing Res.* 6:241–46.

———. 1986a. Afferent and efferent innervation patterns of the superior olivary nucleus of the leopard frog. *Brain. Res.* 364:167–71.

———. 1986b. Afferent and efferent innervation patterns of the cochlear nucleus (dorsal medullary nucleus) of the leopard frog. *Brain. Res.* 367:183–91

Feng, A. S., and R. R. Capranica. 1976. Sound localization in anurans. I. Evidence of binaural interaction in dorsal medullary nucleus of bullfrog (*Rana catesbeiana*). *J. Neurophysiol.* 39:871–81.

———. 1978. Sound localization in anurans. II. Binaural interaction in superior olivary nucleus of the green tree frog (*Hyla cinerea*). *J. Neurophysiol.* 41:43–54.

Feng, A. S., H. C. Gerhardt, and R. R. Capranica. 1976. Sound localization behavior of the green treefrog (*Hyla cinerea*) and the barking treefrog (*Hyla gratiosa*). *J. Comp. Physiol.* 107:241–52.

Feng, A. S., J. C. Hall, and S. S. Siddique. 1991. Coding of temporal parameters of complex sounds by frog auditory nerve fibers. *J. Neurophysiol.* 65:424–45.

Feng, A. S., and P. M. Narins. 1991. Unusual mating behavior of Malaysian treefrogs, *Polypedates leucomystax. Naturwissenschaften* 78:362–65.

Feng, A. S., P. M. Narins, and R. R. Capranica. 1975. Three populations of primary auditory fibers in the bullfrog (*Rana catesbeiana*): Their peripheral origins and frequency sensitivities. *J. Comp. Physiol.* 100:221–29.

Feng, A. S., P. M. Narins, and C.-H. Xu. 2002. Vocal acrobatics in a Chinese frog, *Amolops tormotus. Naturwissenschaften* 89:352–56.

Fenolio, D. B., G. O. Graening, B. A. Collier, and J. F. Stout. 2006. Coprophagy in a cave-adapted salamander: The importance of bat guano examined through nutritional and stable isotope analyses. *Proc. Roy. Soc. London* B 273:439–43.

Fenton, M. B., C. M. Swanepoel, R. M. Brigham, J. Cebek, and M. B. C. Hickey. 1990. Foraging behavior and prey selection by large slit-faced bats (*Nycteris grandis;* Chiroptera: Nycteridae). *Biotropica* 22:2–8.

Fenton, M. B., D. W. Thomas, and R. Sasseen. 1981. *Nycteris grandis* (Nycteridae): An African carnivorous bat. *J. Zool., London* 194:461–65.

Feral, D., M. A. Camann, and H. H. Welsh, Jr. 2005. *Dicamptodon tenebrosus* larvae within hyporheic zones of intermittent streams in California. *Herp. Rev.* 36:16–27.

Ferguson, D. E. 1961. The geographic variation of *Ambystoma macrodactylum* Baird, with the description of two new subspecies. *Am. Midl. Nat.* 65:311–38.

———. 1966. Sun-compass orientation in anurans, pp. 21–34. In *Animal orientation and navigation,* ed. R. E. Storm, 21–34. Corvallis: Oregon State University Press.

———. 1971. The sensory basis of orientation in amphibians. *Ann. New York Acad. Sci.* 188:30–36.

Ferguson, D. E., and H. F. Landreth. 1966. Celestial orientation of Fowler's toad, *Bufo fowleri. Behaviour* 26:105–23.

Ferguson, D. E., H. F. Landreth, and J. P. McKeown. 1967. Sun compass orientation of the northern cricket frog, *Acris crepitans. Anim. Behav.* 15:45–53.

Ferguson, D. E., H. F. Landreth, and M. R. Turnipseed. 1965. Astronomical orientation of the southern cricket frog, *Acris gryllus. Copeia* 1965:58–66.

Ferguson, D. E., J. P. McKeown, O. S. Bosarge, and H. F. Landreth. 1968. Sun-compass orientation in bullfrogs. *Copeia* 1968:230–35.

Ferguson, H. M. 2000. Larval colonization and recruitment in the Pacific giant salamander (*Dicamptodon tenebrosus*) in British Columbia. *Can. J. Zool.* 78:1238–42.

Ferhat-Akat, S. 1939. Untersuchungen über den Gehorsinn der Amphibien. *Z. Vergleich. Physiol.* 26:253–81.

Fermin, P. H. 1765. *Développement parfait du mystère de la generation du fameux crapaud de Surinam, nomme Pipa.* The Netyherlands: Maastricht.

Fernández, K. 1926. Sobre la biología y reproducción de Batracios argentinos. Segunda parte. *Boln Acad. nac. Cien. Córdoba* 29:271–320.

Fernandez, K., and M. Fernandez. 1921. Sobre la biología y reproducción de algunos batracios argentinos. I. Cystignathidae. *Ann. Soc. Cient. Argent.* 91:97–140.

Fernandez, P. J., and J. P. Collins. 1988. Effect of environment and ontogeny on color pattern variation in Arizona tiger salamanders (*Ambystoma tigrinum nebulosum* Hallowell). *Copeia* 1988:928–38.

Ferner, J. W. 1979. A review of marking techniques for amphibians and reptiles. Society for the Study of Amphibians and Reptiles, *Herpetological Circular* no. 9:1–41.

Ferracin, A., M. Lunadei, and N. Falcone. 1980. An ecological note on *Triturus alpestris apuanus* (Bonaparte) and *Triturus* cr*i*tatus *carnifex* (Laurenti) in the Garfagnana (Lucca, Central Italy). *Boll. Zool.* 47:143–47.

Ferraz, S. F. B., C. A. Vettorazzi, D. M. Therobald, and M. V. R. Ballester. 2005. Landscape dynamics of Amazonian deforestation between 1984 and 2002 in central Rondônia, Brazil: Assessment and future scenarios. *Forest Ecol. Manage.* 204:67–83.

Ferreira, H. G., and C. H. Jesus. 1973. Salt adaptation in *Bufo bufo*. *J. Physiol.* 228:583–600.

Ferretti, D. F., J. B. Miller, J. W. C. White, D. M. Etheridge, K. R. Lassay, D. C. Lowe, C. M. MacFarling Meure, M. F. Dreier, C. M. Trudinger, T. D. van Ommen, and R. L. Langenfelds. 2005. Unexpected changes to the global methane budget over the past 2000 years. *Science* 309:1714–17.

Ficetola, G. F., and F. De Bernardi. 2005. Interspecific social interactions and breeding success of the frog *Rana latastei:* A field study. *Ethology* 111:764–74.

Ficken, R. W., M. S. Ficken, and J. P. Hailman. 1974. Temporal pattern shifts to avoid acoustic interference in singing birds. *Science* 183:762–63.

Figge, F. H. J. 1934. The effect of ligation of the pulmonary arch on amphibian metamorphosis. *Physiol. Zool.* 7:149–77.

———. 1936. The differential reaction of the blood vessels of a branchial arch of *Ambystoma tigrinum* (Colorado axolotl). I. The reaction to adrenaline, oxygen and $CO_2$. *Physiol. Zool.* 9:79–101.

Figiel, C. R., Jr., and R. D. Semlitsch. 1990. Population variation in survival and metamorphosis of larval salamanders (*Ambystoma maculatum*) in the presence and absence of fish predation. *Copeia* 1990:818–26.

———. 1991. Effects of nonlethal injury and habitat complexity on predation in tadpole populations. *Can. J. Zool.* 69:830–34.

———. 1995. Experimental determination of oviposition site selection in the marbled salamander, *Ambystoma opacum. J. Herpetol.* 29:452–54.

Filippi, F. 1861. Sulla larva del *Triton alpestris. Arch. Zool. Anat. Fisiol., Genova* 1:206.

Findeis, E. K., and W. E. Bemis. 1990. Functional morphology of tongue projection in *Taricha torosa* (Urodela: Salamandridae). *Zool. J. Linn. Soc.* 99:129–57.

Findlay, C. S., and J. Houlahan. 1997. Anthropogenic correlates of species richness in southeastern Ontario wetlands. *Cons. Biol.* 11:1000–1009.

Findlay, C. S., J. Lenton, and L. G. Zheng. 2001. Land-use correlates of anuran community richness and composition in southeastern Ontario wetlands. *Ecoscience* 8:336–43.

Finke, O. M., J. K. Waage, and W. D. Koenig. 1997. Natural and sexual selection components of odonate mating systems. In *The evolution of mating systems in insects and arachnids,* ed. J. C. Choe and B. J. Crespi, 58–74. New York: Cambridge University Press.

Finkler, M. S., and K. A. Cullum. 2002. Sex-related differences in metabolic rate and energy reserves in spring-breeding small-mouthed salamanders (*Ambystoma texanum*). *Copeia* 2002:824–29.

Fiorito de Lopez, L. E., and D. D. Echeverria. 1989. Microanatomy and histogenesis of the oral apparatus of *Bufo arenarum* tadpoles (Anura, Bufonidae). *Cuadernos Herpetol.* 4:4–10.

Fiorito, S., P. Cazzanti, B. Rolando, S. Castellano, A. Rolando, and C. Giacoma. 1994. Post-breeding dispersal by adult common frogs *Rana temporaria* studied by radio-tracking. *Stud. Trentini Sci. Nat. Acta Biol.* 71:183–89.

Fischer, J. H., M. J. Freake, S. C. Borland, and J. B. Phillips. 2001. Evidence for the use of magnetic map information by an amphibian. *Anim. Behav.* 62:1–10.

Fischer, H. 1969. Massentod von Erdkröten und Fadenmolchen auf der Autobahn und der Staatsstrasse von Kaiseraugst nach Rheinfelden. *Schweiz. Naturschutz* 35:66–69.

Fishbeck, D. W., and J. C. Underhill. 1971. Distribution of stripe polymorphism in wood frogs, *Rana sylvatica* LeConte, from Minnesota. *Copeia* 1971:253–59.

Fisher, R. A. 1930. *The genetical theory of natural selection.* New York: Oxford University Press. (2nd ed., 1958, New York: Dover Publications).

Fisher, R. N., and H. B. Shaffer. 1996. The decline of amphibians in California's Great Central Valley. *Cons. Biol.* 10:1387–97.

Fishwald, T. G., R. A. Schemidt, K. M. Jankens, K. A. Berven, G. J. Gamboa, and C. M. Richards. 1990. Sibling recognition by larval frogs (*Rana pipiens, R. sylvatica,* and *Pseudacris crucifer*). *J. Herpetol.* 24:40–44.

Fitch, H. S. 1941. The feeding habits of California garter snakes. *California Fish and Game* 27:1–32.

———. 1956. A field study of the Kansas ant-eating frog, *Gastrophryne olivacea. Univ. Kansas Publ. Mus. Nat. Hist.* 42:1–60.

———. 1958. Home ranges, territories, and seasonal movements of vertebrates on the Natural History Reservation. *Univ. Kansas Publ. Mus. Nat. Hist.* 11:63–326.

———. 1960. Autecology of the copperhead. *Univ. Kansas Publ. Mus. Nat. Hist.* 13:85–288.

———. 1963. Natural history of the racer *Coluber constrictor. Univ. Kansas Publ. Mus. Nat. Hist.* 15:351–468.

———. 1965. An ecological study of the garter snake, *Thamnophis sirtalis. Univ. Kansas Publ. Mus. Nat. Hist.* 15:493–564.

———. 1975. A demographic study of the ringneck snake (*Diadophis punctatus*) in Kansas. *Univ. Kansas Mus. Nat. Hist. Misc. Publ.* (62):1–53.

———. 1999. *A Kansas snake community: Composition and change over 50 years.* Malabar, FL: Krieger Publishing Company.

Fitch, H. S., P. Goodrum, and C. Newman. 1952. The armadillo in the southeastern United States. *J. Mammal.* 33:21–37.

Fitch, K. L. 1959. Observations on the nesting habits of the mudpuppy, *Necturus maculosus* Rafinesque. *Copeia* 1959:339–40.

Fite, K. V., A. R. Blaustein, L. Bengston, and H. E. Hewitt. 1998. Evidence of retinal light damage in *Rana cascadae,* a declining amphibian species. *Copeia* 1998:906–14.

Fitzgerald, G. J., and J. R. Bider. 1974a. Influence of moon phase and weather factors on locomotor activity in *Bufo americanus. Oikos* 25:338–40.

———. 1974b. Seasonal activity of the toad *Bufo americanus* in southern Quebec as revealed by a sand-transect technique. *Can. J. Zool.* 52:1–5.

Fitzgerald, K. T., L. J. Guillette, Jr., and D. Duvall. 1979. Notes on birth, development and care of *Gastrotheca riobambae* tadpoles in the laboratory (Amphibia, Anura, Hylidae). *J. Herpetol.* 13:457–60.

Fitzpatrick, L. C. 1971. Influence of sex and reproductive condition on metabolic rates in the Allegheny mountain salamander *Desmognathus ochrophaeus. Comp. Biochem. Physiol.* 40A:603–8.

Fitzpatrick, L. C. 1973a. Energy allocation in the Allegheny Mountain salamander, *Desmognathus ochrophaeus. Ecol. Monogr.* 43:43–58.

———. 1973b. Influence of seasonal temperatures on the energy budget and metabolic rate of the northern two-lined salamander, *Eurycea bislineata bislineata. Comp. Biochem. Physiol.* 45A:807–18.

———. 1976. Life history patterns of storage and utilization of lipids for energy in amphibians. *Amer. Zool.* 16:725–32.

Fitzpatrick, L. C., and M. Y. Atebara. 1974. Effects of acclimation to seasonal temperatures on energy metabolism in the toad *Bufo woodhousei*. *Physiol. Zool.* 47:119–29.

Fitzpatrick, L. C., J. R. Bristol, and R. M. Stokes. 1971. Thermal acclimation and metabolism in the Allegheny Mountain salamander *Desmognathus ochrophaeus*. *Comp. Biochem. Physiol.* 40A:681–88.

Fitzpatrick, L. C., J. R. Bristol, and R. M. Stokes. 1972. Thermal acclimation and metabolic rates in the dusky salamander *Desmognathus fuscus*. *Comp. Biochem. Physiol.* 41A:89–96.

Fitzpatrick, L. C., and A. V. Brown. 1975. Metabolic compensation to temperature in the salamander *Desmognathus ochrophaeus* from a high elevation population. *Comp. Biochem. Physiol.* 50A:733–37.

Flageole, S., and R. Leclair, Jr. 1992. Étude démographique d'une population de salamandres (*Ambystoma maculatum*) a l'aide de la méthode squeletto-chronologique. *Can. J. Zool.* 70:740–49.

Flanigan, J. E., and M. Guppy. 1997. Metabolic depression and sodium-potassium ATPase in the aestivating frog, *Neobatrachus kunapalari*. *J. Comp. Physiol.* B 167:135–45.

Flanigan, J. E., P. C. Withers, C. J. Fuery, and M. Guppy. 1993. Metabolic depression and Na+/K+ gradients in the aestivating Australian goldfields frog, *Neobatrachus wilsmorei*. *J. Comp. Physiol.* B 163:587–93.

Flanigan, J. E., P. C. Withers, and M. Guppy. 1991. In vitro metabolic depression of tissues from the aestivating frog *Neobatrachus pelabatoides*. *J. Exp. Biol.* 161:273–83.

Flanigan, J. E., P. C. Withers, K. Storey, and M. Guppy. 1990. Changes in enzyme binding and activity during aestivation in the frog *Neobatrachus pelobatoides*. *Comp. Biochem. Physiol.* 96B:67–71.

Fleck, J. 1982. Erst-Nachzucht des Turkischen Bergbachmolches *Neurergus strauchii* (Amphibia: Caudata: Salamandridae). *Salamandra* 18:138–49.

Flecker, A. S., B. P. Feifarek, and B. W. Taylor. 1999. Ecosystem engineering by a tropical tadpole: density-dependent effects on habitat structure and larval growth rates. *Copeia* 1999:495–500.

Fleet, R. R., and H. S. Fitch. 1974. Food habits of *Basiliscus basiliscus* in Costa Rica. *J. Herpetol.* 8:260–62.

Fleharty, E. D. 1967. Comparative ecology of *Thamnophis elegans, T. cyrtopsis,* and *T. rufipunctatus* in New Mexico. *Southwest. Nat.* 12:207–29.

Fleming, T. H. 1975. The role of small mammals in tropical ecosystems. *Small mammals: Their productivity and population dynamics,* ed. F. B. Golley, K. Petrusewicz, and L. Ryszkowski, 269–98. Cambridge, UK: Cambridge University Press.

Fleming, T. H. 1979. Life-history strategies. In *Ecology of small mammals,* ed. D. M. Stoddart, 1–61. London: Chapman and Hall.

Flowers, M. A., and B. M. Graves. 1997. Juvenile toads avoid chemical cues from snake predators. *Anim. Behav.* 53: 641–46.

Floyd, R.B. 1984. Variation in temperature preference with stage of development of *Bufo marinus* larvae. *J. Herpetol.* 18:153–58.

Fogarty, J. H., and F. J. Vilella. 2003. Use of native forest and *Eucalyptus* plantations by *Eleutherodactylus* frogs. *J. Wildl. Manage.* 67:186–95.

Fogleman, J. C., P. S. Corn, and D. Pettus. 1980. The genetic basis of a dorsal colour polymorphism in *Rana pipiens*. *J. Hered.* 71:439–40.

Foley, J. A., R. DeFries, G. P. Asner, C. Barford, G. Bonan, S. R. Carpenter, F. S. Chapin, M. T. Coe, G. C. Daily, H. K. Gibbs, J. H. Helkowski, T. Holloway, E. A. Howard, C. J. Kucharik, C. Monfreda, J. A. Patz, I. C. Prentice, N. Ramankutty, and P. K. Snyder. 2005. Global consequences of land use. *Science* 309:570–74.

Fontenot, C. L., Jr. 1999. Reproductive biology of the aquatic salamander *Amphiuma tridactylum* in Louisiana. *J. Herpetol.* 33:100–105.

Fontenot, L. W., G. P. Noblet, J. M. Akins, M. D. Stephens, and G. P. Cobb. 2000. Bioaccumulation of polychlorinated biphenyls in ranid frogs and northern water snakes from a hazardous waste site and a contaminated watershed. *Chemosphere* 40:803–9.

Fontenot, L. W., G. P. Noblet, and S. G. Platt. 1996. A survey of herpetofauna inhabiting polychlorinated biphenyl contaminated and reference watersheds in Pickens County, South Carolina. *J. Elisha Mitchell Sci. Soc.* 112:20–30.

Forchhammer, M. C., E. Post, and N. C. Stenseth. 1998. Breeding phenology and climate. *Nature* 391:29–30.

Ford, L. S. 1993. The phylogenetic position of the dart-poison frogs (Dendrobatidae) among anurans: An examination of the competing hypotheses and their characters. *Ethol. Ecol. Evol.* 5:219–31.

Ford, L. S., and D. C. Cannatella. 1993. The major clades of frogs. *Herpetol. Monogr.* 7:94–117.

Ford, W. M., B. R. Chapman, M. A. Menzel, and R. H. Odom. 2002. Stand age and habitat influences on salamanders in Appalachian cove hardwood forests. *Forest Ecol. Manage.* 155:131–41.

Ford, W. M., M. A. Menzel, D. W. McGill, J. Laerm, and T. S. McCay. 1999. Effects of a community restoration fire on small mammals and herpetofauna in the southern Appalachians. *Forest Ecol. Manage.* 14:233–43.

Forester, D. C. 1973. Mating call as a reproductive isolating mechanism between *Scaphiopus bombifrons* and *S. hammondii*. *Copeia* 1973:60–67.

———. 1977. Comments on the female reproductive cycle and philopatry by *Desmognathus ochrophaeus* (Amphibia, Urodela, Plethodontidae). *J. Herpetol.* 11:311–16.

———. 1978. Laboratory encounters between attending *Desmognathus ochrophaeus* (Amphibia, Urodela, Plethodontidae) females and potential predators. *J. Herpetol.* 12:537–41.

———. 1979a. The adaptiveness of parental care in *Desmognathus ochrophaeus*. *Copeia* 1979:332–41.

———. 1979b. Homing to the nest by female mountain dusky salamanders (*Desmognathus ochrophaeus*) with comments on the sensory modalities essential to clutch recognition. *Herpetologica* 35:330–35.

———. 1979c. Observations of ant predation upon brooding salamanders. *Herp. Rev.* 10:3.

———. 1981. Parental care in the salamander *Desmognathus ochrophaeus:* Female activity pattern and trophic behavior. *J. Herpetol.* 15:29–34.

———. 1983. Duration of the brooding period in the mountain dusky salamander (*Desmognathus ochrophaeus*) and its influences on aggression toward conspecifics. *Copeia* 1983:1098–1101.

———. 1984. Brooding behavior by the mountain dusky salamander: Can the female's presence reduce clutch desiccation? *Herpetologica* 40:105–9.

Forester, D. C., J. Cover, and A. Wisnieski. 1993. The influence of time of residency on the tenacity of territorial defense by the dart-poison frog *Dendrobates pumilio. Herpetologica* 49:94–99.

Forester, D. C., and R. Czarnowsky. 1985. Sexual selection in the spring peeper, *Hyla crucifer* (Amphibia, Anura): Role of the advertisement call. *Behaviour* 92:112–28.

Forester, D. C., and R. Daniel. 1986. Observations on the social behavior of the southern cricket frog, *Acris gryllus. Brimleyana* 12:5–11.

Forester, D. C., and W. K. Harrison. 1987. The significance of antiphonal vocalisation by the spring peeper, *Pseudacris crucifer* (Amphibia, Anura). *Behaviour* 103:1–15.

Forester, D. C., K. Harrison, and L. McCall. 1983. The effects of isolation, the duration of brooding, and non-egg olfactory cues on clutch recognition by the salamander, *Desmognathus ochrophaeus. J. Herpetol.* 17:308–14.

Forester, D. C., and D. V. Lykens. 1986. Significance of satellite males in a population of spring peepers (*Hyla crucifer*). *Copeia* 1986:719–24.

———. 1988. The ability of wood frog eggs to withstand prolonged terrestrial stranding: an empirical study. *Can. J. Zool.* 66:1733–35.

———. 1991. Age structure in a population of red-spotted newts from the Allegheny Plateau of Maryland. *J. Herpetol.* 25:373–76.

Forester, D. C., D. V. Lykens, and W. K. Harrison. 1989. The significance of persistent vocalization by the spring peeper, *Pseudacris crucifer* (Anura: Hylidae). Behaviour 108:197–208.

Forester, D. C., and K. J. Thompson. 1998.Gauntlet behaviour as a male sexual tactic in the American toad (Amphibia: Bufonidae). *Behaviour* 135:99–119.

Forester, D. C., and A. Wisnieski. 1991. The significance of airborne olfactory cues to the recognition of home area by the dart-poison frog *Dendrobates pumilio. J. Herpetol.* 25:502–4.

Forey, P. 1986. Relationships of lungfishes. *J. Morphol.*, Suppl. 1:75–91.

———. 1988. Golden jubilee for the coelacanth *Latimeria chalumnae. Nature* 336:727–32.

———. 1991. Blood lines of the coelacanth. *Nature* 351:347–48.

Forey, P. L., B. G. Gardiner, and C. Patterson. 1991. The lungfish, the coelacanth, and the cow revisited. In *Origins of the higher groups of tetrapods,* ed. H.-P. Schultze and L. Trueb, 145–72. Ithaca, NY: Cornell University Press.

Forge, P., and R. Barbault. 1978. Observations sur le régime alimentaire de deux amphibiens sympatriques du Sahel sénégalais: *Bufo pentoni* et *Tomopterna delalandi. Bull. IFAN* 40:674–84.

Formanowicz, D. R., Jr. 1986. Anuran tadpole/aquatic insect predator-prey interactions: tadpole size and predator capture success. *Herpetologica* 42:367–73.

———. 1987. Foraging tactics of *Dytiscus verticalis* larvae (Coleoptera: Dytiscidae): Prey detection, reactive distance and predator size. *J. Kansas Ent. Soc.* 60:92–99.

Formanowicz, D. R., Jr., and E. D. Brodie, Jr. 1979. Palatability and antipredator behavior of selected *Rana* to the shrew *Blarina. Am. Midl. Nat.* 101:456–58.

———. 1982. Relative palatabilities of members of a larval amphibian community. *Copeia* 1982:91–97.

———. 1993. Size-mediated predation pressure in a salamander community. *Herpetologica* 49:265–70.

Formanowicz, D. R., Jr., M. M. Stewart, K. Townsend, F. H. Pough, and P. F. Brussard. 1981. Predation by giant crab spiders on the Puerto Rican frog *Eleutherodactylus coqui. Herpetologica* 37:125–29.

Formas, J. R. 1976. Descriptions of *Batrachyla* (Amphibia, Anura, Leptodactylidae) tadpoles. *J. Herpetol.* 10:221–25.

———. A new species of *Eupsophus* (Amphibia: Anura: Leptodactylidae) from southern Chile. *Proc. Biol. Soc. Washington* 102:568–76.

Formas, J. R., N. F. Diaz, and J. Valencia. 1980. The tadpole of the Chilean frog *Inseutophrynus acarpicus. Herpetologica* 36:316–18.

Formas, J. R., and V. Poblete. 1996. *Eupsophus emiliopugini* (NCN). Aggressive behavior. *Herp. Rev.* 27:139–40.

Formas, J. R., and E. Pugin. 1971. Reproducción y desarrollo de *Batrachyla antartandica* Barrio (Anura, Leptodactylidae). *Bol. Mus. Hist. Nat.* 32:201–13.

———. 1978a. Tadpoles of *Eupsophus roseus* and *Bufo variegatus* (Amphibia, Anura) in southern Chile. *J. Herpetol.* 12:243–46.

———. 1978b. Tadpoles of *Hylorina sylvatica, Eupsophus vittatus,* and *Bufo rubropunctatus* in southern Chile. *Herpetologica* 34:355–58.

Formas, J. R., E. Pugin, and B. Jorquera. 1975. La identidad del batracio chileno *Heminectes rufus* Philippi, 1902. *Physis (Buenos Aires)* 34:147–57.

Formas, J. R., and M. A. Vera. 1980. Reproductive patterns of *Eupsophus roseus* and *E. vittatus. J. Herpetol.* 14:11–14.

Formicki, G., W. Zamachowski, and R. Stawarz. 2003. Effects of UV-A and UV-B on oxygen consumption in common toad (*Bufo bufo*) tadpoles. *J. Zool., London* 259:317–26.

Forrest, T. G. 1994. From sender to receiver: propagation and environmental effects on acoustic signals. *Amer. Zool.* 34:644–54.

Forrest, T. G., G. L. Miller, and J. R. Zagar. 1993. Sound propagation in shallow water: implications for acoustic communication by aquatic organisms. *Bioacoustics* 4:259–70.

Forrest, T. G., and R. Raspet. 1994. Models of female choice in acoustic communication. *Behav. Ecol.* 5:293–303.

Forselius, S. 1963. Distribution and reproductive behaviour of *Rana esculenta* L. in the coastal area of N. Uppland, C. Sweden. *Zool. Bidr. Uppsala* 35:517–28.

Fossi, M. C., and L. Marsili. 2003. Effects of endocrine disruptors in aquatic mammals. *Pure Appl. Chem.* 75:2235–47.

Foster, M. S., and R. W. McDiarmid. 1982. Study of aggregative behavior of *Rhinophrynus dorsalis* tadpoles: design and analysis. *Herpetologica* 38:395–404.

Foster, P. 2001. The potential negative impacts of global climate change on tropical montane cloud forests. *Earth-Sci. Rev.* 55:73–106.

Foster, W. A. 1967. Chorus structure and vocal responses in the Pacific treefrog, *Hyla regilla. Herpetologica* 23:100–104.

Fouquette, M. J., Jr. 1954. Food competition among four sympatric species of garter snakes, genus *Thamnophis. Texas J. Sci.* 6:172–89.

———. 1960. Isolating mechanisms in three sympatric treefrogs in the Canal Zone. *Evolution* 14:484–97.

———. 1975. Speciation in chorus frogs. I. Reproductive character displacement in the *Pseudacris nigrita* complex. *Syst. Zool.* 24:16–22.

Fouquette, M. J., Jr. and M. J. Littlejohn. 1960. Patterns of oviposition in two species of hylid frogs. *Southwest. Nat.* 5:92–96.

Fowler, J. A. 1946. The eggs of *Pseudotriton montanus montanus. Copeia* 1946:105.

Fox, H. 1983. The skin of *Ichthyophis* (Amphibia: Caecilia): An ultrastructural study. *J. Zool. (London)* 199:223–48.

Fox, H. 1984. *Amphibian morphogenesis.* Clifton, NJ: Humana Press.
———. 1985. The tentacles of *Ichthyophis* (Amphibia: Caecilia) with special reference to the skin. *J. Zool., London* (A) 205:223–34.
———. 1986a. Dermal glands. In *Biology of the integument, vol 2. Vertebrates. III. The skin of Amphibia,* ed. J. Bereiter-Hahn, A. G. Matoltsy, and K. S. Richards, 116–35. New York: Springer-Verlag.
———. 1986b. Dermis. In *Biology of the integument, vol 2. Vertebrates. III. The skin of Amphibia,* ed. J. Bereiter-Hahn, A. G. Matoltsy, and K. S. Richards, 111–15. New York: Springer-Verlag.
———. 1986c. Epidermis. In *Biology of the integument, vol 2. Vertebrates. III. The skin of Amphibia,* ed. J. Bereiter-Hahn, A. G. Matoltsy, and K. S. Richards, 78–110. New York: Springer-Verlag.
———. 1994. The structure of the integument. In *Amphibian biology. Vol. 1. The integument,* ed. H. Heatwole and G. T. Barthalmus, 1–32. Chipping Norton, New South Wales, Australia: Surrey Beatty and Sons.
———. 1995. Morphological correlates of auditory sensitivity in anuran amphibians. *Brain Behav. Evol.* 45:327–38.
Fox, W. 1952. Notes on feeding habits of Pacific Coast garter snakes. *Herpetologica* 8:4–8.
Foxon, G. E. H. 1964. Blood and respiration. In *Physiology of the Amphibia,* vol. 1, ed. J. A. Moore, 151–209. New York: Academic Press.
França, L. F., K. G. Facure, and A. A. Giaretta. 2004. Trophic and spatial niches of two large-sized species of *Leptodactylus* (Anura) in southeastern Brazil. *Stud. Neotrop. Fauna Env.* 39:243–48.
Franca e Souza, I., W. vaz Silva, P. V. C. Mathias, and N. J. da Silva, Jr. 2003. *Leptodactylus ocellatus* (Rã Manteiga). Predation. *Herp. Rev.* 34:232.
Francillon, H. 1979. Étude expérimentale des marqués de croissance sur les humerus et les femurs de Tritons crêtes (*Triturus cristatus cristatus* Laurenti) en relatión avec la détermination de l'âgé individuel. *Acta Zool. (Stockholm)* 60:223–32.
Francillon-Viellot, H., J. W. Arntzen, and J. Gerardie. 1990. Age, growth and longevity of sympatric *Triturus cristatus, T. marmoratus* and their hybrids (Amphibia, Urodela): a skeletochronological comparison. *J. Herpetol.* 24:13–22.
Frangioni, G., and G. Borgioli. 1989. Relationships between spleen and respiration in the newt. *J. Exp. Zool.* 252:118–25.
———. 1991a. Effect of spleen congestion and decongestion on newt blood. *J. Zool., London* 223:15–25.
———. 1991b. Effect of temperature on the splenic mechanism of respiratory compensation in the newt. *J. Exp. Zool.* 259:316–23.
———. 1993a. The role of the liver in the cutaneous respiratory compensation of the frog (*Rana esculenta*). *J. Zool., London* 230:483–93.
———. 1993b. Splenic respiratory compensation and blood volume in the newt. *J. Zool., London* 229:505–14.
———. 1994. Hepatic respiratory compensation and blood volume in the frog (*Rana esculenta*). *J. Zool., London* 234:601–11.
Franz, R. 1964. The eggs of the long-tailed salamander from a Maryland cave. *Herpetologica* 20:216.
Franz, R., C. K. Dodd, and C. Jones. 1988. *Rana areolata aesopus* (Florida gopher frog). Movement. *Herp. Rev.* 19:33.
Fraser, D. F. 1976a. Coexistence of salamanders of the genus *Plethodon:* A variation on the Santa Rosalia theme. *Ecology* 57:238–51.
———. 1976b. Empirical evaluation of the hypothesis of food competition in salamander of the genus *Plethodon. Ecology* 57:459–71.
———. 1980. On the environmental control of oocyte maturation in a plethodontid salamander. *Oecologia* 46:302–307.
Frazer, D. 1983. *Reptiles and amphibians in Britain.* London: Collins.
Freake, M. J., S. C. Borland, and J. B. Phillips. 2002. Use of a magnetic compass for y-axis orientation in larval bullfrogs. *Copeia* 2002:466–71.
Freake, M. J., and J. B. Phillips. 2005. Light-dependent shift in bullfrog tadpole magnetic compass orientation: Evidence for a common magnetoreception mechanism in anuran and urodels amphibians. *Ethology* 111:241–54.
Freda, J. 1986. The influence of acidic pond water in amphibians: A review. *Water Air Soil Poll.* 30:439–50.
———. 1991. Effects of acidification on amphibians. In *Biological effects of changes in surface water acid-base chemistry. State-of-Science/Technology Report 13,* ed. J. P. Baker, D. P. Bernard, S. W. Christensen, M. J. Sale, J. Freda, K. Heltcher, D. Marmorek, L. Rowe, P. Scanlon, G. Suter, W. Warren-Hicks, and P. Welbourn 114–29. Washington, DC: National Acid Precipitation Assessment Program.
Freda, J., and W. A. Dunson. 1984. Sodium balance of amphibian larvae exposed to low environmental pH. *Physiol. Zool.* 57:435–43.
———. 1985a. Field and laboratory studies of ion balance and growth rates of ranid tadpoles chronically exposed to low pH. *Copeia* 1985:415–23.
———. 1985b. The influence of external cation concentration on hatching of amphibian embryos in water of low pH. *Can. J. Zool.* 63:2649–56.
———. 1986a. The effect of prior exposure on sodium uptake in tadpoles exposed to low pH water. *J. Comp. Physiol.* B 156:649–54.
———. 1986b. Effects of low pH and other chemical variables on the local distribution of amphibians. *Copeia* 1986:454–66.
Freda, J., and R. J. Gonzalez. 1986. Daily movements of the treefrog, *Hyla andersoni. J. Herpetol.* 20:469–71.
Freda, J., and D. H. Taylor. 1992. Behavioral responses of amphibian larvae to acidic water. *J. Herpetol.* 26:429–33.
Fredericks, L. P., and J. R. Dankert. 2000. Antibacterial and hemolytic activity of the skin of the terrestrial salamander, *Plethodon cinereus. J. Exp. Zool.* 287:340–45.
Freed, A. N. 1980. An adaptive advantage of basking behavior in an anuran amphibian. *Physiol. Zool.* 53:433–44.
Freeland, W. J., and K. C. Martin. 1985. The rate of range expansion by *Bufo marinus* in Northern Australia, 1980–1984. Aust. *Wildl. Res.* 12:555–59.
Freeman, L. G. 1981. The fat of the land: Notes on paleolithic diet in Iberia. In *Omnivorous primates: Gathering and hunting in human evolution,* ed. R. S. O. Harding and G. Teleki, 104–65. New York: Columbia University Press.
Freeman, M. 1977. Sung. In *Food in Chinese culture: Anthropological and historical perspectives,* ed. K. C. Chang, 143–76. New Haven, CT: Yale University Press.
Freeman, S. L., and R. C. Bruce. 2001. Larval period and metamorphosis of the three-lined salamander, *Eurycea guttolineata* (Amphibia: Plethodontidae), in the Chattooga River Watershed. *Am. Midl. Nat.* 145:194–200.

Freiburg, R. E. 1951. An ecological study of the narrow-mouthed toad (*Microhyla*) in northeastern Kansas. *Trans. Kansas Acad. Sci.* 54:374–86.

Freidenburg, L. K. 2003. *Spatial ecology of the wood frog, Rana sylvatica.* Unpublished PhD dissertation. Storrs: University of Connecticut.

Freidenburg, L. K., and D. K. Skelly. 2004. Microgeographical variation in themal preference by an amphibian. *Ecol. Lett.* 7:369–73.

French, N. R., D. M. Stoddart, and B. Bobek. 1975. Patterns of demography in small mammal populations. In *Small mammals: Their productivity and population dynamics,* ed. F. B. Golley, K. Petrusewicz, and L. Ryszkowski, 73–102. Cambridge, UK: Cambridge University Press.

Frétey, T., E. Cam, B. Le Garff, and J-Y. Monnat. 2004. Adult survival and temporary emigration in the common toad. *Can. J. Zool.* 82:859–72.

Frieden, E., and J. J. Just. 1970. Hormonal responses in amphibian metamorphosis. In *Biochemical actions of hormones,* vol. 1, ed. G. Litwack, 1–52. New York: Academic Press.

Friedl, T. W. P., and G. M. Klump. 1997. Some aspects of population biology in the European treefrog, *Hyla arborea. Herpetologica* 53:321–30.

———. 2002. The vocal behaviour of male European treefrogs (*Hyla arborea*): Implications for inter- and intrasexual selection. *Behaviour* 139:113–36.

Frisbie, M. P., J. P. Constanzo, and R. E. Lee, Jr. 2000. Physiological and ecological aspects of low-temperature tolerance in embryos of the wood frog, *Rana sylvatica. Can. J. Zool.* 78:1032–41.

Frisbie, M. P., and R. L. Wyman. 1991. The effects of soil pH on sodium balance in the red-backed salamander, *Plethodon cinereus,* and three other terrestrial salamanders. *Physiol. Zool.* 64:1050–68.

———. 1992. The effect of soil chemistry on sodium balance in the red-backed salamander: A comparison of two forest types. *J. Herpetol.* 26:434–42.

Frischkopf, L. S., R. R. Capranica, and M. H. Goldstein, Jr. 1968. Neural coding in the bullfrog's auditory system—A teleological approach. *Proc. I.E.E.E.* 56:969–80.

Fritz, G., A. S. Rand, and C. W. de Phaniphilis. 1981. The aposematically colored frog, *Dendrobates pumilio,* is distasteful to the large, predatory ant, *Paraponera clavata. Biotropica* 13:158–59.

Fritzsch, B. 1981. The pattern of lateral-line afferents in urodeles. A horseradish-peroxidase study. *Cell Tissue Res.* 218:581–94.

Fritzsch, B., M. J. Ryan, W. Wilczynski, T. E. Hetherington, and W. Walkowiak, eds. 1988. *The evolution of the amphibian auditory system.* New York: Wiley.

Fritzsch, B., and U. Wahnschaffe. 1983. The electroreceptive ampullary organs of urodeles. *Cell Tissue Res.* 229:483–503.

Fritzsch, B., U. Wahnschaffe, M. D. Crapon de Caprona, and W. Himstedt. 1985. Anatomical evidence for electroreception in larval *Ichthyophis kohtaoensis. Naturwissenschaften* 72:102–4.

Fromm, P. O., and R. E. Johnson. 1955. The respiratory metabolism of frogs as related to season. *J. Cell. Comp. Physiol.* 45:343–59.

Frommhold, E. 1956. Über die Wierbeltiernährung von *Rana esculenta* (Linne). *Deutsch. Aquar. Terr. Z.* 9:159–61.

Frost, D. R. 2004. *Amphibian species of the world: An online reference. Version 3.0 (22 August, 2004).* Electronic Database accessible at http://research.amnh.org/herpetology/amphibia/index.html. New York: American Museum of Natural History.

Frost, D. R., T. Grant, J. Faivovich, R. H. Bain, A. Haas, C. F. B. Haddad, R. O. De Sa, A. Channing, M. Wilkinson, S. C. Donnellan, C. J. Raxworthy, J. A. Campbell, B. L. Blotto, P. Moler, R. C. Drewes, R. A. Nussbaum, J. D. Lynch, D. M. Green, and W. C. Wheeler. 2006. The amphibian tree of life. *Bull. Amer. Mus. Nat. Hist.* 297:1–370.

Fry, A. E., and J. C. Kaltenbach. 1999. Histology and lectin-binding patterns in the digestive tract of the carnivorous larvae of the anuran, *Ceratophrys ornata. J. Morphol.* 241:19–32.

Frye, B. E., P. S. Brown, and B. W. Snyder. 1972. Effects of prolactin and somatotropin on growth and metamorphosis of amphibians. *Gen. Comp. Endocrinol.,* Suppl. 3:209–20.

Fu, C.-Y., X.-F. Sun, X.-Y. Wang, J.-K. Chen, and J.-H. Wu. 2003. Effects of size class and group size structure on cannibalism in basin-raising larval salamanders, *Hynobius amjiensis. Zool. Res.* 6:407–12 (in Chinese, with English abstract).

Fuery, C. J., P. C. Withers, A. A. Hobbs, and M. Guppy. 1998. The role of protein synthesis during metabolic depression in the Australian desert frog *Neobatrachus centralis. Comp. Biochem. Physiol.* 119A:469–76.

Fuhrman, F. A., G. J. Fuhrman, and H. S. Mosher. 1969. Toxin from skin of frogs of the genus *Atelopus:* differentiation from dendrobatid toxins. *Science* 165:1376–77.

Fujioka, M., and S. J. Lane. 1997. The impact of changing irrigation practices in rice fields on frog populations of the Kanto Plain, central Japan. *Ecol. Res.* 12:101–8.

Fukuyama, K. 1991. Spawning behaviour and male mating tactics of a foam-nesting treefrog, *Rhacophorus schlegelii. Anim. Behav.* 42:193–99.

Fukuyama, K., and T. Kusano. 1989. Sexual size dimorphism in a Japanese stream-breeding frog, *Buergeria buergeri* (Rhacophoridae, Amphibia). In *Current herpetology in East Asia,* ed. M. Matsui, T. Hikida, and R. C. Goris, 306–13. Kyoto: Herpetological Society of Japan.

———. 1992. Factors affecting breeding activity in a stream-breeding frog, *Buergeria buergeri. J. Herpetol.* 26:88–91.

Fukuyama, K., T. Kusano, and M. Nakane. 1988. A radio-tracking study of the behaviour of females of the frog *Buergeria buergeri* (Rhacophoridae, Amphibia) in a breeding stream in Japan. *Jap. J. Herpetol.* 12:102–7.

Full, R. J. 1986. Locomotion without lungs: Energetics and performance of a lungless salamander. *Am. J. Physiol.* 251:R775–R780.

Full, R. J., B. D. Anderson, C. M. Finnerty, and M. E. Feder. 1988. Exercising with and without lungs. I. The effects of metabolic cost, maximal oxygen transport and body size on terrestrial locomotion in salamander species. *J. Exp. Biol.* 138:471–85.

Fuller, D. O., T. C. Jessup, and A. Salim. 2004. Loss of forest cover in Kalimantan, Indonesia, since the 1997–1998 El Niño. *Cons. Biol.* 18:249–54.

Funk, W. C., M. S. Blouin, P. S. Corn, B. A. Maxell, D. S. Pilliod, S. Amish, and F. W. Allendorf. 2005. Population structure of Columbia spotted frogs (*Rana luteiventris*) is strongly affected by the landscape. *Mol. Ecol.* 14:483–96.

Funk, W. C., and W. W. Dunlap. 1999. Colonization of high-elevation lakes by long-toed salamanders (*Ambystoma macrodactylum*) after the extinction of introduced trout populations. *Can. J. Zool.* 77:1759–67.

Funk, W. C., G. Fletcher-Lazo, F. Nogales-Sornosa, and

D. Almeidea-Reinoso. 2004. First description of a clutch and nest site for the genus *Caecilia* (Gymnophiona: Caeciliidae). *Herp. Rev.* 35:128–30.

Funk, W. C., and L. S. Mills. 2003. Potential causes of population declines in forest fragments in an Amazonian frog. *Biol. Cons.* 111:205–14.

Funkhouser, D., and L. Goldstein. 1973. Urea response to pure osmotic stress in the aquatic toad, *Xenopus laevis*. *Am. J. Physiol.* 224:524–29.

Furst, P. 1976. *Hallucinogens and culture.* San Francisco: Chandler and Sharp.

Fuzessery, Z. M. 1988. Frequency tuning in the anuran central auditory system. In *The evolution of the amphibian auditory system,* ed. B. Fritzsch, M. J. Ryan, W. Wilczynski, T. E. Hetherington, and W. Walkowiak, 253–73. New York: Wiley.

Fuzessery, Z. M., and A. S. Feng. 1981. Frequency representation in the dorsal medullary nucleus of the leopard frog, *Rana p. pipiens*. *J. Comp. Physiol. A* 143:339–47.

———. 1982. Frequency selectivity in the anuran auditory midbrain: single unit responses to single and multiple tone stimulation. *J. Comp. Physiol.* 146:471–84.

———. 1983a. Frequency selectivity in the anuran auditory medulla: Excitatory and inhibitory tuning properties of single neurons in the dorsal medullary and superior olivary nuclei. *J. Comp. Physiol.* 150:107–19.

———. 1983b. Mating call selectivity in the thalamus and midbrain of the leopard frog (*Rana p. pipiens*): Single and multiunit activity. *J. Comp. Physiol.* 150:333–44.

Gabor, C. R. 1995. Correlational test of Mathis' hypothesis that bigger salamanders have better territories. *Copeia* 1995:729–35.

Gabor, C. R., and T. R. Halliday. 1997. Sequential mate choice by multiply mating smooth newts: Females become more choosy. *Behav. Ecol.* 8:162–66.

Gabor, C. R., and R. G. Jaeger. 1995. Resource quality affects the agonistic behaviour of territorial salamanders. *Anim. Behav.* 49:71–79.

———. 1999. When salamanders misrepresent threat signals. *Copeia* 1999:1123–26.

Gabor, C. R., J. D. Krenz, and R. G. Jaeger. 2000. Female choice, male interference, and sperm precedence in the red-spotted newt. *Behav. Ecol.* 11:115–24.

Gaborieau, O., and G. J. Measey. 2004. Termitivore or detritivore? A quantitative investigation into the diet of the East African caecilian *Boulengerula taitanus* (Amphibia: Gymnophiona: Caeciliidae). *Anim. Biol.* 54:45–56.

Gabrion, J., P. Sentein, and C. Gabrion. 1977. Les populations neoteniques des *Triturus helveticus* des causses et du Bas-Languedoc. I. Repartition et charactéristiques. *Terre et Vie* 31:489–506.

Gabrion, J., P. Sentein, and C. Gabrion. 1978. Les populations neoteniques des *Triturus helveticus* des causses et du Bas-Languedoc. II. Écologie. *Terre et Vie* 32:577–610.

Gahlenbeck, H., and H. Bartels. 1968. Temperaturadaptation der Säuerstoff-affinitat des Blutes von *Rana esculenta* L. *Z. Vergl. Physiol.* 59:232–40.

———. 1970. Blood gas transport properties in gill and lung forms of the axolotl (*Ambystoma mexicanum*). *Respir. Physiol.* 9:175–82.

Gal, J. M., and R. W. Blake. 1987. Hydrodynamic drag of two frog species: *Hymenochirus boettgeri* and *Rana pipiens*. *Can. J. Zool.* 65:1085–90.

———. 1988a. Biomechanics of frog swimming. I. Estimation of the propulsive force generated by *Hymenochirus boettgeri*. *J. Exp. Biol.* 138:399–411.

———. 1988b. Biomechanics of frog swimming. II. Mechanics of the limb-beat cycle in *Hymenochirus boettgeri*. *J. Exp. Biol.* 138:413–29.

Galat-Luong, A. 1991. Proies inhabituelles pour le patas d'Afrique de l'ouest (*Erythrocebus patas patas*). *Rev. Ecol. (Terre Vie)* 46:83–84.

Galatti, U. 1992. Population biology of the frog *Leptodactylus pentadactylus* in a central Amazonian rainforest. *J. Herpetol.* 26:23–31.

Gallardo, J. M. 1958. Observaciones biológicas sobre *Leptodactylus prognathus* Boulenger. *Cienc. Invest.* 14:460–65.

———. 1964. Consideraciones sobre *Leptodactylus ocellatus* (L.) (Amphibia, Anura) y especies aliadas. *Physis (B. Aires)* 24:373–84.

Gallie, J. A., R. L. Mumme, and S. A. Wissinger. 2001. Experience has no effect on the development of chemosensory recognition of predators by tadpoles of the American toad, *Bufo americanus*. *Herpetologica* 57:376–83.

Gallien, L. 1940. Recherches sur la physiologie hypophysaire dans ces relations avec les gonades et le cycle sexuel, chez la grenouille rousse, *Rana temporaria* L. *Bull. Biol. France et Belgique* 74:1–42.

Galton, V. A. 1992. The role of thyroid hormone in amphibian metamorphosis. *Trends Endocrinol. Metab.* 3:96–100.

Gambs, R. D., and M. J. Littlejohn. 1979. Acoustic behavior of males of the Rio Grande leopard frog (*Rana berlandieri*): An experimental analysis through field playback trials. *Copeia* 1979:643–50.

Gamradt, S. C., and L. B. Kats. 1996. Effect of introduced crayfish and mosquitofish on California newts. *Cons. Biol.* 10:1155–62.

Gamradt, S. C., L. B. Kats, and C. B. Anzalone. 1997. Aggression by non-native crayfish deters breeding in California newts. *Cons. Biol.* 11:793–96.

Gans, C. 1961. The first record of egg laying in *Siphonops paulensis* Boettger. *Copeia* 1961:490–91.

———. 1973. Sound production in the Salientia: Mechanism and evolution of the emitter. *Amer. Zool.* 13:1179–94.

———. 1974. *Biomechanics.* Philadelphia: J. B. Lippincott.

Gans, C. and G. C. Gorniak. 1982a. Functional morphology of lingual protrusion in marine *toads (Bufo marinus)*. *Am. J. Anat.* 163:195–222.

———. 1982b. How does the toad flip its tongue? Test of two hypotheses. *Science* 216:1335–37.

Gans, C., and G. de Gueldre. 1992. Striated muscle: physiology and functional morphology. In *Environmental physiology of the amphibians,* ed. M. E. Feder and W. W. Burggren, 277–313. Chicago:University of Chicago Press.

Gao, K.-Q., and N. H. Shubin. 2001. Late Jurassic salamanders from northern China. *Nature* 410:574–76.

———. 2003. Earliest known crown-group salamanders. *Nature* 422:424–28.

Gao, K.-Q., and Y. Wang. 2001. Mesozoic anurans from Liaoning Province, China, and phylogenetic relationships of archaeobatrachian anuran clades. *J. Vert. Paleontol.* 21:460–76.

Garagna, S., M. Zuccotti, M. L. Vecchi, P. G. Rubini, E. Capanna, and C. A. Redi. 2001. Human-dominated ecosystems and restoration ecology: Seveso today. *Chemosphere* 43:577–85.

García, A. 2006. Using ecological niche modelling to identify diversity hotspots for the herpetofauna of Pacific lowlands and adjacent interior valleys of Mexico. *Biol. Cons.* 130:25–46.

Garcia, C. M., and H. Drummond. 1988. Seasonal and ontogenetic variation in the diet of the Mexican garter snake, *Thamnophis eques,* in Lake Tecocomulco, Hidalgo. *J. Herpetol.* 22:129–34.

Garcia, T. S., J. Stacy, and A. Sih. 2004. Larval salamander responses to UV radiation and predation risk: color change and microhabitat use. *Ecol. Appl.* 14:1055–64.

García-París, M., M. Alcobendas, D. Buckley, and D. B. Wake. 2003. Dispersal of viviparity across contact zones in Iberian populations of fire salamanders (*Salamandra*) inferred from discordance of genetic and morphological traits. *Evolution* 57:129–43.

García-París, M., D. R. Buchholz, and G. Parra-Olea. 2003. Phylogenetic relationships of Pelabatoidea re-examined using mtDNA. *Mol. Phylogen. Evol.* 28:12–23.

García-París, M., and S. M. Deban. 1995. A novel antipredator mechanism in salamanders: Rolling escape in *Hydromantes platycephalus. J. Herpetol.* 29:149–51.

García-París, M., D. A. Good, G. Parra-Olea, and D. B. Wake. 2000. Biodiversity of Costa Rican salamanders: Implications of high levels of genetic differentiation and phylogeographic structure for species formation. *Proc. Natl. Acad. Sci. USA* 97:1640–47.

Garcia-Rutledge, E. J., and P. M. Narins. 2001. Shared acoustic resources in an Old World frog community. *Herpetologica* 57:104–16.

Garda, A. A., G. R. Colli, O. Aguiar-Junior, S. M. Recco-Pimentel, and S. N. Báo. 2002. The ultrastructure of the spermatozoa of *Epipedobates flavopictus* (Amphibia, Anura, Dendrobatidae), with comments on its evolutionary significance. *Tissue and Cell* 34:356–64.

Gardner, J. D. 2003. The fossil salamander *Proamphiuma cretacea* Estes (Caudata: Amphiumidae) and relationships within the Amphiumidae. *J. Vert. Paleontol.* 23:769–82.

Gardner, M. N., A. W. Smits, and N. J. Smatresk. 2000. The ventilatory responses of the caecilian *Typhlonectes natans* to hypoxia and hypercapnia. *Physiol. Biochem. Zool.* 73:23–29.

Gardner, R., M. R. Morris, and C. E. Nelson. 1987. Conditional evolutionarily stable strategies. *Anim. Behav.* 35:507–17.

Garland, T., Jr. 1984. Physiological correlates of locomotor performance in a lizard: An allometric approach. *Am. J. Physiol.* 247:R806–R815.

———. 1985. Ontogenetic and individual variation in size, shape and speed in the Australian agamid lizard *Amphibolurus nuchalis. J. Zool., London* A 297:425–39.

———. 1988. Genetic basis of activity metabolism. I. Inheritance of speed, stamina and antipredator displays in the garter snake *Thamnophis sirtalis. Evolution* 42:425–39.

Garland, T., Jr., and S. C. Adolph. 1994. Why not to do two-species comparative studies: limitations on inferring adaptation. *Physiol. Zool.* 67:797–828.

Garland, T., Jr., and P. L. Else. 1987. Seasonal, sexual and individual variation in endurance and activity metabolism in lizards. *Am. J. Physiol.* 252:R439–R449.

Garland, T., Jr., E. Hankins, and R. B. Huey. 1990. Locomotor capacity and social dominance in male lizards. *Funct. Ecol.* 4:243–50.

Garlick, R. L., B. J. Davis, M. F. Farmer, J. F. Hans, E. Unni, R. W. Noble, D. A. Powers, A. Riggs, and R. E. Weber. 1978. A fetal-maternal shift in the oxygen equilibrium of hemoglobin from the viviparous caecilian, *Typhlonectes compressicauda. Comp. Biochem. Physiol.* 63A:239–44.

Garner, T. W. J., S. Angelone, and P. B. Pearman. 2003. Genetic depletion in Swiss populations of *Rana latastei:* conservation implications. *Biol. Cons.* 114:371–76.

Garner, T. W. J., and B. R. Schmidt. 2003. Relatedness, body size and paternity in the alpine newt, *Triturus alpestris. Proc. Roy. Soc. London B* 270:619–24.

Garrick, L. D., R. L. Sutton, and J. W. Lang. 1985. Observations on the largest Jamaican tree frog, *Calyptahyla crucialis. Carib. J. Sci.* 21:159–62.

Garrido, O., E. Pugin, and B. Jorquera. 1975. Correspondance ultrastructurale entre la bourse gutturale du *Rhinoderma darwinii* et le tegument des larves. *Boll. Zool.* 42:133–44.

Garson, P. J., and M. L. Hunter. 1979. Effects of temperature and time of year on the singing behavior of wrens *Troglodytes troglodytes* and great tits *Parus major. Ibis* 121:481–87.

Garstang, W. 1985. *Larval forms with other zoological verses.* Chicago: University of Chicago Press.

Garton, J. D., and H. R. Mushinsky. 1979. Integumentary toxicity and unpalatability as an antipredator mechanism in the narrow mouthed toad, *Gastrophryne carolinensis. Can. J. Zool.* 57:1965–73.

Garton, J. S. 1972. Courtship of the small mouthed salamander, *Ambystoma texanum,* in southern Illinois. *Herpetologica* 28:41–45.

Garton, J. S., and R. A. Brandon. 1975. Reproductive ecology of the green treefrog, *Hyla cinerea,* in southern Illinois (Anura: Hylidae). *Herpetologica* 31:150–61.

Gartshore, M. E. 1986. The status of the montane herpetofauna of the Cameroon Highlands. In *Conservation of Cameroon montane forests,* ed. S. N. Stuart, 204–40. Cambridge, UK: International Council for Bird Preservation.

Gartside, D. F. 1980. Analysis of a hybrid zone between chorus frogs of the *Pseudacris nigrita* complex in the southern United States. *Copeia* 1980:56–66.

Gasche, P. 1939. Beiträge zur Kenntnis der Entwicklungsgeschichte von *Salamandra salamandra* L. mit besonderer Berücksichtigung der Winterpass, der Metamorphose und des Verhaltens der Schildruse. *Rev. Suisse Zool.* 46:403–548.

Gascon, C. 1991. Population- and community-level analyses of species occurrences of central Amazonian rainforest tadpoles. *Ecology* 72:1731–46.

———. 1992a. Aquatic predators and tadpole prey in central Amazonia: Field data and experimental manipulations. *Ecology* 73:971–80.

———. 1992b. The effects of reproductive phenology on larval performance traits in a three-species assemblage of central Amazonian tadpoles. *Oikos* 65:307–13

———. 1992c. Spatial distribution of *Osteocephalus taurinus* and *Pipa arrabali* in a central Amazonian forest. *Copeia* 1992:894–97.

Gascon, C. 1993. Breeding-habitat use by five Amazonian frogs at forest edge. *Biodiv. Cons.* 2:438.

———. 1995. Tropical larval anuran fitness in the absence of direct effects of predation and competition. *Ecology* 76:2222–29.

Gascon, C., T. E. Lovejoy, R. O. Bierregaard, J. R. Malcolm, P. C. Stouffer, H. L. Vasconcelos, W. F. Laurance, B. Zimmerman, M. Tocher, and S. Borges. 1999. Matrix habitat and species richness in tropical forest remnants. *Biol. Cons.* 91:223–29.

Gascon, C., and J. Travis. 1992. Does the spatial scale of experimentation matter? A test with tadpoles and dragonflies. *Ecology* 73:2237–43.

Gasser, F., and J. M. J. Joly. 1972. Existence d'un cycle sexuel biennal chez la femelle de *Salamandra salamandra fastuosa* Schreiber (Urodele, Salamandridae) a différentes altitudes dans les Pyrénées centrales: influence des facteurs génétiques et climatiques. *Ann. Sci. Nat. Zool. Biol. Anim.* 14:427–44.

Gasser, K. W., and B. T. Miller. 1986. Osmoregulation of larval blotched tiger salamanders, *Ambystoma tigrinum melanostictum* in saline environments. *Physiol. Zool.* 59:643–48.

Gaston, A. J., H. G. Gilchrist, and J. M. Hipfner. 2005. Climate changes, ice conditions and reproduction in an Arctic nesting marine bird: Brunnich's guillemot (*Uria lomvia* L.). *J. Anim. Ecol.* 74:832–41.

Gates, J. E., R. H. Stouffer, Jr., J. R. Stauffer, Jr., and C. H. Hocutt. 1985. Dispersal patterns of translocated *Cryptobranchus alleganiensis* in a Maryland stream. *J. Herpetol.* 19:436–38.

Gatten, R. E., Jr. 1985. The uses of anaerobiosis by amphibians and reptiles. *Amer. Zool.* 25:945–54.

———. 1987a. Activity metabolism of anuran amphibians: tolerance to dehydration. *Physiol. Zool.* 60:576–85.

———. 1987b. Cardiovascular and other physiological correlates of hibernation in aquatic and terrestrial turtles. *Amer. Zool.* 27:59–68.

Gatten, R. E., Jr., J. P. Caldwell, and M. E. Stockard. 1984. Anaerobic metabolism during intense swimming by anuran larvae. *Herpetologica* 40:164–69.

Gatten, R. E., Jr., and R. M. Clark. 1989. Locomotor performance of hydrated and dehydrated frogs: Recovery following exhaustive exercise. *Copeia* 1989:451–55.

Gatten, R. E., Jr., and C. J. Hill. 1984. Social influences on thermal selection by *Hyla crucifer. J. Herpetol.* 18:87–88.

Gatten, R. E., Jr., K. Miller, and R. J. Full. 1992. Energetics at rest and during locomotion. In *Environmental physiology of the amphibians*, ed. M. E. Feder and W. W. Burggren, 314–77. Chicago:University of Chicago Press.

Gatz, A. J. 1973. Algal entry into the eggs of *Ambystoma maculatum. J. Herpetol.* 7:137–38.

Gatz, J. A., Jr. 1981a. Non-random mating by size in American toads, *Bufo americanus. Anim. Behav.* 29:1004–12.

———. 1981b. Size selective mating in *Hyla versicolor* and *Hyla crucifer. J. Herpetol.* 15:114–16.

Gatz, R. N., E. C. Crawford, and J. Piiper. 1975. Kinetics of inert gas equilibration in an exclusively skin-breathing salamander, *Desmognathus fuscus. Respir. Physiol.* 24:15–29.

Gaudin, A. J. 1965. Larval development of the tree frogs *Hyla regilla* and *Hyla californiae. Hereptologica* 21:117–30.

Gauss, G. H. 1961. Ein Beitrag zur Kenntnis des Balzverhaltens einheimischer Molche. *Z. Tierpsychol.* 18:60–66.

Gautier, P., and C. Miaud. 2003. Feacal pellets used as an economic territorial marker in two terrestrial alpine salamanders. *Ecoscience* 10:134–39.

Gautier, P., J. P. Léna, and C. Miaud. 2004. Responses to conspecific scent marks and the ontogeny of territorial marking in immature terrestrial salamanders. *Behav. Ecol. Sociobiol.* 55:447–53.

Gautier, P., K. Olgun, N. Uzum, and C. Miaud. 2006. Gregarious behavior in a salamander: attraction to conspecific chemical cues in burrow choice. *Behav. Ecol. Sociobiol.* 59:836–41.

Gaymer, R. 1971. New method of locomotion in limbless terrestrial vertebrates. *Nature* 234:150–52.

Gayou, D. C. 1984. Effects of temperature on the mating call of *Hyla versicolor. Copeia* 1984:733–38.

Gee, J. H., and R. C. Waldick. 1995. Ontogenetic buoyancy changes and hydrostatic control in larval amphibians. *Copeia* 1995:861–70.

Geelen, J. F. M., J. J. van Gelder, and H. A. M. M. Sax. 1970. Insekten als voedsel van der groene kikker (*Rana esculenta* L.). *Entomol. Ber. (Amsterdam)* 30:171–78.

Geertsema, A. A. 1985. Aspects of the ecology of the serval *Leptailurus serval* in the Ngorongoro Crater, Tanzania. *Netherlands J. Zool.* 65:1035–38.

Geffeney, S., E. D. Brodie, Jr., P. C. Ruben, and E. D. Brodie III. 2002. Mechanisms of adaptation in a predator-prey arms race: TTX-resistant sodium channels. *Science* 297:1336–39.

Geffeney, S. L., E. Fujimoto, E. D. Brodie III, E. D. Brodie, Jr., and P. C. Ruben. 2005. Evolutionary diversification of TTX-resistant sodium channels in a predator-prey interaction. *Nature* 434:759–63.

Gehlbach, F. R. 1967. Evolution of tiger salamanders (*Ambystoma tigrinum*) on the Grand Canyon rims, Arizona. *Yearbook Amer. Phil. Soc.* 1967:266–69.

———. 1969. Determination of the relationships of tiger salamander larval populations to different stages of pond succession at the Grand Canyon, Arizona. *Yearbook Amer. Phil. Soc.* 1969:299–302.

Gehlbach, F. R., R. Gordon, and W. A. Weems. 1973. Aestivation of the salamander, *Siren intermedia. Am. Midl. Nat.* 89:455–63.

Gehlbach, F. R., and S. E. Kennedy. 1978. Population ecology of a highly productive aquatic salamander (*Siren intermedia*). *Southwest. Nat.* 23:423–30.

Gehlbach, F. R., J. R. Kimmel, and W. A. Weems. 1969. Aggregations and body water relations in tiger salamanders (*Ambystoma tigrinum*) from the Grand Canyon rims, Arizona. *Physiol. Zool.* 42:173–82.

Geise, W., and K. E. Linsenmair. 1986. Adaptations of the reed frog *Hyperolius viridiflavus* (Amphibia, Anura, Hyperoliidae) to its arid environment. II. Some aspects of the water economy of *Hyperolius viridiflavus nitidulus* under wet and dry season conditions. *Oecologia* 68:542–48.

———. 1988. Adaptations of the reed frog *Hyperolius viridiflavus* (Amphibia, Anura, Hyperoliidae) to its arid environment. IV. Ecological significance of water economy with comments on thermoregulation and energy allocation. *Oecologia* 77:327–38.

Geisselmann, B., R. Flindt, and H. Hemmer. 1971. Studien zur Biologie, Ökologie und Merkmalsvariabilität der beiden Braunfroscharten *Rana temporaria* L. und *Rana dalmatina* Bonaparte. *Zool. Jb. Syst.* 98:521–68.

Gergits, W. F., and R. G. Jaeger. 1990a. Field observations of the behavior of the red-backed salamander (*Plethodon cinereus*): Courtship and agonistic interactions. *J. Herpetol.* 24:93–95.

———. 1990b. Site attachment by the red-backed salamander, *Plethodon cinereus. J. Herpetol.* 24:91–93.

Gergus, E. W. A., K. B. Malmos, and B. K. Sullivan. 1999. Natural hybridization among distantly related toads (*Bufo alvarius, Bufo cognatus, Bufo woodhousii*) in Central Arizona. *Copeia* 1999:281–86

Gerhardt, H. C. 1970. *Selected aspects of the reproductive biology of some southeastern United States hylid frogs.* Unpublished PhD dissertation. Austin: University of Texas.

———. 1973. Reproductive interactions between *Hyla crucifer* and *Pseudacris ornata. Am. Midl. Nat.* 89:81–88.

———. 1974a. Behavioral isolation of the treefrogs, *Hyla cinerea* and *Hyla andersonii. Am. Midl. Nat.* 91:424–33.

———. 1974b. The significance of some spectral features in mating call recognition in the green treefrog (*Hyla cinerea*). *J. Exp. Biol.* 61:229–41.

———. 1974c. The vocalizations of some hybrid treefrogs: Acoustic and behavioral analyses. *Behaviour* 49:130–51.

———. 1975. Sound pressure levels and sound radiation patterns of the vocalizations of some North American frogs and toads. *J. Comp. Physiol.* 102:1–12.

———. 1976. Significance of two frequency bands in long distance vocal communication in the green treefrog. *Nature* 261:692–94.

———. 1978a. Discrimination of intermediate sounds in a synthetic call continuum by female green tree frogs. *Science* 199:1089–91.

———. 1978b. Mating call recognition in the green treefrog (*Hyla cinerea*): The significance of some fine-temporal properties. *J. Exp. Biol.* 74:59–73

———. 1978c. Temperature coupling in the vocal communication system of the gray treefrog *Hyla versicolor. Science* 199:992–94.

———. 1981a. Mating call recognition in the barking treefrog (*Hyla gratiosa*): Responses to synthetic calls and comparisons with the green treefrog (*Hyla cinerea*). *J. Comp. Physiol.* 144:17–25.

———. 1981b. Mating call recognition in the green treefrog (*Hyla cinerea*): Importance of two frequency bands as a function of sound pressure level. *J. Comp. Physiol.* 144:9–16.

———. 1982. Sound pattern recognition in some North American treefrogs (Anura: Hylidae): Implications for mate choice. *Amer. Zool.* 22:581–95.

———. 1983. Communication and the environment. In *Communication,* ed. T. R. Halliday and P. J. B. Slater, 82–113. San Francisco: Freeman.

———. 1987. Evolutionary and neurobiological implications of selective phonotaxis in the green treefrog, *Hyla cinerea. Anim. Behav.* 35:1479–89.

———. 1988. Acoustic properties used in call recognition in frogs and toads. In *The evolution of the amphibian auditory system,* ed. B. Fritzsch, M. J. Ryan, W. Wilczynski, T. E. Hetherington, and W. Walkowiak, 455–83. New York: Wiley.

———. 1991. Female mate choice in treefrogs: static and dynamic acoustic criteria. *Anim. Behav.* 42:615–35.

———. 1992a. Conducting playback experiments and interpreting their results. In *Playback and studies of animal communication,* ed. P. K. McGregor, 59–77. New York: Plenum.

———. 1992b. Multiple messages in acoustic signals. *Sem. Neurosci.* 4:391–400.

———. 1994a. The evolution of vocalization in frogs and toads. *Ann. Rev. Ecol. Syst.* 25:293–324.

———. 1994b. Reproductive character displacement of female mate choice in the grey treefrog, *Hyla chrysoscelis. Anim. Behav.* 47:959–69.

———. 1994c. Selective responsiveness to long-range acoustic signals in insects and anurans. *Amer. Zool.* 34:706–14.

———. 2001. Acoustic communication in two groups of closely related treefrogs. *Adv. Stud. Behav.* 30:99–167.

———. 2005. Advertisement-call preferences in diploid-tetraploid treefrogs (*Hyla chrysoscelis* and *Hyla versicolor*): Implications for mate choice and the evolution of communication systems. *Evolution* 59:395–408.

Gerhardt, H. C., S. Allan, and J. J. Schwartz. 1990. Female green treefrogs (*Hyla cinerea*) do not selectively respond to signals with a harmonic structure in noise. *J. Comp. Physiol. A* 166:791–94.

Gerhardt, H. C., R. E. Daniel, S. A. Perrill, and S. Schramm. 1987. Mating behaviour and male mating success in the green treefrog. *Anim. Behav.* 35:1490–1503.

Gerhardt, H. C., and M. S. Davis. 1988. Variation in the coding of species identity in the advertisement call of *Litoria verreauxi* (Anura: Hylidae). *Evolution* 42:556–65.

Gerhardt, H. C., B. Diekamp, and M. Ptacek. 1989. Inter-male spacing in choruses of the spring peeper, *Pseudacris* (*Hyla*) *crucifer. Anim. Behav.* 38:1012–24.

Gerhardt, H. C., and J. A. Doherty. 1988. Acoustic communication in the gray treefrog, *Hyla versicolor:* Evolutionary and neurobiological implications. *J. Comp. Physiol. A* 162:261–78.

Gerhardt, H. C., M. L. Dyson, and S. D. Tanner. 1996. Dynamic properties of the advertisement calls of gray tree frogs: Patterns of variability and female choice. *Behav. Ecol.* 7:7–18.

Gerhardt, H. C., M. L. Dyson, S. D. Tanner, and C. G. Murphy. 1994. Female treefrogs do not avoid heterospecific calls as they approach conspecific calls: Implications for mechanisms of mate choice. *Anim. Behav.* 47:1323–32.

Gerhardt, H. C., S. I. Guttman, and A. A. Karlin. 1980. Natural hybrids between *Hyla cinerea* and *Hyla gratiosa:* Morphology, vocalization and electrophoretic analysis. *Copeia* 1980:577–84.

Gerhardt, H. C., and G. Höbel. 2005. Mid-frequency suppression in the green treefrog (*Hyla cinerea*): Mechanisms and implications for the evolution of acoustic communication. *J. Comp. Physiol. A* 191:707–14.

Gerhardt, H. C., and F. Huber. 2002. *Acoustic communication in insects and frogs: Common problems and diverse solutions.* Chicago: University of Chicago Press.

Gerhardt, H. C., and G. M. Klump. 1988a. Masking of acoustic signals by the chorus background noise in the green tree frog: A limitation on mate choice. *Anim. Behav.* 36:1247–49.

———. 1988b. Phonotactic responses and selectivity of barking treefrogs (*Hyla gratiosa*) to chorus sounds. *J. Comp. Physiol. A* 163:795–802.

Gerhardt, H. C., and K. M. Mudry. 1980. Temperature effects on frequency preferences and mating call frequencies in the green treefrog, *Hyla cinerea* (Anura: Hylidae). *J. Comp. Physiol.* 137:1–6.

Gerhardt, H. C., M. B. Ptacek, L. Barnett, and K. Torke. 1994. Hybridization in the diploid-tetraploid treefrogs, *Hyla chrysoscelis* and *Hyla versicolor. Copeia* 1994:51–59.

Gerhardt, H. C., and J. Rheinlaender. 1980. Accuracy of sound localization in a miniature dendrobatid frog. *Naturwissenschaften* 67:362–63.

———. 1982. Localization of an elevated sound source by the green tree frog. *Science* 217:663–64.

Gerhardt, H. C., J. D. Roberts, M. A. Bee, and J. J. Schwartz. 2000. Call matching in the quacking frog (*Crinia georgiana*). *Behav. Ecol. Sociobiol.* 48:243–51.

Gerhardt, H. C., and H. Schneider. 1980. Mating call discrimination by females of the treefrog *Hyla meridionalis* on Tenerife. *Behav. Proc.* 5:143–49.

Gerhardt, H. C., and J. Shul. 1999. A quantitative analysis of behavioral selectivity for pulse rise-time in the gray treefrog, *Hyla versicolor. J. Comp. Physiol. A* 185:33–40.

Gerhardt, H. C., and J. J. Schwartz. 1995. Interspecific interactions in anuran courtship. In *Amphibian biology, vol. 2. Social behaviour,* ed. H. Heatwole and B. K. Sullivan, 603–32.

Chipping Norton, New South Wales, Australia: Surrey Beatty and Sons.

———. 2001. Auditory tuning and frequency preferences in anurans. In *Anuran communication,* ed. M. J. Ryan, 73–85. Washington, DC: Smithsonian Institution Press.

Gerlach, J., and J. Willi. 2002. A new species of frog, genus *Sooglossus* (Anura, Sooglossidae) from Silhouette Island, Seychelles. *Amphibia-Reptilia* 23:445–58

Gerlanc, N. M., and G. A. Kaufman. 2003. Use of bison wallows by anurans on Konza Prairie. *Am. Midl. Nat.* 150:158–68.

Gerell, R. 1968. Food habits of the mink, *Mustela vison* Schreb., in Sweden. *Viltrevy* 5:119–211.

Gershman, S. N., and P. A. Verrell. 2002. To persuade or be persuaded: which sex controls mating in a plethodontid salamander? *Behaviour* 139:447–62.

Gherardi, F., B. Renai, and C. Corti. 2001. Crayfish predation on tadpoles: A comparison between a native (*Austropotamobius pallipes*) and an alien species (*Procambarus clarkii*). *Bull. Fr. Peche Piscicul.* (361):659–68.

Giacoma, C. 1988. The ecology and distribution of newts in Italy. *Annuar. Ist. Mus. Zool. Napoli* 26:49–84.

Giacoma, C., and S. Castellano. 2001. Advertisement call variation and speciation in the *Bufo viridis* complex. In *Anuran communication,* ed. M. J. Ryan, 205–19. Washington, DC: Smithsonian Institution Press.

Giacoma, C., and N. Crusco. 1987. Courtship and male interference in the Italian newt: A field study. *Monit. Zool. Ital.* (n.s.) 21:190–91.

Giacoma, C., and M. Sparreboom. 1987. On the sexual behaviour of the Italian newt, *Triturus italicus* (Caudata: Salamandridae). *Bijdr. Dierk.* 57:19–30.

Giacoma, C., T. Kozar, and I. Pavignano. 1993. Ethological aspects of the biology of *Hyla arborea.* In *Ecology and conservation of the European tree frog,* ed. A. H. P. Stumpel and U. Tester, 21–28. Wageningen, The Netherlands: DLO Institute for Forestry and Nature Research.

Giaretta, A. A., W. C. A. Bokermann, and C. F. B. Haddad. 1993. A review of the genus *Megaelosia* (Anura: Leptodactylidae) with a description of a new species. *J. Herpetol.* 27:276–85.

Giaretta, A. A., and K. G. Facure. 2003. *Cyclorhamphus boraciensis* (Flattened Waterfall Frog). Clutch attendance. *Herp. Rev.* 34:50.

Giaretta, A. A., and M. Menin. 2004. Reproduction, phenology and mortality sources of a species of *Physalaemus* (Anura: Leptodactylidae). *J. Nat. Hist.* 38:1711–22.

Giaretta, A. A., and R. J. Sawaya. 1998. Second species of *Psyllophryne* (Anura: Brachycephalidae). *Copeia* 1998:985–87.

Gibbons, J. W. 2003a. Societal values and attitudes. Their history and sociological influences on *Amphibian Conservation* problems. In *Amphibian conservation,* ed. R. D. Semlitsch, 214–227. Washington, DC: Smithsonian Books.

———. 2003b. Terrestrial habitat: A vital component for herpetofauna of isolated wetlands. *Wetlands* 23:630–35.

Gibbons, J. W., and D. H. Bennett. 1974. Determination of anuran terrestrial activity patterns by a drift fence method. *Copeia* 1974:236–43.

Gibbons, J. W., D. E. Scott, T. J. Ryan, K. A. Buhlmann, T. D. Tuberville, B. S. Metts, J. L. Greene, T. Mills, Y. Leiden, S. Poppy, and C. T. Winne. 2000. The global decline of reptiles, déjà vu amphibians. *BioScience* 50:653–66.

Gibbons, M. M., and T. K. McCarthy. 1983. Age determination of frogs and toads (Amphibia, Anura) from north-western Europe. *Zool. Scripta* 12:145–51.

———. 1984. Growth, maturation and survival of frogs *Rana temporaria* L. *Holarc. Ecol.* 7:419–27.

———. 1986. The reproductive output of frogs *Rana temporaria* (L.) with particular reference to body size and age. *J. Zool., London* (A) 209:579–93.

Gibbs, J. P. 1993. Importance of small wetlands for the persistence of local populations of wetland-associated animals. *Wetlands* 13:25–31.

———. 1998a. Amphibian movements in response to forest edges, roads, and streambeds in southern New England. *J. Wildl. Manage.* 62:584–89.

———. 1998b. Distribution of woodland amphibians along a forest fragmentation gradient. *Landsc. Ecol.* 13:263–68.

———. 1998c. Genetic structure of redback salamander *Plethodon cinereus* populations in continuous and fragmented forests. *Biol. Cons.* 86:77–81.

———. 2000. Wetland loss and biodiversity conservation. *Cons. Biol.* 14:314–17.

Gibbs, J. P., and A. R. Breisch. 2001. Climate warming and calling phenology of frogs near Ithaca, New York, 1900–1999. *Cons. Biol.* 15:1175–78.

Gibbs, J. P., and W. G. Shriver. 2005. Can road mortality limit populations of pool-breeding amphibians? *Wetlands Ecol. Manage.* 13:281–89.

Gibson, R. C., and K. R. Buley. 2004. Maternal care and obligatory oophagy in *Leptodactylus fallax:* a new reproductive mode in frogs. *Copeia* 2004:128–35.

Gibson, R. C., and M. Freeman. 1997. Conservation at home: Recovery programme for the agile frog *Rana dalmatina* in Jersey. *Dodo* 33:91–104.

Giddings, S. 1984. An observation of a display of aggressive behaviour between two male tusked frogs (*Atelotus brevis*). *Herpetofauna (Australia)* 15:50.

Gies, A. 2003. Government view of endocrine disruption in wildlife. *Pure Appl. Chem.* 75:2563–74.

Gilbert, F. F., and R. Allwine. 1991. Terrestrial amphibian communities in the Oregon Cascade Range. In *Wildlife and vegetation in unmanaged Douglas-fir forests. USDA For. Ser. Gen. Tech. Rep. PNW-GTR-285,* ed. L. F. Ruggiero, K. B. Aubry, A. B. Carey, and M. H. Huff, 341–50. Washington DC: Government Printing Office.

Gilbert, L. I., and E. Frieden, eds. 1981. *Metamorphosis: A problem in developmental biology.* 2nd ed. New York: Plenum.

Gilbert, P. W. 1941. Eggs and nests of *Hemidactylium scutatum* in the Ithaca region. *Copeia* 1941:47.

Gill, D. E. 1978a. Effective population size and interdemic migration rates in a metapopulation of the red-spotted newt, *Notophthalmus viridescens* (Rafinesque). *Evolution* 32:839–49.

———. 1978b. The metapopulation ecology of the red-spotted newt, *Notophthalmus viridescens* (Rafinesque). *Ecol. Monogr.* 48:145–66.

———. 1979. Density dependence and homing behavior in adult red-spotted newts *Notophthalmus viridescens* (Rafinesque). *Ecology* 60:800–13.

———. 1985. Interpreting breeding patterns from census data: A solution to the Husting dilemma. *Ecology* 66:344–54.

Gill, D. E., and B. A. Mock. 1985. Ecological and evolutionary dynamics of parasites: The case of *Trypanosoma diemyctyli* in the red-spotted newt *Notophthalmus viridescens.* In *Ecology and genetics of host-parasite interactions,* ed. D. Rollinson and R. M. Anderson, 157–83. New York: Academic Press.

Gillespie, G., and H. Hines. 1999. Status of temperate riverine frogs in south-eastern Australia. In *Declines and disappearances of Australian frogs,* ed. A. Campbell, 109–30. Canberra, Australia: Environment Australia.

Gillespie, G., S. Howard, D. Lockie, M. Scroggie, and Boeadi. 2005. Herpetofauna richness and community structure on offshore islands of Sulawesi, Indonesia. *Biotropica* 37:279–90.

Gillespie, G. R. 2001. The role of introduced trout in the decline of the spotted tree frog (*Litoria spenceri*) in south-eastern Australia. *Biol. Cons.* 100:187–98.

———. 2002. Impacts of sediment loads, tadpole density, and food type on the growth and development of tadpoles of the spotted tree frog *Litoria spenceri:* An in-stream experiment. *Biol. Cons.* 106:141–50.

Gillespie, G. R., D. Lockie, M. P. Scroggie, and D. T. Iskandar. 2004. Habitat use by stream-breeding frogs in south-east Sulawesi, with some preliminary observations on community organization. *J. Trop. Biol.* 20:439–48.

Gillette, J. R. 2002. Odor discrimination in the California slender salamander, *Batrachoseps attenuatus:* evidence for self-recognition. *Herpetologica* 58:165–70.

Gillette, J. R., S. E. Kolb, J. A. Smith, and R. G. Jaeger. 2000. Pheromonal attractions to particular males by female redback salamanders (*Plethodon cinereus*). In *The biology of plethodontid salamanders,* ed. R. C. Bruce, R. G. Jaeger, and L. D. Houck, 431–40. New York: Kluwer Academic/Plenum.

Gillis, G. B., and A. A. Biewener. 2000. Hindlimb extensor muscle function during jumping and swimming in the toad (*Bufo marinus*). *J. Exp. Biol.* 203:3547–63.

Gillis, G. B., and G. V. Lauder. 1994. Aquatic prey transport and the comparative kinematics of *Ambystoma tigrinum* feeding behaviors. *J. Exp. Biol.* 187:159–79.

Gillis, J. E., and W. J. Breuer. 1984. A comparison of rates of evaporative water loss and tolerance to dehydration in the red-eft and newt of *Notophthalmus viridescens. J. Herpetol.* 18:81–82.

Gilmore, R. J., and F. H. J. Figge. 1929. Morphology of the blood vascular system of the tiger salamander (*Ambystoma tigrinum*): Heart and aortic arches. *Colorado College Publications, General Series* no. 161:1–14.

Gipouloux, J. D. 1986. Phoresie et marsupialisme. In *Traite de zoologie, vol. 14, fasc. IB. Amphibiens,* ed. P.-P. Grasse and M. Delsol, 457–69. Paris: Masson.

Girgenrath, M., and R. L. Marsh. 1997. In vivo performance of trunk muscles in tree frogs during calling. *J. Exp. Biol.* 200:3101–8.

———. 1999. Power output of sound-producing muscles in the tree frogs *Hyla versicolor* and *Hyla chrysoscelis. J. Exp. Biol.* 202:3225–37.

———. 2003. Season and testosterone affect contractile properties of fast calling muscles in the gray tree frog *Hyla chrysoscelis. Am. J. Physiol.* 84:R1513–R1520.

Girish, S., and S. K. Saidapur. 1999a. The effects of kinship and density on growth and metamorphosis of the bronze frog (*Rana temporalis*) tadpoles. *Acta Ethol.* 2:61–66.

———. 1999b. Mating and nesting behaviour, and early development in the tree frog *Polypedates maculatus. Curr. Sci.* 76:91–92.

———. 2000. Interrelationship between food availability, fat body, and ovarian cycles in the frog, *Rana tigrina,* with a discussion on the role of fat body in anuran reproduction. *J. Exp. Zool.* 286:487–93.

———. 2003. Density-dependent growth and metamorphosis in the larval bronze frog *Rana temporalis* is influenced by genetic relatedness of the cohort. *J. Biosci.* 28:489–96.

Gish, S. L., and E. S. Morton. 1981. Structural adaptations to local habitat acoustics in Carolina wren songs. *Z. Tierpsychol.* 56:74–84.

Gislén, T., and H. Kauri. 1959. Zoogeography of the Swedish amphibians and reptiles. *Acta Vert.* 1:193–397.

Gittins, S. P. 1983a. The breeding migration of the common toad (*Bufo bufo*) to a pond in mid-Wales. *J. Zool., London* 199:555–62.

———. 1983b. Diurnal activity of the common toad (*Bufo bufo*) during the breeding migration to a pond in mid-Wales. *Br. J. Herpetol.* 6:292–94.

———. 1983c. Population dynamics of the common toad (*Bufo bufo*) at a lake in mid-Wales. *J. Anim. Ecol.* 52:981–88.

Gittins, S. P., A. G. Parker, and F. M. Slater. 1980. Mate assortment in the common toad (*Bufo bufo*). *J. Nat. Hist.* 14:663–68.

Gittins, S. P., J. E. Steeds, and R. Williams. 1982. Population age-structure of the common toad (*Bufo bufo*) at a lake in mid-Wales determined from annual growth rings in the phalanges. *British J. Herpetol.* 6:249–52.

Given, M. F. 1987. Vocalizations and acoustic interactions of the carpenter frog, *Rana virgatipes. Herpetologica* 43:467–81.

———. 1988a. Growth rate and the cost of calling activity in male carpenter frogs, *Rana virgatipes. Behav. Ecol. Sociobiol.* 22:153–60.

———. 1988b. Territoriality and aggressive interactions of male carpenter frogs, *Rana virgatipes. Copeia* 1988:411–21.

———. 1990. Spatial distribution and vocal interaction in *Rana clamitans* and *R. virgatipes. J. Herpetol.* 24:377–82.

———. 1993a. Male response to female vocalizations in the carpenter frog, *Rana virgatipes. Anim. Behav.* 46:1139–49.

———. 1993b. Vocal interactions in *Bufo woodhousii fowleri. J. Herpetol.* 27:447–52.

———. 1999. Frequency alteration of the advertisement call in the carpenter frog, *Rana virgatipes. Herpetologica* 55:304–17.

———. 2002. Interrelationships among calling effort, growth rate, and chorus tenure in *Bufo fowleri. Copeia* 2002:979–87.

———. 2005. Vocalizations and reproductive behavior of male pickeral frogs, *Rana palustris. J. Herpetol.* 39:223–33.

Given, M. F., and D. M. McKay. 1990. Variation in the citrate synthase activity in calling muscles of carpenter frogs, *Rana virgatipes. Copeia* 1990:863–67.

Glandt, D. 1983. Experimentelle Untersuchugen zum Beute-Räuber-Verhältnis zwischen Stichlingen, *Gasterosteus aculeatus* (L.) und *Pungitius pungitius* (L.) (Teleostei), und Grassfroschlarven, *Rana temporaria* (L.) (Amphibia). *Zool. Anz.* 211:277–84.

———. 1984. Laborexperiment zum Beute-Räuber-Verhältnis zwischen Dreistacheligen Stichlingen, *Gasterosteus aculeatus* (L.) (Teleostei), und Erdkrötenlarven, *Bufo bufo* (L.) (Amphibia). *Zool. Anz.* 213:12–16.

———. 1985. Kaulquappen-Fressen durch Goldfische *Carassius a. auratus* and Rotfedern *Scardinius erythrophthalmus. Salamandra* 21:180–85.

———. 1986. Die saisonalen Wanderungen der mitteleuropäischen Amphibien. *Bonn. Zool. Beitr.* 37:211–28.

Glandt, D., and R. Podloucky, eds. 1987. Der Moorfrosch. *Natursch. Landsch. Niedersachesen Beih.* 19:1–161.

Glasby, C. J., G. J. B. Ross, and P. L. Beesley, eds. 1993. *Fauna of Australia. Vol. 2A. Amphibia and Reptilia.* Canberra, Australia: Australian Govt. Publ. Service.

Glass, B. P. 1951. Age at maturity of neotenic *Ambystoma tigrinum mavortium. Am. Midl. Nat.* 46:391–94.

Glaw, F., and J. Köhler. 1998. Amphibian species diversity exceeds that of mammals. *Herp. Rev.* 29:11–12.

Glaw, F., and M. Vences. 1994. *A field guide to the amphibians and reptiles of Madagascar.* 2nd ed. Cologne, Germany: M. Vences and F. Glaw GbR.

Glaw, F., M. Vences, and W. Böhme. 1998. Systematic revision of the genus *Aglyptodactylus,* Boulenger, 1919 (Amphibia: Ranidae), and analysis of its phylogenetic relationships to other Madagascan ranid genera (*Tomopterna, Boophis, Mantidactylus,* and *Mantella*). *J. Zool. Syst. Evol. Res.* 36:17–37.

Gleason, R. L., and T. H. Craig. 1979. Food habits of burrowing owls in southeastern Idaho. *Great Basin Nat.* 39:274–76.

Glennemeier, K. A., and L. J. Begnoche. 2002. Impact of organochlorine contamination on amphibian populations in southwestern Michigan. *J. Herpetol.* 36:233–44.

Glennemeier, K. A., and R. J. Denver. 2002. Role of corticoids in mediating the response of *Rana pipiens* tadpoles to intraspecific competition. *J. Exp. Zool.* 292:32–40.

Glos, J., T. U. Grafe, M.-O. Rödel, and K. E. Linsenmair. 2003. Geographic variation in pH tolerance of two populations of the European common frog, *Rana temporaria. Copeia* 2003:650–56.

Goater, C. P. 1994. Growth and survival of postmetamorphic toads: Interactions among larval history, density, and parasitism. *Ecology* 75:2264–74.

Goater, C. P., R. D. Semlitsch, and M. V. Baernasconi. 1993. Effects of body size and parasite infection on the locomotory performance of juvenile toads, *Bufo bufo. Oikos* 66:129–36.

Goater, C. P., and R. E. Vandenbos. 1997. Effects of larval history and lungworm infection on the growth and survival of juvenile wood frogs (*Rana sylvatica*). *Herpetologica* 53:331–38.

Goater, C. P., and P. I. Ward. 1992. Negative effects of *Rhabdias bufonis* (Nematoda) on the growth and survival of toads (*Bufo bufo*). *Oecologia* 89:161–65.

Goater, T. M., G. W. Esch, and A. O. Bush. 1987. Helminth parasites of sympatric salamanders: ecological concepts at infracommunity, component and compound community levels. *Am. Midl. Nat.* 118:289–300.

Gochfeld, M. 1978. Intraspecific social stimulation and temporal displacement of song of the lesser skylark *Alauda gugula. Z. Tierpsychol.* 48: 337–44.

Godley, J. S. 1983. Observations on the courtship, nests and young of *Siren intermedia* in southern Florida. *Am. Midl. Nat.* 110:215–19.

Godreau, V., G. Bornette, B. Frochot, C. Amoros, E. Castella, B. Oertli, F. Chambaud, D. Oberti, and E. Craney. 1999. Biodiversity in the floodplain of Saône: A global approach. *Biodivers. Cons.* 8:839–64.

Godwin, G. J., and S. M. Roble. 1983. Mating success in male treefrogs, *Hyla chrysoscelis* (Anura: Hylidae). *Herpetologica* 39:141–46.

Goeldi, E. A. 1899. Über die Entwicklung von *Siphonops annulatus. Zool. Jb. Syst.* 12:170–73.

———. 1907. Description of *Hyla resinifictrix* Goeldi, a new Amazonian tree-frog peculiar for its breeding habits. *Proc. Zool. Soc. London* 1907:135–40.

Goetsch, W. 1924. Lebensraum und Körpergrösse. *Biol. Zentralbl.* 44:529–60.

Goicoechea, O., O. Garrido, and B. Jorquera. 1986. Evidence for a trophic paternal-larval relationship in the frog *Rhinoderma darwinii. J. Herpetol.* 20:168–78.

Goin, C. J. 1943. The lower vertebrate fauna of the water hyacinth community. *Proc. Florida Acad. Sci.* 6:143–54.

———. 1947. Studies on the life history of *Eleutherodactylus ricordii planirostris* (Cope) in Florida. *Univ. Florida Publ. Biol. Sci.* 4:1–66.

———. 1949. The peep order of peepers: A swamp water serenade. *Q. J. Florida Acad. Sci.* 11:59–61.

———. 1950. Color pattern inheritance in some frogs of the genus *Eleutherodactylus. Bull. Chicago Acad. Sci.* 9:1–15.

———. 1951. Notes on the eggs and early larvae of three more Florida salamanders. *Ann. Carnegie Mus.* 32:253–60.

———. 1954. Remarks on evolution of color pattern in the *gossei* group of the frog genus *Eleutherodactylus. Ann. Carnegie Mus.* 33:185–95.

———. 1960a. Amphibians, pioneers of terrestrial breeding habits. *Smithsonian Rep.* 1959:427–55.

———. 1960b. Pattern variation in the frog *Eleutherodactylus nubicola* Dunn. *Bull. Florida State Mus.* 5:243–258.

Goin, C. J., and L. H. Ogren. 1956. Parasitic copepods (Argulidae) on amphibians. *J. Parasit.* 42:154.

Goin, O. B., and C. J. Goin. 1962. Amphibian eggs and the montane environment. *Evolution* 16:364–71.

Golay, N., and H. Durrer. 1994. Inflammation due to toe clipping in natterjack toads (*Bufo calamita*). *Amphibia-Reptilia* 15:81–83.

Goldenberg, S., and M. R. Warburg. 1977. Osmoregulatory effect of prolactin during ontogenesis in two anurans. *Comp. Biochem. Physiol.* 56A:137–43.

———. 1983. Water balance of five amphibian species at different stages and phases, as affected by hypophysial hormones. *Comp. Biochem. Physiol.* 75A:447–55.

Goldstein, J. K. Hoff, and S. D. Hillyard. 2003. Comparison of dehydration and angiotensin II-stimulated cutaneous drinking in toads, *Bufo punctatus. Comp. Biochem. Physiol.* A 136:557–63.

Gollmann, B., and G. Gollmann. 1991. Embryonic development of the myobatrachine frogs *Geocrinia laevis, Geocrinia victoriana,* and their natural hybrids. *Amphibia-Reptilia* 12:103–10.

———. 1993. A laboratory experiment on interspecific competition between tadpoles of *Geocrinia victoriana* and *Pseudophryne semimarmorata* (Anura, Myobatrachidae). *Amphibia-Reptilia* 14:349–56.

Gomes, F. R., C. R. Bevier, and C. A. Navas. 2002. Enviromental and physiological factors influence antipredator behavior in *Scinax hiemalis* (Anura: Hylidae). *Copeia* 2002:994–1005.

Gomes, F. R., J. G. Chauí-Berlinck, J. E. P. Bicudo, and C. A. Navas. 2004. Intraspecific relationships between resting and activity metabolism in anuran amphibians: Influence of ecology and behavior. *Physiol. Biochem. Zool.* 77:197–208.

Gomes, M. R., and O. L. Peixoto. 1997. *Hyla callipygia* (Bocaina Treefrog). *Herp. Rev.* 28:92–93.

Gomez-Mestre, I., and C. Keller. 2003. Experimental assessment of turtle predation on larval anurans. *Copeia* 2003:349–56.

Gomez-Mestre, I., and M. Tejedo. 2003. Local adaptation of an anuran amphibian to osmotically stressful environments. *Evolution* 57:1889–99.

———. 2004. Constrasting patterns of quantitative and neutral genetic variation in locally adapted populations of the natterjack toad, *Bufo calamita. Evolution* 58:2343–52.

Gomez-Mestre, I., M. Tejedo, E. Ramayo, and J. Estepa. 2004. Developmental alterations and osmoregulatory physiology of a larval anuran under osmotic stress. *Physiol. Biochem. Zool.* 77:267–74.

Gompper, M. E., and H. M. Hackett. 2005. The long-term, range-wide decline of a once common carnivore: The eastern spotted skunk (*Spilogale putorius*). *Anim. Cons.* 8:195–201.

Gonçalves, A. A., and P. Sawaya. 1978. Oxygen uptake by *Typhlonectes compressicaudus* related to the body weight. *Comp. Biochem. Physiol.* 61:141–43.

Goniakowska-Witalinska, L. 1978. Ultrastructural and morphometric study of the lung of the European salamander *Salamandra salamandra* L. *Cell Tiss. Res.* 191:343–56.

———. 1980a. Scanning and transmission electron microscopic study of the lung of the newt *Triturus alpestris* Laur. *Cell Tiss. Res.* 205:133–45.

———. 1980b. Ultrastructural and morphometric changes in the lungs of the newt *Triturus cristatus carnifex* Laur. during ontogeny. *J. Anat.* 130:571–85.

———. 1985. The structure of the lung of the treefrog *Hyla arborea* L. *Fortschr. Zool.* 30:375–77.

———. 1986. Lung of the tree frog *Hyla arborea* L. A scanning and transmission electron microscopic study. *Anat. Embryol.* 174:379–89.

Gonser, R. A., and L. L. Woolbright. 1995. Homing behavior of the Puerto Rican frog, *Eleutherodactylus coqui. J. Herpetol.* 29:481–84.

Good, D. A., and D. B. Wake. 1992. Geographic variation and speciation in the torrent salamanders of the genus *Rhyacotriton* (Caudata: Rhyacotritonidae). *Univ. California Publ. Zool.* 126:1–91.

———. 1993. Systematic studies of the Costa Rican moss salamanders, genus *Nototriton,* with descriptions of three new species. *Herpetol. Monogr.* 7:131–59.

Goodman, D. E. 1971. Territorial behavior in a Neotropical frog, *Dendrobates granuliferus. Copeia* 1971:365–70.

Goodsell, J. A., and L. B. Kats. 1999. Effect of introduced mosquitofish on Pacific treefrogs and the role of alternative prey. *Cons. Biol.* 13:921–24.

Goodwin, D. 1976. *Crows of the world.* Ithaca, NY: Cornell University Press.

Goodyear, C. P. 1971. Y-axis orientation of the oak toad, *Bufo quercicus. Herpetologica* 27:320–23.

Goodyear, C. P., and R. Altig. 1971. Orientation of bullfrogs (*Rana catesbeiana*) during metamorphosis. *Copeia* 1971:362–64.

Gooler, D. M., C. J. Condon, J. H. Xu, and A. S. Feng. 1993. Sound direction influences the frequency-tuning characteristics of neurons in the frog inferior colliculus. *J. Neurophysiol.* 69:1018–30.

Gorbman, A. 1964. Endocrinology of the Amphibia. In *Physiology of the Amphibia,* vol. 1., ed. J. A. Moore, 371–425. New York: Academic Press.

Gordon, D. R. 1998. Effects of invasive, non-indigenous plant species on ecosystem processes: lessons from Florida. *Ecol. Appl.* 8:975–89.

Gordon, M. S. 1962. Osmotic regulation in the green toad (*Bufo viridis*). *J. Exp. Biol.* 39:261–70.

———. 1965. Intracellular osmoregulation in skeletal muscle during salanity adaptation in two species of toads. *Biol. Bull.* 128:218–29.

Gordon, M. S., K. Schmidt-Nielsen, and H. M. Kelly. 1961. Osmotic regulation in the crab-eating frog (*Rana cancrivora*). *J. Exp. Biol.* 38:659–78.

Gordon, M. S., and V. A. Tucker. 1965. Osmotic regulation in the tadpoles of the crab-eating frog (*Rana cancrivora*). *J. Exp. Biol.* 42:437–45.

———. 1968. Further observations on the physiology of salinity adaptation in the crab eating frog (*Rana cancrivora*). *J. Exp. Biol.* 49:185–93.

Gordon, N. M. 2004. The effect of supplemental feeding on the territorial behavior of the green frog (*Rana clamitans*). *Amphibia-Reptilia* 25:55–62.

Gordon, R. E. 1952. A contribution to the life history and ecology of the plethodontid salamander *Aneides aeneus* (Cope and Packard). *Am. Midl. Nat.* 47:666–701.

———. 1961. The movement of displaced green salamanders. *Ecology* 42:200–2.

Gore, A. P., and V. R. Prayag. 1993. Should harvesting of frogs be banned? In *Environmental problems and prospects in India,* ed. M. Balakrishna, 399–416. Oxford: Oxford University Press.

Gorman, J. 1956. Reproduction in the plethodontids of the genus *Hydromantes. Herpetologica* 12:249–59.

Gorman, R. R., and D. E. Ferguson. 1970. Sun-compass orientation in the western toad, *Bufo boreas. Herpetologica* 26:34–45.

Gorman, W. L. 1986. Patterns of color polymorphism in the cricket frog, *Acris crepitans,* in Kansas. *Copeia* 1986:995–99.

Gorr, T., and T. Kleinschmidt. 1993. Evolutionary relationships of the coelacanth. *Amer. Sci.* 81:72–82.

Gorr, T., T. Kleinschmidt, and H. Fricke. 1991. Close tetrapod relationships of the coelacanth *Latimeria* indicate by haemoglobin sequences. *Nature* 351:394–97.

Gorzula, S. 1977. Foam nesting in leptodactylids: a possible function. Brit. *J. Herpetol.* 5:657–59.

Gorzula, S. J. 1978. An ecological study of *Caiman crocodilus crocodilus* inhabiting savanna lagoons in the Venezuelan Guayana. *Oecologia* 35:21–34.

Gorzula, S., J. Morales, and L. Hernandez. 1983. Cuido materno en la rana *Stefania scalae. Mem. Soc. Cienc. Nat. La Salle, Caracas* 43:127–28.

Gosá, A. 2003. *Bufo bufo* (common toad), *Alytes obstetricans* (midwife toad), and *Rana temporaria* (common frog). Tree climbing. *Herp. Rev.* 34:355.

Gosling, L. M. 1986. The evolution of male mating strategies in male antelopes. In *Ecological aspects of social evolution,* ed. D. I. Rubenstein and R. W. Wrangham, 244–81. Princeton, NJ: Princeton University Press.

Gosling, L. M., and M. Petrie. 1990. Lekking in topi: A consequence of satellite behaviour by small males at hotspots. *Anim. Behav.* 40:272–87.

Gosner, K. L. 1960. A simplified table for staging anuran embryos and larvae with notes on identification. *Herpetologica* 16:183–90.

Gosner, K. L., and D. A. Rossman. 1959. Observations on the reproductive cycle of the swamp chorus frog, *Pseudacris nigrita. Copeia* 1959:263–66.

———. 1960. Eggs and larval development of the treefrogs *Hyla crucifer* and *Hyla ocularis. Herpetologica* 16:225–32.

Gosse, P. H. 1860. *The romance of natural history.* London: J. Nisbet.

Gottlander, K. 1987. Variation in the song rate of male pied flycatchers *Ficedula hypoleuca:* causes and consequences. *Anim. Behav.* 35:1037–43.

Gottlieb, G., and D. C. Jackson. 1976. Importance of pulmonary ventilation in respiratory control in the bullfrog. *Am. J. Physiol.* 230:608–13.

Gould, A. A. 1853. *The naturalist's library.* Boston: Phillips, Sampson, & Co.

Gould, S. J. 1977. *Ontogeny and phylogeny.* Cambridge, MA: Harvard University Press.

Gower, D. J., A. Kupfer, O. V. Oommen, W. Himstedt, R. A. Nussbaum, S. P. Loader, B. Presswell, H. Müller, S. B. Krishna, R. Boistel, and M. Wilkinson. 2002. A molecular phylogeny of ichthyophiid caecilians (Amphibia: Gymnophiona: Ichthyophiidae): Out of India or out of South East Asia? *Proc. Roy. Soc. London B* 269:1563–69.

Gower, D. J., S. P. Loader, and C. B. Moncrieff, and M. Wilkinson. 2004. Niche separation and comparative abundance of *Boulengerula boulengeri* and *Scolecomorphus vittatus* (Amphibia: Gymnophiona) in an East Usambara Forest, Tanzania. *Afr. J. Herpetol.* 53:183–90.

Gower, D. J., J. B. Rasmussen, S. P. Loader, and M. Wilkinson. 2004. The caecilian amphibian *Scoleomorphus kirkii* Boulenger as prey of the burrowing asp *Atractaspis aterrima* Günther: trophic relationships of fossorial vertebrates. *Afr. J. Ecol.* 42:83–87.

Gower, D. J., and M. Wilkinson. 2005. Conservation biology of caecilian amphibians. *Cons. Biol.* 19:45–55.

Gradwell, N. 1968. The jaw and hyoidean mechanism of the bullfrog tadpole during aqueous ventilation. *Can. J. Zool.* 46:1041–52.

———. 1970. The function of the ventral velum during gill irrigation in *Rana catesbeiana. Can. J. Zool.* 48:1179–86.

———. 1971. *Ascaphus* tadpole: Experiments on the suction and gill irrigation mechanism. *Can. J. Zool.* 49:307–32.

———. 1972a. Gill irrigation in *Rana catesbeiana*. Part I. On the anatomical basis. *Can. J. Zool.* 50:481–99.

———. 1972b. Gill irrigation in *Rana catesbeiana*. Part II. On the musculoskeletal mechanism. *Can. J. Zool.* 50:501–21.

———. 1973. On the functional morphology of suction and gill irrigation in the tadpole of *Ascaphus,* and notes on hibernation. *Herpetologica* 29:84–93.

———. 1975. Experiments on oral suction and gill breathing in five species of Australian tadpole (Anura: Hylidae and Leptodactylidae). *J. Zool., London* 177:81–98.

Graeff, D., and R. Schülte. 1980. Neue Erkenntnisse zur Brutbiologie von *Dendrobates pumilio. Herpetofauna* 2:17–22.

Graf, J.-D., and M. Polls Pelaz. 1989. Evolutionary genetics of the *Rana esculenta* complex. In *Evolution and ecology of unisexual vertebrates,* ed. R. M. Dawley and J. P Bogart, 289–302. Albany: New York State Museum Bull. 466.

Grafe, T. U. 1995. Graded aggressive calls in the African painted reed frog *Hyperolius marmoratus* (Hyperoliidae). *Ethology* 101:67–81.

———. 1996a. Energetics of vocalization in the African reed frog (*Hyperolius marmoratus*). *Comp. Biochem. Physiol.* 114A:235–43.

———. 1996b. The function of call alternation in the African reed frog (*Hyperolius marmoratus*): Precise call timing prevents auditory masking. *Behav. Ecol. Sociobiol.* 38:149–58.

———. 1999. A function of synchronous chorusing and a novel female preference shift in an anuran. *Proc. Roy. Soc. London B* 266:2331–36.

———. 2003. Synchronized interdigitated calling in the Kavangu running frog, *Kassina kuvangensis. Anim. Behav.* 66:127–36.

Grafe, T. U., S. K. Kaminsky, J.H. Bitz, H. Lüssow, and K. E. Linsenmair. 2004. Demographic dynamics of the Afro-tropical pig-nosed frog, *Hemisus marmoratus:* Effects of climate and predation on survival and recruitment. *Oecologia* 141:40–46.

Grafe, T. U., S. K. Kaminsky, and K. E. Linsenmair. 2005. Terrestrial larval development and nitrogen excretion in the Afro-tropical pig-nosed frog, *Hemisus marmoratus. J. Trop. Ecol.* 21:219–22.

Grafe, T. U., R. Schmuck, and K. E. Linsenmair. 1992. Reproductive energetics of the African reed frogs, *Hyperolius viridiflavus* and *Hyperolius marmoratus. Physiol. Zool.* 65:153–71.

Grafe, T. U., M. Spieler, and B. König. 1999. Soziobiolgische Erklärungsansätze des Verhaltens von Amphibien und Reptilien: Aktuelle Theorian und offene Fragen. *Z. Feldherpetol.* 6:1–42.

Grafe, T. U., J. O. Steffan, and C. Stoll. 2000. Vocal repertoire and effect of advertisement call intensity on calling behaviour in the West African tree frog, *Leptopelis viridis. Amphibia-Reptilia* 21:13–23.

Grafe, T. U., and J. Thein. 2001. Energetics of calling and metabolic substrate use during prolonged exercise in the European treefrog *Hyla arborea. J. Comp. Physiol.* B 171:69–76.

Grafe, U. 1997a. Costs and benefits of mate choice in the lek-breeding reed frog, *Hyperolius marmoratus. Anim. Behav.* 53:1103–17.

———. 1997b. Use of metabolic substrates in the gray treefrog *Hyla versicolor:* Implications for calling behavior. *Copeia* 1997:356–62.

Grafen, A. 1988. On the uses of data on lifetime reproductive success. In *Reproductive success,* ed. T. H. Clutton-Brock, 454–71. Chicago: University of Chicago Press.

———. 1990a. Biological signals as handicaps. *J. Theor. Biol.* 144:517–46.

———. 1990b. Do animals really recognize kin? *Anim. Behav.* 39:42–54.

———. 1991a. Kin vision? A reply to Stuart. *Anim. Behav.* 41:1095–96.

———. 1991b. A reply to Blaustein et al. *Anim. Behav.* 41:1085–87.

———. 1991c. A reply to Byers and Beckoff. *Anim. Behav.* 41:1091–92.

Grafen, A., and R. Sibly. 1978. A model of mate desertion. *Anim. Behav.* 26:645–52.

Graham, C. H., S. R. Ron, J. C. Santos, C. J. Schneider, and C. Moritz. 2004. Integrating phylogenetics and environmental niche models to explore speciation mechanisms in dendrobatid frogs. *Evolution* 58:1781–93.

Graham, M. D., and R. D. Vinebrooke. 1998. Trade-offs between herbivore resistance and competitiveness in periphyton of acidified lakes. *Can. J. Fish. Aquat. Sci.* 55:806–14.

Gramapurohit, N. P., B. A. Shanbhag, and S. K. Saidapur. 2004a. Growth, sexual maturation and body size dimorphism in the Indian bullfrog, *Hoplobatrachus tigerinus* (Daud.). *Herpetologica* 60:414–19.

———. 2004b. Postmetamorphic growth and sexual maturation in the bicolored frog, *Rana curtipes* (Jerdon) exhibiting delayed metamorphosis. *Amphibia-Reptilia* 25:445–51.

Gramapurohit, N. P., D. K. Veeranacoudar, B. A. Shanbhag, and S. K. Saidapur. 2004. Relative influence of kinship and density on metamorphic traits of *Tomopterna breviceps. J. Herpetol.* 38:594–99.

Grandison, A. G. C. 1978. The occurrence of *Nectophrynoides* (Anura: Bufonidae) in Ethiopia. A new concept of the genus with a description of a new species. *Monit. Zool. Italiano* (n.s.) Suppl. 11:119–72.

———. 1980. Aspects of breeding morphology in *Mertensophryne micranotis* (Anura: Bufonidae): Secondary sexual characters, eggs and tadpole. *Bull. Br. Mus. Nat. Hist (Zool.)* 39:299–304.

———. 1981. Morphology and phylogenetic position of the West African *Didynamipus sjoestedti* Andersson, 1903 (Anura, Bufonidae). *Monit. Zool. Ital.* (n.s.) Suppl. 15:187–215.

Grandison, A. G. C., and S. Ashe. 1983. The distribution, behavioural ecology and breeding strategy of the pygmy toad, *Mertensophryne micranotis* (Lov.). *Bull. Br. Mus. Nat. Hist. (Zool.)* 45:85–93.

Grant, D., O. Anderson, and V. C. Twitty. 1968. Homing orientation by olfaction in newts (*Taricha rivularis*). *Science* 160:1354–55.

Grant, K. P., and L. E. Licht. 1993. Acid tolerance of anuran embryos and larvae from central Ontario. *J. Herpetol.* 27:1–6.

———. 1995. Effects of ultraviolet radiation on life-history stages of anurans from Ontario, Canada. *Can. J. Zool.* 73:2292–2301.

Grant, T., W. Bolívar-G., and F. Castro. 1998. The advertisement call of *Centrolene geckoideum*. *J. Herpetol.* 32:452–55.

Grant, T., and F. Castro. 1998. The cloud forest *Colostethus* (Anura, Dendrobatidae) of a region of the Cordillera Occidental of Colombia. *J. Herpetol.* 32:378–92.

Grant, T., E. C. Humphrey, and C. W. Myers. 1997. The median lingual process of frogs: A bizarre character of Old World ranoids discovered in South American dendrobatids. *Amer. Mus. Novitates* (3212):1–40.

Grant, W. C. 1955. Territorialism in two species of salamanders. Science 121:137.

Graves, B. M. 1994. The role of nasolabial grooves and the vomeronasal system in recognition of home area by red-backed salamanders. *Anim. Behav.* 47:1216–19.

Graves, B. M., and V. S. Quinn. 2000. Temporal persistence of alarm pheromones in skin secretions of the salamander, *Plethodon cinereus*. *J. Herpetol.* 34:287–91.

Gray, H. M., and J. H. Christy. 2000. Predation by the grapsid crab, *Armases angustum* (Smith, 1870) on tadpoles of the green poison frog, *Dendrobates auratus* Girard, 1855. *Crustaceana* 73:1023–25.

Gray, H. M., and D. M. Green. 2000. *Dendrobtes auratus* (green poison frog). Refuge use. *Herp. Rev.* 31:169–70.

Gray, H. M., D. M. Green, and M. J. Peters. 1999. *Physalaemus pustulosus* (Tungara frog). Predation. *Herp. Rev.* 30:93.

Gray, L. A., and K. C. Nishikawa. 1995. Feeding kinematics of phyllomedusine tree frogs. *J. Exp. Biol.* 198:457–63.

Gray, L. A., J. C. O'Reilly, and K. C. Nishikawa. 1997. Evolution of forelimb movement patterns for prey manipulation in anurans. *J. Exp. Zool.* 277:417–24.

Gray, M. J., L. M. Smith, and R. Brenes. 2004. Effects of agricultural cultivation on demographics of southern High Plains amphibians. *Cons. Biol.* 18:1368–77.

Gray, P., and D. C. Cannatella. 1985. A new species of *Atelopus* (Anura, Bufonidae) from the Andes of northern Peru. *Copeia* 1985:910–17.

Gray, R. H. 1971. Fall activity and overwintering of the cricket frog (*Acris crepitans*) in central Illinois. *Copeia* 1971:748–50.

———. 1977. Lack of physiological differentiation in three color morphs of the cricket frog (*Acris crepitans*) in Illinois. *Trans. Illinois Acad. Sci.* 70:73–79.

———. 1978. Nondifferential predation susceptibility and behavioral selection in three color morphs of Illinois cricket frogs, *Acris crepitans. Trans. Illinois Acad. Sci.* 71:356–60.

———. 1983. Seasonal, annual and geographic variation in color morph frequencies of the cricket frog, *Acris crepitans,* in Illinois. *Copeia* 1983:300–11.

———. 1984. Effective breeding size and the adaptive significance of color polymorphism in the cricket frog (*Acris crepitans*) in Illinois, U.S. A. *Amphibia-Reptilia* 5:101–7.

Gray, R. H., and L. E. Brown. 2005. Decline of northern cricket frogs (*Acris crepitans*). In *Amphibian declines: The conservation status of United States species,* ed. M. J. Lannoo, 47–54. Berkeley: University of California Press.

Graybeal, A. 1997. Phylogenetic relationships of bufonid frogs and tests of alternate macroevolutionary hypotheses characterizing their radiation. *Zool. J. Linn. Soc.* 119:297–338.

Graybeal, A., and D. C. Cannatella. 1995. A new taxon of Bufonidae from Peru, with description of two new species and a review of the phylogenetic status of supraspecific bufonid taxa. *Herpetologica* 51:105–31.

Graybeal, A., and K. de Queiroz. 1992. Inguinal amplexus in *Bufo fastidiosus,* with comments on the systematics of bufonid frogs. *J. Herpetol.* 26:84–87.

Grayson, K. L., L. W. Cook, M. J. Todd, D. Pierce, W. A. Hopkins, R. E. Gatten, Jr., and M. E. Dorcas. 2005. Effects of prey type of specific dynamic action, growth, and mass conversion efficiencies in the horned frog, *Ceratophrys cranwelli. Comp. Biochem. Phsyiol.* A 141:298–304.

Graziadei, P. P. C., and Monti Graziadei, G. A. 1976. Olfactory epithelium of *Necturus maculosus* and *Ambystoma tigrinum. J. Neurocytol.* 5:11–32.

Green, A. J. 1989. The sexual behaviour of the great crested newt, *Triturus cristatus* (Amphibia: Salamandridae). *Ethology* 83:129–53.

———. 1990. Determinants of chorus participation and the effects of size, weight and competition on advertisement calling in the túngara frog, *Physalaemus pustulosus* (Leptodactylidae). *Anim. Behav.* 39:620–38.

———. 1991a. Competition and energetic constraints in the courting great crested newt, *Triturus cristatus* (Amphibia: Salamandridae). *Ethology* 87:66–78.

———. 1991b. Large male crests, an honest indicator of condition, are preferred by female smooth newts, *Triturus vulgaris* (Salamandridae) at the spermatophore transfer stage. *Anim. Behav.* 41:367–69.

Green, D. E., K. A. Converse, and A. K. Schrader. 2002. Epizootiology of sixty-four amphibian morbidity and mortality events in the USA, 1996–2001. *Ann. N. Y. Acad. Sci.* 969:323–39.

Green, D. E., and C. Kagarise Sherman. 2001. Diagnostic histological findings in Yosemite toads (*Bufo canorus*) from a die-off in the 1970s. *J. Herpetol.* 35:92–103.

Green, D. M. 1979. Treefrog toe pads: Comparative surface morphology using scanning electron microscopy. *Can. J. Zool.* 57:2033–46.

———. 1981. Adhesion and the toe-pads of treefrogs. *Copeia* 1981:790–96.

———. 1983. Allozyme variation through a clinal hybrid zone between the toads *Bufo americanus* and *B. hemiophrys* in southeastern Manitoba. *Herpetologica* 39:28–40.

———. 1984. Sympatric hybridization and allozyme variation in the toads *Bufo americanus* and *B. fowleri* in southern Ontario. *Copeia* 1984:18–26.

———. 1997a., ed. *Amphibians in decline: Canadian studies of a global problem, herpetological conservation, volume 1*. St Louis: Society for the Study of Amphibians and Reptiles.

———. 1997b. Perspectives on amphibian population declines: Defining the problem and searching for answers. In *Amphibians in decline: Canadian studies of a global problem, herpetological conservation, volume 1,* ed D. M. Green, 291–308. St Louis: Society for the Study of Amphibians and Reptiles.

———. 1997c. Temporal variation in abundance and age structure in Fowler's toads, *Bufo fowleri,* at Long Point, Ontario. In *Amphibians in decline: Canadian studies of a global problem, herpetological conservation, volume 1,* ed D. M. Green, 45–56. St. Louis, MO: Society for the Study of Amphibians and Reptiles.

———. 2003. The ecology of extinction: Population fluctuation and decline in amphibians. *Biol. Cons.* 111:331–43.

———. 2005a. The biology of amphibian declines. In *Amphibian declines: The conservation status of United States species,* ed. M. J. Lannoo, 28–33. Berkeley: University of California Press.

———. 2005b. *Bufo fowleri* Hinckley, 1882. Fowler's toad. In *Amphibian declines: The conservation status of United States species,* ed. M. J. Lannoo, 408–12. Berkeley: University of California Press.

Green, D. M., and J. Carson. 1988. The adhesion of treefrog toe pads to glass: Cryogenic examination of a capillary adhesion system. *J. Nat. Hist.* 22:131–35.

Green, D. M., and C. Parent. 2003. Variable and asymmetric introgression in a hybrid zone in the toads, *Bufo americanus* and *Bufo fowleri. Copeia* 2003:34–43.

Green, D. M., and M. P. Simon. 1986. Digital microstructure in ecologically diverse sympatric microhylid frogs, genera *Cophixalus* and *Sphenophryne* (Amphibia: Anura) from Papua New Guinea. *Aust. J. Zool.* 34:135–45.

Green, S., and P. Marler. 1979. The analysis of animal communication. In *Handbook of behavioral neurobiology, vol. 3,* ed. P. Marler and J. G. Vandenbergh, 73–158. New York: Plenum.

Greenbank, J. 1945. Limnological conditions in ice-covered lakes, especially as related to winter-kill of fish. *Ecol. Monogr.* 15:343–92.

Greenberg, B. 1942. Some effects of testosterone on the sexual pigments and other sex characters of the cricket frog (*Acris gryllus*). *J. Exp. Zool.* 91:435–47.

Greenberg, C. H., and G. W. Tanner. 2004. Breeding pond selection and movement patterns by eastern spadefoot toads (*Scaphiopus holbrookii*) in relation to weather and edaphic conditions. *J. Herpetol.* 38:569–77.

———. 2005. Spatial and temporal ecology of eastern spadefoot toads on a Florida landscape. *Herpetologica* 61:20–28.

Greene, H. W. 1983. Dietary correlates of the origin and radiation of snakes. *Amer. Zool.* 23:431–41.

———. 1984. Feeding behavior and diet of the eastern coral snake, *Micrurus fulvius*. In *Vertebrate ecology and systematics. A tribute to Henry S. Fitch,* ed. R. A. Seigel, L. E. Hunt, J. L. Knight, L. Malaret, and N. L. Zuschlag, 147–62. Lawrence: Univ. Kansas Mus. Nat. Hist. Spec. Publ. 10.

———. 1988. Species richness in tropical predators. In *Diversity and conservation of tropical rainforests,* ed. F. Almeda and C. M. Pringle, 259–80. San Francisco: California Academy of Sciences and Pacific Division, American Association for the Advancement of Science.

———. 1992. The ecological and behavioral context for pitviper evolution. In *Biology of the pitvipers,* ed. J. A. Campbell and E. D. Brodie, Jr., 107–17. Tyler, TX: Selva.

———. 1997. *Snakes. The evolution of mystery in nature.* Berkeley: University of California Press.

Greene, H. W., and J. A. Campell. 1972. Notes on the use of caudal lures by arboreal green pit vipers. *Herpetologica* 28:32–34.

Greenfield, M. D. 1983. Unsynchronized chorusing in the cone-headed katydid *Neoconocephalus affinis* (Beauvois). *Anim. Behav.* 31:102–12.

———. 1988. Interspecific acoustic interactions among katydids *Neoconocephalus:* Inhibition-induced shifts in diel periodicity. *Anim. Behav.* 36:684–95.

———. 1990. Evolution of acoustic communication in the genus *Neoconocephalus:* Discontinuous songs, synchrony, and interspecific interactions. In *The tettigoniidae: Biology, systematics and evolution,* ed. W. J. Bailey and D. C. F. Rentz, 71–97. New York: Springer-Verlag.

———. 1993. Inhibition of male calling by heterospecific signals: Artifact of chorusing or abstinence during suppression of female phonotaxis? *Naturwissenschaften* 80:570–72.

———. 1994a. Cooperation and conflict in the evolution of signal interactions. *Ann. Rev. Ecol. Syst.* 25:97–126.

———. 1994b. Synchronous and alternating choruses in insects and anurans: Common mechanisms and diverse functions. *Amer. Zool.* 34:605–15.

———. 1997. Acoustic communication in Orthoptera. In *The bionomics of grasshoppers, katydids and their kin,* ed. S. K. Gangwere, M. C. Muralirangan, and M. Muralirangan, 197–230. New York: CAB International.

———. 2002. *Signalers and Receivers. Mechanisms and evolution of arthropod communication.* New York: Oxford University Press.

Greenfield, M. D., and R. L. Minkley. 1993. Acoustic dueling in tarbush grasshoppers: Settlement of territorial contests via alternation of reliable signals. *Ethology* 95:302–26.

Greenfield, M. D., and A. S. Rand. 2000. Frogs have rules: Selective attention algorithms regulate chorusing in *Physalaemus pustulosus* (Leptodactylidae). *Ethology* 106:331–47.

Greenfield, M. D., and I. Roizen. 1993. Katydid synchronous chorusing is an evolutionary stable outcome of female choice. *Nature* 364:618–20.

Greenfield, M. D., and K. C. Shaw. 1983. Adaptive significance of chorusing with special reference to the Orthoptera. In *Orthopteran mating systems: Sexual competition in a diverse group of insects,* ed. G. K. Morris and D. T. Gwynne, 1–27. Boulder, CO: Westview Press.

Greenfield, M. D., M. K. Tourtellot, and W. A. Snedden. 1997. Precedence effects and the evolution of chorusing. *Proc. Roy. Soc. London B* 264:1355–61.

Greenwald, L. 1972. Sodium balance in amphibians from different habitats. *Physiol. Zool.* 45:229–37.

Greenwood, P. J. 1980. Mating systems, philopatry and dispersal in birds and mammals. *Anim. Behav.* 28:1140–62.

Greenwood, P. J. 1983. Mating systems and the evolutionary consequences of dispersal. In *The ecology of animal movement,* ed. I. R. Swingland and P. J. Greenwood, 116–31. New York: Oxford University Press.

Greer, B. J., and K. D. Wells. 1980. Territorial and reproductive behavior of the tropical American frog *Centrolenella fleischmanni. Herpetologica* 36:318–26.

Greer, K. R. 1955. Yearly food habits of the river otter in the

Thompson Lakes region, northwestern Montana, as indicated by scat analysis. *Am. Midl. Nat.* 54:299–313.

Gregg, J. R. 1960. Respiratory regulation in amphibian development. *Biol. Bull.* 119:428–39.

———. 1962. Anaerobic glycolysis in amphibian development. *Biol. Bull.* 123:555–61.

Gregory, P. T. 1978. Feeding habits and diet overlap of three species of garter snakes (*Thamnophis*) on Vancouver Island. *Can. J. Zool.* 56:1967–74.

———. 1979. Predator avoidance behavior of the red-legged frog (*Rana aurora*). *Herpetologica* 35:175–84.

Gregory, P. T., and L. A. Isaac. 2004. Food habits of the grass snake in southeastern England: Is *Natrix natrix* a generalist predator? *J. Herpetol.* 38:88–95.

Gregory, P. T., and K. J. Nelson. 1991. Predation on fish and intersite variation in the diet of common garter snakes, *Thamnophis sirtalis,* on Vancouver Island. *Can. J. Zool.* 69:988–94.

Grelle, C. E., M. A. S. Alves, H. G. Bergallo, L. Geise, C. F. D. Rocha, M. Van Sluys, and U. Caramaschi. 2005. Prediction of threatened tetrapods based on the species-area relationship in Atlantic Forest, Brazil. *J. Zool., London* 265:359–64.

Grenot, C. J., L. Garcin, J. Dao, J.-P. Hérold, B. Fahys, and H. Tséré-Pagès. 2000. How does the European common lizard, *Lacerta vivipara,* survive the cold of winter? *Comp. Biochem. Physiol.* A 127:71–80.

Gressler, S. 1997. Biotopverbund fur Amphibien: Trittsteinbiotope, die neue Naturschutzstrategie. In *Populationsbiologie von amphibien. Eine langzeitstudie auf der wiener donauinsel.* Stapfia 51, ed. W. Hödl, R. Jehle, and G. Gollmann, 235–49. Linz, Austria: Oberösterreich Landesmuseum.

Greven, H. 1976. Notizen zum Geburtsvorgang beim Feuersalamander, *Salamandra salamandra* (L.). *Salamandra* 12:87–93.

———. 1977. Comparative ultrastructural investigations of the uterine epithelium in the viviparous *Salamandra atra* Laur. and the ovoviviparous *Salamandra salamandra* (L.) (Amphibia, Urodela). *Cell Tissue Res.* 181:215–38.

———. 1998. Survey of the oviduct of salamandrids with special reference to the viviparous species. *J. Exp. Zool.* 282:507–25.

———. 2003a. Larviparity and pueriparity. In *Reproductive biology and phylogeny of urodela,* ed. D. M. Sever, 447–75. Enfield: NH: Science Publishers.

———. 2003b. Oviduct and egg jelly. In *Reproductive biology and phylogeny of urodela,* ed. D. M. Sever, 151–81. Enfield: NH: Science Publishers.

Greven, H., and G.-D. Guex. 1994. Structural and physiological aspects of viviparity in *Salamandra salamandra. Mertensiella* 4:139–60.

Grialou, J. A., S. D. West, and R. N. Wilkins. 2000. The effect of forest clearcut harvesting and thinning on terrestrial salamanders. *J. Wildl. Manage.* 64:105–13.

Griffin, P. C., and V. A. Solkin. 1995. Ecology and conservation of *Onychodactylus fischeri* (Caudata, Hynobiidae) in the Russian Far East. *Asiatic Herp. Res.* 6:53–61.

Griffis, M. R., and R. G. Jaeger. 1998. Competition leads to an extinction-prone species of salamander: interspecific territoriality in a metapopulation. *Ecology* 79:2494–2502.

Griffith, E. 1831. *The animal kingdom arranged in conformity with its organization, by the Baron Cuvier, vol. 9.* London: Whittaker, Treacher.

Griffiths, I. 1961. The form and function of the fore-gut in anuran larvae (Amphibia, Salientia) with particular reference to the *manicotto glandulare. Proc. Zool. Soc. London* 137:249–83.

Griffiths, R. A. 1984. Seasonal behaviour and intrahabitat movements in an urban population of smooth newts, *Triturus vulgaris* (Amphibia: Salamandridae). *J. Zool., London* 203:241–51.

———. 1986. Feeding niche overlap and food selection in smooth and palmate newts, *Triturus vulgaris* and *T. helveticus,* at a pond in mid-Wales. *J. Anim. Ecol.* 55:201–14.

———. 1987. Microhabitat and seasonal niche dynamics of smooth and palmate newts, *Triturus vulgaris* and *T. helveticus,* at a pond in mid-Wales. *J. Anim. Ecol.* 56:441–51.

———. 1991. Competition between common frog, *Rana temporaria,* and natterjack toad, *Bufo calamita,* tadpoles: The effect of competitor density and interaction level on tadpole development. *Oikos* 61:187–96.

———. 1995. Determining competition mechanisms in tadpole assemblages. *Herpetol. J.* 5:208–10.

———. 1996. *Newts and salamanders of Europe.* San Diego: Academic Press.

———. 1997. Temporary ponds as amphibian habitats. *Aquat. Cons. Mar. Freshw. Ecosyst.* 7:119–26.

Griffiths, R. A., and J. Denton. 1992. Interspecific associations in tadpoles. *Anim. Behav.* 44:1153–57.

Griffiths, R. A., J. Denton, and A. L.-C. Wong. 1993. The effect of food level on competition in tadpoles: interference mediated by protothecan algae? *J. Anim. Ecol.* 62:274–79.

Griffiths, R. A., P. W. Edgar, and A. L.-C. Wong. 1991. Interspecific competition in tadpoles: growth inhibition and growth retrieval in natterjack toads, *Bufo calamita. J. Anim. Ecol.* 60:1065–76.

Griffiths, R. A., P. de Wijer, and R. T. May. 1994. Predation and competition within an assemblage of larval newts (*Triturus*). *Ecography* 17:176–81.

Griffiths, R. A., and J. P. Foster. 1998. The effect of social interactions on tadpole activity and growth in the British anuran amphibians (*Bufo bufo, B. calamita,* and *Rana temporaria*). *J. Zool., London* 245:431–37.

Griffiths, R. A., and V. J. Mylotte. 1987. Microhabitat selection and feeding relations of smooth and warty newts, *Triturus vulgaris* and *T. cristatus,* at an upland pond in mid-Wales. *Holarc. Ecol.* 10:1–7.

Griffiths, R. A., L. Schley, P. E. Sharp, J. L. Dennis, and A. Roman. 1998. Behavioural responses of Mallorcan midwife toad tadpoles to natural and unnatural snake predators. *Anim. Behav.* 55:207–14.

Grimaldi, D. 1994. Description and immature stages of *Hirtodrosophila batracida* sp. n. (Diptera, Drosophilidae), a predator of frog embryos. Entomol. Scand. 25:129–36.

Grinnell, J., and T. I. Storer. 1924. *Animal life in the Yosemite.* Berkeley: University of California Press.

Grobler, J. H. 1972. Observations on the amphibian *Pyxicephalus adspersus* Tschudi in Rhodesia. *Anoldia (Rhodesia)* 6 (3):1–4.

Gromko, M. H., F. S. Mason, and S. J. Smith-Gill. 1973. Analysis of the crowding effect in *Rana pipiens* tadpoles. *J. Exp. Zool.* 186:63–72.

Grosjean, S. 2003. A redescription of the external and buccopharyngeal morphology of the tadpole of *Ophryophryne microstoma* Boulenger, 1903 (Megophryidae). *Alytes* 21:45–58.

Grosjean, S., M. Vences, and A. Dubois. 2004. Evolutionary significance of oral morphology in the carnivorous tadpoles of tiger frogs, genus *Hoplobatrachus* (Ranidae). *Biol. J. Linn. Soc.* 81:171–81.

Gross, M. R., and R. C. Sargent. 1985. The evolution of male and female parental care in fishes. *Amer. Zool.* 25:807–22.

Gross, M. R., and R. Shine. 1981. Parental care and mode of fertilization in ectothermic vertebrates. *Evolution* 35:775–93.

Grover, M. C. 1998. Influence of cover and moisture on abundance of the terrestrial salamanders *Plethodon cinereus* and *Plethodon glutinosus*. *J. Herpetol.* 32:489–97.

———. 2000. Determinants of salamander distributions along moisture gradients. *Copeia* 2000:156–68.

Grover, M. C., and H. M. Wilbur. 2002. Ecology of ecotones: Interactions between salamanders on a complex environmental gradient. *Ecology* 83:2112–23.

Groves, J. D. 1980. Mass predation on a population of the American toad, *Bufo americanus*. *Am. Midl. Nat.* 103:202–3.

Grow, D. 1977. Reproduction of *Dendrobates auratus* at the SCZ. *Kansas Herp. Soc. Newsletter* (17):9–12.

Grubb, J. C. 1970. Orientation in post-reproductive Mexican toads, *Bufo valliceps*. *Copeia* 1970:674–68.

———. 1972. Differential predation by *Gambusia affinis* on the eggs of seven species of anuran amphibians. *Am. Midl. Natur.* 88:102–8.

———. 1973a. Olfactory orientation in breeding Mexican toads, *Bufo valliceps*. *Copeia* 1973:490–97.

———. 1973b. Olfactory orientation in *Bufo woodhousei fowleri, Pseudacris clarki,* and *Pseudacris streckeri*. *Anim. Behav.* 21:726–32.

———. 1973c. Orientation in newly metamorphosed Mexican toads, *Bufo valliceps*. *Herpetologica* 29:95–100.

———. 1975. Olfactory orientation in southern leopard frogs, *Rana utricularia*. *Herpetologica* 31:219–21.

Gruberg, E. R., and R. V. Stirling. 1972. Observations on the burrowing habits of the tiger salamander (*Ambystoma tigrinum*). *Herp. Rev.* 4:85–89.

Gruberg, E. R., and S. B. Udin. 1978. Topographic projections between the nucleus isthnmi and the tectum of the frog *Rana pipiens*. *J. Comp. Neurol.* 179:487–500.

Guarino, F. M., S. Lunardi, M. Carlomagno, and S. Massotti. 2003. A skeletochronological study of growth, longevity, and age at sexual maturity in a population of *Rana latastei* (Amphibia, Anura). *J. Biosci.* 28:775–82.

Guenther, K. 1931. *A naturalist in Brazil.* Boston: Houghton Mifflin.

Guerry, A. D., and M. L. Hunter, Jr. 2002. Amphibian distributions in a landscape of forests and agriculture: An examination of landscape composition and configuration. *Cons. Biol.* 16:745–54.

Guex, G.-D. 1994. Zur Fortpflanzungsbiologie von *Salamandra atra aurorae* Trevisan, 1982, *S. salamandra bernadezi* Wolterstorff, 1928, *S. Salamandra fastulosa* Schreiber, 1992 und *Mertensiella luschani helversenii* Piper, 1963. *Abh. Ber. Naturk. Magdeburg* 17:143–50.

Guex, G.-D., and P. S. Chen. 1986. Epitheliophagy: Intrauterine cell nourishment in the viviparous alpine salamander, *Salamandra atra* (Laur.). *Experientia* 42:1205–18.

Guex, G.-D., and H. Greven. 1994. Structural and physiological aspects of viviparity in *Salamandra atra*. *Mertensiella* 4:139–60.

Guffey, C., J. G. MacKinster, and R. G. Jaeger. 1998. Familiarity affects interactions between potentially courting territorial salamanders. *Copeia* 1998:205–8.

Guha, K., C. B. Jørgensen, and L. O. Larsen. 1980. Relationship between nutritional state and testes function, together with observations on patterns of feeding, in the toad, *Bufo bufo bufo*. *J. Zool., London* 192:147–55.

Guibe, J., and M. Lamotte. 1958. Morphologie et reproduction par développement direct d'un anoure du Mont Nimba, *Arthroleptis crusculum* Angel. *Bull. Mus. Natl. Hist. Nat., Paris,* 2nd ser., 30:125–33.

Guidali, F., S. Scali, and A. Carettoni. 2000. Diet and trophic niche overlap of two ranid species in northern Italy. *Ital. J. Zool.* 67:67–72.

Guillaume, O. 1999. Does the Pyrenean salamander *Euproctus asper* use chemical cues for sex identification and mating behaviour? *Behav. Proc.* 46:57–62.

———. 2000a. Role of chemical communication and behavioural interactions among conspecifics in the choice of shelters by the cave-dwelling salamander *Proteus anguinus* (Caudata, Proteidae). *Can. J. Zool.* 78:167–73.

———. 2000b. The ventral skin glands, new additional cloacal glands in *Proteus anguinus* (Caudata, Proteidae). I. Female. *Acta Zool.* 81:213–21.

———. 2000c. The ventral skin glands, new additional cloacal glands in *Proteus anguinus* (Caudata, Proteidae). I. Male. *Acta Zool.* 81:223–34.

———. 2002. The importance of chemical communication in the social behaviour of cave salamanders. Comparison between a strict (*Proteus anguinus* L., Proteidae) and a facultative (*Euproctus asper* D., Salamandridae) cave dweller. *Bull. Soc. Zool. France* 127:263–72.

Guillette, L. J., Jr. 2000. Contaminant-associated endocrine disruption in reptiles. In *Ecotoxicology of amphibians and reptiles,* ed. D. W. Sparling, G. Linder, and C. A. Bishop, 595–615. Pensacola, FL: SETAC Press.

Guimond, R. W., and V. H. Hutchison. 1968. The effect of temperature and photoperiod on gas exchange in the leopard frog, *Rana pipiens*. *Comp. Biochem. Physiol.* 27:17–195.

———. 1972. Pulmonary, branchial and cutaneous gas exchange in the mud puppy, *Necturus maculosus maculosus* (Rafinesque). *Comp. Biochem. Physiol.* 42A:367–92.

———. 1973a. Aquatic respiration: An unusual strategy in the hellbender, *Cryptobranchus alleganiensis alleganiensis* (Daudin). *Science* 182:1263–65.

———. 1973b. Trimodal gas exchange in the large aquatic salamander, *Siren lacertina* (Linnaeus). *Comp. Biochem. Physiol.* 46A:249–68.

———. 1974. Aerial and aquatic respiration in the congo eel, *Amphiuma means means* (Garden). *Respir. Physiol.* 20:147–59.

———. 1976. Gas exchange in the giant salamanders of North America. In *Respiration of amphibious vertebrates,* ed. G. M. Hughes, 313–38. New York: Academic Press.

Guix, J. C. 1993. Predación de ejemplares adultos de *Bufo* spp. por mamíferos, en el sudeste de Brasil. *Bol. Asoc. Herpetol. Española* 4:23–25.

Guix, J. C., A. Montori, G. A. Llorente, M. A. Carretero, and X. Santos. 1998. Natural history and conservation of bufonids in four Atlantic rainforest areas of southeastern Brazil. *Herpetol. Nat. Hist.* 6:1–12.

Günther, R. 1969. Paarungsrufe und reproduktive Isolationsmechanismen bei europäischen Anuren der Gattung *Rana* (Amphibia). *Forma et Functio* 1:263–84.

Gunzburger, M. S. 2001. Mysterious salamanders of the mud. Florida Wildlife, May/June 2001:22–24.

———. 2003. Evaluation of the hatching trigger and larval ecology of the salamander *Amphiuma means*. *Herpetologica* 59:459–68.

Gunzburger, M. S., and J. Travis. 2004. Evaluating predation pressure on green treefrog larvae across a habitat gradient. *Oecologia* 140:422–29.

———. 2005. Effects of multiple predator species on green treefrog (*Hyla cinerea*) tadpoles. *Can. J. Zool.* 83:996–1002.

Guppy, M. 1988. Limiting carbohydrate stores: Alleviating the problem in the marathon runner, a hibernating lizard, and the neonate brain. *Can. J. Zool.* 66:1090–97.

Guppy, M., C. J. Fuery, and J. E. Flanigan. 1994. Biochemical principles of metabolic depression. *Comp. Biochem. Physiol.* 109B:175–89.

Gupta, B. K. 1998. Declining amphibians. *Curr. Sci.* 75:81–84.

Gustafson, M. P. 1993. Intraguild predation among larval plethodontid salamanders: A field experiment in artificial stream pools. *Oecologia* 96:271–75.

———. 1994. Size-specific interactions among larvae of the plethodontid salamanders *Gyrinophilus porphyriticus* and *Eurycea cirrigera*. *J. Herpetol.* 28:470–76.

Gutiérrez, G., and H. Lüddecke. 2002. Mating pattern and hatching success in a population of the Andean frog *Hyla labialis*. *Amphibia-Reptilia* 23:281–92.

Guttman, S. I., and K. G. Wilson. 1973. Genetic variation in the genus *Bufo*. I. An extreme degree of transferrin and albumin polymorphism in a population of the American toad (*Bufo americanus*). *Biochem. Gen.* 8:329–34.

Guyétant, R. 1966. Observations écologiques sur les pontes de *Rana temporaria* L. dans la region de Besançon. *Ann. Sci. Univ. Besançon, ser. 3. Physiol. Biol. Anim.* 2:12–18.

———. 1969. Influence du facteur température sur le développement embryonaire de *Rana temporaria* L. et *Rana dalmatina* B. *Vie Milieu* 20:231–42.

Guyétant, R., W. Temmermans, and J. N. Avrillier. 1999. Phenologie de la reproduction chez *Pelodytes punctatus* Daudin, 1802 (Amphibia, Anura). *Amphibia-Reptilia* 20:149–60.

Guzik, M., and L. Shimscheiner. 1985. Changes in weight of the alimentary canal and liver in the fire-bellied toad, *Bombina bomina* (L.) during active life. *Acta Biol. Cracova Ser. Zool.* 27:19–29.

Gwynne, D. T. 1985. Role-reversal in katydids: Habitat influences reproductive behaviour (Orthoptera: Tettigoniidae, *Metaballus* sp.). *Behav. Ecol. Sociobiol.* 16:355–61.

———. 1991. Sexual competition among females: What causes courtship role reversal? *Trends Ecol. Evol.* 6:118–91.

Gwynne, D. T., and L. W. Simmons. 1990. Experimental reversal of courtship roles in an insect. *Nature* 346:172–74.

Haapanen, A. 1970. Site tenacity of the common frog (*Rana temporaria* L.) and the moor frog (*R. arvalis* Nilss.). *Ann. Zool. Fennici* 7:61–66.

———. 1974. Site tenacity of the common toad, *Bufo bufo* (L.). *Ann. Zool. Fennici* 11:251–52.

Haas, A. 1995. Cranial features of dendrobatid larvae (Amphibia: Anura: Dendrobatidae). *J. Morphol.* 224:222–41.

———. 1996. Non-feeding and feeding tadpoles in hemiphractine frogs: Larval head morphology, heterochrony, and systematics of *Flectonotus goeldii* (Amphibia: Anura: Hylidae). *J. Zoo. Syst. Evol. Research* 34:163–71.

———. 2001. Mandibular arch musculature of anuran tadpoles, with comments on homologies of amphibian jaw muscles. *J. Morphol.* 247:1–33.

———. 2003. Phylogeny of frogs as inferred from primarily larval characters (Amphibia: Anura). *Cladistics* 19:23–89.

Haas, A., S. Hertwig, and I. Das. 2006. Extreme tadpoles: the morphology of the fossorial megophryid larva, *Leptobrachella mjobergi*. *Zoology* 109:26–42.

Haas, A., and S. J. Richards. 1998. Correlations of cranial morphology, ecology, and evolution in Australian suctorial tadpoles of the genera *Litoria* and *Nyctimystes* (Amphibia: Anura: Hylidae: Pelodryadinae). *J. Morphol.* 238:109–41.

Haase, A., and H. Pröhl. 2002. Female activity patterns and aggressiveness in the strawberry poison frog *Dendrobates pumilio* (Anura: Dendrobatidae). *Amphibia-Reptilia* 23:129–40.

Habermehl, G. 1965. Über eine alkaloidartige Substanz aus der Haut von *Pseudophryne corroboree*. *Z. Naturf.* 20B:1129.

Hackwood, F. H. 1911. *Good cheer. The romance of food and feasting*. London: T. Fisher Unwin.

Haddad, C. F. B. 1991. Satellite behavior in the Neotropical treefrog *Hyla minuta*. *J. Herpetol.* 25:226–29.

Haddad, C. F. B., and R. P. Bastos. 1997. Predation on the toad *Bufo crucifer* during reproduction (Anura: Bufonidae). *Amphibia-Reptilia* 18:295–98.

Haddad, C. F. B., A. J. Cardoso, and L. M. Castanho. 1990. Hibridação natural entre *Bufo ictericus* e *Bufo crucifer* (Amphibia: Anura). *Rev. Brasil. Biol.* 50:739–44.

Haddad, C. F. B., J. Faivovich, and P. C. A. Garcia. 2005. The specialized reproductive mode of the treefrog *Aplastodiscus perviridis* (Anura: Hylidae). *Amphibia-Reptilia* 26:87–92.

Haddad, C. F. B., and A. A. Giaretta. 1999. Visual and acoustic communication in the Brazilian torrent frog, *Hylodes asper* (Anura: Leptodactylidae). *Herpetologica* 55:324–33.

Haddad, C. F. B., and W. Hödl. 1997. New reproductive mode in anurans: bubble nest in *Chiasmocleis leucosticta* (Microhylidae). *Copeia* 1997:585–88.

Haddad, C. F. B., and M. Martins. 1994. Four species of Brazilian poison frogs related to *Epipedobates pictus* (Dendrobatidae): Taxonomy and natural history observations. *Herpetologica* 50:282–95.

Haddad, C. F. B., and J. P. Pombal. 1998. Redescription of *Physalaemus signifer* (Anura: Leptodactylidae) and description of two new reproductive modes. *J. Herpetol.* 32:557–65.

Haddad, C. F. B., J. P. Pombal, and R. F. Batistic. 1994. Natural hybridization between diploid and tetraploid species of leaf-frogs, genus *Phyllomedusa* (Amphibia). *J. Herpetol.* 28:425–30.

———. 1996. New species of *Hylodes* from the Atlantic Forest of Brazil (Amphibia: Leptodactylidae). *Copeia* 1996:965–69.

Haddad, C. F. B., J. P. Pombal, Jr., and M. Gordo. 1990. Foam nesting in a hylid frog (Amphibia, Anura). *J. Herpetol.* 24:225–26.

Haddad, C. F. B., and C. P. A. Prado. 2005. Reproductive modes in frogs and their unexpected diversity in the Atlantic Forest of Brazil. *BioScience* 55:207–17.

Haddad, C. F. B., and R. J. Sawaya. 2000. Reproductive modes of Atlantic Forest hylid frogs: a general overview and the description of a new mode. *Biotropica* 32:862–71.

Haddad, C. F. B., and I. Sazima. 1992. Anfibios anuros da Serra do Japi. In *Historia natural da Serro do Japi*, ed. L. P. C. Morellato, 188–211. Campinas, Brazil: Editoria da Unicamp.

Hadjioloff, A. 1930. Recherches sur le tissue adipeux chez les poissons et la grenouille. *Bull. Histol. Appl., Paris* 7:8–20.

Häfeli, H. P. 1971. Zur Fortpflanzungsbiologie des Alpensalamanders (*Salamandra atra* Laur.). *Rev. Suisse Zool.* 78:235–93.

Haffner, P. 1997. Assessment of recent introductions of amphibians and reptiles in inland aquatic systems in Metropolitan France. *Bull. Fr. Peche et Piscicul.* 344–345:155–63.

Hagman, M., and A. Forsman. 2003. Correlated evolution of conspicuous coloration and body size in poison frogs (Dendrobatidae). *Evolution* 57:2904–10.

Hagström, T. 1970. Några svenska amphibiearters överwintring. *Göteborgs Naturhist. Mus. Årstryck* 1970:17–22.

———. 1977. Growth studies and ageing methods for adult *Triturus vulgaris* L. and *T. cristatus* Laurenti (Urodela, Salamandridae). *Zool. Scr.* 6:61–68.

———. 1979. Population ecology of *Triturus cristatus* and *T. vulgaris* (Urodela) in SW Sweden. *Holarc. Ecol.* 2:108–14.

———. 1982. Winter habitat selection by some northern European amphibians. *British J. Herpetol.* 6:276–77.

Hailman, J. P., and R. G. Jaeger. 1976. A model of phototaxis and its evaluation with anuran amphibians. *Behaviour* 56:215–49.

Hairston, N. G. 1949. The local distribution and ecology of the plethodontid salamanders of the southern Appalachians. *Ecol. Monogr.* 19:47–73.

———. 1951. Interspecies competition and its probable influence upon the vertical distribution of Appalachian salamanders of the genus *Plethodon*. *Ecology* 32:266–74.

———. 1973. Ecology, selection, and systematics. *Breviora* (414):1–21.

———. 1980a. Evolution under interspecific competition: Field experiments on terrestrial salamanders. *Evolution* 34:409–20.

———. 1980b. The experimental test of an analysis of field distributions: Competition in terrestrial salamanders. *Ecology* 61:817–26.

———. 1980c. Species packing in the salamander genus *Desmognathus:* What are the interspecific interactions involved? *Am. Nat.* 115:354–66.

———. 1981. An experimental test of a guild: Salamander competition. *Ecology* 62:65–72.

———. 1983a. Alpha selection in competing salamanders: Experimental verification of an a priori hypothesis. *Am. Nat.* 122:105–13.

———. 1983b. Growth, survival and reproduction of *Plethodon jordani:* Trade-offs between selective pressures. *Copeia* 1983:1024–35.

———. 1984. Inferences and experimental results in guild structure. In *Ecological communities: Conceptual issues and the evidence,* ed. D. R. Strong, et al., 19–27. Princeton, NJ: Princeton University Press.

———. 1986. Species packing in *Desmognathus salamanders:* Experimental demonstration of predation and competition. *Am. Nat.* 127:266–91.

Hairston, N. G., Sr. 1987. *Community ecology and salamander guilds.* New York: Cambridge University Press.

———. 1989a. *Ecological experiments. Purpose, designs, and execution.* New York: Cambridge University Press.

———. 1989b. Hard choices in ecological experiments. *Herpetologica* 45:119–22.

———. 1993. On the validity of the name *teyahalee* as applied to a member of the *Plethodon glutinosus* complex (Caudata: Plethodontidae): A new name. *Brimleyana* 18:65–69.

———. 1996. Predation and competition in salamander communities. In *Long-term studies of vertebrate communities,* ed. M. L. Cody and J. A. Smallwood, 161–89. San Diego: Academic Press.

Hairston, N. G., Sr., K. C. Nishikawa, and S. L. Stenhouse. 1987. The evolution of competing species of terrestrial salamanders: Niche partitioning or interference? *Evol. Ecol.* 1:247–62.

Hairston, N. G., Sr., and R. H. Wiley. 1993. No decline in salamander (Amphibia: Caudata) populations: A twenty-year study in the southern Appalachians. *Brimleyana* 18:59–64.

Håkansson, P., and J. Loman. 2004. Communal spawning in the common frog *Rana temporaria*—Egg temperature and predation consequences. *Ethology* 110:665–80.

Häkkinen, J., S. Pasanen, and J. V. K. Kukonen. 2001. The effects of solar UV-B radiation on embryonic mortality and development in three boreal anurans (*Rana temporaria, Rana arvalis,* and *Bufo bufo*). *Chemosphere* 44:441–46.

Hall, B. K., and M. H. Wake, eds. 1999. *The origin and evolution of larval forms.* San Diego: Academic Press.

Hall, F. G. 1922. The vital limit of exsiccation of certain animals. *Biol. Bull.* 42:31–51

———. 1966. Hemoglobin function in blood of *Bufo marinus. J. Cell. Physiol.* 68:69–73.

Hall, J. A., J. H. Larsen, Jr., and R. E. Fitzner. 1997. Postembryonic ontogeny of the spadefoot toad, *Scaphiopus intermontanus* (Anura: Pelobatidae): External morphology. *Herpetol. Monogr.* 11:124–78.

Hall, J. A., J. H. Larsen, Jr., D. E. Miller, and R. E. Fitzner. 1995. Discrimination of kin- and diet-based cues by larval spadefoot toads, *Scaphiopus intermontanus* (Anura: Pelobatidae), under laboratory conditions. *J. Herpetol.* 29:233–43.

Hall, J. C. 1994. Central processing of communication sounds in the anuran auditory system. *Amer. Zool.* 34:670–84.

Hall, J. C., and A. S. Feng. 1986. Neural analysis of temporally patterned sounds in the frog's thalamus: processing of pulse duration and pulse repetition rate. *Neurosci. Lett.* 63:215–20.

———. 1987. Evidence for parallel processing in the frog's auditory thalamus. *J. Comp. Neurol.* 258:407–419.

Halley, J. M., R. S. Oldham, and J. W. Arntzen. 1996. Predicting the persistence of amphibian populations with the help of a spatial model. *J. Appl. Ecol.* 33:455–70.

Halliday, T. 2005. Diverse phenomena influencing amphibian population declines. In *Amphibian declines: The conservation status of United States species,* ed. M. J. Lannoo, 3–6. Berkeley: University of California Press.

Halliday, T., and B. Arano. 1991. Resolving the phylogeny of the European newts. *Trends Ecol. Evol.* 6:113–17.

Halliday, T., and M. Tejedo. 1995. Intrasexual selection and alternative mating behaviour. In *Amphibian biology. Vol. 2: Social behaviour,* ed. H. Heatwole and B. K. Sullivan, 419–68. Chipping Norton, New South Wales, Australia: Surrey Beatty and Sons.

Halliday, T. R. 1974. Sexual behavior of the smooth newt *Triturus vulgaris* (Urodela, Salamandridae). *J. Herpetol.* 8:277–92.

———. 1975a. An observational and experimental study of sexual behaviour in the smooth newt, *Triturus vulgaris* (Amphibia, Salamandridae). *Anim. Behav.* 23:291–322.

———. 1975b. On the biological significance of certain morphological characters in males of the smooth newt *Triturus vulgaris* and of the palmate newt *Triturus helveticus* (Urodela: Salamandridae). *Zool. J. Linn. Soc.* 56:291–300.

———. 1976. The libidinous newt. An analysis of variations in the sexual behaviour of the male smooth newt, *Triturus vulgaris. Anim. Behav.* 25:39–45.

———. 1977a. The courtship of European newts: an evolutionary perspective. In *The reproductive biology of amphibians,* ed. D. H. Taylor and S. I. Guttman, 185–232. New York: Plenum.

———. 1977b. The effects of experimental manipulation of breathing behaviour on the sexual behaviour of the smooth newt, *Triturus vulgaris. Anim. Behav.* 25:39–45.

———. 1983a. Do frogs and toads choose their mates? *Nature* 305:226–27.

———. 1983b. The study of mate choice. In *Mate choice,* ed. P. Bateson, 3–32. New York: Cambridge University Press.

———. 1987. Physiological constraints on sexual selection, pp. 247–264. In *Sexual selection: Testing the alternatives,* ed. J. W. Bradbury and M. B. Andersson, 247–64. New York: Wiley.

———. 1990. The evolution of courtship behavior in newts and salamanders. *Adv. Stud. Behav.* 19:137–69.

———. 1998. Sperm competition in amphibians. In *Sperm competition and sexual selection,* ed. T. R. Birkhead and A. P. Moller, 465–502. San Diego: Academic Press.

Halliday, T. R., and K. Adler. 1987. *The encyclopedia of reptiles and amphibians.* New York: Facts on File.

Halliday, T. R., and A. I. Houston. 1991. How long will newts wait? An experiment to test an assumption of a causal model of the courtship of the male smooth newt, *Triturus v. vulgaris. Behaviour* 116:278–91.

Halliday, T. R., and H. P. A. Sweatman. 1976. To breathe or not to breathe: The newt's problem. *Anim. Behav.* 24:551–61.

Halliday, T. R., and P. A. Verrell. 1984. Sperm competition in amphibians. In *Sperm competition and the evolution of animal mating systems,* ed. R. L. Smith, 487–508. New York: Academic Press.

———. 1986. Sexual selection and body size in amphibians. *Herpetol. J.* 1:86–92.

———. 1988. Body size and age in amphibians and reptiles. *J. Herpetol.* 22:253–65.

Halliday, T. R., and A. Worsnop. 1977. Correlation between activity and breathing rate in the smooth newt, *Triturus vulgaris. J. Herpetol.* 11:244–46.

Halloy, M., and R. E. Espinoza. 1999–2000. Territorial encounters and threat displays in the Neotropical frog *Phyllomedusa sauvagii* (Anura: Hylidae). *Herp. Nat. Hist.* 7:175–80.

Halloy, M., and J. M. Fiaño. 2000. Oviposition site selection in *Pleurodema borellii* (Anura: Leptodactylidae) may be influenced by tadpole presence. *Copeia* 2000:606–9.

Halverson, M. A., D. K. Skelly, J. M. Kiesecker, and L. K. Freidenburg. 2003. Forest mediated light regime linked to amphibian distribution and performance. *Oecologia* 134:360–64.

Hamada, K., Y. S Ukai, K. Tsushima, and R. Shukuya. 1966. Biochemical metamorphosis of hemoglobin in *Rana catesbeiana.* III. Molecular change of hemoglobin during spontaneous metamorphosis. *J. Biochem. (Tokyo)* 60:37–41.

Hamblin, N. L. 1984. *Animal use by the Cozumel Maya.* Tucson: University of Arizona Press.

Hamer, A. J., S. J. Lane, and M. J. Mahony. 2002a. Management of freshwater wetlands for the endangered green and golden bell frog (*Litoria aurea*): Roles of habitat determinants and space. *Biol. Cons.* 106:413–24.

Hamer, A. J., S. J. Lane, and M. J. Mahony. 2002b. The role of introduced mosquitofish (*Gambusia holbrooki*) in excluding the native green and golden bell frog (*Litoria aurea*) from original habitats in south-eastern Australia. *Oecologia* 132:445–52.

Hamer, A. J., J. A. Makings, S. L. Lane, and M. J. Mahony. 2004. Amphibian declines and fertilizers used on agricultural land in south-eastern Australia. *Agric. Ecos. Environ.* 102:299–305.

Hamerstrom, F. N., Jr., and F. Hamerstrom. 1951. Food of young raptors on the Edwin S. George Reserve. *Wilson Bull.* 63:16–25.

Hamilton, W. D. 1971. Geometry for the selfish herd. *J. Theor. Biol.* 31:295–311.

Hamilton, W. J. 1930. The food of the Soricidae. *J. Mammal.* 11:26–39.

———. 1948. The egg-laying process in the tiger salamander. *Copeia* 1948:212.

———. 1951. The food and feeding behavior of the garter snake in New York State. *Am. Midl. Nat.* 46:385–90.

Hancock, J., and J. Kushlan. 1984. *The herons handbook.* New York: Harper and Row.

Hanifin, C. T., M. Yotsu-Yamashita, T. Yasumoto, and E. D. Brodie III, and E. D. Brodie, Jr. 1999. Toxicity of dangerous prey: Variation of tetrodotoxin levels within and among populations of the newt *Taricha granulosa. J. Chem. Ecol.* 25:2161–75.

Hanken, J. 1979. Egg development time and clutch size in two Neotropical salamanders. *Copeia* 1979:741–44.

———. 1982. Appendicular skeletal morphology in minute salamanders, genus *Thorius* (Amphibia: Plethodontidae): Growth regulation, adult size determination, and natural variation. *J. Morphol.* 174:57–77.

———. Miniaturization and its effects on cranial morphology in plethodontid salamanders, genus *Thorius* (Amphibia, Plethodontidae): II. The fate of the brain and sense organs and their role in skull morphogenesis and evolution. *J. Morphol.* 177:255–68.

———. 1984. Miniaturization and its effects on cranial morphology in plethodontid salamanders, genus *Thorius* (Amphibia: Plethodontidae). I. Osteological variation. *Biol. J. Linn. Soc.* 23:55–75.

———. 1989. Development and evolution in amphibians. *Amer. Sci.* 77:336–43.

———. 1993. Adaptation of bone growth to miniaturization of body size. In *Bone. Vol. 7. Bone growth B,* ed. B. K. Hall, 79–104. Boca Raton, FL: CRC Press.

———. 1999a. Larvae in amphibian development and evolution. In *The origin and evolution of larval forms,* ed. B. K. Hall and M. H. Wake, 61–108. San Diego: Academic Press.

———. 1999b. Why are there so many new amphibian species when amphibians are declining? *Trends Ecol. Evol.* 14:7–8.

Hanken, J., D. H. Jennings, and L. Olsson. 1997. Mechanistic basis of life-history evolution in anuran amphibians: direct development. *Amer. Zool.* 37:160–71.

Hanken, J., and D. B. Wake. 1993. Miniaturization of body size: Organismal consequences and evolutionary significance. *Ann. Rev. Ecol. Syst.* 24:501–19.

———. 1998. Biology of tiny animals: Systematics of the minute salamanders (*Thorius:* Plethodontidae) from Veracruz and Puebla, Mexico, with descriptions of five new species. *Copeia* 1998:312–45.

Hanken, J., D. B. Wake, and J. M. Savage. 2005. A solution to the large black salamander problem (genus *Bolitoglossa*) in Costa Rica and Panama. *Copeia* 2005:227–45.

Hanlin, H. G., and R. H. Mount. 1978. Reproduction and activity of the greater siren, *Siren lacertina* (Amphibia: Sirenidae) in Alabama. *J. Alabama Acad. Sci.* 49:31–39.

Hanna, G., and W. J. P. Barnes. 1991. Adhesion and detachment of the toe pads of tree frogs. *J. Exp. Biol.* 155:103–25.

Hanselmann, R., A. Rodríguez, M. Lampo, L. Fajardo-Ramos, A. A. Aguirre, A. M. Kilpatrick, J. P. Rodríguez, and P. Daszak. 2004. Presence of an emerging pathogen of amphibians in in-

troduced bullfrogs *Rana catesbeiana* in Venezuela. *Biol. Cons.* 120:115–19.
Hansen, K. L. 1957. Movements, area of activity, and growth of *Rana hecksheri. Copeia* 1957:274–77.
Hansen, L. J., D. L. Fabacher, and R. Calfee. 2002. The role of the egg jelly coat in protecting *Hyla regilla* and *Bufo canorus* embryos from ultraviolet B radiation during development. *Environ. Sci. Poll. Res.* 9:412–16.
Hanson, J. A., and J. L. Vial. 1956. Defensive behavior and effects of toxins in *Bufo alvarius. Herpetologica* 12:141–49.
Haramura, T. 2005. *Buergeria japonica* (Ryukyu Kajika Frog). Oviposition behavior. *Herp. Rev.* 36:429–30.
Härdling, R., H. Kokko, and R. W. Elwood. 2004. Priority versus brute force: When should males begin guarding resources? *Am. Nat.* 163:240–52.
Hardy, D. F. 1959. Chorus structure in the striped chorus frog, *Pseudacris nigrita. Herpetologica* 15:14–16.
Hardy, J. D., Jr. 1969. A summary of recent studies on the salamander, *Ambystoma mabeei. Chesapeake Biol. Lab., Ref. no.* 69 20:1–3.
Hardy, L. M., and L. R. Raymond. 1980. The breeding migration of the mole salamander, *Ambystoma talpoideum,* in Louisiana. *J. Herpetol.* 14:327–35.
———. 1991. Observations on the activity of the pickeral frog, *Rana palustris* (Anura: Ranidae), in northern Louisiana. *J. Herpetol.* 25:220–22.
Hardy, M. P., and J. N. Dent. 1986. Regulation of motility of sperm of the red-spotted newt. *J. Exp. Zool.* 240:385–96.
Harestad, A. S. 1985. *Scaphiopus intermontanus* (Great Basin spadefoot toad). Mortality. *Herp. Rev.* 16:24.
Harestad, A. S., and F. L. Bunnell. 1979. Home range and body size—A reevaluation. *Ecology* 60:389–402.
Harkey, G. A., and R. D. Semlitsch. 1988. Effects of temperature on growth, development, and color polymorphism in the ornate chorus frog *Pseudacris ornata. Copeia* 1988:1001–7.
Harlan, R. A., and R. F. Wilkinson. 1981. The effects of progressive hypoxia and rocking activity on blood oxygen tension for hellbenders, *Cryptobranchus alleganiensis. J. Herpetol.* 15:383–88.
Harlow, H. J. 1977. Seasonal oxygen metabolism and cutaneous osmoregulation in the California newt, *Taricha torosa. Physiol. Zool.* 50:231–36.
———. 1978. Seasonal aerobic and anaerobic metabolism at rest and during activity in the salamander *Taricha torosa. Comp. Biochem. Physiol.* 61A:177–82.
Harpaz, S., and J. E. Steiner. 1990. Analysis of betaine-induced feeding behavior in the prawn *Macrobrachium rosenbergi* (de Man, 1879) (Decopoda, Caridea). *Crustaceana* 58:175–85.
Harper, C. A., and D. C. Guynn, Jr. 1999. Factors affecting salamander density and distribution within four forest types in the Southern Appalachian Mountains. *Forest Ecol. Manage.* 114:245–52.
Harpole, D. N., and C. A. Haas. 1999. Effects of seven silvicultural treatments on terrestrial salamanders. *Forest Ecol. Manage.* 114:349–56.
Harri, M. N. E. 1973. The rate of metabolic temperature acclimation in the frog, *Rana temporaria. Physiol. Zool.* 46:148–56.
Harri, M. N. E., and R. Hedenstam. 1972. Calorigenic effect of adrenaline and noradrenaline in the frog, *Rana temporaria. Comp. Biochem. Physiol.* 41A:409–19.
Harris, M. L., C. A. Bishop, J. Struger, B. Ripley, and J. P. Bogart. 1998. The functional integrity of northern leopard frog (*Rana pipiens*) and green frog (*Rana clamitans*) populations in orchard wetlands. II. Effects of pesticides and eutrophic conditions on early life stage development. *Env. Tox. Chem.* 17:1351–63.
Harris, M. L., C. A. Bishop, J. Struger, M. R. van den Heuvel, G. van der Kraak, D. G. Dixon, B. Ripley, and J. P. Bogart. 1998. The functional integrity of northern leopard frog (*Rana pipiens*) and green frog (*Rana clamitans*) populations in orchard wetlands. I. Genetics, physiology, and biochemistry of breeding adults and young-of-the-year. *Env. Tox. Chem.* 17:1338–50.
Harris, P. M. 1995. Are autecologically similar species also functionally similar? A test in pond communities. *Ecology* 76:544–52.
Harris, R. N. 1980. The consequences of within-year timing of breeding in *Ambystoma maculatum. Copeia* 1980:719–22.
———. 1984. Transplant experiments with *Ambystoma* larvae. *Copeia* 1984:161–69.
———. 1987a. Density-dependent paedomorphosis in the salamander *Notophthalmus viridescens dorsalis. Ecology* 68:705–12.
———. 1987b. An experimental study of population regulation in the salamander, *Notophthalmus viridescens dorsalis* (Urodela: Salamandridae). *Oecologia* 71:280–85.
———. 1999. The anuran tadpole. Evolution and maintenance. In *Tadpoles. The biology of anuran larvae,* ed. R. W. McDiarmid and R. Altig, 279–94. Chicago: University of Chicago Press.
Harris, R. N., R. A. Alford, and H. M. Wilbur. 1988. Density and phenology of *Notophthalmus viridescens dorsalis* in a natural pond. *Herpetologica* 44:234–42.
Harris, R. N., and D. E. Gill. 1980. Communal nesting, brooding behavior, and embryonic survival of the four-toed salamander *Hemidactylium scutatum. Herpetologica* 36:141–44.
Harris, R. N., W. W. Hames, I. T. Knight, C. A. Careno, and T. J. Vess. 1995. An experimental analysis of joint nesting in the salamander *Hemidactylium scutatum* (Caudata: Plethodontidae): the effects of population density. *Anim. Behav.* 50:1309–16.
Harris, R. N., and P. M. Ludwig. 2004. Resource level and reproductive frequency in female four-toed salamanders, *Hemidactylium scutatum. Ethology* 85:1585–90.
Harris, R. N., R. D. Semlitsch, H. M. Wilbur, and J. E. Fauth. 1990. Local variation in the genetic basis of paedomorphosis in the salamander *Ambystoma talpoideum. Evolution* 44:1588–1603.
Harris, R.N., T. J. Vess, J. I. Hammond, and C. J. Lindermuth. 2003. Context-dependent kin discrimination in larval four-toed salamanders *Hemidactylium scutatum* (Caudata: Plethodontidae). *Herpetologica* 59:164–77.
Harris, R. T. 1975. Seasonal activity and microhabitat utilization in *Hyla cadaverina. Herpetologica* 31:236–39.
Harris, W. E., and J. R. Lucas. 2002. A state-based model of sperm competition in a group-breeding salamander. *Behav. Ecol.* 13:705–12.
Harrison, J. A., M. Burger, L. R. Minter, A. L. de Villiers, E. H. W. Baard, E. Scott, P. J. Bishop, and S. Ellis, eds. 2001. *Conservation assessment and mangement plan for southern African frogs. Final Report.* Apple Valley, MN: SSC Conservation Specialist Group,.
Harrison, J. D., S. P. Gittins, and F. M. Slater. 1983. The breeding migrations of smooth and palmate newts (*Triturus vulgaris*

and T. *helveticus*) at a pond in mid-Wales. *J. Zool., London* 199:249–58.

Harrison, J. D., S. P. Gittins, and F. M. Slater. 1984. Morphometric observations of smooth and palmate newts in mid-Wales. *British J. Herpetol.* 6:410–13.

Harrison, J. R. 1967. Observations on the life history, ecology and distribution of *Desmognathus aeneus aeneus* Brown and Bishop. *Am. Midl. Nat.* 77:356–70.

Harrison, P. A. 1987. *Vocal behaviour in the south-eastern Australian tree frogs, Litoria ewingi and L. verreauxi (Anura: Hylidae).* Unpublished M.S. thesis. Melbourne, Australia: University of Melborne.

Harrison, P. A., and M. J. Littlejohn. 1985. Diphasy in the advertisement calls of *Geocrinia laevis* (Anura: Leptodactylidae): Vocal responses of males during field playback experiments. *Behav. Ecol. Sociobiol.* 18:67–73.

Harte, J., and E. Hoffman. 1989. Possible effects of acidic deposition on a Rocky Mountain population of the tiger salamander *Ambystoma tigrinum. Cons. Biol.* 3:149–58.

Hartel, T. 2003. The breeding biology of the frog *Rana dalmatina* in Târnava Mare Valley, Romania. *Russ. J. Herpetol.* 10:169–75.

Hartley, S. 1990. Spawning and rearing the parsley frog (*Pelodytes punctatus*) in captivity. *British Herp. Soc. Bull.* 31:19–25.

Hartmann, M. T., P. A. Hartmann, and C. F. B. Haddad. 2004. Visual signaling and reproductive biology in a nocturnal treefrog, genus *Hyla* (Anura: Hylidae). *Amphibia-Reptilia* 25:395–406.

Hartmann, P. A., and O. A. V. Marques. 2005. Diet and habitat use of two sympatric species of *Philodryas* (Colubridae), in south Brazil. *Amphibia-Reptilia* 26:25–31.

Hartwig, H., and E. Rotmann. 1940. Experimentelle Untersuchungen an einem Massenduftreten von neotenen *Triton taeniatus. W. Roux Archiv Entwickslungsmech.* 140:195–251.

Harvell, C. D., C. E. Mitchell, J. R. Ward, S. Altizer, A. P. Dobson, R. S. Ostfeld, and M. D. Samuel. 2002. Climate warming and disease risks for terrestrial and marine biota. *Science* 296:2158–62.

Harvey, L. A., and C. R. Propper. 1997. Effects of androgens on male sexual behavior and secondary sex characteristics in the explosively breeding spadefoot toad, *Scaphiopus couchii. Horm. Behav.* 31:89–96.

Harvey, L. A., C. R. Propper, S. K. Woodley, and M. C. Moore. 1997. Reproductive endocrinology of the explosively breeding desert spadefoot toad, *Scaphiopus couchii. Gen. Comp. Endocrinol.* 105:102–13.

Harvey, M. A., C. Morrison, and J.-M. Hero. 1999. *Litoria lesueri* (Stony Creek Frog). Predation. *Herp. Rev.* 30:162–63.

Harvey, P. H., and M. D. Pagel. 1991. *The comparative method in evolutionary biology.* New York: Oxford University Press.

Harvey, P. H., R. K. Colwell, J. S. Silvertown, and R. M. May. 1983. Null models in ecology. *Ann. Rev. Ecol. Syst.* 14:189–211.

Hass, C. A., J. F. Dunski, L. R. Maxson, and M. S. Hoogmoed. 1995. Divergent lineages within the *Bufo margaritifera* complex (Amphibia: Anura, Bufonidae) revealed by albumin immunology. *Biotropica* 27:238–49.

Hass, C. A., R. A. Nussbaum, and L. R. Maxson. 1993. Immunological insights into the evolutionary history of caecilians (Amphibia: Gymnophiona): Relationships of the Seychellean caecilians and a preliminary report on family-level relationships. *Herpetol. Monogr.* 7:56–63.

Hassinger, D. D. 1970. Notes on the thermal properties of frog eggs. *Herpetologica* 26:49–51.

Hassinger, D. D., and J. D. Anderson. 1970. The effect of lunar eclipse on nocturnal stratification of larval *Ambystoma opacum. Copeia* 1970:178–79.

Hassinger, D. D., J. D. Anderson, and G. H. Dalrymple. 1970. The early life history and ecology of *Ambystoma tigrinum* and *Ambystoma opacum* in New Jersey. *Am. Midl. Nat.* 84:474–95.

Hastings, D., and W. Burggren. 1995. Developmental changes in oxygen consumption regulation in larvae of the South African clawed frog *Xenopus laevis. J. Exp. Biol.* 198:2465–75.

Hasumi, M. 1994. Reproductive behavior of the salamander *Hynobius nigrescens:* monopoly of egg sacs during scramble competition. *J. Herpetol.* 28:264–67.

Hasumi, M., and H. Iwasawa. 1990. Seasonal changes in body shape and mass in the salamander, *Hynobius nigrescens. J. Herpetol.* 24:113–18.

———. 1992. Wandering behavior and cutaneous changes in winter-dormant male salamanders (*Hynobius nigrescens*). *Herpetologica* 48:279–87.

Hatch, A. C., and A. R. Blaustein. 2003. Combined effedts of UV-B radiation and nitrate fertilizer on larval amphibians. *Ecol. Appl.* 13:1083–93.

Hattingh, J., and H. Bartels. 1973. The oxygen affinity of axolotl blood and hemoglobin before and after metamorphosis. *Respir. Physiol.* 18:1–13.

Hauser, M. D., and D. A. Nelson. 1991. 'Intentional' signaling in animal communication. *Trends Ecol. Evol.* 6:186–89.

Hausfater, G., H. C. Gerhardt, and G. M. Klump. 1990. Parasites and mate choice in gray treefrogs, *Hyla versicolor. Amer. Zool.* 30:299–311.

Haverschmidt, F. von. 1962. Beobachtungen an der Schleiereule, *Tyto alba,* in Surinam. *J. für Ornithol.* 103:236–42.

Hay, J. M., I. Ruvinsky, S. B. Hedges, and L. R. Maxson. 1995. Phylogenetic relationships of amphibian families inferred from DNA sequences of mitochondrial 12S and 16S ribosomal RNA genes. *Mol. Biol. Evol.* 12:928–37.

Hay, R. 1998. Blanchard's cricket frogs in Wisconsin: A status report. In *Status and conservation of Midwestern amphibians,* ed. M. J. Lannoo, 79–82. Iowa City: University of Iowa Press.

Hayden, B. 1981. Subsistence and ecological adaptations of modern hunter/gatherers. In *Omnivorous primates: gathering and hunting in human evolution,* ed. R. S. O. Harding and G. Teleki, 344–421. New York: Columbia University Press.

Hayes, F. E. 1989. Antipredator behavior of recently metamorphosed toads (*Bufo a. americanus*) during encounters with garter snakes (*Thamnophis s. sirtalis*). *Copeia* 1989:1011–15.

Hayes, J. P., and T. Garland, Jr. 1995. The evolution of endothermy: testing the aerobic capacity model. *Evolution* 49:836–47.

Hayes, M. P. 1983. Predation on the adults and prehatching stages of glass frogs (Centrolenidae). *Biotropica* 15:74–76.

———. 1985. Nest structure and attendance in the stream-dwelling frog, *Eleutherodactylus angelicus. J. Herpetol.* 19:168–69.

Hayes, M. P. 1991. *A study of clutch attendance in the Neotropical frog Centrolenella fleischmanni (Anura: Centrolenidae).* Unpublished PhD dissertation. Coral Gables, FL: University of Miami.

Hayes, M. P., and M. R. Jennings. 1986. Decline of ranid frog species in western North America: Are bullfrogs (*Rana catesbeiana*) responsible? *J. Herpetol.* 20:490–509.

Hayes, M. P., and D. M. Krempels. 1986. Vocal sac variation among frogs of the genus *Rana* from western North America. *Copeia* 1986:927–36.

Hayes, M. P., and M. M. Miyamoto. 1984. Biochemical, behavioral, and body size differences between the red-legged frogs, *Rana aurora aurora* and *R. a. draytoni*. *Copeia* 1984:1010–22.

Hayes, M. P., and C. J. Rombough. 2004. *Rana aurora* (Northern Red-legged Frog). Predation. *Herp. Rev.* 35:375–76.

Hayes, M. P., C. J. Rombough, C. B. Hayes, and J. D. Engler. 2005. *Rana pretiosa* (Oregon Spotted Frog). Predation. *Herp. Rev.* 36:307.

Hayes, T., K. Haston, M. Tsui, A. Hoang, C. Haeffele, and A. Vonk. 2002. Feminization of male frogs in the wild. *Nature* 419:895–96.

———. 2003. Atrazine-induced hermaphroditism at 0.1 ppb in American leopard frogs (*Rana pipiens*): Laboratory and field evidence. *Env. Health Perspect.* 111:568–75.

Hayes, T., and P. Licht. 1992. Gonadal involvement in sexual size dimorphism in the African bullfrog (*Pyxicephalus adspersus*). *J. Exp. Zool.* 264:130–35.

Hayes, T. B. 1995. Interdependence of corticosterone and thyroid hormones in larval toads (*Bufo boreas*). I. Thyroid hormone-dependent and independent effects of corticosterone on growth and development. *J. Exp. Zool.* 27:95–102.

———. 1997. Steroids as potential modulators of thyroid hormone activity in anuran metamorphosis. *Amer. Zool.* 37:185–94.

———. 2000. Endocrine disruption in amphibians. In *Ecotoxicology of amphibians and reptiles,* ed. D. W. Sparling, G. Linder, and C. A. Bishop, 573–93. Pensacola, FL: SETAC Press,.

———. 2004. There is no denying this: Defusing the confusion about atrazine. *BioScience* 54:1138–49.

———. 2005. Welcome to the revolution: Integrative biology and assessing the impact of endocrine disruptors on environmental and public health. *Integr. Comp. Biol.* 45:321–29.

Hayes, T. B., A. Collins, M. Lee, M. Mendoza, N. Noriega, A. A. Stuart, and A. Vonk. 2002. Hermaphroditic, demasculinized frogs after exposure to the herbicide atrazine at low ecologically relevant doses. *Proc. Natl. Acad. Sci. USA* 99:5476–80.

Hayes, T. B., and K. P. Menendez. 1999. The effect of sex steroids on primary and secondary sex differentiation in the sexually dichromatic reedfrog (*Hyperolius argus:* Hyperoliidae) from the Arabuko Sokoke Forest of Kenya. *Gen. Comp. Endocrinol.* 115:188–99.

Hayes, T. B., and T. H. Wu. 1995. Interdependence of corticosterone and thyroid hormones in toad larvae (*Bufo boreas*). II. Regulation of corticosterone and thyroid hormones. *J. Exp. Zool.* 27:103–11.

Hayes-Odum, L. A. 1990. Observations on reproduction and embryonic development in *Syrrhophus cystignathoides campi* (Anura: Leptodactylidae). *Southwest. Nat.* 35:358–61.

Hayne, D. W. 1949. Calculation of size of home range. *J. Mammal.* 30:1–18.

Hazard, E. S. III, and V. H. Hutchison. 1978. Ontogenetic changes in erythrocyte organic phosphates in the bullfrog, *Rana catesbeiana*. *J. Exp. Zool.* 206:109–18.

Hazell, D., J-M. Hero, D. Lindenmayer, and R. Cunningham. 2004. A comparison of constructed and natural habitat for frog conservation in an Australian agricultural landscape. *Biol. Cons.* 119:61–71.

Hazell, D., W. Osborne, and D. Lindenmayer. 2003. Impact of post-European stream change on frog habitat: southeastern Australia. *Biodivers. Cons.* 12:301–20.

Hazelwood, E. 1969. A study of a breeding colony of frogs at the Canon Slade Grammar School, near Polton, Lancs. *British J. Herpetol.* 4:96–103.

Healey, M., D. Thompson, and A. Robertson. 1997. Amphibian communities associated with billabong habitats of the Murrumbidgee floodplain, Australia. *Aust. J. Ecol.* 22:270–78.

Healy, W. R. 1970. Reduction of neoteny in Massachusetts populations of *Notophthalmus viridescens*. *Copeia* 1970:578–81.

———. 1973. Life history variation and growth of juvenile *Notophthalmus viridescens* from Massachusetts. *Copeia* 1973:641–47.

———. 1974. Population consequences of alternative life histories in *Notophthalmus v. viridescens*. *Copeia* 1974:221–29.

———. 1975. Terrestrial activity and home range in efts of *Notophthalmus viridescens*. *Am. Midl. Nat.* 93:131–38.

Heang, K. B. 1984. Conservation status of the Malaysian fauna. III. Amphibians. *Malaysian Nat.* 37 (4):6–10.

Hearne, S. 1795. *A journey from Prince of Wales Fort, in Hudson's Bay, to the Northern Ocean in the years 1769, 1770, 1771, and 1772*. London: A. Strahan and T. Cadell (reprint edition: The Champlain Society, Toronto, Canada, 1911).

Heath, A. G. 1975. Behavioral thermoregulation in high altitude tiger salamanders, *Ambystoma tigrinum*. *Herpetologica* 31:84–93.

———. 1976. Respiratory responses to hypoxia by *Ambystoma tigrinum* larvae, paedomorphs, and metamorphosed adults. *Comp. Biochem. Physiol.* 55A:45–49.

Heatwole, H. 1960. Burrowing ability and behavioral responses to desiccation of the salamander, *Plethodon cinereus*. *Ecology* 41:661–68.

———. Habitat selection and activity of the wood frog, *Rana sylvatica* Le Conte. *Am. Midl. Nat.* 66:301–13.

———. 1962a. Contributions to the natural history of *Eleutherodactylus cornutus maussi*. *Stahlia* 2:1–7.

———. 1962b. Environmental factors influencing local distribution and activity of the salamander, *Plethodon cinereus*. *Ecology* 43:460–72.

———. 1963. Ecologic segregation of two species of tropical frogs of genus *Eleutherodactylus*. *Carib. J. Sci.* 3:17–23.

———. 1982. A review of structuring in herpetofaunal assemblages. In *Herpetological communities. Wildl. res. report 13,* ed. N. J. Scott, Jr., 1–19. Washington, DC: U.S. Fish and Wildlife Service.

———. 1984. Adaptations of amphibians to aridity. In *Arid Australia,* ed. H. G. Cogger and E. E. Cameron, 177–222. Sydney: Australian Museum.

Heatwole, H., and G. T. Barthalmus, eds. 1994. *Amphibian biology. Vol. 1. The integument.* Chipping Norton, New South Wales, Australia: Surrey Beatty and Sons.

Heatwole, H., E. Cameron, and G. J. W. Webb. 1971. Studies on anuran water balance—II. Desiccation in the Australian frog, *Notaden bennetti*. *Herpetologica* 27:365–78.

Heatwole, H., and R. L. Carroll, eds. 2000. *Amphibian biology. Vol. 4. Palaeontology.* Chipping Norton, New South Wales, Australia: Surrey Beatty and Sons.

Heatwole, H., and E. Davison. 1976. A review of caudal luring in snakes with notes on its occurrence in the Saharan sand viper, *Cerastes vipera*. *Herpetologica* 32:332–36.

Heatwole, H., and K. Lim. 1961. Relation of substrate moisture to absorption and loss of water by the salamander, *Plethodon cinereus*. *Ecology* 42:814–19.

Heatwole, H., and R. C. Newby. 1972. Interaction of interanal rhythm and loss of body water in influencing activity levels of amphibians. *Herpetologica* 28:156–62.

Heatwole, H., N. Poran, and P. King. 1999. Ontogenetic changes in the resistance of bullfrogs (*Rana catesbeiana*) to the venom of copperheads (*Agkistrodon contortrix*) and cottonmouths (*Agkistrodon piscivorus*). *Copeia* 1999:808–14.

Heatwole, H., and O. J. Sexton. 1966. Herpetofaunal comparisons between two climatic zones in Panama. *Am. Midl. Nat.* 75:45–60.

Heatwole, H., and F. H. Test. 1961. Cannibalism in the salamander *Plethodon cinereus*. *Herpetologica* 17:143.

Heatwole, H., F. Torres, S. B. de Austin, and A. Heatwole. 1969. Studies on anuran water balance—I. Dynamics of evaporative water loss by the coqui, *Eleutherodactylus portoricensis*. *Comp. Biochem. Physiol.* 28:245–69.

Hecht, G. 1931. Winterschlaf und Paarungsdaten deutscher Amphibien. *Sitz. ber. Ges. Naturforch. Freunde Berlin* 1931:316–29.

Hecnar, S. J. 1995. Acute and chronic toxicity of ammonium-nitrate fertilizer to amphibians from southern Ontario. *Env. Tox. Chem.* 14:2131–37.

Hecnar, S. J., and R. T. M'Closkey. 1996. Regional dynamics and the status of amphibians. *Ecology* 77:2091–97.

———. 1997a. Changes in the composition of a ranid frog community following bullfrog extinction. *Am. Midl. Nat.* 137:145–50.

———. 1997b. The effects of predatory fish on amphibian species richness and distribution. *Biol. Cons.* 79:123–31.

———. 1997c. Spatial scale and determination of species status of the green frog. *Cons. Biol.* 11:670–82.

———. 1998. Species richness patterns of amphibians in southwestern Ontario ponds. *J. Biogeogr.* 25:763–72.

Hedeen, S. E. 1971. Body temperatures of the mink frog, *Rana septentrionalis*. *J. Herpetol.* 5:211–12.

———. 1972. Escape behavior and causes of death of the mink frog, *Rana septentrionalis*. *Herpetologica* 28:261–62.

Hedgecock, D. 1978. Population subdivision and genetic divergence in the red-bellied newt, *Taricha rivularis*. *Evolution* 32:271–86.

Hedges, S. B. 1986. An electrophoretic study of Holarctic hylid frog evolution. *Syst. Zool.* 35:1–21.

———. 1987. Vocalization and habitat preference of the Jamaican treefrog, *Hyla marianae* (Anura, Hylidae). *Carib. J. Sci.* 23:380–84.

———. 1989. Evolution and biogeography of West Indian frogs of the genus *Eleutherodactylus:* Slow-evolving loci and the major groups. In *Biogeography of the West Indies: Past, present, and future,* ed. C. Woods, 305–70. Gainesville, FL: Sandhill Crane Press.

———. 1993. Global amphibian declines: A perspective from the Caribbean. *Biodivers. Cons.* 2:290–303.

———. 1996. The origin of West Indian amphibians and reptiles. In *Contributions to West Indian herpetology,* ed. R. Powell and R. W. Henderson, 95–128. Ithaca, NY: Society for the Study of Amphibians and Reptiles.

Hedges, S. B., and L. R. Maxson. 1993. A molecular perspective on lissamphibian phylogeny. *Herpetol. Monogr.* 7:27–42.

Hedges, S. B., R. A. Nussbaum, and L. R. Maxson. 1993. Caecilian phylogeny and biogeography inferred from mitochondrial DNA sequences of the 12S rRNA and 16S rRNA genes (Amphibia: Gymnophiona). *Herpetol. Monogr.* 7:64–76.

Hedges, S. B., and R. Thomas. 1987. A new burrowing frog from Hispaniola with comments on the *inoptatus* group of the genus *Eleutherodactylus* (Anura: Leptodactylidae). *Herpetologica* 43:269–79.

Hedges, S. B., R. Thomas, and R. Franz. 1987. A new species of *Eleutherodactylus* (Anura:Leptodactylidae) from the Massif de la Hotte, Haiti. *Copeia* 1987:943–49.

Hedlund, L. 1990a. Courtship display in a natural population of crested newts, *Triturus cristatus*. *Ethology* 85:279–88.

———. 1990b. Factors affecting differential mating success in male crested newts, *Triturus cristatus*. *J. Zool., London* 220:33–40.

Hedlund, L., and J. G. M. Robertson. 1989. Lekking behaviour in cested newts, *Triturus cristatus*. *Ethology* 80:111–19.

Hedrick, P. W. 1986. Genetic polymorphism in heterogeneous environments: A decade later. *Ann. Rev. Ecol. Syst.* 17:535–66.

Hedrick, P. W., M. E. Ginevan, and E. P. Ewing. 1976. Genetic polymorphism in heterogeneous environments. *Ann. Rev. Ecol. Syst.* 7:1–32.

Heinen, J. T. 1993a. Aggregations of newly metamorphosed *Bufo americanus:* Tests of two hypotheses. *Can. J. Zool.* 71:324–38.

———. 1993b. Substrate choice and predation risk in newly metamorphosed American toads *Bufo americanus:* An experimental analysis. *Am. Midl. Nat.* 130:184–92.

Heinen, J. T., and G. Hammond. 1997. Antipredator behaviors of newly metamorphosed green frogs (*Rana clamitans*) and leopard frogs (*R. pipiens*) in encounters with eastern garter snakes (*Thamnophis s. sirtalis*). *Am. Midl. Nat.* 137:136–44.

Heinzmann, U. 1970. Untersuchungen zur Bio-Akustik und Ökologie der Geburtshelferkröte, *Alytes obstetricans* (Laur.). *Oecologia* 5:19–55.

Heisler, N. 1986. Comparative aspects of acid-base regulation. In *Acid-base regulation in animals,* ed. N. Heisler, 397–450. New York: Elsevier.

Heisler, N., G. Forcht, G. Ultsch, and J. F. Anderson. 1982. Acid-base regulation in response to environmental hypercapnia in two aquatic salamanders, *Siren lacertina* and *Amphiuma means*. *Respir. Physiol.* 49:141–58.

Helder-Jose and K. K. de Andrade. 1997. Food and feeding habits of the Neotropical river otter *Lontra longicaudis* (Carnivora, Mustelidae). *Mammalia* 61:193–203.

Hellbernd, L. 1987. Zweijährige Untersuchungen am Laichplatz vom Moorfrosch (*Rana arvalis* Nilsson) und vom Grasfrosch (*Rana temporaria* L.). *Beih. Schriftenr. Naturschutz Landschaftspfl. Niedersachsen* 19:119–30.

Heller, H. 1965. Osmoregulation in Amphibia. *Arch. Anat. Microsc. Morph. Exp.* 54:471–90

———. 1974. History of neurohypophysial research. In *Handbook of physiology, sec. 7: Endocrinology, vol. 4,* ed. E. Knobil and W. H. Sawyer, 103–17. Washington, DC: American Physiological Society.

Heller, H., and P. J. Bentley. 1965. Phylogenetic distribution of the effects of neurohypophysial hormones on water and salt metabolism. *Gen. Comp. Endocrinol.* 5:96–108.

Heller, H., and B. T. Pickering. 1961. Neurohypophysial hormones of non-mammalian vertebrates. *J. Physiol.* 155:98–114.

Hellmich, W. 1932. Zur Analyse des Farbkleides von *Pleurodema bibroni* Tschudi. *Biol. Zentralbl.* 52:513–22.

Hellriegel, B. 2000. Single- or multistage regulation in complex life cycles: Does it make a difference? *Oikos* 88:239–49.

Hellriegel, B., and H.-U. Reyer. 2000. Factors influencing the composition of mixed populations of a hemiclonal hybrid and its sexual host. *J. Evol. Biol.* 13:906–18.

Hels, T. 2002. Population dynamics in a Danish metapopulation of spadefoot toads *Pelobates fuscus*. *Ecography* 25:303–13.

Hels, T., and E. Buchwald. 2001. The effect of road kills on amphibian populations. *Biol. Cons.* 99:331–40.

Hels, T., and G. Nachman. 2002. Simulating viability of a spadefoot toad *Pelobates fuscus* metapopulation in a landscape fragmented by a road. *Ecography* 25:730–44.

Hemelaar, A. 1988. Age, growth and other population characteristics of *Bufo bufo* from different latitudes and altitudes. *J. Herpetol.* 22:369–88.

Hemelaar, A. S. M. 1981. Age determination of male *Bufo bufo* (Amphibia, Anura) from the Netherlands, based on year rings in phalanges. *Amphibia-Reptilia* 1:223–33.

———. 1983. Age of *Bufo bufo* in amplexus over the spawning period. *Oikos* 40:1–5.

———. 1985. An improved method to estimate the number of year rings resorbed in phalanges of *Bufo bufo* (L.) and its application to populations from different latitudes and altitudes. *Amphibia-Reptilia* 6:323–41.

Hemelaar, A. S. M., and J. J. van Gelder. 1980. Annual growth rings in phalanges of *Bufo bufo* (Anura, Amphibia) from the Netherlands and their use for age determination. *Netherlands J. Zool.* 30:129–35.

Hemmer, H., and H. Jakobs. 1974. Funktionelle Korrelate der Korpersproportionsunterschiede und Grabverhalten der Kreuzkröte (*Bufo calamita* Laur.) und der Wechselkröte (*Bufo viridis* Laur.) (Amphibia, Bufonidae). *Forma et Functio* 7:1–5.

Hemmer, H., and K. Kadel. 1971. Untersuchungen zur Laichgrösse nebst Bemerkungen zur Populationsdynamik der Kreuzkröte (*Bufo calamita* Laur.) und der Wechselkröte (*Bufo viridis* Laur.). *Bonn. Zool. Beitr.* 17:327–36.

Henderson, B. A. 1973. The specialized feeding of *Ambystoma gracile* in Marion Lake, British Columbia. *Can. Field-Nat.* 87:131–34.

Henderson, R. W. 1984. The diets of Hispaniolan colubrid snakes. I. Introduction and prey genera. *Oecologia* 62:234–39.

Henderson, R. W., and B. I. Crother. 1989. Biogeographic patterns of predation in West Indian colubrid snakes. In *Biogeography of the West Indies Past, present and future,* ed. C. W. Woods, 479–517. Gainesville, FL: Sandhill Crane Press.

Henderson, R. W., and L. G. Hoevers. 1977. The seasonal incidence of snakes at a locality in northern Belize. *Copeia* 1977:349–55.

Henderson, R. W., M. A. Nickerson, and L. G. Hoevers. 1977. Observations and comments on the feeding behavior of *Leptophis* (Reptilia, Serpentes, Colubridae). *J. Herpetol.* 11:231–32.

Henderson, R. W., T. Noeske-Hallin, B. Crother, and A. Schwartz. 1988. The diets of Hispaniolan colubrid snakes. II. Prey species, prey size, and phylogeny. *Herpetologica* 44:55–70.

Henderson, R. W., and R. Powell. 1999. West Indian herpetoecology. In *Caribbean amphibians and reptiles,* ed. B. I. Crother, 223–68. San Diego: Academic Press.

Henderson, R. W., and R. A. Sajdak. 1996. Diets of West Indian racers (Colubridae: *Alsophis*): Composition and biogeographic implications. In *Contributions to West Indian herpetology,* ed. R. Powell and R. W. Henderson, 327–338. Ithaca, New York: Society for the Study of Amphibians and Reptiles.

Henderson, R. W., and A. Schwartz. 1986. The diet of the Hispaniolan colubrid snake, *Darlingtonia haeiana. Copeia* 1986:529–31.

Henderson, R. W., A. Schwartz, and T. A. Noeske-Hallin. 1987. Food habits of three colubrid tree snakes (genus *Uromacer*) on Hispaniola. *Herpetologica* 43:241–48.

Hendrickson, J. R. 1954. Ecology and systematics of salamanders of the genus *Batrachoseps. Univ. Calif. Publ. Zool.* 54:1–46.

Henle, K. 1999. Winteraktivität des Moorfrosches (*Rana arvalis*). *Z. Feldherpetol.* 6:221–22.

———. 2001. Pit-falls in data analysis of amphibian breeding cycles and mortality patterns: A comment on Schmidt and Anholt (1999). *Amphibia-Reptilia* 22:379–86.

———. 2005. Analysis of recapture data from breeding populations of amphibians: On temporary emigration, model assumptions, bias, and common toads. *Amphibia-Reptilia* 26:7–16.

Henrich, T. W. 1968. Morphological evidence of secondary intergradation between *Bufo hemiophrys* Cope and *Bufo americanus* Holbrook in eastern South Dakota. *Herpetologica* 24:1–13.

Henrici, A. C. 1994. *Tephrodytes brassicarvalis,* new genus and species (Anura: Pelodytidae), from the Arikareean Cabbage Patch Beds of Montana, USA, and pelotydid-pelobatid relationships. *Ann. Carnegie Mus. Nat. Hist.* 63:155–83.

Henrikson, B.-I. 1990. Predation on amphibian eggs and tadpoles by common predators in acidified lakes. *Holarc. Ecol.* 13:201–6.

Henry, C. 1983. Position trophique du blaireau européen (*Meles meles* L.) dans une fôret de centre de la France. *Acta Oecol./Oecol. Gener.* 4:345–58.

———. 1984a. Adaptation comportement du blaireau européen (*Meles meles* L.) a la prédation d'une espèce-proie venimeuse, le crapaud commun (*Bufo bufo* L.). *Rev. Ecol. (Terre Vie)* 39:291–96.

———. 1984b. Eco-éthologie de l'alimentation du blaireau européen (*Meles meles* L.) dans une fôret du centre de la France. *Mammalia* 48:489–503.

Henry, W. V., and V. C. Twitty. 1940. Contributions to the life histories of *Dicamptodon ensatus* and *Ambystoma gracile. Copeia* 1940:247–50.

Hensel, J. L., Jr. and E. D. Brodie, Jr. 1976. An experimental study of aposematic coloration in the salamander *Plethodon jordani. Copeia* 1976:59–65.

Hensley, F. R. 1993. Ontogenetic loss of phenotypic plasticity of age at metamorphosis in tadpoles. *Ecology* 74:2405–12.

Hentschel, B. T. 1999. Complex life cycles in a variable environment: Predicting when the timing of metamorphosis shifts from resource dependent to developmental fixed. *Am. Nat.* 154:549–58.

Hentschel, B. T., and R. B. Emlet. 2000. Metamorphosis of barnacle nauplii: Effects of food variability and a comparison with amphibian models. *Ecology* 81:3495–3508.

Henzi, S. P., M. L. Dyson, S. E. Piper, N. I. Passmore, and P. Bishop. 1995. Chorus attendance by male and female painted reed frogs (*Hyperolius marmoratus*): environmental factors and selection pressures. *Funct. Ecol.* 9:485–91.

Herbeck, L. A., and D. R. Larsen. 1999. Plethodontid salamander response to silvicultural practices in Missouri Ozark forests. *Cons. Biol.* 13:623–32.

Hermans, K., R. Pinxten, and M. Eens. 2002. Territorial and vocal behaviour in a captive dart-poison frog, *Epipedobates tricolor* Boulenger, 1899 (Anura: Dendrobatidae). *Belgian J. Zool.* 132:105–9.

Hermes, D. J., A. Aertsen, P. Johannesma, and J. J. Eggermont. 1981. Spectrotemporal characteristics of single units in the auditory midbrain of the lightly anesthetised grass frog (*Rana temporaria* L.) investigated with noise stimuli. *Hearing Res.* 5:147–78.

Hermes, D. J., J. J. Eggermont, A. Aertsen, and P. Johannesma. 1982. Spectrotemporal characteristics of single units in the auditory midbrain of the lightly anesthesthetised grass frog (*Rana temporaria* L.) investigated with tonal stimuli. *Hearing Res.* 6:103–26.

Hermy, M., and J. Cornelis. 2000. Towards a monitoring method and a number of multifaceted and hierarchical biodiversity indicators for urban and suburban parks. *Landsc. Urban Plan.* 49:149–62.

Hernández, A. 1995. Depredación sobre anfibios, reptiles y aves por tres especies de alcaudones *Lanius* spp. en el noroeste de la Peninsula Iberica. *Ecologia (Madrid)* 9:409–15.

Hero, J.-M., and U. Galatti. 1990. Characteristics distinguishing *Leptodactylus pentadactylus* and *Leptodactylus knudseni* in the Central Amazon Rainforest. *J. Herpetol.* 24:227–28.

Hero, J.-M., C. Gascon, and W. E. Magnusson. 1998. Direct and indirect effects of predation on tadpole community structure in the Amazonian rainforest. *Aust. J. Ecol.* 23:474–82.

Hero, J.-M., and G. R. Gillespie. 1997. Epidemic disease and amphibian declines in Australia. *Cons. Biol.* 11:1023–25.

Hero, J.-M., M. Littlejohn, and G. Marantelli. 1991. *Frogwatch field guide to Victorian frogs.* Melbourne, Australia: Victoria Dept. of Conservation and Environment.

Hero, J.-M., and L. P. Shoo. 2003. Conservation of amphibians in the Old World tropics: Defining unique problems associated with regional fauna. In *Amphibian conservation,* ed. R. D. Semlitsch, 70–84. Washingrton, DC: Smithsonian Books.

Herreid, C. F., II, and S. Kinney. 1966. Survival of Alaskan woodfrog (*Rana sylvatica*) larvae. *Ecology* 47:1039–41.

Herreid, C.F., and S. Kinney. 1967. Temperature and development of the wood frog, *Rana sylvatica* in Alaska. *Ecology* 48:579–90.

Herrera, R. A., M. M. Steciow, and G. S. Natale. 2005. Chytrid fungus parasitizing the wild amphibian *Leptodactylus ocellatus* (Anura: Leptodactylidae) in Argentina. *Dis. Aquat. Org.* 64:247–52.

Herring, K., and P. Verrell. 1996. Sexual incompatibility and geographical variation in mate recognition systems: Tests in the salamander *Desmognathus ochrophaeus. Anim. Behav.* 52:279–87.

Herrington, R. E., and J. H. Larsen. 1985. Current status, habitat requirements and management of the Larch Mountain salamander, *Plethodon larselli* (Burns). *Biol. Cons.* 34:169–79.

Herrmann, H.-J. 1988a. Einige Bemerkungen über Haltung und Verhalten von *Pleurodema thaul* (Lesson, 1826) (Anura: Leptodactylidae). *Sauria* 10:27–28.

———. 1988b. Über Haltung und Verhalten von *Melanophyrniscus stelzneri* (Weyenbergh 1875) (Anura: Bufonidae). *Sauria* 10:23–24.

———. 1989. Über Haltung und Verhalten von *Phrynomerus microps* (Peters 1875) (Anura: Microhylidae). *Sauria* 11:25–26.

———. 1991. Die Mikromorphologie der Haftzehen einiger aborealer Anuren als taxonomisches Mittel. In*Amphibienforschung,* ed. H.-J. Herrmann, 118–25. Schleusingen, Germany: Naturhistorisches Museum Schloss Bertholdsburg.

Herrmann, H.-J., U. Koepernik, W. Matz, R. Bech, R. Gassel, and V. Utjeshov. 1987. Systematik, Verbreitung, Biologie, Ökologie und Terranienhaltung der Discoglossidae Gunther, 1859, (Amphibia, Anura). *Veröff. Naturhist. Mus. Schleusingen* 2:2–30.

Hershey, J. L., and D. C. Forester. 1980. Sensory orientation in *Notophthalmus v. viridescens* (Amphibia: Salamandridae). *Can. J. Zool.* 58:266–76.

Hertwig, I., and U. Sinsch. 1995. Comparative toe pad morphology in marsupial frogs (genus *Gastrotheca*): arboreal versus ground-dwelling species. *Copeia* 1995:38–47.

Hertwig, S., R. O. da Sa, and A. Haas. 2004. Phylogenetic signal and the utility of 12S and 16S RNA in frog phylogeny. *J. Zool. Syst. Evol. Res.* 42:2–18.

Heselhaus, R. 1984. *Pfeilgiftfrösche.* Essen, Germany: Alfred Kernen Verlag.

Hetherington, T. E. 1985. Role of the opercularis muscle in seismic sensitivity in the bullfrog, *Rana catesbeiana. J. Exp. Zool.* 235:27–34.

———. 1987. Physiological features of the opercularis muscle and their effects on vibratory sensitivity in the bullfrog *Rana catesbeiana. J. Exp. Biol.* 131:1–16.

———. 1988a. Biomechanics of vibration reception in the bullfrog, *Rana catesbeiana. J. Comp. Physiol. A* 163:43–52.

———. 1988b. Metamorphic changes in the middle ear. In *The evolution of the amphibian auditory system,* ed. B. Fritzsch, M. J. Ryan, W. Wilczynski, T. E. Hetherington, and W. Walkowiak, 339–57. New York: Wiley.

———. 1992a. The effects of body size on functional properties of middle ear systems of anuran amphibians. *Brain Behav. Evol.* 39:133–42.

———. 1992b. The effects of body size on the evolution of the amphibian middle ear. In *The evolutionary biology of hearing,* ed. D. B. Webster, R. R. Fay, and A. N. Popper, 421–37. New York: Springer-Verlag.

———. 1994. Sexual differences in the tympanic frequency responses of the American bullfrog (*Rana catesbeiana*). *J. Acoust. Soc. Am.* 96:1186–88.

———. 2001. Laser vibrometric studies of sound-induced motion of the body walls and lungs of salamanders and lizards: implications for lung-based hearing. *J. Comp. Physiol. A* 187:499–507.

Hetherington, T. E., A. P. Jaslow, and R. E. Lombard. 1986. Comparative morphology of the amphibian opercularis system. I. General design features and functional interpretation. *J. Morphol.* 190:43–61.

Hetherington, T. E., and E. D. Lindquist. 1999. Lung-based hearing in an "earless" anuran amphibian. *J. Comp. Physiol. A* 184:395–401.

Hetherington, T. E., and R. E. Lombard. 1982. Biophysics of underwater hearing in anuran amphibians. *J. Exp. Biol.* 98:49–66.

———. 1983. Electromyography of the opercularis muscle of the bullfrog *Rana catesbeiana:* an amphibian tonic muscle. *J. Morphol.* 175:17–26.

Hetherington, T. E., and M. H. Wake. 1979. The lateral line system in larval *Ichthyophis* (Amphibia: Gymnophiona). *Zoomorphology* 93:209–25.

Hettyey, A., and P. B. Pearman. 2003. Social environment and reproductive interference affect reproductive success in the frog *Rana latastei. Behav. Ecol.* 14:294–300.

Hettyey, A., J. Török, and G. Hévizi. 2005. Male mate choice lacking in the agile frog, *Rana dalmatina. Copeia* 2005:403–8.

Heusser, H. 1958a. Über die Beziehungen der Erdkröte (*Bufo bufo* L.) zu ihrem Laichplatz I. *Behaviour* 12:208–32.

———. 1958b. Zum Hautungsverhalten von Amphibien. *Rev. Suisse Zool.* 65:793–23.

———. 1960. Über die Beziehungen der Erdkröte (*Bufo bufo* L.) zu ihrem Laichplatz II. *Behaviour* 16:93–109.

———. 1961. Die Bedeutung der ausseren Situation im Verhalten einiger Amphibienarten. *Rev. Suisse Zool.* 68:1–39.

———. 1963. Die Ovulation des Erdkrötenweibchens im Rahmen der Verhaltensorganisation von *Bufo bufo* L. *Rev. Suisse Zool.* 68:1–39.

———. 1964. Zur Laichplatzorientierung der Erdkröte, *Bufo bufo* L. *Mitt. Naturforsch. Ges. Schaffhausen* 28:101–12.

———. 1968a. Die Lebensweise der Erdkröte, *Bufo bufo* (L.): Der Magenfüllungsgrad in Abhängigkeit von Jagdstimmung und Wetter. *Sitzber. Ges. Naturforsh. Fr. Berlin (N. F.)* 8:148–56.

———. 1968b. Die Lebensweise der Erdkröte, *Bufo bufo* L.: Grossenfrequenzen und Populationsdynamik. *Mitt. Naturforsch. Ges. Schaffhausen* 29:1–29.

———. 1968c. Die Lebensweise der Erdkröte, *Bufo bufo* L.: Laichzeit: Umstimmung, Ovulation, Verhalten. *Vierteljahresschr. Naturforsch. Ges. Zürich* 113:257–89.

———. 1968d. Die Lebensweise der Erdkröte, *Bufo bufo* (L.): Wanderungen und Sommerquartiere. *Rev. Suisse Zool.* 75:927–82.

———. 1969a. Der rudimentäre Ruf der männlichen Erdkröte (*Bufo bufo*). *Salamandra* 5:46–56.

———. 1969b. Die Lebensweise der Erdkröte, *Bufo bufo* (L.): Nährungsaufnahme und Pigmentierung der Daumenschwielen im Jahres-lauf. *Biol. Zentralbl.* 88:457–67.

———. 1969c. Die Lebensweise der Erdkröte, *Bufo bufo* (L.): Das Orienticrungsproblem. *Rev. Suisse Zool.* 76:443–517.

———. 1969c. Ethologische Bedingungen für das Vorkommen von Territorialität bei Anuren. *Salamandra* 5:95–104.

———. 1970. Laich-Fresen durch Kaulquappen als mögliche Ursache spezifischer Biotoppräferenzen und kurzer Laichzeiten bei europäischen Froschlurchen (Amphibia, Anura). *Oecologia* 4:83–88.

———. 1971a. Differenzierendes Kaulquappen-Fressen durch Molche. *Experientia* 27:475

———. 1971b. Laich-Räuben und Kannibalismus bei sympatrichen Anuran-Kaulquappen. *Experientia* 27:474–75.

———. 1972a. Intra- un interspezifische Crowding-effekte bei Kaulquappen einheimische Anuren-Arten. *Vierteljahresschr. Naturforsch. Ges. Zürich* 117:121–28.

———. 1972b. Intra- und interspezifische Crowding-effekte be Kaulquappen der Kreuzkröte, *Bufo calamita* Laur. *Oecologia* 10:93–98.

Heusser, H., and K. Miesterhans. 1969. Zur Populationsdynamik der Kreuzkröte, *Bufo calamita* Laur. *Vierteljahresschr. Naturforsch. Ges. Zürich* 114:269–77.

Hews, D. K. 1988. Alarm response in larval western toads, *Bufo boreas:* Release of larval chemicals by a natural predator and its effect on predator capture efficiency. *Anim. Behav.* 36:125–33.

Hews, D. K., and A. R. Blaustein. 1985. An investigation of the alarm response in *Bufo boreas* and *Rana cascadae* tadpoles. *Behav. Neur. Biol.* 43:47–57.

Heyer, H. R. 1969. The adaptive ecology of the genus *Leptodactylus* (Amphibia, Leptodactylidae). *Evolution* 23:421–28.

Heyer, W. R. 1973. Ecological interactions of frog larvae at a seasonal tropical location in Thailand. *J. Herpetol.* 7:337–61.

———. 1974. Niche measurements of frog larvae from a seasonal tropical location in Thailand. *Ecology* 55:651–56.

———. 1975. A preliminary analysis of the intergeneric relationships of the frog family Leptodactylidae. *Smithsonian Contrib. Zool.* (199):1–55.

———. 1976. Studies in larval amphibian habitat partitioning. *Smithsonian Contrib. Zool.* (242):1–27.

———. 1979a. Annual variation in larval amphibian populations within a temperate pond. *J. Washington Acad. Sci.* 69:65–74.

———. 1979b. Systematics of the *pentadactylus* species group of the frog genus *Leptodactylus* (Amphibia: Leptodactylidae). *Smithsonian Contrib. Zool.* (301):1–43.

———. 1983. Variation and systematics of frogs of the genus *Cycloramphus* (Amphibia, Leptodactylidae). *Arq. Zool., S. Paulo* 30:235–339.

———. 1984. Variation, systematics, and zoogeography of *Eleutherodactylus guentheri* and closely related species (Amphibia: Anura: Leptodactylidae). *Smithsonian Contrib. Zool.* (402):1–42.

———. 1985. Taxonomic and natural history notes on frogs of the genus Centrolenella (Amphibia: Centrolenidae) from southeastern Brazil and adjacent Argentina. *Pap. Avul. Zool.* 36:1–21.

Heyer, W. R., and M. S. Bellin. 1973. Ecological notes on five sympatric *Leptodactylus* (Amphibia, Leptodactylidae) from Ecuador. *Herpetologica* 29:66–72.

Heyer, W. R., and K. A. Berven. 1973. Species diversity of herpetofaunal samples from similar microhabitats at two tropical sites. *Ecology* 54:642–45.

Heyer, W. R., and R. I. Crombie. 1979. Natural history notes on *Craspedoglossa stejnegeri* and *Thoropa petropolitana* (Amphibia: Salientia, Leptodactylidae). *J. Washington Acad. Sci.* 69:17–20.

Heyer, W. R., and L. R. Maxson. 1983. Relationships, zoogeography, and speciation mechanisms of frogs of the genus *Cycloramphus* (Amphibia, Leptodactylidae). *Arq. Zool., Sao Paulo* 30:341–73.

Heyer, W. R., R. W. McDiarmid, and D. L. Weigmann. 1975. Tadpoles, predation and pond habitats in the tropics. *Biotropica* 7:100–11.

Heyer, W. R., and J. B. Murphy. 2005. Declining amphibian task force. In *Amphibian declines: The conservation status of United States species,* ed. M. J. Lannoo, 34–46. Berkeley, CA: University of California Press.

Heyer, W. R., and A. S. Rand. 1977. Foam nest construction in the leptodactylid frogs *Leptodactylus* pentadactylus and *Physalaemus pustulosus* (Amphibia, Anura, Leptodactylidae). *J. Herpetol.* 11:225–28.

Heyer, W. R., A. S. Rand, C. A. Goncalvez da Cruz, and O. L. Peixoto. 1988. Decimations, extinctions, and colonizations of frog populations in southeast Brazil and their evolutionary implications. *Biotropica* 20:230–35.

Heyer, W. R., A. S. Rand, C. A. Goncalvez da Cruz, O. L. Peixoto, and C. E. Nelson. 1990. Frogs of Boracéia. *Arq. Zool., Sao Paulo.* 31:231–410.

Heyer, W. R., and P. A. Silverstone. 1969. The larva of the frog *Leptodactylus hylaedactylus* (Leptodactylidae). *Fieldiana: Zool.* 51:141–45.

Heying, H. E. 2001. Social and reproductive behaviour in the Madagascan poison frog, *Mantella laevigata,* with comparisons to the dendrobatids. *Anim. Behav.* 61:567–77.

———. 2004. Reproductive limitation by oviposition site in a treehole breeding Madagascan poison frog (*Mantella laevigata*). In *Ecology and evolution of phytotelm-breeding anurans. Misc. Publ. Mus. Zool. Univ. Michigan,* no. 193., ed. R. M. Lehtinen, 23–30.

Heym, W. D. 1974. Studien zur Verbreitung, Ökologie und Ethologie der Grunfrösche in der mittleren und nordlichen Niederlausitz. *Mitt. Zool. Mus. Berlin* 50:263–85.

Heymann, E., C. Knogge, and E. R. Tirado Herrera. 2000. Vertebrate predation by sympatric tamarins, *Saguinus mystax* and *Saguinus fuscicollis. Amer. J. Primatol.* 51:153–58.

Hickling, R., D. B. Roy, J. K. Hill, R. Fox, and C. D. Thomas. 2005. A northward shift of range margins in British Odonata. *Global Change Biol.* 11:502–6.

———. 2006. The distributions of a wide range of taxonomic groups are expanding polewards. *Global Change Biol.* 12:450–55.

Hickman, C. R., M. D. Stone, and A. Mathis. 2004. Priority use of chemical over visual cues for detection of predators by graybelly salamanders, *Eurycea multiplicata griseogaster. Herpetologica* 60:203–10.

Hicks, N. G., and S. M. Pearson. 2003. Salamander diversity and abundance in forests with alternative land use histories in the southern Blue Ridge Mountains. *Forest Ecol. Manage.* 177:117–30.

Hidalgo-Vila, J., N. Pérez-Santigosa, and C. Díaz-Paniagua. 2002. The sexual behaviour of the pygmy newt, *Triturus pygmaeus. Amphibia-Reptilia* 23:393–405.

Higgins, R. M. 1979. Temperature-related variation in the duration of morning song of the song thrush *Turdus ericetorum. Ibis* 121:333–35.

Highton, R. 1956. The life history of the slimy salamander, *Plethodon glutinosus* in Florida. *Copeia* 1956:75–93.

———. 1962. Geographic variation in the life history of the slimy salamander. *Copeia* 1962:597–613.

———. 1970. Evolutionary interactions between species of North American salamanders of the genus *Plethodon.* Part 1. Genetic and ecological relationships of *Plethodon jordani* and *P. glutinosus* in the southern Appalachian Mountains. *Evol. Biol.* 4:211–41.

———. 1971. Distributional interactions among eastern North American salamanders of the genus *Plethodon.* In *The distributional history of the biota of the southern Appalachians. Res. Div. Monogr.* 4., ed. P. C. Holt, 139–88. Blacksburg: Virginia Polytechnic Institute and State University.

———. 1977. Comparison of microgeographic variation in morphology and electrophoretic traits. *Evol. Biol.* 10:397–436.

———. 1983. A new species of woodland salamander of the *Plethodon glutinosus* group from the Southern Appalachian Mountains. *Brimleyana* 9:1–20.

———. 1989. Biochemical evolution in the slimy salamanders of the *Plethodon glutinosus* complex in the eastern United States. I. Geographic protein variation. *Illinois Biol. Monogr.* 57:1–78.

———. 1990. Taxonomic treatment of genetically differentiated populations. *Herpetologica* 46:114–21.

———. 1995. Speciation in eastern North American salamanders of the genus *Plethodon. Ann. Rev. Ecol. Syst.* 26:579–600.

———. 1997. Geographic protein variation and speciation in the *Plethodon dorsalis* complex. *Herpetologica* 53:345–56.

———. 1998. Is *Ensatina* a ring-species? *Herpetologica* 54:254–78.

———. 2005. Declines in eastern North American woodland salamanders (*Plethodon*). In *Amphibian declines: The conservation status of United States species,* ed. M. J. Lannoo, 34–46. Berkeley: University of California Press.

Highton, R., and S. A. Henry. 1970. Evolutionary interactions between species of North American salamanders of the genus *Plethodon. Evol. Biol.* 4:211–56.

Highton, R., and A. Larson. 1979. The genetic relationships of salamanders of the genus *Plethodon. Syst. Zool.* 28:579–99.

Highton, R., G. C. Maha, and L. R. Maxson. 1989. Biochemical evolution in the slimy salamanders of the *Plethodon glutinosus* complex in the eastern United States. Univ. *Illinois Biol. Monogr.* 57:1–153.

Highton, R., and R. B. Peabody. 2000. Geographic protein variation and speciation in salamanders of the *Plethodon jordani* and *Plethodon glutinosus* complexes in the southern Appalachian Mountains with the description of four new species. In *The biology of plethodontid salamanders,* ed. R. C. Bruce, R. G. Jaeger, and L. D. Houck, 31–94. New York: Kluwer Academic/Plenum.

Highton, R., and T. Savage. 1961. Functions of brooding behavior in the female red-backed salamander, *Plethodon cinereus. Copeia* 1961:95–98.

Highton, R., and T. P. Webster. 1976. Geographic protein variation and divergence in populations of the salamander *Plethodon cinereus. Evolution* 30:33–45.

Hileman, K. S., and E. D. Brodie, Jr. 1994. Survival strategies of the salamander *Desmognathus ochrophaeus:* interaction of predator-avoidance and anti-predator mechanisms. *Anim. Behav.* 47:1–6.

Hill, K., and K. Hawkes. 1983. Neotropical hunting among the Ache of Eastern Paraguay. In *Adaptive responses of Native Amazonians,* ed. R. B. Hames and W. T. Vickers, 139–88. New York: Academic Press.

Hill, K. G., and J. G. M. Robertson. 1981. The auditory system of the atympanate frog *Uperoleia rugosa. Proc. Aust. Phys. Pharm. Soc.* 12:25.

Hillery, C. M., and P. M. Narins. 1984. Neurophysiological evidence for a traveling wave in the amphibian inner ear. *Science* 225:1037–39.

Hillis, D. M. 1982. Morphological differentiation and adaptation of the larvae of *Rana berlandieri* and *Rana sphenocephala* (*Rana pipiens* complex) in sympatry. *Copeia* 1982:168–74.

———. 1988. Systematics of the *Rana pipiens* complex: puzzle and paradigm. *Ann. Rev. Ecol. Syst.* 19:39–63.

———. 1991. The phylogeny of amphibians: current knowledge and the role of cytogenetics. In *Amphibian cytogenetics and evolution,* ed. D. M. Green and S. K. Sessions, 7–31. New York: Academic Press.

Hillis, D. M., L. K, Ammermann, M. T. Dixon, and R. O. da Sá. 1993. Ribosomal DNA and the phylogeny of frogs. *Herpetol. Monogr.* 7:118–31.

Hillis, D. M., and T. P. Wilcox. 2005. Phylogeny of the New World true frogs (*Rana*). *Mol. Phylogen. Evol.* 34:299–314.

Hillis, R. E., and E. D. Bellis. 1971. Some aspects of the ecology of the hellbender, *Cryptobranchus alleganiensis alleganiensis* in a Pennsylvania stream. *J. Herpetol.* 5:121–26.

Hillman, S. S. 1974. The effect of arginine vasopressin on water and sodium balance in the urodele amphibian *Aneides lugubris. Gen. Comp. Endocrinol.*24:74–82.

———. 1976. Cardiovascular correlates of maximal oxygen consumption rates in anuran amphibians. *J. Comp. Physiol.* 109:199–207.

———. 1978a. Some effects of dehydration on internal distribution of water and solutes in *Xenopus laevis. Comp. Biochem. Physiol.* 61A:303–7.

———. 1978b. The roles of oxygen delivery and electrolyte levels in the dehydrational death of *Xenopus laevis. J. Comp. Physiol.* B 128:169–75.

———. 1980a. The effect of anemia on metabolic performance in the frog, *Rana pipiens*. *J. Exp. Zool.* 211:107–11.

———. 1980b. Physiological correlates of differential dehydration tolerance in anuran amphibians. *Copeia* 1980:125–29.

———. 1982a. The effect of in vivo and in vitro hyperosmolality on skeletal muscle performance in the amphibians *Rana pipiens* and *Scaphiopus couchii*. *Comp. Biochem. Physiol.* 73A:709–12.

———. 1982b. Effects of DL-propranolol on exercise heart rate and maximal rates of oxygen consumption in *Scaphiopus intermontanus*. *Experientia* 38:940–41.

———. 1984. Inotropic influence of dehydration and hyperosmolal solutions on amphibian cardiac muscle. *J. Comp. Physiol.* 154:325–28.

———. 1987. Dehydrational effects on cardiovascular and metabolic capacity in two amphibians. *Physiol. Zool.* 60:608–13.

———. 1988. Dehydrational effects on brain and cerebrospinal fluid electrolytes in two amphibians. *Physiol. Zool.* 61:254–59.

Hillman, S. S., M. S. Hedrick, P. C. Withers, and R. C. Drewes. 2004. Lymph pools in the basement, sump pumps in the attic: the anuran dilemma for lymph movement. *Physiol. Biochem. Zool.* 77:161–73.

Hillman, S. S., and M. S. Lea. 1983. Aerial activity oxygen consumption during metamorphosis of the bullfrog, *Rana catesbeiana*. *Copeia* 1983:407–10.

Hillman, S. S., V. H. Shoemaker, R. Putnam, and P. C. Withers. 1979. Reassessment of aerobic metabolism in amphibians during activity. *J. Comp. Physiol.* 129:309–13.

Hillman, S. S., and P. C. Withers. 1979. An analysis of respiratory surface area as a limit to activity metabolic in anurans. *Can. J. Zool.* 57:2100–2105.

———. 1981. Aerobic contributions to sustained activity metabolism in *Xenopus laevis*. *Comp. Biochem. Physiol.* 69A:605–6.

Hillman, S. S., P. C. Withers, and R. C. Drewes. 2000. Correlation of ventricle mass and dehydration tolerance in amphibians. *Herpetologica* 56:413–20.

Hillman, S. S., P. C. Withers, M. S. Hedrick, and P. B. Kimmel. 1985. The effects of erythrocythemia on blood viscosity, maximal systemic oxygen transport capacity and maximal rates of oxygen consumption in an amphibian. *J. Comp. Physiol.* B 155:577–81.

Hillman, S. S., A. Zygmunt, and M. Baustian. 1987. Transcapillary fluid forces during dehydration in two amphibians. *Physiol. Zool.* 60:339–45.

Hillyard, S. D. 1975. The role of antidiuretic hormones in the water economy of the spadefoot toad, *Scaphiopus couchi*. *Physiol. Zool.* 48:242–51.

———. 1976a. The movement of soil water across the isolated amphibian skin. *Copeia* 1976:314–20.

———. 1976b. Variation in the effects of antidiuretic hormone on the isolated skin of the toad, *Scaphiopus couchii*. *J. Exp. Zool.* 195:199–206.

———. 1979. The effect of isoproterenol on the anuran water balance response. *Comp. Biochem. Physiol.* 62C:93–95.

———. 1999. Behavioral, molecular and integrative mechanisms of amphibian osmoregulation. *J. Exp. Zool.* 283:662–74.

Hillyard, S. D., J. Goldstein, W. Tuttle, K. Hoff. 2004. Transcellular and paracellular elements of salt chemosensation in toad skin. *Chem. Sens.* 29:755–62.

Hillyard, S. D., K. von Seckendorff Hoff, and C. Propper. 1998. The water absorption response: A behavioral assay for physiological processes in terrestrial amphibians. *Physiol. Zool.* 71:127–38.

Himes, J. G. 2003. Diet composition of *Nerodia sipedon* (Serpentes: Colubridae) and its dietary overlap with, and chemical recognition of *Agkistrodon piscivorus* (Serpentes: Viperidae). *Amphibia-Reptilia* 24:181–88.

———. 2004. The non-fish, vertebrate diet of sympatric populations of the cottonmouth (*Agkistrodon piscivorus*) and northern watersnake (*Nerodia sipedon*). *Herp. Rev.* 35:123–28.

Himstedt, W. 1965. Beobachtungen zum Paarungsverhalten des Feuersalamanders (*Salamandra salamandra* L.). *Zool. Anz.* 175:295–300.

———. 1991. Zur Biologie und Larvenentwicklung der thailändischen Blindwühle *Ichthyophis kohtaoensis* (Amphibia: Gymnophiona: Ichthyophiidae). *Veröff. Naturhist. Mus. Schleusingen* 6:16–24.

———. 1994. Sensory systems and orientation in *Salamandra salamandra*. *Mertensiella* 4:225–39.

———. 1995. Structure and function of the eyes in the caecilian *Ichthyophis kohtaoensis* (Amphibia, Gymnophiona). *Zoology* 99:81–94.

———. 1996. *Die Blindwühlen*. Magdeburg, Germany: Westarp Wissenshaften.

Himstedt, W., and B. Fritzsch. 1990. Behavioural evidence for electroreception in larvae of the caecilian *Ichthyophis kohtaoensis* (Amphibia, Gymnophiona). *Zool. Jb. Physiol.* 94:486–92.

Himstedt, W., and D. Simon. 1995. Sensory basis of foraging behaviour in caecilians (Amphibia, Gymnophiona). *Herpetol. J.* 5:266–70.

Hines, H., M. Mahony, and K. McDonald. 1999. An assessment of frog declines in wet subtropical Australia. In *Declines and disappearances of Australian frogs*, ed. A. Campbell, 44–63. Canberra: Environment Australia.

Hinsche, G. 1926. Vergleichende untersuchungen zum sogenannten Unken-reflex. *Biol. Zentralbl.* 46:296–305.

———. 1928. Kampfreaktionen bei einheimischen Anuren. *Biol. Zentralbl.* 48:577–617.

Hinshaw, S. H., and B. K. Sullivan. 1990. Predation on *Hyla versicolor* and *Pseudacris crucifer* during reproduction. *J. Herpetol.* 24:196–97.

Hirai, T. 2004. Diet composition of introduced bullfrogs, *Rana catesbeiana*, in the Mizorogaike Pond of Kyoto, Japan. *Ecol. Res.* 19:375–80.

Hirai, T., and K. Hidaka. 2002. Anuran-dependent predation by giant water bug, *Lethocerus deyrollei* (Hemiptera: Belostomatidae), in rice fields of Japan. *Ecol. Res.* 17:655–61.

Hirai, T., and M. Matsui. 1999. Feeding habits of the pond frog, *Rana nigromaculata*, inhabiting rice fields in Kyoto, Japan. *Copeia* 1999:940–47.

Hirai, T., and M. Matsui. 2000. Myrmecophagy in a ranid frog *Rana rugosa*: Specialization or weak avoidance to ant eating? *Zool. Sci.* 17:459–66.

Hirai, T., and M. Matsui. 2001. Food habits of an endangered Japanese frog, *Rana porosa brevipoda*. *Ecol. Res.* 16:737–43.

Hirano, M., and L. C. Rome. 1984. Jumping performance of frogs (*Rana pipiens*) as a function of muscle temperature. *J. Exp. Biol.* 108:429–39.

Hitchings, S. P., and T. J. C. Beebee. 1997. Genetic substructuring as a result of barriers to gene flow in urban common frog (*Rana temporaria*) populations: Implications for biodiversity conservation. *Heredity* 79:117–27.

———. 1998. Loss of genetic diversity and fitness in common toad (*Bufo bufo*) populations isolated by inimical habitat. *J. Evol. Biol.* 11:269–83.

Ho, C. T., A. Lathrop, R. W. Murphy, and N. Orlov. 1999. A redescription of *Vibrissophora ailaonica* with a new record in Vietnam. *Russian J. Herpetol.* 6:48–54.

Höbel, G. 1999a. Facultative nest construction in the gladiator frog *Hyla rosenbergi* (Anura: Hylidae). *Copeia* 1999:797–801.

———. 1999b. Notes on the natural history and habitat use of *Eleutherodactylus fitzingeri* (Anura: Leptodactylidae). *Amphibia-Reptilia* 20:65–72.

———. 2000. Reproductive ecology of *Hyla rosenbergi* in Costa Rica. *Herpetologica* 56:446–54.

Höbel, G., and H. C. Gerhardt. 2003. Reproductive character displacement in the acoustic communication system of green tree frogs (*Hyla cinerea*). *Evolution* 57:894–904.

Hobson, E. S. 1978. Aggregating as a defense against predators in aquatic and terrestrial environments. In *Contrasts in behavior,* ed. E. S. Reese and F. J. Lighter, 219–34. New York: Wiley.

Hochachka, P. W., and G. N. Somero. 1984. *Biochemical adaptation.* Princeton, NJ: Princeton University Press.

Hock, R. J. 1964. Animals in high altitudes: reptiles and amphibians. In *Handbook of physiology. Section 4: Adaptation to the environment,* ed. D. B. Dill, 841–42. Washington, DC: American Physiological Society.

Hock, R. J. 1967. Temperature effect on breeding of the toad, *Bufo variegatus* in southern Chile. *Copeia* 1967:227–30.

Hödl, W. 1975. Die Entwicklung der spektralen Empfindlichkeit der Netzhaut von *Bombina, Hyla, Pelobates* und *Rana. Zool. Jb. Physiol.* 79:173–203.

———. 1977. Call differences and calling site segregation in anuran species from central Amazonian floating meadows. *Oecologia* 28:351–63.

———. 1983. *Phyllobates femoralis* (Dendrobatidae): Rufverhalten und akustische Orientierung der Männchen (Freilandaufnahmen). *Wiss. Film (Wien)* 30:12–19.

———. 1986. Foam nest construction in South American leptodactylid frogs. In *Studies in herpetology,* ed. Z. Roček, 565–70. Prague: Charles University.

———. 1990a. An analysis of foam nest construction in the Neotropical frog *Physalaemus ephippifer* (Leptodactylidae). *Copeia* 1990:547–54.

———. 1990b. Reproductive diversity in Amazonian lowland frogs. *Fortschr. Zool.* 38:41–60.

———. 1991a. Arboreal oviposition in the Neotropical treefrogs *Hyla brevifrons* and *Hyla sarayacuensis* (Anura: Hylidae). *Wiss. Film (Wien)* 42:53–62.

———. 1991b. *Phrynohyas* resinifictrix (Hylidae, Anura) calling behaviour. *Wiss. Film (Wien)* 42:63–70.

———. 1992. Reproductive behaviour in the Neotropical foam-nesting frog *Pleurodema diplolistris* (Leptodactylidae). *Amphibia-Reptilia* 13:263–74.

Hödl, W., and A. Amezquita. 2001. Visual signaling in anuran amphibians. In *Anuran communication,* ed. M. J. Ryan, 121–41. Washington, DC: Smithsonian Institution Press.

Hödl, W., A. Amezquita, and P. M. Narins. 2004. The role of call frequency and the auditory papillae in phonotactic behavior in male dart-poison frogs *Epipedobates femoralis* (Dendrobatidae). *J. Comp. Physiol. A* 190:823–29.

Hödl, W., and G. Gollmann. 1986. Distress calls in Neotropical frogs. *Amphibia-Reptilia* 7:11–21.

Hödl, W., R. Jehle, and G. Gollmann. 1997. Populationsbiologie von Amphibien. Eine Langzeitstudie auf der Wiener Donauinsel. *Stapfia* 51:1–270.

Hoegg, S., M. Vences, H. Brinkmann, and A. Meyer. 2004. Phylogeny and comparative substitution rates of frogs inferred from sequences of three nuclear genes. *Mol. Biol. Evol.* 21:1188–1200.

Hofer, R., and C. Mokri. 2000. Photoprotection in tadpoles of the common frog, *Rana temporaria. J. Photochem. Photobiol.* B 59:48–53.

Hofer, U., L.-F. Bersier, and D. Borcard. 1999. Spatial organization of a herpetofauna on an elevational gradient revealed by null model tests. *Ecology* 80:976–88.

———. 2000. Ecotones and gradient as determinants of herpetofaunal community structure in the primary forest of Mount Kupe, Cameroon. *J. Trop. Ecol.* 16:517–33.

Hofer, V. R. 1972. Einfluss von Temperatur, Photoperiode und Jahreszeiten auf Verdauung und Atmung zweier Froscharten: *Rana ridibunda* (bzw. *R. esculenta*) und *Rana temporaria. Zool. Jb. Physiol.* 76:507–30.

Hoff, J. G., and S. A. Moss. 1974. A distress call in the bullfrog, *Rana catesbeiana. Copeia* 1974:533–34.

Hoff, K. S., A. R. Blaustein, R. W. McDiarmid, and R. Altig. 1999. Behavior: Interactions and their consequences. In *Tadpoles. The biology of anuran larvae,* ed. R. W. McDiarmid and R. Altig, 215–39. Chicago: University of Chicago Press.

Hoff, K. S., and S. D. Hillyard. 1991. Angiotensin II stimulates cutaneous drinking in the toad *Bufo punctatus. Physiol. Zool.* 64:1165–72.

———. 1993. Inhibition of cutaneous water absorption in dehydrated toads by saralasin is associated with changes in barometric pressure. *Physiol. Zool.* 66:89–98.

Hoff, K. S., N. Huq, V. A. King, and R. J. Wassersug. 1989. The kinematics of larval salamander swimming (Ambystomatidae: Caudata). *Can. J. Zool.* 67:2756–61.

Hoff, K. S., M. J. Lannoo, and R. J. Wassersug. 1985. Kinematics of midwater prey capture by *Ambystoma* (Caudata: Ambystomatidae) larvae. *Copeia* 1985:247–51.

Hoff, K. S., and R. J. Wassersug. 1986. The kinematics of swimming in larvae of the clawed frog, *Xenopus laevis. J. Exp. Biol.* 122:1–12.

———. 2000. Tadpole locomotion: Axial movement and tail functions in a largely vertebraeless vertebrate. *Amer. Zool.* 40:62–76.

Hoffman, E. A., and M. S. Blouin. 2000. A review of colour and pattern polymorphism in anurans. *Biol. J. Linn. Soc.* 70:633–65.

Hoffman, E. A., F. W. Schueler, and M. S. Blouin. 2004. Effective population sizes and the temporal stability of genetic structure in *Rana pipiens,* the northern leopard frog. *Evolution* 58:2536–45.

Hoffman, J., and U. Katz. 1989. The ecological significance of burrowing behaviour in the toad (*Bufo viridis*). *Oecologia* 81:510–13.

———. 1991. Tissue osmolytes and body fluid compartments in the toad *Bufo viridis* under simulated terrestrial conditions. *J. Comp. Physiol.* B 161:433–39.

———. 1994. Urea production and accumulation in the green toad, *Bufo viridis. J. Zool., London* 233:591–3.

———. 1997. Salt and water balance in the toad *Bufo viridis* during recovery from two different osmotically stressful conditions. *Comp. Biochem. Physiol.* 117A:147–54.

———. 1999. Elevated plasma osmotic concentration stimulates water absorption response in a toad. *J. Exp. Zool.* 284:168–73.

Hoffman, J., V. Eliath, and U. Katz. 1988. *Bufo viridis:* physio-

logical responses to osmotic stress (burrowing and salt acclimation). *Israel J. Zool.* 35:103–4.
Hoffman, J., U. Katz, and V. Eliath. 1990. Urea accumulation in response to water restriction in burrowing toads (*Bufo viridis*). *Comp. Biochem. Physiol.* 97 A:423–46.
Hoffman, R. L., G. L. Larson, and B. J. Brokes. 2003. Habitat segregation of *Ambystoma gracile* and *Ambystoma macrodactylum* in mountain ponds and lakes, Mount Rainier National Park, Washington, USA. *J. Herpetol.* 37:24–34.
Hoffman, R. L., G. L. Larson, and B. Samora. 2004. Responses of *Ambystoma gracile* to the removal of introduced nonnative fish from a mountain lake. *J. Herpetol.* 38:578–85.
Höglund, J., and R. V. Alatalo. 1995. *Leks.* Princeton, NJ: Princeton University Press.
Höglund, J., and J. G. M. Robertson. 1987. Random mating by size in a population of common toads (*Bufo bufo*). *Amphibia-Reptilia* 8:321–30.
———. 1988. Chorusing behaviour, a density-dependent alternative mating strategy in male common toads (*Bufo bufo*). *Ethology* 79:324–32.
Hokit, D. G., and A. R. Blaustein. 1994. The effects of kinship on growth and development in tadpoles of *Rana cascadae. Evolution* 48:1383–88.
———. 1995. Predator avoidance and alarm-response behaviour in kin-discriminating tadpoles (*Rana cascadae*). *Ethology* 101:280–90.
———. 1997. The effects of kinship on interactions between tadpoles of *Rana cascadae. Ecology* 78:1722–35.
Hokit, D. G., S. C. Walls, and A. R. Blaustein. 1996. Context-dependent kin discrimination in larvae of the marbled salamander *Ambystoma opacum. Anim. Behav.* 52:17–31.
Holenweg, A.-K., and H.-U. Reyer. 2000. Hibernation behavior of *Rana lessonae* and *R. esculenta* in their natural habitat. *Oecologia* 123:41–47.
Holenweg Peter, A.-K. 2001. Suvival in adults of the water frog *Rana lessonae* and its hybridogenetic associate *Rana esculenta. Can. J. Zool.* 79:652–61.
Hölldobler, B., and E. O. Wilson. 1990. *The ants.* Cambridge, MA: Harvard University Press.
Hollenbeck, R. R. 1974. Growth rates and movements within a population of *Rana pretiosa pretiosa* Baird and Girard in south central Montana. *Diss. Abstr. B* 35:1985.
Holman, J. A. 1995. *Pleistocene amphibians and reptiles in North America.* New York: Oxford University Press.
———. 1998a. Amphibian recolonization of midwestern states in the postglacial Pleistocene. In *Status and conservation of Midwestern amphibians,* ed. M. J. Lannoo, 9–15. Iowa City: University of Iowa Press.
———. 1998b. *Pleistocene amphibians and reptiles in Britain and Europe.* New York: Oxford University Press.
Holmes, R. 2000. Paleozoic temnospondyls. In *Amphibian biology. Vol. 4. Palaeontology,* ed. H. Heatwole and R. L. Carroll, 1081–1120. Chipping Norton, New South Wales, Australia: Surrey Beatty and Sons.
Holloszy, J. O., and F. W. Booth. 1976. Biochemical adaptations to endurance exercise in muscle. *Ann. Rev. Physiol.* 38:273–91.
Holomuzki, J. 1998. Grazing effects by green frog tadpoles (*Rana clamitans*) in a woodland pond. *J. Freshw. Ecol.* 13:1–8.
———. 1980. Synchronous foraging and dietary overlap of three species of plethodontid salamanders. *Herpetologica* 36:109–15.
———. 1982. Homing behavior of *Desmognathus ochrophaeus* along a stream. *J. Herpetol.* 16:307–9.
———. 1986a. Intraspecific predation and habitat use by tiger salamanders (*Ambystoma tigrinum nebulosum*). *J. Herpetol.* 20:439–41.
———. 1986b. Predator avoidance and diel patterns of microhabitat use by larval tiger salamanders. *Ecology* 67:737–48.
———. 1989a. Predation risk and macroalga use by the stream-dwelling salamander *Ambystoma texanum. Copeia* 1989:22–28.
———. 1989b. Salamander predation and vertical distributions of zooplankton. *Freshw. Biol.* 21:461–72.
———. 1995. Oviposition sites and fish-deterrent mechanisms of two stream anurans. *Copeia* 1995:607–13.
Holomuzki, J. R., and J. P. Collins. 1983. Diel movement of larvae of the tiger salamander, *Ambystoma tigrinum nebulosum. J. Herpetol.* 17:276–78.
———. 1987. Trophic dynamics of a top predator, *Ambystoma tigrinum nebulosum* (Caudata: Ambystomatidae) in a lentic community. *Copeia* 1987:949–57.
Holomuzki, J. R., J. P. Collins, and P. E. Brunkow. 1994. Trophic control of fishless ponds by tiger salamander larvae. *Oikos* 71:55–64.
Holt, R. D. 1993. Ecology at the mesoscale: The influence of regional processes on local communities. In *Species diversity in ecological communities. Historical and geographical perspectives,* ed. R. E. Ricklefs and D. Schluter, 77–88. Chicago: University of Chicago Press.
Holyoake, A., B. Waldman, and N. J. Gemmell. 2001. Determining the species status of one of the world's rarest frogs: a conservation dilemma. *Anim. Cons.* 4:29–35.
Holzapfel, R. A. 1937. The cyclic character of hibernation in frogs. *Q. Rev. Biol.* 12:65–84.
Holzman, N., and J. J. McManus. 1973. Effects of acclimation on metabolic rate and thermal tolerance in the carpenter frog, *Rana virgatipes. Comp. Biochem. Physiol.* 45A:833–42.
Hom, C. L. 1987. Reproductive ecology of female dusky salamanders, *Desmognathus fuscus* (Plethodontidae), in the southern Appalachians. *Copeia* 1987:768–77.
———. 1988. Optimal reproductive allocation in female dusky salamanders: A quantitative test. *Am. Nat.* 131:71–90.
Hom, C. L., R. G. Jaeger, and N. H. Willits. 1997. Courtship behaviour in male red-backed salamanders: The ESS dating game. *Anim. Behav.* 54:715–24.
Hom, C. L., N. H. Willits, and C. W. Clark. 1990. Fitness consequences of nest defense in plethodontid salamanders: Predictions of a dynamic optimization model. *Herpetologica* 46:304–19.
Homan, W. P. 1968. Osmotic regulation in *Rana pipiens. Copeia* 1968:876–78.
Hong, S. K. 1957. Effects of pituitrin and cold on water exchanges of frogs. *Am. J. Physiol.* 188:439–42.
Honnegger, R. E. 1981. *Threatened amphibians and reptiles in Europe.* Munich, Germany: Akademie Verlagsgesellschaft.
———. 1984. The frog crisis—Observations on the trade in Ranidae, their introduction and influence on foreign environments and some remarks on possible consequences. *ASRA J.* 2:50–62.
Hoogmoed, M. S. 1969. Notes on the herpetofauna of Surinam III.—A new species of *Dendrobates* (Amphibia, Salientia, Dendrobatidae) from Surinam. *Zool. Med. (Leiden)* 44:133–41.
———. 1972. Frösche, die ihre Kaulquappen huckepack tragen. *Dendrobates azureus* und andere Baumsteigerfrösche. *Aquar. Mag.* 1972:288–93.

———. 1986. Ergänzende Beobachtungen an *Lithodytes lineatus* (Schneider, 1799) (Salientia: Leptodactylidae). *Salamandra* 22:215–17.

———. 1987. New *Osornophryne* (Amphibia: Anura: Bufonidae) from the Atlantic versant of the Andes in Ecuador. *Zool. Med. (Leiden)* 61:209–42.

———. 1990a. Biosystematics of South American Bufonidae, with special reference to the Bufo "typhonius" group. In *Vertebrates in the tropics,* ed.G. Peters and R. Hutterer, 113–23. Bonn, Germany: Museum Alexander Koening.

———. 1990b. Resurrection of *Hyla wavrini* Parker (Amphibia: Anura: Hylidae), a gladiator frog from northern South America. *Zool. Med. (Leiden)* 64:71–93.

Hoogmoed, M. S., and J. E. Cadle. 1991. Natural history and distribution of *Agalychnis craspedopus* (Funkhouser, 1957) (Amphibia: Anura: Hylidae). *Zool. Med. (Leiden)* 65:129–42.

Hopey, M. E., and J. W. Petranka. 1994. Restriction of wood frogs to fish-free habitats: How important is adult choice? *Copeia* 1994:1023–25.

Hopkins, C. D. 1983. Sensory mechanisms in animal communication. In *Animal behaviour, vol. 2. Communication,* ed. T. R. Halliday and P. J. B. Slater, 114–55. San Francisco: W. H. Freeman.

Hopkins, S., and A. Channing. 2003. Chytrid fungus in Northern and Western Cape frog populations, South Africa. *Herp. Rev.* 34:334–36.

Hopkins, W. A., J. Congdon, and J. K. Ray. 2000. Incidences and impact of axial malformations in larval bullfrogs (*Rana catesbeiana*) developing in sites polluted by a coal-burning power plant. *Environ. Toxicol. Chem.* 19:862–68.

Hoppe, D. M. 1978. Thermal tolerance in tadpoles of the chorus frog, *Pseudacris triseriata. Herpetologica* 34:318–21.

Hoppe, D. M. 2005. Malformed frogs in Minnesota: history and interspecific differences. In *Amphibian declines: The conservation status of United States species,* ed. M. J. Lannoo, 103–8. Berkeley: University of California Press.

Hoppe, D. M., and D. Pettus. 1984. Developmental features influencing color polymorphism in chorus frogs. *J. Herpetol.* 18:113–20.

Hoppeler, H., O. Mathieu, R. Krauer, H. Claassen, R. B. Armstrong, and E. R. Weibel. 1981. Design of the *Mammalia*n respiratory system. VI. Distribution of mitochondria and capillaries in various muscles. *Respir. Physiol.* 44:87–111.

Hora, S. L. 1922. Animal life in torrential streams. *J. Bombay Nat. Hist. Soc.* 32:111–26

———. 1930. Ecology, bionomics and evolution of the torrential fauna, with special reference to the organs of attachment. *Phil. Trans. Roy. Soc. London B* 218:171–88.

———. 1932. Development and probable evolution of the suctorial disc in the tadpoles of *Rana afghana* Günther. *Trans. Roy. Soc. Edinburgh* 57, part II:469–72.

Horat, P., and R. D. Semlitsch. 1994. Effects of predation risk and hunger on the behaviour of two species of tadpoles. *Behav. Ecol. Sociobiol.* 34:393–401.

Horn, A. G., M. L. Leonard, and D. M. Weary. 1995. Oxygen consumption during crowing by roosters: Talk is cheap. *Anim. Behav.* 50:1171–75.

Horn, H. S. 1983. Some theories about dispersal. In *The ecology of animal movement,* ed. I. R. Swingland and P. J. Greenwood, 54–62. New York: Oxford University Press.

Horn, H. S., and D. S. Rubenstein. 1984. Behavioural adaptations and life history. In *Behavioural ecology: An evolutionary approach* (2nd. ed.), ed. J. R. Krebs and N. B. Davies, 279–98. Sunderland, MA: Sinauer Associates,.

Horn, S., J. L. Hanula, M. D. Ulyshen, and J. C. Kilgo. 2005. Abundance of green tree frogs and insects in artificial canopy gaps in a bottomland hardwood forest. *Am. Midl. Nat.* 153:321–26.

Horne, E. A. 1988. Aggressive behavior of female red-backed salamanders. *Herpetologica* 44:203–9.

Horne, E. A., and R. G. Jaeger. 1988. Territorial pheromones of female red-backed salamanders. *Ethology* 78:143–52.

Horner, H. A., and H. C. MacGregor. 1983. C-value and cell volume: Their significance in the evolution and development of amphibians. *J. Cell Sci.* 63:135–46.

Horton, P. 1982. Diversity and systematic significance of anuran tongue musculature. *Copeia* 1982:595–602.

Hoskin, C. J. 2004. Australian microhylid frogs (*Cophixalus* and *Austrochaperina*): phylogeny, taxonomy, calls, distribution and breeding biology. *Aust. J. Zool.* 52:237–69.

Hossack, B. R., P. S. Corn, and D. S. Pilliod. 2005. Lack of significant changes in the herpetofauna of Theodore Roosevelt National Park, North Dakota, since the 1920s. *Am. Midl. Nat.* 154:423–32.

Hossack, B. R., S. A. Diamond, and P. S. Corn. 2006. Distribution of boreal toad populations in relation to estimated UV-B dose in Glacier National Park, Montana, USA. *Can. J. Zool.* 84:98–107.

Hostache, G., and J. H. Mol. 1998. Reproductive biology of the Neotropical armoured catfish *Hoplosternum littorale* (Siluriformes—Callichthyidae): A synthesis stressing the role of the floating bubble nest. *Aquat. Liv. Res.* 11:173–85.

Hotz, H. 1970. Zur Laichplatzökologie von *Bufo bufo spinosus* Daudin (Amphibia, Salientia) im tyrrhenischen Ligurien. *Vierteljahresschr. Naturforsch. Ges. Zürich* 115:239–54.

Hotz, H., R. D. Semlitsch, E. Gutmann, G.-D. Guex, and P. Beerli. 1999. Spontaneous heterosis in larval life-history traits of hemiclonal frog hybrids. *Proc. Natl. Acad. Sci. USA* 96:2171–76.

Hou, P.-C., and S.-P. Huang. 1999. Metabolic and ventilatory responses to hypoxia in two altitudinal populations of the toad, *Bufo bankorensis. Comp. Biochem. Physiol.* A 124:4413–21.

Houck, L. D. 1977a. Reproductive biology of a Neotropical salamander, *Bolitoglossa rostrata. Copeia* 1977:70–83.

———. 1977b. Life history patterns and reproductive biology of Neotropical salamanders. In *The reproductive biology of amphibians,* ed. D. H. Taylor and S. I. Guttman, 43–72. New York: Plenum.

———. 1986. The evolution of salamander courtship pheromones. In *Chemical signals in vertebrates 4. Ecology, evolution, and comparative biology,* ed. D. Duvall, D. Müller-Schwarze, and R. M. Silverstein, 173–90. New York: Plenum.

———. 1988. The effect of body size on male courtship success in a plethodontid salamander. *Anim. Behav.* 36:837–42.

Houck, L. D., and S. J. Arnold. 2003. Courtship and mating behavior. In *Reproductive biology and phylogeny of urodela,* ed. D. M. Sever, 383–424. Enfield, NH: Science Publishers.

Houck, L. D., S. J. Arnold, and A. R. Hickman. 1988. Tests for sexual isolation in plethodontid salamanders (genus *Desmognathus*). *J. Herpetol.* 22:186–91.

Houck, L. D., A. M. Bell, N. L. Reagan-Wallin, and R. C. Feldhoff. 1998. Effects of experimental delivery of male courtship pheromones on the timing of courtship in a terrestrial salamander, *Plethodon jordani* (Caudata: Plethodontidae). *Copeia* 1998:214–19.

Houck, L. D., and H. Francillon-Viellot. 1988. Tests for age and size effects on male mating success in a plethodontid salamander. *Amphibia-Reptilia* 9:135–44.
Houck, L. D., and N. L. Reagan. 1990. Male courtship pheromones increase female receptivity in a plethodontid salamander. *Anim. Behav.* 39:729–34.
Houck, L. D., and K. Schwenk. 1984. The potential for long-term sperm competition in a plethodontid salamander. *Herpetologica* 40:410–15.
Houck, L. D., and D. M. Sever. 1994. Role of the skin in reproduction and behavior. In. *Amphibian biology. Vol. 1. The integument,* ed. H. Heatwole and G. T. Barthalmus, 351–81.
Houck, L. D., S. G. Tilley, and S. J. Arnold. 1985. Sperm competition in a plethodontid salamander: Preliminary results. *J. Herpetol.* 19:420–23.
Houck, L. D., and P. A. Verrell. 1993. Studies of courtship behavior in plethodontid salamanders: a review. *Herpetologica* 49:175–84.
Houck, L. D., and S. K. Woodley. 1995. Field studies of steroid hormones and male reproductive behaviour in amphibians, pp. 677–703. In *Amphibian biology, Vol. 2. Social behaviour,* ed. H. Heatwole and B. K. Sullivan, 677–703. Chipping Norton, New South Wales, Australia: Surrey Beatty and Sons.
Houck, M. A., and E. D. Bellis. 1972. Comparative tolerance to desiccation in the salamanders *Desmognathus f. fuscus* and *Desmognathus o. ochrophaeus. J. Herpetol.* 6:209–15.
Houlahan, J. E., and C. S. Findlay. 2003. The effects of adjacent land use on wetland amphibian species richness and community composition. *Can. J. Fish. Aquat. Sci.* 60:1078–94.
Hourdry, J., and A. Beaumont. 1985. *Les métamorphoses des amphibiens.* Paris: Masson.
Hourdry, J., A. l'Hermite, and R. Ferrand. 1996. Changes in the digestive tract and feeding behavior of anuran amphibians during metamorphosis. *Physiol. Zool.* 69:219–51.
Houston, W. W. K. 1973. The food of the common frog, *Rana temporaria,* in high moorland in northern England. *J. Zool., London* 171:153–65.
Howard, A. D. 1951. A river mussel parasitic on a salamander. *Nat. Hist. Misc.* 77:1–6.
Howard, A. K., J. D. Forester, J. M. Ruder, and R. Powell. 1997. Diets of two syntopic frogs: *Eleutherodactylus abbotti* and *E. armstrongi* (Leptodactylidae) from the Sierra de Baoruco, Hispaniola. *Herp. Nat. Hist.* 5:66–72.
Howard, J. H., R. L. Wallace, and J. R. Stauffer, Jr. 1983. Critical thermal maxima in populations of *Ambystoma macrodactylum* from different elevations. *J. Herpetol.* 17:400–2.
Howard, R. D. 1978a. The evolution of mating strategies in bullfrogs, *Rana catesbeiana. Evolution* 32:850–71.
———. 1978b. The influence of male-defended oviposition sites on early embryo mortality in bullfrogs. *Ecology* 59:789–98.
———. 1979. Estimating reproductive success in natural populations. *Am. Nat.* 114:221–31.
———. 1980. Mating behaviour and mating success in woodfrogs, *Rana sylvatica. Anim. Behav.* 28:705–16.
———. 1981a. Male age-size distribution and male mating success in bullfrogs. In *Natural selection and social behavior,* ed. R. D. Alexander and D. W. Tinkle, 61–77. New York: Chiron.
———. 1981b. Sexual dimorphism in bullfrogs. *Ecology* 62:303–10.
———. 1983. Sexual selection and variation in reproductive success in a long-lived organism. *Am. Nat.* 122:301–25.
———. 1984. Alternative mating behaviors of young male bullfrogs. *Amer. Zool.* 24:397–406.
———. 1988a. Reproductive success in two species of anurans. In *Reproductive success,* ed. T. H. Clutton-Brock, 93–113. Chicago: University of Chicago Press.
———. 1988b. Sexual selection on male body size and mating behaviour in American toads, *Bufo americanus. Anim. Behav.* 36:1796–1808.
Howard, R. D., and A. G. Kluge. 1985. Proximate mechanisms of sexual selection in woodfrogs. *Evolution* 39:260–77.
Howard, R. D., R. S. Moorman, and H. H. Whiteman. 1997. Differential effects of mate competition and mate choice on eastern tiger salamanders. *Anim. Behav.* 53:1345–56.
Howard, R. D., and J. G. Palmer. 1995. Female choice in *Bufo americanus:* Effects of dominant frequency and call order. *Copeia* 1996:212–17.
Howard, R. D., H. H. Whiteman, and T. I. Schueller. 1994. Sexual selection in American toads: a test of a good-genes hypothesis. *Evolution* 48:1286–1300.
Howard, R. D., and J. R. Young. 1998. Variation in male vocal traits and female mating preferences in *Bufo americanus. Anim. Behav.* 55:1165–79.
Howard, R. R. 1971. Avoidance learning of spotted salamanders, *Ambystoma maculatum,* by domestic chickens. *Amer. Zool.* 11:637.
Howard, R. R., and E. D. Brodie, Jr. 1971. Experimental study of mimicry in salamanders involving *Notophthalmus viridescens* and *Pseudotriton ruber schencki. Nature* 233:277.
———. 1973. A Batesian mimetic complex in salamanders: responses of avian predators. *Herpetologica* 29:33–41.
Howell, B. J. 1970. Acid-base balance in transition from water-breathing to air-breathing. *Fed. Proc.* 29:1130–34.
Hrbâçek, J. 1950. On the fright reaction of tadpoles of the common toad caused by chemical substances. *Experientia* 6:100–102.
Hsu, M. J., and G. Agoramoorthy. 1997. Wildlife conservation in Taiwan. *Cons. Biol.* 11: 834–36.
Hsu, M. Y., Y. C. Kam, and G. M. Fellers. 2005. Effectiveness of amphibian monitoring techniques in a Taiwanese subtropical forest. *Herpetol. J.* 15: 73–79.
Huang, C., and A. Sih. 1991. An experimental study on the effects of salamander larvae on isopods in stream pools. *Freshw. Biol.* 25:451–59.
Huang, C.-H., M.-F. Liang, and Y.-C. Kam. 2003. The fatty acid composition of oophagous tadpoles (*Chirixalus eiffingeri*) fed conspecific or chicken egg yolk. *Comp. Biochem. Physiol.* A 135:329–36.
Huang, H., and D. D. Brown. 2000. Prolactin is not a juvenile hormone in *Xenopus* laevis metamorphosis. *Proc. Natl. Acad. Sci. USA* 97:195–99.
Hubl, L., R. Mohneke, and H. Schneider. 1977. Temperature dependence of auditory thresholds in two central European anurans, *Bombina variegata variegata* (L.) and *Rana ridibunda ridibunda* Pall. (Amphibia), and its relation to calling. *Behav. Proc.* 2:305–14.
Hubl, L., and H. Schneider. 1979. Temperature and auditory thresholds: Bioacoustic studies of the frogs *Rana r. ridibunda, Hyla a. arborea* and *Hyla a. savignyi* (Anura, Amphibia). *J. Comp. Physiol.* 130:17–27.
Hucko, J. A., and P. V. Cupp, Jr. 2001. Avoidance of chemical alarm cues released from autotomized tails of ravine salamanders (*Plethodon richmondi*). *Ethology* 107:315–22.

Hudson, N. J., and C. E. Franklin. 2002a. Effect of aestivation on muscle characteristics and locomotor performance in the green-striped burrowing frog *Cyclorana alboguttata*. *J. Comp. Physiol.* B 172:177–82.

———. 2002b. Maintaining muscle mass during extended disuse: Aestivating frogs as a model species. *J. Exp. Biol.* 205:2297–2303.

Hudson, W. H. 1903. *The naturalist in La Plata*. London: J. M. Dent.

———. 1919. *The book of a naturalist*. New York: George H. Doran.

Huey, R. B. 1980. Sprint velocity of tadpoles (*Bufo boreas*) through metamorphosis. *Copeia* 1980:537–40.

———. 1982. Temperature, physiology, and the ecology of reptiles. In *Biology of the Reptilia*, vol. 12, ed. C. Gans and F. H. Pough, 25–91. New York: Academic Press.

———. 1987. Phylogeny, history, and the comparative method. In *New directions in ecological physiology*, ed. M. E. Feder, W. W. Burggren, A. F. Bennett, and R. B. Huey, 76–101. New York: Cambridge University Press.

Huey, R. B., and M. Slatkin. 1976. Costs and benefits of lizard thermoregulation. *Q. Rev. Biol.* 51:363–84.

Huey, R. B. and R. D. Stevenson. 1979. Integrating thermal physiology and ecology of ectotherms: a discussion of approaches. *Amer. Zool.* 19:357–366.

Hughes, G. M. 1970. Ultrastructure of the air-breathing organs of some lower vertebrates. *7th Cong. Int. Micro. Electron.* 3:599–600.

Huheey, J. E. 1960. Mimicry in the color pattern of certain Appalachian salamanders. *J. Elisha Mitchell Sci. Soc.* 76:246–51.

———. 1966. Studies in warning coloration and mimicry. V. Red-cheeked dusky salamanders in North Carolina. *J. Elisha Mitchell Sci. Soc.* 82:126–31.

———. 1976. Studies in warning coloration and mimicry. VII. Evolutionary consequences of a Batesian-Müllerian spectrum: a model for Müllerian mimicry. *Evolution* 30:86–93.

———. 1980. Batesian and Müllerian mimicry: Semantic and substantive differences of opinion. *Evolution* 34:1212–15.

———. 1988. Mathematical models of mimicry. *Am. Nat.* 131:S22–S41.

Huheey, J. E., and R. A. Brandon. 1973. Rock-face populations of the mountain salamander, *Desmognathus ochrophaeus*, in North Carolina. *Ecol. Monogr.* 43:59–77.

———. 1974. Studies in warning coloration and mimicry. VI. Comments on the warning coloration of red efts and their presumed mimicry by red salamanders. *Herpetologica* 30:149–55.

Hulse, A. C. 1974. Food habits and feeding behavior in *Kinosternon sonoriense* (Chelonia: Kinosternidae). *J. Herpetol.* 8:195–99.

Hultsch, H., and D. Todt. 1982. Temporal performance roles during vocal interactions in nightingales (*Luscinia megarhynchos* B.). *Behav. Ecol. Sociobiol.* 11:253–60.

Humphrey, R. R. 1977. Factors influencing ovulation in the Mexican axolotl as revealed by induced spawning. *J. Exp. Zool.* 199:209–14.

———. 1979. *Dynamics of a breeding frog community*. Unpublished PhD dissertation. Canberra: The Australian National University,.

Humphries, R. L. 1956. An unusual aggregation of *Plethodon glutinosus* and remarks on its sub-specfic status. *Copeia* 1956:122–23.

Hunn, E. S. 1977. *Tzeltal folk zoology. The classification of discontinuities in nature*. New York: Academic Press.

Hunt, R. H. 1980. Toad sanctuary in a tarantula burrow. *Natural History* 89 (3):48–53.

Hunter, M. L., Jr., and J. R. Krebs. 1979. Geographical variation in the song of the Great Tit (*Parus major*) in relation to ecological factors. *J. Anim. Ecol.* 48:759–85.

Huntly, J. 1985. Samango monkeys, tree frogs and Lengwe National Park. *Afr. Wildl.* 39:43.

Hurlbert, S. H. 1969. The breeding migrations and interhabitat wandering of the vermillion-spotted newt, *Notophthalmus viridescens* (Rafinesque). *Ecol. Monogr.* 39:465–88.

———. 1970. Predator responses to the vermillion-spotted newt (*Notophthalmus viridescens*). *J. Herpetol.* 4:47–55.

Hussain, S. A., and B. C. Choudhury. 1998. Feeding ecology of the smooth-coated otter, *Lutra perspicillata* in the National Chambal Sanctuary, India. In *Behaviour and ecology of riparian mammals*, ed. N. Dunstone and M. Gorman, 229–49. Cambridge: Cambridge University Press.

Husting, E. L. 1965. Survival and breeding structure in a population of *Ambystoma maculatum*. *Copeia* 1965:352–62.

Hutcherson, J. E., C. L. Peterson, and R. F. Wilkinson. 1989. Reproductive and larval biology of *Ambystoma annulatum*. *J. Herpetol.* 23:181–83.

Hutchinson, G. E., G. E. Pickford, and J. F. M. Schnurman. 1932. A contribution to the hydrobiology of pans and other inland waters of South Africa. *Arch. Hydrobiol.* 24:1–154.

Hutchinson, J. M. C., J. M. McNamara, and I. C. Cuthill. 1993. Song, sexual selection, starvation and strategic handicaps. *Anim. Behav.* 45:1153–77.

Hutchison, V. H. 1958. The distribution and ecology of the cave salamander, *Eurycea lucifuga*. *Ecol. Monogr.* 28:1–20.

———. 1959. Aggressive behavior in *Plethodon jordani*. *Copeia* 1958:72–73.

———. 1961. Critical thermal maxima in salamanders. *Physiol. Zool.* 34:92–125.

Hutchison, V. H., and M. J. Dady. 1964. The viability of *Rana pipiens* and *Bufo terrestris* submerged at different temperatures. *Herpetologica* 20:149–62.

Hutchison, V. H., and R. K. Dupré. 1992. Thermoregulation. In *Environmental physiology of the amphibians*, ed. M. E. Feder and W. W. Burggren, 206–49. Chicago: University of Chicago Press.

Hutchison, V. H., H. B. Haines, and G. Engbretson. 1976. Aquatic life at high altitude: respiratory adaptations in the Lake Titicaca frog, *Telmatobius culeus*. *Respir. Physiol.* 27:115–29.

Hutchison, V. H., and C. S. Hammen. 1958. Oxygen utilization in the symbiosis of embryos of the salamander, *Ambystoma maculatum* and the alga, *Oophila amblystomatis*. *Biol. Bull.* 115:483–89.

Hutchison, V. H., and L. G. Hill. 1976. Thermal selection in the hellbender, *Cryptobranchus alleganiensis*, and the mudpuppy, *Necturus maculosus*. *Herpetologica* 32:327–31.

———. 1978. Thermal selection of bullfrog tadpoles (*Rana catesbeiana*) at different stages of development and acclimation temperatures. *J. Therm. Biol.* 3:57–60.

Hutchison, V. H., and M. A. Kohl. 1971. The effect of photoperiod on daily rhythms of oxygen consumption in the tropical toad, *Bufo marinus*. *Z. Vergl. Physiol.* 75:367–82.

Hutchison, V. H., and J. D. Maness. 1979. The role of behavior in temperature acclimation and tolerance in ectotherms. *Amer. Zool.* 19:367–84.

Hutchison, V. H., and K. Miller. 1979a. Aerobic and anaerobic contributions to sustained activity in *Xenopus laevis*. *Respir. Physiol.* 38:93–103.

———. 1979b. Anaerobic capacity of amphibians. *Comp. Biochem. Physiol.* 63A:213–16.

Hutchison, V. H., and K. K. Spriestersbach. 1986. Diel and seasonal cycles of activity and behavioral thermoregulation in the salamander *Necturus maculosus*. *Copeia* 1986:612–18.

Hutchison, V. H., L. D. Turney, and R. K. Gratz. 1977. Aerobic and anaerobic metabolism during activity in the salamander *Ambystoma tigrinum*. *Physiol. Zool.* 50:189–202.

Hutchison, V. H., and W. G. Whitford. 1966. Survival and underwater buccal movements in submerged anurans. *Herpetologica* 22:122–27.

Hutchison, V. H., W. G. Whitford, and M. A. Kohl. 1968. Relation of body size and surface area to gas exchange in anurans. *Physiol. Zool.* 41:65–85.

Huxley, Julian. 1927. *Essays in popular science*. New York: Alfred A. Knopf.

Hyatt, A. D., A. R. Gould, Z. Zupanovic, A. A. Cunningham, S. Hengstberger, R. J. Whittington, J. Kattenbelt, and B. E. H. Coupar. 2000. Comparative studies of piscine and amphibian iridoviruses. *Arch. Virol.* 145:301–31.

Ibáñez, C., and J. I. Ibáñez. 1980. Alimentación del tejón (*Meles meles* L.) en el Rasillo de Cameros (Logroño, España). *I Reunión Iberoamer. Zool. Vert., La Rabida* 1977:517–27.

Ibáñez, R. 1991. *Synchronized calling in Centrolenella granulosa and Smilisca sila (Amphibia, Anura)*. Unpublished PhD dissertation. Storrs: University of Connecticut.

———. 1993. Female phonotaxis and call overlap in the Neotropical glassfrog *Centrolenella granulosa*. *Copeia* 1993:846–50.

Ibáñez, R., and E. M. Smith. 1995. Systematic status of *Colostethus flotator* and *C. nubicola* (Anura: Dendrobatidae) in Panama. *Copeia* 1995:446–56.

Ideker, J. 1976. Tadpole thermoregulatory behavior facilitates grackle predation. *Texas J. Sci.* 27:244–45.

Ifft, J. D. 1942. The effect of environmental factors on the sperm cycle of *Triturus viridescens*. *Biol. Bull.* 83:111–28.

Iga, T., and J. T. Bagnara. 1975. An analysis of color change phenomena in the leaf frog, *Agalychnis dacnicolor*. *J. Exp. Zool.* 192:331–42.

Ihara, S. 1999. Site selection for hibernation by the tree frog, *Rhacophorus schlegelii*. *Jap. J. Herpetol.* 18:39–44.

Ikeda, S. 1897. Notes on the breeding habits and development of *Rhacophorus schlegelii* Gunther. *Ann. Zool. Jap.* 1:113–22.

Inger, R. F. 1947. Preliminary survey of the amphibians of the Riukiu Islands. *Fieldiana: Zool.* 32:297–352.

———. 1954. Systematics and zoogeography of Philippine Amphibia. *Fieldiana: Zool.* 33:181–531.

———. 1956. Morphology and development of the vocal sac apparatus in the African frog *Rana* (*Ptychadena*) *porosissima* Steindachner. *J. Morphol.* 99:57–72.

———. 1960. Notes on toads of the genus *Pelophryne*. *Fieldiana: Zool.* 39:415–18.

———. 1966. The systematics and zoogeography of the Amphibia of Borneo. *Fieldiana: Zool.* 52:1–402.

———. 1969. Organization of communities of frogs along small rain forest streams in Sarawak. *J. Anim. Ecol.* 38:123–48.

———. 1980. Densities of floor-dwelling frogs and lizards in lowland forests of southeast Asia and Central America. *Am. Nat.* 115:761–70.

———. 1985. Tadpoles of the forested regions of Borneo. *Fieldiana: Zool.*, n.s. 26:1–89.

———. 1986. Diets of tadpoles living in a Bornean rain forest. *Alytes* 5:153–64.

———. 1992. Variation of apomorphic characters in stream-dwelling tadpoles of the bufonid genus *Ansonia* (Amphibia: Anura). *Zool. J. Linn. Soc.* 105:225–37.

———. 1996. Commentary on a proposed classification of the family Ranidae. *Herpetologica* 52:241–46.

———. 1999. Distribution of amphibians in Southern Asia and adjacent islands. In *Patterns of distribution of amphibians. A global perspective*, ed. W. E. Duellman, 445–82. Baltimore: Johns Hopkins University Press.

———. 2001. *Amphibian population declines: Asia* (abstract). *2001* Indianapolis, IN: Joint Annual Meetings, Herpetologists' League and the Society for the Study of Amphibians and Reptiles.

Inger, R. F., and J. P. Bacon. 1968. Annual reproduction and clutch size in rain forest frogs from Sarawak. *Copeia* 1968:602–6.

Inger, R. F., and R. K. Colwell. 1977. Organization of contiguous communities of amphibians and reptiles in Thailand. *Ecol. Monogr.* 47:229–53.

Inger, R. F., and B. Greenberg. 1963. The annual reproductive pattern of the frog *Rana erythraea* in Sarawak. *Physiol. Zool.* 36:21–33.

———. 1966. Ecological and competitive relations among three species of frogs (genus *Rana*). *Ecology* 47:746–59.

Inger, R. F., and M. Kottelat. 1998. A new species of ranid frog from Laos. *Raffles Bull. Zool.* 46:29–34.

Inger, R. F., and H. Marx. 1961. The food of amphibians. *Exploration du Parc. National du l'Upemba. Mission G. F. de Witte*, Fasc. 64:1–86.

Inger, R. F., N. Orlov, and I. Darevsky. 1999. Frogs of Vietnam: A report on new collections. *Fieldiana: Zool.* (n.s.) 92:1–46.

Inger, R. F., H. B. Shaffer, M. Koshy, and R. Bakde. 1987. Ecological structure of a herpetological assemblage in South India. *Amphibia-Reptilia* 8:189–202.

Inger, R. F., and R. B. Stuebing. 1989. *Frogs of Sabah*. Sabah Parks Publ. no. 10, Sabah, Malaysia.

———. 1997. *A field guide to the frogs of Borneo*. Kota Kinabalu, Sabah, Malaysia: Natural History Publications (Borneo).

Inger, R. F., and H. K. Voris. 1988. Taxonomic status and reproductive biology of Bornean tadpole-carrying frogs. *Copeia* 1988:1060–62.

———. 1993. A comparison of amphibian communities through time and from place to place in Bornean forests. *J. Trop. Ecol.* 9:409–33.

Inger, R. F., H. K. Voris, and K. J. Frogner. 1986. Organization of a community of tadpoles in rain forest streams in Borneo. *J. Trop. Ecol.* 2:193–205.

Inger, R. F., H. K. Voris, and H. H. Voris. 1974. Genetic variation and population ecology of some southeast Asian frogs of the genera *Bufo* and *Rana*. *Biochem. Gen.* 12:121–45.

Inger, R. F., H. K. Voris, and P. Walker. 1986. Larval transport in a Bornean ranid frog. *Copeia* 1986:523–25.

Ingram, G. J., M. Anstis, and C. J. Corben. 1975. Observations on the Australian leptodactylid frog *Assa darlingtoni*. *Herpetologica* 31:425–29.

Ingram, W. M., and E. C. Raney. 1943. Additional studies of the movement of tagged bullfros, *Rana catesbeiana* Shaw. *Am. Midl. Nat.* 29:239–41.

Inoue, T. 1979. On the territorial behaviors of a Japanese pond frog, *Rana brevipoda. Jap. J. Ecol.* 29:149–61 (in Japanese).

Ireland, M. P. 1973. Studies on the adaptation of *Xenopus laevis* to hyperosmotic media *Comp. Biochem. Physiol.* 46A:469–76.

Ireland, P. H. 1974. Reproduction and larval development of the dark-sided salamander, *Eurycea longicauda melanopleura* (Green). *Herpetologica* 30:338–43.

———. 1976. Reproduction and larval development of the gray-bellied salamander *Eurycea multiplicata griseogaster. Herpetologica* 32:233–38.

Irschick, D. J., and H. B. Shaffer. 1997. The polytypic species revisited: morphological differentiation among tiger salamanders (*Ambystoma tigrinum*) (Amphibia: Caudata). *Herpetologica* 53:30–49.

Irwin, J. T. 2005. Overwintering in northern cricket frogs (*Acris crepitans*). In *Amphibian declines: The conservation status of United States species,* ed. M. J. Lannoo, 55–58. Berkeley: University of California Press.

Irwin, J. T., J. P. Costanzo, and R. E. Lee, Jr. 1999. Terrestrial hibernation in the northern cricket frog, *Acris crepitans. Can. J. Zool.* 77:1240–46.

———. 2003. Postfreeze reduction of locomotor endurance in the freeze-tolerant wood frog, *Rana sylvatica. Physiol. Biochem. Zool.* 76:331–38.

Irwin, J. T., and R. E. Lee, Jr. 2003. Geographic variation in energy storage and physiological responses to freezing in the gray treefrogs *Hyla versicolor* and *H. chrysoscelis. J. Exp. Biol.* 206:2859–67.

Isaac, G. L., and D. C. Crader. 1981. To what extent were early hominids carnivorous? An archaeological perspective. In *Omnivorous primates: Gathering and hunting in human evolution,* ed. R. S. O. Harding and G. Teleki, 37–103. New York: Columbia University Press.

Isaacs, J. S. 1971. Temporal stability of vertebral strip color in a cricket frog population. *Copeia* 1971:551–52.

Isacch, J. P., and M. Barg. 2002. Are bufonid toads specialized ant-feeders? A case test from the Argentinian flooding pampa. *J. Nat. Hist.* 36:2005–12.

Ishchenko, V. G. and E. L. Shchupak. 1974. Ecological differences of individual genotypes in a frog population. *Sov. J. Ecol.* 5:379–80.

Ishii, S., H. Yoneyama, M. Inoue, K. Yamamoto, and S. Kikuyama. 1989. Changes in plasma and pituitary levels of prolactin in the toad, *Bufo japonicus,* throughout the year with special reference to the breeding migration. *Gen. Comp. Endocrinol.* 74:365–72.

Iskandar, D. T. 1998. *The amphibians of Java and Bali.* Jakarta, Indonesia: Research and Development Center for Biology.

Issacs, J. S. 1971. Temporal stability of vertebral stripe color in cricket frog populations. *Copeia* 1971:551–52.

Istenic, L., and B. Bulog. 1984. Some evidence for the ampullary organs in the European cave salamander *Proteus anguineus* (Urodela: Amphibia). *Cell Tissue Res.* 235:393–402

Istock, C. A. 1967. The evolution of complex life cycle phenomena: An ecological perspective. *Evolution* 21:592–605.

Itämies, J., and P. Koskela. 1970. On the diet of the common frog (*Rana temporaria* L.). *Aquilo (Zool.)* 10:53–60.

Izaurralde, R. C., A. M. Thomson, N. J. Rosenberg, and R. A. Brown. 2005. Climate change impacts for the conterminous USA: an integrated assessment. Part 6. Distribution and productivity of unmanaged ecosystems. *Climatic Change* 69:107–26.

Iwasawa, H., and K. Saito. 1989. Adaptive characteristics related to a torrential habitat in *Bufo torrenticola* larvae: comparison with those of the still water-breeding species *Bufo japonicus formosus. Sci. Rep. Niigata Univ., ser. B (Biol.)* 26:13–25.

Izawa, K. 1978. Frog-eating behavior of wild black-capped capuchin (*Cebus apella*). *Primates* 19:633–42.

Izecksohn, E. 1971. Novo genero e nova especie de Brachycephalidae do Estado do Rio de Janeiro, Brasil (Amphibia, Anura). *Bol. Mus. Nac. Rio de Janeiro* (280):1–12.

Izecksohn, E., J. Jim, S. Tenorio de Albuquerque, and W. Furtado de Mendoca. 1971. Observações sobre o desenvolimento d os hábitos de *Myersiella subnigra* (Miranda-Ribeiro). *Arq. Mus. Nac. Rio de Janeiro* 54:69–72

Jackman, R., S. Nowicki, D. J. Aneshansley, and T. Eisner. 1983. Predatory capture of toads by fly larvae. *Science* 222:515–16.

Jackman, T. R., and D. B. Wake. 1994. Evolutionary and historical analysis of protein variation in the blotched forms of salamanders of the *Ensatina* complex. *Evolution* 48:876–97.

Jackson, D. C., and B. A. Braun. 1979. Respiratory control in bullfrogs: Cutaneous versus pulmonary response to selective $CO_2$ exposure. *J. Comp. Physiol.* B 129:339–42.

Jackson, D. R., and E. G. Milstrey. 1989. The fauna of gopher tortoise burrows. In *Gopher tortoise relocation symposium proceedings,* ed. J. E. Diemer, D. R. Jackson, J. L. Landers, J. N. Layne, and D. A. Wood, 86–98. Tallahassee, Florida: *Florida Game and Fresh Water Fish Commission Nongame Wildlife Program Technical Report no. 5.*

Jackson, M. E., D. E. Scott, and R. A. Estes. 1989. Determinants of nest success in the marbled salamander (*Ambystoma opacum*). *Can. J. Zool.* 67:2277–81.

Jackson, M. E., and R. D. Semlitsch. 1993. Paedomorphosis in the salamander *Ambystoma talpoideum:* Effects of a fish predator. *Ecology* 74:342–50.

Jacob, C., C. Miaud, A. J. Crivelli, and M. Veith. 2003. How to cope with periods of drought? Age at maturity, longevity, and growth of marbled newts (*Triturus marmoratus*) in Mediterranean temporary ponds. *Can. J. Zool.* 81:1905–11.

Jacobs, A. J., and D. H. Taylor. 1992. Chemical communication between *Desmognathus quadramaculatus* and *Desmognathus monticola. J. Herpetol.* 26:93–95.

Jacobs, J. F. 1987. A preliminary investigation of geographical genetic variation and systematics of the two-lined salamander, *Eurycea bislineata* (Green). *Herpetologica* 43:423–46.

Jacobson, C. M. 1963. Observations on distribution, behaviour and development in the Australian toad genus *Pseudophryne* Fitzinger. *Proc. Linn. Soc. New South Wales* 88:41–46.

Jacobson, S. K. 1985. Reproductive behavior and male mating success in two species of glass frogs (Centrolenidae). *Herpetologica* 41:396–404.

Jacobson, S. K., and J. J. Vandenberg. 1991. Reproductive ecology of the endangered golden toad (*Bufo periglenes*). *J. Herpetol.* 25:321–27.

Jacques, M. 1999. Bengali frog legs export and uses of pesticides. *TED Case Studies* no. 508 (an online journal available at: http://www.american.edu/projects/mandala/TED/bangfrog.htm).

Jaeger, J.R., B.R. Riddle, R.D. Jennings and D.F. Bradford. 2001. Rediscovering *Rana onca:* Evidence for phylogenetically distinct leopard frogs from the border region of Nevada, Utah, and Arizona. *Copeia* 2000:339–54.

Jaeger, R. G. 1970. Potential extinction through competition between two species of terrestrial salamanders. *Evolution* 24:632–42.

———. 1971a. Competitive exclusion as a factor influencing the distribution of two species of terrestrial salamanders. *Ecology* 52:632–37.

———. 1971b. Moisture as a factor influencing the distributions of two species of terrestrial salamanders. *Oecologia* 6:191–207.

———. 1971c. Toxic reaction to skin secretions of the frog, *Phrynomerus bifasciatus*. *Copeia* 1971:160–61.

———. 1972. Food as a limited resource in competition between two species of terrestrial salamanders. *Ecology* 53:535–46.

———. 1974. Interference or exploitation? A second look at competition between salamanders. *J. Herpetol.* 8:191–94.

———. 1976. A possible prey-call window in anuran auditory processing. *Copeia* 1976:833–34.

———. 1977. Ecological niche dimensions and sensory functions in amphibians. In *Sensory ecology. Review and perspectives*, ed.M. A. Ali, 169–96. New York: Plenum.

———. 1978. Plant climbing by salamanders: Periodic availability of plant-dwelling prey. *Copeia* 1978:686–91.

———. 1979. Seasonal spatial distributions of the terrestrial salamander *Plethodon cinereus*. *Herpetologica* 35:90–93.

———. 1980a. Density-dependent and density-independent causes of extinction of a salamander population. *Evolution* 34:617–21.

———. 1980b. Fluctuations in prey availability and food limitation for a terrestrial salamander. *Oecologia* 44:335–41.

———. 1980c. Microhabitats of a terrestrial forest salamander. *Copeia* 1980:265–68.

———. 1981a. Birds as inefficient predators on terrestrial salamanders. *Am. Nat.* 117:835–37.

———. 1981b. Dear enemy recognition and the costs of aggression between salamanders. *Am. Nat.* 117:962–74.

———. 1981c. Foraging in optimum light as a niche dimension for Neotropical frogs. *Natl. Geogr. Res. Rep.* 13:297–302.

———. 1984. Agonistic behavior of the red-backed salamander. *Copeia* 1984:309–14.

———. 1986. Pheromonal markers as territorial advertisement by terrestrial salamanders. In *Chemical signals in vertebrates 4*, ed. D. Duvall, D. Müller-Schwarze, and R. M. Silverstein, 191–203. New York: Plenum.

———. 1988. A comparison of territorial and non-territorial behaviour in two species of salamanders. *Anim. Behav.* 36:307–10.

Jaeger, R. G., and D. E. Barnard. 1981. Foraging tactics of a terrestrial salamander: Choice of diet in structurally simple environments. *Am. Nat.* 117:639–64.

Jaeger, R. G., D. E. Barnard, and R. G. Joseph. 1982. Foraging tactics of a terrestrial salamander: Assessing prey density. *Am. Nat.* 119:885–90.

Jaeger, R. G., and D. C. Forester. 1993. Social behavior of plethodontid salamanders. *Herpetologica* 49:163–75.

Jaeger, R. G., D. Fortune, G. Hill, A. Palen, and G. Risher. 1993. Salamander homing behavior and territorial pheromones: Alternative hypotheses. *J. Herpetol.* 27:236–39.

Jaeger, R. G., and C. R. Gabor. 1993. Intraspecific chemical communication by a territorial salamander via the postcloacal gland. *Copeia* 1993:1171–74.

Jaeger, R. G., C. R. Gabor, and H. M. Wilbur. 1998. An assemblage of salamanders in the southern Appalachian Mountains: competitive and predatory behavior. *Behaviour* 135:795–821.

Jaeger, R. G., and W. F. Gergits. 1979. Intra- and interspecific communication in salamanders through chemical signals on the substrate. *Anim. Behav.* 27:150–56.

Jaeger, R. G., J. M. Goy, M. Tarver, and C. E. Marquez. 1986. Salamander territoriality: Pheromonal markers as advertisement by males. *Anim. Behav.* 34:860–64.

Jaeger, R. G., and J. P. Hailman. 1973. Effects of intensity on the phototactic responses of adult anuran amphibians: a comparative survey. *Z. Tierpsychol.* 33:352–407.

———. 1981. Activity of Neotropical frogs in relation to ambient light. *Biotropica* 13:59–65.

Jaeger, R. G., J. P. Hailman, and L. S. Jaeger. 1976. Bimodal diel activity of a Panamanian dendrobatid frog, *Colostethus nubicola*, in relation to light. *Herpetologica* 32:77–81.

Jaeger, R. G., R. G. Joseph, and D. E. Barnard. 1981. Foraging tactics of a terrestrial salamander: Sustained yield in territories. *Anim. Behav.* 29:1100–1105.

Jaeger, R. G., D. Kalvarsky, and N. Shimizu. 1982. Territorial behaviour of the red-backed salamander: Expulsion of intruders. *Anim. Behav.* 30:490–96.

Jaeger, R. G., K. C. B. Nishikawa, and D. E. Barnard. 1983. Foraging tactics of a terrestrial salamander: costs of territorial defence. *Anim. Behav.* 3l:191–98.

Jaeger, R. G., and M. E. Peterson. 2002. Familiarity affects agonistic interactions between female red-backed salamanders. *Copeia* 2002:865–69.

Jaeger, R. G., M. E. Peterson, and J. R. Gillette. 2000. A model of alternative mating strategies in the redback salamander, *Plethodon cinereus*. In *The biology of plethodontid salamanders*, ed. R. C. Bruce, R. G. Jaeger, and L. D. Houck, 441–50. New York: Kluwer Academic/Plenum.

Jaeger, R. G., M. E. Peterson, G. Gollman, B. Gollman, and V. R. Townsend, Jr. 2001. Salamander social strategies: Living together in female-male pairs. *J. Herpetol.* 35:335–38.

Jaeger, R. G., E. D. Prosen, and D. C. Adams. 2002. Character displacement and aggression in two species of terrestrial salamanders. *Copeia* 2002:391–401.

Jaeger, R. G., and A. M. Rubin. 1982. Foraging tactics of a terrestrial salamander: Judging prey profitability. *J. Anim. Ecol.* 51:167–76.

Jaeger, R. G., and J. K. Schwarz. 1991. Gradational threat postures by the red-backed salamander. *J. Herpetol.* 25:112–14.

Jaeger, R. G., J. Schwarz, and S. E. Wise. 1995. Territorial male salamanders have foraging tactics attractive to gravid females. *Anim. Behav.* 49:633–39.

Jaeger, R. G., and S. C. Walls. 1989. On salamander guilds and ecological methodology. *Herpetologica* 45:111–19.

Jaeger, R. G., J. A. Wicknick, M. R. Griffis, and C. D. Anthony. 1995. Socioecology of a terrestrial salamander: Juveniles enter adult territories during stressful foraging periods. *Ecology* 76:533–43.

Jaeger, R. G., and S. E. Wise. 1991. A reexamination of the male salamander "sexy faeces" hypothesis. *J. Herpetol.* 25:370–73.

Jakob, C., A. Seitz, A. J. Crivelli, and C. Miaud. 2002. Growth cycle of the marbled newt (*Triturus marmoratus*) in the Mediterranean region assessed by skeletochronology. *Amphibia-Reptilia* 23:407–18.

Jakob, C., G. Poizat, M. Veith, A. Seitz, and A. J. Crivelli. 2003. Breeding phenology and larval distribution of amphibians in a Mediterranean pond network with unpredictable hydrology. *Hydrobiologia* 499:51–61.

James, R. S., R. S. Wilson, J. E. de Carvalho, T. Kohlsdorf, F. R. Gomes, and C. A. Navas. 2005. Interindividual differences in leg muscle mass and pyruvate kinase activity correlate with interindividual differences in jumping performance of *Hyla multilineata. Physiol. Biochem. Zool.* 78:857–67.

Jameson, D. L. 1950. Development of *Eleutherodactylus latrans. Copeia* 1950:44–46.

———. 1954. Social patterns in the leptodactylid frogs *Syrrhophus* and *Eleutherodactylus. Copeia* 1954:36–38.

———. 1955a. Evolutionary trends in the courtship and mating behavior of Salientia. *Syst. Zool.* 4:105–19.

———. 1955b. The population dynamics of the cliff frog, *Syrrhophus marnocki. Am. Midl. Nat.* 54:342–81.

———. 1956a. Growth, dispersal and survival of the Pacific tree frog. *Copeia* 1956:25–29.

———. 1956b. Survival of some central Texas frogs under natural conditions. *Copeia* 1956:55–57.

———. 1957a. Life history and phylogeny in the Salientia. *Syst. Zool.* 6:75–80.

———. 1957b. Population structure and homing responses in the Pacific treefrog. *Copeia* 1957:221–28.

Jameson, D. L., and S. Pequegnat. 1971. Estimation of relative viability and fecundity of color polymorphisms in anurans. *Evolution* 25:180–94.

Jamieson, B. G. M. 2003. *Reproductive biology and phylogeny of Anura.* Enfield, NH: Science Publishers.

Jancovich, J, K., E. W. Davidson, J. F. Morado, B. L. Jacobs, and J. P. Collins. 1997. Isolation of a lethal virus from the endangered tiger salamander *Ambystoma tigrinum stebbensi. Dis. Aquat. Org.* 31:161–67.

Jancovich, J. K., E. W. Davidson, N. Parameswaran, M. V. G. Chinchar, J. P. Collins, B. L. Jacobs, and A. Storfer. 2005. Evidence for emergence of an amphibian iridoviral disease because of human-enhanced spread. *Mol. Ecol.* 14:213–24.

Janetos, A. C. 1980. Strategies of female mate choice: a theoretical analysis. *Behav. Ecol. Sociobiol.* 7:107–12.

Janetsky, W., and E. Vareschi. 1993. Phytotelmata in bromeliads as microhabitats for limnetic organisms. In *Animal-plant interactions in tropical environments,* ed. W. Barthlott, C. M. Naumann, K. Schmid-Loske, and K.-L. Schuchmann, e199–209. Bonn, Germany: Zoologisches Forschungsinstitut,

Jansen, A., and M. Healey. 2003. Frog communities and wetland condition: Relationships with grazing by domestic livestock along an Australian floodplain river. *Biol. Cons.* 109:207–19.

Janssens, P. A., and P. P. Cohen. 1968. Biosynthesis of urea in the estivating African lungfish and in *Xenopus laevis* under conditions of water shortage. *Comp. Biochem. Physiol.* 24:887–98.

Janzen, F. J., and E. D. Brodie, III. 1989. Tall tails and sexy males: Sexual behavior of rough-skinned newts (*Taricha granulosa*) in a natural breeding pond. *Copeia* 1989:1068–71.

Jara, F., and M. G. Perotti. 2004. *Pleurodema bufoninum* (NCN) and *Bufo spinulosus papillosus* (NCN). Predation. *Herp. Rev.* 35:161.

Jaramillo, F. E., C. A. Jaramillo, and R. Ibáñez D. 1997. Renacuajo de la rana de cristal *Hyalinobatrachium colymbiphyllum* (Anura: Centrolenidae). *Rev. Biol. Trop.* 45:867–70.

Jaramillo, C., A. S. Rand, R. Ibáñez, and R. Dudley. 1997. Elastic structures in the vocalization apparatus of the túngara frog *Physalaemus pustulosus* (Leptodactylidae). *J. Morphol.* 233:287–95.

Jared, C., M. M. Antoniazzi, E. Katchburian, R. C. Toledo, and E. Freymüller. 1999. Some aspects of the natural history of the casque-headed tree frog *Corythomantis greeningi* Boulenger (Hylidae). *Ann. Sci. Nat.* 3:105–15.

Jared, C., C. A. Navas, and R. C. Toledo. 1999. An appreciation of the physiology and morphology of the caecilians (Amphibia: Gymnophiona). *Comp. Biochem. Physiol.* 123A:313–28.

Jaroensutasinee, M. and K. Jaroensutasinee. 2001. Bubble nest habitat characteristics of wild Siamese fighting fish. *J. Fish Biol.* 58:1311–19.

Jarvik, E. 1996. The Devonian tetrapod *Ichthyostega. Fossils and Strata* 40:1–213.

Jasienski, M. 1988. Kinship ecology of competition: Size hierarchies in kin and nonkin laboratory cohorts of tadpoles. *Oecologia* 77:407–13.

Jaslow, A. P. 1979. Vocalization and aggression in *Atelopus chiriquiensis* (Amphibia, Anura, Bufonidae). *J. Herpetol.* 13:141–45.

Jaslow, A. P., T. E. Hetherington, and R. E. Lombard. 1988. Structure and function of the amphibian middle ear. In *The evolution of the amphibian auditory system,* ed. B. Fritzsch, M. J. Ryan, W. Wilczynski, T. E. Hetherington, and W. Walkowiak, 69–91. New York: Wiley.

Jaslow, A. P., and R. E. Lombard. 1996. Hearing in the Neotropical frog, *Atelopus chiriquiensis. Copeia* 1996:428–32.

Jastrzebski, M. 1968. Morphological changes in the reproductive organs of the female marsh frog (*Rana arvalis* Nilss.) in the yearly cycle. *Acta Biol. Cracov. (Zool.),* 11:9–19.

Jędrzejewska, B., and M. Brzezinski. 1993. Winter habitat selection and feeding habits of polecats (*Mustela putorius*) in the Białowieża National Park, Poland. *Z. Säugetierkunde* 58:75–83.

Jędrzejewska, B., M. Brzeziński, and W. Jęrzejewski. 2003. Seasonal dynamics and breeding of amphibians in pristine forests (Białowieża National Park, E. Poland) in dry years. *Folia Zool.* 52:77–86.

Jędrzejewska, B., V. E. Sidorovich, M. M. Pikulik, and W. Jędrzejewski. 2001. Feeding habits of otter and the American mink in Białowieża Primeval Forest (Poland) compared to other Eurasian populations. *Ecography* 24:165–80.

Jędrzejewski, W., B. Jędrzejewska, and A. Szymura. 1989. Food niche overlaps in a winter community of predators in the Białowieża Primeval Forest, Poland. *Acta Theriol.* 34:487–96.

Jeffrey, B. M., J. H. Waddle, and A. J. Maskell. 2004. *Hyla cinerea* (Green Treefrog). Predation. *Herp. Rev.* 35:158

Jehle, R., and A. Arak. 1998. Graded call variation in the Asian cricket frog *Rana nicobariensis. Bioacoustics* 9:35–48.

Jehle, R., and J. W. Arntzen. 2000. Post-breeding migrations of newts (*Triturus cristatus* and *T. marmoratus*) with contrasting ecological requirements. *J. Zool., London* 251:297–306.

Jehle, R., W. Hödl, and A. Thonke. 1995. Structure and dynamics of central European amphibian populations: a comparison between *Triturus dobrogicus* (Amphibia, Urodela) and *Pelobates fuscus* (Amphibia, Anura). *Aust. J. Ecol.* 20:362–66.

Jehle, R., G. A. Wilson, W. Arntzen, and T. Burke. 2005. Contemporary gene flow and the spatio-temporal genetic structure of subdivided newt populations (*Triturus cristatus, T. marmoratus*). *J. Evol. Biol.* 18:619–28.

Jenkins, D., and R. J. Harper. 1980. Ecology of otters in northern Scotland. II. Analyses of otter (*Lutra lutra*) and mink (*Mustela vison*) feces from Deeside, N. E. Scotland in 1977–78. *J. Anim. Ecol.* 49:737–54.

Jenkins, F. A., and D. M. Walsh. 1993. An early Jurassic caecilian with limbs. *Nature* 365:246–50.

Jenkins, J. L., and D. L. Swanson. 2005. Liver glycogen, glucose mobilization and freezing survival in chorus frogs, *Pseudacris triseriata. J. Therm. Biol.* 30:485–94.

Jenni, D. A. 1969. A study of the ecology of four species of herons during the breeding season at Lake Alice, Alachoa County, Florida. *Ecol. Monogr.* 39:245–70

———. 1973. Regional variation in the food of nestling cattle egrets. *Auk* 90:821–26.

Jennings, D. H., and J. Hanken. 1998. Mechanistic basis of life history evolution in anuran amphibians: Thyroid gland development in the direct-developing frog, *Eleutherodactylus coqui. Gen. Comp. Endocrinol.* 111:225–32.

Jennings, M. R., and D. Cook. 1998. *Taricha torosa* (California newt). Predation. *Herp. Rev.* 29:230.

Jennings, M. R., and M. P. Hayes. 1985. Pre-1900 overharvest of California red-legged frogs (*Rana aurora draytonii*): The inducement for bullfrog (*Rana catesbeiana*) introduction. *Herpetologica* 41:94–103.

———. 1994. Decline of native ranid frogs in the desert Southwest. In *Herpetology of the North American deserts. Spec. Publ. 5.*, ed. P. R. Brown and J. W. Wright, 183–212. Van Nuys, CA: Southwestern Herpetologists Society.

Jennings, R. D. 2005. *Rana fisheri* Stejneger, 1893. Vegas Valley leopard frog. In *Amphibian declines: The conservation status of United States species,* ed. M. J. Lannoo, 554–55. Berkeley: University of California Press.

Jennings, R. D., and N. J. Scott, Jr. 1993. Ecologically correlated morphological variation in tadpoles of the leopard frog, *Rana chiricahuensis. J. Herpetol.* 27:285–93.

Jennings, W. B., D. F. Bradford, and D. F. Johnson. 1992. Dependence of the garter snake *Thamnophis elegans* on amphibians in the Sierra Nevada of California. *J. Herpetol.* 26:503–5.

Jennions, M. D. 1992. *Breeding behaviour of the African frog, Chiromantis xerampelina. Sperm competition and polyandry.* Unpublished M. S. thesis. Johannesburg, South Africa: University of the Witwatersrand.

Jennions, M. D., and P. R. Y. Backwell. 1992. Chorus size influences on the anti-predator response of a Neotropical frog. *Anim. Behav.* 44:990–92.

Jennions, M. D., P. R. Y. Backwell, and N. I. Passmore. 1992. Breeding behaviour of the African frog, *Chiromantis xerampelina:* Multiple spawning and polyandry. *Anim. Behav.* 44:1091–1100.

Jennions, M. D., P. R. Y. Backwell, and N. I. Passmore. 1995. Repeatability of mate choice: the effect of size in the African painted reed frog, *Hyperolius marmoratus. Anim. Behav.* 49:181–86.

Jennions, M. D., P. J. Bishop, P. R. Y. Backwell, and N. I. Passmore. 1995. Call rate variability and female choice in the African frog, *Hyperolius marmoratus. Behaviour* 132:709–20.

Jennions, M. D., and N. I. Passmore. 1993. Sperm competition in frogs: Testis size and a 'sterile male' experiment on *Chiromantis xerampelina* (Rhacophoridae). *Biol. J. Linn. Soc.* 50:211–20.

Jennions, M. D., and M. Petrie. 1997. Variation in mate choice and mating preferences: A review of causes and consequences. *Biol. Rev.* 72:283–327.

Jennrich, R. I., and F. B. Turner. 1969. Measurement of non-circular home range. *J. Theor. Biol.* 22:227–37.

Jensen, J. B. 2003. *Ambystoma tigrinum* (tiger salamander). Predation. *Herp. Rev.* 34:132–33.

Jensen, J. B. and C. D. Camp. 2003. Human exploitation of amphibians. Direct and indirect impacts. In *Amphibian conservation,* ed. R. D. Semlitsch, 199–213. Washington, DC: Smithsonian Books.

Jensen, J. B., and S. C. Richter. 2005. *Rana capito* LeConte, 1855. Gopher frog. In *Amphibian declines: The conservation status of United States species,* ed. M. J. Lannoo, 536–38. Berkeley: University of California Press.

Jenssen, T. A. 1968. Some morphological and behavioral characteristics of an intergrade population of the green frog, *Rana clamitans,* in southern Illinois. *Trans. Illinois State Acad. Sci.* 61:252–59.

Jenssen, T. A. 1972. Seasonal organ weights of the green frog, *Rana clamitans* (Anura, Ranidae), under natural conditions. *Trans. Illinois State Acad. Sci.* 65:15–24.

Jenssen, T. A., and W. D. Klimstra. 1966. Food habits of the green frog *Rana clamitans* in southern Illinois. *Am. Midl. Nat.* 76:169–82.

Jepson, P., J. K. Jarvie, K. MacKinnon, and K. A. Monk. 2001. The end for Indonesia's lowland forests? *Science* 292:859–61.

Jeschke, J. M., and D. L. Strayer. 2005. Invasion success of vertebrates in Europe and North America. *Proc. Natl. Acad. Sci. USA* 102:7198–7202.

Jiang, S., and D. L. Claussen. 1992. A bioenergetic budget for overwintering newts (*Notophthalmus viridescens*) from southern Ohio: Their fat reserves and aerobic metabolic rates in water. *Comp. Biochem. Physiol.* 101A:743–50.

Jiang, S., and D. L. Claussen. 1993. The effects of temperature on food passage time through the digestive tract in *Notophthalmus viridescens. J. Herpetol.* 27:414–19.

Jockusch, E. L. 1997. An evolutionary correlate of genome size change in plethodontid salamanders. *Proc. Roy. Soc. London B* 264:597–604.

Jockusch, E. L., and M. J. Mahoney. 1997. Communal oviposition and lack of parental care in *Batrachoseps nigriventris* (Caudata: Plethodontidae) with a discussion of the evolution of breeding behavior in plethodontid salamanders. *Copeia* 1997:697–705.

Joglar, R. L. 1989. Phylogenetic relationships of the West Indian frogs of the genus *Eleutherodactylus:* a morphological analysis. In *Biogeography of the West Indies. Past, present and future,* ed. C. A. Woods, 371–408. Gainesville, FL: Sandhill Crane Press.

———. 1998. *Los Coquies de Puerto Rico. Su Historia Natural y Conservacion.* San Juan, Puerto Rico: Editorial de la Universidad de Puerto Rico.

Joglar, R. L.,. and P. A. Burrowes. 1996. Declining amphibian populations in Puerto Rico. In *Contributions to West Indian herpetology,* ed. R. Powell and R. W. Henderson, 371–80. Ithaca, NY: Society for the Study of Amphibians and Reptiles.

Joglar, R. L., P. A. Burrowes, D. Dávila, A. Rodríguez, A. López, A. Longo, and P. Medina. 2005. *Eleutherodactylus wightmanae* (Coquí Melodioso). Reproduction. *Herp. Rev.* 36:433–34.

Joglar, R. L., P. A. Burrowes, and N. Rios. 1996. Biology of the Puerto Rican cave-dwelling frog, *Eleutherodactylus cooki,* with some recommendations for conservation. In *Contributions to West Indian herpetology,* ed. R. Powell and R. W. Henderson, 251–58. Ithaca, New York: Society for the Study of Amphibians and Reptiles.

Johansen, K. 1963. Cardiovascular dynamics in the amphibian *Amphiuma tridactylum. Acta Physiol. Scand.* 60 (Suppl.):1–82.

———. 1985. A phylogenetic overview of cardiovascular shunts. In *Cardiovascular shunts,* ed. K. Johansen and W. W. Burggren, 17–32. Copenhagen, Denmark: Munksgaard.

Johansen, K., A. S. Abe, and R. E. Weber. 1980. Respiratory properties of whole blood and hemoglobin from the burrowing reptile, *Amphisbaenia alba. J. Exp. Zool.* 214:71–77.

Johansen, K., and A. S. F. Ditadi. 1966. Double circulation in the giant toad, *Bufo paracnemis. Physiol. Zool.* 39:140–50.

Johansen, K., and C. Lenfant. 1972. A comparative approach to the adaptability of $O_2$-HB affinity. In *Oxygen affinity of hemoglobin and red cell acid-base status,* ed. M. Rorth and P. Astrup, 750–83. New York: Academic Press.

Johansen, K., C. Lenfant, and D. Hanson. 1970. Phylogenetic development of pulmonary circulation. *Fed. Proc.* 29:1135–40.

Johansen, K., G. Lykkeboe, S. Kornerup, and G. M. O. Maloiy. 1980. Temperature insensitive $O_2$ binding in blood of the tree frog, *Chiromantis petersi. J. Comp. Physiol.* 136:71–76.

Johansen, K., and R. E. Weber. 1976. On the adaptability of hemoglobin function to environmental conditions. In *Perspectives in experimental biology, zoology and notany,* vol.1, ed. P. Spencer Davis, 219–34. New York: Pergamon.

Johansson, M., C. R. Primmer, J. Sahlsten, and J. Merilä. 2005. The influence of landscape structure on occurrence, abundance and genetic diversity of the common frog, *Rana temporaria. Global Change Biol.* 11:1664–79.

John, K. R., and D. Fenster. 1975. The effects of partitions on the growth rate of crowded *Rana pipiens* tadpoles. *Am. Midl. Nat.* 93:123–30.

John-Alder, H. B., M. C. Barnhart, and A. F. Bennett. 1989. Thermal sensitivity of swimming performance and muscle contraction in northern and southern populations of tree frogs (*Hyla crucifer*). *J. Exp. Biol.* 142:357–72.

John-Alder, H. B., and P. J. Morin. 1990. Effects of larval density on jumping ability and stamina in newly metamorphosed *Bufo woodhousii fowleri. Copeia* 1990:856–60.

John-Alder, H. B., P. J. Morin, and S. J. Lawler. 1988. Thermal physiology, phenology, and distribution of treefrogs. *Am. Nat.* 132:506–20.

Johnson, C. R. 1969a. Aggregation as a means of water conservation in juvenile *Limnodynastes* from Australia. *Herpetologica* 25:275–76.

———. 1969b. Water absorption response of some Australian anurans. *Herpetologica* 25:171–172.

———. 1970. Observations on body temperatures, critical thermal maxima and tolerance to water loss in the Australian hylid, *Hyla caerulea* (White). *Proc. Roy. Soc. Queensland* 82:47–50.

Johnson, C. R., and C. B. Schreck. 1969. Food and feeding of larval *Dicamptodon ensatus* from California. *Am. Midl. Nat.* 81:280–81.

Johnson, J. A., and E. D. Brodie, Jr. 1975. The selective advantage of the defensive posture of the newt *Taricha granulosa. Am. Midl. Nat.* 93:139–48.

Johnson, J. B., D. Saenz, C. K. Adams, and R. N. Conner. 2003. The influence of predator threat on the timing of a life-history switch point: Predator-induced hatching in the southern leopard frog (*Rana sphenocephala*). *Can. J. Zool.* 81:1608–13.

Johnson, M. L., and R. Speare. 2003. Survival of *Batrachochytrium dendrobatidis* in water: Quarantine and disease control implications. *Emerg. Infec. Dis.* 9:922–25.

Johnson, P. T. J., and J. M. Chase. 2004. Parasites in the food web: Linking amphibian malformations and aquatic eutrophication. *Ecol. Lett.* 7:521–26.

Johnson, P. T. J., and K. B. Lunde. 2005. Parasite infection and limb malformations: A growing problem in *Amphibian Conservation.* In *Amphibian declines: The conservation status of United States species,* ed. M. J. Lannoo, 124–38. Berkeley: University of California Press.

Johnson, P. T. J., K. B. Lunde, E. C. Ritchie, and A. E. Launer. 1999. The effect of trematode infection on amphibian limb development and survivorship. *Science* 284:802–4.

Johnson, P. T. J., K. B. Lunde, R. W. Haight, J. Bowerman, and A. R. Blaustein. 2001. *Ribeiroia ondatrae* (Trematoda: Digenea) infection induces severe limb malformations in wetern toads (*Bufo boreas*). *Can. J. Zool.* 79:370–79.

Johnson, P. T. J., K. B. Lunde, E. G. Ritchie, J. K. Reaser, and A. E. Launder. 2001. Morphological abnormality patterns in a California amphibian community. *Herpetologica* 57:336–52.

Johnson, P. T. J., K. B. Lunde, E. M. Thurman, E. G. Ritchie, S. N. Wray, D. R. Sutherland, J. M. Kapfer, T. J. Frest, J. Bowerman, and A. R. Blaustein. 2002. Parasite (*Ribeiroia ondatrae*) infection linked to amphibian malformations in the western United States. *Ecol. Monogr.* 72:151–68.

Johnson, P. T. J., K. B. Lunde, D. A. Zelmer, and J. K. Werner. 2003. Limb deformities as an emerging parasitic disease in amphibians: evidence from museum specimens and resurvey data. *Cons. Biol.* 17:1724–37.

Johnson, P. T. J., and D. R. Sutherland. 2003. Amphibian deformities and *Ribeiroia* infection: An emerging helminthiasis. *Trends Parasit.* 19:332–35.

Johnson, P. T. J., D. R. Sutherland, J. M. Kinsella, and K. L. Lunde. 2004. Review of the trematode genus *Ribeiroia* (Psilostomidae): Ecology, life history, and pathogenesis with special emphasis on the amphibian malformation problem. *Adv. Parasitol.* 57:191–254.

Johnson, S. A. 1996. *Hyla femoralis* (Pine woods treefrog). Predation. *Herp. Rev.* 27:140.

———. 2003. Orientation and migration distances of a pond-breeding salamander (*Notophthalmus perstriatus,* Salamandridae). *Alytes* 21:3–22.

Johnson, W. E., and C. R. Propper. 2000. Effects of dehydration on plasma osmolality, thirst-related behavior, and plasma and brain angiotensin concentrations in Couch's spadefoot toad, *Scaphiopus couchii. J. Exp. Zool.* 286:572–84.

Johnston, B., and L. Frid. 2002. Clearcut logging restricts the movements of terrestrial Pacific giant salamanders (*Dicamptodon tenbrosus* Good). *Can. J. Zool.* 80:2170–77.

Johnston, G. R., and S. J. Richards. 1993. Observations on the breeding biology of a microhylid frog (genus *Oreophryne*) from New Guinea. *Trans. Royal Soc. South Australia* 117:105–7

Johnston, I. A. 1987. Respiratory characteristics of muscle fibres in a fish (*Chaenocephalus aceratus*) that lacks haem pigments. *J. Exp. Biol.* 133:415–28.

Johnston, R. F., and G. A. Schad. 1959. Natural history of the salamander, *Aneides hardii. Univ. Kansas Publ. Mus. Nat. Hist.* 10:573–85.

Jokumsen, A., and R. E. Weber. 1980. Hemoglobin oxygen binding properties in the blood of *Xenopus laevis,* with special reference to the influences of aestivation and of temperature and salinity acclimation. *J. Exp. Biol.* 86:19–39.

Joly, J. 1960. La conservation des spermatozoides et les particularites histophysiologiques du receptacle seminale chez la salamandre *Salamandra salamandra taeniata. C. R. Acad. Sci. Paris* 250:2269–71.

———. 1963. La sédentarité et le retour au gîte chez la Salamandre tachetée, *Salamandra salamandra quadri-virgata. C. R. Acad. Sci. Paris* 256:3510–12.

———. 1966. Sur l'éthologie sexuelle de *Salamandra salamandra* (L.). *Z. Tierpsychol.* 23:8–27.

———. 1968. Données écologiques sur la salamandre tachetee *Salamandra salamandra* (L.). *Ann. Sci. Nat., Zool. (Paris)*, ser. 12, 10:301–66.

———. 1971. Les cycles sexuels de *Salamandra* (L.). I. *Ann. Sci. Nat., Zool. (Paris)*, ser. 12, 13:451–504.

———. 1986. La reproduction de la salamandre terrestre (*Salamandra salamandra* L.). In *Traite de zoologie, vol. 14, fasc. IB. Amphibiens,* ed. P.-P. Grasse and M. Delsol, 471–86.Paris: Masson.

Joly, J., F. Chesnel, and D. Boujard. 1994. Biological adaptations and reproductive strategies in the genus *Salamandra. Mertensiella* 4:255–69.

Joly, P., and C. Giacoma. 1992. Limitation of similarity and feeding habits in three syntopic species of newts (*Triturus,* Amphibia). *Ecography* 15:401–11.

Joly, P., and C. Miaud. 1989. Fidelity to the breeding site in the alpine newt *Triturus alpestris. Behav. Proc.* 19:47–56.

———. 1993. How does a newt find its pond—The role of chemical cues in migrating newts (*Triturus alpestris*). *Ethol. Ecol. Evol.* 5:447–55.

Joly, P., C. Miaud, A. Lehmann, and O. Grolet. 2001. Habitat matrix effects on pond occupancy in newts. *Cons. Biol.* 15:239–48.

Joly, P., C, Morand, and A. Cohas. 2003. Habitat fragmentation and amphibian conservation: building a tool for assessing landscape matrix connectivity. *C. R. Biol.* 326:S132–S139.

Jones, A. G., E. M. Adams, and S. J. Arnold. 2002. Topping off: a mechanism of first-male sperm precedence in a vertebrate. *Proc. Natl. Acad. Sci. USA* 99:2078–81.

Jones, A. G., J. R. Arguello, and S. J. Arnold. 2004. Molecular parentage analysis in experimental newt populations: the response of mating system measures to variation in the operational sex ratio. *Am. Nat.* 164:444–56.

Jones, D. R. 1967. Oxygen consumption and heart rate of several species of anuran Amphibia during submergence. *Comp. Biochem. Physiol.* 20:691–707.

———. 1972. Anaerobiosis and the oxygen debt in an anuran amphibian, *Rana esculenta* (L.). *J. Comp. Physiol.* 77:356–82.

Jones, G. P., M. I. McCormick, M. Srinivasan, and J. V. Eagle. 2004. Coral decline threatens fish biodiversity in marine reserves. *Proc. Natl. Acad. Sci. USA* 101:8251–53.

Jones, J. M. 1973. Effects of thirty years hybridization on the toads *Bufo americanus* and *Bufo woodhousii fowleri* at Bloomington, Indiana. *Evolution* 27:435–48.

Jones, J. S., B. H. Leith, and P. Rawlings. 1977. Polymorphism in *Cepaea:* A problem with too many solutions? *Ann. Rev. Ecol. Syst.* 8:109–43.

Jones, K. L. 1982. Prey patterns and trophic niche overlap in four species of Caribbean frogs. In *Herpetological communities.* Wildlife Res. Report 13, ed. N. J. Scott, 49–55. Washington, DC: U.S. Fish and Wildlife Service.

Jones, L. L. C., R. B. Bury, and P. S. Corn. 1990. Field observation of the development of a clutch of Pacific giant salamander (*Dicamptodon tenebrosus*) eggs. *Northwest. Nat.* 71:93–94

Jones, M. S., J. P. Goetti, and L. J. Livo. 1999. *Bufo boreas* (boreal toad). Predation. *Herp. Rev.* 30:91.

Jones, P. D., T. J. Osborn, and K. R. Briffa. 2001. The evolution of climate over the last millenium. *Science* 292:662–66.

Jones, R. L. 1986. Reproductive biology of *Desmognathus fuscus* and *Desmognathus santeetlah* in the Unicoi Mountains. *Herpetologica* 42:323–34.

Jones, R. M. 1980. Metabolic consequences of accelerated urea synthesis during seasonal dormancy of spadefoot toads, *Scaphiopus couchi* and *Scaphiopus intermontanus. J. Exp. Zool.* 212:255–67.

———. 1982. How toads breathe: Control of air flow to and from the lungs by the nares in *Bufo marinus. Respir. Physiol.* 49:251–65.

Jones, R. M., and S. S. Hillman. 1978. Salinity adaptation in the salamander *Batrachoseps. J. Exp. Biol.* 76:1–10.

Jones, T. H., J. S. T. Gorman, R. R. Snelling, J. H. C. Delabie, M. S. Blum, H. M. Garraffo, P. Jain, J. W. Daly, and T. F. Spande. 1999. Further alkaloids common to ants and frogs: decahydroquinolines and quinolizidine. *J. Chem. Ecol.* 25:1179–93.

Jordan, E. O. 1891. The spermatophore of *Diemyctylus. J. Morphol.* 5:263–70.

Jordan, F., and D. A. Arrington. 2001. Weak trophic interactions between large predatory fishes and herpetofauna in the channelized Kissimmee River, Florida, USA. *Wetlands* 21:155–59.

Jordan, O. R., W. W. Byrd, and D. E. Ferguson. 1968. Sun-compass orientation in *Rana pipiens. Herpetologica* 24:335–36.

Jørgensen, C. B. 1950a. Amphibian water economy with special regard to the effect of neurohypophyseal extracts. *Acta Physiol. Scan.* 22, Suppl. 79:1–79.

———. 1950b. Osmotic regulation in the frog, *Rana esculenta* (L.), at low temperatures. *Acta Physiol. Scand.* 20:46–55.

———. 1973a. Mechanisms regulating ovarian function in amphibians (toads). In *The development and maturation of the ovary and Its functions,* ed. H. Peters, 133–51. Amsterdam: Exerpta Medica,.

———. 1973b. Pattern of recruitment of oocytes to second growth phase in normal toads and in hypophysectomized toads, *Bufo bufo bufo* (L.), treated with gonadotropin (HCG). *Gen. Comp. Endocrinol.* 21:152–59.

———. 1974a. Mechanisms regulating ovarian cycle in the toad *Bufo bufo bufo* (L.): Role of presence of second growth phase oocytes in controlling recruitment from pool of first growth phase oocytes. *Gen. Comp. Endocrinol.* 23:170–77.

———. 1974b. Relations between hibernation and ovarian function in a temperate zone frog, *Rana temporaria. Acta Zool. (Stockholm)* 65:239–47.

———. 1975. Factors controlling the annual ovarian cycle in the toad *Bufo bufo bufo* (L.). *Gen. Comp. Endocrinol.* 25:264–73.

———. 1981. Ovarian cycle in a temperate zone frog, *Rana temporaria,* with special reference to factors determining number of size of eggs. *J. Zool., London* 195:449–58.

———. 1982. Factors controlling the ovarian cycle in a temperate zone anuran, the toad *Bufo bufo:* Food uptake, nutritional state, and gonadotropin. *J. Exp. Zool.* 224:437–43.

———. 1984a. Dynamics of oogenesis in a lower vertebrate, the toad *Bufo bufo. Acta Zool. (Stockholm)* 65:179–85.

———. 1984b. Ovarian functional patterns in Baltic and Mediterranean populations of a temperate zone anuran, the toad *Bufo viridis. Oikos* 43:309–21

———. 1986. External and internal control of patterns of feeding, growth and gonadal function in a temperate zone anuran, the toad *Bufo bufo. J. Zool., London* (A) 210:211–41.

———. 1989. Pattern of growth and fattening in young toads, *Bufo bufo,* fed mealworms: effects of growth hormone and feeding regimen. *Copeia* 1989:124–28.

———. 1991. Water and salt balance at low temperature in a cold temperate zone anuran, the toad *Bufo bufo. Comp. Biochem. Physiol.* 100 A:377–84.

———. 1992a. Effects of arginine vasotocin, cortisol and adrenergic factors on water balance in the toad *Bufo bufo:* Physiology or pharmacology? *Comp. Biochem. Physiol.* 101 A:709–16.

———. 1992b. Growth and reproduction. In *Environmental physiology of the amphibians,* ed. M. E. Feder and W. W. Burggren, 439–66. Chicago:University of Chicago Press.

———. 1994a. Robert Townson's observations on amphibian water economy revisited. *Comp. Biochem. Physiol.* 109A:325–34.

———. 1994b. Water economy in a terrestrial toad (*Bufo bufo*), with special reference to cutaneous drinking and urinary bladder function. *Comp. Biochem. Physiol.* 109A:311–24.

———. 1997a. 200 years of amphibian water economy: From Robert Townson to the present. *Biol. Rev.* 72:153–237.

———. 1997b. Urea and amphibian water economy. *Comp. Biochem. Physiol.* 117A:161–70.

———. 2000. Amphibian respiration and olfaction and their relationships: From Robert Townson (1794) to the present. *Biol. Rev.* 75:297–345.

Jørgensen, C. B., K. Brems, and P. Geckler. 1978. Volume and osmotic regulation in the toad *Bufo bufo* (L.) at low temperature, with special reference to amphibian hibernation. In *Osmotic and volume regulation. Alfred Benzon Symposium XI,* ed. C. B. Jørgensen and E. Skadhauge, 62–79. Copenhagen: Munksgaard

Jørgensen, C. B., K.-E. Hede, and L. O. Larsen. 1978. Environmental control of annual ovarian cycle in the toad *Bufo bufo bufo* L.: role of temperature. In *Environmental endocrinology,* ed. I. Assenmacher and D. S. Farner, 28–36. New York: Springer-Verlag.

Jørgensen, C. B., and L. O. Larsen. 1960. Hormonal control of moulting in amphibians. *Nature* 185:244–45.

Jørgensen, C. B., and L. O. Larsen. 1961. Molting and its hormonal control in toads. *Gen. Comp. Endocrinol.* 4:389–400.

Jørgensen, C. B., L. O. Larsen, and B. Lofts. 1979. Annual cycles of fat bodies and gonads in the toad *Bufo bufo bufo* (L.) compared with cycles in other temperate zone anurans. *Biol. Skrift. K. Danske Videns. Selskab* 22:1–37.

Jørgensen, C. B., K. Shakuntala, and S. Vijayakumar. 1986. Body size, reproduction and growth in a tropical toad, *Bufo melanostictus,* with a comparison of ovarian cycles in tropical and temperate zone anurans. *Oikos* 46:379–89.

Jørgensen, C. B., and H. Wind-Larsen. 1987. Energetics of growth in a temperate zone toad, *Bufo bufo:* effects of growth hormone. *Acta Zool. (Stockholm)* 68:107–13.

Jørgensen, M. B. 1991. Comparative studies of the biophysics of directional hearing in anurans. *J. Comp. Physiol. A* 169:591–98.

Jørgensen, M. B., and J. Christensen-Dalsgaard. 1991. Peripheral origins and functional characteristics of vibration-sensitive VIIIth nerve fibers in the frog *Rana temporaria. J. Comp. Physiol. A* 169:341–47.

Jørgensen, M. B., and H. C. Gerhardt. 1991. Directional hearing in the gray tree frog *Hyla versicolor:* eardrum vibrations and phonotaxis. *J. Comp. Physiol. A* 169:177–83.

Jørgensen, M. B., and M. Kanneworff. 1998. Middle ear transmission in the grass frog, *Rana temporaria. J. Comp. Physiol. A* 182:59–64.

Jørgensen, M. B., B. Schmitz, and J. Christensen-Dalsgaard. 1991. Biophysics of directional hearing in the frog *Eleutherodactylus coqui. J. Comp. Physiol. A* 168:223–32.

Jorquera, B., O. Garrido, and E. Pugin. 1982. Comparative studies of the digestive tract development between *Rhinoderma darwinii* and *R. rufum. J. Herpetol.* 16:204–14.

Jorquera, B., E. Pugin, and O. Goicoechea. 1972. Table de desarrollo normal de *Rhinoderma darwinii. Arch. Med. Vet.* 4:1–15.

Joss, J. M. P. 1998. Are extant lungfish neotenic? *Clin. Exp. Pharm. Physiol.* 25:733–35.

Joss, J. M. P., P. S. Rajasekar, R. A. Raj-Prasad, and K. Ruitenberg. 1997. Developmental endocrinology of the dipnoan, *Neoceratodus forsteri. Amer. Zool.* 37:461–69.

Jowers, M. J., and J. R. Downie. 2005. Tadpole deposition behaviour in male stream frogs *Mannophryne trinitatis* (Anura: Dendrobatidae). *J. Nat. Hist.* 39:3013–27.

Judge, K. A., and R. J. Brooks. 2001. Chorus participation by male bullfrogs, *Rana catesbeiana:* a test of the energetic constraint hypothesis. *Anim. Behav.* 62:849–61.

Judge, K. A., S. J. Swanson, and R. J. Brooks. 2000. *Rana catesbeiana* (Bullfrog). Female vocalization. *Herp. Rev.* 31:236–37.

Junca, F. A. 1996. Parental care and egg mortality in *Colostethus stepheni. J. Herpetol.* 30:292–94.

———. 1998. Reproductive biology of *Colostethus stepheni* and *Colostethus marchesianus* (Dendrobatidae), with the description of a new anuran mating behavior. *Herpetologica* 54:377–87.

Junca, F. A., R. Altig, and C. Gascon. 1994. Breeding biology of *Colostethus stepheni,* a dendrobatid frog with a nontransported nidicolous tadpole. *Copeia* 1994:747–50.

Jung, R. E., W. L. Ward, C. O. King, and L. A. Weir. 2000. *Plethodon cinereus* (redback salamander). Predation. *Herp. Rev.* 31:98–99.

Jungfer, K.-H. 1985. Beitrag zur Kenntnis von *Dendrobates speciosus* O. Schmidt, 1857 (Salientia: Dendrobatidae). *Salamandra* 21:263–80.

———. 1989. Pfeilgiftfrösche der Gattung *Epipedobates* mit rot grauliertem Rucken aus dem Oriente von Ecuador and Peru. *Salamandra* 25:81–98.

———. 1996. Reproduction and parental care of the coronated treefrog, *Anotheca spinosa* (Steindachner, 1864) (Anura: Hylidae). *Herpetologica* 52:25–32.

Jungfer, K.-H., and W. Boehme. 1991. The backpack strategy of parental care in frogs, with notes on froglet-carrying in *Stefania evansi* (Boulenger, 1904) (Anura: Hylidae: Hemiphractinae). *Rev. Fr. Aquariol.* 18:91–96.

Jungfer, K.-H., S. Ron, R. Seipp, and A. Almendáriz. 2000. Two new species of hylid frogs, genus *Osteocephalus,* from Amazonian Ecuador. *Amphibia-Reptilia* 21:327–40.

Jungfer, K.-H., and L. C. Schiesari. 1995. Description of a central Amazonian and Guianan tree frog, genus *Osteocephalus* (Anura, Hylidae), with oophagous tadpoles. *Alytes* 13:1–13.

Jungfer, K.-H., and P. Weygoldt. 1994. The reproductive biology of the leaf frog *Phyllomedusa lemur* Boulenger, 1882, and a comparison with other members of the Phyllomedusinae (Anura: Hylidae). *Rev. Fr. Aquariol.* 21:57–64.

Jungfer, K.-H., and P. Weygoldt. 1999. Biparental care in the tadpole-feeding Amazonian treefrog *Osetocephalus oophagus*. *Amphibia-Reptilia* 20:235–49.
Jungfer, W. 1943. Beitrage zur Biologie der Erdkröte (*Bufo bufo* L.) mit besonderer Berücksichtigung der Wanderung zu den Laichgewässern. *Z. Morph. Ökol. Tiere* 40:117–57.
Jungreis, A. J. 1971. Seasonal effects of hyperosmotic sodium chloride on urea production in the frog *Rana pipiens*. *J. Exp. Zool.* 178:403–14.
———. 1976. Partition of excretory nitrogen in Amphibia. *Comp. Biochem. Physiol.* 53 A:133–41.
Junk, W. J. 1970. Investigations on the ecology and production-biology of the "floating meadows" (Paspalum-Echinochloetum) on the Middle Amazon. Part I: The floating vegetation and its ecology. *Amazoniana* 2:449–95.
Junk, W. J. 1983. Ecology of swamps on the Middle Amazon. In *Ecosystems of the world 4B. Mires: Swamp, bog, fen, and moor*, ed. A. J. P. Gore, 269–94. New York: Elsevier.
Junk, W. J., and M. T. F. Piedade. 2004. Status of knowledge, ongoing research, and research needs in Amazonian wetlands. *Wetlands Ecol. Manage.* 12:597–609.
Juraske, N., and U. Sinsch. 1992. Direct development in the marsupial frog, *Gastrotheca griswoldi* Shreve, 1941. In *Proc. Sixth Ord. Gen. Meeting, Soc. Eur. Herpetol.*, ed. Z. Korsos and I. Kiss, 245–49. Budapest, Hungary.
Jurd, R. D. 1985. Haematological and immunological 'metamorphosis' in neotenous urodeles. In *Metamorphosis*, ed. M. Balls and M. Bownes, 313–31. Oxford, UK: Clarendon Press.
Jurgens, J. D. 1971. The morphology of the nasal region of Amphibia and its bearing on the phylogeny of the group. *Ann. Univ. Stellenbosch* 46A2:1–146.
Just, J. J., and B. G. Atkinson. 1972. Hemoglobin transitions in the bullfrog, *Rana catesbeiana*, during spontaneous and induced metamorphosis. *J. Exp. Zool.* 182:271–80.
Just, J. J., J. Kraus-Just, and P. A. Check. 1981. Survey of chordate metamorphosis. In *Metamorphosis: A problem in developmental biology.* 2nd ed., eds. L. I. Gilbert and E. Frieden, 265–326. New York: Plenum.
Justis, C. S., and D. H. Taylor. 1976. Extraocular photoreception and compass orientation in larval bullfrogs, *Rana catesbeiana*. *Copeia* 1976:98–105.
Juszczyk, W. 1950. The food of the aquatic frog *Rana esculenta* L. in the natural water reservoirs and in the artificial fish ponds. *Bull. Acad. Polon. Sci.*, Ser. B, 2:31–80.
Juszczyk, W., S. Krawczyk, M. Zakrzewski, W. Zamachowski, and A. Zysk. 1984. Morphometric structure of population of *Rana temporaria* (L.) hibernating together with other amphibians in aquatic environment. *Acta Biol. Cracov. (Zool.)* 26:39–50.
Juszczyk, W., K. Obrzut, and W. Zamachowski. 1966. Morphological changes in the alimentary canal of the common frog (*Rana temporaria* L.) in the annual cycle. *Acta Biol. Cracov. (Zool.)* 9:239–46.
Juszczyk, W., and W. Zamachowski. 1973. Morphological changes in the reproductive organs of the female edible frog (*Rana esculenta* L.) in the annual cycle. *Acta Biol. Cracov. (Zool.)* 16:167–78.
Juterbock, J. E. 1978. Sexual dimorphism and maturity characteristics of three species of *Desmognathus* (Amphibia, Urodela, Plethodontidae). *J. Herpetol.* 12:217–30.
Juterbock, J. E. 1986. The nesting behavior of the dusky salamander, *Desmognathus fuscus*. I. Nesting phenology. *Herpetologica* 42:457–71.
———. 1987. The nesting behavior of the dusky salamander, *Desmognathus fuscus*. II. Nest site tenacity and disturbance. *Herpetologica* 43:361–68.
———. 1990. Variation in larval growth and metamorphosis in the salamander *Desmognathus fuscus*. *Herpetologica* 46:291–303.
Kabisch, K. 1971a. Ein Winterquartier des Feuersalamander, *Salamandra salamandra terrestris*, bei Wernigerode. *Hercynia* 8:344–46.
———. 1971b. Ein Winterquartier von *Triturus vulgaris, Triturus cristatus, Bombina bombina,* und *Pelobates fuscus*. *Aquar. Terr.* 18:381.
Kabisch, K., and H. Belter. 1968. Das Verzehren von Amphibien und Reptilien durch Vögel. *Zool. Abh. Mus. Tierk. Dresden* 29:191–227.
Kabisch, K., and W.-E. Engelmann. 1971. Zur Überwinterung von *Rana temporaria*. *Hercynia* 8:347–48.
Kabisch, K., and J. Weiss. 1968. *Rana temporaria* L. als Winternährung von *Salmo trutta fario* L. *Zool. Abh. Mus. Tierk. Dresden* 29:289–91.
Kadadevaru, C. G., and R. D. Kanamadi. 2000. Courtship and nesting behaviour of the Malabar gliding frog, *Rhacophorus malabaricus* (Jerdon, 1870). *Curr. Sci.* 79:377–80.
Kadel, K. 1975. Freilandstudien zur Überlebensrate von Kreuzkrötenlarven (*Bufo calamita* Laur.). *Rev. Suisse Zool.* 82:237–44.
Kagarise Sherman, C. 1980. *A comparison of the natural history and mating system of two anurans: Yosemite toads (Bufo canorus) and black toads (Bufo exsul)*. Unpublished PhD dissertation. Ann Arbor: University of Michigan.
Kagarise Sherman, C., and M. L. Morton. 1993. Population declines of Yosemite toads in the eastern Sierra Nevada of California. *J. Herpetol.* 27:186–98.
Kaiser, H. 1997. Origins and introductions of the Caribbean frog, *Eleutherodactylus johnstonei* (Leptodactylidae): management and conservation concerns. *Biodivers. Cons.* 6:1391–1407.
Kaiser, H., and R. Altig. 1994. The atypical tadpole of the dendrobatid frog, *Colostethus chalcopis*, from Martinique, French Antilles. *J. Herpetol.* 28:374–78.
Kaiser, H., L. A. Coloma, and H. M. Gray. 1994. A new species of *Colostethus* (Anura: Dendrobatidae) from Martinique, French Antilles. *Herpetologica* 50:23–32.
Kaiser, H., D. M. Green, and M. Schmid. 1995. Systematics and biogeography of Eastern Caribbean frogs (*Eleutherodactylus:* Leptodactylidae), with the description of a new species from Dominica. *Can. J. Zool.* 72:2217–37.
Kaiser, H., and R. W. Henderson. 1994. The conservation status of Lesser Antillean frogs. *Herp. Nat. Hist.* 2:41–56.
Kaiser, H., C. Steinlein, W. Feichtinger, and M. Schmid. 2003. Chromosome banding of six dendrobatid frogs (*Colostethus, Mannophryne*). *Herpetologica* 59:203–18.
Kalb, H. J., and G. R. Zug. 1990. Age estimation for a population of American toads, *Bufo americanus* (Salientia: Bufonidae) in northern Virginia. *Brimleyana* 16:79–86.
Kalezić, M. L., D. Cvetović, A. Djorović, and G. Džukić. 1994. Paedomorphosis and differences in life-history traits of two neighboring crested newt (*Triturus carnifex*) populations. *Herpetol. J.* 4:151–59.
Kalezić, M. L., D. Cvetović, A. Djorović, and G. Džukić. 1996. Alternative life-history pathways: paedomorphosis and adult

fitness in European newts (*Triturus vulgaris* and *T. alpestris*). *J. Zool. Syst. Evol. Res.* 34:1–7.

Kalezić, M. L., and G. Džukić. 1985. Ecological aspects of the smooth newt (*Triturus vulgaris*) paedomorphosis from Montenegro. *Arh. Biol. Nauka, Beograd* 37:43–50.

Kalezić, M. L., and G. Džukić. 1986. The frequent occurrence of paedomorphosis in the smooth newt (*Triturus vulgaris*) population from the submediterranean area of Yugoslavia. *Amphibia-Reptilia* 7:86–89.

Kaltenbach, J. C. 1996. Endocrinology of amphibian metamorphosis. In *Postembryonic programming of gene expression in amphibian and insect cells,* ed. L. I. Gilbert, J. R. Tata, and B. G. Atkinson, eds. Metamorphosis, 403–31. San Diego: Academic Press.

Kam, Y.-C., and T.-C. Chen. 2000. Abundance and movement of a riparian frog (*Rana swinhoana*) in a subtropical forest of Guandau Stream, Taiwan. *Zool. Stud.* 39:67–76.

Kam, Y.-C., Y.-H. Chen, T.-C. Chen, and I.-R. Tsai. 2000. Maternal brood care of an arboreal breeder, *Chirixalus eiffingeri* (Anura: Rhacophoridae) from Taiwan. *Behaviour* 137:137–51.

Kam, Y.-C., Y.-H. Chen, Z.-S. Chuang, and T.-S. Huang. 1997. Growth and development of oophagous tadpoles in relation to brood care of an arboreal breeder, *Chirixalus eiffingeri* (Rhacophoridae). *Zool. Stud.* 36:186–93.

Kam, Y.-C., T.-C. Chen, J.-T. Yang, F.-C. Yu, and K.-M. Yu. 1998. Seasonal activity, reproduction, and diet of a riparian frog (*Rana swinhoana*) from a subtropical forest in Taiwan. *J. Herpetol.* 32:447–52.

Kam, Y.-C., Z.-S. Chuang, and C.-F. Yen. 1996. Reproduction, oviposition-site selection, and tadpole oophagy of an arboreal nester, *Chirixalus eiffingeri* (Rhacophoridae), from Taiwan. *J. Herpetol.* 30:52–59.

Kam, Y.-C., C.-F. Lin, Y.-S. Lin, and Y.-F. Tsal. 1998. Density effects of oophagous tadpoles of *Chirixalus eiffingeri* (Anura: Rhacophoridae): Importance of maternal brood care. *Herpetologica* 54:425–33.

Kam, Y.-C., Y.-J. Su, J.-L. Liu, and Y.-S. Lin. 2001. Intraspecific interactions among oophagous tadpoles (*Chirixalus eiffingeri:* Rhacophoridae) living in bamboo stumps in Taiwan. *J. Zool., London* 255:519–24.

Kam, Y.C., and H.-W. Yang. 2002. Female-offspring communication in a Taiwanese tree frog, *Chirxalus eiffingeri* (Anura: Rhacophoridae). *Anim. Behav.* 64:881–86.

Kamel, S., J. E. Marsden, and F. H. Pough. 1985. Diploid and tetraploid grey treefrogs (*Hyla chrysoscelis* and *Hyla versicolor*) have similar metabolic rates. *Comp. Biochem. Physiol.* 82A:217–20.

Kaminsky, S. K., K. E. Linsenmair, and T. U. Grafe. 1999. Reproductive timing, nest construction and tadpole guidance in the African pig-nosed frog, *Hemisus marmoratus*. *J. Herpetol.* 33:119–23.

Kanamadi, R. D., C. R. Hiremath, and H. Schneider. 1995. Vocalization and territoriality of the Indian frog *Rana limnocharis*. *Proc. Indian Natl. Acad. Sci.* B61:9–14.

Kanamadi, R. D., and C. S. Jirankali. 1991. Ovarian activity in the tree frog *Polypedates maculatus:* A qualitative and quantitative study of the ovarian cycles and its relation to oviduct and fat body cycle. *Zool. Anz.* 226:149–62.

———. 1992. Testicular activity *Polypedates maculatus* (Rhacophoridae): seasonal changes in spermatogenesis and fat bodies. *J. Herpetol.* 26:329–35.

Kanamadi, R. D., and S. K. Saidapur. 1982. Effect of estradiol-17B, estradiol-17β + homoplastic pars distalis extract, estradiol-17β + testosterone and testosterone on spermatogenesis, Leydig cells and thumb pads in the toad *Bufo melanostictus* (Schn.). *Indian J. Exp. Biol.* 20:209–15.

———. 1988. The effect of fat body ablation on spermatogenesis in the Indian skipper frog *Rana cyanophlyctis* (Schn.). *Zool. Anz.* 220:79–81.

Kanamadi, R. D., S. K. Saidapur, N. U. Bhuttewadkar, and S. M. Yamakanamardi. 1989. Annual changes in fat body of the male toad, *Bufo melanostictus* (Schn.) inhabiting the tropical zone of south India. *Proc. Indian Natl. Sci. Acad.* 55:261–64.

Kanamadi, R. D., H. Schneider, C. R. Hiremath, and C. S. Jirankali. 1993. Vocalization of the treefrog *Polypedates maculatus* (Rhacophoridae). *J. Biosci.* 18:239–45.

Kanou, M., and T. Shimozawa. 1983. The elicitation of the predatory labial strike of dragonfly larvae in response to a purely mechanical stimulus. *J. Exp. Biol.* 107:391–404.

Kao, Y. H., P. S. Alexander, V. V. C. Yang, and J. Y. L. Yu. 1994. Annual patterns of nuptial pad and vocal sac development in the male Chinese bullfrog (*Rana rugulosa* Wiegmann). *Zool. Stud.* 33:153–59.

Kaplan, D. L. 1977. Exploitative competition in salamanders: test of a hypothesis. *Copeia* 1977:234–37.

Kaplan, M. 1994. A new species of frog of the genus *Hyla* from the Cordillera Oriental in northern Colombia with comments on the taxonomy of *Hyla minuta*. *J. Herpetol.* 28:79–87.

———. 1997. Internal and external anatomy of the abdominal disc of *Atelopus* (Bufonidae) larvae. *Caldasia* 19:61–69.

Kaplan, R. H. 1980. The implications of ovum size variability for offspring fitness and clutch size within several populations of salamanders (*Ambystoma*). *Evolution* 34:51–64.

———. 1985. Maternal influences on offspring development in the California newt, *Taricha torosa*. *Copeia* 1985:1028–35.

———. 1987. Developmental plasticity and maternal effects of reproductive characteristics in the frog, *Bombina orientalis*. *Oecologia* 71:273–79.

———. 1989. Ovum size plasticity and maternal effects on the early development of the frog, *Bombina orientalis* Boulenger, in a field population in Korea. *Funct. Ecol.* 3:597–604.

Kaplan, R. H., and W. S. Cooper. 1984. The evolution of developmental plasticity in reproductive characteristics: An application of the "adaptive coin-flipping" principle. *Am. Nat.* 123:393–410.

Kaplan, R. H., and M. L. Crump. 1978. The non-cost of brooding in *Ambystoma* opacum. *Copeia* 1978:99–103.

Kaplan, R. H., and E. G. King. 1997. Egg size is a developmentally plastic trait: Evidence from long term studies in the frog *Bombina orientalis*. *Herpetologica* 53:149–65.

Kaplan, R. H., and S. N. Salthe. 1979. The allometry of reproduction: an empirical view in salamanders. *Am. Nat.* 113:671–89.

Kaplan, R. H., and P. W. Sherman. 1980. Intraspecific oophagy in California newts. *J. Herpetol.* 14:183–85.

Kardong, K. V. 1979. "Protovipers" and the evolution of snake fangs. *Evolution* 33:433–43.

Karlsson, J., A. Jonsson, and M. Jonsson. 2005. Productivity of high-latitude lakes: Climate effect inferred from altitude gradient. *Global Change Biol.* 11:710–15.

Karlstrom, E. L. 1957. The use of $CO^{60}$ as a tag for recovering amphibians in the field. *Ecology* 38:187–95.

———. 1962. The toad genus *Bufo* in the Sierra Nevada of California: Ecological and systematic relationships. *Univ. California Publ. Zool.* 62:1–104.

Karns, D. R. 1992. Effects of acidic bog habitats on amphibian reproduction in a northern Minnesota peatland. *J. Herpetol.* 26:401–12.

Karraker, N. E. 1999. *Rhyacotriton variegatus* (southern torrent salamander). Nest site. *Herp. Rev.* 30:160–61.

Karthaus, G. 1985. Schutzmassnahmen für wanderde Amphibien vor einer Gefahrdung durch den Strassenverkehr—Beobachtungen und Erfahrungen. *Natur Landschaft* 60:242–47.

Karuzas, J. M., J. C. Maerz, and D. M. Madison. 2004. An alternative hypothesis for the primary function of a proposed mate assessment behaviour in red-backed salamanders. *Anim. Behav.* 68:489–94.

Kasbohm, P. 1967. Der Einfluss des Lichtes auf die Temperaturadaptation bei *Rana temporaria. Helgolander Wiss. Meer.* 16:157–78.

Kasinathan, S., A. Gunasing, and S. L. Basu. 1978. Fat bodies and spermatogenesis in south Indian green frog *Rana hexadactyla* Lesson. *Boll. Zool.* 45:15–22.

Kastle, W. 1986. Rival combats in *Salamandra salamandra.* In *Studies in herpetology,* ed. Z. Roček, 535–28. Prague: Charles University.

Kasuya, E., M. Hirota, and H. Shigehara. 1996. Reproductive behavior of the Japanese treefrog, *Rhacophorus arboreus* (Anura: Rhacophoridae). *Res. Pop. Ecol.* 38:1–10.

Kasuya, E., T. Kobayashi, M. Ootaki, N. Oota, and A. Takada. 1997. Female preference for temporal features of vocalization in the Japanese treefrog, *Rhacophorus arboreus. J. Ethol.* 15:103–8.

Kasuya, E., T. Kumaki, and T. Saito. 1992. Vocal repertoire of the Japanese treefrog, *Rhacophorus arboreus* (Anura: Rhacophoridae). *Zool. Sci. (Tokyo)* 9:469–73.

Kasuya, E., H. Shigehara, and M. Hirota. 1987. Mating aggregation in the Japanese treefrog, *Rhacophorus arboreus* (Anura: Rhacophoridae): A test of cooperation hypothesis. *Zool. Sci. (Tokyo)* 4:693–97.

Kato, K. 1956. Ecological notes on the green frogs during the breeding season. II. Breeding habits and others. *Jap. J. Ecol.* 6:57–61 (in Japanese, with English abstract).

Kats, L. B., S. A. Elliott, and J. Currens. 1992. Intraspecific oophagy in stream-breeding California newts (*Taricha torosa*). *Herp. Rev.* 23:7–8.

Kats, L. B., and R. P. Ferrer. 2003. Alien predators and amphibian declines: Review of two decades of science and the transition to conservation. *Divers. Distr.* 9:99–110.

Kats, L. B., J. W. Petranka, and A. Sih. 1988. Antipredator defenses and the persistence of amphibian larvae with fishes. *Ecology* 69:1865–70.

Kats, L. B. and A. Sih. 1991. Effects of refuge availability on the responses of salamander larvae to chemical cues from predatory green sunfish. *Anim. Behav.* 42:330–32

———. 1992. Oviposition site selection and avoidance of fish by streamside salamanders (*Ambystoma barbouri*). *Copeia* 1992:468–73.

Kats, L. B., and R. G. van Dragt. 1986. Background color-matching in the spring peeper, *Hyla crucifer. Copeia* 1986:109–15.

Katsikaros, K., and R. Shine. 1997. Sexual dimorphism in the tusked frog, *Adelotus brevis* (Anura: Myobatrachidae): The roles of natural and sexual selection. *Biol. J. Linn. Soc.* 60:39–51.

Katsuren, S., S. Tanaka, and S. Ikehara. 1977. A brief observation on the breeding site and eggs of a frog, *Rana ishikawae* (Stejneger) in Okinawa Island. *Ecological Studies of Nature Conservation of the Ryukyu Islands* 3:49–54.

Katz, L. C., M. J. Potel, and R. J. Wassersug. 1981. Structure and mechanisms of schooling in tadpoles of the clawed frog, *Xenopus laevis. Anim. Behav.* 29:20–33.

Katz, U. 1973. Studies on the adaptation of the toad *Bufo viridis* to high salinities: Oxygen consumption, plasma concentration and water content of the tissues. *J. Exp. Biol.* 58:785–96.

———. 1975. NaCl adaptation in *Rana ridibunda* and a comparison with the euryhaline toad *Bufo viridis. J. Exp. Biol.* 63:763–73.

———. 1978. Ionic and volume regulation in selected tissues of the euryhaline toad *Bufo viridis.* In *Osmotic and volume regulation. Alfred Benzon Symposium XI,* ed. C. B. Jorgensen and E. Skadhauge, 457–69. Copenhagen: Munksgaard.

———. 1986. The role of amphibian epidermis in osmoregulation and its adaptive response to changing environment. In *Biology of the integument, vol. 2. Vertebrates,* ed. J. Bereiter-Hahn, A. G. Matoltsy, and K. S. Richards, 472–98. New York: Springer-Verlag.

———. 1989. Strategies of adaptation to osmotic stress in anuran Amphibia under salt and burrowing conditions. *Comp. Biochem. Physiol.* 93 A:499–503.

Katz, U., and S. Gabbay. 1984. Acclimation of the euryhaline toad *Bufo viridis* to hyperosmotic solution (NaCl, urea and mannitol). *J. Exp. Biol.* 108:403–9.

———. 1986. Water retention and plasma and urine composition in toads (*Bufo viridis* Laur.) under burrowing conditions. *J. Comp. Physiol.* B 156:735–40.

Katz, U., and N. Gil. 1997. Different temperature relations of two species of toads that coexist at the border of their geographic distributions. *Amphibia-Reptilia* 18:259–68.

Katz, U., and R. Graham. 1980. Water relations in the toad (*Bufo viridis*) and a comparison with the frog (*Rana ridibunda*). *Comp. Biochem. Physiol.* 67 A:245–51.

Katz, U., and J. Hoffman. 1990. Changing plasma osmolality—a strategy of adaptation of anuran amphibia to water scarcity under burrowing conditions. *Fortschr. Zool.* 38:163–68.

Katz, U., J. Hoffman, and N. Gil. 1997. What is the ecological significance of laboratory temperature selection in anuran Amphibia? *Alytes* 15:91–98.

Katz, U., and W. Nagel. 1994. Biophysics of ion transport across amphibian skin. In *Amphibian biology. Vol. 1. The integument,* ed. H. Heatwole and G. T. Barthalmus, 98–119. Chipping Norton, New South Wales, Australia: Surrey Beatty and Sons.

Kaufmann, J. H. 1962. Ecology and social behavior of the coati, *Nasua narica,* on Barro Colorado Island, Panama. *Univ. California Publ. Zool.* 60:95–222.

Kauhala, K. 1996. Habitat use of raccoon dogs, *Nyctereutes procyonides,* in southern Finland. *Z. Säugetierkunde* 61:269–75.

Kauhala, K., and M. Auniola. 2001. Diet of raccoon dogs in summer in the Finnish archipelago. *Ecography* 24:151–56.

Kauhala, K., P. Laukikanen, and I. von Rege. 1998. Summer food composition and food niche overlap of the raccoon dog, red fox and badger in Finland. *Ecography* 21:457–63.

Kaul, R., and V. H. Shoemaker. 1989. Control of thermoregulatory evaporation in the waterproof treefrog *Chiromantis xerampelina. J. Comp. Physiol.* B 158:643–49.

Kaung, H. L. C. 1975. Development of beaks of *Rana pipiens* larvae. *Anat. Rec.* 182:401–14.

Kaushal, S. S., P. M. Groffman, G. E. Likens, K. T. Belt, W. P.

Stack, V. R. Reilly, L. E. Band, and G. T. Fisher. 2005. Increased salinization of fresh water in the northeastern United States. *Proc. Natl. Acad. Sci. USA* 102:13517–20.

Kauta, V. 1971. William Frederic Edwards. *Dictionary of Scientific Biography* 4:285–86.

Kawai, S., M Kobayashi, and H. Kaneko. 2003. Effects of endocrine active substances in wildlife species: genetic, biochemical, and physiological factors in variable susceptibility to endrocrine disruptors. *Pure Appl. Chem.* 75:2335–41.

Kawamichi, T., and H. Ueda. 1998. Spawning at nests of extra-large males in the giant salamander *Andrias japonicus*. *J. Herpetol.* 32:133–36.

Kawamura, K., K. Yamamoto, and S. Kikuyama. 1986. Effects of thyroid hormone, stalk secretion, and transplantation of the pituitary gland on plasma prolactin levels at metamorphic climax in *Rana catesbeiana*. *Gen. Comp. Endocrinol.* 64:129–35.

Kawamura, T., and S. Sawada. 1959. On the sexual isolation among different species and local races of Japanese newts. *J. Sci. Hiroshima Univ. Ser. B* 18:17–31.

Kc, P., D. G. Bernard, M. Gardner, and N. J. Smatresk. 2000. Ventilatory mechanics and the effects of water depth on breathing pattern in the aquatic caecilian *Typhlonectes natans*. *J. Exp. Biol.* 203:263–72.

Keddy-Hector, A. C., W. Wilczynski, and M. J. Ryan. 1992. Call patterns and basilar papilla tuning in cricket frogs. II. Intrapopulation variation and allometry. *Brain Behav. Evol.* 39:238–46.

Keen, W. H. 1975. Breeding and larval development of three species of *Ambystoma* in central Kentucky (Amphibia: Urodela). *Herpetologica* 31:18–21.

———. 1979. Feeding and activity patterns in the salamander *Desmognathus ochrophaeus* (Amphibia, Urodela, Plethodontidae). *J. Herpetol.* 13:461–67.

———. 1982. Habitat selection and interspecific competition in two species of plethodontid salamanders. *Ecology* 63:94–102.

———. 1984. Influence of moisture on the activity of a plethodontid salamander. *Copeia* 1984:684–88.

———. 1985. Habitat selection by two streamside plethodontid salamanders. *Oecologia* 66:437–42.

Keen, W. H., M. G. McManus, and M. Wohltman. 1987. Cover site recognition and sex differences in cover site use by the salamander, *Desmognathus fuscus*. *J. Herpetol.* 21:363–65.

Keen, W. H., and R. W. Reed. 1985. Territorial defence of space and feeding sites by a plethodontid salamander. *Anim. Behav.* 33:1119–23.

Keen, W. H., and E. E. Schroeder. 1975. Temperature selection and tolerance in three species of *Ambystoma* larvae. *Copeia* 1975:523–30.

Keen, W. H., and S. Sharp. 1984. Responses of a plethodontid salamander to conspecific and congeneric intruders. *Anim. Behav.* 32:58–65.

Keen, W. H., J. Travis, and J. Julianna. 1984. Larval growth in three sympatric *Ambystoma* salamander species: Species differences and the effects of temperature. *Can. J. Zool.* 62:1043–47.

Keenlyne, K. D., and J. R. Beer. 1973. Food habits of *Sistrurus catenatus catenatus*. *J. Herpetol.* 7:382–84.

Kehr, A. I., and J. A. Schnack. 1991. Predator-prey relationship between giant water bugs (*Belostoma oxyurum*) and larval anurans (*Bufo arenarum*). *Alytes* 9:61–69.

Keister, A. R. 1977. Communication in amphibians and reptiles. In *How animals communicate,* ed. T. A. Sebeok 519–44. Bloomington: Indiana University Press.

Kelleher, K. E., and J. R. Tester. 1969. Homing and survival in the Manitoba toad, *Bufo hemiophrys,* in Minnesota. *Ecology* 50:1040–48.

Keller, M. J., and H. C. Gerhardt. 2001. Polyploidy alters advertisement call structure in gray treefrogs. *Proc. Roy. Soc. London B* 268:341–45.

Keller, O. 1913. *Die Antike Tierwelt*. 2 vols. Georg Olms Verlagsbuchhandlung, Hildesheim, Germany (1963 reprint).

Kelley, D. B. 1986. Neuroeffectors for vocalization in *Xenopus laevis:* Hormonal regulation of sexual dimorphism. *J. Neurobiol.* 17:231–48.

———. 1996. Sexual differentiation in *Xenopus laevis*. In *The biology of Xenopus,* ed. R. C. Tinsley and H. R. Kobel, 143–76. New York: Oxford University Press.

Kelley, D. B., D. Sassoon, N. Segil, and M. Scudder. 1989. Development and hormone regulation of androgen receptor levels in the sexually dimorphic larynx of *Xenopus laevis*. *Dev. Biol.* 131:111–18.

Kelley, D. B., and M. L. Tobias. 1989. The genesis of courtship song: cellular and molecular control of a sexually differentiated behavior. In *Perspectives in neural systems and behavior,* ed. T. J. Carew and D. B. Kelley, 175–94. New York: Alan Liss.

———. 1999. Vocal communication in *Xenopus laevis*. In *Neural mechanisms of communication,* ed. M. Hauser and M. Konishi, 9–35. Cambridge: MIT Press.

Kelley, D. B., M. L. Tobias, and S. Horng. 2001. Producing and receiving frog songs: dissecting the neural basis for vocal behaviors in *Xenopus laevis*. In *Anuran communication,* ed. M. J. Ryan, 156–66. Washington, DC: Smithsonian Institution Press.

Kennedy, A. B. 1982. Ecce *Bufo:* The toad in nature and Olmec iconography. *Curr. Anthro.* 23:272–90.

Kenny, J. S. 1966. Nest building in *Phyllomedusa trinitatis* Mertens. *Carib. J. Sci.* 6:15–22.

———. 1969a. The Amphibia of Trinidad. *Studies on the Fauna of Curaçao and Other Caribbean Islands* (108):1–78.

———. 1969b. Feeding mechanisms in anuran larvae. *J. Zool., London* 157:225–46.

———. 1969c. Pharyngeal mucous secreting epithelia of anuran larvae. *Acta Zool. (Stockholm)* 50:143–53.

Kephart, D. G. 1982. Microgeographic variation in the diets of garter snakes. *Oecologia* 52:287–91.

Kephart, D. G., and S. J. Arnold. 1982. Garter snake diets in a fluctuating environment: a seven year study. *Ecology* 63:1232–36.

Kerbert, C. 1904. Zur Fortpflanzung von *Megalobatrachus maximus* Schlegel. *Zool. Anz.* 27:305–16.

Kerby, J. L., and L. B. Kats. 1998. Modified interactions between salamander life stages caused by wildfire-induced sedimentation. *Ecology* 79:740–45.

Kessel, E. L., and B. B. Kessel. 1942. An egg cluster of *Aneides lugubris lugubris* (Hallowell). *Wasmann Coll.* 5:71.

Kezer, J. 1952. Thyroxin-induced metamorphosis of the neotenic salamanders *Eurycea tynerensis* and *Eurycea neotenes*. *Copeia* 1952:234–37.

Kezer, J., and D. S. Farner. 1955. Life history patterns of the salamander *Ambystoma macrodactylum* in the high Cascade Mountains of southern Oregon. *Copeia* 1955:127–31.

Khan, M. S. 2005. *Amphibians and reptiles of Pakistan*. Malabar, FL: Krieger.

Khonsue, W., M. Matsui, T. Hirai, and Y. Misawa. 2001a. Age determination of wrinkled frog, *Rana rugosa* with special reference to high variation in postmetamorphic body size (Amphibia: Ranidae). *Zool. Sci. (Tokyo)* 18:605–12.

———. 2001b. A comparison of age structure in two populations of a pond frog *Rana nigromaculata* (Amphibia: Anura). *Zool. Sci. (Tokyo)* 18:597–603.

Khonsue, W., M. Matsui, and Y. Misawa. 2002. Age determination of Daruma pond frog, *Rana porosa brevipoda* from Japan towards its conservation (Amphibia: Anura). *Amphibia-Reptilia* 23:259–68.

Kiesecker, J. 1996. pH-mediated predator-prey interactions between *Ambystoma tigrinum* and *Pseudacris triseriata*. *Ecol. Appl.* 6:1325–31.

Kiesecker, J. M. 2003. Invasive species as a global problem. Toward understanding the worldwide decline of amphibians. In *Amphibian conservation,* ed. R. D. Semlitsch, 113–26. Washington, DC: Smithsonian Books.

Kiesecker, J. M., L. K. Belden, K. Shea, and M. J. Rubbo. 2004. Amphibian decline and emerging disease. *Am. Sci.* 92:138–47.

Kiesecker, J. M., and A. R. Blaustein. 1995. Synergism between UV-B radiation and a pathogen magnifies amphibian embryo mortality in nature. *Proc. Natl. Acad. Sci. USA* 92:11049–52.

———. 1997a. Influences of egg laying behavior on pathogenic infection of amphibian eggs. *Cons. Biol.* 11:214–20.

———. 1997b. Population differences in responses of red-legged frogs (*Rana aurora*) to introduced bullfrogs. *Ecology* 78:1752–1760.

———. 1998. Effects of introduced bullfrogs and smallmouth bass on microhabitat use, growth, and survival of native red legged frogs (*Rana aurora*). *Cons. Biol.* 12:776–87.

———. 1999. Pathogen reverses competition between larval amphibians. *Ecology* 80:2442–48.

Kiesecker, J. M., A. R. Blaustein, and L. K. Belden. 2001. Complex causes of amphibian population declines. *Nature* 410:681–84.

Kiesecker, J. M., A. R. Blaustein, and C. L. Miller. 2001. Potential mechanisms underlying the displacement of native red-legged frogs by introduced bullfrogs. *Ecology* 82:1964–70.

Kiesecker, J. M., D. P. Chivers, and A. R. Blaustein. 1996. The use of chemical cues in predator recognition by western toad tadpoles. *Anim. Behav.* 52:1237–45.

Kiesecker, J. M., D. P. Chivers, A. Marco, C. Quilchano, M. T. Anderson, and A. R. Blaustein. 1999. Identification of a disturbance signal in larval red-legged frogs, *Rana aurora*. *Anim. Behav.* 57:1295–1300.

Kiesecker, J. M., C. L. Miller, and A. R. Blaustein. 2001. The transfer of a pathogen from fish to amphibians. *Cons. Biol.* 15:1064–70.

Kiesecker, J. M., and D. K. Skelly. 2000. Choice of oviposition site by gray treefrogs: The role of potential parasitic infection. *Ecology* 81:2939–43.

———. 2001. Effects of disease and pond drying on gray tree frog growth, development, and survival. *Ecology* 82:1956–63.

Kiester, A. R. 1971. Species density of North American amphibians and reptiles. *Syst. Zool.* 20:127–37.

Kiffney, P. M., and J. S. Richardson. 2001. Interactions among nutrients, periphyton, and invertebrate and vertebrate (*Ascaphus truei*) grazers in experimental channels. *Copeia* 2001:422–29.

Kikuyama, S., K. Kawamura, S. Tanaka, and K. Yamamoto. 1993. Aspects of amphibian metamorphosis: Hormonal control. *Int. Rev. Cytol.* 145:105–48.

Kikuyama, S., F. Toyoda, Y. Ohmiya, K. Matsuda, S. Tanaka, and H. Hayashi. 1995. Sodefrin: a female attracting peptide pheromone in newt cloacal glands. *Science* 267:1643–45.

Kikuyama, S., K. Yamamoto, T. Iwata, and F. Toyoda. 2002. Peptide and protein pheromones in amphibians. *Comp. Biochem. Physiol.* B 132:69–74.

Kilpi, M., and P. Byholm. 1995. Diet of courtship feeding herring gulls in the Gulf of Finland. *Ornis Fennici* 72:135–37.

Kim, J. 2005. A projection of the effects of climate change induced by increased $CO_2$ on extreme hydrologic events in the western U.S. *Climatic Change* 68:153–68.

Kim, Y. H., G. B. Brown, H. S. Mosher, and F. A. Fuhrman. 1975. Tetrodotoxin: Occurrence in atelopid frogs of Costa Rica. *Science* 189:151–52.

Kime, D. E. 1998. *Endocrine disruptors in fish*. New York: Kluwer Academic.

Kime, N. M., S. S. Burmeister, and M. J. Ryan. 2004. Female preferences for socially variable call characters in the cricket frog, *Acris crepitans*. *Anim. Behav.* 68:1391–99.

Kime, N. M., A. S. Rand, M. Kapfer, and M. J. Ryan. 1998. Consistency of female choice in the túngara frog: a permissive preference for complex characters. *Anim. Behav.* 55:641–49.

Kime, N. M., W. R. Turner, and M. J. Ryan. 2000. The transmission of advertisement calls in Central American frogs. *Behav. Ecol.* 11:71–83.

King, D., and B. Green. 1979. Notes on diet and reproduction of the sand goanna, *Varanus gouldii rosenbergi*. *Copeia* 1979:64–70.

King, P. A., M. N. Rosholt, and K. B. Storey. 1995. Seasonal changes in plasma membrane glucose transporters enhance cryoprotectant distribution in the freeze-tolerant wood frog. *Can. J. Zool.* 73:1–9.

King, R. B., S. Hauff, and J. B. Phillips. 1994. Physiological color change in the green treefrog: responses to background brightness and temperature. *Copeia* 1994:422–32.

Kingsbury, B. F. 1895. The spermatheca and methods of fertilization in some American newts and salamanders. *Trans. Amer. Microsc. Soc.* 17:115–46.

Kingsland, S. E. 1985. *Modeling nature*. Chicago: University of Chicago Press.

Kinkead, R. 1997. Episodic breathing in frogs: converging hypotheses on neural control of respiration in air breathing vertebrates. *Amer. Zool.* 37:31–40.

Kirberger, C. 1953. Temperatur Adaptation der Säuerstoffbindung des Blutes von *Rana esculenta* L. *Z. Vergl. Physiol.* 35:153–58.

Kirby, A. C. 1983. Physiology of the sternoradialis muscle: Sexual dimorphism and role in amplexus in the leopard frog (*Rana pipiens*). *Comp. Biochem. Physiol.* 74 A:705–9.

Kirkland, G. L., H. W. Snoddy, and T. L. Amster. 1996. Impact of fire on small mammals and amphibians in central Appalachian deciduous forest. *Am. Midl. Nat.* 135:253–60.

Kirkpatrick, M. 1982. Sexual selection and the evolution of female choice. *Evolution* 36:1–12.

Kirkpatrick, M., and M. J. Ryan. 1991. The evolution of mating preferences and the paradox of the lek. *Nature* 350:33–38.

Kirschner, L. B. 1983. Sodium chloride absorption across the body surface: Frog skins and other epithelia. *Am. J. Physiol.* 244:R429–R443.

Kirschner, L. B., T. Kerstetter, D. Porter, and R. H. Alvarado. 1971. Adaptation of larval *Ambystoma tigrinum* to concentrated environments. *Am. J. Physiol.* 220:1814–19.

Kirtisinghe, P. 1946. The presence in Ceylon of a frog with direct development on land. *Ceylon J. Sci. (B)* 23:109–12.

Kishida, O., and K. Nishimura. 2004. Bulgy tadpoles: inducible defense morph. *Oecologia* 140:414–21.

Kitchener, A. 1991. *The natural history of the wild cats.* Ithaca, NY: Cornell University Press.

Klauber, L. M. 1972. *Rattlesnakes: Their habits, life histories, and influence on mankind.* Berkeley: University of California Press.

Kleeberger, S. R. 1984. A test of competition in two sympatric populations of desmognathine salamanders. *Ecology* 65:1846–56.

———. 1985. Influence of intraspecific density and cover on home range of a plethodontid salamander. *Oecologia* 66:404–10.

Kleeberger, S. R., and J. K. Werner. 1982. Home range and homing behavior of *Plethodon cinereus* in northern Michigan. *Copeia* 1982:409–15.

———. 1983. Post-breeding migration and summer movement of *Ambystoma maculatum. J. Herpetol.* 17:176–77.

Kleinenberg, S. E., and E. M. Smirina. 1969. A contribution to the method of age determination in amphibians. *Zool. Zh.* 48:1090–94.

Klemens, M. W. 1993. *Amphibians and reptiles of Connecticut and adjacent states. State Geological and Natural History Survey of Connecticut Bull. 112.* Hartford, Connecticut.

Klemens, M. W., ed. 2000. *Turtle conservation.* Washington, DC: Smithsonian Institution Press.

Klimstra, W. D. 1959a. Food habits of the cottonmouth in southern Illinois. *Chicago Acad. Sci. Nat. Hist. Misc.* no. 168:1–8.

———. 1959b. Food habits of the yellow-bellied king snake in southern Illinois. *Herpetologica* 15:1–5.

———. 1959c. Foods of the racer, *Coluber constrictor* in southern Illinois. *Copeia* 1959:210–14.

Kling, K. B., J. P. Costanzo, and R. E. Lee, Jr. 1994. Post-freeze recovery of peripheral nerve function in the freeze-tolerant wood frog, *Rana sylvatica. J. Comp. Physiol.* B 164:316–20.

Kloas, W. 2002. Amphibians as a model for the study of endocrine disruptors. *Int. Rev. Cytol.* 216:1–57.

Klöckner, P. 1982. Beiträge zur Fortpflanzungsbiologie einheimischer Grunfrösche der Gattung *Rana* (Amphibia: Ranidae). *Braunschw. Naturk. Schr.* 1:341–66.

Kluge, A. G. 1979. The gladiator frogs of Middle America and Colombia—A reevaluation of their systematics (Anura: Hylidae). *Occ. Pap. Mus. Zool. Univ. Michigan* (688):1–24.

———. 1981. The life history, social organization, and parental behavior of *Hyla rosenbergi* Boulenger, a nest-building gladiator frog. *Misc. Publ. Mus. Zool. Univ. Michigan* (160):1–170.

Klump, G. M., J. H. Bendix, Jr., H. C. Gerhardt, and P. M. Narins. 2004. AM representation in green treefrog auditory nerve fibers: neuroethological implications for pattern recognition and sound localization. *J. Comp. Physiol. A* 190:1011–21.

Klump, G. M., and H. C. Gerhardt. 1987. Use of non-arbitrary acoustic criteria in mate choice by female gray tree frogs. *Nature* 326:286–88.

———. 1989. Sound localization in the barking treefrog. *Naturwissenschaften* 76:35–37.

———. 1992. Mechanisms and function of call-timing in male-male interactions in frogs. In *Playback and studies of animal communication,* ed. P. K. McGregor, 153–74. New York: Plenum.

Knapp, R. A. 2005. Effects of nonnative fish and habitat characteristics on lentic herpetofauna in Yosemite National Park, USA. *Biol. Cons.* 121:265–79.

Knapp, R. A., P. S. Corn, and D. E. Schindler. 2001. The introduction of nonnative fish into wilderness lakes: good intentions, conflicting madates, and unintended consequences. *Ecosystems* 4:275–78.

Knapp, R. A., and K. R. Matthews. 2000. Non-native fish introductions and the decline of the mountain yellow-legged frog from within protected areas. *Cons. Biol.* 14:428–38.

Knapp, R. A., K. R. Matthews, H. K. Preisler, and R. Jellison. 2003. Developing probabilistic models to predict amphibian site occupancy in a patchy landscape. *Ecol. Appl.* 13:1069–82.

Knapp, R. A., K. R. Matthews, and O. Sarnelle. 2001. Resistance and resilience of alpine lake fauna to fish introductions. *Ecol. Monogr.* 71:401–21.

Knapp, S. M., C. A. Haas, D. N. Harpole, and R. L. Kirkpatrick. 2003. Initial effects of clearcutting and alternative silvicultural practices on terrestrial salamander abundance. *Cons. Biol.* 17:752–62.

Knapp, W. 1974. Die jahreszeitliche Steurung der Atmung in abhängigkeit von Akklimationstemperatur und Experimentaltemperatur bei *Triturus alpestris* Laur. und *Salamandra atra* Laur. (Amphibia). *Oecologia* 15:353–74.

Kneitz, S. 1998. *Untersuchungen zur Populationsdynamik und zum Ausbreitungsverhalten von Amphibien in der Agrarlandschaft.* Bochum, Germany: Laurenti Verlag.

Kneller, M. 1982. Die Fortpflanzung von *Dendrobates reticulatus* im naturlichen Lebensraum und im Terrarium. *Das Aquarium* 153:148–51.

Knoeppffler, L.-P. 1962. Contribution a l'étude du genre *Discoglossus* (Amphibiens, Anoures). *Vie Milieu* 13:1–94.

———. 1976. Food habits of *Aubria subsigillata* in Gaboon (Anura: Ranidae). *Zool. Afr.* 11:369–71.

———. 1978–79. *Triops cancriformis* (Bosc.), crustacé phyllopode prédateur de têtards et de jeunes amphibiens. *Vie Milieu* 28–29:117–21.

Knorr, A. 1976. Central control of mating call production and spawning in the tree frog *Hyla arborea savignyi* (Audouin): Results of electrical stimulation of the brain. *Behav. Proc.* 1:295–317.

Knowles, T. W., and P. D. Weigl. 1990. Thermal dependence of anuran burst locomotor performance. *Copeia* 1990:796–802.

Knowlton, N. 1982. Parental care and sex role reversal. In *Current problems in sociobiology,* ed. King's College Sociobiology Study Group, 203–22. New York: Cambridge University Press.

Knudsen, J. W. 1960. The courtship and egg mass of *Ambystoma gracile* and *Ambystoma macrodactylum. Copeia* 1960:44–46.

Knutson, M. G., J. R. Sauer, D. A. Olsen, M. J. Mossman, L. M. Hemesath, and M. J. Lannoo. 1999. Effects of landscape composition and wetland fragmentation on frog and toad abundance and species richness in Iowa and Wisconsin, U.S.A. *Cons. Biol.* 13:1437–46.

Knutson, M. G., W. B. Richardson, D. M. Reineke, B. R. Gray, J. R. Parmelee, and S. E. Weick. 2004. Agricultural ponds support amphibian populations. *Ecol. Appl.* 14:669–84.

Kobelt, F. and K. E. Linsenmair. 1986. Adaptations of the reed frog *Hyperolius viridiflavus* (Amphibia, Anura, Hyperoliidae) to its arid environment. I. The skin of *Hyperolius viridiflavus nitidulus* in wet and dry season conditions. *Oecologia* 68:533–41.

Kobelt, F., and K. E. Linsenmair. 1992. Adaptations of the reed frog *Hyperolius viridiflavus* (Amphibia: Anura: Hyperoliidae) to its arid environment. VI. The iridophores in the skin as radiation reflectors. *J. Comp. Physiol.* B 162:314–26.

Kodandapani, N., M. A. Cochrane, and R. Sukumar. 2004. Conservation threat of increasing fire frequencies in the Western Ghats, India. *Cons. Biol.* 18:1553–61.

Kodric-Brown, A., and J. H. Brown. 1979. Competition between distantly related taxa in the coevolution of plants and pollinators. *Amer. Zool.* 19:1115–28.

Koenig, W. D., S. S Albano, and J. L. Dickinson. 1991. A comparison of methods to partition selection acting via components of fitness: Do larger male bullfrogs have greater hatching success? *J. Evol. Biol.* 4:309–20.

Koestler, A. 1971. *The case of the midwife toad.* New York: Random House.

Kofron, C. P. 1978. Foods and habits of aquatic snakes (Reptilia, Serpentes) in a Louisiana swamp. *J. Herpetol.* 12:543–54.

Kohmatsu, Y., S. Nakano, and N. Yamamura. 2001. Effects of head shape variation on growth, metamorphosis and survivorship in larval salamanders (*Hynobius retardatus*). *Ecol. Res.* 16:73–83.

Kohno, S., M. Fujime, Y. Kamishima, and T. Iguchi. 2004. Sexually dimorphic basal water absorption at the isolated pelvic patch of Japanese tree frog, *Hyla japonica. J. Exp. Zool., part A: Comp. Exp. Biol.* 301A:428–38.

Kok, D., L. H. du Preez, and A. Channing. 1989. Channel construction by the African bullfrog: another anuran parental care strategy. *J. Herpetol.* 23:435–37.

Kokubum, M. N. D. C., and A. A. Giaretta. 2005. Reproductive ecology and behaviour of a species of *Adenomera* (Anura, Leptodactylidae) with endotrophic tadpoles: Systematic implications. *J. Nat. Hist.* 39:1745–58.

Kokubum, M. N. D. C., and A. P. Rodrigues. 2005. *Leptodactylus ocellatus* (Rã-Manteiga). Cannibalism. *Herp. Rev.* 36:303.

Kokubum, M. N. D. C., and W. Zacca. 2003. *Physalaemus* cf. *fuscomaculatus* (NCN). Predation. *Herp. Rev.* 34:232–33.

Kolbe, J. J., B. E. Smith, and D. M. Browning. 2002. Burrow use by tiger salamanders (*Ambystoma tigrinum*) at a black-tailed prairie dog (*Cynomys ludovicianus*) town in southwestern South Dakota. *Herp. Rev.* 33:95–99.

Kollros, J. J. 1961. Mechanisms of amphibian metamorphosis: Hormones. *Amer. Zool.* 1:107–14.

Kolozsvary, M. B., and R. K. Swihart. 1999. Habitat fragmentation and the distribution of amphibians: patch and landscape correlates in farmland. *Can. J. Zool.* 77:1288–99.

König, C., and A. Schlüter. 1991. Nachzucht der Balearen-Geburtshelferkröte *Alytes muletensis* (Sanchiz et Adrover 1979) im Rahmen eines Artenschutzprogrammes (Amphibia: Discoglossidae). *Jh. Ges. Naturkde. Württemburg* 146:193–205.

Kopij, G. 1997. Breeding ecology of the African spoonbill *Platalea alba* in the Free State, South Africa. *Ostrich* 68:77–79.

———. 2003. Diet of cattle egret *Bubulcus ibis* chicks in an intensively managed farmland in South Africa. *Acta Ornithologica* 38:155–57.

Kopij, G., O. B. Kok, and Z. N. Roos. 1996. Food of sacred ibis *Threkiornis aethiopicus* nestlings in the Free State province, South Africa. *Ostrich* 67:138–43.

Kopp, K., M. Wachlevski, and P. C. Eterovick. 2006. Environmental complexity reduces tadpoles predation by water bugs. *Can. J. Zool.* 84:136–40.

Korshgen, L. J., and D. L. Moyle. 1955. Food habits of the bullfrog in central Missouri farm ponds. *Am. Midl. Nat.* 54:332–41.

Korshgen, L. J., and T. S. Baskett. 1963. Foods of impoundment and stream-dwelling bullfrogs in Missouri. *Herpetologica* 19:89–99.

Koskela, P., and S. Pasanen. 1974. The wintering of the common frog, *Rana temporaria* L., in northern Finland. *Aquilo (Zool.)* 15:1–17.

———. 1975. The reproductive biology of the female common frog, *Rana temporaria* L., in northern Finland. *Aquilo (Zool.)* 16:1–12.

Kostinsky, S. J., K. Miller, and C. N. Stewart. 2000. Sodium detection during the water absorption response in *Rana pipiens. Physiol. Behav.* 68:373–76.

Kowalewski, L. 1974. Observations on the phenology and ecology of Amphibia in the region of Czestochowa. *Acta Zool. Cracov.* 19:399–457.

Koyama, H., E. R. Lewis, E. L. Leverenz, and R. A. Baird. 1982. Acute seismic sensitivity in the bullfrog ear. *Brain Res.* 250:168–72.

Koyama, H., T. Nagai, H. Takeuchi, and S. D. Hillyard. 2001. The spinal nerves innervate putative chemosensory cells in the ventral skin of desert toads, *Bufo alvarius. Cell Tiss. Res.* 304:185–92.

Kozak, K. H. 2003. Sexual isolation and courtship behavior in salamanders of the *Eurycea bislineata* species complex, with comments on the evolution of the mental gland and pheromone delivery behavior in the Plethodontidae. *Southeast. Nat.* 2:281–92.

Kozak, K. H., A. Larson, R. M. Bonett, and L. J. Harmon. 2005. Phylogenetic analysis of ecomophological divergence, community structure, and diversification rates in dusky salamanders (Plethodontidae: *Desmognathus*). *Evolution* 59:2000–2016.

Kozłowska, M. 1971. Differences in the reproductive biology of mountain and lowland common frogs, *Rana temporaria* L. *Acta Biol. Cracov. (Zool.)* 14:17–32.

Kraft, P. G., R. S. Wilson, and C. E. Franklin. 2005. Predator-mediated phenotypic plasticity in tadpoles of the striped marsh frog, *Limnodynastes peronii. Aust. Ecol.* 30:558–63.

Krajik, K. 2006. The lost world of the Kihansi toad. *Science* 311:1230–32.

Krakauer, T. 1970. Tolerance limits of the toad, *Bufo marinus,* in south Florida. *Comp. Biochem. Physiol.* 33:15–26.

Kramer, A., A. Kupfer, and W. Himstedt. 2001. Haltung und Zucht der thailändischen Blindwühle *Ichthyophis kohtaoensis* (Amphibia: Gymnophiona: Ichthyophiidae). *Salamandra* 37:1–10.

Kramer, D. C. 1973. Movements of western chorus frogs, *Pseudacris triseriata triseriata* tagged with $CO^{60}$. *J. Herpetol.* 7:231–35.

———. 1974. Home range of the western chorus frog, *Pseudacris triseriata triseriata. J. Herpetol.* 8:245–46.

Kramer, P., N. Reichenbach, M. Hayslett, and P. Sattler. 1993. Population dynamics and conservation of the Peaks of Otter salamander, *Plethodon hubrichti. J. Herpetol.* 27:431–35.

Kraus, F., and A. Allison. 2001. A review of the endemic New Guinea microhylid frog genus *Choerophryne. Herpetologica* 57:214–32.

Kraus, F., and E. W. Campbell III. 2002. Human-mediated escalation of a formerly eradicable problem: the invasion of Caribbean frogs in the Hawaiian Islands. *Biol. Invas.* 4:327–32.

Kraus, F., and J. W. Petranka. 1989. A new sibling species of *Ambystoma* from the Ohio River drainage. *Copeia* 1989:94–110.

Krawczyk, S. 1971. Changes in lipid and water content in some organs of the common frog, *Rana temporaria* L. *Acta Biol. Cracov. (Zool.)* 14:211–37.

Krebs, H. A. 1975. The August Krogh Principle: "For many problems there is an animal on which it can be most conveniently studied." *J. Exp. Zool.* 194:221–26.

Krebs, J. 1978. Optimal foraging: decision rules for predators. In *Behavioural ecology: An evolutionary approach*, ed. J. Krebs and N. B. Davies, 23–63. Sunderland, MA: Sinauer Associates.

Krebs, S. L., and R. A. Brandon. 1984. A new species of salamander (family Ambystomatidae) from Michoacán, Mexico. *Herpetologica* 40:238–45.

Krenz, J. D., and D. E. Scott. 1994. Terrestrial courtship affects mating locations in *Ambystoma opacum. Herpetologica* 50:46–50.

Krishna, S., and K. K. Vijayalaxmi. 2004. *Bufo parietalis* (Forest Toad). Predation. *Herp. Rev.* 35:156–57.

Krishna, S. N., S. B. Krishna, and K. K. Vijayalaxmi. 2004. Breeding ecology of a rare microhylid, *Ramanella montana,* in the forests of Western Ghats, India. *Curr. Sci.* 87:80–82.

Krishnamurthy, S. V. 1999. Amphibian diversity in a few selected environs of Western Ghats. In *Biodiversity of the Western Ghats complex of Karnataka. Resource potential and sustainable utilization,* ed. S. A. Hussein and K. P. Achar, 107–18. Mangalore, India: Biodiversity Initiative Trust.

———. 2003. Amphibian assemblages in undisturbed and disturbed areas of Kudremukh National Park, central Western Ghats, India. *Environ. Cons.* 30:274–82.

Krishnamurthy, S. V., and S. A. Hussein. 2000. Amphibian fauna of Kudremukh National Park, Western Ghats, India. *J. Bombay Nat. Hist. Soc.* 97:436–37.

Krogh, A. 1904. On the cutaneous and pulmonary respiration of the frog. *Skand. Arch. Physiol.* 15:328–419.

———. 1937. Osmotic regulation in the frog (*Rana esculenta*) by active absorption of chloride ions. *Skand. Arch. Physiol.* 76:60–74.

———. 1938. The active absorption of ions in some freshwater animals. *Z. Vergl. Physiol.* 25:335–50.

———. 1939. *Osmotic regulation in aquatic animals.* Cambridge: Cambridge University Press.

———. 1941. *The comparative physiology of respiratory mechanisms.* Philadelphia: University of Pennsylvania Press.

Krügel, P., and S. Richter. 1995. *Syncope antenori*—A bromeliad breeding frog with free-swimming, nonfeeding tadpoles (Anura, Microhylidae). *Copeia* 1995:955–63.

Kruhoffer, M., M. L. Glass, A. S. Abe, and K. Johansen. 1987. Control of breathing in an amphibian *Bufo paracnemis:* effects of temperature and hypoxia. *Respir. Physiol.* 69:267–75.

Krupa, J. J. 1986. Multiple clutch production in the Great Plains toad. *Prairie Nat.* 18:151–52.

———. 1988. Fertilization efficiency in the Great Plains toad (*Bufo cognatus*) *Copeia* 1988:800–802.

———. 1989. Alternative mating tactics in the Great Plains toad. *Anim. Behav.* 37:1035–43.

———. 1990. Advertisement call variation in the Great Plains toad. *Copeia* 1990:884–86.

———. 1994. Breeding biology of the Great Plains toad in Oklahoma. *J. Herpetol.* 28:217–24.

———. 1995. How likely is male mate choice among anurans? *Behaviour* 132:643–64.

Kruse, K. 1990. Male backspace availability in the giant waterbug (*Belostoma flumineum* Say). *Behav. Ecol. Sociobiol.* 26:281–89.

Kruse, K. C. 1981a. Mating success, fertilization potential, and male body size in the American toad (*Bufo americanus*). *Herpetologica* 37:228–33.

———. 1981b. Phonotactic responses of female northern leopard frogs (*Rana pipiens*) to *Rana blairi,* a presumed hybrid, and conspecific mating trills. *J. Herpetol.* 15:145–50.

———. 1983. Optimal foraging by predaceous diving beetle larvae on toad tadpoles. *Oecologia* 58:383–88.

Kruse, K. C., and M. G. Francis. 1977. A predation deterrent in larvae of the bullfrog, *Rana catesbeiana. Trans. Am. Fish. Soc.* 106:248–52.

Kruse, K. C., and B. M. Stone. 1984. Largemouth bass (*Micropterus salmoides*) learn to avoid feeding on toad (*Bufo*) tadpoles. *Anim. Behav.* 32:1035–39.

Krutch, J. W. 1949. *The twelve seasons.* New York: William Sloane Associates.

Kruuk, H. 1989. *The social badger.* New York: Oxford University Press.

Kruuk, H., B. Kanchanasaka, S. O'Sullivan, S. Wanghongsa. 1994. Niche separation in three sympatric otters, *Lutra perspicillata, L. lutra,* and *Aonyx cinerea,* in Huai-Kha-Khaeng, Thailand. *Biol. Cons.* 69:115–20.

Kruuk, L. E. B., J. S. Gilchrist, and N. H. Barton. 1999. Hybrid disfunction in fire-bellied toads (*Bombina*). *Evolution* 53:1611–16.

Krzysik, A. J. 1979. Resource allocation, coexistence, and the niche structure of a streambank salamander community. *Ecol. Monogr.* 49:173–94.

———. 1980. Trophic aspects of brooding behavior in *Desmognathus fuscus fuscus. J. Herpetol.* 14:426–28.

Krzysik, A. J., and E. B. Miller. 1979. Substrate selection by three species of desmognathine salamanders from southwestern Pennsylvania: An experimental approach. *Ann. Carnegie Mus.* 48:111–17.

Kuchta, S. R. 2005. Experimental support for aposematic coloration in the salamander *Ensatina eschscholtzii xanthoptica:* Implications for mimicry of Pacific newts. *Copeia* 2005:265–71.

Kuhn, B., and H. Schneider. 1984. Mating and territorial calls of *Rana ridibunda* and their temperature-dependent variability. *Zool. Anz.* 212:273–305.

Kuhn, J. 1987. Strassentod der Erdkröte (*Bufo bufo* L.): Verlustquoten und Verkehrsaufkommen, Verhalten auf der Strasse. *Beih. Veröff. Naturschutz u. Landschaftspflege in Baden-Württemberg* 41:175–86.

———. 1994a. Lebensgeschichte und Demographie von Erdkrotenweibchen *Bufo bufo bufo* (L.). *Z. Feldherpetol.* 1:3–87.

———. 1994b. Methoden der Anuren-Markierung für Freilandstudien: Übersicht—Knie Ringetiketten—Erfahrungen mit der Phalangenamputation. *Z. Feldherpetol.* 1:177–92.

Kulzer, E. 1954. Untersuchungen über die Schreckreaktion der Erdkrötenkaulquappen (*Bufo bufo* L.). *Z. Vergl. Physiol.* 36:443–63.

Kumbar, S. M., and K. Pancharatna. 2001. Determination of age, longevity and age at reproduction of the frog *Microhyla ornata* by skeletochronology. *J. Biosci.* 26:265–70.

Kunde, F. 1857. Über Wasserentziehung und Bildung vourbegehender Katarakte. *Z. Wiss. Zool.* 8:466–86.

Kunitomo, K. 1910. Über die Entwicklungsgeschichte des *Hynobius nebulosus. Anat. Hefte* 40:194–283.

Kunte, K. 2004. Natural history and reproductive behavior of *Nyctibatrachus cf. humayuni* (Anura: Ranidae). *Herp. Rev.* 35:137–40.

Kunte, K. J. 1996. Grey musk shrews (*Suncus murinus*) feeding on toad. *J. Bombay Nat. Hist. Soc.* 93:581.

Kupfer, A., D. J. Gower, and W. Himstedt. 2003. Field observations of the predation of the caecilian amphibian, genus *Ichthyophis* (Fitzinger, 1826), by the red-tailed pipe snake *Cylindrophis rufus* (Laurenti, 1768). *Amphibia-Reptilia* 24:201–34.

Kupfer, A., J. Nabhitabhata, and W. Himstedt. 2004. Reproductive ecology of female caecilian amphibians (genus *Ichthyophis*). A baseline study. *Biol. J. Linn. Soc.* 83:207–17.

Kupfer, A., J. Nabhitabhata, and W. Himstedt. 2005. Life history of amphibians in the seasonal tropics: Habitat, community and population ecology of a caecilian (genus *Ichthyophis*). *J. Zool., London* 266:237–47.

Kupferberg, S. J. 1997a. Bullfrog (*Rana catesbeiana*) invasion of a California river: The role of larval competition. *Ecology* 78:1736–51.

———. 1997b. Facilitation of periphyton production by tadpole grazing: Functional differences between species. *Freshwater Biol.* 37:427–39.

———. 1997c. The role of larval diet in anuran metamorphosis. *Amer. Zool.* 37:146–59

———. 1998. Predator mediated patch use by tadpoles (*Hyla regilla*): risk balancing or consequence of motionlessness? *J. Herpetol.* 32:84–92.

Kupferberg, S. J., J. C. Marks, and M. E. Power. 1994. Effects of variation in natural algal and detrital diets on larval anuran (*Hyla regilla*) life-history traits. *Copeia* 1994:446–57.

Kuramoto, M. 1973. The amphibians of Iriomote of the Ryukyu Islands: Ecological and zoogeographical notes. *Bull. Fukuoka Univ. Educ.* 22:139–51.

———. 1975. Embryonic temperature adaptation in development rate of frogs. *Physiol. Zool.* 48:360–66.

———. 1978a. Correlation of quantitative parameters of fecundity in amphibians. *Evolution* 32:287–96.

———. 1978b. Thermal tolerance of frog embryos as a function of developmental stage. *Herpetologica* 34:417–22.

———. 1985. A new frog (genus *Rana*) from the Yaeyama group of the Ryukyu Islands. *Herpetologica* 41:150–58.

———. 1986. Call structure of the rhacophorid frogs from Taiwan. *Sci. Rep. Lab. Amphib. Biol. Hiroshima Univ.* 8:45–68.

Kuramoto, M., and C. Wang. 1987. A new rhacophorid treefrog from Taiwan, with comparisons to *Chirixalus eiffingeri* (Anura, Rhacophoridae). *Copeia* 1987:931–42.

Kuramoto, M., C.-S. Wang, and H.-T. Yu. 1984. Breeding, larval morphology and experimental hybridization of Taiwanese brown frogs, *Rana longicrus* and *R. sauteri*. *J. Herpetol.* 18:387–95.

Kurlansky, M. 1998. *Cod. A biography of the fish that changed the world.* New York: Penguin Books.

Kursar, T. A., S. J. Wright, and R. Radulovich. 1995. The effects of the rainy season and irrigation on soil water and oxygen in a seasonal forest in Panama. *J. Trop. Ecol.* 11:497–516.

Kurzava, L. M., and P. J. Morin. 1994. Consequences and causes of geographic variation in the body size of a keystone predator, *Notophthalmus viridescens*. *Oecologia* 99:271–80.

———. 1998. Test of functional equivalence: complementary roles of salamanders and fish in community organization. *Ecology* 79:477–89.

Kusano, T. 1980. Breeding and egg survival of a population of a salamander, *Hynobius nebulosus tokyoensis* Tago. *Res. Pop. Ecol.* 21:181–96.

———. 1981. Growth and survival rate of the larvae of *Hynobius nebulosus* tokyoensis (Amphibia, Hynobiidae). *Res. Pop. Ecol.* 23:360–78.

———. 1998. A radio-tracking study of post-breeding dispersal of the treefrog, *Rhacophorus arboreus* (Amphibia: Rhacophoridae). *Jap. J. Herpetol.* 17:98–106.

Kusano, T., and K. Fukuyama. 1989. Breeding activity of a stream-breeding frog (*Rana* sp.). In *Current herpetology in East Asia,* ed. M. Matsui, T. Hikida, and R. C. Goris, 314–22. Kyoto, Japan: Herpetological Society of Japan.

Kusano, T., K. Fukuyama, and N. Miyashita. 1995. Age determination of the stream frog, *Rana sakuraii,* by skeletochronology. *J. Herpetol.* 29:625–28.

Kusano, T., H. Kusano, and K. Miyashita. 1985. Size-related cannibalism among larval *Hynobius nebulosus*. *Copeia* 1985:472–76.

Kusano, T., K. Maruyama, and S. Kaneko. 1995. Post-breeding dispersal of the Japanese toad, *Bufo japonicus formosus*. *J. Herpetol.* 29:633–38.

———. 1999. Breeding site fidelity in the Japanese toad, *Bufo japonicus formosus*. *Herpetol. J.* 9:9–13.

Kusano, T., and K. Miyashita. 1984. Dispersal of the salamander, *Hynobius nebulosus tokyoensis*. *J. Herpetol.* 18:349–53.

Kusano, T., M. Toda, and K. Fukuyama. 1991. Testes size and breeding systems in Japanese anurans with special reference to large testes in the treefrog, *Rhacophorus arboreus* (Amphibia, Rhacophoridae). *Behav. Ecol. Sociobiol.* 29:27–31.

Kushlan, J. A. 1978. Feeding ecology of wading birds. In *Wading birds,* ed. A. Sprunt IV, J. C. Ogden, and S. Winckles, 249–97. New York: National Audubon Society.

Kutrup, B., U. Bulbul, and N. Yilmaz. 2005. Age structure in two populations of *Triturus vittatus ophryticus* at different altitudes. *Amphibia-Reptilia* 26:49–54.

Kuwabara, K., N. Suzuki, F. Wakabayashi, H. Ashikaga, I. Inoue, and J. Kobara. 1989. Breeding the Japanese giant salamander *Andrias japonicus*. *Int. Zoo Yb.* 28:22–31.

Kuyt, E. 1991. A communal overwintering site for the Canadian toad, *Bufo americanus hemiophrys,* in the Northwest Territories. *Can. Field-Nat.* 105:119–21.

Kuzmin, S. L. 1995. The problem of food competition in amphibians. *Herpetol. J.* 5:252–56.

———. 1999. *The amphibians of the former Soviet Union.* Sofia, Bulgaria: Pensoft.

Kuzmin, S. L., and I. V. Maslova. 2003. *The amphibians of the Russian Far East. Advances in amphibian research in the former Soviet Union 8.* Sofia, Bulgaria: Pensoft.

Kwet, A. 1996a. Zu den natürlichen Feinden des Laiches von Froschlurchen. *Salamandra* 32:31–44.

———. 1996b. Zur Überwinterung juveniler Erdkröten und anderer Amphibien im Naturschutzgebiet "Federsee". *Salamandra* 32:45–54.

Kyriakopoulou-Sklavounou, P., and M. E. Kattoulas. 1990. Contribution to the reproductive biology of *Rana ridibunda* Pallas (Anura, Ranidae). *Amphibia-Reptilia* 11:23–30.

Kyriakopoulou-Sklavounou, P., and N. Loumbourdis. 1990. Annual ovarian cycle in the frog, *Rana ridibunda,* in northern Greece. *J. Herpetol.* 24:185–91.

Laan, R., and B. Verboom. 1990. Effects of pool size and isolation on amphibian communities. *Biol. Cons.* 54:251–62.

Labanick, G. M. 1976. Prey availability, consumption and selection in the cricket frog, *Acris crepitans* (Amphibia, Anura, Hylidae). *J. Herpetol.* 10:293–98.

———. 1983. Inheritance of the red-leg and red-cheek traits in the salamander *Desmognathus ochrophaeus. Herpetologica* 39:114–20.

———. 1984. Anti-predator effectiveness of autotomized tails of the salamander *Desmognathus ochrophaeus. Herpetologica* 40:110–18.

———. 1988. Non-random association of red-leg and red-cheek coloration in the salamander *Desmognathus ochrophaeus. Herpetologica* 44:185–89.

Labanick, G. M., and R. A. Brandon. 1981. An experimental study of Batesian mimicry between the salamanders *Plethodon jordani* and *Desmognathus ochrophaeus. J. Herpetol.* 15:275–81.

Labanick, G. M., and G. T. Davis. 1978. The spermatophore of the small-mouthed salamander, *Ambystoma texanum* (Amphibia, Urodela, Ambystomatidae). *J. Herpetol.* 12:111–14.

Lacey, L. A. 1979. Predaçao en girinos por uma vespa e outras associacoes de insectos com minhos de duas espécies de ras da Amazônia. *Acta Amazônica* 9:755–62.

Laessle, A. M. 1961. A micro-limnological study of Jamaican bromeliads. *Ecology* 42:499–517.

LaFiandra, E. M., and K. J. Babbitt. 2004. Predator induced phenotypic plasticity in the pinewoods tree frog, *Hyla femoralis:* Necessary cues and the cost of development. *Oecologia* 138:350–59.

Lagerspetz, K. Y. H. 1977. Interactions of season and temperature acclimation in the control of metabolism in Amphibia. *J. Therm. Biol.* 2:223–31.

Lagler, K. F. 1943. Food habits and economic relations of the turtles of Michigan with special reference to fish management. *Am. Midl. Nat.* 29:257–312.

Lagler, K. F., and J. C. Salyer, II. 1945a. Food and habits of the common watersnake, *Natrix s. sipedon,* in Michigan. *Pap. Michigan Acad. Sci. Art. Lett.* 31:169–80.

———. 1945b. Influence of availability on the feeding habits of the common garter snake. *Copeia* 1945:159–62.

Lai, S.-J., Y-C. Kam, and Y.-S. Lin. 2003. Elevational variation in reproductive and life history traits of Sauter's frog, *Rana sauteri* Boulenger, 1909 in Taiwan. *Zool. Stud.* 42:193–202.

Lajmanovich, R. C. 1995. Relaciones tróficas de bufonidos (Anura, Bufonidae) en ambientes del Río Paraná, Argentina. *Alytes* 13:87–103.

Lamar, W. W., and E. R. Wild. 1995. Comments on the natural history of *Lithodytes lineatus* (Anura: Leptodactylidae), with a description of the tadpole. *Herpetol. Nat. Hist.* 3:135–42.

La Marca, E. 1986. Description of the tadpole of *Ceratophrys calcarata. J. Herpetol.* 20:459–61.

———. 1992. Catálogo taxonómico, biogeográfico y bibliográfico de las ranas de Venezuela. *Cuad. Geográf. Univ. Los Andes, Mérida* 9:1–197.

———. 1994. Descripción de un género nuevo de ranas (Amphibia: Dendrobatidae) de la Cordillera de Mérida, Venezuela. *Anuario de Investigacion 1991, I.G.C.R.N., Universidad de los Andes, Mérida* 1991:39–41.

———. 1995. Biological and systematic synopsis of a genus of frogs from northern mountains of South America (Anura: Dendrobatidae: *Mannophryne*). *Bull. Maryland Herpetol. Soc.* 31:40–77.

La Marca, E., K. R. Lips, S. Lötters, R. Puschendorf, R. Ibáñez, J. V. Rueda-Almonacid, R. Schulte, C. Marty, F. Castro, J. Manzanilla-Puppo, J. E. García-Pérez, F. Bolaños, G. Chaves, J. A. Pounds, E. Toral, and B. E. Young. 2005. Catastrophic population declines and extinctions in Neotropical harlequin frogs (Bufonidae: *Atelopus*). *Biotropica* 37:190–201.

Lamb, T. 1984a. Amplexus displacement in the southern toad, *Bufo terrestris. Copeia* 1984:1023–25.

———. 1984b. The influence of sex and breeding condition on microhabitat selection and diet in the pig frog, *Rana grylio. Am. Midl. Nat.* 111:311–18.

———. 1987. Call site selection in a hybrid population of treefrogs. *Anim. Behav.* 35:1140–44.

Lamb, T., and J. C. Avise. 1986. Directional introgression of mitochondrial DNA in a hybrid population of tree frogs: the influence of mating behavior. *Proc. Natl. Acad. Sci. USA* 83:2526–30.

Lamberti, G. A., S. V. Gregory, C. P. Hawkins, R. C. Wildman, L. R. Ashkenas, and D. M. Denicola. 1992. Plant-herbivore interactions in streams near Mt. St. Helens. *Freshw. Biol.* 27:237–47.

Lambiris, A. J. L. 1971. A note on "parental care" in the bullfrog, *Pyxicephalus a. adspersus. J. Herp. Assoc. Africa* 8:6.

Lambrinos, J. G., and C. C. Kleier. 2003. Thermoregulation of juvenile Andean toads (*Bufo spinulosus*) at 4300 m. *J. Therm. Biol.* 28:15–19.

Lamoreux, J. F., J. C. Morrison, T. H. Ricketts, D. M. Olson, E. Dinerstein, M. W. McKnight, and H. H. Shugart. 2006. Global tests of biodiversity concordance and the importance of endemism. *Nature* 440:212–14.

Lamotte, M. 1954. Le développement larvaire de deux espèces rhéophiles, *Astylosternus diadematus* et *Petropedetes natator. Bull. IFAN* 16, ser. A:1222–33.

———. 1967. Les Batraciens de la region de Gpakobo (Côte d'Ivoire). *Bull. IFAN* 29 (A):218–54.

———. 1983. Amphibians in savanna ecosystems. In *Tropical savannas,* ed. F. Bourliere, 313–23. Amsterdam: Elsevier.

Lamotte, M., and J. Lescure. 1977. Tendances adaptatives a l'affranchissement du milieu aquatique chez les amphibiens anoures. *Terre et Vie* 31:225–311.

———. 1989a. Les têtards rhéophiles et hygropétriques de l'Ancien et du Nouveau Monde. *Ann. Sci. Natur., Zool.,* 13th ser., 10:111–22.

———. 1989b. Les têtards rhéophiles et hygropétriques de l'Ancien et du Nouveau Monde (suite). *Ann. Sci. Natur., Zool.,* 13th ser., 10:125–44.

Lamotte, M., and J.-L. Perret. 1961. Les formes larvaires de quelques espèces de *Leptopelis: L. aubryi, L. viridis, L. anchiaeta, L. ocellata* et *L. calcaratus. Bull. IFAN* 23, ser. A:855–85.

———. 1963a. Contribution a l'étude des Batraciens de l'Ouest Africain. XIV. Le développement larvaire de *Chiromantis rufescens* Gunther. *Bull. IFAN* 25, ser. A:265–76.

———. 1963b. Contribution a l'étude des Batraciens de l'Ouest Africain. XV. Le développement direct de l'espèce *Arthroleptis poecinolotus* Peters. *Bull. IFAN* 25, ser. A:277–84.

———. 1968. Revision du genre *Conraua* Nieden. *Bull. IFAN* 30, ser. A:1603–44.

Lamotte, M., J.-L. Perret, and S. Dzieduszycka. 1959. Les formes larvaires de *Petropedetes palmipes, Conraua goliath,* et *Acanthixalus spinosus. Bull. IFAN* 21, ser. A:762–76.

Lamotte, M., and C. Sanchez-Lamotte. 1999. Adaptation aux particularités climatiques du cycle biologique d'un anoure tropical, *Nectophrynoides occidentalis* Angel, 1943 (Bufonidae). *Alytes* 16:111–22.

Lamotte, M., and F. Xavier. 1972. Les amphibiens anoures a développement direct d'Afrique. Observations sur la biologie de *Nectophrynoides tornieri* (Roux). *Bull. Soc. Zool. France* 97:413–18.

Lamoureux, V. S., and D. M. Madison. 1999. Overwintering habitats of radio-implanted green frogs, *Rana clamitans. J. Herpetol.* 33:430–35.

Lampert, K. P., and K. E. Linsenmair. 2002. Alternative life cycle strategies in the West African reed frog *Hyperolius nitidulus:* The answer to an unpredictable environment? *Oecologia* 130:364–72.

Lampert, K. P., A. S. Rand, U. G. Mueller, and M. J. Ryan. 2003. Fine-scale genetic pattern and evidence for sex-biased dispersal in the túngara frog, *Physalaemus pustulosus. Mol. Ecol.* 12:3325–34.

Lampo, M., and V. Medialdea. 1996. Energy allocation patterns in *Bufo marinus* from two habitats in Venezuela. *J. Trop. Ecol.* 12:321–31.

Lancaster, D. L., and R. G. Jaeger. 1995. Rules of engagement for adult salamanders in territorial conflicts with heterospecific juveniles. *Behav. Ecol. Sociobiol.* 37:25–29.

Lancaster, D. L., and S. E. Wise. 1996. Differential response by the ringneck snake, *Diadophis punctatus,* to odors of tail-autotomizing prey. *Herpetologica* 52:98–108.

Lance, S. L., and K. D. Wells. 1993. Are spring peeper satellite males physiologically inferior to calling males? *Copeia* 1993:1162–66.

Lanctot, R. B., and P. J. Weatherhead. 1997. Ephemeral lekking behavior in the buff-breasted sandpiper, *Tryngites subruficollis. Behav. Ecol.* 8:268–78.

Lande, R. 1981. Models of speciation by sexual selection of polygenic traits. *Proc. Natl. Acad. Sci. USA* 78:3721–25.

Landreth, H. F., and D. E. Ferguson. 1966. Evidence of sun-compass orientation in the chorus frog, *Pseudacris triseriata. Herpetologica* 22:106–12.

———. 1967a. Movements and orientation of the tailed frog, *Ascaphus truei. Herpetologica* 23:81–93.

———. 1967b. Newt orientation by sun-compass. *Nature* 215:516–18.

———. 1967c. Newts: Sun-compass orientation. *Science* 158:1459–61.

———. 1968. The sun compass of Fowler's toad, *Bufo woodhousei fowleri. Behaviour* 30:27–43.

Landreth, H. F., and M. T. Christensen. 1971. Orientation of the plains spadefoot toad, *Scaphiopus bombifrons,* to solar cues. *Herpetologica* 27:454–61.

Landry, S. O., Jr. 1970. The Rodentia as omnivores. *Q. Rev. Biol.* 45:351–72.

Lane, S. J., and M. J. Mahony. 2002. Larval anurans with synchronous and asynchronous developmental periods: Contrasting responses to water reduction and predator presence. *J. Anim. Ecol.*71:780–92.

Lang, C. 1995. Size-fecundity relationships among stream-breeding hylid frogs. *Herpetol. Nat. Hist.* 3:193–97.

Lang, C., and R. G. Jaeger. 2000. Defense of territories by male-female pairs in the red-backed salamander (*Plethodon cinereus*). *Copeia* 2000:169–77.

Langbein, J., and S. J. Thirgood. 1989. Variation in mating systems of fallow deer (*Dama dama*) in relation to ecology. *Ethology* 83:195–214.

Langhelle, A., M. J. Lindell, and P. Nyström. 1999. Effects of ultraviolet radiation on amphian embryonic and larval development. *J. Herpetol.* 33:449–56.

Langlois, J. P., and J. Pellegrin. 1902. De la deshydration chez le crapaud et des variations correlatives de la densite du sang. *C. R. Soc. Biol.* 54:1377–79.

Langone, J. A. 1993. Notas sobre *Phyllomedusa iheringii* Boulenger, 1885 (Amphibia, Anura, Hylidae). *Com. Zool. Mus. Hist. Nat. Montevideo* (179):1–7.

Langone, J. A., C. M. Prigioni, and L. Venturino. 1985. Informe preliminar sobre el comportamiento reproductor y otros aspectos de la biología de *Phyllomedusa iheringi,* Boulenger, 1885 (Anura, Hylidae). *Com. Zool. Mus. Hist. Nat. Montevideo* (152):1–12.

Langton, T. E. S., ed. 1989. *Amphibians and roads.* Shefford, Bedfordshire, UK: ACO Polymer Products.

Lannoo, M. J. 1996. *Okoboji wetlands: A lesson in natural history.*

———. 1998a, ed. *Status and conservation of Midwestern amphibians.* Iowa City: University of Iowa Press.

———. 1998b. Amphibian conservation and wetland management in the Upper Midwest: A Catch-22 for the cricket frog? In *Status and conservation of Midwestern amphibians,* ed. M. J. Lannoo, 330–39. Iowa City: University of Iowa Press.

———. 1998c. Introduction. In *Status and conservation of Midwestern amphibians,* ed. M. J. Lannoo, xi–xviii. Iowa City: University of Iowa Press.

———. 1999. Integration: Nervous and sensory systems. In *Tadpoles. The biology of anuran larvae,* ed. R. W. McDiarmid and R. Altig, 149–69. Chicago: University of Chicago Press.

———. 2005, ed. *Amphibian declines: The conservation status of United States species.* Berkeley: University of California Press.

Lannoo, M. J., and M. D. Bachmann. 1984a. Aspects of cannibalistic morphs in a population of *Ambystoma t. tigrinum* larvae. *Am. Midl. Nat.* 112:103–10.

———. 1984b. On flotation and air breathing in *Ambystoma tigrinum* larvae: stimuli for and the relationship between these behaviors. *Can. J. Zool.* 62:15–18.

Lannoo, M. J., K. Lang, T. Waltz, and G. S. Phillips. 1994. An altered amphibian assemblage: Dickinson County, Iowa, 70 years after Frank Blanchard's survey. *Am. Midl. Nat.* 131:311–19.

Lannoo, M. J., L. Lowcock, and J. P. Bogart. 1989. Sibling cannibalism in noncannibal morph *Ambystoma tigrinum* larvae and its correlation with high growth rates and early metamorphosis. *Can. J. Zool.* 67:1911–14.

Lannoo, M. J., D. R. Sutherland, P. Jones, D. Rosenberry, R. W. Klaver, D. M. Hoppe, P. T. J. Johnson, K. B. Lunde, C. Facemire, and J. P. Kapfer. 2003. Multiple causes for the malformed frog phenomenon. In *Multiple stressor effects in relation to declining amphibian populations,* ed. G. Linder, S. Krest, D. Sparling, and E. Little, 233–62. West Conshoschocken, PA: American Society for Testing Materials International.

Lannoo, M. J., D. S. Townsend, and R. J. Wassersug. 1987. Larval life in the leaves: Arboreal tadpole types, with special attention to the morphology, ecology, and behavior of the oophagous *Osteopilus brunneus* (Hylidae) larva. *Fieldiana: Zool.* 38:1–31.

Lanza, B., J. M. Cei, and E. G. Crespo. 1976. Further immunological evidence for the validity of the family Bombinatoridae (Amphibia, Salientia). *Monit. Zool. Ital.* (n.s.) 10:311–14.

Laposata, M. M., and W. A. Dunson. 2000. Effects of spray-irrigated wastewater effluent on temporary pond-breeding amphibians. *Ecotox. Env. Safety* 46:192–201.

Lardner, B. 1995. Larval ecology of *Rana arvalis:* an allopatric island population compared with a sympatric mainland population. *Amphibia-Reptilia* 16:101–11.

———. 1998. Plasticity or fixed adaptive traits? Strategies for predation avoidance in *Rana arvalis* tadpoles. *Oecologia* 117:119–26.

———. 2000. Morphological and life history responses to predators in larvae of seven anurans. *Oikos* 88:169–80.

Lardner, B., and M. B. Lakim. 2002. Tree-hole frogs exploit resonance effects. *Nature* 420:475.

———. 2004. Female call preferences in tree-hole frogs: Why are there so many unattractive males? *Anim. Behav.* 68:265–72.

Lardner, B., and J. Loman. 2003. Growth or reproduction? Resource allocation by female frogs *Rana temporaria. Oecologia* 442:541–46.

Largen, M. J. 1977. The status of the genus *Leptopelis* (Amphibia: Anura: Hyperoliidae) in Ethiopia, including descriptions of two new species. *Monit. Zool. Ital.,* n.s. suppl. 9:85–136.

Larsen, J. H., Jr., and D. J. Guthrie. 1975. The feeding system of terrestrial tiger salamanders (*Ambystoma tigrinum melanostictum* Baird). *J. Morphol.* 147:137–54.

Larson, A., and W. W. Dimmick. 1993. Phylogenetic relationships of the salamander families: An analysis of congruence among morphological and molecular characters. *Herpetol. Monogr.* 7:77–93.

Larson, A., and R. Highton. 1978. Geographic protein variation and divergence in salamanders of the *Plethodon welleri* group (Amphibia, Plethodontidae). *Syst. Zool.* 27:431–48.

Larson, A., D. W. Weisrock, and K. H. Kozak. 2003. Phylogenetic systematics of salamanders (Amphibia: Urodela), a review. In *Reproductive biology and phylogeny of Urodela,* ed. D. M. Sever, 31–108. Enfield, NH: Science Publishers.

Larson, D. W. 1968. The occurrence of neotenic salamanders, *Ambystoma tigrinum diaboli,* in Devil's Lake, North Dakota. *Copeia* 1968:620–21.

———. 1971. Desiccation tolerance in three races of *Ambystoma tigrinum* in North Dakota. *Copeia* 1971:743–47.

Larson, K. 2004. Advertisement call complexity in northern leopard frogs, *Rana pipiens. Copeia* 2004:676–82.

Larson, K. L., W. Duffy, E. Johnson, M. F. Donovan, and M. J. Lannoo. 1999. "Paedocannibal" morph barred tiger salamanders (*Ambystoma tigrinum mavortium*) from eastern South Dakota. *Am. Midl. Nat.* 141:124–39.

Larson, P. M., and S. M. Reilly. 2002. Functional morphology of feeding and gill irrigation in the anuran tadpole: Electromyography and muscle function in larval *Rana catesbeiana. J. Morphol.* 255:202–14.

Lasiewski, R. C., and W. R. Dawson. 1967. A re-examination of the relation between standard metabolic rate and body weight in birds. *Condor* 69:13–23.

Lathrop, A. 1997. Taxonomic review of the megophryid frogs (Anura: Pelobatoidea). *Asiatic Herpetol. Res.* 7:68–79.

Lathrop, A., R. W. Murphy, N. L. Orlov, and C. T. Ho. 1998. Two new species of *Leptobrachium* (Anura: Megophryidae) from the central highlands of Vietnam with a redescription of *Leptobrachium chapaense. Russ. J. Herpetol.* 5:51–60.

Latimer, W. 1981. Acoustic competition in bush crickets. *Ecol. Entomol.* 6:35–45.

Latimer, W., and W. B. Broughton. 1984. Acoustic interference in bush crickets: A factor in the evolution of singing insects? *J. Nat. Hist.* 18:599–616.

Lau, M. W. N. and D. Dudgeon. 1999. Composition and distribution of Hong Kong amphibian fauna. *Mem. Hong Kong Nat. Hist. Soc.* 22:1–79.

Lauck, B., R. Swain, and L. Barmuta. 2005. Impacts of shading on larval traits of the frog *Litoria ewingii* in a commercial forest, Tasmania, Australia. *J. Herpetol.* 39:478–86.

Lauder, G. V. 1985. Aquatic feeding in lower vertebrates. In *Functional vertebrate morphology,* ed. M. Hildebrand, D. M. Bramble, K. F. Liem, and D. B. Wake, 210–29. Cambridge, MA: Harvard University Press.

Lauder, G. V., and S. M. Reilly. 1988. Functional design of the feeding mechanism in salamanders: Causal basis of ontogenetic changes in function. *J. Exp. Biol.* 134:219–33.

———. 1990. Metamorphosis of the feeding mechanism in tiger salamanders (*Ambystoma tigrinum*): the ontogeny of cranial muscle mass. *J. Zool., London* 222:59–74.

Lauder, G. V., and H. B. Shaffer. 1985. Functional morphology of the feeding mechanism in aquatic ambystomatid salamanders. *J. Morphol.* 185:297–326.

———. 1986. Functional design of the feeding mechanism in lower vertebrates: Unidirectional and bidirectional flow systems in the tiger salamander. *Zool. J. Linn. Soc.* 88:277–90.

———. 1988. Ontogeny of functional design in tiger salamanders (*Ambystoma tigrinum*): Are motor patterns conserved during major morphological transformations? *J. Morphol.* 197:249–68.

———. 1993. Design of feeding systems in aquatic vertebrates: major patterns and their evolutionary interpretations. In *The skull,* vol. 3. *Functional and evolutionary mechanisms,* eds. J. Hanken and B. K. Hall, 113–49. Chicago: University of Chicago Press.

Laugen, A. T., A. Laurila, and J. Merilä. 2002. Maternal and genetic contributions to geographical variation in *Rana temporaria* larval life-history traits. *Biol. J. Linn. Soc.* 76:61–70.

Laugen, A. T., A. Laurila, K. Räsänen, and J. Merilä. 2003. Latitudinal countergradient variation in the common frog (*Rana temporaria*) development rates—Evidence for local adaptation. *J. Evol. Biol.* 16:996–1005.

Laurance, W. F. 1996. Catastrophic declines of Australian rainforest frogs: Is unusual weather responsible? *Biol. Cons.* 77:203–12.

Laurance, W. F., K. R. McDonald, and R. Speare. 1996. Epidemic disease and the catastrophic decline of Australian rain forest frogs. *Cons. Biol.* 10:406–13.

Laurent, R. F. 1965. The geographical variation of the frog *Hyperolius marmoratus* (Family Hyperoliidae) in Rhodesia, Nyasaland, and Tanganyika. *Breviora* 216:1–9.

———. 1986. Sous classe des lissamphibiens. Systematique. In *Traite de zoologie,* vol. 14, 1B, ed. P.-P. Grasse and M. Delsol, 594–797. Paris: Masson.

Laurila, A. 1998. Breeding habitat selection and larval performance of two anurans in freshwater rock-pools. *Ecography* 21:484–94.

———. 2000a. Behavioural responses to predator chemical cues and local variation in antipredator performance in *Rana temporaria* tadpoles. *Oikos* 88:159–68.

———. 2000b. Competitive ability and the coexistence of anuran larvae in freshwater rock-pools. *Freshwater Biol.* 43:161–74.

Laurila, A., and T. Aho. 1997. Do female common frogs choose their breeding habitats to avoid predation on tadpoles? *Oikos* 78:585–91.

Laurila, A., P.-A. Crochet, and J. Merilä. 2001. Predation-induced effects on hatchling morphology in the common frog (*Rana temporaria*). *Can. J. Zool.* 79:926–30.

Laurila, A., M. Järvi-Laturi, S. Pakkasmaa, and J. Merilä. 2004. Temporal variation in predation risk: Stage-dependency, graded responses and fitness costs in tadpole antipredator defences. *Oikos* 107:90–99.

Laurila, A., S. Karttunen, and J. Merilä. 2002. Adaptive phenotypic plasticity and genetics of larval life histories in two *Rana temporaria* populations. *Evolution* 56:617–27.

Laurila, A., and J. Kujasalo. 1999. Habitat duration, predation risk and phenotypic plasticity in common frog (*Rana temporaria*) tadpoles. *J. Anim. Ecol.*68:1123–32.

Laurila, A., J. Kujasalo, and E. Ranta. 1997. Different antipredator behavior in two anuran tadpoles: Effects of predator diet. *Behav. Ecol. Sociobiol.* 40:329–36.

Laurila, A., J. Kujasalo, and E. Ranta. 1998. Predator-induced changes in life history in two anuran tadpoles: effects of predator diet. *Oikos* 83:307–17.

Laurila, A., and P. Seppa. 1998. Multiple paternity in the common frog (*Rana temporaria*): Genetic evidence from tadpole kin groups. *Biol. J. Linn. Soc.* 63:221–32.

Laurin, M. 2000. Seymouriamorphs. In *Amphibian biology. Vol. 4. Palaeontology,* ed. H. Heatwole and R. L. Carroll, 1064–80. Chipping Norton, New South Wales, Australia: Surrey Beatty and Sons.

———. 2002. Tetrapod phylogeny, amphibian origins, and the definition of the name Tetrapoda. *Syst. Biol.* 51:364–69.

Laurin, M., and R. R. Reisz. 1997. A new perspective on tetrapod phylogeny. In *Amniote origins: Completing the transition to land,* ed. S. S. Sumida and K. L. M. Martin, 9–59. San Diego: Academic Press.

Lawler, K. L., and J.-M. Hero. 1997. Palatability of *Bufo marinus* tadpoles to a predatory fish decreases with development. *Wildl. Res.* 24:327–34.

Lawler, S. P. 1989. Behavioural responses to predators and predation risk in four species of larval anurans. *Anim. Behav.* 38:1039–47.

Lawler, S. P., D. Dritz, T. Strange, and M. Holyoak. 1999. Effects of introduced mosquitofish and bullfrogs on the threatened California red-legged frog. *Cons. Biol.* 13:613–22.

Lawler, S. P., and P. J. Morin. 1993. Temporal overlap, competition, and priority effects in larval anurans. *Ecology* 74:174–82.

Lawson, D. P. 1993. The amphibians and reptiles of Korup National Park Project, Cameroon. *Herpetol. Nat. Hist.* 1:27–93.

Lawton, J. H. 1984. Non-competitive populations, non-convergent communities, and vacant niches: The herbivores of bracken. In *Ecological communities,* ed. D. R. Strong, Jr., D. Simberloff, L. G. Abele, and A. B. Thistle, 67–99. Princeton, NJ: Princeton University Press.

Lawton, R. O., U.S. Nair, R. A. Pielke, Sr., and R. M. Welch. 2001. Climatic impact of tropical lowland deforestation on nearby montane cloud forests. *Science* 294:584–87.

Layne, J. R., Jr. 1995. Seasonal variation in the cryobiology of *Rana sylvatica* from Pennsylvania. *J. Therm. Biol.* 20:349–53.

———. 1999. Freeze tolerance and cryoprotectant mobilization in the gray treefrog (*Hyla versicolor*). *J. Exp. Zool.* 283:221–25.

Layne, J. R., Jr., and D. L. Claussen. 1982. Seasonal variation in the thermal acclimation of critical thermal maxima ($CT_{Max}$) and minima ($CT_{Min}$) in the salamander *Eurycea bislineata. J. Therm. Biol.* 7:29–34.

Layne, J. R., Jr., J. P. Costanzo, and R. E. Lee, Jr. 1998. Freeze duration influences postfreeze survival in the frog *Rana sylvatica. J. Exp. Zool.* 280:197–201.

Layne, J. R., Jr., and M. C. First. 1991. Resumption of physiological function in the wood frog (*Rana sylvatica*) after freezing. *Am. J. Physiol.* 261:R134–R137.

Layne, J. R., Jr., and R. E. Lee, Jr. 1987. Freeze tolerance and the dynamics of ice formation in wood frogs (*Rana sylvatica*) from southern Ohio. *Can. J. Zool.* 65:2062–65.

———. 1989. Seasonal variation in freeze tolerance and ice content of the tree frog *Hyla versicolor. J. Exp. Zool.* 249:133–37.

Layne, J. R., Jr., R. E. Lee, Jr., and M. M. Cutwa. 1996. Post-hibernation excretion of glucose in urine of the freeze tolerant frog *Rana sylvatica. J. Herpetol.* 30:85–87.

Layne, J. R., Jr., R. E. Lee, Jr., and T. L. Heil. 1989. Freezing-induced changes in the heart rate of wood frogs (*Rana sylvatica*). *Am. J. Physiol.* 257:R1046–R1049.

Layne, J. R., Jr., R. E. Lee, Jr., and L. L. Huang. 1990. Inoculation triggers freezing at high subzero temperatures in a freeze-tolerant frog (*Rana sylvatica*) and insect (*Eurosta solidaginis*). *Can. J. Zool.* 68:506–10.

Layne, J. R., Jr., M. A. Romano, and S. I. Guttman. 1989. Responses to desiccation of the treefrogs *Hyla cinerea* and *H. gratiosa* and their natural hybrids. *Am. Midl. Nat.* 121:61–67.

Lazell, J. D., Jr. and R. A. Brandon. 1962. A new stygian salamander from the southern Cumberland Plateau. *Copeia* 1962:300–6.

Le Quang Trong, Y. 1971. Etude de la peau et des glandes cutanees de quelques amphibiens du genre *Phrynobatrachus. Bull. IFAN* 33:987–1025.

———. 1975a. Etude de la peau et des glandes cutanees de quelques amphibiens du genre *Ptychadena. Ann. Univ. Abidjan (E)* 8:31–52.

———. 1975b. La peau et les glandes cutanees de *Dicroglossus occipitalis* Gunther. *Ann. Univ. Abidjan (E)* 8:15–29.

Lea, J., M. Dyson, and T. Halliday. 2001. Calling by male midwife toads stimulates females to maintain reproductive condition. *Anim. Behav.* 61:373–77.

Lea, J., T. Halliday, and M. Dyson. 2000. Reproductive stage and history affect the phonotactic preferences of female midwife toads, *Alytes muletensis. Anim. Behav.* 60:423–27.

———. 2003. The mating strategy of *Alytes muletensis:* Some males are less ready to mate than females. *Amphibia-Reptilia* 24:169–80.

Lea, J. M., L. Luiselli, and E. Politano. 2005. Are there shifts in amphibian faunal composition in Nigerian landscapes undergoing long-term degradation? A case study from a montane environment. *Rev. Ecol. (Terre et Vie)* 60:65–76.

Lea, J. M., E. Politano, and L. Luiselli. 2003. Changes in the herpetofauna of a fresh water river in southern Nigeria, after 20 years of development. *Russ. J. Herpetol.* 10:191–98.

Leary, C. J., D. J. Fox, D;. B. Shepard, and A. M. Garcia. 2005. Body size, age, growth and alternative mating tactics in toads: Satellite males are smaller but not younger than calling males. *Anim. Behav.* 70:663–71.

Leary, C. J., T. S. Jessop, A. M. Garcia, and R. Knapp. 2004. Steroid hormone profiles and relative body condition of calling and satellite toads: Implications for proximate regulation of behavior in anurans. *Behav. Ecol.* 15:313–20.

Leary, C. J., and A. A. Parker. 1998. Egg deposition site characteristics of the red-eyed leaf frog, *Agalychnis callidryas* Cope (Hylidae). *Herpetol. Nat. Hist.* 6:55–59.

Leary, C. J., and V. R. Razafindratsita. 1998. Attempted predation on a hylid frog, *Phrynohyas venulosa,* by an indigo snake, *Drymarchon corais,* and the response of conspecific frogs to distress calls. *Amphibia-Reptilia* 19:442–46.

Lechner, A. J. 1978. The scaling of maximal oxygen consumption and pulmonary dimensions in small mammals. *Respir. Physiol.* 35:439–49.

Lecis, R., and K. Norris. 2003. Habitat correlates of distribution and local population decline of the endemic Sardinian newt *Euproctus platycephalus. Biol. Cons.* 115:303–17.

———. 2004. Population genetic diversity of the endemic Sardinian newt *Euproctus platycephalus:* Implications for conservation. *Biol. Cons.* 119:263–70.

Leclair, R., Jr., Y. Alarie, and J.-P. Bourassa. 1986. Prey choice in larval *Dytiscus harrisii* Kirby and *D. verticalis* Say (Coleoptera, Dytiscidae). *Entomologica Brasiliensia* 11:337–42.

Leclair, R., Jr. and J.-P. Bourassa. 1981. Observation et analyse de la prédation des oeufs d'*Ambystoma maculatum* (Shaw) (Amphibia, Urodela) par les larves de dipteres chironomides, dans la region de Trois-Rivieres (Canada). *Can. J. Zool.* 59:1339–43.

Leclair, R., Jr., J.-P. Bourassa, and D. Gagnon. 1983. Exploitation de pontes d'amphibien par des larves de dipteres chironomides: Etude de du coïncidence phénologique. *Bull. Soc. Entomol. France* 88:268–71.

Leclair, R., Jr., and J. Castanet. 1987. A skeletochronological assessment of age and growth in the frog *Rana pipiens* Schreber (Amphibia, Anura) from southwestern Quebec. *Copeia* 1987:361–69.

Leclair, R., and G. Laurin. 1996. Growth and body size in populations of mink frogs *Rana septentrionalis* from two latitudes. *Ecography* 19:296–304.

Lee, A. K. 1967. Studies in Australian Amphibia II. Taxonomy, ecology and evolution of the genus *Heleioporus* Gray (Anura: Leptodactylidae). *Aust. J. Zool.* 15:367–439.

———. 1968. Water economy of the burrowing frog, *Heleioporus eyrei* (Gray). *Copeia* 1968:741–45.

Lee, A. K., and E. H. Mercer. 1967. Cocoon surrounding desert-dwelling frogs. *Science* 157:87–88.

Lee, A. R., M. Silove, U. Katz, and J. B. Balinsky. 1982. Urea cycle enzymes and glutamate dehydrogenase in *Xenopus laevis* and *Bufo viridis* adapted to high salinity. *J. Exp. Zool.* 221:169–72.

Lee, D. S. 1968. Herpetofauna associated with central Florida mammals. *Herpetologica* 24:83–84.

———. 1969. Floridian herpetofauna associated with cabbage palms. *Herpetologica* 25:70–71.

Lee, J. C. 1980. An ecogeographic analysis of the herpetofauna of the Yucutan Peninsula. *Misc. Publ. Mus. Nat. Hist. Univ. Kansas* (67):1–75.

———. 1986. Is the large-male mating advantage in anurans an epiphenomenon? *Oecologia* 69:207–12.

———. 1993. Geographic variation in size and shape of Neotropical frogs: A precipitation gradient analysis. *Occ. Pap. Mus. Nat. Hist. Univ. Kansas* (163):1–20.

———. 1996. *The amphibians and reptiles of the Yucatan Peninsula.* Ithaca, NY: Cornell University Press.

———. 2001. Evolution of a secondary sexual dimorphism in the toad, *Bufo marinus. Copeia* 2001:928–35.

Lee, J. C., and A. D. Corrales. 2002. Sexual dimorphism in hind-limb muscle mass is associated with male reproductive success in *Bufo marinus. J. Herpetol.* 36:502–5.

Lee, J. C., and M. L. Crump. 1981. Morphological correlates of male mating success in *Triprion petasatus* and *Hyla marmorata* (Anura: Hylidae). *Oecologia* 50:153–57.

Lee, J. S. F., and B. Waldman. 2002. Communication by fecal chemosignals in an archaic frog, *Leiopelma hamiltoni. Copeia* 2002:679–86.

Lee, M. S. Y., and B. G. M. Jamieson. 1992. The ultrastructure of the spermatozoa of three species of myobatrachid frogs (Anura, Amphibia) with phylogenetic considerations. *Acta Zool. (Stockholm)* 73:213–22.

Lee, R. 1855. *Anecdotes of the habits and instincts of birds, reptiles, and fishes.* Philadelphia, PA: Lindsay and Blakiston.

Lee, R. E., Jr. 1988. Freeze tolerance and adaptation to low temperature in insects and frogs. In *Low temperature biotechnology:Emerging applications and engineering contributions,* ed. J. J. McGrath and K. R. Diller, 81–90. New York: American Society of Mechanical Engineers.

———. 1990. Principles of insect low temperature tolerance. In *Insects at low temperatures,* ed. R. E. Lee and D. L. Denlinger, 17–46. New York: Chapman and Hall.

Lee, R. E., Jr., and J. P. Costanzo. 1993. Integrated physiological responses promoting anuran freeze tolerance. In *Living in the cold III: Ecological, physiological, and molecular mechanisms,* ed. C. Carey, G. L. Florant, B. A. Wunder, and B. Horwitz, 501–10. Boulder: CO: Westview Press.

———. 1998. Biological ice nucleation and ice distribution in cold-hardy ectothermic animals. *Ann. Rev. Physiol.* 60:55–72.

Lee, R. E., Jr., J. P. Costanzo, E. C. Davidson, and J. R. Layne, Jr. 1992. Dynamics of body water during freezing and thawing in a freeze tolerant frog (*Rana sylvatica*). *J. Therm. Biol.* 17:263–66.

Lee, W. H., Y. Li, R. Lai, S. Li, Y. Zhang, and W. Wang. 2005. Variety of antimicrobial peptides in the *Bombina maxima* toad and evidence of their rapid diversification. *Euro. J. Immunol.* 35:1220–29.

Lefcourt, H. 1996. Adaptive, chemically mediated fright response in tadpoles of the southern leopard frog, *Rana utricularia. Copeia* 1996:455–59.

———. 1998. Chemically mediated fright response in southern toad (*Bufo terrestris*) tadpoles. *Copeia* 1998:445–50.

Lefcourt, H., and A. R. Blaustein. 1995. Disease, predator avoidance, and vulnerability to predation in tadpoles. *Oikos* 74:469–74.

Lefcourt, H., and S. M. Eiger. 1993. Antipredatory behaviour of feverish tadpoles—Implications for pathogen transmission. *Behaviour* 126:13–27.

Lefcourt, H., K. A. Hancock, K. M. Maur, and D. C. Rostal. 1997. The effects of used motor oil, silt, and the water mold *Saprolegnia parasitica* on the growth and survival of mole salamanders (genus *Ambystoma*). *Arch. Environ. Contam. Toxicol.* 32:383–88.

Lefcourt, H., R. A. Meguire, L. H. Wilson, and W. F. Ettinger. 1998. Heavy metals alter the survival, growth, metamorphosis, and antipredatory behavior of Columbia spotted frog (*Rana luteiventris*) tadpoles. *Arch. Environ. Contam. Toxicol.* 35:447–56.

Leff, L. G., and M. D. Bachmann. 1986. Ontogenetic changes in predatory behavior of larval tiger salamanders (*Ambystoma tigrinum*). *Can. J. Zool.* 64:1337–44.

Le Garff, B. 1998. Relations between meteorological factors and laying in the common frog *Rana temporaria* L. (Amphibia, Anura, Ranidae), in west of France (Forest of Rennes). *Bull. Soc. Zool. France* 123:61–71.

Legler, J. M., and A. Georges. 1993. Family Chelidae. In *Fauna of Australia. Vol. 2A: Amphibia and Reptilia,* ed. C. J. Glasby, G. J. B. Ross, and P. L. Beesley, 142–52. Canberra: Australian Government Publishing Service.

Lehr, E., C. Aguilar, and M. Lundberg. 2004. A new species of *Phyllonastes* from Peru (Amphibia, Anura, Leptodactylidae). *J. Herpetol.* 38:214–18.

Lehr, E., G. Fritzsch, and A. Müller. 2005. Analysis of Andes frogs (*Phrynopus,* Leptodactylidae, Anura) phylogeny based on 12S and 16S mitochondrial rDNA sequences. *Zool. Scripta* 34:593–603.

Lehtinen, R. M. 2003. Parental care and reproduction in two species of *Mantidactylus* (Anura: Mantellidae). *J. Herpetol.* 37:766–68.

Lehtinen, R. M., ed. 2004. Ecology and evolution of phytotelm-breeding anurans. *Misc. Publ. Mus. Zool. Univ. Michigan* (193):1–73.

Lehtinen, R. M., and S. A. Galatowitsch. 2001. Colonization of restored wetlands by amphibians in Minnesota. *Am. Midl. Nat.* 145:388–96.

Lehtinen, R. M., S. A. Galatowitsch, and J. R. Tester. 1999. Consequences of habitat loss and fragmentation for wetland amphibian assemblages. *Wetlands* 19:1–12.

Lehtinen, R. M., M. J. Lannoo, and R. J. Wassersug. 2004. Phytotelm-breeding anurans: Past, present, and future research. In *Ecology and evolution of phytotelm-breeding anurans,* ed. R. M. Lehtinen, 1–9. *Misc. Publ. Mus. Zool. Univ. Michigan,* no. 193.

Lehtinen, R. M., and R. A. Nussbaum. 2003. Parental care: A phylogenetic perspective. In *Reproductive biology and phylogeny of Anura,* ed. B. G. M. Jamieson, 343–86. Enfield, NH: Science Publishers.

Lehtinen, R. M., J.-B. Ramanamanjato, and J. G. Raveloarison. 2003. Edge effects and extinction proneness in a herpetofauna from Madagascar. *Biodiver. Cons.* 12:1357–70.

Lehtinen, R. M., C. M. Richards, and R. A. Nussbaum. 2004. Origin of a complex reproductive trait: phytotelm breeding in mantelline frogs. In *Ecology and evolution of phytotelm-breeding anurans,* ed. R. M. Lehtinen, 45–54. *Misc. Publ. Mus. Zool. Univ. Michigan,* no. 193.

Leibold, M. A., and H. M. Wilbur. 1992. Interactions between food-web structure and nutrients on pond organisms. *Nature* 360:341–43.

Leidy, R. A., and P. B. Moyle. 1998. Conservation status of the world's fish fauna: An overview. In *Conservation biology for the coming decade* (2nd ed.), ed. P. L. Fiedler and P. M. Kareiva, 187–227. New York: Chapman and Hall.

Leips, J., and J. Travis. 1994. Metamorphic responses to changing food levels in two species of hylid tadpoles. *Ecology* 75:1345–56.

Leips, J., M. G. McManus, and J. Travis. 2000. Response of treefrog larvae to drying ponds: comparing temporary and permanent pond breeders. *Ecology* 81:2997–3008.

Leja, W. T. 1998. Aquatic habitats in the Midwest: Waiting for amphibian conservation initiatives. In *Status and conservation of Midwestern amphibians,* ed. M. J. Lannoo, 345–53. Iowa City: University of Iowa Press.

Lemckert, F. 1999. Impacts of selective logging on frogs in a forested area of northern New South Wales. *Biol. Cons.* 89:321–28.

Lemckert, F. L., and T. Brassil. 2003. Movements and habitat use by the giant burrowing frog, *Heleioporus australiacus. Amphibia-Reptilia* 24:207–11.

Lemckert, F. L., and R. Morse. 1999. Frogs in the timber production forests of the Dorrigo Escarpment in nothern NSW: An inventory of species present and the conservation of threatened species. In *Declines and disappearances of Australian frogs,* ed. A. Campbell, 72–80. Canberra: Environment Australia.

Lemckert, F. L., and R. Shine. 1993. Costs of reproduction in a population of the frog *Crinia signifera* (Anura: Myobatrachidae) from southeastern Australia. *J. Herpetol.* 27:420–25.

Lemon, R. E. 1971. Vocal communication of the frog *Eleutherodactylus martinicensis. Can. J. Zool.* 49:211–17.

Lemon, R. E., and J. Struger. 1980. Acoustic entrainment to randomly generated calls by the frog *Hyla crucifer. J. Acoust. Soc. Am.* 67:2090–95.

Lenfant, C., and K. Johansen. 1967. Respiratory adaptations in selected amphibians. *Respir. Physiol.* 2:247–60.

Lengvenus, W., and J. Parzefall. 1992. The role of the visual reaction in the behaviour of an epigean and a cave living population of *Euproctus asper* Duges (Salamandridae, Urodela). *Mem. Biospeol.* 19:111–18.

Leong, A. S.-Y., M. J. Tyler, and D. J. C. Shearman. 1986. Gastric brooding: A new form in a recently discovered Australian frog of the genus *Rheobatrachus. Aust. J. Zool.* 34:205–9.

Leong, T. M., and L. M. Chou. 1999. Larval diversity and development in the Singapore Anura (Amphibia). *Raffles Bull. Zool.* 47:81–137.

Lesbarrères, D., and T. Lodé. 2002. Variations in male calls and response to an unfamiliar advertisement call in a territorial breeding anuran, *Rana dalmatina:* evidence for the "dear enemy" effect. *Ethol. Ecol. Evol.* 14:287–95.

Lesbarrères, D., A. Pagano, and T. Lodé. 2003. Inbreeding and road effect zone in a Ranidae: The case of agile frog, *Rana dalmatina* Bonaparte, 1840. *C. R. Biol.* 326:S68–S72.

Lescure, J. 1964. L'alimentation du crapaud commun *Bufo bufo* Linnaeus 1758. *Vie Mileu* 15:757–64.

———. 1968. Le comportement social des Batraciens. *Rev. Comp. Anim.* 2:1–33.

———. 1971. L'alimentation due crapaud *Bufo regularis* Reuss et de la Grenouille *Dicroglossus occipitalis* au Senegal. *Bull. IFAN* 33 (A):446–66.

———. 1973. Contribution a l'étude des amphibiens de Guyane Francaise. II. *Leptodactylus fuscus* (Schneider). Observations écologiques et éthologiques. *Ann. Mus. Hist. Nat. Nice* 1:91–100.

———. 1975a. Contribution a l'étude des amphibiens de Guyane francaise. III. Une novelle éspéce de *Colostethus* (Dendrobatidae): *Colostethus degranvillei* nov. sp. *Bull. Mus. Natl. Hist. Natur. Paris,* 3rd ser., (293):413–20.

———. 1975b. Contribution a l'étude des amphibiens de Guyane Francaise. IV. Reproduction de *Phyllomedusa tomopterna* (Cope) (Hylidae). *Bull. Soc. Zool. France* 100:117–25.

———. 1976. Étude de deux têtards de *Phyllobates* (Dendrobatidae): *P. femoralis* (Boulenger) et *P. pictus* (Bibron). *Bull. Soc. Zool. France* 101:299–306.

———. 1977. Le comportement de defense chez les Amphibiens. *Bull. Soc. Zool. France*102, Suppl. 2:27–36.

———. 1979. Étude taxinomique et éco-éthologique d'un amphibien des Petites Antilles: *Leptodactylus fallax* Muller, 1926 (Leptodactylidae). *Bull. Mus. Natl. Hist. Natur. Paris,* 4th ser., 1 (sec. A):757–74.

———. 1980. La surveillance de la ponte chez un amphibien anoure a développement direct: *Eleutherodactylus johnstonei* Barbour (Leptodactylidae). *Rev. Ecol. (Terre Vie)* 34:469–70.

———. 1981. Contribution a l'étude des Amphibiens de Guyane francaise IX. Le têtard gastromyzophore d'*Atelopus flavescens* Dumeril et Bibron (Anura, Bufonidae). *Amphibia-Reptilia* 2:209–15.

———. 1984. Las larvas de Dendrobatidae. *II Reunión Iberoamer. Cons. Zool. Vert.,* pp. 37–45.

———. 1986. Modes particuliers de reproduction, pp. 429–456. In: P.-P. Grasse and M. Delsol, eds. *Traite de Zoologie, vol. 14, fasc. IB. Amphibiens.* Masson, Paris, France.

Lescure, J., and R. Bechter. 1982. Le comportement de reproduction en captivité et le polymorphisme de *Dendrobates quinquevittatus* Steindachner (Amphibia, Anura, Dendrobatidae). *Rev. Fr. Aquariol.* 8:107–18.

Lescure, J., F. Grenand, and P. Grenand. 1980. Les amphibiens dans l'univers Wayapi. *J. d'Agric. Trop. Bot. Appl.* 27:247–61.

Lescure, J., and F. Letellier. 1983. Reproduction en captivité de *Leptodactylus fallax* Muller, 1926 (Amphibia, Leptodactylidae). *Rev. Fr. Aquariol.* 10:61–64.

Lescure, J., and C. Marty. 2000. *Atlas des amphibiens de Guyane.* Paris: Muséum National d'Histoire Naturelle.

Lesser, M. P., S. L. Turtle, J. H. Farrell, and C. W. Walker. 2001. Exposure to ultraviolet radiation (290–400 nm) causes oxidation stress, DNA damage, and expression of p53/p73 in laboratory experiments on embryos of the spotted salamander, *Ambystoma maculatum. Physiol. Biochem. Zool.* 74:733–41.

Lever, C. 2001. *The cane toad. The history and ecology of a successful colonist.* Otley, West Yorkshire, UK: Westbury Academic and Scientific Publishing.

Levings, S. C., and D. M. Windsor. 1982. Seasonal and annual variation in litter arthropod populations. In *The ecology of a tropical forest: Seasonal rhythms and long-term changes,* ed. E. G. Leigh, A. S. Rand, and D. M. Windsor, 355–87. Washington, DC: Smithsonian Institution Press.

Lewis, E. R. 1976. Surface morphology of the bullfrog amphibian papilla. *Brain Behav. Evol.* 13:196–215.

———. 1981. Suggested evolution of tonotopic organization in the frog amphibian papilla. *Neurosci. Lett.* 21:131–36.

———. 1983. Dual acoustic sensitivity in the bullfrog ear. In *Hearing—Physiological bases and psychoacoustics,* ed. R. Klinke and R. Hartmann, 61–69. Berlin: Springer-Verlag.

Lewis, E. R., R. A. Baird, E. L. Leverenz, and H. Koyama. 1982. Inner ear: Dye injection reveals peripheral origins of specific sensitivities. *Science* 215:1641–43.

Lewis, E. R., E. I. Hecht, and P. M. Narins. 1992. Diversity of form in the amphibian papilla of Puerto Rican frogs. *J. Comp. Physiol. A* 171:421–35.

Lewis, E. R., E. L. Leverenz, and H. Koyama. 1980. Mapping functionally identified auditory afferents from the peripheral origins to their central terminations. *Brain Res.* 197:223–29.

Lewis, E. R., E. L. Leverenz, and H. Koyama. 1982. The tonotopic organization of the bullfrog amphibian papilla, an auditory organ lacking a basilar membrane. *J. Comp. Physiol.* 145:437–45.

Lewis, E. R., and R. E. Lombard. 1988. The amphibian inner ear. In *The evolution of the amphibian auditory system,* ed. B. Fritzsch, M. J. Ryan, W. Wilczynski, T. E. Hetherington, and W. Walkowiak, 93–123. New York: Wiley.

Lewis, E. R., and P. M. Narins. 1985. Do frogs communicate with seismic signals? *Science* 227:187–89.

———. 1999. The acoustic periphery of amphibians: Anatomy and physiology. In *Comparative hearing: Fish and amphibians,* ed. R. R. Fay and A. N. Popper, 101–54. New York: Springer-Verlag.

Lewis, E. R., P. M. Narins, K. A. Cortopassi, W. M. Yamada, E. H. Poinar, S. W. Moore, and X.-L. Yu. 2001. Do male white-lipped frogs use seismic signals for intraspecific communication? *Amer. Zool.* 41:1185–99.

Lewis, H. L., and F. L. Rose. 1969. Effects of fat body fatty acids on ovarian and liver metabolism of *Ambystoma tigrinum. Comp. Biochem. Physiol.* 30:1055–60.

Lewis, W. M., G. E. Gunning, E. Lyles, and W. L. Bridges. 1961. Food choice of largemouth bass as a function of availability and vulnerability of food items. *Trans. Am. Fish. Soc.* 90:277–80.

Lewis, W. M., and D. R. Helms. 1964. Vulnerability of forage organisms to largemouth bass. *Trans. Am. Fish. Soc.* 93:315–18.

Lewis, T. R. 2002. Threats facing endemic herpetofauna in the cloud forest reserves of Ecuador. *Herpetol. Bull.* (79):18–26

Li, D., S. Fan, A. He, and F. Yin. 2004. Forest resources and environment in China. *J. For. Res.* 9:307–12.

Liang, M.-F., C.-H. Huang, and Y.-C. Kam. 2002. Effects of intermittent feeding on the growth of oophagous (*Chirxalus eiffingeri*) and hervivorous (*Chirixalus idiootocus*) tadpoles from Taiwan. *J. Zool., London* 256:207–13.

Licht, L. E. 1967. Growth inhibition in crowded tadpoles: Intraspecific and interspecific effects. *Ecology* 48:736–45.

———. 1968. Unpalatability and toxicity of toad eggs. *Herpetologica* 24:93–98.

———. 1969a. Comparative breeding behavior of the red-legged frog *Rana aurora aurora* and the western spotted frog (*Rana pretiosa pretiosa*) in southwestern British Columbia. *Can. J. Zool.* 47:1287–99.

———. 1969b. Observations on the courtship behavior of *Ambystoma gracile. Herpetologica* 25:49–52.

———. 1969c. Palatability of *Rana* and *Hyla* eggs. *Am. Midl. Nat.* 88:296–99.

———. 1969d. Unusual aspects of anuran sexual behavior as seen in the red-legged frog, *Rana aurora aurora. Can. J. Zool.* 47:505–9.

———. 1971. Breeding habits and embryonic thermal requirements of the frogs, *Rana aurora aurora* and *Rana pretiosa pretiosa,* in the Pacific Northwest. *Ecology* 52:116–24.

———. 1973. Behavior and sound production by the northwestern salamander *Ambystoma gracile. Can. J. Zool.* 51:1055–56.

———. 1974. Survival of embryos, tadpoles, and adults of the frogs *Rana aurora aurora* and *Rana pretiosa pretiosa* sympatric in southwestern British Columbia. *Can. J. Zool.* 52:613–27.

———. 1975. Comparative life history features of the western spotted frog, *Rana pretiosa,* from low- and high-elevation populations. *Can. J. Zool.* 53:1254–57.

———. 1976. Sexual selection in toads (*Bufo americanus*). *Can. J. Zool.* 54:1277–84.

———. 1986a. Comparative escape behavior of sympatric *Rana aurora* and *Rana pretiosa. Am. Midl. Nat.* 115:239–47.

———. 1986b. Food and feeding behavior of sympatric red-legged frogs, *Rana aurora,* and spotted frogs, *Rana pretiosa,* in southwestern British Columbia. *Can. Field-Nat.* 100:22–31.

———. 1989. Reproductive parameters of unisexual *Ambystoma* on Pelee Island, Ontario. In *Evolution and ecology of unisexual vertebrates.* Bulletin 466, ed. R. M. Dawley and J. P. Bogart, 209–17. Albany: New York State Museum.

———. 1991. Habitat selection of *Rana pipiens* and *Rana sylvatica* during exposure to warm and cold temperatures. *Am. Midl. Nat.* 125:259–68.

———. 1996. Amphibian declines still a puzzle. *BioScience* 46:172–73.

———. 2003a. Response by Licht. *BioScience* 53:1029.

———. 2003b. Shedding light on ultraviolet radiation and amphibian embryos. *BioScience* 53:551–61.

Licht, L. E., and K. P. Grant. 1997. The effects of ultraviolet radiation on the biology of amphibians. *Amer. Zool.* 37:137–45.

Licht, P., and A. G. Brown. 1967. Behavioral thermoregulation and its role in the ecology of the red-bellied newt *Taricha rivularis. Ecology* 47:598–611.

Licht, P., M. E. Feder, and S. Bledsoe. 1975. Salinity tolerance and osmoregulation in the salamander *Batrachoseps. J. Comp. Physiol.* 102:123–34.

Lieberman, S. S. 1986. Ecology of the leaf litter herpetofauna of a Neotropical rain forest: La Selva, Costa Rica. *Acta Zool. Mex.*, ns. 15:1–71.

Liem, D. S. 1970. The morphology, systematics, and evolution of the Old World treefrogs (Rhacophoridae and Hyperoliidae). *Fieldiana: Zool.* 57:1–145.

Liem, K. 1961. On the taxonomic status and the granular patches of the Javanese frog *Rana chalconota* Schegel. *Herpetologica* 17:69–71.

Liggins, G. W., and G. C. Grigg. 1985. Osmoregulation of the cane toad, *Bufo marinus,* in salt water. *Comp. Biochem. Physiol.* 82A:613–19.

Lighton, J. R. B. 1987. The cost of tokking: The energetics of substrate communication in the tok-tok beetle, *Psammodes striatus. J. Comp. Physiol. B* 157:11–20.

Lillo, R. S. 1980. Heart rate and blood pressure in bullfrogs during prolonged maintenance in water at low temperature. *Comp. Biochem. Physiol.* 65A:251–53.

Lillywhite, H. B. 1970. Behavioral thermoregulation in the bullfrog, *Rana catesbeiana. Copeia* 1970:158 68.

———. 1971a. Temperature selection by the bullfrog, *Rana catesbeiana. Comp. Biochem. Physiol.* 40A:213–27

———. 1971b. Thermal modulation of cutaneous mucus discharge as a determinant of evaporative water loss in the frog, *Rana catesbeiana. Z. Vergl. Physiol.* 73:84–104.

———. 1975. Physiological correlates of basking in amphibians. *Comp. Biochem. Physiol.* 52A:323–30.

———. 2006. Water relations of tetrapod integument. *J. Exp. Biol.* 209:202–26.

Lillywhite, H. B., and P. Licht. 1974. Movement of water over toad skin: Functional role of epidermal sculpturing. *Copeia* 1974:165–71.

———. 1975. A comparative study of integumentary mucous secretions in amphibians. *Comp. Biochem. Physiol.* 51A:937–41.

Lillywhite, H. B., P. Licht, and P. Chelgren. 1973. The role of behavioral thermoregulation in the growth energetics of the toad, *Bufo boreas. Ecology* 54:375–83.

Lillywhite, H. B., and P. F. A. Maderson. 1988. The structure and permeability of integument. *Amer. Zool.* 28:945–62.

Lillywhite, H. B., A. K. Mittal, T. K. Garg, and N. Agrawal. 1997a. Integumentary structure and its relationship to wiping behaviour in the common Indian tree frog, *Polypedates maculatus. J. Zool., London* 243:675–87.

———. 1997b. Wiping behavior and its ecophysiological significance in the Indian tree frog *Polypedates maculatus. Copeia* 1997:88–100.

Lillywhite, H. B., A. K. Mittal, T. K. Garc, and I. Das. 1998. Basking behavior, sweating and thermal ecology of the Indian tree frog, *Polypedates maculatus. J. Herpetol.* 32:169–75.

Lillywhite, H. B. and R. J. Wassersug. 1974. Comments on a postmetamorphic aggregate of *Bufo boreas. Copeia* 1974:984–86.

Lima, A. P. 1998. The effects of size on the diets of six sympatric species of postmetamorphic litter anurans in Central Amazonia. *J. Herpetol.* 32:393–99.

Lima, A. P., and J. P. Caldwell. 2001. A new Amazonian species of *Colostethus* with sky blue digits. *Herpetologica* 57:180–89.

Lima, A. P., J. P. Caldwell, and G. M. Biavati. 2002. Territorial and reproductive behavioir of an Amazonian dendrobatid frog, *Colostethus caeruleodactylus. Copeia* 2002:44–51.

Lima, A. P., and C. Keller. 2003. Reproductive characteristics of *Colostethus marchesianus* from its type locality in Amazonas, Brazil. *J. Herpetol.* 37:754–57.

Lima, A. P., W. E. Magnusson, and D. G. Williams. 2000. Differences in diet among frogs and lizards coexisting in subtropical forests of Australia. *J. Herpetol.* 34:40–46.

Lima, A. P., and G. Moreira. 1993. Effects of prey size and foraging mode on the ontogenetic change in feeding niche of *Colostethus stepheni* (Anura: Dendrobatidae). *Oecologia* 95:93–102.

Lima, V., J. W. Arntzen, and N. M. Ferrand. 2000. Age structure and growth pattern in two populations of the golden-striped salamander *Chioglossa lusitanica* (Caudata, Salamandridae). *Amphibia-Reptilia* 22:55–68.

Limbaugh, B. A., and E. P. Volpe. 1957. Early development of the Gulf Coast toad, *Bufo valliceps* Wiegmann. *Amer. Mus. Novitates* (1842):1–32.

Limerick, S. 1980. Courtship behavior and oviposition of the poison-arrow frog *Dendrobates pumilio. Herpetologica* 36:69–71.

Lin, S.-S., and K.-Y. Lue. 1999. *The infestation on foam nests of treefrogs (Rhacophoridae) by frogflies in Taiwan, Abstract.* State College, Pennsylvania State University: 1999 Joint Meeting of American Society of Ichthyologists and Herpetologists, American Elasmobranch Society, Herpetologists' League, and the Society for the Study of Amphibians and Reptiles.

Lind, A. J., and H. H. Welsh, Jr. 1990. Predation by *Thamnophis couchii* on *Dicamptodon ensatus. J. Herpetol.* 24:104–6.

Lindemann, B., and C. Voute. 1976. Structure and function of the epidermis. In *Frog neurobiology,* ed. R. Llanas and W. Precht, 169–210. New York: Springer-Verlag.

Linder, G., S. K. Krest, and D. W. Sparling, eds. 2003. *Amphibian decline: An Integrated analysis of multiple stressor effects.* Pensacola, FL: SETAC Press.

Lindgren, B., and A. Laurila. 2005. Proximate causes of adapative growth rates: Growth efficiency variation among latitudinal populations of *Rana temporaria. J. Evol. Biol.* 18:820–28.

Lindquist, E. D., and T. E. Hetherington. 1996. Field studies on visual and acoustic signaling in the "earless" Panamanian golden frog, *Atelopus zeteki. J. Herpetol.* 30:347–54.

———. 1998. Semaphoring in an earless frog: the origin of a novel visual signal. *Anim. Cogn.* 1:83–87.

Lindquist, E. D., T. E. Hetherington, and S. F. Volman. 1998. Biomechanical and neurophysiological studies on audition in eared and earless harlequin frogs (*Atelopus*). *J. Comp. Physiol. A* 183:265–71.

Ling, J. K. 1972. Adaptive functions of vertebrate molting cycles. *Amer. Zool.* 12:77–93.

Linke, R., G. Roth, and B. Rottluff. 1985. Comparative studies on the eye morphology in lungless salamanders, family Plethodontidae, and the effect of miniaturization. *J. Morphol.* 189:131–43.

Linnaeus, C. 1758. *Systema naturae.* (10th ed.) Uppsala, Sweden.

Linnenbach, M. 1985. Zum Feinbau der Haftscheiben von *Hyla cinerea* (Schneider, 1799) (Salientia: Hylidae). *Salamandra* 21:81–85.

Lippuner, M. 2003. Das Wildschwein (*Sus scrofa*) als Laichrauber des Springfroches (*Rana dalmatina*). *Z. Feldherpetol.* 10:261–63.

Lips, K. R. 1991. Vertebrates associated with tortoise (*Gopherus polyphemus*) burrows in south-central Florida. *J. Herpetol.* 25:477–81.

———. 1996. New treefrog from the Cordillera de Talamanca of Central America with a discussion of systematic relationships in the *Hyla lancasteri* group. *Copeia* 1996:615–26.

———. 1998. Decline of a tropical montane amphibian fauna. *Cons. Biol.* 12:106–17.

———. 1999. Mass mortality and population declines of anurans at an upland site in western Panama. *Cons. Biol.* 13:117–25.

———. 2001. Reproductive trade-offs and bet-hedging in *Hyla calypsa*, a Neotropical treefrog. *Oecologia* 128:509–18.

Lips, K. R., F. Brem, R. Brenes, J. D. Reeve, R. A. Alford, J. Voyles, C. Carey, L. Livo, A. P. Pessier, and J. P. Collins. 2006. Emerging infectious disease and the loss of biodiversity in a Neotropical amphibian community. *Proc. Natl. Acad. Sci. USA* 103:3165–70.

Lips, K. R., P. A. Burrowes, J. R. Mendelson III, and G. Para-Olea. 2005. Amphibian population declines in Latin America: A synthesis. *Biotropica* 37:222–26.

Lips, K. R., and M. A. Donnelly. 2005. Lessons from the tropics. In *Amphibian declines: The conservation status of United States species,* ed. M. J. Lannoo, 198–205. Berkeley: University of California Press.

Lips, K. R., D. E. Green, and R. Papendick. 2003. Chytridiomycosis in wild frogs from southern Costa Rica. *J. Herpetol.* 37:215–18.

Lips, K. R., and D. M. Krempels. 1995. Eggs and tadpoles of *Bufo fastidiosus* Cope, with comments on reproductive behavior. *Copeia* 1995:741–46.

Lips, K. R., J. R. Mendelson III, A. Muñoz-Alonso, L. Canseco-Márquez, and D. G. Mulcahy. 2004. Amphibian population declines in montane southern Mexico: Resurveys of historical locations. *Biol. Cons.* 119:555–64.

Lips, K. R., J. K. Reaser, B. E. Young, and R. Ibáñez. 2001. *Amphibian monitoring in Latin America: A protocol manual.* Salt Lake City, Utah: Society for the Study of Amphibians and Reptiles.

Little, C. 1990. *The terrestrial invasion.* New York: Cambridge University Press.

Little, E. E., R. D. Calfee, D. L. Fabacher, C. Carey, V. S. Blazer, and E. M. Middleton. 2003. Effects of ultraviolet radiation on toad early life stages. *Environ. Sci. Poll. Res.* 10:167–72.

Littleford, R. A., W. F. Keller, and N. E. Phillips. 1947. Studies on the vital limits of water loss in the plethodont salamanders. *Ecology* 28:440–47.

Littlejohn, M. J. 1959. Call differentiation in a complex of seven species of *Crinia* (Anura, Leptodactylidae). *Evolution* 13:452–68.

———. 1960. Call discrimination and potential reproductive isolation in *Pseudacris triseriata* females from Oklahoma. *Copeia* 1960:370–71.

———. 1961. Mating call discrimination by females of the spotted chorus frog (*Pseudacris clarki*). *Tex. J. Sci.* 13:49–50.

———. 1963. The breeding biology of the baw baw frog, *Philoria frosti* Spencer. *Proc. Linn. Soc. New South Wales* 88:273–276.

———. 1965. Premating isolation in the *Hyla ewingi* complex (Anura: Hylidae). *Evolution* 19:234–243.

———. 1969. The systematic significance of isolating mechanisms. In *Systematic Biology: Proceedings of an International Conference,* 459–482. Washington, DC: National Academy of Sciences.

———. 1976. The *Litoria ewingi* complex (Anura: Hylidae) in south-eastern Australia. IV. Variation in mating call structure across a narrow hybrid zone between *L. ewingi* and *L. paraewingi. Aust. J. Zool.* 24:283–93.

———. 1977. Long-range acoustic communication in anurans: an integrated and evolutionary approach. In. *The reproductive biology of amphibians,* ed. D. H. Taylor and S. I. Guttman 263–94. New York: Plenum.

———. 1981. Reproductive isolation: a critical review. In *Evolution and speciation. Essays in honor of M. J. D. White,* ed. W. R. Atchley and D. S. Woodruff, 298–334. New York: Cambridge University Press.

———. 1988. Frog calls and speciation. The retrograde evolution of homogamic acoustic signaling systems in hybrid zones. In *The evolution of the amphibian auditory system,* ed. B. Fritzsch, M. J. Ryan, W. Wilczynski, T. E. Hetherington, and W. Walkowiak, 613–35. New York: Wiley.

———. 1993. Homogamy and speciation: a reappraisal. *Oxford Surveys Evol. Biol.* 9:135–65.

———. 1999. Variation in advertisement calls of anurans across zonal interactions. The evolution and breakdown of homogamy. In *Geographic variation in behavior. Perspectives on evolutionary mechanisms,* ed. S. A. Foster and J. A. Endler, 209–33. New York: Oxford University Press.

———. 2001. Patterns of differentiation in temporal properties of acoustic signals of anurans. In *Anuran communcation,* ed. M. J. Ryan, 102–20. Washington, DC: Smithsonian Institution Press.

Littlejohn, M. J., M. J. Fouquette, Jr. and C. Johnson. 1960. Call discrimination by female frogs of the *Hyla versicolor* complex. *Copeia* 1960:47–49.

Littlejohn, M. J., and P. A. Harrison. 1985. The functional significance of the diphasic advertisement call of *Geocrinia victoriana* (Anura: Leptodactylidae). *Behav. Ecol. Sociobiol.* 16:363–73.

Littlejohn, M. J., P. A. Harrison, and R. C. Mac Nally. 1985. Interspecific acoustic interactions in sympatric populations of *Ranidella signifera* and R. *parinsignifera* (Anura: Leptodactylidae). In *Biology of Australasian frogs and reptiles,* ed. G. Grigg, R. Shine, and H. Ehmann, 287–96. Chipping Norton, New South Wales, Australia: Surrey Beatty and Sons.

Littlejohn, M. J., and J. J. Loftus-Hills. 1968. An experimental evaluation of premating isolation in the *Hyla ewingi* complex (Anura: Hylidae). *Evolution* 22:659–63.

Littlejohn, M. J. and A. A. Martin. 1965a. Mating call structure in three sympatric species of *Limnodynastes* (Anura, Leptodactylidae). *Copeia* 1965:509–11.

———. 1965b. A new species of *Crinia* (Anura: Leptodactylidae) from South Australia. *Copeia* 1965:319–24.

———. 1969. Acoustic interaction between two species of leptodactylid frogs. *Anim. Behav.* 17:785–91.

Littlejohn, M. J., and T. C. Michaud. 1959. Mating call discrimination by females of Strecker's chorus frog (*Pseudacris streckeri*). *Tex. J. Sci.* 11:86–92.

Littlejohn, M. J., and R. S. Oldham. 1968. *Rana pipiens* complex: mating call structure and taxonomy. *Science* 162:1003–5.

Littlejohn, M. J., and J. D. Roberts. 1975. Acoustic analysis of an intergrade zone between two call races of the *Limnodynastes tasmaniensis* complex (Anura: Leptodactylidae) in south-eastern Australia. *Aust. J. Zool.* 23:113–22.

Littlejohn, M. J., J. D. Roberts, G. F. Watson, and M. Davies. 1993. Family Myobatrachidae. In *Fauna of Australia. Vol. 2A: Amphibia and reptilia,* ed. C. J. Glasby, G. J. B. Ross, and P. L. Beesley, 41–57. Canberra: Australian Government Publishing Service.

Littlejohn, M. J., and J. W. Wainer. 1978. Carabid beetle preying on frogs. *Victorian Nat.* 95:251–52.

Littlejohn, M. J., and G. F. Watson. 1973. Mating-call variation across a narrow hybrid zone between *Crinia laevis* and *C. victoriana* (Anura: Leptodactylidae). *Aust. J. Zool.* 21:277–84.

———. 1974. Mating call discrimination and phonotaxis by females of the *Crinia laevis* complex (Anura: Leptodactylidae). *Copeia* 1974:171–75.

———. 1976. Effectiveness of a hybrid mating call in eliciting phonotaxis by females of the *Geocrinia laevis* complex (Anura: Leptodactylidae). *Copeia* 1976:76–79.

———. 1983. The *Litoria ewingi* complex (Anura: Hylidae) in south-eastern Australia VII. Mating-call structure and genetic compatibility across a narrow hybrid zone between *L. ewingi* and *L. paraewingi. Aust. J. Zool.* 30:103–13.

Littlejohn, M. J., G. F. Watson, and J. J. Loftus-Hills. 1971. Contact hybridization in the *Crinia laevis* complex (Anura: Leptodactylidae). *Aust. J. Zool.* 19:85–100.

Liu, C. C. 1931. Sexual behavior in the Siberian toad, *Bufo raddei* and the pond frog, *Rana nigromaculata. Peking Nat. Hist. Bull.* 6:43–60.

———. 1935a. *Rana boulengeri,* with a discussion of the allied species in China. *Peking Nat. Hist. Bull.* 10:55–60.

———. 1935b. Types of vocal sac in the Salientia. *Proc. Boston Soc. Nat. Hist.* 41:19–40.

———. 1936. Secondary sex characters in Chinese frogs and toads. *Field Mus. Nat. Hist., Zool. Ser.* 22:115–56.

———. 1950. Amphibians of Western China. *Fieldiana Zool. Mem.* 2:1–400.

Liu, H., R. J. Wassersug, and K. Kawachi. 1996. A computational fluid dynamics study of tadpole swimming. *J. Exp. Biol.* 199:1245–60.

———. 1997. A three dimensional computational fluid dynamics analysis of flow around a swimming tadpole. *J. Exp. Biol.* 200:2807–19.

Liu, W., A. Lathrop, J. Fu, D. Yang, and R. W. Murphy. 2000. Phylogeny of East Asian bufonids inferred from mitochondrial DNA sequences (Anura: Amphibia). *Mol. Phylogen. Evol.* 14:423–35.

Livezey, R. L. 1952. Some observations on *Pseudacris nigrita triseriata* (Wied) in Texas. *Am. Midl. Nat.* 47:372–81.

———. 1959. The egg mass and larvae of *Plethodon elongatus* Van Denburgh. *Herpetologica* 15:41–42.

Livingstone, D. 1858. *Missionary travels and researches in South Africa.* New York: Harper and Brothers.

Lizana, M., R. Márquez, and R. Martin-Sanchez. 1994. Reproductive biology of *Pelobates cultripes* (Anura: Pelobatidae) in central Spain. *J. Herpetol.* 28:19–27.

Lizana, M., and E. M. Pedraza. 1998. The effects of UV-B radiation on toad mortality in mountainous areas of Central Spain. *Cons. Biol.* 12:703–7.

Lizana, M., and V. Pérez-Mellado. 1990. Depredación por la nutria (*Lutra lutra*) del sapo de la sierra de Gredos (*Bufo bufo gredosicola*). *Doñana Acta Vert.* 17:109–12.

Lizana, M., V. Perez-Mellado, and M. J. Ciudad. 1987. Biometry and relation with the ecology of *Rana iberica* and *Rand perezi* in the Sistema Central (Spain). *Proc. 4th Ord. Gen. Meeting, S. E. H., Nijmegen,* pp. 253–58.

———. 1990. Analysis of the structure of an amphibian community in the central system of Spain. *Herpetol. J.* 1:435–46.

Lizana Avia, M., M. J. Ciudad Pizarro, and V. Pérez-Mellado. 1986. Uso de los recursos troficos en una communidad Iberica de anfibios. *Rev. Esp. Herptol.* 1:209–71.

Lloyd, M., R. F. Inger, and F. W. King. 1968. On the diversity of reptile and amphibian species in a Bornean rain forest. *Am. Nat.* 102:497–515.

Loader, S. P., D. J. Gower, K. M. Howell, N. Doggart, M.-O. Rödel, B. T. Clarke, R. O. da Sa, B. L. Cohen, and M. Wilkinson. 2004. Phylogenetic relationships of African microhylid frogs inferred from DNA sequences of mitochondrial 12S and 16S rRNA genes. *Org. Divers. Evol.* 4:227–35.

Lobos, G., and F. M. Jaksic. 2005. The ongoing invasion of African clawed frogs (*Xenopus laevis*) in Chile: Causes of concern. *Biodiver. Cons.* 14:429–39.

Lodé, T. 1991. Évolution annuelle du régime alimentaire du putois *Mustela putorius* L. en fonction de la disponibilité des proies. *Bull. Ecol.* 22:337–42.

———. 1993. Diet composition and habitat use of sympatric polecat and American mink in western France. *Acta Theriol.* 38:161–68.

———. 1994. Environmental factors influencing habitat exploitation by the polecat *Mustela putorius* in western France. *J. Zool., London* 234:75–88.

———. 1995. Activity pattern of polecats *Mustela putorius* L. in relation to food habits and prey activity. *Ethology* 100:295–308.

———. 1996. Polecat predation on frogs and toads at breeding sites in western France. *Ethol. Ecol. Evol.* 8:115–24.

———. 1997. Trophic status and feeding habits of the European polecat *Mustela putorius* L. 1758. *Mammal Rev.* 27:177–84.

. 2000. Functional response and area related search in a predator: Seasonal exploitation of anurans by the European polecat, *Mustela putorius. Aust. Ecol.* 25:223–31.

Lodé, T., M.-J. Holveck, and D. Lesbarrères. 2005. Asynchronous arrival pattern, operational sex ratio and occurrence of multiple paternities in a territorial breeding anuran, *Rana dalmatina. Biol. J. Linn. Soc.* 86:191–200.

Lodé, T, M.-J. Holveck, D. Lesbarreres, and A. Pagano. 2004. Sex-biased predation by polecats influences the mating system of frogs. *Biol. Lett., Proc. Roy. Soc. London B* (Suppl.) 271: S399–S401.

Lodé, T., and D. Le Jacques. 2003. Influence of advertisement calls on reproductive success in the male midwife toad *Alytes obstetricans. Behaviour* 140:885–98.

Lodé, T., and D. Lesbarrères. 2004. Multiple paternity in *Rana dalmatina,* a monogamous territorial breeding anuran. *Naturwissenschaften* 91:44–47.

Lodi, G., M. Biciotti, and B. Viotto. 1982. Cutaneous osmoregulation in *Triturus cristatus carnifex* (Laur.) (Urodela). *Gen. Comp. Endocrinol.* 46:452–57.

Loeb, M. L. G., J. P. Collins, and T. J. Maret. 1994. The role of prey in controlling expression of a trophic polymorphism in *Ambystoma tigrinum nebulosum. Funct. Ecol.* 8:151–58.

Lofts, B. 1974. Reproduction. In *Physiology of the Amphibia,* ed. B. Lofts, 107–218. vol. 2. New York: Academic Press.

———. 1984. Amphibians. In *Marshall's physiology of reproduc-*

*tion, vol. 1. Reproductive cycles of vertebrates* (4th ed.), ed. G. E. Lamming, 127–205. Edinburgh: Churchill Livingstone.

Loftus-Hills, J. J. 1971. Neural correlates of acoustic behaviour in the Australian bullfrog, *Limnodynastes dorsalis* (Anura: Leptodactylidae). *Z. Vergl. Physiol.* 74:140–52.

———. 1973a. Comparative aspects of auditory function in Australian anurans. *Aust. J. Zool.* 21:353–67.

———. 1973b. Neural mechanisms underlying acoustic behaviour of the frog *Pseudophryne semimarmorata* (Anura: Leptodactylidae). *Anim. Behav.* 21:781–87.

———. 1974. Analysis of an acoustic pacemaker in Strecker's chorus frog, *Pseudacris streckeri* (Anura: Hylidae). *J. Comp. Physiol.* 90:75–87.

Loftus-Hills, J. J., and M. J. Littlejohn. 1971. Pulse repetition rate as the basis for mating call discrimination by two sympatric species of *Hyla. Copeia* 1971:154–56.

Löfvenhaft, K., S. Runborg, and P. Sjögren-Gulve. 2004. Biotope patterns and amphibian distribution as assessment tools in urban landscape planning. *Landsc. Urban Plan.* 68:403–27.

Loh, J., ed. 2000. *Living planet report 2000*. Gland, Switzerland: WWF International.

Lohmann, K. J., J. T. Hester, and C. M. F. Lohmann. 1999. Long-distance navigation in sea turtles. *Ethol. Ecol. Evol.* 11:1–23.

Lohmann, K. J., B. E. Witherington, C. M. F. Lohmann, and M. Salmon. 1997. Orientation, navigation, and natal beach homing in sea turtles. In *The biology of sea turtles,* ed. P. L. Lutz and J. A. Musick, 107–35. Boca Raton, FL: CRC Press.

Loiselle, S. A., S. Bastianoni, L. Bracchini, and C. Rossi. 2004. Neotropical wetlands: New instruments in ecosystem management. *Wetlands Ecol. Manage.* 12:587–96.

Loman, J. 1978. Macro- and microhabitat distribution in *Rana arvalis* and *R. temporaria* during summer. *J. Herpetol.* 12:29–33.

———. 1988. Breeding by *Rana temporaria:* The importance of pond size and isolation. *Mem. Soc. Fauna Flora Fennici* 64:113–15.

———. 1999. Early metamorphosis in common frog, *Rana temporaria* tadpoles at risk of drying: An experimental demonstration. *Amphibia-Reptilia* 20:421–30.

———. 2002a. Microevolution and maternal effects on tadpole *Rana temporaria* growth and development rate. *J. Zool., London* 257:93–99.

———. 2002b. Temperature, genetic and hydroperiod effects on metamorphosis of brown frogs *Rana arvalis* and *R. temporaria* in the field. *J. Zool., London* 258:115–29.

———. 2002c. When crowded tadpoles (*Rana arvalis* and *R. temporaria*) fail to metamorphose early and thus fail to escape drying ponds. *Herpetol. J.* 12:21–28.

Loman, J., and D. Claesson. 2003. Plastic response to pond drying in tadpoles *Rana temporaria:* Tests of cost models. *Evol. Ecol. Res.* 5:179–94.

Loman, J., and T. Madsen. 1986. Reproductive tactics of large and small male toads *Bufo bufo. Oikos* 46:57–61.

Lombard, R. E., R. R. Fay, and Y. L. Werner. 1981. Underwater hearing in the frog, *Rana catesbeiana. J. Exp. Biol.* 91:57–71.

Lombard, R. E., and I. R. Straughan. 1974. Functional aspects of anuran middle ear structures. *J. Exp. Biol.* 61:71–93.

Lombard, R. E., and D. B. Wake. 1976. Tongue evolution in the lungless salamanders, family Plethodontidae. I. Introduction, theory and a general model of dynamics. *J. Morphol.* 148:265–86.

———. 1977. Tongue evolution in the lungless salamanders, family Plethodontidae. II. Function and evolutionary diversity. *J. Morphol.* 153:39–80.

———. 1986. Tongue evolution in the lungless salamanders, Family Plethodontidae. IV. Phylogeny of plethodontid salamanders and the evolution of feeding dynamics. *Syst. Zool.* 35:532–51.

Londos, P. L., and R. J. Brooks. 1988. Effect of temperature acclimation on locomotory performance curves in the toad, *Bufo woodhousii woodhousii. Copeia* 1988:26–32.

Long, C. H., and C. A. Killingley. 1983. *The badgers of the world.* Springfield, IL: Charles C. Thomas.

Long, D. R. 1987a. A comparison of energy substrates and reproductive patterns of two anurans, *Acris crepitans* and *Bufo woodhousei. Comp. Biochem. Physiol.* 87A:81–91.

———. 1987b. Reproductive and lipid patterns of a semiarid-adapted anuran, *Bufo cognatus. Texas J. Sci.* 39:3–13.

———. 1989. Energetics and reproduction in female *Scaphiopus multiplicatus* from western Texas. *J. Herpetol.* 23:176–79.

Long, L. E., L. S. Saylor, and M. E. Soulé. 1995. A pH/UV-B synergism in amphibians. *Cons. Biol.* 9:1301–3.

Longcore, J. E., A. P. Pessier, and D. K. Nichols. 1999. *Batrachochytrium dendrobatidis* gen et sp. nov., a chytrid pathogenic to amphibians. *Mycologia* 91:219–27.

Longphre, M., and R. E. Gatten, Jr. 1994. Individual variability in sustained performance, aerobic metabolism, oxygen transport, and enzyme activity in toads (*Bufo marinus*). *Copeia* 1994:887–96.

Lopez, C. H., and E. D. Brodie, Jr. 1977. The function of costal grooves in salamanders (Amphibia, Urodela). *J. Herpetol.* 11:372–74.

Lopez, M. S. 2003. *Philodryas patagoniensis.* Diet. *Herp. Rev.* 34:71–72.

Lopez, M. S., A. R. Giraudo, and V. Arzmendia. 2003. *Leptophis ahaetulla marginatus* (Southern green parrot-snake). Diet. *Herp. Rev.* 34:68–69.

Lopez, P. T., and P. M. Narins. 1991. Mate choice in the Neotropical frog, *Eleutherodactylus coqui. Anim. Behav.* 41:757–72.

Lopez, P. T., P. M. Narins, E. R. Lewis, and S. W. Moore. 1988. Acoustically induced call modification in the white-lipped frog, *Leptodactylus albilabris. Anim. Behav.* 36:1295–1308.

Lopez-Jurado, L. F., M. R. Caballero, and L. Dos-Santos Freitas. 1979. Biologia de la reproduccion de *Alytes cisternasii* Bosca 1879. *Doñana Acta Vert.* 6:6–17.

López-Nieves, P., and J. A. Hernando. 1984. Food habits of the otter in Central Sierra Moreno (Córdoba, Spain). *Acta Theriol.* 29:383–401.

Lörcher, K. 1969. Vergleichende bioakustische Untersuchungen an der Rot- und Gelbbauchunke *Bombina bombina* (L.) und *Bombina variegata* (L.). *Oecologia* 3:84–124.

Loredo, I., D. Van Vuren, and M. L. Morrison. 1996. Habitat use and migration behavior of the California tiger salamander. *J. Herpetol.* 30:282–85.

Löschenkohl, A. 1986. Niche partitioning and competition in tadpoles. In *Studies in herpetology,* ed. Z. Roček, 399–402. Prague: Charles University Press.

Losos, J. B. 1999. Uncertainty in the reconstruction of ancestral character states and limitations on the use of phylogenetic comparative methods. *Anim. Behav.* 58:1319–24.

Lotshaw, D. P. 1977. Temperature adaptation and effects of thermal acclimation in *Rana sylvatica* and *Rana catesbeiana. Comp. Biochem. Physiol.* 56A:287–94.

Lotter, F., and N. J. Scott, Jr. 1977. Correlation between climate and the distribution of the color morphs of the salamander *Plethodon cinereus*. *Copeia* 1977:681–90.

Lötters, S. 1996. *The Neotropical toad genus Atelopus*. Cologne, Germany: M. Vences and F. Glaw Verlags.

Lötters, S., K.-H. Jungfer, and A. Widmer. 2000. A new genus of aposematic poison frogs (Amphibia: Anura: Dendrobatidae) from the Upper Amazon Basin, with notes on its reproductive behaviour and tadpole morphology. *Jh. Ges. Naturkde. Württemberg* 156:233–43.

Lötters, S., V. R. Morales, and C. Proy. 2003. Another new riparian dendrobatid frog species from the Upper Amazon Basin of Peru. *J. Herpetol.* 37:707–13.

Lötters, S, S. Schick, K. Scheelke, P. Teege, J. Kosuch, D. Rotich, and M. Veith. 2004. Bio-sketches and partitioning of sympatric reed frogs, genus *Hyperolius* (Amphibia: Hyperoliidae), in two humid tropical African forest regions. *J. Nat. Hist.* 38:1969–97.

Lougheed, S. C., C. Gascon, D. A. Jones, P. P Bogart, and P. T. Boag. 1999. Ridges and rivers: A testing of competing hypotheses of Amazonian diversification using a dart-poison frog (*Epipedobates femoralis*). *Proc. Roy. Soc. London B* 266:1829–35.

Loumbourdis, N. S., and P. Kyriakopoulou-Sklavounou. 1991. Reproductive and lipid cycles in the male frog *Rana ridibunda* in northern Greece. *Comp. Biochem. Physiol.* 99A:577–83.

Loveridge, A. 1947. History and habits of the E. African bullfrog. *J. E. Afr. Nat. Hist. Soc.* 19:253–55.

Loveridge, J. P. 1970. Observations on nitrogenous excretion and water relations of *Chiromantis xerampelina* (Amphibia, Anura). *Arnoldia* 5:1–6.

———. 1976. Strategies of water conservation in southern African frogs. *Zool. Afr.* 11:319–33.

Loveridge, J. P., and G. Craye. 1979. Cocoon formation in two species of southern African frogs. *S. Afr. J. Sci.* 75:18–20.

Loveridge, J. P., and P. C. Withers. 1981. Metabolism and water balance of active and cocooned African bullfrogs *Pyxicephalus adspersus*. *Physiol. Zool.* 54:203–14.

Lovgren, S. 2004. Huge, freed pet pythons invade Florida Everglades. *National Geographic News* (http://news.nationalgeographic.com/news/2004/06/0603_040603_invasivespecies.html).

Løvtrop, S., and A. Pigón. 1968. Observations on the gill development in frogs. *Ann. Embryol. Morphol.* 2:3–13.

Low, B. S. 1976. The evolution of amphibian life histories in the desert. In *Evolution of desert biota,* ed. D. W. Goodall, 149–95. Austin: University of Texas Press.

Lowcock, L. A. 1989. Biogeography of hybrid complexes of *Ambystoma:* Interpreting unisexual-bisexual genetic data in space and time. In *Evolution and ecology of unisexual vertebrates. Bulletin 466,* ed. R. M. Dawley and J. P. Bogart, 180–208. Albany: New York State Museum.

Lowe, W. H., and D. T. Bolger. 2002. Local and landscape-scale predictors of salamander abundance in New Hampshire headwater streams. *Cons. Biol.* 183–93.

Lowe, W. H., K. H. Nislow, and D. T. Bolger. 2004. Stage-specific and interactive effects of sedimentation and trout on a headwater stream salamander. *Ecol. Appl.* 14:164–72.

Lowry, G. R. 1966. Production and food of cutthroat trout in three Oregon coastal streams. *J. Wildl. Manage.* 30:654–767.

Lu, X. 2004. Annual cycle of nutrional organ mass in a temperate-zone anuran, *Rana chinensis,* from northern China. *Herpetol. J.* 14:9–12.

Lucas, E. A., and A. W. Reynolds. 1967. Temperature selection by amphibian larvae. *Physiol. Zool.* 40:159–71.

Lucas, J. R., and R. D. Howard. 1995. On alternative reproductive tactics in anurans: Dynamic games with density and frequency dependence. *Am. Nat.* 146:365–97.

Lucas, J. R., R. D. Howard, and J. G. Palmer. 1996. Callers and satellites: Chorus behaviour in anurans as a stochastic dynamic game. *Anim. Behav.* 51:501–18.

Lucas, J. R., A. Schraeder, and C. Jackson. 1999. Carolina chickadee (Aves, Paridae, *Poecile carolinensis*) vocalization rates: effects of body mass and food availability under aviary conditions. *Ethology* 105:503–20.

Luckenbill, L. 1965. Morphogenesis of the horny jaws of *Rana pipiens* larvae. *Devel. Biol.* 11:25–49.

Lüddecke, H. 1974. *Ethologische Untersuchungen zur Fortpflanzung von Phyllobates palmatus (Amphibia, Ranidae)*. Unpubl. PhD dissertation. Mainz, Germany: Johannes Gutenberg-Universität.

———. 1976. Einige Ergebnisse aus Feldbeobachtungen an *Phyllobates palmatus* (Amphibia, Ranidae) in Kolumbien. *Mitt. Inst. Colombo-Aleman Invest. Cient.* 8:157–63.

———. 1995a. Fortpflanzungsleistung im Terrarium von *Hyla labialis*—Weibchen aus dem tropischen Hochgebirge Kolumbiens. *Salamandra* 31:193–208.

———. 1995b. Intra- and interpopulational comparison of temperatures selected by *Hyla labialis* (Anura), pp. 192–196. In *Scientia herpetologica,* ed. G. A. Llorente, A. Montori, X. Santos, and M. A. Carretero, 192–96. Barcelona, Spain: Asociación Herpetológica Española.

———. 1996. Site fidelity and homing ability in *Hyla labialis* (Anura, Hylidae). *Alytes* 13:167–78.

———. 1997. Besiedlungsgeschichte der kolumbianischen Ostanden durch Anuren: Hinweise aus naturgeschlichtlichen Daten von *Hyla labialis. Salamandra* 33:111–32.

———. 1999. Behavioral aspects of the reproductive biology of the Andean frog *Colostethus palmatus* (Amphibia: Dendrobatidae). *Rev. Acad. Colombiana Cienc.* 23 (Suplemento especial):303–16.

———. 2001. Variation in mating pattern in a population of the Andean frog *Hyla labialis. Amphibia-Reptilia* 22:199–207.

———. 2002a. Male and female responses to call playback in the Andean frog *Colostethus subpunctatus. Amphibia-Reptilia* 23:141–50.

———. 2002b. Variation and trade-off in reproductive output of the Andean frog *Hyla labialis. Oecologia* 130:403–10.

Lüddecke, H., and A. Amézquita. 1999. Assessment of disc clipping on the survival and behavior of the Andean frog *Hyla labialis. Copeia* 1999:824–30.

Lüddecke, H. and O. R. Sánchez. 2002. Are tropical highland frog calls colds-adapted? The case of the Andean frog *Hyla labialis. Biotropica* 34:281–288.

Ludwig, D., and L. Rowe. 1990. Life history strategies for energy gain and predator avoidance under time constraints. *Am. Nat.* 135:686–707.

Luiselli, L. 2003. Do snakes exhibit shifts in feeding ecology associated with the presence or absence of potential competitors? A case study from tropical Africa. *Can. J. Zool.* 81:228–36.

Luiselli, L., G. C. Akani, F. M. Angelici, E. Politano, L. Ude, and S. M. Wariboko. 2004. Aspects of the ecology of a population of *Causus maculatus* (Reptilia: Viperidae) from southern Nigeria. *Amphibia-Reptilia* 25:99–104.

Luiselli, L., G. C. Akani, and D. Capizzi. 1998. Food resource partitioning of a community of snakes in a swamp rainforest in south-eastern Nigeria. *J. Zool., London* 246:125–33.

Luiselli, L., F. M. Angelici, and G. C. Akani. 2002. Comparative feeding strategies and dietary plasticity of the sympatric cobras *Naja melanoleuca* and *Naja nigricollis* in three diverging Afrotropical habitats. *Can. J. Zool.* 80:55–63.

Luiselli, L., and M. Capula. 1997. Food habits, growth rates, and reproductive biology of grass snakes, *Natrix natrix* (Colubridae) in the Italian Alps. *J. Zool., London* 241:371–80.

Luiselli, L., and L. Rugiero. 1991. Food niche partitioning by water snakes (genus *Natrix*) at a freshwater environment in central Italy. *J. Freshw. Ecol.* 6:439–44.

Lukose, R. L., and H. K. Reinert. 1998. Absence of cold acclimation responses in gray treefrogs, *Hyla chrysoscelis* and *Hyla versicolor. J. Herpetol.* 32:283–85.

Lüling, K. H. 1971. Der Färberfrosch *Phyllobates bicolor* Bibron der Cordillera Azul (Peru). *Bonn. Zool. Beitr.* 22:161–74.

Lutschinger, G. 1988. Zur Überwinterung der Wasserfrösche in den Donau- und Marchauen (Österreich). *Jb. Feldherpetol.* 1:153–60.

Lutterschmidt, W. I., G. A. Marvin, and V. H. Hutchison. 1994. Alarm response by a plethodontid salamander (*Desmognathus ochrophaeus*): Conspecific and heterospecific "Shreckstoff." *J. Chem. Ecol.* 20:2751–59.

Lutz, A., and B. Lutz. 1939. Notes on the genus *Phyllomedusa* Wagler. Observations on small *Phyllomedusa* without vomerine teeth and conspicuous parotids found in the region of Rio de Janeiro. *Ann. Acad. Brasil.* 11:219–63.

Lutz, B. 1944. Biología et taxonomía de *Zachaenus parvulus.* Bol. *Mus. Nac. Zool., Rio de Janeiro* 17:1–66.

———. 1947. Trends toward non-aquatic and direct development in frogs. *Copeia* 1947:242–52.

———. 1954. Anfibios anuros do Distrito Federal. *Mem. Inst. Oswaldo Cruz* 52:155–238

———. 1960a. The clay nest of *Hyla pardalis* Spix. *Copeia* 1960:364–66.

———. 1960b. Fighting and an incipient notion of territory in male treefrogs. *Copeia* 1960:61–63.

———. 1966. Biological significance of cutaneous secretions in toads and frogs. *Mems. Inst. Butantan (Simp. int.)* 33:55–59.

———. 1971. Venomous toads and frogs. In *Venomous animals and their venoms,* vol. II, ed. W. Bucherl and E. E. Buckley, 423–73. New York: Academic Press.

———. 1973. *Brazilian species of Hyla.* Austin: University of Texas Press.

Lutz, B., and G. Orton. 1946. *Hyla claresignata* A. Lutz and B. Lutz 1939. Aspects of the life history and description of the rhyacophilous tadpole (Salientia: Hylidae). *Bol. Mus. Nac. Zool. (Brazil)* 70:1–20.

Lutz, G. J., and L. C. Rome. 1994. Built for jumping: The design of the frog muscular system. *Science* 263:370–72.

Lydeard, C., R. H. Cowie, W. F. Ponder, A. E. Bogan, P. Bouchet, S. A. Clark, E. S. Cummings, T. J. Frest, O. Gargominy, D. G. Herbert, R. Herschler, K. E. Perez, B. Roth, M. Seddon, E. E. Strong, and F. G. Thompson. 2004. The global decline of nonmarine mollusks. *BioScience* 54:321–30.

Lydeard, C., and R. L. Mayden. 1995. A diverse and endangered aquatic ecosystem of the Southeast United States. *Cons. Biol.* 9:800–805.

Lykens, D. V., and D. C. Forester. 1987. Age structure in the spring peeper: Do males advertise longevity? *Herpetologica* 43:216–23.

Lynch, J. D. 1966. Multiple morphotypy and parallel polymorphism in some Neotropical frogs. *Syst. Zool.* 15:18–23.

———. 1971. Evolutionary relationships, osteology, and zoogeography of leptodactyloid frogs. *Misc. Publ. Mus. Nat. Hist. Univ. Kansas* (53):1–238.

———. 1973. The transition from archaic to advanced frogs. In *Evolutionary biology of the anurans,* ed. J. L. Vial, 133–82. Columbia: University of Missouri Press.

———. 1978. A re-assessment of the telmatobiine leptodactylid frogs of Patagonia. *Occ. Pap. Mus. Nat. Hist. Univ. Kansas* 72:1–57.

———. 1985. Mimetic and non-mimetic populations of *Eleutherodactylus gaigeae* (Dunn) in lower Central America and Colombia (Amphibia: Anura, Leptodactylidae). *Stud. Neotrop. Fauna Environ.* 20:195–202.

———. 2001. A small amphibian fauna from a previously unexplored paramo of the Cordillera Occidental in western Colombia. *J. Herpetol.* 35:226–31.

Lynch, J. D., and W. E. Duellman. 1997. Frogs of the genus *Eleutherodactylus* in Western Ecuador. Systematics, Ecology, and Biogeography. *Mus. Nat. Hist. Univ. Kansas Spec. Publ.* 23:1–36.

Lynch, J. D., and T. Grant. 1998. Dying frogs in western Colombia: Catastrophe or trivial observation? *Rev. Acad. Colomb. Cienc.* 22:149–52.

Lynch, J. D., and C. W. Myers. 1983. Frogs of the *fitzingeri* group of *Eleutherodactylus* in eastern Panama and Chocoan South America (Leptodactylidae). *Bull. Amer. Mus. Nat. Hist.* 175:481–572.

Lynch, J. D., P. M Ruiz, and J. V. Rueda. 1983. Notes on the distribution and reproductive biology of *Centrolene geckoideum* Jimenez de la Espada in Colombia and Ecuador (Amphibia: Centrolenidae). *Stud. Neotropical Fauna Environ.* 18:239–43.

Lynch, J. F. 1985. The feeding ecology of *Aneides flavipunctatus* and sympatric plethodontid salamanders in northwestern California. *J. Herpetol.* 19:328–52.

Lynch, L. C., and D. G. Blackburn. 1995. Effects of testosterone administration and gonadectomy on nuptial pad morphology in overwintering male leopard frogs, *Rana pipiens. Amphibia-Reptilia* 16:113–21.

Lynn, S. G., and C. Lindle. 2002. The effect of anthropogenic habitat modification on habitat use by *Afrana angolensis* along the Dodwe River, Tanzania. *Afr. J. Herpetol.* 51:69–73.

Lynn, W. G. 1942. Embryology of *Eleutherodactylus nubicola,* an anuran which has no tadpole stage. *Carnegie Inst. Washington, Contr. Embryol.* 190:27–62.

Lynn, W. G., and J. N. Dent. 1943. Notes on Jamaican amphibians. *Copeia* 1943:234–42.

Lynn, W. G., and A. Edelman. 1936. Crowding and metamorphosis in the tadpole. *Ecology* 17:104–9.

Lynn, W. G., and C. Grant. 1940. The herpetology of Jamaica. *Bull. Inst. Jamaica, Sci. Ser.* 1:1–148.

Lynn, W. G., and B. Lutz. 1946a. The development of *Eleutherodactylus guentheri* Stdnr. 1864. *Bol. Mus. Nac. Rio de Janeiro, N. S. Zool.,* no. 71:1–46.

———. 1946b. The development of *Eleutherodactylus nasutus* Lutz. I. The external embryology of *Eleutherodactylus nasutus* Lutz. *Bol. Mus. Nac. Zool., Rio de Janeiro,* n.s. 79:1–30.

Mabry, M., and P. Verrell. 2003. All are one and one is all: Sexual uniformity among widely separated populations of the North American seal salamander, *Desmognathus monticola. Biol. J. Linn. Soc.* 78:1–10.

MacCracken, J. G. 2002. Response of forest floor vertebrates to riparian hardwood conversion along the Bear River, southwest Washington. *Forest Sci.* 48:299–308.

Mac Nally, R. C. 1979. Social organisation and interspecific interactions in two sympatric species of *Ranidella* (Anura). *Oecologia* 42:293–306.

———. 1981. On the reproductive energetics of chorusing males: Energy depletion profiles, restoration and growth in two sympatric species of *Ranidella* (Anura). *Oecologia* 51:181–88.

———. 1982. *Ecological, behavioural, and energy dynamics of two sympatric species of Ranidella (Anura).* Unpublished PhD dissertation. Melbourne, Australia: University of Melbourne.

———. 1983. Trophic relationships of two sympatric species of *Ranidella* (Anura). *Herpetologica* 39:130–40.

———. 1984a. Chorus dynamics of two sympatric species of *Ranidella* (Anura): Within-year and between-year variability in organization and their determination. *Z. Tierpsychol.* 65:134–51.

———. 1984b. On the reproductive energetics of chorusing males: Costs and patterns of call production in two sympatric species of *Ranidella* (Anura). *Oikos* 42:82–91.

Mac Nally, R. C., and D. Young. 1981. Song energetics of the bladder cicada *Cystosoma saundersii. J. Exp. Biol.* 90:185–96.

Macan, T. T. 1966. The influence of predation on the fauna of a moorland fishpond. *Arch. Hydrobiol.* 61:432–52.

MacArthur, D. L., and J. W. T. Dandy. 1982. Physiological aspects of overwintering in the boreal chorus frog (*Pseudacris triseriata maculata*). *Comp. Biochem. Physiol.* 72A:137–41.

Macartney, J. M., and P. T. Gregory. 1981. Differential susceptibility of sympatric garter snake species to amphibian skin secretions. *Am. Midl. Nat.* 106:271–81.

MacCallum, C. J., B. D. Nürnberger, N. H. Barton, and J. M. Szymura. 1998. Habitat preference in the *Bombina* hybrid zone in Croatia. *Evolution* 52:2227–39.

MacCracken, J. G. 2004. *Rhyacotriton cascadae* (Cascade Torrent Salamander). Nest. *Herp. Rev.* 35:367.

MacGregor, H. C., S. K. Sessions, and J. W. Arntzen. 1990. An integrative analysis of phylogenetic relationships among newts of the genus *Triturus* (family Salamandridae), using comparative biochemistry, cytogenetics and reproductive interactions. *J. Evol. Biol.* 3:329–73.

Machado, L. L., and C. A. B. Galdino. 2005. *Hyla alvarengai* (NCN). Predation. *Herp. Rev.* 36:301.

Macias Garcia, C., and H. Drummond. 1988. Seasonal and ontogenetic variation in the diet of the Mexican garter snake, *Thamnophis eques,* in Lake Tecocomulco, Hidalgo. *J. Herpetol.* 22:129–34.

Macintyre, D. H., and D. P. Toews. 1976. The mechanics of lung ventilation and the effects of hypercapnia on respiration in *Bufo marinus. Can. J. Zool.* 54:1364–74.

MacIver, D. C., and E. Wheaton. 2005. Tomorrow's forests: Adapting to a changing climate. *Climatic Change* 70:273–82.

MacKenzie, J. A., and D. C. Jackson. 1978. The effect of temperature on cutaneous $CO_2$ loss and conductance in the bullfrog. *Respir. Physiol.* 32:313–23.

Maclean, N., and R. D. Jurd. 1971. Electrophoretic analysis of the haemoglobins of *Ambystoma mexicanum. Comp. Biochem. Physiol.* 40B:751–55.

———. 1972. The control of hemoglobin synthesis. *Biol. Rev.* 47:393–437.

Madej, R.F. 1998. Discovery of green salamanders in Indiana and a distributional survey. In *Status and conservation of Midwestern amphibians,* ed. M. J. Lannoo, 55–60. Iowa City: University of Iowa Press.

Madison, D. M. 1969. Homing behavior of the red-cheeked salamander, *Plethodon jordani. Anim. Behav.* 17:25–39.

———. 1972. Homing and orientation in salamanders: a mechanism involving chemical cues. In *Animal orientation and navigation,* ed. S. R. Galler, K. Schmidt-Koenig, G. J. Jacobs, and R. E. Belleville, 485–98. Washington, DC: NASA.

———. 1975. Intraspecific odor preference between salamanders of the same sex: Dependence on season and proximity of residence. *Can. J. Zool.* 53:1256–1361.

———. 1977. Chemical communication in amphibians and reptiles. In *Chemical signals in vertebrates,* ed. D. Müller-Schwartz and M. M. Mozell, 135–68. New York: Plenum.

———. 1997. The emigration of radio-implanted spotted salamanders, *Ambystoma maculatum. J. Herpetol.* 31:542–51.

———. 1998. Habitat-contingent reproductive behaviour in radio-implanted salamanders: A model and test. *Anim. Behav.* 55:1203–10.

Madison, D. M., and L. Farrand III. 1998. Habitat use during breeding and emigration in radio-implanted tiger salamanders, *Ambystoma tigrinum. Copeia* 1998:402–10.

Madison, D. M., J. C. Maerz, and J. H. McDarby. 1999a. Chemosensory responses of salamanders to snake odors: Flight, freeze and dissociation. In *Chemical signals in vertebrates,* ed. D. Müller-Schwartz and M. M. Mozell, 505–16. New York: Plenum.

———. 1999b. Optimization of predator avoidance by salamanders using chemical cues: Diet and diel effects. *Ethology* 105:1073–86.

Madison, D. M., and C. R. Shoop. 1970. Homing behavior, orientation, and home range of salamanders tagged with Tantalum-182. *Science* 168:1484–87

Maeda, N., and M. Matsui. 1989. *Frogs and toads of Japan.* Tokyo, Japan: Bun-ichi Sogo Shuppan (in Japanese).

Maerz, J. C., and J. Karuzas. 2003. *Plethodon cinereus* (Eastern red-backed salamander). Cannibalism. *Herp. Rev.* 34:46–47.

Maerz, J. C., and D. M. Madison. 2000. Environmental variation and territorial behavior in a terrestrial salamander. In *The biology of plethodontid salamanders,* ed. R. C. Bruce, R. G. Jaeger, and L. D. Houck, 395–406. New York: Kluwer Academic/Plenum.

Maglia, A. M. 1998. Phylogenetic relationships of extant pelobatid frogs (Anura: Pelobatoidea): Evidence from adult morphology. *Sci. Pap. Nat. Hist. Mus. Univ. Kansas* 10:1–19.

Maglia, A. M., L. A. Púgener, and L. Trueb. 2001. Comparative development of frogs: using phylogeny to understand ontogeny. *Am. Zool.* 41:538–51.

Magnusson, W. E., and J.-M. Hero. 1991. Predation and the evolution of complex oviposition behavior in Amazonian rainforest frogs. *Oecologia* 86:310–18.

Magnusson, W. E., A. P. Lima, J.-M. Hero, and M. Carmozina de Araújo. 1999. The rise and fall of a population of *Hyla boans:* Reproduction in a Neotropical gladiator frog. *J. Herpetol.* 33:647–56.

Magnusson, W. E., E. Vieira da Silva, and A. P. Lima. 1987. Diets of Amazonian crocodilians. *J. Herpetol.* 21:85–95.

Magurran, A. E., and J. A. Bendelow. 1990. Conflict and cooperation in White Cloud Mountain minnow schools. *J. Fish Biol.* 37:77–83.

Mahaney, P. A. 1994. Effects of fresh-watrer petroleum contamination on amphibian hatching and metamorphosis. *Environ. Toxicol. Chem.* 13:259–65.

Mahmoud, I. Y. 1968. Feeding behavior in kinosternid turtles. *Herpetologica* 24:300–5.

Mahoney, M. J. 2001. Molecular systematics of *Plethodon* and *Aneides* (Caudata: Plethodontidae: Plethodontini): Phylogenetic analysis of an old and rapid radiation. *Mol. Phylogen. Evol.* 18:174–88.

Main, A. R. 1964. A new species of *Pseudophryne* (Anura: Leptodactylidae) from north-western Australia. *West. Aust. Nat.* 9:66–72.

———. 1968. Ecology, systematics, and evolution of Australian frogs. *Adv. Ecol. Res.* 5:37–86.

Main, A. R., and P. J. Bentley. 1964. Water relations of Australian burrowing frogs and tree frogs. *Ecology* 45:379–82.

Main, A. R., M. J. Littlejohn, and A. K. Lee. 1959. Ecology of Australian frogs. In *Biogeography and ecology in Australia,* ed. A. Keast and R. L. Crocker [*Mon. Biol.*vol. 8], 396–411. The Hague: Dr. W. Junk.

Main, B. Y., and A. R. Main. 1956. Spider predator on a vertebrate. *West. Aust. Natur.* 5:139.

Maina, J. N. 1989. The morphology of the lung of the East African tree frog *Chiromantis petersi* with observations on the skin and the buccal cavity as secondary gas exchange organs. A TEM and SEM study. *J. Anat.* 165:29–43.

Maiorana, V. C. 1977. Tail autotomy, functional conflicts and their resolution by a salamander. *Nature* 265:533–35.

———. 1978a. Behavior of an unobservable species: Diet selection by a salamander. *Copeia* 1978:664–72.

———. 1978b. Difference in diet as an epiphenomenon: Space regulates salamanders. *Can. J. Zool.* 56:1017–25.

Majecki, J., and K. Majecka. 1996. Predation by *Oligotricha striata* caddis larvae on amphibian eggs: Effects of a high quality food on growth rate. *Netherlands J. Aquat. Ecol.* 30:21–25.

———. 1998. Predation of *Oligotricha striata* (Trichoptera, Phryganeidae) larvae on amphibian eggs. *Amphibia-Reptilia* 19:230–33.

Major, P. F. 1977. Predator-prey interactions in schooling fishes during periods of twilight: A study of the silverside *Pranesus insularum* in Hawaii. *Fish. Bull.* 75:415–26.

———. 1978. Predator-prey interactions in two schooling fishes, *Caranx ignobilis* and *Stolephorus purpureus. Anim. Behav.* 26:760–77.

Maksimowich, D. S., and A. Mathis. 2000. Parasitized salamanders are inferior competitors for territories and food resources. *Ethology* 106:319–29.

———. 2001. Pheromonal markers as indicators of parasite load: parasite-mediated behavior in salamanders (*Plethodon angusticlavius*). *Acta Ethol.* 3:83–87.

Maksymovitch, E., and P. Verrell. 1992. The courtship behavior of the Santeetlah dusky salamander, *Desmognathus santeetlah* Tilley (Amphbia: Caudata: Plethodontidae). *Ethology* 90:236–46.

———. 1993. Divergence of mate recognition systems among conspecific populations of the plethodontid salamander *Desmognathus santeetlah. Biol. J. Linn. Soc.* 49:19–29.

Malacarne, G., L. Bottoni, R. Massa, and C. Vellano. 1984. The abdominal gland of the crested newt: A possible source of courtship pheromones. Preliminary ethological and biochemical data. *Monit. Zool. Ital.,* n.s. 18:33–39.

Malacarne, G., and R. Cortassa. 1983. Sexual selection in the crested newt. *Anim. Behav.* 31:1256–57.

Malacarne, G., and C. Giacoma. 1986. Chemical signals in European newt courtship. *Boll. Zool.* 53:79–83.

Malacarne, G., and C. Vellano. 1987. Behavioral evidence of a courtship pheromone in the crested newt, *Triturus cristatus carnifex* Laurenti. *Copeia* 1987:245–47.

Maleek, R., P. Sullivan, K. V. Hoff, B. Baula, and D. D. Hillyard. 1999. Salt sensitivity and hydration behavior of the toad *Bufo marinus. Physiol. Behav.* 67:739–45.

Mallick, P. K. 1986. Mating behaviour of *Rana limnocharis. Biol. Bull. India* 8:89–94.

Mallick, P. K., and S. C. Mallick. 1982. Foam-nest formation in *Polypedates* (= *Rhacophorus*) *maculatus* Gray. *J. Bengal Nat. Hist. Soc.,* n.s. 1:52–62.

Mallory M. A., and J. S. Richardson. 2005. Complex interactions of light, nutrients and consumer density in a stream periphyton-grazer (tailed frog tadpoles) system. *J. Anim. Ecol.* 74:1020–28.

Malmgren, J. C. 2002. How does a newt find its way from a pond? Migration patterns after breeding and metamorphosis in great crested newts (*Triturus cristatus*) and smooth newts (*T. vulgaris*). *Herpetol. J.* 12:29–35.

Malone, B. S. 1985. Mortality during the early life history stages of the baw baw frog, *Philoria frosti* (Anura: Myobatrachidae). In *Biology of Australasian frogs and reptiles,* ed. G. Grigg, R. Shine, and H. Ehmann, 1–5. Chipping Norton, New South Wales, Australia: Surrey Beatty and Sons.

Malone, J. H. 2004. Reproduction in three species of *Smilisca* from Costa Rica. *J. Herpetol.* 38:27–35.

Malvin, G. M. 1988. Microvascular regulation of cutaneous gas exchange in amphibians. *Amer. Zool.* 28:999–1007.

———. 1989. Gill structure and function. Amphibian larvae. In *Comparative pulmonary physiology: Current concepts, vol. 39. Lung biology in health and disease,* ed. S. C. Wood, 121–51. New York: Marcel Dekker.

———. 1994. Cardiovascular regulation of cutaneous gas exchange. In *Amphibian biology, vol. 1. The integument,* ed. H. Heatwole and G. T. Barthalmus, 147–67. Chipping Norton, New South Wales, Australia: Surrey Beatty and Sons.

Malvin, G. M., and R. G. Boutilier. 1985. Ventilation-perfusion relationships in Amphibia. In *Circulation, respiration, and metabolism,* ed. R. Gilles, 114–24. New York: Springer-Verlag.

Malvin, G. M., and N. Heisler. 1988. Blood flow patterns in the salamander, *Ambystoma tigrinum,* before during and after metamorphosis. *J. Exp. Biol.* 137:53–74.

Malvin, G. M., and M. P. Hlastala. 1986a. Effects of intrapulmonary $O_2$, $CO_2$, and volume on cutaneous gas exchange in the frog. *Am. J. Physiol.* 251:R941–R946.

———. 1986b. Regulation of cutaneous gas exchange by environmental oxygen and carbon dioxide in the frog. *Respir. Physiol.* 65:99–111.

———. 1989. Effects of environmental $O_2$ on cutaneous blood flow and diffusing capacity in amphibians. *Respir. Physiol.* 76:229–42.

Malvin, G. M., and S. C. Wood. 1991. Behavioral thermoregulation of the toad, *Bufo marinus:* Effects of air humidity. *J. Exp. Zool.* 258:322–26.

Manamendra-Arachchi, K. and R. Pethiyagoda. 2001. *Ramanella nagaoi,* a new tree-hole frog (Microhylidae) from southern Sri Lanka. *J. South Asian Nat. Hist. Soc.* 5:121–33.

Mangoni, M. L., R. Miele, T. G. Renda, D. Barra, and M. Simmaco. 2001. The synthesis of antimicrobial peptides in the skin of *Rana esculenta* is stimulated by microorganisms. *FASEB J.* 15:U432–U445.

Maniero, G. D., and C. Carey. 1997. Changes in selected aspects of immune function in the leopard frog, *Rana pipiens*, associted with exposure to cold. *J. Comp. Physiol. B* 167:256–63.

Manion, J. J., and L. Cory. 1952. Winter kill of *Rana pipiens* in shallow ponds. *Herpetologica* 8:32

Manjarrez, J., and C. Macias Garcia. 1991. Feeding ecology of *Nerodia rhombifera* in a Veracruz swamp. *J. Herpetol.* 25:499–502.

Mann, K. H., and M. J. Tyler. 1963. Leeches as endoparasites of frogs. *Nature* 197:1224–25.

Mann, T. 1984. Spermatophores. Development, structure, bio chemical attributes, and role in the transfer of spermatozoa. In. *Zoophysiology*, vol. 15, ed. D. S. Farner, 1–217. New York: Springer-Verlag.

Manteifel, U. 1995. Chemically-mediated avoidance of predators by *Rana temporaria* tadpoles. *J. Herpetol.* 29:461–63.

Manteuffel, G., O. Wess, and W. Himstedt. 1977. Messugen am dioptrischen Apparat von Amphibienaugen und Berechnung der Sehschärfe in Wasser und Luft. *Zool. Jb. Physiol.* 81:395–406.

Manz, R. 1975. Die Fusionsfrequenzen der Kehlkopfmuskeln und eines Beinmuskels in Abhängigkeit von der Temperatur bei europäischen Froschlurchen (Anura). *Zool. Jb. Physiol.* 79:221–45.

Manzanilla, J. M., I. Martínez-Solano, M. García-París, and D. Buckley. 2005. *Bufo marinus* (Cane Toad). Predation. *Herp. Rev.* 36:298–99.

Mao, S. 1970. Food of the common venomous snakes of Taiwan. *Herpetologica* 26:45–48.

Maracchi, G., O. Sirotenko, and M. Bindi. 2005. Impacts of present and future climate variability on agriculture and forestry in temperate regions: Europe. *Climatic Change* 70:117–35.

Marangio, M. S. 1978. The occurrence of neotenic rough-skinned newts (*Taricha granulosa*) in montane lakes of southern Oregon. *Northwest Sci.* 52:343–50.

Marangio, M. S., and J. D. Anderson. 1977. Soil moisture preference and water relations of the marbled salmander, *Ambystoma opacum* (Amphibia, Urodela, Ambystomatidae). *J. Herpetol.* 11:169–76.

Marchisin, A., and J. D. Anderson. 1978. Strategies employed by frogs and toads (Amphibia, Anura) to avoid predation by snakes (Reptilia, Serpentes). *J. Herpetol.* 12:151–55.

Marco, A., and A. R. Blaustein. 1998. Egg gelatinous matrix protects *Ambystoma gracile* embryos from prolonged exposure to air. *Herpetol. J.* 8:207–11.

Marco, A., D. Cash, L. K. Belden, and A. R. Blaustein. 2001. Sensitivity to urea fertilization in three amphibian species. *Arch. Enviorn. Contam. Toxicol.* 40:406–9.

Marco, A., J. M. Kiesecker, D. P. Chivers, and A. R. Blaustein. 1998. Sex recognition and mate choice by male western toads, *Bufo boreas. Anim. Behav.* 55:1631–35.

Marco, A., and M. Lizana. 2002. The absence of species and sex recognition during mate search by male common toads, *Bufo bufo. Ethol. Ecol. Evol.* 14:1–8.

Marco, A., M. Lizana, A. Alvarez, and A. R. Blaustein. 2001. Egg-wrapping behaviour protects newt embryos from UV radiation. *Anim. Behav.* 61:639–44.

Marco, A., C. Quilchano, and A. R. Blaustein. 1999. Sensitivity to nitrate and nitrite in pond-breeding amphibians from the Pacific Northwest, USA. *Env. Tox. Chem.* 18:2836–39.

Maret, T. J., and J. P. Collins. 1996. Effect of prey vulnerability on population size structure of a gape-limited predator. *Ecology* 77:320–24.

———. 1997. Ecological origin of morphological diversity: A study of alternative trophic phenotypes in larval salamanders. *Ecology* 51:898–905.

Marimuthu, G., J. Habersetzer, and D. Leippert. 1995. Active acoustic gleaning from the water-surface by the Indian false vampire bat, *Megaderma lyra. Ethology* 99:61–74.

Marinelli, M., and D. Vagnetti. 1988. Morphology of the oral disc of *Bufo bufo* (Salientia: Bufonidae). *J. Morphol.* 195:71–81.

Marker, G. M., and R. E. Gatten, Jr. 1993. Individual variability in sprint performance, lactate production, and enzyme activity in frogs (*Rana pipiens*). *J. Herpetol.* 27:294–99.

Markow, J. T. 1997. *The presence of Fowler's toads, Bufo fowleri, in brackish ponds on Cape Cod, Massachusetts.* Unpublished M. S. thesis. Storrs: University of Connecticut.

Marks, E. N. 1960. Mosquitoes biting frogs. *Aust. J. Sci.* 23:89.

Marks, S. B., and A. Collazo. 1998. Direct development in *Desmognathus aeneus* (Caudata: Plethodontidae): A staging table. *Copeia* 1998:637–48.

Marks, S. B., and L. D. Houck. 1989. Partial cost of spermatophore production in the salamander, *Desmognathus ochrophaeus. J. Herpetol.* 23:81–84.

Marler, C. A., J. Chu, and W. Wilczynski. 1995. Arginine vasotocin injection increases probability of calling in cricket frogs, but causes call changes characteristic of less aggressive males. *Horm. Behav.* 29:554–70.

Marler, C. A., and M. J. Ryan. 1996. Energetic constraints and steroid hormone correlates of male calling behaviour in the túngara frog. *J. Zool., London* 240:397–409.

Márquez, R. 1992. Terrestrial parental care and short breeding seasons: Reproductive phenology of the midwife toads *Alytes obstetricans* and *A. cisternasii. Ecography* 15:279–88.

———. 1993. Male reproductive success in two midwife toads, *Alytes obstetricans* and *A. cisternasii. Behav. Ecol. Sociobiol.* 32:283–91.

———. 1995. Female choice in the midwife toads (*Alytes obstetricans* and *A. cisternasii*). *Behaviour* 132:151–61.

———. 1996. Egg mass and size of tadpoles at hatching in the midwife toads *Alytes obstetricans* and *Alytes cisternasii:* Implication for female choice. *Copeia* 1996:824–31.

Márquez, R., and J. Bosch. 1997a. Female preference in complex acoustical environments in the midwife toads *Alytes obstetricans* and *Alytes cisternasii. Behav. Ecol.* 8:588–94.

———. 1997b. Male advertisement call and female preference in sympatric and allopatric midwife toads. *Anim. Behav.* 54:1333–45.

———. 2001. Communication and mating in the midwife toads (*Alytes obstetricans* and *Alytes cisternassii*). In *Anuran communcation,* ed. M. J. Ryan, 220–31. Washington, DC: Smithsonian Institution Press.

Márquez, R., M. Esteban, and J. Castanet. 1997. Sexual size dimorphism and age in the midwife toads *Alytes obstetricans* and *A. cisternasii. J. Herpetol.* 31:52–59.

Márquez, R, J. M Pargana, and E. G. Crespo. 2001. Acoustic competition in male *Pelodytes ibericus* (Anura: Pelodytidae): Interactive playback tests. *Copeia* 2001:1142–50.

Márquez, R., and P. A. Verrell. 1991. The courtship and mating of the Iberian midwife toad *Alytes cisternasii* (Amphibia: Anura: Discoglossidae). *J. Zool., London* 225:125–39.

Márquez-M. de Orense, R. and M. Tejedo-Madueno. 1990. Size-based mating pattern in the treefrog *Hyla arborea. Herpetologica* 46:176–82.

Marquis, R. J., M. A. Donnelly, and C. Guyer. 1986. Aggregations of calling males of *Agalychnis calcarifer* Boulenger (Anura: Hylidae) in a Costa Rican lowland wet forest. *Biotropica* 18:173–75.

Marra, R. V., F. H. Hatano, L. Boquimpani-Freitas, R. V. Marques, M. van Sluys, and C. F. D. Rocha. 2003. *Scinax altera* (NCN). Predation. *Herp. Rev.* 34:55.

Marsh, D. M. 2000. Variable response to rainfall by breeding túngara frogs. *Copeia* 2000:1104–8.

———. 2001. Behavioral and demographic responses of túngara frogs to variation in pond density. *Ecology* 82:1283–1292.

Marsh, D. M., and B. J. Borrell. 2001. Flexible oviposition strategies in túngara frogs and their implications for tadpole spatial distributions. *Oikos* 93:101–9.

Marsh, D. M., E. H. Fegraus, and S. Harrison. 1999. Effects of breeding pond isolation on the spatial and temporal dynamics of pond use by the túngara frog, *Physalaemus pustulosus. J. Anim. Ecol.*68:804–14.

Marsh, D. M., G. S. Milam, N. P. Gorham, and N. E. Beckman. 2005. Forest roads as partial barriers to terrestrial salamander movement. *Cons. Biol.* 19:2004–8.

Marsh, D. M., and P. B. Pearman. 1997. Effects of habitat fragmentation on the abundance of two species of leptodactylid frogs in an Andean montane forest. *Cons. Biol.* 11:1323–28.

Marsh, D. M., A. S. Rand, and M. J. Ryan. 2000. Effects of interpond distance on the breeding ecology of túngara frogs. *Oecologia* 122:505–13.

Marsh, D. M., K. A. Thakur, K. C. Bulka, and L. B. Clarke. 2004. Dispersal and colonization through open fields by a terrestrial woodland salamander. *Ecology* 85:3396–3405.

Marsh, D. M., and P. C. Trenham. 2001. Metapopulation dynamics and amphibian conservation. *Cons. Biol.* 15:40–49.

Marsh, R. L. 1981. Catabolic enzyme activities in relation to premigratory fattening and muscle hypertrophy in the gray catbird (*Dumetella corolinensis*). *J. Comp. Physiol. B* 141:417–23.

———. 1994. Jumping ability of anuran amphibians. In *Comparative vertebrate exercise physiology,* ed. J. H. Jones, 51–111. New York: Academic Press.

———. 1999. Contractile properties of muscles used in sound production and locomotion in two species of gray tree frog. *J. Exp. Biol.* 202:3215–23.

Marsh, R. L., and H. B. John-Alder. 1994. Jumping performance of hylid frogs measured with high-speed cine film. *J. Exp. Biol.* 188:131–41.

Marsh, R. L., and T. L. Taigen. 1987. Properties enhancing aerobic capacity of calling muscles in gray tree frogs *Hyla versicolor. Am. J. Physiol.* 252:R786–R793.

Marshall, C. J., L. S. Doyle, and R. H. Kaplan. 1990. Intraspecific and sex-specific oophagy in a salamander and a frog: Reproductive convergence of *Taricha torosa* and *Bombina orientalis. Herpetologica* 46:395–99.

Marshall, E., and G. C. Grigg. 1980. Lack of metabolic acclimation to different thermal histories by tadpoles of *Limnodynastes peronii* (Anura: Myobatrachidae). *Aust. Zool.* 20:447–55.

Marshall, J. L. 1996. *Eurycea cirrigera* (southern two-lined salamander). Nest site. *Herp. Rev.* 27:75–76.

Marshall, J. L., C. D. Camp, and R. G. Jaeger. 2004. Potential interference competition between a patchily distributed salamander (*Plethodon petraeus*) and a sympatric congener (*Plethodon glutinosus*). *Copeia* 2004:488–95.

Marshall, V. T., S. C. Humfeld, and M. A. Bee. 2003. Plasticity of aggressive signalling and its evolution in male spring peepers, *Pseudacris crucifer. Anim. Behav.* 65:1223–34.

Martin del Campo, R. 1940. Una observación del desarollo de *Tomodactylus nitidus. Ann. Ins. Biol.* 11:745–46.

Martin, A. A. 1967. Australian anuran life histories: some evolutionary and ecological aspects. In *Australian inland waters and their fauna,* ed. A. H. Weatherly, 175–91. Canberra: Australian National University Press.

———. 1970. Parallel evolution in the adaptive ecology of leptodactylid frogs of South America and Australia. *Evolution* 24:643–44.

———. 1972. Studies in Australian Amphibia. III. The *Limnodynastes dorsalis* complex (Anura: Leptodactylidae). *Aust. J. Zool.* 20:165–211.

Martin, A. A., and A. K. Cooper. 1972. The ecology of terrestrial anuran eggs, genus *Crinia* (Leptodactylidae). *Copeia* 1972:162–68.

Martin, A. A., and G. F. Watson. 1971. Life history as an aid to generic delimitation in the family Hylidae. *Copeia* 1971:78–89.

Martin, C. C., and R. Gordon. 1995. Differentiation trees, a junk DNA molecular clock, and the evolution of neoteny in salamanders. *J. Evol. Biol.* 8:339–54.

Martin, D. L., R. G. Jaeger, and C. P. Labat. 1986. Territoriality in an *Ambystoma* salamander? Support for the null hypothesis. *Copeia* 1986:725–30.

Martín, J., and P. López. 1990. Amphibians and reptiles as prey of birds in southwestern Europe. *Smithsonian Herpetological Information Service,* no. 82.

Martin, K. M., and V. H. Hutchison. 1979. Ventilatory activity in *Amphiuma tridactylum* and *Siren lacertina* (Amphibia, Caudata). *J. Herpetol.* 13:427–34.

Martin, S. B., R. G. Jaeger, and E. D. Prosen. 2005. Territorial red-backed salamanders can detect volatile pheromones from intruders. *Herpetologica* 61:29–35.

Martin, T. E., and D. M. Finch, eds. 1995. *Ecology and management of neotropical migratory birds : A synthesis and review of critical issues.* New York: Oxford University Press.

Martin, W. F. 1971. Mechanics of sound production in the genus *Bufo:* Passive elements. *J. Exp. Zool.* 176:273–94.

———. 1972. Evolution of vocalization in the genus *Bufo.* In *Evolution in the genus Bufo,* 279–308. W. F. Blair, ed. Austin: University of Texas Press.

Martin, W. F., and C. Gans. 1972. Muscular control of the vocal tract during release signaling in the toad *Bufo valliceps. J. Morphol.* 137:1–28.

Martínez-Solano, I. 2000. *Pelobates cultripes* (Iberian spadefoot toad). Predation. *Herp. Rev.* 31:235.

Martínez-Solano, I., L. J. Barbadillo, and M. Lapena. 2003. Effect of introduced fish on amphibian species richness and densities in a montane assemblage in the Sierra de Neila, Spain. *Herpetol. J.* 13:167–73.

Martínez-Solano, I., J. Bosch, and M. García-París. 2003. Demographic trends and community stability in a montane amphibian assemblage. *Cons. Biol.* 17:238–44.

Martins, E. P. 1994. Phylogenetic perspectives on the evolution of lizard territoriality. In *Lizard ecology: Historical and experimental perspectives,* ed. L. J. Vitt and E. R. Pianka, 117–44. Princeton, NJ: Princeton University Press.

Martins, I. 2001. Parental care behaviour in *Leptodactylus podicipinus* (Cope, 1862) (Anura, Leptodactylidae). *Herpetol. J.* 11:29–32.

Martins, M. 1988. Biologia reprodutiva de *Leptodactylus fuscus* em Boa Vista, Roraima (Amphibia: Anura). *Rev. Brasil. Biol.* 48:969–77.

———. 1989. Deimatic behavior in *Pleurodema brachyops*. *J. Herpetol.* 23:305–7.

———. 1993a. Observations on nest dynamics and embryonic and larval development in the nest building gladiator frog, *Hyla faber*. *Amphibia-Reptilia* 14:411–21.

———. 1993b. Observations on the reproductive behaviour of the smith frog, *Hyla faber*. *Herpetol. J.* 3:31–34.

Martins, M., and M. Gordo. 1993. *Bothrops atrox* (common lancehead). Diet. *Herp. Rev.* 24:151–52.

Martins, M., and C. F. B. Haddad. 1988. Vocalizations and reproductive behaviour in the smith frog, *Hyla faber* Wied (Amphibia: Hylidae). *Amphibia-Reptilia* 9:49–60.

Martins, M., and G. Moreira. 1991. The nest and tadpole of *Hyla wavrini* Parker (Amphibia, Anura). *Mem. Inst. Butantan* 53:197–204.

Martins, M., J. P. Pombal, and C. F. B. Haddad. 1998. Escalated aggressive behaviour and facultative parental care in the nest building gladiator frog, *Hyla faber*. *Amphibia-Reptilia* 19:65–73.

Martins, M., I. Sazima, and S. G. Egler. 1993. Predators of the nest building gladiator frog, *Hyla faber*, in southeastern Brazil. *Amphibia-Reptilia* 14:307–9.

Martof, B. S. 1953a. Home range and movement of the green frog, *Rana clamitans*. *Ecology* 34:529–43.

———. 1953b. Territoriality in the green frog, *Rana clamitans*. *Ecology* 34:165–75.

———. 1956a. Factors influencing size and composition of populations of *Rana clamitans*. *Am. Midl. Nat.* 56:224–45.

———. 1956b. Growth and development of the green frog, *Rana clamitans* under natural conditions. *Am. Midl. Nat.* 55:101–17.

———. 1961. Vocalization as an isolating mechanism in frogs. *Am. Midl. Nat.* 65:118–26.

———. 1962a. Some aspects of the life history and ecology of the salamander *Leurognathus*. *Am. Midl. Nat.* 67:1–35.

———. 1962b. Some observations on the role of olfaction among salientian Amphibia. *Physiol. Zool.* 35:270–72.

Martof, B. S., and R. L. Humphries. 1959. Geographic variation in the wood frog *Rana sylvatica*. *Am. Midl. Nat.* 61:350–89.

Martof, B. S., and F. L. Rose. 1963. Geographic variation in southern populations of *Desmognathus ochrophaeus*. *Am. Midl. Nat.* 69:376–425.

Martof, B. S., and D. C. Scott. 1957. The food of the salamander *Leurognathus*. *Ecology* 38:494–501.

Martof, B. S., and E. F. Thompson, Jr. 1958. Reproductive behavior of the chorus frog *Pseudacris nigrita*. *Behaviour* 13:243–57.

———. 1964. A behavioral analysis of the mating call of the chorus frog, *Pseudacris triseriata*. *Am. Midl. Nat.* 71:198–209.

Marunouchi, J., T. Kusano, and H. Ueda. 2002. Fluctuation in abundance and age structure of a breeding population of the Japanese brown frog, *Rana japonica* Günther (Amphibia, Anura). *Zool. Sci.* 19:343–50.

Marunouchi, J., H. Ueda, and O. Ochi. 2000. Variation in age and size among breeding populations at different altitudes in the Japanese newts, *Cynops pyrrhogaster*. *Amphibia-Reptilia* 21:381–96.

Marvin, G. A. 1998a. Interspecific aggression and spatial relationships in the salamanders *Plethodon kentucki* and *Plethodon glutinosus:* Evidence of interspecific interference competition. *Can. J. Zool.* 76:94–103.

———. 1998b. Territorial behavior of the plethodontid salamander *Plethodon kentucki:* Influence of habitat structure and population density. *Oecologia* 114:133–44.

———. 2001. Age, growth, and long-term site fidelity in the terrestrial plethodontid salamander *Plethodon kentucki*. *Copeia* 2001:108–17.

Marvin, G. A., and V. H. Hutchison. 1995. Avoidance response by adult newts (*Cynops pyrrhogaster* and *Notophthalmus viridescens*) to chemical alarm cues. *Behaviour* 132:95–105.

———. 1996. Courtship behavior of the Cumberland Plateau woodland salamander, *Plethodon kentucki* (Amphibia: Plethodontidae), with a review of courtship in the genus *Plethodon*. *Ethology* 102:285–303.

Marvin, G. A., R. R. Whitekiller, and V. H. Hutchison. 2004. Avoidance of alarm chemicals by plethodontid salamanders (genus *Eurycea*): Importance of phylogeny, ecology, and methodology. *Herpetologica* 60:24–33.

Marynick, S. P. 1971. Long term storage of sperm in *Desmognathus fuscus* from Louisiana. *Copeia* 1971:345–47.

Maskey, T., H. H. Schleich, and W. Kästle. 2002. Nepal's herpetofauna on a razor's edge between threat and conservation. In *Amphibians and reptiles of Nepal. Biology, Systematics, Field Guide*, ed. H. H. Schleich and W. Kästle, 59–77. Ruggell, Germany: A. R. G. Gantner Verlag.

Maslin, T. P. 1950. The production of sound in caudate Amphibia. *Univ. Colorado Stud. Ser. Biol.* 1:29–45.

Maslin, T. P., Jr. 1939. Egg-laying of the slender salamander (*Batrachoseps attenuatus*). *Copeia* 1939:209–12.

Maslova, I. V. 2000. On the influence of extreme weather conditions on some amphibian species of the Primorye Region. *Adv. Amphib. Res. Former Soviet Union* 5:227–32.

Mason, C. F., and S. M. Macdonald. 1986. *Otters: Ecology and conservation*. New York: Cambridge University Press.

Mason, M. J. and P. M. Narins. 2002. Vibrometric studies of the middle ear of the bullfrog, *Rana catesbeiana*. II. The operculum. *J. Exp. Biol.* 205:3167–76.

Mason, J. R., M. D. Rabin, and D. A. Stevens. 1982. Conditioned taste aversions: Skin secretions used for defense by tiger salamanders, *Ambystoma tigrinum*. *Copeia* 1982:667–71.

Mason, M. C., and N. I. Passmore. 1998. *Hyperolius tuberilinguis* (tinker reed frog). Oviposition. *Herp. Rev.* 29:231–32.

Massey, A. 1988. Sexual interactions in red-spotted newt populations. *Anim. Behav.* 36:205–10.

Mather, J. G., and R. R. Baker. 1981. Magnetic sense of direction in woodmice for route-based navigation. *Nature* 291:152–55.

Mathis, A. 1989. Do seasonal spatial distributions in a terrestrial salamander reflect reproductive behavior or territoriality? *Copeia* 1989:788–91.

———. 1990a. Territoriality in a terrestrial salamander: the influence of resource quality and body size. *Behaviour* 112:162–75.

———. 1990b. Territorial salamanders assess sexual and competitive information using chemical signals. *Anim. Behav.* 40:953–62.

———. 1991a. Large male advantage for access to females: Evidence of male-male competition and female discrimination in a territorial salamander. *Behav. Ecol. Sociobiol.* 29:133–38.

———. 1991b. Territories of male and female terrestrial salamanders: Costs, benefits and intersexual spatial associations. *Oecologia* 86:433–40.

Mathis, A., and E. Britzke. 1999. The roles of body size and experience in agonistic displays of the Ozark zigzig salamander, *Plethodon angusticlavius*. *Herpetologica* 55:344–52.

Mathis, A., K. Deckard, and C. Duer. 1998. Laboratory evidence for territorial behavior by the southern red-backed salamander, *Plethodon serratus:* Influence of residency status and pheromonal advertisement. *Southwest. Nat.* 43:1–5.

Mathis, A., R. G. Jaeger, W. H. Keen, P. K. Ducey, S. C. Walls, and B. W. Buchanan. 1995. Aggression and territoriality by salamanders and a comparison with the behaviour of frogs. In *Amphibian biology, vol. 2. Social behaviour,* ed. H. Heatwole and B. K. Sullivan, 633–76. Chipping Norton, New South Wales, Australia: Surrey Beatty and Sons.

Mathis, A., and D. Lancaster. 1998. Responses of terrestrial salamanders to chemical stimuli from distressed conspecifics. *Amphibia-Reptilia* 19:1–6.

Mathis, A., K. L. Murray, and C. R. Hickman. 2003. Do experience and body size play a role in responses of larval ringed salamanders, *Ambystoma annulatum,* to predator kairomones? Laboratory and field assays. *Ethology* 109:159–70.

Mathis, A., D. W. Schmidt, and K. A. Medley. 2000. The influence of residency status on agonistic behavior of male and female Ozark zigzag salamanders *Plethodon angusticlavius. Am. Midl. Nat.* 143:245–49.

Mathis, A., and R. R. Simons. 1994. Size-dependent responses of resident male red-backed salamanders to chemical stimuli from conspecifics. *Herpetologica* 50:335–44.

Mathis, A., and F. Vincent. 2000. Differential use of visual and chemical cues in predator recognition and threat-sensitive predator-avoidance responses by larval newts (*Notophthalmus viridescens*). *Can. J. Zool.* 78:1646–52.

Matsui, M. 1989. Breeding strategy in the Japanese common toad, *Bufo japonicus.* In *Current herpetology in East Asia,* ed. M. Matsui, H. Hikida, and R. C. Goris, 332–41. Kyoto: Herpetological Society of Japan,

———. 1995. Calls produced by a "voiceless" frog, *Rana blythi* Boulenger 1920, from Peninsular Malaysia (Amphibia, Anura). *Trop. Zool.* 8:325–31.

———. 2000. Batrachology of Japan and adjacent regions—A systematic review. *Comp. Biochem. Physiol.* B 126:247–56.

Matsui, M., Y. Misawa, K. Nishikawa, and S. Tanabe. 2000. Allozymic variation of *Hynobius kimurae* Dunn (Amphibia, Caudata). *Comp. Biochem. Physiol.* B 125:115–25.

Matsui, M., K. Nishikawa, S. Tanabe, and Y. Misawa. 2001. Systematic status of *Hynobius tokyoensis* (Amphibia: Urodela) from Aichi Prefecture, Japan: A biochemical survey. *Comp. Biochem. Physiol.* B 130:181–89.

Matsui, M., and N. Orlov. 2004. A new species of *Chirixalus* from Vietnam (Anura: Rhacophoridae). *Zool. Sci.* 21:671–76.

Matsui, M., and G.-F. Wu. 1994. Acoustic characteristics of treefrogs from Sichuan, China, with comments on systematic relationships of *Polypedates* and *Rhacophorus* (Anura: Rhacophoridae). *Zool. Sci.* 11:485–90.

Matsui, T., and M. Matsui. 1990. A new brown frog (genus *Rana*) from Honshu, Japan. *Herpetologica* 46:78–85.

Matsumoto, Y., T. Matsumoto, and K. Miyashita. 1984. Feeding habits of the marine toad, *Bufo marinus,* in the Bonin Islands, Japan. *Jap. J. Ecol.* 34:289–97.

Matsushima, T., M. Satou, and K. Ueda. 1986. Glossopharyngeal and tectal influences on tongue-muscle motoneurons in the Japanese toad. *Brain Res.* 365:198–203.

———. 1987. Direct contact between glossopharyngeal afferent terminals and hypoglossal motoneurons revealed by double labeling with cobaltic lysine and horseradish peroxidase in the Japanese toad. *Neurosci. Lett.* 80:241–45.

———. 1988. Neuronal pathways for the lingual reflex in the Japanese toad. *J. Comp. Physiol. A* 164:173–93.

———. 1989. Medullary reticular neurons in the Japanese toad: Morphologies and excitatory inputs from the optic tectum. *J. Comp. Physiol. A* 166:7–22.

Matthes, E. 1934. Bau und Funktion der Lippensaume wasserlebender Urodelen. *Z. Morph. Ökol. Tiere* 28:155–69.

Matthews, K. R. 2003. Response of mountain yellow-legged frogs, *Rana muscosa,* to short distance translocation. *J. Herpetol.* 37:621–26.

Matthews, K. R., R. A. Knapp, and K. L. Pope. 2002. Garter snake distribution in high-elevation aquatic ecosystems: Is there a link with decining amphibian populations and nonnative trout introductions? *J. Herpetol.* 36:16–22.

Matthews, K. R., and K. L. Pope. 1999. A telemetric study of the movement patterns and habitat use of *Rana muscosa,* the mountain yellow-legged frog, in a high-elevation basin in Kings Canyon National Park, California. *J. Herpetol.* 33:615–24.

Matthews, K. R., K. L. Pope, H. K. Preisler, and R. A. Knapp. 2001. Effects of nonnative trout on Pacific treefrogs (*Hyla regilla*) in the Sierra Nevada. *Copeia* 2001:1130–37.

Matthews, T. C. 1971. Genetic changes in a population of boreal chorus frogs (*Pseudacris triseriata*) polymorphic for color. *Am. Midl. Nat.* 85:208–21.

———. 1975. Biochemical polymorphism in populations of the Argentine toad, *Bufo arenarum. Copeia* 1975:454–65.

Matthews, T. C., and D. Pettus. 1966. Color inheritance in *Pseudacris triseriata. Herpetologica* 22:269–75.

Mattison, C. 1987. *Frogs and toads of the world.* New York: Facts on File.

Matutte, B., K. B. Storey, F. C. Knoop, and J. M. Conlon. 2000. Induction of synthesis of an antimicrobial peptide in the skin of the freeze-tolerant frog, *Rana sylvatica,* in response to environmental stimuli. *FEBS Lett.* 483:135–38.

Maurer, E. F., and A. Sih. 1996. Ephemeral habitats and variation in behavior and life history comparisons of sibling salamander species. *Oikos* 76:337–49.

Mautz, W. J. 1980. Factors influencing evaporative water loss in lizards. *Comp. Biochem. Physiol.* 67A:429–37.

Maxson, L. R., and J. M. Szymura. 1984. Relationships among discoglossid frogs: An albumin perspective. *Amphibia-Reptilia* 5:245–52.

Maxson, L. R., A. Song, and R. Lopata. 1981. Phylogenetic relationships among North American toads, genus *Bufo. Biochem. Syst. Ecol.* 9:347–50.

Maxwell, S. S. 1913. On absorption of water by the skin of the frog. *Am. J. Physiol.* 32:286–94.

May, R. M. 1984. An overview: Real and apparent patterns in community structure. In *Ecological communities,* ed. D. R. Strong, Jr., D. Simberloff, L. G. Abele, and A. B. Thistle, 3–16. Princeton, NJ: Princeton University Press.

Mayhew, W. W. 1965. Adaptations of the amphibian, *Scaphiopus couchi,* to desert conditions. *Am. Midl. Nat.* 74:95–109.

———. 1968. Biology of desert amphibians and reptiles. In *Desert biology,* ed. G. W. Brown, 195–356. New York: Academic Press.

Maynard Smith, J. 1977. Parental investment: A prospective analysis. *Anim. Behav.* 25:1–9.

———. 1978. *The evolution of sex.* New York: Cambridge University Press.

Maynard Smith, J., and G. A. Parker. 1976. The logic of asymmetric contests. *Anim. Behav.* 24:159–75.

Mazerolle, M. J. 2001. Amphibian activity, movement patterns, and body size in fragmented peat bogs. *J. Herpetol.* 35:13–20.

———. 2003. Detrimental effects of peat mining on amphibian abudance and species richness in bogs. *Biol. Cons.* 113:215–23.

Mazerolle, M. J., and M. Cormier. 2003. Effects of peat mining intensity on green frog (*Rana clamitans*) occurrence in bog ponds. *Wetlands* 23:709–16.

Mazur, T. 1967. Seasonal variation in the energy reserves of *Bufo bufo* (L.) and *Rana arvalis* (Nilss.) (Anura) in Poland. *Ekol. Pol., Ser. A.*, 15:607–13.

Mazzoni, R. A. A. Cunningham, P. Daszak, A. Apolo, E. Perdomo, and G. Speranza. 2003. Emerging pathogen of wild amphibians in frogs (*Rana catesbeiana*) farmed for international trade. *Emerg. Infect. Diseases* 9:995–98.

Mazzotti, S. 1993. Competizione fra fauna ittica e batracofauna in laghii dell'Appennino Settentrionale. *Quad. Civica Staz. Idrobiol. Milano* 20:67–74.

McAfee, R. D. 1972. Survival of *Rana pipiens* in deionized water. *Science* 178:183–85.

McAlister, W. H. 1959. The vocal structures and method of call production in the genus *Scaphiopus*. *Texas J. Sci.* 11:60–77.

McAlpine, D. F. 1997. Helminth communities in bullfrogs (*Rana catesbeiana*), green frogs (*Rana clamitans*), and leopard frogs (*Rana pipiens*) from New Brunswick, Canada. *Can. J. Zool.* 75:1883–90.

McAlpine, D. F., and M. D. B. Burt. 1998. Helminths of bullfrogs, *Rana catesbeiana*, green frogs, *R. clamitans*, and leopard frogs, *R. pipiens* in New Brunswick. *Can. Field-Nat.* 112:50–68.

McAlpine, D. F., and T. G. Dilworth. 1989. Microhabitat and prey size among three species of *Rana* (Anura: Ranidae) sympatric in eastern Canada. *Can. J. Zool.* 67:2244–52.

McBean, R., and L. Goldstein. 1970a. Renal function during osmotic stress in the aquatic toad *Xenopus laevis*. *Am. J. Physiol.* 219:1115–23.

McBean, R., and L. Goldstein. 1970b. Accelerated synthesis of urea in *Xenopus laevis* during osmotic stress. *Am. J. Physiol.* 219:1124–30.

McCallum, H. 2005. Inconclusiveness of chytridiomycosis as the agent in widespread frog declines. *Cons. Biol.* 19:1421–30.

McCallum, M. L., T. L. Klotz, and S. E. Trauth. 2003. *Rana sylvatica* (Wood frog). Phonotactic stalking. *Herp. Rev.* 34:54.

McCallum, M. L., and S. E. Trauth. 2003. *Acris crepitans* (Northern cricket frog). Communal hibernaculum. *Herp. Rev.* 34:228.

McCann, C. 1932. Biology of frogs in the genus *Rana*. *J. Bombay Nat. Hist. Soc.* 36:152–80.

McCauley, S. J., S. S. Bouchard, B. J. Farina, K. Isvaran, S. Quader, D. W. Wood, and C. M. St. Mary. 2000. Energetic dynamics and anuran breeding phenology: Insights from a dynamic game. *Behav. Ecol.* 11:429–36.

McCay, M. G. 2003. Winds under the rain forest canopy: The aerodynamic environment of gliding tree frogs. *Biotropica* 35:94–102.

McClanahan, L. L. 1964. Osmotic tolerance of the muscles of two desert inhabiting toads, *Bufo cognatus* and *Scaphiopus couchi*. *Comp. Biochem. Physiol.* 12:501–8.

———. 1967. Adaptations of the spadefoot toad, *Scaphiopus couchi*, to desert environments. *Comp. Biochem. Physiol.* 20:73–99.

———. 1972. Changes in body fluids of burrowed spadefoot toads as a function of soil water potential. *Copeia* 1972:209–16.

McClanahan, L. L., and R. Baldwin. 1969. Rate of water uptake through the integument of the desert toad *Bufo punctatus*. *Comp. Biochem. Physiol.* 28:381–89.

McClanahan, L. L., R. Ruibal, and V. H. Shoemaker. 1983. Rate of cocoon formation and its physiological correlates in a cerotophryd frog. *Physiol. Zool.* 56:430–35.

McClanahan, L. L., and V. H. Shoemaker. 1987. Behavior and thermal relations of the arboreal frog *Phyllomedusa sauvagei*. *Nat. Geogr. Res.* 3:11–21.

McClanahan, L. L., V. H. Shoemaker, and R. Ruibal. 1976. Structure and function of the cocoon of a ceratophryd frog. *Copeia* 1976:179–85.

McClanahan, L. L., J. N. Stinner, and V. H. Shoemaker. 1978. Skin lipids, water loss, and energy metabolism in a South American tree frog (*Phyllomedsa sauvagei*). *Physiol. Zool.* 51:179–87.

McClelland, B. E., and W. Wilczynski. 1989. Sexually dimorphic laryngeal morphology in *Rana pipiens*. *J. Morphol.* 201:293–99.

McClelland, B. E., W. Wilczynski, and A. S. Rand. 1997. Sexual dimorphism and species differences in the neurophysiology and morphology of the acoustic communication system of two Neotropical hylids. *J. Comp. Physiol. A* 180:451–62.

McClelland, B. E., W. Wilczynski, and M. J. Ryan. 1990. Sexually dimorphic laryngeal morphology in cricket frogs (*Acris crepitans*). *Soc. Neurosci. Abstracts* 18:882.

McClelland, B. E., W. Wilczynski, and M. J. Ryan. 1998. Intraspecific variation in laryngeal and ear morphology in male cricket frogs (*Acris crepitans*). *Biol. J. Linn. Soc.* 63:51–67.

McCoid, M. J. 1995. Non-native reptiles and amphibians. In *Our living resources. A report to the nation on the distribution, abundance, and health of U.S. plants, animals, and ecosystems,* ed. E. T. LaRoe, G. S. Farris, C. E. Puckett, P. D. Doran, and M. J. Mac, 433–37. Washington, DC: U.S. Dept. of Interior, National Biological Service.

McCoid, M. J., and T. H. Fritts. 1980. Notes on the diet of a feral population of *Xenopus laevis* (Pipidae) in California. *Southwest. Nat.* 25:272–75.

———. 1993. Speculations on colonizing success of the African clawed frog, *Xenopus laevis* (Pipidae), in California. *S. Afr. J. Zool.* 28:59–61.

McCoid, M. J., R. A. Hensley, and G. J. Witteman. 1994. Factors in the decline of *Varanus indicus* on Guam. *Herp. Rev.* 25:60–61.

McCollum, S. A., and J. D. Leimberger. 1997. Predator-induced morphological changes in an amphibian: Predation by dragonflies affects tadpoles shape and color. *Oecologia* 109:615–21.

McCollum, S. A., and J. Van Buskirk. 1996. Costs and benefits of a predator-induced polyphenism in the gray treefrog *Hyla chrysoscelis*. *Evolution* 50:583–93.

McComb, W. C., and R. E. Noble. 1981. Herpetofaunal use of natural tree cavities and nest boxes. *Wildlife Soc. Bull.* 9:261–67.

McCormick, S., and G. A. Polis. 1982. Arthropods that prey on vertebrates. *Biol. Rev.* 57:29–58.

McCoy, C. J. 1982. Amphibians and reptiles in Pennsylvania. *Spec. Publ. Carnegie Mus. Nat. Hist.* 6:1–91.

McCoy, C. J., and E. J. Censky. 1992. Biology of the Yucatan hognosed pitviper, *Porthidium yucatanicum*. In *Biology of the pitvipers,* ed. J. A. Campbell and E. D. Brodie, Jr., 217–22. Tyler, TX: Selva.

McCoy, M. W. 2003. *Bufo terrestris* (Southern toad). Predation. *Herp. Rev.* 34:135–36.

McCrady, E. 1954. A new species of *Gyrinophilus* (Plethodontidae) from Tennessee caves. *Copeia* 1954:200–206.

McCranie, J. R., and L. D. Wilson. 1992. *Nototriton barbouri* (NCN). Reproduction. *Herp. Rev.* 23:115–16.

———. 2002. *Amphibians of Honduras.* Ithaca, NY: Society for the Study of Amphibians and Reptiles.

McCutcheon, F. H. 1936. Hemoglobin function during the life history of the bullfrog. *J. Cell. Comp. Physiol.* 8:63–81.

McCutcheon, F. H., and F. G. Hall. 1937. Hemoglobin in the Amphibia. *J. Cell. Comp. Physiol.* 9:191–97.

McDarby, J. H., D. M. Madison, and J. C. Maerz. 1999. Chemosensory avoidance of predators by red-backed salamanders, *Plethodon cinereus.* In *Chemical signals in vertebrates,* ed. R. Johnston, D. Müller-Schwarze, and S. Sorensen, 489–95. New York: Plenum.

McDiarmid, R. W. 1975. Glass frog romance along a tropical stream. *Terra* 13:14–18.

———. 1978. Evolution of parental care in frogs. In *The development of behavior:Comparative and evolutionary aspects,* ed.G. M. Burghardt and M. Beckoff, 127–47. New York: Garland.

McDiarmid, R. W., and K. Adler. 1974. Notes on territorial and vocal behavior of Neotropical frogs of the genus *Centrolenella. Herpetologica* 30:75–78.

McDiarmid, R. W., and R. Altig, eds. 1999a. *Tadpoles: The biology of anuran larvae.* Chicago: University of Chicago Press.

———. 1999b. Research materials and techniques. In *Tadpoles: The biology of anuran larvae,* ed. R. W. McDiarmid and R. Altig, 7–23. Chicago: University of Chicago Press.

McDiarmid, R. W., and M. S. Foster. 1987. Cocoon formation in another hylid frog, *Smilisca baudinii. J. Herpetol.* 21:352–55.

McDiarmid, R. W., and S. Gorzula. 1989. Aspects of the reproductive ecology and behavior of the Tepui toads, genus *Oreophrynella* (Anura, Bufonidae). *Copeia* 1989:445–51.

McDiarmid, R. W., and R. D. Worthington. 1970. Concerning the reproductive habits of tropical plethodontid salamanders. *Herpetologica* 26:57–70.

McDonald, D. G., R. G. Boutilier, and D. P. Toews. 1980. The effects of enforced activity on ventilation, circulation and blood acid-base balance in the semi-terrestrial anuran, *Bufo marinus. J. Exp. Biol.* 84:273–87.

McDonald, D. G., J. L. Ozog, and B. P. Simons. 1984. The influence of low pH environments on ion regulation the larval stages of the anuran amphibian, *Rana clamitans. Can. J. Zool.* 62:2171–77.

McDonald, K. R. 1990. *Rheobatrachus* Liem and *Taudactylus* Straughan and Lee (Anura: Leptodactylidae) in Eungella National Park, Queensland: distribution and decline. *Trans. Roy. Soc. South Australia* 114:187–94.

McDonald, K. R., and R. A. Alford. 1999. A review of declining frogs in northern Queensland. In *Declines and disappearances of Australian frogs,* ed. A.Campbell, 14–22. Canberra: Environment Australia.

McDonald, K. R., and M. Davies. 1990. Morphology and biology of the Australian tree frog *Litoria pearsoniana* (Copland) (Anura: Hylidae). *Trans. Roy. Soc. South Australia* 114:145–156.

McDonald, K. R., and M. J. Tyler. 1984. Evidence of gastric brooding in the Australian leptodactylid frog *Rheobatrachus vitellinus. Trans. Roy. Soc. South Australia* 108:226.

McDonnell, L. J., D. F. Gartside, and M. J. Littlejohn. 1978. Analysis of a narrow hybrid zone between two species of *Pseudophryne* (Anura: Leptodactylidae) in south-eastern Australia. *Evolution* 32:602–12.

McEwan, L. C., and D. H. Hirth. 1980. Food habits of the bald eagle in north-central Florida. *Condor* 82:229–31.

McGavin, M. 1978. Recognition of conspecific odors by the salamander *Plethodon cinereus. Copeia* 1978:356–58.

McGinley, M. A., D. H. Temme, and M. A. Geber. 1987. Parental investment in offspring in variable environments: Theoretical and empirical considerations. *Am. Nat.* 130:370–98.

McGregor, J. H., and W. R. Teska. 1989. Olfaction as an orientation mechanism in migrating *Ambystoma maculatum. Copeia* 1989:779–81.

McIndoe, R., and D. G. Smith. 1984a. Functional anatomy of the internal gills of the tadpole of *Litoria ewingii* (Anura, Hylidae). *Zoomorphology* 104:280–91.

———. 1984b. Functional morphology of gills in larval amphibians. In *Respiration and metabolism of embryonic vertebrates,* ed. R. S. Seymour, 55–69. Boston: Dr. W. Junk.

McIntyre, P. B. 2003. *Physalaemus pustulosus* (Tungara frog). Habitat. *Herp. Rev.* 34:52–53.

McIntyre, P. B., S. Baldwin, and A. S. Flecker. 2004. Effects of behavioral and morphological plasticity on risk of predation in a Neotropical tadpole. *Oecologia* 141:130–38.

McKamie, J. A., and G. A. Heidt. 1974. A comparison of spring food habits of the bullfrog, *Rana catesbeiana,* in three habitats of central Arkansas. *Southwest. Nat.* 19:107–11.

McKay, D. M. 1989. *Seasonal variation in calling behavior and muscle biochemistry in spring peepers (Hyla crucifer).* Unpublished M.S. thesis. Storrs: University of Connecticut.

McKeever, S. 1977. Observations of *Corethrella* feeding on tree frogs (*Hyla*). *Mosquito News* 37:522–23.

McKeever, S., and F. E. French. 1991. *Corethrella* (Diptera: Corethrellidae) of eastern North America: Laboratory life history and field responses to anuran calls. *Ann. Entomol. Soc. Amer.* 84:493–97.

McKeon, C. S. 2005. *Hyla acreana* (NCN). Larval behavior. *Herp. Rev.* 36:300–301.

McKeown, J. P. 1968. *The ontogenetic development of Y-axis orientation in four species of anurans.* Unpublished PhD dissertation. Mississippi State: Mississippi State University.

McKeown, K. C. 1942–43. Vertebrates captured by Australian spiders. *Proc. Roy. Zool. Soc. New South Wales* 1942–43:17–30.

———. 1952. *Australian spiders: Their lives and habits.* Sydney, Australia: Angus and Robertson.

McKnight, A. D. C., D. R. DiBona, and A. Leaf. 1980. Sodium transport across toad urinary bladder: A model "tight" epithelium. *Physiol. Rev.* 60:615–713.

McLaren, I. A., and J. M. Cooley. 1972. Temperature adaptation of embryonic development rate among frogs. *Physiol. Zool.* 45:223–28.

McLean, I. F. G., A. D. Wight, and G. Williams. 1999. The role of legislation in conserving Europe's threatened species. *Cons. Biol.* 13:966–69.

McLeod, D. S. 2005. Nebraska's declining amphibians. In *Amphibian declines: The conservation status of United States species,* ed. M. J. Lannoo, 292–94. Berkeley: University of California Press.

McLeod, R. F., and J. E. Gates. 1998. Response of herpetofaunal communities to forest cutting and burning at Chesapeake Farms, Maryland. *Am. Midl. Nat.* 139:164–77.

McLister, J. D. 2001. Physical factors affecting the cost and efficiency of sound production in the treefrog *Hyla versicolor. J. Exp. Biol.* 203:69–80.

———. 2003. The metabolic cost of amplexus in the grey tree frog (*Hyla versicolor*): Assessing the energetics of male mating success. *Can. J. Zool.* 81:388–94.

McLister, J. D., E. D. Stevens, and J. P. Bogart. 1995. Comparative contractile dynamics of calling and locomotor muscles in three hylid frogs. *J. Exp. Biol.* 198:1527–38.

McMillan, N. F. 1963. Toads continuing to migrate for spawning to a now vanished pond. *Brit. J. Herp.* 3:88.

McNab, B. K. 1988. Complications inherent in scaling the basal rate of metabolism in mammals. *Q. Rev. Biol.* 63:25–54.

McNamara, J. M., R. H. Mace, and A. I. Houston. 1987. Optimal daily routines of singing and foraging in a bird singing to attract a mate. *Behav. Ecol. Sociobiol.* 20:399–405.

McVey, M. E., R. G. Zahary, D. Perry, and J. MacDougal. 1981. Territoriality and homing behavior in the poison dart frog (*Dendrobates pumilio*). *Copeia* 1981:1–8.

McWilliams, S. R. 1992. Courtship behavior of the small-mouthed salamander (*Ambystoma texanum*): The effects of conspecific males on male mating tactics. *Behaviour* 121:1–19.

McWilliams, S. R. and M. D. Bachmann. 1989a. Foraging ecology and prey preference of pond-form larval small-mouthed salamanders, *Ambystoma texanum*. *Copeia* 1989:948–61.

———. 1989b. Predatory behavior of larval small-mouthed salamanders (*Ambystoma texanum*). *Herpetologica* 45:459–67.

Mead, L. S., and P. A. Verrell. 2002. Evolution of courtship behaviour patterns and reproductive isolation in the *Desmognathus ochrophaeus* complex. *Ethology* 108:403–27.

Means, D. B. 1975. Competitive exclusion along a habitat gradient between two species of salamanders (*Desmognathus*) in western Florida. *J. Biogeogr.* 2:253–63.

Means, D. B. 2005. *Hyla andersonii* Baird 1854. Pine barrens treefrog. In *Amphibian declines: The conservation status of United States species,* ed. M. J. Lannoo, 445–47. Berkeley: University of California Press.

Means, D. B., and H. W. Campbell. 1981. Effects of prescribed burning on amphibians and reptiles. In *Prescribed fire and wildfire in southern forests: Symposium proceedings,* ed. G. W. Wood, 89–97. Georgetown, SC: Belle W. Baruch Forest Science Institute.

Means, D. B., C. K. Dodd, Jr., S. A. Johnson, and J. G. Palis. 2004. Amphibians and fire in longleaf pine ecosystems: Response to Shurbon and Fauth. *Cons. Biol.* 18:1149–53.

Means, D. B., and A. A. Karlin. 1989. A new species of *Desmognathus* from the eastern Gulf coastal plain. *Herpetologica* 45:37–46.

Means, D. B., J. G. Palis, and M. Baggett. 1996. Effects of slash pine silviculture on a Florida population of flatwoods salamander. *Cons. Biol.* 10:426–37.

Measey, G. J. 1998. Diet of feral *Xenopus laevis* (Daudin) in South Wales, UK. *J. Zool., London* 246:287–98.

———. 2004. Are caecilians rare? An East African perspective. *J. East Afr. Nat. Hist.* 93:97–117.

Measey, G. J., and M. Di-Bernardo. 2003. Estimating juvenile abundance in a population of the semiaquatic caecilian *Chthonerpeton indistinctum* (Amphibia: Gymnophiona: Typhlonectidae), in southern Brazil. *J. Herpetol.* 37:371–73.

Measey, G. J., D. J. Gower, O. V. Oommen, and M. Wilkinson. 2003a. A mark-recapture study of the caecilian amphibian *Gegeneophis ramaswamii* (Amphibia: Gymnophiona: Caeciliidae) in southern India. *J. Zool., London* 261:129–33.

———. 2003b. Quantitative surveying of endogeic limbless vertebrates—A case study of *Gegeneophis ramaswamii* (Amphibia: Gymnophiona: Caeciliidae) in southern India. *Appl. Soil Ecol.* 23:43–53.

———. 2004. A subterranean generalist predator: Diet of the soil-dwelling caecilian *Gegeneophis ramaswamii* (Amphibia; Gymnophiona; Caeciliidae) in southern India. *C. R. Biol.* 327:65–76.

Meban, C. 1980. Thickness of the air-blood barriers in vertebrate lungs. *J. Anat.* 131:299–307.

Mecham, J. S. 1960. Introgressive hybridization between two southeastern treefrogs. *Evolution* 14:445–57.

———. 1961. Isolating mechanisms in anuran amphibians. In *Vertebrate speciation,* ed. W. F. Blair, 24–61. Austin: University of Texas Press.

———. 1965. Genetic relationships and reproductive isolation in southeastern frogs of the genera *Pseudacris* and *Hyla*. *Am. Midl. Nat.* 74:269–308.

———. 1971. Vocalizations of the leopard frog, *Rana pipiens,* and three related Mexican species. *Copeia* 1971:505–16.

Meede, U. 1980. Beobactungen an *Dendrobates quinquevittatus* und *Phyllobates femoralis* (Amphibia: Salientia: Dendrobatidae). *Salamandra* 16:38–51.

Meegaskumbura, M., F. Bossuyt, R. Pethiyagoda, K. Manamendra-Arachchi, M. Bahir, M. C. Milinkovitch, and C. J. Schneider. 2002. Sri Lanka: An amphibian hot spot. *Science* 298:379.

Megela, A. L. and R. R. Capranica. 1982. Differential patterns of physiological masking in the anuran auditory nerve. *J. Acoust. Soc. Am.* 71:641–45.

Megela, A. L., and R. R. Capranica. 1983. A neural and behavioral study of auditory habituation in the bullfrog, *Rana catesbeiana*. *J. Comp. Physiol.* 151:423–34.

Megela Simmons, A. 1984. Behavioral and vocal response thresholds to mating calls in the bullfrog *Rana catesbeiana*. *J. Acoust. Soc. Am.* 76:676–81.

———. 1988. Masking patterns in the bullfrog (*Rana catesbeiana*). I.: Behavioral effects. *J. Acoust. Soc. Am.* 83:1087–92.

Megela Simmons, A., R. C. Buxbaum, and M. P. Mirin. 1993. Perception of complex sounds by the green treefrog, *Hyla cinerea:* Envelope and fine-structure cues. *J. Comp. Physiol. A* 173:321–27.

Megela Simmons, A., C. F. Moss, and K. M. Daniel. 1985. Behavioral audiograms of the bullfrog (*Rana catesbeiana*) and the green tree frog (*Hyla cinerea*). *J. Acoust. Soc. Am.* 78:1236–44.

Meirelles, N. C., M. L. C. Viera, L. P. S. Airoldi, and A. Focesi, Jr. 1979. Some larval properties of *Pipa carvalhoi* adult hemoglobins. *Comp. Biochem. Physiol.* 62A:859–62.

Meisterhans, K., and H. Heusser. 1970. *Lucilia* Befall an 4 Anuren Arten (Dipt. Tachinidae). *Mitt. Schweiz. Entom. Ges.* 43:41–44.

Melichna, J., E. Gutmann, A. Herbrychova, and J. Stichova. 1972. Sexual dimorphism in contraction properties and fibre pattern of the flexor carpi radialis muscle of the frog (*Rana temporaria* L.). *Experientia* 28:89–91.

Mellanby, K. 1971. *The mole.* London: Collins.

Melquist, W. E., and M. G. Hornocker. 1983. Ecology of river otters, *Lutra canadensis,* in west central Idaho., USA. *Wildl. Monogr.* (83): 1–60.

Mendelson, J. R., III, E. D. Brodie, J. H. Malone, M. E. Acevedo, M. A. Baker, N. J. Smatresk, and J. A. Campbell. 2004. Factors associated with the catastrophic decline of a cloudforest frog fauna in Guatemala. *Rev. Biol. Trop.* 52:991–1000.

Mendelson, J. R., III, H. R. da Silva, and A. M. Maglia. 2000. Phylogenetic relationships among marsupial frog genera (Anura: Hylidae: Hemiphractinae) based on evidence from morphology and natural history. *Zool. J. Linn. Soc.* 128:125–48.

Mendes, E. G. 1945. Contribuição para a fisiologica dos sistemas respiratório e circulatório de *Siphonops annulatus* (Amphibia—Gymnophiona). *Bol. Fac. Filos., Cienc. e Letras, Univ. Sao Paulo, Ser. Zool.* 9:25–64.

Mendes Castanho, L. 1996. *Phyllomedusa distincta* (leaf-frog). Predation. *Herp. Rev.* 27:141.

Mendiola, P., J. de Costa, M. T. Lozano, and B. Aguilleiro. 1991. Histochemical determination of muscle fiber types in locomotor muscles of anuran amphibians. *Comp. Biochem. Physiol.* 99A:365–69.

Mendiola, P., R. Molina, and J. de Costa. 1991. Comparative study in glycolytic capacity in anuran amphibians. *Comp. Biochem. Physiol.* 98B:209–13.

Mendonça, M. T., P. Licht, M. J. Ryan, and R. Barnes. 1985. Changes in hormone level in relation to breeding behavior in male bullfrogs (*Rana catesbeiana*) at the individual and population levels. *Gen. Comp. Endocrinol.* 58:270–79.

Menin, M. 2005. *Bufo proboscideus* (NCN). Predation. *Herp. Rev.* 36:299.

Menin, M., and A. A. Giaretta. 2003. Predation on foam nests of leptodactyline frogs (Anura: Leptodactylidae) by larvae of *Beckeriella niger* (Diptera: Ephydridae). *J. Zool., London* 261:239–43.

Menzies, J. I. 1963. The climate of Bo, Sierra Leone, and the breeding behavior of the toad *Bufo regularis*. *J. W. Afr. Sci. Assoc.* 8:60–73.

———. 1976. *Handbook of common New Guinea frogs.* Wau, Papua New Guinea: Wau Ecology Institute.

———. 1999. A study of *Albericus* (Anura: Microhylidae) of New Guinea. *Aust. J. Zool.* 47:327–60.

Menzies, J. I., and M. J. Tyler. 1977. The systematics and adaptations of some Papuan microhylid frogs which live underground. *J. Zool., London* 183:431–64.

Merchant, H. C. 1966. *Estimated population density and size of home range for two sympatric, closely related species of terrestrial salamanders.* Unpublished Master's thesis. College Park: University of Maryland.

———. 1970. *Estimated energy budget of the red-backed salamander Plethodon cinereus.* Unpublished PhD dissertation. New Brunswick, NJ: Rutgers University.

———. 1972. Estimated population size and home range of the salamanders *Plethodon jordani* and *Plethodon glutinosus*. *J. Washington Acad. Sci.* 62:248–57.

Merilä, J., A. Laurila, A. T. Laugen, K. Räsänen, and M. Pahkala. 2000. Plasticity in age and size at metamorphosis in *Rana temporaria:* A comparison of high and low latitude populations. *Ecography* 23:457–65.

Merilä, J., A. Laurila, and M. Pahkala. 2000. Effects of ambient UV-B radiation on early development of the common frog (*Rana temporaria*) embryos in the subarctic. *Ann. Zool. Fennici* 37:51–58.

Merilä, J., M. Pahkala, and U. Johanson. 2000. Increased ultraviolet-B radiation, climate change and latitudinal adaptation—A frog perspective. *Ann. Zool. Fennici* 37:129–34.

Merilä, J., and M. Sterner. 2002. Medicinal leeches (*Hirudo medicinalis*) attacking and killing adult amphibians. *Ann. Zool. Fennici* 39:343–46.

Merkle, S., and H. Hanke. 1988. Long-term starvation in *Xenopus laevis* Daudin. I. Effects on general metabolism. *Comp. Biochem. Physiol.* 89B:719–30.

Merrell, D. J. 1965. The distribution of the dominant *burnsi* gene in the leopard frog, *Rana pipiens*. *Evolution* 19:69–85.

———. 1969. Limits on heterozygous advantage as an explanation of polymorphism. *J. Hered.* 60:180–82.

———. 1972. Laboratory studies bearing on pigment pattern polymorphism in wild populations of *Rana pipiens*. *Genetics* 70:141–61.

———. 1973. Ecological genetics of anurans as exemplified by *Rana pipiens*. In *Evolutionary biology of the anurans,* ed. J. L. Vial, 329–35. Columbia: University of Missouri Press.

Merrell, D. J., and C. F. Rodell. 1968. Seasonal selection in the leopard frog, *Rana pipiens*. *Evolution* 22:284–88.

Meshaka, W. E. 1996. Retreat use by the Cuban treefrog (*Osteopilus septentrionalis*): Implications for successful colonization in Florida. *J. Herpetol.* 30:443–45.

———. 2001. *The Cuban treefrog in Florida. Life history of a successful colonizing species.* Gainesville: University Press of Florida.

———. 2005. Exotic species. In *Amphibian declines: Conservation status of United States species,* ed. M. J. Lannoo, 271–74. Berkeley: University of California Press.

Meshaka, W. E., B. P. Butterfield, and J. B. Hauge. 2004. *The exotic amphibians and reptiles of Florida.* Malabar, FL: Krieger Publishing Company.

Messere, M., and P. Ducey. 1998. Forest floor distribution of northern redback salamanders, *Plethodon cinereus,* in relation to canopy gaps: First year following selective logging. *Forest Ecol. Manage.* 107:319–24.

Merritt, R. B., W. H. Kroon, D. A. Wienski, and K. A. Vincent. 1984. Genetic structure of natural populations of the red-spotted newt, *Notophthalmus viridescens*. *Biochem. Genet.* 22:669–86.

Metter, D. E. 1963. Stomach contents of Idaho larval *Dicamptodon*. *Copeia* 1963:435–36.

———. 1964. On breeding and sperm retention in *Ascaphus*. *Copeia* 1964:710–11

———. 1967. Variation in the ribbed frog *Ascaphus truei* Stejneger. *Copeia* 1967:634–49.

Meyer, A., and A. C. Wilson. 1990. Origin of tetrapods inferred from their mitochondrial DNA affiliation to lungfish. *J. Mol. Evol.* 31:359–64.

Meyer, A. H., B. R. Schmidt, and K. Grossenbacher. 1998. Analysis of three amphibian populations with quarter-century long time-series. *Proc. Roy. Soc. London B* 265:523–28.

Meyer, F., C. Kutzscher, and W.-R. Grosse. 1999. Diet composition and feeding ecology of the natterjack toad (*Bufo calamita*) in Central European secondary habitats. *Z. Feldherpetol.* 6:65–77.

Meyer-Aurich, A., P. Zander, A. Werner, and R. Roth. 1998. Developing agricultural land use strategies appropriate to nature conservation goals and environmental protection. *Landsc. Urban Plan.* 41:119–27.

Meyers, J. J., J. C. O'Reilly, J. A. Monroy, and K. C. Nishikawa. 2004. Mechanism of tongue protraction in microhylid frogs. *J. Exp. Biol.* 207:21–31.

Meyers, R. S., and J. Felsher. 1976. Continuous recording and quantification of selective blood distribution in the amphibian *Rana catesbeiana*. *Comp. Biochem. Physiol.* 54A:359–63.

Miaud, C. 1993. Predation on newt eggs (*Triturus alpestris* and *T. helveticus*): identification of predators and protective role of oviposition behaviour. *J. Zool., London* 231:575–82.

———. 1994. Role of wrapping behavior on egg survival in three species of *Triturus* (Amphibia: Urodela). *Copeia* 1994:535–37.

———. 1995. Oviposition site selection in three species of European newts (Salamandridae) genus *Triturus. Amphibia-Reptilia* 16:265–72.

Miaud, C., and O. Guillaume. 2005. Variation in age, body size and growth among surface and cave-dwelling populations of the Pyrenean newt, *Euproctus asper* (Amphibia: Urodela). *Herpetologica* 61:241–49.

Miaud, C., and R. Guyétant. 1998. Plasticité et selection sur les traits de vie d'un organisme à cycle vital complexe, la grenouille rousee *Rana temporaria* (Amphibien: Anoure). *Bull. Soc. Zool. France* 123:325–44.

Miaud, C., R. Guyétant, and J. Elmberg. 1999. Variations in life-history traits in the common frog *Rana temporaria* (Amphibia: Anura): A literature review and new data from the French Alps. *J. Zool., London* 249:61–73.

Miaud, C., R. Guyétant, and H. Faber. 2000. Age, size, and growth of the alpine newt, *Triturus alpestris* (Urodela: Salamandridae), at high altitude and a review of life-history trait variation throughout its range. *Herpetologica* 56:135–44.

Miaud, C., P. Joly, and J. Castanet. 1993. Variation in age structure in a subdivided population of *Triturus cristatus. Can. J. Zool.* 71:1874–79.

Miaud, C., D. Sanuy, and J.-N. Avrillier. 2000. Terrestrial movments of the natterjack toad *Bufo calamita* (Amphibia, Anura) in a semi-arid, agricultural landscape. *Amphibia-Reptilia* 21:367–69.

Michael, S. F. 1996. Courtship calls of three species of *Eleutherodactylus* from Puerto Rico (Anura: Leptodactylidae). *Herpetologica* 52:116–20.

———. 1997a. Captive breeding of *Eleutherodactylus antillensis* (Anura: Leptodactylidae) from Puerto Rico with notes on behavior in captivity. *Herp. Rev.* 28:141–43.

———. 1997b. Vocalization and diurnal retreat defense in the Puerto Rican frog *Eleutherodactylus cochranae. J. Herpetol.* 31:453–56.

Michalak, P. 1996. Repeatability of mating behaviour in Montandon's newt, *Triturus montandoni* (Caudata: Salamandridae). *Ethol. Ecol. Evol.* 8:19–27.

Michalak, P., J. Grzesik, and J. Rafínski. 1997. Tests for sexual incompatibility between two newt species, *Triturus vulgaris* and *Triturus montandoni:* No-choice mating design. *Evolution* 51:2045–50.

Michalak, P., and J. Rafínski. 1999. Sexual isolation between two newt species, *Triturus vulgaris* and *T. montandoni* (Amphibia, Urodela, Salamandridae). *Biol. J. Linn. Soc.* 67:343–52.

Michaud, E. J., and J. R. Dixon. 1989. Prey items of 20 species of the Neotropical colubrid snake genus *Liophis. Herp. Rev.* 20:39–41.

Michaud, T. C. 1962. Call discrimination by females of the chorus frogs, *Pseudacris clarki* and *Pseudacris nigrita. Copeia* 1962:213–15.

Michelsen, A. 1978. Sound reception in different environments. In *Sensory ecology: Review and perspectives,* ed. M. A. Ali, 345–73. New York: Plenum.

Michelsen, A., M. Jørgensen, and J. Christensen-Dalsgaard. 1986. Directional hearing of awake, unrestrained treefrogs. *Naturwissenschaften* 73:682–83.

Michimae, H., and M. Wakahara. 2001. Factors which affect the occurrence of cannibalism and the broad-headed "cannibal" morph in larvae of the salamander *Hynobius retardatus. Behav. Ecol. Sociobiol.* 50:339–45.

Michimae, H. and M. Wakahara. 2002a. A tadpole-induced polyphenism in the salamander *Hynobius retardatus. Evolution* 56:2029–38.

———. 2002b. Variation in cannibalistic polyphenism between populations in the salamander *Hynobius retardatus. Zool. Sci.* 19:703–7.

Middler, S. A., C. R. Kleeman, and E. Edwards. 1968. The role of the urinary bladder in salt and water metabolism of the toad, *Bufo marinus. Comp. Biochem. Physiol.* 26:57–68.

Middleton, E. M., J. R. Herman, E. R. Celarier, J. W. Wilkinson, C. Carey, and R. J. Rusin. 2001. Evaluating ultraviolet radiation exposure with satellite data at sites of amphibian declines in Central and South America. *Cons. Biol.* 15:914–29.

Mijares-Urrutia, A. 1990. The tadpole of *Centrolenella andina* (Anura: Centrolenidae). *J. Herpetol.* 24:410–12.

Mijares-Urrutia, A., and A. Arends. 1999. A new *Mannophryne* (Anura: Dendrobatidae) from western Venezuela, with comments on the generic allocation of *Colostethus laradinus. Herpetologica* 55:106–14.

Mijares-Urrutia, A., A. Arends, and J.-M. Hero. 1997. *Hyla crepitans* (NCN). Predation. *Herp. Rev.* 29:84.

Milanovich, J. R., S. E. Trauth, and M. N. Mary. 2005. *Diadophis punctatus stictogenys* (Mississippi Ring-necked Snake). Diet. *Herp. Rev.* 36:323–24.

Milinski, M. 1977. Experiments on the selection by predators against spatial oddity of their prey. *Z. Tierpsychol.* 43:311–45.

Miller, A. B., E.S. Bryant, and R. W. Birnie. 1998. An analysis of land cover changes in the northern forest of New England using multitemporal landsat MSS data. *Int. J. Rem. Sens.* 19:245–65.

Miller, B. T., and J. H. Larsen, Jr. 1989. Feeding performance in aquatic postmetamorphic newts (Urodela: Salamandridae): Are bidirectional flow systems necessarily inefficient? *Can. J. Zool.* 67:2414–21.

Miller, D. A., M. L. Standish, and A. E. Thurman. 1968. Effects of temperature on water and electrolyte balance in the frog. *Physiol. Zool.* 41:500–506.

Miller, G. L., ed. 2000. *Nature's fading chorus. Classic and contemporary writings on amphibians.* Washington, DC: Island Press.

Miller, K. 1982. Effect of temperature on sprint performance in the frog *Xenopus laevis* and the salamander *Necturus maculosus. Copeia* 1982:695–98.

Miller, K., and J. J. Camilliere. 1981. Physical training improves swimming performance of the African clawed frog *Xenopus laevis. Herpetologica* 37:1–10.

Miller, K., and V. H. Hutchison. 1979. Activity metabolism in the mudpuppy, *Necturus maculosus. Physiol. Zool.* 52:22–37.

———. 1980. Aerobic and anaerobic scope for activity in the giant toad, *Bufo marinus. Physiol. Zool.* 53:170–75.

Miller, K., and G. C. Packard. 1977. An altitudinal cline in critical thermal maxima of chorus frogs (*Pseudacris triseriata*). *Am. Nat.* 111:267–77.

Miller, K., and J. L. Sabol. 1989. The role of phosphocreatine breakdown in the metabolic support of locomotion by clawed frogs, *Xenopus laevis. Comp. Biochem. Physiol.* 93B:251–54.

Miller, K., and G. M. Zoghby. 1986. Thermal acclimation of locomotor performance in anuran amphibians. *Can. J. Zool.* 64:1956–60.

Miller, L. 1944. Notes on the eggs and larvae of *Aneides lugubris. Copeia* 1944:224–30.

Mills, N. E., and M. C. Barnhart. 1999. Effects of hypoxia on embryonic development in two *Ambystoma* and two *Rana* species. *Physiol. Biochem. Zool.* 72:179–88.

Mills, N. E., M. C. Barnhart, and R. D. Semlitsch. 2001. Effects of hypoxia on egg capsule conductance in *Ambystoma* (Class Amphibia, Order Caudata). *J. Exp. Biol.* 204:3747–53.

Milner, A. R. 1982. Small temnospondyl amphibians from the Middle Pennsylvanian of Illinois. *Palaeontology* 25:635–64.

———. 1988. The relationships and origin of living amphibians. In *The phylogeny and classification of the tetrapods. Vol. 1: Amphibians, reptiles, birds,* ed. M. J. Benton, 59–102. New York: Oxford University Press.

———. 1994. Late Triassic and Jurassic amphibians: Fossil record and phylogeny. In *In the shadow of the dinosaurs,* ed. N. C. Fraser and H.-D. Sues, 5–22. New York: Cambridge University Press.

———. 2000. Mesozoic and Tertiary Caudata and Albanerpetonidae. In *Amphibian biology. Vol. 4. Palaeontology,* ed. H. Heatwole and R. L. Carroll, 1412–44. Chipping Norton, New South Wales, Australia: Surrey Beatty and Sons.

Milstead, W. W., A. S. Rand, and M. M. Stewart. 1974. Polymorphism in cricket frogs: an hypothesis. *Evolution* 28:489–91.

Min, M. S., S. Y. Yang, R. M. Bonett, D. R. Vietes, R. A. Brandon, and D. B. Wake. 2005. Discovery of the first Asian plethodontid salamander. *Nature* 435:87–90.

Minter, L. R. 1997. Advertisement call structure and morphology of *Breviceps mossambicus* Peters and *B. poweri* Parker (Anura: Microhylidae) from northern Mozambique. *Ann. Natal Mus.* 38:5–19.

Minter, L. R., M. Burger, J. A. Harrison, H. Braack, P. J. Bishop, and D. Kloepfer, eds. 2004. *Atlas and red data book of the frogs of South African, Lesotho and Swaziland.* Washington, DC: Smithsonian Institution Monitoring and Assessment of Biodiversity Program.

Miranda, L. A., J. M. Affani, and D. A. Paz. 2000. Corticotropin-releasing factor accelerates metamorphosis in *Bufo arenarum:* effect on pituitary ACTH and TSH cells. *J. Exp. Zool.* 286:473–80.

Mitchell, B. L. 1946. A naturalist in Nyasaland. *Nyasaland Agric. Quart. J.* 6:1–47.

Mitchell, J. C. 1990. *Pseudacris feriarum* (Upland chorus frog). Predation. *Herp. Rev.* 21:89–90.

Mitchell, J. C., T. K. Pauley, D. I. Withers, S. M. Roble, B. T. Miller, A. L. Braswell, P. V. Cupp Jr., and C. S. Hobson. 1999. Conservation of the southern Appalachian herpetofauna. *Virginia J. Sci.* 50:1–35.

Mitchell, J. C., S. C. Rinehart, J. F. Pagels, K. A. Buhlmann, and C. A. Pague. 1997. Factors influencing amphibian and small mammal assemblages in central Appalachian forests. *Forest Ecol. Manage.* 95:65–76.

Mitchell, J. C., J. A. Wicknick, and C. D. Anthony. 1996. Effects of timber harvesting practices on Peaks of Otter salamander (*Plethodon hubrichti*) populations. *Amphib. Rept. Cons.* 1:15–19.

Mitchell, N., and R. Swain. 1996. Terrestrial development in the Tasmanian frog *Bryobatrachus nimbus* (Anura, Myobatrachidae), larval development, and a field staging table. *Pap. Proc. Roy. Soc. Tasmania* 130:75–80.

Mitchell, N. J. 2001. Males call more from wetter nests: effects of substrate water potential on reproductive behaviours of terrestrial toadlets. *Proc. Roy. Soc. London B* 268:87–93.

———. 2002a. Low tolerance of embryonic desiccation in the terrestrial nesting frog *Bryobatrachus nimbus* (Anura: Myobatrachinae). *Copeia* 2002:364–73.

———. 2002b. Nest-site selection in a terrestrially breeding frog with protracted development. *Aust. J. Zool.* 50:225–35.

Mitchell, N. J., and R. S. Seymour. 2000. Effects of temperature on energy cost and timing of embryonic and larval development of the terrestrially breeding moss frog, *Bryobatrachus nimbus. Physiol. Biochem. Zool.* 73:829–40.

———. 2003. The effects of nest temperature, nest substrate, and clutch size on the oxygenation of embryos and larvae of the Australian moss frog, *Bryobatrachus nimbus. Physiol. Biochem. Zool.* 76:60–71.

Mitchell, S. L. 1990. The mating system genetically affects offspring performance in Woodhouse's toad (*Bufo woodhousei*). *Evolution* 44:502–19.

Mitchell, S. L. and R. Altig. 1983. The feeding ecology of *Sooglossus gardineri. J. Herpetol.* 17:281–82.

Mitchell, S. L., and G. L. Miller. 1991. Intermale spacing and calling site characteristics in a southern Mississippi chorus of *Hyla cinerea. Copeia* 1991:521–24.

Mitton, J. B., C. Carey, and T. D. Kocher. 1986. The relation of enzyme heterozygosity to standard and active oxygen consumption and body size of tiger salamanders, *Ambystoma tigrinum. Physiol. Zool.* 59:574–82.

Miyamoto, M. M. 1982. Vertical habitat use by *Eleutherodactylus* frogs (Leptodactylidae) at two Costa Rican localities. *Biotropica* 14:141–44.

Miyamoto, M. M., and J. H. Cane. 1980. Behavioral observations of noncalling male Costa Rican *Hyla ebraccata. Biotropica* 12:225–27.

Mizell, S. 1964. Seasonal differences in spermatogenesis and oogenesis in *Rana pipiens. Nature* 202:875–76.

———. 1965. Seasonal changes in energy reserves in the common frog, *Rana pipiens. J. Cell. Comp. Physiol.* 66:251–58.

Mjelstad, H., and M. Sactersdal. 1989. Mallards (*Anas platyrhynchos*) utilizing common frogs (*Rana temporaria*) as a protein source during breeding. *Fauna Norvegica Ser. C* 12:47–48.

Moalli, R., R. S. Meyers, D. C. Jackson, and R. W. Millard. 1980. Skin circulation in the frog, *Rana catesbeiana:* distribution and dynamics. *Respir. Physiol.* 40:137–48.

Mock, B. A., and D. E. Gill. 1984. The infrapopulation dynamics of trypanosomes in red-spotted newts. *Parasitology* 88:267–82.

Moermond, T. C. 1983. Competition between *Anolis* lizards and birds: a reassessment. In. *Advances in herpetology and evolutionary biology,* ed. A. Rhodin and K. Miyata, 506–20. Cambridge, MA: Museum of Comparative Zoology.

Moffat, A. J. M., and R. R. Capranica. 1976. Auditory sensitivity of the saccule in the American toad (*Bufo americanus*). *J. Comp. Physiol.* 105:1–8.

———. 1978. Middle ear sensitivity in anurans and reptiles measured by light scattering spectroscopy. *J. Comp. Physiol.* 127:97–107.

Mohneke, R., and H. Schneider. 1979. Effect of temperature upon auditory thresholds in two anuran species, *Bombina v. variegata* and *Alytes o. obstetricans* (Amphibia, Discoglossidae). *J. Comp. Physiol.* 130:9–16.

Mohr, C. E. 1931. Observations on the early breeding habits of *Ambystoma jeffersonianum* in central Pennsylvania. *Copeia* 1931:102–4.

———. 1943. The eggs of the long-tailed salamander, *Eurycea longicauda* (Green). *Proc. Pennsylvania Acad. Sci.* 17:86.

Mohr, C. O. 1947. Table of equivalent populations of North American small mammals. *Am. Midl. Nat.* 24:581–84.

Mokany, A., and R. Shine. 2002. Pond attributes influence competitive interactions between tadpoles and mosquito larvae. *Aust. Ecol.* 27:396–404.

———. 2003a. Competition between tadpoles and mosquito larvae. *Oecologia* 135:615–20.

———. 2003b. Oviposition site selection by mosquitoes is affected by cues from conspecific larvae and anuran tadpoles. *Aust. Ecol.* 28:33–37.

Moler, P. E., and J. Kezer. 1993. Karyology and systematics of the salamander genus *Pseudobranchus* (Sirenidae). *Copeia* 1993:39–47.

Molina, R., P. Mendiola, and J. de Costa. 1987. Seasonal changes in the activities of glycolytic enzymes from liver and skeletal muscle of *Rana perezi. Rev. Esp. Fisiol.* 43:497–502.

Molur, S., and S. Walker. 1998. Conservation assessment of the herpetofauna of India—An overview. *Hamadryad* 23:169–78.

Monello, R. J., and R. G. Wright. 1999. Amphibian habitat preferences among artificial ponds in the Palouse region of northern Idaho. *J. Herpetol.* 33:298–303.

———. 2001. Predation by goldfish (*Carassius auratus*) on eggs and larvae of the eastern long-toed salamander (*Ambystoma macrodactylum columbianum*). *J. Herpetol.* 35:350–53.

Monnet, J.-M. and M. I. Cherry. 2002. Sexual size dimorphism in anurans. *Proc. Roy. Soc. London B* 269:2301–7.

Monney, J.-C. 1993. Predation of lizards and frogs by adult vipers, *Vipera aspis,* in the Bernese Prealpine region (West Switzerland). *Amphibia-Reptilia* 14:93–95.

Monsen, K. J., and M. S. Blouin. 2003. Genetic structure in a montane ranid frog: Restricted gene flow and nuclear-mitochondrial discordance. *Mol. Ecol.* 12:3275–86.

Monsen, K. J., and M. S. Blouin. 2004. Extreme isolation by distance in a montane frog *Rana cascadae. Cons. Gen.* 6:827–35.

Montori, A. 1990. Skeletochronological results in the Pyrenean newt *Euproctus asper* (Dugés, 1852) from one prepyrenean population. *Ann. Sci. Nat., Zool. Biol. Anim.*, Ser. 13, 11:209 11.

Moodie, G. E. E. 1978. Observations on the life history of the caecilian *Typhlonectes compressicaudus* (Dumeril and Bibron) in the Amazon basin. *Can. J. Zool.* 56:1005–8.

Moore, F. L. 1994. Complexity in the control of amphibian reproduction: Reproductive behavior and physiology of a salamander. In *Captive management and conservation of amphibians and reptiles,* ed. J. B. Murphy, K. Adler, and J. T. Collins, 125–31. Ithaca, NY: Society for the Study of Amphibians and Reptiles.

Moore, F. L., and L. J. Miller. 1983. Arginine vasotocin induces sexual behavior of newts by acting on cells in the brain. *Peptides* 4:97–102.

Moore, F. L., and R. T. Zoeller. 1979. Endocrine control of amphibian sexual behavior: Evidence for a neurohormone-androgen interaction. *Horm. Behav.* 13:207–13.

Moore, F. R., and R. E. Gatten, Jr. 1989. Locomotor performance of hydrated, dehydrated, and osmotically stressed anuran amphibians. *Herpetologica* 45:101–10.

Moore, H. J. 1954. Some observations on the migration of the common toad (*Bufo b. bufo*). *Brit. J. Herp.* 1:194–224.

Moore, J. A. 1939. Temperature tolerance and rates of development in the eggs of Amphibia. *Ecology* 20:459–78.

———. 1940. Adaptive differences in the egg membranes of frogs. *Am. Nat.* 74:89–93.

———. 1942. The role of temperature in the speciation of frogs. *Biol. Symp.* 6:189–213.

———. 1944. Geographic variation in *Rana pipiens* Schreber of eastern North America. *Bull. Amer. Mus. Nat. Hist.* 82:345–70.

———. 1949a. Geographic variation of adaptive characters in *Rana pipiens* Schreber. *Evolution* 3:1–24.

———. 1949b. Patterns of evolution in the genus *Rana.* In *Genetics, paleontology, and evolution,* ed. G. L. Jepsen, E. Mayr, and G. G. Simpson, 315–38. Princeton, NJ: Princeton University Press.

———. 1961. The frogs of eastern New South Wales. *Bull. Amer. Mus. Nat. Hist.* 121:149–386.

Moore, M. K., and V. R. Townsend. 1998. The interactions of temperature, dissolved oxygen and predation pressure in an aquatic predator-prey system. *Oikos* 81:329–36.

Moore, R. D., R. A. Griffiths, C. M. O'Brien, A. Murphy, and D. Jay. 2004. Induced defenses in an endangered amphibian in response to an introduced snake predator. *Oecologia* 141:139–47.

Moore, R. D., R. A. Griffiths, and A. Román. 2004. Distribution of the Mallorcan midwife toad (*Alytes muletensis*) in relation to landscape topography and introduced predators. *Biol. Cons.* 116:327–32.

Moore, R. D., B. Newton, and A. Sih. 1996. Delayed hatching as a response of streamside salamander eggs to chemical cues from predatory fish. *Oikos* 77:331–35.

Moore, S. W., E. R. Lewis, P. M. Narins, and P. M. Lopez. 1989. The call-timing algorithm of the white lipped frog, *Leptodactylus albilabris. J. Comp. Physiol. A* 164:309–19.

Moore, W. G., and A. Burn. 1968. Lethal oxygen thresholds for certain temporary pond invertebrates and their applicability to field situations. *Ecology* 49:349–51.

Morehouse, E. A., T. Y. James, A. R. D. Ganley, R. Vilgalys, L. Berger, P. J. Murphy, and J. E. Longcore. 2003. Multilocus sequence typing suggests the chytrid pathogen of amphibians is a recently emerged clone. *Mol. Ecol.* 12:395–403.

Moreira, G., and L. Barreto. 1997. Seasonal variation in nocturnal calling activity of a savanna anuran community in central Brazil. *Amphibia-Reptilia* 18:49–57.

Moreira, G., and A. P. Lima. 1991. Seasonal patterns of juvenile recruitment and reproduction in four species of leaf litter frogs in Central Amazonia. *Herpetologica* 47:295–300.

Morellato, L. P. C., and C. F. B. Haddad. 2000. Introduction: The Brazilian Atlantic Forest. *Biotropica* 32:786–92.

Morey, S., and D. Reznick. 2000. A comparative analysis of plasticity in larval development in three species of spadefoot toads. *Ecology* 81:1736–49.

———. 2001. Effects of larval density on postmetamorphic spadefoot toads (*Spea hammondii*). *Ecology* 82:510–22.

Morey, S. R. 1990. Microhabitat selection and predation in the Pacific treefrog, *Pseudacris regilla. J. Herpetol.* 24:292–96.

Morgan, A. H., and M. C. Grierson. 1932. Winter habits and yearly food consumption of adult spotted newts, *Triturus viridescens. Ecology* 13:54–62.

Morgan, L. A., and W. A. Buttemer. 1996. Predation by the non-native fish *Gambusia holbrooki* on small *Litoria aurea* and *L. dentata* tadpoles. *Aust. Zool.* 30:143–49.

Mori, A., and H. Moriguchi. 1988. Food habits of the snakes in Japan: a critical review. *The Snake* 20:98–113.

Morin, P. J. 1981. Predatory salamanders reverse the outcome of competition among three species of anuran tadpoles. *Science* 212:1284–86.

———. 1983a. Competitive and predatory interactions in natural and experimental populations of *Notophthalmus viridescens dorsalis* and *Ambystoma tigrinum*. *Copeia* 1983:628–39.

———. 1983b. Predation, competition, and the composition of larval anuran guilds. *Ecol. Monogr.* 53:119–38.

———. 1986. Interactions between intraspecific competition and predation in an amphibian predator-prey system. *Ecology* 67:713–20.

———. 1987a. Predation, breeding asynchrony, and the outcome of competition among treefrog tadpoles. *Ecology* 68:675–83.

———. 1987b. Salamander predation, prey facilitation, and seasonal succession in microcrustacean communities. In *Predation: Direct and indirect impacts on aquatic communities,* ed. W. C. Kerfoot and A. Sih, 174–87. Hanover, NH: University Press of New England.

———. 1989. New directions in amphibian community ecology. *Herpetologica* 45:124–28

———. 1995. Functional redundancy, non-additive interactions, and supply-side dynamics in experimental pond communities. *Ecology* 76:133–49.

———. 1998. Realism, precision, and generality in experimental ecology. In *Experimental ecology. Issues and perspectives,* ed. W. J. Resetarits, Jr. and J. Bernardo, 50–70. New York: Oxford University Press.

Morin, P. J., and E. A. Johnson. 1988. Experimental studies of asymmetric competition among anurans. *Oikos* 53:398–407.

Morin, P. J., S. P. Lawler, and E. A. Johnson. 1988. Competition between aquatic insects and vertebrates: Interaction strength and higher order interactions. *Ecology* 69:1401–9.

Morin, P. J., H. M. Wilbur, and R. N. Harris. 1983. Salamander predation and the structure of experimental communities: responses of *Notophthalmus* and microcrustaceans. *Ecology* 64:1430–36.

Moriwaki, T. 1953. The inheritance of the dorsal median stripe in *Rana limnocharis* Wiegmann. *J. Sci. Hiroshima Univ., Ser. B, Div. 1,* 14:159–64.

Moriya, K. 1952. Genetical study of the pond frog, *Rana nigromaculata.* I. Two types of *Rana nigr. nigromaculata* found in Takata district. *J. Sci. Hiroshima Univ., Ser. B, Div. 1,* 13:189–97.

Moriya, T. 1983. The effect of temperature on the activity of thyroid hormones and prolactin in larvae of the salamander *Hynobius retardatus. Gen. Comp. Endocrin.* 49:1–7.

Morpurgo, G., P. A. Battaglia, and T. Leggio. 1970. Negative Bohr effect in newt hemolysates and its regulation. *Nature* 225:76–77.

Morneault, A. El, B. J. Naylor, L. S. Schaeffer, and D. C. Othmer. 2004. The effect of shelterwood harvesting and site preparation on eastern red-backed salamanders in white pine stands. *Forest Ecol. Manage.* 199:1–10.

Morris, G. K., G. E. Kerr, and J. H. Fullard. 1978. Phonotactic preferences of female meadow katydids (Orthoptera: Tettigoniidae: *Conocephalus nigropleurum*). *Can. J. Zool.* 56:1479–87.

Morris, G. K., A. C. Mason, P. Wall, and J. J. Bellwood. 1994. High ultrasonic and tremulation signals in Neotropical katydids (Orthoptera: Tettigoniidae). *J. Zool., London* 233:129–63.

Morris, M. R. 1989. Female choice of large males in the treefrog *Hyla chrysoscelis:* the importance of identifying the scale of choice. *Behav. Ecol. Sociobiol.* 25:275–81.

———. 1991. Female choice of large males in the treefrog *Hyla ebraccata. J. Zool., London* 223:371–78.

Morris, M. R., and S. L. Yoon. 1989. A mechanism or female choice of large males in the treefrog *Hyla chrysoscelis. Behav. Ecol. Sociobiol.* 25:65–71.

Morris, R. L., and W. W. Tanner. 1969. The ecology of the spotted frog *Rana pretiosa* in Utah. *Great Basin Nat.* 29:45–81.

Morrison, C., and J.-M. Hero. 2002. Geographic variation in life history characteristics of amphibians in mid-eastern Australia: Reproductive traits. In *Frogs in the community. Proceedings of the Brisbane symposium 13–14 February 1999,* ed. A. E. O. Nattrass, 40–48. East Brisbane, Queensland, Australia: The Queensland Frog Society Inc.

Morrison, C., and J.-M. Hero. 2003a. Altitudinal variation in growth and development rates of tadpoles of *Litoris chloris* and *Litoria pearsoniana* in southeast Queensland, Australia. *J. Herpetol.* 37:59–64.

———. 2003b. Geographic variation in life-history characteristics of amphibians: A review. *J. Anim. Ecol.* 72:270–279.

Morrison, C., J.-M. Hero, and J. Browning. 2004. Altitudinal variation in the age at maturity, longevity, and reproductive lifespan of anurans in subtropical Queensland. *Herpetologica* 60:34–44.

Morrison, C., J.-M. Hero, and W. P. Smith. 2001. Mate selection in *Litoria chloris* and *Litoria xanthomera:* Females prefer smaller males. *Aust. Ecol.* 26:223–32.

Morton, E. S. 1975. Ecological sources of selection on avian sounds. *Am. Nat.* 109:17–34.

Morton, M. L. 1981. Seasonal changes in total body lipid and liver weight in the Yosemite toad. *Copeia* 1981:234–38.

Mosconi, G., O. Carnevali, M. F. Franzoni, E. Cottone, I. Lutz, W. Kloas, K. Yamamoto, S. Kikuyama, and A. M. Polzonetti-Magni. 2002. Environmental estrogens and reproductive biology in amphibians. *Gen. Comp. Endocrinol.* 126:125–29.

Moseley, K. R., S. B. Castleberry, and W. M. Ford. 2004. Coarse woody debris and pine litter manipulation effects on movement and microhabitat use of *Ambystoma talpoideum* in a *Pinus taeda* stand. *Forest Ecol. Manage.* 191:387–96.

Moseley, K. R., S. B. Castleberry, J. L. Hanula, and W. M. Ford. 2005. Diet of southern toads (*Bufo terrestris*) in loblolly pine (*Pinus taeda*) stands subject to coarse woody debris manipulations. *Am. Midl. Nat.* 153:327–37.

Moseley, K. R., S. B. Castleberry, and S. H. Schweitzer. 2003. Effects of prescribed fire on herpetofauna in bottomland hardwood forests. *Southeast. Nat.* 2:475–86.

Mosher, J. A., and P. F. Matray. 1974. Size dimorphism: A factor in energy savings for broad-winged hawks. *Auk* 91:325–41.

Mosimann, J. E., and T. M. Uzzell, Jr. 1952. Description of the eggs of the mole salamander, *Ambystoma talpoideum* (Holbrook). *Chicago Acad. Sci., Nat. Hist. Misc.* (104):1–3.

Moss, C. F., and A. Megela Simmons. 1986. Frequency selectivity of hearing in the green treefrog, *Hyla cinerea. J. Comp. Physiol. A* 159:257–66.

Motha, R. P., and W. Baier. 2005. Impacts of present and future climate change and climatic variability on agriculture in the temperate regions: North America. *Climatic Change* 70:137–64.

Motychak, J. E., E. D. Brodie, Jr., and E. D. Brodie III. 1999. Evolutionary response of predators to dangerous prey: Preadaptation and the evolution of tetrodotoxin resistance in garter snakes. *Evolution* 53:1528–35.

Moyle, P. B. 1973. Effects of introduced bullfrogs, *Rana catesbeiana* on the native frogs of the San Joaquin Valley, California. *Copeia* 1973:18–22.

Mudrack, W. 1969. Pflege und Zucht eines Blattsteigerfrosches der Gattung *Phyllobates* aus Ecuador. *Salamandra* 5:81–84.

Mudry, K. M. 1978. *A comparative study of the response properties of higher auditory nuclei in anurans: Correlation with species-specific vocalizations.* Unpublished PhD dissertation. Ithaca, NY: Cornell University.

Mudry, K. M., and R. R. Capranica. 1980. Evoked auditory activity within the telencephalon of the bullfrog (*Rana catesbeiana*). *Brain Res.* 182:303–11.

Mudry, K. M., M. Constantine-Paton, and R. R. Capranica. 1977. Auditory sensitivity of the diencephalon of the leopard frog *Rana p. pipiens. J. Comp. Physiol.* 114:1–13.

Mueller, H. C. 1971. Oddity and specific searching image more important than conspicuousness in prey selection. *Nature* 233:345–46.

———. 1975. Hawks select odd prey. *Science* 188:953–54.

Mueller, N. S. 1980. Mallards capture and eat American toads. *Wilson Bull.* 92:523–24.

Mueller, R. L., J. R. Macey, M. Jaekel, D. B. Wake, and J. L. Boore. 2004. Morphological homoplasy, life history evolution, and historical biogeography of plethodontid salamanders inferred from complete mitochondrial genomes. *Proc. Natl. Acad. Sci. USA* 101:13820–25.

Mukherjee, A. K. 1969. Food-habits of water-birds of the Sundarban, 24–Parganas District, West Bengal, India. *J. Bombay Nat. Hist. Soc.* 66:345–60.

———. 1971a. Food-habits of water-birds of the Sundarban, 24–Parganas District, West Bengal, India. II. Herons and bitterns. *J. Bombay Nat. Hist. Soc.* 68:37–64.

———. 1971b. Food-habits of water-birds of the Sundarban, 24–Parganas District, West Bengal, India. III. Egrets. *J. Bombay Nat. Hist. Soc.* 68:691–716.

———. 1974. Food-habits of water-birds of the Sundarban, 24–Parganas District, West Bengal, India. IV. Stork, teal, moorhen and coot. *J. Bombay Nat. Hist. Soc.* 71:188–200.

Mullally, D. P. 1953. Observations on the ecology of the toad *Bufo canorus. Copeia* 1953:182–83.

Mullally, D. P., and J. P. Cunningham. 1956. Aspects of the thermal ecology of the Yosemite toad. *Herpetologica* 12:57–67.

Mullen, T. L., and R. H. Alvarado. 1976. Osmotic and ionic regulation in amphibians. *Physiol. Zool.* 49:11–23.

Muller, E. R. A., G. Galavazi, and J. A. Szirmai. 1969. Effect of castration and testosterone treatment on fiber width of the flexor carpi radialis muscle in the male frog (*Rana temporaria* L.). *Gen. Comp. Endocrinol.* 13:275–84.

Münch, D., and T. Schröer. 1994. Gefährdung und Vernetzung von Kreuzkrötenpopulationen in der Grosstadt Dortmund. *Ber. Land. Umweltsch. Sachsen-Anhalt (Halle)* 14:69–74.

Muniz, K. P. R. and W. R. Silva. 2005. *Leptodactylus labyrinthicus* (South American Pepper Frog). Predation. *Herp. Rev.* 36:302–3.

Munsey, L. D. 1972. Salinity tolerance of the African pipid frog, *Xenopus laevis. Copeia* 1972:584–86.

Muntz, L., J. E. Hornby, and M. R. A. Dalooi. 1989. A comparison of the distribution of muscle type in the tadpole tails of *Xenopus laevis* and *Rana temporaria:* An histological and ultrastructural study. *Tiss. Cell* 21:773–81.

Münz, H., B. Claas, and B. Fritzsch. 1982. Electrophysiological evidence of electroreception in the axolotl *Siredon mexicanum. Neurosci. Lett.* 28:107–11.

Münz, H., B. Claas, and B. Fritzsch. 1984. Electroreceptive and mechanoreceptive units in the lateral-line of the axolotl *Ambystoma mexicanum. J. Comp. Physiol.* 154:33–44.

Murdoch, W. W., and A. Oaten. 1975. Predation and population stability. *Adv. Ecol. Res.* 9:1–131.

Murphy, C. G. 1994a. Chorus tenure of male barking treefrogs, *Hyla gratiosa. Anim. Behav.* 48:763–77.

———. 1994b. Determinants of chorus tenure in barking treefrogs (*Hyla gratiosa*). *Behav. Ecol. Sociobiol.* 34:285–94.

———. 2003. The cause of correlations between nightly numbers of male and female barking treefrogs (*Hyla gratiosa*) attending choruses. *Behav. Ecol.* 14:274–81.

———. 1999. Nightly timing of chorusing by male barking treefrogs (*Hyla gratiosa*): The influence of female arrival and energy. *Copeia* 1999:333–47.

Murphy, C. G., and S. B. Floyd. 2005. The effect of call amplitude on male spacing in choruses of barking treefrogs, *Hyla gratiosa. Anim. Behav.* 69:419–26.

Murphy, C. G., and H. C. Gerhardt. 1996. Evaluating the design of mate-choice experiments: The effect of amplexus on mate choice by female barking treefrogs, *Hyla gratiosa. Anim. Behav.* 51:881–90.

———. 2000. Mating preference functions of individual female barking treefrogs, *Hyla gratiosa,* for two properties of male advertisement calls. *Evolution* 54:660–69.

———. 2002. Mate sampling by female barking treefrogs (*Hyla gratiosa*). *Behav. Ecol.* 13:472–80.

Murphy, J. C. 1997. *Amphibians and reptiles of Trinidad and Tobago.* Malabar, FL: Krieger Publishing Company.

Murphy, J. E., C. A. Phillips, and V. R. Beasley. 2000. Aspects of amphibian ecology. In *Ecotoxicology of amphibians and reptiles,* ed. D. W. Sparling, G. Linder, and C. A. Bishop, 141–78. Pensavola, FL: SETAC Press.

Murphy, J. F., E. T. Simandle, and D. E. Becker. 2003. Population status and conservation of the black toad, *Bufo exsul. Southwest. Nat.* 48:54–60.

Murphy, P. J. 2003. Does reproductive site choice in a Neotropical frog mirror variable risks facing offspring? *Ecol. Monogr.* 73:45–67.

Murphy, R. W., J. Z. Fu, D. E. Upton, T. De Lama, and E. M. Zhao. 2000. Genetic variation among endangered Chinese giant salamanders, *Andrias davidianus. Mol. Ecol.* 9:1539–47.

Murphy, T. D. 1961. Predation on eggs of the salamander, *Ambystoma maculatum* by caddisfly larvae. *Copeia* 1961:495–96.

Murray, B. R., C. R. Fonseca, and M. Westoby. 1998. The macroecology of Australian frogs. *J. Anim. Ecol.* 67:567–79.

Murray, B. R., and G. C. Hose. 2005. Life-history and ecological correlates of decline and extinction in the endemic Australian frog fauna. *Aust. Ecol.* 30:564–71.

Murray, D. L., and C. L. Jenkins. 1999. Perceived predation risk as a function of predator dietary cues in terrestrial salamanders. *Anim. Behav.* 57:33–39.

Murray, D. L., J. D. Roth, and A. J. Wirsing. 2004. Predation risk avoidance by terrestrial amphibians: The role of prey experience and vulnerability to native and exotic predators. *Ethology* 110:635–47.

Mushinsky, H. R. 1987. Foraging ecology. In *Snakes: Ecology and evolutionary biology,* ed. R. A. Seigel, J. T. Collins, and S. S. Novak, 302–34. New York: Macmillan.

Mushinsky, H. R., and E. D. Brodie, Jr. 1975. Selection of substrate pH by salamanders. *Am. Midl. Nat.* 93:440–43.

Mushinsky, H. R., and J. J. Hebrard. 1977. Food partitioning by five species of water snakes in Louisiana. *Herpetologica* 33:162–66.

Mushinsky, H. R., J. J. Hebrard, and D. S. Vodopich. 1982. Ontogeny of water snake foraging ecology. *Ecology* 63:1624–29.

Mushinsky, H. R., and K. H. Lotz. 1980. Chemoreceptive responses of two sympatric water snakes to commonly ingested prey species: Ontogenetic and ecological considerations. *J. Chem. Ecol.* 6:523–35.

Muskhelishvili, T. A. 1964. New findings about reproduction of the Caucasus salamander (*Mertensiella caucasica*). *Soobshch. Akad. Nauk. Gruz. SSR* 36:183–85.

Muths, E. 2003. Home range and movements of boreal toads in undisturbed habitat. *Copeia* 2003:160–65.

Muths, E., D. H. Campbell, and P. S. Corn. 2003. Hatching success in salamanders and chorus frogs at two sites in Colorado, USA: Effects of acidic deposition and climate. *Amphibia-Reptilia* 24:27–36.

Muths, E., and P. S. Corn. 1997. Basking by adult boreal toads (*Bufo boreas boreas*) during the breeding season. *J. Herpetol.* 31:426–28.

Muths, E., P. S. Corn, A. P. Pessier, and D. E. Green. 2003. Evidence for disease-related amphibian decline in Colorado. *Biol. Cons.* 110:357–65.

Muths, E., and P. Nanjappa. 2005. *Bufo boreas* Baird and Girard, 1852. Western toad. In *Amphibian declines. The conservation status of United States species,* ed. M. J. Lannoo, 392–96. Berkeley: University of California Press.

Myers, C. W. 1965. Biology of the ringneck snake, *Diadophis punctatus,* in Florida. *Bull. Florida State Mus., Biol. Sci.* 10:43–90.

———. 1967. The pine woods snake, *Rhadinaea flavilata* (Cope). *Bull. Florida State Mus., Biol. Sci.* 11:47–97.

———. 1969. The ecological geography of cloud forest in Panama. *Amer. Mus. Novitates* (2396):1–52.

———. 1974. The systematics of *Rhadinaea* (Colubridae), a genus of New World snakes. *Bull. Amer. Mus. Nat. Hist.* 153:1–262.

———. 1982. Spotted poison frogs: descriptions of three new *Dendrobates* from western Amazonia, and resurrection of a lost species from "Chiriqui." *Amer. Mus. Novitates* (2721):1–23.

———. 1987. New generic names for some Neotropical poison frogs (Dendrobatidae). *Pap. Avul. Zool., S. Paulo* 36:301–6.

Myers, C. W., and P. A. Burrowes. 1987. A new poison frog (*Dendrobates*) from Andean Colombia, with notes on a lowland relative. *Amer. Mus. Novitates* (2899):1–17.

Myers, C. W., and J. W. Daly. 1976. Preliminary evaluation of skin toxins and vocalizations in taxonomic and evolutionary studies of poison-dart frogs (Dendrobatidae). *Bull. Amer. Mus. Nat. Hist.* 157:173–262.

———. 1979. A name for the poison frog of Cordillera Azul, eastern Peru, with notes on its biology and skin toxins (Dendrobatidae). *Amer. Mus. Novitates* (2674):1–24.

———. 1980. Taxonomy and ecology of *Dendrobates bombetes,* a new Andean poison frog with new skin toxins. *Amer. Mus. Novitates* (2692):1–13.

———. 1983. Dart-poison frogs. *Sci. Amer.* 248:120–33.

Myers, C. W., J. W. Daly, and B. Malkin. 1978. A dangerously toxic new frog (*Phyllobates*) used by Embera Indians of western Colombia, with discussion of blowgun fabrication and dart poisoning. *Bull. Amer. Mus. Nat. Hist.* 161:307–66.

Myers, C. W., A. Paolillo O., and J. W. Daly. 1991. Discovery of a defensively malodorous and nocturnal frog in the family Dendrobatidae: Phylogenetic significance of a new genus and species from the Venezuelan Andes. *Amer. Mus. Novitates* (3002):1–33.

Myers, E. M., and K. R. Zamudio. 2004. Multiple paternity in an aggregative breeding amphibian: The effect of reproductive skew on estimates of male reproductive success. *Mol. Ecol.* 13:1951–63.

Myers, G. S. 1943. Rediscovery of the Philippine discoglossid frog, *Barbourula busduangensis. Copeia* 1943:148–50.

Myers, J. G. 1931. The present position of the mongoose in the West Indies. *Trop. Agricul., Trinidad* 8:94–95.

Myers, N. 1988. Threatened biotas: "Hot spots" in tropical forests. *The Environmentalist* 8:187–208.

———. 1990. The biodiversity challenge: expanded hot-spots analysis. *The Environmentalist* 10:243–56.

———. 1996. The rich diversity of biodiversity issues. In *Biodiversity II: Understanding and protecting our natural resources,* ed. M. L. Reaka-Kudla, D. W. Wilson, and E. O. Wilson, 125–38. Washington, DC: National Academy Press.

Nagai, T., H. Koyama, K. V. Hoff, and S. D. Hillyard. 1999. Desert toads discriminate salt taste with chemosensory function of the ventral skin. *J. Comp. Neurol.* 408:125–36.

Nagl, A. M., and R. Hofer. 1997. Effects of ultraviolet radiation on early larval stages of the alpine newt, *Triturus alpestris,* under natural and laboratory conditions. *Oecologia* 110:514–19.

Narins, P. M. 1982a. Behavioral refractory period in Neotropical treefrogs. *J. Comp. Physiol.* 148:337–44.

———. 1982b. Effects of masking noise on evoked calling in the Puerto Rican coqui (Anura: Leptodactylidae). *J. Comp. Physiol.* 147:439–46.

———. 1983. Synchronous vocal response mediated by the amphibian papilla in a Neotropical treefrog: behavioural evidence. *J. Exp. Biol.* 105:95–105.

———. 1987. Coding of signals in noise by amphibian auditory nerve fibers. *Hearing Res.* 26:145–54.

———. 1990. Seismic communication in anuran amphibians. *BioScience* 40:268–74.

———. 1992. Evolution of anuran chorus behavior: Neural and behavioral constraints. *Am. Nat.* 139:S90–S104.

———. 1995a. Comparative aspects of interactive communication. In *Active hearing,* ed. A. Flock, D. Ottosen, and M. Ulfendahl, 363–72. London: Elsevier Science.

———. 1995b. Frog communication. *Sci. Amer.* 273:62–67.

———. 1995c. Temperature dependence of auditory functions in the frog. In *Advances in hearing research,* ed. G. A. Manley, G. M. Klump, C. Koppel, H. Fastl, and H. Oeckinghaus, 198–206. Singapore: World Scientific Publishers.

———. 2001a. Ectothermy's last stand: Hearing in the heat and cold. In *Anuran communication,* ed. M. J. Ryan, 61–70. Washington, DC: Smithsonian Institution Press.

———. 2001b. Vibrational communication in vertebrates. In *Ecology of sensing,* ed. F. Barth and A. Schmidt, 127–48. Berlin: Springer-Verlag.

Narins, P. M., and R. R. Capranica. 1976. Sexual differences in the auditory system of the tree frog *Eleutherodactylus coqui. Science* 192:378–80.

———. 1978. Communicative significance of the two-note call of the treefrog *Eleutherodactylus coqui. J. Comp. Physiol.* 127:1–9.

———. 1980. Neural adaptations for processing the two-note call of the Puerto Rican treefrog, *Eleutherodactylus coqui. Brain Behav. Evol.* 17:48–66.

Narins, P. M., G. Ehret, and J. Tautz. 1988. Accessory pathway for sound transfer in a Neotropical frog. *Proc. Natl. Acad. Sci. USA* 85:1508–12.

Narins, P. M., and A. S. Feng, eds. 2007. *Hearing and sound*

*communication in amphibians. Springer handbook of auditory research.* New York: Springer Verlag.
Narins, P. M., A. S. Feng, H.-S. Yong, and J. Christensen-Dalsgaard. 1998. Morphological, behavioral, and genetic divergence of sympatric morphotypes of the treefrog *Polypedates leucomystax* in Peninsular Malaysia. *Herpetologica* 54:129–42.
Narins, P. M., D. S. Grabul, K. K. Soma, P. Gaucher, and W. Hödl. 2005. Cross-modal integration in a dart-poison frog. *Proc. Natl. Acad. Sci. USA USA* 102:2425–29.
Narins, P. M., W. Hödl, and D. S. Grabul. 2003. Biomodal signal requisite for agonistic behavior in a dart-poison frog, *Epipedobates femoralis. Proc. Natl. Acad. Sci. USA USA* 100:577–80.
Narins, P. M., and D. D. Hurley. 1982. The relationship between call intensity and function in the Puerto Rican coqui (Leptodactylidae). *Herpetologica* 38:287–95.
Narins, P. M., and E. R. Lewis. 1984. The vertebrate ear as an exquisite seismic sensor. *J. Acoust. Soc. Amer.* 76:1384–87.
Narins, P. M., E. R. Lewis, and B. E. McClelland. 2000. Hyperextended call note repertoire of the endemic Madagascar treefrog *Boophis madagascariensis* (Rhacophoridae). *J. Zool., London* 250:283–98.
Narins, P. M., E. R. Lewis, A. P. Purgue, P. J. Bishop, L. R. Minter, and D. P. Lawson. 2001. Functional consequences of a novel middle ear adaptation in the central African frog, *Petropedetes parkeri* (Ranidae). *J. Exp. Biol.* 204:1223–32.
Narins, P. M., and S. L. Smith. 1986. Clinal variation in anuran advertisement calls: Basis for acoustic isolation? *Behav. Ecol. Sociobiol.* 19:135–41.
Narins, P. M., and R. D. Zelick. 1988. The effects of noise on auditory processing and behavior in amphibians. In *The evolution of the amphibian auditory system,* ed. B. Fritzsch, M. J. Ryan, W. Wilczynski, T. E. Hetherington, and W. Walkowiak, 511–36. New York: Wiley.
Nascimento, A. C. C., W. Fontes, A. Sebben, and M. S. Castro. 2003. Antimicrobial peptides from anurans skin secretions. *Prot. Pep. Lett.* 10:227–38.
Naughton, G. P., C. B. Henderson, K. R. Foresman, and R. L. McGraw. 2000. Long-toed salamanders in harvested and intact Douglas-fir forests of western Montana. *Ecol. Appl.* 10:1681–89.
Naugle, D. E., K. F. Higgins, R. R. Johnson, T. D. Fischer, and F. R. Quamen. 2005. Landscape ecology. In *Amphibian declines: The conservation status of United States species,* ed. M. J. Lannoo, 185–92. Berkeley: University of California Press.
Nauwelaerts, S., J. Scholliers, and P. Aerts. 2004. A functional analysis of how frogs jump out of water. *Biol. J. Linn. Soc.* 83:413–20.
Navas, C. A. 1996a. The effect of temperature on the vocal activity of anurans: A comparison of high and low-elevation species. *J. Herpetol.* 30:488–97.
———. 1996b. Implications of microhabitat selection and patterns of activity on the thermal ecology of high elevation Neotropical anurans. *Oecologia* 108:617–26.
———. 1996c. Metabolic physiology, locomotor performance, and thermal niche breadth in Neotropical anurans. *Physiol. Zool.* 69:1481–1501.
———. 1996d. Thermal dependency of field locomotor and vocal performance of high-elevation anurans in the tropical Andes. *J. Herpetol.* 30:478–87.
———. 1997. Thermal extremes at high elevations in the Andes: Physiological ecology of frogs. *J. Therm. Biol.* 22:467–77.
———. 2003. Herpetological diversity along Andean elevational gradients: Links with physiological ecology and evolutionary physiology. *Comp. Biochem. Physiol. A* 133:469–85.
———. 2006. Patterns of distribution of anurans in high Andean tropical elevations: Insights from integrating biogeography and evolutionary physiology. *Integr. Comp. Biol.* 46:82–91.
Navas, C. A., R. S. James, J. M. Wakeling, K. M. Kemp, and I. A. Johnston. 1999. An integrative study of the temperature dependence of whole animal and muscle performance during jumping and swimming in the frog *Rana temporaria. J. Comp. Physiol. B* 169:588–96.
Navas, C. A., C. Jared, and M. M. Antoniazzi. 2002. Water economy in the casque-headed tree-frog *Corythomantis greeningi* (Hylidae): Role of behaviour, skin, and skull co-ossification. *J. Zool., London* 257:525–32.
Naya, D. E., and F. Bozinovic. 2004. Digestive phenotype flexibility in post-metamorphic amphibians: studies on a model organism. *Biol. Res. (Chile)* 37:365–70.
Naya, D. E., G. Farfán, P. Sabat, M. A. Méndez, and F. Bozinovic. 2005. Digestive morphology and enzyme activity in the Andean toad *Bufo spinulosus:* Hard-wired or flexible physiology? *Comp. Biochem. Physiol. A* 140:165–70.
Naya, D. E., R. Maneyro, A. Camargo, A. Canavero, and I. de Rosa. 2003. Seasonal changes in gut length in the South American common frog *Leptodactylus ocellatus* (Amphibia, Anura). *Biociências* 11:47–52.
Naylor, B. G. 1978. The earliest known *Necturus* (Amphibia, Urodela), from the Paleocene Ravenscrag formation of Saskatchewan. *J. Herpetol.* 12:565–69.
Naylor, B. G., and R. A. Nussbaum. 1980. The trunk musculature of caecilians (Amphibia: Gymnophiona). *J. Morphol.* 166:259–73.
Neal, E., and C. Cheeseman. 1996. *Badgers.* London: T. and A. D. Poyser.
Near, J. C., D. P. Easton, P. S. Rutledge, D. P. Dickinson, and J. R. Spotila. 1990. Heat shock protein 70 gene expression in intact salamanders (*Eurycea bislineata*) in response to calibrated heat shocks and to high temperatures encountered in the field. *J. Exp. Zool.* 256:303–14.
Neary, T. J. 1974. Diencephalic efferents of the torus semicircularis in the bullfrog, *Rana catesbeiana. Anat. Rec.* 178:425.
———. 1988. Forebrain auditory pathways in ranid frogs. In *The evolution of the amphibian auditory system,* ed. B. Fritzsch, M. J. Ryan, W. Wilczynski, T. E. Hetherington, and W. Walkowiak, 233–52. New York: Wiley.
Neary, T. J., and W. Wilczynski. 1986. Auditory pathways to the hypothalamus in ranid frogs. *Neurosci. Lett.* 71:142–46.
Neckel-Oliveira, S. 2004. Effects of landscape change on clutches of *Phyllomedusa tarsius,* a Neotropical treefrog. *Biol. Cons.* 118:109–16.
Neckel-Oliveira, S., W. E. Magnusson, A. P. Lima, and A. L. K. Albernaz. 2000. Diversity and distribution of frogs in an Amazonian savanna in Brazil. *Amphibia-Reptilia* 21:317–26.
Neckel-Oliveira, S., and M. Wachlevski. 2004. Predation on the arboreal eggs of three species of *Phyllomedusa* in Central Amazonia. *J. Herpetol.* 38:244–48.
Negovetic, S., B. R. Anholt, R. D. Semlitsch, and H.-U. Reyer. 2001. Specific responses of sexual and hybridogenetic European waterfrog tadpoles to temperature. *Ecology* 82:766–74.
Neill, W. E. 1968. Predation on *Bufo valliceps* tadpoles by the

predaceous diving beetle *Acilius semisulcatus*. *Bull. Ecol. Soc. Amer.* 49:169.

Neill, W. E., and J. C. Grubb. 1971. Arboreal habits of *Bufo valliceps* in central Texas. *Copeia* 1971:347–48.

Neill, W. T. 1946. An unusual habitat for frogs and lizards. *Copeia* 1946:258.

———. 1948a. Hibernation of amphibians and reptiles in Richmond County, Georgia. *Herpetologica* 4:107–14.

———. 1948b. Spiders preying on reptiles and amphibians. *Herpetologica* 4:158.

———. 1951. A bromeliad herpetofauna in Florida. *Ecology* 32:140–43.

———. 1952. Burrowing habits of *Hyla gratiosa*. *Copeia* 1952:196.

———. 1958. The occurrence of amphibians and reptiles in saltwater areas, and a bibliography. *Bull. Mar. Sci. Gulf Carib.* 8:1–97.

———. 1960. The caudal lure of various juvenile snakes. *Q. J. Florida Acad. Sci.* 23:173–200.

Neish, I. C. 1971. Comparison of size, structure, and distributional patterns of two salamander populations in Marion Lake, British Columbia. *J. Fish. Res. Bd. Canada* 28:49–58.

Nelson, C. E. 1973. Mating calls of the Microhylinae: Descriptions and phylogenetic and ecological considerations. *Herpetologica* 29:163–76.

Nelson, C. E., and R. R. Humphrey. 1972. Artificial interspecific hybridization among *Ambystoma*. *Herpetologica* 28:27–32.

Nelson, C. E., and J. Lescure. 1975. The taxonomy and distribution of *Myersiella* and *Synapturanus* (Anura: Microhylidae). *Herpetologica* 31:389–97.

Nelson, C. E., and G. A. Miller. 1971. A possible case of mimicry in frogs. *Herp. Rev.* 3:109.

Nelson, D. A. 1984. Communication of intentions in agonistic contexts by the pigeon guillemot, *Cepphus columba*. *Behaviour* 88:145–89.

Nelson, R. T., B. J. Cochrane, P. R. Delis, and D. TeStrake. 2002. *Basidioboliasis* in anurans in Florida. *J. Wildl. Dis.* 38:463–67.

Neuwirth, M., J. W. Daly, C. W. Myers, and L. W. Tice. 1979. Morphology of the granular secretory glands in skin of poison-dart frogs (Dendrobatidae). *Tissue and Cell* 11:755–71.

Neveu, A. 1997. The introduction of allochtonous green frog species into France, two dissimilar cases: *R. catesbeiana* and the foreign taxa of the *esculenta* complex. *Bull Fr. Peche et Piscicul.* 344–345:165–71.

Nevo, E. 1973. Adaptive color polymorphism in cricket frogs. *Evolution* 27:353–67.

———. 1976a. Adaptive strategies of genetic systems in constant and varying environments. In *Population genetics and ecology*, ed. S. Karlin and E. Nevo, 141–58. New York: Academic Press.

———. 1976b. Genetic variation in constant environments. *Experientia* 32:858–59.

———. 1991. Genetic diversity and ecological heterogeneity in amphibian evolution. *Copeia* 1991:565–92.

Nevo, E., A. Beiles, and R. Ben-Shlomo. 1984. The evolutionary significance of genetic diversity: ecological, demographic and life history correlates. In *Evolutionary dynamics of genetic diversity*, ed. G. S. Mani, 174–206. New York: Springer-Verlag.

Nevo, E., and R. R. Capranica. 1985. Evolutionary origin of ethological reproductive isolation in cricket frogs, *Acris*. *Evol. Biol.* 19:147–214.

Nevo, E., H. C. Dessauer, and K. C. Chauang. 1974. Genetic variation as a test of natural selection. *Proc. Natl. Acad. Sci. USA USA* 72:2145–49.

Nevo, E., and H. Schneider. 1976. Mating call pattern of green toads in Israel and its ecological correlates. *J. Zool., London* 178:133–45.

Nevo, E., and S. Y. Yang. 1979. Genetic diversity and climatic determinants of tree frogs in Israel. *Oecologia* 41:47–63.

———. 1982. Genetic diversity and ecological relationships of marsh frog populations in Israel. *Theor. Appl. Genet.* 63:317–33.

Newman, D. G. 1990. Activity, dispersion, and population densities of Hamilton's frog (*Leiopelma hamiltoni*) on Maud and Stephens Islands, New Zealand. *Herpetologica* 46:319–30.

Newman, R. A. 1987. Effects of density and predation on *Scaphiopus couchi* tadpoles in desert ponds. *Oecologia* 71:301–7.

———. 1988a. Adaptive plasticity in development of *Scaphiopus couchii* tadpoles in desert ponds. *Evolution* 42:774–83.

———. 1988b. Genetic variation for larval anuran (*Scaphiopus couchii*) development time in an uncertain environment. *Evolution* 42:763–73.

———. 1989. Developmental plasticity of *Scaphiopus couchii* tadpoles in an unpredictable environment. *Ecology* 70:1775–87.

———. 1992. Adaptive plasticity in amphibian metamorphosis. *BioScience* 42:671–78.

———. 1994. Effects of changing density and food level on metamorphosis of a desert amphibian, *Scaphiopus couchii*. *Ecology* 75:1085–96.

———. 1998. Ecological constraints on amphibian metamorphosis: Interactions of temperature and larval density with responses to changing food level. *Oecologia* 115:9–16.

Newman, R. A., and A. E. Dunham. 1994. Size at metamorphosis and water loss in a desert anuran (*Scaphiopus couchii*). *Copeia* 1994:372–81.

Newsholme, E. A., and J. M. Paul. 1983. The use of in vitro enzyme activities to indicate the changes in metabolic pathways during acclimatisation. In *Cellular acclimatisation to environmental change*, ed. A. R. Cossins and P. Sheterline, 81–101. New York: Cambridge University Press.

Ng, M. Y., and H. M. Wilbur. 1995. The cost of brooding in *Plethodon cinereus*. *Herpetologica* 51:1–8.

Nguyen, Q. T. 1998. The decline of amphibian diversity in exploited forests. *Froglog* (29):1.

———. 2000. Amphibian uses in Vietnam. *Froglog* (38):1–2.

Nicieza, A. G. 1999. Context-dependent aggregation in common frog *Rana temporaria* tadpoles: Influence of developmental stage, predation risk and social environment. *Funct. Ecol.* 13:852–58.

———. 2000. Interacting effects of predation risk and food availability on larval anuran behaviour and development. *Oecologia* 123:497–505.

Nickerson, M. A. and C. E. Mays. 1973a. The hellbenders: North American "giant salamanders." *Milwaukee Publ. Mus. Publ. Biol. Geol.* (1):1–106.

———. 1973b. A study of the Ozark hellbender *Cryptobranchus alleganiensis bishopi*. *Ecology* 54:1164–65.

Nicolas, P., D. Vanhoye, and M. Amiche. 2003. Molecular strategies in biological evolution of antimicrobial peptides. *Peptides* 24:1669–80.

Nie, M., J. D. Crim, and G. R. Ultsch. 1999. Dissolved oxygen, temperature, and habitat selection by bullfrog (*Rana catesbeiana*) tadpoles. *Copeia* 1999:153–62.

Niekisch, M. 1982. Beitrag zu Biologie und Schutz der Kreuzkröte (*Bufo calamita* Laur.). *Decheniana* 135:88–103.

Nielsen, H. I. 1979. Chromatophores and adaptation to coloured backgrounds in two colour types of the edible frog, *Rana esculenta*. *J. Zool., London* 189:349–58.

———. 1980. Color and color adaptation of the European tree frog, *Hyla arborea*. *J. Exp. Zool.* 211:143–51.

Nielsen, H. I., and J. Dyck. 1978. Adaptation of the tree frog, *Hyla cinerea*, to colored backgrounds and the role of the three chromatophore types. *J. Exp. Zool.* 205:79–94.

Nielsen, K. H., and C. B. Jørgensen. 1990. Salt and water balance during hibernation in anurans. *Fortschr. Zool.* 38:333–49.

Nielson, M., K. Lohman, and J. Sullivan. 2001. Phylogeography of the tailed frog (*Ascaphus truei*): Implications for the biogeography of the Pacific Northwest. *Evolution* 55:147–60.

Nijhuis, M. J., and R. H. Kaplan. 1998. Movement patterns and life history characteristics in a population of the Cascade torrent salamander (*Rhyacotriton cascadae*) in the Columbia River Gorge, Oregon. *J. Herpetol.* 32:301–4.

Nishihara, A. 1996. Effects of density on growth of head size in larvae of the salamander *Hynobius retardatus*. *Copeia* 1996:478–83.

Nishikawa, K., M. Matsui, S. Tanabe, and S. Sato. 2001. Geographic enzyme variation in a Japanese salamander *Hynobius boulengeri* Thompson (Amphibia: Caudata). *Herpetologica* 57:281–94.

Nishikawa, K. C. 1985. Competition and the evolution of aggressive behavior in two species of terrestrial salamanders. *Evolution* 39:1282–94.

———. 1987. Interspecific aggressive behaviour in salamanders: Species-specific interference or misidentification? *Anim. Behav.* 35:263–70.

———. 1990. Intraspecific spatial relationships of two species of terrestrial salamanders. *Copeia* 1990:418–26.

———. 1999. Neuromuscular control of prey capture in frogs. *Phil. Trans. Roy. Soc. London B* 354:941–54.

———. 2000. Feeding in frogs. In *Feeding: Form, function, and evolution in tetrapod vertebrates*, ed. K. Schwenk, 117–47. San Diego: Academic Press.

Nishikawa, K. C., C. W. Anderson, S. M. Deban, and J. C. O'Reilly. 1992. The evolution of neural circuits controlling feeding behavior in frogs. *Brain Behav. Evol.* 40:125–40.

Nishikawa, K. C., and D. C. Cannatella. 1991. Kinematics of prey capture in the tailed frog *Ascaphus truei* (Anura: Ascaphidae). *Zool. J. Linn. Soc.* 103:289–307.

Nishikawa, K. C., and C. Gans. 1992. The role of hypoglossal sensory feedback during feeding in the marine toad, *Bufo marinus*. *J. Exp. Zool.* 264:245–52.

Nishikawa, K. C., W. M. Kier, and K. K. Smith. 1999. Morphology and mechanics of tongue movement in the African pig-nosed frog *Hemisus marmoratum*: Auscular hydrostatic model. *J. Exp. Biol.* 202:771–80.

Nishikawa, K. C., and G. Roth. 1991. The mechanism of tongue protraction during prey capture in the frog *Discoglossus pictus*. *J. Exp. Biol.* 159:217–34.

Nishikawa, K. C., and P. M. Service. 1988. A fluorescent marking technique for individual recognition of terrestrial salamanders. *J. Herpetol.* 22:351–53.

Noble, G. K. 1924. The "retrograde metamorphosis" of the Sirenidae: experiments on the functional activity of the thyroid of the perennibranchs. *Anat. Rec.* 29:100.

———. 1925. The integumentary, pulmonary, and cardiac modifications correlated with increased cutaneous respiration in Amphibia: A solution of the "hairy frog" problem. *J. Morphol. Physiol.* 40:341–416.

———. 1926a. The hatching process in *Alytes, Eleutherodactylus* and other amphibians. *Amer. Mus. Novitates* (229):1–7.

———. 1926b. The Long Island newt: A contribution to the life-history of *Triturus viridescens*. *Amer. Mus. Novitates* (228):1–11.

———. 1927. The value of life history data in the study of the evolution of the Amphibia. *Ann. New York Acad. Sci.* 30:31–128.

———. 1929a. The adaptive modifications of the arboreal tadpoles of *Hoplophryne* and the torrent tadpoles of *Staurois*. *Bull. Amer. Mus. Nat. Hist.* 83:201–336.

———. 1929b. Further observations on the life-history of the newt, *Triturus viridescens*. *Amer. Mus. Novitates* (348):1–22.

———. 1929c. The relation of courtship to the secondary sex characters of the two-lined salamander, *Eurycea bislineata* (Green). *Amer. Mus. Novitates* (362):1–5.

———. 1931. *Biology of the Amphibia*. New York: McGraw Hill.

Noble, G. K., and L. R. Aronson. 1942. The sexual behavior of Anura. 1. The normal mating pattern of *Rana pipiens*. *Bull. Amer. Mus. Nat. Hist.* 80:127–42.

Noble, G. K., and M. K. Brady. 1933. Observations on the life history of the marbled salamander, *Ambystoma opacum* Gravenhorst. *Zoologica* 11:89–132.

Noble, G. K., and B. C. Marshall. 1929. The breeding habits of two salamanders. *Amer. Mus. Novitates* (347):1–12.

———. 1932. The validity of *Siren intermedia* LeConte, with observations on its life history. *Amer. Mus. Novitates* (532):1–17.

Noble, G. K., and R. C. Noble. 1923. The Anderson treefrog (*Hyla andersonii* Baird). Observations on its habits and life history. *Zoologica* 2:417–55.

Noble, G. K., and P. G. Putnam. 1931. Observations on the life history of *Ascaphus truei* Stejneger. *Copeia* 1931:97–101.

Noble, G. K., and L. B. Richards. 1932. Experiments on the egg-laying of salamanders. *Amer. Mus. Novitates* (513):1–25.

Noble, G. K., and J. A. Weber. 1929. The spermatophores of *Desmognathus* and other plethodontid salamanders. *Amer. Mus. Novitates* (351):1–15.

Nodzenski, E., R. J. Wassersug, and R. F. Inger. 1989. Developmental differences in visceral morphology of megophryine pelobatid tadpoles in relation to their body form and mode of life. *Biol. J. Linn. Soc.* 38:369–88.

Nogueira, C., R. J. Sawaya, and M. Martins. 2003. Ecology of the pitviper, *Bothrops moojeni*, in the Brazilian Cerrado. *J. Herpetol.* 37:653–59.

Noland, R., and G. R. Ultsch. 1981. The roles of temperature and dissolved oxygen in microhabitat selection by the tadpoles of a frog (*Rana pipiens*) and a toad (*Bufo terrestris*). *Copeia* 1981:645–52.

Nöllert, A. 1984. *Die Knoblauchkröte*. Witteberg Lutherstadt, Germany: A. Ziemsen Verlag.

Noonan, B. P., and R. M. Bonett. 2003. A new species of *Hyalinobatrachium* (Anura: Centrolenidae) from the highlands of Guyana. *J. Herpetol.* 37:91–97.

Norberg, R. A. 1991. The flappet lark *Mirafra rufocinnamomea* doubles its wingbeat rate to 24 Hz in wing-clap display flight: A sexually selected feat. *J. Exp. Biol.* 159:515–23.

Norberg, U. M. and M. B. Fenton. 1988. Carnivorous bats? *Biol. J. Linn. Soc.* 33:383–94.

Norris, K. M., and C. A. Hosie. 2005. A quantitative ethogram for oviposition in *Triturus* newts: Description and comparison of *T. helveticus* and *T. vulgaris*. *Ethology* 111:357–66.

Norris, K. S., and C. H. Lowe. 1964. An analysis of background color-matching in amphibians and reptiles. *Ecology* 45:565–80.

Novacek , M. J., ed. 2001. *The Biodiversity crisis: Losing what counts*. New York: New Press.

Novak, R. M., and D. C. Robinson. 1975. Observations on the reproduction and ecology of the tropical montane toad, *Bufo holdridgei* Taylor in Costa Rica. *Rev. Biol. Trop.* 23:213–37.

Novo, J., A. R. Estrada, and L. V. Moreno. 1987. Apuntes sobre la reproduccion de *Eleutherodactylus atkinsi* (Anura: Leptodactylidae). *Misc. Zool. (Acad. Cien. Cuba)* 29:2–3.

Nowak, R. M., and J. L. Paradiso. 1983. *Walker's mammals of the world*. 2 vols. (4th ed.) Baltimore: Johns Hopkins University Press.

Nowak, R. T., and E. D. Brodie, Jr. 1978. Rib penetration and associated antipredator adaptations in the salamander *Pleurodeles waltl* (Salamandridae). *Copeia* 1978:424–29.

Nudds, R. L., and D. M. Bryant. 2000. The energetic cost of short flights in birds. *J. Exp. Biol.* 203:1561–72.

Nuin, P. A. S., and F. C. do Val. 2005. Phylogenetic analysis of the subfamily Hylodinae (Anura: Leptodactylidae) based on morphological characters. *Amphibia-Reptilia* 26:139–47.

Nuland, G. J., and P. F. H. Claus. 1981. The development of a radio tracking system for anuran species. *Amphibia-Reptilia* 2:107–16.

Nunes, V. da Silva. 1988a. Feeding asymmetry affects territorial disputes between males of *Plethodon cinereus*. *Herpetologica* 44:386–391.

———. 1988b. Vocalizations of treefrogs (*Smilisca sila*) in response to bat predation. *Herpetologica* 44:8–10.

Nunes, V. da Silva, and R. G. Jaeger. 1989. Salamander aggressiveness increases with length of territorial ownership. *Copeia* 1989:712–18.

Nürnberger, B. D., N. H. Barton, C. MacCullum, J. Gilchrist, and M. Appleby. 1995. Natural selection on quantitative traits in the *Bombina* hybrid zone. *Evolution* 49:1224–38.

Nussbaum, R. A. 1969a. Nests and eggs of the Pacific giant salamander, *Dicamptodon ensatus* (Eschscholtz). *Herpetologica* 25:257–62.

———. 1969b. A nest site of the Olympic salamander *Rhyacotriton olympicus* (Gaige). *Herpetologica* 25:277–78.

———. 1970. *Dicamptodon copei*, n. sp., from the Pacific northwest, U.S.A. (Amphibia: Caudata: Ambystomatidae). *Copeia* 1970:506–14.

———. 1976. Geographic variation and systematics of salamanders of the genus *Dicamptodon* Strauch (Ambystomatidae). *Misc. Publ. Mus. Zool. Univ. Michigan* (149):1–94.

———. 1980. Phylogenetic implications of amplectant behavior in sooglossid frogs. *Herpetologica* 36:1–5.

———. 1983. The evolution of a unique dual jaw-closing mechanism in caecilians (Amphibia: Gymnophiona) and its bearing on caecilian ancestry. *J. Zool., London* 199:545–54.

———. 1984. Amphibians of the Seychelles. In *Biogeography and ecology of the Seychelles Islands*, ed. D. R. Stoddart, 379–415. The Hague: Dr. W. Junk.

———. 1985. The evolution of parental care in salamanders. *Misc. Publ. Mus. Zool. Univ. Michigan* (169):1–50.

———. 1986. *Chthonerpeton onorei*, a new caecilian (Amphibia: Gymnophiona: Typhlonectidae) from Ecuador. *Rev. Suisse Zool.* 93:911–18.

———. 2003. Parental care. In *Reproductive biology and phylogeny of Urodela*, ed. D. M. Sever, 527–612. Enfield, NH: Science Publishers.

Nussbaum, R. A., and E. D. Brodie, Jr. 1982. Partitioning of the salamandrid genus *Tylototriton* Anderson (Amphibia: Caudata) with a description of a new genus. *Herpetologica* 38:320–32.

Nussbaum, R. A., E. D. Brodie, Jr., and R. M. Storm. 1983. *Amphibians and reptiles of the Pacific Northwest*. Moscow: University Press of Idaho.

Nussbaum, R. A., and G. W. Clothier. 1973. Population structure, growth, and size of larval *Dicamptodon ensatus* (Eschscholtz). *Northwest Sci.* 47:218–27.

Nussbaum, R. A., A. P. Jaslow, and J. Watson. 1982. Vocalization in frogs of the family Sooglossidae. *J. Herpetol.* 16:198–204.

Nussbaum, R. A., and C. Maser. 1969. Observations of *Sorex palustris* preying on *Dicamptodon ensatus*. *Murrelet* 50:23–24.

Nussbaum, R. A., and B. G. Naylor. 1982. Variation in the trunk musculature of caecilians (Amphibia: Gymnophiona) *J. Zool., London* 198:383–98.

Nussbaum, R. A., and C. K. Tait. 1977. Aspects of the life history and ecology of the Olympic salamander, *Rhyacotriton olympicus* (Gaige). *Am. Midl. Nat.* 98:176–99.

Nussbaum, R. A., and M. Wilkinson. 1989. On the classification and phylogeny of caecilians (Amphibia: Gymnophiona): A critical review. *Herp. Monogr.* 2:1–42.

———. 1995. A new genus of lungless tetrapod: A radically divergent caecilian (Amphibia: Gymnophiona). *Proc. Roy. Soc. London B* 261:331–35.

Nutphund, W. 2001. *Amphibians of Thailand*. Bangkok, Thailand: Amarin Printing and Publishing Public Company, Ltd.

Nyman, S. 1991. Ecological aspects of syntopic larvae of *Ambystoma maculatum* and the *A. laterale-jeffersonianum* complex in two New Jersey ponds. *J. Herpetol.* 25:505–9.

Nyman, S., R. F. Wilkinson, and J. E. Hutcherson. 1993. Cannibalism and size relations in a cohort of larval ringed salamanders (*Ambystoma annulatum*). *J. Herpetol.* 27:78–84.

Nyström, P., and K. Åbjörnsson. 2000. Effects of fish chemical cues on the interactions between tadpoles and crayfish. *Oikos* 88:181–90.

Nyström, P., L. Birkedal, C. Dahlberg, and C. Brönmark. 2002. The declining spadefoot toad *Pelobates fuscus:* Calling site choice and conservation. *Ecography* 25:488–98.

Obert, H.-J. 1973. Untersuchungen zur homonalen Steuerung der Ruf- und Paarungsaktivität bei der Rot- und Gelbbauchunke, *Bombina bombina* und *Bombina v. variegata*. *Zool. Jb. Physiol.* 77:166–98.

———. 1974. Untersuchungen zur homonalen Steuerung der Rufaktivität von Fröschen und Kröten dere Familien Ranidae, Discoglossidae, Hylidae, und Bufonidae. *Zool. Jb. Physiol.* 78:219–41.

———. 1975. The dependence of calling activity in *Rana esculenta* Linne 1758 and *Rana ridibunda* Pallas 1771 upon exongenous factors (Ranidae, Anura). *Oecologia* 18:317–28.

———. 1977. Hormonal influences on calling and reproductive behavior in anurans. In *The reproductive biology of amphibians*, ed. D. H. Taylor and S. I. Guttman, 357–66. New York: Plenum.

Oberweger, K., and F. Goller. 2001. The metabolic cost of birdsong production. *J. Exp. Biol.* 204:3379–88.

O'Brien, J. A. 2002. *Territory acquisition and maintenance in a Neotropical frog, Eleutherodactylus coqui*. Unpublished PhD dissertation. Storrs: University of Connecticut.

Odendaal, F. J., and C. M. Bull. 1980. Influence of water speed on tadpoles of *Ranidella signifera* and *R. riparia* (Anura: Leptodactylidae). *Aust. J. Zool.* 28:79–82.

———. 1982. A parapatric boundary between *Ranidella signifera* and *R. riparia* (Anura: Leptodactylidae) in South Australia. *Aust. J. Zool.* 30:49–57.

———. 1983. Water movements, tadpole competition and limits to the distribution of the frogs *Ranidella riparia* and *R. signifera*. *Oecologia* 57:361–67.

Odendaal, F. J., C. M. Bull, and R. C. Nias. 1982. Habitat selection in tadpoles of *Ranidella signifera* and *R. riparia* (Anura: Leptodactylidae). *Oecologia* 52:411–14.

Odendaal, F. J., C. M. Bull, and S. J. Richards. 1984. Interactions during feeding between tadpoles of *Ranidella signifera* and *R. riparia*. *J. Herpetol.* 18:489–92.

Odendaal, F. J., C. M. Bull, and S. R. Telford. 1983. The vocabulary of calls of *Ranidella riparia* (Anura: Leptodactylidae). *Copeia* 1983:534–37.

———. 1986. Influence of the acoustic environment on the distribution of the frog *Ranidella riparia*. *Anim. Behav.* 34:1836–43.

Odum, R. A., and P. S. Corn. 2005. *Bufo baxteri* Porter, 1968. Wyoming toad. In *Amphibian declines: The conservation status of United States species,* ed. M. J. Lannoo, 390–92. Berkeley: University of California Press.

Ogurtsov, S. V. and V. A. Bastakov. 2001. Imprinting on native pond odour in the pool frog, *Rana lessonae* Cam. In *Chemical signals in vertebrates 9,* ed. A. Marchlewska-Koj, J. J. Lepri, and D. Müller-Schwarze, 433–38. New York: Kluwer Academic/Plenum.

O'Hara, R. K., and A. R. Blaustein. 1981. An investigation of sibling recognition in *Rana cascadae* tadpoles. *Anim. Behav.* 29:1121–26.

———. 1982. Kin preference behavior in *Bufo boreas* tadpoles. *Behav. Ecol. Sociobiol.* 11:43–49.

———. 1985. *Rana cascadae* tadpoles aggregate with siblings: an experimental field study. *Oecologia* 67:44–51.

———. 1988. *Hyla regilla* and *Rana pretiosa* tadpoles fail to display kin recognition behaviour. *Anim. Behav.* 36:946–48.

Ohdachi, S. 1994. Growth, metamorphosis, and gape-limited cannibalism and predation on tadpoles in larvae of salamander *Hynobius retardatus*. *Zool. Sci.* 11:127–31.

Ohler, A. 2003a. Comments on a new book on the Amphibia of Thailand, with a tentative allocation of the figured species. *Alytes* 21:100–102.

———. 2003b. Revision of the genus *Ophryophryne* Boulenger, 1903 (Megophryidae) with description of two new species. *Alytes* 21:23–44.

Ohler, A., S. R. Swan, and J. C. Daltry. 2002. A recent survey of the amphibian fauna of the Cardomom Mountains, Southwest Cambodia, with descriptions of three new species. *Raffles Bull. Zool.* 50:465–81.

Oka, Y., R. Ohtani, M. Satou, and K. Ueda. 1984. Sexually dimorphic muscles in the forelimb of the Japanese toad, *Bufo japonicus*. *J. Morphol.* 180:297–308.

Okada, Y. 1927. Frogs in Japan. *Copeia* 1927:161–66.

———. 1966. *Fauna Japonica. Anura (Amphibia)*. Tokyo: Biogeographical Society of Japan.

Okada, Y., S. Ishiko, S. Daido, J. Kim, and S. Ikeda. 1962. Comparative morphology of the lung with special reference to the alveolar cells. I. Lung of the Amphibia. *Acta Tuberculosa Jap.* 11:63–72.

Okuno, R. 1984. Studies on the natural history of the Japanese toad, *Bufo japonicus japonicus*. I. Habitat groups and the interactions among them. *Jap. J. Ecol.* 34:113–21 (in Japanese, with English summary).

———. 1985. Studies on the natural history of the Japanese toad, *Bufo japonicus japonicus*. VII. Home range and movement of the adult toad. *Jap. J. Ecol.* 35:357–63 (in Japanese, with English summary).

———. 1986. Studies on the natural history of the Japanese toad, *Bufo japonicus japonicus*. X. Mating and spawning behaviors. *Jap. J. Ecol.* 36:11–18 (in Japanese, with English summary).

O'Laughlin, B. E., and R. N. Harris. 2000. Models of metamorphic timing: An experimental evaluation with the pond-dwelling salamander *Hemidactylium scutatum* (Caudata: Plethodontidae). *Oecologia* 124:343–50.

Oldham, R. S. 1963. Homing behavior in *Rana temporaria* Linn. *Brit. J. Herp.* 3:116–27.

———. 1966. Spring movements in the American toad, *Bufo americanus*. *Can. J. Zool.* 44:63–100.

———. 1967. Orienting mechanisms of the green frog, *Rana clamitans*. *Ecology* 48:477–91.

———. 1969. Initiation of breeding behavior in the American toad, *Bufo americanus*. *Can. J. Zool.* 47:1083–86.

———. 1977. Terrestrial locomotion in two species of amphibian larvae. *J. Zool., London* 181:285–95.

———. 1999. Amphibians and agriculture: double jeopardy. In *Aquatic life cycle strategies. Survival in a variable environment,* ed. M. Whitfield, J. Matthews, and C. Reynolds, 105–24. Plymouth, UK: Marine Biological Association.

Oldham, R. S., and H. C. Gerhardt. 1975. Behavioral isolating mechanisms of the treefrogs *Hyla cinerea* and *H. gratiosa*. *Copeia* 1975:223–31.

Oldham, R. S., D. M. Latham, D. Hilton-Brown, M. Towns, A. S. Cooke, and A. Burn. 1997. The effect of ammonium nitrate fertiliser on frog (*Rana temporaria*) survival. *Agric. Ecos. Environ.* 61:69–74.

Oldham, R. S., and M. J. S. Swan. 1992. Effects of ingested radio transmitters on *Bufo bufo* and *Rana temporaria*. *Herpetol. J.* 2:82–85.

Olenick, R. J., and J. H. Gee. 1981. Tiger salamanders (*Ambystoma tigrinum*) and stocked rainbow trout (*Salmo gairdneri*): Potential competitors for food in Manitoba pothole lakes. *Can. Field-Nat.* 95:129–32.

Olgun, K, C. Miaud, and P. Gautier. 2001. Age, growth, and survivorship in the viviparous salamander *Mertensiella luschani* from southwestern Turkey. *Can. J. Zool.* 79:1559–67.

Olgun, K., N. Uzum, A. Avci, and C. Miaud. 2005. Age, size and growth of the southern crested newt *Triturus karelinii* (Strauch 1870) in a population from Bozdag (Western Turkey). *Amphibia-Reptilia* 26:223–30.

Oliveira, M. E., M. Paillette, H. D. Rosa, and E. G. Crespo. 1991. A natural hybrid between *Hyla arborea* and *Hyla meridionalis* detected by mating calls. *Amphibia-Reptilia* 12:15–20.

Oliver, J. 1951. Gliding in amphibians and reptiles, with a remark on an arboreal adaptation in the lizard, *Anolis carolinensis carolinensis* Voigt. *Am. Nat.* 85:171–76.

Olson, D. H. 1989. Predation on breeding western toads (*Bufo boreas*). *Copeia* 1989:391–97.

———. 1992. Ecological susceptibility of amphibians to population declines. In *Proceeding of symposium on biodiversity of Northwestern California, October 28–30, 1991, Santa Rosa, CA.*, R. R. Harris and D. E. Erman, technical coordinators and ed. H. M. Kerner, 55–62. Berkeley: University of California Wildland Resources Center Report 29.

Olson, D. H., A. R. Blaustein, and R. K. O'Hara. 1986. Mating pattern variability among western toad (*Bufo boreas*) populations. *Oecologia* 70:351–56.

Olsson, M., and T. Uller. 2002. Developmental stability and genetic architechture: A comparison within and across thermal regimes in tadpoles. *J. Evol. Biol.* 15:625–33.

———. 2003. Thermal environment, survival and local adaptation in the common frog, *Rana temporaria. Evol. Ecol. Res.* 5:431–37.

Omland, K. S. 1998. Orientation based on ambient directional information by red-spotted newts (*Notophthalmus viridescens*). *Behaviour* 135:757–75.

Oommen, O. V., G. J. Measey, D. J. Gower, and M. Wilkinson. 2000. The distribution and abundance of the caecilian *Gegeneophis ramaswamii* (Amphibia: Gymnophiona) in southern Kerala. *Curr. Sci.* 79:1386–89.

O'Reilly, J. C. 2000. Feeding in caecilians. In. *Feeding: Form, function, and evolution in tetrapod vertebrates,* ed. K. Schwenk, 149–66. San Diego:Academic Press.

O'Reilly, J. C., S. M. Deban, and K. C. Nishikawa. 2002. Derived life history characteristics constrain the evolution of aquatic feeding behavior in adult amphibians. In *Topics in functional and ecological vertebrate morphology,* ed. P. Aerts, K. D'Août, A. Herrel, and R. Van Damme, 153–90. Maastricht, The Netherlands: Shaker Publishing.

O'Reilly, J. C., D. A. Ritter, and D. R. Carrier. 1997. Hydrostatic locomotion in a limbless tetrapod. *Nature* 386:269–72.

O'Reilly, J. C., A. P. Summers, and D. A. Ritter. 2000. The evolution of the functional role of trunk muscles during locomotioin in adult amphibians. *Amer. Zool.* 40:123–35.

O'Reilly, S. R., and K. C. Nishikawa. 1995. Mechanism of tongue protraction during prey capture in the spadefoot toad *Spea multiplicata* (Anura: Pelobatidae). *J. Exp. Zool.* 273:282–96.

Organ, J. A. 1958. The courtship and spermatophore of *Plethodon jordani metcalfi. Copeia* 1958:251–59.

———. 1960a. The courtship and spermatophore of the salamander *Plethodon glutinosus. Copeia* 1960:34–40.

———. 1960b. Studies on the life history of the salamander *Plethodon welleri. Copeia* 1960:287–97.

———. 1961a. The eggs and young of the spring salamander, *Pseudotriton porphyriticus. Herpetologica* 17:53–56.

———. 1961b. Life history of the pygmy salamander, *Desmognathus wrighti,* in Virginia. *Am. Midl. Nat.* 66:384–90.

———. 1961c. Studies on the local distribution, life history, and population dynamics of the salamander genus *Desmognathus* in Virginia. *Ecol. Monogr.* 31:189–220.

———. 1968. Courtship behavior and spermatophore of the cave salamander *Eurycea lucifuga* (Rafinesque). *Copeia* 1968:576–80.

Organ, J. A., and L. A. Lowenthal. 1963. Comparative studies of macroscopic and microscopic features of spermatophores of some plethodontid salamanders. *Copeia* 1963:659–65.

Organ, J. A., and D. J. Organ. 1968. Courtship behavior of the red salamander, *Pseudotriton ruber. Copeia* 1968:217–23.

Orizaola, G. and F. Braña. 2003a. Oviposition behaviour and vulnerability of eggs to predation in four newt species (genus *Triturus*). *Herpetol. J.* 13:121–24.

———. 2003b. Responses of predator-naïve newt larvae to food and predator presence. *Can. J. Zool.* 81:1845–50.

Orizaola, G., and C. R. del Valle. 2000. *Triturus marmoratus* (marbled newt). Predation. *Herp. Rev.* 31:233.

Orlando, K., and A. W. Pinder. 1995. Larval cardiorespiratory ontogeny and allometry in *Xenopus laevis. Physiol. Zool.* 68:63–75.

Orlov, N. 1997. Breeding behavior and nest construction in a Vietnam frog related to *Rana blythi. Copeia* 1997:464–65.

Orr, L. P. 1967. Feeding experiments with a supposed mimetic complex in salamanders. *Am. Midl. Nat.* 77:147–55.

———. 1968. The relative abundance of mimics and models in a supposed mimetic complex in salamanders. *J. Elisha Mitchell Sci. Soc.* 84:303–4.

Ortiz, J. C., H. Ibarra-Videl, and J. R. Formas. 1989. A new species of *Eupsophus* (Anura: Leptodactylidae) from Contulmo, Nahuelbuta Range, southern Chile. *Proc. Biol. Soc. Washington* 102:1031–35.

Osborne, P. L., and A. M. McLachlan. 1985. The effect of tadpoles on algal growth in temporary, rain-filled rock pools. *Freshw. Biol.* 15:77–87.

Osborne, W., D. Hunter, and G. Hollis. 1999. Population declines and range contraction in Australian alpine frogs. In *Declines and disappearances of Australian frogs,* ed. A. Campbell, 145–57. Canberra: Environment Australia.

Osborne, W. S. 1989. Distribution, relative abundance and conservation status of the Corroboree frogs, *Pseudophryne corroboree* Moore (Anura: Myobatrachidae). *Austr. Wildl. Res.* 16:537–47.

Osikowski, A., and J. Rafínski. 2001. Multiple insemination increases reproductive success of female Montandon's newt (*Triturus montandoni,* Caudata, Salamandridae). *Behav. Ecol. Sociobiol.* 49:145–49.

Österdahl, L., and R. Olsson. 1963. The sexual behaviour of *Hymenochirus boettgeri. Oikos* 14:35–43.

Ota, H. 2000. Current status of the threatened amphibians and reptiles of Japan. *Pop. Ecol.* 42:5–9.

Otte, D. 1977. Communication in Orthoptera. In *How animals communicate,* ed. T. A. Sebeok, 334–61. Bloomington: Indiana University Press.

Otte, D., and J. J. Loftus-Hills. 1979. Chorusing in *Syrbula* (Orthoptera: Acrididae). Cooperation, interference competition, or concealment? *Entomol. News* 90:159–65.

Ouellet, M. 2000. Amphibian deformities: Current state of knowledge. In *Ecotoxicology of amphibians and reptiles,* ed. D. W. Sparling, G. Linder, and C. A. Bishop, 617–61. Pensacola, FL: SETAC Press,.

Ouellet, M., I. Mikaelian, B. D. Pauli, J. Rodrigue, and D. M. Green. 2005. Historical evidence of widespread chytrid infection in North American amphibian populations. *Cons. Biol.* 19:1431–40.

Ovaska, K. 1987. Seasonal changes in agonistic behaviour of the western red-backed salamander, *Plethodon vehiculum. Anim. Behav.* 35:67–74.

———. 1988a. Recognition of conspecific odors by the western red-backed salamander, *Plethodon vehiculum. Can. J. Zool.* 66:1293–96.

———. 1988b. Spacing and movements of the salamander *Plethodon vehiculum. Herpetologica* 44:377–86.

———. 1991. Reproductive phenology, population structure, and habitat use of the frog *Eleutherodactylus johnstonei* in Barbados, West Indies. *J. Herpetol.* 25:424–30.

———. 1992. Short- and long-term movements of the frog *Eleutherodactylus johnstonei* in Barbados, West Indies. *Copeia* 1992:569–73.

———. 1993. Aggression and territoriality among sympatric western plethodontid salamanders. *Can. J. Zool.* 71:901–7.

Ovaska, K., and T. M. Davis. 1992. Faecal pellets as burrow markers: intra- and interspecific odour recognition by western plethodontid salamanders. *Anim. Behav.* 43:931–39.

Ovaska, K., and A. R. Estrada. 2003. *Eleutherodactylus antillensis* (CoquíChurí). Reproduction. *Herp. Rev.* 34:229.

Ovaska, K., and M. A. Smith. 1988. Predatory behavior of two species of ground beetles (Coleoptera: Carabidae) towards juvenile salamanders (*Plethodon vehiculum*). *Can. J. Zool* 66:599–604.

Ovaska, K. E., and J. Caldbeck. 1997a. Courtship behavior and vocalizations of the frogs *Eleutherodactylus antillensis* and *E. cochranae* on the British Virgin Islands. *J. Herpetol.* 31:149–55.

———. 1997b. Vocal behaviour of the frog *Eleutherodactylus antillensis* on the British Virgin Islands. *Anim. Behav.* 54:181–88.

———. 1999. Courtship call of the frog *Eleutherodactylus schwartzi* from the British Virgin Islands. *J. Herpetol.* 33:501–4.

Ovaska, K. E., and W. Hunte. 1992. Male mating behavior of the frog *Eleutherodactylus johnstonei* (Leptodactylidae) in Barbados, West Indies. *Herpetologica* 48:40–49.

Overton, E. 1904. Neunenddreissig Thesen uber die Wasserokonomie der Amphibien und osmotischen Eigenschaften der Amphibienhaut. *Verh. Phys.-Med. Ges. Wurzburg* 36:277–95.

Owen, P. C. 2003. *The structure, function, and evolution of aggressive signals in anuran amphibians.* Unpublished PhD dissertation. Storrs: University of Connecticut.

Owen, P. C., and N. M. Gordon. 2005. The effect of perceived intruder proximity and resident body size on the aggressive responses of male green frogs, *Rana clamitans* (Anura: Ranidae). *Behav. Ecol. Sociobiol.* 58:446–55.

Owen, P. C., and S. A. Perrill. 1998. Habituation in the green frog, *Rana clamitans. Behav. Ecol. Sociobiol.* 44:209–13.

———. 2000. Habituation and neighbor-stranger discrimination in green frogs: A reply to Bee and Schachtman. *Behav. Ecol. Sociobiol.* 48:169–71.

Owen, P. C., and J. K. Tucker. 2006. Courtship calls and behavior in two species of chorus frogs, genus *Pseudacris* (Anura: Hylidae). *Copeia* 2006:137–44.

Owen, R. D., and S. A. Johnson. 1997. *Rana curtipes* (bicolored frog). Predation. *Herp. Rev.* 28:200.

Oyama, J. 1929. Some embryological notes on *Hynobius naevius. Copeia* 1929:92–94.

Oza, G. M. 1990. Ecological effects of the frog's legs trade. *The Environmentalist* 10:39–42.

Özeti, N. 1967. The morphology of the salamander *Mertensiella luschani* (Steindachner) and the relationships of *Mertensiella* and *Salamandra. Copeia* 1967:287–98.

———. 1973. A study on the reproduction biology of the salamander *Mertensiella luschani atifi* Basoglu (Urodela, Amphibia). *Sci. Rep. Fac. Sci. Ege Univ.* 164:1–17.

———. 1979. Reproductive biology of the salamander *Mertensiella luschani antalyana. Herpetologica* 35:193–97.

Özeti, N., and D. B. Wake. 1969. The morphology and evolution of the tongue and associated structures in salamanders and newts (Family Salamandridae). *Copeia* 1969:91–123.

Pacala, S. W., and J. Roughgarden. 1985. Population experiments with the *Anolis* lizards of St. Maarten and St. Eustatius. *Ecology* 66:129–41.

Pace, A. E. 1974. Systematic and behavioral studies of the leopard frogs (*Rana pipiens* complex) of the United States. *Misc. Publ. Mus. Zool. Univ. Mich.* 148:1–140.

Packard, G. C. 1972. Inverse compensation for temperature in oxygen consumption of the hylid frog, *Pseudacris triseriata. Physiol. Zool.* 45:270–75.

Packard, G. C., and T. J. Boardman. 1986. The misuse of ratios to scale physiological data that vary allometrically with body size. In *New directions in ecological physiology,* ed. M. E. Feder, A. F. Bennett, W. W. Burggren, and R. B. Huey, 216–39. New York: Cambridge University Press.

Packard, G. C., J. W. Lang, L. D. Lohmiller, and M. J. Packard. 1997. Cold tolerance in hatchling painted turtles (*Chrysemys picta*): Supercooling or tolerance for freezing? *Physiol. Zool.* 70:670–78.

Packard, G. C., J. W. Lang, L. S. Lohmiller, and M. J. Packard. 1999. Resistance to freezing in hatchling painted turtles (*Chrysemys picta*). *Can. J. Zool.* 77:795–801.

Packard, G. C., and M. J. Packard. 1995. The basis for cold tolerance in hatchling painted turtles (*Chrysemys picta*). *Physiol. Zool.* 68:129–48.

Packard G.C., J. K. Tucker, and L. D. Lohmiller. 1998. Distribution of Strecker's chorus frogs (*Pseudacris streckeri*) in relation to their tolerance for freezing. *J. Herpetol.* 32: 437–40.

Packer, W. C. 1960. Bioclimatic influences on the breeding migration of *Taricha rivularis. Ecology* 41:509–17.

———. 1963a. Dehydration, hydration, and burrowing behavior in *Heleioporus eyre* (Gray) (Leptodactylidae). *Ecology* 44:643–51.

———. 1963b. Observations on the breeding migration of *Taricha rivularis. Copeia* 1963:378–82.

Padgett-Flohr, G. E. and J. E. Longcore. 2005. *Ambystoma californiense* (California tiger salamander). Fungal infection. *Herp. Rev.* 36:50–51.

Páez, V. P., B. C. Bock, and A. S. Rand. 1993. Inhibition of evoked calling of *Dendrobates pumilio* due to acoustic interference from cicada calling. *Biotropica* 25:242–45.

Pagano, A., P. A. Crochet, J-D. Graf, P. Joly, and T. Lodé. 2001. Distribution and habitat use of water frog hybrid complexes in France. *Glob. Ecol. Biogeogr.* 10:433–41.

Pagano, A., A. Dubois, D. Lesbarrères, and T. Lodé. 2003. Frog alien species: a way for genetic invasion? *C. R. Biol.* 326:S85–S92.

Pagano, A., P. Joly, S. Plenet, A. Lehman, and O Grolet. 2001. Breeding habitat partitioning in the *Rana esculenta* complex: The intermediate niche hypothesis supported. *Ecoscience* 9:294–300.

Pagano, A., T. Lodé, and P. A. Crochet. 2001. New contact zone and assemblages among water frogs of southern France. *J. Zool. Syst. Evol. Res.* 39:63–67.

Page, R. A., and M. J. Ryan. 2005. Flexibility in assessment of prey cues: Frog-eating bats and frog calls. *Proc. Roy. Soc. London B* 272:841–47.

Pahkala, M., A. Laurila, L. O. Björn, and J. Merilä. 2001. Effects of ultraviolet-B radiation and pH on early development of the moor frog *Rana arvalis. J. Appl. Ecol.* 38:628–36.

Pahkala, M., A. Laurila, and J. Merilä. 2000. Ambient ultraviolet-B radiation reduces hatchling size in the common frog *Rana temporaria. Ecography* 23:531–38.

———. 2001. Carry-over effects of ultraviolet-B radiation on larval fitness in *Rana temporaria. Proc. Roy. Soc. London B* 268:1699–1706.

Pahkala, M., J. Merilä, I. Ots, and A. Laurila. 2003. Effects of ultraviolet-B radiation on metamorphic traits in the common frog *Rana temporaria. J. Zool., London* 259:57–62.

Pahkala, M., K. Räsänen, A. Laurila, U. Johanson, L. O. Björn, and J. Merilä. 2002. Lethal and sublethal effects of UV-B/pH synergism on common frog embryos. *Cons. Biol.* 16:1063–73.

Paillette, M. 1967. Rythme d'activité acoustique des *Hyla arborea* (Linne) et *Hyla meridionalis* Boettger (Amphibiens-Anoures). *C. R. Soc. Biol.* 161:986–89.

———. 1970a. Conditions biophysiques de déclenchement du signal sonore chez *Hyla meridionalis* (Amphibien Anoure). *Terre et Vie* 1970:251–300.

———. 1970b. La notion de territoire chez les amphibiens anoures, et plus particulièrement la valeur des émissions sonores dans le comportement territorial des hylídes: *Hyla arborea* et *Hyla meridionalis*. In *Territoire et domaine vital,* ed.G. Richard, 35–48. Paris: Masson.

———. 1971. Communication acoustique chez les Amphibiens anoures. *J. de Psychol.* 68:327–51.

Pailley, M., and P. Pailley. 1995. La prédation des batraciens par la chouette affraie (*Tyto alba*) en Maine-et-Loire. *Bull. Soc. d'Études Sci. de l'Anjou* (n.s.) 15:135–41.

Paine, R. T. 1974. Intertidal community structure: Experimental studies on the relationship between a dominant competitor and its principal predator. *Oecologia* 15:93–120.

Painter, C. W., N. J. Scott, Jr., and M. J. Altenbach. 1999. *Thamnophis elegans vagrans* (wandering garter snake). Diet. *Herp. Rev.* 30:48.

Pakkasmaa, S., and S. Aikio. 2003. Relatedness and competitive asymmetry—The growth and development of common frog tadpoles. *Oikos* 100:55–64.

Palen, W. J., D. E. Schindler, M. J. Adams, C. A. Pearl, R. B. Bury, and S. A. Diamond. 2002. Optical characteristics of natural waters protect amphibians from UV-B in the U.S. Pacific Northwest. *Ecology* 83:2951–57.

Palen, W. J., D. E. Schindler, M. J. Adams, C. A. Pearl, R. B. Bury, and S. A. Diamond. 2004. Optical characteristics of natural waters protect amphibians from UV-B in the U.S. Pacific Northwest: Reply. *Ecology* 85:1754–59.

Palen, W. J., C. E. Williamson, A. A. Clauser, and D. E. Schindler. 2005. Impact of UV-B exposure on amphibian embryos: Linking species physiology and oviposition behaviour. *Proc. Roy. Soc. London B* 272:1227–34.

Palis, J. G. 1997. Distribution, habitat, and status of the flatwoods salamander (*Ambystoma cingulatum*) in Florida, USA. *Herpetol. Nat. Hist.* 5:53–65.

———. 1998. Breeding biology of the gopher frog, *Rana capito,* in western Florida. *J. Herpetol.* 32:217–23.

———. 2005. *Scaphiopus holbrookii*. Eastern spadefoot toad. In *Amphibian declines: The conservation status of United States species,* ed. M. J. Lannoo, 511–12. Berkeley: University of California Press.

Palis, J. G., and D. B. Means. 2005. *Ambystoma cingulatum* Cope, 1868. Flatwoods salamander. In *Amphibian declines: The conservation status of United States species,* ed. M. J. Lannoo, 608–9. Berkeley: University of California Press.

Pallett, J. R., and N. I. Passmore. 1988. The significance of multinote advertisement calls in the reed frog, *Hyperolius tuberilinguis. Bioacoustics* 1:13–24.

Palmer, C. A., R. A. Watts, R. G. Gregg, M. A. McCall, L. D. Houck, R. Highton, and S. J. Arnold. 2005. Lineage-specific differences in evolutionary mode in a salamander courtship pheromone. *Mol. Bio. Evol.* 22:2243–56.

Palo, J. U., D. Lesbarrères, D. S. Schmeller, C. R. Primmer, and J Merilä. 2004. Microsatellite marker data suggest sex-biased dispersal in the common frog *Rana temporaria. Mol. Ecol.* 13:2865–69.

Palomares, F. 1993. Opportunistic feeding of the Egyptian mongoose, *Herpestes ichneumon,* (L.) in southwestern Spain. *Rev. Ecol. (Terre Vie)* 48:295–304.

Pancharatna, M., and S. K. Saidapur. 1985. Ovarian cycle in the frog *Rana cyanophlyctis:* A quantitative study of follicular kinetics in relation to body mass, oviduct and fat body cycles. *J. Morphol.* 186:135–47.

Panchen, A. L., and T. R. Smithson. 1987. Character diagnosis, fossils and the origin of tetrapods. *Biol. Rev.* 62:341–438.

———. 1988. The relationships of the earliest tetrapods. In *The phylogeny and classification of the tetrapods. Vol. 1: Amphibians, reptiles, birds,* ed. M. J. Benton, 1–32. New York: Oxford University Press.

Panchen, A. L., ed. 1980. *The terrestrial environment and the origin of land vertebrates.* New York: Academic Press.

Pandian, T. J., and M. P. Marian. 1985a. Predicting anuran metamorphosis and energetics. *Physiol. Zool.* 58:538–52.

———. 1985b. Time and energy costs of metamorphosis in the Indian bullfrog, *Rana tigrina. Copeia* 1985:653–62.

Pang, P. K. T. 1977. Osmoregulatory functions of neurohypophysial hormones in fishes and amphibians. *Amer. Zool.* 17:739–49.

Pang, P. K. T., M. Uchiyama, and W. H. Sawyer. 1982. Endocrine and neural control of amphibian renal functions. *Fed. Proc. Am. Soc. Exp. Biol.* 41:2365–70.

Papi, F. 1990. Homing phenomena: mechanisms and classification. *Ethol. Ecol. Evol.* 2:3–10.

Papp, M. C., an C. O. G. Papp. 2000. Decline in a population of the treefrog *Phyllodytes lutelolus* after fire. *Herp. Rev.* 31:93–95.

Para-Olea, G., E. Martínez-Meyer, and G. Pérez-Ponce de León. 2005. Forecasting climate change effects on salamander distribution in the highlands of Central Mexico. *Biotropica* 37:202–8.

Pardini, R. 1998. Feeding ecology of the Neotropical river otter *Lontra longicaudis* in an Atlantic Forest stream, south-eastern Brazil. *J. Zool., London* 245:385–91.

Parichy, D. M., and R. H. Kaplan. 1992. Developmental consequences of tail injury on larvae of the Oriental fire-bellied toad, *Bombina orientalis. Copeia* 1992:129–37.

Park, D., J. M. McGuire, A. L. Majchrzak, J. M. Ziobro, and H. L. Eisthen. 2004. Discrimination of conspecific sex and reproductive condition using chemical cues in axolotls (*Ambystoma mexicanum*). *J. Comp. Physiol. A* 190:415–27.

Park, D., and S.-R. Park. 2000. Multiple insemination and reproductive biology of *Hynobius leechii. J. Herpetol.* 34:594–98.

Park, D., and C. R. Propper. 2001. Repellent function of male pheromones in the red-spotted newt. *J. Exp. Zool.* 289:404–8.

Park, S.-R., D. S. Park, and S. Y. Yang. 1996. Courtship, fighting behaviors and sexual dimorphism of the salamander *Hynobius leechii. Korean J. Zool.* 39:437–46.

Parker, G. A. 1974. Assessment strategy and the evolution of fighting behaviour. *J. Theor. Biol.* 47:223–43.

———. 1983. Mate quality and mating decisions. In *Mate choice,* ed. P. Bateson, 141–64. New York: Cambridge University Press.

Parker, G. A., and L. W. Simmons. 1996. Parental investment and the control of sexual selection: Predicting the direction of sexual competition. *Proc. Roy. Soc. London B* 263:315–21.

Parker, H. W. 1936. The amphibians of the Mamfe Division, Cameroons. I. Zoogeography and systematics. *Proc. Zool. Soc. London* 1936:135–63.

———. 1940. The Australian frogs of the family Leptodactylidae. *Novit. Zool.* 42:1–106.

Parker, M. S. 1993. Predation by Pacific giant salamander larvae on juvenile steelhead trout. *Northwest. Nat.* 74:77–81.

———. 1994. Feeding ecology of stream-dwelling Pacific giant salamander larvae (*Dicamptodon tenebrosus*). *Copeia* 1994:705–18.

Parker, W. S., and M. V. Plummer. 1987. Population ecology. In *Snakes: Ecology and evolutionary biology,* ed. R. A. Seigel, J. T. Collins, and S. S. Novak, 253–301. New York: Macmillan.

Parmelee, J. R. 1993. Microhabitat segregation and spatial relationships among four species of mole salamanders (genus *Ambystoma*). *Occ. Pap. Mus. Nat. Hist. Univ. Kansas* 160:1–33.

———. 1999. Trophic ecology of a tropical anuran assemblage. *Sci. Pap. Nat. Hist. Mus. Univ. Kansas* 11:1–59.

Parmesan, C., and G. Yohe. 2003. A globally coherent fingerprint of climate change across natural systems. *Nature* 421:37–42.

Parra-Olea, G., M. García-París, and D. B. Wake. 2004. Molecular diversification of salamanders of the tropical American genus *Bolitoglossa* (Caudata: Plethodontidae) and its evolutionary and biogeographical implications. *Biol. J. Linn. Soc.* 81:325–46.

Parra-Olea, G., E. Martínez-Meyer, and G. Pérez-Ponce de Leon. 2005. Forecasting climate change effects on salamander distribution in the highlands of central Mexico. *Biotropica* 37:202–8.

Parra-Olea, G., and D. B. Wake. 2001. Extreme morphological and ecological homoplasy in tropical salamanders. *Proc. Natl. Acad. Sci. USA* 98:7888–91.

Parris, K. M. 2001. Distribution, habitat requirements and conservation of the cascade treefrog (*Litoria pearsoniana,* Anura: Hylidae). *Biol. Cons.* 99:285–92.

———. 2002. More bang for your buck: the effect of caller position, habitat and chorus noise on the efficiency of calling in the spring peeper. *Ecol. Model.* 156:213–24.

Parris, K. M., and M. A. McCarthy. 1999. What influences the structure of frog assemblages at forest streams? *Aust. J. Ecol.* 24:495–502.

Parris, M. J. 1998. Terrestrial burrowing ecology of newly metamorphosed frogs (*Rana pipiens* complex). *Can. J. Zool.* 76:2124–29.

———. 2000. Experimental analysis of hybridization in leopard frogs (Anura: Ranidae) larval performance in desiccating environments. *Copeia* 2000:11–19.

———. 2004. Hybrid responses to pathogen infection in interspecific crosses between two amphibian species (Anura: Ranidae). *Evol. Ecol. Res.* 6:457–71.

Parris, M. J., and J. G. Beaudoin. 2004. Chytridiomycosis impacts predator-prey interactions in larval amphibian communities. *Oecologia* 140:626–32.

Parris, M. J., and T. O. Cornelius. 2004. Fungal pathogen causes competitive and developmental stress in larval amphibian communities. *Ecology* 85:3385–95.

Parris, M. J., E. Reese, and A. Storfer. 2006. Antipredator behavior of chytridiomycosis-infected northern leopard frog (*Rana pipiens*) tadpoles. *Can. J. Zool.* 84:58–65.

Parris, M. J., and R. D. Semlitsch. 1998. Asymmetric competition in larval amphibian communities: Conservation implications for the northern crawfish frog, *Rana areolata circulosa. Oecologia* 116:219–26.

Parris, M. J., A. Storfer, J. P. Collins, and E. W. Davidson. 2005. Life-history responses to pathogens in tiger salamander (*Ambystoma tigrinum*) larvae. *J. Herpetol.* 39:366–72.

Parry, C. R., and R. Cavill. 1978. A note on cocoon formation and structure in *Pyxicephalus adspersus* Tschudi (Anura: Ranidae). *Trans. Rhodesian Sci. Assoc.* 58:55–58.

Parsons, R. H. 1994. Effects of skin circulation on water exchange. In *Amphibian biology. Vol. 1. The integument,* ed. H. Heatwole and G. T. Barthalmus, 132–46. Chipping Norton, New South Wales, Australia: Surrey Beatty and Sons.

Parsons, R. H., R. Brady, B. Goddard, and D. Kernan. 1978. Seasonal effects on osmotic permeability of *Rana pipiens* in vivo and in vitro. *J. Exp. Zool.* 204:347–52.

Parsons, R. H., V. McDevitt, V. Aggerwall, T. LeBlang, K. Manley, N. Kim, J. Lopez, and A. Kennedy. 1993. Regulation of pelvic patch water flow in *Bufo marinus:* role of bladder volume and angiotensin II. *Am. J. Physiol.* 264:R1260–R1265.

Partridge, B., and J. R. Krebs. 1978. Tree frog choruses: a mixed evolutionarily stable strategy? *Anim. Behav.* 26:959–63.

Partridge, L., and V. French. 1996. Thermal evolution of ectotherm body size: Why get big in the cold? In *Animals and temperature*, ed. I. A. Johnston and A. F. Bennett, 265–92. New York: Cambridge University Press.

Parzefall, J. 1976. Die Rolle der chemischen Information im Verhalten des Grottenolms *Proteus anguinus* Laur. (Proteidae, Urodela). *Z. Tierpsychol.* 42:29–49.

Parzefall, J., J. P. Durand, and B. Richard. 1980. Chemical communication in *Necturus maculosus* and his cave-living relative *Proteus anguinus* (Proteidae, Urodela). *Z. Tierpsychol.* 53:133–38.

Parzefall, J., J. P. Durand, and B. Richard. 1981. Aggressive behaviour of the European cave salamander *Proteus anguinus* (Proteidae, Urodela). *Proc. Int. Cong. Speleol.* 8:415–19.

Pasanen, S., and M. Karhapää. 1997. Can boreal common frog (*Rana temporaria* L.) survive in frost? *Ann. Zool. Fennici* 34:247–50.

Pasanen, S., and P. Koskela. 1974. Seasonal and age variation in the metabolism of the common frog, *Rana temporaria* L. in northern Finland. *Comp. Biochem. Physiol.* 47A:635–54.

Pasanen, S., and J. Sorjonen. 1994. Partial terrestrial wintering in a northern common frog population (*Rana temporaria* L.). *Ann. Zool. Fennici* 31:275–78.

Pascual, M., P. Macchi, J. Urbanski, F. Marcos, C. R. Rossi, M. Novara, and P. Dell'Arciprete. 2002. Evaluating potential effects of exotic freshwater fish from incomplete species presence-absence data. *Biol. Invas.* 4:101–13.

Passamani, M., and A. B. Rylands. 2000. Feeding behvior of Geoffroy's marmoset (*Callithrix geoffroyi*) in an Atlantic forest fragment of south-eastern Brazil. *Primates* 41:27–38.

Passmore, N. I. 1977. Mating calls and other vocalizations of five species of *Ptychadena* (Anura: Ranidae). *S. Afr. J. Sci.* 73:212–14.

———. 1978. *The vocalizations and aspects of the reproductive behaviour of the genus Ptychadena in South Africa.* Unpublished PhD dissertation. Johannesburg, South Africa: University of the Witwatersrand.

———. 1981. The relevance of the specific mate recognition concept to anuran reproductive biology. *Monit. Zool. Ital.* (N. S.) Suppl. 15:93–108.

Passmore, N. I., P. J. Bishop, and N. Caithness. 1992. Calling behaviour influences mating success in male painted reed frogs, *Hyperolius marmoratus. Ethology* 92:227–41.

Passmore, N. I., R. R. Capranica, S. R. Telford, and P. J. Bishop. 1984. Phonotaxis in the painted reed frog (*Hyperolius marmoratus*): The localization of elevated sound sources. *J. Comp. Physiol. A* 154:189–97.

Passmore, N. I., and V. C. Carruthers. 1979. *South African frogs.* Johannesburg, South Africa: Witwatersrand University Press.

———. 1995. *South African frogs. A complete guide.* Johannesburg, South Africa: University of the Witwatersrand Press.

Passmore, N. I., and H. Malherbe. 1985. High nocturnal body temperature in the painted reed frog (*Hyperolius marmoratus*). *Herpetologica* 41:212–15.

Passmore, N. I., and S. R. Telford. 1981. The effect of chorus organization on mate localization in the painted reed frog (*Hyperolius marmoratus*). *Behav. Ecol. Sociobiol.* 9:291–93.

———. 1983. Random mating by size and age of males in the painted reed frog, *Hyperolius marmoratus. S. Afr. J. Sci.* 79:353–55.

Patel, T. 1993. French may eat Indonesia out of frogs. *New Scientist* 138:7.

Paterson, H. E. H. 1978. More evidence against speciation by reinforcement. *S. Afr. J. Sci.* 74:369–71.

———. 1980. A comment on "Mate Recognition Systems." *Evolution* 34:330–31.

———. 1981. The continuing search for the unknown and unknowable: Critique of contemporary ideas on speciation. *S. Afr. J. Sci.* 77:113–19.

———. 1982. Perspective on speciation by reinforcement. *S. Afr. J. Sci.* 78:53–57.

Paterson, H. E. H. 1985. The recognition concept of species. In *Species and speciation,* ed. E. Vrba, 21–29. Pretoria, South Africa: Transvaal Museum Monograph 4.

Patla, D.A., and C.R. Peterson. 1999. Are amphibians declining in Yellowstone National Park? *Yellowstone Science* 7:2–11.

Paton, P. W., and W. B. Crouch III. 2002. Using the phenology of pond-breeding amphibians to develop conservation strategies. *Cons. Biol.* 16:194–204.

Paton, P. W. C. 2005. A review of vertebrate community composition in seasonal forest pools of the northeastern United States. *Wetlands Ecol. Manage.* 13:235–46.

Patterson, K. K. 1978. Life history aspects of paedogenic populations of the mole salamander, *Ambystoma talpoideum. Copeia* 1978:649–55.

Pauley, T. K. 1978. Moisture as a factor regulating habitat partitioning between two sympatric species of *Plethodon* (Amphibia, Urodela, Plethodontidae). *J. Herpetol.* 12:491–93.

Pauley, T. K., and J. W. Barron. 1995. Natural history and ecology of anurans. In *Mitigated wetland restoration: Environmental effects at Green Bottom Wildlife Management Area, West Virginia,* ed. D. K. Evans and H. H. Allen, 158–71. Technical Report, U.S. Army Corps of Engineers Waterways Experiment Station, Vicksburg, Mississippi. WRP-RE-10.

Pauley, T. K., B. A. Pauley, and M. B. Watson. 2005. *Plethodon nettingi* Green 1938. Cheat Mountain Salamander. In *Amphibian declines: The conservation status of United States species,* ed. M. J. Lannoo, 829–31. Berkeley: University of California Press.

Pauley, T. K. and M. B. Watson. 2005. *Aneides aeneus* Cope and Packard, 1881. Green salamander. In *Amphibian declines: The conservation status of United States species,* ed. M. J. Lannoo, 656–58. Berkeley: University of California Press.

Pauli, B. D., D. R. Coulson, and M. Berrill. 1999. Sensitivity of amphibian embryos and tadpoles to Mimic ® 240 LV insecticide following single or double exposures. *Env. Tox. Chem.* 18:2538–44.

Paulian, R., and A. Vilardebo. 1946. Observation sur la régime alimentaire des batraciens en basse Côte-d'Ivoire. *Bull. Soc. Zool. Fr.* 71:129–32.

Paulsen, K. 1965. Beiträge zur Anatomie und Physiologie des Froschkehlkopfes (mit besonderer Berücksichtingung der Stimmbildung). *Z. Wiss. Zool.* 172:1–16.

Pauly, G. B., D. M. Hillis, and D. C. Cannatella. 2004. The history of a nearctic colonization: Molecular phylogenetics and biogeography of the Nearctic toads (*Bufo*). *Evolution* 58:2517–35.

Pauwels, O. S. G. 2002. *Limnonectes limnocharis* (Paddy frog). Predation. *Herp. Rev.* 33:126.

Pavignano, I. 1990. Niche overlap in tadpole populations of *Pelobates fuscus insubricus* and *Hyla arborea* at a pond in northwestern Italy. *Boll. Zool.* 57:83–87.

Pawar, S. S., G. S. Rawat, and B. C. Choudhury. 2004. Recovery of frog and lizard communities following primary habitat alteration in Mizoram, Northeast India. *BMC Ecology* 4:10 (http://www.biomedcentral.com/1472-6785/4/10).

Payne, R. R. 1984. Sexual selection, lek and arena behavior, and sexual size dimorphism in birds. *Ornithol. Monogr.* 33:1–52.

Peacor, S. D., and E. E. Werner. 1997. Trait-mediated indirect interactions in a simple aquatic food web. *Ecology* 78:1146–56.

Pearl, C. A., M. J. Adams, R. B. Bury, and B. McCreary. 2004. Asymmetrical effects of introduced bullfrogs (*Rana catesbeiana*) on native ranid frogs in Oregon. *Copeia* 2004:11–20.

Pearl, C. A., M. J. Adams, N. Leuthold, and R. B. Bury. 2005. Amphibian occurrence and aquatic invaders in a changing landscape: Implications for wetlands mitigation in the Willamette Valley, Oregon, USA. *Wetlands* 25:76–88.

Pearl, C. A., M. Cervantes, M. Chan, U. Ho, R. Shoji, and E. O. Thomas. 2000. Evidence for a mate-attracting chemosignal in the dwarf African clawed frog *Hymenochirus. Horm. Behav.* 38:67–74.

Pearl, C. A., and M. P. Hayes. 2002. Predation by Oregon spotted frogs (*Rana pretiosa*) on western toads (*Bufo boreas*) in Oregon. *Am. Midl. Nat.* 147:145–52.

———. 2005. *Rana pretiosa* Baird and Girard, 1853. Oregon spotted frog. In *Amphibian declines: The conservation status of United States species,* ed. M. J. Lannoo, 577–80. Berkeley: University of California Press

Pearman, P. B. 1997. Correlates of amphibian diversity in an altered landscape of Amazonian Ecuador. *Cons. Biol.* 11:1211–25.

———. 2002. Interactions between *Ambystoma* salamander larvae: Evidence for competitive asymmetry. *Herpetologica* 58:156–65.

Pearman, P. B., and T. W. J. Garner. 2005. Susceptibility of Italian agile frog populations to an emerging strain of *Ranavirus* parallels population genetic diversity. *Ecol. Lett.* 8:401–8.

Pearson, O. P., and D. F. Bradford. 1976. Thermoregulation of lizards and toads at high altitudes in southern Peru. *Copeia* 1976:155–70.

Pearson, P. G. 1955. Population ecology of the spadefoot toad, *Scaphiopus h. holbrooki* (Harlan). *Ecol. Monogr.* 25:233–67.

———. 1957. Further notes on the population ecology of the spadefoot toad. *Ecology* 38:580–86.

Pechmann, J. H. K. 1989. Influence of wetland hydroperiod on diversity and abundance of metamorphosing juvenile amphibians. *Wetlands Ecol. Manage.* 1:3–11.

———. 2003. Natural population fluctuations and human influences: Null models and interactions. In *Amphibian conservation,* ed. R. D. Semlitsch, 85–93. Washington, DC: Smithsonian Books.

Pechmann, J. H. K., R. A. Estes, D. E. Scott, and J. W. Gibbons. 2001. Amphibian colonization and use of ponds created for trial mitigation of wetland loss. *Wetlands* 21:93–111.

Pechmann, J. H. K., and R. D. Semlitsch. 1986. Diel activity patterns in the breeding migrations of winter-breeding anurans. *Can. J. Zool.* 64:1116–1120.

Pechmann, J. H. K., D. E. Scott, J. W. Gibbons, and R. D. Semlitsch. 1989. Influence of wetland hydroperiod on diversity and abundance of metamorphosing juvenile amphibians. *Wetlands Ecol. Manage.* 1:3–11.

Pechmann, J. H. K., D. E. Scott, R. D. Semlitsch, J. P. Caldwell, L. J. Vitt and J. W. Gibbons. 1991. Declining amphibian populations: The problem of separating human impacts from natural fluctuations. *Science* 253:892–95.

Pechmann, J., and H. Wilbur. 1994. Putting declining amphibian populations in perspective: Natural fluctuations and human impacts. *Herpetologica* 50:65–84.

Pecio, A., and J. Rafinski. 1985. Sexual behavior of the Montandon's newt, *Triturus montandoni* (Boulenger) (Caudata: Salamandridae). *Amphibia-Reptilia* 6:11–22.

Peck, S. B., and B. L. Richardson. 1976. Feeding ecology of the salamander *Eurycea lucifuga* in the entrance, twilight zone, and dark zone of caves. *Ann. Speleol.* 31:175–82.

Peckarsky, B. L., B. W. Taylor, A. R. McIntosh, M. A. McPeek, and D. A. Lytle. 2001. Variation in mayfly size at metamorphosis as a developmental response to risk of predation. *Ecology* 82:740–57.

Pedersen, S. C. 1991. Dental morphology of the cannibal morph in the tiger salamander, *Ambystoma tigrinum. Amphibia-Reptilia* 12:1–14.

Pefaur, J. E., and W. E. Duellman. 1980. Community structure in high Andean herpetofaunas. *Trans. Kansas Acad. Sci.* 83:45–65.

Pehek, E. L. 1995. Competition, pH, and the ecology of larval *Hyla andersonii. Ecology* 76:1786–93.

Peixoto, O. L., U. Caramaschi, and E. M. X. Freire. 2003. Two new species of *Phyllodytes* (Anura: Hylidae) from the state of Alagoas, northeastern Brazil. *Herpetologica* 59:235–46.

Peltzer, P. M., and R. C. Lajmanovich. 2003. *Hyla puchella* (NCN). Predation. *Herp. Rev.* 34:231.

Peltzer, P. M., R. C. Lajmanovich, and A. H. Beltzer. 2003. The effects of habitat fragmentation on amphibian species richness in the floodplain of the Middle Parana River, Argentina. *Herpetol. J.* 13:95–98.

Pengilley, R. K. 1971a. Calling and associated behavior of some species of *Pseudophryne* (Anura: Leptodactylidae). *J. Zool., London* 163:73–92.

———. 1971b. The food of some Australian anurans (Amphibia). *J. Zool., London* 163:93–103.

Penna, M. 1997. Selectivity of evoked vocal responses in the time domain by frogs of the genus *Batrachyla. J. Herpetol.* 31:202–17.

Penna, M., W.-Y. Lin, and A. S. Feng. 1997. Temporal selectivity for complex signals by single neurons in the torus semicircularis of *Pleurodema thaul* (Amphibia: Leptodactylidae). *J. Comp. Physiol. A* 180:313–28.

Penna, M., P. M. Narins, and A. S. Feng. 2005. Thresholds for evoked vocal responses of *Eupsophus emiliopugini* (Amphibia, Leptodactylidae). *Herpetologica* 61:1–8.

Penna, M., H. Pottstock, and N. Velasquez. 2005. Effect of natural and synthetic noise on evoked vocal responses in a frog of the temperate austral forest. *Anim. Behav.* 70:639–51.

Penna, M., and R. Solis. 1996. Influence of burrow acoustics on sound reception by frogs *Eupsophus* (Leptodactylidae). *Anim. Behav.* 51:255–63.

———. 1998. Frog call intensities and sound propagation in the South American temperate forest region. *Behav. Ecol. Sociobiol.* 42:371–81.

———. 1999. Extent and variation of sound enhancement inside burrows of the frog *Eupsophus emiliopugini* (Leptodactylidae). *Behav. Ecol. Sociobiol.* 47:94–103.

Penna, M., and A. Veloso. 1981. Acoustical signals related to reproduction in the *spinulosus* species group of *Bufo* (Amphibia, Bufonidae). *Can. J. Zool.* 59:54–60.

———. 1982. The warning vibration of *Pleurodema thaul. J. Herpetol.* 16:408–10.

———. 1987. Vocalizations by Andean frogs of the genus *Telmatobius* (Leptodactylidae). *Herpetologica* 43:208–16.

———. 1990. Vocal diversity in frogs of the South American temperate forest. *J. Herpetol.* 24:23–33.

Penna, M., A. Veloso, and S. Contreras. 1983. Acoustical repertoires and morphological differences in the ear of two *Alsodes* species (Amphibia, Leptodactylidae). *Can. J. Zool.* 61:2369–76.

Pérez-Rivera, R. A. 1997. The importance of vertebrates in the diet of tanagers. *J. Field Ornithol.* 68:178–82.

Perkins, J. F., Jr. 1964. Historical development of respiratory physiology. In *Handbook of physiology. Section 3: Respiration,* ed. W. O. Fenn and H. Rahn, 1–63. Washington, DC: American Physiological Society.

Perks, A. M. 1987. The neurohypophysis: The place of comparative studies in biology and medicine. In *Vertebrate endocrinology: Fundamentals and biomedical implications, vol. 2. Regulation of water and electrolytes,* ed. P. K. T. Pang and M. P. Schreibman, 9–44. New York: Academic Press.

Perotti, M. G. 1997. Modos reproductivos y variables reproductivas cuantitatives de un ensamble de anuros del Chaco semiárido, Salta, Argentina. *Rev. Chil. Hist. Nat.* 70:277–88.

Perret, J.-L. 1962. La biologie d'*Acanthixalus spinosus* (Amphibia, Salientia). *Rech. Étud. Cameroun* 1:90–101.

———. 1966. Les amphibiens du Cameroun. *Zool. Jb. Syst.* 93:289–464.

Perret, N., R. Pradel, C. Miaud, O. Grolet, and P. Joly. 2003. Transience, dispersal and survival rates in newt patchy populations. *J. Anim. Ecol.* 72:567–75.

Perrill, S. A. 1984. Male mating behavior in *Hyla regilla. Copeia* 1984:727–32.

Perrill, S. A., and R. E. Daniel. 1983. Multiple egg clutches in *Hyla regilla, H. cinerea,* and *H. gratiosa. Copeia* 1983:513–16.

Perrill, S. A., H. C. Gerhardt, and R. Daniel. 1978. Sexual parasitism in the green tree frog (*Hyla cinerea*). *Science* 200:1179–80.

———. 1982. Mating strategy shifts in male green treefrogs (*Hyla cinerea*): An experimental study. *Anim. Behav.* 30:43–48.

Perrill, S. A., and M. Magier. 1988. Male mating behavior in *Acris crepitans. Copeia* 1988:245–48.

Perrill, S. A., and W. J. Shepherd. 1989. Spatial distribution and male-male communication in the northern cricket frog, *Acris crepitans blanchardi. J. Herpetol.* 23:237–43.

Perrins, C. M., and A. L. A. Middleton, eds. 1985. *The encyclopedia of birds.* New York: Facts on File.

Perrone, M., and T. M. Zaret. 1979. Parental care patterns in fishes. *Am. Nat.* 113:351–61.

Perry, A. L., P. J. Low, J. R. Ellis, and J. D. Reynolds. 2005. Climate change and distribution shifts in marine fishes. *Science* 308:1912–15.

Peters, S. E., and D. A. Aulner. 2000. Sexual dimorphism in forelimb muscles of the bullfrog, *Rana catesbeiana:* A functional analysis of isometric contractile properties. *J. Exp. Biol.* 203:3639–54.

Peters, S. E., L. T. Kamel, and D. P. Bashor. 1996. Hopping and swimming in the leopard frog, *Rana pipiens.* I. Step cycles and kinematics. *J. Morphol.* 230:1–16.

Peters, S. E., and K. C. Nishikawa. 1999. Comparison of isometric contractile properties of the tongue muscles in three species of frogs, *Litoria caerulea, Dyscophus guinetti,* and *Bufo marinus. J. Morphol.* 242:107–24.

Peterson, A. G., C. M. Bull, and L. M. Wheeler. 1992. Habitat choice and predator avoidance in tadpoles. *J. Herpetol.* 26:142–46.

Peterson, C. G., and A. J. Boulton. 1999. Stream permanence influences microalgal food availability to grazing tadpoles in arid-zone springs. *Oecologia* 118:340–52.

Peterson, C. H., and S. V. Andre. 1980. An experimental analysis of interspecific competition among marine filter feeders in a soft-sediment environment. *Ecology* 61:129–39.

Peterson, C. L., R. F. Wilkinson, D. Moll, and T. Holder. 1991. Premetamorphic survival of *Ambystoma annulatum. Herpetologica* 47:96–100.

Petersson, C. L., R. F. Wilkinson, Jr., M. S. Topping, and D. E. Metter. 1983. Age and growth of the Ozark hellbender (*Cryptobranchus alleganiensis bishopi*). *Copeia* 1983:225–31.

Peterson, G. S., L. B. Johnson, R. Axler, and S. A. Diamond. 2002. Assessment of the risk of solar ultraviolet radiation to amphibians. II. In situ characterization of exposure in amphibian habitats. *Environ. Sci. Technol.* 36:2859–65.

Peterson, J. A., and A. R. Blaustein. 1991. Unpalatability in anuran larvae as a defense against natural salamander predators. *Ethol. Ecol. Evol.* 3:63–72.

———. 1992. Relative palatabilities of anuran larvae to natural aquatic insect predators. *Copeia* 1992:577–84.

Pethiyagoda, R., and K. Manamendra-Arachchi. 1998. Evaluating Sri Lanka's amphibian diversity. *Occ. Pap. Wild. Her. Trust* (2):1–18.

Petranka, J., and L. Hayes. 1998. Chemically mediated avoidance of a predatory odonate (*Anax junius*) by American toad (*Bufo americanus*) and wood frog (*Rana sylvatica*) tadpoles. *Behav. Ecol. Sociobiol.* 42:263–71.

Petranka, J. W. 1979. The effects of severe winter weather on *Plethodon dorsalis* and *Plethodon richmondi* populations in central Kentucky. *J. Herpetol.* 13:369–71.

———. 1982a. Courtship behavior of the small-mouthed salamander (*Ambystoma texanum*) in central Kentucky. *Herpetologica* 38:333–36.

———. 1982b. Geographic variation in the mode of reproduction and larval characteristics of the small-mouthed salamander (*Ambystoma texanum* in the east-central United States. *Herpetologica* 38:475–85.

———. 1983. Fish predation: A factor affecting the spatial distribution of a stream-breeding salamander. *Copeia* 1983:624–28.

———. 1984a. Breeding migration, breeding season, clutch size, and oviposition of stream-breeding *Ambystoma texanum. J. Herpetol.* 18:106–12.

———. 1984b. Incubation, larval growth, and embryonic and larval survivorship of smallmouth salamanders (*Ambystoma texanum*) in streams. *Copeia* 1984:862–68.

———. 1984c. Ontogeny of the diet and feeding behavior of *Eurycea bislineata* larvae. *J. Herpetol.* 18:48–55.

———. 1984d. Sources of interpopulational variation in growth responses of larval salamanders. *Ecology* 65:1857–65.

———. 1985. Does age-specific mortality decrease with age in amphibian larvae? *Copeia* 1985:1080–83.

———. 1989a. Chemical interference competition in tadpoles: Does it occur outside laboratory aquaria? *Copeia* 1989:921–30.

———. 1989b. Density-dependent growth and survival of larval *Ambystoma:* Evidence from whole-pond manipulations. *Ecology* 70:1752–67.

———. 1989c. Response of toad tadpoles to conflicting chemical stimuli: Predator avoidance versus "optimal" foraging. *Herpetologica* 45:283–92.

———. 1990. Observations on nest site selection, nest desertion, and embryonic survival in marbled salamanders. *J. Herpetol.* 24:229–34.

———. 1994. Responses to impact of timber harvesting on salamanders. *Cons. Biol.* 8:302–4.

———. 1995. Interference competition in tadpoles: Are multiple agents involved? *Herpetol. J.* 5:206–7.

———. 1998. *Salamanders of the United States and Canada.* Washington, DC: Smithsonian Institution Press.

Petranka, J. W., M. P. Brannon, M. E. Hopey, and C. K. Smith. 1994. Effects of timber harvesting on low elevation populations of Southern Appalachian salamanders. *Forest Ecol. Manage.* 67:135–147.

Petranka, J. W., M. E. Eldridge, and K. E. Haley. 1993. Effects of timber harvesting on southern Appalachian salamanders. *Cons. Biol.* 7:363–70.

Petranka, J. W., M. E. Hopey, B. T. Jennings, S. D. Baird, and S. J. Boone. 1994. Breeding habitat segregation of wood frogs and American toads: The role of interspecific tadpole predation and adult choice. *Copeia* 1994:691–97.

Petranka, J. W., J. W. Just, and E. C. Crawford. 1982. Hatching of amphibian embryos: The physiological trigger. *Science* 217:257–59.

Petranka, J. W., L. B. Kats, and A. Sih. 1987. Predator-prey interactions among fish and larval amphibians: Use of chemical cues to detect predatory fish. *Anim. Behav.* 35:420–25.

Petranka, J. W., and C. A. Kennedy. 1999. Pond tadpoles with generalized morphology: Is it time to reconsider their functional role in aquatic communities? *Oecologia* 120:621–31.

Petranka, J. W., and S. S. Murray. 2001. Effectiveness of removal sampling for determining salamander density and biomass: A case study in an Appalachian streamside community. *J. Herpetol.* 35:36–44.

Petranka, J. W., S. S. Murray, and C. A. Kennedy. 2003. Responses of amphibians to restoration of a southern Appalachian wetland: perturbations confound post-restoration assessment. *Wetlands* 23:278–90.

Petranka, J. W., and J. G. Petranka. 1980. Selected aspects of the larval ecology of the marbled salamander *Ambystoma opacum* in the southern portion of its range. *Am. Midl. Nat.* 104:352–62.

———. 1981. On the evolution of nest site selection in the marbled salamander, *Ambystoma opacum. Copeia* 1981:387–91.

Petranka, J. W., A. W. Rushlow, and M. E. Hopey. 1998. Predation by tadpoles of *Rana sylvatica* on embryos of *Ambystoma maculatum:* Implications of ecological role reversals by *Rana* (predator) and *Ambystoma* (prey). *Herpetologica* 54:1–13.

Petranka, J. W., and A. Sih. 1986. Environmental instability, competition, and density-dependent growth and survivorship of a stream-dwelling salamander. *Ecology* 67:729–36.

———. 1987. Habitat duration, length of the larval period and the evolution of a complex life cycle in an amphibian. *Evolution* 41:1347–56.

Petranka, J. W., A. Sih, L. B. Kats, and J. R. Holomuzki. 1987. Stream drift, size-specific predation, and the evolution of ovum size in an amphibian. *Oecologia* 71:624–30.

Petranka, J. W., and C. K. Smith. 2005. A functional analysis of streamside habitat use by southern Appalachian salamanders: implications for riparian forest management. *Forest Ecol. Manage.* 210:443–54.

Petranka, J. W., and D. A. G. Thomas. 1995. Explosive breeding reduces egg and tadpole cannibalism in the wood frog, *Rana sylvatica. Anim. Behav.* 50:731–39.

Pettus, D., and G. M. Angleton. 1967. Comparative reproductive biology of montane and piedmont chorus frogs. *Evolution* 21:138–42.

Peuhkur, N., E. Ranta, and P. Seppä. 1997. Size-assortative schooling in free-ranging sticklebacks. *Ethology* 103:318–24.

Peuhkuri, N. 1997. Size-assortative shoaling in fish: The effect of oddity on foraging behaviour. *Anim. Behav.* 54:271–78.

Pezzo, F., and F. Morimando. 1995. Food habits of the barn owl, *Tyto alba,* in a Mediterranean rural area: Comparison with the diet of two sympatric carnivores. *Boll. Zool.* 62:369–73.

Pfeiffer, W. 1966. Die Verbreitung der Schreckreaktion bei Kaulquappen und die Herkunft des Schreckstoffes. *Z. Vergl. Physiol.* 52:79–98.

———. 1974. Pheromones in fish and amphibia. In *Pheromones,* ed. M. C. Birch, 269–96. New York: North Holland Publishing Company.

Pfennig, D. W. 1990a. The adaptive significance of an environmentally-cued developmental switch in an anuran tadpole. *Oecologia* 85:101–7.

———. 1990b. "Kin recognition" among spadefoot toad tadpoles: A side-effect of habitat selection. *Evolution* 44:785–98.

———. 1992a. Polyphenism in spadefoot toad tadpoles as a locally adjusted evolutionarily stable strategy. *Evolution* 46:1408–20.

———. 1992b. Proximate and functional causes of polyphenism in an anuran tadpole. *Funct. Ecol.* 6:167–74.

———. 1997. Kinship and cannibalism. *BioScience* 47:667–75.

———. 1999. Cannibalistic tadpoles that pose the greatest threat to kin are most likely to discriminate kin. *Proc. Roy. Soc. London B* 266:57–61.

Pfennig, D. W., and J. P. Collins. 1993. Kinship affects morphogenesis in cannibalistic salamanders. *Nature* 362:836–38.

Pfennig, D. W., J. P. Collins, and R. E. Ziemba. 1999. A test of alternative hypotheses for kin recognition in cannibalistic tiger salamanders. *Behav. Ecol.* 10:436–43

Pfennig, D. W., and W. A. Frankino. 1997. Kin-mediated morphogenesis in facultatively cannibalistic tadpoles. *Evolution* 51:1993–99.

Pfennig, D. W., S. G. Ho, and E. A. Hoffman. 1998. Pathogen transmission as a selective force against cannibalism. *Anim. Behav.* 55:1255–61.

Pfennig, D. W., M. L. G. Loeb, and J. P. Collins. 1991. Pathogens as a factor limiting the spread of cannibalism in tiger salamanders. *Oecologia* 88:161–66.

Pfennig, D. W., A. Mabry, and D. Orange. 1991. Environmental causes of correlations between age and size at metamorphosis in *Scaphiopus multiplicatus. Ecology* 72:2240–48.

Pfennig, D. W., and P. J. Murphy. 2000. Character displacement in polyphenic tadpoles. *Evolution* 54:1738–49.

———. 2002. How fluctuating competition and phenotypic plasticity mediate species divergence. *Evolution* 56:1217–28.

Pfennig, D. W., H. K. Reeve, and P. W. Sherman. 1993. Kin recognition and cannibalism in spadefoot toad tadpoles. *Anim. Behav.* 46:87–94.

Pfennig, D. W., P. W. Sherman, and J. P. Collins. 1994. Kin recognition and cannibalism in polyphenic salamanders. *Behav. Ecol.* 5:225–32.

Pfennig, K. S. 1998. The evolution of mate choice and the potential for conflict between species and mate-quality recognition. *Proc. Roy. Soc. London B* 265:1743–48.

———. 2000. Female spadefoot toads compromise on mate quality to ensure conspecific matings. *Behav. Ecol.* 11:220–27.

———. 2003. A test of alternative hypotheses for the evolution of reproductive isolation between spadefoot toads: Support for the reinforcement hypothesis. *Evolution* 57:2842–51.

Pfennig, K. S., and M. A. Simovich. 2002. Differential selection to avoid hybridization in two toad species. *Evolution* 56:1840–48.

Pfennig, K. S., and R. C. Tinsley. 2002. Different mate preferences by parasitized and unparasitized females potentially reduces sexual selection. *J. Evol. Biol.* 15:399–406.

Pfingsten, R. A., and F. L. Downs, eds. 1989. Salamanders of Ohio. *Ohio Biol. Survey Bull.* n.s. 7 (2):1–315.

Pflaumer, C. 1936. Observaciónes biológicas acerca de la *Rhinoderma darwinii* D. & B. *Rev. Chilena Hist. Nat. Pura Apl.* 39:28–30.

Philibosian, R., R. Ruibal, V. H. Shoemaker, and L. McClanahan. 1974. Nesting behavior and early larval life of the frog *Leptodactylus bufonius. Herpetologica* 30:381–86.

Phillips, B. L., G. P. Brown, and R. Shine. 2003. Assessing the potential impact of cane toads on Australian snakes. *Cons. Biol.* 17:1738–47.

Phillips, B. L., and R. Shine. 2004. Adapting to an invasive species: Toxic cane toads induce morphological change in Australian snakes. *Proc. Natl. Acad. Sci. USA USA* 101:17150–55.

Phillips, C. A., and O. J. Sexton. 1989. Orientation and sexual differences during breeding migrations of the spotted salamander, *Ambystoma maculatum. Copeia* 1989:17–22.

Phillips, J. B. 1977. Use of the earth's magnetic field by orienting cave salamanders (*Eurycea lucifuga*). *J. Comp. Physiol.* 121:273–88.

———. 1986a. Magnetic compass orientation in the eastern red-spotted newt. *J. Comp. Physiol. A* 158:103–9.

———. 1986b. Two magnetoreception pathways in a migratory salamander. *Science* 233:765–67.

———. 1987. Laboratory studies of homing orientation in the eastern red-spotted newt, *Notophthalmus viridescens. J. Exp. Biol.* 131:215–29.

———. 1996. Magnetic navigation. *J. Theor. Biol.* 180:309–19.

Phillips, J. B., and K. Adler. 1978. Directional and discriminatory responses of salamanders to weak magnetic fields. In *Animal migration, navigation, and homing,* ed. K. Schmidt-Koenig and W. T. Keeton, 325–33. New York: Springer-Verlag.

Phillips, J. B., K. Adler, and S. C. Borland. 1995. True navigation by an amphibian. *Anim. Behav.* 50:855–58.

Phillips, J. B. and S. C. Borland. 1992a. Behavioral evidence for use of a light-dependent magnetoreception mechanism by a vertebrate. *Nature* 359:142–44.

———. 1992b. Magnetic compass orientation is eliminated under near-infrared light in the eastern red-spotted newt *Notophthalmus viridescens. Anim. Behav.* 44:796–97.

———. 1992c. Wavelength specific effects of light on magnetic compass orientation of the eastern red-spotted newt *Notophthalmus viridescens. Ethol. Ecol. Evol.* 4:33–42.

———. 1994. Use of a specialized magnetoreception system for homing by the eastern red-spotted newt *Notophthalmus viridescens. J. Exp. Biol.* 188:275–91.

Phillips, J. B., S. C. Borland, M. J. Freake, J. Brassart, and J. L. Kirschvink. 2002. 'Fixed-axis' magnetic orientation by an amphibian: Non-shoreward-directed compass orientation, misdirected homing or positioning a magnetic-based map detector in a consistent alignment relative to the magnetic field? *J. Exp. Biol.* 205:3903–14.

Phillips, J. B., M. E. Deutschlander, M. J. Freake, and S. C. Borland. 2001. The role of extraocular photoreceptors in newt magnetic compass orientation: Parallels between light-dependent magnetoreception and polarized light detection in vertebrates. *J. Exp. Biol.* 204:2543–52.

Phillips, K. 1994. *Tracking the vanishing frogs.* New York: St. Martin's Press.

Phoenix, G. K., W. K. Hicks, S. Cinderby, J. C. I. Kulenstierna, W. D. Stock, F. J. Dentener, K. E. Giller, A. T. Austin, R. D. B. Lefroy, B. S. Gimeno, M. R. Ashmore, and P. Ineson. 2006. Atmospheric nitrogen deposition in world biodiversity hotspots: The need for a greater global perspective in assessing deposition impacts. *Global Change Biol.* 12:470–76.

Picado, T. C. 1913. Les bromeliaces epiphytes considereis comme une milieu biologique. *Bull. Sci. Fr. Belg.* 5:215–360.

Picard, A. L. 2005. Courtship in the zig-zag salamander (*Plethodon dorsalis*): insights into a transition in pheromone-delivery behavior. *Ethology* 111:799–809.

Picariello, O. 1993. Dati preliminari riguardanti l'impatto della captazioni idriche e la batracofauna alloctona sulle popu-lazioni di anfibi in Campania. *Quad. Civica Staz. Idrobiol. Milano* 20:95–100.

Picker, M. 1983. Territorial behaviour in the bullfrog. *Afr. Wildl.* 37:238–39.

Picker, M. D. 1980. *Xenopus laevis* (Anura: Pipidae) mating systems—A preliminary synthesis with some data on the female phonoresponse. *S. Afr. J. Zool.* 15:150–58.

———. 1983. Hormonal induction of the aquatic phonotactic response of *Xenopus. Behaviour* 84:74–90.

Picker, M. D., J. A. Harrison, and D. Wallace. 1996. Natural hybridization between *Xenopus laevis laevis* and *X. gilli* in the south-western Cape Province, South Africa. In *The biology of Xenopus,* ed. R. C. Tinsley and H. R. Kobel, 61–71. New York: Oxford University Press.

Picker, M. D., and A. L. de Villiers. 1989. The distribution and conservation status of *Xenopus gilli* (Anura: Pipidae). *Biol. Cons.* 49:169–83.

Picker, M. D., C. J. McKenzie, and P. Fielding. 1993. Embryonic tolerance of *Xenopus* (Anura) to acidic blackwater. *Copeia* 1993:1072–81.

Pierce, B. A. 1985. Acid tolerance in amphibians. *BioScience* 35:239–43.

Pierce, B. A., J. B. Mitton, L. Jacobson, and F. L. Rose. 1983. Head shape and size in cannibal and noncannibal larvae of the tiger salamander from West Texas. *Copeia* 1983:1006–12.

Pierce, B. A., J. B. Mitton, and F. L. Rose. 1981. Allozyme variation among large, small and cannibal morphs of the tiger salamander inhabiting the Llano Estacado of West Texas. *Copeia* 1981:590–95.

Pierce, B. A., and H. M. Smith. 1979. Neoteny or paedogenesis? *J. Herpetol.* 13:119–21.

———. 1980. On the use of the term paedomorphosis. *J. Herpetol.* 14:81–82.

Pierce, B. A., and D. K. Wooten. 1992. Genetic variation in tolerance of amphibians to low pH. *J. Herpetol.* 26:422–29.

Pierce, J. R., and D. B. Ralin. 1972. Vocalizations and behavior of the males of three species in the *Hyla versicolor* complex. *Herpetologica* 28:329–37.

Pietrewicz, A. T., and A. C. Kamil. 1981. Search images and the detection of cryptic prey: An operant approach. In *Foraging behavior,* ed. A. C. Kamil and T.D. Sargent, 311–31. New York: Garland STPM Press.

Piiper, J. 1988. Models for cutaneous gas exchange and transport. *Amer. Zool.* 28:963–72

Piiper, J., R. N. Gatz, and E. C. Crawford. 1976. Gas transport characteristics in an exclusively skin-breathing salamander, *Desmognathus fuscus.* In *Respiration in amphibious vertebrates,* ed. G. M. Hughes, 339–56. New York: Academic Press.

Pikulik, M. M., V. E. Sidorovich, B. Jędrzejewska, and W. Jędrzejewski. 2001. Summer abundance and habitat distribution of frogs (*Rana temporaria, R. arvalis, R. kl. esculenta*) and toads (*Bufo bufo*) in the Białowieża Primeval Forest, E. Poland. *Folia Zool.* 50:65–73.

Pilliod, D. S. 1999. *Rana luteiventris* (Columbia spotted frog). Cannibalism. *Herp. Rev.* 30:93.

Pilliod, D. S., R. B. Bury, E. J. Hyde, C. A. Pearl, and P. S. Corn. 2003. Fire and amphibians in North America. *Forest Ecol. Manage.* 178:163–81.

Pilliod, D. S., and C. R. Peterson. 2001. Local and landscape effects of introduced trout on amphibians in historically fishless watersheds. *Ecosystems* 4:322–33.

Pilliod, D. S., C. R. Peterson, and P. I. Ritson. 2002. Seasonal migration of Columbia spotted frogs (*Rana luteiventris*) among complementary resources in a high mountain basin. *Can. J. Zool.* 80:1849–62.

Pimenta, B. V. S., C. F. B. Haddad, L. B. Nascimento, C. A. Gonçalves Cruz, and J. P. Pombal, Jr. 2005. Comment on "Status and trends of amphibian declines worldwide." Science 309:1999b.

Pimentel, D. 1955. Biology of the Indian mongoose in Puerto Rico. *J. Mammal.* 36:62–68.

Pinder, A. C., and A. R. Palmer. 1983. Mechanical properties of the frog ear: Vibration measurements under free and closed-field acoustic conditions. *Proc. Roy. Soc. London B* 219:371–96.

Pinder, A. W. 1987. Cutaneous diffusing capacity increases during hypoxia in cold, submerged bullfrogs (*Rana catesbeiana*). *Respir. Physiol.* 70:85–95.

Pinder, A. W., and W. W. Burggren. 1983. Respiration during chronic hypoxia and hyperoxia in larval and adult bullfrogs (*Rana catesbeiana*). II. Changes in respiratory properties of whole blood. *J. Exp. Biol.* 105:205–13.

———. 1986. Ventilation and partitioning of oxygen uptake in the frog *Rana pipiens:* Effects of hypoxia and activity. *J. Exp. Biol.* 126:453–68.

Pinder, A. W., and M. E. Feder. 1990. Effect of boundary layers on cutaneous gas exchange. *J. Exp. Biol.* 154:67–80.

Pinder, A. W., and S. C. Friet. 1994. Oxygen transport in egg masses of the amphibians *Rana sylvatica* and *Ambystoma maculatum:* Convection, diffusion and oxygen production by algae. *J. Exp. Biol.* 197:17–30.

Pinder, A. W., K. B. Storey, and G. R. Ultsch. 1992. Estivation and hibernation. In *Environmental physiology of the amphibians,* ed. M. E. Feder and W. W. Burggren, 250–74. Chicago: University of Chicago Press.

Pineda, E. 2003. *Hyla miotympanum* (small-eared treefrog). Predation. *Herp. Rev.* 34:136–37.

Pineda, E., and G. Halffter. 2004. Species diversity and habitat fragmentation: Frogs in a tropical montane landscape in Mexico. *Biol. Cons.* 117:499–508.

Pineda, E., C. Moreno, F. Escobar, and G. Halffter. 2005. Frog, bat, and dung beetle diversity in the cloud forests and coffee agrosystems of Veracruz, Mexico. *Cons. Biol.* 19:400–10.

Pisanó, A. 1957a. Factores de especialización endocrino-sexual en Anfibios de alta montaña y de otras regiones del Norte Argentino. *An. Dep. Invest. Cient. Univ. Nac. Cuyo, Mendoza* 2:1–27.

———. 1957b. Studio comparativo della correlazione endocrina in Anfibi argentini di alta montagna e pianura con spermatogenesi continua. *Arch. Zool. Ital.* 42:161–84.

———. 1958. Características histológicas y correlaciones endócrinas en la hipófisis, tiroides y gónades de *Pleurodema tucumana. Arch. Biol. Quim. Farm. Tucumán* 8:261–70.

———. 1962. La espuma de los nidos de *Leptodactylus bufonius* y su acción inhibidora sobre el desarrollo de las larvas. *Arch. Biol. Quim. Farm. Tucumán* 10:65–77.

Pitcher, T. J., and J. K. Parrish. 1993. Functions of shoaling behaviour in teleosts. In *The behavior of teleost fishes* (2nd ed.), ed. T. J. Pitcher, 363–439. New York: Chapman and Hall.

Pitcher, T. J., A. E. Magurran, and J. R. Allan. 1986. Size-segregative behaviour in minnow shoals. *J. Fish. Biol.* (Suppl. A) 29:83–95.

Pitcher, T. J., A. E. Magurran, and J. I. Edwards. 1985. Schooling mackerel and herring choose neighbors of similar size. *Mar. Biol.* 86:319–22.

Plasa, L. 1979. Heimfindeverhalten bei *Salamandra salamandra* (L.). *Z. Tierpsychol.* 51:113–25.

Platt, D. R. 1969. Natural history of the hognose snakes *Heterodon platyrhinos* and *Heterodon nasicus. Univ. Kansas Publ. Mus. Nat. Hist.* 18:253–420.

Platz, J. E. 1976. Biochemical and morphological variation of leopard frogs in Arizona. *Copeia* 1976:660–72.

———. 1993. *Rana subaquavocalis,* a remarkable new species of leopard frog (*Rana pipiens* complex) from southeastern Arizona that calls under water. *J. Herpetol.* 27:154–62.

Platz, J. E., A. Lathrop, L. Hofbauer, M. Vradenburg. 1997. Age distribution and longevity in the Ramsey Canyon leopard frog, *Rana subaquavocalis. J. Herpetol.* 31:552–57.

Plénet, S., A. Pagano, P. Joly, and P. Fouillet. 2000. Variation of plastic responses to oxygen availability within the hybridogenetic *Rana esculenta* complex. *J. Evol. Biol.* 13:20–28.

Plytycz, B., and J. Bigaj. 1984. Preliminary studies on the growth and movements of the yellow-bellied toad, *Bombina variegata* (Anura: Discoglossidae). *Amphibia-Reptilia* 5:81–86.

———. 1993. Studies on the growth and longevity of the yellow-bellied toad, *Bombina bombina,* in natural environments. *Amphibia-Reptilia* 14:35–44.

Płytycz, B., J. Dulak, and A. Pecio. 1984. Genetic control of length of the larval period in *Rana temporaria. Folia Biol.* 32:155–66.

Poczopko, P. 1957. Further investigations on the cutaneous vasomotor reflexes in the edible frog in connection with the problem of regulation of the cutaneous respiration in frogs. *Zool. Pol.* 8:161–75.

Podloucky, R. 1987. Zur Verbreitung und Bestandssituation des Moorfrosches (*Rana arvalis* Nilsson) in Niedersachsen. *Beih. Shriftenr. Naturschutz Landsschaftspfl. Niedersachsen* 19:15–27.

———. 1989. Protection of amphibians on roads—Examples and experiences from Lower Saxony. In *Amphibians and roads,* ed. T. E. S. Langton, 15–28. Shefford, Bedfordshire, England: ACO Polymer Products.

Poff, N. L., M. M. Brinson, and J. W. Day, Jr. 2002. *Aquatic ecosystems and global climate change. Potential impacts on inland freshwater and coastal wetland ecosystems in the United States.* Report Prepared for the Pew Center on Global Climate Change. Arlington, VA: Pew Center.

Polakow, D. A., P. R. Y. Backwell, N. Caithness, and M. D. Jennions. 1995. Stabilizing or directional selection in signaling systems: investigations in a population of painted reed frogs, *Hyperolius marmoratus. S. Afr. J. Sci.* 91:270–73.

Polder, W. N. 1974a. Dendrobatidae IV. Over verzorging en voortplanting in gevangenshap van *Dendrobates azureus* en enkele andere Dendrobatidae. *Het Aquar.* 45:122–28.

———. 1974b. Pflege und Fortpflanzung von *Dendrobates azureus* und anderer Dendrobatiden. II. *Deutsch. Aquar. Terr. Z.* 27:28–32.

———. 1974c. Pflege und Fortpflanzung von *Dendrobates azureus* und anderer Dendrobatiden. III. *Deutsch. Aquar. Terr. Z.* 27:244–49.

———. 1975. Dendrobatidae V. Nogmaals *Dendrobates histrionicus. Het Aquar.* 45:304–7.

———. 1976a. *Dendrobates, Phyllobates* en *Colostethus. Het Aquar.* 46:260–66.

———. 1976b. Dendrobatidae VII (slot). *Het Aquar.* 46:318–23.

Polis, G. A. 1981. The evolution and dynamics of intraspecific predation. *Ann. Rev. Ecol. Syst.* 12:225–51.

Polis, G. A., and C. A. Myers. 1985. A survey of intraspecific predation among reptiles and amphibians. *J. Herpetol.* 19:99–107.

Pollet, I., and L. I. Bendell-Young. 2000. Amphibians as indicators of wetland quality in wetlands formed from oil sands effluent. *Environ. Toxicol. Chem.* 13:2589–97.

Pombal, J. P., Jr. 1999. Oviposição e desenvolviment o de *Brachycephals ephippium* (Anura, Brachycephalidae). *Rev. Bras. Zool.* 16:967–76.

Pombal, J. P., Jr., and C. F. B. Haddad. 1993. *Hyla luctuosa,* a new treefrog from southeastern Brazil (Amphibia: Hylidae). *Herpetologica* 49:16–21.

———. 1999. Frogs of the genus *Paratelmatobius* (Anura: Leptodactylidae) with descriptions of two new species. *Copeia* 1999:1014–26.

Pombal, J. P., Jr., I. Sazima, and C. F. B. Haddad. 1994. Breeding behavior of the pumpkin toadlet, *Brachycephalus ephippium* (Brachycephalidae). *J. Herpetol.* 28:516–19.

Pomeroy, L. V. 1981. *Characteristics and determinants of polymorphism in the larvae of the spadefoot toad, Scaphiopus spp.* Unpublished PhD dissertation. Riverside: University of California.

Ponssa, M. L. 2001. Cuidado parental y comportamiento de cardumen de larvas en *Leptodactylus insuralum* (Anuran, Leptodactylidae). *Alytes* 19:183–95.

Pool, T. B., and T. R. Hoage. 1973. The ultrastructure of secretion in the spermatheca of the salamander, *Manculus quadridigitatus* (Holbrook). *Tissue and Cell* 5:303–13.

Pool, T. R., and J. N. Dent. 1977. The ultrastructure and the hormonal control of product sysnthesis in the hedonic glands of the red-spotted newt *Notophthalmus viridescens. J. Exp. Zool.* 201:177–202.

Pool, T. R., J. N. Dent, and K. Kemphues. 1977. Neural regulation of product discharge from the hedonic glands of the red-spotted newt, *Notophthalmus viridescens. J. Exp. Zool.* 201:203–20.

Pope, C. H. 1924. Notes on North Carolina salamanders with especial reference to the egg-laying habits of *Leurognathus* and *Desmognathus. Amer. Mus. Novitates* (153):1–15.

———. 1931. Notes on amphibians from Fukien, Hainan, and other parts of China. *Bull. Amer. Mus. Nat. Hist.* 61:397–611.

———. 1935. *The reptiles of China.* New York: American Museum of Natural History.

Pope, K. L. 1999. *Rana muscosa* (Mountain Yellow-legged Frog). Diet. *Herp. Rev.* 30:163–64.

Pope, P. H. 1924. The life-history of the common water newt (*Notophthalmus viridescens*), together with observations on the sense of smell. *Ann. Carnegie Mus.* 15:305–62.

Pope, S. E., L. Fahrig, and H. G. Merriam. 2000. Landscape complementation and metapopulation effects on leopard frog populations. *Ecology* 81:2498–2508.

Popp, J. W., and R. W. Ficken. 1987. Effects of non-specific singing on the song of the ovenbird. *Bird Behav.* 7:22–26.

Popp, J. W., R. W. Ficken, and J. A. Reinartz. 1985. Short-term temporal avoidance of interspecific acoustic interference among forest birds. *Auk* 102:744–48.

Porej, D., M, Micacchion, and T. E. Hetherington. 2004. Core terrestrial habitat for conservation of local populations of salamanders and wood frogs in agricultural landscapes. *Biol. Cons.* 120:403–13.

Porter, K. R. 1972. *Herpetology.* W. B. Philadelphia: Saunders.

Portner, H. O., L. M. MacLatchy, and D. P. Toews. 1991. Metabolic responses of the toad *Bufo marinus* to environmental hypoxia: An analysis of the critical $PO_2$. *Physiol. Zool.* 64:836–49.

Portnoy, J. W. and W. E. Dodge. 1979. Red-shouldered hawk nesting ecology and behavior. *Wilson Bull.* 91:104–17.

Potter, F. E., and G. Rabb. 1960. Thyroxin-induced metamorphosis in a neotenic salamander, *Eurycea nana* Bishop. *Zool. Ser., A. & M. College of Texas* 1:1–12.

Potter, F. E., Jr., and S. S. Sweet. 1981. Generic boundaries in Texas cave salamanders, and a redescription of *Typhlomolge robusta* (Amphibia: Plethodontidae). *Copeia* 1981:64–75.

Potts, W. T. W. 1954. The energetics of osmotic regulation in brackish and freshwater animals. *J. Exp. Biol.* 31:618–30.

Pough, F. H. 1973. Lizard energetics and diet. *Ecology* 54:837–44.

———. 1974a. Comments on the presumed mimicry of red efts (*Notophthalmus*) by red salamanders (*Pseudotriton*). *Herpetologica* 30:24–27.

———. 1974b. Natural daily temperature acclimation of eastern red efts, *Notophthalmus v. viridescens* (Rafinesque) (Amphibia, Caudata). *Comp. Biochem. Physiol.* 47A:71–78.

———. 1976. Acid precipitation and embryonic mortality of spotted salamanders, *Ambystoma maculatum. Science* 192:68–70.

———. 1980. The advantages of ectothermy for tetrapods. *Am. Nat.* 115:92–112.

———. 1983. Amphibians and reptiles and low-energy systems. In *Behavioral energetics: The cost of survival in vertebrates,* ed. W. P. Aspey and S. I. Lustick, 141–88. Columbus: Ohio State University Press.

———. 1988. Mimicry of vertebrates: Are the rules different? *Am. Nat.* 131:S67–S102.

———. 1989. Organismal performance and Darwinian fitness: Approaches and interpretations. *Physiol. Zool.* 62:199–236.

Pough, F. H., R. M. Andrews, J. E. Cadle, M. L. Crump, A. H. Savitzky, and K. D. Wells. 2004. *Herpetology* (3rd ed.). Upper Saddle River, NJ: Prentice Hall.

Pough, F. H., and R. E. Gatten, Jr. 1984. The use of anaerobic metabolism by frogs in a breeding chorus. *Comp. Biochem. Physiol.* 78A:337–40.

Pough, F. H., and J. D. Groves. 1983. Specialization of the body form and food habits of snakes. *Amer. Zool.* 23:443–54.

Pough, F. H., J. B. Heiser, and W. N. McFarland. 1996. *Vertebrate life.* (4th ed.) Upper Saddle River, NJ: Prentice Hall.

Pough, F. H., C. M. Janis, and J. B. Heiser. 2001. *Vertebrate life.* (6th ed.). Upper Saddle River, NJ: Prentice Hall.

Pough, F. H., and S. Kamel. 1984. Post-metamorphic change in activity metabolism in relation to the life histories of anurans. *Oecologia* 65:138–44.

Pough, F. H., W. E. Magnusson, M. J. Ryan, K. D. Wells, and T. L. Taigen 1992. Behavioral energetics. In *Environmental physiology of the amphibians,* ed. M. E. Feder and W. W. Burggren, 395–436. Chicago: University of Chicago Press.

Pough, F. H., E. M. Smith, D. H. Rhodes, and A. Collazo. 1987. The abundance of salamanders in forest stands with different histories of disturbance. *Forest Ecol. Manage.* 20:1–9.

Pough, F. H., M. M. Stewart, and R. G. Thomas. 1977. Physiological basis of habitat partitioning in Jamaican *Eleutherodactylus. Oecologia* 27:285–93.

Pough, F. H., and T. L. Taigen. 1990. Metabolic correlates of the foraging and social behaviour of dart-poison frogs. *Anim. Behav.* 39:145–55.

Pough, F. H., T. L. Taigen, M. M. Stewart, and P. F. Brussard. 1983. Behavioral modification of evaporative water loss by a Puerto Rican frog. *Ecology* 64:244–52.

Pough, F. H., and R. E. Wilson. 1970. Natural daily temperature stress, dehydration, and acclimation in juvenile *Ambystoma maculatum* (Shaw) (Amphibia: Caudata). *Physiol. Zool.* 43:194–205.

———. 1977. Acid precipitation and reproductive success of *Ambystoma* salamanders. *Water Air Soil Pollut.* 7:531–44.

Poulin, B., G. Lefebvre, R. Ibáñez, C. Jaramillo, C. Hernández, and A. S. Rand. 2001. Avian predation upon lizards and frogs in a Neotropical forest understory. *J. Trop. Ecol.* 17:21–40.

Pounds, J. A. 2001. Climate and amphibian declines. *Nature* 410:639–40.

Pounds, J. A., M. R. Bustamante, L. A. Coloma, J. A. Consuegra, M. P. L. Fogden, P. N. Foster, E. La Marca, K. L. Masters, A. Merion-Viteri, R. Puschendorf, S. R. Ron, G. A. Sánchez-Azofeifa, C. J. Still, and B. E. Young. 2006. Widespread amphibian extinctions from epidemic disease driven by global warming. *Nature* 439:161–67.

Pounds, J. A., and M. L. Crump. 1987. Harlequin frogs along a tropical montane stream: Aggregation and the risk of predation by frog-eating flies. *Biotropica* 19:306–9.

———. 1994. Amphibian declines and climate disturbance: The

case of the golden toad and the harlequin frog. *Cons. Biol.* 8:72–85.

Pounds, J. A., M. P. Fogden, and J. H. Campbell. 1999. Biological responses to climate change on a tropical mountain. *Nature* 398:611–15.

Pounds, J. A., M. P. Fogden, J. M. Savage and G. C. Gorman. 1997. Tests of null models for amphibian declines on a tropical mountain. *Cons. Biol.* 11:1307–22.

Pounds, J. A., and R. Puschendorf. 2004. Clouded futures. *Nature* 427:107–9.

Pourbagher, N. 1969. Etude expérimentale de l'effet de groupe chez les têtards de batraciens. *Rev. Comport. Anim.* 3:75–119.

Powders, V. N. 1973. Cannibalism by the slimy salamander, *Plethodon glutinosus* in eastern Tennessee. *J. Herpetol.* 7:139–40.

Powders, V. N., and W. L. Tietjen. 1974. The comparative food habits of sympatric and allopatric salamanders, *Plethodon glutinosus* and *Plethodon jordani* in eastern Tennessee and adjacent areas. *Herpetologica* 30:167–75.

Powell, M. K., J. Mansfield-Jones, and R. E. Gatten. 1999. Specific dynamic effect in the horned frog *Ceratophrys cranwelli*. *Copeia* 1999:710–17.

Powell, R., L. W. Little, and D. D. Smith. 1984. Eine Wohngemeinschaft von *Physalaemus pustulosus* (Cope, 1864) (Salientia: Leptodactylidae) mit einer bodenbewohnenden Vogelspinne. *Salamandra* 20:273–74.

Powers, J. H. 1903. The causes of acceleration and retardation in the metamorphosis of *Ambystoma tigrinum*, a preliminary report. *Am. Nat.* 37:385–410.

———. 1907. Morphological variation and its causes in *Ambystoma tigrinum*. *Stud. Univ. Nebraska* 7:197–274.

Poynton, J. C. 1957. Bullfrog guardians. *Afr. Wildlife* 11:80.

———. 1964a. The Amphibia of southern Africa. *Ann. Natal Mus.* 17:1–334.

———. 1964b. Relationships between habitat and terrestrial breeding in amphibians. *Evolution* 18:131.

———. 1998. Tanzanian bufonid diversity: preliminary findings. *Herpetol. J.* 8:3–6.

———. 1999. Distribution patterns of amphibians in sub-Saharan Africa, Madagascar, and Seychelles. In *Patterns of distribution of amphibians. A global perspective,* ed. W. E. Duellman, 483–539. Baltimore: Johns Hopkins University Press.

———. 2003. Altitudinal species turnover in southern Tanzania shown by anurans: Some zoogeographical considerations. *Syst. Biodiver.* 1:117–26.

Poynton, J. C., and S. Pritchard. 1976. Notes on the biology of *Breviceps* (Anura: Microhylidae). *Zool. Afr.* 11:313–18.

Praderio, M. J., and M. D. Robinson. 1990. Reproduction in the toad *Colostethus trinitatus* (Anura: Dendrobatidae) in a northern Venezuela seasonal environment. *J. Trop. Ecol.* 6:333–41.

Prado, C. P. A., and C. F. B. Haddad. 2003. Testes size in leptodactylid frogs and occurrence of multimale spawning in the genus *Leptodactylus* in Brazil. *J. Herpetol.* 37:354–62.

Prado, C. P. A., M. Uetanabaro, and C. F. B. Haddad. 2002. Description of a new reproductive mode in *Leptodactylus* (Anura, Leptodactylidae), with a review of the reproductive specializations toward terrestriality in the genus. *Copeia* 2002:1128–33.

———. 2005. Breeding activity patterns, reproductive modes, and habitat use by anurans (Amphibia) in a seasonal environment in the Pantanal, Brazil. *Amphibia-Reptilia* 26:211–21.

Prado, C. P. A., M. Uetanabaro, and F. S. Lopes. 2000. Reproductive strategies of *Leptodactylus chaquensis* and *L. podicipinus* in the Pantanal, Brazil. *J. Herpetol.* 34:135–39.

Prado, G. M., and J. H. Borgo. 2003. *Scinax altera* (NCN). Predation. *Herp. Rev.* 34:238–39.

Preest, M. R., D. G. Brust, and M. L. Wygoda. 1992. Cutaneous water loss and the effects of temperature and hydration state on aerobic metabolism of canyon treefrogs, *Hyla arenicolor. Herpetologica* 48:210–19.

Preest, M. R., and F. H. Pough. 2003. Effects of body temperature and hydration state on organismal performance of toads, *Bufo americanus*. *Physiol. Biochem. Zool.* 76:229–39.

Premo, D. B., and A. H. Atmowidjojo. 1987. Dietary patterns of the "crab-eating frog," *Rana cancrivora,* in west Java. *Herpetologica* 43:1–6.

Prendergast, J. R., R. M. Quinn, J. H. Lawton, B. C. Eversham, and D. W. Gibbons. 1993. Rare species, the coincidence of diversity hotspots and conservation strategies. *Nature* 365:335–37.

Prescott, D. R. C. 1985. Probable predation by a tufted titmouse on a salamander. *J. Field Ornithol.* 56:418–19.

Preslar, A. J., III, and V. H. Hutchison. 1978. Energetics for activity in the salamander *Amphiuma tridactylum*. *J. Comp. Physiol. B* 128:139–46.

Presswell, B., D. J. Gower, O. V. Oomen, G. J. Measey, and M. Wilkinson. 2002. Scolecophidian snakes in the diets of south Asian caecilian amphibians. *Herpetol. J.* 12: 123–26.

Prestwich, K. N. 1994. The energetics of acoustic signaling in anurans and insects. *Amer. Zool.* 34:625–43.

Prestwich, K. N., K. E. Brugger, and M. Topping. 1989. Energy and communication in three species of hylid frogs: Power input, power output, and efficiency. *J. Exp. Biol.* 144:53–80.

Prestwich, K. N., and K. O'Sullivan. 2005. Simultaneous measurement of metabolic and acoustic power and the efficiency of sound production in two mole cricket species (Orthoptera: Gryllotalpidae). *J. Exp. Biol.* 208:1495–1512.

Preusser, H.-J., G. Habermehl, M. Sablofski, and D. Schmall-Haury. 1975. Antimicrobial activity of alkaloids from amphibian venoms and effects on the ultrastructure of yeast cells. *Toxicon* 13:285–89.

Pritchard, G. 1965. Prey capture by dragonfly larvae (Odonata: Anisoptera). *Can. J. Zool.* 43:271–89.

———. 1966. On the morphology of the compound eyes of dragonflies (Odonata: Anisoptera), with special reference to their role in prey capture. *Proc. Roy. Ent. Soc. London A* 41:1–8.

Pröhl, H. 1997a. Reproductive pattern in *Dendrobates pumilio* (Anura: Dendrobatidae). *Rev. Biol. Trop.* 45:1669–74.

———. 1997b. Territorial behavior of the strawberry poison-dart frog, *Dendrobates pumilio*. *Amphibia-Reptilia* 18:437–42.

———. 2002. Population differences in female resource abundance, adult sex ratio, and male mating success in *Dendrobates pumilio*. *Behav. Ecol.* 13:175–81.

———. 2003. Variation in male calling behaviour and relation to male mating success in the strawberry poison frog (*Dendrobates pumilio*). *Ethology* 109:273–90.

———. 2005a. Clutch loss affects the operational sex ratio in the strawberry posion frog *Dendrobates pumilio*. *Behav. Ecol. Sociobiol.* 58:310–15.

———. 2005b. Territorial behavior in dendrobatid frogs. *J. Herpetol.* 39:354–65.

Pröhl, H., and O. Berke. 2001. Spatial distribution of male and female strawberry poison frogs and their relation to female reproctive success. *Oecologia* 129:534–42.

Pröhl, H., and W. Hödl. 1999. Parental investment, potential reproductive rates, and mating system in the strawberry dart-poison frog, *Dendrobates pumilio. Behav. Ecol. Sociobiol.* 46:215–20.

Promislow, D. E. L. 1987. Courtship behavior of a plethodontid salamander, *Desmognathus aeneus. J. Herpetol.* 21:298–306.

Pronych, S., and R. Wassersug. 1994. Lung use and development in *Xenopus laevis. Can. J. Zool.* 72:738–43.

Propper, C. R. 1991. Courtship in the rough-skinned newt *Taricha granulosa. Anim. Behav.* 41:547–54.

Propper, C. R., and W. E. Johnson. 1994. Angiotensin II induces water absorption behavior in two species of desert anurans. *Horm. Behav.* 28:41–52.

Propper, C. R., S. D. Hillyard, and W. E. Johnson. 1995. Central angiotensin II induces thirst-related responses in an amphibian. *Horm. Behav.* 29:74–84.

Prudhoe, S., and R. A. Bray. 1982. *Platyhelminth parasites of the Amphibia.* New York: Oxford University Press.

Pruett, S. J., D. F. Hoyt, and D. F. Stiffler. 1991. The allometry of osmotic and ionic regulation in Amphibia with emphasis on intraspecific scaling in larval *Ambystoma tigrinum. Physiol. Zool.* 64:1173–99.

Pryor, G. S., D. P. German, and K. A. Bjorndal. 2006. Gastrointestinal fermentation in greater sirens (*Siren lacertina*). *J. Herpetol.* 40:112–17.

Ptacek, M. B. 1992. Calling sites used by male gray treefrogs, *Hyla versicolor* and *Hyla chrysoscelis,* in sympatry and allopatry in Missouri. *Herpetologica* 48:373–82.

Ptacek, M. B., H. C. Gerhardt, and R. D. Sage. 1994. Speciation by polyploidy in treefrogs: multiple origins of the tetraploid, *Hyla versicolor. Evolution* 48:898–908.

Púgener, L. A., A. M. Maglia, and L. Trueb. 2003. Revisiting the contribution of larval characters to an analysis of phylogenetic relationships of basal anurans. *Zool. J. Linn. Soc.* 139:129–55.

Pulliainen, E., and P. Ollinmäki. 1996a. Common frogs, *Rana temporaria* L., in the winter diet of the pine marten, *Martes martes* L., from the standpoint of their attempts to winter terrestrially. *Aquilo Ser. Zool.* 29:49–53.

———. 1996b. A long-term study of the winter food niche of the pine marten *Martes martes* in northern boreal Finland. *Acta Theriol.* 41:337–52.

Pulliam, H. R., and T. Caraco. 1984. Living in groups: Is there an optimal group size? In *Behavioural ecology: An evolutionary approach* (2nd ed.), ed. J. R. Krebs and N. B. Davies, 122–47. Sunderland, MA: Sinauer Associates.

Punzo, F. 1974. A qualitative and quantitative study of the food items of the yellow mud turtle, *Kinosternon flavescens* (Agassiz). *J. Herpetol.* 8:269–71.

———. 1975. Studies on the feeding behavior, diet, nesting habits and temperature relationships of *Chelydra serpentina osceola* (Chelonia: Chelydridae). *J. Herpetol.* 9:207–10.

———. 1991. Group learning in tadpoles of *Rana heckscheri* (Anura: Ranidae). *J. Herpetol.* 25:214–17.

———. 1992. Socially facilitated behavior in tadpoles of *Rana heckscheri* (Anura: Ranidae). *J. Herpetol.* 26:219–22.

Purgue, A. P. 1997. Tympanic sound radiation in the bullfrog *Rana catesbeiana. J. Comp. Physiol. A* 181:438–45.

Purgue, A. P., and P. M. Narins. 2000. Mechanics of the inner ear of the bullfrog (*Rana catesbeiana*): The contact membranes and the periotic canal. *J. Comp. Physiol. A* 186:481–88.

Putnam, R. W. 1979. The role of lactic acid accumulation in muscle fatigue of two species of anurans, *Xenopus laevis* and *Rana pipiens. J. Exp. Biol.* 82:35–51.

Putnam, R. W., and A. F. Bennett. 1981. Thermal dependence of behavioural performance of anuran amphibians. *Anim. Behav.* 29:502–9.

———. 1983. Histochemical, enzymatic, and contractile properties of skeletal muscles of three anuran amphibians. *Am. J. Physiol.* 224:R558–R567.

Putnam, R. W., and S. S. Hillman. 1977. Activity responses of anurans to dehydration. *Copeia* 1977:746–49.

Puttlitz, M. H., D. P. Chivers, J. M. Kiesecker, and A. R. Blaustein. 1999. Threat-sensitive predator avoidance by larval Pacific treefrogs (Amphibia, Hylidae). *Ethology* 105:449–56.

Pyburn, W. F. 1958. Size and movements of a local population of cricket frogs (*Acris crepitans*). *Texas J. Sci.* 10:325–42.

———. 1961a. The inheritance and distribution of vertebral stripe color in the cricket frog. In *Vertebrate speciation,* ed. W. F. Blair, 235–61. Austin: University of Texas Press.

———. 1961b. Inheritance of the green vertebral stripe in *Acris crepitans. Southwest. Nat.* 6:164–67.

———. 1963. Observations on the life history of the treefrog *Phyllomedusa callidryas* (Cope). *Texas J. Sci.* 15:155–70.

———. 1964. Breeding behavior of the leaf-frog, *Phyllomedusa callidryas,* in southern Veracruz. *Yrbk. Amer. Phil. Soc.* 1964:291–94.

———. 1966. Breeding activity, larvae and relationship of the treefrog *Hyla phaeota cyanosticta. Southwest. Nat.* 11:1–18.

———. 1967. Breeding and larval development of the hylid frog *Phrynohyas spilomma* in southern Veracruz, Mexico. *Herpetologica* 23:184–94.

———. 1970. Breeding behavior of the leaf frogs *Phyllomedusa callidryas* and *Phyllomedusa dacnicolor* in Mexico. *Copeia* 1970:209–18.

———. 1975. A new species of microhylid frog of the genus *Synapturanus* from southeastern Colombia. *Herpetologica* 31:439–43.

———. 1976. A new fossorial frog from the Colombian rain forest (Anura: Microhylidae). *Herpetologica* 32:367–70.

——— 1980a. The function of eggless capsules and leaf in nests of the frog *Phyllomedusa hypochondrialis* (Anura: Hylidae). *Proc. Biol. Soc. Washington* 93:153–67.

———. 1980b. An unusual anuran larva from the Vaupes region of southeastern Colombia. *Pap. Avul. Zool. Sao Paulo* 33:231–38.

Pyburn, W. F., and J. R. Glidewell. 1971. Notes and breeding behavior of *Phyllomedusa hypochondrialis* in Colombia. *J. Herpetol.* 5:49–52.

Pycraft, W. P. 1913. *The courtship of animals.* London: Hutchinson.

Pyke, C. R., and J. Marty. 2005. Cattle grazing mediates climate change impacts on ephemeral wetlands. *Cons. Biol.* 19:1619–25.

Pyke, G. H. 1983. Animal movements: an optimal foraging approach. In *The ecology of animal movement,* ed. I. R. Swingland and P. J. Greenwood, 7–31. New York: Oxford University Press.

Pyke, G. H., and A. Miehs. 2001. Predation by water skinks (*Eulamprus quoyii*) on tadpoles and metamorphs of the green and golden bell frog (*Litoria aurea*). *Herpetofauna* 31:99–101.

Pyke, G. H., and A. W. White. 1996. Habitat requirements for the green and golden bell frog *Litoria aurea* (Anura: Hylidae). *Aust. Zool.* 30:224–32.

———. 2000. Factors influencing predation on eggs and tadpoles of the endangered green and golden bell frog *Litoria aurea* by the introduced plague minnow *Gambusia holbrooki*. *Aust. Zool.* 31:496–505.

Quijano, F. M., G. Santos-Barrera, and J. Pacheco-Rodríguez. 2002. *Eleutherodactylus fitzingeri* (common rain frog). Clutch size and parental care. *Herp. Rev.* 33:125.

Quinlan, R., M. S. V. Douglas, and J. P. Smol. 2005. Food web changes in arctic ecosystems related to climate warming. *Global Change Biol.* 11:1381–86.

Quinn, D., and W. W. Burggren. 1983. Lactate production, tissue distribution, and elimination following exhaustive exercise in larval and adult bullfrogs, *Rana catesbeiana*. *Physiol. Zool.* 56:597–613.

Quinn, J. F., and A. E. Dunham. 1983. On hypothesis testing in ecology and evolution. *Am. Nat.* 122:602–17.

Quinn, T., J. Gallie, and D. P. Volsen. 2001. Amphibian occurrence in artificial and natural wetlands of the Teanaway and Lower Swauk River drainages of Kittitas County, Washington. *Northwest Sci.* 75:84–89.

Quinn, V. S., and B. M. Graves. 1999. Space use in response to conspecifics by the red-backed salamander (*Plethdon cinereus*, Plethodontidae, Caudata). *Ethology* 105:993–1002.

Rabb, G. B. 1956. Some observations on the salamander *Stereochilus marginatum*. *Copeia* 1956:119.

———. 1960. On the unique sound production of the Surinam toad (*Pipa pipa*). *Copeia* 1960:368–69.

———. 1969. Fighting frogs. *Brookfield Banderlog* 37:4–5.

———. 1973. Evolutionary aspects of the reproductive behavior of frogs. In *Evolutionary biology of the anurans*, ed. J. L. Vial, 213–27. Columbia: University of Missouri Press.

Rabb, G. B., and M. S. Rabb. 1960. On the mating and egg-laying behavior of the Surinam toad, *Pipa pipa*. *Copeia* 1960:171–76.

———. 1963a. Additional observations on breeding behavior of the Surinam toad, *Pipa pipa*. *Copeia* 1963:636–42.

———. 1963b. On the behavior and breeding biology of the African pipid frog *Hymenochirus boettgeri*. *Z. Tierpsychol.* 20:215–41.

Rabb, G. B., and R. Snedigar. 1960. Observations on breeding and development of the Surinam toad, *Pipa pipa*. *Copeia* 1960:40–44.

Rachowicz, L. J., J.-M. Hero, R. A. Alford, J. W. Taylor, J. A. T. Morgan, V. T. Vredenburg, J. P. Collins, and C. J. Briggs. 2005. The novel and endemic pathogen hypothesis: competing explanations for the origin of emerging infectious diseases of wildlife. *Cons. Biol.* 19:1441–48.

Rachowicz, L. J., and V. T. Vredenburg. 2004. Transmission of *Batrachochytrium dendrobatidis* within and between amphibian life stages. *Dis. Aquat. Org.* 61:75–83.

Rader, R. B., M. C. Belk, and M. J. Keleher. 2003. The introduction of an invasive snail (*Melanoides tuberculata*) to spring ecosystems in the Bonneville Basin, Utah. *J. Freshw. Ecol.* 18:647–57.

Radesäter, T., S. Jakobsson, N. Andbjer, A. Bylin, and K. Nyström. 1987. Song rate and pair formation in the willow warbler, *Phylloscopus trochilus*. *Anim. Behav.* 35:1645–51.

Radojičić, J. M., D. D. Cvetković, L. M. Tomović, G. V. Džukić, and M. L. Kalezić. 2002. Sexual dimorphism in fire-bellied toads *Bombina* spp. from the central Balkans. *Folia Zool.* 51:129–40.

Radovanović, M. 1951. A new race of the alpine newt from Yugoslavia. *Brit. J. Herpetol.* 1:93–97.

Radwan, N. M. M., and H. Schneider. 1988. Social behaviour, call repertory and variation in the calls of the pool frog, *Rana lessonae* (Anura: Ranidae). *Amphibia-Reptilia* 9:329–51.

Raehmel, C.-A. 1983. *Pelodytes punctatus* (Daudin), der Westliche Schlammtaucher—seine Pflege und seine Zucht. *Herpetofauna* 27:22–24.

Rafinska, A. 1991. Reproductive biology of the fire-bellied toads, *Bombina bombina* and *B. variegata* (Anura: Discoglossidae): Egg size, clutch size and larval period length differences. *Biol. J. Linn. Soc.* 43:197–210.

Rafiński, J., and A. Czaja. 1984. Sexual behaviour of the Bosca's newt, *Triturus boscai*. *Amphibia-Reptilia* 5:197.

Rafiński, J., and A. Osikowski. 2002. Sperm mixing in the alpine newt (*Triturus alpestris*). *Can. J. Zool.* 80:1293–98.

Rafiński, J. N. 1981. Multiple paternity in a natural population of the alpine newt, *Triturus alpestris* (Laur.) *Amphibia-Reptilia* 2:282.

Raimondo, S. M., C. L. Rowe, and J. D. Congdon. 1998. Exposure to coal ash impacts swimming performance and predator avoidance in larval bullfrogs (*Rana catesbeiana*). *J. Herpetol.* 32:289–92.

Ralin, D. B. 1977. Evolutionary aspects of mating call variation in a diploid-tetraploid species complex of treefrogs (Anura). *Evolution* 31:721–36.

Ralin, D. B., and J. S. Rogers. 1972. Aspects of tolerance to desiccation in *Acris crepitans* and *Pseudacris streckeri*. *Copeia* 1972:519–25.

Ralls, K. 1976. Mammals in which females are larger than males. *Q. Rev. Biol.* 51:245–76.

———. 1977. Sexual dimorphism in mammals: avian models and unanswered questions. *Am. Nat.* 111:917–38.

Ralph, C. L. 1957. A diurnal activity rhythm in *Plethodon cinereus* and its modification by an influence having a lunar frequency. *Biol. Bull.* 113:188–97.

Ramer, J. D., T. A. Jenssen, and C. J. Hurst. 1983. Size-related variation in the advertisement call of *Rana clamitans* (Anura: Ranidae), and its effect on conspecific males. *Copeia* 1983:141–55.

Ramirez, J., R. C. Vogt, and J.-L. Villarreal-Benitez. 1998. Population biology of a Neotropical frog (*Rana vaillanti*). *J. Herpetol.* 32:338–44.

Rand, A. S. 1950. Leopard frogs in caves in winter. *Copeia* 1950:324.

———. 1952. Jumping ability of certain anurans, with notes on endurance. *Copeia* 1952:15–20.

———. 1983. *Physalaemus pustulosus*. In *Costa Rican natural history*, ed. D. H. Janzen, 412–15. Chicago: University of Chicago Press.

———. 1988. An overview of anuran acoustic communication. In *The evolution of the amphibian auditory system*, ed. B. Fritzsch, M. J. Ryan, W. Wilczynski, T. E. Hetherington, and W. Walkowiak, 415–31. New York: Wiley.

———. 2001. A history of frog call studies 405 BC to 1980. In *Anuran communcation*, ed. M. J. Ryan, 8–19. Washington, DC: Smithsonian Institution Press.

Rand, A. S., and R. Dudley. 1993. Frogs in helium: The anuran vocal sac is not a cavity resonator. *Physiol. Zool.* 66:793–806.

Rand, A. S., and C. W. Myers. 1990. The herpetofauna of Barro Colorado Island, Panama: an ecological summary. In *Four neotropical rainforests*, ed. A. H. Gentry, 386–409. New Haven, CT: Yale University Press.

Rand, A. S., and M. J. Ryan. 1981. The adaptive significance of a complex vocal repertoire in a Neotropical frog. *Z. Tierpsychol.* 57:209–14.

Rand, A. S., M. J. Ryan, and W. Wilczynski. 1992. Signal redundancy and receiver permissiveness in acoustic mate recognition by the túngara frog, *Physalaemus pustulosus. Amer. Zool.* 32:81–90.

Randall, R. J., W. W. Burggren, A. P. Farrell, and M. S. Haswell. 1981. *The evolution of air-breathing in vertebrates.* New York: Cambridge University Press.

Raney, E. C. 1940. Summer movements of the bullfrog *Rana catesbeiana* Shaw as determined by the jaw-tag method. *Am. Midl. Nat.* 23:733–45.

Raney, E. C., and W. M. Ingram. 1941. Growth of tagged frogs (*Rana catesbeiana* Shaw and *Rana clamitans* Daudin) under natural conditions. *Am. Midl. Nat.* 26:201–6.

Rankin, J. S. 1937. An ecological study of parasites of some North American salamanders. *Ecol. Monogr.* 7:169–269.

Ranta, E., and K. Lindström. 1990. Assortative schooling in three-spined sticklebacks? *Ann. Zool. Fennici* 27:67–75.

Ranta, E., K. Lindström, and N. Peuhkuri. 1992. Size matters when three-spined sticklebacks go to school. *Anim. Behav.* 43:160–62.

Ranta, E., and V. Nuutinen. 1985. Foraging by the smooth newt (*Triturus vulgaris*) on zooplankton: functional response and diet choice. *J. Anim. Ecol.*54:275–93.

Ranvestal, A. W., K. R. Lips, C. M. Pringle, M. R. Whiles, and R. J. Bixby. 2004. Neotropical tadpoles influence stream benthos: Evidence for the ecological cosequences of decline in amphibian populations. *Freshw. Biol.* 49:274–85.

Rao, D.-Q., and D.-T. Yang. 1997. The karyotypes of Megophryinae (Pelobatidae) with a discussion on their classification and phylogenetic relationships. *Asiatic Herpetol. Res.* 7:93–102.

Räsänen, K., A. Laurila, and J. Merilä. 2003a. Geographic variation in acid stress tolerance of the moor frog, *Rana arvalis.* I. Local adaptation. *Evolution* 57:352–62.

Räsänen, K., A. Laurila, and J. Merilä. 2003b. Geographic variation in acid stress tolerance of the moor frog, *Rana arvalis.* II. Adaptive maternal effects. *Evolution* 57:363–71.

Räsänen, K., M. Pahkala, A. Laurila, and J. Merilä. 2003. Does jelly envelope protect the common frog *Rana temporaria* embryos from UV-B radiation? *Herpetologica* 59:293–300.

Rastogi, R. K., and G. Chieffi. 1971. Effect of an antiandrogen, cyproterone acetate, on the pars distalis of pituitary and thumb pad of the male green frog, *Rana esculenta* L. *Steroidologia* 2:276–82.

Rastogi, R. K., I. Izzo-Vitielo, M. Di Meglio, L. Di Matteo, R. Franzese M. G. Di Costanza, S. Minucci, L. Iela, and G. Chieffi. 1983. Ovarian activity and reproduction in the frog, *Rana esculenta. J. Zool., London* 200:233–47.

Rautio, S. A., E. A. Bura, K. A. Berven, and G. J. Gamboa. 1991. Kin recognition in wood frog tadpoles (*Rana sylvatica*): Factors affecting spatial proximity to siblings. *Can. J. Zool.* 69:2569–71.

Raxworthy, C. J. 1989. Courtship, fighting and sexual dimorphism of the banded newt, *Triturus vittatus ophryticus. Ethology* 81:148–70.

———. 1990. Non-random mating by size in the midwife toad *Alytes obstetricans:* bigger males carry more eggs. *Amphibia-Reptilia* 11:247–52.

Raxworthy, C. J., and R. A. Nussbaum. 2000. Extinction and extinction vulnerability of amphibians and reptiles in Madagascar. *Amph. Rept. Cons.* 2:15–23.

Ray, C. 1958. Vital limits and rates of desiccation in salamanders. *Ecology* 39:75–83.

Ray, J. C., and M. E. Sunquist. 2001. Trophic relations in a community of African rainforest carnivores. *Oecologia* 127:395–408.

Raymond, L. R., and L. M. Hardy. 1990. Demography of a population of *Ambystoma talpoideum* (Caudata: Ambystomatidae) in northwestern Louisiana. *Herpetologica* 46:371–82.

Read, K., J. S. Keogh, I. A. W. Scott, J. D. Roberts, and P. Doughty. 2001. Molecular phylogeny of the Australian frog genera *Crinia, Geocrinia,* and allied taxa (Anura: Myobatrachidae). *Mol. Phylogen. Evol.* 21:294–308.

Read, J. L. 1999. Abundance and recruitment patterns of the trilling frog (*Neobatrachus centralis*) in the Australian arid zone. *Aust. J. Zool.* 47:393–404.

Reading, C. J. 1986. Egg production in the common toad, *Bufo bufo. J. Zool., London* (A) 208:99–107.

———. 1990. Palmate newt predation on common frog, *Rana temporaria,* and common toad, *Bufo bufo,* tadpoles. *Herpetol. J.* 1:462–65.

———. 1998. The effect of winter temperatures on the timing of breeding activity in the common toad *Bufo bufo. Oecologia* 117:469–75.

———. 2001. Non-random pairing with respect to past breeding experience in the common toad (*Bufo bufo*). *J. Zool., London* 255:511–18.

———. 2003. The effects of variation in climatic temperature (1980–2001) on breeding activity and tadpole stage duration in the common toad, *Bufo bufo. Sci. Total Envir.* 310:231–36.

Reading, C. J., and R. T. Clarke. 1983. Male breeding behaviour and mate acquisition in the common toad, *Bufo bufo. J. Zool., London* 201:237–46.

———. 1988. Multiple clutches, egg mortality and mate choice in the mid-wife toad, *Alytes obstetricans. Amphibia-Reptilia* 9:357–64.

———. 1995. The effects of density, rainfall and environmental temperature on body condition and fecundity in the common toad, *Bufo bufo. Oecologia* 102:453–59.

———. 1999. Impacts of climate and density on the duration of the tadpole stage of the common toad *Bufo bufo. Oecologia* 121:310–15.

Reading, C. J., and J. L. Davies. 1996. Predation by grass snakes (*Natrix natrix*) at a site in southern England. *J. Zool., London* 239:73–82.

Reading, C. J., and G. M. Jofré. 2003. Reproduction in the nest building Vizcacheras frog *Leptodactylus bufonius* in central Argentina. *Amphibia-Reptilia* 24:415–27.

Reading, C. J., J. Loman, and T. Madsen. 1991. Breeding pond fidelity in the common toad, *Bufo bufo. J. Zool., London* 225:201–11.

Reagan, D. P. 1984. Ecology of the Puerto Rican boa (*Epicrates inornatus*) in the Luquillo Mountains of Puerto Rico. *Carib. J. Sci.* 20:119–27.

Reagan, D. P., G. R. Camilo, and R. B. Waide. 1996. The community food web: Major properties and patterns of organization. In *The food web of a tropical rain forest,* ed. D. P. Reagan and R. B. Waide, 462–510. Chicago: University of Chicago Press.

Reagen, N. L., and P. A. Verrell. 1991. The evolution of plethodontid salamanders: Did terrestrial mating facilitate lunglessness? *Am. Nat.* 138:1307–13.

Reaser, J. K. 1997. Batrachoseps attenuatus (*California slender salamander*). Predation. *Herp. Rev.* 28:81.

Rebelo, R., and M. H. Leclair. 2003a. Differences in size at birth and brood size among Portuguese populations of the fire salamander, *Salamandra salamandra*. *Herpetol. J.* 13:179–87.

———. 2003b. Site tenacity in the terrestrial salamandrid *Salamandra salamandra*. *J. Herpetol.* 37:440–45.

Redfield, A. C. 1933. The evolution of the respiratory function of the blood. *Q. Rev. Biol.* 8:31–59.

Redford, K. H., and J. F. Eisenberg. 1992. *Mammals of the Neotropics. The southern cone, vol. 2. Chile, Argentina, Uruguay, Paraguay.* Chicago: University of Chicago Press.

Redmer, M. 1998. *Hyla avivoca* (bird-voiced treefrog). Amplexus and oviposition. *Herp. Rev.* 29:230–31.

Reeder, A. L., G. L. Foley, D. K. Nichols, L. G. Hansen, B. Wikoff, S. Fach, J. Eisold, M. B. Wheeler, R. Warner, J. E. Murphy, and V. R. Beasley. 1998. Forms and prevalence of intersexuality and effects of environmental contaminants on sexuality in cricket frogs (*Acris crepitans*). *Env. Health Perspect.* 106:261–66.

Reeder, A. L., M. O. Ruiz, A. Pessier, L. E. Brown, J. M. Levengood, C. A. Phillips, M. B. Wheeler, R. E. Warner, and V. B. Beasley. 2005. Intersexuality and the cricket frog decline: historic and geographic trends. *Env. Health Perspect.* 113:261–65.

Regal, P. J. 1966. Feeding specialization and the classification of terrestrial salamanders *Evolution* 20:392–407.

———. 1978. Behavioral differences between reptiles and mammals: An analysis of activity and mental capabilities. In *Behavior and neurology of lizards,* ed. N. Greenberg and P. D. MacLean, 183–202. Rockville, MD: NIMH.

Regal, P. J., and C. Gans. 1976. Functional aspects of the evolution of frog tongues. *Evolution* 30:718–34.

Regester, K. J., K. R. Lips, and M. R. Whiles. 2006. Energy flow and subsidies associated with the complex life cycle of ambystomatid salamanders in ponds and adjacent forest in southern Illinois. *Oecologia* 147:303–14.

Regnault, V., and J. Reiset. 1849. Recherches chimiques sur la respiration des animaux des diverses classes. *Ann. Chim. Physiq.*, ser. 3, 26:299–519.

Regös, J., and A. Schlüter. 1984. Erste Ergebnisse zur Fortpflanzungsbiologie von *Lithodytes lineatus* (Schneider, 1799) (Amphibia: Leptodactylidae). *Salamandra* 20:252–61.

Regosin, J. V., B. S. Windmiller, and J. M. Reed. 2003a. Influence of abundance of small-mammal burrows and conspecifics on the density and distribution of spotted salamanders (*Ambystoma maculatum*) in terrestrial habitats. *Can. J. Zool.* 81:596–605.

———. 2003b. Terrestrial habitat use and winter densities of the wood frog (*Rana sylvatica*). *J. Herpetol.* 37:390–94.

———. 2004. Effects of conspecifics on the burrow occupancy behavior of spotted salamanders (*Ambystoma maculatum*). *Copeia* 2004:152–58.

Reh, W., and A. Seitz. 1990. The influence of land use on the genetic structure of populations of the common frog *Rana temporaria*. *Biol. Cons.* 54:239–49.

Reid, D. G., T. E. Code, A. K. H. Reid, and S. M. Herrero. 1994. Food habits of the river otter in a boreal ecosystem. *Can. J. Zool.* 72:1306–13.

Reid, M. L. 1987. Costliness and reliability in the singing vigor of Ipswich sparrows. *Anim. Behav.* 35:1735–43.

Reig, S., and W. Jędrzejewski. 1988. Winter and early spring food of some carnivores in the Białowieża National Park, eastern Poland. *Acta Theriol.* 33:57–65.

Reilly, S. M. 1986. Ontogeny of cranial ossification in the eastern newt, *Notophthalmus viridescens* (Caudata: Salamandridae), and its relationship to metamorphosis and neoteny. *J. Morphol.* 188:315–26.

———. 1987. Ontogeny of the hypobranchial apparatus in the salamanders *Ambystoma talpoideum* (Ambystomatidae) and *Notophthalmus viridescens* (Salamandridae): the ecological morphology of two neotenic strategies. *J. Morphol.* 191:205–14.

———. 1994. The ecological morphology of metamorphosis: Heterochrony and the evolution of feeding mechanisms in salamanders. In *Ecological morphology: Integrative organismal biology,* ed. P. C. Wainwright and S. M. Reilly, 319–38. Chicago: University of Chicago Press.

———. 1995. The ontogeny of aquatic feeding behavior in *Salamandra salamandra:* Stereotypy and isometry in feeding kinematics. *J. Exp. Biol.* 198:701–8.

———. 1996. The metamorphosis of feeding kinematics in *Salamandra salamandra* and the evolution of terrestrial feeding behavior. *J. Exp. Biol.* 199:1219–27.

Reilly, S. M., and R. N. Brandon. 1994. Partial paedomorphosis in the Mexican stream ambystomatids and the taxonomic status of the genus *Rhyacosiredon* Dunn. *Copeia* 1994:656–62.

Reilly, S. M., and G. V. Lauder. 1988. Ontogeny of aquatic feeding performance in the eastern newt, *Notophthalmus viridescens* (Salamandridae). *Copeia* 1988:87–91.

———. 1989. Physiological bases of feeding behaviour in salamanders: Do motor patterns vary with prey type? *J. Exp. Biol.* 141:343–58.

Reilly, S. M., G. V. Lauder, and J. P. Collins. 1992. Performance consequences of a trophic polymorphism: Feeding behavior in typical and cannibal phenotypes of *Ambystoma tigrinum*. *Copeia* 1992:672–79.

Reilly, S. M., E. O. Wiley, and D. J. Meinhardt. 1997. An integrative approach to heterochrony: The distinction between interspecific and intraspecific phenomena. *Biol. J. Linn. Soc.* 60:119–43.

Reimer, W. J. 1958. Variation and systematic relationships within the salamander genus *Taricha*. *Univ. California Publ. Zool.* 56:301–90.

Reinhold, K. 1999. Energetically costly behaviour and the evolution of resting metabolic rate in insects. *Funct. Ecol.* 13:217–24.

Reinhold, K., M. D. Greenfield, Y. Jang, and A. Broce. 1998. Energetic cost of sexual attractiveness: ultrasonic advertisement in wax moths. *Anim. Behav.* 55:905–13.

Reiss, J. O. 2002. The phylogeny of amphibian metamorphosis. *Zoology* 105:85–96.

Relyea, R. A. 2000. Trait-mediated indirect effects in larval anurans: Reversing competition with the threat of predation. *Ecology* 81:2278–89.

———. 2001a. Morphological and behavioral plasticity of larval anurans in response to different predators. *Ecology* 82:523–40.

———. 2001b. The relationship between predation risk and antipredator responses in larval anurans. *Ecology* 82:541–54.

———. 2002a. Costs of phenotypic plasticity. *Am. Nat.* 159:272–82.

———. 2002b. Local population differences in phenotypic plasticity: Predator-induced changes in wood frog tadpoles. *Ecol. Monogr.* 72:77–93.

———. 2002c. The many faces of predation: How induction, selection, and thinning combine to alter prey phenotypes. *Ecology* 83:1953–64.

———. 2003a. How prey respond to combined predators: A review and an empirical test. *Ecology* 84:1827–39.

———. 2003b. Predator cues and pesticides: A double dose of danger for amphibians. *Ecol. Appl.* 13:1515–21.

———. 2003c. Predators come and predators go: The reversibility of predator-induced traits. *Ecology* 84:1840–48.

———. 2004a. Growth and survival of five amphibian species exposed to combinations of pesticides. *Env. Tox. Chem.* 23:1737–42.

———. 2004b. Synergistic impacts of malathion and predatory stress on six species of North American tadpoles. *Env. Tox. Chem.* 23:1080–84.

———. 2005a. The heritability of inducible defenses in tadpoles. *J. Evol. Biol.* 18:856–66.

———. 2005b. The impact of insectides and herbicides on the biodiversity and productivity of aquatic communities. *Ecol. Appl.* 15:618–27.

———. 2005c. The lethal impact of Roundup and predatory stress on six species of North American tadpoles. *Arch. Environ. Contam. Toxicol.* 48:351–57.

Relyea, R. A., and N. Mills. 2001. Predator-induced stress makes the pesticide carbaryl more deadly to grey treefrog tadpoles (*Hyla versicolor*). *Proc. Natl. Acad. Sci. USA* 98:2491–96.

Relyea, R. A., and E. E. Werner. 1999. Quantifying the relation between predator-induced behavior and growth performance in larval anurans. *Ecology* 80:2117–24.

———. 2000. Morphological plasticity in four larval anurans distributed along an environmental gradient. *Copeia* 2000:178–90.

Renaud, J. M., and E. D. Stevens. 1981. Effects of acclimation temperature and pH on the isometric contraction of toad sartorius muscle. *J. Comp. Physiol.* 145:67–71.

———. 1983a. A comparison between field habits and contractile performance of frog and toad sartorius muscle. *J. Comp. Physiol. B* 151:127–31.

———. 1983b. The extent of long-term temperature compensation for jumping distance in the frog, *Rana pipiens,* and the toad, *Bufo americanus. Can. J. Zool.* 61:1284–87.

———. 1984. The extent of short-term and long-term compensation to temperature shown frog and toad sartorius muscle. *J. Exp. Biol.* 108:57–75.

Rengel, D. 1950. Acción de la temperatura elevada sobre la espermatogenesis de dos formas de *Leptodactylus ocellatus* L. *Acta Zool. Lilloana* 9:425–38.

Reno, H. W., F. R. Gehlbach, and R. A. Turner. 1972. Skin and aestivational cocoon of the aquatic amphibian, *Siren intermedia* Le Conte. *Copeia* 1972:625–31.

Reques, R., and M. Tejedo. 1995. Negative correlation between length of larval period and metamorphic size in natural populations of natterjack toads (*Bufo calamita*). *J. Herpetol.* 29:311–14.

———. 1996. Intraspecific aggressive behaviour in fire salamander larvae (*Salamandra salamandra*): The effects of density and body size. *Herpetol. J.* 6:15–19.

Resetarits, W. J., Jr. 1986. The ecology of cave use by the frog *Rana palustris. Am. Midl. Nat.* 116:256–66.

———. 1991. Ecological interactions among predators in experimental stream communities. *Ecology* 72:1782–93.

———. 1995. Competitive asymmetry and coexistence in size-structured populations of brook trout and spring salamanders. *Oikos* 73:188–98.

———. 1997. Differences in an ensemble of streamside salamanders (Plethodontidae) above and below a barrier to brook trout. *Amphibia-Reptilia* 18:15–25.

———. 1998. Differential vulnerability of *Hyla chrysoscelis* eggs and hatchlings to larval insect predators. *J. Herpetol.* 32:440–43.

Resetarits, W. J., Jr., and J. Bernardo, eds. 1998. *Experimental ecology. Issues and perspectives.* New York: Oxford University Press.

Resetarits, W. J., Jr., and J. E. Fauth. 1998. From cattle tanks to Carolina bays. The utility of model systems for understanding natural communities. In *Experimental ecology. Issues and perspectives,* ed. W. J. Resetarits, Jr. and J. Bernardo, 133–51. New York: Oxford University Press.

Resetarits, W. J., Jr., and H. M. Wilbur. 1989. Choice of oviposition site by *Hyla chrysoscelis:* Role of predators and competitors. *Ecology* 70:220–28.

———. 1991. Calling site choice by *Hyla chrysoscelis:* Effect of predators, competitors, and oviposition sites. *Ecology* 72:778–86.

Reshetnikov, A. N. 2003. The introduced fish, rotan (*Percottus glenii*), depresses populations of aquatic animals (macroinvertebrates, amphibians, and a fish). *Hydrobiologia* 510:83–90.

Reshetnikov, A. N., and Y. B. Manteifel. 1997. Newt-fish interactions in Moscow Province: A new predatory fish colonizer, *Percottus glehni,* transforms metapopulations of newts, *Triturus vulgaris* and *T. cristatus. Adv. Amphib. Res. Former Soviet Union* 2:1–12.

Resnick, L. E., and D. L. Jameson. 1963. Color polymorphisms in Pacific tree frogs. *Science* 142:1081–83.

Ressel, S. J. 1993. *A morphometric analysis of anuran skeletal muscle ultrastructure: implications for the functional design of muscle in ectotherms.* Unpublished PhD dissertation. Storrs: University of Connecticut.

———. 1996. Ultrastructural properties of muscles used for call production in Neotropical frogs. *Physiol. Zool.* 69:952–73.

———. 2001. Ultrastructural design of anuran muscles used for call production in relation to the thermal environment of a species. *J. Exp. Biol.* 204:1445–57.

Retallick, R. W. R., H. McCallum, and R. Speare. 2004. Endemic infection of the amphibian chytrid fungus in a frog community post-decline. *PLOS Biology* 2 (11): e351.

Revilla, E., and F. Palomares. 2002. Does local feeding specialization exist in European badgers? *Can. J. Zool.* 80:83–93.

Rey, J. M., M. Esteban, and B. Sanchiz. 1994. Registros corologícos de anfibios españoles fundamentados en ega gropilas de lechuza commun. *Bol. Asoc. Herpetol. Española* 5:4–8.

Reyer, H.-U., G. Frei, and C. Som. 1999. Cryptic female choice: frogs reduce clutch size when amplexed by undesired males. *Proc. Roy. Soc. London B* 266:2101–7.

Reyer, H.-U., B. Niederer, and A. Hettyey. 2003. Variation in fertilisation abilities between hemiclonal hybrid and sexual parental males of sympatric water frogs (*Rana lessonae, R. esculenta, R. ridibunda*). *Behav. Ecol. Sociobiol.* 54:274–84.

Reynolds, R. P., and M. S. Foster. 1992. Four new species of frogs and one new species of snake from the Chapare region of Bolivia, with notes on other species. *Herpetol. Monogr.* 6:83–104.

Reznick, D. N. 1990. Plasticity in age and size at maturity in male guppies (*Poecilia reticulata*): An experimental evaluation of alternative models of development. *J. Evol. Biol.* 3:185–203.

Rheinlaender, J., H. C. Gerhardt, D. Yager, and R. R. Capranica. 1979. Accuracy of phonotaxis by the green treefrog (*Hyla cinerea*). *J. Comp. Physiol.* 133:247–55.

Rheinlaender, J., and G. Klump. 1988. Behavioral aspects of sound localization. In *The evolution of the amphibian audi-*

*tory system,* ed. B. Fritzsch, M. J. Ryan, W. Wilczynski, T. E. Hetherington, and W. Walkowiak, 297–305. New York: Wiley.

Rheinlaender, J., W. Walkowiak, and H. C. Gerhardt. 1981. Directional hearing in the green treefrog: A variable mechanism? *Naturwissenschaften* 67:430–31.

Ribeiro Matos, R. H., G. V. Andrade, and A. Hass. 2000. Reproductive biology and territoriality of *Phyllomedusa hypochondrialis* in northeastern Brazil. *Herp. Rev.* 31:84–86.

Ribuoli, P., and M. Robbiani. 1990. *Frogs. Art, legend, history.* Boston: Little Brown.

Ricciardi, A., R. J. Neves, and J. B. Rasmussen. 1999. Extinction rates of North American freshwater fauna. *Cons. Biol.* 13:1220–22.

Rice, A. N., and P. S. Lobel. 2002. Enzyme activities of pharyngeal jaw musculature in the cichlid *Tramitichromis intermedius:* Implications for sound production in cichlid fishes. *J. Exp. Biol.* 205:3519–23.

Richards, C. M. 1958. The inhibition of growth in crowded *Rana pipiens* tadpoles. *Physiol. Zool.* 31:138–51.

———. 1962. The control of tadpole growth by alga-like cells. *Physiol. Zool.* 35:285–96.

———. 1976. The development of color dimorphism in *Hyperolius v. viridiflavus* a reed frog from Kenya. *Copeia* 1976:65–70.

———. 1977. Reproductive potential under laboratory conditions of *Hyperolius viridiflavus* (Amphibia, Anura, Hyperoliidae), a Kenyan reed frog. *J. Herpetol.* 11:426–28.

———. 1982. The alteration of chromatophore expression by sex hormones in the Kenyan reed frog, *Hyperolius viridiflavus. Gen. Comp. Endocrin.* 46:59–67.

Richards, C. M., B. Carlson, T. Connelly, S. Rogers, and E. Ashcraft. 1977. A scanning electron microscopic study of differentiation of the digital pad in regenerating digits of the Kenyan reed frog, *Hyperolius viridiflavus ferniquei. J. Morphol.* 153:387–96.

Richards, C. M., and W. S. Moore. 1996. A phylogeny for the African treefrog family Hyperoliidae based on mitochondrial rDNA. *Mol. Phylogen. Evol.* 5:522–32.

———. 1998. A molecular phylogenetic study of the Old World treefrog family Rhacophoridae. *Herpetol. J.* 8:41–46.

Richards, C. M., R. A. Nussbaum, and C. J. Raxworthy. 2000. Phylogenetic relationships within the Madagascan boophids and mantellids as elucidated by mitochondrial ribosomal genes. *Afr. J. Herpetol.* 49:23–32.

Richards, D. G., and R. H. Wiley. 1980. Reverberations and amplitude fluctuations in the propagation of sound in a forest: Implications for animal communication. *Am. Nat.* 115:181–399.

Richards, S. J. 1992. Foot-flagging displays of some Australian frogs. *Mem. Queensland Mus.* 32:302.

———. 1993. Functional significance of nest construction by an Australian rainforest frog: A preliminary analysis. *Mem. Queensland Mus.* 34:89–93.

———. 2002. Influence of flow regime on habitat selection by tadpoles in an Australian rainforest stream. *J. Zool., London* 257:273–79.

Richards, S. J., and R. A. Alford. 1992. Nest construction by an Australian rainforest frog of the *Litoria lesueuri* complex (Anura: Hylidae). *Copeia* 1992:1120–23.

Richards, S. J., and C. M. Bull. 1990a. Non-visual detection of anuran tadpoles by odonate larvae. *J. Herpetol.* 24:311–13.

———. 1990b. Size-limited predation on tadpoles of three Australian frogs. *Copeia* 1990:1041–46.

Richards, S. J., K. R. McDonald, and R. A. Alford. 1993. Declines in populations of Australia's endemic tropical rainforest frogs. *Pacific Cons. Biol.* 1:66–77.

Richardson, J. M. L. 1994. Shoaling in White Cloud Mountain minnows, *Tanichthys albonubes:* effects of predation risk and prey hunger. *Anim. Behav.* 48:727–30.

———. 2001. The relative roles of adaptation and phylogeny in determination of larval traits in diversifying anuran lineages. *Am. Nat.* 157:282–99.

———. 2002a. Burst swim speed in tadpoles inhabiting ponds with different top predators. *Evol. Ecol. Res.* 4:627–42.

———. 2002b. A comparative study of phenotypic traits related to resource utilization in anuran communities. *Evol. Ecol.* 16:101–22.

Richmond, N. D. 1945. Nesting of the two-lined salamander on the coastal plain. *Copeia* 1945:170.

———. 1947. Life history of *Scaphiopus holbrooki* (Harlan). Part I. Larval development and behavior. *Ecology* 28:53–67.

Richter, S. C. 2000. Larval caddisfly predation on the eggs and embryos of *Rana capito* and *Rana sphenocephala. J. Herpetol.* 34:590–93.

Richter, S. C., and J. B. Jensen. 2005. *Rana sevosa* Goin and Netting 1940. Dark gopher frog. In *Amphibian declines: The conservation status of United States species,* ed. M. J. Lannoo, 584–86. Berkeley: University of California Press.

Richter, S. C., and R. A. Seigel. 2002. Annual variation in the population ecology of the endangered gopher frog, *Rana sevosa* Goin and Netting. *Copeia* 2002:962–72.

Richter, S. C., J. E. Young, G. N. Johnson, and R. A. Seigel. 2003. Stochastic variation in reproductive success of a rare frog, *Rana sevosa:* implications for conservation and for monitoring amphibian populations. *Biol. Cons.* 111:171–77.

Richter, S. C., J. E. Young, R. A. Seigel, and G. N. Johnson. 2001. Postbreeding movements of the dark gopher frog, *Rana sevosa* Goin and Netting: Implications for conservation and management. *J. Herpetol.* 35:316–21.

Ricklefs, R. E. 1987. Community diversity: Relative roles of local and regional processes. *Science* 235:167–71.

Ricklefs, R. E., and D. Schluter. 1993a. *Species diversity in ecological communities. Historical and geographical perspectives.* Chicago: University of Chicago Press.

———. 1993b. Species diversity: Regional and historical influences. In *Species diversity in ecological communities. Historical and geographical perspectives,* ed. R. E. Ricklefs and D. Schluter, 350–63. Chicago: University of Chicago Press.

Rico, M., C. F. D. Rocha, V. N. T. Borges, Jr., and M. Van Sluys. 2004. Breeding ecology of *Scinax trapicheiroi* (Anura, Hylidae) at a creek in the Atlantic Rainforest of Ilha Grande, southeastern Brazil. *Amphibia-Reptilia* 25:277–86.

Riddle, O. 1909. The rate of digestion in cold-blooded vertebrates—The influence of season and temperature. *Am. J. Physiol.* 24:447–58.

Ridenhour, B. J., E. D. Brodie, Jr., and E. D. Brodie III. 1999. Effects of repeated injection of tetrodotoxin on growth and resistance to tetrodotoxin in the garter snake *Thamnophis sirtalis. Copeia* 1999:531–35.

Ridewood, W. G. 1897. On the structure and development of the hyobranchial skeleton and larynx in *Xenopus* and *Pipa,* with remarks on the affinities of the Aglossa. *Zool. J. Linn. Soc.* 26:53–128.

———. 1900. On the hyobranchial skeleton and larynx of a new aglossal toad, *Hymenochirus boettgeri. Zool. J. Linn. Soc.* 27:454–60.

Ridley, M. 1978. Paternal care. *Anim. Behav.* 26:904–32.

Ridpath, M. G. 1977. Predation on frogs and small birds by *Hierodula werneri* (Giglio-Tos) (Mantidae) in tropical Australia. *J. Aust. Ent. Soc.* 16:153–54.

Riechert, S. E. 1978. Games spiders play: Behavioral variability in territorial disputes. *Behav. Ecol. Sociobiol.* 3:135–62.

Riggs, A. 1951. The metamorphosis of hemoglobin in the bullfrog. *J. Gen. Physiol.* 35:23–40.

Riha, V., and K. A. Berven. 1991. An analysis of latitudinal variation in the larval development of the wood frog (*Rana sylvatica*). *Copeia* 1991:209–21.

Riley, J., and F. W. Huchzermeyer. 2000. Diet and lung parasites of swamp forest dwarf crocodiles (*Osteolaemus tetraspis osborni*) in the northern Congo Republic. *Copeia* 2000:582–86.

Riley, S. P. D., H. B. Schaffer, S. R. Vos, and B. M. Fitzpatrick. 2003. Hybridization between a rare, native tiger salamander (*Ambystoma californiense*) and its introduced congener. *Ecol. Appl.* 13:1263–75.

Rinaldi, A. C. 2002. Antimicrobial peptides from amphibian skin: An expanding scenario. *Curr. Op. Chem. Biol.* 6:799–804.

Rincón-Franco, F., and F. Castro-H. 1998. Aspectos ecologicos de una comunidad de *Eleutherodactylus* (Anura: Leptodactylidae) en un bosque de niebla del occidente de Colombia. *Caldasia* 20:193–202.

Rising, J. D., and F. W. Schueler. 1980. Screech owl eats fish and salamander in winter. *Wilson Bull.* 92:250–51.

Rissler, L. J., A. M. Barber, and H. M. Wilbur. 2000. Spatial and behavioral interactions between a native and introduced salamander species. *Behav. Ecol. Sociobiol.* 48:61–68.

Rissler, L. J., H. M. Wilbur, and D. R. Taylor. 2004. The influence of ecology and genetics on behavioral variation in salamander populations across the Eastern Continental Divide. *Am. Nat.* 164:201–13.

Rist, L., R. D. Semlitsch, H. Hotz, and H.-U. Reyer. 1997. Feeding behaviour, food consumption, and growth efficiency of hemiclonal and parental tadpoles of the *Rana esculenta* complex. *Funct. Ecol.* 11:735–42.

Ritke, M. E., and J. G. Babb. 1991. Behavior of the gray treefrog (*Hyla chrysoscelis*) during the non-breeding season. *Herp. Rev.* 22:5–8.

Ritke, M. E., J. G. Babb, and M. K. Ritke. 1990. Life history of the gray treefrog (*Hyla chrysoscelis*) in western Tennessee. *J. Herpetol.* 24:135–41.

———. 1991. Breeding-site specificity in the gray treefrog (*Hyla chrysoscelis*). *J. Herpetol.* 25:123–25.

———. 1992. Temporal patterns of reproductive activity in the gray treefrog (*Hyla chrysoscelis*). *J. Herpetol.* 26:107–11.

Ritke, M. E., and C. A. Lessman. 1994. Longitudinal study of ovarian dynamics in female gray treefrogs (*Hyla chrysoscelis*). *Copeia* 1994:1014–22.

Ritke, M. E., and R. L. Mumme. 1993. Choice of calling sites and oviposition sites by gray treefrogs (*Hyla chrysoscelis*)—a comment. *Ecology* 74:623–26.

Ritke, M. E., and R. D. Semlitsch. 1991. Mating behavior and determinants of male mating success in the gray treefrog, *Hyla chrysoscelis*. *Can. J. Zool.* 69:246–50.

Ritter, D., and K. C. Nishikawa. 1995. The kinematics and mechanism of prey capture in the African pig-nosed frog (*Hemisus marmoratum*): Description of a radically divergent anuran tongue. *J. Exp. Biol.* 198:2025–40.

Rivero, J. A. 1978. *Los anfibios y reptiles de Puerto Rico* (2nd ed.). San Juan: Editorial de la Universidad de Puerto Rico.

———. 1969. Observations on the agonistic and breeding behavior of *Leptodactylus pentadactylus* and other amphibian species in Venezuela. *Breviora* (321):1–14.

Roberts, D., S. Conroy, and K. Williams. 1999. Conservation status of frogs in Western Australia. In *Declines and disappearances of Australian frogs*, ed. A. Campbell, 177–84. 1999. Canberra: Environment Australia.

Roberts, J. D. 1981. Terrestrial breeding in the Australian leptodactylid frog *Myobatrachus gouldii* (Gray). *Austr. Wildl. Res.* 8:451–62.

———. 1984. Terrestrial egg deposition and direct development in *Arenophryne rotunda* Tyler, a myobatrachid frog from coastal sand dunes at Shark Bay, W. A. *Austr. Wildl. Res.* 11:191–200.

———. 1993. Natural history of the Anura. In *Fauna of Australia. Vol. 2A. Amphibia and reptilia*, ed. C. J. Glasby, G. J. B. Ross, and P. L. Beasley, 28–34. Canberra: Australian Government Publishing Service.

———. 1997. Call evolution in *Neobatrachus* (Anura: Myobatrachidae): speculations on tetraploid origins. *Copeia* 1997:791–801.

Roberts, J. D., P. Horwitz, G. Wardell-Johnson, L. R. Maxson, and M. J. Mahony. 1997. Taxonomy, relationships and conservation of a new genus and species of myobatrachid frog from the high rainfall region of southwestern Australia. *Copeia* 1997:373–81.

Roberts, J. D., and R. S. Seymour. 1989. Non-foamy egg masses in *Limnodynastes tasmaniensis* (Anura: Myobatrachidae) from South Australia. *Copeia* 1989:488–92.

Roberts, J. D., R. J. Standish, P. G. Byrne, and P. Doughty. 1999. Synchronous polyandry and multiple paternity in the frog *Crinia georgiana* (Anura: Myobatrachidae). *Anim. Behav.* 57:721–26.

Roberts, W. E. 1994. Explosive breeding aggregations and parachuting in a Neotropical frog, *Agalychnis saltator* (Hylidae). *J. Herpetol.* 28:193–99.

Robertson, D. R., H. P. A. Sweatman, G. A. Fletcher, and M. G. Cleland. 1976. Schooling as a means of circumventing the territoriality of competitors. *Ecology* 57:1208–20.

Robertson, J. G. M. 1984. Acoustic spacing by breeding males of *Uperoleia rugosa* (Anura: Leptodactylidae). *Z. Tierpsychol.* 64:283–97.

———. 1986a. Female choice, male strategies and the role of vocalizations in the Australian frog *Uperoleia rugosa*. *Anim. Behav.* 34:773–84.

———. 1986b. Territoriality, fighting and assessment of fighting ability in the Australian frog *Uperoleia rugosa*. *Anim. Behav.* 34:763–72.

———. 1989. Predation by a carabid beetle on Australian frogs. *Herp. Rev.* 20:7–8.

———. 1990. Female choice increases fertilization success in the Australian frog, *Uperoleia laevigata*. *Anim. Behav.* 39:639–45.

Robertson, K. M., and T. E. Ostertag. 2004. Problems with Shurbon and Fauth's test of effects of prescribed burning on amphibian diversity. *Cons. Biol.* 18:1154–55.

Robertson, S. L., and E. N. Smith. 1982. Evaporative water loss in the spiny soft-shelled turtle, *Trionyx spiniferus*. *Physiol. Zool.* 55:124–29.

Robineau, D. 1987. Sur la significatión phylogenetique de quelques caracteres anatomiques remarquables du Coelacanthe *Latimeria chalumnae* Smith, 1939. *Ann. Sci. Natur., Zool.* 8:43–60.

Robinson, D. C. 1961. The identity of the tadpole of *Anotheca coronata* (Stejneger). *Copeia* 1961:495.

———. 1976. A new dwarf salamander of the genus *Bolitoglossa* (Plethodontidae) from Costa Rica. *Proc. Biol. Soc. Washington* 89:289–94.

Robinson, J., R. A. Griffiths, and P. Jeffries. 2003. Susceptibility of frog (*Rana temporaria*) and toad (*Bufo bufo*) eggs to invasion by *Saprolegnia. Amphibia-Reptilia* 24:261–68.

Robinson, R. A., and W. J. Sutherland. 2002. Post-war changes in arable farming and biodiversity in Great Britain. *J. Appl. Ecol.* 39:157–76.

Roble, S. M. 1979. Dispersal movements and plant associations of juvenile gray treefrogs, *Hyla versicolor* Le Conte. *Trans. Kansas Acad. Sci.* 82:235–45.

———. 1985. Observations on satellite males in *Hyla chrysoscelis, Hyla picta,* and *Pseudacris triseriata. J. Herpetol.* 19:432–36.

Roček, Z. 1974. Beitrag zur Erkennung der Neotenie des Alpenmolches *Triturus alpetris* (Laurenti, 1768). *Vest. Českosl. Spol. Zool.* 38:285–94.

———. 1980. Cranial anatomy of frogs of the family Pelobatidae Stannius, 1856, with outlines of their phylogeny and systematics. *Acta Univ. Carol.* 1980:1–164.

———. 2000. Mesozoic anurans. In *Amphibian biology. Vol. 4. Palaeontology,* ed. H. Heatwole and R. L. Carroll, 1295–1331. Chipping Norton, New South Wales, Australia: Surrey Beatty and Sons.

Roček, Z., and J.-C. Rage. 2000a. Anatomical transformations in the transition from temnospondyl to proanuran stages. In *Amphibian biology. Vol. 4. Palaeontology,* ed. H. Heatwole and R. L. Carroll, 1274–82. Chipping Norton, New South Wales, Australia: Surrey Beatty and Sons.

———. 2000b. Proanuran stages (*Triadobatrachus, Czatkobatrahus*). In *Amphibian biology. Vol. 4. Palaeontology,* ed. H. Heatwole and R. L. Carroll, 1283–94. Chipping Norton, New South Wales, Australia: Surrey Beatty and Sons.

Rodda, G. H. 1984. Orientation and navigation of juvenile alligators: Evidence for magnetic sensitivity. *J. Comp. Physiol. A* 154:649–58.

Rodda, G. H., and J. B. Phillips. 1992. Navigational systems develop along similar lines in amphibians, reptiles, and birds. *Ethol. Ecol. Evol.* 4:43–51.

Rödel, M.-O. 1996. *Amphibien der Westafrikanischen Savanne.* Frankfurt am Main, Germany: Edition Chimaira.

———. 1998. A reproductive mode so far unknown in African ranids: *Phrynobatrachus guineensis* Guibé & Lamotte, 1961 breeds in tree holes (Anura: Ranidae). *Herpetozoa* 11:19–26.

———. 1999. Predation on tadpoles by hatchlings of the freshwater turtle *Pelomedusa subrufa. Amphibia-Reptilia* 20:173–83.

———. 2000. *Herpetofauna of West Africa. Vol. 1. Amphibians of the West African savanna.* Frankfurt am Main, Germany: Edition Chimaira.

Rödel, M.-O., and U. Braun. 1999. Association between anurans and ants in a West African savanna (Anura: Microhylidae, Hyperoliidae, and Hymenoptera: Formicidae). *Biotropica* 31:178–83.

Rödel, M.-O., and R. Ernst. 2002. A new reproductive mode for the genus *Phyrnobatrachus: Phrynobatrachus alticola* has nonfeeding, nonhatching tadpoles. *J. Herpetol.* 36:121–25.

Rödel, M.-O., V. H. W. Frohschammer, and K. E. Linsenmair. 2004. Life history of a West African tree-hole breeding frog, *Phyrnobatrachus guineensis,* Guibé & Lamotte, 1961 (Amphibia: Anura: Petropedetidae). In *Ecology and evolution of phytotelm-breeding anurans,* ed. R. M. Lehtinen, 31–44. *Misc. Publ. Mus. Zool. Univ. Michigan,* no. 193.

Rödel, M.-O., J. Kosuch, M. Veith, and R. Ernst. 2003. First record of the genus *Acanthixalus* Laurent, 1944 from the Upper Guinean rain forest, West Africa, with the description of a new species. *J. Herpetol.* 37:43–52.

Rödel, M.-O., and K. E. Linsenmair. 1997. Predator-induced swarms in the tadpoles of an African savanna frog, *Phrynomantis microps. Ethology* 103:902–14.

Rödel, M.-O., F. Range, J.-T. Seppänen, and R. Noë. 2002. Caviar in the rain forest: Monkeys as frog-spawn predators in Taï National Park, Ivory Coast. *J. Trop. Ecol.* 18:289–94.

Rödel, M.-O., G. Schorr, and R. Ernst. 2001. Zur Biologie von *Cardioglossa leucomystax* (Boulenger, 1903) im Taï-Nationalpark, Elfenbeinküste. *Salamandra* 37:239–60.

Rödel, M.-O., M. Spieler, K. Grabow, and C. Böckheler. 1995. *Hemisus marmoratus* (Peters, 1854) (Anura: Hemisotidae), Fortpflanzungsstrategien eines Savannenfrosches. *Bonn. Zool. Beitr.* 45:191–207.

Rodrigues, D. J., T. S. F. de Souza, and P. L Filho. 2003. *Phyllomedusa sauvagii* (Painted-belly leaf frog). Physical combat. *Herp. Rev.* 34:232.

Rodrigues, D. J., M. Uetanabaro, and F. S. Lopes. 2005. Reproductive patterns of *Trachycephalus venulosus* (Laurenti, 1768) and *Scinax fuscovarius* (Lutz, 1925) from the Cerrado, Central Brazil. *J. Nat. Hist.* 39:3217–26.

Rodriguez, L., and C. W. Myers. 1993. A new poison frog from Manu National Park, southeastern Peru (Dendrobatidae, *Epipedobates*). *Amer. Mus. Novitates* (3068):1–15.

Rodríguez, J. P. 2001. Exotic species introductions into South America: an underestimated threat? *Biodiver. Cons.* 10:1983–96.

Rodríguez, M. A., J. A. Belmontes, and B. A. Hawkins. 2005. Energy, water and large-scale patterns of reptile and amphibian species richness in Europe. *Acta Oecol.* 28:65–70.

Rodríguez-Prieto, I., and E. Fernández-Juricic. 2005. Effects of human disturbance on the endemic Iberian frog *Rana iberica* at individual and population levels. *Biol. Cons.* 123:1–9.

Rodriguez-Robles, J. A., D. C. Mulcahy, and H. W. Greene. 1999. Feeding ecology of the desert nightsnake, *Hypsiglena torquata* (Colubridae). *Copeia* 1999:93–100.

Roelants, K., and F. Bossuyt. 2005. Archaeobatrachian paraphyly and Pangean diversification of crown-group frogs. *Syst. Biol.* 54:111–26.

Roesli, M., and H.-U. Reyer. 2000. Male vocalization and female choice in the hybridogenetic *Rana lessonae*/*Rana esculenta* complex. *Anim. Behav.* 60:745–55.

Roff, D. A. 1992. *The evolution of life histories.* New York: Chapman and Hall.

Rogers, C. P. 1996. *Rana catesbeiana* (bullfrog). Predation. *Herp. Rev.* 27:75.

Rogers, K. D., F. Seebacher, and M. B. Thompson. 2004. Biochemical acclimation of metabolic enzymes in response to lowered temperature in tadpoles of *Limnodynastes peronii. Comp. Biochem. Physiol. A* 137:731–38.

Rogers, P. H., and M. Cox. 1988. Underwater sound as a biological stimulus. In *Sensory biology of aquatic animals,* ed. J. Atema, R. R. Fay, A. N. Popper, and W. N. Tavolga, 131–49. New York: Springer-Verlag.

Rogoff, J. L. 1927. The hedonic glands of *Triturus viridescens* a structural and morphological study. *Anat. Rec.* 34:132–33.

Rogowitz, G. L., C. L. Candelaria, L. E. Denizard, and L. J.

Meléndez. 2001. Seasonal reproduction of a Neotropical frog, the cave coquí (*Eleutherodactylus cooki*). *Copeia* 2001:542–47.

Rogowitz, G. L., M. Cortés-Rivera, and K. Nieves-Puigdoller. 1999. Water loss, cutaneous resistance, and effects of dehydration on locomotion in *Eleutherodactylus coqui. J. Comp. Physiol. B* 169:179–86.

Rogowitz, G. L., and J. Sánchez-Rivoleda. 1999. Locomotor performance and aerobic capacity of the cave coquí, *Eleutherodactylus cooki. Copeia* 1999:40–48.

Rohr, J. R., and D. M. Madison. 2003. Dryness increases predation risk in efts: Support for an amphibian decline hypothesis. *Oecologia* 135:657–64.

Rohr, J. R., D. M. Madison, and A. M. Sullivan. 2002a. The ontogeny of chemically-mediated antipredator behaviours in newts (*Notophthalmus viridescens*): responses to injured and non-injured conspecifics. *Behaviour* 139:1043–60.

———. 2002b. Sex differences and seasonal trade-off in responses to injured and non-injured conspecifics in red-spotted newts, *Notophthalmus viridescens. Behav. Ecol. Sociobiol.* 52:385–93.

———. 2003. On temporal variation and conflicting selection pressures: a test of theory using newts. *Ecology* 84:1816–26.

Rohr, J. R., D. Park, A. M. Sullivan, M. McKenna, C. R. Propper, and D. M. Madison. 2005. Operational sex ratio in newts: Field responses and characterization of a constituent chemical cue. *Behav. Ecol.* 16:286–93.

Rohr, J. R., and B. D. Palmer. 2005. Aquatic herbicide exposure increases salamander desiccation risk eight months later in a terrestrial environment. *Environ. Toxicol. Chem.* 24:1253–58.

Roithmair, M. E. 1992. Territoriality and male mating success in the dart-poison frog, *Epipedobates femoralis* (Dendrobatidae, Anura). *Ethology* 92:331–43.

———. 1994. Male territoriality and female mate selection in the dart-poison frog *Epipedobates trivittatus* (Dendrobatidae, Anura). *Copeia* 1994:107–15.

Rolle, F. J. 1963. Life history of the red-legged thrush (*Mimocichla plumbea ardodiacea*) in Puerto Rico. *Stud. Fauna Curacao and Other Caribb. Islands* 14:1–40.

Rollins-Smith, L. A. 1998. Metamorphosis and the amphibian immune system. *Immunol. Rev.* 166:221–30.

———. 2001. Neuroendocrine-immune system interactions in amphibians. *Immunol. Res.* 23:273–80.

Rollins-Smith, L. A., C. Carey, J. E. Longcore, J. K. Doersam, A. Boutte, J. E. Brzgal, and J. M. Conlon. 2002. Activity of antimicrobial skin peptides from ranid frogs against *Batrachochytrium dendrobatidis,* the chytrid fungus associated with gobal amphibian declines. *Dev. Comp. Immunol.* 26:471–79.

Rollins-Smith, L. A., C. Carey, J. M. Conlon, L. K. Reinert, J. K. Doersam, T. Bergman, J. Silberring, H. Lankinen, and D. Wade. 2003. Activities of temporin family peptides against the chytrid fungus (*Batrachochytrium dendrobatidis*) associated with global amphibian declines. *Anitmicrob. Agents Chemother.* 47:1157–60.

Rollins-Smith, L. A., and J. M. Conlon. 2005. Antimicrobial peptide defense against chytridiomycosis, an emerging infectious disease of amphibian populations. *Dev. Comp. Immunol.* 29:589–98.

Rollins-Smith, L. A., J. K. Doersam, J. E. Longcore, S. K. Taylor, J. C. Shamblin, C. Carey, and M. A. Zasloff. 2002. Antimicrobial peptide defenses against pathogens associated with global amphibian declines. *Dev. Comp. Immunol.* 26:63–72.

Rollins-Smith, L. A., L. K. Reinert, V. Miera, and J. M. Conlon. 2002. Antimicrobial peptide defenses of the Tarahumara frog, *Rana tarahumarae. Biochem. Biophys. Res. Comm.* 297:361–67.

Rollins-Smith, L. A., L. K. Reinert, C. J. O'Leary, L. E. Houston, and D. G. Woodhams. 2005. Antimicrobial peptide defenses in amphibian skin. *Integr. Comp. Biol.* 45:137–42.

Rollmann, S. M., L. D. Houck, and R. C. Feldhoff. 1999. Proteinaceous pheromone affecting female receptivity in a terrestrial salamander. *Science* 285:1907–9.

———. 2000. Population variation in salamander courtship pheromones. *J. Chem. Ecol.* 26:2713–24.

———. 2003. Conspecific and heterospecific pheromone effects on female receptivity. *Anim. Behav.* 66:857–61.

Rombough, C. J., D. J. Jordan, and C. A. Pearl. 2003. *Rana cascadae* (Cascade frog). Cannibalism. *Herp. Rev.* 34:138.

Rome, L. C. 1983. The effect of long-term exposure to different temperatures on the mechanical performance of frog muscle. *Physiol. Zool.* 56:33–40.

Rome, L. C., P. T. Loughna, and G. Goldspink. 1985. Temperature acclimation improves sustained swimming performance at low temperatures in carp. *Science* 228:194–96.

Rome, L. C., E. D. Stevens, and H. B. John-Alder. 1992. The influence of temperature and thermal acclimation on physiological function. In *Environmental physiology of the amphibians,* ed. M. E. Feder and W. W. Burggren, 183–205. Chicago: University of Chicago Press.

Romer, A. S. 1959. *The vertebrate story* (4th ed.). Chicago: University of Chicago Press.

Romer, J. D. 1951. Observations on the habits and life-history of the Chinese newt, *Cynops chinensis* Gray. *Copeia* 1951:213–19.

Romspert, A. P. 1976. Osmoregulation of the African clawed frog, *Xenopus laevis,* in hypersaline media. *Comp. Biochem. Physiol.* 54A:207–10.

Romspert, A. P., and L. L. McClanahan. 1981. Osmoregulation of the terrestrial salamander, *Ambystoma tigrinum,* in hypersaline media. *Copeia* 1981:400–405.

Ron, S. R. 2005. Predicting the distribution of the amphibian pathogen *Batrachochytrium dendrobatidis* in the New World. *Biotropica* 37:209–21.

Ron, S. R., and A. Merino. 2000. Amphibian declines in Ecuador: Overview and first report of chytridiomycosis from South America. *Froglog* (42):4.

Ron, S. R., J. C. Santos, and D. C. Cannatella. 2006. Phylogeny of the tungara frog genus *Engystomops* (= *Physalaemus pustulosus* species group; Anura: Leptodactylidae). *Mol. Phylogen. Evol.* 39:392–403.

Rooney, T. P., C. Antolik, and M. D. Moran. 2000. The impact of salmander predation on collembola abundance. *Proc. Entomol. Soc. Washington* 102:308–12.

Roonwal, M. L., and M. B. Kripalani. 1961. New frog *Philautus cherrapunjiae* from Assam, India, with field observations on its behaviour and metamorphosis. *Rec. Indian Mus.* 59:325–33.

Root, T. L., J. T. Price, K. R. Hall, S. H. Schneider, C. Rosenzweig, and J. A. Pounds. 2003. Fingerprints of global warming on wild animals and plants. *Nature* 421:57–60.

Roper, T. J. 1994. The European badger *Meles meles:* food specialist or generalist? *J. Zool., London* 234:437–52.

Roper, T. J., and E. Mickevicius. 1995. Badger *Meles meles* diet: A review of literature from the former Soviet Union. *Mammal Rev.* 25:117–29.

Rorabaugh, J. C. 2005. *Rana pipiens* Schreber, 1782. Northern leopard frog. In *Amphibian declines: The conservation status of United States species,* ed. M. J. Lannoo, 570–77. Berkeley: University of California Press.

Rorabaugh, J. C. and S. F. Hale. 2005. *Rana tarahumarae* Boulenger, 1917. Tarahumara frog. In *Amphibian declines: The conservation status of United States species,* ed. M. J. Lannoo, 593–95. Berkeley: University of California Press.

Rose, B. L. 1982. Lizard home ranges: Methodology and functions. *J. Herpetol.* 16:253–69.

Rose, C. S. 1999. Hormonal control in larval development and evolution—Amphibians. In*The origin and evolution of larval forms,* ed. B. K. Hall and M. H. Wake, 167–216. San Diego: Academic Press.

———. 2005. Ingetrating ecology and developmental biology to explain the timing of frog metamorphosis. *Trends Ecol. Evol.* 20:129–35.

Rose, F. L. 1966. Homing to nests by the salamander *Desmognathus auriculatus. Copeia* 1966:251–53.

———. 1967. Seasonal changes in lipid levels of the salamander *Amphiuma means. Copeia* 1967:662–66.

Rose, F. L., and D. Armentrout. 1976. Adaptive strategies of *Ambystoma tigrinum* Green inhabiting the Llano Estacado of West Texas. *J. Anim. Ecol.*45:713–29.

Rose, F. L., D. Armentrout, and P. Roper. 1971. Physiological responses of paedogenic *Ambystoma tigrinum* to acute anoxia. *Herpetologica* 27:101–7.

Rose, F. L., and R. B. Drotman. 1967. Anaerobiosis in a frog, *Rana pipiens. J. Exp. Zool.* 166:427–32.

Rose, F. L., and H. L. Lewis. 1968. Changes in weight and free fatty acid concentration of fat bodies of paedogenic *Ambystoma tigrinum* during vitellogenesis. *Comp. Biochem. Physiol.* 26:149–54.

Rose, G. J. 1986. A temporal processing mechanism for all species? *Brain Behav. Evol.* 28:134–44.

Rose, G. J., and E. A. Brenowitz. 1991. Aggressive thresholds of male Pacific treefrogs for advertisement calls vary with amplitude of neighbors' calls. *Ethology* 89:244–52.

———. 1997. Plasticity of aggressive thresholds in *Hyla regilla:* discrete accommodation to encounter calls. *Anim. Behav.* 53:353–61.

———. 2002. Pacific treefrogs use temporal integration to differentiate advertisement from encounter calls. *Anim. Behav.* 63:1183–90.

Rose, G. J., E. A. Brenowitz, and R. R. Capranica. 1985. Species specificity and temperature dependency of temporal processing by the auditory midbrain of two species of treefrogs. *J. Comp. Physiol. A* 157:763–69.

Rose, G. J., and R. R. Capranica. 1983. Temporal selectivity in the central auditory system of the leopard frog. *Science* 219:1087–89.

———. 1984. Processing amplitude-modulated sounds by the auditory midbrain of two species of toads: matched temporal filters. *J. Comp. Physiol. A* 154:211–19.

———. 1985. Sensitivity to amplitude modulated sounds in the anuran auditory nervous system. *J. Neurophysiol.* 53:446–65.

Rose, G. J., and A. S. Rand. 1991. Differential response to frequency modulation rate and direction by the Neotropical frog, *Physalaemus pustulosus. Anim. Behav.* 42:413–21.

Rose, G. J., R. Zelick, and A. S. Rand. 1988. Auditory processing of temporal information in a Neotropical frog is independent of signal intensity. *Ethology* 77:330–36.

Rose, S. M. 1960. A feedback mechanism of growth control in tadpoles. *Ecology* 41:188–99.

Rose, S. M., and F. C. Rose. 1961. Growth controlling exudates of tadpoles. *Symp. Soc. Exp. Biol.* 15:207–18.

———. 1965. The control of growth and reproduction in freshwater organisms by specific products. *Mitt. Int. Ver. Limnol.* 13:21–35.

Rose, W. 1956. Parental care in batrachians. *Afr. Wildlife* 10:257.

———. 1962. *The reptiles and amphibians of Southern Africa.* Capetown, South Africa: Maskew Miller.

Roseghini, M., V. Erspamer, and R. Endean. 1976. Indoleimidazole- and phenylalkylamines in the skin of one hundred amphibian species from Australia and New Guinea. *Comp. Biochem. Physiol.* 54C:31–43.

Rosen, D. E., P. L. Forey, B. G. Gardiner, and C. Patterson. 1981. Lungfishes, tetrapods, paleontology, and plesiomorphy. *Bull. Amer. Mus. Nat. Hist.* 167:159–276.

Rosen, M., and R. E. Lemon. 1974. The vocal behavior of spring peepers, *Hyla crucifer. Copeia* 1974:940–50.

Rosen, P. C., and C. R. Schwalbe. 1995. Bullfrogs: introduced predators in southwestern wetlands. In *Our living resources. A report to the nation on the distribution, abundance, and health of U.S. plants, animals, and ecosystems,* ed. E. T. LaRoe, G. S. Farris, C. E. Puckett, P. D. Doran, and M. J. Mac, 452–54. Washington, DC: U.S. Deparment of Interior, National Biological Service.

Rosen, P. C., C. R. Schwalbe, D. A. Parizek, Jr., P. A. Holm, and C. H. Lowe. 1995. Introduced aquatic vertebrates in the Chiricahua region: Effects on declining native ranid frogs. In *Biodiversity and management of the Madrean Archipelago: The Sky Islands of Southwestern United States and Northwestern Mexico,* ed. L. F DeBano, 251–61. USDA Forest Service General Technical Report RM-GTR-264. Fort Collins, CO: Rocky Mountain Forest and Range Experiment Station, U.S. Department of Agriculture.

Rosenfield, R. N., M. W. Gratson, and L. B. Carson. 1984. Food brought by broad-winged hawks to a Wisconsin nest. *J. Field Ornithol.* 55:246–47.

Rosenkilde, P. 1979. The thyroid hormones in Amphibia. In *Hormones and evolution, vol. 1,* ed. E. J. W. Barrington 437–91. New York: Academic Press.

Rosenkilde, P., and A. P. Ussing. 1996. What mechanisms control neoteny and regulate induced metamorphosis in urodeles? *Int. J. Develop. Biol.* 40:665–73.

Rosenthal, G. G., A. S. Rand, and M. J. Ryan. 2004. The vocal sac as a visual cue in anuran communication: an experiemental analysis using video playback. *Anim. Behav.* 68:55–58.

Rosenthal, G. M. 1957. The role of moisture and temperature in the local distribution of the plethodontid salamander *Aneides lugubris. Univ. California Publ. Zool.* 54:371–420.

Rosenzweig, M. L., and Z. Abramsky. 1993. How are diversity and productivity related? In *Species diversity in ecological communities. Historical and geographical perspectives,* ed. R. E. Ricklefs and D. Schluter, 52–65. Chicago: University of Chicago Press.

Ross, B., T. Fredericksen, E. Ross, W. Hoffman, M. L. Morrison, J. Beyea, M. B. Lester, B. N. Johnson, and N. J. Fredericksen. 2000. Relative abundance and species richness of herpetofauna in forest stands in Pennsylvania. *Forest Sci.* 46:139–46.

Ross, D. A. 1989. Amphibians and reptiles in the diets of North American raptors. *Wisconsin Endangered Resources Report 59.* Madison: Wisconsin Department of Natural Resources.

Ross, J. P. 1992. *Crocodile conservation action.* Gland, Switzerland; IUCN-World Conservation Union.

Ross, R. J., and J. J. B. Smith. 1978. Detection of substrate vibrations by salamanders: inner ear organ activity. *Can. J. Zool.* 56:1156–62.

Ross, R. J., and J. J. B. Smith. 1979. Detection of substrate vibrations by salamanders: Eighth cranial nerve activity. *Can. J. Zool.* 57:368–74.

Rosset, S. D., and L. Alcalde. 2004. Distribution of the burrows of *Bufo fernandezae* (Anura: Bufonidae) outside of the breeding season. *Phyllomedusa* 3:95–99.

Rossman, D. A., N. B. Ford, and R. A. Seigel. 1996. *The garter snakes. Evolution and ecology.* Norman: University of Oklahoma Press.

Rotenberry, J. T. 1980. Dietary relationships among shrub-steppe passerine birds: competition or opportunism in a variable environment. *Ecol. Monogr.* 50:93–110.

Roth, A., and P. Schlegel. 1988. Behavioral evidence and supporting electrophysiological observations for electroreception in the blind cave salamander *Proteus anguineus* (Urodela). *Brain Behav. Evol.* 32:277–80.

Roth, A. H., and J. F. Jackson. 1987. The effect of pool size on recruitment of predatory insects and on mortality in a larval anuran. *Herpetologica* 43:224–32.

Roth, E. D., P. G. May, and T. M. Farrell. 1999. Pigmy rattlesnakes use frog-derived chemical cues to select foraging sites. *Copeia* 1999:772–74.

Roth, G. 1976. Experimental analysis of the prey catching behavior *of Hydromantes italicus* Dunn (Amphibia, Plethodontidae). *J. Comp. Physiol.* 109:47–58.

———. 1987. *Visual behaviour in salamanders.* New York: Springer-Verlag.

Roth, G., J. Blanke, and M. Ohle. 1995. Brain size and morphology in miniaturized plethodontid salamanders. *Brain Behav. Evol.* 45:84–95.

Roth, G., J. Blanke, and D. B. Wake. 1994. Cell size predicts morphological complexity in the brains of frogs and salamanders. *Proc. Natl. Acad. Sci. USA USA* 91:4796–4800.

Roth, G., U. Dicke, and K. Nishikawa. 1992. How do ontogeny, morphology, and physiology of sensory systems constrain and direct the evolution of amphibians? *Am. Nat.* 139:S105–S124.

Roth, G., K. C. Nishikawa, C. Naujoks-Manteuffel, A. Schmidt, and D. B. Wake. 1993. Paedomorphosis and simplification in the nervous system of salamanders. *Brain Behav. Evol.* 42:137–70.

Roth, G., K. C. Nishikawa, and D. B. Wake. 1997. Genome size, secondary simplification, and the evolution of the brain in salamanders. *Brain Behav. Evol.* 50:50–59.

Roth, G., and D. B. Wake. 1985a. Trends in the functional morphology and sensorimotor control of feeding behavior in salamanders: An example of internal dynamics in evolution. *Acta Biotheor.* 34:175–92.

———. 1985b. The structure of the brainstem and cervical spinal cord in lungless salamanders (family Plethodontidae) and its relation to feeding. *J. Comp. Neurol.* 241:99–110.

———. 1989. Conservation and innovation in the evolution of feeding in vertebrates, pp. 7–21. In *Complex organismal functions: Integration and evolution in vertebrate,*ed. D. B. Wake and G. Roth, 7–21. New York: Wiley.

Roth, J. J. 1973. Vascular supply to the ventral pelvic region of anurans as related to water balance. *J. Morphol.* 140:443–60.

Rothermel, B. B. 2004. Migratory success of juveniles: A potential constraint on connectivity for pond-breeding amphibians. *Ecol. Appl.* 14:1535–46.

Rothermel, B. B., and R. D. Semlitsch. 2002. An experimental investigation of landscape resistance of forest versus old-field habitats to emerging juvenile amphibians. *Cons. Biol.* 16:1324–32.

Roudebush, R. E., and D. H. Taylor. 1987a. Behavioral interactions between two desmognathine salamander species: Importance of competition and predation. *Ecology* 68:1453–58.

———. 1987b. Chemical communication between two species of desmognathine salamanders. *Copeia* 1987:744–48.

Roughgarden, J. 1983. Competition and theory in community ecology. *Am. Nat.* 122:583–601.

Rouillé, Y., G. Michel, M. T. Chauvet, J. Chauvet, and R. Archer. 1989. Hydrins, hydroosmotic neurohypophysial peptides: Osmoregulatory adaptation in amphibians through vasotocin precursor processing. *Proc. Natl. Acad. Sci. USA* 86:5272–75.

Rounsevell, D. E., D. Ziegeler, P. B. Brown, M. Davies, and M. J. Littlejohn. 1994. A new genus and species of frog (Anura: Leptodactylidae: Myobatrachidae) from southern Tasmania. *Trans. Roy. Soc. South Australia* 118:171–85.

Rouse, J. D., C. A. Bishop, and J. Struger. 1999. Nitrogen pollution: An assessment of its threat to amphibian survival. *Env. Health Perspect.* 107:799–803.

Rowe, C. L., and W. A. Dunson. 1993. Relationships among abiotic parameters and breeding effort by three amphibians in temporary wetlands of central Pennsylvania. *Wetlands* 13:237–46.

———. 1995. Impacts of hydroperiod on growth and survival of larval amphibians in temporary ponds of Central Pennsylvania, USA. *Oecologia* 102:397–403.

Rowe, C. L., and J. Freda. 2000. Effects of acidification on amphibians at multiple levels of biological organization. In *Ecotoxicology of amphibians and reptiles,* ed. D. W. Sparling, G. Linder, and C. A. Bishop, 545–71. Pensacola, FL: SETAC Press.

Rowe, C. L., and W. A. Hopkins. 2003. Anthropogenic activities producing sink habitats for amphibians in the local landscape: A case study of lethal and sublethal effects of coal combustion residues in the aquatic environment. In *Amphibian decline: An integrated analysis of multiple stressor effects,* ed. G. Linder, S. K. Krest, and D. W. Sparling, 271–82. Pensacola, FL: SETAC Press.

Rowe, C. L., W. A. Hopkins, and C. M. Bridges. 2003. Physiological ecology of amphibians in relation to susceptibility to natural and anthropogenic factors. In *Amphibian decline: An integrated analysis of multiple stressor effects,* ed. G. Linder, S. K. Krest, and D. W. Sparling, 9–57. Pensacola, FL: SETAC Press.

Rowe, C. L., O. M. Kinney, R. D. Nagle, and J. D. Congdon. 1998. Elevated maintenance costs in an anuran (*Rana catesbeiana*) exposed to a mixture of trace elements during the embryonic and early larval periods. *Physiol. Zool.* 71:27–35.

Rowe, C. L., W. J. Sadinski, and W. A. Dunson. 1994. Predation on larval and embryonic amphibians by acid-tolerant caddisfly larvae (*Ptilostomis postica*). *J. Herpetol.* 28:357–64.

Rowe, G., and T. J. C. Beebee. 2003. Population on the verge of a mutational meltdown? Fitness costs of genetic load for an amphibian in the wild. *Evolution* 57:177–81.

———. 2005. Intraspecific competition disadvantages inbred natterjack toad (*Bufo calamita*) genotypes over outbred ones in a shared pond environment. *J. Anim. Ecol.*74:71–76.

Rowe, G., T. J. C. Beebee, and T. Burke. 1999. Microsatellite heterozygosity, fitness, and demography in natterjack toads *Bufo calamita. Anim. Cons.*2:85–92.

Rowe, L., and D. Ludwig. 1991. Size and timing of metamorphosis in complex life cycles: Time constraints and variation. *Ecology* 72:413–27.

Rowe-Rowe, D. T. 1977. Food ecology of otters in Natal. *Oikos* 28:210–19.

Rowe-Rowe, D. T., and M. J. Somers. 1998. Diet, foraging behaviour and coexistence of African otters and the water mongoose. In *Behaviour and ecology of riparian mammals,* ed. N. Dunstone and M. Gorman, 215–27. Cambridge: Cambridge University Press.

Rowland, W. J., C. C. Robb, and S. A. Cortwright. 1990. Chemically-mediated mate choice in red-spotted newts: do males select or just detect females? *Anim. Behav.* 39:811–13.

Roy, D., B. Borah, and A. Sarma. 1995. Analysis and significance of female reciprocal call in frogs. *Curr. Sci.* 69:265–70.

Rubbo, M. J., and J. M. Kiesecker. 2004. Leaf litter composition and community structure: translating regional species change into local dynamics. *Ecology* 85:2519–25.

———. 2005. Amphibian breeding distribution in an urbanized landscape. *Cons. Biol.* 19:504–11.

Rubbo, M. J., V. R. Townsend, Jr., S. D. Smyers, and R. G. Jaeger. 2003. An experimental assessment of invertebrate/vertebrate predation: The interaction between wolf spiders (*Gladicosa pulchra*) and terrestrial salamanders (*Ambystoma maculatum*). *J. Zool., London* 261:1–5.

Ruben, J. A., and A. J. Boucot. 1989. The origin of the lungless salamanders (Amphibia: Plethodontidae). *Am. Nat.* 134:161–69.

Ruben, J. A., N. L. Reagan, P. A. Verrell, and A. J. Boucot. 1993. Plethodontid salamander origins: a response to Beachy and Bruce. *Am. Nat.* 142:1038–51.

Rubenstein, D. I. 1980. On the evolution of alternative mating strategies. In *Limits to action: The allocation of individual behavior,* ed. J. E. R. Staddon, 65–99. New York: Academic Press.

Rubenstein, N. A., S. D. Erulkar, and G. T. Schneider. 1983. Sexual dimorphism in the fibers of a "clasp" muscle of *Xenopus laevis*. *Exp. Neurol.* 82:424–31.

Rubinoff, D. 1996. *Aneides lugubris* (arboreal salamander). Predation. *Herp. Rev.* 27:135.

Rubinsky, B., S. T. S. Wong, J.-S. Hong, J. Gilbert, M. Roos, and K. B. Storey. 1994. $^{1}$H magnetic resonance imaging of freezing and thawing in freeze-tolerant frogs. *Am. J. Physiol.* 266:R1771–R1777.

Rugh, R. 1934. The space factor in the growth rate of tadpoles. *Ecology* 15:407–11.

Rühmekorf, E. 1958a. Beitrage zur Ökologie Mitteleuropäischer Salientia. I. Abhängigkeit der Laichabgabe von Aussenfaktoren. *Z. Morph. Ökol. Tiere* 47:1–19.

———. 1958b. Beitrage zur Ökologie Mitteleuropäischer Salientia. II. Temperaturwahl der Larven. *Z. Morph. Ökol. Tiere* 47:20–36.

Ruibal, R. 1959. The ecology of a brackish water population of *Rana pipiens*. *Copeia* 1959:315–22.

———. 1962. The adaptive value of bladder water in the toad, *Bufo cognatus*. *Physiol. Zool.* 35:218–23.

Ruibal, R., and S. Hillman. 1981. Cocoon structure and function in the burrowing hylid frog, *Pternohyla fodiens*. *J. Herpetol.* 15:403–8.

Ruibal, R., and V. Shoemaker. 1984. Osteoderms in anurans. *J. Herpetol.* 18:313–28.

Ruibal, R., and E. Thomas. 1988. The obligate carnivorous larvae of the frog *Lepidobatrachus laevis* (Leptodactylidae). *Copeia* 1988:591–604.

Ruibal, R., L. Tevis, Jr., and V. Roig. 1969. The terrestrial ecology of the spadefoot toad *Scaphiopus hammondii*. *Copeia* 1969:571–84.

Ruiz-Carranza, P. M., and J. I. Hernandez-Camacho. 1976. *Osornophryne,* genero nuevo de anfibios bufonidos de Colombia y Ecuador. *Caldasia* 11:93–148.

Ruiz-Carranza, P. M., and J. D. Lynch. 1991a. Ranas Centrolenidae de Colombia. I. Propuesta de una nueva clasificación genérica. *Lozania (Acta Zool. Colombiana)* (57):1–30.

———. 1991b. Centrolenidae de Colombia II. Nuevas especies de *Centrolene* de la Cordillera Oriental y Sierra Nevada de Santa Marta. *Lozania (Acta Zool. Colombiana)* (58):1–26.

———. 1991c. Centrolenidae de Colombia III. Nuevas especies de *Cochranella* del grupo *granulosa*. *Lozania (Acta Zool. Colombiana)* (59):1–18.

Ruiz-Carranza, P. M., and M. P. Ramirez-Pinilla. 1992. Una nueva especie de *Minyobates* (Anura: Dendrobatidae) de Colombia. *Lozania (Acta Zool. Colombiana)* (61):1–16.

Ruiz-Olmo, J. 1998. Influence of altitude on the distribution, abundance and ecology of the otter (*Lutra lutra*). In *Behaviour and ecology of riparian mammals,* ed. N. Dunstone and M. Gorman, 159–76. Cambridge: Cambridge University Press.

Rundio, D. E., and D. H. Olson. 2003. Antipredator defenses of larval Pacific giant salamanders (*Dicamptodon tenbrosus*) against cutthroat trout (*Oncorhynchus clarki*). *Copeia* 2003:402–7.

Runkle, L. S., K. D. Wells, C. C. Robb, and S. L. Lance. 1994. Individual, nightly, and seasonal variation in calling behavior of the gray treefrog, *Hyla versicolor:* Implications for energy expenditure. *Behav. Ecol.* 5:318–25.

Runkova, G. G., Z. L. Stepanova, and L. A. Koval'chuck. 1974. Organ specificity in the action of metabolites of amphibian larvae on their endogenous metabolism under condition of increased population density. The role of protein in the regulating influence of metabolites. *Akad. Nauk. SSSR Doklady Biol. Sci. Sec.* 217:328–30.

Ruschenberger, W. S. W. 1856. *Elements of herpetology, and of ichthyology: Prepared for the use of schools and colleges.* Philadelphia: J. B. Lippincott.

Russell, K. R., A. A. Gonyaw, J. D. Strom, K. E. Diemer, and K. C. Murk. 2002. Three new nests of the Columbia torrent salamander, *Rhyacotriton kezeri,* in Oregon with observations of nesting behavior. *Northwest. Nat.* 83:19–22.

Russell, K. R., D. C. Guynn, Jr., and H. G. Hanlin. 2002. Importance of small isolated wetlands for herpetofaunal diversity in managed, young growth forests in the Coastal Plain of South Carolina. *Forest Ecol. Manage.* 163:43–59.

Russell, K. R., T. J. Mabee, and M. B. Cole. 2004. Distribution and habitat of Columbia torrent salamanders at multiple spatial scales in managed forests of northwestern Oregon. *J. Wildl. Manage.* 68:405–17.

Russell, K. R., D. H. Van Lear, and D. C. Guynn. 1999. Prescribed fire effects on herpetofauna: review and management implications. *Wildl. Soc. Bull.* 27:374–84.

Russell, L. D., R. A. Brandon, E. J. Zalisko, and J. Martan. 1981. Spermatophore of the salamander *Ambystoma texanum*. *Tissue and Cell* 13:609–21.

Ruta, M., M. I. Coates, and D. L. J. Quicke. 2003. Early tetrapod relationships revisted. *Biol. Rev.* 78:251–345.

Ruth, S. B. 1988. The life history and current status of the Santa Cruz long-toed salamander, *Ambystoma macrodactylum croceum.* In *Conference on California herpetology proceedings.* Spec. Publ. 4. H. F. De Lisle, P. R. Brown, B. Kaufmann, and B. M. McGurty, 89–110. Van Nuys, CA: Southwestern Herpetologists Society.

Rutledge, P. S., J. R. Spotila, and D. P. Easton. 1987a. Heat hardening in response to two types of heat shock in the lungless salamanders *Eurycea bislineata* and *Desmognathus ochrophaeus. J. Therm. Biol.* 12:235–41.

———. 1987b. Heat shock proteins from the lungless salamanders *Eurycea bislineata* and *Desmognathus ochrophaeus. Comp. Biochem. Physiol.* 88B:13–18.

Ruvinsky, I., and L. R. Maxson. 1996. Phylogenetic relationships among bufonoid frogs (Anura: Neobatrachia) inferred from mitochondrial DNA sequences. *Mol. Phylogen. Evol.* 3:533–47.

Ryan, M. J. 1977. A thermal property of the *Rana catesbeiana* egg mass. *J. Herpetol.* 12:247–48.

Ryan, M. J. 1980a. Female mate choice in a Neotropical frog. *Science* 209:523–25.

———. 1980b. The reproductive behavior of the bullfrog (*Rana catesbeiana*). *Copeia* 1980:108–14.

———. 1983a. Frequency modulated calls and species recognition in a Neotropical frog. *J. Comp. Physiol.* 150:217–21.

———. 1983b. Sexual selection and communication in a Neotropical frog, *Physalaemus pustulosus. Evolution* 37:261–72.

———. 1985a. Energetic efficiency of vocalization by the frog *Physalaemus pustulosus. J. Exp. Biol.* 116:47–52.

———. 1985b. *The Túngara frog. A study in sexual selection and communication.* Chicago: University of Chicago Press.

———. 1986a. Neuroanatomy influences speciation rates in anurans. *Proc. Natl. Acad. Sci. USA* 83:1379–82.

———. 1986b. Synchronized calling in a treefrog (*Smilisca sila*): Short behavioral latencies and implications for neural pathways involved in call perception and production. *Brain Behav. Evol.* 29:196–206.

———. 1988a. Constraints and patterns in the evolution of anuran acoustic communication. In *The evolution of the amphibian auditory system,* ed. B. Fritzsch, M. J. Ryan, W. Wilczynski, T. E. Hetherington, and W. Walkowiak, 637–77. New York: Wiley.

———. 1988b. Energy, calling, and selection. *Amer. Zool.* 28:885–98.

———. 1990. Sexual selection, sensory systems and sensory exploitation. *Oxford Surveys in Evolutionary Biology* 7:159–95.

———. 1991. Sexual selection and communication in frogs. *Trends Ecol. Evol.* 6:351–55.

———. 2001. *Anuran communication,* ed. Washington, DC: Smithsonian Institution Press.

Ryan, M. J., G. A. Bartholomew, and A. S. Rand. 1983. Energetics of reproduction in a Neotropical frog, *Physalaemus pustulosus. Ecology* 64:1456–62.

Ryan, M. J., and E. A. Brenowitz. 1985. The role of body size, phylogeny, and ambient noise in the evolution of bird song. *Am. Nat.* 126:87–100.

Ryan, M. J., R. B. Cocroft, and W. Wilczynski. 1990. The role of environmental selection in intraspecific divergence of mate cognition signals in the cricket frog, *Acris crepitans. Evolution* 44:1869–72.

Ryan, M. J., and R. C. Drewes. 1990. Vocal morphology of *Physalaemus pustulosus* species group (Leptodactylidae): Morphological response to sexual selection for complex calls. *Biol. J. Linn. Soc.* 40:37–52.

Ryan, M. J., J. H. Fox, W. Wilczynski, and A. S. Rand. 1990. Sexual selection for sensory exploitation in the frog *Physalaemus pustulosus. Nature* 343:66–67.

Ryan, M. J., and A. Keddy-Hector. 1992. Directional patterns of female mate choice and the role of sensory biases. *Am. Nat.* 139:S4–S35.

Ryan, M. J., S. A. Perrill, and W. Wilczynski. 1992. Auditory tuning and call frequency predict population-based mating preferences in the cricket frog, *Acris crepitans. Am. Nat.* 139:1370–83.

Ryan, M. J., and A. S. Rand. 1990. The sensory basis of sexual selection for complex calls in the túngara frog, *Physalaemus pustulosus* (sexual selection for sensory exploitation). *Evolution* 44:305–14.

———. 1993a. Phylogenetic patterns of behavioral mate recognition systems in the *Physalaemus pustulosus* species group (Anura: Leptodactylidae): The role of ancestral and derived characters and sensory exploitation. In *Evolutionary pattern and processes. Linnean Soc. Symp. Series 14,* ed. D. R. Lees and D. Edwacrds, 251–67. Academic Press, New York.

———. 1993b. Species recognition and sexual selection as a unitary problem in animal communication. *Evolution* 47:647–57.

———. 1993c. Sexual selection and signal evolution: The ghost of biases past. *Phil. Trans. Roy. Soc. London B* 340:187–95.

———. 1995. Female responses to ancestral advertisement calls in túngara frogs. *Science* 269:390–92.

———. 1999a. Phylogenetic inference and the evolution of communication in túngara frogs. In *The design of animal communication,* ed. M. D. Hauser and M. Konishi, 535–57. Cambridge, MA: MIT Press.

———. 1999b. Phylogenetic influence on mating call preferences in female túngara frogs, *Physalaemus pustulosus. Anim. Behav.* 57:945–56.

———. 2001. Feature weighting in signal recognition and discrimination by túngara frogs. In *Anuran communication,* ed. M. J. Ryan, 86–101. Washington, DC: Smithsonian Institution Press.

Ryan, M. J., W. Rand, P. L. Hurd, S. M. Phelps, and A. S. Rand. 2003. Generalization in response to mate recognition signals. *Am. Nat.* 161:380–94.

Ryan, M. J., and B. K. Sullivan. 1989. Transmission effects on temporal structure in the advertisement calls of two toads, *Bufo woodhousii* and *Bufo valliceps. Ethology* 80:182–89.

Ryan, M. J., M. D. Tuttle, and A. S. Rand. 1982. Bat predation and sexual advertisement in a Neotropical frog. *Am. Nat.* 119:136–39.

Ryan, M. J., M. D. Tuttle, and L. K. Taft. 1981. The costs and benefits of frog chorusing behavior. *Behav. Ecol. Sociobiol.* 8:273–78.

Ryan, M. J., and W. Wilczynski. 1988. Coevolution of sender and receiver: Effect on local mate preference in cricket frogs. *Science* 240:1786–88.

———. 1991. Evolution of intraspecific variation in the advertisement call of a cricket frog (*Acris crepitans,* Hylidae). *Biol. J. Linn. Soc.* 44:249–71.

Ryan, R. A. 1953. Growth rates of some ranids under natural conditions. *Copeia* 1953:73–80.

Ryan, T. J., and G. R. Plague. 2004. Hatching asynchrony, survival, and the fitness of alternative adult morphs in *Ambystoma talpoideum. Oecologia* 140:46–51.

Ryan, T. J., and R. D. Semlitsch. 1998. Intraspecific heterochrony and life history evolution: Decoupling somatic and sexual development in a facultatively paedomorphic salamander. *Proc. Natl. Acad. Sci. USA USA* 95:5643–48.

Ryan, T. J., and C. T. Winne. 2001. Effects of hydroperiod on metamorphosis in *Rana sphenocephala*. *Am. Midl. Nat.* 145:46–53.

Ryberg, W. A., and G. H. Dayton. 2004. *Scaphiopus couchii* (Couch's spadefoot). Predation. *Herp. Rev.* 35:263.

Ryser, J. 1986. Altersstruktur, Geschlechterverhaltnis und Dynamik einer Grasfrosch-Population (*Rana temporaria* L.) aus der Schweiz. *Zool. Anz.* 217:234–51.

———. 1988a. Clutch parameters in a Swiss population of *Rana temporaria*. *Herpetol. J.* 1:310–11.

———. 1988b. Determination of growth and maturation in the common frog, *Rana temporaria*, by skeletochronology. *J. Zool., London* 216:673–85.

———. 1989a. The breeding migration and mating system of a Swiss population of the common frog *Rana temporaria*. *Amphibia-Reptilia* 10:13–21.

———. 1989b. Weight loss, reproductive output, and the cost of reproduction in the common frog, *Rana temporaria*. *Oecologia* 78:264–68.

Rzoska, J. 1961. Observations on tropical rainpools and general remarks on temporary waters. *Hydrobiologia* 17:265–80.

———. 1974. The upper Nile swamps, a tropical wet land study. *Freshwater Biol.* 4:1–30.

Saatchi, S., D. Agosti, K. Alger, J. Delabie, and J. Musinky. 2001. Examining fragmentation and loss of primary forest in the southern Bahian Atlantic Forest of Brazil with radar imagery. *Cons. Biol.* 15:867–75.

Sabater-Pi, J. 1985. Contribution to the biology of the giant frog (*Conraua goliath*, Boulenger). *Amphibia-Reptilia* 6:143–53.

Sabath, M. D., W. C. Boughton, and S. Easteal. 1981. Expansion of the range of the introduced toad (*Bufo marinus*) in Australia from 1935 to 1974. *Copeia* 1981:677–80.

Sacher, P. 1986. Zur Entwicklung und Lebensweise von Kreuzkrötenlarven (Doñana Laur.). *Zool. Abh. Mus. Tierk. Dresden* 42:107–24.

Sadinski, W. J., and W. A. Dunson. 1992. A multilevel study of the effects of low pH on amphibians of temporary ponds. *J. Herpetol.* 26:413–22.

Safi, R., A. Begue, C. Hanni, D. Stehelin, J. R. Tata, and V. Lauder. 1997. Thyroid hormone receptor genes of neotenic amphibians. *J. Mol. Evol.* 44:595–604.

Sage, R. D., and R. K. Selander. 1979. Hybridization between species of the *Rana pipiens* complex in central Texas. *Evolution* 33:1069–88.

Sagor, E. S., M. Ouellet, E. Barten, and D. M. Green. 1998. Skeletochronology and geographic variation in age structure in the wood frog, *Rana sylvatica*. *J. Herpetol.* 32:469–74.

Saidapur, S. K., and S. Girish. 2000. The ontogeny of kin recognition in tadpoles of the toad *Bufo melanostictus* (Anura; Bufonidae). *J. Biosci.* 25:267–73.

———. 2001. Growth and metamorphosis of *Bufo melanostictus* tadpoles: effects of kinship and density. *J. Herpetol.* 35:249–54.

Saidapur, S. K., R. D. Kanamadi, and N. U. Bhuttewadkar. 1989. Variations in fat body mass in the male frog, *Rana cyanophlyctis*. *J. Herpetol.* 23:461–63.

Saint Girons, H. 1980. Modification sélective du régime des viperes (Reptilia: Viperidae), lors de la croissance. *Amphibia-Reptilia* 1:127–36.

Saito, T. 1987. Mortality of eggs and site selection for egg laying of the salamander, *Hynobius nigrescens* Stejneger. *Bull. Gumma Pref. Women's College* 1:111–19.

Salas, A. W., G. Tello V., W. Arizabal A., P. J. Sehgelmeble, and D. Neyra H. 1999. Monitoring leaf-litter amphibians in Manu National Park. In *Forest biodiversity in North, Central and South America, and the Caribbean: research and monitoring*, ed. F. Dallmeier and J. A. Comiskey, 615–31. Washington DC: Parthenon.

Salducci, M.-D., C. Marty, R. Chappaz, and A. Gilles. 2002. Molecular phylogeny of French Guiana Hylinae: Implications for the systematics and biodiversity of the Neotropical frogs. *C. R. Biol.* 325:141–53.

Sale, P. F. 1977. Maintenance of high diversity in coral reef fish communities. *Am. Nat.* 111:337–59.

———. 1984. The structure of communities of fish on coral reefs and the merit of a hypothesis-testing manipulative approach to ecology. In *Ecological communities*, ed. D. R. Strong, Jr., D. Simberloff, L. G. Abele, and A. B. Thistle, 478–90. Princeton, NJ: Princeton University Press.

Salinger, M. J. 2005. Climate variability and change: Past, present, and future—An overview. *Climatic Change* 70:9–29.

Salinger, M. J., M. V. K. Sivakumar, and R. Motha. 2005. Reducing vulnerability of agriculture and forestry to climate variability and change: Workshop summary and recommendations. *Climatic Change* 70:341–62.

Salthe, S. N. 1963. The egg capsules in the Amphibia. *J. Morphol.* 113:161–71.

———. 1967. Courtship patterns and the phylogeny of the urodeles. *Copeia* 1967:100–17.

———. 1969. Reproductive modes and the numbers and sizes of ova in the urodeles. *Am. Midl. Nat.* 81:467–90.

Salthe, S. N., and W. E. Duellman. 1973. Quantitative constraints associated with reproductive mode in anurans. In *Evolutionary biology of the anurans*, ed. J. L. Vial, 229–49. Columbia: University of Missouri Press.

Salthe, S. N., and J. S. Mecham. 1974. Reproduction and courtship patterns. In *Physiology of the Amphibia, vol. 2*, ed. B. Lofts, 309–521. New York: Academic Press.

Samollow, P. B., and M. E. Soule. 1983. A case of stress-related heterozygote superiority in nature. *Evolution* 37:646–49.

Sampson, L. V. 1900. Unusual modes of breeding and development among Anura. *Am. Nat.* 34:687–715.

———. 1904. A contribution to the embryology of *Hylodes martinicensis*. *Am. J. Anat.* 3:473–504.

Samways, M. J. 1977a. Bush cricket interspecific acoustic interactions in the field (Orthoptera, Tettigoniidae). *J. Nat. Hist.* 11:155–68.

———. 1977b. Song modification in the Orthoptera. IV. The *Platycleis intermedia*/*P. affinis* interaction quantified. *Physiol. Entomol.* 2:301–15.

Samways, M. J., and W. B. Broughton. 1976. Song modification in the Orthoptera. II. Types of acoustic interaction between *Platycleis intermedia* and other species of the genus (Tettigoniidae). *Physiol. Entomol.* 1:287–97.

Sánchez-Herráiz, M. J., L. J. Barbadillo, A. Machordom, and B. Sanchiz. 2000. A new species of pelodytid frog from the Iberian Peninsula. *Herpetologica* 56:105–18.

Sanchiz, B., and Z. Roček. 1996. An overview of the anuran fossil record. In *The biology of Xenopus*, ed. R. C. Tinsley and H. R. Kobel, 317–28. New York: Oxford University Press.

Sanders, J., and M. Davies. 1984. Burrowing behavior and associated hindlimb myology in some Australian hylid and leptodactylid frogs. *Aust. Zool.* 21:123–42.

Sanderson, I. T. 1936. The amphibians of the Mamfe Division, Cameroons. II. Ecology of the frogs. *Proc. Zool. Soc. London* 1936:165–208.

Sanderson, N., J. M. Szymura, and N. H. Barton. 1992. Variation in mating call across the hybrid zone between the fire-bellied toads *Bombina bombina* and *B. variegata*. *Evolution* 46:595–607.

Sanderson, S. L., and S. J. Kupferberg. 1999. Development and evolution of aquatic larval feeding mechanisms. In *The origin and evolution of larval forms,* ed. B. K. Hall and M. H. Wake, 301–77. San Diego: Academic Press.

San Mauro, D., D. J. Gower, O. V. Oommen, M. Wilkinson, and R. Zardoya. 2004. Phylogeny of caecilian amphibians (Gymnophiona) based on complete mitochondrial genomes and nuclear RAG1. *Mol. Phylogen. Evol.* 33:413–27.

San Mauro, D., M. Vences, M. Alcobendas, R. Zardoya, and A. Meyer. 2005. Initial diversification of living amphibians predated the breakup of Pangaea. *Am. Nat.* 165:590–99.

Santana, C. E., and S. A. Temple. 1988. Breeding biology and diet of red-tailed hawks in Puerto Rico. *Biotropica* 20:151–60.

Santee, W. R., and G. S. Bakken. 1987. Social displays in red-winged blackbirds (*Agelaius phoeniceus*): sensitivity to thermoregulatory costs. *Auk* 104:413–20.

Santos, J. C., L. A. Coloma, and D. C. Cannatella. 2003. Multiple, recurring origins of aposematism and diet specialization in poison frogs. *Proc. Natl. Acad. Sci. USA* 100:12792–97.

Santos, X., and G. A. Llorente. 1998. Sexual and size-related differences in the diet of the snake *Natrix maura* from the Ebro Delta, Spain. *Herpetol. J.* 8:161–65.

Saporito, R. A., M. A. Donnelly, R. L. Hoffman, H. M. Garraffo, and J. W. Daly. 2003. A siphonotid millipede (*Rhinotus*) as the source of spiropyrrolizidine oximes of dendrobatid frogs. *J. Chem. Ecol.* 29:2781–86.

Saporito, R. A., H. M. Garraffo, M. A. Donnelly, A. L. Edwards, J. T. Longino, and J. W. Daly. 2004. Formacine ants: An arthropod source for the pumiliotoxin alkaloids of dendrobatid poison frogs. *Proc. Natl. Acad. Sci. USA* 101:8045–50.

Sarasin, P., and F. Sarasin. 1887–1890. *Ergebnisse naturwissenschaftlichen Forschungen auf Ceylon in den Jahren 1884–1886. Zur Entwicklungsgeschichte u. Anat. der Ceylonische Blindwühle Ichthyophis glutinosus.* Wiesbaden, Germany: C. W. Kreidel's Verlag.

Sargent, T. D. 1981. Antipredator adaptations of underwing moths. In *Foraging behavior,* ed. A. C. Kamil and T. D. Sargent, 259–84. New York: Garland STPM Press.

Sasaki, M. 1924. On a Japanese salamander in Lake Kuttarush, which propagates like the axolotl. *J. Coll. Agr. Hokkaido Imp. Univ.* 15:1–36.

Sassoon, D., and D. Kelley. 1986. The sexually dimorphic larynx of *Xenopus laevis:* Development and androgen regulation. *Am. J. Anat.* 177:457–72.

Sassoon, D., G. Gray, and D. Kelley. 1987. Androgen regulation of muscle fiber type in the sexually dimorphic larynx of *Xenopus laevis. Devel. Biol.* 113:135–40.

Sato, T. 1989. Breeding environment and spawning of a salamander, *Hynobius retardatus,* at the foot of Hidaka Mountains, Hokkaido, Japan. In *Current herpetology in East Asia,* ed. M. Matsui, T. Hikida, and R. C. Goris, 292–304. Kyoto: Herpetological Society of Japan.

———. 1990. Temperature and velocity of water at breeding sites of *Hynobius retardatus. Jap. J. Herpetol.* 13:131–35.

———. 1992. Reproductive behavior in the Japanese salamander *Hynobius retardatus. Jap. J. Herpetol.* 14:184–90.

Sattler, P., and N. Reichenbach. 1998. The effects of timbering on *Plethodon hubrichti:* Short-term effects. *J. Herpetol.* 32:399–404.

Sattler, P. W. 1985. Introgressive hybridization between the spadefoot toads *Scaphiopus bombifrons* and *S. mulitplicatus* (Salientia: Pelobatidae). *Copeia* 1985:324–32.

Saunders, D. L., J. J. Meeuwig, and C. J. Vincent. 2002. Freshwater protected areas: Strategies for conservation. *Cons. Biol.* 16:30–41.

Savage, J. M. 1966. An extraordinary new toad (*Bufo*) from Costa Rica. *Rev. Biol. Trop.* 14:153–67.

———. 1973. The geographic distribution of frogs: patterns and predictions. In *Evolutionary biology of the anurans,* ed. J. L. Vial, 351–445. Columbia: University of Missouri Press.

———. 1987. Systematics and distribution of the Mexican and Central American rainfrogs of the *Eleutherodactylus gollmeri* group (Amphibia: Leptodactylidae). *Fieldiana: Zool.* n.s., no. 33:1–57.

———. 2002. *The amphibians and reptiles of Costa Rica. A herpetofauna between two continents, between two seas.* Chicago: University of Chicago Press.

Savage, J. M., and S. B. Emerson. 1970. Central American frogs allied to *Eleutherodactylus bransfordii* (Cope): a problem of polymorphism. *Copeia* 1970:623–44.

Savage, R. M. 1932. The spawning, voice and sexual behaviour of *Bombina variegata variegata. Proc. Zool. Soc. London* 1932:889–98.

———. 1934. The breeding behaviour of the common frog, *Rana temporaria,* and of the common toad, *Bufo bufo bufo. Proc. Zool. Soc. London* 1934:55–74.

———. 1935a. The ecology of young tadpoles, with special reference to some adaptations to the habit of mass spawning in *Rana temporaria temporaria* L. *Proc. Zool. Soc. London* 1935:605–10.

———. 1935b. The influence of external factors on the spawning date and migration of the common frog, *Rana temporaria temporaria* Linn. *Proc. Zool. Soc. London* 1935:49–98.

———. 1942. The burrowing and emergence of the spade-foot toad, *Pelobates fuscus* Wagler. *Proc. Zool. Soc. London* (Ser. A) 112:21–35.

———. 1952. Ecological, physiological and anatomical observations on some species of anuran tadpoles. *Proc. Zool. Soc. London* 122:467–514.

———. 1961. *The ecology and life history of the common frog.* London: Sir Isaac Pitman and Sons.

Savile, D. 1962. Gliding and flight in the vertebrates. *Am. Zool.* 2:161–66.

Savitsky, A. H. 1983. Coadapted character complexes among snakes: Fossoriality, piscivory, and durophagy. *Amer. Zool.* 23:397–409.

Sawada, S. 1963. Studies on the local races of the Japanese newt, *Triturus pyrrhogaster* Boie. II. Sexual isolation mechanisms. *J. Sci. Hiroshima Univ. Ser. B* 21:167–80.

Sawaya, P. 1947. Metabolismo respiratório de Anfibio Gymnophiona, *Typhlonectes compressicauda* (Dum. et Bibr.). *Bol. Fac. Filos., Cien. Let., Univ. Sao Paulo, Ser. Zool.* 11:51–56.

Sawyer, W. H. 1960a. Evidence for the indentity of natriferin, the frog water balance principle, and arginine vasotocin. *Nature* 187:1030–31.

———. 1960b. Increased water permeability of the bullfrog (*Rana catesbeiana*) bladder in vitro in response to synthetic oxytocin and arginine vasotocin and to neurohypophysial extracts from non-mammalian vertebrates. *Endocrinology* 66:112–20.

Sawyer, W. H., R. A. Munsick, and H. B. van Dyke. 1959. Pharmacological evidence for the presence of argenine vasotocin and oxytocin in neurohypophysial extracts from cold-blooded vertebrates. *Nature* 184:1464–65.

Sawyer, W. H., and P. K. T. Pang. 1975. Endocrine adaptation to osmotic requirements of the environment: Endocrine factors in osmoregulation by lungfishes and amphibians. *Gen. Comp. Endocrinol.* 25:224–29

———. 1979. Responses of vertebrates to neurohypophysial principles. In *Hormones and evolution, vol. 2,* ed. E. J. W. Barrington, 493–523. New York: Academic Press.

Sayler, A. 1966. The reproductive ecology of the red-backed salamander, *Plethodon cinereus,* in Maryland. *Copeia* 1966:183–93.

Sazima, I. 1974. Experimental predation on the leaf-frog *Phyllomedusa rohdei* by the water snake *Liophis miliaris. J. Herpetol.* 8:376–77.

———. 1975. Distress calls in newly metamorphosed smith frog, *Hyla faber* Wied. *Herpetologica* 31:471–72.

———. 1991. Caudal luring in two Neotropical pitvipers, *Bothrops jaracara* and *B. jararacussu. Copeia* 1991:245–46.

———. 1992. Natural history of the jararaca pitviper, *Bothrops jararaca,* in southeastern Brazil. In *Biology of the pitvipers,* ed. J. A. Campbell and E. D. Brodie, Jr., 199–216. Tyler, TX: Selva.

Sazima, I., S. A. Chini, and C. R. C. Souza. 1992. *Rhadinaea bilineata* (NCN). Diet. *Herp. Rev.* 23:120–21.

Schaaf, R. T., and J. S. Garton. 1970. Raccoon predation on the American toad, *Bufo americanus. Herpetologica* 26:334–35.

Schabetsberger, R. 1994. Gastric evacuation rates of adults and larval alpine newts (*Triturus alpestris*) under laboratory and field conditions. *Freshwater Biol.* 31:143–51.

Schabetsberger, R., and C. D. Jersabek. 1995. Alpine newts (*Triturus alpestris*) as top predators in a high-altitude Karst lake: Daily food consumption and impact on the copepod, *Arctodiaptomus alpinus. Freshwater Biol.* 31:47–61.

Schafer, E. H. 1977. T'ang. In *Food in Chinese culture: Anthropological and historical perspectives,* ed. K. C. Chang, 87–140. New Haven, CT: Yale University Press.

Schalk, G., M. R. Forbes, and P. J. Weatherhead. 2002. Developmental plasticity and growth rates of green frog (*Rana clamitans*) embryos and tadpoles in relation to a leech (*Macrobdella decora*) predator. *Copeia* 2002:445–49.

Schardier, B. J., and J. A. Jackson. 1982. Killdeers feeding on frogs. *Wilson Bull.* 94:85–87.

Schatral, A. 1990 Interspecific acoustic interactions in bushcrickets. In *The tettigoniidae: Biology, systematics and evolution,* ed. W. J. Bailey and D. C. F. Rentz, 152–65. New York: Springer-Verlag.

Schatral, A., and P. Yeoh. 1990. Spatial distribution, calling, and interspecific acoustic interactions in two species of the Australian tettigoniid genus *Tympanophora* (Orthoptera, Tettigoniidae). *J. Zool., London* 221:375–90.

Scheel, J. J. 1970. Notes on the biology of the African tree-toad, *Nectophryne afra* Buchholz and Peters, 1875 (Bufonidae, Anura) from Fernando Póo. *Rev. Zool. Bot. Afr.* 81:225–36.

Scheid, P., and J. Piiper. 1986. Control of breathing in birds. In *Handbook of physiology. Section 3: The respiratory system, vol. 2,* ed. A. P. Fishman, N. S. Cherniack, and J. G. Widdicombe, 815–32. Bethesda, MD: American Physiological Society.

Scheltinga, D. M., and B. G. M. Jamieson. 2003. Spermatogenesis and the mature spermatozoon: form, function and phylogenetic implications. In *Reproductive biology and phylogeny of anura,* ed. B. G. M. Jamieson, 119–251. Enfield, NH: Science Publishers.

Scherff-Norris, K. L., and L. J. Livo. 1999. *Bufo boreas* (Boreal Toad). Phoretic host. *Herp. Rev.* 30:162.

Schiesari, L. C., F. Acuna Junca, and G. de Mattos Accacio. 1995. *Hylodes phyllodes* (NCN). Predation. *Herp. Rev.* 26:30–31.

Schiesari, L. C., M. Gordo, and W. Hödl. 2003. Treeholes as calling, breeding, and development sites for the Amazonian canopy frog, *Phrynohyas resinifictrix* (Hylidae). *Copeia* 2003:263–72.

Schiesari, L. C., B. Grillitsch, and C. Vogl. 1996. Comparative morphology of phytotelmonous and pond-dwelling larvae of four Neotropical treefrog species (Anura, Hylidae, *Osteocephalus oophagus, Osteocephalus taurinus, Phrynohyas resinifictrix, Phrynohyas venulosa*). *Alytes* 13:109–39.

Schindelmeiser, J., and H. Greven. 1981. Nitrogen excretion in intra- and extrauterine larvae of the ovoviviparous salamander, *Salamandra salamandra* (L.) (Amphibia, Urodela). *Comp. Biochem. Physiol.* 70A:563–65.

Schiøtz, A. 1963. The amphibians of Nigeria. *Vidensk. Meddr. Dansk Naturh. Foren.* 125:1–92.

———. 1967. The treefrogs (Rhacophoridae) of West Africa. *Spolia Zool. Musei Hauniensis (Copenhagen)* 25:1–346.

———. 1971. The superspecies *Hyperolius viridiflavus* (Anura). *Vidensk. Meddr. Dansk Naturh. Foren.* 134:1–17.

———. 1973. Evolution of anuran mating calls: Ecological aspects. In *Evolutionary biology of the anurans,* ed. J. L. Vial, 311–19. Columbia: University of Missouri Press.

———. 1999. *Treefrogs of Africa.* Frankfurt am Main, Germany: Edition Chimaira.

Schaefer, H.-C., M. Vences, and M. Veith. 2002. Molecular phylogeny of Malagasy poison frogs, genus *Mantella* (Anura: Mantellidae): Homoplastic evolution of colour pattern in aposematic amphibians. *Org. Divers. Evol.* 2:97–105.

Schlaepfer, M. A., and R. Figeroa-Sandí. 1998. Female reciprocal calling in a Costa Rican leaf-litter frog, *Eleutherodactylus podiciferus. Copeia* 1998:1076–80.

Schlaepfer, M. A., and T. A. Gavin. 2001. Edge effects on lizards and frogs in tropical forest fragments. *Cons. Biol.* 15:1079–90.

Schlaghecke, R., and V. Blüm. 1978a. Seasonal variation in fat body metabolism of the green frog, *Rana esculenta* (L.). *Experientia* 34:1019–20.

———. 1978b. Seasonal variation in liver metabolism of the green frog, *Rana esculenta* (L.). *Experientia* 34:456–57.

Schlefer, E. K., M. A. Romano, S. I. Guttman, and S. B. Ruth. 1986. Effects of twenty years of hybridization in a disturbed habitat on *Hyla cinerea* and *Hyla gratiosa. J. Herpetol.* 20:210–21.

Schleich, H. H., C. Anders, and W. Kästle. 2002. Species list with annotations. In *Amphibians and reptiles of Nepal. Biology, systematics, field guide,* ed. H. H. Schleich and W. Kästle, 78–92. Ruggell, Germany: A. R. G. Gantner Verlag.

Schleich, H. H., and W. Kästle, eds. 2002. *Amphibians and reptiles of Nepal. Biology, systematics, field guide.* Ruggell, Germany: A. R. G. Gantner Verlag.

Schleidt, W. M. 1973. Tonic communication: continual effects of discrete signs in animal communication systems. *J. Theor. Biol.* 42:359–386.

Schley, L., and R. A. Griffiths. 1998. Midwife toads (*Alytes muletensis*) avoid chemical cues from snakes (*Natrix maura*). *J. Herpetol.* 32:572–74.

Schlisio, V. W., K. Jurss, and L. Spannhof. 1973. Osmo- und Ionenregulation von *Xenopus laevis* Daud. nach Adaptatio in verschiedenen osmotisch wirksamen Losungen. I. Toleranz und Wasseraushalt. *Zool. Jb. Zool. Physiol. Tiere* 77:275–90.

Schlupp, I. and R. Podloucky. 1994. Changes in breeding site fidelity: A combined study of conservation and behaviour in the common toad *Bufo bufo*. *Biol. Cons.* 69:285–91.

Schlupp, I., M. Kietz, R. Podloucky, and F. M. Stolz. 1989. Pilot project Braken: Preliminary results from the resettlement of adult toads to a substitute breeding site. In *Amphibians and roads,* ed. T. E. S. Langton, 127–33. Shefford, Bedfordshire, England: ACO Polymer Products.

Schlüter, A. 1979. Bio-akustische Untersuchungen an Hyliden in einem begrenzten Gebeit des tropischen Regenwaldes von Peru (Amphibia: Salientia: Hylidae). *Salamandra* 15:211–36.

———. 1980a. Bio-akustische Untersuchungen an Dendrobatiden in einem begrenzten Gebiet des tropischen Regenwaldes von Peru (Amphibia: Salientia: Dendrobatidae). *Salamandra* 16:149–61.

———. 1980b. Bio-akustische Untersuchungen an Leptodactyliden in einem begrenzten Gebiet des tropischen Regenwaldes von Peru (Amphibia: Salientia: Leptodactylidae). *Salamandra* 16:227–47.

———. 1980c. Bio-akustische Untersuchungen an Microhyliden in einem begrenzten Gebiet des tropischen Regenwaldes von Peru (Amphibia: Salientia: Microhylidae). *Salamandra* 16:114–31.

———. 1981. Bio-akustische Untersuchungen an Bufoniden in einem begrenzten Gebiet des tropischen Regenwaldes von Peru (Amphibia: Salientia: Bufonidae). *Salamandra* 17:99–105.

———. 1984. *Ökologische Untersuchungen an einem Stillgewässer im tropischen Regenwald von Peru unter besonderer Berucksichtung der Amphibien.* Unpublished PhD dissertation. Hamburg, Germany: University of Hamburg.

———. 1990. Reproduction and tadpole of *Edalorhina perezi* (Amphibia, Leptodactylidae). *Stud. Neotrop. Fauna Envir.* 25:49–56.

Schlüter, A., and J. Regös. 1981. *Lithodytes lineatus* (Schneider, 1799) (Amphibia: Leptodactylidae) as a dweller in nests of the leaf cutting ant *Atta cephalotes* (Linnaeus, 1758) (Hymenoptera: Attini). *Amphibia-Reptilia* 2:117–21.

Schlüter, A., and A. W. Salas. 1991. Reproduction, tadpoles, and ecological aspects of three syntopic microhylid species from Peru (Amphibia: Microhylidae). *Stutt. Beitr. Naturk.*, ser. A, (458):1–17.

Schmajuk, N. A., and E. T. Segura. 1982. Behavioral regulation of water balance in the toad *Bufo arenarum*. *Herpetologica* 38:296–301.

Schmid, E. 1977. Der Rufapparat der Rot- und Gelbbauchunke und der Geburtshelferkröte *Bombina bombina* (L.), *Bombina v. variegata* (L.), *Alytes o. obstetricans* (Laur.) (Anura, Discoglossidae). *Zool. Jb. Anat.* 98:171–80.

———. 1978. Beitrag zur Morphologie und zur Histologie des Stimmbandes einheimischer Froschlurche (Anura, Amphibia). *Zool. Jb. Anat.* 99:133–50.

Schmid, W. D. 1965. Some aspects of the water economy of nine species of amphibians. *Ecology* 46:261–69.

———. 1969. Physiological specializations of amphibians to habitats of varying aridity. In *Physiological systems in semiarid environments,* ed. C. C. Hoff and M. L. Riedesel, 135–42. Albuquerque: University of New Mexico Press.

———. 1982. Survival of frogs at low temperature. *Science* 215:697–98.

Schmid, W. D., and R. E. Barden. 1965. Water permeability and lipid content of amphibian skin. *Comp. Biochem. Physiol.* 15:423–27.

Schmidt, A., C. Naujoks-Manteuffel, and G. Roth. 1988. Olfactory and vomeronasal projections and the pathway of the nervus terminalis in ten species of salamanders. *Cell Tissue Res.* 251:45–50.

Schmidt, A., and M. H. Wake. 1990. Olfactory and vomeronasal systems of caecilians (Amphibia: Gymnophiona). *J. Morphol.* 205:255–68.

Schmidt, B. R. 2003. Count data, detection probabilities, and the demography, dynamics, distribution, and decline of amphibians. *C. R. Biol.* 326:S119–S124.

Schmidt, B. R., and B. R. Anholt. 1999. Analysis of survival probabilities of female common toads, *Bufo bufo*. *Amphibia-Reptilia* 20:97–108.

Schmidt, B. R., R. Feldmann, and M. Schaub. 2005. Demographic processes underlying population growth and decline in *Salamandra salamandra*. *Cons. Biol.* 19:1149–56.

Schmidt, B. R., M. Schaub, and B. R. Anholt. 2002. Why you should use capture-recapture methods when estimating survival and breeding probabilities: On bias, temporary emigration, overdispersion, and common toads. *Amphibia-Reptilia* 23:375–88.

———. 2005. On the analysis of mark-recapture data: a reply to Henle (2005). *Amphibia-Reptilia* 26:269.

Schmidt, B. R., and J. Van Buskirk. 2005. A comparative analysis of predator-induced plasticity in larval *Triturus* newts. *J. Evol. Biol.* 18:415–25.

Schmidt, K. P. 1936. Guatemalan salamanders of the genus *Oedipus*. *Zool. Ser. Field Mus.* 20:135–66.

Schmidt, K. P., and R. F. Inger. 1959. Amphibians. *Explor. Parc Nat. Upemba* 56:1–264.

Schmidt, R. S. 1965. Larynx control and call production in frogs. *Copeia* 1965:143–47.

———. 1966. Central mechanisms of frog calling. *Behaviour* 26:251–85.

———. 1968a. Chuckle calls of the leopard frog (*Rana pipiens*). *Copeia* 1968:561–69.

———. 1968b. Preoptic activation of frog mating behavior. *Behaviour* 30:239–57.

———. 1969. Preoptic activation of mating call orientation in female anurans. *Behaviour* 35:114–27.

———. 1971. Neural mechanisms of mating call orientation in female toads (*Bufo americanus*). *Copeia* 1971:545–48.

———. 1972a. Action of intrinsic laryngeal muscles during release calling in leopard frog. *J. Exp. Zool.* 181:233–44.

———. 1972b. Release calling and inflating movements in anurans. *Copeia* 1972:240–45.

———. 1973a. Central mechanisms of frog calling. *Amer. Zool.* 13:1169–77.

———. 1973b. Vocal cord mechanisms of release calling in northern leopard frog. *Copeia* 1973:624–27.

———. 1974. Neural correlates of frog calling: trigeminal tegmentum. *J. Comp. Physiol.* 92:229–54.

———. 1976. Release-call differences between two members of the *Rana pipiens* complex. *Copeia* 1976:721–26.

———. 1984. Neural correlates of frog calling: Preoptic area trigger of "mating call." *J. Comp. Physiol. A* 154:847–53.

Schmidt, R. S., C. P. Kemnitz, and W. R. Hudson. 1990. Origin of anuran calling: description of toad releasing. *J. Exp. Zool.* 254:338–42.

Schmidt-Nielsen, B., and R. P. Forster. 1954. The effect of dehydration and low temperature on renal function in the bullfrog. *J. Cell. Comp. Physiol.* 44:233–46.

Schmidt-Nielsen, K., and P. Lee. 1962. Kidney function in the crab-eating frog (*Rana cancrivora*). *J. Exp. Biol.* 39:167–77.

Schmitz, B., T. D. White, and P. M. Narins. 1992. Directionality of phase locking in auditory nerve fibers of the leopard frog *Rana pipiens pipiens*. *J. Comp. Physiol. A* 170:589–604.

Schmuck, R., W. Geise, and K. E. Linsenmair. 1994. Life cycle strategies and physiological adjustments of reedfrog tadpoles (Amphibia: Anura: Hyperoliidae) in relation to environmental conditions. *Copeia* 1994:996–1007.

Schmuck, R., F. Kobelt, and K. E. Linsenmair. 1988. Adaptations of the reed frog *Hyperolius viridiflavus* (Amphibia, Anura, Hyperoliidae) to its arid environment. V. Iridiophores and nitrogen metabolism. *J. Comp. Physiol. B* 158:537–46.

Schmuck, R., and K. E. Linsenmair. 1988. Adaptations of the reed frog *Hyperolius viridiflavus* (Amphibia, Anura, Hyperoliidae) to its arid environment. III. Aspects of nitrogen metabolism and osmoregulation in the reed frog *Hyperolius viridiflavus taeniatus*, with special reference to the role of iridophores. *Oecologia* 75:354–61.

Schneichel, W., and H. Schneider. 1988. Hearing and calls of the banana frog, *Afrixalus fornasinii* (Bianconi) (Anura: Rhacophoridae). *Amphibia-Reptilia* 9:251–64.

Schneider, C. J., M. Cunningham, and C. Moritz. 1998. Comparative phylogeography and the history of endemic vertebrates in the wet tropics rainforests of Australia. *Mol. Ecol.* 7:487–98.

Schneider, H. 1966. Bio-akustik der Froschlurche. *Stutt. Beitr. Naturkd.* 152:1–16.

———. 1967. Rufe und Rufverhalten des Laubfrosches, *Hyla arborea arborea* (L.). *Z. Vergl. Physiol.* 57:174–89.

———. 1968. Bio-akustische Untersuchungen am Mittelmeerlaubfrosch. *Z. Vergl. Physiol.* 61:369–85.

———. 1970. Morphologie des Larynx von *Hyla a. arborea* (L.) und *Hyla meridionalis* Boettger (Amphibia, Anura). *Z. Morph. Ökol. Tiere* 66:299–309.

———. 1977. Acoustic behavior and physiology of vocalization in the European tree frog, Hyla arborea L. In *The reproductive biology of the amphibians*, ed. D. H. Taylor and S. I. Guttman, 295–335. New York: Plenum.

———. 1986. Strukturmerkmale und Funktion des Rufapparates der Froschlurche. *Zool. Jb. Physiol.* 90:317–28.

———. 1988. Peripheral and central mechanisms of vocalization. In *The evolution of the amphibian auditory system*, ed. B. Fritzsch, M. J. Ryan, W. Wilczynski, T. E. Hetherington, and W. Walkowiak, 537–58. New York: Wiley.

Schneider, H., and J. Brzoska. 1981. Die Befreiungsrufe der mitteleuropäischen Wasserfrösche. *Zool. Anz.* 206:189–202.

Schneider, H., F. Hussein, and M. S. A. Akef. 1986. Comparative bioacoustic studies in the yellow-bellied toad, *Bombina variegata* (L.), and relationships of European and Asian species and subspecies of the genus *Bombina* (Anura, Amphibia). *Bonn. Zool. Beitr.* 37:49–67.

Schneider, H., and G. Joermann. 1988. Mating calls of water frogs (Ranidae) from Lake Skutari, Yugoslavia, and the relationship to water frogs of other regions. *Z. Zool. Syst. Evolut.-Forsch.* 26:261–75.

Schneider, H., G. Joermann, and W. Hödl. 1988. Calling and antiphonal calling in four Neotropical anuran species of the family Leptodactylidae. *Zool. Jb. Physiol.* 92:77–103.

Schneider, H., and E. Nevo. 1972. Bio-acoustic study of the yellow-lemon treefrog, *Hyla arborea savignyi* Audouin. *Zool. Jb. Physiol.* 76:497–506.

Schneider, H., T. S. Sofianidou, and P. Kyriakopoulou-Sklavounou. 1988. Calling behavior and calls of *Rana dalmatina* (Anura, Ranidae) in Greece. *Zool. Jb. Physiol.* 92:231–43.

Schnurre, O. 1956. Ernährungsbiologische Studien an Raubvögeln und Eulen der Darsshalbinse (Meckl.). *Beitr. z. Vogelk.* 4:211–45.

———. 1957. Beobachtungen an einem Berliner Waldkauze (*Strix aluco* L.). *Vogelring* 26:41–49.

Schoch, R. R. 1992. Comparative ontogeny of early Permian branchiosaurid amphibians from southwestern Germany. *Palaeontographica* A 222:43–84.

Schoener, T. W. 1968. Sizes of feeding territories among birds. *Ecology* 43:113–41.

———. 1974. Resource partitioning in ecological communities. *Science* 185:27–39.

———. 1982. The controversy over interspecific competition. *Am. Sci.* 70:586–95.

———. 1983. Field experiments on interspecific competition. *Am. Nat.* 122:240–85.

Scholander, P. F., R. Hock, V. Walters, and L. Irving. 1950. Heat regulation in some arctic and tropical mammals and birds. *Biol. Bull.* 99:237–58.

Schotthoefer, A. M., A. V. Koehler, C. U. Meteyer, and R. A. Cole. 2003. Influences of *Ribeiroia ondantrae* (Trematoda: Digenea) infection on limb development and survival in northern leopard frogs (*Rana pipiens*): Effects of host state and parasite-exposure level. *Can. J. Zool.* 81:1144–53.

Schreiber, A. M., and J. L. Specker. 1998. Metamorphosis in the summer flounder (*Paralichthys dentatus*): Stage-specific developmental response to altered thyroid status. *Gen. Comp. Endocrinol.* 111:156–66.

Schroeder, E. E. 1976. Dispersal and movement of newly transformed green frogs, *Rana clamitans*. *Am. Midl. Nat.* 95:471–74.

Schroeder, E. E., and T. E. Baskett. 1968. Age estimation, growth rates, and population structure in Missouri bullfrogs. *Copeia* 1968:583–92.

Schueler, F. W., and F. R. Cook. 1980. Distribution of the middorsal stripe dimorphism in the wood frog, *Rana sylvatica*, in eastern North America. *Can. J. Zool.* 58:1643–51.

Schuett, G. W. 1984. *Calloselasma rhodostoma* (Malayan pit viper). Feeding mimicry. *Herp. Rev.* 15:112.

Schuierer, F. W. 1958. Factors affecting neoteny in the salamander *Dicamptodon ensatus* (Eschscholtz). *Bull. South. California Acad. Sci.* 57:119–21.

———. 1962. Remarks on the natural history of *Bufo exsul* Myers, the endemic toad of Deep Springs Valley, Inyo County, California. *Herpetologica* 17:260–66.

Schul, J., and S. L. Bush. 2002. Non-parallel coevolution of

sender and receiver in the acoustic communication system of treefrogs. *Proc. Roy. Soc. London B* 269:1847–52.
Schülte, R. 1981a. *Dendrobates bassleri.* Freilandbeobachtungen, Haltung und Zucht. *Herpetofauna* 12:23–28.
———. 1981b. *Dendrobates quinquevittatus.* Ökologie, Haltung und Zucht. *Herpetofauna* 10:24–28.
Schultschik, G. 1994. Zur Fortpflanzungsbiologie von *Mertensiella caucasica* (Waga, 1876) (Urodela: Salamandridae). *Abh. Ber. Naturk. Magdeburg* 17:163–75.
Schultze, H.-P. 1991. A comparison of controversial hypotheses on the origin of tetrapods. In *Origins of the higher groups of tetrapods,* ed. H.-P. Schultze and L. Trueb, 29–67. Ithaca, NY: Cornell University Press.
Schurbon, J. M., and J. E. Fauth. 2003. Effects of prescribed burning on amphibian diversity in a southeastern U.S. National Forest. *Cons. Biol.* 17:1338–49.
Schurbon, J. M. and J. E. Fauth. 2004. Fire as friend and foe of amphibians: a reply. *Cons. Biol.* 18:1156–59.
Schwalbe, C. R., and P. C. Rosen. 1988. Preliminary report on effect of bullfrogs on wetland herpetofaunas in southeastern Arizona. In *Management of amphibians, reptiles, and small mammals in North America,* ed. R. C. Szaro, K. E. Severson, and D. R. Patton, 166–73. USDA Forest Service General Technical Report RM-166. Fort Collins, CO: Rocky Mountain Forest and Range Experiment Station, U.S. Department of Agriculture.
Schwalm, P. A., P. H. Starrett, and R. W. McDiarmid. 1977. Infrared reflectance in leaf-sitting Neotropical frogs. *Science* 196:1225–27.
Schwartz, A. 1955. A clutch of eggs of *Aneides hardyi* (Taylor). *Herpetologica* 11:70
Schwartz, A., and D. C. Fowler. 1973. The anura of Jamaica: a progress report. *Stud. Fauna Curaçao Carib. Islands* 43:50–142.
Schwartz, A., and R. Etheridge. 1954. New and additional herpetological records from the North Carolina coastal plain. *Herpetologica* 10:167–71.
Schwartz, J. J. 1986. Male calling behavior and female choice in the Neotropical treefrog *Hyla microcephala. Ethology* 73:116–27.
———. 1987a. The function of call alternation in anuran amphibians: A test of three hypotheses. *Evolution* 41:461–471.
———. 1987b. The importance of spectral and temporal properties in species and call recognition in a Neotropical treefrog with a complex vocal repertoire. *Anim. Behav.* 35:340–47.
———. 1989. Graded aggressive calls of the spring peeper, *Pseudacris crucifer. Herpetologica* 45:172–81.
———. 1991. Why stop calling? A study of unison bout singing in a Neotropical treefrog. *Anim. Behav.* 42:565–77.
———. 1993. Male calling behavior, female discrimination and acoustic interference in the Neotropical treefrog *Hyla microcephala* under realistic acoustic conditions. *Behav. Ecol. Sociobiol.* 32:401–14.
———. 1994. Male advertisement and female choice in frogs: Recent findings and new approaches to the study of communication in a dynamic acoustic environment. *Amer. Zool.* 34:616–24.
———. 2001. Call monitoring and interactive playback systems in the study of acoustic interactions among male anurans. In *Anuran communcation,* ed. M. J. Ryan, 183–204. Washington, DC: Smithsonian Institution Press.
Schwartz, J. J., B. Buchanan, and H. C. Gerhardt. 2001. Female mate choice in gray treefrogs (*Hyla versiolor*) in three experimental environments. *Behav. Ecol. Sociobiol.* 49:443–55.
———. 2002. Acoustic interactions among gray treefrogs, *Hyla versicolor,* in a chorus setting. *Behav. Ecol. Sociobiol.* 53:9–19.
Schwartz, J. J., and H. C. Gerhardt. 1989. Spatially mediated release from auditory masking in an anuran amphibian. *J. Comp. Physiol. A* 166:37–41.
———. 1995. Directionality of the auditory system and call pattern recognition during acoustic interference in the gray tree frog, *Hyla versicolor. Aud. Neurosci.* 1:195–206.
———. 1998. The neuroethology of frequency preferences in the spring peeper. *Anim. Behav.* 56:55–69.
Schwartz, J. J., K. Huth, and T. Hutchin. 2004. How long do females really listen? Assessment time for female mate choice in the grey treefrog, *Hyla versicolor. Anim. Behav.* 68:533–40.
Schwartz, J. J., and A. S. Rand. 1991. The consequences for communication of call overlap in the túngara frog, a Neotropical anuran with a frequency-modulated call. *Ethology* 89:73–83
Schwartz, J. J., S. J. Ressel, and C. R. Bevier. 1995. Carbohydrates and calling: depletion of muscle glycogen and the chorusing dynamics of the Neotropical treefrog *Hyla microcephala. Behav. Ecol. Sociobiol.* 37:125–35.
Schwartz, J. J., and K. D. Wells. 1983a. An experimental study of acoustic interference between two species of Neotropical treefrogs. *Anim. Behav.* 31:181–90.
———. 1983b. The influence of background noise on the behavior of a Neotropical treefrog, *Hyla ebraccata. Herpetologica* 39:121–29.
———. 1984a. Interspecific acoustic interactions of the Neotropical treefrog *Hyla ebraccata. Behav. Ecol. Sociobiol.* 14:211–24.
———. 1984b. Vocal behavior of the Neotropical treefrog *Hyla phlebodes. Herpetologica* 40:452–63.
———. 1985. Intra- and interspecific vocal behavior of the Neotropical treefrog *Hyla microcephala. Copeia* 1985:27–38.
Schwarzkopf, L., and R. A. Alford. 1996. Desiccation and shelter-site use in a tropical amphibian: Comparing toads with physical models. *Funct. Ecol.* 10:193–200.
———. 2002. Nomadic movement in tropical toads. *Oikos* 96:492–506.
Schwenk, K., and D. B. Wake. 1993. Prey processing in *Leurognathus marmoratus* and the evolution of form and function in desmognathine salamanders (Plethodontidae). *Biol. J. Linn. Soc.* 49:141–62.
Scott, D. E. 1990. Effects of larval density in *Ambystoma opacum:* An experiment in large-scale field enclosures. *Ecology* 71:296–306.
———. 1993. Timing of reproduction of paedomorphic and metamorphic *Ambystoma talpoideum. Am. Midl. Nat.* 129:397–402.
———. 1994. The effect of larval density on adult demographic traits in *Ambystoma opacum. Ecology* 75:1383–96.
Scott, D. E., and M. R. Fore. 1995. The effect of food limitation on lipid levels, growth, and reproduction in the marbled salamander, *Ambystoma opacum. Herpetologica* 51:472–87.
Scott, N. J., Jr. 1976. The abundance and diversity of the herpetofauna of tropical forest litter. *Biotropica* 8:41–58.
———. 1982. The herpetofauna of forest litter plots from Cameroon, Africa, pp. 145–150. In *Herpetological communities, Wildl. Res. Report 13,* ed. N. J. Scott, Jr., ed. Washington, DC: U.S. Fish and Wildlife Service.
———. 1983. *Leptodactylus pentadactylus.* In *Costa Rican natural history,* ed. D. H. Janzen, 405–6. Chicago: University of Chicago Press.

Scott, N. J., Jr., and A. L. Aquino. 2005. It's a frog-eat-frog world in the Paraguayan Chaco: Food habits, anatomy, and behavior of the frog-eating anurans. In *Ecology and evolution in the tropics. A herpetological perspective,* ed. M. A. Donnelly, B. I Crother, C. Guyer, M. H. Wake, and M. E. White, 243–59. Chicago: University of Chicago Press.

Scott, N. J., Jr., and A. Starrett. 1974. An unusual breeding aggregation of frogs, with notes on the ecology of *Agalychnis spurrelli* (Anura: Hylidae). *Bull. S. California Acad. Sci.* 73:86–94.

Scott, N. J., Jr., and R. D. Jennings. 1985. The tadpoles of five species of New Mexican leopard frogs. *Occ. Pap. Mus. Southwest. Biol., Univ. New Mexico* 3:1–21.

Scow, R. O. 1965. Hormonal control of postnatal growth of skeletal muscle. In *Proceedings of 2nd International Congress of Endocrinology, London, 196,* 772–78. Amsterdam: Exerpta Medica Foundation,.

Scribner, K. T., J. W. Arntzen, N. Cruddace, R. S. Oldham, and T. Burke. 2001. Environmental correlates of toad abundance and population genetic diversity. *Biol. Cons.* 98:201–10.

Scrocchi, G. J., and E. O. Lavilla. 1990. *Elachistocleis bicolor.* Amplexus behavior. *Herp. Rev.* 21:18.

Scroggie, M. P., and M. J. Littlejohn. 2005. Territorial vocal behavior in hybrid smooth froglets, *Geocrinia laevis* complex (Anura: Myobatrachidae). *Behav. Ecol. Sociobiol.* 58:72–79.

Scroggin, J. B., and W. B. Davis. 1956. Food habits of the Texas dwarf siren. *Herpetologica* 12:231–37.

Seale, D. B. 1980. Influence of amphibian larvae on primary production, nutrient flux, and competition in a pond ecosystem. *Ecology* 61:1531–50.

———. 1982a. Obligate and facultative suspension feeding in anuran larvae: Feeding regulation in *Xenopus* and *Rana. Biol. Bull.* 162:214–31.

———. 1982b. Physical factors influencing oviposition by the woodfrog, *Rana sylvatica,* in Pennsylvania. *Copeia* 1982:627–35.

———. 1987. Amphibia. In *Animal energetics, vol. 2,* ed. T. J. Pandian and F. J. Vernberg, 467–552. New York: Academic Press.

Seale, D. B., and N. Beckvar. 1980. The comparative ability of anuran larvae (genera: *Hyla, Bufo,* and *Rana*) to ingest suspended blue-green algae. *Copeia* 1980:495–503.

Seale, D. B., K. Hoff, and R. J. Wassersug. 1982. *Xenopus laevis* larvae (Amphibia, Anura) as model suspension feeders. *Hydrobiologia* 87:161–69.

Seale, D. B., and R. J. Wassersug. 1979. Suspension feeding dynamics of anuran larvae related to their functional morphology. *Oecologia* 39:259–72.

Seale, D. B., E. Rodgers, and M. E. Boraas. 1975. Effects of suspension-feeding frog larvae on limnological variables and community structure. *Verh. Int. Verein. Theor. Ang. Limnol.* 19:3179–84.

Searcy, W. A. 1979. Sexual selection and body size in male red-winged blackbirds. *Evolution* 33:649–61.

Searcy, W. A., and K. Yasukawa. 1995. *Polygyny and sexual selection in red-winged blackbirds.* Princeton, NJ: Princeton University Press.

Seber, G. A. F. 1973. *The estimation of animal abundance and related parameters.* London: Griffin.

Secki, J. A., and A. Queral-Regil. 1997. *Gyrinophilus porphyriticus porphyriticus* (northern spring salamander). Prey. *Herp. Rev.* 28:82.

Secondi, J., W. Haerty, and T. Lodé. 2005. Female attraction to conspecific chemical cues in the palmate newt *Triturus helveticus. Ethology* 111:726–35.

Secor, S. M. 2001. Regulation of digestive performance: A proposed adaptive response. *Comp. Biochem. Physiol. A* 128:565–77.

———. 2005a. Evolutionary and cellular mechanisms regulating intestinal performance of amphibians and reptiles. *Integr. Comp. Biol.* 45:282–94.

———. 2005b. Physiological responses to feeding, fasting and estivation for anurans. *J. Exp. Biol.* 208:2595–2608.

Secor, S. M., and J. M. Diamond. 2000. Evolution of regulatory responses to feeding in snakes. *Physiol. Biochem. Zool.* 73:123–41.

Secor, S. M., and A. C. Faulkner. 2002. Effects of meal size, meal type, body temperature, and body size on the specific dynamic action of the marine toad, *Bufo marinus. Physiol. Biochem. Zool.* 75:557–71.

Sedra, S. N., and M. I. Michael. 1957. The development of the skull, visceral arches, larynx and visceral muscles of the South African clawed toad, *Xenopus laevis* (Daudin) during the process of metamorphosis (from stage 55 to stage 66). *Verh. Koninkl. Nederl. Akad. Wetensch.* 51:1–80.

Seebacher, F., and R. A. Alford. 1999. Movement and microhabitat use of a terrestrial amphibian (*Bufo marinus*) on a tropical island: Seasonal variation and environmental correlates. *J. Herpetol.* 33:208–14.

———. 2002. Shelter microhabitats determine body temperature and dehydration rates of a terrestrial amphibian (*Bufo marinus*). *J. Herpetol.* 36:69–75.

Seib, R. L. 1984. Prey use in three syntopic Neotropical racers. *J. Herpetol.* 18:412–20.

———. 1985. Europhagy in a tropical snake, *Coniophanes fissidens. Biotropica* 17:57–64.

Seibert, E. A., H. B. Lillywhite, and R. J. Wassersug. 1974. Cranial coossification in frogs: Relationship to rate of evaporative water loss. *Physiol. Zool.* 47:261–65.

Seidel, B. 1989. Phoresis of *Cyclocypris ovum* (Jurine) (Ostracoda, Podocopida, Cyprididae) on *Bombina variegata variegata* (L.) (Anura, Amphibia) and *Triturus vulgaris* (L.) (Urodela, Amphibia). *Crustaceana* 57:171–76.

———. 1990. Amphibien als Tranporteure limnischer Muschelkrebse: Ein parameter zur Analyse der Verteilung von *Bombina bombina. Amphibia-Reptilia* 11:253–61

———. 1999. Water-wave communication between territorial male *Bombina variegata. J. Herpetol.* 33:457–62.

Seifert, R. P. 1984. Does competition structure communities? Field studies on Neotropical *Heliconia* insect communities. In *Ecological communities,* ed. D. R. Strong, Jr., D. Simberloff, L. G. Abele, and A. B. Thistle, 54–63. Princeton, NJ: Princeton University Press.

Seigel, R. A. 1983. Natural survival of eggs and tadpoles of the wood frog, *Rana sylvatica. Copeia* 1983:1096–98.

Seiter, P., H. Schultheiss, and W. Hanke. 1978. Osmotic stress and excretion of ammonia and urea in *Xenopus laevis. Comp. Biochem. Physiol.* 61A:571–76.

Sekar, A. G. 1992. A study of the food habits of six anuran tadpoles. *J. Bombay Nat. Hist. Soc.* 89:9–16.

Selander, R. K. 1972. Sexual selection and dimorphism in birds. In *Sexual selection and the descent of man,* ed. B. G. Campbell, 180–230. Chicago: Aldine Press.

Selby, M. F., S. C. Winkel, and J. W. Petranka. 1996. Geographic uniformity in agonistic behaviors of Jordan's salamander. *Herpetologica* 52:108–15.

Semb-Johannson, A. 1991. Declining populations of the common toad (*Bufo bufo* L.) on two islands in Oslofjord, Norway. *Amphibia-Reptilia* 13:409–412.

Semlitsch, R. D. 1981a. Effects of implanted tantalum-182 tags on the mole salamander, *Ambystoma talpoideum. Copeia* 1981:735–37.

———. 1981b. Terrestrial activity and summer home range of the mole salamander (*Ambystoma talpoideum*). *Can. J. Zool.* 59:315–322.

———. 1983a. Burrowing ability and behavior of salmanders of the genus *Ambystoma. Can. J. Zool.* 61:616–20.

———. 1983b. Structure and dynamics of two breeding populations of the eastern tiger salamander, *Ambystoma tigrinum. Copeia* 1983:608–16.

———. 1985a. Analysis of climatic factors influencing migrations of the salamander *Ambystoma talpoideum. Copeia* 1985:477–89.

———. 1985b. Reproductive strategy of a facultatively paedomorphic salamander *Ambystoma talpoideum. Oecologia* 65:305–13.

———. 1987a. Density-dependent growth and fecundity in the paedomorphic salamander *Ambystoma talpoideum. Ecology* 68:1003–8.

———. 1987b. Interactions between fish and salamander larvae: cost of predator avoidance or competition? *Oecologia* 72:481–86.

———. 1987c. Paedomorphosis in *Ambystoma talpoideum:* effects of density, food, and pond drying. *Ecology* 68:994–1002.

———. 1987d. Relationship of pond drying to the reproductive success of the salamander *Ambystoma talpoideum. Copeia* 1987:61–69.

———. 1988. Allopatric distribution of two salamanders: Effects of fish predation and competitive interactions. *Copeia* 1988:290–98.

———. 1990. Paedomorphosis in *Ambystoma talpoideum:* Maintenance of population variation and alternative life-history pathways. *Evolution* 44:1604–13

———. 1991. Asymmetric competition in mixed populations of tadpoles of the hybridogenetic *Rana esculenta* complex. *Evolution* 47:510–19.

———. 1993. Adaptive genetic variation in growth and development of tadpoles of the hybridogenetic *Rana esculenta* complex. *Evolution* 47:1805–18.

———. 1994. Evolutionary consequences of non-random mating: Do large males increase offspring fitness in the anuran *Bufo bufo*? *Behav. Ecol. Sociobiol.* 34:19–24.

———. 1998. Biological delineation of terrestrial buffer zones for pond-breeding salamanders. *Cons. Biol.* 12:1113–19.

———. 2000. Principles for management of aquatic breeding amphibians. *J. Wildl. Manage.* 64:615–31.

———. 2002. Critical elements for biologically based recovery plans of aquatic-breeding amphibians. *Cons. Biol.* 16:619–29.

———. 2003a. *Amphibian conservation,* ed. Washington, DC: Smithsonian Books.

———. 2003b. Conservation of pond-breeding amphibians. In *Amphibian conservation,* ed. R. D. Semlitsch, 8–23. Washington, DC: Smithsonian Books.

———. 2003c. Introduction: General threats to amphibians. In *Amphibian conservation,* ed. R. D. Semlitsch, 1–7. Washington, DC: Smithsonian Books.

Semlitsch, R.D., and J.R. Bodie. 1998. Are small, isolated wetlands expendable? *Cons. Biol.* 12:1129–33.

———. 2003. Biological criteria for buffer zones around wetland and riparian habitats for amphibians and reptiles. *Cons. Biol.* 17:1219–28.

Semlitsch, R. D., and J. P. Caldwell. 1982. Effects of density on growth, metamorphosis, and survivorship in tadpole of *Scaphiopus holbrooki. Ecology* 63:905–11.

Semlitsch, R. D., and S. Gavasso. 1992. Behavioral responses of *Bufo bufo* and *Bufo calamita* tadpoles to chemical cues of vertebrate and invertebrate predators. *Ethol. Ecol. Evol.* 4:165–73.

Semlitsch, R. D., and J. W. Gibbons. 1985. Phenotypic variation in metamorphosis and paedomorphosis in the salamander *Ambystoma talpoideum. Ecology* 66:1123–30.

———. 1988. Fish predation in size-structured populations of treefrog tadpoles. *Oecologia* 75:321–26.

———. 1990. Effects of egg size on success of larval salamanders in complex aquatic environments. *Ecology* 71:1789–95.

Semlitsch, R. D., J. W. Gibbons, and T. D. Turberville. 1995. Timing of reproduction and metamorphosis in the Carolina gopher frog (*Rana capito capito*) in South Carolina. *J. Herpetol.* 29:612–14.

Semlitsch, R. D., R. N. Harris, and H. M. Wilbur. 1990. Paedomorphosis in *Ambystoma talpoideum:* Maintenance of population variation and alternative life-history pathways. *Evolution* 44:1604–13.

Semlitsch, R. D., H. Hotz, and G.-D. Guex. 1997. Competition among tadpoles of coexisting hemiclones of hybridogenetic *Rana esculenta:* Support for the frozen niche variation model. *Evolution* 51:1249–61.

Semlitsch, R. D., and M. A. McMillan. 1980. Breeding migration, population size structure, and reproduction of the dwarf salamander, *Eurycea quadridigitata* in South Carolina. *Brimleyana* (3):97–105.

Semlitsch, R. D., and J. H. K. Pechmann. 1985. Diel pattern of migratory activity for several species of pond-breeding salamanders. *Copeia* 1985:86–91.

Semlitsch, R. D., J. Pickle, M. J. Parris, and R. D. Sage. 1999. Jumping performance and short-term repeatability of newly metamorphosed hybrid and parental leopard frogs (*Rana sphenocephala* and *Rana blairi*). *Can. J. Zool.* 77:748–54.

Semlitsch, R. D., and S. B. Reichling. 1989. Density-dependent injury in larval salamanders. *Oecologia* 81:100–103.

Semlitsch, R. D., and H.-U. Reyer. 1992a. Modification of anti-predator behaviour in tadpoles by environmental conditions. *J. Anim. Ecol.* 61:353–60.

———. 1992b. Performance of tadpoles from the hybridogenetic *Rana esculenta* complex: interactions with pond drying and interspecific competition. *Evolution* 46:665–76.

Semlitsch, R. D., and B. B. Rothermel. 2003. A foundation for conservation and management of amphibians. In *Amphbibian conservation,* ed. R. D. Semlitsch, 242–59. Washingtom, DC: Smithsonian Institution Press.

Semlitsch, R. D., D. E. Scott, and J. H. K. Pechmann. 1988. Time and size at metamorphosis related to adult fitness in *Ambystoma talpoideum. Ecology* 69:184–92.

Semlitsch, R. D., D. E. Scott, J. H. K. Pechmann, and J. W. Gibbons. 1996. Structure and dynamics of an amphibian community. Evidence from a 16–year study of a natural pond. In *Long-term studies of vertebrate communities,* ed. M. L. Cody and J. A. Smallwood, 217–48. San Diego: Academic Press.

Semlitsch, R. D., and S. C. Walls. 1993. Competition in two species of larval salamanders: a test of geographic variation in competitive ability. *Copeia* 1993:587–95.
Semlitsch, R. D., and C. A. West. 1983. Aspects of the life history and ecology of Webster's salamander, *Plethodon websteri. Copeia* 1983:339–46.
Semlitsch, R. D., and H. M. Wilbur. 1988. Effects of pond drying time on metamorphosis and survival in the salamander *Ambystoma talpoideum. Copeia* 1988:978–83.
———. 1989. Artificial selection for paedomorphosis in the salamander *Ambystoma talpoideum. Evolution* 43:105–12.
Senfft, W. 1936. Das Brutgeschaft des Baumsteigerfrosches (*Dendrobates auratus* Girard) in Gefangenschaft. *Zool. Gart.* 8:122–36.
Senning, W. C. 1940. A study of age determination and growth of *Necturus maculosus,* based on the parasphenoid bone. *Am. J. Anat.* 66:483–94.
Seppä, P., and A. Laurila. 1999. Genetic structure of island populations of the anurans *Rana temporaria* and *Bufo bufo. Heredity* 82:309–17.
Sequeira, F., N. Ferrand, and E. G. Crespo. 2003. Reproductive cycle of the golden-striped salamander *Chioglossa lusitanica* (Caudata, Salamandridae) in NW Portugal. *Amphibia-Reptilia* 24:1–12.
Serbionova, I. A., and V. A. Solkin. 1995. Reprodutive behavior in the long-tailed salamander (*Onychodactylus fischeri* Boulenger). *Asiatic Herpetol. Res.* 6:114–19.
Serra-Cobo, J., G. Lacroix, and S. White. 1998. Comparison between the ecology of the new European frog *Rana pyrenaica* and that of four Pyrenean amphibians. *J. Zool., London* 246:147–54.
Seshaiya, R. V. 1941. Tadpoles as hosts for the glochidia of the fresh-water mussel. *Curr. Sci.* 10:535–36.
———. 1969. Some observations on the life-histories of South Indian freshwater mussels. *Malacologia* 9:286–87.
Sessions, S. K. 2003. What is causing deformed amphibians? In *Amphibian conservation,* ed. R. D. Semlitsch, 168–86. Washington, DC: Smithsonian Books.
Sessions, S. K., R. A. Franssen, and V. L. Horner. 1999. Morphological clues from multilegged frogs: Are retinoids to blame? *Science* 284:800–802.
Sessions, S. K., and A. Larson. 1987. Developmental correlates of genome size in plethodontid salamanders and their implications for genome evolution. *Evolution* 41:1239–51.
Sessions, S. K., and S. B. Ruth. 1990. Explanation for naturally occurring supernumery limbs in amphibians. *J. Exp. Zool.* 254:38–47.
Sever, D. M. 1975a. Morphology and seasonal variation of the mental hedonic glands of the dwarf salamander, *Eurycea quadradigitata* (Holbrook). *Herpetologica* 31:241–51.
———. 1975b. Morphology and seasonal variation of the nasolabial glands of *Eurycea quadradigitata* (Holbrook). *J. Herpetol.* 9:337–48.
———. 1976. Morphology of mental hedonic gland clusters of plethodontid salamanders (Amphibia, Urodela, Plethodontidae). *J. Herpetol.* 10:227–39.
———. 1978. Male cloacal glands of *Plethodon cinereus* and *Plethodon dorsalis* (Amphibia, Plethodontidae). *Herpetologica* 34:1–20.
———. 1980. Cloacal anatomy of male brook salamanders (*Eurycea*). *Herpetologica* 36:51–60.
———. 1981. Cloacal anatomy of male salamanders in the families Ambystomatidae, Salamandridae, and Plethodontidae. *Herpetologica* 37:142–55.
———. 1983. Cloacal anatomy of male salamanders in the plethodontid subfamily Desmognathinae. *Herpetologica* 39:16–27.
———. 1985. Sexually dimorphic glands of *Eurycea nana, Eurycea neotenes* and *Typhlomolge rathbuni* (Amphibia: Plethodontidae). *Herpetologica* 41:71–84.
———. 1989. Caudal hedonic glands in salamanders of the *Eurycea bislineata* complex (Amphibia: Plethodontidae). *Herpetologica* 45:322–29.
———. 1991a. Comparative anatomy and phylogeny of the cloacae of salamanders (Amphibia: Caudata). I. Evolution at the family level. *Herpetologica* 47:165–93.
———. 1991b. Sperm storage and degradation in the spermathecae of the salamander *Eurycea cirrigera. J. Morphol.* 210:71–84.
———. 1992a. Comparative anatomy and phylogeny of the cloacae of salamanders (Amphibia: Caudata). V. Proteidae. *Herpetologica* 48:318–29.
———. 1992b. Spermiophagy by the spermathecal epithelium of the salamander *Eurycea cirrigera. J. Morphol.* 212:281–90.
———. 1994. Comparative anatomy and phylogeny of the cloacae of salamanders (Amphibia: Caudata). VII. Plethodontidae. *Herpetol. Monogr.* 8:276–337.
———. 1995. Spermathecae of *Ambystoma tigrinum* (Amphibia: Caudata): Development and a role for the secretion. *J. Herpetol.* 29:243–55.
———. 1997. Sperm storage in the spermathecae of the red-back salamander, *Plethodon cinereus* (Amphibia: Plethodontidae). *J. Morphol.* 234:131–46.
———. 2002. Female sperm storage in amphibians. *J. Exp. Zool.* 292:165–79.
———. 2003a. Courtship and mating glands. In *Reproductive biology and phylogeny of Urodela,* ed. D. M. Sever, 323–81. Enfield, NH: Science Publishers.
———. 2003b. *Reproductive biology and phylogeny of Urodela,* ed., 323–81. Enfield, NH: Science Publishers.
Sever, D. M., and H. L. Bart, Jr. 1996. Ultrastructure of the spermathecae of *Necturus beyeri* (Amphibia: Proteidae) in relation to its breeding season. *Copeia* 1996:927–37.
Sever, D. M., and R. Brizzi. 1998. Comparative biology of sperm storage in female salamanders. *J. Exp. Zool.* 282:460–76.
Sever, D. M., J. S. Doody, C. A. Reddish, M. M. Wenner, and D. R. Church. 1996. Sperm storage in spermathecae of the great lamper eel, *Amphiuma tridactylum* (Caudata: Amphiumidae). *J. Morphol.* 230:79–97.
Sever, D. M., T. Halliday, V. Waights, J. Brown, H. A. Davies, and E. C. Moriarty. 1999. Sperm storage in females of the smooth newt (*Triturus v. vulgaris* L.): I. Ultrastructure of the spermathecae during the breeding season. *J. Exp. Zool.* 283:51–70.
Sever, D. M., and W. C. Hamlett. 1998. Sperm aggregations in the spermatheca of female desmognathine salamanders (Amphibia: Urodela: Plethodontidae). *J. Morphol.* 238:143–55.
Sever, D. M., W. C. Hamlett, R. Slabach, B. Stephenson, and P. A. Verrell. 2003. Internal fertilization in the Anura with special reference to mating and female sperm storage in *Ascaphus.* In *Reproductive biology and phylogeny of anura,* ed. B. G. M. Jamieson, 319–41. Enfield, NH: Science Publishers.
Sever, D. M., and L. D. Houck. 1985. Spermatophore formation in *Desmognathus ochrophaeus* (Amphibia: Plethodontidae). *Copeia* 1985:394–402.
Sever, D. M., and N. M. Kloepfer. 1993. Spermathecal cytology

of *Ambystoma opacum* (Amphibia: Ambystomatidae) and the phylogeny of sperm storage organs in female salamanders. *J. Morphol.* 217:115–27.

Sever, D. M., J. D. Krenz, K. M. Johnson, and L. C. Rania. 1995. Morphology and evolutionary implications of the annual cycle of secretion and sperm storage in spermathecae of the salamander *Ambystoma opacum* (Amphibia: Ambystomatidae). *J. Morphol.* 223:35–46.

Sever, D. M., E. C. Moriarity, L. C. Rania, and W. C. Hamlett. 2001. Sperm storage in the oviduct of the internal fertilizing frog *Ascaphus truei*. *J. Morphol.* 248:1–21.

Sever, D. M., L. C. Rania, and J. D. Krenz. 1996. The annual cycle of sperm storage in the spermathecae of the red-spotted newt, *Notophthalmus viridescens* (Amphibia: Caudata). *J. Morphol.* 227:155–70.

Sever, D. M., M. Sparreboom, and G. Schultschik. 1997. The dorsal tail tubercle of *Mertensiella caucasica* and *M. luschani* (Amphibia: Salamandridae). *J. Morphol.* 232:93–105.

Sever, D. M., C. K. Tait, L. V. Diller, and L. Burkholder. 2004a. Sperm aggregation in the spermathecae of southern torrent salamanders, *Rhyacotriton variegatus*. *J. Herpetol.* 38:61–72.

———. 2004b. Ultrastructure of the annual cycle of female sperm storage in spermathecae of the torrent salamander, *Rhyacotriton variegatus* (Amphibia: Rhyacotritonidae). *J. Morphol.* 261:1–17.

Sever, D. M., P. A. Verrell, T. R. Halliday, M. Griffiths, and V. Waights. 1990. The cloaca and cloacal glands of the male smooth newt, *Triturus vulgaris vulgaris* (Linnaeus), with especial emphasis on the dorsal gland. *Herpetologica* 46:160–68.

Sexton, O. J. 1956–57. The distribution of *Bothrops atrox* in relation to food supply. *Bol. Mus Cienc. Nat.* 2–3:47–54.

———. 1960. Some aspects of the behavior and of the territory of a dendrobatid frog, *Prostherapis trinitatis*. *Ecology* 41:107–15.

———. 1962. Apparent territorialism in *Leptodactylus insularum*. *Herpetologica* 18:212–14.

Sexton, O. J., and J. R. Bizer. 1978. Life history patterns of *Ambystoma tigrinum* in montane Colorado. *Am. Midl. Nat.* 99:101–18.

Sexton, O. J., J. Bizer, D. C. Gayou, P. Freiling, and M. Moutseous. 1986. Field studies of breeding spotted salamanders, *Ambystoma maculatum,* in eatern Missouri, USA. *Milwaukee Pub. Mus. Contrib. Biol. Geol.* 67:1–19.

Sexton, O. J., and E. P. Ortleb. 1966. Some cues used by the leptodactylid frog, *Engystomops pustulosus,* in selection of the oviposition site. *Copeia* 1966:225–30.

Sexton, O. J., and C. Phillips. 1986. A qualitative study of the fish-amphibian interactions in three Missouri ponds. *Trans. Missouri Acad. Sci.* 20:25–35.

Seymour, C. L., H. M. De Klerk, A. Channing, and T. M. Crowe. 2001. The biogeography of the Anura of sub-equatorial Africa and the prioritization of areas for their conservation. *Biodiver. Cons.* 10:2045–76.

Seymour, R. S. 1972. Behavioral thermoregulation by juvenile green toads, *Bufo debilis*. *Copeia* 1972:572–75.

———. 1973a. Energy metabolism of dormant spadefoot toads (*Scaphiopus*). *Copeia* 1973:435–45.

———. 1973b. Gas exchange in spadefoot toads beneath the ground. *Copeia* 1973:452–60.

———. 1973c. Physiological correlates of forced activity and burrowing in the spadefoot toad, *Scaphiopus hammondii*. *Copeia* 1973:103–15.

———. 1982. Physiological adaptations to aquatic life. In *Biology of the reptilia, vol. 13,* ed. C. Gans and F. H. Pough, 1–51. New York: Academic Press.

———. 1994. Oxygen diffusion through the jelly capsules of amphibian eggs. *Israel J. Zool.* 40:493–506.

———. 1995. Oxygen uptake by embryos in gelatinous egg masses of *Rana sylvatica:* The roles of diffusion and convection. *Copeia* 1995:626–35.

———. 1999. Respiration of aquatic and terrestrial amphibian embryos. *Amer. Zool.* 39:261–70.

Seymour, R. S., and D. F. Bradford. 1987. Gas exchange through the jelly capsule of the terrestrial eggs of the frog, *Pseudophryne bibroni*. *J. Comp. Physiol. B* 157:477–81.

———. 1995. Respiration of amphibian eggs. *Physiol. Zool.* 68:1–25.

Seymour, R. S., F. Geiser, and D. F. Bradford. 1991a. Gas conductance of the jelly capsule of terrestrial frog eggs correlates with embryonic stage, not metabolic demand or ambient $PO_2$. *Physiol. Zool.* 64:673–87.

———. 1991b. Metabolic cost of development in terrestrial frog eggs (*Pseudophryne bibronii*). *Physiol. Zool.* 64:688–96.

Seymour, R. S., and J. P. Loveridge. 1994. Embryonic and larval respiration in the arboreal foam nests of the African frog *Chiromantis xerampelina*. *J. Exp. Biol.* 197:31–46.

Seymour, R. S., M. J. Mahony, and R. Knowles. 1995. Respiration of embryos and larvae of the terrestrially breeding frog *Kyarranus loveridgei*. *Herpetologica* 51:369–76.

Seymour, R. S., and J. D. Roberts. 1991. Embryonic respiration and oxygen distribution in foamy and nonfoamy egg masses of the frog *Limnodynastes tasmaniensis*. *Physiol. Zool.* 64:1322–40.

———. 1995. Oxygen uptake by the aquatic eggs of the Australian frog *Crinia georgiana*. *Physiol. Zool.* 68:206–22.

Seymour, R. S., J. D. Roberts, N. J. Mitchell, and A. J. Blaylock. 2000. Influence of environmental oxygen on development and hatching of aquatic eggs of the Australian frog, *Crinia georgiana*. *Physiol. Biochem. Zool.* 73:501–7.

Shaffer, H. B. 1984a. Evolution in a paedomorphic lineage. I. An electrophoretic analysis of the Mexican ambystomatid salamanders. *Evolution* 38:1194–1206.

———. 1984b. Evolution in a paedomorphic lineage. II. Allometry and form in the Mexican ambystomatid salamanders. *Evolution* 38:1207–18.

———. 1993. Phylogenetics of model organisms: The laboratory axolotl, *Ambystoma mexicanum*. *Syst. Biol.* 42:508–22.

Shaffer, H. B., G. M. Fellers, S. R. Vos, J. C. Oliver, and G. B. Pauly. 2004. Species boundaries, phylogeography and conservation genetics of the red-legged frog (*Rana aurora/draytonii*) complex. *Mol. Ecol.* 13:2667–77.

Shaffer, H. B., and G. V. Lauder. 1985a. Aquatic prey capture in ambystomatid salamanders: Patterns of variation in muscle activity. *J. Morphol.* 183:273–84.

———. 1985b. Patterns of variation in aquatic ambystomatid salamanders: Kinematics of the feeding mechanism. *Evolution* 39:83–92.

———. 1988. The ontogeny of functional design: Metamorphosis of feeding behaviour in the tiger salamander *(Ambystoma tigrinum)*. *J. Zool., London* 216:437–54.

Shaffer, H. B., and M. McKnight. 1996. The polytypic species revisited: Genetic differentiation and molecular phylogenetics of the tiger salamander (*Ambystoma tigrinum*) (Amphibia: Caudata) complex. *Evolution* 50:417–33.

Shaffer, H. B., and S. R. Voss. 1996. Phylogenetic and mechanistic analysis of a developmentally integrated character complex: Alternate life history modes in ambystomatid salamanders. *Amer. Zool.* 36:24–35.

Shannon, P., and D. L. Kramer. 1988. Water depth alters respiratory behaviour of *Xenopus laevis*. *J. Exp. Biol.* 137:597–602.

Sharell, R. 1966. *The tuatara, lizards and frogs of New Zealand.* London: Collins.

Sharitz, R. R. 2003. Carolina Bay wetlands: Unique habitats of the southeastern United States. *Wetlands* 23:550–62.

Sharma, B., and T. Sharma. 1977. Lycosid spiders feeding on juveniles of the skipper frog *Rana cyanophlyctis* Schneider. *J. Bombay Nat. Hist. Soc.* 74:189.

Shea, G. M. 1987. Bibliography of herpetological references in Australian ornithological journals. Smithsonian Herpetological Information Service, no. 73.

Shealy, R. M. 1975. Factors influencing activity in the salamanders *Desmognathus ochrophaeus* and *D. monticola* (Plethodontidae). *Herpetologica* 31:94–102.

Sheil, C. A., J. R. Mendelson III, and H. R. da Silva. 2001. Phylogenetic relationships of the species of Neotropical horned frogs, genus *Hemiphractus* (Anura: Hylidae: Hemiphractinae), based on evidence from morphology. *Herpetologica* 57:203–14.

Shelly, T. E., and T. S. Whittier. 1997. Lek behavior of insects. In *The evolution of mating systems in insects and arachnids,* ed. J. C. Choe and B. J. Crespi, 273–93. New York: Cambridge University Press.

Shelton, G. 1970. The effect of lung ventilation on blood flow to the lungs and body of the amphibian, *Xenopus laevis*. *Respir. Physiol.* 9:183–96.

———. 1976. Gas exchange, pulmonary blood supply, and the partially divided amphibian heart. In *Perspectives in experimental biology. Vol. I. Zoology,* ed. P. Spencer Davies, 247–59. Oxford: Pergamon.

———. 1985. Functional and evolutionary significance of cardiovascular shunts in the Amphibia. In *Cardiovascular shunts,* ed. K. Johansen and W. W. Burggren, 100–120. Copenhagen, Denmark: Munksgaard.

Shelton, G., and R. G. Boutilier. 1982. Apnoea in amphibians and reptiles. *J. Exp. Biol.* 100:245–73.

Shelton, G., D. R. Jones, and W. K. Milsom. 1986. Control of breathing in ectothermic vertebrates. In *Handbook of Physiology, Section 3: The respiratory system, vol. 2,* ed. A. P. Fishman, N. S. Cherniak, and J. G. Widdicombe, 857–909. Bethesda, MD: American Physiological Society.

Shepard, D. B., and J. P. Caldwell. 2005. From foam to free-living: Ecology of larval *Leptodactylus labyrinthicus*. *Copeia* 2005:803–11.

Shepherdson, D. J., T. J. Roper, and P. Lüps. 1990. Diet, food availability and foraging behaviour of badgers (*Meles meles* L.) in southern England. *Z. Säugetierkunde* 55:81–93.

Sheppard, P. M., and J. R. G. Turner. 1977. The existence of Müllerian mimicry. *Evolution* 31:452–53.

Sherk, T. E. 1977. Development of the compound eyes of dragonflies (Odonata). I. Larval compound eyes. *J. Exp. Zool.* 201:391–416.

Sherratt, T. N., and I. F. Harvey. 1989. Predation by larvae of *Pantala flavescens* (Odonata) on tadpoles of *Phyllomedusa trinitatis* and *Physalaemus pustulosus:* The influence of absolute and relative density of prey on predator choice. *Oikos* 56:170–76.

Shi, Y.-B. 2000. *Amphibian metamorphosis. From morphology to molecular biology.* New York: Wiley-Liss.

Shield, J. W., and P. J. Bentley. 1973. Respiration of some urodele and anuran Amphibia. I. In water, role of the skin and gills. *Comp. Biochem. Physiol.* 46A:17–28.

Shields, W. M. 1982. *Philopatry, inbreeding, and the evolution of sex.* Albany: State University of New York Press.

———. 1983. Optimal inbreeding and the evolution of philopatry. In *The ecology of animal movement,* ed. I. R. Swingland and P. J. Greenwood, 132–59. New York: Oxford University Press.

Shillington, C., and P. Verrell. 1996. Multiple matings by females is not dependent on body size in the salamander *Desmognathus ochrophaeus*. *Amphibia-Reptilia* 17:33–38.

Shimoyama, R. 1982. Preliminary report on male territoriality in the pond frog, *Rana nigromaculata,* in the breeding season. *Jap. J. Herpetol.* 9:99–102.

———. 1986. Maturity and clutch frequency of female *Rana porosa brevipoda* in the northern Ina Basin, Nagon Prefecture, Japan. *Jap. J. Herpetol.* 11:167–72.

———. 1989. Breeding ecology of a Japanese pond frog, *Rana p. porosa*. In *Current herpetology in East Asia,* ed. M. Matsui, T. Hikida, and R. C. Goris, 323–31. Kyoto: Herpetological Society of Japan.

———. 1993. Chorus organization and male mating behavior in the Japanese pond frog, *Rana porosa brevipoda*. *J. Ethol.* 11:91–97.

———. 1996. Sympatric and synchronous breeding by two pond frogs, *Rana porosa brevipoda* and *Rana nigromaculata*. *Jap. J. Herpetol.* 16:87–93.

———. 1999. Interspecific interactions between two Japanese pond frogs, *Rana porosa brevipoda* and *Rana nigromaculata*. *Jap. J. Herpetol.* 18:7–15.

Shine, R. 1977. Habitats, diets and sympatry in snakes: a study from Australia. *Can. J. Zool.* 55:1118–28.

———. 1978. Sexual size dimorphism and male combat in snakes. *Oecologia* 33:269–77.

———. 1979. Sexual selection and sexual dimorphism in the Amphibia. *Copeia* 1979: 297–306.

———. 1980a. Comparative ecology of three Australian snake species of the genus *Cacophis* (Serpentes: Elapidae). *Copeia* 1980:831–38.

———. 1980b. Ecology of eastern Australian whipsnakes of the genus *Demansia*. *J. Herpetol.* 14:381–89.

———. 1980c. Ecology of the Australian death adder *Acanthophis antarcticus* (Elapidae): Evidence for convergence with the Viperidae. *Herpetologica* 36:281–89.

———. 1981. Venomous snakes in cold climates: Ecology of the Australian genus *Drysdalia* (Serpentes: Elapidae). *Copeia* 1981:14–25.

———. 1982. Ecology of the Australian elapid snake *Echiopsis curta*. *J. Herpetol.* 16:388–93.

———. 1983. Food habits and reproductive biology of Australian elapid snakes of the genus *Denisonia*. *J. Herpetol.* 17:171–75.

———. 1986a. Food habits, habitats and reproductive biology of four sympatric species of varanid lizards in tropical Australia. *Herpetologica* 42:346–60.

———. 1986b. Natural history of two monotypic snake genera of southwestern Australia, *Elapognathus* and *Rhinoplocephalus* (Elapidae). *J. Herpetol.* 20:436–39.

———. 1987a. Ecological comparisons of island and mainland populations of Australian tigersnakes (*Notechis:* Elapidae). *Herpetologica* 43:233–40.

———. 1987b. Ecological ramifications of prey size: Food habits and reproductive biology of Australian copperhead snakes (*Austerlaps,* Elapidae). *J. Herpetol.* 21:21–28.

———. 1987c. Food habits and reproductive biology of Australian snakes of the genus *Hemiaspis* (Elapidae). *J. Herpetol.* 21:71–74.

Shine, R., W. R. Branch, P. S. Harlow, and J. K. Webb. 1996. Sexual dimorphism, reproductive biology, and food habits of two species of African filesnakes (*Mehelya,* Colubridae). *J. Zool., London* 240:327–40.

Shine, R., and N. Charles. 1982. Ecology of the Australian elapid snake *Tropidechis carinatus. J. Herpetol.* 16:383–87.

Shirose, L. J., and R. J. Brooks. 1995a. Age structure, mortality, and longevity in syntopic populations of three species of ranid frogs in central Ontario. *Can. J. Zool.* 73:1878–86.

———. 1995b. Growth rate and age at maturity in syntopic populations of *Rana clamitans* and *Rana septentrionalis* in central Ontario. *Can. J. Zool.* 73:1468–73.

Shirose, L. J., R. J. Brooks, J. R. Barta, and S. S. Desser. 1993. Intersexual differences in growth, mortality, and size at maturity in bullfrogs in central Ontario. *Can. J. Zool.* 71:2363–69.

Shoemaker, V. H. 1964. The effects of dehydration on electrolyte concentrations in a toad, *Bufo marinus. Comp. Biochem. Physiol.* 13:261–71.

———. 1965. The stimulus for the water-balance response to dehydration in toads. *Comp. Biochem. Physiol.* 15:81–88.

———. 1975. Adaptations to aridity in amphibians and reptiles. In *Physiological adaptation to the environment,* ed. F. J. Vernberg, 143–54. New York: Intext Educational.

———. 1987. Osmoregulation in amphibians. In *Comparative physiology: Life in water and on land.* FIDIA Research Series, vol. 9, ed. P. Dejours, L. Bolis, C. R. Taylor, and E. R. Weibel, 109–20. Padova, Italy: Liviana Press, Springer Verlag.

———. 1988. Physiological ecology of amphibians in arid environments. *J. Arid Environ.* 14:145–53.

Shoemaker, V. H., M. A. Baker, and J. P. Loveridge. 1989. Effect of water balance on thermoregulation in waterproof frogs (*Chiromantis* and *Phyllomedusa*). *Physiol. Zool.* 62:133–46.

Shoemaker, V. H., D. Balding, R. Ruibal, and L. L. McClanahan, Jr. 1972. Uricotelism and low evaporative water loss in a South American frog. *Science* 175:1018–20.

Shoemaker, V. H., and P. E. Bickler. 1979. Kidney and bladder function in a urioctelic treefrog (*Phyllomedusa sauvagei*). *J. Comp. Physiol.* 133:211–18.

Shoemaker, V. H., S. S. Hillman, S. D. Hillyard, D. C. Jackson, L. L. McClanahan, P. C. Withers, and M. Wygoda. 1992. Exchange of water, ions, and respiratory gases in terrestrial amphibians. In *Environmental physiology of the amphibians,* ed. M. E. Feder and W. W. Burggren, 125–50. Chicago: University of Chicago Press.

Shoemaker, V. H., and L. L. McClanahan. 1973. Nitrogen excretion in the larvae of a land-nesting frog (*Leptodactylus bufonius*). *Comp. Biochem. Physiol.* 44A:1149–56.

———. 1975. Evaporative water loss, nitrogen excretion and osmoregulation in phyllomedusine frogs. *J. Comp. Physiol.* 100:331–45.

———. 1980. Nitrogen excretion and water balance in amphibians of Borneo. *Copeia* 1980:446–51.

———. 1982. Enzymatic correlates and ontogeny of uricotelism in tree frogs of the genus *Phyllomedusa. J. Exp. Zool.* 220:163–69.

Shoemaker, V. H., L. L. McClanahan, and R. Ruibal. 1969. Seasonal changes in body fluids in a field population of spadefoot toads. *Copeia* 1969:585–91.

Shoemaker, V. H., L. L. McClanahan, P. C. Withers, S. S. Hillman, and R. C. Drewes. 1987. Thermoregulatory response to heat in the waterproof frogs *Phyllomedusa* and *Chiromantis. Physiol. Zool.* 60:365–72.

Shoemaker, V. H., and K. A. Nagy. 1977. Osmoregulation in amphibians and reptiles. *Ann. Rev. Physiol.* 39:449–71.

Shoemaker, V. H., and H. Waring. 1968. Effect of hypothalamic lesions on the water-balance response of a toad (*Bufo marinus*). *Comp. Biochem. Physiol.* 24:47–54.

Shofner, W. P. 1988. Postmetamorphic changes in the auditory system. In *The evolution of the amphibian auditory system,* ed. B. Fritzsch, M. J. Ryan, W. Wilczynski, T. E. Hetherington, and W. Walkowiak, 377–92. New York: Wiley.

Shofner, W. P., and A. S. Feng. 1981. Post-metamorphic development of the frequency selectivities and sensitivities of the peripheral auditory system of the bullfrog, *Rana catesbeiana. J. Exp. Biol.* 93:181–96.

———. 1983. A quantitative light microscopic study of the bullfrog amphibian papilla tectorum: Correlation with the tonotopic organization. *Hearing Res.* 11:103–16.

———. 1984. Quantitative light and scanning electron microscopic study of the developing auditory organs in the bullfrog: Implications on their functional characteristics. *J. Comp. Neurol.* 224:141–54.

Shoop, C. R. 1960. The breeding habits of the mole salamander, *Ambystoma talpoideum* (Holbrook), in southeastern Louisiana. *Tulane Stud. Zool.* 8:65–82.

———. 1965. Aspects of reproduction in Louisiana *Necturus* populations. *Am. Midl. Nat.* 74:357–67.

———. 1968. Migratory orientation of *Ambystoma maculatum:* movements near breeding ponds and displacements of migrating individuals. *Biol. Bull.* 135:230–38.

———. 1974. Yearly variation in larval survival of *Ambystoma maculatum. Ecology* 55:440–44.

Shoop, C. R., and T. L. Doty. 1972. Migratory orientation by marbled salamanders (*Ambystoma opacum*) near a breeding area. *Behav. Biol.* 7:131–36.

Shpun, S., J. Hoffman, and U. Katz. 1992. Anuran Amphibia which are not acclimable to high salt, tolerate high plama urea. *Comp. Biochem. Physiol.* 103 A:473–77.

Shpun, S., J. Hoffman, E. Nevo, and U. Katz. 1993. Is the distribution of *Pelobates syriacus* related to its limited osmoregulatory capacity? *Comp. Biochem. Physiol.* 105 A:135–39.

Shpun, S., and U. Katz. 1995. Renal function at steady state in a toad (*Bufo viridis*) acclimated in hyperosmotic NaCl and urea solutions. *J. Comp. Physiol. B* 164:646–52.

———. 1999. Renal response of euryhaline toad (*Bufo viridis*) to acute immersion in tap water, NaCl, or urea solutions. *Physiol. Biochem. Zool.* 72:227–37.

Shubin, N. H., E. B. Daeschler, and M. I. Coates. 2004. The early evolution of the tetrapod humerus. *Science* 304:90–93.

Shubin, N. H., E. B. Daeschler, and F. A. Jenkins, Jr. 2006. The pectoral fin of *Tiktaalik roseae* and the origin of the tetrapod limb. *Nature* 440:764–71.

Shubin, N. H., and F. A. Jenkins, Jr. 1995. An early Jurassic jumping frog. *Nature* 377:49–52.

Shumway, W. 1940. Stages in the normal development of *Rana pipiens.* I. External form. *Anat. Rec.* 78:139–44.

Shvarts, S. S., and D. A. Pyastolova. 1970a. Regulators of growth and development of amphibian larvae. I. Specificity of effects. *Sov. J. Ecol.* 1:58–62.

———. 1970b. Regulators of growth and development of amphibian larvae. II. Diversity of effects. *Sov. J. Ecol.* 1:122–34.

Shy, E. 1983. The relation of geographical variation in song to habitat characteristic and body size in North American tanagers (Thraupinae: *Piranga*). *Behav. Ecol. Sociobiol.* 12:71–76.

Siddiqi, A., T. W. Cronin, E. R. Loew, M. Vorobyev, and K. Summers. 2004. Interspecific and intraspecific views of color signals in the strawberry poison frog *Dendrobates pumilio. J. Exp. Biol.* 207:2471–85.

Sidorovich, V., H. Kruuk, D. W. Macdonald, and T. Maran. 1998. Diets of semi-aquatic carnivores in northern Belarus, with implications for population changes. In *Behaviour and ecology of riparian mammals,* ed. N. Dunstone and M. Gorman, 177–89. Cambridge: Cambridge University Press.

Sidorovich, V. E., and M. M. Pikulik. 1997. Toads *Bufo* spp. in the diets of mustelid predators in Belarus. *Acta Theriol.* 42:105–8.

Siedlecki, M. 1909. Zur Kenntnis des Javanischen Flugfrosches. *Biol. Zentralbl.* 29:704–37.

Siegert, F., G. Ruecker, A. Hinrichs, and A. A. Hoffmann. 2001. Increased damage from fires in logged forests during droughts caused by El Niño. *Nature* 414:437–40.

Siegfried, W. R. 1971. The food of the cattle egret. *J. Appl. Ecol.* 8:447–68.

Sievert, L. M., and J. K. Bailey. 2000. Specific dynamic action in the toad, *Bufo woodhousii. Copeia* 2000:1076–78.

Sih, A., P. Crowley, M. McPeek, J. Petranka, and K. Strohmeier. 1985. Predation, competition, and prey communities: A review of field experiments. *Ann. Rev. Ecol. Syst.* 16:269–311.

Sih, A., G. Englund, and D. Wooster. 1998. Emergent impacts of multiple predators on prey. *Trends Ecol. Evol.* 13:350–55.

Sih, A., and L. B. Kats. 1991. Effects of refuge availability on the responses of salamander larvae to chemical cues from predatory green sunfish. *Anim. Behav.* 42:330–32.

———. 1994. Age, experience, and the response of streamside salamander hatchlings to chemical cues from predatory sunfish. *Ethology* 96:253–59.

Sih, A., L. B. Kats, and E. F. Maurer. 2000. Does phylogenetic inertia explain the evolution of ineffective antipredator behavior in a sunfish-salamander system? *Behav. Ecol. Sociobiol.* 49:48–56.

Sih, A., L. B. Kats, and R. D. Moore. 1992. Effects of predatory sunfish on the density, drift, and refuge use of stream salamander larvae. *Ecology* 73:1418–30.

Sih, A., and E. Maurer. 1992. Effects of cryptic oviposition on egg survival for stream-breeding, streamside salamanders. *J. Herpetol.* 26:114–16.

Sih, A., and R. D. Moore. 1993. Delayed hatching of salamander eggs in response to enhanced larval predation risk. *Am. Nat.* 142:947–60.

Sih, A., J. W. Petranka, and L. B. Kats. 1988. The dynamics of prey refuge use: A model and tests with sunfish and salamander larvae. *Am. Nat.* 132:463–83.

Silva, W. R., A. A. Giaretta, and K. G. Facure. 2005. On the natural history of the South American pepper frog, *Leptodactylus labyrinthicus* (Spix, 1824) (Anura: Leptodactylidae). *J. Nat. Hist.* 39:555–66.

Silverin, B., and C. Andren. 1992. The ovarian cycle in the natterjack toad, *Bufo calamita,* and its relation to breeding behavior. *Amphibia-Reptilia* 13:177–92.

Silverstone, P. A. 1973. Observations on the behavior and ecology of a Colombian poison-arrow frog, the Kokoe-pa (*Dendrobates histrionicus* Berthold). *Herpetologica* 29:295–301.

———. 1975. A revision of the poison-arrow frogs of the genus *Dendrobates* Wagler. *Nat. Hist. Mus. Los Angeles Co. Sci. Bull.* 21:1–55.

———. 1976. A revision of the poison-arrow frogs of the genus *Phyllobates* Bibron in Sagra (Family Dendrobatidae). *Nat. Hist. Mus. Los Angeles Co. Sci. Bull.* 27:1–53.

Sima, P. 2003. Nonmammalian vertebrate antibiotic peptides. *Folia Microbiol.* 48:709–24.

Simberloff, D. 1983. Competition theory, hypothesis testing, and other community ecology buzzwords. *Am. Nat.* 122:626–35.

Simmaco, M., M. L Mangoni, A. Boman, D. Barra, and H. G. Boman. 1998. Experimental infections of *Rana esculenta* with *Aeromonas hydrophila:* A molecular mechanism for the control of the normal flora. *Scan. J. Immunol.* 48:357–63.

Simmons, D. 1976. A naturally metamorphosed *Gyrinophilus palleucus* (Amphibia, Urodela, Plethodontidae). *J. Herpetol.* 10:255–57.

Simmons, D. D., C. Bertolotto, and P. M. Narins. 1994. Morphological gradients in sensory hair cells of the amphibian papilla of the frog, *Rana pipiens pipiens. Hearing Res.* 80:71–78.

Simmons, L. W. 1992. Quantification of role reversal in relative parental investment in a bush cricket. *Nature* 358:61–63.

Simmons, L. W., R. J. Teale, M. Maier, R. J. Standish, W. J. Bailey, and P. C. Withers. 1992. Some costs of reproduction for male bushcrickets, *Requena verticalis* (Orthoptera. Tettigoniidae): Allocating resources to mate attraction and nuptial feeding. *Behav. Ecol. Sociobiol.* 31:57–62.

Simon, G. S., and D. M. Madison. 1984. Individual recognition in salamanders: Cloacal odours. *Anim. Behav.* 32:1017–20.

Simon, M. P. 1982. *The ecology and evolution of parental care in a terrestrial breeding frog from New Guinea.* Unpublished PhD dissertation. Davis: University of California.

———. 1983. The ecology of parental care in a terrestrial breeding frog from New Guinea. *Behav. Ecol. Sociobiol.* 14:61–67.

———. 1984. The influence of conspecifics on egg and larval mortality in amphibians. In *Infanticide: Comparative and evolutionary perspectives,* ed. G. Hausfater and S. Blaffer Hrdy, 65–86. New York: Aldine Publishing.

Simon, M. P., and C. A. Toft. 1991. Diet specialization in small vertebrates: Mite-eating in frogs. *Oikos* 61:263–78.

Simons, R. R., and B. E. Felgenhauer. 1992. Identifying areas of chemical signal production in the red-backed salamander, *Plethodon cinereus. Copeia* 1992:776–81.

Simons, R. R., B. E. Felgenhauer, and R. G. Jaeger. 1994. Salamander scent marks: Site of production and their role in territorial defence. *Anim. Behav.* 48:97–103.

Simons, R. R., R. G. Jaeger, and B. E. Felgenhauer. 1995. Juvenile terrestrial salamanders have active post-cloacal glands. *Copeia* 1995:481–83.

———. 1997. Competitor assessment and area defense by territorial salamanders. *Copeia* 1997:70–76.

Simons, R. S., and E. L. Brainerd. 1999. Morphological variation of the hypaxial musculature in salamanders. *J. Morphol.* 241:153–64.

Singhas, C. A., and J. N. Dent. 1975. Hormonal control of the tail fin and of the nuptial pads in the male red-spotted newt. *Gen. Comp. Endocrinol.* 26:382–93.

Sinsch, U. 1984. Thermal influences on the habitat preferences and the diurnal activity in three European *Rana species*. *Oecologia* 64:125–31.

———. 1985. Reproduktionsbiologie eines bachbewohnenden Frosches, *Telmatobius jelskii*, in einem Andenhoctal Zentralperus. *Verh. Dtsch. Zool. Ges.* 78:265.

———. 1987a. Migratory behaviour of the toad *Bufo bufo* within its home range and after displacement. In *Proc. 4th Ord. Gen. Meeting S. E. H.*, ed. J. J. van Gelder, H. Strijbosch, and P. J. M. Bergers, 361–64. Nijmegen, The Netherlands.

———. 1987b. Modifikationen im Wasserhaushalt von drei *Rana*-Arten (Amphibia: Anura) als Anpassungen an ihre spezifischen Habitate. *Verh. Ges. Okol.* 15:365–72.

———. 1987c. Orientation behaviour of toads (*Bufo bufo*) displaced from the breeding site. *J. Comp. Physiol. A* 161:715–27.

———. 1988a. Einfluss von Temperatur und Ernährung auf die diurnale Rufaktivität des Beutelfrosches, *Gastrotheca marsupiata. Verh. Dtsch. Zool. Ges.* 167:197.

———. 1988b. El sapo andino *Bufo spinulosus:* Analisis preliminar de su orientacion hacia sus lagares de reproduccion. *Bol. Lima* 57:83–91.

———. 1988c. Seasonal changes in the migratory behaviour of the toad *Bufo bufo:* Direction and magnitude of movements. *Oecologia* 76:390–398.

———. 1988d. Temporal spacing of breeding activity in the natterjack toad, *Bufo calamita. Oecologia* 76:399–407.

———. 1989a. Behavioural thermoregulation of the Andean toad (*Bufo spinulosus*) at high altitudes. *Oecologia* 80:32–38.

———. 1989b. Migratory behaviour of the common toad *Bufo bufo* and the natterjack toad *Bufo calamita*. In *Amphibians and roads,* ed. T. E. S. Langton, 113–25. Shefford, Bedfordshire, UK: ACO Polymer Products.

———. 1989c. Sommer- und Winterquartiere der Herpetofauna in Auskiesungen. *Salamandra* 25:104–8.

———. 1990. Migration and orientation in anuran amphibians. *Ethol. Ecol. Evol.* 2:65–79.

———. 1991a. Cold acclimation in frogs (*Rana*): Microhabitat choice, osmoregulation, and hydromineral balance. *Comp. Biochem. Physiol.* 98A:469–77.

———. 1991b. The orientation behaviour of amphibians. *Herpetol. J.* 1:541–44.

———. 1991c. Reabsorption of water and electrolytes in the urinary bladder of intact frogs (genus *Rana*). *Comp. Biochem. Physiol.* 99A:559–65.

———. 1992a. Amphibians. In *Animal homing,* ed. F. Papi, 213–33. New York: Chapman and Hall.

———. 1992b. Sex-biased site fidelity and orientation behaviour in reproductive natterjack toads (*Bufo calamita*). *Ethol. Ecol. Evol.* 4:15–32.

———. 1992c. Structure and dynamic of a natterjack toad metapopulation (*Bufo calamita*). *Oecologia* 90:489–92.

———. 1992d. Zwei neue Markierungsmethoden zur individuellen Identifkation von Amphibien in langfristigen Freilanduntersuchungen: Erste Erfahrungen bei Kreuzkröten. *Salamandra* 28:116–28.

———. 1997. Postmetamorphic dispersal and recruitment of first breeders in a *Bufo calamita* metapopulation. *Oecologia* 112:42–47.

———. 1998. *Biologie und Ökologie der Kreuzkröte.* Bochum, Germany: Laurenti Verlag.

Sinsch, U., and D. Seidel. 1995. Dynamics of local and temporal breeding assemblages in a *Bufo calamita* metapopulation. *Aust. J. Ecol.* 20:351–61.

Sinsch, U., R. Seine, and N. Sherif. 1992. Seasonal changes in the tolerance of osmotic stress in natterjack toads (*Bufo calamita*). *Comp. Biochem. Physiol.* 101A:353–60.

Sinsch, U., and N. Sherif. 1989. Osmoregulation of natterjack toads (*Bufo calamita*) in tap water, urea and NaCl solutions. *Verh. Dtsch. Zool. Ges.* 82:307–8.

Sites, J. W., Jr. 1978. The foraging strategy of the dusky salamander, *Desmognathus fuscus* (Amphibia, Urodela, Plethodontidae): An empirical approach to predation theory. *J. Herpetol.* 12:373–83.

Sivak, J. G., and M. R. Warburg. 1980. Optical metamorphosis of the eye of *Salamandra salamandra. Can. J. Zool.* 58:2059–64.

———. 1983. Changes in the optical properties of the eye during metamorphosis of an anuran, *Pelobates syriacus. J. Comp. Physiol. A* 150:329–32.

Sivakumar, M. V. K., H. P. Das, and O. Brunini. 2005. Impacts of present and future climate variability and change on agriculture and forestry in the arid and sem-arid tropics. *Climatic Change* 70:31–72.

Sjögren, P. 1991a. Extinction and isolation gradients in metapopulations: the case of the pool frog (*Rana lessonae*). *Biol. J. Linn. Soc.* 42:135–47.

———. 1991b. Genetic variation in relation to demography of peripheral pool frog populations (*Rana lessonae*). *Evol. Ecol.* 5:248–71.

Sjögren, P., J. Elmberg, and S.-A. Berglind. 1988. Thermal preference in the pool frog *Rana lessonae:* Impact on the reproductive behaviour of a northern fringe population. *Holarc. Ecol.* 11:178–84.

Sjögren Gulve, P. 1994. Distribution and extinction patterns within a northern metapopulation of the pool frog, *Rana lessonae. Ecology* 75:1357–67.

———. 1998a. Spatial movement patterns in frogs: Differences between three *Rana* species. *Ecoscience* 5:148–55.

———. 1998b. Spatial movement patterns in frogs: Target-oriented dispersal in the pool frog, *Rana lessonae. Ecoscience* 5:31–38.

Skelly, D. K. 1992. Field evidence for a cost of behavioral antipredator response in a larval amphibian. *Ecology* 73:704–8.

———. 1994. Activity level and susceptibility of anuran larvae to predation. *Anim. Behav.* 47:465–68.

———. 1995a. A behavioral trade-off and its consequences for the distribution of *Pseudacris* treefrog larvae. *Ecology* 76:150–64.

———. 1995b. Competition and the distribution of spring peeper larvae. *Oecologia* 103:203–7.

———. 1996. Pond drying, predators, and the distribution of *Pseudacris* tadpoles. *Copeia* 1996:599–605.

———. 1997. Tadpole communities. *Amer. Sci.* 85:36–45.

———. 2001. Distribution of pond-breeding anurans: an overview of mechanisms. *Israel J. Zool.* 47:313–32.

———. 2002. Experimental venue and estimation of interaction strength. *Ecology* 83:2097–2101.

———. 2004. Microgeographic countergradient variation in the wood frog, *Rana sylvatica. Evolution* 58:160–65.

Skelly, D. K., and L. K. Friedenburg. 2000. Effects of beaver on the thermal biology of an amphibian. *Ecol. Lett.* 3:483–86.

Skelly, D. K., L. K. Friedenburg, and J. M. Kiesecker. 2002. Forest canopy and the performance of larval amphibians. *Ecology* 83:983–92.

Skelly, D. K., M. A. Halverson, L. K. Freidenburg, and M. C. Urban. 2005. Canopy closure and amphibian diversity in forested wetlands. *Wetlands Ecol. Manage.* 13:261–68.

Skelly, D. K., and J. M. Kiesecker. 2001. Venue and outcome in ecological experiments: Density manipulations of larval anurans. *Oikos* 94:198–208.

Skelly, D. K., and E. E. Werner. 1990. Behavioral and life-historical responses of larval American toads to an odonate predator. *Ecology* 71:2313–22.

Skelly, D. K., E. E. Werner, and S. A. Cortwright. 1999. Long-term distributional dynamics of a Michigan amphibian assemblage. *Ecology* 80:2326–37.

Skelly, D. K., K. L. Yurewicz, E. E. Werner, and R. A. Relyea. 2003. Estimating decline and distributional change in amphibians. *Cons. Biol.* 17: 744–51

Skidds, D. E., and F. C. Golet. 2005. Estimating hydroperiod suitability for breeding amphibians in southern Rhode Island seasonal forest ponds. *Wetlands Ecol. Manage.* 13:349–66.

Slade, N. A., and R. J. Wassersug. 1975. On the evolution of complex life cycles. *Evolution* 29:568–71.

Slater, J. R. 1931. The mating of *Ascaphus truei* Stejneger. *Copeia* 1931:62–63.

Slater, P., and A. R. Main. 1963. Notes on the biology of *Notaden nichollsi* (Anura: Leptodactylidae). *West. Aust. Nat.* 8:163–66.

Slevin, J. R. 1955. Notes on Australian amphibians. *Proc. California Acad. Sci.* 28:355–92.

Slivkoff, M. D. and S. J. Warburton. 2001. Angiotensin II alters blood flow distribution in amphibians. *Physiol. Biochem. Zool.* 74:576–83.

Smallcombe, W. A. 1934. Tawny owl eating newts. *British Birds* 28:205.

Smalley, K. N., and G. W. Nace. 1983. Vitellogenic cycles in laboratory-maintained females of the leopard frog, *Rana pipiens*. *J. Exp. Zool.* 226:211–19.

Smatresk, N. J. 1994. Respiratory control in the transition from water to air breathing in vertebrates. *Amer. Zool.* 34:264–78

Smatresk, N. J., and A. W. Smits. 1991. Effects of central and peripheral chemoreceptor stimulation on ventilation in the marine toad, *Bufo marinus*. *Respir. Physiol.* 83:223–38.

Smirina, E. M., and Z. Roček. 1976. On the possibility of using annual bone layers of Alpine newts, *Triturus alpestris* (Amphibia: Urodela), for their age determination. *Vest. Cs. Spol. Zool.* 40:232–37.

Smith, A. C. 2000. Interspecific differences in prey captured by associating saddleback (*Saguinus fusciocollis*) and moustached (*Saguinus mystax*) tamarins. *J. Zool., London* 251:315–24.

Smith, B. G. 1907. The life history and habits of *Cryptobranchus alleganiensis*. *Biol. Bull.* 13:5–39.

———. 1910. The structure of the spermatophore of *Ambystoma punctatum*. *Biol. Bull.* 4:204–11.

Smith, B. P., M. J. Tyler, T. Kaneko, H. M. Garraffo, T. F. Spande, and J. W. Daly. 2002. Evidence for biosynthesis of pseudophyrnamine alkaloids by an Australian myobatrachid frog (*Pseudophryne*) and for sequestration of dietary pumiliotoxins. *J. Nat. Prod.* 65:439–47.

Smith, C. K. 1990. Effects of variation in body size on intraspecific competition among larval salamanders. *Ecology* 71:1777–88.

Smith, C. K., and J. W. Petranka. 1987. Prey size-distributions and size-specific forgaing success of *Ambystoma* larvae. *Oecologia* 71:239–44.

Smith, C. L. 1950. Seasonal changes in blood sugar, fat body, liver glycogen and gonads in the common frog, *Rana temporaria*. *J. Exp. Biol.* 26:412–29.

Smith, D. C. 1983. Factors controlling tadpole populations of the chorus frog (*Pseudacris triseriata*) on Isle Royale, Michigan. *Ecology* 64:501–10.

———. 1987. Adult recruitment in chorus frogs: Effects of size and date at metamorphosis. *Ecology* 68:344–50.

———. 1990. Population structure and competition among kin in the chorus frog (*Pseudacris triseriata*). *Evolution* 44:1529–41.

Smith, D. C., and J. Van Buskirk. 1995. Phenotypic design, plasticity, and ecological performance in two tadpole species. *Am. Nat.* 145:211–33.

Smith, D. G., and L. Rapson. 1977. Differences in pulmonary microvascular anatomy between *Bufo marinus* and *Xenopus laevis*. *Cell Tiss. Res.* 178:1–15.

Smith, E. M., and F. H. Pough. 1994. Intergeneric aggression in salamanders. *J. Herpetol.* 28:41–45.

Smith, G. C. 1976. Ecological energetics of three species of ectothermic vertebrates. *Ecology* 57:252–64.

Smith, G. R. 1999. Among family variation in tadpole (*Rana catesbeiana*) responses to density. *J. Herpetol.* 33:167–69.

Smith, G. R., M. A. Waters, and J. E. Rettig. 2000. Consequences of embryonic UV-B exposure for embryos and tadpoles of the Plains leopard frog. *Cons. Biol.* 14:1903–7.

Smith, H. H. 1887. On oviposition and nursing in the batrachian genus *Dendrobates*. *Am. Nat.* 21:307–11.

Smith, H. M. 1956. Handbook of amphibians and reptiles of Kansas. *Univ. Kansas Mus. Nat. Hist. Misc. Publ.* (9):1–356.

Smith, H. M., and R. B. Smith. 1971. *Synopsis of the herpetofauna of Mexico, vol. 1. analysis of the literature on the Mexican Axolotl*. Augusta, WV: Eric Lundberg.

Smith, H. M., and F. N. White. 1955. Adrenal enlargement and its significance in the hognose snakes (*Heterodon*). *Herpetologica* 11:137–44.

Smith, K. C. 2005. Effects of nonindigenous tapoles on native tadpoles in Florida: Evidence of competition. *Biol. Cons.* 123:433–41.

Smith, M. 1969. *The British amphibians and reptiles* (4th ed.). London: Collins.

Smith, M. A. 1926. The function of the "funnel" mouth of the tadpole of *Megophrys,* with a note on *M. aceras* Boulenger. *Proc. Zool. Soc. London* 4:983–88.

Smith, M. A., M. Berrill, and C. M. Kapron. 2002. Photolyase activity of the embryo and the ultraviolet absorbance of embryo jelly for several Ontario amphibian species. *Can. J. Zool.* 80:1109–16.

Smith, M. A., and D. M. Green. 2005. Dispersal and the metapopulation paradigm in amphibian ecology and conservation: Are all amphibian populations metapopulations? *Ecography* 28:110–28.

Smith, M. A., C. M. Kapron, and M. Berrill. 2000. Induction of photolyase activity in wood frog (*Rana sylvatica*) embryos. *Photochem. Photobiol.* 72:575–78.

Smith, M. J., and J. D. Roberts. 2003a. Call structure may affect male mating success in the quacking frog *Crinia georgiana* (Anura: Myobatrachidae). *Behav. Ecol. Sociobiol.* 53:221–26.

———. 2003b. An experimental examination of female preference patterns for components of the male advertisement call in the quacking frog, *Crinia georgiana*. *Behav. Ecol. Sociobiol.* 55:144–50.

Smith, M. J., P. C. Withers, and J. D. Roberts. 2003. Reproductive energetics and behavior of an Australian myobatrachid frog *Crinia georgiana*. *Copeia* 2003:248–54.

Smith, P. W. 1948. Food habits of cave dwelling amphibians. *Herpetologica* 4:205–8.

———. 1961. The amphibians and reptiles of Illinois. *Illinois Nat. Hist. Sur. Bull.* 28:1–298.

Smith, R. E. 1941. The spermatophores of *Triturus torosus* and *T. rivularis*. *Proc. Natl. Acad. Sci. USA* 27:261–64.

Smith, R. L. 1979. Paternity assurance and altered roles in the mating behaviour of a giant water bug, *Abedus herberti* (Heteroptera: Belostomatidae). *Anim. Behav.* 27:716–25.

Smith, R. S., and W. K. Ovalle, Jr. 1973. Varieties of fast and slow extrafusal muscle fibers in amphibian hindlimb muscles. *J. Anat.* 116:1–24.

Smith, W. J. 1977. *The behavior of communicating.* Cambridge, MA: Harvard University Press.

Smith-Gill, S. J. 1983. Developmental plasticity: Developmental conversion versus phenotypic modulation. *Amer. Zool.* 23:47–55.

Smith-Gill, S. J., and K. A. Berven. 1979. Predicting amphibian metamorphosis. *Am. Nat.* 113:563–85.

Smith-Gill, S. J., and D. E. Gill. 1978. Curvilinearities in the competition equations: An experiment with ranid tadpoles. *Am. Nat.* 112:557–70.

Smithson, T. 2000. Anthracosaurs. In *Amphibian biology. Vol. 4. Palaeontology,* ed. H. Heatwole and R. L. Carroll, 1053–63. Chipping Norton, New South Wales, Australia: Surrey Beatty and Sons.

Smits, A. W. 1984. Activity patterns and thermal biology of the toad *Bufo boreas halophilus*. *Copeia* 1984:689–96.

Smits, A. W., and D. L. Crawford. 1984. Emergence of toads to activity: A statistical analysis of contributing cues. *Copeia* 1984:696–701.

Smits, A. W., and J. I. Flanagin. 1994. Bimodal respiration in aquatic and terrestrial apodan amphibians. *Amer. Zool.* 34:247–63.

Snell, C., J. Tetter, and I. H. Evans. 2005. Phylogeography of the pool frog (*Rana lessonae* Camerano) in Europe: Evidence for native status in Great Britain and for an unusual postglacial colonization route. *Biol. J. Linn. Soc.* 85:41–51.

Smotherman, M., and P. M. Narins. 1999. The electrical properties of auditory hair cells in the frog amphibian papilla. *J. Neurosci.* 19:5275–92.

———. 2004. Evolution of the amphibian ear. In *Evolution of the vertebrate auditory system. Springer handbook of auditory research,* ed. G.A. Manley, A.N. Popper, and R.R. Fay, 164–99. New York: Springer-Verlag.

Snodgrass, J. W., J. W. Ackerman, A. L. Bryan, Jr., and J. Burger. 1999. Influence of hydroperiod, isolation, and heterospecifics on the distribution of aquatic salamanders (*Siren* and *Amphiuma*) among depression wetlands. *Copeia* 1999:107–13.

Snodgrass, J. W., A. L. Bryan, Jr., and J. Burger. 2000. Development of expectations of larval amphibian assemblage structure in southeastern depression wetlands. *Ecol. Appl.* 10:1219–29.

Snodgrass, J. W., M. J. Komorowski, A. L. Bryan, Jr., and J. Burger. 2000. Relationships among isolated wetland size, hydroperiod, and amphibian species richness: Implications for wetlands regulations. *Cons. Biol.* 14:414–19.

Snyder, D. H. 1971. *The function of brooding behavior in the plethodontid salamander, Aneides aeneus: A field study.* Unpublished PhD dissertation. South Bend, IN: University of Notre Dame.

Snyder, G. K., and W. W. Weathers. 1975. Temperature adaptations in amphibians. *Am. Nat.* 109:93–101.

Snyder, N. F. R. 1974. Breeding biology of swallow-tailed kites in Florida. *Living Bird* 13:73–97.

Snyder, N. F. R., J. W. Wiley, and C. B. Kepler. 1987. *The parrots of Luquillo: Natural history and conservation of the Puerto Rican parrot.* Los Angeles: Western Foundation of Vertebrate Zoology.

Snyder, R.C. 1956. Comparative features of the life histories of *Ambystoma gracile* (Baird) from populations at low and high altitudes. *Copeia* 1956:41–50.

Snyder, W. F., and D. L. Jameson. 1965. Multivariate geographic variation of mating call in populations of the Pacific tree frog (*Hyla regilla*). *Copeia* 1965:129–42.

Sofianidou, T. S., and P. Kyriakopoulou-Sklavounou. 1983. Studies on the biology of the frog *Rana dalmatina* Bonaparte during the breeding season in Greece (Amphibia: Anura: Ranidae). *Amphibia-Reptilia* 4:125–36.

Sofianidou, T. S., and H. Schneider. 1985. The reproductive behavior of the giant toad *Bufo b. spinosus* (Amphibia, Anura) in northern Greece. *Zool. Anz.* 214:209–21.

Sokol, A. 1984. Plasticity in the fine timing of metamorphosis in tadpoles of the hylid frog, *Litoria ewingi*. *Copeia* 1984:868–73.

Sokol, O. M. 1962. The tadpole of *Hymenochirus boettgeri*. *Copeia* 1969:272–84.

———. 1975. The phylogeny of anuran larvae: a new look. *Copeia* 1975:1–23.

Solbakken, J. S., B. Norberg, K. Watanabe, and K. Pittman. 1999. Thyroxine as a mediator of metamorphosis of Atlantic halibut, *Hippoglossus hippoglossus*. *Env. Biol. Fish.* 56:53–65.

Solé, M., and A. Kwet. 2003. *Liophis jaegeri* (Jaeger's ground snake). Diet. *Herp. Rev.* 34:69.

Solis, R., and M. Penna. 1997. Testosterone levels and evoked vocal responses in a natural population of the frog *Batrachyla taeniata*. *Horm. Behav.* 31:101–9.

Som, C., B. R. Anholt, and H.-U. Reyer. 2000. The effect of assortative mating on the coexistence of a hybridogenetic waterfrog and its sexual host. *Am. Nat.* 156:34–46

Somers, M. J., and M. G. Purves. 1996. Trophic overlap between three syntopic semi-aquatic carnivores: Cape clawless otters, spotted-necked otters, and water mongooses. *Afr. J. Ecol.* 34:158–66.

Sorensen, K. 2004. Population characteristics of *Siren lacertina* and *Amphiuma means* in North Florida. *Southeast. Nat.* 3:249–58.

Souder, W. 2000. *Plague of frogs.* New York: Hyperion.

Southerland, M. T. 1986a. Behavioral interactions among four species of the salamander genus *Desmognathus*. *Ecology* 67:175–81.

———. 1986b. Coexistence of three congeneric salamanders: The importance of habitat and body size. *Ecology* 67:721–28.

Southwood, T. R. E. 1978. *Ecological methods, with particular reference to the study of insect populations* (2nd ed.). London: Methuen.

Spallanzani, L. 1803. *Mémoire sur la respiration.* Translated by J. Senebier. Geneva, Switzerland: Paschaud.

Spande, T. F., P. Jain, H. M. Garraffo, L. K. Pannell, H. J. C. Yeh, J. W. Daly, S. Fukumoto, K. Imamura, T. Tokuyama, J. A. Torres, R. R. Snelling, and T. H. Jones. 1999. Occurrence and significance of decahyroquinolines from dendrobatid poison frogs and a myrmicine ant: use of H-1 and C-13 NMR in their conformation analysis. *J. Nat. Prod.* 62:5–21.

Sparling, D. W., and D. Cowman. 2003. Amphibians and pesticides in pristine areas. In *Amphibian decline: An integrated analysis of multiple stressor effects,* ed. G. Linder, S. K. Krest, and D. W. Sparling, 257–64. Pensacola, FL: SETAC Press.

Sparling, D. W., G. M. Fellers, and L. L. McConnell. 2001. Pesticides and amphibian population declines in California, USA. *Env. Tox. Chem.* 20:1591–95.

Sparling, D. W., G. Linder, and C. A. Bishop, eds. 2000. *Ecotoxicology of amphibians and reptiles.* Pensacola, FL: SETAC Press.

Sparreboom, M. 1979. Eieren van *Batrachuperus mustersi. Lacerta* 37:83–88.

———. 1983. On the sexual behaviour of *Paramesotriton caudopunctatus* (Liu & Hu) (Amphibia: Caudata: Salamandridae). *Amphibia-Reptilia* 4:25–33.

———. 1984a. Gedrag en voortplanting van de wrattensalamanders *Paramesotriton hongkongensis* en *P. chinesis* (I). *Lacerta* 43:7–14.

———. 1984b. Gedrag en voortplanting van de wrattensalamanders *Paramesotriton hongkongensis* en *P. chinesis* (II). *Lacerta* 43:28–35.

———. 1984c Gedrag en voortplanting van de wrattensalamanders *Paramesotriton hongkongensis* en *P. chinensis* (III). *Lacerta* 43:42–49.

———. 1993. On the sexual behaviour of *Paramestriton caudopunctatus* (Liu & Hu (Amphibia: Caudata: Salamandridae). *Amphibia-Reptilia* 4:25–33.

———. 1994. On the sexual behaviour of the sword-tailed newt, *Cynops ensicauda* (Halowell, 1860). *Abh. Ber. Naturk. Magdeburg* 17:151–61.

———. 1996. Sexual interference in the sword-tailed newt, *Cynops ensicauda popei* (Amphibia: Salamandridae). *Ethology* 102:672–85.

Sparreboom, M., and M. M. Faria. 1997. Sexual behaviour of the Chinese fire-bellied newt, *Cynops orientalis. Amphibia-Reptilia* 18:27–38.

Sparreboom, M., S. Sreinfartz, and G. Schultschik. 2000. Courtship behaviour of *Neurergus* (Caudata: Salamandridae). *Amphibia-Reptilia* 21:1–11.

Sparreboom, M., and B. Teunis. 1990. The courtship display of the marbled newt, *Triturus m. marmoratus. Amphibia-Reptilia* 11:351–61.

Sparreboom, M., and B. Thiesmeier. 1999. Courtship behaviour of *Pachytriton labiatus* (Caudata: Salamandridae). *Amphibia-Reptilia* 20:339–44.

Sparreboom, M., F. Xie, and F. Liang. 2001. Reproductive behaviour of the Chinhai salamander (*Echiontriton chinhaiensis*) (Caudata: Salamandridae). *Amphibia-Reptilia* 22:309–20.

Spearman, R. I. C. 1968. Epidermal keratinization in the salmander and a comparison with other amphibia. *J. Morphol.* 125:129–44.

Speck, F. G., and E. S. Dodge. 1945. Amphibian and reptile lore of the Six Nations Cayuga. *J. Amer. Folklore* 58:306–9.

Spellerberg, I. F. 2002. *Amphibians and reptiles of north-west Europe. Their natural history, ecology and conservation.* Enfield, NH: Science Publishers.

Spencer, A. W. 1964. *The relationship of dispersal and migration to gene flow in the Boreal chorus frog.* Unpublished PhD dissertation, Fort Collins: Colorado State University.

Spencer, B. 1896. *Report on the work of the Horn scientific expedition to central Australia. Part II. Zoology.* London: Dulau.

———. 1928. *Wanderings in wild Australia.* 2 vols. London: MacMillan.

Sperry, C. C. 1941. Burrowing owls eat spadefoot toads. *Wilson Bull.* 53:45.

Spieler, M. 1997. Anpassungen ausgewählter Froschlurche an Trockenstress und Räuberdruck in einer westafrikanischen Savanne. *Salamandra* 33:133–52.

———. 2000. Developmental plasticity and behavioural adaptations of two West African anurans living in an unpredictable environment (Amphibia, Anura). *Bonn. Zool. Monogr.* 46:109–20.

———. 2003. Risk of predation affects aggregation size: a study with tadpoles of *Phrynomantis microps* (Anura: Microhylidae). *Anim. Behav.* 65:179–84.

Spieler, M., and K. E. Linsenmair. 1997. Choice of optimal oviposition site by *Hoplobatrachus occipitalis* (Anura: Ranidae) in an unpredictable and patchy environment. *Oecologia* 109:184–99.

———. 1998. Migration patterns and diurnal use of shelter in a ranid frog of a West African savannah: a telemetric study. *Amphibia-Reptilia* 19:43–64.

———. 1999. Aggregation behaviour of *Bufo maculatus* tadpoles as an antipredator mechanism. *Ethology* 105:665–86.

Spight, T. M. 1967a. The water economy of salamanders: Exchange of water with soil. *Biol. Bull.* 132:126–32.

———. 1967b. The water economy of salamanders: Water uptake after dehydration. *Comp. Biochem. Physiol.* 20:767–71.

———. 1968. The water economy of salamanders: Evaporative water loss. *Physiol. Zool.* 41:195–203.

Spiller, D. A., and T. A. Schoener. 1988. An experimental study of the effect of lizards on web-spider communities. *Ecol. Monogr.* 58:57–77.

———. 1990. Lizards reduce food consumption by spiders: Mechanisms and consequences. *Oecologia* 83:150–61.

Spinelli, G. R., E. F. Schaefer, and A. I. Kehr. 2002. First record of biting midges (Diptera: Ceratopogonidae) attacking frogs in the Neotropical region. *Proc. Entomol. Soc. Washington* 104:527–28.

Spohr, S. M. 2005. *Ambystoma maculatum* (Spotted Salamander). Predation. *Herp. Rev.* 36:294–95.

Spotila, J. R. 1972. Role of temperature and water in the ecology of lungless salamanders. *Ecol. Monogr.* 42:95–125.

Spotila, J. R., and E. N. Berman. 1976. Determination of skin resistance and the role of the skin in controlling water loss in amphibians and reptiles. *Comp. Biochem. Physiol.* 55A:407–11.

Spotila, J. R., and R. J. Beumer. 1970. The breeding habits of the ringed salamander, *Ambystoma annulatum* (Cope) in northeastern Arkansas. *Am. Midl. Nat.* 84:77–89.

Spotila, J. R., and P. H. Ireland. 1970. Notes on the eggs of the grey-bellied salamander, *Eurycea multiplicata* griseogaster. *Southwest. Nat.* 14:366–68.

Spotila, J. R., M. P. O'Connor, and G. S. Bakken. 1992. Biophysics of heat and mass transfer. In *Environmental physiology of the amphibians,* ed. M. E. Feder and W. W. Burggren, 59–80. Chicago: University of Chicago Press.

Sprules, W. G. 1972. Effects of size-selective predation and food competition on high altitude zooplankton communities. *Ecology* 53:375–86.

———. 1974a. The adaptive significance of paedogenesis in North American species of *Ambystoma* (Amphibia: Caudata): an hypothesis. *Can. J. Zool.* 52:393–400.

———. 1974b. Environmental factors and the incidence of neoteny in *Ambystoma gracile* (Baird) (Amphibia: Caudata). *Can. J. Zool.* 52:1545–52.

Spurway, H., and J. B. S. Haldane. 1953. The comparative ethology of vertebrate breathing. I. Breathing in newts, with a general survey. *Behaviour* 6:8–34.

Sredl, M. J., and J. P. Collins. 1991. The effect of ontogeny on interspecific interactions in larval amphibians. *Ecology* 72:2232–39.

Ståhlberg, F., M. Olsson, and T. Uller. 2001. Population divergence of developmental thermal optima in Swedish common frogs, *Rana temporaria. J. Evol. Biol.* 14:755–62.

Stallard, R. F. 2001. Possible environmental factors underlying amphibian decline in eastern Puerto Rico: analysis of U.S. Government data archives. *Cons. Biol.* 15:943–53.

Stamps, J. A. 1977. Social behavior and spacing patterns in lizards. In *Biology of the reptilia, vol. 7,* ed. C. Gans and D. W. Tinkle, 265–334. New York: Academic Press.

———. 1993. Sexual size dimorphism in species with asymptotic growth after maturity. *Biol. J. Linn. Soc.* 50:123–45.

Stangel, P. W. 1983. Least sandpiper predation on *Bufo americanus* and *Ambystoma maculatum* larvae. *Herp. Rev.* 14:112.

———. 1988. Premetamorphic survival of the salamander *Ambystoma maculatum,* in eastern Massachusetts. *J. Herpetol.* 22:345–47.

Stangel, P. W., and R. D. Semlitsch. 1987. Experimental analysis of predation on the diel vertical migrations of a larval salamander. *Can. J. Zool.* 65:1554–58.

Starnes, S. M., C. A. Kennedy, and J. W. Petranka. 2000. Sensitivity of embryos of southern Appalachian amphibians to ambient solar UV-B radiation. *Cons. Biol.* 14:277–82.

Starrett, P. 1960. Descriptions of tadpoles of Middle American frogs. *Misc. Publ. Mus. Zool. Univ. Michigan,* no. 110:1–37.

———. 1967. Observations on the life-history of frogs of the family Atelopodidae. *Herpetologica* 23:195–204.

Staub, N. L. 1993. Intraspecific agonistic behavior of the salamander *Aneides flavipunctatus* (Amphibia: Plethodontidae) with comparisons to other plethodontid species. *Herpetologica* 49:271–82.

Staub, N. L., C. W. Brown, and D. B. Wake. 1995. Patterns of growth and movements in a population of *Ensatina eschscholtzii platensis* (Caudata: Plethodontidae) in the Sierra Nevada, California. *J. Herpetol.* 29:593–99.

Staub, N. L., and J. Paladin. 1997. The presence of modified granular glands in male and female *Aneides lugubris* (Amphibia: Plethodontidae). *Herpetologica* 53:339–44.

Stauffer, H.-P., and R. D. Semlitsch. 1993. Effects of visual, chemical and tactile cues of fish on the behavioural responses of tadpoles. *Anim. Behav.* 46:355–64.

Stauffer, J. R., Jr., J. E. Gates, and W. L. Goodfellow. 1983. Preferred temperature of two sympatric *Ambystoma* larvae: a proximate factor in niche segregation? *Copeia* 1983:1001–5.

Stearns, S. C. 1989. Trade-offs in life-history evolution. *Funct. Ecol.* 3:259–68.

———. 1992. *The evolution of life histories.* New York: Oxford University Press.

Stebbins, R. C. 1949. Observations on laying, development, and hatching of the eggs of *Batrachoseps wrighti. Copeia* 1949:161–68.

———. 1951. *Amphibians of western North America.* Berkeley: University of California Press.

———. 1954a. *Amphibians and reptiles of western North America.* New York: McGraw Hill.

———. 1954b. Natural history of the salamanders of the plethodontid genus *Ensatina. Univ. California Publ. Zool.* 54:47–123.

———. 1985. *A field guide to western reptiles and amphibians* (2nd ed.). Boston: Houghton Mifflin.

Stebbins, R. C., and N. W. Cohen. 1995. *A natural history of amphibians.* Princeton, NJ: Princeton University Press.

Stebbins, R. C., and J. R. Hendrickson. 1959. Field studies of amphibians in Colombia, South America. *Univ. California Publ. Zool.* 56:497–540.

Steele, C. A., E. D. Brodie, Jr., and J. G. McCracken. 2002. Influence of forest age on densities of Cope's and Pacific giant salamanders. *Northwest Sci.* 76:347–52.

———. 2003a. *Dicamptodon copei* (Cope's giant salamander). Reproduction. *Herp. Rev.* 34:227–28.

———. 2003b. Relationships between abundance of Cascade torrent salamanders and forest age. *J. Wildl. Manage.* 67:447–53.

Steele, C. A., B. C. Carstens, A. Storfer, and J. Sullivan. 2005. Testing hypotheses of speciation timing in *Dicamptodon copei* and *Dicamptodon aterrimus* (Caudata: Dicamptodontidae). *Mol. Phylogen. Evol.* 36:90–100.

Steele, W. K., and G. N. Louw. 1988. Caecilians exhibit cutaneous respiration and high evaporative water loss. *S. Afr. J. Zool.* 23:134–35.

Stefanski, M., R. E. Gatten, Jr., and F. H. Pough. 1989. Activity metabolism of salamanders: tolerance to dehydration. *J. Herpetol.* 23:45–50.

Stegen, J. C., C. M. Gienger, and L. Sun. 2004. The control of color change in the Pacific tree frog, *Hyla regilla. Can. J. Zool.* 82:889–96.

Steggerda, F. R. 1937. Comparative study of water metabolism in amphibians injected with pituitrin. *Proc. Soc. Exp. Biol. Med.* 36:103–6.

Steinfartz, S., U. W. Hwang, D. Tautz, M. Öz, and M. Veith. 2002. Molecular phylogeny of the salamandrid genus *Neurergus:* Evidence for an intrageneric switch of reproductive biology. *Amphibia-Reptilia* 23:419–31.

Steinwarz, D. 1990. Morphology of the larynx of the banana frog *Afrixalus fornasinii* (Bianconi) (Anura). *Amphibia-Reptilia* 11:53–59.

Steinwascher, K. 1978a. The effect of coprophagy on the growth of *Rana catesbeiana* tadpoles. *Copeia* 1978:130–34.

———. 1978b. Interference and exploitation competition among tadpoles of *Rana utricularia. Ecology* 59:1039–46.

———. 1979a. Competitive interactions among tadpoles: Responses to resource levels. *Ecology* 60:1172–83.

———. 1979b. Host-parasite interaction as a potential population-regulating mechanism. *Ecology* 60:884–90.

Steinwascher, K., and J. Travis. 1983. Influence of food quality and quantity on early larval growth of two anurans. *Copeia* 1983:238–42.

Steiof, K., and I. Tietje. 1985. Stockente (*Anas platyrhynos*) tötet Braunfrosch (*Rana* spec.). *Ornithol. Ber. Berl. (West)* 10:48–49.

Stejneger, L. 1907. Herpetology of Japan and adjacent territory. *Bull. U.S. Nat. Mus.* 58:47–161.

Stenhouse, S. L. 1985a. Interdemic variation in predation on salamander larvae. *Ecology* 66:1706–17.

———. 1985b. Migratory orientation and homing in *Ambystoma maculatum* and *Ambystoma opacum. Copeia* 1985:631–37.

———. 1987. Embryo mortality and recruitment of juveniles of *Ambystoma maculatum* and *Ambystoma opacum* in North Carolina. *Herpetologica* 43:496–501.

Stenhouse, S. L., N. G. Hairston, and A. E. Cobey. 1983. Predation and competition in *Ambystoma* larvae: field and laboratory experiments. *J. Herpetol.* 17:210–20.

Stenseth, N. C. 1983. Causes and consequences of dispersal in small mammals. In *The ecology of animal movement,* ed. I. R. Swingland and P. J. Greenwood, 63–101. New York: Oxford University Press.

Stepanova, Z. C. 1974. The chemical nature of the products of metabolism of amphibian larvae, excreted into the water. *Soviet J. Ecol.* 5:148–49.

Stephen, R. O., and J. C. Hartley. 1991. The transmission of bush-cricket calls in natural environments. *J. Exp. Biol.* 155:227–44.

Stephenson, B., and P. Verrell. 2003. Courtship and mating of the tailed frog (*Ascaphus truei*). *J. Zool., London* 259:15–22.

Stephenson, E. M., and N. G. Stephenson. 1957. Field observations on the New Zealand frog *Leiopelma* Fitzinger. *Trans. Roy. Soc. New Zealand* 84:867–82.

Stephenson, N. G. 1951. Observations on the development of the amphicoelus frogs, *Leiopelma* and *Ascaphus. J. Linn. Soc. (Zool.)* 42:18–28.

———. 1955. On the development of the frog *Leiopelma hochstetteri* Fitzinger. *Proc. Zool. Soc. London* 124:785–95.

Stephenson, N. G., and E. M. Thomas. 1945. A note concerning the occurrence and life-history of *Leiopelma* Fitzinger. *Trans. and Proc. Roy. Soc. New Zealand* 75:319–20.

Stevens, E. D. 1988a. The effect of acclimation temperature on feeding performance in the toad, *Bufo americanus. Comp. Biochem. Physiol.* 89A:131–35.

———. 1988b. Feeding performance of toads at different acclimation temperatures. *Can. J. Zool.* 66:537–39.

Stevens, R. A. 1971. A new tree-frog from Malawi (Hyperoliinae, Amphibia). *Zool. Afr.* 6:313–20.

Stevens, V. M., E. Pollus, R. A. Wresselingh, N. Schtickzelle, and M. Baguette. 2004. Quantifying functional connectivity: Experimental evidence for patch-specific resistance in the natterjack toad (*Bufo calamaita*). *Landsc. Ecol.* 19:839–42.

Stevenson, D. J., and K. J. Dyer. 2002. *Rana capito capito* (Carolina gopher frog). Refugia. *Herp. Rev.* 33:128–29.

Steward, J. W. 1969. *The tailed amphibians of Europe.* Newton Abbot, UK: David & Charles.

———. 1971. *The snakes of Europe.* Newton Abbot, UK: David & Charles,.

Stewart, E. R., S. A. Reese, and G. R. Ultsch. 2004. The physiology of hibernation in Canadian leopard frogs (*Rana pipiens*) and bullfrogs (*Rana catesbeiana*). *Physiol. Biochem. Zool.* 77:65–73.

Stewart, M. M. 1967. *Amphibians of Malawi.* Albany: State University of New York Press.

———. 1974. Parallel pattern polymorphism in the genus *Phrynobatrachus* (Amphibia: Ranidae). *Copeia* 1974:823–32.

———. 1979. The role of introduced species in a Jamaican frog community. In *Proceedings of the IV Symposium on Tropical Ecology,* ed. H. Wolda, 113–46. Panama City, Panama.

———. 1985. Arboreal habitat use and parachuting by a subtropical forest frog. *J. Herpetol.* 19:391–401.

———. 1995. Climate driven population fluctuations in rain forest frogs. *J. Herpetol.* 29:437–46.

Stewart, M. M., and P. J. Bishop. 1994. Effects of increased sound level of advertisement calls on calling male frogs, *Eleutherodactylus coqui. J. Herpetol.* 28:46–53.

Stewart, M. M., and G. E. Martin. 1980. Coconut husk-piles—A unique habitat for Jamaican terrestrial frogs. *Biotropica* 12:107–16.

Stewart, M. M., and F. H. Pough. 1983. Population density of tropical forest frogs: Relation to retreat sites. *Science* 221:570–72.

Stewart, M. M., and A. S. Rand. 1991. Vocalizations and the defense of retreat sites by male and female frogs, *Eleutherodactylus coqui. Copeia* 1991:1013–24.

———. 1992. Diel variation in the use of aggressive calls by the frog *Eleutherodactylus coqui. Herpetologica* 48:49–56.

Stewart, M. M., and P. Sandison. 1972. Comparative food habits of sympatric mink frogs, bullfrogs, and green frogs. *J. Herpetol.* 6:241–44.

Stewart, M. M., and L. L. Woolbright. 1996. Amphibians. In *The food web of a tropical rain forest,* ed. D. P. Reagan and R. B. Waide, 274–320. Chicago: University of Chicago Press.

Stewart, R. E. 1949. Ecology of a nesting red-shouldered hawk population. *Wilson Bull.* 61:26–35.

Stickel, L. F. 1950. Population and home range relationships of the box turtle, *Terrepene c. carolina. Ecol. Monogr.* 20:351–78.

Stiebler, I. B., and P. M. Narins. 1990. Temperature-dependence of auditory nerve response properties in the frog. *Hearing Res.* 46:63–82.

Stiffler, D. F. 1988. Cutaneous exchange of ions in lower vertebrates. *Amer. Zool.* 28:1019–29.

Stiffler, D. F., and R. H. Alvarado. 1974. Renal function in response to reduced osmotic load in larval salamanders. *Am. J. Physiol.* 226:1243–49.

———. 1980. Renal regulation of electrolytes in two species of ambystomatid salamanders. *Copeia* 1980:918–21.

Stiffler, D. F., M. L. DeRuyter, and C. R. Talbot. 1990. Osmotic and ionic regulation in the aquatic caecilian *Typhlonectes compressicauda* and the terrestrial caecilian *Ichthyophis kohtaoensis. Physiol. Zool.* 63:649–68.

Stiffler, D. F., and T. Manokham. 1994. Partitioning of nitrogen excretion between urea and ammonia and between skin and kidneys in the aquatic caecilian *Typhlonectes natans. Physiol. Zool.* 67:1077–86.

Stiffler, D. F., S. C. Ryan, and R. A. Mushkot. 1987. Interactions between acid-base balance and cutaneous ion transport in larval *Ambystoma tigrinum* (Amphibia: Caudata) in response to hypercapnia. *J. Exp. Biol.* 130:389 404.

Stiffler, D. F., and C. R. Talbot. 2000. Exchanges of oxygen, carbon dioxide, nitrogen and water in the caecilian *Dermophis mexicanus. J. Comp. Physiol. B* 170:505–9.

Stiffler, D. F., B. L. Tufts, and D. P. Toews. 1983. Acid-base and ionic balance in *Ambystoma tigrinum* and *Necturus maculosus* during hypercapnia. *Am. J. Physiol.* 245:R689–R694.

Stille, W. T. 1952. The nocturnal amphibian fauana of the southern Lake Michigan beach. *Ecology* 33:149–62.

———. 1958. The water absorption resonse of an anuran. *Copeia* 1958:217–18.

Stinner, J., N. Zarlinga, and S. Orcutt. 1994. Overwintering behavior of adult bullfrog, *Rana catesbeiana,* in northern Ohio. *Ohio J. Sci.* 1:8–13.

Stöck, M. 1998. Mating call differences between diploid and tetraploid green toads (*Bufo viridis* complex) in Middle Asia. *Amphibia-Reptilia* 19:29–42.

Stocks, R. 1987. Frogs in paradise. *Paradise* (64):17–20.

Stopper, G. F., L. Hecker, R. A. Franssen, and S. K. Sessions. 2002. How trematodes cause limb deformities in amphibians. *J. Exp. Zool. (Mol. Dev. Evol.)* 294:252–63.

Storer, T. I. 1925. A synopsis of the Amphibia of California. *Univ. California Publ. Zool.* 27:1–342.

Storey, J. M., and K. B. Storey. 1985a. Adaptations of metabolism for freeze tolerance in the gray tree frog, *Hyla versicolor. Can. J. Zool.* 63:49–54.

———. 1985b. Triggering of cryoprotectant synthesis by the initiation of ice nucleation in the freeze tolerant frog, *Rana sylvatica*. *J. Comp. Physiol. B* 156:191–95.

Storey, K. B. 1984. Freeze tolerance in the frog, *Rana sylvatica*. *Experientia* 40:1261–62.

———. 1986. Freeze tolerance in vertebrates. Biochemical adaptation of terrestrially hibernating frogs. In *Living in the cold: Physiological and biochemical adaptations,* ed. H. C. Heller, X. J. Musacchia, and L. C. H. Lang, 131–38. New York: Elsevier.

———. 1987a. Glycolysis and the regulation of cryoprotectant synthesis in liver of the freeze tolerant wood frog. *J. Comp. Physiol. B* 157:373–80.

———. 1987b. Organ-specific metabolism during freezing and thawing in freeze tolerant frog. *Am. J. Physiol.* 253:R292–R297.

———. 1990. Life in a frozen state: Adaptive strategies for natural freeze tolerance in amphibians and reptiles. *Am. J. Physiol.* 258:R559–R568.

———. 1997. Organic solutes in freezing tolerance. *Comp. Biochem. Physiol.* 117A:319–26.

———. 2002. Life in the slow lane: molecular mechanisms of estivation. *Comp. Biochem. Physiol. A* 133:733–54.

Storey, K. B., and J. M. Storey. 1984. Biochemical adaptation for freezing tolerance in the wood frog, *Rana sylvatica*. *J. Comp. Physiol.* 155:29–36.

———. 1986a. Freeze tolerance and intolerance as strategies of winter survival in terrestrially-hibernating amphibians. *Comp. Biochem. Physiol.* 83A:613–17.

———. 1986b. Freeze tolerant frogs: cryoprotectants and tissue metabolism during freeze/thaw cycles. *Can. J. Zool.* 64:49–56.

———. 1987. Persistence of freeze tolerance in terrestrially hibernating frogs after spring emergence. *Copeia* 1987:720–26.

———. 1988. Freeze tolerance in animals. *Physiol. Rev.* 68:27–84.

———. 1992. Natural freeze tolerance in ectothermic vertebrates. *Ann. Rev. Physiol.* 54:619–37.

Storey, K. B, J. M. Storey, S. P. J. Brooks, T. A. Churchill, and R. J. Brooks. 1988. Hatchling turtles survive freezing during winter hibernation. *Proc. Natl. Acad. Sci. USA* 85:8350–54.

Storez, R. A. 1969. Observations on the courtship of *Ambystoma laterale*. *J. Herpetol.* 3:87–95.

Storfer, A. 1999. Gene flow and local adaptation in a sunfish-salamander system. *Behav. Ecol. Sociobiol.* 46:273–79.

———. 2003. Amphibian declines: future directions. *Divers. Distr.* 9:151–63.

Storfer, A., J. Cross, V. Rush, and J. Caruso. 1999. Adaptive coloration and gene flow as a constraint to local adaptation in the streamside salamander, *Ambystoma barbouri*. *Evolution* 53:889–98.

Storfer, A., and A. Sih. 1998. Gene flow and ineffective antipredator behavior in a stream-breeding salamander. *Evolution* 52:558–65.

Storfer, A., and C. White. 2004. Phenotypically plastic responses of larval tiger salamanders, *Ambystoma tigrinum,* to different predators. *J. Herpetol.* 38:612–15.

Storm, R. M. 1947. Eggs and young of *Aneides ferreus*. *Herpetologica* 4:60–62.

Storz, B. L. 2004. Reassessment of the environmental mechanisms controlling developmental polyphenism in spadefoot toad tadpoles. *Oecologia* 141:402–10.

Stouffer, R. H., Jr., J. E. Gates, C. H. Hocutt, and J. R. Stauffer, Jr. 1983. Surgical implantation of a transmitter package for radio-tracking endangered hellbenders. *Wildl. Soc. Bull.* 11:384–86.

Stout, B. M. III, K. K. Stout, and C. W. Stihler. 1992. Predation by the caddisfly *Banksiola dossuaria* on egg masses of the spotted salamander *Ambystoma maculatum*. *Am. Midl. Nat.* 127:368–72.

Strain, J. G., and R. L. Mumme. 1988. Effects of food supplementation, song playback, and temperature on vocal territorial behavior of Carolina wrens. *Auk* 105:11–16.

Strathmann, R. 1978. The evolution and loss of feeding larval stages of marine invertebrates. *Evolution* 32:894–906.

Straub, J. O. 1984. Aspects of the cranial anatomy of *Grandisonia diminutiva* Taylor. *Mem. Soc. Zool. France* 43:55–63.

Straub, M. 1957. Weitere Untersuchungen zur Temperatur Adaptation der Säuerstoff Bindug des Blutes von *Rana esculenta* L. *Z. Vergl. Physiol.* 39:506–23.

Straughan, I. R. 1973. Evolution of anuran mating calls: bioacoustical aspects. In *Evolutionary biology of the anurans,* ed. J. L. Vial, 321–27. Columbia: University of Missouri Press.

———. 1975. An analysis of the mechanism of mating call discrimination in the frogs *Hyla regilla* and *H. cadaverina*. *Copeia* 1975:415–24.

Straughan, I. R., and W. R. Heyer. 1976. A functional analysis of the mating calls of the Neotropical frog genera of the *Leptodactylus* complex (Amphibia, Leptodactylidae). *Pap. Avul. Zool.* 29:221–45.

Straughan, I. R., and A. R. Main. 1966. Speciation and polymorphism in the genus *Crinia* Tschudi (Anura, Leptodactylidae) in Queensland. *Proc. Roy. Soc. Queensland* 78:11–28.

Strawinski, S. 1956. Vascularization of respiratory surfaces in ontogeny of the edible frog, *Rana esculenta* L. *Zool. Pol.* 7:327–65.

Streilein, K. E. 1982. Behavior, ecology, and distribution of South American marsupials. In *Mammalian biology in South America.* Special Publ. Ser. 6, ed. M. A. Mares and H. H. Genoways, 231–50. Linesville, PA: Pymatuning Laboratory of Ecology.

Strijbosch, H. 1980. Habitat selection by amphibians during their terrestrial phase. *Brit. J. Herpetol.* 6:93–98.

Stroganov, S. U. 1969. *Carnivorous mammals of Siberia.* Jerusalem: Israel Program for Scientific Translations.

Strong, D. R., Jr. 1983. Natural variability and the manifold mechanisms of ecological communities. *Am. Nat.* 122:636–60.

———. 1984. Exorcising the ghost of competition past: phytophagous insects. In *Ecological communities,* ed. D. R. Strong, Jr., D. Simberloff, L. G. Abele, and A. B. Thistle, 28–41. Princeton, NJ: Princeton University Press.

Strong, D. R., Jr., D. Simberloff, L. G. Abele, and A. B. Thistle, eds. 1984. *Ecological communities.* Princeton, NJ: Princeton University Press.

Strübing, H. 1954. Über Vorzugstemperaturen von Amphibien. *Z. Morph. Ökol. Tiere* 43:357–86.

Strüssman, C., M. B. Ribeiro do Vale, M. H. Meneghini, and W. E. Magnusson. 1984. Diet and foraging mode of *Bufo marinus* and *Leptodactylus ocellatus*. *J. Herpetol.* 18:138–46.

Stuart, B. L. 2005. New frog records from Laos. *Herp. Rev.* 36:473–79.

Stuart, J. N., and C. W. Painter. 1993. *Rana catesbeiana* (bullfrog). Cannibalism. *Herp. Rev.* 24:103.

Stuart, L. C. 1951. The distributional implications of temperature tolerances and hemoglobin values in the toads *Bufo marinus* (Linnaeus) and *Bufo bocourti* Brocchi. *Copeia* 1951:220–29.

———. 1961. Some observations on the natural history of tadpoles of *Rhinophrynus dorsalis* Dumeril and Bibron. *Herpetologica* 17:73–79.

Stuart, R. J. 1991. Kin recognition as a functional concept. *Anim. Behav.* 41:1093–94.

Stuart, S. N., J. S. Chanson, N. A. Cox, B. E. Young, A. S. L. Rodrigues, D. L. Fischman, and R. W. Waller. 2004. Status and trends of amphibian declines and extinctions worldwide. *Science* 306:1783–86 (supplemental material at http://sciencemag.org/cgi/content/full/1103538/DC1).

———. 2005. Responses to comment on "status and trends of amphibian declines and extinctions worldwide." Science 309:1999c.

Stugren, B. 1966. Geographic variation and distribution of the moor frog, *Rana arvalis* Nilss. *Ann. Zool. Fennici* 3:29–39.

Stumpel, A. H. P. 1990. On hibernation sites in the tree frog, *Hyla arborea*. *Amphibia-Reptilia* 11:304–6.

———. 1992. Successful reproduction of introduced bullfrogs *Rana catesbeiana* in northwestern Europe: a potential threat to indigenous amphibians. *Biol. Cons.* 60:61–62.

Stumpel, A. H. P., and G. Hanekamp. 1986. Habitat and ecology of *Hyla arborea* in the Netherlands. In *Studies in herpetology,* ed. Z. Roček, 409–12. Prague: Charles University.

Stumpel, A. H. P., and U. Tester, eds. 1993. *Ecology and conservation of the European tree frog*. Wageningen, The Netherlands: DLO Institute for Forestry and Nature Research.

Stumpel, A. H. P., and H. van der Voet. 1998. Characterising the suitability of new ponds for amphibians. *Amphibia-Reptilia* 19:125–42.

Su, T., and W. Yu-lan. 1963. L'hibernation, facteur déterminant de la maturation ovulaire chez le crapaud (*Bufo bufo asiaticus*). *Sciencia Sinica* 12:1161–64.

Suggs, D. N., and A. M. Simmons. 2005. Information theory analysis of patterns of modulation in the advertisement call of the male bullfrog, *Rana catesbeiana*. *J. Acoust. Soc. Amer.* 117:2330–37.

Sullivan, A. M., D. M. Madison, and J. R. Rohr. 2004. Variation in the antipredator responses of three sympatric plethodontid salamanders to predator-diet cues. *Herpetologica* 60:401–8.

Sullivan, A. M., A. L. Picard, and D. M. Madison. 2005. To avoid or not to avoid? Factors influencing the discrimination of predator diet cues by a terrestrial salamander. *Anim. Behav.* 69:1425–33.

Sullivan, B. 1974. Amphibian hemoglobins. In *Chemical zoology, vol. 9,* ed. M. Florkin and B. T. Scheer, 77–122. New York: Academic Press.

Sullivan, B. D. 1991. Additional vertebrate prey items of the red squirrel (*Tamiasciurus hudsonicus*). *Can. Field-Nat.* 105:398–99.

Sullivan, B. K. 1982a. Male mating behaviour in the Great Plains toad (*Bufo cognatus*). *Anim. Behav.* 30:939–40.

———. 1982b. Sexual selection in Woodhouse's toad (*Bufo woodhousei*). I. Chorus organization. *Anim. Behav.* 30:680–86.

———. 1982c. Significance of size, temperature and call attributes to sexual selection in *Bufo woodhousei australis*. *J. Herpetol.* 16:103–6.

———. 1983a. Sexual selection in the Great Plains toad (*Bufo cognatus*). *Behaviour* 84:258–64.

———. 1983b. Sexual selection in Woodhouse's toad (*Bufo woodhousei*). II. Female Choice. *Anim. Behav.* 31:1011–17.

———. 1984. Size dimorphism in anurans: a comment. *Am. Nat.* 123:721–24.

———. 1985a. Male calling behavior in response to playback of conspecific advertisement calls in two bufonids. *J. Herpetol.* 19:78–83.

———. 1985b. Sexual selection and mating system variation in pelobatids and bufonids of the Arizona Sonoran Desert. *Great Basin Nat.*45:688–96.

———. 1986a. Hybridization between the toads *Bufo microscaphus* and *Bufo woodhousii* in Arizona: Morphological variation. *J. Herpetol.* 20:11–21.

———. 1986b. Intra-populational variation in the intensity of sexual selection in breeding aggregations of Woodhouse's toad (*Bufo woodhousei*). *J. Herpetol.* 20:88–90.

———. 1987. Sexual selection in Woodhouse's toad (*Bufo woodhousei*). III. Seasonal variation in male mating success. *Anim. Behav.* 35:912–19.

———. 1989a. Desert environments and the structure of anuran mating systems. *J. Arid Environ.* 17:175–83.

———. 1989b. Passive and active female choice: a comment. *Anim. Behav.* 37:692–94.

———. 1992. Sexual selection and calling behavior in the American toad (*Bufo americanus*). *Copeia* 1992:1–7.

———. 1995. Temporal stability in hybridization between *Bufo microscaphus* and *Bufo woodhousii* (Anura: Bufonidae): Behavior and morphology. *J. Evol. Biol.* 8:233–47.

Sullivan, B. K., and S. H. Hinshaw. 1990. Variation in the advertisement calls and male calling behavior in the spring peeper (*Pseudacris crucifer*). *Copeia* 1990:1146–50.

———. 1992. Female choice and selection on male calling behaviour in the grey treefrog *Hyla versicolor*. *Anim. Behav.* 44:733–44.

Sullivan, B. K., and T. Lamb. 1988. Hybridization between the toads *Bufo microscaphus* and Bufo *woodhousii* in Arizona: Variation in release calls and allozymes. *Herpetologica* 44:325–53.

Sullivan, B. K., and M. R. Leek. 1986. Acoustic communication in Woodhouse's toad (*Bufo woodhousei*). I. Response of calling males to variation in spectral and temporal components of advertisement calls. *Behaviour* 98:305–19.

———. 1987. Acoustic communication in Woodhouse's toad (*Bufo woodhousei*). II. Responses of females to variation in spectral and temporal components of advertisement calls. *Behaviour* 103:16–26.

Sullivan, B. K., M. J. Ryan, and P. A. Verrell. 1995. Female choice and mating system structure. In *Amphibian biology, vol. 2. Social behaviour,* ed. H. Heatwole and B. K. Sullivan, 469–517. Chipping Norton, New South Wales, Australia: Surrey Beatty and Sons.

Sullivan, B. K., and E. A. Sullivan. 1985. Variation in advertisement calls and male mating success of *Scaphiopus bombifrons, S. couchi,* and *S. multiplicatus* (Pelobatidae). *Southwest. Nat.* 30:349–55.

Sullivan, B. K., and G. E. Walsberg. 1985. Call rate and aerobic capacity in Woodhouse's toad (*Bufo woodhousei*). *Herpetologica* 41:404–7.

Sullivan, R. M. 1980. Paedomorphosis or neoteny (*sensu lato*)? *J. Herpetol.* 14:79–80.

Summers, A. P., and J. C. O'Reilly. 1997. A comparative study of locomotion in the caecilians *Dermophis mexicanus* and *Typhlonectes natans* (Amphibia: Gymnophiona). *Zool. J. Linn. Soc.* 121:65–76.

Summers, K. 1989. Sexual selection and intra-female competition in the green poison-dart frog, *Dendrobates auratus*. *Anim. Behav.* 37:797–805.

———. 1990. Paternal care and the cost of polygyny in the green dart-poison frog. *Behav. Ecol. Sociobiol.* 27:307–13.

———. 1992. Mating strategies in two species of dart-poison frogs: A comparative study. *Anim. Behav.* 43:907–19.

———. 1999a. *Dendrobates auratus* (Green Poison Frog). Predation. *Herp. Rev.* 30:91.

———. 1999b. The effects of cannibalism on Amazonian poison frog egg and tadpole deposition and survivorship in *Heliconia* axil pools. *Oecologia* 119:557–64.

———. 2000. Mating and aggressive behaviour in dendrobatid frogs from Corcovado National Park, Costa Rica: A comparative study. *Behaviour* 137:7–24.

———. 2003. Convergent evolution of bright coloration and toxicity in frogs. *Proc. Natl. Acad. Sci. USA* 100:12533–34.

Summers, K., and W. Amos. 1997. Behavioral, ecological, and molecular genetic analyses of reproductive strategies in the Amazonian dart-poison frog, *Dendrobates ventrimaculatus. Behav. Ecol.* 8:260–67.

Summers, K., and M. Clough. 2001. The evolution of coloration and toxicity in the poison frog family (Dendrobatidae). *Proc. Natl. Acad. Sci. USA* 98:6227–32.

Summers, K., T. W. Cronin, and T. Kennedy. 2003. Variation in spectral reflectance among populations of *Dendrobates pumilio,* the strawberry poison frog, in the Bocas del Toro Archipelago, Panama. *J. Biogeogr.* 30:35–53.

———. 2004. Cross-breeding of distinct color morphs of the strawberry poison frog (*Dendrobates pumilio*) from the Bocas del Toro Archipelago, Panama. *J. Herpetol.* 38:1–8.

Summers, K., and D. J. D. Earn. 1999. The cost of polygyny and the evolution of female care in poison frogs. *Biol. J. Linn. Soc.* 66:515–38.

Summers, K., and C. S. McKeon. 2004. The evolutionary ecology of phytotelmata use in Neotropical poison frogs. In *Ecology and evolution of phytotelm-breeding anurans,* ed. R. M. Lehtinen, 55–73. *Misc. Publ. Mus. Zool. Univ. Michigan,* no. 193.

Summers, K., and R. Symula. 2001. Cannibalism and kin discrimination in tadpoiles of the Amazonian poison frog, *Dendrobates ventrimaculatus,* in the field. *Herpetol. J.* 11:17–21.

Summers, K., R. Symula, M. Clough, and T. Cronin. 1999. Visual mate choice in poison frogs. *Proc. Roy. Soc. London B* 266:2141–45.

Summers, K., L. A. Weight, P. Boag, and E. Bermingham. 1999. The evolution of female parental care in poison frogs of the genus *Dendrobates:* Evidence from mitochondrial DNA sequences. *Herpetologica* 55:254–70.

Sun, L., W. Wilczynski, A. S. Rand, and M. J. Ryan. 2000. Trade-off in short- and long-distance communication in túngara (*Physalaemus pustulosus*) and cricket (*Acris crepitans*) frogs. *Behav. Ecol.* 11:102–9.

Sunquist, M. E., F. Sunquist, and D. E. Paneke. 1989. Ecological separation in a Venezuelan Llanos carnivore community. In *Advances in neotropical mammalogy,* ed. K. H. Redford and J. F. Eisenberg, 197–233. Gainesville, FL: Sandhill Crane Press.

Sutherland, D. 2005. Parasites of North American frogs. In *Amphibian declines. The conservation status of United States species,* ed. M. J. Lannoo, 109–23. Berkeley: University of California Press.

Sutherland, K. P., J. W. Porter, and C. Torres. 2004. Disease and immunity in Caribbean and Indo-Pacific zooxanthellae corals. *Mar. Ecol. Prog. Ser.* 266:273–302.

Sutherland, W. J. 1985. Measures of sexual selection. *Oxford Surveys in Evol. Biol.* 1:90–101.

Sutter, J., W. E. Martínez A., F. Oliva T. and N. Oswaldo J. 2001. Diet and hunting behavior of the crane hawk in Tikal National Park, Guatemala. *Condor* 103:70–7.

Suzuki, S., D. A. Hill, T. Maruhashi, and T. Tsukahara. 1990. Frog- and lizard-eating behaviour of wild Japanese macaques in Yakushima, Japan. *Primates* 31:421–26.

Swanson, D. L., and B. M. Graves. 1995. Supercooling and freeze intolerance in overwintering juvenile spadefoot toads (*Scaphiopus bombifrons*). *J. Herpetol.* 29:280–85.

Swanson, D. L., B. M. Graves, and K. L. Koster. 1996. Freezing tolerance/intolerance and cryoprotectant synthesis in terrestrially overwintering anurans in the Great Plains, USA. *J. Comp. Physiol. B* 166:110–19.

Sweet, S. S. 1977a. *Eurycea tridentifera. Cat. Amer. Amphib. Rept.* 199.1–199.2.

———. 1977b. Natural metamorphosis in *Eurycea neotenes* and the generic allocation of the Texas *Eurycea* (Amphibia: Plethodontidae). *Herpetologica* 33:364–75.

———. 1982. A distributional analysis of epigean populations of *Eurycea neotenes* in central Texas, with comments on the origin of troglobitic populations. *Herpetologica* 38:430–44.

Swingland, I. R. 1983. Intraspecific differences in movement. In *The ecology of animal movement,* ed. I. R. Swingland and P. J. Greenwood, 102–15. New York: Oxford University Press.

Swingland, I. R., and P. J. Greenwood. 1983. *The ecology of animal movement.* New York: Oxford University Press.

Symula, R., R. Schulte, and K. Summers. 2001. Molecular phylogenetic evidence for a mimetic radiation in Peruvian poison frogs supports a Müllerian mimicry hypothesis. *Proc. Roy. Soc. London B* 268:2415–21.

———. 2003. Molecular systematics and phylogeography of Amazonian poison frogs of the genus *Dendrobates. Mol. Phylogen. Evol.* 26:452–75.

Szarski, H. 1955. La vascularisation des surfaces respiratoires chez les amphibiens. *C. R. Ass. Anat.* 42:1267–73.

———. 1957. The origin of the larva and metamorphosis in Amphibia. *Am. Nat.* 91:283–301.

———. 1959. Some reactions of skin capillaries of the frog. *Proc. 15th Int. Congr. Zool., London,* pp. 531–33.

———. 1964. The structure of respiratory organs in relation to body size in Amphibia. *Evolution* 18:118–26.

Szelistowski, W. A. 1985. Unpalatability of the poison arrow frog *Dendrobates pumilio* to the ctenid spider *Cupiennius coccineus. Biotropica* 17:345–46.

Sztatecsny, M., and R. Schabetsberger. 2005. Into thin air: vertical migrations, body condition, and quality of terrestrial habitats of alpine common toads, *Bufo bufo. Can. J. Zool.* 83:788–96.

Szymura, J. M. 1993. Analysis of hybrid zones with *Bombina.* In *Hybrid zones and the evolutionary process,* ed. R. G. Harrison, 261–89. New York: Oxford University Press.

Szymura, J. M., and N. H. Barton. 1986. Genetic analysis of a hybrid zone in the fire-bellied toads *Bombina bombina* and *B. variegata,* near Cracow in southern Poland. *Evolution* 40:1141–59.

Tabachnick, W. J. 1977. Geographic variation of five biochemical polymorphisms in *Notophthalmus viridescens. J. Hered.* 68:117–22.

Tabachnick, W. J., and D. K. Underhill. 1972. Serum transferrin and serum albumin polymorphism in two populations of the red spotted newt *Notophthalmus viridescens viridescens. Copeia* 1972:525–28.

Taber, C. A., R. F. Wilkinson, Jr., and M. S. Topping. 1975. Age and growth of hellbenders in the Niangua River, Missouri. *Copeia* 1975:633–39.

Taborsky, M. 1994. Sneakers, satellites, and helpers: Parasitic and cooperative behavior in fish reproduction. *Adv. Stud. Behav.* 23:1–100.

Tachi, C. 1963. Seasonal variations in regulation of fat mobilization in the Japanese newt, *Triturus pyrrhogaster. Annot. Zool. Jap.* 36:7–13.

Tago, K. 1929. Notes on the habits and life history of *Megalobatrachus japonicus. 10th Int. Congr. Zool., Budapest* 1:828–38.

———. 1931. *Newts and salamanders.* Tokyo: Maruzen. (in Japanese)

Taigen, T. L. 1983. Activity metabolism of anuran amphibians: Implications for the origin of endothermy. *Am. Nat.* 121:94–109.

Taigen, T. L., and C. A. Beuchat. 1984. Anaerobic threshold of anuran amphibians. *Physiol. Zool.* 57:641–47.

Taigen, T. L., S. B. Emerson, and F. H. Pough. 1982. Ecological correlates of anuran exercise physiology. *Oecologia* 52:49–56.

Taigen, T. L., and F. H. Pough. 1981. Activity metabolism of the toad (*Bufo americanus*): ecological consequences of ontogenetic change. *J. Comp. Physiol. B* 144:247–52.

———. 1983. Prey preference, foraging behavior, and metabolic characteristics of frogs. *Am. Nat.* 122:509–20.

———. 1985. Metabolic correlates of anuran behavior. *Amer. Zool.* 25:987–97.

Taigen, T. L., F. H. Pough, and M. M. Stewart. 1984. Water balance of terrestrial anuran eggs (*Eleutherodactylus coqui*): Importance of parental care. *Ecology* 65:248–55.

Taigen, T. L., and K. D. Wells. 1985. Energetics of vocalization by an anuran amphibian (*Hyla versicolor*). *J. Comp. Physiol. B* 155:163–70.

Taigen, T. L., K. D. Wells, and R. L. Marsh. 1985. The enzymatic basis of high metabolic rates in calling frogs. *Physiol. Zool.* 58:719–26.

Takahashi, H., and H. Iwasawa. 1988a. Interpopulation variations in clutch size and egg size in the Japanese salamander, *Hynobius nigrescens. Zool. Sci. (Tokyo)* 5:1073–81.

———. 1988b. Intraclutch egg size variability in *Hynobius nigrescens* and *Hynobius lichenatus* (Urodela: Hynobiidae). *Sci. Rep. Niigata Univ., ser. D (Biology)* 25:19–29.

———. 1989a. Clutch size and egg size variation in salamanders: traits of the variations in some levels of biological organization. In *Current herpetology in East Asia,* ed. M. Matsui, H. Hikida, and R. C. Goris, 282–91. Kyoto: Herpetological Society of Japan.

———. 1989b. Geographic variation in egg sac transparency in the Japanese salamander *Hynobius nigrescens. Sci. Rep. Niigata Univ., ser. D (Biology)* 26:27–34.

Taketa, F., and M. A. Nickerson. 1973. Comparative studies on the hemoglobins of representative salamanders of the families Cryptobranchidae, Proteidae, and Hynobiidae. *Comp. Biochem. Physiol.* 45B:549–56.

Talbot, C. R., and M. E. Feder. 1992. Relationships among cutaneous surface area, cutaneous mass, and body mass in frogs: A reappraisal. *Physiol. Zool.* 65:1135–47.

Tamsitt, J. R. 1962. Notes on a population of the Manitoba toad (*Bufo hemiophrys*) in the delta marsh region of Lake Manitoba, Canada. *Ecology* 43:147–50.

Tanabe, S. 2002. *Hynobius naevius* (Amphibia). In *Red data book of Kochi prefecture (animals,*164–65. Kochi, Japan: Editorial Committee of Red Data Book of Kochi Prefecture (in Japanese).

Tanaka, K. 1986. The territorial behavior of *Hynobius takedai* in the breeding season (Amphibia: Hynobiidae). *Jap. J. Herpetol.* 11:173–81 (in Japanese with English abstract).

———. 1987. Body size and territorial behaviour of male *Hynobius takedai* in breeding season (Amphibia: Hynobiidae). *Jap. J. Herpetol.* 12:45–49 (in Japanese, with English abstract).

———. 1989. Mating strategy of male *Hynobius nebulosus* (Amphibia: Hynobiidae). In *Current herpetology in East Asia,* ed. M. Matsui, T. Hikida, and R. C. Goris, 437–48. Kyoto: Herpetological Society of Japan.

Tandler, B., T. Nagato, and C. J. Phillips. 1997. Ultrastructure of the unusual accessory submandibular gland in the fringe-lipped bat, *Trachops cirrhosus. Anat. Rec.* 248:164–75.

Tandler, B., C. J. Phillips, and T. Nagato. 1996. Histological convergent evolution of the accessory submandibular glands in four species of frog-eating bats. *Eur. J. Morphol.* 34:163–68.

Tandy, M., and R. Keith. 1972. *Bufo* of Africa. In *Evolution in the genus Bufo,* ed. W. F. Blair, 119–70. Austin: University of Texas Press.

Tandy, M., and J. Tandy. 1976. Evolution of acoustic behavior of African *Bufo. Zool. Afr.* 11:349–68.

Tannahill, R. 1973. *Food in history.* New York: Stein and Day.

Tao, F., M. Yokozawa, Y. Hayashi, and E. Lin. 2005. A perspective on water resources in China: Interactions between climate change and soil degradation. *Climatic Change* 68:169–97.

Tárano, Z. 1998. Cover and ambient light influence nesting preferences in the túngara frog *Physalaemus pustulosus. Copeia* 1998:250–51.

———. 2001. Variation in male advertisement calls in the Neotropical frog *Physalaemus enesefae. Copeia* 2001:1064–72.

Tárano, Z., and E. A. Herrera. 2003. Female preferences for call traits and mating mating success in the Neotropical frog *Physalaemus enesefae. Ethology* 109:121–34.

Tárano, Z., and M. J. Ryan. 2002. No pre-existing biases for heterospecific call traits in the frog *Physalaemus enesefae. Anim. Behav.* 64:599–607.

Tarkhnishvili, D. N. 1994. Breeding of the toad *Bufo verrucosissimus:* Sexual dimorphism and shifting spawning sites. *Amphibia-Reptilia* 15:191–98.

Tarkhnishvili, D. N., and R. K. Gokhelashvili. 1994. Preliminary data of the age structure of a *Mertensiella caucasica* population. *Mertensiella* 4:327–34.

———. 1999. The amphibians of the Caucasus. *Adv. Amphib. Res. Former Soviet Union* 4:1–240.

Tarkhnishvili, D. N., and I. A. Serbinova. 1993. The ecology of the Caucasian salamander (*Mertensiella caucasica* Waga) in a local population. *Asiatic Herpetol. Res.* 5:147–65.

Tashian, R. E., and C. Ray. 1957. The relation of oxygen consumption to temperature in some tropical, temperate, and boreal anuran amphibians. *Zoologica* 42:63–68.

Tata, J. R. 1996. Hormonal interplay and thyroid hormone receptor expression during amphibian metamorphosis. In *Metamorphosis. Postembryonic programming of gene expression in amphibian and insect cells,* ed. L. I. Gilbert, J. R. Tata, and B. G. Atkinson, 465–503. San Diego: Academic Press.

———. 2005. Hormonal control of metamorphosis. In *Amphibian biology. Vol. 6. Endocrinology,* ed. H. Heatwole, 2208–27. Chipping Norton, New South Wales, Australia: Surrey Beatty and Sons.

Tattersall, G. J., and R. G. Boutilier. 1997. Balancing hypoxia and hypothermia in cold-submerged frogs. *J. Exp. Biol.* 200:1031–38.

———. 1999. Behavioural oxy-regulation by cold-submerged frogs in heterogeneous oxygen environments. *Can. J. Zool.* 77:843–50.

Taub, F. B. 1961. The distribution of the red-backed salamander, *Plethodon c. cinereus,* within the soil. *Ecology* 42:681–98.

Taurog, A. 1974. Effect of TSH and long-acting thyroid stimulator on thyroid 131 I-metabolism and metamorphosis of the Mexican axolotl (*Ambystoma mexicanum*). *Gen. Comp. Endocrinol.* 24:257–66.

Taylor, A. C., and J. J. Kollros. 1946. Stages in the normal development of *Rana pipiens* larvae. *Anat. Rec.* 94:7–23.

Taylor, B. E., and D. E. Scott. 1997. Effects of larval density dependence on population dynamics of *Ambystoma opacum. Herpetologica* 53:132–45.

Taylor, B. E., R. A. Estes, J. H. K. Pechmann, and R. D. Semlitsch. 1988. Trophic relations in a temporary pond: Larval salamanders and their microinvertebrate prey. *Can. J. Zool.* 66:2191–98.

Taylor, C. L., and R. Altig. 1995. Oral disc kinematics of four rheophilous anuran tadpoles. *Herpetol. Nat. Hist.* 3:101–6.

Taylor, D. H. 1972. Extra-optic photoreception and compass orientation in larval and adult salamanders (*Ambystoma tigrinum*). *Anim. Behav.* 20:233–36.

Taylor, D. H., and K. Adler. 1973. Spatial orientation by salamanders using plane-polarized light. *Science* 181:285–87.

———. 1978. The pineal body: Site of extraocular perception of celestial cues for orientation in the tiger salamander (*Ambystoma tigrinum*). *J. Comp. Physiol.* 124:357–61.

Taylor, D. H., and J. S. Auburn. 1978. Orientation of amphibians by linearly polarized light. In *Animal migration, navigation, and homing,* ed. K. Schmidt-Koenig and W. T. Keeton, 334–46. New York: Springer-Verlag.

Taylor, D. H., and D. E. Ferguson. 1969. Solar cues and shoreline learning in the southern cricket frog, *Acris gryllus. Herpetologica* 25:147–49.

———. 1970. Extraoptic orientation in the southern cricket frog, *Acris gryllus. Science* 168:390–92.

Taylor, E. H. 1954. Frog-egg-eating tadpoles of *Anotheca coronata* (Stejneger) (Salientia, Hylidae). *Univ. Kansas Sci. Bull.* 36:589–96.

———. 1968. *The caecilians of the world.* Lawrence: University of Kansas Press.

Taylor, G. M., E. Nol, and D. Boire. 1995. Brain regions and encephalization in anurans: adaptation or stability? *Brain Behav. Evol.* 45:96–109.

Taylor, J. 1983a. Orientation and flight responses of a neotenic salamander (*Ambystoma gracile*) in Oregon. *Am. Midl. Nat.* 109:40–49.

———. 1983b. Size-specific associations of larval and neotenic northwestern salamanders, *Ambystoma gracile. J. Herpetol.* 17:203–9.

Taylor, S., and D. W. Ewer. 1956. Moulting in the Anura: The normal moulting cycle of *Bufo regularis* Reusse. *Proc. Zool. Soc. London* 127:461–78.

Taylor, S. K., E. S. Williams, and K. W. Mills. 1999. Effects of malathion on disease susceptibility in Woodhouse's toads. *J. Wildl. Dis.* 35:536–41.

Tejedo, M. 1988. Fighting for females in the toad *Bufo calamita* is affected by the operational sex ratio. *Anim. Behav.* 36:1765–69.

———. 1991. Effect of predation by two species of sympatric tadpoles on embryo survival in the natterjack toad (*Bufo calamita*). *Herpetologica* 47:322–27.

———. 1992a. Absence of the trade-off between the size and number of offspring in the natterjack toad (*Bufo calamita*). *Oecologia* 90:294–96.

———. 1992b. Effects of body size and timing of reproduction on reproductive success in female natterjack toads (*Bufo calamita*). *J. Zool., London* 228:545–55.

———. 1992c. Large male mating advantage in natterjack toads, *Bufo calamita:* sexual selection or energetic constraints? *Anim. Behav.* 44:557–69.

———. 1993a. Do male natterjack toads join larger breeding choruses to increase mating success? *Copeia* 1993:75–80.

———. 1993b. Size-dependent vulnerability and behavioral responses of tadpoles of two anuran species to beetle larvae predators. *Herpetologica* 49:287–94.

Tejedo, M., and R. Reques. 1992. Effects of egg size and density on metamorphic traits in tadpoles of the natterjack toad (*Bufo calamita*). *J. Herpetol.* 26:146–52.

———. 1994a. Does larval growth history determine timing of metamorphosis in anurans? A field experiment. *Herpetologica* 50:113–18.

———. 1994b. Plasticity in metamorphic traits of natterjack tadpoles: The interactive effects of density and pond duration. *Oikos* 71:295–304.

Tejedo, M., R. D. Semlitsch, and H. Hotz. 2000. Covariation of morphology and jumping performance in newly metamorphosed water frogs: Effects of larval growth history. *Copeia* 2000:448–58.

Telford, S. R. 1982. *Aspects of mate recognition and social behaviour in a sub-tropical frog community.* Unpublished PhD dissertation. Johannesburg, South Africa: University of the Witwatersrand.

———. 1985. Mechanisms and evolution of inter-male spacing in the painted reedfrog (*Hyperolius marmoratus*). *Anim. Behav.* 33:1353–61.

Telford, S. R., and M. L. Dyson. 1988. Some determinants of the mating system in a population of painted reed frogs (*Hyperolius marmoratus*). *Behaviour* 106:265–78.

———. 1990. The effect of rainfall on interclutch interval in painted reed frogs (*Hyperolius marmoratus*). *Copeia* 1990:644–48.

Telford, S. R., M. L. Dyson, and N. I. Passmore. 1989. Mate choice occurs only in small choruses of painted reed frogs *Hyperolius marmoratus. Bioacoustics* 2:47–53.

Telford, S. R., and N. I. Passmore. 1981. Selective phonotaxis of four sympatric species of African reed frogs (Genus *Hyperolius*). *Herpetologica* 37:29–32.

Telford, S. R., and J. van Sickle. 1989. Sexual selection in an African toad (*Bufo gutteralis*): The roles of morphology, amplexus displacement and chorus participation. *Behaviour* 110:62–75.

Ten Eyck, G. R. 2005. Arginine vasotocin activates advertisement calling and movement in the territorial Puerto Rican frog, *Eleutherodactylus coqui. Horm. Behav.* 47:223–29.

Tennessen, J. A., and K. A. Zamudio. 2003. Early male reproductive advantage, multiple paternity and sperm storage in an amphibian aggregate breeder. *Mol. Ecol.* 12:1567–76.

Tenney, S. M., and D. F. Boggs. 1986. Comparative mammalian respiratory control. In *Handbook of physiology. Section 3: The respiratory system, vol. 2,* ed. A. P. Fishman, N. S. Cherniack, and J. G. Widdicombe, 833–35. Bethesda, MD: American Physiological Society.

Teplitsky, C., S. Plénet, and P. Joly. 2003. Tadpoles' response to risk of fish introduction. *Oecologia* 134:270–77.

———. 2005. Costs and limits of dosage response to predation risk: To what extent can tadpoles invest in anti-predator morphology? *Oecologia* 145:364–70.

Teplitsky, C., S. Plénet, J.-P. Léna, N. Mermet, E. Malet, and P. Joly. 2005. Escape behaviour and ultimate causes of specific induced defenses in an anuran tadpole. *J. Evol. Biol.* 18:180–90.

Terborgh, J. 1983. *Five new world primates. A study in comparative ecology.* Princeton, NJ: Princeton University Press.

Terhivuo, J. 1988. Phenology of spawning for the common frog (*Rana temporaria* L.) in Finland from 1846 to 1986. *Ann. Zool. Fennici* 25:165–75.

Terres, J. K. 1968. Kingfishers eating bullfrog tadpoles. *Auk* 85:140.

Tessier, C., D. Slaven, and D. M. Green. 1991. Population density and daily movement patterns of Hochstetter's frogs, *Leiopelma hochstetteri,* in a New Zealand mountain stream. *J. Herpetol.* 25:213–14.

Test, F. H. 1946. Relations of the red-backed salamander (*Plethodon cinereus*) to light and contact. *Ecology* 27:246–54.

———. 1954. Social aggressiveness in an amphibian. *Science* 120:140–41.

———. 1955. Seasonal differences in populations of the red-backed salamander in southeastern Michigan. *Pap. Mich. Acad. Sci.* 40:137–53.

Test, F. H., and R. G. McCann. 1976. Foraging behavior of *Bufo americanus* tadpoles in response to high densities of micro-organisms. *Copeia* 1976:576–78.

Test, F. H., O. J. Sexton, and H. Heatwole. 1966. Reptiles of Rancho Grande and vicinity, Estado Aragua, Venezuela. *Misc. Publ. Mus. Zool. Univ. Michigan* 128:1–63.

Tester, J. R., and W. J. Brekenridge. 1964. Winter behavior patterns of the Manitoba toad, *Bufo hemiophrys,* in northern Minnesota. *Ann. Acad. Sci. Fennicae, Ser. A, IV Biologica* 71:423–31.

Tevis, L., Jr. 1966. Unsuccessful breeding by desert toads (*Bufo punctatus*) at the limit of their ecological tolerance. *Ecology* 47:766–75.

Teyssedre, C., and T. Halliday. 1986. Cumulative effect of male's display in the sexual behaviour of the smooth newt *Triturus vulgaris* (Urodela, Salamandridae). *Ethology* 71:89–102.

Thesleff, S., and K. Schmidt-Nielsen. 1962. Osmotic tolerance of the muscles of the crab-eating frog *Rana cancrivora. J. Cell. Comp. Physiol.* 59:31–34.

Thibaudeau, G., and R. Altig. 1999. Endotrophic anurans. Development and evolution. In *Tadpoles. The biology of anuran larvae,* ed. R. W. McDiarmid and R. A. Altig, 170–88. Chicago: University of Chicago Press.

Thibert, P., and M. Nicolet. 1972. Dimorphismé sexuel dans le muscles de la grenouille. *Ann. Sci. Univ. Besançon* 8:37–45.

Thielcke, G., C.-P. Herrn, C.-P. Hutter, and R. L. Schreiber. 1985. *Rettet die Frösche.* Stuttgart, Germany: Pro Natur Verlag.

Thiemann, G. W., and R. J. Wassersug. 2000. Patterns and consequences of behavioural responses to predators and parasites in *Rana* tadpoles. *Biol. J. Linn. Soc.* 71:513–28.

Thiesmeier, B. 1990. Die Larven des Feuersalamanders (*Salamandra salamandra terrestris*). Leitorganismen der fischfreien Mittelgebirgs-Quellbache. *Dtsch. Ges. Limnol., Erweiterte Zusammenfassungen der Jahresstagung* 1990:365–69.

Thiesmeier, B., and C. Hornberg. 1986. *Euproctus asper* (Dugès, 1852): Beobachtungen im Freiland und Angaben zur Haltung und Fortpflanzung im Terrarium (Caudata: Salamandridae). *Salamandra* 22:196–210.

———. 1990. Zur Fortpflanzung sowie zum Paarungsverhalten der Gebirgsmolche, Gattung *Euproctus* (Gené), im Terrarium, unter besonderer Berücksichtingung von *Euproctus asper* (Dugès, 1852). *Salamandra* 26:63–82.

———. 1992. First evidence of parental care in *Pachytriton brevipes* (Amphibia, Salamandridae). *Acta Biol. Benrodis, Düsseldorf* 4:163–64.

———. 1997. Paarung, Fortpflanzung und Larvalentwicklung von *Pacytriton* sp. (*Pachytriton* A) nebst Bemerkungen zur Taxonomie der Gattung. *Salamandra* 33:97–110.

———. 1998. Zur Fortpflanzung von *Pachytriton labiatus*—Ein weiterer Hinweis auf Brutpflegeverhalten in der Gattung *Pachytriton. Salamandra* 34:77–80.

Thiollay, J. M. 1976. Besoins alimentaires quantitatifs de quelques oiseaux tropicaux. *Terre et Vie* 30:229–45.

Thirgood, S., J. Langbein, and R. J. Putnam. 1999. Intraspecific variation in ungulate mating strategies: The case of the flexible fallow deer. *Adv. Stud. Behav.* 28:333–61.

Thomas, C. D., A. Cameron, R. E. Green, M. Bakkenes, L. J. Beaumont, Y. C. Collingham, B. F. N. Erasmus, M. Ferreria de Siqueira, A. Grainger, L. Hannah, L. Hughes, B. Huntley, A. S. van Jaarsveld, G. F. Midgley, L. Miles, M. A. Ortega-Huerta, A. T. Peterson, O. L. Phillips, and S. E. Williams. 2004. Extinction risk from climate change. *Nature* 427:145–48.

Thomas, J. S., R. G. Jaeger, and E. A. Horne. 1989. Are all females welcome? Agonistic behavior of male red backed salamanders. *Copeia* 1989:915–20.

Thomas, E. O., L. Tsang, and P. Licht. 1993. Comparative histochemistry of the sexually dimorphic skin glands of anuran amphibians. *Copeia* 1993:133–43.

Thomas, R. 1966. New species of Antillean *Eleutherodactylus. Q. J. Florida Acad. Sci.* 28:375–91.

Thomas, R. G., and A. G. Kessler. 1996. Nonanoline reptiles. In *The food web of a tropical rain forest,* ed. D. P. Reagan and R. B. Waide, 348–62. Chicago. University of Chicago Press.

Thomas, R. J., I. C. Cuthill, A. R. Goldsmith, D. F. Cosgrove, J. C. Lidgate, and S. L. Burdett Proctor. 2003. The trade-off between singing and mass gain in a daytime-singing bird the European robin. *Behaviour* 140:387–404.

Thomson, A. M., R. A. Brown, N. J. Rosenberg, R. Srinivasan, and R. C. Izaurralde. 2005. Climate change for the conterminous USA: An integrated assessment. Part 4: Water resources. *Climatic Change* 69:67–88.

Thompson, R. L. 1996. Larval habitat, ecology, and parental investment of *Osteopilus brunneus* (Hylidae). In *Contributions to West Indian herpetology,* ed. R. Powell and R. W. Henderson, 259–69. Ithaca, NY: Society for the Study of Amphibians and Reptiles.

Thompson, R. R., and F. L. Moore. 2000. Vasotocin stimulates appetitive responses to the visual and pheromonal stimuli used by male roughskin newts during courtship. *Horm. Behav.* 38:75–85.

Thompson, R. R., Z. Tokar, D. Pistohl, and F. L. Moore. 1999. Behavioral evidence for a sex-pheromone in female rougthskin newts, *Taricha granulosa.* In *Advances in chemical signals in vertebrates,* ed. R. E. Johnston, D. Müller-Schwarze, and P. W. Sorensen, 421–30. New York: Plenum.

Thompson, W. 1843. Notes on a species of toad. *Zoologist* 1:321–22.

Thorn, R. 1962. Protection of the brood by a male salamander, *Hynobius nebulosus. Copeia* 1962:638–40.

———. 1963. Contribution a l'étude d'une salamandre japonaise l'*Hynobius nebulosus* (Schlegel). Comportement et reproduction en captivité. *Arch. Inst. Grand-Duc. Luxembourg, Sect. Sci.*, n.s. 29:201–15.

———. 1967. Nouvelles observations sur l'éthologie sexuelle de *Hynobius nebulosus* (Temminck et Schlegel) (Caudata, Hynobiidae). *Arch. Inst. Grand-Duc. Luxembourg, Sect. Sci.*, n.s. 32:267–71.

Thornhill, N. W., ed. 1993. *The natural history of inbreeding and outbreeding. Theoretical and empirical perspectives.* Chicago: University of Chicago Press.

Thornhill, R., and J. Alcock. 1983. *The evolution of insect mating systems.* Cambridge, MA: Harvard University Press.

Thornton, W. A. 1955. Interspecific hybridization in *Bufo woodhousei* and *Bufo valliceps. Evolution* 9:455–68.

Thorson, T., and A. Svihla. 1943. Correlation of the habitats of amphibians with their ability to survive the loss of water. *Ecology* 24:374–81.

Thorson, T. B. 1955. The relationship of water economy to terrestrialism in amphibians. Ecology 36:100–16.

Thumm, K., and M. Mahony. 1999. Loss and degradation of red-crowned toadlet habitat in the Sydney region. In *Declines and disappearances of Australian frogs,* ed. A. Campbell, 99–108. Canberra: Environment Australia.

———. 2002a. Evidence for continuous interoparity in a temperate-zone frog, the red-crowned toadlet, *Pseudophryne australis* (Anura: Myobatrachidae). *Aust. J. Zool.* 50:151–67.

———. 2002b. Hatching dynamics and bet-hedging in a temperate frog, *Pseudophryne australis* (Anura: Myobatrachidae). *Amphibia-Reptilia* 23:433–44.

———. 2005. Is variable egg size the proximate cause of diversified bet-hedging in the hatching dynamics of the red-crowned toadlet (*Pseudophryne australis*) (Anura: Myobatrachidae)? *Herpetologica* 61:9–19.

Thurley, T., and B. D. Bell. 1994. Habitat distribution and predation on a western population of terrestrial *Leiopelma* (Anura, Leiopelmatidae) in the Northern King Country, New Zealand. *New Zealand J. Zool.* 21:431–36.

Thurow, G. 1976. Aggression and competition in eastern *Plethodon* (Amphibia, Urodela, Plethodontidae). *J. Herpetol.* 10:277–92.

Tietge, J. E., S. A. Diamond, G. T. Ankley, D. L. DeFoe, G. W. Holcombe, K. M. Jensen, S. J. Degitz, G. E. Elonen, and E. Hammer. 2001. Ambient solar UV radiation causes mortality in larvae of three species of *Rana* under controlled exposure conditions. *Photochem. Photobiol.* 74:261–68.

Tihen, J. A. 1960. Comments on the origin of the amniote egg. *Evolution* 14:528–31.

———. 1965. Evolutionary trends in frogs. *Amer. Zool.* 5:309–18.

Tilley, S. G. 1964. A quantitative study of the shrinkage in the digestive tract of the tiger salamander (*Ambystoma tigrinum* Green) during metamorphosis. *J. Ohio Herpetol. Soc.* 4:81–85.

———. 1968. Size-fecundity relationships and their evolutionary implications in five desmognathine salamanders. *Evolution* 22:806–16.

———. 1972. Aspects of parental care and embryonic development in *Desmognathus ochrophaeus. Copeia* 1972:532–40.

———. 1977. Studies of life histories and reproduction in North American plethodontid salamanders. In *The reproductive biology of amphibians,* ed. D. H. Taylor and S. I. Guttman, 1–41. New York: Plenum.

———. 1980. Life histories and comparative demography of two salamander populations. *Copeia* 1980:806–21.

———. 1981. A new species of *Desmognathus* (Amphibia: Caudata: Plethodontidae) from the southern Appalachian Mountains. *Occ. Pap. Mus. Zool. Univ. Michigan* 695:1–23.

———. 1985. *Desmognathus imitator.* Imitator salamander. *Cat. Am. Amphib. Rept.* 359.1–359.2.

———. 1997. Patterns of genetic differentiation in Appalachian desmognathine salamanders. *J. Hered.* 88:305–15.

———. 2000. The systematics of *Desmognathus imitator.* In *The biology of plethodontid salamanders,* ed. R. C. Bruce, R. G. Jaeger, and L. D. Houck, 121–47. New York: Kluwer Academic/Plenum.

Tilley, S. G., and J. Bernardo. 1993. Life history evolution in plethodontid salamanders. *Herpetologica* 49:154–63.

Tilley, S. G., and J. S. Hausman. 1976. Allozymic variation and occurrence of multiple inseminations in populations of the salamander *Desmognathus ochrophaeus. Copeia* 1976:734–41.

Tilley, S. G., B. L. Lundrigan, and L. P. Brower. 1982. Erythrism and mimicry in the salamander *Plethodon cinereus. Herpetologica* 38:409–17.

Tilley, S. G., and M. J. Mahoney. 1996. Patterns of genetic differentiation in salamanders of the *Desmognathus ochrophaeus* complex (Amphibia: Plethodontidae). *Herpetol. Monogr.* 10:1–42.

Tilley, S. G., R. B. Merritt, B. Wu, and R. Highton. 1978. Genetic differentiation in salamanders of the *Desmognathus ochrophaeus* complex (Plethodontidae). *Evolution* 32:93–115.

Tilley, S. G., and P. M. Schwerdtfeger. 1981. Electrophoretic variation in Appalachian populations of the *Desmognathus fuscus* complex (Amphibia, Plethodontidae). *Copeia* 1981:109–18.

Tilley, S. G., P. A. Verrell, and S. J. Arnold. 1990. Correspondence between sexual isolation and allozyme differentiation: a test in the salamander *Desmognathus ochrophaeus. Proc. Natl. Acad. Sci. USA* 87:2715–19.

Ting, H.-P., and A. M. Boring. 1940. The seasonal cycle in the reproductive organs of the Chinese toad *Bufo bufo* and the pond frog *Rana nigromaculata. Peking Nat. Hist. Bull.* 14:49–80.

Tinsley, R. C. 1973. Studies on the ecology and systematics of a new species of clawed toad, the genus *Xenopus,* from western Uganda. *J. Zool., London* 169:1–27.

———. 1989. The effects of host sex on transmission success. *Parasitology* Today 5:190–95.

———. 1990. The influence of parasite infection on mating success in spadefoot toads, *Scaphiopus couchii. Amer. Zool.* 30:313–24.

Tinsley, R. C., and C. M. Earle. 1983. Invasion of vertebrate lungs by the polysomatid monogeneans *Pseudodiplorchis americanus* and *Neodiplorchis scaphiopodis. Parasitology* 86:501–17.

Tinsley, R. C., and H. C. Jackson. 1986. Intestinal migration in the life-cycle of *Pseudodiplorchis americanus* (Monogenea). *Parasitology* 93:451–69.

Tinsley, R. C., and H. C. Jackson. 1988. Pulsed transmission of *Pseudodiplorchis americanus* (Monogenea) between desert hosts (*Scaphiopus couchii*). *Parasitology* 97:437–52.

Tinsley, R. C., C. Loumont, and H. R. Kobel. 1996. Geographical distribution and ecology. In *The biology of Xenopus,* ed. R. C. Tinsley and H. R. Kobel, 35–59. New York: Oxford University Press.

Tinsley, R. C., and K. Tocque. 1995. The population dynamics of a desert anuran, *Scaphiopus couchii*. *Aust. J. Ecol.* 20:376–84.

Titus, T. A., and A. Larson. 1995. A molecular phylogenetic perspective on the evolutionary radiation of the salamander family Salamandridae. *Syst. Biol.* 44:125–51.

———. 1996. Molecular phylogenetics of desmognathine salamanders (Caudata: Plethodontidae): A reevaluation of evolution in ecology, life history, and morphology. *Syst. Biol.* 45:451–72.

Tobias, M. L., and D. B. Kelley. 1987. Vocalizations by a sexually dimorphic isolated larynx: peripheral constraints on behavioral expression. *J. Neurosci.* 7:3191–97.

Tobias, M. L., S. S. Viswanathan, and D. B. Kelley. 1998. Rapping, a female receptive call, initiates male-female duets in the South African clawed frog. *Proc. Natl. Acad. Sci. USA* 95:1870–75.

Tocher, M. D., C. Gascon, and B. L. Zimmerman. 1997. Fragmentation effects on a central Amazonian frog community: a ten-year study. In *Tropical forest remnants,* ed. W. F. Laurance and R. O. Bierregaard, Jr., 124–27. Chicago: University of Chicago Press.

Tocher, M. D., C. Gascon, and J. Meyer. 2002. Community composition and breeding success of Amazonian frogs in continuous forest and matrix habitat aquatic sites. In *Lessons from Amazonia. The ecology and conservation of a fragmented forest,* ed. R. O. Bierregard, Jr., C. Gascon, T. E. Lovejoy, and R. Mesquita, 235–47. New Haven, CT: Yale University Press.

Tockner, K., F. Schiemer, and J. V. Ward. 1998. Conservation by restoration: the management concept for a river-floodplain system on the Danube River in Austria. *Aquat. Cons. Mar. Freshw. Ecosyst.* 8:71–86.

Toda, M., M. Matsui, M. Nishida, and H. Ota. 1998. Genetic divergence among southeast and east Asian populations of *Rana limnocharis* (Amphibia: Anura), with special reference to sympatric cryptic species in Java. *Zool. Sci.* 15:607–13.

Todt, D. 1970. Gesangliche Reaktionen der Amsel *Turdus merula* auf ihren experimentell reproduzierten Eigengesang. *Z. Vergl. Physiol.* 66:294–317

Toews, D. P. 1971. Factors affecting the onset and termination of respiration in the salamander *Amphiuma tridactylum*. *Can. J. Zool.* 49:1231–37.

Toews, D. P., and R. G. Boutilier. 1986. Acid-base regulation in the Amphibia. In *Acid-base regulation in animals,* ed. N. Heisler, 265–308. New York: Elsevier.

Toews, D. P., and N. Heisler. 1982. The effects of hypercapnia on intracellular and extracellular acid-base status in the toad *Bufo marinus*. *J. Exp. Biol.* 97:79–86.

Toews, D. P., and D. Macintyre. 1977. Blood respiratory properties of a viviparous amphibian. *Nature* 266:464–65.

———. 1978. Respiration and circulation in an apodan amphibian. *Can. J. Zool.* 56:998–1004.

Toews, D. P., and A. McRae. 1974. Respiratory mechanisms in the aquatic salamander, *Amphiuma tridactylum*. *Copeia* 1974:917–20.

Toews, D. P., and D. F. Stiffler. 1989. The role of the amphibian kidney and bladder in the regulation of acid-base relevant ions. *Can. J. Zool.* 67:3064–69.

———. 1990. Compensation of progressive hypercapnia in the toad (*Bufo* marinus) and the bullfrog (*Rana catesbeiana*). *J. Exp. Biol.* 148:293–302.

Toft, C. A. 1980a. Feeding ecology of thirteen syntopic species of anurans in a seasonal tropical environment. *Oecologia* 45:131–41.

———. 1980b. Seasonal variation in populations of Panamanian litter frogs and their prey: A comparison of wetter and drier sites. *Oecologia* 47:34–38.

———. 1981. Feeding ecology of Panamanian litter anurans: Patterns in diet and foraging mode. *J. Herpetol.* 15:130–44.

———. 1982. Community structure of litter anurans in a tropical forest, Makokou, Gabon: A preliminary analysis in the minor dry season. *Terre Vie* 36:223–32.

———. 1985. Resource partitioning in amphibians and reptiles. *Copeia* 1985:1–21.

———. 1995. Evolution of diet specialization in poison-dart frogs (Dendrobatidae). *Herpetologica* 51:202–16.

Toft, C. A., and W. E. Duellman. 1979. Anurans of the lower Rio Llullapichis, Amazonian Peru: A preliminary analysis of community structure. *Herpetologica* 35:71–77.

Toft, C. A., A. S. Rand, and M. Clark. 1982. Population dynamics and seasonal recruitment in *Bufo typhonius* and *Colostethus nubicola* (Anura). In *The ecology of a tropical forest: Seasonal rhythms and long-term changes,* ed. E. G. Leigh, Jr., A. S. Rand, and D. M. Windsor, 397–403. Washington, DC: Smithsonian Institution Press.

Toledo, L. F. 2003. Predation on seven South American anuran species by water bugs (Belostomatidae). *Phyllomedusa* 2:105–8.

———. 2004a. *Bufo* cf. *crucifer* (Sapo Cururu). Defensive behavior. *Herp. Rev.* 35:370–71.

———. 2004b. *Elachistocleis* cf. *ovalis* (Common Oval Frog). Death feigning. *Herp. Rev.* 35:371–72.

———. 2005. Predation of juvenile and adult amphibians by invertebrates: Current knowledge and perspectives. *Herp. Rev.* 36:395–400.

Toledo, L. F., and C. F. B. Haddad. 2005. Reproductive biology of *Scinax fuscomarginatus* (Anura, Hylidae) in south-eastern Brazil. *J. Nat. Hist.* 39:3029–37.

Toledo, R. C., and C. Jared. 1993. Cutaneous adaptations to water balance in amphibians. *Comp. Biochem. Physiol.* 105A:593–608.

———. 1995. Cutaneous granular glands and amphibian venoms. *Comp. Biochem. Physiol.* 111A:1–29.

Toll, S. J., R. G. Jaeger, and J. R. Gillette. 2000. Socioecology of a territorial salamander: females and males as territorial residents and intruders. *Copeia* 2000:276–81.

Tome, M. A. and F. H. Pough. 1982. Responses of amphibians to acid precipitation. In *Acid rain/fisheries. Proceedings of an international symposium on acidic precipitation and fishery impacts in northeastern North America,* ed. R. E. Johnson, 245–54. Bethesda, MD: American Fisheries Society.

Tomiałojć, L., and T. Wesołowski. 2004. Diversity of the Białowieża Forest avifauna in space and time. *J. Ornithol.* 145:81–92.

Tominaga, A., M. Matsui, K. Nishikawa, and S. Sato. 2003. Occurrence of two types of *Hynobius naevius* in northern Kyushu, Japan (Amphibia: Urodela). *Zool. Sci.* 20:1467–76.

Tomson, O. H., and D. E. Ferguson. 1972. Y-axis orientation in larvae and juveniles of three species of *Ambystoma*. *Herpetologica* 28:6–9.

Topsell, E. 1658. *The history of serpents.* London: E. Cotes, London (reprint edition (1967): New York: Da Capo Press, New York).

Toral C. E., P. Feinsinger, and M. L. Crump. 2002. Frogs and cloud-forest edge in Ecuador. *Cons. Biol.* 16:735–44.

Tordoff, W. III. 1980. Selective predation of gray jays, *Perisoreus canadensis,* upon boreal chorus frogs, *Pseudacris triseriata. Evolution* 34:1004–8.

Tordoff, W., III, and D. Pettus. 1977. Temporal stability of phenotypic frequencies in *Pseudacris triseriata* (Amphibia, Anura, Hylidae). *J. Herpetol.* 11:161–68.

Tordoff, W. III, D. Pettus, and T. C. Matthews. 1976. Microgeographic variation in gene frequencies in *Pseudacris triseriata* (Amphibia, Anura, Hylidae). *J. Herpetol.* 10:35–40.

Towns, D. R., and C. H. Daugherty. 1994. Patterns of range contractions and extinctions in the New Zealand herpetofauna following human colonization. *New Zealand J. Zool.* 21:325–39.

Townsend, D. S. 1986. The costs of male parental care and its evolution in a Neotropical frog. *Behav. Ecol. Sociobiol.* 19:187–95.

Townsend, D. S. 1989a. The consequences of microhabitat choice for male reproductive success in a tropical frog (*Eleutherodactylus coqui*). *Herpetologica* 45:451–58.

———. 1989b. Sexual selection, natural selection, and a fitness trade-off in a tropical frog with male parental care. *Am. Nat.* 133:266–72.

———. 1996. Patterns of parental care in frogs of the genus *Eleutherodactylus.* In *Contributions to West Indian herpetology,* ed. R. Powell and R. W. Henderson, 229–39. Ithaca, NY: Society for the Study of Amphibians and Reptiles.

Townsend, D. S., and M. M. Stewart. 1985. Direct development in *Eleutherodactylus coqui* (Anura: Leptodactylidae). *Copeia* 1985:423–36.

———. 1986. Courtship and mating behavior of a Puerto Rican frog, *Eleutherodactylus coqui. Herpetologica* 42:165–70.

———. 1994. Reproductive ecology of the Puerto Rican frog *Eleutherodactylus coqui. J. Herpetol.* 28:34–40.

Townsend, D. S., M. M. Stewart, and F. H. Pough. 1984. Male parental care and its adaptive significance in a Neotropical frog. *Anim. Behav.* 32:421–31.

Townsend, D. S., M. M. Stewart, F. H. Pough, and P. F. Brussard. 1981. Internal fertilization in an oviparous frog. *Science* 212:469–71.

Townsend, V. R., Jr., and R. G. Jaeger. 1998. Territorial conflicts over prey: Domination by large male salamanders. *Copeia* 1998:725–29.

Townson, R. 1794–1795. *Observationes physiologicae de amphibiis.* Gottingen, Germany: I. C. Dieterich.

Townson, R. 1799. *Tracts and observations in natural history and physiology.* London: J. White.

Toxopeus, A. G. 1985. Die Eiblage bei *Pelodytes punctatus* (Daudin, 1802) (Salientia: Pelobatidae). *Salamandra* 21:222–24.

Toxopeus, A. G., M. Ohm, and J. W. Arntzen. 1993. Reproductive biology of the parsley frog, *Pelodytes punctatus,* at the northernmost part of its range. *Amphibia-Reptilia* 14:131–47.

Tozetti, A. M., and L. F. Toledo. 2005. Short-term movement and retreat sites of *Leptodactylus labyrinthicus* (Anura: Leptodactylidae) during the breeding season: A spool-and-line tracking study. *J. Herpetol.* 39:640–44.

Tracy, C. R. 1971. Evidence for the use of celestial cues by dispersing immature California toads (*Bufo boreas*). *Copeia* 1971:145–47.

———. 1975. Water and energy relations of terrestrial amphibians: Insights from mechanistic modeling. In *Perspectives in biophysical ecology,* ed. D. M. Gates and R. Schmerl, 325–45. New York: Springer-Verlag.

———. 1976. A model of the dynamic exchanges of water and energy between a terrestrial amphibian and its environment. *Ecol. Monogr.* 46:293–326.

———. 1979. Further thoughts on anuran thermoregulation: Discussion. In *The behavioral significance of color,* ed. E. H. Burtt, 63–68. New York: Garland STPM Press.

Tracy, C. R., C. R. Christian, M. P. O'Connor, and C. R. Tracy. 1993. Behavioral thermoregulation by *Bufo americanus:* the importance of the hydric environment. *Herpetologica* 49:375–82.

Tracy, C. R., and J. W. Dole. 1969a. Evidence of celestial orientation by California toads (*Bufo boreas*) during breeding migration. *Bull. So. California Acad. Sci.* 68:10–18.

———. 1969b. Orientation of displaced California toads, *Bufo boreas,* to their breeding sites. *Copeia* 1969:693–700.

Tracy, C. R., C. R. Tracy, and J. S. Turner. 1992. Contrasting physiological abilities for heating and cooling in an amphibian (*Rana pipiens*) and a reptile (*Sauromalus obesus*). *Herpetologica* 48:57–60.

Tran, D.-Y., K. S. Hoff, and S. D. Hillyard. 1992. Effects of angiotensin II and bladder condition on hydration behavior and water uptake in the toad, *Bufo woodhousei. Comp. Biochem. Physiol.* 103 A:127–30.

Trauth, S. E., and C. T. McAllister. 1995. Vertebrate prey of selected Arkansas snakes. *Proc. Arkansas Acad. Sci.* 49:190–94.

Travis, J. 1980a. Genetic variation for larval specific growth rate in the frog *Hyla gratiosa. Growth* 44:167–81.

———. 1980b. Phenotypic variation and the outcome of interspecific competition in hylid tadpoles. *Evolution* 34:40–50.

———. 1981. Control of larval growth variation in a population of *Pseudacris triseriata* (Anura: Hylidae). *Evolution* 35:423–32.

———. 1983a. Variation in development patterns of larval anurans in temporary ponds. I. Persistent variation within a *Hyla gratiosa* population. *Evolution* 37:496–512.

———. 1983b. Variation in growth and survival of *Hyla gratiosa* larvae in experimental enclosures. *Copeia* 1983:232–37.

———. 1984. Anuran size at metamorphosis: Experimental test of a model based on intraspecific competition. *Ecology* 65:1155–60.

Travis, J., S. B. Emerson, and M. Blouin. 1987. A quantitative-genetic analysis of larval life-history traits in *Hyla crucifer. Evolution* 41:145–56.

Travis, J., W. H. Keen, and J. Juilianna. 1985a. The effects of multiple factors on viability selection in *Hyla gratiosa* tadpoles. *Evolution* 39:1087–99.

———. 1985b. The role of relative body size in a predator-prey relationship between dragonfly naiads and larval anurans. *Oikos* 45:59–65.

Travis, J., and J. C. Trexler. 1984. Investigations on the control of the color polymorphism in *Pseudacris ornata. Herpetologica* 40:252–57.

———. 1986. Interactions among factors affecting growth, development and survival in experimental populations of *Bufo terrestris* (Anura: Bufonidae). *Oecologia* 69:110–16.

Travis, J. M. J. 2003. Climate change and habitat destruction: A deadly anthropogenic cocktail. *Proc. Roy. Soc. London B* 270:467–74.

Treherne, J. E., and W. A. Foster. 1981. Group transmission of predator avoidance behaviour in a marine insect: the Trafalgar Effect. *Anim. Behav.* 29:911–17.

Trenerry, M. P., W. F. Laurance, and K. R. McDonald. 1994. Further evidence for the precipitous decline of endemic rainforest frogs in tropical Australia. *Pacific Cons. Biol.* 1:150–53.

Trenham, P. C. 2001. Terrestrial habitat use by adult California tiger salamanders. *J. Herpetol.* 35: 343–346.

Trenham, P. C., W. D. Koenig, M. J. Mossman, S. L. Stark, and L. A. Jagger. 2003. Regional dynamics of wetland-breeding frogs and toads: Turnover and synchrony. *Ecol. Appl.* 13:1522–32.

Trenham, P. C., H. B. Shaffer, W. D. Koenig, and M. R. Stromberg. 2000. Life history and demographic variation in the California tiger salamander (*Ambystoma californiense*). *Copeia* 2000:365–77.

Trewavas, E. 1933. The hyoid and larynx of the Anura. *Phil. Trans. Roy. Soc. London (ser. B)* 222:401–527.

Trimnell, H. C. 1945. Little owls feeding young on newts. *British Birds* 38:174.

Tripp, J. B., and S. Lustick. 1974. The effects of heating and cooling in water on the heart rate of the bullfrog. *Comp. Biochem. Physiol.* 48A:547–54.

Tristram, D. A. 1977. Intraspecific olfactory communication in the terrestrial salamander *Plethodon cinereus. Copeia* 1977:597–600.

Trivers, R. L. 1972. Parental investment and sexual selection. In *Sexual selection and the descent of man,* ed. B. G. Campbell, 136–79. Chicago: Aldine Press.

———. 1974. Parent-offspring conflict. *Amer. Zool.* 14:249–64.

Trueb, L. 1970. Evolutionary relationships of casque-headed tree frogs with co-ossified skulls (family Hylidae). *Univ. Kansas Publ. Mus. Nat. Hist.* 18:547–716.

. 1974. Systematic relationships of Neotropical horned frogs, genus *Hemiphractus* (Anura: Hylidae). *Occ. Pap. Mus. Nat. Hist. Univ. Kansas,* no. 29:1–60.

———. 1979. Leptodactylid frogs of the genus *Telmatobius* in Ecuador with the description of a new species. *Copeia* 1979:714–33.

———. 1993. Patterns of cranial diversity among the Lissamphibia. In *The skull, vol. 2,* ed. J. Hanken and B. K. Hall, 255–343. Chicago: University of Chicago Press.

Trueb, L. 1996. Historical constraints and morphological novelties in the evolution of the skeletal system of pipid frogs (Anura: Pipidae). In *The biology of Xenopus,* ed. R. C. Tinsley and H. R. Kobel, 349–77. New York: Oxford University Press.

Trueb, L., and P. Alberch. 1985. Miniaturization and the anuran skull: A case study of heterochrony. In *Functional morphology in vertebrates,* ed. H.-R. Duncker and G. Fleischer, 113–21. Stuttgart: Gustav Fischer Verlag.

Trueb, L., and D. C. Cannatella. 1982. The cranial osteology and hyolaryngeal apparatus of *Rhinophrynus dorsalis* (Anura: Rhinophrynidae) with comparisons to recent pipid frogs. *J. Morphol.* 171:11–40.

———. 1986. Systematics, morphology, and phylogeny of genus *Pipa* (Anura: Pipidae). *Herpetologica* 42:412–49.

Trueb, L., and R. Cloutier. 1991a. A phylogenetic investigation of the inter- and intrarelationships of the Lissamphibia (Amphibia: Temnospondyli). In *Origins of the higher groups of tetrapods,* ed. H.-P. Schultze and L. Trueb, 223–313. Ithaca, NY: Cornell University Press.

———. 1991b. Toward an understanding of the amphibians: Two centuries of systematic history. In *Origins of the higher groups of tetrapods,* ed. H.-P. Schultze and L. Trueb, 175–93. Ithaca, NY: Cornell University Press.

Trueb, L., and C. Gans. 1983. Feeding specializations of the Mexican burrowing toad, *Rhinophrynus dorsalis* (Anura: Rhinophrynidae). *J. Zool., London* 199:189–208.

Tryjanowski, P., M. Rybacki, and T. Sparks. 2003. Changes in the first spawning dates of common frogs and common toads in western Poland in 1978–2002. *Ann. Zool. Fennici* 40:459–64.

Tsiora, A., and P. Kyriakopoulou-Sklavounou. 2001. Male reproductive cycle of the water frog *Rana epeirotica* in northwestern Greece. *Amphibia-Reptilia* 22:291–302.

Tsuji, H. 2004. Reproductive ecology and mating success of male *Limnodynastes kuhlii,* a fanged frog from Taiwan. *Herpetologica* 60:155–67.

Tsuji, H., and T. Kawamichi. 1996a. Breeding activity of a stream-breeding toad, *Bufo torrenticola. Jap. J. Herpetol.* 16:117–28.

———. 1996b. Breeding habitats of a stream-breeding toad, *Bufo torrenticola,* in an Asian mountain torrent. *J. Herpetol.* 30:451–54.

———. 1998. Field observations of the spawning behavior of stream toads, *Bufo torrenticola. J. Herpetol.* 32:34–40.

Tsuji, H., and K.-Y. Lue. 1998. Temporal aspects of the amplexus and oviposition behavior of the fanged frog, *Rana kuhlii* from Taiwan. *Copeia* 1998:769–73.

———. 2000. The reproductive ecology of female *Rana* (*Limnonectes*) *kuhlii,* a fanged frog of Taiwan, with particular emphasis on multiple clutches. *Herpetologica* 56:153–65.

Tsuji, H., and M. Matsui. 2002. Male-male combat and head morphology in a fanged frog (*Rana kuhlii*) from Taiwan. *J. Herpetol.* 36:520–26.

Tsutsui, Y. 1931. Notes on the behaviour of the common Japanese newt, *Diemictylus pyrrhogaster* Boie. I. Breeding habit. *Mem. Coll. Sci. Kyoto Univ., Ser. B,* 7:159 78.

Tucić, N., M. Kalezić, and G. Džukić. 1985. Morphometric variability in a *Triturus vulgaris* population with facultative paedomorphosis (Amphibia). *Zool. Anz.* 215:102–8.

Tucker, J. K. 1998. Status of Illinois chorus frogs in southern Illinois. In *Status and conservation of Midwestern amphibians,* ed. M. J. Lannoo, 102–12. Iowa City: University of Iowa Press.

Tufts, B. L., and D. P. Toews. 1986. Renal function and acid-base balance in the toad *Bufo marinus* during short-term dehydration. *Can. J. Zool.* 64:1054–57.

Tumlison, R., G. R. Cline, and P. Zwank. 1990. Morphological discrimination between the Oklahoma salamander (*Eurycea tynerensis*) and the graybelly salamander (*Eurycea mutiplicata griseogaster*). *Copeia* 1990:242–46.

Tumlison, R., and M. Karnes. 1987. Seasonal changes in food habits of river otters in southwestern Arkansas beaver swamps. *Mammalia* 51:225–31.

Tunner, H. 1996. Ein Künstler erforscht die Welt der Frösche. Das Leben und das naturwissenschaftliche Werk des Malers und Kupferstechers August Johann Rösel von Rosenhof (1705–1759). In *Frösche Kröten Unken. Aus der Welt der Amphibien. Stapfia* 47, ed. W. Hödl and G. Aubrecht, 13–51. Linz, Austria: Oberösterreich Landesmuseum.

Tunner, H. G. 1976. Aggressives Verhalten bei *Rana ridibunda, Rana lessonae,* und der hybriden *Rana esculenta. Zool. Anz.* 200:386–90.

———. 1992. Locomotory behaviour in water frogs from Neusiedlersee (Austria, Hungary). 15 km migration of *Rana lessonae* and its hybridogenetic associate *R. esculenta. Proc. 6th Ord. Gen. Meeting S. E. H., Budapest 1991*:449–52.

Tunner, H. G., and W. Hödl. 1978. Warn- oder Alarmruf und Shreckruf bein *Rana ridibunda ridibunda* (Salientia, Ranidae). *Zool. Anz.* 200:386–90.

Tunner, H. G., and H. Nopp. 1979. Heterosis in the common European water frog. *Naturwissenschaften* 78:32–34.

Turbott, E. G. 1949. Discovery of the breeding habits of *Leiopelma hochstetteri* Fitzinger. *Rec. Aukland Inst. Mus.* 3:373–76.

Turner, F. B. 1958. Life history of the western spotted frog in Yellowstone National Park. *Herpetologica* 14:96–100.

———. 1959. Some features of the ecology of *Bufo punctatus* in Death Valley, California. *Ecology* 40:175–81.

———. 1960a. Population structure and dynamics of the western spotted frog, *Rana p. pretiosa* Baird & Girard, in Yellowstone Park, Wyoming. *Ecol. Monogr.* 30:251–78.

———. 1960b. Some features of the ecology of *Bufo punctatus* in Death Valley, California. *Ecology* 40:175–81.

———. 1962. The demography of frogs and toads. *Q. Rev. Biol.* 37:303–14.

———. 1977. The dynamics of populations of squamates, crocodilians and rhynchocephalians. In *Biology of the reptilia, vol. 7,* ed. C. Gans and D. W. Tinkle, 157–264. New York: Academic Press.

Turner, F. B., R. I. Jennrich, and J. D. Weintraub. 1969. Home ranges and the body sizes of lizards. *Ecology* 50:1076–81.

Turner, J. R. G. 1984a. Darwin's coffin and Doctor Pangloss—Do adaptationist models explain mimicry? In *Evolutionary ecology,* ed. B. Shorrocks, 313–61. Oxford, UK: Blackwell.

———. 1984b. Mimicry: The palatability spectrum and the origins of mimicry. *Biol. J. Linn. Soc.* 23:247–68.

Turney, L. D., and V. H. Hutchison. 1974. Metabolic scope, oxygen debt, and the diurnal oxygen consumption cycle of the leopard frog, *Rana pipiens. Comp. Biochem. Physiol.* 49A:583–601.

Turtle, S. L. 2000. Embryonic survivorship of the spotted salamander (*Ambystoma maculatum*) in roadside and woodland vernal pools in southeastern New Hampshire. *J. Herpetol.* 34:60–67.

Tuttle, M. D., and M. J. Ryan. 1981. Bat predation and the evolution of frog vocalizations in the Neotropics. *Science* 214:677–78.

———. 1982. The role of synchronized calling, ambient light, and ambient noise in anti-bat-predator behavior of a treefrog. *Behav. Ecol. Sociobiol.* 11:125–31.

Tuttle, M. D., L. K. Taft, and M. J. Ryan. 1981. Acoustical location of calling frogs by *Philander* opossums. *Biotropica* 13:233–34.

———. 1982. Evasive behavior of a frog in response to bat predation. *Anim. Behav.* 30:393–97.

Twitty, V. C. 1942. The species of California *Triturus. Copeia* 1942:65–76.

———. 1955. Field experiments on the biology and genetic relationships of the California species of *Triturus. J. Exp. Biol.* 129:129–48.

———. 1959. Migration and speciation in newts. *Science* 139:1735–43.

———. 1961. Experiments on homing behavior and speciation in *Taricha.* In *Vertebrate speciation,* ed. W. F. Blair, 415–59. Austin: University of Texas Press.

———. 1966. *Of scientists and salamanders.* San Francisco: Freeman.

Twitty, V. C., D. Grant, and O. Anderson. 1964. Long distance homing in the newt *Taricha rivularis. Proc. Natl. Acad. Sci. USA* 51:51–58.

———. 1967. Home range in relation to homing in the newt *Taricha rivularis* (Amphibia: Caudata). *Copeia* 1967:649–53.

Twombly, S. 1996. Timing of metamorphosis in a freshwater crustacean: Comparison with anuran models. *Ecology* 77:1855–66.

Tyler, J. D. 1983. Notes on burrowing owl *Athene cunicularia* food habits in Oklahoma USA. *Southwest. Nat.* 28:100–102.

Tyler, M. J. 1962. A record of parasitism of New Guinea frogs' eggs by dipterous larvae. *West. Austral. Nat.* 8:102–3.

———. 1963a. A taxonomic study of amphibians and reptiles of the central highlands of New Guinea, with notes on their ecology and biology. I. Anura: Microhylidae. *Trans. Roy. Soc. South Australia* 86:11–29.

———. 1963b. A taxonomic study of amphibians and reptiles of the central highlands of New Guinea, with notes on their ecology and biology. II. Anura: Ranidae and Hylidae. *Trans. Roy. Soc. South Australia* 86:105–30.

———. 1967. Microhylid frogs of New Britain. *Trans. Roy. Soc. South Australia* 91:187–90.

———. 1971. The phylogenetic significance of vocal sac structure in hylid frogs. *Univ. Kansas Publ. Mus. Nat. Hist.* 19:319–60.

———. 1972. Superficial mandibular musculature, vocal sacs and the phylogeny of Australo-Papuan leptodactylid frogs. *Rec. South Australian Mus.* 16 (9):1–20.

———. 1974. Superficial mandibular musculature and vocal sac structure of the Mexican burrowing toad, *Rhinophrynus dorsalis. Herpetologica* 30:313–16.

———. 1976. *Frogs.* Sydney, Australia: Collins.

———. 1980. The evolutionary significance of vocal sac of the European discoglossid frog *Bombina bombina* (L.) (Amphibia, Salientia). *Acta Biol. Cracov. (Zool.)* 22:99–104.

———. 1983. *The gastric brooding frog,* ed. London: Croom Helm.

———. 1985a. Phylogenetic significance of the superficial mandibular musculature and vocal sac structure of sooglossid frogs. *Herpetologica* 41:173–76.

———. 1985b. Reproductive modes in Australian Amphibia, pp. 265–267. In: G. Grigg, R. Shine, and H. Ehmann, eds. *Biology of Australasian Frogs and Reptiles.* Surrey Beatty and Sons, Chipping Norton, New South Wales, Australia.

———. 1998. *Australian frogs. A natural history.* Ithaca, NY: Cornell University Press.

———. 1999. Distribution patterns of amphibians in the Australo-Papuan Region. In *Patterns of distribution of amphibians. A global perspective,* ed. W. E. Duellman, 541–63. Baltimore: Johns Hopkins University Press.

Tyler, M. J., and D. B. Carter. 1981. Oral birth of the young of the gastric brooding frog *Rheobatrachus silus. Anim. Behav.* 29:280–82.

Tyler, M. J., and M. Davies. 1979. Foam nest construction by Australian leptodactylid frogs (Amphibia, Anura, Leptodactylidae). *J. Herpetol.* 13:509–10.

———. 1993. Family Hylidae. In *Fauna of Australia. Vol. 2A. Amphibia and reptilia,* ed.C. J. Glasby, G. J. B. Ross, and P. L. Beesley, 58–63. Canberra: Australian Government Publishing Service.

Tyler, M. J., D. J. Shearman, R. Franco, P. O'Brien, R. Seamark, and R. Kelly. 1983. Inhibition of gastric acid secretion in the gastric brooding frog, *Rheobatrachus silus. Science* 220:609–10.

Tyler, M. J., G. F. Watson, and A. A. Martin. 1981. The Amphibia: Diversity and distribution. In *Ecological biogeography of Australia,* ed. A. Keast, 1277–1301. The Hague: Dr. W. Junk.

Tyler, T., W. J. Liss, L. M. Ganio, G. L. Larson, R. Hoffman, E. Deimling, and G. Lomnicky. 1998. Interactions between introduced trout and larval salamanders (*Ambystoma macrodactylum*) in high-elevation lakes. *Cons. Biol.* 12:94–105.

Tyler, T. J., W. J. Liss, R. L. Hoffman, and L. M. Ganio. 1998. Experimental analysis of trout effects on survival, growth, and habitat use of two species of ambystomatid salamanders. *J. Herpetol.* 32:345–49.

Tymowska, J. 1991. Polyploidy and cytogenetic variation in frogs of the genus *Xenopus*. In *Amphibian cytogenetics and evolution,* ed. D. M. Green and S. K. Sessions, 259–97. San Diego: Academic Press.

Ueck, M. 1967. Der Manicotto glandulare ("Drusenmagen") der Anurenlarve in Bau, Funktion und Beziehung zur Gesamtlänge des Darmes. Eine mikroskopish-anatomische, histochemische und elektronenoptische Studie an der omnivoren und mikrophagen Larve von *Xenopus laevis* und der carnivoren und makrophagen Larve von *Hymenochirus boettgeri* (Anura, Pipidae). *Z. Wiss. Zool.* 176:173–270.

Ueda, H. 1986. Reproduction in *Chirixalus eiffingeri* (Boettger). *Sci. Rep. Lab. Amphibian Biol., Hiroshima Univ.* 8:109–16.

Uhler, F. M., C. Cottam, and T. E. Clarke. 1939. Food of snakes of the George Washington National Forest, Virginia. *Trans. N. Amer. Wildl. Conf.* 4:605–22.

Ultsch, G. R. 1971. The relationship of dissolved carbon dioxide and oxygen to microhabitat selection in *Pseudobranchus striatus*. *Copeia* 1971:247–52.

———. 1973a. The effects of water hyacinths (*Eichhornia crassipes*) on the microenvironment of aquatic communities. *Arch. Hydrobiol.* 72:460–73.

———. 1973b. A theoretical and experimental investigation of the relationships between metabolic rate, body size, and oxygen exchange capacity. *Respir. Physiol.* 18:143–60.

———. 1974. Gas exchange and metabolism in the Sirenidae (Amphibia: Caudata). I. Oxygen consumption of submerged sirenids as a function of body size and respiratory surface area. *Comp. Biochem. Physiol.* 47A:485–98.

———. 1976. Eco-physiological studies of some metabolic and respiratory adaptations of sirenid salamanders. In *Respiration of amphibious vertebrates,* ed. G. M. Hughes, 287–312. New York: Academic Press.

———. 1989. Ecology and physiology of hibernation and overwintering among freshwater fishes, turtles, and snakes. *Biol. Rev.* 64:435–516.

Ultsch, G. R., and S. J. Arceneaux. 1988. Gill loss in larval *Amphiuma tridactylum*. *J. Herpetol.* 22:347–48.

Ultsch, G. R., D. F. Bradford, and J. Freda. 1999. Physiology. Coping with the Environment. In *Tadpoles. The biology of anuran larvae,* ed. R. W. McDiarmid and R. A. Altig, 189–214. Chicago: University of Chicago Press.

Ultsch, G. R., and J. T. Duke. 1990. Gas exchange and habitat selection in the aquatic salamanders *Necturus maculosus* and *Cryptobranchus alleganiensis*. *Oecologia* 83:250–58.

Ultsch, G. R., T. E. Graham, and C. E. Crocker. 2000. An aggregation of overwintering leopard frogs, *Rana pipiens,* and common map turtles, *Graptemys geographica,* in northern Vermont. *Can. Field-Nat.* 114:314–15.

Ultsch, G. R., S. A. Reese, and E. R. Stewart. 2004. Physiology of hibernation in *Rana pipiens:* Metabolic rate, critical oxygen tension, and the effects of hypoxia on several plasma variables. *J. Exp. Zool.* 301A:169–76.

Urano, A., and A. Gorbman. 1981. Effects of pituitary hormonal treatment on responsiveness of anterior preoptic neurons in male leopard frogs, *Rana pipiens*. *J. Comp. Physiol.* 141:163–71.

Ussing, H. H. 1949. The active ion transport through the isolated frog skin. *Acta Physiol. Scand.* 17:1–37.

Utsunomiya, Y., and T. Utsunomiya. 1977. On the development of *Tylototriton andersoni*. *J. Fac. Fish. Anim. Husb., Hiroshima Univ.* 16:65–76 (in Japanese, with English summary).

———. 1983. On the eggs and larvae of frog species occurring in the Ryukyu Archipelago. *J. Fac. App. Biol. Sci. Hiroshima Univ.* 22:255–70 (in Japanese, with English summary).

Utsunomiya, Y., T. Utsunomiya, and S. Katsuren. 1979. Some ecological observations of *Rana ishikawae,* a rare frog endemic to the Ryukyu Islands. *Proc. Japan Acad.* 55:233–37.

Utsunomiya, Y., T. Utsunomiya, S. Katsuren, and M. Toyama. 1983. Habitat segregation observed in the breeding of five frog species dwelling in a mountain stream of Okinawa Island. *Ann. Zool. Jap.* 56:149–53.

Utsunomiya, Y., T. Utsunomiya, and S. Kawachi. 1978. Some ecological observations of *Tylototriton andersoni,* a terrestrial salamander occurring in the Tokunoshima Island. *Proc. Japan Acad.* 54 B:341–46.

Uttendörfer, O. 1939. *Die Ernährung der Deutschen Raubvogel und Eulen und ihr Bedeutung in der Heimischen Natur.* Berlin:J. Neumann-Neudamm.

Uwa, H., Y. Watanabe, Y. Sakakura, S. Asai, and H. Shiono. 1981. Estimation of the spring emergence of the toad *Bufo bufo* using underground temperature. *Zool. Mag. (Tokyo)* 90:157–63.

Uzendoski, K. and P. Verrell. 1993. Sexual incompatibility and mate-recognition systems: a study of two species of sympatric salamanders (Plethodontidae). *Anim. Behav.* 46:267–78.

Uzzell, T. M., Jr. 1964. Relations of the diploid and triploid species of the *Ambystoma jeffersonianum* complex (Amphibia, Caudata). *Copeia* 1964:257–300.

———. 1969. Notes on spermatophore production by salamanders of the *Ambystoma jeffersonianum* complex. *Copeia* 1969:602–12.

Vaira, M. 1997. *Leptodactylus bolivianus* (NCN). Behavior. *Herp. Rev.* 28:200.

———. 2001. Breeding biology of the leaf frog, *Phyllomedusa boliviana* (Anura, Hylidae). *Amphibia-Reptilia* 22:421–29.

———. 2005. Annual variation of breeding patterns of the toad, *Melanophyniscus rubiventris* (Vellard, 1947). *Amphibia-Reptilia* 26:193–99.

Valdez, C. M., and K. C. Nishikawa. 1997. Sensory modulation and behavioral choice during feeding in the Australian frog, *Cyclorana novaehollandiae*. *J. Comp. Physiol. A* 180:187–202.

Valdivieso, D., and J. R. Tamsitt. 1974. Thermal relations of the Neotropical frog *Hyla labialis* (Anura: Hylidae). *Roy. Ontario Mus. Life Sci. Occ. Pap.* 26:1–10.

Valentine, B. D. 1963. Notes on the early life history of the Alabama salamander, *Desmognathus aeneus chermocki* Bishop and Valentine. *Am. Midl. Nat.* 69:182–88.

Valentine, B. D., and D. M. Dennis. 1964. A comparison of the gill-arch system and fins of three genera of larval salamanders, *Rhyacotriton, Gyrinophilus* and *Ambystoma*. *Copeia* 1964:196–201.

Valentine, J. M., S. R. Walther, K. M. M. Cartney, and H. H. Ivy. 1972. Alligator diets on the Sabine National Wildlife Refuge, Louisiana. *J. Wildl. Manage.* 36:809–15.

Valett, B. B., and D. L. Jameson. 1961. The embryology of *Eleutherodactylus augusti latrans*. *Copeia* 1961:103–9.

Vallan, D. 2000. Influence of forest fragmentation on amphibian diversity in the nature reserve of Ambohitantely, highland Madagascar. *Biol. Cons.* 96:31–43.

———. 2002. Effects of anthropogenic environmental changes on amphibian diversity in the rain forest of eastern Madagascar. *J. Trop. Ecol.* 18:725–42.

Van Bergeijk, W. A. 1959. Hydrostatic balancing mechanism of *Xenopus larvae*. *J. Acoust. Soc. Amer.* 31:1340–47.

Van Berkum, F., F. H. Pough, M. M. Stewart, and B. F. Brussard. 1982. Altitudinal and interspecific differences in the rehydration abilities of Puerto Rican frogs (*Eleutherodactylus*). *Physiol. Zool.* 55:130–36.

Van Beurden, E. K. 1980. Energy metabolism of dormant Australian water-holding frogs (*Cyclorana platycephalus*). *Copeia* 1980:787–99.

———. 1984. Survival strategies of the Australian water-holding frog, *Cyclorana platycephalus*. In *Arid Australia,* ed. H. G. Cogger and E. E. Cameron, 223–34. Sydney: Australian Museum.

Van Beurden, E. K., and G. C. Grigg. 1980. An isolated and expanding population of the introduced toad *Bufo marinus* in New South Wales. *Austr. Wildl. Res.* 7:305–10.

Van Buskirk, J. 1988. Interactive effects of dragonfly predation in experimental pond communities. *Ecology* 69:857–67.

———. 2000. The costs of an inducible defense in anuran larvae. *Ecology* 81:2813–21.

———. 2001. Specific induced responses to different predator species in anuran larvae. *J. Evol. Biol.* 14:482–89.

———. 2002a. A comparative test of the adaptive plasticity hypothesis: Relationships between habitat and phenotype in anuran larvae. *Am. Nat.* 160:87–102.

———. 2002b. Phenotypic lability and the evolution of predator-induced plasticity in tadpoles. *Evolution* 56:361–70.

Van Buskirk, J., P. Anderwald, S. Lüpold, L. Reinhardt, and H. Schuler. 2003. The lure effect, tadpole tail shape, and the target of dragonfly strikes. *J. Herpetol.* 37:420–24.

Van Buskirk, J., and M. Arioli. 2002. Dosage response of an induced defense: How sensitive are tadpoles to predation risk? *Ecology* 83:1580–85.

———. 2005. Habitat specializtion and adaptive phenotypic divergence of anuran populations. *J. Evol. Biol.* 18:596–608.

Van Buskirk, J., J. Aschwanden, I. Buckelmüller, S. Reolon, and S. Rüttiman. 2004. Bold tail coloration protects tadpoles from dragonfly strikes. *Copeia* 2004:599–602.

Van Buskirk, J., and S. A. McCollum. 1999. Plasticity and selection explain variation in tadpole phenotype between ponds with different predator composition. *Oikos* 85:31–39.

———. 2000a. Functional mechanisms of an inducible defence in tadpoles: Morphology and behaviour influence risk from predation. *J. Evol. Biol.* 13:336–47.

———. 2000b. Influence of tail shape on tadpole swimming performance. *J. Exp. Biol.* 203:2149–58.

Van Buskirk, J., C. Müller, A. Portmann, and M. Surbeck. 2002. A test of the risk allocation hypothesis: Tadpole responses to temporal change in predation risk. *Behav. Ecol.* 13:526–30.

Van Buskirk, J., and R. A. Relyea. 1998. Selection for phenotypic plasticity in *Rana sylvatica* tadpoles. *Biol. J. Linn. Soc.* 65:301–28.

Van Buskirk, J., and G. Saxer. 2001. Delayed costs of an induced defense in tadpoles? Morphology, hopping, and development rate at metamorphosis. *Evolution* 55:821–29.

Van Buskirk, J., and B. R. Schmidt. 2000. Predator-induced phenotypic plasticity in larval newts: Trade-offs, selection, and variation in nature. *Ecology* 81:3009–28.

Van Buskirk, J., and D. C. Smith. 1991. Density-dependent population regulation in a salamander. *Ecology* 72:1747–56.

Van Buskirk, J., and K. L. Yurewicz. 1998. Effects of predators on prey growth rate: Relative contributions of thinning and reduced activity. *Oikos* 82:20–28.

Van de Mortel, T., W. Buttemer, P. Hoffman, J. Hays, and A. Blaustein. 1998. A comparison of photolyase activity in three Australian tree frogs. *Oecologia* 1998:366–69.

Van der Meijden, A., M. Vences, S. Hoegg, and A. Meyer. 2005. A previously unrecognized radiation of ranid frogs in Southern Africa revealed by nuclear and mitochondrial DNA sequences. *Mol. Phylogen. Evol.* 37:674–85.

Van Dijk, D. E. 1971. Anuran ecology in relation particularly to oviposition and development out of water. *Zool. Afr.* 6:119–32.

———. 1972. The behaviour of southern African anuran tadpoles with particular reference to their ecology and related external morphological features. *Zool. Afr.* 7:49–55.

van Gelder, J. J. 1973. Ecological observations on Amphibia in the Netherlands. II. *Triturus helveticus* Razoumowski: migration, hibernation and neoteny. *Neth. J. Zool.* 23:86–108.

van Gelder, J. J., H. M. J. Aarts, and H.-J. W. M. Staal. 1986. Routes and speed of migrating toads (*Bufo bufo* L.): a telemetric study. *Herpetol. J.* 1:111–14.

van Gelder, J. J., and H. C. M. Hoedemaekers. 1971. Sound activity and migration during the breeding period of *Rana temporaria* L., *R. arvalis* Nilsson, *Pelobates fuscus* Laur. and *Rana esculenta* L. *J. Anim. Ecol.* 40:559–68.

van Gelder, J. J., J. V. J. Olders, J. W. G. Bosch, and P. W. Starmans. 1986. Behaviour and body temperature of hibernating common toads, *Bufo bufo*. *Holarc. Ecol.* 9:225–28.

van Gelder, J. J., and H. Strijbosch. 1996. Marking amphibians: effects of toe clipping on *Bufo bufo* (Anura: Bufonidae). *Amphibia-Reptilia* 17:169–74.

Vanhoye, D., F. Bruston, P. Nicolas, and M. Amiche. 2003. Antimicrobial peptides from hylid and ranin frogs originated from a 150–million-year-old ancestral precursor with a conserved signal peptide but a hypermutable antimicrobial domain. *Eur. J. Biochem.* 270:2068–81.

Van Hyning, O. C. 1932. Food of some Florida snakes. *Copeia* 1932:37.

Van Meeuwen, H. M. 1977. De Trinidadse beekkikker *Colostethus trinitatus*. *Lacerta* 36:3–11.

van Oordt, G. J., and P. G. W. J. van Oordt. 1955. The regulation of spermatogenesis in the frog. *Mem. Soc. Endocrin.* 4:25–38.

van Oordt, P. G. W. J. 1960. The influence of internal and external factors in the regulation of the spermatogenetic cycle in Amphibia. *Symp. Zool. Soc. London* 2: 29–52.

Van Sluys, M., C. F. D. Rocha, and M. B. Souza. 2001. Diet, reproduction, and density of the leptodactylid litter frog *Zachaenus parvulus* in an Atlantic rain forest of southeastern Brazil. *J. Herpetol.* 35:322–25.

van Wijngaarden, R., and F. Bolanos. 1992. Parental care in *Dendrobates granuliferus* (Anura: Dendrobatidae), with a description of the tadpole. *J. Herpetol.* 26:102–5.

van Wijngaarden, R., and S. van Gool. 1994. Site fidelity and territoriality in the dendrobatid frog *Dendrobates granuliferus*. *Amphibia-Reptilia* 15:171–81.

van Winkle, W. 1975. Comparison of several probabilistic home-range models. *J. Wildl. Manage.* 39:118–23.

Vandel, A., and M. Bouillon. 1959. La reproduction du protee (*Proteus anguineus*). *C. R. Acad. Sci. Paris* 248:1267–72.

Vasconcelos, D., and A. J. K. Calhoun. 2004. Movement patterns of adult and juvenile *Rana sylvatica* (LeConte) and *Ambystoma maculatum* (Shaw) in three restored seasonal pools in Maine. *J. Herpetol.* 38:551–61.

Vasudevan, K., A. Kumar, and R. Chellam. 2001. Structure and composition of rainforest floor amphibian communities in Kalakad-Mundanthurai Tiger Reserve. *Curr. Sci.* 80:406–12.

Vasvari, N. 1927/28. Beitrage zur Ernahrungsokologie von *Botaurus stellaris* L. und *Ardetta minuta* L. *Aquila* 34/35:342–74.

———. 1929/30. Studien über die Ernährung des Purpureihers (*Ardea purpurea* L.). *Aquila* 36/37:231–93.

———. 1935/38. Die Ernährungsökologie des Nacht- und Rallenreihers. *Aquila* 42/45:592–613.

Vaughan, A. 2003. *Rana vaillanti* (Web-footed frog). Diet. *Herp. Rev.* 34:238.

Vaz-Ferreira, R., and A. Gehrau. 1974. Proteccion de la prole en leptodactylidos. *Rev. Biol. Uruguay* 2:59–62.

———. 1975. Comportamiento epimeletico de la rana comun, *Leptodactylus ocellatus* (L.) (Amphibia, Leptodactylidae). I. Atencion de la cria y actividades alimentarias y agresivas relacionadas. *Physis (Buenos Aires) sec. B* 34:1–14.

Vaz-Silva, W., and J. G. da Frota. 2004. *Bufo marinus* (Marine Toad). Defensive behavior. *Herp. Rev.* 35:371.

Vaz-Silva, W., H. L. R. Silva, and N. J. da Silva, Jr. 2003a. *Bufo ocellatus* (Ocellated Toad). Diet. *Herp. Rev.* 34:356–57.

———. 2003b. *Bufo paracnemis* (Cururu Toad). Diet. *Herp Rev.* 34:357.

———. 2003c. *Leptodactylus labyrinthicus* (Labyrinth Frog). Diet. *Herp. Rev.* 34:359.

Vehrencamp, S. L., J. W. Bradbury, and R. M. Gibson. 1989. The energetic cost of display in male sage grouse. *Anim. Behav.* 38:885–96.

Veith, M. 1993. Zur Koexistenz von Larven des Feuersalamanders und dekapoden Krebsen—Beobactungen aus der Matra (Ungarn). *Salamandra* 29:82–85.

Veith, M., J. Kosuch, R. Feldmann, H. Martens, and A. Seitz. 2000. A test for correct species declaration of frog legs imports from Indonesia into the European Union. *Biodiver. Cons.* 9:333–41.

Veith, M., J. Kosuch, and M. Vences. 2003. Climatic oscillations triggered post-Messinian speciation of Western Palearctic brown frogs (Amphibia, Ranidae). *Mol. Phylogen. Evol.* 26:310–27.

Veith, M., S. Steinfartz, R. Zardoya, A. Seitz, and A. Meyer. 1998. A molecular phylogeny of "true" salamanders (family Salamandridae) and the evolution of terrestriality of reproductive modes. *J. Zool. Syst. Evol. Res.* 36:7–16.

Veloso, A. 1977. Aggressive behavior and the generic relationships of *Caudiverbera caudiverbera* (Amphibia: Leptodactylidae). *Herpetologica* 33:434–42.

Vences, M., F. Andreone, F. Glaw, J. Kosuch, A. Meyer, H.-C. Schaefer, and M. Veith. 2002. Exploring the potential of life-history key innovation: Brook breeding in the radiation of the Malagasy treefrog genus *Boophis*. *Mol. Ecol.* 11:1453–63.

Vences, M., and I. de la Riva. 2005. *Mantidactylus majori* (NCN). Male egg-guarding. *Herp. Rev.* 36:435–36.

Vences, M., P. Galán, A. Palanca, D. R. Vietes, S. Nieto, and J. Rey. 2000. Summer microhabitat use and diel activity cycles in a high altitude Pyrenean population of *Rana temporaria*. *Herpetol. J.* 10:49–56.

Vences, M., P. Galán, D. R. Vieites, M. Puente, K. Oetter, and S. Wanke. 2002. Field body temperatures and heating rates in a montane frog population: The importance of black dorsal pattern for thermoregulation. *Ann. Zool. Fennici* 39:209–20.

Vences, M., and F. Glaw. 1991. Revision der Gattung *Stumpffia* Boettger 1881 aus Madagaskar, mit Beschreibung von zwei neue Arten. *Acta Biol. Benrodis* 3:203–19.

———. 2001. When molecules claim for taxonomic change: New proposals on the classification of Old World treefrogs. *Spixiana* 24:85–92.

Vences, M., F. Glaw, F. Andreone, R. Jesu, and G. Schimmenti. 2002. Systematic revision of the enigmatic Malagasy broad-headed frogs (*Laurentomantis* Dubois, 1980), and their phylogenetic position within the endemic mantellid radiation of Madagascar. *Contr. Zool.* 70:191–212.

Vences, M., F. Glaw, and W. Böhme. 1997/98. Evolutionary correlates of microphagy in alkaloid-containing frogs (Amphibia: Anura). *Zool. Anz.* 236:217–30.

Vences, M., K. Grossenbacher, M. Puente, A. Palanca, and D. R. Vieites. 2003. The Cambalès fairy tale: Elevational limits of *Rana temporaria* (Amphibia: Ranidae) and other European amphibians revisited. *Folia Zool.* 52:189–202.

Vences, M., J. Kosuch, S. Lötters, A. Widmer, K.-H. Jungfer, J. Köhler, and M. Veith. 2000. Phylogeny and classification of poison frogs (Amphibia: Dendrobatidae), based on mitochondrial 16S and 12S ribosomal RNA gene sequences. *Mol. Phylogen. Evol.* 15:34–40.

Vences, M., J. Kosuch, R. Boistel, C. F. B. Haddad, E. La Marca, S. Lötters, and M. Veith. 2003. Convergent evolution of aposematic coloration in Neotropical poison frogs: a molecular phylogenetic perspective. *Org. Divers. Evol.* 3:215–26.

Vences, M., C. J. Raxworthy, R. A. Nussbaum, and F. Glaw. 2003. New microhylid frog (*Plethodontohyla*) from Madagascar, with semiarboreal habits and possible parental care. *J. Herpetol.* 37:629–36.

Vences, M., D. R. Vieites, F. Glaw, H. Brinkmann, J. Kosuch, M. Veith, and A. Meyer. 2003. Mutliple overseas dispersal in amphibians. *Proc. Roy. Soc. London B* 270:2435–42.

Verdade, V. K., L. C. Schiesari, and J. A. Bertoluci. 2000. Diet of juvenile aquatic caecilians, *Typhlonectes compressicauda*. *J. Herpetol.* 34:291–93.

Vernberg, F. J. 1952. The oxygen consumption of two species of salamanders at different seasons of the year. *Physiol. Zool.* 25:243–49.

———. 1953. Hibernation studies of two species of salamanders, *Plethodon cinereus cinereus* and *Eurycea bislineata bislineata*. *Ecology* 34:55–62.

———. 1955. Correlation of physiological and behavioral indexes of activity in the study of *Plethodon cinereus* (Green) and *Plethodon glutinosus* (Green). *Am. Midl. Nat.* 54:382–93.

Verrell, P. 1999. Bracketing the extremes: courtship behaviour of the smallest- and largest-bodied species in the salamander genus *Desmognathus* (Plethodontidae: Desmognathinae). *J. Zool., London* 247:105–11.

———. 2003. Population and species divergence of chemical cues that influence male recognition of females in desmognathine salamanders. *Ethology* 109:577–86.

Verrell, P., and A. Donovan. 1991. Male-male aggression in the plethodontid salamander *Desmognathus ochrophaeus*. *J. Zool., London* 223:203–12.

Verrell, P., and M. Mabry. 2000. The courtship of plethodontid salamanders: Form, function, and phylogeny. In *The biology of plethodontid salamanders,* ed. R. C. Bruce, R. G. Jaeger, and L. D. Houck, 371–80. New York: Kluwer Academic/Plenum.

———. 2003. Sexual behaviour of the Black Mountain dusky salamander (*Desmognathus welteri*), and the evolutionary history of courtship in the Desmognathinae. *J. Zool., London* 260:367–76.

Verrell, P., and J. Pelton. 1996. The sexual strategy of the central long-toed salamander, *Ambystoma macrodactylum columbianum,* in south-eastern Washington. *J. Zool., London* 240:37–50.

Verrell, P., N. Strand, and E. Hanson. 2001. Sexual dimorphism of mate location in the long-toed salamander *Ambystoma macrodactylum columbianum. Ethology* 107:677–83.

Verrell, P. A. 1982. The sexual behaviour of the red-spotted newt, *Notophthalmus viridescens* (Amphibia: Urodela: Salamandridae). *Anim. Behav.* 30:1224–36.

———. 1983. The influence of the ambient sex ratio and intermale competition on the sexual behavior of the red-spotted newt, *Notophthalmus viridescens* (Amphibia: Urodela: Salamandridae). *Behav. Ecol. Sociobiol.* 13:307–13.

———. 1984a. Responses to different densities of males in the smooth newt, *Triturus vulgaris:* "One at a time, please." *J. Herpetol.* 18:482–84.

———. 1984b. Sexual interference and sexual defense in the smooth newt, *Triturus vulgaris* (Amphibia, Urodela, Salamandridae). *Z. Tierpsychol.* 66:242–54.

———. 1985a. Feeding in adult smooth newts (*Triturus vulgaris*), with particular emphasis on prey changes in the aquatic phase. *Amphibia-Reptilia* 6:133–36.

———. 1985b. Female availability and multiple courtship in male red-spotted newts, *Notophthalmus viridescens* (Amphibia): Decisions that maximize male mating success. *Behaviour* 94:244–53.

———. 1986a. Limited mating capacity in the smooth newt, *Triturus vulgaris vulgaris* (Amphibia). *J. Comp. Psychol.* 100:291–95.

———. 1986b. Male discrimination of larger, more fecund females in the smooth newt, *Triturus vulgaris. J. Herpetol.* 20:416–22.

———. 1986c. Wrestling in the red-spotted newt (*Notophthalmus viridescens*): Resource value and contestant asymmetry determine contest duration and outcome. *Anim. Behav.* 34:398–402.

———. 1988a. Intrinsic male mating capacity in limited in the plethodontid salamander, *Desmognathus ochrophaeus. J. Herpetol.* 22:394–400.

———. 1988b. Mating and female sexual responsiveness in the salamander *Desmognathus ochrophaeus. Herpetologica* 44:334–37.

———. 1998c. Sexual interference in the alpine newt, *Triturus alpestris* (Amphibia, Urodela, Salamandridae). *Zool. Sci.* 5:159–64.

———. 1989a. An experimental study of the behavioral basis of sexual isolation between two sympatric plethodontid salamanders, *Desmognathus imitator* and *D. ochrophaeus. Ethology* 80:274–82.

———. 1989b. Male mate choice for fecund females in a plethodontid salamander. *Anim. Behav.* 38:1086–87.

———. 1989c. The sexual strategies of natural populations of newts and salamanders. *Herpetologica* 45:265–82.

———. 1990a. A note on the courtship of the broken-striped newt, *Notophthalmus dorsalis* (Harlan). *J. Herpetol.* 24:215–17.

———. 1990b. Sexual compatibility among plethodontid salamanders: Tests between *Desmognathus apalachicolae,* and *D. ochrophaeus* and *D. fuscus. Herpetologica* 46:415–22.

———. 1991a. Insemination temporarily inhibits sexual responsiveness in female salamanders (*Desmognathus ochrophaeus*). *Behaviour* 119:51–64.

———. 1991b. Male mating success in the mountain dusky salamander, *Desmognathus ochrophaeus:* Are small, young, inexperienced males at a disadvantage? *Ethology* 88:277–86.

———. 1994a. The courtship behaviour of the Apalachicola dusky salamander, *Desmognathus apalachicolae* (Amphibia: Caudata: Plethodontidae). *Ethol. Ecol. Evol.* 6:497–506.

———. 1994b. Courtship of the salamander *Desmognathus imitator* (Amphibia: Caudata: Plethodontidae). *Amphibia-Reptilia* 15:135–42.

———. 1994c. Evidence against a role for experience in the maintenance of sexual incompatibility between sympatric salamanders. *Herpetologica* 50:475–79.

———. 1994d. Males may choose larger females as mates in the salamander *Desmognathus fuscus. Anim. Behav.* 47:1465–67.

———. 1995. The courtship behaviour of the spotted dusky salamander, *Desmognathus fuscus conanti* (Amphibia: Caudata: Plethodontidae). *J. Zool., London* 235:515–23.

———. 1997. Courtship in desmognathine salamanders: The southern dusky salamander, *Desmognathus auriculatus. J. Herpetol.* 31:273–77.

Verrell, P. A., and L. E. Brown. 1993. Competition among females for mates in a species with male parental care the midwife toad *Alytes obstetricans. Ethology* 93:247–57.

Verrell, P. A. and T. R. Halliday. 1985a. Autumnal migration and aquatic overwintering in the common frog, *Rana temporaria. Brit. J. Herpetol.* 6:433–34.

———. 1985b. Reproductive dynamics of a population of smooth newts, *Triturus vulgaris,* in southern England. *Herpetologica* 41:386–95.

Verrell, P. A., T. R. Halliday, and M. L. Griffiths. 1986. The annual reproductive cycle of the smooth newt (*Triturus vulgaris*) in England. *J. Zool., London* (A) 210:101–19.

Verrell, P. A., and J. D. Krenz. 1998. Competition for mates in the mole salamander, *Ambystoma talpoideum:* Tactics that may maximize male mating success. *Behaviour* 135:121–38.

Verrell, P. A., and N. McCabe. 1986. Mating balls in the common toad, *Bufo bufo. British Herp. Soc. Bull.* 16:28–29.

———. 1988. Field observations of the sexual behaviour of the smooth newt, *Triturus vulgaris vulgaris* (Amphibia: Salamandridae). *J. Zool., London* 214:533–45.

Vertucci, F. A., and P. S. Corn. 1996. Evaluation of episodic acidifications and amphibian declines in the Rocky Mountains. *Ecol. Appl.* 6:449–57.

Vesely, D. G., and W. C. McComb. 2002. Salamander abundance and amphibian species richness in riparian buffer strips in the Oregon Coast Range. *Forest Sci.* 48:291–97.

Vestjens, W. J. M., and L. S. Hall. 1977. Stomach contents of forty-two species of bats from the Australasian region. *Austr. Wildl. Res.* 4:25–35.

Vial, J. L. 1968. The ecology of the tropical salamander *Bolitoglossa subpalmata* in Costa Rica. *Rev. Biol. Trop.* 15:117–21.

Vial, J. L., and F. B. Preib. 1966. Antibiotic assay of dermal secretions from the salamander *Plethodon cinereus. Trans. Missouri Acad. Sci.* 1:37–40.

Viborg, A. L., and P. Rosenkilde. 2001. Angiotensin II elicits water seeking behavior and the water absorption response in the toad *Bufo bufo. Horm. Behav.* 39:225–31.

———. 2004. Water potential receptors in the skin regulate blood perfusion in the ventral pelvic patch of toads. *Physiol. Biochem. Zool.* 77:39–49.

Vieites, D. R., S. Nieto-Roman, M. Barluenga, A. Palanca, M. Vences, and A. Meyer. 2004. Post-mating clutch piracy in an amphibian. *Nature* 431:305–8.

Vielliard, J. M. E., and A. J. Cardoso. 1996. Adaptações de sinais sonoros de anfibios e aves a ambientes de riachos com corredeiras. In *Herpetologia neotropical: Actas del II congreso Latinamericano de herpetología,* Vol. II, ed. J. E. Pefaur, 97–119. Merida, Venezuela: Universidad de los Andes.

Viertel, B. 1980. Überlebensraten und Mortalität bei Erdkrötenlarven (*Bufo bufo* L.) im Freiland (Amphibia: Salientia: Bufonidae). *Salamandra* 16:19–37.

———. 1982. The oral cavities of central European anuran larvae (Amphibia). Morphology, ontogenesis and generic diagnosis. *Amphibia-Reptilia* 4:327–60.

———. 1984a. Filtration, eine Strategie der Nährungsaufnahme der Larven von *Xenopus laevis, Rana temporaria,* und *Bufo calamita* (Amphibia, Anura). *Verh. Ges. Ökol.* 12:563–73.

———. 1984b. Suspension feeding of the larvae of *Balaeophryne muletensis* Sanchiz and Adrover 1977 (Amphibia: Anura: Discoglossidae). In *História biológica del ferreret,* ed. H. Hemmer and J. A. Alcover, 153–61. Palma de Mallorca, Spain: Editorial Moll.

———. 1985. The filter apparatus of *Rana temporaria* and *Bufo bufo* larvae (Amphibia, Anura). *Zoomorphology* 105:345–55.

———. 1987. The filter apparatus of *Xenopus laevis, Bombina variegata,* and *Bufo calamita* (Amphibia, Anura): A comparison of different larval types. *Zool. Jb. Anat.* 115:425–52.

———. 1990. Suspension feeding of anuran larvae at low concentrations of *Chlorella* algae (Amphibia, Anura). *Oecologia* 85:167–77.

———. 1991. The ontogeny of the filter apparatus of anuran larvae (Amphibia, Anura). *Zoomorphology* 110:239–66.

———. 1992. Functional response of suspension feeding anuran larvae to different particle sizes at low concentrations (Amphibia). *Hydrobiologia* 234:151–73.

———. 1999. Salt tolerance of *Rana temporaria:* spawning site selection and survival during embryonic development (Amphibia, Anura). *Amphibia-Reptilia* 20:161–71.

Viertel, B., and S. Richter. 1999. Anatomy. Viscera and endocrines. In *Tadpoles. The biology of anuran larvae,* ed. R. W. McDiarmid and R. A. Altig, 92–148. Chicago: University of Chicago Press.

Vigny, C. 1979. The mating calls of 12 species and sub-species of the genus *Xenopus* (Amphibia: Anura). *J. Zool., London* 188:103–22.

Villa, J. 1969. Comportamiento defensivo de la "Rana Ternero," *Leptodactylus pentadactylus. Rev. Biol. Trop.* 15:323–29.

———. 1977. A symbiotic relationship between frog and fly larvae. *J. Herpetol.* 11:317–22.

———. 1979a. Synopsis of the biology of the Middle American highland frog *Rana maculata* Brocchi. *Milwaukee Pub. Mus. Contr. Biol. Geol.* (21):1–16.

———. 1979b. Two fungi lethal to frog eggs in Central America. *Copeia* 1979:650–55.

———. 1980. "Frogflies" from Central and South America with notes on other organisms of the amphibian egg microhabitat. *Brenesia* 17:49–68.

———. 1984. Biology of a Neotropical glass frog, *Centrolenella fleischmanni* (Boettger), with special reference to its frogfly associates. *Milwaukee Pub. Mus. Contr. Biol. Geol.* (55):1–60.

Villa, J., R. W. McDiarmid, and J. M. Gallardo. 1982. Arthropod predators of leptodactylid frog foam nests. *Brenesia* 19/20:577–90.

Villa, J., and D. S. Townsend. 1983. Viable frog eggs eaten by phorid fly larvae. *J. Herpetol.* 17:278–81.

Villa, J., and C. E. Valerio. 1982. Red, white and brown: Preliminary observations on the color of the centrolenid tadpole (Amphibia: Anura: Centrolenidae). *Brenesia* 19/20:1–16.

Vilter, V. 1986. La reproduction de la salamandre noire (*Salamandra atra*). In *Traite de zoologie, vol. 14, fasc. IB. Amphibiens,* ed. P.-P. Grasse and M. Delsol, 487–95. Paris: Masson.

Vilter, V., and A. Lugand. 1959. Trophismé intra-uterin et croissance embryonnaire chez *Nectophrynoides occidentalis* Ang., crapaud totalement vivipare du Mont Nimba (Haute Guinee). *C. R. Soc. Biol.* 153:29–32.

Vilter, V., and A. Vilter. 1960. Sur la gestation de la salamandre noire des Alpes, la *Salamandra atra* Laur. *C. R. Soc. Biol.* (Paris) 154:290–94.

———. 1962. Role de l'oviducte das l'uniparite uterine de la salamandre noire des Alpes orientales (*Salamandra atra* Laur.). *C. R. Soc. Biol.* (Paris) 156:49–51.

———. 1964. Sur l'évolution des corps jaunes ovariens chez *Salamandra atra* Laur. Des Alpes vaudoises. *C. R. Soc. Biol.* (Paris) 158:457–61.

Vincent, A., I. Ahnesjo, A. Berglund, and G. Rosenqvist. 1992. Pipefishes and seahorses: Are they all sex role reversed? *Trends Ecol. Evol.* 7:237–41.

Vinegar, A., and V. H. Hutchison. 1965. Pulmonary and cutaneous gas exchange in the green frog, *Rana clamitans. Zoologica* 50:47–53.

Vines, T. H., S. C. Köhler, M. Thiel, I. Ghira, T. R. Sands, C. J. MacCallum, N. H. Barton, and B. Nürnberger. 2003. The maintenance of reproductive isolation in a mosaic hybrid zone between the fire-bellied toads *Bombinia bombina* and *B. variegata. Evolution* 57:1876–88.

Vinnedge, B., and P. Verrell. 1998. Variance in male mating success and female choice for persuasive courtship displays. *Anim. Behav.* 56:443–48.

Vinton, K. W. 1951. Observations on the life history of *Leptodactylus pentadactylus. Herpetologica* 7:73–75.

Viparina, S., and J. J. Just. 1975. The life period, growth, and differentiation of *Rana catesbeiana* larvae occurring in nature. *Copeia* 1975:103–9.

Virgos, E., M. Llorente, and Y. Cortes. 1999. Geographical variation in genet (*Genetta genetta* L.) diet: A literature review. *Mammal Rev.* 29:119–28.

Viro, P., and H. Mikkola. 1981. Food composition of the raccoon dog *Nyctereutes procyonoides* Gray, 1834 in Finland. *Z. Säugetierkunde* 46:20–26.

Visser, J. 1971. Hunting the eggs of the ghost frog. *African Wild Life* 25:22–24.

Vitalis, T. Z. 1990. Pulmonary and cutaneous oxygen uptake and oxygen consumption of isolated skin in the frog, *Rana pipiens. Respir. Physiol.* 81:91–100.

Vitanen, P. 1967. Hibernation and seasonal movements of the viper, *Vipera berus berus* (L.), in southern Finland. *Ann. Zool. Fennici* 4:472–546.

Vitt, L. J. 1983. Ecology of an anuran-eating guild of terrestrial tropical snakes. *Herpetologica* 39:52–66.

———. 1987. Communities. In *Snakes: Ecology and evolutionary biology,* ed. R. A. Seigel, J. T. Collins, and S. A. Novak, 335–65. New York: Macmillan.

———. 1992. Mimicry of millipedes and centipedes by elongate terrestrial vertebrates. *Nat. Geog. Res. Expl.* 8:76–95.

———. 1996. Ecological observations on the tropical colubrid snake *Leptodeira annulata. Herpetol. Nat. Hist.* 4:69–76.

Vitt, L. J., and J. P. Caldwell. 1994. Resource utilization and guild structure of small vertebrates in the Amazon forest leaf litter. *J. Zool., London* 234:463–76.

———. 2001. Effects of logging on reptiles and amphibians of tropical forests. In *The effects of logging on tropical forests,* ed. R. A. Fimbel, A. Grajal, and J. G. Robinson, 239–59. New York: Columbia University Press.

Vitt, L. J., J. P. Caldwell, H. M. Wilbur, and D. C. Smith. 1990. Amphibians as harbingers of decay. *BioScience* 40:418.

Vitt, L. J., and L. D. Vangilder. 1983. Ecology of a snake community in northeastern Brazil. *Amphibia-Reptilia* 4:273–96.

Vlaming, M. S. M. G., A. M. H. J. Aertsen, and W. J. M. Epping. 1984. Directional hearing in the grassfrog (*Rana temporaria* L.). I. Mechanical vibrations of tympanic membrane. *Hearing Res.* 14:191–201.

Vleck, C. M., and D. F. Hoyt. 1991. Metabolism and energetics of reptilian and avian embryos, pp. 285–306. In *Egg incubation: Its effects on embryonic development in birds and reptiles,* ed. D. C. Deeming and M. J. W. Ferguson, eds. New York: Cambridge University Press.

Vogt, R. C., and S. G. Guzman. 1988. Food partitioning in a Neotropical freshwater turtle community. *Copeia* 1988:37–47.

Voituron, Y., M. Eugene, and H. Barre. 2003. Survival and metabolic responses to freezing by the water frog (*Rana ridibunda*). *J. Exp. Zool.* 299A:118–26.

Voituron, Y., J. M. Storey, C. Grenot, and K. B. Storey. 2002. Freezing survival, body ice content and blood composition of the freeze-tolerant European common lizard, *Lacerta vivpara. J. Comp. Physiol. B* 172:71–76.

Volpe, E. P. 1953. Embryonic temperature adaptations and relationships in toads. *Physiol. Zool.* 26:344–59.

———. 1957. Embryonic temperature tolerance and rate of development in *Bufo valliceps. Physiol. Zool.* 30:164–76.

———. 1960. Evolutionary consequences of hybrid sterility and vigor in toads. *Evolution* 14:181–93.

———. 1961. Polymorphism in anuran populations. In *Vertebrate speciation,* ed. W. F. Blair, 221–34. Austin: University of Texas Press.

Vonesh, J. R. 2000. Dipteran predation on the arboreal eggs of four *Hyperolius* frog species in western Uganda. *Copeia* 2000:560–66.

———. 2005a. Egg predation and predator-induced hatching plasticity in the African reed frog, *Hyperolius spinigularis. Oikos* 110:241–52.

———. 2005b. Sequential predator effects across three life stages of the African tree frog, *Hyperolius spinigularis. Oecologia* 143:280–90.

Vonesh, J. R., and C. Osenberg. 2003. Multi-predator effects across life-history stages: Non-additivity of egg- and larval-stage predation in an African treefrog. *Ecol. Lett.* 6:503–8.

Vorburger, C. 2001. Non-hybrid offspring from matings between hemiclonal hybrid waterfrogs suggest occasional recombination between clonal genomes. *Ecol. Lett.* 4:628–36.

Vorburger, C., and H.-U. Reyer. 2003. A genetic mechanism of species replacement in European waterfrogs? *Cons. Gen.* 4:141–55.

Voris, H. K. 1977. Comparison of herpetofaunal diversity in tree buttresses of evergreen tropical forests. *Herpetologica* 33:375–80.

Voris, H. K., and J. P. Bacon. 1966. Differential predation on tadpoles. *Copeia* 1966:594–98.

Voris, H. K., and R. F. Inger. 1995. Frog abundance along streams in Bornean forests. *Cons. Biol.* 9:679–83.

Voris, H. K., and D. R. Karns. 1996. Habitat utilization, movements, and activity patterns of *Enhydris plumbea* (Serpentes: Homalopsinae) in a rice paddy wetland in Borneo. *Herpetol. Nat. Hist.* 4:111–26.

Vorndran, I. C., E. Reichwaldt, and B. Nürnberger. 2002. Does differential susceptibility to predation in tadpoles stabilize the *Bombina* hybrid zone? *Ecology* 83:1648–59.

Vorobyeva, E., and H.-P. Schultze. 1991. Description and systematics of panderichthyid fishes with comments on their relationship to tetrapods. In *Origins of the higher groups of tetrapods,* ed. H.-P. Schultze and L. Trueb, 68–109. Ithaca, NY: Cornell University Press.

Vos, C. C., and J. P. Chardon. 1998. Effects of habitat fragmentation and road density on the distribution pattern of the moor frog *Rana arvalis. J. Appl. Ecol.* 35:44–56.

Vos, C. C., and A. H. P. Stumpel. 1996. Comparison of habitat-isolation parameters in relation to fragmented distribution patterns in the tree frog (*Hyla arborea*). *Landsc. Ecol.* 11:203–14.

Voss, R. S., J. L. Sliva L., and J. A. Valdeo L. 1982. Feeding behavior and diets of Neotropical water rats, genus *Ichthyomys* Thomas, 1893. *Z. Säugetierkunde* 47:364–69.

Voss, S. R. 1993. Relationship between stream order and length of larval period in the salamander *Eurycea wilderae. Copeia* 1993:736–42.

Voss, S. R., and H. B. Shaffer. 1997. Adaptive evolution via a major gene effect: Paedomorphosis in the Mexican axolotl. *Proc. Natl. Acad. Sci. USA* 94:14185–89.

Vredenburg, V. 2000. *Rana muscosa* (mountain yellow-legged frog). Egg predation. *Herp. Rev.* 31:170–171.

Vredenburg, V. 2004. Reversing introduced species effects: experimental removal of introduced fish leads to rapid recovery of a declining frog. *Proc. Natl. Acad. Sci. USA* 101:7646–50.

Wabnitz, P. A., J. H. Bowie, M. J. Tyler, J. C. Wallace, and B. P. Smith. 1999. Aquatic sex pheromone from a male tree frog. *Nature* 401:444–45.

Wabnitz, P. A., J. H. Bowie, M. J. Tyler, J. C. Wallace, and B. P. Smith. 2000. Differences in the skin peptides of the male and female Australian tree frog *Litoria splendida.* The discovery of the aquatic male sex pheromone spledipherin, together with Phe8 caerulein and a new antibiotic peptide caerin 1.10. *Eur. J. Biochem.* 267:269–75.

Wacholder, K., and J. F. von Ledebur. 1930. Untersuchungen über tonische und nicht tonische Wirbeltiermuskeln. *Pflügers Arch.* 225:627–42.

Wada, M., and A. Gorbman. 1977. Mate calling induced by electrical stimulation in freely moving leopard frogs, *Rana pipiens. Horm. Behav.* 9:141–49.

Waddington, L. F. G. 1952. Toads hibernating under water. *Brit. J. Herpetol.* 1:112–13.

Wade, M., and L. Buhlman. 2000. Lateralization and effects of adult androgen in a sexually dimorphic neuromuscular system controlling song in zebra finches. *J. Comp. Neurol.* 426:154–64.

Wade, M. J. 1979. Sexual selection and variance in reproductive success. *Am. Nat.* 114:742–47.

Wade, M. J., and S. J. Arnold. 1980. The intensity of sexual selection in relation to male sexual behaviour, female choice, and sperm precedence. *Anim. Behav.* 28:446–61.

Wager, V. A. 1926. Breeding habits and life histories of some Transvaal Amphibia. *Trans. Roy. Soc. South Africa* 13:163–74.

———. 1956. Parental care in batrachians. *Afr. Wildlife* 10:341–43.

———. 1965. *The frogs of South Africa.* Capetown, South Africa: Purnell and Sons.

Wagner, W. E., Jr. 1989a. Fighting, assessment, and frequency alteration in Blanchard's cricket frog. *Behav. Ecol. Sociobiol.* 25:429–36.

———. 1989b. Graded aggressive calls in Blanchard's cricket frog: Vocal responses to opponent proximity and size. *Anim. Behav.* 38:1025–38.

———. 1989c. Social correlates of variation in male calling behavior in Blanchard's cricket frog, *Acris crepitans blanchardi. Ethology* 82:27–45.

———. 1992. Deceptive or honest signalling of fighting ability? A test of alternative hypotheses for the function of changes in dominant frequency by male cricket frogs. *Anim. Behav.* 44:449–62.

———. 1998. Measuring female mating preferences. *Anim. Behav.* 55:1029–42.

Wagner, W. E., Jr., and B. K. Sullivan. 1992. Chorus organization in the Gulf Coast toad (*Bufo valliceps*): Male and female behavior and the opportunity for sexual selection. *Copeia* 1992:647–58.

———. 1995. Sexual selection in the Gulf Coast toad, *Bufo valliceps:* Female choice based on variable characters. *Anim. Behav.* 49:305–19.

Wahbe, T. R., and F. L. Bunnell. 2003. Relations among larval tailed frogs, forest harvesting, stream microhabitat, and site parameters in southwestern British Columbia. *Can. J. For. Res.* 33:1256–66.

Wahl, M. 1969. Untersuchungen zur Bio-Akustik des Wasserfrosches *Rana esculenta* (L.). *Oecologia* 3:14–55.

Wahlert, G. von. 1953. Eileiter, Laich und Kloake der Salamandriden. *Zool. Jb.* 73:276–310.

Wahnschaffe, U., B. Fritzsch, and W. Himstedt. 1985. The fine structure of the lateral-line organs of larval *Ichthyophis* (Amphibia: Gymnophiona). *J. Morphol.* 186:369–77.

Waide, R. B., and D. P. Reagan. 1996. The rain forest setting. In *The food web of a tropical rain forest,* ed. D. P. Reagan and R. B. Waide, 1–16. Chicago: University of Chicago Press.

Wainwright, P. C., C. P. Sanford, S. M. Reilly, and G. V. Lauder. 1989. Evolution of motor patterns: aquatic feeding in salamanders and ray-finned fishes. *Brain Behav. Evol.* 34:329–41.

Wakahara, M. 1995. Cannibalism and the resulting dimorphism in larvae of a salamander *Hynobius retardatus,* inhabited in Hokkaido, Japan. *Zool. Sci.* 12:467–73.

Wakano, J. Y., Y. Kohmatsu, and N. Yamamura. 2002. Evolutionary dynamics of frequency-dependent growth strategy in cannibalistic amphibians. *Evol. Ecol. Res.* 4:719–36.

Wake, D. B. 1966. Comparative osteology and evolution of the lungless salamanders, Family Plethodontidae. *Mem. Southern California Acad. Sci.* 4:1–111.

———. 1970. The abundance and diversity of tropical salamanders. *Am. Nat.* 104:211–13.

———. 1980. Paedomorphosis. *J. Herpetol.* 14:80–81.

———. 1982. Functional and developmental constraints and opportunities in the evolution of feeding systems in urodeles. In *Environmental adaptation and evolution,* ed. D. Mossakowski and G. Roth, 51–66. Stuttgart, Germany: Gustav Fischer Verlag.

———. 1987. Adaptive radiation of salamanders in middle American cloud forests. *Ann. Missouri Bot. Gard.* 74:242–64.

———. 1991. Homoplasy: The result of natural selection, or evidence of design limitations? *Am. Nat.* 138:543–67.

———. 1992. An integrated approach to evolutionary studies of salamanders. In *Herpetology: Current research on the biology of amphibians and reptiles. Proceedings of the first world congress of herpetology,* ed. K. Adler, 163–77. Oxford, OH: Society for the Study of Amphibians and Reptiles.

———. 1997. Incipient species formation in salamanders of the *Ensatina* complex. *Proc. Natl. Acad. Sci. USA* 94:7761–67.

Wake, D. B., and A. H. Brame, Jr. 1969. Systematics and evolution of Neotropical salamanders of the *Bolitoglossa helmrichi* group. *Nat. Hist. Mus. Los Angeles County Contrib. Sci.* (175):1–40.

Wake, D. B., and J. Castanet. 1995. A skeletochronological study of growth and age in relation to adult size in *Batrachoseps attenuatus. J. Herpetol.* 29:60–65.

Wake, D. B., and S. M. Deban. 2000. Terrestrial feeding in salamanders. In *Feeding: Form, function, and evolution in tetrapod vertebrates,* ed. K. Schwenk, 95–116. San Diego: Academic Press.

Wake, D. B., and I. G. Dresner. 1967. Functional morphology and evolution of tail autotomy in salamanders. *J. Morphol.* 122:265–306.

Wake, D. B., and P. Elias. 1983. New genera and a new species of Central American salamanders, with a review of the tropical genera (Amphibia, Caudata, Plethodontidae). *Contrib. Sci. Los Angeles Co. Mus. Nat. Hist.* (345):1–19.

Wake, D. B., and J. Hanken. 1996. Direct development in the lungless salamanders: What are the consequences for developmental biology, evolution and phylogenesis? *Int. J. Devel. Biol.* 40:859–69.

Wake, D. B. and E. L. Jockusch. 2000. Detecting species borders using diverse data sets: examples from plethodontid salamanders in California. In *The biology of plethodontid salamanders,* ed. R. C. Bruce, R. G. Jaeger, and L. D. Houck, 95–119. New York: Kluwer Academic/Plenum.

Wake, D. B., and A. Larson. 1987. Multidimensional analysis of an evolving lineage. *Science* 238:42–48.

Wake, D. B., and J. F. Lynch. 1976. The distribution, ecology, and evolutionary history of plethodontid salamanders in tropical America. *Sci. Bull. Los Angeles Co. Mus. Nat. Hist.* (25):1–65.

Wake, D. B., and S. B. Marks. 1993. Development and evolution of plethodontid salamanders: A review of prior studies and a prospectus for future research. *Herpetologica* 49:194–203.

Wake, D. B., and G. Roth. 1989. The linkage between ontogeny and phylogeny in the evolution of complex systems. In *Complex organismal functions: Integration and evolution in vertebrates,* ed. D. B. Wake and G. Roth, 361–77. New York: Wiley.

Wake, D. B., and C. J. Schneider. 1998. Taxonomy of the plethodontid salamander genus *Ensatina. Herpetologica* 54:279–98.

Wake, D. B., and N. Shubin. 1994. Urodele limb development in relation to phylogeny and life history. *J. Morphol.* 220:407–8.

Wake, M. H. 1972. Evolutionary morphology of the caecilian urogenital system. Part IV. The cloaca. *J. Morphol.* 136:353–66.

———. 1974. The comparative morphology of the caecilian lung. *Anat. Rec.* 178:483.

———. 1977a. Fetal maintenance and its evolutionary significance in the Amphibia: Gymnophiona. *J. Herpetol.* 11:379–86.

———. 1977b. The reproductive biology of caecilians: an evolutionary perspective. In *The reproductive biology of amphibians,* ed. D. H. Taylor and S. I. Guttman, 73–101. New York: Plenum.

———. 1978. The reproductive biology of *Eleutherodactylus jasperi* (Amphibia, Anura, Leptodactylidae), with comments on the evolution of live-bearing systems. *J. Herpetol.* 12:121–33.

———. 1980a. Fetal tooth development and adult replacement in *Dermophis mexicanus* (Amphibia: Gymnophiona): fields versus clones. *J. Morphol.* 166:203–16.

———. 1980b. Reproduction, growth, and population structure of *Dermophis mexicanus* (Amphibia: Gymnophiona). *Herpetologica* 36:244–56.

———. 1980c. The reproductive biology of *Nectophrynoides malcolmi* (Amphibia: Bufonidae), with comments on the evolution of reproductive modes in the genus *Nectophrynoides. Copeia* 1980:193–209.

———. 1981. Structure and function of the male Müllerian gland in caecilians (Amphibia: Gymnophiona), with comments on its evolutionary significance. *J. Herpetol.* 15:17–22.

———. 1982. Diversity within a framework of constraints. Amphibian reproductive modes. In *Environmental adaptation and evolution,* ed. D. Mossakowski and G. Roth, 87–106. New York: Gustav Fischer.

———. 1985. The comparative morphology and evolution of the eyes of caecilians (Amphibia, Gymnophiona). *Zoomorphology* 105:277–95.

———. 1986a. The morphology of *Idiocranium russeli* (Amphibia: Gymnophiona), with comments on miniaturization through heterochrony. *J. Morphol.* 189:1–16.

———. 1986b. A perspective on the systematics and morphology of the Gymnophiona (Amphibia). *Mem. Soc. Zool. France* 43:21–38.

———. 1989a. Metamorphosis of the hyobranchial apparatus in *Epicrionops* (Amphibia: Gymnophiona: Rhinatrematidae): replacement of bone by cartilage. *Ann. Sci. Nat. Zoologie,* 13th ser. 10:171–82.

———. 1989b. Phylogenesis of direct development and viviparity in vertebrates. In *Complex organismal function: Integration and evolution in vertebrates,* ed. D. B. Wake and G. Roth, 235–50. New York: Wiley.

———. 1992. Reproduction in caecilians. In *Reproductive biology of South American vertebrates,* ed. W. C. Hamlett, 112–20. New York: Springer-Verlag.

———. 1993a. Non-traditional characters in the assessment of caecilian phylogenetic relationships. *Herp. Monogr.* 7:42–55.

———. 1993b. The skull as a locomotor organ. In *The skull, vol. 3,* ed. J. Hanken and B. K. Hall, 197–240. Chicago: University of Chicago Press.

Wake, M. H., and J. Hanken. 1982. Development of the skull of *Dermophis mexicanus* (Amphibia: Gymnophiona), with comments on skull kinesis and amphibian relationships. *J. Morphol.* 173:203–23.

Wakeman, J. M., and G. R. Ultsch. 1976. The effects of dissolved $O_2$ and $CO_2$ on metabolism and gas-exchange partitioning in aquatic salamanders. *Physiol. Zool.* 48:348–59.

Waldick, R. C., B. Freeman, and R. J. Wassersug. 1999. The consequences for amphibians of the conversion of natural, mixed-species forests to conifer plantations in southern New Brunswick. *Can. Field-Nat.* 113:408–18.

Waldman, B. 1981. Sibling recognition in toad tadpoles: The role of experience. *Z. Tierpsychol.* 56:341–58.

———. 1982a. Adaptive significance of communal oviposition in wood frogs (*Rana sylvatica*). *Behav. Ecol. Sociobiol.* 10:169–74.

———. 1982b. Sibling association among schooling toad tadpoles: Field evidence and implications. *Anim. Behav.* 30:700–13.

———. 1984. Kin recognition and sibling association among wood frog (*Rana sylvatica*) tadpoles. *Behav. Ecol. Sociobiol.* 14:171–80.

———. 1985a. Olfactory basis of kin recognition in toad tadpoles. *J. Comp. Physiol.* 156:565–77.

———. 1985b. Sibling recognition in toad tadpoles: are kinship labels transferred among individuals? *Z. Tierpsychol.* 68:41–57.

———. 1986a. Chemical ecology of kin recognition in anuran amphibians. In *Chemical signals in vertebrates IV: Ecology, evolution and comparative biology,* ed. D. Duvall, D. Müller-Schwarze, and R. M. Silverstein, 225–41. New York: Plenum.

———. 1986b. Preference for unfamiliar siblings over familiar non-siblings in American toad (*Bufo americanus*) tadpoles. *Anim. Behav.* 34:48–53.

———. 1989. Do anuran larvae retain kin recognition abilities following metamorphosis? *Anim. Behav.* 37:1055–58.

———. 1991. Kin recognition in amphibians. In *Kin recognition,* ed. P. G. Hepper, 162–219. New York: Cambridge University Press.

———. 2001. Kin recognition, sexual selection, and mate choice in toads. In *Anuran communication,* ed. M. J. Ryan, 232–44. Washington, DC: Smithsonian Institution Press.

Waldman, B., and K. Adler. 1979. Toad tadpoles associate preferentially with siblings. *Nature* 282:611–13.

Waldman, B., and P. J. Bishop. 2004. Chemical communication in an archaic anuran amphibian. *Behav. Ecol.* 15:88–93.

Waldman, B., and J. S. McKinnon. 1993. Inbreeding and outbreeding in fishes, amphibians, and reptiles. In *The natural history of inbreeding and outbreeding. Theoretical and empirical perspectives,* ed. N. W. Thornhill, 250–82. Chicago: University of Chicago Press.

Waldman, B., J. E. Rice, and R. L. Honeycutt. 1992. Kin recognition and incest avoidance in toads. *Amer. Zool.* 32:18–30.

Waldman, B., and M. J. Ryan. 1983. Thermal advantages of communal egg mass deposition in wood frogs (*Rana sylvatica*). *J. Herpetol.* 17:70–72.

Waldman, B., K van de Wolfshaar, J. D. Klena, V. Andjic, P. J. Bishop, and R. J. de B. Norman. 2001. Chytridiomycosis in New Zealand frogs. *Surveillance* 28:9–11.

Walker, R. F., and W. G. Whitford. 1970. Soil water absorption capabilities in selected species of anurans. *Herpetologica* 26:411–18.

Walker, S. L. 1989. *The effect of substrate depletion on the calling behavior of the gray treefrog.* Unpublished M. S. thesis, Storrs: University of Connecticut.

Walker, T. J. 1983. Mating modes and female choice in short-tailed crickets (*Anurogryllus arboreus*). In *Orthopteran mating systems: Sexual competition in a diverse group of insects*, ed. D. T. Gwynne and G. K. Morris, 240–67. Boulder, CO: Westview Press.

Walkowiak, W. 1980a. The coding of auditory signals in the torus semicircularis of the fire-bellied toad and the grass frog: Responses to simple stimuli and to conspecific calls. *J. Comp. Physiol.* 138:141–48.

———. 1980b. Sensitivity, range and temperature dependence of hearing in the grass frog and fire-bellied toad. *Behav. Proc.* 5:363–72.

———. 1984. Neuronal correlates of the recognition of pulsed sound signals in the grass frog. *J. Comp. Physiol. A* 155:57–66.

———. 1988a. Central temporal encoding. In *The evolution of the amphibian auditory system,* ed. B. Fritzsch, M. J. Ryan,

W. Wilczynski, T. E. Hetherington, and W. Walkowiak, 275–94. New York: Wiley.

———. 1988b. Neuroethology of anuran call recognition. In *The evolution of the amphibian auditory system,* ed. B. Fritzsch, M. J. Ryan, W. Wilczynski, T. E. Hetherington, and W. Walkowiak, 485–509. New York: Wiley.

———. 1992. Acoustic communication in the fire-bellied toad: An integrative neurobiological approach. *Ethol. Ecol. Evol.* 4:63–74.

Walkowiak, W., and J. Brzoska. 1982. Significance of spectral and temporal call parameters in the auditory communication of male grass frogs. *Behav. Ecol. Sociobiol.* 11:247–52.

Walkowiak, W., R. R. Capranica, and H. Schneider. 1981. A comparative study of auditory sensitivity in the genus *Bufo* (Amphibia). *Behav. Proc.* 6:223–37.

Walkowiak, W., and H. Luksch. 1994. Sensory motor interfacing in acoustic behavior of anurans. *Amer. Zool.* 34:685–95.

Walkowiak, W., and H. Munz. 1985. The significance of water surface-waves in the communication of fire-bellied toads. *Naturwissenschaften* 72:49–51.

Wallace, A. R. 1869. *The Malay archipelago.* New York: Harper and Brothers.

Wallace, R. L., and L. V. Diller. 1998. Length of the larval cycle of *Ascaphus truei* in coastal streams of the redwood region, northern California. *J. Herpetol.* 32:404–9.

Walls, S. C. 1990. Interference competition in postmetamorphic salamanders: Interspecific differences in aggression by coexisting species. *Ecology* 71:307–14.

———. 1991. Ontogenetic shifts in the recognition of siblings and neighbors by juvenile salamanders. *Anim. Behav.* 42:423–34.

———. 1995. Differential vulnerability to predation and refuge use in competing larval salamanders. *Oecologia* 101:86–93.

———. 1998. Density dependence in larval salamanders: the effects of interference and food limitation. *Copeia* 1998:926–35.

Walls, S. C., and R. Altig. 1986. Female reproductive biology and larval life history of *Ambystoma* salamanders: A comparison of egg size, hatchling size, and larval growth. *Herpetologica* 42:334–45.

Walls, S. C., J. J. Beatty, B.N. Tissot, D. G. Hokit, and A. R. Blaustein. 1993. Morphological variation and cannibalism in a larval salamander (*Ambystoma macrodactylum columbianum*). *Can. J. Zool.* 71:1543–51.

Walls, S. C., and A. R. Blaustein. 1994. Does kinship influence density dependence in a larval salamander? *Oikos* 71:459–68.

———. 1995. Larval marbled salamanders, *Ambystoma opacum,* eat their kin. *Anim. Behav.* 50:537–45.

Walls, S. C., and R. G. Jaeger. 1987. Aggression and exploitation and mechanisms of competition in larval salamanders. *Can. J. Zool.* 65:2938–44.

Walls, S. C., A. Mathis, R. G. Jaeger, and W. F. Gergits. 1989. Male salamanders with high-quality diets have faeces attractive to females. *Anim. Behav.* 38:546–47.

Walls, S. C., and R. E. Roudebush. 1991. Reduced aggression toward siblings as evidence of kin recognition in cannibalistic salamanders. *Am. Nat.* 138:1027–38.

Walls, S. C., and R. D. Semlitsch. 1991. Visual and movement displays function as agonistic behavior in larval salamanders. *Copeia* 1991:936–42.

Walls, S. C., and M. G. Williams. 2001. The effect of community composition on persistence of prey with their predators in an assemblage of pond-breeding amphibians. *Oecologia* 128:134–41.

Walsberg, G. E. 1986. Comparison of two techniques for estimating the maximum aerobic capacity of amphibians. *Herpetologica* 42:389–94.

Walsberg, G. E., M. S. Lea, and S. S. Hillman. 1986. Individual variation in maximum aerobic capacity: cardiovascular and enzymatic correlates in *Rana catesbeiana. J. Exp. Zool.* 239:1–5.

Walters, B. 1975. Studies of interspecific predation within an amphibian community. *J. Herpetol.* 9:267–79.

Walters, P. J., and L. Greenwald. 1977. Physiological adaptations of aquatic newts (*Notophthalmus viridescens*) to a terrestrial environment. *Physiol. Zool.* 50:88–98.

Walther, G.-R., E. Post, P. Convey, A. Menzel, C. Parmesan, T. J. C. Beebee, J.-M. Fromentin, O. Hoegh-Guldberg, and F. Bairlein. 2002. Ecological responses to recent climate change. *Nature* 416:389–95.

Walton, B. M. 1988. Relationships among metabolic, locomotory, and field measures of organismal performance in the Fowler's toad (*Bufo woodhousei fowleri*). *Physiol. Zool.* 61:107–18.

———. 1993. Physiology and phylogeny: the evolution of locomotor energetics in hylid frogs. *Am. Nat.* 141:26–50.

Walton, B. M., and B. D. Anderson. 1988. The aerobic cost of saltatory locomotion in the Fowler's toad (*Bufo woodhousei fowleri*). *J. Exp. Biol.* 136:273–88.

Walton, B. M., C. C. Peterson, and A. F. Bennett. 1994. Is walking costly for anurans? The energetic cost of walking in the northern toad *Bufo boreas halophilus. J. Exp. Biol.* 197:165–78.

Walvoord, M. E. 2003. Cricket frogs maintain body hydration and temperature near levels allowing maximum jump performance. *Physiol. Biochem. Zool.* 76:825–35.

Waltz, E. C. 1982. Alternative mating tactics and the law of diminishing returns: The satellite threshold model. *Behav. Ecol. Sociobiol.* 10:75–83.

Waltz, E. C., and L. L. Wolf. 1984. By jove! Why do alternative mating tactics assume so many different forms? *Amer. Zool.* 24:333–43.

Wambreuse, P., and V. Bels. 1984. Analyse qualitative et quantitative de la parade sexuelle du triton palmé *Triturus helveticus* (Razoumowsky 1798). *Cahiers d'Ethologie Appl.* 4:193–218.

Wang, J., T. A. Ludwig, and P. M. Narins. 1996. Spatial and spectral dependence of the auditory periphery in the northern leopard frog. *J. Comp. Physiol. A* 178:159–72.

Wang, T., M. Busk, and J. Overgaard. 2001. The respiratory costs of feeding in amphibians and reptiles. *Comp. Biochem. Physiol. A* 128:535–49.

Warbeck, A., I. Breiter, and J. Parzefall. 1996. Evidence for chemical communication in the aquatic caecilian *Typhlonectes natans* (Typhlonectidae, Gymnophiona). *Mém. Biospéol.* 23:37–41.

Warbeck, A., and J. Parzefall. 2001. Mate recognition via waterborne chemical cues in the viviparous caecilian *Typhlonectes natans* (Amphibia: Gymnophiona). In *Chemical signals in vertebrates 9,* ed. A. Marchlewska-Koj, J. J. Lepri, and D. Müller-Schwarze, 263–68. New York: Kluwer Academic/Plenum.

Warburg, M. R. 1965. Studies on the water economy of some Australian frogs. *Aust. J. Zool.* 13:317–30.

———. 1967. On thermal and water balance of three central Australian frogs. *Comp. Biochem. Physiol.* 20:27–43.

———. 1971a. On the water economy of Israel amphibians: The anurans. *Comp. Biochem. Physiol.* 40 A:911–24.

———. 1971b. The water economy of Israel amphibians: The

urodeles *Triturus vittatus* (Jenyns) and *Salamandra salamandra* (L.). *Comp. Biochem. Physiol.* 40 A:1055–63.

———. 1972. Water economy and thermal balance on Israeli and Australian Amphibia from xeric habitats. In *Comparative physiology of desert animals. Symp. Zool. Soc. London, no. 31,* ed. G. M. O. Maloiy, 79–111. New York: Academic Press.

———. 1992. Breeding patterns in a fringe population of fire salamanders, *Salamandra salamandra. Herpetol. J.* 2:54–58.

———. 1994. Reproductive ecology, breeding activity, longevity, and reproductive strategies of *Salamandra salamandra* during an 18–year long study of an isolated population on Mt. Carmel, Israel. *Mertensiella* 4:399–421.

———. 1997. *Ecophysiology of amphibians inhabiting xeric environments.* New York: Springer-Verlag.

Warburg, M. R., and G. Degani. 1979. Evaporative water loss and uptake in juvenile and adult *Salamandra salamandra* (L.) (Amphibia: Urodela). *Comp. Biochem. Physiol.* 62 A:1071–75.

Warburg, M. R., and S. Goldenberg. 1978. The changes in osmoregulatory effects of prolactin during the life cycle of two urodeles. *Comp. Biochem. Physiol.* 61 A:321–24.

Warburg, M. R., D. Lewinson, and M. Rosenberg. 1994. Ontogenesis of amphibian epidermis. In *Amphibian biology. Vol. 1. The integument,* ed. H. Heatwole and G. T. Barthalmus, 33–63. Chipping Norton, New South Wales, Australia: Surrey Beatty and Sons.

Ward, D., and O. J. Sexton. 1981. Anti-predator role of salamander egg membranes. *Copeia* 1981:724–26.

Ward, S., H. M. Lampe, and P. J. B. Slater. 2004. Singing is not energetically demanding for pied flycatchers, *Ficedula hypoleuca. Behav. Ecol.* 15:477–84.

Ward, S., and P. J. B. Slater. 2005. Heat transfer and the energetic cost of singing by canaries *Serinus canaria. J. Comp. Physiol. A* 191:953–64.

Ward, S., J. R. Speakman, and P. J. B. Slater. 2003. The energy of song in the canary *Serinus canaria. Anim. Behav.* 66:893–902.

Waringer-Löschenkohl, A. 1988. An experimental study of microhabitat selection and microhabitat shifts in European tadpoles. *Amphibia-Reptilia* 9:219–36.

Warkentin, K. M. 1995. Adaptive plasticity in hatching age: A response to predation risk trade-offs. *Proc. Natl. Acad. Sci. USA* 92:3507–10.

———. 1999a. The development of behavioral defenses: A mechanistic analysis of vulnerability in red-eyed tree frog hatchlings. *Behav. Ecol.* 10:251–62.

———. 1999b. Effects of hatching age of development and hatchling morphology in the red-eyed treefrog, *Agalychnis callidryas. Biol. J. Linn. Soc.* 68:443–70.

———. 2000a. Environmental and developmental effects on external gill loss in the red-eyed tree frog, *Agalychnis callidryas. Physiol. Biochem. Zool.* 73:557–65.

———. 2000b. Wasp predation and wasp-induced hatching of red-eyed treefrog eggs. *Anim. Behav.* 60:503–10.

Warkentin, K. M., C. R. Currie, and S. A. Rehner. 2001. Egg-killing fungus induces early hatching of red-eyed treefrog eggs. *Ecology* 82:2860–69.

Warner, S. C., W. A. Dunson, and J. Travis. 1991. Interaction of pH, density, and priority effects on the survivorship and growth of two species of hylid tadpoles. *Oecologia* 88:331–39.

Warner, S. C., J. Travis, and W. A. Dunson. 1993. Effect of pH variation on interspecific competition between two species of hylid tadpoles. *Ecology* 74:183–94.

Warren, A. 2000. Secondarily aquatic temnospondyls of the Upper Permian and Mesozoic. In *Amphibian biology. Vol. 4. Palaeontology,* ed. H. Heatwole and R. L. Carroll, 1121–49. Chipping Norton, New South Wales, Australia: Surrey Beatty and Sons.

Warren, M. L., B. M. Burr, S. J. Walsh, H. L. Bart, R. C. Cashner, D. A. Etnier, B. J. Freeman, B. R. Kuhajda, R. L. Mayden, H. W. Robison, S. T. Ross, and W. C. Starnes. 2000. Diversity, distribution, and conservation status of the native freshwater fishes of the southern United States. *Fisheries* 25:7–31.

Wassen, H. 1934. The frog in Indian mythology and imaginative world. *Anthropos* 29:613–58.

Wasserman, A. O. 1957. Factors affecting interbreeding in sympatric species of spadefoots (*Scaphiopus*). *Evolution* 11:320–38.

Wasserman, F. E. 1977. Intraspecific acoustical interference in the white-throated sparrow (*Zonotrichia albicollis*). *Anim. Behav.* 25:949–52.

Wassersug, R. J. 1971. On the comparative palatability of some dry-season tadpoles from Costa Rica. *Am. Midl. Nat.* 86:101–9.

———. 1972. The mechanism of ultraplanktonic entrapment in anuran larvae. *J. Morphol.* 137:279–88.

———. 1973. Aspects of social behavior in anuran larvae. In *Evolutionary biology of the anurans,* ed. J. L. Vial, 273–97. Columbia: University of Missouri Press.

———. 1975. The adaptive significance of the tadpole stage with comments on the maintenance of complex life cycles in anurans. *Amer. Zool.* 15:405–17.

———. 1976. Internal oral features in *Hyla regilla* (Anura: Hylidae) larvae: An ontogenetic study. *Occ. Pap. Mus. Nat. Hist. Univ. Kansas* 49:1–24.

———. 1980. Internal oral features of larvae from eight anuran families: Functional systematic, evolutionary and ecological considerations. *Univ. Kansas Mus. Nat. Hist. Misc. Publ.* (68):1–146.

———. 1989a. Locomotion in amphibian larvae (or "Why aren't tadpoles built like fishes?"). *Amer. Zool.* 29:65–84.

———. 1989b. What, if anything, is a microhylid (Orton type II) tadpole? *Forsch. Zool.* 35:534–38.

———. 1996. The biology of *Xenopus* tadpoles. In *The biology of Xenopus,* ed. R. C. Tinsley and H. R. Kobel, 195–211. New York: Oxford University Press.

———. 1997. Where the tadpole meets the world—Observations and speculations on biomechanical and biochemical factors that influence metamorphosis in anurans. *Amer. Zool.* 37:124–36.

Wassersug, R. J., and W. E. Duellman. 1984. Oral structures and their development in egg-brooding hylid frog embryos and larvae: Evolutionary and ecological implications. *J. Morphol.* 182:1–37.

Wassersug, R. J., and M. E. Feder. 1983. The effects of aquatic oxygen concentration, body size, and respiratory behaviours on the stamina of obligate aquatic (*Bufo americanus*) and facultative air-breathing (*Xenopus laevis* and *Rana berlandieri*) anuran larvae. *J. Exp. Biol.* 105:173–90.

Wassersug, R. J., K. J. Frogner, and R. F. Inger. 1981. Adaptations for life in tree holes by rhacophorid tadpoles from Thailand. *J. Herpetol.* 15:41–52.

Wassersug, R. J., and C. M. Hessler. 1971. Tadpole behaviour: aggegation in larval *Xenopus laevis. Anim. Behav.* 19:386–89.

Wassersug, R. J., and W. R. Heyer. 1983. Morphological correlates of subaerial existence in leptodactylid tadpole associated with flowing water. *Can. J. Zool.* 61:761–69.

———. 1988. A survey of internal oral features of leptodactyloid larvae (Amphibia: Anura). *Smithsonian Contrib. Zool.* (457):1–99.

Wassersug, R. J., and K. Hoff. 1979. A comparative study of the buccal pumping mechanism of tadpoles. *Biol. J. Linn. Soc.* 12:225–59.

———. 1982. Developmental changes in the orientation of the anuran jaw suspension. *Evol. Biol.* 15:223–46.

———. 1985. The kinematics of swimming in anuran larvae. *J. Exp. Biol.* 119:1–30.

Wassersug, R. J., A. M. Lum, and M. J. Potel. 1981. An analysis of school structure for tadpoles (Anura: Amphibia). *Behav. Ecol. Sociobiol.* 9:15–22.

Wassersug, R. J., and A. M. Murphy. 1987. Aerial respiration facilitates growth in suspension-feeding anuran larvae (*Xenopus laevis*). *J. Exp. Biol.* 46:141–47.

Wassersug, R. J., and W. F. Pyburn. 1987. The biology of the Pe-ret' Toad, *Otophryne robusta* (Microhylidae), with special consideration of its fossorial larva and systematic relationships. *Zool. J. Linn. Soc.* 91:137–69.

Wassersug, R. J., and K. Rosenberg. 1979. Surface anatomy of the secretory epithelium in the branchial food traps of tadpoles: a comparative study. *J. Morphol.* 159:393–425.

Wassersug, R. J., and E. A. Seibert. 1975. Behavioral responses of amphibian larvae to variation in dissolved oxygen. *Copeia* 1975:86–103.

Wassersug, R. J., and D. G. Sperry. 1977. The relationship of locomotion to differential predation on *Pseudacris triseriata* (Anura: Hylidae). *Ecology* 58:830–39.

Wassersug, R. J., and D. B. Wake. 1995. Fossil tadpoles from the Miocene of Turkey. *Alytes* 12:145–57.

Wassersug, R. J., and M. Yamashita. 2001. Plasticity and constraints on feeding kinematics in anuran larvae. *Comp. Biochem. Physiol. A* 131:183–95.

Watanabe, S., N. Nakanishi, and M. Izawa. 2005. Seasonal abundance in the floor-dwelling frog fauna on Iriomote Island of the Ryukyu Archipelago, Japan. *J. Trop. Ecol.* 21:85–91.

Watkins, T. B. 1996. Predator-mediated selection on burst swimming performance in tadpoles of the Pacific tree frog, *Pseudacris regilla. Physiol. Zool.* 69:154–67.

———. 2000. The effects of acute and developmental temperature on burst swimming speed and myofibrillar ATPase activity in tadpoles of the Pacific tree frog, *Hyla regilla. Physiol. Biochem. Zool.* 73:356–64.

Watling, J. I., and M. A. Donnelly. 2002. Seasonal patterns of reproduction and abundance of leaf litter frogs in a Central American rainforest. *J. Zool., London* 258:269–76.

Watson, G. F., and A. A. Martin. 1973. Life history, larval morphology and relationships of Australian leptodactylid frogs. *Trans. Roy. Soc. South Australia* 97:33–45.

Watson, J. W., K. R. McAllister, and D. J. Pierce. 2003. Home ranges, movements, and habitat selection of Oregon spotted frogs (*Rana pretiosa*). *J. Herpetol.* 37:292–300.

Watson, R. T., A. Mathis, and R. Thompson. 2004. Influence of physical stress, distress cues, and predator kairomones on the foraging behavior of Ozark zigzag salamanders, *Plethodon angusticlavius. Behav. Proc.* 65:201–9.

Watt, K. W. K., and A. Riggs. 1975. Hemoglobins of the tadpoles of the bullfrog, *Rana catesbeiana.* Structure and function of the isolated components. *J. Biol. Chem.* 250:5934–44.

Watt, P. J., and P. Jarvis. 1997. Survival analysis in palmate newts exposed to ammonium nitrate agricultural fertilizer. *Ecotoxicology* 6:355–62.

Watt, P. J., S. F. Nottingham, and S. Young. 1997. Toad tadpole aggregation behaviour: Evidence for a predator avoidance function. *Anim. Behav.* 54:865–72.

Watts, R. A., C. A. Palmer, R. C. Feldhoff, P. W. Feldhoff, L. D. Houck, A. G. Jones, M. E. Pfrender, S. M. Rollmann, and S. J. Arnold. 2004. Stabilizing selection on behavior and morphology masks positive selection on the signal in a salamander pheromone signaling complex. *Mol. Biol. Evol.* 21:1032–41.

Watters, G. T. 1997. Glochidial metamorphosis of the freshwater mussel *Lampsilis cardim* (Bivalvia: Unionidae) on larval tiger salamanders, *Ambystoma tigrinum* ssp. (Amphibia: Ambystomatidae). *Can. J. Zool.* 75:505–8.

Watters, G. T. and S. H. O'Dee. 1998. Metamorphosis of freshwater mussel glochidia (Bivalvia: Unionidae) on amphibians and exotic fishes. *Am. Midl. Nat.* 139:49–57.

Waugh, F. W. 1916. Iroquois foods and food preparation. *Canada Geological Survey Memoir 86. Anthropological Series no. 12.* Ottawa, Ontario: Government Printing Bureau,

Weathers, W. W. 1975. Circulatory responses of *Rana catesbeiana* to temperature, season and previous thermal history. *Comp. Biochem. Physiol.* 51A:43–52.

———. 1976. Influence of temperature acclimation on oxygen consumption, haemodynamics and oxygen transport in bullfrogs. *Aust. J. Zool.* 24:321–30.

Weathers, W. W., and G. K. Snyder. 1977. Relation of oxygen consumption to temperature and time of day in tropical anuran amphibians. *Aust. J. Zool.* 25:19–24.

Webb, G., S. C. Manolis, and P. J. Whitehead. 1987. *Wildlife management: Crocodiles and alligators.* Chipping Norton, New South Wales, Australia: Surrey Beatty and Sons.

Webb, G. J. W., S. C. Manolis, and R. Buckworth. 1982. *Crocodylus johnstoni* in the McKinley River area, N. T. 1. Variation in the diet and a new method of assessing the relative importance of prey. *Aust. J. Zool.* 30:877–99.

Webb, J. K., R. Shine, and K. A. Christian. 2005. Does intraspecific niche partitioning in a native predator influence its response to an invasion by a toxic prey species? *Aust. Ecol.* 30:201–9.

Webb, R. G., and W. L. Roueche. 1971. Life history aspects of the tiger salamander (*Ambystoma tigrinum mavortium*) in the Chihuahuan desert. *Great Basin Nat.* 31:193–212.

Weber, D. 1989a. The diet of polecats (*Mustela putorius* L.) in Switzerland. *Z. Säugetierkunde* 54:157–71.

———. 1989b. Foraging in polecats (*Mustela putorius* L.) of Switzerland: The case of a specialist anuran predator. *Z. Säugetierkunde* 54:377–92.

Weber, E. 1974. Vergleichende Untersuchungen zur Bioakustik von *Discoglossus pictus,* Otth 1837, und *Discoglossus sardus,* Tschudi 1837 (Discoglossidae, Anura). *Zool. Jb. Physiol.* 78:40–84.

———. 1975a. Die Veränderung der Befreiungsrufe bei sechs europäischen Anuren nach Ausschaltung von Kehlkopfmuskeln (Amphibia). *Zool. Jb. Physiol.* 79:311–20.

———. 1975b. Schreckrufe von *Rana r. ridibunda* Pallas (Amphibia, Anura). *Z. Naturforsch.* 3:695.

———. 1976. Die Veränderung der Paarungs- und Revierrufe von *Hyla arborea savignyi* Audouin (Anura) nach Ausschaltung von Kehlkopfmuskeln. *Bonn. Zool. Beitr.* 27:87–97.

———. 1977. Effects of extirpation of laryngeal muscles and vocal cords on vocalizations of *Bufo viridis* (Amphibia, Anura). *Israel J. Zool.* 26:230–47.

———. 1978a. Die Veränderungen der Befreiungsrufe nach Ausschaltungen an Kehlkopfmuskel und Stimmbandern bei *Rana ridibunda* (Amphibia, Anura). *Zool. Jb. Physiol.* 82:103–12.

———. 1978b. Distress calls of *Bufo calamita* and *B. viridis* (Amphibia: Anura). *Copeia* 1978:354–56.

Weber, J. A. 1944. Observations on the life history of *Amphiuma means*. *Copeia* 1944:61–62.

Weber, J.-M. 1988. Design of exogenous fuel supply systems: Adaptive strategies for endurance locomotion. *Can. J. Zool.* 66:1116–21.

———. 1990. Seasonal exploitation of amphibians by otters (*Lutra lutra*) in north-east Scotland. *J. Zool., London* 220:641–51.

———. 1992. Pathways for oxidative fuel provision to working muscle: ecological consequences of maximal supply limitations. *Experientia* 48:557–64.

Weber, M. 1972. Frosch. In *Reallexikon fur Antike und Christentum,* vol. 8, ed. T. Klauser, 524–38. Stuttgart, Germany: Anton Hiersemann,.

Weber, N. A. 1938. The food of the giant toad, *Bufo marinus* (L.), in Trinidad and British Guiana with special reference to ants. *Ann. Entomol. Soc. Amer.* 31:499–503.

Wederkinch, E. 1988. Population size, migration barriers, and other features of *Rana dalmatina* populations near Koge, Zealand, Denmark. *Mem. Soc. Fauna Flora Fennici* 64:101–3.

Wehner, R. 1962. Ein Beitrag zur Ernährungsbiologie des Waldkauzes (*Strix aluco*). *Orn. Mitt.* 14:90–94.

Weibel, E. R. 1972. Morphometric estimation of pulmonary diffusion capacity. V. Comparative morphometry of alveolar lungs. *Respir. Physiol.* 14:26–43.

Weick, S. E., M. G. Knutson, B. G. Knights, and B. C. Pember. 2005. A comparison of internal and external radio transmitters with northern leopard frogs (*Rana pipiens*). *Herp. Rev.* 36:415–21.

Weigmann, D. C., and R. Altig. 1975. Anoxic tolerances of three species of salamander larvae. *Comp. Biochem. Physiol.* 50A:681–84.

Weintraub, J. D. 1974. Movement patterns of the red-spotted toad, *Bufo punctatus*. *Herpetologica* 30:212–15.

Weir, L. A., and M. J. Mossman. 2005. North American Amphibian Monitoring Program (NAAMP). In *Amphibian declines: The conservation status of United States species,* ed. M. J. Lannoo, 307–13. Berkeley: University of California Press.

Weismann, A. 1878. On the change of the Mexican axolotl to an *Amblystoma*. *Smithsonian Report 1877*:349–75.

Weisrock, D. W., L. J. Harmon, and A. Larson. 2005. Resolving deep phylogenetic relationships in salamanders: Analysis of mitochondrial and nuclear genomic data. *Syst. Biol.* 54:758–77.

Weisrock, D. W., K. H. Kozak, and A. Larson. 2005. Phylogenetic analysis of mitochondrial gene flow and introgression in the salamander, *Plethodon shermani*. *Mol. Ecol.* 14:1457–72.

Weisrock, D. W., J. R. Macey, I. H. Ugurtas, A. Larson, and T. J. Papenfuss. 2001. Molecular phylogenetics and historical biogeography among salamandrids of the "true" salamander clade: Rapid branching of numerous highly divergent lineages in *Mertensiella luschani* associated with the rise of Anatolia. *Mol. Phylogen. Evol.* 18:434–48.

Weitere, M., D. Tautz, D. Neumann, and S. Steinfartz. 2004. Adaptive divergence vs. environmental plascticity: Tracing local genetic adaptation of metamorphosis traits in salamanders. *Mol. Ecol.* 13:1665–77.

Welch, A. M. 2003. Genetic benefits of a female mating preference in gray tree frogs are context-dependent. *Evolution* 57:883–93.

Welch, A. M., R. D. Semlitsch, and H. C. Gerhardt. 1998. Call duration as an indicator of genetic quality in male gray tree frogs. *Science* 280:1928–30.

Welch, N. E., and J. A. MacMahon. 2005. Identifying habitat variables important to the rare Columbia spotted frog in Utah (U.S.A.): An information-theoretic approach. *Cons. Biol.* 19:473–81.

Weldon, C., and L. H. du Preez. 2004. Decline of the Kihansi spray toad, *Nectophrynoides asperginis,* from the Udzungwa Mountains, Tanzania. *Froglog* (62):2–3.

Weldon, C., L. H. du Preez, A. D. Hyatt, R. Muller, and R. Speare. 2004. Origin of the amphibian chytrid fungus. *Emerg. Infect. Dis.* 10:2100–5.

Wellborn, G. A., D. K. Skelly, and E. E. Werner. 1996. Mechanisms creating community structure across a freshwater habitat gradient. *Ann. Rev. Ecol. Syst.* 27:337–63.

Weller, W. F. and D. M. Green. 1997. Checklist and current status of Canadian amphibians. In *Amphibians in decline: Canadian studies of a global problem,* ed. D. M. Green, 309–28. St. Louis: Society for the Study of Amphibians and Reptiles.

Wells, C. S., and R. N. Harris. 2001. Activity level and the trade-off between growth and survival in the salamanders *Ambystoma jeffersonianum* and *Hemidactylium scutatum*. *Herpetologica* 57:116–27.

Wells, H. G., J. Huxley, and G. P. Wells. 1929. *The science of life*. New York: Literary Guild.

Wells, K. D. 1976a. Multiple egg clutches in the green frog (*Rana clamitans*). *Herpetologica* 32:85–87.

———. 1976b. *Territorial behavior of the green frog, Rana clamitans*. Unpublished PhD dissertation. Ithaca, NY: Cornell University.

———. 1977a. The courtship of frogs, pp. 233–262. In *The Reproductive biology of amphibians,* ed. D. Taylor and S. Guttman, 233–62. New York: Plenum.

———. 1977b. The social behaviour of anuran amphibians. *Anim. Behav.* 25:666–93.

———. 1977c. Territoriality and male mating success in the green frog (*Rana clamitans*). Ecology 58:750–62.

———. 1978ba. Courtship and parental behavior in a Panamanian poison-arrow frog, *Dendrobates auratus*. *Herpetologica* 34:148–55.

———. 1978b. Territoriality in the green frog (*Rana clamitans*): Vocalizations and aggressive behaviour. *Anim. Behav.* 26:1051–63.

———. 1979. Reproductive behavior and male mating success in a Neotropical toad, *Bufo typhonius*. *Biotropica* 11:301–7.

———. 1980a. Behavioral ecology and social organization of a dendrobatid frog (*Colostethus inguinalis*). *Behav. Ecol. Sociobiol.* 6:199–209.

———. 1980b. Evidence for growth of tadpoles during parental transport in *Colostethus inguinalis*. *J. Herpetol.* 14:428–30.

———. 1980c. Social behavior and communication of a dendrobatid frog (*Colostethus trinitatis*). *Herpetologica* 36:189–99.

———. 1981a. Parental behavior of male and female frogs. In *Natural selection and social behavior,* ed. R. D. Alexander and D. W. Tinkle, 184–97. New York: Chiron Press.

———. 1981b. Territorial behavior of the frog *Eleutherodactylus urichi* in Trinidad. *Copeia* 1981:726–28.

———. 1988. The effect of social interactions on anuran vocal behavior. In *The evolution of the amphibian auditory system,* ed. B. Fritzsch, M. J. Ryan, W. Wilczynski, T. E. Hetherington, and W. Walkowiak, 433–54. New York: Wiley.

———. 1989. Vocal communication in a Neotropical treefrog,

*Hyla ebraccata:* responses of males to graded aggressive calls. *Copeia* 1989:461–66.

———. 2001a. The energetics of calling in frogs. In *Anuran communication,* ed. M. J. Ryan, 45–60. Washington, DC: Smithsonian Institution Press.

———. 2001b. Mating systems and sexual selection. In *Herpetology* (2nd ed.), ed. F. H. Pough, R. M. Andrews, J. E. Cadle, M. L. Crump, A. H. Savitsky, and K. D. Wells, 398–430. Upper Saddle River, NJ: Prentice Hall.

Wells, K. D., and K. M. Bard. 1987. Vocal communication in a Neotropical treefrog, *Hyla ebraccata:* responses of females to advertisement and aggressive calls. *Behaviour* 101:200–210.

———. 1988. Parental behavior of an aquatic-breeding tropical frog, *Leptodactylus bolivianus. J. Herpetol.* 22:361–64.

Wells, K. D., and C. R. Bevier. 1997. Contrasting patterns of energy substrate use in two species of frogs that breed in cold weather. *Herpetologica* 53:70–80.

Wells, K. D., and B. J. Greer. 1981. Vocal responses to conspecific calls in a Neotropical hylid frog, *Hyla ebraccata. Copeia* 1981:615–24.

Wells, K. D., and J. J. Schwartz. 1982. The effect of vegetation on the propagation of calls in the Neotropical frog *Centrolenella* fleischmanni. *Herpetologica* 38:449–55.

———. 1984a. Vocal communication in a Neotropical treefrog, *Hyla ebraccata:* Advertisement calls. *Anim. Behav.* 32:405–20.

———. 1984b. Vocal communication in a Neotropical treefrog, *Hyla ebraccata:* Aggressive calls. *Behaviour* 91:128–45.

———. 2007. The behavioral ecology of anuran communication. In *Hearing and sound communication in amphibians. Springer handbook of auditory research,* ed. P. M. Narins and A. S. Feng, 44–86. New York: Springer Verlag.

Wells, K. D., and T. L. Taigen. 1984. Reproductive behavior and aerobic capacities of male American toads (*Bufo americanus*): Is behavior constrained by physiology? *Herpetologica* 40:292–98.

———. 1986. The effect of social interactions on calling energetics in the gray treefrog (*Hyla versicolor*). *Behav. Ecol. Sociobiol.* 19:9–18.

———. 1989. Calling energetics of a Neotropical treefrog, *Hyla microcephala. Behav. Ecol. Sociobiol.* 25:13–22.

Wells, K. D., T. L. Taigen, and J. A. O'Brien. 1996. The effect of temperature on calling energetics of the spring peeper (*Pseudacris crucifer*). *Amphibia-Reptilia* 17:149–58.

Wells, K. D., T. L. Taigen, S. W. Rusch, and C. C. Robb. 1995. Seasonal and nightly variation in glycogen reserves of calling gray treefrogs (*Hyla versicolor*). *Herpetologica* 51:359–68.

Wells, K. D., and R. A. Wells. 1976. Patterns of movement in a population of the slimy salamander, *Plethodon glutinosus,* with observations on aggregations. *Herpetologica* 32:156–62.

Wells, R. M. G. 1990. Hemoglobin physiology in vertebrate animals: a cautionary approach to adaptationist thinking. In *Vertebrate gas exchange: From environment to cell,* ed. R. G. Boutilier, 143–61. New York: Springer-Verlag.

Welsch, U., M. Muller, and C. Schubert. 1977. Electronenmikroskopische und histochemische Beobachtungen zur Fortpflanzungsbiologie viviparer Gymnophionen (*Chthonerpeton indistinctum*). *Zool. Jb. Anat.* 97:532–49.

Welsch, U., V. Storch, and W. Fuchs. 1974. The fine structure of the digital pads of rhacophorid tree frogs. *Cell Tissue Res.* 148:407–16.

Welsh, H. H., Jr. 1990. Relictual amphibians and old-growth forests. *Cons. Biol.* 4:309–19.

Welsh, H. H., Jr., and S. Droege. 2001. A case for using plethodontid salamanders for monitoring biodiversity and ecosystem integrity of North American forests. *Cons. Biol.* 15:558–69.

Welsh, H. H., Jr., and A. J. Lind. 1995. Habitat correlates of the Del Norte salamander, *Plethodon elongatus* (Caudata: Plethodontidae), in northwestern California. *J. Herpetol.* 29:198–210.

Welsh, H. H., Jr., and L. A. Ollivier. 1998. Stream amphibians as indicators of ecosystem stress: a case study from California's redwoods. *Ecol. Appl.* 8:1118–32.

Wendland, V. 1984. The influence of prey fluctuations on the breeding success of the tawny owl *Strix aluco. Ibis* 126:284–95.

Weng, S. P., J. G. He, X. H. Wang, L. Lu, M. Deng, and S. M. Chan. 2002. Outbreaks of an iridovirus disease in cultured tiger frog, *Rana tigerina rugulosa,* in southern China. *J. Fish. Dis.* 25:423–27.

Wente, W. H., and J. B. Phillips. 2003. Fixed green and brown color morphs and a novel color-changing morph of the Pacific tree frog *Hyla regilla. Am. Nat* 162:461–73.

———. 2005. Seasonal color change in a population of Pacific treefrogs (*Pseudacris regilla*). *J. Herpetol.* 39:161–65.

Wentzell, L. A., S. A. McNeil, and D. P. Toews. 1993. The role of the lymphatic system in water balance processes in the toad *Bufo marinus* (L.). *Physiol. Zool.* 66:307–21.

Werner, E. E. 1986. Amphibian metamorphosis: growth rate, predation risk, and the optimal size at transformation. *Am. Nat.* 128:319–41.

———. 1988. Size, scaling, and the evolution of complex life cycles. In *Size-structured populations: Ecology and evolution,* ed. B. Ebenman and L. Persson, 60–81. New York: Springer-Verlag.

———. 1991. Nonlethal effects of a predator on competitive interactions between two anuran larvae. *Ecology* 72:1709–20.

———. 1992. Competitive interactions between wood frog and northern leopard frog larvae: The influence of size and activity. *Copeia* 1992:26–35.

———. 1994. Ontogenetic scaling of competitive relations: Size dependent effects and responses in two anuran larvae. *Ecology* 75:197–213.

———. 1998. Ecological experiments and a research program in community ecology. In *Experimental ecology. Issues and perspectives,* ed. W. J. Resetarits, Jr. and J. Bernardo, 3–26. New York: Oxford University Press.

Werner, E. E., and B. R. Anholt. 1993. Ecological consequences of the tradeoff between growth and mortality rates mediated by foraging activity. *Am. Nat.* 142:242–72.

———. 1996. Predator-induced behavioral indirect effects: Consequences to comparative interactions in anuran larvae. *Ecology* 77:157–69.

Werner, E. E., and J. F. Gilliam. 1984. The ontogenetic niche and species interactions in size-structured populations. *Ann. Rev. Ecol. Syst.* 15:393–425.

Werner, E. E., and K. S. Glennemeier. 1999. Influence of forest canopy cover on the breeding pond distributions of several amphibian species. *Copeia* 1999:1–12.

Werner, E. E., and D. J. Hall. 1977. Competition and habitat shift in two sunfishes (Centrarchidae). *Ecology* 58:869–76.

Werner, E. E., and M. A. McPeek. 1994. Direct and indirect effects of predators on two anuran species along an environmental gradient. *Ecology* 75:1368–82.

Werner, E. E., G. A. Wellborn, and M. A. McPeek. 1995. Diet composition of postmetamorphic bullfrogs and green frogs:

Implications for interspecific predation and competition. *J. Herpetol.* 29:600–607.

Werner, J. K. 1969. Temperature-photoperiod effects on spermatogenesis in the salamander *Plethodon cinereus. Copeia* 1969:592–602.

Werren, J. H., M. R. Gross, and R. Shine. 1980. Paternity and the evolution of male parental care. *J. Theor. Biol.* 82:619–31.

Werschkul, D. F., and M. T. Christensen. 1977. Differential predation by *Lepomis macrochirus* on the eggs and tadpoles of *Rana. Herpetologica* 33:237–41.

West, L. B. 1960. The nature of growth inhibiting material from crowded *Rana pipiens* tadpoles. *Physiol. Zool.* 33:232–39.

West, N. H., and W. W. Burggren. 1982. Gill and lung ventilatory responses to steady-state aquatic hypoxia and hyperoxia in the bullfrog tadpole. *Respir. Physiol.* 47:165–76.

———. 1984. Factors influencing pulmonary and cutaneous arterial blood flow in the toad, *Bufo marinus. Am. J. Physiol.* 247:R884–R894.

West, N. H., and D. R. Jones. 1975. Breathing movements in the frog *Rana pipiens.* I. The mechanical events associated with lung and buccal ventilation. *Can. J. Zool.* 53:332–44.

West Eberhard, M. J. 1983. Sexual selection, social competition, and speciation. *Q. Rev. Biol.* 58:155–83.

Westerman, A. G., A. J. Wigginton, D. J. Price, G. Linder, and W. J. Birge. 2003. Integrating amphibians into ecological risk assessment strategies. In *Amphibian decline: An integrated analysis of multiple stressor effects,* ed. G. Linder, S. K. Krest, and D. W. Sparling, 283–313. Pensacola, FL: SETAC Press.

Wetmore, A. 1916. *Birds of Puerto Rico. U.S. Dept. of Agriculture Bulletin 326.* Washington, DC: U.S. Department of Agriculture.

Wetzel, D. M., U. L. Haerter, and D. B. Kelley. 1985. A proposed neural pathway for vocalization in South African clawed frogs *Xenopus laevis. J. Comp. Physiol. A* 157:749–61.

Wever, E. G. 1973. The ear and hearing in the frog *Rana pipiens. J. Morphol.* 141:461–78.

———. 1978. Sound transmission in the salamander ear. *Proc. Natl. Acad. Sci. USA* 75:529–30.

———. 1979. Middle ear muscles of the frog. *Proc. Natl. Acad. Sci. USA* 76:3031–33

———. 1985. *The amphibian ear.* Princeton, NJ: Princeton University Press.

Weygoldt, P. 1976a. Beobachtungen zur Biologie und Ethologie von *Pipa (Hemipipa) calvalhoi* Mir. RiB. 1937 (Anura, Pipidae). *Z. Tierpsychol.* 40:80–99.

———. 1976b. Beobachtungen zur Fortpflanzungsbiologie der Wabenkröte *Pipa carvalhoi* Miranda Ribeiro. *Z. Kölner Zoo* 19:77–84.

———. 1980a. Complex brood care and reproductive behavior in captive poison-arrow frogs, *Dendrobates pumilio* O. Schmidt. *Behav. Ecol. Sociobiol.* 7:329–32.

———. 1980b. Zur Fortpflanzung des Erdbeerfrosches *Dendrobates pumilio. Aquar. Mag.* 1980:460–62.

———. 1980c. Zur Fortpflanzungsbiologie von *Phyllobates femoralis* (Boulenger) im Terrarium. *Salamandra* 16:215–26.

———. 1981a. Beobachtungen zur Fortpflanzungsbiologie von *Phyllodytes luteolus* (Wied 1824) im Terrarium. *Salamandra* 17:1–11.

———. 1981b. Durch Nachzucht erhalten: Der Makifrosh *Phyllomedusa tomopterna. Aquar. Mag.* 15:160–65.

———. 1982. Durch Nachzucht erhalten: der Färberfrosch *Dendrobates tinctorius Aquar. Mag.* 1982:6–13.

———. 1983. Durch Nachzucht erhalten: Blattsteigerfrösche. Drei Arten aus der *Phyllobates pictus*-Gruppe. *Aquar. Mag.* 1983:566–72.

———. 1984a. Beobachtungen zur Fortpflanzungsbiologie von *Dendrobates pumilio* Schmidt, 1857 im Terrarium (Salientia: Dendrobatidae). *Salamandra* 20:112–20.

———. 1984b. Die Eiblage von *Phyllomedusa marginata* Izecksohn & da Cruz, 1976 (Anura: Hylidae) im Terrarium. *Salamandra* 20:248–51.

———. 1985. Ein Kleinod unter den Makifröschen: *Phyllomedusa marginata. Aquar. Mag.* 19:163–67.

———. 1986. Beobachtungen zur Ökologie und Biologie von Fröschen an einem neotropischen Bergbach. *Zool. Jb. Syst.* 113:429–54.

———. 1987. Evolution of parental care in dart poison frogs (Amphibia: Anura: Dendrobatidae). *Z. Zool. Syst. Evolut.-Forsch.* 25:51–67.

———. 1989a. Changes in the composition of mountain stream frog communities in the Atlantic mountains of Brazil: Frogs as indicators of environmental deteriorations? *Stud. Neotrop. Fauna Environ.* 243:249–55.

———. 1989b. Feeding behavior of the larvae of *Fritziana goeldii* (Anura, Hylidae). *Amphibia-Reptilia* 10:419–22.

———. 1991. Zur Biologie und zum Verhalten von *Phyllomedusa marginata* Izecksohn & da Cruz, 1976, im Terrarium. *Salamandra* 27:83–96.

Weygoldt, P., and S. P. Carvalho e Silva. 1991. Observations on mating, oviposition, egg sac formation and development in the egg-brooding frog, *Fritziana goeldii. Amphibia-Reptilia* 12:67–80.

———. 1992. Mating and oviposition in the hylodine frog *Crossodactylus gaudichaudii* (Anura: Leptodactylidae). *Amphibia-Reptilia* 13:35–45.

Wheeler, C. A., H. H. Welsh, Jr., and L. L. Heise. 2003. *Rana boylii* (Foothill Yellow-Legged Frog). Oviposition behavior. *Herp. Rev.* 34:234.

Wheater, C. P. 1986. Prey-size and parasite relationships in the common toad *Bufo bufo. Herpetol. J.* 1:62–66.

Whiles, M. R., K. R. Lips, C. M. Pringle, S. S. Kilham, R. J. Bixby, R. Brenes, S. Connelly, J. Checo Colon-Gaud, M. Hunte-Brown, A. D. Huryn, C. Montgomery, and S. Peterson. 2006. The effects of amphibian population declines on the structure and function of Neotropical stream ecosystems. *Front. Ecol. Env.* 4:27–34.

Whitaker, J. O., Jr., and D. C. Rubin. 1971. Food habits of *Plethodon jordani metcalfi* and *Plethodon jordani shermani* from North Carolina. *Herpetologica* 27:81–86.

Whitaker, J. O., Jr., D. Rubin, and J. R. Munsee. 1977. Observations on food habits of four species of spadefoot toads, genus *Scaphiopus. Herpetologica* 33:468–75.

White, B. A., and C. S. Nicoll. 1981. Hormonal control of amphibian metamorphosis. In *Metamorphosis: A problem in developmental biology* (2nd. ed.), ed. L. I. Gilbert and E. Frieden, 363–96. New York: Plenum.

White, M., and J. A. Kolb. 1974. A preliminary study of *Thamnophis* near Sagehen Creek, California. *Copeia* 1974:126–36.

White, T. D., B. Schmitz, and P. M. Narins. 1992. Directional dependence of auditory sensitivity and frequency selectivity in the leopard frog. *J. Acoust. Soc. Amer.* 92:1953–61.

Whitear, M. 1977. A functional comparison between the epidermis of fish and of amphibians. In *Comparative biology of skin. Symp. Zool. Soc. London, no. 39,* ed. R. I. C. Spearman, 291–313. New York: Academic Press.

Whitehead, P. J., J. T. Puckridge, C. M. Leigh, and R. S. Seymour. 1989. Effect of temperature on jump performance of the frog *Limnodynastes tasmaniensis. Physiol. Zool.* 62:937–49.

Whiteman, H. H. 1994. Evolution of facultative paedomorphosis in salamanders. *Q. Rev. Biol.* 69:205–21.

———. 1997. Maintenance of polymorphism promoted by sex-specific fitness payoffs. *Evolution* 51:2039–44.

Whiteman, H. H., J. P. Sheen, E. B. Johnson, A. VanDeusen, R. Cargille, and T. W. Sacco. 2003. Heterospecific prey and trophic polyphenism in larval tiger salamanders. *Copeia* 2003:56–67.

Whiteman, H. H., and S. A. Wissinger. 1991. Differences in the antipredator behavior of three plethodontid salamanders to snake attack. *J. Herpetol.* 25:352–55.

Whiteman, H. H., S. A. Wissinger, and W. S. Brown. 1996. Growth and foraging consequences of facultative paedomorphosis in the tiger salamander, *Ambystoma tigrinum nebulosum. Evol. Ecol.* 10:433–46.

Whitfield, S. M., and M. S. F. Pierce. 2003. *Eleutherodactylus noblei* (NCN). Reproduction and parental care. *Herp. Rev.* 34:357–58.

Whitford, W. G. 1968. Physiological responses to temperature and desiccation in the endemic New Mexico plethodontids, *Plethodon neomexicanus* and *Aneides hardii. Copeia* 1968:247–51.

———. 1973. The effects of temperature on respiration in the Amphibia. *Amer. Zool.* 13:505–12.

Whitford, W. G., and V. H. Hutchison. 1963. Cutaneous and pulmonary gas exchange in the spotted salamander, *Ambystoma maculatum. Biol. Bull.* 124:344–54.

———. 1965a. Effect of photoperiod on pulmonary and cutaneous respiration in the spotted salamander, *Ambystoma maculatum. Copeia* 1965:53–58.

———. 1965b. Gas exchange in salamanders. *Physiol. Zool.* 38:228–42.

———. 1966. Cutaneous and pulmonary gas exchange in ambystomatid salamanders. *Copeia* 1966:573–77.

———. 1967. Body size and metabolic rate in salamanders. *Physiol. Zool.* 40:127–33.

Whitford, W. G., and K. H. Meltzer. 1976. Changes in $O_2$ consumption, body water and lipid in burrowed desert juvenile anurans. *Herpetologica* 32:23–25.

Whitford, W. G., and M. Massey. 1970. Responses of a population of *Ambystoma tigrinum* to thermal and oxygen gradients. *Herpetologica* 26:372–76.

Whitford, W. G., and R. E. Sherman. 1968. Aerial and aquatic respiration in axolotl and transformed *Ambystoma tigrinum. Herpetologica* 24:233–37.

Whitford, W. G., and A. Vinegar. 1966. Homing, survivorship and overwintering of larvae in spotted salamanders, *Ambystoma maculatum. Copeia* 1966:515–19.

Whitney, C. L. 1980. The role of the "encounter" call in spacing of Pacific tree frogs, *Hyla regilla. Can. J. Zool.* 58:75–78.

Whitney, C. L., and J. R. Krebs. 1975a. Mate selection in Pacific tree frogs, *Hyla regilla.* Nature 255:325–26.

———. 1975b. Spacing and calling in Pacific tree frogs, *Hyla regilla. Can. J. Zool.* 53:1519–27.

Wickler, W. 1968. *Mimicry in plants and animals.* New York: McGraw Hill.

———. 1972. Duettieren zwischen artverscheidenen Vögeln im Freiland. Z. *Tierpsychol.* 31:98–103.

Wickler, W., and U. Seibt. 1974. Rufen und Antworten bei *Kassina senegalensis, Bufo regularis* und anderen Anuren. *Z. Tierpsychol.* 34:524–37.

Wickramasinghe, D. D., K. L. Oseen, S. W, Kotagama, and R. J. Wassersug. 2005. Raising semi-terrestrial tadpoles of *Nannophrys ceylonensis* (Anura, Ranidae) in the laboratory. *Herp. Rev.* 36:23–25.

Widmer, A., S. Lötters, and K.-H. Jungfer. 2000. A molecular phylogenetic analysis of the Neotropical dart-poison frog genus *Phyllobates* (Amphibia: Dendrobatidae). *Naturwissenschaften* 87:559–62.

Wieczorek, A. M., A. Channing, and R. C. Drewes. 1998. A review of the taxonomy of the *Hyperolius viridiflavus* complex. *Herpetol. J.* 8:29–34.

Wieczorek, A. M., R. C. Drewes , and A. Channing. 2001. Phylogenetic relationships within the *Hyperolius viridiflavus* complex (Anura: Hyperoliidae), and comments on taxonomic status. *Amphibia-Reptilia* 22:155–66.

Wiederholt, M., W. J. Sullivan, and G. Giebisch. 1971. Transport of potassium and sodium across single distal tubules of Amphiuma. *J. Gen. Physiol.* 57:495–529.

Wiens, J. A. 1973. Pattern and process in grassland bird communities. *Ecol. Monogr.* 43:237–70.

———. 1976. Population responses to patchy environments. *Ann. Rev. Ecol. Syst.* 7:81–120.

———. 1977. On competition and variable environments. *Amer. Sci.* 65:590–97.

———. 1981. Single-sample surveys of communities: Are the revealed patterns real? *Am. Nat.* 117:90–98.

———. 1984. On understanding a non-equilibrium world: Myth and reality in community patterns and processes. In *Ecological communities,* ed. D. R. Strong, Jr., D. Simberloff, L. G. Abele, and A. B. Thistle, 439–57. Princeton, NJ: Princeton University Press.

Wiens, J. A., and J. T. Rotenberry. 1979. Diet niche relationships among North American grassland and shrubsteppe birds. *Oecologia* 42:253–92.

———. 1980. Patterns of morphology and ecology in grassland and shrubsteppe bird populations. *Ecol. Monogr.* 50:287–308.

Wiens, J. J. 1993. Systematics of the leptodactylid frog genus *Telmatobius* in the Andes of northern Peru. *Occ. Pap. Mus. Nat. Hist. Univ. Kansas* (162):1–76.

Wiens, J. J., R. M. Bonett, and P. T. Chippindale. 2005. Ontogeny discombobulates phylogeny: Paedomorphosis and higher-level salamander relationships. *Syst. Biol.* 54:91–110.

Wiens, J. J., P. T. Chippindale, and D. M. Hillis. 2003. When are phylogenetic analyses misled by convergence? A case study in Texas cave salamanders. *Syst. Biol.* 52:501–14.

Wiens, J. J., J. W. Fetzner, Jr., C. L Parkinson, and T. W. Reeder. 2005. Hylid frog phylogeny and sampling strategies for speciose clades. *Syst. Biol.* 54:719–48.

Wiest, J. A., Jr. 1982. Anuran succession at temporary ponds in a post oak-savanna region of Texas. In *Herpetological communities. U.S. Fish and Wildlife Service Wildlife Report 13,* ed. N. J. Scott, Jr., 39–47. U.S. Fish and Wildlife Service: Washington, DC.

Wiewandt, T. A. 1969. Vocalization, aggressive behavior, and territoriality in the bullfrog, *Rana catesbeiana. Copeia* 1969:276–85.

———. 1971. Breeding biology of the Mexican leaf-frog. *Fauna* (2):29–34.

Wijesinghe, M. R., and M. de L. Brooke. 2004. What causes the vulnerability of endemic animals? A case study from Sri Lanka. *J. Zool., London* 263:135–40.

Wijeyamohan, S., M. da Silva, and C. Santiapillai. 1994. Conservation of amphibian diversity. *Loris* 20:154–57.

Wikelski, M., C. Carbone, and F. Trillmich. 1996. Lekking in marine iguanas: Female grouping and male reproductive strategies. *Anim. Behav.* 52:581–96.

Wikelski, M., and F. Trillmich. 1997. Body size an sexual size dimorphism in marine iguanas fluctuate as a result of opposing natural and sexual selection: An island comparison. *Evolution* 51:922–36.

Wilbur, H. M. 1971. The ecological relationship of the salamander *Ambystoma laterale* to its all-female, gynogenetic associate. *Evolution* 25:168–79.

———. 1972. Competition, predation, and the structure of the *Ambystoma-Rana sylvatica* community. *Ecology* 53:3–21.

———. 1976. Density-dependent aspects of metamorphosis in *Ambystoma* and *Rana sylvatica. Ecology* 57:1289–96.

———. 1977a. Density-dependent aspects of growth and metamorphosis in *Bufo americanus. Ecology* 58:196–200.

———. 1977b. Interactions of food level and population density in *Rana sylvatica. Ecology* 58:206–9.

———. 1977c. Propagule size, number, and dispersion patterns in *Ambystoma* and *Asclepias. Am. Nat.* 111:43–68.

———. 1980. Complex life cycles. *Ann. Rev. Ecol. Syst.* 11:67–93.

———. 1982. Competition between tadpoles of *Hyla femoralis* and *Hyla gratiosa* in laboratory experiments. *Ecology* 63:278–82.

———. 1984. Complex life cycles and community organization in amphibians. In *A new ecology: Novel approaches to interactive systems,* ed. P. W. Price, C. N. Slobodchikoff, and W. S. Gaud, 196–224. New York: Wiley.

———. 1987. Regulation of structure in complex systems: Experimental temporary pond communities. *Ecology* 68:1437–52.

———. In defense of cattle tanks. *Herpetologica* 45:122–23.

———. 1997. Experimental ecology of food webs: Complex systems in temporary ponds. *Ecology* 78:2279–2302.

Wilbur, H. M., and R. A. Alford. 1985. Priority effects in experimental pond communities: responses of *Hyla* to *Bufo* and *Rana. Ecology* 66:1106–14.

Wilbur, H. M., and J. P. Collins. 1973. Ecological aspects of amphibian metamorphosis. *Science* 182:1305–14.

Wilbur, H. M., and J. E. Fauth. 1990. Experimental aquatic food webs: Interactions between two predators and two prey. *Am. Nat.* 135:176–204.

Wilbur, H. M., P. J. Morin, and R. N. Harris. 1983. Salamander predation and the structure of experimental communities: anuran responses. *Ecology* 64:1423–29.

Wilbur, H. M., D. I. Rubenstein, and L. Fairchild. 1978. Sexual selection in toads: The roles of female choice and male body size. *Evolution* 32:264–70.

Wilbur, H. M., and R. D. Semlitsch. 1990. Ecological consequences of tail injury in *Rana* tadpoles. *Copeia* 1990:18–24.

Wilbur, H. M., and J. Travis. 1984. An experimental approach to understanding pattern in natural communities. In *Ecological communities: Conceptual issues and the evidence,* ed. D. R. Strong, D. Simberloff, L. G. Abele, and A. B. Thistle, 113–22. Princeton, NJ: Princeton University Press.

Wilczynski, W. 1981. Afferents to the midbrain auditory center in the bullfrog, *Rana catesbeiana. J. Comp. Neurol.* 198:421–33.

———. 1988. Brainstem auditory pathways in anuran amphibians. In *The evolution of the amphibian auditory system,* ed. B. Fritzsch, M. J. Ryan, W. Wilczynski, T. E. Hetherington, and W. Walkowiak, 209–31. New York: Wiley.

———. 1992. Auditory and endocrine inputs to forebrain centers in anuran amphibians. *Ethol. Ecol. Evol.* 4:75–87.

Wilczynski, W., and J. D. Allison. 1989. Acoustic modulation of neural activity in the hypothalamus of the leopard frog. *Brain Behav. Evol.* 33:317–24.

Wilczynski, W., J. D. Allison, and C. A. Marler. 1993. Sensory pathways linking social and environmental cues to endocrine control regions of amphibian forebrains. *Brain Behav. Evol.* 42:252–64.

Wilczynski, W., and E. A. Brenowitz. 1988. Acoustic cues mediate inter-male spacing in a Neotropical frog. *Anim. Behav.* 36:1054–63.

Wilczynski, W., and R. R. Capranica. 1984. The auditory system of anuran amphibians. *Progr. Neurobiol.* 22:1–38.

Wilczynski, W., and J. Chu. 2001. Acoustic communication, endocrine control, and the neurochemical systems of the brain. In *Anuran communication,* ed. M. J. Ryan, 23–35. Washington, DC: Smithsonian Institution Press.

Wilczynski, W., A. C. Keddy-Hector, and M. J. Ryan. 1992. Call patterns and basilar papilla tuning in cricket frogs. I. Differences among populations and between sexes. *Brain Behav. Evol.* 39:229–37.

Wilczynski, W., B. E. McClelland, and A. S. Rand. 1993. Acoustic, auditory, and morphological divergence in three species of Neotropical frog. *J. Comp. Physiol. A* 172:425–38.

Wilczynski, W., and R. G. Northcutt. 1977. Afferents to the optic tectum of the leopard frog: an HRP study. *J. Comp. Neurol.* 173:219–30.

———. 1983. Connections of the bullfrog striatum: Afferent organization. *J. Comp. Neurol.* 214:321–32.

Wilczynski, W., A. S. Rand, and M. J. Ryan. 1995. The processing of spectral cues by the call analysis system of the túngara frog, *Physalaemus pustulosus. Anim. Behav.* 49:911–29.

Wilczynski, W., A. S. Rand, and M. J. Ryan. 1999. Female preferences for temporal order of call components in the túngara frog: A Bayesian analysis. *Anim. Behav.* 58:841–51.

Wilczynski, W., A. S. Rand, and M. J. Ryan. 2001. Evolution of calls and auditory tuning in the *Physalaemus pustulosus* species group. *Brain Behav. Evol.* 58:137–51.

Wilczynski, W., C. Resler, and R. R. Capranica. 1987. Tympanic and extratympanic sound transmission in the leopard frog. *J. Comp. Physiol. A* 161:659–69.

Wilczynski, W., H. H. Zakon, and E. A. Brenowitz. 1984. Acoustic communication in spring peepers. Call characteristics and neurophysiological aspects. *J. Comp. Physiol. A* 155:577–84.

Wild, E. R. 1995. New genus and species of Amazonian microhylid frog with a phylogenetic analysis of New World genera. *Copeia* 1995:837–49.

———. 1996. Natural history and resource use of four Amazonian tadpole assemblages. *Occas. Pap. Nat. Hist. Mus. Univ. Kansas* (176):1–59.

Wilder, I. 1923. Spermatophores of *Desmognathus fusca. Copeia* 1923:88–92.

Wilder, I., L. Whipple and E. R. Dunn. 1920. The correlation of lunglessness in salamanders with a mountain brook habitat. *Copeia* 1920:63–68.

Wilder, I. W. 1913. The life history of *Desmognathus fusca. Biol. Bull.* 24:251–343.

———. 1924. The development history of *Eurycea bislineata* in western Massachusetts. *Copeia* 1924:77–80.

———. 1925. *The Morphology of Amphibian Metamorphosis.* Northampton, MA: Smith College.

Wildy, E. L., and A. R. Blaustein. 2001. Learned recognition of intraspecific predators in larval long-toed salamanders *Ambystoma macrodactylum. Ethology* 107:479–93.

Wildy, E. L., D. P. Chivers, and A. R. Blaustein. 1999. Shifts in life-history traits as a response to cannibalism in larval long-toed salmanders (*Ambystoma macrodactylum*). *J. Chem. Ecol.* 25:2337–46.

Wildy, E. L., D. P. Chivers, J. M. Kiesecker, and A. R. Blaustein. 1998. Cannibalism enhances growth in larval long-toed salamanders (*Ambystoma macrodactylum*). *J. Herpetol.* 32:286–89.

Wildy, E. L., D. P. Chivers, J. M. Kiesecker, and A. R. Blaustein. 2001. The effects of food level and conspecific density on biting and cannibalism in larval long-toed salamanders, *Ambystoma macrodactylum. Oecologia* 128:202–9.

Wiley, R. H., and D. G. Richards. 1978. Physical constraints on acoustic communication in the atmosphere: Implications for the evolution of animal vocalizations. *Behav. Ecol. Sociobiol.* 3:69–94.

———. 1982. Adaptations for acoustic communication in birds: sound transmission and signal detection. In *Acoustic communication in birds.* Vol. 1, ed. D. E. Kroodsma and E. H. Miller, 132–81. New York: Academic Press.

Wilhelm, K. H., H. Comtesse, and W. Phlumm. 1980. Zur Abhängigket des Gesangs von Nährungsangebot beim Gelbbauchnektarvogel (*Nectarina venusta*). *Z. Tierpsychol.* 54:185–202.

Wilkins, R. N., and N. P. Peterson. 2000. Factors related to amphibian occurrence and abundance in headwater streams draining second-growth Douglas-fir forests in southwestern Washington. *Forest Ecol. Manage.* 139:79–91.

Wilkinson, J. A., R. C. Drewes, and O. L. Tatum. 2002. A molecular phylogenetic analysis of the family Rhacophoridae with emphasis on the Asian and African genera. *Mol. Phylogen. Evol.* 24:265–73.

Wilkinson, M. 1992. On the life history of the caecilian genus *Uraeotyphlus* (Amphibia: Gymnophiona). *Herpetol. J.* 2:121–24.

———. 1997. Characters, congruence and quality: A study of neuroanatomical and traditional data in caecilian phylogeny. *Biol. Rev.* 72:423–70.

Wilkinson, M., and R. A. Nussbaum. 1996. On the phylogenetic position of the Uraeotyphlidae (Amphibia: Gymnophiona). *Copeia* 1996:550–62.

———. 1997. Comparative morphology and evolution of the lungless caecilian *Aretochoana eiselti* (Taylor) (Amphibia: Gymnophiona: Typhlonectidae). *Biol. J. Linn. Soc.* 62:39–109.

———. 1998. Caecilian viviparity and amniote origins. *J. Nat. Hist.* 32:1403–9.

———. 1999. Evolutionary relationships of the lungless caecilian *Aretochoana eiselti* (Amphibia: Gymnophiona: Typhlonectidae). *Zool. J. Linn. Soc.* 126:191–223.

Wilkinson, M., A. Sebben, E. N. F. Schwartz, and C. Schwartz. 1998. The largest lungless tetrapod: report on a second specimen of *Aretochoana eiselti* (Amphibia: Gymnophiona: Typhlonectidae) from Brazil. *J. Nat. Hist.* 32:617–27.

Wilkinson, M., J. A. Sheps, O. V. Oommen, and B. L. Cohen. 2002. Phylogenetic relationships of Indian caecilians (Amphibia: Gymnophiona) inferred from mitochondrial rDNA gene sequences. *Mol. Phylogen. Evol.* 23:401–7.

Will, U. 1988. Organization and projections of the area octavolateralis in amphibians. In *The evolution of the amphibian auditory system,* ed. B. Fritzsch, M. J. Ryan, W. Wilczynski, T. E. Hetherington, and W. Walkowiak, 185–208. New York: Wiley.

Will, U., and B. Fritzsch. 1988. The eighth nerve of amphibians: peripheral and central distribution. In *The evolution of the amphibian auditory system,* ed. B. Fritzsch, M. J. Ryan, W. Wilczynski, T. E. Hetherington, and W. Walkowiak, 159–83. New York: Wiley.

Williams, B. L., E. D. Brodie, Jr., and E. D. Brodie III. 2003. Coevolution of deadly toxins and predator resistance: Self-assessment of resistance by garter snakes leads to behavioral rejection of toxic newt prey. *Herpetologica* 59:155–63.

Williams, C. B. 1918. Mongoose. The food habits of the mongoose in Trinidad. *Bull. Dept. Agricul. Trinidad and Tobago* 17:167–86.

Williams, F. 1936. Two water beetles that lay their eggs in the frothy egg masses of a frog or tree toad. *Pan-Pacific Entomol.* 12:6–7.

Williams, G. C. 1966. *Adaptation and natural selection.* Princeton, NJ: Princeton University Press.

———. 1975. *Sex and evolution.* Princeton, NJ: Princeton University Press,

Williams, P. K. 1973. *Seasonal movements and population dynamics of four sympatric mole salamanders, genus Ambystoma.* Unpublished. PhD dissertation. Bloomington: Indiana University.

Williams, S. E., E. E. Bolitho, and S. Fox. 2003. Climate change in Australian tropical rainforests: an impending environmental catastrophe. *Proc. Roy. Soc. London B* 270:1887–92.

Williams, S. E., and J.-M. Hero. 1998. Rainforest frogs of the Australian wet tropics: guild classification and the ecological similarity of declining species. *Proc. Roy. Soc. London B* 265:597–602.

Williamson, I. 1999. Competition between the larvae of the introduced cane toad *Bufo marinus* (Anura: Bufonidae) and native anurans from the Darling Downs area of southern Queensland. *Aust. J. Ecol.* 24:636–43.

Williamson, I., and C. M. Bull. 1989. Life history variation in a population of the Australian frog *Ranidella signifera:* Egg size and early development. *Copeia* 1989:349–56.

———. 1994. Population ecology of the Australian frog *Crinia signifera:* Egg-laying patterns and egg mortality. *Wildl. Res.* 21:621–32.

———. 1996. Population ecology of the Australian frog *Crinia signifera:* Adults and juveniles. *Wildl. Res.* 23:249–66.

———. 1999. Population ecology of the Australian frog *Crinia signifera:* Larvae. *Wildl. Res.* 26:81–99.

Willis, Y. I., D. L. Moyle, and T. S. Baskett. 1956. Emergence, breeding, hibernation, movements and transformation of the bullfrog, *Rana catesbeiana,* in Missouri. *Copeia* 1956:30–41.

Wills, I. A. 1936. The respiratory rate of developing amphibians with special reference to sex differentiation. *J. Exp. Zool.* 73:481–510.

Wilnow, I. 1964. Die Lunge von *Amphiuma means means.* I. Mitteilung: Morphologie. *Zool. Beitr. Berlin* 10:29–85.

———. 1968. Die Lunge von *Amphiuma means means.* II. Mitteilung: Diskussion. *Zool. Beitr. Berlin* 14:359–86.

Wilson, A. G., and E. M. Simon. 1985. *Plethodon vandykei idahoensis* (Coeur d'Alene salamander). Predation. *Herp. Rev.* 16:111.

Wilson, D. J., and H. Lefcourt. 1993. The effect of predator diet on the alarm responses of red-legged frog, *Rana aurora,* tadpoles. *Anim. Behav.* 46:1017–19.

Wilson, D. S. 1980. *The natural selection of populations and communities.* Menlo Park, CA: Benjamin/Cummings.

Wilson, E. O. 1975. *Sociobiology: The new synthesis.* Cambridge, MA: Harvard University Press.

Wilson, L. D., and J. R. McCranie. 2003. The conservation status of the herpetofauna of Honduras. *Amphib. Rept. Cons.* 3:6–33.

Wilson, R. S. 2001. Geographic variation in thermal sensitivity of jumping performance in the frog *Limnodynastes peronii. J. Exp. Biol.* 204:4227–36.

Wilson, R. S., and C. E. Franklin. 1999. Thermal acclimation of locomotor performance in tadpoles of the frog *Limnodynastes peronii. J. Comp. Physiol. B* 169:445–51.

Wilson, R. S. and C. E. Franklin. 2000a. Effect of ontogenetic increases in body size on burst swimming performance in tadpoles of the striped marsh frog, *Limnodynastes peronii. Physiol. Biochem. Zool.* 73:142–52.

———. 2000b. Inability of adult *Limnodynastes peronii* (Amphibia: Anura) to thermally acclimate locomotor performance. *Comp. Biochem. Physiol. A* 127:21–28.

Wilson, R. S., R. S. James, and I. A. Johnston. 2000. Thermal acclimation of locomotor performance in tadpoles and adults of the aquatic frog *Xenopus laevis. J. Comp. Physiol. B* 170:117–24.

Wilson, R. S., R. S. James, T. Kohlsdorf, and V. M. Cox. 2004. Interindividual variation of isolated muscle performance and fibre-type composition in the toad *Bufo viridis. J. Comp. Physiol. B* 174:453–59.

Wilson, R. S., P. G. Kraft, and R. Van Damme. 2005. Predator-specific changes in the morphology and swimming performance of larval *Rana lessonae. Funct. Ecol.* 19:238–44.

Wiltenmuth, E. B. 1996. Agonistic and sensory behaviour of the salamander *Ensatina eschscholtzii* during asymmetrical contests. *Anim. Behav.* 52:841–50.

Wiltenmuth, E. B., and K. C. Nishikawa. 1998. Geographical variation in agonistic behaviour in a ring species of salamander, *Ensatina eschscholtzii. Anim. Behav.* 55:1595–1606.

Wiltschko, R., and W. Wiltschko. 1995. *Magnetic orientation in animals.* New York: Springer-Verlag.

Wiltshire, D. J., and C. M. Bull. 1977. Potential competitive interactions between larvae of *Pseudophryne bibroni* and *P. semimarmorata* (Anura: Leptodactylidae). *Aust. J. Zool.* 25:449–54.

Wind-Larsen, H., and C. B. Jørgensen. 1987. Hormonal control of seasonal growth in a temperate-zone toad *Bufo bufo. Acta Zool. (Stockholm)* 68:49–56.

Windmiller, B. 1996. Tracking techniques useful for field studies of anuran orientation and movement. *Herp. Rev.* 27:13–15.

Wine, R. N., and R. E. Gatten, Jr. 1992. Glycolytic support for locomotion in *Anolis carolinensis* and *Rana pipiens:* Simultaneous effects of temperature and duration of exercise. *Physiol. Zool.* 65:724–41.

Winemiller, K. O., and K. A. Rose. 1993. Why do most fish produce so many tiny offspring? *Am. Nat.* 142:585–603.

Winokur, R. M., and S. D. Hillyard. 1992. Pelvic cutaneous musculature in toads of the genus *Bufo. Copeia* 1992:760–69.

Winterbourn, M. J. 1971a. An ecological study of *Banksiola crotchi* Banks (Trichoptera, Phryganeidae) in Marion Lake, British Columbia. *Can. J. Zool.* 49:636–45.

———. 1971b. The life histories and trophic relationships of the Trichoptera of Marion Lake, British Columbia. *Can. J. Zool.* 49:623–35.

Wirsig-Wiechmann, C. R., L. D. Houck, P. W. Feldhoff, and R. C. Feldhoff. 2002. Pheromonal activation of vomeronasal neurons in plethodontid salamanders. *Brain Res.* 952:335–44.

Wirsing, A. J., J. D. Roth, and D. L. Murray. 2005. Can prey use dietary cues to distinguish predators? A test involving three terrestrial amphibians. *Herpetologica* 61:104–10.

Wirtz, W. O., D. H. Austin, and G. W. Deklee. 1985. Food habits of the common long-nosed armadillo *Dasypus novemcinctus* in Florida, 1960–61. In *The evolution and ecology of armadillos, sloths, and vermilinguas,* ed. G. G. Montgomery, 439–51. Washington, DC: Smithsonian Institution Press.

Wise, D. H. 1984. The role of competition in spider communities: Insights from field experiments with a model organism. In *Ecological communities,* ed. D. R. Strong, Jr., D. Simberloff, L. G. Abele, and A. B. Thistle, 42–53. Princeton, NJ: Princeton University Press.

Wise, M. H., I. J. Linn, and C. R. Kennedy. 1981. A comparison of the feeding biology of mink (*Mustela vison*) and otter (*Lutra lutra*). *J. Zool., London* 195:181–213.

Wise, S. E., and R. G. Jaeger. 1998. The influence of tail autotomy on agonistic behaviour in a territorial salamander. *Anim. Behav.* 55:1707–16.

Wise, S. E., K. S. Siex, K. M. Brown, and R. G. Jaeger. 1993. Recognition influences social interactions in red-spotted newts. *J. Herpetol.* 27:149–54.

Wisniewski, P. J., L. M. Paull, and F. M. Slater. 1981. The effects of temperature on the breeding migration and spawning of the common toad (*Bufo bufo*). *Brit. J. Herp.* 6:119–21.

Wissinger, S. A., and H. Whiteman. 1992. Fluctuation in a Rocky Mountain population of salamanders: anthropogenic acidification or natural variation? *J. Herpetol.* 26:377–91.

Wissinger, S. A., H. H. Whiteman, G. B. Sparks, G. L Rouse, and W. S. Brown. 1999. Foraging trade-offs along a predator-permanence gradient in subalpine wetlands. *Ecology* 80:2102–16.

Withers, P. C. 1980. Oxygen consumption of plethodontid salamanders during rest, activity, and recovery. *Copeia* 1980:781–86.

———. 1981. Physiological correlates of limblessness and fossoriality in scincid lizards. *Copeia* 1981:197–204.

———. 1993. Metabolic depression during aestivation in the Australian frogs *Neobatrachus* and *Cyclorana. Aust. J. Zool.* 41:467–73.

———. 1995. Cocoon formation and structure in the aestivating Australian desert frogs, *Neobatrachus* and *Cyclorana. Aust. J. Zool.* 43:429–41.

———. 1998. Evaporative water loss and the role of cocoon formation in Australian frogs. *Aust. J. Zool.* 46:405–18.

Withers, P. C., and M. Guppy. 1996. Do Australian desert frogs co-accumulate counteracting solutes with urea during aestivation? *J. Exp. Biol.* 199:1809–16.

Withers, P. C., and S. S. Hillman. 1981. Oxygen consumption of *Amphiuma means* during forced activity and recovery. *Comp. Biochem. Physiol.* 69A:141–44.

———. 1983. The effects of hypoxia on pulmonary function and maximal rates of oxygen consumption in two anuran amphibians. *J. Comp. Physiol.* 152:125–29.

———. 1988. A steady-state model of maximal oxygen and carbon dioxide transport in anuran amphibians. *J. Appl. Physiol.* 64:860–68.

Withers, P. C., S. S. Hillman, and R. C. Drewes. 1984. Evaporative water loss and skin lipids of anuran amphibians. *J. Exp. Zool.* 232:11–17.

Withers, P. C., S. S. Hillman, R. C. Drewes, and O. M. Sokol. 1982. Water loss and nitrogen excretion in sharp-nosed reed frogs (*Hyperolius nasutus:* Anura, Hyperoliidae). *J. Exp. Biol.* 97:335–43.

Withers, P. C., S. S. Hillman, L. A. Simmons, and A. C. Zygmunt. 1988. Cardiovascular adjustments to enforced activity in the anuran amphibian, *Bufo marinus. Comp. Biochem. Physiol.* 89A:45–49.

Withers, P. C., M. Lea, T. C. Solberg, M. Baustian, and M. Hedrick. 1988. Metabolic fates of lactate during recovery from activity in an anuran amphibian, *Bufo americanus. J. Exp. Zool.* 246:236–43.

Withers, P., G. Louw, and S. Nicholson. 1982. Water loss, oxygen consumption and colour change in 'waterproof' reed frogs (*Hyperolius*). *S. Afr. J. Sci.* 78:30–32.

Withers, P. C., and J. D. Roberts. 1993. Water balance response in the terrestrial Western Australian frog, *Myobatrachus gouldii* (Anura: Myobatrachidae). *J. Roy. Soc. West. Australia* 76:49–52.

Witte, K., K.-C. Chen, W. Wilczynski, and M. J. Ryan. 2000. Influence of amplexus on phonotaxis in the cricket frog *Acris crepitans blanchardi. Copeia* 2000:257–61.

Witte, K., H. E. Farris, M. J. Ryan, and W. Wilczynski. 2005. How cricket frog females deal with a noisy world: habitat-related differences in auditory tuning. *Behav. Ecol.* 16:571–79.

Witte, K., M. J. Ryan, and W. Wilczynski. 2001. Changes in the frequency structure of a mating call decreases its attractiveness to females in the cricket frog *Acris crepitans blanchardi. Ethology* 107:685–99.

Wittenberger, J. F. 1981. *Animal social behavior.* Boston: Duxbury Press.

———. 1983. Tactics of mate choice. In *Mate choice,* ed. P. Bateson, 435–47. New York: Cambridge University Press.

Witters, L. R., and L. Sievert. 2001. Feeding causes thermophily in the Woodhouse's toad (*Bufo woodhousii*). *J. Therm. Biol.* 26:205–8.

Witz, B. W., D. S. Wilson, and M. D. Palmer. 1991. Distribution of *Gopherus polyphemus* and its vertebrate symbionts in three burrow categories. *Am. Midl. Nat.* 126:152–58.

Wogel, H., P. A. Abrunhosa, and J. P. Pombal, Jr. 2004. Vocalizations and aggressive behavior of *Phyllomedusa rohdei* (Anura: Hylidae). *Herp. Rev.* 35:239–43.

Wogel, H., P. A. Abrunhosa, and J. P. Pombal. 2005. Breeding behaviour and mating success of *Phyllomedusa rohdei* (Anura, Hylidae) in south-eastern Brazil. *J. Nat. Hist.* 39:2035–45.

Wolf, J. H. D., and C. J. F. Konings. 2001. Toward the sustainable harvesting of epiphytic bromeliads: A pilot study from the highlands of Chiapas, Mexico. *Biol. Cons.* 101:23–31.

Wolf, N. G. 1985. Odd fish abandon mixed-species groups when threatened. *Behav. Ecol. Sociobiol.* 17:47–52.

———. 1976. The winter food of the European water-shrew. *Acta Theriol.* 21:117–29.

Wollerman, L. 1998. Stabilizing and directional preferences of female *Hyla ebraccata* for calls differing in static properties. *Anim. Behav.* 55:1619–30.

———. 1999. Acoustic interference limits call detection in a Neotropical frog *Hyla ebraccata. Anim. Behav.* 57:529–36.

Wollerman, L., and R. H. Wiley. 2002a. Background noise from a natural chorus alters female discrimination of male calls in a Neotropical frog. *Anim. Behav.* 63:15–22.

———. 2002b. Possibilities for error during communication by Neotropical frogs in a complex acoustic environment. *Behav. Ecol. Sociobiol.* 52:465–73.

Wollmuth, L. P., and L. I. Crawshaw. 1988. The effect of development and season on temperature selection in bullfrog tadpoles. *Physiol. Zool.* 6l:461–69.

Wollmuth, L. P., L. I. Crawshaw, R. B. Forbes, and D. A. Grahn. 1987. Temperature selection during development in a montane anuran species, *Rana cascadae. Physiol. Zool.* 60:472–80.

Wolterstorff, W. 1923. Überwintern Tritonen im Wasser und aus welchen Ursachen? *Bl. Aquar. Terrariek. (Stuttgart)* 34:10.

Wong, A. L.-C., and T. J. C. Beebee. 1994. Identification of a unicellular nonpigmented alga that mediates growth inhibition in tadpoles: A new member of the genus *Prototheca* (Chlorophyceae: Chlorococcales). *Hydrobiologia* 277:85–96.

Wong, B. B. M., A. N. N. Cowling, R. B. Cunningham, C. F. Donnelly, and P. D. Cooper. 2004. Do temperature and social environment interact to affect call rate in frogs (*Crinia signifera*)? *Aust. Ecol.* 29:209–14.

Wong, S., and D. T. Booth. 1994. Hypoxia induces surfacing behaviour in brown-striped frog (*Limnodynastes peronii*) larvae. *Comp. Biochem. Physiol.* 109A:437–45.

Wood, J. T. 1950. Eggs of the two-lined salamander, *Eurycea b. bislineata. Virginia J. Sci.* 1:348.

———. 1951. Protective behavior and photic orientation in aquatic adult and larval two-lined salamanders, *Eurycea* b. *bislineata* X *cirrigera. Virginia J. Sci.,* n.s. 2:113–21.

———. 1953a. The nesting of the two-lined salamander, *Eurycea bislineata,* on the Virginia coastal plain. *Nat. Hist. Misc. Chicago Acad. Nat. Sci.* (122):7.

———. 1953b. Observations on the complements of ova and nesting of the four-toed salamander in Virginia. *Am. Nat.* 87:77–86.

Wood, J. T., and R. F. Clarke. 1955. The dusky salamander: Oophagy in nesting sites. *Herpetologica* 11:150–51.

Wood, J. T., and H. N. McCutcheon. 1954. Ovarian egg complements and nests of the two-lined salamander, *Eurycea b. bislineata* X *cirrigera,* from southeastern Virginia. *Am. Midl. Nat.* 52:433–36.

Wood, J. T., and R. H. Rageot. 1963. The nesting of the many-lined salamander in the Dismal Swamp. *Virginia J. Sci.,* n.s. 14:121–25.

Wood, K. V., J. D. Nichols, H. F. Percival, and J. E. Hines. 1998. Size-sex variation in survival rates and abundance of pig frogs, *Rana grylio,* in northern Florida wetlands. *J. Herpetol.* 32:527–35.

Wood, S. C. 1971. Effect of metamorphosis on blood respiratory properties and erythrocyte adenosine triphosphate level of the salamander, *Dicamptodon ensatus. Respir. Physiol.* 12:53–66.

———. 1980. Adaptation of red blood cell function to hypoxia and temperature in ectothermic vertebrates. *Amer. Zool.* 20:163–72.

Wood, S. C., R. W. Hoyt, and W. W. Burggren. 1982. Control of hemoglobin function in the salamander, *Ambystoma tigrinum. Mol. Physiol.* 2:263–72.

Wood, S. C., and C. Lenfant. 1987. Phylogeny of the gas-exchange system: Red cell function. In *Handbook of physiol-*

*ogy. Section 3: The respiratory system, vol. 4*, ed. A. P. Fishman, L. E. Farhi, and S. M. Tenney, 131–46. Bethesda, MD: American Physiological Society.

Wood, S. C., R. E. Weber, G. M. O. Maloiy, and K. Johansen. 1975. Oxygen uptake and blood respiratory properties of the caecilian *Boulengerula taitanus*. *Respir. Physiol.* 24:355–63.

Woodford, J. E., and M. W. Meyer. 2003. Impact of lakeshore development on green frog abundance. *Biol. Cons.* 110:277–84.

Woodhams, D. C., and R. A. Alford. 2005. Ecology of chytridiomycosis in rainforest stream frog assemblages of tropical Queensland. *Cons. Biol.* 19:1449–59.

Woodhams, D. C., L. A. Rollins-Smith, C. Carey, L. Reinert, M. J. Tyler, and R. A. Alford. 2006. Population trends associated with skin peptide defenses against chytridiomycosis in Australian frogs. *Oecologia* 146:531–40.

Woodhams, D. C., R. A. Alford, and G. Marantelli. 2003. Emerging disease of amphibians cured by elevated body temperature. *Dis. Aquat. Organ.* 55:65–67.

Woods, M., R. A. McDonald, and S. Harris. 2003. Predation of wildlife by domestic cats *Felis catus* in Great Britain. *Mammal Rev.* 33:174–88.

Woodley, S. K. 1994. Plasma androgen levels, spermatogenesis, and secondary sexual characteristics in two species of plethodontid salamanders with dissociated reproductive patterns. *Gen. Comp. Endocrinol.* 96:206–14.

Woodruff, D. S. 1973. Natural hybridization and hybrid zones. *Syst. Zool.* 22:213–18.

———. 1976a. Courtship, reproductive rates, and mating systems in three Australian *Pseudophryne* (Amphibia, Anura, Leptodactylidae). *J. Herpetol.* 10:313–18.

———. 1976b. Embryonic mortality in *Pseudophryne* (Anura: Leptodactylidae). *Copeia* 1976:445–49.

———. 1977. Male postmating brooding behavior in three Australian *Pseudophryne* (Anura: Leptodactylidae). *Herpetologica* 33:296–303.

———. 1978. Hybridization between two species of *Pseudophryne* (Anura: Leptodactylidae) in the Sydney Basin, Australia. *Proc. Linn. Soc. New South Wales* 102:131–47.

Woodward, B. D. 1982a. Local intraspecific variation in clutch parameters in the spotted salamander (*Ambystoma maculatum*). *Copeia* 1982:157–60.

———. 1982b. Male persistence and mating success in Woodhouse's toad (*Bufo woodhousei*). *Ecology* 63:583–85.

———. 1982c. Sexual selection and nonrandom mating patterns in desert anurans (*Bufo woodhousei, Scaphiopus couchi, S. multiplicatus*, and *S. bombifrons*). *Copeia* 1982:351–55.

———. 1982d. Tadpole competition in a desert anuran community. *Oecologia* 54:96–100.

———. 1983. Predator-prey interactions and breeding-pond use of temporary-pond species in a desert anuran community. *Ecology* 64:1549–55.

———. 1986. Paternal effects on juvenile growth in *Scaphiopus multiplicatus* (the New Mexico spadefoot toad). *Am. Nat.* 128:58–65.

———. 1987. Interactions between Woodhouse's toad tadpoles (*Bufo woodhousii*) of mixed sizes. *Copeia* 1987:380–86.

Woodward, B. D., J. Travis, and S. Mitchell. 1988. The effects of the mating system on progeny performance in *Hyla crucifer* (Anura: Hylidae). *Evolution* 42:784–94.

Woody, D. R., and A. Mathis. 1997. Avoidance of areas labeled with chemical stimuli from damaged conspecifics by adult newts, *Notophthalmus viridescens*, in a natural habitat. *J. Herpetol.* 31:316–18.

———. 1998. Acquired recognition of chemical stimuli from an unfamiliar predator: Associative learning by adult newts, *Notophthalmus viridescens*. *Copeia* 1998:1027–31.

Woolard, P., W. J. M. Vestjens, and L. MacLean. 1978. The ecology of the eastern water rat *Hydromys chrysogaster* at Griffith, N. S. W.: Food and feeding habits. *Austral. Wildl. Res.* 5:59–73.

Woolbright, L. L. 1983. Sexual selection and size dimorphism in anuran Amphibia. *Am. Nat.* 121:110–19.

———. 1985a. Patterns of nocturnal movement and calling by the tropical frog *Eleutherodactylus coqui*. *Herpetologica* 41:1–9.

———. 1985b. *Sexual dimorphism in body size of the subtropical frog, Eleutherodactylus coqui.* Unpublished PhD dissertation. Albany: State University of New York.

———. 1989. Sexual dimorphism in *Eleutherodactylus coqui:* Selection pressures and growth rates. *Herpetologica* 45:68–74.

———. 1991. The impact of Hurrican Hugo on forest frogs in Puerto Rico. *Biotropica* 23:462–67.

———. 1996. Disturbance influences long-term population patterns in the Puerto Rican frog *Eleutherodactylus coqui* (Anura: Leptodactylidae). *Biotropica* 28:493–501.

———. 1997. Local extinctions of anuran amphibians in the Luquillo Experimental Forest of northeastern Puerto Rico. *J. Herpetol.* 31:572–76.

Woolbright, L. L., E. J. Greene, and G. C. Rapp. 1990. Density-dependent mate searching strategies of male wood frogs. *Anim. Behav.* 40:135–42.

Woolbright, L. L., A. H. Hara, C. M. Jacobsen, W. J. Mautz, and F. L. Benevides, Jr. 2006. Population densities of the coquí, *Eleutherodactylus coqui* (Anura: Leptodactylidae) in newly invaded Hawaii and in native Puerto Rico. *J. Herpetol.* 40:122–26.

Woolbright, L. L., and M. M. Stewart. 1987. Foraging success of the tropical frog, *Eleutherodactylus coqui:* the cost of calling. *Copeia* 1987:69–75.

Workman, G., and K. C. Fisher. 1941. Temperature selection and the effect of temperature on movement in frog tadpoles. *Amer. J. Physiol.* 133:P499–P500.

Worrell, E. 1963. *Reptiles of Australia: Crocodiles, turtles, tortoises, lizards, snakes.* Sydney, Australia: Angus and Robertson.

Worster, D. 1977. *Nature's economy.* New York: Cambridge University Press.

Worthington, R. D. 1968. Observations on the relative sizes of three species of salamander larvae in a Maryland pond. *Herpetologica* 24:242–46.

———. 1969. Additional observations on sympatric species of salamander larvae in a Maryland pond. *Herpetologica* 25:227–29.

Worthington, S., and E. B. Worthington, 1933. *Inland waters of Africa.* London: Macmillan.

Worthylake, K. M., and P. Hovingh. 1989. Mass mortality of salamanders (*Ambystoma tigrinum*) by bacteria (*Acinetobacter*) in an oligotrophic seepage mountain lake. *Great Basin Nat.* 49:364–72.

Wright, A. H. 1914. *Life-Histories of the anura of Ithaca, New York.* Washington, DC: Carnegie Institution of Washington.

———. 1920. *Frogs: Their natural history and utilization. Appendix VI. Report of the U.S. Commissioner of Fish and Fisheries for 1919.* Washington, DC.

———. 1932. *Life-histories of the frogs of Okefenokee Swamp, Georgia.* New York: Macmillan.

Wright, A. H., and A. A. Allen. 1908. Notes on the breeding habits of the swamp cricket frog, *Chorophilus triseriatus* Wied. *Am. Nat.* 42:39–42.

Wright, A. H., and A. A. Allen. 1949. *Handbook of frogs and toads.* Ithaca, NY: Comstock.

———. 1957. *Handbook of snakes.* 2 vols. Ithaca, NY: Comstock.

Wright, P. M., and P. A. Wright. 1996. Nitrogen metabolism and excretion in bullfrog (*Rana catesbeiana*) tadpoles and adults exposed to elevated environmental ammonia levels. *Physiol. Zool.* 69:1057–78.

Wright, S. J. 1979. Competition between insectivorous lizards and birds in central Panama. *Am. Zool.* 19:1145–56.

———. 1981. Extinction-mediated competition: The *Anolis* lizards and insectivorous birds of the West Indies. *Am. Nat.* 117:181–92.

Wrobel, D. J., W. F. Gergits, and R. G. Jaeger. 1980. An experimental study of interference competition among terrestrial salamanders. *Ecology* 61:1034–39.

Wu, C.-S., and Y.-C. Kam. 2005. Thermal tolerance and thermoregulation by Taiwanese rhacophorid tadpoles (*Buergeria japonica*) living in geothermal hot springs and streams. *Herpetologica* 61:35–46.

Wunder, W. 1932. Nestbau und Brutpflege bei Amphibien. *Ergebn. Biol.* 8:180–220.

Wyatt, J. L., and E. E. Forys. 2004. Conservation implications of predation by Cuban treefrogs (*Osteopilus septentrionalis*) on native hylids in Florida. *Southeast. Nat.* 3:695–700.

Wycherley, J., S. Doran, and T. J. C. Beebee. 2002. Male advertisement call characters as phylogeographical indicators in European water frogs. *Biol. J. Linn. Soc.* 77:355–65.

Wygoda, M. L. 1984. Low cutaneous evaporative water loss in arboreal frogs. *Physiol. Zool.* 57:329–37.

———. 1988. Adaptive control of water loss resistance in an arboreal frog. *Herpetologica* 44:251–57.

———. 1989. A comparative study of heating rates in arboreal and nonarboreal frogs. *J. Herpetol.* 23:141–45.

Wygoda, M. L., and R. H. Garman. 1993. Ontogeny of resistance to water loss in green tree frogs, *Hyla cinerea. Herpetologica* 49:365–74.

Wygoda, M. L., R. H. Garman, and C. E. Howard. 1987. Cutaneous and subcutaneous adipose tissue in anuran amphibians. *Copeia* 1987:1031–35.

Wygoda, M. L., and A. A. Williams. 1991. Body temperatures in free-ranging green tree frogs (*Hyla cinerea*): a comparison with "typical" frogs. *Herpetologica* 47:328–35.

Wyman, R. L. 1988. Soil acidity and moisture and the distribution of amphibians in five forests of southcentral New York. *Copeia* 1988:394–99.

———. 1990. What's happening to the amphibians? *Cons. Biol.* 4:350–52.

———. 1998. Experimental assessment of salamanders as predators on detrital food webs: Effects on invertebrates, decomposition and the carbon cycle. *Biodivers. Cons.* 7:641–50.

———. 2003. Conservation of terrestrial salamanders with direct development. In *Amphibian conservation,* ed. R. D. Semlitsch, 37–52. Washington, DC: Smithsonian Books.

Wyman, R. L. and D. S. Hawksley-Lescault. 1987. Soil acidity affects distribution, behavior, and physiology of the salamander *Plethodon cinereus. Ecology* 68:1819–27.

Wyman, R. L., and J. Jancola. 1992. Degree and scale of terrestrial acidification and amphibian community structure. *J. Herpetol.* 26:392–401.

Wyman, R. L., and J. H. Thrall. 1972. Sound production by the spotted salamander, *Ambystoma maculatum. Herpetologica* 28:210–12.

Wynne-Edwards, V. C. 1952. Freshwater vertebrates of the Arctic and Subarctic. *Bull. Fish. Res. Bd. Canada* 94:1–28.

Xavier, F. 1977. An exceptional reproductive strategy in Anura: *Nectophrynoides occidentalis* Angel (Bufonidae), an example of adaptation to terrestrial life by viviparity. In *Major patterns in vertebrate evolution,* ed. M. K. Hecht, P. C. Goody, and B. M. Hecht, 545–52. New York: Plenum.

Xavier, F. 1986. La reproduction des *Nectophrynoides.* In *Traite de zoologie, vol. 14, fas. IB. Amphibiens,* ed. P.-P. Grasse and M. Delsol, 497–513. Paris: Masson.

Xia, X. 1995. Body temperature, rate of biosynthesis, and evolution of genome size. *Mol. Biol. Evol.* 12:834–42.

Xie, F., L. Fei, C. Ye, C. Cai, Z. Wang, and M. Sparreboom. 2000. Breeding migration and oviposition of the Chinhai salamander, *Echinotriton chihaiensis. Herpetol. J.* 10:111–18.

Xu, J., D. M. Gooler, and A. S. Feng. 1994. Single neurons in the frog inferior colliculus exhibit direction-dependent frequency selectivity to isointensity tone bursts. *J. Acoust. Soc. Amer.* 95:2160–70.

———. 1996. Effects of sound direction on the processing of amplitude-modulated signals in the frog inferior colliculus. *J. Comp. Physiol. A* 178:435–45.

Xu, Q., and R. S. Oldham. 1997. Lethal and sublethal effects of nitrogen fertilizer ammonium nitrate on common toad (Bufo bufo) tadpoles. *Arch. Environ. Contam. Toxicol.* 32:298–303.

Yager, D. D. 1992a. Underwater acoustic communication in the African pipid frog *Xenopus borealis. Bioacoustics* 4:1–24.

———. 1992b. A unique sound production mechanism in the pipid anuran *Xenopus borealis. Zool. J. Linn. Soc.* 104:351–75.

———. 1996. Sound production and acoustic communication in *Xenopus borealis.* In *The biology of Xenopus,* ed. R. C. Tinsley and H. R. Kobel, 121–41. New York: Oxford University Press.

Yamano, K., and S. Miwa. 1998. Differential gene expression of thyroid hormone receptor alpha and beta in fish development. *Gen. Comp. Endocrinol.* 109:75–85.

Yan, X., K. Yaci, H. Akiyama, and H. Akimoto. 2005. Statistical analysis of the major variables controlling methane emission from rice fields. *Global Change Biol.* 11:1131–41.

Yano, M. 1978. Ecological studies of *Bufo japonicus formosus* Schlegel. III. Movements. *Rep. Inst. Nat. Stud.* 8:107–19 (in Japanese).

Ye, C., L. Fei, and S. Hu. 1993. *Rare and economic amphibians of China.* Chengdu, China: Sichuan House of Science and Technology (in Chinese).

Yeh, J. 2002. The effect of miniaturized body size on skeletal morphology in frogs. *Evolution* 56:628–41.

Yekta, N., and D. G. Blackburn. 1992. Sexual dimorphism in mass and protein content of the forelimb muscles of the northern leopard frog, *Rana pipiens. Can. J. Zool.* 70:670–74.

Yesilada, E., M. Ozmen, and O. Yesilada. 1999. Studies on the toxic and genotoxic effect of olive oil mill wastewater. *Fresenius Environ. Bull.* 8:732–39.

Yiming, L., and D. S. Wilcove. 2005. Threats to vertebrate species in China and the United States. *BioScience* 55:147–53.

Yip, J. Y., R. T. Corlett, and D. Dudgeon. 2004. A fine-scale gap analysis of the existing protected area system in Hong Kong, China. *Biodiver. Cons.* 13:943–57.

Yoder, H. R., and J. R. Coggins. 1996. Helminth communities in the northern spring peeper, *Pseudacris c. crucifer* Weid, and the wood frog, *Rana sylvatica* LeConte, from southeastern Wisconsin. *J. Helminthol. Soc. Washington* 63:211–14.

Yokota, S. D., S. Benyajati, adn W. H. Dantzler. 1985. Comparative aspects of glomerular filtration in vertebrates. *Renal Physiol.* 8:193–221.

Yokota, S. D., and S. S. Hillman. 1984. Adrenergic control of the anuran cutaneous hydroosmotic response. *Gen. Comp. Endocrinol.* 53:309–14.

Yom-Tov, Y., and J. Yom-Tov. 2005. Global warming, Bergmann's rule and body size in the masked shrew Sorex cinereus Kerr in Alaska. *J. Anim. Ecol.*74:803–8.

Yorio, T., and P. J. Bentley. 1977. Asymmentrical permeability of the integument of tree frogs (Hylidae). *J. Exp. Biol.* 67:197–204.

———. 1978. The permeability of the skin of the aquatic anuran *Xenopus laevis* (Pipidae). *J. Exp. Biol.* 72:285–89.

Yorke, C. D. 1983. Survival of embryos and larvae of the frog *Polypedates leucomystax* in Malaysia. *J. Herpetol.* 17:235–41.

Young, B. E., K. R. Lips, J. K. Reaser, R. Ibáñez, A. W. Salas, J. R. Cedeño, L. A. Coloma, S. Ron, E. La Marca, J. R. Meyer, A. Muñoz, F. Bolaños, G. Chaves, and D. Romo. 2001. Population declines and priorities for amphibian conservation in Latin America. *Cons. Biol.* 15:1213–23.

Young, B. E., S. N. Stuart, J. S. Chanson, N. A. Cox, and T. M. Boucher. 2004. *Disappearing jewels. The status of new world amphibians.* Arlington, VA: NatureServe.

Youson, J. H. 1997. Is lamprey metamorphosis regulated by thyroid hormones? *Amer. Zool.* 37:441–60.

Yu, X., E. R. Lewis, and D. Feld. 1991. Seismic and auditory tuning curves from bullfrog saccular and amphibian papillar axons. *J. Comp. Physiol. A* 169:241–48.

Yu, Z., R. Dickstein, W. E. Magee, and J. R. Spotila. 1998. Heat shock response in the salamanders *Plethodon jordani* and *Plethodon cinereus*. *J. Therm. Biol.* 23:259–65.

Yu, Z., W. E. Magee, and J. R. Spotila. 1994. Monoclonal antibody ELISA test indicatee that there are large amounts of hsp-70 present in salamanders, turtles and fish. *J. Therm. Biol.* 19:41–53.

Yurewicz, K. L., and H. M. Wilbur. 2004. Resource availability and costs of reproduction in the salamander *Plethodon cinereus*. *Copeia* 2004:28–36.

Zaga, A., E. E. Little, C. F. Rabeni, and M. R. Ellersieck. 1998. Photo-enhanced toxicity of a carbamate insecticide to early life stage anuran amphibians. *Env. Tox. Chem.* 17:2543–53.

Zahl, P. A. 1967. In quest of the world's largest frog. *Nat. Geogr.* 132:146–52.

Zakon, H. H., and W. Wilczynski. 1988. The physiology of the anuran eighth nerve. In *The evolution of the amphibian auditory system,* ed. B. Fritzsch, M. J. Ryan, W. Wilczynski, T. E. Hetherington, and W. Walkowiak, 125–55. New York: Wiley.

Zalisko, E. J., R. A. Brandon, and J. Martan. 1984. Microstructure and histochemistry of salamander spermatophores (Ambystomatidae, Salamandridae, and Plethodontidae). *Copeia* 1984:739–47.

Zamachowski, W. 1968. Changes in the water content in the organism of the common frog (*Rana temporaria* L.) and the water frog (*Rana esculenta* L.) in the annual cycle. *Acta Biol. Cracov. Ser. Zool.* 11:213–25.

———. 1977. The water economy in some European species of anuran amphibians during the annual cycle. I. Water content of the organism. *Acta Biol. Cracov. Ser. Zool.* 20:181–89.

Zardoya, R., Y. Cao, M. Hasegawa, and A. Meyer. 1998. Searching for the closest living relative(s) of tetrapods through evolutionary analysis of mitochondrial and nuclear data. *Mol. Biol. Evol.* 15:506–17.

Zardoya, R., and A. Meyer. 1997. Molecular phylogenetic information on the identity of the closest living relative(s) of land vertebrates. *Naturwissenschaften* 84:389–97.

———. 2001. On the origin of and phylogenetic relationships among living amphibians. *Proc. Natl. Acad. Sci. USA* 98:7380–83.

Zasloff, M. 1987. Magainins, a class of antimicrobial peptides from *Xenopus* skin: Isolation, characterization of two active forms, and partial cDNA sequence of a precursor. *Proc. Natl. Acad. Sci. USA* 84:5449–53.

Zedler, J. B., and S. Kercher. 2004. Causes and consequences of invasive plants in wetlands: Opportunities, opportunists, and outcomes. *Crit. Rev. Plant Sci.* 23:431–52.

Zedler, P. H. 2003. Vernal pools and the concept of "isolated wetlands." *Wetlands* 23:597–607.

Zelick, R. D., and P. M. Narins. 1982. Analysis of acoustically evoked call suppression behaviour in a Neotropical treefrog. *Anim. Behav.* 30:728–33.

———. 1983. Intensity discrimination and the precision of call timing in two species of Neotropical treefrogs. *J. Comp. Physiol. A* 153:403–12.

———. 1985a. Characterization of the advertisement call oscillator in the frog *Eleutherodactylus coqui*. *J. Comp. Physiol. A* 156:223–29.

———. 1985b. Temporary threshold shifts, adaptation, and recovery characteristics of frog auditory nerve fibers. *Hearing Res.* 17:161–76.

Zerba, K. E., and J. P. Collins. 1992. Spatial heterogeneity and individual variation in diet of an aquatic top predator. *Ecology* 73:268–79.

Zhang, P., H. Zhou, Y-Q. Chen, Y.-F. Liu, and L-H. Qu. 2005. Mitogenomic perspectives on the origin and phylogeny of living amphibians. *Syst. Biol.* 54:391–400.

Zhao, E.-M., ed. 1998. *China red data book of endangered animals. Amphibia & reptilia.* Beijing: Science Books (in Chinese).

Zhao, E.-M. 1999. Distribution patterns of amphibians in temperate eastern Asia. In *Patterns of distribution of amphibians,* ed. W. E. Duellman, 421–43. Baltimore: Johns Hopkins University Press.

Zhao, E.-M., and K. Adler. 1993. *Herpetology of China.* Oxford, OH: Society for the Study of Amphibians and Reptiles.

Zhao, E.-M., R. F. Inger, G.-F. Wu, and H. B. Shaffer. 1994. Morphological variation and ecological distribution of co-occurring larval forms of *Oreolalax* (Anura: Pelobatidae). *Amphibia-Reptilia* 15:109–21.

Zhao, E.-M., G. Wu, and R. F. Inger. 1989. Ecological and geographic distribution of the amphibians of Sichuan, China. *Copeia* 1989:549–57.

Zhao, Y., C. Wang, S. Wang, and L. V. Tibig. 2005. Impacts of present and future climate variability on agriculture and forestry in the humid and sub-humid tropics. *Climatic Change* 70:73–116.

Zhou, L., R. E. Dickinson, Y. Tian, J. Fang, Q. Li, R. K. Kaufmann, C. J. Tucker, and R. B. Myneni. 2004. Evidence for a significant urbanization effect on climate in China. *Proc. Natl. Acad. Sci. USA* 101:9540–44.

Zielinski, W. J., and G. T. Barthalmus. 1989. African clawed frog skin compounds: Antipredatory effects on African and North American water snakes. *Anim. Behav.* 38:1083–86.

Ziemba, E. E., and J. P. Collins. 1999. Development of size structure in tiger salamanders: The role of intraspecific interference. *Oecologia* 120:524–29.

Zimmerman, B. L. 1983. A comparison of structural features of calls of open and forest habitat frog species in the central Amazon. *Herpetologica* 39:235–46.

Zimmerman, B. L., and J. P. Bogart. 1984. Vocalizations of primary forest frog species in the central Amazon. *Acta Amazonica* 14:473–519.

———. 1988. Ecology and calls of four species of Amazonian forest frogs. *J. Herpetol.* 22:97–108.

Zimmerman, B. L., and W. Hödl. 1983. Distinction of *Phrynohyas resinifictrix* (Goeldi, 1907) from *Phrynohyas venulosa* (Laurenti, 1786) based on acoustical and behavioural parameters (Amphibia, Anura, Hylidae). *Zool. Anz.* 211:341–52.

Zimmerman, B. L., and M. T. Rodrigues. 1990. Frogs, snakes, and lizards of the INPA-WWF Reserves near Manaus, Brazil. In *Four neotropical rainforests,* ed. A. H. Gentry, 426–54. New Haven, CT: Yale University Press.

Zimmerman, B. L., and D. Simberloff. 1996. An historical interpretation of habitat use by frogs in a Central Amazonian forest. *J. Biogeogr.* 23:27–46.

Zimmermann, E. 1983. *Das Züchten von Terrarientieren.* Stuttgart, Germany: Gesellschaft der Naturfreunde Franckh'sche Verlagshandlung.

———. 1990. Behavioral signals and reproduction modes in the Neotropical frog family Dendrobatidae. *Fortschr. Zool.* 38:61–73.

Zimmermann, E., and H. Zimmermann. 1982. Soziale Interaktionen, Brutpflege und Zucht des Pfeilgiftfrosches *Dendrobates histrionicus* (Amphibia: Salientia: Dendrobatidae). *Salamandra* 18:150–67.

———. 1985. Brutpflegestrategien bei Pfeilgiftfroschen (Dendrobatidae). *Verh. Dtsch. Zool. Ges.* 78:220.

Zimmermann, H. 1974. Die Aufzucht des Goldbaumsteigers *Dendrobates auratus. Aquar. Mag.* 1974:526–31.

———. 1978. Verhaltensbeobachtungen an Färberfröschen. *Aquar. Mag.* 1978:458–63.

———. 1982. Durch Nachzucht erhalten: Blattsteigerfrosche *Phyllobates vittatus* und *P. lugubris. Aquar. Mag.* 1982:109–12.

———. 1983. Durch Nachzucht erhalten: *Phyllobates tricolor. Aquar. Mag.* 1983:17–22.

———. 1989. Conservation studies of the dart-poison frogs Dendrobatidae in the field and in captivity. *Int. Zoo Ybk.* 28:31–44.

———. 1992. Nachzucht und Schutz von *Mantella crocea, Mantella viridis,* und vom madagassischen Goldfroschen *Mantella aurantiaca. Z. Kölner Zoo* 35:165–72.

Zimmermann, H., and E. Zimmermann. 1980a. Durch Nachzucht erhalten: Der Baumsteiger *Dendrobates leucomelas. Aquar. Mag.* 1980:211–17.

———. 1980b. Durch Nachzuct erhalten: *Dendrobates histrionicus* und *D. lehmanni. Aquar. Mag.* 1980:562–69.

———. 1981. Sozialverhalten, Fortpflanzungsverhalten und Zucht der Färberfrösche *Dendrobates histrionicus* und *D. lehmanni* sowie einiger anderer Dendrobatiden. *Z. Kölner Zoo* 24:83–99.

———. 1983. Das Phänomen der Eierfresser. *Aquar. Mag.* 1983:44.

———. 1984. Durch Nachzucht erhalten: Baumsteigerfrösche *Dendrobates quinquevittatus* und *D. reticulatus. Aquar. Mag.* 1984:35–41.

———. 1985. Zur Fortpflanzungsstrategie des Pfeilgiftfrosches *Phyllobates terribilis* Myers, Daly & Malkinm, 1978 (Salientia: Dendrobatidae). *Salamandra* 21:281–97.

———. 1988. Etho-Taxonomie und zoogeographische Artensgruppenbildung bei Pfeilgiftfröschen. *Salamandra* 24:125–60.

———. 1990. Behavioral systematics and zoogeography of the formation of species groups in dart-poison frogs (Anura, Dendrobatidae). *Newsletter of the International Society for the Study of Dendrobatid Frogs* 3:75–112 (English translation of H. and E. Zimmermann 1988).

Zimmitti, S. J. 1999. Individual variation in morphological, physiological, and biochemical features associated with calling in spring peepers (*Pseudacris crucifer*). *Physiol. Biochem. Zool.* 72:666–76.

Zippelius, H. M. 1948. Untersuchungen über das Balzverhalten ein heimischer Molche. *Verh. Dtsch. Zool. Ges. Kiel* 1948:127–30.

Zschokke, F. 1900. Die Tierwelt der Hochgebirgsseen. *Neue Denkschr. Allg. Schweiz. Gesellsch.* 37:1–400.

Zug, G. R. 1972. Anuran locomotion: structure and function. I. Preliminary observation on relation between jumping and osteometrics of appendicular and postaxial skeleton. *Copeia* 1972:613–24.

———. 1978. Anuran locomotion—Structure and function. 2: Jumping performance of semiaquatic, terrestrial, and arboreal frogs. *Smithsonian Contrib. Zool.* (276):1–31.

———. 1985. Anuran locomotion: fatigue and jumping performance. *Herpetologica* 41:188–94.

Zug, G. R., L. J. Vitt, and J. P. Caldwell. 2001. *Herpetology* (2nd ed.). San Diego: Academic Press.

Zug, G. R., and P. B. Zug. 1979. The marine toad, *Bufo marinus:* A natural history resume of native populations. *Smithsonian Contrib. Zool.* (284):1–58.

Zuiderwijk, A., and M. Sparreboom. 1986. Territorial behaviour in crested newt *Triturus cristatus* and marbled newt *T. marmoratus* (Amphibia, Urodela). *Bijdr. Dierk.* 56:205–13.

Zuk, M., and L. H. Simmons. 1997. Reproductive strategies of crickets (Orthoptera: Gryllidae). In *The evolution of mating systems in insects and arachnids,* ed. J. C. Choe and B. J. Crespi, 89–109. New York: Cambridge University Press.

Zvirgzds, J., M. Stašuls, and V. Vilnītis. 1995. Reintroduction of the European tree trog (*Hyla arborea*) in Latvia. *Mem. Soc. Fauna Flora Fennica* 71: 139–42.

Zweifel, R. G. 1955. Ecology, distribution, and systematics of frogs of the *Rana boylei* group. *Univ. California Publ. Zool.* 54:207–92.

———. 1956. Results of the Archbold expeditions. No. 72. Microhylid frogs from New Guinea, with descriptions of new species. *Amer. Mus. Novitates* (1766):1–49.

———. 1958. Results of the Archbold Expedition no. 78. Frogs of the Papuan hylid genus *Nyctimystes. Amer. Mus. Novitates* (1896):1–51.

———. 1968a. Effects of temperature, body size, and hybridization on mating calls of toads, *Bufo a. americanus* and *Bufo woodhousii fowleri. Copeia* 1968:269–85.

———. 1968b. Reproductive biology of anurans of the arid

southwest, with emphasis on adaptation of embryos to temperature. *Bull. Amer. Mus. Nat. Hist.* 140:1–64.

———. 1969. Frogs of the genus *Platymantis* (Ranidae) in New Guinea, with the description of a new species. *Amer. Mus. Novitates* (2374):1–19.

———. 1972. Results of the Archbold Expeditions. No. 97. A revision of the frogs of the subfamily Asterophryinae, family Microhylidae. *Bull. Amer. Mus. Nat. Hist.* 148:411–546.

———. 1975. Two new frogs of the genus *Platymantis* (Ranidae) from New Britain. *Amer. Mus. Novitates* (2582):1–7.

———. 1977. Upper thermal tolerances of anuran embryos in relation to stages of development and breeding habits. *Amer. Mus. Novitates* (2617):1–21.

———. 1985. Australian frogs of the family Microhylidae. *Bull. Amer. Mus. Nat. Hist.* 182:265–388.

———. 1986. A new genus and species of microhylid frog from the Cerro de al Neblina region of Venezuela and a discussion of relationships among New World microhylid genera. *Amer. Mus. Novitates* (2863):1–12.

———. 2000. Partition of the Australopapuan microhylid frog genus *Sphenophryne* with descriptions of new species. *Bull. Amer. Mus. Nat. Hist.* 253:1–130.

Zweifel, R. G., and A. Allison. 1982. A new montane microhylid frog from Papua New Guinea, and comments on the status of the genus *Aphantophryne. Amer. Mus. Novitates* (2723):1–14.

Zweifel, R. G., and F. Parker. 1989. New species of microhylid frogs from the Owen Stanley Mountains of Papua New Guinea and resurrection of the genus *Aphantophryne. Amer. Mus. Novitates,* (2954):1–20.

Zweifel, R. G., and M. J. Tyler. 1982. Amphibia of New Guinea. *Monogr. Biol.* 42:759–801.

Zylberberg, L., J. Castanet, and A. de Ricqles. 1980. Structure of the dermal scales in Gymnophiona (Amphibia). *J. Morphol.* 165:41–51.

Zylerberg, L., and M. H. Wake. 1990. Structure of the scales of *Dermophis* and *Microcaecilia* (Amphibia: Gymnophiona), and a comparison to dermal ossifications of other vertebrates. *J. Morphol.* 206:25–43.

# Index

Full or abbreviated chapter titles serve as main subject headings in the index (**boldface**), with major topics indexed under those main headings. Topics in Chapter 1 are indexed under the three major groups of amphibians, Anura (frogs), Gymnophiona (caecilians), and Urodela (salamanders), along with other topics specific to those groups. Family synopses are indexed under family names. Amphibian species are indexed by Latin name only, because many species mentioned in this book do not have standardized English common names. All non-amphibian organisms are indexed by Latin name, followed by a descriptive common name in parentheses. Author names are not indexed, except for historically important figures (e.g., Charles Darwin).